Unit Conversion Factors

Length

1 m = 100 cm = 1000 mm = 10^6 μm = 10^9 nm

1 km = 1000 m = 0.6214 mi

1 m = 3.281 ft = 39.37 in

1 cm = 0.3937 in

1 in = 2.54 cm (exactly)

1 ft = 30.48 cm (exactly)

1 yd = 91.44 cm (exactly)

1 mi = 5280 ft = 1.609344 km (exactly)

1 Angstrom = 10^{-10} m = 10^{-8} cm = 0.1 nm

1 nautical mile = 6076 ft = 1.151 mi

1 light-year = $9.461 \cdot 10^{15}$ m

Area

1 m^2 = 10^4 cm^2 = 10.76 ft^2

1 cm^2 = 0.155 in^2

1 in^2 = 6.452 cm^2

1 ft^2 = 144 in^2 = 0.0929 m^2

1 hectare = 2.471 acres = 10,000 m^2

1 acre = 0.4047 hectare = 43,560 ft^2

1 mi^2 = 640 acres (exactly)

1 yd^2 = 0.8361 m^2

Volume

1 liter = 1000 cm^3 = 10^{-3} m^3 = 0.03531 ft^3 = 61.02 in^3 = 33.81 fluid ounce

1 ft^3 = 0.02832 m^3 = 28.32 liter = 7.481 gallon

1 gallon = 3.785 liters

1 quart = 0.9464 liter

Time

1 min = 60 s

1 h = 3600 s

1 day = 86,400 s

1 week = 604,800 s

1 year = $3.156 \cdot 10^7$ s

Angle

1 rad = 57.30° = 180°/π

1° = 0.01745 rad = (π/180) rad

1 rev = 360° = 2π rad

1 rev/min (rpm) = 0.1047 rad/s = 6°/s

Speed

1 mile per hour (mph) = 0.4470 m/s = 1.467 ft/s = 1.609 km/h

1 m/s = 2.237 mph = 3.281 ft/s

1 km/h = 0.2778 m/s = 0.6214 mph

1 ft/s = 0.3048 m/s

1 knot = 1.151 mph = 0.5144 m/s

Acceleration

1 m/s^2 = 100 cm/s^2 = 3.281 ft/s^2

1 cm/s^2 = 0.01 m/s^2 = 0.03281 ft/s^2

1 ft/s^2 = 0.3048 m/s^2 = 30.48 cm/s^2

Mass

1 kg = 1000 g = 0.0685 slug

1 slug = 14.59 kg

1 kg = 2.2046 lb

1 lb = 0.4536 kg

Force

1 N = 0.2248 lb

1 lb = 4.448 N

1 stone = 14 lb = 62.28 N

Pressure

1 Pa = 1 N/m^2 = $1.450 \cdot 10^{-4}$ lb/in^2 = 0.0209 lb/ft^2

1 atm = $1.013 \cdot 10^5$ Pa = 101.3 kPa = 14.7 lb/in^2 = 2117 lb/ft^2 = 760 mm Hg = 29.92 in Hg

1 lb/in^2 = 6895 Pa

1 lb/ft^2 = 47.88 Pa

1 mm Hg = 1 Torr = 133.3 Pa

1 bar = 10^5 Pa = 100 kPa

Energy

1 J = 0.239 cal

1 cal = 4.186 J

1 Btu = 1055 J = 252 cal

1 kW·h = $3.600 \cdot 10^6$ J

1 ft·lb = 1.356 J

1 eV = $1.602 \cdot 10^{-19}$ J

Power

1 W = 1 J s

1 hp = 746 W = 0.746 kW = 550 ft·lb/s

1 Btu/h = 0.293 W

1 GW = 1000 MW = $1.0 \cdot 10^9$ W

1 kW = 1.34 hp

Temperature

Fahrenheit to Celsius: $T_C = \frac{5}{9}(T_F - 32\ °F)$

Celsius to Fahrenheit: $T_F = \frac{9}{5}T_C + 32\ °C$

Celsius to Kelvin: $T_K = T_C + 273.15\ °C$

Kelvin to Celsius: $T_C = T_K - 273.15\ K$

Second Edition

University
Physics

with
Modern Physics

Wolfgang Bauer
Michigan State University

Gary D. Westfall
Michigan State University

Connect
Learn
Succeed™

UNIVERSITY PHYSICS WITH MODERN PHYSICS, SECOND EDITION

Published by McGraw-Hill, a business unit of The McGraw-Hill Companies, Inc., 1221 Avenue of the Americas, New York, NY 10020.
Copyright © 2014 by The McGraw-Hill Companies, Inc. All rights reserved. Printed in the United States of America. Previous edition © 2011.
No part of this publication may be reproduced or distributed in any form or by any means, or stored in a database or retrieval system, without
the prior written consent of The McGraw-Hill Companies, Inc., including, but not limited to, in any network or other electronic storage or
transmission, or broadcast for distance learning.

Some ancillaries, including electronic and print components, may not be available to customers outside the United States.

This book is printed on acid-free paper.

1 2 3 4 5 6 7 8 9 0 DOW/DOW 1 0 9 8 7 6 5 4 3

ISBN 978–0–07–351388–1
MHID 0–07–351388–1

Senior Vice President, Products & Markets: *Kurt L. Strand*
Vice President, General Manager, Products & Markets: *Marty Lange*
Vice President, Content Production & Technology Services: *Kimberly Meriwether David*
Managing Director: *Thomas Timp*
Director: *Michael Lange*
Director of Development: *Rose Koos*
Developmental Editor: *Eve L. Lipton*
Marketing Manager: *Bill Welsh*
Senior Project Manager: *Jayne L. Klein*
Senior Buyer: *Sandy Ludovissy*
Lead Media Project Manager: *Judi David*
Senior Designer: *David W. Hash*
Cover/Interior Designer: *John Joran*
Cover Image: *Abstract orbital lines on white background, ©imagewerks/Getty Images, Inc. RF; NIF Target Chamber, ©Lawrence Livermore
National Laboratory; Time-averaged surface temperature of the Earth in June 1992, ©MHHE; The temperature of the cosmic microwave
background radiation everywhere in the universe, ©MHHE; Large wind turbine, ©MHHE; Niagara Falls, ©W. Bauer and G.D. Westfall; Solar
panels, ©Rob Atkins/Photographer's Choice/Getty Images, Inc.; Bose-Einstein condensate, image courtesy NIST/JILA/CU-Boulder; Scaled
chrysophyte, Mallomanas lychenensis, SEM, ©Dr. Peter Siver/Visuals Unlimited/Getty Images, Inc; Snow boarder, ©W. Bauer and G.D. Westfall.*
Lead Content Licensing Specialist: *Carrie K. Burger*
Photo Research: *Danny Meldung/PhotoAffairs, Inc.*
Compositor: *Precision Graphics*
Typeface: *10/12 Minion Pro*
Printer: *R. R. Donnelley*

All credits appearing on page or at the end of the book are considered to be an extension of the copyright page.

Library of Congress Cataloging-in-Publication Data

Bauer, W. (Wolfgang), 1959-
  University physics with modern physics / Wolfgang Bauer, Michigan State University, Gary Westfall, Michigan State University. – Second
edition.
      pages cm
  Includes index.
  ISBN 978–0–07–351388–1 — ISBN 0–07–351388–1 (hard copy : acid-free paper); ISBN 978–0–07–740962–3 — ISBN 0–07–740962–0
(standard version : acid-free paper); ISBN 978–0–07–740963–0 — ISBN 0–07–740963–9 (v. 1 : acid-free paper); ISBN 978–0–07–740960–9 —
ISBN 0–07–740960–4 (v. 2 : acid-free paper) 1. Physics–Textbooks. I. Westfall, Gary D. II. Title.
  QC23.2.B38 2014
  530–dc23
                                                                                                                    2012031018

The Internet addresses listed in the text were accurate at the time of publication. The inclusion of a website does not indicate an endorsement
by the authors or McGraw-Hill, and McGraw-Hill does not guarantee the accuracy of the information presented at these sites.

www.mhhe.com

Brief Contents

About the Authors

Wolfgang Bauer was born in Germany and obtained his Ph.D. in theoretical nuclear physics from the University of Giessen in 1987. After a post-doctoral fellowship at the California Institute of Technology, he joined the faculty at Michigan State University in 1988, with a dual appointment at the National Superconducting Cyclotron Laboratory (NSCL). He has worked on a large variety of topics in theoretical and computational physics, from high-temperature superconductivity to supernova explosions, but has been especially interested in relativistic nuclear collisions. He is probably best known for his work on phase transitions of nuclear matter in heavy ion collisions. In recent years, Dr. Bauer has focused much of his research and teaching on issues concerning energy, including fossil fuel resources, ways to use energy more efficiently, and, in particular, alternative and carbon-neutral energy resources. In 2009, he founded the Institute for Cyber-Enabled Research and served as its first director until 2013. He presently serves as chairperson of the Department of Physics and Astronomy and is a University Distinguished Professor at Michigan State University.

Gary D. Westfall started his career at the Center for Nuclear Studies at the University of Texas at Austin, where he completed his Ph.D. in experimental nuclear physics in 1975. From there he went to Lawrence Berkeley National Laboratory (LBNL) in Berkeley, California, to conduct his post-doctoral work in high-energy nuclear physics and then stayed on as a staff scientist. While he was at LBNL, Dr. Westfall became internationally known for his work on the nuclear fireball model and the use of fragmentation to produce nuclei far from stability. In 1981, Dr. Westfall joined the National Superconducting Cyclotron Laboratory (NSCL) at Michigan State University (MSU) as a research professor; there he conceived, constructed, and ran the MSU 4π Detector. His research using the 4π Detector produced information concerning the response of nuclear matter as it is compressed in a supernova collapse. In 1987, Dr. Westfall joined the Department of Physics and Astronomy at MSU while continuing to carry out his research at NSCL. In 1994, Dr. Westfall joined the STAR Collaboration, which is carrying out experiments at the Relativistic Heavy Ion Collider (RHIC) at Brookhaven National Laboratory on Long Island, New York. In 2003, he was named University Distinguished Professor at Michigan State University.

The Westfall/Bauer Partnership Drs. Bauer and Westfall have collaborated on nuclear physics research and on physics education research for more than two decades. The partnership started in 1988, when both authors were speaking at the same conference and decided to go downhill skiing together after the session. On this occasion, Westfall recruited Bauer to join the faculty at Michigan State University (in part by threatening to push him off the ski lift if he declined). They obtained NSF funding to develop novel teaching and laboratory techniques, authored multimedia physics CDs for their students at the Lyman Briggs School, and co-authored a textbook on CD-ROM, called *cliXX Physik*. In 1992, they became early adopters of the Internet for teaching and learning by developing the first version of their online homework system. In subsequent years, they were instrumental in creating the Learning*Online* Network with CAPA, which is now used at more than 70 universities and colleges in the United States and around the world. Since 2008, Bauer and Westfall have been part of a team of instructors, engineers, and physicists, who investigate the use of peer-assisted learning in the introductory physics curriculum. This project has received funding from the NSF STEM Talent Expansion Program, and its best practices have been incorporated into this textbook.

Dedication This book is dedicated to our families. Without their patience, encouragement, and support, we could never have completed it.

A Note from the Authors

We are excited to introduce the second edition of our textbook, *University Physics*. Physics is a thriving science, alive with intellectual challenge and presenting innumerable research problems on topics ranging from the largest galaxies to the smallest subatomic particles. Physicists have managed to bring understanding, order, consistency, and predictability to our universe and will continue that endeavor into the exciting future.

However, when we open most current introductory physics textbooks, we find that a different story is being told. Physics is painted as a completed science in which the major advances happened at the time of Newton, or perhaps early in the 20th century. Only toward the end of the standard textbooks is "modern" physics covered, and even that coverage often includes only discoveries made through the 1960s.

Our main motivation in writing this book is to change this perception by weaving exciting, contemporary physics throughout the text. Physics is an amazingly dynamic discipline—continuously on the verge of new discoveries and life-changing applications. In order to help students see this, we need to tell the full, absorbing story of our science by integrating contemporary physics into the first-year calculus-based course. Even the very first semester offers many opportunities to do this by weaving recent results from nonlinear dynamics, chaos, complexity, and high-energy physics research into the introductory curriculum. Because we are actively carrying out research in these fields, we know that many of the cutting-edge results are accessible in their essence to the first-year student.

Recent results involving renewable energy, the environment, engineering, medicine, and technology show physics as an exciting, thriving, and intellectually alive subject motivating students, invigorating classrooms, and making the instructor's job easier and more enjoyable. In particular, we believe that talking about the broad topic of energy provides a great opening gambit to capture students' interest. Concepts of energy sources (fossil, renewable, nuclear, and so forth), energy efficiency, energy storage, alternative energy sources, and environmental effects of energy supply choices (global warming and ocean acidification, for example) are very much accessible on the introductory physics level. We find that discussions of energy spark our students' interest like no other current topic, and we have addressed different aspects of energy throughout our book.

In addition to being exposed to the exciting world of physics, students benefit greatly from gaining the ability to **problem solve and think logically about a situation.** Physics is based on a core set of ideas that is fundamental to all of science. We acknowledge this and provide a useful problem-solving method (outlined in Chapter 1) which is used throughout the entire book. This problem-solving method involves a multistep format that we have developed with students in our classes. But mastery of concepts also involves actively applying them. To this end, we have asked more than a dozen contributors from some of the leading universities across the country to share their best work in the end-of-chapter exercises. New to this edition are approximately 400 multi-version exercises, which allow students to address the same problem from different perspectives.

In 2012, the National Research Council published a framework for K-12 science education, which covers the essential science and engineering practices, the concepts that have application across fields, and the core ideas in four disciplinary areas (in physics, these are matter and its interactions, motion and stability, energy, and waves and their applications in information transfer). We have structured the second edition of this textbook to tie the undergraduate physics experience to this framework and have provided concept checks and self-test opportunities in each chapter.

With all of this in mind, along with the desire to write a captivating textbook, we have created what we hope will be a tool to engage students' imaginations and to better prepare them for future courses in their chosen fields (admittedly, hoping we can convert at least a few students to physics majors along the way). Having feedback from more than 400 people, including a board of advisors, several contributors, manuscript reviewers, and focus group participants, assisted greatly in this enormous undertaking, as did field testing our ideas with approximately 6000 students in our introductory physics classes at Michigan State University. We thank you all!

—*Wolfgang Bauer and Gary D. Westfall*

Contents

PART 2: EXTENDED OBJECTS, MATTER, AND CIRCULAR MOTION

PART 3: OSCILLATIONS AND WAVES

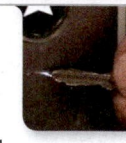

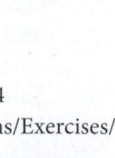

How to Use This Book

Problem-Solving Skills: Learning to Think Like a Scientist

Perhaps one of the greatest skills students can take from their physics course is the ability to **problem solve and think critically about a situation.** Physics is based on a core set of fundamental ideas that can be applied to various situations and problems. *University Physics* by Bauer and Westfall acknowledges this and provides a problem-solving method that has been class-tested by the authors, which is used throughout the text. The text's problem-solving method has a multistep format.

Problem-Solving Method

Solved Problems

The book's numbered **Solved Problems** are fully worked problems, each consistently following the seven-step method described in Section 1.5. Each Solved Problem begins with the problem statement and then provides a complete solution. The seven-step method is also used in Connect Physics. The familiar seven steps are outlined in the guided solutions, with additional help where you need it.

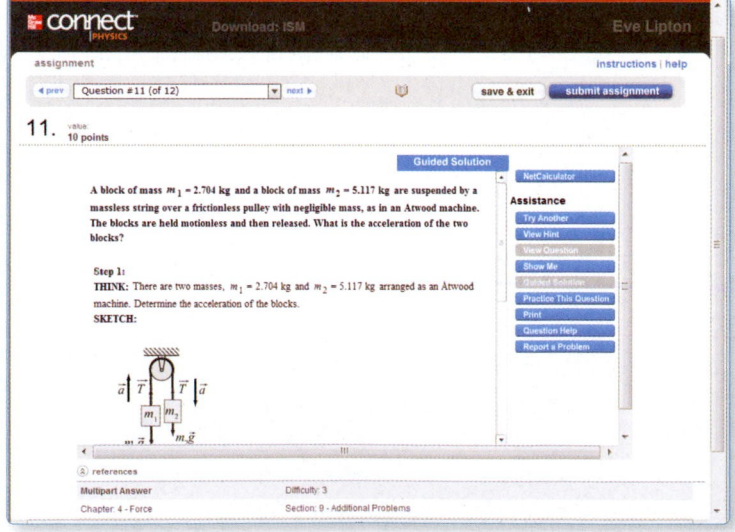

SOLVED PROBLEM 13.2 | **Weighing Earth's Atmosphere**

The Earth's atmosphere is composed (by volume) of 78.08% nitrogen (N_2), 20.95% oxygen (O_2), 0.93% argon (Ar), 0.25% water vapor (H_2O), and traces of other gases, most importantly, carbon dioxide (CO_2). The CO_2 content of the atmosphere is currently around 0.039% = 390 ppm (parts per million), but it varies with the seasons by about 6–7 ppm and has been rising since the start of the Industrial Revolution, mainly as a result of the burning of fossil fuels. Approximately 2 ppm of CO_2 are being added to the atmosphere each year.

PROBLEM
What is the mass of the Earth's atmosphere, and what is the mass of 1 ppm of atmospheric CO_2?

SOLUTION
THINK At first glance, this problem seems rather daunting, because very little information is given. However, we know that the atmospheric pressure is $1.01 \cdot 10^5$ Pa and that pressure is force per area.

SKETCH The sketch in Figure 13.13 shows a column of air with weight mg above an area A of Earth's surface. This air exerts a pressure, p, on the surface.

RESEARCH We start with the relationship between pressure and force, $p = F/A$, where the area is the surface area of Earth, $A = 4\pi R^2$, and $R = 6370$ km is the radius of Earth. For the force, we can use the atmospheric weight, $F = mg$, where m is the mass of the atmosphere.

SIMPLIFY We combine the equations just mentioned
$$p = \frac{F}{A} = \frac{mg}{4\pi R^2}$$
and solve for the mass of the atmosphere
$$m = \frac{4\pi R^2 p}{g}.$$

CALCULATE We substitute the numerical values:
$$m = 4\pi(6.37 \cdot 10^6 \text{ m})^2(1.01 \cdot 10^5 \text{ Pa})/(9.81 \text{ m/s}^2) = 5.24978 \cdot 10^{18} \text{ kg}.$$

ROUND We round to three significant figures and obtain
$$m = 5.25 \cdot 10^{18} \text{ kg}.$$

DOUBLE-CHECK In order to obtain the mass of 1 ppm of CO_2 in the atmosphere, we have to realize that the molar mass of CO_2 is $12 + (2 \cdot 16) = 44$ g. The average mass of a mole of the atmosphere is approximately $0.78(2 \cdot 14) + 0.21(2 \cdot 16) + 0.01(40) = 28.96$ g. The mass of 1 ppm of CO_2 in the atmosphere is therefore
$$m_{1 \text{ ppm } CO_2} = 10^{-6} \cdot m \frac{44}{28.96} = 7.97 \cdot 10^{12} \text{ kg} = 8.0 \text{ billion tons}.$$

Humans add approximately 2 ppm of CO_2 to the atmosphere each year by burning fossil fuels, which amounts to approximately 16 billion tons of CO_2, a scary number. It is not easy to double-check the orders of magnitude for this calculation. However, data published by the U.S. Energy Information Administration show that total carbon dioxide emissions from burning fossil fuels are currently approximately 30 billion tons per year, higher than our result by a factor of 2. Where does the other half of the CO_2 go? Mainly, it dissolves in the Earth's oceans.

Examples

Briefer **Examples** (problem statement and solution only) focus on a specific point or concept. The Examples also serve as a bridge between fully worked-out Solved Problems (with all seven steps) and the homework problems.

EXAMPLE 18.9 **Estimate of Earth's Internal Thermal Energy**

Since Earth's core and mantle are at very high temperatures relative to its surface, there must be a lot of thermal energy available inside Earth.

PROBLEM

What is the thermal energy stored in Earth's interior?

SOLUTION

Obviously, we can make only a rough estimate, because the exact radial temperature profile of Earth is not known. Let's assume an average temperature of 3000 K, which is approximately half of the difference between the surface and core temperatures.

The specific heats (see Table 18.1) for the materials in the Earth's interior range from 0.45 kJ/(kg K) for iron to 0.92 kJ/(kg K) for rocks in the crust. In order to make our estimate, we will use an average value of 0.7 kJ/(kg K). The total mass of Earth is (see Table 12.1) $5.97 \cdot 10^{24}$ kg.

Inserting the numbers into equation 18.12, we find

$$Q_{\text{Earth}} = m_{\text{Earth}} c \Delta T = \left(6 \cdot 10^{24}\,\text{kg}\right)\left[0.7\,\text{kJ}/\left(\text{kg K}\right)\right]\left(3000\,\text{K}\right) = 10^{31}\,\text{J}.$$

Does it matter that some part of Earth's core is liquid and not solid? Should we account for the latent heat of fusion in our estimate? The answer is yes, in principle, but since the latent heat of fusion for metals is typically on the order of a few hundred kilojoules per kilogram, it would contribute only 10–20% of what the specific heat does in this case. For our order-of-magnitude estimate, we can safely neglect this contribution.

Problem-Solving Guidelines

Located before the end-of-chapter exercise sets, **Problem-Solving Guidelines** summarize important skills or techniques that can help you solve problems related to the material in the chapter. Acknowledging that physics is based on a core set of fundamental ideas that can be applied to various situations and problems, *University Physics* emphasizes that there is no single way to solve every problem and helps you think critically about the most effective problem-solving method before beginning to work on a solution.

PROBLEM-SOLVING GUIDELINES: NEWTON'S LAWS

Analyzing a situation in terms of forces and motion is a vital skill in physics. One of the most important techniques is the proper application of Newton's laws. The following guidelines can help you solve mechanics problems in terms of Newton's three laws. These are part of the seven-step strategy for solving all types of physics problems and are most relevant to the Sketch, Think, and Research steps.

1. An overall sketch can help you visualize the situation and identify the concepts involved, but you also need a separate free-body diagram for each object to identify which forces act on that particular object and no others. Drawing correct free-body diagrams is the key to solving all problems in mechanics, whether they involve static (nonmoving) objects or kinetic (moving) ones. Remember that the $m\vec{a}$ from Newton's Second Law should not be included as a force in any free-body diagram.

2. Choosing the coordinate system is important—often the choice of coordinate system makes the difference between very simple equations and very difficult ones. Placing an axis along the same direction as an object's acceleration, if there is any, is often very helpful. In a statics problem, orienting an axis along a surface, whether horizontal or inclined, is often useful. Choosing the most advantageous coordinate system is an acquired skill gained through experience as you work many problems.

3. Once you have chosen your coordinate directions, determine whether the situation involves acceleration in either direction. If no acceleration occurs in the y-direction, for example, then Newton's First Law applies in that direction, and the sum of forces (the net force) equals zero. If acceleration does occur in a given direction, for example, the x-direction,

then Newton's Second Law applies in that direction, and the net force equals the object's mass times its acceleration.

4. When you decompose a force vector into components along the coordinate directions, be careful about which direction involves the sine of a given angle and which direction involves the cosine. Do not generalize from past problems and think that all components in the x-direction involve the cosine; you will find problems where the x-component involves the sine. Rely instead on clear definitions of angles and coordinate directions and the geometry of the given situation. Often the same angle appears at different points and between different lines in a problem. This usually results in similar triangles, often involving right angles. If you create a sketch of a problem with a general angle θ, try to use an angle that is not close to 45°, because it is hard to distinguish between such an angle and its complement in your sketch.

5. Always check your final answer. Do the units make sense? Are the magnitudes reasonable? If you change a variable to approach some limiting value, does your answer make a valid prediction about what happens? Sometimes you can estimate the answer to a problem by using order-of-magnitude approximations, as discussed in Chapter 1; such an estimate can often reveal whether you made an arithmetical mistake or wrote down an incorrect formula.

6. The friction force always opposes the direction of motion and acts parallel to the contact surface; the static friction force opposes the direction in which the object would move, if the friction force were not present. Note that the kinetic friction force is *equal* to the product of the coefficient of friction and the normal force, whereas the static friction force is *less than or equal* to that product.

End-of-Chapter Questions and Exercise Sets

Along with providing problem-solving guidelines, examples, and strategies, *University Physics* also offers a **wide variety of end-of-chapter Questions and Exercises.** Included in each chapter are Multiple-Choice Questions, Conceptual Questions, Exercises (by section), Additional Exercises (no section "clue"), and Multi-Version Exercises. One bullet identifies slightly more challenging Exercises, and two bullets identify the most challenging Exercises.

Calculus Primer

Since this course is typically taken in the first year of study at universities, this book assumes knowledge of high school physics and mathematics. It is preferable that students have had a course in calculus before they start this course, but calculus can also be taken in parallel. To facilitate this, the text contains a short calculus primer in an appendix, giving the main results of calculus without the rigorous derivations.

Building Conceptual Understanding

Chapter Opening Outline

At the beginning of each chapter is an outline presenting the section heads within the chapter. The outline also includes the titles of the Examples and Solved Problems found in the chapter. At a quick glance, you will know if a desired topic, example, or problem is in the chapter.

What We Will Learn / What We Have Learned

Each chapter of *University Physics* is organized like a good research seminar. It was once said, "Tell them what you will tell them, then tell them, and then tell them what you told them!" Each chapter starts with **What We Will Learn**—a quick summary of the main points, without any equations. And at the end of each chapter, **What We Have Learned/Exam Study Guide** contains key concepts, including major equations.

WHAT WE WILL LEARN

- An electric field represents the electric force at different points in space.
- Electric field lines represent the net force vectors exerted on a unit positive electric charge. They originate on positive charges and terminate on negative charges.
- The electric field of a point charge is radial, proportional to the charge, and inversely proportional to the square of the distance from the charge.
- An electric dipole consists of a positive charge and a negative charge of equal magnitude.
- The electric flux is the electric field component normal to an area times the area.
- Gauss's Law states that the electric flux through a closed surface is proportional to the net electric charge

enclosed within the surface. This law provides simple ways to solve seemingly complicated electric field problems.
- The electric field inside a conductor is zero.
- The magnitude of the electric field due to a uniformly charged, infinitely long wire varies as the inverse of the perpendicular distance from the wire.
- The electric field due to an infinite sheet of charge does not depend on the distance from the sheet.
- The electric field outside a spherical distribution of charge is the same as the field of a point charge with the same total charge located at the sphere's center.

Conceptual Introductions

Conceptual explanations are provided in the text prior to any mathematical explanations, formulas, or derivations in order to establish why the concept or quantity is needed, why it is useful, and why it must be defined accurately. The authors then move from the conceptual explanation and definition to a formula and exact terms.

Self-Test Opportunity 5.3

A block is hanging vertically from a spring at the equilibrium displacement. The block is then pulled down a bit and released from rest. Draw the free-body diagram for the block in each of the following cases:

a) The block is at the equilibrium displacement.

b) The block is at its highest vertical point.

c) The block is at its lowest vertical point.

Self-Test Opportunities

In each chapter, a series of questions focus on major concepts within the text to encourage students to develop an internal dialogue. These questions will help students think critically about what they have just read, decide whether they have a grasp of the concept, and develop a list of follow-up questions to ask in class. The answers to the Self-Tests are found at the end of each chapter.

Concept Checks

Concept Checks are designed to be used with personal response system technology. They will appear in the text so that you may begin contemplating the concepts. Answers will only be available to instructors.

Concept Check 25.8

Three light bulbs are connected in series with a battery that delivers a constant potential difference, V_{emf}. When a wire is connected across light bulb 2 as shown in the figure, light bulbs 1 and 3

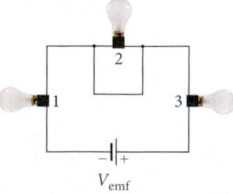

V_{emf}

a) burn just as brightly as they did before the wire was connected.

b) burn more brightly than they did before the wire was connected.

c) burn less brightly than they did before the wire was connected.

d) go out.

Student Solutions Manual

The *Student Solutions Manual* contains answers and worked-out solutions to selected end-of-chapter Questions and Exercises (those indicated by a blue number). Worked-out solutions for all items in Chapters 1 through 13 follow the complete seven-step problem-solving method introduced in Section 1.5. Chapters 14 through 40 continue to use the seven-step method for challenging (one bullet) and most challenging (two bullet) exercises, but present more abbreviated solutions for the less challenging (no bullet) exercises.

Seeing the Big Picture

Contemporary Examples

The authors have included recent physics research results throughout the text. Results involving renewable energy, the environment, aerospace, engineering, medicine, and technology demonstrate that physics is an exciting, thriving, and intellectually stimulating field. Available online at www.mhhe.com/bauerwestfall2e, the student resource center provides a number of items to enhance your understanding and help you prepare for lectures, labs, and tests.

ConnectPlus eBook

Linked to multimedia assets—including author videos, applets that allow you to explore fundamental physics principles, and images—the eBook allows you to take notes, highlight, and even search for specific words or phrases. All of the textbook figures, videos, and interactive content are also listed in line and by chapter, so you can navigate directly to the resource you need. Links to the ConnectPlus eBook are included in the online homework and LearnSmart assignments, so if you are having trouble with an exercise or concept, you can navigate directly to the relevant portion of the text.

Visual Program

Familiarity with graphics and animation on the Internet and in video games has raised the bar for the graphical presentations in textbooks, which must now be more sophisticated to excite both students and faculty. Here are some of the techniques and ideas implemented in *University Physics*:

- Line drawings are superimposed on photographs to connect sometimes very abstract physics concepts to students' realities and everyday experiences.

- A three-dimensional look for line drawings adds plasticity to the presentations. Mathematically accurate graphs and plots were created by the authors in software programs such as Mathematica and then used by the graphic artists to ensure complete accuracy as well as a visually appealing style.

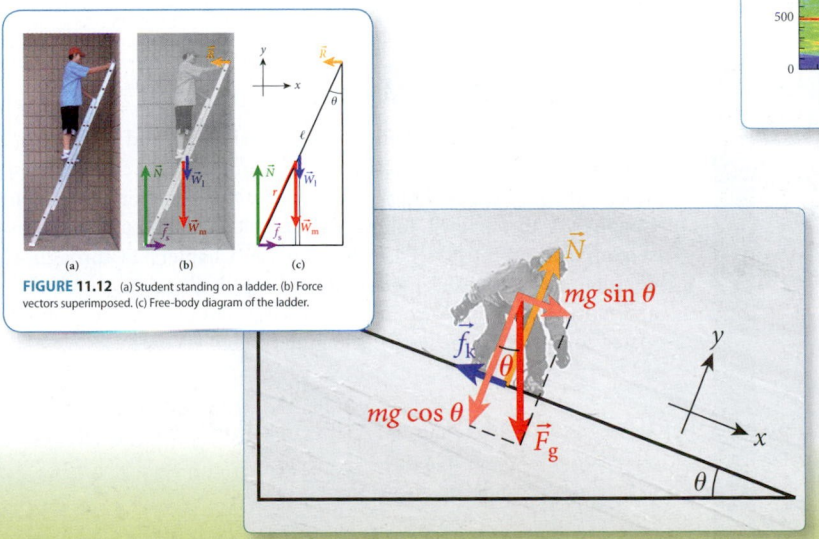

FIGURE 11.12 (a) Student standing on a ladder. (b) Force vectors superimposed. (c) Free-body diagram of the ladder.

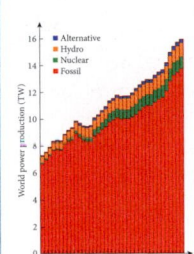
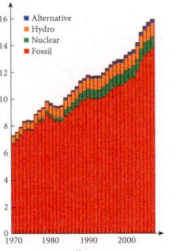

SOLVED PROBLEM 5.3 Wind Power

The total power consumption of all humans combined is approximately 16 TW ($1.6 \cdot 10^{13}$ W), and it is expected to double during the next 15 to 20 years. Almost 90% of the power produced comes from fossil fuels; see Figure 5.20. Since the burning of fossil fuels is currently adding more than 10 billion tons of carbon dioxide to Earth's atmosphere per year, it is not clear how much longer this mode of power generation is sustainable. Other sources of power, such as wind, have to be considered. Some huge wind farms have been constructed (see Figure 5.21), and many more are under development.

PROBLEM
How much average power is contained in wind blowing at 10.0 m/s across the rotor of a large wind turbine, such as the Enercon E-126, which has a hub height of 135 m and a rotor radius of 63 m?

SOLUTION
THINK Since the wind speed is given, we can calculate the kinetic energy of the amount of air blowing across the rotor's surface. If we can calculate how much air moves across the rotor per unit of time, then we can calculate the power as the ratio of the kinetic energy of the air to the time interval.

SKETCH The rotor surface is a circle, and we can assume that the wind blows perpendicular to it, because the turbines in wind farms are oriented so that that is the case. Indicated in the sketch (Figure 5.22) is the cylindrical volume of air moving across the rotor per time interval.

FIGURE 5.20 Worldwide power production as a function of time for different power sources.

RESEARCH Earlier in this chapter, we learned that the kinetic energy is given by $E = \frac{1}{2}mv^2$; here, m is the mass of air, and v is the wind speed. A very handy rule of thumb is that 1 m^3 of air has a mass of 1.20 kg at sea level and room temperature. The average power is given by $P = W/\Delta t$, and the work is related to the change in kinetic energy through the work–kinetic energy theorem $W = \Delta K$.

We can thus write, for the average power of the wind moving across the rotor of the wind turbine,

$$ P = \frac{W}{\Delta t} = \frac{\Delta K}{\Delta t} = \frac{\Delta\left(\frac{1}{2}mv^2\right)}{\Delta t} = \frac{1}{2}v^2\frac{\Delta m}{\Delta t}. $$

In the last step, we have assumed that the wind speed is constant and does not change.

What is Δm? We know that density is mass/volume, and so we can write $\Delta m = \rho \Delta V$, where $\rho = 1.20$ kg/m^3 is the air density and ΔV is the volume of air moved across the rotor per unit of time. Here ΔV is a cylinder with length $l = v\Delta t$ and base area A = area of the rotor (see Figure 5.22), v is again the wind speed, and the area is the area of a circle, $A = \pi R^2$.

FIGURE 5.21 Large-scale wind farm producing power.

SIMPLIFY Now we are ready to insert our expressions for Δm and ΔV into our equation for the average power:

$$ P = \frac{1}{2}v^2\frac{\Delta m}{\Delta t} = \frac{1}{2}v^2\frac{\rho\Delta V}{\Delta t} = \frac{1}{2}v^2\frac{\rho Al}{\Delta t} = \frac{1}{2}v^2\frac{\rho(\pi R^2)(v\Delta t)}{\Delta t} = \frac{1}{2}v^3\rho\pi R^2. $$

We see that the average wind power is proportional to the cube of the wind speed!

CALCULATE Inserting the given numbers for the rotor's radius, the wind speed, and the air density yields

$$ P = \frac{1}{2}(10.0 \text{ m/s})^3(1.2 \text{ kg/m}^3)\pi(63 \text{ m})^2 = 7.481389 \cdot 10^6 \text{ kg m}^2/\text{s}^3. $$

ROUND Since the rotor's radius was given to only two significant figures, we round the final result to the same number of significant figures. So our answer is 7.5 MW.

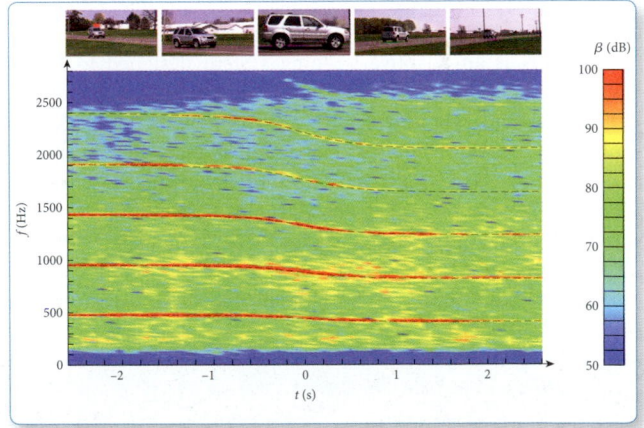

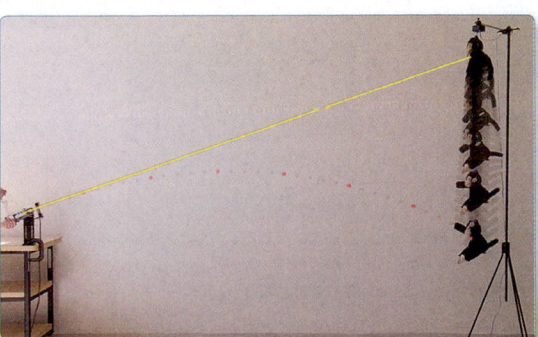

Student Resources

Online Resources

Available online at www.mhhe.com/bauerwestfall2e, the student resource center provides a number of items to enhance your study. A full suite of author-provided applets is available to help you visualize physics concepts that are presented throughout the book—from vectors and kinematics to quantum and nuclear physics. Interactive simulations allow you to simulate real experiments, while viewing data in real time, thereby linking concepts and principles you have just learned to real, quantifiable results. Videos illustrating important results are also available, giving you another opportunity to see the dynamics of physical situations.

McGraw-Hill LearnSmart™

McGraw-Hill LearnSmart™ is available as an integrated feature of McGraw-Hill Connect® Physics or as a stand-alone product. LearnSmart is an adaptive learning system designed to help you learn faster, study more efficiently, and retain more knowledge for greater success. LearnSmart assesses your knowledge of the course content through a series of adaptive questions, meaning that you will not waste time studying topics you already know. It is designed to pinpoint concepts you do not understand and map out a personalized study plan for success. LearnSmart gives you a tool to help guide your studies, so you can study more effectively and retain more knowledge. Visit the following site for a demonstration: www.mhlearnsmart.com.

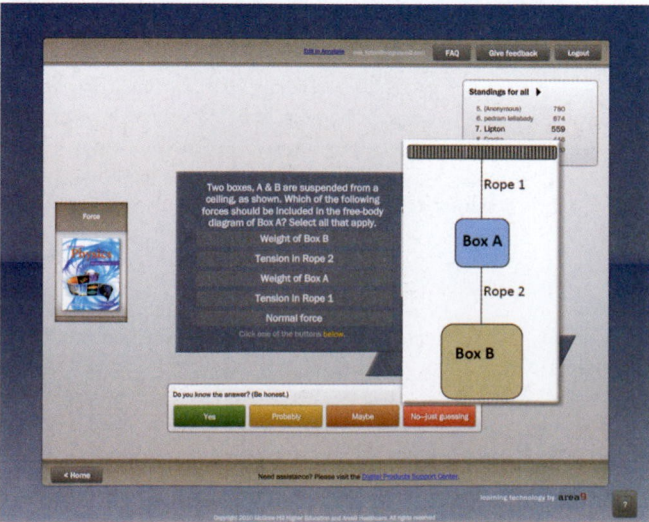

McGraw-Hill Connect® Physics

McGraw-Hill Connect® Physics provides a single online source for your class. In addition to online homework, you can find resources on the site including study modules, communication from instructors, and course information. LearnSmart study modules are available for each chapter to help guide your studying and review for quizzes and exams.

McGraw-Hill ConnectPlus® Physics

McGraw-Hill ConnectPlus® Physics provides all of the benefits of Connect, plus a fully interactive eBook. Links to applicable student resources—including applets, video, and assessments—appear in line so you can follow them without losing your place in the text. All of the textbook figures, videos, and interactive content are also listed by chapter, so you can navigate directly to the resource you need. You can highlight, take notes, and even search the eBook for specific words or phrases. Links to the ConnectPlus eBook are included in the online homework and LearnSmart assignments, so if you are having trouble with an exercise or concept, you can navigate directly to the relevant portion of the text.

Student Solutions Manual

The *Student Solutions Manual* contains answers and worked-out solutions to selected end-of-chapter Questions and Exercises (those indicated by a blue number). Worked-out solutions for all items in Chapters 1 through 13 follow the complete seven-step problem-solving method introduced in Section 1.5. Chapters 14 through 40 continue to use the seven-step method for challenging (one bullet) and most challenging (two bullet) exercises, but present more abbreviated solutions for the less challenging (no bullet) exercises.

Resources for Instructors

Online Tools

A collection of online tools—including photos, artwork, an *Instructor's Solutions Manual,* and other media—can be accessed from the *University Physics* website at www.mhhe.com/bauerwestfall2e. These tools provide content for novice and experienced instructors who teach in a variety of styles. Included in the collection are PowerPoint® slides containing full-color digital files of all illustrations in the text, a collection of digital files of photographs from the text, libraries of all the solved problems, examples, tables, and numbered equations from the text, and ready-made PowerPoint lecture outlines that include art, lecture notes, and additional examples for each section of the text. An *Instructor's Solutions Manual* with complete worked-out solutions to all of the end-of-chapter Questions and Exercises is available in document and PDF formats. The latest research in physics education shows that in-class use of personal response systems (or "clickers") improves student learning, so a full set of clicker questions based on the Concept Checks from the text is available on the companion website.

Computerized Test Bank Online

Over 2300 text questions in multiple-choice format, written by exceptional instructors and fully updated to reflect the content of the second edition, are included on the Connect website. These questions are organized by test section and represent a variety of difficulty levels. By working within the ConnectPhysics online platform, you can select questions from multiple McGraw-Hill test banks or create your own to better reflect the interests and needs of your students. This test bank allows you to easily create paper and online tests or quizzes from anywhere at any time, without installing any software.

McGraw-Hill LearnSmart™

McGraw-Hill LearnSmart™ is available as an integrated feature of McGraw-Hill Connect® Physics or as a stand-alone product for instructors who choose not to use online homework. Written by experienced professors and informed by the latest in physics education research, LearnSmart is an adaptive learning system designed to help students learn faster, study more efficiently, and retain more knowledge for greater success. LearnSmart assesses a student's knowledge of course content through a series of adaptive questions. It is designed to address common student misconceptions, pinpointing concepts a student does not understand and mapping out a personalized study plan for success. This innovative study tool also has features that allow instructors to see exactly what students have accomplished and a built-in assessment tool for graded assignments. Visit the following site for a demonstration: www.mhlearnsmart.com.

McGraw-Hill ConnectPlus® Physics

McGraw-Hill Connect® Physics provides online presentation, assignment, and assessment tools. It connects your students with the resources they'll need to achieve success. With Connect Physics, you can deliver assignments, quizzes, and tests online. A full set of online homework problems, test bank questions, LearnSmart study modules, clicker questions, animations, interactive applets, and videos are available. Each online homework problem in Connect Physics uses the text problem-solving methodology and is tailored with specific hints and feedback for all questions. As an instructor, you can edit existing questions or problems and author entirely new ones. Track individual student performance—by question, assignment, or in relation to the class overall—with detailed grade reports. Integrate grade reports easily with Learning Management Systems (LMS), such as WebCT, Desire2Learn, and Blackboard, and much more. ConnectPlus Physics provides students with all the advantages of Connect Physics, plus 24/7 online access to an eBook. This media-rich version of the book is available through the McGraw-Hill Connect platform and allows seamless integration of text, media, and assessments. To learn more, visit www.mcgrawhillconnect.com.

McGraw-Hill Create™

With McGraw-Hill Create™, you can easily rearrange chapters, combine material from other content sources, and quickly upload content you have written, such as a course syllabus or your teaching notes. Find the content you need in Create by searching through thousands of leading McGraw-Hill textbooks. Arrange your book to fit your teaching style. Create even allows you to personalize the book's appearance by selecting the cover and adding your name, school, and course information. Order a Create book and you'll receive a complimentary print review copy in 3–5 business days or a complimentary electronic review copy (eComp) via e-mail in minutes. Go to www.mcgrawhillcreate.com today and register to experience how McGraw-Hill Create empowers you to teach your students your way.

My Lectures—Tegrity®

McGraw-Hill Tegrity® records and distributes a class lecture with just a click of a button. Students can view a lecture anytime and anywhere via computer, iPod, or mobile device. Tegrity indexes as it records your PowerPoint® presentations and anything shown on your computer, so students can use keywords to find exactly what they want to study or review. Tegrity is available as an integrated feature of McGraw-Hill Connect and as a stand-alone.

Acknowledgments

A book like the one that you are holding in your hands is impossible to produce without tremendous work by an incredible number of dedicated individuals. First and foremost, we would like to thank the talented marketing and editorial team from McGraw-Hill who worked on the first two editions of the text: Marty Lange, Kent Peterson, Thomas Timp, Ryan Blankenship, Mary Hurley, Liz Recker, Daryl Bruflodt, Lisa Nicks, Dan Wallace, and, in particular, Michael Lange, Eve Lipton, Bill Welsh, Curt Reynolds, and Deb Hash helped us in innumerable ways and managed to reignite our enthusiasm after each revision. Their team spirit, good humor, and unbending optimism kept us on track and always made it fun for us to put in the seemingly endless hours it took to produce the manuscript.

The developmental editors, Richard Heinz, David Chelton, Mary Hurley, and Eve Lipton, helped us work through the near infinite number of comments and suggestions for improvements from our reviewers. They, as well as the reviewers and our board of advisors, deserve a large share of the credit for improving the quality of the final manuscript. Our colleagues on the faculty of the Department of Physics and Astronomy at Michigan State University—Alexandra Gade, Alex Brown, Bernard Pope, Carl Schmidt, Chong-Yu Ruan, C.P. Yuan, Dan Stump, Ed Brown, Hendrik Schatz, Kris Starosta, Lisa Lapidus, Michael Harrison, Michael Moore, Reinhard Schwienhorst, Salemeh Ahmad, S.B. Mahanti, Scott Pratt, Stan Schriber, Tibor Nagy, and Thomas Duguet—helped us in innumerable ways as well, teaching their classes and sections with the materials developed by us and in the process providing invaluable feedback on what worked and what needed additional refinement. We thank all of them. We also thank all of our problem contributors for sharing some of their best work with us, in particular, Richard Hallstein, who took on the task of organizing and processing all the contributions.

At the point when we turned in the final manuscript to the publisher, a whole new army of professionals took over and added another layer of refinement, which transformed a manuscript into a book. Kurt Norlin and the team at LaurelTech (a division of diacriTech) worked through each and every homework problem, each exercise, and each number and equation we wrote down. The photo researchers, in particular, Danny Meldung, improved the quality of the images used in the book immensely, and they made the selection process fun for us. The team at Precision Graphics used our original drawings but improved their quality substantially, while at the same time remaining true to our original calculations that went into producing the drawings. Our copyeditor, Jane Hoover, pulled it all together in the end, deciphered our scribbling, and made sure that the final product is as readable as possible. McGraw-Hill's design and production team of Jayne Klein, David Hash, Carrie Burger, Sandy Ludovissy, Tammy Juran, Judi David, Mary Powers, Sherry Padden, and Annette Doerr expertly guided the book and its ancillary materials through to publication. All of them deserve our tremendous gratitude.

Finally, we could not have made it through the last decade of effort without the support of our families, who had to put up with us working on the book through untold evenings, weekends, and even during many vacations. We hope that all of their patience and encouragement has paid off, and we thank them from the bottom of our hearts for sticking with us during the completion of this book.

—*Wolfgang Bauer*
—*Gary D. Westfall*

Reviewers, Advisors, and Contributors for the Second Edition

Nina Abramzon, *California State Polytechnic University–Pomona*
Royal Albridge, *Vanderbilt University*
Deepthi Amarasuriya, *Northwest College*
Michael G. Anderson, *University of California–San Diego*
Roy Arunava, *Young Harris College*
Arlette Baljon, *San Diego State University*
Charles Benesh, *Wesleyan College*
Ward Beyermann, *University of California–Riverside*
Angela Biselli, *Fairfield University*
Guy Blaylock, *University of Massachusetts*
Jeff Bodart, *Chipola College*
Kenneth Bolland, *The Ohio State University*
Bruce Bolon, *Hamline University*
Luca Bombelli, *University of Mississippi*
Gary D. Branum, *Friends University*
Charles Burkhardt, *St. Louis Community College–Florissant Valley*
Duncan Carlsmith, *University of Wisconsin–Madison*
Ryan Case, *Elizabethtown Community and Technical College*
George Caviris, *Farmingdale State College*
Soumitra Chattopadhyay, *Georgia Highlands College*
Xidong Chen, *Cedarville University*
Scott Crittenden, *University of South Carolina*
Danielle Dalafave, *The College of New Jersey*
Daniel Dale, *University of Wyoming*
Dipangkar Dutta, *Mississippi State University*
Shalom Fisher, *George Mason University*
Juhan Frank, *Louisiana State University*
Stuart Gazes, *University of Chicago*
Richard Gelderman, *Western Kentucky University*
Margaret Geppert, *William Rainey Harper College*
Awad Gerges, *University of North Carolina–Charlotte*

James Gering, *Florida Institute of Technology*
Patrick C. Gibbons, *Washington University in St. Louis*
James Gilbert, *Rose State College*
Christopher Gould, *University of Southern California*
Edwin E. Hach, III, *Rochester Institute of Technology*
Martin Hackworth, *Idaho State University*
Rob Hagood, *Washtenaw Community College*
Jim Hamm, *Big Bend Community College*
Athula Herat, *Slippery Rock University of Pennsylvania*
Scott Hildreth, *Chabot College*
Barbara Hoeling, *California State Polytechnic University–Pomona*
Richard Holland, *Southeastern Illinois College and Southern Illinois University*
George Igo, *University of California–Los Angeles*
Sai Iyer, *Washington University in St. Louis*
Howard Jackson, *University of Cincinnati*
Amin Jazaeri, *George Mason University*
Craig Jensen, *Northern Virginia Community College*
Robert Johnson, *University of California–Santa Cruz*
Scott Kennedy, *Anderson University*
Timothy Kidd, *University of Northern Iowa*
Yangsoo Kim, *Virginia Polytechnic Institute and State University*
Yeong E. Kim, *Purdue University*
Brian Koberlein, *Rochester Institute of Technology*
Zlatko Koinov, *University of Texas–San Antonio*
Ichishiro Konno, *University of Texas–San Antonio*
Ilya Kravchenko, *University of Nebraska–Lincoln*
Ratnappuli Kulasiri, *Southern Polytechnic State University*
Allen Landers, *Auburn University*
Kwan Lee, *St. Louis Community College–Meramec*
Todd R. Leif, *Cloud County Community College*
Pedram Leilabady, *University of North Carolina at Charlotte*
Steve Liebling, *Long Island University–C.W. Post*
Mike Lisa, *The Ohio State University*
Jorge Lopez, *University of Texas–El Paso*
Hong Luo, *State University of New York–Buffalo*
Kingshuk Majumdar, *Grand Valley State University*
Lloyd Makarowitz, *Farmingdale State College*
Pete Markowitz, *Florida International University*
Bruce Mason, *University of Oklahoma–Norman*
Krista McBride, *Nashville State Community College*
Arthur Meyers, *Iowa State University*
Rudi Michalak, *University of Wyoming*
Jonathan Morris, *St. Louis Community College–Forest Park*
L. Kent Morrison, *University of New Mexico*
Richard Mowat, *North Carolina State University*
Rajamani Narayanan, *Florida International University*
Peter D. Persans, *Rensselaer Polytechnic Institute*
Robert Philbin, *Trinidad State Junior College*
Michael Politano, *Marquette University*
Amy Pope, *Clemson University*
G. Albert Popson, Jr., *West Virginia Wesleyan College*
Tommy Ragland, *University of New Mexico–Taos*
Roberto Ramos, *Drexel University*
Bea Rasmussen, *University of Texas at Dallas*
David Reid, *University of Chicago*
Adam Rengstorf, *Purdue University–Calumet*
Mark Rupright, *Birmingham-Southern College*
Gina Sorci, *Hinds Community College*
Ben Shaevitz, *Slippery Rock University of Pennsylvania*

Ananda Shastri, *Minnesota State University–Moorhead*
Marllin Simon, *Auburn University*
David Slimmer, *Lander University*
Leigh Smith, *University of Cincinnati*
Steffen Strauch, *University of South Carolina*
Eugene Surdutovich, *Oakland University*
John T. Tarvin, *Samford University*
Michael Thackston, *Southern Polytechnic State University*
Cynthia Trevisan, *California Maritime Academy of the California State University*
Marian Tzolov, *Lock Haven University of Pennsylvania*
Christos Velissaris, *University of Central Florida*
David Verdonck, *Elmira College*
Margaret Wessling, *L.A. Pierce College*
Joseph Wiest, *West Virginia Wesleyan College*
Weldon J. Wilson, *University of Central Oklahoma*
Matthew Wood, *Florida Institute of Technology*
Scott Yost, *The Citadel*
Todd Young, *Wayne State College*

Advisors, Contributors, Testers, and Reviewers for the First Edition

El Hassan El Aaoud, *University of Hail, Hail KSA*
Mohamed S. Abdelmonem, *King Fahd University of Petroleum and Minerals, Dhahran, Saudi Arabia*
Nina Abramzon, *California State Polytechnic University–Pomona*
Edward Adelson, *Ohio State University*
Mohan Aggarwal, *Alabama A&M University*
Salemeh Ahmad, *Michigan State University*
Albert Altman, *UMASS Lowell*
Paul Avery, *University of Florida*
David T. Bannon, *Oregon State University*
Marco Battaglia, *UC Berkeley and LBNL*
Rene Bellweid, *Wayne State University*
Douglas R. Bergman, *Rutgers, the State University of New Jersey*
Carlos Bertulani, *Texas A&M University–Commerce*
Luca Bertello, *University of California–Los Angeles*
Peter Beyersdorf, *San Jose State University*
Sudeb Bhattacharya, *Saha Institute of Nuclear Physics, Kolkata, India*
Helmut Biritz, *Georgia Institute of Technology*
Ken Thomas Bolland, *Ohio State University*
Richard Bone, *Florida International University*
Dieter Brill, *University of Maryland–College Park*
Alex Brown, *Michigan State University*
Ed Brown, *Michigan State University*
Jason Brown, *Clemson University*
Ronald Brown, *California Polytechnic University–San Luis Obispo*
Branton J. Campbell, *Brigham Young University*
Duncan Carlsmith, *University of Wisconsin–Madison*
Neal Cason, *University of Notre Dame*
John Cerne, *State University of New York–Buffalo*
Ralph Chamberlain, *Arizona State University*
K. Kelvin Cheng, *Texas Tech University*
Chris Church, *Miami University of Ohio–Oxford*
Eugenia Ciocan, *Clemson University*
Robert Clare, *University of California–Riverside*
Roy Clarke, *University of Michigan*
J. M. Collins, *Marquette University*

Brent A. Corbin, *University of California–Los Angeles*
Stephane Coutu, *The Pennsylvania State University*
William Dawicke, *Milwaukee School of Engineering*
Mike Dennin, *University of California–Irvine*
John Devlin, *University of Michigan–Dearborn*
John DiNardo, *Drexel University*
Fivos R. Drymiotis, *Clemson University*
Mike Dubson, *University of Colorado–Boulder*
Thomas Duguet, *Michigan State University*
Michael DuVernois, *University of Hawaii–Manoa*
David Elmore, *Purdue University*
Robert Endorf, *University of Cincinnati*
David Ermer, *Mississippi State University*
Harold Evensen, *University of Wisconsin–Platteville*
Lisa L. Everett, *University of Wisconsin–Madison*
Gus Evrard, *University of Michigan*
Michael Famiano, *Western Michigan University*
Frank Ferrone, *Drexel University*
Leonard Finegold, *Drexel University*
Ray Frey, *University of Oregon*
Alexandra Gade, *Michigan State University*
J. William Gary, *University of California–Riverside*
Stuart Gazes, *University of Chicago*
Chris Gould, *University of Southern California*
Benjamin Grinstein, *University of California–San Diego*
John B. Gruber, *San Jose State University*
Shi-Jian Gu, *The Chinese University of Hong Kong, Shatin, N.T., Hong Kong*
Edwin E. Hach, III, *Rochester Institute of Technology*
Nasser M. Hamdan, *The American University of Sharjah*
John Hardy, *Texas A&M University*
Kathleen A. Harper, *Denison University*
David Harrison, *University of Toronto*
Michael Harrison, *Michigan State University*
Richard Heinz, *Indiana University*
Satoshi Hinata, *Auburn University*
Laurent Hodges, *Iowa State University*
John Hopkins, *The Pennsylvania State University*
George K. Horton, *Rutgers University*
T. William Houk, *Miami University–Ohio*
Eric Hudson, *Massachusetts Institute of Technology*
Yung Huh, *South Dakota State University*
Moustafa Hussein, *Arab Academy for Science & Engineering, Egypt*
A. K. Hyder, *University of Notre Dame*
David C. Ingram, *Ohio University–Athens*
Diane Jacobs, *Eastern Michigan University*
A.K. Jain, *I.I.T. Roorkee*
Rongying Jin, *The University of Tennessee–Knoxville*
Kate L. Jones, *University of Tennessee*
Steven E. Jones, *Brigham Young University*
Teruki Kamon, *Texas A&M University*
Lev Kaplan, *Tulane University*
Joseph Kapusta, *University of Minnesota*
Kathleen Kash, *Case Western Reserve*
Sanford Kern, *Colorado State University*
Eric Kincanon, *Gonzaga University*
Elaine Kirkpatrick, *Rose-Hulman Institute of Technology*
Carsten Knudsen, *Technical University of Denmark*
Brian D. Koberlein, *Rochester Institute of Technology*
W. David Kulp, III, *Georgia Institute of Technology*
Ajal Kumar, *The University of the South Pacific–Fiji*

Fred Kuttner, *University of California–Santa Cruz*
David Lamp, *Texas Tech University*
Lisa Lapidus, *Michigan State University*
André LeClair, *Cornell University*
Patrick R. LeClair, *University of Alabama*
Luis Lehner, *Louisiana State University–Baton Rouge*
Pedram Leilabady, *University of North Carolina–Charlotte*
Michael Lisa, *The Ohio State University*
M.A.K. Lodhi, *Texas Tech University*
Samuel E. Lofland, *Rowan University*
Jerome Long, *Virginia Tech*
S. B. Mahanti, *Michigan State University*
A. James Mallmann, *Milwaukee School of Engineering*
Pete Markowitz, *Florida International University*
Daniel Marlow, *Princeton University*
Bruce Mason, *Oklahoma University*
Martin McHugh, *Loyola University*
Michael McInerney, *Rose-Hulman Institute of Technology*
David McIntyre, *Oregon State University*
Bruce Mellado, *University of Wisconsin–Madison*
Marina Milner-Bolotin, *Ryerson University–Toronto*
C. Fred Moore, *University of Texas–Austin*
Michael Moore, *Michigan State University*
Jeffrey Morgan, *University of Northern Iowa*
Kieran Mullen, *University of Oklahoma*
Nazir Mustapha, *Al-Imam University*
Charley Myles, *Texas Tech University*
Tibor Nagy, *Michigan State University*
Reza Nejat, *McMaster University*
Curt Nelson, *Gonzaga University*
Mark Neubauer, *University of Illinois at Urbana–Champaign*
Cindy Neyer, *Tri-State University*
Craig Ogilvie, *Iowa State University*
Bradford G. Orr, *The University of Michigan*
Karur Padmanabhan, *Wayne State University*
Kiumars Parvin, *San Jose State University*
Jacqueline Pau, *University of California–Los Angeles*
Todd Pedlar, *Luther College*
Leo Piilonen, *Virginia Tech*
Amy Pope, *Clemson University*
Bernard Pope, *Michigan State University*
K. Porsezian, *Pondicherry University–Puducherry*
Scott Pratt, *Michigan State University*
Earl Prohofsky, *Purdue University*
Claude Pruneau, *Wayne State University*
Wang Qing-hai, *National University of Singapore*
Corneliu Rablau, *Kettering University*
Johann Rafelski, *University of Arizona*
Kenneth J. Ragan, *McGill University*
Roberto Ramos, *Drexel University*
Ian Redmount, *Saint Louis University*
Lawrence B. Rees, *Brigham Young University*
Andrew J. Rivers, *Northwestern University*
James W. Rohlf, *Boston University*
Philip Roos, *University of Maryland*
Chong-Yu Ruan, *Michigan State University*
Dubravka Rupnik, *Louisiana State University*
Homeyra Sadaghiani, *California State Polytechnic University–Pomona*
Ertan Salik, *California State Polytechnic University–Pomona*
Otto Sankey, *Arizona State University*
Sergey Savrasov, *University of California–Davis*

Hendrik Schatz, *Michigan State University*
Carl Schmidt, *Michigan State University*
Stan Schriber, *Michigan State University*
John Schroeder, *Rensselaer Polytechnic Institute*
Reinhard Schwienhorst, *Michigan State University*
Kunnat Sebastian, *University of Massachusetts–Lowell*
Bjoern Seipel, *Portland State University*
Jerry Shakov, *Tulane University*
Ralph Shiell, *Trent University*
Irfan Siddiqi, *University of California–Berkeley*
Ravindra Kumar Sinha, *Delhi College of Engineering*
Marllin L. Simon, *Auburn University*
Alex Small, *California State Polytechnic University–Pomona*
Leigh Smith, *University of Cincinnati*
Todd Smith, *University of Dayton*
Xian-Ning Song, *Richland College*
Jeff Sonier, *Simon Fraser University–Surrey Central*
Chad E. Sosolik, *Clemson University*
Kris Starosta, *Michigan State University*
Donna W. Stokes, *University of Houston*
James Stone, *Boston University*
Michael G. Strauss, *University of Oklahoma*
Dan Stump, *Michigan State University*
Yang Sun, *University of Notre Dame*

Stephen Swingle, *City College of San Francisco*
Maarij Syed, *Rose-Hulman Institute of Technology*
Gregory Tarlé, *University of Michigan*
Marshall Thomsen, *Eastern Michigan University*
Douglas C. Tussey, *The Pennsylvania State University*
Somdev Tyagi, *Drexel University*
Prem Vaishnava, *Kettering University*
Erich W. Varnes, *University of Arizona*
John Vasut, *Baylor University*
Gautam Vemuri, *Indiana University-Purdue University–Indianapolis*
Thad Walker, *University of Wisconsin–Madison*
Fuqiang Wang, *Purdue University*
David J. Webb, *University of California–Davis*
Kurt Wiesenfeld, *Georgia Tech*
Fred Wietfeldt, *Tulane University*
Gary Williams, *University of California–Los Angeles*
Sun Yang, *University of Notre Dame*
L. You, *Georgia Tech*
Billy Younger, *College of the Albemarle*
C. P. Yuan, *Michigan State University*
Andrew Zangwill, *Georgia Institute of Technology*
Jens Zorn, *University of Michigan–Ann Arbor*
Michael Zudov, *University of Minnesota*

New to the Second Edition

General Changes

The content of the second edition of *University Physics* has been completely updated to reflect new research in physics and physics education. The changes range from the inclusion of newly discovered elements in the periodic table to the addition of new in-text Concept Checks and end-of-chapter exercises. New Multi-Version Exercises have been added to the end of every chapter. These consist of groups of related exercises that use a common problem setup but ask you to solve for a different quantity. The Multi-Version Exercises will help you to build conceptual understanding, learn how different physical quantities are related to one another, and recognize related problems when you see them again.

Chapter-Specific Changes

Chapter 1

Two new Concept Checks and one new Self-Test Opportunity have been added, as well as Example 1.4, "Greenhouse Gas Production." There are six new Multiple-Choice Questions and thirteen new end-of-chapter exercises comprising section-specific, general, and Multi-Version Exercises.

Chapter 2

Five new Concept Checks have been added. There are four new Multiple-Choice Questions and eleven new end-of-chapter exercises comprising section-specific, general, and Multi-Version Exercises.

Chapter 3

Five new Concept Checks have been added. There are three new Multiple-Choice Questions and seven new Multi-Version Exercises.

Chapter 4

One new Concept Check and two new Self-Test Opportunities have been added. Solved Problem 4.2, "Two Blocks Connected by a Rope," has replaced the related example from the first edition. There are four new Multiple-Choice Questions and ten new Multi-Version Exercises.

Chapter 5

Two new Concept Checks have been added. Solved Problems 5.3, "Wind Power," and 5.4, "Riding a Bicycle," are new. There are four new Multiple-Choice Questions and fourteen new end-of-chapter exercises comprising general and Multi-Version Exercises.

Chapter 6

Three new Concept Checks have been added and Example 6.1, "Weightlifting," is new. There are four new Multiple-Choice Questions and nine new Multi-Version Exercises.

Chapter 7

One new Concept Check has been added. There are four new Multiple-Choice Questions and thirteen new end-of-chapter exercises comprising general and Multi-Version Exercises.

Chapter 8

Three new Concept Checks have been added. There are five new Multiple-Choice Questions and eleven new Multi-Version Exercises.

Chapter 9

Four new Concept Checks have been added. There are four new Multiple-Choice Questions and seven new Multi-Version Exercises.

Chapter 10

Three new Concept Checks have been added. Solved Problem 10.5, "Bullet Hitting a Pole," is new. There are three new Multiple-Choice Questions and ten new end-of-chapter exercises comprising Conceptual Questions and Multi-Version Exercises.

Chapter 11

Three new Concept Checks have been added. Example 11.3, "Standing on a Board," is new. There are four new Multiple-Choice Questions and eight new Multi-Version Exercises.

Chapter 12

Three new Concept Checks have been added. There are four new Multiple-Choice Questions and eight new Multi-Version Exercises.

Chapter 13

Four new Concept Checks and one new Self-Test Opportunity have been added. Example 13.7, "Betz Limit," and Solved Problem 13.2, "Weighing Earth's Atmosphere," are new. There are four new Multiple-Choice Questions and nine new Multi-Version Exercises.

Chapter 14

Five new Concept Checks and one new Self-Test Opportunity have been added. Solved Problem 14.4, "Grandfather Clock," is new. There are four new Multiple-Choice Questions and eight new Multi-Version Exercises.

Chapter 15

Two new Concept Checks have been added. There are three new Multiple-Choice Questions and nine new Multi-Version Exercises.

Chapter 16

Three new Concept Checks have been added. Solved Problem 16.2, "Tuning a Violin," is new. There are four new Multiple-Choice Questions and eight new Multi-Version Exercises. New subsections discuss sound attenuation, sound diffraction, and sound localization.

Chapter 17

Solved Problem 17.1, "Temperature Conversion," and a new Concept Check have been added. There are four new Multiple-Choice Questions and nine new Multi-Version Exercises.

Chapter 18

Five new Concept Checks have been added. Example 18.9, "Estimate of Earth's Internal Thermal Energy," and Solved Problem 18.5, "Enhanced Geothermal System," are new. A subsection on geothermal power resources has been added. There are six new Multiple-Choice Questions and six new Multi-Version Exercises.

Chapter 19

Two new Self-Test Opportunities have been added. Solved Problem 19.2, "Home Energy Storage," is new. A new section on real gases and a new subsection on compressed air have been added. There are four new Multiple-Choice Questions and six new Multi-Version Exercises.

Chapter 20

There are four new Multiple-Choice Questions and nine new Multi-Version Exercises.

Chapter 21

One new Concept Check, four new Multiple-Choice Questions, and six new Multi-Version Exercises have been added. A new subsection discusses triboelectric charging.

Chapter 22

Six new Concept Checks have been added. There are two new Multiple-Choice Questions and six new Multi-Version Exercises.

Chapter 23

Two new Concept Checks have been added. Solved Problem 23.4, "Charged Disk," is new. A new subsection covers the case of a dipole in a constant electric field. There are four new Multiple-Choice Questions and six new Multi-Version Exercises.

Chapter 24

In addition to a new subsection on electrolytic capacitors, there are four new Multiple-Choice Questions and six new Multi-Version Exercises.

Chapter 25

Two new Concept Checks and a new Self-Test Opportunity have been added. There are two new Multiple-Choice Questions and eight new Multi-Version Exercises.

Chapter 26

Three new Concept Checks have been added. There are four new Multiple-Choice Questions and nine new Multi-Version Exercises.

Chapter 27

Three new Concept Checks have been added. There are four new Multiple-Choice Questions and six new Multi-Version Exercises.

Chapter 28

There are four new Multiple-Choice Questions and six new Multi-Version Exercises.

Chapter 29

Three new Concept Checks have been added. There are six new Multiple-Choice Questions and six new Multi-Version Exercises.

Chapter 30

There are two new Multiple-Choice Questions and five new end-of-chapter exercises comprising section-specific and Multi-Version Exercises.

Chapter 31

There are two new Multiple-Choice Questions and six new Multi-Version Exercises.

Chapter 32

Six new Concept Checks and a new Self-Test Opportunity have been added. There are six new Multiple-Choice Questions and eight new Multi-Version Exercises. A new subsection discusses concentrating solar power.

Chapter 33

Seven new Concept Checks and a new Self-Test Opportunity have been added. There are six new Multiple-Choice Questions and six new Multi-Version Exercises. The section on converging and diverging lenses has been extensively revised, and a new subsection on lens equations has been added.

Chapter 34

Four new Concept Checks have been added. There are four new Multiple-Choice Questions and eight new Multi-Version Exercises.

Chapter 35

One new Concept Check has been added. There are six new Multiple-Choice Questions and six new Multi-Version Exercises.

Chapter 36

An example on Compton scattering has been replaced by Solved Problem 36.1, "Compton Scattering," with a complete seven-step solution. There are eight new Multi-Version Exercises.

Chapter 37

Improved notation for quantum operators and new material on wave functions have been added. An example concerning a finite potential well has been replaced by Solved Problem 37.1, "Finite Potential Well." There are seven new Multi-Version Exercises.

Chapter 38

New Problem-Solving Guidelines and five new Multi-Version Exercises have been added.

Chapter 39

New Problem-Solving Guidelines, one new end-of-chapter exercise, and six new Multi-Version Exercises have been added. The text discussion of fundamental particles has been revised to include the 2012 discovery of a Higgs boson at CERN.

Chapter 40

There is a new subsection on the thorium fission cycle. Three new Multi-Version Exercises have been added.

The Big Picture

Frontiers of Modern Physics

This book will give you insight into some of the astounding recent progress made in physics and show you how this progress aids practically all other areas of science and engineering. Examples from advanced areas of research are accessible with the knowledge available at the introductory level. At many universities, freshmen and sophomores are already involved in cutting-edge physics research. Often, this participation requires nothing more than the tools developed in this book, a few days or weeks of additional reading, and the curiosity and willingness necessary to learn new facts and skills.

The following pages present some of the amazing frontiers of current physics research and describe some of the results that have been obtained during the last few years. This introduction stays at a qualitative level, skipping all mathematical and other technical details. Chapter references indicate where more in-depth explorations of the topics can be found.

Energy and Power

Probably the greatest problem facing humanity in this century is how to satisfy the ever-increasing demand for energy and power. Energy is a very basic physics topic, and power is the rate at which energy is converted; we will devote two entire chapters to these key concepts (Chapters 5 and 6). In 2010 and 2011, the twin disasters of the explosion of the Deepwater Horizon oil rig in the Gulf of Mexico and the destruction of the Fukushima Daiichi nuclear reactors by a tsunami following a very strong earthquake off the Pacific Coast of Japan (Figure 1) made it abundantly clear that satisfying our energy needs can carry huge risks.

(a)

(b)

FIGURE 1 Two environmental disasters: (a) The burning Deepwater Horizon oil rig in the Gulf of Mexico, source of the biggest accidental ocean oil spill of all time. (b) The heavily damaged reactor buildings at the Fukushima Daiichi nuclear power plant, source of the largest nuclear radiation contamination to occur in the last quarter century.

FIGURE 2 An offshore wind farm between Malmö, Sweden, and Copenhagen, Denmark, that delivers a power output comparable to that of a large coal-fired power plant.

FIGURE 3 Target chamber of the National Ignition Facility, which is used to study nuclear fusion.

FIGURE 4 Physicists from the University of Bonn, Germany, observing their newly created "super-photons," or Bose-Einstein condensate consisting of light, which they discovered in 2010.

Physics is at the center of a large number of interdisciplinary collaborations seeking ways to secure our future energy supply in the face of strongly rising demand. Wind (Figure 2), water, biomass, and sunlight are all possible alternative sources of energy. The coming decades will see increased emphasis on exploiting these resources, and physicists will work on optimizing these processes. Some of these technologies and their physical foundations are highlighted in many chapters in this book.

Seventy years ago, nuclear physicist Hans Bethe and his colleagues figured out how nuclear fusion in the Sun produces the light that makes life on Earth possible. Today, nuclear physicists are working on how to utilize nuclear fusion on Earth to produce nearly limitless energy. Both the International Thermonuclear Experimental Reactor (ITER) (Chapter 40), presently under construction in southern France through a collaborative effort involving many industrialized countries, and the National Ignition Facility (Chapters 38 and 40, Figure 3) at Lawrence Livermore National Laboratory, completed in 2009, will help researchers investigate many of the important questions that need to be resolved before the use of fusion is technically feasible and commercially viable.

Quantum Physics

The year 2005 marked the 100th anniversary of Albert Einstein's landmark papers on Brownian motion (proving that atoms are real; see Chapters 13 and 38), on the theory of relativity (Chapter 35), and on the photoelectric effect (Chapter 36). This last paper introduced one of the basic ideas underlying quantum mechanics, the physics of matter on the scale of atoms and molecules. Quantum mechanics is a product of the 20th century that led, for example, to the invention of lasers, which are now routinely used in CD, DVD, and Blu-ray players, in price scanners, and even in eye surgery, among many other applications. Quantum mechanics has also provided a more fundamental understanding of chemistry: Physicists are using ultrashort laser pulses less than 10^{-15} s in duration to gain an understanding of how chemical bonds develop. The quantum revolution has included exotic discoveries such as antimatter, and there is no end in sight. During the last decade, groups of atoms called *Bose-Einstein condensates* have been formed in electromagnetic traps; this work has opened an entirely new realm of research in atomic and quantum physics (see Chapters 36 through 38, Figure 4).

Condensed Matter Physics and Electronics

Physics innovations created and continue to drive high-tech industry. Only slightly more than 50 years ago, the first transistor was invented at Bell Labs, ushering in the electronic age. The central processing unit (CPU) of a typical desktop or laptop computer contains more than 100 million transistor elements. The incredible growth in the capabilities and the scope of applications of computers over the last few decades has been made possible by research in condensed matter physics. Gordon Moore, cofounder of Intel, famously observed that computer processing power doubles every 2 years, a trend that is predicted to continue for at least another decade or more.

Computer storage capacity grows even faster than processing power, with a doubling time of 12 months. The 2007 Nobel Prize in Physics was awarded to Albert Fert and Peter Grünberg for their

1988 discovery of *giant magnetoresistance.* It took only a decade for this discovery to be applied in computer hard disks, enabling storage capacities of hundreds of gigabytes (1 gigabyte = 1 billion pieces of information) and even terabytes (1 terabyte = 1 trillion pieces of information).

Network capacity and bandwidth doubles every 9 months. You can now go to almost any country on Earth and find wireless access points, from which you can connect your laptop or WiFi-enabled smartphone to the Internet. Yet it is only a couple of decades since the conception of the World Wide Web by Tim Berners-Lee, who was then working at the particle physics laboratory CERN in Switzerland and who developed this new medium to facilitate collaboration among particle physicists in different parts of the world.

Cell phones and other powerful communication devices have found their way into just about everybody's hands. Modern physics research enables a progressive miniaturization of consumer electronics devices. This process drives a digital convergence, making it possible to equip cell phones with digital cameras, video recorders, e-mail capability, Web browsers, and global positioning system receivers. More functionality is added continuously, while prices continue to fall. Less than 50 years after the first Moon landing smart phones now pack more computing power than the Apollo spaceship used for that trip to the Moon.

Quantum Computing

Physics researchers are still pushing the limits of computing. At present, many groups are investigating ways to build a quantum computer. Theoretically, a quantum computer consisting of N processors would be able to execute 2^N instructions simultaneously, whereas a conventional computer consisting of N processors can execute only N instructions at the same time. Thus, a quantum computer consisting of 100 processors would exceed the combined computing power of all currently existing supercomputers. Although many complex problems have to be solved before this vision can become reality, remember that 50 years ago it seemed utterly impossible to pack 100 million transistors onto a computer chip the size of a thumbnail.

Computational Physics

The interaction between physics and computers works both ways. Traditionally, physics investigations were either experimental or theoretical in nature. Textbooks appear to favor the theoretical side, because they analyze the conceptual ideas that are encapsulated in the main formulas of physics. On the other hand, much research originates on the experimental side, when newly observed phenomena seem to defy theoretical description. However, the rise of computers has made possible a third branch of physics: computational physics. Most physicists now rely on computers to process data, visualize data, solve large sets of coupled equations, or study systems for which simple analytical formulations are not known.

The emerging field of chaos and nonlinear dynamics is the prime example of such study. Arguably, MIT atmospheric physicist Edward Lorenz first simulated chaotic behavior with the aid of a computer in 1963, when he solved three coupled equations for a simple model of weather and detected a sensitive dependence on the initial conditions—even the smallest differences in the beginning of the simulation resulted in very large deviations later in time. This phenomenon is now sometimes called the *butterfly effect,* from the idea that a butterfly flapping its wings in China could change the weather in the United States a few weeks later. This sensitivity to initial conditions implies that long-term deterministic weather prediction is impossible.

Complexity and Chaos

Systems of many constituents often exhibit very complex behavior, even if the individual constituents follow simple rules of nonlinear dynamics. Physicists have started to address complexity in many systems, including simple sand piles, traffic jams, the stock market, biological evolution, fractals, and self-assembly of molecules and nanostructures. The science of complexity is another field that emerged only during the last decade and is experiencing rapid growth. Chaos and nonlinear dynamics are discussed in Chapter 7 on momentum and

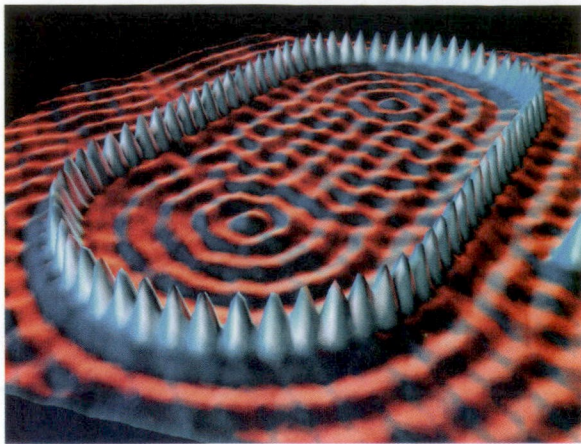

FIGURE 5 Individual iron atoms, arranged in the shape of a stadium on a copper surface. The ripples inside the "stadium" are the result of standing waves formed by electron density distributions. This arrangement was created and then imaged by using a scanning tunneling microscope (Chapter 37).

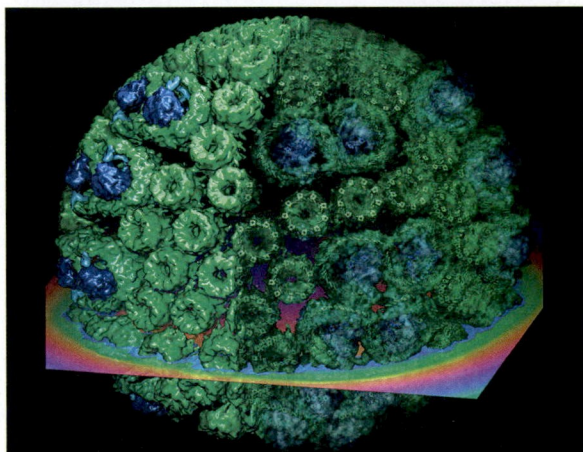

FIGURE 6 Computer simulation of the approximately 200 proteins in a chromatophore vesicle, which channels the energy of sunlight into the synthesis of adenosine triphosphate.

FIGURE 7 Aerial view of Geneva, Switzerland, with the location of the underground tunnel of the Large Hadron Collider superimposed in red.

in Chapter 14 on oscillations. The models are often quite straightforward, and first-year physics students can make valuable contributions. However, contributing generally requires some computer programming skills. Programming expertise will enable you to contribute to many advanced physics research projects.

Nanotechnology

Physicists are beginning to acquire the knowledge and skills needed to manipulate matter one atom at a time. During the last two decades, scanning, tunneling, and atomic force microscopes have allowed researchers to see individual atoms (Figure 5) and, in some cases, to move them around in controlled ways. Nanoscience and nanotechnology are devoted to these types of challenges, whose solution holds promise for great technological advances, from even more miniaturized and thus more powerful electronics to the design of new drugs, or even the manipulation of DNA to cure some diseases.

Biophysics

Just as physicists moved into the domain of chemists during the 20th century, a rapid interdisciplinary convergence of physics and molecular biology is taking place in the 21st century. Researchers are already able to use laser tweezers (Chapter 33) to move individual biomolecules. X-ray diffraction (Chapter 34) has become sophisticated enough that researchers can obtain pictures of the three-dimensional structures of very complicated proteins. In addition, theoretical biophysicists are beginning to be successful in predicting the spatial structure and the associated functionality of these molecules from the sequences of amino acids they contain. Researchers are beginning to obtain a microscopic understanding of biological entities, and some groups are making progress toward molecular-level simulation of biological processes (Figure 6).

High-Energy/Particle Physics

Nuclear and particle physicists are probing deeper and deeper into the smallest constituents of matter (Chapters 39 and 40). For example, the Large Hadron Collider (LHC) started its physics program as the highest-energy accelerator in the world in March 2010 at the CERN laboratory in Geneva, Switzerland. The LHC is located in a circular underground tunnel with a circumference of 27 km (17 mi), indicated by the red circle in Figure 7. This new instrument is the most expensive research facility ever built, at a cost of more than \$8 billion. Particle physicists use this facility to collide protons at the highest energies ever produced to try to find out what causes different elementary particles to have different masses, to probe what the true elementary constituents of the universe are, and perhaps to look for hidden extra dimensions or other exotic phenomena. Nuclear physicists at CERN smash lead nuclei into each other at very high energies in order to recreate the state of the universe a small fraction of a second after its beginning, the *Big Bang*. The ALICE (A Large Ion Collider Experiment) detector (Figure 8a) generates computerized images of the tracks (Figure 8b) of the several thousand subatomic particles produced by such a collision. Chapters 27 and 28 on magnetism and the magnetic field will explain how these tracks are analyzed in order to discover the properties of the particles that produced them.

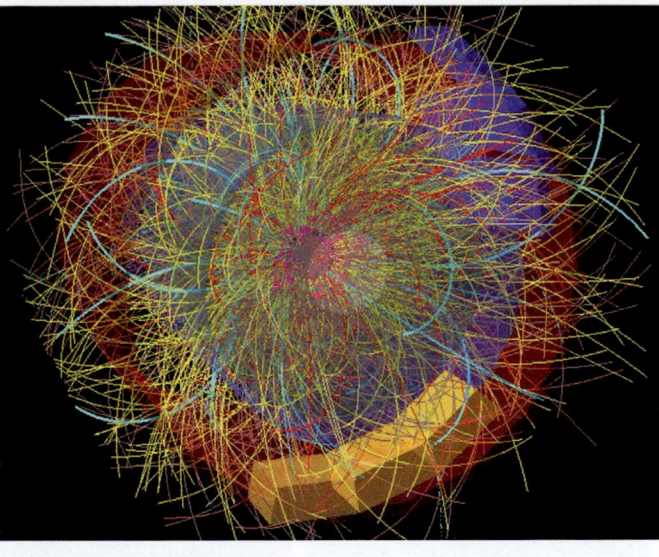

(a)   **(b)**

FIGURE 8 (a) The ALICE detector at the Large Hadron Collider during its construction. (b) Electronically reconstructed tracks of thousands of charged subatomic particles produced inside the ALICE detector by a high-energy collision of two lead nuclei.

String Theory

Particle physics has a standard model of all particles and their interactions (Chapter 39), but why this model works so well is not yet understood. String theory is currently thought to be the most likely candidate for a framework that will eventually provide this explanation. Sometimes string theory is hubristically called the *Theory of Everything*. It predicts extra spatial dimensions, which sounds at first like science fiction, but many physicists are trying to find ways in which to test this theory experimentally.

Astrophysics

Physics and astronomy have extensive interdisciplinary overlap in the areas of investigating the history of the early universe, modeling the evolution of stars, and studying the origin of gravitational waves or cosmic rays of the highest energies. Ever more precise and sophisticated observatories, such as the James Webb Space Telescope (Figure 9), are being built to study these phenomena.

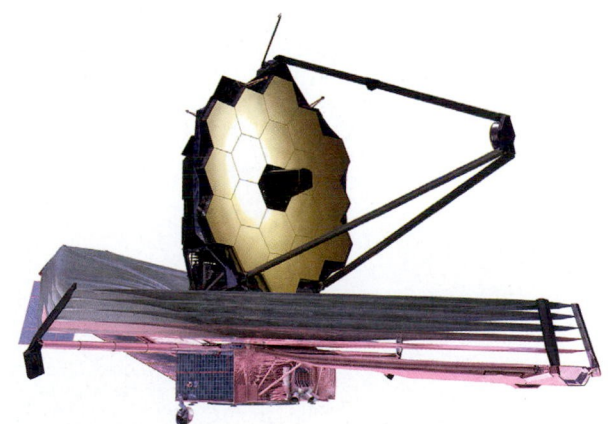

Astrophysicists continue to make astounding discoveries that reshape our understanding of the universe. Only during the last few years has it been discovered that most of the matter in the universe is not contained in stars. The composition of this *dark matter* is still unknown (Chapter 12), but its effects are revealed through *gravitational lensing,* as shown in Figure 10 by the arcs observed in the galaxy cluster Abell 2218, which is 2 billion light-years away from Earth, in the constellation Draco. These arcs are images of even more distant galaxies, distorted by the presence of large quantities of dark matter. This phenomenon is discussed in more detail in Chapter 35 on relativity.

FIGURE 9 The James Webb Space Telescope.

Symmetry, Simplicity, and Elegance

From the smallest subatomic particles to the universe at large, physical laws govern all structures and dynamics from atomic nuclei to black holes. Physicists have discovered a tremendous amount, but each new discovery opens more exciting unknown territory. Thus, we continue to construct theories to explain any and all physical phenomena. The development of these theories is guided by the need to match experimental facts, as well as by the

FIGURE 10 Galaxy cluster Abell 2218, with the arcs that are created by gravitational lensing due to dark matter.

conviction that symmetry, simplicity, and elegance are key design principles. The fact that laws of nature can be formulated in simple mathematical equations ($F = ma$, $E = mc^2$, and many other, less famous ones) is stunning.

This introduction has tried to convey a bit about the frontiers of modern physics research and their relevance to progress in other fields, from biology and medicine to engineering. This textbook should help you build a foundation for appreciating, understanding, and perhaps even participating in this vibrant research enterprise, which continues to refine and even to reshape our understanding of the world around us.

1

Overview

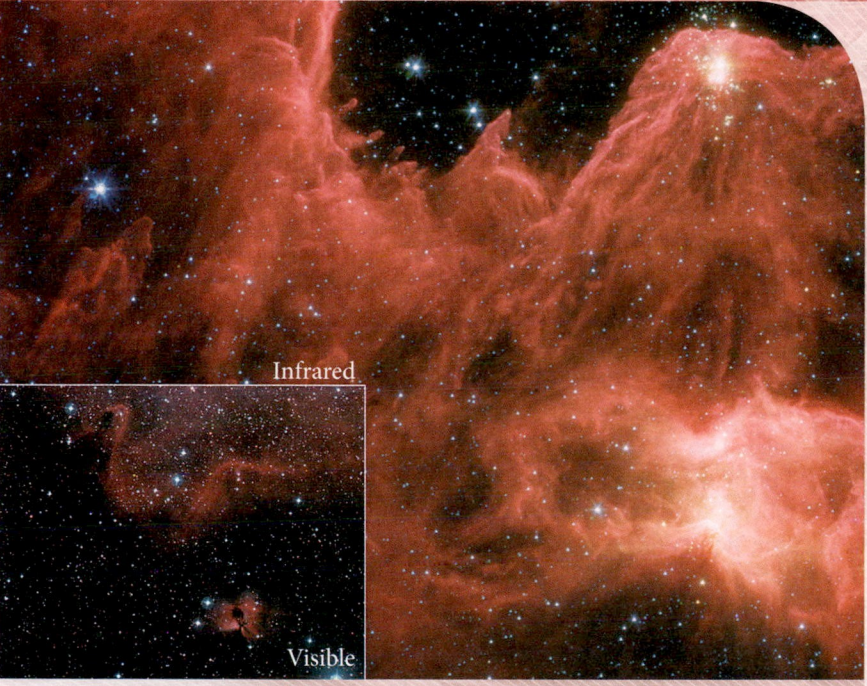

Infrared

Visible

FIGURE 1.1   An image of the W5 star-forming region taken by the Spitzer Space Telescope using infrared light.

The dramatic image in Figure 1.1 could be showing any of several things: a colored liquid spreading out in a glass of water, or perhaps biological activity in some organism, or maybe even an artist's idea of mountains on some unknown planet. If we said the view was 70 wide, would that help you decide what the picture shows? Probably not—you need to know if we mean, for example, 70 meters or 70 millionths of an inch or 70 thousand miles.

In fact, this infrared image taken by the Spitzer Space Telescope shows huge clouds of gas and dust about 70 light-years across. (A light-year is the distance traveled by light in 1 year, about 10 quadrillion meters.) These clouds are about 6500 light-years away from Earth and contain newly formed stars embedded in the glowing regions. The technology that enables us to see images such as this one is at the forefront of contemporary astronomy, but it depends in a real way on the basic ideas of numbers, units, and vectors presented in this chapter.

The ideas described in this chapter are not necessarily principles of physics, but they help us to formulate and communicate physical ideas and observations. We will use the concepts of units, scientific notation, significant figures, and vector quantities throughout the course. Once you have understood these concepts, we can go on to discuss physical descriptions of motion and its causes.

WHAT WE WILL LEARN

- The use of scientific notation and the appropriate number of significant figures is important in physics.
- We will become familiar with the international unit system and the definitions of the base units as well as methods of converting among other unit systems.
- We will use available length, mass, and time scales to establish reference points for grasping the vast diversity of systems in physics.

- We will apply a problem-solving strategy that will be useful in analyzing and understanding problems throughout this course and in science and engineering applications.
- We will work with vectors: vector addition and subtraction, multiplication of vectors, unit vectors, and length and direction of vectors.

1.1 Why Study Physics?

Perhaps your reason for studying physics can be quickly summed up as "Because it is required for my major!" While this motivation is certainly compelling, the study of science, and particularly physics, offers a few additional benefits.

Physics is the science on which all other natural and engineering sciences are built. All modern technological advances—from laser surgery to television, from computers to refrigerators, from cars to airplanes—trace back directly to basic physics. A good grasp of essential physics concepts gives you a solid foundation on which to construct advanced knowledge in all sciences. For example, the conservation laws and symmetry principles of physics also hold true for all scientific phenomena and many aspects of everyday life.

The study of physics will help you grasp the scales of distance, mass, and time, from the smallest constituents inside the nuclei of atoms to the galaxies that make up our universe. All natural systems follow the same basic laws of physics, providing a unifying concept for understanding how we fit into the overall scheme of the universe.

Physics is intimately connected with mathematics because it brings to life the abstract concepts used in trigonometry, algebra, and calculus. The analytical thinking and general techniques for problem solving that you learn here will remain useful for the rest of your life.

Science, especially physics, helps remove irrationality from our explanations of the world around us. Prescientific thinking resorted to mythology to explain natural phenomena. For example, the old Germanic tribes believed the god Thor using his hammer caused thunder. You may smile when you read this account, knowing that thunder and lightning come from electric discharges in the atmosphere. However, if you read the daily news, you will find that some misconceptions from prescientific thinking persist even today. You may not find the answer to the meaning of life in this course, but at the very least you will come away with some of the intellectual tools that enable you to weed out inconsistent, logically flawed theories and misconceptions that contradict experimentally verifiable facts. Scientific progress over the last millennium has provided a rational explanation for most of what occurs in the natural world surrounding us.

Through consistent theories and well-designed experiments, physics has helped us obtain a deeper understanding of our surroundings and has given us greater ability to control them. In a time when the consequences of air and water pollution, limited energy resources, and global warming threaten the continued existence of huge portions of life on Earth, the need to understand the results of our interactions with the environment has never been greater. Much of environmental science is based on fundamental physics, and physics drives much of the technology essential to progress in chemistry and the life sciences. You may well be called upon to help decide public policy in these areas, whether as a scientist, an engineer, or simply as a citizen. Having an objective understanding of basic scientific issues is of vital importance in making such decisions. Thus, you need to acquire scientific literacy, an essential tool for every citizen in our technology-driven society.

You cannot become scientifically literate without command of the necessary elementary tools, just as it is impossible to make music without the ability to play an instrument. This is the main purpose of this text: to properly equip you to make sound contributions to the important discussions and decisions of our time. You will emerge from reading and working with this text with a deeper appreciation for the fundamental laws that govern our universe and for the tools that humanity has developed to uncover them, tools that transcend cultures and historic eras.

1.2   Working with Numbers

Scientists have established logical rules to govern how they communicate quantitative information to one another. If you want to report the result of a measurement—for example, the distance between two cities, your own weight, or the length of a lecture—you have to specify this result in multiples of a standard unit. Thus, a measurement is the combination of a number and a unit.

At first thought, writing down numbers doesn't seem very difficult. However, in physics, we need to deal with two complications: how to deal with very big or very small numbers, and how to specify precision.

Scientific Notation

If you want to report a really big number, it becomes tedious to write it out. For example, the human body contains approximately 7,000,000,000,000,000,000,000,000,000 atoms. If you used this number often, you would surely like to have a more compact notation for it. This is exactly what **scientific notation** is. It represents a number as the product of a number greater than or equal to 1 and less than 10 (called the *mantissa*) and a power (or exponent) of 10:

$$\text{number} = \text{mantissa} \cdot 10^{\text{exponent}}. \tag{1.1}$$

The number of atoms in the human body can thus be written compactly as $7 \cdot 10^{27}$, where 7 is the mantissa and 27 is the exponent.

Another advantage of scientific notation is that it makes it easy to multiply and divide large numbers. To multiply two numbers in scientific notation, we multiply their mantissas and then add their exponents. If we wanted to estimate, for example, how many atoms are contained in the bodies of all the people on Earth, we could do this calculation rather easily. Earth hosts approximately 7 billion ($=7 \cdot 10^9$) humans. All we have to do to find our answer is to multiply $7 \cdot 10^{27}$ by $7 \cdot 10^9$. We do this by multiplying the two mantissas and adding the exponents:

$$(7 \cdot 10^{27}) \cdot (7 \cdot 10^9) = (7 \cdot 7) \cdot 10^{27+9} = 49 \cdot 10^{36} = 4.9 \cdot 10^{37}. \tag{1.2}$$

In the last step, we follow the common convention of keeping only one digit in front of the decimal point of the mantissa and adjusting the exponent accordingly. (But be advised that we will have to further adjust this answer—read on!)

Division with scientific notation is equally straightforward: If we want to calculate A / B, we divide the mantissa of A by the mantissa of B and subtract the exponent of B from the exponent of A.

Significant Figures

When we specified the number of atoms in the average human body as $7 \cdot 10^{27}$, we meant to indicate that we know it is at least $6.5 \cdot 10^{27}$ but smaller than $7.5 \cdot 10^{27}$. However, if we had written $7.0 \cdot 10^{27}$, we would have implied that we know the true number is somewhere between $6.95 \cdot 10^{27}$ and $7.05 \cdot 10^{27}$. This statement is more precise than the previous statement.

As a general rule, the number of digits you write in the mantissa specifies how precisely you claim to know it. The more digits specified, the more precision is implied (see Figure 1.2). We call the number of digits in the mantissa the number of **significant figures.** Here are some rules about using significant figures followed in each case by an example:

- The number of significant figures is the number of reliably known digits. For example, 1.62 has three significant figures; 1.6 has two significant figures.

Concept Check 1.1

The total surface area of Earth is $A = 4\pi R^2 = 4\pi (6370 \text{ km})^2 = 5.099 \cdot 10^{14} \text{ m}^2$. Assuming there are 7.0 billion humans on the planet, what is the available surface area per person?

a) $7.3 \cdot 10^4 \text{ m}^2$ c) $3.6 \cdot 10^{24} \text{ m}^2$

b) $7.3 \cdot 10^{24} \text{ m}^2$ d) $3.6 \cdot 10^4 \text{ m}^2$

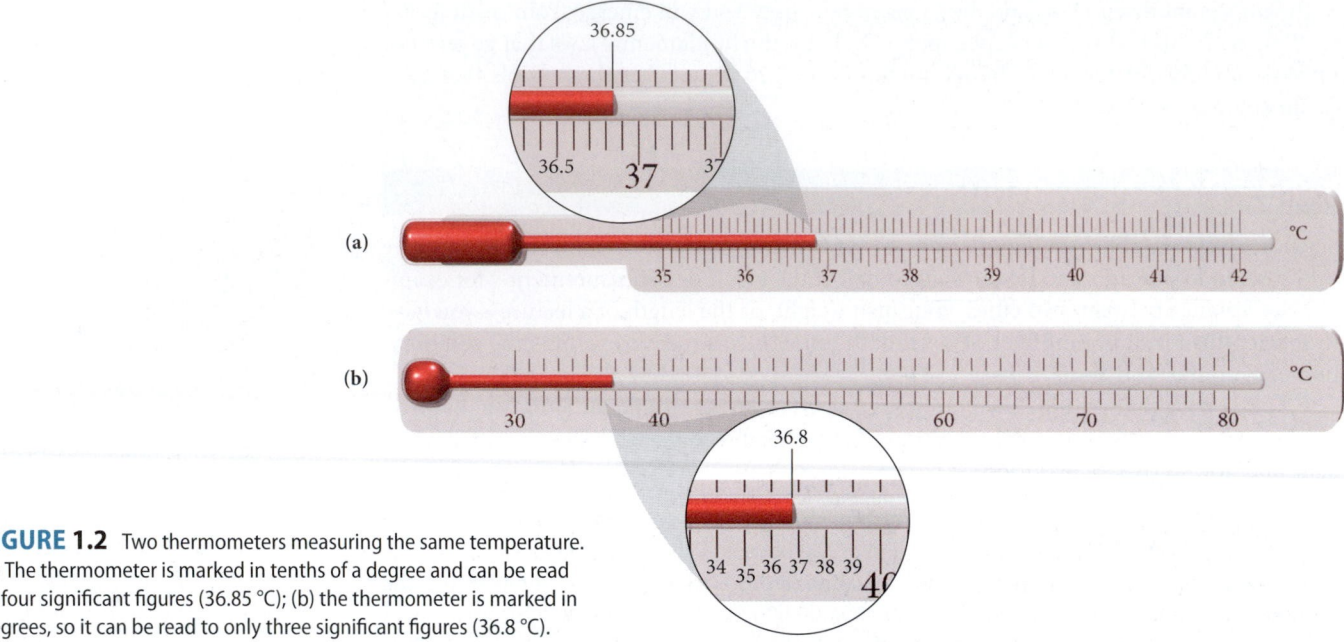

FIGURE 1.2 Two thermometers measuring the same temperature. (a) The thermometer is marked in tenths of a degree and can be read to four significant figures (36.85 °C); (b) the thermometer is marked in degrees, so it can be read to only three significant figures (36.8 °C).

Concept Check 1.2

How many significant figures are in each of the following numbers?

a) 2.150   d) 0.215000

b) 0.000215   e) 0.215 + 0.21

c) 215.00

Concept Check 1.3

For the two numbers $x = 0.43$ and $y = 3.53$, which of the following has the greatest number of significant figures?

a) the sum, $x + y$ d) the number x

b) the product, xy e) the number y

c) the difference, $x - y$

- If you give a number as an integer, you specify it with infinite precision. For example, if someone says that he or she has 3 children, this means exactly 3, no less and no more.

- Leading zeros do not count as significant digits. The number 1.62 has the same number of significant digits as 0.00162. There are three significant figures in both numbers. We start counting significant digits from the left at the first nonzero digit.

- Trailing zeros, on the other hand, do count as significant digits. The number 1.620 has four significant figures. Writing a trailing zero implies greater precision!

- Numbers in scientific notation have as many significant figures as their mantissa. For example, the number $9.11 \cdot 10^{-31}$ has three significant figures because that's what the mantissa (9.11) has. The size of the exponent has no influence.

- You can never have more significant figures in a result than you start with in any of the factors of a multiplication or division. For example, 1.23 / 3.4461 is not equal to 0.3569252. Your calculator may give you that answer, but calculators do not automatically display the correct number of significant figures. Correctly, 1.23 / 3.4461 = 0.357. You must round a calculator result to the proper number of significant figures—in this case, three, which is the number of significant figures in the numerator.

- You can only add or subtract when there are significant figures for that place in every number. For example, 1.23 + 3.4461 = 4.68, and not 4.6761 as you may think. This rule, in particular, requires some getting used to.

To finish this discussion of significant figures, let's reconsider the total number of atoms contained in the bodies of all people on Earth. We started with two quantities that were given to only one significant figure. Therefore, the result of our multiplication needs to be properly rounded to one significant digit. The combined number of atoms in all human bodies is thus correctly stated as $5 \cdot 10^{37}$.

1.3 SI Unit System

In high school, you may have been introduced to the international system of units and compared it to the British system of units in common use in the United States. You may have driven on a freeway on which the distances are posted both in miles and in kilometers or purchased food where the price was quoted per pound and per kilogram.

Table 1.1	Unit Names and Abbreviations for the Base Units of the SI System of Units	
Unit	**Abbreviation**	**Base Unit for**
meter	m	length
kilogram	kg	mass
second	s	time
ampere	A	current
kelvin	K	temperature
mole	mol	amount of a substance
candela	cd	luminous intensity

The international system of units is often abbreviated as SI (for Système International). Sometimes the units in this system are called *metric units*. The **SI unit system** is the standard used for scientific work around the world. The seven base units for the SI system are given in Table 1.1.

The first letters of the first four base units provide another commonly used name for the SI system: the MKSA system. We will only use the first three units (meter, kilogram, and second) in the entire first part of this book and in all of mechanics. The current definitions of these base units are as follows:

- 1 meter (m) is the distance that a light beam in vacuum travels in 1/299,792,458 of a second. Originally, the meter was related to the size of the Earth (Figure 1.3).

- 1 kilogram (kg) is defined as the mass of the international prototype of the kilogram. This prototype, shown in its elaborate storage container in Figure 1.4, is kept just outside Paris, France, under carefully controlled environmental conditions.

- 1 second (s) is the time interval during which 9,192,631,770 oscillations of the electromagnetic wave (see Chapter 31) that corresponds to the transition between two specific states of the cesium-133 atom. Until 1967, the standard for the second was 1/86,400 of a mean solar day. However, the atomic definition is more precise and more reliably reproducible.

FIGURE 1.3 Originally, the meter was defined as 1 ten-millionth of the length of the meridian through Paris from the North Pole to the Equator.

Notation Convention: It is common practice to use roman letters for unit abbreviations and italic letters for physical quantities. We follow this convention in this book. For example, m stands for the unit meter, while *m* is used for the physical quantity mass. Thus, the expression $m = 17.2$ kg specifies that the mass of an object is 17.2 kilograms.

Units for all other physical quantities can be derived from the seven base units of Table 1.1. The unit for area, for example, is m^2. The units for volume and mass density are m^3 and kg/m^3, respectively. The units for velocity and acceleration are m/s and m/s^2, respectively. Some derived units were used so often that it became convenient to give them their own names and symbols. Often the name is that of a famous physicist. Table 1.2 lists the 20 derived SI units with special names. In the two rightmost columns of the table, the named unit is listed in terms of other named units and then in terms of SI base units. Also included in this table are the radian and steradian, the dimensionless units of angle and solid angle, respectively.

You can obtain SI-recognized multiples of the base units and derived units by multiplying them by various factors of 10. These factors have universally accepted letter abbreviations that are used as prefixes, shown in Table 1.3. The use of standard prefixes (factors of 10) makes it easy to determine, for example, how many centimeters (cm) are in a kilometer (km):

$$1 \text{ km} = 10^3 \text{ m} = 10^3 \text{ m} \cdot (10^2 \text{ cm / m}) = 10^5 \text{ cm}. \tag{1.3}$$

In comparison, note how tedious it is to figure out how many inches are in a mile:

$$1 \text{ mile} = (5,280 \text{ feet/mile}) \cdot (12 \text{ inches/foot}) = 63,360 \text{ inches}. \tag{1.4}$$

As you can see, not only do you have to memorize particular conversion factors in the British system, but calculations also become more complicated. For calculations in the SI system, you only have to know the standard prefixes shown in Table 1.3 and how to add or subtract integers in the powers of 10.

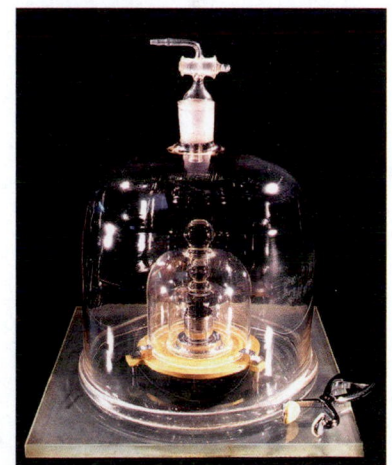

FIGURE 1.4 Prototype of the kilogram, stored near Paris, France.

Table 1.2	Common SI Derived Units			
Derived or Dimensionless Unit	**Name**	**Symbol**	**Equivalent**	**Expressions**
Absorbed dose	gray	Gy	J/kg	$m^2\,s^{-2}$
Angle	radian	rad	—	—
Capacitance	farad	F	C/V	$m^{-2}\,kg^{-1}\,s^4\,A^2$
Catalytic activity	katal	kat	—	$s^{-1}\,mol$
Dose equivalent	sievert	Sv	J/kg	$m^2\,s^{-2}$
Electric charge	coulomb	C	—	s A
Electric conductance	siemens	S	A/V	$m^{-2}\,kg^{-1}\,s^3\,A^2$
Electric potential	volt	V	W/A	$m^2\,kg\,s^{-3}\,A^{-1}$
Electric resistance	ohm	Ω	V/A	$m^2\,kg\,s^{-3}\,A^{-2}$
Energy	joule	J	N m	$m^2\,kg\,s^{-2}$
Force	newton	N	—	$m\,kg\,s^{-2}$
Frequency	hertz	Hz	—	s^{-1}
Illuminance	lux	lx	lm/m^2	$m^{-2}\,cd$
Inductance	henry	H	Wb/A	$m^2\,kg\,s^{-2}\,A^{-2}$
Luminous flux	lumen	lm	cd sr	cd
Magnetic flux	weber	Wb	V s	$m^2\,kg\,s^{-2}\,A^{-1}$
Magnetic field	tesla	T	Wb/m^2	$kg\,s^{-2}\,A^{-1}$
Power	watt	W	J/s	$m^2\,kg\,s^{-3}$
Pressure	pascal	Pa	N/m^2	$m^{-1}\,kg\,s^{-2}$
Radioactivity	becquerel	Bq	—	s^{-1}
Solid angle	steradian	sr	—	—
Temperature	degree Celsius	°C	—	K

Table 1.3	SI Standard Prefixes				
Factor	**Prefix**	**Symbol**	**Factor**	**Prefix**	**Symbol**
10^{24}	yotta-	Y	10^{-24}	yocto-	y
10^{21}	zetta-	Z	10^{-21}	zepto-	z
10^{18}	exa-	E	10^{-18}	atto-	a
10^{15}	peta-	P	10^{-15}	femto-	f
10^{12}	tera-	T	10^{-12}	pico-	p
10^{9}	giga-	G	10^{-9}	nano-	n
10^{6}	mega-	M	10^{-6}	micro-	μ
10^{3}	kilo-	k	10^{-3}	milli-	m
10^{2}	hecto-	h	10^{-2}	centi-	c
10^{1}	deka-	da	10^{-1}	deci-	d

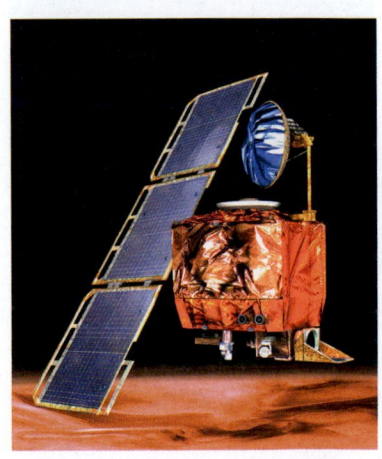

FIGURE 1.5 The *Mars Climate Orbiter,* a victim of faulty unit conversion.

The international system of units was adopted by France in 1799 and is now used daily in almost all countries of the world, the one notable exception being the United States. In the United States, we buy milk and gasoline in gallons, not in liters. Our cars show speed in miles per hour, not meters per second. When we go to the lumberyard, we buy wood as two-by-fours (actually 1.5 inches by 3.5 inches, but that's a different story). Since we use British units in our daily lives, this book will indicate British units where appropriate, to establish connections with everyday experiences. But we will use the SI unit system in all calculations to avoid having to work with the British unit conversion factors. However, we will sometimes provide equivalents in British units. This is a compromise that will be necessary only until the United States decides to adopt SI units as well.

The use of British units can be costly. The cost can range from a small expense, such as that incurred by car mechanics who need to purchase two sets of wrench socket sets, one metric and one British, to the very expensive loss of the *Mars Climate Orbiter* spacecraft (Figure 1.5) in September 1999. The crash of this spacecraft has been blamed on the fact that one of the engineering teams used British units and the other one SI units. The two teams relied on each other's numbers, without realizing that the units were not the same.

The use of powers of 10 is not completely consistent even within the SI system itself. The notable exception is in the time units, which are not factors of 10 times the base unit (second):

- 365 days form a year,
- a day has 24 hours,
- an hour contains 60 minutes, and
- a minute consists of 60 seconds.

Early metric pioneers tried to establish a completely consistent set of metric time units, but these attempts failed. The not-exactly-metric nature of time units extends to some derived units. For example, a European sedan's speedometer does not show speeds in meters per second, but in kilometers per hour.

EXAMPLE 1.1   Units of Land Area

The unit of land area used in countries that use the SI system is the hectare, defined as 10,000 m^2. In the United States, land area is given in acres; an acre is defined as 43,560 ft^2.

PROBLEM
You just bought a plot of land with dimensions 2.00 km by 4.00 km. What is the area of your new purchase in hectares and acres?

SOLUTION
The area A is given by

$$A = \text{length} \cdot \text{width} = (2.00 \text{ km})(4.00 \text{ km}) = (2.00 \cdot 10^3 \text{ m})(4.00 \cdot 10^3 \text{ m})$$

$$A = 8.00 \text{ km}^2 = 8.00 \cdot 10^6 \text{ m}^2.$$

The area of this plot of land in hectares is then

$$A = 8.00 \cdot 10^6 \text{ m}^2 \frac{1 \text{ hectare}}{10,000 \text{ m}^2} = 8.00 \cdot 10^2 \text{ hectares} = 800. \text{ hectares}.$$

To find the area of the land in acres, we need the length and width in British units:

$$\text{length} = 2.00 \text{ km} \frac{1 \text{ mi}}{1.609 \text{ km}} = 1.24 \text{ mi} \frac{5,280 \text{ ft}}{1 \text{ mi}} = 6,563 \text{ ft}$$

$$\text{width} = 4.00 \text{ km} \frac{1 \text{ mi}}{1.609 \text{ km}} = 2.49 \text{ mi} \frac{5,280 \text{ ft}}{1 \text{ mi}} = 13,130 \text{ ft}.$$

The area is then

$$A = \text{length} \cdot \text{width} = (1.24 \text{ mi})(2.49 \text{ mi}) = (6,563 \text{ ft})(13,130 \text{ ft})$$

$$A = 3.09 \text{ mi}^2 = 8.61 \cdot 10^7 \text{ ft}^2.$$

In acres, this is

$$A = 8.61 \cdot 10^7 \text{ ft}^2 \frac{1 \text{ acre}}{43,560 \text{ ft}^2} = 1980 \text{ acres}.$$

Metrology: Research on Measures and Standards

The work of defining the standards for the base units of the SI system is by no means complete. A great deal of research is devoted to refining measurement technologies and pushing them to greater precision. This field of research is called **metrology.** In the United States, the laboratory that has the primary responsibility for this work is the National Institute of Standards and Technology (NIST). NIST works in collaboration with similar institutes in other countries to refine accepted standards for the SI base units.

One current research project is to find a definition of the kilogram based on reproducible quantities in nature. This definition would replace the current definition of the

FIGURE 1.6 Setup of the world's most precise clock at NIST.

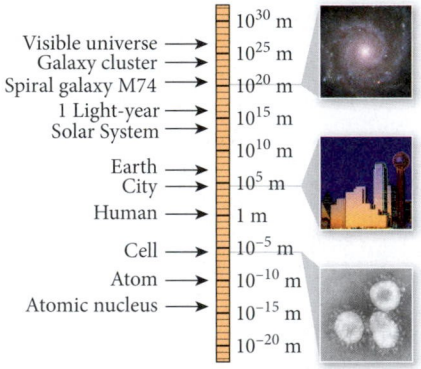

FIGURE 1.7 Range of length scales for physical systems. The pictures top to bottom are the spiral galaxy M74, the Dallas skyline, and the SARS virus.

kilogram, which is based on the mass of a standard object kept in Sèvres on the outskirts of Paris (see Figure 1.4). The most promising effort in this direction seems to be Project Avogadro, which attempts to define the kilogram using highly purified silicon crystals.

Research on keeping time ever more precisely is one of the major tasks of NIST and similar institutions. Figure 1.6 shows what is currently the most accurate clock at NIST in Boulder, Colorado. It is accurate to ±1 second in 3.7 billion years!

Greater precision in timekeeping is needed for many applications in our information-based society, where signals can travel around the world in less than 0.2 second. The Global Positioning System (GPS) is one example of technology that would be impossible to realize without the precision of atomic clocks and the physics research that enters into their construction. The GPS system also relies on Einstein's theory of relativity, which we will study in Chapter 35.

1.4 The Scales of Our World

The most amazing fact about physics is that its laws govern every object, from the smallest to the largest. The scales of the systems for which physics holds predictive power span many orders of magnitude (powers of 10), as we'll see in this section.

Nomenclature: In the following, you will read "on the order of" several times. This phrase means "within a factor of 2 or 3."

Length Scales

Length is defined as the distance measurement between two points in space. Figure 1.7 shows some length scales for common objects and systems that span over 40 orders of magnitude.

Let's start with ourselves. On average, in the United States, a woman is 1.62 m (5 ft 4 in) tall, and a man measures 1.75 m (5 ft 9 in). Thus, human height is on the order of a meter. If you reduce the length scale for a human body by a factor of a million, you arrive at a micrometer. This is the typical diameter of a cell in your body or a bacterium.

If you reduce the length of your measuring stick by another factor of 10,000, you are at a scale of 10^{-10} m, the typical diameter of an individual atom. This is the smallest size we can resolve with the aid of the most advanced microscopes.

Inside the atom is its nucleus, with a diameter about 1/10,000 that of the atom, on the order of 10^{-14} m. The individual protons and neutrons that make up the atomic nucleus have a diameter of approximately 10^{-15} m $= 1$ fm (a femtometer).

Considering objects larger than ourselves, we can look at the scale of a typical city, on the order of kilometers. The diameter of Earth is just a little bigger than 10,000 km (12,760 km, to be more precise). As discussed earlier, the definition of the meter is now stated in terms of the speed of light. However, the meter was originally defined as 1 ten-millionth of the length of the meridian through Paris from the North Pole to the Equator. If a quarter circle has an arc length of 10 million meters (= 10,000 km), then the circumference of the entire circle would be exactly 40,000 km. Using the modern definition of the meter, the equatorial circumference of the Earth is 40,075 km, and the circumference along the meridian is 40,008 km.

The distance from the Earth to the Moon is 384,000 km, and the distance from the Earth to the Sun is greater by a factor of approximately 400, or about 150 million km. This distance is called an *astronomical unit* and has the symbol AU. Astronomers used this unit before the distance from Earth to Sun became known with accuracy, but it is still convenient today. In SI units, an astronomical unit is

$$1 \text{ AU} = 1.495\,98 \cdot 10^{11} \text{ m}. \tag{1.5}$$

The diameter of our Solar System is conventionally stated as approximately 10^{13} m, or 60 AU.

We have already remarked that light travels in vacuum at a speed of approximately 300,000 km/s. Therefore, the distance between Earth and Moon is covered by light in just over 1 second, and light from the Sun takes approximately 8 minutes to reach Earth. In order to cover distance scales outside our Solar System, astronomers have introduced the (non-SI, but handy) unit of the light-year, the distance that light travels in 1 year in vacuum:

$$1 \text{ light-year} = 9.46 \cdot 10^{15} \text{ m.} \tag{1.6}$$

The nearest star to our Sun is just over 4 light-years away. The Andromeda Galaxy, the sister galaxy of our Milky Way, is about 2.5 million light-years $= 2 \cdot 10^{22}$ m away.

Finally, the radius of the visible universe is approximately 14 billion light-years $= 1.5 \cdot 10^{26}$ m. Thus, about 41 orders of magnitude span between the size of an individual proton and that of the entire visible universe.

Mass Scales

Mass is the amount of matter in an object. When you consider the range of masses of physical objects, you obtain an even more awesome span of orders of magnitude (Figure 1.8) than for lengths.

Atoms and their parts have incredibly small masses. The mass of an electron is only $9.11 \cdot 10^{-31}$ kg. A proton's mass is $1.67 \cdot 10^{-27}$ kg, roughly a factor of 2000 more than the mass of an electron. An individual atom of lead has a mass of $3.46 \cdot 10^{-25}$ kg.

The mass of a single cell in the human body is on the order of 10^{-13} kg to 10^{-12} kg. Even a fly has more than 100 million times the mass of a cell, at approximately 10^{-4} kg.

A car's mass is on the order of 10^3 kg, and that of a passenger plane is on the order of 10^5 kg.

A mountain typically has a mass of 10^{12} kg to 10^{14} kg, and the combined mass of all the water in all of the Earth's oceans is estimated to be on the order of 10^{21} kg.

The mass of the entire Earth can be specified fairly precisely at $6.0 \cdot 10^{24}$ kg. The Sun has a mass of $2.0 \cdot 10^{30}$ kg, or about 300,000 times the mass of the Earth. Our entire galaxy, the Milky Way, is estimated to have 200 billion stars in it and thus has a mass around $3 \cdot 10^{41}$ kg. Finally, the entire universe contains billions of galaxies. Depending on the assumptions about dark matter, a currently active research topic (see Chapter 12), the mass of the universe as a whole is roughly 10^{51} kg. However, you should recognize that this number is an estimate and may be off by a factor of up to 100.

Interestingly, some objects have no mass. For example, photons, the "particles" that light is made of, have zero mass.

FIGURE 1.8 Range of mass scales for physical systems.

Time Scales

Time is the duration between two events. Human time scales lie in the range from a second (the typical duration of a human heartbeat) to a century (about the life expectancy of a person born now). Incidentally, human life expectancy is increasing at an ever faster rate. During the Roman Empire, 2000 years ago, a person could expect to live only 25 years. In 1850, actuary tables listed the mean lifetime of a human as 39 years. Now that number is 80 years. Thus, it took almost 2000 years to add 50% to human life expectancy, but in the last 150 years, life expectancy has doubled again. This is perhaps the most direct evidence that science has basic benefits for all of us. Physics contributes to this progress by aiding the development of more sophisticated medical imaging and treatment equipment, and today's fundamental research will enter clinical practice tomorrow. Laser surgery, cancer radiation therapy, magnetic resonance imaging, and positron emission tomography are just a few examples of technological advances that have helped increase life expectancy.

In their research, the authors of this book study ultrarelativistic heavy-ion collisions. These collisions occur during time intervals on the order of 10^{-22} s, more than a million times shorter than the time intervals we can measure directly. During this course, you will learn that the time scale for the oscillation of visible light is 10^{-15} s, and that of audible sound is 10^{-3} s.

The longest time span we can measure *indirectly* or *infer* is the age of the universe. Current research puts this number at 13.7 billion years, but with an uncertainty of up to 0.2 billion years.

We cannot leave this topic without mentioning one interesting fact to ponder during your next class lecture. Lectures typically last 50 minutes at most universities. A century, by comparison, has $100 \cdot 365 \cdot 24 \cdot 60 \approx 50{,}000{,}000$ minutes. So a lecture lasts about 1 millionth of a century, leading to a handy (non-SI) time unit, the microcentury = duration of one lecture.

1.5 General Problem-Solving Strategy

Physics involves more than solving problems—but problem solving is a big part of it. At times, while you are laboring over your homework assignments, it may seem that's all you do. However, repetition and practice are important parts of learning.

A basketball player spends hours practicing the fundamentals of free throw shooting. Many repetitions of the same action enable a player to become very reliable at this task. You need to develop the same philosophy toward solving mathematics and physics problems: You have to practice good problem-solving techniques. This work will pay huge dividends, not just during the remainder of this physics course, not just during exams, not even just in your other science classes, but also throughout your entire career.

What constitutes a good problem-solving strategy? Everybody develops his or her own routines, procedures, and shortcuts. However, here is a general blueprint that should help get you started:

1. **THINK** Read the problem carefully. Ask yourself what quantities are known, what quantities might be useful but are unknown, and what quantities are asked for in the solution. Write down these quantities and represent them with their commonly used symbols. Convert into SI units, if necessary.

2. **SKETCH** Make a sketch of the physical situation to help you visualize the problem. For many learning styles, a visual or graphical representation is essential, and it is often essential for defining the variables.

3. **RESEARCH** Write down the physical principles or laws that apply to the problem. Use equations representing these principles to connect the known and unknown quantities to each other. In some cases, you will immediately see an equation that has only the quantities that you know and the one unknown that you are supposed to calculate, and nothing else. More often you may have to do a bit of deriving, combining two or more known equations into the one that you need. This requires some experience, more than any of the other steps listed here. To the beginner, the task of deriving a new equation may look daunting, but you will get better at it the more you practice.

4. **SIMPLIFY** Do not plug numbers into your equation yet! Instead, simplify your result algebraically as much as possible. For example, if your result is expressed as a ratio, cancel out common factors in the numerator and the denominator. This step is particularly helpful if you need to calculate more than one quantity.

5. **CALCULATE** Put the numbers with units into the equation and get to work with a calculator. Typically, you will obtain a number and a physical unit as your answer.

6. **ROUND** Determine the number of significant figures that you want to have in your result. As a rule of thumb, a result obtained by multiplying or dividing should be rounded to the same number of significant figures as in the input quantity that is given with the least number of significant figures. You should not round in intermediate steps, as rounding too early might give you a wrong solution.

7. **DOUBLE-CHECK** Step back and look at the result. Judge for yourself if the answer (both the number and the units) seems realistic. You can often avoid handing in a wrong solution by making this final check. Sometimes the units of your answer are simply wrong, and you know you must have made an error. Or sometimes the order of magnitude comes out totally wrong. For example, if your task is to calculate the mass of the Sun (we will do this later in this book), and your answer comes out near 10^6 kg (only a few thousand tons), you know you must have made a mistake somewhere.

Let's put this strategy to work in the following example.

SOLVED PROBLEM 1.1 / Volume of a Cylinder

PROBLEM
Nuclear waste material in a physics laboratory is stored in a cylinder of height $4\frac{3}{16}$ inches and circumference $8\frac{3}{16}$ inches. What is the volume of this cylinder, measured in metric units?

SOLUTION
In order to practice problem-solving skills, we will go through each of the steps of the strategy outlined above.

THINK From the question, we know that the height of the cylinder, converted to centimeters, is

$$h = 4\tfrac{13}{16} \text{ in} = 4.8125 \text{ in}$$
$$= (4.8125 \text{ in}) \cdot (2.54 \text{ cm/in})$$
$$= 12.22375 \text{ cm}.$$

Also, the circumference of the cylinder is specified as

$$c = 8\tfrac{3}{16} \text{ in} = 8.1875 \text{ in}$$
$$= (8.1875 \text{ in}) \cdot (2.54 \text{ cm/in})$$
$$= 20.79625 \text{ cm}.$$

Even though it is clear that the number of significant digits in our SI-converted values for h and c are clearly too high, we keep them for our intermediate calculations, and we only round our final answer to the proper number of significant digits.

SKETCH Next, we produce a sketch, something like Figure 1.9. Note that the given quantities are shown with their symbolic representations, not with their numerical values. The circumference is represented by the thicker circle (oval, actually, in this projection).

RESEARCH Now we have to find the volume of the cylinder in terms of its height and its circumference. This relationship is not commonly listed in collections of geometric formulas. Instead, the volume of a cylinder is given as the product of base area and height:

$$V = \pi r^2 h.$$

Once we find a way to connect the radius and the circumference, we'll have the formula we need. The top and bottom areas of a cylinder are circles, and for a circle we know that

$$c = 2\pi r.$$

SIMPLIFY Remember: We do not plug in numbers yet! To simplify our numerical task, we can solve the second equation for r and insert this result into the first equation:

$$c = 2\pi r \Rightarrow r = \frac{c}{2\pi}$$

$$V = \pi r^2 h = \pi \left(\frac{c}{2\pi}\right)^2 h = \frac{c^2 h}{4\pi}.$$

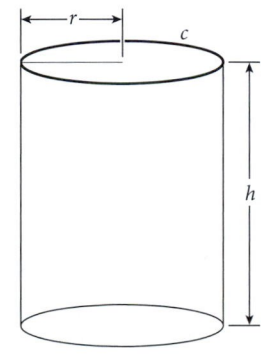

FIGURE 1.9 Sketch of right cylinder.

– Continued

CALCULATE Now it's time to get out the calculator and put in the numbers:

$$V = \frac{c^2 h}{4\pi}$$

$$= \frac{(20.79625 \text{ cm})^2 \cdot (12.22375 \text{ cm})}{4\pi}$$

$$= 420.69239 \text{ cm}^3.$$

ROUND The output of the calculator has again made our result look much more precise than we can claim realistically. We need to round. Since the input quantities are given to only three significant figures, our result needs to be rounded to three significant figures. Our final answer is $V = 421.$ cm^3.

DOUBLE-CHECK Our last step is to check that the answer is reasonable. First, we look at the unit we got for our result. Cubic centimeters are a unit for volume, so our result passes its first test. Now let's look at the magnitude of our result. You might recognize that the height and circumference given for the cylinder are close to the corresponding dimensions of a soda can. If you look on a can of your favorite soft drink, it will list the contents as 12 fluid ounces and also give you the information that this is 355 mL. Because 1 mL $= 1$ cm^3, our answer is reasonably close to the volume of the soda can. Note that this does *not* tell us that our calculation is correct, but it shows that we are not way off.

Suppose the researchers decide that a soda can is not large enough for holding the waste in the lab and replace it with a larger cylindrical container with a height of 44.6 cm and a circumference of 62.5 cm. If we want to calculate the volume of this replacement cylinder, we don't need to do all of Solved Problem 1.1 over again. Instead, we can go directly to the algebraic formula we derived in the Simplify step and substitute our new data into that, ending up with a volume of 13,900 cm^3 when rounded to three significant figures. This example illustrates the value of waiting to substitute in numbers until algebraic simplification has been completed.

In Solved Problem 1.1, you can see that we followed the seven steps outlined in our general strategy. It is tremendously helpful to train your brain to follow a certain procedure in attacking all kinds of problems. This is not unlike following the same routine whenever you are shooting free throws in basketball, where frequent repetition helps you build the muscle memory essential for consistent success, even when the game is on the line.

Perhaps more than anything, an introductory physics class should enable you to develop methods to come up with your own solutions to a variety of problems, eliminating the need to accept "authoritative" answers uncritically. The method we used in Solved Problem 1.1 is extremely useful, and we will practice it again and again in this book. However, to make a simple point, one that does not require the full set of steps used for a solved problem, we'll sometimes use an illustrative example.

EXAMPLE 1.2 Volume of a Barrel of Oil

PROBLEM
The volume of a barrel of oil is 159 L. We need to design a cylindrical container that will hold this volume of oil. The container needs to have a height of 1.00 m to fit in a transportation container. What is the required circumference of the cylindrical container?

SOLUTION
Starting with the equation we derived in the Simplify step in Solved Problem 1.1, we can relate the circumference, c, and the height, h, of the container to the volume, V, of the container:

$$V = \frac{c^2 h}{4\pi}.$$

Solving for the circumference, we get

$$c = \sqrt{\frac{4\pi V}{h}}.$$

The volume in SI units is

$$V = 159 \text{ L} \frac{1000 \text{ mL}}{\text{L}} \frac{1 \text{ cm}^3}{1 \text{ mL}} \frac{1 \text{ m}^3}{10^6 \text{ cm}^3} = 0.159 \text{ m}^3.$$

The required circumference is then

$$c = \sqrt{\frac{4\pi V}{h}} = \sqrt{\frac{4\pi \left(0.159 \text{ m}^3\right)}{1.00 \text{ m}}} = 1.41 \text{ m}.$$

As you may have already realized from the preceding problem and example, a good command of algebra is essential for success in an introductory physics class. For engineers and scientists, most universities and colleges also require calculus, but at many schools an introductory physics class and a calculus class can be taken concurrently. This first chapter does not contain any calculus, and subsequent chapters will review the relevant calculus concepts as you need them. However, there is another field of mathematics that is used extensively in introductory physics: trigonometry. Virtually every chapter of this book uses right triangles in some way. Therefore, it is a good idea to review the formulas for sine, cosine, and the like, as well as the indispensable Pythagorean Theorem. Let's look at another solved problem, which makes use of trigonometric concepts.

SOLVED PROBLEM 1.2 / View from the Willis Tower

PROBLEM
It goes without saying that one can see farther from a tower than from ground level; the higher the tower, the farther one can see. The Willis Tower (formerly named Sears Tower) in Chicago has an observation deck, which is 412 m above ground. How far can one see out over Lake Michigan from this observation deck under perfect weather conditions? (Assume eye level is at 413 m above the level of the lake.)

SOLUTION
THINK As we have stressed before, this is the most important step in the problem-solving process. A little preparation at this stage can save a lot of work at a later stage. Perfect weather conditions are specified, so fog or haze is not a limiting factor. What else could determine how far one can see? If the air is clear, one can see mountains that are quite far away. Why mountains? Because they are very tall. But the landscape around Chicago is flat. What then could limit the viewing range? Nothing, really; one can see all the way to the horizon. And what is the deciding factor for where the horizon is? It is the curvature of the Earth. Let's make a sketch to make this a little clearer.

SKETCH Our sketch does not have to be elaborate, but it needs to show a simple version of the Willis Tower on the surface of the Earth. It is not important that the sketch be to scale, and we elect to greatly exaggerate the height of the tower relative to the size of the Earth. See Figure 1.10.

It seems obvious from this sketch that the farthest point (point C) that one can see from the top of the Willis Tower (point B) is where the line of sight just touches the surface of the Earth tangentially. Any point on Earth's surface farther away from the Willis Tower is hidden from view (below the dashed line segment). The viewing range is then given by the distance r between that surface point C and the observation deck (point B) on top of the tower, at height h. Included in the sketch is also a line from the center of Earth (point A) to the foot of the Willis Tower. It has length R, which is the radius of Earth. Another line of the same length, R, is drawn to the point where the line of sight touches the Earth's surface tangentially.

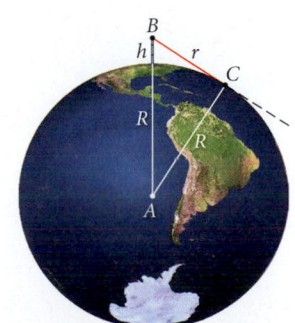

FIGURE 1.10 Distance from the top of the Willis Tower (B) to the horizon (C).

– Continued

RESEARCH As you can see from the sketch, a line drawn from the center of the Earth to the point where the line of sight touches the surface (A to C) will form a right angle with that line of sight (B to C); that is, the three points A, B, and C form the corners of a right triangle. This is the key insight, which enables us to use trigonometry and the Pythagorean Theorem to attack the solution of this problem. Examining the sketch in Figure 1.10, we find

$$r^2 + R^2 = (R+h)^2.$$

SIMPLIFY Remember, we want to find the distance to the horizon, for which we used the symbol r in the previous equation. Isolating that variable on one side of our equation gives

$$r^2 = (R+h)^2 - R^2.$$

Now we can simplify the square and obtain

$$r^2 = R^2 + 2hR + h^2 - R^2 = 2hR + h^2.$$

Finally, we take the square root and obtain our final algebraic answer:

$$r = \sqrt{2hR + h^2}.$$

CALCULATE Now we are ready to insert numbers. The accepted value for the radius of Earth is $R = 6.37 \cdot 10^6$ m, and $h = 413$ m $= 4.13 \cdot 10^2$ m was given in the problem. This leads to

$$r = \sqrt{2(4.13 \cdot 10^2 \, \text{m})(6.37 \cdot 10^6 \, \text{m}) + (4.13 \cdot 10^2 \, \text{m})^2} = 7.25382 \cdot 10^4 \, \text{m}.$$

ROUND The Earth's radius was given to three-digit precision, as was the elevation of the eye level of the observer. So we round to three digits and give our final result as

$$r = 7.25 \cdot 10^4 \, \text{m} = 72.5 \, \text{km}.$$

DOUBLE-CHECK Always check the units first. Since the problem asked "how far," the answer needs to be a distance, which has the dimension of length and thus the base unit meter. Our answer passes this first check. What about the magnitude of our answer? Since the Willis Tower is almost 0.5 kilometer high, we expect the viewing range to be at least several kilometers; so a multikilometer range for the answer seems reasonable. Lake Michigan is slightly more than 80 km wide, if you look toward the east from Chicago. Our answer then implies that you cannot see the Michigan shore of Lake Michigan if you stand on top of the Willis Tower. Experience shows that this is correct, which gives us additional confidence in our answer.

Problem-Solving Guidelines: Limits

In Solved Problem 1.2, we found a handy formula, $r = \sqrt{2hR + h^2}$, for how far one can see on the surface of Earth from an elevation h, where R is the radius of Earth. There is another test that we can perform to check the validity of this formula. We did not include it in the Double-Check step, because it deserves seperate consideration. This general problem-solving technique is examining the limits of an equation.

What does "examining the limits" mean? In terms of Solved Problem 1.2, it means that instead of just inserting the given number for h into our formula and computing the solution, we can also step back and think about what should happen to the distance r one can see if h becomes very large or very small. Obviously, the smallest that h can become is zero. In this case, r will also approach zero. This is expected, of course; if your eye level is at ground level, you cannot see very far. On the other hand, we can ponder what happens if h becomes large compared to the radius of Earth (see Figure 1.11). (Yes, it is impossible to build a tower that tall but h could also stand for the altitude of a satellite above ground.) In that case, we expect that the viewing range will eventually simply be the height h. Our formula also bears out this expectation, because as h becomes large compared to R, the first term in the square root can be neglected, and we find $\lim_{h \to \infty} \sqrt{2hR + h^2} = h$.

What we have illustrated by this example is a general guideline: If you derive a formula, you can check its validity by substituting extreme values of the variables in the formula and checking if these limits agree with common sense. Often the formula simplifies drastically

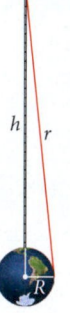

FIGURE 1.11 Viewing range in the limit of very large h.

at a limit. If your formula has a limiting behavior that is correct, it does not necessarily mean that the formula itself is correct, but it gives you additional confidence in its validity.

Problem-Solving Guidelines: Ratios

Another very common class of physics problems asks what happens to a quantity that depends on a certain parameter if that parameter changes by a given factor. These problems provide excellent insight into physical concepts and take almost no time to do. This is true, in general, *if* two conditions are met: First, you have to know what formula to use; and second, you have to know how to solve this general class of problems. But that is a big *if*. Studying will equip your memory with the correct formulas, but you need to acquire the skill of solving problems of this general type.

Here is the trick: Write down the formula that connects the dependent quantity to the parameter that changes. Write it twice, once with the dependent quantity and the parameters indexed (or labeled) with 1 and once with them indexed with 2. Then form ratios of the indexed quantities by dividing the right-hand sides and the left-hand sides of the two equations. Next, insert the factor of change for the parameter (expressed as a ratio) and do the calculation to find the factor of change for the dependent quantity (also expressed as a ratio).

Here is an example that demonstrates this method.

EXAMPLE 1.3 Change in Volume

PROBLEM

If the radius of a cylinder increases by a factor of 2.73, by what factor does the volume change? Assume that the height of the cylinder stays the same.

SOLUTION

The formula that connects the volume of a cylinder, V, and its radius, r, is

$$V = \pi r^2 h.$$

The way the problem is phrased, V is the dependent quantity and r is the parameter it depends on. The height of the cylinder, h, also appears in the equation but remains constant, according to the problem statement.

Following the general problem-solving guideline, we write the equation twice, once with 1 as indexes and once with 2:

$$V_1 = \pi r_1^2 h$$

$$V_2 = \pi r_2^2 h.$$

Now we divide the second equation by the first, obtaining

$$\frac{V_2}{V_1} = \frac{\pi r_2^2 h}{\pi r_1^2 h} = \left(\frac{r_2}{r_1}\right)^2.$$

As you can see, h did not receive an index because it stayed constant in this problem; it canceled out in the division.

The problem states that the change in radius is given by:

$$r_2 = 2.73 r_1.$$

We substitute for r_2 in our ratio:

$$\frac{V_2}{V_1} = \left(\frac{r_2}{r_1}\right)^2 = \left(\frac{2.73 r_1}{r_1}\right)^2 = 2.73^2 = 7.4529,$$

or

$$V_2 = 7.45 V_1,$$

where we have rounded the solution to the three significant digits that the quantity given in the problem had. Thus, the answer is that the volume of the cylinder increases by a factor of 7.45 when you increase its radius by a factor of 2.73.

Problem-Solving Guidelines: Estimation

Sometimes you don't need to solve a physics problem exactly. When an estimate is all that is asked for, knowing the order of magnitude of some quantity is enough. For example, an answer of $1.24 \cdot 10^{20}$ km is for most purposes not much different from $1 \cdot 10^{20}$ km. In such cases, you can round off all the numbers in a problem to the nearest power of 10 and carry out the necessary arithmetic. For example, the calculation in Solved Problem 1.1 reduces to

$$\frac{(20.8 \text{ cm})^2 \cdot (12.2 \text{ cm})}{4\pi} \approx \frac{(2 \cdot 10^1 \text{ cm})^2 \cdot (10 \text{ cm})}{10} = \frac{4 \cdot 10^3 \text{ cm}^3}{10} = 400 \text{ cm}^3,$$

which is pretty close to our answer of 420. cm^3. Even an answer of 100 cm^3 (rounding 20.8 cm to 10 cm) has the correct order of magnitude for the volume. Notice that you can often round the number π to 3 or round π^2 to 10. With practice, you can find more tricks of approximation like these that can make estimations simpler and faster.

The technique of gaining useful results through careful estimation was made famous by the 20th-century physicist Enrico Fermi (1901–1954), who estimated the energy released by the Trinity nuclear explosion on July 16, 1945, near Socorro, New Mexico, by observing how far a piece of paper was blown by the wind from the blast. There is a class of estimation problems called *Fermi problems* that can yield interesting results when reasonable assumptions are made about quantities that are not known exactly.

Estimates are useful to gain insight into a problem before turning to more complicated methods of calculating a precise answer. For example, one could estimate how many tacos people eat and how many taco stands there are in town before investing in a complete business plan to construct a taco stand. In order to practice estimation skills, let's estimate the amount of carbon dioxide that is added to Earth's atmosphere annually by humans breathing.

EXAMPLE 1.4 Greenhouse Gas Production

PROBLEM

The concentration of greenhouse gases, including carbon dioxide (CO_2), in the Earth's atmosphere is increasing. Estimate how much CO_2 is added to the atmosphere each year by humans breathing.

SOLUTION

Since we are asked to estimate, we have to come up with the order of magnitude of the amount. The precise number does not matter so much. Let's start with the amount of CO_2 in one breath, estimate how many breaths each of us takes per year, and then multiply by the number of humans on the planet.

When we breathe, we take in air that is a mixture of 21% oxygen and 78% nitrogen (plus traces of other gases). We exhale air that has approximately 16% oxygen and 5% CO_2. Even though our lung capacity is approximately 3 to 5 L, we only use about 10% of that capacity in normal breathing. So, let's say that one breath of air is approximately 0.4 L. Then, 5% of 0.4 L is $2 \cdot 10^{-2}$ L. You may remember from high school science that 22.4 L of gas comprise 1 mole, and that 1 mole of CO_2 has a mass of $2 \cdot 16 \text{ g} + 12 \text{ g} = 44$ g. This means that in one breath we produce

$$m_1 = \frac{(2 \cdot 10^{-2} \text{ L})(44 \text{ g})}{22.4 \text{ L}} \approx 4 \cdot 10^{-2} \text{ g}$$

of CO_2.

We take a breath of air about once every 4 seconds. (You can use a stopwatch to convince yourself that this is true, or you can count the number of breaths you take in a minute.) This means that we breathe about 1000 times per hour, and since a year has about 10,000 hours, we take $N = 10^7$ breaths per year.

Now, we can put all this together and get our estimate. Humanity (~7 billion, or $7 \cdot 10^9$, humans) produces about

$$M = Nm_1 N_{\text{humans}} = 10^7 (4 \cdot 10^{-2} \text{ g})(7 \cdot 10^9) \approx 3 \cdot 10^{15} \text{ g} = 3 \cdot 10^{12} \text{ kg}$$

of CO_2.

Concept Check 1.5

Estimate the number of gallons of gasoline consumed each day in the United States by commuters driving to work.

a) 10,000 gallons

b) 100,000 gallons

c) 1,000,000 gallons

d) 10,000,000 gallons

e) 100,000,000 gallons

In other words, our estimate indicates that humans add approximately 3 billion metric tons of CO_2 to Earth's atmosphere each year just by breathing. Remember, this is just an estimate; we cannot be sure of the mantissa, but we can be reasonably confident of the exponent. Our answer certainly is billions of tons, but we cannot be sure if it is 1 or 3 billion.

For comparison, measurements indicate that the amount of CO_2 in the atmosphere increases by approximately 15 billion metric tons per year. This amount is clearly much higher than our estimate, but is mainly due to burning of fossil fuels (more on this topic in Chapter 18).

1.6 Vectors

Vectors are mathematical descriptions of quantities that have both magnitude and direction. The magnitude of a vector is a nonnegative number, often combined with a physical unit. Many vector quantities are important in physics and, indeed, in all of science. Therefore, before we start this study of physics, you need to be familiar with vectors and some basic vector operations.

Vectors have a starting point and an ending point. For example, consider a flight from Seattle to New York. To represent the change of the plane's position, we can draw an arrow from the plane's departure point to its destination (Figure 1.12). (Real flight paths are not exactly straight lines due to the fact that the Earth is a sphere and due to airspace restrictions and air traffic regulations, but a straight line is a reasonable approximation for our purpose.) This arrow represents a *displacement vector*, which always goes from somewhere to somewhere else. Any vector quantity has a magnitude and a direction. If the vector represents a physical quantity, such as displacement, it will also have a physical unit. A quantity that can be represented without giving a direction is called a **scalar.** A scalar quantity has just a magnitude and possibly a physical unit. Examples of scalar quantities are time and temperature.

This book denotes a vector quantity by a letter with a small horizontal arrow pointing to the right above it. For example, in the drawing of the trip from Seattle to New York (Figure 1.12), the displacement vector has the symbol $\vec{C}$. In the rest of this section, you will learn how to work with vectors: how to add and subtract them and how to multiply them. In order to perform these operations, it is very useful to introduce a coordinate system in which to represent vectors.

Cartesian Coordinate System

A **Cartesian coordinate system** is defined as a set of two or more axes with angles of 90° between each pair. These axes are said to be orthogonal to each other. In a two-dimensional space, the coordinate axes are typically labeled x and y. We can then uniquely specify any point P in the two-dimensional space by giving its coordinates P_x and P_y along the two coordinate axes, as shown in Figure 1.13. We will use the notation (P_x, P_y) to specify a

FIGURE 1.12 Flight path from Seattle to New York as an example of a vector.

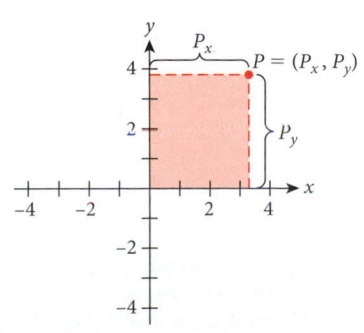

FIGURE 1.13 Representation of a point P in two-dimensional space in terms of its Cartesian coordinates.

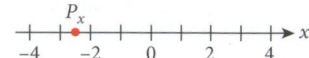

FIGURE 1.14 Representation of a point *P* in a one-dimensional Cartesian coordinate system.

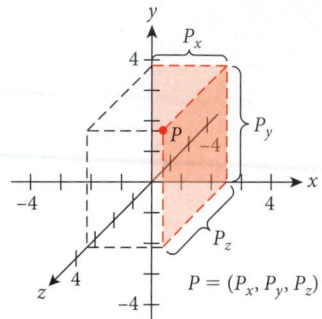

FIGURE 1.15 Representation of a point *P* in a three-dimensional space in terms of its Cartesian coordinates.

point in terms of its coordinates. In Figure 1.13, for example, the point *P* has the position (3.3,3.8), because its *x*-coordinate has a value of 3.3 and its *y*-coordinate has a value of 3.8. Note that each coordinate is a number and can have a positive or negative value or be zero.

We can also define a one-dimensional coordinate system, for which any point is located on a single straight line, conventionally called the *x*-axis. Any point in this one-dimensional space is then uniquely defined by specifying one number, the value of the *x*-coordinate, which again can be negative, zero, or positive (Figure 1.14). The point *P* in Figure 1.14 has the *x*-coordinate $P_x = -2.5$.

Clearly, one- and two-dimensional coordinate systems are easy to draw, because the surface of paper has two dimensions. In a three-dimensional coordinate system, the third coordinate axis is perpendicular to the other two; thus, to be represented accurately, it would have to stick straight out of the plane of the page. In order to draw a three-dimensional coordinate system, we have to rely on conventions that make use of the techniques for perspective drawings. We represent the third axis by a line that is at a 45° angle with the other two (Figure 1.15).

In a three-dimensional space, we have to specify three numbers to uniquely determine the coordinates of a point. We use the notation $P = (P_x, P_y, P_z)$ to accomplish this. It is possible to construct Cartesian coordinate systems with more than three orthogonal axes, although they are almost impossible to visualize. Modern string theories (described in Chapter 39), for example, are usually constructed in 10-dimensional spaces. However, for the purposes of this book and for almost all of physics, three dimensions are sufficient. As a matter of fact, for most applications, the essential mathematical and physical understanding can be obtained from two-dimensional representations.

Cartesian Representation of Vectors

The example of the flight from Seattle to New York established that vectors are characterized by two points: start and finish, represented by the tail and head of an arrow, respectively. Using the Cartesian representation of points, we can define the Cartesian representation of a displacement vector as the difference in the coordinates of the end point and the starting point. Since the difference between the two points for a vector is all that matters, we can shift the vector around in space as much as we like. As long as we do not change the length or direction of the arrow, the vector remains mathematically the same. Consider the two vectors in Figure 1.16.

Figure 1.16a shows the displacement vector $\vec{A}$ that points from point $P = (-2, -3)$ to point $Q = (3, 1)$. With the notation just introduced, the **components** of $\vec{A}$ are the coordinates of point *Q* minus those of point *P*, $\vec{A} = (3-(-2), 1-(-3)) = (5, 4)$. Figure 1.16b shows another vector from point $R = (-3, -1)$ to point $S = (2, 3)$. The difference between these coordinates is $(2-(-3), 3-(-1)) = (5, 4)$, which is the same as the vector $\vec{A}$ pointing from *P* to *Q*.

For simplicity, we can shift the beginning of a vector to the origin of the coordinate system, and the components of the vector will be the same as the coordinates of its end point (Figure 1.17). As a result, we see that we can represent a vector in Cartesian coordinates as

$$\vec{A} = (A_x, A_y) \text{ in two-dimensional space} \tag{1.7}$$

$$\vec{A} = (A_x, A_y, A_z) \text{ in three-dimensional space} \tag{1.8}$$

where A_x, A_y, and A_z are numbers. Note that the notation for a point in Cartesian coordinates is similar to the notation for a vector in Cartesian coordinates. Whether the notation specifies a point or a vector will be clear from the context of the reference.

Graphical Vector Addition and Subtraction

Suppose that the direct flight from Seattle to New York shown in Figure 1.12 was not available, and you had to make a connection through Dallas (Figure 1.18). Then the displacement vector $\vec{C}$ for the flight from Seattle to New York is the sum of a displacement vector $\vec{A}$ from Seattle to Dallas and a displacement vector $\vec{B}$ from Dallas to New York:

$$\vec{C} = \vec{A} + \vec{B}. \tag{1.9}$$

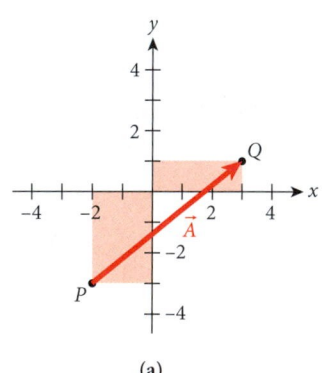

(a)

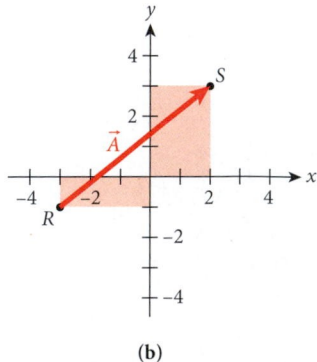

(b)

FIGURE 1.16 Cartesian representations of a vector $\vec{A}$. (a) Displacement vector from *P* to *Q*; (b) displacement vector from *R* to *S*.

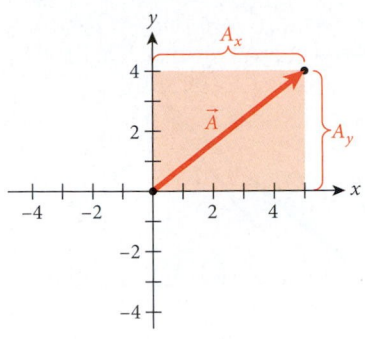

FIGURE **1.17** Cartesian components of vector $\vec{A}$ in two dimensions.

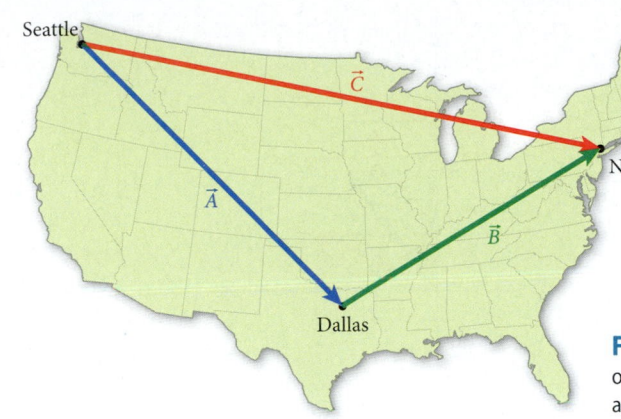

FIGURE **1.18** Direct flight versus one-stop flight as an example of vector addition.

This example shows the general procedure for vector addition in a graphical way: Move the tail of vector $\vec{B}$ to the head of vector $\vec{A}$; then the vector from the tail of vector $\vec{A}$ to the head of vector $\vec{B}$ is the sum vector, or **resultant,** of the two.

If you add two real numbers, the order does not matter: $3+5=5+3$. This property is called the *commutative property of addition*. Vector addition is also commutative:

$$\vec{A} + \vec{B} = \vec{B} + \vec{A}. \tag{1.10}$$

Figure 1.19 demonstrates this commutative property of vector addition graphically. It shows the same vectors as in Figure 1.18, but also shows the beginning of vector $\vec{A}$ moved to the tip of vector $\vec{B}$ (dashed arrows)—note that the resultant vector is the same as before.

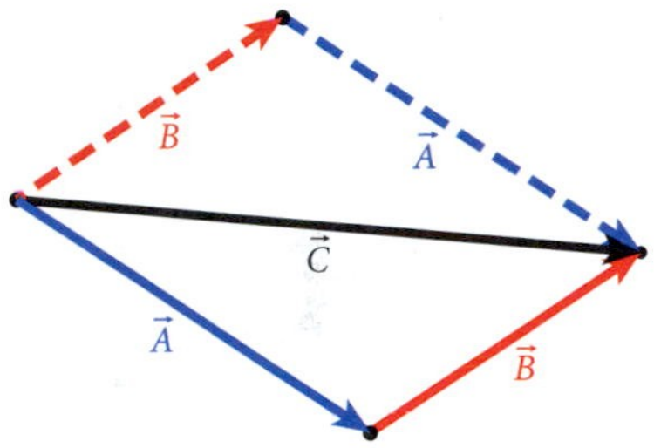

Next, the inverse (or reverse or negative) vector, $-\vec{C}$, of the vector $\vec{C}$ is a vector with the same length as $\vec{C}$ but pointing in the opposite direction (Figure 1.20). For the vector representing the flight from Seattle to New York, for example, the inverse vector is the return trip. Clearly, if you add $\vec{C}$ and its inverse vector, $-\vec{C}$, you end up at the point you started from. Thus, we find

$$\vec{C} + (-\vec{C}) = \vec{C} - \vec{C} = (0,0,0), \tag{1.11}$$

and the magnitude is zero, $|\vec{C} - \vec{C}| = 0$. This seemingly simple identity shows that we can treat vector subtraction as vector addition, by simply adding the inverse vector. For example, the vector $\vec{B}$ in Figure 1.19 can be obtained as $\vec{B} = \vec{C} - \vec{A}$. Therefore, vector addition and subtraction follow exactly the same rules as the addition and subtraction of real numbers.

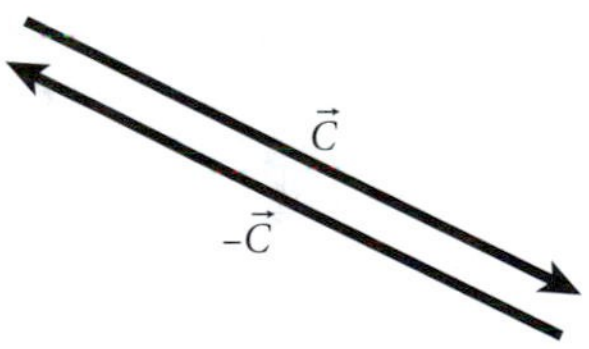

FIGURE **1.19** Commutative property of vector addition.

FIGURE **1.20** Inverse vector $-\vec{C}$ of a vector $\vec{C}$.

Vector Addition Using Components

Graphical vector addition illustrates the concepts very well, but for practical purposes the component method of vector addition is much more useful. (This is because calculators are easier to use and much more precise than rulers and graph paper.) Let's consider the component method for addition of three-dimensional vectors. The equations for two-dimensional vectors are special cases that arise by neglecting the z-components. Similarly, the one-dimensional equation can be obtained by neglecting all y- and z-components.

If you add two three-dimensional vectors, $\vec{A}=(A_x, A_y, A_z)$ and $\vec{B}=(B_x, B_y, B_z)$, the resulting vector is

$$\vec{C} = \vec{A} + \vec{B} = (A_x, A_y, A_z) + (B_x, B_y, B_z) = (A_x + B_x, A_y + B_y, A_z + B_z). \tag{1.12}$$

In other words, the components of the sum vector are the sums of the components of the individual vectors:

$$\begin{aligned} C_x &= A_x + B_x \\ C_y &= A_y + B_y \\ C_z &= A_z + B_z. \end{aligned} \tag{1.13}$$

The relationship between graphical and component methods is illustrated in Figure 1.21. Figure 1.21a shows two vectors $\vec{A} = (4, 2)$ and $\vec{B} = (3, 4)$ in two-dimensional space, and

FIGURE 1.21 Vector addition by components. (a) Components of vectors $\vec{A}$ and $\vec{B}$; (b) the components of the resultant vector are the sums of the components of the individual vectors.

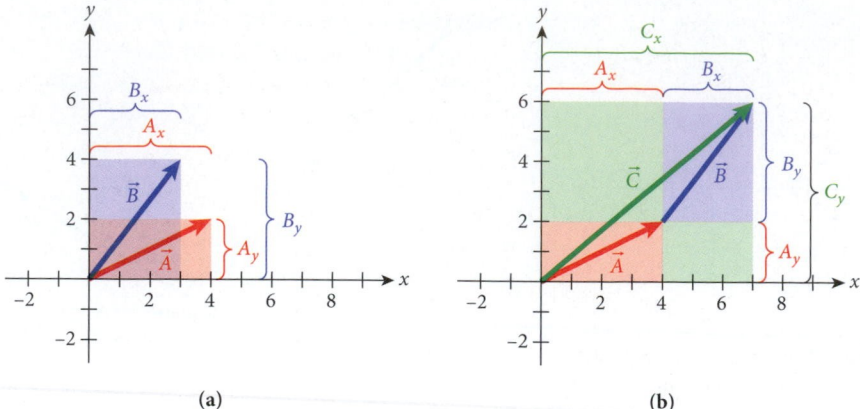

(a)   (b)

Figure 1.21b displays their sum vector $\vec{C} = (4+3, 2+4) = (7,6)$. Figure 1.21b clearly shows that $C_x = A_x + B_x$ because the whole is equal to the sum of its parts.

In the same way, we can take the difference $\vec{D} = \vec{A} - \vec{B}$, and the Cartesian components of the difference vector are given by

$$D_x = A_x - B_x$$
$$D_y = A_y - B_y \tag{1.14}$$
$$D_z = A_z - B_z .$$

Multiplication of a Vector with a Scalar

What is $\vec{A} + \vec{A} + \vec{A}$? If your answer to this question is $3\vec{A}$, you already understand multiplying a vector with a scalar. The vector that results from multiplying the vector $\vec{A}$ with the scalar 3 is a vector that points in the same direction as the original vector $\vec{A}$ but is 3 times as long.

Multiplication of a vector with an arbitrary positive scalar—that is, a positive number—results in another vector that points in the same direction but has a magnitude that is the product of the magnitude of the original vector and the value of the scalar. Multiplication of a vector by a negative scalar results in a vector pointing in the opposite direction to the original with a magnitude that is the product of the magnitude of the original vector and the magnitude of the scalar.

Again, the component notation is useful. For the multiplication of a vector $\vec{A}$ with a scalar s, we obtain:

$$\vec{E} = s\vec{A} = s(A_x, A_y, A_z) = (sA_x, sA_y, sA_z). \tag{1.15}$$

In other words, each component of the vector $\vec{A}$ is multiplied by the scalar in order to arrive at the components of the product vector:

$$E_x = sA_x$$
$$E_y = sA_y \tag{1.16}$$
$$E_z = sA_z .$$

Unit Vectors

There is a set of special vectors that make much of the math associated with vectors easier. Called **unit vectors,** they are vectors of magnitude 1 directed along the main coordinate axes of the coordinate system. In two dimensions, these vectors point in the positive x-direction and the positive y-direction. In three dimensions, a third unit vector points in the positive z-direction. In order to distinguish these as unit vectors, we give them the symbols $\hat{x}$, $\hat{y}$, and $\hat{z}$. Their component representation is

$$\hat{x} = (1,0,0)$$
$$\hat{y} = (0,1,0) \tag{1.17}$$
$$\hat{z} = (0,0,1).$$

Figure 1.22a shows the unit vectors in two dimensions, and Figure 1.22b shows the unit vectors in three dimensions.

What is the advantage of unit vectors? We can write any vector as a sum of these unit vectors instead of using the component notation; each unit vector is multiplied by the corresponding Cartesian component of the vector:

$$
\begin{aligned}
\vec{A} &= (A_x, A_y, A_z) \\
&= (A_x, 0, 0) + (0, A_y, 0) + (0, 0, A_z) \\
&= A_x(1,0,0) + A_y(0,1,0) + A_z(0,0,1) \\
&= A_x\hat{x} + A_y\hat{y} + A_z\hat{z}.
\end{aligned}
\tag{1.18}
$$

In two dimensions, we have

$$
\vec{A} = A_x\hat{x} + A_y\hat{y}.
\tag{1.19}
$$

This unit vector representation of a general vector will be particularly useful for multiplying two vectors.

Vector Length and Direction

If we know the component representation of a vector, how can we find its length (magnitude) and the direction it is pointing in? Let's look at the most important case: a vector in two dimensions. In two dimensions, a vector $\vec{A}$ can be specified uniquely by giving the two Cartesian components, A_x and A_y. We can also specify the same vector by giving two other numbers: its length A and its angle θ with respect to the positive x-axis.

Let's take a look at Figure 1.23 to see how we can determine A and θ from A_x and A_y. Figure 1.23a shows the graphical representation of equation 1.19. The vector $\vec{A}$ is the sum of the vectors $A_x\hat{x}$ and $A_y\hat{y}$. Since the unit vectors $\hat{x}$ and $\hat{y}$ are by definition orthogonal to each other, these vectors form a 90° angle. Thus, the three vectors $\vec{A}$, $A_x\hat{x}$, and $A_y\hat{y}$ form a right triangle with side lengths A, A_x, and A_y, as shown in Figure 1.23b.

Now we can employ basic trigonometry to find θ and A. Using the Pythagorean Theorem results in

$$
A = \sqrt{A_x^2 + A_y^2}.
\tag{1.20}
$$

We can find the angle θ from the definition of the tangent function

$$
\theta = \tan^{-1}\frac{A_y}{A_x}.
\tag{1.21}
$$

In using equation 1.21, you must be careful that θ is in the correct quadrant. We can also invert equations 1.20 and 1.21 to obtain the Cartesian components of a vector of given length and direction:

$$
A_x = A\cos\theta
\tag{1.22}
$$

$$
A_y = A\sin\theta.
\tag{1.23}
$$

You will encounter these trigonometric relations again and again throughout introductory physics. If you need to refamiliarize yourself with trigonometry, consult the mathematics primer provided in Appendix A.

Scalar Product of Vectors

Above we saw how to multiply a vector with a scalar. Now we will define one way of multiplying a vector with a vector and obtain the **scalar product.** The scalar product of two vectors $\vec{A}$ and $\vec{B}$ is defined as

$$
\vec{A}\cdot\vec{B} = \left|\vec{A}\right|\left|\vec{B}\right|\cos\alpha,
\tag{1.24}
$$

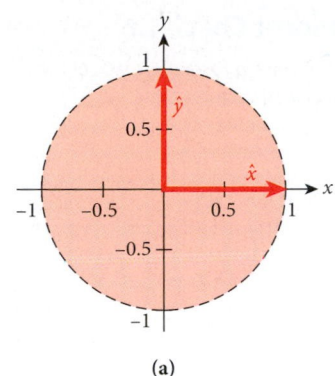

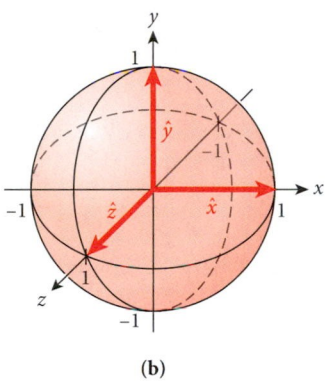

FIGURE 1.22 Cartesian unit vectors in (a) two and (b) three dimensions.

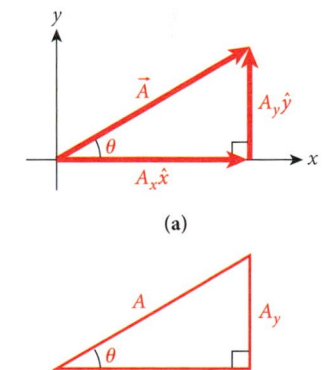

FIGURE 1.23 Length and direction of a vector. (a) Cartesian components A_x and A_y; (b) length A and angle θ.

Concept Check 1.6

Into which quadrant do each of the following vectors point?

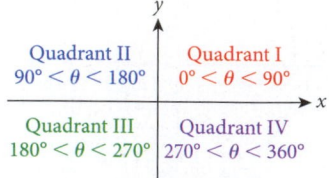

a) $A = (A_x, A_y)$ with $A_x = 1.5$ cm, $A_y = -1.0$ cm

b) a vector with length 2.3 cm and direction angle 131°

c) the inverse vector of $B = (0.5$ cm, 1.0 cm)

d) the sum of the unit vectors in the x- and y-directions

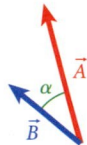

FIGURE 1.24 Two vectors $\vec{A}$ and $\vec{B}$ and the angle α between them.

where α is the angle between the vectors $\vec{A}$ and $\vec{B}$, as shown in Figure 1.24. Note the use of the larger dot (•) as the multiplication sign for the scalar product between vectors, in contrast to the smaller dot (·) that is used for the multiplication of scalars. Because of the dot, the scalar product is often referred to as the *dot product*.

If two vectors form a 90° angle, then the scalar product has the value zero. In this case, the two vectors are orthogonal to each other. The scalar product of a pair of orthogonal vectors is zero.

If $\vec{A}$ and $\vec{B}$ are given in Cartesian coordinates as $\vec{A} = (A_x, A_y, A_z)$ and $\vec{B} = (B_x, B_y, B_z)$, then their scalar product can be shown to be equal to:

$$\vec{A} \bullet \vec{B} = (A_x, A_y, A_z) \bullet (B_x, B_y, B_z) = A_x B_x + A_y B_y + A_z B_z. \tag{1.25}$$

From equation 1.25, we can see that the scalar product has the commutative property:

$$\vec{A} \bullet \vec{B} = \vec{B} \bullet \vec{A}. \tag{1.26}$$

This result is not surprising, since the commutative property also holds for the multiplication of two scalars.

For the scalar product of any vector with itself, we have, in component notation, $\vec{A} \bullet \vec{A} = A_x^2 + A_y^2 + A_z^2$. Then, from equation 1.24, we find $\vec{A} \bullet \vec{A} = |\vec{A}| \, |\vec{A}| \cos\alpha = |\vec{A}| \, |\vec{A}| = |\vec{A}|^2$ (because the angle between the vector $\vec{A}$ and itself is zero, and the cosine of that angle has the value 1). Combining these two equations, we obtain the expression for the length of a vector that was introduced in the previous subsection:

$$|\vec{A}| = \sqrt{A_x^2 + A_y^2 + A_z^2}. \tag{1.27}$$

We can also use the definition of the scalar product to compute the angle between two arbitrary vectors in three-dimensional space:

$$\vec{A} \bullet \vec{B} = |\vec{A}| \, |\vec{B}| \cos\alpha \Rightarrow \cos\alpha = \frac{\vec{A} \bullet \vec{B}}{|\vec{A}| \, |\vec{B}|} \Rightarrow \alpha = \cos^{-1}\left(\frac{\vec{A} \bullet \vec{B}}{|\vec{A}| \, |\vec{B}|}\right). \tag{1.28}$$

For the scalar product, the same distributive property that is valid for the conventional multiplication of numbers holds:

$$\vec{A} \bullet (\vec{B} + \vec{C}) = \vec{A} \bullet \vec{B} + \vec{A} \bullet \vec{C}. \tag{1.29}$$

The following example puts the scalar product to use.

EXAMPLE 1.5 | **Angle Between Two Position Vectors**

PROBLEM
What is the angle α between the two position vectors shown in Figure 1.25, $\vec{A} = (4.00, 2.00, 5.00)$ cm and $\vec{B} = (4.50, 4.00, 3.00)$ cm?

SOLUTION
To solve this problem, we have to put the numbers for the components of each of the two vectors into equation 1.27 and equation 1.25 then use equation 1.28:

$$|\vec{A}| = \sqrt{4.00^2 + 2.00^2 + 5.00^2} \text{ cm} = 6.71 \text{ cm}$$

$$|\vec{B}| = \sqrt{4.50^2 + 4.00^2 + 3.00^2} \text{ cm} = 6.73 \text{ cm}$$

$$\vec{A} \bullet \vec{B} = A_x B_x + A_y B_y + A_z B_z = (4.00 \cdot 4.50 + 2.00 \cdot 4.00 + 5.00 \cdot 3.00) \text{ cm}^2 = 41.0 \text{ cm}^2$$

$$\Rightarrow \alpha = \cos^{-1}\left(\frac{41.0 \text{ cm}^2}{6.71 \text{ cm} \cdot 6.73 \text{ cm}}\right) = 24.7°.$$

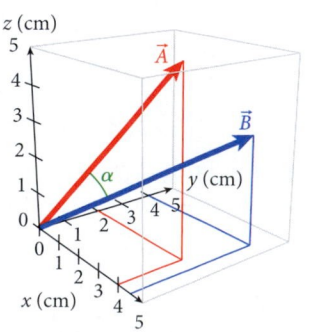

FIGURE 1.25 Calculating the angle between two position vectors.

Scalar Product for Unit Vectors.   On page 26 we introduced unit vectors in the three-dimensional Cartesian coordinate system: $\hat{x} = (1,0,0)$, $\hat{y} = (0,1,0)$, and $\hat{z} = (0,0,1)$. With our definition (1.25) of the scalar product, we find

$$\hat{x} \bullet \hat{x} = \hat{y} \bullet \hat{y} = \hat{z} \bullet \hat{z} = 1 \tag{1.30}$$

and

$$\hat{x} \bullet \hat{y} = \hat{x} \bullet \hat{z} = \hat{y} \bullet \hat{z} = 0$$
$$\hat{y} \bullet \hat{x} = \hat{z} \bullet \hat{x} = \hat{z} \bullet \hat{y} = 0. \tag{1.31}$$

Self-Test Opportunity 1.1

Show that equations 1.30 and 1.31 are correct by using equation 1.25 and the definitions of the unit vectors.

Now we see why the unit vectors are called that: Their scalar products with themselves have the value 1. Thus, the unit vectors have length 1, or unit length, according to equation 1.27. In addition, any pair of different unit vectors has a scalar product that is zero, meaning that these vectors are orthogonal to each other. Equations 1.30 and 1.31 thus state that the unit vectors $\hat{x}$, $\hat{y}$, and $\hat{z}$ form an orthonormal set of vectors, which makes them extremely useful for the description of physical systems.

Geometrical Interpretation of the Scalar Product.   In the definition of the scalar product $\vec{A} \bullet \vec{B} = |\vec{A}| \, |\vec{B}| \cos\alpha$ (equation 1.24), we can interpret $|\vec{A}|\cos\alpha$ as the projection of the vector $\vec{A}$ onto the vector $\vec{B}$ (Figure 1.26a). In this drawing, the line $|\vec{A}|\cos\alpha$ is rotated by 90° to show the geometrical interpretation of the scalar product as the area of a rectangle with sides $|\vec{A}|\cos\alpha$ and $|\vec{B}|$. In the same way, we can interpret $|\vec{B}|\cos\alpha$ as the projection of the vector $\vec{B}$ onto the vector $\vec{A}$ and construct a rectangle with side lengths $|\vec{B}|\cos\alpha$ and $|\vec{A}|$ (Figure 1.26b). The areas of the two yellow rectangles in Figure 1.26 are identical and are equal to the scalar product of the two vectors $\vec{A}$ and $\vec{B}$.

Finally, if we substitute from equation 1.28 for the cosine of the angle between the two vectors, the projection $|\vec{A}|\cos\alpha$ of the vector $\vec{A}$ onto the vector $\vec{B}$ can be written as

$$|\vec{A}|\cos\alpha = |\vec{A}| \frac{\vec{A} \bullet \vec{B}}{|\vec{A}| \, |\vec{B}|} = \frac{\vec{A} \bullet \vec{B}}{|\vec{B}|},$$

and the projection $|\vec{B}|\cos\alpha$ of the vector $\vec{B}$ onto the vector $\vec{A}$ can be expressed as

$$|\vec{B}|\cos\alpha = \frac{\vec{A} \bullet \vec{B}}{|\vec{A}|}.$$

Vector Product

The **vector product** (or cross product) between two vectors $\vec{A} = (A_x, A_y, A_z)$ and $\vec{B} = (B_x, B_y, B_z)$ is defined as

$$\vec{C} = \vec{A} \times \vec{B}$$
$$C_x = A_y B_z - A_z B_y$$
$$C_y = A_z B_x - A_x B_z \tag{1.32}$$
$$C_z = A_x B_y - A_y B_x.$$

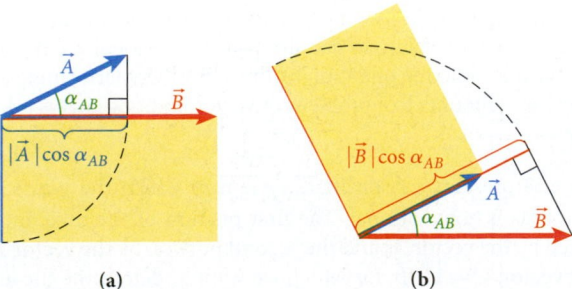

(a)   (b)

FIGURE 1.26 Geometrical interpretation of the scalar product as an area. (a) The projection of $\vec{A}$ onto $\vec{B}$. (b) The projection of $\vec{B}$ onto $\vec{A}$.

In particular, for the vector products of the Cartesian unit vectors, this definition implies

$$\hat{x} \times \hat{y} = \hat{z}$$
$$\hat{y} \times \hat{z} = \hat{x} \qquad (1.33)$$
$$\hat{z} \times \hat{x} = \hat{y}.$$

The absolute magnitude of the vector $\vec{C}$ is given by

$$|\vec{C}| = |\vec{A}||\vec{B}| \sin \theta. \qquad (1.34)$$

Here θ is the angle between $\vec{A}$ and $\vec{B}$, as shown in Figure 1.27. This result implies that the magnitude of the vector product of two vectors is at its maximum when $\vec{A} \perp \vec{B}$ and is zero when $\vec{A} \parallel \vec{B}$. We can also interpret the right-hand side of this equation as either the product of the magnitude of vector $\vec{A}$ times the component of $\vec{B}$ perpendicular to $\vec{A}$ or the product of the magnitude of $\vec{B}$ times the component of $\vec{A}$ perpendicular to $\vec{B}$: $|\vec{C}| = |\vec{A}| B_{\perp A} = |\vec{B}| A_{\perp B}$. Either interpretation is valid.

The direction of the vector $\vec{C}$ can be found using the right-hand rule: If vector $\vec{A}$ points along the direction of the thumb and vector $\vec{B}$ points along the direction of the index finger, then the vector product is perpendicular to *both* vectors and points along the direction of the middle finger, as shown in Figure 1.27.

It is important to realize that for the vector product, the order of the factors matters:

$$\vec{B} \times \vec{A} = -\vec{A} \times \vec{B}. \qquad (1.35)$$

Thus, the vector product differs from both regular multiplication of scalars and multiplication of vectors to form a scalar product.

We'll see immediately from the definition of the vector product that for any vector $\vec{A}$, the vector product with itself is always zero:

$$\vec{A} \times \vec{A} = 0. \qquad (1.36)$$

Finally, there is a handy rule for a double vector product involving three vectors: The vector product of the vector $\vec{A}$ with the vector product of the vectors $\vec{B}$ and $\vec{C}$ is the sum of two vectors, one pointing in the direction of the vector $\vec{B}$ and multiplied by the scalar product $\vec{A} \cdot \vec{C}$ and another one pointing in the direction of the vector $\vec{C}$ and multiplied by $-\vec{A} \cdot \vec{B}$:

$$\vec{A} \times (\vec{B} \times \vec{C}) = \vec{B}(\vec{A} \cdot \vec{C}) - \vec{C}(\vec{A} \cdot \vec{B}). \qquad (1.37)$$

This *BAC-CAB rule* is straightforward to prove using the Cartesian components in the definitions of the vector product and the scalar product, but the proof is cumbersome and thus omitted here. However, the rule occasionally comes in handy, in particular when dealing with torque and angular momentum. And the mnemonic BAC-CAB makes it fairly easy to remember.

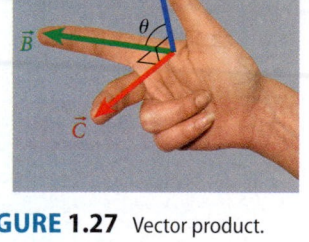

FIGURE 1.27 Vector product.

SOLVED PROBLEM 1.3 / Hiking

PROBLEM
You are hiking in the Florida Everglades heading southwest from your base camp, for 1.72 km. You reach a river that is too deep to cross; so you make a 90° right turn and hike another 3.12 km to a bridge. How far away are you from your base camp?

SOLUTION
THINK If you are hiking, you are moving in a two-dimensional plane: the surface of Earth (because the Everglades are flat). Thus, we can use two-dimensional vectors to characterize the various segments of the hike. Making one straight-line hike, then performing a turn, followed by another straight-line hike amounts to a problem of vector addition that is asking for the length of the resultant vector.

SKETCH Figure 1.28 presents a coordinate system in which the y-axis points north and the x-axis points east, as is conventional. The first portion of the hike, in the southwestern direction, is indicated by the vector $\vec{A}$, and the second portion by the vector $\vec{B}$. The figure also shows the resultant vector, $\vec{C} = \vec{A} + \vec{B}$, for which we want to determine the length.

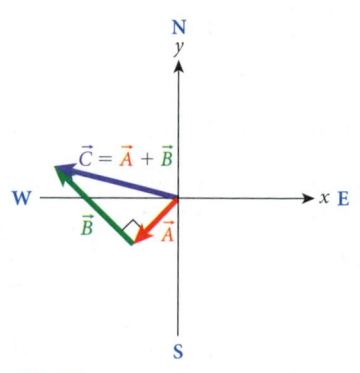

FIGURE 1.28 Hike with a 90° turn.

RESEARCH If you have drawn the sketch with sufficient accuracy, making the lengths of the vectors in your drawing to be proportional to the lengths of the segments of the hike (as was done in Figure 1.28), then you can measure the length of the vector $\vec{C}$ to determine the distance from your base camp at the end of the second segment of the hike. However, the given distances are specified to three significant digits, so the answer should also have three significant digits. Thus, we cannot rely on the graphical method but must use the component method of vector addition.

In order to calculate the components of the vectors, we need to know their angles relative to the positive x-axis. For the vector $\vec{A}$, which points southwest, this angle is $\theta_A = 225°$, as shown in Figure 1.29. The vector $\vec{B}$ has an angle of 90° relative to $\vec{A}$, and thus $\theta_B = 135°$ relative to the positive x-axis. To make this point clearer, the starting point of $\vec{B}$ has been moved to the origin of the coordinate system in Figure 1.29. (Remember: We can move vectors around at will. As long as we leave the direction and length of a vector the same, the vector remains unchanged.)

Now we have everything in place to start our calculation. We have the lengths and directions of both vectors, allowing us to calculate their Cartesian components. Then, we will add their components to calculate the components of the vector $\vec{C}$, from which we can calculate the length of this vector.

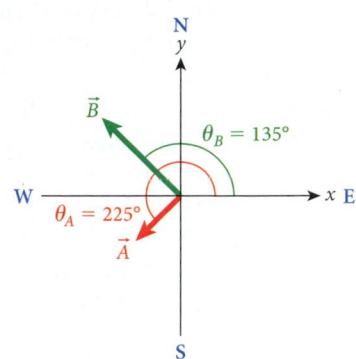

FIGURE 1.29 Angles of the two hike segments.

SIMPLIFY The components of the vector $\vec{C}$ are:

$$C_x = A_x + B_x = A\cos\theta_A + B\cos\theta_B$$
$$C_y = A_y + B_y = A\sin\theta_A + B\sin\theta_B.$$

Thus, the length of the vector $\vec{C}$ is (compare with equation 1.20)

$$C = \sqrt{C_x^2 + C_y^2} = \sqrt{(A_x + B_x)^2 + (A_y + B_y)^2}$$
$$= \sqrt{(A\cos\theta_A + B\cos\theta_B)^2 + (A\sin\theta_A + B\sin\theta_B)^2}.$$

CALCULATE Now all that is left is to put in the numbers to obtain the vector length:

$$C = \sqrt{\left((1.72\text{ km})\cos225° + (3.12\text{ km})\cos135°\right)^2 + \left((1.72\text{ km})\sin225° + (3.12\text{ km})\sin135°\right)^2}$$
$$= \sqrt{\left(1.72\cdot(-\sqrt{1/2}) + 3.12\cdot(-\sqrt{1/2})\right)^2 + \left((1.72\cdot(-\sqrt{1/2}) + 3.12\cdot\sqrt{1/2}\right)^2}\text{ km}.$$

Entering these numbers into a calculator, we obtain:

$$C = 3.562695609\text{ km}.$$

ROUND Because the initial distances were given to three significant figures, our final answer should also have (at most) the same precision. Rounding to three significant figures yields our final answer:

$$C = 3.56\text{ km}.$$

DOUBLE-CHECK This problem was intended to provide practice with vector concepts. However, if you forget for a moment that the displacements are vectors and note that they form a right triangle, you can immediately calculate the length of side C from the Pythagorean Theorem as follows:

$$C = \sqrt{A^2 + B^2} = \sqrt{1.72^2 + 3.12^2}\text{ km} = 3.56\text{ km}.$$

Here we also rounded our result to three significant figures, and we see that it agrees with the answer obtained using the longer procedure of vector addition.

WHAT WE HAVE LEARNED | EXAM STUDY GUIDE

- Large and small numbers can be represented using scientific notation, consisting of a mantissa and a power of ten.

- Physical systems are described by the SI system of units. These units are based on reproducible standards and provide convenient methods of scaling and calculation. The base units of the SI system include meter (m), kilogram (kg), second (s), and ampere (A).

- Physical systems have widely varying sizes, masses, and time scales, but the same physical laws govern all of them.

- A number (with a specific number of significant figures) or a set of numbers (such as components of a vector) must be combined with a unit or units to describe physical quantities.

- Vectors in three dimensions can be specified by their three Cartesian components, $\vec{A} = (A_x, A_y, A_z)$. Each of these Cartesian components is a number.

- Vectors can be added or subtracted. In Cartesian components, $\vec{C} = \vec{A} + \vec{B} = (A_x, A_y, A_z) + (B_x, B_y, B_z)$

  $= (A_x + B_x, A_y + B_y, A_z + B_z)$.

- Multiplication of a vector with a scalar results in another vector in the same or opposite direction but of different magnitude, $\vec{E} = s\vec{A} = s(A_x, A_y, A_z) = (sA_x, sA_y, sA_z)$.

- Unit vectors are vectors of length 1. The unit vectors in Cartesian coordinate systems are denoted by $\hat{x}$, $\hat{y}$, and $\hat{z}$.

- The length and direction of a two-dimensional vector can be determined from its Cartesian components: $A = \sqrt{A_x^2 + A_y^2}$ and $\theta = \tan^{-1}(A_y / A_x)$.

- The Cartesian components of a two-dimensional vector can be calculated from the vector's length and angle with respect to the x-axis: $A_x = A\cos\theta$ and $A_y = A\sin\theta$.

- The scalar product, or dot product, of two vectors yields a scalar quantity and is defined as $\vec{A} \cdot \vec{B} = A_x B_x + A_y B_y + A_z B_z$.

- The vector product, or cross product, of two vectors yields another vector and is defined as $\vec{A} \times \vec{B} = \vec{C} = (A_y B_z - A_z B_y, A_z B_x - A_x B_z, A_x B_y - A_y B_x)$.

ANSWERS TO SELF-TEST OPPORTUNITIES

1.1 Equation 1.30

$\hat{x} \cdot \hat{x} = (1,0,0) \cdot (1,0,0) = 1 \cdot 1 + 0 \cdot 0 + 0 \cdot 0 = 1$

$\hat{y} \cdot \hat{y} = (0,1,0) \cdot (0,1,0) = 0 \cdot 0 + 1 \cdot 1 + 0 \cdot 0 = 1$

$\hat{z} \cdot \hat{z} = (0,0,1) \cdot (0,0,1) = 0 \cdot 0 + 0 \cdot 0 + 1 \cdot 1 = 1$

Equation 1.31

$\hat{x} \cdot \hat{y} = (1,0,0) \cdot (0,1,0) = 1 \cdot 0 + 0 \cdot 1 + 0 \cdot 0 = 0$

$\hat{x} \cdot \hat{z} = (1,0,0) \cdot (0,0,1) = 1 \cdot 0 + 0 \cdot 0 + 0 \cdot 1 = 0$

$\hat{y} \cdot \hat{z} = (0,1,0) \cdot (0,0,1) = 0 \cdot 0 + 1 \cdot 0 + 0 \cdot 1 = 0$

$\hat{y} \cdot \hat{x} = (0,1,0) \cdot (1,0,0) = 0 \cdot 1 + 1 \cdot 0 + 0 \cdot 0 = 0$

$\hat{z} \cdot \hat{x} = (0,0,1) \cdot (1,0,0) = 0 \cdot 1 + 0 \cdot 0 + 1 \cdot 0 = 0$

$\hat{z} \cdot \hat{y} = (0,0,1) \cdot (0,1,0) = 0 \cdot 0 + 0 \cdot 1 + 1 \cdot 0 = 0$

PROBLEM-SOLVING GUIDELINES: NUMBERS, UNITS, AND VECTORS

1. Try to use our seven-step strategy for problem solving, even if you have no idea how to arrive at the final solution. Sometimes the process of making a sketch can give you a hint about what to do next.

2. In general, you should try to convert all given units to SI units before you start working with numbers. Working with quantities in SI units makes computations easier.

3. In most situations, the number of significant figures your final solution should be rounded to is the number of significant figures in the least precisely given quantity.

4. Estimating a solution for a problem can be very useful for obtaining an idea of the order of magnitude of the solution. Often, estimating can be used as a double-check.

5. In working with vectors, you should generally use the Cartesian coordinate system. Make use of your knowledge of trigonometry when solving vector problems!

6. The graphical method for vector addition and subtraction is useful for making sketches. But the component method is more precise, and thus preferred if you have to arrive at a numerical answer.

MULTIPLE-CHOICE QUESTIONS

1.1 Which of the following is the frequency of the musical note C5?

a) 376 g b) 483 m/s c) 523 Hz d) 26.5 J

1.2 If $\vec{A}$ and $\vec{B}$ are vectors and $\vec{B} = -\vec{A}$, which of the following is true?

a) The magnitude of $\vec{B}$ is equal to the negative of the magnitude of $\vec{A}$.

b) $\vec{A}$ and $\vec{B}$ are perpendicular.

c) The direction angle of $\vec{B}$ is equal to the direction angle of $\vec{A}$ plus 180°.

d) $\vec{A} + \vec{B} = 2\vec{A}$.

1.3 Compare three SI units: millimeter, kilogram, and microsecond. Which is the largest?

a) millimeter c) microsecond

b) kilogram d) The units are not comparable.

1.4 What is(are) the difference(s) between 3.0 and 3.0000?

a) 3.0000 could be the result from an intermediate step in a calculation; 3.0 has to result from a final step.

b) 3.0000 represents a quantity that is known more precisely than 3.0.

c) There is no difference.

d) They convey the same information, but 3.0 is preferred for ease of writing.

1.5 A speed of 7 mm/μs is equal to _____.

a) 7000 m/s b) 70 m/s c) 7 m/s d) 0.07 m/s

1.6 A hockey puck, whose diameter is approximately 3 inches, is to be used to determine the value of π to three significant figures by carefully measuring its diameter and its circumference. For this calculation to be done properly, the measurements must be made to the nearest _____.

a) hundredth of a mm c) mm e) in

b) tenth of a mm d) cm

1.7 What is the sum of $5.786 \cdot 10^3$ m and $3.19 \cdot 10^4$ m?

a) $6.02 \cdot 10^{23}$ m c) $8.976 \cdot 10^3$ m

b) $3.77 \cdot 10^4$ m d) $8.98 \cdot 10^3$ m

1.8 What is the number of carbon atoms in 0.5 nanomoles of carbon? One mole contains $6.02 \cdot 10^{23}$ atoms.

a) $3.2 \cdot 10^{14}$ atoms e) $3.19 \cdot 10^{17}$ atoms

b) $3.19 \cdot 10^{14}$ atoms f) $3. \cdot 10^{17}$ atoms

c) $3. \cdot 10^{14}$ atoms

d) $3.2 \cdot 10^{17}$ atoms

1.9 The resultant of the two-dimensional vectors (1.5 m, 0.7 m), (−3.2 m, 1.7 m), and (1.2 m, −3.3 m) lies in quadrant _____.

a) I b) II c) III d) IV

1.10 By how much does the volume of a cylinder change if the radius is halved and the height is doubled?

a) The volume is quartered. d) The volume doubles.

b) The volume is cut in half. e) The volume quadruples.

c) There is no change in the volume.

1.11 How is the number 0.009834 expressed in scientific notation?

a) $9.834 \cdot 10^4$ c) $9.834 \cdot 10^3$

b) $9.834 \cdot 10^{-4}$ d) $9.834 \cdot 10^{-3}$

1.12 How many significant figures does the number 0.4560 have?

a) five c) three e) one

b) four d) two

1.13 How many watts are in 1 gigawatt (GW)?

a) 10^3 c) 10^9 e) 10^{15}

b) 10^6 d) 10^{12}

1.14 What is the limit of $\gamma = 1/\sqrt{1-(v/c)^2}$, where c is a constant and $v \rightarrow 0$?

a) $\gamma = 1$ c) $\gamma = 2$ e) $\gamma = v/2$

b) $\gamma = 0$ d) $\gamma = v$

1.15 For the two vectors $\vec{A} = (2,1,0)$ and $\vec{B} = (0,1,2)$, what is their scalar product, $\vec{A} \bullet \vec{B}$?

a) 3 b) 6 c) 2 d) 0 e) 1

1.16 For the two vectors $\vec{A} = (2,1,0)$ and $\vec{B} = (0,1,2)$, what is their vector product, $\vec{A} \times \vec{B}$?

a) (2, −4, 2) c) (2, 0, 2) e) (0, 0, 0)

b) (1, 0, 1) d) (3, −2, 1)

CONCEPTUAL QUESTIONS

1.17 In Europe, cars' gas consumption is measured in liters per 100 kilometers. In the United States, the unit used is miles per gallon.

a) How are these units related?

b) How many miles per gallon does your car get if it consumes 12.2 liters per 100 kilometers?

c) What is your car's gas consumption in liters per 100 kilometers if it gets 27.4 miles per gallon?

d) Can you draw a curve plotting miles per gallon versus liters per 100 kilometers? If yes, draw the curve.

1.18 If you draw a vector on a sheet of paper, how many components are required to describe it? How many components does a vector in real space have? How many components would a vector have in a four-dimensional world?

1.19 Since vectors in general have more than one component and thus more than one number is used to describe them, they are obviously more difficult to add and subtract than single numbers. Why then work with vectors at all?

1.20 If $\vec{A}$ and $\vec{B}$ are vectors specified in magnitude-direction form, and $\vec{C} = \vec{A} + \vec{B}$ is to be found and to be expressed in magnitude-direction form, how is this done? That is, what is the procedure for adding vectors that are given in magnitude-direction form?

1.21 Suppose you solve a problem and your calculator's display reads 0.0000000036. Why not just write this down? Is there any advantage to using the scientific notation?

1.22 Since the British system of units is more familiar to most people in the United States, why is the international (SI) system of units used for scientific work in the United States?

1.23 Is it possible to add three equal-length vectors and obtain a vector sum of zero? If so, sketch the arrangement of the three vectors. If not, explain why not.

1.24 Is mass a vector quantity? Why or why not?

1.25 Two flies sit exactly opposite each other on the surface of a spherical balloon. If the balloon's volume doubles, by what factor does the distance between the flies change?

1.26 What is the ratio of the volume of a cube of side r to that of a sphere of radius r? Does your answer depend on the particular value of r?

1.27 Consider a sphere of radius r. What is the length of a side of a cube that has the same surface area as the sphere?

1.28 The mass of the Sun is $2 \cdot 10^{30}$ kg, and the Sun contains more than 99% of all the mass in the Solar System. Astronomers estimate there are approximately 100 billion stars in the Milky Way and approximately 100 billion galaxies in the universe. The Sun and other stars are predominantly composed of hydrogen; a hydrogen atom has a mass of approximately $2 \cdot 10^{-27}$ kg.

a) Assuming that the Sun is an average star and the Milky Way is an average galaxy, what is the total mass of the universe?

b) Since the universe consists mainly of hydrogen, can you estimate the total number of atoms in the universe?

1.29 A futile task is proverbially said to be "like trying to empty the ocean with a teaspoon." Just how futile is such a task? Estimate the number of teaspoonfuls of water in the Earth's oceans.

1.30 The world's population passed 6.5 billion in 2006. Estimate the amount of land area required if each person were to stand in such a way as to be unable to touch another person. Compare this area to the land area of the United States, 3.5 million square miles, and to the land area of your home state (or country).

1.31 Advances in the field of nanotechnology have made it possible to construct chains of single metal atoms linked one to the next. Physicists are particularly interested in the ability of such chains to conduct electricity with little resistance. Estimate how many gold atoms would be required to make such a chain long enough to wear as a necklace. How many would be required to make a chain that encircled the Earth? If 1 mole of a substance is equivalent to roughly $6.022 \cdot 10^{23}$ atoms, how many moles of gold are required for each necklace?

1.32 One of the standard clichés in physics courses is to talk about approximating a cow as a sphere. How large a sphere makes the best approximation to an average dairy cow? That is, estimate the radius of a sphere that has the same mass and density as a dairy cow.

1.33 Estimate the mass of your head. Assume that its density is that of water, 1000 kg/m^3.

1.34 Estimate the number of hairs on your head.

EXERCISES

A blue problem number indicates a worked-out solution is available in the Student Solutions Manual. One • and two •• indicate increasing level of problem difficulty.

Section 1.2

1.35 How many significant figures are in each of the following numbers?

a) 4.01   c) 4   e) 0.00001   g) $7.01 \cdot 3.1415$

b) 4.010   d) 2.00001   f) $2.1 - 1.10042$

1.36 Two different forces, acting on the same object, are measured. One force is 2.0031 N and the other force, in the same direction, is 3.12 N. These are the only forces acting on the object. Find the total force on the object *to the correct number of significant figures*.

1.37 Three quantities, the results of measurements, are to be added. They are 2.0600, 3.163, and 1.12. What is their sum *to the correct number of significant figures*?

1.38 Given the equation $w = xyz$, and $x = 1.1 \cdot 10^3$, $y = 2.48 \cdot 10^{-2}$, and $z = 6.000$, what is w, in scientific notation and with the correct number of significant figures?

1.39 Write this quantity in scientific notation: one ten-millionth of a centimeter.

1.40 Write this number in scientific notation: one hundred fifty-three million.

Section 1.3

1.41 How many inches are in 30.7484 miles?

1.42 What metric prefixes correspond to the following powers of 10?

a) 10^3 b) 10^{-2} c) 10^{-3}

1.43 How many millimeters in a kilometer?

1.44 A hectare is a hundred ares, and an are is a hundred square meters. How many hectares are there in a square kilometer?

1.45 The unit of pressure in the SI system is the pascal. What is the SI name for 1 one-thousandth of a pascal?

1.46 The masses of four sugar cubes are measured to be 25.3 g, 24.7 g, 26.0 g, and 25.8 g. Express the answers to the following questions in scientific notation, with standard SI units and an appropriate number of significant figures.

a) If the four sugar cubes were crushed and all the sugar collected, what would be the total mass, in kilograms, of the sugar?

b) What is the average mass, in kilograms, of these four sugar cubes?

•**1.47** What is the surface area of a right cylinder of height 20.5 cm and radius 11.9 cm?

Section 1.4

1.48 You step on your brand-new digital bathroom scale, and it reads 125.4 pounds. What is your mass in kilograms?

1.49 The distance from the center of the Moon to the center of the Earth ranges from approximately 356,000 km to 407,000 km.

a) What is the minimum distance to the Moon in miles?

b) What is the maximum distance to the Moon in miles?

1.50 In Major League baseball, the pitcher delivers his pitches from a distance of 60 feet, 6 inches from home plate. What is the distance in meters?

1.51 A flea hops in a straight path along a meter stick, starting at 0.7 cm and making successive jumps, which are measured to be 3.2 cm, 6.5 cm, 8.3 cm, 10.0 cm, 11.5 cm, and 15.5 cm. Express the answers to the following questions in scientific notation, with units of meters and an appropriate number of significant figures. What is the total distance covered by the flea in these six hops? What is the average distance covered by the flea in a single hop?

•**1.52** One cubic centimeter of water has a mass of 1 gram. A milliliter is equal to a cubic centimeter. What is the mass, in kilograms, of a liter of water? A metric ton is a thousand kilograms. How many cubic centimeters of water are in a metric ton of water? If a metric ton of water were held in a thin-walled cubical tank, how long (in meters) would each side of the tank be?

•**1.53** The speed limit on a particular stretch of road is 45 miles per hour. Express this speed limit in millifurlongs per microfortnight. A furlong is $\frac{1}{8}$ mile, and a fortnight is a period of 2 weeks.

•**1.54** According to one mnemonic rhyme, "A pint's a pound, the world around." Investigate this statement of equivalence by calculating the weight of a pint of water, assuming that the density of water is 1000. kg/m^3 and that the weight of 1.00 kg of a substance is 2.21 pounds. The volume of 1.00 fluid ounce is 29.6 mL.

Section 1.5

1.55 If the radius of a planet is larger than that of Earth by a factor of 8.7, how much bigger is the surface area of the planet than Earth's?

1.56 If the radius of a planet is larger than that of Earth by a factor of 5.8, how much bigger is the volume of the planet than Earth's?

1.57 What is the maximum distance from which a sailor on top of the mast of ship 1, at 34 m above the ocean's surface, can see another sailor on top of the mast of ship 2, at 26 m above the ocean's surface?

1.58 You are flying in a jetliner at an altitude of 35,000 ft. How far away is the horizon?

1.59 How many cubic inches are in 1.56 barrels of oil?

1.60 A car's gasoline tank has the shape of a right rectangular box with a square base whose sides measure 62 cm. Its capacity is 52 L. If the tank has only 1.5 L remaining, how deep is the gasoline in the tank, assuming the car is parked on level ground?

•**1.61** The volume of a sphere is given by the formula $\frac{4}{3}\pi r^3$, where r is the radius of the sphere. The average density of an object is simply the ratio of its mass to its volume. Using the numerical data found in Table 12.1, express the answers to the following questions in scientific notation, with SI units and an appropriate number of significant figures.

a) What is the volume of the Sun?

b) What is the volume of the Earth?

c) What is the average density of the Sun?

d) What is the average density of the Earth?

•**1.62** A tank is in the shape of an inverted cone, having height $h = 2.5$ m and base radius $r = 0.75$ m. If water is poured into the tank at a rate of 15 L/s, how long will it take to fill the tank?

•**1.63** Water flows into a cubical tank at a rate of 15 L/s. If the top surface of the water in the tank is rising by 1.5 cm every second, what is the length of each side of the tank?

••**1.64** The atmosphere has a weight that is, effectively, about 15 pounds for every square inch of Earth's surface. The average density of air at the Earth's surface is about 1.275 kg/m³. If the atmosphere were uniformly dense (it is not—the density varies quite significantly with altitude), how thick would it be?

Section 1.6

1.65 A position vector has a length of 40.0 m and is at an angle of 57.0° above the x-axis. Find the vector's components.

1.66 In the triangle shown in the figure, the side lengths are $a = 6.6$ cm, $b = 13.7$ cm, and $c = 9.2$ cm. What is the value of the angle γ? (*Hint:* See Appendix A for the law of cosines.)

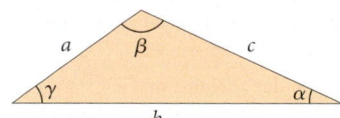

1.67 Find the components of the vectors $\vec{A}$, $\vec{B}$, $\vec{C}$, and $\vec{D}$, if their lengths are given by $A = 75.0$, $B = 60.0$, $C = 25.0$, $D = 90.0$ and their direction angles are as shown in the figure. Write the vectors in terms of unit vectors.

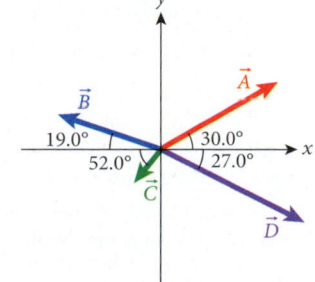

•**1.68** Use the components of the vectors from Problem 1.67 to find

a) the sum $\vec{A} + \vec{B} + \vec{C} + \vec{D}$ in terms of its components

b) the magnitude and direction of the sum $\vec{A} - \vec{B} + \vec{D}$

•**1.69** The Bonneville Salt Flats, located in Utah near the border with Nevada, not far from Interstate I80, cover an area of over 30,000 acres. A race car driver on the Flats first heads north for 4.47 km, then makes a sharp turn and heads southwest for 2.49 km, then makes another turn and heads east for 3.59 km. How far is she from where she started?

•**1.70** A map in a pirate's log gives directions to the location of a buried treasure. The starting location is an old oak tree. According to the map, the treasure's location is found by proceeding 20 paces north from the oak tree and then 30 paces northwest. At this location, an iron pin is sunk in the ground. From the iron pin, walk 10 paces south and dig. How far (in paces) from the oak tree is the spot at which digging occurs?

••**1.71** The next page of the pirate's log contains a set of directions that differ from those on the map in Problem 1.70. These say the treasure's location is found by proceeding 20 paces north from the old oak tree and then 30 paces northwest. After finding the iron pin, one should "walk 12 paces nor'ward and dig downward 3 paces to the treasure box." What is the vector that points from the base of the old oak tree to the treasure box? What is the length of this vector?

••**1.72** The Earth's orbit has a radius of $1.5 \cdot 10^{11}$ m, and that of Venus has a radius of $1.1 \cdot 10^{11}$ m. Consider these two orbits to be perfect circles (though in reality they are ellipses with slight eccentricity). Write the direction and length of a vector from Earth to Venus (take the direction from Earth to Sun to be 0°) when Venus is at the maximum angular separation in the sky relative to the Sun.

••**1.73** A friend walks away from you a distance of 550 m, and then turns (as if on a dime) an unknown angle, and walks an additional 178 m in the new direction. You use a laser range-finder to find out that his final distance from you is 432 m. What is the angle between his initial departure direction and the direction to his final location? Through what angle did he turn? (There are two possibilities.)

Additional Exercises

1.74 The radius of Earth is 6378. km. What is its circumference to three significant figures?

1.75 Estimate the product of 4,308,229 and 44 to one significant figure (show your work and do not use a calculator), and express the result in standard scientific notation.

1.76 Find the vector $\vec{C}$ that satisfies the equation $3\hat{x} + 6\hat{y} - 10\hat{z} + \vec{C} = -7\hat{x} + 14\hat{y}$.

1.77 A position vector has components $x = 34.6$ m and $y = -53.5$ m. Find the vector's length and angle with the x-axis.

1.78 For the planet Mars, calculate the distance around the Equator, the surface area, and the volume. The radius of Mars is $3.39 \cdot 10^6$ m.

•**1.79** Find the magnitude and direction of (a) $9\vec{B} - 3\vec{A}$ and (b) $-5\vec{A} + 8\vec{B}$, where $\vec{A} = (23.0, 59.0)$, $\vec{B} = (90.0, -150.0)$.

•**1.80** Express the vectors $\vec{A} = (A_x, A_y) = (-30.0$ m, -50.0 m$)$ and $\vec{B} = (B_x, B_y) = (30.0$ m, 50.0 m$)$ by giving their magnitude and direction as measured from the positive x-axis.

•**1.81** The force F that a spring exerts on you is directly proportional to the distance x that you stretch it beyond its resting length. Suppose that when you stretch a spring 8.00 cm, it exerts a force of 200. N on you. How much force will it exert on you if you stretch it 40.0 cm?

•**1.82** The distance a freely falling object drops, starting from rest, is proportional to the square of the time it has been falling. By what factor will the distance fallen change if the time of falling is three times as long?

•**1.83** A pilot decides to take his small plane for a Sunday afternoon excursion. He first flies north for 155.3 miles, then makes a 90° turn to his right and flies on a straight line for 62.5 miles, then makes another 90° turn to his right and flies 47.5 miles on a straight line.

a) How far away from his home airport is he at this point?

b) In which direction does he need to fly from this point on to make it home in a straight line?

c) What was the farthest distance from the home airport that he reached during the trip?

•**1.84** As the photo shows, during a total eclipse, the Sun and the Moon appear to the observer to be almost exactly the same size. The radii of the Sun and Moon are $r_S = 6.96 \cdot 10^8$ m and $r_M = 1.74 \cdot 10^6$ m, respectively. The distance between the Earth and the Moon is $d_{EM} = 3.84 \cdot 10^8$ m.

Total solar eclipse.

a) Determine the distance from the Earth to the Sun at the moment of the eclipse.

b) In part (a), the implicit assumption is that the distance from the observer to the Moon's center is equal to the distance between the centers of the Earth and the Moon. By how much is this assumption incorrect, if the observer of the eclipse is on the Equator at noon? [*Hint:* Express this quantitatively, by calculating the relative error as a ratio: (assumed observer-to-Moon distance – actual observer-to-Moon distance)/(actual observer-to-Moon distance).]

c) Use the corrected observer-to-Moon distance to determine a corrected distance from Earth to the Sun.

•**1.85** A hiker travels 1.50 km north and turns to a heading of 20.0° north of west, traveling another 1.50 km along that heading. Subsequently, she then turns north again and travels another 1.50 km. How far is she from her original point of departure, and what is the heading relative to that initial point?

•**1.86** Assuming that 1 mole ($6.02 \cdot 10^{23}$ molecules) of an ideal gas has a volume of 22.4 L at standard temperature and pressure (STP) and that nitrogen, which makes up 78% of the air we breathe, is an ideal gas, how many nitrogen molecules are there in an average 0.5-L breath at STP?

•**1.87** On August 27, 2003, Mars approached as close to Earth as it will for over 50,000 years. If its angular size (the planet's diameter, measured by the angle the radius subtends) on that day was measured by an astronomer

to be 24.9 seconds of arc, and its diameter is known to be 6784 km, how close was the approach distance? Be sure to use an appropriate number of significant figures in your answer.

•**1.88** A football field's length is exactly 100 yards, and its width is $53\frac{1}{3}$ yards. A quarterback stands at the exact center of the field and throws a pass to a receiver standing at one corner of the field. Let the origin of coordinates be at the center of the football field and the x-axis point along the longer side of the field, with the y-direction parallel to the shorter side of the field.

a) Write the direction and length of a vector pointing from the quarterback to the receiver.

b) Consider the other three possibilities for the location of the receiver at corners of the field. Repeat part (a) for each.

•**1.89** The circumference of the Cornell Electron Storage Ring is 768.4 m. Express the diameter in inches, to the proper number of significant figures.

••**1.90** Roughly 4% to 5% of what you exhale is carbon dioxide. Assume that 22.4 L is the volume of 1 mole ($6.02 \cdot 10^{23}$ molecules) of carbon dioxide and that you exhale 0.5 L per breath.

a) Estimate how many carbon dioxide molecules you breathe out each day.

b) If each mole of carbon dioxide has a mass of 44.0 g, how many kilograms of carbon dioxide do you exhale in a year?

••**1.91** The Earth's orbit has a radius of $1.5 \cdot 10^{11}$ m, and that of Mercury has a radius of $4.6 \cdot 10^{10}$ m. Consider these orbits to be perfect circles (though in reality they are ellipses with slight eccentricity). Write down the direction and length of a vector from Earth to Mercury (take the direction from Earth to Sun to be 0°) when Mercury is at the maximum angular separation in the sky relative to the Sun.

••**1.92** The star (other than the Sun) that is closest to Earth is Proxima Centauri. Its distance from Earth can be measured using parallax. *Parallax* is half the apparent angular shift of the star when observed from Earth at points on opposite sides of the Sun. The parallax of Proxima Centauri is 769 milliarcseconds. How far away is Proxima Centauri? (Give your answer in light-years with the correct number of significant digits.)

MULTI-VERSION EXERCISES

1.93 Write the vectors $\vec{A}$, $\vec{B}$, and $\vec{C}$ in Cartesian coordinates.

1.94 Calculate the length and direction of the vectors $\vec{A}$, $\vec{B}$, and $\vec{C}$.

1.95 Add the three vectors $\vec{A}$, $\vec{B}$, and $\vec{C}$ graphically.

1.96 Determine the difference vector $\vec{E} = \vec{B} - \vec{A}$ graphically.

1.97 Add the three vectors $\vec{A}$, $\vec{B}$, and $\vec{C}$ using the component method, and find their sum vector $\vec{D}$.

1.98 Use the component method to determine the length of the vector $\vec{F} = \vec{C} - \vec{A} - \vec{B}$.

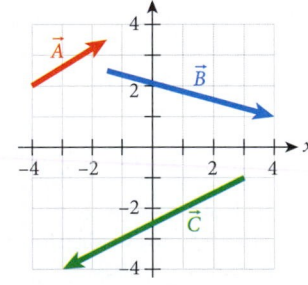

Figure for Problems 1.93 through 1.98

1.99 Sketch the vectors with the components $\vec{A} = (A_x, A_y) = (30.0$ m, -50.0 m) and $\vec{B} = (B_x, B_y) = (-30.0$ m,50.0 m), and find the magnitudes of these vectors.

1.100 What angle does $\vec{A} = (A_x, A_y) = (30.0$ m,-50.0 m) make with the positive x-axis? What angle does it make with the negative y-axis?

1.101 Sketch the vectors with the components $\vec{A} = (A_x, A_y) = (-30.0$ m, -50.0 m) and $\vec{B} = (B_x, B_y) = (30.0$ m, 50.0 m), and find the magnitudes of these vectors.

1.102 What angle does $\vec{B} = (B_x, B_y) = (30.0$ m,50.0 m) make with the positive x-axis? What angle does it make with the positive y-axis?

1.103 Find the magnitude and direction of each of the following vectors, which are given in terms of their x- and y-components: $\vec{A} = (23.0, 59.0)$, and $\vec{B} = (90.0, -150.0)$.

1.104 Find the magnitude and direction of $-\vec{A} + \vec{B}$, where $\vec{A} = (23.0, 59.0)$, $\vec{B} = (90.0, -150.0)$.

1.105 Find the magnitude and direction of $-5\vec{A} + \vec{B}$, where $\vec{A} = (23.0, 59.0)$, $\vec{B} = (90.0, -150.0)$.

1.106 Find the magnitude and direction of $-7\vec{B} + 3\vec{A}$, where $\vec{A} = (23.0, 59.0)$, $\vec{B} = (90.0, -150.0)$.

1.107 Which of the six cases shown in the figure has the largest absolute value of the scalar product of the vectors $\vec{A}$ and $\vec{B}$?

1.108 Which of the six cases shown in the figure has the smallest absolute value of the scalar product of the vectors $\vec{A}$ and $\vec{B}$?

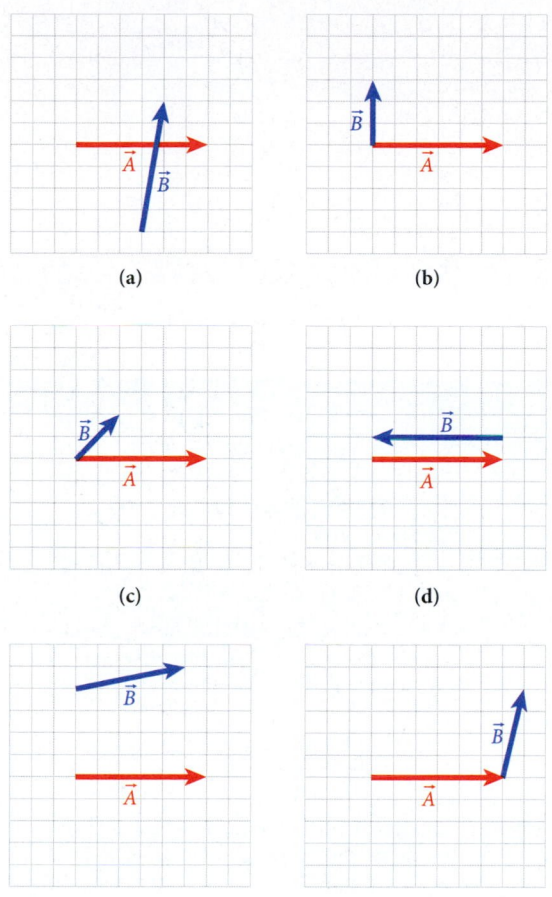

(a)   (b)

(c)   (d)

(e)   (f)

Figure for Problems 1.107 to 1.112

1.109 Which of the six cases shown in the figure has the largest absolute value of the vector product of the vectors $\vec{A}$ and $\vec{B}$?

1.110 Which of the six cases shown in the figure has the smallest absolute value of the vector product of the vectors $\vec{A}$ and $\vec{B}$?

•**1.111** Rank order the six cases shown in the figure from the smallest absolute value to the largest absolute value of the scalar product of the vectors $\vec{A}$ and $\vec{B}$.

•**1.112** Rank order the six cases shown in the figure from the smallest absolute value to the largest absolute value of the vector product of the vectors $\vec{A}$ and $\vec{B}$.

1.113 Toward the end of their lives many stars become much bigger. Assume that they remain spherical in shape and that their masses do not change in this process. If the radius of a star increases by a factor of 11.4, by what factors do the following change:

a) its surface area,

b) its circumference,

c) its volume?

1.114 Toward the end of their lives many stars become much bigger. Assume that they remain spherical in shape and that their masses do not change in this process. If the circumference of a star increases by a factor of 12.5, by what factors do the following change:

a) its surface area,

b) its radius,

c) its volume?

1.115 Toward the end of their lives many stars become much bigger. Assume that they remain spherical in shape and that their masses do not change in this process. If the volume of a star increases by a factor of 872, by what factors do the following change:

a) its surface area,

b) its circumference,

c) its diameter?

•**1.116** Toward the end of their lives many stars become much bigger. Assume that they remain spherical in shape and that their masses do not change in this process. If the surface area of a star increases by a factor of 274, by what factors do the following change:

a) its radius,

b) its volume,

c) its density?

Motion in a Straight Line

2

FIGURE 2.1 A fast-moving train passes a railroad crossing.

You can see that the train in Figure 2.1 is moving very fast by noticing that its image is blurred compared with the stationary crossing signal and telephone pole. But can you tell if the train is speeding up, slowing down, or zipping by at constant speed? A photograph can convey an object's speed because the object moves during the exposure time, but a photograph cannot show a change in speed, referred to as acceleration. Yet acceleration is extremely important in physics, at least as important as speed itself.

In this chapter, we look at the terms used in physics to describe an object's motion: displacement, velocity, and acceleration. We examine motion along a straight line (one-dimensional motion) in this chapter and motion on a curved path (motion in a plane, or two-dimensional motion) in the next chapter. One of the greatest advantages of physics is that its laws are universal, so the same general terms and ideas apply to a wide range of situations. Thus, we can use the same equations to describe the flight of a baseball and the lift-off of a rocket into space from Earth to Mars. In this chapter, we will use some of the problem-solving techniques discussed in Chapter 1 along with some new ones.

As you continue in this course, you will see that almost everything moves relative to other objects on some scale or other, whether it is a comet plunging through space at several kilometers per second or the atoms in a seemingly stationary object vibrating millions of times per second. The terms we introduce in this chapter will be part of your study for the rest of the course and afterward.

WHAT WE WILL LEARN

- We will learn to describe the motion of an object traveling in a straight line or in one dimension.

- We will learn to define position, displacement, and distance.

- We will learn to describe motion of an object in a straight line with constant acceleration.

- We will see that an object can be in free fall in one dimension, where it undergoes constant acceleration due to gravity.

- We will define the concepts of instantaneous velocity and average velocity.

- We will define the concepts of instantaneous acceleration and average acceleration.

- We will learn to calculate the position, velocity, and acceleration of an object moving in a straight line.

2.1 Introduction to Kinematics

The study of physics is divided into several large parts, one of which is mechanics. **Mechanics,** or the study of motion and its causes, is usually subdivided. In this chapter and the next, we examine the kinematics aspect of mechanics. **Kinematics** is the study of the motion of objects. These objects may be, for example, cars, baseballs, people, planets, or atoms. For now, we will set aside the question of what causes this motion. We will return to that question when we study forces.

We will also not consider rotation in this chapter, but concentrate on only translational motion (motion without rotation). Furthermore, we will neglect all internal structure of a moving object and consider it to be a point particle, or pointlike object. That is, to determine the equations of motion for an object, we imagine it to be located at a single point in space at each instant of time. What point of an object should we choose to represent its location? Initially, we will simply use the geometric center, the middle. (Chapter 8, on systems of particles and extended objects, will give a more precise definition for the point location of an object, called the *center of mass*.)

Concept Check 2.1

The train in Figure 2.1 is

a) speeding up.

b) slowing down.

c) traveling at a constant speed.

d) moving at a rate that can't be determined from the photo.

2.2 Position Vector, Displacement Vector, and Distance

The simplest motion we can investigate is that of an object moving in a straight line. Examples of this motion include a person running the 100-m dash, a car driving on a straight segment of road, and a stone falling straight down off a cliff. In later chapters, we will consider motion in two or more dimensions and will see that the concepts we derive here for one-dimensional motion still apply.

If an object is located on a particular point on a line, we can denote this point with its **position vector,** as described in Section 1.6. Throughout this book, we use the symbol $\vec{r}$ to denote the position vector. Since we are working with motion in only one dimension in this chapter, the position vector has only one component. If the motion is in the horizontal direction, this one component is the x-component. (For motion in the vertical direction, we will use the y-component; see Section 2.7.) One number, the x-coordinate or x-component of the position vector (with a corresponding unit), uniquely specifies the position vector in one-dimensional motion. Some valid ways of writing a position are $x = 4.3$ m, $x = 7\frac{3}{8}$ inches, and $x = -2.04$ km; it is understood that these specifications refer to the x-component of the position vector. Note that a position vector's x-component can have a positive or a negative value, depending on the location of the point and the axis direction that we choose to be positive. The value of the x-component also depends on where we define the origin of the coordinate system—the zero of the straight line.

The position of an object can change as a function of time, t; that is, the object can move. We can therefore formally write the position vector in function notation: $\vec{r} = \vec{r}(t)$. In one dimension, this means that the x-component of the vector is a function of time, $x = x(t)$. If we want to specify the position at some specific time t_1, we use the notation $x_1 \equiv x(t_1)$.

Position Graphs

Before we go any further, let's graph an object's position as a function of time. Figure 2.2a illustrates the principle involved by showing several frames of a video of a car driving down a road. The video frames were taken at time intervals of $\frac{1}{3}$ second.

We are free to choose the origins of our time measurements and of our coordinate system. In this case, we choose the time of the second frame to be $t = \frac{1}{3}$ s and the position of the center of the car in the second frame as $x = 0$. We can now draw our coordinate axes and graph over the frames (Figure 2.2b). The position of the car as a function of time lies on a straight line. Again, keep in mind that we are representing the car by a single point.

When drawing graphs, it is customary to plot the independent variable—in this case, the time t—on the horizontal axis and to plot x, which is called the dependent variable because its value depends on the value of t, on the vertical axis. Figure 2.3 is a graph of the car's position as a function of time drawn in this customary way. (Note that if Figure 2.2b were rotated 90° counterclockwise and the pictures of the car removed, the two graphs would be the same.)

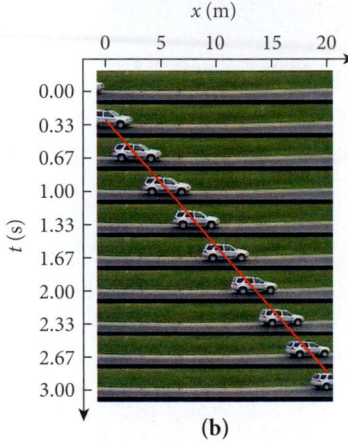

FIGURE 2.2 (a) Series of video frames of a moving car, taken every $\frac{1}{3}$ second; (b) same series, but with a coordinate system and a red line connecting the center of the car in successive frames.

Displacement

Now that we have specified the position vector, let's go one step further and define displacement. **Displacement** is simply the difference between the final position vector, $\vec{r}_2 \equiv \vec{r}(t_2)$, at the end of a motion and the initial position vector, $\vec{r}_1 \equiv \vec{r}(t_1)$. We write the displacement vector as

$$\Delta\vec{r} = \vec{r}_2 - \vec{r}_1. \tag{2.1}$$

We use the notation $\Delta\vec{r}$ for the displacement vector to indicate that it is a difference between two position vectors. Note that the displacement vector is independent of the location of the origin of the coordinate system. Why? Any shift of the coordinate system will add to the position vector $\vec{r}_2$ the same amount that it adds to the position vector $\vec{r}_1$; thus, the difference between the position vectors, $\Delta\vec{r}$, will not change.

Just like the position vector, the displacement vector in one dimension has only an x-component, which is the difference between the x-components of the final and initial position vectors:

$$\Delta x = x_2 - x_1. \tag{2.2}$$

Also just like position vectors, displacement vectors can be positive or negative. In particular, the displacement vector $\Delta\vec{r}_{ba}$ for going from point a to point b is exactly the negative of $\Delta\vec{r}_{ab}$ going from point b to point a:

$$\Delta\vec{r}_{ba} = \vec{r}_b - \vec{r}_a = -(\vec{r}_a - \vec{r}_b) = -\Delta\vec{r}_{ab}. \tag{2.3}$$

And it is probably obvious to you at this point that this relationship also holds for the x-component of the displacement vector, $\Delta x_{ba} = x_b - x_a = -(x_a - x_b) = -\Delta x_{ab}$.

Distance

For motion on a straight line without changing directions, the **distance,** ℓ, that a moving object travels is the absolute value of the displacement vector:

$$\ell = \left|\Delta\vec{r}\right|. \tag{2.4}$$

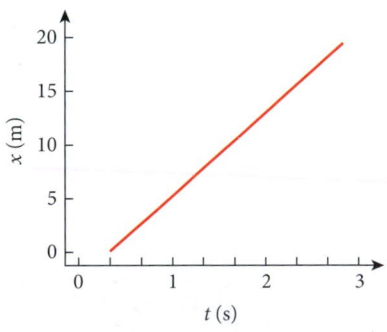

FIGURE 2.3 Same graph as Figure 2.2b, but rotated so that the time axis is horizontal, and without the pictures of the car.

For one-dimensional motion, this distance is also the absolute value of the x-component of the displacement vector, $\ell = \left|\Delta x\right|$. (For multidimensional motion, we calculate the length of the displacement vector as shown in Chapter 1.) The distance is always greater than or equal to zero and is measured in the same units as position and displacement. However, distance is a scalar quantity, not a vector. If the displacement is not in a straight line or if it is not all in the same direction, the displacement must be broken up into segments that are approximately straight and unidirectional, and then the distances for the various segments are added to get the total distance. The following solved problem illustrates the difference between distance and displacement.

SOLVED PROBLEM 2.1 | Trip Segments

The distance between Des Moines and Iowa City is 170.5 km (106.0 miles) along Interstate 80, and as you can see from the map (Figure 2.4), the route is a straight line to a good approximation. Approximately halfway between the two cities, where I80 crosses highway US63, is the city of Malcom, 89.9 km (55.9 miles) from Des Moines.

FIGURE 2.4 Route I80 between Des Moines and Iowa City.

PROBLEM

If we drive from Malcom to Des Moines and then go to Iowa City, what are the total distance and total displacement for this trip?

SOLUTION

THINK Distance and displacement are not identical. If the trip consisted of one segment in one direction, the distance would just be the absolute value of the displacement, according to equation 2.4. However, this trip is composed of segments with a direction change, so we need to be careful. We'll treat each segment individually and then add up the segments in the end.

SKETCH Because I80 is almost a straight line, it is sufficient to draw a straight horizontal line and make this our coordinate axis. We enter the positions of the three cities as x_I (Iowa City), x_M (Malcom), and x_D (Des Moines). We always have the freedom to define the origin of our coordinate system, so we elect to put it at Des Moines, thus setting $x_D = 0$. As is conventional, we define the positive direction to the right, in the eastward direction. See Figure 2.5.

We also draw arrows for the displacements of the two segments of the trip. We represent segment 1 from Malcom to Des Moines by a red arrow, and segment 2 from Des Moines to Iowa City by a blue arrow. Finally, we draw a diagram for the total trip as the sum of the two trips.

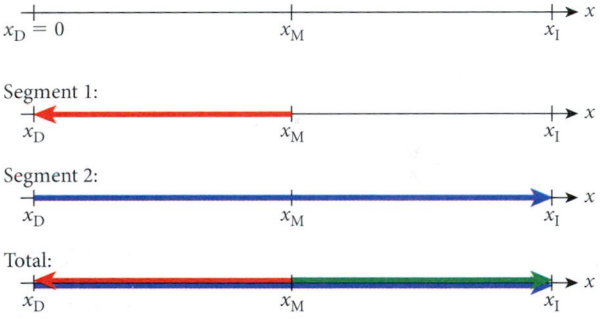

FIGURE 2.5 Coordinate system and trip segments for the Malcom to Des Moines to Iowa City trip.

RESEARCH With our assignment of $x_D = 0$, Des Moines is the origin of the coordinate system. According to the information given us, Malcom is then at $x_M = +89.9$ km and Iowa City is at $x_I = +170.5$ km. Note that we write a plus sign in front of the numbers for x_M and x_I to remind us that these are components of position vectors and can have positive or negative values.

For the first segment, the *displacement* is given by

$$\Delta x_1 = x_D - x_M.$$

Thus, the distance driven for this segment is

$$\ell_1 = \left| \Delta x_1 \right| = \left| x_D - x_M \right|.$$

In the same way, the displacement and distance for the second segment are

$$\Delta x_2 = x_I - x_D$$
$$\ell_2 = \left| \Delta x_2 \right| = \left| x_I - x_D \right|.$$

– Continued

Concept Check 2.2

Your dorm room is located 0.25 mile from the Dairy Store. You walk from your room to the Dairy Store and back. Which of the following statements about your trip is true?

a) The distance is 0.50 mile, and the displacement is 0.50 mile.

b) The distance is 0.50 mile, and the displacement is 0.00 mile.

c) The distance is 0.00 mile, and the displacement is 0.50 mile.

d) The distance is 0.00 mile, and the displacement is 0.00 mile.

For the sum of the two segments, the total trip, we use simple addition to find the displacement,

$$\Delta x_{\text{total}} = \Delta x_1 + \Delta x_2 \, ,$$

and the total distance,

$$\ell_{\text{total}} = \ell_1 + \ell_2 .$$

SIMPLIFY We can simplify the equation for the total displacement a little bit by inserting the expressions for the displacements for the two segments:

$$\begin{aligned}\Delta x_{\text{total}} &= \Delta x_1 + \Delta x_2 \\ &= (x_D - x_M) + (x_I - x_D) \\ &= x_I - x_M.\end{aligned}$$

This is an interesting result—for the total displacement of the entire trip, it does not matter at all that we went to Des Moines. All that matters is where the trip started and where it ended. The total displacement is a result of a one-dimensional vector addition, as indicated in the bottom part of Figure 2.5 by the green arrow.

CALCULATE Now we can insert the numbers for the positions of the three cities in our coordinate system. We then obtain for the net displacement in our trip

$$\Delta x_{\text{total}} = x_I - x_M = (+170.5 \text{ km}) - (+89.9 \text{ km}) = +80.6 \text{ km}.$$

For the total distance driven, we get

$$\ell_{\text{total}} = |89.9 \text{ km}| + |170.5 \text{ km}| = 260.4 \text{ km}.$$

(Remember, the distance between Des Moines and Malcom, or Δx_1, and that between Des Moines and Iowa City, or Δx_2, were given in the problem; so we do not have to calculate them again from the differences in the position vectors of the cities.)

ROUND The numbers for the distances were initially given to a tenth of a kilometer. Since our entire calculation only amounted to adding or subtracting these numbers, it is not surprising that we end up with numbers that are also accurate to a tenth of a kilometer. No further rounding is needed.

DOUBLE-CHECK As is customary, we first make sure that the units of our answer came out properly. Since we are looking for quantities with the dimension of length, it is comforting that our answers have the units of kilometers. At first sight, it may be surprising that the net displacement for the trip is only 80.6 km, much smaller than the total distance traveled. This is a good time to recall that the relationship between the absolute value of the displacement and the distance (equation 2.4) is valid only if the moving object does *not* change direction (but it did in this example).

This discrepancy is even more apparent for a round trip. In that case, the total distance driven is twice the distance between the two cities, but the total displacement is zero, because the starting point and the end point of the trip are identical.

Self-Test Opportunity 2.1

Suppose we had chosen to put the origin of the coordinate system in Solved Problem 2.1 at Malcom instead of Des Moines. Would the final result of our calculation change? If yes, how? If no, why not?

This result is a general one: If the initial and final positions are the same, the total displacement is 0. As straightforward as this seems for the trip example, it is a potential pitfall in many exam questions. You need to remember that displacement is a vector, whereas distance is a positive scalar.

2.3 Velocity Vector, Average Velocity, and Speed

Not only do distance (a scalar) and displacement (a vector) mean different things in physics, but their rates of change with time are also different. Although the words "speed" and "velocity" are often used interchangeably in everyday speech, in physics "speed" refers to a scalar and "velocity" to a vector.

We define v_x, the x-component of the velocity vector, as the change in position (i.e., the displacement component) in a given time interval divided by that time interval, $\Delta x/\Delta t$. Velocity can change from moment to moment. The velocity calculated by taking the ratio of displacement per time interval is the average of the velocity over this time interval, or the x-component of the **average velocity,** $\bar{v}_x$:

$$\bar{v}_x = \frac{\Delta x}{\Delta t}. \tag{2.5}$$

Notation: A bar above a symbol is the notation for averaging over a finite time interval.

In calculus, a time derivative is obtained by taking a limit as the time interval approaches zero. We use the same concept here to define the **instantaneous velocity,** usually referred to simply as the **velocity,** as the time derivative of the displacement. For the x-component of the velocity vector, this implies

$$v_x = \lim_{\Delta t \to 0} \bar{v}_x = \lim_{\Delta t \to 0} \frac{\Delta x}{\Delta t} \equiv \frac{dx}{dt}. \tag{2.6}$$

We can now introduce the velocity vector, $\vec{v}$, as the vector for which each component is the time derivative of the corresponding component of the position vector,

$$\vec{v} = \frac{d\vec{r}}{dt}, \tag{2.7}$$

with the understanding that the derivative operation applies to each of the components of the vector. In the one-dimensional case, this velocity vector $\vec{v}$ has only an x-component, v_x, and the velocity is equivalent to a single velocity component in the spatial x-direction.

Figure 2.6 presents three graphs of the position of an object with respect to time. Figure 2.6a shows that we can calculate the average velocity of the object by finding the change in position of the object between two points and dividing by the time it takes to go from x_1 to x_2. That is, the average velocity is given by the displacement, Δx_1, divided by the time interval, Δt_1, or $\bar{v}_1 = \Delta x_1/\Delta t_1$. In Figure 2.6b, the average velocity, $\bar{v}_2 = \Delta x_2/\Delta t_2$, is determined over a smaller time interval, Δt_2. In Figure 2.6c, the instantaneous velocity, $v(t_3) = dx/dt|_{t=t_3}$, is represented by the slope of the blue line tangent to the red curve at $t = t_3$.

Velocity is a vector, pointing in the same direction as the vector of the infinitesimal displacement, dx. Because the position $x(t)$ and the displacement $\Delta x(t)$ are functions of time, so is the velocity. Because the velocity vector is defined as the time derivative of the displacement vector, all the rules of differentiation introduced in calculus hold. If you need a refresher, consult Appendix A.

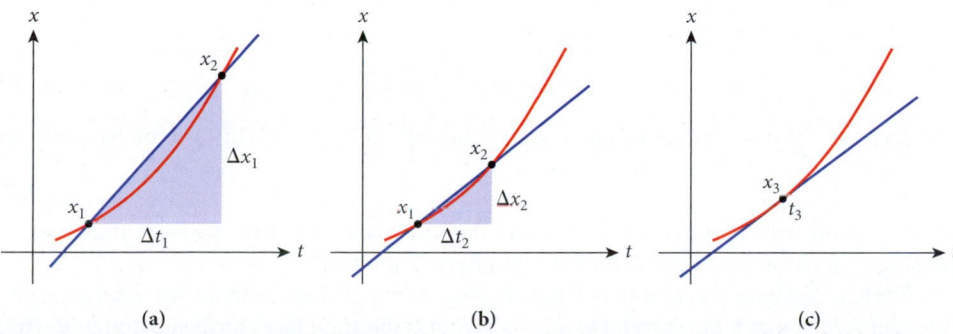

FIGURE 2.6 Instantaneous velocity as the limit of the ratio of displacement to time interval: (a) an average velocity over a large time interval; (b) an average velocity over a smaller time interval; and (c) the instantaneous velocity at a specific time, t_3.

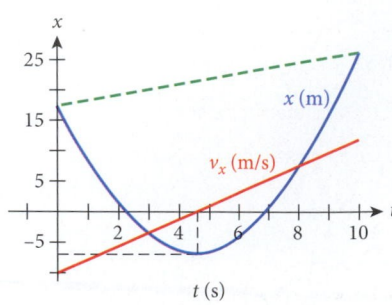

FIGURE 2.7 Graph of the position x and velocity v_x as a function of the time t. The slope of the dashed line represents the average velocity for the time interval from 0 to 10 s.

EXAMPLE 2.1 | Time Dependence of Velocity

PROBLEM
During the time interval from 0.0 to 10.0 s, the position vector of a car on a road is given by $x(t) = a + bt + ct^2$, with $a = 17.2$ m, $b = -10.1$ m/s, and $c = 1.10$ m/s². What is the car's velocity as a function of time? What is the car's average velocity during this interval?

SOLUTION
According to the definition of velocity in equation 2.6, we simply take the time derivative of the position vector function to arrive at our solution:

$$v_x = \frac{dx}{dt} = \frac{d}{dt}(a + bt + ct^2) = b + 2ct = -10.1 \text{ m/s} + 2 \cdot (1.10 \text{ m/s}^2)t.$$

It is instructive to graph this solution. In Figure 2.7, the position as a function of time is shown in blue, and the velocity as a function of time is shown in red. Initially, the velocity has a value of −10.1 m/s, and at $t = 10$ s, the velocity has a value of +11.9 m/s.

Note that the velocity is initially negative, is zero at 4.59 s (indicated by the vertical dashed line in Figure 2.7), and then is positive after 4.59 s. At $t = 4.59$ s, the position graph $x(t)$ shows an extremum (a minimum in this case), just as expected from calculus, since

$$\frac{dx}{dt} = b + 2ct_0 = 0 \Rightarrow t_0 = -\frac{b}{2c} = -\frac{-10.1 \text{ m/s}}{2.20 \text{ m/s}^2} = 4.59 \text{ s}.$$

From the definition of average velocity, we know that to determine the average velocity during a time interval, we need to subtract the position at the beginning of the interval from the position at the end of the interval. By inserting $t = 0$ and $t = 10$ s into the equation for the position vector as a function of time, we obtain $x(t = 0) = 17.2$ m and $x(t = 10\text{ s}) = 26.2$ m. Therefore,

$$\Delta x = x(t = 10) - x(t = 0) = 26.2 \text{ m} - 17.2 \text{ m} = 9.0 \text{ m}.$$

We then obtain for the average velocity over this time interval:

$$\bar{v}_x = \frac{\Delta x}{\Delta t} = \frac{9.0 \text{ m}}{10 \text{ s}} = 0.90 \text{ m/s}.$$

The slope of the green dashed line in Figure 2.7 is the average velocity over this time interval.

Speed

Speed is the absolute value of the velocity vector. For a moving object, speed is always positive. "Speed" and "velocity" are used interchangeably in everyday contexts, but in physical terms they are very different. Velocity is a vector, which has a direction. For one-dimensional motion, the velocity vector can point in either the positive or the negative direction; in other words, its component can have either sign. Speed is the absolute magnitude of the velocity vector and is thus a scalar quantity:

$$\text{speed} \equiv v = |\vec{v}| = |v_x|. \tag{2.8}$$

The last part of this equation makes use of the fact that the velocity vector has only an x-component for one-dimensional motion.

In everyday experience, we recognize that speed can never be negative: Speed limits are always posted as positive numbers, and the radar monitors that measure the speed of passing cars also always display positive numbers (Figure 2.8).

Earlier, distance was defined as the absolute value of displacement for each straight-line segment *in which the movement does not reverse direction* (see the discussion following equation 2.4). The average speed when a distance ℓ is traveled during a time interval Δt is

$$\text{average speed} \equiv \bar{v} = \frac{\ell}{\Delta t}. \tag{2.9}$$

FIGURE 2.8 Measuring the speeds of passing cars.

EXAMPLE 2.2 | Speed and Velocity

Suppose a swimmer completes the first 50 m of the 100-m freestyle in 38.2 s. Once she reaches the far side of the 50-m-long pool, she turns around and swims back to the start in 42.5 s.

PROBLEM

What are the swimmer's average velocity and average speed for (a) the leg from the start to the far side of the pool, (b) the return leg, and (c) the total lap?

SOLUTION

We start by defining our coordinate system, as shown in Figure 2.9. The positive x-axis points toward the bottom of the page.

(a) *First leg of the swim:*
The swimmer starts at $x_1 = 0$ and swims to $x_2 = 50$ m. It takes her $\Delta t = 38.2$ s to accomplish this leg. Her average velocity for leg 1 then, according to our definition, is

$$\bar{v}_{x1} = \frac{x_2 - x_1}{\Delta t} = \frac{50 \text{ m} - 0 \text{ m}}{38.2 \text{ s}} = \frac{50}{38.2} \text{ m/s} = 1.31 \text{ m/s}.$$

Her average speed is the distance divided by time interval, which, in this case, is the same as the absolute value of her average velocity, or $|\bar{v}_{x1}| = 1.31$ m/s.

(b) *Second leg of the swim:*
We use the same coordinate system for leg 2 as for leg 1. This choice means that the swimmer starts at $x_1 = 50$ m and finishes at $x_2 = 0$, and it takes $\Delta t = 42.5$ s to complete this leg. Her average velocity for this leg is

$$\bar{v}_{x2} = \frac{x_2 - x_1}{\Delta t} = \frac{0 \text{ m} - 50 \text{ m}}{42.5 \text{ s}} = \frac{-50}{42.5} \text{ m/s} = -1.18 \text{ m/s}.$$

Note the negative sign for the average velocity for this leg. The average speed is again the absolute magnitude of the average velocity, or $|\bar{v}_{x2}| = |-1.18 \text{ m/s}| = 1.18$ m/s.

(c) *The entire lap:*
We can find the average velocity in two ways, demonstrating that they result in the same answer. First, because the swimmer started at $x_1 = 0$ and finished at $x_2 = 0$, the difference is 0. Thus, the net displacement is 0, and consequently the average velocity is also 0.

We can also find the average velocity for the whole lap by taking the time-weighted sum of the components of the average velocities of the individual legs:

$$\bar{v}_x = \frac{\bar{v}_{x1} \cdot \Delta t_1 + \bar{v}_{x2} \cdot \Delta t_2}{\Delta t_1 + \Delta t_2} = \frac{(1.31 \text{ m/s})(38.2 \text{ s}) + (-1.18 \text{ m/s})(42.5 \text{ s})}{(38.2 \text{ s}) + (42.5 \text{ s})} = 0.$$

What do we find for the average speed? The average speed, according to our definition, is the total distance divided by the total time. The total distance is 100 m and the total time is 38.2 s plus 42.5 s, or 80.7 s. Thus,

$$\bar{v} = \frac{\ell}{\Delta t} = \frac{100 \text{ m}}{80.7 \text{ s}} = 1.24 \text{ m/s}.$$

We can also use the time-weighted sum of the average speeds, leading to the same result. Note that the average speed for the entire lap is between that for leg 1 and that for leg 2. It is not exactly halfway between these two values, but is closer to the lower value because the swimmer spent more time completing leg 2.

FIGURE 2.9 Choosing an x-axis in a swimming pool.

2.4 Acceleration Vector

Just as the average velocity is defined as the displacement per time interval, the x-component of **average acceleration** is defined as the velocity change per time interval:

$$\bar{a}_x = \frac{\Delta v_x}{\Delta t}. \tag{2.10}$$

Concept Check **2.4**

Average acceleration is defined as the

a) displacement change per time interval.

b) position change per time interval.

c) velocity change per time interval.

d) speed change per time interval.

Concept Check **2.5**

When you're driving a car along a straight road, you may be traveling in the positive or negative direction and you may have a positive acceleration or a negative acceleration. Match the following combinations of velocity and acceleration with the list of outcomes.

a) positive velocity, positive acceleration

b) positive velocity, negative acceleration

c) negative velocity, positive acceleration

d) negative velocity, negative acceleration

1) slowing down in positive direction

2) speeding up in negative direction

3) speeding up in positive direction

4) slowing down in negative direction

Concept Check **2.6**

An example of one-dimensional motion with constant acceleration is

a) the motion of a car during a NASCAR race.

b) the Earth orbiting the Sun.

c) an object in free fall.

d) None of the above describe one-dimensional motion with constant acceleration.

Similarly, the x-component of the **instantaneous acceleration** is defined as the limit of the average acceleration as the time interval approaches 0:

$$a_x = \lim_{\Delta t \to 0} \bar{a}_x = \lim_{\Delta t \to 0} \frac{\Delta v_x}{\Delta t} \equiv \frac{dv_x}{dt}. \tag{2.11}$$

We can now define the acceleration vector as

$$\vec{a} = \frac{d\vec{v}}{dt}, \tag{2.12}$$

where again the derivative operation is understood to act component-wise, just as in the definition of the velocity vector.

Figure 2.10 illustrates this relationship among velocity, time interval, average acceleration, and instantaneous acceleration as the limit of the average acceleration (for a decreasing time interval). In Figure 2.10a, the average acceleration is given by the velocity change, Δv_1, divided by the time interval Δt_1: $\bar{a}_1 = \Delta v_1/\Delta t_1$. In Figure 2.10b, the average acceleration is determined over a smaller time interval, Δt_2. In Figure 2.10c, the instantaneous acceleration, $a(t_3) = dv/dt|_{t=t_3}$, is represented by the slope of the blue line tangent to the red curve at $t = t_3$. Figure 2.10 looks very similar to Figure 2.6, and this is not a coincidence. The similarity emphasizes that the mathematical operations and physical relationships that connect the velocity and acceleration vectors are the same as those that connect the position and velocity vectors.

The acceleration is the time derivative of the velocity, and the velocity is the time derivative of the displacement. The acceleration is therefore the second derivative of the displacement:

$$a_x = \frac{d}{dt} v_x = \frac{d}{dt} \left(\frac{d}{dt} x \right) = \frac{d^2}{dt^2} x. \tag{2.13}$$

There is no word in everyday language for the absolute value of the acceleration.

Note that we often refer to the *deceleration* of an object as a decrease in the speed of the object over time, which corresponds to acceleration in the opposite direction of the motion of the object.

In one-dimensional motion, an acceleration, which is a change in velocity, necessarily entails a change in the magnitude of the velocity—that is, the speed. However, in the next chapter, we will consider motion in more than one spatial dimension, where the velocity vector can also change its direction, not just its magnitude. In Chapter 9, we will examine motion in a circle with constant speed; in that case, there is a constant acceleration that keeps the object on a circular path but leaves the speed constant.

As Concept Check 2.5 shows, even in one dimension, a positive acceleration does not necessarily mean speeding up and a negative acceleration does not mean that an object must be slowing down. Rather, the combination of velocity and acceleration determines the motion. *If the velocity and acceleration are in the same direction, the object moves faster; if they are in opposite directions, it slows down.* We will examine this relationship further in the next chapter.

FIGURE 2.10 Instantaneous acceleration as the limit of the ratio of velocity change to time interval: (a) average acceleration over a large time interval; (b) average acceleration over a smaller time interval; and (c) instantaneous acceleration in the limit as the time interval goes to zero.

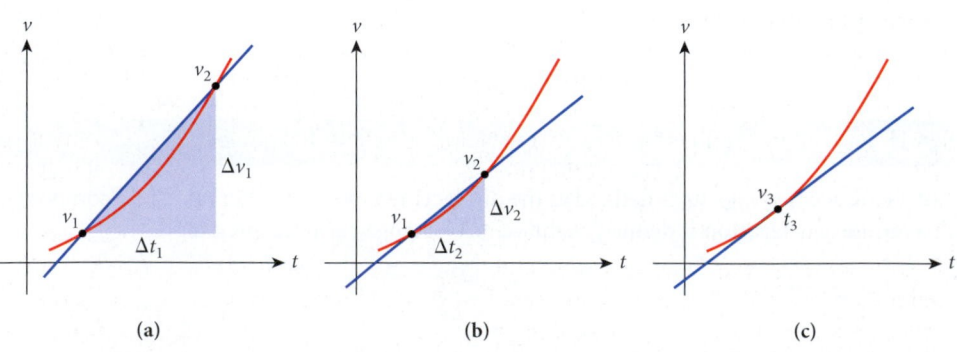

2.5 Computer Solutions and Difference Formulas

In some situations, the acceleration changes as a function of time, but the exact functional form is not known beforehand. However, we can still calculate velocity and acceleration even if the position is known only at certain points in time. The following example illustrates this procedure.

EXAMPLE 2.3 / World Record for the 100-m Dash

In the 1991 Track and Field World Championships in Tokyo, Japan, Carl Lewis of the United States set a new world record in the 100-m dash. Figure 2.11 lists the times at which he arrived at the 10-m mark, the 20-m mark, and so on, as well as values for his average velocity and average acceleration, calculated from the formulas in equations 2.5 and 2.10. From Figure 2.11, it is clear that after about 3 s, Lewis reached an approximately constant average velocity between 11 and 12 m/s.

Figure 2.11 also indicates how the values for average velocity and average acceleration were obtained. Take, for example, the upper two green boxes, which contain the times and positions for two measurements. From these we get $\Delta t = 2.96$ s – 1.88 s = 1.08 s and $\Delta x =$ 20 m – 10 m = 10 m. The average velocity in this time interval is then $\bar{v} = \Delta x / \Delta t$ = 10 m/1.08 s = 9.26 m/s. We have rounded this result to three significant digits, because the times were given to that accuracy. The accuracy for the distances can be assumed to be even better, because these data were extracted from video analysis that showed the times at which Lewis crossed marks on the ground.

In Figure 2.11 the calculated average velocity is positioned halfway between the lines for time and distance, indicating that it is a good approximation for the instantaneous velocity in the middle of the time interval.

The average velocities for other time intervals were obtained in the same way. Using the numbers in the second and third green boxes in Figure 2.11, we obtained an average velocity of 10.87 m/s for the time interval from 2.96 s to 3.88 s. With two velocity values, we can use the difference formula for the acceleration to calculate the average acceleration. Here we assume that the instantaneous velocity at a time corresponding to halfway between the first two green boxes (2.42 s) is equal to the average velocity during the interval between the first two green boxes, or 9.26 m/s. Similarly, we take the instantaneous velocity at 3.42 s (midway between the second and third boxes) to be 10.87 m/s. Then, the average acceleration between 2.42 s and 3.42 s is

$$\bar{a}_x = \Delta v_x / \Delta t = (10.87 \text{ m/s} - 9.26 \text{ m/s})/(3.42 \text{ s} - 2.42 \text{ s}) = 1.61 \text{ m/s}^2.$$

From the entries in Figure 2.11 obtained in this way, we can see that Lewis did most of his accelerating between the start of the race and the 30-m mark, where he attained his maximum velocity between 11 and 12 m/s. He then ran with about that velocity until he reached the finish line. This result is clearer in a graphical display of his position versus time during the race (Figure 2.12a). The red dots represent the data points from Figure 2.11, and the green straight line represents a constant velocity of 11.58 m/s. In Figure 2.12b, Lewis's velocity is plotted as a function of time. The green line again represents a constant velocity of 11.58 m/s, fitted to the last six points, where Lewis is no longer accelerating but running at a constant velocity.

t(s)	x(m)	$\bar{v}_x$(m/s)	$\bar{a}_x$(m/s²)
0.00	0		5.66
		5.32	
1.88	10		2.66
		9.26	
2.96	20		1.61
		10.87	
3.88	30		0.40
		11.24	
4.77	40		0.77
		11.90	
5.61	50		−0.17
		11.76	
6.46	60		0.17
		11.90	
7.30	70		0.17
		12.05	
8.13	80		−0.65
		11.49	
9.00	90		0.00
		11.49	
9.87	100		

FIGURE 2.11 Time, position, average velocity, and average acceleration during Carl Lewis's world record 100-m dash.

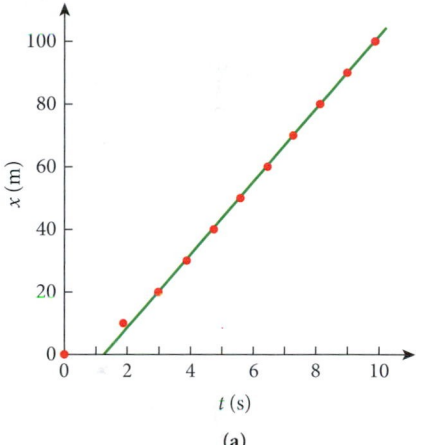

(a)

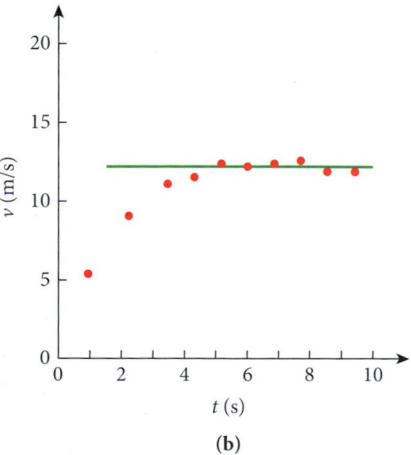

(b)

FIGURE 2.12 Analysis of Carl Lewis's 100-m dash in 1991: (a) his position as a function of time; (b) his velocity as a function of time.

The type of numerical analysis that treats average velocities and accelerations as approximations for the instantaneous values of these quantities is very common in all types of scientific and engineering applications. It is indispensable in situations where the precise functional dependencies on time are not known and researchers must rely on numerical approximations for derivatives obtained via the difference formulas. Most practical solutions of scientific and engineering problems found with the aid of computers make use of difference formulas such as those introduced here.

The entire field of numerical analysis is devoted to finding better numerical approximations that will enable more precise and faster computer calculations and simulations of natural processes. Difference formulas similar to those introduced here are as important for the everyday work of scientists and engineers as the calculus-based analytic expressions.

This importance is a consequence of the computer revolution in science and technology, which, however, does not make the contents of this textbook any less important. In order to devise a valid solution to an engineering or science problem, you have to understand the basic underlying physical principles, no matter what calculation techniques you use. This fact is well recognized by cutting-edge movie animators and creators of special digital effects, who have to take basic physics classes to ensure that the products of their computer simulations look realistic to the audience.

2.6 Finding Displacement and Velocity from Acceleration

The fact that integration is the inverse operation of differentiation is known as the *Fundamental Theorem of Calculus*. It allows us to reverse the differentiation process leading from displacement to velocity to acceleration and instead integrate the equation for velocity (2.6) to obtain displacement and the equation for acceleration (2.13) to obtain velocity. Let's start with the equation for the x-component of the velocity:

$$v_x(t) = \frac{dx(t)}{dt} \Rightarrow$$

$$\int_{t_0}^{t} v_x(t')dt' = \int_{t_0}^{t} \frac{dx(t')}{dt'}dt' = x(t) - x(t_0) \Rightarrow$$

$$x(t) = x_0 + \int_{t_0}^{t} v_x(t')dt'. \tag{2.14}$$

Notation: Here again we have used the convention that $x(t_0) = x_0$, the initial position. Further, we have used the notation t' in the definite integrals in equation 2.14. This prime notation reminds us that the integration variable is a dummy variable, which serves to identify the physical quantity we want to integrate. Throughout this book, we will reserve the prime notation for dummy integration variables in definite integrals. (Note that some books use a prime to denote a spatial derivative, but to avoid possible confusion, this book will not do so.)

In the same way, we integrate equation 2.13 for the x-component of the acceleration to obtain an expression for the x-component of the velocity:

$$a_x(t) = \frac{dv_x(t)}{dt} \Rightarrow$$

$$\int_{t_0}^{t} a_x(t')dt' = \int_{t_0}^{t} \frac{dv_x(t')}{dt'}dt' = v_x(t) - v_x(t_0) \Rightarrow$$

$$v_x(t) = v_{x0} + \int_{t_0}^{t} a_x(t')dt'. \tag{2.15}$$

Here $v_x(t_0) = v_{x0}$ is the initial velocity component in the x-direction. Just like the derivative operation, integration is understood to act component-wise, so we can write the integral relationships for the vectors from those for the components in equations 2.14 and 2.15. Formally, we then have

$$\vec{r}(t) = \vec{r}_0 + \int_{t_0}^{t} \vec{v}(t')dt' \tag{2.16}$$

and

$$\vec{v}(t) = \vec{v}_0 + \int_{t_0}^{t} \vec{a}(t')dt'. \tag{2.17}$$

This result means that for any given time dependence of the acceleration vector, we can calculate the velocity vector, provided we are given the initial value of the velocity vector. We can also calculate the displacement vector, if we know its initial value and the time dependence of the velocity vector.

In calculus, you probably learned that the geometrical interpretation of the definite integral is an area under a curve. This is true for equations 2.14 and 2.15. We can interpret the area under the curve of $v_x(t)$ between t_0 and t as the difference in the position between these two times, as shown in Figure 2.13a. Figure 2.13b shows that the area under the curve of $a_x(t)$ in the time interval between t_0 and t is the velocity difference between these two times.

2.7 Motion with Constant Acceleration

In many physical situations, the acceleration experienced by an object is approximately, or perhaps even exactly, constant. We can derive useful equations for these special cases of motion with constant acceleration. If the acceleration, a_x, is a constant, then the time integral used to obtain the velocity in equation 2.15 results in

$$v_x(t) = v_{x0} + \int_0^t a_x \, dt' = v_{x0} + a_x \int_0^t dt' \Rightarrow$$

$$v_x(t) = v_{x0} + a_x t, \qquad (2.18)$$

where we have taken the lower limit of the integral to be $t_0 = 0$ for simplicity. This means that the velocity is a linear function of time.

$$x = x_0 + \int_0^t v_x(t') dt' = x_0 + \int_0^t (v_{x0} + a_x t') dt'$$

$$= x_0 + v_{x0} \int_0^t dt' + a_x \int_0^t t' dt' \Rightarrow$$

$$x(t) = x_0 + v_{x0} t + \tfrac{1}{2} a_x t^2. \qquad (2.19)$$

Thus, with a constant acceleration, the velocity is always a linear function of time, and the position is a quadratic function of time. Three other useful equations can be derived using equations 2.18 and 2.19 as starting points. After listing these three equations, we'll work through their derivations.

The average velocity in the time interval from 0 to t is the average of the velocities at the beginning and end of the time interval:

$$\overline{v}_x = \tfrac{1}{2}(v_{x0} + v_x). \qquad (2.20)$$

The average velocity from equation 2.20 leads to an alternative way to express the position:

$$x = x_0 + \overline{v}_x t. \qquad (2.21)$$

Finally, we can write an equation for the square of the velocity that does not contain the time explicitly:

$$v_x^2 = v_{x0}^2 + 2a_x(x - x_0). \qquad (2.22)$$

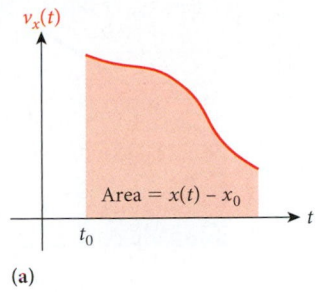

(a)

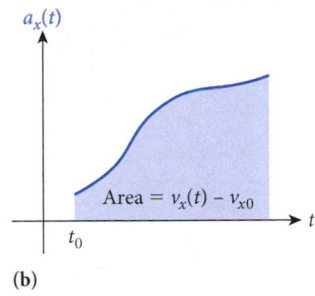

(b)

FIGURE 2.13 Geometrical interpretation of the integrals of (a) velocity and (b) acceleration with respect to time.

DERIVATION 2.1

Mathematically, to obtain the time average of a quantity over a certain interval Δt, we have to integrate this quantity over the time interval and then divide by the time interval:

$$\overline{v}_x = \frac{1}{t} \int_0^t v_x(t') dt' = \frac{1}{t} \int_0^t (v_{x0} + a_x t') dt'$$

$$= \frac{v_{x0}}{t} \int_0^t dt' + \frac{a_x}{t} \int_0^t t' dt' = v_{x0} + \tfrac{1}{2} a_x t$$

$$= \tfrac{1}{2} v_{x0} + \tfrac{1}{2}(v_{x0} + a_x t)$$

$$= \tfrac{1}{2}(v_{x0} + v_x). \qquad \text{– Continued}$$

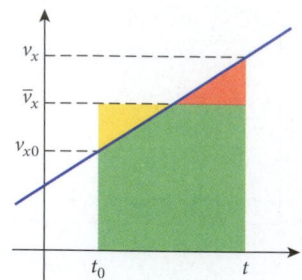

FIGURE 2.14 Graph of velocity versus time for motion with constant acceleration.

This averaging procedure for the time interval t_0 to t is illustrated in Figure 2.14. You can see that the area of the trapezoid formed by the blue line representing $v(t)$ and the two vertical lines at t_0 and t is equal to the area of the square formed by the horizontal line to $\bar{v}_x$ and the two vertical lines. The base line for both areas is the horizontal t-axis. It is more apparent that these two areas are equal if you note that the yellow triangle (part of the square) and the orange triangle (part of the trapezoid) are equal in size. [Algebraically, the area of the square is $\bar{v}_x(t - t_0)$, and the area of the trapezoid is $\frac{1}{2}(v_{x0} + v_x)(t - t_0)$. Setting these two areas equal to each other gives us equation 2.20 again.]

To derive the equation for the position, we take $t_0 = 0$, use the expression $\bar{v}_x = v_{x0} + \frac{1}{2}a_x t$, and multiply both sides by the time:

$$\bar{v}_x = v_{x0} + \tfrac{1}{2}a_x t$$
$$\Rightarrow \bar{v}_x t = v_{x0}t + \tfrac{1}{2}a_x t^2.$$

Now we compare this result to the expression we already obtained for x (equation 2.19) and find:

$$x = x_0 + v_{x0}t + \tfrac{1}{2}a_x t^2 = x_0 + \bar{v}_x t.$$

For the derivation of equation 2.22 for the square of the velocity, we solve $v_x = v_{x0} + a_x t$ for the time, getting $t = (v_x - v_{x0})/a_x$. We then substitute into the expression for the position, which is equation 2.19:

$$x = x_0 + v_{x0}t + \tfrac{1}{2}a_x t^2$$
$$= x_0 + v_{x0}\left(\frac{v_x - v_{x0}}{a_x}\right) + \tfrac{1}{2}a_x \left(\frac{v_x - v_{x0}}{a_x}\right)^2$$
$$= x_0 + \frac{v_x v_{x0} - v_{x0}^2}{a_x} + \tfrac{1}{2}\frac{v_x^2 + v_{x0}^2 - 2v_x v_{x0}}{a_x}.$$

Next, we subtract x_0 from both sides of the equation and then multiply by a_x:

$$a_x(x - x_0) = v_x v_{x0} - v_{x0}^2 + \tfrac{1}{2}(v_x^2 + v_{x0}^2 - 2v_x v_{x0})$$
$$\Rightarrow a_x(x - x_0) = \tfrac{1}{2}v_x^2 - \tfrac{1}{2}v_{x0}^2$$
$$\Rightarrow v_x^2 = v_{x0}^2 + 2a_x(x - x_0).$$

Here are the five kinematical equations we have obtained for the *special case of motion with constant acceleration* (where the initial time when $x = x_0$ and $v = v_0$ is taken to be 0):

$$
\begin{array}{lll}
\text{(i)} & x = x_0 + v_{x0}t + \tfrac{1}{2}a_x t^2 & \\
\text{(ii)} & x = x_0 + \bar{v}_x t & \\
\text{(iii)} & v_x = v_{x0} + a_x t & \text{(2.23)} \\
\text{(iv)} & \bar{v}_x = \tfrac{1}{2}(v_x + v_{x0}) & \\
\text{(v)} & v_x^2 = v_{x0}^2 + 2a_x(x - x_0) & \\
\end{array}
$$

These five equations allow us to solve many kinds of problems for motion in one dimension with constant acceleration. However, remember that if the acceleration is not constant, these equations will not give the correct solutions.

Many real-life problems involve motion along a straight line with constant acceleration. In these situations, equations 2.23 provide the template for answering any question about the motion. The following solved problem and example will illustrate how useful these kinematical equations are. However, keep in mind that physics is not simply about finding an appropriate equation and plugging in numbers, but is instead about understanding concepts. Only if you understand the underlying ideas will you be able to extrapolate from specific examples and become skilled at solving more general problems.

SOLVED PROBLEM 2.2 | Airplane Takeoff

As an airplane rolls down a runway to reach takeoff speed, it is accelerated by its jet engines. On one particular flight, one of the authors of this book measured the acceleration produced by the plane's jet engines. Figure 2.15 shows the measurements.

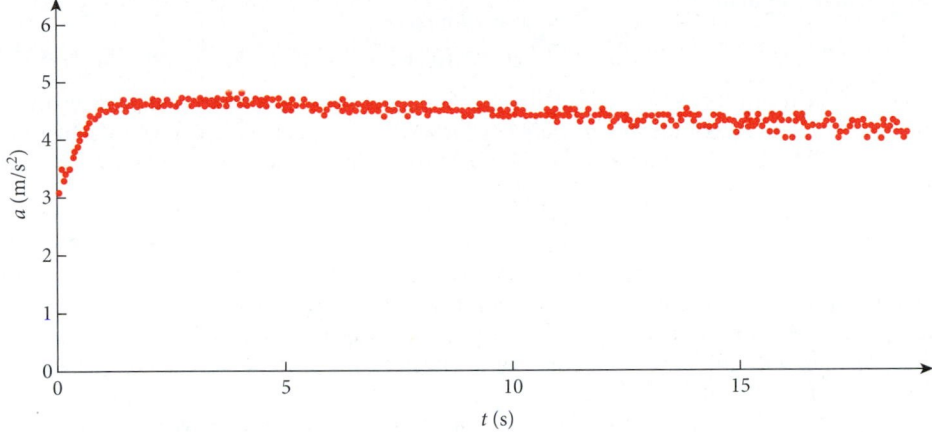

FIGURE 2.15 Data for the acceleration of a jet plane before takeoff.

You can see that the assumption of constant acceleration is not quite correct in this case. However, an average acceleration of $a_x = 4.3 \text{ m/s}^2$ over the 18.4 s (measured with a stopwatch) it took the airplane to take off is a good approximation.

PROBLEM
Assuming a constant acceleration of $a_x = 4.3 \text{ m/s}^2$ starting from rest, what is the airplane's takeoff velocity after 18.4 s? How far down the runway has the plane moved by the time it takes off?

SOLUTION
THINK An airplane moving along a runway prior to takeoff is a nearly perfect example of one-dimensional accelerated motion. Because we are assuming constant acceleration, we know that the velocity increases linearly with time, and the displacement increases as the second power of time. Since the plane starts from rest, the initial value of the velocity is 0. As usual, we can define the origin of our coordinate system at any location; it is convenient to locate it at the point of the plane's standing start.

SKETCH The sketch in Figure 2.16 shows how we expect the velocity and displacement to increase for this case of constant acceleration, where the initial conditions are set at $v_{x0} = 0$ and $x_0 = 0$. Note that no scales have been placed on the axes, because displacement, velocity, and acceleration are measured in different units. Thus, the points at which the three curves intersect are completely arbitrary.

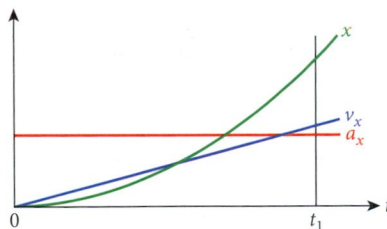

FIGURE 2.16 The acceleration, velocity, and displacement of the plane before takeoff.

RESEARCH Finding the takeoff velocity is actually a straightforward application of equation 2.23(iii):

$$v_x = v_{x0} + a_x t.$$

Similarly, the distance down the runway that the airplane moves before taking off can be obtained from equation 2.23(i):

$$x = x_0 + v_{x0} t + \tfrac{1}{2} a_x t^2.$$

SIMPLIFY The airplane accelerates from a standing start, so the initial velocity is $v_{x0} = 0$, and by our choice of coordinate system origin, we have set $x_0 = 0$. Therefore, the equations for takeoff velocity and distance simplify to

$$v_x = a_x t$$
$$x = \tfrac{1}{2} a_x t^2.$$

– Continued

CALCULATE The only thing left to do is to put in the numbers:

$$v_x = (4.3 \text{ m/s}^2)(18.4 \text{ s}) = 79.12 \text{ m/s}$$

$$x = \tfrac{1}{2}(4.3 \text{ m/s}^2)(18.4 \text{ s})^2 = 727.904 \text{ m}.$$

ROUND The acceleration was specified to two significant digits, and the time to three. Multiplying these two numbers must result in an answer that has two significant digits. Our final answers are therefore

$$v_x = 79 \text{ m/s}$$

$$x = 7.3 \cdot 10^2 \text{ m}.$$

Note that the measured time to takeoff of 18.4 s was probably not actually that precise. If you have ever tried to determine the moment at which a plane starts to accelerate down the runway, you will have noticed that it is almost impossible to determine that point in time with an accuracy to 0.1 s.

DOUBLE-CHECK As this book has repeatedly stressed, the most straightforward check of any answer to a physics problem is to make sure the units fit the situation. This is the case here, because we obtained displacement in units of meters and velocity in units of meters per second. Solved problems in the rest of this book may sometimes skip this simple test; however, if you want to do a quick check for algebraic errors in your calculations, it can be valuable to first look at the units of the answer.

Now let's see if our answers have the appropriate orders of magnitude. A takeoff displacement of 730 m (~0.5 mi) is reasonable, because it is on the order of the length of an airport runway. A takeoff velocity of $v_x = 79$ m/s translates into

$$(79 \text{ m/s})(1 \text{ mi}/1609 \text{ m})(3600 \text{ s}/1 \text{ h}) \approx 180 \text{ mph}.$$

This answer also appears to be in the right ballpark.

ALTERNATIVE SOLUTION

Many problems in physics can be solved in several ways, because it is often possible to use more than one relationship between the known and unknown quantities. In this case, once we obtain the final velocity, we could use this information and solve kinematical equation 2.23(v) for x. This alternative approach results in

$$v_x^2 = v_{x0}^2 + 2a_x(x - x_0) \Rightarrow$$

$$x = x_0 + \frac{v_x^2 - v_{x0}^2}{2a_x} = 0 + \frac{(79 \text{ m/s})^2}{2(4.3 \text{ m/s}^2)} = 7.3 \cdot 10^2 \text{ m}.$$

Thus, we arrive at the same answer for the distance in a different way, giving us additional confidence that our solution makes sense.

Solved Problem 2.2 was a fairly easy one; solving it amounted to little more than plugging in numbers. Nevertheless, it shows that the kinematical equations we derived can be applied to real-world situations and lead to answers that have physical meaning. The following short example, this time from motor sports, addresses the same concepts of velocity and acceleration, but in a slightly different light.

EXAMPLE 2.4 Top Fuel Racing

Accelerating from rest, a top fuel race car (Figure 2.17) can reach 333.2 mph (= 148.9 m/s), a record established in 2003, at the end of a quarter mile (= 402.3 m). For this example, we will assume constant acceleration.

PROBLEM 1

What is the value of the race car's constant acceleration?

SOLUTION 1

Since the initial and final values of the velocity are given and the distance is known, we are looking for a relationship between these three quantities and the acceleration, the unknown.

In this case, it is most convenient to use kinematical equation 2.23(v) and solve for the acceleration, a_x:

$$v_x^2 = v_{x0}^2 + 2a_x(x - x_0) \Rightarrow a_x = \frac{v_x^2 - v_{x0}^2}{2(x - x_0)} = \frac{(148.9 \text{ m/s})^2}{2(402.3 \text{ m})} = 27.6 \text{ m/s}^2.$$

PROBLEM 2

How long does it take the race car to complete a quarter-mile run from a standing start?

SOLUTION 2

Because the final velocity is 148.9 m/s, the average velocity is [using equation 2.23(iv)]: $\bar{v}_x = \frac{1}{2}(148.9 \text{ m/s} + 0) = 74.45$ m/s. Relating this average velocity to the displacement and time using equation 2.23(ii), we obtain:

$$x = x_0 + \bar{v}_x t \Rightarrow t = \frac{x - x_0}{\bar{v}_x} = \frac{402.3 \text{ m}}{74.45 \text{ m/s}} = 5.40 \text{ s}.$$

Note that we could have obtained the same result by using kinematical equation 2.23(iii), because we already calculated the acceleration in Solution 1.

If you are a fan of top fuel racing, however, you know that the real record time for the quarter mile is slightly lower than 4.5 s. The reason our calculated answer is somewhat higher is that our assumption of constant acceleration is not quite correct. The acceleration of the car at the beginning of the race is actually higher than the value we calculated above, and the actual acceleration toward the end of the race is lower than our value.

From what you have seen up to now, it may appear that almost all problems involving constant acceleration can be solved by using equations 2.23. But it's important to remember that you can't apply these equations blindly. The following two solved problems illustrate this point.

SOLVED PROBLEM 2.3 | Racing with a Head Start

Cheri has a new Dodge Charger with a Hemi engine and has challenged Vince, who owns a tuned VW GTI, to a race at a local track. Vince knows that Cheri's Charger is rated to go from 0 to 60 mph in 5.3 s, whereas his VW needs 7.0 s. Vince asks for a head start and Cheri agrees to give him exactly 1.0 s.

PROBLEM

How far down the track is Vince before Cheri gets to start the race? At what time does Cheri catch Vince? How far away from the start are they when this happens? (Assume constant acceleration for each car during the race.)

SOLUTION

THINK This race is a good example of one-dimensional motion with constant acceleration. The temptation is to look over the kinematical equations 2.23 and see which one we can apply. However, we have to be a bit more careful here, because the time delay between Vince's start and Cheri's start adds a complication. In fact, if you try to solve this problem using the kinematical equations directly, you will not get the right answer. Instead, this problem requires some careful definition of the time coordinates for each car.

SKETCH For our sketch, we plot time on the horizontal axis and position on the vertical axis. Both cars move with constant acceleration from a standing start, so we expect simple parabolas for their paths in this diagram.

Since Cheri's car has the greater acceleration, its parabola (blue curve in Figure 2.18) has the larger curvature and thus the steeper rise. Therefore, it is clear that Cheri will catch Vince at some point, but it is not yet clear where this point is.

– Continued

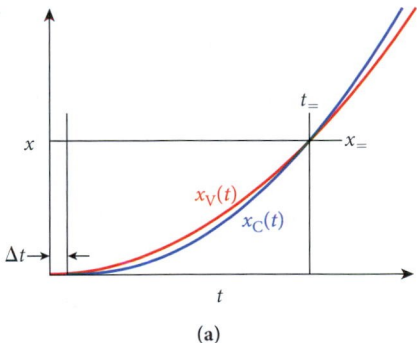

(a)

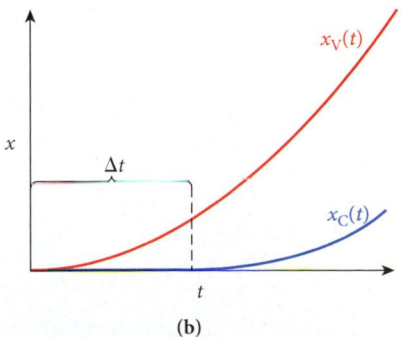

(b)

FIGURE 2.18 Position versus time for the race between Cheri and Vince. Part (a) shows the entire duration of the race, and part (b) shows a magnified view of the start.

RESEARCH We set up the problem quantitatively. We call the time delay before Cheri can start Δt, and we use the indices (subscripts) C for Cheri's Charger and V for Vince's VW. We put the coordinate system's origin at the starting line. Thus, both cars have initial position $x_C(t=0) = x_V(t=0) = 0$. Since both cars are at rest at the start, their initial velocities are zero. The equation of motion for Vince's VW is

$$x_V(t) = \tfrac{1}{2}a_V t^2.$$

Here we use the symbol a_V for the acceleration of the VW. We can calculate its value from the 0 to 60 mph time given by the problem statement, but we will postpone this step until it is time to put in the numbers.

To obtain the equation of motion for Cheri's Charger, we have to be careful, because she is forced to wait Δt after Vince has taken off. We can account for this delay with a shifted time: $t' = t - \Delta t$. Once t reaches the value of Δt, the time t' has the value 0 and then Cheri can take off. So her Charger's equation of motion is

$$x_C(t) = \tfrac{1}{2}a_C t'^2 = \tfrac{1}{2}a_C(t - \Delta t)^2 \qquad \text{for } t \geq \Delta t.$$

Just like a_V, the constant acceleration a_C for Cheri's Charger will be evaluated below.

SIMPLIFY When Cheri catches Vince, their coordinates will have the same value. We will call the time at which this happens $t_=$, and the coordinate where it happens $x_= \equiv x(t_=)$. Because the two coordinates are the same, we have

$$x_= = \tfrac{1}{2}a_V t_=^2 = \tfrac{1}{2}a_C(t_= - \Delta t)^2.$$

We can solve this equation for $t_=$ by dividing out the common factor of $\tfrac{1}{2}$ and then taking the square root of both sides of the equation:

$$\sqrt{a_V}\, t_= = \sqrt{a_C}\,(t_= - \Delta t) \Rightarrow$$
$$t_=\left(\sqrt{a_C} - \sqrt{a_V}\right) = \Delta t \sqrt{a_C} \Rightarrow$$
$$t_= = \frac{\Delta t \sqrt{a_C}}{\sqrt{a_C} - \sqrt{a_V}}.$$

Why do we use the positive root and discard the negative root here? The negative root would lead to a physically impossible solution: We are interested in the time when the two cars meet *after* they have left the start, not in a negative value that would imply a time before they left.

CALCULATE Now we can get a numerical answer for each of the questions that were asked. First, let's figure out the values of the cars' constant accelerations from the 0 to 60 mph specifications given. We use $a = (v_x - v_{x0})/t$ and get

$$a_V = \frac{60 \text{ mph}}{7.0 \text{ s}} = \frac{26.8167 \text{ m/s}}{7.0 \text{ s}} = 3.83095 \text{ m/s}^2$$

$$a_C = \frac{60 \text{ mph}}{5.3 \text{ s}} = \frac{26.8167 \text{ m/s}}{5.3 \text{ s}} = 5.05975 \text{ m/s}^2.$$

Again, we postpone rounding our results until we have completed all steps in our calculations. However, with the values for the accelerations, we can immediately calculate how far Vince travels during the time $\Delta t = 1.0$ s:

$$x_V(1.0 \text{ s}) = \tfrac{1}{2}(3.83095 \text{ m/s}^2)(1.0 \text{ s})^2 = 1.91548 \text{ m}.$$

Now we can calculate the time when Cheri catches up to Vince:

$$t_= = \frac{\Delta t \sqrt{a_C}}{\sqrt{a_C} - \sqrt{a_V}} = \frac{(1.0 \text{ s})\sqrt{5.05975 \text{ m/s}^2}}{\sqrt{5.05975 \text{ m/s}^2} - \sqrt{3.83095 \text{ m/s}^2}} = 7.70055 \text{ s}.$$

At this time, both cars have traveled to the same point. Thus, we can insert this value into either equation of motion to find this position:

$$x_= = \tfrac{1}{2}a_V t_=^2 = \tfrac{1}{2}(3.83095 \text{ m/s}^2)(7.70055 \text{ s})^2 = 113.585 \text{ m}.$$

ROUND   The initial data were specified to a precision of only two significant digits. Rounding our results to the same precision, we finally arrive at our answers: Vince receives a head start of 1.9 m, and Cheri will catch him after 7.7 s. At that time, they will be $1.1 \cdot 10^2$ m into their race.

DOUBLE-CHECK   It may seem strange to you that Vince's car was able to move only 1.9 m, or approximately half the length of the car, during the first second. Have we done anything wrong in our calculation? The answer is no; from a standing start, cars move only a comparatively short distance during the first second of acceleration. The following solved problem contains visible proof for this statement.

Figure 2.19 shows a plot of the equations of motion for both cars, this time with the proper units.

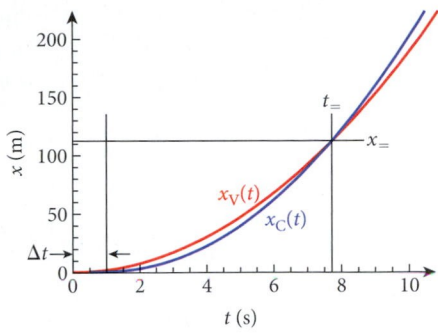

FIGURE 2.19 Plot of the parameters and equations of motion for the race between Cheri and Vince.

SOLVED PROBLEM 2.4 | Accelerating Car

PROBLEM
You are given the image sequence shown in Figure 2.20, and you are told that there is a time interval of 0.333 s between successive frames. Can you determine how fast this car (whose length is 174.9 inches) was accelerating from rest? Also, can you give an estimate for the time it takes this car to go from 0 to 60 mph?

SOLUTION
THINK   Acceleration is measured in dimensions of length per time per time. To find a number for the value of the acceleration, we need to know the time and length scales in Figure 2.20. The time scale is straightforward because we were given the information that 0.333 s passes between successive frames. We can get the length scale from the specified vehicle dimensions. For example, if we focus on the length of the car and compare it to the overall width of the frame, we can find the distance the car covered between the first and the last frame (which are 3.000 s apart).

SKETCH   We draw vertical yellow lines over Figure 2.20, as shown in Figure 2.21. We put the center of the car at the line between the front and rear windows (the exact location is irrelevant, as long as we are consistent). Now we can use a ruler and measure the perpendicular distance between the two yellow lines, indicated by the yellow double-headed arrow in the figure. We can also measure the length of the car, indicated by the red double-headed arrow.

RESEARCH   Dividing the measured length of the yellow double-headed arrow by that of the red one gives us a ratio of 3.474. Since the two vertical yellow lines mark the position of the center of the car at 0.000 s and 3.000 s, we know that the car covered a distance of 3.474 car lengths in this time interval. The car length was given as 174.9 in = 4.442 m. Thus, the total distance covered is $d = 3.474 \cdot \ell_{car} = 3.474 \cdot 4.442$ m = 15.4315 m (remember that we round to the proper number of significant digits at the end).

SIMPLIFY   We have two choices on how to proceed. The first one is more complicated: We could measure the position of the car in each frame and then use difference formulas like those in Example 2.3. The other and much faster way to proceed is to assume constant acceleration and then use the measurements of the car's positions in only the first and last frames. We'll use the second way, but in the end we'll need to double-check that our assumption of constant acceleration is justified.

For a constant acceleration from a standing start, we simply have

$$x = x_0 + \tfrac{1}{2}a_x t^2 \Rightarrow$$
$$d = x - x_0 = \tfrac{1}{2}a_x t^2 \Rightarrow$$
$$a_x = \frac{2d}{t^2}.$$

– Continued

FIGURE 2.20 Video sequence of a car accelerating from a standing start.

FIGURE 2.21 Determination of the length scale for Figure 2.20.

This is the acceleration we want to find. Once we have the acceleration, we can give an estimate for the time from 0 to 60 mph by using $v_x = v_{x0} + at \Rightarrow t = (v_x - v_{x0})/a_x$. A standing start means $v_{x0} = 0$, and so we have

$$t(0-60 \text{ mph}) = \frac{60 \text{ mph}}{a_x}.$$

CALCULATE We insert the numbers to find the acceleration:

$$a_x = \frac{2d}{t^2} = \frac{2 \cdot (15.4315 \text{ m})}{(3.000 \text{ s})^2} = 3.42922 \text{ m/s}^2.$$

Then, for the time from 0 to 60 mph, we obtain

$$t(0-60 \text{ mph}) = \frac{60 \text{ mph}}{a_x} = \frac{(60 \text{ mph})(1609 \text{ m/mi})(1 \text{ h}/3600 \text{ s})}{3.42922 \text{ m/s}^2} = 7.82004 \text{ s}.$$

ROUND The length of the car sets our length scale, and it was given to four significant figures. The time was given to three significant figures. Are we entitled to quote our results to three significant digits? The answer is no, because we also performed measurements on Figure 2.21, which are probably accurate to only two digits at best. In addition, you can see field-of-vision and lens distortions in the image sequence: In the first few frames, you see a little bit of the front of the car, and in the last few frames, a little bit of the back. Taking all this into account, our results should be quoted to two significant figures. So our final answer for the acceleration is

$$a_x = 3.4 \text{ m/s}^2.$$

For the time from 0 to 60 mph, we give

$$t(0-60 \text{ mph}) = 7.8 \text{ s}.$$

DOUBLE-CHECK The numbers we have found for the acceleration and the time from 0 to 60 mph are fairly typical for cars or small SUVs; see also Solved Problem 2.3. Thus, we have confidence that we are not off by orders of magnitude.

What we must also double-check, however, is the assumption of constant acceleration. For constant acceleration from a standing start, the points $x(t)$ should fall on a parabola $x(t) = \frac{1}{2}a_x t^2$. Therefore, if we plot x on the horizontal axis and t on the vertical axis as in Figure 2.22, the points $t(x)$ should follow a square root dependence: $t(x) = \sqrt{2x/a_x}$. This functional dependence is shown by the red curve in Figure 2.22. We can see that the same point of the car is hit by the curve in every frame, giving us confidence that the assumption of constant acceleration is reasonable.

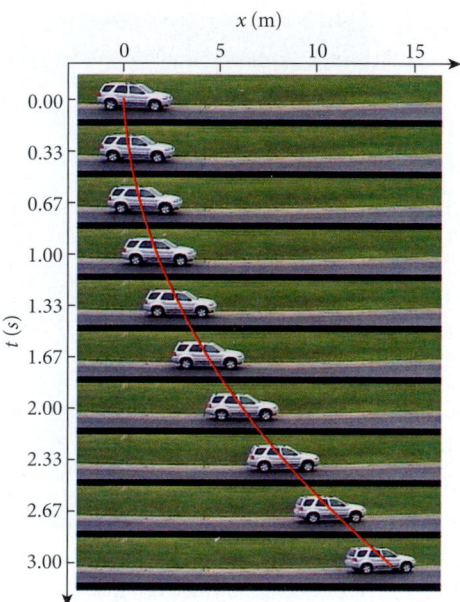

FIGURE 2.22 Graphical analysis of the accelerating car problem.

2.8 Free Fall

The acceleration due to the gravitational force is constant, to a good approximation, near the surface of Earth. If this statement is true, it must have observable consequences. Let's assume it is true and work out the consequences for the motion of objects under the influence of Earth's gravitational attraction. Then we'll compare our results with experimental observations and see if constant acceleration due to gravity makes sense.

The acceleration due to gravity near the surface of the Earth has the value $g = 9.81 \text{ m/s}^2$. We call the vertical axis the y-axis and define the positive direction as up. Then the acceleration vector $\vec{a}$ has only a nonzero y-component, which is given by

$$a_y = -g. \tag{2.24}$$

This situation is a specific application of motion with constant acceleration, which we discussed earlier in this section. We modify equations 2.23 by substituting for acceleration from equation 2.24. We also use y instead of x to indicate that the displacement takes place in the y-direction. We obtain:

$$\text{(i)} \qquad y = y_0 + v_{y0}t - \tfrac{1}{2}gt^2$$

$$\text{(ii)} \qquad y = y_0 + \bar{v}_y t$$

$$\text{(iii)} \qquad v_y = v_{y0} - gt \qquad\qquad (2.25)$$

$$\text{(iv)} \qquad \bar{v}_y = \tfrac{1}{2}(v_y + v_{y0})$$

$$\text{(v)} \qquad v_y^2 = v_{y0}^2 - 2g(y - y_0)$$

Motion under the sole influence of a gravitational acceleration is called **free fall**, and equations 2.25 allow us to solve problems for objects in free fall.

Now let's consider an experiment that tested the assumption of constant gravitational acceleration. The authors went to the top of a building of height 12.7 m and dropped a computer from rest ($v_{y0} = 0$) under controlled conditions. The computer's fall was recorded by a digital video camera. Because the camera records at 30 frames per second, we know the time information. Figure 2.23 displays 14 frames, equally spaced in time, from this experiment, with the time after release marked on the horizontal axis for each frame. The yellow curve superimposed on the frames has the form

$$y = 12.7 \text{ m} - \tfrac{1}{2}(9.81 \text{ m/s}^2)t^2,$$

which is what we expect for initial conditions $y_0 = 12.7$ m, $v_{y0} = 0$ and the assumption of a constant acceleration, $a_y = -9.81$ m/s^2. As you can see, the computer's fall follows this curve almost perfectly. This agreement is, of course, not a conclusive proof, but it is a strong indication that the gravitational acceleration is constant near the surface of Earth, and that it has the stated value.

In addition, the value of the gravitational acceleration is the same for all objects. This is by no means a trivial statement. Objects of different sizes and masses, if released from the same height, should hit the ground at the same time. Is this consistent with our everyday experience? Well, not quite! In a common lecture demonstration, a feather and a coin are dropped from the same height. It is easy to observe that the coin reaches the floor first, while the feather slowly floats down. This difference is due to air resistance. If this experiment is done in an evacuated glass tube, the coin and the feather fall at the same rate. We will return to air resistance in Chapter 4, but for now we can conclude that the gravitational acceleration near the surface of Earth is constant, has the absolute value of $g = 9.81$ m/s^2, and is the same for all objects provided we can neglect air resistance. In Chapter 4, we will examine the conditions under which the assumption of zero air resistance is justified.

To help you understand the answer to Concept Check 2.7, consider throwing a ball straight up, as illustrated in Figure 2.24. In Figure 2.24a, the ball is thrown upward with a

Concept Check 2.7

Throwing a ball straight up into the air provides an example of free-fall motion. At the instant the ball reaches its maximum height, which of the following statements is true?

a) The ball's acceleration vector points down, and its velocity vector points up.

b) The ball's acceleration is zero, and its velocity vector points up.

c) The ball's acceleration vector points up, and its velocity vector points up.

d) The ball's acceleration vector points down, and its velocity is zero.

e) The ball's acceleration vector points up, and its velocity is zero.

f) The ball's acceleration is zero, and its velocity vector points down.

Concept Check 2.8

A ball is thrown upward with a speed v_1, as shown in Figure 2.24. The ball reaches a maximum height of $y = h$. What is the ratio of the speed of the ball, v_2, at $y = h/2$ in Figure 2.24b, to the initial upward speed of the ball, v_1, at $y = 0$ in Figure 2.24a?

a) $v_2/v_1 = 0$

b) $v_2/v_1 = 0.50$

c) $v_2/v_1 = 0.71$

d) $v_2/v_1 = 0.75$

e) $v_2/v_1 = 0.90$

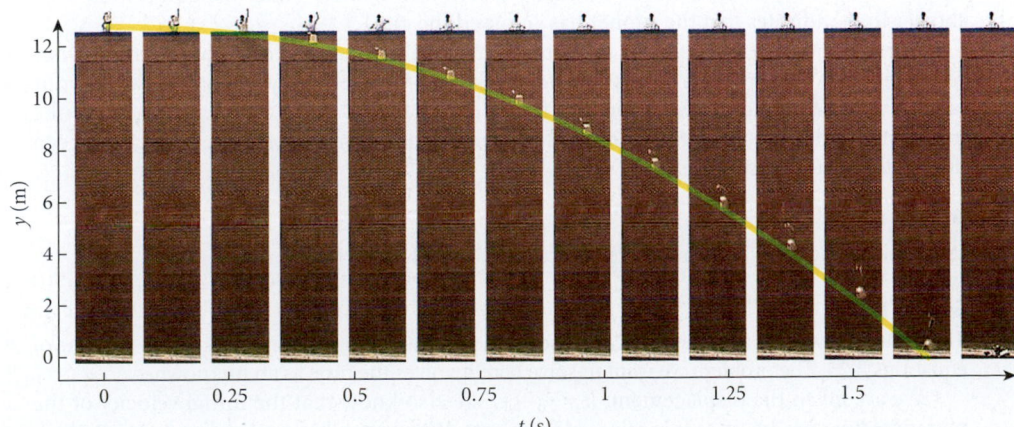

FIGURE 2.23 Free-fall experiment: dropping a computer off the top of a building.

FIGURE 2.24 The velocity vector and acceleration vector of a ball thrown straight up in the air. (a) The ball is initially thrown upward at $y = 0$. (b) The ball going upward at a height of $y = h/2$. (c) The ball at its maximum height of $y = h$. (d) The ball coming down at $y = h/2$. (e) The ball back at $y = 0$ going downward.

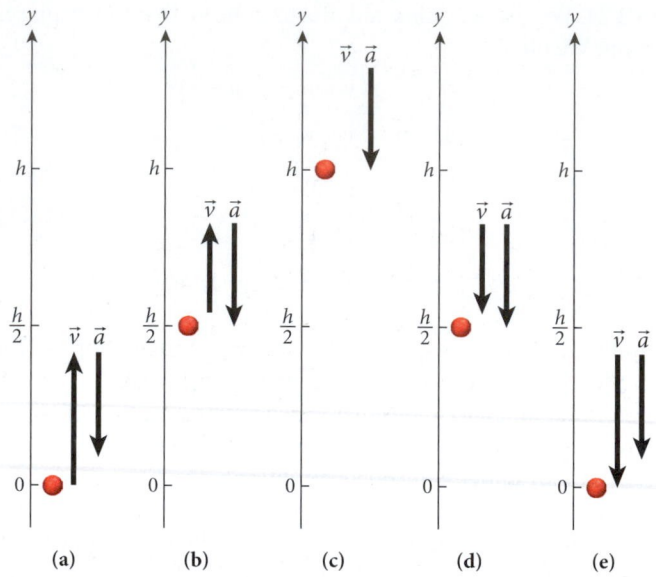

(a)   (b)   (c)   (d)   (e)

velocity $\vec{v}$. When the ball is released, it experiences only the force of gravity and thus accelerates downward with the acceleration due to gravity, which is given by $\vec{a} = -g\hat{y}$. As the ball travels upward, the acceleration due to gravity acts to slow the ball down. In Figure 2.24b, the ball is moving upward and has reached half of its maximum height, h. The ball has slowed down, but its acceleration is still the same. The ball reaches its maximum height in Figure 2.24c. Here the velocity of the ball is zero, but the acceleration is still $\vec{a} = -g\hat{y}$. The ball now begins to move downward, and the acceleration due to gravity acts to speed it up. In Figure 2.24d, the velocity of the ball is downward and the acceleration is still the same. Finally, the ball returns to $y = 0$ in Figure 2.24e. The velocity of the ball now has the same magnitude as when the ball was initially thrown upward, but its direction is now downward. The acceleration is still the same. Note that the acceleration remains constant and downward even though the velocity changes from upward to zero to downward.

EXAMPLE 2.5   Reaction Time

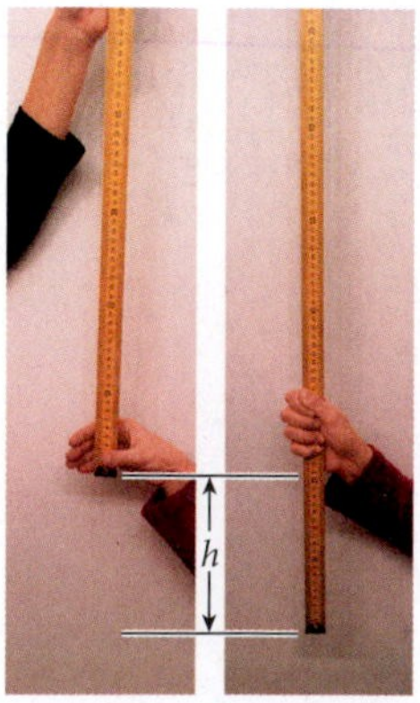

It takes time for a person to react to any external stimulus. For example, at the beginning of a 100-m dash in a track-and-field meet, a gun is fired by the starter. A slight time delay occurs before the runners come out of the starting blocks, due to their nonzero reaction time. In fact, it counts as a false start if a runner leaves the blocks less than 0.1 s after the gun is fired. Any shorter time indicates that the runner has "jumped the gun."

There is a simple test, shown in Figure 2.25, that you can perform to determine your reaction time. Your partner holds a meter stick, and you get ready to catch it when your partner releases it, as shown in the left frame of the figure. From the distance h that the meter stick falls after it is released until you grab it (shown in the right frame), you can determine your reaction time.

FIGURE 2.25 Simple experiment to measure reaction time.

PROBLEM

If the meter stick falls 0.20 m before you catch it, what is your reaction time?

SOLUTION

This situation is a free-fall scenario. For these problems, the solution invariably comes from one of equations 2.25. The problem we want to solve here involves the time as an unknown.

We are given the displacement, $h = y_0 - y$. We also know that the initial velocity of the meter stick is zero because it is released from rest. We can use kinematical equation 2.25(i): $y = y_0 + v_{y0}t - \frac{1}{2}gt^2$. With $h = y_0 - y$ and $v_0 = 0$, this equation becomes

$$y = y_0 - \tfrac{1}{2}gt^2$$
$$\Rightarrow h = \tfrac{1}{2}gt^2$$
$$\Rightarrow t = \sqrt{\frac{2h}{g}} = \sqrt{\frac{2 \cdot 0.20 \text{ m}}{9.81 \text{ m/s}^2}} = 0.20 \text{ s}.$$

Your reaction time was 0.20 s. This reaction time is typical. For comparison, when Usain Bolt established a world record of 9.69 s for the 100-m dash in August 2008, his reaction time was measured to be 0.165 s. (One year later Bolt ran the 100 m in 9.58 s, which is the current world record.)

Let's consider one more free-fall scenario, this time with two moving objects.

SOLVED PROBLEM 2.5 | Melon Drop

Suppose you decide to drop a melon from rest from the first observation platform of the Eiffel Tower. The initial height h from which the melon is released is 58.3 m above the head of your French friend Pierre, who is standing on the ground right below you. At the same instant you release the melon, Pierre shoots an arrow straight up with an initial velocity of 25.1 m/s. (Of course, Pierre makes sure the area around him is cleared and gets out of the way quickly after he shoots his arrow.)

PROBLEM
(a) How long after you drop the melon will the arrow hit it? (b) At what height above Pierre's head does this collision occur?

SOLUTION
THINK At first sight, this problem looks complicated. We will solve it using the full set of steps and then examine a shortcut we could have taken. Obviously, the dropped melon is in free fall. However, because the arrow is shot straight up, the arrow is also in free fall, only with an upward initial velocity.

SKETCH We set up our coordinate system with the y-axis pointing vertically up, as is conventional, and we locate the origin of the coordinate system at Pierre's head (Figure 2.26). Thus, the arrow is released from an initial position $y = 0$, and the melon from $y = h$.

RESEARCH We use the subscripts m for "melon" and a for "arrow." We start with the general free-fall equation, $y = y_0 + v_{y0}t - \tfrac{1}{2}gt^2$, and use the initial conditions given for the melon ($v_{y0} = 0$, $y_0 = h = 58.3$ m) and for the arrow ($v_{y0} \equiv v_{a0} = 25.1$ m/s, $y_0 = 0$) to set up the two equations of free-fall motion:

$$y_m(t) = h - \tfrac{1}{2}gt^2$$
$$y_a(t) = v_{a0}t - \tfrac{1}{2}gt^2.$$

The key insight is that at t_c, the moment when the melon and arrow collide, their coordinates are identical:

$$y_a(t_c) = y_m(t_c).$$

SIMPLIFY Inserting t_c into the two equations of motion and setting them equal results in

$$h - \tfrac{1}{2}gt_c^2 = v_{a0}t_c - \tfrac{1}{2}gt_c^2 \Rightarrow$$
$$h = v_{a0}t_c \Rightarrow$$
$$t_c = \frac{h}{v_{a0}}.$$

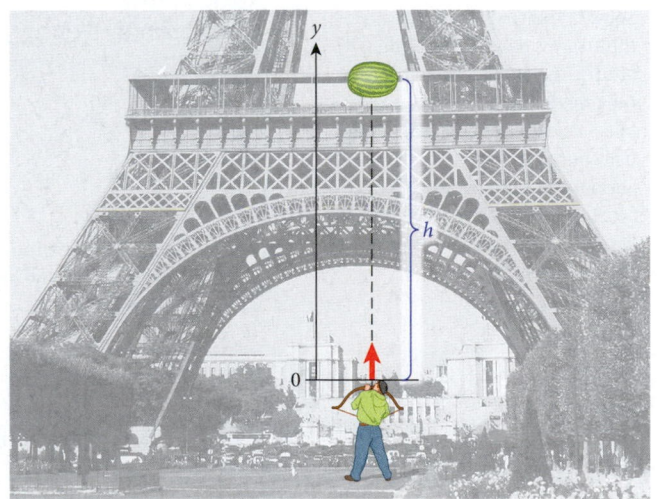

FIGURE 2.26 The melon drop (melon and person are not drawn to scale!).

– Continued

We can now insert this value for the time of collision in either of the two free-fall equations and obtain the height above Pierre's head at which the collision occurs. We select the equation for the melon:

$$y_m(t_c) = h - \tfrac{1}{2} g t_c^2 .$$

CALCULATE   (a) All that is left to do is to insert the numbers given for the height of release of the melon and the initial velocity of the arrow, which results in

$$t_c = \frac{58.3 \text{ m}}{25.1 \text{ m/s}} = 2.32271 \text{ s}$$

for the time of impact.

(b) Using the number we obtained for the time, we find the position at which the collision occurs:

$$y_m(t_c) = 58.3 \text{ m} - \tfrac{1}{2}(9.81 \text{ m/s}^2)(2.32271 \text{ s})^2 = 31.8376 \text{ m}.$$

ROUND   Since the initial values of the release height and the arrow velocity were given to three significant figures, we have to limit our final answers to three digits. Thus, the arrow will hit the melon after 2.32 s, and this will occur at a position 31.8 m above Pierre's head.

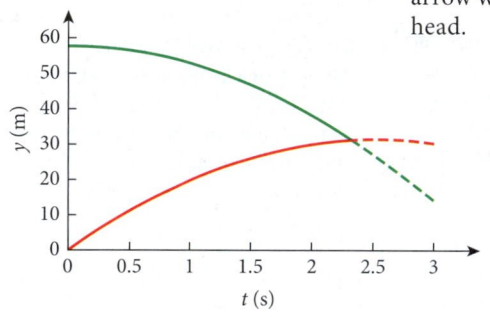

FIGURE 2.27 Position as a function of time for the arrow (red curve) and the melon (green curve).

DOUBLE-CHECK   Could we have obtained the answers in an easier way? Yes, if we had realized that both melon and arrow fall under the influence of the same gravitational acceleration, and thus their free-fall motion does not influence the distance between them. This means that the time it takes them to meet is simply the initial distance between them divided by their initial velocity difference. With this realization, we could have written $t = h/v_{a0}$ right away and been done. However, thinking in terms of relative motion in this way takes some practice, and we will return to it in more detail in the next chapter.

Figure 2.27 shows the complete graph of the positions of arrow and melon as functions of time. The dashed portions of both graphs indicate where the arrow and melon would have gone, had they not collided.

ADDITIONAL QUESTION

What are the velocities of melon and arrow at the moment of the collision?

SOLUTION

We obtain the velocity by taking the time derivative of the position. For arrow and melon, we get

$$y_m(t) = h - \tfrac{1}{2} g t^2 \Rightarrow v_m(t) = \frac{dy_m(t)}{dt} = -gt$$

$$y_a(t) = v_{a0} t - \tfrac{1}{2} g t^2 \Rightarrow v_a(t) = \frac{dy_a(t)}{dt} = v_{a0} - gt.$$

Now, inserting the time of the collision, 2.32 s, will produce the answers. Note that, unlike the positions of the arrow and the melon, the velocities of the two objects are not the same right before contact!

$$v_m(t_c) = -(9.81 \text{ m/s}^2)(2.32 \text{ s}) = -22.8 \text{ m/s}$$

$$v_a(t_c) = (25.1 \text{ m/s}) - (9.81 \text{ m/s}^2)(2.32 \text{ s}) = 2.34 \text{ m/s}.$$

Furthermore, you should note that the difference between the two velocities is still 25.1 m/s, just as it was at the beginning of the trajectories.

We finish this problem by plotting (Figure 2.28) the velocities as a function of time. You can see that the arrow starts out with a velocity that is 25.1 m/s greater than that of the melon. As time progresses, arrow and melon experience the same change in velocity under the influence of gravity, meaning that their velocities maintain the initial difference.

Self-Test Opportunity 2.3

As you can see from the answer to Solved Problem 2.5, the velocity of the arrow is only 2.34 m/s when it hits the melon. This means that by the time the arrow hits the melon, its initial velocity has dropped substantially because of the effect of gravity. Suppose that the initial velocity of the arrow was smaller by 5.0 m/s. What would change? Would the arrow still hit the melon?

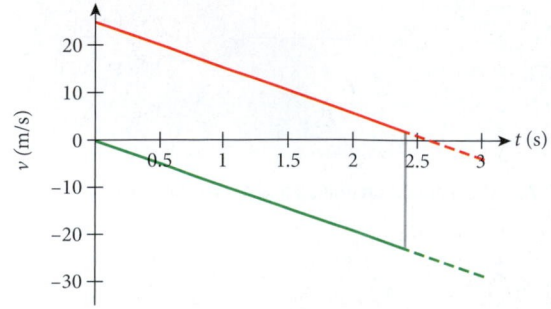

FIGURE 2.28 Velocities of the arrow (red curve) and melon (green curve) as a function of time.

2.9 Reducing Motion in More Than One Dimension to One Dimension

Motion in one spatial dimension is not all there is to kinematics. We can also investigate more general cases, in which objects move in two or three spatial dimensions. We will do this in the next few chapters. However, in some cases, motion in more than one dimension can be reduced to one-dimensional motion. Let's consider a very interesting case of motion in two dimensions for which each segment can be described by motion in a straight line.

EXAMPLE 2.6 Aquathlon

The triathlon is a sporting competition that was invented by the San Diego Track Club in the 1970s and became an event in the Sydney Olympic Games in 2000. Typically, it consists of a 1.5-km swim, followed by a 40-km bike race, and finishing with a 10-km foot race. To be competitive, athletes must be able to swim the 1.5-km distance in less than 20 minutes, do the 40-km bike race in less than 70 minutes, and run the 10-km race in less than 35 minutes.

However, for this example, we'll consider a competition where thinking is rewarded in addition to athletic prowess. The competition consists of only two legs: a swim followed by a run. (This competition is sometimes called an *aquathlon*.) Athletes start a distance $b = 1.5$ km from shore, and the finish line is a distance $a = 3$ km along the shoreline, as shown in Figure 2.29. Suppose you can swim with a speed of $v_1 = 3.5$ km/h and can run across the sand with a speed of $v_2 = 14$ km/h.

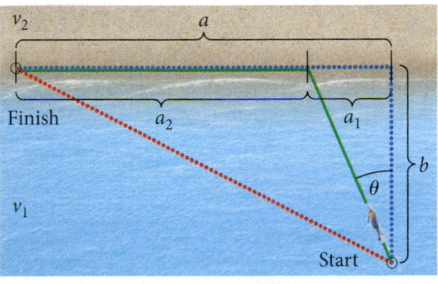

FIGURE 2.29 Geometry of the aquathlon course.

PROBLEM
What angle θ will result in the shortest finish time under these conditions?

SOLUTION
Clearly, the dotted red line marks the shortest distance between start and finish. This distance is $\sqrt{a^2 + b^2} = \sqrt{1.5^2 + 3^2}$ km = 3.354 km. Because this entire path is in water, it couldn't be used for the aquathlon, but the time it would take to swim this distance is

$$t_{red} = \frac{\sqrt{a^2 + b^2}}{v_1} = \frac{3.354 \text{ km}}{3.5 \text{ km/h}} = 0.958 \text{ h}.$$

Because you can run faster than you can swim, we can also try the approach indicated by the dotted blue line: swim straight to shore and then run. This takes

$$t_{blue} = \frac{b}{v_1} + \frac{a}{v_2} = \frac{1.5 \text{ km}}{3.5 \text{ km/h}} + \frac{3 \text{ km}}{14 \text{ km/h}} = 0.643 \text{ h}.$$

Thus, the blue path is better than the red one. But is it the best? To answer this question, we need to search the angle interval from 0 (blue path) to $\tan^{-1}(3/1.5) = 63.43°$ (red path). Let's consider the green path, with an arbitrary angle θ with respect to the straight line to shore (that is, the normal to the shoreline). On the green path, you have to swim a distance of $\sqrt{a_1^2 + b^2}$ and then run a distance of a_2, as indicated in Figure 2.29. The total time for this path is

$$t = \frac{\sqrt{a_1^2 + b^2}}{v_1} + \frac{a_2}{v_2}.$$

To find the minimum time, we can express the time in terms of the distance a_1 only, take the derivative of that time with respect to that distance, set that derivative equal to zero, and solve for the distance. Using a_1, we can then calculate the angle θ that the athlete must swim.

We can express the distance a_2 in terms of the given distance a and a_1:

$$a_2 = a - a_1.$$

– Continued

We can then express the time to complete the race in terms of a_1:

$$t(a_1) = \frac{\sqrt{a_1^2 + b^2}}{v_1} + \frac{a - a_1}{v_2}. \tag{2.26}$$

Taking the derivative with respect to a_1 and setting that result equal to zero gives us

$$\frac{dt(a_1)}{da_1} = \frac{a_1}{v_1\sqrt{a_1^2 + b^2}} - \frac{1}{v_2} = 0.$$

Solving for a_1 gives us

$$a_1 = \frac{bv_1}{\sqrt{v_2^2 - v_1^2}}.$$

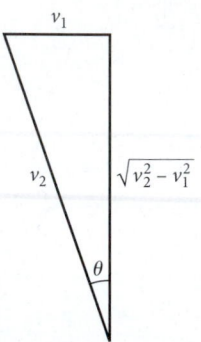

v_1

v_2 $\sqrt{v_2^2 - v_1^2}$

θ

FIGURE 2.30 Relation between v_1, v_2, and θ.

In Figure 2.29, we can see that $\tan\theta = a_1/b$, so we can write

$$\tan\theta = \frac{a_1}{b} = \frac{\dfrac{bv_1}{\sqrt{v_2^2 - v_1^2}}}{b} = \frac{v_1}{\sqrt{v_2^2 - v_1^2}}.$$

We can simplify this result by looking at the triangle for the angle θ in Figure 2.30. The Pythagorean Theorem tells us that the hypotenuse of the triangle is

$$\sqrt{v_1^2 + v_2^2 - v_1^2} = v_2.$$

We can then write

$$\sin\theta = \frac{v_1}{v_2}.$$

This result is very interesting because the distances a and b do not appear in it at all! Instead the sine of the optimum angle is simply the ratio of the speeds in water and on land. For the given values of the two speeds, this optimum angle is

$$\theta_m = \sin^{-1}\frac{3.5}{14} = 14.48°.$$

Inserting $a_1 = b\tan\theta_m$ into equation 2.26, we find $t(\theta_m) = 0.629$ h. This time is approximately 49 seconds faster than swimming straight to shore and then running (blue path).

Strictly speaking, we have not quite shown that this angle results in the minimum time. To accomplish this, we also need to show that the second derivative of the time with respect to the angle is larger than zero. However, since we did find one extremum, and since its value is smaller than those of the boundaries, we know that this extremum is a true minimum.

Finally, Figure 2.31 plots the time, in hours, needed to complete the race for all angles between 0° and 63.43°, indicated by the green curve. This plot is obtained by substituting $a_1 = b\tan\theta$ into equation 2.26 and using the identity $\tan^2\theta + 1 = \sec^2\theta$, which gives us

$$t(\theta) = \frac{\sqrt{(b\tan\theta)^2 + b^2}}{v_1} + \frac{a - b\tan\theta}{v_2} = \frac{b\sec\theta}{v_1} + \frac{a - b\tan\theta}{v_2}.$$

A vertical red line marks the maximum angle, corresponding to a straight-line swim from start to finish. The vertical blue line marks the optimum angle that we calculated, and the horizontal blue line marks the duration of the race for this angle.

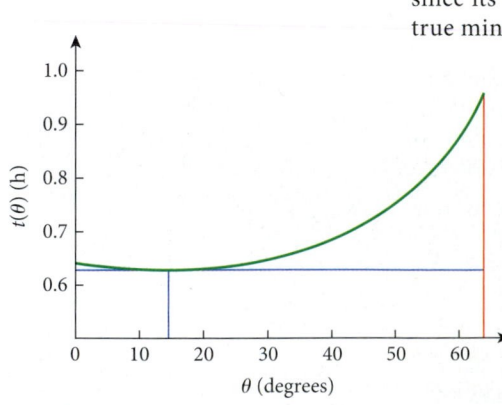

$t(\theta)$ (h)

θ (degrees)

FIGURE 2.31 Duration of the race as a function of initial angle.

Having completed Example 2.6, we can address a more complicated question: If the finish line is not at the shoreline, but at a perpendicular distance b away from the shoreline, as shown in Figure 2.32, what are the angles θ_1 and θ_2 that a competitor needs to select to achieve the minimum time?

We proceed in a fashion very similar to our approach in Example 2.6. However, now we have to realize that the time depends on two angles, θ_1 and θ_2. These two angles are not independent of each other. We can better see the relationship between θ_1 and θ_2 by changing the orientation of the lower triangle in Figure 2.32, as shown in Figure 2.33.

Now we see that the two right triangles a_1bc_1 and a_2bc_2 have a common side b, which helps us relate the two angles to each other. We can express the time to complete the race as

$$t = \frac{\sqrt{a_1^2 + b^2}}{v_1} + \frac{\sqrt{a_2^2 + b^2}}{v_2}.$$

Again realizing that $a_2 = a - a_1$, we can write

$$t(a_1) = \frac{\sqrt{a_1^2 + b^2}}{v_1} + \frac{\sqrt{(a - a_1)^2 + b^2}}{v_2}.$$

Taking the derivative of the time to complete the race with respect to a_1 and setting that result equal to zero, we obtain

$$\frac{dt(a_1)}{da_1} = \frac{a_1}{v_1 \sqrt{a_1^2 + b^2}} - \frac{a - a_1}{v_2 \sqrt{(a - a_1)^2 + b^2}} = 0.$$

We can rearrange this equation to get

$$\frac{a_1}{v_1 \sqrt{a_1^2 + b^2}} = \frac{a - a_1}{v_2 \sqrt{(a - a_1)^2 + b^2}} = \frac{a_2}{v_2 \sqrt{a_2^2 + b^2}}.$$

Looking at Figure 2.33 and referring to our previous result from Figure 2.30, we can see that

$$\sin \theta_1 = \frac{a_1}{\sqrt{a_1^2 + b^2}} \qquad \text{and} \qquad \sin \theta_2 = \frac{a_2}{\sqrt{a_2^2 + b^2}}.$$

We can insert these two results into the previous equation and finally find that to finish the race in the shortest time, a competitor should select angles that satisfy

$$\frac{\sin \theta_1}{v_1} = \frac{\sin \theta_2}{v_2}. \tag{2.27}$$

We can now see that our previous result, where we forced the running to take place along the beach, is a special case of this more general result, equation 2.27, with $\theta_2 = 90°$.

Just as for that special case, we find that the relationship between the angles does not depend on the values of the displacements a and b, but depends only on the speeds with which the competitor can move in the water and on land. The angles are still related to a and b by the overall constraint that the competitor has to get from the start to the finish. However, for the path of minimum time, the change in direction at the boundary between water and land, as expressed by the two angles θ_1 and θ_2, is determined exclusively by the ratio of the speeds v_1 and v_2.

The initial condition specified that the perpendicular distance b from the starting point to the shoreline be the same as the perpendicular distance between the shoreline and the finish. We did this to keep the algebra relatively brief. However, in the final formula you see that there are no more references to b; it has canceled out. Equation 2.27 is even valid in the case where the two perpendicular distances have different values. The ratio of angles for the path of minimum time is exclusively determined by the two speeds in the different media.

Interestingly, we will encounter the same relationship between two angles and two speeds when we study light and see that it changes direction at the interface between two media

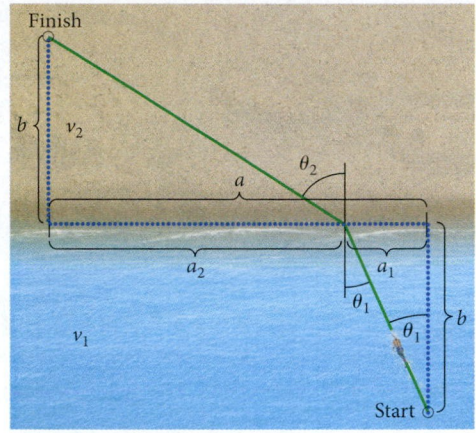

FIGURE 2.32 Modified aquathlon, with finish away from the shoreline.

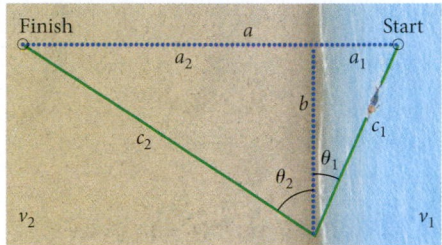

FIGURE 2.33 Same as previous figure, but with lower triangle reflected along the horizontal axis.

through which it moves at different speeds. In Chapter 32, we will discover that light also moves along the path of minimum time and that the result obtained in equation 2.27 is known as Snell's Law.

Finally, we make an observation that may appear trivial, but is not: If a competitor started at the point marked "Finish" in Figure 2.32 and ended up at the point marked "Start," he or she would have to take exactly the same path as the one we just calculated for the reverse direction. Snell's Law holds in both directions.

WHAT WE HAVE LEARNED | EXAM STUDY GUIDE

- x is the x-component of the position vector. Displacement is the change in position: $\Delta x = x_2 - x_1$.

- Distance is the absolute value of displacement, $\ell = |\Delta x|$, and is a positive scalar for motion in one direction.

- The average velocity of an object in a given time interval is given by $\bar{v}_x = \dfrac{\Delta x}{\Delta t}$.

- The x-component of the (instantaneous) velocity *vector* is the derivative of the x-component of the position *vector* as a function of time, $v_x = \dfrac{dx}{dt}$.

- Speed is the absolute value of the velocity: $v = |v_x|$.

- The x-component of the (instantaneous) acceleration *vector* is the derivative of the x-component of the velocity *vector* as a function of time, $a_x = \dfrac{dv_x}{dt}$.

- For constant acceleration, five kinematical equations describe motion in one dimension:
  - (i) $x = x_0 + v_{x0}t + \frac{1}{2}a_x t^2$
  - (ii) $x = x_0 + \bar{v}_x t$
  - (iii) $v_x = v_{x0} + a_x t$
  - (iv) $\bar{v}_x = \frac{1}{2}(v_x + v_{x0})$
  - (v) $v_x^2 = v_{x0}^2 + 2a_x(x - x_0)$

  where x_0 is the initial position, v_{x0} is the initial velocity, and the initial time t_0 is set to zero.

- For situations involving free fall (constant acceleration), we replace the acceleration a with $-g$ and x with y in the above kinematical equations to obtain
  - (i) $y = y_0 + v_{y0}t - \frac{1}{2}gt^2$
  - (ii) $y = y_0 + \bar{v}_y t$
  - (iii) $v_y = v_{y0} - gt$
  - (iv) $\bar{v}_y = \frac{1}{2}(v_y + v_{y0})$
  - (v) $v_y^2 = v_{y0}^2 - 2g(y - y_0)$

  where y_0 is the initial position, v_{y0} is the initial velocity, and the y-axis points up.

ANSWERS TO SELF-TEST OPPORTUNITIES

2.1 The result would not change, because a shift in the origin of the coordinate system has no influence on the net displacements or distances.

2.2

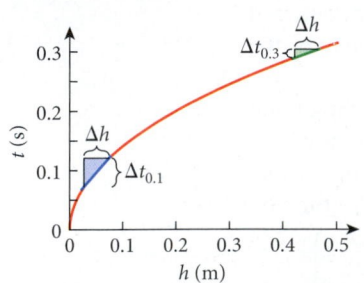

This method is more precise for longer reaction times, because the slope of the curve decreases as a function of

the height, h. (In the graph, the slope at 0.1 s is indicated by the blue line, and that at 0.3 s by the green line.) So, a given uncertainty, Δh, in measuring the height results in a smaller uncertainty, Δt, in the value for the reaction time for longer reaction times.

2.3 The arrow would still hit the melon, but at the time of collision the arrow would already have negative velocity and thus be moving downward again. Thus, the melon would catch the arrow as it was falling. The collision would occur a bit later, after $t = 58.3$ m/(20.1 m/s) = 2.90 s. The altitude of the collision would be quite a bit lower, at $y_m(t_c) = 58.3$ m $- \frac{1}{2}(9.81$ m/s$^2)(2.90$ s$)^2 = 17.0$ m above Pierre's head.

PROBLEM-SOLVING GUIDELINES: ONE-DIMENSIONAL KINEMATICS

1. In one-dimensional motion, displacement and velocity are vectors and can have positive and negative values; distance and speed are always nonnegative scalar quantities. It is important to keep this distinction in mind.

2. For problems involving one-dimensional motion, first determine if the acceleration is constant or not. The five kinematical equations 2.23 *apply only to motion with constant acceleration.*

3. If the acceleration is constant, first check whether any of equations 2.23 can help. Although you shouldn't blindly apply these equations, there is no point to reinventing the wheel.

4. Free-fall problems simply involve special cases of motion with constant acceleration, where the value of the acceleration is 9.81 m/s^2 in the downward direction.

MULTIPLE-CHOICE QUESTIONS

2.1 Two athletes jump straight up. Upon leaving the ground, Adam has half the initial speed of Bob. Compared to Adam, Bob jumps

a) 0.50 times as high.

b) 1.41 times as high.

c) twice as high.

d) three times as high.

e) four times as high.

2.2 Two athletes jump straight up. Upon leaving the ground, Adam has half the initial speed of Bob. Compared to Adam, Bob is in the air

a) 0.50 times as long.

b) 1.41 times as long.

c) twice as long.

d) three times as long.

e) four times as long.

2.3 A car is traveling due west at 20.0 m/s. Find the velocity of the car after 3.00 s if its acceleration is 1.0 m/s^2 due west. Assume that the acceleration remains constant.

a) 17.0 m/s west

b) 17.0 m/s east

c) 23.0 m/s west

d) 23.0 m/s east

e) 11.0 m/s south

2.4 A car is traveling due west at 20.0 m/s. Find the velocity of the car after 37.00 s if its acceleration is 1.0 m/s^2 due east. Assume that the acceleration remains constant.

a) 17.0 m/s west

b) 17.0 m/s east

c) 23.0 m/s west

d) 23.0 m/s east

e) 11.0 m/s south

2.5 An electron, starting from rest and moving with a constant acceleration, travels 1.0 cm in 2.0 ms. What is the magnitude of this acceleration?

a) 25 km/s^2

b) 20 km/s^2

c) 15 km/s^2

d) 10 km/s^2

e) 5.0 km/s^2

2.6 A car travels at 22.0 m/s north for 30.0 min and then reverses direction and travels at 28.0 m/s for 15.0 min. What is the car's total displacement?

a) 1.44·10^4 m   b) 6.48·10^4 m   c) 3.96·10^4 m   d) 9.98·10^4 m

2.7 Which of these statements is (are) true?

1. An object can have zero acceleration and be at rest.

2. An object can have nonzero acceleration and be at rest.

3. An object can have zero acceleration and be in motion.

a) 1 only   b) 1 and 3   c) 1 and 2   d) 1, 2, and 3

2.8 A car moving at 60 km/h comes to a stop in 4.0 s. What was its average deceleration?

a) 2.4 m/s^2 b) 15 m/s^2 c) 4.2 m/s^2 d) 41 m/s^2

2.9 You drop a rock from a cliff. If air resistance is neglected, which of the following statements is (are) true?

1. The speed of the rock will increase.

2. The speed of the rock will decrease.

3. The acceleration of the rock will increase.

4. The acceleration of the rock will decrease.

a) 1   b) 1 and 4   c) 2   d) 2 and 3

2.10 A car travels at 22.0 mph for 15.0 min and at 35.0 mph for 30.0 min. How far does it travel overall?

a) 23.0 m   b) 3.70·10^4 m   c) 1.38·10^3 m   d) 3.30·10^2 m

2.11 If the melon in Solved Problem 2.5 is thrown straight up with an initial velocity of 5.00 m/s at the same time that the arrow is shot upward, how long does it take before the collision occurs?

a) 2.32 s

b) 2.90 s

c) 1.94 s

d) They do not collide before the melon hits the ground.

2.12 The figure describes the position of an object as a function of time. Which one of the following statements is true?

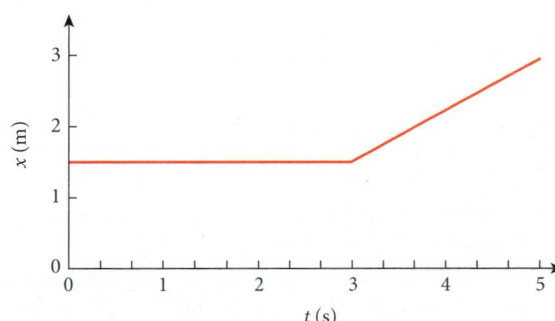

a) The position of the object is constant.

b) The velocity of the object is constant.

c) The object moves in the positive x-direction until $t = 3$ s, and then the object is at rest.

d) The object's position is constant until $t = 3$ s, and then the object begins to move in the positive x-direction.

e) The object moves in the positive x-direction from $t = 0$ to $t = 3$ s and then moves in the negative x-direction from $t = 3$ s to $t = 5$ s.

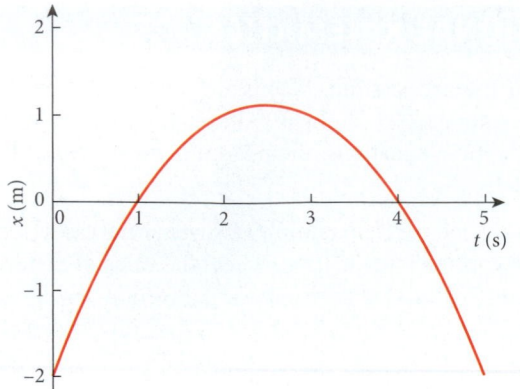

This figure describes the position of an object as a function of time. Refer to it to answer Questions 2.13–2.16.

2.13 Which one of the following statements is true at $t = 1$ s?

a) The x-component of the velocity of the object is zero.

b) The x-component of the acceleration of the object is zero.

c) The x-component of the velocity of the object is positive.

d) The x-component of the velocity of the object is negative.

2.14 Which one of the following statements is true at $t = 4$ s?

a) The x-component of the velocity of the object is zero.

b) The x-component of the acceleration of the object is zero.

c) The x-component of the velocity of the object is positive.

d) The x-component of the velocity of the object is negative.

2.15 Which one of the following statements is true at $t = 2.5$ s?

a) The x-component of the velocity of the object is zero.

b) The x-component of the acceleration of the object is zero.

c) The x-component of the velocity of the object is positive.

d) The x-component of the velocity of the object is negative.

2.16 Which one of the following statements is true at $t = 2.5$ s?

a) The x-component of the acceleration of the object is zero.

b) The x-component of the acceleration of the object is positive.

c) The x-component of the acceleration of the object is negative.

d) The acceleration of the object at that time can't be determined from the figure.

CONCEPTUAL QUESTIONS

2.17 Consider three ice skaters: Anna moves in the positive x-direction without reversing. Bertha moves in the negative x-direction without reversing. Christine moves in the positive x-direction and then reverses the direction of her motion. For which of these skaters is the magnitude of the average velocity smaller than the average speed over some time interval?

2.18 You toss a small ball vertically up in the air. How are the velocity and acceleration vectors of the ball oriented with respect to one another during the ball's flight up and down?

2.19 After you apply the brakes, the acceleration of your car is in the opposite direction to its velocity. If the acceleration of your car remains constant, describe the motion of your car.

2.20 Two cars are traveling at the same speed, and the drivers hit the brakes at the same time. The deceleration of one car is double that of the other. By what factor do the times required for the two cars to come to a stop differ?

2.21 If the acceleration of an object is zero and its velocity is nonzero, what can you say about the motion of the object? Sketch graphs of velocity versus time and acceleration versus time to support your explanation.

2.22 Can an object's acceleration be in the opposite direction to its motion? Explain.

2.23 You and a friend are standing at the edge of a snow-covered cliff. At the same time, you both drop a snowball over the edge of the cliff. Your snowball is twice as heavy as your friend's. Neglect air resistance. (a) Which snowball will hit the ground first? (b) Which snowball will have the greater speed?

2.24 You and a friend are standing at the edge of a snow-covered cliff. At the same time, you throw a snowball straight upward with a speed of 8.0 m/s over the edge of the cliff and your friend throws a snowball straight downward over the edge of the cliff with the same speed. Your snowball is twice as heavy as your friend's. Neglecting air resistance, which snowball will hit the ground first, and which will have the greater speed?

2.25 A car is slowing down and comes to a complete stop. The figure shows an image sequence of this process. The time between successive frames is 0.333 s, and the car is the same as the one in Solved Problem 2.4. Assuming constant acceleration, what is its value? Can you give some estimate of the error in your answer? How well justified is the assumption of constant acceleration?

2.26 A car moves along a road with a constant velocity. Starting at time $t = 2.5$ s, the driver accelerates with constant acceleration. The resulting position of the car as a function of time is shown by the blue curve in the figure.

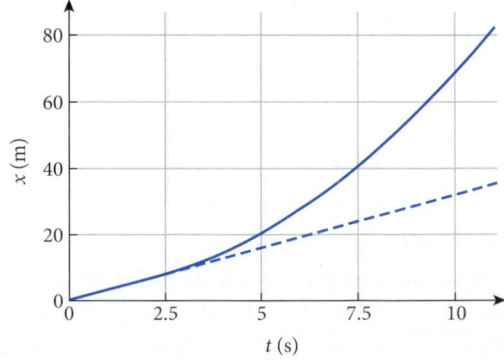

a) What is the value of the constant velocity of the car before 2.5 s? (*Hint:* The dashed blue line is the path the car would take in the absence of the acceleration.)

b) What is the velocity of the car at $t = 7.5$ s? Use a graphical technique (i.e., draw a slope).

c) What is the value of the constant acceleration?

2.27 You drop a rock over the edge of a cliff from a height h. Your friend throws a rock over the edge from the same height with a speed v_0 vertically downward, at some time t after you drop your rock. Both rocks hit the ground at the same time. How long after you dropped your rock did your friend throw hers? Express your answer in terms of v_0, g, and h.

2.28 A wrench is thrown vertically upward with speed v_0. How long after its release is it halfway to its maximum height?

EXERCISES

A blue problem number indicates a worked-out solution is available in the Student Solutions Manual. One • and two •• indicate increasing level of problem difficulty.

Section 2.2

2.29 A car travels north at 30.0 m/s for 10.0 min. It then travels south at 40.0 m/s for 20.0 min. What are the total distance the car travels and its displacement?

2.30 You ride your bike along a straight line from your house to a store 1000. m away. On your way back, you stop at a friend's house which is halfway between your house and the store.

a) What is your displacement?

b) What is the distance you have traveled?

c) After talking to your friend, you continue to your house. When you arrive back at your house, what is your displacement?

d) What is the total distance you have traveled?

Section 2.3

2.31 Running on a 50-m by 40-m rectangular track, you complete one lap in 100 s. What is your average velocity for the lap?

2.32 An electron moves in the positive x-direction a distance of 2.42 m in $2.91 \cdot 10^{-8}$ s, bounces off a moving proton, and then moves in the opposite direction a distance of 1.69 m in $3.43 \cdot 10^{-8}$ s.

a) What is the average velocity of the electron over the entire time interval?

b) What is the average speed of the electron over the entire time interval?

2.33 The graph describes the position of a particle in one dimension as a function of time.

a) In which time interval does the particle have its maximum speed? What is that speed?

b) What is the average velocity in the time interval between −5 s and +5 s?

c) What is the average speed in the time interval between −5 s and +5 s?

d) What is the ratio of the velocity in the interval between 2 s and 3 s to the velocity in the interval between 3 s and 4 s?

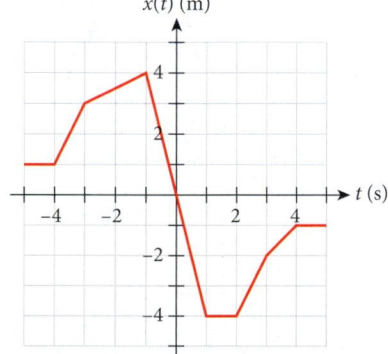

e) At what time(s) is the particle's velocity zero?

2.34 The position of a particle moving along the x-axis is given by $x = (11 + 14t - 2.0t^2)$, where t is in seconds and x is in meters. What is the average velocity during the time interval from $t = 1.0$ s to $t = 4.0$ s?

•2.35 The position of a particle moving along the x-axis is given by $x = 3.0t^2 - 2.0t^3$, where x is in meters and t is in seconds. What is the position of the particle when it achieves its maximum speed in the positive x-direction?

2.36 The rate of continental drift is on the order of 10.0 mm/yr. Approximately how long did it take North America and Europe to reach their current separation of about 3000 mi?

•2.37 The position of an object as a function of time is given as $x = At^3 + Bt^2 + Ct + D$. The constants are $A = 2.10$ m/s^3, $B = 1.00$ m/s^2, $C = -4.10$ m/s, and $D = 3.00$ m.

a) What is the velocity of the object at $t = 10.0$ s?

b) At what time(s) is the object at rest?

c) What is the acceleration of the object at $t = 0.50$ s?

d) Plot the acceleration as a function of time for the time interval from $t = -10.0$ s to $t = 10.0$ s.

•2.38 The trajectory of an object is given by the equation

$$x(t) = (4.35 \text{ m}) + (25.9 \text{ m/s})t - (11.79 \text{ m/s}^2)t^2$$

a) For which time t is the displacement $x(t)$ at its maximum?

b) What is this maximum value?

Section 2.4

2.39 A bank robber in a getaway car approaches an intersection at a speed of 45 mph. Just as he passes the intersection, he realizes that he needed to turn. So he steps on the brakes, comes to a complete stop, and then accelerates driving straight backward. He reaches a speed of 22.5 mph moving backward. Altogether his deceleration and re-acceleration in the opposite direction take 12.4 s. What is the average acceleration during this time?

2.40 A car is traveling west at 22.0 m/s. After 10.0 s, its velocity is 17.0 m/s in the same direction. Find the magnitude and direction of the car's average acceleration.

2.41 Your friend's car starts from rest and travels 0.500 km in 10.0 s. What is the magnitude of the constant acceleration required to do this?

2.42 A fellow student found in the performance data for his new car the velocity-versus-time graph shown in the figure.

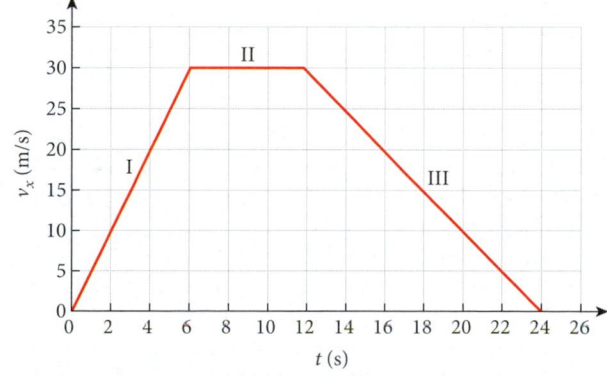

a) Find the average acceleration of the car during each of the segments I, II, and III.

b) What is the total distance traveled by the car from $t = 0$ s to $t = 24$ s?

•2.43 The velocity of a particle moving along the x-axis is given, for $t > 0$, by $v_x = (50.0t - 2.0t^3)$ m/s, where t is in seconds. What is the acceleration of the particle when (after $t = 0$) it achieves its maximum displacement in the positive x-direction?

•**2.44** The 2007 world record for the men's 100-m dash was 9.77 s. The third-place runner crossed the finish line in 10.07 s. When the winner crossed the finish line, how far was the third-place runner behind him?

a) Compute an answer that assumes that each runner ran at his average speed for the entire race.

b) Compute another answer that uses the result of Example 2.3, that a world-class sprinter runs at a speed of 12 m/s after an initial acceleration phase. If both runners in this race reach this speed, how far behind is the third-place runner when the winner finishes?

Section 2.5

2.45 On October 30th, 2011, in an NFL football game between the Detroit Lions and the Denver Broncos, Lions left cornerback Chris Johnson intercepted a pass 1 yard behind the goal line, ran down the entire length of the football field on a straight line, and scored a touchdown. Video analysis of this scoring play showed the following approximate times for him to cross the yard lines painted on the field.

Yard	−1	0	5	10	15	20	25	30	35	40	45
Time	0.00	0.23	1.16	1.80	2.33	2.87	3.37	3.87	4.33	4.80	5.27

50	45	40	35	30	25	20	15	10	5	0	−1
5.73	6.20	6.67	7.17	7.64	8.14	8.67	9.20	9.71	10.34	11.47	12.01

a) What was his average speed from the time he caught the ball until he reached midfield?

b) What was his average speed from the time he crossed midfield until he came to a stop 1 yard behind the opposite goal line?

c) What was the average acceleration during his entire run?

••**2.46** An F-14 Tomcat fighter jet is taking off from the deck of the USS *Nimitz* aircraft carrier with the assistance of a steam-powered catapult. The jet's location along the flight deck is measured at intervals of 0.20 s. These measurements are tabulated as follows:

t (s)	0.00	0.20	0.40	0.60	0.80	1.00	1.20	1.40	1.60	1.80	2.00
x (m)	0.0	0.70	3.0	6.6	11.8	18.5	26.6	36.2	47.3	59.9	73.9

Use difference formulas to calculate the jet's average velocity and average acceleration for each time interval. After completing this analysis, can you say whether the F-14 Tomcat accelerated with approximately constant acceleration?

Section 2.6

2.47 A particle starts from rest at $x = 0.00$ and moves for 20.0 s with an acceleration of +2.00 cm/s². For the next 40.0 s, the acceleration of the particle is −4.00 cm/s². What is the position of the particle at the end of this motion?

2.48 A car moving in the x-direction has an acceleration a_x that varies with time as shown in the figure. At the moment $t = 0.0$ s, the car is located at $x = 12$ m and has a velocity of 6.0 m/s in the positive x-direction. What is the velocity of the car at $t = 5.0$ s?

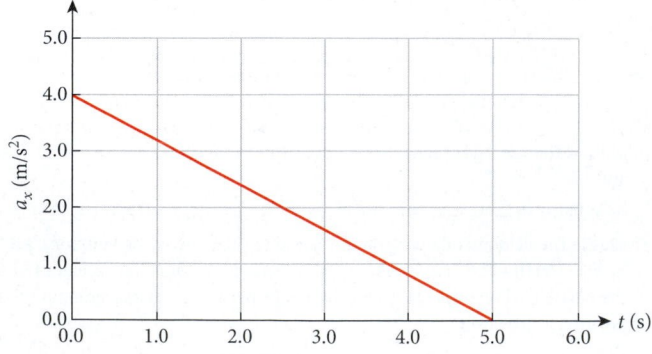

2.49 The velocity as a function of time for a car on an amusement park ride is given as $v = At^2 + Bt$ with constants $A = 2.0$ m/s³ and $B = 1.0$ m/s². If the car starts at the origin, what is its position at $t = 3.0$ s?

2.50 An object starts from rest and has an acceleration given by $a = Bt^2 - \frac{1}{2}Ct$, where $B = 2.0$ m/s⁴ and $C = -4.0$ m/s³.

a) What is the object's velocity after 5.0 s?

b) How far has the object moved after $t = 5.0$ s?

•**2.51** A car is moving along the x-axis and its velocity, v_x, varies with time as shown in the figure. If $x_0 = 2.0$ m at $t_0 = 2.0$ s, what is the position of the car at $t = 10.0$ s?

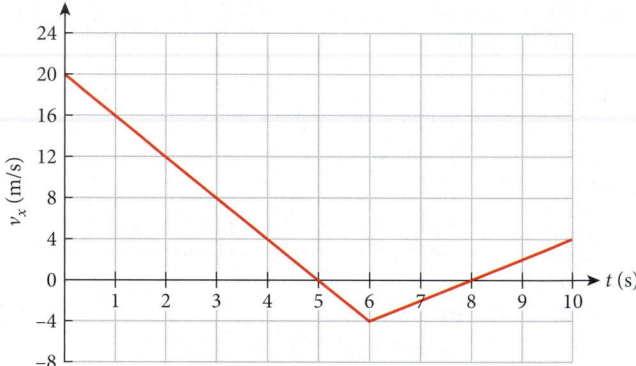

•**2.52** A car is moving along the x-axis and its velocity, v_x, varies with time as shown in the figure. What is the displacement, Δx, of the car from $t = 4$ s to $t = 9$ s ?

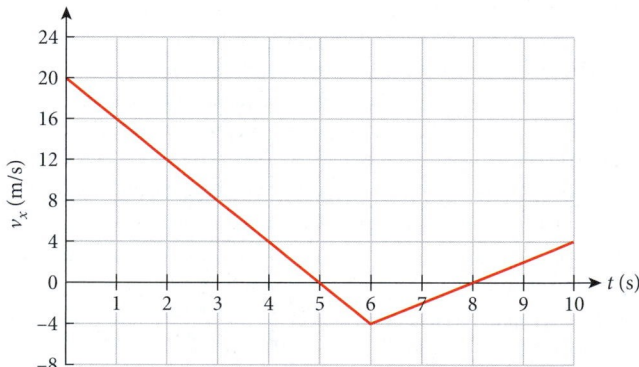

•**2.53** A motorcycle starts from rest and accelerates as shown in the figure. Determine (a) the motorcycle's speed at $t = 4.00$ s and at $t = 14.0$ s and (b) the distance traveled in the first 14.0 s.

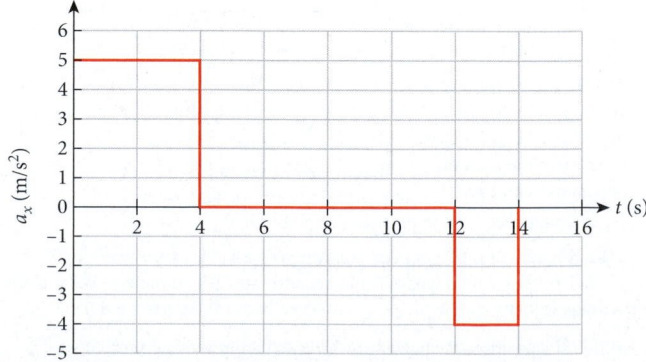

Section 2.7

2.54 How much time does it take for a car to accelerate from a standing start to 22.2 m/s if the acceleration is constant and the car covers 243 m during the acceleration?

2.55 A car slows down from a speed of 31.0 m/s to a speed of 12.0 m/s over a distance of 380. m.

a) How long does this take, assuming constant acceleration?

b) What is the value of this acceleration?

2.56 A runner of mass 57.5 kg starts from rest and accelerates with a constant acceleration of 1.25 m/s^2 until she reaches a velocity of 6.3 m/s. She then continues running with this constant velocity.

a) How far has she run after 59.7 s?

b) What is the velocity of the runner at this point?

2.57 A fighter jet lands on the deck of an aircraft carrier. It touches down with a speed of 70.4 m/s and comes to a complete stop over a distance of 197.4 m. If this process happens with constant deceleration, what is the speed of the jet 44.2 m before its final stopping location?

2.58 A bullet is fired through a board 10.0 cm thick, with a line of motion perpendicular to the face of the board. If the bullet enters with a speed of 400. m/s and emerges with a speed of 200. m/s, what is its acceleration as it passes through the board?

2.59 A car starts from rest and accelerates at 10.0 m/s^2. How far does it travel in 2.00 s?

2.60 An airplane starts from rest and accelerates at 12.1 m/s^2. What is its speed at the end of a 500.-m runway?

2.61 Starting from rest, a boat increases its speed to 5.00 m/s with constant acceleration.

a) What is the boat's average speed?

b) If it takes the boat 4.00 s to reach this speed, how far has it traveled?

•**2.62** Runner 1 is standing still on a straight running track. Runner 2 passes him, running with a constant speed of 5.1 m/s. Just as runner 2 passes, runner 1 accelerates with a constant acceleration of 0.89 m/s^2. How far down the track does runner 1 catch up with runner 2?

•**2.63** A girl is riding her bicycle. When she gets to a corner, she stops to get a drink from her water bottle. At that time, a friend passes by her, traveling at a constant speed of 8.0 m/s.

a) After 20 s, the girl gets back on her bike and travels with a constant acceleration of 2.2 m/s^2. How long does it take for her to catch up with her friend?

b) If the girl had been on her bike and rolling along at a speed of 1.2 m/s when her friend passed, what constant acceleration would she need to catch up with her friend in the same amount of time?

•**2.64** A speeding motorcyclist is traveling at a constant speed of 36.0 m/s when he passes a police car parked on the side of the road. The radar, positioned in the police car's rear window, measures the speed of the motorcycle. At the instant the motorcycle passes the police car, the police officer starts to chase the motorcyclist with a constant acceleration of 4.0 m/s^2.

a) How long will it take the police officer to catch the motorcyclist?

b) What is the speed of the police car when it catches up to the motorcycle?

c) How far will the police car be from its original position?

•**2.65** Two train cars are on a straight, horizontal track. One car starts at rest and is put in motion with a constant acceleration of 2.00 m/s^2. This car moves toward a second car that is 30.0 m away. The second car is moving away from the first car and is traveling at a constant speed of 4.00 m/s.

a) Where will the cars collide?

b) How long will it take for the cars to collide?

Section 2.8

2.66 A ball is tossed vertically upward with an initial speed of 26.4 m/s. How long does it take before the ball is back on the ground?

2.67 A stone is thrown upward, from ground level, with an initial velocity of 10.0 m/s.

a) What is the velocity of the stone after 0.50 s?

b) How high above ground level is the stone after 0.50 s?

2.68 A stone is thrown downward with an initial velocity of 10.0 m/s. The acceleration of the stone is constant and has the value of the free-fall acceleration, 9.81 m/s^2. What is the velocity of the stone after 0.500 s?

2.69 A ball is thrown directly downward, with an initial speed of 10.0 m/s, from a height of 50.0 m. After what time interval does the ball strike the ground?

2.70 An object is thrown vertically upward and has a speed of 20.0 m/s when it reaches two thirds of its maximum height above the launch point. Determine its maximum height.

2.71 What is the velocity at the midway point of a ball able to reach a height y when thrown upward with an initial velocity v_0?

2.72 On August 2, 1971, Astronaut David Scott, while standing on the surface of the Moon, dropped a 1.3-kg hammer and a 0.030-kg falcon feather from a height of 1.6 m. Both objects hit the Moon's surface 1.4 s after being released. What is the acceleration due to gravity on the surface of the Moon?

•**2.73** Bill Jones has a bad night in his bowling league. Feeling disgusted when he gets home, he drops his bowling ball out the window of his apartment, from a height of 63.17 m above the ground. John Smith sees the bowling ball pass by his window when it is 40.95 m above the ground. How much time passes from the time when John Smith sees the bowling ball pass his window to when it hits the ground?

•**2.74** Picture yourself in the castle of Helm's Deep from the *Lord of the Rings*. You are on top of the castle wall and are dropping rocks on assorted monsters that are 18.35 m below you. Just when you release a rock, an archer located exactly below you shoots an arrow straight up toward you with an initial velocity of 47.4 m/s. The arrow hits the rock in midair. How long after you release the rock does this happen?

•**2.75** An object is thrown vertically and has an upward velocity of 25 m/s when it reaches one fourth of its maximum height above its launch point. What is the initial (launch) speed of the object?

•**2.76** In a fancy hotel, the back of the elevator is made of glass so that you can enjoy a lovely view on your ride. The elevator travels at an average speed of 1.75 m/s. A boy on the 15th floor, 80.0 m above the ground level, drops a rock at the same instant the elevator starts its ascent from the 1st to the 5th floor. Assume the elevator travels at its average speed for the entire trip and neglect the dimensions of the elevator.

a) How long after it was dropped do you see the rock?

b) How long does it take for the rock to reach ground level?

••**2.77** You drop a water balloon straight down from your dormitory window 80.0 m above your friend's head. At 2.00 s after you drop the balloon, your friend (not aware that the balloon has water in it) fires a dart from a gun, which is at the same height as his head, directly upward toward the balloon with an initial velocity of 20.0 m/s.

a) How long after you drop the balloon will the dart burst the balloon?

b) How long after the dart hits the balloon will your friend have to move out of the way of the falling water? Assume the balloon breaks instantaneously at the touch of the dart.

Additional Exercises

2.78 A runner of mass 56.1 kg starts from rest and accelerates with a constant acceleration of 1.23 m/s^2 until she reaches a velocity of 5.10 m/s. She then continues running at this constant velocity. How long does the runner take to travel 173 m?

2.79 A jet touches down on a runway with a speed of 142.4 mph. After 12.4 s, the jet comes to a complete stop. Assuming constant acceleration of the jet, how far down the runway from where it touched down does the jet stand?

2.80 On the graph of position as a function of time, mark the points where the velocity is zero and the points where the acceleration is zero.

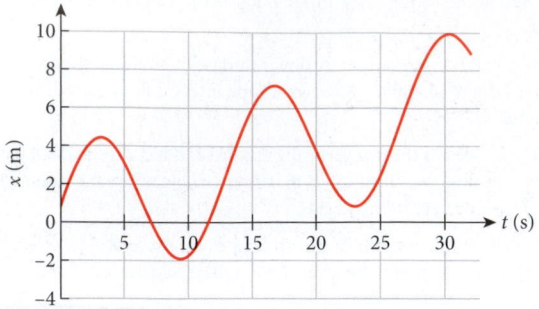

2.81 A car accelerates from a standing start to 60.0 mph in 4.20 s. Assume that its acceleration is constant.

a) What is the acceleration?

b) How far does the car travel?

2.82 The minimum distance necessary for a car to brake to a stop from a speed of 100.0 km/h is 40.00 m on a dry pavement. What is the minimum distance necessary for this car to brake to a stop from a speed of 130.0 km/h on dry pavement?

2.83 A car moving at 60.0 km/h comes to a stop in $t = 4.00$ s. Assume uniform deceleration.

a) How far does the car travel while stopping?

b) What is its deceleration?

2.84 You are driving at 29.1 m/s when the truck ahead of you comes to a halt 200.0 m away from your bumper. Your brakes are in poor condition, and you decelerate at a constant rate of 2.4 m/s^2.

a) How close do you come to the bumper of the truck?

b) How long does it take you to come to a stop?

2.85 A train traveling at 40.0 m/s is headed straight toward another train, which is at rest on the same track. The moving train decelerates at 6.0 m/s^2, and the stationary train is 100.0 m away. How far from the stationary train will the moving train be when it comes to a stop?

2.86 The driver of a car traveling at 25.0 m/s applies the brakes, and the car decelerates uniformly at a rate of 1.2 m/s^2.

a) How far does it travel in 3.0 s?

b) What is its velocity at the end of this time interval?

c) How long does it take for the car to come to a stop?

d) What distance does the car travel before coming to a stop?

2.87 The fastest speed in NASCAR racing history was 212.809 mph (reached by Bill Elliott in 1987 at Talladega). If the race car decelerated from that speed at a rate of 8.0 m/s^2, how far would it travel before coming to a stop?

2.88 You are flying on a commercial airline on your way from Houston, Texas, to Oklahoma City, Oklahoma. Your pilot announces that the plane is directly over Austin, Texas, traveling at a constant speed of 245 mph, and will be flying directly over Dallas, Texas, 362 km away. How long will it be before you are directly over Dallas, Texas?

2.89 The position of a rocket sled on a straight track is given as $x = at^3 + bt^2 + c$, where $a = 2.0$ m/s^3, $b = 2.0$ m/s^2, and $c = 3.0$ m.

a) What is the sled's position between $t = 4.0$ s and $t = 9.0$ s?

b) What is the average speed between $t = 4.0$ s and $t = 9.0$ s?

2.90 A girl is standing at the edge of a cliff 100. m above the ground. She reaches out over the edge of the cliff and throws a rock straight upward with a speed 8.00 m/s.

a) How long does it take the rock to hit the ground?

b) What is the speed of the rock the instant before it hits the ground?

•**2.91** A double speed trap is set up on a freeway. One police cruiser is hidden behind a billboard, and another is some distance away under a bridge. As a sedan passes by the first cruiser, its speed is measured as 105.9 mph. Since the driver has a radar detector, he is alerted to the fact that his speed has been measured, and he tries to slow his car down gradually without stepping on the brakes and alerting the police that he knew he was going too fast. Just taking the foot off the gas leads to a constant deceleration. Exactly 7.05 s later, the sedan passes the second police cruiser. Now its speed is measured as only 67.1 mph, just below the local freeway speed limit.

a) What is the value of the deceleration?

b) How far apart are the two cruisers?

•**2.92** During a test run on an airport runway, a new race car reaches a speed of 258.4 mph from a standing start. The car accelerates with constant acceleration and reaches this speed mark at a distance of 612.5 m from where it started. What was its speed after one-fourth, one-half, and three-fourths of this distance?

•**2.93** The vertical position of a ball suspended by a rubber band is given by the equation

$$y(t) = (3.8 \text{ m})\sin(0.46\,t/\text{s} - 0.31) - (0.2 \text{ m/s})t + 5.0 \text{ m}$$

a) What are the equations for the velocity and acceleration of this ball?

b) For what times between 0 and 30 s is the acceleration zero?

•**2.94** The position of a particle moving along the x-axis varies with time according to the expression $x = 4t^2$, where x is in meters and t is in seconds.

a) Evaluate the particle's position at $t = 2.00$ s.

b) Evaluate its position at 2.00 s $+ \Delta t$.

c) Evaluate the limit of $\Delta x/\Delta t$ as Δt approaches zero, to find the velocity at $t = 2.00$ s.

•**2.95** In 2005, Hurricane Rita hit several states in the southern United States. In the panic to escape her wrath, thousands of people tried to flee Houston, Texas, by car. One car full of college students traveling to Tyler, Texas, 199 miles north of Houston, moved at an average speed of 3.0 m/s for one-fourth of the time, then at 4.5 m/s for another one-fourth of the time, and at 6.0 m/s for the remainder of the trip.

a) How long did it take the students to reach their destination?

b) Sketch a graph of position versus time for the trip.

•**2.96** A ball is thrown straight upward in the air at a speed of 15.0 m/s. Ignore air resistance.

a) What is the maximum height the ball will reach?

b) What is the speed of the ball when it reaches 5.00 m?

c) How long will the ball take to reach 5.00 m above its initial position on the way up?

d) How long will the ball take to reach 5.00 m above its initial position on its way down?

•**2.97** The Bellagio Hotel in Las Vegas, Nevada, is well known for its Musical Fountains, which use 192 HyperShooters to fire water hundreds of feet into the air to the rhythm of music. One of the HyperShooters fires water straight upward to a height of 240. ft.

a) What is the initial speed of the water?

b) What is the speed of the water when it is at half this height on its way down?

c) How long will it take for the water to fall back to its original height from half its maximum height?

•**2.98** You are trying to improve your shooting skills by shooting at a can on top of a fence post. You miss the can, and the bullet, moving at 200. m/s, is embedded 1.5 cm into the post when it comes to a stop. If constant acceleration is assumed, how long does it take for the bullet to stop?

•**2.99** You drive with a constant speed of 13.5 m/s for 30.0 s. You then steadily accelerate for 10.0 s to a speed of 22.0 m/s. You then slow smoothly to a stop in 10.0 s. How far have you traveled?

•**2.100** A ball is dropped from the roof of a building. It hits the ground and is caught at its original height 5.0 s later.

a) What was the speed of the ball just before it hit the ground?

b) How tall was the building?

c) You are watching from a window 2.5 m above the ground. The window opening is 1.2 m from the top to the bottom. At what time after the ball was dropped did you first see the ball in the window?

2.101 The position of a particle as a function of time is given as $x(t) = \frac{1}{4} x_0 e^{3\alpha t}$, where α is a positive constant.

a) At what time is the particle at $2x_0$?

b) What is the speed of the particle as a function of time?

c) What is the acceleration of the particle as a function of time?

d) What are the SI units for α?

2.102 The position of an object as a function of time is given by $x = At^4 - Bt^3 + C$.

a) What is the instantaneous velocity as a function of time?

b) What is the instantaneous acceleration as a function of time?

MULTI-VERSION EXERCISES

2.103 An object is thrown upward with a speed of 28.0 m/s. How long does it take it to reach its maximum height?

2.104 An object is thrown upward with a speed of 28.0 m/s. How high above the projection point is it after 1.00 s?

2.105 An object is thrown upward with a speed of 28.0 m/s. What maximum height above the projection point does it reach?

2.106 The acceleration due to gravity on the surface of Mars at the Equator is 3.699 m/s². How long does it take for a rock dropped from a height of 1.013 m to hit the surface?

2.107 The acceleration due to gravity on the surface of Mars at the Equator is 3.699 m/s². A rock released from rest takes 0.8198 s to reach the surface. From what height was the rock dropped?

2.108 A steel ball is dropped from a height of 12.37 m above the ground. What is its speed when it reaches 2.345 m above the ground?

2.109 A steel ball is dropped from a height of 13.51 m above the ground. How high above the ground is the ball when it has a speed of 14.787 m/s?

2.110 When a dropped steel ball reaches a height of 2.387 m above the ground, it has a speed of 15.524 m/s. From what height was the ball dropped?

Motion in Two and Three Dimensions

3

FIGURE 3.1 Multiple exposure sequence of a bouncing ball.

Everyone has seen a bouncing ball, but have you ever looked closely at the path it takes? If you could slow the ball down, as in the photo in Figure 3.1, you would see the symmetrical arch of each bounce, which gets smaller until the ball stops. This path is characteristic of a kind of two-dimensional motion known as *projectile motion*. You can see the same parabolic shape in water fountains, fireworks, basketball shots—any kind of isolated motion where the force of gravity is relatively constant and the moving object is dense enough that air resistance (a force that tends to slow down objects moving though air) can be ignored.

This chapter extends Chapter 2's discussion of displacement, velocity, and acceleration to two-dimensional motion. The definitions of these vectors in two dimensions are very similar to the one-dimensional definitions, but we can apply them to a greater variety of real-life situations. Two-dimensional motion is still more restricted than general motion in three dimensions, but it applies to a wide range of common and important motions that we will consider throughout this course.

WHAT WE WILL LEARN

- We will learn to handle motion in two and three dimensions using methods developed for one-dimensional motion.
- We will determine the parabolic path of ideal projectile motion.
- We will calculate the maximum height and maximum range of an ideal projectile trajectory in terms of the initial velocity vector and the initial position.

- We will learn to describe the velocity vector of a projectile at any time during its flight.
- We will study qualitatively how realistic trajectories of objects like baseballs are affected by air friction and how such trajectories deviate from parabolic ones.
- We will learn to transform velocity vectors from one reference frame to another.

3.1 Three-Dimensional Coordinate Systems

Having studied motion in one dimension, we next tackle more complicated problems in two and three spatial dimensions. To describe this motion, we will work in Cartesian coordinates. In a three-dimensional Cartesian coordinate system, we choose the x- and y-axes to lie in the horizontal plane and the z-axis to point vertically upward (Figure 3.2). The three coordinate axes are at 90° (orthogonal) to one another, as required for a Cartesian coordinate system.

The convention followed without exception in this book is that the Cartesian coordinate system is **right-handed.** This convention means that you can obtain the relative orientation of the three coordinate axes using your right hand. To determine the positive directions of the three axes, hold your right hand with the thumb sticking straight up and the index finger pointing straight out; they will naturally have a 90° angle relative to each other. Then stick out your middle finger so that it is at a right angle with both the index finger and the thumb (Figure 3.3). The three axes are assigned to the fingers as shown in Figure 3.3: thumb is x, index finger is y, and middle finger is z. You can rotate your right hand in any direction, but the relative orientation of thumb and fingers stays the same. Your right hand can always be orientated in three-dimensional space in such a way that the axes assignments on your fingers can be brought into alignment with the schematic coordinate axes shown in Figure 3.2.

With this set of Cartesian coordinates, a position vector can be written in component form as

$$\vec{r} = (x, y, z) = x\hat{x} + y\hat{y} + z\hat{z}. \tag{3.1}$$

A velocity vector is

$$\vec{v} = (v_x, v_y, v_y) = v_x\hat{x} + v_y\hat{y} + v_z\hat{z}. \tag{3.2}$$

For one-dimensional vectors, the time derivative of the position vector defines the velocity vector. This is also the case for more than one dimension:

$$\vec{v} = \frac{d\vec{r}}{dt} = \frac{d}{dt}(x\hat{x} + y\hat{y} + z\hat{z}) = \frac{dx}{dt}\hat{x} + \frac{dy}{dt}\hat{y} + \frac{dz}{dt}\hat{z}. \tag{3.3}$$

In the last step of this equation, we used the sum and product rules of differentiation, as well as the fact that the unit vectors are constant vectors (fixed directions along the coordinate axes and constant magnitude of 1). Comparing equations 3.2 and 3.3, we see that

$$v_x = \frac{dx}{dt}, \quad v_y = \frac{dy}{dt}, \quad v_z = \frac{dz}{dt}. \tag{3.4}$$

The same procedure leads us from the velocity vector to the acceleration vector by taking the time derivative of the former:

$$\vec{a} = \frac{d\vec{v}}{dt} = \frac{dv_x}{dt}\hat{x} + \frac{dv_y}{dt}\hat{y} + \frac{dv_z}{dt}\hat{z}. \tag{3.5}$$

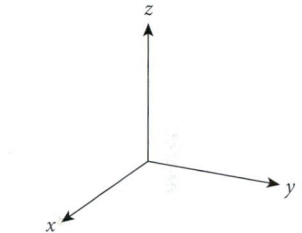

FIGURE 3.2 A right-handed xyz Cartesian coordinate system.

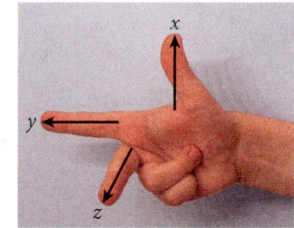

FIGURE 3.3 A right-handed Cartesian coordinate system.

We can therefore write the Cartesian components of the acceleration vector:

$$a_x = \frac{dv_x}{dt}, \qquad a_y = \frac{dv_y}{dt}, \qquad a_z = \frac{dv_z}{dt}. \tag{3.6}$$

3.2 Velocity and Acceleration in Two or Three Dimensions

The most striking difference between velocity along a line and velocity in two or more dimensions is that the latter can change direction even in cases where the speed stays constant. Because acceleration is defined as a change in velocity—any change in velocity—divided by a time interval, there can be acceleration even when the magnitude of the velocity does not change.

Consider, for example, a particle moving in two dimensions (that is, in a plane). At time t_1, the particle has velocity $\vec{v}_1$, and at a later time t_2, the particle has velocity $\vec{v}_2$. The change in velocity of the particle is $\Delta\vec{v} = \vec{v}_2 - \vec{v}_1$. The average acceleration, $\vec{a}_{ave}$, for the time interval $\Delta t = t_2 - t_1$ is given by

$$\vec{a}_{ave} = \frac{\Delta\vec{v}}{\Delta t} = \frac{\vec{v}_2 - \vec{v}_1}{t_2 - t_1}. \tag{3.7}$$

Figure 3.4 shows three different cases for the change in velocity of a particle moving in two dimensions over a given time interval. Figure 3.4a shows the initial and final velocities of the particle having the same direction, but the magnitude of the final velocity is greater than the magnitude of the initial velocity. The resulting change in velocity and the average acceleration are in the same direction as the velocities. Figure 3.4b again shows the initial and final velocities pointing in the same direction, but the magnitude of the final velocity is less than the magnitude of the initial velocity. The resulting change in velocity and the average acceleration are in the opposite direction from the velocities. Figure 3.4c illustrates the case when the initial and final velocities have the same magnitude but the direction of the final velocity vector is different from the direction of the initial velocity vector. Even though the magnitudes of the initial and final velocity vectors are the same, the change in velocity and the average acceleration are not zero and can be in a direction not obviously related to the initial or final velocity directions.

Concept Check 3.1

In all of the cases shown below, the velocity vectors $\vec{v}_1$ and $\vec{v}_2$ have the same length. In which case does $\Delta\vec{v} = \vec{v}_2 - \vec{v}_1$ have the *largest* absolute value?

a)   b)

c)   d)

e) All of the cases are identical.

Concept Check 3.2

In all of the cases shown in Concept Check 3.1, the velocity vectors $\vec{v}_1$ and $\vec{v}_2$ have the same length. In which case does the acceleration $\vec{a} = \Delta\vec{v} / \Delta t$ have the *smallest* absolute value?

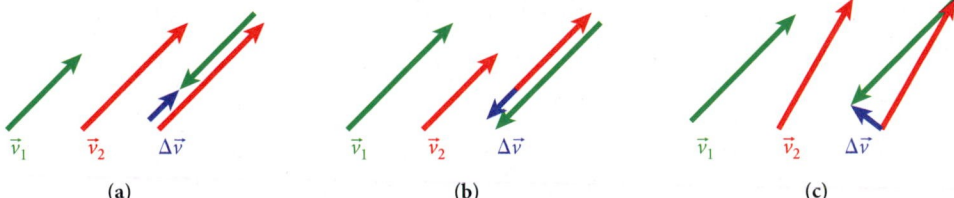

FIGURE 3.4 At time t_1, a particle has a velocity $\vec{v}_1$. At a later time t_2, the particle has a velocity $\vec{v}_2$. The average acceleration is given by $\vec{a}_{ave} = \Delta\vec{v} / \Delta t = (\vec{v}_2 - \vec{v}_1) / (t_2 - t_1)$. (a) A time interval corresponding to $|\vec{v}_2| > |\vec{v}_1|$, with $\vec{v}_2$ and $\vec{v}_1$ in the same direction. (b) A time interval corresponding to $|\vec{v}_2| < |\vec{v}_1|$, with $\vec{v}_2$ and $\vec{v}_1$ in the same direction. (c) A time interval with $|\vec{v}_2| = |\vec{v}_1|$, but with $\vec{v}_2$ in a different direction from $\vec{v}_1$.

In general, an acceleration vector arises if an object's velocity vector changes in magnitude or direction. Any time an object travels along a curved path, in two or three dimensions, it must have acceleration. We will examine the components of acceleration in more detail in Chapter 9, when we discuss circular motion.

3.3 Ideal Projectile Motion

In some special cases of three-dimensional motion, the horizontal projection of the trajectory, or flight path, is a straight line. This situation occurs whenever the accelerations in the horizontal xy-plane are zero, so the object has constant velocity components, v_x and v_y, in the horizontal plane. Such a case is shown in Figure 3.5 for a baseball tossed in the air. In this case, we can assign new coordinate axes such that the x-axis points along the horizontal projection of the trajectory and the y-axis is the vertical axis. In this special case, the motion in three dimensions can in effect be described as a motion in two dimensions. A large class of real-life problems falls into this category, especially problems that involve ideal projectile motion.

An **ideal projectile** is any object that is released with some initial velocity and then moves only under the influence of gravitational acceleration, which is assumed to be constant and in the vertical downward direction. A basketball free throw (Figure 3.6) is a good example of ideal projectile motion, as is the flight of a bullet or the trajectory of a car that becomes airborne. **Ideal projectile motion** neglects air resistance and wind speed, spin of the projectile, and other effects influencing the flight of real-life projectiles. For realistic situations in which a golf ball, tennis ball, or baseball moves in air, the actual trajectory is not well described by ideal projectile motion and requires a more sophisticated analysis. We will discuss these effects in Section 3.5, but will not go into quantitative detail.

Let's begin with ideal projectile motion, with no effects due to air resistance or any other forces besides gravity. We work with two Cartesian components: x in the horizontal direction and y in the vertical (upward) direction. Therefore, the position vector for projectile motion is

$$\vec{r} = (x,y) = x\hat{x} + y\hat{y}, \tag{3.8}$$

and the velocity vector is

$$\vec{v} = (v_x, v_y) = v_x\hat{x} + v_x\hat{y} = \left(\frac{dx}{dt}, \frac{dy}{dt}\right) = \frac{dx}{dt}\hat{x} + \frac{dy}{dt}\hat{y}. \tag{3.9}$$

Given our choice of coordinate system, with a vertical y-axis, the acceleration due to gravity acts downward, in the negative y-direction; there is no acceleration in the horizontal direction:

$$\vec{a} = (0,-g) = -g\hat{y}. \tag{3.10}$$

For this special case of a constant acceleration only in the y-direction and with zero acceleration in the x-direction, we have a free-fall problem in the vertical direction and motion with constant velocity in the horizontal direction. The kinematical equations for the x-direction are those for an object moving with constant velocity:

$$x = x_0 + v_{x0}t \tag{3.11}$$

$$v_x = v_{x0}. \tag{3.12}$$

Just as in Chapter 2, we use the notation $v_{x0} \equiv v_x(t=0)$ for the initial value of the x-component of the velocity. The kinematical equations for the y-direction are those for free-fall motion in one dimension:

$$y = y_0 + v_{y0}t - \tfrac{1}{2}gt^2 \tag{3.13}$$

$$y = y_0 + \bar{v}_y t \tag{3.14}$$

$$v_y = v_{y0} - gt \tag{3.15}$$

$$\bar{v}_y = \tfrac{1}{2}(v_y + v_{y0}) \tag{3.16}$$

$$v_y^2 = v_{y0}^2 - 2g(y - y_0). \tag{3.17}$$

For consistency, we write $v_{y0} \equiv v_y(t=0)$. With these seven equations for the x- and y-components, we can solve any problem involving an ideal projectile. Notice that since two-dimensional motion can be split into separate one-dimensional motions, these equations are written in component form, without the use of unit vectors.

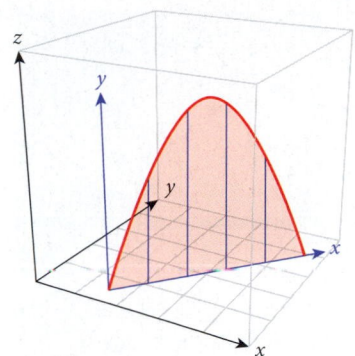

FIGURE 3.5 Trajectory in three dimensions reduced to a trajectory in two dimensions.

FIGURE 3.6 Photograph of a free throw with the parabolic trajectory of the basketball superimposed. This image was generated by superposition of 13 frames, with a time of 1/12 s between each two successive frames. Note that the ball moves the same horizontal distance (black vertical lines) between each two frames!

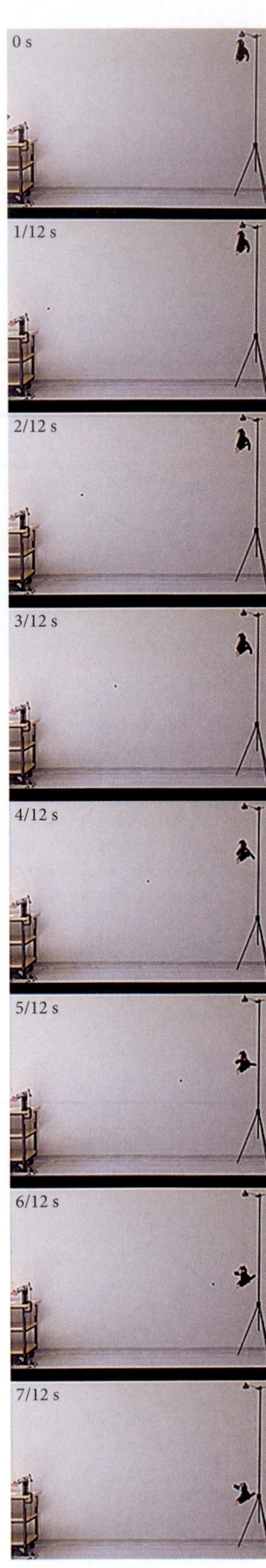

EXAMPLE 3.1 | Shoot the Monkey

Many lecture demonstrations illustrate that motion in the *x*-direction and motion in the *y*-direction are indeed independent of each other, as assumed in the derivation of the equations for projectile motion. One popular demonstration, called "shoot the monkey," is shown in Figure 3.7. The demonstration is motivated by a story. A monkey has escaped from the zoo and has climbed a tree. The zookeeper wants to shoot the monkey with a tranquilizer dart in order to recapture it, but she knows that the monkey will let go of the branch it is holding onto at the sound of the gun firing. Her challenge is therefore to hit the monkey in the air as it is falling.

PROBLEM

Where does the zookeeper need to aim to hit the falling monkey?

SOLUTION

The zookeeper must aim directly at the monkey, as shown in Figure 3.7, assuming that the time for the sound of the gun firing to reach the monkey is negligible and the speed of the dart is fast enough to cover the horizontal distance to the tree. As soon as the dart leaves the gun, it is in free fall, just like the monkey. Because both the monkey and the dart are in free fall, they fall with the same acceleration, independent of the dart's motion in the *x*-direction and of the dart's initial velocity. The dart and the monkey will meet at a point directly below the point from which the monkey dropped.

DISCUSSION

Any sharpshooter can tell you that, for a fixed target, you need to correct your gun sight for the free-fall motion of the projectile on the way to the target. As you can infer from Figure 3.7, even a bullet fired from a high-powered rifle will not fly in a straight line but will drop under the influence of gravitational acceleration. Only in a situation like the shoot-the-monkey demonstration, where the target is in free fall as soon as the projectile leaves the muzzle, can one aim directly at the target without making corrections for the free-fall motion of the projectile.

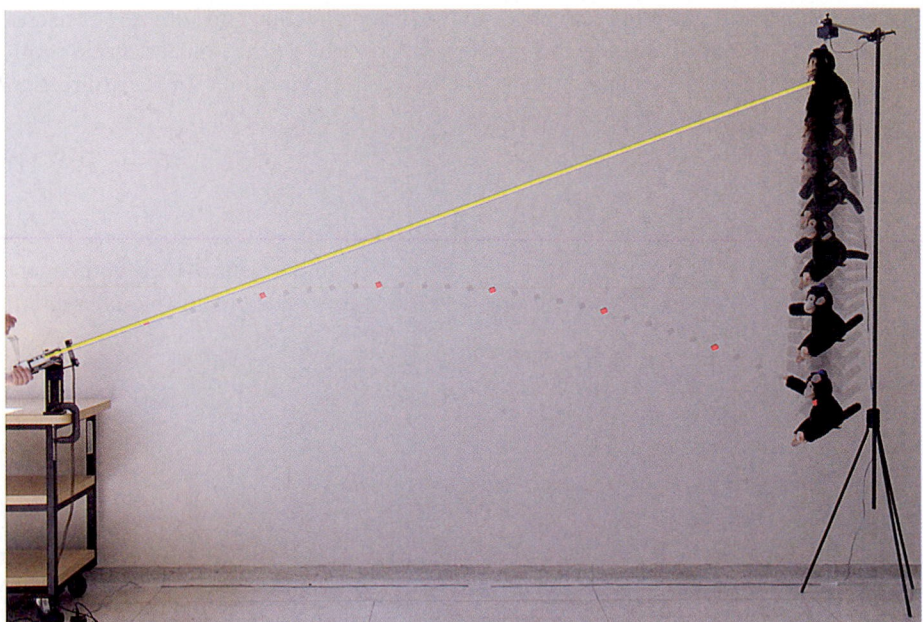

FIGURE 3.7　The shoot-the-monkey lecture demonstration. On the left are some of the individual frames of the video, with information on their timing in the upper-left corners. On the right, these frames have been combined into a single image with a superimposed yellow line indicating the initial aim of the projectile launcher.

Shape of a Projectile's Trajectory

Let's now examine the **trajectory** of a projectile in two dimensions. To find y as a function of x, we solve the equation $x = x_0 + v_{x0}t$ for the time, $t = (x - x_0)/v_{x0}$, and then substitute for t in the equation $y = y_0 + v_{y0}t - \frac{1}{2}gt^2$:

$$y = y_0 + v_{y0}t - \frac{1}{2}gt^2 \Rightarrow$$

$$y = y_0 + v_{y0}\frac{x - x_0}{v_{x0}} - \frac{1}{2}g\left(\frac{x - x_0}{v_{x0}}\right)^2 \Rightarrow$$

$$y = \left(y_0 - \frac{v_{y0}x_0}{v_{x0}} - \frac{gx_0^2}{2v_{x0}^2}\right) + \left(\frac{v_{y0}}{v_{x0}} + \frac{gx_0}{v_{x0}^2}\right)x - \frac{g}{2v_{x0}^2}x^2. \tag{3.18}$$

Thus, the trajectory follows an equation of the general form $y = c + bx + ax^2$, with constants a, b, and c. This is the form of an equation for a parabola in the xy-plane. It is customary to set the x-component of the initial point of the parabola equal to zero: $x_0 = 0$. In this case, the equation for the parabola becomes

$$y = y_0 + \frac{v_{y0}}{v_{x0}}x - \frac{g}{2v_{x0}^2}x^2. \tag{3.19}$$

The trajectory of the projectile is completely determined by three input constants. These constants are the initial height of the release of the projectile, y_0, and the x- and y-components of the initial velocity vector, v_{x0} and v_{y0}, as shown in Figure 3.8.

We can also express the initial velocity vector $\vec{v}_0$, in terms of its magnitude, v_0, and direction, θ_0. Expressing $\vec{v}_0$ in this manner involves the transformation

$$v_0 = \sqrt{v_{x0}^2 + v_{y0}^2}$$
$$\theta_0 = \tan^{-1}\frac{v_{y0}}{v_{x0}}. \tag{3.20}$$

In Chapter 1, we discussed this transformation from Cartesian coordinates to length and angle of the vector, as well as the inverse transformation:

$$v_{x0} = v_0\cos\theta_0$$
$$v_{y0} = v_0\sin\theta_0. \tag{3.21}$$

Expressed in terms of the magnitude and direction of the initial velocity vector, the equation for the path of the projectile becomes

$$y = y_0 + (\tan\theta_0)x - \frac{g}{2v_0^2\cos^2\theta_0}x^2. \tag{3.22}$$

The fountain shown in Figure 3.9 is in the Detroit Metropolitan Wayne County (DTW) airport. You can clearly see that the water shot out of many pipes traces almost perfect parabolic trajectories.

Note that because a parabola is symmetric, a projectile takes the same amount of time and travels the same distance from its launch point to the top of its trajectory as from the top of its trajectory back to launch level. Also, the speed of a projectile at a given height on its way up to the top of its trajectory is the same as its speed at that same height going back down.

Time Dependence of the Velocity Vector

From equation 3.12, we know that the x-component of the velocity is constant in time: $v_x = v_{x0}$. This result means that a projectile will cover the same horizontal distance in each time interval of the same duration. Thus, in a video of projectile motion, such as a basketball player shooting a free throw as in Figure 3.6, or the path of the dart in the shoot-the-monkey demonstration in Figure 3.7, the horizontal displacement of the projectile from one frame of the video to the next will be constant.

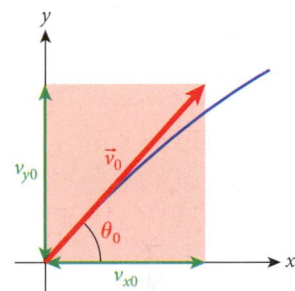

FIGURE 3.8 Initial velocity vector $\vec{v}_0$ and its components, v_{x0} and v_{y0}.

FIGURE 3.9 A fountain with water following parabolic trajectories.

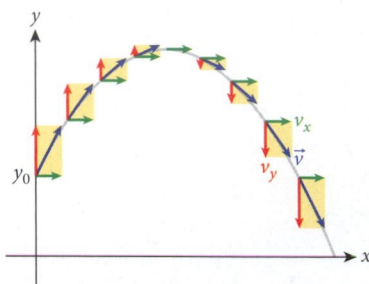

FIGURE 3.10 Graph of a parabolic trajectory with the velocity vector and its Cartesian components shown at constant time intervals.

The y-component of the velocity vector changes according to equation 3.15, $v_y = v_{y0} - gt$; that is, the projectile falls with constant acceleration. Typically, projectile motion starts with a positive value, v_{y0}. The apex (highest point) of the trajectory is reached at the point where $v_y = 0$ and the projectile is moving only in the horizontal direction. At the apex, the y-component of the velocity is momentarily zero as it changes sign from positive to negative.

We can indicate the instantaneous values of the x- and y-components of the velocity vector on a plot of y versus x for the flight path of a projectile (Figure 3.10). The x-components, v_x, of the velocity vector are shown by green arrows, and the y-components, v_y, by red arrows. Note the identical lengths of the green arrows, demonstrating the fact that v_x remains constant. Each blue arrow is the vector sum of the x- and y-velocity components and depicts the instantaneous velocity vector along the path. Note that the direction of the velocity vector is always tangential to the trajectory. This is because the slope of the velocity vector is

$$\frac{v_y}{v_x} = \frac{dy/dt}{dx/dt} = \frac{dy}{dx},$$

which is also the local slope of the flight path. At the top of the trajectory, the green and blue arrows are identical because the velocity vector has only an x-component—that is, it points in the horizontal direction.

Although the vertical component of the velocity vector is equal to zero at the top of the trajectory, the gravitational acceleration has the same constant value as on any other part of the trajectory. Beware of the common misconception that the gravitational acceleration is equal to zero at the top of the trajectory. The gravitational acceleration has the same constant value everywhere along the trajectory.

Finally, let's explore the functional dependence of the absolute value of the velocity vector on time and/or the y-coordinate. We start with the dependence of $|\vec{v}|$ on y. We use the fact that the absolute value of a vector is given as the square root of the sum of the squares of the components. Then we use kinematical equation 3.12 for the x-component and kinematical equation 3.17 for the y-component. We obtain

$$|\vec{v}| = \sqrt{v_x^2 + v_y^2} = \sqrt{v_{x0}^2 + v_{y0}^2 - 2g(y - y_0)} = \sqrt{v_0^2 - 2g(y - y_0)}. \tag{3.23}$$

Note that the initial launch angle does not appear in this equation. The absolute value of the velocity—the speed—depends only on the initial value of the speed and the difference between the y-coordinate and the initial launch height. Thus, if we release a projectile from a certain height above ground and want to know the speed with which it hits the ground, it does not matter if the projectile is shot straight up, or horizontally, or straight down. Chapter 5 will discuss the concept of kinetic energy, and then the reason for this seemingly strange fact will become more apparent.

Concept Check 3.3

At the top of the trajectory of any projectile, which of the following statement, if any, is (are) true?

a) The acceleration is zero.

b) The x-component of the acceleration is zero.

c) The y-component of the acceleration is zero.

d) The speed is zero.

e) The x-component of the velocity is zero.

f) The y-component of the velocity is zero.

Self-Test Opportunity 3.1

What is the dependence of $|\vec{v}|$ on the x-coordinate?

3.4   Maximum Height and Range of a Projectile

When launching a projectile, for example, throwing a ball, we are often interested in the **range** (R), or how far the projectile will travel horizontally before returning to its original vertical position, and the **maximum height** (H) it will reach. These quantities R and H are illustrated in Figure 3.11. We find that the maximum height reached by the projectile is

$$H = y_0 + \frac{v_{y0}^2}{2g}. \tag{3.24}$$

We'll derive this equation below. We'll also derive this equation for the range:

$$R = \frac{v_0^2}{g} \sin 2\theta_0, \tag{3.25}$$

where v_0 is the absolute value of the initial velocity vector and θ_0 is the launch angle. The maximum range, for a given fixed value of v_0, is reached when $\theta_0 = 45°$.

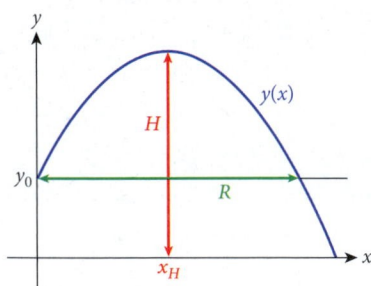

FIGURE 3.11 The maximum height (red) and range (green) of a projectile.

DERIVATION 3.1

Let's investigate the maximum height of the projectile first. To determine its value, we obtain an expression for the height, differentiate it, set the result equal to zero, and solve for the maximum height. Suppose v_0 is the initial speed and θ_0 is the launch angle. We take the derivative of the path function $y(x)$, equation 3.22, with respect to x:

$$\frac{dy}{dx} = \frac{d}{dx}\left(y_0 + (\tan\theta_0)x - \frac{g}{2v_0^2\cos^2\theta_0}x^2\right) = \tan\theta_0 - \frac{g}{v_0^2\cos^2\theta_0}x.$$

Now we look for the point x_H where the derivative is zero:

$$0 = \tan\theta_0 - \frac{g}{v_0^2\cos^2\theta_0}x_H$$

$$\Rightarrow x_H = \frac{v_0^2\cos^2\theta_0\tan\theta_0}{g} = \frac{v_0^2}{g}\sin\theta_0\cos\theta_0 = \frac{v_0^2}{2g}\sin 2\theta_0.$$

In the second line above, we used the trigonometric identities $\tan\theta = \sin\theta/\cos\theta$ and $2\sin\theta\cos\theta = \sin 2\theta$. Now we insert this value for x into equation 3.22 and obtain the maximum height, H:

$$H \equiv y(x_H) = y_0 + x_H\tan\theta_0 - \frac{g}{2v_0^2\cos^2\theta_0}x_H^2$$

$$= y_0 + \frac{v_0^2}{2g}\sin 2\theta_0\tan\theta_0 - \frac{g}{2v_0^2\cos^2\theta_0}\left(\frac{v_0^2}{2g}\sin 2\theta_0\right)^2$$

$$= y_0 + \frac{v_0^2}{g}\sin^2\theta_0 - \frac{v_0^2}{2g}\sin^2\theta_0$$

$$= y_0 + \frac{v_0^2}{2g}\sin^2\theta_0.$$

Because $v_{y0} = v_0\sin\theta_0$, we can also write

$$H = y_0 + \frac{v_{y0}^2}{2g},$$

which is equation 3.24.

The range, R, of a projectile is defined as the horizontal distance between the launching point and the point where the projectile reaches the same height from which it started, $y(R) = y_0$. Inserting $x = R$ into equation 3.22:

$$y_0 = y_0 + R\tan\theta_0 - \frac{g}{2v_0^2\cos^2\theta_0}R^2$$

$$\Rightarrow \tan\theta_0 = \frac{g}{2v_0^2\cos^2\theta_0}R$$

$$\Rightarrow R = \frac{2v_0^2}{g}\sin\theta_0\cos\theta_0 = \frac{v_0^2}{g}\sin 2\theta_0,$$

which is equation 3.25.

Note that the range, R, is twice the value of the x-coordinate, x_H, at which the trajectory reached its maximum height: $R = 2x_H$.

Finally, we consider how to maximize the range of the projectile. One way to maximize the range is to maximize the initial speed, v_0. Given a specific initial speed, what is the dependence of the range on the launch angle θ_0? To answer this question, we take the derivative of the range (equation 3.25) with respect to the launch angle:

$$\frac{dR}{d\theta_0} = \frac{d}{d\theta_0}\left(\frac{v_0^2}{g}\sin 2\theta_0\right) = 2\frac{v_0^2}{g}\cos 2\theta_0.$$

Then we set this derivative equal to zero and find the angle for which the maximum value is achieved. The angle between 0° and 90° for which $\cos 2\theta_0 = 0$ is 45°. So the maximum range of an ideal projectile is given by

$$R_{\max} = \frac{v_0^2}{g}. \tag{3.26}$$

We could have obtained this result directly from the formula for the range because, according to that formula (equation 3.25), the range is at a maximum when $\sin 2\theta_0$ has its maximum value of 1, and it has this maximum when $2\theta_0 = 90°$, or $\theta_0 = 45°$.

Most sports involving balls provide numerous examples of projectile motion. We next consider a few examples where the effects of air resistance and spin do not dominate the motion, and so the findings are reasonably close to what happens in reality. In the next section, we'll look at what effects air resistance and spin can have on a projectile.

SOLVED PROBLEM 3.1 / Throwing a Baseball

When listening to a radio broadcast of a baseball game, you often hear the phrase "line drive" or "frozen rope" for a ball hit really hard and at a low angle with respect to the ground. Some announcers even use "frozen rope" to describe a particularly strong throw from second or third base to first base. This figure of speech implies movement on a straight line—but we know that the ball's actual trajectory is a parabola.

PROBLEM
What is the maximum height that a baseball reaches if it is thrown from second base to first base or from third base to first base, and in either case is released from a height of 6.0 ft, with a speed of 90. mph, and caught at the same height?

SOLUTION
THINK The dimensions of a baseball infield are shown in Figure 3.12. (In this problem, we'll need to perform lots of unit conversions. Generally, this book uses SI units, but baseball is full of British units.) The baseball infield is a square with sides 90. ft long. This is the distance between second and first base, and we get $d_{12} = 90.$ ft $= 90. \cdot 0.3048$ m $= 27.432$ m. The distance from third to first base is the length of the diagonal of the infield square: $d_{13} = d_{12}\sqrt{2} = 38.795$ m.

A speed of 90. mph (the speed of a good Major League fastball) translates into

$$v_0 = 90.\text{ mph} = 90. \cdot 0.44704 \text{ m/s} = 40.2336 \text{ m/s}.$$

As with most trajectory problems, there are many ways to solve this problem. The most straightforward approach follows from our considerations of range and maximum height. We can equate the base-to-base distance with the range of the projectile because the ball is released and caught at the same height, $y_0 = 6.0$ ft $= 6.0 \cdot 0.3048$ m $= 1.8288$ m.

SKETCH

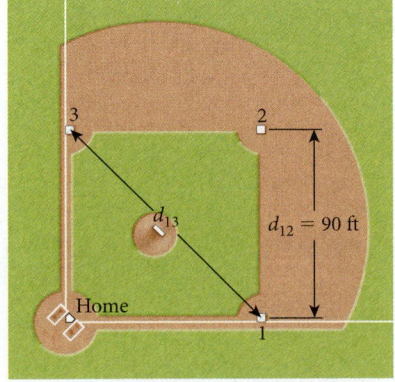

FIGURE 3.12 Dimensions of a baseball infield.

RESEARCH   In order to obtain the initial launch angle of the ball, we use equation 3.25, setting the range equal to the distance between first and second base:

$$d_{12} = \frac{v_0^2}{g} \sin 2\theta_0 \Rightarrow \theta_0 = \frac{1}{2} \sin^{-1}\left(\frac{d_{12} g}{v_0^2}\right).$$

Furthermore, we already have an equation for the maximum height:

$$H = y_0 + \frac{v_0^2 \sin^2 \theta_0}{2g}.$$

SIMPLIFY   Substituting our expression for the launch angle into the equation for the maximum height results in

$$H = y_0 + \frac{v_0^2 \sin^2\left(\frac{1}{2} \sin^{-1}\left(\frac{d_{12} g}{v_0^2}\right)\right)}{2g}.$$

CALCULATE   We are ready to insert numbers:

$$H = 1.8288 \text{ m} + \frac{(40.2336 \text{ m/s})^2 \sin^2\left(\frac{1}{2} \sin^{-1}\left(\frac{(27.432 \text{ m})(9.81 \text{ m/s}^2)}{(40.2336 \text{ m/s})^2}\right)\right)}{2(9.81 \text{ m/s}^2)} = 2.40285 \text{ m}.$$

ROUND   The precision of the given values was two significant digits. So we round our final result to

$$H = 2.4 \text{ m}.$$

Thus, a 90.-mph throw from second to first base is 2.40 m – 1.83 m = 0.57 m—that is, almost 2 ft—above a straight line at the middle of its trajectory. This number is even bigger for the throw from third to first base, for which we find an initial angle of 6.8° and a maximum height of 3.0 m, or 1.2 m (almost 4 ft) above the straight line connecting the points of release and catch (see Figure 3.13).

FIGURE 3.13   Trajectory of a baseball thrown from third to first base.

DOUBLE-CHECK   Common sense says that the longer throw from third to first needs to have a greater maximum height than the throw from second to first, and our answers agree with that. If you watch a baseball game from the stands or on television, these calculated heights may seem too large. However, if you watch a game from ground level, you'll see that the other infielders really do have to get some height on the ball to make a good throw to first base.

Let's consider one more example from baseball and calculate the trajectory of a batted ball (see Figure 3.14).

EXAMPLE 3.2 | **Batting a Baseball**

During the flight of a batted baseball, in particular, a home run, air resistance has a quite noticeable impact. For now though, we want to neglect it. Section 3.5 will discuss the effect of air resistance.

PROBLEM

If the ball comes off the bat with a launch angle of 35.0° and an initial speed of 110 mph, how far will the ball fly? How long will it be in the air? What will its speed be at the top of its trajectory? What will its speed be when it lands?

– Continued

FIGURE 3.14   Batting a baseball.

SOLUTION

Again, we need to convert to SI units first: $v_0 = 110$ mph $= 49.2$ m/s. We first find the range:

$$R = \frac{v_0^2}{g} \sin 2\theta_0 = \frac{(49.2 \text{ m/s})^2}{9.81 \text{ m/s}^2} \sin 70° = 231.6 \text{ m}.$$

This distance is about 760 feet, which would be a home run in even the biggest ballpark. However, this calculation does not take air resistance into account. If we took friction due to air resistance into account, the distance would be reduced to approximately 400 feet. (See Section 3.5 on realistic projectile motion.)

In order to find the baseball's time in the air, we can divide the range by the horizontal component of the velocity, assuming the ball is hit at about ground level.

$$t = \frac{R}{v_0 \cos\theta_0} = \frac{231.5 \text{ m}}{(49.2 \text{ m/s})(\cos 35°)} = 5.75 \text{ s}.$$

Now we calculate the speeds at the top of the trajectory and at landing. At the top of the trajectory, the velocity has only a horizontal component, which is $v_0 \cos\theta_0 = 40.3$ m/s. When the ball lands, we can calculate its speed using equation 3.23: $|\vec{v}| = \sqrt{v_0^2 - 2g(y - y_0)}$. Because we assume that the altitude at which it lands is the same as the one from which it was launched, we see that the speed is the same at the landing point as at the launching point, 49.2 m/s.

A real baseball would not quite follow the trajectory calculated here. If instead we launched a small steel ball of the size of the baseball with the same angle and speed, neglecting air resistance would have led to a very good approximation, and the trajectory parameters just found would be verified in such an experiment. The reason we can comfortably neglect air resistance for the steel ball bearing is that it has a much higher mass density and smaller surface area than a baseball, so drag effects (which depend on cross-sectional area) are small compared to gravitational effects.

Baseball is not the only sport that provides examples of projectile motion. Let's consider an example from football.

SOLVED PROBLEM 3.2 | Hang Time

When a football team is forced to punt the ball away to the opponent, it is very important to kick the ball as far as possible but also to attain a sufficiently long hang time—that is, the ball should remain in the air long enough that the punt-coverage team has time to run downfield and tackle the receiver right after the catch.

PROBLEM
What are the initial angle and speed with which a football has to be punted so that its hang time is 4.41 s and it travels a distance of 49.8 m (= 54.5 yd)?

SOLUTION
THINK A punt is a special case of projectile motion for which the initial and final values of the vertical coordinate are both zero. If we know the range of the projectile, we can figure out the hang time from the fact that the horizontal component of the velocity vector remains at a constant value; thus, the hang time must simply be the range divided by this horizontal component of the velocity vector. The equations for hang time and range give us two equations in the two unknown quantities that we are looking for: v_0 and θ_0.

SKETCH This is one of the few cases in which a sketch does not seem to provide additional information.

RESEARCH We have already seen (equation 3.25) that the range of a projectile is given by

$$R = \frac{v_0^2}{g} \sin 2\theta_0.$$

As already mentioned, the hang time can be most easily computed by dividing the range by the horizontal component of the velocity:

$$t = \frac{R}{v_0 \cos\theta_0}.$$

Thus, we have two equations in the two unknowns, v_0 and θ_0. (Remember, R and t were given in the problem statement.)

SIMPLIFY We solve both equations for v_0^2 and set them equal:

$$R = \frac{v_0^2}{g}\sin 2\theta_0 \Rightarrow v_0^2 = \frac{gR}{\sin 2\theta_0}$$

$$t = \frac{R}{v_0 \cos\theta_0} \Rightarrow v_0^2 = \frac{R^2}{t^2 \cos^2\theta_0}$$

$$\frac{gR}{\sin 2\theta_0} = \frac{R^2}{t^2 \cos^2\theta_0}.$$

Now, we can solve for θ_0. Using $\sin 2\theta_0 = 2\sin\theta_0\cos\theta_0$, we find

$$\frac{g}{2\sin\theta_0\cos\theta_0} = \frac{R}{t^2 \cos^2\theta_0}$$

$$\Rightarrow \tan\theta_0 = \frac{gt^2}{2R}$$

$$\Rightarrow \theta_0 = \tan^{-1}\left(\frac{gt^2}{2R}\right).$$

Next, we substitute this expression in either of the two equations we started with. We select the equation for hang time and solve it for v_0:

$$t = \frac{R}{v_0 \cos\theta_0} \Rightarrow v_0 = \frac{R}{t\cos\theta_0}.$$

CALCULATE All that remains is to insert numbers into the equations we have obtained:

$$\theta_0 = \tan^{-1}\left(\frac{(9.81\text{ m/s}^2)(4.41\text{ s})^2}{2(49.8\text{ m})}\right) = 62.4331°$$

$$v_0 = \frac{49.8\text{ m}}{(4.41\text{ s})(\cos 62.4331°)} = 24.4013\text{ m/s}.$$

ROUND The range and hang time were specified to three significant figures, so we state our final results to this precision:

$$\theta_0 = 62.4°$$

and

$$v_0 = 24.4\text{ m/s}.$$

DOUBLE-CHECK We know that the maximum range is reached with a launch angle of 45°. The punted ball here is launched at an initial angle that is significantly steeper, at 62.4°. Thus, the ball does not travel as far as it could go with the value of the initial speed that we computed. Instead, it travels higher and thus maximizes the hang time. If you watch good college or pro punters practice their skills during football games, you'll see that they try to kick the ball with an initial angle larger than 45°, in agreement with what we found in our calculations.

Concept Check 3.5

The same range as in Solved Problem 3.2 could be achieved with the same initial speed of 24.4 m/s but a launch angle different from 62.4°. What is the value of this angle?

a) 12.4°   c) 45.0°

b) 27.6°   d) 55.2°

Concept Check 3.6

What is the hang time for that other launch angle found in Concept Check 3.5?

a) 2.30 s   c) 4.41 s

b) 3.14 s   d) 5.14 s

SOLVED PROBLEM 3.3 | Time of Flight

You may have participated in Science Olympiad during middle school or high school. In one of the events in Science Olympiad, the goal is to hit a horizontal target at a fixed distance with a golf ball launched by a trebuchet. Competing teams build their own trebuchets. Your team has constructed a trebuchet that is able to launch the golf ball with an initial speed of 17.2 m/s, according to extensive tests performed before the competition.

PROBLEM

If the target is located at the same height as the elevation from which the golf ball is released and at a horizontal distance of 22.42 m away, how long will the golf ball be in the air before it hits the target?

SOLUTION

THINK Let's first eliminate what does not work. We cannot simply divide the distance between trebuchet and target by the initial speed, because this would imply that the initial velocity vector

– Continued

is in the horizontal direction. Since the projectile is in free fall in the vertical direction during its flight, it would certainly miss the target. So we have to aim the golf ball with an angle larger than zero relative to the horizontal. But at what angle do we need to aim?

If the golf ball, as stated, is released from the same height as the height of the target, then the horizontal distance between the trebuchet and the target is equal to the range. Because we also know the initial speed, we can calculate the release angle. Knowing the release angle and the initial speed lets us determine the horizontal component of the velocity vector. Since this horizontal component does not change in time, the flight time is simply given by the range divided by the horizontal component of the velocity.

SKETCH We don't need a sketch at this point because it would simply show a parabola, as for all projectile motion. However, we do not know the initial angle yet, so we will need a sketch later.

RESEARCH The range of a projectile is given by equation 3.25:

$$R = \frac{v_0^2}{g} \sin 2\theta_0.$$

If we know the value of this range and the initial speed, we can find the angle:

$$\sin 2\theta_0 = \frac{gR}{v_0^2}.$$

Once we have the value for the angle, we can use it to calculate the horizontal component of the initial velocity:

$$v_{x0} = v_0 \cos\theta_0.$$

Finally, as noted previously, we obtain the flight time as the ratio of the range and the horizontal component of the velocity:

$$t = \frac{R}{v_{x0}}.$$

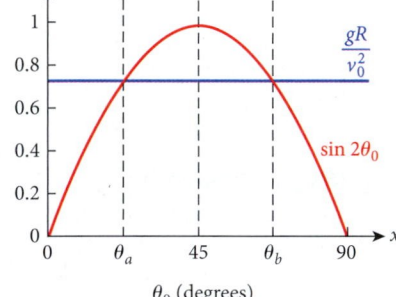

FIGURE 3.15 Two solutions for the initial angle.

SIMPLIFY If we solve the equation for the angle, $\sin 2\theta_0 = Rg/v_0^2$, we see that it has two solutions: one for an angle of less than 45° and one for an angle of more than 45°. Figure 3.15 plots the function $\sin 2\theta_0$ (in red) for all possible values of the initial angle θ_0 and shows where that curve crosses the plot of gR/v_0^2 (blue horizontal line). We call the two solutions θ_a and θ_b.

Algebraically, these solutions are given as

$$\theta_{a,b} = \tfrac{1}{2}\sin^{-1}\left(\frac{Rg}{v_0^2}\right).$$

Substituting this result into the formula for the horizontal component of the velocity results in

$$t = \frac{R}{v_{x0}} = \frac{R}{v_0 \cos\theta_0} = \frac{R}{v_0 \cos\left(\frac{1}{2}\sin^{-1}\left(\frac{Rg}{v_0^2}\right)\right)}.$$

CALCULATE Inserting numbers, we find:

$$\theta_{a,b} = \tfrac{1}{2}\sin^{-1}\left(\frac{(22.42\text{ m})(9.81\text{ m/s}^2)}{(17.2\text{ m/s})^2}\right) = 24.0128° \text{ or } 65.9872°$$

$$t_a = \frac{R}{v_0 \cos\theta_a} = \frac{22.42\text{ m}}{(17.2\text{ m/s})(\cos 24.0128°)} = 1.42699\text{ s}$$

$$t_b = \frac{R}{v_0 \cos\theta_b} = \frac{22.42\text{ m}}{(17.2\text{ m/s})(\cos 65.9872°)} = 3.20314\text{ s}.$$

ROUND The range was specified to four significant figures, and the initial speed to three. Therefore, we also state our final results to three significant figures:

$$t_a = 1.43\text{ s}, \quad t_b = 3.20\text{ s}.$$

Note that both solutions are valid in this case, and the team can select either one.

DOUBLE-CHECK Back to the approach that does not work: simply taking the distance from the trebuchet to the target and dividing it by the speed. This incorrect procedure leads to $t_{min} = d/v_0 = 1.30\text{ s}$. We use t_{min} to symbolize this value to indicate that it is some lower boundary representing the case in which the initial velocity vector points horizontally and in which we neglect the free-fall motion of the projectile. Thus, t_{min} serves as an absolute lower boundary, and it is reassuring to note that the shorter time we obtained above is a little larger than this lowest possible, but physically unrealistic, value.

3.5 Realistic Projectile Motion

If you are familiar with tennis or golf or baseball, you know that the parabolic model for the motion of a projectile is only a fairly crude approximation to the actual trajectory of any real ball. However, by ignoring some factors that affect real projectiles, we were able to focus on the physical principles that are most important in projectile motion. This is a common technique in science: Ignore some factors involved in a real situation in order to work with fewer variables and come to an understanding of the basic concept. Then go back and consider how the ignored factors affect the model. Let's briefly consider the most important factors that affect real projectile motion: air resistance, spin, and surface properties of the projectile.

The first modifying effect that we need to take into account is air resistance. Typically, we can parameterize air resistance as a velocity-dependent acceleration. The general analysis exceeds the scope of this book; however, the resulting trajectories are called *ballistic curves*.

Figure 3.16 shows the trajectories of baseballs launched at an initial angle of 35° with respect to the horizontal at initial speeds of 90 and 110 mph. Compare the trajectory shown for the launch speed of 110 mph with the result we calculated in Example 3.2: The real range of this ball is only slightly more than 400 ft, whereas we found 760 ft when we neglected air resistance. Obviously, for a long fly ball, neglecting air resistance is not valid.

Another important effect that the parabolic model neglects is the spin of the projectile as it moves through the air. When a quarterback throws a "spiral" in football, for example, the spin is important for the stability of the flight motion and prevents the ball from rotating end-over-end. In tennis, a ball with topspin drops much faster than a ball without noticeable spin, given the same initial values of speed and launch angle. Conversely, a tennis ball with underspin, or backspin, "floats" deeper into the court. In golf, backspin is sometimes desired, because it causes a steeper landing angle and thus helps the ball come to rest closer to its landing point than a ball hit without backspin. Depending on the magnitude and direction of rotation, sidespin of a golf ball can cause a deviation from a straight-line path along the ground (draws and fades for good players or hooks and slices for the rest of us).

In baseball, sidespin is what enables a pitcher to throw a curveball. By the way, there is no such thing as a "rising fastball" in baseball. However, balls thrown with severe backspin do not drop as fast as the batter expects and are thus sometimes perceived as rising—an optical illusion. In the graph of ballistic baseball trajectories in Figure 3.16, an initial backspin of 2000 rpm was assumed.

Curving and practically all other effects of spin on the trajectory of a moving ball are a result of the air molecules bouncing with higher speeds off the side of the ball (and the boundary layer of air molecules) that is rotating in the direction of the flight motion (and thus has a higher velocity relative to the incoming air molecules) than off the side of the ball rotating against the flight direction. We will return to this topic in Chapter 13 when we discuss fluid motion.

The surface properties of projectiles also have significant effects on their trajectories. Golf balls have dimples to make them fly farther. Balls that are otherwise identical to typical golf balls but have a smooth surface can be driven only about half as far. This surface effect is also the reason why sandpaper found in a pitcher's glove leads to ejection of that player from the game, because a baseball that is roughened on parts of its surface moves differently from one that is not.

Concept Check 3.7

Looking at Figure 3.16, what can you say about the ratio R_{real}/R_{ideal}, the real projectile range divided by the range calculated for ideal projectile motion?

a) For the launch angle of 35°, the ratio *increases* with the launch speed.

b) For the launch angle of 35°, the ratio *decreases* with the launch speed.

c) The ratio is independent of the launch speed.

d) For all launch angles, the ratio *increases* with the launch speed.

e) For all launch angles, the ratio *decreases* with the launch speed.

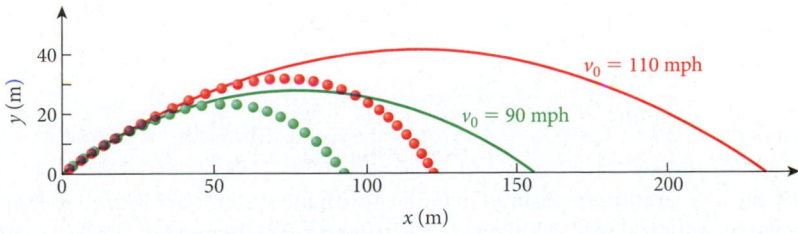

FIGURE 3.16 Trajectories of baseballs initially launched at an angle of 35° above the horizontal at speeds of 90 mph (green) and 110 mph (red). Solid curves neglect air resistance and backspin; dotted curves reflect air resistance and backspin.

3.6   Relative Motion

To study motion, we have allowed ourselves to shift the origin of the coordinate system by properly choosing values for x_0 and y_0. In general, x_0 and y_0 are constants that can be chosen freely. If this choice is made intelligently, it can help make a problem more manageable. For example, when we calculated the path of the projectile, $y(x)$, we set $x_0 = 0$ to simplify our calculations. The freedom to select values for x_0 and y_0 arises from the fact that our ability to describe any kind of motion does not depend on the location of the origin of the coordinate system.

So far, we have examined physical situations where we have kept the origin of the coordinate system at a fixed location during the motion of the object we wanted to consider. However, in some physical situations, it is impractical to choose a reference system with a fixed origin. Consider, for example, a jet plane landing on an aircraft carrier that is going forward at full throttle at the same time. You want to describe the plane's motion in a coordinate system fixed to the carrier, even though the carrier is moving. The reason why this is important is that the plane needs to come to rest *relative* to the carrier at some fixed location on the deck. The reference frame from which we view motion makes a big difference in how we describe the motion, producing an effect known as **relative velocity.**

Another example of a situation for which we cannot neglect relative motion is a transatlantic flight from Detroit, Michigan, to Frankfurt, Germany, which takes 8 h and 10 min. Using the same aircraft and going in the reverse direction, from Frankfurt to Detroit, takes 9 h and 10 min, a full hour longer. The primary reason for this difference is that the prevailing wind at high altitudes, the jet stream, tends to blow from west to east at speeds as high as 67 m/s (150 mph). Even though the airplane's speed relative to the air around it is the same in both directions, that air is moving with its own speed. Thus, the relationship of the coordinate system of the air inside the jet stream to the coordinate system in which the locations of Detroit and Frankfurt remain fixed is important in understanding the difference in flight times.

For a more easily analyzed example of a moving coordinate system, let's consider motion on a moving walkway, as is typically found in airport terminals. This system is an example of one-dimensional relative motion. Suppose that the walkway surface moves with a certain velocity, v_{wt}, relative to the terminal. We use the subscripts w for walkway and t for terminal. Then a coordinate system that is fixed to the walkway surface has exactly velocity v_{wt} relative to a coordinate system attached to the terminal. The man shown in Figure 3.17 is walking with a velocity v_{mw} as measured in a coordinate system on the walkway, and he has a velocity $v_{mt} = v_{mw} + v_{wt}$ with respect to the terminal. The two velocities v_{mw} and v_{wt} add as vectors since the corresponding displacements add as vectors. (We will show this explicitly when we generalize to three dimensions.) For example, if the walkway moves with $v_{wt} = 1.5$ m/s and the man moves with $v_{mw} = 2.0$ m/s, then he will progress through the terminal with a velocity of $v_{mt} = v_{mw} + v_{wt} = 2.0$ m/s + 1.5 m/s = 3.5 m/s.

One can achieve a state of no motion relative to the terminal by walking in the direction opposite of the motion of the walkway with a velocity that is exactly the negative of the walkway velocity. Children often try to do this. If a child were to walk with $v_{mw} = -1.5$ m/s on this walkway, her velocity would be zero relative to the terminal.

It is essential for this discussion of relative motion that the two coordinate systems have a velocity relative to each other that is constant in time. In this case, we can show that the accelerations measured in both coordinate systems are identical: $v_{wt} = \text{const.} \Rightarrow dv_{wt}/dt = 0$. From $v_{mt} = v_{mw} + v_{wt}$, we then obtain:

$$\frac{dv_{mt}}{dt} = \frac{d(v_{mw} + v_{wt})}{dt} = \frac{dv_{mw}}{dt} + \frac{dv_{wt}}{dt} = \frac{dv_{mw}}{dt} + 0$$

$$\Rightarrow a_{mt} = a_{mw}. \tag{3.27}$$

Therefore, the accelerations measured in both coordinate systems are indeed the same. This type of velocity addition is also known as a **Galilean transformation.** Before we go on to the two- and three-dimensional cases, note that this type of transformation is valid only for speeds that are small compared to the speed of light. Once the speed approaches the speed

FIGURE 3.17 Man walking on a moving walkway, demonstrating one-dimensional relative motion.

of light, we must use a different transformation, which we discuss in detail in Chapter 35 on the theory of relativity.

Now let's generalize this result to more than one spatial dimension. We assume that we have two coordinate systems: x_l, y_l, z_l and x_m, y_m, z_m. (Here we use the subscripts l for the coordinate system that is at rest in the laboratory and m for the one that is moving.) At time $t = 0$, suppose the origins of both coordinate systems are located at the same point, with their axes exactly parallel to one another. As indicated in Figure 3.18, the origin of the moving $x_m y_m z_m$ coordinate system moves with a constant translational velocity $\vec{v}_{ml}$ (blue arrow) relative to the origin of the laboratory $x_l y_l z_l$ coordinate system. After a time t, the origin of the moving $x_m y_m z_m$ coordinate system is thus located at the point $\vec{r}_{ml} = \vec{v}_{ml} t$.

We can now describe the motion of any object in either coordinate system. If the object is located at coordinate $\vec{r}_l$ in the $x_l y_l z_l$ coordinate system and at coordinate $\vec{r}_m$ in the $x_m y_m z_m$ coordinate system, then the position vectors are related to each other via simple vector addition:

$$\vec{r}_l = \vec{r}_m + \vec{r}_{ml} = \vec{r}_m + \vec{v}_{ml} t. \tag{3.28}$$

A similar relationship holds for the object's velocities, as measured in the two coordinate systems. If the object has velocity $\vec{v}_{ol}$ in the $x_l y_l z_l$ coordinate system and velocity $\vec{v}_{om}$ in the $x_m y_m z_m$ coordinate system, these two velocities are related via:

$$\vec{v}_{ol} = \vec{v}_{om} + \vec{v}_{ml}. \tag{3.29}$$

This equation can be obtained by taking the time derivative of equation 3.28, because $\vec{v}_{ml}$ is constant. Note that the two inner subscripts on the right-hand side of this equation are the same (and will be in any application of this equation). This makes the equation understandable on an intuitive level, because it says that the velocity of the object in the *lab* frame (subscript "ol") is equal to the sum of the velocity with which the *object* moves relative to the *moving* frame (subscript "om") and the velocity with which the *moving* frame moves relative to the *lab* frame (subscript "ml").

Taking another time derivative produces the accelerations. Again, because $\vec{v}_{ml}$ is constant and thus has a derivative equal to zero, we obtain, just as in the one-dimensional case,

$$\vec{a}_{ol} = \vec{a}_{om}. \tag{3.30}$$

The magnitude and direction of the acceleration for an object is the same in both coordinate systems.

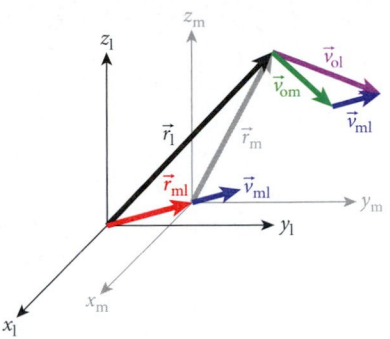

FIGURE 3.18 Reference frame transformation of a velocity vector and a position vector at some particular time.

EXAMPLE 3.3 | Airplane in a Crosswind

Airplanes move relative to the air that surrounds them. Suppose a pilot points his plane in the northeast direction. The airplane moves with a speed of 160. m/s relative to the wind, and the wind is blowing at 32.0 m/s in a direction from east to west (measured by an instrument at a fixed point on the ground).

PROBLEM
What is the velocity vector—speed and direction—of the airplane relative to the ground? How far off course does the wind blow this plane in 2.0 h?

SOLUTION
Figure 3.19 shows a vector diagram of the velocities. The airplane heads in the northeast direction, and the yellow arrow represents its velocity vector relative to the wind. The velocity vector of the wind is represented in orange and points due west. Graphical vector addition results in the green arrow that represents the velocity of the plane relative to the ground. To solve this problem, we apply the basic transformation of equation 3.29 embodied in the equation

$$\vec{v}_{pg} = \vec{v}_{pw} + \vec{v}_{wg}.$$

Here $\vec{v}_{pw}$ is the velocity of the plane with respect to the wind and has these components:

$$v_{pw,x} = v_{pw} \cos\theta = 160 \text{ m/s} \cdot \cos 45° = 113 \text{ m/s}$$

$$v_{pw,y} = v_{pw} \sin\theta = 160 \text{ m/s} \cdot \sin 45° = 113 \text{ m/s}.$$

– Continued

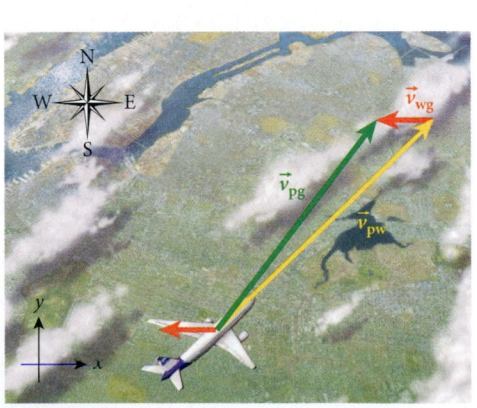

FIGURE 3.19 Velocity of an airplane with respect to the wind (yellow), the velocity of the wind with respect to the ground (orange), and the resultant velocity of the airplane with respect to the ground (green).

The velocity of the wind with respect to the ground, $\vec{v}_{wg}$, has these components:

$$v_{wg,x} = -32 \text{ m/s}$$

$$v_{wg,y} = 0.$$

We next obtain the components of the airplane's velocity relative to a coordinate system fixed to the ground, $\vec{v}_{pg}$:

$$v_{pg,x} = v_{pw,x} + v_{wg,x} = 113 \text{ m/s} - 32 \text{ m/s} = 81 \text{ m/s}$$

$$v_{pg,y} = v_{pw,y} + v_{pw,y} = 113 \text{ m/s}.$$

The absolute value of the velocity vector and its direction in the ground-based coordinate system are therefore

$$v_{pg} = \sqrt{v_{pg,x}^2 + v_{pg,y}^2} = 139 \text{ m/s}$$

$$\theta = \tan^{-1}\left(\frac{v_{pg,y}}{v_{pg,x}}\right) = 54.4°.$$

Now we need to find the course deviation due to the wind. To find this quantity, we can multiply the plane's velocity vectors in each coordinate system by the elapsed time of 2 h = 7200 s, then take the vector difference, and finally obtain the magnitude of the vector difference. The answer can be obtained more easily if we use equation 3.29 multiplied by the elapsed time to reflect that the course deviation, $\vec{r}_T$, due to the wind is the wind velocity, $\vec{v}_{wg}$, times 7200 s:

$$\left|\vec{r}_T\right| = \left|\vec{v}_{wg}\right| t = 32.0 \text{ m/s} \cdot 7200 \text{ s} = 230.4 \text{ km}.$$

DISCUSSION

The Earth itself moves a considerable amount in 2 h, as a result of its own rotation and its motion around the Sun, and you might think we have to take these motions into account. That the Earth moves is true, but is irrelevant for the present example: The airplane, the air, and the ground all participate in this rotation and orbital motion, which is superimposed on the relative motion of the objects described in the problem. Thus, we simply perform our calculations in a coordinate system in which the Earth is at rest and not rotating.

Another interesting consequence of relative motion can be seen when observing rain while in a moving car. You may have wondered why the rain always seems to come almost straight at you as you are driving. The following example answers this question.

Concept Check 3.8

It is raining, and there is practically no wind. While driving through the rain, you speed up. What happens to the angle of the rain relative to the horizontal that you observe from inside the car?

a) It *increases*.

b) It *decreases*.

c) It stays the same.

d) It can increase or decrease, depending on the direction in which you are driving.

EXAMPLE 3.4   Driving through Rain

Let's supppose rain is falling straight down on a car, as indicated by the white lines in Figure 3.20. A stationary observer outside the car would be able to measure the velocities of the rain (blue arrow) and of the moving car (red arrow).

However, if you are sitting inside the moving car, the outside world of the stationary observer (including the street, as well as the rain) moves with a relative velocity of $\vec{v} = -\vec{v}_{car}$. The velocity of this relative motion has to be added to all outside events as observed from inside the moving car. This motion results in a velocity vector $\vec{v}'_{rain}$ for the rain as observed from inside the moving car (Figure 3.21); mathematically, this vector is a sum, $\vec{v}'_{rain} = \vec{v}_{rain} - \vec{v}_{car}$, where $\vec{v}_{rain}$ and $\vec{v}_{car}$ are the velocity vectors of the rain and the car as observed by the stationary observer.

FIGURE 3.20 The velocity vectors of a moving car and of rain falling straight down on the car, as viewed by a stationary observer.

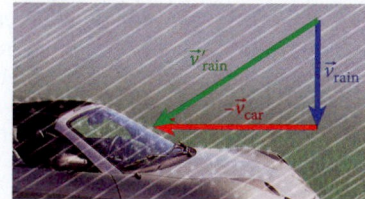

FIGURE 3.21 The velocity vector $\vec{v}'_{rain}$ of rain, as observed from inside the moving car.

SOLVED PROBLEM 3.4 | Moving Deer

The zookeeper who captured the monkey in Example 3.1 now has to capture a deer. We found that she needed to aim directly at the monkey for that earlier capture. She decides to fire directly at her target again, indicated by the bull's-eye in Figure 3.22.

PROBLEM
Where will the tranquilizer dart hit if the deer is $d = 25$ m away from the zookeeper and running from her right to her left with a speed of $v_d = 3.0$ m/s? The tranquilizer dart leaves her rifle horizontally with a speed of $v_0 = 90.$ m/s.

FIGURE 3.22 The red arrow indicates the velocity of the deer in the zookeeper's reference frame.

SOLUTION
THINK The deer is moving at the same time as the dart is falling, which introduces two complications. It is easiest to think about this problem in the moving reference frame of the deer. In that frame, the sideways horizontal component of the dart's motion has a constant velocity of $-\vec{v}_d$. The vertical component of the motion is a free-fall motion. The total displacement of the dart is then the vector sum of the displacements caused by both of these motions.

SKETCH We draw the two displacements in the reference frame of the deer (Figure 3.23). The blue arrow is the displacement due to the free-fall motion, and the red arrow is the sideways horizontal motion of the dart in the reference frame of the deer. The advantage of drawing the displacements in this moving reference frame is that the bull's-eye is attached to the deer and is moving with it.

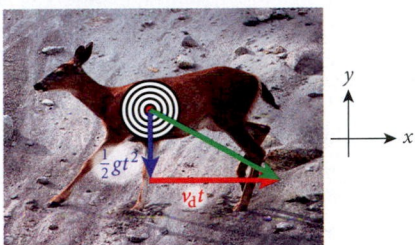

FIGURE 3.23 Displacement of the tranquilizer dart in the deer's reference frame.

RESEARCH First, we need to calculate the time it takes the tranquilizer dart to move 25 m in the direct line of sight from the gun to the deer. Because the dart leaves the rifle in the horizontal direction, the initial forward horizontal component of the dart's velocity vector is 90. m/s. For projectile motion, the horizontal velocity component is constant. Therefore, for the time the dart takes to cross the 25-m distance, we have

$$t = \frac{d}{v_0}.$$

During this time, the dart falls under the influence of gravity, and this vertical displacement is

$$\Delta y = -\tfrac{1}{2}gt^2.$$

Also, during this time, the deer has a sideways horizontal displacement in the reference frame of the zookeeper of $x = -v_d t$ (the deer moves to the left, hence the negative value of the horizontal velocity component). Therefore, the displacement of the dart in the reference frame of the deer is (see Figure 3.23)

$$\Delta x = v_d t.$$

SIMPLIFY Substituting the expression for the time into the equations for the two displacements results in

$$\Delta x = v_d \frac{d}{v_0} = \frac{v_d}{v_0} d$$

$$\Delta y = -\tfrac{1}{2}gt^2 = -\frac{d^2 g}{2v_0^2}.$$

CALCULATE We are now ready to put in the numbers:

$$\Delta x = \frac{(3.0 \text{ m/s})}{(90. \text{ m/s})}(25 \text{ m}) = 0.833333 \text{ m}$$

$$\Delta y = -\frac{(25 \text{ m})^2(9.81 \text{ m/s}^2)}{2(90. \text{ m/s})^2} = -0.378472 \text{ m}.$$

ROUND Rounding our results to two significant figures gives:

$$\Delta x = 0.83 \text{ m}$$

$$\Delta y = -0.38 \text{ m}.$$

The net effect is the vector sum of the sideways horizontal and vertical displacements, as indicated by the green diagonal arrow in Figure 3.23: The dart will miss the deer and hit the ground behind the deer.

– Continued

DOUBLE-CHECK Where should the zookeeper aim? If she wants to hit the running deer, she has to aim approximately 0.38 m above and 0.83 m to the left of her intended target. A dart fired in this direction will hit the deer, but not in the center of the bull's-eye. Why? With this aim, the initial velocity vector does not point in the horizontal direction. This lengthens the flight time, as we saw in Solved Problem 3.3. A longer flight time translates into a larger displacement in both x- and y-directions. This correction is small, but calculating it is a bit too involved to show here.

WHAT WE HAVE LEARNED | EXAM STUDY GUIDE

- In two or three dimensions, any change in the magnitude or direction of an object's velocity corresponds to acceleration.

- Projectile motion of an object can be separated into motion in the x-direction, described by the equations

    (1) $x = x_0 + v_{x0}t$

    (2) $v_x = v_{x0}$

    and motion in the y-direction, described by

    (3) $y = y_0 + v_{y0}t - \frac{1}{2}gt^2$

    (4) $y = y_0 + \bar{v}_y t$

    (5) $v_y = v_{y0} - gt$

    (6) $\bar{v}_y = \frac{1}{2}(v_y + v_{y0})$

    (7) $v_y^2 = v_{y0}^2 - 2g(y - y_0)$

- The relationship between the x- and y-coordinates for ideal projectile motion can be described by a parabola given by the formula $y = y_0 + (\tan\theta_0)x - \dfrac{g}{2v_0^2\cos^2\theta_0}x^2$, where y_0 is the initial vertical position, v_0 is the initial speed

- of the projectile, and θ_0 is the initial angle with respect to the horizontal at which the projectile is launched.

- The range R of an ideal projectile is given by

$$R = \frac{v_0^2}{g}\sin 2\theta_0.$$

- The maximum height H reached by an ideal projectile is given by $H = y_0 + \dfrac{v_{y0}^2}{2g}$, where v_{y0} is the vertical component of the initial velocity.

- Projectile trajectories are not parabolas when air resistance is taken into account. In general, the trajectories of realistic projectiles do not reach the maximum predicted height, and they have a significantly shorter range.

- The velocity $\vec{v}_{ol}$ of an object with respect to a stationary laboratory reference frame can be calculated using a Galilean transformation of the velocity, $\vec{v}_{ol} = \vec{v}_{om} + \vec{v}_{ml}$, where $\vec{v}_{om}$ is the velocity of the object with respect to a moving reference frame and $\vec{v}_{ml}$ is the constant velocity of the moving reference frame with respect to the laboratory frame.

ANSWERS TO SELF-TEST OPPORTUNITIES

3.1 Use equation 3.23 and $t = (x - x_0)/v_{x0} = (x - x_0)/(v_0\cos\theta_0)$ to find

$$|\vec{v}| = \sqrt{v_0^2 - 2g(x - x_0)(\tan\theta_0) + g^2(x - x_0)^2/(v_0\cos\theta_0)^2}$$

3.2 The time to reach the top is given by $v_y = v_{y0} - gt_{top} = 0 \Rightarrow t_{top} = v_{y0}/g = v_0\sin\theta_0/g$. The total flight time is $t_{total} = 2t_{top}$ because of the symmetry of the parabolic projectile trajectory. The range is the product of the total flight time and the horizontal velocity component: $R = t_{total}v_{x0} = 2t_{top}v_0\cos\theta_0 = 2(v_0\sin\theta_0/g)v_0\cos\theta_0 = v_0^2\sin(2\theta_0)/g$.

PROBLEM-SOLVING GUIDELINES

1. In all problems involving moving reference frames, it is important to clearly distinguish which object has what motion in which frame and relative to what. It is convenient to use subscripts consisting of two letters, where the first letter stands for a particular object and the second letter for the object it is moving relative to. The moving walkway situation discussed in Section 3.6 provides a good example of this use of subscripts.

2. In all problems concerning ideal projectile motion, the motion in the x-direction is independent of that in the y-direction. To solve these, you can often use the seven kinematical equations (3.11 through 3.17), which describe motion with constant velocity in the horizontal direction and free-fall motion with constant acceleration in the vertical direction. In general, you should avoid cookie-cutter–style application of formulas, but in exam situations, these seven kinematical equations can be your first line of defense. Keep in mind, however, that these equations work only in situations in which the horizontal acceleration component is zero and the vertical acceleration component is constant.

MULTIPLE-CHOICE QUESTIONS

3.1 An arrow is shot horizontally with a speed of 20. m/s from the top of a tower 60. m high. The time to reach the ground will be

a) 8.9 s.

b) 7.1 s.

c) 3.5 s.

d) 2.6 s.

e) 1.0 s.

3.2 A projectile is launched from the top of a building with an initial velocity of 30.0 m/s at an angle of 60.0° above the horizontal. The magnitude of its velocity at $t = 5.00$ s after the launch is

a) −23.0 m/s.

b) 7.3 m/s.

c) 15.0 m/s.

d) 27.5 m/s.

e) 50.4 m/s.

3.3 A ball is thrown at an angle between 0° and 90° with respect to the horizontal. Its velocity and acceleration vectors are parallel to each other at a launch angle of

a) 0°.

b) 45°.

c) 60°.

d) 90°.

e) none of the above.

3.4 During practice two baseball outfielders throw a ball to the shortstop. In both cases the distance is 40.0 m. Outfielder 1 throws the ball with an initial speed of 20.0 m/s, outfielder 2 throws the ball with an initial speed of 30.0 m/s. In both cases the balls are thrown and caught at the same height above ground.

a) Ball 1 is in the air for a shorter time than ball 2.

b) Ball 2 is in the air for a shorter time than ball 1.

c) Both balls are in the air for the same duration.

d) The answer cannot be decided from the information given.

3.5 A 50-g ball rolls off a countertop and lands 2 m from the base of the counter. A 100-g ball rolls off the same countertop with the same speed. It lands _____ from the base of the counter.

a) less than 1 m

b) 1 m

c) 2 m

d) 4 m

e) more than 4 m

3.6 For a given initial speed of an ideal projectile, there is (are) _____ launch angle(s) for which the range of the projectile is the same.

a) only one

b) two different

c) more than two but a finite number of

d) only one if the angle is 45° but otherwise two different

e) an infinite number of

3.7 A cruise ship moves southward in still water at a speed of 20.0 km/h, while a passenger on the deck of the ship walks toward the east at a speed of 5.0 km/h. The passenger's velocity with respect to Earth is

a) 20.6 km/h, at an angle of 14.04° east of south.

b) 20.6 km/h, at an angle of 14.04° south of east.

c) 25.0 km/h, south.

d) 25.0 km/h, east.

e) 20.6 km/h, south.

3.8 Two cannonballs are shot from different cannons at angles $\theta_{01} = 20°$ and $\theta_{02} = 30°$, respectively. Assuming ideal projectile motion, the ratio of the launching speeds, v_{02}/v_{01}, for which the two cannonballs achieve the same range is

a) 0.742.

b) 0.862.

c) 1.212.

d) 1.093.

e) 2.222.

3.9 The acceleration due to gravity on the Moon is 1.62 m/s², approximately a sixth of the value on Earth. For a given initial velocity v_0 and a given launch angle θ_0, the ratio of the range of an ideal projectile on the Moon to the range of the same projectile on Earth, R_{Moon}/R_{Earth}, will be approximately

a) 6.

b) 3.

c) 12.

d) 5.

e) 1.

3.10 A baseball is launched from the bat at an angle $\theta_0 = 30.0°$ with respect to the positive x-axis and with an initial speed of 40.0 m/s, and it is caught at the same height from which it was hit. Assuming ideal projectile motion (positive y-axis upward), the velocity of the ball when it is caught is

a) $(20.00\,\hat{x} + 34.64\,\hat{y})$ m/s.

b) $(-20.00\,\hat{x} + 34.64\,\hat{y})$ m/s.

c) $(34.64\,\hat{x} - 20.00\,\hat{y})$ m/s.

d) $(34.64\,\hat{x} + 20.00\,\hat{y})$ m/s.

3.11 In ideal projectile motion, the velocity and acceleration of the projectile at its maximum height are, respectively,

a) horizontal, vertical downward.

b) horizontal, zero.

c) zero, zero.

d) zero, vertical downward.

e) zero, horizontal.

3.12 In ideal projectile motion, when the positive y-axis is chosen to be vertically upward, the y-component of the acceleration of the object during the ascending part of the motion and the y-component of the acceleration during the descending part of the motion are, respectively,

a) positive, negative.

b) negative, positive.

c) positive, positive.

d) negative, negative.

3.13 In ideal projectile motion, when the positive y-axis is chosen to be vertically upward, the y-component of the velocity of the object during the ascending part of the motion and the y-component of the velocity during the descending part of the motion are, respectively,

a) positive, negative.

b) negative, positive.

c) positive, positive.

d) negative, negative.

3.14 A projectile is launched from a height $y_0 = 0$. For a given launch angle, if the launch speed is doubled, what will happen to the range R and the maximum height H of the projectile?

a) R and H will both double.

b) R and H will both quadruple.

c) R will double, and H will stay the same.

d) R will quadruple, and H will double.

e) R will double, and H will quadruple.

3.15 A projectile is launched twice from a height $y_0 = 0$ at a given launch speed, v_0. The first launch angle is 30.0°; the second angle is 60.0°. What can you say about the range R of the projectile in these two cases?

a) R is the same for both cases.

b) R is larger for a launch angle of 30.0°.

c) R is larger for a launch angle of 60.0°.

d) None of the preceding statements is true.

CONCEPTUAL QUESTIONS

3.16 A ball is thrown from ground at an angle between 0° and 90°. Which of the following remain constant: x, y, v_x, v_y, a_x, a_y?

3.17 A ball is thrown straight up by a passenger in a train that is moving with a constant velocity. Where would the ball land—back in his

hands, in front of him, or behind him? Does your answer change if the train is accelerating in the forward direction? If yes, how?

3.18 A rock is thrown at an angle 45° below the horizontal from the top of a building. Immediately after release will its acceleration be greater than, equal to, or less than the acceleration due to gravity?

3.19 Three balls of different masses are thrown horizontally from the same height with different initial speeds, as shown in the figure. Rank in order, from the shortest to the longest, the times the balls take to hit the ground.

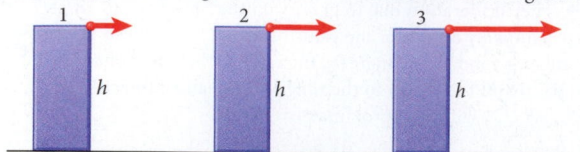

3.20 To attain maximum height for the trajectory of a projectile, what angle would you choose between 0° and 90°, assuming that you can launch the projectile with the same initial speed independent of the launch angle. Explain your reasoning.

3.21 An airplane is traveling at a constant horizontal speed v, at an altitude h above a lake, when a trapdoor at the bottom of the airplane opens and a package is released (falls) from the plane. The airplane continues horizontally at the same altitude and velocity. Neglect air resistance.

a) What is the distance between the package and the plane when the package hits the surface of the lake?

b) What is the horizontal component of the velocity vector of the package when it hits the lake?

c) What is the speed of the package when it hits the lake?

3.22 Two cannonballs are shot in sequence from a cannon, into the air, with the same muzzle velocity, at the same launch angle. Based on their trajectory and range, how can you tell which one is made of lead and which one is made of wood. If the same cannonballs were launched in vacuum, what would your answer be?

3.23 One should never jump off a moving vehicle (train, car, bus, etc.). Assuming, however, that one does perform such a jump, from a physics standpoint, what would be the best direction to jump in order to minimize the impact of the landing? Explain.

3.24 A boat travels at a speed of v_{BW} relative to the water in a river of width D. The speed at which the water is flowing is v_W.

a) Prove that the time required to cross the river to a point exactly opposite the starting point and then to return is $T_1 = 2D / \sqrt{v_{BW}^2 - v_W^2}$.

b) Prove that the time for the boat to travel a distance D downstream and then return is $T_1 = 2Dv_{BW} / (v_{BW}^2 - v_W^2)$.

3.25 A rocket-powered hockey puck is moving on a (frictionless) horizontal air-hockey table. The x- and y-components of its velocity as a function of time are presented in the graphs below. Assuming that at $t = 0$ the puck is at $(x_0, y_0) = (1, 2)$, draw a detailed graph of the trajectory $y(x)$.

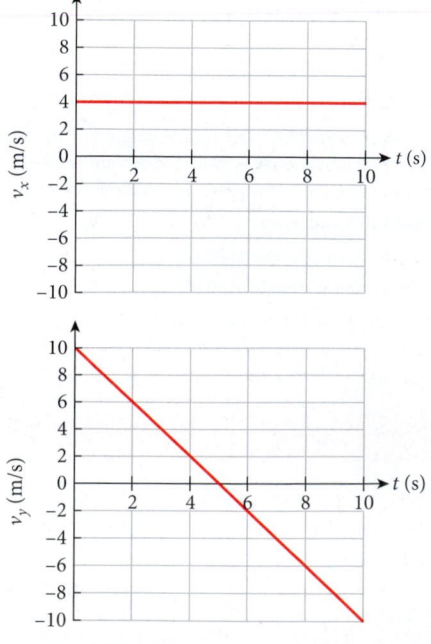

3.26 For a certain object in three-dimensional motion, the x-, y-, and z-coordinates as a function of time are given by

$$x(t) = \frac{\sqrt{2}}{2}t, \quad y(t) = \frac{\sqrt{2}}{2}t, \quad \text{and} \quad z(t) = -4.9t^2 + \sqrt{3}t.$$

Describe the motion and the trajectory of the object in an xyz coordinate system.

3.27 An object moves in the xy-plane. The x- and y-coordinates of the object as a function of time are given by the following equations: $x(t) = 4.9t^2 + 2t + 1$ and $y(t) = 3t + 2$. What is the velocity vector of the object as a function of time? What is its acceleration vector at the time $t = 2$ s?

3.28 A particle's motion is described by the following two parametric equations:
$$x(t) = 5\cos(2\pi t)$$
$$y(t) = 5\sin(2\pi t)$$
where the displacements are in meters and t is the time, in seconds.

a) Draw a graph of the particle's trajectory (that is, a graph of y versus x).

b) Determine the equations that describe the x- and y-components of the velocity, v_x and v_y, as functions of time.

c) Draw a graph of the particle's speed as a function of time.

3.29 In a proof-of-concept experiment for an antiballistic missile defense system, a missile is fired from the ground of a shooting range toward a stationary target on the ground. The system detects the missile by radar, analyzes in real time its parabolic motion, and determines that it was fired from a distance $x_0 = 5.00$ km, with an initial speed of 0.600 km/s in the xz-plane such that the x-component of the initial velocity is in the negative x-direction, and with a launch angle $\theta_0 = 20.0°$. The defense system then calculates the required time delay measured from the launch of the missile and fires a small rocket situated at $y_0 = -1.00$ km with an initial velocity of v_r m/s at a launch angle $\alpha_0 = 60.0°$ in the yz-plane, to intercept the missile. Determine the initial speed v_r of the intercept rocket and the required time delay.

3.30 A projectile is launched at an angle of 45.0° above the horizontal. What is the ratio of its horizontal range to its maximum height? How does the answer change if the initial speed of the projectile is doubled?

3.31 In a projectile motion, the horizontal range and the maximum height attained by the projectile are equal.

a) What is the launch angle?

b) If everything else stays the same, how should the launch angle, θ_0, of the projectile be changed for the range to be halved?

3.32 An air-hockey puck has a model rocket rigidly attached to it. The puck is pushed from one corner along the long side of the 2.00-m-long air-hockey table, with the rocket pointing along the short side of the table, and at the same time the rocket is fired. If the rocket thrust imparts an acceleration of 2.00 m/s^2 to the puck, and the table is 1.00 m wide, with what minimum initial velocity should the puck be pushed to make it to the opposite short side of the table without bouncing off either long side? Draw the trajectory of the puck for three initial velocities: $v < v_{min}$, $v = v_{min}$, and $v > v_{min}$. Neglect friction and air resistance.

3.33 On a battlefield, a cannon fires a cannonball up a slope, from ground level, with an initial velocity v_0 at an angle θ_0 above the horizontal. The ground itself makes an angle α above the horizontal ($\alpha < \theta_0$). What is the range R of the cannonball, measured along the inclined ground? Compare your result with the equation for the range on horizontal ground (equation 3.25).

3.34 Two swimmers with a soft spot for physics engage in a peculiar race that models a famous optics experiment: the Michelson-Morley experiment. The race takes place in a river 50.0 m wide that is flowing at a steady rate of 3.00 m/s. Both swimmers start at the same point on one bank and swim at the same speed of 5.00 m/s *with respect to the stream*. One of the swimmers swims directly across the river to the closest point on the opposite bank and then turns around and swims back to the starting point. The other swimmer swims *along* the river bank, first upstream a distance exactly equal to the width of the river and then downstream back to the starting point. Who gets back to the starting point first?

EXERCISES

A blue problem number indicates a worked-out solution is available in the Student Solutions Manual. One • and two •• indicate increasing level of problem difficulty.

Section 3.2

3.35 What is the magnitude of an object's average velocity if the object moves from a point with coordinates $x = 2.0$ m, $y = -3.0$ m to a point with coordinates $x = 5.0$ m, $y = -9.0$ m in a time interval of 2.4 s?

3.36 A man in search of his dog drives first 10.0 mi northeast, then 12.0 mi straight south, and finally 8.0 mi in a direction 30.0° north of west. What are the magnitude and direction of his resultant displacement?

3.37 During a jaunt on your sailboat, you sail 2.00 km east, then 4.00 km southeast, and finally an additional distance in an unknown direction. Your final position is 6.00 km directly east of the starting point. Find the magnitude and direction of the third leg of your journey.

3.38 A truck travels 3.02 km north and then makes a 90.0° left turn and drives another 4.30 km. The whole trip takes 5.00 min.

a) With respect to a two-dimensional coordinate system on the surface of Earth such that the y-axis points north, what is the net displacement vector of the truck for this trip?

b) What is the magnitude of the average velocity for this trip?

•3.39 A rabbit runs in a garden such that the x- and y-components of its displacement as functions of time are given by $x(t) = -0.45t^2 - 6.5t + 25$ and $y(t) = 0.35t^2 + 8.3t + 34$. (Both x and y are in meters and t is in seconds.)

a) Calculate the rabbit's position (magnitude and direction) at $t = 10.0$ s.

b) Calculate the rabbit's velocity at $t = 10.0$ s.

c) Determine the acceleration vector at $t = 10.0$ s.

••3.40 Some rental cars have a GPS unit installed, which allows the rental car company to check where you are at all times and thus also know your speed at any time. One of these rental cars is driven by an employee in the company's lot and, during the time interval from 0 to 10.0 s, is found to have a position vector as a function of time given by

$$\vec{r}(t) = \left((24.4 \text{ m}) - t(12.3 \text{ m/s}) + t^2(2.43 \text{ m/s}^2), \right.$$
$$\left. (74.4 \text{ m}) + t^2(1.80 \text{ m/s}^2) - t^3(0.130 \text{ m/s}^3)\right)$$

a) What is the distance of this car from the origin of the coordinate system at $t = 5.00$ s?

b) What is the velocity vector as a function of time?

c) What is the speed at $t = 5.00$ s?

Extra credit: Can you produce a plot of the trajectory of the car in the xy-plane?

Section 3.3

3.41 A skier launches off a ski jump with a horizontal velocity of 30.0 m/s (and no vertical velocity component). What are the magnitudes of the horizontal and vertical components of her velocity the instant before she lands 2.00 s later?

3.42 An archer shoots an arrow from a height of 1.14 m above ground with an initial speed of 47.5 m/s and a launch angle of 35.2° above the horizontal. At what time after the release of the arrow from the bow will the arrow be flying exactly horizontally?

3.43 A football is punted with an initial speed of 27.5 m/s and a launch angle of 56.7°. What is its hang time (the time until it hits the ground again)?

3.44 You serve a tennis ball from a height of 1.80 m above the ground. The ball leaves your racket with a speed of 18.0 m/s at an angle of 7.00° above the horizontal. The horizontal distance from the court's baseline to the net is 11.83 m, and the net is 1.07 m high. Neglect any spin imparted on the ball as well as air resistance effects. Does the ball clear the net? If yes, by how much? If not, by how much did it miss?

3.45 Stones are thrown horizontally with the same velocity from two buildings. One stone lands twice as far away from its building as the other stone. Determine the ratio of the heights of the two buildings.

3.46 You are practicing throwing darts in your dorm. You stand 3.00 m from the wall on which the board hangs. The dart leaves your hand with a horizontal velocity at a point 2.00 m above the ground. The dart strikes the board at a point 1.65 m from the ground. Calculate:

a) the time of flight of the dart;

b) the initial speed of the dart;

c) the velocity of the dart when it hits the board.

•3.47 A football player kicks a ball with a speed of 22.4 m/s at an angle of 49.0° above the horizontal from a distance of 39.0 m from the goalpost.

a) By how much does the ball clear or fall short of clearing the crossbar of the goalpost if that bar is 3.05 m high?

b) What is the vertical velocity of the ball at the time it reaches the goalpost?

•3.48 An object fired at an angle of 35.0° above the horizontal takes 1.50 s to travel the last 15.0 m of its vertical distance and the last 10.0 m of its horizontal distance. With what speed was the object launched? (Note: The problem does not specify that the initial and final elevation of the object are the same!)

•3.49 A conveyor belt is used to move sand from one place to another in a factory. The conveyor is tilted at an angle of 14.0° above the horizontal and the sand is moved without slipping at the rate of 7.00 m/s. The sand is collected in a big drum 3.00 m below the end of the conveyor belt. Determine the horizontal distance between the end of the conveyor belt and the middle of the collecting drum.

•3.50 Your friend's car is parked on a cliff overlooking the ocean on an incline that makes an angle of 17.0° below the horizontal. The brakes fail, and the car rolls from rest down the incline for a distance of 29.0 m to the edge of the cliff, which is 55.0 m above the ocean, and, unfortunately, continues over the edge and lands in the ocean.

a) Find the car's position relative to the base of the cliff when the car lands in the ocean.

b) Find the length of time the car is in the air.

•3.51 An object is launched at a speed of 20.0 m/s from the top of a tall tower. The height y of the object as a function of the time t elapsed from launch is $y(t) = -4.90t^2 + 19.32t + 60.0$, where h is in meters and t is in seconds. Determine:

a) the height H of the tower;

b) the launch angle;

c) the horizontal distance traveled by the object before it hits the ground.

•3.52 A projectile is launched at a 60.0° angle above the horizontal on level ground. The change in its velocity between launch and just before landing is found to be $\Delta \vec{v} \equiv \vec{v}_{\text{landing}} - \vec{v}_{\text{launch}} = -20.0 \hat{y}$ m/s. What is the initial velocity of the projectile? What is its final velocity just before landing?

••**3.53** The figure shows the paths of a tennis ball your friend drops from the window of her apartment and of the rock you throw from the ground at the same instant. The rock and the ball collide at $x = 50.0$ m, $y = 10.0$ m and $t = 3.00$ s. If the ball was dropped from a height of 54.1 m, determine the velocity of the rock initially and at the time of its collision with the ball.

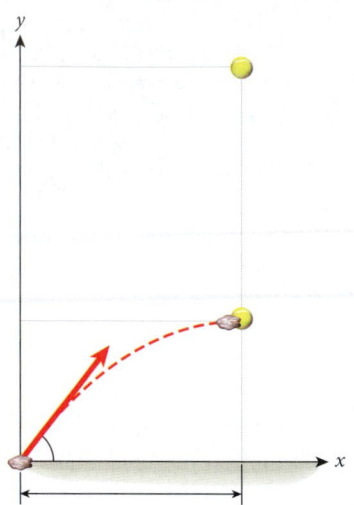

Section 3.4

3.54 For a science fair competition, a group of high school students build a kicker-machine that can launch a golf ball from the origin with a velocity of 11.2 m/s and a launch angle of 31.5° with respect to the horizontal.

a) Where will the golf ball fall back to the ground?

b) How high will it be at the highest point of its trajectory?

c) What is the ball's velocity vector (in Cartesian components) at the highest point of its trajectory?

d) What is the ball's acceleration vector (in Cartesian components) at the highest point of its trajectory?

3.55 If you want to use a catapult to throw rocks and the maximum range you need these projectiles to have is 0.67 km, what initial speed do the rocks have to have as they leave the catapult?

3.56 What is the maximum height above ground that a projectile of mass 0.790 kg, launched from ground level, can achieve if you are able to give it an initial speed of 80.3 m/s?

•**3.57** During one of the games, you were asked to punt for your football team. You kicked the ball at an angle of 35.0° with a velocity of 25.0 m/s. If your punt goes straight down the field, determine the average speed at which the running back of the opposing team standing at 70.0 m from you must run to catch the ball at the same height as you released it. Assume that the running back starts running as the ball leaves your foot and that the air resistance is negligible.

•**3.58** By trial and error, a frog learns that it can leap a maximum horizontal distance of 1.30 m. If, in the course of an hour, the frog spends 20.0% of the time resting and 80.0% of the time performing identical jumps of that maximum length, in a straight line, what is the distance traveled by the frog?

•**3.59** A circus juggler performs an act with balls that she tosses with her right hand and catches with her left hand. Each ball is launched at an angle of 75.0° and reaches a maximum height of 90.0 cm above the launching height. If it takes the juggler 0.200 s to catch a ball with her left hand, pass it to her right hand and toss it back into the air, what is the maximum number of balls she can juggle?

••**3.60** In an arcade game, a ball is launched from the corner of a smooth inclined plane. The inclined plane makes a 30.0° angle with the horizontal and has a width of $w = 50.0$ cm. The spring-loaded launcher makes an angle of 45.0° with the lower edge of the inclined plane. The goal is to

get the ball into a small hole at the opposite corner of the inclined plane. With what initial speed should you launch the ball to achieve this goal? (*Hint:* If the hole is small, the ball should enter it with zero vertical velocity component.)

••**3.61** A copy-cat daredevil tries to reenact Evel Knievel's 1974 attempt to jump the Snake River Canyon in a rocket-powered motorcycle. The canyon is $L = 400.$ m wide, with the opposite rims at the same height. The height of the launch ramp at one rim of the canyon is $h = 8.00$ m above the rim, and the angle of the end of the ramp is 45.0° with the horizontal.

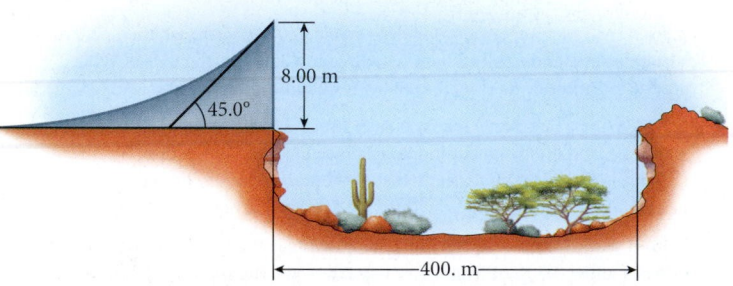

a) What is the minimum launch speed required for the daredevil to make it across the canyon? Neglect the air resistance and wind.

b) Famous after his successful first jump, but still recovering from the injuries sustained in the crash caused by a strong bounce upon landing, the daredevil decides to jump again but to add a landing ramp with a slope that will match the angle of his velocity at landing. If the height of the landing ramp at the opposite rim is 3.00 m, what is the new required launch speed?

Section 3.5

3.62 A golf ball is hit with an initial angle of 35.5° with respect to the horizontal and an initial velocity of 83.3 mph. It lands a distance of 86.8 m away from where it was hit. By how much did the effects of wind resistance, spin, and so forth reduce the range of the golf ball from the ideal value?

Section 3.6

3.63 You are walking on a moving walkway in an airport. The length of the walkway is 59.1 m. If your velocity relative to the walkway is 2.35 m/s and the walkway moves with a velocity of 1.77 m/s, how long will it take you to reach the other end of the walkway?

3.64 The captain of a boat wants to travel directly across a river that flows due east with a speed of 1.00 m/s. He starts from the south bank of the river and heads toward the north bank. The boat has a speed of 6.10 m/s with respect to the water. In what direction (in degrees) should the captain steer the boat? Note that 90° is east, 180° is south, 270° is west, and 360° is north.

3.65 The captain of a boat wants to travel directly across a river that flows due east. He starts from the south bank of the river and heads toward the north bank. The boat has a speed of 5.57 m/s with respect to the water. The captain steers the boat in the direction 315°. How fast is the water flowing? Note that 90° is east, 180° is south, 270° is west, and 360° is north.

•**3.66** The air-speed indicator of a plane that took off from Detroit reads 350. km/h and the compass indicates that it is heading due east to Boston. A steady wind is blowing due north at 40.0 km/h. Calculate the velocity of the plane with reference to the ground. If the pilot wishes to fly directly to Boston (due east) what must the compass read?

•**3.67** You want to cross a straight section of a river that has a uniform current of 5.33 m/s and is 127. m wide. Your motorboat has an engine that can generate a speed of 17.5 m/s for your boat. Assume that you reach top speed right away (that is, neglect the time it takes to accelerate the boat to top speed).

a) If you want to go directly across the river with a 90.0° angle relative to the riverbank, at what angle relative to the riverbank should you point your boat?

b) How long will it take to cross the river in this way?

c) In which direction should you aim your boat to achieve minimum crossing time?

d) What is the minimum time to cross the river?

e) What is the minimum speed of your boat that will still enable you to cross the river with a 90.0° angle relative to the riverbank?

•**3.68** During a long airport layover, a physicist father and his 8-year-old daughter try a game that involves a moving walkway. They have measured the walkway to be 42.5 m long. The father has a stopwatch and times his daughter. First, the daughter walks with a constant speed in the same direction as the conveyor. It takes 15.2 s to reach the end of the walkway. Then, she turns around and walks with the same speed relative to the conveyor as before, just this time in the opposite direction. The return leg takes 70.8 s. What is the speed of the walkway conveyor relative to the terminal, and with what speed was the girl walking?

•**3.69** An airplane has an air speed of 126.2 m/s and is flying due north, but the wind blows from the northeast to the southwest at 55.5 m/s. What is the plane's actual ground speed?

Additional Exercises

3.70 A cannon is fired from a hill 116.7 m high at an angle of 22.7° with respect to the horizontal. If the muzzle velocity is 36.1 m/s, what is the speed of a 4.35-kg cannonball when it hits the ground 116.7 m below?

3.71 A baseball is thrown with a velocity of 31.1 m/s at an angle of $\theta = 33.4°$ above horizontal. What is the horizontal component of the ball's velocity at the highest point of the ball's trajectory?

3.72 A rock is thrown horizontally from the top of a building with an initial speed of $v = 10.1$ m/s. If it lands $d = 57.1$ m from the base of the building, how high is the building?

3.73 A car is moving at a constant 19.3 m/s, and rain is falling at 8.90 m/s straight down. What angle θ (in degrees) does the rain make with respect to the horizontal as observed by the driver?

3.74 You tried to pass the salt and pepper shakers to your friend at the other end of a table of height 0.850 m by sliding them across the table. They both slid off the table, with velocities of 5.00 m/s and 2.50 m/s, respectively.

a) Compare the times it takes the shakers to hit the floor.

b) Compare the horizontal distance that each shaker travels from the edge of the table to the point it hits the floor.

3.75 A box containing food supplies for a refugee camp was dropped from a helicopter flying horizontally at a constant elevation of 500. m. If the box hit the ground at a distance of 150. m horizontally from the point of its release, what was the speed of the helicopter? With what speed did the box hit the ground?

3.76 A car is driven straight off the edge of a cliff that is 60.0 m high. The police at the scene of the accident note that the point of impact is 150. m from the base of the cliff. How fast was the car traveling when it went over the cliff?

3.77 At the end of the spring term, a high school physics class celebrates by shooting a bundle of exam papers into the town landfill with a homemade catapult. They aim for a point that is 30.0 m away and at the same height from which the catapult releases the bundle. The initial horizontal velocity component is 3.90 m/s. What is the initial velocity component in the vertical direction? What is the launch angle?

3.78 Salmon often jump upstream through waterfalls to reach their breeding grounds. One salmon came across a waterfall 1.05 m in height, which she jumped in 2.10 s at an angle of 35.0° above the horizontal to continue upstream. What was the initial speed of her jump?

3.79 A firefighter, 60.0 m away from a burning building, directs a stream of water from a ground-level fire hose at an angle of 37.0° above the horizontal. If the water leaves the hose at 40.3 m/s, which floor of the building will the stream of water strike? Each floor is 4.00 m high.

3.80 A projectile leaves ground level at an angle of 68.0° above the horizontal. As it reaches its maximum height, H, it has traveled a horizontal distance, d, in the same amount of time. What is the ratio H/d?

3.81 The McNamara Delta terminal at the Metro Detroit Airport has moving walkways for the convenience of the passengers. Robert walks beside one walkway and takes 30.0 s to cover its length. John simply stands on the walkway and covers the same distance in 13.0 s. Kathy walks on the walkway with the same speed as Robert's. How long does Kathy take to complete her stroll?

3.82 Rain is falling vertically at a constant speed of 7.00 m/s. At what angle from the vertical do the raindrops appear to be falling to the driver of a car traveling on a straight road with a speed of 60.0 km/h?

3.83 To determine the gravitational acceleration at the surface of a newly discovered planet, scientists perform a projectile motion experiment. They launch a small model rocket at an initial speed of 50.0 m/s and an angle of 30.0° above the horizontal and measure the (horizontal) range on flat ground to be 2165 m. Determine the value of g for the planet.

3.84 A diver jumps from a 40.0-m-high cliff into the sea. Rocks stick out of the water for a horizontal distance of 7.00 m from the foot of the cliff. With what minimum horizontal speed must the diver jump off the cliff in order to clear the rocks and land safely in the sea?

3.85 An outfielder throws a baseball with an initial speed of 32.0 m/s at an angle of 23.0° to the horizontal. The ball leaves his hand from a height of 1.83 m. How long is the ball in the air before it hits the ground?

•**3.86** A rock is tossed off the top of a cliff of height 34.9 m. Its initial speed is 29.3 m/s, and the launch angle is 29.9° above the horizontal. What is the speed with which the rock hits the ground at the bottom of the cliff?

•**3.87** During the 2004 Olympic Games, a shot putter threw a shot put with a speed of 13.0 m/s at an angle of 43.0° above the horizontal. She released the shot put from a height of 2.00 m above the ground.

a) How far did the shot put travel in the horizontal direction?

b) How long was it until the shot put hit the ground?

•**3.88** A salesman is standing on the Golden Gate Bridge in a traffic jam. He is at a height of 71.8 m above the water below. He receives a call on his cell phone that makes him so mad that he throws his phone horizontally off the bridge with a speed of 23.7 m/s.

a) How far does the cell phone travel horizontally before hitting the water?

b) What is the speed with which the phone hits the water?

•**3.89** A security guard is chasing a burglar across a rooftop, both running at 4.20 m/s. Before the burglar reaches the edge of the roof, he has to decide whether or not to try jumping to the roof of the next building, which is 5.50 m away and 4.00 m lower. If he decides to jump horizontally to get away from the guard, can he make it? Explain your answer.

•**3.90** A blimp is ascending at the rate of 7.50 m/s at a height of 80.0 m above the ground when a package is thrown from its cockpit horizontally with a speed of 4.70 m/s.

a) How long does it take for the package to reach the ground?

b) With what velocity (magnitude and direction) does it hit the ground?

•**3.91** Wild geese are known for their lack of manners. One goose is flying northward at a level altitude of $h_g = 30.0$ m above a north-south highway, when it sees a car ahead in the distance moving in the southbound lane and decides to deliver (drop) an "egg." The goose is flying at a speed of $v_g = 15.0$ m/s, and the car is moving at a speed of $v_c = 100.0$ km/h.

a) Given the details in the figure, where the separation between the goose and the car's windshield, $d = 104.0$ m, is specified at the instant when the goose takes action, will the driver have to wash the windshield after this encounter? (The center of the windshield is $h_c = 1.00$ m off the ground.)

b) If the delivery is completed, what is the relative velocity of the "egg" with respect to the car at the moment of the impact?

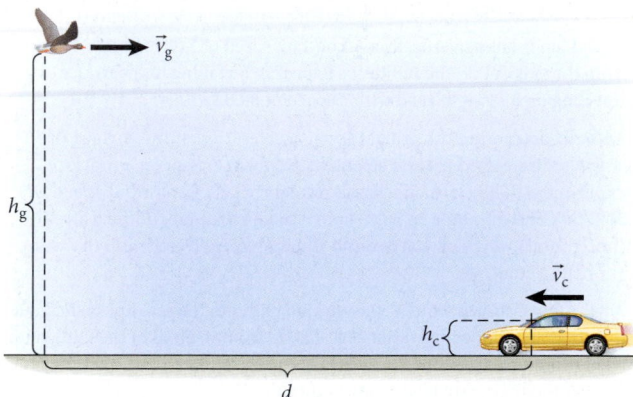

•**3.92** You are at the mall on the top step of a down escalator when you lean over laterally to see your 1.80-m-tall physics professor on the bottom step of the adjacent up escalator. Unfortunately, the ice cream you hold in your hand falls out of its cone as you lean. The two escalators have identical angles of 40.0° with the horizontal, a vertical height of 10.0 m, and move at the same speed of 0.400 m/s. Will the ice cream land on your professor's head? Explain. If it does land on his head, at what time and at what vertical height does that happen? What is the relative speed of the ice cream with respect to the head at the time of impact?

•**3.93** A basketball player practices shooting three-pointers from a distance of 7.50 m from the hoop, releasing the ball at a height of 2.00 m above the floor. A standard basketball hoop's rim top is 3.05 m above the floor. The player shoots the ball at an angle of 48.0° with the horizontal. At what initial speed must she shoot to make the basket?

•**3.94** Wanting to invite Juliet to his party, Romeo is throwing pebbles at her window with a launch angle of 37.0° from the horizontal. He is launching the pebble 7.00 m below her window and 10.0 m from the base of the wall. What is the initial speed of the pebbles?

•**3.95** An airplane flies horizontally above the flat surface of a desert at an altitude of 5.00 km and a speed of 1000. km/h. If the airplane is to drop a bomb that is supposed to hit a target on the ground, where should the plane be with respect to the target when the bomb is released? If the target covers a circular area with a diameter of 50.0 m, what is the "window of opportunity" (or margin of error allowed) for the release time?

•**3.96** A plane diving with constant speed, at an angle of 49.0° with the vertical, releases a package at an altitude of 600. m. The package hits the ground 3.50 s after release. How far horizontally does the package travel?

••**3.97** At a time of 10.0 s after being fired, a cannonball strikes a point 500. m horizontally from and 100. m vertically above the point of launch.

a) With what initial velocity was the cannonball launched?

b) What maximum height was attained by the ball?

c) What is the magnitude and direction of the ball's velocity just before it strikes the given point?

••**3.98** Neglect air resistance for the following. A soccer ball is kicked from the ground into the air. When the ball is at a height of 12.5 m, its velocity is $(5.60\hat{x} + 4.10\hat{y})$ m/s.

a) To what maximum height will the ball rise?

b) What horizontal distance will be traveled by the ball?

c) With what velocity (magnitude and direction) will it hit the ground?

MULTI-VERSION EXERCISES

3.99 For a Science Olympiad competition, a group of middle school students build a trebuchet that can fire a tennis ball from a height of 1.55 m with a velocity of 10.5 m/s and a launch angle of 35.0° above the horizontal. What horizontal distance will the tennis ball cover before it hits the ground?

3.100 For a Science Olympiad competition, a group of middle school students build a trebuchet that can fire a tennis ball from a height of 1.55 m with a velocity of 10.5 m/s and a launch angle of 35.0° above the horizontal. What is the x-component of the velocity of the tennis ball just before it hits the ground?

3.101 For a Science Olympiad competition, a group of middle school students build a trebuchet that can fire a tennis ball from a height of 1.55 m with a velocity of 10.5 m/s and a launch angle of 35.0° above the horizontal. What is the y-component of the velocity of the tennis ball just before it hits the ground?

3.102 For a Science Olympiad competition, a group of middle school students build a trebuchet that can fire a tennis ball from a height of 1.55 m with a velocity of 10.5 m/s and a launch angle of 35.0° above the horizontal. What is the speed of the tennis ball just before it hits the ground?

3.103 A pilot flies her airplane from its initial position to a position 200.0 km north of that spot. The airplane flies with a speed of 250.0 km/h with respect to the air. The wind is blowing west to east with a speed of 45.0 km/h. In what direction should the pilot steer her plane to accomplish this trip? (Express your answer in degrees noting that east is 90°, south is 180°, west is 270°, and north is 360°.)

3.104 A pilot flies her airplane from its initial position to a position 200.0 km north of that spot. The airplane flies with a speed of 250.0 km/h with respect to the air. The wind is blowing west to east with a speed of 45.0 km/h. What is the speed of the plane with respect to the ground?

3.105 A pilot flies her airplane from its initial position to a position 200.0 km north of that spot. The airplane flies with a speed of 250.0 km/h with respect to the air. The wind is blowing west to east with a speed of 45.0 km/h. How long will it take to make the trip?

4

Force

FIGURE 4.1 The Space Shuttle *Columbia* lifts off from the Kennedy Space Center.

The launch of a space shuttle was an awesome sight. Huge clouds of smoke obscured the shuttle until it rose up high enough to be seen above them, with brilliant exhaust flames exiting the main engines. The boosters provided a force of over 30 meganewtons, enough to shake the ground for miles around. This tremendous force accelerated the shuttle (whose mass was over 2 million kilograms) sufficiently for lift-off.

The space shuttle has been called one of the greatest technological achievements of the 20th century, but the basic principles of force, mass, and acceleration that govern its operation have been known for over 300 years. First stated by Isaac Newton in 1687, the laws of motion apply to all interactions between objects. Just as kinematics describes how objects move, Newton's laws of motion are the foundation of **dynamics,** which describes what makes objects move. We will study dynamics for the next several chapters.

In this chapter, we examine Newton's laws of motion and explore the various kinds of forces they describe. The process of identifying the forces that act on an object, determining the motion caused by those forces, and interpreting the overall vector result is one of the most common and important types of analysis in physics, and we will use it numerous times throughout this book. Many of the kinds of forces introduced in this chapter, such as contact forces, friction forces, and weight, will play a role in many of the concepts and principles discussed later.

WHAT WE WILL LEARN

- A force is a vector quantity that is a measure of how an object interacts with other objects.

- Fundamental forces include gravitational attraction and electromagnetic attraction and repulsion. In daily experience, important forces include tension and normal, friction, and spring forces.

- Multiple forces acting on an object sum to a net force.

- Free-body diagrams are valuable aids in working problems.

- Newton's three laws of motion govern the motion of objects under the influence of forces.

  a) The first law deals with objects for which external forces are balanced.

  b) The second law describes those cases for which external forces are not balanced.

  c) The third law addresses equal (in magnitude) and opposite (in direction) forces that two bodies exert on each other.

- The gravitational mass and the inertial mass of an object are equivalent.

- Kinetic friction opposes the motion of moving objects; static friction opposes the impending motion of objects at rest.

- Friction is important to the understanding of real-world motion, but its causes and exact mechanisms are still under investigation.

- Newton's laws of motion apply to situations involving multiple objects, multiple forces, and friction; applying the laws to analyze a situation is among the most important problem-solving techniques in physics.

4.1 Types of Forces

You are probably sitting on a chair as you are reading this page. The chair exerts a force on you, which prevents you from falling to the ground. You can feel this force from the chair on the underside of your legs and your backside. Conversely, you exert a force on the chair.

If you pull on a string, you exert a force on the string, and that string, in turn, can exert a force on something tied to its other end. This force, as well as the force you exert on a chair you're sitting in, is an example of a **contact force,** in which one object has to be in contact with another to exert a force on it. If you push or pull on an object, you exert a contact force on it. Pulling on an object, such as a rope or a string, gives rise to the contact force called **tension.** Pushing on an object causes the contact force called **compression.** The force that acts on you when you sit on a chair is called a **normal force,** where the word *normal* means "perpendicular to the surface." We will examine normal forces in more detail a little later in this chapter.

The **friction force** is another important contact force that we will study in more detail in this chapter. If you push a glass across the surface of a table, it comes to rest rather quickly. The force that causes the glass's motion to stop is the friction force, sometimes also simply called *friction*. Interestingly, the exact nature and microscopic origin of the friction force are still under intense investigation, as we'll see.

A force is needed to compress a spring as well as to extend it. The **spring force** has the special property of being linearly dependent on the change in length of the spring. Chapter 5 will introduce the spring force and describe some of its properties. Chapter 14 will focus on oscillations, a special kind of motion resulting from the action of spring forces.

Contact forces, friction forces, and spring forces are the results of the **fundamental forces** of nature acting between the constituents of objects. The **gravitational force,** often simply called *gravity,* is one example of a fundamental force. If you hold an object in your hand and let go of it, it falls downward. We know what causes this effect: the gravitational attraction between the Earth and the object. Gravitational acceleration was introduced in Chapter 2, and this chapter describes how it is related to the gravitational force. Gravity is also responsible for holding the Moon in orbit around the Earth and the Earth in orbit around the Sun. In a famous story (which may even be true!), Isaac Newton was reported to have had this insight in the 17th century, after sitting under an apple tree and being hit by an apple falling off the tree: The same type of gravitational force acts between celestial objects as operates between terrestrial objects. However, keep in mind that the gravitational force discussed in this chapter is a limited instance, valid only near the surface of Earth, of the more general gravitational force. Near the surface of Earth, a constant gravitational force acts on all objects, which is sufficient to solve practically all trajectory problems of the kind covered in Chapter 3. The more general form of the gravitational interaction, however, is inversely proportional to the square of the distance between the two objects exerting the gravitational force on each other. Chapter 12 is devoted to this force.

Another fundamental force that can act at a distance is the **electromagnetic force,** which, like the gravitational force, is inversely proportional to the square of the distance over which it acts. The most apparent manifestation of this force is the attraction or repulsion between two magnets, depending on their relative orientation. The entire Earth also acts as a huge magnet, which makes compass needles orient themselves toward the North Pole. The electromagnetic force was the big physics discovery of the 19th century, and its refinement during the 20th century led to many of the high-tech conveniences (basically everything that plugs into an electrical outlet or uses batteries) we enjoy today. Chapters 21 through 31 will provide an extensive tour of the electromagnetic force and its many manifestations.

In particular, we will see that all of the contact forces listed above (normal force, tension, friction, spring force) are fundamentally consequences of the electromagnetic force. Why then study these contact forces in the first place? The answer is that phrasing a problem in terms of contact forces gives us great insight and allows us to formulate simple solutions to real-world problems whose solutions would otherwise require the use of supercomputers if we tried to analyze them in terms of the electromagnetic interactions between atoms.

The other two fundamental forces—called the **strong nuclear force** and the **weak nuclear force**—act only on the length scales of atomic nuclei and between elementary particles. These forces between elementary particles will be discussed in Chapter 39 on particle physics and Chapter 40 on nuclear physics. In general, forces can be defined as the means for objects to influence each other (Figure 4.2).

Most of the forces mentioned here have been known for hundreds of years. However, the ways in which scientists and engineers use forces continue to evolve as new materials and new designs are created. For example, the idea of a bridge to cross a river or deep ravine has been used for thousands of years, starting with simple forms such as a log dropped across a stream or a series of ropes strung across a gorge. Over time, engineers developed the idea of an arch bridge that can support a heavy roadway and a load of traffic using compressive forces. Many such bridges were built from stone or steel, materials that can support compression well (Figure 4.3a). In the late 19th and 20th centuries, bridges were built with

(a)

(b)

(c)

FIGURE 4.2 Some common types of forces. (a) A grinding wheel works by using the force of friction to remove the outer surface of an object. (b) Springs are often used as shock absorbers in cars to reduce the force transmitted to the wheels by the ground. (c) Some dams are among the largest structures ever built. They are designed to resist the force exerted by the water they hold back.

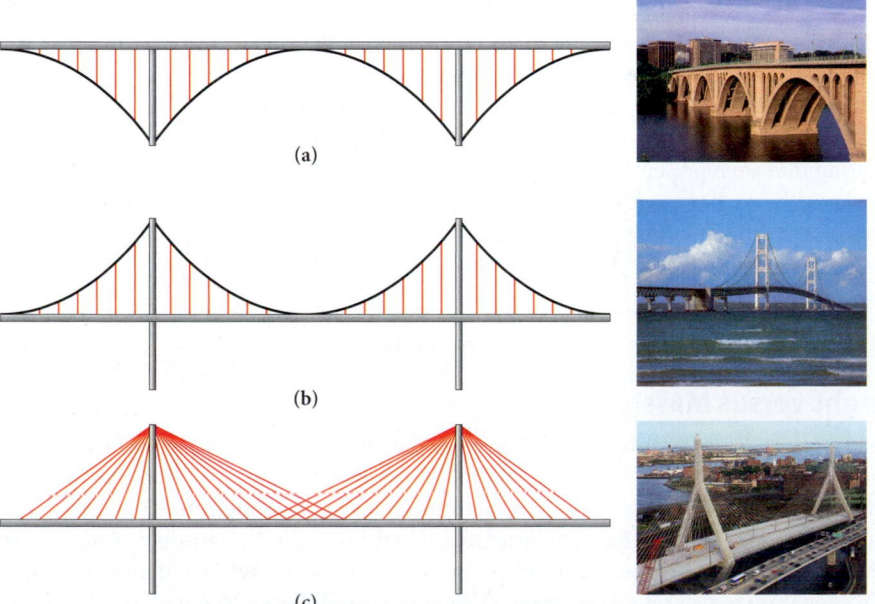

(a)

(b)

(c)

FIGURE 4.3 Different ways to use forces. (a) Arch bridges (such as Francis Scott Key Bridge in Washington, DC) support a road by compressive forces, with each end of the arch anchored in place. (b) Suspension bridges (such as Mackinac Bridge in Michigan) support the roadway by tension forces in the cables, which are in turn supported by compressive forces in the tall piers sunk into the ground below the water. (c) Cable-stayed bridges (such as Zakim Bridge in Boston) also use tension forces in cables to support the roadway, but the load is distributed over many more cables, which do not need to be as strong as in suspension bridges.

the roadway suspended from steel cables supported by tall piers (Figure 4.3b). The cables supported tension, and these bridges could be lighter as well as longer than previous bridge designs. In the late 20th century, cable-stayed bridges began to appear, with the roadway supported by cables attached directly to the piers (Figure 4.3c). These bridges are not generally as long as suspension bridges but are less expensive and time-consuming to build.

4.2   Gravitational Force Vector, Weight, and Mass

Having had a general introduction to forces, it is time to get more quantitative. Let's start with an obvious fact: Forces have a direction. For example, if you are holding a laptop computer in your hand, you can easily tell that the gravitational force acting on the computer points downward. This direction is the direction of the **gravitational force vector** (Figure 4.4). Again, to characterize a quantity as a vector quantity throughout this book, a small right-pointing arrow appears above the symbol for the quantity. Thus, the gravitational force vector acting on the laptop is denoted by $\vec{F}_g$ in the figure.

Figure 4.4 also shows a convenient Cartesian coordinate system, which follows the convention introduced in Chapter 3 but has been rotated so that up is the positive y-direction (and down the negative y-direction). The x- and z-directions then lie in the horizontal plane, as shown. As always, we use a right-handed coordinate system. Also, we restrict ourselves to two-dimensional coordinate systems with x- and y-axes wherever possible.

In the coordinate system in Figure 4.4, the force vector of the gravitational force acting on the laptop is pointing in the negative y-direction:

$$\vec{F}_g = -F_g\hat{y}. \tag{4.1}$$

Here we see that the force vector is the product of its magnitude, F_g, and its direction, $-\hat{y}$. The magnitude F_g is called the **weight** of the object.

Near the surface of the Earth (within a few hundred meters above the ground), the magnitude of the gravitational force acting on an object is given by the product of the mass of the object, m, and the Earth's gravitational acceleration, g:

$$F_g = mg. \tag{4.2}$$

We have used the magnitude of the Earth's gravitational acceleration in previous chapters: It has the value $g = 9.81 \text{ m/s}^2$. Note that this constant value is valid only up to a few hundred meters above the ground, as we will see in Chapter 12.

With equation 4.2, we find that the unit of force is the product of the unit of mass (kg) and the unit of acceleration (m/s²), which makes the unit of force kg m/s². (Perhaps it is worth repeating that we represent units with roman letters and physical quantities with italic letters. Thus, m is the unit of length; m stands for the physical quantity of mass.) Since dealing with forces is so common in physics, the unit of force has received its own name, the newton (N), after Sir Isaac Newton, the British physicist who made a key contribution to the analysis of forces.

$$1 \text{ N} \equiv 1 \text{ kg m/s}^2. \tag{4.3}$$

Weight versus Mass

Before discussing forces in greater detail, we need to clarify the concept of mass. Under the influence of gravity, an object has a weight that is proportional to its **mass,** which is (intuitively) the amount of matter in the object. This weight is the magnitude of a force that acts on an object due to its gravitational interaction with the Earth (or another object). Near the surface of Earth, the magnitude of this force is $F_g = mg$, as stated in equation 4.2. The mass in this equation is also called the **gravitational mass** to indicate that it is responsible for the gravitational interaction. However, mass has a role in dynamics as well.

Newton's laws of motion, which will be introduced later in this chapter, deal with inertial mass. To understand the concept of inertial mass, consider the following examples: It is a lot easier to throw a tennis ball than a shot put. It is also easier to pull open a door made of lightweight materials like foam-core with wood veneer than one made of a heavy material

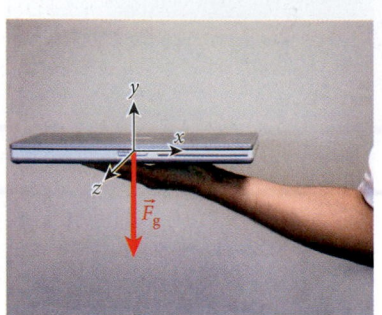

FIGURE 4.4 Force vector of gravity acting on a laptop computer, in relation to the conventional right-handed Cartesian coordinate system.

like iron. The more massive objects seem to resist being put into motion more than the less massive ones do. This property of an object is referred to as its **inertial mass.** However, the gravitational mass and the inertial mass are identical, so most of the time we refer simply to the mass of an object.

For a laptop computer with mass $m = 3.00$ kg, for example, the magnitude of the gravitational force is $F_g = mg = (3.00 \text{ kg})(9.81 \text{ m/s}^2) = 29.4$ kg m/s^2 = 29.4 N. Now we can write an equation for the force vector that contains both the magnitude and the direction of the gravitational force acting on the laptop computer (see Figure 4.4):

$$\vec{F}_g = -mg\hat{y}. \tag{4.4}$$

To summarize, the mass of an object is measured in kilograms and the weight of an object is measured in newtons. The mass and the weight of an object are related to each other by multiplying the mass (in kilograms) by the gravitational acceleration constant, $g = 9.81$ m/s^2, to arrive at the weight (in newtons). For example, if your mass is 70.0 kg, then your weight is 687 N.

Orders of Magnitude of Forces

The concept of a force is a central theme of this book, and we will return to it again and again. For this reason, it is instructive to look at the orders of magnitude that different forces can have. Figure 4.5 gives an overview of magnitudes of some typical forces with the aid of a logarithmic scale, similar to those used in Chapter 1 for length and mass.

A human's body weight is in the range between 100 and 1000 N and is represented by the young soccer player in Figure 4.5. The ear to the player's left in Figure 4.5 symbolizes the force exerted by sound on our eardrums, which can be as large as 10^{-4} N, but is still detectable when it is as small as 10^{-13} N. (Chapter 16 will focus on sound.) A single electron is kept in orbit around a proton by an electrostatic force of approximately 10^{-7} N, which will be introduced in Chapter 21 on electrostatics. Forces as small as 10^{-15} N ≡ 1 fN can be measured in the lab; these forces are typical of those needed to stretch the double-helical DNA molecule.

The Earth's atmosphere exerts quite a sizable force on our bodies, on the order of 10^5 N, which is approximately 100 times the average body weight. Chapter 13 on solids and fluids will expand on this topic and will also show how to calculate the force of water on a dam. For example, the Hoover Dam (shown in Figure 4.5) has to withstand a force close to 10^{11} N, a huge force, more than 30 times the weight of the Empire State Building. But this force pales, of course, in comparison to the gravitational force that the Sun exerts on Earth, which is $3.5 \cdot 10^{22}$ N. (Chapter 12 on gravity will describe how to calculate this force.)

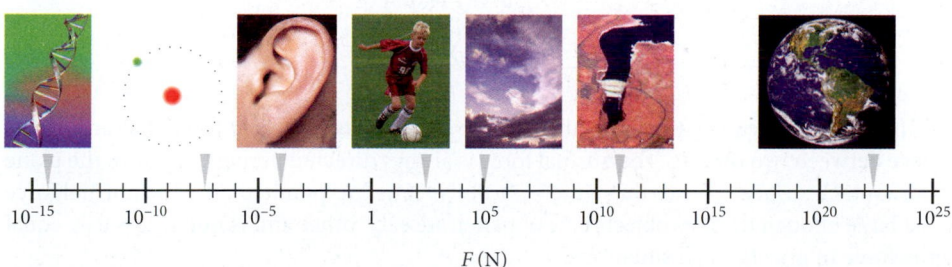

FIGURE 4.5 Typical magnitudes for different forces.

Higgs Particle

As far as our studies are concerned, mass is an intrinsic, or given, property of an object. The origin of mass is still under intense study in nuclear and particle physics. Different elementary particles have been observed to vary widely in mass—in fact, some are several thousand times more massive than others. Why? We don't really know. In recent years, particle physicists have theorized that the so-called **Higgs particle** (named after Scottish physicist Peter Higgs, who first proposed it) may be responsible for imparting mass to all other particles, with the mass of a particular type of particle depending on how it interacts with the Higgs particle. On July 4,

Concept Check 4.1

Consider two steel balls. Ball 1 has a weight of 1.5 N. Ball 2 has a weight of 15 N. Both balls are released simultaneously from a height of 2.0 m. Which ball hits the ground first?

a)　Ball 1 hits the ground first.

b)　Ball 2 hits the ground first.

c)　Both balls hit the ground at the same time.

d)　It can't be determined from the information provided.

2012, the ATLAS and CMS Collaborations at the Large Hadron Collider at CERN in Geneva, Switzerland, announced the discovery of the Higgs particle, which was the last undiscovered particle predicted by the standard model of particle physics.

4.3 Net Force

Because forces are vectors, we must add them as vectors, using the methods developed in Chapter 1. We define the **net force** as the vector sum of all force vectors that act on an object:

$$\vec{F}_{net} = \sum_{i=1}^{n} \vec{F}_i = \vec{F}_1 + \vec{F}_2 + \cdots + \vec{F}_n. \tag{4.5}$$

Following the rules for the addition of vectors using components, we can write the Cartesian components of the net force:

$$F_{net,x} = \sum_{i=1}^{n} F_{i,x} = F_{1,x} + F_{2,x} + \cdots + F_{n,x}$$

$$F_{net,y} = \sum_{i=1}^{n} F_{i,y} = F_{1,y} + F_{2,y} + \cdots + F_{n,y} \tag{4.6}$$

$$F_{net,z} = \sum_{i=1}^{n} F_{i,z} = F_{1,z} + F_{2,z} + \cdots + F_{n,z}.$$

To explore the concept of the net force, let's return again to the example of the laptop held up by a hand.

Normal Force

So far we have only looked at the gravitational force acting on the laptop computer. However, other forces are also acting on it. What are they?

In Figure 4.6, the force exerted on the laptop computer by the hand is represented by the yellow arrow labeled $\vec{N}$. (Careful—the magnitude of the normal force is represented by the italic letter N, whereas the force unit, the newton, is represented by the roman letter N.) Note in the figure that the magnitude of the vector $\vec{N}$ is exactly equal to that of the vector $\vec{F}_g$ and that the two vectors point in opposite directions, or $\vec{N} = -\vec{F}_g$. This situation is not an accident. We will see shortly that there is no net force on an object at rest. If we calculate the net force acting on the laptop computer, we obtain

$$\vec{F}_{net} = \sum_{i=1}^{n} \vec{F}_i = \vec{F}_g + \vec{N} = \vec{F}_g - \vec{F}_g = 0.$$

In general, we can characterize the *normal force, $\vec{N}$,* as a contact force that acts at the surface between two objects. The normal force is always directed perpendicular to the plane of the contact surface. (Hence the name—*normal* means "perpendicular.") The normal force is just large enough that the objects do not penetrate each other and is not necessarily equal to the force of gravity in all situations.

For the hand holding the laptop computer, the contact surface between the hand and the computer is the bottom surface of the computer, which is aligned with the horizontal plane. By definition, the normal force has to point perpendicular to this plane, or vertically upward in this case.

Free-body diagrams greatly ease the task of determining net forces on objects.

Free-Body Diagrams

We have represented the entire effect that the hand has in holding up the laptop computer by the force vector $\vec{N}$. We do not need to consider the influence of the arm, the person to whom the arm belongs, or the entire rest of the world when we want to consider the forces acting on the laptop computer. We can just eliminate them from our consideration,

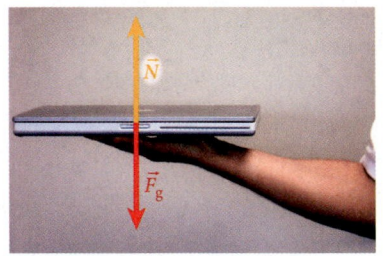

FIGURE 4.6 Force of gravity acting downward and normal force acting upward exerted by the hand holding the laptop computer.

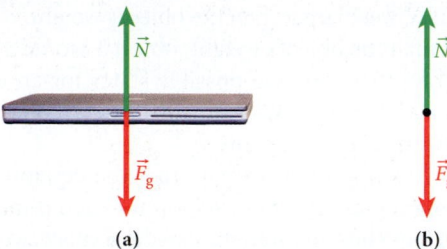

FIGURE 4.7 (a) Forces acting on a real object, a laptop computer; (b) abstraction of the object as a free body being acted on by two forces.

as illustrated in Figure 4.7a, where everything but the laptop computer and the two force vectors has been removed. For that matter, a realistic representation of the laptop computer is not necessary either; it can be shown as a dot, as in Figure 4.7b. This type of drawing of an object, in which all connections to the rest of the world are ignored and only the force vectors that act on it are drawn, is called a **free-body diagram**.

Self-Test Opportunity 4.1

Draw the free-body diagrams for a golf ball resting on a tee, your car parked on the street, and you sitting on a chair.

4.4 Newton's Laws

So far this chapter has introduced several types of forces without really explaining how they work and how we can deal with them. The key to working with forces involves understanding Newton's laws. We discuss these laws in this section and then present several examples showing how they apply to practical situations.

Sir Isaac Newton (1642–1727) was perhaps the most influential scientist who ever lived. He is generally credited with being the founder of modern mechanics, as well as calculus (along with the German mathematician Gottfried Leibniz). The first few chapters of this book are basically about Newtonian mechanics. Although he formulated his three famous laws in the 17th century, these laws are still the foundation of our understanding of forces. To begin this discussion, we simply list Newton's three laws, published in 1687.

Newton's First Law:

If the net force on an object is equal to zero, the object will remain at rest if it was at rest. If it was moving, it will remain in motion in a straight line with the same constant velocity.

Newton's Second Law:

If a net external force, $\vec{F}_{net}$, acts on an object with mass m, the force will cause an acceleration, $\vec{a}$, in the same direction as the force:

$$\vec{F}_{net} = m\vec{a}.$$

Newton's Third Law:

The forces that two interacting objects exert on each other are always exactly equal in magnitude and opposite in direction:

$$\vec{F}_{1\to 2} = -\vec{F}_{2\to 1}.$$

Newton's First Law

The earlier discussion of net force mentioned that zero net external force is the necessary condition for an object to be at rest. We can use this condition to find the magnitude and direction of any unknown forces in a problem. That is, if we know that an object is at rest and we know its weight force, then we can use the condition $\vec{F}_{net} = 0$ to solve for other forces acting on the object. This kind of analysis led to the magnitude and direction of the force $\vec{N}$ in the example of the laptop computer being held at rest.

We can use this way of thinking as a general principle: If object 1 rests on object 2, then the normal force $\vec{N}$ equal to the object's weight keeps object 1 at rest, and therefore the net force on object 1 is zero. If $\vec{N}$ were larger than the object's weight, object 1 would lift off into the air. If $\vec{N}$ were smaller than the object's weight, object 1 would sink into object 2.

Newton's First Law says there are two possible states for an object with no net force on it: An object at rest is said to be in **static equilibrium.** An object moving with constant velocity is said to be in **dynamic equilibrium.**

Before we move on, it is important to state that the equation $\vec{F}_{net} = 0$ as a condition for static equilibrium really represents one equation for each dimension of the coordinate space that we are considering. Thus, in three-dimensional space, we have three independent equilibrium conditions:

$$F_{net,x} = \sum_{i=1}^{n} F_{i,x} = F_{1,x} + F_{2,x} + \cdots + F_{n,x} = 0$$

$$F_{net,y} = \sum_{i=1}^{n} F_{i,y} = F_{1,y} + F_{2,y} + \cdots + F_{n,y} = 0$$

$$F_{net,z} = \sum_{i=1}^{n} F_{i,z} = F_{1,z} + F_{2,z} + \cdots + F_{n,z} = 0.$$

However, Newton's First Law also addresses the case when an object is already in motion with respect to some particular reference frame. For this case, the law specifies that the acceleration is zero, provided the net external force is zero. Newton's abstraction claims something that seemed at the time in conflict with everyday experience. Today, however, we have the benefit of having seen videos or movies showing objects floating in a spaceship, moving with unchanged velocities until an astronaut pushes them and thus exerts a force on them. This visual experience is in complete accord with what Newton's First Law claims, but no one in Newton's time had seen it.

Consider a car that is out of gas and needs to be pushed to the nearest gas station on a horizontal street. As long as you push the car, you can make it move. However, as soon as you stop pushing, the car slows down and comes to a stop. It seems that as long as you push the car, it moves at constant velocity, but as soon as you stop exerting a force on it, it stops moving. This idea that a constant force is required to move something with a constant speed was the Aristotelian view, which originated from the ancient Greek philosopher Aristotle (384–322 BC) and his students. Galileo (1564–1642) proposed a law of inertia and theorized that moving objects slowed down because of friction. Newton's First Law builds on this law of inertia.

What about the car that slows down once you stop pushing? This situation is not a case of zero net force. Instead, a force *is* acting on the car to slow it down—the force of friction. Because the friction force acts as a nonzero net force, the example of the car slowing down turns out to be an example not of Newton's First Law, but of Newton's Second Law. We will work more with friction later in this chapter.

Newton's First Law is sometimes also called the *law of inertia.* Inertial mass was defined earlier (Section 4.2), and the definition implied that inertia is an object's resistance to a change in its motion. This is exactly what Newton's First Law says: To change an object's motion, you need to apply an external net force—the motion won't change by itself, neither in magnitude nor in direction.

Newton's Second Law

The second law relates the concept of acceleration, for which we use the symbol $\vec{a}$, to the force. We have already considered acceleration as the time derivative of the velocity and the second time derivative of the position. Newton's Second Law tells us what causes acceleration.

Newton's Second Law:

If a net external force, $\vec{F}_{net}$, acts on an object with mass m, the force will cause an acceleration, $\vec{a}$, in the same direction as the force:

$$\vec{F}_{net} = m\vec{a}. \tag{4.7}$$

This formula, $F = ma$, is arguably the second-most famous equation in all of physics. (We will encounter the most famous, $E = mc^2$, later in the book.) Equation 4.7 tells us that the magnitude of the acceleration of an object is proportional to the magnitude of the net external force acting on it. It also tells us that for a given external force, the magnitude of the acceleration is inversely proportional to the mass of the object. All things being equal, more massive objects are harder to accelerate than less massive ones.

However, equation 4.7 tells us even more, because it is a vector equation. It says that the acceleration vector experienced by the object with mass m is in the same direction as the net external force vector that is acting on the object to cause this acceleration. Because it is a vector equation, we can immediately write the equations for the three spatial components:

$$F_{\text{net},x} = ma_x, \qquad F_{\text{net},y} = ma_y, \qquad F_{\text{net},z} = ma_z.$$

This result means that $F = ma$ holds independently for each Cartesian component of the force and acceleration vectors.

Newton's Third Law

If you have ever ridden a skateboard, you must have made the following observation: If you are standing at rest on the skateboard, and you step off over the front or back, the skateboard shoots off into the opposite direction. In the process of stepping off, the skateboard exerts a force on your foot, and your foot exerts a force onto the skateboard. This experience seems to suggest that these forces point in opposite directions, and it provides one example of a general truth, quantified in Newton's Third Law.

Newton's Third Law:

The forces that two interacting objects exert on each other are always exactly equal in magnitude and opposite in direction:

$$\vec{F}_{1 \to 2} = -\vec{F}_{2 \to 1}. \tag{4.8}$$

Note that these two forces do not act on the same body but are the forces with which two bodies act on each other.

Newton's Third Law seems to present a paradox. For example, if a horse pulls forward on a wagon with the same force with which the wagon pulls backward on the horse, then how do horse and wagon move anywhere? The answer is that these forces act on different objects in the system. The wagon experiences the pull from the horse and moves forward. The horse feels the pull from the wagon and pushes hard enough against the ground to overcome this force and move forward. A free-body diagram of an object can show only half of such an action-reaction pair of forces.

Newton's Third Law is a consequence of the requirement that internal forces—that is, forces that act between different components of the same system—must add to zero; otherwise, their sum would contribute to a net external force and cause an acceleration, according to Newton's Second Law. No object or group of objects can accelerate itself or themselves without interacting with external objects. The story of Baron Münchhausen, who claimed to have pulled himself out of a swamp by simply pulling very hard on his own hair, is unmasked by Newton's Third Law as complete fiction.

We'll consider some examples of the use of Newton's laws to solve problems, after we discuss how ropes and pulleys convey forces. Many problems involving Newton's laws involve forces on a rope (or a string), often one wrapped around a pulley.

4.5 Ropes and Pulleys

Problems that involve ropes and pulleys are very common. In this chapter, we consider only massless (idealized) ropes and pulleys. Whenever a rope is involved, the direction of the force due to the pull on the rope acts exactly in the direction along the rope. The force with which we pull on the massless rope is transmitted through the entire rope unchanged. The magnitude of this force is referred to as the *tension* in the rope. Every rope can withstand only a certain maximum force, but for now we will assume that all applied forces are below this limit. Ropes cannot support a *compression* force.

FIGURE 4.8 A rope passing over a pulley with force measurement devices attached, showing that the magnitude of the force is constant throughout the rope.

If a rope is guided over a pulley, the direction of the force is changed, but the magnitude of the force is still the same everywhere inside the rope. In Figure 4.8, the right end of the green rope was tied down and someone pulled on the other end with a certain force, 11.5 N, as indicated by the inserted force measurement devices. As you can clearly see, the magnitude of the force on both sides of the pulley is the same. (The weight of the force measurement devices themselves is a small real-world complication, but enough pulling force was used that it is reasonably safe to neglect this effect.)

EXAMPLE 4.1 | Modified Tug-of-War

In a tug-of-war competition, two teams try to pull each other across a line. If neither team is moving, then the two teams exert equal and opposite forces on a rope. This is an immediate consequence of Newton's Third Law. That is, if the team shown in Figure 4.9 pulls on the rope with a force of magnitude F, the other team necessarily has to pull on the rope with a force of the same magnitude but in the opposite direction.

FIGURE 4.9 Men competing in a tug-of-war competition at Braemar Games Highland Gathering, Scotland, UK.

PROBLEM

Now let's consider the situation where three ropes are tied together at one point, with a team pulling on each rope. Suppose team 1 is pulling due west with a force of 2750 N, and team 2 is pulling due north with a force of 3630 N. Can a third team pull in such a way that the three-team tug-of-war ends at a standstill, that is, no team is able to move the rope? If yes, what is the magnitude and direction of the force needed to accomplish this?

SOLUTION

The answer to the first question is yes, no matter with what force and in what direction teams 1 and 2 pull. This is the case because the two forces will always add up to a combined force, and all that team 3 has to do is pull with a force equal to and opposite in direction to that combined force. Then all three forces will add up to zero, and by Newton's First Law, the system has achieved static equilibrium. Nothing will accelerate, so if the ropes start at rest, nothing will move.

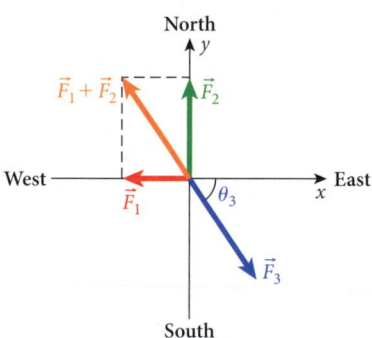

FIGURE 4.10 Addition of force vectors in the three-team tug-of-war.

Figure 4.10 represents this physical situation. The vector addition of forces exerted by teams 1 and 2 is particularly simple, because the two forces are perpendicular to each other. We choose a conventional coordinate system with its origin at the point where all the ropes meet, and we designate north to be in the positive y-direction and west to be in the negative x-direction. Thus, the force vector for team 1, $\vec{F}_1$, points in the negative x-direction and the force vector for team 2, $\vec{F}_2$, points in the positive y-direction. We can now write the two force vectors and their sum as follows:

$$\vec{F}_1 = -(2750 \text{ N})\hat{x}$$
$$\vec{F}_2 = (3630 \text{ N})\hat{y}$$
$$\vec{F}_1 + \vec{F}_2 = -(2750 \text{ N})\hat{x} + (3630 \text{ N})\hat{y}.$$

The addition was made easier by the fact that the two forces pointed along the chosen coordinate axes. However, more general cases of two forces can be added in terms of their components. Because the sum of all three forces has to be zero for a standstill, we obtain the force that the third team has to exert:

$$0 = \vec{F}_1 + \vec{F}_2 + \vec{F}_3$$
$$\Leftrightarrow \vec{F}_3 = -(\vec{F}_1 + \vec{F}_2)$$
$$= (2750 \text{ N})\hat{x} - (3630 \text{ N})\hat{y}.$$

This force vector is also shown in Figure 4.10. Having the Cartesian components of the force vector we were looking for, we can get the magnitude and direction by using trigonometry:

$$F_3 = \sqrt{F_{3,x}^2 + F_{3,y}^2} = \sqrt{(2750 \text{ N})^2 + (-3630 \text{ N})^2} = 4554 \text{ N}$$

$$\theta_3 = \tan^{-1}\left(\frac{F_{3,y}}{F_{3,x}}\right) = \tan^{-1}\left(\frac{-3630 \text{ N}}{2750 \text{ N}}\right) = -52.9°.$$

These results complete our answer.

Because this type of problem occurs frequently, let's work through another example.

EXAMPLE 4.2 / Still Rings

A gymnast of mass 55 kg hangs vertically from a pair of parallel rings (Figure 4.11a).

PROBLEM 1

If the ropes supporting the rings are vertical and attached to the ceiling directly above, what is the tension in each rope?

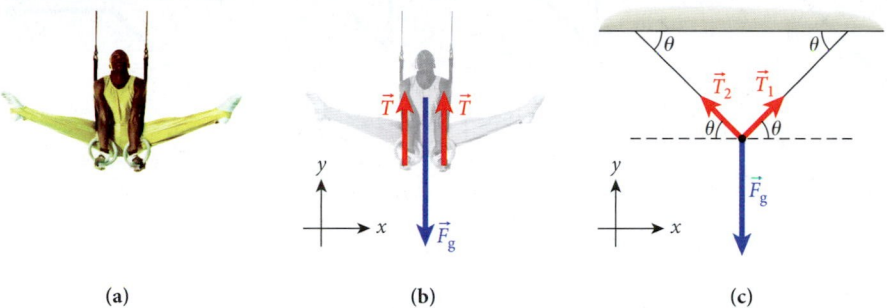

(a) (b) (c)

FIGURE 4.11 (a) Still rings in men's gymnastics. (b) Free-body diagram for problem 1. (c) Free-body diagram for problem 2.

SOLUTION 1

In this example, we define the x-direction to be horizontal and the y-direction to be vertical. The free-body diagram is shown Figure 4.11b. For now, there are no forces in the x-direction. In the y-direction, we have $\sum_i F_{y,i} = T_1 + T_2 - mg = 0$. Because both ropes support the gymnast equally, the tension has to be the same in both ropes, $T_1 = T_2 \equiv T$, and we get

$$T + T - mg = 0$$
$$\Rightarrow T = \tfrac{1}{2}mg = \tfrac{1}{2}(55 \text{ kg})(9.81 \text{ m/s}^2) = 270 \text{ N}.$$

PROBLEM 2

If the ropes are attached so that they make an angle $\theta = 45°$ with the ceiling (Figure 4.11c), what is the tension in each rope?

SOLUTION 2

In this part, forces do occur in both x- and y-directions. We will work in terms of a general angle and then plug in the specific angle, $\theta = 45°$, at the end.
In the x-direction, we have for our equilibrium condition:

$$\sum_i F_{x,i} = T_1 \cos\theta - T_2 \cos\theta = 0.$$

In the y-direction, our equilibrium condition is

$$\sum_i F_{y,i} = T_1 \sin\theta + T_2 \sin\theta - mg = 0.$$

From the equation for the x-direction, we again get $T_1 = T_2 \equiv T$, and from the equation for the y-direction, we then obtain:

$$2T \sin\theta - mg = 0 \Rightarrow T = \frac{mg}{2\sin\theta}.$$

Putting in the numbers, we obtain the tension in each rope:

$$T = \frac{(55 \text{ kg})(9.81 \text{ m/s}^2)}{2\sin 45°} = 382 \text{ N}.$$

PROBLEM 3

How does the tension in the ropes change as the angle θ between the ceiling and the ropes becomes smaller and smaller?

SOLUTION 3

As the angle θ between the ceiling and the ropes becomes smaller, the tension in the ropes, $T = mg/2\sin\theta$, gets larger. As θ approaches zero, the tension becomes infinitely large. In reality, of course, the gymnast has only finite strength and cannot hold his position with small angles.

Concept Check 4.2

Choose the set of three coplanar vectors that sum to a net force of zero: $\vec{F}_1 + \vec{F}_2 + \vec{F}_3 = 0$.

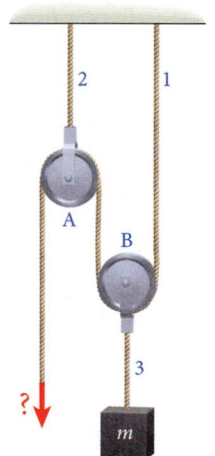

FIGURE 4.12 Rope guided over two pulleys.

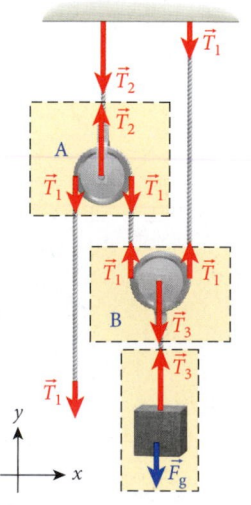

FIGURE 4.13 Free-body diagrams for the two pulleys and the mass to be lifted.

Force Multiplier

Ropes and pulleys can be combined to lift objects that are too heavy to lift otherwise. In order to see how this can be done, consider Figure 4.12. The system shown consists of rope 1, which is tied to the ceiling (upper right) and then guided over pulleys B and A. Pulley A is also tied to the ceiling with rope 2. Pulley B is free to move vertically and is attached to rope 3. The object of mass m, which we want to lift, is hanging from the other end of rope 3. We assume that the two pulleys have negligible mass and that rope 1 can glide across the pulleys without friction.

What force do we need to apply to the free end of rope 1 to keep the system in static equilibrium? We will call the tension force in rope 1, $\vec{T}_1$, that in rope 2, $\vec{T}_2$, and that in rope 3, $\vec{T}_3$. Again, the key idea is that the magnitude of this tension force is the same everywhere in a given rope.

Figure 4.13 again shows the system in Figure 4.12, but with dashed lines and shaded areas, indicating the free-body diagrams of the two pulleys and the object of mass m. We start with the mass m. For the condition of zero net force to be fulfilled, we need

$$\vec{T}_3 + \vec{F}_g = 0$$

or

$$F_g = mg = T_3.$$

From the free-body diagram of pulley B, we see that the tension force applied from rope 1 acts on both sides of pulley B. This tension has to balance the tension from rope 3, giving us

$$2T_1 = T_3.$$

Combining the last two equations, we see that

$$T_1 = \tfrac{1}{2}mg.$$

This result means that the force we need to apply to suspend the object of mass m in this way is only half as large as the force we would have to use to simply hold it up with a rope, without pulleys. This change in force is why a pulley is called a *force multiplier*.

Even greater force multiplication is achieved if rope 1 passes a total of n times over the same two pulleys. In this case, the force needed to suspend the object of mass m is

$$T = \frac{1}{2n}mg. \tag{4.9}$$

Figure 4.14 shows the situation for the lower pulley in Figure 4.13 with $n = 3$. This arrangement results in $2n = 6$ force arrows of magnitude T pointing up, able to balance a downward force of magnitude $6T$, as expressed by equation 4.9.

Concept Check 4.3

Using a pulley pair with two loops, we can lift a weight of 440 N. If we add two loops to the pulley, with the same force, we can lift

a) half the weight.

b) twice the weight.

c) one-fourth the weight.

d) four times the weight.

e) the same weight.

FIGURE 4.14 Pulley with three loops.

4.6 Applying Newton's Laws

Now let's look at how Newton's laws allow us to solve various kinds of problems involving force, mass, and acceleration. We will make frequent use of free-body diagrams and will assume massless ropes and pulleys. We will also neglect friction for now but will consider it in Section 4.7.

EXAMPLE 4.3 | Two Books on a Table

We have considered the simple situation of one object (a laptop computer) supported from below and held at rest. Now let's look at two objects at rest: two books on a table (Figure 4.15a).

PROBLEM

What is the magnitude of the force that the table exerts on the lower book?

SOLUTION

We start with a free-body diagram of the book on top, book 1 (Figure 4.15b). This situation is the same as that of the laptop computer held steady by the hand. The gravitational force due to the Earth's attraction that acts on the upper book is indicated by $\vec{F}_1$. It has the magnitude m_1g, where m_1 is the mass of the upper book, and it points straight down. The magnitude of the normal force, $\vec{N}_1$, that the lower book exerts on the upper book from below is then $N_1 = F_1 = m_1g$, from the condition of zero net force on the upper book (Newton's First Law). The force $\vec{N}_1$ points straight up, as shown in the free-body diagram, $\vec{N}_1 = -\vec{F}_1$.

Newton's Third Law now allows us to calculate the force that the upper book exerts on the lower book. This force is equal in magnitude and opposite in direction to the force that the lower book exerts on the upper one:

$$\vec{F}_{1\to2} = -\vec{N}_1 = -(-\vec{F}_1) = \vec{F}_1.$$

This relationship says that the force that the upper book exerts on the lower one is exactly equal to the gravitational force acting on the upper book—that is, its weight. You may find this result trivial at this point, but the application of this general principle allows us to analyze and do calculations for complicated situations.

Now consider the free-body diagram of the lower book, book 2 (Figure 4.15c). This free-body diagram allows us to calculate the normal force that the table exerts on the lower book. We sum up all the forces acting on this book:

$$\vec{F}_{1\to2} + \vec{N}_2 + \vec{F}_2 = 0 \Rightarrow \vec{N}_2 = -(\vec{F}_{1\to2} + \vec{F}_2) = -(\vec{F}_1 + \vec{F}_2),$$

where $\vec{N}_2$ is the normal force exerted by the table on the lower book, $\vec{F}_{1\to2}$ is the force exerted by the upper book on the lower book, and $\vec{F}_2$ is the gravitational force on the lower book. In the last step, we used the result we obtained from the free-body diagram of book 1. This result means that the force that the table exerts on the lower book is exactly equal in magnitude and opposite in direction to the sum of the weights of the two books.

(a)

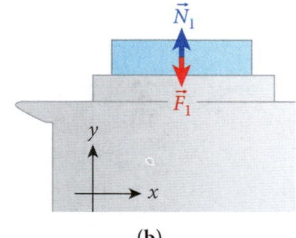

(b)

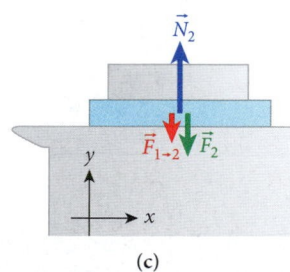

(c)

FIGURE 4.15 (a) Two books on top of a table. (b) Free-body diagram for book 1. (c) Free-body diagram for book 2.

The use of Newton's Second Law enables us to perform a wide range of calculations involving motion and acceleration. The following problem is a classic example: Consider an object of mass m located on a plane that is inclined at an angle θ relative to the horizontal. Assume that there is no friction force between the plane and the object. What can Newton's Second Law tell us about this situation?

SOLVED PROBLEM 4.1 | Snowboarding

PROBLEM

A snowboarder (mass 72.9 kg, height 1.79 m) glides down a slope with an angle of 22° with respect to the horizontal (Figure 4.16a). If we can neglect friction, what is his acceleration?

SOLUTION

THINK The motion is restricted to motion along the slope (or inclined plane) because the snowboarder cannot sink into the snow, and he cannot lift off from its surface. (At least not without jumping!) It is always advisable to start with a free-body diagram. Figure 4.16b shows the force vectors for gravity, $\vec{F}_g$, and the normal force, $\vec{N}$. Note that the normal force vector is directed perpendicular to the contact surface, as required by the definition of the normal force. Also observe that the normal force and the force of gravity do not point in exactly opposite directions and thus do not cancel each other out completely.

(a)

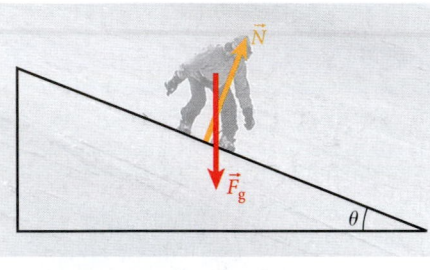

(b)

SKETCH Now we pick a convenient coordinate system. As shown in Figure 4.16c, we choose a coordinate system with the x-axis along the direction of the inclined plane. This choice ensures that the acceleration is only in the x-direction. Another advantage of this choice of coordinate system is that the normal force is pointing exactly in the y-direction. The price we pay for this convenience is that the gravitational force vector does not point along one of the major axes in our coordinate system but has an x- and a y-component. The red arrows in the figure indicate the two components of the gravitational force vector. Note that the angle of inclination of the plane, θ, also appears in the rectangle constructed from the two components of the gravity force vector, which is the diagonal of that rectangle. You can see this relationship by considering the similar triangles with sides abc and ABC in Figure 4.16d. Because a is perpendicular to C and c is perpendicular to A, it follows that the angle between a and c is the same as the angle between A and C.

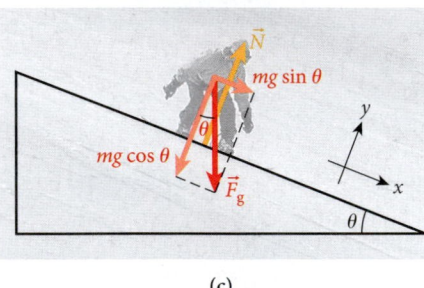

(c)

RESEARCH The x- and y-components of the gravitational force vector are found from trigonometry:

$$F_{g,x} = F_g \sin\theta = mg \sin\theta$$
$$F_{g,y} = -F_g \cos\theta = -mg \cos\theta.$$

SIMPLIFY Now we do the math in a straightforward way, separating calculations by components.

First, there is no motion in the y-direction, which means that, according to Newton's First Law, all the external force components in the y-direction have to add up to zero:

$$F_{g,y} + N = 0 \Rightarrow$$
$$-mg \cos\theta + N = 0 \Rightarrow$$
$$N = mg \cos\theta.$$

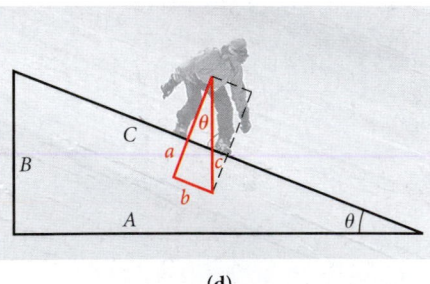

(d)

FIGURE 4.16 (a) Snowboarding as an example of motion on an inclined plane. (b) Free-body diagram of the snowboarder on the inclined plane. (c) Free-body diagram of the snowboarder, with a coordinate system added. (d) Similar triangles in the inclined-plane problem.

Our analysis of the motion in the y-direction has given us the magnitude of the normal force, which balances the component of the weight of the snowboarder perpendicular to the slope. This is a very typical result. The normal force almost always balances the net force perpendicular to the contact surface that is contributed by all other forces. Thus, objects do not sink into or lift off from surfaces.

The information we are interested in comes from looking at the x-direction. In this direction, there is only one force component, the x-component of the gravitational force. Therefore, according to Newton's Second Law, we obtain

$$F_{g,x} = mg \sin\theta = ma_x \Rightarrow$$
$$a_x = g \sin\theta.$$

Thus, we now have the acceleration vector in the specified coordinate system:

$$\vec{a} = (g \sin\theta)\hat{x}.$$

Note that the mass, m, dropped out of our answer. The acceleration does not depend on the mass of the snowboarder; it depends only on the angle of inclination of the plane. Thus, the mass of the snowboarder given in the problem statement turns out to be just as irrelevant as his height.

CALCULATE Putting in the given value for the angle leads to

$$a_x = (9.81 \text{ m/s}^2)(\sin 22°) = 3.67489 \text{ m/s}^2.$$

ROUND Because the angle of the slope was given to only two-digit accuracy, it makes no sense to give our result to a greater precision. The final answer is

$$a_x = 3.7 \text{ m/s}^2.$$

DOUBLE-CHECK The units of our answer, m/s², are those of acceleration. The number we obtained is positive, which means a positive acceleration down the slope in the coordinate system we have chosen. Also the number is less than 9.81, which is comforting. It means that our calculated acceleration is less than that for free fall. As a final step, let's check for consistency of our answer, $a_x = g \sin \theta$, in limiting cases. In the case where $\theta \to 0°$, the sine also converges to zero, and the acceleration vanishes. This result is consistent because we expect no acceleration of the snowboarder if he rests on a horizontal surface. As $\theta \to 90°$, the sine approaches 1, and the acceleration is the acceleration due to gravity, as we expect as well. In this limiting case, the snowboarder would be in free fall.

Inclined-plane problems, like the one we have just solved, are very common and provide practice with component decomposition of forces. Another common type of problem involves redirection of forces via pulleys and ropes. The next solved problem shows how to proceed in a simple case.

SOLVED PROBLEM 4.2 Two Blocks Connected by a Rope

In this classic problem, a hanging mass causes the acceleration of a second mass that is resting on a horizontal surface (Figure 4.17a). Block 1, of mass $m_1 = 3.00$ kg, rests on a horizontal frictionless surface and is connected via a massless rope (for simplicity, oriented in the horizontal direction) running over a massless pulley to block 2, of mass $m_2 = 1.30$ kg.

PROBLEM
What is the acceleration of block 1 and of block 2?

SOLUTION
THINK From the figure, it is clear that block 1 can only move horizontally and that block 2 can only move vertically. Since the two blocks are connected by a rope, they exert a force on each other via the tension in the rope. Newton's Third Law tells us that the magnitude of the tension acting on either block is the same. We also assume that the rope does not stretch appreciably. This means that any displacement of block 1 results in a displacement of equal magnitude of block 2. This also means that the blocks have the same speed at any time and thus their accelerations have the same magnitude, a. Finally, let's think about the sign of the acceleration: If block 1 moves to the right, then block 2 moves down. It is important to note this consequence, because otherwise sign errors can creep into our calculations.

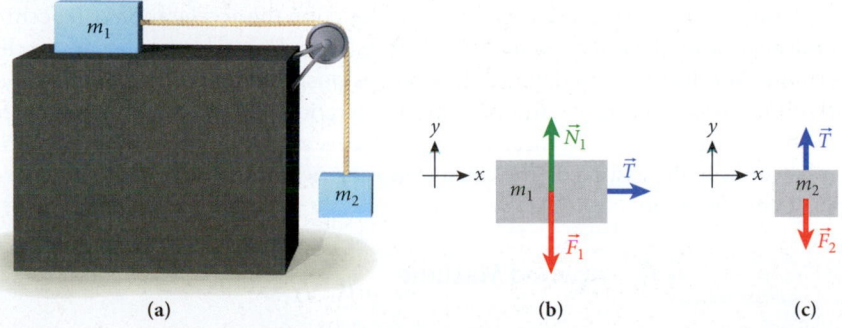

| (a) | (b) | (c) |

FIGURE 4.17 (a) Block 2 hanging vertically from a rope that runs over a pulley and is connected to block 1, resting on a horizontal frictionless surface. (b) Free-body diagram for block 1. (c) Free-body diagram for block 2.

– Continued

SKETCH Again, we start with a free-body diagram for each object. For block 1, the free-body diagram is shown in Figure 4.17b. The gravitational force vector, $\vec{F}_1$, points straight down. The tension force in the rope, $\vec{T}$, acts along the direction of the rope and thus is in the horizontal direction, which we have chosen as the x-direction. The normal force, $\vec{N}_1$, acting on block 1 is perpendicular to the contact surface between that block and the supporting surface. Because the surface is horizontal, $\vec{N}_1$ acts in the vertical direction. Figure 4.17c shows the free-body diagram for block 2. The tension force, $\vec{T}$, acts in the upward direction, and the gravitational force vector, $\vec{F}_2$, acting on this block points straight down, just like $\vec{F}_1$.

RESEARCH Let's start with the forces acting on block 1 and write the equations (based on Newton's Second Law) for the x- and y-components:

$$T = m_1 a$$

$$N_1 - m_1 g = 0.$$

The second equation just tells us that the normal force acting on block 1 is equal to the weight of block 1. However, the first equation contains the acceleration, a, which we want to find. All we have to do now is to find the tension, T.

The free-body diagram for block 2 allows us to write one more equation. In the horizontal direction, there are no forces acting on it, but in the vertical direction, from Newton's Second Law, we have

$$T - m_2 g = -m_2 a .$$

Note the negative sign on the right side! As block 1 accelerates in the positive x-direction (to the right), block 2 has to accelerate in the negative y-direction (downward).

SIMPLIFY We substitute from $T = m_1 a$ into $T - m_2 g = -m_2 a$ and find

$$T - m_2 g = m_1 a - m_2 g = -m_2 a \Rightarrow m_1 a + m_2 a = m_2 g \Rightarrow$$

$$a = g \frac{m_2}{m_1 + m_2}.$$

CALCULATE Now all we have to do is insert the values for the masses:

$$a = (9.81 \text{ m/s}^2) \frac{1.30 \text{ kg}}{3.00 \text{ kg} + 1.30 \text{ kg}} = 2.96581 \text{ m/s}^2.$$

ROUND Since the gravitational acceleration was given to only three significant digits and the masses were specified to the same precision, we round our result to $a = 2.97 \text{ m/s}^2$.

DOUBLE-CHECK This result makes sense: In the limit where m_1 is very large compared to m_2, there will be almost no acceleration, whereas if m_1 is very small compared to m_2, then m_2 will accelerate with almost the full acceleration due to gravity, as if m_1 were not there.

Finally, we can calculate the magnitude of the tension by substituting our result for the acceleration into one of the two equations we obtained using Newton's Second Law:

$$T = m_1 a = g \frac{m_1 m_2}{m_1 + m_2}.$$

In Solved Problem 4.2, it is clear in which direction the acceleration will occur. In more complicated cases, the direction in which objects begin to accelerate may not be clear at the beginning. You just have to define a direction as positive and use this assumption consistently throughout your calculations. If the acceleration value you obtain in the end turns out to be negative, this result means that the objects accelerate in the direction opposite to the one you initially assumed. The calculated value will remain correct. Example 4.4 illustrates such a situation.

EXAMPLE 4.4 | Atwood Machine

The Atwood machine consists of two hanging weights (with masses m_1 and m_2) connected via a rope running over a pulley. For now, we consider a friction-free case, where the pulley does not move, and the rope glides over it. (In Chapter 10 on rotation, we will return to this problem and solve it with friction present, which causes the pulley to rotate.) We also assume that $m_1 > m_2$. In this case, the acceleration is as shown in Figure 4.18a. (The formula derived in the following is correct for any case. If $m_1 < m_2$, then the value of the acceleration, a, will have a negative sign, which will mean that the acceleration direction is opposite to what we have assumed in working the problem.)

We start with free-body diagrams for m_1 and m_2, as shown in Figure 4.18b and Figure 4.18c. For both free-body diagrams, we elect to point the positive y-axis upward, and both diagrams show our choice for the direction of the acceleration. The rope exerts a tension T, of magnitude still to be determined, upward on both m_1 and m_2. With our choices of coordinate system and acceleration direction, the downward acceleration of m_1 is acceleration in a negative direction. This leads to an equation that can be solved for T:

$$T - m_1 g = -m_1 a \;\Rightarrow\; T = m_1 g - m_1 a = m_1\left(g - a\right).$$

From the free-body diagram for m_2 and the assumption that the upward acceleration of m_2 corresponds to acceleration in a positive direction, we get

$$T - m_2 g = m_2 a \;\Rightarrow\; T = m_2 g + m_2 a = m_2\left(g + a\right).$$

Equating the two expressions for T, we obtain

$$m_1\left(g - a\right) = m_2\left(g + a\right),$$

which leads to an expression for the acceleration:

$$(m_1 - m_2)g = (m_1 + m_2)a \Rightarrow$$

$$a = g\left(\frac{m_1 - m_2}{m_1 + m_2}\right).$$

From this equation, you can see that the magnitude of the acceleration, a, is always smaller than g in this situation. If the masses are equal, we obtain the expected result of no acceleration. By selecting the proper combination of masses, we can generate any value of acceleration between zero and g that we desire.

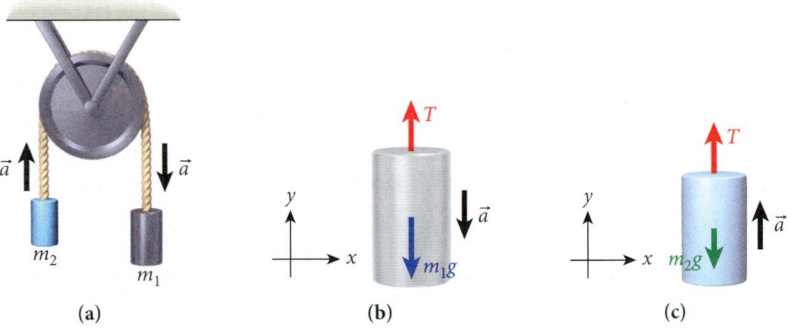

FIGURE 4.18 (a) Atwood machine with the direction of positive acceleration assumed to be as indicated. (b) Free-body diagram for the weight on the right side of the Atwood machine. (c) Free-body diagram for the weight on the left side of the Atwood machine.

Self-Test Opportunity 4.2

What is the acceleration of the masses on the Atwood machine at the limits where m_1 approaches infinity, m_1 approaches zero, and $m_1 = m_2$?

Self-Test Opportunity 4.3

For the Atwood machine, can you write a formula for the magnitude of the tension in the rope?

Concept Check 4.4

If you double both masses in an Atwood machine, the resulting acceleration will be

a) twice as large.

b) half as large.

c) the same.

d) one-quarter as large.

e) four times as large.

EXAMPLE 4.5 Collision of Two Vehicles

Suppose that an SUV of mass $m = 3260$ kg has a head-on collision with a subcompact car of mass $m = 1194$ kg and exerts a force of magnitude $2.9 \cdot 10^5$ N on the subcompact.

PROBLEM

What is the magnitude of the force that the subcompact exerts on the SUV in the collision?

SOLUTION

As paradoxical as it may seem at first, the little subcompact exerts just as much force on the SUV as the SUV exerts on it. This equality is a straightforward consequence of Newton's Third Law, equation 4.8. So, the answer is $2.9 \cdot 10^5$ N.

– Continued

Concept Check 4.5

For the collision in Example 4.5, if we call the absolute value of the acceleration experienced by the SUV a_{SUV} and that for the subcompact car a_{car}, we find that approximately

a) $a_{SUV} \approx \frac{1}{9} a_{car}$.

b) $a_{SUV} \approx \frac{1}{3} a_{car}$.

c) $a_{SUV} \approx a_{car}$.

d) $a_{SUV} \approx 3 a_{car}$.

e) $a_{SUV} \approx 9 a_{car}$.

DISCUSSION

The answer may be straightforward, but it is by no means intuitive. The subcompact will usually sustain much more damage in such a collision, and its passengers will have a much bigger chance of getting hurt. However, this difference is due to Newton's Second Law, which says that the same force applied to a less massive object yields a higher acceleration than if applied to a more massive one. Even in a head-on collision between a mosquito and a car on the freeway, the forces exerted on each body are equal; the difference in damage to the car (none) and to the mosquito (obliteration) is due to their different resulting accelerations and the fact that the mosquito's body is made of organic matter and thus much more easily damaged than the car's windshield. We will revisit collisions in Chapter 7.

4.7 Friction Force

So far, we have neglected the force of friction and considered only frictionless approximations. However, in general, we have to include friction in most of our calculations when we want to describe physically realistic situations.

We could conduct a series of very simple experiments to learn about the basic characteristics of friction. Here are the findings we would obtain:

- If an object is at rest, it takes an external force with a certain threshold magnitude and acting parallel to the contact surface between the object and the surface to overcome the friction force and make the object move.

- The friction force that has to be overcome to make an object at rest move is larger than the friction force that has to be overcome to keep the object moving at a constant velocity.

- The magnitude of the friction force acting on a moving object is proportional to the magnitude of the normal force.

- The friction force is independent of the size of the contact area between object and surface.

- The friction force depends on the roughness of the surfaces; that is, a smoother interface generally provides less friction force than a rougher one.

- The friction force is independent of the velocity of the object.

These statements about friction are not principles in the same way as Newton's laws. Instead, they are general observations based on experiments. For example, you might think that the contact of two extremely smooth surfaces would yield very low friction. However, in some cases, extremely smooth surfaces actually fuse together as a cold weld. Investigations into the nature and causes of friction continue, as we discuss later in this section.

From these findings, it is clear that we need to distinguish between the case where an object is at rest relative to its supporting surface (static friction) and the case where an object is moving across the surface (kinetic friction). The case in which an object is moving across a surface is easier to treat, and so we consider kinetic friction first.

Kinetic Friction

The above general observations can be summarized in the following approximate formula for the magnitude of the kinetic friction force, f_k:

$$f_k = \mu_k N. \tag{4.10}$$

Here N is the magnitude of the normal force and μ_k is the **coefficient of kinetic friction.** This coefficient is always equal to or greater than zero. (The case where $\mu_k = 0$ corresponds to a frictionless approximation. In practice, however, it can never be reached perfectly.) In almost all cases, μ_k is also less than 1. (Some special tire surfaces used for car racing, though, have a coefficient of friction with the road that can significantly exceed 1.) Some representative coefficients of kinetic friction are shown in Table 4.1.

Table 4.1	Typical Coefficients of Both Static and Kinetic Friction Between Material 1 and Material 2*		
Material 1	**Material 2**	μ_s	μ_k
rubber	dry concrete	1	0.8
rubber	wet concrete	0.7	0.5
steel	steel	0.7	0.6
wood	wood	0.5	0.3
waxed ski	snow	0.1	0.05
steel	oiled steel	0.12	0.07
Teflon	steel	0.04	0.04
curling stone	ice		0.017

*Note that these values are approximate and depend strongly on the condition of the contact surface between the two materials.

The direction of the kinetic friction force is *always opposite to the direction of motion* of the object relative to the surface it moves on.

If you push an object with an external force parallel to the contact surface, and the force has a magnitude exactly equal to that of the force of kinetic friction on the object, then the total net external force is zero, because the external force and the friction force cancel each other. In that case, according to Newton's First Law, the object will continue to slide across the surface with constant velocity.

Static Friction

If an object is at rest, it takes a certain threshold amount of external force to set it in motion. For example, if you push lightly against a refrigerator, it will not move. As you push harder and harder, you reach a point where the refrigerator finally slides across the kitchen floor.

For any external force acting on an object that remains at rest, the friction force is exactly equal in magnitude and opposite in direction to the component of that external force that acts along the contact surface between the object and its supporting surface. However, the magnitude of the static friction force has a maximum value: $f_s \leq f_{s,max}$. This maximum magnitude of the static friction force is proportional to the normal force, but with a different proportionality constant than the coefficient of kinetic friction: $f_{s,max} = \mu_s N$. For the magnitude of the force of static friction, we can write

$$f_s \leq \mu_s N = f_{s,max}, \tag{4.11}$$

where μ_s is called the **coefficient of static friction.** Some typical coefficients of static friction are shown in Table 4.1. In general, for any object on any supporting surface, the maximum static friction force is greater than the force of kinetic friction. You may have experienced this when trying to slide a heavy object across a surface: As soon as the object starts moving, a lot less force is required to keep it in constant sliding motion. We can express this finding as a mathematical inequality between the two coefficients:

$$\mu_s > \mu_k. \tag{4.12}$$

Figure 4.19 presents a graph showing how the friction force depends on an external force, F_{ext}, applied to an object. If the object is initially at rest, a small external force results in a small force of friction, rising linearly with the external force until it reaches a value of $\mu_s N$. Then it drops rather quickly to a value of $\mu_k N$ when the object is set in motion. At this point, the external force has a value of $F_{ext} = \mu_s N$, resulting in a sudden acceleration of the object. This dependence of the friction force on the external force is shown in Figure 4.19 as a red line.

On the other hand, if we start with a large external force and the object is already in motion, then we can reduce the external force below a value of $\mu_s N$, but still above $\mu_k N$, and the object will keep moving and accelerating. Thus, the friction coefficient retains a value of μ_k until the external force is reduced to a value of $\mu_k N$. At this point (and only at this point!), the object will move with a constant velocity, because the external force and the friction force are equal in magnitude. If we reduce the external force further, the object decelerates (horizontal segment of the blue line left of the red

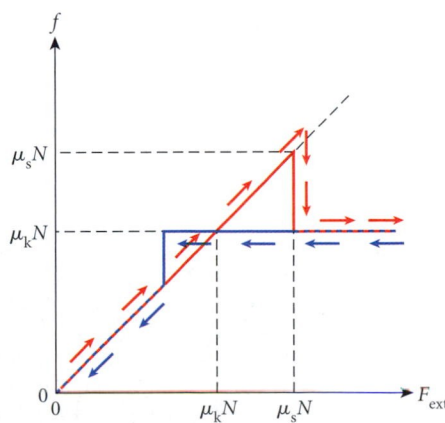

FIGURE 4.19 Magnitudes of the forces of friction as a function of the magnitude of an external force. The red line represents the forces for an object initially at rest, with the external force increasing from zero. The blue line represents the forces for an object initially in motion under the influence of an external force that is larger than the friction force but is then gradually reduced.

diagonal in Figure 4.19), because the kinetic friction force is bigger than the external force. Eventually, the object comes to rest because of the kinetic friction, and the external force is not sufficient to move it anymore. Then static friction takes over, and the friction force is reduced proportionally to the external force until both reach zero. The blue line in Figure 4.19 illustrates this dependence of the friction force on the external force. Where the blue line and the red line overlap, this is indicated by alternating blue and red dashes. The most interesting feature of Figure 4.19 is that the blue and red lines do not coincide between $\mu_k N$ and $\mu_s N$.

Let's return to the attempt to move a refrigerator across the kitchen floor. Initially, the refrigerator sits on the floor, and the static friction force resists your effort to move it. Once you push hard enough, the refrigerator jars into motion. In this process, the friction force follows the red path in Figure 4.19. Once the refrigerator moves, you can push less hard and still keep it moving. If you push with less force so that it moves with constant velocity, the external force you apply follows the blue path in Figure 4.19 until it is reduced to $F_{ext} = \mu_k N$. Then the friction force and the force you apply to the fridge add up to zero, and there is no net force acting on the refrigerator, allowing it to move with constant velocity.

EXAMPLE 4.6 | Realistic Snowboarding

Let's reconsider the snowboarding situation from Solved Problem 4.1, but now include friction. A snowboarder moves down a slope for which $\theta = 22°$. Suppose the coefficient of kinetic friction between his board and the snow is 0.21, and his velocity, which is along the direction of the slope, is measured as 8.3 m/s at a given instant.

PROBLEM 1
Assuming a constant slope, what will be the speed of the snowboarder along the direction of the slope when he is 100 m farther down the slope?

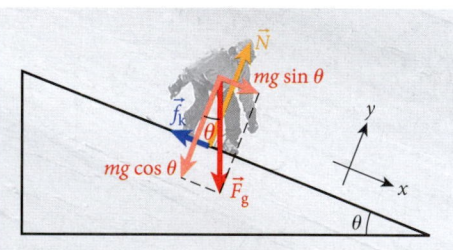

FIGURE 4.20 Free-body diagram of a snowboarder, including the friction force.

SOLUTION 1
Figure 4.20 shows a free-body diagram for this problem. The gravitational force points downward and has magnitude mg, where m is the mass of the snowboarder and his equipment. We choose convenient x- and y-axes parallel and perpendicular to the slope, respectively, as indicated in Figure 4.20. The angle θ that the slope makes with the horizontal (22° in this case) also appears in the decomposition of the components of the gravitational force parallel and perpendicular to the slope. (This analysis is a general feature of any inclined-plane problem.) The force component along the plane is then $mg \sin\theta$, as shown in Figure 4.20. The normal force is given by $N = mg \cos\theta$, and the force of kinetic friction is $f_k = -\mu_k mg \cos\theta$, with the minus sign indicating that the force is acting in the negative x-direction, in our chosen coordinate system.

We thus get for the total force component in the x-direction:

$$mg \sin\theta - \mu_k mg \cos\theta = ma_x \Rightarrow$$
$$a_x = g(\sin\theta - \mu_k \cos\theta).$$

Here we have used Newton's Second Law, $F_x = ma_x$, in the first line. The mass of the snowboarder drops out, and the acceleration, a_x, along the slope is constant. Inserting the numbers given in the problem statement, we obtain

$$a \equiv a_x = (9.81 \text{ m/s}^2)(\sin 22° - 0.21\cos 22°) = 1.76 \text{ m/s}^2.$$

Thus, we see that this is a problem of motion in a straight line in one direction with constant acceleration. We can apply the relationship between the squares of the initial and final velocities and the acceleration that we have derived for one-dimensional motion with constant acceleration:

$$v^2 = v_0^2 + 2a(x - x_0).$$

With $v_0 = 8.3$ m/s and $x - x_0 = 100$ m, we calculate the final speed:

$$v = \sqrt{v_0^2 + 2a(x - x_0)}$$
$$= \sqrt{(8.3 \text{ m/s})^2 + 2(1.76 \text{ m/s}^2)(100 \text{ m})}$$
$$= 20.5 \text{ m/s}.$$

PROBLEM 2

How long does it take the snowboarder to reach this speed?

SOLUTION 2

Since we now know the acceleration and the final speed and were given the initial speed, we use

$$v = v_0 + at \Rightarrow t = \frac{v - v_0}{a} = \frac{(20.5 - 8.3) \text{ m/s}}{1.76 \text{ m/s}^2} = 6.9 \text{ s.}$$

PROBLEM 3

Given the same coefficient of friction, what would the angle of the slope have to be for the snowboarder to glide with constant velocity?

SOLUTION 3

Motion with constant velocity implies zero acceleration. We have already derived an equation for the acceleration as a function of the slope angle. We set this expression equal to zero and solve the resulting equation for the angle θ:

$$a_x = g(\sin\theta - \mu_k \cos\theta) = 0$$
$$\Rightarrow \sin\theta = \mu_k \cos\theta$$
$$\Rightarrow \tan\theta = \mu_k$$
$$\Rightarrow \theta = \tan^{-1} \mu_k$$

Because $\mu_k = 0.21$ was given, the angle is $\theta = \tan^{-1} 0.21 = 12°$. With a steeper slope, the snowboarder will accelerate, and with a shallower slope, the snowboarder will slow down until he comes to a stop.

Air Resistance

So far we have ignored the friction due to moving through the air. Unlike the force of kinetic friction that acts when you drag or push one object across the surface of another, air resistance increases as speed increases. Thus, we need to express the friction force as a function of the velocity of the object relative to the medium it moves through. The direction of the force of air resistance is opposite to the direction of the velocity vector.

In general, the magnitude of the friction force due to air resistance, or **drag force,** can be expressed as $F_{\text{frict}} = K_0 + K_1 v + K_2 v^2 + ...$, with the constants $K_0, K_1, K_2, ...$ determined experimentally. For the drag force on macroscopic objects moving at relatively high speeds, we can neglect the linear term in the velocity. The magnitude of the drag force is then approximately

$$F_{\text{drag}} = Kv^2. \tag{4.13}$$

This equation means that the force due to air resistance is proportional to the square of the speed.

When an object falls through air, the force from air resistance increases as the object accelerates until it reaches a **terminal speed.** At this point, the upward force of air resistance and the downward force due to gravity equal each other. Thus, the net force is zero, and there is no more acceleration. Because there is no more acceleration, the falling object has constant terminal speed:

$$F_g = F_{\text{drag}} \Rightarrow mg = Kv^2.$$

Solving this for the terminal speed, we obtain

$$v = \sqrt{\frac{mg}{K}}. \tag{4.14}$$

Note that the terminal speed depends on the mass of the object, whereas the mass of the object does not affect the object's motion when air resistance is neglected. In the absence of air resistance, all objects fall at the same rate, but the presence of air resistance explains why heavy objects fall faster than light ones that have the same (drag) constant K.

To compute the terminal speed for a falling object, we need to know the value of the constant K. This constant depends on many variables, including the size of the cross-sectional area, A, exposed to the air stream. In general terms, the bigger the area, the bigger is the constant K. K also depends linearly on the air density, ρ. All other dependences on the shape of the object, on its inclination relative to the direction of motion, on air viscosity, and compressibility are usually collected in a drag coefficient, c_d:

$$K = \tfrac{1}{2} c_d A \rho. \qquad (4.15)$$

Equation 4.15 has the factor $\tfrac{1}{2}$ to simplify calculations involving the energy of objects undergoing free fall with air resistance. We will return to this subject when we discuss kinetic energy in Chapter 5.

Creating a low drag coefficient is an important consideration in automotive design, because it has a strong influence on the maximum speed of a car and its fuel consumption. Numerical computations are useful, but the drag coefficient is usually optimized experimentally by putting car prototypes into wind tunnels and testing the air resistance at different speeds. The same wind tunnel tests are also used to optimize the performance of equipment and athletes in events such as downhill ski racing and bicycle racing.

For motion in very viscous media or at low velocities, the linear velocity term of the friction force cannot be neglected. In such cases, the friction force can be approximated by the form $F_{frict} = K_1 v$. This form applies to most biological processes, including motion of large biomolecules or even microorganisms such as bacteria through liquids. This approximation of the friction force is also useful when analyzing the sinking of an object in a fluid, for example, a small stone or fossil shell in water.

Concept Check 4.6

An unused coffee filter reaches its terminal speed very quickly if you let it fall. Suppose you release a single coffee filter from a height of 1 m. From what height do you have to release a stack of two coffee filters at the same instant so that they will hit the ground at the same time as the single coffee filter? (You can safely neglect the time needed to reach terminal speed.)

a) 0.5 m

b) 0.7 m

c) 1 m

d) 1.4 m

e) 2 m

EXAMPLE 4.7 Sky Diving

An 80.0-kg skydiver falls through air with a density of 1.15 kg/m^3. Assume that his drag coefficient is $c_d = 0.570$. When he falls in the spread-eagle position, as shown in Figure 4.21a, his body presents an area $A_1 = 0.940$ m^2 to the wind, whereas when he dives head first, with arms close to the body and legs together, as shown in Figure 4.21b, his area is reduced to $A_2 = 0.210$ m^2.

PROBLEM
What are the terminal speeds in both cases?

SOLUTION
We use equation 4.14 for the terminal speed, substitute the expression for the air resistance constant from equation 4.15, and insert the given numbers:

$$v = \sqrt{\frac{mg}{K}} = \sqrt{\frac{mg}{\tfrac{1}{2} c_d A \rho}}$$

$$v_1 = \sqrt{\frac{(80.0 \text{ kg})(9.81 \text{ m/s}^2)}{\tfrac{1}{2} 0.570(0.940 \text{ m}^2)(1.15 \text{ kg/m}^3)}} = 50.5 \text{ m/s}$$

$$v_2 = \sqrt{\frac{(80.0 \text{ kg})(9.81 \text{ m/s}^2)}{\tfrac{1}{2} 0.570(0.210 \text{ m}^2)(1.15 \text{ kg/m}^3)}} = 107 \text{ m/s}.$$

These results show that, by diving head first, the skydiver can reach higher velocities during free fall than when he uses the spread-eagle position. Therefore, it is possible to catch up to a person who has fallen out of an airplane, assuming that the person is not diving head first, too. However, in general, this technique cannot be used to save such a person because it would be nearly impossible to hold onto him or her during the sudden deceleration shock caused by the rescuer's parachute opening.

(a)

(b)

FIGURE 4.21 (a) Skydiver in the high-resistance position. (b) Skydiver in low-resistance position.

Tribology

What causes friction? The answer to this question is not at all easy or obvious. When surfaces rub against each other, different atoms (more on atoms in Chapter 13) from the two surfaces make contact with each other in different ways. Atoms get dislocated in the process of dragging surfaces across each other. Electrostatic interactions (more on these in Chapter 21) between the atoms on the surfaces cause additional static friction. A true microscopic understanding of friction is beyond the scope of this book and is currently the focus of great research activity.

The science of friction has a name: **tribology.** The laws of friction we have discussed were already known 300 years ago. Their discovery is generally credited to Guillaume Amontons and Charles Augustin de Coulomb, but even Leonardo da Vinci may have known them. Yet amazing things are still being discovered about friction, lubrication, and wear.

Perhaps the most interesting advance in tribology that occurred in the last two decades was the development of atomic and friction force microscopes. The basic principle that these microscopes employ is the dragging of a very sharp tip across a surface with analysis by cutting-edge computer and sensor technology. Such friction force microscopes can measure friction forces as small as 10 pN = 10^{-11} N. Figure 4.22a shows a diagram of an atomic force microscope (AFM); Figure 4.22b is a close-up photograph (magnification 1000×) of its cantilever tip. State-of-the-art microscopic simulations of friction are still not able to completely explain it, and so this research area is of great interest in the field of nanotechnology.

Friction is responsible for the breaking off of small particles from surfaces that rub against each other, causing wear. This phenomenon is of particular importance in high-performance car engines, which require specially formulated lubricants. Understanding the influence of small surface impurities on the friction force is of great interest in this context. Research into lubricants continues to try to find ways to reduce the coefficient of kinetic friction, μ_k, to a value as close to zero as possible. For example, modern lubricants include *buckyballs*—molecules consisting of 60 carbon atoms arranged in the shape of a soccer ball, which were discovered in 1985. These molecules act like microscopic ball bearings.

Solving problems involving friction is also important to car racing. In the Formula 1 circuit, using the right tires that provide optimally high friction is essential for winning races. While friction coefficients are normally in the range between 0 and 1, it is not unusual for top fuel race cars to have tires that have friction coefficients with the track surface of 3 or even larger.

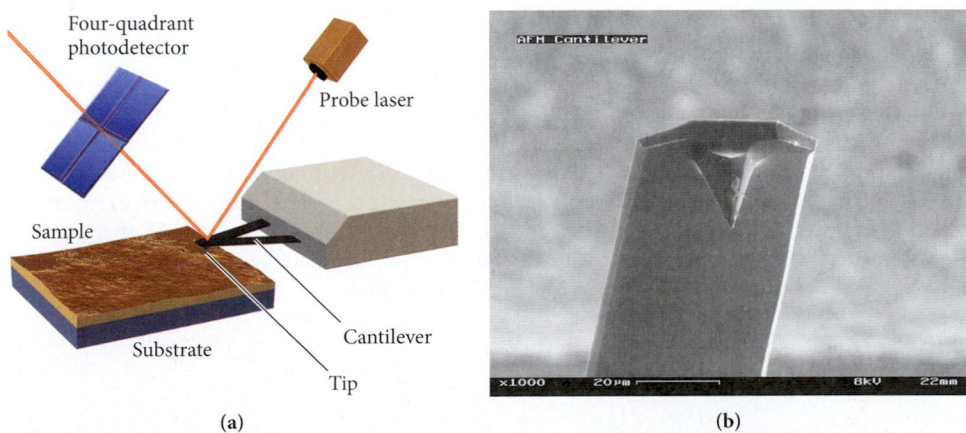

FIGURE 4.22 Atomic force microscope (AFM): (a) schematic diagram of the parts of the microscope, (b) photograph (at 1000×) of the tip of the cantilever, which is dragged across the sample.

4.8 Applications of the Friction Force

With Newton's three laws, we can solve a huge class of problems. Knowing about static and kinetic friction allows us to approximate real-world situations and come to meaningful conclusions. Because it is helpful to see various applications of Newton's laws, we will solve several practice problems. These examples are designed to demonstrate a range of techniques that are useful in the solution of many kinds of problems.

EXAMPLE 4.8   Two Blocks Connected by a Rope—with Friction

We solved this problem in Solved Problem 4.2, with the assumptions that block 1 slides without friction across the horizontal support surface and that the rope slides without friction across the pulley. Here we will allow for friction between block 1 and the surface it slides across. For now, we will still assume that the rope slides without friction across the pulley. (Chapter 10 will present techniques that let us deal with the pulley being set into rotational motion by the rope moving across it.)

PROBLEM 1
Let the coefficient of static friction between block 1 (mass m_1 = 2.3 kg) and its support surface have a value of 0.73 and the coefficient of kinetic friction have a value of 0.60. (Refer back to Figure 4.17.) If block 2 has mass m_2 = 1.9 kg, will block 1 accelerate from rest?

SOLUTION 1
All the force considerations from Solved Problem 4.2 remain the same, except that the free-body diagram for block 1 (Figure 4.23) now has a force arrow corresponding to the friction force, f. Keep in mind that in order to draw the direction of the friction force, you need to know in which direction movement would occur in the absence of friction. Because we have already solved the frictionless case, we know that block 1 would move to the right. Because the friction force is directed opposite to the movement, the friction vector thus points to the left.

The equation we derived in Solved Problem 4.2 by applying Newton's Second Law to block 1 changes from $m_1 a = T$ to

$$m_1 a = T - f.$$

Combining this with the equation we obtained in Solved Problem 4.2 via application of Newton's Second Law to m_2, $T - m_2 g = -m_2 a$, and again eliminating T gives us

$$m_1 a + f = T = m_2 g - m_2 a \Rightarrow$$

$$a = \frac{m_2 g - f}{m_1 + m_2}.$$

So far, we have avoided specifying any further details about the friction force. We now do so by first calculating the maximum magnitude of the static friction force, $f_{s,max} = \mu_s N_1$. For the magnitude of the normal force, we already found $N_1 = m_1 g$, so we have for the maximum static friction force:

$$f_{s,max} = \mu_s N_1 = \mu_s m_1 g = (0.73)(2.3 \text{ kg})(9.81 \text{ m/s}^2) = 16.5 \text{ N}.$$

We need to compare this value to that of $m_2 g$ in the numerator of our equation for the acceleration, $a = (m_2 g - f)/(m_1 + m_2)$. If $f_{s,max} \geq m_2 g$, then the static friction force will assume a value exactly equal to $m_2 g$, causing the acceleration to be zero. In other words, there will be no motion, because the pull due to block 2 hanging from the rope is not sufficient to overcome the force of static friction between block 1 and its supporting surface. If $f_{s,max} < m_2 g$, then there will be positive acceleration, and the two blocks will start moving. In the present case, because $m_2 g = (1.9 \text{ kg})(9.81 \text{ m/s}^2) = 18.6 \text{ N}$, the blocks will start moving.

PROBLEM 2
What is the value of the acceleration?

SOLUTION 2
As soon as the static friction force is overcome, kinetic friction takes over. We can then use our equation for the acceleration, $a = (m_2 g - f)/(m_1 + m_2)$, substitute $f = \mu_k N_1 = \mu_k m_1 g$, and obtain

$$a = \frac{m_2 g - \mu_k m_1 g}{m_1 + m_2} = g\left(\frac{m_2 - \mu_k m_1}{m_1 + m_2}\right).$$

Inserting the numbers, we find

$$a = (9.81 \text{ m/s}^2)\left|\frac{(1.9 \text{ kg}) - 0.6 \cdot (2.3 \text{ kg})}{(2.3 \text{ kg}) + (1.9 \text{ kg})}\right| = 1.21 \text{ m/s}^2.$$

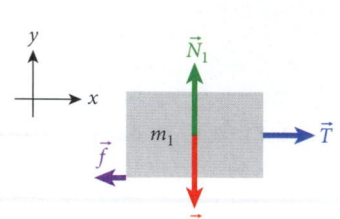

FIGURE 4.23 Free-body diagram for block 1, including the force of friction.

Self-Test Opportunity 4.4

(a) What is the maximum mass m_2 for which the system of these two blocks does not move? (b) What is the value of the friction force if m_2 is smaller than this value?

SOLVED PROBLEM 4.3 / Wedge

A wedge of mass $m = 37.7$ kg is held in place on a fixed plane that is inclined by an angle $\theta = 20.5°$ with respect to the horizontal. A force $F = 309.3$ N in the horizontal direction pushes on the wedge, as shown in Figure 4.24a. The coefficient of kinetic friction between the wedge and the plane is $\mu_k = 0.171$. Assume that the coefficient of static friction is low enough that the net force will move the wedge.

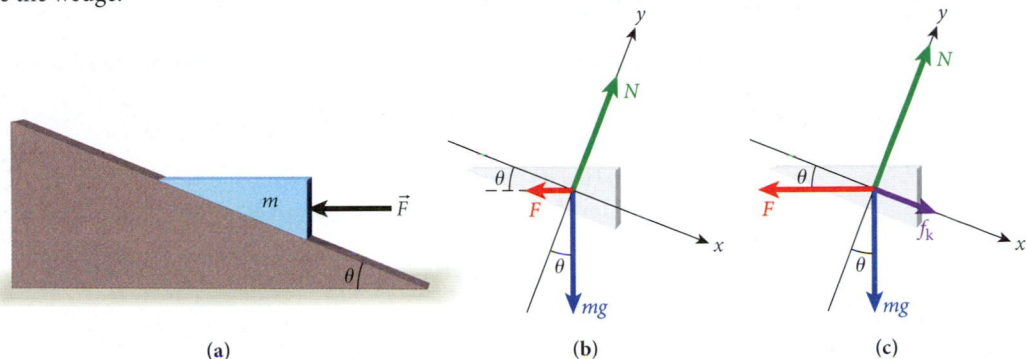

(a) (b) (c)

FIGURE 4.24 (a) A wedge-shaped block being pushed on an inclined plane. (b) Free-body diagram of the wedge, including the external force, the force of gravity, and the normal force. (c) Free-body diagram including the external force, the force of gravity, the normal force, and the friction force.

PROBLEM
What is the acceleration of the wedge along the plane when it is released and free to move?

SOLUTION
THINK We want to know the acceleration a of the wedge of mass m along the plane, which requires us to determine the component of the net force that acts on the wedge parallel to the surface of the inclined plane. Also, we need to find the component of the net force that acts on the wedge perpendicular to the plane, to allow us to determine the force of kinetic friction.

The forces acting on the wedge are gravity, the normal force, the force of kinetic friction f_k, and the external force F. The coefficient of kinetic friction, μ_k, is given, so we can calculate the friction force once we determine the normal force. Before we can continue with our analysis of the forces, we must determine in which direction the wedge will move after it is released as a result of force F. Once we know in which direction the wedge will go, we can determine the direction of the friction force and complete our analysis.

To determine the net force before the wedge begins to move, we need a free-body diagram with just the forces F, N, and mg. Once we determine the direction of motion, we can determine the direction of the friction force, using a second free-body diagram with the friction force added.

SKETCH A free-body diagram showing the forces acting on the wedge before it is released is presented in Figure 4.24b. We have defined a coordinate system in which the x-axis is parallel to the surface of the inclined plane, with the positive x-direction pointing down the plane. The sum of the forces in the x-direction is

$$mg\sin\theta - F\cos\theta = ma.$$

We need to determine if the mass will move to the right (positive x-direction, or down the plane) or to the left (negative x-direction, or up the plane). We can see from the equation that the quantity $mg\sin\theta - F\cos\theta$ will determine the direction of the motion. With the given numerical values, we have

$$mg\sin\theta - F\cos\theta = (37.7 \text{ kg})(9.81 \text{ m/s}^2)(\sin 20.5°) - (309.3 \text{ N})(\cos 20.5°)$$

$$= -160.193 \text{ N}.$$

Thus, the mass will move up the plane (to the left, or in the negative x-direction). Now we can redraw the free-body diagram as shown in Figure 4.24c by inserting the arrow for the force of

– Continued

kinetic friction, f_k, pointing down the plane (in the positive x-direction), because the friction force always opposes the direction of motion.

RESEARCH Now we can write the components of the forces in the x- and y-directions based on the final free-body diagram. For the x-direction, we have

$$mg\sin\theta - F\cos\theta + f_k = ma. \tag{i}$$

For the y-direction we have

$$N - mg\cos\theta - F\sin\theta = 0.$$

From this equation, we can get the normal force N that we need to calculate the friction force:

$$f_k = \mu_k N = \mu_k (mg\cos\theta + F\sin\theta). \tag{ii}$$

SIMPLIFY Having related all the known and unknown quantities to each other, we can get an expression for the acceleration of the mass using equations i and ii:

$$mg\sin\theta - F\cos\theta + \mu_k (mg\cos\theta + F\sin\theta) = ma.$$

We can rearrange the left side:

$$mg\sin\theta - F\cos\theta + \mu_k mg\cos\theta + \mu_k F\sin\theta = ma$$

$$(mg + \mu_k F)\sin\theta + (\mu_k mg - F)\cos\theta = ma,$$

and then solve for the acceleration:

$$a = \frac{(mg + \mu_k F)\sin\theta + (\mu_k mg - F)\cos\theta}{m}. \tag{iii}$$

CALCULATE Now we put in the numbers and get a numerical result. The first term in the numerator of equation iii is

$$\big((37.7\text{ kg})(9.81\text{ m/s}^2) + (0.171)(309.3\text{ N})\big)(\sin 20.5°) = 148.042\text{ N}.$$

Note that we have not rounded this result yet. The second term in the numerator of equation iii is

$$\big((0.171)(37.7\text{ kg})(9.81\text{ m/s}^2) - (309.3\text{ N})\big)(\cos 20.5°) = -230.476\text{ N}.$$

Again we have not yet rounded the result. Now we calculate the acceleration using equation iii:

$$a = \frac{(148.042\text{ N}) + (-230.476\text{ N})}{37.7\text{ kg}} = -2.1866\text{ m/s}^2.$$

ROUND Because all of the numerical values were initially given with three significant figures, we report our final result as

$$a = -2.19\text{ m/s}^2.$$

DOUBLE-CHECK Looking at our answer, we see that the acceleration is negative, which means it is in the negative x-direction. We had determined that the mass would move to the left (up the plane, or in the negative x-direction), which agrees with the sign of the acceleration in our final result. The magnitude of the acceleration is a fraction of the acceleration of gravity (9.81 m/s^2), which makes physical sense.

(a)

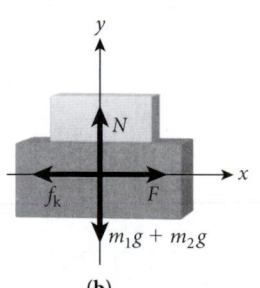

(b)

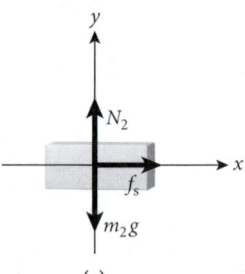

(c)

FIGURE 4.25 (a) Two stacked blocks being pulled to the right. (b) Free-body diagram for the two blocks moving together. (c) Free-body diagram for the upper block.

SOLVED PROBLEM 4.4 Two Blocks

Two rectangular blocks are stacked on a table as shown in Figure 4.25a. The upper block has a mass of 3.40 kg, and the lower block has a mass of 38.6 kg. The coefficient of kinetic friction between the lower block and the table is 0.260. The coefficient of static friction between the blocks is 0.551. A string is attached to the lower block, and an external force $\vec{F}$ is applied horizontally, pulling on the string as shown.

PROBLEM
What is the maximum force that can be applied to the string without having the upper block slide off?

SOLUTION

THINK To begin this problem, we note that as long as the force of static friction between the two blocks is not overcome, the two blocks will travel together. Thus, if we pull gently on the lower block, the upper block will stay in place on top of it, and the two blocks will slide as one. However, if we pull hard on the lower block, the force of static friction between the blocks will not be sufficient to keep the upper block in place and it will begin to slide off the lower block.

The forces acting in this problem are the external force F pulling on the string, the force of kinetic friction f_k between the lower block and the surface on which the blocks are sliding, the weight m_1g of the lower block, the weight m_2g of the upper block, and the force of static friction f_s between the blocks and the normal forces.

SKETCH We start with a free-body diagram of the two blocks moving together (Figure 4.25b), because we will treat the two blocks as one system for the first part of this analysis. We define the x-direction to be parallel to the surface on which the blocks are sliding and parallel to the external force pulling on the string, with the positive direction to the right, in the direction of the external force. The sum of the forces in the x-direction is

$$F - f_k = (m_1 + m_2)a. \tag{i}$$

The sum of the forces in the y-direction is

$$N - (m_1g + m_2g) = 0. \tag{ii}$$

Equations i and ii describe the motion of the two blocks together.

Now we need a second free-body diagram to describe the forces acting on the upper block. The forces in the free-body diagram for the upper block (Figure 4.25c) are the normal force N_2 exerted by the lower block, the weight $m_2\,g$, and the force of static friction f_s. The sum of the forces in the x-direction is

$$f_s = m_2a. \tag{iii}$$

The sum of the forces in the y-direction is

$$N_2 - m_2g = 0. \tag{iv}$$

RESEARCH The maximum value of the force of static friction between the upper and lower blocks is given by

$$f_s = \mu_s N_2 = \mu_s(m_2g).$$

where we have used equations iii and iv. Thus, the maximum acceleration that the upper block can have without sliding is

$$a_{max} = \frac{f_s}{m_2} = \frac{\mu_s m_2 g}{m_2} = \mu_s g. \tag{v}$$

This maximum acceleration for the upper block is also the maximum acceleration for both blocks together. From equation ii, we get the normal force between the lower block and the sliding surface:

$$N = m_1g + m_2g. \tag{vi}$$

The force of kinetic friction between the lower block and the sliding surface is then

$$f_k = \mu_k(m_1g + m_2g). \tag{vii}$$

SIMPLIFY We can now relate the maximum acceleration to the maximum force, F_{max}, that can be exerted without the upper block sliding off, using equations v and vii:

$$F_{max} - \mu_k(m_1g + m_2g) = (m_1 + m_2)\mu_s g.$$

We solve this for the maximum force to obtain

$$F_{max} = \mu_k(m_1g + m_2g) + (m_1 + m_2)\mu_s g = g(m_1 + m_2)(\mu_k + \mu_s).$$

CALCULATE Putting in the given numerical values, we get

$$F_{max} = (9.81 \text{ m/s}^2)(38.6 \text{ kg} + 3.40 \text{ kg})(0.260 + 0.551) = 334.148 \text{ N}.$$

ROUND All of the numerical values were given to three significant figures, so we report our answer as

$$F_{max} = 334 \text{ N}.$$

DOUBLE-CHECK The answer is a positive value, implying a force to the right, which agrees with the free-body diagram in Figure 4.25b.

The maximum acceleration is

$$a_{max} = \mu_s g = (0.551)(9.81 \text{ m/s}^2) = 5.41 \text{ m/s}^2,$$

– Continued

which is a fraction of the acceleration due to gravity, which seems reasonable. If there were no friction between the lower block and the surface on which it slides, the force required to accelerate both blocks would be

$$F = (m_1 + m_2)a_{max} = (38.6 \text{ kg} + 3.40 \text{ kg})(5.41 \text{ m/s}^2) = 227 \text{ N}.$$

Thus, our answer of 334 N for the maximum force seems reasonable because it is higher than the force calculated when there is no friction.

EXAMPLE 4.9 | Pulling a Sled

Suppose you are pulling a sled across a level snow-covered surface by exerting constant force on a rope, at an angle θ relative to the ground.

PROBLEM 1
If the sled, including its load, has a mass of 15.3 kg, the coefficients of friction between the sled and the snow are $\mu_s = 0.076$ and $\mu_k = 0.070$, and you pull with a force of 25.3 N on the rope at an angle of 24.5° relative to the horizontal ground, what is the sled's acceleration?

SOLUTION 1
Figure 4.26 shows the free-body diagram for the sled, with all the forces acting on it. The directions of the force vectors are correct, but the magnitudes are not necessarily drawn to scale. The acceleration of the sled will, if it occurs at all, be directed along the horizontal, in the x-direction. In terms of components, Newton's Second Law gives:

$$x\text{-component: } ma = T\cos\theta - f$$

and

$$y\text{-component: } 0 = T\sin\theta - mg + N.$$

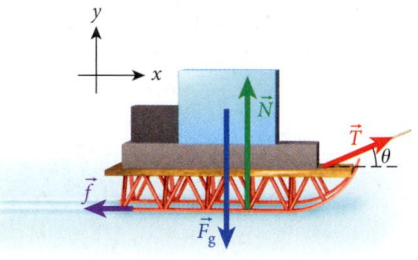

FIGURE 4.26 Free-body diagram of the sled and its load.

For the friction force, we will use the form $f = \mu N$ for now, without specifying whether it is kinetic or static friction, but in the end we will have to return to this point. The normal force can be calculated from the above equation for the y-component and then substituted into the equation for the x-component:

$$N = mg - T\sin\theta$$

$$ma = T\cos\theta - \mu(mg - T\sin\theta) \Rightarrow$$

$$a = \frac{T}{m}(\cos\theta + \mu\sin\theta) - \mu g.$$

We see that the normal force is less than the weight of the sled, because the force pulling on the rope has an upward y-component. The vertical component of the force pulling on the rope also contributes to the acceleration of the sled since it affects the normal force and hence the horizontal frictional force.

When putting in the numbers, we first use the value of the coefficient of static friction to see if enough force is applied by pulling on the rope to generate a positive acceleration. If the resulting value for a turns out to be negative, this means that there is not enough pulling force to overcome the static friction force. With the given value of μ_s (0.076), we obtain

$$a' = \frac{25.3 \text{ N}}{15.3 \text{ kg}}(\cos 24.5° + 0.076\sin 24.5°) - 0.076(9.81 \text{ m/s}^2) = 0.81 \text{ m/s}^2.$$

Because this calculation results in a positive value for a', we know that the force is strong enough to overcome the friction force. We now use the given value for the coefficient of kinetic friction to calculate the actual acceleration of the sled:

$$a = \frac{25.3 \text{ N}}{15.3 \text{ kg}}(\cos 24.5° + 0.070\sin 24.5°) - 0.070(9.81 \text{ m/s}^2) = 0.87 \text{ m/s}^2.$$

PROBLEM 2
What angle of the rope with the horizontal will produce the maximum acceleration of the sled for the given value of the magnitude of the pulling force, T? What is that maximum value of a?

SOLUTION 2
In calculus, to find the extremum of a function, we take the first derivative and find the value of the independent variable for which that derivative is zero:

$$\frac{d}{d\theta}a = \frac{d}{d\theta}\left(\frac{T}{m}(\cos\theta + \mu\sin\theta) - \mu g\right) = \frac{T}{m}(-\sin\theta + \mu\cos\theta).$$

Searching for the root of this equation results in

$$\frac{da}{d\theta}\bigg|_{\theta=\theta_{max}} = \frac{T}{m}(-\sin\theta_{max} + \mu\cos\theta_{max}) = 0$$

$$\Rightarrow \sin\theta_{max} = \mu\cos\theta_{max} \Rightarrow$$

$$\theta_{max} = \tan^{-1}\mu.$$

Inserting the given value for the coefficient of kinetic friction, 0.070, into this equation results in $\theta_{max} = 4.0°$. This means that the rope should be oriented almost horizontally. The resulting value of the acceleration can be obtained by inserting the numbers into the equation for a we used in Solution 1:

$$a_{max} \equiv a(\theta_{max}) = 0.97 \text{ m/s}^2.$$

Note: A zero first derivative is only a necessary condition for a maximum, not a sufficient one. You can convince yourself that we have indeed found the maximum by first realizing that we obtained only one root of the first derivative, meaning that the function $a(\theta)$ has only one extremum. Also, because the value of the acceleration we calculated at this point is bigger than the value we previously obtained for 24.5°, we are assured that our single extremum is indeed a maximum. Alternatively, we could have taken the second derivative and found that it is negative at the point $\theta_{max} = 4.0°$; then, we could have compared the value of the acceleration obtained at that point with those at $\theta = 0°$ and $\theta = 90°$.

Self-Test Opportunity 4.5

What is the value of the second derivative if the angle of the sled's rope relative to the ground is 4.0°?

WHAT WE HAVE LEARNED | EXAM STUDY GUIDE

- The net force on an object is the vector sum of the forces acting on the object: $\vec{F}_{net} = \sum_{i=1}^{n}\vec{F}_i$.

- Mass is an intrinsic quality of an object that quantifies both the object's ability to resist acceleration and the gravitational force on the object.

- A free-body diagram is an abstraction showing all forces acting on an isolated object.

- Newton's three laws are as follows:

  Newton's First Law. In the absence of a net force on an object, the object will remain at rest if it was at rest. If it was moving, it will remain in motion in a straight line with the same velocity.

  Newton's Second Law. If a net external force, $\vec{F}_{net}$, acts on an object with mass m, the force will cause an acceleration, $\vec{a}$, in the same direction as the force: $\vec{F}_{net} = m\vec{a}$.

  Newton's Third Law. The forces that two interacting objects exert on each other are always exactly equal in magnitude and opposite in direction: $\vec{F}_{1\to2} = -\vec{F}_{2\to1}$.

- Two types of friction occur: static and kinetic friction. Both types of friction are proportional to the normal force, N.

  Static friction describes the force of friction between an object at rest on a surface in terms of the coefficient of static friction, μ_s. The static friction force, f_s, opposes a force trying to move an object and has a maximum value, $f_{s,max}$, such that $f_s \le \mu_s N = f_{s,max}$.

  Kinetic friction describes the force of friction between a moving object and a surface in terms of the coefficient of kinetic friction, μ_k. Kinetic friction is given by $f_k = \mu_k N$.

  In general, $\mu_s > \mu_k$.

ANSWERS TO SELF-TEST OPPORTUNITIES

4.1

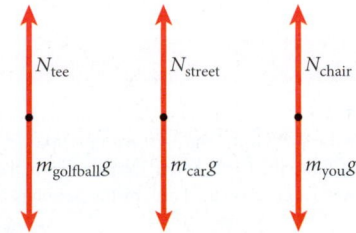

4.2 We have found for the acceleration of the two masses the general result $a = g(m_1 - m_2)/(m_1 + m_2)$. If m_1 approaches infinity, we can neglect m_2 compared to m_1. This means

that the numerator and denominator of the fraction both approach 1, and we have $a = g$. For $m_1 = 0$, we have $a = -g$, and for $m_1 = m_2$, we have $a = 0$.

4.3 Using $T = m_2(g + a)$ and inserting the value for the acceleration, $a = g\left(\dfrac{m_1 - m_2}{m_1 + m_2}\right)$, we find

$$T = m_2(g + a) = m_2\left(g + g\frac{m_1 - m_2}{m_1 + m_2}\right) = m_2 g\left(\frac{m_1 + m_2}{m_1 + m_2} + \frac{m_1 - m_2}{m_1 + m_2}\right)$$

$$= 2g\frac{m_1 m_2}{m_1 + m_2}.$$

4.4 (a) From $a = (m_2 g - f)/(m_1 + m_2)$ and the condition that $f \leq f_{s,max} = \mu_s m_1 g$, we see that the largest value m_2 can have is $m_2 = f_{s,max}/g = \mu_s m_1$. (b) For $m_2 < \mu_s m_1$, nothing moves; so the acceleration has to be zero, which means that the numerator of the expression for the acceleration must be zero. Therefore, in this case, $f = m_2 g$.

4.5 Taking the second derivative, we find

$$\frac{d^2 a}{d\theta^2} = \frac{d}{d\theta}\left(\frac{T}{m}(-\sin\theta + \mu\cos\theta)\right) = \frac{T}{m}(-\cos\theta - \mu\sin\theta) = -a.$$

Since we have found $a(\theta_{max}) = 0.97 \text{ m/s}^2$, the value of the second derivative at $4.00°$ is -0.97 m/s^2, which is smaller than zero, confirming that our extremum is indeed a maximum.

PROBLEM-SOLVING GUIDELINES: NEWTON'S LAWS

Analyzing a situation in terms of forces and motion is a vital skill in physics. One of the most important techniques is the proper application of Newton's laws. The following guidelines can help you solve mechanics problems in terms of Newton's three laws. These are part of the seven-step strategy for solving all types of physics problems and are most relevant to the Sketch, Think, and Research steps.

1. An overall sketch can help you visualize the situation and identify the concepts involved, but you also need a separate free-body diagram for each object to identify which forces act on that particular object and no others. Drawing correct free-body diagrams is the key to solving all problems in mechanics, whether they involve static (nonmoving) objects or kinetic (moving) ones. Remember that the $m\vec{a}$ from Newton's Second Law should not be included as a force in any free-body diagram.

2. Choosing the coordinate system is important—often the choice of coordinate system makes the difference between very simple equations and very difficult ones. Placing an axis along the same direction as an object's acceleration, if there is any, is often very helpful. In a statics problem, orienting an axis along a surface, whether horizontal or inclined, is often useful. Choosing the most advantageous coordinate system is an acquired skill gained through experience as you work many problems.

3. Once you have chosen your coordinate directions, determine whether the situation involves acceleration in either direction. If no acceleration occurs in the y-direction, for example, then Newton's First Law applies in that direction, and the sum of forces (the net force) equals zero. If acceleration does occur in a given direction, for example, the x-direction,

then Newton's Second Law applies in that direction, and the net force equals the object's mass times its acceleration.

4. When you decompose a force vector into components along the coordinate directions, be careful about which direction involves the sine of a given angle and which direction involves the cosine. Do not generalize from past problems and think that all components in the x-direction involve the cosine; you will find problems where the x-component involves the sine. Rely instead on clear definitions of angles and coordinate directions and the geometry of the given situation. Often the same angle appears at different points and between different lines in a problem. This usually results in similar triangles, often involving right angles. If you create a sketch of a problem with a general angle θ, try to use an angle that is not close to $45°$, because it is hard to distinguish between such an angle and its complement in your sketch.

5. Always check your final answer. Do the units make sense? Are the magnitudes reasonable? If you change a variable to approach some limiting value, does your answer make a valid prediction about what happens? Sometimes you can estimate the answer to a problem by using order-of-magnitude approximations, as discussed in Chapter 1; such an estimate can often reveal whether you made an arithmetical mistake or wrote down an incorrect formula.

6. The friction force always opposes the direction of motion and acts parallel to the contact surface; the static friction force opposes the direction in which the object would move, if the friction force were not present. Note that the kinetic friction force is *equal* to the product of the coefficient of friction and the normal force, whereas the static friction force is *less than or equal* to that product.

MULTIPLE-CHOICE QUESTIONS

4.1 A car of mass M travels in a straight line at constant speed along a level road with a coefficient of friction between the tires and the road of μ and a drag force of D. The magnitude of the net force on the car is

a) μMg.

b) $\mu Mg + D$.

c) $\sqrt{(\mu Mg)^2 + D^2}$.

d) zero.

4.2 A person stands on the surface of the Earth. The mass of the person is m, and the mass of the Earth is M. The person jumps upward, reaching a maximum height h above the Earth. When the person is at this height h, the magnitude of the force exerted on the Earth by the person is

a) mg.

b) Mg.

c) $M^2 g/m$.

d) $m^2 g/M$.

e) zero.

4.3 Leonardo da Vinci discovered that the magnitude of the friction force is simply proportional to the magnitude of the normal force only; that is, the friction force does not depend on the width or length of the contact area. Thus, the main reason to use wide tires on a race car is that they

a) look cool.

b) have more apparent contact area.

c) cost more.

d) can be made of softer materials.

4.4 The Tornado is a carnival ride that consists of a hollow vertical cylinder that rotates rapidly about its vertical axis. As the Tornado rotates, the riders are pressed against the inside wall of the cylinder by the rotation, and the floor of the cylinder drops away. The force that points upward, preventing the riders from falling downward, is

a) friction force.   c) gravity.

b) a normal force.   d) a tension force.

4.5 When a bus makes a sudden stop, passengers tend to jerk forward. Which of Newton's laws can explain this?

a) Newton's First Law

b) Newton's Second Law

c) Newton's Third Law

d) It cannot be explained by Newton's laws.

4.6 Only two forces, $\vec{F}_1$ and $\vec{F}_2$, are acting on a block. Which of the following can be the magnitude of the net force, $\vec{F}$, acting on the block (indicate all possibilities)?

a) $F > F_1 + F_2$ c) $F < F_1 + F_2$

b) $F = F_1 + F_2$ d) none of the above

4.7 Which of the following observations about the friction force is (are) incorrect?

a) The magnitude of the kinetic friction force is always proportional to the normal force.

b) The magnitude of the static friction force is always proportional to the normal force.

c) The magnitude of the static friction force is always proportional to the external applied force.

d) The direction of the kinetic friction force is always opposite the direction of the relative motion of the object with respect to the surface the object moves on.

e) The direction of the static friction force is always opposite that of the impending motion of the object relative to the surface it rests on.

f) All of the above are correct.

4.8 A horizontal force equal to the object's weight is applied to an object resting on a table. What is the acceleration of the moving object when the coefficient of kinetic friction between the object and floor is 1 (assuming the object is moving in the direction of the applied force)?

a) zero

b) 1 m/s^2

c) Not enough information is given to find the acceleration.

4.9 Two blocks of equal mass are connected by a massless horizontal rope and resting on a frictionless table. When one of the blocks is pulled away by a horizontal external force $\vec{F}$, what is the ratio of the net forces acting on the blocks?

a) 1:1   c) 1:2

b) 1:1.41   d) none of the above

4.10 If a cart sits motionless on level ground, there are no forces acting on the cart.

a) true   b) false   c) maybe

4.11 An object whose mass is 0.092 kg is initially at rest and then attains a speed of 75.0 m/s in 0.028 s. What average net force acted on the object during this time interval?

a) $1.2 \cdot 10^2 \text{ N}$ c) $2.8 \cdot 10^2 \text{ N}$

b) $2.5 \cdot 10^2 \text{ N}$ d) $4.9 \cdot 10^2 \text{ N}$

4.12 You push a large crate across the floor at constant speed, exerting a horizontal force F on the crate. There is friction between the floor and the crate. The force of friction has a magnitude that is

a) zero.   d) less than F.

b) F.   e) impossible to quantify without further

c) greater than F.   information.

4.13 Which of the following fundamental forces are not apparent to us in our everyday lives?

a) gravitational force   c) strong nuclear force

b) electromagnetic force   d) weak nuclear force

4.14 An SUV of mass 3250 kg has a head-on collision with a 1250-kg subcompact. Identify all the statements that are incorrect.

a) The SUV exerts a larger force on the subcompact than the subcompact exerts on the SUV.

b) The subcompact exerts a larger force on the SUV than the SUV exerts on the subcompact.

c) The subcompact experiences a larger acceleration than the SUV.

d) The SUV experiences a larger acceleration that the subcompact.

4.15 Which one of the following statements is correct?

a) The gravitational force on an object is always directed upward.

b) The gravitational force on an object is always directed downward.

c) The gravitational force on an object depends on the vertical speed of the object.

d) The gravitational force on an object depends on the horizontal speed of the object.

4.16 A normal force is a contact force that acts at the surface between two objects. Which of the following statements concerning the normal force is not correct?

a) The normal force is always equal to the force of gravity.

b) The normal force is just large enough to keep the two objects from penetrating each other.

c) The normal force is not necessarily equal to the force of gravity.

d) The normal force is perpendicular to the plane of the contact surface between the two objects.

CONCEPTUAL QUESTIONS

4.17 You are at the shoe store to buy a pair of basketball shoes that have the greatest traction on a specific type of hardwood. To determine the coefficient of static friction, μ, you place each shoe on a plank of the wood and tilt the plank to an angle θ, at which the shoe just starts to slide. Obtain an expression for μ as a function of θ.

4.18 A heavy wooden ball is hanging from a ceiling by a piece of string that is attached to the ceiling and to the top of the ball. A similar piece of string

is attached to the bottom of the ball. If the loose end of the lower string is pulled down sharply, which is the string that is most likely to break?

4.19 A car pulls a trailer down the highway. Let F_t be the magnitude of the force on the trailer due to the car, and let F_c be the magnitude of the force on the car due to the trailer. If the car and trailer are moving at a constant velocity across level ground, then $F_t = F_c$. If the car and trailer are accelerating up a hill, what is the relationship between the two forces?

4.20 A car accelerates down a level highway. What is the force in the direction of motion that accelerates the car?

4.21 If the forces that two interacting objects exert on each other are always exactly equal in magnitude and opposite in direction, how is it possible for an object to accelerate?

4.22 True or false: A physics book on a table will not move at all if and only if the net force is zero.

4.23 A mass slides on a ramp that is at an angle of θ above the horizontal. The coefficient of friction between the mass and the ramp is μ.

a) Find an expression for the magnitude and direction of the acceleration of the mass as it slides up the ramp.

b) Repeat part (a) to find an expression for the magnitude and direction of the acceleration of the mass as it slides down the ramp.

4.24 A shipping crate that weighs 340 N is initially stationary on a loading dock. A forklift arrives and lifts the crate with an upward force of 500 N, accelerating the crate upward. What is the magnitude of the force due to gravity acting on the shipping crate while it is accelerating upward?

4.25 A block is sliding on a (near) frictionless slope with an incline of 30.0°. Which force is greater in magnitude, the net force acting on the block or the normal force acting on the block?

4.26 A tow truck of mass M is using a cable to pull a shipping container of mass m across a horizontal surface as shown in the figure. The cable is attached to the container at the front bottom corner and makes an angle θ with the vertical as shown. The coefficient of kinetic friction between the surface and the crate is μ.

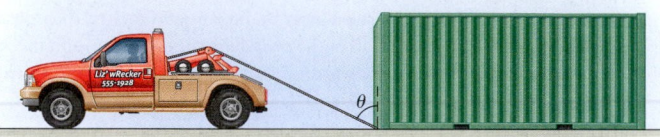

a) Draw a free-body diagram for the container.

b) Assuming that the truck pulls the container at a constant speed, write an equation for the magnitude T of the string tension in the cable.

EXERCISES

A blue problem number indicates a worked-out solution is available in the Student Solutions Manual. One • and two •• indicate increasing level of problem difficulty.

Section 4.2

4.27 The gravitational acceleration on the Moon is a sixth of that on Earth. The weight of an apple is 1.00 N on Earth.

a) What is the weight of the apple on the Moon?

b) What is the mass of the apple?

Section 4.4

4.28 A 423.5-N force accelerates a go-cart and its driver from 10.4 m/s to 17.9 m/s in 5.00 s. What is the mass of the go-cart plus driver?

4.29 You have just joined an exclusive health club, located on the top floor of a skyscraper. You reach the facility by using an express elevator. The elevator has a precision scale installed so that members can weigh themselves before and after their workouts. A member steps into the elevator and gets on the scale before the elevator doors close. The scale shows a weight of 183.7 lb. Then the elevator accelerates upward with an acceleration of 2.43 m/s², while the member is still standing on the scale. What is the weight shown by the scale's display while the elevator is accelerating?

4.30 An elevator cabin has a mass of 358.1 kg, and the combined mass of the people inside the cabin is 169.2 kg. The cabin is pulled upward by a cable, with a constant acceleration of 4.11 m/s². What is the tension in the cable?

4.31 An elevator cabin has a mass of 363.7 kg, and the combined mass of the people inside the cabin is 177.0 kg. The cabin is pulled upward by a cable, in which there is a tension force of 7638 N. What is the acceleration of the elevator?

4.32 Two blocks are in contact on a frictionless, horizontal tabletop. An external force, $\vec{F}$, is applied to block 1, and the two blocks are moving with a constant acceleration of 2.45 m/s². Use $M_1 = 3.20$ kg and $M_2 = 5.70$ kg.

a) What is the magnitude, F, of the applied force?

b) What is the contact force between the blocks?

c) What is the net force acting on block 1?

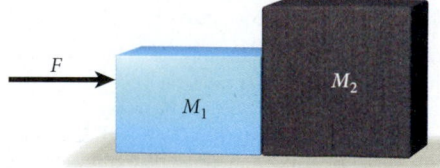

•**4.33** The density (mass per unit volume) of ice is 917 kg/m³, and the density of seawater is 1024 kg/m³. Only 10.45% of the volume of an iceberg is above the water's surface. If the volume of a particular iceberg that is above water is 4205.3 m³, what is the magnitude of the force that the seawater exerts on this iceberg?

•**4.34** In a physics laboratory class, three massless ropes are tied together at a point. A pulling force is applied along each rope: $F_1 = 150.$ N at 60.0°, $F_2 = 200.$ N at 100.°, $F_3 = 100.$ N at 190.°. What is the magnitude of a fourth force and the angle at which it acts to keep the point at the center of the system stationary? (All angles are measured from the positive x-axis.)

Section 4.5

4.35 Four weights, of masses $m_1 = 6.50$ kg, $m_2 = 3.80$ kg, $m_3 = 10.70$ kg, and $m_4 = 4.20$ kg, are hanging from a ceiling as shown in the figure. They are connected with ropes. What is the tension in the rope connecting masses m_1 and m_2?

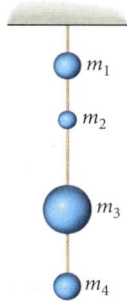

4.36 A hanging mass, $M_1 = 0.500$ kg, is attached by a light string that runs over a frictionless pulley to a mass $M_2 = 1.50$ kg that is initially at rest on a frictionless table. Find the magnitude of the acceleration, a, of M_2.

•**4.37** A hanging mass, $M_1 = 0.500$ kg, is attached by a light string that runs over a frictionless pulley to the front of a mass $M_2 = 1.50$ kg that is initially at rest on a frictionless table. A third mass $M_3 = 2.50$ kg, which is also initially at rest on the same frictionless table, is attached to the back of M_2 by a light string.

a) Find the magnitude of the acceleration, a, of mass M_3.

b) Find the tension in the string between masses M_1 and M_2.

•**4.38** A hanging mass, $M_1 = 0.400$ kg, is attached by a light string that runs over a frictionless pulley to a mass $M_2 = 1.20$ kg that is initially at rest on a frictionless ramp. The ramp is at an angle of $\theta = 30.0°$ above the horizontal, and the pulley is at the top of the ramp. Find the magnitude and direction of the acceleration, a_2, of M_2.

•**4.39** A force table is a circular table with a small ring that is to be balanced in the center of the table. The ring is attached to three hanging masses by strings of negligible mass that pass over frictionless pulleys mounted on the edge of the table. The magnitude and direction of each of the three horizontal forces acting on the ring can be adjusted by changing the amount of each hanging mass and the position of each pulley, respectively. Given a mass $m_1 = 0.0400$ kg pulling in the positive x-direction and a mass $m_2 = 0.0300$ kg pulling in the positive y-direction, find the mass (m_3) and the angle (θ, counterclockwise from the positive x-axis) that will balance the ring in the center of the table.

•**4.40** A monkey is sitting on a wood plate attached to a rope whose other end is passed over a tree branch, as shown in the figure. The monkey holds the rope and tries to pull it down. The combined mass of the monkey and the wood plate is 100. kg. Assume that you can neglect the friction between the rope and the branch.

a) What is the minimum force the monkey has to apply to lift itself and the plate off the ground?

b) What applied force is needed to move the monkey with an upward acceleration of 2.45 m/s²?

c) Explain how the answers would change if a second monkey on the ground pulled on the rope instead.

Section 4.6

4.41 A bosun's chair is a device used by a boatswain to lift himself to the top of the mainsail of a ship. A simplified device consists of a chair, a rope of negligible mass, and a frictionless pulley attached to the top of the mainsail. The rope goes over the pulley, with one end attached to the chair, and the boatswain pulls on the other end, lifting himself upward. The chair and boatswain have a total mass $M = 90.0$ kg.

a) If the boatswain is pulling himself up at a constant speed, with what magnitude of force must he pull on the rope?

b) If, instead, the boatswain moves in a jerky fashion, accelerating upward with a maximum acceleration of magnitude $a = 2.00$ m/s², with what maximum magnitude of force must he pull on the rope?

4.42 A granite block of mass 3311 kg is suspended from a pulley system as shown in the figure. The rope is wound around the pulleys 6 times. What is the force with which you would have to pull on the rope to hold the granite block in equilibrium?

4.43 Arriving on a newly discovered planet, the captain of a spaceship performed the following experiment to calculate the gravitational acceleration for the planet: She placed masses of 100.0 g and 200.0 g on an Atwood device made of massless string and a frictionless pulley and measured that it took 1.52 s for each mass to travel 1.00 m from rest.

a) What is the gravitational acceleration for the planet?

b) What is the tension in the string?

•**4.44** A store sign of mass 4.25 kg is hung by two wires that each make an angle of $\theta = 42.4°$ with the ceiling. What is the tension in each wire?

•**4.45** A crate of oranges slides down an inclined plane without friction. If it is released from rest and reaches a speed of 5.832 m/s after sliding a distance of 2.29 m, what is the angle of inclination of the plane with respect to the horizontal?

•**4.46** A load of bricks of mass $M = 200.0$ kg is attached to a crane by a cable of negligible mass and length $L = 3.00$ m. Initially, when the cable hangs vertically downward, the bricks are a horizontal distance $D = 1.50$ m from the wall where the bricks are to be placed. What is the magnitude of the horizontal force that must be applied to the load of bricks (without moving the crane) so that the bricks will rest directly above the wall?

•**4.47** A large ice block of mass $M = 80.0$ kg is held stationary on a frictionless ramp. The ramp is at an angle of $\theta = 36.9°$ above the horizontal.

a) If the ice block is held in place by a tangential force along the surface of the ramp (at angle θ above the horizontal), find the magnitude of this force.

b) If, instead, the ice block is held in place by a horizontal force, directed horizontally toward the center of the ice block, find the magnitude of this force.

•**4.48** A mass, $m_1 = 20.0$ kg, on a frictionless ramp is attached to a light string. The string passes over a frictionless pulley and is attached to a hanging mass, m_2. The ramp is at an angle of $\theta = 30.0°$ above the horizontal. The mass m_1 moves up the ramp uniformly (at constant speed). Find the value of m_2.

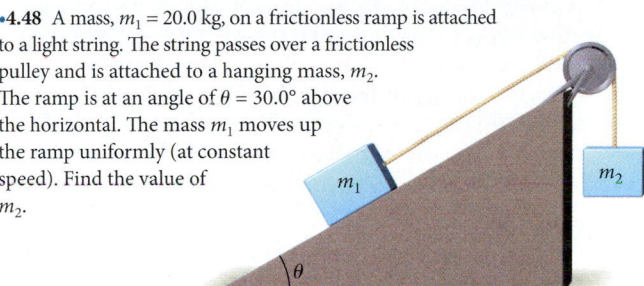

•**4.49** A piñata of mass $M = 8.00$ kg is attached to a rope of negligible mass that is strung between the tops of two vertical poles. The horizontal distance between the poles is $D = 2.00$ m, and the top of the right pole is a vertical distance $h = 0.500$ m higher than the top of the left pole. The piñata is attached to the rope at a horizontal position halfway between the two poles and at a vertical distance $s = 1.00$ m below the top of the left pole. Find the tension in each part of the rope due to the weight of the piñata.

••**4.50** A piñata of mass $M = 12.0$ kg hangs on a rope of negligible mass that is strung between the tops of two vertical poles. The horizontal distance between the poles is $D = 2.00$ m, the top of the right pole is a vertical distance $h = 0.500$ m higher than the top of the left pole, and the total length of the rope between the poles is $L = 3.00$ m. The piñata is attached to a ring, with the rope passing through the center of the ring. The ring is frictionless, so that it can slide freely along the rope until the piñata comes to a point of static equilibrium.

a) Determine the distance from the top of the left (lower) pole to the ring when the piñata is in static equilibrium.

b) What is the tension in the rope when the piñata is at this point of static equilibrium?

••**4.51** Three objects with masses $m_1 = 36.5$ kg, $m_2 = 19.2$ kg, and $m_3 = 12.5$ kg are hanging from ropes that run over pulleys. What is the acceleration of m_1?

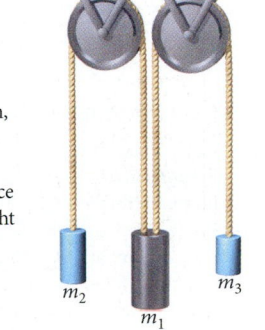

••**4.52** A rectangular block of width $w = 116.5$ cm, depth $d = 164.8$ cm, and height $h = 105.1$ cm is cut diagonally from one upper corner to the opposing lower corners so that a triangular surface is generated, as shown in the figure. A paperweight of mass $m = 16.93$ kg is sliding down the incline without friction. What is the magnitude of the acceleration that the paperweight experiences?

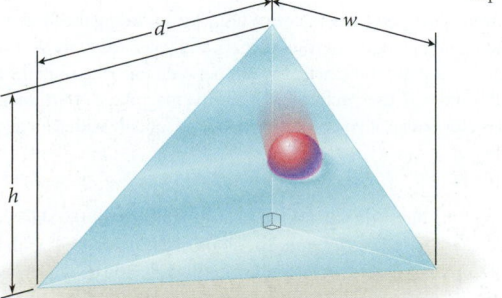

••**4.53** A large cubical block of ice of mass $M = 64.0$ kg and sides of length $L = 0.400$ m is held stationary on a frictionless ramp. The ramp is at an angle of $\theta = 26.0°$ above the horizontal. The ice cube is held in place by a rope of negligible mass and length $l = 1.60$ m. The rope is attached to the surface of the ramp and to the upper edge of the ice cube, a distance L above the surface of the ramp. Find the tension in the rope.

••**4.54** A bowling ball of mass $M_1 = 6.00$ kg is initially at rest on the sloped side of a wedge of mass $M_2 = 9.00$ kg that is on a frictionless horizontal floor. The side of the wedge is sloped at an angle of $\theta = 36.9°$ above the horizontal.

a) What is the magnitude of the horizontal force that should be exerted on the bowling ball to keep it at a constant height on the slope?

b) What is the magnitude of the acceleration of the wedge, if no external force is applied?

Section 4.7

4.55 A skydiver of mass 82.3 kg (including outfit and equipment) floats downward suspended from her parachute, having reached terminal speed. The drag coefficient is 0.533, and the area of her parachute is 20.11 m². The density of air is 1.14 kg/m³. What is the air's drag force on her?

4.56 The elapsed time for a top fuel dragster to start from rest and travel in a straight line a distance of $\frac{1}{4}$ mile (402 m) is 4.441 s. Find the minimum coefficient of friction between the tires and the track needed to achieve this result. (Note that the minimum coefficient of friction is found from the simplifying assumption that the dragster accelerates with constant acceleration. For this problem we neglect the downward forces from spoilers and the exhaust pipes.)

4.57 An engine block of mass M is on the flatbed of a pickup truck that is traveling in a straight line down a level road with an initial speed of 30.0 m/s. The coefficient of static friction between the block and the bed is $\mu_s = 0.540$. Find the minimum distance in which the truck can come to a stop without the engine block sliding toward the cab.

•**4.58** A box of books is initially at rest a distance $D = 0.540$ m from the end of a wooden board. The coefficient of static friction between the box and the board is $\mu_s = 0.320$, and the coefficient of kinetic friction is $\mu_k = 0.250$. The angle of the board is increased slowly, until the box just begins to slide; then the board is held at this angle. Find the speed of the box as it reaches the end of the board.

•**4.59** A block of mass $M_1 = 0.640$ kg is initially at rest on a cart of mass $M_2 = 0.320$ kg with the cart initially at rest on a level air track. The coefficient of static friction between the block and the cart is $\mu_s = 0.620$, but there is essentially no friction between the air track and the cart. The cart is accelerated by a force of magnitude F parallel to the air track. Find the maximum value of F that allows the block to accelerate with the cart, without sliding on top of the cart.

Section 4.8

4.60 Coffee filters behave like small parachutes, with a drag force that is proportional to the velocity squared, $F_{drag} = Kv^2$. A single coffee filter, when dropped from a height of 2.00 m, reaches the ground in a time of 3.00 s. When a second coffee filter is nestled within the first, the drag coefficient remains the same, but the weight is doubled. Find the time for the combined filters to reach the ground. (Neglect the brief period when the filters are accelerating up to their terminal speed.)

4.61 Your refrigerator has a mass of 112.2 kg, including the food in it. It is standing in the middle of your kitchen, and you need to move it. The coefficients of static and kinetic friction between the fridge and the tile floor are 0.460 and 0.370, respectively. What is the magnitude of the force of friction acting on the fridge, if you push against it horizontally with a force of each magnitude?

a) 300.0 N   b) 500.0 N   c) 700.0 N

•**4.62** On the bunny hill at a ski resort, a towrope pulls the skiers up the hill with constant speed of 1.74 m/s. The slope of the hill is 12.4° with respect to the horizontal. A child is being pulled up the hill. The coefficients of static

and kinetic friction between the child's skis and the snow are 0.152 and 0.104, respectively, and the child's mass is 62.4 kg, including clothing and equipment. What is the force with which the towrope has to pull on the child?

•**4.63** A skier starts with a speed of 2.00 m/s and skis straight down a slope with an angle of 15.0° relative to the horizontal. The coefficient of kinetic friction between her skis and the snow is 0.100. What is her speed after 10.0 s?

••**4.64** A block of mass $m_1 = 21.9$ kg is at rest on a plane inclined at $\theta = 30.0°$ above the horizontal. The block is connected via a rope and massless pulley system to another block of mass $m_2 = 25.1$ kg, as shown in the figure. The coefficients of static and kinetic friction between block 1 and the inclined plane are $\mu_s = 0.109$ and $\mu_k = 0.086$, respectively. If the blocks are released from rest, what is the displacement of block 2 in the vertical direction after 1.51 s? Use positive numbers for the upward direction and negative numbers for the downward direction.

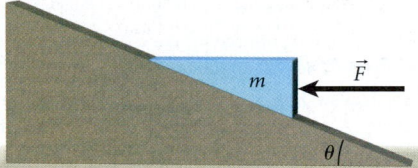

••**4.65** A wedge of mass $m = 36.1$ kg is located on a plane that is inclined by an angle $\theta = 21.3°$ with respect to the horizontal. A force $F = 302.3$ N in the horizontal direction pushes on the wedge, as shown in the figure. The coefficient of kinetic friction between the wedge and the plane is 0.159. What is the acceleration of the wedge along the plane?

••**4.66** A chair of mass M rests on a level floor, with a coefficient of static friction $\mu_s = 0.560$ between the chair and the floor. A person wishes to push the chair across the floor. He pushes downward on the chair with a force F at an angle θ relative to the horizontal. What is the minimum value of θ for which the chair will not start to move across the floor, no matter how large F gets?

••**4.67** As shown in the figure, blocks of masses $m_1 = 250.0$ g and $m_2 = 500.0$ g are attached by a massless string over a frictionless and massless pulley. The coefficients of static and kinetic friction between the block and inclined plane are 0.250 and 0.123, respectively. The angle of the incline is $\theta = 30.0°$, and the blocks are at rest initially.

a) In which direction do the blocks move?

b) What is the acceleration of the blocks?

••**4.68** A block of mass $M = 500.0$ g sits on a horizontal tabletop. The coefficients of static and kinetic friction are 0.530 and 0.410, respectively, at the contact surface between table and block. The block is pushed on with a 10.0 N external force at an angle θ with the horizontal.

a) What angle will lead to the maximum acceleration of the block for a given pushing force?

b) What is the maximum acceleration?

Additional Exercises

4.69 A car without ABS (antilock brake system) was moving at 15.0 m/s when the driver slammed on the brakes to make a sudden stop. The coefficients of static and kinetic friction between the tires and the road are 0.550 and 0.430, respectively.

a) What was the acceleration of the car during the interval between braking and stopping?

b) How far did the car travel before it stopped?

4.70 A 2.00-kg block (M_1) and a 6.00-kg block (M_2) are connected by a massless string. Applied forces, $F_1 = 10.0$ N and $F_2 = 5.00$ N, act on the blocks, as shown in the figure.

a) What is the acceleration of the blocks?

b) What is the tension in the string?

c) What is the net force acting on M_1? (Neglect friction between the blocks and the table.)

4.71 An elevator contains two masses: $M_1 = 2.00$ kg is attached by a string (string 1) to the ceiling of the elevator, and $M_2 = 4.00$ kg is attached by a similar string (string 2) to the bottom of mass 1.

a) Find the tension in string 1 (T_1) if the elevator is moving upward at a constant velocity of $v = 3.00$ m/s.

b) Find T_1 if the elevator is accelerating upward with an acceleration of $a = 3.00$ m/s^2.

4.72 What coefficient of friction is required to stop a hockey puck sliding at 12.5 m/s initially over a distance of 60.5 m?

4.73 A spring of negligible mass is attached to the ceiling of an elevator. When the elevator is stopped at the first floor, a mass M is attached to the spring, stretching the spring a distance D until the mass is in equilibrium. As the elevator starts upward toward the second floor, the spring stretches an additional distance $D/4$. What is the magnitude of the acceleration of the elevator? Assume the force provided by the spring is linearly proportional to the distance stretched by the spring.

4.74 A crane of mass $M = 1.00 \cdot 10^4$ kg lifts a wrecking ball of mass $m = 1200.$ kg directly upward.

a) Find the magnitude of the normal force exerted on the crane by the ground while the wrecking ball is moving upward at a constant speed of $v = 1.00$ m/s.

b) Find the magnitude of the normal force if the wrecking ball's upward motion slows at a constant rate from its initial speed $v = 1.00$ m/s to a stop over a distance $D = 0.250$ m.

4.75 A block of mass 20.0 kg supported by a vertical massless cable is initially at rest. The block is then pulled upward with a constant acceleration of 2.32 m/s^2.

a) What is the tension in the cable?

b) What is the net force acting on the mass?

c) What is the speed of the block after it has traveled 2.00 m?

4.76 Three identical blocks, A, B, and C, are on a horizontal frictionless table. The blocks are connected by strings of negligible mass, with block B between the other two blocks. If block C is pulled horizontally by a force of magnitude $F = 12.0$ N, find the tension in the string between blocks B and C.

•4.77 A block of mass $m_1 = 3.00$ kg and a block of mass $m_2 = 4.00$ kg are suspended by a massless string over a frictionless pulley with negligible mass, as in an Atwood machine. The blocks are held motionless and then released. What is the acceleration of the two blocks?

•4.78 Two blocks of masses m_1 and m_2 are suspended by a massless string over a frictionless pulley with negligible mass, as in an Atwood machine. The blocks are held motionless and then released. If $m_1 = 3.50$ kg, what value does m_2 have to have for the system to experience an acceleration of $a = 0.400\ g$? (*Hint:* There are two solutions to this problem.)

•4.79 A tractor pulls a sled of mass $M = 1000.$ kg across level ground. The coefficient of kinetic friction between the sled and the ground is $\mu_k = 0.600$. The tractor pulls the sled by a rope that connects to the sled at an angle of $\theta = 30.0°$ above the horizontal. What magnitude of tension in the rope is necessary to move the sled horizontally with an acceleration $a = 2.00$ m/s^2?

•4.80 A 2.00-kg block is on a plane inclined at 20.0° with respect to the horizontal. The coefficient of static friction between the block and the plane is 0.600.

a) How many forces are acting on the block?

b) What is the normal force?

c) Is this block moving? Explain.

•4.81 A block of mass 5.00 kg is sliding at a constant velocity down an inclined plane that makes an angle of 37.0° with respect to the horizontal.

a) What is the friction force?

b) What is the coefficient of kinetic friction?

•4.82 A skydiver of mass 83.7 kg (including outfit and equipment) falls in the spread-eagle position, having reached terminal speed. Her drag coefficient is 0.587, and her surface area that is exposed to the air stream is 1.035 m^2. How long does it take her to fall a vertical distance of 296.7 m? (The density of air is 1.14 kg/m^3.)

•4.83 A 0.500-kg physics textbook is hanging from two massless wires of equal length attached to a ceiling. The tension on each wire is measured as 15.4 N. What is the angle of the wires with the horizontal?

•4.84 In the figure, an external force $\vec{F}$ is holding a bob of mass 500. g in a stationary position. The angle that the massless rope makes with the vertical is $\theta = 30.0°$.

a) What is the magnitude, F, of the force needed to maintain equilibrium?

b) What is the tension in the rope?

•4.85 In a physics class, a 2.70-g ping-pong ball was suspended from a massless string. The string makes an angle of $\theta = 15.0°$ with the vertical when air is blown horizontally at the ball at a speed of 20.5 m/s. Assume that the friction force is proportional to the squared speed of the air stream.

a) What is the proportionality constant in this experiment?

b) What is the tension in the string?

•4.86 A nanowire is a (nearly) one-dimensional structure with a diameter on the order of a few nanometers. Suppose a 100.0-nm-long nanowire made of pure silicon (density of Si = 2.33 g/cm^3) has a diameter of 5.00 nm. This nanowire is attached at the top and hanging down vertically due to the force of gravity.

a) What is the tension at the top?

b) What is the tension in the middle?

(*Hint:* Treat the nanowire as a cylinder of diameter 5.00 nm and length 100.0 nm, made of silicon.)

•4.87 Two blocks are stacked on a frictionless table, and a horizontal force F is applied to the top block (block 1). Their masses are $m_1 = 2.50$ kg and $m_2 = 3.75$ kg. The coefficients of static and kinetic friction between the blocks are 0.456 and 0.380, respectively.

a) What is the maximum applied force F for which m_1 will not slide off m_2?

b) What are the accelerations of m_1 and m_2 when the force $F = 24.5$ N is applied to m_1?

•4.88 Two blocks ($m_1 = 1.23$ kg and $m_2 = 2.46$ kg) are glued together and are moving downward on an inclined plane having an angle of 40.0° with respect to the horizontal. Both blocks are lying flat on the surface of the inclined plane. The coefficients of kinetic friction are 0.23 for m_1 and 0.35 for m_2. What is the acceleration of the blocks?

•4.89 A marble block of mass $m_1 = 567.1$ kg and a granite block of mass $m_2 = 266.4$ kg are connected to each other by a rope that runs over

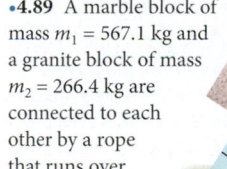
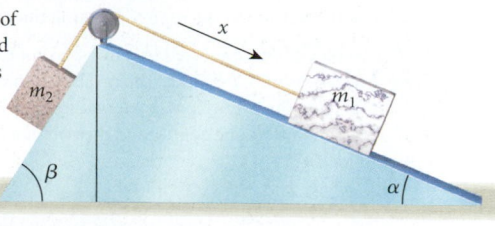

a pulley, as shown in the figure. Both blocks are located on inclined planes, with angles $\alpha = 39.3°$ and $\beta = 53.2°$. Both blocks move without friction, and the rope glides over the pulley without friction. What is the acceleration of the marble block? Note that the positive x-direction is indicated in the figure.

••**4.90** A marble block of mass $m_1 = 559.1$ kg and a granite block of mass $m_2 = 128.4$ kg are connected to each other by a rope that runs over a pulley as shown in the figure with Problem 4.89. Both blocks are located on inclined planes with angles $\alpha = 38.3°$ and $\beta = 57.2°$. The rope glides over the pulley without friction, but the coefficient of friction between block 1 and the inclined plane is $\mu_1 = 0.130$, and that between block 2 and the inclined plane is $\mu_2 = 0.310$. (For simplicity, assume that the coefficients of static and kinetic friction are the same in each case.) What is the acceleration of the marble block? Note that the positive x-direction is indicated in the figure.

••**4.91** As shown in the figure, two masses, $m_1 = 3.50$ kg and $m_2 = 5.00$ kg, are on a frictionless tabletop and mass $m_3 = 7.60$ kg is hanging from m_1. The coefficients of static and kinetic friction between m_1 and m_2 are 0.600 and 0.500, respectively.

a) What are the accelerations of m_1 and m_2?

b) What is the tension in the string between m_1 and m_3?

••**4.92** A block of mass $m_1 = 2.30$ kg is placed in front of a block of mass $m_2 = 5.20$ kg, as shown in the figure. The coefficient of static friction between m_1 and m_2 is 0.65, and there is negligible friction between the larger block and the tabletop.

a) What forces are acting on m_1?

b) What is the minimum external force F that can be applied to m_2 so that m_1 does not fall?

c) What is the contact force between m_1 and m_2?

d) What is the net force acting on m_2 when the force found in part (b) is applied?

••**4.93** A suitcase of weight $Mg = 450.$ N is being pulled by a small strap across a level floor. The coefficient of kinetic friction between the suitcase and the floor is $\mu_k = 0.640$.

a) Find the optimal angle of the strap above the horizontal. (The optimal angle minimizes the force necessary to pull the suitcase at constant speed.)

b) Find the minimum tension in the strap needed to pull the suitcase at constant speed.

••**4.94** As shown in the figure, a block of mass $M_1 = 0.450$ kg is initially at rest on a slab of mass $M_2 = 0.820$ kg, and the slab is initially at rest on a level table. A string of negligible mass is connected to the slab, runs over a frictionless pulley on the edge of the table, and is attached to a hanging mass M_3. The block rests on the slab but is not tied to the string, so friction provides the only horizontal force on the block. The slab has a coefficient of kinetic friction $\mu_k = 0.340$ and a coefficient of static friction $\mu_s = 0.560$ with both the table and the block. When released, M_3 pulls on the string and accelerates the slab, which accelerates the block. Find the maximum mass of M_3 that allows the block to accelerate with the slab, without sliding on top of the slab.

••**4.95** As shown in the figure with Problem 4.94, a block of mass $M_1 = 0.250$ kg is initially at rest on a slab of mass $M_2 = 0.420$ kg, and the slab is initially at rest on a level table. A string of negligible mass is connected to the slab, runs over a frictionless pulley on the edge of the table, and is attached to a hanging mass $M_3 = 1.80$ kg. The block rests on the slab but is not tied to the string, so friction provides the only horizontal force on the block. The slab has a coefficient of kinetic friction $\mu_k = 0.340$ with both the table and the block. When released, M_3 pulls on the string, which accelerates the slab so quickly that the block starts to slide on the slab. Before the block slides off the top of the slab:

a) Find the magnitude of the acceleration of the block.

b) Find the magnitude of the acceleration of the slab.

MULTI-VERSION EXERCISES

4.96 Two blocks are connected by a massless rope, as shown in the figure. Block 1 has mass $m_1 = 1.267$ kg, and block 2 has mass $m_2 = 3.557$ kg. The two blocks move on a frictionless, horizontal tabletop. A horizontal external force, $F = 12.61$ N, acts on block 2. What is the tension in the rope connecting the two blocks?

4.97 Two blocks are connected by a massless rope as shown in the figure. Block 2 has mass $m_2 = 3.577$ kg. The two blocks move on a frictionless, horizontal tabletop. A horizontal external force, $F = 13.89$ N, acts on block 2. The tension in the rope connecting the two blocks is 4.094 N. What is the mass of block 1?

4.98 Two blocks are connected by a massless rope as shown in the figure. Block 1 has mass $m_1 = 1.725$ kg. The two blocks move on a frictionless, horizontal tabletop. A horizontal external force, $F = 15.17$ N, acts on block 2. The tension in the rope connecting the two blocks is 4.915 N. What is the mass of block 2?

4.99 Two blocks are connected by a massless rope as shown in the figure. Block 1 has mass $m_1 = 1.955$ kg, and block 2 has mass $m_2 = 3.619$ kg. The two blocks move on a frictionless, horizontal tabletop. The tension in the rope connecting the two blocks is 5.777 N. What is the magnitude of the horizontal external force F that is acting on block 2?

Pulley 1   Pulley 2

m_1 m_2

4.100 Two blocks are connected by a massless rope running over two frictionless pulleys, as shown in the figure. Block 1 has mass $m_1 = 1.183$ kg, and block 2 has mass $m_2 = 3.639$ kg. The blocks are released from rest. What is the acceleration of block 1?

4.101 Two blocks are connected by a massless rope running over two frictionless pulleys, as shown in the figure. Block 1 has mass $m_1 = 1.411$ kg. The blocks are released from rest. Block 1 accelerates upward with an acceleration of 4.352 m/s². What is the mass of block 2?

4.102 Two blocks are connected by a massless rope running over two frictionless pulleys, as shown in the figure. Block 2 has mass $m_2 = 3.681$ kg. The blocks are released from rest. Block 2 accelerates downward with an acceleration of 3.760 m/s². What is the mass of block 1?

4.103 A curling stone of mass 19.00 kg is released with an initial speed v_0 and slides on level ice. The coefficient of kinetic friction between the curling stone and the ice is 0.01869. The curling stone travels a distance of 36.01 m before it stops. What is the initial speed of the curling stone?

4.104 A curling stone of mass 19.00 kg is released with an initial speed $v_0 = 2.788$ m/s and slides on level ice. The coefficient of kinetic friction between the curling stone and the ice is 0.01097. How far does the curling stone travel before it stops?

4.105 A curling stone of mass 19.00 kg is released with an initial speed $v_0 = 3.070$ m/s and slides on level ice. The curling stone travels 36.21 m before it stops. What is the coefficient of kinetic friction between the curling stone and the ice?

5 Kinetic Energy, Work, and Power

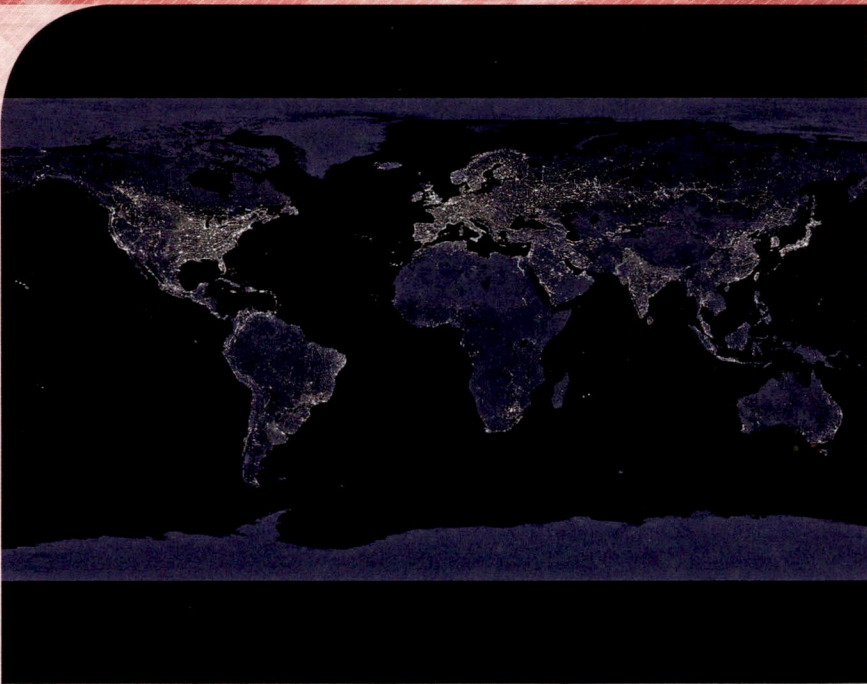

FIGURE 5.1 A composite image of NASA satellite photographs taken at night. Photos were taken from November 1994 through March 1995.

Figure 5.1 is a composite image of satellite photographs taken at night, showing which parts of the world use the most energy for nighttime illumination. Not surprisingly, the United States, Western Europe, and Japan stand out. The amount of light emitted by a region during the night is a good measure of the amount of energy that region consumes.

In physics, energy has a fundamental significance: Practically no physical activity takes place without the expenditure or transformation of energy. Calculations involving the energy of a system are of primary importance in all of science and engineering. As we'll see in this chapter, problem-solving methods involving energy provide an alternative to working with Newton's laws and are often simpler and easier to use.

This chapter presents the concepts of kinetic energy, work, and power and introduces some techniques that use these ideas, such as the work–kinetic energy theorem, to solve several types of problems. Chapter 6 will introduce additional types of energy and expand the work–kinetic energy theorem to cover these; it will also discuss one of the great ideas in physics and, indeed, in all of science: the law of conservation of energy.

WHAT WE WILL LEARN

- Kinetic energy is the energy associated with the motion of an object.

- Work is energy transferred to an object or from an object as the result of the action of an external force. Positive work transfers energy to the object, and negative work transfers energy from the object.

- Work is the scalar product of the force vector and the displacement vector.

- The change in kinetic energy due to applied forces is equal to the work done by the forces.

- Power is the rate at which work is done.

- The power provided by a force acting on an object is the scalar product of the velocity vector for that object and the force vector.

5.1 Energy in Our Daily Lives

No physical quantity has a greater importance in our daily lives than energy. Energy consumption, energy efficiency, and energy "production" are of the utmost economic importance and are the focus of heated discussions about national policies and international agreements. (The word *production* is in quotes because energy is not produced but rather is converted from a less usable form to a more usable form.) Energy also has an important role in each individual's daily routine: energy intake through food calories and energy consumption through cellular processes, activities, work, and exercise. Weight loss or weight gain is ultimately due to an imbalance between energy intake and use.

Energy has many forms and requires several different approaches to cover completely. Thus, energy is a recurring theme throughout this book. We start in this chapter and the next by investigating forms of mechanical energy: kinetic energy and potential energy. But as we progress through the topics in this book you will see that other forms of energy are playing very important roles as well. Thermal energy is one of the central pillars of thermodynamics. Chemical energy is stored in chemical compounds, and chemical reactions can either consume energy from the environment (endothermic reactions) or yield usable energy to the surroundings (exothermic reactions). Our petroleum economy makes use of chemical energy and its conversion to mechanical energy and heat, which is another form of energy (or energy transfer).

In Chapter 31, we will see that electromagnetic radiation contains energy. This energy is the basis for one renewable form of energy—solar energy. Almost all other renewable energy sources on Earth can be traced back to solar energy. Solar energy is responsible for the wind that drives large wind turbines (Figure 5.2). The Sun's radiation is also responsible for evaporating water from the Earth's surface and moving it into the clouds, from which it falls down as rain and eventually joins rivers that can be dammed (Figure 5.3) to extract energy. Biomass, another renewable energy resource, depends on the ability of plants and animals to store solar energy during their metabolic and growth processes.

FIGURE 5.2 Wind farms harvest renewable energy.

(a)   (b)   (c)

FIGURE 5.3 Dams provide renewable electrical energy. (a) The Grand Coulee Dam on the Columbia River in Washington. (b) The Itaipú Dam on the Paraná River in Brazil and Paraguay. (c) The Three Gorges Dam on the Yangtze River in China.

(a)

(b)

FIGURE 5.4 (a) Solar farm with an adjustable array of mirrors; (b) solar panel.

In fact, the energy radiated onto the surface of Earth by the Sun exceeds the energy needs of the entire human population by a factor of more than 10,000. It is possible to convert solar energy directly into electrical energy by using photovoltaic cells (Figure 5.4b). Currently, intense research efforts are focused on increasing the efficiency and reliability of these photocells while reducing their cost. Versions of solar cells are already being used for some practical purposes, for example, in patio and garden lights. Experimental solar farms like the one in Figure 5.4a are in operation as well. The chapter on quantum physics (Chapter 36) will discuss in detail how photocells work. Problems with using solar energy are that it is not available at night, has seasonal variations, and is strongly reduced in cloudy conditions or bad weather. Depending on the installation and conversion methods used, present solar devices convert only 10–15% of solar energy into electrical energy; increasing this fraction is a key goal of research activity. Materials with a 30% or higher yield of electrical energy from solar energy have been developed in the laboratory but are still not deployed on an industrial scale. Using biomass to generate electricity, in comparison, has much lower efficiencies of solar energy capture, on the order of 1% or less.

The Earth itself contains useful energy in the form of heat, which we can harvest using geothermal power plants. Iceland satisfies about half of its energy needs from this resource. The world's largest geothermal power plant complex, the Geysers, is located in Northern California and provides approximately 60% of that region's electricity. Also, the currents, tides, and waves of Earth's oceans can be exploited to extract useful energy. These and other alternative energy resources are the focus of intensive research, and the near future will see impressive developments.

In Chapter 35, on relativity, we will see that energy and mass are not totally separate concepts but are related to each other via Einstein's famous formula $E = mc^2$. When we study nuclear physics (Chapter 40), we will find that splitting massive atomic nuclei (such as uranium or plutonium) liberates energy. Conventional nuclear power plants are based on this physical principle, called *nuclear fission.* We can also obtain useful energy by merging atomic nuclei with very small masses (hydrogen, for example) to form more massive nuclei, a process called *nuclear fusion.* The Sun and all other stars in the universe use nuclear fusion to generate energy.

The energy from nuclear fusion is thought by many to be the most likely means of satisfying the long-term energy needs of modern industrialized society. Perhaps the most likely approach to achieving progress toward controlled fusion reactions is the proposed international nuclear fusion reactor facility ITER ("the way" in Latin), which will be constructed in France. But there are other promising approaches to solving the problem of how to use nuclear fusion, for example, the National Ignition Facility (NIF) opened in May 2009 at Lawrence Livermore National Laboratory in California. We will discuss these technologies in greater detail in Chapter 40.

Related to energy are work and power. We all use these words informally, but this chapter will explain how these quantities relate to energy in precise physical and mathematical terms.

You can see that energy occupies an essential place in our lives. One of the goals of this book is to give you a solid grounding in the fundamentals of energy science. Then you will be able to participate in some of the most important policy discussions of our time in an informed manner.

A final question remains: What is energy? In many textbooks, energy is defined as the ability to do work. However, this definition only shifts the mystery without giving a deeper explanation. And the truth is that there is no deeper explanation. In his famous *Feynman Lectures on Physics,* the Nobel laureate and physics folk hero Richard Feynman wrote in 1963: "It is important to realize that in physics today, we have no knowledge of what energy *is.* We do not have a picture that energy comes in little blobs of a definite amount. It is not that way. However, there are formulas for calculating some numerical quantity, and when we add it all together it gives '28'—always the same number. It is an abstract thing in that it does not tell us the mechanism or the *reasons* for the various formulas." Five decades later, this has not changed. The concept of energy and, in particular, the law of energy conservation (see Chapter 6), are extremely useful tools for figuring out the behavior of systems. But no one has yet given an explanation as to the true nature of energy.

5.2 Kinetic Energy

The first kind of energy we'll consider is the energy associated with the motion of a moving object: **kinetic energy.** Kinetic energy is defined as one-half the product of a moving object's mass and the square of its speed:

$$K = \tfrac{1}{2}mv^2. \tag{5.1}$$

Note that, by definition, kinetic energy is always positive or equal to zero, and it is only zero for an object at rest. Also note that kinetic energy, like all forms of energy, is a scalar, not a vector, quantity. Because it is the product of mass (kg) and speed squared (m/s·m/s), the units of kinetic energy are kg m²/s². Because energy is such an important quantity, it has its own SI unit, the **joule (J).** The SI force unit, the newton, is 1 N = 1 kg m/s², and we can make a useful conversion:

$$\text{Energy unit:} \quad 1\,\text{J} = 1\,\text{N m} = 1\,\text{kg m}^2/\text{s}^2. \tag{5.2}$$

Let's look at a few sample energy values to get a feeling for the size of the joule. A car of mass 1310 kg being driven at the speed limit of 55 mph (24.6 m/s) has a kinetic energy of

$$K_{car} = \tfrac{1}{2}mv^2 = \tfrac{1}{2}(1310\,\text{kg})(24.6\,\text{m/s})^2 = 4.0 \cdot 10^5 \,\text{J}.$$

The mass of the Earth is $6.0 \cdot 10^{24}$ kg, and it orbits the Sun with a speed of $3.0 \cdot 10^4$ m/s. The kinetic energy associated with this motion is $2.7 \cdot 10^{33}$ J. A person of mass 64.8 kg jogging at 3.50 m/s has a kinetic energy of 397 J, and a baseball (mass of "5 ounces avoirdupois" = 0.142 kg) thrown at 80. mph (35.8 m/s) has a kinetic energy of 91 J. On the atomic scale, the average kinetic energy of an air molecule is $6.1 \cdot 10^{-21}$ J, as we will see in Chapter 19. The typical magnitudes of kinetic energies of some moving objects are presented in Figure 5.5. You can see from these examples that the range of energies involved in physical processes is tremendously large.

Some other frequently used energy units are the electron-volt (eV), the food calorie (Cal), and the mega-ton of TNT (Mt):

$$1\,\text{eV} = 1.602 \cdot 10^{-19}\,\text{J}$$

$$1\,\text{Cal} = 4186\,\text{J}$$

$$1\,\text{Mt} = 4.18 \cdot 10^{15}\,\text{J}.$$

On the atomic scale, 1 electron-volt (eV) is the kinetic energy that an electron gains when accelerated by an electric potential of 1 volt. The energy content of the food we eat is usually (and mistakenly) given in terms of calories but should be given in food calories. As we'll see when we study thermodynamics, 1 food calorie is equal to 1 kilocalorie; a nice round number to remember is that about 10 MJ (~2500 food calories) of energy is stored in the food we eat each day. On a larger scale, 1 Mt is the energy released by exploding 1 million metric tons of the explosive TNT, an energy release achieved only by nuclear weapons or by catastrophic natural events such as the impact of a large asteroid. For comparison, in 2007,

FIGURE 5.5 Range of kinetic energies displayed on a logarithmic scale. The kinetic energies (left to right) of an air molecule, a red blood cell traveling through the aorta, a mosquito in flight, a thrown baseball, a moving car, and the Earth orbiting the Sun are compared with the energy released from a 15-Mt nuclear explosion and by a supernova, which emits particles with a total kinetic energy of approximately 10^{46} J.

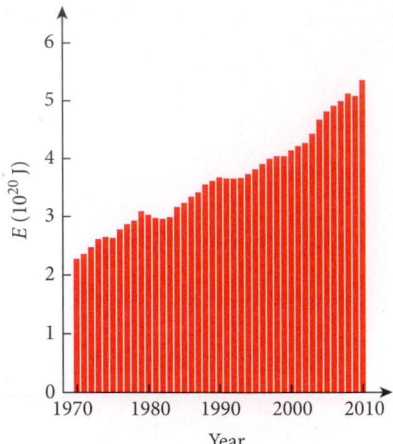

FIGURE 5.6 Total annual global energy consumption by humans from 1970 to 2010. (Data sources: US Energy Information Agency and yearbook.enerdata.net).

the annual energy consumption by all humans on Earth reached $5 \cdot 10^{20}$ J (see Figure 5.6). (All of these concepts will be discussed further in subsequent chapters.)

For motion in more than one dimension, we can write the total kinetic energy as the sum of the kinetic energies associated with the components of velocity in each spatial direction. To show this, we start with the definition of kinetic energy (equation 5.1) and then use $v^2 = v_x^2 + v_y^2 + v_z^2$:

$$K = \tfrac{1}{2}mv^2 = \tfrac{1}{2}m\left(v_x^2 + v_y^2 + v_z^2\right) = \tfrac{1}{2}mv_x^2 + \tfrac{1}{2}mv_y^2 + \tfrac{1}{2}mv_z^2. \tag{5.3}$$

(*Note:* Kinetic energy is a scalar, so these components are not added like vectors but simply by taking their algebraic sum.) Thus, we can think of kinetic energy as the sum of the kinetic energies associated with the motion in the x-direction, y-direction, and z-direction. This concept is particularly useful for ideal projectile problems, where the motion consists of free fall in the vertical direction (y-direction) and motion with constant velocity in the horizontal direction (x-direction).

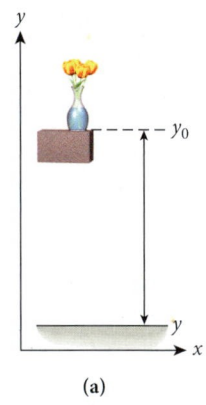

EXAMPLE 5.1 **Falling Vase**

PROBLEM

A crystal vase (mass = 2.40 kg) is dropped from a height of 1.30 m and falls to the floor, as shown in Figure 5.7. What is its kinetic energy just before impact? (Neglect air resistance for now.)

SOLUTION

Once we know the velocity of the vase just before impact, we can put it into the equation defining kinetic energy. To obtain this velocity, we recall the kinematics of free-falling objects. In this case, it is most straightforward to use the relationship between the initial and final velocities and heights that we derived in Chapter 2 for free-fall motion:

$$v_y^2 = v_{y0}^2 - 2g(y - y_0).$$

(Remember that the y-axis must be pointing up to use this equation.) Because the vase is released from rest, the initial velocity components are $v_{x0} = v_{y0} = 0$. Because there is no acceleration in the x-direction, the x-component of velocity remains zero during the fall of the vase: $v_x = 0$. Therefore, we have

$$v^2 = v_x^2 + v_y^2 = 0 + v_y^2 = v_y^2.$$

We then obtain

$$v^2 = v_y^2 = 2g(y_0 - y).$$

We use this result in equation 5.1:

$$K = \tfrac{1}{2}mv^2 = \tfrac{1}{2}m\left(2g(y_0 - y)\right) = mg(y_0 - y).$$

Inserting the numbers given in the problem statement gives us the answer:

$$K = (2.40 \text{ kg})(9.81 \text{ m/s}^2)(1.30 \text{ m}) = 30.6 \text{ J}.$$

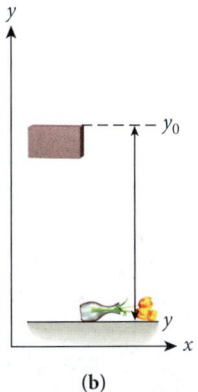

FIGURE 5.7 (a) A vase is released from rest at a height of y_0. (b) The vase falls to the floor, which has a height of y.

5.3 Work

In Example 5.1, the vase started out with zero kinetic energy, just before it was released. After falling a distance of 1.30 m, it had acquired a kinetic energy of 30.6 J. The greater the height from which the vase is released, the greater the speed the vase will attain (ignoring air resistance), and therefore the greater its kinetic energy becomes. In fact, as we found in Example 5.1, the kinetic energy of the vase depends linearly on the height from which it falls: $K = mg(y_0 - y)$.

The gravitational force, $\vec{F}_g = -mg\hat{y}$, accelerates the vase and therefore gives it its kinetic energy. We can see from the equation we used in Example 5.1 that the kinetic energy also depends linearly on the magnitude of the gravitational force. Doubling the mass of the vase would double the gravitational force acting on it and thus double its kinetic energy.

Because the speed of an object can be increased or decreased by accelerating or decelerating it, respectively, its kinetic energy also changes in this process. For the vase, we have just seen that the force of gravity is responsible for this change. We account for a change in the kinetic energy of an object caused by a force with the concept of work, W.

> **Definition**
>
> **Work** is the energy transferred to or from an object as the result of the action of a force. Positive work is a transfer of energy to the object, and negative work is a transfer of energy from the object.

The vase gained kinetic energy from positive work done by the gravitational force and so $W_g = mg(y_0 - y)$.

Note that this definition is not restricted to kinetic energy. The relationship between work and energy described in this definition holds in general for different forms of energy besides kinetic energy. This definition of work is not exactly the same as the meaning attached to the word *work* in everyday language. The work being considered in this chapter is mechanical work in connection with energy transfer. However, the work, physical as well as mental, that we commonly speak of does not necessarily involve the transfer of energy.

5.4 Work Done by a Constant Force

Suppose we let the vase of Example 5.1 slide, from rest, along an inclined plane that has an angle θ with respect to the horizontal (Figure 5.8). For now, we neglect the friction force, but we will come back to it later. As we showed in Chapter 4, in the absence of friction, the acceleration along the plane is given by $a = g \sin \theta = g \cos \alpha$. (Here the angle $\alpha = 90° - \theta$ is the angle between the gravitational force vector and the displacement vector; see Figure 5.8.)

We can determine the kinetic energy the vase has in this situation as a function of the displacement, $\Delta \vec{r}$. Most conveniently, we can perform this calculation by using the relationship between the squares of initial and final velocities, the displacement, and the acceleration, which we obtained for one-dimensional motion in Chapter 2:

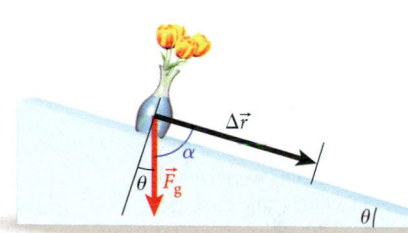

FIGURE 5.8 Vase sliding without friction on an inclined plane.

$$v^2 = v_0^2 + 2a\Delta r.$$

We set $v_0 = 0$ because we are again assuming that the vase is released from rest, that is, with zero kinetic energy. Then we use the expression for the acceleration, $a = g \cos \alpha$, that we just obtained. Now we have

$$v^2 = (2g\cos\alpha)\Delta r \Rightarrow K = \tfrac{1}{2}mv^2 = mg\Delta r\cos\alpha.$$

The kinetic energy transferred to the vase was the result of positive work done by the gravitational force, and so

$$\Delta K = mg\Delta r\cos\alpha = W_g. \qquad (5.4)$$

Let's look at two limiting cases for equation 5.4:

- For $\alpha = 0$, both the gravitational force and the displacement are in the negative y-direction. Thus, these vectors are parallel, and we have the result we already derived for the case of the vase falling under the influence of gravity, $W_g = mg\Delta r$.

- For $\alpha = 90°$, the gravitational force is still in the negative y-direction, but the vase cannot move in the negative y-direction because it is sitting on the horizontal surface of the plane. Hence, there is no change in the kinetic energy of the vase, and there is no work done by the gravitational force on the vase; that is, $W_g = 0$. The work done on the vase by the gravitational force is also zero if the vase moves at a constant speed along the surface of the plane.

Self-Test Opportunity 5.1

Draw the free-body diagram for the vase that is sliding down the inclined plane.

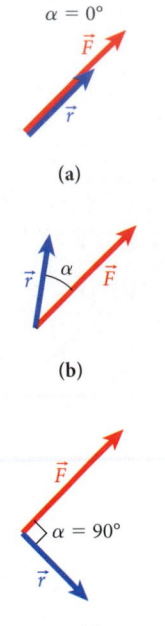

$\alpha = 0°$

(a)

(b)

$\alpha = 90°$

(c)

FIGURE 5.9 (a) $\vec{F}$ is parallel to $\vec{r}$ and $W = |\vec{F}||\vec{r}|$. (b) The angle between $\vec{F}$ and $\vec{r}$ is α and $W = |\vec{F}||\vec{r}|\cos\alpha$. (c) $\vec{F}$ is perpendicular to $\vec{r}$ and $W = 0$.

Concept Check 5.1

Consider an object undergoing a displacement $\Delta\vec{r}$ and experiencing a force $\vec{F}$. In which of the three cases shown below is the work done by the force on the object zero?

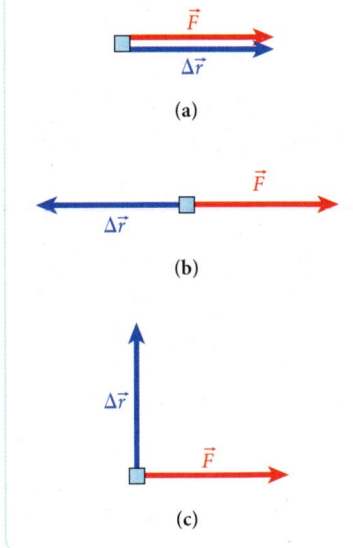

(a)

(b)

(c)

Because $mg = |\vec{F}_g|$ and $\Delta r = |\Delta\vec{r}|$, we can write the work done on the vase as $W = |\vec{F}||\Delta\vec{r}|\cos\alpha$. From the two limiting cases we have just discussed, we gain confidence that we can use the equation we have just derived for motion on an inclined plane as the definition of the work done by a constant force:

$$W = |\vec{F}||\Delta\vec{r}|\cos\alpha, \quad \text{where } \alpha \text{ is the angle between } \vec{F} \text{ and } \Delta\vec{r}.$$

This equation for the work done by a constant force acting over some spatial displacement holds for all constant force vectors, arbitrary displacement vectors, and angles between the two. Figure 5.9 shows three cases for the work done by a force $\vec{F}$ acting over a displacement $\vec{r}$. In Figure 5.9a, the maximum work is done because $\alpha = 0$ and $\vec{F}$ and $\vec{r}$ are in the same direction. In Figure 5.9b, $\vec{F}$ is at an arbitrary angle α with respect to $\vec{r}$. In Figure 5.9c, no work is done because $\vec{F}$ is perpendicular to $\vec{r}$.

Using a scalar product (see Section 1.6), we can write the work done by a constant force as

$$W = \vec{F} \cdot \Delta\vec{r}. \tag{5.5}$$

This equation is the main result of this section. It says that the work done by a constant force $\vec{F}$ in displacing an object by $\Delta\vec{r}$ is the scalar product of the two vectors. In particular, if the displacement is perpendicular to the force, the scalar product is zero, and no work is done.

Note that we can use any force vector and any displacement vector in equation 5.5. If there is more than one force acting on an object, the equation holds for any of the individual forces, and it holds for the net force. The mathematical reason for this generalization lies in the distributive property of the scalar product. To verify this statement, we can look at a constant net force that is the sum of individual constant forces, $\vec{F}_{net} = \sum_i \vec{F}_i$. According to equation 5.5, the work done by this net force is

$$W_{net} = \vec{F}_{net} \cdot \Delta\vec{r} = \left(\sum_i \vec{F}_i\right) \cdot \Delta\vec{r} = \sum_i \left(\vec{F}_i \cdot \Delta\vec{r}\right) = \sum_i W_i.$$

In other words, the net work done by the net force is equal to the sum of the work done by the individual forces. We have demonstrated this additive property of work only for constant forces, but it is also valid for variable forces (or, strictly speaking, only for conservative forces, which we'll encounter in Chapter 6). But to repeat the main point: Equation 5.5 is valid for each individual force as well as the net force. We will typically consider the net force when calculating the work done on an object, but we will omit the index "net" to simplify the notation.

One-Dimensional Case

In all cases of motion in one dimension, the work done to produce the motion is given by

$$\begin{aligned} W &= \vec{F} \cdot \Delta\vec{r} \\ &= \pm F_x \cdot |\Delta\vec{r}| = F_x \Delta x \\ &= F_x (x - x_0). \end{aligned} \tag{5.6}$$

The force $\vec{F}$ and displacement $\Delta\vec{r}$ can point in the same direction, $\alpha = 0 \Rightarrow \cos\alpha = 1$, resulting in positive work, or they can point in opposite directions, $\alpha = 180° \Rightarrow \cos\alpha = -1$, resulting in negative work.

Work–Kinetic Energy Theorem

The relationship between kinetic energy of an object and the work done by the force(s) acting on it, called the **work–kinetic energy theorem**, is expressed formally as

$$\Delta K \equiv K - K_0 = W. \tag{5.7}$$

Here, K is the kinetic energy that an object has after work W has been done on it and K_0 is the kinetic energy before the work is done. The definitions of W and K are such that equation 5.7 is equivalent to Newton's Second Law. To see this equivalence, consider a constant force acting in one dimension on an object of mass m. Newton's Second Law is then $F_x = ma_x$, and the (also constant!) acceleration, a_x, of the object is related to the difference in the squares of its initial and final velocities via $v_x^2 - v_{x0}^2 = 2a_x(x - x_0)$, which is one of the five kinematical equations we derived in Chapter 2. Multiplication of both sides of this equation by $\frac{1}{2}m$ yields

$$\frac{1}{2}mv_x^2 - \frac{1}{2}mv_{x0}^2 = ma_x(x - x_0) = F_x\Delta x = W. \tag{5.8}$$

Thus, we see that, for this one-dimensional case, the work–kinetic energy theorem is equivalent to Newton's Second Law.

Because of the equivalence we have just established, if more than one force is acting on an object, we can use the net force to calculate the work done. Alternatively, and more commonly in energy problems, if more than one force is acting on an object, we can calculate the work done by each force, and then W in equation 5.7 represents their sum.

The work–kinetic energy theorem specifies that the change in kinetic energy of an object is equal to the work done on the object by the forces acting on it. We can rewrite equation 5.7 to solve for K or K_0:

$$K = K_0 + W$$

or

$$K_0 = K - W.$$

By definition, the kinetic energy cannot be less than zero; so, if an object has $K_0 = 0$, the work–kinetic energy theorem implies that $K = K_0 + W = W \geq 0$.

While we have only verified the work–kinetic energy theorem for a constant force, it is also valid for variable forces, as we will see below. Is it valid for all kinds of forces? The short answer is no! Friction forces are one kind of force that violate the work–kinetic energy theorem. We will discuss this point further in Chapter 6.

Work Done by the Gravitational Force

With the work–kinetic energy theorem at our disposal, we can now take another look at the problem of an object falling under the influence of the gravitational force, as in Example 5.1. On the way down, the work done by the gravitational force on the object is

$$W_g = +mgh, \tag{5.9}$$

where $h = |y - y_0| = |\Delta\vec{r}| > 0$. The displacement $\Delta\vec{r}$ and the force of gravity $\vec{F}_g$ point in the same direction, resulting in a positive scalar product and therefore positive work. This situation is illustrated in Figure 5.10a. Since the work is positive, the gravitational force increases the kinetic energy of the object.

We can reverse this situation and toss the object vertically upward, making it a projectile and giving it an initial kinetic energy. This kinetic energy will decrease until the projectile reaches the top of its trajectory. During this time, the displacement vector $\Delta\vec{r}$ points up, in the opposite direction to the force of gravity (Figure 5.10b). Thus, the work done by the gravitational force during the object's upward motion is

$$W_g = -mgh. \tag{5.10}$$

Therefore, the work done by the gravitational force reduces the kinetic energy of the object during its upward motion. This conclusion is consistent with the general formula for work done by a constant force, $W = \vec{F} \cdot \Delta\vec{r}$, because the displacement (pointing upward) of the object and the gravitational force (pointing downward) are in opposite directions.

Work Done in Lifting and Lowering an Object

Now let's consider the situation in which a vertical external force is applied to an object—for example, by attaching the object to a rope and lifting it up or lowering it down. The

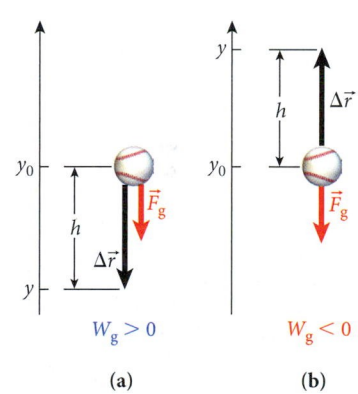

$W_g > 0$ $W_g < 0$

(a)   (b)

FIGURE 5.10 Work done by the gravitational force. (a) The object during free fall. (b) Tossing an object upward.

work–kinetic energy theorem now has to include the work done by the gravitational force, W_g, and the work done by the external force, W_F:

$$K - K_0 = W_g + W_F.$$

For the case where the object is at rest both initially, $K_0 = 0$, and finally, $K = 0$, we have

$$W_F = -W_g.$$

The work done by force in lifting or lowering the object is then

$$W_F = -W_g = mgh \text{ (for lifting) or } W_F = -W_g = -mgh \text{ (for lowering).} \tag{5.11}$$

EXAMPLE 5.2 | Weightlifting

In the sport of weightlifting, the task is to pick up a very large mass, lift it over your head, and hold it there at rest for a moment. This action is an example of doing work by lifting and lowering a mass.

PROBLEM 1

The German lifter Ronny Weller won the silver medal at the Olympic Games in Sydney, Australia, in 2000. He lifted 257.5 kg in the "clean and jerk" competition. Assuming he lifted the mass to a height of 1.83 m and held it there, what was the work he did in this process?

SOLUTION 1

This problem is an application of equation 5.11 for the work done against the gravitational force. The work Weller did was

$$W = mgh = (257.5 \text{ kg})(9.81 \text{ m/s}^2)(1.83 \text{ m}) = 4.62 \text{ kJ}.$$

PROBLEM 2

Once Weller successfully completed the lift and was holding the mass with outstretched arms above his head, what was the work done by him in lowering the weight slowly (with negligible kinetic energy) back down to the ground?

SOLUTION 2

This calculation is the same as that in Solution 1 except the sign of the displacement changes. Thus, the answer is that –4.62 kJ of work is done in bringing the weight back down—exactly the opposite of what we obtained for Problem 1!

Now is a good time to remember that we are dealing with strictly mechanical work. Every lifter knows that you can feel the muscles "burn" just as much when holding the mass overhead or lowering the mass (in a controlled way) as when lifting it. (In Olympic competitions, by the way, the weightlifters just drop the mass after the successful lift.) However, this physiological effect is not mechanical work, which is what we are presently interested in. Instead it is the conversion of chemical energy, stored in different molecules such as sugars, into the energy needed to contract the muscles.

You may think that Olympic weightlifting is not the best example to consider, because the force used to lift the mass is not constant. This is true, but as we discussed previously, the work–kinetic energy theorem applies to nonconstant forces. Additionally, even when a crane lifts a mass very slowly and with constant speed, the lifting force is still not exactly constant, because a slight initial acceleration is needed to get the mass from zero speed to a finite value and a deceleration occurs at the end of the lifting process.

Lifting with Pulleys

When we studied pulleys and ropes in Chapter 4, we learned that pulleys act as force multipliers. For example, with the setup shown in Figure 5.11, the force needed to lift a pallet of bricks of mass m by pulling on the rope is only half the gravitational force, $T = \frac{1}{2}mg$. How does the work done in pulling up the pallet of bricks with ropes and pulleys compare to the work of lifting it without such mechanical aids?

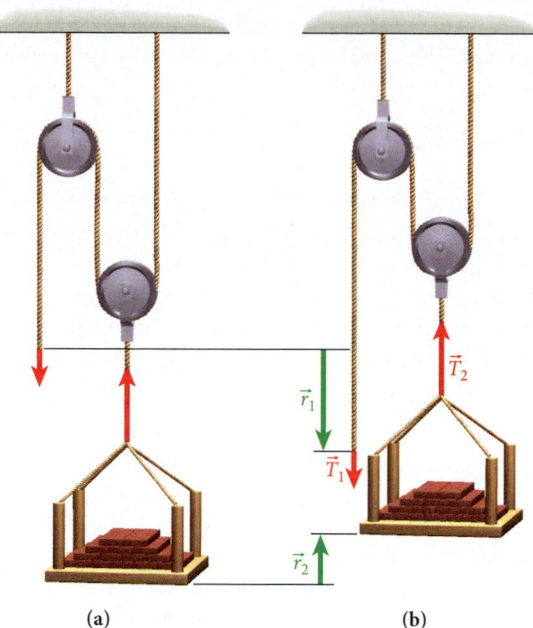

FIGURE 5.11   Forces and displacements for the process of lifting a pallet of bricks at a work site with the aid of a rope-and-pulley mechanism. (a) Pallet in the initial position. (b) Pallet in the final position.

Figure 5.11 shows the initial and final positions of the pallet of bricks and the ropes and pulleys used to lift it. Lifting it without mechanical aids would require the force $\vec{T}_2$, as indicated, whose magnitude is given by $T_2 = mg$. The work done by force $\vec{T}_2$ in this case is $W_2 = \vec{T}_2 \cdot \vec{r}_2 = T_2 r_2 = mgr_2$. Pulling on the rope with force $\vec{T}_1$ of magnitude $T_1 = \frac{1}{2}T_2 = \frac{1}{2}mg$ accomplishes the same thing. However, the displacement is then twice as long, $r_1 = 2r_2$, as you can see by examining Figure 5.11. Thus, the work done in this case is $W_1 = \vec{T}_1 \cdot \vec{r}_1 = (\frac{1}{2}T_2)(2r_2) = mgr_2 = W_2$.

The same amount of work is done in both cases. It is necessary to compensate for the reduced force by pulling the rope through a longer distance. This result is general for the use of pulleys or lever arms or any other mechanical force multiplier: The total work done is the same as it would be if the mechanical aid were not used. Any reduction in the force is always going to be compensated for by a proportional lengthening of the displacement.

We end this section with a solved problem, which we solve with kinetic energy concepts of Section 5.2. As a double-check step we use the concepts of work for a constant force, which we developed in the present section, showing that both paths lead to the same answer.

Concept Check 5.2

If you lift an object a distance h with the aid of a rope and n pulleys and do work W_h in the process, how much work will be required to lift the same object a distance $2h$?

a) W_h d) nW_h

b) $2W_h$ e) $2W_h/n$

c) $0.5W_h$

SOLVED PROBLEM 5.1   Shot Put

PROBLEM

Shot put competitions use metal balls with a mass of 16 lb (7.26 kg). A competitor throws the shot at an angle of 43.3° and releases it from a height of 1.82 m above where it lands, and it lands a horizontal distance of 17.7 m from the point of release. What is the kinetic energy of the shot as it leaves the thrower's hand?

SOLUTION

THINK   We are given the horizontal distance, $x_s = 17.7$ m, the height of release, $y_0 = 1.82$ m, and the angle of the initial velocity, $\theta_0 = 43.3°$, but not the initial speed, v_0. If we can figure out the initial speed from the given data, then calculating the initial kinetic energy will be straightforward because we also know the mass of the shot: $m = 7.26$ kg.

Because the shot is very heavy, air resistance can be safely ignored. This situation is an excellent realization of ideal projectile motion. After the shot leaves the thrower's hand, the

– Continued

only force on the shot is the force of gravity, and the shot will follow a parabolic trajectory until it lands on the ground. Thus, we'll solve this problem by application of the rules about ideal projectile motion.

SKETCH The trajectory of the shot is shown in Figure 5.12.

RESEARCH The initial kinetic energy K of the shot of mass m is given by

$$K = \tfrac{1}{2}mv_0^2.$$

Now we need to decide how to obtain v_0. We are given the distance, x_s, to where the shot hits the ground, but this is *not* equal to the range, R (for which we obtained a formula in Chapter 3), because the range formula assumes that the heights of the start and end of the trajectory are equal. Here the initial height of the shot is y_0, and the final height is zero. Therefore, we have to use the full expression for the trajectory of an ideal projectile from Chapter 3:

$$y = y_0 + x\tan\theta_0 - \frac{x^2 g}{2v_0^2\cos^2\theta_0}.$$

This equation describes the y-component of the trajectory as a function of the x-component.

In this problem, we know that $y(x = x_s) = 0$, that is, that the shot touches the ground at $x = x_s$. Substituting for x when $y = 0$ in the equation for the trajectory results in

$$0 = y_0 + x_s\tan\theta_0 - x_s^2 \frac{g}{2v_0^2\cos^2\theta_0}.$$

SIMPLIFY We solve this equation for v_0^2:

$$y_0 + x_s\tan\theta_0 = \frac{x_s^2 g}{2v_0^2\cos^2\theta_0} \Rightarrow$$

$$2v_0^2\cos^2\theta_0 = \frac{x_s^2 g}{y_0 + x_s\tan\theta_0} \Rightarrow$$

$$v_0^2 = \frac{x_s^2 g}{2\cos^2\theta_0\left(y_0 + x_s\tan\theta_0\right)}.$$

Now, substituting for v_0^2 in the expression for the initial kinetic energy gives us

$$K = \tfrac{1}{2}mv_0^2 = \frac{mx_s^2 g}{4\cos^2\theta_0\left(y_0 + x_s\tan\theta_0\right)}.$$

CALCULATE Putting in the given numerical values, we get

$$K = \frac{(7.26\text{ kg})(17.7\text{ m})^2(9.81\text{ m/s}^2)}{4\left(\cos^2 43.3°\right)\left[1.82\text{ m} + (17.7\text{ m})(\tan 43.3°)\right]} = 569.295\text{ J}.$$

ROUND All of the numerical values given for this problem had three significant figures, so we report our answer as

$$K = 569\text{ J}.$$

DOUBLE-CHECK Since we have an expression for the initial speed, $v_0^2 = x_s^2 g / [2\cos^2\theta_0(y_0 + x_s\tan\theta_0)]$, we can find the horizontal and vertical components of the initial velocity vector:

$$v_{x0} = v_0\cos\theta_0 = 9.11\text{ m/s}$$

$$v_{y0} = v_0\sin\theta_0 = 8.59\text{ m/s}.$$

As we discussed in Section 5.2, we can split up the total kinetic energy in ideal projectile motion into contributions from the motion in horizontal and vertical directions (see equation 5.3). The kinetic energy due to the motion in the x-direction remains constant. The kinetic energy due to the motion in the y-direction is initially

$$\tfrac{1}{2}mv_{y0}^2 = 268\text{ J}.$$

At the top of the shot's trajectory, the vertical velocity component is zero, as in all projectile motion. This also means that the kinetic energy associated with the vertical motion is zero at this point. All 268 J of the initial kinetic energy due to the y-component of the motion has been used up to do work against the force of gravity. This work is (refer to equation 5.10) –268 J = –mgh, where $h = y_{max} - y_0$ is the maximum height of the trajectory. We thus find the value of h:

$$h = \frac{268\text{ J}}{mg} = \frac{268\text{ J}}{(7.26\text{ kg})(9.81\text{ m/s}^2)} = 3.76\text{ m}.$$

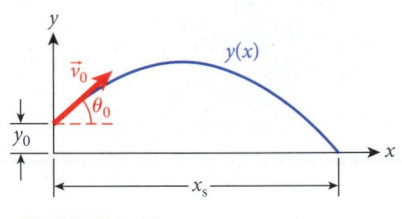

y

$\vec{v}_0$

θ_0

$y(x)$

y_0

x_s

x

FIGURE 5.12 Parabolic trajectory of a thrown shot.

Let's use known concepts of projectile motion to find the maximum height for the initial velocity we have determined. In Section 3.4, the maximum height H of an object in projectile motion was shown to be

$$H = y_0 + \frac{v_{y0}^2}{2g}.$$

Putting in the numbers gives $v_{y0}^2 / 2g = 3.76$ m. This value is the same as that obtained by applying energy considerations.

5.5 Work Done by a Variable Force

Suppose the force acting on an object is not constant. What is the work done by such a force? In a case of motion in one dimension with a variable x-component of force, $F_x(x)$, the work is

$$W = \int_{x_0}^{x} F_x(x')dx'. \tag{5.12}$$

(The integrand has x' as a dummy variable to distinguish it from the integral limits.) Equation 5.12 shows that the work W is the area under the curve of $F_x(x)$ (see Figure 5.13 in the following derivation).

DERIVATION 5.1

If you have already taken integral calculus, you can skip this section. If equation 5.12 is your first exposure to integrals, the following derivation is a useful introduction. We'll derive the one-dimensional case and use our result for the constant force as a starting point.

In the case of a constant force, we can think of the work as the area under the horizontal line that plots the value of the constant force in the interval between x_0 and x. For a variable force, the work is the area under the curve $F_x(x)$, but that area is no longer a simple rectangle. In the case of a variable force, we need to divide the interval from x_0 to x into many small equal intervals. Then we approximate the area under the curve $F_x(x)$ by a series of rectangles and add their areas to approximate the work. As you can see from Figure 5.13a, the area of the rectangle between x_i and x_{i+1} is given by $F_x(x_i)(x_{i+1} - x_i) = F_x(x_i)\Delta x$. We obtain an approximation for the work by summing over all rectangles:

$$W \approx \sum_i W_i = \sum_i F_x(x_i)\Delta x.$$

Now we space the points x_i closer and closer by using more and more of them. This method makes Δx smaller and causes the total area of the series of rectangles to be a better approximation of the area under the curve $F_x(x)$, as shown in Figure 5.13b. In the limit as $\Delta x \to 0$, the sum approaches the exact expression for the work:

$$W = \lim_{\Delta x \to 0}\left(\sum_i F_x(x_i)\Delta x\right).$$

This limit of the sum of the areas is exactly how the integral is defined:

$$W = \int_{x_0}^{x} F_x(x')dx'.$$

We have derived this result for the case of one-dimensional motion. The derivation of the three-dimensional case proceeds along similar lines but is more involved in terms of algebra.

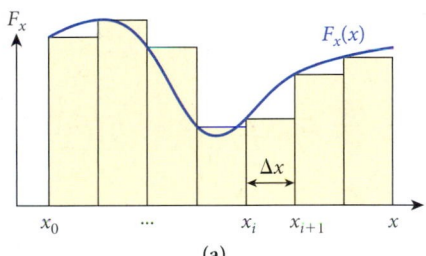

(a)

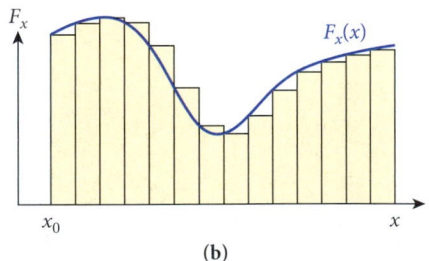

(b)

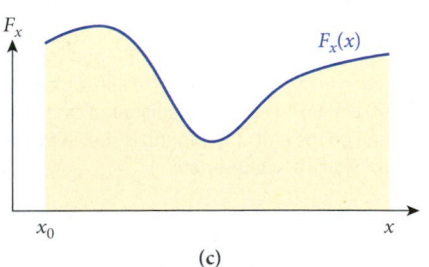

(c)

FIGURE 5.13 (a) A series of rectangles approximates the area under the curve obtained by plotting the force as a function of the displacement; (b) a better approximation using rectangles of smaller width; (c) the exact area under the curve.

As promised earlier, we can verify that the work–kinetic energy theorem (equation 5.7) is valid when the force is variable. We show this result for one-dimensional motion for simplicity, but the work–kinetic energy theorem also holds for variable forces and displacements in more than one dimension. We assume a variable force in the x-direction, $F_x(x)$, as in equation 5.12, which we can express as

$$F_x(x) = ma,$$

using Newton's Second Law. We use the chain rule of calculus to obtain

$$a = \frac{dv}{dt} = \frac{dv}{dx}\frac{dx}{dt}.$$

We can then use equation 5.12 and integrate over the displacement to get the work done:

$$W = \int_{x_0}^{x} F_x(x')dx' = \int_{x_0}^{x} ma\,dx' = \int_{x_0}^{x} m\frac{dv}{dx'}\frac{dx'}{dt}dx'.$$

We now change the variable of integration from displacement (x) to velocity (v):

$$W = \int_{x_0}^{x} m\frac{dx'}{dt}\frac{dv}{dx'}dx' = \int_{v_0}^{v} mv'\,dv' = m\int_{v_0}^{v} v'\,dv',$$

where v' is a dummy variable of integration. We carry out the integration and obtain the promised result:

$$W = m\int_{v_0}^{v} v'\,dv' = m\left[\frac{v'^2}{2}\right]_{v_0}^{v} = \frac{1}{2}mv^2 - \frac{1}{2}mv_0^2 = K - K_0 = \Delta K.$$

5.6 Spring Force

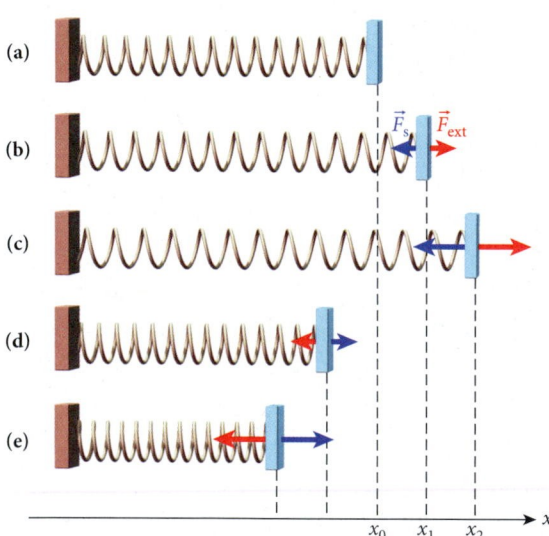

FIGURE 5.14 Spring force. The spring is in its equilibrium position in (a), is stretched in (b) and (c), and is compressed in (d) and (e). In each nonequilibrium case, the external force acting on the end of the spring is shown as a red arrow, and the spring force as a blue arrow.

Let's examine the force that is needed to stretch or compress a spring. We start with a spring that is neither stretched nor compressed from its normal length and take the end of the spring in this condition to be located at the equilibrium position, x_0, as shown in Figure 5.14a. If we pull the end of this spring a bit toward the right using an external force, $\vec{F}_{ext}$, the spring gets longer. In the stretching process, the spring generates a force directed to the left, that is, pointing toward the equilibrium position, and increasing in magnitude with increasing length of the spring. This force is conventionally called the **spring force, $\vec{F}_s$.**

Pulling with an external force of a given magnitude stretches the spring to a certain displacement from equilibrium, at which point the spring force is equal in magnitude to the external force (Figure 5.14b). Doubling this external force doubles the displacement from equilibrium (Figure 5.14c). Conversely, pushing with an external force toward the left compresses the spring from its equilibrium length, and the resulting spring force points to the right, again toward the equilibrium position (Figure 5.14d). Doubling the amount of compression (Figure 5.14e) also doubles the spring force, just as with stretching.

We can summarize these observations by noting that the magnitude of the spring force is proportional to the magnitude of the displacement of the end of the spring from its equilibrium position, and that the spring force always points toward the equilibrium position and thus is in the direction opposite to the displacement vector:

$$\vec{F}_s = -k(\vec{x} - \vec{x}_0). \tag{5.13}$$

As usual, this vector equation can be written in terms of components; in particular, for the x-component, we can write

$$F_s = -k\left(x - x_0\right). \tag{5.14}$$

The constant k is by definition always positive. The negative sign in front of k indicates that the spring force is always directed opposite to the direction of the displacement from the equilibrium position. We can choose the equilibrium position to be $x_0 = 0$, allowing us to write

$$F_s = -kx. \tag{5.15}$$

This simple force law is called **Hooke's Law,** after the British physicist Robert Hooke (1635–1703), a contemporary of Newton and the Curator of Experiments for the Royal Society.

Note that for a displacement $x > 0$, the spring force points in the negative direction, and $F_s < 0$. The converse is also true; if $x < 0$, then $F_s > 0$. Thus, in all cases, the spring force points toward the equilibrium position, $x = 0$. At exactly the equilibrium position, the spring force is zero, $F_s(x = 0) = 0$. As a reminder from Chapter 4, zero force is one of the defining conditions for equilibrium. The proportionality constant, k, that appears in Hooke's Law is called the **spring constant** and has units of N/m = kg/s^2. The spring force is an important example of a **restoring force:** It always acts to restore the end of the spring to its equilibrium position.

Linear restoring forces that follow Hooke's Law can be found in many systems in nature. Examples are the forces on an atom that has moved slightly out of equilibrium in a crystal lattice, the forces due to shape deformations in atomic nuclei, and any other force that leads to oscillations in a physical system (discussed in further detail in Chapters 14 through 16). In Chapter 6, we will see that we can usually approximate the force in many physical situations by a force that follows Hooke's Law.

Of course, Hooke's Law is not valid for all spring displacements. Everyone who has played with a spring knows that if it is stretched too much, it will deform and then not return to its equilibrium length when released. If stretched even further, it will eventually break into two parts. Every spring has an elastic limit—a maximum deformation—below which Hooke's Law is still valid; however, where exactly this limit lies depends on the material characteristics of the particular spring. For our considerations in this chapter, we assume that springs are always inside the elastic limit.

EXAMPLE 5.3 / Spring Constant

PROBLEM 1
A spring has a length of 15.4 cm and is hanging vertically from a support point above it (Figure 5.15a). A weight with a mass of 0.200 kg is attached to the spring, causing it to extend to a length of 28.6 cm (Figure 5.15b). What is the value of the spring constant?

SOLUTION 1
We place the origin of our coordinate system at the top of the spring, with the positive direction upward, as is customary. Then, $x_0 = -15.4$ cm and $x = -28.6$ cm. According to Hooke's Law, the spring force is

$$F_s = -k(x - x_0).$$

Also, we know the force exerted on the spring was provided by the weight of the 0.200-kg mass: $F = -mg = -(0.200 \text{ kg})(9.81 \text{ m/s}^2) = -1.962$ N. Again, the negative sign indicates the direction. Now we can solve the force equation for the spring constant:

$$k = -\frac{F_s}{x - x_0} = -\frac{1.962 \text{ N}}{(-0.286 \text{ m}) - (-0.154 \text{ m})} = 14.9 \text{ N/m}.$$

Note that we would have obtained exactly the same result if we had put the origin of the coordinate system at another point or if we had elected to designate the downward direction as positive.

PROBLEM 2
How much force is needed to hold the weight at a position 4.6 cm above -28.6 cm (Figure 5.15c)?

SOLUTION 2
At first sight, this problem might appear to require a complicated calculation. However, remember that the mass has stretched the spring to a new equilibrium position. To move the mass from that position takes an external force. If the external force moves the mass up 4.6 cm, then it has to be exactly equal in magnitude and opposite in direction to the spring force resulting from a displacement of 4.6 cm. Thus, all we have to do to find the external force is to use Hooke's Law for the spring force (choosing new equilibrium position to be at $x_0 = 0$):

$$F_{ext} + F_s = 0 \Rightarrow F_{ext} = -F_s = kx = (0.046 \text{ m})(14.9 \text{ N/m}) = 0.68 \text{ N}.$$

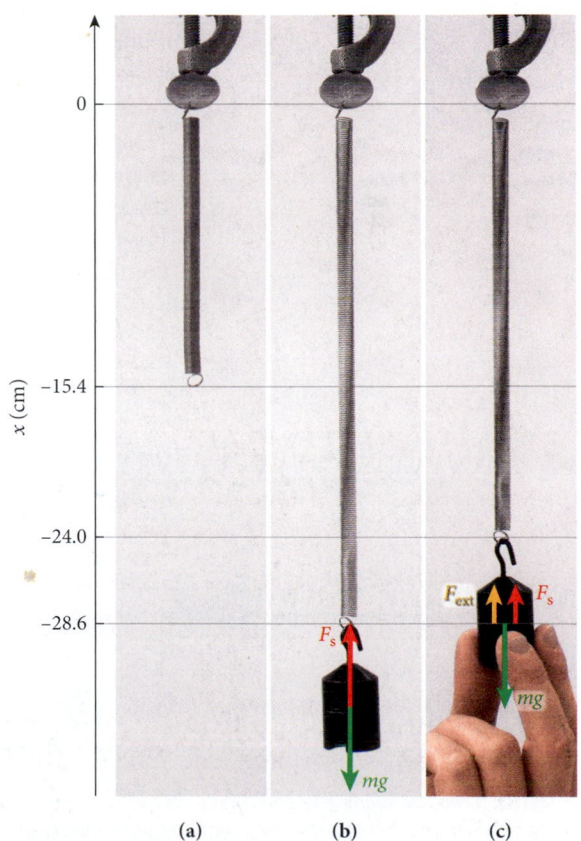

FIGURE 5.15 Mass on a spring. (a) The spring without any mass attached. (b) The spring with the mass hanging freely. (c) The mass pushed upward by an external force.

Self-Test Opportunity 5.3

A block is hanging vertically from a spring at the equilibrium displacement. The block is then pulled down a bit and released from rest. Draw the free-body diagram for the block in each of the following cases:

a) The block is at the equilibrium displacement.

b) The block is at its highest vertical point.

c) The block is at its lowest vertical point.

At this point, it is worthwhile to generalize the observations made in Example 5.3: Adding a constant force—for example, by suspending a mass from the spring—only shifts the equilibrium position. (This generalization is true for all forces that depend linearly on displacement.) Moving the mass, up or down, away from the new equilibrium position then results in a force that is linearly proportional to the displacement from the new equilibrium position. Adding another mass will only cause an additional shift to a new equilibrium position. Of course, adding more mass cannot be continued without limit. At some point, the addition of more and more mass will overstretch the spring. Then the spring will not return to its original length once the mass is removed, and Hooke's Law is no longer valid.

Work Done by the Spring Force

The displacement of a spring is a case of motion in one spatial dimension. Thus, we can apply the one-dimensional integral of equation 5.12 to find the work done by the spring force in moving from x_0 to x. The result is

$$W_s = \int_{x_0}^{x} F_s(x')dx' = \int_{x_0}^{x} (-kx')dx' = -k\int_{x_0}^{x} x'dx'.$$

The work done by the spring force is then

$$W_s = -k\int_{x_0}^{x} x'dx' = -\tfrac{1}{2}kx^2 + \tfrac{1}{2}kx_0^2. \tag{5.16}$$

If we set $x_0 = 0$ and start at the equilibrium position, as we did in arriving at Hooke's Law (equation 5.15), the second term on the right side in equation 5.16 becomes zero and we obtain

$$W_s = -\tfrac{1}{2}kx^2. \tag{5.17}$$

Note that because the spring constant is always positive, the work done by the spring force is always negative for displacements from equilibrium. Equation 5.16 shows that the work done by the spring force is positive if the starting spring displacement is farther from equilibrium than the ending displacement. External work of magnitude $\tfrac{1}{2}kx^2$ will stretch or compress it out of its equilibrium position.

SOLVED PROBLEM 5.2 | Compressing a Spring

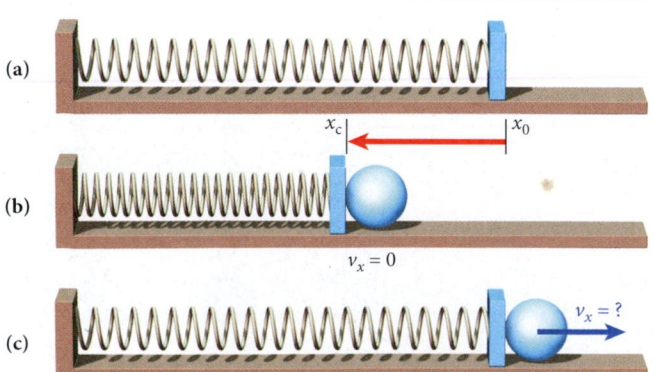

(a)

x_c | x_0

(b)

$v_x = 0$

(c)

$v_x = ?$

FIGURE 5.16 (a) Spring in its equilibrium position; (b) compressing the spring; (c) relaxing the compression and accelerating the steel ball.

A massless spring located on a smooth horizontal surface is compressed by a force of 63.5 N, which results in a displacement of 4.35 cm from the initial equilibrium position. As shown in Figure 5.16, a steel ball of mass 0.075 kg is then placed in front of the spring and the spring is released.

PROBLEM
What is the speed of the steel ball when it is shot off by the spring, that is, right after it loses contact with the spring? (Assume there is no friction between the surface and the steel ball; the steel ball will then simply slide across the surface and will not roll.)

SOLUTION
THINK If we compress a spring with an external force, we do work against the spring force. Releasing the spring by withdrawing the external force enables the spring to do work on the steel ball, which acquires kinetic energy in this process. Calculating the initial work done against the spring force enables us to figure out the kinetic energy that the steel ball will have and thus will lead us to the speed of the ball.

SKETCH We draw a free-body diagram at the instant before the external force is removed (see Figure 5.17). At this instant, the steel ball is at rest in equilibrium, because the external force and the spring force exactly balance each other. Note that the diagram also includes the support surface and shows two more forces acting on the ball: the force of gravity, $\vec{F}_g$, and the normal force from the support surface, $\vec{N}$. These two forces cancel each other out and thus do not enter into our calculations, but it is worthwhile to note the complete set of forces that act on the ball.

We set the x-coordinate of the ball at its left edge, which is where the ball touches the spring. This is the physically relevant location, because it measures the elongation of the spring from its equilibrium position.

RESEARCH The motion of the steel ball starts once the external force is removed. Without the blue arrow in Figure 5.17, the spring force is the only unbalanced force in the situation, and it accelerates the ball. This acceleration is not constant over time (as is the case for free-fall motion, for example), but rather changes in time. However, the beauty of applying energy considerations is that we do not need to know the acceleration to calculate the final speed.

As usual, we are free to choose the origin of the coordinate system, and we put it at x_0, the equilibrium position of the spring. This implies that we set $x_0 = 0$. The relation between the x-component of the spring force at the moment of release and the initial compression of the spring x_c is

$$F_s(x_c) = -kx_c.$$

Because $F_s(x_c) = -F_{ext}$, we find

$$kx_c = F_{ext}.$$

The magnitude of this external force, as well as the value of the displacement, was given, and so we can calculate the value of the spring constant from this equation. Note that with our choice of the coordinate system, $F_{ext} < 0$, because its vector arrow points in the negative x-direction. In addition, $x_c < 0$, because the displacement from equilibrium is in the negative direction.

We can now calculate the work W needed to compress this spring. Since the force that the ball exerts on the spring is always equal and opposite to the force that the spring exerts on the ball, the definition of work allows us to set

$$W = -W_s = \tfrac{1}{2}kx_c^2.$$

According to the work–kinetic energy theorem, this work is related to the change in the kinetic energy of the steel ball via

$$K = K_0 + W = 0 + W = \tfrac{1}{2}kx_c^2.$$

Finally, the ball's kinetic energy is, by definition,

$$K = \tfrac{1}{2}mv_x^2,$$

which allows us to determine the ball's speed.

SIMPLIFY We solve the equation for the kinetic energy for the speed, v_x, and then use the $K = \tfrac{1}{2}kx_c^2$ to obtain

$$v_x = \sqrt{\frac{2K}{m}} = \sqrt{\frac{2(\tfrac{1}{2}kx_c^2)}{m}} = \sqrt{\frac{kx_c^2}{m}} = \sqrt{\frac{F_{ext}x_c}{m}}.$$

(In the third step, we canceled out the factors 2 and $\tfrac{1}{2}$, and in the fourth step, we used $kx_c = F_{ext}$.)

CALCULATE Now we are ready to insert the numbers: $x_c = -0.0435$ m, $m = 0.075$ kg, and $F_{ext} = -63.5$ N. Our result is

$$v_x = \sqrt{\frac{(-63.5 \text{ N})(-0.0435 \text{ m})}{0.075 \text{ kg}}} = 6.06877 \text{ m/s}.$$

Note that we choose the positive root for the x-component of the ball's velocity. By examining Figure 5.16, you can see that this is the appropriate choice, because the ball will move in the positive x-direction after the spring is released.

ROUND Rounding to the two-digit accuracy to which the mass was specified, we state our result as

$$v_x = 6.1 \text{ m/s}.$$

DOUBLE-CHECK We are limited in the checking we can perform to verify that our answer makes sense until we study motion under the influence of the spring force in more detail in Chapter 14. However, our answer passes the minimum requirements in that it has the proper units and the order of magnitude seems in line with typical velocities for balls propelled from spring-loaded toy guns.

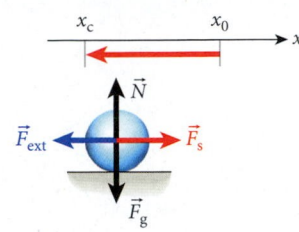

FIGURE 5.17 Free-body diagram of the steel ball before the external force is removed.

Concept Check 5.3

If you compress a spring a distance h from its equilibrium position and do work W_h in the process, how much work will be required to compress the same spring a distance $2h$?

a) W_h

b) $2W_h$

c) $0.5W_h$

d) $4W_h$

e) $0.25W_h$

5.7 Power

We can now readily calculate the amount of work required to accelerate a 1550-kg (3420-lb) car from a standing start to a speed of 26.8 m/s (60.0 mph). The work done is simply the difference between the final and initial kinetic energies. The initial kinetic energy is zero, and the final kinetic energy is

$$K = \tfrac{1}{2}mv^2 = \tfrac{1}{2}(1550 \text{ kg})(26.8 \text{ m/s})^2 = 557 \text{ kJ},$$

which is also the amount of work required. However, the work requirement is not that interesting to most of us—we'd be more interested in how quickly the car is able to reach 60 mph. That is, we'd like to know the rate at which the car can do this work.

Power is the rate at which work is done. Mathematically, this means that the power, P, is the time derivative of the work, W:

$$P = \frac{dW}{dt}. \tag{5.18}$$

It is also useful to define the average power, $\overline{P}$ as

$$\overline{P} = \frac{W}{\Delta t}. \tag{5.19}$$

The SI unit of power is the **watt** (W). [Beware of confusing the symbol for work, W (*italicized*), and the abbreviation for the unit of power, W (nonitalicized).]

$$1 \text{ W} = 1 \text{ J/s} = 1 \text{ kg m}^2/\text{s}^3. \tag{5.20}$$

Conversely, one joule is also one watt times one second. This relationship is reflected in a very common unit of energy (not power!), the **kilowatt-hour** (kWh):

$$1 \text{ kWh} = (1000 \text{ W})(3600 \text{ s}) = 3.6 \cdot 10^6 \text{ J} = 3.6 \text{ MJ}.$$

The unit kWh appears on utility bills and quantifies the amount of electrical energy that has been consumed. Kilowatt-hours can be used to measure any kind of energy. Thus, the kinetic energy of the 1550-kg car moving with a speed of 26.8 m/s, which we calculated as 557 kJ, can be expressed with equal validity as

$$(557{,}000 \text{ J})(1 \text{ kWh}/3.6 \cdot 10^6 \text{ J}) = 0.155 \text{ kWh}.$$

The two most common non-SI power units are the horsepower (hp) and the foot-pound per second (ft lb/s): 1 hp = 550 ft lb/s = 746 W.

It is instructive to look at a logarithmic scale of power consumption of different devices. This is done in the lower row of Figure 5.18. Wristwatches consume about a μW of power, and a green laser pointer is rated up to 5 mW. Among the items in your home a blow dryer has one of the highest power consumption rates at 1 to 2 kW. Cars have engines rated at approximately 100 kW, and a large plane (Boeing 747, Airbus 380) uses on the order of 100 MW to become airborne. The average power consumption of the entire United States is approximately 3.5 TW. Since there are a little more than 300 million people living in the

Concept Check 5.4

Is each of the following statements true or false?

a) Work cannot be done in the absence of motion.

b) More power is required to lift a box slowly than to lift a box quickly.

c) A force is required to do work.

FIGURE 5.18 Logarithmic display of some power sources (upper row of pictures) and power consumption (lower row).

United States, this means an average power use of approximately 10 kW per person. For comparison, the average power a human needs to stay alive is approximately 100 W (= 10 MJ of food intake per day divided by the number of seconds in a day).

The upper row of Figure 5.18 lists power sources. On the left we mark 1 horsepower. A typical wind turbine generates on the order of 1 MW, a typical nuclear power plant 1 GW. The largest power producer in the world is the Chinese Three Gorges Dam, completed in 2012, with a peak power production of more than 22 GW. The entire power that the Sun radiates onto the surface of Earth is 175 PW, more than 10,000 times the present combined power consumption of all humans.

EXAMPLE 5.4 / Accelerating a Car

PROBLEM
Returning to the example of an accelerating car, let's assume that the car, of mass 1550 kg, can reach a speed of 60 mph (26.8 m/s) in 7.1 s. What is the average power needed to accomplish this?

SOLUTION
We already found that the car's kinetic energy at 60 mph is

$$K = \tfrac{1}{2}mv^2 = \tfrac{1}{2}(1550 \text{ kg})(26.8 \text{ m/s})^2 = 557 \text{ kJ}.$$

The work to get the car to the speed of 60 mph is then

$$W = \Delta K = K - K_0 = 557 \text{ kJ}.$$

The average power needed to get to 60 mph in 7.1 s is therefore

$$\overline{P} = \frac{W}{\Delta t} = \frac{5.57 \cdot 10^5 \text{ J}}{7.1 \text{ s}} = 78.4 \text{ kW} = 105 \text{ hp}.$$

If you own a car with a mass of at least 1550 kg that has an engine with 105 hp, you know that it cannot possibly reach 60 mph in 7.1 s. An engine with at least 180 hp is needed to accelerate a car of mass 1550 kg (including the driver, of course) to 60 mph in that time interval.

Our calculation in Example 5.4 is not quite correct for several reasons. First, not all of the power output of the engine is available to do useful work such as accelerating the car. Second, friction and air resistance forces act on a moving car, but were ignored in Example 5.4. Chapter 6 will address work and energy in the presence of friction forces (rolling friction and air resistance in this case). Finally, a car's rated horsepower is a peak specification, realizable only at the most beneficial rpm-domain of the engine. As you accelerate the car from rest, this peak output of the engine is not maintainable as you shift through the gears.

The average mass, power, and fuel efficiency (for city driving) of mid-sized cars sold in the United States from 1975 to 2010 are shown in Figure 5.19. The mass of a car is important in city driving because of the many instances of acceleration in stop-and-go conditions. We can combine the work–kinetic energy theorem (equation 5.17) and the definition of average power (equation 5.19) to get

$$\overline{P} = \frac{W}{\Delta t} = \frac{\Delta K}{\Delta t} = \frac{\tfrac{1}{2}mv^2}{\Delta t} = \frac{mv^2}{2\Delta t}. \tag{5.21}$$

You can see that the average power required to accelerate a car from rest to a speed v in a given time interval, Δt, is proportional to the mass of the car. The energy consumed by the car is equal to the average power times the time interval. Thus, the larger the mass of a car, the more energy is required to accelerate it in a given amount of time.

Following the 1973 oil embargo, the average mass of mid-sized cars decreased from 2100 kg to 1500 kg between 1975 and 1982. During that same period, the average power decreased from 160 hp to 110 hp, and the fuel

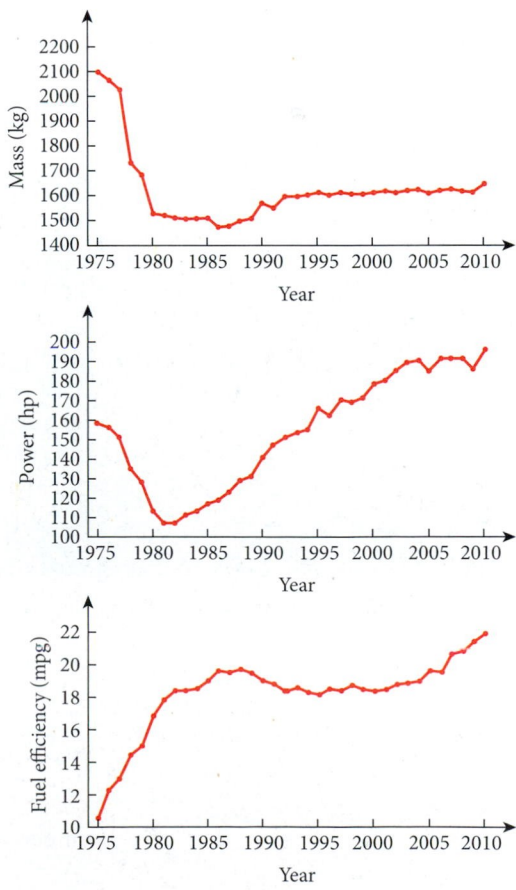

FIGURE 5.19 The mass, power, and fuel efficiency of mid-sized cars sold in the United States from 1975 to 2010. The fuel efficiency is that for typical city driving.

efficiency increased from 10 to 18 mpg. From 1982 to 2010, however, the average mass and fuel efficiency of mid-sized cars stayed roughly constant, while the power increased steadily. Apparently, buyers of mid-sized cars in the United States have valued increased power over increased efficiency.

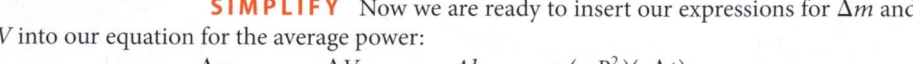

SOLVED PROBLEM 5.3 Wind Power

The total power consumption of all humans combined is approximately 16 TW ($1.6 \cdot 10^{13}$ W), and it is expected to double during the next 15 to 20 years. Almost 90% of the power produced comes from fossil fuels; see Figure 5.20. Since the burning of fossil fuels is currently adding more than 10 billion tons of carbon dioxide to Earth's atmosphere per year, it is not clear how much longer this mode of power generation is sustainable. Other sources of power, such as wind, have to be considered. Some huge wind farms have been constructed (see Figure 5.21), and many more are under development.

PROBLEM
How much average power is contained in wind blowing at 10.0 m/s across the rotor of a large wind turbine, such as the Enercon E-126, which has a hub height of 135 m and a rotor radius of 63 m?

SOLUTION
THINK   Since the wind speed is given, we can calculate the kinetic energy of the amount of air blowing across the rotor's surface. If we can calculate how much air moves across the rotor per unit of time, then we can calculate the power as the ratio of the kinetic energy of the air to the time interval.

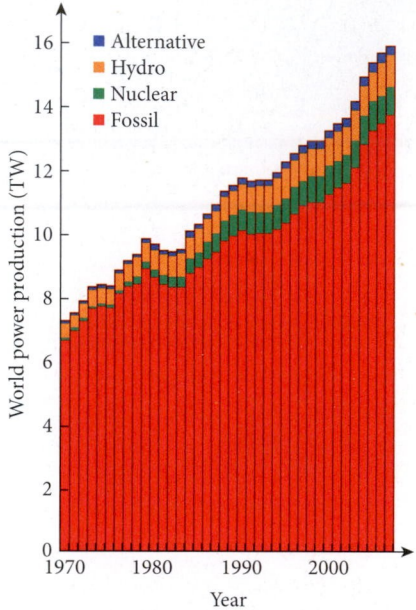

FIGURE 5.20 Worldwide power production as a function of time for different power sources.

SKETCH   The rotor surface is a circle, and we can assume that the wind blows perpendicular to it, because the turbines in wind farms are oriented so that that is the case. Indicated in the sketch (Figure 5.22) is the cylindrical volume of air moving across the rotor per time interval.

RESEARCH   Earlier in this chapter, we learned that the kinetic energy is given by $E = \frac{1}{2}mv^2$; here, m is the mass of air, and v is the wind speed. A very handy rule of thumb is that 1 m^3 of air has a mass of 1.20 kg at sea level and room temperature. The average power is given by $P = W/\Delta t$, and the work is related to the change in kinetic energy through the work–kinetic energy theorem $W = \Delta K$.

We can thus write, for the average power of the wind moving across the rotor of the wind turbine,

$$P = \frac{W}{\Delta t} = \frac{\Delta K}{\Delta t} = \frac{\Delta\left(\frac{1}{2}mv^2\right)}{\Delta t} = \frac{1}{2}v^2\frac{\Delta m}{\Delta t}.$$

In the last step, we have assumed that the wind speed is constant and does not change.

What is Δm? We know that density is mass/volume, and so we can write $\Delta m = \rho\Delta V$, where $\rho = 1.20$ kg/m^3 is the air density and ΔV is the volume of air moved across the rotor per unit of time. Here ΔV is a cylinder with length $l = v\Delta t$ and base area A = area of the rotor (see Figure 5.22), v is again the wind speed, and the area is the area of a circle, $A = \pi R^2$.

FIGURE 5.21 Large-scale wind farm producing power.

SIMPLIFY   Now we are ready to insert our expressions for Δm and ΔV into our equation for the average power:

$$P = \frac{1}{2}v^2\frac{\Delta m}{\Delta t} = \frac{1}{2}v^2\frac{\rho\Delta V}{\Delta t} = \frac{1}{2}v^2\frac{\rho Al}{\Delta t} = \frac{1}{2}v^2\frac{\rho(\pi R^2)(v\Delta t)}{\Delta t} = \frac{1}{2}v^3\rho\pi R^2.$$

We see that the average wind power is proportional to the cube of the wind speed!

CALCULATE   Inserting the given numbers for the rotor's radius, the wind speed, and the air density yields

$$P = \frac{1}{2}(10.0 \text{ m/s})^3(1.2 \text{ kg/m}^3)\pi(63 \text{ m})^2 = 7.481389 \cdot 10^6 \text{ kg m}^2/\text{s}^3$$

ROUND   Since the rotor's radius was given to only two significant figures, we round the final result to the same number of significant figures. So our answer is 7.5 MW.

DOUBLE-CHECK In stating our final result, we replaced kg m²/s³ with W, which is correct by definition and also means that our answer is dimensionally correct. Although 7.5 MW seems to be a very large amount of power, enough to satisfy the average electricity needs of approximately 5000 households, this is a gigantic wind turbine. If you look it up on the Internet, you will see that it is rated close to our result. In Example 13.7 we will return to wind turbines and determine what maximum fraction of this total power contained in the wind one can extract.

Obviously, since the wind power is proportional to the cube of the wind speed, it matters *a lot* where wind turbines are located. Figure 5.23 shows a map of average wind speed in the United States. You can see that winds up to 10 m/s are realistic in the central part of the country, from North Dakota to the Texas panhandle, which makes these states prime locations for wind farms.

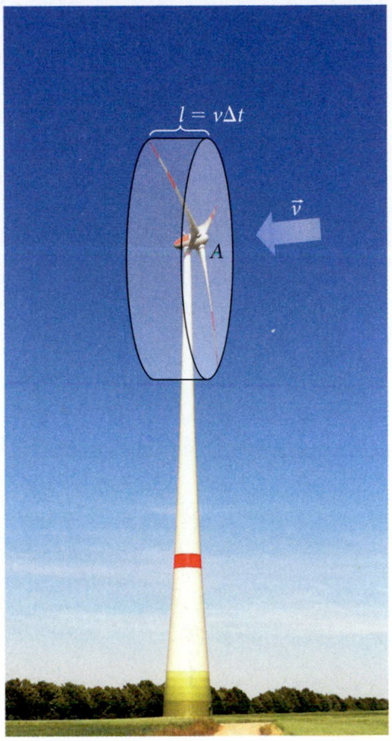

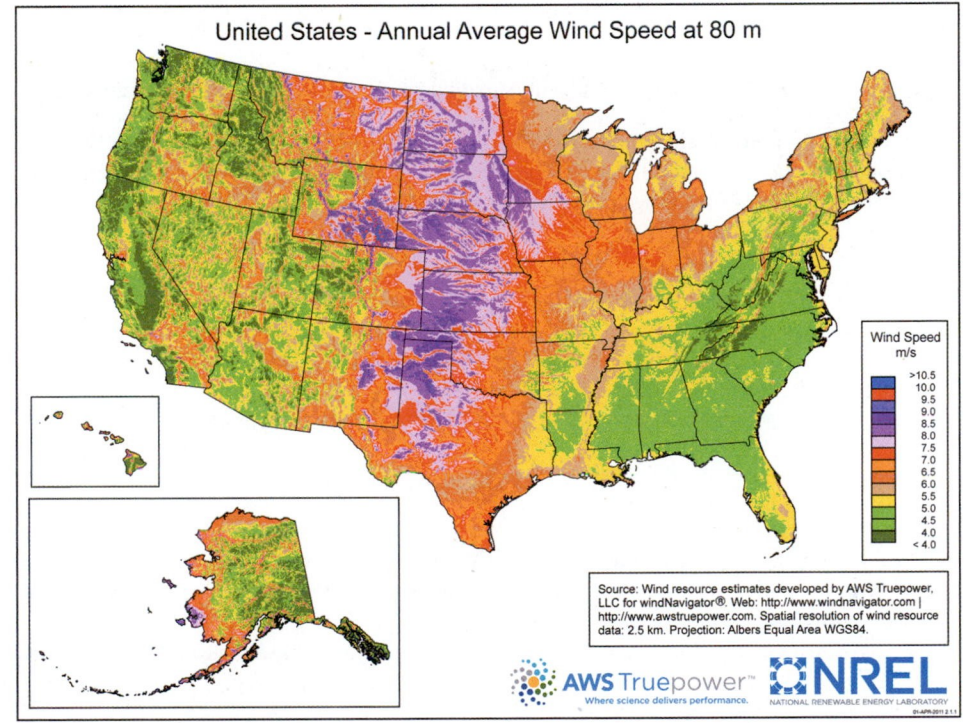

FIGURE 5.23 Map of the United States, showing average wind speed at an altitude of 80 m above ground.

FIGURE 5.22 Sketch for finding the wind power for a large wind turbine.

Power, Force, and Velocity

For a constant force, the work is given by $W = \vec{F} \cdot \Delta\vec{r}$ and the differential work as $dW = \vec{F} \cdot d\vec{r}$. In this case, the time derivative is

$$P = \frac{dW}{dt} = \frac{\vec{F} \cdot d\vec{r}}{dt} = \vec{F} \cdot \vec{v} = Fv\cos\alpha, \tag{5.22}$$

where α is the angle between the force vector and the velocity vector. Therefore, the power is the scalar product of the force vector and the velocity vector. While we show this only for a constant force, equation 5.22 also holds for a nonconstant force.

SOLVED PROBLEM 5.4 | **Riding a Bicycle**

PROBLEM

A bicyclist coasts down a 4.2° slope at a steady speed of 5.1 m/s. Assuming a total mass of 82.2 kg (bicycle plus rider), what power output must the cyclist expend to pedal up the same slope at the same speed?

– Continued

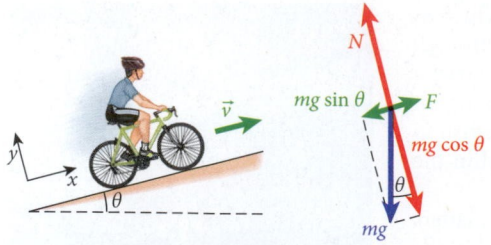

FIGURE 5.24 Sketch of the bicycle moving up the slope (left) and the free-body diagram (right).

SOLUTION

THINK If you ride your bike on a horizontal surface and stop pedaling, you slow down to a stop. The net force that causes you to stop is the combination of friction in the mechanical components of the bike and air resistance. In the problem statement, we learned that the bicycle rolls down the hill at a constant speed, which means that the net force acting on it is zero (Newton's First Law!). The force directed along the slope is $mg \sin \theta$ (see Figure 5.24). For the net force to be zero, this force has to be balanced by the forces of friction and air resistance, which act opposite to the direction of motion, or up the slope. So in this case the forces of friction and air resistance have exactly the same magnitude as the component of the gravitational force along the slope. But if the cyclist is pedaling up the same slope, gravity and the forces of air resistance and friction point in the same direction. Thus, we can calculate the total work done against all forces in this case (and only in this case!) by simply calculating the work done against gravity and then multiplying by a factor of 2.

SKETCH Figure 5.24 shows a sketch of the situation where the cyclist is pedaling against gravity.

RESEARCH We can calculate the work done against gravity, and then multiply it by 2. The component of the gravitational force along the slope is $mg \sin \theta$. F is the force exerted by the bicyclist and Power = Fv. Using Newton's Second Law, we have

$$\sum F_x = ma_x = 0 \Rightarrow F - mg \sin \theta = 0 \Rightarrow F = mg \sin \theta.$$

SIMPLIFY We find the total power the cyclist has to expend by inserting the expression for the force and using the factor of 2, as discussed above: $P = 2Fv = 2(mg \sin \theta)v$.

CALCULATE Inserting the given numbers, we get

$$P = 2(82.2 \text{ kg})(9.81 \text{ m/s}^2)\sin(4.2°)(5.1 \text{ m/s}) = 602.391 \text{ W}.$$

ROUND The slope and the speed are given to two significant figures, which is what we round our final answer to: $P = 0.60$ kW.

DOUBLE-CHECK Our result is $P = 0.60$ kW (1 hp/0.746 kW) = 0.81 hp. Thus, going up a 4.2° slope at 5.1 m/s requires approximately 0.8 hp, which is what a good cyclist can expend for quite some time. (But it's hard!)

WHAT WE HAVE LEARNED | EXAM STUDY GUIDE

- Kinetic energy is the energy associated with the motion of an object, $K = \frac{1}{2}mv^2$.

- The SI unit of work and energy is the joule: $1 \text{ J} = 1 \text{ kg m}^2/\text{s}^2$.

- Work is the energy transferred to an object or transferred from an object as the result of the action of a force. Positive work is a transfer of energy to the object, and negative work is a transfer of energy from the object.

- Work done by a constant force is $W = |\vec{F}||\Delta\vec{r}|\cos\alpha$, where α is the angle between $\vec{F}$ and $\Delta\vec{r}$.

- Work done by a variable force in one dimension is
$$W = \int_{x_0}^{x} F_x(x')dx'.$$

- Work done by the gravitational force in the process of lifting an object is $W_g = -mgh < 0$, where $h = |y - y_0|$; the work done by the gravitational force in lowering an object is $W_g = +mgh > 0$.

- The spring force is given by Hooke's Law: $F_s = -kx$.

- Work done by the spring force is
$$W = -k \int_{x_0}^{x} x'dx' = -\tfrac{1}{2}kx^2 + \tfrac{1}{2}kx_0^2.$$

- The work–kinetic energy theorem is $\Delta K \equiv K - K_0 = W$.

- Power, P, is the time derivative of the work, W: $P = \dfrac{dW}{dt}$.

- The average power, $\bar{P}$, is $\bar{P} = \dfrac{W}{\Delta t}$.

- The SI unit of power is the watt (W): 1 W = 1 J/s.

- The relationship between power, force, and velocity is
$$P = \frac{dW}{dt} = \frac{\vec{F} \cdot d\vec{r}}{dt} = \vec{F} \cdot \vec{v} = Fv\cos\alpha, \text{ where } \alpha \text{ is the angle}$$
between the force vector and the velocity vector.

ANSWERS TO SELF-TEST OPPORTUNITIES

5.1

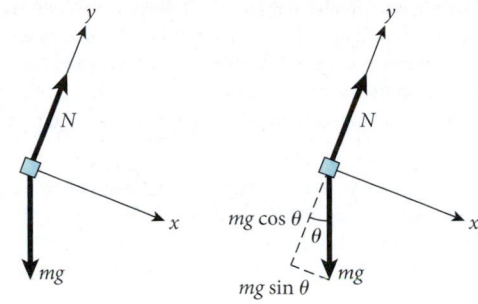

5.2 $\vec{F} = m\vec{a}$ can be rewritten as

$F_x = ma_x$

$F_y = ma_y$

$F_z = ma_z$

for each component

$v_x^2 - v_{x0}^2 = 2a_x(x - x_0)$

$v_y^2 - v_{y0}^2 = 2a_y(y - y_0)$

$v_z^2 - v_{z0}^2 = 2a_z(z - z_0)$

multiply by $\frac{1}{2}m$

$\frac{1}{2}mv_x^2 - \frac{1}{2}mv_{x0}^2 = ma_x(x - x_0)$

$\frac{1}{2}mv_y^2 - \frac{1}{2}mv_{y0}^2 = ma_y(y - y_0)$

$\frac{1}{2}mv_z^2 - \frac{1}{2}mv_{z0}^2 = ma_z(z - z_0)$

add the three equations

$\frac{1}{2}m\left(v_x^2 + v_y^2 + v_z^2\right) - \frac{1}{2}m\left(v_{x0}^2 + v_{y0}^2 + v_{z0}^2\right) =$
$ma_x(x - x_0) + ma_y(y - y_0) + ma_z(z - z_0)$

$K = \frac{1}{2}m\left(v_x^2 + v_y^2 + v_z^2\right) = \frac{1}{2}mv^2$

$K_0 = \frac{1}{2}m\left(v_{x0}^2 + v_{y0}^2 + v_{z0}^2\right) = \frac{1}{2}mv_0^2$

$\Delta\vec{r} = (x - x_0)\hat{x} + (y - y_0)\hat{y} + (z - z_0)\hat{z}$

$\vec{F} = ma_x\hat{x} + ma_y\hat{y} + ma_z\hat{z}$

$K - K_0 = \Delta K = \vec{F} \cdot \Delta\vec{r} = W$

5.3

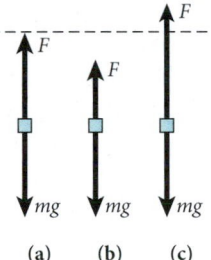

(a)   (b)   (c)

PROBLEM-SOLVING GUIDELINES: KINETIC ENERGY, WORK, AND POWER

1. In all problems involving energy, the first step is to clearly identify the system and the changes in its conditions. If an object undergoes a displacement, check that the displacement is always measured from the same point on the object, such as the front edge or the center of the object. If the speed of the object changes, identify the initial and final speeds at specific points. A diagram is often helpful to show the position and the speed of the object at two different times of interest.

2. Be careful to identify the force that is doing work. Also note whether forces doing work are constant forces or variable forces, because they need to be treated differently.

3. You can calculate the sum of the work done by individual forces acting on an object or the work done by the net force acting on an object; the result should be the same. (You can use this as a way to check your calculations.)

4. Remember that the direction of the restoring force exerted by a spring is always opposite the direction of the displacement of the spring from its equilibrium point.

5. The formula for power, $P = \vec{F} \cdot \vec{v}$, is very useful, if information on the velocity is available. When using the more general definition of power, be sure to distinguish between the average power, $\overline{P} = \dfrac{W}{\Delta t}$, and the instantaneous value of the power, $P = \dfrac{dW}{dt}$.

MULTIPLE-CHOICE QUESTIONS

5.1 Which of the following is a correct unit of energy?

a) $kg\ m/s^2$

b) $kg\ m^2/s$

c) $kg\ m^2/s^2$

d) $kg^2\ m/s^2$

e) $kg^2\ m^2/s^2$

5.2 An 800-N box is pushed up an inclined plane that is 4.0 m long. It requires 3200 J of work to get the box to the top of the plane, which is 2.0 m above the base. What is the magnitude of the average friction force on the box? (Assume the box starts at rest and ends at rest.)

a) zero

b) not zero but less than 400 N

c) greater than 400 N

d) 400 N

e) 800 N

5.3 An engine pumps water continuously through a hose. If the speed with which the water passes through the hose nozzle is v and if k is the mass per unit length of the water jet as it leaves the nozzle, what is the power being imparted to the water?

a) $\frac{1}{2}kv^3$

b) $\frac{1}{2}kv^2$

c) $\frac{1}{2}kv$

d) $\frac{1}{2}v^2/k$

e) $\frac{1}{2}v^3/k$

5.4 A 1500-kg car accelerates from 0 to 25 m/s in 7.0 s. What is the average power delivered by the engine (1 hp = 746 W)?

a) 60 hp

b) 70 hp

c) 80 hp

d) 90 hp

e) 180 hp

5.5 Which of the following is a correct unit of power?

a) $kg\ m/s^2$

b) N

c) J

d) m/s^2

e) W

5.6 How much work is done when a 75.0-kg person climbs a flight of stairs 10.0 m high at constant speed?

a) $7.36 \cdot 10^5$ J

b) 750 J

c) 75 J

d) 7500 J

e) 7360 J

5.7 How much work do movers do (horizontally) in pushing a 150-kg crate 12.3 m across a floor at constant speed if the coefficient of friction is 0.70?

a) 1300 J

b) 1845 J

c) $1.3 \cdot 10^4$ J

d) $1.8 \cdot 10^4$ J

e) 130 J

5.8 Eight books, each 4.6 cm thick and of mass 1.8 kg, lie on a flat table. How much work is required to stack them on top of one another?

a) 141 J

b) 23 J

c) 230 J

d) 0.81 J

e) 14 J

5.9 A particle moves parallel to the x-axis. The net force on the particle increases with x according to the formula $F_x = (120\ N/m)x$, where the force is in newtons when x is in meters. How much work does this force do on the particle as it moves from $x = 0$ to $x = 0.50$ m?

a) 7.5 J

b) 15 J

c) 30 J

d) 60 J

e) 120 J

5.10 A skydiver is subject to two forces: gravity and air resistance. Falling vertically, she reaches a constant terminal speed at some time after jumping from a plane. Since she is moving at a constant velocity from that time until her chute opens, we conclude from the work–kinetic energy theorem that, over that time interval,

a) the work done by gravity is zero.

b) the work done by air resistance is zero.

c) the work done by gravity equals the negative of the work done by air resistance.

d) the work done by gravity equals the work done by air resistance.

e) her kinetic energy increases.

5.11 Jack is holding a box that has a mass of m kg. He walks a distance of d m at a constant speed of v m/s. How much work, in joules, has Jack done on the box?

a) mgd

b) $-mgd$

c) $\frac{1}{2}mv^2$

d) $-\frac{1}{2}mv^2$

e) zero

5.12 If negative work is being done by an object, which one of the following statements is true?

a) An object is moving in the negative x-direction.

b) An object has negative kinetic energy.

c) Energy is being transferred from an object.

d) Energy is being transferred to an object.

5.13 The work–kinetic energy theorem is equivalent to

a) Newton's First Law.

b) Newton's Second Law.

c) Newton's Third Law.

d) Newton's Fourth Law.

e) none of Newton's laws.

5.14 Kathleen climbs a flight of stairs. What can we say about the work done by gravity on her?

a) Gravity does negative work on her.

b) Gravity does positive work on her.

c) Gravity does no work on her.

d) We can't tell what work gravity does on her.

CONCEPTUAL QUESTIONS

5.15 If the net work done on a particle is zero, what can be said about the particle's speed?

5.16 Paul and Kathleen start from rest at the same time at height h at the top of two differently configured water slides. The slides are nearly frictionless. a) Which slider arrives first at the bottom? b) Which slider is traveling faster at the bottom? What physical principle did you use to answer this?

Paul

Kathleen

h

5.17 Does the Earth do any work on the Moon as the Moon moves in its orbit?

5.18 A car, of mass m, traveling at a speed v_1 can brake to a stop within a distance d. If the car speeds up by a factor of 2, so that $v_2 = 2v_1$, by what factor is its stopping distance increased, assuming that the braking force F is approximately independent of the car's speed?

EXERCISES

A blue problem number indicates a worked-out solution is available in the Student Solutions Manual. One • and two •• indicate increasing level of problem difficulty.

Section 5.2

5.19 The damage done by a projectile on impact is correlated with its kinetic energy. Calculate and compare the kinetic energies of these three projectiles:

a) a 10.0-kg stone at 30.0 m/s

b) a 100.0-g baseball at 60.0 m/s

c) a 20.0-g bullet at 300. m/s

5.20 A limo is moving at a speed of 100. km/h. If the mass of the limo, including passengers, is 1900. kg, what is its kinetic energy?

5.21 Two railroad cars, each of mass 7000. kg and traveling at 90.0 km/h, collide head on and come to rest. How much mechanical energy is lost in this collision?

5.22 Think about the answers to these questions next time you are driving a car:

a) What is the kinetic energy of a 1500.-kg car moving at 15.0 m/s?

b) If the car changed its speed to 30.0 m/s, how would the value of its kinetic energy change?

5.23 A 200.-kg moving tiger has a kinetic energy of 14,400 J. What is the speed of the tiger?

•5.24 Two cars are moving. The first car has twice the mass of the second car but only half as much kinetic energy. When both cars increase their speed by 5.00 m/s, they then have the same kinetic energy. Calculate the original speeds of the two cars.

•5.25 What is the kinetic energy of an ideal projectile of mass 20.1 kg at the apex (highest point) of its trajectory, if it was launched with an initial speed of 27.3 m/s and at an initial angle of 46.9° with respect to the horizontal?

Section 5.4

5.26 A force of 5.00 N acts over a distance of 12.0 m in the direction of the force. Find the work done.

5.27 Two baseballs are thrown off the top of a building that is 7.25 m high. Both are thrown with initial speed of 63.5 mph. Ball 1 is thrown horizontally, and ball 2 is thrown straight down. What is the difference in the speeds of the two balls when they touch the ground? (Neglect air resistance.)

5.28 A 95.0-kg refrigerator rests on the floor. How much work is required to move it at constant speed for 4.00 m along the floor against a friction force of 180. N?

5.29 A hammerhead of mass $m = 2.00$ kg is allowed to fall onto a nail from a height $h = 0.400$ m. Calculate the maximum amount of work it could do on the nail.

5.30 You push your couch a distance of 4.00 m across the living room floor with a horizontal force of 200.0 N. The force of friction is 150.0 N. What is the work done by you, by the friction force, by gravity, and by the net force?

•5.31 Suppose you pull a sled with a rope that makes an angle of 30.0° to the horizontal. How much work do you do if you pull with 25.0 N of force and the sled moves 25.0 m?

•5.32 A father pulls his son, whose mass is 25.0 kg and who is sitting on a swing with ropes of length 3.00 m, backward until the ropes make an angle of 33.6° with respect to the vertical. He then releases his son from rest. What is the speed of the son at the bottom of the swinging motion?

•5.33 A constant force, $\vec{F} = (4.79, -3.79, 2.09)$ N, acts on an object of mass 18.0 kg, causing a displacement of that object by $\vec{r} = (4.25, 3.69, -2.45)$ m. What is the total work done by this force?

•5.34 A mother pulls her daughter, whose mass is 20.0 kg and who is sitting on a swing with ropes of length 3.50 m, backward until the ropes make an angle of 35.0° with respect to the vertical. She then releases her daughter from rest. What is the speed of the daughter when the ropes make an angle of 15.0° with respect to the vertical?

•5.35 A ski jumper glides down a 30.0° slope for 80.0 ft before taking off from a negligibly short horizontal ramp. If the jumper's takeoff speed is 45.0 ft/s, what is the coefficient of kinetic friction between skis and slope? Would the value of the coefficient of friction be different if expressed in SI units? If yes, by how much would it differ?

•5.36 At sea level, a nitrogen molecule in the air has an average kinetic energy of $6.2 \cdot 10^{-21}$ J. Its mass is $4.7 \cdot 10^{-26}$ kg. If the molecule could shoot straight up without colliding with other molecules, how high would it rise? What percentage of the Earth's radius is this height? What is the molecule's initial speed? (Assume that you can use $g = 9.81$ m/s^2; although we'll see in Chapter 12 that this assumption may not be justified for this situation.)

••5.37 A bullet moving at a speed of 153 m/s passes through a plank of wood. After passing through the plank, its speed is 130. m/s. Another bullet, of the same mass and size but moving at 92.0 m/s, passes through an identical plank. What will this second bullet's speed be after passing through the plank? Assume that the resistance offered by the plank is independent of the speed of the bullet.

Section 5.5

•5.38 A particle of mass m is subjected to a force acting in the x-direction. $F_x = (3.00 + 0.500x)$ N. Find the work done by the force as the particle moves from $x = 0.00$ to $x = 4.00$ m.

•5.39 A force has the dependence $F_x(x) = -kx^4$ on the displacement x, where the constant $k = 20.3$ N/m^4. How much work does it take to change the displacement, working against the force, from 0.730 m to 1.35 m?

•5.40 A body of mass m moves along a trajectory $\vec{r}(t)$ in three-dimensional space with constant kinetic energy. What geometric relationship has to exist between the body's velocity vector, $\vec{v}(t)$, and its acceleration vector, $\vec{a}(t)$, in order to accomplish this?

•5.41 A force given by $\vec{F}(x) = 5x^3\hat{x}$ (in N/m^3) acts on a 1.00-kg mass moving on a frictionless surface. The mass moves from $x = 2.00$ m to $x = 6.00$ m.

a) How much work is done by the force?

b) If the mass has a speed of 2.00 m/s at $x = 2.00$ m, what is its speed at $x = 6.00$ m?

Section 5.6

5.42 An ideal spring has the spring constant $k = 440.$ N/m. Calculate the distance this spring must be stretched from its equilibrium position for 25.0 J of work to be done.

5.43 A spring is stretched 5.00 cm from its equilibrium position. If this stretching requires 30.0 J of work, what is the spring constant?

5.44 A spring with spring constant k is initially compressed a distance x_0 from its equilibrium length. After returning to its equilibrium position, the spring is then stretched a distance x_0 from that position. What is the ratio of the work that needs to be done on the spring in the stretching to the work done in the compressing?

•5.45 A spring with a spring constant of 238.5 N/m is compressed by 0.231 m. Then a steel ball bearing of mass 0.0413 kg is put against the end of the spring, and the spring is released. What is the speed of the ball bearing right after it loses contact with the spring? (The ball bearing will come off the spring exactly as the spring returns to its equilibrium position. Assume that the mass of the spring can be neglected.)

Section 5.7

5.46 A horse draws a sled horizontally across a snow-covered field. The coefficient of friction between the sled and the snow is 0.195, and the mass of the sled, including the load, is 202.3 kg. If the horse moves the sled at a constant speed of 1.785 m/s, what is the power needed to accomplish this?

5.47 A horse draws a sled horizontally on snow at constant speed. The horse can produce a power of 1.060 hp. The coefficient of friction between the sled and the snow is 0.115, and the mass of the sled, including the load, is 204.7 kg. What is the speed with which the sled moves across the snow?

5.48 While a boat is being towed at a speed of 12.0 m/s, the tension in the towline is 6.00 kN. What is the power supplied to the boat through the towline?

5.49 A car of mass 1214.5 kg is moving at a speed of 62.5 mph when it misses a curve in the road and hits a bridge piling. If the car comes to rest in 0.236 s, how much average power (in watts) is expended in this interval?

5.50 An engine expends 40.0 hp in moving a car along a level track at a speed of 15.0 m/s. How large is the total force acting on the car in the direction opposite to the motion of the car?

•**5.51** A car of mass 942.4 kg accelerates from rest with a constant power output of 140.5 hp. Neglecting air resistance, what is the speed of the car after 4.55 s?

•**5.52** A bicyclist coasts down a 7.0° slope at a steady speed of 5.0 m/s. Assuming a total mass of 75 kg (bicycle plus rider), what must the cyclist's power output be to pedal up the same slope at the same speed?

•**5.53** A small blimp is used for advertising purposes at a football game. It has a mass of 93.5 kg and is attached by a towrope to a truck on the ground. The towrope makes an angle of 53.3° downward from the horizontal, and the blimp hovers at a constant height of 19.5 m above the ground. The truck moves on a straight line for 840.5 m on the level surface of the stadium parking lot at a constant velocity of 8.90 m/s. If the drag coefficient (K in $F = Kv^2$) is 0.500 kg/m, how much work is done by the truck in pulling the blimp (assuming there is no wind)?

••**5.54** A car of mass m accelerates from rest along a level straight track, not at a constant acceleration but with constant engine power, P. Assume that air resistance is negligible.

a) Find the car's velocity as a function of time.

b) A second car starts from rest alongside the first car on the same track, but maintains a constant acceleration. Which car takes the initial lead? Does the other car overtake it? If yes, write a formula for the distance from the starting point at which this happens.

c) You are in a drag race, on a straight level track, with an opponent whose car maintains a constant acceleration of 12.0 m/s². Both cars have identical masses of 1000. kg. The cars start together from rest. Air resistance is assumed to be negligible. Calculate the minimum power your engine needs for you to win the race, assuming the power output is constant and the distance to the finish line is 0.250 mi.

Additional Exercises

5.55 At the 2004 Olympic Games in Athens, Greece, the Iranian athlete Hossein Rezazadeh won the super-heavyweight class gold medal in weightlifting. He lifted 472.5 kg (1042 lb) combined in his two best lifts in the competition. Assuming that he lifted the weights a height of 196.7 cm, what work did he do?

5.56 How much work is done against gravity in lifting a 6.00-kg weight through a distance of 20.0 cm?

5.57 A certain tractor is capable of pulling with a steady force of 14.0 kN while moving at a speed of 3.00 m/s. How much power in kilowatts and in horsepower is the tractor delivering under these conditions?

5.58 A shot-putter accelerates a 7.30-kg shot from rest to 14.0 m/s. If this motion takes 2.00 s, what average power was supplied?

5.59 An advertisement claims that a certain 1200.-kg car can accelerate from rest to a speed of 25.0 m/s in 8.00 s. What average power must the motor supply in order to cause this acceleration? Ignore losses due to friction.

5.60 A car of mass $m = 1250$ kg is traveling at a speed of $v_0 = 105$ km/h (29.2 m/s). Calculate the work that must be done by the brakes to completely stop the car.

5.61 An arrow of mass $m = 88.0$ g (0.0880 kg) is fired from a bow. The bowstring exerts an average force of $F = 110.$ N on the arrow over a distance $d = 78.0$ cm (0.780 m). Calculate the speed of the arrow as it leaves the bow.

5.62 The mass of a physics textbook is 3.40 kg. You pick the book up off a table and lift it 0.470 m at a constant speed of 0.270 m/s.

a) What is the work done by gravity on the book?

b) What is the power you supplied to accomplish this task?

5.63 A sled, with mass m, is given a shove up a frictionless incline, which makes a 28.0° angle with the horizontal. Eventually, the sled comes to a stop at a height of 1.35 m above where it started. Calculate its initial speed.

5.64 A man throws a rock of mass $m = 0.325$ kg straight up into the air. In this process, his arm does a total amount of work $W_{net} = 115$ J on the rock. Calculate the maximum distance, h, above the man's throwing hand that the rock will travel. Neglect air resistance.

5.65 A car does the work $W_{car} = 7.00 \cdot 10^4$ J in traveling a distance $x = 2.80$ km at constant speed. Calculate the average force F (from all sources) acting on the car in this process.

•**5.66** A softball, of mass $m = 0.250$ kg, is pitched at a speed $v_0 = 26.4$ m/s. Due to air resistance, by the time it reaches home plate, it has slowed by 10.0%. The distance between the plate and the pitcher is $d = 15.0$ m. Calculate the average force of air resistance, F_{air}, that is exerted on the ball during its movement from the pitcher to the plate.

•**5.67** A flatbed truck is loaded with a stack of sacks of cement whose combined mass is 1143.5 kg. The coefficient of static friction between the bed of the truck and the bottom sack in the stack is 0.372, and the sacks are not tied down but held in place by the force of friction between the bed and the bottom sack. The truck accelerates uniformly from rest to 56.6 mph in 22.9 s. The stack of sacks is 1 m from the end of the truck bed. Does the stack slide on the truck bed? The coefficient of kinetic friction between the bottom sack and the truck bed is 0.257. What is the work done on the stack by the force of friction between the stack and the bed of the truck?

•**5.68** A driver notices that her 1000.-kg car slows from $v_0 = 90.0$ km/h (25.0 m/s) to $v = 70.0$ km/h (19.4 m/s) in $t = 6.00$ s moving on level ground in neutral gear. Calculate the power needed to keep the car moving at a constant speed, $v_{ave} = 80.0$ km/h (22.2 m/s). Assume that energy is lost at a constant rate during the deceleration.

•**5.69** The 125-kg cart in the figure starts from rest and rolls with negligible friction. It is pulled by three ropes as shown. It moves 100. m horizontally. Find the final velocity of the cart.

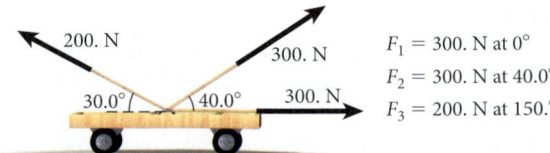

$F_1 = 300.$ N at 0°
$F_2 = 300.$ N at 40.0°
$F_3 = 200.$ N at 150.°

•**5.70** Calculate the power required to propel a 1000.0-kg car at 25.0 m/s up a straight slope inclined 5.00° above the horizontal. Neglect friction and air resistance.

•**5.71** A grandfather pulls his granddaughter, whose mass is 21.0 kg and who is sitting on a swing with ropes of length 2.50 m, backward and releases her from rest. The speed of the granddaughter at the bottom of the swinging motion is 3.00 m/s. What is the angle (in degrees, measured relative to the vertical) from which she is released?

•**5.72** A 65-kg hiker climbs to the second base camp on Nanga Parbat in Pakistan, at an altitude of 3900 m, starting from the first base camp at 2200 m. The climb is made in 5.0 h. Calculate (a) the work done against gravity, (b) the average power output, and (c) the rate of energy input required, assuming the energy conversion efficiency of the human body is 15%.

5.73 An x-component of a force has the dependence $F_x(x) = -cx^3$ on the displacement x, where the constant $c = 19.1$ N/m^3. How much work does it take to oppose this force and change the displacement from 0.810 m to 1.39 m?

5.74 A massless spring lying on a smooth horizontal surface is compressed by a force of 63.5 N, which results in a displacement of 4.35 cm from the initial equilibrium position. How much work will it take to compress the spring from 4.35 cm to 8.15 cm?

•**5.75** A car is traveling at a constant speed of 26.8 m/s. It has a drag coefficient $c_d = 0.333$ and a cross-sectional area of 3.25 m^2. How much power is required just to overcome air resistance and keep the car traveling at this constant speed? Assume the density of air is 1.15 kg/m^3.

MULTI-VERSION EXERCISES

5.76 A variable force is given by $F(x) = Ax^6$, where $A = 11.45$ N/m^6. This force acts on an object of mass 2.735 kg that moves on a frictionless surface. Starting from rest, the object moves from $x = 1.093$ m to $x = 4.429$ m. How much does the kinetic energy of the object change?

5.77 A variable force is given by $F(x) = Ax^6$, where $A = 13.75$ N/m^6. This force acts on an object of mass 3.433 kg that moves on a frictionless surface. Starting from rest, the object moves from $x = 1.105$ m to a new position, x. The object gains $5.662 \cdot 10^3$ J of kinetic energy. What is the new position x?

5.78 A variable force is given by $F(x) = Ax^6$, where $A = 16.05$ N/m^6. This force acts on an object of mass 3.127 kg that moves on a frictionless surface. Starting from rest, the object moves from a position x_0 to a new position, $x = 3.313$ m. The object gains $1.00396 \cdot 10^4$ J of kinetic energy. What is the initial position x_0?

5.79 Santa's reindeer pull his sleigh through the snow at a speed of 3.333 m/s. The mass of the sleigh, including Santa and the presents, is 537.3 kg. Assuming that the coefficient of kinetic friction between the runners of the sleigh and the snow is 0.1337, what is the total power (in hp) that the reindeer are providing?

5.80 Santa's reindeer pull his sleigh through the snow at a speed of 2.561 m/s. The mass of the sleigh, including Santa and the presents, is 540.3 kg. Assuming that the reindeer can provide a total power of 2.666 hp, what is the coefficient of friction between the runners of the sleigh and the snow?

5.81 Santa's reindeer pull his sleigh through the snow at a speed of 2.791 m/s. Assuming that the reindeer can provide a total power of 3.182 hp and the coefficient of friction between the runners of the sleigh and the snow is 0.1595, what is the mass of the sleigh, including Santa and the presents?

5.82 A horizontal spring with spring constant $k = 15.19$ N/m is compressed 23.11 cm from its equilibrium position. A hockey puck with mass $m = 170.0$ g is placed against the end of the spring. The spring is released, and the puck slides on horizontal ice, with a coefficient of kinetic friction of 0.02221 between the puck and the ice. How far does the hockey puck travel on the ice after it leaves the spring?

5.83 A horizontal spring with spring constant $k = 17.49$ N/m is compressed 23.31 cm from its equilibrium position. A hockey puck with mass $m = 170.0$ g is placed against the end of the spring. The spring is released, and the puck slides on horizontal ice a distance of 12.13 m after it leaves the spring. What is the coefficient of kinetic friction between the puck and the ice?

5.84 A load of bricks at a construction site has a mass of 75.0 kg. A crane raises this load from the ground to a height of 45.0 m in 52.0 s at a low constant speed. What is the average power of the crane?

5.85 A load of bricks at a construction site has a mass of 75.0 kg. A crane with 725 W of power raises this load from the ground to a height of 45.0 m at a low constant speed. How long does it take to raise the load?

5.86 A load of bricks at a construction site has a mass of 75.0 kg. A crane with 815 W of power raises this load from the ground to a certain height in 52.0 s at a low constant speed. What is the final height of the load?

6

Potential Energy and Energy Conservation

FIGURE 6.1 Niagara Falls.

Niagara Falls is one of the most spectacular sights in the world, with about 3000 cubic meters of water dropping 49 m (161 ft) every *second!* Horseshoe Falls on the Canadian border, shown in Figure 6.1, has a length of 790. m (2592 ft); American Falls on the United States side extends another 305 m (1001 ft). Together, they are one of the great tourist attractions of North America. However, Niagara Falls is more than a scenic wonder. It is also one of the largest sources of electric power in the world, producing over 2500 megawatts (see Solved Problem 6.1). Humans have exploited the energy of falling water since ancient times, using it to turn large paddlewheels for mills and factories. Today, the conversion of energy in falling water to electrical energy by hydroelectric dams is a major source of energy throughout the world.

As we saw in Chapter 5, energy is a fundamental concept in physics that governs many of the interactions involving forces and motions of objects. In this chapter, we continue our study of energy, introducing several new forms of energy and new laws that govern its use. We will return to the laws of energy in the chapters on thermodynamics, building on much of the material presented here. Until then, we will continue our study of mechanics, relying heavily on the ideas discussed here.

WHAT WE WILL LEARN

- Potential energy, U, is the energy stored in the configuration of a system of objects that exert forces on one another.

- When a conservative force does work on an object that travels on a path and returns to where it started (a closed path), the total work is zero. A force that does not fulfill this requirement is a nonconservative force.

- A potential energy can be associated with any conservative force. The change in potential energy due to some spatial rearrangement of a system is equal to the negative of the work done by the conservative force during this spatial rearrangement.

- The mechanical energy, E, is the sum of kinetic energy and potential energy.

- The total mechanical energy is conserved (it remains constant over time) for any mechanical process that occurs within an isolated system and involves only conservative forces.

- In an isolated system, the total energy—that is, the sum of all forms of energy, mechanical or other—is always conserved. This holds with both conservative and nonconservative forces.

- Small perturbations about a stable equilibrium point result in small oscillations around the equilibrium point; for an unstable equilibrium point, small perturbations result in an accelerated movement away from the equilibrium point.

6.1 Potential Energy

Chapter 5 examined in detail the relationship between kinetic energy and work, and one of the main points was that work and kinetic energy can be converted into one another. Now, this section introduces another kind of energy, called *potential energy*.

Potential energy, U, is the energy stored in the configuration of a system of objects that exert forces on one another. For example, we have seen that work is done by an external force in lifting a load against the force of gravity, and this work is given by $W = mgh$, where m is the mass of the load and $h = y - y_0$ is the height to which the load is lifted above its initial position. (In this chapter, we will assume the y-axis points upward unless specified differently.) This lifting can be accomplished without changing the kinetic energy, as in the case of a weightlifter who lifts a mass above his head and holds it there. There is energy stored in holding the mass above the head. If the weightlifter lets go of the mass, this energy can be converted back into kinetic energy as the mass accelerates and falls to the ground. We can express the gravitational potential energy as

$$U_g = mgy. \tag{6.1}$$

The change in the gravitational potential energy of the mass is then

$$\Delta U_g \equiv U_g(y) - U_g(y_0) = mg(y - y_0) = mgh. \tag{6.2}$$

(Equation 6.1 is valid only near the surface of the Earth, where $F_g = mg$, and in the limit that Earth is infinitely massive relative to the object. We will encounter a more general expression for U_g in Chapter 12.) In Chapter 5, we found that the work done by the gravitational force on an object that is lifted through a height h is $W_g = -mgh$. From this, we see that the work done by the gravitational force and the gravitational potential energy for an object lifted from rest to a height h are related by

$$\Delta U_g = -W_g. \tag{6.3}$$

EXAMPLE 6.1 | Weightlifting

PROBLEM

Let's consider the gravitational potential energy in a specific situation: a weightlifter lifting a barbell of mass m. What is the gravitational potential energy and the work done during the different phases of lifting the barbell?

– Continued

SOLUTION

The weightlifter starts with the barbell on the floor, as shown in Figure 6.2a. At $y = 0$, the gravitational potential energy can be defined to be $U_g = 0$. The weightlifter then picks up the barbell, lifts it to a height of $y = h/2$, and holds it there, as shown in Figure 6.2b. The gravitational potential energy is now $U_g = mgh/2$, and the work done by gravity on the barbell is $W_g = -mgh/2$. The weightlifter next lifts the barbell over his head to a height of $y = h$, as shown in Figure 6.2c. The gravitational potential energy is now $U_g = mgh$, and the work done by gravity during this part of the lift is again $W_g = -mgh/2$. Having completed the lift, the weightlifter lets go of the barbell, and it falls to the floor, as illustrated in Figure 6.2d. The gravitational potential energy of the barbell on the floor is again $U_g = 0$, and the work done by gravity during the fall is $W_g = mgh$.

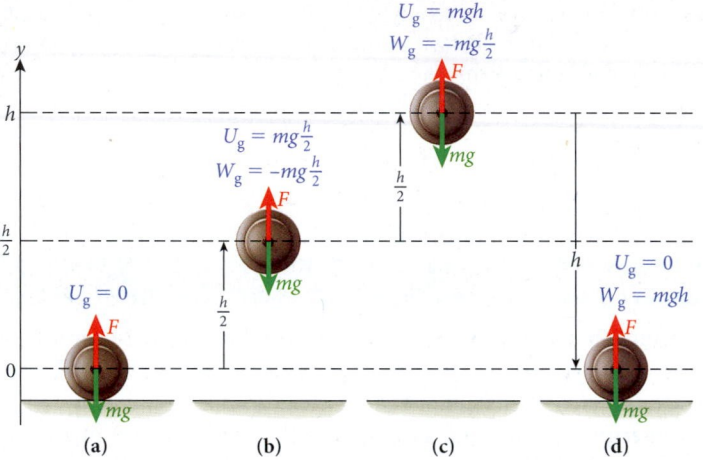

FIGURE 6.2 Lifting a barbell and potential energy (the diagram shows the barbell in side view and omits the weightlifter). The weight of barbell is mg, and the normal force exerted by the floor or the weightlifter to hold the weight up is F. (a) The barbell is initially on the floor. (b) The weightlifter lifts the barbell of mass m to a height of $h/2$ and holds it there. (c) The weightlifter lifts the barbell an additional distance $h/2$, to a height of h, and holds it there. (d) The weightlifter lets the barbell drop to the floor.

Equation 6.3 is true even for complicated paths involving horizontal as well as vertical motion of the object, because the gravitational force does no work during horizontal segments of the motion. In horizontal motion, the displacement is perpendicular to the force of gravity (which always points vertically down), and thus the scalar product between the force and displacement vectors is zero; hence, no work is done.

Lifting any mass to a higher elevation involves doing work against the force of gravity and generates an increase in gravitational potential energy of the mass. This energy can be stored for later use. This principle is employed, for example, at many hydroelectric dams. The excess electricity generated by the turbines is used to pump water to a reservoir at a higher elevation. There it constitutes a reserve that can be tapped in times of high energy demand and/or low water supply. Stated in general terms, if ΔU_g is positive, there exists the potential (hence the name *potential energy*) to allow ΔU_g to be negative in the future, thereby extracting positive work, since $W_g = -\Delta U_g$.

SOLVED PROBLEM 6.1 | Power Produced by Niagara Falls

PROBLEM

The Niagara River delivers an average of 5520 m³ of water per second to the top of Niagara Falls, where it drops 49.0 m. If all the potential energy of that water could be converted to electrical energy, how much electrical power could Niagara Falls generate?

SOLUTION

THINK The mass of 1 m³ of water is 1000 kg. The work done by the falling water is equal to the change in its gravitational potential energy. The average power is the work per unit time.

SKETCH A sketch of a vertical coordinate axis is superimposed on a photo of Niagara Falls in Figure 6.3.

RESEARCH The average power is given by the work per unit time:

$$\bar{P} = \frac{W}{t}.$$

The work that is done by the water going over Niagara Falls is equal to the change in gravitational potential energy,

$$\Delta U = W.$$

The change in gravitational potential energy of a given mass m of water falling a distance h is given by

$$\Delta U = mgh.$$

SIMPLIFY We can combine the preceding three equations to obtain

$$\bar{P} = \frac{W}{t} = \frac{mgh}{t} = \left(\frac{m}{t}\right)gh.$$

CALCULATE We first calculate the mass of water moving over the falls per unit time from the given volume of water per unit time, using the density of water:

$$\frac{m}{t} = \left(5520\,\frac{\text{m}^3}{\text{s}}\right)\left(\frac{1000\,\text{kg}}{1\,\text{m}^3}\right) = 5.52\cdot10^6\ \text{kg/s}.$$

The average power is then

$$\bar{P} = \left(5.52\cdot10^6\ \text{kg/s}\right)\left(9.81\ \text{m/s}^2\right)\left(49.0\ \text{m}\right) = 2653.4088\ \text{MW}.$$

ROUND We round to three significant figures:

$$\bar{P} = 2.65\ \text{GW}.$$

DOUBLE-CHECK Our result is comparable to the output of nuclear power plants, on the order of 1000 MW (1 GW). The combined power generation capability of all of the hydroelectric power stations at Niagara Falls has a peak of 4.4 GW during the high water season in the spring, which is close to our answer. However, you may ask how the water produces power by simply falling over Niagara Falls. The answer is that it doesn't. Instead, a large fraction of the water of the Niagara River is diverted upstream from the falls and sent through tunnels, where it drives power generators. The water that makes it across the falls during the daytime and in the summer tourist season is only about 50% of the flow of the Niagara River. This flow is reduced even further, down to 10%, and more water is diverted for power generation during the nighttime and in the winter.

FIGURE 6.3 Niagara Falls, showing an elevation of h for the drop of the water going over the falls.

6.2 Conservative and Nonconservative Forces

Before we can calculate the potential energy from a given force, we have to ask: Can all kinds of forces be used to store potential energy for later retrieval? If not, what kinds of forces can we use? To answer this question, we need to consider what happens to the work done by a force when the direction of the path taken by an object is reversed. In the case of the gravitational force, we have already seen what happens. As shown in Figure 6.4, the work done by F_g when an object of mass m is lifted from elevation y_A to y_B has the same magnitude as, but the opposite sign to, the work done by F_g when lowering the same object from elevation y_B to y_A. This means that the total work done by F_g in lifting the object from some elevation to a different one and then returning it to the same elevation is zero. This fact is the basis for the definition of a conservative force (refer to Figure 6.5a).

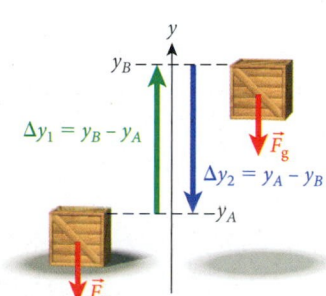

FIGURE 6.4 Gravitational force vectors and displacement for lifting and lowering a box.

Definition

A **conservative force** is any force for which the work done over any closed path is zero. A force that does not fulfill this requirement is called a **nonconservative force**.

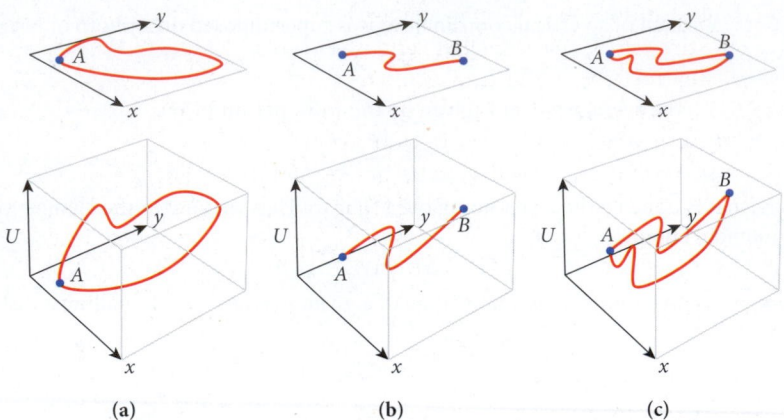

FIGURE 6.5 Various paths for the potential energy related to a conservative force as a function of positions *x* and *y*, with *U* proportional to *y*. The two-dimensional plots are projections of the three-dimensional plots onto the *xy*-plane. (a) Closed loop. (b) A path from point *A* to point *B*. (c) Two different paths between points *A* and *B*.

For conservative forces, we can immediately state two consequences of this definition:

1. If we know the work, $W_{A \to B}$, done by a conservative force on an object as the object moves along a path from point *A* to point *B*, then we also know the work, $W_{B \to A}$, that the same force does on the object as it moves along the path in the reverse direction, from point *B* to point *A* (see Figure 6.5b):

$$W_{B \to A} = -W_{A \to B} \text{ (for conservative forces).} \tag{6.4}$$

The proof of this statement is obtained from the condition of zero work over a closed loop. Because the path from *A* to *B* to *A* forms a closed loop, the sum of the work contributions from the loop has to equal zero. In other words,

$$W_{A \to B} + W_{B \to A} = 0,$$

from which equation 6.4 follows immediately.

2. If we know the work, $W_{A \to B, \text{path 1}}$, done by a conservative force on an object moving along path 1 from point *A* to point *B*, then we also know the work, $W_{A \to B, \text{path 2}}$, done by the same force on the object when it uses any other path (path 2) to go from point *A* to point *B* (see Figure 6.5c). The work is the same; the work done by a conservative force is independent of the path taken by the object:

$$W_{A \to B, \text{path 2}} = W_{A \to B, \text{path 1}} \tag{6.5}$$

(for arbitrary paths 1 and 2, for conservative forces).

This statement is also easy to prove from the definition of a conservative force as a force for which the work done over any closed path is zero. The path from point *A* to point *B* on path 1 and then back from *B* to *A* on path 2 is a closed loop; therefore, $W_{A \to B, \text{path 2}} + W_{B \to A, \text{path 1}} = 0$. Now we use equation 6.4 for reversal of the path direction, $W_{B \to A, \text{path 1}} = -W_{A \to B, \text{path 1}}$. Combining these two results gives us $W_{A \to B, \text{path 2}} - W_{A \to B, \text{path 1}} = 0$, from which equation 6.5 follows.

One physical application of the mathematical results just given involves riding a bicycle from one point, such as your home, to another point, such as the swimming pool. Assuming that your home is located at the foot of a hill and the pool at the top, we can use the Figure 6.5c to illustrate this example, with point *A* representing your home and point *B* the pool. What the preceding statements regarding conservative forces mean is that you do the same amount of work riding your bike from home to the pool, independent of the route you select. You can take a shorter and steeper route or a flatter and longer route; you can even take a route that goes up and down between points *A* and *B*. The total work will be the same. As with almost all real-world examples, however, there are some complications here: It matters whether you use the handbrakes; there is air resistance and tire friction to consider; and your body also performs other metabolic functions during the ride, in addition to moving your mass and that of the bicycle from point A to B. But this example can help you develop a mental picture of the concepts of path-independence of work and conservative forces.

The gravitational force, as we have seen, is an example of a conservative force. Another example of a conservative force is the spring force. Not all forces are conservative, however. Which forces are nonconservative?

Friction Forces

Let's consider what happens in sliding a box across a horizontal surface, from point A to point B and then back to point A, if the coefficient of kinetic friction between the box and the surface is μ_k (Figure 6.6). As we have learned, the friction force is given by $f = \mu_k N = \mu_k mg$ and always points in the direction opposite to that of the motion. Let's use results from Chapter 5 to find the work done by this friction force. Since the friction force is constant, the amount of work it does is found by simply taking the scalar product between the friction force and displacement vectors.

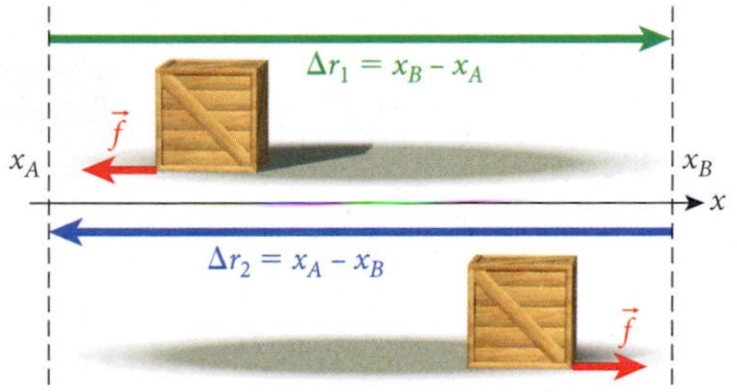

For the motion from A to B, we use the general scalar product formula for work done by a constant force:

$$W_{f1} = \vec{f} \bullet \Delta\vec{r}_1 = -f \cdot (x_B - x_A) = -\mu_k mg \cdot (x_B - x_A).$$

We have assumed that the positive x-axis is pointing to the right, as is conventional, so the friction force points in the negative x-direction. For the motion from B back to A, then, the friction force points in the positive x-direction. Therefore, the work done for this part of the path is

$$W_{f2} = \vec{f} \bullet \Delta\vec{r}_2 = f \cdot (x_A - x_B) = \mu_k mg \cdot (x_A - x_B).$$

This result leads us to conclude that the total work done by the friction force while the box slides across the surface on the closed path from point A to point B and back to point A is not zero, but instead

$$W_f = W_{f1} + W_{f2} = -2\mu_k mg(x_B - x_A) < 0. \tag{6.6}$$

There appears to be a contradiction between this result and the work–kinetic energy theorem. The box starts with zero kinetic energy and at a certain position, and it ends up with zero kinetic energy and at the same position. According to the work–kinetic energy theorem, the total work done should be zero. This leads us to conclude that the friction force does not do work in the way that a conservative force does. Instead, the friction force converts kinetic and/or potential energy into internal excitation energy of the two objects that exert friction on each other (the box and the support surface, in this case). This internal excitation energy can take the form of vibrations or thermal energy or even chemical or electrical energy. The main point is that the conversion of kinetic and/or potential energy to internal excitation energy is not reversible; that is, the internal excitation energy cannot be fully converted back into kinetic and/or potential energy.

Thus, we see that the friction force is an example of a nonconservative force. Because the friction force always acts in a direction opposite to the displacement, the dissipation of energy due to the friction force always reduces the total mechanical energy, whether or not the path is closed. Work done by a conservative force, W, can be positive or negative, but the dissipation from the friction force, W_f, is always negative, withdrawing kinetic and/or potential energy and converting it into internal excitation energy. Using the symbol W_f for this dissipated energy is a reminder that we use the same procedures to calculate it as to calculate work for conservative forces.

The decisive fact is that the friction force switches direction as a function of the direction of motion and causes dissipation. The friction force vector is always antiparallel to the velocity vector; any force with this property cannot be conservative. Dissipation

FIGURE 6.6 Friction force vector and displacement vector for the process of sliding a box back and forth across a surface with friction.

Concept Check 6.1

A person pushes a box of mass m a distance d across a floor. The coefficient of kinetic friction between the box and the floor is μ_k. The person then picks up the box, raises it to a height h, carries it back to the starting point, and puts it back down on the floor. How much work has the person done on the box?

a) zero

b) $\mu_k mgd$

c) $\mu_k mgd + 2mgh$

d) $\mu_k mgd - 2mgh$

e) $2\mu_k mgd + 2mgh$

converts kinetic energy into internal energy of the object, which is another important characteristic of a nonconservative force. In Section 6.7, we will examine this point in more detail.

Another example of a nonconservative force is the force of air resistance. It is also velocity dependent and always points in the direction opposite to the velocity vector, just like the force of kinetic friction. Yet another example of a nonconservative force is the damping force (discussed in Chapter 14). It, too, is velocity dependent and opposite in direction to the velocity.

6.3 Work and Potential Energy

In considering the work done by the gravitational force and its relationship to gravitational potential energy in Section 6.1, we found that the change in potential energy is equal to the negative of the work done by the force, $\Delta U_g = -W_g$. This relationship is true for all conservative forces. In fact, we can use it to define the concept of potential energy.

For any conservative force, the change in potential energy due to some spatial rearrangement of a system is equal to the negative of the work done by the conservative force during this spatial rearrangement:

$$\Delta U = -W. \tag{6.7}$$

We have already seen that work is given by

$$W = \int_{x_0}^{x} F_x\left(x'\right) dx'. \tag{6.8}$$

Combining equations 6.7 and 6.8 gives us the relationship between the conservative force and the potential energy:

$$\Delta U = U(x) - U(x_0) = -\int_{x_0}^{x} F_x\left(x'\right) dx'. \tag{6.9}$$

We could use equation 6.9 to calculate the potential energy change due to the action of any given conservative force. Why should we bother with the concept of potential energy when we can deal directly with the conservative force itself? The answer is that the change in potential energy depends only on the beginning and final states of the system and is independent of the path taken to get to the final state. Often, we have a simple expression for the potential energy (and thus its change) prior to working a problem! In contrast, evaluating the integral on the right-hand side of equation 6.9 could be quite complicated. And, in fact, the computational savings is not the only rationale, as the use of energy considerations is based on an underlying physical law (the law of energy conservation, to be introduced in Section 6.5).

In Chapter 5, we evaluated the integral in equation 6.9 for the force of gravity and for the spring force. The result for the gravitational force is

$$\Delta U_g = U_g(y) - U_g(y_0) = -\int_{y_0}^{y} (-mg) dy' = mg \int_{y_0}^{y} dy' = mgy - mgy_0. \tag{6.10}$$

This is in accord with the result we found in Section 6.1. Consequently, the gravitational potential energy is

$$U_g(y) = mgy + \text{constant}. \tag{6.11}$$

Note that we are able to determine the potential energy at coordinate y only to within an additive constant. The only physically observable quantity, the work done, is related to the *difference* in the potential energy. If we add an arbitrary constant to the value of the potential energy everywhere, the difference in the potential energies remains unchanged.

In the same way, we find for the spring force that

$$\Delta U_s = U_s(x) - U_s(x_0)$$

$$= -\int_{x_0}^{x} F_s(x')dx'$$

$$= -\int_{x_0}^{x} (-kx')dx'$$

$$= k\int_{x_0}^{x} x'dx'$$

$$\Delta U_s = \tfrac{1}{2}kx^2 - \tfrac{1}{2}kx_0^2. \tag{6.12}$$

Thus, the potential energy associated with elongating a spring from its equilibrium position, at $x = 0$, is

$$U_s(x) = \tfrac{1}{2}kx^2 + \text{constant.} \tag{6.13}$$

Again, the potential energy is determined only to within an additive constant. However, keep in mind that the physical situation will often force a choice of this additive constant.

6.4 Potential Energy and Force

How can we find the conservative force when we have information on the corresponding potential energy? In calculus, taking the derivative is the inverse operation of integrating, and integration is used in equation 6.9 for the change in potential energy. Therefore, we take the derivative of that expression to obtain the force from the potential energy:

$$F_x(x) = -\frac{dU(x)}{dx}. \tag{6.14}$$

Equation 6.14 is an expression for the force from the potential energy for the case of motion in one dimension. As you can see from this expression, any constant that you add to the potential energy will not have any influence on the result you obtain for the force, because taking the derivative of a constant term results in zero. This is further evidence that the potential energy can be determined only within an additive constant.

We will not consider motion in three-dimensional situations until later in this book. However, for completeness, we can state the expression for the force from the potential energy for the case of three-dimensional motion:

$$\vec{F}(\vec{r}) = -\left(\frac{\partial U(\vec{r})}{\partial x}\hat{x} + \frac{\partial U(\vec{r})}{\partial y}\hat{y} + \frac{\partial U(\vec{r})}{\partial z}\hat{z} \right). \tag{6.15}$$

Here, the force components are given as partial derivatives with respect to the corresponding coordinates. If you major in engineering or science, you will encounter partial derivatives in many situations, certainly in higher-level calculus classes during your undergraduate studies. Basically, to take a partial derivative with respect to x, you follow the usual calculus process for finding a derivative and treat the variables y and z as constants. To take a partial derivative with respect to y, you treat x and z as constants, and to take a partial derivative with respect to z, you treat x and y as constants. We will revisit multidimensional potential energy surfaces in Chapter 11, when we talk about stability.

Concept Check 6.2

The potential energy, $U(x)$, is shown as a function of position, x, in the figure. In which region is the magnitude of the force the highest?

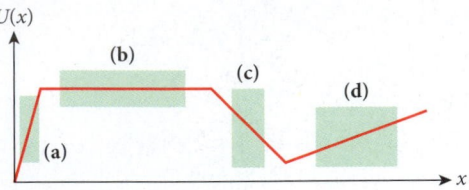

Lennard-Jones Potential

Empirically, the potential energy associated with the interaction of two atoms in a molecule as a function of the separation of the atoms has a form that is called the *Lennard-Jones potential*. This potential energy as a function of the separation, x, is given by

$$U(x) = 4U_0\left(\left(\frac{x_0}{x}\right)^{12} - \left(\frac{x_0}{x}\right)^6\right). \tag{6.16}$$

Here U_0 is a constant energy and x_0 is a constant distance. The Lennard-Jones potential is one of the most important concepts in atomic physics and is used for most numerical simulations of molecular systems.

EXAMPLE 6.2 Molecular Force

PROBLEM
What is the force resulting from the Lennard-Jones potential?

SOLUTION
We simply take the negative of the derivative of the potential energy with respect to x:

$$F_x(x) = -\frac{dU(x)}{dx}$$

$$= -\frac{d}{dx}\left(4U_0\left(\left(\frac{x_0}{x}\right)^{12} - \left(\frac{x_0}{x}\right)^6\right)\right)$$

$$= -4U_0x_0^{12}\frac{d}{dx}\left(\frac{1}{x^{12}}\right) + 4U_0x_0^6\frac{d}{dx}\left(\frac{1}{x^6}\right)$$

$$= 48U_0x_0^{12}\frac{1}{x^{13}} - 24U_0x_0^6\frac{1}{x^7}$$

$$= \frac{24U_0}{x_0}\left(2\left(\frac{x_0}{x}\right)^{13} - \left(\frac{x_0}{x}\right)^7\right).$$

PROBLEM
At what value of x does the Lennard-Jones potential have its minimum?

SOLUTION
Since we just found that the force is the derivative of the potential energy function, all we have to do is find the point(s) where $F(x) = 0$. This leads to

$$F_x(x)\big|_{x=x_{min}} = \frac{24U_0}{x_0}\left(2\left(\frac{x_0}{x_{min}}\right)^{13} - \left(\frac{x_0}{x_{min}}\right)^7\right) = 0.$$

This condition can be fulfilled only if the expression in the larger parentheses is zero; thus

$$2\left(\frac{x_0}{x_{min}}\right)^{13} = \left(\frac{x_0}{x_{min}}\right)^7.$$

Multiplying both sides by $x_{\min}^{13} x_0^{-7}$ then yields

$$2x_0^6 = x_{\min}^6$$

or

$$x_{\min} = 2^{1/6} x_0 \approx 1.1225 x_0.$$

Mathematically, it is not enough to show that the derivative is zero to establish that the potential indeed has a minimum at this coordinate. We should also make sure that the second derivative is positive. You can do this as an exercise.

Figure 6.7a shows the shape of the Lennard-Jones potential, plotted from equation 6.16, with $x_0 = 0.34$ nm and $U_0 = 1.70 \cdot 10^{-21}$ J, for the interaction of two argon atoms as a function of the separation of the centers of the two atoms. Figure 6.7b plots the corresponding molecular force, using the expression we found in Example 6.2. The vertical gray dashed line marks the coordinate where the potential has a minimum and where consequently the force is zero. Also note that close to the minimum point of the potential (within ±0.1 nm), the force can be closely approximated by a linear function, $F_x(x) \approx -k(x - x_{\min})$. This means that close to the minimum, the molecular force due to the Lennard-Jones potential behaves like a spring force.

Chapter 5 mentioned that forces similar to the spring force appear in many physical systems, and the connection between potential energy and force just described tells us why. Look, for example, at the skateboarder in the half-pipe in Figure 6.8. The curved surface of the half-pipe approximates the shape of the Lennard-Jones potential close to the minimum. If the skateboarder is at $x = x_{\min}$, he can remain there at rest. If he is to the left of the minimum, where $x < x_{\min}$, then the half-pipe exerts a force on him, which points to the right, $F_x > 0$; the further to the left he moves, the bigger the force becomes. On the right side of the half-pipe, for $x > x_{\min}$, the force points to the left, that is, $F_x < 0$. Again, these observations can be summarized with a force expression that approximately follows Hooke's Law: $F_x(x) = -k(x - x_{\min})$.

In addition, we can reach this same conclusion mathematically, by writing a Taylor expansion for $F_x(x)$ around $x_{\min}$:

$$F_x(x) = F_x(x_{\min}) + \left(\frac{dF_x}{dx}\right)_{x=x_{\min}} \cdot (x - x_{\min}) + \frac{1}{2}\left(\frac{d^2 F_x}{dx^2}\right)_{x=x_{\min}} \cdot (x - x_{\min})^2 + \cdots.$$

Since we are expanding around the potential energy minimum and since we have just shown that the force is zero there, we have $F_x(x_{\min}) = 0$. If there is a potential minimum at $x = x_{\min}$, then the second derivative of the potential must be positive. Since, according to equation 6.14, the force is $F_x(x) = -dU(x)/dx$, this means that the derivative of the force is $dF_x(x)/dx = -d^2 U(x)/dx^2$. At the minimum of the potential, we thus have $(dF_x/dx)_{x=x_{\min}} < 0$.

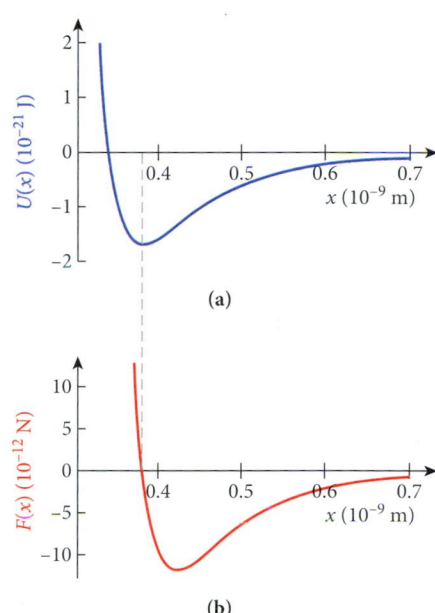

FIGURE 6.7 (a) Dependence of the potential energy on the x-coordinate of the potential energy function in equation 6.16. (b) Dependence of the force on the x-coordinate of the potential energy function in equation 6.16.

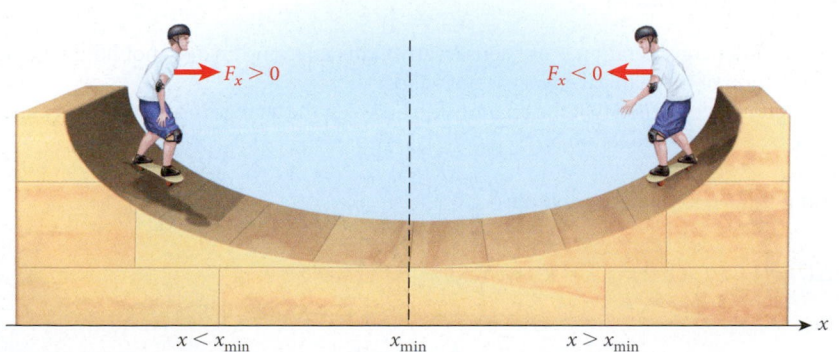

FIGURE 6.8 Skateboarder in a half-pipe.

Expressing the value of the first derivative of the force at coordinate x_{min} as some constant, $(dF_x/dx)_{x = x_{min}} = -k$ (with $k > 0$), we find $F_x(x) = -k(x - x_{min})$, if we are sufficiently close to x_{min} that we can neglect terms proportional to $(x - x_{min})^2$ and higher powers.

These physical and mathematical arguments establish why it is important to study Hooke's Law and the resulting equations of motion in detail. In this chapter, we study the work done by the spring force. In Chapter 14 on oscillations, we will analyze the motion of an object under the influence of the spring force.

6.5 Conservation of Mechanical Energy

We have defined potential energy in reference to a system of objects. We will examine different kinds of general systems in later chapters, but here we focus on one particular kind of system: an **isolated system,** which by definition is a system of objects that exert forces on one another but for which no force external to the system causes energy changes within the system. This means that no energy is transferred into or out of the system. This very common situation is highly important in science and engineering and has been extensively studied. One of the fundamental concepts of physics involves energy within an isolated system.

To investigate this concept, we begin with a definition of **mechanical energy,** E, as the sum of kinetic energy and potential energy:

$$E = K + U. \tag{6.17}$$

(Later, when we move beyond mechanics, we will add other kinds of energy to this sum and call it the *total energy*.)

For any mechanical process that occurs inside an isolated system and involves only conservative forces, the total mechanical energy is conserved. This means that the total mechanical energy remains constant in time:

$$\Delta E = \Delta K + \Delta U = 0. \tag{6.18}$$

An alternative way of writing this result (which we'll derive below) is

$$K + U = K_0 + U_0, \tag{6.19}$$

where K_0 and U_0 are the initial kinetic energy and potential energy, respectively. This relationship, which is called the law of **conservation of mechanical energy,** does not imply that the kinetic energy of the system cannot change, or that the potential energy alone remains constant. Rather it states that their changes are exactly compensating and thus offset each other. It is worth repeating that conservation of mechanical energy is valid only for conservative forces and for an isolated system, for which the influence of external forces can be neglected.

DERIVATION 6.1

As we have already seen in equation 6.7, if a conservative force does work, then the work causes a change in potential energy:

$$\Delta U = -W.$$

(If the force under consideration is not conservative, this relationship does not hold in general, and conservation of mechanical energy is not valid.)

In Chapter 5, we learned that the relationship between the change in kinetic energy and the work done by a force is (equation 5.7):

$$\Delta K = W.$$

Combining these two results, we obtain

$$\Delta U = -\Delta K \Rightarrow \Delta U + \Delta K = 0.$$

Using $\Delta U = U - U_0$ and $\Delta K = K - K_0$, we find

$$0 = \Delta U + \Delta K = U - U_0 + K - K_0 = U + K - (U_0 + K_0) \Rightarrow$$

$$U + K = U_0 + K_0.$$

Note that Derivation 6.1 did not make any reference to the particular path along which the force did the work that caused the rearrangement. In fact, you do not need to know any detail about the work or the force, other than that the force is conservative. Nor do you need to know how many conservative forces are acting. If more than one conservative force is present, you interpret ΔU as the sum of all the potential energy changes and W as the total work done by all of the conservative forces, and the derivation is still valid.

The law of energy conservation enables us to easily solve a huge number of problems that involve only conservative forces, problems that would have been very hard to solve without this law. Later in this chapter, the more general work-energy theorem for mechanics, which includes nonconservative forces, will be presented. This law will enable us to solve an even wider range of problems, including those involving friction.

Equation 6.19 introduces our first conservation law, the law of conservation of mechanical energy. Chapters 18 and 20 will extend this law to include thermal energy (heat) as well. Chapter 7 will present a conservation law for linear momentum. When we discuss rotation in Chapter 10, we will encounter a conservation law for angular momentum. In studying electricity and magnetism, we will find a conservation law for net charge (Chapter 21), and in looking at elementary particle physics (Chapter 39), we will find conservation laws for several other quantities. This list is intended to give you a flavor of a central theme of physics—the discovery of conservation laws and their use in determining the dynamics of various systems.

Before we solve a sample problem, one more remark on the concept of an isolated system is in order. In situations that involve the motion of objects under the influence of the Earth's gravitational force, the isolated system to which we apply the law of conservation of energy actually consists of the moving object plus the entire Earth. However, in using the approximation that the gravitational force is a constant, we assume that the Earth is infinitely massive and thus immobile (and that the moving object is close to the surface of Earth). Therefore, no change in the kinetic energy of the Earth can result from the rearrangement of the system. Thus, we calculate all changes in kinetic energy and potential energy only for the "junior partner"—the object moving under the influence of the gravitational force. This force is conservative and internal to the system consisting of Earth plus moving object, so all the conditions for the utilization of the law of energy conservation are fulfilled.

Specific examples of situations that involve objects moving under the influence of the gravitational force are projectile motion and pendulum motion occurring near the Earth's surface.

SOLVED PROBLEM 6.2 The Catapult Defense

Your task is to defend Neuschwanstein Castle from attackers (Figure 6.9). You have a catapult with which you can lob a rock with a launch speed of 14.2 m/s from the courtyard over the castle walls onto the attackers' camp in front of the castle at an elevation 7.20 m below that of the courtyard.

PROBLEM
What is the speed with which a rock will hit the ground at the attackers' camp? (Neglect air resistance.)

SOLUTION
THINK We can solve this problem by applying the conservation of mechanical energy. Once the catapult launches a rock, only the conservative force of gravity is acting on the rock. Thus, the total mechanical energy is conserved, which means the sum of the kinetic and potential energies of the rock always equals the total mechanical energy.

SKETCH The trajectory of the rock is shown in Figure 6.10, where the initial speed of the rock is v_0, the initial kinetic energy K_0, the initial potential energy U_0, and the initial height y_0. The final speed is v, the final kinetic energy K, the final potential energy U, and the final height y.

– Continued

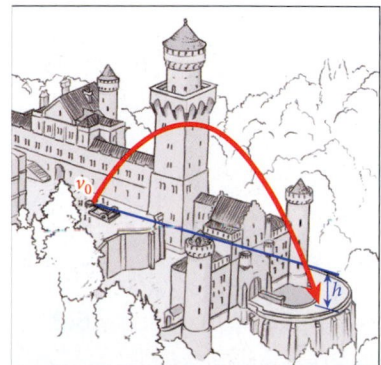

FIGURE 6.9 Illustration for a possible projectile path (red parabola) from the courtyard to the camp below and in front of the castle gate. The blue line indicates the horizontal.

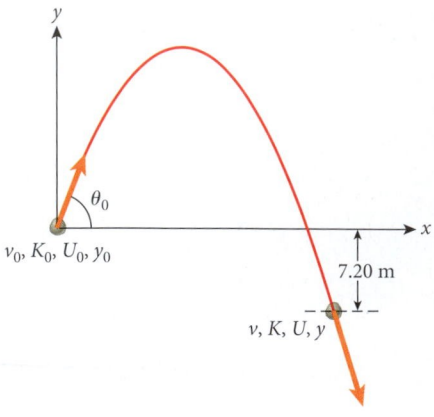

FIGURE 6.10 Trajectory of the rock launched by the catapult.

RESEARCH   We can use conservation of mechanical energy to write
$$E = K + U = K_0 + U_0,$$
where E is the total mechanical energy. The kinetic energy of the projectile can be expressed as
$$K = \tfrac{1}{2}mv^2,$$
where m is the mass of the projectile and v is its speed when it hits the ground. The potential energy of the projectile can be expressed as
$$U = mgy,$$
where y is the vertical component of the position vector of the projectile when it hits the ground.

SIMPLIFY   We substitute for K and U in $E = K + U$ to get
$$E = \tfrac{1}{2}mv^2 + mgy = \tfrac{1}{2}mv_0^2 + mgy_0.$$
The mass of the rock, m, cancels out, and we are left with
$$\tfrac{1}{2}v^2 + gy = \tfrac{1}{2}v_0^2 + gy_0.$$
We solve this for the speed:
$$v = \sqrt{v_0^2 + 2g(y_0 - y)}. \tag{6.20}$$

CALCULATE   According to the problem statement, $y_0 - y = 7.20$ m and $v_0 = 14.2$ m/s. Thus, for the final speed, we find
$$v = \sqrt{(14.2 \text{ m/s})^2 + 2(9.81 \text{ m/s}^2)(7.20 \text{ m})} = 18.51766724 \text{ m/s}.$$

ROUND   The relative height was given to three significant figures, so we report our final answer as
$$v = 18.5 \text{ m/s}.$$

DOUBLE-CHECK   Our answer for the speed of the rock when it hits the ground in front of the castle is 18.5 m/s, compared with the initial launch speed of 14.2 m/s, which seems reasonable. This speed has to be bigger due to the gain from the difference in gravitational potential energy, and it is comforting that our answer passes this simple test.

Because we were only interested in the speed at impact, we did not even need to know the initial launch angle θ_0 to solve the problem. All launch angles will give the same result (for a given launch speed), which is a somewhat surprising finding. (Of course, if you were in this situation, you would obviously want to aim high enough to clear the castle wall and accurately enough to strike the attackers' camp.)

We can also solve this problem using the concepts of projectile motion, which is a way to double-check our answer and to show the power of applying the concept of energy conservation. We start by writing the components of the initial velocity vector $\vec{v}_0$:
$$v_{x0} = v_0 \cos\theta_0$$
and
$$v_{y0} = v_0 \sin\theta_0.$$
The final x-component of the velocity, v_x, is equal to the initial x-component of the initial velocity v_{x0},
$$v_x = v_{x0} = v_0 \cos\theta_0.$$
The final component of the velocity in the y-direction can be obtained from a result of the analysis of projectile motion in Chapter 3:
$$v_y^2 = v_{y0}^2 - 2g(y - y_0).$$
Therefore, the final speed of the rock as it hits the ground is
$$v = \sqrt{v_x^2 + v_y^2}$$
$$= \sqrt{(v_0 \cos\theta_0)^2 + (v_{y0}^2 - 2g(y - y_0))}$$
$$= \sqrt{v_0^2 \cos^2\theta_0 + v_0^2 \sin^2\theta_0 - 2g(y - y_0)}.$$
Remembering that $\sin^2\theta + \cos^2\theta = 1$, we can further simplify and get:
$$v = \sqrt{v_0^2(\cos^2\theta_0 + \sin^2\theta_0) - 2g(y - y_0)} = \sqrt{v_0^2 - 2g(y - y_0)} = \sqrt{v_0^2 + 2g(y_0 - y)}.$$
This is the same as equation 6.20, which we obtained using energy conservation. Even though the final result is the same, the solution process based on energy conservation was by far easier than that based on kinematics.

Self-Test Opportunity 6.2

In Solved Problem 6.2, we neglected air resistance. Discuss qualitatively how our final answer would have changed if we had included the effects of air resistance.

y

y_0

FIGURE 6.11 Race of two balls down different inclines of the same height.

As you can see from Solved Problem 6.2, applying the conservation of mechanical energy provides us with a powerful technique for solving problems that seem rather complicated at first sight.

In general, we can determine final speed as a function of the elevation in situations where the gravitational force is at work. For instance, consider the image sequence in Figure 6.11. Two balls are released at the same time from the same height at the top of two ramps with different shapes. At the bottom end of the ramps, both balls reach the same lower elevation. Therefore, in both cases, the height difference between the initial and final points is the same. Both balls also experience normal forces in addition to the gravitational force; however, the normal forces do no work because they are perpendicular to the contact surface, by definition, and the motion is parallel to the surface. Thus, the scalar product of the normal force and displacement vectors is zero. (There is a small friction force, but it is negligible in this case.) Energy conservation considerations (see equation 6.20 in Solved Problem 6.2) tell us that the speed of both balls at the bottom end of the ramps has to be the same:

$$v = \sqrt{2g(y_0 - y)}.$$

This equation is a special case of equation 6.20 with $v_0 = 0$. Note that, depending on the curve of the bottom ramp, this result could be rather difficult to obtain using Newton's Second Law. However, even though the velocities at the top and the bottom of the ramps are the same for both balls, you cannot conclude from this result that both balls arrive at the bottom at the same time. The image sequence clearly shows that this is not the case.

Self-Test Opportunity 6.3

Why does the lighter-colored ball arrive at the bottom in Figure 6.11 before the other ball?

SOLVED PROBLEM 6.3 / Trapeze Artist

PROBLEM
A circus trapeze artist starts her motion with the trapeze at rest at an angle of 45.0° relative to the vertical. The trapeze ropes have a length of 5.00 m. What is her speed at the lowest point in her trajectory?

SOLUTION
THINK Initially, the trapeze artist has only gravitational potential energy. We can choose a coordinate system such that $y = 0$ is at her trajectory's lowest point, so the potential energy is zero at that lowest point. When the trapeze artist is at the lowest point, her kinetic energy will be a maximum. We can then equate the initial gravitational potential energy to the final kinetic energy of the trapeze artist.

SKETCH We represent the trapeze artist in Figure 6.12 as an object of mass m suspended by a rope of length ℓ. We indicate the position of the trapeze artist at a given value of the angle θ by the blue disk. The lowest point of the trajectory is reached at $\theta = 0$, and we indicate this in Figure 6.12 by a gray disk.

The figure shows that the trapeze artist is at a distance ℓ (the length of the rope) below the ceiling at the lowest point and at a distance $\ell \cos\theta$ below the ceiling for all other values of θ. This means that she is at a height $h = \ell - \ell\cos\theta = \ell(1 - \cos\theta)$ above the lowest point in the trajectory when the trapeze forms an angle θ with the vertical.

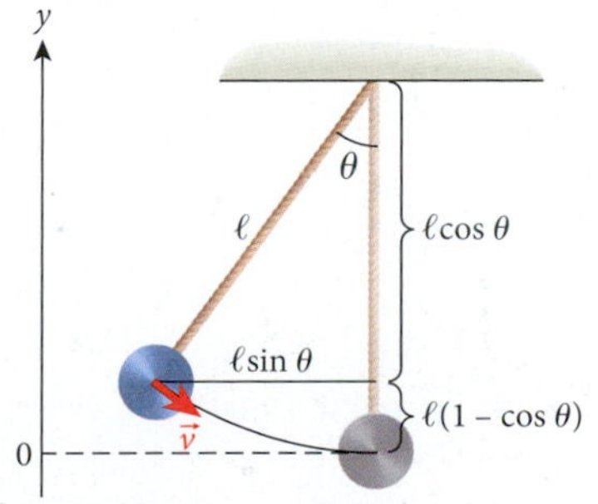

RESEARCH The trapeze is pulled back to an initial angle θ_0 relative to the vertical, and thus the trapeze artist is at a height $h = \ell(1 - \cos\theta_0)$ above the lowest point in the trajectory, according to our analysis of Figure 6.12. The potential energy at this maximum deflection, θ_0, is therefore

$$E = K + U = 0 + U = mg\ell(1 - \cos\theta_0).$$

– Continued

y

θ

ℓ

$\ell\cos\theta$

$\ell\sin\theta$

$\ell(1 - \cos\theta)$

$\vec{v}$

0

FIGURE 6.12 Geometry of a trapeze artist's swing or trajectory.

This is also the value for the total mechanical energy, because the trapeze artist has zero kinetic energy at the point of maximum deflection. For any other deflection, the energy is the sum of kinetic and potential energies:

$$E = mg\ell(1 - \cos\theta) + \tfrac{1}{2}mv^2.$$

SIMPLIFY Solving the preceding equation for the speed, we obtain

$$mg\ell(1 - \cos\theta_0) = mg\ell(1 - \cos\theta) + \tfrac{1}{2}mv^2 \Rightarrow$$

$$mg\ell(\cos\theta - \cos\theta_0) = \tfrac{1}{2}mv^2 \Rightarrow$$

$$|v| = \sqrt{2g\ell(\cos\theta - \cos\theta_0)}.$$

Here, we are interested in the speed for $|v(\theta = 0)|$, which is

$$|v(\theta = 0)| = \sqrt{2g\ell(\cos 0 - \cos\theta_0)} = \sqrt{2g\ell(1 - \cos\theta_0)}.$$

CALCULATE The initial condition is $\theta_0 = 45.0°$. Inserting the numbers, we find

$$|v(0°)| = \sqrt{2(9.81 \text{ m/s}^2)(5.00 \text{ m})(1 - \cos 45.0°)} = 5.360300809 \text{ m/s}.$$

ROUND All of the numerical values were specified to three significant figures, so we report our answer as

$$|v(0°)| = 5.36 \text{ m/s}.$$

DOUBLE-CHECK First, the obvious check of units: m/s is the SI unit for velocity and speed. The speed of the trapeze artist at the lowest point is 5.36 m/s (12 mph), which seems in line with what we see in the circus.

We can perform another check on the formula $|v(\theta = 0)| = \sqrt{2g\ell(1 - \cos\theta_0)}$ by considering the limiting cases for the initial angle θ_0 to see if they yield reasonable results. In this situation, the limiting values for θ_0 are 90°, where the trapeze starts out horizontal, and 0°, where it starts out vertical. If we use $\theta_0 = 0$, the trapeze is just hanging at rest, and we expect zero speed, an expectation borne out by our formula. On the other hand, if we use $\theta_0 = 90°$, or $\cos\theta_0 = \cos 90° = 0$, we obtain the limiting result $\sqrt{2g\ell}$, which is the same result as a free fall from the ceiling to the bottom of the trapeze swing. Again, this limit is as expected, which gives us additional confidence in our solution.

6.6 Work and Energy for the Spring Force

In Section 6.3, we found that the potential energy stored in a spring is $U_s = \tfrac{1}{2}kx^2 + \text{constant}$, where k is the spring constant and x is the displacement from the equilibrium position. Here we choose the additive constant to be zero, corresponding to having $U_s = 0$ at $x = 0$. Using the principle of energy conservation, we can find the velocity v as a function of the position. First, we can write, in general, for the total mechanical energy:

$$E = K + U_s = \tfrac{1}{2}mv^2 + \tfrac{1}{2}kx^2. \tag{6.21}$$

Once we know the total mechanical energy, we can solve this equation for the velocity. What is the total mechanical energy? The point of maximum elongation of a spring from the equilibrium position is called the **amplitude,** A. When the displacement reaches the amplitude, the velocity is briefly zero. At this point, the total mechanical energy of an object oscillating on a spring is

$$E = \tfrac{1}{2}kA^2.$$

However, conservation of mechanical energy means that this is the value of the energy for any point in the spring's oscillation. Inserting the above expression for E into equation 6.21 yields

$$\tfrac{1}{2}kA^2 = \tfrac{1}{2}mv^2 + \tfrac{1}{2}kx^2. \tag{6.22}$$

From equation 6.22, we can get an expression for the speed as a function of the position:

$$v = \sqrt{(A^2 - x^2)\frac{k}{m}}. \tag{6.23}$$

Note that we did not rely on kinematics to get this result, as that approach is rather challenging—another piece of evidence that using principles of conservation (in this case, conservation of mechanical energy) can yield powerful results. We will return to the equation of motion for a mass on a spring in Chapter 14.

SOLVED PROBLEM 6.4 | Human Cannonball

In a favorite circus act, called the "human cannonball," a person is shot from a long barrel, usually with a lot of smoke and a loud bang added for theatrical effect. Before the Italian Zacchini brothers invented the compressed air cannon for shooting human cannonballs in the 1920s, the Englishman George Farini used a spring-loaded cannon for this purpose in the 1870s.

Suppose someone wants to recreate Farini's spring-loaded human cannonball act with a spring inside a barrel. Assume the barrel is 4.00 m long, with a spring that extends the entire length of the barrel. Further, the barrel is upright, so it points vertically toward the ceiling of the circus tent. The human cannonball is lowered into the barrel and compresses the spring to some degree. An external force is added to compress the spring even further, to a length of only 0.70 m. At a height of 7.50 m above the top of the barrel is a spot on the tent that the human cannonball, of height 1.75 m and mass 68.4 kg, is supposed to touch at the top of his trajectory. Removing the external force releases the spring and fires the human cannonball vertically upward.

PROBLEM 1

What is the value of the spring constant needed to accomplish this stunt?

SOLUTION 1

THINK Let's apply energy conservation considerations to solve this problem. Potential energy is stored in the spring initially and then converted to gravitational potential energy at the top of the human cannonball's flight. As a reference point for our calculations, we select the top of the barrel and place the origin of our coordinate system there. To accomplish the stunt, enough energy has to be provided, through compressing the spring, that the top of the head of the human cannonball is elevated to a height of 7.50 m above the zero point we have chosen. Since the person has a height of 1.75 m, his feet need to be elevated by only $h = 7.50 \text{ m} - 1.75 \text{ m} = 5.75 \text{ m}$. We can specify all position values for the human cannonball on the y-coordinate as the position of the bottom of his feet.

SKETCH To clarify this problem, let's apply energy conservation at different instants of time. Figure 6.13a shows the initial equilibrium position of the spring. In Figure 6.13b, the external force $\vec{F}$ and the weight of the human cannonball compress the spring by 3.30 m to a length

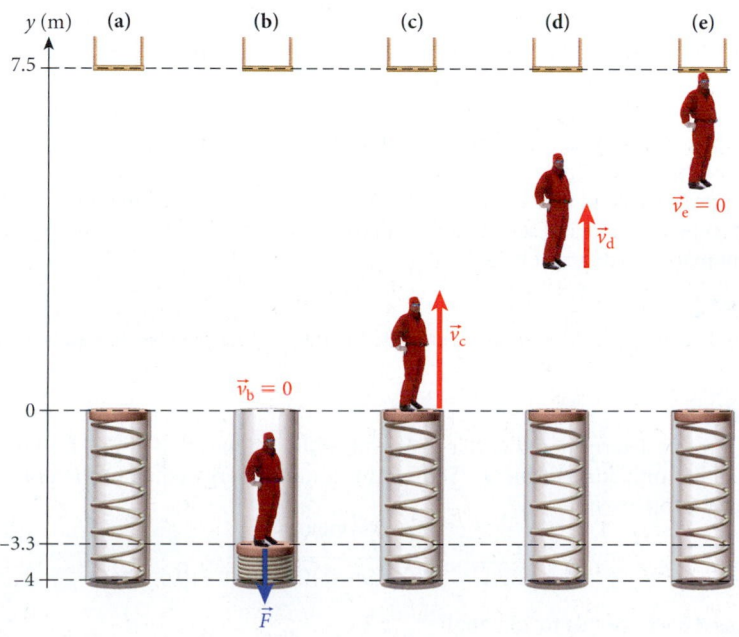

FIGURE 6.13 The human cannonball stunt at five different instants of time.   – *Continued*

Concept Check 6.3

A spring with spring constant k is oriented vertically and compressed downward a distance x from its equilibrium position. An object of mass m is placed on the upper end of the spring, and the spring is released. The object rises a distance h (with $h \gg x$) above the equilibrium position of the spring. If the spring is then compressed downward by the same distance x and an object of mass $3m$ is placed on it, how high will the object rise when the spring is released?

a) h

d) h^3

b) $3h$

e) $h^{1/2}$

c) $h/3$

Concept Check 6.4

A ball of mass m is thrown vertically into the air with an initial speed v. Which of the following equations correctly describes the maximum height, h, of the ball?

a) $h = \sqrt{\dfrac{v}{2g}}$

d) $h = \dfrac{mv^2}{g}$

b) $h = \dfrac{g}{\frac{1}{2}v^2}$

e) $h = \dfrac{v^2}{2g}$

c) $h = \dfrac{2mv}{g}$

Self-Test Opportunity 6.4

Graph the potential and kinetic energies of the human cannonball of Solved Problem 6.4 as a function of the y-coordinate. For what value of the displacement is the speed of the human cannonball at a maximum? (*Hint:* This occurs not exactly at $y = 0$ but at a value of $y < 0$.)

---

of 0.70 m. When the spring is released, the cannonball accelerates and has a velocity $\vec{v}_c$ as he passes the spring's equilibrium position (see Figure 6.13c). From this position, he has to rise 5.75 m and arrive at the spot (Figure 6.13e) with zero velocity.

RESEARCH   We are free to choose the zero point for the gravitational potential energy arbitrarily. We elect to set the gravitational potential to zero at the equilibrium position of the spring without a load, as shown in Figure 6.13a.

At the instant depicted in Figure 6.13b, the human cannonball has zero kinetic energy and potential energies from the spring force and gravity. Therefore, the total energy at this instant is
$$E = \tfrac{1}{2}ky_b^2 + mgy_b.$$
At the instant shown in Figure 6.13c, the human cannonball has only kinetic energy and zero potential energy:
$$E = \tfrac{1}{2}mv_c^2.$$

Right after this instant, the human cannonball leaves the spring, flies through the air as shown in Figure 6.13d, and finally reaches the top (Figure 6.13e). At the top, he has only gravitational potential energy and no kinetic energy (because the spring is designed to allow him to reach the top with no residual speed):
$$E = mgy_e.$$

SIMPLIFY   Energy conservation requires that the total energy remain the same. Setting the first and third expressions written above for E equal, we obtain
$$\tfrac{1}{2}ky_b^2 + mgy_b = mgy_e.$$
We can rearrange this equation to obtain the spring constant:
$$k = 2mg\frac{y_e - y_b}{y_b^2}.$$

CALCULATE   According to the given information and the origin of the coordinate system we selected, $y_b = -3.30$ m and $y_e = 5.75$ m. Thus, we find that the necessary value of the spring constant is
$$k = 2(68.4\text{ kg})(9.81\text{ m/s}^2)\frac{5.75\text{ m} - (-3.30\text{ m})}{(3.30\text{ m})^2} = 1115.26\text{ N/m}.$$

ROUND   All of the numerical values used in the calculation have three significant figures, so our final answer is
$$k = 1.12 \cdot 10^3\text{ N/m}.$$

DOUBLE-CHECK   When the spring is compressed initially, the potential energy stored in it is
$$U = \tfrac{1}{2}ky_b^2 = \tfrac{1}{2}\left(1.12 \cdot 10^3\text{ N/m}\right)(3.30\text{ m})^2 = 6.07\text{ kJ}.$$
The gravitational potential energy gained by the human cannonball is
$$U = mg\Delta y = (68.4\text{ kg})(9.81\text{ m/s}^2)(9.05\text{ m}) = 6.07\text{ kJ},$$
which is the same as the energy stored in the spring initially. Our calculated value for the spring constant makes sense.

Note that the mass of the human cannonball enters into the equation for the spring constant. We can turn this around and state that the same cannon with the same spring will shoot people of different masses to different heights.

PROBLEM 2

What is the speed that the human cannonball reaches as he passes the equilibrium position of the spring?

SOLUTION 2

We have already determined that our choice of origin implies that at this instant the human cannonball has only kinetic energy. Setting this kinetic energy equal to the potential energy reached at the top, we find
$$\tfrac{1}{2}mv_c^2 = mgy_e \Rightarrow$$
$$v_c = \sqrt{2gy_e} = \sqrt{2(9.81\text{ m/s}^2)(5.75\text{ m})} = 10.6\text{ m/s}.$$
This speed corresponds to 23.8 mph.

EXAMPLE 6.3   Bungee Jumper

A bungee jumper locates a suitable bridge that is 75.0 m above the river below, as shown in Figure 6.14. The jumper has a mass of $m = 80.0$ kg and a height of $L_{jumper} = 1.85$ m. We can think of a bungee cord as a spring. The spring constant of the bungee cord is $k = 50.0$ N/m. Assume that the mass of the bungee cord is negligible compared with the jumper's mass.

PROBLEM
The jumper wants to know the maximum length of bungee cord he can safely use for this jump.

SOLUTION
We are looking for the unstretched length of the bungee cord, L_0, as the jumper would measure it standing on the bridge. The distance from the bridge to the water is $L_{max} = 75.0$ m. Energy conservation tells us that the gravitational potential energy that the jumper has, as he dives off the bridge, will be converted to potential energy stored in the bungee cord. The jumper's gravitational potential energy on the bridge is

$$U_g = mgy = mgL_{max},$$

assuming that gravitational potential energy is zero at the level of the water. Before he starts his jump, he has zero kinetic energy, and so his total energy when he is on top of the bridge is

$$E_{top} = mgL_{max}.$$

At the bottom of the jump, where the jumper's head just touches the water, the potential energy stored in the bungee cord is

$$U_s = \tfrac{1}{2}ky^2 = \tfrac{1}{2}k\left(L_{max} - L_{jumper} - L_0\right)^2,$$

where $L_{max} - L_{jumper} - L_0$ is the length that the bungee cord stretches beyond its unstretched length. (Here we have to subtract the jumper's height from the height of the bridge to obtain the maximum length, $L_{max} - L_{jumper}$, to which the bungee cord is allowed to stretch, assuming that it is tied around his ankles.) Since the bungee jumper is momentarily at rest at this lowest point of his jump, the kinetic energy is zero at that point, and the total energy is then

$$E_{bottom} = \tfrac{1}{2}k\left(L_{max} - L_{jumper} - L_0\right)^2.$$

From the conservation of mechanical energy, we know that $E_{top} = E_{bottom}$, and so we find

$$mgL_{max} = \tfrac{1}{2}k\left(L_{max} - L_{jumper} - L_0\right)^2.$$

Solving for the required unstretched length of bungee cord gives us

$$L_0 = L_{max} - L_{jumper} - \sqrt{\frac{2mgL_{max}}{k}}.$$

Putting in the given numbers, we get

$$L_0 = \left(75.0 \text{ m}\right) - \left(1.85 \text{ m}\right) - \sqrt{\frac{2\left(80.0 \text{ kg}\right)\left(9.81 \text{ m/s}^2\right)\left(75.0 \text{ m}\right)}{50.0 \text{ N/m}}} = 24.6 \text{ m}.$$

For safety, the jumper would be wise to use a bungee cord shorter than this and to test it with a dummy mass similar to his.

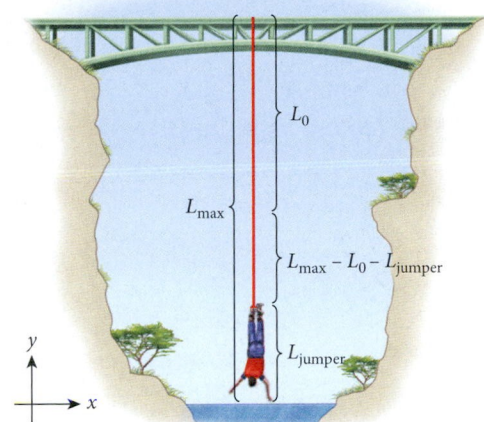

FIGURE 6.14 A bungee jumper needs to calculate how long a bungee cord he can safely use.

Concept Check 6.5

At the moment of maximum stretching of the bungee cord in Example 6.3, what is the net acceleration that the jumper experiences (in terms of $g = 9.81$ m/s^2)?

a) $0g$

b) $1.0g$, directed downward

c) $1.0g$, directed upward

d) $2.1g$, directed downward

e) $2.1g$, directed upward

Self-Test Opportunity 6.5

Can you derive an expression for the acceleration that the bungee jumper experiences at the maximum stretching of the bungee cord? How does this acceleration depend on the spring constant of the bungee cord?

Potential Energy of an Object Hanging from a Spring

We saw in Solved Problem 6.4 that the initial potential energy of the human cannonball has contributions from the spring force and the gravitational force. In Example 5.3 and the discussion following it, we established that hanging an object of mass m from a spring with spring constant k shifts the equilibrium position of the spring from zero to y_0, given by the equilibrium condition,

$$ky_0 = -mg \Rightarrow y_0 = -\frac{mg}{k}. \tag{6.24}$$

Figure 6.15 shows the forces that act on an object suspended from a spring when it is in different positions. This figure shows two different choices for the origin of the vertical coordinate

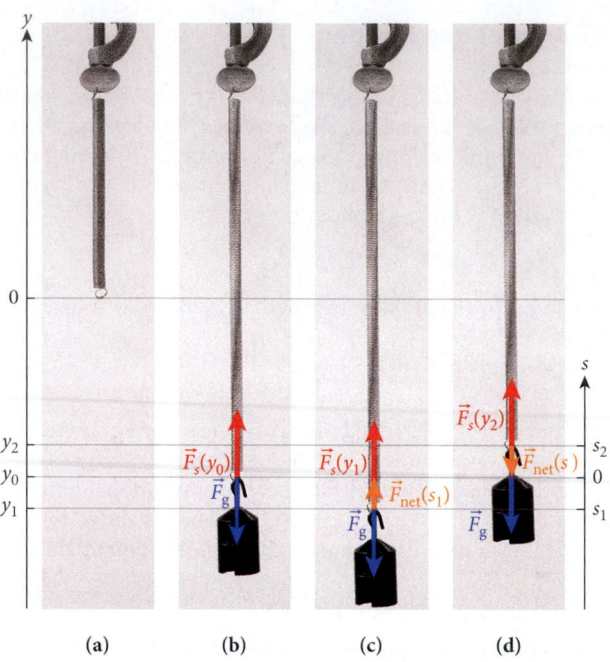

FIGURE 6.15   (a) A spring is hanging vertically with its end in the equilibrium position at $y = 0$. (b) An object of mass m is hanging at rest from the same spring, with the end of the spring now at $y = y_0$ or $s = 0$. (c) The end of the spring with the object attached is at $y = y_1$ or $s = s_1$. (d) The end of the spring with the object attached is at $y = y_2$ or $s = s_2$.

axis: In Figure 6.15a, the vertical coordinate is called y and has zero at the equilibrium position of the end of the spring without the mass hanging from it; in Figure 6.15b the new equilibrium point, y_0, with the object suspended from the spring, is calculated according to equation 6.24. This new equilibrium point is the origin of the axis and the vertical coordinate is called s. The end of the spring is located at $s = 0$. The system is in equilibrium because the force exerted by the spring on the object balances the gravitational force acting on the object:

$$\vec{F}_s(y_0) + \vec{F}_g = 0.$$

In Figure 6.15c, the object has been displaced downward away from the new equilibrium position, so $y = y_1$ and $s = s_1$. Now there is a net upward force tending to restore the object to the new equilibrium position:

$$\vec{F}_{net}(s_1) = \vec{F}_s(y_1) + \vec{F}_g.$$

If instead, the object is displaced upward, above the new equilibrium position, as shown in Figure 6.15d, there is a net downward force that tends to restore the object to the new equilibrium position:

$$\vec{F}_{net}(s_2) = \vec{F}_s(y_2) + \vec{F}_g.$$

We can calculate the potential energy of the object and the spring for these two choices of the coordinate system and show that they differ by only a constant. We start by defining the potential energy of the object connected to the spring, taking y as the variable and assuming that the potential energy is zero at $y = 0$:

$$U(y) = \tfrac{1}{2}ky^2 + mgy.$$

Using the relation $y = s + y_0$, we can express this potential energy in terms of the variable s:

$$U(s) = \tfrac{1}{2}k(s + y_0)^2 + mg(s + y_0).$$

Rearranging gives us

$$U(s) = \tfrac{1}{2}ks^2 + ksy_0 + \tfrac{1}{2}ky_0^2 + mgs + mgy_0.$$

Substituting $ky_0 = -mg$, from equation 6.24, into this equation, we get

$$U(s) = \tfrac{1}{2}ks^2 - (mg)s + \tfrac{1}{2}(mg)y_0 + mgs - mgy_0.$$

Thus, we find that the potential energy in terms of s is

$$U(s) = \tfrac{1}{2}ks^2 - \tfrac{1}{2}mgy_0. \tag{6.25}$$

Figure 6.16 shows the potential energy functions for these two coordinate axes. The blue curve in Figure 6.16 (a) shows the potential energy as a function of the vertical coordinate y, with the choice of zero potential energy at $y = 0$ corresponding to the spring hanging vertically without the object connected to it. The new equilibrium position, y_0, is determined by the displacement that occurs when an object of mass m is attached to the spring, as calculated using equation 6.24. The red curve in Figure 6.16 (b) represents the potential energy as a function of the vertical coordinate s, with the equilibrium position chosen to be $s = 0$. The potential energy curves $U(y)$ and $U(s)$ are both parabolas, which are offset from each other by a simple constant.

Thus, we can express the potential energy of an object of mass m hanging from a vertical spring in terms of the displacement s about an equilibrium point as

$$U(s) = \tfrac{1}{2}ks^2 + C,$$

where C is a constant. For many problems, we can choose zero as the value of this constant, allowing us to write

$$U(s) = \tfrac{1}{2}ks^2.$$

This result allows us to use the same spring force potential for different masses attached to the end of a spring by simply shifting the origin to the new equilibrium position. (Of course, this only works if we do not attach too much mass to the end of the spring and overstretch it beyond its elastic limit.)

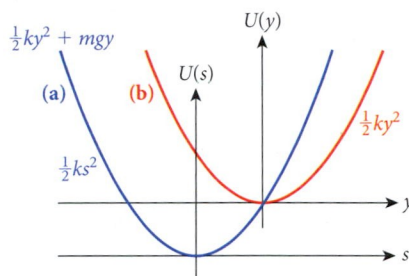

FIGURE 6.16 Potential energy functions for the two vertical coordinate axes used in Figure 6.15.

6.7 Nonconservative Forces and the Work-Energy Theorem

With the introduction of the potential energy we can extend and augment the work–kinetic energy theorem of Chapter 5 for conservative forces. By including the potential energy as well, we find the **work-energy theorem**

$$W = \Delta E = \Delta K + \Delta U \tag{6.26}$$

where W is the work done by an *external* force, ΔK is the change of kinetic energy, and ΔU is the change in potential energy. This relationship means that external work done to a system can change the total energy of the system.

Is energy conservation violated in the presence of nonconservative forces? The word *nonconservative* seems to imply that it is violated, and, indeed, the total *mechanical* energy is not conserved. Where, then, does the energy go? Section 6.2 showed that the friction force docs not do work but instead dissipates mechanical energy into internal excitation energy, which can be vibration energy, deformation energy, chemical energy, or electrical energy, depending on the material of which the object is made and on the particular form of the friction force. In Section 6.2, we found that W_f is the total energy dissipated by nonconservative forces into internal energy and then into other energy forms besides mechanical energy. If we add this type of energy to the total mechanical energy, we obtain the **total energy:**

$$E_{\text{total}} = E_{\text{mechanical}} + E_{\text{other}} = K + U + E_{\text{other}}. \tag{6.27}$$

Here E_{other} stands for all other forms of energy that are not kinetic or potential energies. The change in the other energy forms is exactly the negative of the energy dissipated by the friction force in going from the initial to the final state of the system:

$$\Delta E_{other} = - W_f.$$

The total energy is conserved—that is, stays constant in time—even for nonconservative forces. This is the most important point in this chapter:

The total energy—the sum of all forms of energy, mechanical or other—is *always* conserved in an isolated system.

We can also write this law of energy conservation in the form that states that the change in the total energy of an isolated system is zero:

$$\Delta E_{total} = 0. \tag{6.28}$$

Since we do not yet know what exactly this internal energy is and how to calculate it, it may seem that we cannot use energy considerations when at least one of the forces acting is nonconservative. However, this is not the case. For the case in which only conservative forces are acting, we found that (see equation 6.18) the total mechanical energy is conserved, or $\Delta E = \Delta K + \Delta U = 0$, where E refers to the total mechanical energy. In the presence of nonconservative forces, combining equations 6.28, 6.27, and 6.26 gives

$$W_f = \Delta K + \Delta U. \tag{6.29}$$

This relationship is a generalization of the work-energy theorem. In the absence of nonconservative forces, $W_f = 0$, and equation 6.29 reduces to the law of the conservation of mechanical energy, equation 6.19. When applying either of these two equations, you must select two times—a beginning and an end. Usually this choice is obvious, but sometimes care must be taken, as demonstrated in the following solved problem.

SOLVED PROBLEM 6.5 | Block Pushed Off a Table

Consider a block on a table. This block is pushed by a spring attached to the wall, slides across the table, and then falls to the ground. The block has a mass $m = 1.35$ kg. The spring constant is $k = 560.$ N/m, and the spring is initially compressed by 0.110 m. The block slides a distance $d = 0.65$ m across the table of height $h = 0.750$ m. The coefficient of kinetic friction between the block and the table is $\mu_k = 0.160$.

PROBLEM
What speed will the block have when it lands on the floor?

SOLUTION
THINK At first sight, this problem does not seem to be one to which we can apply mechanical energy conservation, because the nonconservative force of friction is in play. However, we can utilize the work-energy theorem, equation 6.29. To be certain that the block actually leaves the table, though, we first calculate the total energy imparted to the block by the spring and make sure that the potential energy stored in the compressed spring is sufficient to overcome the friction force.

SKETCH Figure 6.17a shows the block of mass m pushed by the spring. The mass slides on the table a distance d and then falls to the floor, which is a distance h below the table.

We choose the origin of our coordinate system such that the block starts at $x = y = 0$, with the x-axis running along the bottom surface of the block and the y-axis running through its center (Figure 6.17b). The origin of the coordinate system can be placed at any point, but it is important to fix an origin, because all potential energies have to be expressed relative to some reference point.

RESEARCH *Step 1:* Let's analyze the problem situation without the friction force. In this case, the block initially has potential energy from the spring and no kinetic energy, since it is

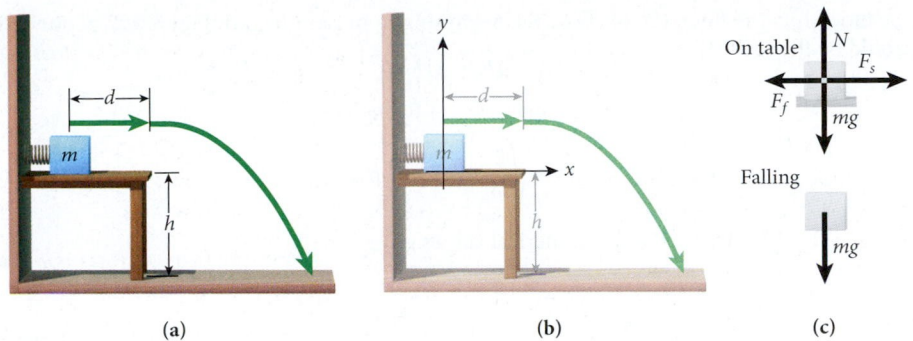

FIGURE 6.17 (a) Block of mass m is pushed off a table by a spring. (b) A coordinate system is superimposed on the block and table. (c) Free-body diagrams of the block while moving on the table and while falling.

at rest. When the block hits the floor, it has kinetic energy and negative gravitational potential energy. Conservation of mechanical energy results in

$$K_0 + U_0 = K + U \Rightarrow$$

$$0 + \tfrac{1}{2}kx_0^2 = \tfrac{1}{2}mv^2 - mgh. \qquad (i)$$

Usually, we would solve this equation for the speed and put in the numbers later. However, because we will need them again, let's evaluate the two expressions for potential energy:

$$\tfrac{1}{2}kx_0^2 = 0.5(560 \text{ N/m})(0.11 \text{ m})^2 = 3.39 \text{ J}$$

$$mgh = (1.35 \text{ kg})(9.81 \text{ m/s}^2)(0.75 \text{ m}) = 9.93 \text{ J}.$$

Now, solving equation (i) for the speed results in

$$v = \sqrt{\frac{2}{m}(\tfrac{1}{2}kx_0^2 + mgh)} = \sqrt{\frac{2}{1.35 \text{ kg}}(3.39 \text{ J} + 9.93 \text{ J})} = 4.44 \text{ m/s}.$$

Step 2: Now we include friction. Our considerations remain almost unchanged, except that we have to include the energy dissipated by the nonconservative force of friction. We find the force of friction using the upper free-body diagram in Figure 6.17c. We can see that the normal force is equal to the weight of the block and write

$$N = mg.$$

The friction force is given by

$$F_k = \mu_k N = \mu_k mg.$$

We can then write the energy dissipated by the friction force as

$$W_f = -\mu_k mgd.$$

In applying the generalization of the work-energy theorem, we choose the initial time to be when the block is about to start moving (see Figure 6.17a) and the final time to be when the block reaches the edge of the table and is about to start the free-fall portion of its path. Let K_{top} be the kinetic energy at the final time, chosen to make sure that the block makes it to the end of the table. Using equation 6.29 and the value we calculated above for the block's initial potential energy, we find:

$$W_f = \Delta K + \Delta U = K_{\text{top}} - \tfrac{1}{2}kx_0^2 = -\mu_k mgd$$

$$K_{\text{top}} = \tfrac{1}{2}kx_0^2 - \mu_k mgd$$

$$= 3.39 \text{ J} - (0.16)(1.35 \text{ kg})(9.81 \text{ m/s}^2)(0.65 \text{ m})$$

$$= 3.39 \text{ J} - 1.38 \text{ J} = 2.01 \text{ J}.$$

Because the kinetic energy $K_{\text{top}} > 0$, the block can overcome friction and slide off the table. Now we can calculate the block's speed when it hits the floor.

SIMPLIFY For this part of the problem, we choose the initial time to be when the block is at the table's edge to exploit the calculations we have already done. The final time is when

– Continued

the block hits the floor. (If we chose the beginning to be as shown in Figure 6.17a, our result would be the same.)

$$W_f = \Delta K + \Delta U = 0$$

$$\tfrac{1}{2}mv^2 - K_{top} + 0 - mgh = 0,$$

$$v = \sqrt{\frac{2}{m}(K_{top} + mgh)}.$$

CALCULATE   Putting in the numerical values gives us

$$v = \sqrt{\frac{2}{1.35 \text{ kg}}(2.01 \text{ J} + 9.93 \text{ J})} = 4.20581608 \text{ m/s}.$$

ROUND   All of the numerical values were given to three significant figures, so we have

$$v = 4.21 \text{ m/s}.$$

DOUBLE-CHECK   As you can see, the main contribution to the speed of the block at impact originates from the free-fall portion of its path. Why did we go through the intermediate step of figuring out the value of K_{top}, instead of simply using the formula $v = \sqrt{2(\tfrac{1}{2}kx_0^2 - \mu_k mgd + mgh)/m}$, which we got from a generalization of the work-energy theorem? We needed to calculate K_{top} first to ensure that it is positive, meaning that the energy imparted to the block by the spring is sufficient to exceed the work to be done against the friction force. If K_{top} had turned out to be negative, the block would have stopped on the table. For example, if we had attempted to solve the same problem with a coefficient of kinetic friction between the block and the table of $\mu_k = 0.50$ instead of $\mu_k = 0.16$, we would have found that

$$K_{top} = 3.39 \text{ J} - 4.30 \text{ J} = -0.91 \text{ J},$$

which is impossible.

Concept Check 6.6

A curling stone of mass m is given an initial velocity v on ice, where the coefficient of kinetic friction is μ_k. The stone travels a distance d. If the initial velocity is doubled, how far will the stone slide?

a) d

b) $2d$

c) d^2

d) $4d$

e) $4d^2$

As Solved Problem 6.5 shows, energy considerations are still a powerful tool for performing otherwise very difficult calculations, even in the presence of nonconservative forces. However, the principle of conservation of mechanical energy cannot be applied quite as straightforwardly when nonconservative forces are present, and you have to account for the energy dissipated by these forces. Let's try one more solved problem to elaborate on these concepts.

SOLVED PROBLEM 6.6   Sledding down a Hill

PROBLEM

A boy on a sled starts from rest and slides down a snow-covered hill. Together the boy and sled have a mass of 23.0 kg. The hill's slope makes an angle $\theta = 35.0°$ with the horizontal. The surface of the hill is 25.0 m long. When the boy and the sled reach the bottom of the hill, they continue sliding on a horizontal snow-covered field. The coefficient of kinetic friction between the sled and the snow is 0.100. How far do the boy and sled move on the horizontal field before stopping?

SOLUTION

THINK   The boy and sled start with zero kinetic energy and finish with zero kinetic energy and have gravitational potential energy at the top of the hill. As boy and sled go down the hill, they gain kinetic energy. At the bottom of the hill, their potential energy is zero, and they have kinetic energy. However, the boy and sled are continuously losing energy to friction. Thus, the change in potential energy will equal the energy lost to friction. We must take into account the fact that the friction force will be different when the sled is on the slope than when it is on the flat field.

SKETCH   A sketch of the boy sledding is shown in Figure 6.18.

RESEARCH   The boy and sled start with zero kinetic energy and finish with zero kinetic energy. We call the length of the slope of the hill d_1, and the distance that the boy and sled

FIGURE 6.18 (a) Sketch of the boy and sled on the slope and on the flat field, showing the angle of incline and distances. (b) Free-body diagram for the boy and sled on the slope. (c) Free-body diagram for the boy and sled on the flat field.

travel on the flat field d_2, as shown in Figure 6.18a. Assuming that the gravitational potential energy of the boy and sled is zero at the bottom of the hill, the change in the gravitational potential energy from the top of the hill to the flat field is

$$\Delta U = -mgh,$$

where m is the mass of the boy and sled together and

$$h = d_1 \sin\theta.$$

The force of friction is different on the slope and on the flat field because the normal force is different. From Figure 6.18b, the force of friction on the slope is

$$f_{k1} = \mu_k N_1 = \mu_k mg \cos\theta.$$

From Figure 6.18c, the force of friction on the flat field is

$$f_{k2} = \mu_k N_2 = \mu_k mg.$$

The energy dissipated by friction, W_f, is equal to the energy dissipated by friction while sliding on the slope, W_1, plus the energy dissipated while sliding on the flat field, W_2:

$$W_f = W_1 + W_2.$$

The energy dissipated by friction on the slope is

$$W_1 = -f_{k1} d_1,$$

and the energy dissipated by friction on the flat field is

$$W_2 = -f_{k2} d_2.$$

SIMPLIFY According to the preceding three equations, the total energy dissipated by friction is given by

$$W_f = -f_{k1} d_1 - f_{k2} d_2.$$

Substituting the two expressions for the friction forces into this equation gives us

$$W_f = -\left(\mu_k mg \cos\theta\right) d_1 - \left(\mu_k mg\right) d_2.$$

The change in potential energy is obtained by combining the equation $\Delta U = -mgh$ with the expression obtained for the height, $h = d_1 \sin\theta$:

$$\Delta U = -mg d_1 \sin\theta.$$

Since the sled is at rest at the top of the hill and at the end of the ride as well, we have $\Delta K = 0$, and so according to equation 6.29, in this case, $\Delta U = W_f$. Now we can equate the change in potential energy with the energy dissipated by friction:

$$mg d_1 \sin\theta = \left(\mu_k mg \cos\theta\right) d_1 + \left(\mu_k mg\right) d_2.$$

Canceling out mg on both sides and solving for the distance the boy and sled travel on the flat field, we get

$$d_2 = \frac{d_1 \left(\sin\theta - \mu_k \cos\theta\right)}{\mu_k}.$$

– *Continued*

CALCULATE   Putting in the given numerical values, we get

$$d_2 = \frac{(25.0 \text{ m})(\sin 35.0° - 0.100 \cdot \cos 35.0°)}{0.100} = 122.9153 \text{ m}.$$

ROUND   All of the numerical values were specified to three significant figures, so we report our answer as

$$d_2 = 123. \text{ m}.$$

DOUBLE-CHECK   The distance that the sled moves on the flat field is a little longer than a football field, which certainly seems possible after coming off a steep hill of length 25 m.

We can double-check our answer by assuming that the friction force between the sled and the snow is the same on the slope of the hill as it is on the flat field,

$$f_k = \mu_k mg.$$

The change in potential energy would then equal the approximate energy dissipated by friction for the entire distance the sled moves:

$$mgd_1 \sin\theta = \mu_k mg(d_1 + d_2).$$

The approximate distance traveled on the flat field would then be

$$d_2 = \frac{d_1(\sin\theta - \mu_k)}{\mu_k} = \frac{(25.0 \text{ m})(\sin 35.0° - 0.100)}{0.100} = 118. \text{ m}.$$

This result is close to but less than our answer of 123 m, which we expect because the friction force on the flat field is higher than the friction force on the slope. Thus, our answer seems reasonable.

6.8   Potential Energy and Stability

Let's return to the relationship between force and potential energy. Perhaps it may help you gain physical insight into this relationship if you visualize the potential energy curve as the track of a roller coaster. This analogy is not a perfect one, because a roller coaster moves in a two-dimensional plane or even in three-dimensional space, not in one dimension, and there is some small amount of friction between the cars and the track. Still, it is a good approximation to assume that there is conservation of mechanical energy. The motion of the roller coaster car can then be described by a potential energy function.

Shown in Figure 6.19 are plots of the potential energy (yellow line following the outline of the track), the total energy (horizontal orange line), and the kinetic energy (difference between these two, the red line) as a function of position for a segment of a roller coaster ride. You can see that the kinetic energy has a minimum at the highest point of the track, where the speed of the cars is smallest, and the speed increases as the cars roll down the incline. All of these effects are a consequence of the conservation of total mechanical energy.

Figure 6.20 shows graphs of a potential energy function (part a) and the corresponding force (part b). Because the potential energy can be determined only within an additive

FIGURE 6.19 Total, potential, and kinetic energy for a roller coaster. The yellow line shows the potential energy and follows the roller coaster track. The red curve shows the kinetic energy, and the orange line at the top is the (constant!) sum of the two.

constant, the zero value of the potential energy in Figure 6.20a is set at the lowest value. However, for all physical considerations, this is irrelevant. On the other hand, the zero value for the force cannot be chosen arbitrarily.

Equilibrium Points

Three special points on the x-coordinate axis of Figure 6.20b are marked by vertical gray lines. These points indicate where the force has a value of zero. Because the force is the derivative of the potential energy with respect to the x-coordinate, the potential energy has an extremum—a maximum or minimum value—at such points. You can clearly see that the potential energies at x_1 and x_3 represent minima, and the potential energy at x_2 is a maximum. At all three points, an object would experience no acceleration, because it is located at an extremum where the force is zero. Because there is no force, Newton's Second Law tells us that there is no acceleration. Thus, these points are equilibrium points.

The equilibrium points in Figure 6.20 represent two different kinds. Points x_1 and x_3 are stable equilibrium points, and x_2 is an unstable equilibrium point. What distinguishes stable and unstable equilibrium points is the response to perturbations (small changes in position around the equilibrium position).

> ### Definition
>
> At **stable equilibrium points,** small perturbations result in small oscillations around the equilibrium point. At **unstable equilibrium points,** small perturbations result in an accelerating movement away from the equilibrium point.

The roller coaster analogy may be helpful here: If you are sitting in a roller coaster car at point x_1 or x_3 and someone gives the car a push, it will just rock back and forth on the track, because you are sitting at a local point of lowest energy. However, if the car gets the same small push while sitting at x_2, it will roll down the slope.

What makes an equilibrium point stable or unstable from a mathematical standpoint is the value of the second derivative of the potential energy function, or the curvature. Negative curvature means a local maximum of the potential energy function, and therefore an unstable equilibrium point; a positive curvature indicates a stable equilibrium point. Of course, there is also the situation between a stable and unstable equilibrium, between a positive and negative curvature. This is a point of metastable equilibrium, with zero local curvature, that is, a value of zero for the second derivative of the potential energy function.

Turning Points

Figure 6.21a shows the same potential energy function as Figure 6.20, but with the addition of horizontal lines for four different values of the total mechanical energy (E_1 through E_4). For each value of this total energy and for each point on the potential energy curve, we can calculate the value of the kinetic energy by simple subtraction. Let's first consider the largest value of the total mechanical energy shown in the figure, E_1 (blue horizontal line):

$$K_1(x) = E_1 - U(x). \tag{6.30}$$

The kinetic energy, $K_1(x)$, is shown in Figure 6.21b by the blue curve, which is clearly an upside-down version of the potential energy curve in Figure 6.21a. However, its absolute height is not arbitrary but results from equation 6.30. As previously mentioned, we can always add an arbitrary additive constant to the potential energy, but then we are forced to add the same additive constant to the total mechanical energy, so that their difference, the kinetic energy, remains unchanged.

For the other values of the total mechanical energy in Figure 6.21, an additional complication arises: the condition that the kinetic energy has to be larger than or equal to zero. This condition means that the kinetic energy is not defined in a region where $E_i - U(x)$ is negative. For the total mechanical energy of E_2, the kinetic energy is greater than zero only

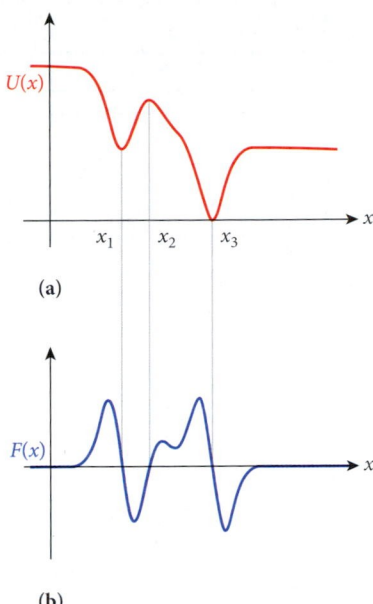

FIGURE 6.20 (a) Potential energy as a function of position; (b) the force corresponding to this potential energy function, as a function of position.

Concept Check 6.7

Which of the four drawings represents a stable equilibrium point for the ball on its supporting surface?

(a)

(b)

(c)

(d)

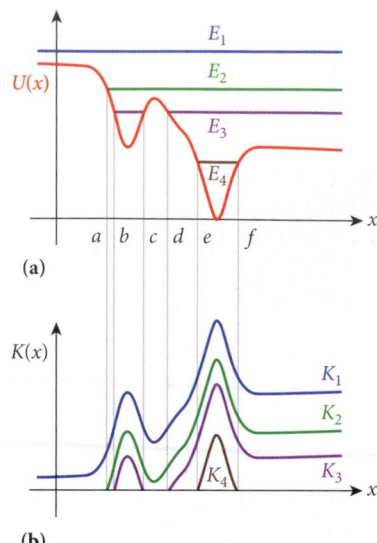

FIGURE 6.21 (a) The same potential energy function as in Figure 6.20. Shown are lines representing four different values of the total energy, E_1 through E_4. (b) The corresponding kinetic energy functions for these four total energies and the potential energy function in the upper part. The gray vertical lines mark the turning points.

for $x \geq a$, as indicated in Figure 6.21b by the green curve. Thus, an object moving with total energy E_2 from right to left in Figure 6.21 will reach the point $x = a$ and have zero velocity there. Referring to Figure 6.20, you see that the force at that point is positive, pushing the object to the right, that is, making it turn around. This is why such a point is called a **turning point.** Moving to the right, this object will pick up kinetic energy and follow the same kinetic energy curve from left to right, making its path reversible. This behavior is a consequence of the conservation of total mechanical energy.

Definition

Turning points are points where the kinetic energy is zero and where a net force moves the object away from the point.

An object with total energy equal to E_4 in Figure 6.21 has two turning points in its path: $x = e$ and $x = f$. The object can move only between these two points. It is trapped in this interval and cannot escape. Perhaps the roller coaster analogy is again helpful: A car released from point $x = e$ will move through the dip in the potential energy curve to the right until it reaches the point $x = f$, where it will reverse direction and move back to $x = e$, never having enough total mechanical energy to escape the dip. The region in which the object is trapped is often referred to as a *potential well.*

Perhaps the most interesting situation is that for which the total energy is E_3. If an object moves in from the right in Figure 6.21 with energy E_3, it will be reflected at the turning point where $x = d$, in complete analogy to the situation at $x = a$ for the object with energy E_2. However, there is another allowed part of the path farther to the left, in the interval $b \leq x \leq c$. If the object starts out in this interval, it remains trapped in a dip, just like the object with energy E_4. Between the allowed intervals $b \leq x \leq c$ and $x \geq d$ is a *forbidden region* that an object with total mechanical energy E_3 cannot cross.

Preview: Atomic Physics

In studying atomic physics, we will again encounter potential energy curves. Particles with energies such as E_4 in Figure 6.21, which are trapped between two turning points, are said to be in *bound states.* One of the most interesting phenomena in atomic and nuclear physics, however, occurs in situations like the one shown in Figure 6.21 for a total mechanical energy of E_3. From our considerations of classical mechanics in this chapter, we expect that an object sitting in a bound state between $b \leq x \leq c$ cannot escape. However, in atomic and nuclear physics applications, a particle in such a bound state has a small probability of escaping out of this potential well and through the classically forbidden region into the region $x \geq d$. This process is called *tunneling.* Depending on the height and width of the barrier, the tunneling probability can be quite large, leading to a fast escape, or quite small, leading to a very slow escape. For example, the isotope ^{235}U of the element uranium, used in nuclear fission power plants and naturally occurring on Earth, has a half-life of over 700 million years, which is the average time that elapses until an alpha particle (a tightly bound cluster of two neutrons and two protons in the nucleus) tunnels through its potential barrier, causing the uranium nucleus to decay. In contrast, the isotope ^{238}U has a half-life of 4500 million years. Thus, much of the original ^{235}U present on Earth has decayed away. The fact that ^{235}U comprises only 0.7% of all naturally occurring uranium means that the first sign that a nation is attempting to use nuclear power, for any purpose, is the acquisition of equipment that can separate ^{235}U from the much more abundant (99.3%) ^{238}U, which is not suitable for nuclear fission power production.

The previous paragraph is included to whet your appetite for things to come. In order to understand the processes of atomic and nuclear physics, you'll need to be familiar with quite a few more concepts. However, the basic considerations of energy introduced here will remain virtually unchanged.

WHAT WE HAVE LEARNED | EXAM STUDY GUIDE

- Potential energy, U, is the energy stored in the configuration of a system of objects that exert forces on one another.

- Gravitational potential energy is defined as $U_g = mgy$.

- The potential energy associated with elongating a spring from its equilibrium position at $x = 0$ is $U_s(x) = \frac{1}{2}kx^2$.

- A conservative force is a force for which the work done over any closed path is zero. A force that does not fulfill this requirement is a nonconservative force.

- For any conservative force, the change in potential energy due to some spatial rearrangement of a system is equal to the negative of the work done by the conservative force during this spatial rearrangement.

- The relationship between a potential energy and the corresponding conservative force is
$$\Delta U = U(x) - U(x_0) = -\int_{x_0}^{x} F_x(x')dx'.$$

- In one-dimensional situations, the force component can be obtained from the potential energy using $F_x(x) = -\dfrac{dU(x)}{dx}$.

- The mechanical energy, E, is the sum of the kinetic energy and the potential energy: $E = K + U$.

- The total mechanical energy is conserved for any mechanical process inside an isolated system that involves only conservative forces: $\Delta E = \Delta K + \Delta U = 0$. An alternative way of expressing this law of conservation of mechanical energy is $K + U = K_0 + U_0$.

- The total energy—the sum of all forms of energy, mechanical or other—is always conserved in an isolated system. This holds for conservative as well as nonconservative forces: $E_{total} = E_{mechanical} + E_{other} = K + U + E_{other} = $ constant.

- Energy problems involving nonconservative forces can be solved using the work-energy theorem: $W_f = \Delta K + \Delta U$.

- At stable equilibrium points, small perturbations result in small oscillations around the equilibrium point; at unstable equilibrium points, small perturbations result in an accelerating movement away from the equilibrium point.

- Turning points are points where the kinetic energy is zero and where a net force moves the object away from the point.

ANSWERS TO SELF-TEST OPPORTUNITIES

6.1 The potential energy is proportional to the inverse of the distance between the two objects. Examples of these forces are the force of gravity (see Chapter 12) and the electrostatic force (see Chapter 21).

6.2 To handle this problem with air resistance included, we would have introduced the work done by air resistance, which can be treated as a friction force. We would have modified our statement of energy conservation to reflect the fact that work, W_f, is done by the friction force:

$$W_f + K + U = K_0 + U_0.$$

The solution would have to be done numerically because the work done by friction in this case would depend on the distance that the rock actually traveled through the air.

6.3 The lighter-colored ball descends to a lower elevation earlier in its motion and thus converts more of its potential energy to kinetic energy early on. Greater kinetic energy means higher speed. Thus, the lighter-colored ball reaches higher speeds earlier and is able to move to the bottom of the track faster, even though its path length is greater.

6.4 The speed is at a maximum where the kinetic energy is at a maximum:

$$K(y) = U(-3.3\text{ m}) - U(y) = (3856\text{ J}) - (671\text{ J/m})y - (557.5\text{ J/m}^2)y^2$$

$$\frac{d}{dy}K(y) = -(671\text{ J/m}) - (1115\text{ J/m}^2)y = 0 \Rightarrow y = -0.602\text{ m}$$

$$v(-0.602\text{ m}) = \sqrt{2K(-0.602\text{ m})/m} = 10.89\text{ m/s}.$$

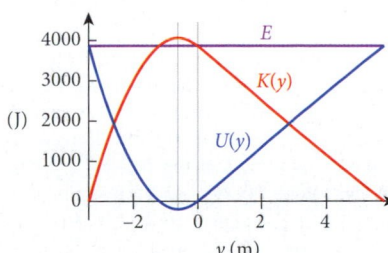

Note that the value at which the speed is maximum is the equilibrium position of the spring once it is loaded with the human cannonball.

6.5 The net force at the maximum stretching is $F = k(L_{max} - L_{jumper} - L_0) - mg$. Therefore, the acceleration at this point is $a = k(L_{max} - L_{jumper} - L_0)/m - g$. Inserting the expression we found for L_0 gives

$$a = \sqrt{\frac{2gL_{max}k}{m}} - g.$$

The maximum acceleration increases with the square root of the spring constant. If one wants to jump from a great height, L_{max}, a very soft bungee cord is needed.

PROBLEM-SOLVING GUIDELINES: CONSERVATION OF ENERGY

1. Many of the problem-solving guidelines given in Chapter 5 apply to problems involving conservation of energy as well. It is important to identify the system and determine the state of the objects in it at different key times, such as the beginning and end of each kind of motion. You should also identify which forces in the situation are conservative or nonconservative, because they affect the system in different ways.

2. Try to keep track of each kind of energy throughout the problem situation. When does the object have kinetic en-

ergy? Does gravitational potential energy increase or decrease? Where is the equilibrium point for a spring?

3. Remember that you can choose where potential energy is zero, so try to determine what choice will simplify the calculations.

4. A sketch is almost always helpful, and often a free-body diagram is useful as well. In some cases, drawing graphs of potential energy, kinetic energy, and total mechanical energy is a good idea.

MULTIPLE-CHOICE QUESTIONS

6.1 A block of mass 5.0 kg slides without friction at a speed of 8.0 m/s on a horizontal table surface until it strikes and sticks to a horizontal spring (with spring constant of $k = 2000$. N/m and very small mass), which in turn is attached to a wall. How far is the spring compressed before the mass comes to rest?

a) 0.40 m   c) 0.30 m   e) 0.67 m

b) 0.54 m   d) 0.020 m

6.2 A pendulum swings in a vertical plane. At the bottom of the swing, the kinetic energy is 8 J and the gravitational potential energy is 4 J. At the highest position of the swing, the kinetic and gravitational potential energies are

a) kinetic energy = 0 J and gravitational potential energy = 4 J.

b) kinetic energy = 12 J and gravitational potential energy = 0 J.

c) kinetic energy = 0 J and gravitational potential energy = 12 J.

d) kinetic energy = 4 J and gravitational potential energy = 8 J.

e) kinetic energy = 8 J and gravitational potential energy = 4 J.

6.3 A ball of mass 0.50 kg is released from rest at point A, which is 5.0 m above the bottom of a tank of oil, as shown in the figure. At point B, which is 2.0 m above the bottom of the tank, the ball has a speed of 6.0 m/s. The work done on the ball by the force of fluid friction is

a) +15 J.   d) –9 J.

b) +9 J.   e) –5.7 J.

c) –15 J.

6.4 A child throws three identical marbles from the same height above the ground so that they land on the *flat* roof of a building. The marbles are launched with the same initial speed. The first marble, marble A, is thrown at an angle of 75° above horizontal, while marbles B and C are thrown with launch angles of 60° and 45°, respectively. Neglecting air resistance, rank the marbles according to the speeds with which they hit the roof.

a) A < B < C

b) C < B < A

c) A and C have the same speed; B has a lower speed.

d) B has the highest speed; A and C have the same speed.

e) A, B, and C all hit the roof with the same speed.

6.5 Which of the following is *not* a valid potential energy function for the spring force $F = -kx$?

a) $(\frac{1}{2})kx^2$ c) $(\frac{1}{2})kx^2 - 10$ J   e) None of the above is valid.

b) $(\frac{1}{2})kx^2 + 10$ J   d) $-(\frac{1}{2})kx^2$

6.6 You use your hand to stretch a spring to a displacement x from its equilibrium position and then slowly bring it back to that position. Which is true?

a) The spring's ΔU is positive.   d) The hand's ΔU is negative.

b) The spring's ΔU is negative.   e) None of the above statements is true.

c) The hand's ΔU is positive.

6.7 In Question 6, what is the work done by the hand?

a) $-(\frac{1}{2})kx^2$ d) zero

b) $+(\frac{1}{2})kx^2$ e) none of the above

c) $(\frac{1}{2})mv^2$, where v is the speed of the hand

6.8 Which of the following is *not* a unit of energy?

a) newton-meter   c) kilowatt-hour   e) All of the above are

b) joule   d) kg m²/ s²   units of energy.

6.9 A spring has a spring constant of 80. N/m. How much potential energy does it store when stretched by 1.0 cm?

a) $4.0 \cdot 10^{-3}$ J   c) 80 J   e) 0.8 J

b) 0.40 J   d) 800 J

6.10 What is the maximum acceleration that the human cannonball of Solved Problem 6.4 experiences?

a) $1.00g$ c) $3.25g$ e) $7.30g$

b) $2.14g$ d) $4.48g$

6.11 For an object sliding on the ground, the friction force

a) always acts in the same direction as the displacement.

b) always acts in a direction perpendicular to the displacement.

c) always acts in a direction opposite to the displacement.

d) acts either in the same direction as the displacement or in the direction opposite to the displacement depending on the value of the coefficient of kinetic friction.

6.12 Some forces in nature vary with the inverse of the distance squared between two objects. For a force like this, how does the potential energy vary with the distance between the two objects?

a) The potential energy varies with the distance.

b) The potential energy varies with the distance squared.

c) The potential energy varies with the inverse of the distance.

d) The potential energy varies with the inverse of the distance squared.

e) The potential energy does not depend on the distance.

6.13 A baseball is dropped from the top of a building. Air resistance acts on the baseball as it drops. Which of the following statements is true?

a) The change in potential energy of the baseball as it falls is equal to the kinetic energy of the baseball just before it strikes the ground.

b) The change in potential energy of the baseball as it falls is greater than the kinetic energy of the baseball just before it strikes the ground.

c) The change in potential energy of the baseball as it falls is less than the kinetic energy of the baseball just before it strikes the ground.

d) The change in potential energy of the baseball is equal to the energy lost due to the friction from the air resistance while the ball is falling.

CONCEPTUAL QUESTIONS

6.14 Can the kinetic energy of an object be negative? Can the potential energy of an object be negative?

6.15 a) If you jump off a table onto the floor, is your mechanical energy conserved? If not, where does it go? b) A car moving down the road smashes into a tree. Is the mechanical energy of the car conserved? If not, where does it go?

6.16 How much work do you do when you hold a bag of groceries while standing still? How much work do you do when carrying the same bag a distance d across the parking lot of the grocery store?

6.17 An arrow is placed on a bow, the bowstring is pulled back, and the arrow is shot straight up into the air; the arrow then comes back down and sticks into the ground. Describe all of the changes in work and energy that occur.

6.18 Two identical billiard balls start at the same height and the same time and roll along different tracks, as shown in the figure.

a) Which ball has the highest speed at the end?

b) Which one will get to the end first?

6.19 A girl of mass 49.0 kg is on a swing, which has a mass of 1.0 kg. Suppose you pull her back until her center of mass is 2.0 m above the ground. Then you let her go, and she swings out and returns to the same point. Are all forces acting on the girl and swing conservative?

6.20 Can a potential energy function be defined for the force of friction?

6.21 Can the potential energy of a spring be negative?

6.22 One end of a rubber band is tied down, and you pull on the other end to trace a complicated *closed* trajectory. If you measured the elastic force F at every point, took its scalar product with the local displacements, $F \cdot \Delta \vec{r}$, and then summed all of these, what would you get?

6.23 Can a unique potential energy function be identified with a particular conservative force?

6.24 In skydiving, the vertical velocity component of the skydiver is typically zero at the moment he or she leaves the plane; the vertical component of the velocity then increases until the skydiver reaches terminal speed (see Chapter 4). For a simplified model of this motion, we assume that the horizontal velocity component is zero and that the vertical velocity component increases linearly, with acceleration $a_y = -g$, until the skydiver reaches terminal velocity, after which it stays constant. Thus, our simplified model assumes free fall without air resistance followed by falling at constant speed. Sketch the kinetic energy, potential energy, and total energy as a function of time for this model.

6.25 A projectile of mass m is launched from the ground at $t = 0$ with a speed v_0 and at an angle θ_0 above the horizontal. Assuming that air resistance is negligible, write the kinetic, potential, and total energies of the projectile as explicit functions of time.

6.26 The energy height, H, of an aircraft of mass m at altitude h and with speed v is defined as its total energy (with the zero of the potential energy taken at ground level) divided by its weight. Thus, the energy height is a quantity with units of length.

a) Derive an expression for the energy height, H, in terms of the quantities m, h, and v.

b) A Boeing 747 jet with mass $3.5 \cdot 10^5$ kg is cruising in level flight at 250.0 m/s at an altitude of 10.0 km. Calculate the value of its energy height.

Note: The energy height is the maximum altitude an aircraft can reach by "zooming" (pulling into a vertical climb without changing the engine thrust). This maneuver is not recommended for a 747, however.

6.27 A body of mass m moves in one dimension under the influence of a force, $F(x)$, which depends only on the body's position.

a) Prove that Newton's Second Law and the law of conservation of energy for this body are exactly equivalent.

b) Explain, then, why the law of conservation of energy is considered to be of greater significance than Newton's Second Law.

6.28 The molecular bonding in a diatomic molecule such as the nitrogen (N_2) molecule can be modeled by the Lennard-Jones potential, which has the form

$$U(x) = 4U_0 \left(\left(\frac{x_0}{x} \right)^{12} - \left(\frac{x_0}{x} \right)^6 \right),$$

where x is the separation distance between the two nuclei and x_0 and U_0 are constants. Determine, in terms of these constants, the following:

a) the corresponding force function;

b) the equilibrium separation x_E, which is the value of x for which the two atoms experience zero force from each other; and

c) the nature of the interaction (repulsive or attractive) for separations larger and smaller than x_E.

6.29 A particle of mass m moving in the xy-plane is confined by a two-dimensional potential function, $U(x, y) = \frac{1}{2}k(x^2 + y^2)$.

a) Derive an expression for the net force, $\vec{F} = F_x \hat{x} + F_y \hat{y}$.

b) Find the equilibrium point on the xy-plane.

c) Describe qualitatively the effect of the net force.

d) What is the magnitude of the net force on the particle at the coordinate (3.00, 4.00) in centimeters if $k = 10.0$ N/cm?

e) What are the turning points if the particle has 10.0 J of total mechanical energy?

6.30 For a rock dropped from rest from a height h, to calculate the speed just before it hits the ground, we use the conservation of mechanical energy and write $mgh = \frac{1}{2}mv^2$. The mass cancels out, and we solve for v. A very common error made by some beginning physics students is to assume, based on the appearance of this equation, that they should set the kinetic energy equal to the potential energy at the same point in space. For example, to calculate the speed v_1 of the rock at some height $y_1 < h$, they often write $mgy_1 = \frac{1}{2}mv_1^2$ and solve for v_1. Explain why this approach is wrong.

EXERCISES

A blue problem number indicates a worked-out solution is available in the Student Solutions Manual. One • and two •• indicate increasing level of problem difficulty.

Section 6.1

6.31 What is the gravitational potential energy of a 2.00-kg book 1.50 m above the floor?

6.32 a) If the gravitational potential energy of a 40.0-kg rock is 500. J relative to a value of zero on the ground, how high is the rock above the ground?

b) If the rock were lifted to twice its original height, how would the value of its gravitational potential energy change?

6.33 A rock of mass 0.773 kg is hanging from a string of length 2.45 m on the Moon, where the gravitational acceleration is a sixth of that on Earth. What is the change in gravitational potential energy of this rock when it is moved so that the angle of the string changes from 3.31° to 14.01°? (Both angles are measured relative to the vertical.)

6.34 A 20.0-kg child is on a swing attached to ropes that are $L = 1.50$ m long. Take the zero of the gravitational potential energy to be at the position of the child when the ropes are horizontal.

a) Determine the child's gravitational potential energy when the child is at the lowest point of the circular trajectory.

b) Determine the child's gravitational potential energy when the ropes make an angle of 45.0° relative to the vertical.

c) Based on these results, which position has the higher potential energy?

Section 6.3

6.35 A $1.50 \cdot 10^3$-kg car travels 2.50 km up an incline at constant velocity. The incline has an angle of 3.00° with respect to the horizontal. What is the change in the car's potential energy? What is the net work done on the car?

6.36 A constant force of 40.0 N is needed to keep a car traveling at constant speed as it moves 5.00 km along a road. How much work is done? Is the work done on or by the car?

6.37 A piñata of mass 3.27 kg is attached to a string tied to a hook in the ceiling. The length of the string is 0.810 m, and the piñata is released from rest from an initial position in which the string makes an angle of 56.5° with the vertical. What is the work done by gravity by the time the string is in a vertical position for the first time?

Section 6.4

•6.38 A particle is moving along the x-axis subject to the potential energy function $U(x) = a(1/x) + bx^2 + cx - d$, where $a = 7.00$ J m, $b = 10.0$ J/m^2, $c = 6.00$ J/m, and $d = 28.0$ J.

a) Express the force felt by the particle as a function of x.

b) Plot this force and the potential energy function.

c) Determine the net force on the particle at the coordinate $x = 2.00$ m.

•6.39 Calculate the force $F(y)$ associated with each of the following potential energies:

a) $U(y) = ay^3 - by^2$ b) $U(y) = U_0 \sin (cy)$

•6.40 The potential energy of a certain particle is given by $U(x,z) = ax^2 + bz^3$, where a and b are constants. Find the force vector exerted on the particle.

Section 6.5

6.41 A ball is thrown up in the air, reaching a height of 5.00 m. Using energy conservation considerations, determine its initial speed.

6.42 A cannonball of mass 5.99 kg is shot from a cannon at an angle of 50.21° relative to the horizontal and with an initial speed of 52.61 m/s. As the cannonball reaches the highest point of its trajectory, what is the gain in its potential energy relative to the point from which it was shot?

6.43 A basketball of mass 0.624 kg is shot from a vertical height of 1.20 m and at a speed of 20.0 m/s. After reaching its maximum height, the ball moves into the hoop on its downward path, at 3.05 m above the ground. Using the principle of energy conservation, determine how fast the ball is moving just before it enters the hoop.

•6.44 A classmate throws a 1.00-kg book from a height of 1.00 m above the ground straight up into the air. The book reaches a maximum height of 3.00 m above the ground and begins to fall back. Assume that 1.00 m above the ground is the reference level for zero gravitational potential energy. Determine

a) the gravitational potential energy of the book when it hits the ground.

b) the velocity of the book just before hitting the ground.

•6.45 Suppose you throw a 0.0520-kg ball with a speed of 10.0 m/s and at an angle of 30.0° above the horizontal from a building 12.0 m high.

a) What will be its kinetic energy when it hits the ground?

b) What will be its speed when it hits the ground?

•6.46 A uniform chain of total mass m is laid out straight on a frictionless table and held stationary so that one-third of its length, $L = 1.00$ m, is hanging vertically over the edge of the table. The chain is then released. Determine the speed of the chain at the instant when only one-third of its length remains on the table.

•6.47 a) If you are at the top of a toboggan run that is 40.0 m high, how fast will you be going at the bottom, provided you can ignore friction between the sled and the track?

b) Does the steepness of the run affect how fast you will be going at the bottom?

c) If you do not ignore the small friction force, does the steepness of the track affect the value of the speed at the bottom?

Section 6.6

6.48 A block of mass 0.773 kg on a spring with spring constant 239.5 N/m oscillates vertically with amplitude 0.551 m. What is the speed of this block at a distance of 0.331 m from the equilibrium position?

6.49 A spring with $k = 10.0$ N/cm is initially stretched 1.00 cm from its equilibrium length.

a) How much more energy is needed to further stretch the spring to 5.00 cm beyond its equilibrium length?

b) From this new position, how much energy is needed to compress the spring to 5.00 cm shorter than its equilibrium position?

•6.50 A 5.00-kg ball of clay is thrown downward from a height of 3.00 m with a speed of 5.00 m/s onto a spring with $k = 1600$. N/m. The clay compresses the spring a certain maximum amount before momentarily stopping.

a) Find the maximum compression of the spring.

b) Find the total work done on the clay during the spring's compression.

•6.51 A slingshot that launches a stone horizontally consists of two light, identical springs (with spring constants of 30.0 N/m) and a light cup that holds a 1.00-kg stone. Each spring has an equilibrium length of 50.0 cm. When the springs are in equilibrium, they line up vertically. Suppose that the cup containing the mass is pulled to $x = 70.0$ cm to the left of the vertical and then released. Determine

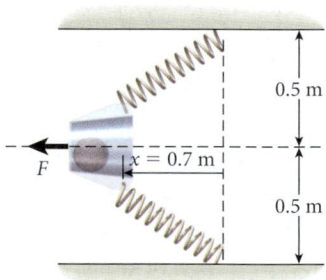

a) the system's total mechanical energy;

b) the speed of the stone at $x = 0$.

•6.52 Suppose the stone in Problem 6.51 is instead launched vertically and the mass is a lot smaller ($m = 0.100$ kg). Take the zero of the gravitational potential energy to be at the equilibrium point.

a) Determine the total mechanical energy of the system.

b) How fast is the stone moving as it passes the equilibrium point?

Section 6.7

6.53 A 80.0-kg fireman slides down a 3.00-m pole by applying a frictional force of 400. N against the pole with his hands. If he slides from rest, how fast is he moving once he reaches the ground?

6.54 A large air-filled 0.100-kg plastic ball is thrown up into the air with an initial speed of 10.0 m/s. At a height of 3.00 m, the ball's speed is 3.00 m/s. What fraction of its original energy has been lost to air friction?

6.55 How much mechanical energy is lost to friction if a 55.0-kg skier slides down a ski slope at constant speed of 14.4 m/s? The slope is 123.5 m long and makes an angle of 14.7° with respect to the horizontal.

•6.56 A truck of mass 10,212 kg moving at a speed of 61.2 mph has lost its brakes. Fortunately, the driver finds a runaway lane, a gravel-covered incline that uses friction to stop a truck in such a situation; see the figure. In this case, the incline makes an angle of $\theta = 40.15°$ with the horizontal, and the gravel has a coefficient of friction of 0.634 with the tires of the truck. How far along the incline (Δx) does the truck travel before it stops?

•6.57 A snowboarder of mass 70.1 kg (including gear and clothing), starting with a speed of 5.10 m/s, slides down a slope at an angle $\theta = 37.1°$ with the horizontal. The coefficient of kinetic friction is 0.116. What is the net work done on the snowboarder in the first 5.72 s of descent?

•6.58 The greenskeepers of golf courses use a stimpmeter to determine how "fast" their greens are. A stimpmeter is a straight aluminum bar with a V-shaped groove on which a golf ball can roll. It is designed to release the golf ball once the angle of the bar with the ground reaches a value of $\theta = 20.0°$. The golf ball (mass = 1.62 oz = 0.0459 kg) rolls 30.0 in down the bar and then continues to roll along the green for several feet. This distance is called the "reading." The test is done on a level part of the green, and stimpmeter readings between 7 and 12 ft are considered acceptable. For a stimpmeter reading of 11.1 ft, what is the coefficient of friction between the ball and the green? (The ball is rolling and not sliding, as we usually assume when considering friction, but this does not change the result in this case.)

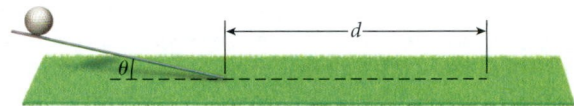

•6.59 A 1.00-kg block is pushed up and down a rough plank of length $L = 2.00$ m, inclined at 30.0° above the horizontal. From the bottom, it is pushed a distance $L/2$ up the plank, then pushed back down a distance $L/4$, and finally pushed back up the plank until it reaches the top end. If the coefficient of kinetic friction between the block and plank is 0.300, determine the work done by the block against friction.

••6.60 A 1.00-kg block initially at rest at the top of a 4.00-m incline with a slope of 45.0° begins to slide down the incline. The upper half of the incline is frictionless, while the lower half is rough, with a coefficient of kinetic friction $\mu_k = 0.300$.

a) How fast is the block moving midway along the incline, before entering the rough section?

b) How fast is the block moving at the bottom of the incline?

••6.61 A spring with a spring constant of 500. N/m is used to propel a 0.500-kg mass up an inclined plane. The spring is compressed 30.0 cm from its equilibrium position and launches the mass from rest across a horizontal surface and onto the plane. The plane has a length of 4.00 m and is inclined at 30.0°. Both the plane and the horizontal surface have a coefficient of kinetic friction with the mass of 0.350. When the spring is compressed, the mass is 1.50 m from the bottom of the plane.

a) What is the speed of the mass as it reaches the bottom of the plane?

b) What is the speed of the mass as it reaches the top of the plane?

c) What is the total work done by friction from the beginning to the end of the mass's motion?

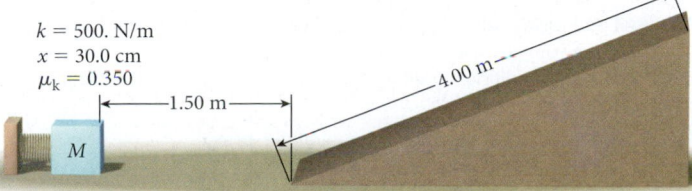

$k = 500.$ N/m
$x = 30.0$ cm
$\mu_k = 0.350$

••6.62 The sled shown in the figure leaves the starting point with a velocity of 20.0 m/s. Use the work-energy theorem to calculate the sled's speed at the end of the track or the maximum height it reaches if it stops before reaching the end.

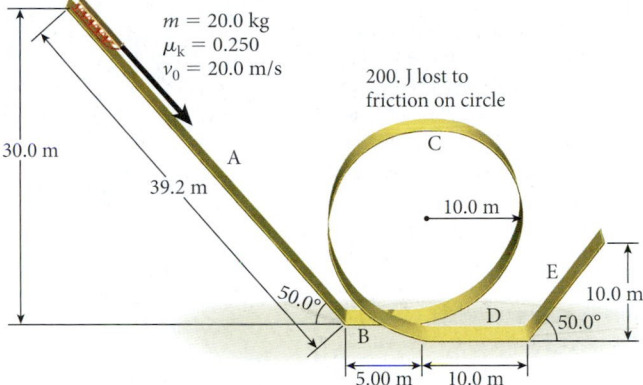

$m = 20.0$ kg
$\mu_k = 0.250$
$v_0 = 20.0$ m/s

200. J lost to friction on circle

Section 6.8

•6.63 On the segment of roller coaster track shown in the figure, a cart of mass 237.5 kg starts at $x = 0$ with a speed of 16.5 m/s. Assuming that the dissipation of energy due to friction is small enough to be ignored, where is the turning point of this trajectory?

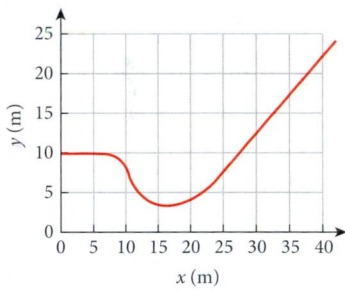

•6.64 A 70.0-kg skier moving horizontally at 4.50 m/s encounters a 20.0° incline.

a) How far up the incline will the skier move before she momentarily stops, ignoring friction?

b) How far up the incline will the skier move if the coefficient of kinetic friction between the skies and snow is 0.100?

•6.65 A 0.200-kg particle is moving along the x-axis, subject to the potential energy function shown in the figure, where $U_A = 50.0$ J, $U_B = 0$ J, $U_C = 25.0$ J, $U_D = 10.0$ J, and $U_E = 60.0$ J along the path. If the particle was initially at $x = 4.00$ m and had a total mechanical energy of 40.0 J, determine:

a) the particle's speed at $x = 3.00$ m,

b) the particle's speed at $x = 4.50$ m, and

c) the particle's turning points.

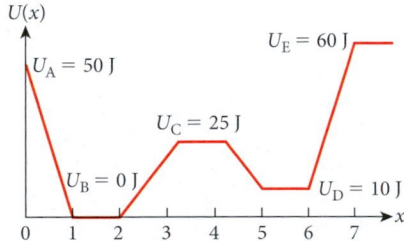

Additional Exercises

6.66 A ball of mass 1.84 kg is dropped from a height $y_1 = 1.49$ m and then bounces back up to a height of $y_2 = 0.87$ m. How much mechanical energy is lost in the bounce? The effect of air resistance has been experimentally found to be negligible in this case, and you can ignore it.

6.67 A car of mass 987 kg is traveling on a horizontal segment of a freeway with a speed of 64.5 mph. Suddenly, the driver has to hit the brakes hard to try to avoid an accident up ahead. The car does not have an ABS (antilock braking system), and the wheels lock, causing the car to slide some distance before it is brought to a stop by the friction force between its tires and the road surface. The coefficient of kinetic friction is 0.301. How much mechanical energy is lost to heat in this process?

6.68 Two masses are connected by a light string that goes over a light, frictionless pulley, as shown in the figure. The 10.0-kg mass is released and falls through a vertical distance of 1.00 m before hitting the ground. Use conservation of mechanical energy to determine:

a) how fast the 5.00-kg mass is moving just before the 10.0-kg mass hits the ground; and

b) the maximum height attained by the 5.00-kg mass.

6.69 In 1896 in Waco, Texas, William George Crush, owner of the K-T (or "Katy") Railroad, parked two locomotives at opposite ends of a 6.4-km-long track, fired them up, tied their throttles open, and then allowed them to crash head-on at full speed in front of 30,000 spectators. Hundreds of people were hurt by flying debris; a few were killed. Assuming that each locomotive weighed $1.2 \cdot 10^6$ N and its acceleration along the track was a constant 0.26 m/s^2, what was the total kinetic energy of the two locomotives just before the collision?

6.70 A baseball pitcher can throw a 5.00-oz baseball with a speed measured by a radar gun to be 90.0 mph. Assuming that the force exerted by the pitcher on the ball acts over a distance of two arm lengths, each 28.0 in, what is the average force exerted by the pitcher on the ball?

6.71 A 1.50-kg ball has a speed of 20.0 m/s when it is 15.0 m above the ground. What is the total energy of the ball?

6.72 If it takes an average force of 5.50 N to push a 4.50-g dart 6.00 cm into a dart gun, assuming that the barrel is frictionless, how fast will the dart exit the gun?

6.73 A high jumper approaches the bar at 9.00 m/s. What is the highest altitude the jumper can reach, if he does not use any additional push off the ground and is moving at 7.00 m/s as he goes over the bar and if he stays upright during the jump?

6.74 A roller coaster is moving at 2.00 m/s at the top of the first hill (h = 40.0 m). Ignoring friction and air resistance, how fast will the roller coaster be moving at the top of a subsequent hill, which is 15.0 m high?

6.75 You are on a swing with a chain 4.00 m long. If your maximum displacement from the vertical is 35.0°, how fast will you be moving at the bottom of the arc?

6.76 A truck is descending a winding mountain road. When the truck is 680. m above sea level and traveling at 15.0 m/s, its brakes fail. What is the maximum possible speed of the truck at the foot of the mountain, 550. m above sea level?

6.77 Tarzan swings on a taut vine from his tree house to a limb on a neighboring tree, which is located a horizontal distance of 10.0 m from and a vertical distance of 4.00 m below his starting point. Amazingly the vine neither stretches nor breaks; Tarzan's trajectory is thus a portion of a circle. If Tarzan starts with zero speed, what is his speed when he reaches the limb?

6.78 The graph shows the component ($F \cos \theta$) of the net force that acts on a 2.00-kg block as it moves along a flat horizontal surface. Find

a) the net work done on the block;

b) the final speed of the block if it starts from rest at s = 0.

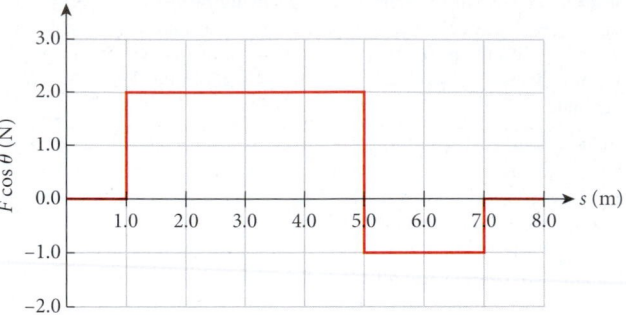

•6.79 A 3.00-kg model rocket is launched vertically upward with sufficient initial speed to reach a height of $1.00 \cdot 10^2$ m, even though air resistance (a nonconservative force) performs $-8.00 \cdot 10^2$ J of work on the rocket. How high would the rocket have gone, if there were no air resistance?

•6.80 A 0.500-kg mass is attached to a horizontal spring with k = 100. N/m. The mass slides across a frictionless surface. The spring is stretched 25.0 cm from equilibrium, and then the mass is released from rest.

a) Find the mechanical energy of the system.

b) Find the speed of the mass when it has moved 5.00 cm.

c) Find the maximum speed of the mass.

•6.81 You have decided to move a refrigerator (mass = 81.3 kg, including all the contents) to the other side of a room. You slide it across the floor on a straight path of length 6.35 m, and the coefficient of kinetic friction between floor and fridge is 0.437. Happy about your accomplishment, you leave the apartment. Your roommate comes home, wonders why the fridge is on the other side of the room, picks it up (you have a strong roommate!), carries it back to where it was originally, and puts it down. How much net mechanical work have the two of you done together?

•6.82 A 1.00-kg block compresses a spring for which k = 100. N/m by 20.0 cm; the spring is then released, and the block moves across a horizontal, frictionless table, where it hits and compresses another spring, for which k = 50.0 N/m. Determine

a) the total mechanical energy of the system,

b) the speed of the mass while moving freely between springs, and

c) the maximum compression of the second spring.

•6.83 A 1.00-kg block is resting against a light, compressed spring at the bottom of a rough plane inclined at an angle of 30.0°; the coefficient of kinetic friction between block and plane is μ_k = 0.100. Suppose the spring is compressed 10.0 cm from its equilibrium length. The spring is then released, and the block separates from the spring and slides up the incline a distance of only 2.00 cm beyond the spring's normal length before stopping. Determine

a) the change in total mechanical energy of the system, and

b) the spring constant k.

•6.84 A 0.100-kg ball is dropped from a height of 1.00 m and lands on a light (approximately massless) cup mounted on top of a light, vertical spring initially at its equilibrium position. The maximum compression of the spring is to be 10.0 cm.

a) What is the required spring constant of the spring?

b) Suppose you ignore the change in the gravitational energy of the ball during the 10.0-cm compression. What is the percentage difference between the calculated spring constant for this case and the answer obtained in part (a)?

•**6.85** A mass of 1.00 kg attached to a spring with a spring constant of 100. N/m oscillates horizontally on a smooth frictionless table with an amplitude of 0.500 m. When the mass is 0.250 m away from equilibrium, determine:

a) its total mechanical energy,

b) the system's potential energy and the mass's kinetic energy, and

c) the mass's kinetic energy when it is at the equilibrium point.

d) Suppose there was friction between the mass and the table so that the amplitude was cut in half after some time. By what factor has the mass's maximum kinetic energy changed?

e) By what factor has the maximum potential energy changed?

•**6.86** Bolo, the human cannonball, is ejected from a 3.50-m long barrel. If Bolo ($m = 80.0$ kg) has a speed of 12.0 m/s at the top of his trajectory, 15.0 m above the ground, what was the average force exerted on him while in the barrel?

•**6.87** A 1.00-kg mass is suspended vertically from a spring with $k = 100$. N/m and oscillates with an amplitude of 0.200 m. At the top of its oscillation, the mass is hit in such a way that it instantaneously moves down with a speed of 1.00 m/s. Determine

a) its total mechanical energy,

b) how fast it is moving as it crosses the equilibrium point, and

c) its new amplitude.

•**6.88** A runner reaches the top of a hill with a speed of 6.50 m/s. He descends 50.0 m and then ascends 28.0 m to the top of the next hill. His speed is now 4.50 m/s. The runner has a mass of 83.0 kg. The total distance that the runner covers is 400. m, and there is a constant resistance to motion of 9.00 N. Use energy considerations to find the work done by the runner over the total distance.

•**6.89** A package is dropped on a horizontal conveyor belt. The mass of the package is m, the speed of the conveyor belt is v, and the coefficient of kinetic friction between the package and the belt is μ_k.

a) How long does it take for the package to stop sliding on the belt?

b) What is the package's displacement during this time?

c) What is the energy dissipated by friction?

d) What is the total work done by the conveyor belt?

•**6.90** A father exerts a $2.40 \cdot 10^2$ N force to pull a sled with his daughter on it (combined mass of 85.0 kg) across a horizontal surface. The rope with which he pulls the sled makes an angle of 20.0° with the horizontal. The coefficient of kinetic friction is 0.200, and the sled moves a distance of 8.00 m. Find

a) the work done by the father,

b) the work done by the friction force, and

c) the total work done by all the forces.

•**6.91** A variable force acting on a 0.100-kg particle moving in the xy-plane is given by $F(x, y) = (x^2\hat{x} + y^2\hat{y})$ N, where x and y are in meters. Suppose that due to this force, the particle moves from the origin, O, to point S, with coordinates (10.0 m, 10.0 m). The coordinates of points P and Q are (0 m, 10.0 m) and (10.0 m, 0 m), respectively. Determine the work performed by the force as the particle moves along each of the following paths:

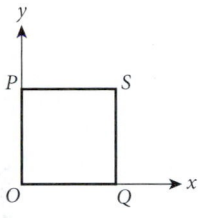

a) *OPS*　　　　　c) *OS*　　　　　e) *OQSPO*

b) *OQS*　　　　　d) *OPSQO*

••**6.92** In the situation in Problem 6.91, suppose there is friction between the 0.100-kg particle and the xy-plane, with $\mu_k = 0.100$. Determine the net work done by all forces on this particle when it takes each of the following paths:

a) *OPS*　　　　　c) *OS*　　　　　e) *OQSPO*

b) *OQS*　　　　　d) *OPSQO*

MULTI-VERSION EXERCISES

6.93 A snowboarder starts from rest, rides 38.09 m down a snow-covered slope that makes an angle of 30.15° with the horizontal, and reaches the flat snow near the lift. How far will she travel along the flat snow if the coefficient of kinetic friction between her board and the snow is 0.02501?

6.94 A snowboarder starts from rest, rides 30.37 m down a snow-covered slope that makes an angle of 30.35° with the horizontal, and reaches the flat snow near the lift. She travels 506.4 m along the flat snow. What is the coefficient of kinetic friction between her board and the snow?

6.95 A snowboarder starts from rest, rides down a snow-covered slope that makes an angle 30.57° with the horizontal, and reaches the flat snow near the lift. She travels 478.0 m along the flat snow. The coefficient of kinetic friction between her board and the snow is 0.03281. How far did she travel down the slope?

6.96 A batter hits a pop-up straight up in the air from a height of 1.397 m. The baseball rises to a height of 7.653 m above the ground. Ignoring air resistance, what is the speed of the baseball when the catcher gloves it 1.757 m above the ground?

6.97 A batter hits a pop-up straight up in the air from a height of 1.581 m. The baseball rises to a height h above the ground. The speed of the baseball when the catcher gloves it 1.859 m above the ground is 10.74 m/s. To what height h did the ball rise?

6.98 A batter hits a pop-up straight up in the air from a height of 1.273 m. The baseball rises to a height of 7.777 m above the ground. The speed of the baseball when the catcher gloves it is 10.73 m/s. At what height above the ground did the catcher glove the ball?

6.99 A ball is thrown horizontally from the top of a building that is 20.27 m high with a speed of 24.89 m/s. Neglecting air resistance, at what angle with respect to the horizontal will the ball strike the ground?

6.100 A ball is thrown horizontally from the top of a building that is 26.01 m high. The ball strikes the ground at an angle of 41.86° with respect to the horizontal. Neglecting air resistance, with what speed was the ball thrown?

6.101 A ball is thrown horizontally from the top of a building with a speed of 25.51 m/s. The ball strikes the ground at an angle of 44.37° with respect to the horizontal. What is the height of the building?

Momentum and Collisions

7

FIGURE 7.1 A supertanker.

Supertankers for transporting oil around the world are the largest ships ever built (Figure 7.1). They can have a mass (including cargo) of up to 650,000 tons and carry over 2 million barrels (84 million gallons = 318 million liters) of oil. However, their large size creates practical problems. Supertankers are too big to enter most shipping ports and have to stop at offshore platforms to offload their oil. In addition, piloting a ship of this size is extremely difficult. For example, when the captain gives the order to reverse engines and come to a stop, the ship can continue to move forward for more than 3 miles!

The physical quantity that makes a large moving object difficult to stop is *momentum*, the subject of this chapter. Momentum is a fundamental property associated with an object's motion, similar to kinetic energy. Moreover, both momentum and energy are subjects of important conservation laws in physics. However, momentum is a vector quantity, whereas energy is a scalar. Thus, working with momentum requires taking account of angles and components, as we did for force in Chapter 4.

The importance of momentum becomes most evident when we deal with collisions between two or more objects. In this chapter, we examine various collisions in one and two dimensions. In later chapters, we will make use of conservation of momentum in many different situations on vastly different scales—from bursts of elementary particles to collisions of galaxies.

WHAT WE WILL LEARN

- The momentum of an object is the product of its velocity and mass. Momentum is a vector quantity and points in the same direction as the velocity vector.

- Newton's Second Law can be phrased more generally as follows: The net force on an object equals the time derivative of the object's momentum.

- A change of momentum, called *impulse,* is the time integral of the net force that causes the momentum change.

- In all collisions, the momentum is conserved.

- Besides conservation of momentum, elastic collisions also have the property that the total kinetic energy is conserved.

- In totally inelastic collisions, the maximum amount of kinetic energy is removed, and the colliding objects stick to each other. Total kinetic energy is not conserved, but momentum is.

- Collisions that are neither elastic nor totally inelastic are partially inelastic, and the change in kinetic energy is proportional to the square of the coefficient of restitution.

- Momentum conservation is applied in such far-reaching fields as particle physics, where it is used in the discovery of new particles, and chaos theory, where it is used to show that it is impossible to predict long-term weather patterns.

7.1   Linear Momentum

For the terms *force, position, velocity,* and *acceleration,* the precise physical definitions are quite close to the words' usage in everyday language. With the term *momentum,* the situation is more analogous to that of *energy,* for which there is only a vague connection between conversational use and precise physical meaning. You sometimes hear that the campaign of a particular political candidate gains momentum or that legislation gains momentum in Congress. Often, sports teams or individual players are said to gain or lose momentum. What these statements imply is that the objects said to gain momentum have become harder to stop. However, Figure 7.2 shows that even objects with large momentum can be stopped!

Definition of Momentum

In physics, **momentum** is defined as the product of an object's mass and its velocity:

$$\vec{p} = m\vec{v}. \tag{7.1}$$

As you can see, the lowercase letter $\vec{p}$ is the symbol for linear momentum. The velocity $\vec{v}$ is a vector and is multiplied by a scalar quantity, the mass m. The product is thus a vector as well. The momentum vector, $\vec{p}$, and the velocity vector, $\vec{v}$, are parallel to each other; that is, they point in the same direction. As a simple consequence of equation 7.1, the magnitude of the momentum is

$$p = mv.$$

The momentum is also referred to as *linear momentum* to distinguish it from angular momentum, a concept we will study in Chapter 10 on rotation. The units of momentum

FIGURE 7.2 Rocket-sled crash test of a fighter plane. Tests like these can be used to improve the design of critical structures such as nuclear reactors so that they can withstand the impact of a plane crash.

are kg m/s. Unlike the unit for energy, the unit for momentum does not have a special name. The magnitude of momentum spans a large range. Momenta of various objects, from a subatomic particle to a planet orbiting the Sun, are given in Table 7.1.

Momentum and Force

Let's take the time derivative of equation 7.1. We use the product rule of differentiation to obtain

$$\frac{d}{dt}\vec{p} = \frac{d}{dt}(m\vec{v}) = m\frac{d\vec{v}}{dt} + \frac{dm}{dt}\vec{v}.$$

For now, we assume that the mass of the object does not change, and therefore the second term is zero. Because the time derivative of the velocity is the acceleration, we have

$$\frac{d}{dt}\vec{p} = m\frac{d\vec{v}}{dt} = m\vec{a} = \vec{F},$$

according to Newton's Second Law. The relationship

$$\vec{F} = \frac{d}{dt}\vec{p} \tag{7.2}$$

is an equivalent form of Newton's Second Law. This form is more general than $\vec{F} = m\vec{a}$ because it also holds in cases where the mass is not constant in time. This distinction will become important when we examine rocket motion in Chapter 8. Because equation 7.2 is a vector equation, we can also write it in Cartesian components:

$$F_x = \frac{dp_x}{dt}; \quad F_y = \frac{dp_y}{dt}; \quad F_z = \frac{dp_z}{dt}.$$

Momentum and Kinetic Energy

In Chapter 5, we established the relationship, $K = \frac{1}{2}mv^2$ (equation 5.1), between the kinetic energy K, the speed v, and the mass m. We can use $p = mv$ to obtain

$$K = \frac{mv^2}{2} = \frac{m^2v^2}{2m} = \frac{p^2}{2m}.$$

This equation gives us an important relationship between kinetic energy, mass, and momentum:

$$K = \frac{p^2}{2m}. \tag{7.3}$$

At this point, you may wonder why we need to reformulate the concepts of force and kinetic energy in terms of momentum. This reformulation is far more than a mathematical game. We will see that momentum is conserved in collisions and disintegrations, and this principle will provide an extremely helpful way to find solutions to complicated problems. These relationships of momentum with force and kinetic energy will be very useful in working such problems. First, though, we need to explore the physics of changing momentum in a little more detail.

Table 7.1	Momenta of Various Objects
Object	**Momentum (kg m/s)**
Alpha (α) particle from ^{238}U decay	$9.53 \cdot 10^{-20}$
90-mph fastball	5.75
Charging rhinoceros	$3 \cdot 10^4$
Car moving on freeway	$5 \cdot 10^4$
Supertanker at cruising speed	$4 \cdot 10^9$
Moon orbiting Earth	$7.54 \cdot 10^{25}$
Earth orbiting Sun	$1.78 \cdot 10^{29}$

7.2 Impulse

The change in momentum is defined as the difference between the final (index f) and initial (index i) momenta:

$$\Delta \vec{p} \equiv \vec{p}_f - \vec{p}_i.$$

To see why this definition is useful, we have to do a bit of math. Let's start by exploring the relationship between force and momentum just a little further. We can integrate each component of the equation $\vec{F} = d\vec{p}/dt$ over time. For the integral of F_x, for example, we obtain:

$$\int_{t_i}^{t_f} F_x \, dt = \int_{t_i}^{t_f} \frac{dp_x}{dt} \, dt = \int_{p_{x,i}}^{p_{x,f}} dp_x = p_{x,f} - p_{x,i} \equiv \Delta p_x.$$

This equation requires some explanation. In the second step, we performed a substitution of variables to transform an integration over time into an integration over momentum. Figure 7.3a illustrates this relationship: The area under the $F_x(t)$ curve is the change in momentum, Δp_x. We can obtain similar equations for the y- and z-components.

Combining all three component equations into one vector equation yields the following result:

$$\int_{t_i}^{t_f} \vec{F} \, dt = \int_{t_i}^{t_f} \frac{d\vec{p}}{dt} \, dt = \int_{\vec{p}_i}^{\vec{p}_f} d\vec{p} = \vec{p}_f - \vec{p}_i \equiv \Delta \vec{p}.$$

The time integral of the force is called the **impulse**, $\vec{J}$:

$$\vec{J} \equiv \int_{t_i}^{t_f} \vec{F} \, dt. \tag{7.4}$$

This definition immediately gives us the relationship between the impulse and the momentum change:

$$\vec{J} = \Delta \vec{p}. \tag{7.5}$$

From equation 7.5, we can calculate the momentum change over some time interval, if we know the time dependence of the force. If the force is constant or has some form that we can integrate, then we can simply evaluate the integral of equation 7.4. However, we can also define an average force,

$$\vec{F}_{ave} = \frac{\int_{t_i}^{t_f} \vec{F} \, dt}{\int_{t_i}^{t_f} dt} = \frac{1}{t_f - t_i} \int_{t_i}^{t_f} \vec{F} \, dt = \frac{1}{\Delta t} \int_{t_i}^{t_f} \vec{F} \, dt. \tag{7.6}$$

This integral gives us

$$\vec{J} = \vec{F}_{ave} \Delta t. \tag{7.7}$$

You may think this transformation is trivial in that it conveys the same information as equation 7.5. After all, the integration is still there, hidden in the definition of the average force. This is true, but sometimes we are only interested in the average force. Measuring the time interval, Δt, over which a force acts as well as the resulting impulse an object receives tells us the average force that the object experiences during that time interval. Figure 7.3b illustrates the relationship between the time-averaged force, the momentum change, and the impulse.

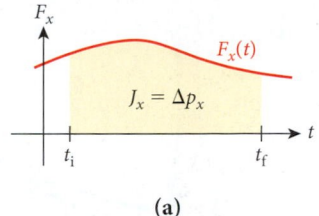

(a)

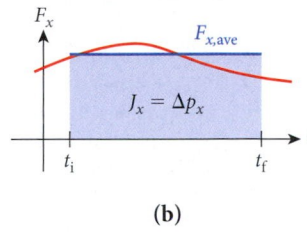

(b)

FIGURE 7.3 (a) The impulse (yellow area) is the time integral of the force; (b) same impulse resulting from an average force.

EXAMPLE 7.1 | Baseball Home Run

A Major League pitcher throws a fastball that crosses home plate with a speed of 90.0 mph (40.23 m/s) and an angle of 5.0° below the horizontal. A batter slugs it for a home run, launching it with a speed of 110.0 mph (49.17 m/s) at an angle of 35.0° above the horizontal (Figure 7.4). The mass of a baseball is required to be between 5 and 5.25 oz; let's say that the mass of the ball hit here is 5.10 oz (0.145 kg).   – Continued

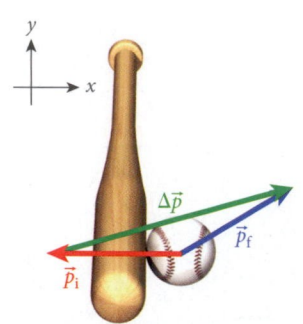

FIGURE 7.4 Baseball being hit by a bat. Initial (red) and final (blue) momentum vectors, as well as the impulse (green) vector (or change in momentum vector) are shown.

PROBLEM 1
What is the magnitude of the impulse the baseball receives from the bat?

SOLUTION 1
The impulse is equal to the momentum change of the baseball. Unfortunately, there is no shortcut; we must calculate $\Delta\vec{v} \equiv \vec{v}_f - \vec{v}_i$ for the x- and y-components separately, add them as vectors, and finally multiply by the mass of the baseball:

$$\Delta v_x = (49.17 \text{ m/s})(\cos 35.0°) - (40.23 \text{ m/s})(\cos 185.0°) = 80.35 \text{ m/s}$$
$$\Delta v_y = (49.17 \text{ m/s})(\sin 35.0°) - (40.23 \text{ m/s})(\sin 185.0°) = 31.71 \text{ m/s}$$
$$\Delta v = \sqrt{\Delta v_x^2 + \Delta v_y^2} = \sqrt{(80.35)^2 + (31.71)^2} \text{ m/s} = 86.38 \text{ m/s}$$
$$\Delta p = m\Delta v = (0.145 \text{ kg})(86.38 \text{ m/s}) = 12.5 \text{ kg m/s}.$$

Avoiding a Common Mistake:

It is tempting to just add the magnitudes of the initial and final momentum vectors, because they point approximately in opposite directions. This method would lead to $\Delta p_{\text{wrong}} = m(v_1 + v_2) = 12.96$ kg m/s. As you can see, this answer is pretty close to the correct one, only about 3% off. It can serve as a first estimate, if you realize that the vectors point in almost opposite directions and that, in such a case, vector subtraction implies an addition of the two magnitudes. However, to get the correct answer, you have to go through the calculations above.

PROBLEM 2
High-speed video shows that the ball-bat contact lasts only about 1 ms (0.001 s). Suppose, for the home run we're considering, that the contact lasted 1.20 ms. What was the magnitude of the average force exerted on the ball by the bat during that time?

SOLUTION 2
The force can be calculated by simply using the formula for the impulse:

$$\Delta\vec{p} = \vec{J} = \vec{F}_{\text{ave}}\Delta t$$
$$\Rightarrow F_{\text{ave}} = \frac{\Delta p}{\Delta t} = \frac{12.5 \text{ kg m/s}}{0.00120 \text{ s}} = 10.4 \text{ kN}.$$

This force is approximately the same as the weight of an entire baseball team! The collision of the bat and the ball results in significant compression of the baseball, as shown in Figure 7.5.

FIGURE 7.5 A baseball being compressed as it is hit by a baseball bat.

Concept Check 7.2

If the baseball in Example 7.1 was hit so that it had the same speed of 110 mph after it left the bat but was launched at an angle of 38° above the horizontal, the impulse the baseball received would have been

a) bigger.   c) the same.

b) smaller.

Self-Test Opportunity 7.1

Can you think of other everyday articles that are designed to minimize the average force for a given impulse?

Some important safety devices, such as air bags and seat belts in cars, make use of equation 7.7 relating impulse, average force, and time. If the car you are driving has a collision with another vehicle or a stationary object, the impulse—that is, the momentum change of your car—is rather large, and it can be delivered over a very short time interval. Equation 7.7 then results in a very large average force:

$$\vec{F}_{\text{ave}} = \frac{\vec{J}}{\Delta t}.$$

If no seat belts or air bags were installed in your car, a sudden stop could cause your head to hit the windshield and experience the impulse during a very short time of only a few milliseconds. This could result in a big average force acting on your head, causing injury or even death. Air bags and seat belts are designed to make the time over which the momentum change occurs as long as possible. Maximizing this time and having the driver's body decelerate in contact with the air bag minimize the force acting on the driver, greatly reducing injuries (see Figure 7.6).

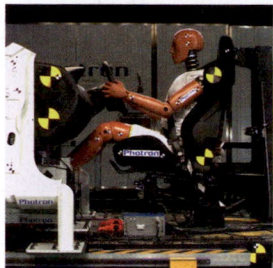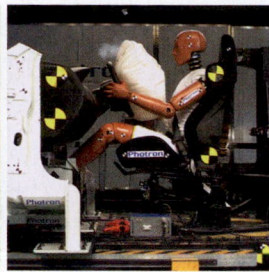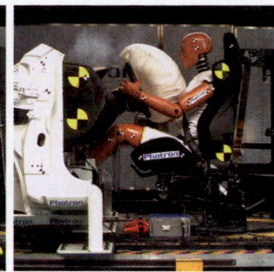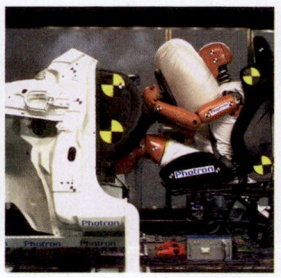

FIGURE 7.6 Time sequence of a crash test, showing the role of air bags, seat belts, and crumple zones in reducing the forces acting on the driver during a crash. The air bag can be seen deploying in the second photograph of the sequence.

Concept Check 7.3

Several cars are designed with active crumple zones in the front that get severely damaged during head-on collisions. The purpose of this design is to

a) reduce the impulse experienced by the driver during the collision.

b) increase the impulse experienced by the driver during the collision.

c) reduce the collision time and thus reduce the force acting on the driver.

d) increase the collision time and thus reduce the force acting on the driver.

e) make the repair as expensive as possible.

SOLVED PROBLEM 7.1 Egg Drop

PROBLEM
An egg in a special container is dropped from a height of 3.70 m. The container and egg together have a mass of 0.144 kg. A net force of 4.42 N will break the egg. What is the minimum time over which the egg/container can come to a stop without breaking the egg?

SOLUTION
THINK   When the egg/container is released, it accelerates with the acceleration due to gravity. When the egg/container strikes the ground, its velocity goes from the final velocity due to the gravitational acceleration to zero. As the egg/container comes to a stop, the force stopping it times the time interval (the impulse) will equal the mass of the egg/container times the change in speed. The time interval over which the velocity change takes place will determine whether the force exerted on the egg by the collision with the floor will break the egg.

SKETCH   The egg/container is dropped from rest from a height of $h = 3.70$ m (Figure 7.7).

RESEARCH   From the discussion of kinematics in Chapter 2, we know the final speed, v_y, of the egg/container resulting from free fall from a height of y_0 to a final height of y, starting with an initial velocity v_{y0}, is given by

$$v_y^2 = v_{y0}^2 - 2g(y - y_0).$$ (i)

We know that $v_{y0} = 0$ because the egg/container was released from rest. We define the final height to be $y = 0$ and the initial height to be $y_0 = h$, as shown in Figure 7.7. Thus, equation (i) for the final speed in the y-direction reduces to

$$v_y = \sqrt{2gh}.$$ (ii)

When the egg/container strikes the ground, the impulse, $\vec{J}$, exerted on it is given by

$$\vec{J} = \Delta \vec{p} = \int_{t_1}^{t_2} \vec{F}\, dt,$$ (iii)

– Continued

y

$m = 0.144$ kg

$h = 3.70$ m

FIGURE 7.7 An egg in a special container is dropped from a height of 3.70 m.

where $\Delta\vec{p}$ is the change in momentum of the egg/container and $\vec{F}$ is the force exerted to stop it. We assume the force is constant, so we can rewrite the integral in equation (iii) as

$$\int_{t_1}^{t_2} \vec{F}\,dt = \vec{F}(t_2 - t_1) = \vec{F}\Delta t.$$

The momentum of the egg/container will change from $p = mv_y$ to $p = 0$ when it strikes the ground, so we can write

$$\Delta p_y = 0 - \left(-mv_y\right) = mv_y = F_y\Delta t, \tag{iv}$$

where the term $-mv_y$ is negative because the velocity of the egg/container just before impact is in the negative y-direction.

SIMPLIFY We can now solve equation (iv) for the time interval and substitute the expression for the final velocity from equation (ii):

$$\Delta t = \frac{mv_y}{F_y} = \frac{m\sqrt{2gh}}{F_y}. \tag{v}$$

CALCULATE Inserting the numerical values, we get

$$\Delta t = \frac{(0.144\text{ kg})\sqrt{2(9.81\text{ m/s}^2)(3.70\text{ m})}}{4.42\text{ N}} = 0.277581543\text{ s}.$$

ROUND All of the numerical values in this problem were given with three significant figures, so we report our answer as

$$\Delta t = 0.278\text{ s}.$$

DOUBLE-CHECK Slowing the egg/container from its final velocity to zero over a time interval of 0.278 s seems reasonable. Looking at equation (v) we see that the force exerted on the egg as it hits the ground is given by

$$F = \frac{mv_y}{\Delta t}.$$

For a given height, we could reduce the force exerted on the egg in several ways. First, we could make Δt larger by making some kind of crumple zone in the container. Second, we could make the egg/container as light as possible. Third, we could construct the container so that it had a large surface area and thus significant air resistance, which would reduce the value of v_y below that for frictionless free fall.

7.3 Conservation of Linear Momentum

Suppose two objects collide with each other. They might then rebound away from each other, like two billiard balls on a billiard table. This kind of collision is called an **elastic collision** (at least it is approximately elastic, as we will see later). Another example of a collision is that of a subcompact car with an 18-wheeler, where the two vehicles stick to each other. This kind of collision is called a **totally inelastic collision.** Before seeing exactly what is meant by the terms *elastic* and *inelastic collisions,* let's look at the momenta, $\vec{p}_1$ and $\vec{p}_2$, of two objects during a collision.

We find that the sum of the two momenta after the collision is the same as the sum of the two momenta before the collision (index i1 indicates the initial value for object 1, just before the collision, and index f1 indicates the final value for the same object):

$$\vec{p}_{f1} + \vec{p}_{f2} = \vec{p}_{i1} + \vec{p}_{i2}. \tag{7.8}$$

This equation is the basic expression of the law of **conservation of total momentum,** the most important result of this chapter and the second conservation law we have encountered (the first being the law of conservation of energy in Chapter 6). Let's first go through its derivation and then consider its consequences.

DERIVATION 7.1

During a collision, object 1 exerts a force on object 2. Let's call this force $\vec{F}_{1\to2}$. Using the definition of impulse and its relationship to the momentum change, we get for the momentum change of object 2 during the collision:

$$\int_{t_i}^{t_f} \vec{F}_{1\to2}\,dt = \Delta\vec{p}_2 = \vec{p}_{f2} - \vec{p}_{i2}.$$

Here we neglect external forces; if they exist, they are usually negligible compared to $\vec{F}_{1\to2}$ during the collision. The initial and final times are selected to bracket the time of the collision process. In addition, the force $\vec{F}_{2\to1}$, which object 2 exerts on object 1, is also present. The same argument as before leads to

$$\int_{t_i}^{t_f} \vec{F}_{2\to1}\,dt = \Delta\vec{p}_1 = \vec{p}_{f1} - \vec{p}_{i1}.$$

Newton's Third Law (see Chapter 4) tells us that these forces are equal and opposite to each other, $\vec{F}_{1\to2} = -\vec{F}_{2\to1}$, or

$$\vec{F}_{1\to2} + \vec{F}_{2\to1} = 0.$$

Integration of this equation results in

$$0 = \int_{t_i}^{t_f} (\vec{F}_{2\to1} + \vec{F}_{1\to2})\,dt = \int_{t_i}^{t_f} \vec{F}_{2\to1}\,dt + \int_{t_i}^{t_f} \vec{F}_{1\to2}\,dt = \vec{p}_{f1} - \vec{p}_{i1} + \vec{p}_{f2} - \vec{p}_{i2}.$$

Collecting the initial momentum vectors on one side and the final momentum vectors on the other gives us equation 7.8:

$$\vec{p}_{f1} + \vec{p}_{f2} = \vec{p}_{i1} + \vec{p}_{i2}.$$

Equation 7.8 expresses the principle of conservation of linear momentum. The sum of the final momentum vectors is exactly equal to the sum of the initial momentum vectors. Note that this equation does not depend on any particular conditions for the collision. It is valid for all two-body collisions, elastic or inelastic.

You may object that other external forces may be present. In a collision of billiard balls, for example, there is a friction force due to each ball rolling or sliding across the table. In a collision of two cars, friction acts between the tires and the road. However, what characterizes a collision is the occurrence of a very large impulse due to a very large contact force during a relatively short time. If you integrate the external forces over the collision time, you obtain only very small or moderate impulses. Thus, these external forces can usually be safely neglected in calculations of collision dynamics, and we can treat two-body collisions as if only internal forces are at work. We will assume we are dealing with an isolated system, which is a system with no external forces.

In addition, the same argument holds if there are more than two objects taking part in the collision or if there is no collision at all. As long as the net external force is zero, the total momentum of the interaction of objects will be conserved:

$$\text{if} \quad \vec{F}_{net} = 0 \quad \text{then} \quad \sum_{k=1}^{n} \vec{p}_k = \text{constant}. \tag{7.9}$$

Equation 7.9 is the general formulation of the law of conservation of momentum. We will return to this general formulation in Chapter 8 when we talk about systems of particles. For the remainder of this chapter, we consider only idealized cases in which the net external force is negligibly small, and thus the total momentum is always conserved in all processes.

7.4   Elastic Collisions in One Dimension

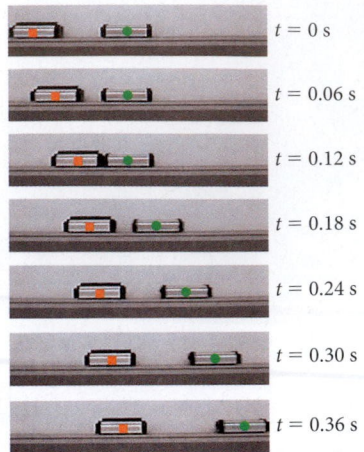

$t = 0$ s

$t = 0.06$ s

$t = 0.12$ s

$t = 0.18$ s

$t = 0.24$ s

$t = 0.30$ s

$t = 0.36$ s

FIGURE 7.8 Video sequence of a collision between two carts of nonequal masses on an air track. The cart with the orange square carries a black metal bar to increase its mass.

Figure 7.8 shows the collision of two carts on an almost frictionless track. The collision was videotaped, and the figure includes seven frames of this video, taken at intervals of 0.06 s. The cart marked with the green circle is initially at rest. The cart marked with the orange square has a larger mass and is approaching from the left. The collision happens in the frame marked with the time $t = 0.12$ s. You can see that after the collision, both carts move to the right, but the lighter cart moves with a significantly higher speed. (The speed is proportional to the horizontal distance between the carts' markings in adjacent video frames.) Next, we'll derive equations that can be used to determine the velocities of the carts after the collision.

What exactly is an elastic collision? As with so many concepts in physics, it is an idealization. In practically all collisions, at least some kinetic energy is converted into other forms of energy that are not conserved. The other forms can be heat or sound or the energy to deform an object, for example. However, an elastic collision is defined as one in which the total kinetic energy of the colliding objects is conserved. This definition does not mean that each object involved in the collision retains its kinetic energy. Kinetic energy can be transferred from one object to the other, but *in an elastic collision, the sum of the kinetic energies has to remain constant.*

We'll consider objects moving in one dimension and use the notations $p_{i1,x}$ for the initial momentum and $p_{f1,x}$ for the final momentum of object 1. (We use the subscript x to remind ourselves that these could equally well be the x-components of the two- or three-dimensional momentum vector.) In the same way, we denote the initial and final momenta of object 2 by $p_{i2,x}$ and $p_{f2,x}$. Because we are restricted to collisions in one dimension, the equation for conservation of kinetic energy can be written as

$$\frac{p_{f1,x}^2}{2m_1} + \frac{p_{f2,x}^2}{2m_2} = \frac{p_{i1,x}^2}{2m_1} + \frac{p_{i2,x}^2}{2m_2}. \qquad (7.10)$$

(For motion in one dimension, the square of the x-component of the vector is also the square of the absolute value of the vector.) The equation for conservation of momentum in the x-direction can be written as

$$p_{f1,x} + p_{f2,x} = p_{i1,x} + p_{i2,x}. \qquad (7.11)$$

(Remember that momentum is conserved in any collision in which the external forces are negligible.)

Let's look more closely at equations 7.10 and 7.11. What is known, and what is unknown? Typically, we know the two masses and components of the initial momentum vectors, and we want to find the final momentum vectors after the collision. This calculation can be done because equations 7.10 and 7.11 give us two equations for two unknowns, $p_{f1,x}$ and $p_{f2,x}$. This is by far the most common use of these equations, but it is also possible, for example, to calculate the two masses if the initial and final momentum vectors are known.

Let's find the components of the final momentum vectors:

$$p_{f1,x} = \left(\frac{m_1 - m_2}{m_1 + m_2}\right) p_{i1,x} + \left(\frac{2m_1}{m_1 + m_2}\right) p_{i2,x}$$

$$\qquad (7.12)$$

$$p_{f2,x} = \left(\frac{2m_2}{m_1 + m_2}\right) p_{i1,x} + \left(\frac{m_2 - m_1}{m_1 + m_2}\right) p_{i2,x}.$$

Derivation 7.2 shows how this result is obtained. It will help you solve similar problems.

DERIVATION 7.2

We start with the equations for energy and momentum conservation and collect all quantities connected with object 1 on the left side and all those connected with object 2 on the right. Equation 7.10 for the (conserved) kinetic energy then becomes:

$$\frac{p_{f1,x}^2}{2m_1} - \frac{p_{i1,x}^2}{2m_1} = \frac{p_{i2,x}^2}{2m_2} - \frac{p_{f2,x}^2}{2m_2}$$

or

$$m_2(p_{f1,x}^2 - p_{i1,x}^2) = m_1(p_{i2,x}^2 - p_{f2,x}^2). \qquad (i)$$

By rearranging equation 7.11 for momentum conservation, we obtain

$$p_{f1,x} - p_{i1,x} = p_{i2,x} - p_{f2,x}. \tag{ii}$$

Next, we divide the left and right sides of equation (i) by the corresponding sides of equation (ii). To do this division, we use the algebraic identity $a^2 - b^2 = (a + b)(a - b)$. This process results in

$$m_2(p_{i1,x} + p_{f1,x}) = m_1(p_{i2,x} + p_{f2,x}). \tag{iii}$$

Now we can solve equation (ii) for $p_{f1,x}$ and substitute the expression $p_{i1,x} + p_{i2,x} - p_{f2,x}$ into equation (iii):

$$m_2(p_{i1,x} + [p_{i1,x} + p_{i2,x} - p_{f2,x}]) = m_1(p_{i2,x} + p_{f2,x})$$

$$2m_2 p_{i1,x} + m_2 p_{i2,x} - m_2 p_{f2,x} = m_1 p_{i2,x} + m_1 p_{f2,x}$$

$$p_{f2,x}(m_1 + m_2) = 2m_2 p_{i1,x} + (m_2 - m_1)p_{i2,x}$$

$$p_{f2,x} = \frac{2m_2 p_{i1,x} + (m_2 - m_1)p_{i2,x}}{m_1 + m_2}.$$

This result is one of the two desired components of equation 7.12. We can obtain the other component by solving equation (ii) for $p_{f2,x}$ and substituting the expression $p_{i1,x} + p_{i2,x} - p_{f1,x}$ into equation (iii). We can also obtain the result for $p_{f1,x}$ from the result for $p_{f2,x}$ that we just derived by exchanging the indices 1 and 2. It is, after all, arbitrary which object is labeled 1 or 2, and so the resulting equations should be symmetric under the exchange of the two labels. Use of this type of symmetry principle is very powerful and very convenient. (But it does take some getting used to at first!)

With the result for the final momenta, we can also obtain expressions for the final velocities by using $p_x = mv_x$:

$$v_{f1,x} = \left(\frac{m_1 - m_2}{m_1 + m_2}\right)v_{i1,x} + \left(\frac{2m_2}{m_1 + m_2}\right)v_{i2,x}$$

$$v_{f2,x} = \left(\frac{2m_1}{m_1 + m_2}\right)v_{i1,x} + \left(\frac{m_2 - m_1}{m_1 + m_2}\right)v_{i2,x}. \tag{7.13}$$

These equations for the final velocities look, at first sight, very similar to those for the final momenta (equation 7.12). However, there is one important difference: In the second term of the right-hand side of the equation for $v_{f1,x}$ the numerator is $2m_2$ instead of $2m_1$; and conversely, the numerator is now $2m_1$ instead of $2m_2$ in the first term of the equation for $v_{f2,x}$.

As a last point in this general discussion, let's find the relative velocity, $v_{f1,x} - v_{f2,x}$, after the collision:

$$v_{f1,x} - v_{f2,x} = \left(\frac{m_1 - m_2 - 2m_1}{m_1 + m_2}\right)v_{i1,x} + \left(\frac{2m_2 - (m_2 - m_1)}{m_1 + m_2}\right)v_{i2,x} \tag{7.14}$$

$$= -v_{i1,x} + v_{i2,x} = -(v_{i1,x} - v_{i2,x}).$$

We see that *in elastic collisions, the relative velocity simply changes sign, $\Delta v_f = -\Delta v_i$*. We will return to this result later in this chapter. You should not try to memorize the general expressions for momentum and velocity in equations 7.13 and 7.14, but instead study the method we used to derive them. Next, we examine two special cases of these general results.

Special Case 1: Equal Masses

If $m_1 = m_2$, the general expressions in equation 7.12 simplify considerably, because the terms proportional to $m_1 - m_2$ are equal to zero and the ratios $2m_1/(m_1 + m_2)$ and $2m_2/(m_1 + m_2)$ become unity. We then obtain the extremely simple result

$$\begin{aligned} p_{f1,x} &= p_{i2,x} \\ p_{f2,x} &= p_{i1,x}. \end{aligned} \qquad \text{(for the special case where } m_1 = m_2) \tag{7.15}$$

This result means that in any elastic collision of two objects of equal mass moving in one dimension, the two objects simply *exchange* their momenta. The initial momentum of object 1 becomes the final momentum of object 2. The same is true for the velocities:

$$v_{f1,x} = v_{i2,x}$$
$$v_{f2,x} = v_{i1,x}.$$ (for the special case where $m_1 = m_2$)   (7.16)

Special Case 2: One Object Initially at Rest

Now suppose the two objects in a collision are not necessarily the same in mass, but one of the two is initially at rest, that is, has zero momentum. Without loss of generality, we can say that object 1 is the one at rest. (Remember that the equations are invariant under exchange of the indices 1 and 2.) By using the general expressions in equation 7.12 and setting $p_{i1,x} = 0$, we get

$$p_{f1,x} = \left(\frac{2m_1}{m_1 + m_2} \right) p_{i2,x}$$
$$p_{f2,x} = \left(\frac{m_2 - m_1}{m_1 + m_2} \right) p_{i2,x}.$$ (for the special case where $p_{i1,x} = 0$)   (7.17)

In the same way, we obtain for the final velocities

$$v_{f1,x} = \left(\frac{2m_2}{m_1 + m_2} \right) v_{i2,x}$$
$$v_{f2,x} = \left(\frac{m_2 - m_1}{m_1 + m_2} \right) v_{i2,x}.$$ (for the special case where $p_{i1,x} = 0$)   (7.18)

If $v_{i2,x} > 0$, object 2 moves from left to right, with the conventional assignment of the positive x-axis pointing to the right. This situation is shown in Figure 7.8. Depending on which mass is larger, the collision can have one of four outcomes:

1. $m_2 > m_1 \Rightarrow (m_2 - m_1)/(m_2 + m_1) > 0$: The final velocity of object 2 points in the same direction but is reduced in magnitude.

2. $m_2 = m_1 \Rightarrow (m_2 - m_1)/(m_2 + m_1) = 0$: Object 2 is at rest, and object 1 moves with the initial velocity of object 2.

3. $m_2 < m_1 \Rightarrow (m_2 - m_1)/(m_2 + m_1) < 0$: Object 2 bounces back; the direction of its velocity vector changes.

4. $m_2 \ll m_1 \Rightarrow (m_2 - m_1)/(m_2 + m_1) \approx -1$ and $2m_2/(m_1 + m_2) \approx 0$: Object 1 remains at rest, and object 2 approximately reverses its velocity. This situation occurs, for example, in the collision of a ball with the ground. In this collision, object 1 is the entire Earth and object 2 is the ball. If the collision is sufficiently elastic, the ball bounces back with the same speed it had right before the collision, but in the opposite direction—up instead of down.

Concept Check 7.4

Suppose an elastic collision occurs in one dimension, like the one shown in Figure 7.8, where the cart marked with the green circle is initially at rest and the cart with the orange square initially has $v_{orange} > 0$, that is, is moving from left to right. What can you say about the masses of the two carts?

a) $m_{orange} < m_{green}$

b) $m_{orange} > m_{green}$

c) $m_{orange} = m_{green}$

Concept Check 7.5

In the situation shown in Figure 7.8, suppose the mass of the cart with the orange square is very much larger than that of the cart with the green circle. What outcome do you expect?

a) The outcome is about the same as the one shown in the figure.

b) The cart with the orange square moves with almost unchanged velocity after the collision, and the cart with the green circle moves with a velocity almost twice as large as the initial velocity of the cart with the orange square.

c) Both carts move with almost the same speed that the cart with the orange square had before the collision.

d) The cart with the orange square stops, and the cart with the green circle moves to the right with the same speed that the cart with the orange square had originally.

Concept Check 7.6

In the situation shown in Figure 7.8, if the mass of the cart with the green circle (originally at rest) is very much larger than that of the cart with the orange square, what outcome do you expect?

a) The outcome is about the same as shown in the figure.

b) The cart with the orange square moves with an almost unchanged velocity after the collision, and the cart with the green circle moves with a velocity almost twice as large as the initial velocity of the cart with the orange square.

c) Both carts move with almost the same speed that the cart with the orange square had before the collision.

d) The cart with the green circle moves with a very low speed slightly to the right, and the cart with the orange square bounces back to the left with almost the same speed it had originally.

EXAMPLE 7.2 Average Force on a Golf Ball

A driver is a golf club used to hit a golf ball a long distance. The head of a driver typically has a mass of 200. g. A skilled golfer can give the club head a speed of around 40.0 m/s. The mass of a golf ball is 45.0 g. The ball stays in contact with the face of the driver for 0.500 ms.

PROBLEM
What is the average force exerted on the golf ball by the driver?

SOLUTION
The golf ball is initially at rest. Because the driver head and the ball are in contact for only a short time, we can consider the collision between them to be an elastic collision. We can use equation 7.18 to calculate the speed of the golf ball, $v_{f1,x}$, after the collision with the driver head

$$v_{f1,x} = \left(\frac{2m_2}{m_1+m_2}\right)v_{i2,x},$$

where m_1 is the mass of the golf ball, m_2 is the mass of the driver head, and $v_{i2,x}$ is the speed of the driver head. The speed of the golf ball leaving the face of the driver head in this case is

$$v_{f1,x} = \frac{2(0.200 \text{ kg})}{0.0450 \text{ kg}+0.200 \text{ kg}}(40.0 \text{ m/s}) = 65.3 \text{ m/s}.$$

Note that if the driver head were much more massive than the golf ball, the golf ball would attain twice the speed of the driver head. However, in that case, the golfer would have a difficult time giving the club head a substantial speed. The momentum change of the golf ball is

$$\Delta p = m\Delta v = mv_{f1,x}.$$

The impulse is then

$$\Delta p = F_{ave}\Delta t,$$

where F_{ave} is the average force exerted by the driver head and Δt is the time that the driver head and golf ball are in contact. The average force is then

$$F_{ave} = \frac{\Delta p}{\Delta t} = \frac{mv_{f1,x}}{\Delta t} = \frac{(0.045 \text{ kg})(65.3 \text{ m/s})}{0.500\cdot10^{-3} \text{ s}} = 5880 \text{ N}.$$

Thus, the driver exerts a very large force on the golf ball. This force compresses the golf ball significantly, as shown in the video sequence in Self-Test Opportunity 7.2. Also, note that the driver does not propel the ball in the horizontal direction and imparts spin to the ball. Thus, an accurate description of striking a golf ball with a driver requires a more detailed analysis.

Self-Test Opportunity 7.2

The figure shows a high-speed video sequence of the collision of a golf club with a golf ball. The ball experiences significant deformation, but this deformation is sufficiently restored before the ball leaves the club's face. Thus, this collision can be approximated as a one-dimensional elastic collision. Discuss the speed of the ball relative to that of the club after the collision and how the cases we have discussed apply to this result.

Collision of a golf club with a golf ball.

7.5 Elastic Collisions in Two or Three Dimensions

Collisions with Walls

To begin our discussion of two- and three-dimensional collisions, we consider the elastic collision of an object with a solid wall. In Chapter 4 on forces, we saw that a solid surface exerts a force on any object that attempts to penetrate the surface. Such forces are normal forces; they are directed perpendicular to the surface (Figure 7.9). If a normal force acts

Concept Check 7.7

Choose the correct statement:

a) In an elastic collision of an object with a wall, energy may or may not be conserved.

b) In an elastic collision of an object with a wall, momentum may or may not be conserved.

c) In an elastic collision of an object with a wall, the incident angle is equal to the final angle.

d) In an elastic collision of an object with a wall, the original momentum vector does not change as a result of the collision.

e) In an elastic collision of an object with a wall, the wall cannot change the momentum of the object because momentum is conserved.

Concept Check 7.8

Choose the correct statement:

a) When a moving object strikes a stationary object, the angle between the velocity vectors of the two objects after the collision is always 90°.

b) For a real-life collision between a moving object and a stationary object, the angle between the velocity vectors of the two objects after the collision is never less than 90°.

c) When a moving object has a head-on collision with a stationary object, the angle between the two velocity vectors after the collision is 90°.

d) When a moving object collides head-on and elastically with a stationary object of the same mass, the object that was moving stops and the other object moves with the original velocity of the moving object.

e) When a moving object collides elastically with a stationary object of the same mass, the angle between the two velocity vectors after the collision cannot be 90°.

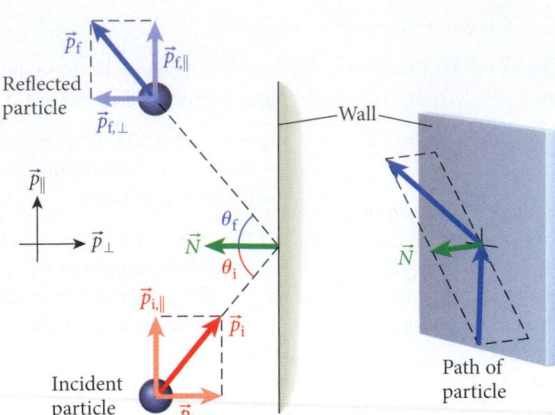

FIGURE 7.9 Elastic collision of an object with a wall. The symbol ⊥ represents the component of the momentum perpendicular to the wall, and the symbol ∥ represents the component of the momentum parallel to the wall.

on an object colliding with a wall, the normal force can only transmit an impulse that is perpendicular to the wall; the normal force has no component parallel to the wall. Thus, the momentum component of the object directed along the wall does not change, $p_{f,\parallel} = p_{i,\parallel}$. In addition, for an elastic collision, we have the condition that the kinetic energy of the object colliding with the wall has to remain the same. This makes sense because the wall stays at rest (it is connected to the Earth and has a much bigger mass than the ball). The kinetic energy of the object is $K = p^2/2m$, so we see that $p_f^2 = p_i^2$.

Because $p_f^2 = p_{f,\parallel}^2 + p_{f,\perp}^2$ and $p_i^2 = p_{i,\parallel}^2 + p_{i,\perp}^2$, we get $p_{f,\perp}^2 = p_{i,\perp}^2$. The only two outcomes possible for the collision are then $p_{f,\perp} = p_{i,\perp}$ and $p_{f,\perp} = -p_{i,\perp}$. Only for the second solution does the perpendicular momentum component point away from the wall after the collision, so it is the only solution that makes sense physically.

To summarize, when an object collides elastically with a wall, the length of the object's momentum vector remains unchanged, as does the momentum component directed along the wall; the momentum component perpendicular to the wall changes sign, but retains the same absolute value. The angle of incidence, θ_i, on the wall (Figure 7.9) is then also the same as the angle of reflection, θ_f:

$$\theta_i = \cos^{-1}\frac{p_{i,\perp}}{p_i} = \cos^{-1}\frac{p_{f,\perp}}{p_f} = \theta_f. \tag{7.19}$$

We will see this same relationship again when we study light and its reflection off a mirror in Chapter 32.

Collisions of Two Objects in Two Dimensions

We have just seen that problems involving elastic collisions in one dimension are always solvable if we have the initial velocity or momentum conditions for the two colliding objects, as well as their masses. Again, this is true because we have two equations for the two unknown quantities, $p_{f1,x}$ and $p_{f2,x}$.

For collisions in two dimensions, each of the final momentum vectors has two components. Thus, this situation gives us four unknown quantities to determine. How many equations do we have at our disposal? Conservation of kinetic energy again provides one of them. Conservation of linear momentum provides independent equations for the x- and y-directions. Therefore, we have only three equations for the four unknown quantities. Unless an additional condition is specified for the collision, there is no unique solution for the final momenta.

For collisions in three dimensions, the situation is even worse. Here we need to determine two vectors with three components each, for a total of six unknown quantities. However, we have only four equations: one from energy conservation and three from the conservation equations for the x-, y-, and z-components of momentum.

Incidentally, this fact is what makes the game of billiards or pool interesting from a physics perspective. The final momenta of two balls after a collision are determined by where on their spherical surfaces they hit each other. Speaking of billiard ball collisions, an interesting observation can be made. Suppose object 2 is initially at rest and both objects have the same mass. Then conservation of momentum results in

$$\vec{p}_{f1} + \vec{p}_{f2} = \vec{p}_{i1}$$
$$(\vec{p}_{f1} + \vec{p}_{f2})^2 = (\vec{p}_{i1})^2$$
$$p_{f1}^2 + p_{f2}^2 + 2\vec{p}_{f1} \cdot \vec{p}_{f2} = p_{i1}^2.$$

Here we squared the equation for momentum conservation and then used the properties of the scalar product. On the other hand, conservation of kinetic energy leads to

$$\frac{p_{f1}^2}{2m} + \frac{p_{f2}^2}{2m} = \frac{p_{i1}^2}{2m}$$
$$p_{f1}^2 + p_{f2}^2 = p_{i1}^2,$$

for $m_1 = m_2 \equiv m$. If we subtract this result from the previous result, we obtain

$$2\vec{p}_{f1} \cdot \vec{p}_{f2} = 0. \tag{7.20}$$

However, the scalar product of two vectors can be zero only if the two vectors are perpendicular to each other or if one of them has length zero. The latter condition is in effect in a head-on collision of two billiard balls, after which the cue ball remains at rest ($\vec{p}_{f1} = 0$) and the other ball moves away with the momentum that the cue ball had initially. After all non–head-on collisions, both balls move, and they move in directions that are perpendicular to each other.

You can do a simple experiment to see if the 90° angle between final velocity vectors works out quantitatively. Put two identical coins on a piece of paper, as shown in Figure 7.10. Mark the position of one of them (the target coin) on the paper by drawing a circle around it. Then flick the other coin with your finger into the target coin (Figure 7.10a). The coins will bounce off each other and slide briefly, before friction forces bring them to rest (Figure 7.10b). Then draw a line from the final position of the target coin back to the circle that you drew, as shown in Figure 7.10c, and thereby deduce the trajectory of the other coin. The images of parts (a) and (b) are superimposed on that in part (c) of the figure to show the motion of the coins before and after the collision, indicated by the red arrows. Measuring the angle between the two black lines in Figure 7.10c results in $\theta = 80°$, so the theoretically derived result of $\theta = 90°$ is not quite correct for this experiment. Why?

What we neglected in our derivation is the fact that—for collisions between coins or billiard balls—some of each object's kinetic energy is associated with rotation and the transfer of energy due to that motion, as well as the fact that such a collision is not quite elastic. However, the 90° rule just derived is a good first approximation for two colliding coins. You can perform a similar experiment of this kind on any billiard table; you will find that the angle of motion between the two billiard balls is not quite 90°, but this approximation will give you a good idea where your cue ball will go after you hit the target ball.

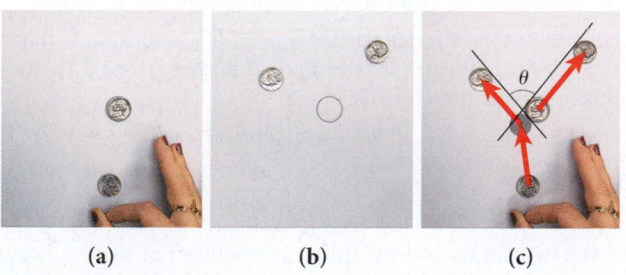

(a)   **(b)**   **(c)**

FIGURE 7.10 Collision of two nickels.

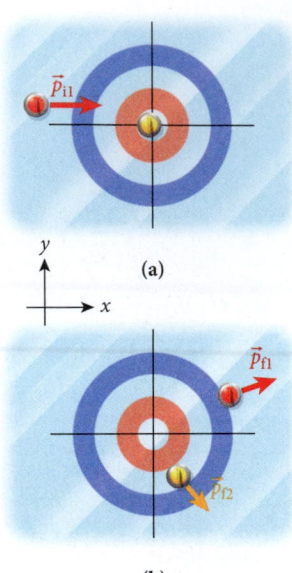

(a)

y
$\uparrow$
$\rightarrow x$

(b)

FIGURE 7.11 Overhead view of a collision of two curling stones: (a) just before the collision; (b) just after the collision.

SOLVED PROBLEM 7.2 / Curling

The sport of curling is all about collisions. A player slides a 19.0-kg (41.9-lb) granite "stone" 35–40 m down the ice into a target area (concentric circles with cross hairs). Teams take turns sliding stones, and the stone closest to the bull's-eye in the end wins. Whenever a stone of one team is closest to the bull's-eye, the other team attempts to knock that stone out of the way, as shown in Figure 7.11.

PROBLEM

The red curling stone shown in Figure 7.11 has an initial velocity of 1.60 m/s in the x-direction and is deflected after colliding with the yellow stone to an angle of 32.0° relative to the x-axis. What are the two final momentum vectors right after this elastic collision, and what is the sum of the stones' kinetic energies?

SOLUTION

THINK Momentum conservation tells us that the sum of the momentum vectors of both stones before the collision is equal to the sum of the momentum vectors of both stones after the collision. Energy conservation tells us that in an elastic collision, the sum of the kinetic energies of both stones before the collision is equal to the sum of the kinetic energies of both stones after the collision. Before the collision, the red stone (stone 1) has momentum and kinetic energy because it is moving, while the yellow stone (stone 2) is at rest and has no momentum or kinetic energy. After the collision, both stones have momentum and kinetic energy. We must calculate the momentum in terms of x- and y-components.

SKETCH A sketch of the momentum vectors of the two stones before and after the collision is shown in Figure 7.12a. The x- and y-components of the momentum vectors after the collision of the two stones are shown in Figure 7.12b.

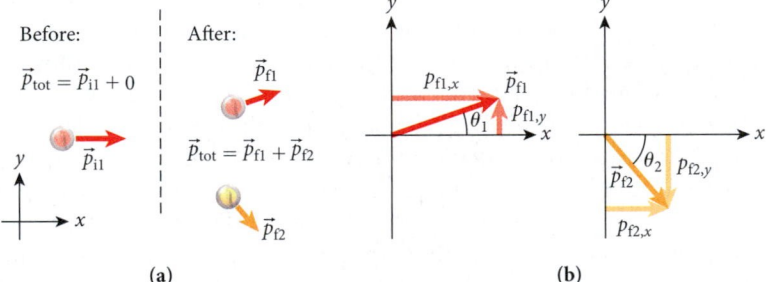

(a)   (b)

FIGURE 7.12 (a) Sketch of momentum vectors before and after the two stones collide. (b) The x- and y-components of the momentum vectors of the two stones after the collision.

RESEARCH Momentum conservation dictates that the sum of the momenta of the two stones before the collision must equal the sum of the momenta of the two stones after the collision. We know the momenta of both stones before the collision, and our task is to calculate their momenta after the collision, based on the given directions of those momenta. For the x-components, we can write

$$p_{i1,x} + 0 = p_{f1,x} + p_{f2,x}.$$

For the y-components, we can write

$$0 + 0 = p_{f1,y} + p_{f2,y}.$$

The problem specifies that stone 1 is deflected at $\theta_1 = 32.0°$. According to the 90°-rule that we derived for perfectly elastic collisions between equal masses, stone 2 has to be deflected at $\theta_2 = -58.0°$. Therefore, in the x-direction we obtain

$$p_{i1,x} = p_{f1,x} + p_{f2,x} = p_{f1}\cos\theta_1 + p_{f2}\cos\theta_2. \qquad (i)$$

And in the y-direction, we have

$$0 = p_{f1,y} + p_{f2,y} = p_{f1}\sin\theta_1 + p_{f2}\sin\theta_2. \qquad (ii)$$

Because we know the two angles and the initial momentum of stone 1, we need to solve a system of two equations for two unknown quantities, which are the magnitudes of the final momenta, p_{f1} and p_{f2}.

SIMPLIFY We solve this system of equations by direct substitution. We can solve the y-component equation (ii) for p_{f1}

$$p_{f1} = -p_{f2} \frac{\sin\theta_2}{\sin\theta_1} \qquad\qquad \text{(iii)}$$

and substitute into the x-component equation (i) to get

$$p_{i1,x} = \left(-p_{f2} \frac{\sin\theta_2}{\sin\theta_1}\right)\cos\theta_1 + p_{f2}\cos\theta_2.$$

We can rearrange this equation to get

$$p_{f2} = \frac{p_{i1,x}}{\cos\theta_2 - \sin\theta_2 \cot\theta_1}.$$

CALCULATE First, we calculate the magnitude of the initial momentum of stone 1:

$$p_{i1,x} = mv_{i1,x} = (19.0\text{ kg})(1.60\text{ m/s}) = 30.4\text{ kg m/s}.$$

We can then calculate the magnitude of the final momentum of stone 2:

$$p_{f2} = \frac{30.4\text{ kg m/s}}{(\cos{-58.0°}) - (\sin{-58.0°})(\cot 32.0°)} = 16.10954563\text{ kg m/s}.$$

The magnitude of the final momentum of stone 1 is

$$p_{f1} = -p_{f2}\frac{\sin(-58.0°)}{\sin(32.0°)} = 25.78066212\text{ kg m/s}.$$

Now we can answer the question concerning the sum of the kinetic energies of the two stones after the collision. Because this collision is elastic, we can simply calculate the initial kinetic energy of the red stone (the yellow one was at rest). Thus, our answer is

$$K = \frac{p_{i1}^2}{2m} = \frac{(30.4\text{ kg m/s})^2}{2(19.0\text{ kg})} = 24.32\text{ J}.$$

ROUND Because all the numerical values were specified to three significant figures, we report the magnitude of the final momentum of the first stone as

$$p_{f1} = 25.8\text{ kg m/s}.$$

The direction of the first stone is +32.0° with respect to the horizontal. We report the magnitude of the final momentum of the second stone as

$$p_{f2} = 16.1\text{ kg m/s}.$$

The direction of the second stone is −58.0° with respect to the horizontal.
The total kinetic energy of the two stones after the collision is

$$K = 24.3\text{ J}.$$

Self-Test Opportunity 7.5

Double-check the results for the final momenta of the two stones in Solved Problem 7.2 by calculating the individual kinetic energies of the two stones after the collision to verify that their sum is indeed equal to the initial kinetic energy.

7.6 Totally Inelastic Collisions

In all collisions that are not completely elastic, the conservation of kinetic energy no longer holds. These collisions are called *inelastic*, because some of the initial kinetic energy gets converted into internal energy of excitation, deformation, vibration, or (eventually) heat. At first sight, this conversion of energy may make the task of calculating the final momentum or velocity vectors of the colliding objects appear more complicated. However, that is not the case; in particular, the algebra becomes considerably easier for the limiting case of totally inelastic collisions.

A totally inelastic collision is one in which the colliding objects stick to each other after colliding. This result implies that both objects have the same velocity vector after the collision: $\vec{v}_{f1} = \vec{v}_{f2} \equiv \vec{v}_f$. (Thus, the *relative velocity* between the two colliding objects is zero after the collision.) Using $\vec{p} = m\vec{v}$ and conservation of momentum, we get the final velocity vector:

$$\vec{v}_f = \frac{m_1\vec{v}_{i1} + m_2\vec{v}_{i2}}{m_1 + m_2}. \qquad\qquad (7.21)$$

This useful formula enables you to solve practically all problems involving totally inelastic collisions. Derivation 7.3 shows how it was obtained.

We start with the conservation law for total momentum (equation 7.8):

$$\vec{p}_{f1} + \vec{p}_{f2} = \vec{p}_{i1} + \vec{p}_{i2}.$$

Now we use $\vec{p} = m\vec{v}$ and get

$$m_1\vec{v}_{f1} + m_2\vec{v}_{f2} = m_1\vec{v}_{i1} + m_2\vec{v}_{i2}.$$

The condition that the collision is totally inelastic implies that the final velocities of the two objects are the same. Thus, we have equation 7.21:

$$m_1\vec{v}_f + m_2\vec{v}_f = m_1\vec{v}_{i1} + m_2\vec{v}_{i2}$$
$$(m_1 + m_2)\vec{v}_f = m_1\vec{v}_{i1} + m_2\vec{v}_{i2}$$
$$\vec{v}_f = \frac{m_1\vec{v}_{i1} + m_2\vec{v}_{i2}}{m_1 + m_2}.$$

Note that the condition of a totally inelastic collision implies only that the final velocities are the same for both objects. In general, the final momentum vectors of the objects can have quite different magnitudes.

We know from Newton's Third Law (see Chapter 4) that the forces two objects exert on each other during a collision are equal in magnitude. However, the changes in velocity—that is, the accelerations that the two objects experience in a totally inelastic collision—can be drastically different. The following example illustrates this phenomenon.

EXAMPLE 7.3 | Head-on Collision

Consider a head-on collision of a full-size SUV, with mass $M = 3023$ kg, and a compact car, with mass $m = 1184$ kg. Each vehicle has an initial speed of $v = 22.35$ m/s (50 mph), and they are moving in opposite directions (Figure 7.13). So, as shown in the figure, we can say v_x is the initial velocity of the compact car and $-v_x$ is the initial velocity of the SUV. The two cars crash into each other and become entangled, a case of a totally inelastic collision.

FIGURE 7.13 Head-on collision of two vehicles with different masses and identical speeds.

PROBLEM
What are the changes of the two cars' velocities in the collision? (Neglect friction between the tires and the road.)

SOLUTION
We first calculate the final velocity that the combined mass has immediately after the collision. To do this calculation, we simply use equation 7.21 and get

$$v_{f,x} = \frac{mv_x - Mv_x}{m + M} = \left(\frac{m - M}{m + M}\right)v_x$$

$$= \frac{1184\ \text{kg} - 3023\ \text{kg}}{1184\ \text{kg} + 3023\ \text{kg}}(22.35\ \text{m/s}) = -9.77\ \text{m/s}.$$

Thus, the velocity change for the SUV turns out to be

$$\Delta v_{\text{SUV},x} = -9.77\ \text{m/s} - (-22.35\ \text{m/s}) = 12.58\ \text{m/s}.$$

However, the velocity change for the compact car is

$$\Delta v_{compact,x} = -9.77 \text{ m/s} - (22.35 \text{ m/s}) = -32.12 \text{ m/s}.$$

We obtain the corresponding average accelerations by dividing the velocity changes by the time interval, Δt, during which the collision takes place. This time interval is obviously the same for both cars, which means that the magnitude of the acceleration experienced by the body of the driver of the compact car is bigger than that experienced by the body of the driver of the SUV by a factor of $32.12/12.58 = 2.55$.

From this result alone, it is clear that it is safer to be in the SUV in this head-on collision than in the compact car. Keep in mind that this result is true even though Newton's Third Law says that the forces exerted by the two vehicles on each other are the same (compare Example 4.5).

Self-Test Opportunity 7.6

In Example 7.3, what is the ratio of the accelerations of the two cars? (*Hint:* Start with Newton's Third Law and make use of the fact that the forces the two cars exert on each other are equal.)

Ballistic Pendulum

A **ballistic pendulum** is a device that can be used to measure muzzle speeds of projectiles shot from firearms. It consists of a block of material into which the bullet is fired. This block is suspended so that it forms a pendulum (Figure 7.14). From the deflection angle of the pendulum and the known masses of the bullet, m, and the block, M, we can calculate the speed of the bullet right before it hits the block.

To obtain an expression for the speed of the bullet in terms of the deflection angle, we have to calculate the speed of the block-plus-bullet combination right after the bullet is embedded in the block. This collision is a prototypical totally inelastic collision, and thus we can apply equation 7.21. Because the pendulum is at rest before the bullet hits it, the speed of the block-plus-bullet combination is

$$v = \frac{m}{m+M} v_b,$$

where v_b is the speed of the bullet before it hits the block and v is the speed of the combined masses right after impact. The kinetic energy of the bullet is $K_b = \frac{1}{2} m v_b^2$ just before it hits the block, whereas right after the collision, the block-plus-bullet combination has the kinetic energy

$$K = \tfrac{1}{2}(m+M)v^2 = \tfrac{1}{2}(m+M)\left(\frac{m}{m+M}v_b\right)^2 = \tfrac{1}{2}mv_b^2 \frac{m}{m+M} = \frac{m}{m+M}K_b. \quad (7.22)$$

Clearly, kinetic energy is not conserved in the process whereby the bullet embeds in the block. (With a real ballistic pendulum, the kinetic energy is transferred into deformation of bullet and block. In this demonstration version, kinetic energy is transferred into frictional work between the bullet and the block.) Equation 7.22 shows that the total kinetic energy (and with it the total mechanical energy) is reduced by a factor of $m/(m+M)$. However, after the collision, the block-plus-bullet combination retains its remaining total energy in the ensuing pendulum motion, converting all of the initial kinetic energy of equation 7.22 into potential energy at the highest point:

$$U_{max} = (m+M)gh = K = \tfrac{1}{2}\left(\frac{m^2}{m+M}\right)v_b^2. \quad (7.23)$$

As you can see from Figure 7.14b, the height h and angle θ are related via $h = \ell(1 - \cos\theta)$, where ℓ is the length of the pendulum. (We found the same relationship in Solved Problem 6.3.) Substituting this result into equation 7.23 yields

$$(m+M)g\ell(1-\cos\theta) = \tfrac{1}{2}\left(\frac{m^2}{m+M}\right)v_b^2 \Rightarrow$$

$$v_b = \frac{m+M}{m}\sqrt{2g\ell(1-\cos\theta)}. \quad (7.24)$$

It is clear from equation 7.24 that practically any bullet speed can be measured with a ballistic pendulum, provided the mass of the block, M, is chosen appropriately. For example, if you shoot a .357 Magnum caliber round ($m = 0.0081$ kg) into a block ($M = 3.00$ kg) suspended by a 1.00 m long rope, the deflection is 21.4°, and equation 7.24 lets you determine that the muzzle speed of this bullet fired from the particular gun that you used is 432 m/s (which is a typical value for this type of ammunition).

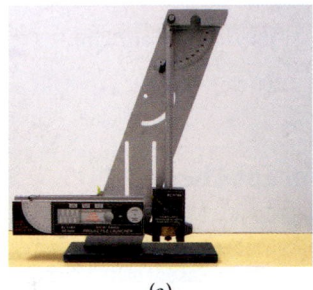

(a)

(b)

FIGURE 7.14 Ballistic pendulum used in an introductory physics laboratory.

Self-Test Opportunity 7.7

If you fire a bullet with half the mass of a .357 Magnum caliber round at the same speed into the ballistic pendulum described in the text, what is the deflection angle?

Kinetic Energy Loss in Totally Inelastic Collisions

As we have just seen, total kinetic energy is not conserved in totally inelastic collisions. How much kinetic energy is lost in the general case? We can find this loss by taking the difference between the total initial kinetic energy, K_i, and the total final kinetic energy, K_f:

$$K_{\text{loss}} = K_i - K_f.$$

The total initial kinetic energy is the sum of the individual kinetic energies of the two objects before the collision:

$$K_i = \frac{p_{i1}^2}{2m_1} + \frac{p_{i2}^2}{2m_2}.$$

The total final kinetic energy for the case in which the two objects stick together and move as one, with the total mass of $m_1 + m_2$ and velocity $\vec{v}_f$, using equation 7.21 is

$$K_f = \frac{1}{2}(m_1 + m_2)v_f^2$$

$$= \frac{1}{2}(m_1 + m_2)\left(\frac{m_1\vec{v}_{i1} + m_2\vec{v}_{i2}}{m_1 + m_2}\right)^2$$

$$= \frac{(m_1\vec{v}_{i1} + m_2\vec{v}_{i2})^2}{2(m_1 + m_2)}.$$

Now we can take the difference between the final and initial kinetic energies and obtain the kinetic energy loss:

$$K_{\text{loss}} = K_i - K_f = \frac{1}{2}\frac{m_1 m_2}{m_1 + m_2}(\vec{v}_{i1} - \vec{v}_{i2})^2. \qquad (7.25)$$

The derivation of this result involves a bit of algebra and is omitted here. What matters, though, is that the difference in the initial velocities—that is, the initial relative velocity—enters into the equation for energy loss. We will explore the significance of this fact in the following section and again in Chapter 8, when we consider center-of-mass motion.

SOLVED PROBLEM 7.3 | Forensic Science

Figure 7.15a shows a sketch of a traffic accident. The white pickup truck (car 1) with mass $m_1 = 2209$ kg was traveling north and hit the westbound red car (car 2) with mass $m_2 = 1474$ kg. When the two vehicles collided, they became entangled (stuck together). Skid marks on the road reveal the exact location of the collision, and the direction in which the two vehicles were sliding immediately afterward. This direction was measured to be 38° relative to the initial direction of the white pickup truck. The white pickup truck had the right of way, because the red car had a stop sign. The driver of the red car, however, claimed that the driver of the white

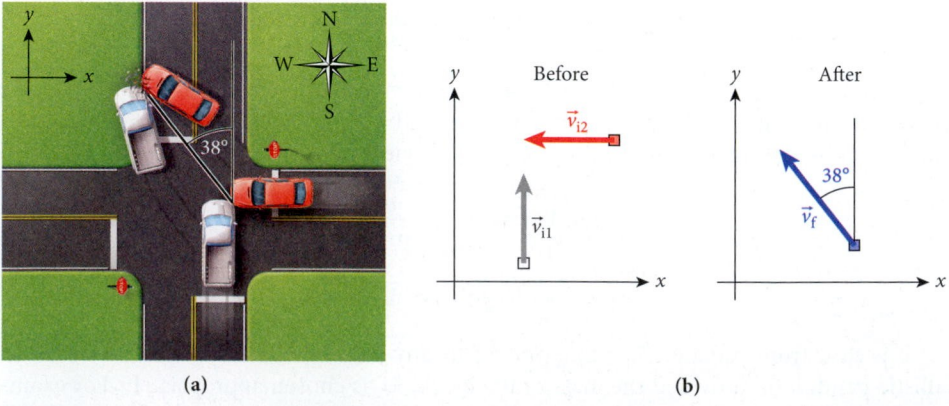

(a)   (b)

FIGURE 7.15 (a) Sketch of the accident scene. (b) Velocity vectors of the two cars before and after the collision.

pickup truck was moving at a speed of at least 50. mph (22 m/s), although the speed limit was 25 mph (11 m/s). Furthermore, the driver of the red car claimed that he had stopped at the stop sign and was then driving through the intersection at a speed of less than 25 mph when the white pickup truck hit him. If the driver of the white pickup truck was speeding, he would legally forfeit the right of way and be declared responsible for the accident.

PROBLEM
Can the version of the accident as described by the driver of the red car be correct?

SOLUTION
THINK This collision is clearly a totally inelastic collision, and so we know that the velocity of the entangled cars after the collision is given by equation 7.21. We are given the angles of the initial velocities of both cars and the angle of the final velocity vector of the entangled cars. However, we are not given the magnitudes of these three velocities. Thus, we have one equation and three unknowns. However, to solve this problem, we only need to determine the ratio of the magnitude of the initial velocity of car 1 to the magnitude of the initial velocity of car 2. Here the initial velocity is the velocity of the car just before the collision occurred.

SKETCH A sketch of the velocity vectors of the two cars before and after the collision is shown in Figure 7.15b. A sketch of the components of the velocity vector of the two stuck together after the collision is shown in Figure 7.16.

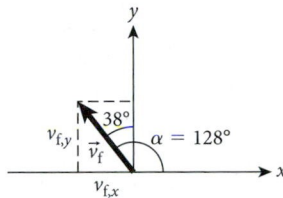
FIGURE 7.16 Components of the velocity vector of the two entangled cars after the collision.

RESEARCH Using the coordinate system shown in Figure 7.15a, the white pickup truck (car 1) has only a y-component for its velocity vector, $\vec{v}_{i1} = v_{i1}\hat{y}$, where v_{i1} is the initial speed of the white pickup. The red car (car 2) has a velocity component only in the negative x-direction, $\vec{v}_{i2} = -v_{i2}\hat{x}$. The final velocity $\vec{v}_f$ of the cars when stuck together after the collision, expressed in terms of the initial velocities, is given by

$$\vec{v}_f = \frac{m_1\vec{v}_{i1} + m_2\vec{v}_{i2}}{m_1 + m_2}.$$

SIMPLIFY Substituting the initial velocities into this equation for the final velocity gives

$$\vec{v}_f = v_{f,x}\hat{x} + v_{f,y}\hat{y} = \frac{-m_2 v_{i2}}{m_1 + m_2}\hat{x} + \frac{m_1 v_{i1}}{m_1 + m_2}\hat{y}.$$

The components of the final velocity vector, $v_{f,x}$ and $v_{f,y}$, are shown in Figure 7.16. From trigonometry, we obtain an expression for the tangent of the angle of the final velocity as the ratio of its y- and x-components:

$$\tan\alpha = \frac{v_{f,y}}{v_{f,x}} = \frac{\dfrac{m_1 v_{i1}}{m_1 + m_2}}{\dfrac{-m_2 v_{i2}}{m_1 + m_2}} = -\frac{m_1 v_{i1}}{m_2 v_{i2}}.$$

Thus, we can find the initial speed of car 1 in terms of the initial speed of car 2:

$$v_{i1} = -\frac{m_2 \tan\alpha}{m_1}v_{i2}.$$

We have to be careful with the value of the angle α. It is *not* 38°, as you might conclude from a casual examination of Figure 7.15b. Instead, it is 38° + 90° = 128°, as shown in Figure 7.16, because angles must be measured relative to the positive x-axis when using the tangent formula, $\tan\alpha = v_{f,y}/v_{f,x}$.

CALCULATE With this result and the known values of the masses of the two cars, we find

$$v_{i1} = -\frac{(1474 \text{ kg})(\tan 128°)}{2209 \text{ kg}}v_{i2} = 0.854066983 v_{i2}.$$

ROUND The angle at which the two cars moved after the collision was specified to two significant digits, so we report our answer to two significant digits:

$$v_{i1} = 0.85 v_{i2}.$$

The white pickup truck (car 1) was moving at a slower speed than the red car (car 2). The story of the driver of the red car is not consistent with the facts. Apparently, the driver of the white pickup truck was not speeding at the time of the collision, and the cause of the accident was the driver of the red car running the stop sign.

Self-Test Opportunity 7.8

To double-check the result for Solved Problem 7.3, estimate what the angle of deflection would have been if the white pickup truck had been traveling with a speed of 50. mph and the red car had been traveling with a speed of 25 mph just before the collision.

Explosions

In totally inelastic collisions, two or more objects merge into one and move in unison with the same total momentum after the collision as before it. The reverse process is also possible. If an object moves with initial momentum $\vec{p}_i$ and then explodes into fragments, the process of the explosion only generates internal forces among the fragments. Because an explosion takes place over a very short time, the impulse due to external forces can usually be neglected. In such a situation, according to Newton's Third Law, the total momentum is conserved. This result implies that the sum of the momentum vectors of the fragments has to add up to the initial momentum vector of the object:

$$\vec{p}_i = \sum_{k=1}^{n} \vec{p}_{fk}. \tag{7.26}$$

This equation relating the momentum of the exploding object just before the explosion to the sum of the momentum vectors of the fragments after the explosion is exactly the same as the equation for a totally inelastic collision except that the indices for the initial and final states are exchanged. In particular, if an object breaks up into two fragments, equation 7.26 matches equation 7.21, with the indices i and f exchanged:

$$\vec{v}_i = \frac{m_1 \vec{v}_{f1} + m_2 \vec{v}_{f2}}{m_1 + m_2}. \tag{7.27}$$

This relationship allows us, for example, to reconstruct the initial velocity if we know the fragments' velocities and masses. Further, we can calculate the energy released in a breakup that yields two fragments from equation 7.25, again with the indices i and f exchanged:

$$K_{\text{release}} = K_f - K_i = \frac{1}{2} \frac{m_1 m_2}{m_1 + m_2} \left(\vec{v}_{f1} - \vec{v}_{f2} \right)^2. \tag{7.28}$$

EXAMPLE 7.4   Decay of a Radon Nucleus

Radon is a gas that is produced by the radioactive decay of naturally occurring heavy nuclei, such as thorium and uranium. Radon gas can be breathed into the lungs, where it can decay further. The nucleus of a radon atom has a mass of 222 u, where u is an atomic mass unit (to be introduced in Chapter 40). Assume that the nucleus is at rest when it decays into a polonium nucleus with mass 218 u and a helium nucleus with mass 4 u (called an *alpha particle*), releasing 5.59 MeV of energy.

PROBLEM

What are the kinetic energies of the polonium nucleus and the alpha particle?

SOLUTION

The polonium nucleus and the alpha particle are emitted in opposite directions. We will say that the alpha particle is emitted with speed v_{x1} in the positive x-direction and the polonium nucleus is emitted with speed v_{x2} in the negative x-direction. The mass of the alpha particle is $m_1 = 4$ u, and the mass of the polonium nucleus is $m_2 = 218$ u. The initial velocity of the radon nucleus is zero, so we can use equation 7.27 to write

$$\vec{v}_i = 0 = \frac{m_1 \vec{v}_1 + m_2 \vec{v}_2}{m_1 + m_2},$$

which gives us

$$m_1 v_{x1} = -m_2 v_{x2}. \tag{i}$$

Using equation 7.28, we can express the kinetic energy released by the decay of the radon nucleus

$$K_{\text{release}} = \frac{1}{2} \frac{m_1 m_2}{m_1 + m_2} \left(v_{x1} - v_{x2} \right)^2. \tag{ii}$$

We can then use equation (i) to express the velocity of the polonium nucleus in terms of the velocity of the alpha particle:

$$v_{x2} = -\frac{m_1}{m_2} v_{x1}.$$

Substituting from equation (i) into equation (ii) gives us

$$K_{\text{release}} = \frac{1}{2}\frac{m_1 m_2}{m_1+m_2}\left(v_{x1}+\frac{m_1}{m_2}v_{x1}\right)^2 = \frac{1}{2}\frac{m_1 m_2 v_{x1}^2}{m_1+m_2}\left(\frac{m_1+m_2}{m_2}\right)^2.$$ (iii)

Rearranging equation (iii) leads to

$$K_{\text{release}} = \frac{1}{2}\frac{m_1(m_1+m_2)v_{x1}^2}{m_2} = \frac{1}{2}m_1 v_{x1}^2\frac{m_1+m_2}{m_2} = K_1\left(\frac{m_1+m_2}{m_2}\right),$$

where K_1 is the kinetic energy of the alpha particle. This kinetic energy is then

$$K_1 = K_{\text{release}}\left(\frac{m_2}{m_1+m_2}\right).$$

Putting in the numerical values, we get the kinetic energy of the alpha particle:

$$K_1 = (5.59 \text{ MeV})\left(\frac{218}{4+218}\right) = 5.49 \text{ MeV}.$$

The kinetic energy, K_2, of the polonium nucleus is then

$$K_2 = K_{\text{release}} - K_1 = 5.59 \text{ MeV} - 5.49 \text{ MeV} = 0.10 \text{ MeV}.$$

The alpha particle gets most of the kinetic energy when the radon nucleus decays, and this is sufficient energy to damage surrounding tissue in the lungs.

EXAMPLE 7.5 / Particle Physics

The conservation laws of momentum and energy are essential for analyzing the products of particle collisions at high energies, such as those produced at Fermilab's Tevatron, near Chicago, Illinois, which was the world's highest-energy proton-antiproton accelerator. (An accelerator called LHC, for Large Hadron Collider, began operation in 2008 at the CERN Laboratory in Geneva, Switzerland, and it is more powerful than Tevatron. However, the LHC is a proton-proton accelerator.)

In the Tevatron accelerator, particle physicists collide protons and antiprotons at total energies of 1.96 TeV (hence the name). Remember that 1 eV = $1.602 \cdot 10^{-19}$ J; so 1.96 TeV = $1.96 \cdot 10^{12}$ eV = $3.1 \cdot 10^{-7}$ J. The Tevatron is set up so that the protons and antiprotons move in the collider ring in opposite directions, with, for all practical purposes, exactly opposite momentum vectors. The main particle detectors, D-Zero and CDF, are located at the interaction regions, where protons and antiprotons collide.

Figure 7.17a shows an example of such a collision. In this computer-generated display of one particular collision event at the D-Zero detector, the proton's initial momentum vector

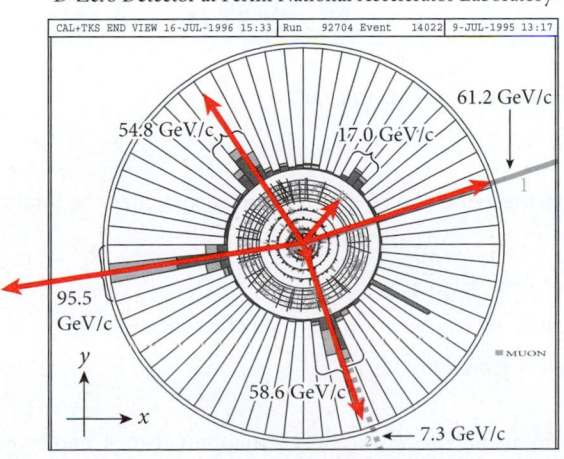
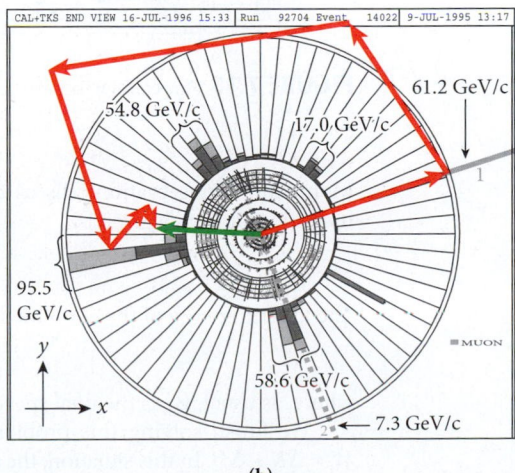

FIGURE 7.17 Event display generated by the D-Zero collaboration and education office at Fermilab, showing a top-quark event. (a) Momentum vectors of the detected particles produced by the event; (b) graphical addition of the momentum vectors, showing that they add up to a nonzero sum, indicated by the thicker green arrow.

– Continued

points straight into the page and that of the antiproton points straight out of the page. Thus, the total initial momentum of the proton-antiproton system is zero. The explosion produced by this collision produces several fragments, almost all of which are registered by the detector. These measurements are indicated in the display (Figure 7.17a). Superimposed on this event display are the momentum vectors of the corresponding parent particles of these fragments, with their lengths and directions based on computer analysis of the detector response. (The momentum unit GeV/c, commonly used in high-energy physics and shown in the figure, is the energy unit GeV divided by the speed of light.) In Figure 7.17b, the momentum vectors have been summed graphically, giving a nonzero vector, indicated by the thicker green arrow.

However, conservation of momentum absolutely requires that the sum of the momentum vectors of all particles produced in this collision must be zero. Thus, conservation of momentum allows us to assign the missing momentum represented by the green arrow to a particle that escaped undetected. With the aid of this missing-momentum analysis, physicists at Fermilab were able to show that the event displayed here was one in which an elusive particle known as a *top quark* was produced.

Self-Test Opportunity 7.9

The length of each vector arrow in Figure 7.17b is proportional to the magnitude of the momentum vector of the particular particle. Can you determine the momentum of the nondetected particle (green arrow)?

SOLVED PROBLEM 7.4 | Collision with a Parked Car

PROBLEM

A moving truck strikes a car parked in the breakdown lane of a highway. During the collision, the vehicles stick together and slide to a stop. The moving truck has a total mass of 1982 kg (including the driver), and the parked car has a total mass of 966 kg. If the vehicles slide 10.5 m before coming to rest, how fast was the truck going? The coefficient of kinetic friction between the tires and the road is 0.350.

SOLUTION

THINK This situation is a totally inelastic collision of a moving truck with a parked car. The kinetic energy of the truck/car combination after the collision is reduced by the energy dissipated by friction while the truck/car combination is sliding. The kinetic energy of the truck/car combination can be related to the initial speed of the truck before the collision.

SKETCH Figure 7.18 is a sketch of the moving truck, m_1, and the parked car, m_2. Before the collision, the truck is moving with an initial speed of $v_{i1,x}$. After the truck collides with the car, the two vehicles slide together with a speed of $v_{f,x}$.

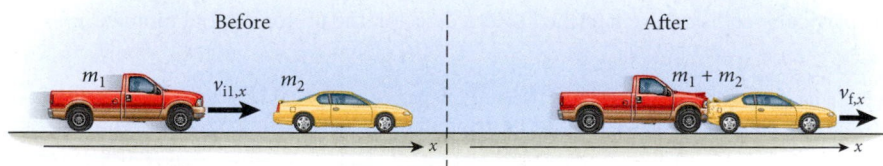

Before m_1 $v_{i1,x}$ m_2 x **After** $m_1 + m_2$ $v_{f,x}$ x

FIGURE 7.18 The collision between a moving truck and a parked car.

RESEARCH

Momentum conservation tells us that the velocity of the two vehicles just after the totally inelastic collision is given by

$$v_{f,x} = \frac{m_1 v_{i1,x}}{m_1 + m_2}.$$

The kinetic energy of the combined truck/car combination just after the collision is

$$K = \tfrac{1}{2}(m_1 + m_2)v_{f,x}^2, \tag{i}$$

where, as usual, $v_{f,x}$ is the final speed.

We finish solving this problem by using the work-energy theorem from Chapter 6, $W_f = \Delta K + \Delta U$. In this situation, the energy dissipated by friction, W_f, on the truck/car system is equal to the change in kinetic energy, ΔK, of the truck/car system, since $\Delta U = 0$. We can thus write

$$W_f = \Delta K.$$

The change in kinetic energy is equal to zero (since the truck and car finally stop) minus the kinetic energy of the truck/car system just after the collision. The truck/car system slides a distance d. The x-component of the frictional force slowing down the truck/car system is given by $f_x = -\mu_k N$, where μ_k is the coefficient of kinetic friction and N is the magnitude of the normal force. The normal force has a magnitude equal to the weight of the truck/car system, or $N = (m_1 + m_2)g$. The energy dissipated is equal to the x-component of the friction force times the distance that the truck/car system slides along the x-axis, so we can write

$$W_f = f_x d = -\mu_k(m_1 + m_2)gd. \tag{ii}$$

SIMPLIFY We can substitute for the final speed in equation (i) for the kinetic energy and obtain

$$K = \tfrac{1}{2}(m_1 + m_2)v_{f,x}^2 = \tfrac{1}{2}(m_1 + m_2)\left(\frac{m_1 v_{i1,x}}{m_1 + m_2}\right)^2 = \frac{(m_1 v_{i1,x})^2}{2(m_1 + m_2)}.$$

Combining this equation with the work-energy theorem and equation (ii) for the energy dissipated by friction, we get

$$\Delta K = W_f = 0 - \frac{(m_1 v_{i1,x})^2}{2(m_1 + m_2)} = -\mu_k(m_1 + m_2)gd.$$

Solving for $v_{i1,x}$ finally leads us to

$$v_{i1,x} = \frac{(m_1 + m_2)}{m_1}\sqrt{2\mu_k gd}.$$

CALCULATE Putting in the numerical values results in

$$v_{i1,x} = \frac{1982 \text{ kg} + 966 \text{ kg}}{1982 \text{ kg}}\sqrt{2(0.350)(9.81 \text{ m/s}^2)(10.5 \text{ m})} = 12.62996079 \text{ m/s}.$$

ROUND All of the numerical values given in this problem were specified to at least three significant figures. Therefore, we report our result as

$$v_{i1,x} = 12.6 \text{ m/s}.$$

DOUBLE-CHECK The initial speed of the truck was 12.6 m/s (28.2 mph), which is within the range of normal speeds for vehicles on highways and thus certainly within the expected magnitude for our result.

7.7 Partially Inelastic Collisions

What happens if a collision is neither elastic nor totally inelastic? Most real collisions are somewhere between these two extremes, as we saw in the coin collision experiment in Figure 7.10. Therefore, it is important to look at partially inelastic collisions in more detail.

We have already seen that the relative velocity of the two objects in a one-dimensional elastic collision simply changes sign. The relative velocity becomes zero in totally inelastic collisions. Thus, it seems logical to define the elasticity of a collision in a way that involves the ratio of initial and final relative velocities.

The **coefficient of restitution,** symbolized by ϵ, is the ratio of the magnitudes of the final and initial relative velocities in a collision:

$$\epsilon = \frac{|\vec{v}_{f1} - \vec{v}_{f2}|}{|\vec{v}_{i1} - \vec{v}_{i2}|}. \tag{7.29}$$

This definition gives a coefficient of restitution of $\epsilon = 1$ for elastic collisions and $\epsilon = 0$ for totally inelastic collisions.

First, let's examine what happens in the limit where one of the two colliding objects is the ground (for all intents and purposes, infinitely massive) and the other one is a ball. We can see from equation 7.29 that if the ground does not move when the ball bounces, $\vec{v}_{i1} = \vec{v}_{f1} = 0$, and we can write for the speed of the ball:

$$v_{f2} = \epsilon v_{i2}.$$

If we release the ball from some height, h_i, we know that it reaches a speed of $v_i = \sqrt{2gh_i}$ immediately before it collides with the ground. If the collision is elastic, the speed of the ball just after the collision is the same, $v_f = v_i = \sqrt{2gh_i}$, and it bounces back to the same height from which it was released. If the collision is totally inelastic, as in the case of a ball of putty that falls to the ground and then just stays there, the final speed is zero. For all cases in between, we can find the coefficient of restitution from the height h_f that the ball returns to:

$$h_f = \frac{v_f^2}{2g} = \frac{\epsilon^2 v_i^2}{2g} = \epsilon^2 h_i \Rightarrow$$

$$\epsilon = \sqrt{h_f / h_i}.$$

Self-Test Opportunity 7.10

The maximum kinetic energy loss is obtained in the limit of a totally inelastic collision. What fraction of this maximum possible energy loss is obtained in the case where $\epsilon = \frac{1}{2}$?

Using this formula to measure the coefficient of restitution, we find $\epsilon = 0.58$ for baseballs, using typical relative velocities that would occur in ball-bat collisions in Major League games.

In general, we can state (without proof) that the kinetic energy loss in partially inelastic collisions is

$$K_{loss} = K_i - K_f = \frac{1}{2} \frac{m_1 m_2}{m_1 + m_2} (1 - \epsilon^2)(\vec{v}_{i1} - \vec{v}_{i2})^2. \tag{7.30}$$

In the limit $\epsilon \rightarrow 1$, we obtain $K_{loss} = 0$, i.e., no loss in kinetic energy, as required for elastic collisions. In addition, in the limit $\epsilon \rightarrow 0$, equation 7.30 matches the energy release for totally inelastic collisions already shown in equation 7.28.

Partially Inelastic Collision with a Wall

If you play racquetball or squash, you know that the ball loses energy when you hit it against the wall. While the angle at which a ball bounces off the wall in an elastic collision is the same as the angle at which it hit the wall, the final angle is not so clear for a partially inelastic collision (Figure 7.19).

The key to obtaining a first approximation to this angle is to consider only the normal force, which acts in a direction perpendicular to the wall. Then the momentum component directed along the wall remains unchanged, just as in an elastic collision. However, the momentum component perpendicular to the wall is not simply inverted but is also reduced in magnitude by the coefficient of restitution: $p_{f,\perp} = -\epsilon p_{i,\perp}$. This approximation gives us an angle of reflection relative to the normal that is larger than the initial angle:

$$\theta_f = \cot^{-1} \frac{p_{f,\perp}}{p_{f,\parallel}} = \cot^{-1} \frac{\epsilon p_{i,\perp}}{p_{i,\parallel}} > \theta_i. \tag{7.31}$$

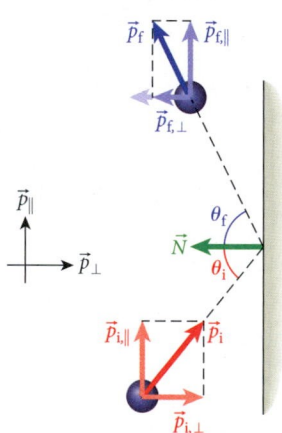

FIGURE 7.19 Partially inelastic collision of a ball with a wall.

The magnitude of the final momentum vector is also changed and reduces to

$$p_f = \sqrt{p_{f,\parallel}^2 + p_{f,\perp}^2} = \sqrt{p_{i,\parallel}^2 + \epsilon^2 p_{i,\perp}^2} < p_i. \tag{7.32}$$

If we want a more accurate description, we need to include the effect of a friction force between ball and wall, acting for the duration of the collision. (This is why squash balls and racquetballs leave marks on the walls.) Further, the collision with the wall also changes the rotation of the ball, and thus additionally alters the direction and kinetic energy of the ball as it bounces off. However, equations 7.31 and 7.32 still provide a very reasonable first approximation to partially inelastic collisions with walls.

7.8 Billiards and Chaos

Let's look at billiards in an abstract way. The abstract billiard system is a rectangular (or even square) billiard table, on which particles can bounce around and have elastic collisions with the sides. Between collisions, these particles move on straight paths without energy loss. When two particles start off close to each other, as in Figure 7.20a, they stay close to each other. The figure shows the paths (red and green lines) of two particles, which start close to each other with the same initial momentum (indicated by the red arrow) and clearly stay close.

The situation becomes qualitatively different when a circular wall is added in the middle of the billiard table. Now each collision with the circle drives the two paths farther apart.

In Figure 7.20b, you can see that one collision with the circle was enough to separate the red and green lines for good. This type of billiard system is called a *Sinai billiard,* named after the Russian academician Yakov Sinai (b. 1935) who first studied it in 1970. The Sinai billiard exhibits **chaotic motion,** or motion that follows the laws of physics—it is not random—but cannot be predicted because it changes significantly with slight changes in conditions, including starting conditions. Surprisingly, Sinai billiard systems are still not fully explored. For example, only in the last two decades have the decay properties of these systems been described. Researchers interested in the physics of chaos are continually gaining new knowledge of these systems.

Here is one example from the authors' own research. If you block off the pockets and cut a hole into the side wall of a conventional billiard table and then measure the time it takes a ball to hit this hole and escape, you obtain a power-law decay time distribution: $N(t) = N(t = 0)t^{-1}$, where $N(t = 0)$ is the number of balls used in the experiment and $N(t)$ is the number of balls remaining after time t. However, if you did the same for the Sinai billiard, you would obtain an exponential time dependence: $N(t) = N(t = 0)e^{-t/T}$.

These types of investigations are not simply theoretical speculations. Try the following experiment: Place a billiard ball on the surface of a table and hold onto the ball. Then hold a second billiard ball as exactly as you can directly above the first one and release it from a height of a few inches (or centimeters). You will see that the upper ball cannot be made to bounce on the lower one more than three or four times before going off in some uncontrollable direction. Even if you could fix the location of the two balls to atomic precision, the upper ball would bounce off the lower one after only ten to fifteen times. This result means that the ability to predict the outcome of this experiment extends to only a few collisions. This limitation of predictability goes to the heart of chaos science. It is one of the main reasons, for example, that exact long-term weather forecasting is impossible. After all, air molecules bounce off each other, too.

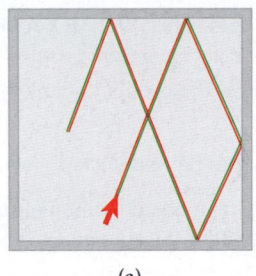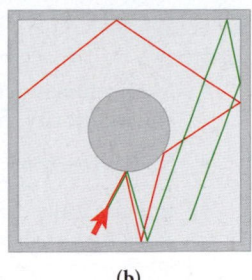

(a) **(b)**

FIGURE 7.20 Collisions of particles with walls for two particles starting out very close to each other and with the same momentum: (a) regular billiard table; (b) Sinai billiard.

Laplace's Demon

Marquis Pierre-Simon Laplace (1749–1827) was an eminent French physicist and mathematician of the 18th century. He lived during the time of the French Revolution and other major social upheavals, characterized by the struggle for self-determination and freedom. No painting symbolizes this struggle better than *Liberty Leading the People* (1830) by Eugène Delacroix (Figure 7.21).

Laplace had an interesting idea, now known as *Laplace's Demon.* He reasoned that everything is made of atoms, and all atoms obey differential equations governed by the forces acting on them. If the initial positions and velocities of all atoms, together with all force laws, were fed into a huge computer (he called this an "intellect"), then "for such an intellect nothing could be uncertain and the future just like the past would be present before its eyes." This reasoning implies that everything is predetermined; we are only cogs in a huge clockwork, and nobody has free will. Ironically, Laplace came up with this idea at a time when quite a few people believed that they could achieve free will, if only they could overthrow those in power.

The downfall of Laplace's Demon comes from the combination of two principles of physics. One is from chaos science, which points out that long-term predictability depends sensitively on knowledge of the initial conditions, as seen in the experiment with the bouncing billiard balls. This principle applies to molecules interacting with one another, as, for example, air molecules do. The other physics principle notes the impossibility of specifying both the position and the momentum of any object exactly at the same time. This is the uncertainty relation in quantum physics (discussed in Chapter 36). Thus, free will is still alive and well, at least from the perspective of physics constraints—the predictability of large or complex systems such as the weather or the human brain over the long term is impossible. The combination of chaos theory and quantum theory ensures that Laplace's "intellect" or any computer cannot possibly calculate and predict our individual decisions.

FIGURE 7.21 *Liberty Leading the People,* Eugène Delacroix (Louvre, Paris, France).

WHAT WE HAVE LEARNED | EXAM STUDY GUIDE

- Momentum is defined as the product of an object's mass and its velocity: $\vec{p} = m\vec{v}$.

- Newton's Second Law can be written as $\vec{F} = d\vec{p}/dt$.

- Impulse is the change in an object's momentum and is equal to the time integral of the applied external force:
$$\vec{J} = \Delta \vec{p} = \int_{t_i}^{t_f} \vec{F}\,dt.$$

- In a collision of two objects, momentum can be exchanged, but the sum of the momenta of the colliding objects remains constant: $\vec{p}_{f1} + \vec{p}_{f2} = \vec{p}_{i1} + \vec{p}_{i2}$. This is the law of conservation of total momentum.

- Collisions can be elastic, totally inelastic, or partially inelastic.

- In elastic collisions, the total kinetic energy also remains constant:
$$\frac{p_{f1}^2}{2m_1} + \frac{p_{f2}^2}{2m_2} = \frac{p_{i1}^2}{2m_1} + \frac{p_{i2}^2}{2m_2}.$$

- For one-dimensional elastic collisions in general, the final velocities of the two colliding objects can be expressed as functions of the initial velocities:
$$v_{f1,x} = \left(\frac{m_1 - m_2}{m_1 + m_2}\right) v_{i1,x} + \left(\frac{2m_2}{m_1 + m_2}\right) v_{i2,x}$$
$$v_{f2,x} = \left(\frac{2m_1}{m_1 + m_2}\right) v_{i1,x} + \left(\frac{m_2 - m_1}{m_1 + m_2}\right) v_{i2,x}.$$

- In totally inelastic collisions, the colliding objects stick together after the collision and have the same velocity: $\vec{v}_f = (m_1\vec{v}_{i1} + m_2\vec{v}_{i2})/(m_1 + m_2)$.

- All partially inelastic collisions are characterized by a coefficient of restitution, defined as the ratio of the magnitudes of the final and the initial relative velocities: $\epsilon = |\vec{v}_{f1} - \vec{v}_{f2}|/|\vec{v}_{i1} - \vec{v}_{i2}|$. The kinetic energy loss in a partially inelastic collision is then given by
$$\Delta K = K_i - K_f = \frac{1}{2}\frac{m_1 m_2}{m_1 + m_2}(1 - \epsilon^2)(\vec{v}_{i1} - \vec{v}_{i2})^2.$$

ANSWERS TO SELF-TEST OPPORTUNITIES

7.1 Examples include: rubber door stop, padded baseball glove, padded dashboard in car, water-filled barrels in front of bridge abutments on highways, pads on basketball goal support, pads on goalpost supports on football field, portable pad for gym floor, padded shoe inserts.

7.2 The collision is that of a stationary golf ball being struck by a moving golf club head. The head of the driver has more mass than the golf ball. If we take the golf ball to be m_1 and the driver head to be m_2, then $m_2 > m_1$ and
$$v_{f1,x} = \left(\frac{2m_2}{m_1 + m_2}\right) v_{i2,x}.$$

So the speed of the golf ball after impact will be a factor of
$$\left(\frac{2m_2}{m_1 + m_2}\right) > 1$$

greater than the speed of the golf club head. If the golf club head is much more massive than the golf ball, then
$$\left(\frac{2m_2}{m_1 + m_2}\right) \approx 2.$$

7.3 Take two coins, each with radius R. Collide with impact parameter b. The angle by which the moving coin is deflected is θ.
$$b = 0 \rightarrow \theta = 90°$$
$$b = 2R \rightarrow \theta = 0°$$

The complete function is
$$\theta = \cos^{-1}\left(\frac{b}{2R}\right).$$

7.4 Suppose the larger coin is at rest and we shoot a smaller coin at the larger coin. For $b = 0$, we get $\theta = 180°$ and for $b = R_1 + R_2$, we get $\theta = 0°$.

For a very small moving coin incident on a very large stationary coin with radius R, we get
$$b = R\cos(\theta/2) \Rightarrow \theta = 2\cos^{-1}\left(\frac{b}{R}\right).$$

Suppose the smaller coin is at rest and we shoot a larger coin at the smaller coin. For $b = 0$, we get $\theta = 0$ because the larger coin continues forward in a head-on collision. For $b = R_1 + R_2$, we get $\theta = 0$.

7.5 The initial kinetic energy is
$$K_i = \tfrac{1}{2}mv^2 = \tfrac{1}{2}(19.0 \text{ kg})(1.60 \text{ m/s})^2 = 24.3 \text{ J}.$$

7.6 $F = ma$, and both experience the same force
$$m_{SUV}a_{SUV} = m_{compact}a_{compact}$$
$$\frac{a_{compact}}{a_{SUV}} = \frac{m_{SUV}}{m_{compact}} = \frac{3023 \text{ kg}}{1184 \text{ kg}} = 2.553.$$

7.7 $v_b = \dfrac{m + M}{m}\sqrt{2g\ell(1 - \cos\theta)} \Rightarrow$
$$\theta = \cos^{-1}\left(1 - \frac{1}{2g\ell}\left(\frac{mv_b}{m + M}\right)^2\right)$$
$$\theta = \cos^{-1}\left(1 - \frac{1}{2(9.81 \text{ m/s}^2)(1.00 \text{ m})}\left(\frac{(0.004 \text{ kg})(432 \text{ m/s})}{0.004 \text{ kg} + 3.00 \text{ kg}}\right)^2\right)$$
$$\theta = 10.5°$$
or slightly less than half the original angle.

7.8 Again we have to be careful with the tangent. The x-component is negative and the y-component is positive. So we must end up in quadrant II with $90° < \alpha < 180°$.

$$\tan\alpha = \frac{-m_1 v_{i1}}{m_2 v_{i2}}$$

$$\alpha = \tan^{-1}\left(\frac{-m_1 v_{i1}}{m_2 v_{i2}}\right) = \tan^{-1}\left(\frac{-2209\ \text{kg}}{1474\ \text{kg}} \frac{50\ \text{mph}}{25\ \text{mph}}\right)$$

$\alpha = 108°$ or $18°$ from the vertical, compared with $38°$ from the vertical in the actual collision.

7.9 The length of the missing momentum vector is 1.4 cm. The length of the 95.5 GeV/c vector is 4.0 cm. The missing momentum is then $(4.0/1.4)95.5$ GeV/c = 33 GeV/c.

7.10 Loss for $\epsilon = 0.5$ is

$$K_{\text{loss},0.5} = K_i - K_f = \frac{1}{2}\frac{m_1 m_2}{m_1 + m_2}(1 - (0.5)^2)(\vec{v}_{i1} - \vec{v}_{i2})^2$$

$$= (0.75)\frac{1}{2}\frac{m_1 m_2}{m_1 + m_2}(\vec{v}_{i1} - \vec{v}_{i2})^2 = 0.75\, K_{\text{loss,max}}.$$

PROBLEM-SOLVING GUIDELINES: CONSERVATION OF MOMENTUM

1. Conservation of momentum applies to isolated systems with no external forces acting on them—always be sure the problem involves a situation that satisfies or approximately satisfies these conditions. Also, be sure that you take account of every interacting part of the system; conservation of momentum applies to the entire system, not just one object.

2. If the situation you're analyzing involves a collision or explosion, identify the momenta immediately before and immediately after the event to use in the conservation of momentum equation. Remember that a change in total momentum equals an impulse, but the impulse can be either an instanta-

neous force acting over an instant of time or an average force acting over a time interval.

3. If the problem involves a collision, you need to recognize what kind of collision it is. If the collision is perfectly elastic, kinetic energy is conserved, but this is not true for other kinds of collisions.

4. Remember that momentum is a vector and is conserved in the x-, y-, and z-directions separately. For a collision in more than one dimension, you may need additional information to analyze momentum changes completely.

MULTIPLE-CHOICE QUESTIONS

7.1 In many old Western movies, a bandit is knocked back 3 m after being shot by a sheriff. Which statement best describes what happened to the sheriff after he fired his gun?

a) He remained in the same position.

b) He was knocked back a step or two.

c) He was knocked back approximately 3 m.

d) He was knocked forward slightly.

e) He was pushed upward.

7.2 A fireworks projectile is traveling upward as shown in the figure on the right just before it explodes. Sets of possible momentum vectors for the shell fragments immediately after the explosion are shown below. Which sets could actually occur?

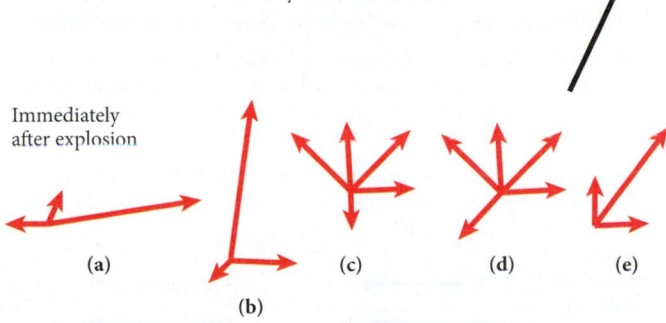

Immediately after explosion

(a)   (b)   (c)   (d)   (e)

Immediately before explosion

7.3 The figure shows sets of possible momentum vectors before and after a collision, with no external forces acting. Which sets could actually occur?

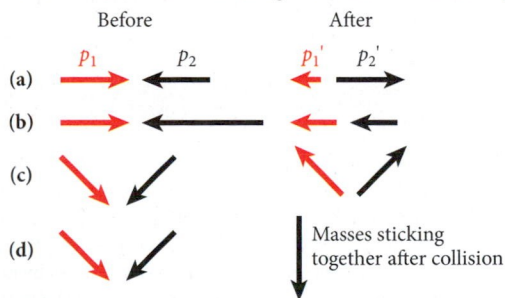

Before   After

p_1 p_2 p_1' p_2'

(a)

(b)

(c)

(d)   Masses sticking together after collision

7.4 The value of the momentum for a system is the same at a later time as at an earlier time if there are no

a) collisions between particles within the system.

b) inelastic collisions between particles within the system.

c) changes of momentum of individual particles within the system.

d) internal forces acting between particles within the system.

e) external forces acting on particles of the system.

7.5 Consider these three situations:

(i) A ball moving to the right at speed v is brought to rest.

(ii) The same ball at rest is projected at speed v toward the left.

(iii) The same ball moving to the left at speed v speeds up to $2v$.

In which situation(s) does the ball undergo the largest change in momentum?

a) situation (i)   d) situations (i) and (ii)

b) situation (ii)   e) all three situations

c) situation (iii)

7.6 Consider two carts, of masses m and $2m$, at rest on a frictionless air track. If you push the lower-mass cart for 35 cm and then the other cart for the same distance and with the same force, which cart undergoes the larger change in momentum?

a) The cart with mass m has the larger change.

b) The cart with mass $2m$ has the larger change.

c) The change in momentum is the same for both carts.

d) It is impossible to tell from the information given.

7.7 Consider two carts, of masses m and $2m$, at rest on a frictionless air track. If you push the lower-mass cart for 3 s and then the other cart for the same length of time and with the same force, which cart undergoes the larger change in momentum?

a) The cart with mass m has the larger change.

b) The cart with mass $2m$ has the larger change.

c) The change in momentum is the same for both carts.

d) It is impossible to tell from the information given.

7.8 Which of the following statements about car collisions is (are) true?

a) The essential safety benefit of crumple zones (parts of the front of a car designed to receive maximum deformation during a head-on collision) is due to their absorbing kinetic energy, converting it into deformation, and lengthening the effective collision time, thus reducing the average force experienced by the driver.

b) If car 1 has mass m and speed v, and car 2 has mass $0.5m$ and speed $1.5v$, then both cars have the same momentum.

c) If two identical cars with identical speeds collide head on, the magnitude of the impulse received by each car and each driver is the same as if one car at the same speed had collided head on with a concrete wall.

d) Car 1 has mass m, and car 2 has mass $2m$. In a head-on collision of these cars while moving at identical speeds in opposite directions, car 1 experiences a bigger acceleration than car 2.

e) Car 1 has mass m, and car 2 has mass $2m$. In a head-on collision of these cars while moving at identical speeds in opposite directions, car 1 receives an impulse of bigger magnitude than that received by car 2.

7.9 A fireworks projectile is launched upward at an angle above a large flat plane. When the projectile reaches the top of its flight, at a height of h above a point that is a horizontal distance D from where it was launched,

the projectile explodes into two equal pieces. One piece reverses its velocity and travels directly back to the launch point. How far from the launch point does the other piece land?

a) D　　　　　　　　c) $3D$

b) $2D$　　　　　　　　d) $4D$

7.10 A red curling stone moving with a speed of 2.0 m/s collides head-on with a yellow curling stone at rest (totally elastic collision). What are the speeds of the two curling stones just after the collision?

a) The red stone is at rest, and the yellow stone is moving with a speed of 2.0 m/s.

b) The red stone and the yellow stone are both moving with a speed of 1.0 m/s.

c) The red stone bounces off the yellow stone and moves with a speed of 2.0 m/s, and the yellow stone remains at rest.

7.11 For a totally elastic collision between two objects, which of the following statements is (are) true?

a) The total mechanical energy is conserved.

b) The total kinetic energy is conserved.

c) The total momentum is conserved.

d) The momentum of each object is conserved.

e) The kinetic energy of each object is conserved.

7.12 For a totally inelastic collision between two objects, which of the following statements is (are) true?

a) The total mechanical energy is conserved.

b) The total kinetic energy is conserved.

c) The total momentum is conserved.

d) The total momentum after the collision is always zero.

e) The total kinetic energy after the collision can never be zero.

7.13 A ballistic pendulum is used to measure the speed of a bullet shot from a gun. The mass of the bullet is 50.0 g, and the mass of the block is 20.0 kg. When the bullet strikes the block, the combined mass rises a vertical distance of 5.00 cm. What was the speed of the bullet as it struck the block?

a) 397 m/s　　　　　　d) 479 m/s

b) 426 m/s　　　　　　e) 503 m/s

c) 457 m/s

CONCEPTUAL QUESTIONS

7.14 An astronaut becomes stranded during a space walk after her jet pack malfunctions. Fortunately, there are two objects close to her that she can push to propel herself back to the International Space Station (ISS). Object A has the same mass as the astronaut, and Object B is 10 times more massive. To achieve a given momentum toward the ISS by pushing one of the objects away from the ISS, which object should she push? That is, which one requires less work to produce the same impulse? Initially, the astronaut and the two objects are at rest with respect to the ISS. (*Hint:* Recall that work is force times distance and think about how the two objects move when they are pushed.)

7.15 Consider a ballistic pendulum (see Section 7.6) in which a bullet strikes a block of wood. The wooden block is hanging from the ceiling and swings up to a maximum height after the bullet strikes it. Typically, the bullet becomes embedded in the block. Given the same bullet, the same initial bullet speed, and the same block, would the maximum height of the block change if the bullet did not get stopped by the block but passed through to the other side? Would the height change if the bullet and its speed were the same but the block was steel and the bullet bounced off it, directly backward?

7.16 A bungee jumper is concerned that his elastic cord might break if it is overstretched and is considering replacing the cord with a high-tensile-strength steel cable. Is this a good idea?

7.17 A ball falls straight down onto a block that is wedge-shaped and is sitting on frictionless ice. The block is initially at rest (see the figure). Assuming the collision is perfectly elastic, is the total momentum of the block/ball system conserved? Is the total kinetic energy of the block/ball system exactly the same before and after the collision? Explain.

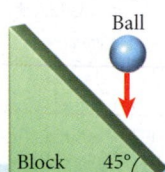

7.18 To solve problems involving projectiles traveling through the air by applying the law of conservation of momentum requires evaluating the momentum of the system *immediately* before and *immediately* after the collision or explosion. Why?

7.19 Two carts are riding on an air track as shown in the figure. At time $t = 0$, cart B is at the origin traveling in the positive x-direction with speed v_B, and cart A is at rest as shown in the diagram. The carts collide, but do not stick.

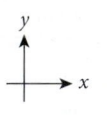

Before

$\vec{v}_B$ ⟶　　　　　Initially at rest

Cart B　　　　　Cart A

Each of the graphs depicts a possible plot of a physical parameter with respect to time. Each graph has two curves, one for each cart, and each curve is labeled with the cart's letter. For each property (a)–(e), specify the graph that could be a plot of the property; if a graph for a property is not shown, choose alternative 9.

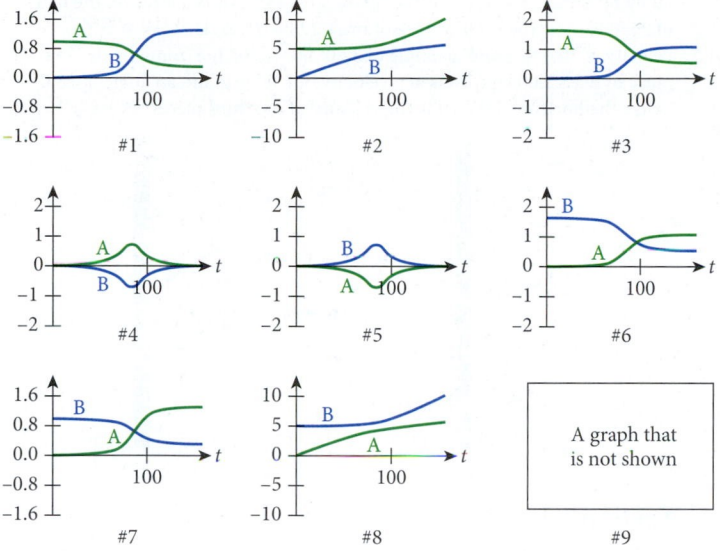

a) the forces *exerted by* the carts

b) the positions of the carts

c) the velocities of the carts

d) the accelerations of the carts

e) the momenta of the carts

7.20 Using momentum and force principles, explain why an air bag reduces injury in an automobile collision.

7.21 A rocket works by expelling gas (fuel) from its nozzles at a high velocity. However, if we take the system to be the rocket and fuel, explain qualitatively why a stationary rocket is able to move.

7.22 When hit in the face, a boxer will "ride the punch"; that is, if he anticipates the punch, he will start moving his head backward before the fist arrives. His head then moves back easily from the blow. From a momentum-impulse standpoint, explain why this is much better than stiffening his neck muscles and bracing himself against the punch.

7.23 An open train car moves with speed v_0 on a flat frictionless railroad track, with no engine pulling it. It begins to rain. The rain falls straight down and begins to fill the train car. Does the speed of the car decrease, increase, or stay the same? Explain.

EXERCISES

A blue problem number indicates a worked-out solution is available in the Student Solutions Manual. One • and two •• indicate increasing level of problem difficulty.

Section 7.1

7.24 Rank the following objects from highest to lowest in terms of momentum and from highest to lowest in terms of energy.

a) an asteroid with mass 10^6 kg and speed 500 m/s

b) a high-speed train with a mass of 180,000 kg and a speed of 300 km/h

c) a 120-kg linebacker with a speed of 10 m/s

d) a 10-kg cannonball with a speed of 120 m/s

e) a proton with a mass of $2 \cdot 10^{-27}$ kg and a speed of $2 \cdot 10^8$ m/s

7.25 A car of mass 1200. kg, moving with a speed of 72.0 mph on a highway, passes a small SUV with a mass $1\frac{1}{2}$ times bigger, moving at 2/3 the speed of the car.

a) What is the ratio of the momentum of the SUV to that of the car?

b) What is the ratio of the kinetic energy of the SUV to that of the car?

7.26 The electron-volt, eV, is a unit of energy (1 eV = $1.602 \cdot 10^{-19}$ J, 1 MeV = $1.602 \cdot 10^{-13}$ J). Since the unit of momentum is an energy unit divided by a velocity unit, nuclear physicists usually specify momenta of nuclei in units of MeV/c, where c is the speed of light ($c = 2.998 \cdot 10^8$ m/s). In the same units, the mass of a proton ($1.673 \cdot 10^{-27}$ kg) is given as 938.3 MeV/c^2. If a proton moves with a speed of 17,400 km/s, what is its momentum in units of MeV/c?

7.27 A soccer ball with a mass of 442 g bounces off the crossbar of a goal and is deflected upward at an angle of 58.0° with respect to horizontal. Immediately after the deflection, the kinetic energy of the ball is 49.5 J. What are the vertical and horizontal components of the ball's momentum immediately after striking the crossbar?

•**7.28** A billiard ball of mass m = 0.250 kg hits the cushion of a billiard table at

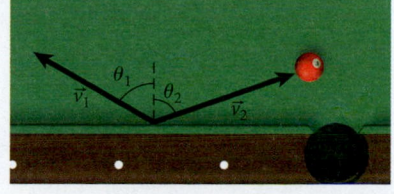

an angle of θ_1 = 60.0° and a speed of v_1 = 27.0 m/s. It bounces off at an angle of θ_2 = 71.0° and a speed of v_2 = 10.0 m/s.

a) What is the magnitude of the change in momentum of the billiard ball?

b) In which direction does the change-of-momentum vector point?

Section 7.2

7.29 In the movie *Superman,* Lois Lane falls from a building and is caught by the diving superhero. Assuming that Lois, with a mass of 50.0 kg, is falling at a terminal velocity of 60.0 m/s, how much average force is exerted on her if it takes 0.100 s to slow her to a stop? If Lois can withstand a maximum acceleration of 7.00 g's, what minimum time should it take Superman to stop her after he begins to slow her down?

7.30 One of the events in the Scottish Highland Games is the sheaf toss, in which a 9.09-kg bag of hay is tossed straight up into the air using a pitchfork. During one throw, the sheaf is launched straight up with an initial speed of 2.70 m/s.

a) What is the impulse exerted on the sheaf by gravity during the upward motion of the sheaf (from launch to maximum height)?

b) Neglecting air resistance, what is the impulse exerted by gravity on the sheaf during its downward motion (from maximum height until it hits the ground)?

c) Using the total impulse produced by gravity, determine how long the sheaf is airborne.

7.31 An 83.0-kg running back leaps straight ahead toward the end zone with a speed of 6.50 m/s. A 115-kg linebacker, keeping his feet on the ground, catches the running back and applies a force of 900. N in the opposite direction for 0.750 s before the running back's feet touch the ground.

a) What is the impulse that the linebacker imparts to the running back?

b) What change in the running back's momentum does the impulse produce?

c) What is the running back's momentum when his feet touch the ground?

d) If the linebacker keeps applying the same force after the running back's feet have touched the ground, is this still the only force acting to change the running back's momentum?

7.32 A baseball pitcher delivers a fastball that crosses the plate at an angle of 7.25° below the horizontal and a speed of 88.5 mph. The ball (of mass 0.149 kg) is hit back over the head of the pitcher at an angle of 35.53° above the horizontal and a speed of 102.7 mph. What is the magnitude of the impulse received by the ball?

•**7.33** Although they don't have mass, photons—traveling at the speed of light—have momentum. Space travel experts have thought of capitalizing on this fact by constructing *solar sails*—large sheets of material that would work by reflecting photons. Since the momentum of the photon would be reversed, an impulse would be exerted on it by the solar sail, and—by Newton's Third Law—an impulse would also be exerted on the sail, providing a force. In space near the Earth, about $3.84 \cdot 10^{21}$ photons are incident per square meter per second. On average, the momentum of each photon is $1.30 \cdot 10^{-27}$ kg m/s. For a 1000.-kg spaceship starting from rest and attached to a square sail 20.0 m wide, how fast could the ship be moving after 1 hour? After 1 week? After 1 month? How long would it take the ship to attain a speed of 8000. m/s, roughly the speed of a space shuttle in orbit?

•**7.34** In a severe storm, 1.00 cm of rain falls on a flat horizontal roof in 30.0 min. If the area of the roof is 100. m^2 and the terminal velocity of the rain is 5.00 m/s, what is the average force exerted on the roof by the rain during the storm?

•**7.35** NASA has taken an increased interest in near-Earth asteroids. These objects, popularized in certain blockbuster movies, can pass very close to Earth on a cosmic scale, sometimes as close as 1 million miles. Most are small—less than 500 m across—and while an impact with one of the smaller ones could be dangerous, experts believe that it may not be catastrophic to the human race. One possible defense system against near-Earth asteroids involves hitting an incoming asteroid with a rocket to divert its course. Assume a relatively small asteroid with a mass of $2.10 \cdot 10^{10}$ kg is traveling toward the Earth at a modest speed of 12.0 km/s.

a) How fast would a large rocket with a mass of $8.00 \cdot 10^4$ kg have to be moving when it hit the asteroid head on in order to stop the asteroid? Assume that the rocket and the asteroid stick together after colliding.

b) An alternative approach would be to divert the asteroid from its path by a small amount and thus cause it to miss Earth. How fast would the rocket of part (a) have to be traveling at impact to divert the asteroid's path by 1.00°? In this case, assume that the rocket hits the asteroid while traveling along a line perpendicular to the asteroid's path.

•**7.36** In nanoscale electronics, electrons can be treated like billiard balls. The figure shows a simple device currently under study in which an electron elastically collides with a rigid wall (a ballistic electron transistor). The green bars represent electrodes that can apply a vertical force of $8.00 \cdot 10^{-13}$ N to the electrons. If an electron initially has velocity components $v_x = 1.00 \cdot 10^5$ m/s and $v_y = 0$ and the wall is at 45.0°, the deflection angle θ_D is 90.0°. How long does the vertical force from the electrodes need to be applied to obtain a deflection angle of 120.°?

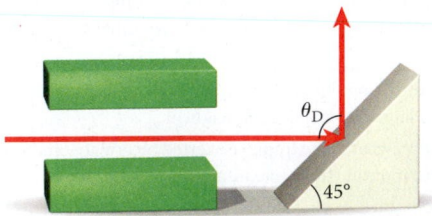

•**7.37** The largest railway gun ever built was called Gustav and was used briefly in World War II. The gun, mount, and train car had a total mass of $1.22 \cdot 10^6$ kg. The gun fired a projectile that was 80.0 cm in diameter and weighed 7502 kg. In the firing illustrated in the figure, the gun has been elevated 20.0° above the horizontal. If the railway gun was at rest before firing and moved to the right at a speed of 4.68 m/s immediately after

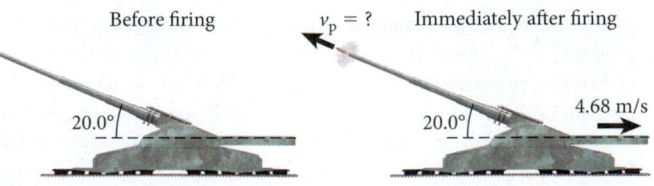

Before firing $v_p = ?$ Immediately after firing

20.0°   20.0°   4.68 m/s

firing, what was the speed of the projectile as it left the barrel (muzzle velocity)? How far will the projectile travel if air resistance is neglected? Assume that the wheel axles are frictionless.

••**7.38** A 6.00-kg clay ball is thrown directly against a perpendicular brick wall at a velocity of 22.0 m/s and shatters into three pieces, which all fly backward, as shown in the figure. The wall exerts a force on the ball of 2640 N for 0.100 s. One piece of mass 2.00 kg travels backward at a velocity of 10.0 m/s and an angle of 32.0° above the horizontal. A second piece of mass 1.00 kg travels at a velocity of 8.00 m/s and an angle of 28.0° below the horizontal. What is the velocity of the third piece?

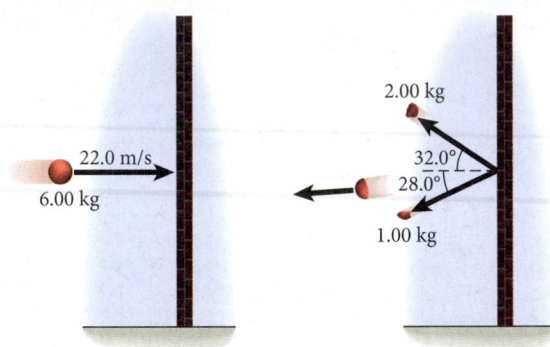

22.0 m/s

6.00 kg

2.00 kg

32.0°
28.0°

1.00 kg

Section 7.3

7.39 A sled initially at rest has a mass of 52.0 kg, including all of its contents. A block with a mass of 13.5 kg is ejected to the left at a speed of 13.6 m/s. What is the speed of the sled and the remaining contents?

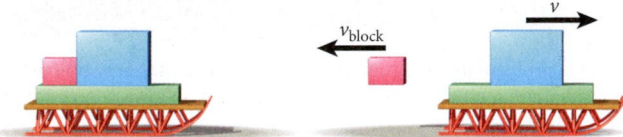

v_{block} v

7.40 Stuck in the middle of a frozen pond with only your physics book, you decide to put physics in action and throw the 5.00-kg book. If your mass is 62.0 kg and you throw the book at 13.0 m/s, how fast do you then slide across the ice? (Assume the absence of friction.)

7.41 Astronauts are playing baseball on the International Space Station. One astronaut with a mass of 50.0 kg, initially at rest, hits a baseball with a bat. The baseball was initially moving toward the astronaut at 35.0 m/s, and after being hit, travels back in the same direction with a speed of 45.0 m/s. The mass of a baseball is 0.140 kg. What is the recoil velocity of the astronaut?

•**7.42** An automobile with a mass of 1450 kg is parked on a moving flatbed railcar; the flatbed is 1.50 m above the ground. The railcar has a mass of 38,500 kg and is moving to the right at a constant speed of 8.70 m/s on a frictionless rail. The automobile then accelerates to the left, leaving the railcar at a speed of 22.0 m/s with respect to the ground. When the

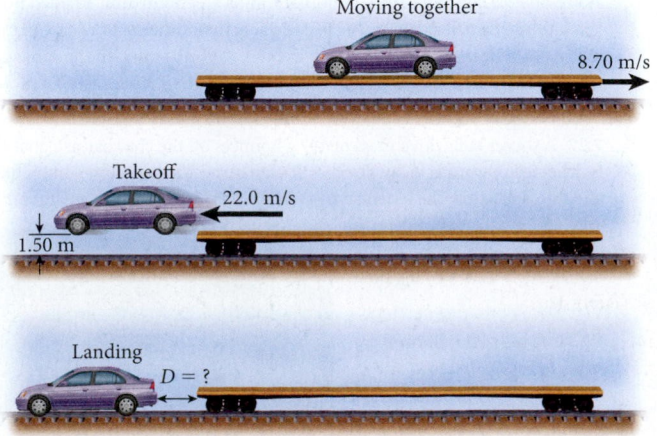

Moving together

8.70 m/s

Takeoff   22.0 m/s

1.50 m

Landing

$D = ?$

automobile lands, what is the distance D between it and the left end of the railcar? See the figure.

•7.43 Three people are floating on a 120.-kg raft in the middle of a pond on a warm summer day. They decide to go swimming, and all jump off the raft at the same time and from evenly spaced positions around the perimeter of the raft. One person, with a mass of 62.0 kg, jumps off the raft at a speed of 12.0 m/s. The second person, of mass 73.0 kg, jumps off the raft at a speed of 8.00 m/s. The third person, of mass 55.0 kg, jumps off the raft at a speed of 11.0 m/s. At what speed does the raft drift from its original position?

•7.44 A missile is shot straight up into the air. At the peak of its trajectory, it breaks up into three pieces of equal mass, all of which move horizontally away from the point of the explosion. One piece travels in a horizontal direction of 28.0° east of north with a speed of 30.0 m/s. The second piece travels in a horizontal direction of 12.0° south of west with a speed of 8.00 m/s. What is the velocity of the remaining piece? Give both speed and angle.

••7.45 Once a favorite playground sport, dodgeball is becoming increasingly popular among adults of all ages who want to stay in shape and who even form organized leagues. Gutball is a less popular variant of dodgeball in which players are allowed to bring their own (typically nonregulation) equipment and in which cheap shots to the face are permitted. In a gutball tournament against people half his age, a physics professor threw his 0.400-kg soccer ball at a kid throwing a 0.600-kg basketball. The balls collided in midair (see the figure), and the basketball flew off with an energy of 95.0 J at an angle of 32.0° relative to its initial path. Before the collision, the energy of the soccer ball was 100. J and the energy of the basketball was 112 J. At what angle and speed did the soccer ball move away from the collision?

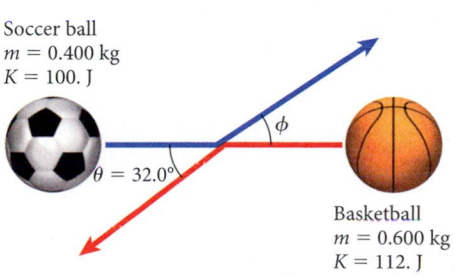

Soccer ball
$m = 0.400$ kg
$K = 100.$ J

ϕ

$\theta = 32.0°$

Basketball
$m = 0.600$ kg
$K = 112.$ J

Section 7.4

7.46 Two bumper cars moving on a frictionless surface collide elastically. The first bumper car is moving to the right with a speed of 20.4 m/s and rear-ends the second bumper car, which is also moving to the right but with a speed of 9.00 m/s. What is the speed of the first bumper car after the collision? The mass of the first bumper car is 188 kg, and the mass of the second bumper car is 143 kg. Assume that the collision takes place in one dimension.

7.47 A satellite with a mass of 274 kg approaches a large planet at a speed $v_{i,1} = 13.5$ km/s. The planet is moving at a speed $v_{i,2} = 10.5$ km/s in the opposite direction. The satellite partially orbits the planet and then moves away from the planet in a direction opposite to its original direction (see the figure). If this interaction is assumed to approximate an elastic collision in one dimension, what is the speed of the satellite after the collision? This so-called *slingshot effect* is often used to accelerate space probes for journeys to distance parts of the Solar System (see Chapter 12).

$v_{i,2}$

$v_{i,1}$

•7.48 You see a pair of shoes tied together by the laces and hanging over a telephone line. You throw a 0.250-kg stone at one of the shoes (mass = 0.370 kg), and it collides elastically with the shoe with a velocity of 2.30 m/s in the horizontal direction. How far up does the shoe move?

•7.49 Block A and block B are forced together, compressing a spring (with spring constant $k = 2500.$ N/m) between them 3.00 cm from its equilibrium length. The spring, which has

$m_A = 1.00$ kg $m_B = 3.00$ kg

$k = 2500.$ N/m

negligible mass, is not fastened to either block and drops to the surface after it has expanded. What are the speeds of block A and block B at this moment? (Assume that the friction between the blocks and the supporting surface is so low that it can be neglected.)

•7.50 An alpha particle (mass = 4.00 u) has a head-on, elastic collision with a nucleus (mass = 166 u) that is initially at rest. What percentage of the kinetic energy of the alpha particle is transferred to the nucleus in the collision?

•7.51 You notice that a shopping cart 20.0 m away is moving with a velocity of 0.700 m/s toward you. You launch an identical cart with a velocity of 1.10 m/s directly at the other cart in order to intercept it. When the two carts collide elastically, they remain in contact for 0.200 s. Graph the position, velocity, and force for both carts as a function of time.

•7.52 A 0.280-kg ball has an elastic, head-on collision with a second ball that is initially at rest. The second ball moves off with half the original speed of the first ball.

a) What is the mass of the second ball?

b) What fraction of the original kinetic energy ($\Delta K/K$) is transferred to the second ball?

••7.53 Cosmic rays from space that strike Earth contain some charged particles with energies billions of times higher than any that can be produced in the biggest accelerator. One model that was proposed to account for these particles is shown schematically in the figure. Two very strong sources of magnetic fields move toward each other and repeatedly reflect the charged particles trapped between them. These magnetic field sources can be approximated as infinitely heavy walls from which charged particles get reflected elastically. The high-energy particles that strike the Earth would have been reflected a large number of times to attain the observed energies. An analogous case with only a few reflections demonstrates this effect. Suppose a particle has an initial velocity of −2.21 km/s (moving in the negative x-direction, to the left), the left wall moves with a velocity of 1.01 km/s to the right, and the right wall moves with a velocity of 2.51 km/s to the left. What is the velocity of the particle after six collisions with the left wall and five collisions with the right wall?

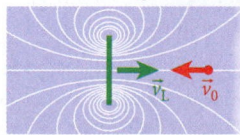
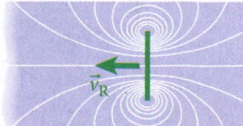

v_L v_0 v_R

••7.54 Here is a popular lecture demonstration that you can perform at home. Place a golf ball on top of a basketball, and drop the pair from rest so they fall to the ground. (For reasons that should become clear once you solve this problem, do not attempt to do this experiment inside, but outdoors instead!) With a little practice, you can achieve the situation pictured here: The golf ball stays on top of the basketball until the basketball hits the floor. The mass of the golf ball is 0.0459 kg, and the mass of the basketball is 0.619 kg.

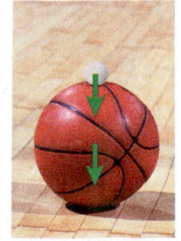

a) If the balls are released from a height where the bottom of the basketball is at 0.701 m above the ground, what is the absolute value of the basketball's momentum just before it hits the ground?

b) What is the absolute value of the momentum of the golf ball at this instant?

c) Treat the collision of the basketball with the floor and the collision of the golf ball with the basketball as totally elastic collisions in one dimension. What is the absolute magnitude of the momentum of the golf ball after these collisions?

d) Now comes the interesting question: How high, measured from the ground, will the golf ball bounce up after its collision with the basketball?

Section 7.5

7.55 A hockey puck with mass 0.170 kg traveling along the blue line (a blue-colored straight line on the ice in a hockey rink) at 1.50 m/s strikes a stationary puck with the same mass. The first puck exits the collision in a direction that is 30.0° away from the blue line at a speed of 0.750 m/s (see the figure). What is the direction and magnitude of the velocity of the second puck after the collision? Is this an elastic collision?

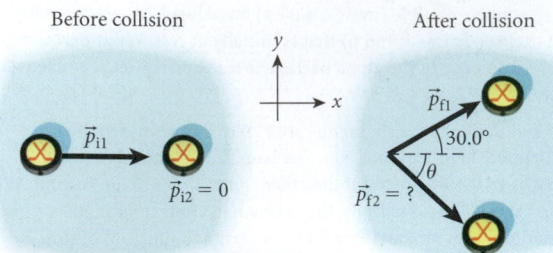

•**7.56** A ball with mass $m = 0.210$ kg and kinetic energy $K_1 = 2.97$ J collides elastically with a second ball of the same mass that is initially at rest. After the collision, the first ball moves away at an angle of $\theta_1 = 30.6°$ with respect to the horizontal, as shown in the figure. What is the kinetic energy of the first ball after the collision?

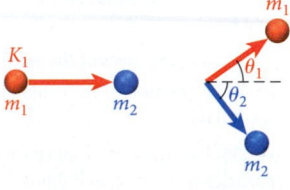

•**7.57** When you open the door to an air-conditioned room, you mix hot gas with cool gas. Saying that a gas is hot or cold actually refers to its average energy; that is, the hot gas molecules have a higher kinetic energy than the cold gas molecules. The difference in kinetic energy in the mixed gases decreases over time as a result of elastic collisions between the gas molecules, which redistribute the energy. Consider a two-dimensional collision between two nitrogen molecules (N_2, molecular weight = 28.0 g/mol). One molecule moves to the right and upward at 30.0° with respect to the horizontal with a velocity of 672 m/s. This molecule collides with a second molecule moving in the negative horizontal direction at 246 m/s. What are the molecules' final velocities if the one that is initially more energetic moves in the positive vertical direction after the collision?

•**7.58** A ball with mass 3.00 kg falls straight down onto a 45°-wedge that is rigidly attached to the ground. The ball is moving a speed of 4.50 m/s when it strikes the wedge. Assuming that the collision is instantaneous and perfectly elastic, what is the recoil momentum that the Earth receives during this collision?

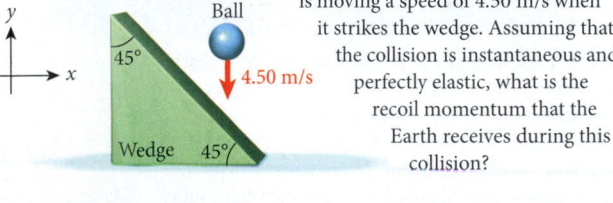

••**7.59** Betty Bodycheck ($m_B = 55.0$ kg, $v_B = 22.0$ km/h in the positive x-direction) and Sally Slasher ($m_S = 45.0$ kg, $v_S = 28.0$ km/h in the positive y-direction) are both racing to get to a hockey puck. Immediately after the collision, Betty is heading in a direction that is 76.0° counterclockwise from her original direction, and Sally is heading back and to her right in a direction that is 12.0° from the x-axis. What are Betty and Sally's final kinetic energies? Is their collision elastic?

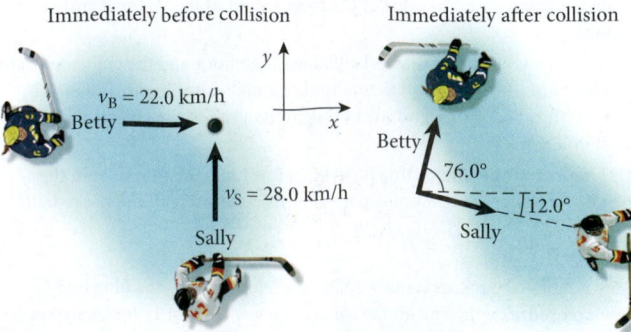

••**7.60** Current measurements and cosmological theories suggest that only about 4.6% of the total mass of the universe is composed of ordinary matter. About 23.% of the mass is composed of *dark matter*, which does not emit or reflect light and can only be observed through its gravitational interaction with its surroundings (see Chapter 12). Suppose a galaxy with mass M_G is moving in a straight line in the x-direction. After it interacts with an invisible clump of dark matter with mass M_{DM}, the galaxy moves with 50.% of its initial speed in a straight line in a direction that is rotated by an angle θ from its initial velocity. Assume that initial and final velocities are given for positions where the galaxy is very far from the clump of dark matter, that the gravitational attraction can be neglected at those positions, and that the dark matter is initially at rest. Determine M_{DM} in terms of M_G, v_0, and θ.

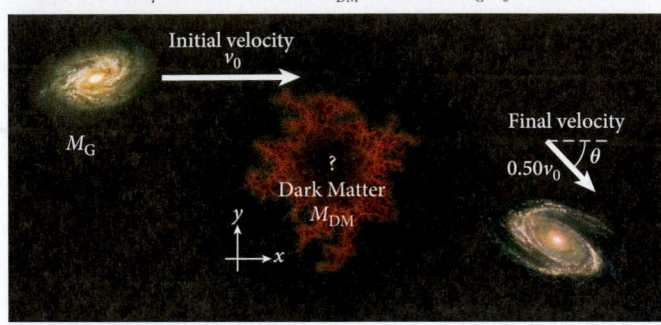

Section 7.6

7.61 A 1439-kg railroad car traveling at a speed of 12.0 m/s strikes an identical car at rest. If the cars lock together as a result of the collision, what is their common speed (in m/s) afterward?

7.62 Bats are extremely adept at catching insects in midair. If a 50.0-g bat flying in one direction at 8.00 m/s catches a 5.00-g insect flying in the opposite direction at 6.00 m/s, what is the speed of the bat immediately after catching the insect?

•**7.63** A small car of mass 1000. kg traveling at a speed of 33.0 m/s collides head on with a large car of mass 3000. kg traveling in the opposite direction at a speed of 30.0 m/s. The two cars stick together. The duration of the collision is 100. ms. What acceleration (in g) do the occupants of the small car experience? What acceleration (in g) do the occupants of the large car experience?

•**7.64** To determine the muzzle velocity of a bullet fired from a rifle, you shoot the 2.00-g bullet into a 2.00-kg wooden block. The block is suspended by wires from the ceiling and is initially at rest. After the bullet is embedded in the block, the block swings up to a maximum height of 0.500 cm above its initial position. What is the velocity of the bullet on leaving the gun's barrel?

•**7.65** A 2000.-kg Cadillac and a 1000.-kg Volkswagen meet at an intersection. The stoplight has just turned green, and the Cadillac, heading north, drives forward into the intersection. The Volkswagen, traveling east, fails to stop. The Volkswagen crashes into the left front fender of the Cadillac; then the cars stick together and slide to a halt. Officer Tom, responding to the accident, sees that the skid marks are directed 55.0° north of east from the point of impact. The driver of the Cadillac, who keeps a close eye on the speedometer, reports that he was traveling at 30.0 m/s when the accident occurred. How fast was the Volkswagen going just before the impact?

•**7.66** Two balls of equal mass collide and stick together as shown in the figure. The initial velocity of ball B is twice that of ball A.

a) Calculate the angle above the horizontal of the motion of mass A + B after the collision.

b) What is the ratio of the final velocity of the mass A + B to the initial velocity of ball A, v_f/v_A?

c) What is the ratio of the final energy of the system to the initial energy of the system, E_f/E_i?

••**7.67** Tarzan swings on a vine from a cliff to rescue Jane, who is standing on the ground surrounded by snakes. His plan is to push off the cliff, grab Jane at the lowest point of his swing, and carry them both to the safety of a nearby tree (see the figure). Tarzan's mass is 80.0 kg, Jane's mass is 40.0 kg, the height of the lowest limb of the target tree is 10.0 m, and Tarzan is initially standing on a cliff of height 20.0 m. The length of the vine is 30.0 m. With what speed should Tarzan push off the cliff if he and Jane are to make it to the tree limb successfully?

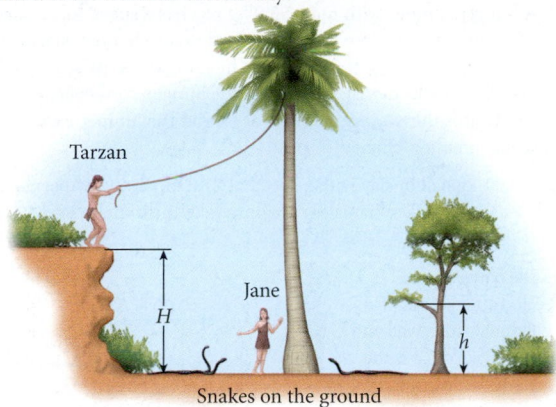

••**7.68** A 3000.-kg Cessna airplane flying north at 75.0 m/s at an altitude of 1600. m over the jungles of Brazil collided with a 7000.-kg cargo plane flying 35.0° north of west with a speed of 100. m/s. As measured from a point on the ground directly below the collision, the Cessna wreckage was found 1000. m away at an angle of 25.0° south of west, as shown in the figure. The cargo plane broke into two pieces. Rescuers located a 4000.-kg piece 1800. m away from the same point at an angle of 22.0° east of north. Where should they look for the other piece of the cargo plane? Give a distance and a direction from the point directly below the collision.

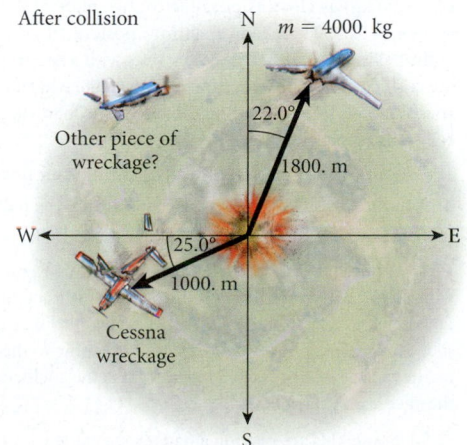

Section 7.7

7.69 The objects listed in the table are dropped from a height of 85 cm. The height they reach after bouncing has been recorded. Determine the coefficient of restitution for each object.

Object	H (cm)	h_1 (cm)	ϵ
Range golf ball	85.0	62.6	
Tennis ball	85.0	43.1	
Billiard ball	85.0	54.9	
Handball	85.0	48.1	
Wooden ball	85.0	30.9	
Steel ball bearing	85.0	30.3	
Glass marble	85.0	36.8	
Ball of rubber bands	85.0	58.3	
Hollow, hard plastic ball	85.0	40.2	

7.70 A golf ball is released from rest from a height of 0.811 m above the ground and has a collision with the ground, for which the coefficient of restitution is 0.601. What is the maximum height reached by this ball as it bounces back up after this collision?

•**7.71** A billiard ball of mass 0.162 kg has a speed of 1.91 m/s and collides with the side of the billiard table at an angle of 35.9° with respect to the normal. For this collision, the coefficient of restitution is 0.841. What is the angle relative to the side (in degrees) at which the ball moves away from the collision?

•**7.72** A soccer ball rolls out of a gym through the center of a doorway into the next room. The adjacent room is 6.00 m by 6.00 m with the 2.00-m wide doorway located at the center of the wall. The ball hits the center of a side wall at 45.0°. If the coefficient of restitution for the soccer ball is 0.850, does the ball bounce back out of the room? (Note that the ball rolls without slipping, so no energy is lost to the floor.)

•**7.73** In a *Tom and Jerry*™ cartoon, Tom the cat is chasing Jerry the mouse outside in a yard. The house is in a neighborhood where all the houses are exactly the same, each with the same size yard and the same 2.00-m-high fence around each yard. In the cartoon, Tom rolls Jerry into a ball and throws him over the fence. Jerry moves like a projectile, bounces at the center of the next yard, and continues to fly toward the next fence, which is 7.50 m away. If Jerry's original height above ground when he was thrown is 5.00 m, his original range is 15.0 m, and his coefficient of restitution is 0.80, does he make it over the next fence?

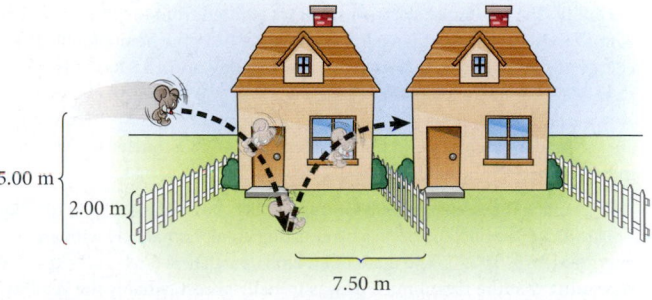

••**7.74** Two Sumo wrestlers are involved in an inelastic collision. The first wrestler, Hakurazan, has a mass of 135 kg and moves forward along the positive x-direction at a speed of 3.50 m/s. The second wrestler, Toyohibiki, has a mass of 173 kg and moves straight toward Hakurazan at a speed of 3.00 m/s. Immediately after the collision, Hakurazan is deflected to his right by 35.0° (see the figure). In the collision, 10.0% of the wrestlers' initial total kinetic energy is lost. What is the angle at which Toyohibiki is moving immediately after the collision?

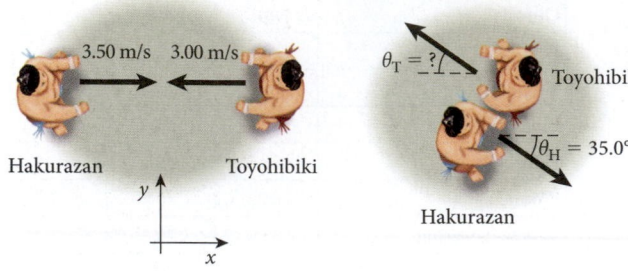

Immediately before collision   Immediately after collision

3.50 m/s   3.00 m/s

θ_T = ?   Toyohibiki

Hakurazan   Toyohibiki

θ_H = 35.0°

Hakurazan

••**7.75** A hockey puck ($m = 170.$ g and $v_0 = 2.00$ m/s) slides without friction on the ice and hits the rink board at 30.0° with respect to the normal. The puck bounces off the board at a 40.0° angle with respect to the normal. What is the coefficient of restitution for the puck? What is the ratio of the puck's final kinetic energy to its initial kinetic energy?

Additional Exercises

7.76 How fast would a 5.00-g fly have to be traveling to slow a 1900.-kg car traveling at 55.0 mph by 5.00 mph if the fly hit the car in a totally inelastic head-on collision?

7.77 Attempting to score a touchdown, an 85.0-kg tailback jumps over his blockers, achieving a horizontal speed of 8.90 m/s. He is met in midair just short of the goal line by a 110.-kg linebacker traveling in the opposite direction at a speed of 8.00 m/s. The linebacker grabs the tailback.

a) What is the speed of the entangled tailback and linebacker just after the collision?

b) Will the tailback score a touchdown (provided that no other player has a chance to get involved, of course)?

7.78 The nucleus of radioactive thorium-228, with a mass of about $3.78 \cdot 10^{-25}$ kg, is known to decay by emitting an alpha particle with a mass of about $6.64 \cdot 10^{-27}$ kg. If the alpha particle is emitted with a speed of $1.80 \cdot 10^7$ m/s, what is the recoil speed of the remaining nucleus (which is the nucleus of a radon atom)?

7.79 A 60.0-kg astronaut inside a 7.00-m-long space capsule of mass 500. kg is floating weightlessly on one end of the capsule. He kicks off the wall at a velocity of 3.50 m/s toward the other end of the capsule. How long does it take the astronaut to reach the far wall?

7.80 Moessbauer spectroscopy is a technique for studying molecules by looking at a particular atom within them. For example, Moessbauer measurements of iron (Fe) inside hemoglobin, the molecule responsible for transporting oxygen in the blood, can be used to determine the hemoglobin's flexibility. The technique starts with X-rays emitted from the nuclei of ^{57}Co atoms. These X-rays are then used to study the Fe in the hemoglobin. The energy and momentum of each X-ray are 14.0 keV and 14.0 keV/c (see Example 7.5 for an explanation of the units). A ^{57}Co nucleus recoils as an X-ray is emitted. A single ^{57}Co nucleus has a mass of $9.52 \cdot 10^{-26}$ kg. What are the final momentum and kinetic energy of the ^{57}Co nucleus? How do these compare to the values for the X-ray?

7.81 Assume the nucleus of a radon atom, ^{222}Rn, has a mass of $3.68 \cdot 10^{-25}$ kg. This radioactive nucleus decays by emitting an alpha particle with an energy of $8.79 \cdot 10^{-13}$ J. The mass of an alpha particle is $6.64 \cdot 10^{-27}$ kg. Assuming that the radon nucleus was initially at rest, what is the velocity of the nucleus that remains after the decay?

7.82 A skateboarder of mass 35.0 kg is riding her 3.50-kg skateboard at a speed of 5.00 m/s. She jumps backward off her skateboard, sending the skateboard forward at a speed of 8.50 m/s. At what speed is the skateboarder moving when her feet hit the ground?

7.83 During an ice-skating extravaganza, *Robin Hood on Ice*, a 50.0-kg archer is standing still on ice skates. Assume that the friction between the ice skates and the ice is negligible. The archer shoots a 0.100-kg arrow horizontally at a speed of 95.0 m/s. At what speed does the archer recoil?

7.84 Astronauts are playing catch on the International Space Station. One 55.0-kg astronaut, initially at rest, throws a baseball of mass 0.145 kg at a speed of 31.3 m/s. At what speed does the astronaut recoil?

7.85 A bungee jumper with mass 55.0 kg reaches a speed of 13.3 m/s moving straight down when the elastic cord tied to her feet starts pulling her back up. After 1.25 s, the jumper is heading back up at a speed of 10.5 m/s. What is the average force that the bungee cord exerts on the jumper? What is the average number of *g*'s that the jumper experiences during this direction change?

7.86 A 3.00-kg ball of clay with a speed of 21.0 m/s is thrown against a wall and sticks to the wall. What is the magnitude of the impulse exerted on the ball?

7.87 The figure shows before and after scenes of a cart colliding with a wall and bouncing back. What is the cart's change of momentum? (Assume that right is the positive direction in the coordinate system.)

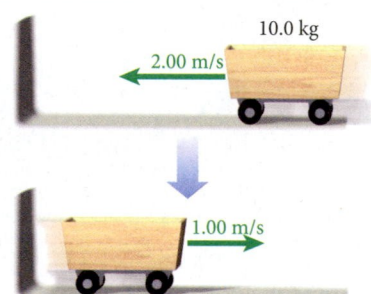

10.0 kg

2.00 m/s

1.00 m/s

7.88 Tennis champion Venus Williams is capable of serving a tennis ball at around 127 mph.

a) Assuming that her racquet is in contact with the 57.0-g ball for 0.250 s, what is the average force of the racquet on the ball?

b) What average force would an opponent's racquet have to exert in order to return Williams's serve at a speed of 50.0 mph, assuming that the opponent's racquet is also in contact with the ball for 0.250 s?

7.89 Three birds are flying in a compact formation. The first bird, with a mass of 100. g, is flying 35.0° east of north at a speed of 8.00 m/s. The second bird, with a mass of 123 g, is flying 2.00° east of north at a speed of 11.0 m/s. The third bird, with a mass of 112 g, is flying 22.0° west of north at a speed of 10.0 m/s. What is the momentum vector of the formation? What would be the speed and direction of a 115-g bird with the same momentum?

7.90 A golf ball of mass 45.0 g moving at a speed of 120. km/h collides head on with a French TGV high-speed train of mass $3.80 \cdot 10^5$ kg that is traveling at 300. km/h. Assuming that the collision is elastic, what is the speed of the golf ball after the collision? (Do not try to conduct this experiment!)

7.91 In bocce, the object of the game is to get your balls (each with mass $M = 1.00$ kg) as close as possible to the small white ball (the *pallina*, mass $m = 0.0450$ kg). Your first throw positioned your ball 2.00 m to the left of the pallina. If your next throw arrives with a speed of $v = 1.00$ m/s and the coefficient of kinetic friction is $\mu_k = 0.200$, what are the final distances of your two balls from the pallina in each of the following cases? Assume that collisions are elastic.

a) You throw your ball from the left, hitting your first ball.

b) You throw your ball from the right, hitting the pallina.

(*Hint:* Use the fact that $m \ll M$, and ignore ball diameters.)

•**7.92** A bored boy shoots a soft pellet from an air gun at a piece of cheese with mass 0.250 kg that sits, keeping cool for dinner guests, on a block of ice. On one particular shot, his 1.20-g pellet gets stuck in the cheese, causing it to slide 25.0 cm before coming to a stop. According to the package the gun came in, the muzzle velocity is 65.0 m/s. What is the coefficient of friction between the cheese and the ice?

•**7.93** Some kids are playing a dangerous game with fireworks. They strap several firecrackers to a toy rocket and launch it into the air at an angle of 60.0° with respect to the ground. At the top of its trajectory, the contraption explodes, and the rocket breaks into two equal pieces. One of the pieces has half the speed that the rocket had before it exploded and travels straight upward with respect to the ground. Determine the speed and direction of the second piece.

•**7.94** A ball with mass 0.265 kg is initially at rest and is kicked at an angle of 20.8° with respect to the horizontal. The ball travels through the air a horizontal distance of 52.8 m after it is kicked. What is the impulse received by the ball during the kick? Assume there is no air resistance.

•**7.95** Tarzan, King of the Jungle (mass = 70.4 kg), grabs a vine of length 14.5 m hanging from a tree branch. The angle of the vine was 25.9° with respect to the vertical when he grabbed it. At the lowest point of his trajectory, he picks up Jane (mass = 43.4 kg) and continues his swinging motion. What angle relative to the vertical will the vine have when Tarzan and Jane reach the highest point of their trajectory?

•**7.96** A bullet with mass 35.5 g is shot horizontally from a gun. The bullet embeds in a 5.90-kg block of wood that is suspended by strings. The combined mass swings upward, gaining a height of 12.85 cm. What was the speed of the bullet as it left the gun? (Air resistance can be ignored here.)

•**7.97** A 170.-g hockey puck moving in the positive x-direction at 30.0 m/s is struck by a stick at time $t = 2.00$ s and moves in the opposite direction at 25.0 m/s. If the puck is in contact with the stick for 0.200 s, plot the momentum and the position of the puck, and the force acting on it as a function of time, from 0 to 5.00 s. Be sure to label the coordinate axes with reasonable numbers.

•**7.98** Balls are sitting on a billiard table as shown in the figure. You are playing stripes, and your opponent is playing solids. Your plan is to hit your target ball by bouncing the white ball off the table bumper.

a) At what angle relative to the normal does the cue ball need to hit the bumper for a purely elastic collision?

b) As an ever observant player, you determine that in fact collisions between billiard balls and bumpers have a coefficient of restitution of 0.600. What angle do you now choose to make the shot?

•**7.99** You have dropped your cell phone behind a very long bookcase and cannot reach it from either the top or the sides. You decide to get the phone out by elastically colliding a set of keys with it so that both will slide out. If the mass of the cell phone is 0.111 kg, the key ring's mass is 0.020 kg, and each key's mass is 0.023 kg, what is the minimum number of keys you need on the ring so that both the keys and the cell phone will come out on the same side of the bookcase? If your key ring has five keys on it and the velocity is 1.21 m/s when it hits the cell phone, what are the final velocities of the cell phone and key ring? Assume the collision is one-dimensional and elastic, and neglect friction.

•**7.100** After several large firecrackers have been inserted into its holes, a bowling ball is projected into the air using a homemade launcher and explodes in midair. During the launch, the 7.00-kg ball is shot into the air with an initial speed of 10.0 m/s at a 40.0° angle; it explodes at the peak of its

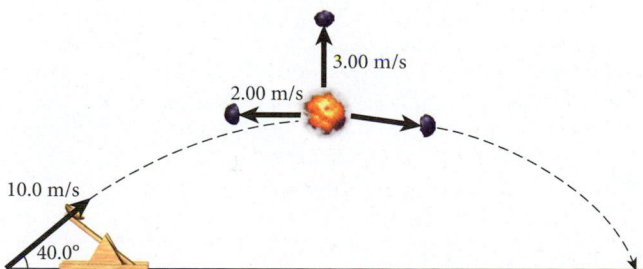

trajectory, breaking into three pieces of equal mass. One piece travels straight up with a speed of 3.00 m/s. Another piece travels straight back with a speed of 2.00 m/s. What is the velocity of the third piece (speed and direction)?

•**7.101** In waterskiing, a "garage sale" occurs when a skier loses control and falls and water skis fly in different directions. In one particular incident, a novice skier was skimming across the surface of the water at 22.0 m/s when he lost control. One ski, with a mass of 1.50 kg, flew off at an angle of 12.0° to the left of the initial direction of the skier with a speed of 25.0 m/s. The other identical ski flew from the crash at an angle of 5.00° to the right with a speed of 21.0 m/s. What was the velocity of the 61.0-kg skier? Give a speed and a direction relative to the initial velocity vector.

•**7.102** An uncovered hopper car from a freight train rolls without friction or air resistance along a level track at a constant speed of 6.70 m/s in the positive x-direction. The mass of the car is $1.18 \cdot 10^5$ kg.

a) As the car rolls, a monsoon rainstorm begins, and the car begins to collect water in its hopper (see the figure). What is the speed of the car after $1.62 \cdot 10^4$ kg of water collects in the car's hopper? Assume that the rain is falling vertically in the negative y-direction.

b) The rain stops, and a valve at the bottom of the hopper is opened to release the water. The speed of the car when the valve is opened is again 6.70 m/s in the positive x-direction (see the figure). The water drains out vertically in the negative y-direction. What is the speed of the car after all the water has drained out?

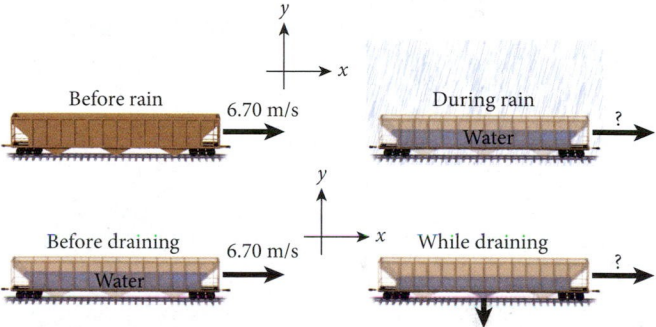

•**7.103** A rare isotope facility produces a beam of $7.25 \cdot 10^5$ nuclei per second of a rare isotope with mass $8.91 \cdot 10^{-26}$ kg. The nuclei are moving with 24.7% of the speed of light when they hit a beam stop (which is a block of materials that slows down the particles in the beam to zero speed). What is the magnitude of the average force that this beam exerts on the beam stop?

•**7.104** A student with a mass of 60.0 kg jumps straight up in the air by using her legs to apply an average force of 770. N to the ground for 0.250 s. Assume that the initial momentum of the student and the Earth are zero. What is the momentum of the student immediately after this impulse? What is the momentum of the Earth after this impulse? What is the speed of the Earth after the impulse? What fraction of the total kinetic energy that the student produces with her legs goes to the Earth (the mass of the Earth is $5.98 \cdot 10^{24}$ kg)? Using conservation of energy, how high does the student jump?

•**7.105** A potato cannon is used to launch a potato on a frozen lake, as shown in the figure. The mass of the cannon, m_c, is 10.0 kg, and the mass of the potato, m_p, is 0.850 kg. The cannon's spring (with spring constant $k_c = 7.06 \cdot 10^3$ N/m) is compressed 2.00 m. Prior to launching the potato, the cannon is at rest. The potato leaves the cannon's muzzle moving horizontally to the right at a speed of $v_p = 175$ m/s. Neglect the effects of the potato spinning. Assume there is no friction between the cannon and the lake's ice or between the cannon barrel and the potato.

a) What are the direction and magnitude of the cannon's velocity, v_c, after the potato leaves the muzzle?

b) What is the total mechanical energy (potential and kinetic) of the potato/cannon system before and after the firing of the potato?

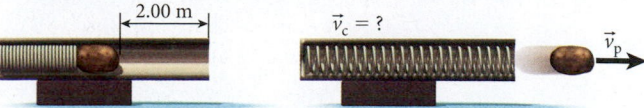

•**7.106** A potato cannon is used to launch a potato on a frozen lake, as in Exercise 7.105. All quantities are the same as in that problem, except the potato has a large diameter and is very rough, causing friction between the cannon barrel and the potato. The rough potato leaves the cannon's muzzle moving horizontally to the right at a speed of $v_p = 165$ m/s. Neglect the effects of the potato spinning.

a) What are the direction and magnitude of the cannon's velocity, v_c, after the rough potato leaves the muzzle?

b) What is the total mechanical energy (potential and kinetic) of the potato/cannon system before and after the firing of the potato?

c) What is the work, W_f, done by the force of friction on the rough potato?

•**7.107** A particle ($M_1 = 1.00$ kg) moving at 30.0° downward from the horizontal with $v_1 = 2.50$ m/s hits a second particle ($M_2 = 2.00$ kg), which is at rest. After the collision, the speed of M_1 is reduced to 0.500 m/s, and it is moving to the left and at an angle of 32.0° downward with respect to the horizontal. You cannot assume that the collision is elastic. What is the speed of M_2 after the collision?

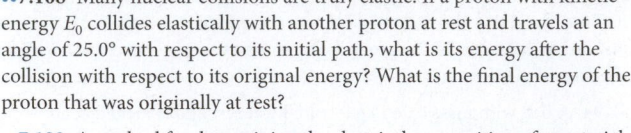

••**7.108** Many nuclear collisions are truly elastic. If a proton with kinetic energy E_0 collides elastically with another proton at rest and travels at an angle of 25.0° with respect to its initial path, what is its energy after the collision with respect to its original energy? What is the final energy of the proton that was originally at rest?

••**7.109** A method for determining the chemical composition of a material is Rutherford backscattering (RBS), named for the scientist who first discovered that an atom contains a high-density positively charged nucleus, rather than having positive charge distributed uniformly throughout (see Chapter 39). In RBS, alpha particles are shot straight at a target material, and the energy of the alpha particles that bounce directly back is measured. An alpha particle has a mass of $6.65 \cdot 10^{-27}$ kg. An alpha particle having an initial kinetic energy of 2.00 MeV collides elastically with atom X. If the backscattered alpha particle's kinetic energy is 1.59 MeV, what is the mass of atom X? Assume that atom X is initially at rest. You will need to find the square root of an expression, which will result in two possible answers (if $a = b^2$, then $b = \pm\sqrt{a}$). Since you know that atom X is more massive than the alpha particle, you can choose the correct root accordingly. What element is atom X? (Check a periodic table of elements, where atomic mass is listed as the mass in grams of 1 mol of atoms, which is $6.02 \cdot 10^{23}$ atoms.)

MULTI-VERSION EXERCISES

7.110 A Super Ball has a coefficient of restitution of 0.8887. If the ball is dropped from a height of 3.853 m above the floor, what maximum height will it reach on its third bounce?

7.111 A Super Ball has a coefficient of restitution of 0.9115. From what height should the ball be dropped so that its maximum height on its third bounce is 2.234 m?

7.112 A Super Ball is dropped from a height of 3.935 m. Its maximum height on its third bounce is 2.621 m. What is the coefficient of restitution of the ball?

7.113 Two gliders are moving on a horizontal frictionless air track. Glider 1 has mass $m_1 = 160.1$ g and is moving to the right (positive x-direction) with a speed of 2.723 m/s. Glider 2 has mass $m_2 = 354.1$ g and is moving to the left (negative x-direction) with a speed of 3.515 m/s. The gliders undergo a totally elastic collision. What is the velocity of glider 1 after the collision?

7.114 Two gliders are moving on a horizontal frictionless air track. Glider 1 has mass $m_1 = 176.3$ g and is moving to the right (positive x-direction) with a speed of 2.199 m/s. Glider 2 is moving to the left (negative x-direction) with a speed of 3.301 m/s. The gliders undergo a totally elastic collision. The velocity of glider 1 after the collision is -4.511 m/s. What is the mass of glider 2?

7.115 Two gliders are moving on a horizontal frictionless air track. Glider 1 is moving to the right (positive x-direction) with a speed of 2.277 m/s. Glider 2 has mass $m_2 = 277.3$ g and is moving to the left (negative x-direction) with a speed of 3.789 m/s. The gliders undergo a totally elastic collision. The velocity of glider 1 after the collision is -4.887 m/s. What is the mass of glider 1?

7.116 A racquetball of mass 41.05 g has a speed of 15.49 m/s and collides with the wall of the court at an angle of 43.53° relative to the normal to the wall. The coefficient of restitution of the racquetball is 0.8199. What is the angle relative to the normal at which the ball leaves the wall?

7.117 A racquetball of mass 41.97 g has a speed of 15.69 m/s and collides with the wall of the court at an angle of 48.67° relative to the normal to the wall. The racquetball leaves the wall at an angle of 55.75° relative to the normal to the wall. What is the coefficient of restitution of the ball?

7.118 A racquetball of mass 38.87 g has a speed of 15.89 m/s and collides with the wall of the court. The ball leaves the wall at an angle of 57.24° relative to the normal to the wall. The coefficient of restitution of the racquetball is 0.8787. What is the initial angle relative to the normal to the wall at which the ball hits the wall?

7.119 A 48.95-kg boy is playing rollerblade dodgeball. His rollerblades are frictionless, and he is initially at rest. A dodgeball with a mass of 511.1 g is thrown directly at him with a speed of 23.63 m/s. He catches the dodgeball. With what speed does the boy move after the catch?

7.120 A 53.53-kg boy is playing rollerblade dodgeball. His rollerblades are frictionless, and he is initially at rest. A dodgeball with a mass of 513.1 g is thrown directly at him, and he catches it. After catching the ball, the boy moves with a speed of 0.2304 m/s. With what speed was the dodgeball thrown?

7.121 A boy is playing rollerblade dodgeball. His rollerblades are frictionless, and he is initially at rest. A dodgeball with a mass of 515.1 g is thrown directly at him with a speed of 24.91 m/s. He catches the dodgeball and then moves with a speed of 0.2188 m/s. What is the mass of the boy?

8

Systems of Particles and Extended Objects

FIGURE 8.1 The International Space Station photographed from the Space Shuttle *Discovery*.

The International Space Station (ISS), shown in Figure 8.1, is a remarkable engineering achievement. It has been continuously inhabited since 2000. It orbits Earth at a speed of over 7.5 km/s, in an orbit ranging from 320 to 350 km above Earth's surface. When engineers track the ISS, they treat it as a point particle, even though it measures roughly 109 m by 73 m by 25 m. Presumably this point represents the center of the ISS, but how exactly do engineers determine where the center is?

Every object has a point where all the mass of the object can be considered to be concentrated. Sometimes this point, called the *center of mass,* is not even within the object. This chapter explains how to calculate the location of the center of mass and shows how to use it to simplify calculations involving conservation of momentum. We have been assuming in earlier chapters that objects could be treated as particles. This chapter shows why that assumption works.

This chapter also discusses changes in momentum for the situation where an object's mass varies as well as its velocity. This occurs with rocket propulsion, where the mass of fuel is often much greater than the mass of the rocket itself.

WHAT WE WILL LEARN

- The center of mass is the point at which we can imagine all the mass of an object to be concentrated.

- The position of the combined center of mass of two or more objects is found by taking the sum of their position vectors, weighted by their individual masses.

- The translational motion of the center of mass of an extended object can be described by Newtonian mechanics.

- The center-of-mass momentum is the sum of the linear momentum vectors of the parts of a system. Its time derivative is equal to the total net external force acting on the system, an extended formulation of Newton's Second Law.

- Analyses of rocket motion have to consider systems of varying mass. This variation leads to a logarithmic dependence of the velocity of the rocket on the ratio of initial to final mass.

- It is possible to calculate the location of the center of mass of an extended object by integrating its mass density over its entire volume, weighted by the coordinate vector, and then dividing by the total mass.

- If an object has a plane of symmetry, the center of mass lies in that plane. If the object has more than one symmetry plane, the center of mass lies on the line or point of intersection of the planes.

8.1 Center of Mass and Center of Gravity

So far, we have represented the location of an object by coordinates of a single point. However, a statement such as "a car is located at $x = 3.2$ m" surely does not mean that the entire car is located at that point. So, what does it mean to give the coordinate of one particular point to represent an extended object? Answers to this question depend on the particular application. In auto racing, for example, a car's location is represented by the coordinate of the frontmost part of the car. When this point crosses the finish line, the race is decided. On the other hand, in soccer, a goal is counted only if the entire ball has crossed the goal line; in this case, it makes sense to represent the soccer ball's location by the coordinates of the rearmost part of the ball. However, these examples are exceptions. In almost all situations, there is a natural choice of a point to represent the location of an extended object. This point is called the *center of mass*.

> **Definition**
>
> The **center of mass** is the point at which we can imagine all the mass of an object to be concentrated.

Thus, the center of mass is also the point at which we can imagine the force of gravity acting on the entire object to be concentrated. If we can imagine all of the mass to be concentrated at this point when calculating the force due to gravity, it is legitimate to call this point the *center of gravity*, a term that can often be used interchangeably with *center of mass*. (To be precise, we should note that these two terms are only equivalent in situations where the gravitational force is constant everywhere throughout the object. In Chapter 12, we will see that this is not the case for very large objects.)

It is appropriate to mention here that if an object's mass density is constant, the center of mass (center of gravity) is located in the geometrical center of the object. Thus, for most objects in everyday experience, it is a reasonable first guess that the center of gravity is the middle of the object. The derivations in this chapter will bear out this conjecture.

Combined Center of Mass for Two Objects

If we have two identical objects of equal mass and want to find the center of mass for the combination of the two, it is reasonable to assume from considerations of symmetry that the combined center of mass of this system lies exactly midway between the individual centers of mass of the two objects. If one of the two objects is more massive, then it is equally reasonable to assume that the center of mass for the combination is closer to that of the more massive one. Thus, we have a general formula for calculating the location of the center of mass, $\vec{R}$, for two masses m_1 and m_2 located at positions $\vec{r}_1$ and $\vec{r}_2$ in an arbitrary coordinate system (Figure 8.2):

$$\vec{R} = \frac{\vec{r}_1 m_1 + \vec{r}_2 m_2}{m_1 + m_2}. \tag{8.1}$$

This equation says that the center-of-mass position vector is an average of the position vectors of the individual objects, weighted by their mass. Such a definition is consistent with the empirical evidence we have just cited. For now, we will use this equation as an operating definition and gradually work out its consequences. Later in this chapter and in the following chapters, we will see additional reasons why this definition makes sense.

Note that we can immediately write vector equation 8.1 in Cartesian coordinates as follows:

$$X = \frac{x_1 m_1 + x_2 m_2}{m_1 + m_2}, \quad Y = \frac{y_1 m_1 + y_2 m_2}{m_1 + m_2}, \quad Z = \frac{z_1 m_1 + z_2 m_2}{m_1 + m_2}. \tag{8.2}$$

In Figure 8.2, the location of the center of mass lies exactly on the straight (dashed black) line that connects the two masses. Is this a general result—does the center of mass always lie on this line? If yes, why? If no, what is the special condition that is needed for this to be the case? The answer is that this is a general result for all two-body systems: The center of mass of such a system always lies on the connecting line between the two objects. To see this, we can place the origin of the coordinate system at one of the two masses in Figure 8.2, say m_1. (As we know, we can always shift the origin of a coordinate system without changing the physics results.) Using equation 8.1, we then see that $\vec{R} = \vec{r}_2 m_2 / (m_1 + m_2)$, because with this choice of coordinate system, we define $\vec{r}_1$ as zero. Thus, the two vectors $\vec{R}$ and $\vec{r}_2$ point in the same direction, but $\vec{R}$ is shorter by a factor of $m_2 / (m_1 + m_2) < 1$. This shows that $\vec{R}$ always lies on the straight line that connects the two masses.

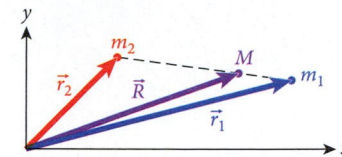

FIGURE 8.2 Location of the center of mass for a system of two masses m_1 and m_2, where $M = m_1 + m_2$.

Concept Check 8.1

In the case shown in Figure 8.2, what are the relative magnitudes of the two masses m_1 and m_2?

a) $m_1 < m_2$

b) $m_1 > m_2$

c) $m_1 = m_2$

d) Based solely on the information given in the figure, it is not possible to decide which of the two masses is larger.

Concept Check 8.2

A cylindrical bottle of oil-and-vinegar salad dressing whose volume is 1/2 vinegar (mass density of 1.01 g/cm³) and 1/2 oil (mass density of 0.910 g/cm³) rests on a table. Initially, the oil and the vinegar are separated, with the oil floating on top of the vinegar. The bottle is shaken so that the oil and vinegar mix uniformly and then returned to the table. How has the height of the center of mass of the salad dressing changed as a result of the mixing?

a) It is higher.

b) It is lower.

c) It is the same.

d) There is not enough information to answer this question.

SOLVED PROBLEM 8.1 Center of Mass of Earth and Moon

The Earth has a mass of $5.97 \cdot 10^{24}$ kg, and the Moon has a mass of $7.36 \cdot 10^{22}$ kg. The Moon orbits the Earth at a distance of 384,000 km; that is, the center of the Moon is a distance of 384,000 km from the center of Earth, as shown in Figure 8.3a.

PROBLEM
How far from the center of the Earth is the center of mass of the Earth-Moon system?

SOLUTION
THINK The center of mass of the Earth-Moon system can be calculated by taking the center of the Earth to be located at $x = 0$ and the center of the Moon to be located at $x = 384,000$ km. The center of mass of the Earth-Moon system will lie along a line connecting the center of the Earth and the center of the Moon (as in Figure 8.3a).

SKETCH A sketch showing Earth and Moon to scale is presented in Figure 8.3b.

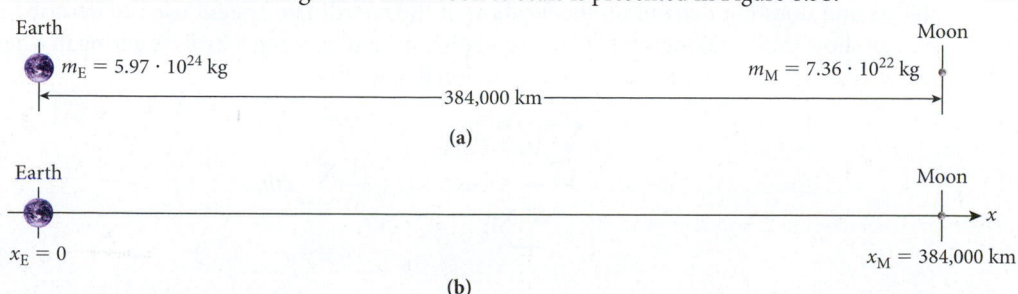

FIGURE 8.3 (a) The Moon orbits the Earth at a distance of 384,000 km (drawing to scale). (b) A sketch showing the Earth at $x_E = 0$ and the Moon at $x_M = 384,000$ km.

– Continued

RESEARCH We define an x-axis and place the Earth at $x_E = 0$ and the Moon at $x_M = 384,000$ km. We can use equation 8.2 to obtain an expression for the x-coordinate of the center of mass of the Earth-Moon system:

$$X = \frac{x_E m_E + x_M m_M}{m_E + m_M}.$$

SIMPLIFY Since we have put the origin of our coordinate system at the center of Earth, we set $x_E = 0$. This results in

$$X = \frac{x_M m_M}{m_E + m_M}.$$

CALCULATE Inserting the numerical values, we get the x-coordinate of the center of mass of the Earth-Moon system:

$$X = \frac{x_M m_M}{m_E + m_M} = \frac{(384,000 \text{ km})(7.36 \cdot 10^{22} \text{ kg})}{5.97 \cdot 10^{24} \text{ kg} + 7.36 \cdot 10^{22} \text{ kg}} = 4676.418 \text{ km}.$$

ROUND All of the numerical values were given to three significant figures, so we report our result as

$$X = 4680 \text{ km}.$$

DOUBLE-CHECK Our result is in kilometers, which is the correct unit for a position. The center of mass of the Earth-Moon system is close to the center of the Earth. This distance is small compared to the distance between the Earth and the Moon, which makes sense because the mass of the Earth is much larger than the mass of the Moon. In fact, this distance is less than the radius of the Earth, $R_E = 6370$ km. Both Earth and Moon actually orbit the common center of mass. Thus, the Earth seems to wobble as the Moon orbits it.

Combined Center of Mass for Several Objects

The definition of the center of mass in equation 8.1 can be generalized to a total of n objects with different masses, m_i, located at different positions, $\vec{r}_i$. In this general case,

$$\vec{R} = \frac{\vec{r}_1 m_1 + \vec{r}_2 m_2 + \cdots + \vec{r}_n m_n}{m_1 + m_2 + \cdots + m_n} = \frac{\sum_{i=1}^{n} \vec{r}_i m_i}{\sum_{i=1}^{n} m_i} = \frac{1}{M} \sum_{i=1}^{n} \vec{r}_i m_i, \tag{8.3}$$

where M represents the combined mass of all n objects:

$$M = \sum_{i=1}^{n} m_i. \tag{8.4}$$

Writing equation 8.3 in Cartesian components, we obtain

$$X = \frac{1}{M} \sum_{i=1}^{n} x_i m_i, \quad Y = \frac{1}{M} \sum_{i=1}^{n} y_i m_i, \quad Z = \frac{1}{M} \sum_{i=1}^{n} z_i m_i. \tag{8.5}$$

The location of the center of mass is a fixed point relative to the object or system of objects and does not depend on the location of the coordinate system used to describe it. We can show this by taking the system of equation 8.3 and moving it by $\vec{r}_0$, resulting in a new center-of-mass position, $\vec{R} + \vec{R}_0$. Using equation 8.3, we find

$$\vec{R} + \vec{R}_0 = \frac{\sum_{i=1}^{n} (\vec{r}_0 + \vec{r}_i) m_i}{\sum_{i=1}^{n} m_i} = \vec{r}_0 + \frac{1}{M} \sum_{i=1}^{n} \vec{r}_i m_i.$$

Thus, $\vec{R}_0 = \vec{r}_0$, and the location of the center of mass does not change relative to the system.

Now we can determine the center of mass of a collection of objects in the following example.

EXAMPLE 8.1 | Shipping Containers

Large freight containers, which can be transported by truck, railroad, or ship, come in standard sizes. One of the most common sizes is the ISO 20' container, which has a length of 6.1 m, a width of 2.4 m, and a height of 2.6 m. This container is allowed to have a mass (including its contents, of course) of up to 30,400 kg.

PROBLEM

The five freight containers shown in Figure 8.4 sit on the deck of a container ship. Each one has a mass of 9,000 kg, except for the red one, which has a mass of 18,000 kg. Assume that each of the containers has an individual center of mass at its geometric center. What are the x-coordinate and the y-coordinate of the containers' combined center of mass? Use the coordinate system shown in the figure to describe the location of this center of mass.

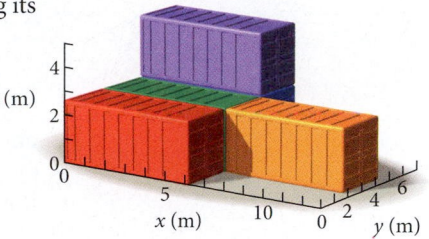

FIGURE 8.4 Freight containers arranged on the deck of a container ship.

SOLUTION

We need to calculate the individual Cartesian components of the center of mass, so we'll use equation 8.5. There does not seem to be a shortcut we can utilize.

Let's call the length of each container ℓ (6.1 m), the width of each container w (2.4 m), and the mass of the green container m_0 (9,000 kg). The mass of the red container is then $2m_0$, and all the others also have a mass of m_0.

First, we need to calculate the combined mass, M. According to equation 8.4, it is

$$M = m_{\text{red}} + m_{\text{green}} + m_{\text{orange}} + m_{\text{blue}} + m_{\text{purple}}$$

$$= 2m_0 + m_0 + m_0 + m_0 + m_0$$

$$= 6m_0.$$

For the x-coordinate of the combined center of mass, we find

$$X = \frac{x_{\text{red}} m_{\text{red}} + x_{\text{green}} m_{\text{green}} + x_{\text{orange}} m_{\text{orange}} + x_{\text{blue}} m_{\text{blue}} + x_{\text{purple}} m_{\text{purple}}}{M}$$

$$= \frac{\frac{1}{2}\ell 2m_0 + \frac{1}{2}\ell m_0 + \frac{3}{2}\ell m_0 + \frac{1}{2}\ell m_0 + \frac{1}{2}\ell m_0}{6m_0}$$

$$= \frac{\ell\left(1 + \frac{1}{2} + \frac{3}{2} + \frac{1}{2} + \frac{1}{2}\right)}{6}$$

$$= \frac{2}{3}\ell = 4.1 \text{ m}.$$

In the last step, we substituted the value of 6.1 m for ℓ.

In the same way, we can calculate the y-coordinate:

$$Y = \frac{y_{\text{red}} m_{\text{red}} + y_{\text{green}} m_{\text{green}} + y_{\text{orange}} m_{\text{orange}} + y_{\text{blue}} m_{\text{blue}} + y_{\text{purple}} m_{\text{purple}}}{M}$$

$$= \frac{\frac{1}{2}w 2m_0 + \frac{3}{2}wm_0 + \frac{3}{2}wm_0 + \frac{5}{2}wm_0 + \frac{5}{2}wm_0}{6m_0}$$

$$= \frac{w\left(1 + \frac{3}{2} + \frac{3}{2} + \frac{5}{2} + \frac{5}{2}\right)}{6}$$

$$= \frac{3}{2}w = 3.6 \text{ m}.$$

Here again we substituted the numerical value of 2.4 m in the last step. (Note that we rounded both center-of-mass coordinates to two significant figures to be consistent with the given values.)

Self-Test Opportunity 8.1

Determine the z-coordinate of the center of mass of the container arrangement in Figure 8.4.

8.2 Center-of-Mass Momentum

Now we can take the time derivative of the position vector of the center of mass to get $\vec{V}$, the velocity vector of the center of mass. We take the time derivative of equation 8.3:

$$\vec{V} \equiv \frac{d}{dt}\vec{R} = \frac{d}{dt}\left(\frac{1}{M}\sum_{i=1}^{n}\vec{r}_i m_i\right) = \frac{1}{M}\sum_{i=1}^{n} m_i \frac{d}{dt}\vec{r}_i = \frac{1}{M}\sum_{i=1}^{n} m_i \vec{v}_i = \frac{1}{M}\sum_{i=1}^{n}\vec{p}_i. \qquad (8.6)$$

For now, we have assumed that the total mass, M, and the masses, m_i, of the individual objects remain constant. (Later in this chapter, we will give up this assumption and study the consequences for rocket motion.) Equation 8.6 is an expression for the velocity vector of the center of mass, $\vec{V}$. Multiplication of both sides of equation 8.6 by M yields

$$\vec{P} = M\vec{V} = \sum_{i=1}^{n} \vec{p}_i. \tag{8.7}$$

We thus find that the center-of-mass momentum, $\vec{P}$, is the product of the total mass, M, and the center-of-mass velocity, $\vec{V}$, and is the sum of all the individual momentum vectors.

Taking the time derivative of both sides of equation 8.7 yields Newton's Second Law for the center of mass:

$$\frac{d}{dt}\vec{P} = \frac{d}{dt}(M\vec{V}) = \frac{d}{dt}\left(\sum_{i=1}^{n} \vec{p}_i\right) = \sum_{i=1}^{n} \frac{d}{dt}\vec{p}_i = \sum_{i=1}^{n} \vec{F}_i. \tag{8.8}$$

In the last step, we used the result from Chapter 7 that the time derivative of the momentum of particle i is equal to the net force, $\vec{F}_i$, acting on it. Note that if the particles (objects) in a system exert forces on one another, those forces do not make a net contribution to the sum of forces in equation 8.8. Why? According to Newton's Third Law, the forces that two objects exert on each other are equal in magnitude and opposite in direction. Therefore, adding them yields zero. Thus, we obtain Newton's Second Law for the center of mass:

$$\frac{d}{dt}\vec{P} = \vec{F}_{\text{net}}, \tag{8.9}$$

where $\vec{F}_{\text{net}}$ is the sum of all *external* forces acting on the system of particles.

The center of mass has the same relationships among position, velocity, momentum, force, and mass that have been established for point particles. It is thus possible to consider the center of mass of an extended object or a group of objects as a point particle. This conclusion justifies the approximation we used in earlier chapters of representing objects as points.

Recoil

When a bullet is fired from a gun, the gun **recoils;** that is, it moves in the direction opposite to that in which the bullet is fired. Another demonstration of the same physical principle occurs if you are sitting in a boat that is at rest and you throw an object off the boat: The boat moves in the direction opposite from that of the object. You also experience the same effect if you stand on a skateboard and toss a (reasonably heavy) ball. This well-known recoil effect is a consequence of Newton's Third Law.

SOLVED PROBLEM 8.2 | Cannon Recoil

Suppose a cannonball of mass 13.7 kg is fired at a target that is 2.30 km away from the cannon, which has a mass of 249.0 kg. The distance 2.30 km is also the maximum range of the cannon. The target and cannon are at the same elevation, and the cannon is resting on a horizontal surface.

PROBLEM
What is the velocity with which the cannon will recoil?

SOLUTION
THINK   First, we realize that the cannon can recoil only in the horizontal direction, because the normal force exerted by the ground will prevent it from acquiring a downward velocity component. We use the fact that the x-component of the center-of-mass momentum of the system (cannon and cannonball) remains unchanged in the process of firing the cannon, because the explosion of the gunpowder inside the cannon, which sets the cannonball in motion, creates only forces internal to the system. No net external force component occurs in the horizontal direction because the two external forces (normal force and gravity) are both

vertical. The y-component of the center-of-mass velocity changes because a net external force component does occur in the y-direction when the normal force increases to prevent the cannon from penetrating the ground. Because the cannonball and cannon are both initially at rest, the center-of-mass momentum of this system is initially zero, and its x-component remains zero after the firing of the cannon.

SKETCH Figure 8.5a is a sketch of the cannon just as the cannonball is fired. Figure 8.5b shows the velocity vector of the cannonball, $\vec{v}_2$, including the x- and y-components.

RESEARCH Using equation 8.7 with index 1 for the cannon and index 2 for the cannonball, we obtain

$$\vec{P} = \vec{p}_1 + \vec{p}_2 = m_1\vec{v}_1 + m_2\vec{v}_2 = 0 \Rightarrow \vec{v}_1 = -\frac{m_2}{m_1}\vec{v}_2.$$

For the horizontal component of the velocity, we then have

$$v_{1,x} = -\frac{m_2}{m_1}v_{2,x}. \qquad (i)$$

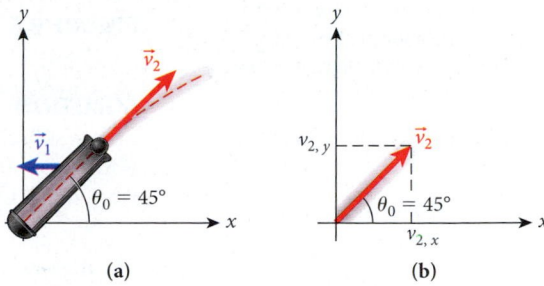

FIGURE 8.5 (a) Cannonball being fired from a cannon. (b) The initial velocity vector of the cannonball.

We can obtain the horizontal component of the cannonball's initial velocity (at firing) from the fact that the range of the cannon is 2.30 km. In Chapter 3, we saw that the range of the cannon is related to the initial velocity via $R = (v_0^2/g)(\sin 2\theta_0)$. The maximum range is reached for $\theta_0 = 45°$ and is $R = v_0^2/g \Rightarrow v_0 = \sqrt{gR}$. For $\theta_0 = 45°$, the initial speed and horizontal velocity component are related via $v_{2,x} = v_0\cos 45° = v_0/\sqrt{2}$. Combining these two results, we can relate the maximum range to the horizontal component of the initial velocity of the cannonball:

$$v_{2,x} = \frac{v_0}{\sqrt{2}} = \sqrt{\frac{gR}{2}}. \qquad (ii)$$

SIMPLIFY Substituting from equation (ii) into equation (i) gives us the result we are looking for

$$v_{1,x} = -\frac{m_2}{m_1}v_{2,x} = -\frac{m_2}{m_1}\sqrt{\frac{gR}{2}}.$$

CALCULATE Inserting the numbers given in the problem statement, we obtain

$$v_{1,x} = -\frac{m_2}{m_1}\sqrt{\frac{gR}{2}} = -\frac{13.7 \text{ kg}}{249 \text{ kg}}\sqrt{\frac{(9.81 \text{ m/s}^2)(2.30\cdot10^3 \text{ m})}{2}} = -5.84392 \text{ m/s}.$$

ROUND Expressing our answer to three significant figures gives

$$v_{1,x} = -5.84 \text{ m/s}.$$

DOUBLE-CHECK The minus sign means that the cannon moves in the opposite direction to that of the cannonball, which is reasonable. The cannonball should have a much larger initial velocity than the cannon because the cannon is much more massive. The initial velocity of the cannonball was

$$v_0 = \sqrt{gR} = \sqrt{(9.81 \text{ m/s}^2)(2.3\cdot10^3 \text{ m})} = 150 \text{ m/s}.$$

The fact that our answer for the velocity of the cannon is much less than the initial velocity of the cannonball also seems reasonable.

Mass can be ejected continuously from a system, producing a continuous recoil. As an example, let's consider the spraying of water from a fire hose.

EXAMPLE 8.2 | Fire Hose

PROBLEM

What is the magnitude of the force, F, that acts on a firefighter holding a fire hose that ejects 360. L of water per minute with a muzzle speed of $v = 39.0$ m/s, as shown in Figure 8.6?

– Continued

Concept Check 8.3

The fire hose of Example 8.2 is used to spray firefighting foam (with a mass density half that of water) at the same flow rate of 360. L/min and at the same muzzle speed of 39.0 m/s. The force acting on a firefighter holding the hose in this case is

a) four times the force found in Example 8.2.

b) twice the force found in Example 8.2.

c) the same as the force found in Example 8.2.

d) half the force found in Example 8.2.

e) a quarter of the force found in Example 8.2.

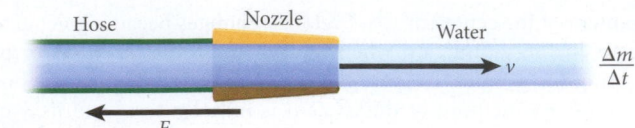

FIGURE 8.6   A fire hose with water leaving at speed v.

SOLUTION

Let's first find the total mass of the water that is being ejected per minute. The mass density of water is $\rho = 1000.$ kg/m^3 = 1.000 kg/L. Because $\Delta V = 360.$ L, we get for the total mass of water ejected in a minute:

$$\Delta m = \Delta V \rho = (360. \text{ L})(1.000 \text{ kg/L}) = 360. \text{ kg}.$$

The momentum of the water is then $\Delta p = v\Delta m$, and, from the definition of the average force, $F = \Delta p / \Delta t$, we have

$$F = \frac{v\Delta m}{\Delta t} = \frac{(39.0 \text{ m/s})(360. \text{ kg})}{60 \text{ s}} = 234 \text{ N}.$$

This force is sizable, which is why it is so dangerous for firefighters to let go of operating fire hoses: They would whip around, potentially causing injury.

General Motion of the Center of Mass

Extended solid objects can have motions that appear, at first sight, rather complicated. One example of such motion is high jumping. During the 1968 Olympic Games in Mexico City, the American track-and-field star Dick Fosbury won a gold medal using a new high-jump technique, which became known as the Fosbury flop (see Figure 8.7). Properly executed, the technique allows the athlete to cross over the bar while his or her center of mass remains below it, thus adding effective height to the jump.

Figure 8.8a shows a wrench twirling through the air, in a multiple-exposure series of images with equal time intervals between sequential frames. While this motion looks complicated, we can use what we know about the center of mass to perform a straightforward analysis of this motion. If we assume that all the mass of the wrench is concentrated at a point, then this point will move on a parabola through the air under the influence of gravity, as discussed in Chapter 3. Superimposed on this motion is a rotation of the wrench about its center of mass. You can see this parabolic trajectory clearly in Figure 8.8b, where a superimposed parabola (green) passes through the location of the center of mass of the wrench in each exposure. In addition, a superimposed black line rotates with a constant rate about the center of mass of the wrench. You can clearly see that the handle of the wrench is always aligned with the black line, indicating that the wrench rotates with constant rate about its center of mass (we will analyze such rotational motion in Chapter 10).

The techniques introduced here allow us to analyze many kinds of complicated problems involving moving solid objects in terms of superposition of the motion of the center of mass and a rotation of the object about the center of mass.

FIGURE 8.7   Dick Fosbury clears the high-jump bar during the finals of the Olympic Games in Mexico City on October 20, 1968.

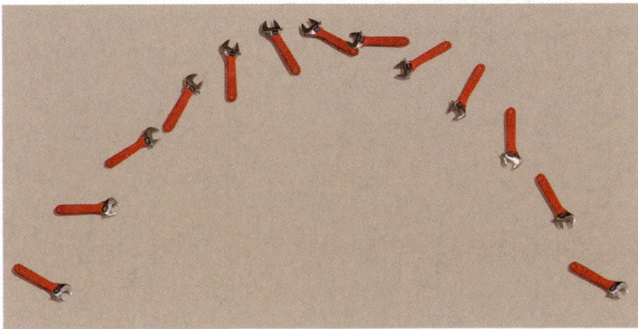

(a)

(b)

FIGURE 8.8   (a) Digitally processed multiple-exposure series of images of a wrench tossed through the air. (b) Same series as in part (a), but with a parabola for the center-of-mass motion superimposed.

8.3 Rocket Motion

Example 8.2 about the fire hose is the first situation we have examined that involves a change in momentum due to a change in mass rather than in velocity. Another important situation in which changing momentum is due to changing mass is rocket motion, where part of the mass of the rocket is ejected through a nozzle or nozzles at the rear (Figure 8.9). Rocket motion is an important case of the recoil effect discussed in Section 8.2. A rocket does not "push against" anything. Instead, its forward thrust is gained from ejecting its propellant from its rear, according to the law of conservation of total momentum.

In order to obtain an expression for the acceleration of a rocket, we'll first consider ejecting discrete amounts of mass out of the rocket. Then we can approach the continuum limit. Let's use a toy model of a rocket that moves in interstellar space, propelling itself forward by shooting cannonballs out its back end (Figure 8.10). (We specify that the rocket is in interstellar space so we can treat it and its components as an isolated system, for which we can neglect outside forces.) Initially, the rocket is at rest. All motion is in the x-direction, so we can use notation for one-dimensional motion, with the signs of the x-components of the velocities (which, for simplicity, we will refer to as velocities) indicating their direction. Each cannonball has a mass of Δm, and the initial mass of the rocket, including all cannonballs, is m_0. Each cannonball is fired with a speed of v_c relative to the common center of mass of the rocket and the cannonballs, resulting in a cannonball momentum of $v_c \Delta m$.

After the first cannonball is fired, the mass of the rocket is reduced to $m_0 - \Delta m$. Firing the cannonball does not change the center-of-mass momentum of the system (rocket plus cannonball). (Remember, this is an isolated system, on which no net external forces act.) Thus, the rocket receives a recoil momentum opposite to that of the cannonball. The momentum of the cannonball is

FIGURE 8.9 A Delta II rocket lifting a GPS satellite into orbit.

$$p_c = v_c \Delta m, \tag{8.10}$$

and the momentum of the rocket is

$$p_r = (m_0 - \Delta m)v_1,$$

where v_1 is the velocity of the rocket after the cannonball is fired. Because momentum is conserved, we can write $p_r + p_c = 0$, and then substitute p_r and p_c from the preceding two expressions:

$$(m_0 - \Delta m)v_1 + v_c \Delta m = 0.$$

We define the change in the velocity, Δv_1, of the rocket after firing one cannonball by

$$v_1 = v_0 + \Delta v = 0 + \Delta v = \Delta v_1, \tag{8.11}$$

where the assumption that the rocket was initially at rest means $v_0 = 0$. This gives us the recoil velocity of the rocket due to the firing of one cannonball:

$$\Delta v_1 = -\frac{v_c \Delta m}{m_0 - \Delta m}. \tag{8.12}$$

In the moving system of the rocket, we can then fire the second cannonball. Firing the second cannonball reduces the mass of the rocket from $m_0 - \Delta m$ to $m_0 - 2\Delta m$, which results in an additional recoil velocity of

$$\Delta v_2 = -\frac{v_c \Delta m}{m_0 - 2\Delta m}.$$

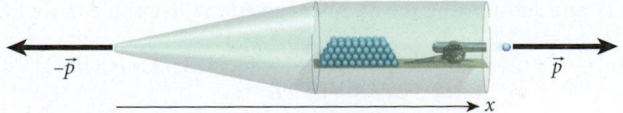

FIGURE 8.10 Toy model for rocket propulsion: firing cannonballs.

The total velocity of the rocket then increases to $v_2 = v_1 + \Delta v_2$. After firing the nth cannonball, the velocity change is

$$\Delta v_n = -\frac{v_c \Delta m}{m_0 - n\Delta m}. \tag{8.13}$$

Thus, the velocity of the rocket after firing the nth cannonball is

$$v_n = v_{n-1} + \Delta v_n.$$

This kind of equation, which defined the nth term of a sequence where each term is expressed as a function of the preceding terms, is called a *recursion relation*. It can be solved in a straightforward manner by using a computer. However, we can use a very helpful approximation for the case where the mass emitted per unit time is constant and small compared to *m,* the overall (time-dependent) mass of the rocket. In this limit, we obtain from equation 8.13

$$\Delta v = -\frac{v_c \Delta m}{m} \Rightarrow \frac{\Delta v}{\Delta m} = -\frac{v_c}{m}. \tag{8.14}$$

Here v_c is the speed with which the cannonball is ejected. In the limit $\Delta m \to 0$, we then obtain the derivative

$$\frac{dv}{dm} = -\frac{v_c}{m}. \tag{8.15}$$

The solution of this differential equation is

$$v(m) = -v_c \int_{m_0}^{m} \frac{1}{m'} dm' = -v_c \ln m' \Big|_{m_0}^{m} = v_c \ln\left(\frac{m_0}{m}\right). \tag{8.16}$$

(You can verify that equation 8.16 is indeed the solution of equation 8.15 by taking the derivative of equation 8.16 with respect to *m.*)

If m_i is the initial value for the total mass at some time t_i and m_f is the final mass at a later time, we can use equation 8.16 to obtain $v_i = v_c \ln(m_0/m_i)$ and $v_f = v_c \ln(m_0/m_f)$ for the initial and final velocities of the rocket. Then, using the property of logarithms, $\ln(a/b) = \ln a - \ln b$, we find the difference in those two velocities:

$$v_f - v_i = v_c \ln\left(\frac{m_0}{m_f}\right) - v_c \ln\left(\frac{m_0}{m_i}\right) = v_c \ln\left(\frac{m_i}{m_f}\right). \tag{8.17}$$

EXAMPLE 8.3 | Rocket Launch to Mars

One proposed scheme for sending astronauts to Mars involves assembling a spaceship in orbit around Earth, thus avoiding the need for the spaceship to overcome most of Earth's gravity at the start. Suppose such a spaceship has a payload of 50,000 kg, carries 2,000,000 kg of fuel, and is able to eject the propellant with a speed of 23.5 km/s. (Current chemical rocket propellants yield a maximum speed of approximately 5 km/s, but electromagnetic rocket propulsion is predicted to yield a speed of perhaps 40 km/s.)

PROBLEM
What is the final speed that this spaceship can reach, relative to the velocity it initially had in its orbit around Earth?

SOLUTION
Using equation 8.17 and substituting the numbers given in this problem, we find

$$v_f - v_i = v_c \ln\left(\frac{m_i}{m_f}\right) = (23.5 \text{ km/s}) \ln\left(\frac{2{,}050{,}000 \text{ kg}}{50{,}000 \text{ kg}}\right) = (23.5 \text{ km/s})(\ln 41) = 87.3 \text{ km/s}.$$

For comparison, the Saturn V multistage rocket that carried astronauts to the Moon in the late 1960s and early 1970s was able to reach a speed of only about 12 km/s.

However, even with advanced technology such as electromagnetic propulsion, it would still take several months for astronauts to reach Mars, even under the most favorable conditions. The *Mars Rover,* for example, took 207 days to travel from Earth to Mars. NASA estimates that astronauts on such a mission would receive approximately 10 to 20 times more radiation than the maximum allowable annual dose for radiation workers, leading to high probabilities of developing cancer and brain damage. No shielding mechanism has yet been proposed that could protect the astronauts from this danger.

Another and perhaps easier way to think of rocket motion is to go back to the definition of momentum as the product of mass and velocity and take the time derivative to obtain the force. However, now the mass of the object can change as well:

$$\vec{F}_{net} = \frac{d}{dt}\vec{p} = \frac{d}{dt}(m\vec{v}) = m\frac{d\vec{v}}{dt} + \vec{v}\frac{dm}{dt}.$$

(The last step in this equation represents the application of the product rule of differentiation from calculus.) If no external force is acting on an object ($\vec{F}_{net} = 0$), then we obtain

$$m\frac{d\vec{v}}{dt} = -\vec{v}\frac{dm}{dt}.$$

In the case of rocket motion (as illustrated in Figure 8.11), the outflow of propellant, dm/dt, is constant and creates the change in mass of the rocket. The propellant moves with a constant velocity, $\vec{v}_c$, relative to the rocket, so we obtain

$$m\frac{d\vec{v}}{dt} = m\vec{a} = -\vec{v}_c\frac{dm}{dt}.$$

The combination $v_c(dm/dt)$ is called the **thrust** of the rocket. It is a force and thus is measured in newtons:

$$\vec{F}_{thrust} = -\vec{v}_c\frac{dm}{dt}. \tag{8.18}$$

The thrust generated by space shuttle rocket engines and boosters was approximately 30.4 MN (30.4 meganewtons, or approximately 7.0 million pounds). The initial total mass of a space shuttle, including payload, fuel tanks, and rocket fuel, was slightly greater than 2.0 million kg; thus, the shuttle's rocket engines and boosters were able to produce an initial acceleration of

$$a = \frac{F_{net}}{m} = \frac{3.04 \cdot 10^7 \text{ N}}{2.0 \cdot 10^6 \text{ kg}} = 15 \text{ m/s}^2.$$

This acceleration is sufficient to lift the shuttle off the launch pad against the acceleration of gravity (-9.81 m/s^2). Once the shuttle rises and its mass decreases, it can generate a larger acceleration. As the fuel is expended, the main engines are throttled back to make sure that the acceleration does not exceed 3g (three times gravitational acceleration) in order to avoid damaging the cargo or injuring the astronauts.

FIGURE 8.11 Rocket motion.

SOLVED PROBLEM 8.3 Thruster Firing

PROBLEM

Suppose a spacecraft has an initial mass of 1,850,000 kg. Without its propellant, the spacecraft has a mass of 50,000 kg. The rocket that powers the spacecraft is designed to eject the propellant with a speed of 25 km/s with respect to the rocket at a constant rate of 15,000 kg/s. The spacecraft is initially at rest in space and travels in a straight line. How far will the spacecraft travel before its rocket uses all the propellant and shuts down?

SOLUTION

THINK   The total mass of propellant is the total mass of the spacecraft minus the mass of the spacecraft after all the propellant is ejected. The rocket ejects the propellant at a fixed rate, so we can calculate the amount of time during which the rocket operates. As the propellant is used up, the mass of the spacecraft decreases and the speed of the spacecraft increases. If the spacecraft starts from rest, the speed $v(t)$ at any time while the rocket is operating can be obtained from equation 8.17, with the final mass of the spacecraft replaced by the mass of the spacecraft at that time. The distance traveled before all the propellant is used is given by the integral of the speed as a function of time.

SKETCH   The flight of the spacecraft is sketched in Figure 8.12.

FIGURE 8.12 The various parameters for the spacecraft as the rocket operates.

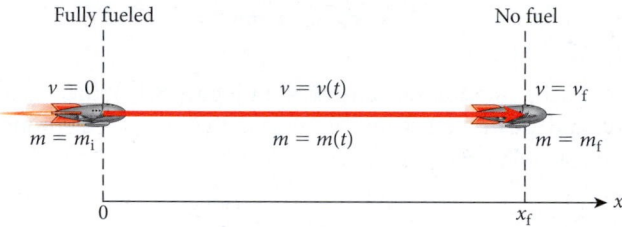

RESEARCH   We symbolize the rate at which the propellant is ejected by r_p. The time t_{max} during which the rocket will operate is then given by

$$t_{max} = \frac{(m_i - m_f)}{r_p},$$

where m_i is the initial mass of the spacecraft and m_f is the mass of the spacecraft after all the propellant is ejected. The total distance the spacecraft travels in this time interval is the integral of the speed over time:

$$x_f = \int_0^{t_{max}} v(t)\,dt. \tag{i}$$

While the rocket is operating, the mass of the spacecraft at a time t is given by

$$m(t) = m_i - r_p t.$$

The speed of the spacecraft at any given time after the rocket starts to operate and before all the propellant is used up is given by (compare to equation 8.17)

$$v(t) = v_c \ln\left(\frac{m_i}{m(t)}\right) = v_c \ln\left(\frac{m_i}{m_i - r_p t}\right) = v_c \ln\left(\frac{1}{1 - r_p t / m_i}\right), \tag{ii}$$

where v_c is the speed of the ejected propellant with respect to the rocket.

SIMPLIFY   Now we substitute from equation (ii) for the time dependence of the speed of the spacecraft into equation (i) and obtain

$$x_f = \int_0^{t_{max}} v(t)\,dt = \int_0^{t_{max}} v_c \ln\left(\frac{1}{1 - r_p t / m_i}\right) dt = -v_c \int_0^{t_{max}} \ln\left(1 - \frac{r_p t}{m_i}\right) dt. \tag{iii}$$

Because $\int \ln(1-ax)\,dx = \dfrac{ax-1}{a}\ln(1-ax)-x$ (you can look up this result in an integral table), the integral evaluates to

$$\int_0^{t_{max}} \ln(1-r_p t/m_i)\,dt = \left[\left(\frac{r_p t/m_i-1}{r_p/m_i}\right)\ln(1-r_p t/m_i)-t\right]_0^{t_{max}}$$

$$= \left(\frac{r_p t_{max}/m_i-1}{r_p/m_i}\right)\ln(1-r_p t_{max}/m_i)-t_{max}$$

$$= \left(t_{max}-m_i/r_p\right)\ln(1-r_p t_{max}/m_i)-t_{max}.$$

The distance traveled is then

$$x_f = -v_c\left[\left(t_{max}-m_i/r_p\right)\ln\left(1-r_p t_{max}/m_i\right)-t_{max}\right].$$

CALCULATE The time during which the rocket is operating is

$$t_{max} = \frac{m_i-m_f}{r_p} = \frac{1{,}850{,}000 \text{ kg}-50{,}000 \text{ kg}}{15{,}000 \text{ kg/s}} = 120 \text{ s}.$$

Putting numerical values into the factor $1-r_p t_{max}/m_i$ gives

$$1-\frac{r_p t_{max}}{m_i} = 1-\frac{15{,}000 \text{ kg/s}\cdot 120 \text{ s}}{1{,}850{,}000 \text{ kg}} = 0.027027.$$

Thus, we find for the distance traveled

$$x_f = -\left(25\cdot 10^3 \text{ m/s}\right)\left[-(120 \text{ s})+\left\{(120 \text{ s})-\left(1.85\cdot 10^6 \text{ kg}\right)/\left(15\cdot 10^3 \text{ kg/s}\right)\right\}\ln(0.027027)\right]$$

$$= 2.69909\cdot 10^6 \text{ m}.$$

ROUND Because the propellant speed was given to only two significant figures, we need to round to that accuracy:

$$x_f = 2.7\cdot 10^6 \text{ m}.$$

DOUBLE-CHECK To double-check our answer for the distance traveled, we use equation 8.17 to calculate the final velocity of the spacecraft:

$$v_f = v_c \ln\left(\frac{m_i}{m_f}\right) = (25 \text{ km/s})\ln\left(\frac{1.85\cdot 10^6 \text{ kg}}{5\cdot 10^4 \text{ kg}}\right) = 90.3 \text{ km/s}.$$

If the spacecraft accelerated at a constant rate, the speed would increase linearly with time, as shown in Figure 8.13, and the average speed during the time the propellant was being ejected would be $\bar v = v_f/2$. Taking this average speed and multiplying by that time gives

$$x_{a\text{-const}} \approx \bar v t_{max} = (v_f/2)t_{max} = (90.3 \text{ km/s})(120 \text{ s})/2 = 5.4\cdot 10^6 \text{ m}.$$

This approximate distance is bigger than our calculated answer, because in the calculation the velocity increases in time until it reaches the value of 90.3 km/s. The approximation is about twice the calculated distance, giving us confidence that our answer at least has the right order of magnitude.

Figure 8.13 shows the exact solution for $v(t)$ (red curve). The distance traveled, x_f, is the area under the red curve. The blue line shows the case where constant acceleration leads to the same final velocity. As you can see, the area under the blue line is approximately twice that under the red curve. Since we just calculated the area under the blue line, $x_{a\text{-const}}$, and found it to be about twice as big as our calculated result, we gain confidence that we interpreted correctly.

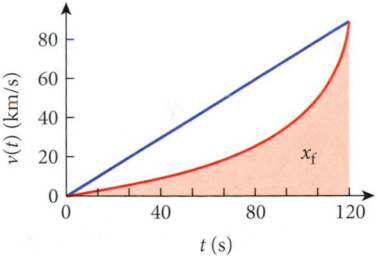

FIGURE 8.13 Comparison of the exact solution for $v(t)$ (red curve) to one for constant acceleration (blue line).

8.4 Calculating the Center of Mass

So far, we have not addressed a key question: How do we calculate the location of the center of mass for an arbitrarily shaped object? To answer this question, let's find the location of the center of mass of the hammer shown in Figure 8.14. To do this, we can represent the

FIGURE 8.14 Calculating the center of mass for a hammer.

Concept Check 8.5

If we have an object with variable mass density $\rho(\vec{r})$ and center-of-mass coordinate $\vec{R}$ and we replace it with an object of exactly the same shape but with a variable mass density that is twice as high as in the first object at every point, then the new center-of-mass coordinate is

a) $\vec{R}$.

b) $2\vec{R}$.

c) $\vec{R}/2$.

d) any of the above, depending on the shape of the objects.

hammer by small identically sized cubes, as shown in the lower part of the figure. The centers of the cubes are their individual centers of mass, marked with red dots. The red arrows are the position vectors of the cubes. If we accept the collection of cubes as a good approximation for the hammer, we can use equation 8.3 to find the center of mass of the collection of cubes and thus that of the hammer.

Note that not all the cubes have the same mass, because the densities of the wooden handle and the iron head are very different. The relationship between mass density (ρ), mass, and volume is given by

$$\rho = \frac{dm}{dV}. \tag{8.19}$$

If the mass density is uniform throughout an object, we simply have

$$\rho = \frac{M}{V} \text{ (for constant } \rho). \tag{8.20}$$

We can then use the mass density and rewrite equation 8.3:

$$\vec{R} = \frac{1}{M}\sum_{i=1}^{n}\vec{r}_i m_i = \frac{1}{M}\sum_{i=1}^{n}\vec{r}_i \rho(\vec{r}_i)V.$$

Here we have assumed that the mass density of each small cube is uniform (but still possibly different from one cube to another) and that each cube has the same (small) volume, V.

We can obtain a better and better approximation by shrinking the volume of each cube and using a larger and larger number of cubes. This procedure should look very familiar to you, because it is exactly what is done in calculus to arrive at the limit for an integral. In this limit, we obtain for the location of the center of mass for an arbitrarily shaped object:

$$\vec{R} = \frac{1}{M}\int_V \vec{r}\rho(\vec{r})\,dV. \tag{8.21}$$

Here the three-dimensional volume integral extends over the entire volume of the object under consideration.

The next question that arises is what coordinate system to choose in order to evaluate this integral. You may have never seen a three-dimensional integral before and may have worked only with one-dimensional integrals of the form $\int f(x)dx$. However, all three-dimensional integrals that we will use in this chapter can be reduced to (at most) three successive one-dimensional integrals, most of which are very straightforward to evaluate, provided one selects an appropriate system of coordinates.

Three-Dimensional Non-Cartesian Coordinate Systems

Chapter 1 introduced a three-dimensional orthogonal coordinate system, the Cartesian coordinate system, with coordinates x, y, and z. However, for some applications, it is mathematically simpler to represent the position vector in another coordinate system. This section briefly introduces two commonly used three-dimensional coordinate systems that can be used to specify a vector in three-dimensional space: spherical coordinates and cylindrical coordinates.

Spherical Coordinates. In **spherical coordinates,** the position vector $\vec{r}$ is represented by giving its length, r; its polar angle relative to the positive z-axis, θ; and the azimuthal angle of the vector's projection onto the xy-plane relative to the positive x-axis, ϕ (Figure 8.15).

We can obtain the Cartesian coordinates of the vector $\vec{r}$ from its spherical coordinates via the transformation

$$x = r\cos\phi\sin\theta$$
$$y = r\sin\phi\sin\theta \tag{8.22}$$
$$z = r\cos\theta.$$

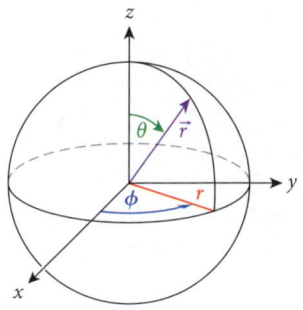

FIGURE 8.15 Three-dimensional spherical coordinate system.

The inverse transformation from Cartesian to spherical coordinates is

$$r = \sqrt{x^2 + y^2 + z^2}$$

$$\theta = \cos^{-1}\left(\frac{z}{\sqrt{x^2 + y^2 + z^2}}\right) \tag{8.23}$$

$$\phi = \tan^{-1}\left(\frac{y}{x}\right).$$

Cylindrical Coordinates. **Cylindrical coordinates** can be thought of as an intermediate between Cartesian and spherical coordinate systems, in the sense that the Cartesian z-coordinate is retained, but the Cartesian coordinates x and y are replaced by the coordinates $r_\perp$ and ϕ (Figure 8.16). Here $r_\perp$ specifies the length of the projection of the position vector $\vec{r}$ onto the xy-plane, so it measures the perpendicular distance to the z-axis. Just as in spherical coordinates, ϕ is the angle of the vector's projection into the xy-plane relative to the positive x-axis.

We obtain the Cartesian coordinates from the cylindrical coordinates via

$$x = r_\perp \cos\phi$$

$$y = r_\perp \sin\phi \tag{8.24}$$

$$z = z.$$

The inverse transformation from Cartesian to cylindrical coordinates is

$$r_\perp = \sqrt{x^2 + y^2}$$

$$\phi = \tan^{-1}\left(\frac{y}{x}\right) \tag{8.25}$$

$$z = z.$$

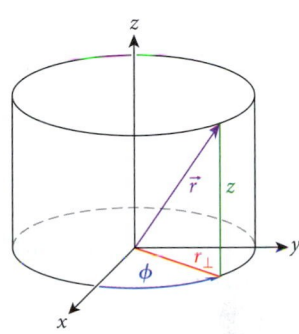

FIGURE 8.16 Cylindrical coordinate system in three dimensions.

As a rule of thumb, you should use a Cartesian coordinate system in your first attempt to describe any physical situation. However, cylindrical and spherical coordinate systems are often preferable when working with objects that have symmetry about a point or a line. Later in this chapter, we will make use of a cylindrical coordinate system to perform a three-dimensional volume integral. Chapter 9 will discuss polar coordinates, which can be thought of as the two-dimensional equivalent of either cylindrical or spherical coordinates. Finally, in Chapter 10, we will again use spherical and cylindrical coordinates to solve slightly more complicated problems requiring integration.

Volume Integrals

Even though calculus is a prerequisite for physics, many universities allow students to take introductory physics and calculus courses concurrently. In general, this approach works well, but when students encounter multidimensional integrals in physics, it is often the first time they have seen this notation. Therefore, let's review the basic procedure for performing these integrations.

If we want to integrate any function over a three-dimensional volume, we need to find an expression for the volume element dV in an appropriate set of coordinates. Unless there is an extremely important reason not to, you should always use orthogonal coordinate systems. The three commonly used three-dimensional orthogonal coordinate systems are the Cartesian, cylindrical, and spherical systems.

It is easiest by far to express the volume element dV in Cartesian coordinates; it is simply the product of the three individual coordinate elements (Figure 8.17). The three-dimensional volume integral written in Cartesian coordinates is

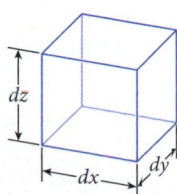

FIGURE 8.17 Volume element in Cartesian coordinates.

$$\int_V f(\vec{r})\,dV = \int_{z_{min}}^{z_{max}}\left(\int_{y_{min}}^{y_{max}}\left(\int_{x_{min}}^{x_{max}} f(\vec{r})\,dx\right)dy\right)dz. \tag{8.26}$$

In this equation, $f(\vec{r})$ can be an arbitrary function of the position. The lower and upper boundaries for the individual coordinates are denoted by $x_{min}, x_{max}, \ldots$. The convention is to solve the innermost integral first and then work outward. For equation 8.26, this means that we first execute the integration over x, then the integration over y, and finally the integration over z. However, any other order is possible. An equally valid way of writing the integral in equation 8.26 is

$$\int_V f(\vec{r})dV = \int_{x_{min}}^{x_{max}} \left(\int_{y_{min}}^{y_{max}} \left(\int_{z_{min}}^{z_{max}} f(\vec{r})dz \right) dy \right) dx, \qquad (8.27)$$

which implies that the order of integration is now z, y, x. Why might the order of integration make a difference? The only time the order of the integration matters is when the integration boundaries in a particular coordinate depend on one or both of the other coordinates. Example 8.4 will consider such a situation.

Because the angle ϕ is one of the coordinates in the cylindrical coordinate system, the volume element is not cube-shaped. For a given differential angle, $d\phi$, the size of the volume element depends on how far away from the z-axis the volume element is located. This size increases linearly with the distance $r_\perp$ from the z-axis (Figure 8.18) and is given by

$$dV = r_\perp dr_\perp d\phi dz. \qquad (8.28)$$

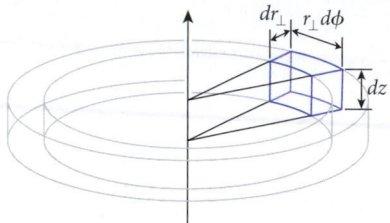

FIGURE 8.18 Volume element in cylindrical coordinates.

The volume integral is then

$$\int_V f(\vec{r})dV = \int_{z_{min}}^{z_{max}} \left(\int_{\phi_{min}}^{\phi_{max}} \left(\int_{r_{\perp min}}^{r_{\perp max}} f(\vec{r})r_\perp dr_\perp \right) d\phi \right) dz. \qquad (8.29)$$

Again the order of integration can be chosen to make the task as simple as possible.

Finally, in spherical coordinates, we use two angular variables, θ and ϕ (Figure 8.19). Here the size of the volume element for a given value of the differential coordinates depends on the distance r to the origin as well as the angle relative to the $\theta = 0$ axis (equivalent to the z-axis in Cartesian or cylindrical coordinates). The differential volume element in spherical coordinates is

$$dV = r^2 dr \sin\theta d\theta d\phi. \qquad (8.30)$$

The volume integral in spherical coordinates is

$$\int_V f(\vec{r})dV = \int_{r_{min}}^{r_{max}} \left(\int_{\phi_{min}}^{\phi_{max}} \left(\int_{\theta_{min}}^{\theta_{max}} f(\vec{r})\sin\theta d\theta \right) d\phi \right) r^2 dr. \qquad (8.31)$$

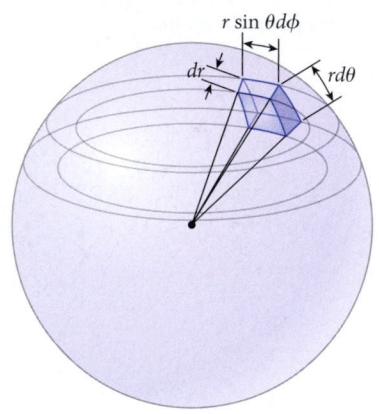

FIGURE 8.19 Volume element in spherical coordinates.

EXAMPLE 8.4   Volume of a Cylinder

To illustrate why it may be simpler to use non-Cartesian coordinates in certain circumstances, let's use volume integrals to find the volume of a cylinder with radius R and height H. We have to integrate the function $f(\vec{r}) = 1$ over the entire cylinder to obtain the volume.

PROBLEM
Use a volume integral to find the volume of a right cylinder of height H and radius R.

SOLUTION
In Cartesian coordinates, we place the origin of our coordinate system at the center of the cylinder's circular base (bottom surface), so that the shape in the xy-plane that we have to integrate over is a circle with radius R (Figure 8.20). The volume integral in Cartesian coordinates is then

$$\int_V dV = \int_0^H \left(\int_{y_{min}}^{y_{max}} \left(\int_{x_{min}(y)}^{x_{max}(y)} dx \right) dy \right) dz. \qquad (i)$$

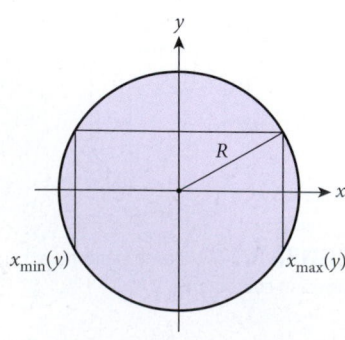

FIGURE 8.20 Bottom surface of a right cylinder of radius R.

The innermost integral has to be done first and is straightforward:

$$\int_{x_{\min}(y)}^{x_{\max}(y)} dx = x_{\max}(y) - x_{\min}(y). \tag{ii}$$

The integration boundaries depend on y: $x_{\max} = \sqrt{R^2 - y^2}$ and $x_{\min} = -\sqrt{R^2 - y^2}$. Thus, the solution of equation (ii) is $x_{\max}(y) - x_{\min}(y) = 2\sqrt{R^2 - y^2}$. We insert this into equation (i) and obtain

$$\int_V dV = \int_0^H \left(\int_{-R}^R 2\sqrt{R^2 - y^2}\, dy \right) dz. \tag{iii}$$

The inner of these two remaining integrals evaluates to

$$\int_{-R}^R 2\sqrt{R^2 - y^2}\, dy = \left(y\sqrt{R^2 - y^2} + R^2 \tan^{-1}\left(\frac{y}{\sqrt{R^2 - y^2}} \right) \right)\Bigg|_{-R}^R = \pi R^2.$$

You can check this result by looking up the definite integral in an integral table. Inserting this result into equation (iii) finally yields our answer:

$$\int_V dV = \int_0^H \pi R^2\, dz = \pi R^2 \int_0^H dz = \pi R^2 H.$$

As you can see, obtaining the volume of the cylinder was rather cumbersome in Cartesian coordinates. What about using cylindrical coordinates? According to equation 8.29, the volume integral is then

$$\int_V f(\vec{r})\, dV = \int_0^H \left(\int_0^{2\pi} \left(\int_0^R r_\perp\, dr_\perp \right) d\phi \right) dz = \int_0^H \left(\int_0^{2\pi} \left(\tfrac{1}{2} R^2 \right) d\phi \right) dz$$

$$= \tfrac{1}{2} R^2 \int_0^H \left(\int_0^{2\pi} d\phi \right) dz = \tfrac{1}{2} R^2 \int_0^H 2\pi\, dz = \pi R^2 \int_0^H dz = \pi R^2 H.$$

In this case, it was much easier to use cylindrical coordinates, a consequence of the geometry of the object over which we had to integrate.

Self-Test Opportunity 8.2

Using spherical coordinates, show that the volume V of a sphere with radius R is $V = \tfrac{4}{3} \pi R^3$.

Now we can return to the problem of calculating the location of an object's center of mass. For the Cartesian components of the position vector, we find, from equation 8.21:

$$X = \frac{1}{M} \int_V x\rho(\vec{r})\, dV, \quad Y = \frac{1}{M} \int_V y\rho(\vec{r})\, dV, \quad Z = \frac{1}{M} \int_V z\rho(\vec{r})\, dV. \tag{8.32}$$

If the mass density for the entire object is constant, $\rho(\vec{r}) \equiv \rho$, we can remove this constant factor from the integral and obtain a special case of equation 8.21 for constant mass density:

$$\vec{R} = \frac{\rho}{M} \int_V \vec{r}\, dV = \frac{1}{V} \int_V \vec{r}\, dV \quad \text{(for constant } \rho\text{)}, \tag{8.33}$$

where we have used equation 8.20 in the last step. In Cartesian components, we obtain for this case:

$$X = \frac{1}{V} \int_V x\, dV, \quad Y = \frac{1}{V} \int_V y\, dV, \quad Z = \frac{1}{V} \int_V z\, dV. \tag{8.34}$$

Equations 8.33 and 8.34 indicate that any object that has a symmetry plane has its center of mass located in that plane. An object having three mutually perpendicular symmetry planes (such as a cylinder, a rectangular solid, or a sphere) has its center of mass where these three planes intersect, which is the geometric center. Example 8.5 develops this idea further.

EXAMPLE 8.5 | Center of Mass for a Half-Sphere

PROBLEM

Consider a solid half-sphere of constant mass density with radius R_0 (Figure 8.21a). Where is its center of mass?

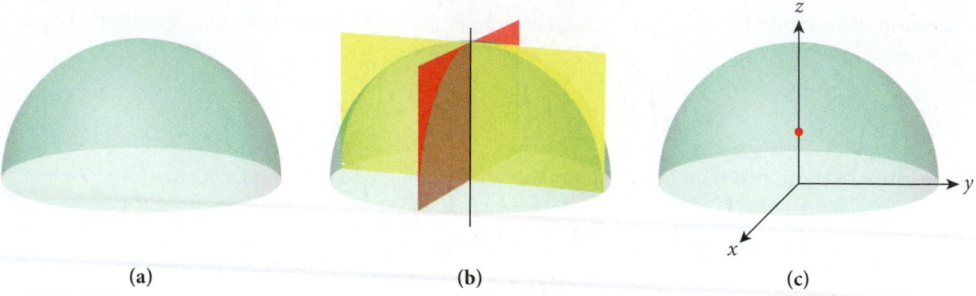

(a)　　　　　　　　**(b)**　　　　　　　　**(c)**

FIGURE 8.21 Determination of the center of mass: (a) half-sphere; (b) symmetry planes and symmetry axis; (c) coordinate system, with location of center of mass marked by red dot.

SOLUTION

As shown in Figure 8.21b, symmetry planes can divide this object into equal, mirror-image halves. Shown are two perpendicular planes in red and yellow, but any plane through the vertical symmetry axis (indicated by the thin black line) is a symmetry plane.

We now position the coordinate system so that one axis (the z-axis, in this case) coincides with this symmetry axis. We are then assured that the center of mass is located exactly on this axis. Because the mass distribution is symmetric and the integrands of equation 8.33 or 8.34 are odd powers of $\vec{r}$, the integral for X or Y has to have the value zero. Specifically,

$$\int_{-a}^{a} x \, dx = 0 \text{ for all values of the constant } a.$$

Positioning the coordinate system so that the z-axis is the symmetry axis ensures that $X = Y = 0$. This is shown in Figure 8.21c, where the origin of the coordinate system is positioned at the center of the half-sphere's circular bottom surface.

Now we have to find the value of the third integral in equation 8.34:

$$Z = \frac{1}{V} \int_{V} z \, dV.$$

The volume of a half-sphere is half the volume of a sphere, or

$$V = \frac{2\pi}{3} R_0^3. \tag{i}$$

To evaluate the integral for Z, we use cylindrical coordinates, in which the differential volume element is given (see equation 8.28) as $dV = r_\perp \, dr_\perp \, d\phi \, dz$. The integral is then evaluated as follows:

$$\int_{V} z \, dV = \int_0^{R_0} \left(\int_0^{\sqrt{R_0^2 - z^2}} \left(\int_0^{2\pi} z r_\perp \, d\phi \right) dr_\perp \right) dz = \int_0^{R_0} z \left(\int_0^{\sqrt{R_0^2 - z^2}} r_\perp \left(\int_0^{2\pi} d\phi \right) dr_\perp \right) dz$$

$$= 2\pi \int_0^{R_0} z \left(\int_0^{\sqrt{R_0^2 - z^2}} r_\perp \, dr_\perp \right) dz = \pi \int_0^{R_0} z (R_0^2 - z^2) \, dz$$

$$= \frac{\pi}{4} R_0^4.$$

Combining this result and the expression for the volume of a half-sphere from equation (i), we obtain the z-coordinate of the center of mass:

$$Z = \frac{1}{V} \int_{V} z \, dV = \frac{3}{2\pi R_0^3} \frac{\pi R_0^4}{4} = \frac{3}{8} R_0.$$

Note that the center of mass of an object does not always have to be located inside the object. Two obvious examples are shown in Figure 8.22. From symmetry considerations, it follows that the center of mass of the donut (Figure 8.22a) is exactly in the center of its hole, at a point outside the donut. The center of mass of the boomerang (Figure 8.22b) lies on the dashed symmetry axis but, again, outside the object.

(a)   (b)

FIGURE 8.22 Objects with a center of mass (indicated by the red dot) outside their mass distribution: (a) donut; (b) boomerang. The symmetry axis of the boomerang is shown by a dashed line.

SOLVED PROBLEM 8.4 | Center of Mass of a Disk with a Hole in It

PROBLEM
Where is the center of mass of a disk with a rectangular hole in it (Figure 8.23)? The height of the disk is $h = 11.0$ cm, and its radius is $R = 11.5$ cm. The rectangular hole has a width $w = 7.0$ cm and a depth $d = 8.0$ cm. The right side of the hole is located so that its midpoint coincides with the central axis of the disk.

SOLUTION
THINK One way to approach this problem is to write mathematical formulas that describe the three-dimensional geometry of the disk with a hole in it and then integrate over that volume to obtain the coordinates of the center of mass. If we did that, we would be faced with several difficult integrals. A simpler way to approach this problem is to think of the disk with a hole in it as a solid disk *minus* a rectangular hole. That is, we treat the hole as a solid object with a negative mass. Using the symmetry of the solid disk and of the hole, we can specify the coordinates of the center of mass of the solid disk and of the center of mass of the hole. We can then combine these coordinates, using equation 8.1, to find the center of mass of the disk with a hole in it.

SKETCH Figure 8.24a shows a top view of the disk with a hole in it, with x- and y-axes assigned.

Figure 8.24b shows the two symmetry planes of the disk with a hole in it. One plane corresponds to the x-y plane, and the second plane is a plane along the x-axis and perpendicular to the x-y plane. The line where the two planes intersect is marked A.

RESEARCH The center of mass must lie along the intersection of the two planes of symmetry. Therefore, we know that the center of mass can only be located along the x-axis. The center of mass for the disk without the hole is at the origin of the coordinate system, at $x_d = 0$, and the volume of the solid disk is $V_d = \pi R^2 h$. If the hole were a solid object with the same dimensions ($h = 11.0$ cm, $w = 7.0$ cm, and $d = 8.0$ cm), that object would have a volume of $V_h = hwd$. If this imagined solid object were located where the hole is, its center of mass would be in the middle of the hole, at $x_h = -3.5$ cm. We now multiply each of the volumes by ρ, the mass density of the material of the disk, to get the corresponding masses, and assign a negative mass to the hole. Then we use equation 8.1 to get the x-coordinate of the center of mass:

$$X = \frac{x_d V_d \rho - x_h V_h \rho}{V_d \rho - V_h \rho}. \tag{i}$$

This method of treating a hole as an object of the same shape and then using its volume in calculations, but with negative mass (or charge), is very common in atomic and subatomic physics. We will encounter it again when we explore atomic physics (Chapter 37) and nuclear and particle physics (Chapters 39 and 40).

SIMPLIFY We can simplify equation (i) by realizing that $x_d = 0$ and that ρ is a common factor:

$$X = \frac{-x_h V_h}{V_d - V_h}.$$

– Continued

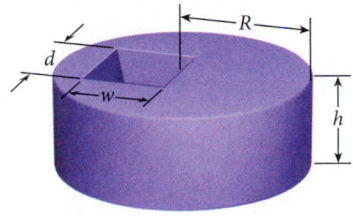

FIGURE 8.23 Three-dimensional view of a disk with a rectangular hole in it.

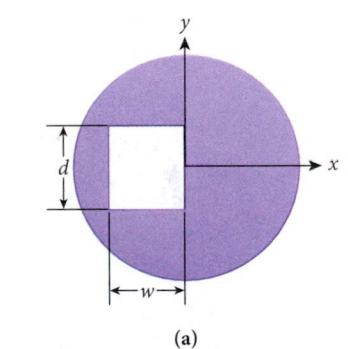

(a)

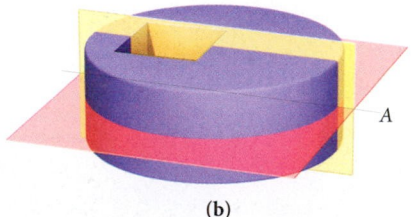

(b)

FIGURE 8.24 (a) Top view of the disk with a hole in it with a coordinate system assigned. (b) Symmetry planes of the disk with a hole in it.

Substituting the expressions we obtained for V_d and V_h, we get

$$X = \frac{-x_h V_h}{V_d - V_h} = \frac{-x_h (hwd)}{\pi R^2 h - hwd} = \frac{-x_h wd}{\pi R^2 - wd}.$$

Defining the area of the disk in the x-y plane to be $A_d = \pi R^2$ and the area of the hole in the x-y plane to be $A_h = wd$, we can write

$$X = \frac{-x_h wd}{\pi R^2 - wd} = \frac{-x_h A_h}{A_d - A_h}.$$

CALCULATE Inserting the given numbers, we find that the area of the disk is

$$A_d = \pi R^2 = \pi (11.5 \text{ cm})^2 = 415.475 \text{ cm}^2,$$

and the area of the hole is

$$A_h = wd = (7.0 \text{ cm})(8.0 \text{ cm}) = 56 \text{ cm}^2.$$

Therefore, the location of the center of mass of the disk with the hole in it (remember that $x_h = -3.5$ cm) is

$$X = \frac{-x_h A_h}{A_d - A_h} = \frac{-(-3.5 \text{ cm})(56 \text{ cm}^2)}{(415.475 \text{ cm}^2) - (56 \text{ cm}^2)} = 0.545239 \text{ cm}.$$

ROUND Expressing our answer with two significant figures, we report the x-coordinate of the center of mass of the disk with a hole in it as

$$X = 0.55 \text{ cm}.$$

DOUBLE-CHECK This point is slightly to the right of the center of the solid disk, by a distance that is a small fraction of the radius of the disk. This result seems reasonable because taking material out of the disk to the left of $x = 0$ should shift the center of gravity to the right, just as we calculated.

Center of Mass for One- and Two-Dimensional Objects

Not all problems involving calculation of a center of mass focus on three-dimensional objects. For example, you may want to calculate the center of mass of a two-dimensional object, such as a flat metal plate. We can write the equations for the center-of-mass coordinates of a two-dimensional object whose area mass density (or mass per unit area) is $\sigma(\vec{r})$ by modifying the expressions for X and Y given in equation 8.32:

$$X = \frac{1}{M} \int_A x\sigma(\vec{r}) dA, \quad Y = \frac{1}{M} \int_A y\sigma(\vec{r}) dA, \tag{8.35}$$

where the mass is

$$M = \int_A \sigma(\vec{r}) dA. \tag{8.36}$$

If the area mass density of the object is constant, then $\sigma = M/A$, and we can rewrite equation 8.35 to give the coordinates of the center of mass of a two-dimensional object in terms of the area, A, and the coordinates x and y:

$$X = \frac{1}{A} \int_A x\, dA, \quad Y = \frac{1}{A} \int_A y\, dA, \tag{8.37}$$

where the total area is obtained from

$$A = \int_A dA. \tag{8.38}$$

If the object is effectively one-dimensional, such as a long, thin rod with length L and linear mass density (or mass per unit length) $\lambda(x)$, the coordinate of the center of mass is given by

$$X = \frac{1}{M} \int_L x\lambda(x)dx, \tag{8.39}$$

where the mass is

$$M = \int_L \lambda(x)dx. \tag{8.40}$$

If the linear mass density of the rod is constant, then clearly the center of mass is located at the geometric center—the middle of the rod—and no further calculation is required.

SOLVED PROBLEM 8.5 | Center of Mass of a Long, Thin Rod

PROBLEM
A long, thin rod lies along the x-axis. One end of the rod is located at $x = 1.00$ m, and the other end of the rod is located at $x = 3.00$ m. The linear mass density of the rod is given by $\lambda(x) = ax^2 + b$, where $a = 0.300$ kg/m^3 and $b = 0.600$ kg/m. What are the mass of the rod and the x-coordinate of its center of mass?

SOLUTION
THINK The linear mass density of the rod is not uniform but depends on the x-coordinate. Therefore, to get the mass, we must integrate the linear mass density over the length of the rod. To get the center of mass, we need to integrate the linear mass density, weighted by the distance in the x-direction, and then divide by the mass of the rod.

SKETCH The long, thin rod oriented along the x-axis is shown in Figure 8.25.

FIGURE 8.25 A long, thin rod oriented along the x-axis.

RESEARCH We obtain the mass of the rod by integrating the linear mass density, λ, over the rod from $x_1 = 1.00$ m to $x_2 = 3.00$ m (see equation 8.40):

$$M = \int_{x_1}^{x_2} \lambda(x)dx = \int_{x_1}^{x_2} \left(ax^2 + b\right)dx = \left[a\frac{x^3}{3} + bx\right]_{x_1}^{x_2}.$$

To find the x-coordinate of the center of mass of the rod, X, we evaluate the integral of the differential mass times x and then divide by the mass, which we have just calculated (see equation 8.39):

$$X = \frac{1}{M}\int_{x_1}^{x_2} \lambda(x)x\,dx = \frac{1}{M}\int_{x_1}^{x_2}\left(ax^2 + b\right)x\,dx = \frac{1}{M}\int_{x_1}^{x_2}\left(ax^3 + bx\right)dx = \frac{1}{M}\left[a\frac{x^4}{4} + b\frac{x^2}{2}\right]_{x_1}^{x_2}.$$

SIMPLIFY Inserting the upper and lower limits, x_2 and x_1, we get the mass of the rod:

$$M = \left[a\frac{x^3}{3} + bx\right]_{x_1}^{x_2} = \left(a\frac{x_2^3}{3} + bx_2\right) - \left(a\frac{x_1^3}{3} + bx_1\right) = \frac{a}{3}\left(x_2^3 - x_1^3\right) + b\left(x_2 - x_1\right).$$

And, in the same way, we find the x-coordinate of the center of mass of the rod:

$$X = \frac{1}{M}\left[a\frac{x^4}{4} + b\frac{x^2}{2}\right]_{x_1}^{x_2} = \frac{1}{M}\left\{\left(a\frac{x_2^4}{4} + b\frac{x_2^2}{2}\right) - \left(a\frac{x_1^4}{4} + b\frac{x_1^2}{2}\right)\right\},$$

which we can further simplify to

$$X = \frac{1}{M}\left\{\frac{a}{4}\left(x_2^4 - x_1^4\right) + \frac{b}{2}\left(x_2^2 - x_1^2\right)\right\}.$$

– Continued

CALCULATE Substituting the given numerical values, we compute the mass of the rod:

$$M = \frac{0.300 \text{ kg/m}^3}{3}\left((3.00 \text{ m})^3 - (1.00 \text{ m})^3\right) + (0.600 \text{ kg/m})(3.00 \text{ m} - 1.00 \text{ m}) = 3.8 \text{ kg}.$$

With the numerical values, the x-coordinate of the rod is

$$X = \frac{1}{3.8 \text{ kg}}\left\{\frac{0.300 \text{ kg/m}^3}{4}\left((3.00 \text{ m})^4 - (1.00 \text{ m})^4\right) + \frac{0.600 \text{ kg/m}}{2}\left((3.00 \text{ m})^2 - (1.00 \text{ m})^2\right)\right\}$$

$$= 2.210526316 \text{ m}.$$

Self-Test Opportunity 8.3

A plate with height h is cut from a thin metal sheet with uniform mass density, as shown in the figure. The lower boundary of the plate is defined by $y = 2x^2$. Show that the center of mass of this plate is located at $x = 0$ and $y = \frac{3}{5}h$.

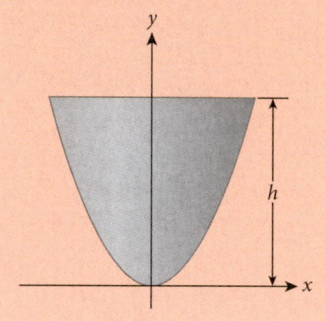

ROUND All of the numerical values in the problem statement were specified to three significant figures, so we report our results as

$$M = 3.80 \text{ kg}$$

and

$$X = 2.21 \text{ m}.$$

DOUBLE-CHECK To double-check our answer for the mass of the rod, let's assume that the rod has a constant linear mass density equal to the linear mass density obtained by setting $x = 2$ m (the middle of the rod) in the expression for λ given in the problem, that is,

$$\lambda = (0.3 \cdot 4 + 0.6) \text{ kg/m} = 1.8 \text{ kg/m}.$$

The mass of the rod is then $m \approx 2 \text{ m} \cdot 1.8 \text{ kg/m} = 3.6 \text{ kg}$, which is reasonably close to our exact calculation of $M = 3.80$ kg.

To double-check the x-coordinate of the center of mass of the rod, we again assume that the linear mass density is constant. Then the center of mass will be located at the middle of the rod, or $X \approx 2$ m. Our calculated answer is $X = 2.21$ m, which is slightly to the right of the middle of the rod. Looking at the function for the linear mass density, we see that the linear mass of the rod increases toward the right, which means that the center of mass of the rod must be to the right of the rod's geometric center. Our result is therefore reasonable.

WHAT WE HAVE LEARNED | EXAM STUDY GUIDE

- The center of mass is the point at which we can imagine all the mass of an object to be concentrated.

- The location of the center of mass for an arbitrarily shaped object is given by $\vec{R} = \frac{1}{M}\int_V \vec{r}\rho(\vec{r})dV$, where the mass density of the object is $\rho = \frac{dm}{dV}$, the integration extends over the entire volume V of the object, and M is its total mass.

- When the mass density is uniform throughout the object, that is, $\rho = \frac{M}{V}$, the center of mass is at $\vec{R} = \frac{1}{V}\int_V \vec{r}\,dV$.

- If an object has a plane of symmetry, the location of the center of mass must be in that plane.

- The location of the center of mass for a combination of several objects can be found by taking the mass-weighted average of the locations of the centers of mass of the individual objects:
$$\vec{R} = \frac{\vec{r}_1 m_1 + \vec{r}_2 m_2 + \cdots + \vec{r}_n m_n}{m_1 + m_2 + \cdots + m_n} = \frac{1}{M}\sum_{i=1}^n \vec{r}_i m_i.$$

- The motion of an extended rigid object can be described by the motion of its center of mass.

- The velocity of the center of mass is given by the derivative of its position vector: $\vec{V} \equiv \frac{d}{dt}\vec{R}$.

- The center-of-mass momentum for a combination of several objects is $\vec{P} = M\vec{V} = \sum_{i=1}^n \vec{p}_i$. This momentum obeys Newton's Second Law:
$$\frac{d}{dt}\vec{P} = \frac{d}{dt}(M\vec{V}) = \sum_{i=1}^n \vec{F}_i = \vec{F}_{net}.$$ Internal forces between the objects do not contribute to the sum that yields the net force (because they always come in action-reaction pairs adding up to zero) and thus do not change the center-of-mass momentum.

- Rocket motion is an example of motion during which the mass of the moving object is not constant. The equation of motion for a rocket in interstellar space is given by $\vec{F}_{thrust} = m\vec{a} = -\vec{v}_c\frac{dm}{dt}$, where $\vec{v}_c$ is the velocity of the propellant relative to the rocket and $\frac{dm}{dt}$ is the rate of change in mass due to outflow of propellant.

- The speed of a rocket as a function of its mass is given by $v_f - v_i = v_c \ln(m_i/m_f)$, where the indices i and f indicate initial and final masses and speeds.

ANSWERS TO SELF-TEST OPPORTUNITIES

8.1

$$Z = \frac{z_{red}m_{red} + z_{green}m_{green} + z_{orange}m_{orange} + z_{blue}m_{blue} + z_{purple}m_{purple}}{M}$$

$$= \frac{\frac{1}{2}h2m_0 + \frac{1}{2}hm_0 + \frac{1}{2}hm_0 + \frac{1}{2}hm_0 + \frac{3}{2}hm_0}{6m_0}$$

$$= \frac{h\left(1 + \frac{1}{2} + \frac{1}{2} + \frac{1}{2} + \frac{3}{2}\right)}{6} = \frac{2}{3}h = 1.7 \text{ m.}$$

8.2 We use spherical coordinates and integrate the angle θ from 0 to π, the angle ϕ from 0 to 2π, and the radial coordinate r from 0 to R.

$$V = \int_0^R \left(\int_0^{2\pi} \left(\int_0^\pi \sin\theta \, d\theta \right) d\phi \right) r^2 \, dr$$

First, evaluate the integral over the azimuthal angle:

$$\int_0^\pi \sin\theta \, d\theta = \left[-\cos\theta\right]_0^\pi = -\left[\cos(\pi) - \cos(0)\right] = 2$$

$$V = 2\int_0^R \left(\int_0^{2\pi} d\phi \right) r^2 \, dr$$

Now evaluate the polar angle integral:

$$\int_0^{2\pi} d\phi = \left[\phi\right]_0^{2\pi} = 2\pi$$

Finally:

$$V = 4\pi \int_0^R r^2 \, dr = 4\pi \left[\frac{r^3}{3}\right]_0^R = \frac{4}{3}\pi R^3.$$

8.3 $dA = 2x(y)dy; y = 2x^2 \Rightarrow x = \sqrt{y/2}$

$$2x(y) = 2\sqrt{y/2} = \sqrt{2y}; dA = \sqrt{2y}\, dy$$

$$Y = \frac{\int_0^h y\sqrt{2y}\, dy}{\int_0^h \sqrt{2y}\, dy} = \frac{\sqrt{2}\int_0^h y^{3/2}\, dy}{\sqrt{2}\int_0^h y^{1/2}\, dy} = \frac{\left[\frac{y^{5/2}}{5/2}\right]_0^h}{\left[\frac{y^{3/2}}{3/2}\right]_0^h}$$

$Y = \frac{3}{5}h$. ($X = 0$, by symmetry).

PROBLEM-SOLVING GUIDELINES: CENTER OF MASS AND ROCKET MOTION

1. The first step in locating the center of mass of an object or a system of particles is to look for planes of symmetry. The center of mass must be located on the plane of symmetry, on the line of intersection of two planes of symmetry, or at the point of intersection of more than two planes.

2. For complicated shapes, break the object down into simpler geometric forms and locate the center of mass for each individual form. Then combine the separate centers of mass into one overall center of mass using the weighted average of distances and masses. Treat holes as objects of negative mass.

3. Any motion of an object can be treated as a superposition of motion of its center of mass (according to Newton's Second Law) and rotation of the object about the center of mass. Collisions can often be conveniently analyzed by considering a reference frame with the origin located at the center of mass.

4. Often, integration is unavoidable when you need to locate the center of mass. In such a case, it is always best to think carefully about the dimensionality of the situation and the choice of the coordinate system (Cartesian, cylindrical, or spherical).

5. In problems involving rocket motion, you cannot simply apply $F = ma$, because the mass of the rocket does not remain constant.

MULTIPLE-CHOICE QUESTIONS

8.1 A man standing on frictionless ice throws a boomerang, which returns to him. Choose the correct statement:

a) Since the momentum of the man-boomerang system is conserved, the man will come to rest holding the boomerang at the same location from which he threw it.

b) It is impossible for the man to throw a boomerang in this situation.

c) It is possible for the man to throw a boomerang, but because he is standing on frictionless ice when he throws it, the boomerang cannot return.

d) The total momentum of the man-boomerang system is not conserved, so the man will be sliding backward holding the boomerang after he catches it.

8.2 When a bismuth-208 nucleus at rest decays, thallium-204 is produced, along with an alpha particle (helium-4 nucleus). The mass numbers of bismuth-208, thallium-204, and helium-4 are 208, 204, and 4, respectively. (The mass number represents the total number of protons and neutrons in the nucleus.) The kinetic energy of the thallium nucleus is

a) equal to that of the alpha particle.

b) less than that of the alpha particle.

c) greater than that of the alpha particle.

8.3 Two objects with masses m_1 and m_2 are moving along the x-axis in the positive direction with speeds v_1 and v_2, respectively, where v_1 is less than v_2. The speed of the center of mass of this system of two bodies is

a) less than v_1.

b) equal to v_1.

c) equal to the average of v_1 and v_2.

d) greater than v_1 and less than v_2.

e) greater than v_2.

8.4 An artillery shell is moving on a parabolic trajectory when it explodes in midair. The shell shatters into a very large number of fragments. Which of the following statements is (are) true (select all that apply)?

a) The force of the explosion will increase the momentum of the system of fragments, and so the momentum of the shell is *not* conserved during the explosion.

b) The force of the explosion is an internal force and thus cannot alter the total momentum of the system.

c) The center of mass of the system of fragments will continue to move on the initial parabolic trajectory until the last fragment touches the ground.

d) The center of mass of the system of fragments will continue to move on the initial parabolic trajectory until the first fragment touches the ground.

e) The center of mass of the system of fragments will have a trajectory that depends on the number of fragments and their velocities right after the explosion.

8.5 An 80-kg astronaut becomes separated from his spaceship. He is 15.0 m away from it and at rest relative to it. In an effort to get back, he throws a 500-g object with a speed of 8.0 m/s in a direction away from the ship. How long does it take him to get back to the ship?

a) 1 s

b) 10 s

c) 20 s

d) 200 s

e) 300 s

8.6 You find yourself in the (realistic?) situation of being stuck on a 300-kg raft (including yourself) in the middle of a pond with nothing but a pile of 7-kg bowling balls and 55-g tennis balls. Using your knowledge of rocket propulsion, you decide to start throwing balls from the raft to move toward shore. Which of the following will allow you to reach the shore faster?

a) throwing the tennis balls at 35 m/s at a rate of 1 tennis ball per second

b) throwing the bowling balls at 0.5 m/s at a rate of 1 bowling ball every 3 s

c) throwing a tennis ball and a bowling ball simultaneously, with the tennis ball moving at 15 m/s and the bowling ball moving at 0.3 m/s, at a rate of 1 tennis ball and 1 bowling ball every 4 s

d) not enough information to decide

8.7 The figures show a high jumper using different techniques to get over the crossbar. Which technique would allow the jumper to clear the highest setting of the bar?

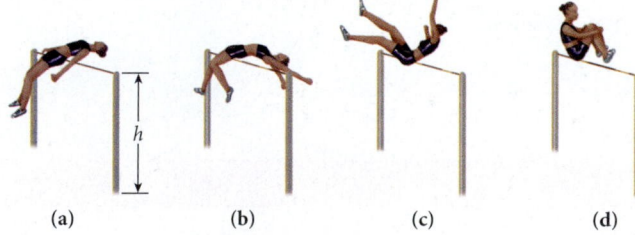

(a)　　　(b)　　　(c)　　　(d)

8.8 The center of mass of an irregular rigid object is *always* located

a) at the geometrical center of the object.

b) somewhere within the object.

c) both of the above.

d) none of the above

8.9 A catapult on a level field tosses a 3-kg stone a horizontal distance of 100 m. A second 3-kg stone tossed in an identical fashion breaks apart in the air into two pieces, one with a mass of 1 kg and one with a mass of 2 kg. Both of the

pieces hit the ground at the same time. If the 1-kg piece lands a distance of 180 m away from the catapult, how far away from the catapult does the 2-kg piece land? Ignore air resistance.

a) 20 m

b) 60 m

c) 100 m

d) 120 m

e) 180 m

8.10 Two point masses are located in the same plane. The distance from mass 1 to the center of mass is 3.0 m. The distance from mass 2 to the center of mass is 1.0 m. What is m_1/m_2, the ratio of mass 1 to mass 2?

a) 3/4

b) 4/3

c) 4/7

d) 7/4

e) 1/3

f) 3/1

8.11 A cylindrical bottle of oil-and-vinegar salad dressing whose volume is 1/3 vinegar ($\rho = 1.01$ g/cm³) and 2/3 oil ($\rho = 0.910$ g/cm³) is at rest on a table. Initially, the oil and the vinegar are separated, with the oil floating on top of the vinegar. The bottle is shaken so that the oil and vinegar mix uniformly, and the bottle is returned to the table. How has the height of the center of mass of the salad dressing changed as a result of the mixing?

a) It is higher.

b) It is lower.

c) It is the same.

d) There is not enough information to answer this question.

8.12 A one-dimensional rod has a linear density that varies with position according to the relationship $\lambda(x) = cx$, where c is a constant and $x = 0$ is the left end of the rod. Where do you expect the center of mass to be located?

a) the middle of the rod

b) to the left of the middle of the rod

c) to the right of the middle of the rod

d) at the right end of the rod

e) at the left end of the rod

8.13 A garden hose is used to fill a 20-L bucket in 1 min. The speed of the water leaving the hose is 1.05 m/s. What force acts on the person holding the hose?

a) 0.35 N

b) 2.1 N

c) 9.8 N

d) 12 N

e) 21 N

8.14 A garden hose is used to fill a 20-L bucket in 1 min. The speed of the water leaving the hose is 2.35 m/s. A force F is acting on the person holding the hose. Another hose can fill the same bucket in 2 min, with the water having the same speed leaving this hose. The force acting on the person holding this hose is

a) F/4

b) F/2

c) F

d) 2F

e) 4F

8.15 Consider a rocket firing in the vacuum of outer space. Which of the following statements is (are) true?

a) The rocket will not produce any thrust because there is no air to push against.

b) The rocket will produce the same thrust in vacuum that it can produce in air.

c) The rocket will produce half the thrust in vacuum that it can produce in air.

d) The rocket will produce twice the thrust in vacuum that it can produce in air.

8.16 The center of mass of the Sun and Jupiter is located

a) exactly at the center of the Sun.

b) near the center of the Sun.

c) exactly at the center of Jupiter.

d) near the center of Jupiter.

e) halfway between the Sun and Jupiter.

CONCEPTUAL QUESTIONS

8.17 A projectile is launched into the air. Part way through its flight, it explodes. How does the explosion affect the motion of the center of mass of the projectile?

8.18 Find the center of mass of the arrangement of uniform identical cubes shown in the figure. The length of the sides of the each cube is d.

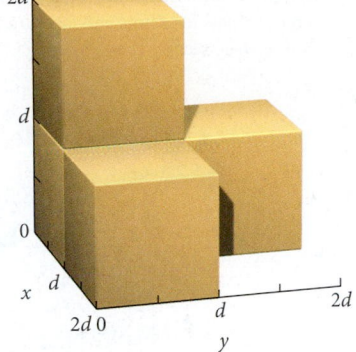

8.19 A model rocket that has a horizontal range of 100 m is fired. A small explosion splits the rocket into two equal parts. What can you say about the points where the fragments land on the ground?

8.20 Can the center of mass of an object be located at a point outside the object, that is, at a point in space where no part of the object is located? Explain.

8.21 Is it possible for two masses to undergo a collision such that the system of two masses has more kinetic energy than the two separate masses had? Explain.

8.22 Prove that the center of mass of a thin metal plate in the shape of an equilateral triangle is located at the intersection of the triangle's altitudes by direct calculation and by physical reasoning.

8.23 A soda can of mass m and height L is filled with soda of mass M. A hole is punched in the bottom of the can to drain out the soda.

a) What is the center of mass of the system consisting of the can and the soda remaining in it when the level of soda in the can is h, where $0 < h < L$?

b) What is the minimum value of the center of mass as the soda drains out?

8.24 An astronaut of mass M is floating in space at a constant distance D from his spaceship when his safety line breaks. He is carrying a toolbox of mass $M/2$ that contains a big sledgehammer of mass $M/4$, for a total mass of $3M/4$. He can throw the items with a speed v relative to his final speed after each item is thrown. He wants to return to the spaceship as soon as possible.

a) To attain the maximum final speed, should the astronaut throw the two items together, or should he throw them one at a time? Explain.

b) To attain the maximum speed, is it best to throw the hammer first or the toolbox first, or does the order make no difference? Explain.

c) Find the maximum speed at which the astronaut can start moving toward the spaceship.

8.25 A metal rod with a length density (mass per unit length) λ is bent into a circular arc of radius R and subtending a total angle of ϕ, as shown in the figure. What is the distance of the center of mass of this arc from O as a function of the angle ϕ? Plot this center-of-mass coordinate as a function of ϕ.

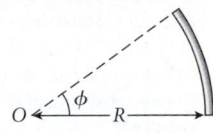

8.26 The carton shown in the figure is filled with a dozen eggs, each of mass m. Initially, the center of mass of the eggs is at the center of the carton, which is the same point as the origin of the Cartesian coordinate system shown. Where is the center of mass of the remaining eggs, in terms of the egg-to-egg distance d, in each of the following situations? Neglect the mass of the carton.

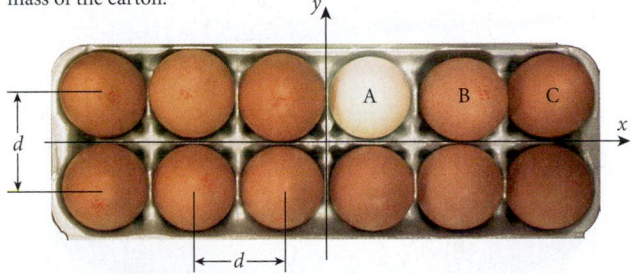

a) Only egg A is removed.

b) Only egg B is removed.

c) Only egg C is removed.

d) Eggs A, B and C are removed.

8.27 A circular pizza of radius R has a circular piece of radius $R/4$ removed from one side, as shown in the figure. Where is the center of mass of the pizza with the hole in it?

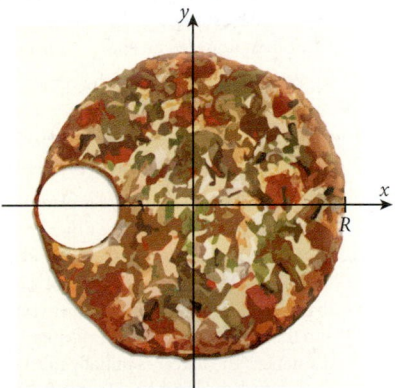

8.28 Suppose you place an old-fashioned hourglass, with sand in the bottom, on a very sensitive analytical balance to determine its mass. You then turn it over (handling it with very clean gloves) and place it back on the balance. You want to predict whether the reading on the balance will be less than, greater than, or the same as before. What do you need to calculate to answer this question? Explain carefully what should be calculated and what the results would imply. You do not need to attempt the calculation.

EXERCISES

A blue problem number indicates a worked-out solution is available in the Student Solutions Manual. One • and two •• indicate increasing level of problem difficulty.

Section 8.1

8.29 Find the following center-of-mass information about objects in the Solar System. You can look up the necessary data on the Internet or in the tables in Chapter 12 of this book. Assume spherically symmetrical mass distributions for all objects under consideration.

a) Determine the distance from the center of mass of the Earth-Moon system to the geometric center of Earth.

b) Determine the distance from the center of mass of the Sun-Jupiter system to the geometric center of the Sun.

•**8.30** The coordinates of the center of mass for the extended object shown in the figure are $(L/4, -L/5)$. What are the coordinates of the 2.00-kg mass?

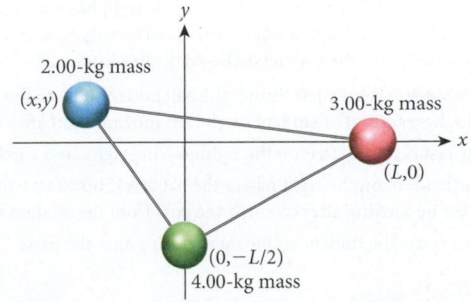

•**8.31** Young acrobats are standing still on a circular horizontal platform suspended at the center. The origin of the two-dimensional Cartesian coordinate system is assumed to be at the center of the platform. A 30.0-kg acrobat is located at (3.00 m, 4.00 m), and a 40.0-kg acrobat is located at (−2.00 m, −2.00 m). Assuming that the acrobats stand still in their positions, where must a 20.0-kg acrobat be located so that the center of mass of the system consisting of the three acrobats is at the origin and the platform is balanced?

Section 8.2

8.32 A man with a mass of 55.0 kg stands up in a 65.0-kg canoe of length 4.00 m floating on water. He walks from a point 0.750 m from the back of the canoe to a point 0.750 m from the front of the canoe. Assume negligible friction between the canoe and the water. How far does the canoe move?

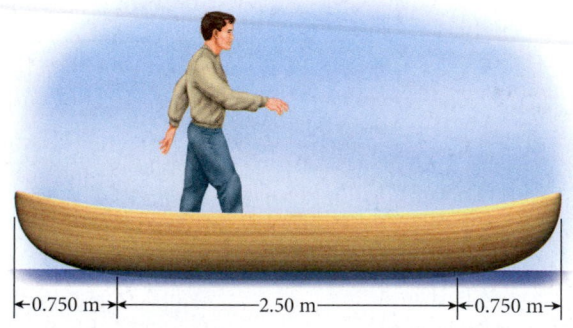

8.33 A toy car of mass 2.00 kg is stationary, and a child rolls a toy truck of mass 3.50 kg straight toward it with a speed of 4.00 m/s. The toy car and truck then undergo an elastic collision.

a) What is the velocity of the center of mass of the system consisting of the two toys?

b) What are the velocities of the truck and the car with respect to the center of mass of the system consisting of the two toys before and after the collision?

8.34 A motorcycle stunt rider plans to start from one end of a railroad flatcar, accelerate toward the other end of the car, and jump from the flatcar to a platform. The motorcycle and rider have a mass of 350. kg and a length of 2.00 m. The flatcar has a mass of 1500. kg and a length of 20.0 m. Assume that there is negligible friction between the flatcar's wheels and the rails and that the motorcycle and rider can move through the air with negligible resistance. The flatcar is initially touching the platform. The promoters of the event have asked you how far the flatcar will be from the platform when the stunt rider reaches the end of the flatcar. What is your answer?

•**8.35** Starting at rest, two students stand on 10.0-kg sleds, which point away from each other on ice, and they pass a 5.00-kg medicine ball back and forth. The student on the left has a mass of 50.0 kg and can throw the ball with a relative speed of 10.0 m/s. The student on the right has a mass of 45.0 kg and can throw the ball with a relative speed of 12.0 m/s. (Assume there is no friction between the ice and the sleds and no air resistance.)

a) If the student on the left throws the ball horizontally to the student on the right, how fast is the student on the left moving right after the throw?

b) How fast is the student on the right moving right after catching the ball?

c) If the student on the right passes the ball back, how fast will the student on the left be moving after catching the pass from the student on the right?

(d) How fast is the student on the right moving after the pass?

•**8.36** Two skiers, Annie and Jack, start skiing from rest at different points on a hill at the same time. Jack, with mass 88.0 kg, skis from the top of the hill down a steeper section with an angle of inclination of 35.0°. Annie, with mass 64.0 kg, starts from a lower point and skis a less steep section, with an angle of inclination of 20.0°. The length of the steeper section is 100. m. Determine the acceleration, velocity, and position vectors of the combined center of mass for Annie and Jack as a function of time before Jack reaches the less steep section.

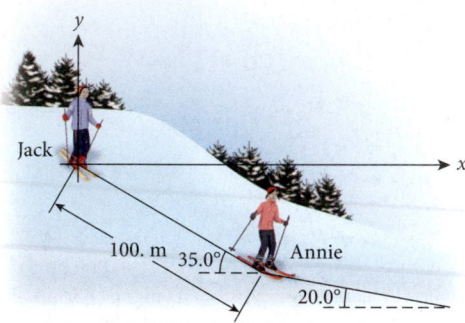

•**8.37** Many nuclear collisions studied in laboratories are analyzed in a frame of reference relative to the laboratory. A proton, with a mass of $1.6726 \cdot 10^{-27}$ kg and traveling at a speed of 70.0% of the speed of light, c, collides with a tin-116 (^{116}Sn) nucleus with a mass of $1.9240 \cdot 10^{-25}$ kg. What is the speed of the center of mass with respect to the laboratory frame? Answer in terms of c, the speed of light.

•**8.38** A system consists of two particles. Particle 1 with mass 2.00 kg is located at (2.00 m, 6.00 m) and has a velocity of (4.00 m/s, 2.00 m/s). Particle 2 with mass 3.00 kg is located at (4.00 m, 1.00 m) and has a velocity of (0, 4.00 m/s).

a) Determine the position and the velocity of the center of mass of the system.

b) Sketch the position and velocity vectors for the individual particles and for the center of mass.

•**8.39** A fire hose 4.00 cm in diameter is capable of spraying water at a velocity of 10.0 m/s. For a continuous horizontal flow of water, what horizontal force should a fireman exert on the hose to keep it stationary?

••**8.40** A block of mass m_b = 1.20 kg slides to the right at a speed of 2.50 m/s on a frictionless horizontal surface, as shown in the figure. It "collides" with a wedge of mass m_w, which moves to the left at a speed of 1.10 m/s. The wedge is shaped so that the block slides seamlessly up the Teflon (frictionless!) surface, as the two come together. Relative to the horizontal surface, block and wedge are moving with a common velocity v_{b+w} at the instant the block stops sliding up the wedge.

a) If the block's center of mass rises by a distance h = 0.370 m, what is the mass of the wedge?

b) What is v_{b+w}?

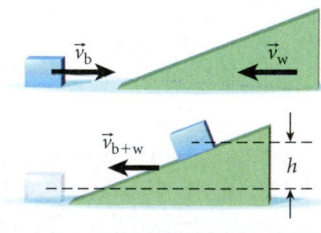

Section 8.3

8.41 One important characteristic of rocket engines is the specific impulse, which is defined as the total impulse (time integral of the thrust) per unit ground weight of fuel/oxidizer expended. (The use of weight, instead of mass, in this definition is due to purely historical reasons.)

a) Consider a rocket engine operating in free space with an exhaust nozzle speed of v. Calculate the specific impulse of this engine.

b) A model rocket engine has a typical exhaust speed of v_{toy} = 800. m/s. The best chemical rocket engines have exhaust speeds of approximately v_{chem} = 4.00 km/s. Evaluate and compare the specific impulse values for these engines.

•8.42 An astronaut is performing a space walk outside the International Space Station. The total mass of the astronaut with her space suit and all her gear is 115 kg. A small leak develops in her propulsion system, and 7.00 g of gas is ejected each second into space with a speed of 800. m/s. She notices the leak 6.00 s after it starts. How much will the gas leak have caused her to move from her original location in space by that time?

•8.43 A rocket in outer space has a payload of 5190.0 kg and $1.551 \cdot 10^5$ kg of fuel. The rocket can expel propellant at a speed of 5.600 km/s. Assume that the rocket starts from rest, accelerates to its final velocity, and then begins its trip. How long will it take the rocket to travel a distance of $3.82 \cdot 10^5$ km (approximately the distance between Earth and Moon)?

••8.44 A uniform chain with a mass of 1.32 kg per meter of length is coiled on a table. One end is pulled upward at a constant rate of 0.470 m/s.

a) Calculate the net force acting on the chain.

b) At the instant when 0.150 m of the chain has been lifted off the table, how much force must be applied to the end being raised?

••8.45 A spacecraft engine creates 53.2 MN of thrust with a propellant velocity of 4.78 km/s.

a) Find the rate (dm/dt) at which the propellant is expelled.

b) If the initial mass is $2.12 \cdot 10^6$ kg and the final mass is $7.04 \cdot 10^4$ kg, find the final speed of the spacecraft (assume the initial speed is zero and any gravitational fields are small enough to be ignored).

c) Find the average acceleration till burnout (the time at which the propellant is used up; assume the mass flow rate is constant until that time).

••8.46 A cart running on a frictionless air track is propelled by a stream of water expelled by a gas-powered pressure washer stationed on the cart. There is a 1.00-m³ water tank on the cart to provide the water for the pressure washer. The mass of the cart, including the operator riding it, the pressure washer with its fuel, and the empty water tank, is 400. kg. The water can be directed, by switching a valve, either backward or forward. In both directions, the pressure washer ejects 200. L of water per min with a muzzle velocity of 25.0 m/s.

a) If the cart starts from rest, after what time should the valve be switched from backward (forward thrust) to forward (backward thrust) for the cart to end up at rest?

b) What is the mass of the cart at that time, and what is its velocity? (*Hint:* It is safe to neglect the decrease in mass due to the gas consumption of the gas-powered pressure washer!)

c) What is the thrust of this "rocket"?

d) What is the acceleration of the cart immediately before the valve is switched?

Section 8.4

8.47 A 32.0-cm-by-32.0-cm checkerboard has a mass of 100. g. There are four 20.0-g checkers located on the checkerboard, as shown in the figure. Relative to the origin located at the bottom left corner of the checkerboard, where is the center of mass of the checkerboard-checkers system?

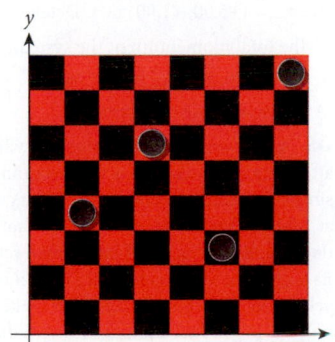

•8.48 A uniform, square metal plate with side $L = 5.70$ cm and mass 0.205 kg is located with its lower left corner at $(x, y) = (0, 0)$, as shown in the figure. A square with side $L/4$ and its lower left edge located at $(x, y) = (0, 0)$ is removed from the plate. What is the distance from the origin of the center of mass of the remaining plate?

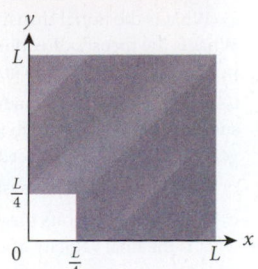

•8.49 Find the x- and y-coordinates of the center of mass of the flat triangular plate of height $H = 17.3$ cm and base $B = 10.0$ cm shown in the figure.

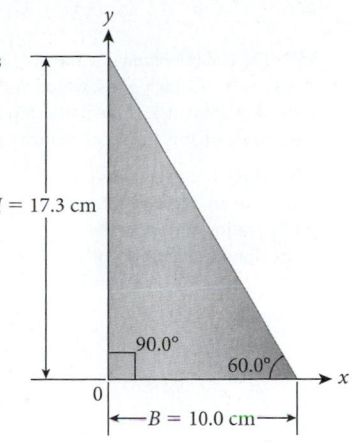

•8.50 The density of a 1.00-m long rod can be described by the linear density function $\lambda(x) = 100.$ g/m $+10.0x$ g/m². One end of the rod is positioned at $x = 0$ and the other at $x = 1.00$ m. Determine (a) the total mass of the rod, and (b) the center-of-mass coordinate.

•8.51 A thin rectangular plate of uniform area density $\sigma_1 = 1.05$ kg/m² has a length $a = 0.600$ m and a width $b = 0.250$ m. The lower left corner is placed at the origin, $(x, y) = (0, 0)$. A circular hole of radius $r = 0.0480$ m with center at $(x, y) = (0.068$ m, 0.068 m$)$ is cut in the plate. The hole is plugged with a disk of the same radius that is composed of another material of uniform area density $\sigma_2 = 5.32$ kg/m². What is the distance from the origin of the resulting plate's center of mass?

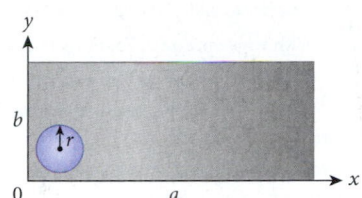

•8.52 A uniform, square metal plate with side $L = 5.70$ cm and mass 0.205 kg is located with its lower left corner at $(x, y) = (0, 0)$, as shown in the figure. Two squares with side length $L/4$ are removed from the plate.

a) What is the x-coordinate of the center of mass?

b) What is the y-coordinate of the center of mass?

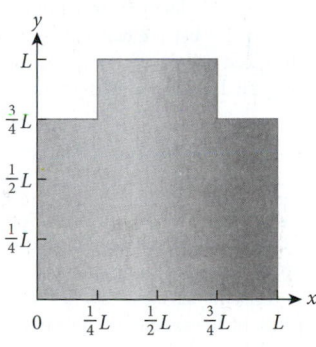

••8.53 The linear mass density, $\lambda(x)$, for a one-dimensional object is plotted in the graph. What is the location of the center of mass for this object?

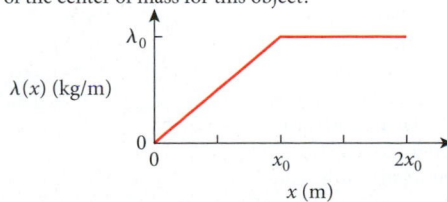

Additional Exercises

8.54 A 750.-kg cannon fires a 15.0-kg projectile with a speed of 250. m/s with respect to the muzzle. The cannon is on wheels and can recoil with negligible friction. Just after the cannon fires the projectile, what is the speed of the projectile with respect to the ground?

8.55 The distance between a carbon atom ($m = 12.0$ u; 1 u = 1 atomic mass unit) and an oxygen atom ($m = 16.0$ u) in a carbon monoxide (CO) molecule is $1.13 \cdot 10^{-10}$ m. How far from the carbon atom is the center of mass of the molecule?

8.56 One method of detecting extrasolar planets involves looking for indirect evidence of a planet in the form of wobbling of its star about the star-planet system's center of mass. Assuming that the Solar System consisted mainly of the Sun and Jupiter, how much would the Sun wobble? That is, what back-and-forth distance would it move due to its rotation about the center of mass of the Sun-Jupiter system? How far from the center of the Sun is that center of mass?

8.57 The USS *Montana* is a massive battleship with a weight of 136,634,000 lb. It has twelve 16-inch guns, which are capable of firing 2700.-lb projectiles at a speed of 2300. ft/s. If the battleship fires three of these guns (in the same direction), what is the recoil velocity of the ship?

8.58 Three identical balls of mass *m* are placed in the configuration shown in the figure. Find the location of the center of mass.

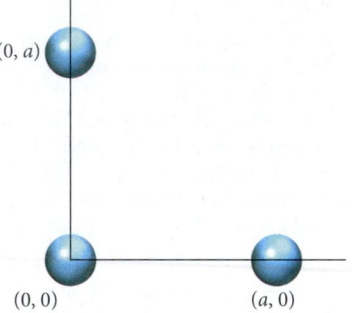

8.59 Sam (61.0 kg) and Alice (44.0 kg) stand on an ice rink, providing them with a nearly frictionless surface to slide on. Sam gives Alice a push, causing her to slide away at a speed (with respect to the rink) of 1.20 m/s.

a) With what speed does Sam recoil?

b) Calculate the change in the kinetic energy of the Sam-Alice system.

c) Energy cannot be created or destroyed. What is the source of the final kinetic energy of this system?

8.60 A baseball player uses a bat with mass m_{bat} to hit a ball with mass m_{ball}. Right before he hits the ball, the bat's initial velocity is 35.0 m/s, and the ball's initial velocity is –30.0 m/s (the positive direction is along the positive *x*-axis). The bat and ball undergo a one-dimensional elastic collision. Find the speed of the ball after the collision. Assume that m_{bat} is much greater than m_{ball}, so the center of mass of the two objects is essentially at the bat.

8.61 A student with a mass of 40.0 kg can throw a 5.00-kg ball with a relative speed of 10.0 m/s. The student is standing at rest on a cart of mass 10.0 kg that can move without friction. If the student throws the ball horizontally, what will the velocity of the ball with respect to the ground be?

•8.62 Find the location of the center of mass of a two-dimensional sheet of constant density σ that has the shape of an isosceles triangle (see the figure).

•8.63 A rocket consists of a payload of 4390.0 kg and $1.761 \cdot 10^5$ kg of fuel. Assume that the rocket starts from rest in outer space, accelerates to its final velocity, and then begins its trip. What is the speed at which the propellant must be expelled to make the trip from the Earth to the Moon, a distance of $3.82 \cdot 10^5$ km, in 7.00 h?

•8.64 A 350.-kg cannon, sliding freely on a frictionless horizontal plane at a speed of 7.50 m/s, shoots a 15.0-kg cannonball at an angle of 55.0° above the horizontal. The velocity of the ball relative to the cannon is such that when the shot occurs, the cannon stops cold. What is the velocity of the ball relative to the cannon?

•8.65 The Saturn V rocket, which was used to launch the *Apollo* spacecraft on their way to the Moon, has an initial mass $M_0 = 2.80 \cdot 10^6$ kg and a final mass $M_1 = 8.00 \cdot 10^5$ kg and burns fuel at a constant rate for 160. s. The speed of the exhaust relative to the rocket is about $v = 2700.$ m/s.

a) Find the upward acceleration of the rocket, as it lifts off the launch pad (while its mass is the initial mass).

b) Find the upward acceleration of the rocket, just as it finishes burning its fuel (when its mass is the final mass).

c) If the same rocket were fired in deep space, where there is negligible gravitational force, what would be the net change in the speed of the rocket during the time it was burning fuel?

•8.66 Find the location of the center of mass for a one-dimensional rod of length *L* and of linear density $\lambda(x) = cx$, where *c* is a constant. (*Hint:* You will need to calculate the mass in terms of *c* and *L*.)

•8.67 Find the center of mass of a rectangular plate of length 20.0 cm and width 10.0 cm. The mass density varies linearly along the length. At one end, it is 5.00 g/cm^2; at the other end, it is 20.0 g/cm^2.

•8.68 A uniform log of length 2.50 m has a mass of 91.0 kg and is floating in water. Standing on this log is a 72.0-kg man, located 22.0 cm from one end. On the other end is his daughter (*m* = 20.0 kg), standing 1.00 m from the end.

a) Find the center of mass of this system.

b) If the father jumps off the log backward away from his daughter (*v* = 3.14 m/s with respect to the water), what is the initial speed of log and child?

8.69 A sculptor has commissioned you to perform an engineering analysis of one of his works, which consists of regularly shaped metal plates of uniform thickness and density, welded together as shown in the figure. Using the intersection of the two axes shown as the origin of the coordinate system, determine the Cartesian coordinates of the center of mass of this piece.

•8.70 A jet aircraft is traveling at 223 m/s in horizontal flight. The engine takes in air at a rate of 80.0 kg/s and burns fuel at a rate of 3.00 kg/s. The exhaust gases are ejected at 600. m/s relative to the speed of the aircraft. Find the thrust of the jet engine.

•8.71 A bucket is mounted on a skateboard, which rolls across a horizontal road with no friction. Rain is falling vertically into the bucket. The bucket is filled with water, and the total mass of the skateboard, bucket, and water is *M* = 10.0 kg. The rain enters the top of the bucket and simultaneously leaks out of a hole at the bottom of the bucket at equal rates of $\lambda = 0.100$ kg/s. Initially, bucket and skateboard are moving at a speed of v_0. How long will it take before the speed is reduced by half?

•8.72 A 1000.-kg cannon shoots a 30.0-kg shell at an angle of 25.0° above the horizontal and a speed of 500. m/s with respect to the ground. What is the recoil velocity of the cannon?

•8.73 Two masses, $m_1 = 2.00$ kg and $m_2 = 3.00$ kg, are moving in the *xy*-plane. The velocity of their common center of mass and the velocity of mass 1 relative to mass 2 are given by the vectors $\vec{v}_{cm} = (-1.00, +2.40)$ m/s and $\vec{v}_{rel} = (+5.00, +1.00)$ m/s. Determine

a) the total momentum of the system,

b) the momentum of mass 1, and

c) the momentum of mass 2.

••8.74 You are piloting a spacecraft whose total mass is 1000. kg and attempting to dock with a space station in deep space. Assume for simplicity that the station is stationary, that your spacecraft is moving at 1.00 m/s toward the station, and that both are perfectly aligned for docking. Your spacecraft has a small retro-rocket at its front end to slow its approach, which can burn fuel at a rate of 1.00 kg/s and with an exhaust velocity of 100. m/s relative to the rocket. Assume that your spacecraft has only 20.0 kg of fuel left and sufficient distance for docking.

a) What is the initial thrust exerted on your spacecraft by the retro-rocket? What is the thrust's direction?

b) For safety in docking, NASA allows a maximum docking speed of 0.0200 m/s. Assuming you fire the retro-rocket from time *t* = 0 in one sustained burst, how much fuel (in kilograms) has to be burned to slow your spacecraft to this speed relative to the space station?

c) How long should you sustain the firing of the retro-rocket?

d) If the space station's mass is $5.00 \cdot 10^5$ kg (close to the value for the ISS), what is the final velocity of the station after the docking of your spacecraft, which arrives with a speed of 0.0200 m/s?

••**8.75** A chain whose mass is 3.00 kg and length is 5.00 m is held at one end so that the bottom end of the chain just touches the floor (see the figure). The top end of the chain is released. What is the force exerted by the chain on the floor just as the last link of the chain lands on the floor?

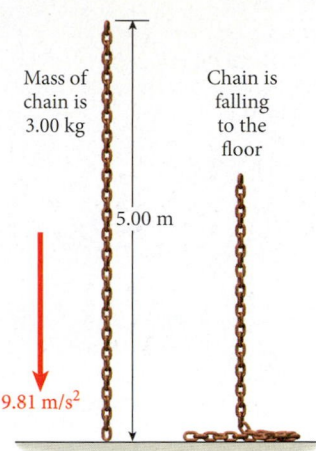

Mass of chain is 3.00 kg

Chain is falling to the floor

5.00 m

9.81 m/s^2

MULTI-VERSION EXERCISES

8.76 An ion thruster mounted in a satellite uses electric forces to eject xenon ions with a speed of 21.45 km/s. The ion thruster produces a thrust of $1.187 \cdot 10^{-2}$ N. What is the rate of fuel consumption of the thruster?

8.77 An ion thruster mounted in a satellite uses electric forces to eject xenon ions with a speed of 23.75 km/s. The rate of fuel consumption of the thruster is $5.082 \cdot 10^{-7}$ kg/s. How much thrust is produced?

8.78 An ion thruster mounted in a satellite uses electric forces to eject xenon ions and produces a thrust of $1.229 \cdot 10^{-2}$ N. The rate of fuel consumption of the thruster is $4.718 \cdot 10^{-7}$ kg/s. With what speed are the xenon ions ejected from the thruster?

8.79 An ion thruster mounted in a satellite with mass 2149 kg (including 23.37 kg of fuel) uses electric forces to eject xenon ions with a speed of 28.33 km/s. If the ion thruster operates continuously while pointed in the same direction until it uses all of the fuel, what is the change in the speed of the satellite?

8.80 An ion thruster mounted in a satellite with mass 2161 kg (including fuel) uses electric forces to eject xenon ions with a speed of 20.61 km/s. The ion thruster operates continuously while pointed in the same direction until it uses all the available fuel. The change in the speed of the satellite is 236.4 m/s. How much fuel was available to the thruster?

8.81 An ion thruster mounted in a satellite uses electric forces to eject xenon ions with a speed of 22.91 km/s. The ion thruster operates continuously while pointed in the same direction until it uses all 25.95 kg of the available fuel. The change in speed of the satellite is 275.0 m/s. What was the mass of the satellite and the fuel before the thruster started operating?

8.82 A 75.19-kg fisherman is sitting in his 28.09-kg fishing boat along with his 13.63-kg tackle box. The boat and its cargo are at rest near a dock. He throws the tackle box toward the dock with a speed of 2.911 m/s relative to the dock. What is the recoil speed of the fisherman and his boat?

8.83 A 77.49-kg fisherman is sitting in his 28.31-kg fishing boat along with his 14.27-kg tackle box. The boat and its cargo are at rest near a dock. He throws the tackle box toward the dock, and he and his boat recoil with a speed of 0.3516 m/s. With what speed, as seen from the dock, did the fishermen throw his tackle box?

8.84 A fisherman is sitting in his 28.51-kg fishing boat along with his 14.91-kg tackle box. The boat and its cargo are at rest near a dock. He throws the tackle box toward the dock, which it approaches with a speed of 3.303 m/s. The fisherman and his boat recoil with a speed of 0.4547 m/s. What is the mass of the fisherman?

8.85 A proton with mass $1.673 \cdot 10^{-27}$ kg is moving with a speed of $1.823 \cdot 10^{6}$ m/s toward an alpha particle with mass $6.645 \cdot 10^{-27}$ kg, which is at rest. What is the speed of the center of mass of the proton–alpha particle system?

8.86 The center of mass of a proton with mass $1.673 \cdot 10^{-27}$ kg and an alpha particle with mass $6.645 \cdot 10^{-27}$ kg is moving with a speed of $5.509 \cdot 10^{5}$ m/s. What is the speed of the proton in the laboratory frame of reference, where the alpha particle is at rest?

9 Circular Motion

FIGURE 9.1 NASA's 20*g* centrifuge subjects astronauts and equipment to *g*-forces similar to those encountered during rocket launches.

In the Ames Research Center at Moffet Field, California, NASA operates a giant centrifuge, which is shown in Figure 9.1. By moving objects on a circular path, this centrifuge creates artificial gravity forces up to twenty times stronger than what we experience on Earth.

In this chapter, we study circular motion and see how force is involved in making turns, which will let us understand how NASA's 20*g* centrifuge works. This discussion relies on concepts of force, velocity, and acceleration presented in Chapters 3 and 4. Chapter 10 will combine these ideas with some of the concepts from Chapters 5 through 8, such as energy and momentum. Much of what you will learn in these two chapters on circular motion and rotation is analogous to earlier material on linear motion, force, and energy. Because most objects do not travel in perfectly straight lines, the concepts of circular motion will be applied many times in later chapters.

WHAT WE WILL LEARN

- The motion of objects traveling in a circle rather than in a straight line can be described using coordinates based on radius and angle rather than Cartesian coordinates.

- There is a relationship between linear motion and circular motion.

- Circular motion can be described in terms of the angular coordinate, angular frequency, and period.

- An object undergoing circular motion can have angular velocity and angular acceleration.

9.1 Polar Coordinates

In Chapter 3, we discussed motion in two dimensions. In this chapter, we examine a special case of motion in a two-dimensional plane: motion of an object along the circumference of a circle. To be precise, we will only study circular motion of objects that we can consider to be point particles. In Chapter 10 on rotation, we will relax this condition and also examine extended bodies.

Circular motion is surprisingly common. Riding on a carousel or many other amusement park rides, such as those shown in Figure 9.2, qualifies as circular motion. Indy-style car racing also involves circular motion, as the cars alternate between moving along straight sections and half-circle segments of the track. CD, DVD, and Blu-ray players also operate with circular motion, although this motion is usually hidden from the eye.

During an object's **circular motion**, its x- and y-coordinates change continuously, but the distance from the object to the center of the circular path stays the same. We can take advantage of this fact by using **polar coordinates** to study circular motion. Shown in Figure 9.3 is the position vector, $\vec{r}$, of an object in circular motion. This vector changes as a function of time, but its tip always moves on the circumference of a circle. We can specify $\vec{r}$ by giving its x- and y-components. However, we can specify the same vector by giving two other numbers: the angle of $\vec{r}$ relative to the x-axis, θ, and the length of $\vec{r}$, $r = |\vec{r}|$ (Figure 9.3).

Trigonometry provides the relationship between the Cartesian coordinates x and y and the polar coordinates θ and r:

$$r = \sqrt{x^2 + y^2} \tag{9.1}$$

$$\theta = \tan^{-1}(y/x). \tag{9.2}$$

The inverse transformation from polar to Cartesian coordinates is given by

$$x = r\cos\theta \tag{9.3}$$

$$y = r\sin\theta. \tag{9.4}$$

The major advantage of using polar coordinates for analyzing circular motion is that r never changes. It remains the same as long as the tip of the vector $\vec{r}$ moves along the circular path. Thus, we can reduce the description of two-dimensional motion on the circumference of a circle to a one-dimensional problem involving the angle θ.

Figure 9.3 also shows the unit vectors in the radial and tangential directions, $\hat{r}$ and $\hat{t}$, respectively. The angle between $\hat{r}$ and $\hat{x}$ is again the angle θ. Therefore, the Cartesian components of the radial unit vector can be written as follows (refer to Figure 9.4):

$$\hat{r} = \frac{x}{r}\hat{x} + \frac{y}{r}\hat{y} = (\cos\theta)\hat{x} + (\sin\theta)\hat{y} \equiv (\cos\theta, \sin\theta). \tag{9.5}$$

Similarly, we obtain the Cartesian components of the tangential unit vector (refer again to Figure 9.4 and note that the tangential and radial unit vectors are always perpendicular to each other; thus, the angle between the tangential unit vector and the y-axis is the same as the angle between the radial unit vector and the x-axis):

$$\hat{t} = \frac{-y}{r}\hat{x} + \frac{x}{r}\hat{y} = (-\sin\theta)\hat{x} + (\cos\theta)\hat{y} \equiv (-\sin\theta, \cos\theta). \tag{9.6}$$

FIGURE 9.2 Circular motion in the horizontal and in the vertical plane.

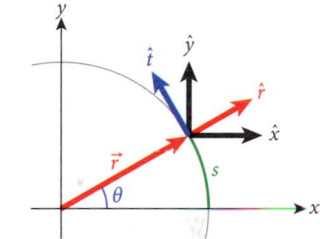

FIGURE 9.3 Polar coordinate system for circular motion.

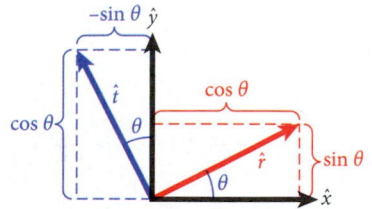

FIGURE 9.4 Relationship between the radial and tangential unit vectors shown in Figure 9.3, the Cartesian unit vectors, and the sine and cosine of the angle.

(Note that the tangential unit vector is always denoted with a caret, $\hat{t}$, and can thus be readily distinguished from the time, t.) It is straightforward to verify that the radial and tangential unit vectors are perpendicular to each other by taking their scalar product:

$$\hat{r} \cdot \hat{t} = (\cos\theta, \sin\theta) \cdot (-\sin\theta, \cos\theta) = -\cos\theta\sin\theta + \sin\theta\cos\theta = 0.$$

We can also verify that these two unit vectors have a length of 1, as required:

$$\hat{r} \cdot \hat{r} = (\cos\theta, \sin\theta) \cdot (\cos\theta, \sin\theta) = \cos^2\theta + \sin^2\theta = 1$$
$$\hat{t} \cdot \hat{t} = (-\sin\theta, \cos\theta) \cdot (-\sin\theta, \cos\theta) = \sin^2\theta + \cos^2\theta = 1.$$

Finally, it is important to stress that, for circular motion, there is a major difference between the unit vectors for polar coordinates and those for Cartesian coordinates: The Cartesian unit vectors stay constant in time, while the radial and tangential unit vectors change their directions during the process of circular motion. This is because both of these vectors depend on the angle θ, which depends on the time for circular motion.

9.2   Angular Coordinates and Angular Displacement

Polar coordinates allow us to describe and analyze circular motion, where the distance to the origin, r, of the object in motion stays constant and the angle θ varies as a function of time, $\theta(t)$. As was already pointed out, the angle θ is measured relative to the positive x-axis. Any point on the positive x-axis has an angle $\theta = 0$. As the definition in equation 9.2 implies, a move in the counterclockwise direction away from the positive x-axis toward the positive y-axis results in positive values for the angle θ. Conversely, a clockwise move away from the positive x-axis toward the negative y-axis results in negative values of θ.

The two most commonly used units for angles are degrees (°) and radians (rad). These units are defined such that the angle measured by one complete circle is 360°, which corresponds to 2π rad. Thus, the unit conversion between the two angular measures is

$$\theta \text{ (degrees)} \frac{\pi}{180} = \theta \text{ (radians)} \Leftrightarrow \theta \text{ (radians)} \frac{180}{\pi} = \theta \text{ (degrees)}$$

$$1 \text{ rad} = \frac{180°}{\pi} \approx 57.3°.$$

Like the linear position x, the angle θ can have positive and negative values. However, θ is periodic; a complete turn around the circle (2π rad, or 360°) returns the coordinate for θ to the same point in space. Just as the linear displacement, Δx, is defined to be the difference between two positions, x_2 and x_1, the angular displacement, $\Delta\theta$, is the difference between two angles:

$$\Delta\theta = \theta_2 - \theta_1.$$

EXAMPLE 9.1   Locating a Point with Cartesian and Polar Coordinates

A point has a location given in Cartesian coordinates as (4,3), as shown in Figure 9.5.

PROBLEM
How do we represent the position of this point in polar coordinates?

SOLUTION
Using equation 9.1, we can calculate the radial coordinate:
$$r = \sqrt{x^2 + y^2} = \sqrt{4^2 + 3^2} = 5.$$
Using equation 9.2, we can calculate the angular coordinate:
$$\theta = \tan^{-1}(y/x) = \tan^{-1}(3/4) = 0.64 \text{ rad} = 37°.$$
Therefore, we can express the position of point P in polar coordinates as $(r, \theta) = (5, 0.64 \text{ rad}) = (5, 37°)$. Note that we can specify the same position by adding (any integral multiple of) 2π rad, or 360°, to θ:
$$(r, \theta) = (5, 0.64 \text{ rad}) = (5, 37°) = (5, 2\pi \text{ rad} + 0.64 \text{ rad}) = (5, 360° + 37°).$$

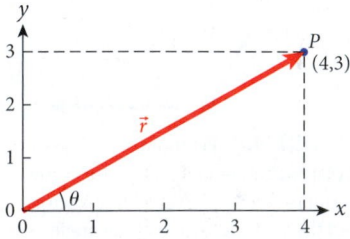

FIGURE 9.5 A point located at (4,3) in a Cartesian coordinate system.

Arc Length

Figure 9.3 also shows (in green) the path on the circumference of the circle traveled by the tip of the vector $\vec{r}$ in going from an angle of zero to θ. This path is called the *arc length, s*. It is related to the radius and angle via

$$s = r\theta. \tag{9.7}$$

For this relationship to work out numerically, the angle has to be measured in radians. The fact that the circumference of a circle is $2\pi r$ is a special case of equation 9.7 with $\theta = 2\pi$ rad, corresponding to one full turn around the circle. The arc length has the same unit as the radius.

For small angles, measuring a degree or less, the sine of the angle is approximately (to four significant figures) equal to the angle measured in radians. Because of this fact as well as the need to use equation 9.7 to solve problems, the preferred unit for angular coordinates is the radian. But the use of degrees is common, and this book uses both units.

Self-Test Opportunity 9.1

Use polar coordinates and calculus to show that the circumference of a circle with radius R is $2\pi R$.

Self-Test Opportunity 9.2

Use polar coordinates and calculus to show that the area of a circle with radius R is πR^2.

EXAMPLE 9.2 | CD Track

The track on a compact disc (CD) is represented in Figure 9.6. The track is a spiral, originating at an inner radius of $r_1 = 25$ mm and terminating at an outer radius of $r_2 = 58$ mm. The spacing between successive loops of the track is a constant, $\Delta r = 1.6$ μm.

PROBLEM
What is the total length of this track?

SOLUTION 1, WITHOUT CALCULUS
At a given radius r between r_1 and r_2, the track is almost perfectly circular. Since the track spacing is $\Delta r = 1.6$ μm, and the distance between the innermost and outermost parts is $r_2 - r_1 = (58 \text{ mm}) - (25 \text{ mm}) = 33$ mm, we find that the track winds around a total of

$$n = \frac{r_2 - r_1}{\Delta r} = \frac{3.3 \cdot 10^{-2} \text{ m}}{1.6 \cdot 10^{-6} \text{ m}} = 20{,}625 \text{ times.}$$

(Note that we did not round this intermediate result!) The radii of these 20,625 circles grow linearly from 25 mm to 58 mm; consequently, the circumferences ($c = 2\pi r$) also grow linearly. The average radius of the circles is then $\bar{r} = \frac{1}{2}(r_2 + r_1) = 41.5$ mm, and the average circumference is

$$\bar{c} = 2\pi\bar{r} = (2\pi)(41.5 \text{ mm}) = 0.2608 \text{ m.}$$

Now we have to multiply this average circumference by the number of circles to obtain our result:

$$L = n\bar{c} = (20{,}625)(0.2608 \text{ m}) = 5.4 \text{ km,}$$

rounded to two significant figures. Thus, the length of a track on a CD is more than 3.3 mi!

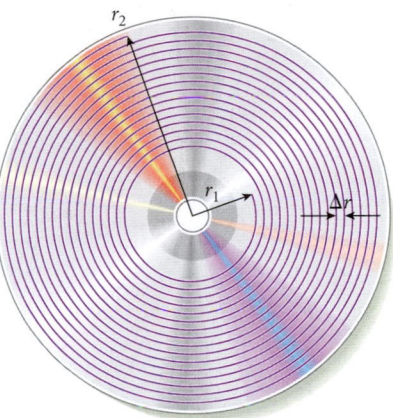

FIGURE 9.6 The track of a compact disc.

SOLUTION 2, WITH CALCULUS
The spacing between successive loops of the track is a constant, $\Delta r = 1.6$ μm; therefore, the track density (that is, the number of times the track is crossed in the radial direction per unit length moving outward from the innermost point) is $1/\Delta r = 625{,}000$ m^{-1}. At a given radius r between r_1 and r_2, the track is almost perfectly circular. This segment of the spiral track has a length of $2\pi r$, but the loops' lengths grow steadily longer moving to the outside. We obtain the overall length of the track by integrating the length of each loop from r_1 to r_2, multiplied by the number of loops per unit length:

$$L = \frac{1}{\Delta r} \int_{r_1}^{r_2} 2\pi r \, dr = \frac{1}{\Delta r} \pi r^2 \bigg|_{r_1}^{r_2} = \frac{1}{\Delta r} \pi \left(r_2^2 - r_1^2 \right)$$

$$= (625{,}000 \text{ m}^{-1})\pi\left((0.058 \text{ m})^2 - (0.025 \text{ m})^2\right) = 5.4 \text{ km.}$$

Satisfyingly, this calculus-based solution agrees with the solution we obtained without the use of calculus. You may want to think about what conditions had to be met to avoid using calculus in this case. (For a DVD, by the way, the track density is higher by a factor of 2.2, resulting in a track that is almost 12 km long.)

Figure 9.7 shows a small portion of the tracks of two CDs, at a magnification of 500 times. Figure 9.7a shows a factory-pressed CD, with the individual aluminum bumps visible. In

– Continued

Figure 9.7b is a read-write CD, which is burned when a laser induces a phase change (Chapter 34 on wave optics will explain this process) in the continuous track. The lower-right portion of this picture shows a portion of the CD on which nothing has yet been written.

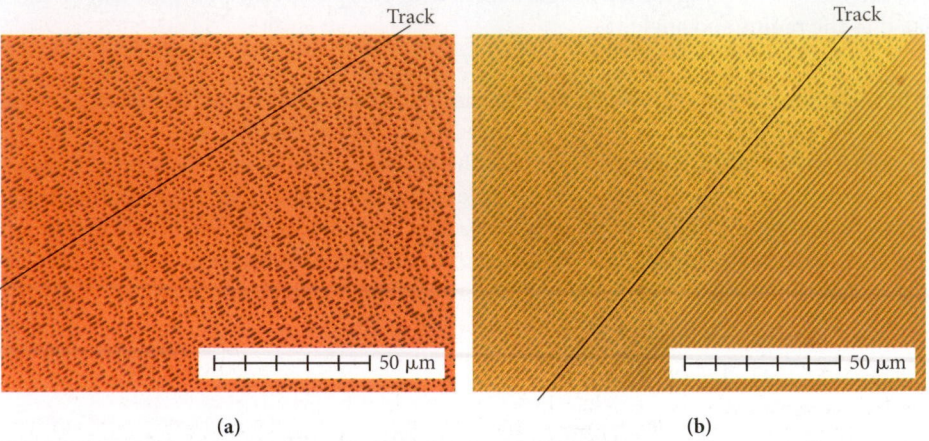

FIGURE 9.7 Microscope pictures (magnification 500×) of (a) a factory-pressed CD and (b) a read-write CD. The solid lines indicate the direction of the tracks.

9.3 Angular Velocity, Angular Frequency, and Period

We have seen that the change of an object's linear coordinates in time is its velocity. Similarly, the change of an object's angular coordinate in time is its **angular velocity.** The average magnitude of the angular velocity is defined as

$$\bar{\omega} = \frac{\theta_2 - \theta_1}{t_2 - t_1} = \frac{\Delta\theta}{\Delta t}.$$

This definition uses the notation $\theta_1 \equiv \theta(t_1)$ and $\theta_2 \equiv \theta(t_2)$. The horizontal bar above the symbol ω for angular velocity again indicates a time average. By taking the limit of this expression as the time interval approaches zero, we find the instantaneous value of the magnitude of the angular velocity:

$$\omega = \lim_{\Delta t \to 0} \bar{\omega} = \lim_{\Delta t \to 0} \frac{\Delta\theta}{\Delta t} \equiv \frac{d\theta}{dt}. \tag{9.8}$$

The most common unit of angular velocity is radians per second (rad/s); degrees per second is not generally used.

Angular velocity is a vector. Its direction is that of an axis through the center of the circular path and perpendicular to the plane of the circle. (This axis is a rotation axis, as we will discuss in more detail in Chapter 10.) This definition allows for two possibilities for the direction in which the vector $\vec{\omega}$ can point: up or down, parallel or antiparallel to the axis of rotation. A right-hand rule helps us decide which is the correct direction: When the fingers point in the direction of rotation along the circle's circumference, the thumb points in the direction of $\vec{\omega}$, as shown in Figure 9.8.

Angular velocity measures how fast the angle θ changes in time. Another quantity also specifies how fast this angle changes in time—the **frequency,** f. For example, the rpm number on the tachometer in your car indicates how many times per minute the engine cycles and thus specifies the frequency of engine revolution. Figure 9.9 shows a tachometer, with the units specified as "1/min × 1000"; the engine hits the red line at 6000 revolutions per minute. Thus, the frequency, f, measures cycles per unit time, instead of radians per unit time as the angular velocity does. The frequency is related to the magnitude of the angular velocity, ω, by

$$f = \frac{\omega}{2\pi} \Leftrightarrow \omega = 2\pi f. \tag{9.9}$$

This relationship makes sense because one complete turn around a circle requires an angle change of 2π rad. (Be careful—both frequency and angular velocity have the same unit of inverse seconds and can be easily confused.)

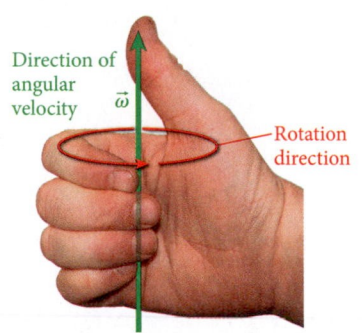

FIGURE 9.8 The right-hand rule for determining the direction of the angular velocity vector.

FIGURE 9.9 The tachometer of a car measures the frequency (in cycles per minute) of revolutions of the engine.

Because the unit inverse second is used so widely, it was given its own name, the **hertz** (Hz), for the German physicist Heinrich Rudolf Hertz (1857–1894): 1 Hz = 1 s^{-1}. The **period of rotation,** T, is defined as the inverse of the frequency:

$$T = \frac{1}{f}. \qquad (9.10)$$

The period measures the time interval between two successive instances where the angle has the same value; that is, the time it takes to pass once around the circle. The unit of the period is the same as that of time, the second (s). Given the relationships between period and frequency and between frequency and angular velocity, we also obtain

$$\omega = 2\pi f = \frac{2\pi}{T}. \qquad (9.11)$$

Angular Velocity and Linear Velocity

If we take the time derivative of the position vector, we obtain the linear velocity vector. To find the angular velocity, it is most convenient to write the radial position vector in Cartesian coordinates and take the time derivatives component by component:

$$\vec{r} = x\hat{x} + y\hat{y} = (x, y) = (r\cos\theta, r\sin\theta) = r(\cos\theta, \sin\theta) = r\hat{r} \Rightarrow$$

$$\vec{v} = \frac{d\vec{r}}{dt} = \frac{d}{dt}(r\cos\theta, r\sin\theta) = \left(\frac{d}{dt}(r\cos\theta), \frac{d}{dt}(r\sin\theta)\right).$$

Now we can use the fact that, for motion along a circle, the distance r to the origin does not change in time but is constant. This results in

$$\vec{v} = \left(\frac{d}{dt}(r\cos\theta), \frac{d}{dt}(r\sin\theta)\right) = \left(r\frac{d}{dt}(\cos\theta), r\frac{d}{dt}(\sin\theta)\right)$$

$$= \left(-r\sin\theta\frac{d\theta}{dt}, r\cos\theta\frac{d\theta}{dt}\right)$$

$$= (-\sin\theta, \cos\theta)r\frac{d\theta}{dt}.$$

Here we used the chain rule of differentiation in the next to last step and then factored out the common factor $r\,d\theta/dt$. We already know that the time derivative of the angle is the angular velocity (see equation 9.8). In addition, we recognize the vector $(-\sin\theta, \cos\theta)$ as the tangential unit vector (see equation 9.6). Thus, we have the relationship between angular and linear velocities for circular motion:

$$\vec{v} = r\omega\hat{t}. \qquad (9.12)$$

(Again, $\hat{t}$ is the symbol for the tangential unit vector and has no connection with the time, t!)

Because the velocity vector points in the direction of the tangent to the trajectory at any given time, it is always tangential to the circle's circumference, pointing in the direction of the motion, as shown in Figure 9.10. Thus, the velocity vector is always perpendicular to the position vector, which points in the radial direction. If two vectors are perpendicular to each other, their scalar product is zero. Thus, for circular motion, we always have

$$\vec{r} \cdot \vec{v} = (r\cos\theta, r\sin\theta) \cdot (-r\omega\sin\theta, r\omega\cos\theta) = 0.$$

If we take the absolute values of the left- and right-hand sides of equation 9.12, we obtain an important relationship between the magnitudes of the linear and angular speeds for circular motion:

$$v = r\omega. \qquad (9.13)$$

Remember that this relationship holds only for the *magnitudes* of the linear and angular velocities. Their vectors point in different directions and, for uniform circular motion, are perpendicular to each other, with $\vec{\omega}$ pointing in the direction of the rotation axis and $\vec{v}$ tangential to the circle.

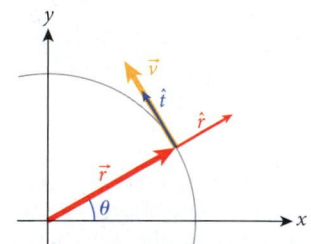

FIGURE 9.10 Linear velocity and coordinate vectors.

Concept Check 9.1

A bicycle's wheels have a radius R. The bicycle is traveling with speed v. Which one of the following expressions describes the angular speed of the front tire?

a) $\omega = \frac{1}{2}Rv^2$ d) $\omega = Rv$

b) $\omega = \frac{1}{2}vR^2$ e) $\omega = v/R$

c) $\omega = R/v$

EXAMPLE 9.3 | **Revolution and Rotation of the Earth**

PROBLEM
The Earth orbits around the Sun and also rotates on its pole-to-pole axis. What are the angular velocities, frequencies, and linear speeds of these motions?

SOLUTION
Any point on the surface of Earth moves in circular motion around the rotation axis (pole-to-pole), with a rotation period of 1 day. Expressed in seconds, this period is

$$T_{Earth} = 1 \text{ day} \cdot \frac{24 \text{ h}}{1 \text{ day}} \cdot \frac{3600 \text{ s}}{1 \text{ h}} = 8.64 \cdot 10^4 \text{ s}.$$

The Earth moves around the Sun on an elliptical path; the path is very close to circular, so let's treat the Earth's orbit as circular motion. The orbital period for the motion of the Earth around the Sun is 1 year. Expressing this period in seconds, we obtain

$$T_{Sun} = 1 \text{ year} \cdot \frac{365 \text{ days}}{1 \text{ year}} \cdot \frac{24 \text{ h}}{1 \text{ day}} \cdot \frac{3600 \text{ s}}{1 \text{ h}} = 3.15 \cdot 10^7 \text{ s}.$$

Both circular motions have constant angular velocity. Thus, we can use $T = 1/f$ and $\omega = 2\pi f$ to obtain the frequencies and angular velocities:

$$f_{Earth} = \frac{1}{T_{Earth}} = 1.16 \cdot 10^{-5} \text{ Hz}; \ \omega_{Earth} = 2\pi f_{Earth} = 7.27 \cdot 10^{-5} \text{ rad/s}$$

$$f_{Sun} = \frac{1}{T_{Sun}} = 3.17 \cdot 10^{-8} \text{ Hz}; \ \omega_{Sun} = 2\pi f_{Sun} = 1.99 \cdot 10^{-7} \text{ rad/s}.$$

Note that the 24-hour period that we used as the length of a day is how long it takes the Sun to reach the same position in the sky. If we wanted to specify the frequencies and angular velocities to greater precision, we would have to use the sidereal day: Because the Earth is also moving around the Sun during each day, the time it actually takes to complete one full rotation and bring the stars into the same position in the night sky is the sidereal day, which consists of 23 h, 56 min, and a little more than 4 s, or 86,164.09 s = (1 – 1/366.242) · 86,400 s. (We have used the fact that it takes a fraction of a day more than 365 to complete an orbit around the Sun—which is why leap years are necessary.)

Now let's find the linear velocity at which the Earth orbits the Sun. Because we are assuming circular motion, the relationship between the orbital speed and angular velocity is given by $v = r\omega$. To get our answer, we need to know the radius of the orbit. The radius of this orbit is the Earth-Sun distance, $r_{Earth-Sun} = 1.49 \cdot 10^{11}$ m. Therefore, the linear orbital speed—the speed with which the Earth moves around the Sun—is

$$v = r\omega = (1.49 \cdot 10^{11} \text{ m})(1.99 \cdot 10^{-7} \text{ s}^{-1}) = 2.97 \cdot 10^4 \text{ m/s}.$$

This is over 66,000 mph!

Now we want to find the speed of a point on the surface of the rotating Earth relative to the center of the Earth. We note that points at different latitudes have different distances to the axis of rotation, as shown in Figure 9.11. At the Equator, the radius of rotation is $r = R_{Earth} = 6380$ km. Away from the Equator, the radius of rotation as a function of the latitude angle is $r = R_{Earth} \cos \vartheta$; see Figure 9.11. (The script theta, ϑ, is used for the latitude angles here in order to avoid confusion with θ, which is used for the motion in the circle.) In general, we obtain the following formula for the speed of rotation:

$$v = \omega r = \omega R_{Earth} \cos \vartheta$$
$$= (7.27 \cdot 10^{-5} \text{ s}^{-1})(6.38 \cdot 10^6 \text{ m})(\cos \vartheta)$$
$$= (464 \text{ m/s})(\cos \vartheta).$$

At the poles, where $\vartheta = 90°$, the rotation speed is zero; at the Equator, where $\vartheta = 0°$, the speed is 464 m/s. Seattle, with $\vartheta = 47.5°$, moves at $v = 313$ m/s, and Miami, with $\vartheta = 25.7°$, has a speed of $v = 418$ m/s.

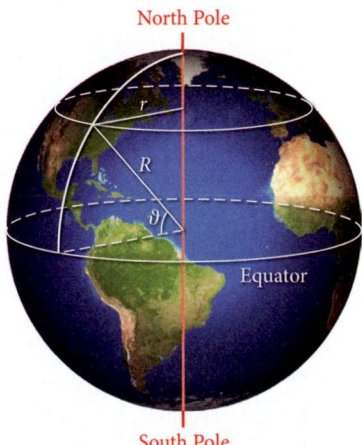

North Pole

South Pole

FIGURE 9.11 Earth's rotation axis is indicated by the vertical line. Points at different latitudes on the Earth's surface move at different speeds.

9.4 Angular and Centripetal Acceleration

The rate of change of an object's angular velocity is its **angular acceleration,** denoted by the Greek letter α. The definition of the magnitude of the angular acceleration is analogous to that for the linear acceleration. Its time average is defined as

$$\bar{\alpha} = \frac{\Delta \omega}{\Delta t}.$$

The instantaneous magnitude of the angular acceleration is obtained in the limit as the time interval approaches zero:

$$\alpha = \lim_{\Delta t \to 0} \bar{\alpha} = \lim_{\Delta t \to 0} \frac{\Delta \omega}{\Delta t} \equiv \frac{d\omega}{dt} = \frac{d^2\theta}{dt^2}. \tag{9.14}$$

Just as we related the linear velocity to the angular velocity, we can also relate the tangential acceleration to the angular acceleration. We start with the definition of the linear acceleration vector as the time derivative of the linear velocity vector. Then we substitute the expression for the linear velocity in circular motion from equation 9.12:

$$\vec{a}(t) = \frac{d}{dt}\vec{v}(t) = \frac{d}{dt}(v\hat{t}) = \left(\frac{dv}{dt}\right)\hat{t} + v\left(\frac{d\hat{t}}{dt}\right). \tag{9.15}$$

In the last step here, we used the product rule of differentiation. Thus, the acceleration in circular motion has two components. The first part arises from the change in the magnitude of the velocity; this is the **tangential acceleration.** The second part is due to the fact that the velocity vector always points in the tangential direction and thus has to change its direction continuously as the tip of the radial position vector moves around the circle; this is the **radial acceleration.**

Let's look at the two components individually. First, we can calculate the time derivative of the linear speed, v, using the relationship between linear speed and angular speed in equation 9.13 and again invoking the product rule:

$$\frac{dv}{dt} = \frac{d}{dt}(r\omega) = \left(\frac{dr}{dt}\right)\omega + r\frac{d\omega}{dt}.$$

Because r is constant for circular motion, $dr/dt = 0$, and the first term in the sum on the right-hand side is zero. From equation 9.14, $d\omega/dt = \alpha$, and so the second term in the sum is equal to $r\alpha$. Thus, the change in speed is related to the angular acceleration by

$$\frac{dv}{dt} = r\alpha. \tag{9.16}$$

However, the acceleration vector of equation 9.15 also has a second component, which is proportional to the time derivative of the tangential unit vector. For this quantity, we find

$$\frac{d}{dt}\hat{t} = \frac{d}{dt}(-\sin\theta, \cos\theta) = \left(\frac{d}{dt}(-\sin\theta), \frac{d}{dt}(\cos\theta)\right)$$

$$= \left(-\cos\theta\frac{d\theta}{dt}, -\sin\theta\frac{d\theta}{dt}\right) = -\frac{d\theta}{dt}(\cos\theta, \sin\theta)$$

$$= -\omega\hat{r}.$$

Therefore, we find that the time derivative of the tangential unit vector points in the direction opposite to that of the radial unit vector. With this result, we can finally write for the linear acceleration vector of equation 9.15:

$$\vec{a}(t) = r\alpha\hat{t} - v\omega\hat{r}. \tag{9.17}$$

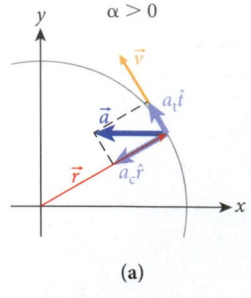

(a)

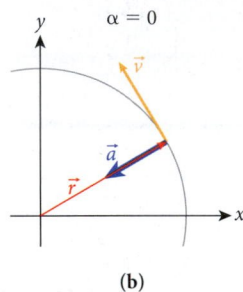

(b)

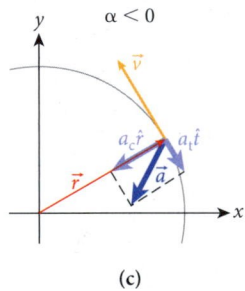

(c)

FIGURE 9.12 Relationships among linear acceleration, centripetal acceleration, and angular acceleration for (a) increasing speed; (b) constant speed; and (c) decreasing speed.

Again, for circular motion, the acceleration vector has two physical components (Figure 9.12): The first results from the change in speed and points in the tangential direction, and the second comes from the continuous change of direction of the velocity vector and points in the negative radial direction, toward the center of the circle. This second component is present even if the circular motion proceeds at constant speed. If the angular velocity is constant, the tangential angular acceleration is zero, but the velocity vector still changes direction continuously as the object moves in its circular path. The acceleration that changes the direction of the velocity vector without changing its magnitude is often called **centripetal acceleration** (*centripetal* means "center-seeking"), and it is directed in the inward radial direction. Thus, we can write equation 9.17 for the acceleration of an object in circular motion as the sum of the tangential acceleration and the centripetal acceleration:

$$\vec{a} = a_t \hat{t} - a_c \hat{r}. \tag{9.18}$$

The magnitude of the centripetal acceleration is

$$a_c = v\omega = \frac{v^2}{r} = \omega^2 r. \tag{9.19}$$

The first expression for the centripetal acceleration in equation 9.19 can be simply read off from equation 9.17 as the coefficient of the unit vector pointing in the negative radial direction. The second and third expressions for the centripetal acceleration then follow from the relationship between linear and angular speeds and the radius (equation 9.13).

For the magnitude of the acceleration in circular motion, we thus have, from equations 9.17 and 9.19,

$$a = \sqrt{a_t^2 + a_c^2} = \sqrt{(r\alpha)^2 + (r\omega^2)^2} = r\sqrt{\alpha^2 + \omega^4}. \tag{9.20}$$

EXAMPLE 9.4   **Ultracentrifuge**

One of the most important pieces of equipment in biomedical labs is the ultracentrifuge (Figure 9.13). It is used for separation of substances (such as colloids or proteins) consisting of *particles* of different masses through the process of sedimentation (more massive particles sink to the bottom). Instead of relying on the acceleration of gravity to accomplish sedimentation, an ultracentrifuge utilizes the centripetal acceleration from rapid rotation to speed up the process. Some ultracentrifuges can reach centripetal acceleration values of up to $10^6 g$ ($g = 9.81 \text{ m/s}^2$).

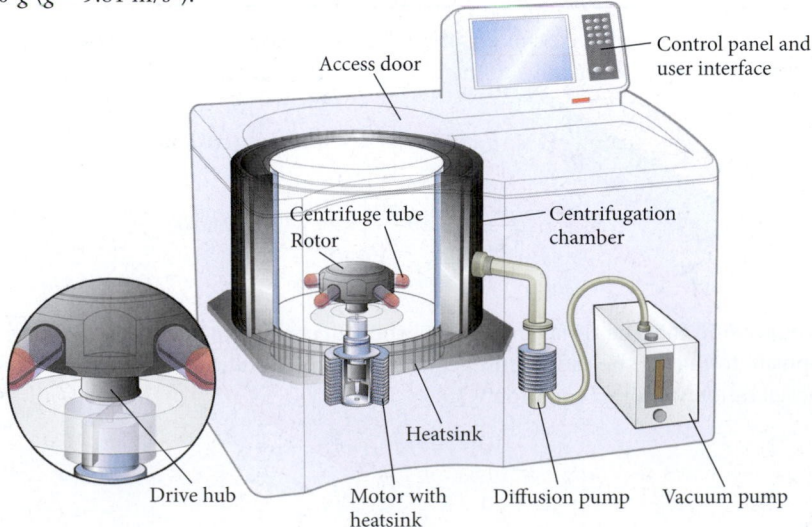

FIGURE 9.13 Cutaway diagram of an ultracentrifuge.

PROBLEM

If you want to generate 840,000g of centripetal acceleration in a sample rotating at a distance of 23.5 cm from the ultracentrifuge's rotation axis, what is the frequency you have to enter into the controls? What is the linear speed with which the sample is then moving?

SOLUTION

The centripetal acceleration is given by $a_c = \omega^2 r$, and the angular velocity is related to the frequency via equation 9.11: $\omega = 2\pi f$. Thus, the relationship between the frequency and the centripetal acceleration is $a_c = (2\pi f)^2 r$, or

$$f = \frac{1}{2\pi}\sqrt{\frac{a_c}{r}} = \frac{1}{2\pi}\sqrt{\frac{(840,000)(9.81 \text{ m/s}^2)}{0.235 \text{ m}}} = 942 \text{ s}^{-1} = 56,500 \text{ rpm.}$$

For the linear speed of the sample inside the centrifuge, we then find

$$v = r\omega = 2\pi rf = 2\pi(0.235 \text{ m})(942 \text{ s}^{-1}) = 1.39 \text{ km/s.}$$

Other types of centrifuges called *gas centrifuges* are used in the process of uranium enrichment. In this process, the isotopes ^{235}U and ^{238}U (discussed in Chapter 40), which differ in mass by only slightly more than 1%, are separated. Natural uranium contains more than 99% of the harmless ^{238}U. But if uranium is enriched to contain more than 90% ^{235}U, it can be used for nuclear weapons. The gas centrifuges used in the enrichment process have to spin at approximately 100,000 rpm, which puts an incredible strain on the mechanisms and materials of these centrifuges and makes them very hard to design and manufacture. In order to impede proliferation of nuclear weapons, the design of these centrifuges is a closely guarded secret.

EXAMPLE 9.5 Centripetal Acceleration Due to Earth's Rotation

Because the Earth rotates, points on its surface move with a rotational velocity, and it is interesting to compute the corresponding centripetal acceleration. This acceleration can slightly alter the commonly stated value of the acceleration due to gravity at the surface of Earth.

We can insert the data for the Earth into equation 9.19 to find the magnitude of the centripetal acceleration:

$$a_c = \omega^2 r = \omega^2 R_{\text{Earth}} \cos \vartheta$$

$$= (7.27 \cdot 10^{-5} \text{ s}^{-1})^2 (6.38 \cdot 10^6 \text{ m})(\cos \vartheta)$$

$$= (0.034 \text{ m/s}^2)(\cos \vartheta).$$

Here we have used the same notation as in Example 9.3, with ϑ indicating the latitude angle relative to the Equator. Our result shows that the centripetal acceleration due to the Earth's rotation changes the effective gravitational acceleration observed on the surface of Earth by a factor ranging between 0.34 percent (at the Equator) and zero (at the poles). Using Seattle and Miami as examples, we obtain a centripetal acceleration of 0.02 m/s^2 for Seattle and 0.03 m/s^2 for Miami. These values are relatively small compared to the quoted value for the acceleration of gravity, 9.81 m/s^2, but are not always negligible.

EXAMPLE 9.6 CD Player

PROBLEM

In Example 9.2, we established that a CD track is 5.4 km long. A music CD can store 74 min of music. What are the angular velocity and the tangential acceleration of the disc as it spins inside a CD player, assuming a constant linear velocity?

SOLUTION

Since the track is 5.4 km long and has to pass the laser that reads it in a time interval of $\Delta t = 74$ min = 4440 s, the speed of the track passing by the reader has to be $v = (5.4 \text{ km})/(4440 \text{ s}) = 1.216$ m/s. From Example 9.2, the track is a spiral with 20,625 loops, starting at an

– Continued

Concept Check 9.2

The rotation of the Earth on its axis creates a centripetal acceleration at the surface of the Earth. Suppose you were standing on the Equator and the Earth stopped rotating. When the Earth stopped, you would

a) feel slightly lighter than before.

b) feel slightly heavier than before.

c) fly off the surface of the Earth.

d) not be able to tell whether the Earth was still rotating.

inner radius $r_1 = 25$ mm and reaching an outer radius $r_2 = 58$ mm. At each value of the radius r, we can approximate the spiral track by a circle, as we did in Example 9.2. Then we can use the relationship between linear and angular speeds expressed in equation 9.13 to solve for the angular velocity as a function of the radius:

$$v = r\omega \Rightarrow \omega = \frac{v}{r}.$$

Inserting the values for v and r, we obtain

$$\omega(r_1) = \frac{1.216 \text{ m/s}}{0.025 \text{ m}} = 48.64 \text{ s}^{-1}$$

$$\omega(r_2) = \frac{1.216 \text{ m/s}}{0.058 \text{ m}} = 20.97 \text{ s}^{-1}.$$

Self-Test Opportunity 9.4

How does the centripetal accelera-tion of a CD rotating in a player com-pare to its tangential acceleration?

This means that a CD player has to slow the disc's rate of rotation during the course of playing it. The average angular acceleration during this process is

$$\alpha = \frac{\omega(r_2) - \omega(r_1)}{\Delta t} = \frac{20.97 \text{ s}^{-1} - 48.64 \text{ s}^{-1}}{4440 \text{ s}} = -6.2 \cdot 10^{-3} \text{ s}^{-2}.$$

9.5 Centripetal Force

Concept Check 9.3

You are sitting on a carousel, which is in motion. Where should you sit so that the largest possible centripetal force is acting on you?

a) close to the outer edge

b) close to the center

c) in the middle

d) The force is the same everywhere.

The centripetal force, $\vec{F}_c$, is not another fundamental force of nature but is simply the net inward force needed to provide the centripetal acceleration necessary for circular motion. It has to point inward, toward the circle's center. Its magnitude is the product of the mass of the object and the centripetal acceleration required to force it onto a circular path:

$$F_c = ma_c = mv\omega = m\frac{v^2}{r} = m\omega^2 r. \qquad (9.21)$$

To arrive at equation 9.21, we simply wrote the centripetal acceleration in terms of the lin-ear velocity v, the angular velocity ω, and the radius r, as in equation 9.19, and multiplied it by the mass of the object forced onto a circular path by the centripetal force.

Figure 9.14 shows a top view of a spinning table with three identical (except for color) poker chips on it. The black chip is located close to the center, the red chip close to the outer edge, and the blue chip in the middle between them. If we spin the table slowly as in part (a), all three chips are in circular motion. In this case, the static friction force between the table and the chips provides the centripetal force required to keep the chips in circular motion. In parts (b), (c), and (d), the table is spinning progressively faster. A higher angular velocity means a larger centripetal force, according to equation 9.21. The chips slide when the friction force is not large enough to provide the centripetal force needed. As you can see, the outermost chip slides off first, and the innermost chip last. This clearly indicates that, for a given angular velocity, the centripetal force increases with the distance from the center. Equation 9.21 in the form $F_c = m\omega^2 r$ can explain this observed behavior. All points on the surface of the spinning table have the same angular velocity, ω, because all of them take the same time to complete one revolution. Thus, for the three poker chips, the centripetal force is proportional to the distance from the center, explaining why the red chip slides off first and the black chip slides off last.

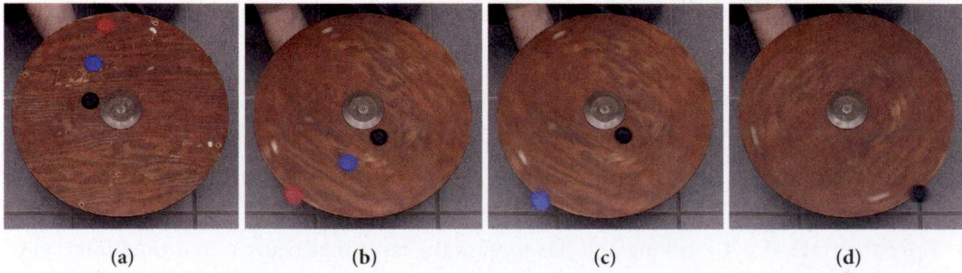

FIGURE 9.14 Poker chips on a spinning table. Shown from left to right are the initial positions of the chips and the moments when the three chips slide off during the process of circular motion.

Conical Pendulum

Figure 9.15 is a picture of the Wave Swinger, a ride at an amusement park. The riders sit in seats that are suspended from a solid disk by long chains. At the beginning of the ride, the chains hang straight down, but as the ride starts to rotate, the chains form an angle φ with the vertical, as you can see. This angle is independent of the mass of the rider and depends only on the angular velocity of the circular motion. How can we find the value of this angle in terms of that velocity?

In order to gain this understanding, we consider a similar, but somewhat simpler situation: a mass suspended from the ceiling by a string of length ℓ and performing circular motion such that the angle between the string and the vertical is φ. The string outlines the surface of a cone, which is why this setup is called a *conical pendulum*; see Figure 9.16.

Figure 9.16c shows a free-body diagram for the mass. There are only two forces acting on it. Acting vertically downward is the force of gravity, $\vec{F}_g$, indicated by the red arrow in the free-body diagram; as usual, its magnitude is mg. The only other force acting on the mass is the string tension, $\vec{T}$, which acts along the direction of the string at an angle φ with the vertical. This string tension is broken down into its x- and y-components ($T_x = T \sin \varphi$, $T_y = T \cos \varphi$). There is no motion in the vertical direction; therefore, we have to have zero net force in that direction, $F_{net,y} = T \cos \varphi - mg = 0$, leading to

$$T \cos \varphi = mg.$$

In the horizontal direction, the horizontal component of the string tension is the only force component; it provides the centripetal force. Newton's Second Law, $F_{net,x} = F_c = ma_c$, then yields

$$T \sin \varphi = mr\omega^2.$$

As you can see from Figure 9.16b, the radius of the circular motion is given by $r = \ell \sin \varphi$. Using this relationship, we find the string tension in terms of the angular velocity:

$$T = m\ell\omega^2. \tag{9.22}$$

We have just seen that $T \cos \varphi = mg$, and we can substitute the expression for T from equation 9.22 into this equation to get

$$(m\ell\omega^2)(\cos\varphi) = mg$$

$$\omega^2 = \frac{g}{\ell \cos\varphi}$$

$$\omega = \sqrt{\frac{g}{\ell \cos\varphi}}. \tag{9.23}$$

The mass cancels out, which explains why all the chains in Figure 9.15 have the same angle with the vertical. Clearly, there is a unique and interesting relationship between the angle of the conical pendulum and its angular velocity. As the angle φ approaches zero, the angular velocity does not approach zero but rather some finite minimum value, $\sqrt{g/\ell}$. (We will gain a deeper understanding of this result in Chapter 14, when we study pendulum motion.) In the limit where the angle φ approaches 90°, ω becomes infinite.

FIGURE 9.15 Wave Swinger at an amusement park.

Self-Test Opportunity 9.5

Sketch a graph of the string tension as a function of the angle φ.

Concept Check 9.4

A certain angular velocity, ω_0, of a conical pendulum results in an angle φ_0. If this conical pendulum were taken to the Moon, where the gravitational acceleration is a sixth of that on Earth, how would one have to adjust the angular velocity to obtain the same angle φ_0?

a) $\omega_{Moon} = 6\omega_0$

b) $\omega_{Moon} = \sqrt{6}\,\omega_0$

c) $\omega_{Moon} = \omega_0$

d) $\omega_{Moon} = \omega_0/\sqrt{6}$

e) $\omega_{Moon} = \omega_0/6$

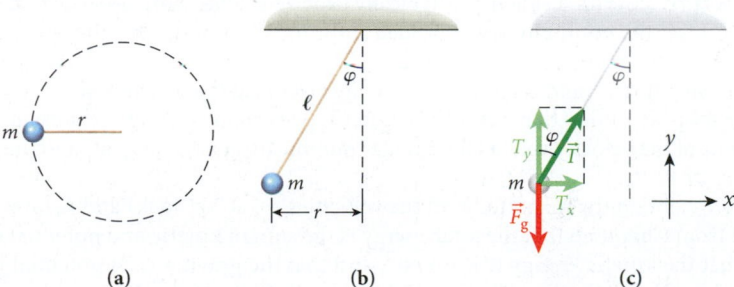

(a)   (b)   (c)

FIGURE 9.16 Conical pendulum: (a) top view, with the circular path of the mass indicated by the dashed line; (b) side view; (c) free-body diagram.

FIGURE 9.17 Modern roller coaster with a vertical loop.

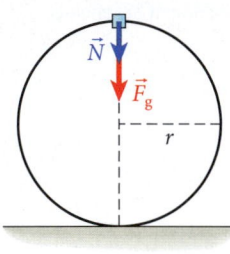

(a)

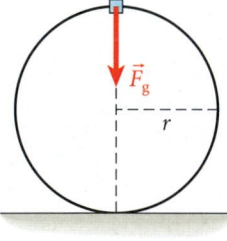

(b)

FIGURE 9.18 (a) Free-body diagram for a passenger at the top of the vertical loop of a roller coaster. (b) Condition for the feeling of weightlessness.

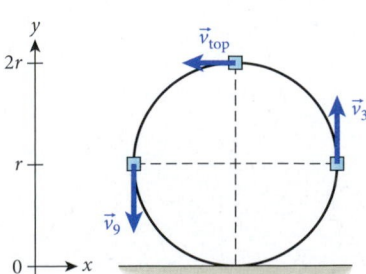

FIGURE 9.19 Directions of the velocity vectors at several points along the vertical roller coaster loop.

SOLVED PROBLEM 9.1 | Analysis of a Roller Coaster

Perhaps the biggest thrill to be had at an amusement park is on a roller coaster with a vertical loop in it (Figure 9.17), where passengers feel almost weightless at the top of the loop.

PROBLEM

Suppose the vertical loop has a radius of 5.00 m. What does the linear speed of the roller coaster have to be at the top of the loop for the passengers to feel weightless? (Assume that friction between roller coaster and rails can be neglected.)

SOLUTION

THINK A person feels weightless when there is no supporting force, from a seat or a restraint, acting to counter his or her weight. For a person to feel weightless at the top of the loop, no normal force can be acting on him or her at this point.

SKETCH The free-body diagrams in Figure 9.18 may help to conceptualize the situation. The force of gravity and the normal force acting on a passenger in the roller coaster at the top of the loop are shown in Figure 9.18a. The sum of these two forces is the net force, which has to equal the centripetal force in circular motion. If the net force (centripetal force here) is equal to the gravitational force, then the normal force is zero, and the passenger feels weightless. This situation is illustrated in Figure 9.18b.

RESEARCH We have just stated that the net force is equal to the centripetal force and that the net force is the sum of the normal force and the force of gravity:

$$\vec{F}_c = \vec{F}_{net} = \vec{F}_g + \vec{N}.$$

For the feeling of weightlessness at the top of the loop, we need $\vec{N} = 0$, and thus

$$\vec{F}_c = \vec{F}_g \Rightarrow F_c = F_g. \tag{i}$$

As always, we have $F_g = mg$. For the magnitude of the centripetal force, we use equation 9.21:

$$F_c = ma_c = m\frac{v^2}{r}.$$

SIMPLIFY After substituting the expressions for the centripetal and the gravitational forces into equation (i), we solve for the linear speed at the top of the loop:

$$F_c = F_g \Rightarrow m\frac{v_{top}^2}{r} = mg \Rightarrow v_{top} = \sqrt{rg}.$$

CALCULATE Using $g = 9.81$ m/s^2 and the value of 5.00 m given for the radius, we obtain

$$v_{top} = \sqrt{(5.00 \text{ m})(9.81 \text{ m/s}^2)} = 7.00357 \text{ m/s}.$$

ROUND Rounding our result to three-digit precision, we have

$$v_{top} = 7.00 \text{ m/s}.$$

DOUBLE-CHECK Obviously, our answer passes the simplest check in that the units are those of speed, meters per second. The formula for the linear speed at the top, $v_{top} = \sqrt{rg}$, indicates that a larger radius necessitates a higher speed, which seems plausible.

Is 7.00 m/s for the speed at the top reasonable? Converting this value gives 15.7 mph, which seems fairly slow for a ride that is typically experienced as tremendously fast. But keep in mind that this is the minimum speed needed at the top of the loop, and the operators of the rides do not want to get too close to this value.

Let's go a step further and calculate the velocity vectors at the 3 o'clock and 9 o'clock positions on the loop, assuming that the roller coaster moves counterclockwise around the loop. The directions of the velocity vectors for circular motion are always tangential to the circle, as shown in Figure 9.19.

How do we obtain the magnitudes of the velocities v_3 (at 3 o'clock) and v_9 (at 9 o'clock)? First, recall from Chapter 6 that the total energy is the sum of kinetic and potential energies, $E = K + U$, that the kinetic energy is $K = \frac{1}{2}mv^2$, and that the gravitational potential energy is proportional to the height above ground, $U = mgy$. In Figure 9.19, the coordinate system is placed so that the zero of the y-axis is at the bottom of the loop. We then can write the equation for the conservation of mechanical energy assuming no nonconservative forces are acting:

$$E = K_3 + U_3 = K_{\text{top}} + U_{\text{top}} = K_9 + U_9 \Rightarrow$$

$$\tfrac{1}{2}mv_3^2 + mgy_3 = \tfrac{1}{2}mv_{\text{top}}^2 + mgy_{\text{top}} = \tfrac{1}{2}mv_9^2 + mgy_9.$$

(ii)

We can see from the figure that the y-coordinates and therefore the potential energies are the same at the 3 o'clock and 9 o'clock positions; therefore, the kinetic energies at both points have to be the same. Consequently, the absolute values of the speeds at both points are the same: $v_3 = v_9$. Solving equation (ii) for v_3, we obtain

$$\tfrac{1}{2}mv_3^2 + mgy_3 = \tfrac{1}{2}mv_{\text{top}}^2 + mgy_{\text{top}} \Rightarrow$$

$$\tfrac{1}{2}v_3^2 + gy_3 = \tfrac{1}{2}v_{\text{top}}^2 + gy_{\text{top}} \Rightarrow$$

$$v_3 = \sqrt{v_{\text{top}}^2 + 2g(y_{\text{top}} - y_3)}.$$

Again, the mass cancels out. Further, only the difference in the y-coordinates enters into this formula for v_3; therefore, the choice of the origin of the coordinate system is irrelevant. The difference in the y-coordinates between the two points is $y_{\text{top}} - y_3 = r$. Inserting the given value of $r = 5.00$ m and the result $v_{\text{top}} = 7.00$ m/s that we found previously, we see that the speed at the 3 o'clock and at the 9 o'clock positions in the loop is

$$v_3 = \sqrt{(7.00 \text{ m/s})^2 + 2(9.81 \text{ m/s}^2)(5.00 \text{ m})} = 12.1 \text{ m/s}.$$

As you can see from this discussion, practically any force can act as the centripetal force. It was the force of static friction for the poker chips on the rotating table and the horizontal component of the tension in the string for the conical pendulum. But it can also be the gravitational force, which forces planets into (almost) circular orbits around the Sun (see Chapter 12), the Coulomb force acting on the electrons in atoms (see Chapter 21), or the normal force from a wall (see the following solved problem).

SOLVED PROBLEM 9.2 Carnival Ride

PROBLEM
One of the rides found at carnivals is a rotating cylinder, as shown in Figure 9.20. The riders step inside the vertical cylinder and stand with their backs against the curved wall. The cylinder spins very rapidly, and at some angular velocity, the floor is pulled away. The thrill-seekers now hang like flies on the wall. (The cylinder in Figure 9.20, is lifted up and tilted after the ride reaches it operating angular velocity, but we will not deal with this additional complication.) If the radius of the cylinder is $r = 2.10$ m, the rotation axis of the cylinder remains vertical, and the coefficient of static friction between the people and the wall is $\mu_s = 0.390$, what is the minimum angular velocity, ω, at which the floor can be withdrawn?

SOLUTION
THINK When the floor drops away, the magnitude of the force of static friction between a rider and the wall of the rotating cylinder must equal the magnitude of the force of gravity acting on the rider. The static friction between the rider and the wall depends on the normal force being exerted on the rider and the coefficient of static friction. As the cylinder spins faster, the normal force (which acts as the centripetal force) being exerted on the rider increases. At a certain angular velocity, the maximum magnitude of the force of static friction will equal the magnitude of the force of gravity. That angular velocity is the minimum angular velocity at which the floor can be withdrawn.

SKETCH A top view of the rotating cylinder is shown in Figure 9.21a. The free-body diagram for one of the riders is shown in Figure 9.21b, where the rotation axis is assumed to be the y-axis. In this sketch, $\vec{f}$ is the force of static friction, $\vec{N}$ is the normal force exerted on the rider of mass m by the wall of the cylinder, and $\vec{F}_g$ is the force of gravity acting on the rider.

– Continued

Concept Check 9.5

When you go through a vertical loop on a high-speed roller coaster, what keeps you in your seat?

a) centrifugal force

b) the normal force from the track

c) the force of gravity

d) the force of friction

e) the force exerted by the seat belt

Self-Test Opportunity 9.6

What speed must the roller coaster of Solved Problem 9.1 have at the top of the loop to accomplish the same feeling of weightlessness if the radius of the loop is doubled?

FIGURE 9.20 A carnival ride consisting of a rotating cylinder.

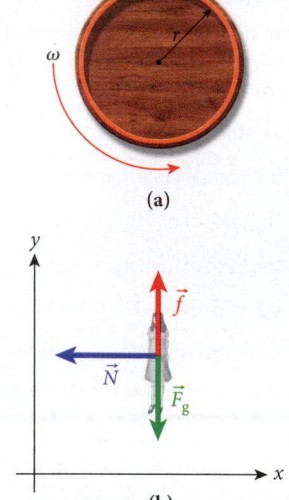

FIGURE 9.21 (a) Top view of the rotating cylinder of a carnival ride. (b) Free-body diagram for one of the riders.

RESEARCH   At the minimum angular velocity required to keep the rider from falling, the magnitude of the force of static friction between the rider and the wall is equal to the magnitude of the force of gravity acting on the rider. To analyze these forces, we start with the free-body diagram shown in Figure 9.21b. In the free-body diagram, the x-direction is along the radius of the cylinder, and the y-direction is vertical. In the x-direction, the normal force exerted by the wall on the rider provides the centripetal force that makes the rider move in a circle:

$$F_c = N. \tag{i}$$

In the y-direction, the thrill-seeking rider sticks to the wall only if the upward force of static friction between the rider and wall balances the downward force of gravity. The force of gravity on the rider is his or her weight, so we can write

$$f = F_g = mg. \tag{ii}$$

We know that the centripetal force is given by

$$F_c = mr\omega^2, \tag{iii}$$

and the force of static friction is given by

$$f \le f_{\max} = \mu_s N. \tag{iv}$$

SIMPLIFY   We can combine equations (ii) and (iv) to obtain

$$mg \le \mu_s N. \tag{v}$$

Substituting for F_c from equation (i) into (iii), we find

$$N = mr\omega^2. \tag{vi}$$

Combining equations (v) and (vi), we have

$$mg \le \mu_s mr\omega^2,$$

which we can solve for ω:

$$\omega \ge \sqrt{\frac{g}{\mu_s r}}.$$

Thus, the minimum value of the angular velocity is given by

$$\omega_{\min} = \sqrt{\frac{g}{\mu_s r}}.$$

Note that the mass of the rider canceled out. This is crucial since people of different masses want to ride at the same time!

CALCULATE   Putting in the numerical values, we find

$$\omega_{\min} = \sqrt{\frac{g}{\mu_s r}} = \sqrt{\frac{9.81 \text{ m/s}^2}{(0.390)(2.10 \text{ m})}} = 3.46093 \text{ rad/s}.$$

ROUND   Expressing our result to the three significant figures that the numbers in the problem statement had gives

$$\omega_{\min} = 3.46 \text{ rad/s}.$$

DOUBLE-CHECK   To double-check, let's express our result for the angular velocity in revolutions per minute (rpm):

$$3.46 \, \frac{\text{rad}}{\text{s}} = \left(3.46 \, \frac{\text{rad}}{\text{s}}\right)\left(\frac{60 \text{ s}}{1 \text{ min}}\right)\left(\frac{1 \text{ rev}}{2\pi \text{ rad}}\right) = 33 \text{ rpm}.$$

An angular velocity of 33 rpm for the rotating cylinder seems reasonable, because this means that it completes almost one full turn every 2 s. If you have ever been on one of these rides or watched one, you know that the answer is in the right ballpark.

Note that the coefficient of friction, μ_s, between the clothing of the rider and the wall is not identical in all cases. Our formula, $\omega_{\min} = \sqrt{g/(\mu_s r)}$, indicates that a smaller coefficient of friction necessitates a larger angular velocity. Designers of this kind of ride need to make sure that they allow for the smallest coefficient of friction that can be expected to occur. Obviously, they want to have a somewhat sticky contact surface on the wall of the ride, just to make sure!

One last point to check: $\omega_{\min} = \sqrt{g/(\mu_s r)}$ indicates that the minimum angular velocity required will decrease as a function of the radius of the cylinder. The example of the poker chips on the spinning table presented earlier in this section established that the centripetal force increases with radial distance, which is consistent with this result.

Is There a Centrifugal Force?

This is a good time to clarify an important point regarding the direction of the force responsible for circular motion. You often hear people talking about centrifugal (or "center-fleeing," in the radial outward direction) acceleration or centrifugal force (mass times acceleration). You can experience the sensation of seemingly being pulled outward on many rotating amusement park rides, like the one considered in Solved Problem 9.2. This sensation is due to your body's inertia, which resists the centripetal acceleration toward the center. Thus, you feel a seemingly outward-pointing force—the centrifugal force. Keep in mind that this perception is due to your body moving in an accelerated reference frame; there is no centrifugal force. The real force that acts on your body and forces it to move on a circular path is the centripetal force, and it points inward. Nevertheless, the effect in the rotating frame is the same as that of a force pointing in the outward direction. This is why NASA's centrifuge, shown in Figure 9.1, can simulate up to 20g of gravitational acceleration for astronauts and equipment. The effect of an outward-pointing force felt by an observer inside the centrifuge is real, but it is caused by the observer's experiencing a continuous acceleration in the inward direction. Chapter 35 on relativity will continue this discussion.

You have also experienced a similar effect in straight-line motion. When you are sitting in your car at rest and then step on the gas pedal, you feel as if you are being pressed back into your seat. This sensation of a force that presses you in the backward direction also comes from the inertia of your body, which is accelerated forward by your car. Both of these sensations of forces acting on your body—the "centrifugal" force and the force "pushing" you into the seat—are the result of your body experiencing an acceleration in the opposite direction and putting up resistance, inertia, against this acceleration.

Concept Check 9.6

At the top of a vertical loop in a roller coaster, what condition must be met for the car to stay on the track?

a) The centrifugal force acting on the car must equal the centripetal force.

b) The normal force exerted by the track on the car must be equal to the force of gravity.

c) The normal force exerted by the track on the car must be in the direction opposite to the force of gravity.

d) The centripetal force required to keep the car moving in a circle must be equal to or greater than the force of gravity.

e) The normal force exerted by the track on the car must be zero.

9.6 Circular and Linear Motion

Table 9.1 summarizes the relationships between linear and angular quantities for circular motion. The relationships shown in the table relate the angular quantities (θ, ω, and α) to the linear quantities (s, v, and a). The radius r of the circular path is constant and provides the connection between the two sets of quantities. (In Chapter 10, the rotational counterparts to mass, kinetic energy, momentum, and force will be added to this list.)

As we have just seen, there is a formal correspondence between motion on a straight line with constant velocity and circular motion with constant angular velocity. However, there is one big difference. As we saw in Section 3.6 on relative motion, you cannot always distinguish between moving with constant velocity in a straight line and being at rest. This is because the origin of the coordinate system can be placed at any point—even a point that moves with constant velocity. The physics of translational motion does not change under this Galilean transformation. In contrast, in circular motion, you are always moving on a circular path with a well-defined center. Experiencing the "centrifugal" force is then a sure sign of circular motion, and the strength of that force is a measure of the magnitude of angular velocity. You may argue that you are in constant circular motion around the center of the Earth, around the center of the Solar System, and around the center of the Milky Way Galaxy but you do not feel the effects of those circular motions. True, but the very small magnitudes of the angular velocities involved in these motions cause their perceived effects to be negligible.

Table 9.1	Comparison of Kinematical Variables for Circular Motion		
Quantity	**Linear**	**Angular**	**Relationship**
Displacement	s	θ	$s = r\theta$
Velocity	v	ω	$v = r\omega$
Acceleration	a	α	$a_t = r\alpha$
			$a_c = r\omega^2$
			$\vec{a} = r\alpha\hat{t} - r\omega^2\hat{r}$

Constant Angular Acceleration

Chapter 2 discussed at some length the special case of constant acceleration. Under this assumption, we derived five equations that proved useful in solving all kinds of problems. For ease of reference, here are those five equations of linear motion with constant acceleration:

$$\text{(i)}\qquad x = x_0 + v_{x0}t + \tfrac{1}{2}a_x t^2$$

$$\text{(ii)}\qquad x = x_0 + \bar{v}_x t$$

$$\text{(iii)}\qquad v_x = v_{x0} + a_x t$$

$$\text{(iv)}\qquad \bar{v}_x = \tfrac{1}{2}(v_x + v_{x0})$$

$$\text{(v)}\qquad v_x^2 = v_{x0}^2 + 2a_x(x - x_0).$$

Now we'll take the same steps as in Chapter 2 to derive the equivalent equations for constant angular acceleration. We start with equation 9.14 and integrate, using the usual notation convention of $\omega_0 \equiv \omega(t_0)$:

$$\alpha(t) = \frac{d\omega}{dt} \Rightarrow$$

$$\int_{t_0}^{t} \alpha(t')\,dt' = \int_{t_0}^{t} \frac{d\omega(t')}{dt'}\,dt' = \omega(t) - \omega(t_0) \Rightarrow$$

$$\omega(t) = \omega_0 + \int_{t_0}^{t} \alpha(t')\,dt'.$$

This relationship is the inverse of equation 9.14 and holds in general. If we assume the angular acceleration, α, is constant in time, we can evaluate the integral and obtain

$$\omega(t) = \omega_0 + \alpha \int_{0}^{t} dt' = \omega_0 + \alpha t. \qquad (9.24)$$

For convenience, we have set $t_0 = 0$, just as we did in Chapter 2 at this juncture. Next, we use equation 9.8, expressing the angular velocity as the derivative of the angle with respect to time and with the notation $\theta_0 = \theta(t = 0)$:

$$\frac{d\theta(t)}{dt} = \omega(t) = \omega_0 + \alpha t \Rightarrow$$

$$\theta(t) = \theta_0 + \int_{0}^{t} \omega(t')\,dt' = \theta_0 + \int_{0}^{t} (\omega_0 + \alpha t')\,dt' \Rightarrow$$

$$= \theta_0 + \omega_0 \int_{0}^{t} dt' + \alpha \int_{0}^{t} t'\,dt' \Rightarrow$$

$$\theta(t) = \theta_0 + \omega_0 t + \tfrac{1}{2}\alpha t^2. \qquad (9.25)$$

Comparing equations 9.24 and 9.25 to equations (iii) and (i) for linear motion reveals that these two equations are the circular motion equivalents of the two kinematical equations for straight-line linear motion in one dimension. With the straightforward substitutions $x \to \theta$, $v_x \to \omega$, and $a_x \to \alpha$, we can write five kinematical equations for circular motion under constant angular acceleration:

$$\text{(i)}\qquad \theta = \theta_0 + \omega_0 t + \tfrac{1}{2}\alpha t^2$$

$$\text{(ii)}\qquad \theta = \theta_0 + \bar{\omega}t$$

$$\text{(iii)}\qquad \omega = \omega_0 + \alpha t \qquad (9.26)$$

$$\text{(iv)}\qquad \bar{\omega} = \tfrac{1}{2}(\omega + \omega_0)$$

$$\text{(v)}\qquad \omega^2 = \omega_0^2 + 2\alpha(\theta - \theta_0).$$

Self-Test Opportunity 9.7

The derivation has been provided for two of these five kinematical equations for circular motion. Can you provide the derivation for the remaining equations? (*Hint:* The chain of reasoning proceeds along exactly the same lines as Derivation 2.1 in Chapter 2.)

EXAMPLE 9.7 | Hammer Throw

One of the most interesting events in track-and-field competitions is the hammer throw. The task is to throw the "hammer," a 12-cm-diameter iron ball attached to a grip by a steel cable, a maximum distance. The hammer's total length is 121.5 cm, and its total mass is 7.26 kg. The athlete has to accomplish the throw from within a circle of radius 2.13 m (7 ft), and the best way to throw the hammer is for the athlete to spin, allowing the hammer to move in a circle around him, before releasing it. At the 1988 Olympic Games in Seoul, the Russian thrower Sergey Litvinov won the gold medal with an Olympic record distance of 84.80 m. He took seven turns before releasing the hammer, and the period to complete each turn was obtained from examining the video recording frame by frame: 1.52 s, 1.08 s, 0.72 s, 0.56 s, 0.44 s, 0.40 s, and 0.36 s.

PROBLEM 1

What was the average angular acceleration during the seven turns? Assume constant angular acceleration for the solution, and then check whether this assumption is justified.

SOLUTION 1

In order to find the average angular acceleration, we add all the time intervals for the seven turns, obtaining the total time:

$$t_{all} = 1.52 \text{ s} + 1.08 \text{ s} + 0.72 \text{ s} + 0.56 \text{ s} + 0.44 \text{ s} + 0.40 \text{ s} + 0.36 \text{ s} = 5.08 \text{ s}.$$

During this time, Litvinov completed seven full turns, resulting in a total angle of

$$\theta_{all} = 7(2\pi \text{ rad}) = 14\pi \text{ rad}.$$

Because we are assuming constant acceleration, we can readily solve for the angular acceleration and insert the given numbers to obtain our answer:

$$\theta = \tfrac{1}{2}\alpha t^2 \Rightarrow \alpha = \frac{2\theta_{all}}{t^2} = 2\frac{14\pi \text{ rad}}{(5.08 \text{ s})^2} = 3.41 \text{ rad/s}^2.$$

DISCUSSION

Because we know how long it took to complete each turn, we can generate a plot of the angle of the hammer in the horizontal plane as a function of time. This plot is shown in Figure 9.22, with the red dots representing the data points. The blue line that is fit to the data points in Figure 9.22 assumes a constant angular acceleration of $\alpha = 3.41 \text{ rad/s}^2$. As you can see, the assumption of constant angular acceleration is almost, but not quite justified.

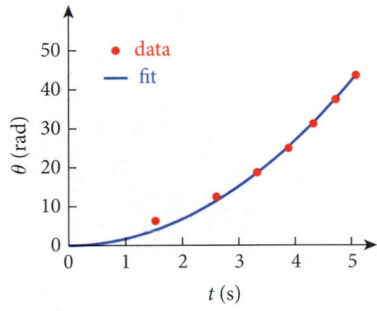

FIGURE 9.22 Angle as a function of time for Sergey Litvinov's 1988 gold-medal–winning hammer throw.

PROBLEM 2

Assuming that the radius of the circle on which the hammer moves is 1.67 m (the length of the hammer plus the arms of the athlete), what is the linear speed with which the hammer is released?

SOLUTION 2

With constant angular acceleration from rest for a period of 5.08 s, the final angular velocity is

$$\omega = \alpha t = (3.41 \text{ rad/s}^2)(5.08 \text{ s}) = 17.3 \text{ rad/s}.$$

Using the relationship between linear and angular velocity, we obtain the linear speed at release:

$$v = r\omega = (1.67 \text{ m})(17.3 \text{ rad/s}) = 28.9 \text{ m/s}.$$

PROBLEM 3

What is the centripetal force that the hammer thrower has to exert on the hammer right before he releases it?

SOLUTION 3

The centripetal acceleration right before release is given by

$$a_c = \omega^2 r = (17.3 \text{ rad/s})^2 (1.67 \text{ m}) = 501. \text{ m/s}^2.$$

With a mass of 7.26 kg for the hammer, the centripetal force required is

$$F_c = ma_c = (7.26 \text{ kg})(501. \text{ m/s}^2) = 3640 \text{ N}.$$

This is an astonishingly large force, equivalent to the weight of an object of mass 371 kg! This is why world-class hammer throwers have to be very strong.

– *Continued*

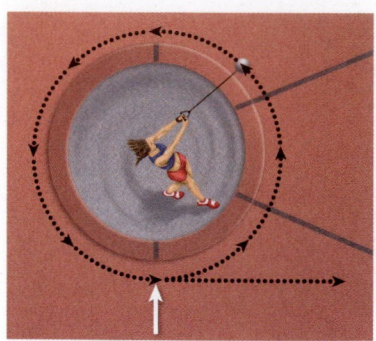

FIGURE 9.23 Overhead view of the trajectory of the hammer (black dots, with the arrows indicating the direction of the velocity vector) during the time the athlete holds on to it (circular path) and after release (straight line). The white arrow marks the point of release.

PROBLEM 4
After release, what is the direction in which the hammer moves?

SOLUTION 4
It is a common misconception that the hammer "spirals" in some sort of circular motion with an ever increasing radius after release. This idea is wrong because there is no horizontal component of force once the athlete releases the hammer. Newton's Second Law tells us that there will be no horizontal component of acceleration and hence no centripetal acceleration. Instead, the hammer moves in a direction that is tangential to the circle at the point of release. If you were to look straight down on the stadium from a blimp, you would see that the hammer moves on a straight line, as shown in Figure 9.23. Viewed from the side, the shape of the hammer's trajectory is a parabola, as shown in Chapter 3.

SOLVED PROBLEM 9.3   Flywheel

PROBLEM
The flywheel of a steam engine starts to rotate from rest with a constant angular acceleration of $\alpha = 1.43$ rad/s^2. The flywheel undergoes this constant angular acceleration for $t = 25.9$ s and then continues to rotate at a constant angular velocity, ω. After the flywheel has been rotating for 59.5 s, what is the total angle through which it has rotated since it started?

SOLUTION

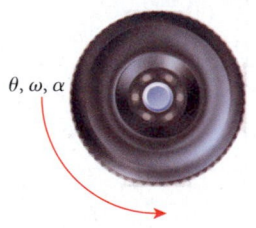
θ, ω, α

FIGURE 9.24 Top view of the rotating flywheel.

THINK Here we are trying to determine the total angular displacement, θ. For the time interval when the flywheel is undergoing angular acceleration, we can use equation 9.26(i) with $\theta_0 = 0$ and $\omega_0 = 0$. When the flywheel is rotating at a constant angular velocity, we use equation 9.26(i) with $\theta_0 = 0$ and $\alpha = 0$. To get the total angular displacement, we add these two angular displacements.

SKETCH A top view of the rotating flywheel is shown in Figure 9.24.

RESEARCH Let's call the time during which the flywheel is undergoing angular acceleration t_a and the total time the flywheel is rotating t_b. Thus, the flywheel rotates at a constant angular velocity for a time interval equal to $t_b - t_a$. The angular displacement, θ_a, that occurs while the flywheel is undergoing angular acceleration is given by

$$\theta_a = \tfrac{1}{2}\alpha t_a^2. \tag{i}$$

The angular displacement, θ_b, that occurs while the flywheel is rotating at the constant angular velocity, ω, is given by

$$\theta_b = \omega\left(t_b - t_a\right). \tag{ii}$$

The angular velocity, ω, reached by the flywheel after undergoing the angular acceleration α for a time t_a is given by

$$\omega = \alpha t_a. \tag{iii}$$

The total angular displacement is given by

$$\theta_{\text{total}} = \theta_a + \theta_b. \tag{iv}$$

SIMPLIFY We can combine equations (ii) and (iii) to obtain the angular displacement while the flywheel is rotating at a constant angular velocity:

$$\theta_b = \left(\alpha t_a\right)\left(t_b - t_a\right) = \alpha t_a t_b - \alpha t_a^2. \tag{v}$$

We can combine equations (v), (iv), and (i) to get the total angular displacement of the flywheel:

$$\theta_{\text{total}} = \theta_a + \theta_b = \tfrac{1}{2}\alpha t_a^2 + \left(\alpha t_a t_b - \alpha t_a^2\right) = \alpha t_a t_b - \tfrac{1}{2}\alpha t_a^2.$$

CALCULATE Putting in the numerical values gives us

$$\theta_{\text{total}} = \alpha t_a t_b - \tfrac{1}{2}\alpha t_a^2 = \left(1.43 \text{ rad/s}^2\right)\left(25.9 \text{ s}\right)\left(59.5 \text{ s}\right) - \tfrac{1}{2}\left(1.43 \text{ rad/s}^2\right)\left(25.9 \text{ s}\right)^2$$

$$= 1724.07 \text{ rad.}$$

ROUND Expressing our result to three significant figures, we have

$$\theta_{\text{total}} = 1720 \text{ rad.}$$

DOUBLE-CHECK It is comforting that our answer has the right unit, the rad. Our formula, $\theta_{\text{total}} = \alpha t_a t_b - \frac{1}{2}\alpha t_a^2 = \alpha t_a (t_b - \frac{1}{2}t_a)$, gives a value that increases linearly with the value of the angular acceleration. It is also always larger than zero, as expected, because $t_b > t_a$.

To perform a further check, let's calculate the angular displacement in two steps. The first step is to calculate the angular displacement while the flywheel is accelerating

$$\theta_a = \tfrac{1}{2}\alpha t_a^2 = \tfrac{1}{2}\left(1.43 \text{ rad/s}^2\right)\left(25.9 \text{ s}\right)^2 = 480 \text{ rad.}$$

The angular velocity of the flywheel after the angular acceleration ends is

$$\omega = \alpha t_a = \left(1.43 \text{ rad/s}^2\right)\left(25.9 \text{ s}\right) = 37.0 \text{ rad/s.}$$

Next, we calculate the angular displacement while the flywheel is rotating at constant velocity:

$$\theta_b = \omega\left(t_b - t_a\right) = \left(37.0 \text{ rad/s}\right)\left(59.5 \text{ s} - 25.9 \text{ s}\right) = 1240 \text{ rad.}$$

The total angular displacement is then

$$\theta_{\text{total}} = \theta_a + \theta_b = 480 \text{ rad} + 1240 \text{ rad} = 1720 \text{ rad,}$$

which agrees with our answer.

9.7 More Examples for Circular Motion

Let's look at another example and solved problem that demonstrate how useful the concepts of circular motion we've just discussed are.

EXAMPLE 9.8 / Formula 1 Racing

If you watch a Formula 1 race, you can see that the race cars approach curves from the outside, cut through to the inside, and then drift again to the outside, as shown by the red path in Figure 9.25a. The blue path is shorter. Why don't the drivers follow the shortest path?

PROBLEM

Suppose that cars move through the U-turn shown in Figure 9.25a at constant speed and that the coefficient of static friction between the tires and the road is $\mu_s = 1.2$. (As was mentioned in Chapter 4, modern race car tires can have coefficients of friction that exceed 1 when they are heated to race temperature and thus are very sticky.) If the radius of the inner curve shown in the figure is $R_B = 10.3$ m and radius of the outer is $R_A = 32.2$ m and the cars move at their maximum speed, how much time will it take to move from point A to A' and from point B to B'?

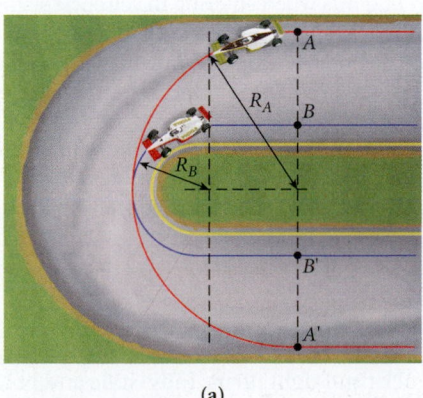

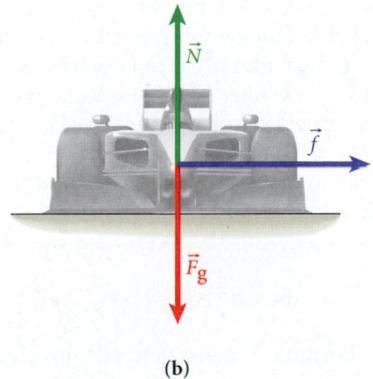

(a)   (b)

FIGURE 9.25 (a) Paths of race cars negotiating a turn on an oval track in two ways. (b) Free-body diagram for a race car in a curve.

– Continued

SOLUTION

We start by drawing a free-body diagram, as shown in Figure 9.25b. The diagram shows all forces that act on the car with the force arrows originating at the center of mass of the car. The force of gravity, acting downward with magnitude $F_g = mg$, is shown in red. This force is exactly balanced by the normal force, which the road exerts on the car, shown in green. As a car makes the turn, a net force is required to change the car's velocity vector and act as the centripetal force that pushes the car onto a circular path. This net force is generated by the friction force (shown in blue) between the car's tires and the road. This force arrow points horizontally and inward, toward the center of the curve. As usual, the magnitude of the friction force is the product of the normal force and the coefficient of friction: $f_{max} = \mu_s mg$. (*Note:* In this case, we use the equals sign, because the race car drivers push their cars and tires to the limit and thus reach the maximum possible static friction force.) The arrow for the friction force is longer than that for the normal force by a factor of 1.2, because $\mu_s = 1.2$.

First, we need to calculate the maximum velocity that a race car can have on each trajectory. For each radius of curvature, R, the resulting centripetal force, $F_c = mv^2/R$, must be provided by the friction force, $f_{max} = \mu_s mg$:

$$m\mu_s g = m\frac{v^2}{R} \Rightarrow v = \sqrt{\mu_s gR}.$$

Therefore, for the red and blue curves, we get

$$v_{red} = \sqrt{\mu_s gR_A} = \sqrt{(1.2)(9.81 \text{ m/s}^2)(32.2 \text{ m})} = 19.5 \text{ m/s}$$

$$v_{blue} = \sqrt{\mu_s gR_B} = \sqrt{(1.2)(9.81 \text{ m/s}^2)(10.3 \text{ m})} = 11.0 \text{ m/s}.$$

These speeds are only about 43.6 mph and 24.6 mph, respectively! However, the curve shown is a very tight one, typically only encountered on a city course like the one in Monaco.

Even though a car can move much faster on the red curve, the blue curve is shorter than the red one. For the path length of the red curve, we simply have the distance along the semicircle, $\ell_{red} = \pi R_A = 101.$ m. For the path length of the blue curve, we have to add the two straight sections and the semicircular curve with the smaller radius:

$$\ell_{blue} = \pi R_B + 2(R_A - R_B) = 76.2 \text{ m}.$$

We then get for the time to go from A to A' on the red path:

$$t_{red} = \frac{\ell_{red}}{v_{red}} = \frac{101. \text{ m}}{19.5 \text{ m/s}} = 5.20 \text{ s}.$$

To travel along the blue curve from B to B' takes

$$t_{blue} = \frac{\ell_{blue}}{v_{blue}} = \frac{76.2 \text{ m}}{11.0 \text{ m/s}} = 6.92 \text{ s}.$$

With the assumption that the cars have to use a constant speed, it is clearly a big advantage to cut through the curve, as shown by the red path.

DISCUSSION

In a race situation, it is unreasonable to expect the car on the blue path to move along the straight-line segments with constant velocity. Instead, the driver will come to point B with the maximum speed that allows him to slow down to 11.0 m/s when entering the circular segment. If we worked this out in detail, we would find that the blue path is still slower, but not by much. In addition, the car following the blue path can reach point B with a slightly higher speed than that at which the red car can reach point A. In other words, we have a tradeoff to consider before declaring a winner in this situation. In real races, cars slow down as they approach a curve and then accelerate as they come out of the curve. The most advantageous path for cutting through a curve is not the red semicircle, but a path that looks closer to elliptical, starting on the outside, cutting to the extreme inside in the middle of the turn, and then drifting to the outside again while accelerating out of the turn.

Formula 1 racing generally involves flat tracks and tight turns. Indy-style and NASCAR races take place on tracks with bigger turning radii as well as banked curves. To study the forces involved in this type of racing situation, we must combine concepts of static equilibrium on an inclined plane with concepts of circular motion.

SOLVED PROBLEM 9.4 NASCAR Racing

As a NASCAR racer moves through a banked curve, the banking helps the driver achieve higher speeds. Let's see how. Figure 9.26 shows a race car on a banked curve.

PROBLEM
If the coefficient of static friction between the track surface and the car's tires is $\mu_s = 0.620$ and the radius of the turn is $R = 110.$ m, what is the maximum speed with which a driver can take a curve banked at $\theta = 21.1°$? (This is a fairly typical banking angle for NASCAR tracks. Indianapolis has only 9° banking, but there are some tracks with banking angles over 30°, including Daytona (31°), Talladega (33°), and Bristol (36°).)

FIGURE 9.26 Race car on a banked curve.

SOLUTION

THINK The three forces acting on the race car are gravity, $\vec{F}_g$, the normal force, $\vec{N}$, and friction, $\vec{f}$. The curve is banked at an angle θ, which is also the angle between the normal to the track surface and the gravity force vector, as shown in Figure 9.27a. To draw the vector for the force of friction, we have assumed that the car has entered the curve at high speed, so the direction of the force of friction is down the incline. In contrast to the situation of static equilibrium, these three forces do not add up to zero, but instead add up to a net force, $\vec{F}_{net}$, as shown in Figure 9.27b. This net force has to provide the centripetal force, $\vec{F}_c$, which forces the car to move in a circle. Thus, the net force has to act in the horizontal direction because this is the direction of the center of the circle in which the car is moving.

SKETCH The free-body diagram for the race car on the banked curve, showing the x- and y-components of the forces, is presented in Figure 9.27c. The orientation for the coordinate system was selected to give a horizontal x-axis and a vertical y-axis.

RESEARCH Like problems involving linear motion, we can solve problems involving circular motion by starting with the familiar Newton's Second Law: $\sum \vec{F} = m\vec{a}$. And just as in the linear case, we can generally work the problems in Cartesian components. From the free-body diagram in Figure 9.27c, we can see that the x-components of the forces acting on the race car are

$$N \sin\theta + f \cos\theta = F_{net}. \tag{i}$$

Similarly, the forces acting in the y-direction are

$$N \cos\theta - F_g - f \sin\theta = 0. \tag{ii}$$

As usual, the maximum friction force is given by the product of the coefficient of friction and the normal force: $f = \mu_s N$. The gravitational force is the product of mass and gravitational acceleration: $F_g = mg$.

The key to solving this problem is to realize that the net force has to be the force that causes the race car to move through the curve, that is, that provides the centripetal force. Therefore, using the expression for the centripetal force from equation 9.21, we have

$$F_{net} = F_c = m\frac{v^2}{R},$$

where R is the radius of the curve.

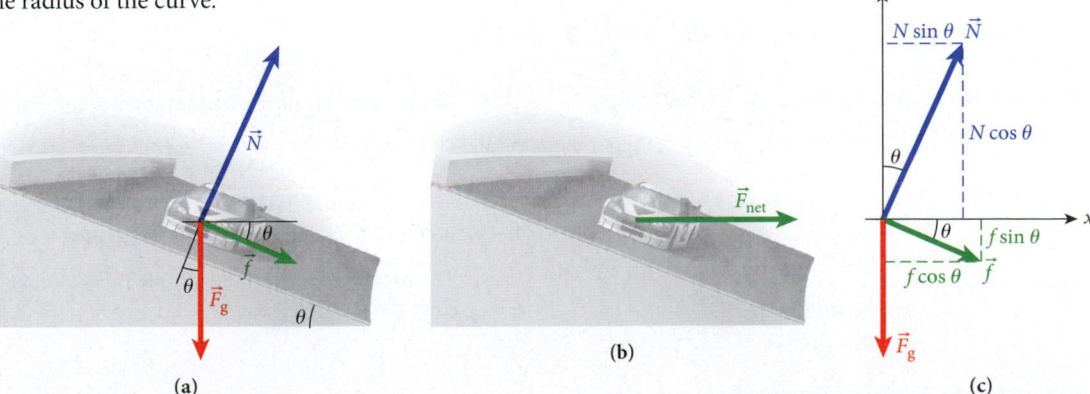

FIGURE 9.27 (a) Forces on a race car going around a banked curve on a racetrack. (b) The net force, the sum of the three forces in part (a). (c) A free-body diagram showing the x- and y-components of the forces acting on the car.

– *Continued*

SIMPLIFY We insert the expressions for the maximum friction force, the gravitational force, and the net force into equations (i) and (ii) for the x- and y-components of the forces:

$$N \sin\theta + \mu_s N \cos\theta = m\frac{v^2}{R} \Rightarrow N(\sin\theta + \mu_s \cos\theta) = m\frac{v^2}{R}$$

$$N \cos\theta - mg - \mu_s N \sin\theta = 0 \Rightarrow N(\cos\theta - \mu_s \sin\theta) = mg.$$

This is a system of two equations for two unknown quantities: the magnitude of the normal force, N, and the speed of the car, v. It is easy to eliminate N by dividing the first equation above by the second:

$$\frac{\sin\theta + \mu_s \cos\theta}{\cos\theta - \mu_s \sin\theta} = \frac{v^2}{gR}.$$

We solve for v:

$$v = \sqrt{\frac{Rg(\sin\theta + \mu_s \cos\theta)}{\cos\theta - \mu_s \sin\theta}}. \tag{iii}$$

Note that the mass of the car, m, canceled out. Thus, what matters in this situation is the coefficient of friction between the tires and the track surface, the radius of the turn, and the angle of banking.

CALCULATE Putting in the numbers, we obtain

$$v = \sqrt{\frac{(110.\ \text{m})(9.81\ \text{m/s}^2)[\sin 21.1° + 0.620(\cos 21.1°)]}{\cos 21.1° - 0.620(\sin 21.1°)}} = 37.7726\ \text{m/s}.$$

ROUND Expressing our result to three significant figures gives us
$$v = 37.8\ \text{m/s}.$$

DOUBLE-CHECK To double-check our result, let's compare the speed for a banked curve to the maximum speed a race car can achieve on a curve with the same radius but without banking. Without banking, the only force keeping the race car on the circular path is the force of friction. Then our result reduces to the equation $v = \sqrt{\mu_s gR}$ that we found in Example 9.8, and we can insert the numerical values given here to obtain the maximum speed around a flat curve of the same radius:

$$v = \sqrt{\mu_s gR} = \sqrt{(0.620)(9.81\ \text{m/s}^2)(110.\ \text{m})} = 25.9\ \text{m/s}.$$

Our result for the maximum speed around a banked curve, 37.8 m/s (84.6 mph), is considerably larger than this result for a flat curve, 25.9 m/s (57.9 mph), which seems reasonable.

Note that the vector for the force of friction in Figure 9.27 points along the surface of the track and toward the inside of the curve, just as the friction force vector for the nonbanked curve does in Figure 9.25b. However, you can see that as the banking angle increases, it reaches a value for which the denominator of the velocity formula, equation (iii), approaches zero. This occurs when $\cot\theta = \mu_s$. For the given value, $\mu_s = 0.620$, this angle is 58.2°. For larger angles, the driver needs to maintain a minimum speed driving through the curve to prevent the car from sliding to the bottom of the incline.

WHAT WE HAVE LEARNED | EXAM STUDY GUIDE

- The conversion between Cartesian coordinates, x and y, and polar coordinates, r and θ, is given by
$$r = \sqrt{x^2 + y^2}$$
$$\theta = \tan^{-1}(y/x).$$

- The conversion between polar coordinates and Cartesian coordinates is given by
$$x = r\cos\theta$$
$$y = r\sin\theta.$$

- For circular motion, the linear displacement, s, is related to the angular displacement, θ, by $s = r\theta$, where r is the radius of the circular path and θ is measured in radians.

- The magnitude of the instantaneous angular velocity, ω, is given by $\omega = \dfrac{d\theta}{dt}$.

- The magnitude of the angular velocity is related to the magnitude of the linear velocity, v, by $v = r\omega$.

- The magnitude of the instantaneous angular acceleration, α, is given by
$$\alpha = \frac{d\omega}{dt} = \frac{d^2\theta}{dt^2}.$$

- The magnitude of the angular acceleration is related to the magnitude of the tangential acceleration, a_t, by $a_t = r\alpha$.

- The magnitude of the centripetal acceleration, a_c, required to keep an object moving in a circle with constant angular velocity is given by $a_c = \omega^2 r = \dfrac{v^2}{r}$.

- The magnitude of the total acceleration of an object in circular motion is $a = \sqrt{a_t^2 + a_c^2} = r\sqrt{\alpha^2 + \omega^4}$.

- The kinematical equations for circular motion are
$$\theta = \theta_0 + \omega_0 t + \tfrac{1}{2}\alpha t^2$$
$$\theta = \theta_0 + \bar{\omega} t$$
$$\omega = \omega_0 + \alpha t$$
$$\bar{\omega} = \tfrac{1}{2}(\omega + \omega_0)$$
$$\omega^2 = \omega_0^2 + 2\alpha(\theta - \theta_0).$$

ANSWERS TO SELF-TEST OPPORTUNITIES

9.1 Differential arc length is $R\,d\theta$ for a circle with radius R; integral arc length around the circle is the circumference C:

$$C = \int_0^{2\pi} r\,d\theta = r\int_0^{2\pi} d\theta = r[\theta]_0^{2\pi} = 2\pi r.$$

9.2 The differential area is shown in the sketch. The differential area is $dA = 2\pi dr$. Area of a circle is

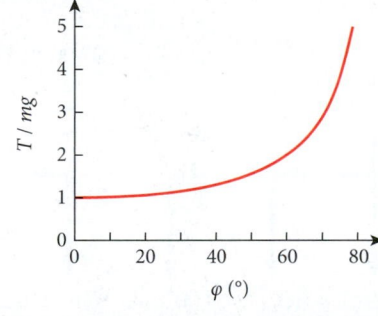

$$\int_0^R 2\pi r\,dr = 2\pi \int_0^R r\,dr$$

$$= 2\pi \left[\frac{r^2}{2}\right]_0^R = \pi R^2.$$

9.3 $\omega^2 r = (2\pi/\text{yr})^2 (1\text{ au}) = 5.9 \cdot 10^{-3}\text{ m/s}^2 \approx g/1700$.

9.4 At $r_1 = 25$ mm, the magnitude of the tangential acceleration is $a_t = \alpha r_1 = 1.6 \cdot 10^{-4}\text{ m/s}^2$, and that of the centripetal acceleration is $a_c = v\omega(r_1) = 59.\text{ m/s}^2$, larger by more than four orders of magnitude. At $r_2 = 58$ mm, the accelerations are $a_t = \alpha r_2 = 3.6 \cdot 10^{-4}\text{ m/s}^2$ and $a_c = v\omega(r_2) = 25.\text{ m/s}^2$.

9.5

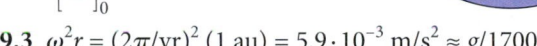

9.6 If the radius is doubled, the speed at the top of the loop has to increase by a factor of $\sqrt{2}$. Thus, the required speed is $(7.00\text{ m/s})(\sqrt{2}) = 9.90\text{ m/s}$.

9.7 (iv) $\bar{\omega} = \dfrac{1}{t}\displaystyle\int_0^t \omega(t')\,dt' = \dfrac{1}{t}\int_0^t (\omega_0 + \alpha t')\,dt'$

$= \dfrac{\omega_0}{t}\displaystyle\int_0^t dt' + \dfrac{\alpha}{t}\int_0^t t'\,dt' = \omega_0 + \tfrac{1}{2}\alpha t$

$= \tfrac{1}{2}\omega_0 + \tfrac{1}{2}(\omega_0 + \alpha t)$

$= \tfrac{1}{2}(\omega_0 + \omega)$

(ii) $\bar{\omega} = \omega_0 + \tfrac{1}{2}\alpha t$

$\Rightarrow \bar{\omega}t = \omega_0 t + \tfrac{1}{2}\alpha t^2$

$\theta = \theta_0 + \omega_0 t + \tfrac{1}{2}\alpha t^2 = \theta_0 + \bar{\omega}t$

(v) $\theta = \theta_0 + \omega_0 t + \tfrac{1}{2}\alpha t^2$

$= \theta_0 + \omega_0\left(\dfrac{\omega - \omega_0}{\alpha}\right) + \tfrac{1}{2}\alpha\left(\dfrac{\omega - \omega_0}{\alpha}\right)^2$

$= \theta_0 + \dfrac{\omega\omega_0 - \omega_0^2}{\alpha} + \tfrac{1}{2}\dfrac{\omega^2 + \omega_0^2 - 2\omega\omega_0}{\alpha}$

Now we subtract θ_0 from both sides of the equation and then multiply by α:

$\alpha(\theta - \theta_0) = \omega\omega_0 - \omega_0^2 + \tfrac{1}{2}(\omega^2 + \omega_0^2 - 2\omega\omega_0)$

$\Rightarrow \alpha(\theta - \theta_0) = \tfrac{1}{2}\omega^2 - \tfrac{1}{2}\omega_0^2$

$\Rightarrow \omega^2 = \omega_0^2 + 2\alpha(\theta - \theta_0)$

PROBLEM-SOLVING GUIDELINES: CIRCULAR MOTION

1. Motion in a circle always involves centripetal force and centripetal acceleration. However, remember that centripetal force is not a new kind of force but is simply the net force causing the motion; it consists of the sum of whatever forces are acting on the moving object. This net force equals the mass times the centripetal acceleration; do not make the common mistake of counting mass times acceleration as a force to be added to the net force on one side of the equation of motion.

2. Be sure you note whether the situation involves angles in degrees or in radians. The radian is not a unit that necessarily has to be carried through a calculation, but check that your result makes sense in terms of the angular units.

3. The equations of motion with constant angular acceleration have the same form as the equations of motion with constant linear acceleration. However, neither set of equations applies if the acceleration is not constant.

MULTIPLE-CHOICE QUESTIONS

9.1 An object is moving in a circular path. If the centripetal force is suddenly removed, how will the object move?

a) It will move radially outward.

b) It will move radially inward.

c) It will move vertically downward.

d) It will move in the direction in which its velocity vector points at the instant the centripetal force vanishes.

9.2 The angular acceleration for an object undergoing circular motion is plotted versus time in the figure. If the object started from rest at $t = 0$, the net angular displacement of the object at $t = t_f$

a) is in the clockwise direction.

b) is in the counterclockwise direction.

c) is zero.

d) cannot be determined.

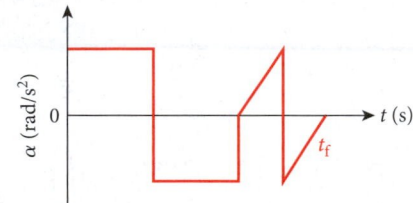

9.3 The latitude of Lubbock, Texas (known as the Hub City of the South Plains), is 33° N. What is its rotational speed, assuming the radius of the Earth at the Equator to be 6380 km?

a) 464 m/s

b) 389 m/s

c) 253 m/s

d) 0.464 m/s

e) 0.389 m/s

9.4 A rock attached to a string moves clockwise in uniform circular motion. In which direction from point A is the rock thrown off when the string is cut?

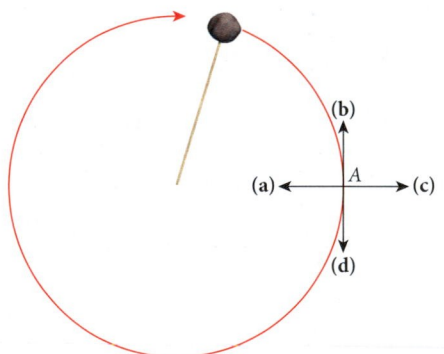

9.5 A Ferris wheel rotates slowly about a horizontal axis. Passengers sit on seats that remain horizontal on the Ferris wheel as it rotates. Which type of force provides the centripetal acceleration on the passengers when they are at the top of the Ferris wheel?

a) centrifugal

b) normal

c) gravity

d) tension

9.6 In a conical pendulum, a bob moves in a horizontal circle, as shown in the figure. The period of the pendulum (the time it takes for the bob to perform a complete revolution) is

a) $T = 2\pi\sqrt{L\cos\theta/g}$.

b) $T = 2\pi\sqrt{g\cos\theta/L}$.

c) $T = 2\pi\sqrt{Lg\sin\theta}$.

d) $T = 2\pi\sqrt{L\sin\theta/g}$.

e) $T = 2\pi\sqrt{L/g}$.

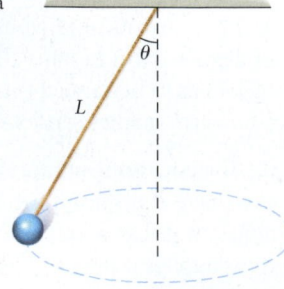

9.7 A ball attached to the end of a string is swung around in a circular path of radius r. If the radius is doubled and the linear speed is kept constant, the centripetal acceleration

a) remains the same.

b) increases by a factor of 2.

c) decreases by a factor of 2.

d) increases by a factor of 4.

e) decreases by a factor of 4.

9.8 The angular speed of the hour hand of a clock (in radians per second) is

a) $\dfrac{\pi}{21,600}$

b) $\dfrac{\pi}{7200}$

c) $\dfrac{\pi}{3600}$

d) $\dfrac{\pi}{1800}$

e) $\dfrac{\pi}{60}$

9.9 You put three identical coins on a turntable at different distances from the center and then turn the motor on. As the turntable speeds up, the outermost coin slides off first, followed by the one at the middle distance, and, finally, when the turntable is going the fastest, the innermost one. Why is this?

a) For greater distances from the center, the centripetal acceleration is higher, and so the force of friction becomes unable to hold the coin in place.

b) The weight of the coin causes the turntable to flex downward, so the coin nearest the edge falls off first.

c) Because of the way the turntable is made, the coefficient of static friction decreases with distance from the center.

d) For smaller distances from the center, the centripetal acceleration is higher.

9.10 A point on a Blu-ray disc is a distance $R/4$ from the axis of rotation. How far from the axis of rotation is a second point that has, at any instant, a linear velocity twice that of the first point?

a) $R/16$

b) $R/8$

c) $R/2$

d) R

9.11 The figure shows a rider stuck to the wall without touching the floor in the Barrel of Fun at a carnival. Which diagram correctly shows the forces acting on the rider?

 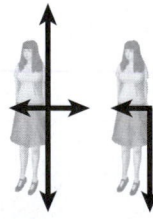  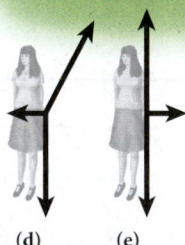

(a)   (b)   (c)   (d)   (e)

9.12 A string is tied to a rock, and the rock is twirled around in a circle at a constant speed. If gravity is ignored and the period of the circular motion is doubled, the tension in the string is

a) reduced to $\frac{1}{4}$ of its original value.

b) reduced to $\frac{1}{2}$ of its original value.

c) unchanged.

d) increased to 2 times its original value.

e) increased to 4 times its original value.

9.13 A bicycle's wheels have a radius of 33.0 cm. The bicycle is traveling at a speed of 6.5 m/s. What is the angular speed of the front tire?

a) 0.197 rad/s

b) 1.24 rad/s

c) 5.08 rad/s

d) 19.7 rad/s

e) 215 rad/s

9.14 The period of rotation of the Earth on its axis is 24 h. At this angular velocity, the centripetal acceleration at the surface of the Earth is small compared with the acceleration due to gravity. What would Earth's period of rotation have to be for the magnitude of the centripetal acceleration at its surface at the Equator to be equal to the magnitude of the acceleration due to gravity? (With this period of rotation, you could levitate just above the Earth's surface!)

a) 0.043 h   d) 1.41 h

b) 0.340 h   e) 3.89 h

c) 0.841 h   f) 12.0 h

9.15 In Solved Problem 9.1, with what speed does the roller coaster car have to enter the bottom of the loop in order to produce the feeling of weightlessness at the top?

a) 7.00 m/s   d) 15.7 m/s

b) 12.1 m/s   e) 21.4 m/s

c) 13.5 m/s

9.16 Figure 9.18a shows the free-body diagram for the forces acting on a passenger in a roller coaster car at the top of the loop, where the normal force exerted by the track has a magnitude smaller than that of the gravitational force. If the speed of the car is 7.00 m/s, what does the radius of the loop have to be for that free-body diagram to be correct?

a) less than 5 m   b) 5 m   c) more than 5 m

CONCEPTUAL QUESTIONS

9.17 A ceiling fan is rotating in clockwise direction (as viewed from below), but it is slowing down. What are the directions of ω and α?

9.18 A hook above a stage is rated to support 150. lb. A 3.00-lb rope is attached to the hook, and a 147-lb actor is going to attempt to swing across the stage on the rope. Will the hook hold the actor up during the swing?

9.19 A popular carnival ride consists of seats attached to a central disk through cables, as shown in the figure. The passengers travel in uniform circular motion. The mass of one of the passengers (including the chair he is sitting on) is 65 kg; the mass of an empty chair on the opposite side of the central disk is 5.0 kg. If θ_1 and θ_2 are the angles that the cables attached to the two chairs make with respect to vertical, how do these two angles compare qualitatively? Is θ_2 larger than, smaller than, or equal to θ_1?

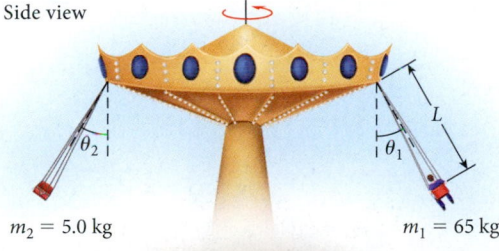

Side view

$m_2 = 5.0$ kg $m_1 = 65$ kg

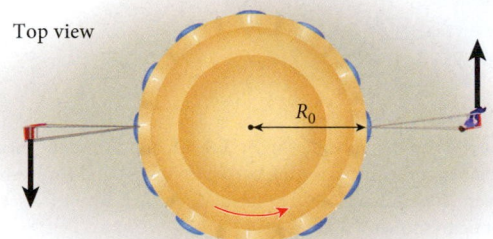

Top view

R_0

9.20 A person rides on a Ferris wheel of radius R, which is rotating at a constant angular velocity ω. Compare the normal force of the seat pushing up on the person at point A to that at point B in the figure. Which force is greater, or are they the same?

9.21 Bicycle tires range in size from about 25. cm in diameter to about 70. cm in diameter. Why is it impractical to make tires much smaller than 25. cm in diameter? (You will learn why bicycle tires can't be too large in Chapter 10.)

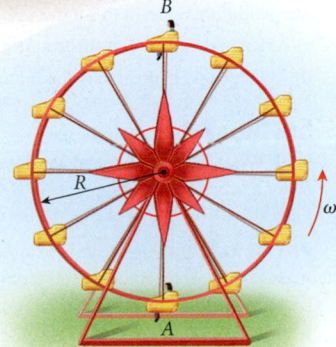

9.22 A CD starts from rest and speeds up to the operating angular frequency of the CD player. Compare the angular velocity and acceleration of a point on the edge of the CD to those of a point halfway between the center and the edge of the CD. Do the same for the linear velocity and acceleration.

9.23 A car is traveling around an unbanked curve at a maximum speed. Which force(s) is(are) responsible for keeping it on the road?

9.24 Two masses hang from two strings of equal length that are attached to the ceiling of a car. One mass is over the driver's seat; the other is over the passenger's seat. As the car makes a sharp turn, both masses swing away from the center of the turn. In their resulting positions, will they be farther apart, closer together, or the same distance apart as they were when the car wasn't turning?

9.25 A point mass m starts sliding from a height h along the frictionless surface shown in the figure. What is the minimum value of h in order for the mass to complete the loop of radius R?

9.26 In a conical pendulum, the bob attached to the string (which can be considered massless) moves in a horizontal circle at constant speed. The string sweeps out a cone as the bob rotates. What forces are acting on the bob?

9.27 Is it possible to swing a mass attached to a string in a perfectly horizontal circle (with the mass and the string parallel to the ground)?

9.28 A small ice block of mass m starts from rest from the top of an inverted bowl in the shape of a hemisphere, as shown in the figure. The hemisphere is fixed to the ground, and the block slides without friction along the surface of the hemisphere. Find the normal force exerted by the block on the sphere when the line between the block and the center of the sphere makes an angle θ with the horizontal. Discuss the result.

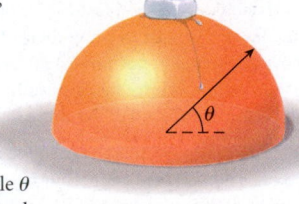

9.29 Suppose you are riding on a roller coaster, which moves through a vertical circular loop. Show that your apparent weight at the bottom of the loop is six times your weight when you experience weightlessness at the top, independent of the size of the loop. Assume that friction is negligible.

9.30 The following event actually occurred on the Sunshine Skyway Bridge near St. Petersburg, Florida, in 1997. Five daredevils tied a 55-m-long cable to the center of the bridge. They hoped to swing back and forth under the bridge at the end of this cable. The five people (total weight = W) attached themselves to the end of the cable, at the same level and 55 m away from where it was attached to the bridge and dropped straight down from the bridge, following the dashed circular path indicated in the figure. Unfortunately, the daredevils were not well versed in the laws of physics, and the cable broke (at the point it was linked to their seats) at the bottom of their swing. Determine how strong the cable (and all the links where the seats and the bridge are attached to it) would have had to be in order to support the five people at the bottom of the swing. Express your result in terms of their total weight, W.

EXERCISES

A blue problem number indicates a worked-out solution is available in the Student Solutions Manual. One • and two •• indicate increasing level of problem difficulty.

Section 9.2

9.31 What is the angle in radians that the Earth sweeps out in its orbit during winter?

9.32 Assuming that the Earth is spherical and recalling that latitudes range from 0° at the Equator to 90° N at the North Pole, how far apart, measured on the Earth's surface, are Dubuque, Iowa (42.50° N latitude), and Guatemala City (14.62° N latitude)? The two cities lie on approximately the same longitude. Do *not* neglect the curvature of the Earth in determining this distance.

•9.33 Refer to the information given in Problem 9.32. If one could burrow through the Earth and dig a straight-line tunnel from Dubuque to Guatemala City, how long would the tunnel be? From the point of view of the digger, at what angle below the horizontal would the tunnel be directed?

Section 9.3

9.34 A baseball is thrown at approximately 88.0 mph and with a spin rate of 110. rpm. If the distance between the pitcher's point of release and the catcher's glove is 60.5 ft, how many full turns does the ball make between release and catch? Neglect any effect of gravity or air resistance on the ball's flight.

9.35 A vinyl record plays at 33.3 rpm. Assume it takes 5.00 s for it to reach this full speed, starting from rest.

a) What is its angular acceleration during the 5.00 s?

b) How many revolutions does the record make before reaching its final angular speed?

9.36 At a county fair, a boy takes his teddy bear on the giant Ferris wheel. Unfortunately, at the top of the ride, he accidentally drops his stuffed buddy. The wheel has a diameter of 12.0 m, the bottom of the wheel is 2.00 m above the ground and its rim is moving at a speed of 1.00 m/s. How far from the base of the Ferris wheel will the teddy bear land?

•9.37 Having developed a taste for experimentation, the boy in Problem 9.36 invites two friends to bring their teddy bears on the same Ferris wheel. The boys are seated in positions 45.0° from each other. When the wheel brings the second boy to the maximum height, they all drop their stuffed animals. How far apart will the three teddy bears land?

•9.38 Mars orbits the Sun at a mean distance of 228 million km, in a period of 687 days. The Earth orbits at a mean distance of 149.6 million km, in a period of 365.26 days.

a) Suppose Earth and Mars are positioned such that Earth lies on a straight line between Mars and the Sun. Exactly 365.26 days later, when the Earth has completed one orbit, what is the angle between the Earth-Sun line and the Mars-Sun line?

b) The initial situation in part (a) is a closest approach of Mars to Earth. What is the time, in days, between two closest approaches? Assume constant speed and circular orbits for both Mars and Earth.

c) Another way of expressing the answer to part (b) is in terms of the angle between the lines drawn through the Sun, Earth, and Mars in the two closest approach situations. What is that angle?

••9.39 Consider a large simple pendulum that is located at a latitude of 55.0° N and is swinging in a north-south direction with points A and B being the northernmost and the southernmost points of the swing, respectively. A stationary (with respect to the fixed stars) observer is looking directly down on the pendulum at the moment shown in the figure. The Earth is rotating once every 23 h and 56 min.

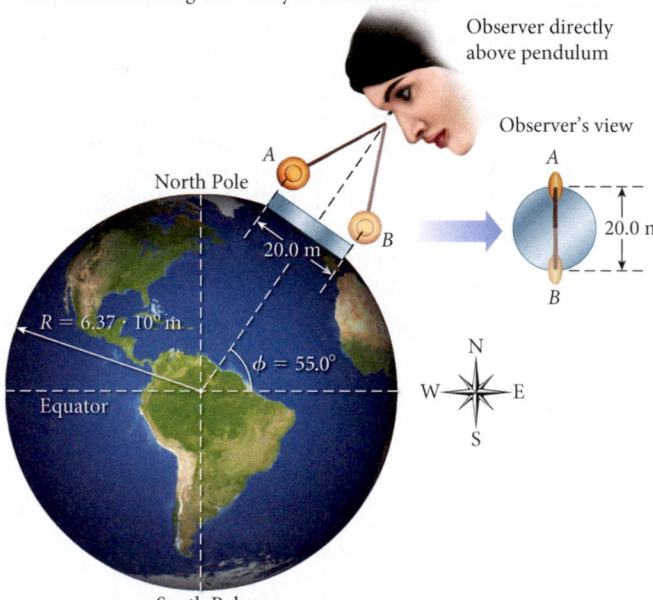

a) What are the directions (in terms of N, E, W, and S) and the magnitudes of the velocities of the surface of the Earth at points A and B as seen by the observer? *Note:* You will need to calculate answers to at least seven significant figures to see a difference.

b) What is the angular speed with which the 20.0-m diameter circle under the pendulum appears to rotate?

c) What is the period of this rotation?

d) What would happen to a pendulum swinging at the Equator?

Section 9.4

9.40 What is the centripetal acceleration of the Moon? The period of the Moon's orbit about the Earth is 27.3 days, measured with respect to the fixed stars. The radius of the Moon's orbit is $R_M = 3.85 \cdot 10^8$ m.

9.41 You are holding the axle of a bicycle wheel with radius 35.0 cm and mass 1.00 kg. You get the wheel spinning at a rate of 75.0 rpm and then stop it by pressing the tire against the pavement. You notice that it takes 1.20 s for the wheel to come to a complete stop. What is the angular acceleration of the wheel?

9.42 Life scientists use ultracentrifuges to separate biological components or to remove molecules from suspension. Samples in a symmetric array of containers are spun rapidly about a central axis. The centrifugal acceleration they experience in their moving reference frame acts as "artificial gravity" to effect a rapid separation. If the sample containers are 10.0 cm from the rotation axis, what rotation frequency is required to produce an acceleration of $1.00 \cdot 10^5 g$?

9.43 A centrifuge in a medical laboratory rotates at an angular speed of 3600. rpm (revolutions per minute). When switched off, it rotates 60.0 times before coming to rest. Find the constant angular acceleration of the centrifuge.

9.44 A discus thrower (with arm length of 1.20 m) starts from rest and begins to rotate counterclockwise with an angular acceleration of 2.50 rad/s^2.

a) How long does it take the discus thrower's speed to get to 4.70 rad/s?

b) How many revolutions does the thrower make to reach the speed of 4.70 rad/s?

c) What is the linear speed of the discus at 4.70 rad/s?

d) What is the linear acceleration of the discus thrower at this point?

e) What is the magnitude of the centripetal acceleration of the discus thrown?

f) What is the magnitude of the discus's total acceleration?

•9.45 In a department store toy display, a small disk (disk 1) of radius 0.100 m is driven by a motor and turns a larger disk (disk 2) of radius 0.500 m. Disk 2, in turn, drives disk 3, whose radius is 1.00 m. The three disks are in contact, and there is no slipping. Disk 3 is observed to sweep through one complete revolution every 30.0 s.

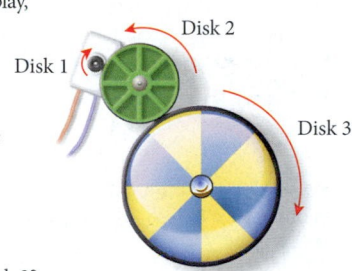

a) What is the angular speed of disk 3?

b) What is the ratio of the tangential velocities of the rims of the three disks?

c) What are the angular speeds of disks 1 and 2?

d) If the motor malfunctions, resulting in an angular acceleration of 0.100 rad/s^2 for disk 1, what are disks 2 and 3's angular accelerations?

•9.46 A particle is moving clockwise in a circle of radius 1.00 m. At a certain instant, the magnitude of its acceleration is $a = |\vec{a}| = 25.0$ m/s^2, and the acceleration vector has an angle of $\theta = 50.0°$ with the position vector, as shown in the figure. At this instant, find the speed, $v = |\vec{v}|$, of this particle.

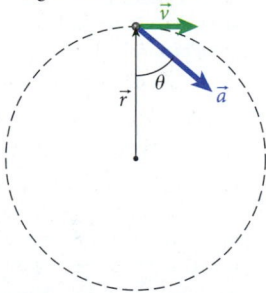

•9.47 In a tape recorder, the magnetic tape moves at a constant linear speed of 5.60 cm/s. To maintain this constant linear speed, the angular speed of the driving spool (the take-up spool) has to change accordingly.

a) What is the angular speed of the take-up spool when it is empty, with radius $r_1 = 0.800$ cm?

b) What is the angular speed when the spool is full, with radius $r_2 = 2.20$ cm?

c) If the total length of the tape is 100.80 m, what is the average angular acceleration of the take-up spool while the tape is being played?

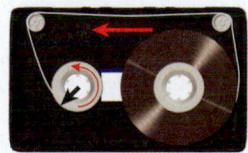 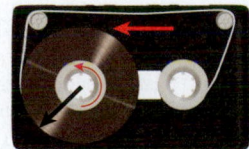

••9.48 A ring is fitted loosely (with no friction) around a long, smooth rod of length $L = 0.500$ m. The rod is fixed at one end, and the other end is spun in a horizontal circle at a constant angular velocity of $\omega = 4.00$ rad/s. The

ring has zero radial velocity at its initial position, a distance of $r_0 = 0.300$ m from the fixed end. Determine the radial velocity of the ring as it reaches the moving end of the rod.

••9.49 A flywheel with a diameter of 1.00 m is initially at rest. Its angular acceleration is plotted versus time in the figure.

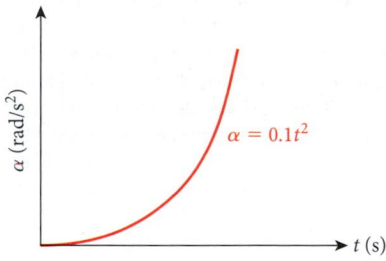

a) What is the angular separation between the initial position of a fixed point on the rim of the flywheel and the point's position 8.00 s after the wheel starts rotating?

b) The point starts its motion at $\theta = 0$. Calculate and sketch the linear position, velocity vector, and acceleration vector 8 s after the wheel starts rotating.

Section 9.5

9.50 Calculate the centripetal force exerted on a vehicle of mass $m = 1500$. kg that is moving at a speed of 15.0 m/s around a curve of radius $R = 400$. m. Which force plays the role of the centripetal force in this case?

9.51 What is the apparent weight of a rider on the roller coaster of Solved Problem 9.1 at the *bottom* of the loop?

9.52 Two skaters, A and B, of equal mass are moving in clockwise uniform circular motion on the ice. Their motions have equal periods, but the radius of skater A's circle is half that of skater B's circle.

a) What is the ratio of the speeds of the skaters?

b) What is the ratio of the magnitudes of the forces acting on each skater?

•9.53 A small block of mass m is in contact with the inner wall of a large hollow cylinder. Assume the coefficient of static friction between the object and the wall of the cylinder is μ_s. Initially, the cylinder is at rest, and the block is held in place by a peg supporting its weight. The cylinder starts rotating about its center axis, as shown in the figure, with an angular acceleration of α. Determine the minimum time interval after the cylinder begins to rotate before the peg can be removed without the block sliding against the wall.

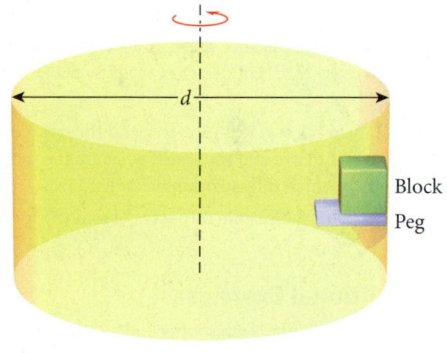

•9.54 A race car is making a U-turn at constant speed. The coefficient of friction between the tires and the track is $\mu_s = 1.20$. If the radius of the curve is 10.0 m, what is the maximum speed at which the car can turn without sliding? Assume that the car is undergoing uniform circular motion.

•9.55 A car speeds over the top of a hill. If the radius of curvature of the hill at the top is 9.00 m, how fast can the car be traveling and maintain constant contact with the ground?

•9.56 A ball of mass $m = 0.200$ kg is attached to a (massless) string of length $L = 1.00$ m and is undergoing circular motion in the horizontal plane, as shown in the figure.

a) Draw a free-body diagram for the ball.

b) Which force plays the role of the centripetal force?

c) What should the speed of the mass be for θ to be 45.0°?

d) What is the tension in the string?

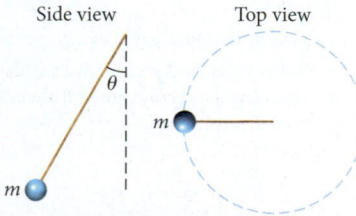

•**9.57** You are flying to Chicago for a weekend away from the books. In your last physics class, you learned that the airflow over the wings of the plane creates a *lift force,* which acts perpendicular to the wings. When the plane is flying level, the upward lift force exactly balances the downward *weight force.* Since O'Hare is one of the busiest airports in the world, you are not surprised when the captain announces that the flight is in a holding pattern due to the heavy traffic. He informs the passengers that the plane will be flying in a circle of radius 7.00 mi at a speed of 360. mph and an altitude of $2.00 \cdot 10^4$ ft. From the safety information card, you know that the total length of the wingspan of the plane is 275 ft. From this information, estimate the banking angle of the plane relative to the horizontal.

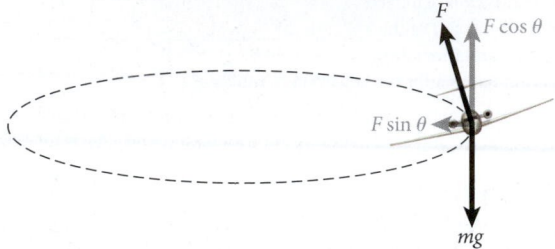

••**9.58** A 20.0-g metal cylinder is placed on a turntable, with its center 80.0 cm from the turntable's center. The coefficient of static friction between the cylinder and the turntable's surface is $\mu_s = 0.800$. A thin, massless string of length 80.0 cm connects the center of the turntable to the cylinder, and *initially,* the string has zero tension in it. Starting from rest, the turntable very slowly attains higher and higher angular velocities, but the turntable and the cylinder can be considered to have uniform circular motion at any instant. Calculate the tension in the string when the angular velocity of the turntable is 60.0 rpm (rotations per minute).

••**9.59** A speedway turn, with radius of curvature R, is banked at an angle θ above the horizontal.

a) What is the optimal speed at which to take the turn if the track's surface is iced over (that is, if there is very little friction between the tires and the track)?

b) If the track surface is ice-free and there is a coefficient of friction μ_s between the tires and the track, what are the maximum and minimum speeds at which this turn can be taken?

c) Evaluate the results of parts (a) and (b) for $R = 400.$ m, $\theta = 45.0°$, and $\mu_s = 0.700$.

Additional Exercises

9.60 A particular Ferris wheel takes riders in a vertical circle of radius 9.00 m once every 12.0 s.

a) Calculate the speed of the riders, assuming it to be constant.

b) Draw a free-body diagram for a rider at a time when she is at the bottom of the circle. Calculate the normal force exerted by the seat on the rider at that point in the ride.

c) Perform the same analysis as in part (b) for a point at the top of the ride.

9.61 A boy is on a Ferris wheel, which takes him in a vertical circle of radius 9.00 m once every 12.0 s.

a) What is the angular speed of the Ferris wheel?

b) Suppose the wheel comes to a stop at a uniform rate during one quarter of a revolution. What is the angular acceleration of the wheel during this time?

c) Calculate the tangential acceleration of the boy during the time interval described in part (b).

9.62 Consider a 53.0-cm-long lawn mower blade rotating about its center at 3400. rpm.

a) Calculate the linear speed of the tip of the blade.

b) If safety regulations require that the blade be stoppable within 3.00 s, what minimum angular acceleration will accomplish this? Assume that the angular acceleration is constant.

9.63 A car accelerates uniformly from rest and reaches a speed of 22.0 m/s in 9.00 s. The diameter of a tire on this car is 58.0 cm.

a) Find the number of revolutions the tire makes during the car's motion, assuming that no slipping occurs.

b) What is the final angular speed of a tire in revolutions per second?

9.64 Gear A, with a mass of 1.00 kg and a radius of 55.0 cm, is in contact with gear B, with a mass of 0.500 kg and a radius of 30.0 cm. The gears do not slip with respect to each other as they rotate. Gear A rotates at 120. rpm and slows to 60.0 rpm in 3.00 s. How many rotations does gear B undergo during this time interval?

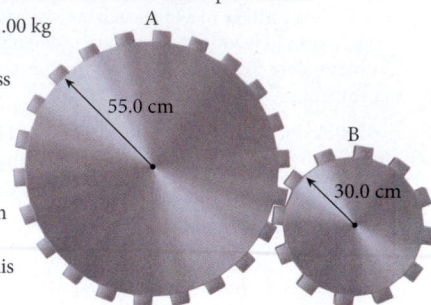

9.65 A top spins for 10.0 min, beginning with an angular speed of 10.0 rev/s. Determine its angular acceleration, assuming it is constant, and its total angular displacement.

9.66 A penny is sitting on the edge of an old phonograph disk that is spinning at 33.0 rpm and has a diameter of 12.0 inches. What is the minimum coefficient of static friction between the penny and the surface of the disk to ensure that the penny doesn't fly off?

9.67 A vinyl record that is initially turning at $33\frac{1}{3}$ rpm slows uniformly to a stop in a time of 15.0 s. How many rotations are made by the record while stopping?

9.68 Determine the linear and angular speeds and accelerations of a speck of dirt located 2.00 cm from the center of a CD rotating inside a CD player at 250. rpm.

9.69 What is the acceleration of the Earth in its orbit? (Assume the orbit is circular.)

9.70 A day on Mars is 24.6 Earth hours long. A year on Mars is 687 Earth days long. How do the angular velocities of Mars's rotation and orbit compare to the angular velocities of Earth's rotation and orbit?

9.71 A monster truck has tires with a diameter of 1.10 m and is traveling at 35.8 m/s. After the brakes are applied, the truck slows uniformly and is brought to rest after the tires rotate through 40.2 turns.

a) What is the initial angular speed of the tires?

b) What is the angular acceleration of the tires?

c) What distance does the truck travel before coming to rest?

•**9.72** The motor of a fan turns a small wheel of radius $r_m = 2.00$ cm. This wheel turns a belt, which is attached to a wheel of radius $r_f = 3.00$ cm that is mounted to the axle of the fan blades. Measured from the center of this axle, the tip of the fan blades are at a distance $r_b = 15.0$ cm. When the fan is in operation, the motor spins at an angular speed of $\omega = 1200.$ rpm. What is the tangential speed of the tips of the fan blades?

•**9.73** A car with a mass of 1000. kg goes over a hill at a constant speed of 60.0 m/s. The top of the hill can be approximated as an arc length of a circle with a radius of curvature of 370. m. What force does the car exert on the hill as it passes over the top?

•**9.74** Unlike a ship, an airplane does not use its rudder to turn. It turns by banking its wings: The lift force, perpendicular to the wings, has a horizontal component, which provides the centripetal acceleration for the turn, and a vertical component, which supports the plane's weight. (The rudder counteracts yaw and thus it keeps the plane pointed in the direction it is moving.) A famous spy plane the SR-71 Blackbird, flying at 4800. km/h, had a turning radius of 290. km. Find its banking angle.

•**9.75** An 80.0-kg pilot in an aircraft moving at a constant speed of 500. m/s pulls out of a vertical dive along an arc of a circle of radius 4000. m.

a) Find the centripetal acceleration and the centripetal force acting on the pilot.

b) What is the pilot's apparent weight at the bottom of the dive?

•**9.76** A ball that has a mass of 1.00 kg is attached to a string 1.00 m long and is whirled in a vertical circle at a constant speed of 10.0 m/s.

a) Determine the tension in the string when the ball is at the top of the circle.

b) Determine the tension in the string when the ball is at the bottom of the circle.

c) Consider the ball at some point other than the top or bottom. What can you say about the tension in the string at this point?

•**9.77** A car starts from rest and accelerates around a flat curve of radius $R = 36.0$ m. The tangential component of the car's acceleration remains constant at $a_t = 3.30$ m/s^2, while the centripetal acceleration increases to keep the car on the curve as long as possible. The coefficient of friction between the tires and the road is $\mu = 0.950$. What distance does the car travel around the curve before it begins to skid? (Be sure to include both the tangential and centripetal components of the acceleration.)

•**9.78** A girl on a merry-go-round platform holds a pendulum in her hand. The pendulum is 6.00 m from the rotation axis of the platform. The rotational speed of the platform is 0.0200 rev/s. It is found that the pendulum hangs at an angle θ to the vertical. Find θ.

••**9.79** A carousel at a carnival has a diameter of 6.00 m. The ride starts from rest and accelerates at a constant angular acceleration to an angular speed of 0.600 rev/s in 8.00 s.

a) What is the value of the angular acceleration?

b) What are the centripetal and angular accelerations of a seat on the carousel that is 2.75 m from the rotation axis?

c) What is the total acceleration, magnitude and direction, 8.00 s after the angular acceleration starts?

••**9.80** A car of weight $W = 10.0$ kN makes a turn on a track that is banked at an angle of $\theta = 20.0°$. Inside the car, hanging from a short string tied to the rear-view mirror, is an ornament. As the car turns, the ornament swings out at an angle of $\varphi = 30.0°$ measured

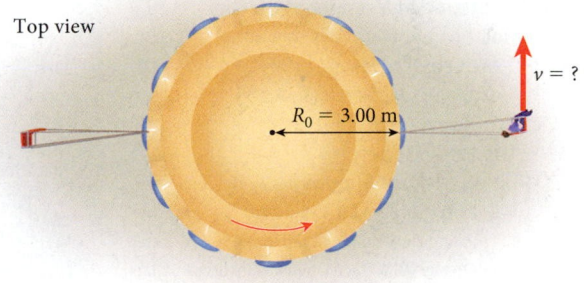

from the vertical inside the car. What is the force of static friction between the car and the road?

••**9.81** A popular carnival ride consists of seats attached to a central disk through cables. The passengers travel in uniform circular motion. As shown in the figure, the radius of the central disk is $R_0 = 3.00$ m, and the length of the cable is $L = 3.20$ m. The mass of one of the passengers (including the chair he is sitting on) is 65.0 kg.

a) If the angle θ that the cable makes with respect to the vertical is 30.0°, what is the speed, v, of this passenger?

b) What is the magnitude of the force exerted by the cable on the chair?

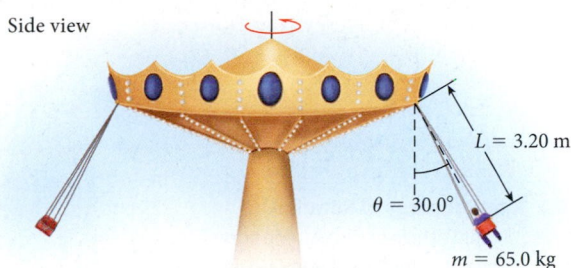

Side view

$L = 3.20$ m

$\theta = 30.0°$

$m = 65.0$ kg

Top view

$v = ?$

$R_0 = 3.00$ m

MULTI-VERSION EXERCISES

9.82 In the Thunder Sphere, a motorcycle moves on the inside of a sphere, traveling in a horizontal circle along the equator of the sphere. The inner radius of the sphere is 12.61 m, and the coefficient of static friction between the tires of the motorcycle and the inner surface of the sphere is 0.4601. What is the minimum speed the motorcycle must maintain to avoid falling?

9.83 In the Thunder Sphere, a motorcycle moves on the inside of a sphere, traveling in a horizontal circle along the equator of the sphere. The inner radius of the sphere is 13.75 m, and the motorcycle maintains a speed of 17.01 m/s. What is the minimum value for the coefficient of static friction between the tires of the motorcycle and the inner surface of the sphere to ensure that the motorcycle does not fall?

9.84 In the Thunder Sphere, a motorcycle moves on the inside of a sphere, traveling in a horizontal circle along the equator of the sphere. The motorcycle maintains a speed of 15.11 m/s, and the coefficient of static friction between the tires of the motorcycle and the inner surface of the sphere is 0.4741. What is the maximum radius that the sphere can have if the motorcycle is not to fall?

9.85 When the tips of the rotating blade of an airplane propeller have a linear speed greater than the speed of sound, the propeller makes a lot of noise, which is undesirable. If the blade turns at 2403 rpm, what is the maximum length it can have without the tips exceeding the speed of sound? Assume that the speed of sound is 343.0 m/s and that the blade rotates about its center.

9.86 When the tips of the rotating blade of an airplane propeller have a linear speed greater than the speed of sound, the propeller makes a lot of noise, which is undesirable. If the tip-to-tip length of the blade is 2.601 m, what is the maximum angular frequency (in revolutions per minute) with which it can rotate? Assume that the speed of sound is 343.0 m/s and that the blade rotates about its center.

9.87 The rear wheels of a sports car have a radius of 46.65 cm. The sports car goes from rest to a speed of 29.13 m/s in 3.945 s with constant acceleration. What is the angular acceleration of the rear wheels? Assume that the wheels roll without slipping.

9.88 The rear wheels of a sports car have a radius of 48.95 cm. The sports car accelerates from rest with constant acceleration for 3.997 s. The angular acceleration of the rear wheels is 14.99 s^{-2}. What is the final linear speed of the car? Assume that the wheels roll without slipping.

9.89 A sports car accelerates from rest to a speed of 29.53 m/s with constant acceleration in 4.047 s. The angular acceleration of the rear wheels is 17.71 s^{-2}. What is the radius of the rear wheels? Assume that the wheels roll without slipping.

9.90 A flywheel of radius 27.01 cm rotates with a frequency of 4949 rpm. What is the centripetal acceleration at a point on the edge of the flywheel?

9.91 A flywheel of radius 31.59 cm is rotating at a constant frequency. The centripetal acceleration at a point on the edge of the flywheel is $8.629 \cdot 10^4$ m/s^2. What is the frequency of rotation of the flywheel (in rpm)?

10 Rotation

FIGURE 10.1 A modern jet engine.

The large fans at the front of modern jet engines, like the one shown in Figure 10.1, pull air into a compression chamber, where the air mixes with fuel and ignites. The explosion propels gases out the back of the engine, producing the thrust that moves the plane forward. These fans rotate at 7000–9000 rpm and must be inspected often—no one wants a broken fan blade at an altitude of 6 mi.

Almost all engines have rotating parts that transfer energy to the output device, which is often rotating as well. In fact, most objects in the universe rotate, from molecules to stars and galaxies. Figure 10.2 shows rotation on very large (a hurricane) and humongous (a spiral galaxy) scales. Furthermore, in 2011, a team including five undergraduates from the University of Michigan analyzed a large sample of galaxies and found evidence that the universe as a whole rotates. The laws that govern rotation are as fundamentally important as any other part of mechanics.

Chapter 9 introduced some basic concepts of circular motion, and this chapter uses some of those same ideas—angular velocity, angular acceleration, and axis of rotation. In this chapter, we complete our comparison of translational and rotational quantities and encounter another conservation law of basic importance: the law of conservation of angular momentum.

WHAT WE WILL LEARN

- The kinetic energy due to an object's rotational motion must be accounted for when considering energy conservation.

- For rotation of a round object about an axis through its center of mass, the moment of inertia is proportional to the product of the object's mass and the square of the largest perpendicular distance from any part of the object to the axis of rotation. The proportionality constant has a value between zero and one and depends on the shape of the object.

- For rotation about an axis parallel to an axis through an object's center of mass, the moment of inertia equals the center-of-mass moment of inertia plus the product of the mass of the object and the square of the distance between the two axes.

- For rolling objects, the kinetic energies of rotation and translation are related.

- Torque is the vector product of the position vector and the force vector.

- Newton's Second Law also applies to rotational motion.

- Angular momentum is defined as the vector product of the position vector and the momentum vector.

- Relationships analogous to those for linear quantities exist among angular momentum, torque, moment of inertia, angular velocity, and angular acceleration.

- Another fundamental conservation law is the law of conservation of angular momentum.

10.1 Kinetic Energy of Rotation

In Chapter 8, we saw that we can describe the motion of an extended object in terms of the path that its center of mass follows and the rotation of the object around its center of mass. However, even though we covered circular motion for point particles in Chapter 9, we have not yet considered the rotation of extended objects. Analyzing this motion is the purpose of this chapter.

Point Particle in Circular Motion

Chapter 9 introduced the kinematical quantities of circular motion. Angular speed, ω, and angular acceleration, α, were defined in terms of the time derivatives of the angular displacement, θ:

$$\omega = \frac{d\theta}{dt}$$

$$\alpha = \frac{d\omega}{dt} = \frac{d^2\theta}{dt^2}.$$

We saw that the angular quantities are related to the linear quantities as follows:

$$s = r\theta, \quad v = r\omega,$$

$$a_t = r\alpha, \quad a_c = \omega^2 r, \quad a = \sqrt{a_c^2 + a_t^2},$$

where s is the arc length, v is the linear speed of the center of mass, a_t is the tangential acceleration, a_c is the centripetal acceleration, and a is the linear acceleration.

The most straightforward way to introduce the physical quantities for the description of rotation is through the kinetic energy of rotation of an extended object. In Chapter 5 on work and energy, the kinetic energy of a moving object was defined as

$$K = \tfrac{1}{2} mv^2. \tag{10.1}$$

If the motion of this object is circular, we can use the relationship between linear and angular velocity to obtain

$$K = \tfrac{1}{2} mv^2 = \tfrac{1}{2} m(r\omega)^2 = \tfrac{1}{2} mr^2 \omega^2, \tag{10.2}$$

which is the **kinetic energy of rotation** for a point particle's motion on the circumference of a circle of radius r about a fixed axis, as illustrated in Figure 10.3.

Several Point Particles in Circular Motion

Just as we proceeded in Chapter 8 in finding the location of the center of mass of a system of particles, we start with a collection of individual rotating objects and then

(a)

(b)

FIGURE 10.2 (a) Rotating air mass forming a hurricane. (b) Rotation on an enormous scale—spiral galaxy M74.

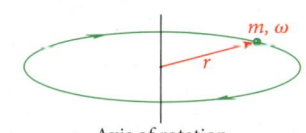

FIGURE 10.3 A point particle moving in a circle about the axis of rotation.

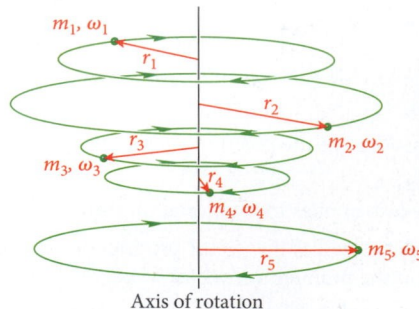

FIGURE 10.4 Five point particles moving in circles about a common axis of rotation.

Concept Check 10.1

Consider two equal masses, m, connected by a thin, massless rod. As shown in the figures, the two masses spin in a horizontal plane around a vertical axis represented by the dashed line. Which system has the highest moment of inertia?

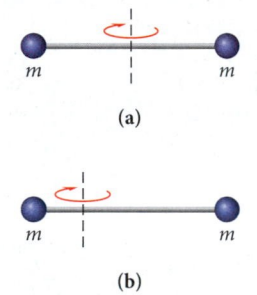

(a)

(b)

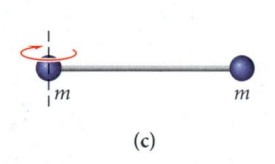

(c)

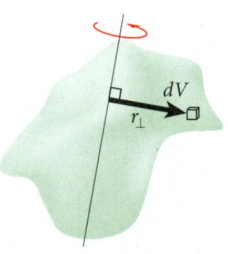

FIGURE 10.5 Definition of $r_\perp$ as the perpendicular distance of an infinitesimal volume element from the axis of rotation.

approach the continuous limit. The kinetic energy of a collection of rotating objects is given by

$$K = \sum_{i=1}^{n} K_i = \frac{1}{2}\sum_{i=1}^{n} m_i v_i^2 = \frac{1}{2}\sum_{i=1}^{n} m_i r_i^2 \omega_i^2.$$

This result is simply a consequence of using equation 10.2 for several point particles and writing the total kinetic energy as the sum of the individual kinetic energies. Here ω_i is the angular velocity of particle i and r_i is the perpendicular distance from i to a fixed axis. This fixed axis is the **axis of rotation** for these particles. An example of a system of five rotating point particles is shown in Figure 10.4.

Now we assume that all of the point particles whose kinetic energies we have summed keep their distances with respect to one another and with respect to the axis of rotation fixed. Then, all of the point particles in the system will undergo circular motion around the common axis of rotation with the same angular velocity. With this constraint, the sum of the particles' kinetic energies becomes

$$K = \frac{1}{2}\sum_{i=1}^{n} m_i r_i^2 \omega^2 = \frac{1}{2}\left(\sum_{i=1}^{n} m_i r_i^2\right)\omega^2 = \frac{1}{2}I\omega^2. \tag{10.3}$$

The quantity I introduced in equation 10.3 is called the **moment of inertia,** also known as the *rotational inertia*. It depends only on the masses of the individual particles and their distances to the axis of rotation:

$$I = \sum_{i=1}^{n} m_i r_i^2. \tag{10.4}$$

In Chapter 9, we saw that all quantities associated with circular motion have equivalents in linear motion. The angular velocity ω and the linear velocity v form such a pair. Comparing the expressions for the kinetic energy of rotation (equation 10.3) and the kinetic energy of linear motion (equation 10.1), we see that the moment of inertia I plays the same role for circular motion as the mass m does for linear motion.

10.2 Calculation of Moment of Inertia

We can use the moment of inertia of several point particles, as given in equation 10.4, as a starting point to find the moment of inertia of an extended object. We'll proceed in the same way as we did in finding the center-of-mass location in Chapter 8. We again represent an extended object by a collection of small identical-sized cubes of volume V and (possibly different) mass density ρ. Then equation 10.4 becomes

$$I = \sum_{i=1}^{n} \rho(\vec{r}_i) r_i^2 V. \tag{10.5}$$

Again, as in Chapter 8, we follow the conventional calculus approach, letting the volume of the cubes approach zero, $V \to 0$. In this limit, the sum in equation 10.5 approaches the integral, which gives an expression for the moment of inertia of an extended object:

$$I = \int_V r_\perp^2 \rho(\vec{r})\,dV. \tag{10.6}$$

The symbol $r_\perp$ represents the perpendicular distance of an infinitesimal volume element from the axis of rotation (Figure 10.5).

We also know that the total mass of an object can be obtained by integrating the mass density over the total volume of the object:

$$M = \int_V \rho(\vec{r})\,dV. \tag{10.7}$$

Equations 10.6 and 10.7 are the most general expressions for the moment of inertia and the mass of an extended object. However, just as with the equations for the center of mass, some of the physically most interesting cases are those where the mass density is constant throughout the volume. In this case, equations 10.6 and 10.7 reduce to

and

$$I = \rho \int_V r_\perp^2 dV \quad \text{(for constant mass density, } \rho\text{)},$$

$$M = \rho \int_V dV = \rho V \quad \text{(for constant mass density, } \rho\text{)}.$$

Thus, the moment of inertia for an object with constant mass density is given by

$$I = \frac{M}{V} \int_V r_\perp^2 dV \quad \text{(for constant mass density, } \rho\text{)}. \tag{10.8}$$

We can now calculate moments of inertia for some objects with particular shapes. First we'll assume that the axis of rotation passes through the center of mass of the object. Then, we'll derive a theorem that connects this special case in a simple way to the general case, for which the axis of rotation does not pass through the center of mass.

Rotation about an Axis through the Center of Mass

For any object with constant mass density, we can use equation 10.8 to calculate the moment of inertia with respect to rotation about a fixed axis that passes through the object's center of mass. For convenience, the location of the center of mass is usually chosen as the origin of the coordinate system. Because the integral in equation 10.8 is a three-dimensional volume integral, the choice of coordinate system is usually extremely important for evaluating the integral with as little computational work as possible.

In this section, we consider two cases: a hollow cylinder and a solid sphere. These cases represent the two most common classes of objects that can roll, and they illustrate the use of two different coordinate systems for the integral.

Figure 10.6a shows a hollow cylinder rotating about its symmetry axis. Its moment of inertia is

$$I = \tfrac{1}{2} M \left(R_1^2 + R_2^2 \right) \quad \text{(hollow cylinder)}. \tag{10.9}$$

This is the general result for the moment of inertia of a hollow cylinder rotating about its symmetry axis, where M is the cylinder's total mass, R_1 is its inner radius, and R_2 is its outer radius.

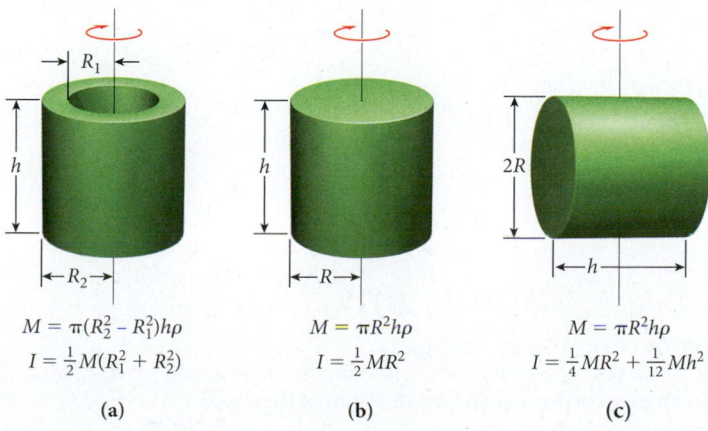

(a) $M = \pi(R_2^2 - R_1^2)h\rho$
$I = \tfrac{1}{2}M(R_1^2 + R_2^2)$

(b) $M = \pi R^2 h\rho$
$I = \tfrac{1}{2}MR^2$

(c) $M = \pi R^2 h\rho$
$I = \tfrac{1}{4}MR^2 + \tfrac{1}{12}Mh^2$

FIGURE 10.6 Moment of inertia for (a) a hollow cylinder and (b) a solid cylinder rotating about the axis of symmetry. (c) Moment of inertia for a cylinder rotating about an axis through its center of mass but perpendicular to its symmetry axis.

Using equation 10.9, we can obtain the moment of inertia for a solid cylinder rotating about its symmetry axis (see Figure 10.6b) by setting $R_1 = R$ and $R_2 = 0$:

$$I = \tfrac{1}{2}MR^2 \quad \text{(solid cylinder).}$$

We can also obtain the limiting case of a thin cylindrical shell or hoop, for which all the mass is concentrated on the circumference, by setting $R_1 = R_2 = R$. In this case, the moment of inertia is

$$I = MR^2 \quad \text{(hoop or cylindrical thin shell).}$$

Finally, the moment of inertia for a solid cylinder of height h rotating about an axis through its center of mass but perpendicular to its symmetry axis (see Figure 10.6c) is given by

$$I = \tfrac{1}{4}MR^2 + \tfrac{1}{12}Mh^2 \quad \text{(solid cylinder, perpendicular to axis of rotation).}$$

If the radius R is very small compared to the height h, as is the case for a long thin rod, the moment of inertia in that limit is given by dropping the first term of the preceding equation:

$$I = \tfrac{1}{12}Mh^2 \quad \text{(thin rod of length h, perpendicular to axis of rotation).}$$

We'll derive the formula for the moment of inertia of the hollow cylinder using a volume integral of the kind introduced in Section 8.4. This integral involves separate one-dimensional integrations over each of the three coordinates. This derivation and the next one show how these integrations are done for cylindrical and spherical coordinates. They are not essential for the physical concepts developed in this chapter, but they may be interesting to you.

DERIVATION 10.1 | Moment of Inertia for a Wheel

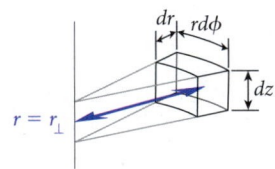

FIGURE 10.7 Volume element in cylindrical coordinates.

In order to derive the moment of inertia of a hollow cylinder of constant density ρ, height h, inner radius R_1, and outer radius R_2, with its symmetry axis as the axis of rotation (see Figure 10.6a), we'll use cylindrical coordinates. For most problems involving cylinders or disks, the coordinate system chosen should generally be cylindrical. In this system (see the discussion of volume integrals in Section 8.4), the volume element is given by (see Figure 10.7).

$$dV = r_\perp \, dr_\perp \, d\phi \, dz.$$

For cylindrical coordinates (and only for cylindrical coordinates!), the perpendicular distance, $r_\perp$, is the same as the radial coordinate, r. With this in mind, we can evaluate the integrals for the hollow cylinder. For the mass, we obtain

$$M = \rho \int_V dV = \rho \int_{R_1}^{R_2} \left(\int_0^{2\pi} \left(\int_{-h/2}^{h/2} dz \right) d\phi \right) r_\perp \, dr_\perp$$

$$= \rho h \int_{R_1}^{R_2} \left(\int_0^{2\pi} d\phi \right) r_\perp \, dr_\perp$$

$$= \rho h 2\pi \int_{R_1}^{R_2} r_\perp \, dr_\perp$$

$$= \rho h 2\pi \left(\tfrac{1}{2}R_2^2 - \tfrac{1}{2}R_1^2 \right)$$

$$= \pi \left(R_2^2 - R_1^2 \right) h \rho.$$

Alternatively, we can express the density as a function of the mass:

$$M = \pi \left(R_2^2 - R_1^2 \right) h \rho \Leftrightarrow \rho = \frac{M}{\pi \left(R_2^2 - R_1^2 \right) h}. \tag{i}$$

The reason for performing this last step may not be entirely obvious right now, but it should become clear after we evaluate the integral for the moment of inertia:

$$I = \rho \int_V r_\perp^2 \, dV = \rho \int_{R_1}^{R_2} \left(\int_0^{2\pi} \left(\int_{-h/2}^{h/2} dz \right) d\phi \right) r_\perp^3 \, dr_\perp$$

$$= \rho h \int_{R_1}^{R_2} \left(\int_0^{2\pi} d\phi \right) r_\perp^3 \, dr_\perp$$

$$= \rho h 2\pi \int_{R_1}^{R_2} r_\perp^3 \, dr_\perp$$

$$= \rho h 2\pi \left(\tfrac{1}{4} R_2^4 - \tfrac{1}{4} R_1^4 \right).$$

Now we can substitute for the density from equation (i):

$$I = \tfrac{1}{2} \rho h \pi \left(R_2^4 - R_1^4 \right) = \frac{M}{\pi \left(R_2^2 - R_1^2 \right) h} \tfrac{1}{2} h \pi \left(R_2^4 - R_1^4 \right).$$

Finally, we make use of the identity $a^4 - b^4 = (a^2 - b^2)(a^2 + b^2)$ to obtain equation 10.9:

$$I = \tfrac{1}{2} M \left(R_1^2 + R_2^2 \right).$$

$$M = \tfrac{4\pi}{3} R^3 \rho$$

$$I = \tfrac{2}{5} M R^2$$

(a)

For objects other than disklike ones, the use of cylindrical coordinates may not be advantageous. The most important such objects are spheres and rectangular blocks. The moment of inertia for a sphere rotating about any axis through its center of mass (see Figure 10.8a) is given by

$$I = \tfrac{2}{5} M R^2 \quad \text{(solid sphere)}. \tag{10.10}$$

The moment of inertia for a thin spherical shell rotating about any axis through its center of mass is

$$I = \tfrac{2}{3} M R^2 \quad \text{(thin spherical shell)}.$$

The moment of inertia for a rectangular block with side lengths a, b, and c rotating about an axis through the center of mass and parallel to side c (see Figure 10.8b) is

$$I = \tfrac{1}{12} M (a^2 + b^2) \quad \text{(rectangular block)}.$$

Again, we'll derive the formula for the solid sphere only in order to illustrate working in a different coordinate system.

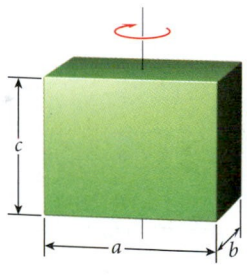

$$M = abc\rho$$

$$I = \tfrac{1}{12} M (a^2 + b^2)$$

(b)

FIGURE 10.8 Moments of inertia of (a) a sphere and (b) a block.

DERIVATION 10.2 ╱ Moment of Inertia for a Sphere

To calculate the moment of inertia for a solid sphere (Figure 10.8a) of constant mass density ρ and radius R rotating about an axis through its center, it is not appropriate to use cylindrical coordinates. Spherical coordinates are the best choice. In spherical coordinates, the volume element is given by (see Figure 10.9)

$$dV = r^2 \sin\theta \, dr \, d\theta \, d\phi.$$

Note that for spherical coordinates, the radial coordinate, r, and the perpendicular distance, $r_\perp$, are not identical. Instead, they are related via (see Figure 10.9)

$$r_\perp = r \sin\theta.$$

– Continued

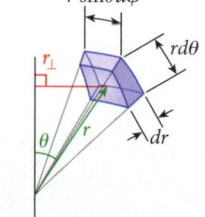

FIGURE 10.9 Volume element in spherical coordinates.

(It is a very common source of error in these kinds of calculations to omit the sine of the angle. You need to keep this in mind when you are using spherical coordinates.)

Again, we first evaluate the integral for the mass:

$$M = \rho \int_V dV = \rho \int_0^R \left(\int_0^{2\pi} \left(\int_0^\pi \sin\theta \, d\theta \right) d\phi \right) r^2 \, dr$$

$$= 2\rho \int_0^R \left(\int_0^{2\pi} d\phi \right) r^2 \, dr = 4\pi\rho \int_0^R r^2 \, dr$$

$$= \frac{4\pi}{3} R^3 \rho.$$

Thus,

$$\rho = \frac{3M}{4\pi R^3}. \tag{i}$$

Then we evaluate the integral for the moment of inertia in a similar way:

$$I = \rho \int_V r_\perp^2 \, dV = \rho \int_V r^2 \sin^2\theta \, dV$$

$$= \rho \int_0^R \left(\int_0^{2\pi} \left(\int_0^\pi \sin^3\theta \, d\theta \right) d\phi \right) r^4 \, dr = \rho \frac{4}{3} \int_0^R \left(\int_0^{2\pi} d\phi \right) r^4 \, dr$$

$$= \rho \frac{8\pi}{3} \int_0^R r^4 \, dr.$$

Thus,

$$I = \rho \frac{8\pi}{15} R^5. \tag{ii}$$

Inserting the expression for the density from equation (i) into equation (ii) we obtain

$$I = \rho \frac{8\pi}{15} R^5 = \frac{3M}{4\pi R^3} \frac{8\pi}{15} R^5 = \tfrac{2}{5} MR^2.$$

Finally, note this important general observation: If R is the largest perpendicular distance of any part of the rotating object from the axis of rotation, then the moment of inertia is always related to the mass of an object by

$$I = cMR^2, \text{ with } 0 < c \le 1. \tag{10.11}$$

The constant c can be calculated from the geometrical configuration of the rotating object and always has a value between zero and one. The more the bulk of the mass is pushed toward the axis of rotation, the smaller the value of the constant c will be. If all the mass is located on the outer edge of the object, as in a hoop, for example, then c approaches the value 1. (Mathematically, this equation is a consequence of the mean value theorem, which you may have encountered in a calculus class.) For a cylinder rotating around its symmetry axis, $c_{\text{cyl}} = \tfrac{1}{2}$, and for a sphere, $c_{\text{sph}} = \tfrac{2}{5}$, as we have seen.

Various objects rotating about an axis that passes through the center of mass are shown in Figure 10.10. Table 10.1 gives the moment of inertia for each object as well as the constant c from equation 10.11, where applicable.

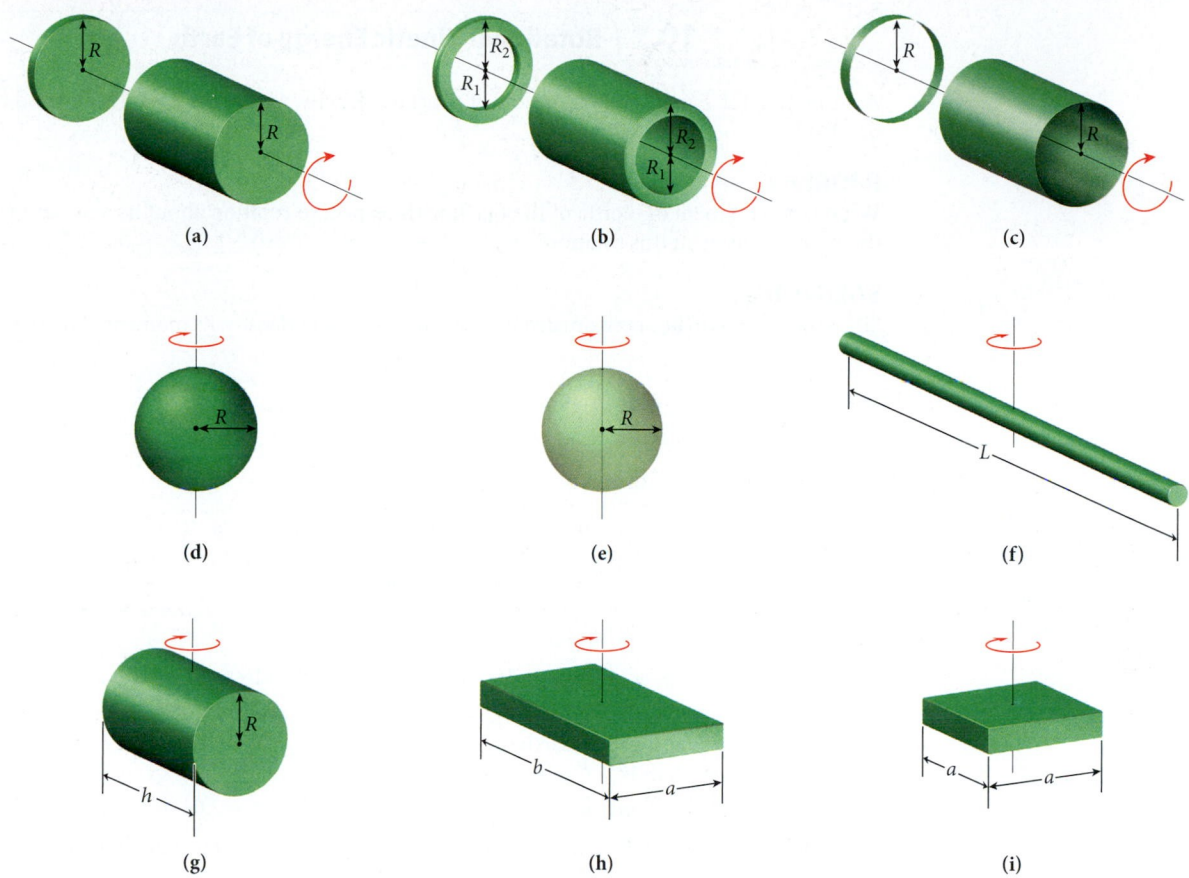

FIGURE 10.10 Orientation of the axis of rotation passing through the center of mass and definition of dimensions for objects listed in Table 10.1.

Table 10.1	The Moment of Inertia and Value of Constant c for the Objects Shown in Figure 10.10. All Objects Have Mass M	
Object	I	c
a) Solid cylinder or disk	$\frac{1}{2}MR^2$	$\frac{1}{2}$
b) Thick hollow cylinder or wheel	$\frac{1}{2}M(R_1^2 + R_2^2)$	
c) Hollow cylinder or hoop	MR^2	1
d) Solid sphere	$\frac{2}{5}MR^2$	$\frac{2}{5}$
e) Hollow sphere	$\frac{2}{3}MR^2$	$\frac{2}{3}$
f) Thin rod	$\frac{1}{12}ML^2$	
g) Solid cylinder perpendicular to symmetry axis	$\frac{1}{4}MR^2 + \frac{1}{12}Mh^2$	
h) Flat rectangular plate	$\frac{1}{12}M(a^2 + b^2)$	
i) Flat square plate	$\frac{1}{6}Ma^2$	

EXAMPLE 10.1 | Rotational Kinetic Energy of Earth

Assume that the Earth is a solid sphere of constant density, with mass $5.98 \cdot 10^{24}$ kg and radius 6370 km.

PROBLEM
What is the moment of inertia of the Earth with respect to rotation about its axis, and what is the kinetic energy of this rotation?

SOLUTION
Since the Earth is to be approximated by a sphere of constant density, its moment of inertia is
$$I = \tfrac{2}{5}MR^2.$$
Inserting the values for the mass and the radius, we obtain
$$I = \tfrac{2}{5}MR^2 = \tfrac{2}{5}(5.98 \cdot 10^{24} \text{ kg})(6.37 \cdot 10^{6} \text{ m})^2 = 9.71 \cdot 10^{37} \text{ kg m}^2.$$
The angular frequency of Earth's rotation is
$$\omega = \frac{2\pi}{1 \text{ day}} = \frac{2\pi}{86,164 \text{ s}} = 7.29 \cdot 10^{-5} \text{ rad/s}.$$
(Note that we used the sidereal day here; see Example 9.3.)

With the results for the moment of inertia and the angular frequency, we can find the kinetic energy of the Earth's rotation:
$$K = \tfrac{1}{2}I\omega^2 = 0.5(9.71 \cdot 10^{37} \text{ kg m}^2)(7.29 \cdot 10^{-5} \text{ rad/s})^2 = 2.58 \cdot 10^{29} \text{ J}.$$
Let's compare this to the kinetic energy of Earth's motion around the Sun. In Chapter 9, we calculated the orbital speed of Earth to be $v = 2.97 \cdot 10^{4}$ m/s. Thus, the kinetic energy of Earth's motion around the Sun is
$$K = \tfrac{1}{2}mv^2 = 0.5(5.98 \cdot 10^{24} \text{ kg})(2.97 \cdot 10^{4} \text{ m/s})^2 = 2.64 \cdot 10^{33} \text{ J},$$
which is larger than the kinetic energy of rotation by a factor greater than 10,000.

Parallel-Axis Theorem

We have determined the moment of inertia for a rotation axis through the center of mass of an object, but what is the moment of inertia for rotation about an axis that does not pass through the center of mass? The **parallel-axis theorem** answers this question. It states that the moment of inertia, $I_{\parallel}$, for rotation of an object of mass M about an axis located a distance d away from the object's center of mass and parallel to an axis through the center of mass, for which the moment of inertia is I_{cm}, is given by

$$I_{\parallel} = I_{cm} + Md^2. \tag{10.12}$$

DERIVATION 10.3 | Parallel-Axis Theorem

For this derivation, consider the object in Figure 10.11. Suppose we have already calculated the moment of inertia of this object for rotation about the z-axis, which passes through the center of mass of this object. The origin of the xyz-coordinate system is at the center of mass, and the z-axis is the rotation axis. Any axis parallel to the rotation axis can then be described by a simple shift in the xy-plane indicated in the figure by the vector $\vec{d}$, with components d_x and d_y.

If we shift the coordinate system in the xy-plane so that the new vertical axis, the z'-axis, coincides with the new rotation axis, then the transformation from the xyz-coordinates to the new $x'y'z'$-coordinates is given by

$$x' = x - d_x, \quad y' = y - d_y, \quad z' = z.$$

To calculate the moment of inertia of the object rotating about the new axis in the new coordinate system, we can simply use equation 10.6, the most general equation, which applies to the case where the mass density is not constant:

$$I_{\parallel} = \int_V (r_{\perp}')^2 \rho \, dV. \tag{i}$$

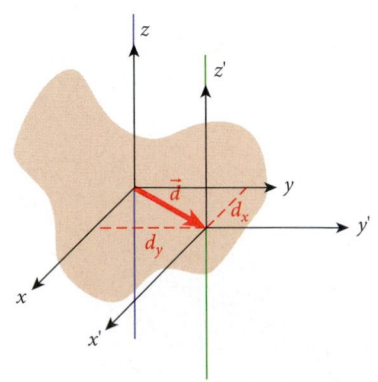

FIGURE 10.11 Coordinates and distances for the parallel-axis theorem.

According to the coordinate transformation,

$$(r_\perp')^2 = (x')^2 + (y')^2 = (x - d_x)^2 + (y - d_y)^2$$
$$= x^2 - 2xd_x + d_x^2 + y^2 - 2yd_y + d_y^2$$
$$= (x^2 + y^2) + (d_x^2 + d_y^2) - 2xd_x - 2yd_y$$
$$= r_\perp^2 + d^2 - 2xd_x - 2yd_y.$$

(Keep in mind that $\vec{r}_\perp$ lies in the xy-plane because of the way we constructed the coordinate system.) Now we substitute this expression for $(r_\perp')^2$ into equation (i) and obtain

$$I_\parallel = \int_V (r_\perp')^2 \, \rho \, dV$$

$$= \int_V r_\perp^2 \rho \, dV + d^2 \int_V \rho \, dV - 2d_x \int_V x\rho \, dV - 2d_y \int_V y\rho \, dV. \tag{ii}$$

The first integral in equation (ii) gives the moment of inertia for rotation about the center of mass, which we already know. The second integral is simply equal to the mass (compare to equation 10.7). The third and fourth integrals were introduced in Chapter 8 and give the locations of the x- and y-coordinates of the center of mass when divided by the mass. However, by construction, they equal zero, because we put the origin of the xyz-coordinate system at the center of mass. Thus, we obtain the parallel-axis theorem:

$$I_\parallel = I_{cm} + d^2 M.$$

Note that, according to equations 10.11 and 10.12, the moment of inertia with respect to rotation about an arbitrary axis parallel to an axis through the center of mass can be written as

$$I = (cR^2 + d^2)M, \text{ with } 0 < c \le 1.$$

Here R is the maximum perpendicular distance of any part of the object from its axis of rotation through the center of mass, and d is the distance of the rotation axis from a parallel axis through the center of mass.

Self-Test Opportunity 10.1

Show that the moment of inertia of a thin rod of mass m and length L rotating around one end is $I = \frac{1}{3}mL^2$.

10.3 Rolling without Slipping

Rolling motion is a special case of rotational motion that is performed by round objects of radius R that move across a surface without slipping. For rolling motion, we can connect the linear and angular quantities by realizing that the linear distance moved by the center of mass is the same as the length of corresponding arc of the object's circumference, as shown in Table 9.1. Thus, the relationship between the linear distance, r, traveled by the center of mass and the rotation angle is

$$r = R\theta.$$

Taking the time derivative and keeping in mind that the radius, R, remains constant, we obtain the relationships between linear and angular speeds and accelerations:

$$v = R\omega$$

and

$$a = R\alpha.$$

The total kinetic energy of an object in rolling motion is the sum of its translational (linear motion of its center of mass) and rotational kinetic energies (rotation about the center of mass):

$$K = K_{trans} + K_{rot} = \frac{1}{2}mv^2 + \frac{1}{2}I\omega^2. \tag{10.13}$$

We can now substitute for ω from $v = R\omega$ and for I from equation 10.11:

$$K = \tfrac{1}{2}mv^2 + \tfrac{1}{2}I\omega^2$$

$$= \tfrac{1}{2}mv^2 + \tfrac{1}{2}(cR^2m)\left(\frac{v}{R}\right)^2$$

$$= \tfrac{1}{2}mv^2 + \tfrac{1}{2}mv^2c \Rightarrow$$

$$K = (1+c)\tfrac{1}{2}mv^2, \tag{10.14}$$

where $0 < c \le 1$ is the constant introduced in equation 10.11. Equation 10.14 implies that the kinetic energy of a rolling object is always greater than that of an object that is sliding, provided they have the same mass and linear velocity.

With an expression for the kinetic energy that includes the contribution due to rotation, we can apply the concept of conservation of total mechanical energy (the sum of kinetic and potential energy) that we applied in Chapter 6.

SOLVED PROBLEM 10.1 | Sphere Rolling Down an Inclined Plane

PROBLEM
A solid sphere with a mass of 5.15 kg and a radius of 0.340 m starts from rest at a height of 2.10 m above the base of an inclined plane and rolls down without sliding under the influence of gravity. What is the linear speed of the center of mass of the sphere just as it leaves the incline and rolls onto a horizontal surface?

SOLUTION

FIGURE 10.12 Sphere rolling down an inclined plane.

THINK At the top of the incline, the sphere is at rest. At that point, the sphere has gravitational potential energy and no kinetic energy. As the sphere starts to roll, it loses potential energy and gains kinetic energy of linear motion and kinetic energy of rotation. At the bottom of the inclined plane, all of the original potential energy is in the form of kinetic energy. The kinetic energy of linear motion is linked to the kinetic energy of rotation through the radius of the sphere.

SKETCH A sketch of the problem situation is shown in Figure 10.12 with the zero of the y-coordinate at the bottom of the incline.

RESEARCH At the top of the incline, the sphere is at rest and has zero kinetic energy. At the top, its energy is therefore its potential energy, mgh:

$$E_{\text{top}} = K_{\text{top}} + U_{\text{top}} = 0 + mgh = mgh,$$

where m is the mass of the sphere, h is the height of the sphere above the horizontal surface, and g is the acceleration due to gravity. At the bottom of the incline, just as the sphere starts rolling onto the horizontal surface, the potential energy is zero. According to equation 10.14, the sphere has a total kinetic energy (sum of translational and rotational kinetic energy) of $(1 + c)\tfrac{1}{2}mv^2$. Thus, the total energy at the bottom of the incline is

$$E_{\text{bottom}} = K_{\text{bottom}} + U_{\text{bottom}} = (1+c)\tfrac{1}{2}mv^2 + 0 = (1+c)\tfrac{1}{2}mv^2.$$

Because the moment of inertia of a sphere is $I = \tfrac{2}{5}mR^2$ (see equation 10.10), the constant c has the value $\tfrac{2}{5}$ in this case.

SIMPLIFY Because of the conservation of energy, the energy at the top of the incline is equal to that at the bottom:

$$mgh = (1+c)\tfrac{1}{2}mv^2.$$

Solving for the linear velocity gives us

$$v = \sqrt{\frac{2gh}{1+c}}.$$

For a sphere, $c = \tfrac{2}{5}$, as noted above, and so the speed of the rolling object is in this case

$$v = \sqrt{\frac{2gh}{1+\tfrac{2}{5}}} = \sqrt{\frac{10}{7}gh}.$$

CALCULATE   Putting in the numerical values gives us

$$v = \sqrt{\frac{10}{7}(9.81 \text{ m/s}^2)(2.10 \text{ m})} = 5.42494 \text{ m/s}.$$

ROUND   Expressing our result with three significant figures leads to

$$v = 5.42 \text{ m/s}.$$

DOUBLE-CHECK   If the sphere did not roll, but instead slid down the inclined plane without friction, the final speed would be

$$v = \sqrt{2gh} = \sqrt{2(9.81 \text{ m/s}^2)(2.10 \text{ m})} = 6.42 \text{ m/s},$$

which is faster than the speed we found for the rolling sphere. It seems reasonable that the final linear velocity of the rolling sphere is somewhat less than the final linear velocity of the sliding sphere, because some of the initial potential energy of the rolling sphere is transformed into rotational kinetic energy. This energy is then not available to become kinetic energy of linear motion of the center of mass of the sphere. Note that we did not need the mass or radius of the sphere in this calculation.

The formula derived in Solved Problem 10.1 for the speed of a rolling sphere at the bottom of the incline,

$$v = \sqrt{\frac{2gh}{1+c}}, \tag{10.15}$$

is a rather general result. It can be applied to various situations in which potential gravitational energy is converted into translational and rotational kinetic energy of a rolling object.

Galileo Galilei famously showed that the acceleration of an object in free fall is independent of its mass. This is also true for an object rolling down an incline, as we have just seen in Solved Problem 10.1, leading to equation 10.15. However, while the total mass of the rolling object does not matter, the distribution of mass within it does matter. Mathematically, this is reflected in equation 10.15 by the fact that the constant c from equation 10.11, which is calculated from the geometric distribution of the mass, appears in the denominator. The following example demonstrates clearly that the distribution of mass in rolling objects matters.

EXAMPLE 10.2   Race Down an Incline

PROBLEM

A solid sphere, a solid cylinder, and a hollow cylinder (a tube), all of the same mass m and the same outer radius R, are released from rest at the top of an incline and start rolling without sliding. In which order do they arrive at the bottom of the incline?

SOLUTION

We can answer this question by using energy considerations only. Since the total mechanical energy is conserved for each of the three objects during its rolling motion, we can write for each object

$$E = K + U = K_0 + U_0.$$

The objects were released from rest, so $K_0 = 0$. For the potential energy, we again use $U = mgh$, and for the kinetic energy, we use equation 10.14. Therefore, we have

$$K_{\text{bottom}} = U_{\text{top}} \Rightarrow (1+c)\tfrac{1}{2}mv^2 = mgh \Rightarrow$$

$$v = \sqrt{\frac{2gh}{1+c}},$$

which is the same formula as equation 10.15.

We already noted that the mass and radius of the object canceled out of this formula. However, we can make an additional important observation. The constant c that is determined by

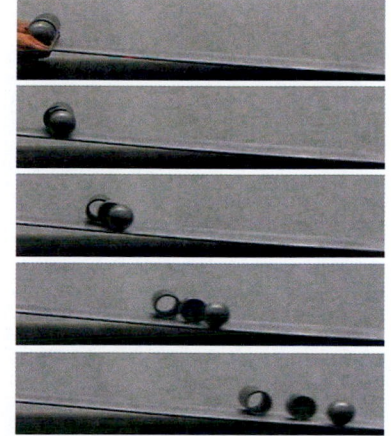

FIGURE 10.13   Race of a sphere, a solid cylinder, and a hollow cylinder of the same mass and radius down an inclined plane. The frames were shot at intervals of 0.5 s.

Concept Check 10.3

Suppose we repeat the race of Example 10.2, but let an unopened can of soda participate. In which place will the can finish?

a) first   c) third

b) second   d) fourth

– Continued

the distribution of the mass appears in the denominator. We already know the value of c for the three rolling objects: $c_{sphere} = \frac{2}{5}$, $c_{cylinder} = \frac{1}{2}$, and $c_{tube} \approx 1$. Since the constant for the sphere is the smallest, the sphere's velocity for any given height h will be the largest, implying that the sphere will win the race. Physically, we see that since all three objects have the same mass and thus have the same change in potential energy, all final kinetic energies are equal. Thus, an object with a higher value of c will have relatively more of its kinetic energy in rotation and therefore a lower translational kinetic energy and a lower linear speed. The solid cylinder will come in second in the race, followed by the tube. Figure 10.13 shows frames from a videotaped experiment that verify our conclusions.

Self-Test Opportunity 10.2

Can you explain why the can of soda finishes the race in the position determined in Concept Check 10.3?

SOLVED PROBLEM 10.2 | Ball Rolling through a Loop

A solid sphere is released from rest and rolls down an incline and then into a circular loop of radius R (see Figure 10.14). Assume that the radius of the sphere is much smaller than the radius of the loop.

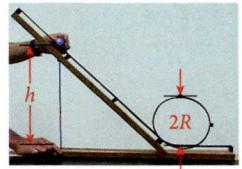 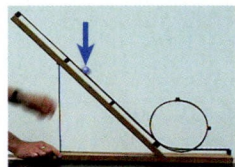 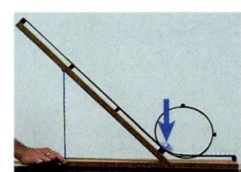 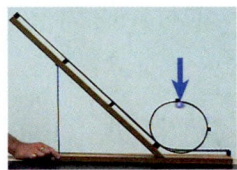 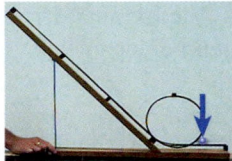

FIGURE 10.14 Sphere rolling down an incline and into a loop. The frames were shot at intervals of 0.25 s.

PROBLEM
What is the minimum height h from which the sphere has to be released so that it does not fall off the track when in the loop?

SOLUTION
THINK When released from rest at a height h on the incline, the sphere has gravitational potential energy but no kinetic energy. As the sphere rolls down the incline and around the loop, gravitational potential energy is converted to kinetic energy. At the top of the loop, the sphere has dropped a distance $h - 2R$. The key to solving this problem is to realize that at the top of the loop, the centripetal acceleration has to be equal to or larger than the acceleration due to gravity. (When these accelerations are equal, "weightlessness" of the object occurs, as shown in Solved Problem 9.1. When the centripetal acceleration is greater, there must be a downward supporting force, which the track can supply. When the centripetal acceleration is less, there must be an upward supporting force, which the track cannot supply, so the sphere falls off the track.) What is unknown here is the speed v at the top of the loop. We can again employ energy conservation considerations to calculate the minimum velocity required and then find the minimum starting height required for the sphere to remain on the track.

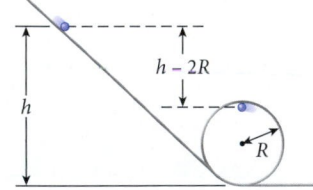

FIGURE 10.15 Sphere rolling on an incline and a loop.

SKETCH A sketch of the sphere rolling on the incline and loop is shown in Figure 10.15.

RESEARCH When the sphere is released, it has potential energy $U_0 = mgh$ and zero kinetic energy. At the top of the loop, the potential energy is $U = mg2R$, and the total kinetic energy is, according to equation 10.14, $K = (1 + c)\frac{1}{2}mv^2$, where $c_{sphere} = \frac{2}{5}$. Thus, conservation of total mechanical energy tells us that

$$E = K + U = (1+c)\tfrac{1}{2}mv^2 + mg2R = K_0 + U_0 = mgh. \tag{i}$$

At the top of the loop, the centripetal acceleration, a_c, has to be equal to or larger than the acceleration due to gravity, g:

$$g \le a_c = \frac{v^2}{R}. \tag{ii}$$

SIMPLIFY We can solve equation (i) for v^2:

$$v^2 = \frac{2g(h-2R)}{1+c}.$$

If we then substitute this expression for v^2 into equation (ii), we find

$$g \le \frac{2g(h-2R)}{R(1+c)}.$$

Multiplying both sides of this equation by the denominator of the fraction on the right side leads to

$$R(1+c) \le 2h - 4R.$$

Therefore, we have

$$h \ge \frac{5+c}{2}R.$$

CALCULATE This result is valid for any rolling object, with the constant c determined from the object's geometry. In this problem, we have a solid sphere, so $c = \frac{2}{5}$. Thus, the result is

$$h \ge \frac{5+\frac{2}{5}}{2}R = \frac{27}{10}R.$$

ROUND This problem situation was described in terms of variables rather than numerical values, so we can quote our result exactly as

$$h \ge 2.7R.$$

DOUBLE-CHECK If the sphere were not rolling but sliding without friction, we could equate the kinetic energy of the sphere at the top of the loop with the change in gravitational potential energy:

$$\tfrac{1}{2}mv^2 = mg(h-2R).$$

We could then express the inequality between gravitational and centripetal acceleration given in equation (ii) as follows:

$$g \le a_c = \frac{v^2}{R} = \frac{2g(h-2R)}{R}.$$

Solving this equation for h gives us

$$h \ge \frac{5}{2}R = 2.5R.$$

This required height for a sliding sphere is slightly lower than the height we found for a rolling sphere. We would expect that less energy, in the form of gravitational potential energy, would be required to keep the sphere on the track if it were not rolling because its kinetic energy would be entirely translational. Thus, our result for keeping a rolling sphere on the loop seems reasonable.

10.4 Torque

So far, in discussing forces, we have seen that a force can cause linear motion of an object, which can be described in terms of the motion of the center of mass of the object. However, we have not addressed one general question: Where are the force vectors acting on an extended object placed in a free-body diagram? A force can be exerted on an extended object at a point away from its center of mass, which can cause the object to rotate as well as move linearly.

Moment Arm

Consider the hand attempting to use a wrench to loosen a bolt in Figure 10.16. It is clear that it will be easiest to turn the bolt in Figure 10.16c, quite a bit harder in Figure 10.16b,

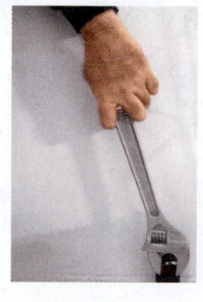

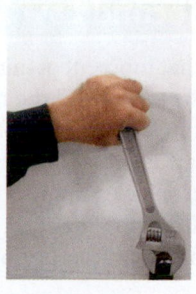
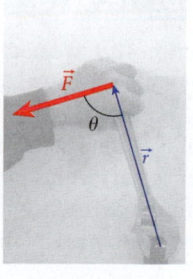

(a)   (b)   (c)   (d)

FIGURE 10.16 (a)–(c) Three ways to use a wrench to loosen a bolt. (d) The force $\vec{F}$ and moment arm r, with the angle θ between them.

and downright impossible in Figure 10.16a. This example shows that the magnitude of the force is not the only relevant quantity. The perpendicular distance from the line of action of the force to the axis of rotation, called the **moment arm,** is also important. In addition, the angle at which the force is applied, relative to the moment arm, matters as well. In parts (b) and (c) of Figure 10.16, this angle is 90°. (An angle of 270° would be just as effective, but then the force would act in the opposite direction.) An angle of 180° or 0° (Figure 10.16a) will not turn the bolt.

These considerations are quantified by the concept of torque, τ. **Torque** (also called *moment*) is the vector product of the force $\vec{F}$ and the position vector $\vec{r}$:

$$\vec{\tau} = \vec{r} \times \vec{F}. \tag{10.16}$$

The position vector $\vec{r}$ is measured with the origin at the axis of rotation. The symbol × denotes the **vector product,** or *cross product*. (We introduced vector products in Chapter 1, which you may want to consult for a review.)

The SI unit of torque is N m, not to be confused with the unit of energy, which is the joule (J = N m)

$$[\tau] = [F] \cdot [r] = N\,m.$$

In English units, torque is often expressed in foot-pounds (ft-lb).

The magnitude of the torque is the product of the magnitude of the force and the distance to the axis of rotation (the magnitude of the position vector, or the moment arm) times the sine of the angle between the force vector and the position vector (see Figure 10.17):

$$\tau = rF\sin\theta. \tag{10.17}$$

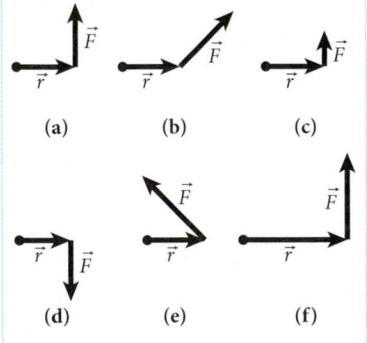

FIGURE 10.17 Right-hand rule for the direction of the torque for a given force and position vector.

Angular quantities can also be vectors, called *axial vectors*. (An axial vector is any vector that points along the rotation axis.) Torque is an example of an axial vector, and its magnitude is given by equation 10.17. The direction of the torque is given by a right-hand rule (Figure 10.17). Note that right-hand rules apply to all vector products! The torque points in a direction perpendicular to the plane spanned by the force and position vectors. Thus, if the position vector points along the thumb and the force vector points along the index finger, then the direction of the axial torque vector is the direction of the middle finger, as shown in Figure 10.17. Note that the torque vector is perpendicular to both the force vector and the position vector.

With the mathematical definition of the torque, its magnitude, and its relation to the force vector, the position vector, and their relative angle, we can understand why the approach shown in part (c) of Figure 10.16 yields maximum torque for a given magnitude of the force, whereas that in part (a) yields zero torque. We see that the magnitude of the torque is the decisive factor in determining how easy or hard it is to loosen (or tighten) a bolt.

Torques around any fixed axis of rotation can be clockwise or counterclockwise. As indicated by the force vector in Figure 10.16d, the torque generated by the hand pulling on the wrench would be counterclockwise. The **net torque** is defined as the difference between the sum of all clockwise torques and the sum of all counterclockwise torques:

$$\tau_{\text{net}} = \sum_i \tau_{\text{counterclockwise},i} - \sum_j \tau_{\text{clockwise},j}$$

Concept Check 10.4

Choose the combination of position vector, $\vec{r}$, and force vector, $\vec{F}$, that produces the torque of highest magnitude around the point indicated by the black dot.

(a)   (b)   (c)

(d)   (e)   (f)

10.5 Newton's Second Law for Rotation

In Section 10.1, we noted that moment of inertia I is the rotational equivalent of mass. From Chapter 4, we know that the product of the mass and the linear acceleration is the net force acting on the object, as expressed by Newton's Second Law, $F_{\text{net}} = ma$. What is the equivalent of Newton's Second Law for rotational motion?

Let's start with a point particle of mass M moving in a circle around an axis at a distance R from the axis. If we multiply the moment of inertia for rotation about an axis parallel to the center of mass by the angular acceleration, we obtain

$$I\alpha = (R^2 M)\alpha = RM(R\alpha) = RMa = RF_{\text{net}}.$$

To obtain this result, we first used equation 10.11 with $c = 1$, then the relationship between angular and linear acceleration for motion on a circle, and then Newton's Second Law. Thus, the product of the moment of inertia and the angular acceleration is proportional to the product of a distance quantity and a force quantity. In Section 10.4, we saw that this product is a torque, τ. Thus, we can write the following form of Newton's Second Law for rotational motion:

$$\tau = I\alpha. \tag{10.18}$$

Combining equations 10.16 and 10.18 gives us

$$\vec{\tau} = \vec{r} \times \vec{F}_{net} = I\vec{\alpha}. \tag{10.19}$$

This equation for rotational motion is analogous to Newton's Second Law, $\vec{F} = m\vec{a}$, for linear motion. Figure 10.18 shows the relationship between force, position, torque, and angular acceleration for a point particle moving around an axis of rotation. Note that, strictly speaking, equation 10.19 holds only for a point particle in circular orbit. It seems plausible that this equation holds for a moment of inertia for an extended object in general, but we have not proven it. Later in this chapter, we will come back to this equation.

Newton's First Law stipulates that in the absence of a net force, an object experiences no acceleration and thus no change in velocity. The equivalent of Newton's First Law for rotational motion is that an object experiencing no net torque has no angular acceleration and thus no change in angular velocity. In particular, this means that for objects to remain stationary, the net torque on them has to be zero. We will return to this statement in Chapter 11, where we investigate static equilibrium.

Now, however, we can use the rotational form of Newton's Second Law (equation 10.19) to solve interesting problems of rotational motion, such as the following one.

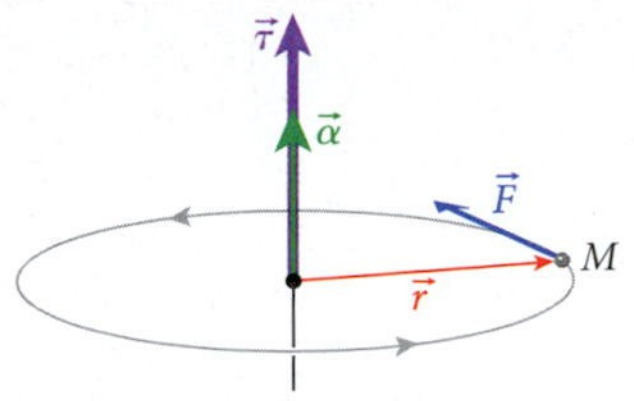

FIGURE 10.18 A force exerted on a point particle creates a torque.

EXAMPLE 10.3 | Toilet Paper

This may have happened to you: You are trying to put a new roll of toilet paper into its holder. However, you drop the roll, managing to hold onto just the first sheet. On its way to the floor, the toilet paper roll unwinds, as Figure 10.19a shows.

PROBLEM
How long does it take the roll of toilet paper to hit the floor, if it was released from a height of 0.73 m? The roll has an inner radius $R_1 = 2.7$ cm, an outer radius $R_2 = 6.1$ cm, and a mass of 274 g.

SOLUTION
For the falling roll of toilet paper, the acceleration is $a_y = -g$, and in Chapter 2, we saw that for free fall from rest, the position as a function of time is given in general as $y = y_0 + v_0 t - \frac{1}{2}gt^2$. In this case, the initial velocity is zero; therefore, $y = y_0 - \frac{1}{2}gt^2$. If we put the origin at floor level, then we have to find the time at which $y = 0$. This implies $y_0 = \frac{1}{2}gt^2$ for the time it takes the roll to hit the floor. Therefore, the time it would take for the roll in free fall to hit the floor is

$$t_{free} = \sqrt{\frac{2y_0}{g}} = \sqrt{\frac{2(0.73 \text{ m})}{9.81 \text{ m/s}^2}} = 0.386 \text{ s}. \tag{i}$$

However, since you hold onto the first sheet and the toilet paper unwinds on the way down, the roll of toilet paper is rolling without slipping (in the sense in which this was defined earlier). Therefore, the acceleration will be different from the free-fall case. Once we know the value of this acceleration, we can use a formula relating the initial height and fall time that is similar to equation (i).

How do we calculate the acceleration that the toilet paper experiences? Again, we start with a free-body diagram. Figure 10.19b shows the toilet paper roll in side view and indicates the force of gravity, $\vec{F}_g = mg(-\hat{y})$, and the tension from the sheet that is held by the hand, $\vec{T} = T\hat{y}$. Newton's Second Law then allows us to relate the net force acting on the toilet paper to the acceleration of the roll:

$$T - mg = ma_y. \tag{ii}$$

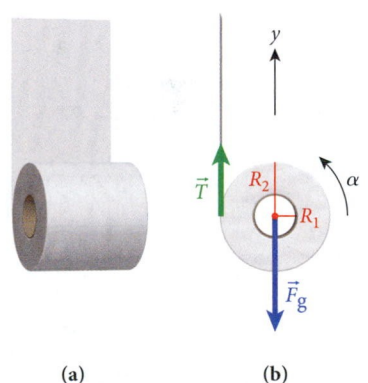

FIGURE 10.19 (a) Unrolling toilet paper. (b) Free-body diagram of the roll of toilet paper.

– Continued

The tension and the acceleration are both unknown, so we need to find a second equation relating these quantities. This second equation can be obtained from the rotational motion of the roll, for which the net torque is the product of the moment of inertia and the angular acceleration: $\tau = I\alpha$. The moment of inertia of the roll of toilet paper is that of a hollow cylinder, $I = \frac{1}{2}m(R_1^2 + R_2^2)$, which is equation 10.9.

We can again relate the linear and angular accelerations via $a_y = R_2\alpha$, where R_2 is the outer radius of the roll of toilet paper. We need to specify which direction is positive for the angular acceleration—otherwise, we could get the sign wrong and thus obtain a false result. To be consistent with the choice of up as the positive y-direction, we need to select counterclockwise rotation as the positive angular direction, as indicated in Figure 10.19b.

For the torque about the axis of symmetry of the roll of toilet paper, we have $\tau = -R_2T$, with the sign convention for positive angular acceleration that we just established. The force of gravity does not contribute to the torque about the axis of symmetry, because its moment arm has a length of zero. Newton's Second Law for rotational motion then leads to

$$\tau = I\alpha$$

$$-R_2T = \left[\tfrac{1}{2}m(R_1^2 + R_2^2)\right]\frac{a_y}{R_2}$$

$$-T = \tfrac{1}{2}m\left(1 + \frac{R_1^2}{R_2^2}\right)a_y. \qquad \text{(iii)}$$

Equations (ii) and (iii) form a set of two equations for the two unknown quantities, T and a_y. Adding them results in

$$-mg = \tfrac{1}{2}m\left(1 + \frac{R_1^2}{R_2^2}\right)a_y + ma_y.$$

The mass of the roll of toilet paper cancels out, and we find for the acceleration

$$a_y = -\frac{g}{\dfrac{3}{2} + \dfrac{R_1^2}{2R_2^2}}.$$

Using the given values for the inner radius, $R_1 = 2.7$ cm, and the outer radius, $R_2 = 6.1$ cm, we find the value of the acceleration:

$$a_y = -\frac{9.81 \text{ m/s}^2}{\dfrac{3}{2} + \dfrac{(2.7 \text{ cm})^2}{2(6.1 \text{ cm})^2}} = -6.14 \text{ m/s}^2.$$

Inserting this value for the acceleration into a formula for the fall time that is analogous to equation (i) gives our answer

$$t = \sqrt{\frac{2y_0}{-a_y}} = \sqrt{\frac{2(0.73 \text{ m})}{6.14 \text{ m/s}^2}} = 0.488 \text{ s}.$$

This is approximately 0.1 s longer than the time for the free-falling roll of toilet paper released from the same height.

DISCUSSION

Note that we assumed that the outer radius of the roll of toilet paper did not change as the paper unwound. For the short distance of less than 1 m, this is justified. However, if we wanted to calculate how long it took for the toilet paper to unwind over a distance of, say, 10 m, we would need to take the change in outer radius into account. Of course, then we would also need to take into account the effect of air resistance!

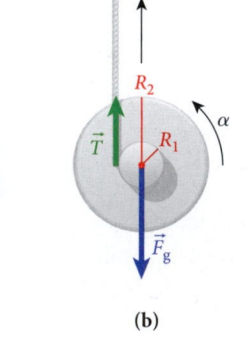

FIGURE 10.20 (a) Yo-yo. (b) Free-body diagram of the yo-yo.

As an extension of Example 10.3, we can consider a yo-yo. A yo-yo consists of two solid disks of radius R_2, with a thin smaller disk of radius R_1 mounted between them, and a string wound around the smaller disk (see Figure 10.20). For the purpose of this analysis, we can consider the moment of inertia of the yo-yo to be that of a solid disk with radius R_2: $I = \frac{1}{2}mR_2^2$. The free-body diagrams of Figure 10.19b and Figure 10.20b are almost identical, except for one detail: For the yo-yo, the string tension acts on the surface at the inner radius R_1, as opposed to the outer radius R_2, as was the case for the roll of toilet paper. This implies that the angular and linear accelerations for the yo-yo are proportional to each other, with proportionality constant R_1 (instead of R_2 as for the roll). Thus, the torque for the yo-yo rolling without slipping along the string is

$$\tau = I\alpha \Rightarrow -TR_1 = \left(\tfrac{1}{2}mR_2^2\right)\frac{a_y}{R_1}$$

$$-T = \tfrac{1}{2}m\frac{R_2^2}{R_1^2}a_y.$$

It is instructive to compare this equation to $-T = \tfrac{1}{2}m(1 + R_1^2/R_2^2)a_y$, which we derived in Example 10.3 for the roll of toilet paper. They look very similar, but the ratio of the radii is different. On the other hand, the equation derived from Newton's Second Law (for translation) is the same in both cases:

$$T - mg = ma_y.$$

Adding this equation and the equation for the torque of the yo-yo yields the acceleration of the yo-yo:

$$-mg = ma_y + \tfrac{1}{2}m\frac{R_2^2}{R_1^2}a_y \Rightarrow$$

$$a_y = -\frac{g}{1 + \tfrac{1}{2}\dfrac{R_2^2}{R_1^2}} = -\frac{2R_1^2}{2R_1^2 + R_2^2}g.$$

For example, if $R_2 = 5R_1$, the acceleration of the yo-yo is

$$a_y = \frac{-g}{1 + \tfrac{1}{2}(25)} = \frac{-g}{13.5} = -0.727 \text{ m/s}^2.$$

Atwood Machine

Chapter 4 introduced an Atwood machine, which consists of two weights with masses m_1 and m_2 connected by a string or rope, which is guided over a pulley. The machines we analyzed in Chapter 4 were subject to the condition that the rope glides without friction over the pulley so that the pulley does not rotate (or, equivalently, that the pulley is massless). In such cases, the common acceleration of the two masses is $a = g(m_1 - m_2)/(m_1 + m_2)$. With the concepts of rotational dynamics, we can take another look at the Atwood machine and consider the case where friction occurs between the rope and the pulley, causing the rope to turn the pulley without slipping.

In Chapter 4, we saw that the magnitude of the tension, T, is the same everywhere in a string. However, now friction forces are involved to keep the string attached to the pulley, and we cannot assume that the tension is constant. Instead, the string tension is separately determined in each segment of the string from which one of the two masses hang. Thus, there are two different string tensions, T_1 and T_2, in the free-body diagrams for the two masses (as shown in Figure 10.21b). Just as in Chapter 4, applying Newton's Second Law individually to each free-body diagram leads to

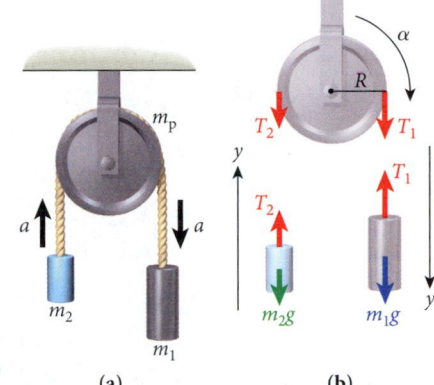

FIGURE 10.21 Atwood machine: (a) physical setup; (b) free-body diagrams.

$$-T_1 + m_1g = m_1a \tag{10.20}$$

$$T_2 - m_2g = m_2a. \tag{10.21}$$

Here again we used the (arbitrary) sign convention that a positive acceleration ($a > 0$) is one for which m_1 moves downward and m_2 upward. This convention is indicated in the free-body diagrams by the direction of the positive y-axis.

Figure 10.21b also shows a free-body diagram for the pulley, but it includes only the forces that can cause a torque: the two string tensions, T_1 and T_2. The downward force of gravity and the upward force of the support structure on the pulley are not shown. The pulley does not have translational motion, so all forces acting on the pulley add up to zero. However, a net torque does act on the pulley. According to equation 10.17, the magnitude of the torque due to the string tensions is given by

$$\tau = \tau_1 - \tau_2 = RT_1\sin 90° - RT_2\sin 90° = R(T_1 - T_2). \tag{10.22}$$

These two torques have opposite signs, because one is acting clockwise and the other counterclockwise. According to equation 10.18, the net torque is related to the moment of inertia of the pulley and its angular acceleration by $\tau = I\alpha$. The moment of inertia of the pulley (mass of m_p) is that of a disk: $I = \frac{1}{2}m_pR^2$. Since the rope moves across the pulley without slipping, the acceleration of the rope (and masses m_1 and m_2) is related to the angular acceleration via $\alpha = a/R$, just like the correspondence established in Chapter 9 between linear and angular acceleration for a point particle moving on the circumference of a circle. Inserting the expressions for the moment of inertia and the angular acceleration results in $\tau = I\alpha = (\frac{1}{2}m_pR^2)(a/R)$. Substituting this expression for the torque into equation 10.22, we find

$$R(T_1 - T_2) = \tau = \left(\tfrac{1}{2}m_pR^2\right)\left(\frac{a}{R}\right) \Rightarrow$$

$$T_1 - T_2 = \tfrac{1}{2}m_pa. \tag{10.23}$$

Equations 10.20, 10.21, and 10.23 form a set of three equations for three unknown quantities: the two values of the string tension, T_1 and T_2, and the acceleration, a. The easiest way to solve this system for the acceleration is to add the equations. We then find

$$m_1g - m_2g = (m_1 + m_2 + \tfrac{1}{2}m_p)a \Rightarrow$$

$$a = \frac{m_1 - m_2}{m_1 + m_2 + \tfrac{1}{2}m_p}g. \tag{10.24}$$

Note that equation 10.24 matches the equation for the case of a massless pulley (or the case in which the rope slides over the pulley without friction), except for the additional term of $\frac{1}{2}m_p$ in the denominator, which represents the contribution of the pulley to the overall inertia of the system. The factor $\frac{1}{2}$ reflects the shape of the pulley, a disk, because $c = \frac{1}{2}$ for a disk in the relationship between moment of inertia, mass, and radius (equation 10.11).

Thus, we have answered the question of what happens when a force is exerted on an extended object some distance away from its center of mass: The force produces a torque as well as linear motion. This torque leads to rotation, which we left out of our original considerations of the result of exerting a force on an object because we assumed that all forces acted on the center of mass of the object.

SOLVED PROBLEM 10.3 / Falling Horizontal Rod

A thin rod of length $L = 2.50$ m and mass $m = 3.50$ kg is suspended horizontally by a pair of vertical strings attached to the ends (Figure 10.22). The string supporting end B is then cut.

PROBLEM
What is the linear acceleration of end B of the rod just after the string is cut?

SOLUTION
THINK Before the string is cut, the rod is at rest. When the string supporting end B is cut, a net torque acts on the rod, with a pivot point at end A. This torque is due to the force of gravity acting on the rod. We can consider the mass of the rod to be concentrated at its center of mass, which is located at $L/2$. The initial torque is then equal to the weight of the rod times the moment arm, which is $L/2$. The resulting initial angular acceleration can be related to the linear acceleration of end B of the rod.

SKETCH Figure 10.23 is a sketch of the rod after the string is cut.

RESEARCH When the string supporting end B is cut, the torque, τ, on the rod is due to the force of gravity, F_g, acting on the rod times the moment arm, $r_\perp = L/2$:

$$\tau = r_\perp F_g = \left(\frac{L}{2}\right)(mg) = \frac{mgL}{2}. \tag{i}$$

The angular acceleration, α, is given by

$$\tau = I\alpha, \tag{ii}$$

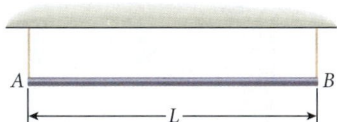

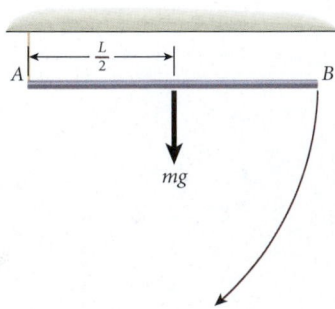

FIGURE 10.22 A thin rod supported in a horizontal position by a vertical string at each end.

FIGURE 10.23 The thin rod just after the string supporting end B is cut.

where the moment of inertia, I, of the thin rod rotating around end A is given by

$$I = \tfrac{1}{3}mL^2.$$ (iii)

The linear acceleration, a, of end B can be related to the angular acceleration through

$$a = L\alpha,$$ (iv)

because end B of the rod is undergoing circular motion as the rod pivots around end A.

SIMPLIFY   We can combine equations (i) and (ii) to get

$$\tau = I\alpha = \frac{mgL}{2}.$$ (v)

Substituting for I and a from equations (iii) and (iv) into equation (v) gives us

$$I\alpha = \left(\tfrac{1}{3}mL^2\right)\left(\frac{a}{L}\right) = \frac{mgL}{2}.$$

Dividing out common factors, we obtain

$$\frac{a}{3} = \frac{g}{2}$$

or

$$a = 1.5g.$$

CALCULATE   Inserting the numerical value for the acceleration of gravity gives us

$$a = 1.5\left(9.81 \text{ m/s}^2\right) = 14.715 \text{ m/s}^2.$$

ROUND   Expressing our result to three significant figures gives

$$a = 14.7 \text{ m/s}^2.$$

DOUBLE-CHECK   Perhaps this answer is somewhat surprising, because you may have assumed that the acceleration cannot exceed the free-fall acceleration, g. If both strings were cut at the same time, the acceleration of the entire rod would be $a = g$. Our result of an initial acceleration of end B of $a = 1.5g$ seems reasonable because the full force of gravity is acting on the rod and end A of the rod remains fixed. Therefore, the acceleration of the moving ends is not just that due to free fall—there is an additional acceleration due to the rotation of the rod.

10.6   Work Done by a Torque

In Chapter 5, we saw that the work W done by a force $\vec{F}$ is given by the integral

$$W = \int_{x_0}^{x} F_x\left(x'\right)dx'.$$

We can now consider the work done by a torque $\vec{\tau}$.

Torque is the angular equivalent of force. The angular equivalent of the linear displacement, $d\vec{r}$, is the angular displacement, $d\vec{\theta}$. Since both the torque and the angular displacement are axial vectors and point in the direction of the axis of rotation, we can write their scalar product as $\vec{\tau} \cdot d\vec{\theta} = \tau d\theta$. Therefore, the work done by a torque is

$$W = \int_{\theta_0}^{\theta} \tau\left(\theta'\right)d\theta'.$$ (10.25)

For the special case where the torque is constant and thus does not depend on θ, the integral of equation 10.25 simply evaluates to

$$W = \tau(\theta - \theta_0).$$ (10.26)

Chapter 5 also presented the first version of the work–kinetic energy theorem: $\Delta K \equiv K - K_0 = W$. The rotational equivalent of this work–kinetic energy relationship can be written with the aid of equation 10.3 as follows:

$$\Delta K \equiv K - K_0 = \tfrac{1}{2}I\omega^2 - \tfrac{1}{2}I\omega_0^2 = W.$$ (10.27)

For the case of a constant torque, we can use equation 10.26 and find the work–kinetic energy theorem for constant torque:

$$\tfrac{1}{2}I\omega^2 - \tfrac{1}{2}I\omega_0^2 = \tau(\theta - \theta_0).$$ (10.28)

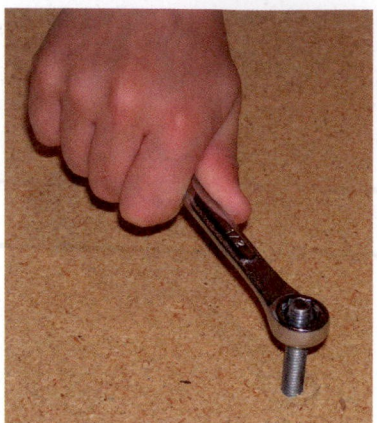

FIGURE 10.24 Tightening a bolt.

EXAMPLE 10.4 | Tightening a Bolt

PROBLEM

What is the total work required to completely tighten the bolt shown in Figure 10.24? The total number of turns is 30.5, the diameter of the bolt is 0.860 cm, and the friction force between the nut and the bolt is a constant 14.5 N.

SOLUTION

Since the friction force is constant and the diameter of the bolt is constant, we can calculate directly the torque needed to turn the nut:

$$\tau = Fr = \tfrac{1}{2}Fd = \tfrac{1}{2}(14.5 \text{ N})(0.860 \text{ cm}) = 0.0623 \text{ N m}.$$

In order to calculate the total work required to completely tighten the bolt, we need to figure out the total angle. Each turn corresponds to an angle of 2π rad, so the total angle in this case is $\Delta\theta = 30.5(2\pi) = 191.6$ rad.

The total work required is then obtained by making use of equation 10.26:

$$W = \tau\Delta\theta = (0.0623 \text{ N m})(191.6) = 11.9 \text{ J}.$$

As you can see, figuring out the work done is not very difficult with a constant torque. However, in many physical situations, the torque cannot be considered to be constant. The next example illustrates such a case.

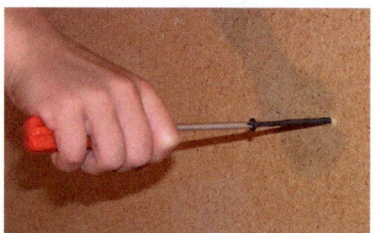

FIGURE 10.25 Driving a drywall screw into a block of wood.

EXAMPLE 10.5 | Driving a Screw

The friction force between a drywall screw and wood is proportional to the contact area between the screw and the wood. Since the drywall screw has a constant diameter, this means that the torque required for turning the screw increases linearly with the depth to which the screw has penetrated into the wood.

PROBLEM

Suppose it takes 27.3 turns to screw a drywall screw completely into a block of wood (Figure 10.25). The torque needed to turn the screw increases linearly from zero at the beginning to a maximum of 12.4 N m at the end. What is the total work required to drive in the screw?

SOLUTION

Clearly, the torque is a function of the angle in this situation and is not constant anymore. Thus, we have to use the integral of equation 10.25 to find our answer. First, let us calculate the total angle, θ_{total}, that the screw turns through: $\theta_{\text{total}} = 27.3(2\pi) = 171.5$ rad. Now we need to find an expression for $\tau(\theta)$. A linear increase with θ from zero to 12.4 N m means

$$\tau(\theta) = \theta\frac{\tau_{\max}}{\theta_{\text{total}}}.$$

Now we can evaluate the integral as follows:

$$W = \int_0^{\theta_{\text{total}}} \tau(\theta')d\theta' = \int_0^{\theta_{\text{total}}} \theta'\frac{\tau_{\max}}{\theta_{\text{total}}}d\theta' = \frac{\tau_{\max}}{\theta_{\text{total}}}\int_0^{\theta_{\text{total}}} \theta'd\theta' = \frac{\tau_{\max}}{\theta_{\text{total}}}\tfrac{1}{2}\theta^2\Big|_0^{\theta_{\text{total}}} = \tfrac{1}{2}\tau_{\max}\theta_{\text{total}}.$$

Inserting the numbers, we obtain

$$W = \tfrac{1}{2}\tau_{\max}\theta_{\text{total}} = \tfrac{1}{2}(12.4 \text{ N m})(171.5 \text{ rad}) = 1.06 \text{ kJ}.$$

Concept Check 10.5

If you want to reduce the torque required to drive in a screw, you can rub soap on the thread beforehand. Suppose the soap reduces the coefficient of friction between the screw and the wood by a factor of 2 and therefore reduces the required torque by a factor of 2. How much does it change the total work required to turn the screw into the wood?

a) It leaves the work the same.

b) It reduces the work by a factor of 2.

c) It reduces the work by a factor of 4.

SOLVED PROBLEM 10.4 | Atwood Machine

PROBLEM

Two weights with masses $m_1 = 3.00$ kg and $m_2 = 1.40$ kg are connected by a very light rope that runs without sliding over a pulley (solid disk) of mass $m_p = 2.30$ kg. The two masses initially hang at the same height and are at rest. Once released, the heavier mass, m_1, descends and lifts up the lighter mass, m_2. What speed will m_2 have at height $h = 0.16$ m?

SOLUTION

THINK We could try to calculate the acceleration of the two masses and then use kinematical equations to relate this acceleration to the vertical displacement. However, we can also use energy considerations, which will lead to a fairly direct solution. Initially, the two hanging masses and the pulley are at rest, so the total kinetic energy is zero. We can choose a coordinate system such that the initial potential energy is zero, and thus the total energy is zero. As one of the masses is lifted, it gains gravitational potential energy, and the other mass loses potential energy. Both masses gain translational kinetic energy, and the pulley gains rotational kinetic energy. Since the kinetic energy is proportional to the square of the speed, we can then use energy conservation to solve for the speed.

SKETCH Figure 10.26a shows the initial state of the Atwood machine with both hanging masses at the same height. We decide to set that height as the origin of the vertical axis, thus ensuring that the initial potential energy, and therefore the total initial energy, is zero. Figure 10.26b shows the Atwood machine with the masses displaced by h.

RESEARCH The gain in gravitational potential energy for m_2 is $U_2 = m_2 gh$. At the same time, m_1 is lowered the same distance, so its potential energy is $U_1 = -m_1 gh$. The kinetic energy of m_1 is $K_1 = \frac{1}{2}m_1 v^2$, and the kinetic energy of m_2 is $K_2 = \frac{1}{2}m_2 v^2$. Note that the same speed, v, is used in the energy expressions for the two masses. This equality is assured, because a rope connects them. (We assume that the rope does not stretch.)

What about the rotational kinetic energy of the pulley? The pulley is a solid disk with a moment of inertia given by $I = \frac{1}{2}m_p R^2$ and a rotational kinetic energy given by $K_r = \frac{1}{2}I\omega^2$. Since the rope runs over the pulley without slipping and also moves with the same speed as the two masses, points on the surface of the pulley also have to move with that same linear speed, v. Just as for a rolling solid disk, the linear speed is related to the angular speed via $\omega R = v$. Thus, the rotational kinetic energy of the pulley is

$$K_r = \tfrac{1}{2}I\omega^2 = \tfrac{1}{2}\left(\tfrac{1}{2}m_p R^2\right)\omega^2 = \tfrac{1}{4}m_p R^2\omega^2 = \tfrac{1}{4}m_p v^2.$$

Now we can express the total energy as the sum of the potential and translational and rotational kinetic energies. This overall sum has to equal zero, because that was the initial value of the total energy and energy conservation applies:

$$0 = U_1 + U_2 + K_1 + K_2 + K_r$$

$$= -m_1 gh + m_2 gh + \tfrac{1}{2}m_1 v^2 + \tfrac{1}{2}m_2 v^2 + \tfrac{1}{4}m_p v^2.$$

SIMPLIFY We can rearrange the preceding equation to isolate the speed, v:

$$(m_1 - m_2)gh = (\tfrac{1}{2}m_1 + \tfrac{1}{2}m_2 + \tfrac{1}{4}m_p)v^2 \Rightarrow$$

$$v = \sqrt{\frac{2(m_1 - m_2)gh}{m_1 + m_2 + \tfrac{1}{2}m_p}}. \tag{i}$$

CALCULATE Now we insert the numbers:

$$v = \sqrt{\frac{2(3.00 \text{ kg} - 1.40 \text{ kg})(9.81 \text{ m/s}^2)(0.16 \text{ m})}{3.00 \text{ kg} + 1.40 \text{ kg} + \tfrac{1}{2}(2.30 \text{ kg})}} = 0.951312 \text{ m/s}.$$

ROUND The displacement h was given with the least precision, to two digits, so we round our result to

$$v = 0.95 \text{ m/s}.$$

– Continued

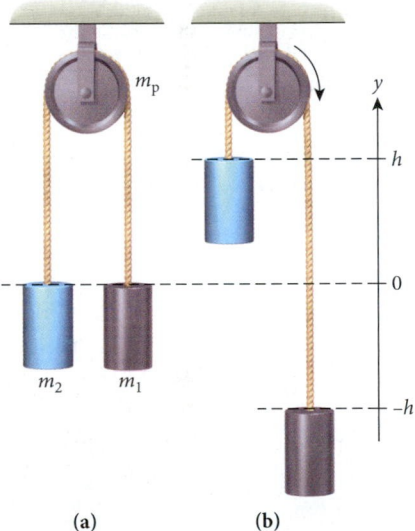

(a)   **(b)**

FIGURE 10.26 One more Atwood machine: (a) initial positions; (b) positions after the weights move a distance h.

Concept Check 10.7

We used the law of energy conservation in Solved Problem 10.4. But the friction force, a nonconservative force, prevents the rope from slipping over the pulley and causes the pulley to rotate. Which of the following statements justifies our use of energy conservation in this situation?

a) Because the friction force acts on both sides of the pulley, it cancels out.

b) Because the friction force points in the same direction as the tension in the rope, it does not contribute to the work done.

c) Because the friction force does not cause a displacement of the rope relative to the pulley, it does not do any work and therefore does not change the total mechanical energy.

d) Since the pulley rotates freely, the friction force does not act on the pulley.

DOUBLE-CHECK The acceleration of the masses is given by equation 10.24, which we developed in Section 10.5 in discussing the Atwood machine:

$$a = \frac{m_1 - m_2}{m_1 + m_2 + \frac{1}{2}m_p} g. \tag{ii}$$

Chapter 2 presented kinematical equations for one-dimensional linear motion. One of these, which relates the initial and final speeds, the displacement, and the acceleration, now comes in handy:

$$v^2 = v_0^2 + 2a(y - y_0).$$

Here $v_0 = 0$ and $y - y_0 = h$. Then, inserting the expression for a from equation (ii) leads to our result

$$v^2 = 2ah = 2\frac{m_1 - m_2}{m_1 + m_2 + \frac{1}{2}m_p} gh \Rightarrow v = \sqrt{\frac{2(m_1 - m_2)gh}{m_1 + m_2 + \frac{1}{2}m_p}}.$$

This equation for the speed is identical to equation (i), which we obtained using energy considerations. The effort it took to develop the equation for the acceleration makes it clear that the energy method is the quicker way to proceed.

10.7   Angular Momentum

Although we have discussed the rotational equivalents of mass (moment of inertia), velocity (angular velocity), acceleration (angular acceleration), and force (torque), we have not yet encountered the rotational analogue to linear momentum. Since linear momentum is the product of an object's velocity and its mass, by analogy, the angular momentum should be the product of angular velocity and moment of inertia. In this section, we will find that this relationship is indeed true for an extended object with a fixed moment of inertia. However, to reach that conclusion, we need to start from a definition of angular momentum for a point particle and proceed from there.

Point Particle

The **angular momentum**, $\vec{L}$, of a point particle is the vector product of its position and momentum vectors:

$$\vec{L} = \vec{r} \times \vec{p}. \tag{10.29}$$

Because the angular momentum is defined as $\vec{L} = \vec{r} \times \vec{p}$ and the torque is defined as $\vec{\tau} = \vec{r} \times \vec{F}$, statements can be made about angular momentum that are similar to those made about torque in Section 10.4. For example, the magnitude of the angular momentum is given by

$$L = rp \sin\theta, \tag{10.30}$$

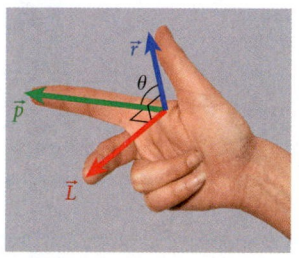

FIGURE 10.27 Right-hand rule for the direction of the angular momentum vector: The thumb is aligned with the position vector and the index finger with the momentum vector; then the angular momentum vector points along the middle finger.

where θ is the angle between the position and momentum vectors. Also, just like the direction of the torque vector, the direction of the angular momentum vector is given by a right-hand rule. Let the thumb of the right hand point along the position vector, $\vec{r}$, of a point particle and the index finger point along the momentum vector, $\vec{p}$, then the middle finger will indicate the direction of the angular momentum vector $\vec{L}$ (Figure 10.27). As an example, the angular momentum vector of a point particle located in the xy-plane is illustrated in Figure 10.28.

Given the definition of the angular momentum in equation 10.29, we can take the time derivative:

$$\frac{d}{dt}\vec{L} = \frac{d}{dt}(\vec{r} \times \vec{p}) = \left[\left(\frac{d}{dt}\vec{r}\right) \times \vec{p}\right] + \left(\vec{r} \times \frac{d}{dt}\vec{p}\right) = (\vec{v} \times \vec{p}) + (\vec{r} \times \vec{F}).$$

To take the derivative of the vector product, we apply the product rule of calculus. The term $\vec{v} \times \vec{p}$ is always zero, because $\vec{v} \parallel \vec{p}$. Also, from equation 10.16, we know that $\vec{r} \times \vec{F} = \vec{\tau}$. Thus, we obtain for the time derivative of the angular momentum vector:

$$\frac{d}{dt}\vec{L} = \vec{\tau}. \tag{10.31}$$

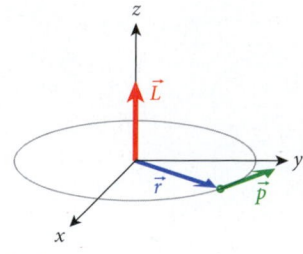

FIGURE 10.28 The angular momentum of a point particle.

The time derivative of the angular momentum vector for a point particle is the torque vector acting on that point particle. This result is again analogous to the linear motion case, where the time derivative of the linear momentum vector is equal to the force vector.

The vector product allows us to revisit the relationship between the linear velocity vector, the coordinate vector, and the angular velocity vector, which was introduced in Chapter 9. (Again, we consider the special case illustrated in Figure 10.28, where the position and momentum vectors are located in a two-dimensional plane.) For circular motion, the magnitudes of those vectors are related via $\omega = v/r$, and the direction of $\vec{\omega}$ is given by a right-hand rule. Using the definition of the vector product, we can write $\vec{\omega}$ as

$$\vec{\omega} = \frac{\vec{r} \times \vec{v}}{r^2}. \tag{10.32}$$

Comparing equations 10.29 and 10.32 reveals that the angular momentum and angular velocity vectors for the point particle are parallel, with

$$\vec{L} = \vec{\omega} \cdot (mr^2). \tag{10.33}$$

The quantity mr^2 is the moment of inertia of a point particle orbiting the axis of rotation at a distance r.

System of Particles

It is straightforward to generalize the concept of angular momentum to a system of n point particles. The total angular momentum of the system of particles is simply the sum of the angular momenta of the individual particles:

$$\vec{L} = \sum_{i=1}^{n} \vec{L}_i = \sum_{i=1}^{n} \vec{r}_i \times \vec{p}_i = \sum_{i=1}^{n} m_i \vec{r}_i \times \vec{v}_i. \tag{10.34}$$

Again, we take the time derivative of this sum of angular momenta in order to obtain the relationship between the total angular momentum of this system and the torque:

$$\frac{d}{dt} \vec{L} = \frac{d}{dt}\left(\sum_{i=1}^{n} \vec{L}_i\right) = \frac{d}{dt}\left(\sum_{i=1}^{n} \vec{r}_i \times \vec{p}_i\right) = \sum_{i=1}^{n} \frac{d}{dt}(\vec{r}_i \times \vec{p}_i)$$

$$= \sum_{i=1}^{n} \left[\underbrace{\left(\frac{d}{dt}\vec{r}_i\right)}_{\text{Equals } \vec{v}_i} \times \vec{p}_i + \vec{r}_i \times \underbrace{\left(\frac{d}{dt}\vec{p}_i\right)}_{\vec{F}_i} \right] = \sum_{i=1}^{n} \vec{r}_i \times \vec{F}_i = \sum_{i=1}^{n} \vec{\tau}_i = \vec{\tau}_{\text{net}}.$$

Equals zero, because $\vec{v}_i \| \vec{p}_i$

As expected, we find that the time derivative of the total angular momentum for a system of particles is given by the total net external torque acting on the system. It is important to keep in mind that this is the net *external* torque due to the *external* forces, $\vec{F}_i$.

Rigid Objects

A rigid object will rotate about a fixed symmetry axis with an angular velocity $\vec{\omega}$ ωthat is the same for every part of the object. In this case, the angular momentum is proportional to the angular velocity, and the proportionality constant is the moment of inertia:

$$\vec{L} = I\vec{\omega}. \tag{10.35}$$

Self-Test Opportunity 10.3

Can you show that internal torques (those due to internal forces between particles in a system) do not contribute to the total net torque? (*Hint:* Use Newton's Third Law, $\vec{F}_{i \to j} = -\vec{F}_{j \to i}$.)

DERIVATION 10.4 / **Angular Momentum of a Rigid Object**

Representing the rigid object by a collection of point particles allows us to use the results of the preceding subsection as a starting point. In order for point particles to represent the rigid object, their relative distances from one another must remain constant (rigid). Then all of these point particles rotate with a constant angular velocity, $\vec{\omega}$, about the common rotation axis.

From equation 10.34, we obtain

$$\vec{L} = \sum_{i=1}^{n} \vec{L}_i = \sum_{i=1}^{n} m_i \vec{r}_i \times \vec{v}_i = \sum_{i=1}^{n} m_i r_{i\perp}^2 \vec{\omega}.$$

– Continued

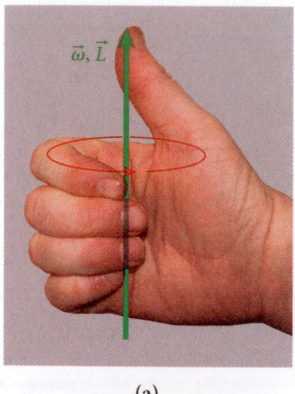

(a)

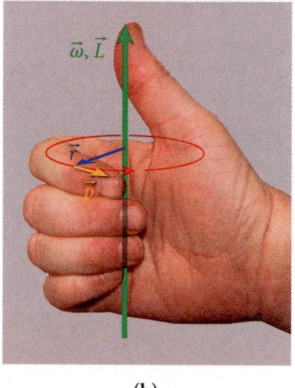

(b)

FIGURE 10.29 (a) Right-hand rule for the direction of the angular momentum (along the thumb) as a function of the direction of rotation (along the fingers). (b) The momentum and position vectors of a point particle in circular motion.

In the last step, we have used the relationship between the angular velocity and the vector product of the position and linear velocity vectors for point particles, equation 10.32, where $r_{i\perp}$ is the orbital radius of the point particle i. Note that the angular velocity vector is the same for all point particles in this rigid object. Therefore, we can move it out of the sum as a common factor:

$$\vec{L} = \vec{\omega} \sum_{i=1}^{n} m_i r_{i\perp}^2 .$$

We can identify this sum as the moment of inertia of a collection of point particles; see equation 10.4. Thus, we finally have our result:

$$\vec{L} = I\vec{\omega}.$$

For rigid objects rotating about a symmetry axis, just as for point particles, the direction of the angular momentum vector is the same as the direction of the angular velocity vector. Figure 10.29 shows the right-hand rule used to determine the direction of the angular momentum vector (arrow along the direction of the thumb) as a function of the sense of rotation (direction of the fingers). What happens if the axis of rotation is not a symmetry axis of the rigid object? In this case, the angular momentum vector does not necessarily point in the direction of the angular velocity vector, and the math becomes quite a bit more complicated, beyond the scope of this book. A common example of this situation occurs when a car's wheel goes out of alignment (for example, after hitting a curb with the rim). The wheel will start to wobble, which can be quite dangerous, because it puts a lot of mechanical stress on both wheel and axle.

EXAMPLE 10.6 | Golf Ball

PROBLEM
What is the magnitude of the angular momentum of a golf ball ($m = 4.59 \cdot 10^{-2}$ kg, $R = 2.13 \cdot 10^{-2}$ m) spinning at 4250 rpm (revolutions per minute) after a good hit with a driver?

SOLUTION
First, we need to find the angular velocity of the golf ball, which involves use of the concepts introduced in Chapter 9:

$$\omega = 2\pi f = 2\pi(4250 \text{ min}^{-1}) = 2\pi(4250/60 \text{ s}^{-1}) = 445.1 \text{ rad/s}.$$

The moment of inertia of the golf ball is

$$I = \tfrac{2}{5} mR^2 = 0.4(4.59 \cdot 10^{-2} \text{ kg})(2.13 \cdot 10^{-2} \text{ m})^2 = 8.33 \cdot 10^{-6} \text{ kg m}^2.$$

The magnitude of the angular momentum of the golf ball is then simply the product of these two numbers:

$$L = (8.33 \cdot 10^{-6} \text{ kg m}^2)(445.1 \text{ s}^{-1}) = 3.71 \cdot 10^{-3} \text{ kg m}^2 \text{ s}^{-1}.$$

Using equation 10.35 for the angular momentum of a rigid object, we can show that the relationship between the rate of change of the angular momentum and the torque is still valid. Taking the time derivative of equation 10.35, and assuming a rigid body whose moment of inertia is constant in time, we obtain

$$\frac{d}{dt}\vec{L} = \frac{d}{dt}(I\vec{\omega}) = I\frac{d}{dt}\vec{\omega} = I\vec{\alpha} = \vec{\tau}_{\text{net}}. \tag{10.36}$$

Note the addition of the index "net" to the symbol for torque, indicating that this equation also holds if different torques are present. Earlier, equation 10.19 was said to be true only for a point particle. However, equation 10.36 clearly shows that equation 10.19 holds for any object with a fixed (constant in time) moment of inertia.

The time derivative of the angular momentum is equal to the torque, just as the time derivative of the linear momentum is equal to the force. Equation 10.31 is another formulation of Newton's Second Law for rotation and is more general than equation 10.19, because it also encompasses the case of a moment of inertia that is not constant in time.

Conservation of Angular Momentum

If the net external torque on a system is zero, then according to equation 10.36, the time derivative of the angular momentum is also zero. However, if the time derivative of a quantity is zero, then the quantity is constant in time. Therefore, we can write the law of **conservation of angular momentum:**

$$\text{If } \vec{\tau}_{net} = 0 \Rightarrow \vec{L} = \text{constant} \Rightarrow \vec{L}(t) = \vec{L}(t_0) \equiv \vec{L}_0. \tag{10.37}$$

This is the third major conservation law that we have encountered—the first two applied to mechanical energy (Chapter 6) and linear momentum (Chapter 7). Like the other conservation laws, this one can be used to solve problems that would otherwise be very hard to attack.

If there are several objects present in a system with zero net external torque, the equation for conservation of angular momentum becomes

$$\sum_i \vec{L}_{\text{initial}} = \sum_i \vec{L}_{\text{final}}. \tag{10.38}$$

For the special case of a rigid body rotating around a fixed axis of rotation, we find (since, in this case, $\vec{L} = I\vec{\omega}$):

$$I\vec{\omega} = I_0\vec{\omega}_0 \quad (\text{for } \vec{\tau}_{net} = 0), \tag{10.39}$$

or, equivalently,

$$\frac{\omega}{\omega_0} = \frac{I_0}{I} \quad (\text{for } \vec{\tau}_{net} = 0).$$

This conservation law is the basis for the functioning of gyroscopes (Figure 10.30). Gyroscopes are objects (usually disks) that spin around a symmetry axis at high angular velocities. The axis of rotation is able to turn on ball bearings, almost without friction, and the suspension system is able to rotate freely in all directions. This freedom of motion ensures that no net external torque can act on the gyroscope. Without torque, the angular momentum of the gyroscope remains constant and thus points in the same direction, no matter what the object carrying the gyroscope does. Both airplanes and satellites rely on gyroscopes for navigation. The Hubble Space Telescope, for example, is equipped with six gyroscopes, at least three of which must work to allow the telescope to orient itself in space.

Equation 10.39 is also of importance in many sports, most notably gymnastics, platform diving (Figure 10.31), and figure skating. In all three sports, the athletes rearrange their bodies and thus adjust their moments of inertia to manipulate their rotation frequencies. The changing moment of inertia of a platform diver is illustrated in Figure 10.31. The platform diver starts her dive stretched out, as shown in Figure 10.31a. She then pulls her legs and arms into a tucked position, decreasing her moment of inertia, as shown in Figure 10.31b. She then completes several rotations as she falls. Before entering the water, she stretches out her arms and legs, increasing her moment of inertia and slowing her rotation, as shown in Figure 10.31c. Pulling her arms and legs into a tucked position reduces the moment of

FIGURE 10.30 Toy gyroscopes.

(a)   (b)   (c)

FIGURE 10.31 Laura Wilkinson at the 2000 Olympic Games in Sydney, Australia. (a) She leaves the diving platform. (b) She holds the tucked position. (c) She stretches out before entering the water.

inertia of her body by some factor $I' = I/k$, where $k > 1$. Conservation of angular momentum then increases the angular velocity by the same factor k: $\omega' = k\omega$. Thus, the diver can control the rate of rotation. Tucking in the arms and legs can increase the rate of rotation relative to that in the stretched-out position by more than a factor of 2.

EXAMPLE 10.7 / Death of a Star

At the end of the life of a massive star more than five times as massive as the Sun, the core of the star consists almost entirely of the metal iron. Once this stage is reached, the core becomes unstable and collapses (as illustrated in Figure 10.32) in a process that lasts only about a second and is the initial phase of a supernova explosion. Among the most powerful energy-releasing events in the universe, supernova explosions are thought to be the source of most of the elements heavier than iron. The explosion throws off debris, including the heavier elements, into outer space, and may leave behind a neutron star, which consists of stellar material that is compressed to a density millions of times higher than the highest densities found on Earth.

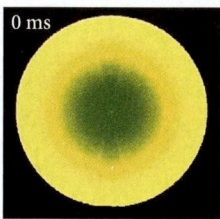 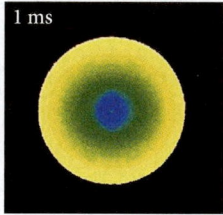 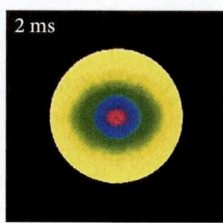 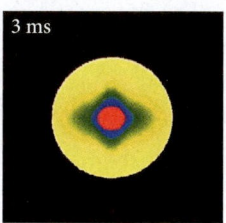

FIGURE 10.32 Computer simulation of the initial stages of the collapse of the core of a massive star. The different colors represent the varying density of the star's core, increasing from yellow through green and blue to red.

PROBLEM

If the iron core initially spins 9.00 revolutions per day and if its radius decreases during the collapse by a factor of 700, what is the angular velocity of the core at the end of the collapse? (The assumption that the iron core has a constant density is not really justified. Computer simulations show that it falls off exponentially in the radial direction. However, the same simulations show that the moment of inertia of the iron core is still approximately proportional to the square of its radius during the collapse process.)

SOLUTION

Because the collapse of the iron core occurs under the influence of its own gravitational pull, no net external torque acts on the core. Thus, according to equation 10.31, angular momentum is conserved. From equation 10.39 we then obtain

$$\frac{\omega}{\omega_0} = \frac{I_0}{I} = \frac{R_0^2}{R^2} = 700^2 = 4.90 \cdot 10^5.$$

With $\omega_0 = 2\pi f_0 = 2\pi[(9 \text{ rev})/(24 \cdot 3600 \text{ s})] = 6.55 \cdot 10^{-4}$ rad/s, we obtain the magnitude of the final angular velocity:

$$\omega = 4.90 \cdot 10^5 \, \omega_0 = 4.90 \cdot 10^5 (6.55 \cdot 10^{-4} \text{ rad/s}) = 321 \text{ rad/s}.$$

Thus, the neutron star that results from this collapse rotates with a rotational frequency of 51.0 rev/s.

DISCUSSION

Astronomers can observe the rotation of neutron stars, which are called *pulsars*. It is estimated that the fastest a pulsar can rotate when formed from a single-star supernova explosion is about 60 rev/s. One of the fastest known rotational frequencies for a pulsar is

$$f = 716 \text{ rev/s},$$

which corresponds to an angular speed of

$$\omega = 2\pi f = 2\pi \left(716 \text{ s}^{-1}\right) = 4500 \text{ rad/s}.$$

After its formation from a stellar collapse, such a fast-rotating pulsar increases its rotational frequency by taking matter from a companion star orbiting nearby.

The next example ends the section with a current cutting-edge engineering application, which ties together the concepts of moment of inertia, rotational kinetic energy, torque, and angular momentum.

EXAMPLE 10.8 Flybrid

The process of braking to slow a car down decreases the car's kinetic energy and dissipates it through the action of the friction force between the brake pads and the drums. Gas-electric hybrid vehicles convert some or most of this kinetic energy into reusable electric energy stored in a large battery. However, there is a way to accomplish this energy storage without the use of a large battery by storing the energy temporarily in a flywheel (Figure 10.33). Flywheel kinetic energy recovery systems were pioneered by Flybrid Systems and are being used in Formula 1 races and endurance races such as Le Mans.

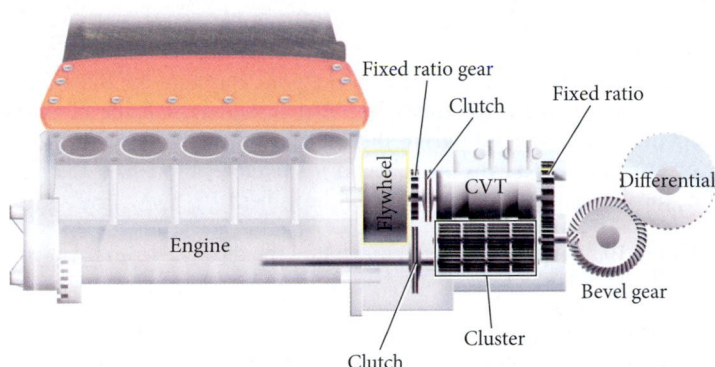

FIGURE 10.33 Diagram of the integration of a flywheel into the drive train of a car. The continuously variable transmission (CVT) is used to store energy in the flywheel and to extract energy from the flywheel.

PROBLEM

A flywheel made of carbon steel has a mass of 5.00 kg, an inner radius of 8.00 cm, and an outer radius of 14.2 cm. If it is supposed to store 400.0 kJ of rotational energy, how fast (in rpm) does it have to rotate? If the rotational energy can be stored or withdrawn in 6.67 s, how much average power and torque can this flywheel deliver during that time?

SOLUTION

The moment of inertia of the flywheel is given by equation 10.9: $I = \frac{1}{2}M(R_1^2 + R_2^2)$. The rotational kinetic energy (equation 10.3) is $K = \frac{1}{2}I\omega^2$. We solve this for the angular speed:

$$\omega = \sqrt{\frac{2K}{I}} = \sqrt{\frac{4K}{M(R_1^2 + R_2^2)}}.$$

So, for the rotational frequency, we find

$$f = \frac{\omega}{2\pi} = \sqrt{\frac{K}{\pi^2 M(R_1^2 + R_2^2)}}$$

$$= \sqrt{\frac{400.0 \text{ kJ}}{\pi^2 (5.00 \text{ kg})\left[(0.0800 \text{ m})^2 + (0.142 \text{ m})^2\right]}}$$

$$= 552 \text{ s}^{-1} = 33{,}100 \text{ rpm}.$$

Since the average power is given by the change in kinetic energy divided by the time (see Chapter 5), we have

$$P = \frac{\Delta K}{\Delta t} = \frac{400.0 \text{ kJ}}{6.67 \text{ s}} = 60.0 \text{ kW}.$$

We find the average torque from equation 10.36 and the fact that the average angular acceleration is the change in angular speed $\Delta\omega$, divided by the time interval, Δt:

$$\tau = I\alpha = I\frac{\Delta\omega}{\Delta t} = \frac{1}{2}M(R_1^2 + R_2^2)\frac{1}{\Delta t}\sqrt{\frac{4K}{M(R_1^2 + R_2^2)}} = \frac{1}{\Delta t}\sqrt{M(R_1^2 + R_2^2)K}$$

$$= \frac{1}{6.67 \text{ s}}\sqrt{(400.0 \text{ kJ})(5.00 \text{ kg})\left[(0.0800 \text{ m})^2 + (0.142 \text{ m})^2\right]}$$

$$= 34.6 \text{ N m}.$$

Concept Check 10.8

The flybrid rotates fastest when the Formula 1 car is moving slowest, in the process of making a tight turn. Knowing that it takes torque to change an angular momentum vector, how would you orient the axis of rotation of the flywheel in order to have the least impact on steering the car through the curve?

a) The flywheel should be aligned with the main axis of the race car.

b) The flywheel should be vertical.

c) The flywheel should be aligned with the wheel axles.

d) It makes no difference; all three orientations are equally problematic.

e) Orientations (a) and (c) are both equally good and better than (b).

SOLVED PROBLEM 10.5 | Bullet Hitting a Pole

PROBLEM

A .22-caliber long rifle bullet of mass m = 2.59 g is shot from a gun during a Boy Scout merit badge exercise and is moving with a speed of 374.5 m/s when it collides with a pole of mass M = 3.00 kg and length ℓ = 2.00 m. The pole is initially at rest, in a vertical position, and pivots about an axis going through its center of mass. The bullet embeds itself in the pole at a point that is $\frac{1}{3}$ of the length of the pole above the pivot point. As a result, the bullet-pole system starts rotating.

a) Find the angular momentum of the bullet-pole system after the collision.

b) What is the rotational kinetic energy of the pole and bullet after the collision?

SOLUTION

THINK   First, a seemingly trivial observation: If the bullet hit the pole right in the center, then there would be no rotation at all, because in that case the bullet has no angular momentum with respect to the pivot point at the center of the pole. But here the bullet hit the pole off-center, and the distance to the center times the bullet's momentum is the angular momentum. Angular momentum is conserved during the collision of the bullet with the pole (because there are no external torques during the collision!). Therefore, all we have to do to answer part (a) is calculate the initial angular momentum of the bullet.

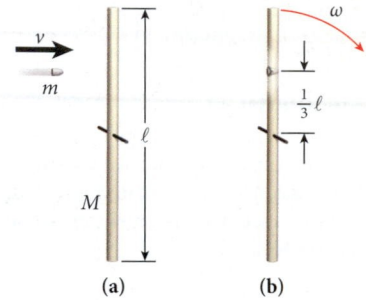

(a)   (b)

FIGURE 10.34   A bullet hits a pole and sticks.

SKETCH   Figure 10.34 includes a sketch indicating the distance from the pivot point to the point where the bullet hits the pole.

RESEARCH   The momentum of the bullet is given by $p = mv$. When it hits the pole at the distance $\ell/3$ from the pole's center of mass, its angular momentum is $p\ell/3$. The moment of inertia of the pole is $M\ell^2/12$ and that of the bullet in the pole is $m(\ell/3)^2$. The kinetic energy of the bullet and pole together is all rotational, $K_r = I\omega^2/2$, and the relationship among angular momentum, moment of inertia, and angular velocity is given by $L = I\omega$.

SIMPLIFY

a) The initial angular momentum is $L_i = mv\ell/3$. The final angular momentum is the same, $L_f = L_i = mv\ell/3$.

b) The total moment of inertia is $I = (M\ell^2/12) + (m\ell^2/9)$. The rotational kinetic energy is

$$K_r = \tfrac{1}{2}I\omega^2 = \tfrac{1}{2}I^2\omega^2/I = \tfrac{1}{2}L^2/I = \frac{L^2}{2((M/12)+(m/9))\ell^2}.$$

CALCULATE

a) $L_f = (2.59\cdot10^{-3}\text{ kg})(374.5\text{ m/s})(2.00\text{ m})/3 = 0.6466367\text{ kg m}^2/\text{s}.$

b) $K_r = \dfrac{\left(0.6466367\text{ kg m}^2/\text{s}\right)^2}{2\left((3.00\text{ kg})/12+\left(2.59\cdot10^{-3}\text{ kg}\right)/9\right)(2.00\text{ m})^2} = 0.2088291\text{ J}.$

ROUND   The least precise input number was specified to three significant figures, to which we also round our final results:

a) $L_f = 0.647\text{ kg m}^2/\text{s}$

b) $K_r = 0.209\text{ J}$

DOUBLE-CHECK   Our results pass the minimum test in that the units work out properly. If you compare the initial translational kinetic energy of the bullet to the rotational kinetic energy just determined, you will find that $K_i = \tfrac{1}{2}mv^2 = \tfrac{1}{2}(2.59\cdot10^{-3}\text{ kg})(374.5\text{ m/s})^2 = 182$ J, a value that is larger than our answer for part (b) by a factor of ~280. This indicates that almost all of the kinetic energy is lost in this inelastic collision. A final issue to ponder: We have stated that the relationship $L = I\omega$ is only valid for an object rotating about an axis through its center of mass, a symmetry axis. Strictly speaking, the relationship is not exactly valid here, because the addition of the bullet to the upper part of the pole shifts the combined center of mass of the bullet-pole system slightly upward. However, since the mass of the bullet is only ~0.1% of that of the pole, this effect is negligible.

10.8 Precession

Spinning tops were popular toys when your parents or grandparents were kids. When put in rapid rotational motion, they stand upright without falling down. What's more, if tilted at an angle relative to the vertical, they still do not fall down. Instead, the rotation axis moves on the surface of a cone as a function of time (Figure 10.35). This motion is called **precession**. What causes it?

First, we note that a spinning top has an angular momentum vector, $\vec{L}$, which is aligned with its symmetry axis, pointing either up or down, depending on whether it is spinning clockwise or counterclockwise (Figure 10.36). Because the top is tilted, its center of mass (marked with a black dot in Figure 10.36) is not located above the contact point with the support surface. The gravitational force acting on the center of mass then results in a torque, $\vec{\tau}$, about the contact point, as indicated in the figure; in this case, the torque vector points straight out of the page. The position vector, $\vec{r}$, of the center of mass, which helps determine the torque, is exactly aligned with the angular momentum vector. The angle of the symmetry axis of the top with respect to the vertical is labeled ϕ in the figure. The angle between the gravitational force vector and the position vector is then $\pi - \phi$ (see Figure 10.36). Because $\sin(\pi - \phi) = \sin\phi$, we can write the magnitude of the torque as a function of the angle ϕ:

$$\tau = rF\sin\phi = rmg\sin\phi.$$

Since $d\vec{L}/dt = \vec{\tau}$, the change in the angular momentum vector, $d\vec{L}$, points in the same direction as the torque and is thus perpendicular to the angular momentum vector. This effect forces the angular momentum vector to sweep along the surface of a cone of angle ϕ as a function of time, with the tip of the angular momentum vector following a circle in the horizontal plane, shown in gray in Figure 10.36.

We can even calculate the magnitude of the angular velocity, ω_p, for this precessional motion. Figure 10.36 indicates that the radius of the circle that the tip of the angular momentum vector sweeps out as a function of time is given by $L\sin\phi$. The magnitude of the differential change in angular momentum, dL, is the arc length of this circle, and it can be calculated as the product of the circle's radius and the differential angle swept out by the radius, $d\theta$:

$$dL = (L\sin\phi)d\theta.$$

Consequently, for the time derivative of the angular momentum, dL/dt, we find

$$\frac{dL}{dt} = (L\sin\phi)\frac{d\theta}{dt}.$$

The time derivative of the deflection angle, θ, is the angular velocity of precession, ω_p. Since $dL/dt = \tau$, we use the preceding equation and the expression for the torque, $\tau = rmg\sin\phi$, to obtain

$$rmg\sin\phi = \tau = \frac{dL}{dt} = (L\sin\phi)\frac{d\theta}{dt} = (L\sin\phi)\omega_p \Rightarrow$$

$$\omega_p = \frac{rmg\sin\phi}{L\sin\phi}.$$

We see that the term $\sin\phi$ cancels out of the last expression, yielding $\omega_p = rmg/L$. The angular frequency of precession is the same for all values of ϕ, the tilt angle of the rotation axis! This result may seem a bit surprising, but experiments verify that it is indeed the case. For the final step, we use the fact that the angular momentum for a rigid object, L, is the product of the moment of inertia, I, and the angular velocity, ω. Thus, substituting $I\omega$ for L in the expression for the precessional angular speed, ω_p, gives us our final result:

$$\omega_p = \frac{rmg}{I\omega}. \tag{10.40}$$

This formula reflects the interesting property that the precessional angular speed is inversely proportional to the angular speed of the spinning top. As the top slows down because of friction, its angular speed is gradually reduced, and therefore the precessional angular speed increases gradually. The faster and faster precession eventually causes the top to wobble and fall down.

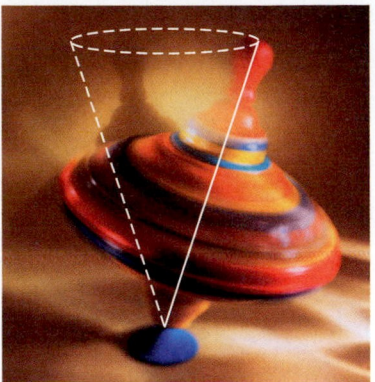

FIGURE 10.35 A spinning top may tilt from the vertical but does not fall down.

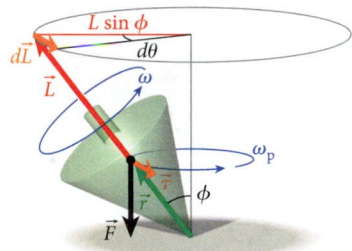

FIGURE 10.36 Precession of a spinning top.

Concept Check 10.10

Estimate the precessional angular speed of the wheel in Figure 10.37.

a) 0.01 rad/s   c) 5 rad/s

b) 0.6 rad/s   d) 10 rad/s

Self-Test Opportunity 10.4

The wheel shown in Figure 10.37 has a mass of 2.5 kg, almost all of which is concentrated on the rim. It has a radius of 22 cm, and the distance between the suspension point and the center of mass is 5.0 cm. Estimate the angular speed with which it is spinning.

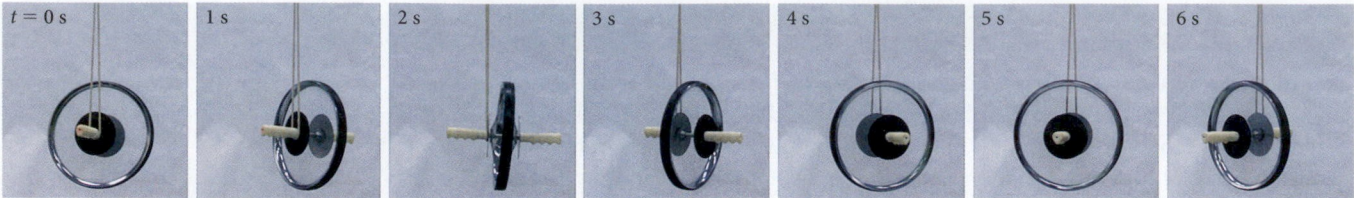

FIGURE 10.37 Precession of a rapidly spinning wheel suspended from a rope.

Precession is impressively demonstrated in the photo sequence of Figure 10.37. Here a rapidly spinning wheel is suspended off-center from a string attached to the ceiling. As you can see, the wheel does not fall down, as one would expect a nonspinning wheel to do in the same situation, but slowly precesses around the suspension point.

10.9 Quantized Angular Momentum

To finish our discussion of angular momentum and rotation, let's consider the smallest quantity of angular momentum that an object can have. From the definition of the angular momentum of a point particle (equation 10.29), $\vec{L} = \vec{r} \times \vec{p}$ or $L = rp \sin \theta$, it would appear that there is no smallest amount of angular momentum, because either the distance to the rotation axis, r, or the momentum, p, can be reduced by a factor between 0 and 1, and the corresponding angular momentum will be reduced by the same factor.

However, for atoms and subatomic particles, the notion of a continuously variable angular momentum doesn't apply. Instead, a quantum of angular momentum is observed. This quantum of angular momentum is called **Planck's constant**, $h = 6.626 \cdot 10^{-34}$ J s. Very often Planck's constant appears in equations divided by the factor 2π, and physicists have given this ratio the symbol $\hbar$: $\hbar \equiv h/2\pi = 1.055 \cdot 10^{-34}$ J s. Chapter 36 will provide a full discussion of the experimental observations that led to the introduction of this fundamental constant. Here we simply note an amazing fact: All elementary particles have an intrinsic angular momentum, often called *spin*, which is either an integer multiple $(0, 1\hbar, 2\hbar, \ldots)$ or a half-integer multiple $(\frac{1}{2}\hbar, \frac{3}{2}\hbar, \ldots)$ of Planck's quantum of angular momentum. Astonishingly, the integer or half-integer spin values of particles make all the difference in the ways they interact with one another. Particles with integer-valued spins include photons, which are the elementary particles of light. Particles with half-integer values of spin include electrons, protons, and neutrons, the particles that constitute the building blocks of matter. We will return to the fundamental importance of angular momentum in atoms and subatomic particles in Chapters 37 through 40.

WHAT WE HAVE LEARNED | EXAM STUDY GUIDE

- An object's kinetic energy of rotation is given by $K = \frac{1}{2}I\omega^2$. This relationship holds for point particles as well as for solid objects.

- The moment of inertia for rotation of an object about an axis through the center of mass is defined as $I = \int_V r_\perp^2 \rho(\vec{r})dV$, where $r_\perp$ is the perpendicular distance of the volume element dV to the axis of rotation and $\rho(\vec{r})$ is the mass density.

- If the mass density is constant, the moment of inertia is $I = \frac{M}{V}\int_V r_\perp^2 dV$, where M is the total mass of the rotating object and V is its volume.

- The moment of inertia of all round objects is $I = cMR^2$, with $c \in [0,1]$.

- The parallel-axis theorem states that the moment of inertia, $I_\parallel$, for rotation about an axis parallel to one through the center of mass is given by $I_\parallel = I_{cm} + Md^2$, where d is the distance between the two axes and I_{cm} is the moment of inertia for rotation about the axis through the center of mass.

- For an object that is rolling without slipping, the center of mass coordinate, r, and the rotation angle, θ, are related by $r = R\theta$, where R is the radius of the object.

- The kinetic energy of a rolling object is the sum of its translational and rotational kinetic energies: $K = K_{trans} + K_{rot} = \frac{1}{2}mv_{cm}^2 + \frac{1}{2}I_{cm}\omega^2 = \frac{1}{2}(1 + c)mv_{cm}^2$, with $c \in [0,1]$ and with c depending on the shape of the object.

- Torque is defined as the vector product of the position vector and the force vector: $\vec{\tau} = \vec{r} \times \vec{F}$.

- The angular momentum of a point particle is defined as $\vec{L} = \vec{r} \times \vec{p}$.

- The rate of change of the angular momentum is equal to the torque: $\dfrac{d}{dt}\vec{L} = \vec{\tau}$. This is the rotational equivalent of Newton's Second Law.

- For rigid objects rotating about their symmetry axis, the angular momentum is $\vec{L} = I\vec{\omega}$, and the torque is $\vec{\tau} = I\vec{\alpha}$.

- In the case of vanishing net external torque, angular momentum is conserved: $I\vec{\omega} = I_0\vec{\omega}_0$ (for $\vec{\tau}_{net} = 0$).

- The equivalent quantities for linear and rotational motion are summarized in the table.

Quantity	Linear	Circular	Relationship
Displacement	$\vec{s}$	$\vec{\theta}$	$\vec{s} = r\vec{\theta}$
Velocity	$\vec{v}$	$\vec{\omega}$	$\vec{\omega} = \vec{r} \times \vec{v}/r^2$
Acceleration	$\vec{a}$	$\vec{\alpha}$	$\vec{a} = r\alpha\,\hat{t} - r\omega^2\hat{r}$ $a_t = r\alpha$ $a_c = \omega^2 r$
Momentum	$\vec{p}$	$\vec{L}$	$\vec{L} = \vec{r} \times \vec{p}$
Mass/moment of inertia	m	I	
Kinetic energy	$\frac{1}{2}mv^2$	$\frac{1}{2}I\omega^2$	
Force/torque	$\vec{F}$	$\vec{\tau}$	$\vec{\tau} = \vec{r} \times \vec{F}$

ANSWERS TO SELF-TEST OPPORTUNITIES

10.1 $I_{\parallel} = \dfrac{1}{12}mL^2 + m\left(\dfrac{L}{2}\right)^2 = mL^2\left(\dfrac{1}{12} + \dfrac{1}{4}\right) = \dfrac{1}{3}mL^2.$

10.2 The full can of soda is not a solid object and thus does not rotate as a solid cylinder. Instead, most of the liquid inside the can does not participate in the rotation even when the can reaches the bottom of the incline. The mass of the can itself is negligible compared to the mass of the liquid inside it. Therefore, a can of soda rolling down an incline approximates a mass sliding down an incline without friction. The constant c used in equation 10.15 is then close to zero, and so the can wins the race.

10.3 Newton's Third Law states that internal forces occur in equal and opposite pairs that act along the line of the separation of each pair of particles. Thus, the torque due to each pair of forces is zero. Summing up the torques from all the internal forces leads to zero net internal torque.

10.4 Since the mass is concentrated on the rim of the wheel, the wheel's moment of inertia is $I = mR^2$. With the aid of equation 10.40, we then obtain

$$\omega_p = \frac{rmg}{mR^2\omega} = \frac{rg}{R^2\omega} = \frac{(0.050\ \text{m})(9.81\ \text{m/s}^2)}{(0.22\ \text{m})^2(0.62\ \text{rad/s})} = 16\ \text{rad/s}.$$

PROBLEM-SOLVING GUIDELINES: ROTATIONAL MOTION

1. Newton's Second Law and the work–kinetic energy theorem are powerful and complementary tools for solving a wide variety of problems in rotational mechanics. As a general guideline, you should try an approach based on Newton's Second Law and free-body diagrams when the problem involves calculating an angular acceleration. An approach based on the work–kinetic energy theorem is more useful when you need to calculate an angular speed.

2. Many concepts of translational motion are equally valid for rotational motion. For example, conservation of linear momentum applies when no external forces are present; conservation of angular momentum applies when no external torques are present. Remember the correspondences between translational and rotational quantities.

3. It is crucial to remember that the shape of an object is important in situations involving rotational motion. Be sure to use the correct formula for moment of inertia, which depends on the location of the axis of rotation as well as on the geometry of the object. Torque also depends on the location of the axis of symmetry; be sure to be consistent in calculating clockwise and counterclockwise torques.

4. Many relationships for rotational motion depend on the geometry of the situation—for example, the relationship of the linear velocity of a hanging weight to the angular velocity of the rope moving over a pulley. Sometimes the geometry of a situation changes in a problem, for example, if there is a different rotational inertia between starting and ending points of a rotation. Be sure you understand what quantities change during the course of any rotational motion.

5. Many physical situations involve rotating objects that roll, with or without slipping. If rolling without slipping is occurring, you can relate linear and angular displacements, velocities, and accelerations to each other at points on the perimeter of the rolling object.

6. The law of conservation of angular momentum is just as important for problems involving circular or rotational motion as the law of conservation of linear momentum is for problems involving straight-line motion. Thinking of a problem situation in terms of conserved angular momentum often provides a straightforward path to a solution, which otherwise would be difficult to obtain. But keep in mind that angular momentum is only conserved if the net external torque is zero.

MULTIPLE-CHOICE QUESTIONS

10.1 A circular object begins from rest and rolls without slipping down an incline, through a vertical distance of 4.0 m. When the object reaches the bottom, its translational velocity is 7.0 m/s. What is the constant c relating the moment of inertia to the mass and radius (see equation 10.11) of this object?

a) 0.80   c) 0.40

b) 0.60   d) 0.20

10.2 Two solid steel balls, one small and one large, are on an inclined plane. The large ball has a diameter twice as large as that of the small ball. Starting from rest, the two balls roll without slipping down the incline until their centers of mass are 1 m below their starting positions. What is the speed of the large ball (v_L) relative to that of the small ball (v_S) after rolling 1 m?

a) $v_L = 4v_S$ c) $v_L = v_S$ e) $v_L = 0.25v_S$

b) $v_L = 2v_S$ d) $v_L = 0.5v_S$

10.3 A generator's flywheel, which is a homogeneous cylinder of radius R and mass M, rotates about its longitudinal axis. The linear velocity of a point on the rim (side) of the flywheel is v. What is the kinetic energy of the flywheel?

a) $K = \frac{1}{2}Mv^2$ d) $K = \frac{1}{2}Mv^2R$

b) $K = \frac{1}{4}Mv^2$ e) not given enough information to

c) $K = \frac{1}{2}Mv^2/R$ answer

10.4 Four hollow spheres, each with a mass of 1 kg and a radius $R = 10$ cm, are connected with massless rods to form a square with sides of length $L = 50$ cm. In case 1, the masses rotate about an axis that bisects two sides of the square. In case 2, the masses rotate about an axis that passes through the diagonal of the square, as shown in the figure. Compute the ratio of the moments of inertia, I_1/I_2, for the two cases.

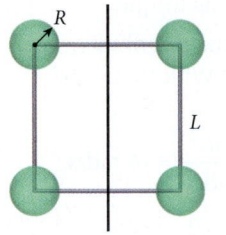

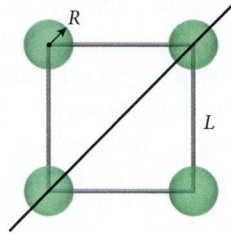

Case 1   Case 2

a) $I_1/I_2 = 8$ c) $I_1/I_2 = 2$ e) $I_1/I_2 = 0.5$

b) $I_1/I_2 = 4$ d) $I_1/I_2 = 1$

10.5 If the hollow spheres of Question 10.4 were replaced by solid spheres of the same mass and radius, the ratio of the moments of inertia for the two cases would

a) increase.   c) stay the same.

b) decrease.   d) be zero.

10.6 An extended object consists of two point masses, m_1 and m_2, connected via a rigid massless rod of length L, as shown in the figure. The object is rotating at a constant angular velocity about an axis perpendicular to the page through the midpoint of the rod. Two time-varying tangential forces, F_1 and F_2, are applied to m_1 and m_2, respectively. After the forces have been applied, what will happen to the angular velocity of the object?

a) It will increase.   c) It will remain unchanged.

b) It will decrease.   d) Not enough information is given to make a determination.

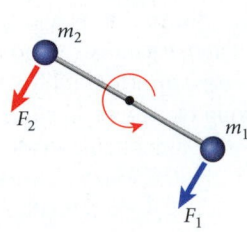

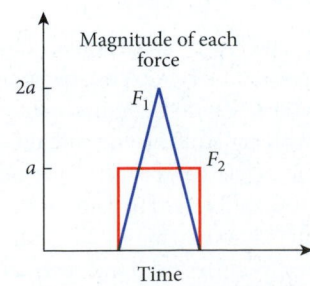

10.7 Consider a cylinder and a hollow cylinder, rotating about an axis going through their centers of mass. If both objects have the same mass and the same radius, which object will have the larger moment of inertia?

a) The moment of inertia will be the same for both objects.

b) The solid cylinder will have the larger moment of inertia because its mass is uniformly distributed.

c) The hollow cylinder will have the larger moment of inertia because its mass is located away from the axis of rotation.

10.8 A basketball of mass 610 g and circumference 76 cm is rolling without slipping across a gymnasium floor. Treating the ball as a hollow sphere, what fraction of its total kinetic energy is associated with its rotational motion?

a) 0.14   d) 0.40

b) 0.19   e) 0.67

c) 0.29

10.9 A solid sphere rolls without slipping down an incline, starting from rest. At the same time, a box starts from rest at the same altitude and slides down the same incline, with negligible friction. Which object arrives at the bottom first?

a) The solid sphere arrives first.

b) The box arrives first.

c) Both arrive at the same time.

d) It is impossible to determine.

10.10 A cylinder is rolling without slipping down a plane, which is inclined by an angle θ relative to the horizontal. What is the work done by the friction force while the cylinder travels a distance s along the plane (μ_s is the coefficient of static friction between the plane and the cylinder)?

a) $+\mu_s mgs \sin\theta$ d) $-mgs \sin\theta$

b) $-\mu_s mgs \sin\theta$ e) No work is done.

c) $+mgs \sin\theta$

10.11 A ball attached to the end of a string is swung in a vertical circle. The angular momentum of the ball at the top of the circular path is

a) greater than the angular momentum at the bottom of the circular path.

b) less than the angular momentum at the bottom of the circular path.

c) the same as the angular momentum at the bottom of the circular path.

10.12 You are unwinding a large spool of cable. As you pull on the cable with a constant tension, what happens to the angular acceleration and angular velocity of the spool, assuming that the radius at which you are extracting the cable remains constant and there is no friction force?

a) Both increase as the spool unwinds.

b) Both decrease as the spool unwinds.

c) Angular acceleration increases, and angular velocity decreases.

d) Angular acceleration decreases, and angular velocity increases.

e) It is impossible to tell.

10.13 A disk of clay is rotating with angular velocity ω. A blob of clay is stuck to the outer rim of the disk, and it has a mass $\frac{1}{10}$ of that of the disk. If the blob detaches and flies off tangent to the outer rim of the disk, what is the angular velocity of the disk after the blob separates?

a) $\frac{5}{6}\omega$ d) $\frac{11}{10}\omega$

b) $\frac{10}{11}\omega$ e) $\frac{6}{5}\omega$

c) ω

10.14 An ice skater spins with her arms extended and then pulls her arms in and spins faster. Which statement is correct?

a) Her kinetic energy of rotation does not change because, by conservation of angular momentum, the fraction by which her angular velocity increases is the same as the fraction by which her rotational inertia decreases.

b) Her kinetic energy of rotation increases because of the work she does to pull her arms in.

c) Her kinetic energy of rotation decreases because of the decrease in her rotational inertia; she loses energy because she gradually gets tired.

10.15 An ice skater rotating on frictionless ice brings her hands into her body so that she rotates faster. Which, if any, of the conservation laws hold?

a) conservation of mechanical energy and conservation of angular momentum

b) conservation of mechanical energy only

c) conservation of angular momentum only

d) neither conservation of mechanical energy nor conservation of angular momentum

10.16 If the iron core of a collapsing star initially spins with a rotational frequency of $f_0 = 3.20 \text{ s}^{-1}$, and if the core's radius decreases during the collapse by a factor of 22.7, what is the rotational frequency of the iron core at the end of the collapse?

a) 10.4 kHz   d) 0.460 kHz

b) 1.66 kHz   e) 5.20 kHz

c) 65.3 kHz

10.17 A bicycle is moving with a speed of 4.02 m/s. If the radius of the front wheel is 0.450 m, how long does it take for that wheel to make a complete revolution?

a) 0.703 s   d) 4.04 s

b) 1.23 s   e) 6.78 s

c) 2.34 s

10.18 Which one of the following statements concerning the moment of inertia of an extended rigid object is correct?

a) The moment of inertia is independent of the axis of rotation.

b) The moment of inertia depends on the axis of rotation.

c) The moment of inertia depends only on the mass of the object.

d) The moment of inertia depends only on the largest perpendicular dimension of the object.

10.19 The London Eye (basically a very large Ferris wheel) can be viewed as 32 pods, each with mass m_p, evenly spaced along the edge of a disk with mass m_d and radius R. Which of the following expressions gives the moment of inertia of the London Eye about the symmetry axis of the disk?

a) $(m_p + m_d)R^2$ d) $(32m_p + m_d)R^2$

b) $(m_p + \frac{1}{2}m_d)R^2$ e) $(16m_p + \frac{1}{2}m_d)R^2$

c) $(32m_p + \frac{1}{2}m_d)R^2$

10.20 A solid cylinder, a hollow cylinder, a solid sphere, and a hollow sphere are rolling without slipping. All four objects have the same mass and radius and are traveling with the same linear speed. Which one of the following statements is true?

a) The solid cylinder has the highest kinetic energy.

b) The hollow cylinder has the highest kinetic energy.

c) The solid sphere has the highest kinetic energy.

d) The hollow sphere has the highest kinetic energy.

e) All four objects have the same kinetic energy.

CONCEPTUAL QUESTIONS

10.21 A uniform solid sphere of radius R, mass M, and moment of inertia $I = \frac{2}{5}MR^2$ is rolling without slipping along a horizontal surface. Its total kinetic energy is the sum of the energies associated with translation of the center of mass and rotation about the center of mass. Find the *fraction* of the sphere's total kinetic energy that is attributable to rotation.

10.22 A thin ring, a solid sphere, a hollow spherical shell, and a disk of uniform thickness are placed side by side on a wide ramp of length ℓ and inclined at angle θ to the horizontal. At time $t = 0$, all four objects are released and roll without slipping on parallel paths down the ramp to the bottom. Friction and air resistance are negligible. Determine the order of finish of the race.

10.23 In another race, a solid sphere and a thin ring roll without slipping from rest down a ramp that makes angle θ with the horizontal. Find the ratio of their accelerations, $a_{\text{ring}}/a_{\text{sphere}}$.

10.24 A uniform solid sphere of mass m and radius r is placed on a ramp inclined at an angle θ to the horizontal. The coefficient of static friction between sphere and ramp is μ_s. Find the maximum value of θ for which the sphere will roll without slipping, starting from rest, in terms of the other quantities.

10.25 A round body of mass M, radius R, and moment of inertia I about its center of mass is struck a sharp horizontal blow along a line at height h above its center (with $0 \le h \le R$, of course). The body rolls away without slipping immediately after being struck. Calculate the ratio $I/(MR^2)$ for this body.

10.26 A projectile of mass m is launched from the origin at speed v_0 and an angle θ_0 above the horizontal. Air resistance is negligible.

a) Calculate the angular momentum of the projectile about the origin.

b) Calculate the rate of change of this angular momentum.

c) Calculate the torque acting on the projectile, about the origin, during its flight.

10.27 A solid sphere of radius R and mass M is placed at a height h_0 on an inclined plane of slope θ. When released, it rolls without slipping to the bottom of the incline. Next, a cylinder of same mass and radius is released on the same incline. From what height h should it be released in order to have the same speed as the sphere at the bottom?

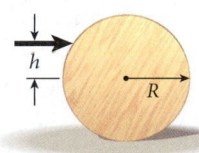

10.28 It is harder to move a door if you lean against it (along the plane of the door) toward the hinge than if you lean against the door perpendicular to its plane. Why is this so?

10.29 A figure skater draws her arms in during a final spin. Since angular momentum is conserved, her angular velocity will increase. Is her rotational kinetic energy conserved during this process? If not, where does the extra energy come from or go to?

10.30 Does a particle traveling in a straight line have an angular momentum? Explain.

10.31 A cylinder with mass M and radius R is rolling without slipping through a distance s along an inclined plane that makes an angle θ with respect to the horizontal. Calculate the work done by (a) gravity, (b) the normal force, and (c) the frictional force.

10.32 Using the conservation of mechanical energy, calculate the final speed and the acceleration of a cylindrical object of mass M and radius R after it rolls a distance s without slipping along an inclined plane that makes an angle θ with respect to the horizontal.

10.33 A *couple* is a set of two forces of equal magnitude and opposite directions, whose lines of action are parallel but not identical. Prove that the net torque of a couple of forces is independent of the pivot point about which the torque is calculated and of the points along their lines of action where the two forces are applied.

10.34 Why does a figure skater pull in her arms while increasing her angular velocity in a tight spin?

10.35 To turn a motorcycle moving at high speed to the right, you momentarily turn the handlebars to the left to initiate the turn. After the turn is begun, you steer and lean to the right to complete the turn. Explain, as precisely as you can, how this countersteering begins the turn in the desired direction. (*Hint:* The wheels of a motorcycle in motion have a great deal of angular momentum.)

10.36 The Moon's tide-producing effect on the Earth is gradually slowing the rotation of the Earth, owing to tidal friction. Studies of corals from the Devonian Period indicate that the year was 400 days long in that period. What, if anything, does this indicate about the angular momentum of the Moon in the Devonian Period relative to its value at present?

10.37 A light rope passes over a light, frictionless pulley. One end is fastened to a bunch of bananas of mass M, and a monkey of the same mass clings to the other end. The monkey climbs the rope in an attempt to reach the bananas. The radius of the pulley is R.

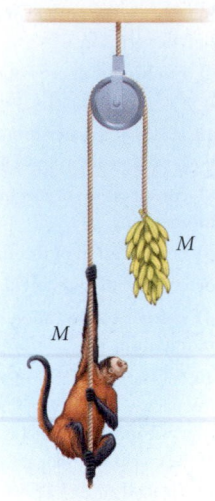

a) Treating the monkey, bananas, rope, and pulley as a system, evaluate the net torque about the pulley axis.

b) Using the result of part (a), determine the total angular momentum about the pulley axis as a function of time.

EXERCISES

A blue problem number indicates a worked-out solution is available in the Student Solutions Manual. One • and two •• indicate increasing level of problem difficulty.

Sections 10.1 and 10.2

10.38 A uniform solid cylinder of mass M = 5.00 kg is rolling without slipping along a horizontal surface. The velocity of its center of mass is 30.0 m/s. Calculate its energy.

10.39 Determine the moment of inertia for three children weighing 60.0 lb, 45.0 lb, and 80.0 lb sitting at different points on the edge of a rotating merry-go-round, which has a radius of 12.0 ft.

•**10.40** A 24.0-cm-long pen is tossed up in the air, reaching a maximum height of 1.20 m above its release point. On the way up, the pen makes 1.80 revolutions. Treating the pen as a thin uniform rod, calculate the ratio between the rotational kinetic energy and the translational kinetic energy at the instant the pen is released. Assume that the rotational speed does not change during the toss.

•**10.41** A solid ball and a hollow ball, each with a mass of 1.00 kg and radius of 0.100 m, start from rest and roll down a ramp of length 3.00 m at an incline of 35.0°. An ice cube of the same mass slides without friction down the same ramp.

a) Which ball will reach the bottom first? Explain!

b) Does the ice cube travel faster or slower than the solid ball at the base of the incline? Explain your reasoning.

c) What is the speed of the solid ball at the bottom of the incline?

•**10.42** A solid ball of mass m and radius r rolls without slipping through a loop of radius R, as shown in the figure. From what height h should the ball be launched in order to make it through the loop without falling off the track?

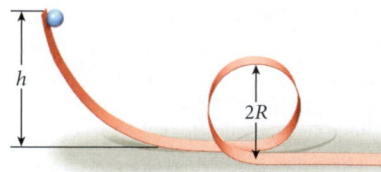

••**10.43** The Crab pulsar ($m \approx 2 \cdot 10^{30}$ kg, R = 12 km) is a neutron star located in the Crab Nebula. The rotation rate of the Crab pulsar is currently about 30 rotations per second, or 60π rad/s. The rotation rate of the pulsar, however, is decreasing; each year, the rotation period increases by 10^{-5} s. Justify the following statement: The loss in rotational energy of the pulsar is equivalent to 100,000 times the power output of the Sun. (The total power radiated by the Sun is about $4 \cdot 10^{26}$ W.)

••**10.44** A block of mass m = 4.00 kg is attached to a spring (k = 32.0 N/m) by a rope that hangs over a pulley of mass M = 8.00 kg and radius R = 5.00 cm, as shown in the figure. Treating the pulley as a solid homogeneous disk, neglecting friction at the axle of the pulley, and assuming the system starts from rest with the spring at its natural length, find (a) the speed of the block after it falls 1.00 m, and (b) the maximum extension of the spring.

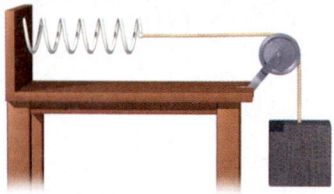

Section 10.3

•**10.45** A small circular object with mass m and radius r has a moment of inertia given by $I = cmr^2$. The object rolls without slipping along the track shown in the figure. The track ends with a ramp of height R = 2.50 m that launches the object vertically. The object starts from a height H = 6.00 m. To what maximum height will it rise after leaving the ramp if c = 0.400?

•**10.46** A uniform solid sphere of mass M and radius R is rolling without sliding along a level plane with a speed v = 3.00 m/s when it encounters a ramp that is at an angle θ = 23.0° above the horizontal. Find the maximum distance that the sphere travels up the ramp in each case:

a) The ramp is frictionless, so the sphere continues to rotate with its initial angular speed until it reaches its maximum height.

b) The ramp provides enough friction to prevent the sphere from sliding, so both the linear and rotational motion stop when the object reaches its maximum height.

Section 10.4

•**10.47** A disk with a mass of 30.0 kg and a radius of 40.0 cm is mounted on a frictionless horizontal axle. A string is wound many times around the disk and then attached to a 70.0-kg block, as shown in the figure. Find the acceleration of the block, assuming that the string does not slip.

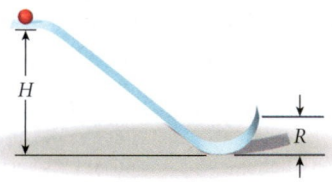

•**10.48** A force, $\vec{F} = (2\hat{x} + 3\hat{y})$ N, is applied to an object at a point whose position vector with respect to the pivot point is $\vec{r} = (4\hat{x} + 4\hat{y} + 4\hat{z})$ m. Calculate the torque created by the force about that pivot point.

••10.49 A disk with a mass of 14.0 kg, a diameter of 30.0 cm, and a thickness of 8.00 cm is mounted on a rough horizontal axle as shown on the left in the

figure. (There is a friction force between the axle and the disk.) The disk is initially at rest. A constant force, $F = 70.0$ N, is applied to the edge of the disk at an angle of 37.0°, as shown on the right in the figure. After 2.00 s, the force is reduced to $F = 24.0$ N, and the disk spins with a constant angular velocity.

a) What is the magnitude of the torque due to friction between the disk and the axle?

b) What is the angular velocity of the disk after 2.00 s?

c) What is the kinetic energy of the disk after 2.00 s?

Section 10.5

10.50 A thin uniform rod (length = 1.00 m, mass = 2.00 kg) is pivoted about a horizontal frictionless pin through one of its ends. The moment of inertia of the rod through this axis is $\frac{1}{3}mL^2$. The rod is released when it is 60.0° below the horizontal. What is the angular acceleration of the rod at the instant it is released?

10.51 An object made of two disk-shaped sections, A and B, as shown in the figure, is rotating about an axis through the center of disk A. The masses and the radii of disks A and B, respectively, are 2.00 kg and 0.200 kg and 25.0 cm and 2.50 cm.

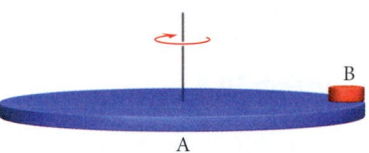

a) Calculate the moment of inertia of the object.

b) If the axial torque due to friction is 0.200 N m, how long will it take for the object to come to a stop if it is rotating with an initial angular velocity of -2π rad/s?

•10.52 You are the technical consultant for an action-adventure film in which a stunt calls for the hero to drop off a 20.0-m-tall building and land on the ground safely at a final vertical speed of 4.00 m/s. At the edge of the building's roof, there is a 100.-kg drum that is wound with a sufficiently long rope (of negligible mass), has a radius of 0.500 m, and is free to rotate about its cylindrical axis with a moment of inertia I_0. The script calls for the 50.0-kg stuntman to tie the rope around his waist and walk off the roof.

a) Determine an expression for the stuntman's linear acceleration in terms of his mass m, the drum's radius r, and the moment of inertia I_0.

b) Determine the required value of the stuntman's acceleration if he is to land safely at a speed of 4.00 m/s, and use this value to calculate the moment of inertia of the drum about its axis.

c) What is the angular acceleration of the drum?

d) How many revolutions does the drum make during the fall?

•10.53 In a tire-throwing competition, a man holding a 23.5-kg car tire quickly swings the tire through three full turns and releases it, much like a discus thrower. The tire starts from rest and is then accelerated in a circular path. The orbital radius r for the tire's center of mass is 1.10 m, and the path is horizontal to the ground.

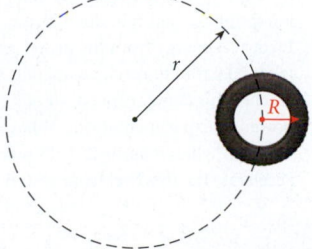

The figure shows a top view of the tire's circular path, and the dot at the center marks the rotation axis. The man applies a constant torque of 20.0 N m to accelerate the tire at a constant angular acceleration. Assume that all of the tire's mass is at a radius $R = 0.350$ m from its center.

a) What is the time, t_{throw}, required for the tire to complete three full revolutions?

b) What is the final linear speed of the tire's center of mass (after three full revolutions)?

c) If, instead of assuming that all of the mass of the tire is at a distance 0.350 m from its center, you treat the tire as a hollow disk of inner radius 0.300 m and outer radius 0.400 m, how does this change your answers to parts (a) and (b)?

•10.54 A 100.-kg barrel with a radius of 50.0 cm has two ropes wrapped around it, as shown in the figure. The barrel is released from rest, causing the ropes to unwind and the barrel to fall spinning toward the ground. What is the speed of the barrel after it has fallen a distance of 10.0 m? What is the tension in each rope? Assume that the barrel's mass is uniformly distributed and that the barrel rotates as a solid cylinder.

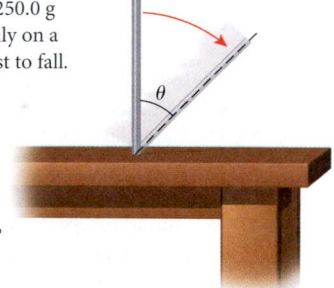

•10.55 A wheel with $c = \frac{4}{9}$, a mass of 40.0 kg, and a rim radius of 30.0 cm is mounted vertically on a horizontal axis. A 2.00-kg mass is suspended from the wheel by a rope wound around the rim. Find the angular acceleration of the wheel when the mass is released.

••10.56 A uniform rod of mass $M = 250.0$ g and length $L = 50.0$ cm stands vertically on a horizontal table. It is released from rest to fall.

a) What forces are acting on the rod?

b) Calculate the angular speed of the rod, the vertical acceleration of the moving end of the rod, and the normal force exerted by the table on the rod as it makes an angle $\theta = 45.0°$ with respect to the vertical.

c) If the rod falls onto the table without slipping, find the linear acceleration of the end point of the rod when it hits the table and compare it with g.

••10.57 A demonstration setup consists of a uniform board of length L, hinged at the bottom end and elevated at an angle θ by means of a support stick. A ball rests at the elevated end, and a light cup is attached to the board at the distance d from the elevated end to catch the ball when the stick supporting the board is suddenly removed. You want to use a thin hinged board 1.00 m long and 10.0 cm wide, and you plan to have the vertical support stick located right at its elevated end.

a) What is the maximum length you can make the support stick so that the ball has a chance to be caught?

b) Assume that you choose to use the longest possible support stick placed at the elevated end of the board. What distance d from that end should the cup be located to ensure that the ball will be caught in the cup?

Section 10.6

•10.58 The flywheel of an old steam engine is a solid homogeneous metal disk of mass $M = 120.$ kg and radius $R = 80.0$ cm. The engine rotates the wheel at 500. rpm. In an emergency, to bring the engine to a stop, the flywheel is disengaged from the engine and a brake pad is applied at the edge to provide a radially inward force $F = 100.$ N. If the coefficient of kinetic friction between the pad and the flywheel is $\mu_k = 0.200$, how many revolutions does the flywheel make before coming to rest? How long does it take for the flywheel to come to rest? Calculate the work done by the torque during this time.

•**10.59** The turbine and associated rotating parts of a jet engine have a total moment of inertia of 25.0 kg m². The turbine is accelerated uniformly from rest to an angular speed of 150. rad/s in a time of 25.0 s. Find

a) the angular acceleration,

b) the net torque required,

c) the angle turned through in 25.0 s,

d) the work done by the net torque, and

e) the kinetic energy of the turbine at the end of the 25.0 s.

Section 10.7

10.60 Two small 6.00-kg masses are joined by a string, which can be assumed to be massless. The string has a tangle in it, as shown in the figure. With the string tangled, the masses are separated by 1.00 m. The two masses are then made to rotate about their center of mass on a frictionless table at a rate of 5.00 rad/s. As they are rotating, the string untangles and lengthens to 1.40 m. What is the angular velocity of the masses after the string untangles?

6.00 kg   6.00 kg

•**10.61** It is sometimes said that if the entire population of China stood on chairs and jumped off simultaneously, it would alter the rotation of the Earth. Fortunately, physics gives us the tools to investigate such speculations.

a) Calculate the moment of inertia of the Earth about its axis. For simplicity, treat the Earth as a uniform sphere of mass $m_E = 5.977 \cdot 10^{24}$ kg and radius 6371 km.

b) Calculate an upper limit for the contribution by the population of China to the Earth's moment of inertia, by assuming that the whole group is at the Equator. Take the population of China to be 1.30 billion people, of average mass 70.0 kg.

c) Calculate the change in the contribution in part (b) associated with a 1.00-m simultaneous change in the radial position of the entire group.

d) Determine the fractional change in the length of the day the change in part (c) would produce.

•**10.62** A bullet of mass $m_B = 1.00 \cdot 10^{-2}$ kg is moving with a speed of 100. m/s when it collides with a rod of mass $m_R = 5.00$ kg and length $L = 1.00$ m (shown in the figure). The rod is initially at rest, in a vertical position, and pivots about an axis going through its center of mass. The bullet embeds itself in the rod at a distance $L/4$ from the pivot point. As a result, the bullet-rod system starts rotating.

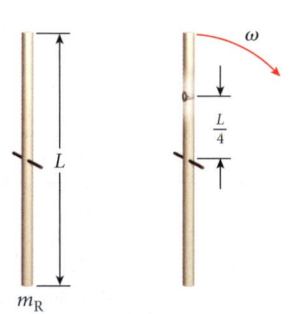

a) Find the angular velocity, ω, of the bullet-rod system after the collision. You can neglect the width of the rod and can treat the bullet as a point mass.

b) How much kinetic energy is lost in the collision?

••**10.63** A uniform solid sphere of radius R and mass M sits on a horizontal tabletop. A horizontally directed impulse with magnitude J is delivered to a spot on the ball a vertical distance h above the tabletop.

a) Determine the angular and translational velocity of the sphere just after the impulse is delivered.

b) Determine the distance h_0 at which the delivered impulse causes the ball to immediately roll without slipping.

•**10.64** A circular platform of radius $R_p = 4.00$ m and mass $M_p = 400.$ kg rotates on frictionless air bearings about its vertical axis at 6.00 rpm. An

80.0-kg man standing at the very center of the platform starts walking (at $t = 0$) radially outward at a speed of 0.500 m/s with respect to the platform. Approximating the man by a vertical cylinder of radius $R_m = 0.200$ m, determine an equation (specific expression) for the angular velocity of the platform as a function of time. What is the angular velocity when the man reaches the edge of the platform?

•**10.65** A 25.0-kg boy stands 2.00 m from the center of a frictionless playground merry-go-round, which has a moment of inertia of 200. kg m². The boy begins to run in a circular path with a speed of 0.600 m/s relative to the ground.

a) Calculate the angular velocity of the merry-go-round.

b) Calculate the speed of the boy relative to the surface of the merry-go-round.

•**10.66** The Earth has an angular speed of $7.272 \cdot 10^{-5}$ rad/s in its rotation. Find the new angular speed if an asteroid ($m = 1.00 \cdot 10^{22}$ kg) hits the Earth while traveling at a speed of $1.40 \cdot 10^3$ m/s (assume the asteroid is a point mass compared to the radius of the Earth) in each of the following cases:

a) The asteroid hits the Earth dead center.

b) The asteroid hits the Earth nearly tangentially in the direction of Earth's rotation.

c) The asteroid hits the Earth nearly tangentially in the direction opposite to Earth's rotation.

Section 10.8

10.67 A demonstration gyroscope consists of a uniform disk with a 40.0-cm radius, mounted at the midpoint of a light 60.0-cm axle. The axle is supported at one end while in a horizontal position. How fast is the gyroscope precessing, in units of rad/s, if the disk is spinning around the axle at 30.0 rev/s?

Additional Exercises

10.68 Most stars maintain an equilibrium size by balancing two forces—an inward gravitational force and an outward force resulting from the star's nuclear reactions. When the star's fuel is spent, there is no counterbalance to the gravitational force. Whatever material is remaining collapses in on itself. Stars about the same size as the Sun become white dwarfs, which glow from leftover heat. Stars that have about three times the mass of the Sun compact into neutron stars. And a star with a mass greater than three times the Sun's mass collapses into a single point, called a *black hole*. In most cases, protons and electrons are fused together to form neutrons— this is the reason for the name *neutron star*. Neutron stars rotate very fast because of the conservation of angular momentum. Imagine a star of mass $5.00 \cdot 10^{30}$ kg and radius $9.50 \cdot 10^8$ m that rotates once in 30.0 days. Suppose this star undergoes gravitational collapse to form a neutron star of radius 10.0 km. Determine its rotation period.

10.69 In experiments at the Princeton Plasma Physics Laboratory, a plasma of hydrogen atoms is heated to over 500 million degrees Celsius (about 25 times hotter than the center of the Sun) and confined for tens of milliseconds by powerful magnetic fields (100,000 times greater than the Earth's magnetic field). For each experimental run, a huge amount of energy is required over a fraction of a second, which translates into a power requirement that would cause a blackout if electricity from the normal grid were to be used to power the experiment. Instead, kinetic energy is stored in a colossal flywheel, which is a spinning solid cylinder with a radius of 3.00 m and mass of $1.18 \cdot 10^6$ kg. Electrical energy from the power grid starts the flywheel spinning, and it takes 10.0 min to reach an angular speed of 1.95 rad/s. Once the flywheel reaches this angular speed, all of its energy can be drawn off very quickly to power an experimental run. What is the mechanical energy stored in the flywheel when it spins at 1.95 rad/s? What is the average torque required to accelerate the flywheel from rest to 1.95 rad/s in 10.0 min?

10.70 A 2.00-kg thin hoop with a 50.0-cm radius rolls down a 30.0° slope without slipping. If the hoop starts from rest at the top of the slope, what is its translational velocity after it rolls 10.0 m along the slope?

10.71 An oxygen molecule (O_2) rotates in the xy-plane about the z-axis. The axis of rotation passes through the center of the molecule, perpendicular to its length. The mass of each oxygen atom is $2.66 \cdot 10^{-26}$ kg, and the average separation between the two atoms is $d = 1.21 \cdot 10^{-10}$ m.

a) Calculate the moment of inertia of the molecule about the z-axis.

b) If the angular speed of the molecule about the z-axis is $4.60 \cdot 10^{12}$ rad/s, what is its rotational kinetic energy?

10.72 A 0.0500-kg bead slides on a wire bent into a circle of radius 0.400 m. You pluck the bead with a force tangent to the circle. What force is needed to give the bead an angular acceleration of 6.00 rad/s²?

10.73 A professor doing a lecture demonstration stands at the center of a frictionless turntable, holding 5.00-kg masses in each hand with arms extended so that each mass is 1.20 m from his centerline. A (carefully selected!) student spins the professor up to a rotational frequency of 1.00 rpm. If he then pulls his arms in by his sides so that each mass is 0.300 m from his centerline, what is his new angular speed? Assume that his rotational inertia without the masses is 2.80 kg m², and neglect the effect on the rotational inertia of the position of his arms, since their mass is small compared to the mass of the body.

•**10.74** The system shown in the figure is held initially at rest. Calculate the angular acceleration of the system as soon as it is released. You can treat M_A (1.00 kg) and M_B (10.0 kg) as point masses located on either end of the rod of mass M_C (20.0 kg) and length L (5.00 m).

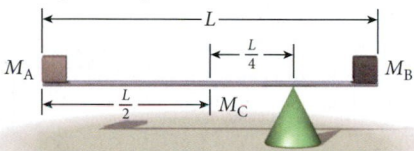

•**10.75** A child builds a simple cart consisting of a 60.0 cm by 1.20 m sheet of plywood of mass 8.00 kg and four wheels, each 20.0 cm in diameter and with a mass of 2.00 kg. It is released from the top of a 15.0° incline that is 30.0 m long. Find the speed at the bottom. Assume that the wheels roll along the incline without slipping and that friction between the wheels and their axles can be neglected.

•**10.76** A CD has a mass of 15.0 g, an inner diameter of 1.50 cm, and an outer diameter of 11.9 cm. Suppose you toss it, causing it to spin at a rate of 4.30 revolutions per second.

a) Determine the moment of inertia of the CD, approximating its density as uniform.

b) If your fingers were in contact with the CD for 0.250 revolutions while it was acquiring its angular velocity and applied a constant torque to it, what was the magnitude of that torque?

•**10.77** A sheet of plywood 1.30 cm thick is used to make a cabinet door 55.0 cm wide by 79.0 cm tall, with hinges mounted on the vertical edge. A small 150.-g handle is mounted 45.0 cm from the lower hinge at the same height as that hinge. If the density of the plywood is 550. kg/m³, what is the moment of inertia of the door about the hinges? Neglect the contribution of hinge components to the moment of inertia.

•**10.78** A machine part is made from a uniform solid disk of radius R and mass M. A hole of radius $R/2$ is drilled into the disk, with the center of the hole at a distance $R/2$ from the center of the disk (the diameter of the hole spans from the center of the disk to its outer edge). What is the moment of inertia of this machine part about the center of the disk in terms of R and M?

•**10.79** A space station is to provide artificial gravity to support long-term habitation by astronauts and cosmonauts. It is designed as a large wheel, with all the compartments in the rim, which is to rotate at a speed that will provide an acceleration similar to that of terrestrial gravity for the astronauts (their feet will be on the inside of the outer wall of the space station and their heads will be pointing toward the hub). After the space station is assembled in orbit, its rotation will be started by the firing of a rocket motor fixed to the outer rim, which fires tangentially to the rim. The radius of the space station is $R = 50.0$ m, and the mass is $M = 2.40 \cdot 10^5$ kg. If the thrust of the rocket motor is $F = 1.40 \cdot 10^2$ N, how long should the motor fire?

•**10.80** Many pulsars radiate radio-frequency or other radiation in a periodic manner and are bound to a companion star in what is known as a *binary pulsar system*. In 2003, a double pulsar system, PSR J0737-3039A and J0737-3039B, was discovered by astronomers at the Jodrell Bank Observatory in the United Kingdom. In this system, *both* stars are pulsars. The pulsar with the faster rotation period rotates once every 0.0230 s, while the other pulsar has a rotation period of 2.80 s. The faster pulsar has a mass 1.337 times that of the Sun, while the slower pulsar has a mass 1.250 times that of the Sun.

a) If each pulsar has a radius of 20.0 km, express the ratio of their rotational kinetic energies. Consider each star to be a uniform sphere with a fixed rotation period.

b) The orbits of the two pulsars about their common center of mass are rather eccentric (highly squashed ellipses), but an estimate of their average translational kinetic energy can be obtained by treating each orbit as circular with a radius equal to the mean distance from the system's center of mass. This radius is equal to $4.23 \cdot 10^8$ m for the larger star, and $4.54 \cdot 10^8$ m for the smaller star. If the orbital period is 2.40 h, calculate the ratio of rotational to translational kinetic energies for each star.

•**10.81** A student of mass 52.0 kg wants to measure the mass of a playground merry-go-round, which consists of a solid metal disk of radius $R = 1.50$ m that is mounted in a horizontal position on a low-friction axle. She tries an experiment: She runs with speed $v = 6.80$ m/s toward the outer rim of the merry-go-round and jumps on to the outer rim, as shown in the figure. The merry-go-round is initially at rest before the student jumps on and rotates at 1.30 rad/s immediately after she jumps on. You may assume that the student's mass is concentrated at a point.

a) What is the mass of the merry-go-round?

b) If it takes 35.0 s for the merry-go-round to come to a stop after the student has jumped on, what is the average torque due to friction in the axle?

c) How many times does the merry-go-round rotate before it stops, assuming that the torque due to friction is constant?

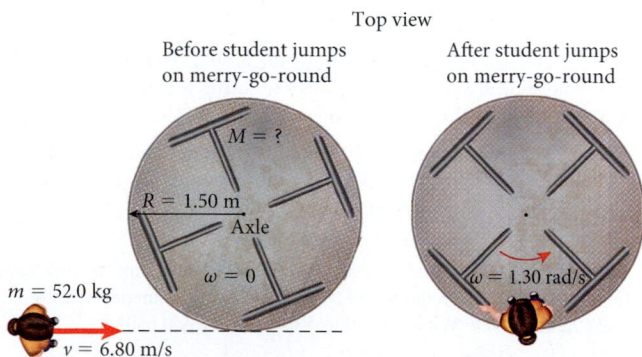

Top view

Before student jumps on merry-go-round

After student jumps on merry-go-round

•**10.82** A ballistic pendulum consists of an arm of mass M and length $L = 0.480$ m. One end of the arm is pivoted so that the arm rotates freely in a vertical plane. Initially, the arm is motionless and hangs vertically from the pivot point. A projectile of the same mass M hits the lower end of the arm with a horizontal velocity of $V = 3.60$ m/s. The projectile remains stuck to the free end of the arm during their subsequent motion. Find the maximum angle to which the arm and attached mass will swing in each case:

a) The arm is treated as an ideal pendulum, with all of its mass concentrated as a point mass at the free end.

b) The arm is treated as a thin rigid rod, with its mass evenly distributed along its length.

••**10.83** A wagon wheel is made entirely of wood. Its components consist of a rim, 12 spokes, and a hub. The rim has mass 5.20 kg, outer radius 0.900 m, and inner radius 0.860 m. The hub is a solid cylinder with mass 3.40 kg and radius 0.120 m. The spokes are thin rods of mass 1.10 kg that extend from the hub to the inner side of the rim. Determine the constant $c = I/MR^2$ for this wagon wheel.

••**10.84** The figure shows a solid, homogeneous ball with radius R. Before falling to the floor, its center of mass is at rest, but it is spinning with angular velocity ω_0 about a horizontal axis through its center. The lowest point of the ball is at a height h above the floor. When released, the ball falls under the influence of gravity, and rebounds to a new height such that its lowest point is ah above the floor. The deformation of the ball and the floor due to the impact can be considered negligible; the impact time, though, is nonzero. The mass of the ball is m, and the coefficient of kinetic friction between the ball and the floor is μ_k. Ignore air resistance.

For the situation where the ball is slipping throughout the impact, find each of the following:

a) $\tan \theta$, where θ is the rebound angle indicated in the diagram,

b) the horizontal distance traveled in flight between the first and second impacts, and

c) the minimum value of ω_0 for this situation.

For the situation where the ball stops slipping before the impact ends, find each of the following:

d) $\tan \theta$, and

e) the horizontal distance traveled in flight between the first and second impacts.

Taking both of the situations into account, sketch the variation of $\tan \theta$ with ω_0.

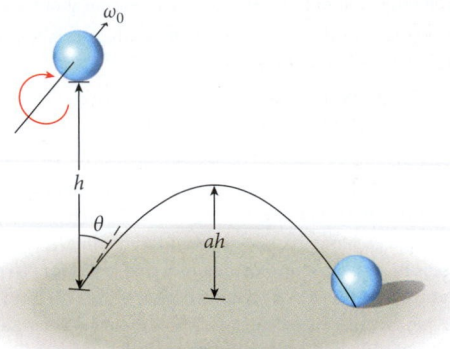

MULTI-VERSION EXERCISES

10.85 The propeller of a light plane has a length of 2.012 m and a mass of 17.36 kg. The propeller is rotating with a frequency of 3280. rpm. What is the rotational kinetic energy of the propeller? You can treat the propeller as a thin rod rotating about its center.

10.86 The propeller of a light plane has a length of 2.092 m and a mass of 17.56 kg. The rotational energy of the propeller is 422.8 kJ. What is the rotational frequency of the propeller (in rpm)? You can treat the propeller as a thin rod rotating about its center.

10.87 The propeller of a light plane has a length of 1.812 m and rotates at 2160. rpm. The rotational kinetic energy of the propeller is 124.3 kJ. What is the mass of the propeller? You can treat the propeller as a thin rod rotating about its center.

10.88 A golf ball with mass 45.90 g and diameter 42.60 mm is struck such that it moves with a speed of 51.85 m/s and rotates with a frequency of 2857 rpm. What is the kinetic energy of the golf ball?

10.89 A golf ball with mass 45.90 g and diameter 42.60 mm is struck such that it moves with a speed of 54.15 m/s while rotating. The golf ball has a kinetic energy of 67.67 J. What is the rotational frequency of the golf ball (in rpm)?

10.90 A golf ball with mass 45.90 g and diameter 42.60 mm is struck such that it rotates with a frequency of 2875 rpm and has a kinetic energy of 73.51 J as it moves. What is the linear speed of the golf ball?

10.91 A string is wrapped many times around a pulley and is connected to a block of mass $m_b = 4.243$ kg, which is hanging vertically. The pulley consists of a wheel of radius 46.21 cm and mass $m_p = 5.907$ kg, with spokes that have negligible mass. What is the magnitude of the acceleration of the block?

10.92 A string is wrapped many times around a pulley and is connected to a block of mass $m_b = 4.701$ kg, which is hanging vertically. The pulley consists of a wheel of radius 47.49 cm, with spokes that have negligible mass. The block accelerates downward at 4.330 m/s². What is the mass of the pulley, m_p?

10.93 A string is wrapped many times around a pulley and is connected to a block that is hanging vertically. The pulley consists of a wheel of radius 48.77 cm and mass $m_p = 5.991$ kg, with spokes that have negligible mass. The block accelerates downward at 4.539 m/s². What is the mass of the block, m_b?

11

Static Equilibrium

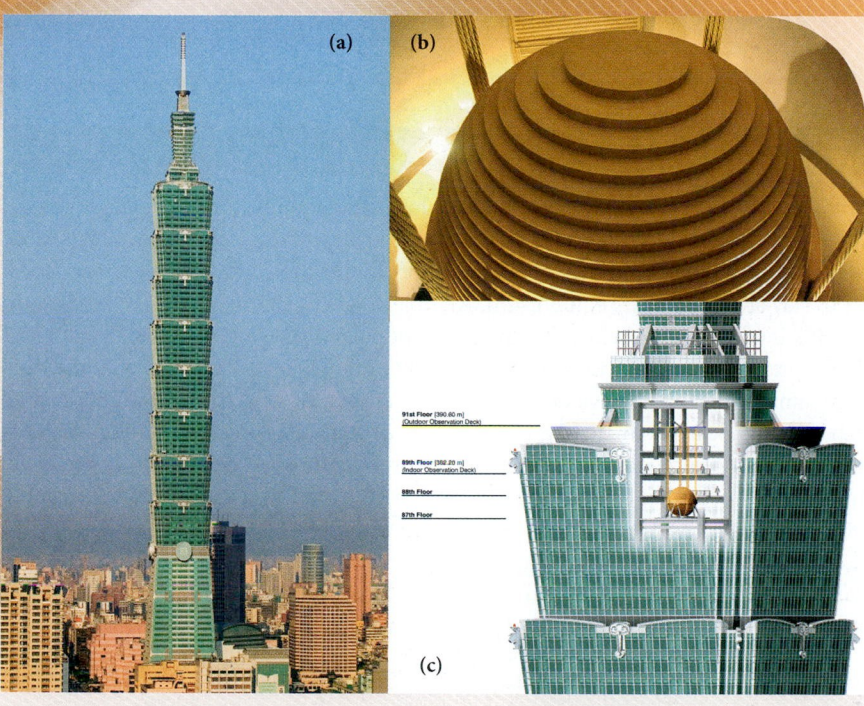

FIGURE 11.1 The tallest building in the world as of 2008, Taipei 101 in Taiwan: (a) view of the tower; (b) view of the sway damper inside the tower; (c) cutaway drawing of the top of the tower showing the location of the damper.

he tallest building in the world as of 2008 was the Taipei 101 tower (Figure 11.1) in Taiwan, at 509 m (1670 ft) tall. Like any skyscraper, this building sways when winds near the top gust at high speeds. To minimize the motion, the Taipei 101 tower contains a mass damper between the 87th and 92nd floors, consisting of a steel ball built from 5-in-thick disks. The damper has a mass of 660 metric tons, enough to reduce the tower's motion by about 40%. Restaurants and observation decks surround the damper, making it a leading tourist attraction.

Stability and safety are of prime importance in the design and construction of any building. In this chapter, we examine the conditions for static equilibrium, which occurs when an object is at rest, and subject to zero net forces and torques. However, as we will see, a structure must be able to resist outside forces that tend to set it in motion. The long-term stability of a large structure—a building, a bridge, or a monument—depends on the builders' ability to judge how strong outside forces might be and to design the structure to withstand these forces.

WHAT WE WILL LEARN

- Static equilibrium is defined as mechanical equilibrium for the special case of an object at rest.
- An object (or a collection of objects) can be in static equilibrium only if the net external force is zero and the net external torque is zero.
- A necessary condition for static equilibrium is that the first derivative of the potential energy function is zero at the equilibrium point.
- Stable equilibrium is achieved at points where the potential energy function has a minimum.

- Unstable equilibrium occurs at points where the potential energy function has a maximum.
- Neutral equilibrium (or indifferent equilibrium or marginally stable equilibrium) exists at points where the first and second derivatives of the potential energy function are both zero.
- Equilibrium considerations are used to find otherwise unknown forces acting on an unmoving object or to find the forces required to prevent an object from moving.

11.1 Equilibrium Conditions

(a)

(b)

FIGURE 11.2 (a) This 420-kg installation created by Alexander Calder hangs from the ceiling at the National Gallery of Art (Washington, DC) in perfect static equilibrium. (b) You can create this installation, consisting of a fork, a spoon, and a toothpick balanced on a glass, at the dinner table.

In Chapter 4, we saw that the necessary condition for static equilibrium is the absence of a net external force. In that case, Newton's First Law stipulates that an object stays at rest or moves with constant velocity. However, we often want to find the conditions necessary for a rigid object to stay at rest, in *static* equilibrium. An object (or collection of objects) is in **static equilibrium** if it is at rest and not experiencing translational or rotational motion. Figure 11.2a shows a famous example of a collection of objects in static equilibrium. Part of what makes this installation so amazing is that the eye does not want to accept that the configuration is stable. Figure 11.2b shows another example of a set of objects in static equilibrium that you can create at home or at a restaurant: The toothpick holds the spoon and fork on the rim of the glass. (For theatrical effect, the end of the toothpick was burnt after the "installation" was balanced.)

The requirement of no translational or rotational motion means that the linear and angular velocities of an object in static equilibrium are always zero. The fact that the linear and angular velocities do not change with time implies that the linear and angular accelerations are also zero at all times. In Chapter 4, we saw that Newton's Second Law,

$$\vec{F}_{net} = m\vec{a}, \tag{11.1}$$

implies that if the linear acceleration, $\vec{a}$, is zero, the external net force, $\vec{F}_{net}$, must be zero. Furthermore, we saw in Chapter 10 that Newton's Second Law for rotation,

$$\vec{\tau}_{net} = I\vec{\alpha}, \tag{11.2}$$

implies that if the angular acceleration, $\vec{\alpha}$, is zero, the external net torque, $\vec{\tau}_{net}$, must be zero. These facts lead to two conditions for static equilibrium.

Static Equilibrium Condition 1

An object can stay in static equilibrium only if the net force acting on it is zero:

$$\vec{F}_{net} = 0. \tag{11.3}$$

Static Equilibrium Condition 2

An object can stay in static equilibrium only if the net torque acting on it is zero:

$$\vec{\tau}_{net} = 0. \tag{11.4}$$

Even if Newton's First Law is satisfied (no net force acts on an object), and an object has no translational motion, it will still rotate if it experiences a net torque.

It is important to remember that torque is always defined with respect to a pivot point (the point where the axis of rotation intersects the plane defined by $\vec{F}$ and $\vec{r}$). When we compute the net torque, the pivot point must be the same for all forces involved in the calculation. If we try to solve a static equilibrium problem, with vanishing net torque, the

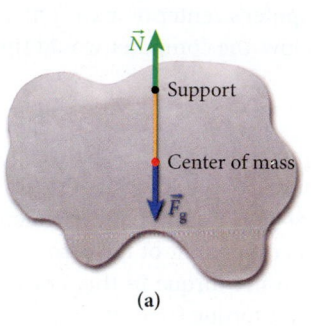 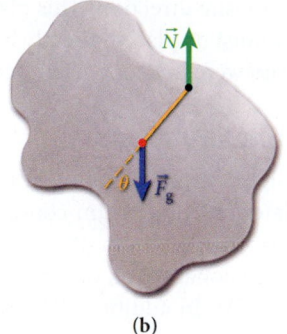

(a)   (b)

FIGURE 11.3 (a) This object experiences zero net torque, because it is supported from a pin located exactly above the center of mass. (b) A net torque results when the center of mass of the same object is at a location not exactly below the support point.

net torque has to be zero for any pivot point chosen. Thus, we have the freedom to select a pivot point that best suits our purpose. A clever selection of a pivot point is often the key to a quick solution. For example, if an unknown force is present in the problem, we can select the point where the force acts as the pivot point. Then, that force will not enter into the torque equation because it has a moment arm of length zero.

If an object is supported from a pin located directly above its center of mass, as in Figure 11.3a (where the red dot marks the center of mass), then the object stays balanced; that is, it does not start to rotate. Why? Because in this case only two forces act on the object—the force of gravity, $\vec{F}_g$ (blue arrow), and the vertical supporting force $\vec{N}$ (green arrow), from the pin—and they lie on the same line (yellow line in Figure 11.3a). The two forces cancel each other out and produce no net torque, resulting in static equilibrium; the object is in balance.

On the other hand, if an object is supported in the same way from a pin but its center of mass is not below the support point, then the situation is that shown in Figure 11.3b. The normal and gravitational force vectors still point in opposite directions; however, a nonzero net torque now acts, because the angle θ between the gravitational force vector, $\vec{F}_g$, and the moment arm (directed along the yellow line) is not zero any more. This torque violates the condition that the net torque must be zero for static equilibrium. However, suspending an object from different points is a practical method for finding the center of mass of the object, even a strangely shaped object like the one in Figure 11.3.

Experimentally Locating the Center of Mass

To locate an object's center of mass experimentally, we can support the object from a pin in such a way that it can rotate freely around the pin and then let it come to rest. Once the object has come to rest, its center of mass is located on the line directly below the pin. We hang a weight (a plumb bob in Figure 11.4) from the same pin used to support the object, and it identifies the line. We mark this line on the object. If we do this for two different support points, the intersection of the two lines will mark the precise location of the center of mass.

You can use another technique to determine the location of the center of mass for many objects (see Figure 11.5). You simply support the object on two fingers placed in such a way that the center of mass is located somewhere between them. (If this is not the case, you will know right away, because the object will fall.) Then slowly slide the fingers closer to each other. At the point where they meet, they are directly below the center of mass, and the object is balanced on top.

Why does this technique work? The finger that is closer to the center of mass exerts a larger normal force on the object. Thus, when moving, this finger exerts a larger friction force on the object than the finger that is farther away. Consequently, if the fingers slide toward each other, the finger that is closer to the center of mass will take the suspended object along with it. This continues until the other finger becomes closer to the center of mass, when the effect is reversed. In this way, the two fingers always keep the center of mass located between them. When the fingers are next to each other, the center of mass is located.

In Figure 4.6, showing a hand holding up a laptop computer, the force vector $\vec{N}$ exerted by the hand on the laptop acted at the laptop's center, just like the gravitational force vector

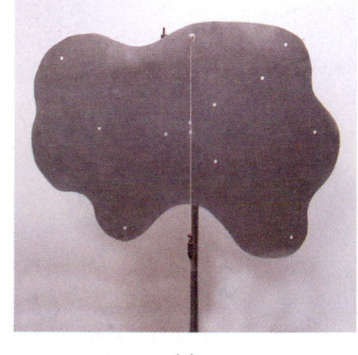

(a)

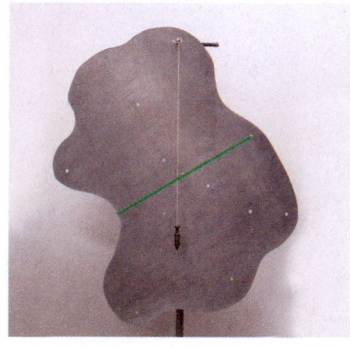

(b)

FIGURE 11.4 Finding the center of mass for an arbitrarily shaped object.

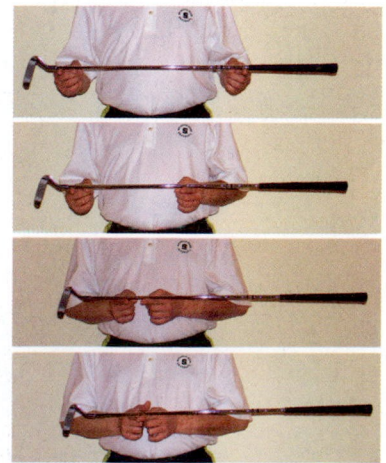

FIGURE 11.5 Determining the center of mass of a golf club experimentally.

but in the opposite direction. This placement is necessary. For a hand to hold up a laptop computer, it must be placed directly below the computer's center of mass. Otherwise, if the center of mass were not supported from directly below, the computer would tip over.

Equilibrium Equations

With a qualitative understanding of the concepts and conditions for static equilibrium, we can formulate the equilibrium conditions for a more quantitative analysis. In Chapter 4, we found that the condition of zero net force translates into three independent equations in three-dimensional space, one for each Cartesian component of the zero net force (refer to equation 11.3). In addition, the condition of zero net torque in three dimensions also implies three equations for the components of the net torque (refer to equation 11.4), representing independent rotations about the three possible axes of rotation, which are all perpendicular to each other. In this chapter, we will not deal with the three-dimensional situations (involving six equations), but rather will concentrate on problems of static equilibrium in two-dimensional space, that is, a plane. In a plane, there are two independent translational degrees of freedom for a rigid body (in the x- and the y-directions) and one possible rotation, either clockwise or counterclockwise around a rotation axis that is perpendicular to the plane. Thus, the two equations for the net force components are

$$F_{\text{net},x} = \sum_{i=1}^{n} F_{i,x} = F_{1,x} + F_{2,x} + \cdots + F_{n,x} = 0 \tag{11.5}$$

$$F_{\text{net},y} = \sum_{i=1}^{n} F_{i,y} = F_{1,y} + F_{2,y} + \cdots + F_{n,y} = 0. \tag{11.6}$$

In Chapter 10, the net torque about a fixed axis of rotation was defined as the difference between the sum of the counterclockwise torques and the sum of the clockwise torques. The static equilibrium condition of zero net torque about each axis of rotation can thus be written as

$$\tau_{\text{net}} = \sum_{i} \tau_{\text{counterclockwise},i} - \sum_{j} \tau_{\text{clockwise},j} = 0. \tag{11.7}$$

These three equations (11.5 through 11.7) form the basis for the quantitative analysis of static equilibrium in the problems in this chapter.

11.2 Examples Involving Static Equilibrium

The two conditions for static equilibrium (zero net force and zero net torque) are all we need to solve a very large class of problems involving static equilibrium. We do not need calculus to solve these problems; all the calculations use only algebra and trigonometry. Let's start with an example for which the answer seems obvious. This will provide practice with the method and show that it leads to the right answer.

EXAMPLE 11.1 | Seesaw

A playground seesaw consists of a pivot and a bar, of mass M, that is placed on the pivot so that the ends can move up and down freely (Figure 11.6a). If an object of mass m_1 is placed on one end of the bar at a distance r_1 from the pivot point, as shown in Figure 11.6b, that end goes down, simply because of the force and torque that the object exerts on it.

PROBLEM 1

Where do we have to place an object of mass m_2 (assumed to be equal to the mass of m_1) to get the seesaw to balance, so the bar is horizontal and neither end touches the ground?

SOLUTION 1

Figure 11.6b is a free-body diagram of the bar showing the forces acting on it and the points where they act. The force that m_1 exerts on the bar is simply $m_1 g$, acting downward as shown in Figure 11.6b. The same is true for the force that m_2 exerts on the bar. In addition, because the bar has a mass M of its own, it experiences a gravitational force, Mg. The gravitational force acts at the center of mass of the bar, right in the middle of the bar. The final force acting on the bar is the normal force, N, exerted by the bar's support. It acts exactly at the axle of the seesaw (marked with an orange dot).

The equilibrium equation for the y-components of the forces leads to an expression for the value of the normal force:

$$F_{\text{net},y} = \sum_i F_{i,y} = -m_1 g - m_2 g - Mg + N = 0$$

$$\Rightarrow N = g(m_1 + m_2 + M).$$

The signs in front of the individual force components indicate whether they act upward (positive) or downward (negative).

Because all forces act in the y-direction, it is not necessary to write equations for the net force components in the x- or z-directions.

Now we can consider the net torque. The selection of the proper pivot point can make our computations simple. For a seesaw, the natural selection is at the axle, the point marked with an orange dot in the center of the bar in Figure 11.6b. Because the normal force, N, and the weight of the bar, Mg, act exactly through this point, their moment arms have length zero. Thus, these two forces do not contribute to the torque equation if this is selected as the pivot point. The forces $F_1 = m_1 g$ and $F_2 = m_2 g$ are the only ones contributing torques: F_1 generates a counterclockwise torque, and F_2 a clockwise torque. The torque equation is then

$$\tau_{\text{net}} = \sum_i \tau_{\text{counterclockwise},i} - \sum_j \tau_{\text{clockwise},j}$$

$$= m_1 g r_1 \sin 90° - m_2 g r_2 \sin 90° = 0$$

$$\Rightarrow m_2 r_2 = m_1 r_1$$

$$\Rightarrow r_2 = r_1 \frac{m_1}{m_2}. \tag{i}$$

Even though they equal 1 and thus have no effect, the factors $\sin 90°$ are included above as a reminder that the angle between force and moment arm usually affects the calculation of the torques.

The question was where to put m_2 for the case that the two masses were the same; the answer is $r_2 = r_1$ in this case. This expected result shows that our systematic way of approaching the solution works in this easily verifiable case.

PROBLEM 2

How big does m_2 need to be to balance m_1 if $r_1 = 3r_2$, that is, if m_2 is three times closer to the pivot point than m_1?

SOLUTION 2

We use the same free-body diagram (Figure 11.6b) and arrive at the same general equation for the masses and distances. Solving equation (i) for m_2 gives

$$m_2 r_2 = m_1 r_1$$

$$\Rightarrow m_2 = m_1 \frac{r_1}{r_2}.$$

Using $r_1 = 3r_2$, we obtain

$$m_2 = m_1 \frac{r_1}{r_2} = m_1 \frac{3r_2}{r_2} = 3m_1.$$

For this case, we find that the mass of m_2 has to be three times that of m_1 to establish static equilibrium.

(a)

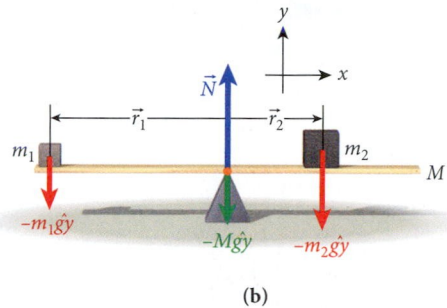

(b)

FIGURE 11.6 (a) A playground seesaw; (b) free-body diagram showing forces and moment arms.

As Example 11.1 shows, a clever choice of a pivot point can often greatly simplify a solution. It is important, however, to realize that one can use *any* pivot point. If the torques are balanced about any pivot point, they are balanced about *all* pivot points. Thus, if we change the pivot point, it can make the calculations more complicated in certain situations, but the end result of the calculation will not change.

Self-Test Opportunity 11.1

Suppose that the pivot point for the seesaw in Problem 1 of Example 11.1 is placed instead below the center of mass of m_2. Show that this leads to the same result.

EXAMPLE 11.2 | **Force on Biceps**

Suppose you are holding a barbell in your hand, as shown in Figure 11.7a. Your biceps supports your forearm. The biceps is attached to the bone of the forearm at a distance $r_b = 2.0$ cm from the elbow, as shown in Figure 11.7b. The mass of your forearm is 0.85 kg. The length of your forearm is 31 cm. Your forearm makes an angle $\theta = 75°$ with the vertical, as shown in Figure 11.7b. The barbell has a mass of 15 kg.

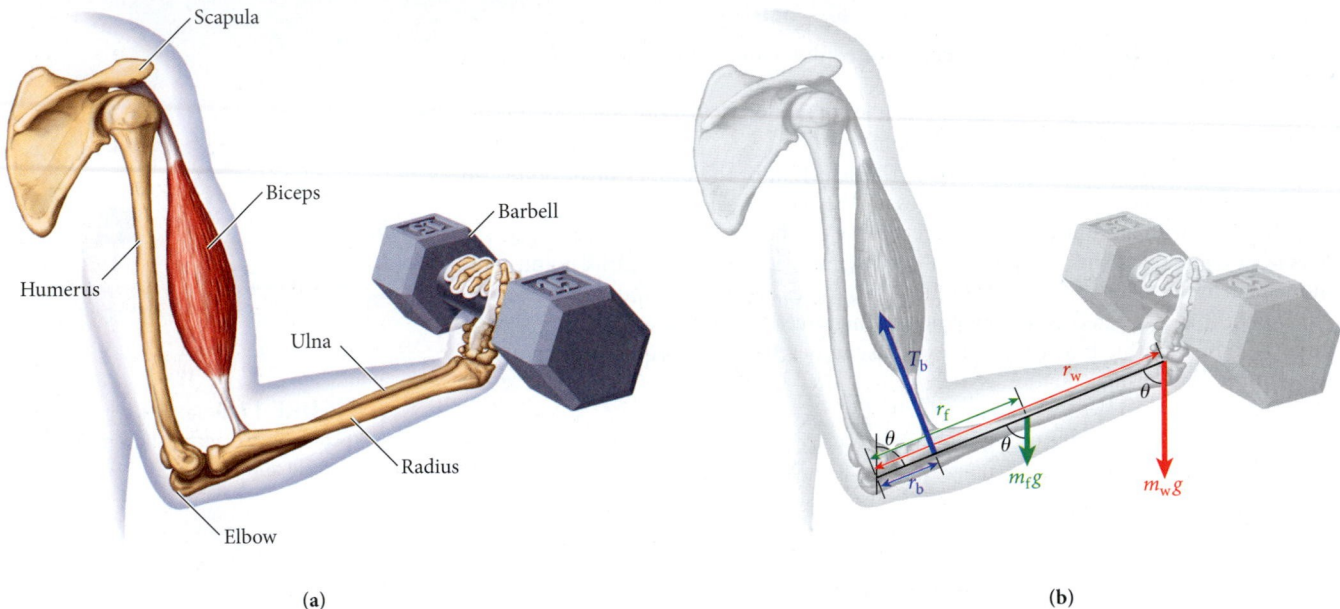

(a)   **(b)**

FIGURE 11.7 (a) A human arm holding a barbell. (b) Forces and moment arms for a human arm holding a barbell.

PROBLEM

What is the force that the biceps must exert to hold up your forearm and the barbell? Assume that the biceps exerts a force perpendicular to the forearm at the point of attachment.

SOLUTION

The pivot point is the elbow. The net torque on your forearm must be zero, so the counterclockwise torque must equal the clockwise torque:

$$\sum_i \tau_{\text{counterclockwise},i} = \sum_j \tau_{\text{clockwise},j}.$$

The counterclockwise torque is provided by the biceps:

$$\sum_i \tau_{\text{counterclockwise},i} = T_b r_b \sin 90° = T_b r_b,$$

where T_b is the force exerted by the biceps and r_b is the moment arm for the force exerted by the biceps. The clockwise torque is the sum of the torque exerted by the weight of the forearm and the torque exerted by the barbell:

$$\sum_j \tau_{\text{clockwise},j} = m_f g r_f \sin\theta + m_w g r_w \sin\theta,$$

where m_f is the mass of the forearm, r_f is the moment arm for the force exerted by the weight of the forearm, m_w is the mass of the barbell, and r_w is the moment arm of the force exerted by the weight of the barbell. We take the r_w to be equal to the length of the forearm and r_f to be half of that length, or $r_w/2$. Equating the counterclockwise torque and the clockwise torque gives us

$$T_b r_b = m_f g r_f \sin\theta + m_w g r_w \sin\theta.$$

Solving for the force exerted by the biceps, we obtain

$$T_b = \frac{m_f g r_f \sin\theta + m_w g r_w \sin\theta}{r_b} = g\sin\theta\left(\frac{m_f r_f + m_w r_w}{r_b}\right).$$

Putting in the given numbers gives the force exerted by the biceps

$$T_b = g\sin\theta\left(\frac{m_f r_f + m_w r_w}{r_b}\right)$$

$$= \left(9.81 \text{ m/s}^2\right)\left(\sin 75°\right)\left(\frac{(0.85 \text{ kg})\left(\frac{0.31 \text{ m}}{2}\right) + (15 \text{ kg})(0.31 \text{ m})}{0.020 \text{ m}}\right)$$

$$= 2300 \text{ N}.$$

You may wonder why evolution gave the biceps such a huge mechanical disadvantage. Apparently, it was more advantageous to be able to swing the arms a long distance (and therefore moving the hands with greater speed!) while the biceps exerted a comparatively huge force than for the biceps to be able to exert a smaller force and move the arms a shorter distance. This is in contrast, by the way, to the jaw muscles, which have evolved the ability to crunch tough food with huge force by attaching to the jaw farther from the pivot point and closer to the point where the force is applied by the teeth.

Self-Test Opportunity 11.2

In Example 11.2, suppose you hold the barbell so that your forearm makes an angle of 180° with the vertical. Why is it that you can still lift the barbell?

The following example for static equilibrium also applies the formulas for computing the center of mass that were introduced in Chapter 8, and at the same time has a very surprising outcome.

EXAMPLE 11.3 / Stacking Blocks

PROBLEM
Consider a collection of identical blocks stacked at the edge of a table (Figure 11.8). How far out can we push the leading edge of the top block without the pile falling off?

SOLUTION
Let's start with one block. If the block has length ℓ and uniform mass density, then its center of mass is located at $\frac{1}{2}\ell$. Clearly, it can stay at rest as long as at least half of it is on the table, with its center of mass supported by the table from below. The block can stick out an infinitesimal amount less than $\frac{1}{2}\ell$ beyond the support, and it will remain at rest.

Next, we consider two identical blocks. If we call the x-coordinate of the center of mass of the upper block x_1 and that of the lower block x_2, we obtain for the x-coordinate of the center of mass of the combined system, according to Section 8.1,

$$x_{12} = \frac{x_1 m_1 + x_2 m_2}{m_1 + m_2}.$$

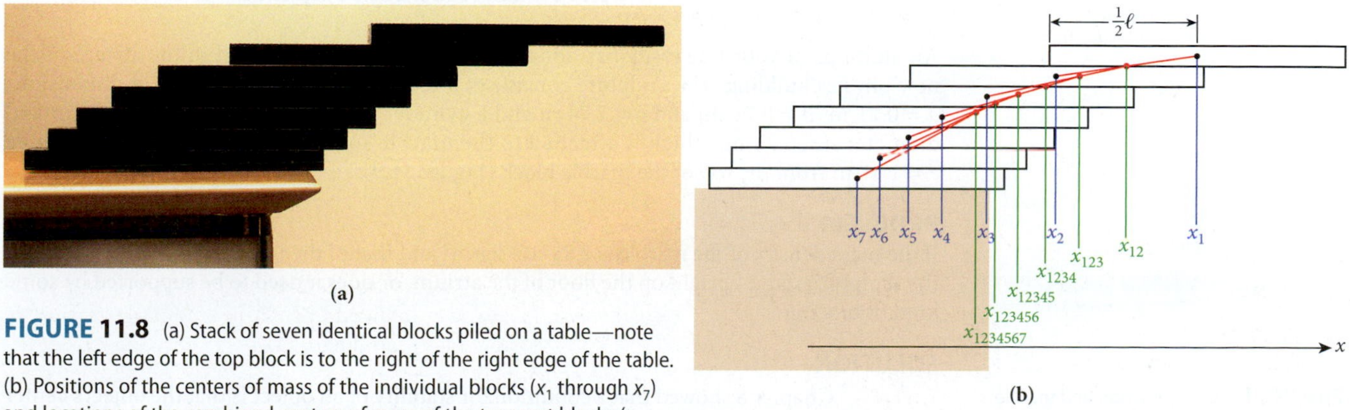

FIGURE 11.8 (a) Stack of seven identical blocks piled on a table—note that the left edge of the top block is to the right of the right edge of the table. (b) Positions of the centers of mass of the individual blocks (x_1 through x_7) and locations of the combined centers of mass of the topmost blocks (x_{12} through $x_{1234567}$).

– *Continued*

For identical blocks, $m_1 = m_2$, which simplifies the expression for x_{12} to

$$x_{12} = \tfrac{1}{2}(x_1 + x_2).$$

Since $x_1 = x_2 + \tfrac{1}{2}\ell$ in the limiting case that the center of mass of the first block is still supported from below by the second block, we obtain

$$x_{12} = \tfrac{1}{2}(x_1 + x_2) = \tfrac{1}{2}((x_2 + \tfrac{1}{2}\ell) + x_2) = x_2 + \tfrac{1}{4}\ell. \tag{i}$$

Now we can go to three blocks. The top two blocks will not topple if the combined center of mass, x_{12}, is supported from below. Shifting x_{12} to the very edge of the third block, we obtain $x_{12} = x_3 + \tfrac{1}{2}\ell$. Combining this with equation (i), we have

$$x_{12} = x_2 + \tfrac{1}{4}\ell = x_3 + \tfrac{1}{2}\ell \Rightarrow x_2 = x_3 + \tfrac{1}{4}\ell.$$

Note that equation (i) is still valid after the shift because we have expressed x_{12} in terms of x_2 and because x_{12} and x_2 change by the same amount when the two blocks move together. We can now calculate the center of mass for the three blocks in the same way as before, by applying the same principle to find the new combined center of mass:

$$x_{123} = \frac{x_{12}(2m) + x_3 m}{2m + m} = \tfrac{2}{3}x_{12} + \tfrac{1}{3}x_3 = \tfrac{2}{3}(x_3 + \tfrac{1}{2}\ell) + \tfrac{1}{3}x_3 = x_3 + \tfrac{1}{3}\ell. \tag{ii}$$

Requiring that the top three blocks are supported by the fourth block from below results in $x_{123} = x_4 + \tfrac{1}{2}\ell$. Combined with equation (ii), this establishes

$$x_{123} = x_3 + \tfrac{1}{3}\ell = x_4 + \tfrac{1}{2}\ell \Rightarrow x_3 = x_4 + \tfrac{1}{6}\ell.$$

You can see how this series continues. If we have $n - 1$ blocks supported in this way by the nth block, then the coordinates of the $(n-1)$st and nth block are related as follows:

$$x_{n-1} = x_n + \frac{\ell}{2n - 2}.$$

We can now add up all the terms and find out how far away x_1 can be from the edge:

$$x_1 = x_2 + \tfrac{1}{2}\ell = x_3 + \tfrac{1}{4}\ell + \tfrac{1}{2}\ell = x_4 + \tfrac{1}{6}\ell + \tfrac{1}{4}\ell + \tfrac{1}{2}\ell = \cdots = x_{n+1} + \tfrac{1}{2}\ell \left(\sum_{i=1}^{n} \frac{1}{i} \right).$$

Self-Test Opportunity 11.3

Suppose you had 10,000 identical blocks of height 4.0 cm and length 15.0 cm. If you arranged them in the way demonstrated in Example 11.3, how far would the right edge of the top block stick out?

You may remember from calculus that the sum $\sum_{i=1}^{n} i^{-1}$ does not converge, that is, does not have an upper limit for $n \to \infty$. This gives the astonishing result that x_1 can move *infinitely* far away from the table's edge, provided there are enough blocks under it and that the edge of the table can support their weight without significant deformation! (But see Self-Test Opportunity 11.3 to put this *infinity* into perspective.) Figure 11.8a shows only 7 blocks stacked on a table, and the left edge of the top block is already to the right of the right edge of the table.

The following problem, involving a composite extended object, serves to review the calculation of the center of mass of such objects, which was introduced in Chapter 8.

SOLVED PROBLEM 11.1 | An Abstract Sculpture

An alumnus of your university has donated a sculpture to be displayed in the atrium of the new physics building. The sculpture consists of a rectangular block of marble of dimensions $a = 0.71$ m, $b = 0.71$ m, and $c = 2.74$ m and a cylinder of wood with length $\ell = 2.84$ m and diameter $d = 0.71$ m, which is attached to the marble so that its upper edge is a distance $e = 1.47$ m from the top of the marble block (Figure 11.9).

PROBLEM

If the mass density of the marble is $2.85 \cdot 10^3$ kg/m^3 and that of the wood is $4.40 \cdot 10^2$ kg/m^3, can the sculpture stand upright on the floor of the atrium, or does it need to be supported by some kind of bracing?

SOLUTION

THINK Chapter 8 showed that a condition for stability of an object is that the object's center of mass needs to be directly supported from below. In order to decide if the sculpture can stand upright without additional support, we therefore need to determine the location of the center

(a)

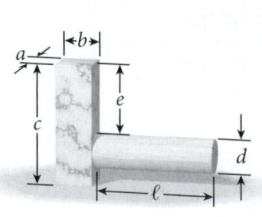

(b)

FIGURE 11.9 (a) A wood and marble sculpture; (b) diagram of the sculpture with dimensions labeled.

of mass of the sculpture and find out if it is located at a point inside the marble block. Since the floor supports the block from below, the sculpture will be able to stand upright if this is the case. If the center of mass of the sculpture is located outside the marble block, the sculpture will need bracing.

SKETCH   We draw a side view of the sculpture (Figure 11.10), representing the marble block and the wooden cylinder by rectangles. The sketch also indicates a coordinate system with a horizontal x-axis and a vertical y-axis and the origin at the right edge of the marble block. (What about the z-coordinate? We can use symmetry arguments in the same way as we did in Chapter 8 and find that the z-component of the center of mass is located in the plane that divides block and cylinder into halves.)

RESEARCH   With the chosen coordinate system, we also do not need to worry about calculating the y-coordinate of the center of mass of the sculpture. The condition for stability depends only on whether the x-coordinate of the center of mass is within the marble block; it does not depend on how high the center of mass is off the ground. Thus, the only task left is to calculate the x-component of the center of mass. According to the general principles for calculating center-of-mass coordinates, developed in Chapter 8, we can write

$$X = \frac{1}{M}\int_V x\rho(\vec{r})dV, \tag{i}$$

where M is the mass of the entire sculpture and V is its volume. Note that in this case the mass density is not homogeneous, since the marble and the wood have different densities.

In order to perform the integration, we split the volume V into convenient parts: $V = V_1 + V_2$, where V_1 is the volume of the marble block and V_2 is the volume of the wooden cylinder. Then equation (i) becomes

$$X = \frac{1}{M}\int_{V_1} x\rho(\vec{r})dV + \frac{1}{M}\int_{V_2} x\rho(\vec{r})dV$$

$$= \frac{1}{M}\int_{V_1} x\rho_1\, dV + \frac{1}{M}\int_{V_2} x\rho_2\, dV.$$

SIMPLIFY   To calculate the location of the x-component of the center of mass of the marble block alone, we can use the equation for constant density from Chapter 8:

$$X_1 = \frac{1}{M_1}\int_{V_1} x\rho_1\, dV = \frac{\rho_1}{M_1}\int_{V_1} x\, dV.$$

(Since the density is constant over this entire volume, we can move it out of the integral.) In the same way, we can find the x-component of the center of mass of the wooden cylinder:

$$X_2 = \frac{\rho_2}{M_2}\int_{V_2} x\, dV.$$

Therefore, the expression for the center of mass of the composite object—that is, the entire sculpture—is

$$X = \frac{\rho_1}{M}\int_{V_1} x\, dV + \frac{\rho_2}{M}\int_{V_2} x\, dV$$

$$= \frac{M_1}{M}\frac{\rho_1}{M_1}\int_{V_1} x\, dV + \frac{M_2}{M}\frac{\rho_2}{M_2}\int_{V_2} x\, dV \tag{ii}$$

$$= \frac{M_1}{M}X_1 + \frac{M_2}{M}X_2.$$

This is a very important general result: Even for extended objects, the combined center of mass can be calculated in the same way as it is for point particles. Since the total mass of the sculpture is the combined mass of its two parts, $M = M_1 + M_2$, equation (ii) becomes

$$X = \frac{M_1}{M_1 + M_2}X_1 + \frac{M_2}{M_1 + M_2}X_2. \tag{iii}$$

It is important to note that this relationship between the coordinate of combined center of mass of a composite object and the individual center-of-mass coordinates is true even in the

– Continued

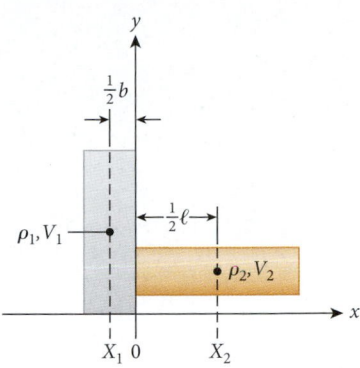

FIGURE 11.10 Sketch for calculating the center of mass of the sculpture.

case where the parts are separate objects. Further, it holds in the case when the density inside a given object is not constant. Formally, this result relies on the fact that a volume integral can always be split into a set of integrals over disjoint subvolumes that add up to the whole. That is, integration is linear, as it is merely addition.

Deriving equation (iii) has simplified a complicated problem greatly, because the center-of-mass coordinates of the two individual objects can be calculated easily. Since the density of each of them is constant, their center-of-mass locations are identical to their geometrical centers. One look at Figure 11.10 is enough to convince us that $X_2 = \frac{1}{2}\ell$ and $X_1 = -\frac{1}{2}b$. (Remember, we chose the origin of the coordinate system as the right edge of the marble block.)

All that is left now is to calculate the masses of the two objects. Since we know their densities, we only need to figure out each object's volume; then the mass is given by $M = \rho V$.

Since the marble block is rectangular, its volume is $V = abc$. Thus, we have

$$M_1 = \rho_1 abc.$$

The horizontal wooden part is a cylinder, so its mass is

$$M_2 = \frac{\rho_2 \ell \pi d^2}{4}.$$

CALCULATE Inserting the numbers given in the problem statement, we obtain for the individual masses

$$M_1 = (2850 \text{ kg/m}^3)(0.71 \text{ m})(0.71 \text{ m})(2.74 \text{ m}) = 3936.52 \text{ kg}$$

$$M_2 = \frac{(440 \text{ kg/m}^3)(2.84 \text{ m})\pi(0.71 \text{ m})^2}{4} = 494.741 \text{ kg}.$$

Thus, the combined mass is

$$M = M_1 + M_2 = 3936.52 \text{ kg} + 494.741 \text{ kg} = 4431.261 \text{ kg}.$$

The location of the x-component of the center of mass of the sculpture is then

$$X = \frac{3936.52 \text{ kg}}{4431.261 \text{ kg}}\left[(-0.5)(0.71 \text{ m})\right] + \frac{494.741 \text{ kg}}{4431.261 \text{ kg}}\left[(0.5)(2.84 \text{ m})\right] = -0.156825 \text{ m}.$$

ROUND The densities were given to three significant figures, but the length dimensions were given to only two significant figures. Rounding thus results in

$$X = -0.16 \text{ m}.$$

Since this number is negative, the center of mass of the sculpture is located to the left of the right edge of the marble block. Thus, it is located above the base of the block and is supported from directly below. The sculpture is stable and can stand without bracing.

DOUBLE-CHECK From Figure 11.9a, it seems hardly possible that this sculpture wouldn't tip over. However, our eyes can deceive us, as the density of the sculpture isn't constant. The ratio of the densities of the two materials used in the sculpture is $\rho_1/\rho_2 = (2850 \text{ kg/m}^3)/(440 \text{ kg/m}^3) = 6.48$. Therefore, we would obtain the same location for the center of mass if the 2.84-m-long cylinder made of wood were replaced by a cylinder of the same length made of marble, with its central axis located at the same place as that of the wooden cylinder, but thinner by a factor of $\sqrt{6.48} = 2.55$. The sculpture shown in Figure 11.11 has the same center-of-mass location as the sculpture in Figure 11.9a. Figure 11.11 should convince you that the sculpture is able to stand in stable equilibrium without bracing.

FIGURE 11.11 Sculpture made entirely of marble that has the same center-of-mass location as the marble-and-wood sculpture in Figure 11.9a.

Concept Check 11.1

In Solved Problem 11.1, the total mass of the sculpture was M, and the x-component of the sculpture's center of mass was found to be X. If a point mass, m_3, were placed at the far right end of the wooden cylinder ($X_3 = \ell$), which of the following would express the criterion for the largest magnitude m_3 could have without making the sculpture unstable?

a) $\dfrac{M + m_3}{X + X_3} = \ell$

b) $\dfrac{MX + m_3 X_3}{M - m_3} = \ell/2$

c) $\dfrac{M + m_3}{X + X_3} = 0$

d) $\dfrac{MX + m_3 X_3}{M + m_3} = 0$

e) $\dfrac{MX - m_3 X_3}{X - X_3} = 0$

The next example considers a situation in which the force of static friction plays an essential role. Static friction forces help maintain many arrangements of objects in equilibrium.

EXAMPLE 11.4 Person Standing on a Ladder

Typically, a ladder stands on a horizontal surface (the floor) and leans against a vertical surface (the wall). Suppose a ladder of length $\ell = 3.04$ m, with mass $m_l = 13.3$ kg, rests against a smooth wall at an angle of $\theta = 24.8°$. A student, who has a mass of $m_m = 62.0$ kg, stands on the ladder (Figure 11.12a). The student is standing on a rung that is $r = 1.43$ m along the ladder, measured from where the ladder touches the ground.

PROBLEM 1

What friction force must act on the bottom of the ladder to keep it from slipping? Neglect the (small) force of friction between the smooth wall and the ladder.

SOLUTION 1

Let's start with the free-body diagram shown in Figure 11.12c. Here $\vec{R} = -R\hat{x}$ is the normal force exerted by the wall on the ladder, $\vec{N} = N\hat{y}$ is the normal force exerted by the floor on the ladder, and $\vec{W}_m = -m_m g\hat{y}$ and $\vec{W}_l = -m_l g\hat{y}$ are the weights of the student and the ladder: $m_m g = (62.0 \text{ kg})(9.81 \text{ m/s}^2) = 608. \text{ N}$ and $m_l g = (13.3 \text{ kg})(9.81 \text{ m/s}^2) = 130. \text{ N}$.

We'll let $\vec{f}_s = f_s\hat{x}$ be the force of static friction between the floor and the bottom of the ladder, which is the answer to the problem. Note that this force vector is directed in the positive x-direction (if the ladder slips, its bottom will slide in the negative x-direction, and the friction force must necessarily oppose that motion). As instructed, we neglect the force of friction between wall and ladder.

The ladder and the student are in translational and rotational equilibrium, so we have the three equilibrium conditions introduced in equations 11.5 through 11.7:

$$\sum_i F_{x,i} = 0, \quad \sum_i F_{y,i} = 0, \quad \sum_i \tau_i = 0.$$

Let's start with the equation for the force components in the horizontal direction:

$$\sum_i F_{x,i} = f_s - R = 0 \Rightarrow R = f_s.$$

From this equation, we learn that the force the wall exerts on the ladder and the friction force between the ladder and the floor have the same magnitude. Next, we write the equation for the force components in the vertical direction:

$$\sum_i F_{y,i} = N - m_m g - m_l g = 0 \Rightarrow N = g(m_m + m_l).$$

The normal force that the floor exerts on the ladder is exactly equal in magnitude to the sum of the weights of ladder and man: $N = 608. \text{ N} + 130. \text{ N} = 738. \text{ N}$. (Again, we neglect the friction force between wall and ladder, which would otherwise have come in here.)

Now we sum the torques, assuming that the pivot point is where the ladder touches the ground. This assumption has the advantage of allowing us to ignore the forces acting at that point, because their moment arms are zero.

$$\sum_i \tau_i = (m_l g)\left(\frac{\ell}{2}\right)\sin\theta + (m_m g)r\sin\theta - R\ell\cos\theta = 0. \tag{i}$$

Note that the torque from the wall's normal force acts counterclockwise, whereas the two torques from the weights of the student and the ladder act clockwise. Also, the angle between the normal force, $\vec{R}$, and its moment arm, $\vec{\ell}$, is $90° - \theta$, and $\sin(90° - \theta) = \cos\theta$. Now we solve equation (i) for R:

$$R = \frac{\frac{1}{2}(m_l g)\ell\sin\theta + (m_m g)r\sin\theta}{\ell\cos\theta} = \left(\frac{1}{2}m_l g + m_m g\frac{r}{\ell}\right)\tan\theta.$$

Numerically, we obtain

$$R = \left(\frac{1}{2}(130. \text{ N}) + (608. \text{ N})\frac{1.43 \text{ m}}{3.04 \text{ m}}\right)(\tan 24.8°) = 162. \text{ N}.$$

However, we already found that $R = f_s$, so our answer is $f_s = 162. \text{ N}$.

PROBLEM 2

Suppose that the coefficient of static friction between ladder and floor is 0.31. Will the ladder slip?

SOLUTION 2

We found the normal force in the first part of this example: $N = g(m_m + m_l) = 738. \text{ N}$. It is related to the maximum static friction force via $f_{s,\text{max}} = \mu_s N$. So, the maximum static friction

– Continued

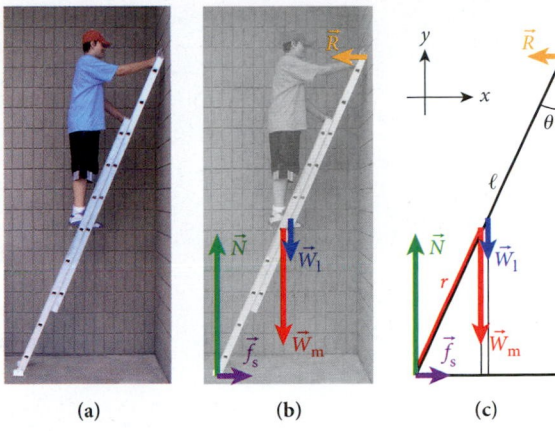

FIGURE 11.12 (a) Student standing on a ladder. (b) Force vectors superimposed. (c) Free-body diagram of the ladder.

Concept Check 11.2

What can the student in Example 11.4 do if he really has to get up just a bit higher than the maximum height allowed by equation (ii) for the given situation?

a) He can increase the angle θ between the wall and the ladder.

b) He can decrease the angle θ between the wall and the ladder.

c) Neither increasing nor decreasing the angle will make any difference.

force is 229. N, well above the 162. N that we just found necessary for static equilibrium. In other words, the ladder will not slip.

In general, the ladder will not slip as long as the force from the wall is smaller than the maximum force of static friction, leading to the condition

$$R = \left(\tfrac{1}{2}m_1 + m_m \frac{r}{\ell}\right)g\tan\theta \le \mu_s(m_1 + m_m)g. \tag{ii}$$

PROBLEM 3
In which direction does the force that the ground exerts on the ladder point?

SOLUTION 3
Figure 11.12 shows the horizontal ($\vec{f_s}$) and vertical ($\vec{N}$) components of the force that the ground exerts on the ladder. Calling the angle between this force and the ground φ, we have

$$\varphi = \tan^{-1}\left(\frac{F_y}{F_x}\right) = \tan^{-1}\left(\frac{N}{f_s}\right) = \tan^{-1}\left(\frac{738.}{162.}\right) = 77.6°.$$

The problem statement says that the ladder makes an angle of 24.8° with the wall. Therefore, it makes an angle of 90° – 24.8° = 65.2° with the ground. Note that this is not the same as φ; the force exerted by the ground on a ladder in a situation like this does *not*, in general, point along the ladder!

PROBLEM 4
What happens as the student climbs higher on the ladder?

SOLUTION 4
From equation (ii), we see that R grows larger with increasing r. Eventually, this force will overcome the maximum force of static friction, and the ladder will slip. You can now understand why it is not a good idea to climb too high on a ladder in this kind of situation.

Concept Check 11.3

The absolute value of the force exerted on the ladder by the floor is given by

a) N.

b) f_s.

c) $N + f_s$.

d) $N - f_s$.

e) $\sqrt{N^2 + f_s^2}$.

SOLVED PROBLEM 11.2 | Hanging a Storefront Sign

It is not uncommon for businesses to hang a sign over the sidewalk, suspended from a building's front wall. They often attach a post to the wall by a hinge and hold it horizontal by a cable that is also attached to the wall: the sign is then suspended from the post. Suppose the mass of the sign in Figure 11.13a is M = 33.1 kg, and the mass of the post is m = 19.7 kg. The length of the post is l = 2.40 m, and the sign is attached to the post as shown at a distance r = 1.95 m from the wall. The cable is attached to the wall a distance d = 1.14 m above the post.

PROBLEM
What is the tension in the cable holding the post? What is the magnitude and direction of the force, $\vec{F}$, that the wall exerts on the post?

SOLUTION
THINK This problem involves static equilibrium, moment arms, and torques. Static equilibrium means vanishing net external force and torque. In order to determine the torques, we

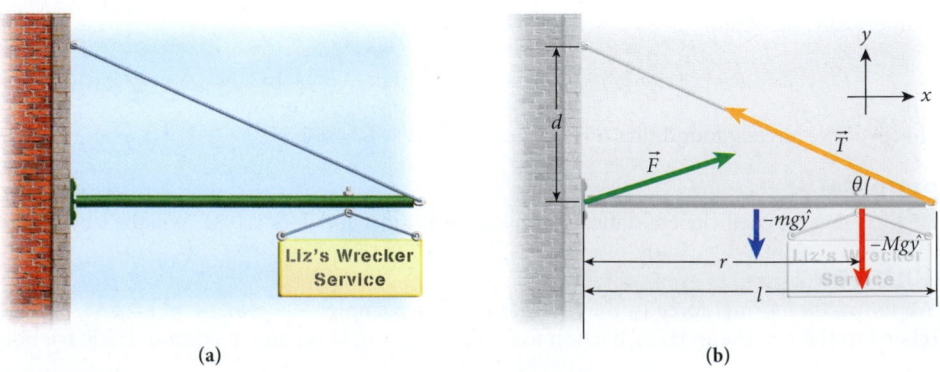

(a)   (b)

FIGURE 11.13 (a) Hanging storefront sign; (b) free-body diagram for the post.

have to pick a pivot point. It seems natural to pick the point where the hinge attaches the post to the wall. Because a hinge is used, the post can rotate about this point. Picking this point also has the advantage that we do not need to pay attention to the force that the wall exerts on the post, because that force will act at the contact point (the hinge) and thus have a moment arm of zero and consequently make no contribution to the torque.

SKETCH   To calculate the net torque, we start with a free-body diagram that shows all the forces acting on the post (Figure 11.13b). We know that the weight of the sign (red arrow) acts at the point where the sign is suspended from the post. The gravitational force acting on the post is represented by the blue arrow, which points downward from the center of mass of the post. Finally, we know that the tension, $\vec{T}$ (yellow arrow), is acting along the direction of the cable.

RESEARCH   The angle θ between the cable and the post (see Figure 11.13b) can be found from the given data:

$$\theta = \tan^{-1}\left(\frac{d}{l}\right).$$

The equation for the torques about the point where the post touches the wall is then

$$mg\frac{l}{2}\sin 90° + Mgr\sin 90° - Tl\sin\theta = 0. \qquad (i)$$

Figure 11.13b also shows a green arrow for the force, $\vec{F}$, that the wall exerts on the post, but the direction and magnitude of this force vector are still to be determined. Equation (i) cannot be used to find this force, because the point where this force acts is the pivot point, and thus the corresponding moment arm has length zero.

On the other hand, once we have found the tension in the cable, we will have determined all the other forces involved in the problem situation, and we know from the condition of static equilibrium that the net force has to be zero. We can thus write separate equations for the horizontal and vertical force components. In the horizontal direction, we have only two force components, from the tension and the force from the wall:

$$F_x - T\cos\theta = 0 \Rightarrow F_x = T\cos\theta.$$

In the vertical direction, we have the weights of the beam and the sign, in addition to the vertical components of the tension and the force from the wall:

$$F_y + T\sin\theta - mg - Mg = 0 \Rightarrow F_y = (m+M)g - T\sin\theta.$$

SIMPLIFY   We solve equation (i) for the tension:

$$T = \frac{(ml + 2Mr)g}{2l\sin\theta}. \qquad (ii)$$

For the magnitude of the force that the wall exerts on the post, we find

$$F = \sqrt{F_x^2 + F_y^2}.$$

The direction of this force is given by

$$\theta_F = \tan^{-1}\left(\frac{F_y}{F_x}\right).$$

CALCULATE   Inserting the numbers given in the problem statement, we find the angle θ:

$$\theta = \tan^{-1}\left(\frac{1.14 \text{ m}}{2.40 \text{ m}}\right) = 25.4077°.$$

We obtain the tension in the cable from equation (ii):

$$T = \frac{\left[(19.7 \text{ kg})(2.40 \text{ m}) + 2(33.1 \text{ kg})(1.95 \text{ m})\right](9.81 \text{ m/s}^2)}{2(2.40 \text{ m})(\sin 25.4077°)} = 840.113 \text{ N}.$$

The magnitudes of the components of the force that the wall exerts on the post are

$$F_x = (840.113 \text{ N})(\cos 25.4077°) = 758.855 \text{ N}$$

$$F_y = (19.7 \text{ kg} + 33.1 \text{ kg})(9.81 \text{ m/s}^2) - (840.113 \text{ N})(\sin 25.4077°) = 157.512 \text{ N}.$$

– Continued

Concept Check 11.4

If all other parameters in Solved Problem 11.2 remain unchanged, but the sign is moved farther away from the wall, toward the end of the post, what happens to the tension, T?

a) It decreases.

b) It stays the same.

c) It increases.

Concept Check 11.5

If all other parameters in Solved Problem 11.2 remain unchanged, but the angle θ between the cable and the post is increased, what happens to the tension, T?

a) It decreases.

b) It stays the same.

c) It increases.

Therefore, the magnitude and direction of that force are given by

$$F = \sqrt{(157.512 \text{ N})^2 + (758.855 \text{ N})^2} = 775.030 \text{ N}$$

$$\theta_F = \tan^{-1}\left(\frac{157.512}{785.855}\right) = 11.726°.$$

ROUND All of the given quantities were specified to three significant figures, and so we round our final answers to three digits: $T = 840.$ N, $F = 775.$ N, and $\theta_F = 11.7°$.

DOUBLE-CHECK The two forces that we calculated have rather large magnitudes, considering that the combined weight of the post and attached sign is only

$$F_g = (m + M)g = (19.7 \text{ kg} + 33.1 \text{ kg})(9.81 \text{ m/s}^2) = 518. \text{ N}.$$

In fact, the sum of the magnitudes of the force from the cable on the post, T, and the force from the wall on the post, F, is larger than the combined weight of the post and the sign by more than a factor of 3. Does this make sense? Yes, because the two force vectors, $\vec{T}$ and $\vec{F}$, have rather large horizontal components that have to cancel each other. When we calculate the magnitudes of these forces, their horizontal components are also included. As the angle θ between the cable and the post approaches zero, the horizontal components of $\vec{T}$ and $\vec{F}$ become larger and larger. Thus, you can see that selecting a distance d in Figure 11.13b that is too small relative to the length of the post will result in a huge tension in the cable and a very stressed suspension system.

11.3 Stability of Structures

For a skyscraper or a bridge, designers and builders need to worry about the ability of the structure to remain standing under the influence of external forces. For example, after standing for 40 years, the bridge carrying Interstate 35W across the Mississippi River in Minneapolis, shown in Figure 11.14, collapsed on August 1, 2007, probably from design-related causes. This bridge collapse and other architectural disasters are painful reminders that the stability of structures is a paramount concern.

Let's try to quantify the concept of stability by looking at Figure 11.15a, which shows a box in static equilibrium, resting on a horizontal surface. Our experience tells us that if we use a finger to push with a small force in the way shown in the figure, the box remains in the same position. The small force we exert on the box is exactly balanced by the force of friction between the box and the supporting surface. The net force is zero, and there is no motion. If we steadily increase the magnitude of the force we apply, there are two possible outcomes: If the friction force is not sufficient to counterbalance the force exerted by the finger, the box begins to slide to the right. Or, if the torque due to the weight of the box acting at its center of gravity is less than the torque due the applied force and the friction force, the box starts to tilt as shown in Figure 11.15b. Thus, the static equilibrium of the box is stable with respect to small external forces, but a sufficiently large external force destroys the equilibrium.

This simple example illustrates the characteristic of **stability.** Engineers need to be able to calculate the maximum external forces and torques that can be present without undermining the stability of a structure.

FIGURE 11.14 The bridge carrying Interstate 35W across the Mississippi River in Minneapolis collapsed on August 1, 2007, during rush hour.

Quantitative Condition for Stability

In order to be able to quantify the stability of an equilibrium situation, we start with the relationship between potential energy and force in one dimension from Chapter 6:

$$F_x(x) = -\frac{dU(x)}{dx}.$$

In three dimensions, this is

$$\vec{F}(\vec{r}) = -\vec{\nabla}U(\vec{r}),$$

where $\vec{\nabla}U(\vec{r}) = \left(\dfrac{\partial U(\vec{r})}{\partial x}\hat{x} + \dfrac{\partial U(\vec{r})}{\partial y}\hat{y} + \dfrac{\partial U(\vec{r})}{\partial z}\hat{z} \right)$ is the first gradient derivative of the potential energy function with respect to the position vector. A vanishing net force is one of the equilibrium conditions, which we can write as $\dfrac{dU(x)}{dx} = 0$ in one dimension or $\vec{\nabla}U(\vec{r}) = 0$ in three dimensions, at a given point in space. So far, the condition of vanishing first derivative adds no new insight. However, we can use the second derivative of the potential energy function to distinguish three different cases, depending on the sign of the second derivative.

Case 1 Stable Equilibrium

$$\text{Stable equilibrium: } \left.\frac{d^2U(x)}{dx^2}\right|_{x=x_0} > 0. \tag{11.8}$$

If the second derivative of the potential energy function with respect to the coordinate is positive at a point, then the potential energy has a local minimum at that point. The system is in **stable equilibrium**. In this case, a small deviation from the equilibrium position creates a restoring force that drives the system back to the equilibrium point. This situation is illustrated in Figure 11.16a: If the red dot is moved away from its equilibrium position at x_0 in either the positive or the negative direction and released, it will return to the equilibrium position.

Case 2 Unstable Equilibrium

$$\text{Unstable equilibrium: } \left.\frac{d^2U(x)}{dx^2}\right|_{x=x_0} < 0. \tag{11.9}$$

If the second derivative of the potential energy function with respect to the coordinate is negative at a point, then the potential energy has a local maximum at that point. The system is in **unstable equilibrium**. In this case, a small deviation from the equilibrium position creates a force that drives the system away from the equilibrium point. This situation is illustrated in Figure 11.16b: If the red dot is moved even slightly away from its equilibrium position at x_0 in either the positive or the negative direction and released, it will move away from the equilibrium position.

Case 3 Neutral Equilibrium

$$\text{Neutral equilibrium: } \left.\frac{d^2U(x)}{dx^2}\right|_{x=x_0} = 0. \tag{11.10}$$

The case in which the sign of the second derivative of the potential energy function with respect to the coordinate is neither positive nor negative at a point is called **neutral equilibrium**, also referred to as *indifferent* or *marginally stable*. This situation is illustrated in Figure 11.16c: If the red dot is displaced by a small amount, it will neither return to nor move away from its original equilibrium position. Instead, it will simply stay in the new position, which is also an equilibrium position.

Multidimensional Surfaces and Saddle Points

The three cases just discussed cover all possible types of stability for one-dimensional systems. They can be generalized to two- and three-dimensional potential energy functions that depend on more than one coordinate. Instead of looking at only the derivative with respect to one coordinate, as in equations 11.8 through 11.10, we have to examine all partial derivatives. For the two-dimensional potential energy function $U(x,y)$, the equilibrium condition is that the first derivative with respect to each of the two coordinates is zero. In addition, stable equilibrium requires that at the point of equilibrium the second derivative of the potential energy function is positive for both coordinates, whereas unstable equilibrium implies that it is negative for both, and neutral equilibrium means that it is zero for both. Parts (a) through (c) of Figure 11.17 show these three cases, respectively.

(a)

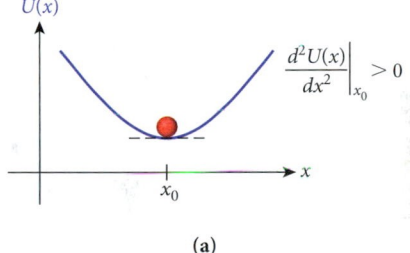

(b)

FIGURE 11.15 (a) Pushing with a small force against the upper edge of a box. (b) Exerting a larger force on the box results in tilting it.

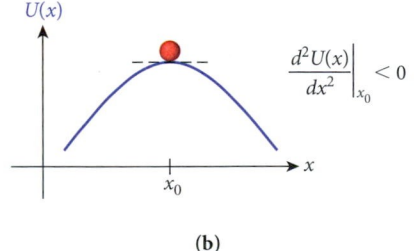

(a)

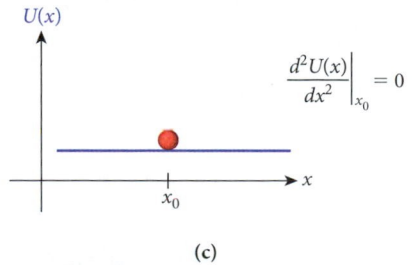

(b)

(c)

FIGURE 11.16 Local shape of the potential energy function at an equilibrium point: (a) stable equilibrium; (b) unstable equilibrium; (c) neutral equilibrium.

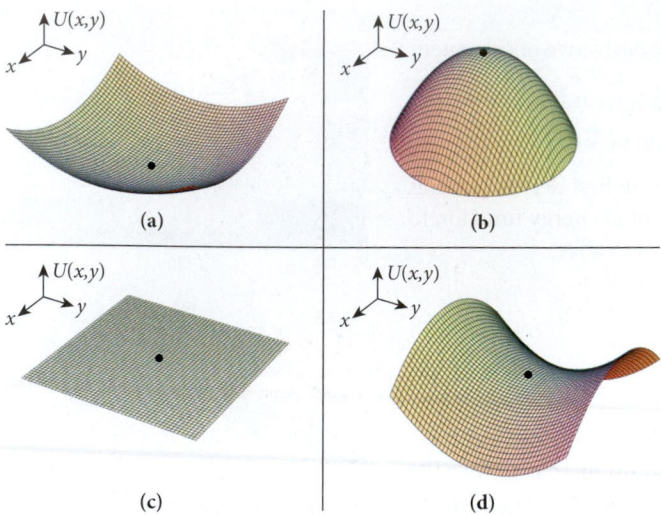

FIGURE 11.17 Different types of equilibrium for a three-dimensional potential energy function.

However, in more than one spatial dimension, the possibility also exists that at one equilibrium point the second derivative with respect to one coordinate is positive, whereas it is negative for the other coordinate. These points are called *saddle points,* because the potential energy function is locally shaped like a saddle. Figure 11.17d shows such a saddle point, where one of the second partial derivatives is negative and one is positive. The equilibrium at this saddle point is stable with respect to small displacements in the *y*-direction, but unstable with respect to small displacements in the *x*-direction.

In a strict mathematical sense, the conditions noted above for the second derivative are sufficient for the existence of maxima and minima, but not necessary. Sometimes, the first derivative of the potential energy function is not continuous, but extrema can still exist, as the following example shows.

Concept Check 11.6

In Figure 11.17, which surface—(a), (b), (c), or (d)—contains equilibrium points other than the one marked with the black dot?

EXAMPLE 11.5 / Pushing a Box

PROBLEM 1
What is the force required to hold the box in Figure 11.15 in equilibrium at a given tilt angle?

SOLUTION 1
Before the finger pushes on the box, the box is resting on a level surface. The only two forces acting on it are the force of gravity and a balancing normal force. There is no net force and no net torque; the box is in equilibrium (Figure 11.18a).

Once the finger starts pushing in the horizontal direction at the very top of the box's left side and the box starts tilting, the normal force vector acts at the contact point (Figure 11.18b). The force of static friction acts at the same point but in the horizontal direction. Since the box does not slip, the friction force vector has exactly the same magnitude as the external force vector due to the pushing by the finger, but acts in the opposite direction.

We can now calculate the torques due to these forces and find the condition for equilibrium, that is, how much force it takes for the finger to hold the box at an angle θ with respect to the vertical. Figure 11.18b also indicates the angle θ_{max}, which is a geometric property of the box that can be calculated from the ratio of the width w and the height h: $\theta_{max} = \tan^{-1}(w/h)$. Of crucial importance is the angle ϕ, which is the difference between these two angles (see Figure 11.18b): $\phi = \theta_{max} - \theta$. The angle ϕ decreases with increasing θ until $\theta = \theta_{max} \Rightarrow \phi = 0$, at which time the box falls over into the horizontal position.

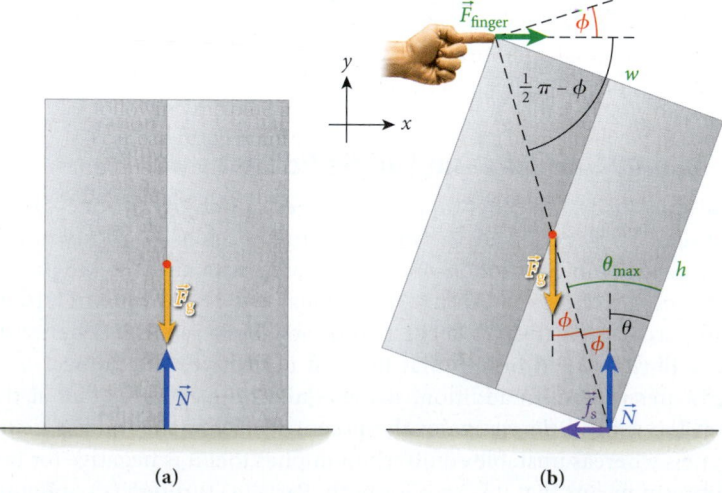

FIGURE 11.18 Free-body diagrams for the box (a) resting on the level surface and (b) being tilted by a finger pushing in the horizontal direction.

Using equation 11.7, we can calculate the net torque. The natural pivot point in this situation is the contact point between the box and the support surface. The friction force and the normal force then have moment arms of zero length and thus do not contribute to the net torque. The only clockwise torque is due to the force from the finger, and the only counterclockwise torque results from the force of gravity. The length of the moment arm for the force from the finger is (see Figure 11.18b) $\ell = \sqrt{h^2 + w^2}$, and the length of the moment arm for the force of gravity is half of this value, or $\ell/2$. This means that equation 11.7 becomes

$$(F_g)(\tfrac{1}{2}\ell)\sin\phi - (F_{finger})(\ell)\sin(\tfrac{1}{2}\pi - \phi) = 0.$$

We can use $\sin(\tfrac{1}{2}\pi - \phi) = \cos\phi$ and $F_g = mg$ and then solve for the force the finger must provide to keep the box at equilibrium at a given angle:

$$F_{finger}(\theta) = \tfrac{1}{2}mg\tan\left[\tan^{-1}\left(\frac{w}{h}\right) - \theta\right]. \tag{i}$$

Figure 11.19 shows a plot of the force from the finger that is needed to hold the box at equilibrium at a given angle, from equation (i), for different ratios of box width to height. The curves reflect values of the angle θ between zero and θ_{max}, which is the point where the force required from the finger is zero, and where the box tips over.

PROBLEM 2
Determine the potential energy function for this box.

SOLUTION 2
Finding the solution to this part of the example is much more straightforward than for the first part. The potential energy is the gravitational potential energy, $U = mgy$, where y is the vertical coordinate of the center of mass of the box. Looking at Figure 11.18b we can see that $\cos\phi = \dfrac{y}{\sqrt{(w/2)^2 + (h/2)^2}}$. Remembering that $\phi = \theta_{max} - \theta$ we can write the potential energy as a function of θ as

$$U = mgy = mg\sqrt{(w/2)^2 + (h/2)^2}\,\cos(\theta_{max} - \theta) = mg\sqrt{(w/2)^2 + (h/2)^2}\,\cos\left[\tan^{-1}\left(\frac{w}{h}\right) - \theta\right].$$

Figure 11.20, shows (red curve) the location of the center of mass of the box for different tipping angles. The curve traces out a segment of a circle with center at the lower right corner of the box. The dashed red line shows the same curve, but for angles $\theta > \theta_{max}$, for which the box tips over into the horizontal position without the finger exerting a force on it. You can clearly see that this potential energy function has a maximum at the point where the box stands on edge and its center of mass is exactly above the contact point with the surface. The curve for the location of the center of mass when the box is tilted to the left is shown in blue in Figure 11.20. You can see that the potential energy function has a minimum when the box rests flat on the table. *Note:* At this equilibrium point, the first derivative does not exist in the mathematical sense, but it is apparent from the figure that the function has a minimum, which is sufficient for stable equilibrium.

Self-Test Opportunity 11.4

What is the minimum value that the coefficient of static friction can have if the box in Example 11.5 is to tip over and the ratio of the width to the height of the box is 0.4?

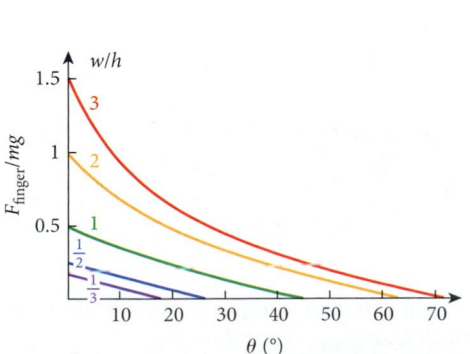

FIGURE 11.19 Ratio of the force of the finger to the weight of the box needed to hold the box in equilibrium as a function of the angle, plotted for different representative values of the ratio of the width to the height of the box.

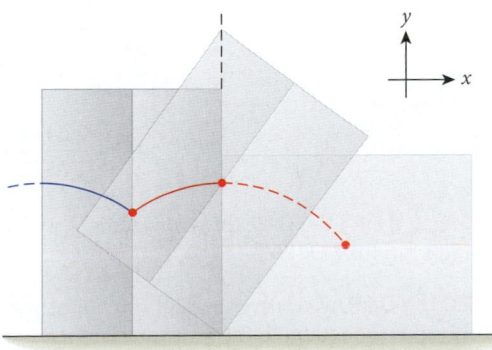

FIGURE 11.20 Location of the center of mass of the box as a function of tipping angle.

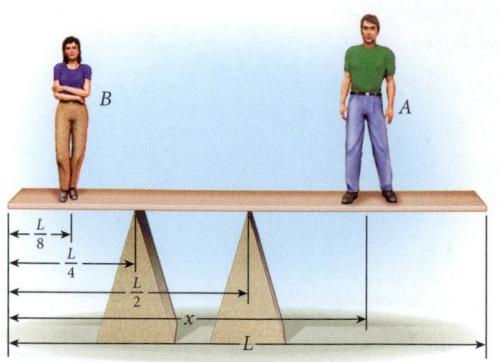

FIGURE 11.21 Two people on a board.

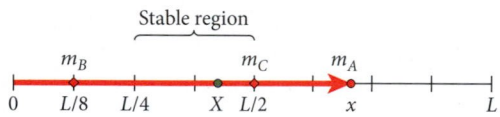

FIGURE 11.22 Coordinate axis showing the location of the combined center of mass for the two people on the board.

SOLVED PROBLEM 11.3 | Standing on a Board

PROBLEM

Persons A and B are standing on a board of uniform linear density that is balanced on two supports, as shown in Figure 11.21. What are the minimum and maximum distances x from the left end of the board at which person A can stand without tipping the board? Treat the people as point masses. The mass of person A is 1.35 times that of person B, and the mass of the board is 0.29 times that of person B.

SOLUTION

THINK This problem asks for the range of values for the position x of person A. If he moves too far to the right, the board may tip in the clockwise direction, and if he moves too far to the left, it may tip in the counterclockwise direction. When tipping in the clockwise direction, the board will pivot about the tip of the right support; and when tipping in the counterclockwise direction, the board will pivot about the tip of the left support. Selecting these two points for the limiting cases of maximum and minimum values of x will make our solution much easier.

SKETCH The sketch in Figure 11.22 consists of a coordinate axis, for which we chose the location of the origin at the left end of the board. The individual centers of mass of person A (with mass m_A), person B (with mass m_B), and the board (with mass m_C) are indicated with red dots on the coordinate axis. The location of the combined center of mass of all three is shown with a blue dot. As long as this blue dot resides between $L/4$ and $L/2$, the locations of the two supports, the overall arrangement of the two persons on the board is stable.

RESEARCH The coordinate of the combined center of mass is given by

$$X = \frac{m_A x + m_B L/8 + m_C L/2}{m_A + m_B + m_C}.$$

Here we have used the fact that person B is located at $L/8$, as specified in Figure 11.21, and that the center of mass of the board is in its middle, at $L/2$. The location x of person A is to be determined. The minimum value of x is obtained when the center-of-mass coordinate is at the left support, $X(x_{min}) = L/4$, and the maximum value of x is obtained when the center-of-mass coordinate is at the right support, $X(x_{max}) = L/2$.

SIMPLIFY The masses are given in terms of m_B: $m_A = 1.35\ m_B$ and $m_C = 0.29\ m_B$. We thus find for the combined center of mass:

$$X = \frac{1.35x + L/8 + 0.29L/2}{1.35 + 1 + 0.29} \Rightarrow$$

$$x = \frac{2.64}{1.35} X - \frac{1/8 + 0.29/2}{1.35} L.$$

We can then express x_{min} as

$$x_{min} = \frac{2.64}{1.35}\left(\frac{L}{4}\right) - \frac{1/8 + 0.29/2}{1.35} L$$

and x_{max} as

$$x_{max} = \frac{2.64}{1.35}\left(\frac{L}{2}\right) - \frac{1/8 + 0.29/2}{1.35} L.$$

CALCULATE The expressions for x_{min} and x_{max} can be simplified to yield $x_{min} = 0.288889\ L$ and $x_{max} = 0.777778\ L$.

ROUND The relative mass of the board was specified to two significant digits, and so we round our answers to the same precision: $x_{min} = 0.29\ L$ and $x_{max} = 0.78\ L$.

DOUBLE-CHECK The answers are fractions of the length of the board, which means they pass the minimum test of dimensional analysis. We also expect the numerical coefficients to be between 0 and 1, and our answers pass this second test as well.

Concept Check 11.7

If the two people in Solved Problem 11.3 traded places, so that person A was standing at $L/8$, how would the region of stability for person B change relative to what it was in Solved Problem 11.3?

a) It would be wider.

b) It would be narrower.

c) It would be the same.

Dynamic Adjustments for Stability

How does the mass damper in the Taipei 101 tower provide stability to the structure? To answer this question, let's first look at how human beings stand up straight. When you stand up straight, your center of mass is located directly above your feet. The gravitational force then exerts no net torque on you, and you can remain standing up straight. If other forces act on you (a strong wind blowing, for example, or a load you have to lift) and provide additional torques, your brain senses this through nerves coupled to the fluid in your inner ears and provides corrective action through slight shifts in the body's mass distribution. You can get a demonstration of your brain's impressive ability to make these dynamic stability adjustments by holding out your backpack (loaded with books, laptop computer, etc.) with outstretched arms in front of your body. This action will not cause you to fall over. However, if you stand straight against a wall, with your heels touching the base of the wall, the same attempt to hold your backpack with your arms outstretched will cause you to fall forward. Why? Because the wall behind you prevents your brain from shifting your body's mass distribution in order to compensate for the torque due to the backpack's weight.

Another way to gain a practical understanding of dynamic stability control is to balance an object like a hockey stick or a golf club with its handle end resting on the palm of your hand. You can accomplish this task with a little practice, in which you train your brain to tell the muscles in your arm to perform small corrective actions that counter the tilting motion of the object you are balancing. The same principles of dynamic stability adjustment are incorporated into the Segway Personal Transporter (Figure 11.23), an innovative system on two wheels, which is electric-powered and can reach speeds of up to 12 mph. Just like your brain, the Segway senses its orientation relative to the vertical. But it uses gyroscopes instead of the fluid in the inner ear. And just like the brain, it counterbalances a net torque by providing a compensating torque in the opposite direction. The Segway accomplishes this by slightly turning its wheels in a clockwise or counterclockwise direction.

Finally, the mass damper at the top of the Taipei 101 tower is used in a similar way, to provide a subtle shift in the building's mass distribution, thus contributing to stability in the presence of net external torques due to the forces of strong winds. But it also dampens the oscillation of the building due to these forces, a topic we will return to in Chapter 14.

FIGURE 11.23 The Segway Personal Transporter provides dynamic stability adjustments.

WHAT WE HAVE LEARNED | EXAM STUDY GUIDE

- Static equilibrium is mechanical equilibrium for the special case where the object in equilibrium is at rest.

- An object (or a collection of objects) can be in static equilibrium only if the net external force is zero and the net external torque is zero:

$$\vec{F}_{\text{net}} = \sum_{i=1}^{n} \vec{F}_i = \vec{F}_1 + \vec{F}_2 + \cdots + \vec{F}_n = 0$$

$$\vec{\tau}_{\text{net}} = \sum_{i} \vec{\tau}_i = 0$$

- The condition for static equilibrium can also be expressed as $\vec{\nabla} U(\vec{r})\big|_{\vec{r}_0} = 0$; that is, the first gradient derivative of the potential energy function with respect to the position vector is zero at the equilibrium point.

- The condition for stable equilibrium is that the potential energy function has a minimum at that point. A sufficient condition for stability is that the second derivative of the potential function with respect to the coordinate at the equilibrium point is positive.

- The condition for unstable equilibrium is that the potential energy function has a maximum at that point. A sufficient condition for instability is that the second derivative of the potential function with respect to the coordinate at the equilibrium point is negative.

- If the second derivative of the potential energy function with respect to the coordinate is zero at the equilibrium point, the equilibrium is neutral (or indifferent or marginally stable).

ANSWERS TO SELF-TEST OPPORTUNITIES

11.1 Choose the pivot point to be at the location of m_2.

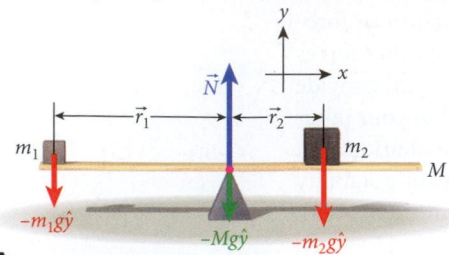

$$F_{\text{net},y} = \sum_i F_{i,y} = -m_1 g - m_2 g - Mg + N = 0$$

$$\Rightarrow N = g(m_1 + m_2 + M) = m_1 g + m_2 g + Mg$$

$$\tau_{\text{net}} = \sum_i \tau_{\text{counterclockwise},i} - \sum_j \tau_{\text{clockwise},j}$$

$$\tau_{\text{net}} = m_1 g(r_1 + r_2) \sin 90° + Mg r_2 \sin 90° - N r_2 \sin 90° = 0$$

$$N r_2 = m_1 g r_1 + m_1 g r_2 + Mg r_2$$

$$(m_1 g + m_2 g + Mg) r_2 = m_1 g r_1 + m_2 g r_2 + Mg r_2 = m_1 g r_1 + m_1 g r_2 + Mg r_2$$

$$\cancel{m_1 g r_2} + m_2 g r_2 + \cancel{Mg r_2} = m_1 g r_1 + \cancel{m_1 g r_2} + \cancel{Mg r_2}$$

$$m_2 \cancel{g} r_2 = m_1 \cancel{g} r_1$$

$$m_2 r_2 = m_1 r_1.$$

11.2 The biceps still has a nonzero moment arm even when the arm is fully extended, because the tendon has to wrap around the elbow joint and is thus never parallel to the radius bone.

11.3 The stack would be $(10,000)(4 \text{ cm}) = 400 \text{ m}$ high, comparable to the tallest skyscrapers. And the top block's right edge would stick out by

$$\frac{1}{2}(15 \text{ cm}) \sum_{i=1}^{10,000} \frac{1}{i} = \frac{1}{2}(15 \text{ cm})(9.78761) = 73.4 \text{ cm}.$$

11.4 From Figure 11.19, we can see that the force required is greatest at the beginning, for $\theta = 0$. If we set $\theta = 0$ in equation (i) of Example 11.5, we find the initial force that the finger needs to apply to displace the box from equilibrium:

$$F_{\text{finger}}(0) = \frac{1}{2} mg \tan\left[\tan^{-1}\left(\frac{w}{h}\right) - 0 \right] = \frac{1}{2} mg\left(\frac{w}{h}\right).$$

A friction coefficient of at least $\mu_s = \frac{1}{2}(w/h)$ is needed to provide the matching friction force that prevents the box from sliding along the supporting surface: $\mu_s = \frac{1}{2}(0.4) = 0.2$.

PROBLEM-SOLVING GUIDELINES: STATIC EQUILIBRIUM

1. Almost all static equilibrium problems involve summing forces in the coordinate directions, summing torques, and setting the sums equal to zero. However, the right choices of coordinate axes and a pivot point for the torques can make the difference between a hard solution and an easy one. Generally, choosing a pivot point that eliminates the moment arm for an unknown force (and often more than one force!) will simplify the equations so that you can solve for some force components.

2. The key step in writing correct equations for situations of static equilibrium is to draw a correct free-body diagram. Be careful about the locations where forces act; because torques are involved, you must represent objects as extended bodies, not point particles, and the point of application of a force makes a difference. Check each force to make sure that it is exerted *on* the object in equilibrium and not *by* the object.

MULTIPLE-CHOICE QUESTIONS

11.1 A 3.00-kg broom is leaning against a coffee table. A woman lifts the broom handle with her arm fully stretched so that her hand is a distance of 0.450 m from her shoulder. What torque is produced on her shoulder by the broom if her arm is at an angle of 50.0° below the horizontal?

a) 7.00 N m
b) 5.80 N m
c) 8.51 N m
d) 10.1 N m

11.2 A uniform beam of mass M and length L is held in static equilibrium, and so the magnitude of the net torque about its center of mass is zero. The magnitude of the net torque on this beam about one of its ends, a distance of $L/2$ from the center of mass, is

a) MgL
b) $MgL/2$
c) zero
d) $2MgL$

11.3 A very light and rigid rod is pivoted at point A and weights m_1 and m_2 are hanging from it, as shown in the figure. The ratio of the weight of m_1

to that of m_2 is 1:2. What is the ratio of L_1 to L_2, the distances from the pivot point to m_1 and m_2, respectively?

a) 1:2
b) 2:1
c) 1:1
d) not given enough information to determine

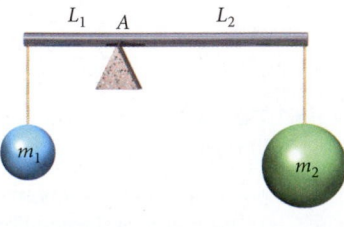

11.4 Which of the following are in static equilibrium?

a) a pendulum at the top of its swing
b) a merry-go-round spinning at constant angular velocity
c) a projectile at the top of its trajectory (with zero velocity)
d) all of the above
e) none of the above

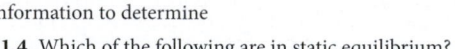

11.5 The object in the figure below is suspended at its center of mass—thus, it is balanced. If the object is cut in two pieces at its center of mass, what is the relation between the two resulting masses?

a) The masses are equal.

b) M_1 is less than M_2.

c) M_2 is less than M_1.

d) It is impossible to tell.

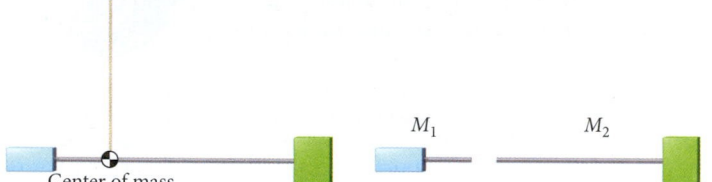

Center of mass

11.6 As shown in the figure, two weights are hanging on a uniform wooden bar that is 60 cm long and has a mass of 100 g. Is this system in equilibrium?

a) yes

b) no

c) cannot be determined

d) depends on the value of the normal force

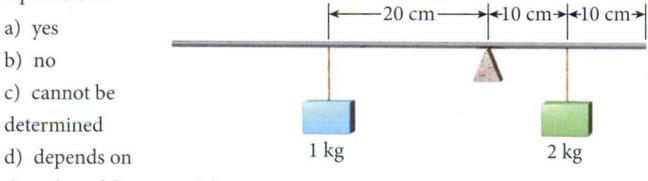

11.7 A 15-kg child sits on a playground seesaw, 2.0 m from the pivot. A second child located 1.0 m on the other side of the pivot would have to have a mass _____ to lift the first child off the ground.

a) greater than 30 kg   c) equal to 30 kg

b) less than 30 kg

11.8 A mobile is constructed from a metal bar and two wooden blocks, as shown in the figure. The metal bar has a mass of 1.0 kg and is 10. cm long. The metal bar has a 3.0-kg wooden block hanging from the left end and a string tied to it at a distance of 3.0 cm from the left end. What mass should the wooden block hanging from the right end of the bar have to keep the bar level?

a) 0.70 kg

b) 0.80 kg

c) 0.90 kg

d) 1.0 kg

e) 1.3 kg

f) 3.0 kg

g) 7.0 kg

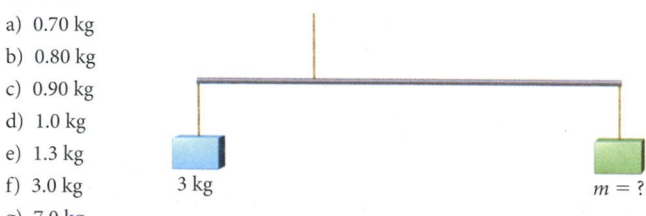

11.9 If a point mass were placed at the right end of the wooden cylinder in the sculpture in Figure 11.9a, how large could this mass be without the sculpture tipping over?

a) 2.4 kg   c) 37.5 kg   e) 1210 kg

b) 29.1 kg   d) 245 kg

11.10 Concerning the function $U(x)$ shown in the figure, which one of the following statements is true at $x = x_0$?

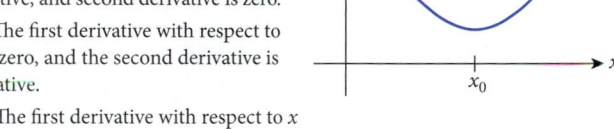

a) The first derivative with respect to x is zero, and the second derivative is positive.

b) The first derivative with respect to x is positive, and second derivative is zero.

c) The first derivative with respect to x is zero, and the second derivative is negative.

d) The first derivative with respect to x is negative, and the second derivative is zero.

e) Both the first derivative and the second derivative with respect to x are zero.

11.11 Which one of the following statements correctly expresses the requirement(s) for a system to be in static equilibrium?

a) The center of mass of the system must be at rest, but the system can be rotating.

b) The center of mass of the system must be at rest or moving with a constant speed.

c) The center of mass of the system must be at rest, and the system must not be rotating.

d) The system must be rotating about its center of mass.

e) The system must be rotating at a constant angular speed around its center of mass, which is at rest.

11.12 In the figure, a cable supports a mass m, and a massless bar makes an angle θ with the horizontal. The bar is hinged to the wall. Which one of the following statements is correct?

a) As θ approaches zero, the tension $\vec{T}$ in the cable approaches zero.

b) As θ approaches 90°, the tension $\vec{T}$ in the cable approaches zero.

c) As θ approaches zero, the tension $\vec{T}$ in the cable approaches mg.

d) As θ approaches 90°, the tension $\vec{T}$ in the cable approaches mg.

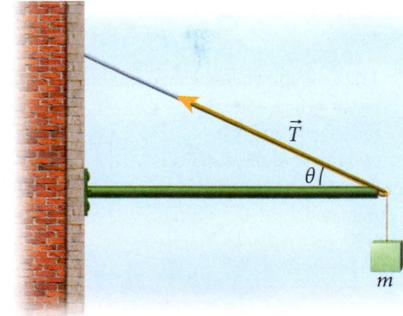

CONCEPTUAL QUESTIONS

11.13 There are three sets of landing gear on an airplane: One main set is located under the centerline of each wing, and the third set is located beneath the nose of the plane. Each set of main landing gear has four tires, and the nose landing gear has two tires. If the load on each of the tires is the same when the plane is at rest, find the center of mass of the plane. Express your result as a fraction of the perpendicular distance between the centerline of the plane's wings to the plane's nose gear. (Assume that the landing gear struts are vertical when the plane is at rest and the dimensions of the landing gear are negligible compared to the dimensions of the plane.)

11.14 A semicircular arch of radius a stands on level ground as shown in the figure. The arch is uniform in cross section and density, with total weight W. By symmetry, at the top of the arch, each of the two legs exerts only horizontal forces on the other; ideally, the stress at that point is uniform compression over the cross section of the arch. What are the

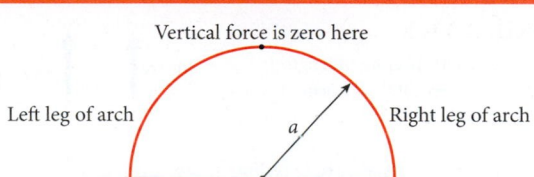

vertical and horizontal force components which must be supplied at the base of each leg to support the arch?

11.15 In the absence of any symmetry or other constraints on the forces involved, how many unknown force components can be determined in a situation of static equilibrium in each of the following cases?

a) All forces and objects lie in a plane.

b) Forces and objects are in three dimensions.

c) Forces act in n spatial dimensions.

11.16 You have a meter stick that balances at the 55-cm mark. Is your meter stick homogeneous?

11.17 You have a meter stick that balances at the 50-cm mark. Is it possible for your meter stick to be inhomogeneous?

11.18 Why does a helicopter with a single main rotor generally have a second small rotor on its tail?

11.19 The system shown in the figure consists of a uniform (homogeneous) rectangular board that is resting on two identical rotating cylinders. The two cylinders rotate in opposite directions at equal angular velocities. Initially, the board is placed perfectly symmetrically relative to the center point between the two cylinders. Is this an equilibrium position for the board? If yes, is it in stable or unstable equilibrium? What happens if the board is given a very slight displacement out of the initial position?

11.20 If the wind is blowing strongly from the east, stable equilibrium for an open umbrella is achieved if its shaft points west. Why is it relatively easy to hold the umbrella directly into the wind (in this case, eastward) but very difficult to hold it perpendicular to the wind?

11.21 A sculptor and his assistant are carrying a wedge-shaped marble slab up a flight of stairs, as shown in the figure. The density of the marble is uniform. Both are lifting straight up as they hold the slab completely stationary for a moment. Does the sculptor have to exert more force than the assistant to keep the slab stationary? Explain.

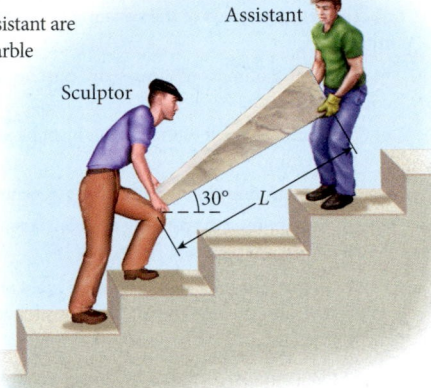

11.22 As shown in the figure, a thin rod of mass M and length L is suspended by two wires—one at the left end and one two-thirds of the distance from the left end to the right end.

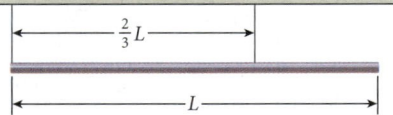

a) What is the tension in each wire?

b) Determine the mass that an object hung by a string attached to the far right-hand end of the rod would have to have for the tension in the left-hand wire to be zero.

11.23 A uniform disk of mass M_1 and radius R_1 has a circular hole of radius R_2 cut out as shown in the figure.

a) Find the center of mass of the resulting object.

b) How many equilibrium positions does this object have when resting on its edge? Which ones are stable, which neutral, and which unstable?

11.24 Consider the system shown in the figure. If a pivot point is placed at a distance $L/2$ from the ends of the rod of length L and mass $5M$, the system will rotate clockwise. Thus, in order for the system to not rotate, the pivot point should be placed away from the center of the rod. In which direction from the center of the rod should the pivot point be placed? How far from the center of the rod should the pivot point be placed in order for the system not to rotate? (Treat masses M and $2M$ as point masses.)

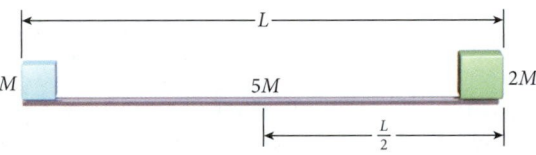

11.25 A child has a set of blocks that are all made from the same type of wood. The blocks come in three shapes: a cube of side L, a piece the size of two cubes, and a piece equivalent in size to three of the cubes placed end to end. The child stacks three blocks as shown in the figure: a cube on the bottom, one of the longest blocks horizontally on top of that, and the medium-sized block placed vertically on top. The centers of each block are initially on a vertical line. How far can the top block be slid along the middle block before the middle block tips?

11.26 Why didn't the ancient Egyptians build their pyramids upside-down? In other words, use force and center-of-mass principles to explain why it is more advantageous to construct buildings with broad bases and narrow tops than the other way around.

EXERCISES

A blue problem number indicates a worked-out solution is available in the Student Solutions Manual. One • and two •• indicate increasing level of problem difficulty.

Section 11.1

11.27 A 1000.-N crate of length L rests on a horizontal platform. It is being pulled up by two vertical ropes. The left rope has a tension of 400. N and is attached a distance of $L/4$ from the left end of the crate. Assuming the platform moves downward and the crate does not move or rotate, what can you say about the tension in the right rope?

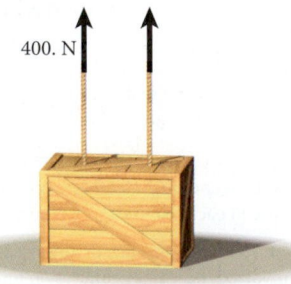

11.28 In preparation for a demonstration on conservation of energy, a professor attaches a 5.00-kg bowling ball to a 4.00-m-long rope. He pulls the ball 20.0° away from the vertical and holds the ball while he discusses the physics principles involved. Assuming that the force he exerts on the ball is entirely in the horizontal direction, find the tension in the rope and the force the professor is exerting on the ball.

11.29 A sculptor and his assistant stop for a break as they carry a marble slab of length $L = 2.00$ m and mass 75.0 kg up the steps, as shown in the figure. The mass of the slab is uniformly distributed along its length. As they rest, both the sculptor and his assistant are pulling *directly up* on each end of the slab, which is at an angle of 30.0° with respect to horizontal. What are the magnitudes of the forces that the sculptor and assistant must exert on the marble slab to keep it stationary during their break?

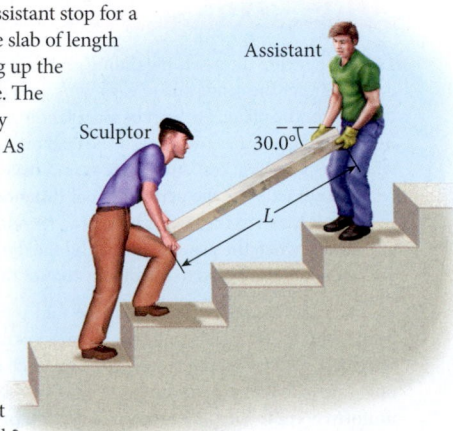

11.30 During a picnic, you and two of your friends decide to have a three-way tug-of-war, so you tie the ends of three ropes into a knot. Roberta pulls to the west with 420. N of force; Michael pulls to the south with 610. N. In what direction and with what magnitude of force should you pull to keep the knot from moving?

11.31 The figure shows a photo of a typical merry-go-round, like those found on many playgrounds, and a diagram giving a top view. Four children are standing on the ground and pulling on the merry-go-round as indicated by the force arrows. The four forces have the magnitudes $F_1 = 104.9$ N, $F_2 = 89.1$ N, $F_3 = 62.8$ N, and $F_4 = 120.7$ N. All the forces act in a tangential direction. With what force, $\vec{F}$, also in a tangential direction and acting at the black point, does a fifth child have to pull in order to prevent the merry-go-round from moving? Specify the magnitude of the force and state whether the force is acting counterclockwise or clockwise.

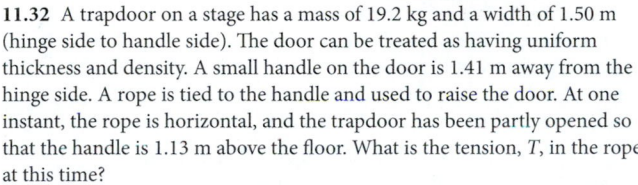

11.32 A trapdoor on a stage has a mass of 19.2 kg and a width of 1.50 m (hinge side to handle side). The door can be treated as having uniform thickness and density. A small handle on the door is 1.41 m away from the hinge side. A rope is tied to the handle and used to raise the door. At one instant, the rope is horizontal, and the trapdoor has been partly opened so that the handle is 1.13 m above the floor. What is the tension, T, in the rope at this time?

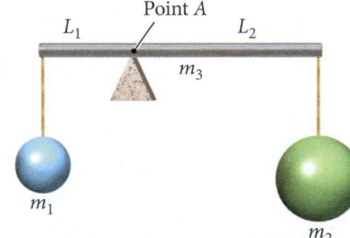

•11.33 A rigid rod of mass m_3 is pivoted at point A, and masses m_1 and m_2 are hanging from it, as shown in the figure.

a) What is the normal force acting on the pivot point?

b) What is the ratio of L_1 to L_2, where these are the distances from the pivot point to m_1 and m_2, respectively? The ratio of the weights of m_1, m_2, and m_3 is 1:2:3.

•11.34 When only the front wheels of an automobile are on a platform scale, the scale balances at 8.00 kN; when only the rear wheels are on the scale, it balances at 6.00 kN. What is the weight of the automobile, and how far is its center of mass behind the front axle? The distance between the axles is 2.80 m.

•11.35 By considering the torques about your shoulder, estimate the force your deltoid muscles (those on top of the shoulder) must exert on the bone of your upper arm, in order to keep your arm extended straight out at shoulder level. Then, estimate the force the muscles must exert to hold a 10.0-lb weight at arm's length. You'll need to estimate the distance from your shoulder pivot point to the point where your deltoid muscles connect to the bone of your upper arm in order to determine the necessary forces. Assume the deltoids are the only contributing muscles.

••11.36 A uniform, equilateral triangle of side length 2.00 m and weight $4.00 \cdot 10^3$ N is placed across a gap. One point is on the north end of the gap, and the opposite side is on the south end. Find the force on each side.

Section 11.2

11.37 A 600.0-N bricklayer is 1.50 m from one end of a uniform scaffold that is 7.00 m long and weighs 800.0 N. A pile of bricks weighing 500.0 N is 3.00 m from the same end of the scaffold. If the scaffold is supported at both ends, calculate the force on each end.

11.38 The uniform rod in the figure is supported by two strings. The string attached to the wall is horizontal, and the string attached to the ceiling makes an angle of ϕ with respect to the vertical. The rod itself is tilted from the vertical by an angle θ. If $\phi = 30.0°$, what is the value of θ?

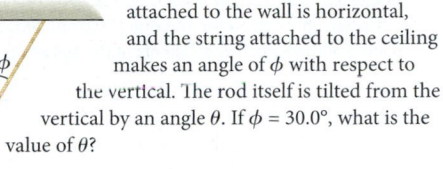

11.39 A construction supervisor of mass $M = 92.1$ kg is standing on a board of mass $m = 27.5$ kg. Two sawhorses at a distance $\ell = 3.70$ m apart support the board, which overhangs each sawhorse by an equal amount. If the man stands a distance $x_1 = 1.07$ m away from the left-hand sawhorse as shown in the figure, what is the force that the board exerts on that sawhorse?

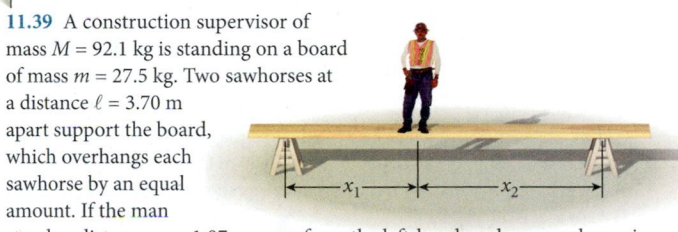

11.40 In a butcher shop, a horizontal steel bar of mass 4.00 kg and length 1.20 m is supported by two vertical wires attached to its ends. The butcher hangs a sausage of mass 2.40 kg from a hook that is at a distance of 0.20 m from the left end of the bar. What are the tensions in the two wires?

•11.41 Two uniform planks, each of mass m and length L, are connected by a hinge at the top and by a chain of negligible mass attached at their centers, as shown in the figure. The assembly will stand upright, in the shape of an A, on a frictionless surface without collapsing. As a function of the length of the chain, find each of the following:

a) the tension in the chain,

b) the force on the hinge of each plank, and

c) the force of the ground on each plank.

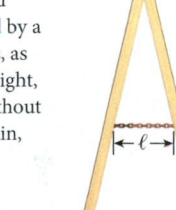

•11.42 Three strings are tied together. They lie on top of a circular table, and the knot is exactly in the center of the table, as shown in the figure (top view). Each string hangs over the edge of the table, with a weight supported from it. The masses $m_1 = 4.30$ kg and $m_2 = 5.40$ kg are known. The angle $\alpha = 74°$ between strings 1 and 2 is also known. What is the angle β between strings 1 and 3?

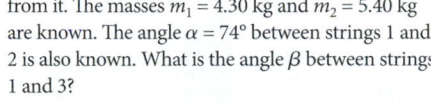

•11.43 A uniform ladder 10.0 m long is leaning against a frictionless wall at an angle of 60.0° above the horizontal. The weight of the ladder is 20.0 lb. A 61.0-lb boy climbs 4.00 m up the ladder. What is the magnitude of the frictional force exerted on the ladder by the floor?

•11.44 Robin is making a mobile to hang over her baby sister's crib. She purchased four stuffed animals: a teddy bear (16.0 g), a lamb (18.0 g), a little pony (22.0 g), and a bird (15.0 g). She also purchased three small wooden dowels, each 15.0 cm long and of mass 5.00 g, and thread of negligible mass. She wants to hang the bear

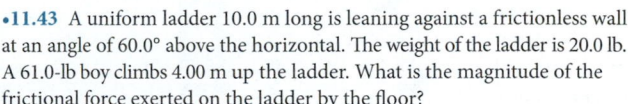

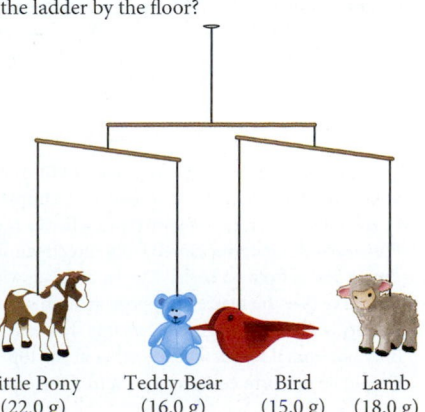

Little Pony (22.0 g)   Teddy Bear (16.0 g)   Bird (15.0 g)   Lamb (18.0 g)

and the pony from the ends of one dowel and the lamb and the bird from the ends of the second dowel. Then, she wants to suspend the two dowels from the ends of the third dowel and hang the whole assembly from the ceiling. Explain where the thread should be attached on each dowel so that the entire assembly will hang level.

•**11.45** A door, essentially a uniform rectangle of height 2.00 m, width 0.800 m, and weight 100.0 N, is supported at one edge by two hinges, one 30.0 cm above the bottom of the door and one 170.0 cm above the bottom of the door. Calculate the horizontal components of the forces on the two hinges.

••**11.46** The figure shows a 20.0-kg, uniform ladder of length L hinged to a horizontal platform at point P_1 and anchored with a steel cable of the same length as the ladder attached at the ladder's midpoint. Calculate the tension in the cable and the forces in the hinge when an 80.0-kg person is standing three-quarters of the way up the ladder.

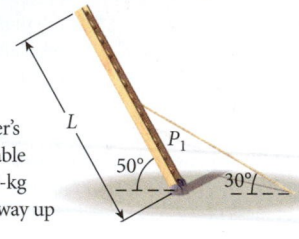

••**11.47** A beam with a length of 8.00 m and a mass of 100. kg is attached by a large bolt to a support at a distance of $d = 3.00$ m from one end. The beam makes an angle $\theta = 30.0°$ with the horizontal, as shown in the figure. A mass $M = 500.$ kg is attached with a rope to one end of the beam, and a second rope is attached at a right angle to the other end of the beam. Find the tension, T, in the second rope and the force exerted on the beam by the bolt.

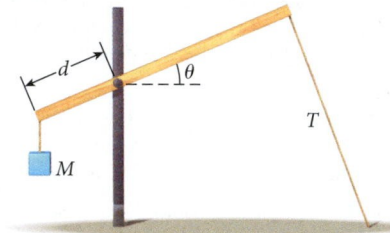

••**11.48** A ball of mass 15.49 kg rests on a table of height 0.720 m. The tabletop is a rectangular glass plate of mass 12.13 kg, which is supported at the corners by thin legs, as shown in the figure. The width of the tabletop is $w = 138.0$ cm, and its depth is $d = 63.8$ cm. If the ball touches the tabletop at a point $(x,y) = (69.0$ cm,16.6 cm) relative to corner 1, what is the force that the tabletop exerts on each leg?

••**11.49** A wooden bridge crossing a canyon consists of a plank with length density $\lambda = 2.00$ kg/m suspended at $h =10.0$ m below a tree branch by two ropes, each with a length $L = 2h$ and a maximum rated tension of 2000. N,

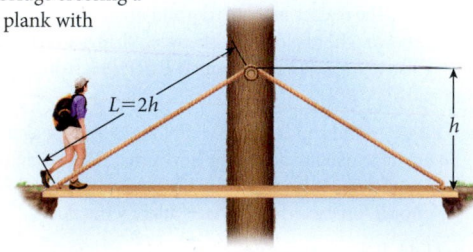

which are attached to the ends of the plank, as shown in the figure. A hiker steps onto the bridge from the left side, causing the bridge to tip to an angle of 25.0° with respect to the horizontal. What is the mass of the hiker?

••**11.50** The famous Gateway Arch in St. Louis, Missouri, is approximately an *inverted catenary*. A simple example of such a curve is given by $y(x) = 2a - a\cosh(x/a)$, where y is vertical height and x is horizontal distance, measured from directly under the top of the curve; thus, x varies from $-a\cosh^{-1}2$ to $+a\cosh^{-1}2$, with a the height of the top of the curve (see the figure). Suppose an arch of uniform cross section and density, with total weight W, has this shape. The two legs of the arch exert only horizontal forces on each other at the top; ideally, the stress there should be uniform compression across the cross section.

a) Calculate the vertical and horizontal force components that are acting at the base of each leg of this arch.

b) At what angle should the bottom face of the legs be oriented?

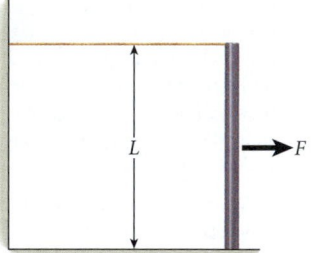

Vertical force is zero here
Left leg of arch
Right leg of arch
a
$2a\cosh^{-1}2$

Section 11.3

11.51 A uniform rectangular bookcase of height H and width $W = H/2$ is to be pushed at a constant velocity across a level floor. The bookcase is pushed horizontally at its top edge, at the distance H above the floor. What is the maximum value the coefficient of kinetic friction between the bookcase and the floor can have if the bookcase is not to tip over while being pushed?

11.52 The system shown in the figure is in static equilibrium. The rod of length L and mass M is held in an upright position. The top of the rod is tied to a fixed vertical surface by a string, and a force F is applied at the midpoint of the rod. The coefficient of static friction between the rod and the horizontal surface is μ_s. What is the maximum force, F, that can be applied and have the rod remain in static equilibrium?

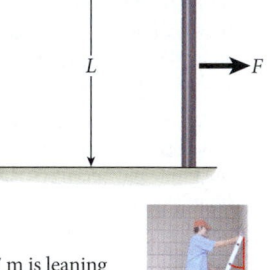

L
F

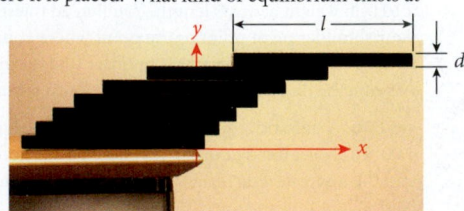

11.53 A ladder of mass 37.7 kg and length 3.07 m is leaning against a wall at an angle θ. The coefficient of static friction between ladder and floor is 0.313; assume that the friction force between ladder and wall is zero. What is the maximum value that θ can have before the ladder starts slipping?

•**11.54** A uniform rigid pole of length L and mass M is to be supported from a vertical wall in a horizontal position, as shown in the figure. The pole is not attached directly to the wall, so the coefficient of static friction, μ_s, between the wall and the pole provides the only vertical force on one end of the pole. The other end of the pole is supported by a light rope that is attached to the wall at a point a distance D directly above the point where the pole contacts the wall. Determine the minimum value of μ_s, as a function of L and D, that will keep the pole horizontal and not allow its end to slide down the wall.

D
M
L

•**11.55** A boy weighing 60.0 lb is playing on a plank. The plank weighs 30.0 lb, is uniform, is 8.00 ft long, and lies on two supports, one 2.00 ft from the left end and the other 2.00 ft from the right end.

a) If the boy is 3.00 ft from the left end, what force is exerted by each support?

b) The boy moves toward the right end. How far can he go before the plank will tip?

•**11.56** A track has a height that is a function of horizontal position x, given by $h(x) = x^3 + 3x^2 - 24x +16$. Find all the positions on the track where a marble will remain where it is placed. What kind of equilibrium exists at each of these positions?

•**11.57** The figure shows a stack of seven identical aluminum blocks, each of length $l = 15.9$ cm and thickness $d = 2.20$ cm, stacked on a table.

y
l
d
x

a) How far to the right of the edge of the table is it possible for the right edge of the top (seventh) block to extend?

b) What is the minimum height of a stack of these blocks for which the left edge of the top block is to the right of the right edge of the table?

•**11.58** You are using a 5.00-m-long ladder to paint the exterior of your house. The point of contact between the ladder and the siding of the house is 4.00 m above the ground. The ladder has a mass of 20.0 kg. If you weigh 60.0 kg and stand three-quarters of the way up the ladder, determine

a) the forces exerted by the side wall and the ground on the ladder and

b) the coefficient of static friction between the ground and the base of the ladder that is necessary to keep the ladder stable.

•**11.59** A ladder of mass M and length $L = 4.00$ m is on a level floor leaning against a vertical wall. The coefficient of static friction between the ladder and the floor is $\mu_s = 0.600$, while the friction between the ladder and the wall is negligible. The ladder is at an angle of $\theta = 50.0°$ above the horizontal. A man of mass $3M$ starts to climb the ladder. To what distance up the ladder can the man climb before the ladder starts to slip on the floor?

•**11.60** A machinist makes the object shown in the figure. The larger-diameter cylinder is made of brass (mass density = 8.60 g/cm^3); the smaller-diameter cylinder is made of aluminum (mass density = 2.70 g/cm^3). The dimensions are $r_1 = 2.00$ cm, $r_2 = 4.00$ cm, $d_1 = 20.0$ cm, and $d_2 = 4.00$ cm.

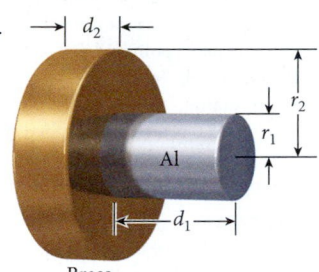

a) Find the location of the center of mass.

b) If the object is on its side, as shown in the figure, is it in equilibrium? If yes, is this stable equilibrium?

•**11.61** An object is restricted to movement in one dimension. Its position is specified along the x-axis. The potential energy of the object as a function of its position is given by $U(x) = a(x^4 - 2b^2x^2)$, where a and b represent positive numbers. Determine the location(s) of any equilibrium point(s), and classify the equilibrium at each point as stable, unstable, or neutral.

•**11.62** A two-dimensional object with uniform mass density is in the shape of a thin square and has a mass of 2.00 kg. The sides of the square are each 20.0 cm long. The coordinate system has its origin at the center of the square. A point mass, m, of $2.00 \cdot 10^2$ g is placed at one corner of the square object, and the assembly is held in equilibrium by positioning the support at position (x,y), as shown in the figure. Find the location of the support.

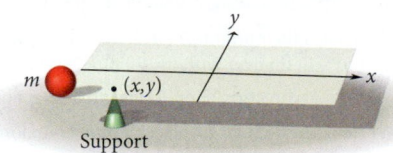

•**11.63** Persons A and B are standing on a board of uniform linear density that is balanced on two supports, as shown in the figure. What is the maximum distance x from the right end of the board at which person A can stand without tipping the board? Treat persons A and B as point masses. The mass of person B is twice that of person A, and the mass of the board is half that of person A. Give your answer in terms of L, the length of the board.

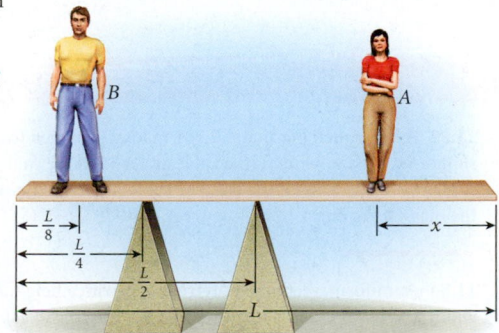

••**11.64** An SUV has a height h and a wheelbase of length b. Its center of mass is midway between the wheels and at a distance αh above the ground, where $0 < \alpha < 1$. The SUV enters a turn at a dangerously high speed, v. The radius of the turn is R ($R \gg b$), and the road is flat. The coefficient of static friction between the road and the properly inflated tires is μ_s. After entering the turn, the SUV will either skid out of the turn or begin to tip.

a) The SUV will skid out of the turn if the friction force reaches its maximum value, $F \rightarrow \mu_s N$. Determine the speed, v_{skid}, for which this will occur. Assume no tipping occurs.

b) The torque keeping the SUV from tipping acts on the outside wheel. The highest value that the force responsible for this torque can have is equal to the entire normal force. Determine the speed, v_{tip}, at which tipping will occur. Assume no skidding occurs.

c) It is safer if the SUV skids out before it tips. This will occur as long as $v_{skid} < v_{tip}$. Apply this condition, and determine the maximum value for α in terms of b, h and μ_s.

Additional Exercises

11.65 A wooden plank with length $L = 8.00$ m and mass $M = 100.$ kg is centered on a granite cube with side $S = 2.00$ m. A person of mass $m = 65.0$ kg begins walking from the center of the plank outward, as shown in the figure. How far from the center of the plank does the person get before the plank starts tipping?

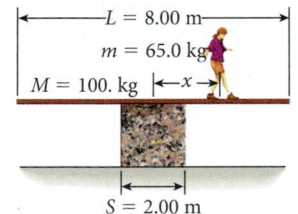

11.66 A board, with a weight $mg = 120.0$ N and a length of 5.00 m, is supported by two vertical ropes, as shown in the figure. Rope A is connected to one end of the board, and rope B is connected at a distance $d = 1.00$ m from the other end of the board. A box with a weight $Mg = 20.0$ N is placed on the board with its center of mass at $d = 1.00$ m from rope A. What are the tensions in the two ropes?

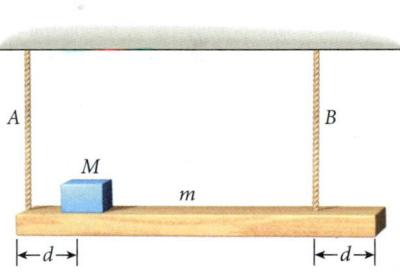

11.67 In a car, which is accelerating at 5.00 m/s^2, an air freshener is hanging from the rear-view mirror, with the string maintaining a constant angle with respect to the vertical. What is this angle?

11.68 Typical weight sets used for bodybuilding consist of disk-shaped weights with holes in the center that can slide onto 2.20-m-long barbells. A barbell is supported by racks located a fifth of its length from each end, as shown in the figure. What is the minimum mass m of the barbell if a bodybuilder is to slide a weight with $M = 22.0$ kg onto the end without the barbell tipping off the rack? Assume that the barbell is a uniform rod.

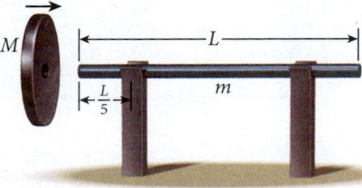

11.69 A 5.00-m-long board of mass 50.0 kg is used as a seesaw. On the left end of the seesaw sits a 45.0-kg girl, and on the right end sits a 60.0-kg boy. Determine the position of the pivot point for static equilibrium.

11.70 A mobile consists of two very lightweight rods of length $l = 0.400$ m connected to each other and the ceiling by vertical strings. (Neglect the masses of the rods and strings.) Three objects are suspended by strings from the rods. The masses of objects 1 and 3 are $m_1 = 6.40$ kg and $m_3 = 3.20$ kg. The distance x shown in the figure is 0.160 m. What is the mass of m_2?

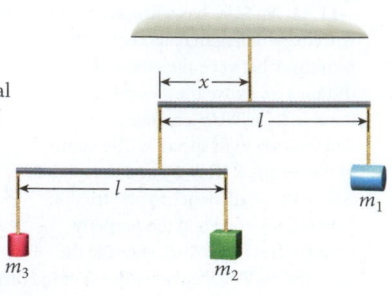

•11.71 In the experimental setup shown in the figure, a beam, B_1, of unknown mass M_1 and length $L_1 = 1.00$ m is pivoted about its lowest point at P_1. A second beam, B_2, of mass $M_2 = 0.200$ kg and length $L_2 = 0.200$ m is suspended (pivoted) from B_1 at a point P_2, which is a horizontal distance $d = 0.550$ m from P_1. To keep the system at equilibrium, a mass $m = 0.500$ kg has to be suspended from a massless string that runs horizontally from P_3, at the top of beam B_1, and passes over a frictionless pulley. The string runs at a vertical distance $y = 0.707$ m above the pivot point P_1. Calculate the mass of beam B_1.

•11.72 An important characteristic of the condition of static equilibrium is the fact that the net torque has to be zero irrespective of the choice of pivot point. For the setup in Problem 11.71, prove that the torque is indeed zero with respect to a pivot point at P_1, P_2, or P_3.

•11.73 One end of a heavy beam of mass $M = 50.0$ kg is hinged to a vertical wall, and the other end is tied to a steel cable of length 3.00 m, as shown in the figure. The other end of the cable is also attached to the wall at a distance of 4.00 m above the hinge. A mass $m = 20.0$ kg is hung from one end of the beam by a rope.

a) Determine the tensions in the cable and the rope.

b) Find the force the hinge exerts on the beam.

•11.74 A 100.-kg uniform bar of length $L = 5.00$ m is attached to a wall by a hinge at point A and supported in a horizontal position by a light cable attached to its other end. The cable is attached to the wall at point B, at a distance $D = 2.00$ m above point A. Find: a) the tension, T, on the cable and b) the horizontal and vertical components of the force acting on the bar at point A.

•11.75 A mobile over a baby's crib displays small colorful shapes. What values for m_1, m_2, and m_3 are needed to keep the mobile balanced (with all rods horizontal)?

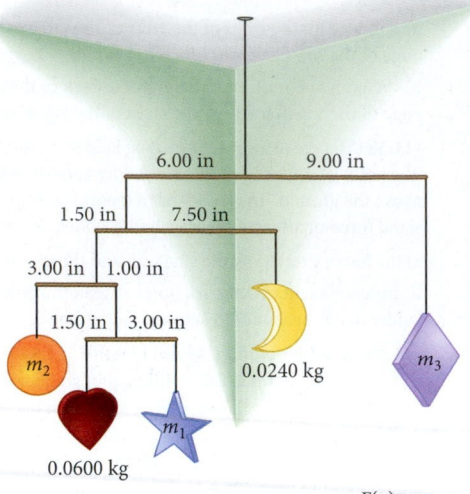

•11.76 Consider the rod of length L shown in the figure. The mass of the rod is $m = 2.00$ kg, and the pivot point is located at the left end (at $x = 0$). In order to prevent the rod from rotating, a variable force given by $F(x) = (15.0 \text{ N})(x/L)^4$ is applied to the rod. At what point x on the rod should the force be applied in order to keep it from rotating?

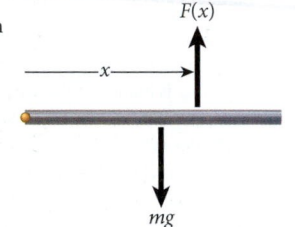

•11.77 A pipe that is 2.20 m long and has a mass of 8.13 kg is suspended horizontally over a stage by two chains, each located 0.20 m from an end. Two 7.89-kg theater lights are clamped onto the pipe, one 0.65 m from the left end and the other 1.14 m from the left end. Calculate the tension in each chain.

•11.78 A 2.00-m-long diving board of mass 12.0 kg is 3.00 m above the water. It has two attachments holding it in place. One is located at the very back end of the board, and the other is 25.0 cm away from that end.

a) Assuming that the board has uniform density, find the forces acting on each attachment (take the downward direction to be positive).

b) If a diver of mass 65.0 kg is standing on the front end, what are the forces acting on the two attachments?

•11.79 A 20.0-kg box with a height of 80.0 cm and a width of 30.0 cm has a handle on the side that is 50.0 cm above the ground. The box is at rest, and the coefficient of static friction between the box and the floor is 0.280.

a) What is the minimum force, F, that can be applied to the handle so that the box will tip over without slipping?

b) In what direction should this force be applied?

••11.80 The angular displacement of a torsional spring is proportional to the applied torque; that is $\tau = \kappa\theta$, where κ is a constant. Suppose that such a spring is mounted to an arm that moves in a vertical plane. The mass of the arm is 45.0 g, and it is 12.0 cm long. The arm-spring system is at equilibrium with the arm at an angular displacement of 17.0° with respect to the horizontal. If a mass of 0.420 kg is hung from the arm 9.00 cm from the axle, what will be the angular displacement in the new equilibrium position (relative to that with the unloaded spring)?

MULTI-VERSION EXERCISES

11.81 A horizontal bar that is 2.141 m long and has a mass of 81.95 kg is hinged to a wall. The bar is supported at its other end by a cable attached to the wall, which makes an angle of $\theta = 38.89°$ with the bar. What is the tension in the cable?

11.82 A horizontal bar that is 2.261 m long and has a mass of 82.45 kg is hinged to a wall. The bar is supported at its other end by a cable attached to the wall. The cable has a tension of 618.8 N. What is the angle between the cable and the bar?

11.83 A horizontal bar that is 2.381 m long is hinged to a wall. The bar is supported at its other end by a cable attached to the wall, which makes an angle of $\theta = 42.75°$ with the bar. The cable has a tension of 599.3 N. What is the mass of the bar?

11.84 A uniform, rectangular refrigerator with a height $h = 2.177$ m and a width $w = 1.247$ m has a mass of 98.57 kg. It is pushed at a constant velocity across a level floor by a force in the horizontal direction that is applied halfway between the floor and the top of the refrigerator. The refrigerator does not tip over while being pushed. What is the maximum value of the coefficient of kinetic friction between the refrigerator and the floor?

11.85 A uniform, rectangular refrigerator with a height $h = 2.187$ m and a width w has a mass of 83.91 kg. It is pushed at a constant velocity across a level floor by a force in the horizontal direction that is applied halfway between the floor and the top of the refrigerator. The refrigerator does not tip over while being pushed. The coefficient of kinetic friction between the refrigerator and the floor is 0.4696. What is the minimum width, w, of the refrigerator?

11.86 A uniform, rectangular refrigerator with a width $w = 1.059$ m and a height h has a mass of 88.27 kg. It is pushed at a constant velocity across a level floor by a force in the horizontal direction that is applied halfway between the floor and the top of the refrigerator. The refrigerator does not

tip over while being pushed. The coefficient of kinetic friction between the refrigerator and the floor is 0.4820. What is the maximum height, h, of the refrigerator?

11.87 A 92.61-kg person is halfway up a uniform ladder of length 3.413 m and mass 23.63 kg that is leaning against a wall. The angle between the ladder and the wall is θ. The coefficient of static friction between the ladder and the floor is 0.2881. Assume that the friction force between the ladder and the wall is zero. What is the maximum value that the angle between the ladder and the wall can have without the ladder slipping?

11.88 A 96.97-kg person is halfway up a uniform ladder of length 3.433 m and mass 24.91 kg that is leaning against a wall. The angle between the ladder and the wall is $\theta = 27.30°$. Assume that the friction force between the ladder and the wall is zero. What is the minimum value of the coefficient of static friction between the ladder and the floor that will keep the ladder from slipping?

12

Gravitation

FIGURE 12.1 Earthset photographed on November 7, 2007, by the Japanese satellite *Kaguya* while orbiting the Moon. The Earth and the Moon orbit around each other and are kept together by their gravitational interaction.

Figure 12.1 shows the Earth setting over the horizon of the Moon, photographed by a satellite orbiting the Moon. We are so used to seeing the Moon in the sky that it's somewhat surprising to see the Earth instead. In fact, astronauts on the Moon do not see earthrise or earthset because the Moon always keeps the same face turned toward Earth. Only astronauts orbiting the Moon can see the Earth seeming to change position and rise or set. However, the image reminds us that, like all forces, the force of gravity is a mutual attraction between two objects—Earth pulls on the Moon, but the Moon also pulls on Earth.

We have examined forces in general terms in previous chapters. In this chapter, we focus on one particular force, the force of gravity, which is one of the four fundamental forces in nature. Gravity is the weakest of these four forces (inside atoms, for example, gravity is negligible relative to the electromagnetic forces), but it operates over all distances and is always a force of attraction between objects with mass (as opposed to the electromagnetic interaction, for which the charges come in positive and negative varieties so resulting forces can be attractive or repulsive and tend to sum to zero for most macroscopic objects). As a result, the gravitational force is of primary importance over the vast distances and for the enormous masses involved in astronomical studies.

WHAT WE WILL LEARN

- The gravitational interaction between two point masses is proportional to the product of their masses and inversely proportional to the square of the distance separating them.

- The gravitational force on an object inside a homogeneous solid sphere rises linearly with the distance that the object is from the center of the solid sphere.

- At the surface of the Earth, the acceleration due to gravity is well approximated by a constant, g. The value of g used for free-fall situations can be verified from the more general gravitational force law.

- A general expression for the gravitational potential energy indicates that it is inversely proportional to the distance between two objects.

- Escape speed is the minimum speed with which a projectile must be launched to escape to an infinite distance from Earth (or another body).

- Kepler's three laws of planetary motion state that planets move in elliptical orbits with the Sun at one focal point, that the radius vector connecting the Sun and a planet sweeps out equal areas in equal times, and that the square of the orbit's period for any planet is proportional to the cube of its semimajor axis.

- The kinetic, potential, and total energies of satellites in orbit have a fixed relationship with each other.

- There is evidence for a great amount of dark matter and dark energy in the universe.

Figure 12.2 is an image of the center of our galaxy obtained with the space-based infrared Spitzer telescope. The galactic center contains a supermassive black hole, and knowledge of the gravitational interaction allows astronomers to calculate that the mass of this black hole is approximately 4.1 million times the mass of the Sun or more than 1 trillion times the mass of the Earth. (Example 12.4 shows how scientists arrived at this conclusion. A black hole is an object so massive and dense that nothing can escape from its surface, not even light.)

FIGURE 12.2 Center of our galaxy, the Milky Way, which contains a supermassive black hole. The size of the region shown here is 890 by 640 light-years. The Solar System is 26,000 light-years away from the center of the galaxy.

12.1 Newton's Law of Gravity

Up to now, we have encountered the gravitational force only in the form of a constant gravitational acceleration, $g = 9.81$ m/s^2, multiplied by the mass of the object on which the force acts. However, from videos of astronauts running and jumping on the Moon (Figure 12.3), we know that the gravitational force is different there. Therefore, the approximation we have used of a constant gravitational force that depends only on the mass of the object the force acts on cannot be correct away from the surface of the Earth.

The general expression for the magnitude of the gravitational interaction between two point masses, m_1 and m_2, at a distance $r = |\vec{r}_2 - \vec{r}_1|$ from each other (Figure 12.4) is

$$F(r) = G\frac{m_1 m_2}{r^2}. \tag{12.1}$$

This relationship, known as **Newton's Law of Gravity,** is an empirical law, deduced from experiments and verified extensively. The proportionality constant G is called the **universal gravitational constant** and has the value (to four significant figures)

$$G = 6.6738 \cdot 10^{-11} \text{ N m}^2/\text{kg}^2. \tag{12.2}$$

Since 1 N = 1 kg m/s^2, we can also write this value as $G = 6.6738 \cdot 10^{-11}$ m^{3}kg^{-1}s^{-2}. Equation 12.1 says that the strength of the gravitational interaction is proportional to each of the two masses involved in the interaction and is inversely proportional to the square of the distance between them. For example, doubling one of the masses will double the strength of the interaction, whereas doubling the distance will reduce the strength of the interaction by a factor of 4.

Because a force is a vector, the direction of the gravitational force must be specified. The gravitational force $\vec{F}_{2\rightarrow1}$ acting from object 2 on object 1 always points toward object 2. We can express this concept in the form of an equation:

$$\vec{F}_{2\rightarrow1} = F(r)\hat{r}_{21} = F(r)\frac{\vec{r}_2 - \vec{r}_1}{|\vec{r}_2 - \vec{r}_1|}.$$

FIGURE 12.3 *Apollo 16* Commander John Young jumps on the surface of the Moon and salutes the U.S. flag on April 21, 1972.

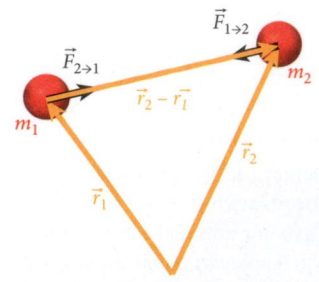

FIGURE 12.4 The gravitational interaction of two point masses.

Combining this result with equation 12.1 results in

$$\vec{F}_{2\rightarrow1} = G\frac{m_1 m_2}{\left|\vec{r}_2 - \vec{r}_1\right|^3}\left(\vec{r}_2 - \vec{r}_1\right). \qquad (12.3)$$

Equation 12.3 is the general form for the gravitational force acting on object 1 due to object 2. It is strictly valid for point particles, as well as for extended spherically symmetrical objects, in which case the position vector is the position of the center of mass. It is also a very good approximation for nonspherical extended objects, represented by their center-of-mass coordinates, provided that the separation between the two objects is large relative to their individual sizes. Note that the center of gravity is identical to the center of mass for spherically symmetrical objects.

Chapter 4 introduced Newton's Third Law: The force $\vec{F}_{1\rightarrow2}$ exerted on object 2 by object 1 has to be of the same magnitude as and in the opposite direction to the force $\vec{F}_{2\rightarrow1}$ exerted on object 1 by object 2:

$$\vec{F}_{1\rightarrow2} = -\vec{F}_{2\rightarrow1}.$$

You can see that the force described by equation 12.3 fulfills the requirements of Newton's Third Law by exchanging the indexes 1 and 2 on all variables (F, m, and r) and observing that the magnitude of the force remains the same, but the sign changes. Equation 12.3 governs the motion of the planets around the Sun, as well as of objects in free fall near the surface of the Earth.

Superposition of Gravitational Forces

If more than one object has a gravitational interaction with object 1, we can compute the total gravitational force on object 1 by using the **superposition principle,** which states that the vector sum of all gravitational forces on a specific object yields the total gravitational force on that object. That is, to find the total gravitational force acting on an object, we simply add the contributions from all other objects:

$$\vec{F}_1 = \vec{F}_{2\rightarrow1} + \vec{F}_{3\rightarrow1} + \cdots + \vec{F}_{n\rightarrow1} = \sum_{i=2}^{n}\vec{F}_{i\rightarrow1}. \qquad (12.4)$$

The individual forces $\vec{F}_{i\rightarrow1}$ can be found from equation 12.3:

$$\vec{F}_{i\rightarrow1} = G\frac{m_i m_1}{\left|\vec{r}_i - \vec{r}_1\right|^3}\left(\vec{r}_i - \vec{r}_1\right).$$

Conversely, the total gravitational force on any one of n objects experiencing mutual gravitational interaction can be written as

$$\vec{F}_j = \sum_{i=1,i\neq j}^{n}\vec{F}_{i\rightarrow j} = G\sum_{i=1,i\neq j}^{n}\frac{m_i m_j}{\left|\vec{r}_i - \vec{r}_j\right|^3}\left(\vec{r}_i - \vec{r}_j\right). \qquad (12.5)$$

The notation "$i \neq j$" below the summation symbol indicates that the sum of forces does not include any object's interaction with itself.

The conceptualization of the superposition of forces is straightforward, but the solution of the resulting equations of motion can become complicated. Even in a system of three approximately equal masses that interact with one another, some initial conditions can lead to regular trajectories, whereas others lead to chaotic motion. The numerical investigation of this kind of system started to become possible only with the advent of computers. Over the last 10 years, the field of many-body physics has developed into one of the most interesting in all of physics, and many intriguing endeavors are likely to be undertaken, such as studying the origin of galaxies from small initial fluctuations in the density of the universe.

DERIVATION 12.1 / Gravitational Force from a Sphere

Earlier, it was claimed that the gravitational interaction of an extended spherically symmetrical object could be treated like that of a point particle with the same mass located at the center of mass of the extended sphere. We can prove this statement with the aid of calculus and some elementary geometry.

To start, we treat a sphere as a collection of concentric and very thin spherical shells. If we can prove that a thin spherical shell has the same gravitational interaction as a point particle located at its center, then the superposition principle implies that a solid sphere does, too.

Figure 12.5 shows a point particle of mass M located outside a spherical shell of mass m at a distance r from the center of the shell. We want to find the x-component of the force on the mass M due to a ring of angular width $d\phi$. This ring has a radius of $a = R \sin \phi$ and thus a circumference of $2\pi R \sin \phi$. It has a width of $R \, d\phi$, as shown in Figure 12.5, and thus a total area of $2\pi R^2 \sin \phi \, d\phi$. Because the mass m is homogeneously distributed over the spherical shell of area $4\pi R^2$, the differential mass of the ring is

$$dm = m \frac{2\pi R^2 \sin\phi d\phi}{4\pi R^2} = \tfrac{1}{2} m \sin\phi d\phi.$$

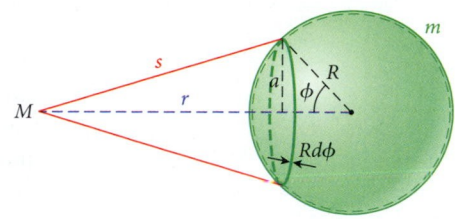

FIGURE 12.5 Gravitational force on a particle of mass M exerted by a spherical shell (hollow sphere) of mass m.

Because the ring is positioned symmetrically around the horizontal axis, there is no net force in the vertical direction from the ring acting on the mass M. The horizontal force component is (Figure 12.6)

$$dF_x = \cos\theta \left(G \frac{M \, dm}{s^2} \right) = \cos\theta \left(G \frac{Mm \sin\phi d\phi}{2s^2} \right). \qquad \text{(i)}$$

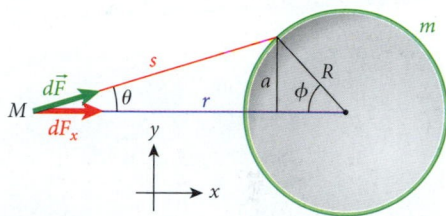

FIGURE 12.6 Cross section through the center of the spherical shell of Figure 12.5.

Now we can relate $\cos\theta$ to s, r, and R via the law of cosines:

$$\cos\theta = \frac{s^2 + r^2 - R^2}{2sr}.$$

In the same way,

$$\cos\phi = \frac{R^2 + r^2 - s^2}{2Rr}. \qquad \text{(ii)}$$

If we differentiate both sides of equation (ii), we obtain

$$-\sin\phi d\phi = -\frac{s}{Rr} ds.$$

Inserting the expressions for $\sin\phi \, d\phi$ and $\cos\theta$ into equation (i) for the differential force component gives

$$dF_x = \frac{s^2 + r^2 - R^2}{2sr} \left(G \frac{Mm}{2s^2} \right) \frac{s}{Rr} ds$$

$$= G \frac{Mm}{r^2} \left(\frac{s^2 + r^2 - R^2}{4s^2 R} \right) ds.$$

Now we can integrate over ds from the minimum value of $s = r - R$ to the maximum value of $s = r + R$:

$$F_x = \int_{r-R}^{r+R} G \frac{Mm}{r^2} \frac{(s^2 + r^2 - R^2)}{4s^2 R} ds = G \frac{Mm}{r^2} \underbrace{\int_{r-R}^{r+R} \frac{s^2 + r^2 - R^2}{4s^2 R} ds}_{\text{Equals 1}} = G \frac{Mm}{r^2}.$$

(It is not obvious that the value of the integral is 1, but you can look this up in an integral table.) Thus, a spherical shell (and by the superposition principle a solid sphere also) exerts the same force on the mass M as a point mass located at the center of the sphere, which is what we set out to prove.

Self-Test Opportunity 12.1

Derivation 12.1 assumes that $r > R$, which implies that the mass M is located outside the spherical shell. What changes if $r < R$?

The Solar System

The Solar System consists of the Sun, which contains the overwhelming majority of the total mass of the Solar System, the four Earthlike inner planets (Mercury, Venus, Earth, and Mars), the asteroid belt between the orbits of Mars and Jupiter, the four gas giants (Jupiter, Saturn, Uranus, and Neptune), a number of dwarf planets (including Ceres, Eris, Haumea, Makemake, and Pluto), and many other minor objects found in the Kuiper Belt. Figure 12.7 shows the orbits and relative sizes of the planets. Table 12.1 provides some physical data for the planets and the Sun.

Not listed as a planet is Pluto, and this omission deserves an explanation. The planet Neptune was discovered in 1846. This discovery had been predicted based on observed small irregularities in the orbit of Uranus, which hinted that another planet's gravitational interaction was the cause. Careful observations of the orbit of Neptune revealed more irregularities, which pointed toward the existence of yet another planet, and Pluto (mass = $1.3 \cdot 10^{22}$ kg) was discovered in 1930. After that, schoolchildren were taught that the Solar System has nine planets. However, in 2003, Sedna (mass = $\sim 3 \cdot 10^{21}$ kg) and, in 2005, Eris (mass = $1.7 \cdot 10^{22}$ kg) were discovered in the Kuiper Belt, a region in which various objects orbit the Sun at distances between 30 and 48 AU. When Eris's moon Dysnomia was discovered in late 2005, it allowed astronomers to calculate that Eris is more massive than Pluto, which began the discussion about what defines a planet. The choice was either to give Sedna, Eris, Ceres (an asteroid that was actually classified as a planet from 1801 to about 1850), and many other Plutolike objects in the Kuiper Belt the status of planet or to reclassify Pluto as a dwarf planet. In August 2006, the International Astronomical Union voted to remove full planetary status from Pluto.

As the story of Pluto shows, the Solar System still holds the potential for many discoveries. For example, almost 400,000 asteroids have been identified, and about 5000 more are discovered each month. The total mass of all asteroids in the asteroid belt is less than 5% of the mass of the Moon. However, more than 200 of these asteroids are known to have a diameter of more than 100 km! Tracking them is very important, considering the damage one could cause if it hit the Earth. Another area of ongoing research is the investigation of the objects in the Kuiper Belt. Some models postulate that the combined mass of all Kuiper Belt objects is up to 30 times the mass of the Earth, but the observed mass so far is smaller than this value by a factor of about 1000.

FIGURE 12.7 The Solar System. On this scale, the sizes of the planets themselves would be too small to be seen. The small yellow dot at the origin of the axis represents the Sun, but it is 30 times bigger than the Sun would appear if it were drawn to scale. The pictures of the planets and of part of the surface of the Sun (lower part of the figure) are all magnified by a factor of 30,000 relative to the scale for the orbits.

Table 12.1	Selected Physical Data for the Solar System						
Planet	Mean Radius (km)	Mass (10^{24} kg)	g (m/s^2)	Escape Speed (km/s)	Mean Orbital Radius (10^6 km)	Eccentricity	Orbital Period (yr)
Mercury	2,440	0.330	3.70	4.3	57.9	0.206	0.241
Venus	6,052	4.87	8.87	10.4	108.2	0.007	0.615
Earth	6,371	5.97	9.81	11.2	149.6	0.017	1
Mars	3,390	0.642	3.71	5.0	227.9	0.094	1.88
Jupiter	69,910	1900.	24.8	59.5	778	0.049	11.9
Saturn	58,230	568	10.4	35.5	1434	0.057	29.5
Uranus	25,360	86.8	8.87	21.3	2872	0.046	84.0
Neptune	24,620	102	11.2	23.5	4495	0.011	165
Sun	696,000	1,990,000	274	618	—	—	—

EXAMPLE 12.1 / Influence of Celestial Objects

Astronomy is the science that focuses on planets, stars, galaxies, and the universe as a whole. The similarly named field of astrology has no scientific basis whatsoever. It may be fun to read the daily horoscope, but constellations of stars and/or alignments of planets have no influence on our lives. The only way stars and planets can interact with us is through the gravitational force. Let's calculate the force of gravity exerted on a person by the Moon, Mars, and the stars in the constellation Gemini.

PROBLEM 1

Suppose you live in the Midwest of the United States. You go outside at 9:00 p.m. on February 16, 2008, and look up in the sky. You see the Moon, the planet Mars, and the constellation Gemini, as shown in Figure 12.8. Your mass is $m = 85$ kg. What is the force of gravity on you due to these celestial objects?

SOLUTION 1

Let's start with the Moon. The mass of the Moon is $7.35 \cdot 10^{22}$ kg, and the distance from the Moon to you is $3.84 \cdot 10^8$ m. The gravitational force the Moon exerts on you is then

$$F_{\text{Moon}} = G\frac{M_{\text{Moon}}m}{r_{\text{Moon}}^2} = \left(6.67 \cdot 10^{-11} \text{ N m}^2/\text{kg}^2\right)\frac{\left(7.35 \cdot 10^{22} \text{ kg}\right)\left(85 \text{ kg}\right)}{\left(3.84 \cdot 10^8 \text{ m}\right)^2} = 0.0028 \text{ N}.$$

The distance between Mars and Earth on February 16, 2008, was 136 million km, and the mass of Mars is $M_{\text{Mars}} = 6.4 \cdot 10^{23}$ kg. Thus, the gravitational force that Mars exerts on you is

$$F_{\text{Mars}} = G\frac{M_{\text{Mars}}m}{r_{\text{Mars}}^2} = \left(6.67 \cdot 10^{-11} \text{ N m}^2/\text{kg}^2\right)\frac{\left(6.4 \cdot 10^{23} \text{ kg}\right)\left(85 \text{ kg}\right)}{\left(1.36 \cdot 10^{11} \text{ m}\right)^2} = 2.0 \cdot 10^{-7} \text{ N}.$$

We can estimate the gravitational force exerted by the constellation Gemini by calculating the force exerted by its three brightest objects: Castor, Pollux, and Alhena (see Figure 12.8). Castor is a triple binary star system with a mass 6.0 times the mass of the Sun and is located a distance of 51.6 light-years from you. Pollux is a star with a mass 1.9 times the mass of the Sun, located 33.7 light-years away. Alhena is a binary star system with a mass 3.8 times the mass of the Sun, and it is 105 light-years away. The mass of the Sun is $2.0 \cdot 10^{30}$ kg, and a light-year corresponds to $9.5 \cdot 10^{15}$ m. The gravitational force exerted on you by the constellation Gemini is then

$$F_{\text{Gemini}} = G\frac{M_{\text{Castor}}m}{r_{\text{Castor}}^2} + G\frac{M_{\text{Pollux}}m}{r_{\text{Pollux}}^2} + G\frac{M_{\text{Alhena}}m}{r_{\text{Alhena}}^2} = Gm\left(\frac{M_{\text{Castor}}}{r_{\text{Castor}}^2} + \frac{M_{\text{Pollux}}}{r_{\text{Pollux}}^2} + \frac{M_{\text{Alhena}}}{r_{\text{Alhena}}^2}\right)$$

$$= \left(6.67 \cdot 10^{-11} \text{ N m}^2/\text{kg}^2\right)\left(85 \text{ kg}\right)\frac{2.0 \cdot 10^{30} \text{ kg}}{\left(9.5 \cdot 10^{15} \text{ m}\right)^2}\left[\frac{6.0}{\left(51.6\right)^2} + \frac{1.9}{\left(33.7\right)^2} + \frac{3.8}{\left(105\right)^2}\right]$$

$$= 5.4 \cdot 10^{-13} \text{ N}.$$

– Continued

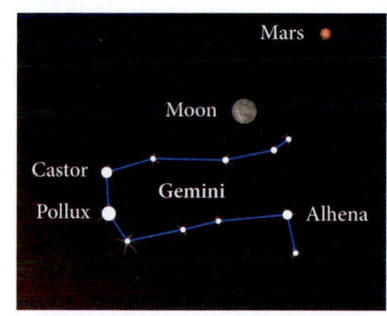

FIGURE 12.8 Relative positions of the Moon, Mars, and the constellation Gemini in the sky over the Midwest on February 16, 2008.

The Moon exerts a very small but measurable force on you, whereas Mars and Gemini exert only negligible forces.

PROBLEM 2

When Mars and Earth are at their minimum separation distance, they are $r_M = 5.6 \cdot 10^{10}$ m apart. How far away from you does a truck of mass 16,000 kg have to be to have the same gravitational interaction with your body as Mars does at this minimum separation distance?

SOLUTION

If the two gravitational forces are equal in magnitude, we can write

$$G\frac{M_M m}{r_M^2} = G\frac{m_T m}{r_T^2},$$

where m is your mass, m_T is the mass of the truck, and r_T is the distance we want to find. Canceling out your mass and the universal gravitational constant and solving for r_T, we have

$$r_T = r_M \sqrt{\frac{m_T}{M_M}}.$$

Inserting the given numerical values leads to

$$r_T = (5.6 \cdot 10^{10} \text{ m})\sqrt{\frac{1.6 \cdot 10^4 \text{ kg}}{6.4 \cdot 10^{23} \text{ kg}}} = 8.9 \text{ m}.$$

This result means that if you get closer to the truck than 8.9 m, it exerts a bigger gravitational pull on you than does Mars at its closest approach.

12.2   Gravitation near the Surface of the Earth

Now we can use the general expression for the gravitational interaction between two masses to reconsider the gravitational force due to the Earth on an object near the surface of the Earth. We can neglect the gravitational interaction of this object with any other objects, because the magnitude of the gravitational interaction with Earth is many orders of magnitudes larger as a result of Earth's very large mass. Since we can represent an extended object by a point particle of the same mass located at the object's center of gravity, any object on the surface of Earth is experiencing a gravitational force directed toward the center of Earth. This direction is straight down anywhere on the surface of Earth, completely in accordance with empirical evidence.

It is more interesting to determine the magnitude of the gravitational force that an object experiences near the surface of the Earth. Inserting the mass of the Earth, M_E, as the mass of one object in equation 12.1 and expressing the altitude h above the surface of the Earth as $h + R_E = r$, where R_E is the radius of the Earth, we find for this special case that

$$F = G\frac{M_E m}{(R_E + h)^2}. \tag{12.6}$$

Because $R_E = 6370$ km, the altitude, h, of the object above the ground can be neglected for many applications. If we make this assumption, we find that $F = mg$, with

$$g = \frac{GM_E}{R_E^2} = \frac{\left(6.67 \cdot 10^{-11} \text{ m}^3\text{kg}^{-1}\text{s}^{-2}\right)\left(5.97 \cdot 10^{24} \text{ kg}\right)}{\left(6.37 \cdot 10^6 \text{ m}\right)^2} = 9.81 \text{ m/s}^2. \tag{12.7}$$

As expected, near the surface of the Earth, the acceleration due to gravity can be approximated by the value of g that was introduced in Chapter 2. We can insert the mass and radius of other planets (see Table 12.1), moons, or stars into equation 12.7 and find their surface gravity as well. For example, the gravitational acceleration at the surface of the Sun is approximately 28 times larger than that at the surface of the Earth.

If we want to find g for altitudes where we cannot safely neglect h, we can start with equation 12.6, divide by the mass m to find the acceleration of that mass,

$$g(h) = \frac{GM_E}{(R_E + h)^2} = \frac{GM_E}{R_E^2}\left(1 + \frac{h}{R_E}\right)^{-2} = g\left(1 + \frac{h}{R_E}\right)^{-2},$$

and then expand in powers of h/R_E to obtain, to the first order,

$$g(h) \approx g\left(1 - 2\frac{h}{R_E} + \cdots\right). \tag{12.8}$$

Equation 12.8 holds for all values of the altitude h that are small compared to the radius of Earth, and it implies that the gravitational acceleration falls off approximately linearly as a function of the altitude above ground. At the top of Mount Everest, the Earth's highest peak, at an altitude of 8850 m, the gravitational acceleration is reduced by 0.28%, or by less than 0.03 m/s². The International Space Station is at an altitude of 365 km, where the gravitational acceleration is reduced by 11.5%, to a value of 8.7 m/s². For higher altitudes, the linear approximation of equation 12.8 should definitely not be used.

However, to obtain a more precise determination of the gravitational acceleration, we need to consider other effects. First, the Earth is not an exact sphere; it has a slightly larger radius at the Equator than at the poles. (The value of 6371 km in Table 12.1 is the mean radius of Earth; the radius varies from 6357 km at the poles to 6378 km at the Equator.) Second, the density of Earth is not uniform, and for a precise determination of the gravitational acceleration, the density of the ground right below the measurement makes a difference. Third, and perhaps most important, there is a systematic variation (as a function of the polar angle θ; see Figure 12.9) of the apparent gravitational acceleration, due to the rotation of the Earth and the associated centripetal acceleration. From Chapter 9 on circular motion, we know that the centripetal acceleration is given by $a_c = \omega^2 r$, where r is the radius of the circular motion. For the rotation of the Earth, this radius is the perpendicular distance to the rotation axis. At the poles, this distance is zero, and there is no contribution from the centripetal acceleration. At the Equator, $r = R_E$, and the maximum value for a_c is

$$a_{c,\text{max}} = \omega^2 R_E = \left(7.29\cdot10^{-5}\text{ s}^{-1}\right)^2 (6378\text{ km}) = 0.034\text{ m/s}^2.$$

Thus, we find that the reduction of the apparent gravitational acceleration at the Equator due to the Earth's rotation is approximately equal to the reduction at the top of Mount Everest.

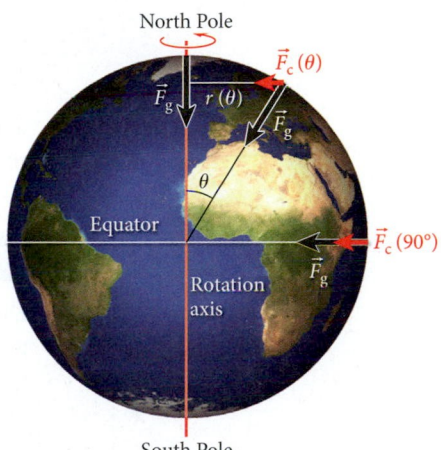

FIGURE 12.9 Variation of the effective force of gravity due to the Earth's rotation. (The lengths of the red arrows representing the centripetal force have been scaled up by a factor of 200 relative to the black arrows representing the gravitational force.)

Self-Test Opportunity 12.2

What is the acceleration due to Earth's gravity at a distance $d = 3R_E$ from the center of the Earth, where R_E is the radius of the Earth?

EXAMPLE 12.2 Gravitational Tear from a Black Hole

A black hole is a very massive and extremely compact object that is so dense that light emitted from its surface cannot escape. (Thus, it appears black.)

PROBLEM
Suppose a black hole has a mass of $6.0\cdot10^{30}$ kg, three times the mass of the Sun. A spaceship of length $h = 85$ m approaches the black hole until the front of the ship is at a distance $R = 13{,}500$ km from the black hole. What is the difference in the acceleration due to gravity between the front and the back of the ship?

SOLUTION
We can determine the gravitational acceleration at the front of the spaceship due to the black hole via

$$g_{\text{bh}} = \frac{GM_{\text{bh}}}{R^2} = \frac{\left(6.67\cdot10^{-11}\text{ m}^3\text{kg}^{-1}\text{s}^{-2}\right)\left(6.0\cdot10^{30}\text{ kg}\right)}{\left(1.35\cdot10^7\text{ m}\right)^2} = 2.2\cdot10^6\text{ m/s}^2.$$

– Continued

Now we can use the linear approximation of equation 12.8 to obtain

$$g_{bh}(h) - g_{bh}(0) = g_{bh}\left(1 - 2\frac{h}{R}\right) - g_{bh}(0) = -2g_{bh}\frac{h}{R},$$

where h is the length of the spaceship. Inserting the numerical values, we find the difference in the gravitational acceleration at the front and the back:

$$g_{bh}(h) - g_{bh}(0) = -2(2.2 \cdot 10^6 \text{ m/s}^2)\frac{85 \text{ m}}{1.35 \cdot 10^7 \text{ m}} = -27.7 \text{ m/s}^2.$$

You can see that in the vicinity of the black hole, the differential acceleration between front and back is so large that the spaceship would have to have tremendous integral strength to avoid being torn apart! (Near a black hole, Newton's Law of Gravity needs to be modified, but this example has ignored that change. In Section 35.8, we will discuss black holes in more detail and see why light is affected by the gravitational interaction.)

12.3 Gravitation inside the Earth

Derivation 12.1 showed that the gravitational interaction of a mass m with a spherically symmetrical mass distribution (where m is located outside the sphere) is not affected if the sphere is replaced with a point particle with the same total mass, located at its center of mass (or center of gravity). Derivation 12.2 now shows that on the inside of a spherical shell of uniform density, the net gravitational force is zero.

DERIVATION 12.2   Force of Gravity inside a Hollow Sphere

We want to show that the gravitational force acting on a point mass inside a hollow homogeneous spherical shell is zero everywhere inside the shell. To do this, we could use calculus and a mathematical law known as Gauss's Law. (Gauss's Law, discussed in Chapter 22, applies to electrostatic interaction and the Coulomb force, which is another force that falls off with the inverse square of the distance between the interacting objects.) However, instead we'll use a geometrical argument that was presented by Newton in 1687 in his book *Philosophiae Naturalis Principia Mathematica,* usually known as the *Principia.*

Consider the (infinitesimally thin) spherical shell shown in Figure 12.10. We mark a point P at an arbitrary location inside the shell and then draw a straight line through this point. This straight line intersects the shell at two points, and the distances between those two intersection points and P are the radii r_1 and r_2. Now we draw cones with tips at P and with small opening angle θ around the straight line. The areas where the two cones intersect the shell are called A_1 and A_2. These areas depend on the angle θ, which is the same for both of them. Area A_1 is also proportional to r_1^2, and area A_2 is proportional to r_2^2 (see Figure 12.10b). Further, since the shell is homogeneous, the mass of any segment of it is

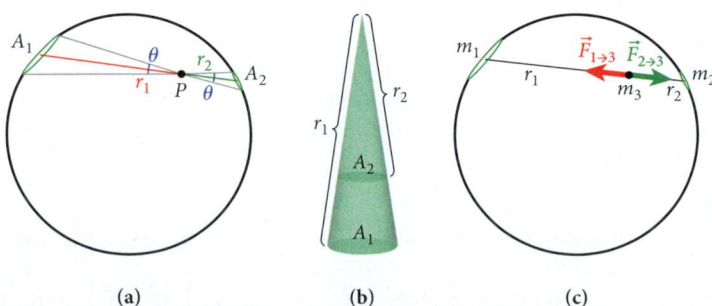

FIGURE 12.10 Gravitational interaction of a point with the surface of a hollow sphere: (a) cones from point P within the sphere to the sphere's surface; (b) details of the cones from P to the surface; (c) the balance of gravitational forces acting on P due to the opposite areas of the spherical shell.

proportional to the segment's area. Therefore, $m_1 = ar_1^2$ and $m_2 = ar_2^2$, with the same proportionality constant a.

The gravitational force $\vec{F}_{1\to3}$ that the mass m_1 of area A_1 exerts on the mass m_3 at point P then points along r_1 toward the center of A_1. The gravitational force $\vec{F}_{2\to3}$ acting on m_3 points in exactly the opposite direction along r_2. We can also find the magnitudes of these two forces:

$$F_{1\to3} = \frac{Gm_1m_3}{r_1^2} = \frac{G(ar_1^2)m_3}{r_1^2} = Gam_3$$

$$F_{2\to3} = \frac{Gm_2m_3}{r_2^2} = \frac{G(ar_2^2)m_3}{r_2^2} = Gam_3.$$

Since the dependence on the distance cancels out, the magnitudes of the two forces are the same. Since their magnitudes are the same, and their directions are opposite, the forces $\vec{F}_{1\to3}$ and $\vec{F}_{2\to3}$ exactly cancel each other (see Figure 12.10c).

Since the location of the point and the orientation of the line drawn through it were arbitrary, the result is true for any point inside the spherical shell. The entire shell can be accounted for with pairs of small opposite areas with no area left over. Thus, the net force of gravity acting on a point mass inside this spherical shell is indeed zero.

We can now gain physical insight into the gravitational force acting inside the Earth. Think of the Earth as composed of many concentric thin spherical shells. Then the gravitational force at a point P inside the Earth at a distance r from the center is due to those shells with a radius less than r. All shells with a greater radius do not contribute to the gravitational force on P. Furthermore, the mass, $M(r)$, of all contributing shells can be imagined to be concentrated in the center, at a distance r away from P. The gravitational force acting on an object of mass m at a distance r from the center of the Earth is then

$$F(r) = G\frac{M(r)m}{r^2}. \tag{12.9}$$

This is Newton's Law of Gravity (equation 12.1), with the mass of the contributing shells to be determined. In order to determine that mass, we make the simplifying assumption of a constant density, ρ_E, inside Earth. Then we obtain

$$M(r) = \rho_E V(r) = \rho_E \tfrac{4}{3}\pi r^3. \tag{12.10}$$

We can calculate the Earth's mass density from its total mass and radius:

$$\rho_E = \frac{M_E}{V_E} = \frac{M_E}{\tfrac{4}{3}\pi R_E^3} = \frac{\left(5.97 \cdot 10^{24}\ \text{kg}\right)}{\tfrac{4}{3}\pi\left(6.37 \cdot 10^6\ \text{m}\right)^3} = 5.5 \cdot 10^3\ \text{kg/m}^3.$$

Substituting the expression for the mass from equation 12.10 into equation 12.9, we obtain an equation for the radial dependence of the gravitational force inside Earth:

$$F(r) = G\frac{M(r)m}{r^2} = G\frac{\rho_E \tfrac{4}{3}\pi r^3 m}{r^2} = \tfrac{4}{3}\pi G\rho_E mr. \tag{12.11}$$

Equation 12.11 states that the gravitational force increases linearly with the distance from the center of the Earth. In particular, if the point is located exactly at the center of the Earth, zero gravitational force acts on it.

Now we can compare the radial dependence of the gravitational acceleration divided by g (for an object acted on by just the force due to the Earth) inside and outside the Earth. Figure 12.11 shows the linear rise of this quantity inside the Earth, and the fall-off in an inverse square fashion outside. At the surface of the Earth, these curves intersect, and the gravitational acceleration has the value of g. Also shown in the figure is the linear approximation (equation 12.8) to the dependence of the gravitational acceleration as a function of the height above the Earth's surface: $g(h) \approx g(1 - 2h/R_E + \cdots) \Rightarrow g(r) \approx g(3 - 2r/R_E + \cdots)$, because $r = h + R_E$. You can clearly see that this approximation is valid to within a few percentage points for altitudes of a few hundred kilometers above the surface of the Earth.

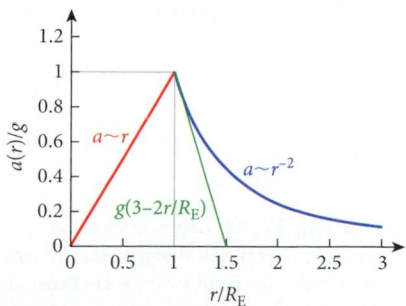

FIGURE 12.11 Dependence of the gravitational acceleration on the radial distance from the center of the Earth.

Concept Check 12.1

The Moon can be considered as a sphere of uniform density with mass M_M and radius R_M. At the center of the Moon, the magnitude of the gravitational force acting on a mass m due to the mass of the Moon is

a) mGM_M/R_M^2.

b) $\frac{1}{2}mGM_M/R_M^2$.

c) $\frac{3}{5}mGM_M/R_M^2$.

d) zero.

e) $2mGM_M/R_M^2$.

Note that the functional form of the force in equation 12.11 is that of a spring force, with a restoring force that increases linearly as a function of displacement from equilibrium at $r = 0$. Equation 12.11 specifies the magnitude of the gravitational force. Because the force always points toward the center of the Earth, we can also write equation 12.11 as a one-dimensional vector equation in terms of x, the displacement from equilibrium:

$$F_x(x) = -\tfrac{4}{3}\pi G\rho_E mx = -\frac{mg}{R_E}x = -kx.$$

This result is Hooke's Law for a spring, which we encountered in Chapter 5. Therefore, the "spring constant" of the gravitational force is

$$k = \tfrac{4}{3}\pi G\rho_E m = \frac{mg}{R_E}.$$

Similar considerations also apply to the gravitational force inside other spherically symmetrical mass distributions, such as planets or stars.

12.4   Gravitational Potential Energy

In Chapter 6, we saw that the gravitational potential energy is given by $U = mgh$, where h is the distance in the y-direction, provided the gravitational force is written $\vec{F} = -mg\hat{y}$ (with the sign convention that positive is straight up). Using Newton's Law of Gravity, we can obtain a more general expression for gravitational potential energy. Integrating equation 12.1 yields an expression for the gravitational potential energy of a system of two masses m_1 and m_2 separated by a distance r:

$$U(r) - U(\infty) = -\int_{\infty}^{r} \vec{F}(\vec{r}\,') \cdot d\vec{r}\,' = \int_{\infty}^{r} F(r')dr' = \int_{\infty}^{r} G\frac{m_1 m_2}{r'^2}dr'$$

$$= Gm_1 m_2 \int_{\infty}^{r} \frac{1}{r'^2}dr' = -Gm_1 m_2 \frac{1}{r'}\Big|_{\infty}^{r} = -G\frac{m_1 m_2}{r}.$$

The first part of this equation is the general relation between force and potential energy. For a gravitational interaction, the force depends only on the radial separation, $\vec{F}(\vec{r}) = \vec{F}(r)$, and is always attractive, $\vec{F}(r) = -F(r)\hat{r}$. Therefore $\vec{F}(\vec{r}) \cdot d\vec{r} = -F(r)\,dr$, which was used in the second part of this equation.

Note that the equation describing gravitational potential energy only tells us the *difference* between the gravitational potential energy at separation r and at separation infinity. We set $U(\infty) = 0$, which implies that the gravitational potential energy vanishes between two objects that are separated by an infinite distance. This choice gives us the following expression for the **gravitational potential energy** as a function of the separation of two masses:

$$U(r) = -G\frac{m_1 m_2}{r}. \tag{12.12}$$

Note that the gravitational potential energy is always less than zero with $U(\infty) = 0$. This dependence of the gravitational potential energy on $1/r$ is illustrated by the red curve in Figure 12.12 for an arbitrary mass near the surface of the Earth.

For interactions among more than two objects, we can write all pairwise interactions between two of them and integrate. The gravitational potential energies from these interactions are simply added to give the total gravitational potential energy. For three point particles, we find, for example,

$$U = U_{12} + U_{13} + U_{23} = -G\frac{m_1 m_2}{|\vec{r}_1 - \vec{r}_2|} - G\frac{m_1 m_3}{|\vec{r}_1 - \vec{r}_3|} - G\frac{m_2 m_3}{|\vec{r}_2 - \vec{r}_3|}.$$

An important special case occurs when one of the two interacting objects is the Earth. For altitudes h that are small compared to the radius of the Earth, we expect to repeat the previous result that the gravitational potential energy is mgh. Because the Earth is many orders of magnitude more massive than any object on the surface of the Earth for which we might want to calculate the gravitational potential energy, the combined center

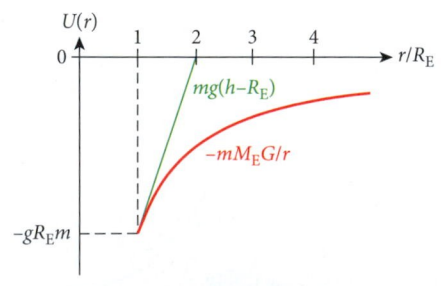

FIGURE 12.12 Dependence of the gravitational potential energy on the distance to the center of the Earth, for distances larger than the radius of the Earth. The red curve represents the exact expression; the green line represents the linear approximation for values of r not much larger than the radius of the Earth, R_E.

of mass of the Earth and the object is practically identical to the center of mass of the Earth, which we then select as the origin of the coordinate system. Using equation 12.12, this results in

$$U(h) = -G\frac{M_E m}{R_E + h} = -G\frac{M_E m}{R_E}\left(1 + \frac{h}{R_E}\right)^{-1}$$

$$\approx -\frac{GM_E m}{R_E} + \frac{GM_E m}{R_E^2}h = -gmR_E + mgh.$$

In the second step here, we used the fact that $h \ll R_E$ and expanded. This result (plotted in green in Figure 12.12) looks almost like the expression $U = mgh$ from Chapter 6, except for the addition of the constant term $-gmR_E$. This constant term is a result of the choice of the integration constant in equation 12.12. However, as was stressed in Chapter 6, we can add any additive constant to the expression for potential energy without changing the physical results for the motion of objects. The only physically relevant quantity is the difference in potential energy between two different locations. Taking the difference between altitude h and altitude zero results in

$$\Delta U = U(h) - U(0) = (-gmR_E + mgh) - (-gmR_E) = mgh.$$

As expected, the additive constant $-gmR_E$ cancels out, and we obtain the same result for low altitudes, h, that we had previously derived: $\Delta U = mgh$.

Escape Speed

With an expression for the gravitational potential energy, we can determine the total mechanical energy for a system consisting of an object of mass m_1 and speed v_1, that has a gravitational interaction with another object of mass m_2 and speed v_2, if the two objects are separated by a distance $r = |\vec{r}_1 - \vec{r}_2|$:

$$E = K + U = \tfrac{1}{2}m_1 v_1^2 + \tfrac{1}{2}m_2 v_2^2 - \frac{Gm_1 m_2}{|\vec{r}_1 - \vec{r}_2|}. \qquad (12.13)$$

Of particular interest is the case where one of the objects is the Earth ($m_1 \equiv M_E$). If we consider the reference frame to be that of the Earth ($v_1 = 0$), the Earth has no kinetic energy in this frame. Again, we put the origin of the coordinate system at the center of the Earth. Then the expression for the total energy of this system is just the kinetic energy of object 2 (but we have omitted the subscript 2), plus the gravitational potential energy:

$$E = \tfrac{1}{2}mv^2 - \frac{GM_E m}{R_E + h},$$

where, as before, h is the altitude above Earth's surface of the object with mass m.

If we want to find out what initial speed a projectile must have to escape to an infinite distance from Earth, called the **escape speed,** v_E (where E stands for Earth), we can use energy conservation. At infinite separation, the gravitational potential energy is zero, and the minimum kinetic energy is also zero. Thus, the minimum total energy with which a projectile can escape from the Earth's gravitational influence is zero. Energy conservation then implies, starting from the Earth's surface:

$$E(h = 0) = \tfrac{1}{2}mv_E^2 - \frac{GM_E m}{R_E} = 0.$$

Solving this for v_E gives an expression for the minimum escape speed:

$$v_E = \sqrt{\frac{2GM_E}{R_E}}. \qquad (12.14)$$

Inserting the numerical values of these constants, we finally obtain

$$v_E = \sqrt{\frac{2\left(6.67 \cdot 10^{-11}\ \text{m}^3\text{kg}^{-1}\text{s}^{-2}\right)\left(5.97 \cdot 10^{24}\ \text{kg}\right)}{6.37 \cdot 10^6\ \text{m}}} = 11.2\ \text{km/s}.$$

This escape speed is approximately equal to 25,000 mph. The same calculation can also be performed for other planets, moons, and stars by inserting the relevant constants (for example, see Table 12.1).

Note that the angle with which the projectile is launched into space does not enter into the expression for the escape speed. Therefore, it does not matter if the projectile is shot straight up or almost in a horizontal direction. However, we neglected air resistance, and the launch angle would make a difference if we accounted for its effect. An even bigger effect results from the Earth's rotation. Since the Earth rotates once around its axis each day, a point on the surface of the Earth located at the Equator has a speed of $v = 2\pi R_E/(1 \text{ day}) \approx 0.46$ km/s, which decreases to zero at the poles. The direction of the corresponding velocity vector points east, tangentially to the Earth's surface. Therefore, the launch angle matters most at the Equator. For a projectile fired in the eastern direction from any location on the Equator, the escape speed is reduced to approximately 10.7 km/s.

Can a projectile launched from the surface of the Earth with a speed of 11.2 km/s escape the Solar System? Doesn't the gravitational potential energy of the projectile due to its interaction with the Sun play a role? At first glance, it would appear not. After all, the gravitational force the Sun exerts on an object located near the surface of the Earth is negligible compared to the force the Earth exerts on that object. As proof, consider that if you jump up in the air, you land at the same place, independent of what time of day it is, that is, where the Sun is in the sky. Thus, we can indeed neglect the Sun's gravitational force near the surface of the Earth.

However, the gravitational force is far different from the gravitational potential energy. In contrast to the force, which falls off as r^{-2}, the potential energy falls off much more slowly, as it is proportional to r^{-1}. It is straightforward to generalize equation 12.14 for the escape speed from any planet or star with mass M, if the object is initially separated a distance R from the center of that planet or star:

$$v = \sqrt{\frac{2GM}{R}}. \tag{12.15}$$

Inserting the mass of the Sun and the size of the orbit of the Earth we find v_S, the speed needed for an object to escape from the gravitational influence of the Sun if it is initially a distance from it equal to the radius of the Earth's orbit:

$$v_S = \sqrt{\frac{2\left(6.67 \cdot 10^{-11} \text{ m}^3\text{kg}^{-1}\text{s}^{-2}\right)\left(1.99 \cdot 10^{30} \text{ kg}\right)}{1.50 \cdot 10^{11} \text{ m}}} = 42 \text{ km/s}.$$

This is quite an astonishing result: The escape speed needed to leave the Solar System from the orbit of the Earth is almost four times larger than the escape speed needed to get away from the gravitational attraction of the Earth.

Launching a projectile from the surface of the Earth with enough speed to leave the Solar System requires overcoming the combined gravitational potential energy of the Earth and the Sun. Since the two potential energies add, the combined escape speed is

$$v_{ES} = \sqrt{v_E^2 + v_S^2} = 43.5 \text{ km/s}.$$

We found that the rotation of the Earth has a nonnegligible, but still small, effect on the escape speed. However, a much bigger effect arises from the orbital motion of the Earth around the Sun. The Earth orbits the Sun with an orbital speed of $v_O = 2\pi R_{ES}/(1 \text{ yr}) = 29.8$ km/s, where R_{ES} is the distance between Earth and Sun (149.6, or approximately 150, million km, according to Table 12.1). A projectile launched in the direction of this orbital velocity vector needs a launch velocity of only $v_{ES,min} = (43.5 - 29.8)$ km/s = 13.7 km/s, whereas one launched in the opposite direction must have $v_{ES,max} = (43.5 + 29.8)$ km/s = 73.3 km/s. Other launch angles produce all values between these two extremes.

When NASA launches a probe to explore the outer planets or to leave the Solar System—for example, *Voyager 2*—the gravity assist technique is used to lower the required launch velocity. This technique is illustrated in Figure 12.13. Part (a) is a sketch of a flyby of Jupiter as seen by an observer at rest relative to Jupiter. Notice that the velocity vector of the spacecraft changes direction but has the same length at the same distance from Jupiter as the spacecraft approaches and as it leaves. This is a consequence of energy conservation. Figure 12.13b is a sketch of the spacecraft's trajectory as seen by an observer at rest relative to the Sun. In this reference frame, Jupiter moves with an orbital velocity of approximately 13 km/s. To transform from Jupiter's reference frame to the Sun's reference frame, we have to add the velocity of Jupiter to the velocities observed in Jupiter's frame (red arrows) to obtain the velocities in the Sun's frame (blue arrows). As you can see from the figure, in the Sun's frame, the length of the final velocity vector is significantly greater than that of the initial velocity vector. This means that the spacecraft has acquired significant additional kinetic energy (and Jupiter has lost this kinetic energy) during the flyby, allowing it to continue escaping the gravitational pull of the Sun.

EXAMPLE 12.3 / Asteroid Impact

One of the most likely causes of the extinction of dinosaurs at the end of the Cretaceous period, about 65 million years ago, was a large asteroid hitting the Earth. Let's look at the energy released during an asteroid impact.

PROBLEM

Suppose a spherical asteroid, with a radius of 1.00 km and a mass density of 4750 kg/m³, enters the Solar System with negligible speed and then collides with Earth in such a way that it hits Earth from a radial direction with respect to the Sun. What kinetic energy will this asteroid have in the Earth's reference frame just before its impact on Earth?

SOLUTION

First, we calculate the mass of the asteroid:

$$m_a = V_a \rho_a = \tfrac{4}{3}\pi r_a^3 \rho_a = \tfrac{4}{3}\pi (1.00 \cdot 10^3 \text{ m})^3 (4750 \text{ kg/m}^3) = 1.99 \cdot 10^{13} \text{ kg.}$$

If the asteroid hits the Earth in a radial direction with respect to the Sun, the Earth's velocity vector will be perpendicular to that of the asteroid at impact, because the Earth moves tangentially around the Sun. Thus, there are three contributions to the kinetic energy of the asteroid as measured in Earth's reference frame: (1) conversion of the gravitational potential energy between Earth and asteroid, (2) conversion of the gravitational potential energy between Sun and asteroid, and (3) kinetic energy of the Earth's motion relative to that of the asteroid.

Because we have already calculated the escape speeds corresponding to the two gravitational potential energy terms and because the asteroid will arrive with the kinetic energy that corresponds to these escape speeds, we can simply write

$$K = \tfrac{1}{2}m_a (v_E^2 + v_S^2 + v_O^2).$$

Now we insert the numerical values:

$$K = 0.5(1.99 \cdot 10^{13} \text{ kg})[(1.1 \cdot 10^4 \text{ m/s})^2 + (4.2 \cdot 10^4 \text{ m/s})^2 + (3.0 \cdot 10^4 \text{ m/s})^2]r$$

$$= 2.8 \cdot 10^{22} \text{ J.}$$

This value is equivalent to the energy released by approximately 300 million nuclear weapons of the magnitude of those used to destroy Hiroshima and Nagasaki in World War II. You can begin to understand the destructive power of an asteroid impact of this magnitude—an event like this could wipe out human life on Earth. A somewhat bigger asteroid, with a diameter of 6 to 10 km, hit Earth near the tip of the Yucatan peninsula in the Gulf of Mexico approximately 65 million years ago. It is believed to be responsible for the K-T (Cretaceous-Tertiary) extinction, which killed off the dinosaurs.

DISCUSSION

Figure 12.14 shows the approximately 1.5 km in diameter and almost 200 m deep Barringer impact crater, which was formed approximately 50,000 years ago when a meteorite of approximately 50-m diameter, with a mass of approximately 300,000 tons (3·10⁸ kg), hit Earth with a speed of approximately 12 km/s. This was a much smaller object than the asteroid described in this example, but the impact still had the destructive power of 250 atomic bombs of the Hiroshima/Nagasaki class.

(a)

(b)

FIGURE 12.13 Gravity assist technique: (a) trajectory of a spacecraft passing Jupiter, as seen in Jupiter's reference frame; (b) the same trajectory as seen in the reference frame of the Sun, in which Jupiter moves with a velocity of approximately 13 km/s.

FIGURE 12.14 Barringer impact crater in central Arizona.

Gravitational Potential

Equation 12.12 states that the gravitational potential energy of any object is proportional to that object's mass. When we employed energy conservation to calculate the escape speed, we saw that the mass of the object canceled out, because both kinetic energy and gravitational potential energy are proportional to an object's mass. Thus, the kinematics are independent of the object's mass. For example, let's consider the gravitational potential energy of a mass m interacting with Earth, $U_E(r) = -GM_E m/r$. The Earth's gravitational potential $V_E(r)$ is defined as the ratio of the gravitational potential energy to the mass of the object, $V_E(r) = U_E(r)/m$, or

$$V_E(r) = -\frac{GM_E}{r}. \tag{12.16}$$

This definition has the advantage of giving information on the gravitational interaction with Earth, independently of the other mass involved. (We will explore the concept of potentials in much more depth in Chapter 23 on electric potentials.)

SOLVED PROBLEM 12.1   Astronaut on a Small Moon

PROBLEM

A small spherical moon has a radius of $6.30 \cdot 10^4$ m and a mass of $8.00 \cdot 10^{18}$ kg. An astronaut standing on the surface of the moon throws a rock straight up. The rock reaches a maximum height of 2.20 km above the surface of the moon before it returns to the surface. What was the initial speed of the rock as it left the hand of the astronaut? (This moon is too small to have an atmosphere.)

SOLUTION

THINK   We know the mass and radius of the moon, allowing us to compute the gravitational potential. Knowing the height that the rock attains, we can use energy conservation to calculate the initial speed of the rock.

SKETCH   Figure 12.15 is a sketch showing the rock at its highest point.

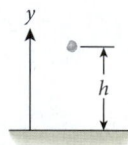

FIGURE 12.15 The rock at the highest point on its straight-line trajectory.

RESEARCH   We denote the mass of the rock as m, the mass of the moon as m_{moon}, and the radius of the moon as R_{moon}. The potential energy of the rock on the surface of the moon is then

$$U(R_{moon}) = -G\frac{m_{moon}}{R_{moon}}m. \tag{i}$$

The potential energy at the top of the rock's trajectory is

$$U(R_{moon} + h) = -G\frac{m_{moon}}{R_{moon} + h}m. \tag{ii}$$

The kinetic energy of the rock as it leaves the hand of the astronaut depends on its mass and the initial speed v:

$$K(R_{moon}) = \tfrac{1}{2}mv^2. \tag{iii}$$

At the top of the rock's trajectory, $K(R_{moon} + h) = 0$. Conservation of total energy now helps us solve the problem:

$$U(R_{moon}) + K(R_{moon}) = U(R_{moon} + h) + K(R_{moon} + h). \tag{iv}$$

SIMPLIFY   We substitute the expressions from equations (i) through (iii) into (iv):

$$-G\frac{m_{moon}}{R_{moon}}m + \tfrac{1}{2}mv^2 = -G\frac{m_{moon}}{R_{moon} + h}m + 0 \Rightarrow$$

$$-G\frac{m_{moon}}{R_{moon}} + \tfrac{1}{2}v^2 = -G\frac{m_{moon}}{R_{moon} + h} \Rightarrow$$

$$v = \sqrt{2Gm_{moon}\left(\frac{1}{R_{moon}} - \frac{1}{R_{moon} + h}\right)}.$$

CALCULATE Putting in the numerical values, we get

$$v = \sqrt{2(6.67 \cdot 10^{-11} \text{ N m}^2/\text{kg}^2)(8.00 \cdot 10^{18} \text{ kg})\left(\frac{1}{6.30 \cdot 10^4 \text{ m}} - \frac{1}{(6.30 \cdot 10^4 \text{ m}) + (2.20 \cdot 10^3 \text{ m})}\right)}$$

$$= 23.9078 \text{ m/s}.$$

ROUND Rounding to three significant figures gives

$$v = 23.9 \text{ m/s}.$$

DOUBLE-CHECK To double-check our result, let's make the simplifying assumption that the force of gravity on the small moon does not change with altitude. We can find the initial speed of the rock using conservation of energy:

$$mg_{\text{moon}}h = \tfrac{1}{2}mv^2, \tag{v}$$

where h is the height attained by the rock, v is the initial speed, m is the mass of the rock, and g_{moon} is the acceleration due to gravity at the surface of the small moon. The acceleration due to gravity is

$$g_{\text{moon}} = G\frac{m_{\text{moon}}}{R_{\text{moon}}^2} = \left(6.67 \cdot 10^{-11} \text{ N m}^2/\text{kg}^2\right)\frac{8.00 \cdot 10^{18} \text{ kg}}{\left(6.30 \cdot 10^4 \text{ m}\right)^2} = 0.134 \text{ m/s}^2.$$

Solving equation (v) for the initial speed of the rock gives us

$$v = \sqrt{2g_{\text{moon}}h}.$$

Putting in the numerical values, we get

$$v = \sqrt{2\left(0.134 \text{ m/s}^2\right)\left(2200 \text{ m}\right)} = 24.3 \text{ m/s}.$$

This result is close to but larger than our answer, 23.9 m/s. The fact that the simple assumption of constant gravitational force leads to a larger initial speed than found for the realistic case where the force decreases with altitude makes sense. The close agreement of the two results also makes sense because the altitude gained (2.20 km) is not too large compared to the radius of the small moon (63.0 km). Thus, our answer seems reasonable.

12.5 Kepler's Laws and Planetary Motion

Johannes Kepler (1571–1630) used empirical observations, mainly from data gathered by Tycho Brahe, and sophisticated calculations to arrive at the famous **Kepler's laws of planetary motion,** published in 1609 and 1619. These laws were published decades before the birth, in 1643, of Isaac Newton, whose law of gravitation would eventually show *why* Kepler's laws were true. What is particularly significant about Kepler's laws is that they challenged the prevailing world view at the time, with the Earth in the center of the universe (a geocentric theory) and the Sun and all the planets and stars orbiting around it, just as the Moon does. Kepler and other pioneers, particularly Nicolaus Copernicus and Galileo Galilei, changed this geocentric view into a heliocentric (Sun-centered) cosmology. Today, space probes have provided direct observations from vantage points outside the Earth's atmosphere and verified that Copernicus, Kepler, and Galilei were correct. However, the simplicity with which the heliocentric model was able to explain astronomical observations won intelligent people over to their view long before external observations were possible.

Kepler's First Law: Orbits
All planets move in elliptical orbits with the Sun at one focal point.

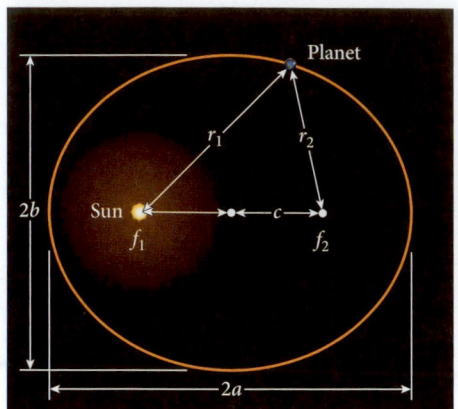

FIGURE 12.16 Parameters used in describing ellipses and elliptical orbits.

An *ellipse* is a closed curve in a two-dimensional plane. It has two focal points, f_1 and f_2, separated by a distance $2c$ (Figure 12.16). For each point on an ellipse, the sum of the distances to the two focal points is a constant:

$$r_1 + r_2 = 2a.$$

The length a is called the semimajor axis of the ellipse (see Figure 12.16). (*Note:* Unfortunately, the standard notation for the semimajor axis of an ellipse uses the same letter, a, as is conventionally used to symbolize acceleration. You should be careful to avoid confusion.) The semiminor axis, b, is related to a and c via

$$b^2 \equiv a^2 - c^2.$$

In terms of the Cartesian coordinates x and y, the points on the ellipse satisfy the equation

$$\frac{x^2}{a^2} + \frac{y^2}{b^2} = 1,$$

where the origin of the coordinate system is at the center of the ellipse. If $a = b$, a circle (a special case of an ellipse) results.

It is useful to introduce the eccentricity, e, of an ellipse, defined as

$$e = \frac{c}{a} = \sqrt{1 - \frac{b^2}{a^2}}.$$

An eccentricity of zero, the smallest possible value, characterizes a circle. The ellipse shown in Figure 12.16 has an eccentricity of 0.5. The eccentricity of the Earth's orbit around the Sun is only 0.017. If you were to plot an ellipse with this value of e, you could not distinguish it from a circle by visual inspection. The length of the semiminor axis of Earth's orbit is approximately 99.98% of the length of the semimajor axis. At its closest approach to the Sun, called the *perihelion*, the Earth is 147.1 million km away from the Sun. The *aphelion*, which is the farthest point from the Sun in Earth's orbit, is 152.1 million km.

It is important to note that the change in seasons is *not* caused primarily by the eccentricity of the Earth's orbit. (The point of closest approach to the Sun is reached in early January each year, in the middle of the cold season in the Northern Hemisphere.) Instead, the seasons are caused by the fact that the Earth's axis of rotation is tilted by an angle of 23.4° relative to the plane of the orbital ellipse. This tilt exposes the Northern Hemisphere to the Sun's rays for longer periods and at a more direct angle in the summer months.

Among the other planetary orbits, Mercury's has the largest eccentricity, 0.206. (Pluto's orbital eccentricity is even larger, at 0.249, but Pluto has been declassified as a planet since August 2006.) Venus's orbit has the smallest eccentricity, 0.007, followed by Neptune with an eccentricity of 0.009.

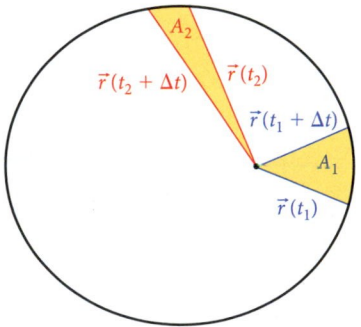

FIGURE 12.17 Kepler's Second Law states that equal areas are swept out in equal time periods, or $A_1 = A_2$.

Kepler's Second Law: Areas

A straight line connecting the center of the Sun and the center of any planet (Figure 12.17) sweeps out an equal area in any given time interval:

$$\frac{dA}{dt} = \text{constant.} \tag{12.17}$$

Kepler's Third Law: Periods

The square of the period of a planet's orbit is proportional to the cube of the semimajor axis of the orbit:

$$\frac{T^2}{a^3} = \text{constant.} \tag{12.18}$$

This proportionality constant can be expressed in terms of the mass of the Sun and the universal gravitational constant

$$\frac{T^2}{a^3} = \frac{4\pi^2}{GM}.$$

(12.19)

DERIVATION 12.3 / Kepler's Laws

The general proof of Kepler's laws uses Newton's Law of Gravity, equation 12.1, and the law of conservation of angular momentum. With quite a bit of algebra and calculus, it is then possible to prove all three of Kepler's laws. Here we will derive Kepler's Second and Third Laws for circular orbits with the Sun in the center, allowing us to put the Sun at the origin of the coordinate system and neglect the motion of the Sun around the common center of mass of the Sun-planet system.

First, we show that circular motion is indeed possible. From Chapter 9, we know that in order to obtain a closed circular orbit, the centripetal force needs to be equal to the gravitational force:

$$m\frac{v^2}{r} = G\frac{Mm}{r^2} \Rightarrow v = \sqrt{\frac{GM}{r}}.$$

This result establishes two important facts: First, the mass of the orbiting object has canceled out, so all objects can have the same orbit, provided that their mass is small compared to the Sun. Second, any given orbital radius, r, has a unique orbital velocity corresponding to it. Also, for a given orbital radius, we obtain a constant value of the angular velocity:

$$\omega = \frac{v}{r} = \sqrt{\frac{GM}{r^3}}.$$

(i)

Next, we examine the area swept out by a radial vector connecting the Sun and the planet. As indicated in Figure 12.18, the area swept out is $dA = \frac{1}{2}rs = \frac{1}{2}r^2 d\theta$. Taking the time derivative results in

$$\frac{dA}{dt} = \frac{1}{2}r^2\frac{d\theta}{dt} = \frac{1}{2}r^2\omega.$$

Because, for a given orbit, ω and r are constant, we have derived Kepler's Second Law, which states that $dA/dt =$ constant.

Finally, for Kepler's Third Law, we use $T = 2\pi/\omega$ and substitute the expression for the angular velocity from equation (i). This results in

$$T = 2\pi\sqrt{\frac{r^3}{GM}}.$$

We can rearrange this equation and obtain

$$\frac{T^2}{r^3} = \frac{4\pi^2}{GM}.$$

This proves Kepler's Third Law and gives the value for the proportionality constant between the square of the orbital period and the cube of the orbital radius.

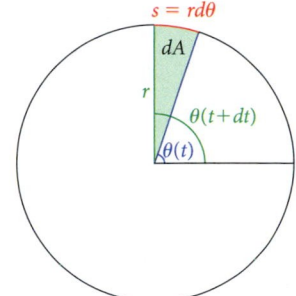

FIGURE 12.18 Angle, arc length, and area as a function of time.

Again, keep in mind that Kepler's laws are valid for elliptic orbits in general, not just for circular orbits. Instead of referring to the radius of the circle, r has to be taken as the semimajor axis of the ellipse for such orbits. Perhaps an even more useful formulation of Kepler's Third Law may be written as

$$\frac{T_1^2}{a_1^3} = \frac{T_2^2}{a_2^3}.$$

(12.20)

With this formula, we can easily find orbital periods and radii for two different orbiting objects.

SOLVED PROBLEM 12.2 | Orbital Period of Sedna

PROBLEM

On November 14, 2003, astronomers discovered a previously unknown object in part of the Kuiper Belt beyond the orbit of Neptune. They named this object Sedna, after the Inuit goddess of the sea. The average distance of Sedna from the Sun is $77.6 \cdot 10^9$ km. How long does it take Sedna to complete one orbit around the Sun?

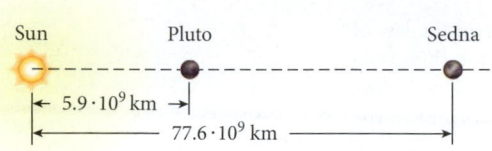

FIGURE 12.19 The distance of Sedna from the Sun compared with the distance of Pluto from the Sun.

SOLUTION

THINK   We can use Kepler's laws to relate Sedna's distance from the Sun to the period of Sedna's orbit around the Sun.

SKETCH   A sketch comparing the average distance of Sedna from the Sun with the average distance of Pluto from the Sun is shown in Figure 12.19.

RESEARCH   We can relate the orbit of Sedna to the known orbit of the Earth using equation 12.20 (a form of Kepler's Third Law):

$$\frac{T_{Earth}^2}{a_{Earth}^3} = \frac{T_{Sedna}^2}{a_{Sedna}^3}, \tag{i}$$

where T_{Earth} is the period of Earth's orbit, a_{Earth} is the radius of Earth's orbit, T_{Sedna} is the period of Sedna's orbit, and a_{Sedna} is the radius of Sedna's orbit.

SIMPLIFY   We can solve equation (i) for the period of Sedna's orbit:

$$T_{Sedna} = T_{Earth} \left(\frac{a_{Sedna}}{a_{Earth}} \right)^{3/2}.$$

CALCULATE   Putting in the numerical values, we get

$$T_{Sedna} = \left(1 \text{ yr}\right) \left(\frac{77.6 \cdot 10^9 \text{ km}}{0.150 \cdot 10^9 \text{ km}} \right)^{3/2} = 12{,}018 \text{ yr.}$$

ROUND   We report our result to two significant figures:

$$T_{Sedna} = 1.18 \cdot 10^4 \text{ yr.}$$

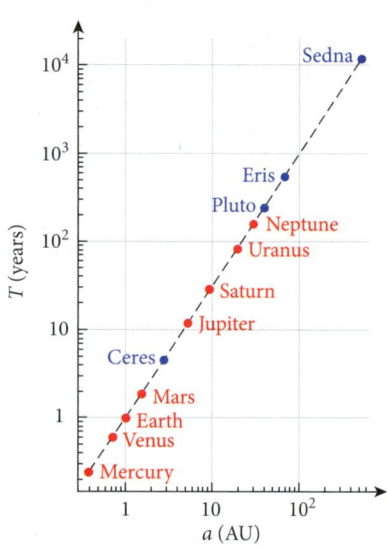

FIGURE 12.20 Orbital period versus length of the semimajor axis for orbits of objects in the Solar System.

DOUBLE-CHECK   We can compare our result for Sedna with the measured values for the semimajor axes of the orbits and orbital periods of the planets and several dwarf planets. As Figure 12.20 shows, our calculated result (dashed line, representing Kepler's Third Law) fits well with the extrapolation of the data from the planets (red dots) and dwarf planets (blue dots).

We can also use Kepler's Third Law to determine the mass of the Sun. We obtain this result by solving equation 12.19 for the mass of the Sun:

$$M = \frac{4\pi^2 a^3}{GT^2}. \tag{12.21}$$

Inserting the data for Earth's orbital period and radius gives

$$M = \frac{4\pi^2 (1.496 \cdot 10^{11} \text{ m})^3}{(6.67 \cdot 10^{-11} \text{ m}^3 \text{kg}^{-1} \text{s}^{-2})(3.16 \cdot 10^7 \text{ s})^2} = 1.98 \cdot 10^{30} \text{ kg.}$$

It is also possible to use Kepler's Third Law to determine the mass of the Earth from the period and radius of the Moon's orbit around Earth. In fact, astronomers can use this law to determine the mass of any astronomical object that has a satellite orbiting it if they know the radius and period of the orbit.

Self-Test Opportunity 12.3

Use the fact that the gravitational interaction between Earth and Sun provides the centripetal force that keeps Earth on its orbit to confirm equation 12.21. (Assume a circular orbit.)

EXAMPLE 12.4 | Black Hole in the Center of the Milky Way

PROBLEM
There is a supermassive black hole in the center of the Milky Way. What is its mass?

SOLUTION
In June 2007, astronomers measured the mass of the center of the Milky Way. Seven stars orbiting near the galactic center had been tracked for 15 years, as shown in Figure 12.21. The periods and semimajor axes extracted by the astronomers are shown in Table 12.2. Using these data and Kepler's Third Law (equation 12.21), we can calculate the mass of the galactic center. The resulting mass of the galactic center is shown in Table 12.2 for each set of star measurements. The average mass of the galactic center is $3.7 \cdot 10^6$ times the mass of the Sun. Thus, astronomers infer that there is a supermassive black hole at the center of the galaxy, because no star is visible at that point.

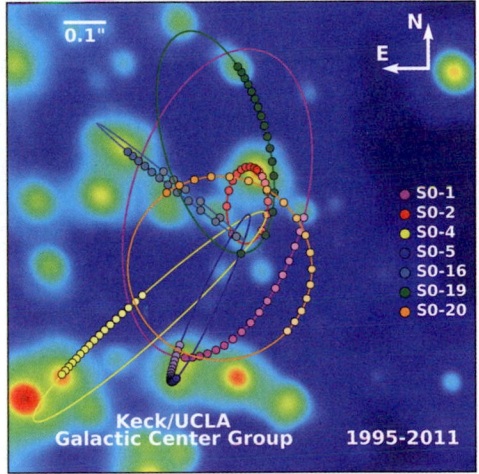

DISCUSSION
If there is a supermassive black hole in the center of the Milky Way, you may ask yourself why isn't Earth being pulled toward it? The answer is the same as the answer to the question of why the Earth does not fall into the Sun: The Earth orbits the Sun, and the Sun orbits the galactic center, a distance of 26,000 light-years away from the Solar System.

FIGURE 12.21 The orbits of seven stars close to the center of the Milky Way as tracked by astronomers from the Keck/UCLA Galactic Center Group from 1995 to 2011. The measured positions, represented by colored dots, are superimposed on a picture of the stars taken at the start of the tracking. The lines represent fits to the measurements that were used to extract the periods and semimajor axes of the stars' orbits. The side of the image is a distance of approximately $\frac{1}{15}$ of a light-year.

Table 12.2	Periods and Semimajor Axes for Stars Orbiting the Center of the Milky Way					
Star	Period (yr)	Semimajor Axis (AU)	Period (10^8 s)	Semimajor Axis (10^{14} m)	Mass of Galactic Center (10^{36} kg)	Equivalent in Solar Masses (10^6)
S0-2	14.43	919	4.55	1.37	7.44	3.74
S0-16	36	1680	113	2.51	7.31	3.67
S0-19	37.2	1720	117	2.57	7.34	3.69
S0-20	43	1900	135	2.84	7.41	3.72
S0-1	190	5100	599	7.63	7.34	3.69
S0-4	2600	30,000	819	44.9	7.98	4.01
S0-5	9900	70,000	3120	105	6.99	3.51
	Average				7.40	3.72

Kepler's Second Law and Conservation of Angular Momentum

Chapter 10 (on rotation) stressed the importance of the concept of angular momentum, in particular, the importance of the conservation of angular momentum. It is quite straightforward to prove the law of conservation of angular momentum for planetary motion and, as a consequence, also derive Kepler's Second Law. Let's work through this proof.

First, we show that the angular momentum, $\vec{L} = \vec{r} \times \vec{p}$, of a point particle is conserved if the particle moves under the influence of a central force. A **central force** is a force that acts only in the radial direction, $\vec{F}_{central} = F\hat{r}$. To prove this statement, we take the time derivative of the angular momentum:

$$\frac{d\vec{L}}{dt} = \frac{d}{dt}(\vec{r} \times \vec{p}) = \frac{d\vec{r}}{dt} \times \vec{p} + \vec{r} \times \frac{d\vec{p}}{dt}.$$

For a point particle, the velocity vector, $\vec{v} = d\vec{r}/dt$, and the momentum vector, $\vec{p}$ are parallel; therefore, their vector product vanishes: $(d\vec{r}/dt) \times \vec{p} = 0$. This leaves only the term

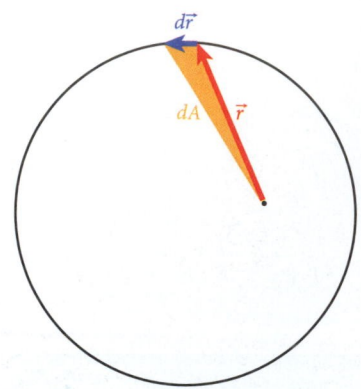

FIGURE 12.22 Area swept out by the radius vector.

$\vec{r} \times (d\vec{p}/dt)$ in the preceding equation. Using Newton's Second Law, we obtain (see Chapter 7 on momentum) $d\vec{p}/dt = \vec{F}$. If this force is a central force, then it is parallel (or antiparallel) to the vector $\vec{r}$. Thus, for a central force, the vector product $\vec{r} \times (d\vec{p}/dt)$ also vanishes:

$$\frac{d\vec{L}}{dt} = \vec{r} \times \frac{d\vec{p}}{dt} = \vec{r} \times \vec{F}_{central} = \vec{r} \times F\hat{r} = 0.$$

Since $d\vec{L}/dt = 0$, we have shown that angular momentum is conserved for a central force. The force of gravity is such a central force, and therefore angular momentum is conserved for any planet moving on an orbit.

How does this general result help in deriving Kepler's Second Law? If we can show that the area dA swept out by the radial vector, $\vec{r}$, during some infinitesimal time, dt, is proportional to the absolute value of the angular momentum, then we are done, because the angular momentum is conserved.

As you can see in Figure 12.22, the infinitesimal area dA swept out by the vector $\vec{r}$ is the triangle spanned by that vector and the differential change in it, $d\vec{r}$:

$$dA = \tfrac{1}{2}\left|\vec{r} \times d\vec{r}\right| = \tfrac{1}{2}\left|\vec{r} \times \frac{d\vec{r}}{dt}dt\right| = \tfrac{1}{2}\left|\vec{r} \times \frac{1}{m}m\frac{d\vec{r}}{dt}dt\right| = \frac{dt}{2m}\left|\vec{r} \times \vec{p}\right| = \frac{dt}{2m}\left|\vec{L}\right|.$$

Therefore, the area swept out in each time interval, dt, is given by

$$\frac{dA}{dt} = \frac{\left|\vec{L}\right|}{2m} = \text{constant},$$

which is exactly what Kepler's Second Law states.

Concept Check 12.3

Which one of the following statements tells how to accurately measure the mass of the Moon?

a) Measure the distance from the Earth to the Moon, and then equate the force of gravity between the Earth and the Moon to the centripetal force required to keep the Moon in orbit around the Earth.

b) Measure the diameter of the Moon, calculate its volume, and then multiply by the density of water.

c) Measure the period of the Moon's orbit around the Earth and the distance from the Earth to the Moon, and then apply Kepler's Third Law.

d) Put a satellite in orbit around the Moon, measure the semimajor axis and the period of the satellite's orbit, and then apply Kepler's Third Law.

e) Measure the effect that the Moon has on the direction of Earth's rotational axis as the Moon orbits the Earth, and then apply the conservation of angular momentum.

12.6 Satellite Orbits

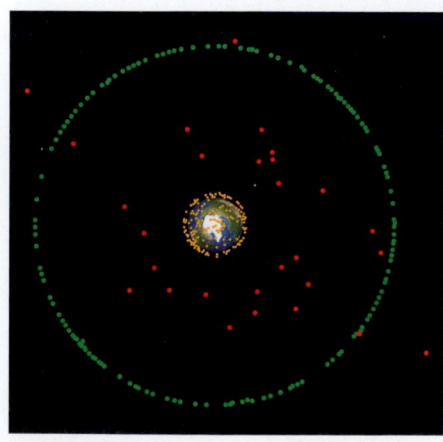

Figure 12.23 shows the positions of many of the several hundreds of satellites in orbit around Earth. Each dot represents the position of a satellite on the afternoon of June 23, 2004. In low orbits, only a few hundred kilometers above sea level, are communication satellites for phone systems, the International Space Station, the Hubble Space Telescope, and other applications (yellow dots). The perfect circle of satellites at a distance of approximately 5.6 Earth radii above the surface (green dots) is composed of **geostationary satellites,** which orbit at the same angular speed as Earth, and so remain above the same spot on the ground. The satellites in between the geostationary and the low-orbit satellites (red dots) are mainly those used for the Global Positioning System, but some carry research instruments.

FIGURE 12.23 Positions of some of the satellites in orbit around Earth on June 23, 2004, looking down on the North Pole. This illustration was produced with data available from NASA.

SOLVED PROBLEM 12.3 | Satellite in Orbit

PROBLEM
A satellite is in a circular orbit around the Earth. The orbit has a radius of 3.75 times the radius of the Earth. What is the linear speed of the satellite?

SOLUTION
THINK   The force of gravity provides the centripetal force that keeps the satellite in its circular orbit around the Earth. We can obtain the satellite's linear speed by equating the centripetal force expressed in terms of the linear speed with the force of gravity between the satellite and the Earth.

SKETCH   A sketch of the problem situation is presented in Figure 12.24.

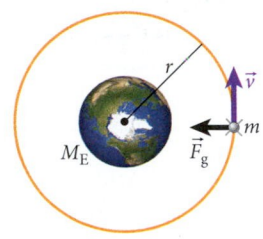

FIGURE 12.24 Satellite in circular orbit around the Earth.

RESEARCH   For a satellite with mass m moving with linear speed v, the centripetal force required to keep the satellite moving in a circle with radius r is

$$F_c = \frac{mv^2}{r}. \tag{i}$$

The gravitational force, F_g, between the satellite and the Earth is

$$F_g = G\frac{M_E m}{r^2}, \tag{ii}$$

where G is the universal gravitational constant and M_E is the mass of the Earth. Equating the forces described by equations (i) and (ii), we get

$$F_c = F_g \Rightarrow$$

$$\frac{mv^2}{r} = G\frac{M_E m}{r^2}.$$

SIMPLIFY   The mass of the satellite cancels out; thus, the orbital speed of a satellite does not depend on its mass. We obtain

$$v^2 = G\frac{M_E}{r} \Rightarrow$$

$$v = \sqrt{\frac{GM_E}{r}}. \tag{iii}$$

CALCULATE   The problem statement specified that the radius of the satellite's orbit is $r = 3.75R_E$, where R_E is the radius of the Earth. Substituting for r in equation (iii) and then inserting the known numerical values gives us

$$v = \sqrt{\frac{GM_E}{3.75R_E}} = \sqrt{\frac{\left(6.67\cdot10^{-11}\ \mathrm{m^3kg^{-1}s^{-2}}\right)\left(5.97\cdot10^{24}\ \mathrm{kg}\right)}{3.75(6.37\cdot10^6\ \mathrm{m})}} = 4082.86\ \mathrm{m/s}.$$

ROUND   Expressing our result with three significant figures gives

$$v = 4080\ \mathrm{m/s} = 4.08\ \mathrm{km/s}.$$

DOUBLE-CHECK   The time it takes this satellite to complete one orbit is

$$T = \frac{2\pi r}{v} = \frac{2\pi(3.75R_E)}{v} = \frac{2\pi\left[3.75(6.37\cdot10^6\ \mathrm{m})\right]}{4080\ \mathrm{m/s}} = 36{,}800\ \mathrm{s} = 10.2\ \mathrm{h},$$

which seems reasonable—communication satellites take 24 h but are at higher altitudes, and the Hubble Space Telescope takes 1.6 h at a lower altitude.

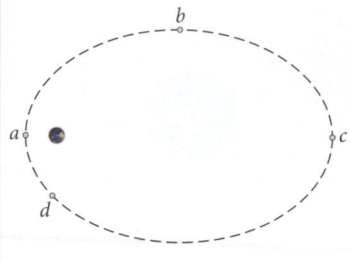
Combining the expression for the orbital speed from Solved Problem 12.3, $v = \sqrt{GM_E/r}$, with equation 12.15 for the escape speed, $v_{esc} = \sqrt{2GM_E/r}$, we find that the orbital velocity of a satellite is always

$$v(r) = \frac{1}{\sqrt{2}} v_{esc}(r). \tag{12.22}$$

The Earth is a satellite of the Sun, and, as we determined in Section 12.4, the escape speed from the Sun starting from the orbital radius of the Earth is 42 km/s. Using equation 12.22, we can predict that the orbital speed of the Earth moving around the Sun is $42/\sqrt{2}$ km/s, or approximately 30 km/s, which matches the value of the orbital speed we found in Chapter 9.

Energy of a Satellite

Having worked through Solved Problem 12.3, we can readily obtain an expression for the kinetic energy of a satellite in orbit around Earth. Multiplying both sides of $mv^2/r = GM_E m/r^2$, which we found by equating the centripetal and gravitational forces, by $r/2$ yields

$$\tfrac{1}{2}mv^2 = \tfrac{1}{2}G\frac{M_E m}{r}.$$

The left-hand side of this equation is the kinetic energy of the satellite. Comparing the right-hand side to the expression for the gravitational potential energy, $U = -GM_E m/r$, we see that this side is equivalent to $-\tfrac{1}{2}U$. Thus, we obtain the kinetic energy of a satellite in circular orbit:

$$K = -\tfrac{1}{2}U. \tag{12.23}$$

The total mechanical energy of the satellite is then

$$E = K + U = -\tfrac{1}{2}U + U = \tfrac{1}{2}U = -\tfrac{1}{2}G\frac{M_E m}{r}. \tag{12.24}$$

Consequently, the total energy is exactly the negative of the satellite's kinetic energy:

$$E = -K. \tag{12.25}$$

It is important to note that equations 12.23 through 12.25 all hold for any orbital radius.

For an elliptical orbit with a semimajor axis a, obtaining the energy of the satellite requires a little more mathematics. The result is very similar to equation 12.24, with the radius r of the circular orbit replaced by the semimajor axis a of the elliptical orbit:

$$E = -\tfrac{1}{2}G\frac{M_E m}{a}.$$

Orbit of Geostationary Satellites

For many applications, a satellite needs to remain at the same point in the sky. For example, a satellite TV dish always points to the same place in the sky, and a satellite has to be located there to ensure that the dish can receive a signal. Satellites that are continuously at the same point in the sky are said to be *geostationary*.

What are the conditions that a satellite must fulfill to be geostationary? First, it has to move in a circle, because this is the only orbit that has a constant angular velocity. Second, the period of rotation must match that of the Earth's, exactly 1 day. And third, the axis of rotation of the satellite's orbit must be exactly aligned with that of the Earth's rotation. Because the center of the Earth must be at the center of a circular orbit for any satellite, the only possible geostationary orbit is one exactly above the Equator. These conditions leave only the radius of the orbit to be determined.

To find the radius, we use Kepler's Third Law in the form of equation 12.19 and solve for r:

$$\frac{T^2}{r^3} = \frac{4\pi^2}{GM} \Rightarrow r = \left(\frac{GMT^2}{4\pi^2}\right)^{1/3}. \tag{12.26}$$

The mass M in this case is that of the Earth. Inserting the numerical values, we find

$$r = \left(\frac{(6.674 \cdot 10^{-11} \ \text{m}^3\text{kg}^{-1}\text{s}^{-2})(5.9736 \cdot 10^{24} \ \text{kg})(86,164 \ \text{s})^2}{4\pi^2} \right)^{1/3} = 42,167 \ \text{km}.$$

Note that we used the best available value of the mass of the Earth and the sidereal day as the correct period of the Earth's rotation (see Chapter 9). The distance of a geostationary satellite above sea level at the Equator is then $42,167 \ \text{km} - R_E$. Taking into account that the Earth is not a perfect sphere, but slightly oblate, this distance is

$$d = r - R_E = 35,790 \ \text{km}.$$

This distance is 5.61 times Earth's radius. This is why the geostationary satellites form an almost perfect circle with a radius of $6.61R_E$ in Figure 12.23.

Figure 12.25 shows a cross section through Earth and the location of a geostationary satellite, with the radius of its orbit drawn to scale. This figure shows the angle ξ relative to the horizontal at which to orient a satellite dish for the best TV reception. Because any geostationary satellite is located in the plane of the Equator, a dish in the Northern Hemisphere should point in a southerly direction.

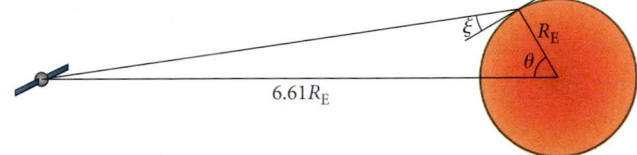

FIGURE 12.25 Angle of a satellite dish, ξ, relative to the local horizontal as a function of the angle of latitude, θ.

There are also geosynchronous satellites in orbit around the Earth. A geosynchronous satellite also has an orbital period of 1 day but does not need to remain at the same point in the sky as viewed from the surface of Earth. For example, NASA's Solar Dynamics Observatory (which was launched in February 2010) has a geosynchronous orbit that is inclined and that traces out a figure 8 in the sky as viewed from the ground. A geostationary orbit is a special case of a geosynchronous orbit.

SOLVED PROBLEM 12.4 | Satellite TV Dish

You just received your new television system, but the company cannot come out to install the dish for you immediately. You want to watch the big game tonight, so you decide to set up the satellite dish yourself.

PROBLEM
Assuming that you live in a location at latitude 42.75° N and that the satellite TV company has a satellite aligned with your longitude, in what direction should you point the satellite dish?

SOLUTION
THINK Satellite TV companies use geostationary satellites to broadcast the signals. Thus, we know you need to point the satellite dish southward, toward the Equator, but you also need to know the angle of inclination of the satellite dish with respect to the horizontal. In Figure 12.25, this is the angle ξ. To determine ξ, we can use the law of cosines, incorporating the distance of a satellite in geostationary orbit, the radius of the Earth, and the latitude of the location of the dish.

SKETCH Figure 12.26 is a sketch of the geometry of the location of the geostationary satellite and the point on the surface of the Earth where the dish is being set up. In this sketch, R_E is the radius of the Earth, R_S is the distance of the satellite from the center of the Earth, d_S is the distance from the satellite to the point on the Earth's surface where the dish

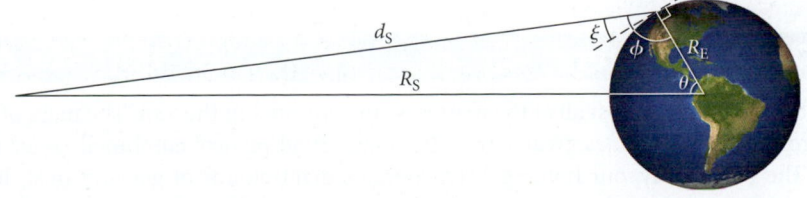

FIGURE 12.26 Geometry of a geostationary satellite in orbit around the Earth.

– Continued

is located, θ is the angle of the latitude of the surface of that location, and ϕ is the angle between d_S and R_E.

RESEARCH   To determine the angle ξ, we first need to determine the angle ϕ. We can see from Figure 12.26 that $\xi = \phi - 90°$ because the dashed line is tangent to the surface of the Earth and thus is perpendicular to a line from the point to the center of the Earth. To determine ϕ, we can apply the law of cosines to the triangle defined by d_S, R_E, and R_S. We will need to apply the law of cosines to this triangle twice. To use the law of cosines to determine ϕ, we need to know the lengths of the sides d_S and R_E. We know R_E but not d_S. We can determine the length d_S using the law of cosines, the angle θ, and the lengths of the two known sides of R_E and R_S:

$$d_S^2 = R_S^2 + R_E^2 - 2R_S R_E \cos\theta. \tag{i}$$

We can now get an equation for the angle ϕ using the law of cosines with the angle ϕ and the two known lengths d_S and R_E:

$$R_S^2 = d_S^2 + R_E^2 - 2d_S R_E \cos\phi. \tag{ii}$$

SIMPLIFY   We know that $R_S = 6.61R_E$ for geostationary satellites. The angle θ corresponds to the latitude, $\theta = 42.75°$. We can substitute these quantities into equation (i):

$$d_S^2 = \left(6.61R_E\right)^2 + R_E^2 - 2\left(6.61R_E\right)R_E\left(\cos 42.75°\right).$$

We can now write an expression for d_S in terms of R_E:

$$d_S^2 = R_E^2\left[6.61^2 + 1 - 2\left(6.61\right)\left(\cos 42.75°\right)\right] = 34.984R_E^2,$$

or

$$d_S = 5.915R_E.$$

We can solve equation (ii) for ϕ:

$$\phi = \cos^{-1}\left(\frac{d_S^2 + R_E^2 - R_S^2}{2d_S R_E}\right). \tag{iii}$$

CALCULATE   Inserting the values we have for d_S and R_S into equation (iii) gives

$$\phi = \cos^{-1}\left(\frac{34.984R_E^2 + R_E^2 - \left(6.61R_E\right)^2}{2\left(5.915R_E\right)R_E}\right) = 130.66°.$$

The angle at which you need to aim the satellite dish with respect to the horizontal is then

$$\xi = \phi - 90° = 130.66° - 90° = 40.66°.$$

ROUND   Expressing our result with three significant figures gives

$$\xi = 40.7°.$$

DOUBLE-CHECK   If the satellite were very far away, the lines d_S and R_S would be parallel to each other, and the sketch of the geometry of the situation would be redrawn as shown in Figure 12.27. We can see from this sketch that $\phi = 180° - \theta$. Remembering that $\xi = \phi - 90°$, we can write

$$\xi = \left(180° - \theta\right) - 90° = 90° - \theta.$$

In this situation, $\theta = 42.75°$, so the estimated angle would be $\xi = 90° - 42.75° = 47.25°$, which is close to our result of $\xi = 40.7°$, but definitely larger, as required. Thus, our answer seems reasonable.

Concept Check 12.7

If a permanent base is established on Mars, it will be necessary to have Martian-stationary satellites in orbit around Mars to facilitate communications. A day on Mars lasts 24 h, 37 min, and 22 s. What is the radius of the orbit for a Martian-stationary satellite?

a) 12,560 km   d) 29,320 km

b) 15,230 km   e) 43,350 km

c) 20,430 km

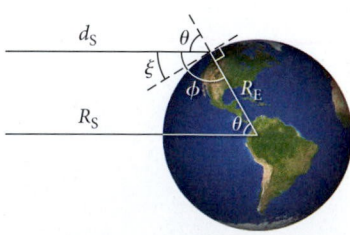

FIGURE 12.27   Geometry for a satellite that is very far away.

12.7   Dark Matter

In the Solar System, almost all of the matter is concentrated in the Sun. The mass of the Sun is approximately 750 times greater than the mass of all planets combined (refer to Table 12.1). The Milky Way, our home galaxy, contains giant clouds of gas and dust, but their combined mass is only about a tenth of that contained in the galaxy's stars. Extrapolating from these facts, you might conclude that the entire universe is composed almost exclusively of luminous matter—that is, stars. Nonluminous matter in the form of dust, asteroids, moons, and planets should contribute only a small fraction of the mass of the universe.

Astronomers know the typical masses of stars and can estimate their numbers in galaxies. Thus, they have obtained fairly accurate estimates of the masses of galaxies and clusters of galaxies. Furthermore, they use modern X-ray telescopes, such as Chandra, to take images of the hot interstellar gas trapped within galaxy clusters. They can deduce the temperature of this gas from the X-ray emissions, and they can also determine how much gravitational force it takes to keep this hot gas from escaping the galaxy clusters. The surprise that has emerged from this research is that the mass contained in luminous matter is too small by a factor of approximately 3 to 5 to provide this gravitational force. This analysis leads to the conclusion that there must be other matter, referred to as *dark matter*, which provides the missing gravitational force.

Some of the observational evidence for dark matter has emerged from measurements of the orbital speeds of stars around the center of their galaxy as a function of their distance from the center, as shown in Figure 12.28 for M 31, the Andromeda galaxy. Even if we assume that there is a supermassive black hole, which is obviously nonluminous, at the center of the galaxy, we would expect star velocity to fall off toward zero for large distances from the center. Experimental evidence, however, indicates that this is not the case. Thus, it is likely that large quantities of dark matter, with a very large radial extent, are present.

Other evidence for dark matter has emerged from observations of gravitational lensing. In Figure 12.29a, the light blue shading represents the distribution of dark matter around a galaxy cluster, as calculated from the observed gravitational lensing. In Figure 12.29b, white circles mark the positions of five images of the same galaxy produced by the gravitational lensing from the unseen dark matter. (Gravitational lensing will be explained in Chapter 35 on relativity. For now, this observation is presented simply as one more empirical fact pointing toward the existence of dark matter.)

Supporting data have also emerged from the WMAP (Wilkinson Microwave Anisotropy Probe) mission, which measured the cosmic background radiation left over from the Big Bang. The best estimate based on this data is that 23% of the universe is composed of dark matter. In addition, the combination of images from the Hubble Space Telescope, the Chandra X-ray Observatory, and the Magellan telescope has yielded a direct empirical proof of the existence of dark matter in the "bullet cluster" (Figure 12.30). The measured temperature of the intergalactic gas in this galaxy cluster is too large for the gas to be contained within the cluster without the presence of dark matter.

Intense speculation and extensive theoretical investigation as to the nature of this dark matter have been ongoing during the last few years, and theories about it are still being modified as new observations emerge. However, all of the proposed theories require a fundamental rethinking of the standard model of the universe and quite possibly of the fundamental models for the

Self-Test Opportunity 12.4

One of the pieces of evidence for dark matter is the velocity curve of stars in rotating galaxies (see, for example, Figure 12.28). Astronomers observe that the velocity of stars in such a galaxy first increases and then remains constant as a function of distance from the center of the galaxy. What would you expect the dependence of the velocity of the stars as a function of distance from the center of the galaxy to be for stars outside the luminous parts of the galaxy assuming that no dark matter were present?

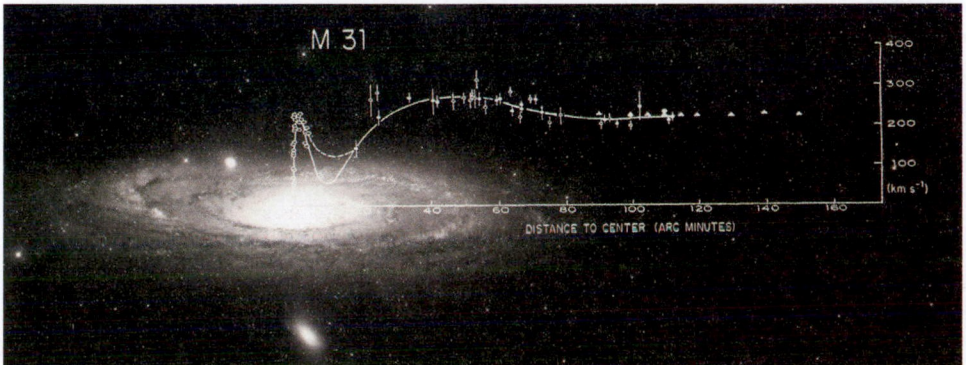

FIGURE 12.28 Image of the Andromeda galaxy, with data on the orbital speeds of stars superimposed. (The triangles represent radio-telescope data from 1975, and the other symbols show optical-wavelength observations from 1970; the solid and dashed lines are simple fits to guide the eye.) The main feature of interest here is that the orbital speeds remain approximately constant well outside the luminous portion of Andromeda, indicating the presence of dark matter.

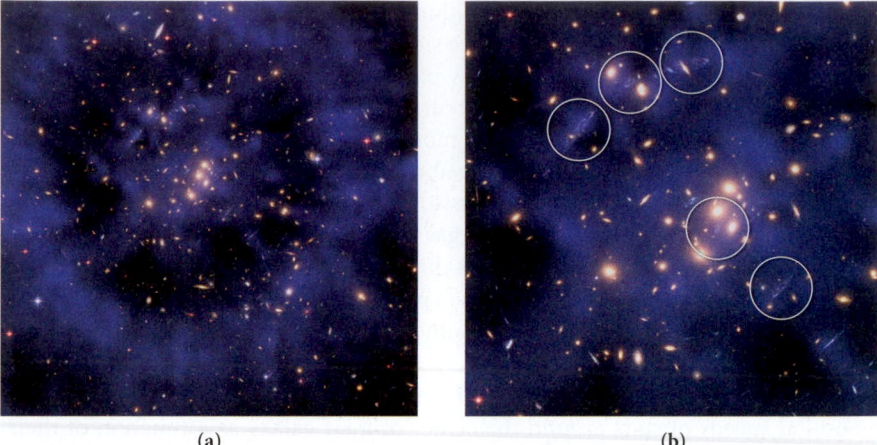

(a)   (b)

FIGURE 12.29 An example of gravitational lensing by dark matter. (a) Photograph of the galaxy cluster Cl 0024+17 taken with the Hubble Space Telescope. The light blue shading represents the distribution of dark matter based on the observed gravitational lensing. (b) Expanded section of the center section of the photograph in part (a), with five images of the same galaxy produced by gravitational lensing marked by circles.

FIGURE 12.30 Superposition of X-ray and optical images of the "bullet cluster," galaxy cluster 1E 0657-56, which contains direct empirical proof of the existence of dark matter.

interaction of particles. Whimsical names have been suggested for the possible constituents of dark matter, such as WIMP (Weakly Interacting Massive Particle) or MACHO (Massive Astrophysical Compact Halo Object). (The physical properties of these postulated constituents are being actively investigated.)

During the last few years, an even stranger phenomenon has been discovered: It seems that, in addition to dark matter, there is also *dark energy*. This dark energy seems to be responsible for an increasing acceleration in the expansion of the universe. A stunning 73% of the mass-energy of the universe is estimated to be dark energy. Together with the 23% estimated to be dark matter, this leaves only 4% of the universe for the stars, planets, moons, gas, and all other objects made of conventional matter.

These are very exciting new areas of research that are sure to change our picture of the universe in the coming decades.

WHAT WE HAVE LEARNED | EXAM STUDY GUIDE

- The gravitational force between two point masses is proportional to the product of their masses and inversely proportional to the distance between them, $F(r) = G\dfrac{m_1 m_2}{r^2}$, with the proportionality constant $G = 6.674 \cdot 10^{-11}$ m³kg⁻¹s⁻², known as the universal gravitational constant.

- In vector form, the equation for the gravitational force can be written as $\vec{F}_{2\to 1} = G\dfrac{m_1 m_2}{|\vec{r}_2 - \vec{r}_1|^3}(\vec{r}_2 - \vec{r}_1)$. This is Newton's Law of Gravity.

- If more than two objects interact gravitationally, the resulting force on one object is given by the vector sum of the forces acting on it due to the other objects.

- Near the surface of the Earth, the gravitational acceleration can be approximated by the function
  $g(h) = g\left(1 - 2\dfrac{h}{R_E} + \cdots\right)$; that is, it falls off linearly with height above the surface.

- The gravitational acceleration at sea level can be derived from Newton's Law of Gravity: $g = \dfrac{GM_E}{R_E^2}$.

- No gravitational force acts on an object inside a massive spherical shell. Because of this, the gravitational force inside a uniform sphere increases linearly with radius: $F(r) = \tfrac{4}{3}\pi G\rho m r$.

- The gravitational potential energy between two objects is given by $U(r) = -G\dfrac{m_1 m_2}{r}$.

- The escape speed from the surface of the Earth is $v_E = \sqrt{\dfrac{2GM_E}{R_E}}$.

- Kepler's laws of planetary motion are as follows

  - All planets move in elliptical orbits with the Sun at one focal point.

- A straight line connecting the center of the Sun and the center of any planet sweeps out an equal area in any given time interval: $\frac{dA}{dt} = \text{constant}$.

- The square of the period of a planet's orbit is proportional to the cube of the semimajor axis of the orbit: $\frac{T^2}{a^3} = \text{constant}$.

- The relationships of the kinetic, potential, and total energy of a satellite in a circular orbit are $K = -\frac{1}{2}U$, $E = K + U = -\frac{1}{2}U + U = \frac{1}{2}U = -\frac{1}{2}G\frac{Mm}{r}$, $E = -K$.

- Geostationary satellites have an orbit that is circular, above the Equator, and with a radius of 42,168 km.

- Evidence points strongly to the existence of dark matter and dark energy, which make up the vast majority of the universe.

ANSWERS TO SELF-TEST OPPORTUNITIES

12.1 The derivation is almost identical, but the integration from $R - r$ to $R + r$ yields zero force. Another proof, based on geometry, is given in Derivation 12.2.

12.2 $g_d = g/9 = 1.09 \text{ m/s}^2$.

12.3 Gravitational attraction between Sun and Earth provides the centripetal force to keep Earth in orbit around the Sun.

$$\frac{m_E v^2}{r} = G\frac{m_E M}{r^2}$$

The mass of Earth cancels. Next, substitute for v from $v = \frac{2\pi r}{T}$; then solve for M:

$$\frac{\left(\frac{2\pi r}{T}\right)^2}{r} = G\frac{M}{r^2} \Rightarrow M = \frac{(2\pi)^2 r^3}{GT^2}.$$

12.4 For stars outside the radius of the visible galaxy, the velocity would decrease proportional to $1/\sqrt{r}$.

PROBLEM-SOLVING GUIDELINES: GRAVITATION

1. This chapter introduced the general form of the equation for the gravitational force, so the approximation $F = mg$ is no longer valid in general. Be sure to keep in mind that in general the acceleration due to gravity is not constant. And this means that you are not able to use the kinematic equations of Chapters 2 and 3 to solve gravitation problems.

2. Energy conservation is essential for many dynamic problems involving gravitation. Be sure to remember that the gravitational potential energy is not given simply by $U = mgh$, as presented in Chapter 6.

3. The principle of superposition of forces is important for situations involving interactions of more than two objects. It allows you to calculate the forces between the objects pairwise and then add them appropriately.

4. For planetary and satellite orbits, Kepler's laws are very useful computational tools, enabling you to connect orbital periods and orbital radii.

MULTIPLE-CHOICE QUESTIONS

12.1 A planet is in a circular orbit about a remote star, far from any other object in the universe. Which of the following statements is true?

a) There is only one force acting on the planet.

b) There are two forces acting on the planet and their resultant is zero.

c) There are two forces acting on the planet and their resultant is not zero.

d) None of the above statements are true.

12.2 Two 30.0-kg masses are held at opposite corners of a square of sides 20.0 cm. If one of the masses is released and allowed to fall toward the other mass, what is the acceleration of the first mass just as it is released? Assume that the only force acting on the mass is the gravitational force of the other mass.

a) $1.52 \cdot 10^{-7} \text{ m/s}^2$

b) $2.50 \cdot 10^{-7} \text{ m/s}^2$

c) $7.50 \cdot 10^{-7} \text{ m/s}^2$

d) $3.73 \cdot 10^{-7} \text{ m/s}^2$

12.3 With the usual assumption that the gravitational potential energy goes to zero at infinite distance, the gravitational potential energy due to the Earth at its center is

a) positive.

b) negative.

c) zero.

d) undetermined.

12.4 A man inside a sturdy box is fired out of a cannon. Which of the following statements regarding the man's sensation of weightlessness is correct?

a) The man senses weightlessness only when he and the box are traveling upward.

b) The man senses weightlessness only when he and the box are traveling downward.

c) The man senses weightlessness when he and the box are traveling both upward and downward.

d) The man does not sense weightlessness at any time of the flight.

12.5 In a binary star system consisting of two stars of equal mass, where is the gravitational potential equal to zero?

a) exactly halfway between the stars

b) along a line bisecting the line connecting the stars

c) infinitely far from the stars

d) none of the above

12.6 Two planets have the same mass, M, but one of them is much denser than the other. Identical objects of mass m are placed on the surfaces of the planets. Which object will have the gravitational potential energy of larger magnitude?

a) Both objects will have the same gravitational potential energy.

b) The object on the surface of the denser planet will have the larger gravitational potential energy.

c) The object on the surface of the less dense planet will have the larger gravitational potential energy.

d) It is impossible to tell.

12.7 Two planets have the same mass, M. Each planet has a constant density, but the density of planet 2 is twice as high as that of planet 1. Identical objects of mass m are placed on the surfaces of the planets. What is the relationship of the gravitational potential energy on planet 1 (U_1) to that on planet 2 (U_2)?

a) $U_1 = U_2$

b) $U_1 = \frac{1}{2}U_2$

c) $U_1 = 2U_2$

d) $U_1 = 1.26U_2$

e) $U_1 = 0.794U_2$

12.8 For two identical satellites in circular motion around the Earth, which statement is true?

a) The one in the lower orbit has less total energy.

b) The one in the higher orbit has more kinetic energy.

c) The one in the lower orbit has more total energy.

d) Both have the same total energy.

12.9 Which condition do all geostationary satellites orbiting the Earth have to fulfill?

a) They have to orbit above the Equator.

b) They have to orbit above the poles.

c) They have to have an orbital radius that locates them less than 30,000 km above the surface.

d) They have to have an orbital radius that locates them more than 42,000 km above the surface.

12.10 An object is placed between the Earth and the Moon, along the straight line that joins their centers. About how far away from the center of the Earth should the object be placed so that the net gravitational force on the object from the Earth and the Moon is zero?

a) halfway to the center of the Moon

b) 60% of the way to the center of the Moon

c) 70% of the way to the center of the Moon

d) 85% of the way to the center of the Moon

e) 90% of the way to the center of the Moon

12.11 A man of mass 100. kg feels a gravitational force, F_m, from a woman of mass 50.0 kg sitting 1 m away. The gravitational force, F_w, experienced by the woman will be _____ that experienced by the man.

a) more than

b) less than

c) the same as

d) not enough information given

12.12 Halfway between Earth's center and its surface, the gravitational acceleration is

a) zero.

b) $g/4$.

c) $g/2$.

d) $2g$.

e) $4g$.

12.13 The best estimate of the orbital period of the Solar System around the center of the Milky Way is between 220 and 250 million years. How much mass (in terms of solar masses) is enclosed by the 26,000 light-years ($1.7 \cdot 10^9$ AU) radius of the Solar System's orbit? (An orbital period of 1 yr for an orbit of radius 1 AU corresponds to 1 solar mass.)

a) 90 billion solar masses

b) 7.2 billion solar masses

c) 52 million solar masses

d) 3.7 million solar masses

e) 432,000 solar masses

12.14 Imagine a large hollow sphere with mass M and outer radius R located in outer space. The hollow sphere has a thickness t, where $t \ll R$. What is the gravitational force on an object with mass m on the outer and inner surfaces of the hollow sphere, respectively?

a) zero, zero

b) mMG/R^2, zero

c) zero, $mMG/(R-t)^2$

d) mMG/R^2, $mMG/(R-t)^2$

e) zero, mMG/R^2

12.15 The ratio of the mass of the Earth to the mass of the Moon is 81. The magnitude of the gravitational force that the Earth exerts on the Moon is

a) the same as the magnitude of the gravitational force that the Moon exerts on the Earth.

b) 81 times the gravitational force exerted by the Moon on the Earth.

c) 1/81 of the gravitational force exerted by the Moon on the Earth.

d) 9 times the gravitational force exerted by the Moon on the Earth.

e) 1/9 of the gravitational force exerted by the Moon on the Earth.

CONCEPTUAL QUESTIONS

12.16 Can the expression for gravitational potential energy $U_g(y) = mgy$ be used to analyze high-altitude motion? Why or why not?

12.17 Even though the Moon does not have an atmosphere, the trajectory of a projectile near its surface is only approximately a parabola. This is because the acceleration due to gravity near the surface of the Moon is only approximately constant. Describe as precisely as you can the *actual* shape of a projectile's path on the Moon, even one that travels a long distance over the surface of the Moon.

12.18 A scientist working for a space agency has noticed that a Russian satellite of mass 250. kg is on collision course with an American satellite of mass 600. kg orbiting at 1000. km above the surface. Both satellites are moving in circular orbits but in opposite directions. If the two satellites collide and stick together, will they continue to orbit or crash to the Earth? Explain.

12.19 Three asteroids, located at points P_1, P_2, and P_3, which are not in a line, and having known masses m_1, m_2, and m_3, interact with one another through their mutual gravitational forces only; they are isolated in space and do not interact with any other bodies. Let σ denote the axis going through the center of mass of the three asteroids, perpendicular to the triangle $P_1P_2P_3$. What conditions should the angular velocity ω of the system (around the axis σ) and the distances

$$P_1P_2 = a_{12}, \qquad P_2P_3 = a_{23}, \qquad P_1P_3 = a_{13},$$

fulfill to allow the shape and size of the triangle $P_1P_2P_3$ to remain unchanged during the motion of the system? That is, under what conditions does the system rotate around the axis σ as a rigid body?

12.20 The more powerful the gravitational force of a planet, the greater its escape speed, v, and the greater the gravitational acceleration, g, at its surface. However, in Table 12.1, the value for v is much greater for Uranus than for Earth—but g is smaller on Uranus than on Earth! How can this be?

12.21 Is the orbital speed of the Earth when it is closest to the Sun greater than, less than, or equal to the orbital speed when it is farthest from the Sun? Explain.

12.22 Point out any flaw in the following physics exam statement: *"Kepler's First Law states that all planets move in elliptical orbits with the Sun at one focal point. It follows that during one complete revolution around the Sun (1 year), the Earth will pass through a closest point to the Sun—the perihelion—as well as through a furthest point from the Sun—the aphelion. This is the main cause of the seasons (summer and winter) on Earth."*

12.23 A comet orbiting the Sun moves in an elliptical orbit. Where is its kinetic energy, and therefore its speed, at a maximum—at perihelion or aphelion? Where is its gravitational potential energy at a maximum?

12.24 Where the International Space Station orbits, the gravitational acceleration is just 11.5% less than its value on the surface of the Earth. Nevertheless, astronauts in the space station float. Why is this so?

12.25 Satellites in low orbit around the Earth lose energy from colliding with the gases of the upper atmosphere, causing them to slowly spiral inward. What happens to their kinetic energy as they fall inward?

12.26 Compare the magnitudes of the gravitational force that the Earth exerts on the Moon and the gravitational force that the Moon exerts on the Earth. Which is larger?

12.27 Imagine that two tunnels are bored completely through the Earth, passing through the center. Tunnel 1 is along the Earth's axis of rotation, and tunnel 2 is in the equatorial plane, with both ends at the Equator. Two identical balls, each with a mass of 5.00 kg, are simultaneously dropped into the tunnels. Neglect air resistance and friction from the tunnel walls. Do the balls reach the center of the Earth (point *C*) at the same time? If not, which ball reaches the center of the Earth first?

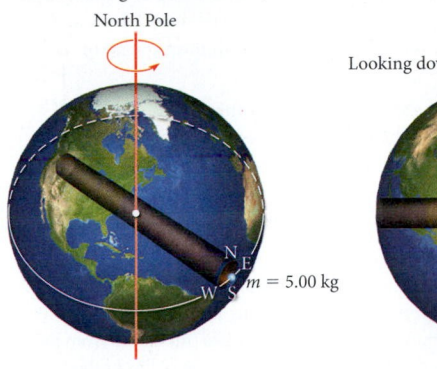

12.28 Imagine that a tunnel is bored in the Earth's equatorial plane, going completely through the center of the Earth with both ends at the Equator. A mass of 5.00 kg is dropped into the tunnel at one end, as shown in the figure. The tunnel has a radius that is slightly larger than that of the mass. The mass is dropped into the center of the tunnel. Neglect air resistance and friction from the tunnel wall. Does the mass ever touch the wall of the tunnel as it falls? If so, which side does it touch first, north, east, south, or west? (*Hint:* The angular momentum of the mass is conserved if the only forces acting on it are radial.)

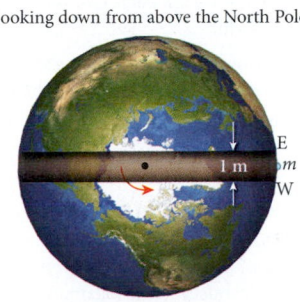

12.29 A plumb bob located at latitude 55.0° N hangs motionlessly with respect to the ground beneath it. A straight line from the string supporting the bob does not go exactly through the Earth's center. Does this line intersect the Earth's axis of rotation south or north of the Earth's center?

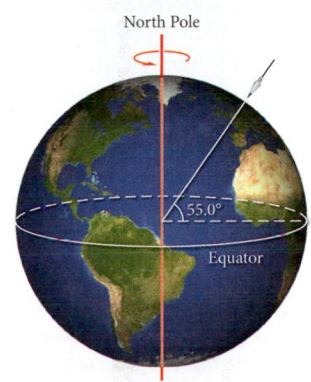

EXERCISES

A blue problem number indicates a worked-out solution is available in the Student Solutions Manual. One • and two •• indicate increasing level of problem difficulty.

Section 12.1

12.30 The Moon causes tides because the gravitational force it exerts differs between the side of the Earth nearer to it and the side farther from it. Find the difference in the accelerations toward the Moon of objects on the nearer and farther sides of the Earth.

12.31 After a spacewalk, a 1.00-kg tool is left 50.0 m from the center of gravity of a 20.0-metric ton space station, orbiting along with it. How much closer to the space station will the tool drift in an hour due to the gravitational attraction of the space station?

•12.32 a) What is the total force on m_1 due to m_2, m_3, and m_4 if all four masses are located at the corners of a square of side a? Let $m_1 = m_2 = m_3 = m_4$.

b) Sketch all the forces acting on m_1.

•12.33 A spaceship of mass m is located between two planets of masses M_1 and M_2; the distance between the two planets is L, as shown in the figure.

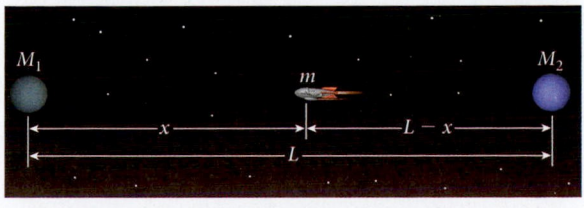

Assume that L is much larger than the radius of either planet. What is the position, x, of the spacecraft (given as a function of L, M_1, and M_2) if the net force on the spacecraft is zero?

•12.34 A carefully designed experiment can measure the gravitational force between masses of 1 kg. Given that the density of iron is 7860 kg/m³, what is the gravitational force between two 1.00-kg iron spheres that are touching?

••12.35 A uniform rod of mass 333 kg is in the shape of a semicircle of radius 5.00 m. Calculate the magnitude of the force on a 77.0-kg point mass placed at the center of the semicircle, as shown in the figure.

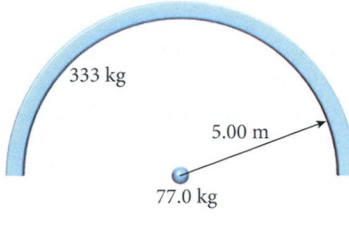

••12.36 The figure shows a system of four masses. The center-to-center distance between any two of the masses is 10.0 cm. The base of the pyramid is in the *xz*-plane and the 20.0-kg mass is on the *y*-axis. What is the magnitude and direction of the gravitational force acting on the 10.0-kg mass? Give direction of the net force with respect to the *xyz*-coordinates shown.

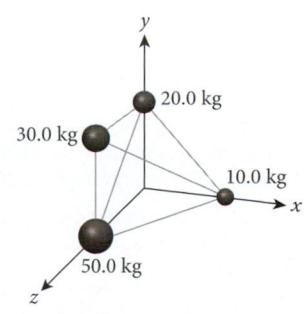

Sections 12.2 and 12.3

12.37 Suppose a new extrasolar planet is discovered. Its mass is double the mass of the Earth, but it has the same density and spherical shape as the Earth. How would the weight of an object at the new planet's surface differ from its weight on Earth?

12.38 What is the magnitude of the free-fall acceleration of a ball (mass m) due to the Earth's gravity at an altitude of $2R$, where R is the radius of the Earth? Ignore the rotation of the Earth.

12.39 Some of the deepest mines in the world are in South Africa and are roughly 3.5 km deep. Consider the Earth to be a uniform sphere of radius 6370 km.

a) How deep would a mine shaft have to be for the gravitational acceleration at the bottom to be reduced by a factor of 2 from its value on the Earth's surface?

b) What is the percentage difference in the gravitational acceleration at the bottom of the 3.5-km-deep shaft relative to that at the Earth's mean radius? That is, what is the value of $(a_{surf} - a_{3.5km})/a_{surf}$?

•12.40 In an experiment performed at the bottom of a very deep vertical mine shaft, a ball is tossed vertically in the air with a known initial velocity of 10.0 m/s, and the maximum height the ball reaches (measured from its launch point) is determined to be 5.113 m. Knowing the radius of the Earth, $R_E = 6370$ km, and the gravitational acceleration at the surface of the Earth, $g(0) = 9.81$ m/s^2, calculate the depth of the shaft.

••12.41 Careful measurements of local variations in the acceleration due to gravity can reveal the locations of oil deposits. Assume that the Earth is a uniform sphere of radius 6370 km and density 5500. kg/m^3, except that there is a spherical region of radius 1.00 km and density 900. kg/m^3, whose center is at a depth of 2.00 km. Suppose you are standing on the surface of the Earth directly above the anomaly with an instrument capable of measuring the acceleration due to gravity with great precision. What is the fractional deviation of the acceleration due to gravity that you measure compared to what you would have measured had the density been 5500. kg/m^3 everywhere? (*Hint:* Think of this as a superposition problem involving two uniform spherical masses, one with a negative density.)

Section 12.4

12.42 A spaceship is launched from the Earth's surface with a speed v. The radius of the Earth is R. What will its speed be when it is very far from the Earth?

12.43 What is the ratio of the escape speed to the orbital speed of a satellite at the surface of the Moon, where the gravitational acceleration is about a sixth of that on Earth?

12.44 Standing on the surface of a small spherical moon whose radius is $6.30 \cdot 10^4$ m and whose mass is $8.00 \cdot 10^{18}$ kg, an astronaut throws a rock of mass 2.00 kg straight upward with an initial speed 40.0 m/s. (This moon is too small to have an atmosphere.) What maximum height above the surface of the moon will the rock reach?

12.45 An object of mass m is launched from the surface of the Earth. Show that the minimum speed required to send the projectile to a height of $4R_E$ above the surface of the Earth is $v_{min} = \sqrt{8GM_E/5R_E}$. M_E is the mass of the Earth and R_E is the radius of the Earth. Neglect air resistance.

12.46 For the satellite in Solved Problem 12.3, orbiting the Earth at a distance of $3.75R_E$ with a speed of 4.08 km/s, with what speed would the satellite hit the Earth's surface if somehow it suddenly stopped and fell to Earth? Ignore air resistance.

•12.47 Estimate the radius of the largest asteroid from which you could escape by jumping. Assume spherical geometry and a uniform density equal to the Earth's average density.

•12.48 Eris, the largest dwarf planet known in the Solar System, has a radius $R = 1200$ km and an acceleration due to gravity on its surface of magnitude $g = 0.77$ m/s^2.

a) Use these numbers to calculate the escape speed from the surface of Eris.

b) If an object is fired directly upward from the surface of Eris with half of this escape speed, to what maximum height above the surface will the object rise? (Assume that Eris has no atmosphere and negligible rotation.)

•12.49 Two identical 20.0-kg spheres of radius 10.0 cm are 30.0 cm apart (center-to-center distance).

a) If they are released from rest and allowed to fall toward one another, what is their speed when they first make contact?

b) If the spheres are initially at rest and just touching, how much energy is required to separate them to 1.00 m apart? Assume that the only force acting on each mass is the gravitational force due to the other mass.

••12.50 Imagine that a tunnel is bored completely through the Earth along its axis of rotation. A ball with a mass of 5.00 kg is dropped from rest into the tunnel at the North Pole, as shown in the figure. Neglect air resistance and friction from the tunnel wall. Calculate the potential energy of the ball as a function of its distance from the center of the Earth. What is the speed of the ball when it arrives at the center of the Earth (point C)?

Section 12.5

12.51 The *Apollo 8* mission in 1968 included a circular orbit at an altitude of 111 km above the Moon's surface. What was the period of this orbit? (You need to look up the mass and radius of the Moon to answer this question!)

•12.52 Halley's comet orbits the Sun with a period of 75.3 yr.

a) Find the semimajor axis of the orbit of Halley's comet in astronomical units (1 AU is equal to the semimajor axis of the Earth's orbit).

b) If Halley's comet is 0.586 AU from the Sun at perihelion, what is its maximum distance from the Sun, and what is the eccentricity of its orbit?

••12.53 A satellite of mass m is in an elliptical orbit (that satisfies Kepler's laws) about a body of mass M, with m negligible compared to M.

a) Find the total energy of the satellite as a function of its speed, v, and distance, r, from the body it is orbiting.

b) At the maximum and minimum distance between the satellite and the body, and only there, the angular momentum is simply related to the speed and distance. Use this relationship and the result of part (a) to eliminate v and obtain a relationship between the extreme distance r and the satellite's energy and angular momentum.

c) Solve the result of part (b) for the maximum and minimum radii of the orbit in terms of the energy and angular momentum per unit mass of the satellite.

d) Transform the results of part (c) into expressions for the semimajor axis, a, and eccentricity of the orbit, e, in terms of the energy and angular momentum per unit mass of the satellite.

••12.54 Consider the Sun to be at the origin of an xy-coordinate system. A telescope spots an asteroid in the xy-plane at a position given by $(2.00 \cdot 10^{11}$ m, $3.00 \cdot 10^{11}$ m$)$ with a velocity given by $(-9.00 \cdot 10^3$ m/s, $-7.00 \cdot 10^3$ m/s$)$. What will the asteroid's speed and distance from the Sun be at closest approach?

Section 12.6

12.55 A spy satellite was launched into a circular orbit with a height of 700. km above the surface of the Earth. Determine its orbital speed and period.

12.56 Express algebraically the ratio of the gravitational force on the Moon due to the Earth to the gravitational force on the Moon due to the Sun. Why, since the ratio is so small, doesn't the Sun pull the Moon away from the Earth?

12.57 A space shuttle is initially in a circular orbit at a radius of $r = 6.60 \cdot 10^6$ m from the center of the Earth. A retrorocket is fired forward, reducing the total energy of the space shuttle by 10.0% (that is, increasing the magnitude of the negative total energy by 10.0%), and the space shuttle moves to a new circular orbit with a radius that is smaller than r. Find the speed of the space shuttle (a) before and (b) after the retrorocket is fired.

•12.58 A 200.-kg satellite is in circular orbit around the Earth and moving at a speed of 5.00 km/s. How much work must be done to move the satellite into another circular orbit that is twice as high above the surface of the Earth?

•12.59 The radius of a black hole is the distance from the black hole's center at which the escape speed is the speed of light.

a) What is the radius of a black hole with a mass twice that of the Sun?

b) At what radius from the center of the black hole in part (a) would the orbital speed be equal to the speed of light?

c) What is the radius of a black hole with the same mass as that of the Earth?

•12.60 A satellite is in a circular orbit around a planet. The ratio of the satellite's kinetic energy to its gravitational potential energy, K/U_g, is a constant whose value is independent of the masses of the satellite and planet and of the radius and velocity of the orbit. Find the value of this constant. (Potential energy is taken to be zero at infinite separation.)

•12.61 Determine the minimum amount of energy that a projectile of mass 100.0 kg must gain to reach a circular orbit 10.00 km above the Earth's surface if launched from (a) the North Pole or (b) the Equator (keep answers to four significant figures). Do not be concerned about the direction of the launch or of the final orbit. Is there an advantage or disadvantage to launching from the Equator? If so, how significant is the difference? Do not neglect the rotation of the Earth when calculating the initial energies. Use $5.974 \cdot 10^{24}$ kg for the mass of the Earth and 6357 km as the radius of the Earth.

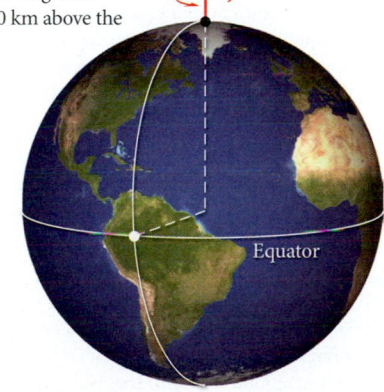

North Pole

Equator

••12.62 A rocket with mass $M = 12.0$ metric tons is moving around the Moon in a circular orbit at the height of $h = 100$. km. The braking engine is activated for a short time to lower the orbital height so that the rocket can make a lunar landing. The velocity of the ejected gases is $u = 1.00 \cdot 10^4$ m/s relative to the rocket's initial velocity. The Moon's radius is $R_M = 1.74 \cdot 10^3$ km; the acceleration of gravity near the Moon's surface is $g_M = 1.62$ m/s^2.

a) What amount of fuel will be used by the braking engine if it is activated at point A of the orbit and the rocket lands on the Moon at point B (see the left part of the figure)?

b) Suppose that, at point A, the rocket is given an impulse directed toward the center of the Moon, to put it on a trajectory that meets the Moon's surface at point C (see the right part of the figure). What amount of fuel is needed in this case?

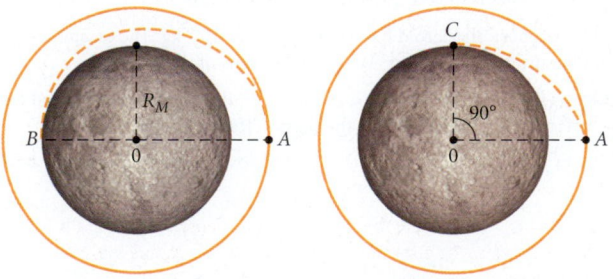

Additional Exercises

12.63 Calculate the magnitudes of the gravitational forces exerted on the Moon by the Sun and by the Earth when the two forces are in direct competition, that is, when the Sun, Moon, and Earth are aligned with the Moon between the Sun and the Earth. (This alignment corresponds to a solar eclipse.) Does the orbit of the Moon ever actually curve away from the Sun, toward the Earth?

12.64 A projectile is shot vertically from the surface of the Earth by means of a very powerful cannon. If the projectile reaches a height of 55.0 km above Earth's surface, what was the speed of the projectile when it left the cannon?

12.65 Newton's Law of Gravity specifies the magnitude of the interaction force between two point masses, m_1 and m_2, separated by a distance r as $F(r) = Gm_1m_2/r^2$. The gravitational constant G can be determined by directly measuring the interaction force (gravitational attraction) between two sets of spheres by using the apparatus constructed in the late 18th century by the English scientist Henry Cavendish. This apparatus was a torsion balance consisting of a 6.00-ft wooden rod suspended from a torsion wire, with a lead sphere having a diameter of 2.00 in and a weight of 1.61 lb attached to each end. Two 12.0-in, 348-lb lead balls were located near the smaller balls, about 9.00 in away, and held in place with a separate suspension system. Today's accepted value for G is $6.674 \cdot 10^{-11}$ m^{3}kg^{-1}s^{-2}. Determine the force of attraction between the larger and smaller balls that had to be measured by this balance. Compare this force to the weight of the small balls.

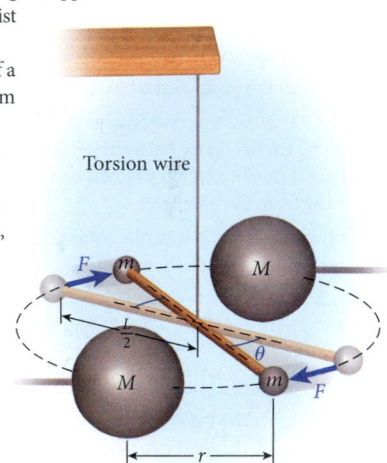

Torsion wire

12.66 Newton was holding an apple of mass 100. g and thinking about the gravitational forces exerted on the apple by himself and by the Sun. Calculate the magnitude of the gravitational force acting on the apple due to (a) Newton, (b) the Sun, and (c) the Earth, assuming that the distance from the apple to Newton's center of mass is 50.0 cm and Newton's mass is 80.0 kg.

12.67 A 1000.-kg communications satellite is released from a space shuttle to initially orbit the Earth at a radius of $7.00 \cdot 10^6$ m. After being deployed, the satellite's rockets are fired to put it into a higher altitude orbit of radius $5.00 \cdot 10^7$ m. What is the minimum mechanical energy supplied by the rockets to effect this change in orbit?

12.68 Consider a 0.300-kg apple (a) attached to a tree and (b) falling. Does the apple exert a gravitational force on the Earth? If so, what is the magnitude of this force?

12.69 At what height h above the Earth will a satellite moving in a circular orbit have half the period of the Earth's rotation about its own axis?

•12.70 In the Earth-Moon system, there is a point where the gravitational forces balance. Assume that the mass of the Moon is $\frac{1}{81}$ that of the Earth.

a) At what point, on a line between the Earth and the Moon, is the gravitational force exerted on an object by the Earth exactly balanced by the gravitational force exerted on the object by the Moon?

b) Is this point one of stable or unstable equilibrium?

c) Calculate the ratio of the force of gravity due to the Sun acting on an object at this point to the force of gravity due to Earth and, separately, to the force of gravity due to the Moon.

••**12.71** Consider a particle on the surface of the Earth, at a position with an angle of latitude $\lambda = 30.0°$ N, as shown in the figure. For this problem, assume that the Earth is a sphere with radius $R = 6.37 \cdot 10^6$ m and that $g = 9.81$ m/s^2. Find (a) the magnitude, and (b) the direction of the effective gravitational force acting on the particle, taking into consideration the rotation of the Earth. (c) What angle λ gives rise to the maximum deviation of the gravitational acceleration?

•**12.72** An asteroid is discovered to have a tiny moon that orbits it in a circular path at a distance of 100. km and with a period of 40.0 h. The asteroid is roughly spherical (unusual for such a small body) with a radius of 20.0 km.

a) Find the acceleration of gravity at the surface of the asteroid.

b) Find the escape speed from the asteroid.

•**12.73** a) By what percentage does the gravitational potential energy of the Earth change between perihelion and aphelion? (Assume that the Earth's potential energy would be zero if it moved to a very large distance away from the Sun.)

b) By what percentage does the kinetic energy of the Earth change between perihelion and aphelion?

•**12.74** A planet with a mass of $7.00 \cdot 10^{21}$ kg is in a circular orbit around a star with a mass of $2.00 \cdot 10^{30}$ kg. The planet has an orbital radius of $3.00 \cdot 10^{10}$ m.

a) What is the linear orbital velocity of the planet?

b) What is the period of the planet's orbit?

c) What is the total mechanical energy of the planet?

•**12.75** The astronomical unit (AU, equal to the mean radius of the Earth's orbit) is $1.4960 \cdot 10^{11}$ m, and a year is $3.1557 \cdot 10^7$ s. Newton's gravitational constant is $G = 6.6738 \cdot 10^{-11}$m^{3}kg^{-1}s^{-2}. Calculate the mass of the Sun in kilograms. (Recalling or looking up the mass of the Sun does not constitute a solution to this problem.)

•**12.76** The distances from the Sun at perihelion and aphelion for Pluto are $4410 \cdot 10^6$ km and $7360 \cdot 10^6$ km, respectively. What is the ratio of Pluto's orbital speed around the Sun at perihelion to that at aphelion?

•**12.77** The weight of a star is usually balanced by two forces: the gravitational force, acting inward, and the force created by nuclear reactions, acting outward. Over a long period of time, the force due to nuclear reactions gets weaker, causing the gravitational collapse of the star and crushing atoms out of existence. Under such extreme conditions, protons and electrons are squeezed to form neutrons, giving birth to a neutron star. Neutron stars are massively heavy—a teaspoon of the substance of a neutron star would weigh 50 million metric tons on the Earth.

a) Consider a neutron star whose mass is twice the mass of the Sun and whose radius is 10.0 km. If it rotates with a period of 1.00 s, what is the speed of a point on the equator of this star? Compare this speed with the speed of a point on Earth's Equator.

b) What is the value of g at the surface of this star?

c) Compare the weight of a 1.00-kg mass on the Earth with its weight on the neutron star.

d) If a satellite is to circle 10.0 km above the surface of such a neutron star, how many revolutions per minute will it make?

e) What is the radius of the geostationary orbit for this neutron star?

••**12.78** You have been sent in a small spacecraft to rendezvous with a space station that is in a circular orbit of radius $2.5000 \cdot 10^4$ km from the Earth's center. Due to a mishandling of units by a technician, you find yourself in the same orbit as the station but exactly halfway around the orbit from it! You do not apply forward thrust in an attempt to chase the station; that would be fatal folly. Instead, you apply a brief braking force against the direction of your motion, to put you into an elliptical orbit, whose highest point is your present position, and whose period is half that of your present orbit. Thus, you will return to your present position when the space station has come halfway around the circle to meet you. Is the minimum radius from the Earth's center—the low point—of your new elliptical orbit greater than the radius of the Earth (6370 km), or have you botched your last physics problem?

••**12.79** If you and the space station are initially in a low orbit—say, with a radius of 6720 km, approximately that of the orbit of the International Space Station—the maneuver of Problem 12.78 will fail unpleasantly. Keeping in mind that the life-support capabilities of your small spacecraft are limited and so time is of the essence, can you perform a similar maneuver that will enable you to rendezvous with the station? Find the perihelion, aphelion, and period of the transfer orbit you should use.

••**12.80** A satellite is placed between the Earth and the Moon, along a straight line that connects their centers of mass. The satellite has an orbital period around the Earth that is the same as that of the Moon, 27.3 days. How far away from the Earth should this satellite be placed?

MULTI-VERSION EXERCISES

12.81 An electromagnetic rail accelerator is used to launch a research probe vertically from the surface of the Moon. The initial speed of the projectile is 114.5 m/s. What height does it reach above the surface of the Moon? Assume that the radius of the Moon is 1737 km and the mass of the Moon is $7.348 \cdot 10^{22}$ kg.

12.82 An electromagnetic rail accelerator is used to launch a research probe vertically from the surface of the Moon. The probe reaches a height of 4.905 km above the surface of the Moon. What was the initial speed of the probe? Assume that the radius of the Moon is 1737 km and the mass of the Moon is $7.348 \cdot 10^{22}$ kg.

12.83 A comet orbits the Sun with a period of 89.17 yr. At perihelion, the comet is 1.331 AU from the Sun. How far from the Sun (in AU) is the comet at aphelion?

12.84 A comet orbits the Sun with a period of 98.11 yr. At aphelion, the comet is 41.19 AU from the Sun. How far from the Sun (in AU) is the comet at perihelion?

12.85 A comet orbits the Sun. The aphelion of its orbit is 31.95 AU from the Sun. The perihelion is 1.373 AU. What is the period (in years) of the comet's orbit?

12.86 A spherical asteroid has a mass of $1.869 \cdot 10^{20}$ kg and a radius of 358.9 km. What is the escape speed from its surface?

12.87 A spherical asteroid has a mass of $1.769 \cdot 10^{20}$ kg. The escape speed from its surface is 273.7 m/s. What is the radius of the asteroid?

12.88 A spherical asteroid has a radius of 365.1 km. The escape speed from its surface is 319.2 m/s. What is the mass of the asteroid?

13

Solids and Fluids

FIGURE 13.1 Horns Rev wind farm in the North Sea. Under the right atmospheric conditions, the wakes of the turbines can become visible when condensation of water vapor in the air creates streams of clouds.

Figure 13.1 shows the Horns Rev offshore wind farm, in the North Sea, 15 km off the west coast of Denmark. When this 80-turbine wind farm went into operation in 2002, it was the largest in the world, with a capacity of 160 MW. Wind turbines can make a great contribution toward satisfying our future needs for electrical power by extracting energy from the wind, which is the motion of the fluid surrounding us—the air in Earth's atmosphere. The turbines in a wind farm cannot be placed too close to one another, because they create downwind wakes, which reduce the efficiency of any turbine in them. A great deal of research effort is currently devoted to studying and modeling these wakes. Normally, the wakes are invisible to the naked eye, but in Figure 13.1 the atmospheric conditions (temperature and water vapor content) were just right for condensation to form cloud streams. Fluid dynamics is currently a major area of research, with applications to all areas of physics from astronomy to nuclear research.

So far, we have studied motion of idealized objects, ignoring factors such as the materials they are made of and the behavior of these materials in response to the forces exerted on them. This chapter considers some of these factors, presenting an overview of the physical characteristics of solids, liquids, and gases.

WHAT WE WILL LEARN

- The atom is the basic building block of macroscopic matter.
- The diameter of an atom is approximately 10^{-10} m.
- Matter can exist as a gas, a liquid, or a solid.
  - A gas is a system in which the atoms move freely through space.
  - A liquid is a system in which the atoms move freely but form a nearly incompressible substance.
  - A solid defines its own size and shape.
- Solids are nearly incompressible.
- Various forms of stress, such as stretching, compression, and shearing, can deform solids. These deformations can be expressed in terms of linear relationships between the applied stress and the resulting deformation.

- Pressure is force per unit area.
- The pressure of the Earth's atmosphere can be measured using a mercury barometer or a similar instrument.
- The pressure of gas can be measured using a mercury manometer.
- Pascal's Principle states that pressure applied to a confined fluid is transmitted to all parts of the fluid.
- Archimedes' Principle states that the buoyant force on an object in a fluid is equal to the weight of the fluid displaced by the object.
- Bernoulli's Principle states that the faster a fluid flows, the less pressure it exerts on its boundary.

13.1 Atoms and the Composition of Matter

During the evolution of physics, scientists have explored ever smaller dimensions, looking deeper into matter in order to examine its elementary building blocks. This general way of learning more about a system by studying its subsystems is called *reductionism* and has proven a fruitful guiding principle during the last four or five centuries of scientific advancements.

Today, we know that **atoms** are the basic building blocks of matter, although they themselves are composite particles. The substructure of atoms, however, can be resolved only with accelerators and other tools of modern nuclear and particle physics. For the purposes of this chapter, it is reasonable to view atoms as the elementary building blocks. In fact, the word *atom* comes from the Greek *atomos*, which means "indivisible." The diameter of an atom is about 10^{-10} m = 0.1 nm. This distance is often called an angstrom (Å).

The simplest atom is hydrogen, composed of a proton and an electron. Hydrogen is the most abundant element in the universe. The next most abundant element is helium. Helium has two protons and two neutrons in its nucleus, along with two electrons surrounding the nucleus. Another common atom is oxygen, with eight protons and eight electrons, as well as (usually) eight neutrons. The heaviest naturally occurring atom is uranium, with 92 protons, 92 electrons, and usually 146 neutrons. So far, 118 different elements have been recognized and classified in the periodic table of the elements.

Essentially all of the hydrogen in the universe, along with some helium, was produced in the Big Bang about 13.7 billion years ago. Heavier elements, up to iron, were and still are produced in stars. Elements with more protons and neutrons than iron (gold, mercury, lead, and so forth) are thought to have been produced by supernova explosions. Most of the atoms of the elements on Earth were produced more than 5 billion years ago, probably in a supernova explosion, and have been recycled ever since. Even our bodies are composed of atoms from the ashes of a dying star. (The production and recycling of the elements will be covered in detail in Chapters 39 and 40.)

Consider the number of atoms in 12 g of the most common isotope of carbon, ^{12}C, where 12 represents the atomic mass number—that is, the total number of protons (six) plus neutrons (six) in the carbon nucleus. This number of atoms has been measured to be $6.022 \cdot 10^{23}$, which is called **Avogadro's number,** N_A:

$$N_A = 6.022 \cdot 10^{23}.$$

To get a feeling for how much carbon corresponds to N_A atoms, consider two forms of carbon: diamond and graphite. Diamonds are composed of carbon atoms arranged in a crystal lattice, whereas the carbon atoms in graphite are arranged in two-dimensional layers (Figure 13.2). In the diamond structure, carbon atoms are bonded in an interlocking lattice, which makes diamond very hard and transparent to light. In graphite, carbon atoms are arranged in layers that can slide over one another, making graphite soft and slippery. It does not reflect light, but appears black instead. Graphite is used in pencils and for lubrication.

The density of diamond is 3.51 g/cm^3, and the density of graphite is 2.20 g/cm^3. Masses of diamonds are usually given in carats, where 1 carat = 200 mg. A 12-g diamond would have N_A atoms, a mass of 60 carats, and a volume of about 3.4 cm^3, about 1.3 times bigger than the Hope Diamond. A typical wedding ring might have a 1-carat diamond containing approximately 10^{22} carbon atoms.

New structures composed entirely of carbon atoms have been produced; these include fullerenes and carbon nanotubes. Fullerenes are composed of 60 carbon atoms arranged in a truncated icosahedron, a shape like a soccer ball, as illustrated in Figure 13.3a. The 1996 Nobel Prize in Chemistry was awarded to Robert Curl, Harold Kroto, and Richard Smalley for their discovery of fullerenes. Fullerenes are also called *buckyballs* and were named after Buckminster Fuller (1895–1983), who invented the geodesic dome, whose geometry resembles that of the fullerenes. Figure 13.3b shows the structure of a carbon nanotube, consisting of interlocking hexagons of carbon atoms. A carbon nanotube can be thought of as a layer of graphite (Figure 13.2c) rolled up into a tube. Fullerenes and carbon nanotubes are early products of an emerging area of research termed *nanotechnology*, referring to the size of the objects under investigation. New materials constructed using nanotechnology may revolutionize materials science. For example, fullerene crystals are harder than diamonds, and fibers composed of carbon nanotubes are lightweight and stronger than steel. And carbon has still other potentially useful forms. In 2004, Andre Geim and Konstantin Novoselov discovered graphene, a two-dimensional sheet of carbon atoms arranged in a hexagonal lattice structure (Figure 13.2e). They received the 2010 Nobel Prize in Physics for their discovery of this material, which holds great promise for electronics applications.

One **mole** (mol) of a substance contains $N_A = 6.022 \cdot 10^{23}$ atoms or molecules. Because the masses of a proton and a neutron are about equal and the mass of either is far greater than that of an electron, the mass of 1 mol of a substance in grams is given by the atomic mass number. Thus, 1 mol of ^{12}C has a mass of 12 g, and 1 mol of ^{4}He has a mass of 4 g. The periodic table of the elements lists the atomic mass number for each element. This atomic mass

FIGURE 13.2 Diamonds and graphite are composed of carbon atoms. (a) Structure of diamond consisting of carbon atoms bonded in a tetrahedral arrangement. (b) Diamonds. (c) Structure of graphite composed of parallel layers of hexagonal structures. (d) A pencil showing the graphite lead. (e) Structure of graphene, a two-dimensional sheet of carbon atoms arranged in hexagons.

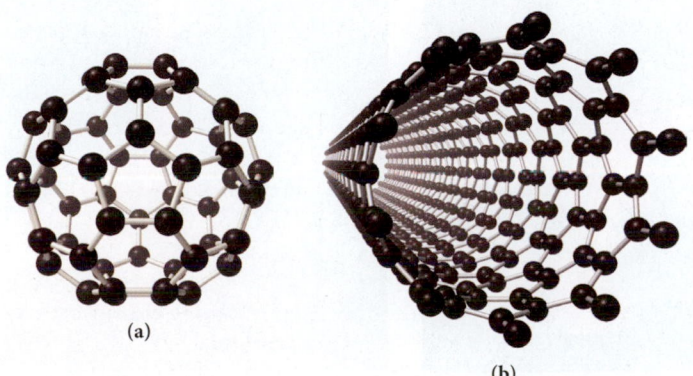

FIGURE 13.3 Nanostructures consisting of arrangements of carbon atoms: (a) fullerene, or buckyball; (b) carbon nanotube.

number is equal to the number of protons and neutrons contained in the nucleus of the atom; it is not an integer because it takes into account the natural isotopic abundances. (Isotopes of an element have varying numbers of neutrons in the nucleus. If a carbon nucleus has seven neutrons, it is the ^{13}C isotope.) For molecules, the molar mass is obtained by adding the mass numbers of all atoms in the molecule. Thus, 1 mol of water, $^{1}H_2{}^{16}O$, has a mass of 18.02 g. (It is not exactly 18 g, because 0.2% of oxygen atoms are the isotope ^{18}O.)

Atoms are electrically neutral. They have the same number of positively charged protons as negatively charged electrons. The chemical properties of an atom are determined by its electronic structure. This structure allows bonding of certain atoms with other atoms to form molecules. For example, water is a molecule containing two hydrogen atoms and one oxygen atom. The electronic structures of atoms and molecules determine a substance's macroscopic properties, such as whether it exists as a gas, liquid, or solid at a given temperature and pressure.

13.2 States of Matter

A **gas** is a system in which each atom or molecule moves through space as a free particle. Occasionally, an atom or molecule collides with another atom or molecule or with the wall of the container. A gas can be treated as a **fluid** because it can flow and exert pressure on the walls of its container. A gas is compressible, which means that the volume of the container can be changed and the gas will still fill the volume, although the pressure it exerts on the walls of the container will change.

In contrast to gases, most liquids are nearly incompressible. If a gas is placed in a container, it will expand to fill the container (Figure 13.4a). When a **liquid** is placed in a container, it fills only the volume corresponding to its initial volume (Figure 13.4b). If the volume of the liquid is less than the volume of the container, the container is only partially filled.

A **solid** does not require a container but instead defines its own shape (Figure 13.4c). Like liquids, solids are nearly incompressible. However, solids can be compressed and deformed slightly.

The categorization of matter into solids, liquids, and gases does not cover the entire range of possibilities. Clearly, which state a certain substance is in depends on its temperature. Water, for example, can be ice (solid), water (liquid), or steam (gas). The same condition holds true for practically all other substances. However, there are states of matter that do not fit into the solid/liquid/gas classification. Matter in stars, for example, is not in any of those three states. Instead, it forms a *plasma*, a system of ionized atoms. On Earth, many beaches are made of sand, a prime example of a *granular medium*. The grains of granular media are solids, but their macroscopic characteristics can be closer to those of liquids (Figure 13.5). For instance, sand can flow like a liquid. *Glasses* seem to be solids at first glance, because they do not change their shape. However, there is also some justification to the view that glass is a type of liquid with

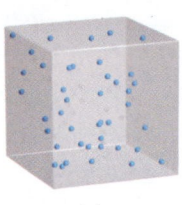

(a)

(b)

(c)

FIGURE 13.4 (a) A cubical container filled with a gas; (b) the same container partially filled with a liquid; (c) a solid, which does not need a container.

(a)

(b)

FIGURE 13.5 (a) Pouring liquid silver (a solid metal at room temperature); (b) pouring sand (a granular medium).

an extremely high viscosity. For the purposes of a classification of matter into states, a glass is neither a solid nor a liquid, but a separate state of matter. *Foams* and *gels* are yet other states of matter, which are currently receiving a lot of interest from researchers. In a foam, the material forms thin membranes around enclosed bubbles of gas of different sizes; thus, some foams are very rigid while having a very low mass density.

Less than two decades ago, the existence of a new form of matter called *Bose-Einstein condensates* was experimentally verified. An understanding of this new state of matter requires some basic concepts of quantum physics. However, in basic terms, at very low temperatures, a gas of certain kinds of atoms can assume an ordered state in which all of the atoms tend to have the same energy and momentum, very similar to the way that light assumes an ordered state in a laser (see Chapter 38).

Finally, the matter in our bodies and in most other biological organisms does not fit into any of these classifications. Biological tissue consists predominantly of water, yet it is able to keep or change its shape, depending on the environmental boundary conditions.

13.3 Tension, Compression, and Shear

Let's examine how solids respond to external forces.

Elasticity of Solids

Many solids are composed of atoms arranged in a three-dimensional crystal lattice in which the atoms have a well-defined equilibrium distance from their neighbors. Atoms in a solid are held in place by interatomic forces that can be modeled as springs. The lattice is very rigid, which implies that the imaginary springs are very stiff. Macroscopic solid objects such as wrenches and spoons are composed of atoms arranged in such a rigid lattice. However, other solid objects, such as rubber balls, are composed of atoms arranged in long chains rather than in a well-defined lattice. Depending on their atomic or molecular structure, solids can be extremely rigid or more easily deformable.

All rigid objects are somewhat elastic, even though they do not appear to be. Compression, pulling, or twisting can deform a rigid object. If a rigid object is deformed a small amount, it will return to its original size and shape when the deforming force is removed. If a rigid object is deformed past a point called its **elastic limit,** it will not return to its original size and shape but will remain permanently deformed. If a rigid object is deformed too far beyond its elastic limit, it will break.

(a)

(b)

Stress and Strain

Deformations of solids are usually classified into three types: stretching (or tension), compression, and shear. Examples of stretching, compression, and shear are shown in Figure 13.6. What these three deformations have in common is that a **stress,** or deforming force per unit area, produces a **strain,** or unit deformation. **Stretching,** or **tension,** is associated with tensile stress. **Compression** can be produced by hydrostatic stress. **Shear** is produced by shearing stress, sometimes also called *deviatory stress*. When a shear force is applied, planes of material parallel to the force and on either side of it remain parallel but shift relative to each other.

Although stress and strain take different forms for the three types of deformation, they are related linearly through a constant called the **modulus of elasticity:**

$$\text{stress} = \text{modulus of elasticity} \cdot \text{strain}. \tag{13.1}$$

This empirical relationship applies as long as the elastic limit of the material is not exceeded.

In the case of tension, a force F is applied to opposite ends of an object of length L and the object stretches to a new length, $L + \Delta L$ (Figure 13.7). The stress for stretching or tension is defined as the force, F, per unit area, A, applied to the end of an object. The strain is defined as the fractional change in length of the object, $\Delta L/L$. The relationship between stress and strain up to the elastic limit is then

$$\frac{F}{A} = Y\frac{\Delta L}{L}, \tag{13.2}$$

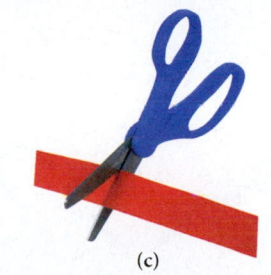

(c)

FIGURE 13.6 Three examples of stress and strain: (a) the stretching of power lines; (b) the compression of the Hoover Dam; (c) shearing by scissors.

Table 13.1 | **Some Typical Values of Young's Modulus** | |
|---|---|
| **Material** | **Young's Modulus** (10^9 N/m^2) |
| Aluminum | 70 |
| Bone | 10–20 |
| Concrete | 20–30 (compression) |
| Diamond | 1000–1200 |
| Glass | 70 |
| Polystyrene | 3 |
| Rubber | 0.01–0.1 |
| Steel | 200 |
| Titanium | 100–120 |
| Tungsten | 400 |
| Wood | 10–15 |

Table 13.2 | **Some Typical Values of the Bulk Modulus** | |
|---|---|
| **Material** | **Bulk Modulus** (10^9 N/m^2) |
| Air | 0.000142 |
| Aluminum | 76 |
| Basalt rock | 50–80 |
| Gasoline | 1.5 |
| Granite rock | 10–50 |
| Mercury | 28.5 |
| Steel | 160 |
| Water | 2.2 |

Table 13.3 | **Some Typical Values of the Shear Modulus** | |
|---|---|
| **Material** | **Shear Modulus** (10^9 N/m^2) |
| Aluminum | 25 |
| Copper | 45 |
| Glass | 26 |
| Polyethylene | 0.12 |
| Rubber | 0.0003 |
| Titanium | 41 |
| Steel | 70–90 |

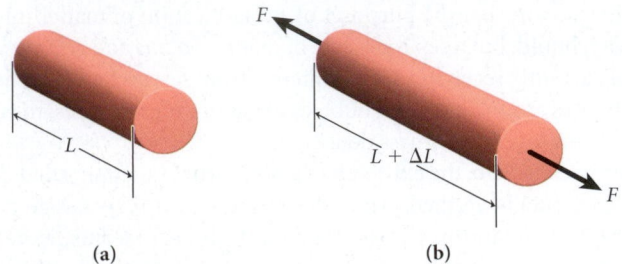

FIGURE 13.7 Tension applied to opposite ends of an object by a pulling force. (a) Object before force is applied. (b) Object after force is applied. *Note:* Tension can also be applied by pushing, with a resulting negative change in length (not shown).

where Y is called **Young's modulus** and depends only on the type of material and not on its size or shape. Some typical values of Young's modulus are given in Table 13.1.

Linear compression can be treated in a manner similar to stretching for most materials, within the elastic limits. However, many materials have different breaking points for stretching and compression. The most notable example is concrete, which resists compression much better than stretching, which is why steel rods are added to it in places where greater tolerance of stretching is required. A steel rod resists stretching much better than compression, under which it can buckle.

The stress related to volume compression is caused by a force per unit area applied to the entire surface area of an object for example, one submerged in a liquid (Figure 13.8). The resulting strain is the fractional change in the volume of the object, $\Delta V/V$. The modulus of elasticity in this case is the **bulk modulus,** B. We can thus write the equation relating stress and strain for volume compression as

$$\frac{F}{A} = B\frac{\Delta V}{V}. \tag{13.3}$$

Some typical values of the bulk modulus are given in Table 13.2. Note the extremely large jump in the bulk modulus from air, which is a gas and can be compressed rather easily, to liquids such as gasoline and water. Solids such as rocks and metals have values for the bulk modulus that are higher than those of liquids by a factor between 5 and 100.

In the case of shearing, the stress is again a force per unit area. However, for shearing, the force is parallel to the area rather than perpendicular to it (Figure 13.9). For shearing, stress is given by the force per unit area, F/A, exerted on the end of the object. The resulting strain is given by the fractional deflection of the object, $\Delta x/L$. The stress is related to the strain through the **shear modulus,** G:

$$\frac{F}{A} = G\frac{\Delta x}{L}. \tag{13.4}$$

Some typical values of the shear modulus are given in Table 13.3.

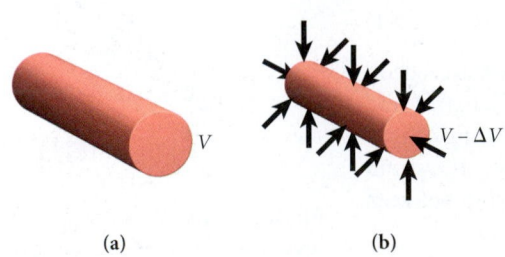

FIGURE 13.8 Compression of an object by fluid pressure; (a) object before compression; (b) object after compression.

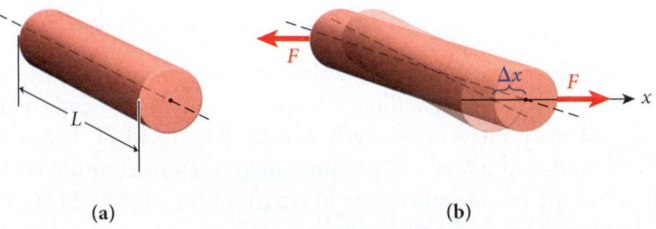

FIGURE 13.9 Shearing of an object caused by a force parallel to the area of the end of the object. (a) Object before shear force is applied. (b) Object after shear force is applied.

EXAMPLE 13.1 | Wall Mount for a Flat-Panel TV

You've just bought a new flat-panel TV (Figure 13.10) and want to mount it on the wall with four bolts, each with a diameter each of 0.50 cm. You cannot mount the TV flush against the wall but have to leave a 10.0-cm gap between wall and TV for air circulation.

PROBLEM 1
If the mass of your new TV is 42.8 kg, what is the shear stress on the bolts?

SOLUTION 1
The combined cross-sectional area of the bolts is

$$A = 4\left(\frac{\pi d^2}{4}\right) = \pi(0.005 \text{ m})^2 = 7.85 \cdot 10^{-5} \text{ m}^2.$$

One force acting on the bolts is $\vec{F}_g$, the force of gravity on the TV, exerted at one end of each bolt. This force is balanced by a force due to the wall, acting on the other end of the bolts. This wall force holds the TV in place; thus, it has exactly the same magnitude as the force of gravity but points in the opposite direction. Therefore, the force entering into equation 13.4 for the stress is

$$F = mg = (42.8 \text{ kg})(9.81 \text{ m/s}^2) = 420. \text{ N}.$$

Thus, we obtain the shear stress on the bolts:

$$\frac{F}{A} = \frac{420. \text{ N}}{7.85 \cdot 10^{-5} \text{ m}^2} = 5.35 \cdot 10^6 \text{ N/m}^2.$$

PROBLEM 2
The shear modulus of the steel used in the bolts is $9.0 \cdot 10^{10}$ N/m^2. What is the resulting vertical deflection of the bolts?

SOLUTION 2
Solving equation 13.4 for the deflection, Δx, we find

$$\Delta x = \left(\frac{F}{A}\right)\frac{L}{G} = (5.35 \cdot 10^6 \text{ N/m}^2)\frac{0.1 \text{ m}}{9.0 \cdot 10^{10} \text{ N/m}^2} = 5.94 \cdot 10^{-6} \text{ m}.$$

Even though the shear stress is more than 5 million N/m^2, the resulting sag of your flat-panel TV is only about 0.006 mm, a distance that is undetectable with the naked eye.

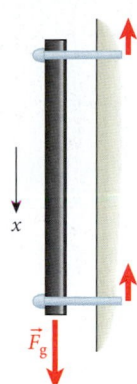

FIGURE 13.10 Forces acting on a wall mount for a flat-panel TV.

SOLVED PROBLEM 13.1 | Stretched Wire

PROBLEM
A 1.50-m-long steel wire with a diameter of 0.400 mm is hanging vertically. A spotlight of mass $m = 1.50$ kg is attached to the wire and released. How much does the wire stretch?

SOLUTION
THINK From the diameter of the wire, we can get its cross-sectional area. The weight of the spotlight provides a downward force. The stress on the wire is the weight divided by the cross-sectional area. The strain is the change in the length of the wire divided by its original length. The stress and strain are related through Young's modulus for steel.

SKETCH Figure 13.11 shows the wire before and after the spotlight is attached.

RESEARCH The cross-sectional area of the wire, A, can be calculated from the diameter, d, of the wire:

$$A = \pi\left(\frac{d}{2}\right)^2 = \frac{\pi d^2}{4}. \tag{i}$$

The force on the wire is the weight of the searchlight,

$$F = mg. \tag{ii}$$

– Continued

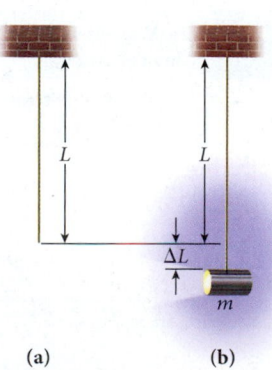

FIGURE 13.11 A wire (a) before and (b) after a spotlight is attached to it.

We can relate the stress and the strain on the wire through Young's modulus, Y, for steel:

$$\frac{F}{A} = Y\frac{\Delta L}{L},\qquad\text{(iii)}$$

where ΔL is the change in the length of the wire and L is its original length.

SIMPLIFY We can combine equations (i), (ii), and (iii) to get

$$\frac{F}{A} = \frac{mg}{\left(\frac{1}{4}\pi d^2\right)} = \frac{4mg}{\pi d^2} = Y\frac{\Delta L}{L}.$$

Solving for the change in the wire's length, we obtain

$$\Delta L = \frac{4mgL}{Y\pi d^2}.$$

CALCULATE Putting in the numerical values gives us

$$\Delta L = \frac{4mgL}{Y\pi d^2} = \frac{4(1.50\text{ kg})(9.81\text{ m/s}^2)(1.50\text{ m})}{\left(200\cdot10^9\text{ N/m}^2\right)\pi\left(0.400\cdot10^{-3}\text{ m}\right)^2} = 0.00087824\text{ m}.$$

ROUND We report our result to three significant figures:

$$\Delta L = 0.000878\text{ m} = 0.878\text{ mm}.$$

DOUBLE-CHECK An easy and important first check of a result is to confirm that the units are right. The change in length, which we have calculated, has the units of millimeters, which makes sense. Next, we examine the magnitude: The wire stretches just less than 1 mm compared with its original length of 1.50 m. This stretch is less than 0.1%, which seems reasonable. Any time a steel wire experiences noticeable stretch, breaking of the wire becomes a concern. This is not the case here, which gives us some confidence that our result is of approximately the right order of magnitude.

As noted earlier, the stress applied to an object is proportional to the strain as long as the elastic limit of the material is not exceeded. Figure 13.12 shows a typical stress-strain diagram for a ductile (easily drawn into a wire) metal under tension. Up to the proportional limit, the ductile metal responds linearly to stress. If the stress is removed, the material will return to its original length. If stress is applied past the proportional limit, the material will continue to lengthen until it reaches its yield point. If stress is applied between the proportional limit and the fracture point and then is removed, the material will not return to its original length but will be permanently deformed. The yield point is the point where the stress causes sudden deformation without any increase in force as can be seen from the flattening of the curve (see Figure 13.12). Additional stress will continue to stretch the material until it reaches its fracture point, where it breaks. This breaking stress is also called the *ultimate stress*, or in the case of tension, the *tensile strength*. Some approximate breaking stresses are given in Table 13.4.

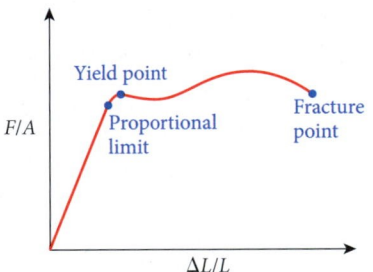

FIGURE 13.12 A typical stress-strain diagram for a ductile metal under tension showing the proportional limit, the yield point, and the fracture point.

Table 13.4	Breaking Stress for Common Materials

Material	Breaking Stress (10^6 N/m^2)
Aluminum	455
Brass	550
Copper	220
Steel	400
Bone	130

Note that these values are approximate.

13.4 Pressure

This section examines the properties of liquids and gases, together termed *fluids*. The properties of stress and strain, discussed in Section 13.3, are most useful when studying solids, because substances that flow, such as liquids and gases, offer little resistance to shear or tension. In this section, we consider the properties of fluids at rest. Later in this chapter, we will discuss the properties of fluids in motion.

The **pressure**, p, is the force per unit area:

$$p = \frac{F}{A}.\qquad\text{(13.5)}$$

Pressure is a scalar quantity. The SI unit of pressure is N/m^2, which has been named the **pascal**, abbreviated Pa:

$$1\text{ Pa} \equiv \frac{1\text{ N}}{1\text{ m}^2}.$$

The average pressure of the Earth's atmosphere at sea level, 1 atm, is a commonly used non-SI unit that is expressed in other units as follows:

$$1 \text{ atm} = 1.01 \cdot 10^5 \text{ Pa} = 760. \text{ torr} = 14.7 \text{ lb/in}^2.$$

Gauges used to measure how much air has been removed from a vessel are often calibrated in *torr*, a unit named after the Italian physicist Evangelista Torricelli (1608–1647). Automobile tire pressures in the United States are often measured in pounds per square inch (lb/in^2, or psi).

SOLVED PROBLEM 13.2 Weighing Earth's Atmosphere

The Earth's atmosphere is composed (by volume) of 78.08% nitrogen (N_2), 20.95% oxygen (O_2), 0.93% argon (Ar), 0.25% water vapor (H_2O), and traces of other gases, most importantly, carbon dioxide (CO_2). The CO_2 content of the atmosphere is currently around 0.039% = 390 ppm (parts per million), but it varies with the seasons by about 6–7 ppm and has been rising since the start of the Industrial Revolution, mainly as a result of the burning of fossil fuels. Approximately 2 ppm of CO_2 are being added to the atmosphere each year.

PROBLEM
What is the mass of the Earth's atmosphere, and what is the mass of 1 ppm of atmospheric CO_2?

SOLUTION
THINK At first glance, this problem seems rather daunting, because very little information is given. However, we know that the atmospheric pressure is $1.01 \cdot 10^5$ Pa and that pressure is force per area.

SKETCH The sketch in Figure 13.13 shows a column of air with weight mg above an area A of Earth's surface. This air exerts a pressure, p, on the surface.

RESEARCH We start with the relationship between pressure and force, $p = F/A$, where the area is the surface area of Earth, $A = 4\pi R^2$, and $R = 6370$ km is the radius of Earth. For the force, we can use the atmospheric weight, $F = mg$, where m is the mass of the atmosphere.

SIMPLIFY We combine the equations just mentioned

$$p = \frac{F}{A} = \frac{mg}{4\pi R^2}$$

and solve for the mass of the atmosphere

$$m = \frac{4\pi R^2 p}{g}.$$

CALCULATE We substitute the numerical values:

$$m = 4\pi (6.37 \cdot 10^6 \text{ m})^2 (1.01 \cdot 10^5 \text{ Pa})/(9.81 \text{ m/s}^2) = 5.24978 \cdot 10^{18} \text{ kg}.$$

ROUND We round to three significant figures and obtain

$$m = 5.25 \cdot 10^{18} \text{ kg}.$$

DOUBLE-CHECK In order to obtain the mass of 1 ppm of CO_2 in the atmosphere, we have to realize that the molar mass of CO_2 is $12 + (2 \cdot 16) = 44$ g. The average mass of a mole of the atmosphere is approximately $0.78(2 \cdot 14) + 0.21(2 \cdot 16) + 0.01(40) = 28.96$ g. The mass of 1 ppm of CO_2 in the atmosphere is therefore

$$m_{1 \text{ ppm CO}_2} = 10^{-6} \cdot m \frac{44}{28.96} = 7.97 \cdot 10^{12} \text{ kg} = 8.0 \text{ billion tons}.$$

Humans add approximately 2 ppm of CO_2 to the atmosphere each year by burning fossil fuels, which amounts to approximately 16 billion tons of CO_2, a scary number. It is not easy to double-check the orders of magnitude for this calculation. However, data published by the U.S. Energy Information Administration show that total carbon dioxide emissions from burning fossil fuels are currently approximately 30 billion tons per year, higher than our result by a factor of 2. Where does the other half of the CO_2 go? Mainly, it dissolves in the Earth's oceans.

Concept Check 13.1

Suppose you have a jar of nuts that has been vacuum packed to preserve the contents. You turn the lid and hear a whoosh. Which of the following statements correctly describes the force exerted by the atmosphere on the lid before and after you turn it?

a) The atmosphere exerts more force on the lid before the lid is turned than after it is turned.

b) The atmosphere exerts less force on the lid after the lid is turned than before it is turned.

c) The atmosphere exerts the same force on the lid before and after the lid is turned.

d) There is no way to tell what happens to the force exerted by the atmosphere on the lid when it is turned.

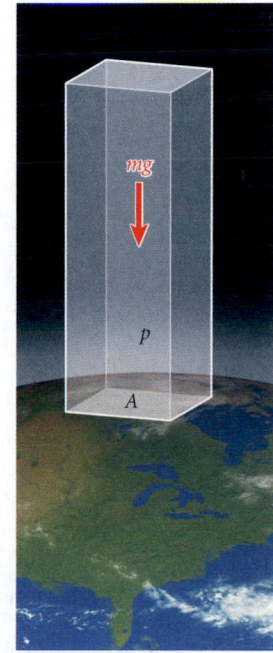

FIGURE 13.13 Weight of the atmosphere.

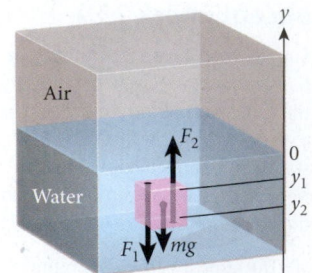

FIGURE 13.14 Cube of water in a tank of water.

Pressure-Depth Relationship

Consider a tank of water open to the Earth's atmosphere, and imagine a cube of the water inside of it (shown in pink in Figure 13.14). Assume that the top surface of the cube is horizontal and at a depth of y_1 and that the bottom surface of the cube is horizontal and at a depth of y_2. The other sides of the cube are oriented vertically. The water pressure acting on the cube produces forces. However, by Newton's First Law, there must be no net force acting on this stationary cube of water. The forces acting on the vertical sides of the cube clearly cancel out. The vertical forces acting on the bottom and top sides of the cube must also add to zero:

$$F_2 - F_1 - mg = 0, \tag{13.6}$$

where F_1 is the force downward on the top of the cube, F_2 is the force upward on the bottom of the cube, and mg is the weight of the cube of water. The pressure at depth y_1 is p_1, and the pressure at depth y_2 is p_2. We can write the forces at these depths in terms of the pressures, assuming that the area of the top and bottom surfaces of the cube is A:

$$F_1 = p_1 A$$
$$F_2 = p_2 A.$$

We can also express the mass, m, of the water in terms of the density of water, ρ (assumed to be constant), and the volume, V, of the cube, $m = \rho V$. Substituting for F_1, F_2, and m in equation 13.6 gives

$$p_2 A - p_1 A - \rho V g = 0.$$

Rearranging this equation and substituting $A(y_1 - y_2)$ for V, we get

$$p_2 A = p_1 A + \rho A(y_1 - y_2)g.$$

Dividing out the area yields an expression for the pressure as a function of depth in a liquid of uniform density ρ:

$$p_2 = p_1 + \rho g(y_1 - y_2). \tag{13.7}$$

A common problem involves the pressure as a function of depth below the surface of a liquid. Starting with equation 13.7, we can define the pressure at the surface of the liquid ($y_1 = 0$) to be p_0 and the pressure at a given depth h ($y_2 = -h$) to be p. These assumptions lead to the following expression for the pressure at a given depth of a liquid with uniform density ρ:

$$p = p_0 + \rho g h. \tag{13.8}$$

Note that in deriving equation 13.8, we made use of the fact that the density of the fluid does not change as a function of depth. This assumption of incompressibility is essential for obtaining this result. Later in this section, relaxing this incompressibility requirement will lead to a different formula relating height and pressure for gases. Further, it is important to note that equation 13.8 specifies the pressure in a liquid at a vertical depth of h, without any dependence on horizontal position. Thus, the equation holds regardless of the shape of the vessel containing the liquid. Figure 13.15, for example, shows three connected columns containing a fluid. You can see that the fluid reaches the same height in each column, independent of the shape or cross-sectional area; this occurs because the bottoms are interconnected and thus have the same pressure.

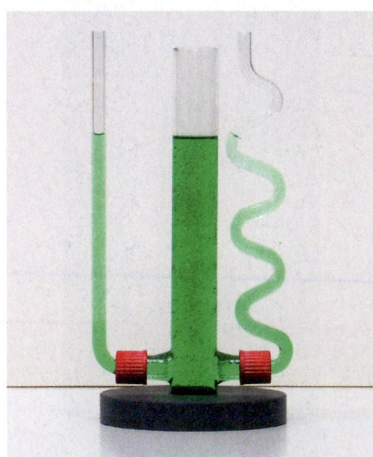

FIGURE 13.15 Three connected columns—the fluid rises to the same height in each.

EXAMPLE 13.2   Submarine

A U.S. Navy submarine of the Los Angeles class is 110 m long and has a hull diameter of 10. m (Figure 13.16). Assume that the submarine has a flat top with an area of $A = 1100$ m^2 and that the density of seawater is 1024 kg/m^3.

PROBLEM

What is the total force pushing down on the top of this submarine at a diving depth of 250. m?

SOLUTION

The pressure inside the submarine is normal atmospheric pressure, p_0. According to equation 13.8, the pressure at a depth of 250. m is given by $p = p_0 + \rho gh$. Therefore, the pressure difference between the inside and the outside of the submarine is

$$\Delta p = \rho gh = (1024 \text{ kg/m}^3)(9.81 \text{ m/s}^2)(250 \text{ m}) = 2.51 \cdot 10^6 \text{ N/m}^2 = 2.51 \text{ MPa},$$

or approximately 25 atm.

Multiplying the area times the pressure gives the total force, according to equation 13.5. Thus, we find a total force of

$$F = \Delta pA = (2.51 \cdot 10^6 \text{ N/m}^2)(1100 \text{ m}^2) = 2.8 \cdot 10^9 \text{ N}.$$

This number is astonishingly large and corresponds to the weight of a mass of $\approx 280{,}000$ metric tons, over 40 times the submarine's own weight!

FIGURE 13.16 Submarine traveling at the surface of the water.

Concept Check 13.2

Three containers of water have different shapes, as shown below. The distance between the top surface of the water and the bottom of each container is the same. Which of the following statements correctly characterizes the pressure at the bottom of the containers?

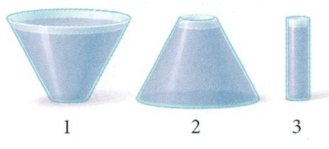

1   2   3

a) The pressure at the bottom of container 1 is the highest.

b) The pressure at the bottom of container 2 is the highest.

c) The pressure at the bottom of container 3 is the highest.

d) The pressure at the bottom of all three containers is the same.

Gauge Pressure and Barometers

The pressure p in equation 13.8 is an **absolute pressure,** which means it includes the pressure of the liquid as well as the pressure of the air above it. The difference between an absolute pressure and the atmospheric air pressure is called a **gauge pressure.** For example, a gauge used to measure the air pressure in a tire is calibrated so that the atmospheric pressure reads zero. When connected to the tire's compressed air, the gauge measures the additional pressure present in the tire. In equation 13.8, the gauge pressure is ρgh.

A simple device used to measure atmospheric pressure is the mercury **barometer** (Figure 13.17). You can construct a mercury barometer by taking a long glass tube, closed at one end, filling it with mercury, and inverting it with the open end in a dish of mercury. The space above the mercury is a vacuum and thus has zero pressure. The difference in height between the top of the mercury in the tube and the top of the mercury in the dish, h, can be related to the atmospheric pressure, p_0, using equation 13.7 with $p_2 = p_1 + \rho g(y_1 - y_2)$, $y_2 - y_1 = h$, $p_2 = 0$, and $p_1 = p_0$:

$$p_0 = \rho gh.$$

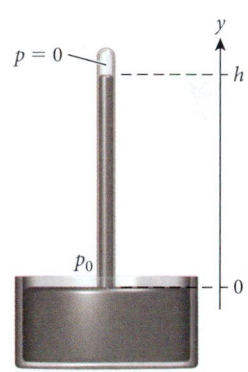

FIGURE 13.17 A mercury barometer measures atmospheric pressure, p_0.

Note that the measurement of atmospheric pressure using a mercury barometer depends on the local value of g. Atmospheric pressure is often expressed in millimeters of mercury (mmHg), corresponding to the height difference h. The torr is equivalent to 1 mmHg, so standard atmospheric pressure is 760. torr, or 29.92 in of mercury (101.325 kPa).

An open-tube **manometer** measures the gauge pressure of a gas. It consists of a U-shaped tube partially filled with a liquid such as mercury (Figure 13.18). The closed end of the manometer is connected to a vessel containing the gas whose gauge pressure, p_g, is being measured, and the other end is open and thus experiences atmospheric pressure, p_0. Using equation 13.7 with $y_1 = 0$, $p_1 = p_0$, $y_2 = -h$, and $p_2 = p$, where p is the absolute pressure of the gas in the vessel, we obtain $p = p_0 + \rho gh$, the pressure-depth relationship derived earlier. The gauge pressure of the gas in the vessel is then

$$p_g = p - p_0 = \rho gh. \tag{13.9}$$

Note that gauge pressure can be positive or negative. The gauge pressure of the air in an inflated automobile tire is positive. The gauge pressure at the end of a straw being used by a person drinking a milkshake is negative.

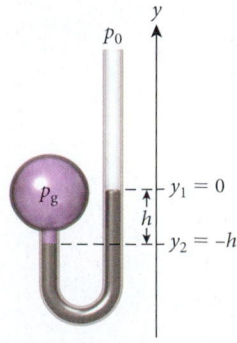

FIGURE 13.18 An open-tube manometer measures the gauge pressure of a gas.

Barometric Altitude Relation for Gases

In deriving equation 13.8, we made use of the incompressibility of liquids. However, if the fluid is a gas, we cannot make this assumption. Let's start again with a thin layer of fluid in a column of the fluid. The pressure difference between the top and bottom surfaces is the negative of the weight of the thin layer of fluid divided by its area:

$$\Delta p = -\frac{F}{A} = -\frac{mg}{A} = -\frac{\rho V g}{A} = -\frac{\rho(\Delta h A)g}{A} = -\rho g \Delta h. \tag{13.10}$$

The negative sign reflects the fact that the pressure decreases with increasing altitude (h), since the weight of the fluid column above the layer will be reduced. So far, nothing is different from the derivation for the incompressible fluid. However, for compressible fluids, the density is proportional to the pressure:

$$\frac{\rho}{\rho_0} = \frac{p}{p_0}. \tag{13.11}$$

Strictly speaking, this relationship is true only for an ideal gas at constant temperature, as we will see in Chapter 19. However, if we combine equations 13.10 and 13.11, we obtain

$$\frac{\Delta p}{\Delta h} = -\frac{g\rho_0}{p_0}p.$$

Taking the limit as $\Delta h \to 0$, we have

$$\frac{dp}{dh} = -\frac{g\rho_0}{p_0}p.$$

This is an example of a differential equation. We need to find a function whose derivative is proportional to the function itself, which leads us to an exponential function:

$$p(h) = p_0 e^{-h\rho_0 g/p_0}. \tag{13.12}$$

It is easy to convince yourself that equation 13.12 is indeed a solution to the preceding differential equation by simply taking the derivative with respect to h. Equation 13.12 is sometimes called the **barometric pressure formula,** and it relates the pressure to the altitude in gases. It applies only as long as the temperature does not change as a function of altitude and as long as the gravitational acceleration can be assumed to be constant. (We will consider the effect of temperature change when we discuss the Ideal Gas Law in Chapter 19.)

We can obtain a formula for the air density, ρ, as a function of altitude by combining equations 13.11 and 13.12:

$$\rho(h) = \rho_0 e^{-h\rho_0 g/p_0}. \tag{13.13}$$

Even though results obtained with this equation are only an approximation, they match actual atmospheric data fairly closely, as shown in Figure 13.19, where $\rho(h)$ was plotted with $g = 9.81$ m/s^2, $p_0 = 1.01 \cdot 10^5$ Pa , the air pressure at sea level ($h = 0$), and $\rho_0 = 1.229$ kg/m^3, the air density at sea level. As you can see, the agreement is very close up to the top of the stratosphere, approximately 50 km above ground.

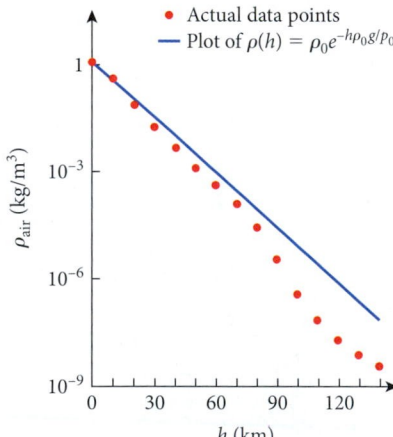

FIGURE 13.19 Comparison of the air density plotted from equation 13.13 (blue line) with actual data for the atmosphere (red dots).

EXAMPLE 13.3 Air Pressure on Mount Everest

As climbers approach the peak of Earth's highest mountain, Mount Everest, they usually have to wear breathing equipment. The reason for this is that the air pressure is very low, too low for climbers' lungs, which are usually accustomed to near sea-level conditions.

PROBLEM
What is the air pressure at the top of Mount Everest (Figure 13.20)?

SOLUTION
This is a situation where we can use the barometric pressure formula (equation 13.12). We first note the constants: $p_0 = 1.01 \cdot 10^5$ Pa, the sea-level air pressure, and $\rho_0 = 1.229$ kg/m^3, the air

density at sea level. Then we find the inverse of the constant part of the exponent in equation 13.12:

$$\frac{p_0}{\rho_0 g} = \frac{1.01 \cdot 10^5 \text{ Pa}}{(1.229 \text{ kg/m}^3)(9.81 \text{ m/s}^2)} = 8377 \text{ m}.$$

We can then rewrite equation 13.12 as

$$p(h) = p_0 e^{-h/(8377 \text{ m})}.$$

The height of Mount Everest is 8850 m. Therefore, we obtain

$$p(8850 \text{ m}) = p_0 e^{-8850/8377} = 0.348 p_0 = 35 \text{ kPa}.$$

The calculated air pressure at the top of Mount Everest is only 35% of the air pressure at sea level. (The actual pressure is slightly lower, mainly because of temperature effects.)

FIGURE 13.20 Mount Everest is Earth's highest peak, at 8850 m (29,035 ft).

You do not have to travel to the top of Mount Everest to appreciate the change in air pressure with altitude. You have probably experienced your ears "popping" while driving in the mountains. This physiological effect of feeling pressure on the eardrums results because of a lag in your body's adjusting the internal pressure to the change in external pressure due to the rapid change in altitude.

Another demonstration of the change of air pressure with altitude is shown in Figure 13.21. A plastic water bottle made from a very soft plastic was emptied in Winter Park, Colorado, and sealed with air in it by tightly closing the lid. Winter Park is the highest incorporated town in the United States, at an altitude of over 3600 m. The bottle was then driven to the Denver airport, at an altitude of 1600 m, with the temperature inside the car kept constant. As you can see, the bottle crumpled, shrinking its volume significantly.

Concept Check 13.3

As you descend into a mine shaft below sea level, the air pressure

a) decreases linearly.

b) decreases exponentially.

c) increases linearly.

d) increases exponentially.

 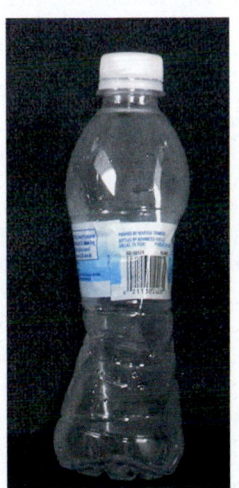

FIGURE 13.21 Demonstration of the change in air pressure with altitude. Left side: an air-filled, sealed plastic bottle at Winter Park, Colorado; right side: the same bottle in Denver, Colorado.

Concept Check 13.4

The air pressure inside the crushed bottle shown on the right in Figure 13.21

a) is significantly lower than the outside air pressure.

b) is significantly higher than the outside air pressure.

c) is approximately the same as the outside air pressure.

d) could be any of the above.

Pascal's Principle

If pressure is exerted on a part of an incompressible fluid, that pressure will be transmitted to all parts of the fluid without loss. This is **Pascal's Principle,** which can be stated as follows:

> *When a change in pressure occurs at any point in a confined fluid, an equal change in pressure occurs at every point in the fluid.*

Pascal's Principle is the basis for many modern hydraulic devices, such as automobile brakes, large earth-moving machines, and car lifts.

Pascal's Principle can be demonstrated by taking a cylinder partially filled with water, placing a piston on top of the column of water, and placing a weight on top of the piston

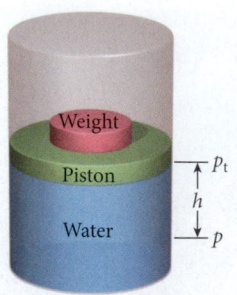

FIGURE 13.22 Cylinder partially filled with water, with a piston placed on top of the water and a weight placed on the piston.

(Figure 13.22). The air pressure and the weight exert a pressure, p_t, on top of the column of water. The pressure, p, at depth h is given by

$$p = p_t + \rho g h.$$

Because water can be considered incompressible, if a second weight is added on top of the piston, the change in pressure, Δp, at depth h is due solely to the change in the pressure, Δp_t, on top of the water. The density of the water does not change and the depth does not change, so we can write

$$\Delta p = \Delta p_t.$$

This result does not depend on h, so it must hold for all positions in the liquid.

Now consider two connected pistons in cylinders filled with oil, as shown in Figure 13.23. One piston has area A_{in} and the other piston has area A_{out}, with $A_{in} < A_{out}$. A force, F_{in}, is exerted on the first piston, producing a change in pressure in the oil. This change in pressure is transmitted to all points in the oil, including points adjacent to the second piston. We can write

$$\Delta p = \frac{F_{in}}{A_{in}} = \frac{F_{out}}{A_{out}},$$

or

$$F_{out} = F_{in} \frac{A_{out}}{A_{in}}. \tag{13.14}$$

Because A_{out} is larger than A_{in}, F_{out} is larger than F_{in}. Thus, the force applied to the first piston is magnified. This phenomenon is the basis of hydraulic devices that produce large output forces with small input forces.

The amount of work done on the first piston is the same as the amount of work done by the second piston. To calculate the work done, we need to calculate the distance over which the forces act. For both pistons, the volume, V, of incompressible oil that is moved is the same:

$$V = h_{in} A_{in} = h_{out} A_{out},$$

where h_{in} is the distance the first piston moves and h_{out} is the distance the second piston moves. We can see that

$$h_{out} = h_{in} \frac{A_{in}}{A_{out}}, \tag{13.15}$$

which means that the second piston moves a smaller distance than the first piston because $A_{in} < A_{out}$. We can find the work done by using the fact that work is force times distance and using equations 13.14 and 13.15:

$$W = F_{in} h_{in} = \left(F_{out} \frac{A_{in}}{A_{out}} \right)\left(h_{out} \frac{A_{out}}{A_{in}} \right) = F_{out} h_{out}.$$

Thus, this hydraulic device transmits a larger force over a smaller distance. However, no additional work is done.

Concept Check 13.5

A car weighing 10,000. N is supported by a hydraulic car lift, as illustrated in Figure 13.23. The large piston supporting the car has a diameter of 25.4 cm. The small piston has a diameter of 2.54 cm. How much force must be exerted on the small piston to support the car?

a) 1.00 N   d) 14.1 N

b) 10.0 N   e) 141 N

c) 100. N

FIGURE 13.23 Application of Pascal's Principle in a hydraulic lift. (The scale of the lift relative to the car is greatly out of proportion in order to show the essential details clearly.)

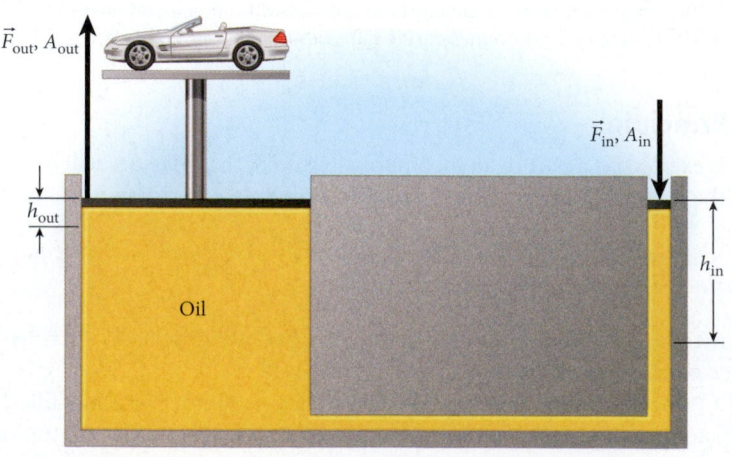

13.5 Archimedes' Principle

Archimedes (287–212 BC) of Syracuse, Sicily, was one of the greatest mathematicians of all time. The king, Hiero of Syracuse, ordered a new crown made and gave the goldsmith the exact amount of gold needed to create the crown. When the crown was finished, it had the correct weight, but Hiero suspected that the goldsmith had used some silver in the crown to replace the more valuable gold. Hiero could not prove this and went to Archimedes for help. According to legend, the answer occurred to Archimedes as he was about to take a bath and noticed the water level rising and his apparent weight diminishing when he got into the bathtub. He then ran naked through the streets of Syracuse to the palace, shouting "Eureka!" (Greek for "I have found it!"). Using his discovery, he was able to demonstrate the silver-for-gold switch and thus prove the theft committed by the goldsmith. Later in this section, we'll examine Archimedes' method for solving this problem, after considering buoyant force and fluid displacement.

Buoyant Force

Figure 13.14 shows a cube of water within a larger volume of water. The weight of the cube of water is supported by the force resulting from the pressure difference between the top and bottom surfaces of the cube, as given by equation 13.6, which we can rewrite as

$$F_2 - F_1 = mg = F_B, \quad (13.16)$$

where F_B is defined as the **buoyant force** acting on the cube of water. For the case of the cube of water, the buoyant force is equal to the weight of the water. In general, the buoyant force acting on a submerged object is given by the weight of the fluid displaced,

$$F_B = m_f g.$$

Now suppose the cube of water is replaced with a cube of steel (Figure 13.24). Because the steel cube has the same volume and is at the same depth as the cube of water, the buoyant force remains the same. However, the steel cube weighs more than the cube of water, so a net y-component of force acts on the steel cube, given by

$$F_{net,y} = F_B - m_{steel}g < 0.$$

This net downward force causes the steel cube to sink.

If the cube of water is replaced by a cube of wood, the weight of the cube of wood is less than that of the cube of water, so the net force will be upward. The wooden cube will rise toward the surface. If an object that is less dense than water is placed in water, it will float. An object of mass m_{object} will sink in water until the weight of the displaced water equals the weight of the object:

$$F_B = m_f g = m_{object} g.$$

A floating body displaces its own weight of fluid. This statement is true, independent of the amount of fluid present. To make this clearer, Figure 13.25 shows a ship in a lock. Part (a) shows the ship in the low position, and part (b) in the high position. In both positions, the ship floats, with the same fraction of the ship below the water level. What matters for the buoyant force is not the total amount of the water in the lock (light blue in the figure), but the amount of water displaced by the underwater portion of the ship (hatched blue and brown in the figure). Clearly, in Figure 13.25a, there is much less water left in the lock than the volume that has been displaced by the ship. The only thing that matters is the weight of the water that *would be where the ship is,* not the weight of the water still in the lock. (In fact, with a container of the right shape, even a single gallon of water could be "enough to float a battleship.")

If an object that has a higher density than water is placed under water, it will experience an upward buoyant force that is less than its weight. Its apparent weight is then given by

$$\text{Actual weight} - \text{buoyant force} = \text{apparent weight}. \quad (13.17)$$

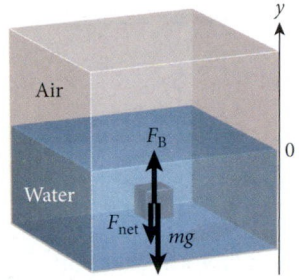

FIGURE 13.24 The weight of a steel cube submerged in water is larger than the buoyant force acting on the cube.

(a)

(b)

FIGURE 13.25 A ship in a lock: (a) low position; (b) high position.

EXAMPLE 13.4 / Floating Iceberg

Icebergs, such as the one shown in Figure 13.26a, pose grave dangers for ocean-going ships. Many ships, the *Titanic* most famously, have sunk after collisions with icebergs. The trouble is that a large fraction of an iceberg's volume is hidden below the waterline and thus practically invisible to sailors, as illustrated in Figure 13.26b.

FIGURE 13.26 (a) An iceberg floating in the ocean. (b) Illustration showing the fraction of the volume of the iceberg above and below the waterline.

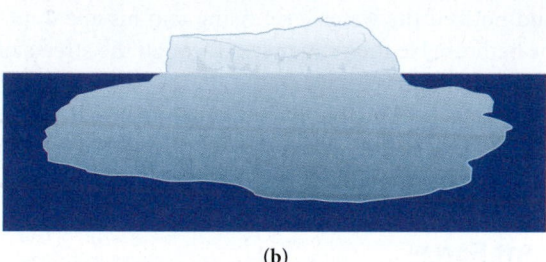

(a)   (b)

PROBLEM
What fraction of the volume of an iceberg floating in seawater is visible above the surface?

SOLUTION
Let V_t be the total volume of the iceberg and V_s be the volume of the iceberg that is submerged. The fraction, f, above water is then

$$f = \frac{V_t - V_s}{V_t} = 1 - \frac{V_s}{V_t}.$$

Because the iceberg is floating, the submerged volume must displace a volume of seawater that has the same weight as the entire iceberg. The mass of the iceberg, m_t, can be calculated from the volume of the iceberg and the density of ice, $\rho_{ice} = 0.917$ g/cm^3. The mass of the displaced seawater can be calculated from the submerged volume and the known density of seawater, $\rho_{seawater} = 1.024$ g/cm^3. We equate the two weights:

$$\rho_{ice} V_t g = \rho_{seawater} V_s g,$$

or

$$\rho_{ice} V_t = \rho_{seawater} V_s.$$

We can rearrange this equation to get

$$\frac{V_s}{V_t} = \frac{\rho_{ice}}{\rho_{seawater}}.$$

We can now find the fraction above water:

$$f = 1 - \frac{V_s}{V_t} = 1 - \frac{\rho_{ice}}{\rho_{seawater}} = 1 - \frac{0.917 \text{ g/cm}^3}{1.024 \text{ g/cm}^3} = 0.104,$$

or about 10%. Figure 13.26b shows an iceberg with approximately 10% of its volume above the waterline.

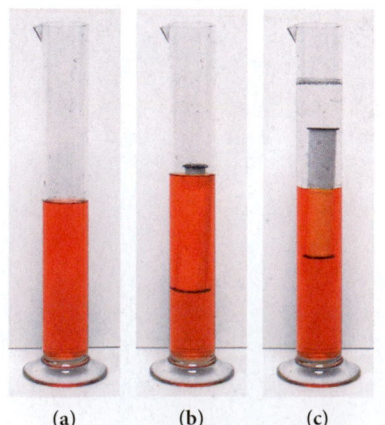

(a)   (b)   (c)

FIGURE 13.27 Buoyancy experiment with two liquids: (a) water with red dye; (b) a swimmer placed in the water; (c) paint thinner added above the water.

An interesting experiment on buoyancy can be performed as follows. You pour water (with red food coloring) into a container (Figure 13.27a) and put a swimmer into it (Figure 13.27b). The swimmer is 90% submerged, so its average density is 90% of that of water, like the iceberg in Example 13.4. Next, you pour paint thinner on top of the water. Paint thinner does not mix with water and has a density equal to 80% of that of water. Thus, it rests on top of the water. If you put the swimmer into a container of paint thinner, the swimmer would sink to the bottom. So, what happens when you pour the paint thinner on top of the water with the swimmer in it? Will the swimmer rise, stay at the same level, or sink?

The answer is shown in Figure 13.27c: The swimmer rises. Why? In the two liquids, two buoyant forces are present—one from the fraction of the swimmer's volume submerged in water, and the other from the fraction of the swimmer's volume submerged in the paint thinner. Therefore, to generate the same buoyant force, less of the volume of the swimmer needs to be submerged in the water when paint thinner is also present than when air and no paint thinner lies on top of the water.

EXAMPLE 13.5 | A Hot-Air Balloon

A typical hot-air balloon has a volume of 2200. m^3. The density of air at a temperature of 20. °C is 1.205 kg/m^3. The density of the hot air inside the balloon at a temperature of 100. °C is 0.946 kg/m^3.

PROBLEM
How much weight can the hot-air balloon shown in Figure 13.28 lift (counting the balloon itself)?

SOLUTION
The weight of the air at 20 °C that the balloon displaces is equal to the buoyant force:
$$F_B = \rho_{20} V g.$$
The weight of the hot air inside the balloon is
$$W_{balloon} = \rho_{100} V g.$$
The weight that can be lifted is
$$W = F_B - W_{balloon} = \rho_{20} V g - \rho_{100} V g$$
$$= V g (\rho_{20} - \rho_{100}) = (2200 \text{ m}^3)(9.81 \text{ m/s}^2)(1.205 \text{ kg/m}^3 - 0.946 \text{ kg/m}^3)$$
$$= 5590 \text{ N}.$$

Note that this weight must include the balloon envelope, the basket, and the fuel, as well as any payload such as the pilot and passengers.

FIGURE 13.28 Hot-air balloon in flight.

Self-Test Opportunity 13.3

Suppose the balloon in Example 13.5 were filled with helium instead of hot air. How much weight could the helium-filled balloon lift? The density of helium is 0.164 kg/m^3.

Determination of Density

Let's return to Archimedes' method of figuring out if the king's new crown was pure gold or a mixture of silver and gold. Since the density of gold is 19.3 g/cm^3 and the density of silver is only 10.5 g/cm^3, Archimedes needed to measure the density of the crown to find out what fraction of it was gold and what fraction was silver. However, how could he measure the density? He needed to submerge the crown in water and determine its volume from the rise in the water level. Weighing the crown gave him the mass, and dividing the mass by the volume yielded the density, the answer the king wanted.

EXAMPLE 13.6 | Finding the Density of an Object

The method shown in Figure 13.29 can be used to determine the density of an object just by weighing it, without the need to measure water levels. This method is usually more precise and thus is preferred. We first weigh a beaker containing water (with green food coloring added) whose density is assumed to be 1000 kg/m^3, as shown in part (a). The mass of the beaker and water is determined to be $m_0 = 0.437$ kg. In part (b), we submerge a metal ball, suspended by a string, in the water, being careful not to let the ball touch the bottom of the beaker. The new total apparent mass of this arrangement is determined in part (c) as $m_1 = 0.458$ kg. Finally, in part (d), we let the ball rest on the bottom of the beaker, without pulling on the string. Now the mass is determined to be $m_2 = 0.596$ kg.

PROBLEM
What is the density of the metal ball?

SOLUTION
Figure 13.30 shows free-body diagrams for the different parts of Figure 13.29. The leftmost diagram shows the ball suspended from the string in static equilibrium, with its weight $-m_b g \hat{y}$,

– Continued

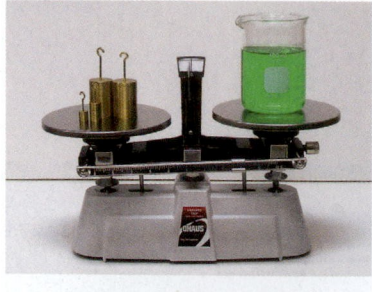

(a)

(b)

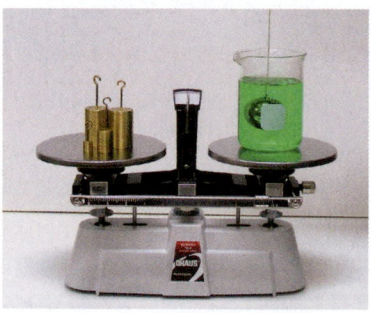

(c)

(d)

FIGURE 13.29 Method for determining the density of a metal object.

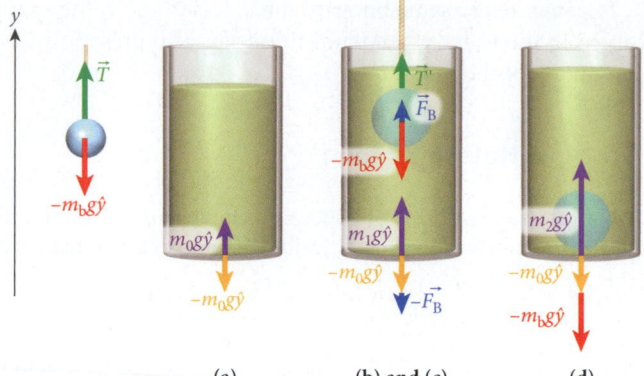

(a)   **(b) and (c)**   **(d)**

FIGURE 13.30 Free-body diagrams for the different parts of the experiment.

balanced by the string tension, $\vec{T}$. The diagram second from the left in Figure 13.30 corresponds to Figure 13.29a and shows the only forces acting on the water-filled container, which are its weight, $-m_0 g\hat{y}$, and the normal force, $m_0 g\hat{y}$. The value of m_0 was determined as the result of the first measurement. The diagram third from the left in Figure 13.30 corresponds to parts (b) and (c) in Figure 13.29 and shows the free-body diagram of the submerged ball as well as the free-body diagram of the container; it illustrates the effect of the buoyant force, $\vec{F}_B$, on the ball and on the liquid-filled container. Note that the ball experiences an upward buoyant force and thus the container must experience a force of equal magnitude and opposite direction (downward), or $-\vec{F}_B$, according to Newton's Third Law. The normal force required to keep the container in equilibrium in this situation is $m_1 g\hat{y}$. The rightmost free-body diagram in Figure 13.30 corresponds to part (d) of Figure 13.29 and shows the normal force as $m_2 g\hat{y}$.

The combined mass of water and beaker, m_0, was determined in part (a) of the method illustrated in Figure 13.30. In part (c), we measured in addition the mass of the water displaced by the (as yet unknown) volume V of the metal ball. The mass of the displaced water is $m_w = \rho_w V_b$, and the total mass measured in part (c) is

$$m_1 = m_0 + \rho_w V_b. \tag{i}$$

Finally, the mass measured in part (d) is the combined mass of the ball, the beaker, and the water. For the mass of the ball, we can again use the product of the density and the volume:

$$m_2 = m_0 + m_b = m_0 + \rho_b V_b. \tag{ii}$$

Combining equations (i) and (ii), we have

$$\frac{m_2 - m_0}{m_1 - m_0} = \frac{\rho_b V_b}{\rho_w V_b} = \frac{\rho_b}{\rho_w}.$$

As you can see, the volume of the ball has canceled out, and we can obtain the density of the ball from the known density of water and our three measurements, m_0, m_1, and m_2:

$$\rho_b = \frac{m_2 - m_0}{m_1 - m_0} \rho_w = \frac{0.596 - 0.437}{0.458 - 0.437}(1000 \text{ kg/m}^3) = 7570 \text{ kg/m}^3.$$

Self-Test Opportunity 13.4

What is the mass of the metal ball in Example 13.6?

Concept Check 13.6

Suppose we use the same setup as in Example 13.6 and a ball of the same size, but half as dense: $\rho_b = 3785 \text{ kg/m}^3$. The measurement of m_1 (with the ball held by a string so that it is under water, but not touching the bottom of the beaker) will yield

a) a lower value.   b) the same value.   c) a higher value.

The mass measurement of m_2 (with the ball resting at the bottom of the beaker) will yield

a) a lower value.   b) the same value.   c) a higher value.

SOLVED PROBLEM 13.3 | Finding the Density of an Unknown Liquid

PROBLEM
A solid aluminum sphere (density = 2700. kg/m³) is suspended in air from a scale (Figure 13.31a). The scale reads 13.06 N. The sphere is then submerged in a liquid with unknown density (Figure 13.31b). The scale then reads 8.706 N. What is the density of the liquid?

SOLUTION
THINK From Newton's Third Law, the buoyant force exerted by the unknown liquid on the aluminum sphere is the difference between the scale reading when the sphere is suspended in air and the scale reading when the sphere is submerged in the liquid.

SKETCH Free-body diagrams of the sphere in each situation are shown in Figure 13.32.

RESEARCH The buoyant force, F_B, exerted by the unknown liquid is

$$F_B = m_l g, \tag{i}$$

where the mass of the displaced liquid, m_l, can be expressed in terms of the volume of the sphere, V, and the density of the liquid, ρ_l:

$$m_l = \rho_l V. \tag{ii}$$

We can obtain the buoyant force exerted by the unknown liquid by subtracting the measured weight of the sphere when it is submerged in the unknown liquid, F_{sub}, from the measured weight when it is suspended in air, F_{air}:

$$F_B = F_{air} - F_{sub}. \tag{iii}$$

The mass of the sphere can be obtained from its weight in air:

$$F_{air} = m_s g. \tag{iv}$$

Using the known density of aluminum, ρ_{Al}, we can express the volume of the aluminum sphere as

$$V = \frac{m_s}{\rho_{Al}}. \tag{v}$$

SIMPLIFY We can combine equations (i), (ii), and (iii) to get

$$F_B = m_l g = \rho_l V g = F_{air} - F_{sub}. \tag{vi}$$

We can use equations (iv) and (v) to get an expression for the volume of the sphere in terms of known quantities:

$$V = \frac{m_s}{\rho_{Al}} = \frac{\left(\dfrac{F_{air}}{g}\right)}{\rho_{Al}} = \frac{F_{air}}{\rho_{Al} g}.$$

Substituting this expression for V into equation (vi) gives us

$$\rho_l V g = \rho_l \left(\frac{F_{air}}{\rho_{Al} g}\right) g = F_{air} \frac{\rho_l}{\rho_{Al}} = F_{air} - F_{sub}.$$

Now, we solve for the density of the unknown liquid:

$$\rho_l = \rho_{Al} \frac{(F_{air} - F_{sub})}{F_{air}}.$$

CALCULATE Putting in the numerical values gives us the density of the unknown liquid:

$$\rho_l = \rho_{Al} \frac{(F_{air} - F_{sub})}{F_{air}} = (2700. \text{ kg/m}^3) \frac{13.06 \text{ N} - 8.706 \text{ N}}{13.06 \text{ N}} = 900.138 \text{ kg/m}^3.$$

ROUND We report our result to four significant figures:

$$\rho_l = 900.1 \text{ kg/m}^3.$$

DOUBLE-CHECK The calculated density of the unknown liquid is 90% of the density of water. Many liquids have such densities, so our answer seems reasonable.

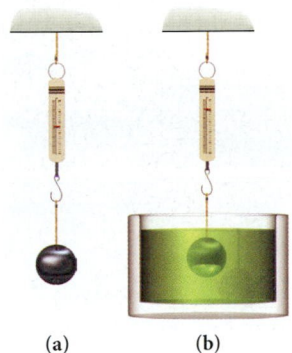

FIGURE 13.31 (a) An aluminum sphere suspended in air. (b) The aluminum sphere submerged in an unknown liquid.

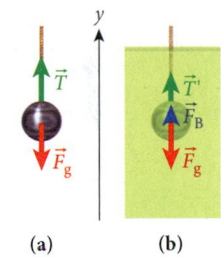

FIGURE 13.32 Free-body diagrams for the sphere (a) suspended in air and (b) submerged in the liquid.

13.6 Ideal Fluid Motion

(a)

(b)

(c)

FIGURE 13.33 (a) Laminar flow of the Firehole River at Yellowstone; (b) transition from laminar to turbulent flow in rising smoke; (c) turbulent flow at the Upper Falls on the Yellowstone River.

FIGURE 13.34 Streamlines of a fluid in laminar flow.

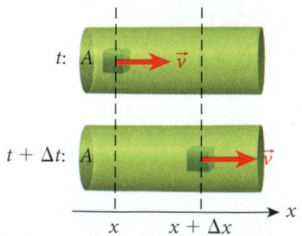

FIGURE 13.35 Blowing air through the gap between two empty soft-drink cans.

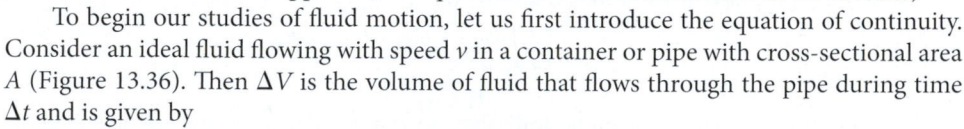

FIGURE 13.36 Ideal fluid flowing through a pipe with a constant cross-sectional area.

We have been examining the behavior of fluids at rest but now turn to fluids in motion. The motion of real-life fluids is complicated and difficult to describe. Numerical techniques and computers are often required to calculate the quantities related to fluid motion. In this section, we consider ideal fluids that can be treated more simply and still yield important and relevant results. To be considered ideal, a fluid must exhibit flow that is laminar, incompressible, nonviscous, and irrotational.

Laminar flow means that the velocity of the moving fluid relative to a fixed point in space does not change with time. A gently flowing river displays laminar flow (Figure 13.33a). Water in a waterfall exhibits nonlaminar flow, or **turbulent flow** (Figure 13.33c). Rising smoke exemplifies a transition from laminar to turbulent flow (Figure 13.33b). The warm smoke initially displays laminar flow as it rises. As the smoke continues to rise, its speed increases until turbulent flow sets in.

Incompressible flow means that the density of the liquid does not change as the liquid flows. A **nonviscous fluid** flows completely freely. Some liquids that do not flow freely are pancake syrup and lava. The viscosity of a fluid has an effect analogous to friction. An object moving in a nonviscous liquid experiences no friction-like force, but the same object moving in a viscous liquid is subject to a drag force that is due to the viscosity and is similar to friction. A flowing viscous fluid loses kinetic energy of motion to thermal energy. We assume that ideal fluids do not lose energy as they flow.

Irrotational flow means that no part of the fluid rotates about its own center of mass. Rotational motion of a small part of the fluid would mean that the rotating part had rotational energy, which is assumed not to occur in ideal fluids.

Laminar flow can be described in terms of streamlines (Figure 13.34). A streamline represents the path that a small element of the fluid takes over time. The velocity, $\vec{v}$, of the small element is always tangent to the streamline. Note that streamlines never cross; if they did, the velocity of flow at the crossing point would have two values.

Bernoulli's Equation

What provides the lift that allows an airplane to fly through the air? To understand this phenomenon, you can do a simple demonstration with two empty soft-drink cans and five drinking straws. Place each empty can on two straws, with a gap of approximately 1 cm between them, as shown in Figure 13.35. In this arrangement, the cans are able to make lateral movements relatively easily. Now, using the fifth straw, blow air through the gap between the cans. What happens? (This question will be answered later in the section.)

To begin our studies of fluid motion, let us first introduce the equation of continuity. Consider an ideal fluid flowing with speed v in a container or pipe with cross-sectional area A (Figure 13.36). Then ΔV is the volume of fluid that flows through the pipe during time Δt and is given by

$$\Delta V = A\Delta x = Av\Delta t.$$

We can write the volume of fluid passing a given point in the pipe per unit time as

$$\frac{\Delta V}{\Delta t} = Av.$$

Now consider an ideal fluid flowing in a pipe that has a cross-sectional area that changes (Figure 13.37). The fluid is initially flowing with speed v_1 through a part of the pipe with cross-sectional area A_1. Downstream, the fluid flows with speed v_2 in a part of the pipe with cross-sectional area A_2. The volume of ideal fluid that enters this section of the pipe per unit time must equal the volume of ideal fluid exiting the pipe per unit time because the fluid is incompressible and the pipe has no leaks. We can express the volume flowing in the first part of the pipe per unit time as

$$\frac{\Delta V}{\Delta t} = A_1 v_1$$

and the volume flowing in the second part of the pipe per unit time as

$$\frac{\Delta V}{\Delta t} = A_2 v_2.$$

The volume of fluid passing any point in the pipe per unit time must be the same in all parts of the pipe, or else fluid would somehow be created or destroyed. Thus, we have

$$A_1 v_1 = A_2 v_2. \tag{13.18}$$

This equation is called the **equation of continuity.**

We can express equation 13.18 as a constant volume flow rate, R_V:

$$R_V = Av.$$

Assuming an ideal fluid whose density does not change, we can also express a constant mass flow rate, R_m:

$$R_m = \rho Av.$$

The SI unit for mass flow rate is kilograms per second (kg/s).

Let's now consider what happens to the pressure in an ideal fluid flowing through a pipe at a steady rate (Figure 13.38). We start by applying the work-energy theorem to the ideal fluid flowing between the lower part and the upper part of the pipe. In the lower left part of the pipe, the flowing fluid of constant density, ρ, is characterized by pressure p_1, speed v_1, and elevation y_1. The same fluid flows through the transition region and into the upper right part of the pipe. Here the fluid is characterized by pressure p_2, speed v_2, and elevation y_2. We will show that the relationship between the pressures and velocities in this situation is given by

$$p_1 + \rho g y_1 + \tfrac{1}{2}\rho v_1^2 = p_2 + \rho g y_2 + \tfrac{1}{2}\rho v_2^2. \tag{13.19}$$

Another way of stating this relationship is

$$p + \rho g y + \tfrac{1}{2}\rho v^2 = \text{constant.} \tag{13.20}$$

This equation is **Bernoulli's Equation.** If there is no flow, $v = 0$ and equation 13.20 is equivalent to equation 13.8.

One major consequence of Bernoulli's Equation becomes evident if $y = 0$, which means that, at constant elevation,

$$p + \tfrac{1}{2}\rho v^2 = \text{constant.} \tag{13.21}$$

From equation 13.21, we can see that if the velocity of a moving fluid is increased, the pressure must decrease. The decrease of the pressure transverse to the fluid flow has many practical applications, including the measurement of the flow rate of a fluid and the creation of a partial vacuum.

Derivation 13.1 develops Bernoulli's Equation mathematically. The only two conditions that enter into this derivation are the law of conservation of energy and the equation of continuity, which simply restates the basic fact that the number of atoms in the ideal fluid is conserved.

Concept Check 13.7

Another demonstration is done by holding two sheets of paper parallel to each other at a separation of approximately an inch and blowing air between them. What do you expect the motion of the sheets of paper will be?

a) They will move away from each other.

b) They will remain at approximately the same separation.

c) They will move toward each other.

FIGURE 13.37 Ideal fluid flowing through a pipe with a changing cross-sectional area.

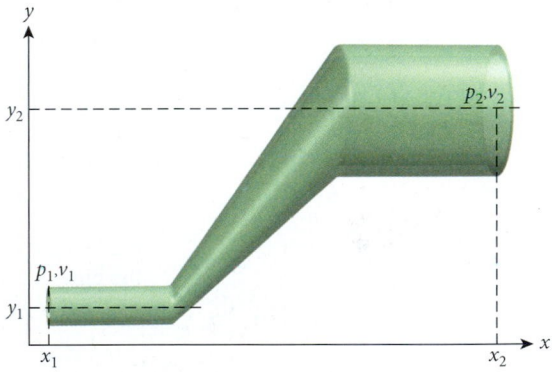

FIGURE 13.38 Ideal fluid flowing through a pipe with changing cross-sectional area and elevation.

DERIVATION 13.1 | Bernoulli's Equation

The work-energy theorem tells us that the net work done on the system, W, is equal to the change in kinetic energy, ΔK, of the fluid flowing between the initial and final cross sections of the pipe in Figure 13.38:

$$W = \Delta K. \tag{i}$$

The change in kinetic energy is given by

$$\Delta K = \tfrac{1}{2}\Delta m v_2^2 - \tfrac{1}{2}\Delta m v_1^2, \tag{ii}$$

where Δm is the amount of mass entering the lower part of the pipe and exiting the upper part of the pipe in time Δt. The flow of mass per unit time is the density of the fluid times the change in volume per unit time:

$$\frac{\Delta m}{\Delta t} = \rho \frac{\Delta V}{\Delta t}.$$

Thus, we can rewrite equation (ii) as

$$\Delta K = \tfrac{1}{2}\rho \Delta V\left(v_2^2 - v_1^2\right).$$

The work done by gravity on the flowing fluid is given by

$$W_g = -\Delta m g\left(y_2 - y_1\right), \tag{iii}$$

where the negative sign arises because negative work is being done by gravity on the fluid when $y_2 > y_1$. We can also rewrite equation (iii) in terms of the volume flow ΔV and the density of the fluid:

$$W_g = -\rho \Delta V g\left(y_2 - y_1\right).$$

The work done by a force, F, acting over a distance, Δx, is given by $W = F\Delta x$, which in this case we can express as

$$W = F\Delta x = \left(pA\right)\Delta x = p\Delta V,$$

because the force arises from the pressure in the fluid. We can then express the work done on the fluid by the pressure forcing the fluid to flow into the pipe as $p_1\Delta V$ and the work done on the exiting fluid as $-p_2\Delta V$, giving the work done as a result of pressure,

$$W_p = \left(p_1 - p_2\right)\Delta V.$$

$p_1\Delta V$ is positive because it arises from fluid to the left of flow exerting a force on fluid entering the pipe. $p_2\Delta V$ is negative because it arises from fluid to the right of flow exerting a force on fluid exiting the pipe. Using equation (i) and $W = W_p + W_g$, we get

$$\left(p_1 - p_2\right)\Delta V - \rho \Delta V g\left(y_2 - y_1\right) = \tfrac{1}{2}\rho \Delta V\left(v_2^2 - v_1^2\right),$$

which we can simplify to

$$p_1 + \rho g y_1 + \tfrac{1}{2}\rho v_1^2 = p_2 + \rho g y_2 + \tfrac{1}{2}\rho v_2^2.$$

This is Bernoulli's Equation.

EXAMPLE 13.7 | Betz Limit

PROBLEM

In Chapter 5, we found that the power contained in wind moving with a speed v through an area A is $P = \tfrac{1}{2}Av^3\rho$, where ρ is the density of air. Now that we know about ideal fluids and energy conservation, we can derive a limit on the fraction of this power that a turbine can actually extract from the wind.

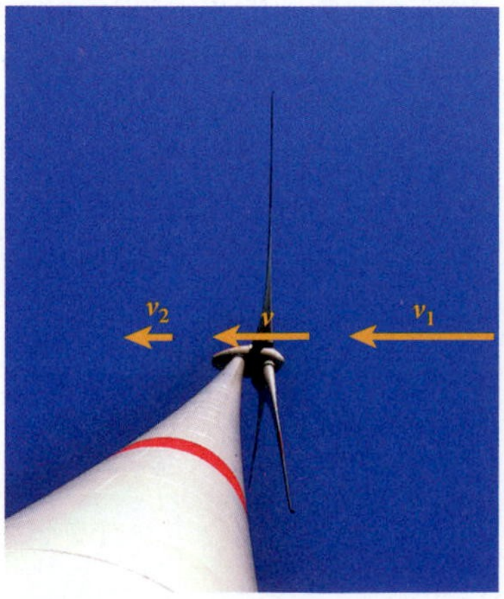

SOLUTION

Obviously, the wind speed has to change during the interaction with the rotor; otherwise, the kinetic energy of the wind would not change, and energy conservation would not allow the turbine to extract energy from the wind. If v_1 is the speed of the wind before it hits the rotor, v_2 is the speed after it passes through the rotor, and v is the speed at the rotor (see Figure 13.39), then v can be assumed to be the average of v_1 and v_2: $v = \tfrac{1}{2}(v_1 + v_2)$.

The average power is the kinetic energy that the wind loses from before hitting the rotor to after doing so:

$$P = \frac{\Delta K}{\Delta t} = \tfrac{1}{2}\frac{\Delta m}{\Delta t}(v_1^2 - v_2^2).$$

FIGURE 13.39 Wind speeds in front of, at, and behind a rotor.

We already calculated the mass flow rate for an incompressible fluid:

$$R_m \equiv \frac{\Delta m}{\Delta t} = \rho A v.$$

Substituting for $\Delta m/\Delta t$ in our expression for the average power and then substituting the expression we obtained above for v yields

$$P = \tfrac{1}{2}\frac{\Delta m}{\Delta t}(v_1^2 - v_2^2) = \tfrac{1}{2}\rho A v(v_1^2 - v_2^2) = \tfrac{1}{2}\rho A \tfrac{1}{2}(v_1 + v_2)(v_1^2 - v_2^2).$$

Rearranging this and using a bit of algebra gives

$$P = \tfrac{1}{2}\rho A v_1^3 \tfrac{1}{2}\Big(1 + (v_2/v_1) - (v_2/v_1)^2 - (v_2/v_1)^3\Big).$$

This is the result that we found in Chapter 5 but with an extra factor, which is the sum of various powers of the ratio $\chi \equiv v_2/v_1$ of the wind speeds behind and in front of the rotor:

$$P = \tfrac{1}{2}\rho A v_1^3 f(\chi), \quad \text{with } f(\chi) = \tfrac{1}{2}\Big(1 + \chi - \chi^2 - \chi^3\Big).$$

To find the maximum extracted power, we have to differentiate the expression for the average power with respect to $\chi \equiv v_2/v_1$:

$$\frac{dP}{d\chi} = \tfrac{1}{2}\rho A v_1^3 \frac{df(\chi)}{d\chi}.$$

Taking the derivative, we obtain

$$\frac{df(\chi)}{d\chi} = \tfrac{1}{2}\Big(1 - 2\chi - 3\chi^2\Big) = 0 \Rightarrow \chi = \tfrac{1}{3} \text{ or } \chi = -1.$$

Since the ratio of the wind speeds has to be between 0 and 1, only $\chi_m = \tfrac{1}{3}$ is a solution. By taking the second derivative, you can convince yourself that this value is indeed the maximum. Inserting it into f, we find $f(\chi_m) = 16/27 \approx 0.593$.

What does this result mean? In order to extract the maximum fraction of the power contained in wind, the rotor of a turbine has to be designed so that the wind speed behind it is $\tfrac{1}{3}$ of the wind speed in front of it, and in this optimum case, approximately 59.3% of the total wind power can be extracted (see Figure 13.40). This limit is known as the *Betz limit*, after Albert Betz, who derived it first, in 1919.

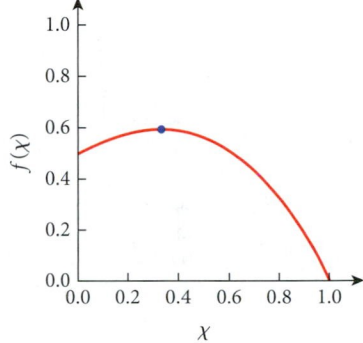

FIGURE 13.40 Dependence of the fraction of power that can be extracted from wind on the ratio of the wind speeds behind and in front of a rotor. The blue dot marks the maximum.

Applications of Bernoulli's Equation

Now that we have derived Bernoulli's Equation, we can return to the demonstration of Figure 13.35. If you performed this demonstration, you found that the two soft-drink cans moved closer to each other. This is the opposite of what most people expect, which is that blowing air in the gap will force the cans apart. Bernoulli's Equation explains this surprising result. Since $p + \tfrac{1}{2}\rho v^2 = \text{constant}$, the large speed with which the air is moving between the cans causes the pressure to decrease. Thus, the air pressure between the cans is smaller than the air pressure on other parts of the cans, causing the cans to be pushed toward each other. This is the *Bernoulli effect*. Another way to show the same effect is to hold two sheets of paper parallel to each other about an inch apart and then blow air between them: The sheets are pushed together.

Truck drivers know about this effect. When two 18-wheelers with the typical rectangular box trailers drive in lanes next to each other at high speed, the drivers have to pay attention not to come too close to each other, because the Bernoulli effect can then push the trailers toward each other.

Race car designers also make use of the Bernoulli effect. The biggest limitation on the acceleration of a race car is the maximum friction force between the tires and the road. This friction force, in turn, is proportional to the normal force. The normal force can be increased by increasing the car's mass, but this defeats the purpose, because a larger mass means a smaller acceleration, according to Newton's Second Law. A much more efficient way to increase the normal force is to develop a large pressure difference between the upper and lower surfaces of the car. According to Bernoulli's Equation, this can be accomplished by causing the air to move faster across the bottom of the car than across the top. (Another way to accomplish this would be to use a wing to deflect the air up and thus create a downward force. However, wings have been ruled out in Formula 1 racing.)

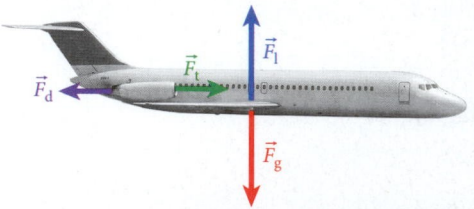

FIGURE 13.41 Forces acting on an airplane in flight.

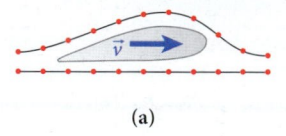

(a)

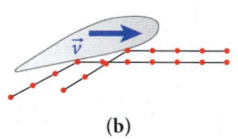

(b)

FIGURE 13.42 Two extreme views of the process that creates lift on the wing of an airplane: (a) the Bernoulli effect; (b) Newton's Third Law.

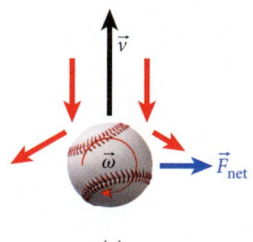

(a)

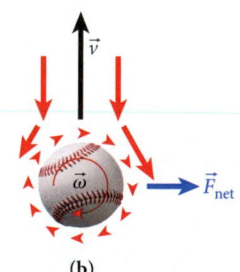

(b)

FIGURE 13.43 Top view of clockwise spinning baseball moving from the bottom of the page toward the top of the page: (a) interpretation using Newton's Third Law; (b) explanation involving a boundary layer.

Let's reconsider the question of what causes an airplane to fly. Figure 13.41 shows the forces acting on an airplane flying with constant velocity and at constant altitude. The thrust, $\vec{F}_t$, is generated by the jet engines taking in air from the front and expelling it out the back. This force is directed toward the front of the plane. The drag force, $\vec{F}_d$, due to air resistance (discussed in Chapter 4) is directed toward the rear of the plane. At constant velocity, these two forces cancel: $\vec{F}_t = \vec{F}_d$. The forces acting in the vertical direction are the weight, $\vec{F}_g$, and the lift, $\vec{F}_l$, which is provided almost exclusively by the wings. If the plane flies at constant altitude, these two forces cancel as well: $\vec{F}_l = \vec{F}_g$. Since a fully loaded and fueled Boeing 747 typically has a mass of 350 metric tons, its weight is 3.4 MN. Thus, the lift provided by the wings to keep the airplane flying must be very large. By comparison, the maximum thrust produced by all four engines on a 747 is 0.9 MN.

The most commonly held notion of what generates the lift that allows a plane to stay airborne is shown in Figure 13.42a: The wing moves through the air and forces the air streamlines that move above it onto a longer path. Thus, the air above the wing needs to move at a higher speed in order to reconnect with the air that moves below the wing. Bernoulli's Equation then implies that the pressure on the top side of the wing is lower than that on the bottom side. Since the combined surface area of the two wings of a Boeing 747 is 511. m², a pressure difference of $\Delta p = F/A = (3.4 \text{ MN})/(511. \text{ m}^2) = 6.65 \text{ kPa}$ (6.6% of the atmospheric pressure at sea level) between the bottom and top of the wings would be required for this notion to work out quantitatively.

An alternative physical interpretation is shown in Figure 13.42b. In this view, air molecules are deflected downward by the lower side of the wing, and according to Newton's Third Law, the wing then experiences an upward force that provides the lift. In order for this idea to work, the bottom of the wing has to have some nonzero angle (the angle of attack) relative to the horizontal. (To attain this position, the nose of the aircraft points slightly upward, so the engine thrust also contributes to the lift. Alternatively, the angle of attack can also be realized by changing the angle of the flaps on the wings.)

Both of these simple ideas contain elements of the truth, and the effects contribute to the lift to different degrees, depending on the type of aircraft and the flight phase. The Newtonian effect is more important during takeoff and landing, and during all flight phases for fighter planes (which have a smaller wing area). However, in general, air does not act as an ideal incompressible fluid as it streams around an airplane's wing. It is compressed at the front edge of the wing and decompressed at its back edge. This compression-decompression effect is stronger on the top surface than on the bottom surface, creating a higher net pressure on the bottom of the wing and providing the lift. The design of aircraft wings is still under intense study, and ongoing research seeks to develop new wing designs that can yield higher fuel efficiency and stability.

We can apply similar considerations to the flight of a curveball in baseball. Curveballs have a sideward spin, and if a left-handed pitcher gives a strong clockwise rotation (as seen from above), the ball will deviate to the right relative to a straight line (as seen by the pitcher). In Figure 13.43, the relative velocity between the air and a point on the left surface of the ball is larger than the relative velocity between the air and a point on the right surface. Application of Bernoulli's Equation might cause us to expect the pressure to be lower on the left side of the ball than on the right, thus causing the ball to move left.

The interpretation using Newton's Third Law (Figure 13.43a) comes closer to explaining correctly why a clockwise spinning baseball experiences a clockwise deflection. The figure shows a top view of a clockwise spinning baseball, moving from the bottom of the page toward the top, with the air molecules that the ball encounters represented by red arrows. As the air molecules collide with the surface of the ball, those on the side where the surface is rotating toward them (left side in Figure 13.43a) receive a stronger sideways hit, or impulse, than those on the other side. By Newton's Third Law, the net recoil force on the baseball thus deflects the baseball in the same direction as its rotation. While the Newtonian explanation gets the direction of the effect right, it is not quite the correct explanation, because the oncoming air molecules do not reach the surface of the baseball undisturbed, as is required for this model to work.

Just like the lift of an airplane, the phenomenon of a curveball has a more complex explanation. A rotating sphere moving through air drags a boundary layer of air along with its surface. The air molecules that encounter this boundary layer get dragged along to some extent. This causes the air molecules on the right side of the ball in Figure 13.43b to be accelerated and those on the left side to be slowed down. The differentially higher speed of air on the right side implies a lower pressure due to the Bernoulli effect and thus causes a deflection to the right. This is known as the *Magnus effect*.

In tennis, topspin causes the ball to dip faster than a ball hit without spin; backspin causes the ball to sail longer. Both kinds of effect occur for the same reason as we just noted for the curveball in baseball. Golfers also use backspin to make their drives carry longer. A well-driven golf ball will have backspin of approximately 4000 rpm. Sidespin in golf causes draws or hooks and fades or slices, depending on the severity and direction of the spin. Incidentally, the dimples on a golf ball are essential for its flight characteristics; they cause turbulence around the golf ball and thus reduce the air resistance. Even the best professionals would not be able to hit a golf ball without dimples more than 200 m.

EXAMPLE 13.8 Spray Bottle

PROBLEM
If you squeeze the handle of a spray bottle (Figure 13.44), you cause air to flow horizontally across the opening of a tube that extends down into the liquid almost to the bottom of the bottle. If the air is moving at 50.0 m/s, what is the pressure difference between the top of the tube and the atmosphere? Assume that the density of air is $\rho = 1.20$ kg/m^3.

SOLUTION
Before you squeeze the handle, the airflow speed is $v_0 = 0$. Using Bernoulli's Equation for a negligible height difference (equation 13.21), we find

$$p + \tfrac{1}{2}\rho v^2 = p_0 + \tfrac{1}{2}\rho v_0^2.$$

Solving this equation for the pressure difference, $p - p_0$, under the condition that $v_0 = 0$, we arrive at

$$p - p_0 = -\tfrac{1}{2}\rho v^2 = -\frac{\left(1.20 \text{ kg/m}^3\right)\left(50.0 \text{ m/s}\right)^2}{2} = -1500 \text{ Pa}.$$

Therefore, we see that the pressure is lowered by 1.50 kPa, causing the liquid to be pushed upward and broken up into small droplets in the air stream, to form a mist.

The same principle is used in old-fashioned carburetors, which mix the fuel with air in older cars. (In newer cars, the carburetor has been replaced by fuel injectors.)

FIGURE 13.44 Spray bottle for dispensing liquids in the form of a fine mist.

SOLVED PROBLEM 13.4 Venturi Tube

PROBLEM
On some light aircraft, a device called a *Venturi tube* is used to create a pressure difference that can be used to drive gyroscope-based instruments for navigation. The Venturi tube is mounted on the outside of the fuselage in an area of free airflow. Suppose a Venturi tube has a circular opening with a diameter of 10.0 cm, narrowing down to a circular opening with a diameter of 2.50 cm, and then opening back up to the original diameter of 10.0 cm. What is the pressure difference between the 10-cm opening and the narrowest region of the Venturi tube, assuming that the aircraft is flying at a constant speed of 38.0 m/s at a low altitude where the density of air can be assumed to be that at sea level ($\rho = 1.30$ kg/m^3) at 5 °C?

SOLUTION
THINK The equation of continuity (equation 13.18) tells us that the product of the area and the velocity of the flow through the Venturi tube is constant. We can then relate the area of the opening, the area of the narrowest region, the velocity of the air entering the Venturi tube, and the velocity of the air in the narrowest region of the tube. Using Bernoulli's Equation, we can then relate the pressure at the opening to the pressure in the narrowest region.

SKETCH A Venturi tube is sketched in Figure 13.45.

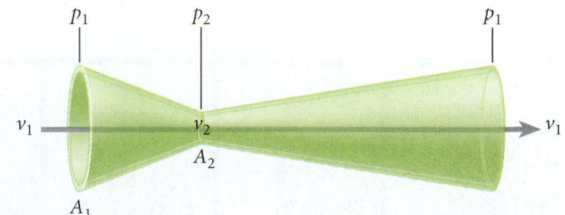

FIGURE 13.45 A Venturi tube with air flowing through it.

– Continued

RESEARCH   The equation of continuity is

$$A_1 v_1 = A_2 v_2.$$

Bernoulli's Equation tells us that

$$p_1 + \tfrac{1}{2}\rho v_1^2 = p_2 + \tfrac{1}{2}\rho v_2^2.$$

SIMPLIFY   The pressure difference between the opening of the Venturi tube and the narrowest area is found by rearranging the Bernoulli Equation:

$$p_1 - p_2 = \Delta p = \tfrac{1}{2}\rho v_2^2 - \tfrac{1}{2}\rho v_1^2 = \tfrac{1}{2}\rho\left(v_2^2 - v_1^2\right).$$

Solving the equation of continuity for v_2 and substituting that result into the rearranged Bernoulli Equation gives us the pressure difference, Δp:

$$\Delta p = \tfrac{1}{2}\rho\left(v_2^2 - v_1^2\right) = \tfrac{1}{2}\rho\left[\left(\frac{A_1}{A_2}v_1\right)^2 - v_1^2\right] = \tfrac{1}{2}\rho v_1^2\left(\frac{A_1^2}{A_2^2} - 1\right).$$

Since both areas are circular, we have $A_1 = \pi r_1^2$ and $A_2 = \pi r_2^2$, and thus $(A_1/A_2)^2 = (\pi r_1^2/\pi r_2^2)^2 = (r_1/r_2)^4$, which leads to

$$\Delta p = \tfrac{1}{2}\rho v_1^2\left[\left(\frac{r_1}{r_2}\right)^4 - 1\right].$$

CALCULATE   Putting in the numerical values, including $\rho = 1.30$ kg/m³ for the density of air, gives us

$$\Delta p = \tfrac{1}{2}\left(1.30 \text{ kg/m}^3\right)\left(38.0 \text{ m/s}\right)^2\left[\left(\frac{10.0/2 \text{ cm}}{2.5/2 \text{ cm}}\right)^4 - 1\right] = 239{,}343.$$

ROUND   We report our result to three significant figures:
$$\Delta p = 239. \text{ kPa.}$$

DOUBLE-CHECK   The pressure difference between the interior of the airplane and the narrowest part of the Venturi tube is 239 kPa, or more than twice normal atmospheric pressure. Connecting a hose between the narrowest part of the tube and the interior of the plane thus allows for a steady stream of air through the tube, which can be used to drive a rapidly rotating gyroscope.

Draining a Tank

Let's analyze another simple experiment. A large container with a small hole at the bottom is filled with water and measurements are made of how long it takes to drain. Figure 13.46 illustrates this experiment. We can analyze this process quantitatively and arrive at a description of the height of the fluid column as a function of time. Since the bottle is open at the top, the atmospheric pressure is the same at the upper surface of the fluid column as it is at the small hole. We thus obtain from Bernoulli's Equation (equation 13.19)

$$\rho g y_1 + \tfrac{1}{2}\rho v_1^2 = \rho g y_2 + \tfrac{1}{2}\rho v_2^2.$$

Canceling out the density and reordering the terms, we obtain

$$v_2^2 - v_1^2 = 2g(y_1 - y_2) = 2gh,$$

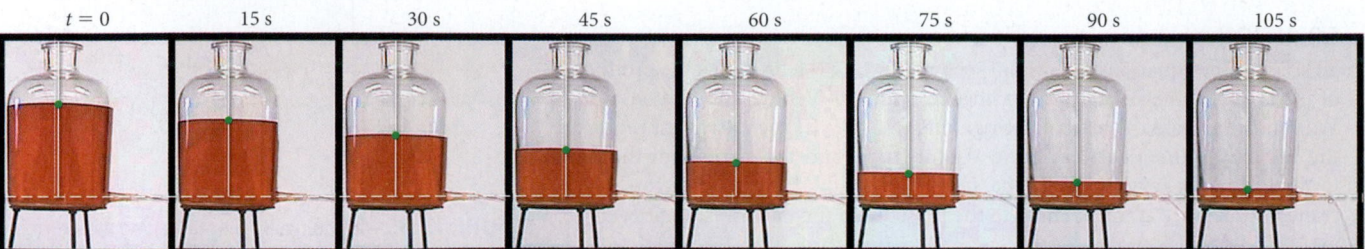

FIGURE 13.46   Draining a bottle of water through a small hole at the base.

where $h = y_1 - y_2$ is the height of the fluid column. We obtained this same result in Chapter 2 for a particle in free fall! Thus, the streaming of an ideal fluid under the influence of gravity proceeds in the same way as free fall of a point particle.

The equation of continuity (equation 13.18) relates the two speeds, v_1 and v_2, to each other via the ratio of their corresponding cross-sectional areas: $A_1 v_1 = A_2 v_2$. Thus, we find the speed with which the fluid flows from the container as a function of the height of the fluid column above the hole:

$$v_1 = v_2 \frac{A_2}{A_1} \Rightarrow v_2^2 - v_1^2 = v_2^2 \left(1 - \frac{A_2^2}{A_1^2}\right) \Rightarrow$$

$$v_2^2 = \frac{2gh}{1 - \frac{A_2^2}{A_1^2}} = \frac{2A_1^2 g}{A_1^2 - A_2^2} h.$$

If A_2 is small compared to A_1, this result simplifies to

$$v_2 = \sqrt{2gh}. \tag{13.22}$$

The speed with which the fluid streams from the container is sometimes called the **speed of efflux,** and equation 13.22 is often called *Torricelli's Theorem.*

How does the height of the fluid column change as a function of time? To answer this, we note that the speed v_1 is the negative of the time derivative of the height of the fluid column, h: $v_1 = -dh/dt$, since the height decreases with time. From the equation of continuity, we then find

$$A_1 v_1 = -A_1 \frac{dh}{dt} = A_2 v_2 = A_2 \sqrt{2gh} \Rightarrow$$

$$\frac{dh}{dt} = -\frac{A_2}{A_1} \sqrt{2gh}. \tag{13.23}$$

Because equation 13.23 relates the height to its derivative, it is a differential equation. Its solution is

$$h(t) = h_0 - \frac{A_2}{A_1} \sqrt{2gh_0}\, t + \frac{g}{2} \frac{A_2^2}{A_1^2} t^2, \tag{13.24}$$

where h_0 is the initial height of the fluid column. You can convince yourself that equation 13.24 is really the solution by taking its derivative and inserting that into equation 13.23. The derivative is

$$\frac{dh}{dt} = -\frac{A_2}{A_1} \sqrt{2gh_0} + g \frac{A_2^2}{A_1^2} t.$$

The point in time at which this derivative reaches zero is the time at which the draining process has finished:

$$\frac{dh}{dt} = 0 \Rightarrow t_f = \frac{A_1}{A_2} \sqrt{\frac{2h_0}{g}}. \tag{13.25}$$

If you substitute this expression for t into equation 13.24, you see that t_f is also the time at which the height of the fluid column reaches zero, which is as expected. Because, according to equation 13.23, the time derivative of the height is proportional to the square root of the height, the height has to be zero when the derivative is zero.

Equation 13.25 also tells us that the time it takes to drain a cylindrical container is proportional to the cross-sectional area of the container and inversely proportional to the area of the hole the liquid is draining through. However, the time is also proportional to the square root of the initial height of the fluid. Thus, if two containers hold the same volume of liquid and drain through holes of the same size, the container that is taller and with a smaller cross-sectional area will drain faster.

Finally, keep in mind that equation 13.25 holds only for an ideal fluid in a container with constant cross-sectional area as a function of height and with a hole whose cross-sectional area is small compared to that of the container.

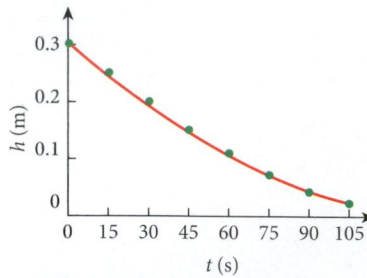

FIGURE 13.47 Height of fluid column as a function of time: Green dots are observed data points; the red line is the calculated solution.

Concept Check 13.8

If the bottle and initial fluid level remain the same as in Example 13.9 but the diameter of the hole is halved, the time required to drain all the fluid will be approximately

a) 36 s.   d) 288 s.

b) 72 s.   e) 576 s.

c) 144 s.

(a)

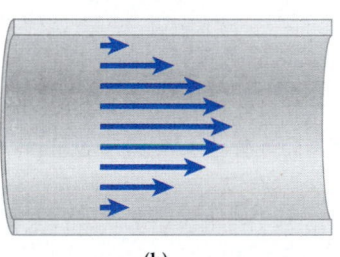

(b)

FIGURE 13.48 Velocity profiles in a cylindrical tube: (a) nonviscous ideal fluid flow; (b) viscous flow.

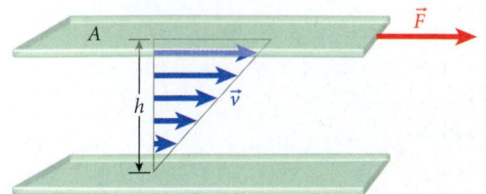

FIGURE 13.49 Measuring the viscosity of a liquid with two parallel plates.

EXAMPLE 13.9 | Draining a Bottle

Figure 13.46 shows a large cylindrical bottle of cross-sectional area $A_1 = 0.100$ m². The liquid (water with red food coloring) drained through a small hole of radius 7.40 mm, or of area $A_2 = 1.72 \cdot 10^{-4}$ m². The frames in the figure represent times at 15-s intervals. The initial height of the fluid column above the hole was $h_0 = 0.300$ m.

PROBLEM

How long did it take to drain this bottle?

SOLUTION

We can simply use equation 13.25 and insert the given values:

$$t_f = \frac{A_1}{A_2}\sqrt{\frac{2h_0}{g}} = \frac{0.100 \text{ m}^2}{1.72 \cdot 10^{-4} \text{ m}^2}\sqrt{\frac{2(0.300 \text{ m})}{9.81 \text{ m/s}^2}} = 144 \text{ s}.$$

While the solution to this problem is straightforward, it is instructive to plot the height as a function of time, using equation 13.24, and compare the plot to the experimental data. In Figure 13.46, the level of the hole is marked with a dashed horizontal line, and the height of the fluid column in each frame is marked with a green dot. Figure 13.47 shows the comparison between the calculated solution (the red line) and the data obtained from the experiment. You can see that the agreement is well within the measurement uncertainty, represented by the size of the green dots.

13.7 Viscosity

If you have ever drifted in a boat on a gentle river, you may have noticed that your boat moves faster in the middle of the river than very close to the banks. Why would this happen? If the water in the river were an ideal fluid in laminar motion, it should make no difference how far away from shore you are. However, water is not quite an ideal fluid. Instead, it has some degree of "stickiness," called **viscosity.** For water, the viscosity is quite low; for heavy motor oil, it is significantly higher, and it is even higher yet for substances like honey, which flow very slowly. Viscosity causes the fluid streamlines at the surface of a river to partially stick to the boundary and neighboring streamlines to partially stick to one another.

The velocity profile for the streamlines in viscous flow in a tube is sketched in Figure 13.48b. The profile is parabolic, with the velocity approaching zero at the walls and reaching its maximum value in the center. This flow is still laminar, with the streamlines all flowing parallel to one another.

How is the viscosity of a fluid measured? The standard procedure is to use two parallel plates of area A and fill the gap of width h between them with the fluid. Then one of the plates is dragged across the other and the force F that is required to do so is measured. The resulting velocity profile of the fluid flow is linear (Figure 13.49). The viscosity, η, is defined as the ratio of the force per unit area divided by the velocity difference between the top and bottom plates over the distance between the plates:

$$\eta = \frac{F/A}{\Delta v/h} = \frac{Fh}{A\Delta v}. \tag{13.26}$$

The unit of viscosity represents pressure (force per unit area) multiplied by time, or pascal seconds (Pa s). This unit is also called a *poiseuille* (Pl). [Care must be taken to avoid confusing this SI unit with the cgi unit poise (P), because 1 P = 0.1 Pa s.]

It is important to realize that the viscosity of any fluid depends strongly on temperature. You can see an example of this temperature dependence in the kitchen. If you store olive oil in the refrigerator and then pour it from the bottle, you can see how slowly it flows. Heat the same olive oil in a pan, and it flows almost as readily as water. Temperature dependence is of great concern for motor oils, and the goal is to have a small temperature dependence. Table 13.5 lists some typical viscosity values for different fluids. All values are those at room temperature (20 °C = 68 °F) except that of blood, whose value is given for the physiologically relevant human body temperature (37 °C = 98.6 °F). Incidentally, the viscosity of blood increases by about 20% during a human's lifetime, and the average value for men is slightly higher than that for women ($4.7 \cdot 10^{-3}$ Pa s vs. $4.3 \cdot 10^{-3}$ Pa s).

The viscosity of a fluid is important in determining how much fluid can flow through a pipe of given radius r and length ℓ. Gotthilf Heinrich Ludwig Hagen (in 1839) and Jean Louis Marie Poiseuille (in 1840) found independently that R_v, the volume of fluid that can flow per unit time, is

$$R_v = \frac{\pi r^4 \Delta p}{8 \eta \ell}. \qquad (13.27)$$

Here Δp is the pressure difference between the two ends of the pipe. As expected, the flow is inversely proportional to the viscosity and the length of the pipe. Most significantly, though, it is proportional to the fourth power of the radius of the pipe. If we consider a blood vessel as a pipe, this relationship helps us understand the problem associated with clogging of the arteries. If cholesterol-induced deposits reduce the diameter of a blood vessel by 50%, then the blood flow through the vessel is reduced to $1/2^4 = 1/16$, or 6.25% of the original rate—a reduction of 93.75%.

Table 13.5	Some Typical Values of Viscosity at Room Temperature	
Material		**Viscosity (Pa s)**
Air		$1.8 \cdot 10^{-5}$
Alcohol (ethanol)		$1.1 \cdot 10^{-3}$
Blood (at body temperature)		$4 \cdot 10^{-3}$
Honey		10
Mercury		$1.5 \cdot 10^{-3}$
Motor oil (SAE 10 to SAE 40)		0.06 to 0.7
Olive oil		0.08
Water		$1.0 \cdot 10^{-3}$

EXAMPLE 13.10 | Hypodermic Needle

For many people, the scariest part of a visit to the doctor is an injection. Learning about the fluid mechanics of the hypodermic needle (Figure 13.50) won't change that, but is interesting nonetheless.

PROBLEM 1
If 2.0 cm³ of water is to be pushed out of a 1.0-cm-diameter syringe through a 3.5-cm-long 15-gauge needle (interior needle diameter = 1.37 mm) in 0.4 s, what force must be applied to the plunger of the syringe?

SOLUTION 1
The Hagen-Poiseuille Law (equation 13.27) relates fluid flow to the pressure difference causing the flow. We can solve that equation for the pressure difference, Δp, between the tip of the needle and the end that is connected to the syringe:

$$\Delta p = \frac{8 \eta \ell R_v}{\pi r^4}.$$

The viscosity of water can be obtained from Table 13.5: $\eta = 1.0 \cdot 10^{-3}$ Pa s. The flow rate, R_v, is just the ratio of volume to time:

$$R_v = \frac{\Delta V}{\Delta t} = \frac{2.0 \cdot 10^{-6} \text{ m}^3}{0.4 \text{ s}} = 5.0 \cdot 10^{-6} \text{ m}^3/\text{s}.$$

The geometric dimensions of the syringe are specified in the problem statement, so we obtain

$$\Delta p = \frac{8(1.0 \cdot 10^{-3} \text{ Pa s})(0.035 \text{ m})(5.0 \cdot 10^{-6} \text{ m}^3/\text{s})}{\pi \left[0.5(1.37 \cdot 10^{-3} \text{ m})\right]^4} = 2024 \text{ Pa}.$$

Because pressure is force per unit area, we can obtain the required force by multiplying the pressure difference we just calculated by the appropriate area. What is this area? The required force is provided by pushing on the plunger, so we use the area of the plunger:

$$A = \pi R^2 = \pi \left[0.5(0.01 \text{ m})\right]^2 = 7.8 \cdot 10^{-5} \text{ m}^2.$$

Thus, the force required to push out 2.0 cm³ of water in 0.4 s is only

$$F = A \Delta p = (7.8 \cdot 10^{-5} \text{ m}^2)(2024 \text{ Pa}) = 0.16 \text{ N}.$$

PROBLEM 2
What is the speed with which the water emerges from the needle of the syringe?

FIGURE 13.50 A hypodermic needle illustrates fluid flow with viscosity.

– Continued

SOLUTION 2

We saw in Section 13.6 that the speed of fluid flow is related to the volume rate of flow by $R_v = Av$, where A is the cross-sectional area—in this case, the area of the needle tip whose diameter is 1.37 mm. Solving this equation for the speed, we find

$$v = \frac{R_v}{A} = \frac{5.0 \cdot 10^{-6} \text{ m}^3/\text{s}}{\pi \left[0.5 \left(1.37 \cdot 10^{-3} \text{ m} \right) \right]^2} = 3.4 \text{ m/s}.$$

DISCUSSION

If you have ever pushed the plunger of a syringe, you know it takes more force than 0.16 N. (A force of 0.16 N is equivalent to the weight of a cheap ballpoint pen.) What gives rise to the higher force requirement? Remember that the plunger has to provide a tight seal with the wall of the syringe. Thus, the main effort in pushing the plunger results from overcoming the friction force between the syringe wall and the plunger. It would be an altogether different story, though, if the syringe were filled with honey. Note also that the 15-gauge needle used in this example is larger than the needle typically used to make an injection, which is a 24- or 25-gauge needle.

13.8 Turbulence and Research Frontiers in Fluid Flow

In laminar flow, the streamlines of a fluid follow smooth paths. In contrast, for a fluid in turbulent flow, vortices form, detach, and propagate (Figure 13.51). We have already seen that ideal laminar flow or viscous laminar flow transitions into turbulent flow when the velocity of flow exceeds a certain value. This transition is clearly illustrated in Figure 13.33b, which shows how rising cigarette smoke undergoes a transition from laminar to turbulent flow. What is the criterion that determines whether flow is laminar or turbulent?

The answer lies in the **Reynolds number,** Re, which is the ratio of the typical inertial force to the viscous force and thus is a pure dimensionless number. The inertial force has to be proportional to the density, ρ, and the typical velocity of the fluid, $\overline{v}$, because $F = dp/dt$, according to Newton's Second Law. The viscous force is proportional to the viscosity, η, and inversely proportional to the characteristic length scale, L, over which the flow varies. For flow through a pipe with a circular cross section, this length scale is the diameter of the pipe, $L = 2r$. Thus, the formula for calculating the Reynolds number is

$$\text{Re} = \frac{\rho \overline{v} L}{\eta}. \tag{13.28}$$

As a rule of thumb, a Reynolds number less than 2000 means laminar flow, and one higher than 4000 means turbulent flow. For Reynolds numbers between 2000 and 4000, the character of the flow depends on many fine details of the exact configuration. Engineers try hard to avoid this interval because of this essential unpredictability.

The true power of the Reynolds number lies in the fact that fluid flows in systems with the same geometry and the same Reynolds number behave similarly. This allows engineers to reduce typical length scales or velocity scales, build scale models of boats or airplanes, and test their performance in water tanks or wind tunnels of relatively modest scale (Figure 13.52).

FIGURE 13.51 Extreme turbulent flow—a tornado is a vortex.

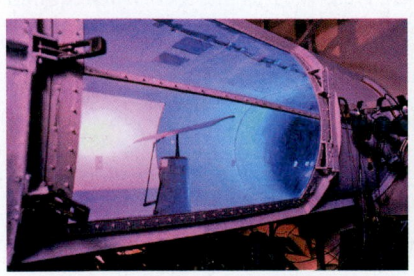

FIGURE 13.52 (a) Wind-tunnel testing of a scale model of a wing. (b) Scale model of fighter jet with interchangeable parts for wind-tunnel testing.

(a)
(b)

Rather than scale models, modern research on fluid flow and turbulence relies on computer models. Hydrodynamic modeling is employed for studying applications of an incredible variety of physical systems, such as the performance and aerodynamics of cars, airplanes, rockets, and boats. The wake effects of wind turbines, which can be seen in Figure 13.1, are a particularly important area of current study. However, hydrodynamic modeling is also utilized in studying the collisions of atomic nuclei at the highest energies attainable in modern accelerators and in the modeling of supernova explosions (Figure 13.53). In 2005, an experimental group working at the Relativistic Heavy Ion Collider in Brookhaven, New York, discovered that gold nuclei show the characteristics of a perfect nonviscous fluid when they smash into each other at the highest attainable energies. [One of the authors (Westfall) had the privilege to announce this discovery at the 2006 annual meeting of the American Physical Society.] Exciting research results on fluid motion will continue to emerge over the next decades, as this is one of the most interesting interdisciplinary areas in the physical sciences.

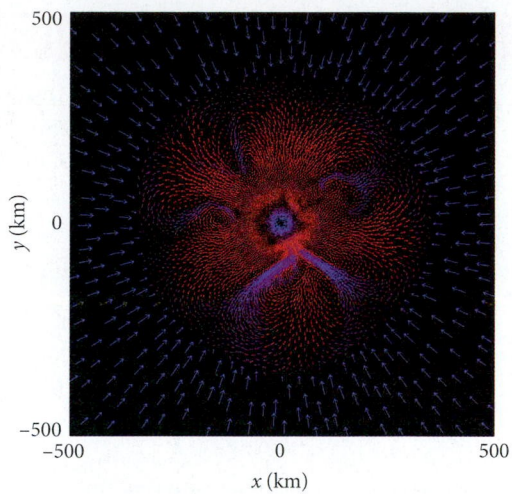

FIGURE 13.53 Hydrodynamic modeling of the collapse of a supernova core. The arrows indicate the flow directions of the fluid elements, and the color indicates the temperature.

WHAT WE HAVE LEARNED | EXAM STUDY GUIDE

- One mole of a material has $N_A = 6.022 \cdot 10^{23}$ atoms or molecules. The mass of 1 mol of a material in grams is given by the sum of the atomic mass numbers of the atoms that make up the material.

- For a solid, stress = modulus·strain, where stress is force per unit area and strain is a unit deformation. There are three types of stress and strain, each with its own modulus:

  - Tension or linear compression leads to a positive or negative change in length: $\frac{F}{A} = Y\frac{\Delta L}{L}$, where Y is Young's modulus.

  - Volume compression leads to a change in volume: $\frac{F}{A} = B\frac{\Delta V}{V}$, where B is the bulk modulus.

  - Shear leads to bending: $\frac{F}{A} = G\frac{\Delta x}{L}$, where G is the shear modulus.

- Pressure is defined as force per unit area: $p = \frac{F}{A}$.

- The absolute pressure, p, at a depth h in a liquid with density ρ and pressure p_0 on its surface is $p = p_0 + \rho g h$.

- Gauge pressure is the difference in pressure between the gas in a container and the pressure of Earth's atmosphere.

- Pascal's Principle states that when a change in pressure occurs at any point in a confined incompressible fluid, an equal change in pressure occurs at every point in the fluid.

- The buoyant force on an object immersed in a fluid is equal to the weight of the fluid displaced: $F_B = m_f g$.

- An ideal fluid is assumed to exhibit flow that is laminar, incompressible, nonviscous, and irrotational.

- The flow of an ideal fluid follows streamlines.

- The equation of continuity for a flowing ideal fluid relates the velocity and the area of the fluid flowing through a container or pipe: $A_1 v_1 = A_2 v_2$.

- Bernoulli's Equation relates the pressure, height, and velocity of an ideal fluid flowing through a container or pipe: $p + \rho g y + \frac{1}{2}\rho v^2 = $ constant.

- For viscous fluids, the viscosity, η, is the ratio of the force per unit area to the velocity difference per unit length: $\eta = \frac{F/A}{\Delta v/h} = \frac{Fh}{A\Delta v}$.

- For viscous fluids, the volume flow rate through a cylindrical pipe of radius r and length ℓ is given by $R_v = \frac{\pi r^4 \Delta p}{8\eta\ell}$, where Δp is the pressure difference between the two ends of the pipe.

- The Reynolds number determines the ratio of inertial force to viscous force and is defined as $\text{Re} = \frac{\rho \bar{v} L}{\eta}$, where $\bar{v}$ is the average fluid velocity and L is the characteristic length scale over which the flow changes. A Reynolds number less than 2000 means laminar flow, and one greater than 4000 means turbulent flow.

ANSWERS TO SELF-TEST OPPORTUNITIES

13.1 The volume of the water in the bottle is 500 cm³. The density of water is 1 g/cm³. Thus, the bottle contains 500 g of water. The mass of 1 mol of water is 18.02 g, so $n = (500 \text{ g})/(18.02 \text{ g/mol}) = 27.7$ mol. Thus, $N = n \cdot N_A = 27.7(6.022 \cdot 10^{23} \text{ mol}^{-1}) = 1.67 \cdot 10^{25}$ molecules.

13.2 $\rho_H = \dfrac{2 \text{ g}}{22.4 \text{ L}} = 0.089 \text{ g/L} = 0.089 \text{ kg/m}^3$

$\rho_{He} = \dfrac{4 \text{ g}}{22.4 \text{ L}} = 0.18 \text{ g/L} = 0.18 \text{ kg/m}^3$

13.3
$$W = F_B - W_{balloon} = \rho_{20}Vg - \rho_{He}Vg$$
$$= Vg(\rho_{20} - \rho_{100}) = (2200 \text{ m}^3)(9.81 \text{ m/s}^2)(1.205 \text{ kg/m}^3 - 0.164 \text{ kg/m}^3)$$
$$= 22{,}500 \text{ N, or } 5050 \text{ lb}_f.$$

13.4 The mass of the ball is the difference between the measurements in part (d) and part (a) of Figure 13.29: $m_b = m_2 - m_0 = 0.159$ kg.

PROBLEM-SOLVING GUIDELINES: SOLIDS AND FLUIDS

1. The three types of stress all have the same relationship with strain: The ratio of stress to strain equals a constant for the material, which can be Young's modulus, the bulk modulus, or the shear modulus. Be sure you understand what kind of stress is acting in a particular situation.

2. Remember that the buoyant force is exerted by a fluid on an object floating or submerged in it. The net force depends on the density of the object as well as the density of the fluid; you will often need to calculate mass and volume

separately and take their ratio to obtain a value for the density.

3. Bernoulli's Equation is derived from the work-energy theorem, and the main problem-solving guidelines for energy problems apply to flow problems as well. In particular, be sure to clearly identify where point 1 and point 2 are located for applying Bernoulli's Equation and to list known values for pressure, height, and velocity of the fluid at each point.

MULTIPLE-CHOICE QUESTIONS

13.1 Salt water has a greater density than freshwater. A boat floats in both freshwater and salt water. The buoyant force on the boat in salt water is _____ that in freshwater.

a) equal to   b) smaller than   c) larger than

13.2 You fill a tall glass with ice and then add water to the level of the glass's rim, so some fraction of the ice floats above the rim. When the ice melts, what happens to the water level? (Neglect evaporation, and assume that the ice and water remain at 0 °C during the melting process.)

a) The water overflows the rim.

b) The water level drops below the rim.

c) The water level stays at the rim.

d) It depends on the difference in density between water and ice.

13.3 The figure shows four identical open-top tanks filled to the brim with water and sitting on a scale. Balls float in tanks (2) and (3), but an object sinks to the bottom in tank (4). Which of the following correctly ranks the weights shown on the scales?

(1)   (2)   (3)   (4)

a) (1) < (2) < (3) < (4)

b) (1) < (2) = (3) < (4)

c) (1) < (2) = (3) = (4)

d) (1) = (2) = (3) < (4)

13.4 You are in a boat filled with large rocks in the middle of a small pond. You begin to drop the rocks into the water. What happens to the water level of the pond?

a) It rises.

b) It falls.

c) It doesn't change.

d) It rises momentarily and then falls when the rocks hit bottom.

e) There is not enough information to say.

13.5 Rank in order, from largest to smallest, the magnitudes of the forces F_1, F_2, and F_3 required for balancing the masses shown in the figure.

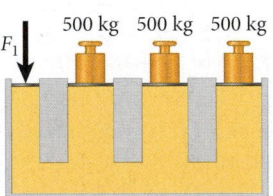 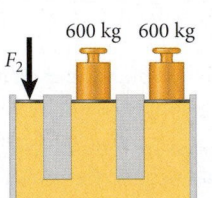 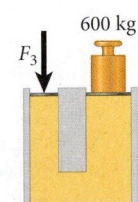

13.6 In a horizontal water pipe that narrows to a smaller radius, the velocity of the water in the section with the smaller radius will be larger. What happens to the pressure?

a) The pressure will be the same in both the wider and narrower sections of the pipe.

b) The pressure will be higher in the narrower section of the pipe.

c) The pressure will be higher in the wider section of the pipe.

d) It is impossible to tell.

13.7 In one of the *Star Wars*™ movies, four of the heroes are trapped in a trash compactor on the Death Star. The compactor's walls begin to close in, and the heroes need to pick an object from the trash to place between the closing walls to stop them. All the objects are the same length and have a circular cross section, but their diameters and compositions are different. Assume that each object is oriented horizontally and does not bend. They have the time and strength to hold up only one object between the walls. Which of the objects shown in the figure will work best—that is, will withstand the greatest force per unit of compression?

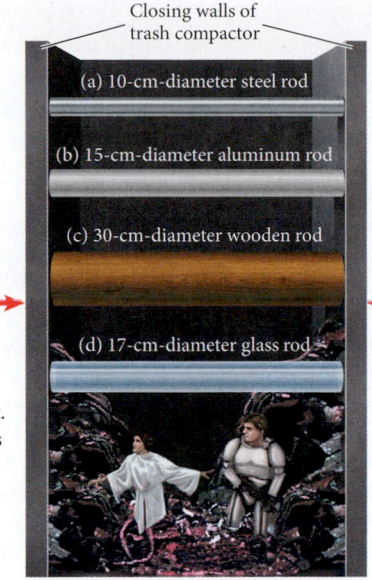

Closing walls of trash compactor

(a) 10-cm-diameter steel rod

(b) 15-cm-diameter aluminum rod

(c) 30-cm-diameter wooden rod

(d) 17-cm-diameter glass rod

13.8 Many altimeters determine altitude changes by measuring changes in the air pressure. An altimeter that is designed to be able to detect altitude changes of 100 m near sea level should be able to detect pressure changes of

a) approximately 1 Pa.
b) approximately 10 Pa.
c) approximately 100 Pa.
d) approximately 1 kPa.
e) approximately 10 kPa.

13.9 Which of the following assumptions is *not* made in the derivation of Bernoulli's Equation?

a) Streamlines do not cross.
b) There is negligible viscosity.
c) There is negligible friction.
d) There is no turbulence.
e) There is negligible gravity.

13.10 A beaker is filled with water to the rim. Gently placing a plastic toy duck in the beaker causes some of the water to spill out. The weight of the beaker with the duck floating in it is

a) greater than the weight before adding the duck.
b) less than the weight before adding the duck.
c) the same as the weight before adding the duck.
d) greater or less than the weight before the duck was added, depending on the weight of the duck.

13.11 A piece of cork (density = 0.33 g/cm³) with a mass of 10 g is held in place under water by a string, as shown in the figure. What is the tension, T, in the string?

a) 0.10 N   c) 0.30 N   e) 200 N
b) 0.20 N   d) 100 N   f) 300 N

13.12 Suppose you pump the air out of a paint can, which is covered by a lid. The can is cylindrical, with a height of 22.4 cm and a diameter of 16.0 cm. How much force does the atmosphere exert on the lid of the evacuated paint can?

a) 9.81 N   c) 2030 N
b) 511 N   d) 8120 N

13.13 A steel sphere with a diameter of 0.250 m is submerged in the ocean to a depth of 500.0 m. What is the percentage change in the volume of the sphere? The bulk modulus of steel is $160 \cdot 10^9$ Pa.

a) 0.0031%   c) 0.33%   e) 1.5%
b) 0.045%   d) 0.55%

13.14 A flowing fluid has a Reynolds number of 1000. Which statement about this flow is true?

a) It is laminar flow.
b) It is turbulent flow.
c) It is frictionless flow.
d) It is perfect flow.
e) It is nonviscous flow.

13.15 A beaker of water is sitting on a scale. A steel ball hanging from a string is lowered into the water until the ball is completely submerged but is not touching the beaker. The weight registered by the scale will

a) increase.   b) decrease.   c) stay the same.

CONCEPTUAL QUESTIONS

13.16 You know from experience that if a car you are riding in suddenly stops, heavy objects in the rear of the car move toward the front. Why does a helium-filled balloon in such a situation move, instead, toward the rear of the car?

13.17 A piece of paper is folded in half and then opened up and placed on a flat table so that it "peaks" up in the middle as shown in the figure. If you blow air between the paper and the table, will the paper move up or down? Explain.

13.18 In what direction does a force due to water flowing from a showerhead act on a shower curtain, inward toward the shower or outward? Explain.

13.19 Point out and discuss any flaws in the following statement: *The hydraulic car lift is a device that operates on the basis of Pascal's Principle. Such a device can produce large output forces with small input forces. Thus, with a small amount of work done by the input force, a much larger amount of work is produced by the output force, and the heavy weight of a car can be lifted.*

13.20 Given two springs of identical size and shape, one made of steel and the other made of aluminum, which has the higher spring constant? Why? Does the difference depend more on the shear modulus or the bulk modulus of the material?

13.21 One material has a higher density than another. Are the individual atoms or molecules of the first material necessarily more massive than those of the second?

13.22 Analytic balances are calibrated to give correct mass values for such items as steel objects of density ρ_s = 8000.00 kg/m³. The calibration compensates for the buoyant force arising because the measurements are made in air, of density ρ_a = 1.205 kg/m³. What compensation must be made to measure the masses of objects of a different material, of density ρ? Does the buoyant force of air matter?

13.23 If you turn on the faucet in the bathroom sink, you will observe that the stream seems to narrow from the point at which it leaves the spigot to the point at which it hits the bottom of the sink. Why does this occur?

13.24 In many problems involving application of Newton's Second Law to the motion of solid objects, friction is neglected for the sake of making the solution easier. The counterpart of friction between solids is viscosity of liquids. Do problems involving fluid flow become simpler if viscosity is neglected? Explain.

13.25 You have two identical silver spheres and two unknown fluids, A and B. You place one sphere in fluid A, and it sinks; you place the other sphere in fluid B, and it floats. What can you conclude about the buoyant force of fluid A versus that of fluid B?

13.26 Water flows from a circular faucet opening of radius r_0, directed vertically downward, at speed v_0. As the stream of water falls, it narrows. Find an expression for the radius of the stream as a function of distance fallen, $r(y)$, where y is measured downward from the opening. Neglect the eventual breakup of the stream into droplets, and any resistance due to drag or viscosity.

EXERCISES

A blue problem number indicates a worked-out solution is available in the Student Solutions Manual. One • and two •• indicate increasing level of problem difficulty.

Sections 13.1 and 13.2

13.27 Air consists of molecules of several types, with an average molar mass of 28.95 g. About how many molecules does an adult who inhales 0.50 L of air at sea level take in?

•13.28 Ordinary table salt (NaCl) consists of sodium and chloride ions arranged in a *face-centered cubic crystal lattice*. That is, a sodium chloride crystal consists of cubic unit cells with a sodium ion on each corner and at the center of each face, and a chloride ion at the center of the cube and at the midpoint of each edge. The density of sodium chloride is $2.165 \cdot 10^3$ kg/m^3. Calculate the spacing between adjacent sodium and chloride ions in the crystal.

Section 13.3

13.29 A 20-kg chandelier is suspended from a ceiling by four vertical steel wires. Each wire has an unloaded length of 1 m and a diameter of 2 mm, and each bears an equal load. When the chandelier is hung, how far do the wires stretch?

13.30 Find the minimum diameter of a 50.0-m-long nylon string that will stretch no more than 1.00 cm when a load of 70.0 kg is suspended from its lower end. Assume that $Y_{\text{nylon}} = 3.51 \cdot 10^9$ N/m^2.

13.31 A 2.00-m-long steel wire in a musical instrument has a radius of 0.300 mm. When the wire is under a tension of 90.0 N, how much does its length change?

•13.32 A rod of length L is attached to a wall. The load on the rod increases linearly (as shown by the arrows in the figure) from zero at the left end to W newtons per unit length at the right end. Find the shear force at

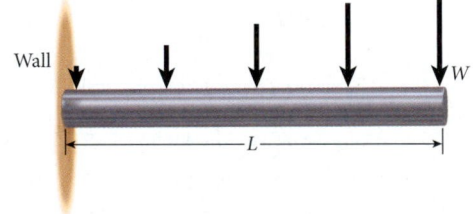

a) the right end,   b) the center, and   c) the left end.

••13.33 Challenger Deep in the Marianas Trench of the Pacific Ocean is the deepest known spot in the Earth's oceans, at 10.922 km below sea level. Taking the density of seawater at atmospheric pressure ($p_0 = 101.3$ kPa) to be 1024 kg/m^3 and its bulk modulus to be $B(p) = B_0 + 6.67(p - p_0)$, with $B_0 = 2.19 \cdot 10^9$ Pa, calculate the pressure and the density of the seawater at the bottom of Challenger Deep. Disregard variations in water temperature and salinity with depth. Is it a good approximation to treat the density of seawater as essentially constant?

Section 13.4

13.34 How much force is the air exerting on the front cover of your textbook? What mass has a weight equivalent to this force?

13.35 Blood pressure is usually reported in millimeters of mercury (mmHg) or the height of a column of mercury producing the same pressure value. Typical values for an adult human are 130/80; the first value is the *systolic* pressure, during the contraction of the ventricles of the heart, and the second is the *diastolic* pressure, during the contraction of the auricles of the heart. The head of an adult male giraffe is 6.0 m above the ground; the giraffe's heart is 2.0 m above the ground. What is the minimum systolic pressure (in mmHg) required at the heart to drive blood to the head (neglect the additional pressure required to overcome the effects of viscosity)? The density of giraffe blood is 1.00 g/cm^3, and that of mercury is 13.6 g/cm^3.

13.36 A scuba diver must decompress after a deep dive to allow excess nitrogen to exit safely from the bloodstream. The length of time required

for decompression depends on the total change in pressure that the diver experienced. Find this total change in pressure for a diver who starts at a depth of $d = 20.0$ m in the ocean (density of seawater = 1024 kg/m^3) and then travels aboard a small plane (with an unpressurized cabin) that rises to an altitude of $h = 5000$. m above sea level.

•13.37 A child loses his balloon, which rises slowly into the sky. If the balloon is 20.0 cm in diameter when the child loses it, what is its diameter at an altitude of (a) 1000. m, (b) 2000. m, and (c) 5000. m? Assume that the balloon is very flexible, allowing surface tension to be neglected.

•13.38 The atmosphere of Mars exerts a pressure of only 600. Pa on the surface and has a density of only 0.0200 kg/m^3.

a) What is the thickness of the Martian atmosphere, assuming the boundary between atmosphere and outer space to be the point where atmospheric pressure drops to 0.0100% of its value at surface level?

b) What is the atmospheric pressure at the bottom of Mars's Hellas Planitia canyon, at a depth of 8.18 km?

c) What is the atmospheric pressure at the top of Mars's Olympus Mons volcano, at a height of 21.3 km?

d) Compare the relative change in air pressure, $\Delta p/p$, between these two points on Mars and between the equivalent extremes on Earth—the Dead Sea shore, at 400. m below sea level, and Mount Everest, at an altitude of 8850 m.

•13.39 The air density at the top of Mount Everest ($h_{\text{Everest}} = 8850$ m) is 34.8% of the air density at sea level. If the air *pressure* at the top of Mount McKinley in Alaska is 47.7% of the air pressure at sea level, calculate the height of Mount McKinley, using only the information given here.

•13.40 A sealed vertical cylinder of radius R and height $h = 0.60$ m is initially filled halfway with water, and the upper half is filled with air. The air is initially at standard atmospheric pressure, $p_0 = 1.01 \cdot 10^5$ Pa. A small valve at the bottom of the cylinder is opened, and water flows out of the cylinder until the reduced pressure of the air in the upper part of the cylinder prevents any further water from escaping. By what distance is the depth of the water lowered? (Assume that the temperature of water and air do not change and that no air leaks into the cylinder.)

••13.41 A temporary above-ground square pool with 100.-m-long sides is created in a concrete parking lot. The walls are concrete, are 50.0 cm thick, and have a density of 2.50 g/cm^3. The coefficient of static friction between the walls and the parking lot is 0.450. What is the maximum possible depth of water in the pool?

••13.42 The calculation of atmospheric pressure at the summit of Mount Everest carried out in Example 13.3 used the model known as the *isothermal atmosphere*, in which gas pressure is proportional to density: $p = \gamma \rho$, with γ constant. Consider a spherical cloud of gas supporting itself under *its own* gravitation and following this model.

a) Write the equation of hydrostatic equilibrium for the cloud, in terms of the gas density as a function of radius, $\rho(r)$.

b) Show that $\rho(r) = \alpha/r^2$ is a solution of this equation, for an appropriate choice of constant α. Explain why this solution is not suitable as a model of a star.

Section 13.5

13.43 A racquetball with a diameter of 5.6 cm and a mass of 42 g is cut in half to make a boat for American pennies made after 1982. The mass and volume of an American penny made after 1982 are 2.5 g and 0.36 cm^3. How many pennies can be placed in the racquetball boat without sinking it?

13.44 A future supertanker filled with oil has a total mass of $10.2 \cdot 10^8$ kg. If the dimensions of the ship are those of a rectangular box 250. m long, 80.0 m wide, and 80.0 m high, determine how far the bottom of the ship is below sea level ($\rho_{\text{sea}} = 1024$ kg/m^3).

13.45 A box with a volume $V = 0.0500$ m^3 lies at the bottom of a lake whose water has a density of $1.00 \cdot 10^3$ kg/m^3. How much force is required to lift the box, if the mass of the box is (a) 1000. kg, (b) 100. kg, and (c) 55.0 kg?

•**13.46** A man of mass 64 kg and density 970 kg/m³ stands in a shallow pool with 32% of the volume of his body below water. Calculate the normal force that the bottom of the pool exerts on his feet.

•**13.47** A block of cherry wood that is 20.0 cm long, 10.0 cm wide, and 2.00 cm thick has a density of 800. kg/m³. What is the volume of a piece of iron that, if glued to the bottom of the block, makes the block float in water with its top just at the surface of the water? The density of iron is 7860 kg/m³, and the density of water is 1000. kg/m³.

•**13.48** The average density of the human body is 985 kg/m³, and the typical density of seawater is about 1024 kg/m³.

a) Draw a free-body diagram of a human body floating in seawater and determine what percentage of the body's volume is submerged.

b) The average density of the human body, after maximum inhalation of air, changes to 945 kg/m³. As a person floating in seawater inhales and exhales slowly, what percentage of his volume moves up out of and down into the water?

c) The Dead Sea (a saltwater lake between Israel and Jordan) is the world's saltiest large body of water. Its average salt content is more than six times that of typical seawater, which explains why there is no plant and animal life in it. Two-thirds of the volume of the body of a person floating in the Dead Sea is observed to be submerged. Determine the density (in kg/m³) of the seawater in the Dead Sea.

•**13.49** A tourist of mass 60.0 kg notices a chest with a short chain attached to it at the bottom of the ocean. Imagining the riches it could contain, he decides to dive for the chest. He inhales fully, thus setting his average body density to 945 kg/m³, jumps into the ocean (with saltwater density = 1024 kg/m³), grabs the chain, and tries to pull the chest to the surface. Unfortunately, the chest is too heavy and will not move. Assume that the man does not touch the bottom.

a) Draw the man's free-body diagram, and determine the tension on the chain.

b) What mass (in kg) has a weight that is equivalent to the tension force in part (a)?

c) After realizing he cannot free the chest, the tourist releases the chain. What is his upward acceleration (assuming that he simply allows the buoyant force to lift him up to the surface)?

•**13.50** A very large balloon with mass M = 10.0 kg is inflated to a volume of 20.0 m³ using a gas of density ρ_{gas} = 0.20 kg/m³. What is the maximum mass m that can be tied to the balloon using a 2.00-kg piece of rope without the balloon falling to the ground? (Assume that the density of air is 1.30 kg/m³ and that the volume of the gas is equal to the volume of the inflated balloon).

•**13.51** The *Hindenburg*, the German zeppelin that caught fire in 1937 while docking in Lakehurst, New Jersey, was a rigid duralumin-frame balloon filled with 2.000·10⁵ m³ of hydrogen. The *Hindenburg's* useful lift (beyond the weight of the zeppelin structure itself) is reported to have been 1.099·10⁶ N (or 247,000 lb). Use ρ_{air} = 1.205 kg/m³, ρ_H = 0.08988 kg/m³, and ρ_{He} = 0.1786 kg/m³.

a) Calculate the weight of the zeppelin structure (without the hydrogen gas).

b) Compare the useful lift of the (highly flammable) hydrogen-filled *Hindenburg* with the useful lift the *Hindenburg* would have had if it had been filled with (nonflammable) helium, as originally planned.

••**13.52** Brass weights are used to weigh an aluminum object on an analytical balance. The weighing is done one time in dry air and another time in humid air. What should the mass of the object be to produce a noticeable difference in the balance readings, provided the balance's sensitivity is m_0 = 0.100 mg? (The density of aluminum is ρ_A = 2.70·10³ kg/m³; the density of brass is ρ_B = 8.50·10³ kg/m³. The density of the dry air is 1.2285 kg/m³, and the density of the humid air is 1.2273 kg/m³.)

Section 13.6

13.53 A fountain sends water to a height of 100. m. What is the difference between the pressure of the water just before it is released upward and the atmospheric pressure?

13.54 Water enters a horizontal pipe with a rectangular cross section at a speed of 1.00 m/s. The width of the pipe remains constant but the height decreases. Twenty meters from the entrance, the height is half of what it is at the entrance. If the water pressure at the entrance is 3000. Pa, what is the pressure 20.0 m downstream?

•**13.55** Water at room temperature flows with a constant speed of 8.00 m/s through a nozzle with a square cross section, as shown in the figure. Water enters the nozzle at point A and exits the nozzle at point B. The lengths of the sides of the square cross section at A and B are 50.0 cm and 20.0 cm, respectively.

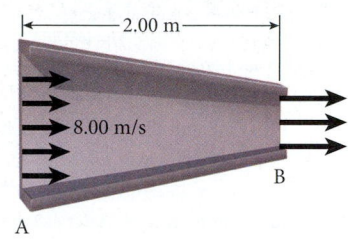

a) What is the volume flow rate at the exit?

b) What is the acceleration at the exit? The length of the nozzle is 2.00 m.

c) If the volume flow rate through the nozzle is increased to 6.00 m³/s, what is the acceleration of the fluid at the exit?

•**13.56** Water is flowing in a pipe as depicted in the figure. What pressure is indicated on the upper pressure gauge?

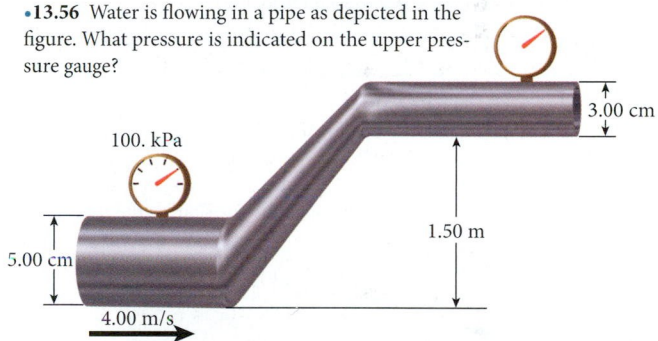

•**13.57** An open-topped tank completely filled with water has a release valve near its bottom. The valve is 1.0 m below the water surface. Water is released from the valve to power a turbine, which generates electricity. The area of the top of the tank, A_T is 10 times the cross-sectional area, A_V, of the valve opening. Calculate the speed of the water as it exits the valve. Neglect friction and viscosity. In addition, calculate the speed of a drop of water released from rest at h = 1.0 m when it reaches the elevation of the valve. Compare the two speeds.

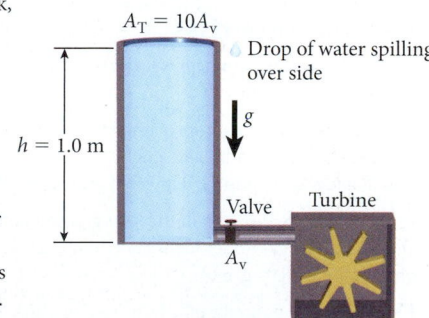

•**13.58** A tank of height H is filled with water and sits on the ground, as shown in the figure. Water squirts from a hole at a height y above the ground and has a range R. For two y values, 0 and H, R is zero. Determine the value of y for which the range will be a maximum.

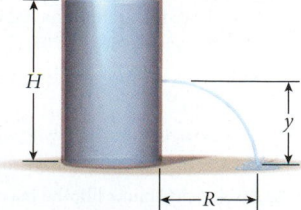

••**13.59** A water-powered backup sump pump uses tap water at a pressure of 3.00 atm ($p_1 = 3p_{atm} = 3.03 \cdot 10^5$ Pa) to pump water out of a well, as shown in the figure ($p_{well} = p_{atm}$). This system allows water to be pumped out of a basement sump well when the electric pump stops working during an electrical power outage. Using water to pump water may sound strange at first, but these pumps are quite efficient, typically pumping out 2.00 L of well water for every 1.00 L of pressurized tap water. The supply water moves to the right in a large pipe with cross-sectional area A_1 at a speed $v_1 = 2.05$ m/s. The water then flows into a pipe of smaller diameter with a cross-sectional area that is ten times smaller ($A_2 = A_1/10$).

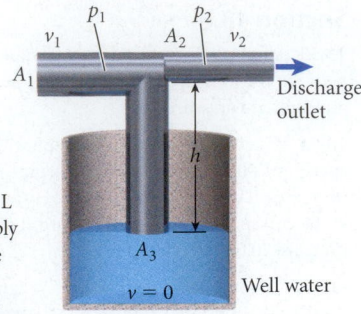

a) What is the speed v_2 of the water in the smaller pipe, with area A_2?

b) What is the pressure p_2 of the water in the smaller pipe, with area A_2?

c) The pump is designed so that the vertical pipe, with cross-sectional area A_3, that leads to the well water also has a pressure of p_2 at its top. What is the maximum height, h, of the column of water that the pump can support (and therefore act on) in the vertical pipe?

Section 13.7

13.60 A basketball of circumference 75.5 cm and mass 598 g is forced to the bottom of a swimming pool and then released. After initially accelerating upward, it rises at a constant velocity.

a) Calculate the buoyant force on the basketball.

b) Calculate the drag force the basketball experiences while it is moving upward at constant velocity.

•**13.61** The cylindrical container shown in the figure has a radius of 1.00 m and contains motor oil with a viscosity of 0.300 Pa s and a density of 670. kg m^{-3}. Oil flows out of the 20.0 cm long, 0.200 cm diameter tube at the bottom of the container. How much oil flows out of the tube in a period of 10.0 s if the container is originally filled to a height of 0.500 m?

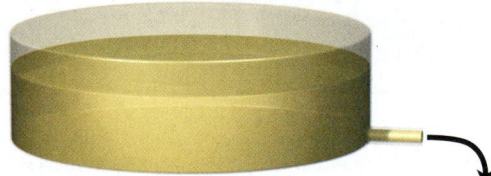

Additional Exercises

13.62 Estimate the atmospheric pressure inside Hurricane Rita. The wind speed was 290 km/h.

13.63 The following data are obtained for a car: The tire pressure is measured as 28.0 psi, and the width and length of the contact surface of each tire are 7.50 in and 8.75 in, respectively. What is the approximate weight of the car?

13.64 Calculate the ratio of the lifting powers of helium (He) gas and hydrogen (H$_2$) gas under identical circumstances. Assume that the molar mass of air is 28.95 g/mol.

13.65 Water is poured into a large barrel whose height is 2.0 m; a cylindrical, 3.0-cm-diameter cork is stuck into the side at a height of 0.50 m above the ground. As the water level in the barrel just reaches maximum height, the cork flies out of the barrel.

a) What was the magnitude of the static friction force between the barrel and the cork?

b) If the barrel were filled with seawater, would the cork have flown out before the barrel's full capacity was reached?

13.66 In a hydraulic lift, the maximum gauge pressure is 17.00 atm. If the diameter of the output line is 22.5 cm, what is the heaviest vehicle that can be lifted?

13.67 A water pipe narrows from a radius of $r_1 = 5.00$ cm to a radius of $r_2 = 2.00$ cm. If the speed of the water in the wider part of the pipe is 2.00 m/s, what is the speed of the water in the narrower part?

13.68 Donald Duck and his nephews manage to sink Uncle Scrooge's yacht ($m = 4500$ kg), which is made of steel ($\rho = 7800$ kg/m^3). In typical comic-book fashion, they decide to raise the yacht by filling it with ping-pong balls. A ping-pong ball has a mass of 2.7 g and a volume of $3.35 \cdot 10^{-5}$ m^3.

a) What is the buoyant force on one ping-pong ball in water?

b) How many balls are required to float the ship?

13.69 A wooden block floating in seawater has two thirds of its volume submerged. When the block is placed in mineral oil, 80.0% of its volume is submerged. Find the density of (a) the wooden block and (b) the mineral oil.

13.70 An approximately round tendon that has an average diameter of 3.5 mm and is 15 cm long is found to stretch 0.37 mm when acted on by a force of 13.4 N. Calculate Young's modulus for the tendon.

•**13.71** The Jovian moon Europa may have oceans (covered by ice, which can be ignored). What would the pressure be 1.00 km below the surface of a Europan ocean? The surface gravity of Europa is 13.5% that of the Earth's.

•**13.72** Two balls of the same volume are released inside a tank filled with water, as shown in the figure. The densities of balls A and B are 0.90 g/cm^3 and 0.80 g/cm^3. Find the acceleration of (a) ball A and (b) ball B. (c) Which ball wins the race to the top?

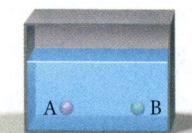

•**13.73** An airplane is moving through the air at a velocity $v = 200.$ m/s. Streamlines just over the top of the wing are compressed to 80.0% of their original cross-sectional area, and those under the wing are not compressed at all.

a) Determine the velocity of the air just over the wing.

b) Find the difference in the pressure of the air just over the wing, P, and that of the air under the wing, P'. The density of the air is 1.30 kg/m^3.

c) Find the net upward force on both wings due to the pressure difference, if the area of the wings is 40.0 m^2.

•**13.74** A cylindrical buoy with hemispherical ends is dropped in seawater of density 1027 kg/m^3, as shown in the figure. The mass of the buoy is 75.0 kg, the radius of the cylinder and the hemispherical caps is $R = 0.200$ m, and the length of the cylindrical center section of the buoy is $L = 0.600$ m. The lower part of the buoy is weighted so that the buoy is upright in the water, as shown in the figure. In calm water, how high (distance h) will the top of the buoy be above the waterline?

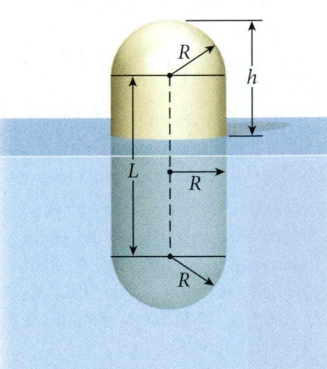

•**13.75** Water of density 998.2 kg/m^3 is moving at negligible speed under a pressure of 101.3 kPa but is then accelerated to a high speed by the blades of a spinning propeller. The vapor pressure of the water at the initial temperature of 20.0 °C is 2.3388 kPa. At what flow speed will the water begin to boil? This effect, known as *cavitation*, limits the performance of propellers in water. (Vapor pressure is the pressure of the vapor resulting from evaporation of a liquid above a sample of the liquid in a closed container.)

•**13.76** An astronaut wishes to measure the atmospheric pressure on Mars using a mercury barometer like that shown in the figure. The calibration of the barometer is the standard calibration for Earth: 760 mmHg corresponds to the pressure due to Earth's atmosphere, 1 atm or 101.325 kPa. How does the barometer need to be recalibrated for use in the atmosphere of

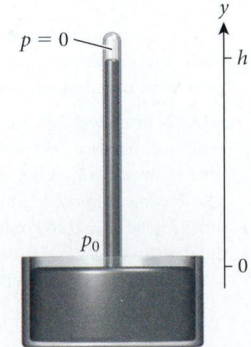

Mars—by what factor does the barometer's scale need to be "stretched"? In her handy table of planetary masses and radii, the astronaut finds that Mars has an average radius of $3.39 \cdot 10^6$ m and a mass of $6.42 \cdot 10^{23}$ kg.

•**13.77** In many locations, such as Lake Washington in Seattle, floating bridges are preferable to conventional bridges. Such a bridge can be constructed out of concrete pontoons, which are essentially concrete boxes filled with air, Styrofoam, or another extremely low-density material. Suppose a floating bridge pontoon is constructed out of concrete and Styrofoam, which have densities of 2200 kg/m^3 and 50.0 kg/m^3. What must the volume ratio of concrete to Styrofoam be if the pontoon is to float with 35.0% of its overall volume above water?

•**13.78** A 1.0-g balloon is filled with helium gas. When a mass of 4.0 g is attached to the balloon, the combined mass hangs in static equilibrium in midair. Assuming that the balloon is spherical, what is its diameter?

•**13.79** A large water tank has an inlet pipe and an outlet pipe. The inlet pipe has a diameter of 2.00 cm and is 1.00 m above the bottom of the tank. The outlet pipe has a diameter of 5.00 cm and is 6.00 m above the bottom of the tank. A volume of 0.300 m^3 of water enters the tank every minute at a gauge pressure of 1.00 atm.

a) What is the velocity of the water in the outlet pipe?

b) What is the gauge pressure in the outlet pipe?

•**13.80** Two spheres of the same diameter (in air), one made of a lead alloy and the other one made of steel, are submerged at a depth $h = 2000.$ m below the surface of the ocean. The ratio of the volumes of the two spheres at this depth is $V_{\text{Steel}}(h)/V_{\text{Lead}}(h) = 1.001206$. Knowing the density of ocean water $\rho = 1024$ kg/m^3 and the bulk modulus of steel, $B_{\text{Steel}} = 160 \cdot 10^9$ N/m^2, calculate the bulk modulus of the lead alloy.

MULTI-VERSION EXERCISES

13.81 A diving bell with interior air pressure equal to atmospheric pressure is submerged in Lake Michigan at a depth of 129.1 m. The diving bell has a flat, transparent, circular viewing port with a diameter of 22.89 cm. What is the magnitude of the net force on the viewing port?

13.82 A diving bell with interior air pressure equal to atmospheric pressure is submerged in Lake Michigan. The diving bell has a flat, transparent, circular viewing port with a diameter of 23.11 cm. The magnitude of the net force on the viewing port is $6.251 \cdot 10^4$ N. At what depth is the diving bell?

13.83 A diving bell with interior air pressure equal to atmospheric pressure is submerged in Lake Michigan at a depth of 174.9 m. The diving bell has a flat, transparent, circular viewing port. The magnitude of the net force on the viewing port is $7.322 \cdot 10^4$ N. What is the diameter of the viewing port?

13.84 A hot-air balloon has a volume of 2979 m^3. The density of the air outside the balloon is 1.205 kg/m^3. The density of the hot air inside the balloon is 0.9441 kg/m^3. How much weight can the balloon lift (including its own weight)?

13.85 A hot-air balloon can lift a weight of 5626 N (including its own weight). The density of the air outside the balloon is 1.205 kg/m^3. The density of the hot air inside the balloon is 0.9449 kg/m^3. What is the volume of the balloon?

13.86 A hot-air balloon has a volume of 2435 m^3. The balloon can lift a weight of 6194 N (including its own weight). The density of the air outside the balloon is 1.205 kg/m^3. What is the density of the hot air inside the balloon?

13.87 A waterproof ball made of rubber with a bulk modulus of $6.309 \cdot 10^7$ N/m^2 is submerged under water to a depth of 55.93 m. What is the fractional change in the volume of the ball?

13.88 A waterproof ball made of rubber with a bulk modulus of $8.141 \cdot 10^7$ N/m^2 is submerged under water. The fractional change in the volume of the ball is $6.925 \cdot 10^{-3}$. To what depth is the ball submerged?

13.89 A waterproof rubber ball is submerged under water to a depth of 59.01 m. The fractional change in the volume of the ball is $2.937 \cdot 10^{-2}$. What is the bulk modulus of the rubber ball?

13.90 One can use turbines to exploit the energy contained in ocean currents, just like one can do it for wind. What is the maximum amount of power that can be extracted from a current flowing with a speed of 1.35 m/s, if one uses a turbine with rotor diameter of 24.5 m? (*Hint 1:* The density of seawater is 1024 kg/m^3. *Hint 2:* The Betz limit applies to any fluid, including seawater.)

13.91 One can use turbines to exploit the energy contained in ocean currents, just like one can do it for wind. If the maximum amount of power is 571.8 kW, which can be extracted from a current flowing with a speed of 1.57 m/s, what is the rotor diameter of the turbine? (*Hint 1:* The density of seawater is 1024 kg/m^3. *Hint 2:* The Betz limit applies to any fluid, including seawater.)

13.92 One can use turbines to exploit the energy contained in ocean currents, just like one can do it for wind. If the maximum amount of power is 918.8 kW, which can be extracted from an ocean current with a turbine of rotor diameter 25.5 m, what is the speed of the ocean current? (*Hint 1:* The density of seawater is 1024 kg/m^3. *Hint 2:* The Betz limit applies to any fluid, including seawater.)

14

Oscillations

(a)   (b)

FIGURE 14.1 Harmonic oscillations used for keeping time. (a) Multiple-exposure picture of an old-fashioned pendulum clock. (b) The world's smallest atomic clock, from the National Institute of Standards and Technology, in which cesium atoms perform 9.2 billion oscillations each second. The clock is the size of a grain of rice and is accurate to 1 part in 10 billion—or off by less than 1 s over a period of 300 yr.

Even when an object appears to be perfectly at rest, its atoms and molecules are rapidly vibrating. Sometimes these vibrations can be put to use; for example, the atoms in a quartz crystal vibrate with a very steady frequency if the crystal is subjected to a periodic electric field. This vibration is used to keep track of time in modern quartz-crystal clocks and wrist watches. Vibrations of cesium atoms are used in atomic clocks (Figure 14.1).

In this chapter, we examine the nature of oscillatory motion. Most of the situations we'll consider involve springs or pendulums, but these are just the simplest examples of oscillators. Later in the book, we will study other kinds of vibrating systems, which can be modeled as a spring or a pendulum for the purpose of analyzing the motion. In this chapter, we also investigate the concept of resonance, which is an important property of all oscillating systems, from the atomic level to bridges and skyscrapers. Chapters 15 and 16 will apply the concepts of oscillations to analyze the nature of waves and sound.

WHAT WE WILL LEARN

- The spring force leads to a sinusoidal oscillation in time referred to as *simple harmonic motion.*

- A similar force law applies to the oscillation of a pendulum swinging through small angles.

- Oscillations can be represented as a projection of circular motion onto one of the two Cartesian coordinate axes.

- In the presence of damping, oscillations slow down exponentially over time. Depending on the strength of the damping, it is possible that no oscillations will occur.

- Periodic external driving of an oscillator leads to sinusoidal motion at the driving frequency, with a maximum amplitude close to the resonant angular speed.

- Plotting the motion of oscillators in terms of the velocity and position shows that the motion of undamped oscillators follows an ellipse and that of damped oscillators spirals in.

- A damped and driven oscillator can exhibit chaotic motion in which the trajectory in time depends sensitively on the initial conditions.

14.1 Simple Harmonic Motion

Repetitive motion, usually called **periodic motion,** is important in science, engineering, and daily life. Common examples of objects in periodic motion are windshield wipers on a car and the pendulum on a grandfather clock. However, periodic motion is also involved in the alternating current that powers the electronic grid of modern cities, atomic vibrations in molecules, and your own heartbeat and circulatory system.

Simple harmonic motion is a particular type of repetitive motion, which is displayed by a pendulum or by a massive object on a spring. Chapter 5 introduced the spring force, which is described by Hooke's Law: The spring force is proportional to the displacement of the spring from equilibrium. The spring force is a restoring force, always pointing toward the equilibrium position, and is thus opposite in direction to the displacement vector:

$$F_x = -kx.$$

The proportionality constant k is called the *spring constant.*

We have encountered the spring force repeatedly in other chapters. The main reason why forces that depend linearly on displacement are so important in many branches of physics was emphasized in Section 6.4, where we saw that for a system at equilibrium, a small displacement from the equilibrium position results in a springlike force with linear dependence on the displacement from the equilibrium position.

Now let's consider the situation in which an object of mass m is attached to a spring that is then stretched or compressed out of its equilibrium position. When the object is released, it oscillates back and forth. This motion is called **simple harmonic motion (SHM),** and it occurs whenever the restoring force is proportional to the displacement. (As we just noted, a linear restoring force is present in all systems close to a stable equilibrium point, so simple harmonic motion is seen in many physical systems.) Figure 14.2 shows frames of a videotape of the vertical oscillation of a weight on a spring. A vertical x-axis is superimposed, along with a horizontal axis that represents time, with each frame 0.06 s from the neighboring ones. The red curve running through this sequence is a sine function.

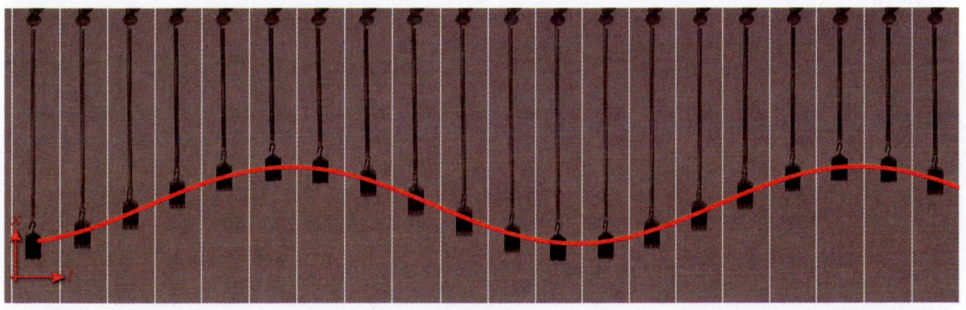

FIGURE 14.2 Consecutive video images of a weight hanging from a spring undergoing simple harmonic motion. A coordinate system and graph of the position as a function of time are superimposed on the images.

With the insight gained from Figure 14.2, we can describe this type of motion mathematically. We start with the force law for the spring force, $F_x = -kx$, and use Newton's Second Law, $F_x = ma$, to obtain

$$ma = -kx.$$

We know that acceleration is the second time derivative of the position: $a = d^2x/dt^2$. Substituting this expression for a into the preceding equation obtained with Newton's Second Law gives us

$$m\frac{d^2x}{dt^2} = -kx,$$

or

$$\frac{d^2x}{dt^2} + \frac{k}{m}x = 0. \tag{14.1}$$

Equation 14.1 includes both the position, x, and its second derivative with respect to time. Both are functions of the time, t. This type of equation is called a *differential equation*. The solution to this particular differential equation is the mathematical description of simple harmonic motion.

From the curve in Figure 14.2, we can see that the solution to the differential equation should be a sine or a cosine function. Let's see if the following will work:

$$x = A\sin(\omega_0 t).$$

The constants A and ω_0 are called the **amplitude** of the oscillation and its **angular speed,** respectively. The amplitude is the maximum displacement away from the equilibrium position, as you learned in Chapter 6. This sinusoidal function works for any value of the amplitude, A. However, the amplitude cannot be arbitrarily large or the spring will be over-stretched. On the other hand, we'll see that *not* all values of ω_0 produce a solution.

Taking the second derivative of the trial sine function results in

$$x = A\sin(\omega_0 t) \Rightarrow$$

$$\frac{dx}{dt} = \omega_0 A\cos(\omega_0 t) \Rightarrow$$

$$\frac{d^2x}{dt^2} = -\omega_0^2 A\sin(\omega_0 t).$$

Inserting this result and the sinusoidal expression for x into equation 14.1 yields

$$\frac{d^2x}{dt^2} + \frac{k}{m}x = -\omega_0^2 A\sin(\omega_0 t) + \frac{k}{m}A\sin(\omega_0 t) = 0.$$

This condition is fulfilled if $\omega_0^2 = k/m$, or

$$\omega_0 = \sqrt{\frac{k}{m}}.$$

We have thus found a valid solution to the differential equation (equation 14.1). In the same way, we can show that the cosine function leads to a solution as well, also with arbitrary amplitude and with the same angular speed. Thus, the complete solution for constants B and C is

$$x(t) = B\sin(\omega_0 t) + C\cos(\omega_0 t), \quad \text{with } \omega_0 = \sqrt{\frac{k}{m}}. \tag{14.2}$$

The units of ω_0 are radians per second (rad/s). When equation 14.2 is applied, $\omega_0 t$ must be expressed in radians, not degrees.

Here is another useful form of equation 14.2:

$$x(t) = A\sin(\omega_0 t + \theta_0), \quad \text{with } \omega_0 = \sqrt{\frac{k}{m}}. \tag{14.3}$$

Self-Test Opportunity 14.1

Show that $x(t) = A\sin(\omega_0 t + \theta_0)$ is a solution to equation 14.1.

Self-Test Opportunity 14.2

Show that $A\sin(\omega_0 t + \theta_0) = B\sin(\omega_0 t) + C\cos(\omega_0 t)$, where the relationships between the constants are given in equations 14.4 and 14.5.

This form allows you to see more readily that the motion is sinusoidal. Instead of having two amplitudes for the sine and cosine functions, equation 14.3 has one amplitude, A, and a *phase angle*, θ_0. These two constants are related to the constants B and C of equation 14.2 via

$$A = \sqrt{B^2 + C^2} \tag{14.4}$$

and

$$\theta_0 = \tan^{-1}\left(\frac{C}{B}\right). \tag{14.5}$$

Initial Conditions

How do we determine the values to use for the constants B and C, the amplitudes of the sine and cosine functions in equation 14.2? The answer is that we need two pieces of information, usually given in the form of the initial position, $x_0 = x(t = 0)$, and initial velocity, $v_0 = v(t = 0) = (dx/dt)|_{t=0}$, as in the following example.

EXAMPLE 14.1 / **Initial Conditions**

PROBLEM 1

A spring with spring constant $k = 56.0$ N/m has a lead weight of mass 1.00 kg attached to its end (Figure 14.3). The weight is pulled +5.5 cm from its equilibrium position and then pushed so that it receives an initial velocity of –0.32 m/s. What is the equation of motion for the resulting oscillation?

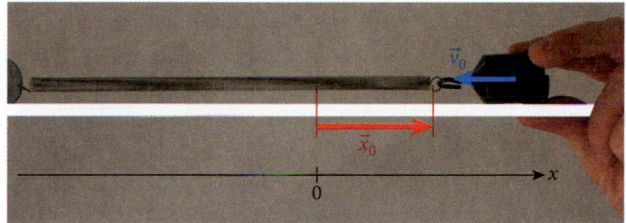

FIGURE 14.3 A weight attached to a spring, with the initial position and velocity vectors shown.

SOLUTION 1

The general equation of motion for this situation is equation 14.2 for simple harmonic motion:

$$x(t) = B\sin(\omega_0 t) + C\cos(\omega_0 t), \quad \text{with } \omega_0 = \sqrt{\frac{k}{m}}.$$

From the data given in the problem statement, we can calculate the angular speed:

$$\omega_0 = \sqrt{\frac{k}{m}} = \sqrt{\frac{56.0 \text{ N/m}}{1.00 \text{ kg}}} = 7.48 \text{ s}^{-1}.$$

Now we must determine the values of the constants B and C.

We take the first derivative of the general equation of motion:

$$x(t) = B\sin(\omega_0 t) + C\cos(\omega_0 t)$$
$$\Rightarrow v(t) = \omega_0 B\cos(\omega_0 t) - \omega_0 C\sin(\omega_0 t).$$

At time $t = 0$, $\sin(0) = 0$ and $\cos(0) = 1$, so these equations reduce to

$$x_0 = x(t = 0) = C$$
$$v_0 = v(t = 0) = \omega_0 B.$$

We were given the initial conditions for the position, $x_0 = 0.055$ m, and velocity, $v_0 = -0.32$ m/s. Thus, we find that $C = x_0 = 0.055$ m and $B = -0.043$ m.

PROBLEM 2

What is the amplitude of this oscillation? What is the phase angle?

– Continued

SOLUTION 2

With the values of the constants B and C, we can calculate the amplitude, A, from equation 14.4:

$$A = \sqrt{B^2 + C^2} = \sqrt{(0.043 \text{ m})^2 + (0.055 \text{ m})^2} = 0.070 \text{ m.}$$

Therefore, the amplitude of this oscillation is 7.0 cm. Note that, as a consequence of the nonzero initial velocity, the amplitude is *not* 5.5 cm, the value of the initial elongation of the spring.

The phase angle is found by application of equation 14.5 taking care to get the correct quadrant for the inverse tangent:

$$\theta_0 = \tan^{-1}\left(\frac{C}{B}\right) = \tan^{-1}\left(-\frac{0.055}{0.043}\right) = 2.234 \text{ rad.}$$

Expressed in degrees, this phase angle is $\theta_0 = 128.0°$.

Position, Velocity, and Acceleration

Let's look again at the relationships of position, velocity, and acceleration. In a form that describes oscillatory motion in terms of amplitude, A, and a phase angle determined by θ_0, they are

$$x(t) = A\sin(\omega_0 t + \theta_0)$$
$$v(t) = \omega_0 A\cos(\omega_0 t + \theta_0) \qquad (14.6)$$
$$a(t) = -\omega_0^2 A\sin(\omega_0 t + \theta_0).$$

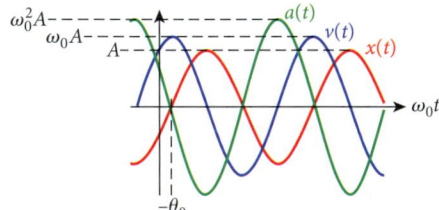

FIGURE 14.4 Graphs of position, velocity, and acceleration for simple harmonic motion as a function of time.

Here the velocity and acceleration are obtained from the position vector by taking successive time derivatives. These equations suggest that the velocity and acceleration vectors have the same phase angle as the position vector, determined by θ_0, but they have an additional phase angle of $\pi/2$ (for the velocity) and π (for the acceleration), which correspond to the phase difference between the sine and cosine functions and between the sine and the negative sine functions, respectively.

The position, velocity, and acceleration, as given by equations 14.6, are plotted in Figure 14.4, with $\omega_0 = 1.25 \text{ s}^{-1}$ and $\theta_0 = -0.5$ rad. Indicated in Figure 14.4 is the phase angle, as well as the three amplitudes of the oscillations: A is the amplitude of the oscillation of the position vector, $\omega_0 A$ (or $1.25A$, in this case) is the amplitude of the oscillation of the velocity vector, and $\omega_0^2 A$ [or $(1.25)^2 A$ here] is the amplitude of the oscillation of the acceleration vector. You can see that whenever the position vector passes through zero, the value of the velocity vector is at a maximum or a minimum, and vice versa. You can also observe that the acceleration (just like the force) is always in the opposite direction to the position vector. When the position passes through zero, so does the acceleration.

Figure 14.5 shows a block connected to a spring and undergoing simple harmonic motion by sliding on a frictionless surface. The velocity and acceleration vectors of the block at eight different positions are shown. In Figure 14.5a, the block is released from $x = A$. The block accelerates to the left as indicated. The block has reached $x = A/\sqrt{2}$ in Figure 14.5b. At this point, the velocity of the block and the acceleration of the block are directed toward the left. In Figure 14.5c, the block has reached the equilibrium position. The block at this position has zero acceleration and its maximum velocity to the left. The block continues through the equilibrium position of the spring and begins to slow down. In Figure 14.5d, the block is located at $x = -A/\sqrt{2}$. The acceleration of the block is now directed toward the right, although the block is still moving to the left. The block reaches $x = -A$ in Figure 14.5e. At this position, the velocity of the block is zero, and the acceleration of the block is directed toward the right. In Figure 14.5f, the block is again located at $x = -A/\sqrt{2}$, but now the velocity of the block and the acceleration are directed toward the right. The block again reaches the equilibrium position in Figure 14.5g with its velocity vector pointing to the right. The block continues through the equilibrium position and reaches $x = A/\sqrt{2}$ in Figure 14.5h, where the velocity is still toward the right but the acceleration is now toward the left. The block returns to its original configuration in Figure 14.5a, and the cycle continues.

Concept Check 14.1

A spring has a mass attached to its end. The mass is pulled +2 cm from the equilibrium position and released. As the mass passes through the equilibrium position, it has a speed of 3.2 m/s. After coming to rest at the equilibrium position, the mass is pulled +4 cm from that position and released. The speed of the mass when it next passes through the equilibrium position will be

a) 0.8 m/s.   d) 4.8 m/s.

b) 1.6 m/s.   e) 6.4 m/s.

c) 3.2 m/s.

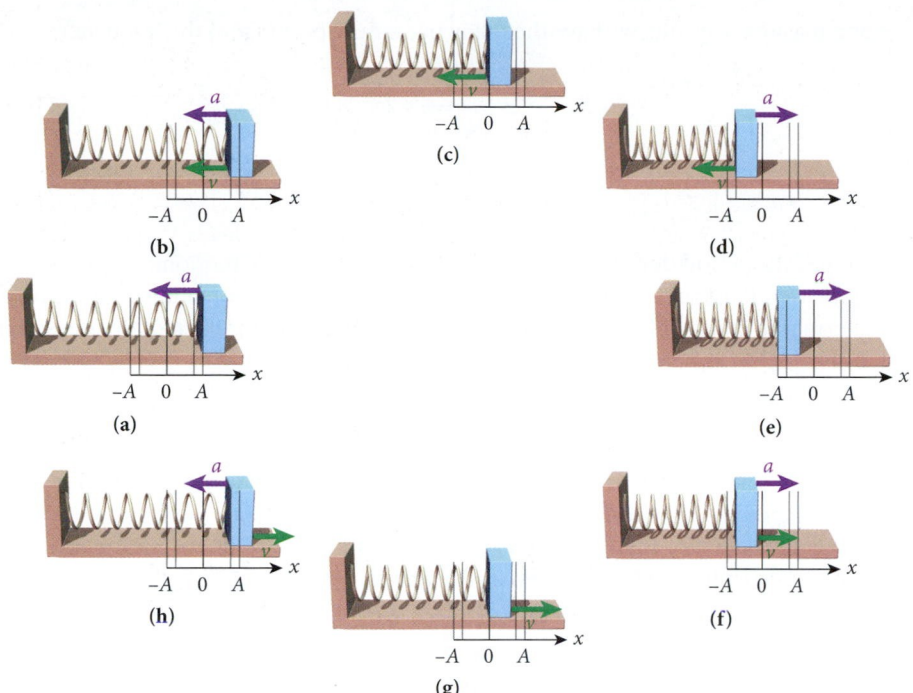

FIGURE 14.5 A block on a spring undergoing simple harmonic motion. The velocity and acceleration vectors are shown at different points in the oscillation: (a) $x = A$; (b) $x = A/\sqrt{2}$; (c) $x = 0$; (d) $x = -A/\sqrt{2}$; (e) $x = -A$; (f) $x = -A/\sqrt{2}$; (g) $x = 0$; (h) $x = A/\sqrt{2}$.

Period and Frequency

As you know, the sine and cosine functions are periodic, with a period of 2π. The position, velocity, and acceleration for the oscillations of simple harmonic motion are described by a sine or cosine function, and adding a multiple of 2π to the argument of such a function does not change its value:

$$\sin(\omega t) = \sin(2\pi + \omega t) = \sin\left[\omega\left(\frac{2\pi}{\omega} + t\right)\right].$$

(To obtain the right-hand side of this equation, we rewrote the expression in the middle by multiplying and dividing 2π by ω and then factoring out the common factor ω.) We have dropped the index 0 on ω, because we are deriving a universal relationship that is valid for all angular speeds, not just for the particular situation of a mass on a spring.

The time interval over which a sinusoidal function repeats itself is the **period,** denoted by T. From the preceding equation for the periodicity of the sine function, we can see that

$$T = \frac{2\pi}{\omega}, \tag{14.7}$$

because $\sin(\omega t) = \sin[\omega(T + t)]$. The same argument works for the cosine function. In other words, replacing t by $t + T$ yields the same position, velocity, and acceleration vectors, as demanded by the definition of the period of simple harmonic motion.

The inverse of the period is the **frequency,** f:

$$f = \frac{1}{T}, \tag{14.8}$$

where f is the number of complete oscillations per unit time. For example, if $T = 0.2$ s, then 5 oscillations occur in 1 s, and $f = 1/T = 1/(0.2 \text{ s}) = 5.0 \text{ s}^{-1} = 5.0$ Hz. Substituting for T from equation 14.7 into equation 14.8 yields an expression for the angular speed in terms of the frequency:

$$f = \frac{1}{T} = \frac{1}{2\pi/\omega} = \frac{\omega}{2\pi}$$

or

$$\omega = 2\pi f. \tag{14.9}$$

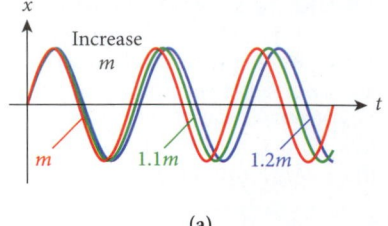

(a)

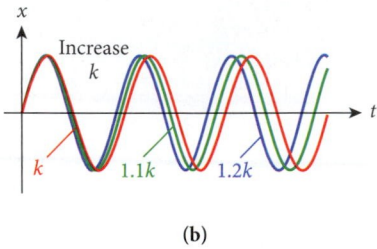

(b)

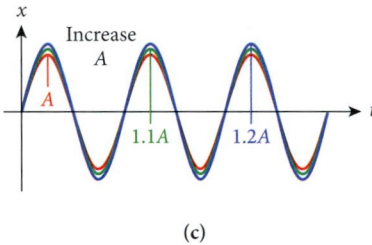

(c)

FIGURE 14.6 The effect on the simple harmonic motion of an object connected to a spring resulting from increasing (a) the mass, m; (b) the spring constant, k; and (c) the amplitude, A.

For a mass on a spring, we have the following for the period and the frequency:

$$T = \frac{2\pi}{\omega_0} = \frac{2\pi}{\sqrt{k/m}} = 2\pi\sqrt{\frac{m}{k}} \qquad (14.10)$$

and

$$f = \frac{\omega_0}{2\pi} = \frac{1}{2\pi}\sqrt{\frac{k}{m}}. \qquad (14.11)$$

Interestingly, the period does not depend on the amplitude of the motion.

Figure 14.6 illustrates the effect of changing the values of the variables affecting the simple harmonic motion of an object on a spring. The simple harmonic motion is described by equation 14.3 with $\theta_0 = 0$:

$$x = A\sin\left(\sqrt{\frac{k}{m}}\,t\right).$$

From Figure 14.6a, you can see that increasing the mass, m, increases the period of the oscillations. Figure 14.6b shows that increasing the spring constant, k, decreases the period of the oscillations. Figure 14.6c reinforces the earlier conclusion that increasing the amplitude, A, does not change the period of the oscillations.

EXAMPLE 14.2   Tunnel through the Moon

Suppose we could drill a tunnel straight through the center of the Moon, from one side to the other.

PROBLEM

If we released a steel ball of mass 5.0 kg from rest at one end of this tunnel, what would its motion be like?

SOLUTION

From Chapter 12, the magnitude of the gravitational force inside a spherical mass distribution of constant density is $F_g = mgr/R$, where m is the mass acted on, r is the distance from the center of the sphere, and R is the radius of the sphere, with $r < R$. This force points toward the center of the sphere, that is, in the direction opposite to the displacement. In other words, the gravitational force inside a homogeneous spherical mass distribution follows Hooke's Law, $F(x) = -kx$, with a "spring constant" of $k = mg/R$, where g is the gravitational acceleration experienced at the surface.

First, we need to calculate the gravitational acceleration at the surface on the Moon. Since the mass of the Moon is $7.35 \cdot 10^{22}$ kg (1.2% of the mass of Earth) and its radius is $1.735 \cdot 10^6$ m (27% of the radius of Earth), we find (see Chapter 12):

$$g_M = \frac{GM_M}{R_M^2} = \frac{\left(6.67 \cdot 10^{-11}\ \mathrm{m^3 kg^{-1} s^{-2}}\right)\left(7.35 \cdot 10^{22}\ \mathrm{kg}\right)}{\left(1.735 \cdot 10^6\ \mathrm{m}\right)^2} = 1.63\ \mathrm{m/s^2}.$$

The acceleration due to gravity at the Moon's surface is approximately a sixth of what it is on the surface of the Earth.

The appropriate equation of motion is equation 14.2:

$$x(t) = B\sin(\omega_0 t) + C\cos(\omega_0 t).$$

Releasing the ball from the surface of the Moon at time $t = 0$ implies that $x(0) = R_M = B\sin(0) + C\cos(0)$, or $C = R_M$. To determine the other initial condition, we use the velocity equation from Example 14.1: $v(t) = \omega_0 B\cos(\omega_0 t) - \omega_0 C\sin(\omega_0 t)$. The ball was released from rest, so $v(0) = 0 = \omega_0 B\cos(0) - \omega_0 C\sin(0) = \omega_0 B$, giving us $B = 0$. Thus, the equation of motion in this case becomes

$$x(t) = R_M\cos(\omega_0 t).$$

The angular speed of the oscillation is

$$\omega_0 = \sqrt{\frac{k}{m}} = \sqrt{\frac{g_M}{R_M}} = \sqrt{\frac{1.63\ \mathrm{m/s^2}}{1.735 \cdot 10^6\ \mathrm{m}}} = 9.69 \cdot 10^{-4}\ \mathrm{s^{-1}}.$$

Note that the mass of the steel ball turns out to be irrelevant. The period of the oscillation is

$$T = \frac{2\pi}{\omega_0} = 6485 \text{ s.}$$

The steel ball would arrive at the surface on the other side of the Moon 3243 s after it was released, and then oscillate back. Passing through the entire Moon in a little less than an hour would characterize an extremely efficient mode of transportation, especially since no power supply would be needed.

The velocity of the steel ball during its oscillation would be

$$v(t) = \frac{dx}{dt} = -\omega_0 R_M \sin(\omega_0 t).$$

The maximum velocity would be reached as the ball crossed the center of the Moon and would have the numerical value

$$v_{max} = \omega_0 R_M = \left(9.69 \cdot 10^{-4} \text{ s}^{-1}\right)\left(1.735 \cdot 10^6 \text{ m}\right) = 1680 \text{ m/s} = 3760 \text{ mph.}$$

If the tunnel were big enough, the same motion could be achieved by a vehicle holding one or more people, providing a very efficient means of transportation to the other side of the Moon, without the need for propulsion. During the entire journey, the people inside the vehicle would feel absolute weightlessness, because they would experience no supporting force from the vehicle! In fact, it would not even be necessary to use a vehicle to make this trip—wearing a space suit, you could just jump into the tunnel.

SOLVED PROBLEM 14.1 | Block on a Spring

PROBLEM

A 1.55-kg block sliding on a frictionless horizontal plane is connected to a horizontal spring with spring constant $k = 2.55$ N/m. The block is pulled to the right a distance $d = 5.75$ cm and released from rest. What is the block's velocity 1.50 s after it is released?

SOLUTION

THINK The block will undergo simple harmonic motion. We can use the given initial conditions to determine the parameters of the motion. With those parameters, we can calculate the velocity of the block at the specified time.

SKETCH Figure 14.7 shows the block attached to a spring and displaced a distance d from the equilibrium position.

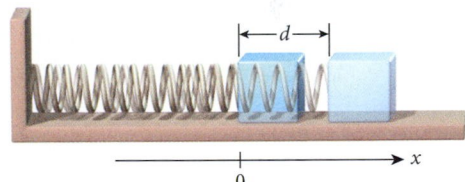

FIGURE 14.7 A block attached to a spring and displaced a distance d from the equilibrium position.

RESEARCH The position of the block undergoing simple harmonic motion is described by equation 14.3. The first initial condition is that at $t = 0$, the position is $x = d$. Thus, we can write

$$x(t = 0) = d = A\sin\left[(\omega_0 \cdot 0) + \theta_0\right] = A\sin\theta_0. \tag{i}$$

We have one equation with two unknowns. To get a second equation, we use the second initial condition: at $t = 0$, the velocity is zero. This leads to

$$v(t = 0) = 0 = \omega_0 A\cos\left[(\omega_0 \cdot 0) + \theta_0\right] = \omega_0 A\cos\theta_0. \tag{ii}$$

We now have two equations and two unknowns.

SIMPLIFY We can simplify equation (ii) to obtain $\cos\theta_0 = 0$, from which we get the phase angle, $\theta_0 = \pi/2$. Substituting this result into equation (i) gives us

$$d = A\sin\theta_0 = A\sin\left(\frac{\pi}{2}\right) = A.$$

Thus, we can write the velocity as a function of time as

$$v(t) = \omega_0 d\cos\left[\omega_0 t + \left(\frac{\pi}{2}\right)\right] = -\omega_0 d\sin(\omega_0 t).$$

– Continued

Since the angular speed is given by $\omega_0 = \sqrt{k/m}$, we obtain

$$v(t) = -\sqrt{\frac{k}{m}}\, d \sin\left(\sqrt{\frac{k}{m}}\, t\right).$$

CALCULATE Putting in the numerical values gives us

$$v(t = 1.50 \text{ s}) = -\sqrt{\frac{2.55 \text{ N/m}}{1.55 \text{ kg}}}(0.0575 \text{ m}) \sin\left[\sqrt{\frac{2.55 \text{ N/m}}{1.55 \text{ kg}}}(1.50 \text{ s})\right] = -0.06920005 \text{ m/s}.$$

ROUND We report our result to three significant figures:

$$v = -0.0692 \text{ m/s} = -6.92 \text{ cm/s}.$$

DOUBLE-CHECK As usual, it is a good idea to verify that the answer has appropriate units. This is the case here, because meters per second are units for velocity. The maximum speed that the block can attain is $v = \omega_0 d = 7.38$ cm/s. The magnitude of our result is less than that maximum speed, so it seems reasonable.

Relationship of Simple Harmonic Motion to Circular Motion

In Chapter 9, we analyzed circular motion along a path of constant radius, r. We saw that the x- and y-coordinates of such motion are given by the equations $x = r \cos\theta$ and $y = \sin\theta$. We also saw that we could express θ as a function of the time t, a constant angular velocity ω, and an initial angle θ_0 as $\theta = \omega t + \theta_0$. We can then express the x- and y-coordinates of circular motion as a function of time as $x(t) = r \cos(\omega t + \theta_0)$ and $y(t) = r \sin(\omega t + \theta_0)$. Figure 14.8a shows how the vector $\vec{r}(t)$ performs circular motion with constant angular velocity as a function of time and with an initial angle of $\theta_0 = 0$. The red arc segment shows the path of the tip of this radius vector. Figure 14.8b shows the projection of the radius vector onto the y-coordinate. You can clearly see that the motion of the y-component of the radius vector traces out a sine function. The motion of the x-component as a function of time is shown in Figure 14.8c, and it traces out a cosine function. These two projections of circular motion of constant angular velocity exhibit simple harmonic oscillations. This observation makes it clear that the frequency,

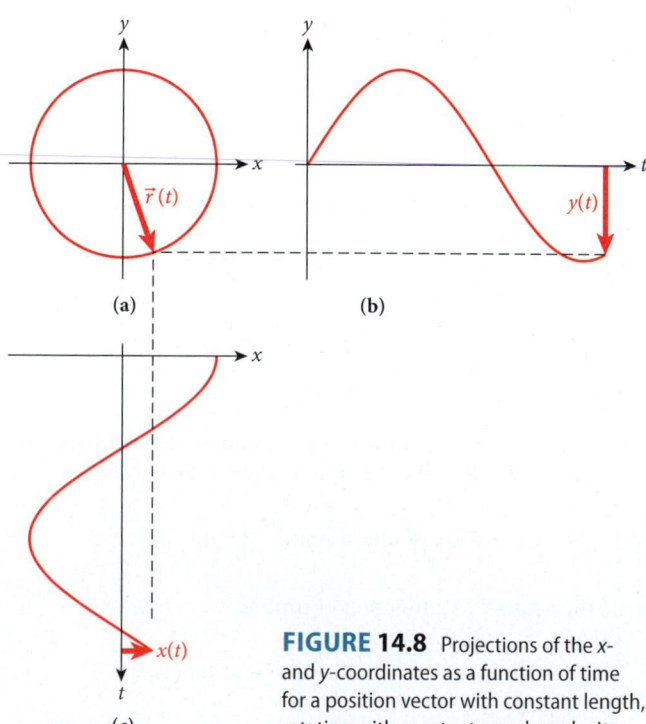

FIGURE 14.8 Projections of the x- and y-coordinates as a function of time for a position vector with constant length, rotating with constant angular velocity.

angular speed, and period defined here for oscillatory motion are identical to the quantities introduced in Chapter 9 for circular motion.

In Figure 14.8, the position vector, $\vec{r}(t)$, originates at $(x,y) = (0,0)$ and rotates with angular velocity ω. In Chapter 9, we saw that for circular motion the linear velocity, $\vec{v}(t)$, is tangent to the circle and the linear acceleration, $\vec{a}(t)$, always points toward the center of the circle. The vectors $\vec{v}(t)$ and $\vec{a}(t)$ can be moved so that they originate at $(x,y) = (0,0)$, as shown in Figure 14.9. Thus, Figure 14.9 shows that, for circular motion, the three vectors $\vec{r}(t)$, $\vec{v}(t)$, and $\vec{a}(t)$ rotate together with angular velocity ω. The linear velocity vector is always 90° out of phase with the position vector, and the linear acceleration vector is always 180° out of phase with the position vector. The projections of the linear velocity and linear acceleration vectors on the x- and y-axes correspond to the velocity and acceleration of an object undergoing simple harmonic motion.

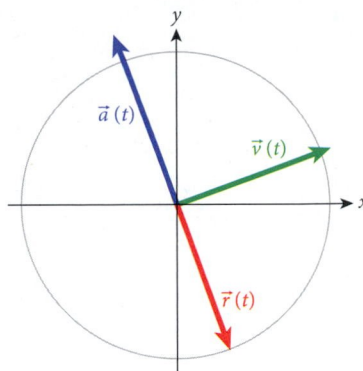

FIGURE 14.9 The position vector, linear velocity vector, and acceleration vector for circular motion.

14.2 Pendulum Motion

We are all familiar with another common oscillating system: the **pendulum.** In its ideal form, a pendulum consists of a thin string attached to a massive object that swings back and forth. The string is assumed to be massless, that is, of such small mass that the mass can be neglected. This assumption is, by the way, a very good approximation to the situation of a person on a swing.

Let's determine the equation of motion for any pendulum-like object. Shown in Figure 14.10 is a ball at the end of a string of length ℓ at an angle θ relative to the vertical. For small angles θ, the differential equation for the motion of the pendulum (which is derived in Derivation 14.1) is

$$\frac{d^2\theta}{dt^2} + \frac{g}{\ell}\theta = 0. \tag{14.12}$$

Equation 14.12 has the solution

$$\theta(t) = B\sin(\omega_0 t) + C\cos(\omega_0 t), \quad \text{with } \omega_0 = \sqrt{\frac{g}{\ell}}. \tag{14.13}$$

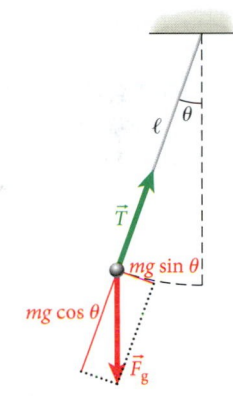

FIGURE 14.10 A pendulum with the force vectors due to gravity and string tension.

DERIVATION 14.1 | Pendulum Motion

The displacement, s, of the pendulum in Figure 14.10 is measured along the circumference of a circle with radius ℓ. This displacement can be obtained from the length of the string and the angle: $s = \ell\theta$. Because the length does not change with time, we can write the second derivative of the displacement as

$$\frac{d^2s}{dt^2} = \frac{d^2(\ell\theta)}{dt^2} = \ell\frac{d^2\theta}{dt^2}.$$

The next task is to find the angular acceleration, $d^2\theta/dt^2$, as a function of time. To do this, we need to determine the force that causes the acceleration. Two forces act on the ball: the force of gravity, $\vec{F}_g$, acting downward, and the force of the tension, $\vec{T}$, acting along the string. Since the string stays taut and does not stretch, the tension must compensate for the component of the gravitational force along the string. However, the tangential component of the gravitational force vector, $mg\sin\theta$, remains uncompensated for, as shown in Figure 14.10. The net force is always in the direction opposite to the displacement, s.

Using $F_{\text{net}} = ma$, we then obtain

$$m\frac{d^2s}{dt^2} = -mg\sin\theta \Rightarrow$$

$$\frac{d^2s}{dt^2} = -g\sin\theta \Rightarrow$$

$$\ell\frac{d^2\theta}{dt^2} + g\sin\theta = 0 \Rightarrow$$

$$\frac{d^2\theta}{dt^2} + \frac{g}{\ell}\sin\theta = 0.$$

– Continued

Self-Test Opportunity 14.3

What is the relationship between the radial component of the gravitational force vector and the string tension?

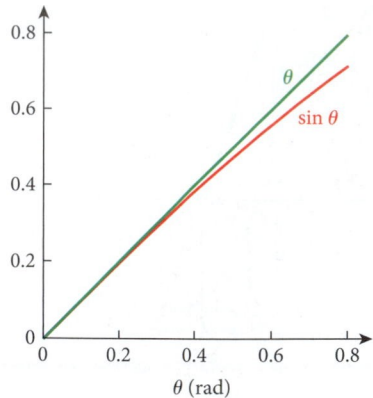

FIGURE 14.11 Graph indicating the error incurred using the small-angle approximation, $\sin\theta \approx \theta$, where θ is measured in radians.

This equation is difficult to solve without using the *small-angle approximation*, $\sin\theta \approx \theta$ (in radians). Doing so yields the desired differential equation 14.12. Figure 14.11 shows that the small-angle approximation introduces little error for $\theta < 0.5$ rad (approximately 30°). For these small angles, the motion of a pendulum is approximately simple harmonic motion because the restoring force is approximately proportional to θ.

To solve equation 14.12, we could go through exactly the same steps that led to the solution of the differential equation for the spring. However, since equations 14.1 and 14.12 are identical in form, we simply take the solution for spring motion and perform the appropriate substitutions: The angle θ takes the place of x, and g/ℓ takes the place of k/m. In this manner, we arrive at the solution without performing a second derivation.

Period and Frequency of a Pendulum

The period and frequency of a pendulum are related to the angular speed just as for a mass on a spring, but with the angular speed given by $\omega_0 = \sqrt{g/\ell}$:

$$T = \frac{2\pi}{\omega_0} = 2\pi\sqrt{\frac{\ell}{g}} \tag{14.14}$$

$$f = \frac{1}{2\pi}\sqrt{\frac{g}{\ell}}. \tag{14.15}$$

Thus, the solution of the equation of motion for the pendulum leads to harmonic motion, just as for the case of a mass on a spring. However, for the pendulum—*unlike* the spring—the frequency is independent of the mass of the oscillating object. This result means that two otherwise identical pendulums with different masses have the same period. The only way to change the period of a pendulum—other than by taking it to another planet or the Moon, where the gravitational acceleration is different—is by varying its length. (Of course, the gravitational acceleration also has small variations on Earth, for example, depending on altitude; so a pendulum's period is not *exactly* the same everywhere on Earth.)

To shorten the period of a pendulum by a factor of 2 requires shortening the length of the string by a factor of 4. This effect is illustrated in Figure 14.12, which shows the motion of two pendulums, with one four times longer than the other. In this sequence, the shorter pendulum completes two oscillations in the same time in which the longer pendulum completes one.

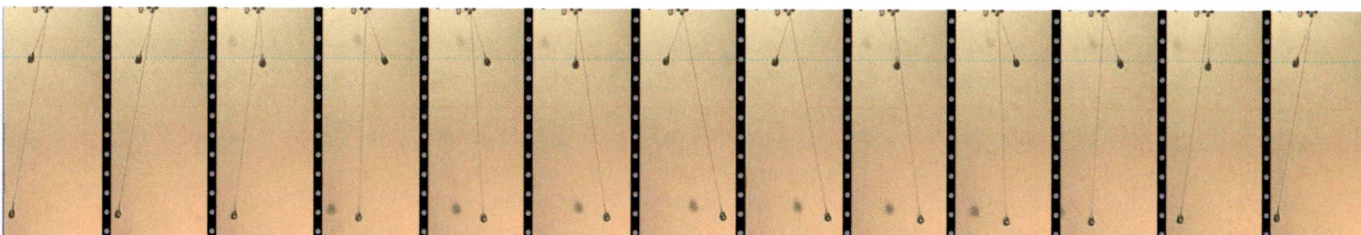

FIGURE 14.12 Video sequence of the oscillation of two pendulums whose lengths have the ratio 4:1.

EXAMPLE 14.3   Restricted Pendulum

PROBLEM

A pendulum of length 45.3 cm is hanging from the ceiling. Its motion is restricted by a peg that is sticking out of the wall 26.6 cm directly below the pivot point (Figure 14.13). What is the period of oscillation? (Note that specifying the mass is not necessary.)

SOLUTION

We must solve this problem separately for motion on the left side and on the right side of the peg. On the left side, the pendulum oscillates with its full length, $\ell_1 = 45.3$ cm. On the right side, the pendulum oscillates with a reduced length, $\ell_2 = 45.3$ cm – 26.6 cm = 18.7 cm. On each side, it performs exactly $\frac{1}{2}$ of a full oscillation. Thus, the overall period of the pendulum is $\frac{1}{2}$ of the sum of the two periods calculated with the different lengths:

$$T = \tfrac{1}{2}(T_1 + T_2) = \frac{2\pi}{2}\left(\sqrt{\frac{\ell_1}{g}} + \sqrt{\frac{\ell_2}{g}}\right) = \frac{\pi}{\sqrt{g}}\left(\sqrt{\ell_1} + \sqrt{\ell_2}\right)$$

$$= \frac{\pi}{\sqrt{9.81 \text{ m/s}^2}}\left(\sqrt{0.453 \text{ m}} + \sqrt{0.187 \text{ m}}\right)$$

$$= 1.11 \text{ s}.$$

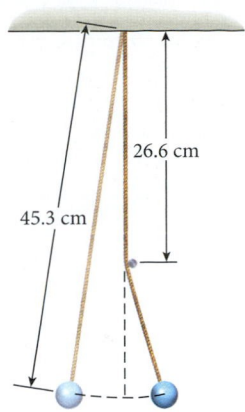

FIGURE 14.13 Restricted pendulum.

SOLVED PROBLEM 14.2 | Oscillation of a Thin Rod

PROBLEM
A long, thin rod swings about a frictionless pivot at one end. The rod has a mass of 2.50 kg and a length of 1.25 m. The bottom of the rod is pulled to the right until the rod makes an angle $\theta = 20.0°$ with respect to the vertical. The rod is then released from rest and oscillates in simple harmonic motion. What is the period of this motion?

SOLUTION
THINK We cannot apply the standard analysis to the motion of this pendulum because we must take into account the moment of inertia of the thin rod. This type of pendulum is called a *physical pendulum*. We can relate the torque due to the weight of the rod and acting at its center of gravity to the angular acceleration. From this relationship, we can get a differential equation that has a form similar to that of the equation we derived for a simple pendulum. By analogy to the simple pendulum, we can find the period of the oscillating thin rod.

SKETCH Figure 14.14 shows the oscillating thin rod.

RESEARCH Looking at Figure 14.14, we can see that the weight, mg, of the thin rod acts as a force producing a torque of magnitude τ, given by

$$\tau = I\alpha = -mgr\sin\theta, \qquad (i)$$

where I is the moment of inertia, α is the resulting angular acceleration, r is the moment arm, and θ is the angle between the force and the moment arm. The problem states that the initial angular displacement is 20°, so we can use the small-angle approximation, $\sin\theta \approx \theta$. We can then rewrite equation (i) as

$$I\alpha + mgr\theta = 0.$$

We now replace the angular acceleration with the second derivative of the angular displacement and divide through by the moment of inertia, which gives us

$$\frac{d^2\theta}{dt^2} + \frac{mgr}{I}\theta = 0.$$

This differential equation has the same form as equation 14.12; therefore, we can write its solution as

$$\theta(t) = B\sin(\omega_0 t) + C\cos(\omega_0 t), \quad \text{with } \omega_0 = \sqrt{\frac{mgr}{I}}.$$

SIMPLIFY We can write the period of oscillation of the thin rod:

$$T = \frac{2\pi}{\omega_0} = 2\pi\sqrt{\frac{I}{mgr}}.$$

This result holds in general for the period of a physical pendulum that has moment of inertia I and center of gravity at a distance r from the pivot point. In this case, $r = \ell/2$ and $I = \frac{1}{3}m\ell^2$, so we have

$$T = 2\pi\sqrt{\frac{\tfrac{1}{3}m\ell^2}{mg(\ell/2)}} = 2\pi\sqrt{\frac{2\ell}{3g}}.$$

– Continued

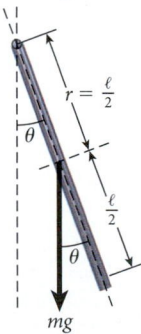

FIGURE 14.14 A physical pendulum consisting of a thin rod swinging from one end.

CALCULATE   Putting in the numerical values gives us

$$T = 2\pi \sqrt{\frac{2(1.25 \text{ m})}{3(9.81 \text{ m/s}^2)}} = 1.83128 \text{ s.}$$

ROUND   We report our result to three significant figures:

$$T = 1.83 \text{ s.}$$

DOUBLE-CHECK   If the mass m of the rod were concentrated at a point a distance ℓ from the pivot point, the rod would be a simple pendulum and the period would be

$$T = 2\pi \sqrt{\frac{\ell}{g}} = 2.24 \text{ s.}$$

The period of the thin rod is less than this value, which is reasonable because the mass of the thin rod is spread out over its length rather than being concentrated at the end.

As a second check, we start with $T = 2\pi\sqrt{I/mgr}$ and substitute the expression for the moment of inertia of a point mass moving around a point at a distance r away: $I = mr^2$. Taking $r = \ell$, we get

$$T = 2\pi \sqrt{\frac{m\ell^2}{mg\ell}} = 2\pi \sqrt{\frac{\ell}{g}},$$

which is the result for the period of a simple pendulum.

14.3   Work and Energy in Harmonic Oscillations

The concepts of work and kinetic energy were introduced in Chapter 5. In Chapter 6, we found the potential energy due to the spring force. In this section, we analyze the energies associated with motion of a mass on a spring. Then we'll see that almost all those results are applicable to a pendulum as well, but that we need to correct for the small-angle approximation used in solving the differential equation for pendulum motion.

Mass on a Spring

In Chapter 6, we did not have the equation for the displacement as a function of time at our disposal. However, we derived the potential energy stored in a spring, U_s, as

$$U_s = \tfrac{1}{2}kx^2,$$

where k is the spring constant and x is the displacement from the equilibrium position. We also saw that the total mechanical energy of a mass on a spring undergoing oscillations of amplitude A is given by

$$E = \tfrac{1}{2}kA^2.$$

Conservation of total mechanical energy means that this expression gives the value of the energy for any point in the oscillation. Using the law of energy conservation, we can write

$$\tfrac{1}{2}kA^2 = \tfrac{1}{2}mv^2 + \tfrac{1}{2}kx^2.$$

Then we can solve for the velocity as a function of the position:

$$v = \sqrt{(A^2 - x^2)\frac{k}{m}}. \tag{14.16}$$

The functions $v(t)$ and $x(t)$ given in equations 14.6 describe the oscillation over time of a mass on a spring. We can use these functions to verify the relationship between position and velocity expressed in equation 14.16, which we obtained from energy conservation. In this way, we can directly test whether this new result is consistent with what we found previously. First, we evaluate $A^2 - x^2$:

$$A^2 - x^2 = A^2 - A^2 \sin^2\left(\omega_0 t + \theta_0\right)$$
$$= A^2 \left[1 - \sin^2\left(\omega_0 t + \theta_0\right)\right]$$
$$= A^2 \cos^2\left(\omega_0 t + \theta_0\right).$$

Multiplying both sides by $\omega_0^2 = k/m$, we obtain

$$\frac{k}{m}\left(A^2 - x^2\right) = \frac{k}{m}A^2 \cos^2\left(\omega_0 t + \theta_0\right) = v^2.$$

By taking the square root of both sides of this equation, we obtain the same equation for the velocity that we got by applying energy considerations, equation 14.16.

Figure 14.15 illustrates the oscillations of the kinetic and potential energies as a function of time for a mass oscillating on a spring. As you can see, even though the potential and kinetic energies oscillate in time, their sum—the total mechanical energy—is constant. The kinetic energy always reaches its maximum wherever the displacement passes through zero, and the potential energy reaches its maximum for maximum elongation of the spring from the equilibrium position.

The position, velocity, and acceleration vectors at several positions for a block on a spring undergoing simple harmonic motion were shown in Figure 14.5. Figure 14.16 reproduces that figure but also shows the potential and kinetic energies (U and K) at each position. In Figure 14.16a, the block is released from rest at position $x = A$. At this point, all the energy in the system exists in the form of potential energy stored in the spring, and the block has no kinetic energy. The block then accelerates to the left. At $x = A/\sqrt{2}$, shown in Figure 14.16b, the potential energy stored in the spring and the kinetic energy of the block are equal. In Figure 14.16c, the block reaches the equilibrium position of the spring, $x = 0$, at which point the potential energy stored in the

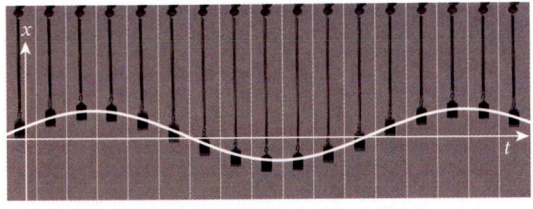

(a)

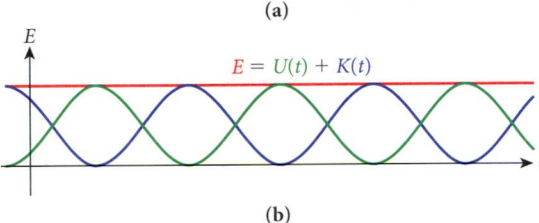

(b)

FIGURE 14.15 Harmonic oscillation of a mass on a spring: (a) displacement as a function of time (same as Figure 14.2); (b) potential and kinetic energies as a function of time on the same time scale.

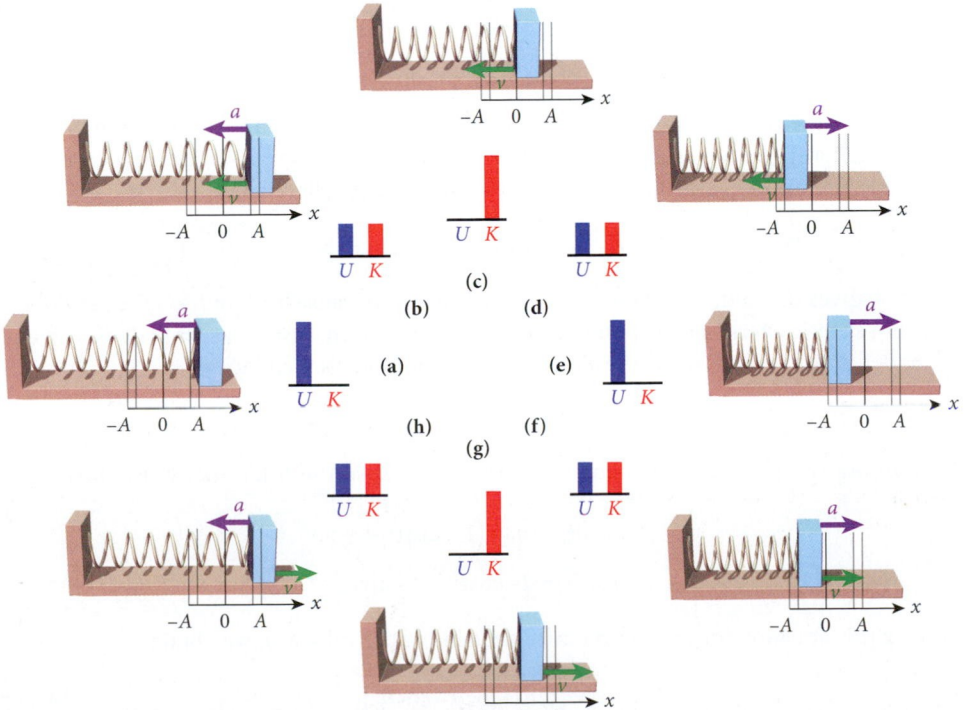

FIGURE 14.16 Position, velocity, and acceleration vectors for a mass on a spring (as in Figure 14.5), with the potential and kinetic energies at each position.

spring is zero and the block has its maximum kinetic energy. The block continues moving, through the equilibrium position, and in Figure 14.16d, it is located at $x = -A/\sqrt{2}$. Again the potential energy in the spring and the kinetic energy of the block are equal. In Figure 14.16e, the block is located at $x = -A$, where the potential energy stored in the spring is at its maximum and the kinetic energy of the block is zero. The block has returned to $x = -A/\sqrt{2}$ in Figure 14.16f, and the potential energy in the spring is again equal to the kinetic energy of the block. In Figure 14.16g, the block is located at $x = 0$, where the potential energy in the spring is zero and the kinetic energy of the block is at its maximum. The block continues through that equilibrium position and reaches $x = A/\sqrt{2}$ in Figure 14.16h, and again the potential energy in the spring is equal to the kinetic energy of the block. The block returns to its original position in Figure 14.16a.

Energy of a Pendulum

In Section 14.2, we saw that the time dependence of the deflection angle of a pendulum is $\theta(t) = \theta_0 \cos\left(\sqrt{g/\ell}\, t\right)$, if the initial conditions of maximum deflection and zero speed at time zero, $\theta(t = 0) = \theta_0$, are known. We can then find the linear velocity at each point in time by taking the derivative, $d\theta/dt = \omega$, and multiplying this angular velocity by the radius of the circle, ℓ:

$$v = \ell\,\frac{d\theta(t)}{dt} = -\theta_0\sqrt{g\ell}\,\sin\left(\sqrt{\frac{g}{\ell}}\,t\right).$$

Because $\sin\alpha = \sqrt{1 - \cos^2\alpha}$, we can insert the expression for the deflection angle as a function of time into this equation for the velocity and obtain the speed of the pendulum as a function of the angle:

$$|v| = |\theta_0|\sqrt{g\ell}\,\left|\sin\left(\sqrt{\frac{g}{\ell}}\,t\right)\right|$$

$$= |\theta_0|\sqrt{g\ell}\,\sqrt{1 - \cos^2\left(\sqrt{\frac{g}{\ell}}\,t\right)}$$

$$= |\theta_0|\sqrt{g\ell}\,\sqrt{1 - \frac{\theta^2}{\theta_0^2}} \Rightarrow$$

$$|v| = \sqrt{g\ell\left(\theta_0^2 - \theta^2\right)}. \tag{14.17}$$

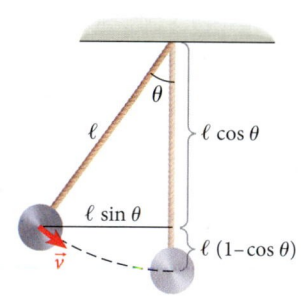

FIGURE 14.17 Geometry of a pendulum.

What is the energy of the pendulum? At time zero, the pendulum has only gravitational potential energy. We can assign the potential energy a value of zero at the lowest point of the arc. From Figure 14.17, the potential energy at maximum deflection, θ_0, is

$$E = K + U = 0 + U = mg\ell\left(1 - \cos\theta_0\right).$$

This also gives the value of the total mechanical energy, because by definition the pendulum has zero kinetic energy at the point of maximum deflection, just as a spring does. For any other deflection, the energy is the sum of kinetic and potential energies:

$$E = mg\ell\left(1 - \cos\theta\right) + \tfrac{1}{2}mv^2.$$

Combining the preceding two equations for E (because the total energy is conserved) leads to

$$mg\ell\left(1 - \cos\theta_0\right) = mg\ell\left(1 - \cos\theta\right) + \tfrac{1}{2}mv^2 \Rightarrow$$

$$mg\ell\left(\cos\theta - \cos\theta_0\right) = \tfrac{1}{2}mv^2.$$

Solving this equation for the speed (absolute value of the velocity), we obtain

$$|v| = \sqrt{2g\ell\left(\cos\theta - \cos\theta_0\right)}. \tag{14.18}$$

Equation 14.18 is the exact expression for the speed of the pendulum at any angle θ, which we obtained quite straightforwardly and without the need to solve a differential equation. This equation, however, does not match equation 14.17, which we obtained from the solution of a differential equation. However, remember that we used the small-angle approximation earlier for solving a differential equation. For small angles, we can approximate $\cos\theta \approx 1 - \frac{1}{2}\theta^2 + \cdots$, and so equation 14.17 is a special case of equation 14.18. Our derivation of equation 14.18 using energy considerations did not make use of the small-angle approximation, and thus this equation is valid for all deflection angles. However, the differences between the values obtained with the two equations are quite small, as you can see from Figure 14.18.

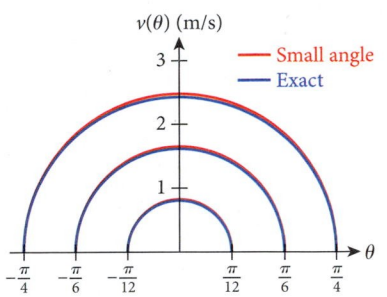

FIGURE 14.18 Comparison of the speeds of a pendulum (of length 1 m) as a function of the deflection angle, calculated with the small-angle approximation (red curves) and the exact formula based on conservation of energy (blue curves). The three sets of curves represent initial angles of 15°, 30°, and 45° (from inside to outside).

EXAMPLE 14.4 · Speed on a Trapeze

PROBLEM
A circus trapeze artist starts her motion on the trapeze from rest with the rope at an angle of 45° relative to the vertical. The rope has a length of 5.00 m. What is her speed at the lowest point in her trajectory?

SOLUTION
The initial condition is $\theta_0 = 45° = \pi/4$ rad. We are interested in finding $v(\theta = 0)$. Applying conservation of energy, we use equation 14.18:

$$v(\theta) = \sqrt{2g\ell(\cos\theta - \cos\theta_0)}.$$

Inserting the numbers, we obtain

$$v(0) = \sqrt{2(9.81 \text{ m/s}^2)(5.00 \text{ m})\left(1 - \sqrt{\tfrac{1}{2}}\right)} = 5.36 \text{ m/s}.$$

With the small-angle approximation, for comparison, we get $v(0) = \theta_0\sqrt{g\ell} = 5.50$ m/s. This is close to the exact result but would not be precise enough for many applications.

14.4 · Damped Harmonic Motion

Springs and pendulums do not go on oscillating forever; after some time interval, they come to rest. Thus, some force must be present that slows them down. This speed-diminishing effect is called **damping**. An example is shown in Figure 14.19, which shows a mass on a spring oscillating in water, which provides resistance to the motion of the mass and damps it. The center of the mass follows the red curve superimposed on the video frames in Figure 14.19b. The red curve is the product of a plot of simple harmonic motion (orange curve) and an exponentially falling function (yellow curve).

In order to quantify the effects of damping, we need to consider the damping force. As we have just seen, a force like the spring force, F_s, which depends linearly on position, does not accomplish damping. However, a force that depends on velocity does. For speeds

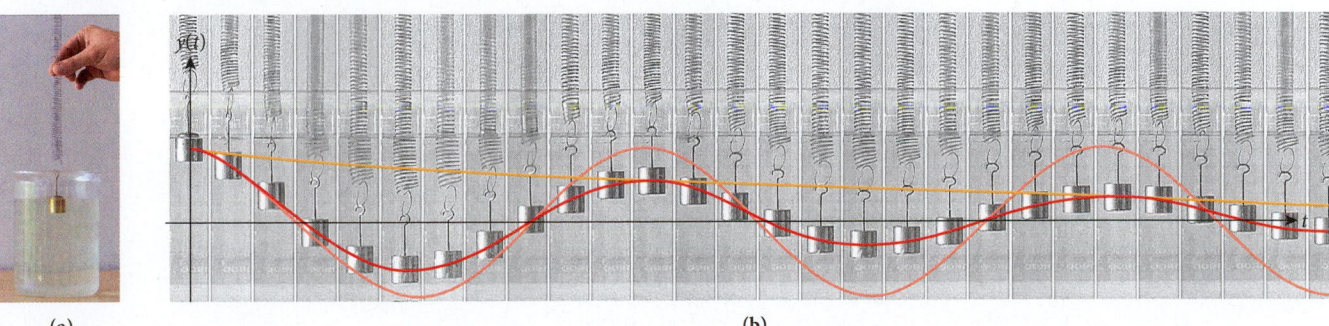

(a)   (b)

FIGURE 14.19 Example of damped harmonic motion: mass on a spring oscillating in water. (a) Initial conditions. (b) Video frames showing the motion of the mass after it is released.

that are not too large, the damping is given by a drag force of the form $F_\gamma = -bv$, where b is a constant, known as the *damping constant*, and $v = dx/dt$ is the velocity. With this damping force, F_γ, we can write the differential equation describing damped harmonic motion:

$$ma = F_\gamma + F_s$$

$$m\frac{d^2x}{dt^2} = -b\frac{dx}{dt} - kx \Rightarrow$$

$$\frac{d^2x}{dt^2} + \frac{b}{m}\frac{dx}{dt} + \frac{k}{m}x = 0. \tag{14.19}$$

The solution of this equation depends on how large the damping force is relative to the linear restoring force responsible for the harmonic motion. This ratio establishes three general cases: small damping, large damping, and the intermediate case.

Small Damping

For small values of the damping constant b ("small" will be specified below), the solution of equation 14.19 is

$$x(t) = Be^{-\omega_\gamma t}\sin(\omega't) + Ce^{-\omega_\gamma t}\cos(\omega't). \tag{14.20}$$

The coefficients B and C are determined by the initial conditions, that is, the position, x_0, and velocity, v_0, at time $t = 0$:

$$B = \frac{v_0 + x_0\omega_\gamma}{\omega'} \quad \text{and} \quad C = x_0.$$

The angular speeds in this solution are given by

$$\omega_\gamma = \frac{b}{2m}$$

$$\omega' = \sqrt{\omega_0^2 - \omega_\gamma^2} = \sqrt{\frac{k}{m} - \left(\frac{b}{2m}\right)^2}.$$

This solution is valid for all values of the damping constant, b, for which the argument of the square root that determines ω' remains positive:

$$b < 2\sqrt{mk}. \tag{14.21}$$

This is the condition for small damping, also known as *underdamping*.

DERIVATION 14.2 / Small Damping

We can show that equation 14.20 satisfies the differential equation for damped harmonic motion (equation 14.19) in the limit of small damping. We'll start with an assumed solution, which mathematicians call an *Ansatz* (the German word for "attempt"). In order to derive the result from the *Ansatz*, no further knowledge of differential equations is needed; all that is required is taking derivatives.

Ansatz:

$$x(t) = Ce^{-\omega_\gamma t}\cos(\omega't)$$

$$\Rightarrow \frac{dx}{dt} = -\omega_\gamma Ce^{-\omega_\gamma t}\cos(\omega't) - \omega'Ce^{-\omega_\gamma t}\sin(\omega't)$$

$$\Rightarrow \frac{d^2x}{dt^2} = \left[\omega_\gamma^2 - (\omega')^2\right]Ce^{-\omega_\gamma t}\cos(\omega't) + 2\omega_\gamma\omega'Ce^{-\omega_\gamma t}\sin(\omega't).$$

We insert these expressions into equation 14.19:

$$\left[\omega_\gamma^2 - (\omega')^2\right]Ce^{-\omega_\gamma t}\cos(\omega't) + 2\omega_\gamma\omega'Ce^{-\omega_\gamma t}\sin(\omega't)$$

$$+ \frac{b}{m}\left[-\omega_\gamma Ce^{-\omega_\gamma t}\cos(\omega't) - \omega'Ce^{-\omega_\gamma t}\sin(\omega't)\right]$$

$$+ \frac{k}{m}\left[Ce^{-\omega_\gamma t}\cos(\omega't)\right] = 0.$$

Now we rearrange terms:

$$\left[\omega_\gamma^2 - (\omega')^2 - \frac{b}{m}\omega_\gamma + \frac{k}{m}\right]Ce^{-\omega_\gamma t}\cos(\omega't) + \left(2\omega_\gamma\omega' - \frac{b}{m}\omega'\right)Ce^{-\omega_\gamma t}\sin(\omega't) = 0.$$

This equation can hold for all times t only if the coefficients in front of the sine and the cosine functions are zero. We thus have two conditions:

$$2\omega_\gamma\omega' - \frac{b}{m}\omega' = 0 \Rightarrow \omega_\gamma = \frac{b}{2m}$$

and

$$\omega_\gamma^2 - (\omega')^2 - \frac{b}{m}\omega_\gamma + \frac{k}{m} = 0.$$

To simplify the second condition, we use the first condition, $\omega_\gamma = b/(2m)$, and the expression we obtained earlier for the angular speed for simple harmonic motion, $\omega_0 = \sqrt{k/m}$. We then obtain for the second condition:

$$\left[-(\omega')^2 - \omega_\gamma^2 + \omega_0^2\right] = 0 \Rightarrow \omega' = \sqrt{\omega_0^2 - \omega_\gamma^2}.$$

We could go through the same steps to show that $Be^{-\omega_\gamma t}\sin(\omega't)$ is also a valid solution and could further show that these two solutions are the only possible solutions (but we will skip these demonstrations).

In the derivations in this chapter that involve differential equations we have proceeded in the same manner. This method is a general approach to the solution of problems involving differential equations: Choose a trial solution, and then adjust parameters and make other changes based on the results obtained when the trial solution is substituted into the differential equation.

Just as for the case of undamped motion, we can also write the solution in terms of one amplitude, A, and a phase shift determined by θ_0 instead of using the coefficients B and C for the sine and cosine functions as in equation 14.20:

$$x(t) = Ae^{-\omega_\gamma t}\sin(\omega't + \theta_0). \tag{14.22}$$

Now let's look at a plot of equation 14.20, which describes weakly damped harmonic motion (Figure 14.20). The sine and cosine functions describe the oscillating behavior. Their combination results in another sine function, with a phase shift. The exponential functions that multiply them can be thought of as reducing the amplitude over time. Thus, the oscillations show exponentially decaying amplitude. The angular speed, ω', of oscillation is reduced relative to the angular speed of oscillation without damping, ω_0. The graph in Figure 14.20 was generated with $k = 11.00$ N/m, $m = 1.800$ kg, and $b = 0.500$ kg/s. These parameters result in an angular speed $\omega' = 2.468$ s^{-1}, compared to an angular speed $\omega_0 = 2.472$ s^{-1} for the case without damping. The amplitude is $A = 5$ cm, and $\theta_0 = 1.6$. The dark blue curve is the plot of the function, and the two light blue curves show the exponential envelope, within which the amplitude is reduced as a function of time.

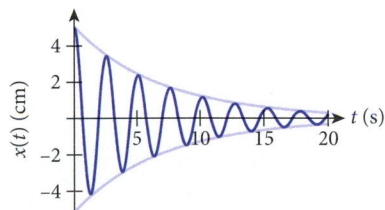

FIGURE 14.20 Position versus time for a weakly damped harmonic oscillator.

EXAMPLE 14.5   Bungee Jumping

A bridge over a deep valley is ideal for bungee jumping. The first part of a bungee jump consists of a free fall of the same length as the unstretched rope. Suppose the height of the bridge is 50.0 m. A 30.0-m-long bungee rope is used and is stretched 5.00 m by the weight of a 70.0-kg person. Thus, the equilibrium length of the bungee rope is 35.0 m. This bungee rope has been found to have a damping angular speed of $\omega_\gamma = 0.300$ s^{-1}.

PROBLEM
Describe the vertical motion of the bungee jumper as a function of time.

– Continued

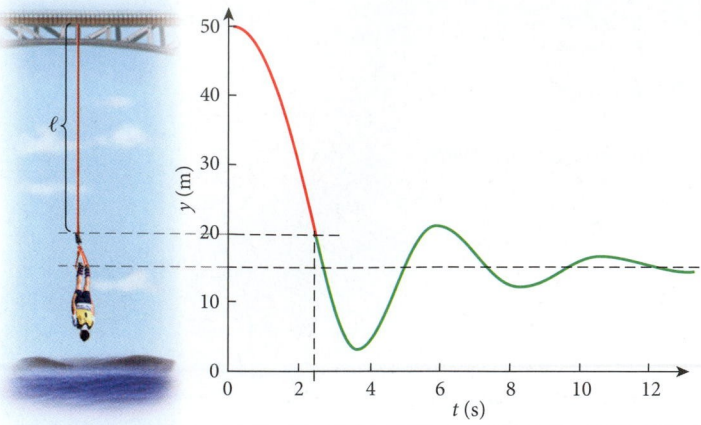

FIGURE 14.21 Idealized vertical motion as a function of time for a bungee jump.

SOLUTION

From the top of the bridge, the jumper experiences free fall for the first 30.0 m of her descent, shown by the red part of the trajectory in Figure 14.21. Once she falls the length of the rope, 30.0 m, she reaches a velocity of $v_0 = -\sqrt{2g\ell} = -24.26$ m/s. This part of the trajectory takes time $t = \sqrt{2\ell/g} = 2.47$ s.

She then enters a damped oscillation $y(t)$ about the equilibrium position, $y_e = 50.0$ m $- 35.0$ m $= 15.0$ m, with the initial displacement $y_0 = 5.00$ m. We can calculate the angular speed of her vertical motion from knowing that the rope is stretched by 5.00 m due to the 70.0-kg mass and using $mg = k(5.00$ m$)$:

$$\omega_0 = \sqrt{\frac{k}{m}} = \sqrt{\frac{g}{\ell_0 - \ell}} = \sqrt{\frac{9.81 \text{ m/s}^2}{5.00 \text{ m}}} = 1.40 \text{ s}^{-1}.$$

Because $\omega_\gamma = 0.300$ s^{-1} was specified, we have $\omega' = \sqrt{\omega_0^2 - \omega_\gamma^2} = 1.37$ s^{-1}. The bungee jumper oscillates according to equation 14.20, with coefficients $B = y_0 = 5.00$ m and $C = (v_0 + y_0\omega_\gamma)/\omega' = -16.6$ m/s. This motion is represented by the green part of the curve in Figure 14.21.

DISCUSSION

The approximation of damped harmonic motion for the final part of this bungee jump is not quite accurate. If you examine Figure 14.21 carefully, you see that the jumper rises above 20 m for approximately 1 s, starting at approximately 5.5 s. During this interval, the bungee cord is not stretched and so this part of the motion is not sinusoidal. Instead, the bungee jumper is again in free-fall motion. However, for the present qualitative discussion, the approximation of damped harmonic motion is acceptable. If you plan to bungee jump (strongly discouraged in light of past fatalities!), you should always test the setup first with an object of weight equal to or greater than yours.

Large Damping

What happens when the condition for small damping, $b < 2\sqrt{mk}$, is no longer fulfilled? The argument of the square root that determines ω' will be smaller than zero, and so the *Ansatz* we used for small damping does not work any more. The solution of the differential equation for damped harmonic oscillations (equation 14.19) for the case where $b > 2\sqrt{mk}$, a situation called **overdamping**, is

$$x(t) = Be^{-\left(\omega_\gamma + \sqrt{\omega_\gamma^2 - \omega_0^2}\right)t} + Ce^{-\left(\omega_\gamma - \sqrt{\omega_\gamma^2 - \omega_0^2}\right)t}, \quad \text{for } b > 2\sqrt{mk}, \tag{14.23}$$

with the amplitudes given by

$$B = \frac{1}{2}x_0 - \frac{x_0\omega_\gamma + v_0}{2\sqrt{\omega_\gamma^2 - \omega_0^2}}, \quad C = \frac{1}{2}x_0 + \frac{x_0\omega_\gamma + v_0}{2\sqrt{\omega_\gamma^2 - \omega_0^2}}.$$

Again, we have for the angular speeds

$$\omega_\gamma = \frac{b}{2m}, \quad \omega_0 = \sqrt{\frac{k}{m}}.$$

The coefficients B and C are determined by the initial conditions.

The solution given by equation 14.23 has no oscillations but instead consists of two exponential terms. The term with the argument $-\omega_\gamma t + \sqrt{\omega_\gamma^2 - \omega_0^2}\,t$ governs the long-time behavior of the system because it decays more slowly than the other term.

An example of the motion of an overdamped oscillator is graphed in Figure 14.22. The graph was generated for $k = 1.0$ N/m, $m = 1.0$ kg, and $b = 3.0$ kg/s, such that $b\left(= 3.0 \text{ kg/s}\right) > 2\sqrt{mk}\left(= 2 \text{ kg/s}\right)$. These parameters result in an angular speed of $\omega_\gamma = 1.5$ s^{-1}, compared to an angular speed of $\omega_0 = 1.0$ s^{-1} for the case without damping. The initial

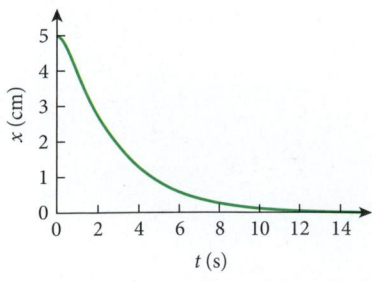

FIGURE 14.22 Position versus time for an overdamped oscillator.

displacement was $x_0 = 5.0$ cm, and the system was assumed to be released from rest. The displacement goes to zero without any oscillations.

The coefficients B and C can have opposite signs, so the value of x determined from equation 14.23 can change sign one time at most. Physically, this sign change corresponds to the situation in which the oscillator receives a large initial velocity toward the equilibrium position. In that case, the oscillator can overshoot the equilibrium position and approach it from the other side.

Critical Damping

Because we have already covered the solutions of equation 14.19 for $b < 2\sqrt{mk}$ and $b > 2\sqrt{mk}$, you might be tempted to think that we can obtain the solution for $b = 2\sqrt{mk}$ via some limit process or some interpolation between the two solutions. This case, however, is not quite so straightforward, and we have to employ a slightly different *Ansatz*. The solution for the case where $b = 2\sqrt{mk}$, the solution known as **critical damping**, is given by

$$x(t) = Be^{-\omega_\gamma t} + tCe^{-\omega_\gamma t}, \quad \text{for } b = 2\sqrt{mk}, \tag{14.24}$$

where the amplitude coefficients have the values

$$B = x_0, \quad C = v_0 + x_0\omega_\gamma.$$

The angular speed in this solution is still $\omega_\gamma = b/(2m)$.

Let's look at an example in which the coefficients B and C are determined from the initial conditions for the damped harmonic motion.

EXAMPLE 14.6 | Damped Harmonic Motion

PROBLEM

A spring with the spring constant $k = 1.00$ N/m has an object of mass $m = 1.00$ kg attached to it, which moves in a medium with damping constant $b = 2.00$ kg/s. The object is released from rest at $x = +5.00$ cm from the equilibrium position. Where will it be after 1.75 s?

SOLUTION

First, we have to decide which of the three types of damping applies in this situation. To find out, we calculate $2\sqrt{mk} = 2\sqrt{(1.00 \text{ kg})(1.00 \text{ N/m})} = 2.00$ kg/s, which happens to be exactly equal to the value that was given for the damping constant, b. Therefore, this motion is critically damped. We can also calculate the damping angular speed: $\omega_\gamma = b/(2m) = 1.00 \text{ s}^{-1}$.

The problem statement gives the initial conditions: $x_0 = +5.00$ cm and $v_0 = 0$. We determine the constants B and C from these initial conditions (expressions for these constants). We use equation 14.24 for critical damping:

$$x(t) = Be^{-\omega_\gamma t} + tCe^{-\omega_\gamma t}$$

$$\Rightarrow x(0) = Be^{-\omega_\gamma 0} + 0 \cdot Ce^{-\omega_\gamma 0} = B.$$

Next, we take the time derivative:

$$v = \frac{dx}{dt} = -\omega_\gamma Be^{-\omega_\gamma t} + C(1 - \omega_\gamma t)e^{-\omega_\gamma t}$$

$$\Rightarrow v(0) = -\omega_\gamma Be^{-\omega_\gamma 0} + C\left[1 - (\omega_\gamma \cdot 0)\right]e^{-\omega_\gamma 0} = -\omega_\gamma B + C.$$

According to the problem statement, $v_0 = 0$, which yields $C = \omega_\gamma B$, and $x_0 = +5.00$ cm, which determines B. We have already calculated $\omega_\gamma = 1.00 \text{ s}^{-1}$. Thus, we have

$$x(t) = (5.00 \text{ cm})(1 + \omega_\gamma t)e^{-\omega_\gamma t}$$

$$= (5.00 \text{ cm})(1 + (1.00 \text{ s}^{-1})t)e^{-(1.00/\text{s})t}.$$

We now calculate the position of the mass after 1.75 s:

$$x(1.75 \text{ s}) = (5.00 \text{ cm})(1 + 1.75)e^{-1.75} = 2.39 \text{ cm}.$$

– Continued

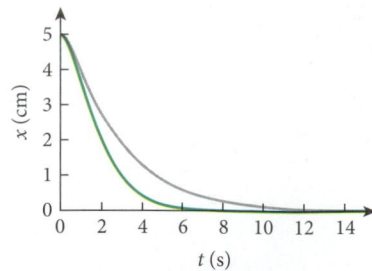

FIGURE 14.23 The displacement of a critically damped oscillator versus time. The gray line reproduces the curve for the overdamped oscillator from Figure 14.22.

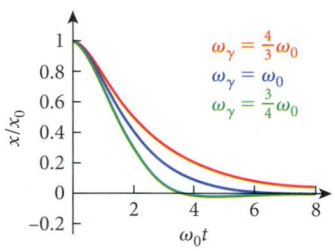

FIGURE 14.24 Position versus time for underdamped motion (green line), overdamped motion (red line), and critically damped motion (blue line).

FIGURE 14.25 Shock absorbers on the front wheel of a motorcycle.

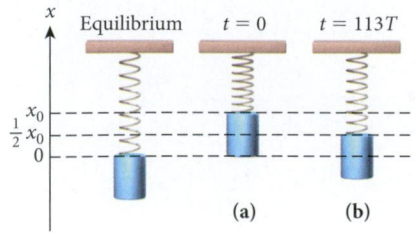

FIGURE 14.26 Position of the oscillating block at (a) $t = 0$ and (b) $t = 113T$.

Figure 14.23 shows a plot of the displacement of this critically damped oscillator as a function of time. It also includes the curve from Figure 14.22 for the overdamped oscillator, which has the same mass and spring constant but a larger damping constant, $b = 3.0$ kg/s. Note that the critically damped oscillator approaches zero displacement more quickly than the overdamped oscillator does.

To compare underdamped, overdamped, and critically damped motion, Figure 14.24 graphs position against angular speed times time for each type of motion. In all three cases, the oscillator started from rest at time $t = 0$. The green line represents the underdamped case; you can see that oscillation continues. The red line reflects large damping above the critical value, and the blue line reflects critical damping.

An engineering solution providing a damped oscillation that returns as fast as possible to the equilibrium position and has no oscillations needs to use the condition of critical damping. Examples of such an application are the shock absorbers in cars, motorcycles (Figure 14.25), and bicycles. For maximum performance, they need to work at the critical damping limit or just slightly below. As they get worn, the damping effect becomes weaker, and the shock absorbers provide only small damping. This results in a "bouncing" sensation for riders, indicating that it is time to change the vehicle's shock absorbers.

SOLVED PROBLEM 14.3 | Determining the Damping Constant

PROBLEM

A 1.75-kg block is connected to a vertical spring with spring constant $k = 3.50$ N/m. The block is pulled upward a distance $d = 7.50$ cm and released from rest. After 113 complete oscillations, the amplitude of the oscillation is half the original amplitude. The damping of the block's motion is proportional to the speed. What is the damping constant, b?

SOLUTION

THINK Because the block executes 113 oscillations and the amplitude decreases by a factor of 2, we know we are dealing with small damping. After an integral number of oscillations, the cosine term will be 1 and the sine term will be zero in equation 14.20. Knowing that the amplitude has decreased by a factor of 2 after 113 oscillations, we can get an expression for ω_γ in terms of the time required to complete 113 oscillations, from which we can obtain the damping constant.

SKETCH Figure 14.26 shows the position of the block at $t = 0$ and at $t = 113T$, where T is the period of oscillation.

RESEARCH Because we know that we are dealing with small damping, we can apply equation 14.20:

$$x(t) = Be^{-\omega_\gamma t}\cos(\omega' t) + Ce^{-\omega_\gamma t}\sin(\omega' t).$$

At $t = 0$, we know that $x = x_0$ and so $B = x_0$. After 113 complete oscillations, the cosine term will be 1, the sine term will be zero, and $x = x_0/2$. Thus, we can write

$$\frac{x_0}{2} = \left(x_0 e^{-\omega_\gamma(113T)} \cdot 1\right) + 0, \tag{i}$$

where $T = 2\pi/\omega'$ is the period of oscillation. We can express equation (i) as

$$0.5 = e^{-\omega_\gamma\left[113(2\pi/\omega')\right]}. \tag{ii}$$

Taking the natural logarithm of both sides of equation (ii) gives us

$$\ln 0.5 = -\omega_\gamma\left[113\left(\frac{2\pi}{\omega'}\right)\right].$$

Remembering that $\omega' = \sqrt{\omega_0^2 - \omega_\gamma^2}$, where $\omega_0 = \sqrt{k/m}$ (k is the spring constant and m is the mass), we can write

$$-\frac{\ln 0.5}{113(2\pi)} = \frac{\omega_\gamma}{\omega'} = \frac{\omega_\gamma}{\sqrt{\omega_0^2 - \omega_\gamma^2}}. \tag{iii}$$

We can square equation (iii) to obtain

$$\frac{\omega_\gamma^2}{\omega_0^2 - \omega_\gamma^2} = \left(-\frac{\ln 0.5}{113(2\pi)}\right)^2. \tag{iv}$$

SIMPLIFY In order to save a little effort in writing the constants on the right-hand side of equation (iv), we use the abbreviation

$$c^2 = \left(-\frac{\ln 0.5}{113(2\pi)}\right)^2.$$

Then we can write

$$\omega_\gamma^2 = \left(\omega_0^2 - \omega_\gamma^2\right)c^2 = \omega_0^2 c^2 - \omega_\gamma^2 c^2.$$

We can now solve for ω_γ:

$$\omega_\gamma = \sqrt{\frac{\omega_0^2 c^2}{1 + c^2}}.$$

Remembering that $\omega_\gamma = b/2m$, we can write an expression for the damping constant, b:

$$b = 2m\sqrt{\frac{\omega_0^2 c^2}{1 + c^2}}.$$

Substituting $\omega_0 = \sqrt{k/m}$, we obtain

$$b = 2m\sqrt{\frac{(k/m)c^2}{1 + c^2}} = 2\sqrt{\frac{mkc^2}{1 + c^2}}.$$

CALCULATE We first put in the numerical values to find c^2:

$$c^2 = \left(-\frac{\ln 0.5}{113(2\pi)}\right)^2 = 9.53091 \cdot 10^{-7}.$$

We can now calculate the damping constant:

$$b = 2\sqrt{\frac{(1.75 \text{ kg})(3.50 \text{ N/m})(9.53091 \cdot 10^{-7})}{1 + 9.53091 \cdot 10^{-7}}} = 0.00483226 \text{ kg/s}.$$

ROUND We report our result to three significant figures:

$$b = 4.83 \cdot 10^{-3} \text{ kg/s}.$$

DOUBLE-CHECK We can check that the calculated damping constant is consistent with small damping by verifying that $b < 2\sqrt{mk}$:

$$4.83 \cdot 10^{-3} \text{ kg/s} < 2\sqrt{(1.75 \text{ kg})(3.50 \text{ N/m})} = 4.95 \text{ kg/s}.$$

Thus, the damping constant is consistent with small damping.

Energy Loss in Damped Oscillations

Damped oscillations are the result of the velocity-dependent (and thus nonconservative) damping force, $F_y = -bv$, and are described by the differential equation (compare to equation 14.19)

$$\frac{d^2 x}{dt^2} + 2\omega_\gamma \frac{dx}{dt} + \omega_0^2 x = 0, \tag{14.25}$$

where $\omega_0 = \sqrt{k/m}$ (k is the spring constant and m is the mass) is the angular speed of the harmonic oscillation without damping and $\omega_\gamma = b/2m$ is the damping angular speed. The nonconservative damping force must result in a loss of total mechanical energy. We'll examine the energy loss for small damping only, because this case is by far of the greatest technological importance. However, large and critical damping can be treated in an analogous manner.

For small damping, we saw that the displacement as a function of time can be written in terms of a single sinusoidal function with a phase shift (see equation 14.22):

$$x(t) = A e^{-\omega_\gamma t} \sin\left(\omega' t + \theta_0\right). \tag{14.26}$$

This solution can be understood as a sinusoidal oscillation in time with angular speed $\omega' = \sqrt{\omega_0^2 - \omega_\gamma^2}$ and with exponentially decreasing amplitude.

We know that the damping force is nonconservative, and so we can be sure that mechanical energy is lost over time. Since the energy of harmonic oscillation is proportional to the square of the amplitude, $E = \frac{1}{2}kA^2$, we might be led to conclude that the energy decreases smoothly and exponentially for a damped oscillation. However, this is not quite the case, which we can demonstrate as follows.

First, we can find the velocity as a function of time by taking the time derivative of equation 14.26:

$$v(t) = \omega' A e^{-\omega_\gamma t} \cos(\omega' t + \theta_0) - \omega_\gamma A e^{-\omega_\gamma t} \sin(\omega' t + \theta_0).$$

Next, we know the potential energy stored in the spring as a function of time, $U = \frac{1}{2}kx^2(t)$, as well as the kinetic energy due to the motion of the mass, $K = \frac{1}{2}mv^2(t)$. We can plot these energies and the displacement and velocity as functions of time, as shown in the graphs in Figure 14.27, where $k = 11.00$ N/m, $m = 1.800$ kg, and $b = 0.500$ kg/s. These parameters result in an angular speed of $\omega' = 2.468$ s^{-1}, compared to an angular speed of $\omega_0 = 2.472$ s^{-1} for the case without damping, and a damping angular speed of $\omega_\gamma = 0.139$ s^{-1}. The amplitude is $A = 5$ cm, and $\theta_0 = 1.6$.

Figure 14.27 shows that the total energy does not decrease in a smooth exponential way but instead declines in a stepwise fashion. This energy loss can be represented in a straightforward way by writing an expression for the power, that is, the rate of energy loss:

$$\frac{dE}{dt} = -bv^2 = vF_\gamma.$$

FIGURE 14.27 (a) Potential energy (red), kinetic energy (green), and total energy (blue). (b) Displacement (red) and velocity (green) for a weakly damped harmonic oscillator. The five thin vertical lines mark one complete oscillation period, T.

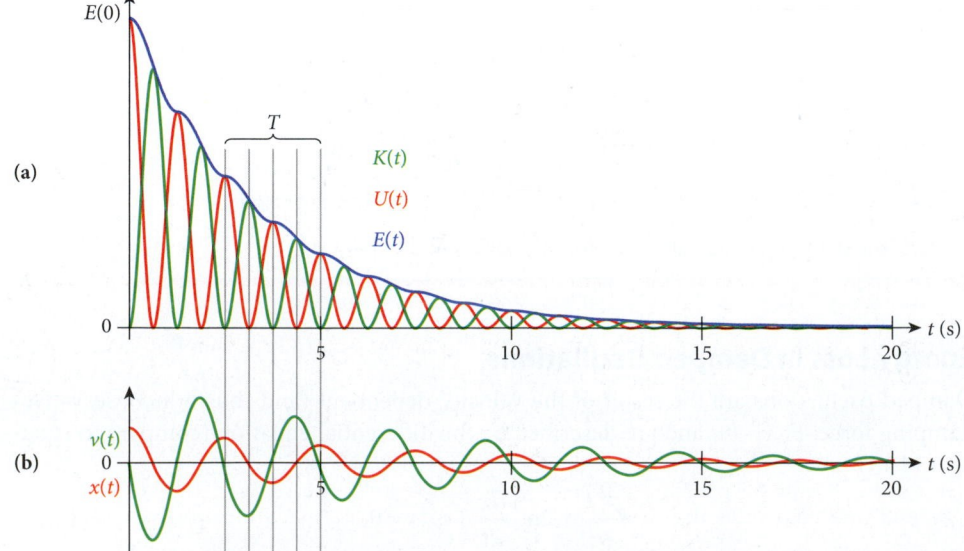

Thus, the rate of energy loss is largest wherever the velocity has the largest absolute value, which happens as the mass passes through the equilibrium position. When the mass reaches the points where the velocity is zero and is about to reverse direction, all energy is potential energy and there is no energy loss at these points.

DERIVATION 14.3 / Energy Loss

We can derive the formula for the energy loss in damped oscillation by taking the time derivatives of the expressions for kinetic and potential energies:

$$\frac{dE}{dt} = \frac{dK}{dt} + \frac{dU}{dt} = \frac{d}{dt}\left(\tfrac{1}{2}mv^2\right) + \frac{d}{dt}\left(\tfrac{1}{2}kx^2\right) = mv\frac{dv}{dt} + kx\frac{dx}{dt}.$$

We use $v = dx/dt$ and factor out the common factor, $m\,dx/dt$:

$$\frac{dE}{dt} = m\frac{dx}{dt}\frac{d^2x}{dt^2} + kx\frac{dx}{dt} = m\frac{dx}{dt}\left(\frac{d^2x}{dt^2} + \frac{k}{m}x\right).$$

Now we can use $\omega_0^2 = k/m$ to obtain

$$\frac{dE}{dt} = m\frac{dx}{dt}\left(\frac{d^2x}{dt^2} + \omega_0^2 x\right). \tag{i}$$

We can reorder equation 14.25 by moving the damping term to the right-hand side:

$$\frac{d^2x}{dt^2} + \omega_0^2 x = -2\omega_\gamma\frac{dx}{dt} = -\frac{b}{m}\frac{dx}{dt}.$$

Because the left-hand side of this differential equation is equal to the term inside the parentheses in equation (i) for the rate of energy loss, we obtain

$$\frac{dE}{dt} = m\frac{dx}{dt}\left(-\frac{b}{m}\frac{dx}{dt}\right) = -b\left(\frac{dx}{dt}\right)^2 = -bv^2.$$

The damping force is $F_\gamma = -bv$, and so we have shown that the rate of energy loss is indeed $dE/dt = vF_\gamma$.

Practical applications require consideration of the **quality** of the oscillator, Q, which specifies the ratio of total energy, E, to the energy loss, ΔE, over one complete oscillation period, T:

$$Q = 2\pi\frac{E}{|\Delta E|}. \tag{14.27}$$

In the limit of zero damping, the oscillator experiences no energy loss, and $Q \to \infty$. In the limit of small damping, the quality of the oscillator can be approximated by

$$Q \approx \frac{\omega_0}{2\omega_\gamma}. \tag{14.28}$$

Combining these two results provides a handy formula for the energy loss during a complete period of weakly damped oscillatory motion:

$$|\Delta E| = E\frac{2\pi}{Q} = E\frac{4\pi\omega_\gamma}{\omega_0}. \tag{14.29}$$

For the harmonic oscillation graphed in Figure 14.27, equation 14.28 results in $Q = 2.472\ \text{s}^{-1}/2(0.139)\ \text{s}^{-1} = 8.89$. Substituting this value of Q into equation 14.29 gives an energy loss of $|\Delta E| = 0.71E$ during each oscillation period. We can compare this to the exact result for Figure 14.27 by measuring the energy at the beginning and end of the interval indicated in the figure. We find for the energy difference divided by the average energy a value of 0.68, in fairly good agreement with the value of 0.71 obtained by using the approximation of equation 14.28.

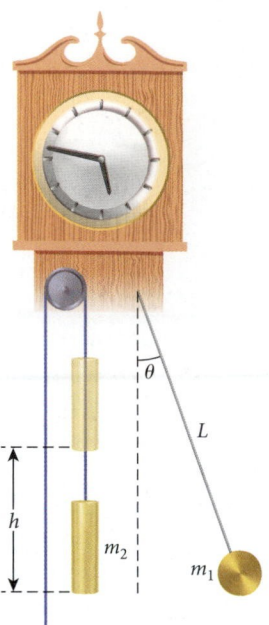

FIGURE 14.28 The workings of a grandfather clock, with the pendulum on the right and the winding mechanism on the left.

SOLVED PROBLEM 14.4 / Grandfather Clock

PROBLEM
A grandfather clock has a pendulum of length 0.994 m that oscillates to a maximum angle of ±6.25° relative to the vertical. The pendulum bob has mass $m_1 = 0.750$ kg and is sufficiently compact to be considered a point mass. A pendulum loses energy because of friction and air resistance and thus has to be supplied with energy to keep going. In a grandfather clock, this is usually done by means of a weight, which hangs from a chain over a pulley, slowly descends in time, and occasionally has to be lifted manually to rewind the clock. For this clock, the weight has mass $m_2 = 1.550$ kg, and it can descend through $h = 1.35$ m before it reaches the ground. If the quality of the pendulum is $Q = 2.95 \cdot 10^3$, what is the maximum time that can pass between successive rewindings of the clock?

SOLUTION
THINK The period, T, is solely a function of the length of the pendulum and can thus be found easily since that information is given. The quality, Q, specifies the ratio of the total energy of an oscillation to the energy lost in one period. Since we are given the angular amplitude and the length and mass of the pendulum, we can calculate the energy lost in one oscillation, ΔE. The energy lost has to be supplied by work done on the mass m_2 by gravity. The maximum total work done on this mass before it hits the ground is m_2gh, which is equal to some number n times the energy lost per oscillation, ΔE. Therefore, at most, a time t equal to n periods of the oscillation can pass before the clock needs rewinding.

SKETCH A sketch of the grandfather clock is shown in Figure 14.28. Usually the fulcrum of the pendulum and the pulley for the winding mechanism are right behind the center of the clock face, but they are separated in the sketch for easier visualization of the various parameters.

RESEARCH The maximum time between successive rewindings of the clock is $t = nT$, and the period is given by $T = 2\pi\sqrt{L/g}$. The number n is the ratio of the total work done on the mass m_2 to the energy lost by the pendulum in one oscillation, $n = m_2gh/\Delta E$. The energy lost during one period is $\Delta E = 2\pi E/Q$ (see equation 14.29), and the energy of the pendulum is the difference in potential energy between the highest and lowest points of the oscillation, $E = m_1gL(1 - \cos\theta_{max})$.

SIMPLIFY We can insert our expressions for T and n into our expression for the time, t, and obtain

$$t = nT = n2\pi\sqrt{\frac{L}{g}} = \frac{m_2gh}{\Delta E}2\pi\sqrt{\frac{L}{g}}.$$

Next, substituting for ΔE and then for E and canceling out $2\pi g$ gives

$$t = \frac{Qm_2gh}{2\pi E}2\pi\sqrt{\frac{L}{g}} = \frac{Qm_2gh}{2\pi m_1gL(1-\cos\theta_{max})}2\pi\sqrt{\frac{L}{g}}$$

$$= \frac{Qm_2h}{m_1L(1-\cos\theta_{max})}\sqrt{\frac{L}{g}}.$$

CALCULATE We can now insert the given numbers:

$$t = \frac{(2.95\cdot10^3)(1.550 \text{ kg})(1.35 \text{ m})}{(0.750 \text{ kg})(0.994 \text{ m})(1-\cos6.25°)}\sqrt{\frac{0.994 \text{ m}}{9.81 \text{ m/s}^2}} = 4.434498\cdot10^5 \text{ s}.$$

ROUND We round our result to three significant digits:

$$t = 4.43\cdot10^5 \text{ s}.$$

DOUBLE-CHECK It's worth repeating occasionally that the first and easiest double-check is to make sure that the units are appropriate. We substituted quite a few quantities with different units into our equation for t, but all the units canceled out except for seconds, which are appropriate for measuring time. What about the order of magnitude of our solution? Since a day has 86,400 s, our solution corresponds to a few days (1 week = $6.048\cdot10^5$ s). If you own a grandfather clock, then you know that it needs rewinding every few days, and our solution is consistent with that experience.

Concept Check 14.5

What could you do to increase the time between successive rewindings of the grandfather clock in Solved Problem 14.4?

a) increase the mass of the pendulum bob

b) increase the mass of the weight on the chain

c) increase the length of the pendulum

d) increase the pendulum's amplitude

e) All of the above would increase the time between rewindings.

f) None of the above would increase the time between rewindings, because Q does not change.

14.5 Forced Harmonic Motion and Resonance

If you are pushing someone sitting on a swing, you give the person periodic pushes, with the general goal of making him or her swing higher—that is, of increasing the amplitude of the oscillation. This situation is an example of **forced harmonic motion.** Periodically forced or driven oscillations arise in many kinds of problems in various areas of physics, including mechanics, acoustics, optics, and electromagnetism. To represent the periodic driving force, we use

$$F(t) = F_d \cos(\omega_d t), \tag{14.30}$$

where F_d and ω_d are constants.

To analyze forced harmonic motion, we start by considering the case of no damping. Using the driving force, $F(t)$, the differential equation for this situation is

$$m\frac{d^2 x}{dt^2} = -kx + F_d \cos(\omega_d t) \Rightarrow$$

$$\frac{d^2 x}{dt^2} + \frac{k}{m}x - \frac{F_d}{m}\cos(\omega_d t) = 0.$$

The solution to this differential equation is

$$x(t) = B\sin(\omega_0 t) + C\cos(\omega_0 t) + A_d \cos(\omega_d t).$$

The coefficients B and C for the part of the motion that oscillates with the intrinsic angular speed, $\omega_0 = \sqrt{k/m}$, can be determined from the initial conditions. However, much more interesting is the part of the motion that oscillates with the driving angular speed. It can be shown that the amplitude of this forced oscillation is given by

$$A_d = \frac{F_d}{m(\omega_0^2 - \omega_d^2)}.$$

Thus, the closer the driving angular speed, ω_d, is to the intrinsic angular speed, ω_0, the larger the amplitude becomes. In the situation of pushing a person on a swing, this amplification is apparent. You can increase the amplitude of the swing's motion only when you push at approximately the same frequency with which the swing is already oscillating. If you push with that frequency, the person on the swing will go higher and higher.

If the driving angular speed is exactly equal to the intrinsic angular speed of the oscillator, the equation $A_d = F_d/[m(\omega_0^2 - \omega_d^2)]$ predicts that the amplitude will grow infinitely big. In real life, this infinite growth does not occur. Some damping is always present in the system. In the presence of damping, the differential equation to solve becomes

$$\frac{d^2 x}{dt^2} + \frac{b}{m}\frac{dx}{dt} + \frac{k}{m}x - \frac{F_d}{m}\cos(\omega_d t) = 0. \tag{14.31}$$

This equation has the steady-state solution

$$x(t) = A_\gamma \cos(\omega_d t - \theta_\gamma), \tag{14.32}$$

where x oscillates at the driving angular speed and the amplitude is

$$A_\gamma = \frac{F_d}{m\sqrt{(\omega_0^2 - \omega_d^2)^2 + 4\omega_d^2\omega_\gamma^2}}. \tag{14.33}$$

As you can see, the amplitude given by equation 14.33 cannot be infinite, because even at $\omega_d = \omega_0$, it still has a finite value, $F_d/(2m\omega_d\omega_\gamma)$, which is also *approximately* its maximum. The driving frequency corresponding to the maximum amplitude can be calculated by

Self-Test Opportunity 14.6

Is a plucked guitar string an example of forced harmonic motion? If no, why not? If yes, what is the driver, and what resonates?

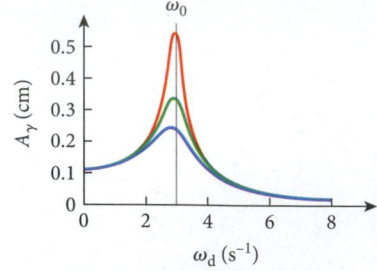

FIGURE 14.29 The amplitude of forced oscillations as a function of the angular speed of the driving force. The three curves represent different damping angular speeds: $\omega_\gamma = 0.3$ s^{-1} (red curve), $\omega_\gamma = 0.5$ s^{-1} (green curve), and $\omega_\gamma = 0.7$ s^{-1} (blue curve).

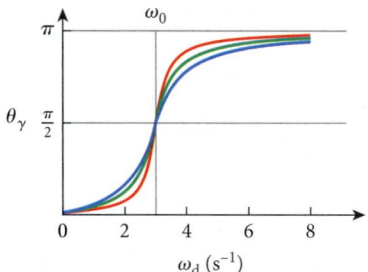

FIGURE 14.30 Phase shift as a function of driving angular speed with the same parameters as in Figure 14.29. The three curves represent different damping angular speeds: $\omega_\gamma = 0.3$ s^{-1} (red curve), $\omega_\gamma = 0.5$ s^{-1} (green curve), and $\omega_\gamma = 0.7$ s^{-1} (blue curve).

taking the derivative of A_γ with respect to ω_d, setting that derivative equal to 0, and solving for ω_d, which gives the result that A_γ is maximum when $\omega_d = \sqrt{\omega_0^2 - 2\omega_\gamma^2}$. Note that for $\omega_\gamma \ll \omega_0$, $\omega_d \approx \omega_0$. The shape of the curve of the amplitude A_γ, as a function of the driving angular speed, ω_d, is called a **resonance shape**, and it is characteristic of all resonance phenomena. When $\omega_d = \sqrt{\omega_0^2 - 2\omega_\gamma^2}$, the driving angular speed is called the **resonant angular speed**; A_γ is close to a maximum at this angular speed. In Figure 14.29, A_γ is plotted from equation 14.33 for $\omega_0 = 3$ s^{-1} and three values of the damping angular speed, ω_γ. As the damping gets weaker, the resonance shape becomes sharper. You can see that in every case, the amplitude curve reaches its maximum at a value of ω_d slightly below $\omega_0 = 3$ s^{-1}.

The phase angle θ_γ in equation 14.32 depends on the driving angular speed as well as on the damping and intrinsic angular speeds:

$$\theta_\gamma = \frac{\pi}{2} - \tan^{-1}\left(\frac{\omega_0^2 - \omega_d^2}{2\omega_d \omega_\gamma}\right). \tag{14.34}$$

The phase shift as a function of the driving angular speed is graphed in Figure 14.30 with the same parameters used in Figure 14.29. You can see that for small driving angular speeds, the forced oscillation follows the driving in phase. As the driving angular speed increases, the phase difference begins to grow and reaches $\pi/2$ at $\omega_d = \omega_0$. For driving angular speeds much larger than the resonant angular speed, the phase shift approaches π. The lower the damping angular speed, ω_γ, the steeper the transition of the phase shift from 0 to π becomes.

A practical note is in order at this point: The solution presented in equation 14.32 is usually only reached after some time. The full solution for the differential equation for damped and driven oscillation (equation 14.31) is given by a combination of equations 14.20 and 14.32. During the transition time, the effect of the particular initial conditions for the system is damped out, and the solution given by equation 14.32 is approached asymptotically.

Figure 14.31 shows motion of a system that is driven with a driving angular speed of $\omega_d = 1.2$ s^{-1} and an acceleration of $F/m = 0.6$ m/s^2. The system has an intrinsic angular speed of $\omega_0 = 2.2$ s^{-1} and a damping angular speed of $\omega_\gamma = 0.4$ s^{-1} and starts at $x_0 = 0$ with a positive initial velocity. The unforced motion of the system follows equation 14.20 and is shown by the green dashed curve. The motion according to equation 14.32 is shown in red. The full solution (blue curve) is the sum of the two. After some complicated-looking oscillations, during which the initial conditions are damped out, the full solution approaches the simple harmonic motion given by equation 14.32.

Driving mechanical systems to resonance has important technical consequences, not all of them desirable. Most worrisome are resonance frequencies in architectural structures. Builders have to be very careful, for example, not to construct high-rises in such a way that earthquakes can drive them into oscillations near resonance. Perhaps the most infamous example of a structure that was destroyed by a resonance phenomenon was the Tacoma Narrows Bridge, known as "Galloping Gertie," in the state of Washington. It collapsed on November 7, 1940, only a few months after construction was finished (Figure 14.32). A 40-mph wind was able to drive the oscillations of the bridge into resonance, and catastrophic mechanical failure occurred.

Columns of soldiers marching across a bridge are often told to fall out of lockstep to avoid creating a resonance due to the periodic stomping of their feet. However, a large number of walkers can get locked into phase through a feedback mechanism. On June 12, 2000, the London Millennium Bridge on the Thames River had to be closed after only three days of operation, because 2000 pedestrians walking over the bridge drove it into resonance and earned it the nickname "wobbly bridge." The lateral swaying of the bridge was such that it provided feedback and locked the pedestrians into a stepping rhythm that amplified the oscillations.

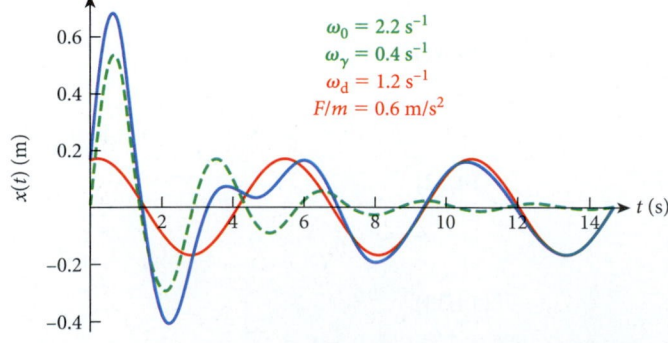

FIGURE 14.31 Position as a function of time for damped and driven harmonic oscillation: The red curve plots equation 14.32, and the green dashed curve shows the motion with damping (from equation 14.20). The blue curve is the sum of the red and green curves, giving the complete picture.

Concept Check **14.6**

The system of Figure 14.31 is driven at four different driving angular speeds, and position versus time for all four cases are plotted in the figures below. Which of the four cases is closest to resonance? (The colors of the curves in each panel are the same as in Figure 14.31.)

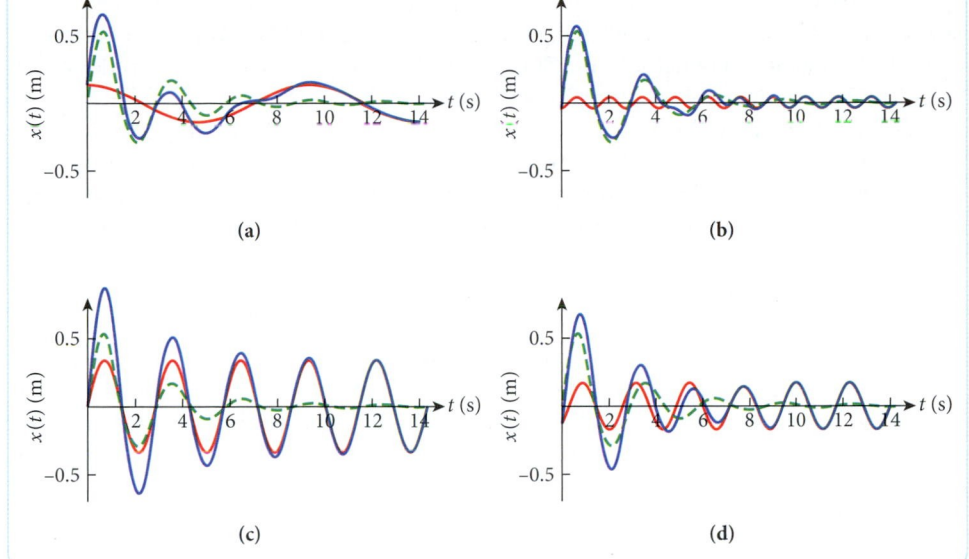

(a)   (b)

(c)   (d)

FIGURE 14.32 Collapse of the midsection of the Tacoma Narrows Bridge on November 7, 1940.

Concept Check **14.7**

Is forced harmonic motion produced by external sinusoidal driving periodic?

a) yes, always

b) only during the initial phase of the motion

c) only asymptotically during the later part of the motion

d) no, never

14.6 Phase Space

We will return repeatedly to the physics of oscillations during the course of this book. As mentioned earlier, the reason for this emphasis is that many systems have restoring forces that drive them back to equilibrium if they are nudged only a small distance from equilibrium. Many problem situations in physics can therefore be modeled in terms of a pendulum or a spring. Before we leave the topic of oscillations for now, let's consider one more interesting way of displaying the motion of oscillating systems. Instead of plotting displacement, velocity, or acceleration as a function of time, we can plot velocity versus displacement. A display of velocity versus displacement is called a **phase space**. Looking at physical motion in this way provides interesting insights.

Let's start with a harmonic oscillation without damping. Plotting an oscillation with time dependence given by $x(t) = A\sin(\omega t)$, amplitude $A = 4$ cm, and angular speed $\omega = 0.7$ s^{-1} in a phase space gives an ellipse, the red curve in Figure 14.33. Other values of the amplitude, between 3 and 7 in steps of 1, yield different ellipses, shown in blue in Figure 14.33.

Consider another example: Figure 14.34 shows a plot in a phase space of the motion of a weakly damped oscillator (with $k = 11$ N/m, $m = 1.8$ kg, $b = 0.5$ kg/s, $A = 5$ cm, and $\theta_0 = 1.6$), the same oscillator whose position was plotted in Figure 14.20 as a function of time. You can see the starting point of the oscillatory motion, at $x(t = 0) = (5$ cm$)\sin(1.6)e^{-0}$ and $v(t = 0) = -1.05$ cm/s. The trajectory in the phase space spirals inward toward the point $(0,0)$. It looks as though the trajectory is attracted to that point. If the oscillation had different initial conditions, the plot would have spiraled inward to the same point, but on a different trajectory. A point to which all trajectories eventually converge is called a *point attractor*.

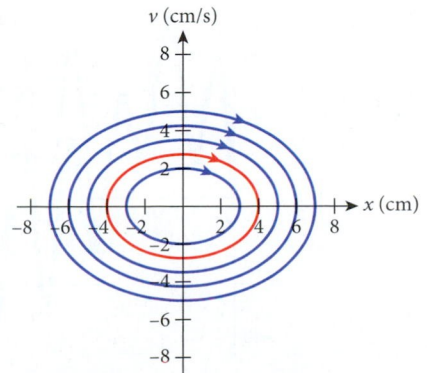

FIGURE 14.33 Velocity versus displacement for different amplitudes of simple harmonic motion.

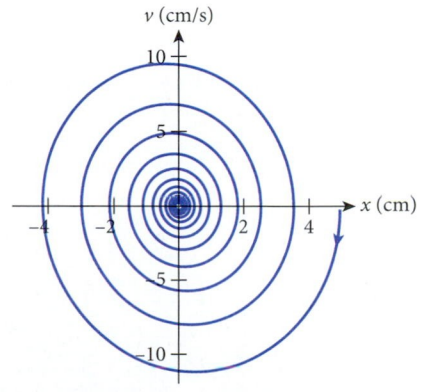

FIGURE 14.34 Velocity versus displacement for a damped oscillator.

14.7 Chaos

One field of research in physics has emerged only since the advent of computers. This field of **nonlinear dynamics,** including those of **chaos,** has yielded interesting and relevant results. If we look hard enough and under certain conditions, we can discover signs of chaotic motion even in the oscillators we have been dealing with in this chapter. In Chapter 7, we looked at chaos and billiards, noting the sensitivity of a system's motion to initial conditions; here we continue this examination in greater detail.

When we analyzed the motion of a pendulum, we used the small-angle approximation (Section 14.2). However, suppose we want to study large deflection angles. We can do this by using a pendulum whose string has been replaced with a thin solid rod. Such a pendulum can move through more than 180°, even through a full 360°. In this case, the approximation $\sin \theta \approx \theta$ does not work, and the sine function must be included in the differential equation. If a periodic external driving force as well as damping is present for this pendulum, we cannot solve the equation of motion in an analytic fashion. However, we can use numerical integration procedures with a computer to obtain the solution. The solution depends on the given initial conditions. Based on what you have learned so far about oscillators, you might expect that similar initial conditions give similar patterns of motion over time. However, for some combinations of driving angular speed and amplitude, damping constant, and pendulum length, this expectation is not borne out.

Figure 14.35 shows plots of the pendulum's deflection angle as a function of time using the same equation of motion and the same value for the initial velocity. However, for the red curve, the initial angle is exactly zero radians. For the green curve, the initial angle is 10^{-3} rad. You can see that these two curves stay close to one another for a few oscillations, but then rapidly diverge. If the initial angle is 10^{-5} rad (blue curve), the curves do not stay close for much longer. Since the blue curve starts 100 times closer to the red curve, it might be expected to stay close to the red curve longer in time by a factor of 100. This is not the case. This behavior is called *sensitive dependence on initial conditions*. If a system is sensitively dependent on the initial conditions, a long-term prediction of its motion is impossible. This sensitivity is one of the hallmarks of chaotic motion.

Precise long-range weather forecasts are impossible because of sensitive dependence on initial conditions. With the advent of computers in the 1950s, meteorologists thought they would be able to describe the future weather accurately if only they generated enough measurements of all the relevant variables—such as temperature and wind direction—at as many points as possible across the country. However, by the 1960s, it had become obvious that the basic differential equations for air masses moving in the atmosphere show sensitive dependence on initial conditions. Thus, the dream of precise long-range weather forecasting was ended.

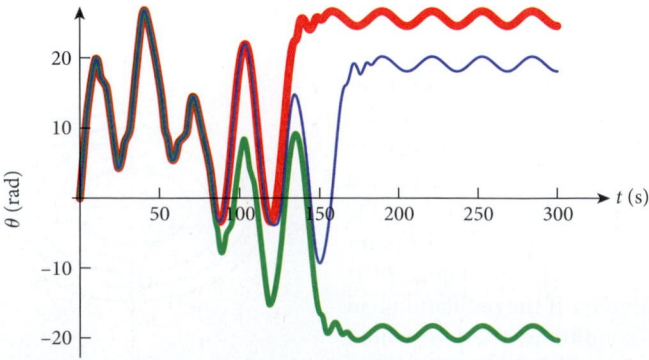

FIGURE 14.35 Deflection angle as a function of time for a driven pendulum with three different initial conditions: The red line resulted from an initial angle of zero. The green line resulted from an initial angle of 10^{-3} rad. The blue line resulted from an initial angle of 10^{-5} rad.

WHAT WE HAVE LEARNED | EXAM STUDY GUIDE

- An object acting under Hooke's Law, $F = -kx$, undergoes simple harmonic motion, with a restoring force that acts in the opposite direction to the displacement from equilibrium.

- The equation of motion for a mass on a spring (with no damping) is $x(t) = B \sin(\omega_0 t) + C \cos(\omega_0 t)$, where $\omega_0 = \sqrt{k/m}$, or, alternatively, $x(t) = A \sin(\omega_0 t + \theta_0)$.

- The equation of motion for a pendulum (with no damping) is $\theta(t) = B \sin(\omega_0 t) + C \cos(\omega_0 t)$ with $\omega_0 = \sqrt{g/\ell}$.

- Velocity and acceleration for simple harmonic oscillations are given by
$$x(t) = A \sin(\omega_0 t + \theta_0)$$
$$\Rightarrow v(t) = \omega_0 A \cos(\omega_0 t + \theta_0)$$
$$\Rightarrow a(t) = -\omega_0^2 A \sin(\omega_0 t + \theta_0).$$

- The period of an oscillation is $T = 2\pi/\omega$.

- The frequency of an oscillation is $f = 1/T = \omega/2\pi$, or $\omega = 2\pi f$.

- The equation of motion for a mass on a spring with small damping $\left(b < 2\sqrt{mk}\right)$ is
$$x(t) = Be^{-\omega_\gamma t} \sin(\omega' t) + Ce^{-\omega_\gamma t} \cos(\omega' t), \text{ with}$$
$$\omega_\gamma = b/2m, \ \omega' = \sqrt{k/m - (b/2m)^2} = \sqrt{\omega_0^2 - \omega_\gamma^2},$$
and $\omega_0 = \sqrt{k/m}$.

- The equation of motion for a mass on a spring with large damping $\left(b > 2\sqrt{mk}\right)$ is
$$x(t) = Be^{-\left(\omega_\gamma + \sqrt{\omega_\gamma^2 - \omega_0^2}\right)t} + Ce^{-\left(\omega_\gamma - \sqrt{\omega_\gamma^2 - \omega_0^2}\right)t}.$$

- The equation of motion for a mass on a spring with critical damping $\left(b = 2\sqrt{mk}\right)$ is $x(t) = (B + tC)e^{-\omega_\gamma t}$.

- For a damped harmonic oscillator, the rate of energy loss is $dE/dt = -bv^2 = vF_\gamma$.

- The quality of an oscillator is related to the energy loss during one complete oscillation period, $Q = 2\pi E/|\Delta E|$.

- If a periodic external driving force, $F(t) = F_d \cos(\omega_d t)$, is applied to an oscillating system, the equation of motion becomes (after some transient time) $x(t) = A_\gamma \cos(\omega_d t - \theta_\gamma)$, with amplitude $A_\gamma = \dfrac{F_d}{m\sqrt{\left(\omega_0^2 - \omega_d^2\right)^2 + 4\omega_d^2 \omega_\gamma^2}}$. The graph of this amplitude displays a resonance shape in which the amplitude is a maximum when the driving angular speed approximately equals the intrinsic angular speed ($\omega_d = \omega_0$).

- A phase space displays a plot of the velocity of an object versus its position. Simple harmonic motion yields an ellipse in such a phase space. A plot of damped harmonic motion spirals in toward the point attractor at (0,0).

- With the right choice of parameters, a damped and driven rod pendulum can show chaotic motion, which is sensitively dependent on the initial conditions. In this case, the system's long-term behavior is not predictable.

ANSWERS TO SELF-TEST OPPORTUNITIES

14.1 $\dfrac{d^2 x}{dt^2} + \dfrac{k}{m} x = 0$. Use the trial function

$x(t) = A \sin(\omega_0 t + \theta_0)$ and take derivatives:

$$\dfrac{dx}{dt} = A\omega_0 \cos(\omega_0 t + \theta_0);$$

$$\dfrac{d^2 x}{dt^2} = -A\omega_0^2 \sin(\omega_0 t + \theta_0).$$

Now insert the latter expression back into the differential equation with

$\omega_0^2 = \dfrac{k}{m}$. We find

$$-A\omega_0^2 \sin(\omega_0 t + \theta_0) + A\omega_0^2 \sin(\omega_0 t + \theta_0) = 0.$$

14.2 Use the trig identity $\sin(\alpha + \beta) = \sin(\alpha) \cos(\beta) + \cos(\alpha) \sin(\beta)$:

$A \sin(\omega_0 t + \theta_0) = A \sin(\omega_0 t) \cos(\theta_0) + A \cos(\omega_0 t) \sin(\theta_0)$

$A \sin(\omega_0 t + \theta_0) = (A \cos(\theta_0)) \sin(\omega_0 t) + (A \sin(\theta_0)) \cos(\omega_0 t) = B \sin(\omega_0 t) + C \cos(\omega_0 t)$

$B = A \cos(\theta_0)$

$C = A \sin(\theta_0).$

Square and add these two equations:

$B^2 + C^2 = A^2 \sin^2(\theta_0) + A^2 \cos^2(\theta_0) = A^2(\sin^2(\theta_0) + \cos^2(\theta_0)) = A^2$

$A = \sqrt{B^2 + C^2}$.

Now divide the same two equations:

$\dfrac{C}{B} = \dfrac{A \sin(\theta_0)}{A \cos(\theta_0)} = \tan(\theta_0)$

$\theta_0 = \tan^{-1}\left(\dfrac{C}{B}\right).$

14.3 The radial component of the gravitational force is $mg \cos\theta$, as indicated in Figure 14.10. Since the mass on the string is forced to move on a circular path, a centripetal force, $F_c = mv^2/\ell$, is required to keep the mass on that path. This force has to be supplied by the string tension, which also has to compensate for the radial component of the gravitational force. Therefore, in general, the relationship between the string tension and the radial component of the gravitational force is $T \geq mg \cos \theta$, where equality exists only at the turning points of the pendulum's motion.

14.4 $T_{Moon} = \sqrt{\dfrac{g_{Earth}}{g_{Moon}}} T = \sqrt{\dfrac{9.81 \text{ m/s}^2}{1.63 \text{ m/s}^2}} T = 2.45T.$

14.5 $\omega_\gamma = \dfrac{\alpha}{2}; \; \omega_0 = \sqrt{\dfrac{g}{\ell}}; \; \omega' = \sqrt{\omega_0^2 - \omega_\gamma^2}.$

14.6 Yes, a plucked guitar string is an example of forced harmonic motion. The oscillating string provides the harmonic driving force, and the air inside the guitar body resonates at the same frequency as the string's oscillation.

PROBLEM-SOLVING GUIDELINES

1. In oscillatory motion, some quantities are characteristic of the motion: position, x; velocity, v; acceleration, a; energy, E; phase angle, θ_0; amplitude, A; and maximum velocity, v_{max}. Other quantities are characteristic of the particular system: mass, m; spring constant, k; pendulum length, ℓ; period, T; frequency, f; and angular frequency, ω. It is helpful to list the quantities you are looking for in a problem and the quantities you know or need to determine to find those unknowns.

2. All harmonic oscillations are produced by forces that depend linearly on the deflection from the equilibrium position and that always point toward the equilibrium position.

Any time the given situation includes such a force, harmonic motion will result.

3. For harmonic oscillations, the kinetic energy is at a maximum when the oscillator is passing through the equilibrium position, and the potential energy is at a maximum (and the kinetic energy is zero) at the maximum displacement from equilibrium.

4. For problems involving damping, it is essential to first determine whether the system is weakly damped, critically damped, or overdamped, because the form of the equation of motion you should use depends on the type of damping.

MULTIPLE-CHOICE QUESTIONS

14.1 Two children are on adjacent playground swings with chains of the same length. They are given pushes by an adult and then left to swing. Assuming that each child on a swing can be treated as a simple pendulum and that friction is negligible, which child takes the longer time for one complete swing (has a longer period)?

a) the bigger child

b) the lighter child

c) neither child

d) the child given the bigger push

14.2 Identical blocks oscillate on the end of a vertical spring, one on Earth and one on the Moon. Where is the period of the oscillations greater?

a) on Earth

b) on the Moon

c) same on both Earth and Moon

d) cannot be determined from the information given

14.3 A mass that can oscillate without friction on a horizontal surface is attached to a horizontal spring that is pulled to the right 10.0 cm and is released from rest. The period of oscillation for the mass is 5.60 s. What is the speed of the mass at $t = 2.50$ s?

a) $-2.61 \cdot 10^{-1}$ m/s

b) $-3.71 \cdot 10^{-2}$ m/s

c) $-3.71 \cdot 10^{-1}$ m/s

d) $-2.01 \cdot 10^{-1}$ m/s

14.4 The spring constant for a spring-mass system undergoing simple harmonic motion is doubled. If the total energy remains unchanged, what will happen to the maximum amplitude of the oscillation? Assume that the system is underdamped.

a) It will remain unchanged.

b) It will be multiplied by 2.

c) It will be multiplied by $\frac{1}{2}$.

d) It will be multiplied by $1/\sqrt{2}$.

14.5 With the right choice of parameters, a damped and driven physical pendulum can show chaotic motion, which is sensitively

dependent on the initial conditions. Which statement about such a pendulum is true?

a) Its long-term behavior can be predicted.

b) Its long-term behavior is not predictable.

c) Its long-term behavior is like that of a simple pendulum of equivalent length.

d) Its long-term behavior is like that of a conical pendulum.

e) None of the above is true.

14.6 A spring is hanging from the ceiling with a mass attached to it. The mass is pulled downward, causing it to oscillate vertically with simple harmonic motion. Which of the following will decrease the frequency of the oscillation?

a) adding a second, identical spring with one end attached to the mass and the other to the ceiling

b) adding a second, identical spring with one end attached to the mass and the other to the floor

c) increasing the mass

d) adding both springs, as described in (a) and (b)

14.7 A child of mass M is swinging on a swing of length L to a maximum deflection angle of θ. A man of mass $4M$ is swinging on a similar swing of length L to a maximum angle of 2θ. Each swing can be treated as a simple pendulum undergoing simple harmonic motion. If the period for the child's motion is T, then the period for the man's motion is

a) T.

b) $2T$.

c) $T/2$.

d) $T/4$.

14.8 Object A is four times heavier than object B. Each object is attached to a spring, and the springs have equal spring constants. The two objects are then pulled from their equilibrium positions and released from rest. What is the ratio of the periods of the two oscillators if the amplitude of A is half that of B?

a) $T_A:T_B = 1:4$

b) $T_A:T_B = 4:1$

c) $T_A:T_B = 2:1$

d) $T_A:T_B = 1:2$

14.9 Rank the simple harmonic oscillators shown in the figure in order of their intrinsic frequencies, from highest to lowest. All the springs have identical spring constants, and all the blocks have identical masses.

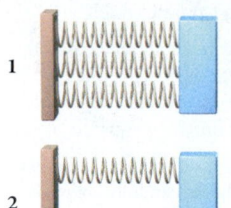

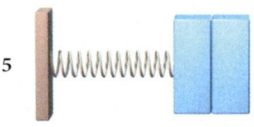
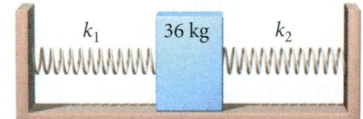

a) 1, 2, 3, 4=5

b) 4=5, 3, 2, 1

c) 1, 2, 3=4, 5

d) 3, 4, 2, 1, 5

e) 3, 2=5, 4, 1

14.10 A pendulum is suspended from the ceiling of an elevator. When the elevator is at rest, the period of the pendulum is T. The elevator accelerates upward, and the period of the pendulum is then

a) still T.   b) less than T.   c) greater than T.

14.11 A 36-kg mass is placed on a horizontal frictionless surface and then connected to walls by two springs with spring constants $k_1 = 3.0$ N/m and $k_2 = 4.0$ N/m, as shown in the figure. What is the period of oscillation for the 36-kg mass if it is displaced slightly to one side?

a) 11 s   d) 20. s

b) 14 s   e) 32 s

c) 17 s   f) 38 s

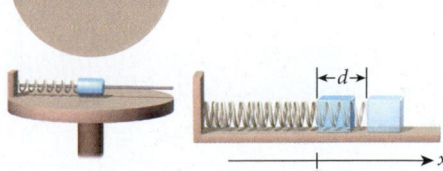

14.12 A spring with $k = 12.0$ N/m has a mass $m = 3.00$ kg attached to its end. The mass is pulled +10.0 cm from the equilibrium position and released from rest. What is the velocity of the mass as it passes the equilibrium position?

a) −0.125 m/s   d) +0.500 m/s

b) +0.750 m/s   e) −0.633 m/s

c) −0.200 m/s

14.13 A grandfather clock keeps time using a pendulum consisting of a light rod connected to a small, heavy mass. How long should the rod be to make the period of the oscillations 1.00 s?

a) 0.0150 m   d) 0.439 m

b) 0.145 m   e) 0.750 m

c) 0.248 m

14.14 An automobile with a mass of 1640 kg is lifted into the air. During the lift, the suspension spring on each wheel lengthens by 30.0 cm. What damping constant is required for the shock absorber on each wheel to produce critical damping?

a) 101 kg/s   d) 2310 kg/s

b) 234 kg/s   e) 4690 kg/s

c) 1230 kg/s

14.15 Suppose you start another weakly damped oscillator with the same initial conditions as in Figure 14.34 and with all parameters unchanged except for a larger mass. How will the trajectory in the phase space change?

a) It remains unchanged.

b) It remains basically unchanged but spirals inward to a different point.

c) It spirals inward to the same point but does so faster; that is, it crosses the x-axis fewer times on its way in.

d) It spirals inward to the same point but does so more slowly; that is, it crosses the x-axis more times on its way in.

CONCEPTUAL QUESTIONS

14.16 You sit in your SUV in a traffic jam. Out of boredom, you rock your body from side to side and then sit still. You notice that your SUV continues rocking for about 2 s and makes three side-to-side oscillations, before subsiding. You estimate that the amplitude of the SUV's motion must be less than 5% of the amplitude of your initial rocking. Knowing that your SUV has a mass of 2200 kg (including the gas, your stuff, and yourself), what can you say about the effective spring constant and damping of your SUV's suspension?

14.17 An astronaut, taking his first flight on a space shuttle, brings along his favorite miniature grandfather clock. At launch, the clock and his digital wristwatch are synchronized, and the grandfather clock is pointed in the direction of the shuttle's nose. During the boost phase, the shuttle has an upward acceleration whose magnitude is several times the value of the gravitational acceleration at the Earth's surface. As the shuttle reaches a constant cruising speed after completion of the boost phase, the astronaut compares his grandfather clock's time to that of his watch. Are the timepieces still synchronized? If not, which is ahead of the other? Explain.

14.18 A small cylinder of mass m can slide without friction on a shaft that is attached to a turntable, as shown in the figure. The shaft also passes through the coils of a spring with spring constant k, which is attached to the turntable at one end and to the cylinder at the other end. The equilibrium length of the spring (unstretched and uncompressed) matches the radius of the turntable; thus, when the turntable is not rotating, the cylinder is at equilibrium at the center of the turntable. The cylinder is given a small initial displacement, and, at the same time, the turntable is set into uniform circular motion with angular speed ω. Calculate the period of the oscillations performed by the cylinder. Discuss the result.

14.19 A door closer on a door allows the door to close by itself, as shown in the figure. The door closer consists of a return spring attached to an oil-filled damping piston. As the spring pulls the piston to the right, teeth on the piston rod engage a gear, which rotates and causes the door to close. Should the system be underdamped, critically damped, or overdamped for the best performance (where the door closes quickly without slamming into the frame)?

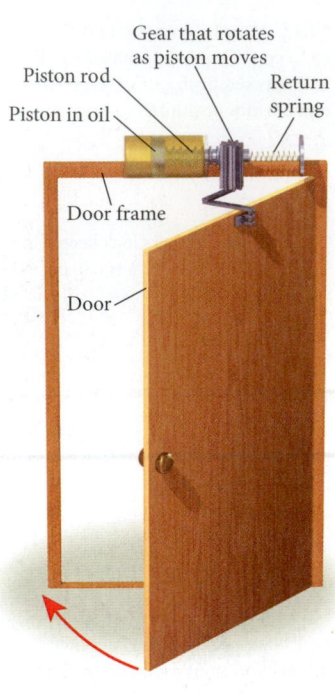

14.20 When the amplitude of the oscillation of a mass on a stretched string is increased, why doesn't the period of oscillation also increase?

14.21 Mass-spring systems and pendulum systems can both be used in mechanical timing devices. What are the advantages of using one type of system rather than the other in a device designed to generate reproducible time measurements over an extended period of time?

14.22 You have a linear (following Hooke's Law) spring with an unknown spring constant, a standard mass, and a timer. Explain carefully how you could most practically use these to measure masses *in the absence of gravity*. Be as quantitative as you can. Regard the mass of the spring as negligible.

14.23 Pendulum A has a bob of mass m hung from a string of length L; pendulum B is identical to A except its bob has mass $2m$. Compare the frequencies of small oscillations of the two pendulums.

14.24 A sharp spike in a plotted curve of frequency can be represented as a sum of sinusoidal functions of all possible frequencies, with equal amplitudes. A bell struck with a hammer rings at its natural frequency, that is, the frequency at which it vibrates as a free oscillator. Explain why, as clearly and concisely as you can.

EXERCISES

A blue problem number indicates a worked-out solution is available in the Student Solutions Manual. One • and two •• indicate increasing level of problem difficulty.

Section 14.1

14.25 A mass $m = 5.00$ kg is suspended from a spring and oscillates according to the equation of motion $x(t) = 0.500 \cos(5.00t + \pi/4)$. What is the spring constant?

14.26 Determine the frequency of oscillation of a 200.-g block that is connected by a spring to a wall and sliding on a frictionless surface, for the three conditions shown in the figure.

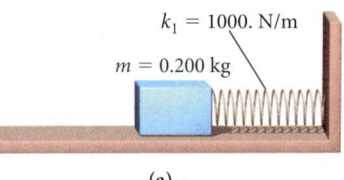

(a)

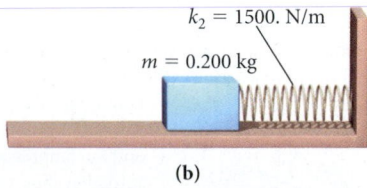

(b)

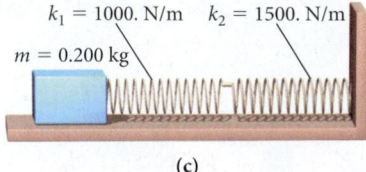

(c)

•**14.27** A mass of 10.0 kg is hanging by a steel wire 1.00 m long and 1.00 mm in diameter. If the mass is pulled down slightly and released, what will be the *frequency* of the resulting oscillations? Young's modulus for steel is $2.0 \cdot 10^{11}$ N/m^2.

•**14.28** A 100.-g block hangs from a spring with $k = 5.00$ N/m. At $t = 0$ s, the block is 20.0 cm below the equilibrium position and moving upward with a speed of 200. cm/s. What is the block's speed when the displacement from equilibrium is 30.0 cm?

•**14.29** A block of wood of mass 55.0 g floats in a swimming pool, oscillating up and down in simple harmonic motion with a frequency of 3.00 Hz.

a) What is the value of the effective spring constant of the water?

b) A partially filled water bottle of almost the same size and shape as the block of wood but with mass 250. g is placed on the water's surface. At what frequency will the bottle bob up and down?

•**14.30** When a mass is attached to a vertical spring, the spring is stretched a distance d. The mass is then pulled down from this position and released. It undergoes 50.0 oscillations in 30.0 s. What was the distance d?

•**14.31** A cube of density ρ_c floats in a liquid of density ρ_l, as shown in the figure. At rest, an amount h of the cube's height is submerged in liquid. If the cube is pushed down, it bobs up and down like a spring and oscillates about its equilibrium position. Show that the frequency of its oscillations is given by $f = (2\pi)^{-1} \sqrt{g/h}$.

•**14.32** A U-shaped glass tube with a uniform cross-sectional area, A, is partly filled with fluid of density ρ. Increased pressure is applied to one of the arms, resulting in a difference in elevation, h, between the two arms of the tube, as shown in the figure. The increased pressure is then removed, and the fluid oscillates in the tube. Determine the period of the oscillations of the fluid column. (You have to determine what the known quantities are.)

•**14.33** The figure shows a mass $m_2 = 20.0$ g resting on top of a mass $m_1 = 20.0$ g, which is attached to a spring with $k = 10.0$ N/m. The coefficient of static friction between the two masses is 0.600. The masses are oscillating with simple harmonic motion on a frictionless surface. What is the maximum amplitude the oscillation can have without m_2 slipping off m_1?

••**14.34** Consider two identical oscillators, each with spring constant k and mass m, in simple harmonic motion. One oscillator is started with initial conditions x_0 and v_0; the other starts with slightly different conditions, $x_0 + \delta x$ and $v_0 + \delta v$.

a) Find the difference in the oscillators' positions, $x_1(t) - x_2(t)$, for all t.

b) This difference is *bounded*; that is, there exists a constant C, independent of time, for which $|x_1(t) - x_2(t)| \le C$ holds for all t. Find an expression for C. What is the *best* bound, that is, the smallest value of C that works? (*Note:* An important characteristic of *chaotic* systems is exponential sensitivity to initial conditions; the difference in position of two such systems with slightly different initial conditions grows exponentially with time. You have just shown that an oscillator in simple harmonic motion is not a chaotic system.)

Section 14.2

14.35 What is the period of a simple pendulum that is 1.00 m long in each situation?

a) in the physics lab

b) in an elevator accelerating at 2.10 m/s^2 upward

c) in an elevator accelerating 2.10 m/s^2 downward

d) in an elevator that is in free fall

14.36 Suppose a simple pendulum is used to measure the acceleration due to gravity at various points on the Earth. If g varies by 0.16% over the various locations where it is sampled, what is the corresponding variation in the period of the pendulum? Assume that the length of the pendulum does not change from one site to another.

•**14.37** A 1.00-kg mass is connected to a 2.00-kg mass by a massless rod 30.0 cm long, as shown in the figure. A hole is then drilled in the rod 10.0 cm away from the 1.00-kg mass, and the rod and masses are free to rotate about this pivot point, P. Calculate the period of oscillation for the masses if they are displaced slightly from the stable equilibrium position.

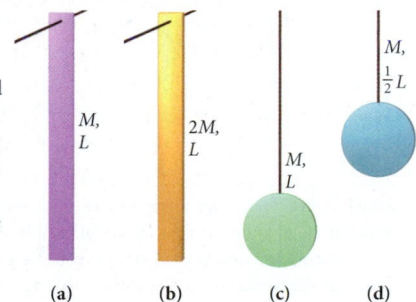

•**14.38** A torsional pendulum is a vertically suspended wire with a mass attached to one end that is free to rotate. The wire resists twisting as well as elongation and compression, so it is subject to a rotational equivalent of Hooke's Law: $\tau = -\kappa\theta$. Use this information to find the frequency of a torsional pendulum.

•**14.39** A physical pendulum consists of a uniform rod of mass M and length L. The pendulum is pivoted at a point that is a distance x from the center of the rod, so the period for oscillation of the pendulum depends on x: $T(x)$.

a) What value of x gives the maximum value for T?

b) What value of x gives the minimum value for T?

•**14.40** Shown in the figure are four different pendulums: (a) a rod has a mass M and a length L; (b) a rod has a mass $2M$ and a length L; (c) a mass M is attached to a massless string of a length L; and (d) a mass M is attached to a massless string of length $\frac{1}{2}L$. Find the period of each pendulum when it is pulled 20° to the right and then released.

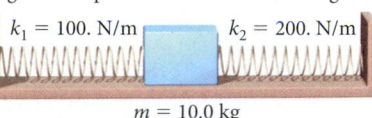

(a) (b) (c) (d)

•**14.41** A grandfather clock uses a physical pendulum to keep time. The pendulum consists of a uniform thin rod of mass M and length L that is pivoted freely about one end, with a solid sphere of the same mass, M, and a radius of $L/2$ centered about the free end of the rod.

a) Obtain an expression for the moment of inertia of the pendulum about its pivot point as a function of M and L.

b) Obtain an expression for the period of the pendulum for small oscillations.

c) Determine the length L that gives a period of $T = 2.0$ s.

•**14.42** A standard CD has a diameter of 12 cm and a hole that is centered on the axis of symmetry and has a diameter of 1.5 cm. The CD's thickness is 2.0 mm. If a (very thin) pin extending through the hole of the CD suspends it in a vertical orientation, as shown in the figure, the CD may oscillate about an axis parallel to the pin, rocking back and forth. Calculate the oscillation frequency.

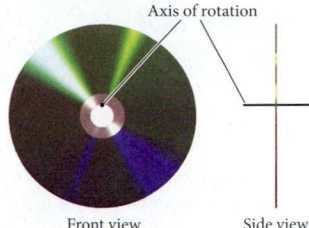

Axis of rotation

Front view   Side view

Section 14.3

14.43 A massive object of $m = 5.00$ kg oscillates with simple harmonic motion. Its position as a function of time varies according to the equation $x(t) = 2\sin([\pi/2]t + \pi/6)$, where x is in meters.

a) What are the position, velocity, and acceleration of the object at $t = 0$ s?

b) What is the kinetic energy of the object as a function of time?

c) At which time after $t = 0$ s is the kinetic energy first at a maximum?

14.44 The mass m shown in the figure is displaced a distance x to the right from its equilibrium position.

a) What is the net force acting on the mass, and what is the effective spring constant?

$k_1 = 100.$ N/m $k_2 = 200.$ N/m

$m = 10.0$ kg

b) What will the frequency of the oscillation be when the mass is released?

c) If $x = 0.100$ m, what is the total energy of the mass-spring system after the mass is released, and what is the maximum velocity of the mass?

•**14.45** A 2.00-kg mass attached to a spring is displaced 8.00 cm from the equilibrium position. It is released and then oscillates with a frequency of 4.00 Hz.

a) What is the energy of the motion when the mass passes through the equilibrium position?

b) What is the speed of the mass when it is 2.00 cm from the equilibrium position?

•**14.46** A Foucault pendulum is designed to demonstrate the effect of the Earth's rotation. A Foucault pendulum displayed in a museum is typically quite long, making the effect easier to see. Consider a Foucault pendulum of length 15.0 m with a 110.-kg brass bob. It is set to swing with an amplitude of 3.50°.

a) What is the period of the pendulum?

b) What is the maximum kinetic energy of the pendulum?

c) What is the maximum speed of the pendulum?

••**14.47** A mass, $m_1 = 8.00$ kg, is at rest on a frictionless horizontal surface and connected to a wall by a spring with $k = 70.0$ N/m, as shown in the figure. A second mass, $m_2 = 5.00$ kg, is moving to the right at $v_0 = 17.0$ m/s. The two masses collide and stick together.

a) What is the maximum compression of the spring?

b) How long will it take after the collision to reach this maximum compression?

m_2 v_0 m_1 k

••**14.48** A mass $M = 0.460$ kg moves with an initial speed $v = 3.20$ m/s on a level frictionless air track. The mass is initially a distance $D = 0.250$ m away from a spring with $k = 840$ N/m, which is mounted rigidly at one end of the air track. The mass compresses the spring a maximum distance d, before reversing direction. After bouncing off the spring, the mass travels with the same speed v, but in the opposite direction.

a) Determine the maximum distance that the spring is compressed.

b) Find the total elapsed time until the mass returns to its starting point. (*Hint:* The mass undergoes a partial cycle of simple harmonic motion while in contact with the spring.)

••**14.49** The relative motion of two atoms in a molecule can be described as the motion of a single body of mass m moving in one dimension, with potential energy $U(r) = A/r^{12} - B/r^6$, where r is the separation between the atoms and A and B are positive constants.

a) Find the equilibrium separation, r_0, of the atoms, in terms of the constants A and B.

b) If moved slightly, the atoms will oscillate about their equilibrium separation. Find the angular frequency of this oscillation, in terms of A, B, and m.

Section 14.4

14.50 A 3.00-kg mass attached to a spring with $k = 140$. N/m is oscillating in a vat of oil, which damps the oscillations.

a) If the damping constant of the oil is $b = 10.0$ kg/s, how long will it take the amplitude of the oscillations to decrease to 1.00% of its original value?

b) What should the damping constant be to reduce the amplitude of the oscillations by 99.0% in 1.00 s?

•**14.51** A vertical spring with a spring constant of 2.00 N/m has a 0.300-kg mass attached to it, and the mass moves in a medium with a damping constant of 0.0250 kg/s. The mass is released from rest at a position 5.00 cm from the equilibrium position. How long will it take for the amplitude to decrease to 2.50 cm?

14.52 A mass of 0.404 kg is attached to a spring with a spring constant of 206.9 N/m. Its oscillation is damped, with damping constant $b = 14.5$ kg/s. What is the frequency of this damped oscillation?

•**14.53** Cars have shock absorbers to damp the oscillations that would otherwise occur when the springs that attach the wheels to the car's frame are compressed or stretched. Ideally, the shock absorbers provide critical damping. If the shock absorbers fail, they provide less damping, resulting in an underdamped motion. You can perform a simple test of your shock absorbers by pushing down on one corner of your car and then quickly releasing it. If this results in an up-and-down oscillation of the car, you know that your shock absorbers need changing. The spring on each wheel of a car has a spring constant of 4005 N/m, and the car has a mass of 851 kg, equally distributed over all four wheels. Its shock absorbers have gone bad and provide only 60.7% of the damping they were initially designed to provide. What will the period of the underdamped oscillation of this car be if the pushing-down shock absorber test is performed?

•**14.54** In a lab, a student measures the unstretched length of a spring as 11.2 cm. When a 100.0-g mass is hung from the spring, its length is 20.7 cm. The mass-spring system is set into oscillatory motion, and the student observes that the amplitude of the oscillation decreases by about a factor of 2 after five complete cycles.

a) Calculate the period of oscillation for this system, assuming no damping.

b) If the student can measure the period to the nearest 0.05 s, will she be able to detect the difference between the period with no damping and the period with damping?

•**14.55** An 80.0-kg bungee jumper is enjoying an afternoon of jumps. The jumper's first oscillation has an amplitude of 10.0 m and a period of 5.00 s. Treating the bungee cord as a spring with no damping, calculate each of the following:

a) the spring constant of the bungee cord,

b) the bungee jumper's maximum speed during the oscillation, and

c) the time for the amplitude to decrease to 2.00 m (with air resistance providing the damping of the oscillations at 7.50 kg/s).

••**14.56** A small mass, $m = 50.0$ g, is attached to the end of a massless rod that is hanging from the ceiling and is free to swing. The rod has length $L = 1.00$ m. The rod is displaced 10.0° from the vertical and released at time $t = 0$. Neglect air resistance. What is the period of the rod's oscillation? Now suppose the entire system is immersed in a fluid with a small damping constant, $b = 0.0100$ kg/s, and the rod is again released from an initial displacement angle of 10.0°. What is the time for the amplitude of the oscillation to reduce to 5.00°? Assume that the damping is small. Also note that since the amplitude of the oscillation is small and all the mass of the pendulum is at the end of the rod, the motion of the mass can be treated as strictly linear, and you can use the substitution $R\theta(t) = x(t)$, where $R = 1.0$ m is the length of the pendulum rod.

Section 14.5

14.57 A 3.00-kg mass is vibrating on a spring. It has a resonant angular speed of 2.40 rad/s and a damping angular speed of 0.140 rad/s. If the sinusoidal driving force has an amplitude of 2.00 N, find the maximum amplitude of the vibration if the driving angular speed is (a) 1.20 rad/s, (b) 2.40 rad/s, and (c) 4.80 rad/s.

•**14.58** A mass of 0.404 kg is attached to a spring with a spring constant of 204.7 N/m and hangs at rest. The spring hangs from a piston. The piston then moves up and down, driven by a force given by (29.4 N) cos $[(17.1$ rad/s$)t]$.

a) What is the maximum displacement from its equilibrium position that the mass can reach?

b) What is the maximum speed that the mass can attain in this motion?

••**14.59** A mass, $M = 1.60$ kg, is attached to a wall by a spring with $k = 578$ N/m. The mass slides on a frictionless floor. The spring and mass are immersed in a fluid with a damping constant of 6.40 kg/s. A horizontal force, $F(t) = F_d \cos(\omega_d t)$, where $F_d = 52.0$ N, is applied to the mass through a knob, causing the mass to oscillate back and forth. Neglect the mass of the spring and of the knob and rod. At what frequency will the amplitude of the mass's oscillation be greatest, and what is the maximum amplitude? If the driving frequency is reduced slightly (but the driving amplitude remains the same), at what frequency will the amplitude of the mass's oscillation be half of the maximum amplitude?

Additional Exercises

14.60 When the displacement of a mass on a spring is half of the amplitude of its oscillation, what fraction of the mass's energy is kinetic energy?

14.61 A mass m is attached to a spring with a spring constant of k and set into simple harmonic motion. When the mass has half of its maximum kinetic energy, how far away from its equilibrium position is it, expressed as a fraction of its maximum displacement?

14.62 If you kick a harmonic oscillator sharply, you impart to it an initial velocity but no initial displacement. For a weakly damped oscillator with mass m, spring constant k, and damping force $F_\gamma = -bv$, find $x(t)$, if the total impulse delivered by the kick is J_0.

14.63 A mass $m = 1.00$ kg in a spring-mass system with $k = 1.00$ N/m is observed to be moving to the right, past its equilibrium position with a speed of 1.00 m/s at time $t = 0$.

a) Ignoring all damping, determine the equation of motion.

b) Suppose the initial conditions are such that at time $t = 0$, the mass is at $x = 0.500$ m and moving to the right with a speed of 1.00 m/s. Determine the new equation of motion. Assume the same spring constant and mass.

14.64 The motion of a block-spring system is described by $x(t) = A \sin(\omega t)$. Find ω, if the potential energy equals the kinetic energy at $t = 1.00$ s.

14.65 A hydrogen gas molecule can be thought of as a pair of protons bound together by a spring. If the mass of a proton is $1.7 \cdot 10^{-27}$ kg and the period of oscillation is $8.0 \cdot 10^{-15}$ s, what is the effective spring constant for the bond in a hydrogen molecule?

14.66 A shock absorber that provides critical damping with $\omega_\gamma = 72.4$ rad/s is compressed by 6.41 cm. How far from the equilibrium position is it after 0.0247 s?

14.67 Imagine you are an astronaut who has landed on another planet and wants to determine the free-fall acceleration on that planet. In one of the experiments you decide to conduct, you use a pendulum 0.500 m long and find that the period of oscillation for this pendulum is 1.50 s. What is the acceleration due to gravity on that planet?

14.68 A horizontal tree branch is directly above another horizontal tree branch. The elevation of the higher branch is 9.65 m above the ground, and the elevation of the lower branch is 5.99 m above the ground. Some children decide to use the two branches to hold a tire swing. One end of the tire swing's rope is tied to the higher tree branch so that the bottom of the tire swing is 0.470 m above the ground. This swing is thus a restricted pendulum. Starting with the complete length of the rope at an initial angle of 14.2° with respect to the vertical, how long does it take a child of mass 29.9 kg to complete one swing back and forth?

•**14.69** Two pendulums with identical lengths of 1.000 m are suspended from the ceiling and begin swinging at the same time. One is at Manila, in the Philippines, where $g = 9.784$ m/s^2, and the other is at Oslo, Norway, where $g = 9.819$ m/s^2. After how many oscillations of the Manila pendulum will the two pendulums be in phase again? How long will it take for them to be in phase again?

•**14.70** Two springs, each with $k = 125$ N/m, are hung vertically, and 1.00-kg masses are attached to their ends. One spring is pulled down 5.00 cm and released at $t = 0$; the other is pulled down 4.00 cm and released at $t = 0.300$ s. Find the phase difference, in degrees, between the oscillations of the two masses and the equations for the vertical displacements of the masses, taking upward to be the positive direction.

•**14.71** A piston that has a small toy car sitting on it is undergoing simple harmonic motion vertically, with amplitude of 5.00 cm, as shown in the figure. When the frequency of oscillation is low, the car stays on the piston. However, when the frequency is increased sufficiently, the car leaves the piston. What is the maximum frequency at which the car will remain in place on the piston?

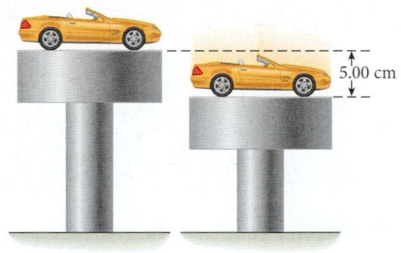

•**14.72** The period of a pendulum is 0.24 s on Earth. The period of the same pendulum is found to be 0.48 s on planet X, whose mass is equal to that of Earth. (a) Calculate the gravitational acceleration at the surface of planet X. (b) Find the radius of planet X in terms of that of Earth.

•**14.73** A grandfather clock uses a pendulum and a weight. The pendulum has a period of 2.00 s, and the mass of the bob is 250. g. The weight slowly falls, providing the energy to overcome the damping of the pendulum due to friction. The weight has a mass of 1.00 kg, and it moves down 25.0 cm every day. Find Q for this clock. Assume that the amplitude of the oscillation of the pendulum is 10.0°.

•**14.74** A cylindrical can of diameter 10.0 cm contains some ballast so that it floats vertically in water. The mass of can and ballast is 800.0 g, and the density of water is 1.00 g/cm^3. The can is lifted 1.00 cm from its equilibrium position and released at $t = 0$. Find its vertical displacement from equilibrium

as a function of time. Determine the period of the motion. Ignore the damping effect due to the viscosity of the water.

•**14.75** The period of oscillation of an object in a frictionless tunnel running through the center of the Moon is $T = 2\pi/\omega_0 = 6485$ s, as shown in Example 14.2. What is the period of oscillation of an object in a similar tunnel through the Earth ($R_E = 6.37 \cdot 10^6$ m; $R_M = 1.74 \cdot 10^6$ m; $M_E = 5.98 \cdot 10^{24}$ kg; $M_M = 7.35 \cdot 10^{22}$ kg)?

•**14.76** The motion of a planet in a circular orbit about a star obeys the equations of simple harmonic motion. If the orbit is observed edge-on, so the planet's motion appears to be one-dimensional, the analogy is quite direct: The motion of the planet looks just like the motion of an object on a spring.

a) Use Kepler's Third Law of planetary motion to determine the "spring constant" for a planet in circular orbit around a star with period T.

b) When the planet is at the extremes of its motion observed edge-on, the analogous "spring" is extended to its largest displacement. Using the "spring" analogy, determine the orbital velocity of the planet.

••**14.77** An object in simple harmonic motion is *isochronous*, meaning that the period of its oscillations is independent of their amplitude. (Contrary to a common assertion, the operation of a pendulum clock is not based on this principle. A pendulum clock operates at fixed, finite amplitude. The gearing of the clock compensates for the anharmonicity of the pendulum.) Consider an oscillator of mass m in one-dimensional motion, with a restoring force $F(x) = -cx^3$, where x is the displacement from equilibrium and c is a constant with appropriate units. The motion of this oscillator is periodic but not isochronous.

a) Write an expression for the period of the undamped oscillations of this oscillator. If your expression involves an integral, it should be a definite integral. You do not need to evaluate the expression.

b) Using the expression of part (a), determine the dependence of the period of oscillation on the amplitude.

c) Generalize the results of parts (a) and (b) to an oscillator of mass m in one-dimensional motion with a restoring force corresponding to the potential energy $U(x) = \gamma |x|^\alpha / \alpha$, where α is any positive value and γ is a constant.

MULTI-VERSION EXERCISES

14.78 A block of mass 1.605 kg is attached to a horizontal spring with spring constant 14.55 N/m and rests on a frictionless surface at the equilibrium position of the spring. The block is then pulled 12.09 cm from the equilibrium position and released. What is the distance of the block from the equilibrium position after $2.834 \cdot 10^{-1}$ s?

14.79 A block of mass 1.833 kg is attached to a horizontal spring with spring constant 14.97 N/m and rests on a frictionless surface at the equilibrium position of the spring. The block is then pulled 13.37 cm from the equilibrium position and released. At what time is the block located 4.990 cm from the equilibrium position?

14.80 A block of mass 1.061 kg is attached to a horizontal spring with spring constant 15.39 N/m and rests on a frictionless surface at the equilibrium position of the spring. The block is then pulled from the equilibrium position and released. After $3.900 \cdot 10^{-1}$ s, the block is located 1.25 cm from the equilibrium position. How far away from that position was the block pulled?

14.81 A pendulum consisting of a small sphere of mass 1.145 kg and a string of length 71.57 cm is suspended from a ceiling. Its motion is restricted by a peg that is sticking out of the wall 40.95 cm directly below the pivot point. What is the period of oscillation?

14.82 A pendulum consisting of a small sphere of mass 1.261 kg and a string of length 72.39 cm is suspended from a ceiling. Its motion is

restricted by a peg that is sticking out of the wall directly below the pivot point. The period of oscillation is 1.404 s. How far below the ceiling is the restricting peg?

14.83 A vertical spring with spring constant 23.31 N/m is hanging from a ceiling. A small object with a mass of 1.375 kg is attached to the lower end of the spring, and the spring stretches to its equilibrium length. The object is then pulled down a distance of 18.51 cm and released. What is the speed of the object when it is 1.849 cm from the equilibrium position?

14.84 A vertical spring with spring constant 23.51 N/m is hanging from a ceiling. A small object is attached to the lower end of the spring, and the spring stretches to its equilibrium length. The object is then pulled down a distance of 19.79 cm and released. The speed of the object a distance 7.417 cm from the equilibrium point is 0.7286 m/s. What is the mass of the object?

14.85 A vertical spring with spring constant 23.73 N/m is hanging from a ceiling. A small object with mass 1.103 kg is attached to the lower end of the spring, and the spring stretches to its equilibrium length. The object is then pulled down and released. The speed of the object when it is 4.985 cm from the equilibrium position is 0.4585 m/s. How far down was the object pulled?

15

Waves

FIGURE 15.1 Surfer riding a huge wave in Australia.

When most people hear the word *waves,* the first thing they think of is ocean waves. Everybody has seen pictures of surfers riding enormous ocean waves in Hawaii or Australia (Figure 15.1) or of the towering waves created by a fierce storm on the open sea. These walls of water are certainly waves, and so are gentle breakers that wash onto a beach. However, waves are far more widespread than disturbances in water. Light and sound are both waves, and wave motion is involved in earthquakes and tsunamis as well as radio and television broadcasts. In fact, later in this book, we'll see that the smallest, most fundamental particles of matter share many properties with waves. In a sense, we are all made of waves as much as particles.

In this chapter, we examine the properties and behavior of waves, building directly on Chapter 14's coverage of oscillations. We'll return to the subject of wave motion when we study light and when we consider quantum mechanics and atomic structure. Whole areas of physics are devoted to wave studies; for example, optics is the science of light waves and acoustics is the study of sound waves (see Chapter 16). Waves are also of fundamental importance in astronomy and in electronics. The concepts in this chapter are just the first steps toward understanding a large part of the physical world of waves around us.

WHAT WE WILL LEARN

- Waves are disturbances (or excitations) that propagate through space or some medium as a function of time. A wave can be understood as a series of individual oscillators coupled to their nearest neighbors.

- Waves can be either longitudinal or transverse.

- The wave equation governs the general motion of waves.

- Waves transport energy across distances; however, even though a wave is propagated by vibrating atoms in a medium, it does not generally transport matter with it.

- The superposition principle states a key property of waves: Two waveforms can be added to form another one.

- Waves can interfere with each other constructively or destructively. Under certain conditions, interference patterns form and can be analyzed.

- A standing wave can be formed by superposition of two traveling waves.

- Boundary conditions on strings determine the possible wavelengths and frequencies of standing waves.

15.1   Wave Motion

If you throw a stone in a still pond, it creates circular ripples that travel along the surface of the water, radially outward (Figure 15.2). If an object is floating on the surface of the pond (a stick, a leaf, or a rubber duck), you can observe that the object moves up and down as the ripples move outward underneath it but does not move noticeably outward with the waves.

If you tie a rope to a wall, hold one end and stretch the rope out, and then rapidly move your arm up and down, this motion will create a wave crest that travels down the rope, as illustrated in Figure 15.3. Again this wave moves, but the material that the wave travels through (the rope) moves up and down but otherwise remains in place.

These two examples illustrate a major characteristic of waves: A **wave** is an excitation that propagates through space or some medium as a function of time but does not generally transport matter along with it. In this fundamental way, wave motion is completely different from the motion we have studied so far. Electromagnetic and gravitational waves do not need a medium in which to propagate; they can move through empty space. Other waves do propagate in a medium; for example, sound waves can travel through a gas (air) or a liquid (water) or a solid (steel rails), but not through a vacuum.

Many other examples of waves surround us in everyday life, most of them with oscillations that are too small for us to observe. Light is an electromagnetic wave, as are the radio waves that FM and AM radios receive and the waves that carry TV signals into our houses—either through coaxial cables or through the air via broadcasts from antenna towers or geostationary satellites. We'll focus on electromagnetic waves in Chapter 31 (after we have studied electricity and magnetism in Chapters 21 through 30).

In this chapter, we study the motion of waves through media. Mechanical vibrations fall in this category, as do sound waves, earthquake waves, and water waves. Since sound is such an important part of human communication, Chapter 16 is devoted exclusively to it.

The stadium wave—called simply "the wave"—is a part of most major sports events: Spectators stand up and raise their arms as soon as they see their neighbor doing it, and then they sit down again. Thus, the wave travels around the stadium, but each spectator stays in place. How fast does this wave travel? Suppose that each person has a time delay comparable to his or her reaction time, 0.1 s, before imitating the adjacent spectator and getting up from the seat. Typically, seats are approximately 0.6 m (2 ft) wide. This means that the wave travels 0.6 m in 0.1 s. From simple kinematics, we know that this corresponds to a velocity of $v = \Delta x/\Delta t = (0.6 \text{ m})/(0.1 \text{ s}) = 6$ m/s. Since a typical football stadium is an oval with a circumference of approximately 400 m, we can estimate that it takes about 1 min for "the wave" to make it around the stadium once. This rather crude estimate is approximately right, within a factor of 2. Empirically, we find that "the wave" moves with a speed of approximately 12 m/s. The reason for this higher speed is that people anticipate the arrival of the wave and jump up when they see the person two or three seats ahead of them jump up.

FIGURE 15.2 Circular ripples on the surface of water.

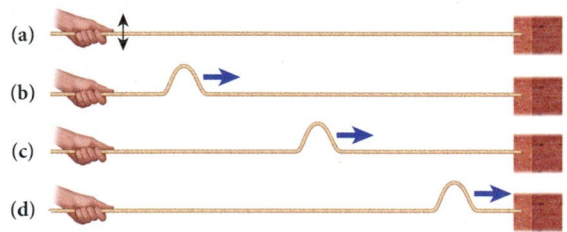

FIGURE 15.3 A rope held by a hand at one end and attached firmly to a wall at the other end. (a) The end of the rope is moved up and down by moving the hand. In (b), (c), and (d), a wave travels along the rope toward the wall.

15.2   Coupled Oscillators

Let's begin our quantitative discussion of waves by examining a series of coupled oscillators—for example, a string of identical massive objects coupled by identical springs. One physical realization is shown in Figure 15.4, where a series of identical metal rods is held in place by a central horizontal bar. The rods can pivot freely about their centers and are connected to their nearest neighbors by identical springs. When the first rod is pushed, this excitation is transmitted from each rod to its neighbor, and the pulse generated in this way travels down the row of rods. Each rod is an oscillator and moves back and forth in the plane of the page. This setup represents a longitudinal coupling between the oscillators, in the sense that the direction of motion and the direction of coupling to the nearest neighbors are the same.

A physical realization of a transverse coupling is shown in Figure 15.5. In part (a) of the figure, all rods are attached to a tension strip in their horizontal equilibrium position; in part (b), rod i is deflected from the horizontal by an angle θ_i. A deflection of this kind has two effects: First, the tension strip provides a restoring force that tries to pull rod i back to the equilibrium position. Second, since the tension strip is now twisted on both sides of rod i, the deflection of that rod at an angle θ_i creates a force that acts to deflect rods $i - 1$ and $i + 1$ toward θ_i. If those rods are allowed to move freely, they will be deflected in that direction, and each rod in turn will generate a force acting on its two nearest neighbors.

Figure 15.6 shows a side view of the arrangement of parallel rods in Figure 15.5. Each circle represents the end of a rod, and the motion of this circle can be characterized by a coordinate y_i. Then y_i is related to the deflection angle, θ_i, via $y_i = \frac{1}{2}l \sin \theta_i$, where l is the length of the rod. The line connecting the circles represents the edge of the tension strip. In Figure 15.6a, all rods are in equilibrium, which implies that all deflections are zero. In Figure 15.6b, rod i is experiencing a deflection, the same as is depicted in Figure 15.5b. The force acting on rod i due to the tension in the strip is a type of spring force given by $F_i = -2ky_i$, where the effective spring constant is determined from the properties of the tension strip. Finally, in Figure 15.6c, all rods are allowed to move. The net force acting on rod i is then the sum of the forces exerted on it by rods $i + 1$ and $i - 1$, as mediated by the tension strip connecting the rods. These forces depend on the difference between the y-coordinates at the beginning and the end of the section of tension strip that connects the three rods:

$$F_+ = -k(y_i - y_{i+1})$$

and

$$F_- = -k(y_i - y_{i-1}).$$

Thus, the net force acting on rod i is the sum of these two forces:

$$F_i = F_+ + F_- = -k(y_i - y_{i+1}) - k(y_i - y_{i-1}) = k(y_{i+1} - 2y_i + y_{i-1}).$$

From Newton's Second Law, $F = ma$, we then find the acceleration, a_i, of rod i:

$$ma_i = k(y_{i+1} - 2y_i + y_{i-1}). \qquad (15.1)$$

With initial conditions for the position and velocity of each oscillator, we could solve the resulting set of equations on a computer. However, we can actually derive a wave equation using equation 15.1, and we'll do so in Section 15.4.

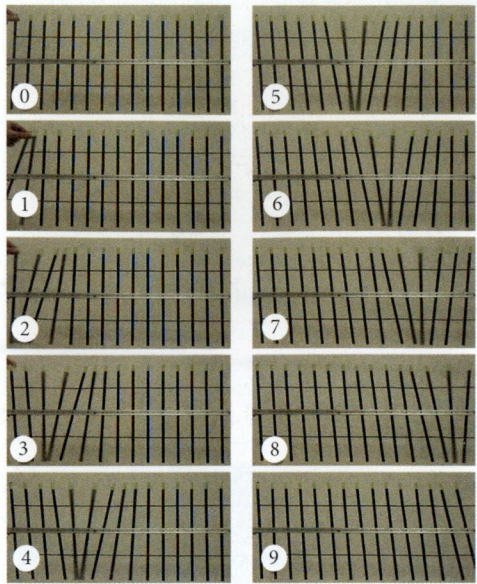

FIGURE 15.4 Series of identical rods coupled to their nearest neighbors by springs. The numbers on the frames indicate the time sequence. Each pair of frames is separated by a time interval of 0.133 s.

FIGURE 15.5 Parallel rods supported by a tension strip as a model for transversely coupled oscillators: (a) initial position, all rods parallel; (b) rod i is deflected by an angle θ_i.

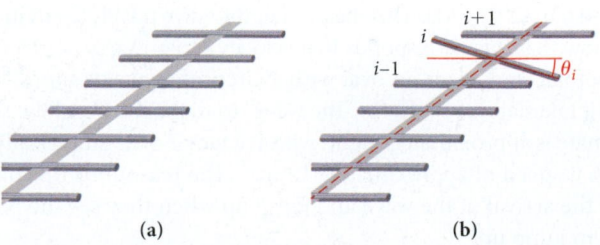

(a)   (b)

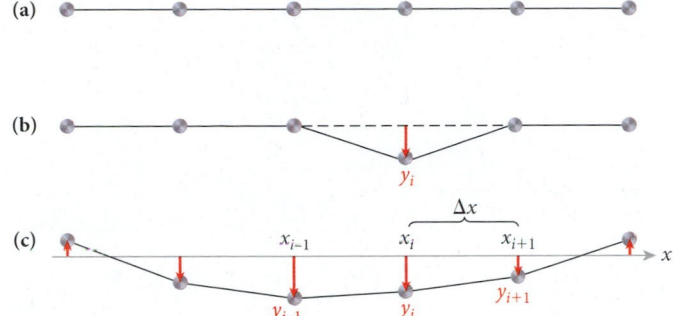

FIGURE 15.6 Side views of coupled rods of Figure 15.5: (a) equilibrium position; (b) one rod is deflected; (c) all rods are allowed to move and become deflected.

Transverse and Longitudinal Waves

We need to emphasize an essential difference between the two systems of coupled oscillators shown in Figure 15.4 and Figure 15.5. In the first case, the oscillators are allowed to move only in the direction of their nearest neighbors, whereas in the second case, the oscillators are restricted to moving in a direction perpendicular to the direction of their nearest neighbors (Figure 15.7). In general, a wave that propagates along the direction in which the oscillators move is called a **longitudinal wave.** In Chapter 16, we'll see that sound waves are prototypical longitudinal waves. A wave that moves in a direction perpendicular to the direction in which the individual oscillators move is called a **transverse wave.** Light waves are transverse waves, as we will discuss in Chapter 31 on electromagnetic waves.

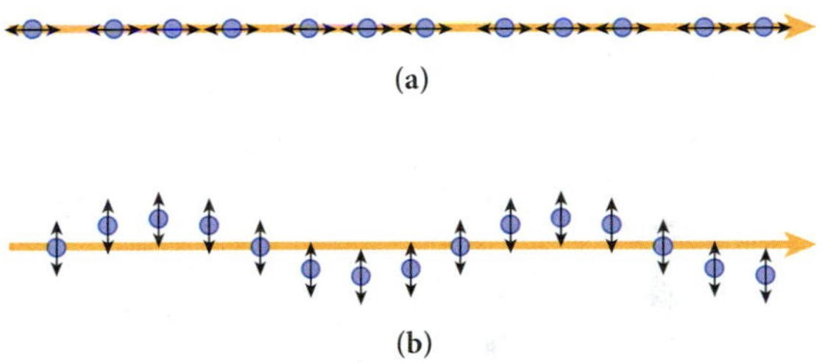

Seismic waves created by earthquakes exist as longitudinal (or compression) waves and as transverse waves. These waves can travel along the surface of the Earth as well as through the Earth and be detected far from the epicenter of the earthquake. The study of seismic waves has revealed detailed information about the composition of the interior of the Earth, as we'll see in Example 15.2.

FIGURE 15.7 Representations of (a) longitudinal wave and (b) transverse wave propagating horizontally to the right. The long yellow arrows indicate the direction of wave propagation. The blue circles represent the individual oscillators, and the black double-headed arrows show the direction of their oscillations.

15.3 Mathematical Description of Waves

So far, we have examined single wave pulses, such as the stadium wave and the wave that results from pushing once on a chain of coupled oscillators. A much more common class of wave phenomena involves a periodic excitation—in particular, a sinusoidal excitation. In the chain of oscillators of Figure 15.4 or Figure 15.5, a continuous wave can be generated by moving the first rod back and forth periodically. This sinusoidal oscillation also travels down the chain of oscillators, just like a single wave pulse.

Period, Wavelength, and Velocity

Figure 15.8a shows the result of a sinusoidal excitation as a function of both time and the horizontal coordinate, in the limit where a very large number of coupled oscillators are very close to each other—that is, in the continuum limit. First, let's examine this wave along the time axis for $x = 0$. This projection is shown in Figure 15.8b, which is a plot of the sinusoidal oscillation of the first oscillator in the chain:

$$y(x = 0, t) = A\sin(\omega t + \theta_0) = A\sin(2\pi ft + \theta_0), \tag{15.2}$$

where ω is the angular frequency (or angular velocity), f is the frequency, A is the amplitude of the oscillation, and θ_0 gives the phase shift. (The values used in Figure 15.8 are $A = 2.0$ cm, $\omega = 0.2$ s^{-1}, and $\theta_0 = 0.5$.) As for all oscillators, the period T is defined as the time interval between two successive maxima and is related to the frequency by

$$T = \frac{1}{f}.$$

FIGURE 15.8 Sinusoidal wave:
(a) the wave as a function of space and time coordinates; (b) dependence of the wave on time at position $x = 0$; (c) dependence of the wave on position at time $t = 0$ (blue curve) and at a later time $t = 1.5$ s (gray dashed curve).

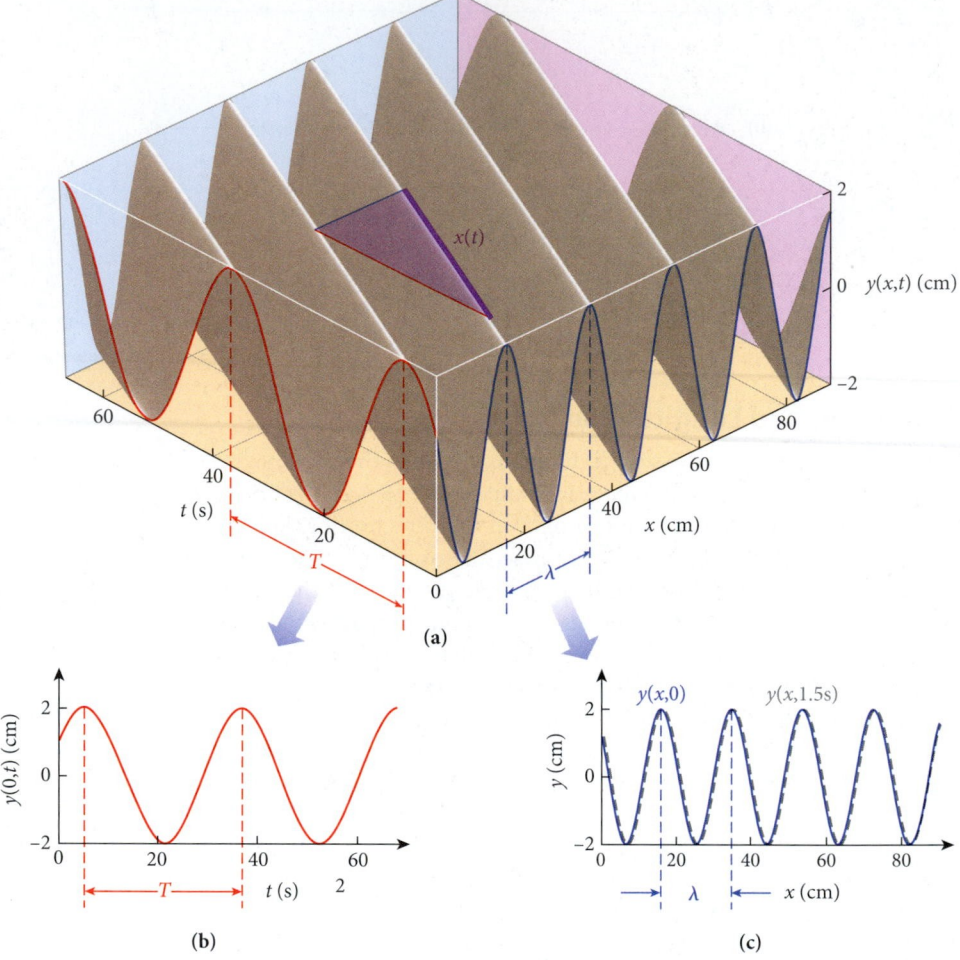

(a)

(b)

(c)

Concept Check 15.1

The sinusoidal wave shown in Figure 15.8 is

a) traveling to the right (toward larger values of x).

b) traveling to the left (toward smaller values of x).

c) traveling neither to the right nor to the left.

d) Since this is just a snapshot of a wave, the direction of motion cannot be inferred.

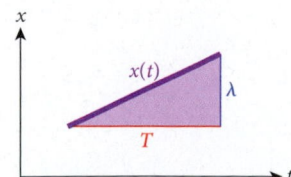

FIGURE 15.9 View of the xt-plane of Figure 15.8a showing that the slope of the line $x(t)$ is equal to the wave velocity defined by the ratio of the wavelength divided by the period, $v = \lambda/T$.

Figure 15.8c shows that for any given time, t, the dependence on the horizontal coordinate, x, of the oscillations is also sinusoidal. This dependence is shown for $t = 0$ (blue curve) and $t = 1.5$ s (gray dashed curve). The distance in space between two consecutive maxima is defined as the **wavelength**, λ.

As you can see in Figure 15.8a, diagonal lines of equal height describe the wave motion in the xt-plane. What is most important to realize about the wave motion depicted in Figure 15.8 and wave motion in general is that during one period, any given point on the wave advances by one wavelength.

Figure 15.9 presents the light purple triangle shown on the top surface of the wave in Figure 15.8a in the xt-plane to further illustrate the relationship among the wave velocity, the period, and the wavelength. Now you can see that the speed of propagation of the wave is in general given by

$$v = \frac{\Delta x}{\Delta t} = \frac{\lambda}{T},$$

or, using the relationship between period and frequency, $T = 1/f$, by

$$v = \lambda f. \tag{15.3}$$

This relationship among wavelength, frequency, and velocity holds for all types of waves and is one of the most important results in this chapter.

Sinusoidal Waveform, Wave Number, and Phase

How do we obtain a mathematical description of a sinusoidal wave as a function of space and time? We start with the motion of the oscillator at $x = 0$ (see equation 15.2), and realize that it takes a time $t = x/v$ for the excitation to move a distance x. Thus, we can simply replace t in equation 15.2 by $t - x/v$ to find the displacement $y(x,t)$ away from the equilibrium position as a function of space and time:

$$y(x,t) = A \sin\left(\omega\left(t - \frac{x}{v}\right) + \theta_0\right).$$

We can then use equation 15.3 for the wave velocity, $v = \lambda f$, and obtain

$$y(x,t) = A \sin\left(2\pi f\left(t - \frac{x}{v}\right) + \theta_0\right)$$

$$= A \sin\left(2\pi f t - \frac{2\pi f x}{\lambda f} + \theta_0\right)$$

$$= A \sin\left(\frac{2\pi}{T}t - \frac{2\pi}{\lambda}x + \theta_0\right).$$

Similar to $\omega = 2\pi/T$, the expression for the **wave number**, κ, is

$$\kappa = \frac{2\pi}{\lambda}. \tag{15.4}$$

Substituting ω and κ into the above expression for $y(x,t)$ then yields the descriptive function we want:

$$y(x,t) = A \sin\left(\omega t - \kappa x + \theta_0\right). \tag{15.5}$$

A sinusoidal waveform of this kind is almost universally applicable when there is only one spatial dimension in which the wave can propagate. Later in this chapter, we'll generalize this waveform to more than one spatial coordinate.

We can also start with a given sinusoidal waveform at some point in space and ask where this waveform will be at a later time. From this perspective, we can write the expression for the waveform as a function of space and time as

$$y(x,t) = A \sin\left(\kappa x - \omega t + \phi_0\right). \tag{15.6}$$

Equations 15.5 and 15.6 are equivalent to each other, because $\sin(-x) = -\sin(x) = \sin(\pi + x)$. The phase shift, ϕ_0, in equation 15.6 is related to the phase angle, θ_0, in equation 15.5 via $\phi_0 = \pi - \theta_0$. We'll use equation 15.6, which is the more common form, from here on.

The argument of the sine function in equation 15.6 is called the wave's **phase**, ϕ:

$$\phi = \kappa x - \omega t + \phi_0. \tag{15.7}$$

In considering more than one wave simultaneously, the relationship of the phases of the waves plays an essential role in the analysis. Later in this chapter, we'll return to this point.

Finally, using $\kappa = 2\pi/\lambda$ and $\omega = 2\pi/T$, we can rewrite the equation for the velocity of the wave front (15.3) in terms of the wave number and the angular frequency:

$$v = \frac{\omega}{\kappa}. \tag{15.8}$$

The angular frequency, ω, counts the number of oscillations in each time interval of 2π s: $\omega = 2\pi/T$, where T is the period. Similarly, the wave number, κ, counts the number of wavelengths, λ, that fit into a distance of 2π m: $\kappa = 2\pi/\lambda$. For example, $\kappa = 0.33$ cm^{-1} in the plot of Figure 15.8.

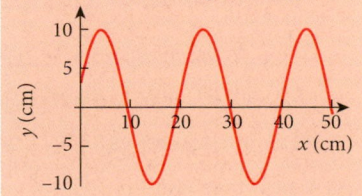

SOLVED PROBLEM 15.1 | Traveling Wave

PROBLEM

A wave on a string is given by the function

$$y(x,t) = (0.00200 \text{ m})\sin\left[\left(78.8 \text{ m}^{-1}\right)x + \left(346 \text{ s}^{-1}\right)t\right].$$

What is the wavelength of this wave? What is its period? What is its velocity?

SOLUTION

THINK From the given wave function, we can identify the wave number and the angular frequency, from which we can extract the wavelength and period. We can then calculate the velocity of the traveling wave from the ratio of the angular frequency to the wave number.

SKETCH Figure 15.10 shows plots of the wave function at $t = 0$ (part a) and at $x = 0$ (part b).

FIGURE 15.10 (a) The wave function at $t = 0$. (b) The wave function as a function of time at $x = 0$.

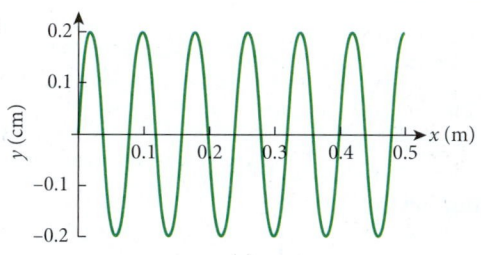

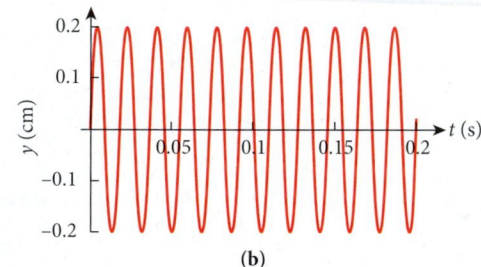

RESEARCH Looking at the wave function given in the problem statement, $y(x,t) = (0.00200 \text{ m})$ $\sin\left[\left(78.8 \text{ s}^{-1}\right)x + \left(346 \text{ s}^{-1}\right)t\right]$, and comparing it with the solution to the wave equation for a wave traveling in the negative x-direction (see equation 15.6),

$$y(x,t) = A\sin(\kappa x + \omega t + \phi_0),$$

we see that $\kappa = 78.8 \text{ m}^{-1}$ and $\omega = 346 \text{ s}^{-1}$.

SIMPLIFY We can obtain the wavelength from the wave number,

$$\lambda = \frac{2\pi}{\kappa},$$

and the period from the angular frequency,

$$T = \frac{2\pi}{\omega}.$$

We can calculate the velocity of the wave from

$$v = \frac{\omega}{\kappa}.$$

CALCULATE Putting in the numerical values, we get the wavelength and the period:

$$\lambda = \frac{2\pi}{78.8 \text{ m}^{-1}} = 0.079736 \text{ m}$$

and

$$T = \frac{2\pi}{346 \text{ s}^{-1}} = 0.018159 \text{ s}.$$

We now calculate the velocity of the traveling wave:

$$v = \frac{346 \text{ s}^{-1}}{78.8 \text{ m}^{-1}} = 4.39086 \text{ m/s}.$$

ROUND We report our results to three significant figures:

$$\lambda = 0.0797 \text{ m}$$
$$T = 0.0182 \text{ s}$$
$$v = 4.39 \text{ m/s}.$$

Concept Check 15.2

The sinusoidal wave in Solved Problem 15.1 is traveling in the negative x-direction. Which of the following waves is traveling in the positive x-direction?

a) $y(x,t) = (-0.002 \text{ m})\sin[(78.8 \text{ m}^{-1})x + (346 \text{ s}^{-1})t]$

b) $y(x,t) = (0.002 \text{ m})\sin[(78.8 \text{ m}^{-1})x + (346 \text{ s}^{-1})t - 1]$

c) $y(x,t) = (0.002 \text{ m})\sin[(-78.8 \text{ m}^{-1})x + (346 \text{ s}^{-1})t]$

d) None of the above.

e) All of the above.

DOUBLE-CHECK To double-check our result for the wavelength, we look at Figure 15.10a. We can see that the wavelength (the distance required to complete one oscillation) is approximately 0.08 m, in agreement with our result. Looking at Figure 15.10b, we can see that the period (the time required to complete one oscillation) is approximately 0.018 s, also in agreement with our result.

15.4 Derivation of the Wave Equation

In this section, we derive a general equation of motion for waves, called simply the **wave equation:**

$$\frac{\partial^2}{\partial t^2} y(x,t) - v^2 \frac{\partial^2}{\partial x^2} y(x,t) = 0, \tag{15.9}$$

where $y(x,t)$ is the displacement from the equilibrium position as a function of a single space coordinate, x, and a time coordinate, t, and v is the speed of propagation of the wave. This wave equation describes all undamped wave motion in one dimension.

DERIVATION 15.1   Wave Equation

Figure 15.11 reproduces Figure 15.6c. The x-axis, pointing in the conventional horizontal direction and to the right, is (as usual) perpendicular to the y-axis, along which the motion of each individual oscillator occurs. The horizontal distance between neighboring oscillators is Δx. The displacements of the oscillators, given by $y(x,t)$, are then functions of time and of position along the x-axis, with $y_i = y(x_i, t)$. The acceleration is defined as the second derivative of the displacement, $y(x,t)$, with respect to time. Since the displacement depends on more than one variable in this case (time as well as the horizontal coordinate), we need to use a partial derivative to express this relationship:

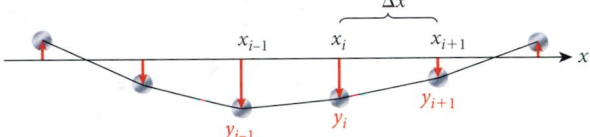

FIGURE 15.11 Side views of coupled rods where all rods are allowed to deflect.

$$a = \frac{\partial^2}{\partial t^2} y(x,t). \tag{i}$$

If we go back to the original definition of the derivative as the limit of the ratio of differences, we can approximate a partial derivative with respect to the x-coordinate as

$$\frac{\partial}{\partial x} y(x,t) = \frac{\Delta y}{\Delta x} = \frac{y_{i+1} - y_i}{\Delta x}.$$

We can then approximate the second derivative as

$$\frac{\partial^2}{\partial x^2} y(x,t) = \frac{\Delta\left(\frac{\Delta y}{\Delta x}\right)}{\Delta x} = \frac{\frac{(y_{i+1} - y_i)}{\Delta x} - \frac{(y_i - y_{i-1})}{\Delta x}}{\Delta x} = \frac{y_{i+1} - 2y_i + y_{i-1}}{(\Delta x)^2}.$$

Both the right-hand side of this equation and the right-hand side of equation 15.1 contain $y_{i+1} - 2y_i + y_{i-1}$. Thus, we can combine these two equations:

$$ma_i = k(\Delta x)^2 \frac{\partial^2}{\partial x^2} y(x,t).$$

Substituting the second partial derivative of position with respect to time from equation (i) for the acceleration, we find

$$m\left[\frac{\partial^2}{\partial t^2} y(x,t)\right] = k(\Delta x)^2 \frac{\partial^2}{\partial x^2} y(x,t).$$

Finally, we divide by the mass of the oscillator and obtain

$$\frac{\partial^2}{\partial t^2} y(x,t) - \frac{k}{m}(\Delta x)^2 \frac{\partial^2}{\partial x^2} y(x,t) = 0. \tag{ii}$$

– *Continued*

This equation relating the second partial derivatives with respect to time and position is a form of the wave equation. The term $\frac{k}{m}(\Delta x)^2$ in equation (ii) is a constant determined by the specific parameters of the coupled oscillator being studied. We can define a constant $a^2 = \frac{k}{m}(\Delta x)^2$ with units $(\text{m/s})^2$ so that equation (ii) becomes the wave equation

$$\frac{\partial^2}{\partial t^2} y(x,t) - a^2 \frac{\partial^2}{\partial x^2} y(x,t) = 0,$$

where a^2 is a constant to be determined.

The factor $\frac{k}{m}(\Delta x)^2$ can be factored into two components, $k(\Delta x)$ and $1/\left(m/(\Delta x)\right)$. In the limit where the oscillators get very close to one another, the first factor remains constant because as a spring is shortened, its spring constant increases. The second term is the inverse of the linear mass density, which is the same for any small section as it is for the overall system. Thus, the solution to the wave equation 15.9 provides the exact solution to the difference equation 15.1. Equation 15.9 is of general importance, and its applicability far exceeds the problem of coupled oscillators. The basic form given in equation 15.9 holds for the propagation of sound waves, light waves, or waves on a string. The difference in physical systems exhibiting wave motion manifests itself only in the way v, the speed of propagation of the wave, is calculated.

In mathematical terms, equation 15.9 is a partial differential equation. As you may know, some semester-long courses are entirely devoted to partial differential equations, and you will encounter many more partial differential equations if you major in the physical sciences or engineering.

Solutions of the Wave Equation

In Section 15.3, we developed equation 15.6, describing a sinusoidal waveform. Let's see if this is a solution of the general wave equation found in Derivation 15.1. To do this, we have to take the partial derivatives of equation 15.6 with respect to the x-coordinate and with respect to time:

$$\frac{\partial}{\partial t} y(x,t) = \frac{\partial}{\partial t}\left(A\sin\left(\kappa x - \omega t + \phi_0\right)\right) = -A\omega\cos\left(\kappa x - \omega t + \phi_0\right)$$

$$\frac{\partial^2}{\partial t^2} y(x,t) = \frac{\partial}{\partial t}\left(-A\omega\cos\left(\kappa x - \omega t + \phi_0\right)\right) = -A\omega^2\sin\left(\kappa x - \omega t + \phi_0\right)$$

$$\frac{\partial}{\partial x} y(x,t) = \frac{\partial}{\partial x}\left(A\sin\left(\kappa x - \omega t + \phi_0\right)\right) = A\kappa\cos\left(\kappa x - \omega t + \phi_0\right)$$

$$\frac{\partial^2}{\partial x^2} y(x,t) = \frac{\partial}{\partial x}\left(A\kappa\cos\left(\kappa x - \omega t + \phi_0\right)\right) = -A\kappa^2\sin\left(\kappa x - \omega t + \phi_0\right).$$

Inserting these partial derivatives into the result from Derivation 15.1, we find

$$\frac{\partial^2}{\partial t^2} y(x,t) - a^2 \frac{\partial^2}{\partial x^2} y(x,t) =$$

$$\left[-A\omega^2\sin\left(\kappa x - \omega t + \phi_0\right)\right] - v^2\left[-A\kappa^2\sin\left(\kappa x - \omega t + \phi_0\right)\right] =$$

$$-A\sin\left(\kappa x - \omega t + \phi_0\right)\left(\omega^2 - a^2\kappa^2\right) = 0. \qquad (15.10)$$

Thus we see that the function given by equation 15.6 is a solution of the wave equation derived in Derivation 15.1 provided that $\omega^2 - a^2\kappa^2 = 0$, or equivalently $a^2 = \omega^2/\kappa^2$. From equation 15.8 we know that $v = \omega/\kappa$ so equation 15.6 is a solution to the wave equation with $a^2 = v^2$, corresponding to the general wave equation stated in equation 15.9.

Is there a larger class of solutions that includes the one we have already found? Yes, any arbitrary, continuously differentiable function Y with an argument containing the same linear combination of x and t as in equation 15.6, $\kappa x - \omega t$, is a solution to the wave equation. The constants ω, κ, and ϕ_0 can have arbitrary values. However, the convention generally followed is that the angular frequency and the wave number are both positive numbers. Thus, there are two general solutions to the wave equation:

$$Y(\kappa x - \omega t + \phi_0) \text{ for a wave moving in the positive } x\text{-direction} \qquad (15.11)$$

and

$$Y(\kappa x + \omega t + \phi_0) \text{ for a wave moving in the negative } x\text{-direction}. \qquad (15.12)$$

This result is perhaps more obvious if we write these functions in terms of arguments that involve the wave speed, $v = \omega/\kappa$. We then see that $Y(x - vt + \phi_0/\kappa)$ is a solution for an arbitrary waveform moving in the positive x-direction and $Y(x + vt + \phi_0/\kappa)$ is a solution for a waveform moving in the negative x-direction.

As previously stated, the functional form of Y does not matter at all. We can see this by looking at the chain rule for differentiation, $df(g(x))/dx = (df/dg)(dg/dx)$. The partial derivatives in space and time contain the common factor df/dg, which cancels out in the same way that the sine function factored out in equation 15.10. Thus, the derivatives of the function $g = \kappa x \pm \omega t + \phi_0$ always lead to the same condition, $\omega^2 - v^2\kappa^2 = 0$, as applied to the solution for the sinusoidal waveform.

Waves on a String

String instruments form a large class of musical instruments. Guitars (Figure 15.12), cellos, violins, mandolins, harps, and others fall into this category. These instruments produce musical sounds when vibrations are induced on their strings.

Suppose a string has a mass M and a length L. In order to find the wave speed, v, on this string, we can think of it as composed of many small individual segments of length Δx and mass m, where

$$m = \frac{M\Delta x}{L} = \mu\Delta x.$$

Here the linear mass density, μ, of the string is its mass per unit length (assumed to be constant):

$$\mu = \frac{M}{L}.$$

We can then treat the motion of a wave on a string as that of coupled oscillators. The restoring force is provided by the string tension force, F. The motion of a wave on a string is described by the wave equation with a wave velocity given by

$$v = \sqrt{\frac{F}{\mu}}. \qquad (15.13)$$

If you are curious about the origin of equation 15.13, you can work through Conceptual Question 15.21.

What does equation 15.13 imply? There are two immediate consequences. First, increasing the linear mass density of the string (that is, using a heavier string) reduces the wave velocity on the string. Second, increasing the string tension (for example, by tightening the tuning knobs at the top of the electric guitar shown in Figure 15.12) increases the wave velocity. If you play a string instrument, you know that tightening the tension increases the pitch of the sound produced by a string. We'll return to the relationship of tension to sound later in this chapter.

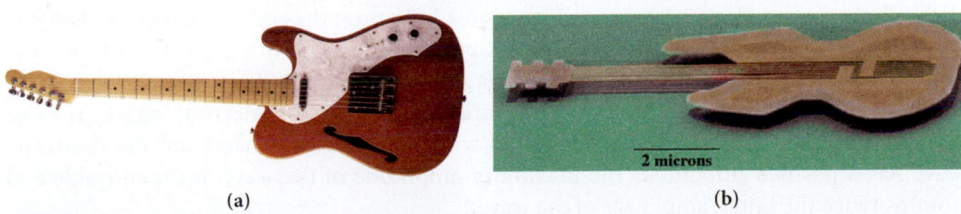

(a)   (b)

FIGURE 15.12 (a) Typical electric guitar used by a rock band. (b) World's smallest guitar, produced in 2004 in the Cornell NanoScale Science and Technology Facility. It is smaller than the guitar in part (a) by a factor of 10^5 and has a length of 10 μm, approximately $\frac{1}{10}$ of the width of a human hair.

EXAMPLE 15.1 | Elevator Cable

An elevator repairman (mass = 73 kg) sits on top of an elevator cabin of mass 655 kg inside a shaft in a skyscraper. A 61-m-long steel cable of mass 38 kg suspends the cabin. The man sends a signal to his colleague at the top of the elevator shaft by tapping the cable with his hammer.

PROBLEM

How long will it take for the wave pulse generated by the hammer to travel up the cable?

SOLUTION

The cable is under tension from the combined weight of the elevator cabin and the repairman:

$$F = mg = (73 \text{ kg} + 655 \text{ kg})(9.81 \text{ m/s}^2) = 7142 \text{ N}.$$

The linear mass density of the steel cable is

$$\mu = \frac{M}{L} = \frac{38 \text{ kg}}{61 \text{ m}} = 0.623 \text{ kg/m}.$$

Therefore, according to equation 15.13, the wave speed on this cable is

$$v = \sqrt{\frac{F}{\mu}} = 107 \text{ m/s}.$$

To travel up the 61-m-long cable, the pulse needs a time of

$$t = \frac{L}{v} = \frac{61 \text{ m}}{107 \text{ m/s}} = 0.57 \text{ s}.$$

DISCUSSION

This solution implies that the string tension is the same everywhere in the cable. Is this true? Not quite! We cannot really neglect the weight of the cable itself in determining the tension. The speed we calculated is correct for the lowest point of the cable, where it attaches to the cabin. However, the tension at the upper end of the cable also has to include the complete weight of the cable. There the tension is

$$F = mg = (73 \text{ kg} + 655 \text{ kg} + 38 \text{ kg})(9.81 \text{ m/s}^2) = 7514 \text{ N},$$

which yields a wave speed of 110 m/s, about 3% greater than what we calculated. If we had to take this effect into account, we would have to calculate how the wave speed depends on the height and then perform the proper integration. However, since the maximum correction is 3%, we can assume that the average correction is about 1.5%, use an average wave speed of 108.5 m/s, and obtain a signal propagation time of 0.56 s. Thus, taking the mass of the cable into account changes the result by only 0.01 s.

Reflection of Waves

What happens when a wave traveling on a string encounters a boundary? The end of a string tied to a support on the wall can be considered a boundary. In Figure 15.3, a wave was initiated on a rope by moving one end of the rope up and down. The wave then traveled along the rope. When the wave reaches the end of the rope that is tied to a fixed support on the wall, the wave reflects as shown in Figure 15.13. Because the end of the rope is fixed, the reflected wave is inverted and returns in the opposite direction with a negative height. The amplitude of the wave at the fixed end point is zero.

If the end of the rope is attached via a frictionless movable connection, such as the one shown in Figure 15.14, the wave is reflected with no change in phase, and the returning wave has a positive amplitude. The maximum amplitude of the wave at the movable end point is twice the initial amplitude of the wave.

In Chapter 34, we'll see that a light wave can be reflected at a boundary between two optical media with or without a phase change, depending on the properties of the two media.

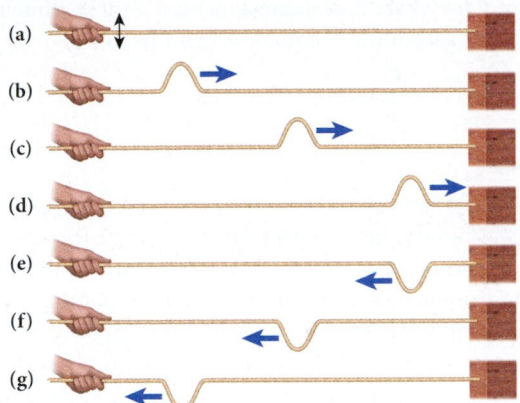

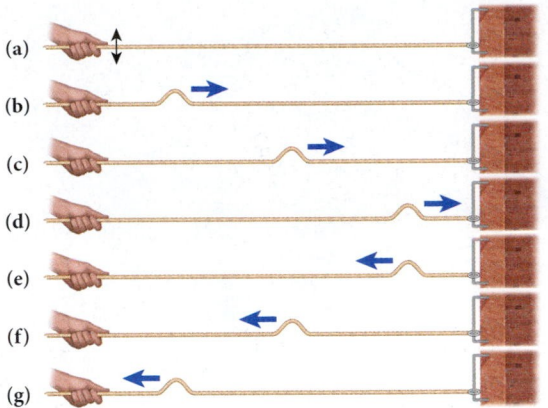

FIGURE 15.13 A rope held by a hand at one end and attached firmly to a wall at the other end. (a) The end of the rope is moved up and down. In (b), (c), and (d), a wave travels along the rope toward the right. In (e), (f), and (g), the wave travels along the rope toward the left after reflecting from the fixed end point at the wall.

FIGURE 15.14 A rope held by a hand at one end and attached with a frictionless movable connection to a wall at the other end. (a) The end of the rope is moved up and down. In (b), (c), and (d), a wave travels along the rope toward the right. In (e), (f), and (g), the wave travels along the rope toward the left after reflecting from the movable end point.

15.5 Waves in Two- and Three-Dimensional Spaces

So far, we have only described one-dimensional waves, those moving on a string and on a linear chain of coupled oscillators. However, waves on the surface of water, like those in Figure 15.2, spread in a two-dimensional space—the plane of the water surface. A light bulb and a loudspeaker also emit waves, and these waves spread in three-dimensions. Thus, a more complete mathematical description of wave motion is needed. A completely general treatment exceeds the scope of this text, but we can examine two special cases that cover most of the important wave phenomena that occur in nature.

Spherical Waves

If you examine Figure 15.2 and Figure 15.15, you see waves generated from one point and spreading as concentric circles. The mathematical idealization of waves emitted from point sources describes the waves as moving outward in concentric circles in two spatial dimensions or in concentric spheres in three spatial dimensions. Sinusoidal waveforms that spread in this way are described mathematically by

$$\psi(\vec{r},t) = A(r)\sin(\kappa r - \omega t + \phi_0). \tag{15.14}$$

Here $\vec{r}$ is the position vector (in two or three dimensions), and $r = |\vec{r}|$ is its absolute value. Note that for the multidimensional case, we no longer use the letter y to represent the wave displacement but instead use the Greek letter ψ (psi). This is necessary because multidimensional wave motion can depend on the y-coordinate as well as on the x-coordinate.

You can convince yourself that equation 15.14 describes waveforms in concentric circles (in two dimensions) or spheres (in three dimensions) if you remember that the circumference of a circle or surface of a sphere has a constant distance to the origin and thus r has a constant value. A constant value of r at some time t implies a constant phase, $\phi = \kappa r - \omega t + \phi_0$. Since all of the points on a given circle or on the surface of a given sphere have the same phase, they have the same state of oscillation. Therefore, equation 15.14 indeed describes circular or spherical waves.

A waveform described by equation 15.14 does not have a constant amplitude, A, but one that depends on the radial distance to the origin of the wave (Figure 15.16). When we talk about energy transport by waves in Section 15.6, it will become clear that radial dependence of the amplitude is necessary for two- and three-dimensional waves and that the amplitude has to be a steadily decreasing function of the distance from the source. We'll use energy conservation to find out exactly how the amplitude varies with distance, but for now

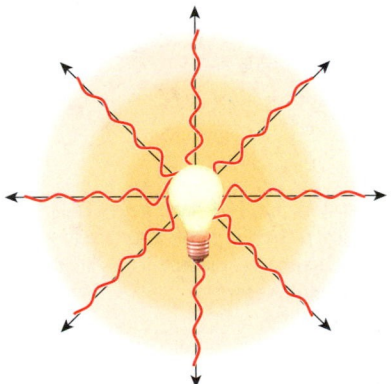

FIGURE 15.15 Waves spreading radially outward from a point source (a light bulb, in this case). The inner and outer surfaces of each concentric colored band are 1 wavelength apart.

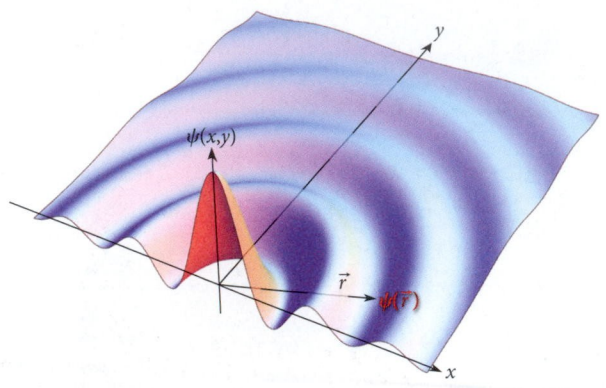

FIGURE 15.16 A spherical wave as a function of two spatial coordinates.

we'll use $A(r) = C/r$ (where C is a constant) for a three-dimensional spherical wave. For such a wave, equation 15.14 becomes

$$\psi(\vec{r},t) = \frac{C}{r}\sin(\kappa r - \omega t + \phi_0). \tag{15.15}$$

Plane Waves

Spherical waves originating from a point source can be approximated as plane waves at points that are very far away from the source. As shown in Figure 15.17, **plane waves** are waves for which the surfaces of constant phase are planes. For example, the light coming from the Sun is in the form of plane waves when it hits the Earth. Why can we use the plane-wave approximation? If we calculate the curvature of a sphere as a function of its radius and let that radius approach infinity, the curvature approaches zero. This means that any surface segment of this sphere locally looks like a plane. For example, the Earth is basically a sphere, but it is so large that locally the surface looks like a plane to us.

For a mathematical description of a plane wave, we can use the same functional form as for a one-dimensional wave:

$$\psi(\vec{r},t) = A\sin(\kappa x - \omega t + \phi_0), \tag{15.16}$$

where the x-axis is locally defined as the axis perpendicular to the plane of the wave. You may wonder why equation 15.16 again includes a constant amplitude. The answer is that very far away from the origin, ψ/r varies very slowly as a function of r, so using a constant amplitude gives the same degree of approximation as approximating the curvature of a sphere as zero. For a perfectly flat planar wave, the amplitude is indeed constant.

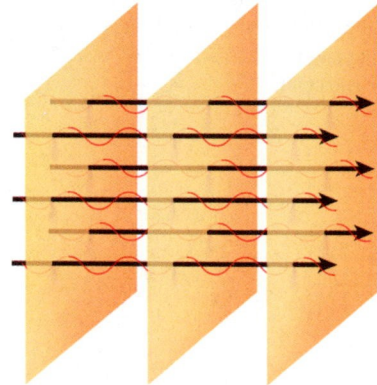

FIGURE 15.17 Plane waves. Each parallel plane represents a surface of constant phase. Adjacent planes are two wavelengths apart.

Surface Waves

On December 26, 2004, an earthquake of magnitude 9.0 under the Indian Ocean off the coast of Sumatra triggered a *tsunami* (a Japanese word meaning "harbor wave"). This tsunami caused incredible destruction in the Aceh province of Indonesia and then raced across the Indian Ocean with the speed of a jetliner and caused many casualties in India and Sri Lanka. A computer simulation of the propagation of the tsunami is shown in Figure 15.18. What is the physical basis of the motion of a tsunami?

When an underwater earthquake occurs, the seafloor is often locally displaced and shifted upward as two plates of the Earth's crust move over each other. Thus, the water in the ocean is displaced, causing a pressure wave. This is one of the rare instances of a wave that

FIGURE 15.18 A computer simulation of the propagation of the tsunami near Sumatra that occurred on December 26, 2004. The four frames cover a time period of several hours.

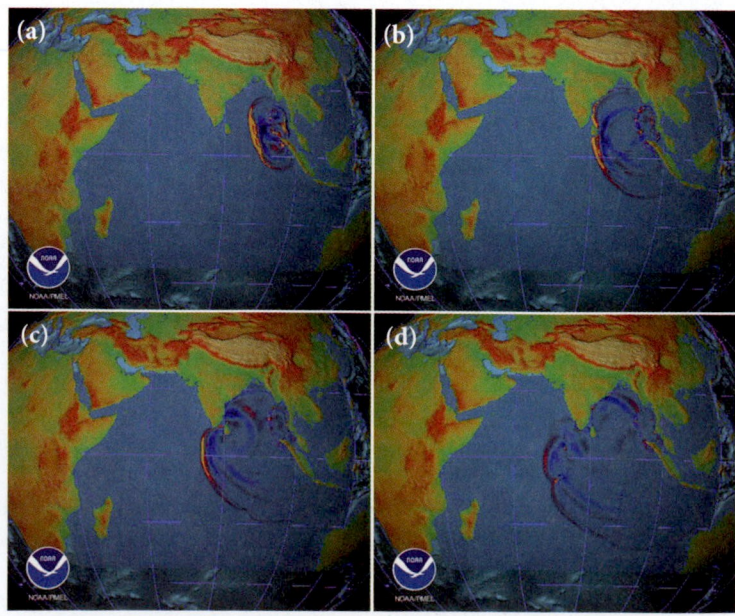

carries matter (in this case, water) along with it; water does crash on the shore. However, the water reaching shore did not originate at the site of the earthquake, so no long-distance transportation of water takes place in a tsunami. We'll consider pressure waves in more detail in Chapter 16. For now, you only need to know that the pressure waves in the ocean have very long wavelengths, up to 100 km or even more. Since this wavelength is much larger than the typical depth of the ocean, which is approximately 4 km, the pressure waves resulting from earthquakes cannot spread spherically, but only as surface waves. For these surface waves, the local wave speed is not constant, but depends on the depth of the ocean:

$$v_{surface} = \sqrt{gd},$$ (15.17)

where $g = 9.81$ m/s^2 is the acceleration due to gravity and d is the depth of the ocean, in meters. For a depth of 4000 m, this results in a wave speed of 200 m/s, or approximately 450 mph, as fast as a jetliner. However, when the leading edge of a tsunami reaches the shallow coastal waters, the wave speed decreases according to equation 15.17. This causes a pileup, as water farther out in the ocean is still traveling at a higher speed and catches up with the water that forms the leading edge of the tsunami. Thus, a tsunami wave crest has a height on the order of only a foot in the open ocean and can pass unnoticed under ships, while near shore it creates a wall of water up to 100 ft high that sweeps away everything in its path.

Seismic Waves

Earthquakes also generate waves that travel through the Earth. Two kinds of seismic waves generated by earthquakes are compression waves and transverse waves. Seismic compression waves are often called *P* (or *primary*) *waves,* and seismic transverse waves are often termed *S* (or *secondary*) *waves.* P waves can travel in solids and liquids, but S waves can travel only in solids. Liquids such as molten iron cannot sustain S waves.

The Earth has a variable composition as a function of distance from the center, as shown in Figure 15.19a. At the center of the Earth is an inner core composed mainly of solid iron and nickel. Next comes the outer core, which is composed mostly of molten iron. The inner and outer mantles are composed of lighter elements and exist in a solid, plastic state. The crust is hard and brittle. Most of the mass of the Earth is concentrated in the mantle.

Seismic waves travel from the epicenter of the earthquake through the Earth and can be detected on the surface far away. Seismic compression (P) waves can traverse the entire volume of the Earth, while seismic transverse (S) waves cannot penetrate the molten core. Thus, the core limits the extent of observation for S waves, as illustrated in Figure 15.19.

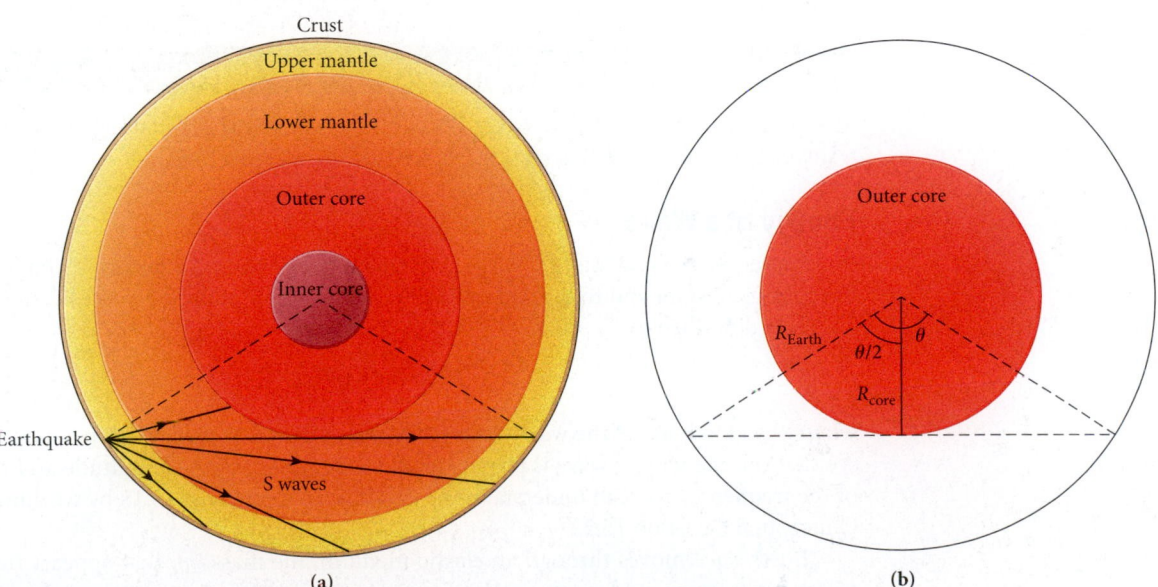

FIGURE 15.19   (a) The composition of the interior of the Earth. An earthquake takes place, creating S waves. The S waves can travel in the crust, the upper mantle, and the lower mantle but cannot travel in the molten outer core. (b) No S waves are observed at points around the Earth at angles greater than θ from the earthquake because the S waves cannot transit the molten outer core.

EXAMPLE 15.2 | **Measuring the Size of the Earth's Molten Core**

Scientists have observed that the maximum angle around the surface of the Earth where S waves from an earthquake are observed is $\theta = 105°$ (see Figure 15.19b). This effect is due to the fact that S waves cannot travel in the Earth's outer core because molten iron cannot support shear forces. The radius of the Earth is 6370 km.

PROBLEM

What is the radius of the molten core of the Earth?

SOLUTION

Looking at Figure 15.19b, we see that we can relate the radius of the molten core, R_{core}, the radius of the Earth, R_{Earth}, and the maximum angle of observation of S waves at the surface of the Earth:

$$\cos\left(\frac{\theta}{2}\right) = \frac{R_{core}}{R_{Earth}}.$$

We can then solve for the radius of the molten core:

$$R_{core} = R_{Earth}\cos\left(\frac{\theta}{2}\right) = (6370 \text{ km})\cos\left(\frac{105°}{2}\right) = 3880 \text{ km}.$$

Note that this result is only approximate. Seismic transverse waves do not travel in exactly straight lines because of variations in density as a function of depth in the Earth. However, the calculated result is reasonably close to the measured value, $R_{core} = 3480$ km, which has been obtained by combining several different methods.

Earthquakes are not the only source of seismic waves. Such waves can also be generated through the underground use of explosives such as dynamite or TNT or even nuclear bombs. Thus, the monitoring of seismic waves is used to verify nuclear test bans and nonproliferation of nuclear weapons. Other means of generating seismic waves are thumper-trucks (which lift a large mass approximately 3 m off the ground and drop it), air guns, plasma sound sources, and hydraulic or electromechanical shakers. Why would anyone want to generate seismic waves? The waves can be used similar to the way S waves were used in Example 15.2 to look for underground deposits of different density. This basic technique of reflection seismology is widely utilized in oil and gas exploration and is thus of great current research interest.

15.6   Energy, Power, and Intensity of Waves

Section 15.1 pointed out that waves do not generally transport matter. However, they do transport energy, and in this section, we will determine how much.

Energy of a Wave

In Chapter 14, we saw that the energy of a mass oscillating on a spring is proportional to the spring constant and to the square of the oscillation amplitude: $E = \frac{1}{2}kA^2$. The energy of a wave is described by

$$E = \tfrac{1}{2}m\omega^2\left[A(r)\right]^2, \tag{15.18}$$

where the amplitude of the wave has a radial dependence, as explained in Section 15.5. We see that the energy of a wave is proportional to the square of the amplitude and the square of the frequency. You can understand the derivation of equation 15.18 by working through Conceptual Question 15.22.

If the wave moves through an elastic medium, the mass, m, that appears in equation 15.18 can be expressed as $m = \rho V$, where ρ is the density and V is the volume through which the wave passes. This volume is the product of the cross-sectional area, $A_\perp$ (the area perpendicular to the velocity vector of the wave), that the wave passes through and the distance that the wave travels in time t, or $l = vt$. (*Note:* The symbol $A_\perp$ indicates the area, whereas the symbol A, without the subscript $_\perp$, is used for the wave amplitude!) Thus, we find for the energy in this case

$$E = \tfrac{1}{2}\rho V \omega^2 \left[A(r)\right]^2 = \tfrac{1}{2}\rho A_\perp l\omega^2 \left[A(r)\right]^2 = \tfrac{1}{2}\rho A_\perp vt\omega^2 \left[A(r)\right]^2. \tag{15.19}$$

If the wave is a spherical wave radiating outward from a point source and if no energy is lost to damping, we can determine the radial dependence of the amplitude from equation 15.19. In such a case, the energy for a given time t is a constant at a fixed value of r, as are the equilibrium density of the medium, the angular frequency, and the wave velocity. Equation 15.19 then yields the condition $A_\perp[A(r)]^2 = $ constant. For a spherical wave in three dimensions, the area through which the wave passes is the surface of a sphere, as shown in Figure 15.15. Thus, in this case, $A_\perp = 4\pi r^2$, and we obtain $r^2 [A(r)]^2 = $ constant $\Rightarrow A(r) \propto 1/r$. This is exactly the result for the radial dependence of the amplitude that was used in equation 15.15.

If the wave is spreading strictly in a two-dimensional space, as is true of surface waves on water, then the cross-sectional area is proportional to the circumference of a circle and is thus proportional to r, not r^2. In this case, we obtain for the radial dependence of the wave amplitude $r[A(r)]^2 = $ constant $\Rightarrow A(r) \propto 1/\sqrt{r}$.

Power and Intensity of a Wave

In Chapter 5, we saw that the average power is the rate of energy transfer per time. With the energy contained in a wave as given by equation 15.19, we then have

$$\overline{P} = \frac{E}{t} = \tfrac{1}{2}\rho A_\perp v\omega^2 \left[A(r)\right]^2. \tag{15.20}$$

Since all quantities on the right-hand side of equation 15.20 are independent of time, the power radiated by a wave is constant in time. Note that this equation applies to both transverse and longitudinal waves.

The intensity, I, of a wave is defined as the power radiated per unit cross-sectional area:

$$I = \frac{\overline{P}}{A_\perp}.$$

Substituting for $\overline{P}$ from equation 15.20 into this definition leads to

$$I = \tfrac{1}{2}\rho v\omega^2 \left[A(r)\right]^2. \tag{15.21}$$

As you can see from this result, the intensity depends on the square of the amplitude, and so the radial dependence of the intensity is due to the dependence of the amplitude on the radial distance from the point of propagation. For a spherical wave in three dimensions, $A(r) \propto 1/r$, and thus

$$I \propto \frac{1}{r^2}.$$

Thus, for two points that are at distances of r_1 and r_2 from the point of propagation of a spherical wave, the intensities $I_1 = I(r_1)$ and $I_2 = I(r_2)$ measured at these two locations are related via

$$\frac{I_1}{I_2} = \left(\frac{r_2}{r_1}\right)^2, \tag{15.22}$$

as shown in Figure 15.20.

Figure 15.21 illustrates the dependence of the intensity of a wave on the distance from the source. If the intensity of a wave at a distance r is I, then the intensity of the wave at a distance of $2r$ is $I/4$. You can see why this relationship is true in Figure 15.21 because the same amount of power is passing through an area four times the size of $A_\perp$. Similarly, at a distance of $3r$, the intensity is $I/9$ because the same amount of power is passing through nine times the area.

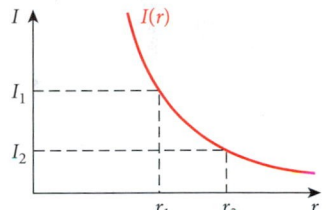

FIGURE 15.20 Inverse square law for the intensities of a spherical wave at two different distances from the source.

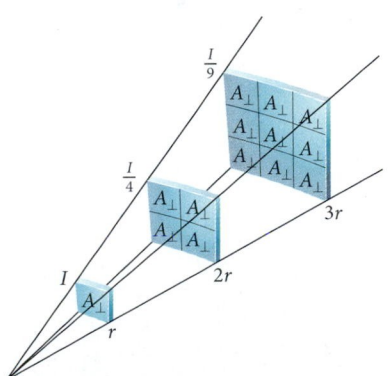

FIGURE 15.21 The intensity of a wave at a distance r is I. At a distance of $2r$, the intensity is $I/4$. At a distance of $3r$, the intensity is $I/9$.

Self-Test Opportunity 15.3

The intensity of electromagnetic waves from the Sun at the Earth is approximately 1400 W/m². The distance of the Earth from the Sun is $149.6 \cdot 10^6$ km. How much power does the Sun generate?

$$y(x,t) = \overbrace{(0.130\ \text{m})}^{A}\cos(\overbrace{(9.00\ \text{m}^{-1})}^{\kappa}x + \overbrace{(72.0\ \text{s}^{-1})}^{\omega}t)$$

FIGURE 15.22 The wave function with the relevant factors identified.

SOLVED PROBLEM 15.2 | Power Transmitted by a Wave on a String

PROBLEM

The transverse displacement from equilibrium of a particular stretched string as a function of position and time is given by $y(x,t) = (0.130\ \text{m})\cos[(9.00\ \text{m}^{-1})x + (72.0\ \text{s}^{-1})t]$. The linear mass density of the string is 0.00677 kg/m. What is the average power transmitted by the string?

SOLUTION

THINK We start with equation 15.20 for the average power transmitted by a wave. We can extract the amplitude, $A(r)$, the wave number, κ, and the angular frequency, ω, from the given function $y(x,t)$ by comparing it with the solution to the wave equation for a wave traveling in the negative x-direction (see equation 15.6). The product $\rho A_\perp$ is given by the linear mass density, μ.

SKETCH Figure 15.22 identifies the relevant factors of the wave function.

RESEARCH The power transmitted by a wave is given by equation 15.20:

$$\overline{P} = \tfrac{1}{2}\rho A_\perp v\omega^2 \left[A(r)\right]^2.$$

For a wave moving on an elastic string, the product of the density, ρ, and the cross-sectional area, $A_\perp$, can be written as

$$\rho A_\perp = \frac{m}{V}A_\perp = \frac{m}{L \cdot A_\perp}A_\perp = \frac{m}{L} = \mu,$$

where $\mu = m/L$ is the linear mass density of the string. For a wave moving on an elastic string, $A(r)$ is constant, and so we can express the transmitted power as

$$\overline{P} = \tfrac{1}{2}\mu v\omega^2 \left[A(r)\right]^2.$$

We can use equation 15.8 to obtain the speed of the wave:

$$v = \frac{\omega}{\kappa}.$$

SIMPLIFY From the wave function specified in the problem statement, we see that $A(r) = 0.130$ m, $\kappa = 9.00$ m^{-1}, and $\omega = 72.0$ s^{-1}. By convention, ω is a positive number, and the specified wave function represents a wave traveling in the negative x-direction. We can then calculate v:

$$v = \frac{\omega}{\kappa} = \frac{72.0\ \text{s}^{-1}}{9.00\ \text{m}^{-1}} = 8.00\ \text{m/s}.$$

CALCULATE Putting the numerical values into equation 15.20, we get

$$\overline{P} = \tfrac{1}{2}\mu v\omega^2 \left[A(r)\right]^2 = \tfrac{1}{2}(0.00677\ \text{kg/m})(8.00\ \text{m/s})(72.0\ \text{s}^{-1})^2(0.130\ \text{m})^2$$

$$= 2.37247\ \text{W}.$$

ROUND We report our result to three significant figures:

$$\overline{P} = 2.37\ \text{W}.$$

DOUBLE-CHECK To double-check our answer, let's assume that the energy of the wave corresponds to the energy of an object whose mass is equal to the mass of a one-wavelength-long section of string. This object is moving in the negative x-direction at the speed of the traveling wave. The power is then the energy transmitted in one period. From the values for κ and ω extracted from the given wave function, we get the wavelength and the period:

$$\lambda = \frac{2\pi}{\kappa} = 0.698\ \text{m}.$$

$$T = \frac{2\pi}{\omega} = 0.0873\ \text{s}.$$

We can write the kinetic energy as

$$K = \tfrac{1}{2}(\mu\lambda)v^2 = 0.151\ \text{J}.$$

The power is then

$$P = \frac{K}{T} = \frac{0.151\ \text{J}}{0.0873\ \text{s}} = 1.73\ \text{W}.$$

This result is comparable to our result for the power transmitted along the string. Thus, our answer seems to be reasonable.

15.7 Superposition Principle and Interference

When two or more waves are simultaneously present at a point, the resultant displacement is simply the sum of the displacements of the component waves. This simple statement, called the **superposition principle**, is one of the most important in wave physics, but how do we know that it is true? Wave equations like equation 15.9 have a very important property: They are linear. What does this mean? If you find two different solutions, $y_1(x,t)$ and $y_2(x,t)$, to a linear differential equation, then any linear combination of those solutions, such as

$$y(x,t) = ay_1(x,t) + by_2(x,t), \tag{15.23}$$

where a and b are arbitrary constants, is also a solution to the same linear differential equation. You can see this linear property in the differential equation, because it contains the function $y(x,t)$ to the first power only [there is no term containing $y(x,t)^2$ or $\sqrt{y(x,t)}$ or any other power of the function].

In physical terms, the linear property means that wave solutions can be added, subtracted, or combined in any other linear combination, and the result is again a wave solution. This physical property is the mathematical basis of the superposition principle:

Two or more wave solutions can be added, resulting in another wave solution.

$$y(x,t) = y_1(x,t) + y_2(x,t). \tag{15.24}$$

Equation 15.24 is a special case of equation 15.23 with $a = b = 1$. Let's look at an example of the superposition of two wave pulses with different amplitudes and different velocities.

EXAMPLE 15.3 | Superposition of Wave Pulses

Two wave pulses, $y_1(x,t) = A_1 e^{-(x-v_1t)^2}$ and $y_2(x,t) = A_2 e^{-(x+v_2t)^2}$, with $A_2 = 1.7A_1$ and $v_1 = 1.6v_2$, are at some distance from each other. Because of their mathematical form, these are called *Gaussian wave packets*. We assume that the two waves are of the same type, such as waves on a rope.

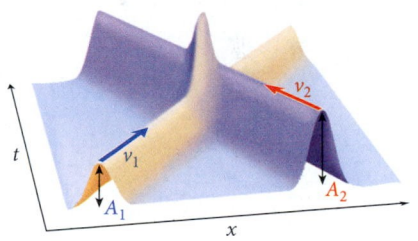

(a)

PROBLEM
What is the amplitude of the resulting wave at maximum overlap of the two pulses, and at what time does this occur?

SOLUTION
Both of these Gaussian wave packets are of the functional form $y(x,t) = Y(x \pm vt)$ and are therefore valid solutions of the one-dimensional wave equation 15.9. Thus, the superposition principle (equation 15.23) holds, and we can simply add the two wave functions to obtain the function for the resulting wave. The two Gaussian wave packets have maximum overlap when their centers are at the same position in coordinate space. The center of a Gaussian wave packet is at the location where the exponent has a value of zero. As you can see from the given functions, both wave packets will be centered at $x = 0$ at time $t = 0$, which is the answer to the question as to when the maximum overlap occurs. The amplitude at maximum overlap is simply $A = A_1 + A_2 = 2.7A_1$ (because $A_2 = 1.7A_1$, as specified in the problem statement).

DISCUSSION
It is instructive to look at plots illustrating these waves and their superposition. The two Gaussian wave packets start at time $t = -3$, in units of the width of a wave packet divided by the speed v_1, and propagate. Figure 15.23a is a three-dimensional plot of the wave function $y(x,t) = y_1(x,t) + y_2(x,t)$ as a function of coordinate x and time t. The velocity vectors $\vec{v}_1$ and $\vec{v}_2$ are shown. Figure 15.23b shows coordinate space plots of the waves at different points in time. The blue curve represents $y_1(x,t)$, the red curve represents $y_2(x,t)$, and the black curve is their sum. At the middle point in time, the two wave packets achieve maximum overlap.

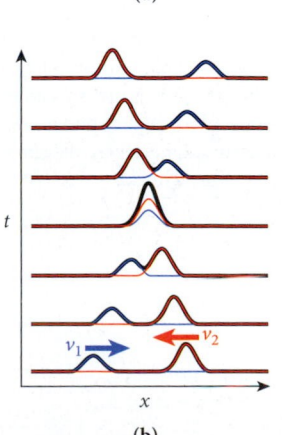

(b)

FIGURE 15.23 Superposition of two Gaussian wave packets: (a) three-dimensional representation; (b) coordinate space plots for different times.

What is essential to remember from the superposition principle is that waves can penetrate each other without changing their frequency, amplitude, speed, or direction. Modern communications technology is absolutely dependent on this fact. In any city, several TV and radio stations broadcast their signals at different frequencies; there are also cell phone and satellite TV transmissions, as well as light waves and sound waves. All of these waves have to be able to penetrate each other without changing; otherwise, everyday communication would be impossible.

Interference of Waves

Interference is one consequence of the superposition principle. If waves pass through each other, their displacements simply add, according to equation 15.24. Interesting cases arise when the wavelengths and frequencies of the two waves are the same, or at least close to each other. Here we'll look at the case of two waves with identical wavelengths and frequencies. The case of waves whose frequencies are close to each other will be discussed in Chapter 16 on sound, when we discuss beats.

First, let's consider two one-dimensional waves with identical amplitude, A, wave number, κ, and angular frequency, ω. The phase shift ϕ_0 is set at zero for wave 1 but allowed to vary for wave 2. The sum of the waveforms is then

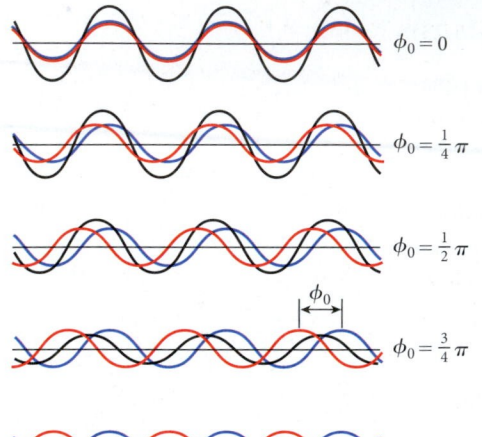

$$y(x,t) = A\sin(\kappa x - \omega t) + A\sin(\kappa x - \omega t + \phi_0). \tag{15.25}$$

For $\phi_0 = 0$, the two sine functions in equation 15.25 have identical arguments, so the sum is simply $y(x,t) = 2A\sin(\kappa x - \omega t)$. In this situation, the two waves add to the maximum extent possible, which is called **constructive interference.**

If we shift the argument of a sine or cosine function by π, we obtain the negative of that function: $\sin(\theta + \pi) = -\sin\theta$. For this reason, the two terms in equation 15.25 add up to exactly zero when $\phi_0 = \pi$. This situation is called **destructive interference.** The interference patterns for these two values of ϕ_0, along with three others, are shown in Figure 15.24 for $t = 0$.

FIGURE 15.24 Interference of one-dimensional waves as a function of their relative phase shifts at $t = 0$. At later times, the patterns move rigidly to the right. Wave 1 is red, wave 2 is blue, and the sum of the two waves is black.

Interesting interference patterns are obtained with two identical circular waves whose sources are some distance from each other. Figure 15.25 shows the schematic interference pattern for two identical waves with $\kappa = 1 \text{ m}^{-1}$ and the same values of ω and A. One of the waves was shifted by Δx to the right and the other by $-\Delta x$ to the left. The figure shows different values of Δx, which should give you a good idea of how the interference pattern develops as a function of the separation of the centers of the waves. If you have ever dropped two stones at a time into a pond, you will recognize these interference patterns as similar to the surface ripples that resulted.

FIGURE 15.25 Schematic interference patterns of two identical two-dimensional periodic waves shifted a distance Δx to either side of the origin, where the unit of Δx is meters. White and black circles mark minima and maxima; gray indicates points where the sum of the waves equals zero.

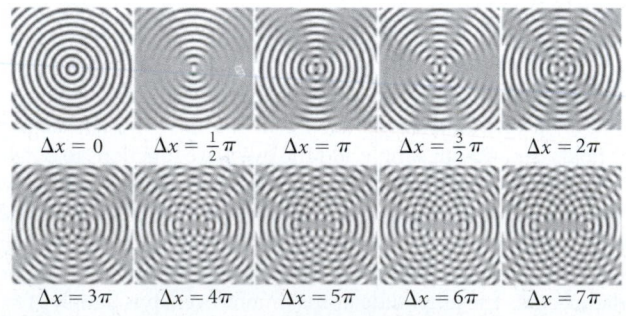

15.8 Standing Waves and Resonance

A special type of superposition occurs for two traveling waves if they are identical except for opposite signs for ω: $y_1(x,t) = A\sin(\kappa x + \omega t)$, $y_2(x,t) = A\sin(\kappa x - t)$. Let's first look at the mathematical result of this superposition (after which we'll gain physical insight):

$$y(x,t) = y_1(x,t) + y_2(x,t)$$
$$= A\sin(\kappa x + \omega t) + A\sin(\kappa x - \omega t) \Rightarrow$$
$$y(x,t) = 2A\sin(\kappa x)\cos(\omega t). \tag{15.26}$$

In the last step to obtain equation 15.26, we used the trigonometric addition formula, $\sin(\alpha \pm \beta) = \sin\alpha \cos\beta \pm \cos\alpha \sin\beta$. Thus, for the superposition of two traveling waves with the same amplitude, same wave number, and same speed but opposite direction of propagation, the dependence on the spatial coordinate and the dependence on the time separate (or factorize) into a function of just x times a function of just t. The result of this superposition is a wave that has **nodes** (where $y = 0$) and **antinodes** (where y can reach its maximum value) at particular points along the x-axis. Each antinode is located midway between neighboring nodes.

This superposition is much easier to visualize in graphical form. Figure 15.26a shows a wave given by $y_1(x,t) = A\sin(\kappa x + \omega t)$, and Figure 15.26b shows a wave given by $y_2(x,t) = A\sin(\kappa x - \omega t)$. Figure 15.26c shows the addition of the two waves. In all three plots, the waveform is shown as a function of the x-coordinate. Each plot also shows the wave at 10 different instants in time, with a time difference of $\pi/10$ between times as if viewing a time-lapse photograph. To compare the same instants of time in each plot, the curves are color-coded starting from red and progressing through orange to yellow. You can see that the traveling waves move toward the left and the right, respectively, but the wave resulting from their superposition oscillates in place, with the nodes and the antinodes remaining at fixed locations on the x-axis. This interference wave is known as a **standing wave** and results from the factorization of the time dependence and the spatial dependence in equation 15.26. The two traveling waves are always out of phase at the nodes.

The nodes of a standing wave always have zero amplitude. According to equation 15.18, the energy contained in a wave is proportional to the square of the amplitude. Thus, no energy is transported across a node by a standing wave. Wave energy gets trapped between the nodes of a standing wave and remains localized there. Also note that even though a standing wave is not moving, the relationship between wavelength, frequency, and wave speed, $v = \lambda f$, still holds. Here, v is the speed of the two traveling waves, which make up the standing wave.

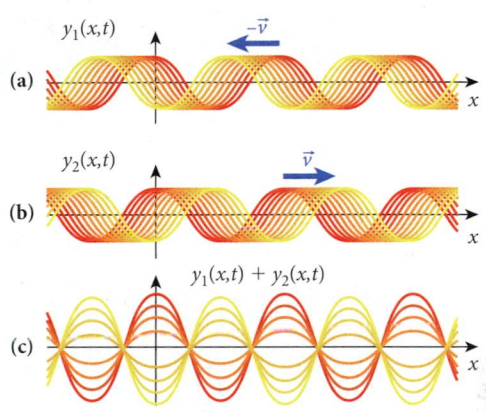

FIGURE **15.26**   (a) and (b) Two traveling waves with velocity vectors in opposite directions. (c) Superposition of the two waves results in a standing wave.

Standing Waves on a String

The basis for the creation of musical sounds by string instruments is the generation of standing waves on strings that are put under tension. Section 15.4 discussed how the velocity of a wave on a string depends on the string tension and the linear mass density of the string (see equation 15.13). In this section, we discuss the basic physics of one-dimensional standing waves.

Let's start with a demonstration. Shown in Figure 15.27 is a string tied to an anchor on the left side and to a piston on the right. The piston oscillates up and down, in sinusoidal fashion, with a variable frequency, f. The up-and-down motion of the piston has only a very small amplitude, so we can consider the string to be fixed at both ends. The frequency of the piston's oscillation is varied, and at a certain frequency, f_0, the string develops a large-amplitude motion with a central antinode (Figure 15.27a). Clearly, this excitation of the string produces a standing wave. This is a resonant excitation of the string in the sense that the amplitude gets large only at a well-defined **resonance frequency**. If the frequency of the piston's motion is increased by 10%, as shown in Figure 15.27b, this resonant oscillation of the string vanishes. Lowering the piston's frequency by 10% has the same effect. Increasing the frequency to twice the value of f_0 results in another resonant excitation, one that has a single node in the center and antinodes at $\frac{1}{4}$ and $\frac{3}{4}$ of the length of the string (Figure 15.27c). Increasing the frequency to triple the value of f_0 results in a standing wave with two nodes and three antinodes (Figure 15.27d). The continuation of this series is clear: A frequency of nf_0 will result in a standing wave with n antinodes and $n-1$ nodes (plus the two nodes at both ends of the string), which are equally spaced along the length of the string.

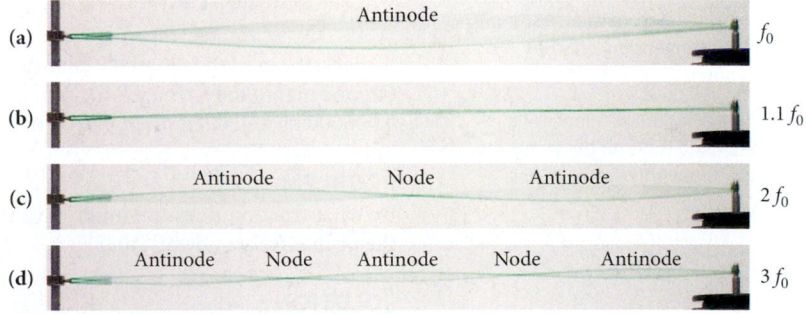

FIGURE **15.27**   Generation of standing waves on a string.

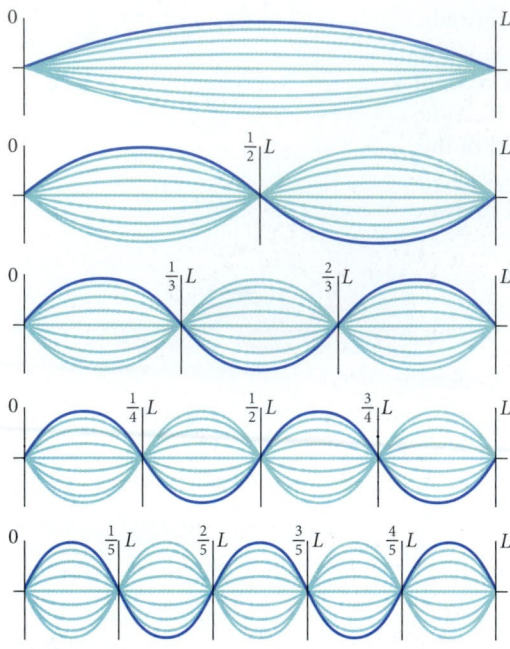

FIGURE 15.28 Lowest five standing-wave excitations of a string.

Figure 15.28 shows that the condition for a standing wave is that an integer multiple, n, of half wavelengths fits exactly into the length of the string L. We give these special wavelengths the index n and write

$$n\frac{\lambda_n}{2} = L, \quad n = 1, 2, 3, \dots .$$

Solving for the wavelengths, we obtain

$$\lambda_n = \frac{2L}{n}, \quad n = 1, 2, 3, \dots . \tag{15.27}$$

The index n on the wavelength (or frequency) indicates the **harmonic.** That is, $n = 1$ identifies the first harmonic (also called the *fundamental frequency*), $n = 2$ labels the second harmonic, $n = 3$ the third, and so on.

As mentioned before, $v = \lambda f$ still holds for a standing wave. Using this relationship, we can find the resonance frequencies of a string from equation 15.27:

$$f_n = \frac{v}{\lambda_n} = n\frac{v}{2L}, \quad n = 1, 2, 3, \dots .$$

Finally, using equation 15.13 for the wave velocity on a string with linear mass density μ under tension F, $v = \sqrt{F/\mu}$, we have

$$f_n = \frac{v}{\lambda_n} = n\frac{\sqrt{F}}{2L\sqrt{\mu}} = n\sqrt{\frac{F}{4L^2\mu}}. \tag{15.28}$$

Equation 15.28 reveals several interesting facts about the construction of string instruments. First, the longer the string, the lower the resonance frequencies. This is the basic reason why a cello, which produces lower notes, is longer than a violin. Second, the resonance frequencies are proportional to the square root of the tension on the string. If an instrument's frequency is too low (it sounds "flat"), you need to increase the tension. Third, the higher the linear mass density, μ, of the string, the lower the frequency. "Fatter" strings produce bass notes. Fourth, the second harmonic is twice as high as the fundamental frequency, the third is three times as high, and so on. When we discuss sound in Chapter 16, we'll see that this fact is the basis of the definition of the octave. For now, you can see that you can produce the second harmonic on the same string that is used for the first harmonic by putting a finger on the exact middle of the string and thus forcing a node to be located there.

Concept Check 15.4

A string of a given length is clamped at both ends and given a certain tension. Which of the following statements about a standing wave on this string is true?

a) The higher the frequency of a standing wave on the string, the closer together are the nodes.

b) Regardless of the frequency of the standing wave, the nodes are always the same distance apart.

c) For a standing wave on a string at a given tension, only one frequency is ever possible.

d) The lower the frequency of a standing wave on the string, the closer together are the nodes.

EXAMPLE 15.4 | Tuning a Piano

A piano tuner's job is to make sure that all keys on the instrument produce the proper tones. On one piano, the string for the middle-A key is under a tension of 2900 N and has a mass of 0.006000 kg and a length of 0.6300 m.

PROBLEM

By what fraction does the tuner have to change the tension on this string in order to obtain the proper frequency of 440.0 Hz?

SOLUTION

First, we calculate the fundamental frequency (or first harmonic), using $n = 1$ in equation 15.28 and the given tension, length, and mass:

$$f_1 = \sqrt{\frac{2900 \text{ N}}{4(0.6300 \text{ m})(0.006000 \text{ kg})}} = 438 \text{ Hz}.$$

This frequency is off by 2 Hz, or too low by 0.5%, so the tension has to be increased. The proper tension can be found by solving equation 15.28 for F and then using the fundamental frequency of 440.0 Hz, the proper frequency for middle A:

$$f_n = n\sqrt{\frac{F}{4L^2(m/L)}} = n\sqrt{\frac{F}{4Lm}} \Rightarrow F = 4Lm\left(\frac{f_n}{n}\right)^2.$$

Inserting the numbers, we find

$$F = 4(0.6300 \text{ m})(0.006000 \text{ kg})(440.0 \text{ Hz})^2 = 2927 \text{ N}.$$

The tension needs to be increased by 27 N, which is 0.93% of 2900 N. The piano tuner needs to increase the tension on the string by 0.93% for it to produce the proper middle-A frequency.

DISCUSSION

We could also have found the required change in the tension by arguing in the following way: The frequency is proportional to the square root of the tension, and $\sqrt{1-x} \approx 1 - \frac{1}{2}x$ for small x. Since a 0.5% change in the frequency is needed, the tension has to be changed by twice that amount, or 1%.

By the way, a piano tuner can detect a frequency that is off by 2 Hz easily, by listening for beats, which we'll discuss in Chapter 16.

Self-Test Opportunity 15.4

Researchers at the Cornell NanoScale Science and Technology Facility, where the guitar in Figure 15.12b was built, have also made an even smaller guitar string, from a single carbon nanotube only 1 to 4 nm in diameter. It is suspended over a trench of width $W = 1.5 \text{ μm}$. In their 2004 paper in *Nature*, the Cornell researchers reported a fundamental resonance frequency of the carbon nanotube at 55 MHz. What is the speed of a wave on this nanotube?

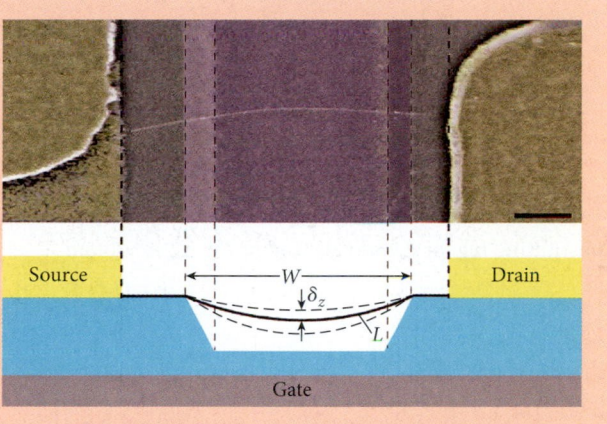

– Continued

Concept Check 15.5

A guitar string is 0.750 m long and has a mass of 5.00 g. The string is tuned to E (660 Hz) when it vibrates at its fundamental frequency. How should the tension in the string be changed to tune it to A (440 Hz)?

a) increase the tension by a factor of $\frac{3}{2}$

b) increase the tension by a factor of $\left(\frac{3}{2}\right)^{1/2}$

c) decrease the tension to $\frac{2}{3}$ of its current value

d) decrease the tension to $\left(\frac{2}{3}\right)^{1/2}$

e) decrease the tension to $\left(\frac{2}{3}\right)^2$

SOLVED PROBLEM 15.3 | Standing Wave on a String

PROBLEM

A mechanical driver is used to set up a standing wave on an elastic string, as shown in Figure 15.29a. Tension is put on the string by running it over a frictionless pulley and hanging a metal block from it (Figure 15.29b). The length of the string from the top of the pulley to the driver is 1.25 m, and the linear mass density of the string is 5.00 g/m. The frequency of the driver is 45.0 Hz. What is the mass of the metal block?

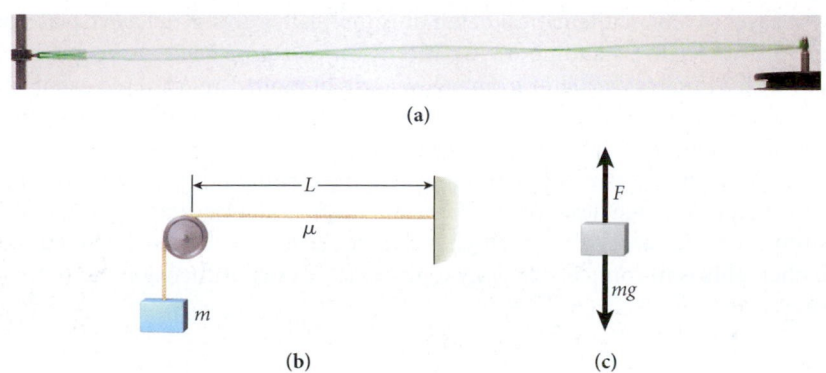

FIGURE 15.29 (a) A standing wave on an elastic string driven by a mechanical driver. (b) The string is put under tension by a hanging mass. (c) Free-body diagram for the hanging block.

SOLUTION

THINK The weight of the metal block is equal to the tension placed on the elastic string. We can see from Figure 15.29a that the string is vibrating at its third harmonic because three antinodes are visible. We can thus use equation 15.28 to relate the tension on the string, the harmonic, the linear mass density of the string, and the frequency. Once we have determined the tension on the string, we can calculate the mass of the metal block.

SKETCH Figure 15.29b shows the elastic string under tension due to a metal block hanging from it. The mechanical driver sets up a standing wave on the string. Here L is the length of the string from the pulley to the wave driver, μ is the linear mass density of the string, and m is the mass of the metal block. Figure 15.29c shows the free-body diagram of the hanging metal block, where F is the tension on the string and mg is the weight of the metal block.

RESEARCH A standing wave with harmonic n and frequency f_n on an elastic string of length L and linear mass density μ satisfies equation 15.28:

$$f_n = n\sqrt{\frac{F}{4L^2\mu}}.$$

From the free-body diagram in Figure 15.29c, and the fact that the metal block does not move, we can write

$$F - mg = 0.$$

Therefore, the tension on the string is $F = mg$.

SIMPLIFY We solve equation 15.28 for the tension on the string:

$$F = 4L^2\mu\frac{f_n^2}{n^2}.$$

Substituting mg for F and rearranging terms, we obtain

$$m = \frac{\mu}{g}\left(\frac{2Lf_n}{n}\right)^2.$$

CALCULATE Putting in the numerical values, we get

$$m = \frac{\mu}{g}\left(\frac{2Lf_n}{n}\right)^2 = \frac{0.00500 \text{ kg/m}}{9.81 \text{ m/s}^2}\left(\frac{2(1.25 \text{ m})(45.0 \text{ Hz})}{3}\right)^2$$
$$= 0.716743 \text{ kg}.$$

ROUND We report our result to three significant figures:

$$m = 0.717 \text{ kg}.$$

DOUBLE-CHECK As a double-check of our result, we use the fact that a half-liter bottle of water has a mass of about 0.5 kg. Even a light string could support this mass without breaking. Thus, our answer seems reasonable.

Standing waves can also be created in two- and three-dimensions. While we will not treat these cases in any mathematical detail, it is interesting to see the wave patterns that can emerge. To visualize standing wave patterns on two-dimensional sheets, the sheets are driven with a tunable oscillator from below and salt grains are scattered evenly on top. The salt grains gather at the wave nodes, where they do not receive large kicks from the oscillator below. Figure 15.30 shows the result for a square plate driven at three different frequencies. You can see that the standing waves have nodal lines, which have different shapes at different frequencies. We won't discuss these nodal lines in quantitative detail, but it is important to realize that the higher the frequency, the closer the nodal lines are to each other. This is in complete analogy to the case of a one-dimensional standing wave on a string.

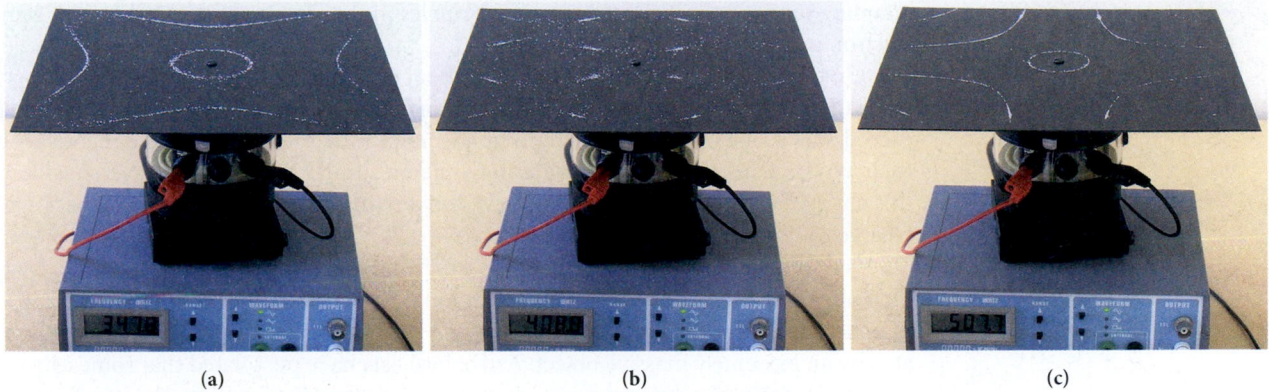

FIGURE 15.30 Standing-wave nodal patterns on a square plate driven at frequencies of (a) 348 Hz, (b) 409 Hz, and (c) 508 Hz.

Concept Check 15.6

The circular plate shown in the figure is driven at three different frequencies. From examining the three parts, rank the three frequencies from lowest to highest.

a) 1, 2, 3

b) 2, 3, 1

c) 3, 1, 2

d) 1, 3, 2

e) 2, 1, 3

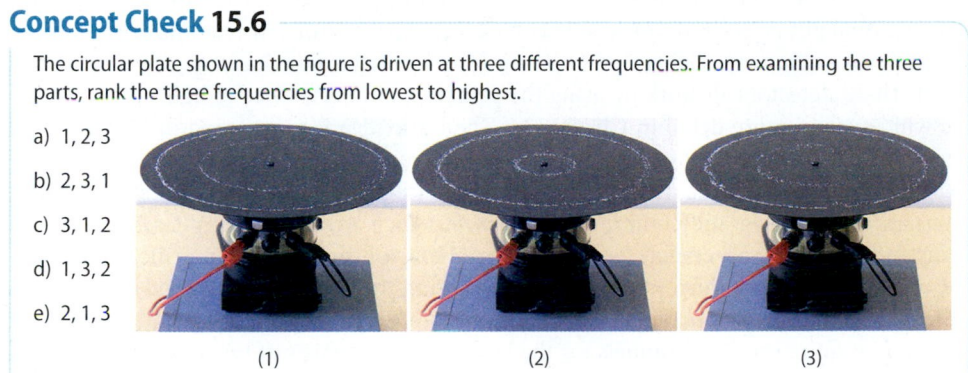

15.9   Research on Waves

This chapter gives only a brief overview of wave phenomena. All physics majors and most engineering majors will take another complete course on waves that goes into more detail. Several other chapters in this book deal with waves. Chapter 16 is devoted to the longitudinal pressure waves called *sound waves*. Chapter 31 focuses on electromagnetic waves, Chapter 34 examines wave optics, and Chapter 36 explores a wave description of matter. This wave character of matter led directly to the development of quantum mechanics, the basis for most of modern physics, including nanoscience and nanotechnology.

Scientists and engineers continue to discover new aspects of wave physics, from fundamental concepts to applications. On the applied side, engineers are studying ways to guide microwaves through different geometries with the aid of finite element methods (which are computational methods that break a large object into a large number of small objects) and very time-intensive computer programs. If these geometries are not spatially symmetric—that is, if they are not exactly round or exactly rectangular—some surprising results can occur. For example, physicists can study quantum chaos with the aid of microwaves that are sent into a metal cavity in the shape of a stadium (a shape that has semicircles at both ends, connected by straight sections). Figure 15.31 shows such a stadium geometry, not for microwaves but for a physical system— a collection of individual iron atoms arranged on a copper surface by using a

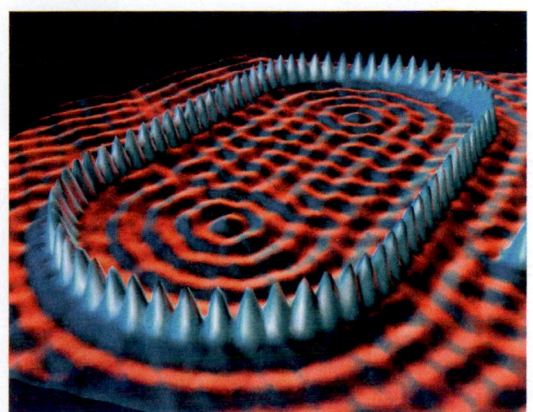

FIGURE 15.31 Iron atoms on a copper surface form a stadium-shaped cavity in which physicists can study electron matter waves.

scanning tunneling microscope. The wave ripples in the interior of the stadium are standing electron matter waves.

On the most advanced level of investigation, particle theorists and mathematicians have come up with new classes of theories called *string theories*. The mathematics of string theories exceeds the scope of this book. However, it is interesting to note that string theorists obtain their physical insights from considering systems similar to those we have considered in this chapter—in particular, standing waves on a string. We'll touch on this subject again in Chapter 39 on particle physics.

Perhaps one of the most interesting experimental investigations into wave physics is the search for gravitational waves. Gravitational waves were predicted by Einstein's theory of general relativity (see Chapter 35) but are very weak and hard to detect unless generated by an extremely massive object. Astrophysicists have postulated that some astronomical objects emit gravitational waves of such great intensity that they should be observable on Earth. Indirect evidence of the existence of such gravitational waves has come from the study of orbital frequencies of neutron star pairs, since the decrease of measured rotational energy in the pair can be explained by the radiation of gravitational waves. However, no direct observation of gravitational waves has yet been made. Groups of American and international physicists have constructed or are in the process of constructing several gravitational wave detectors, such as LIGO (Laser Interferometer Gravitational-Wave Observatory). These detectors all work by using the principle of interference of light waves, which we will cover in more detail in Chapter 34. The basic idea is to suspend in a vacuum by very thin wires several heavy test masses with mirrored sides, placed at the ends of two perpendicular arms of the instrument. As a gravitational wave passes by, it should cause alternating stretching and compressing of space, which will be out of phase for the two arms and should be observable as minute movements of the mirrors. In principle, bouncing a laser light off the mirrors and then recombining the beams would reveal interference of light due to the motion, detectable as an interference pattern. These gravitational wave detectors require very long tunnels for the lasers. LIGO consists of two observatories, one in Hanford, Washington, and the other in Livingston, Louisiana (see Figure 15.32), with a distance of 3002 km between them. Each observatory has two tunnels, each 4 km long,

FIGURE 15.32 (a) Aerial view of the Livingston site of LIGO (Laser Interferometer Gravitational-Wave Observatory); (b) visualization of a gravitational wave; (c) a scientist adjusts the interferometer; (d) map of LIGO.

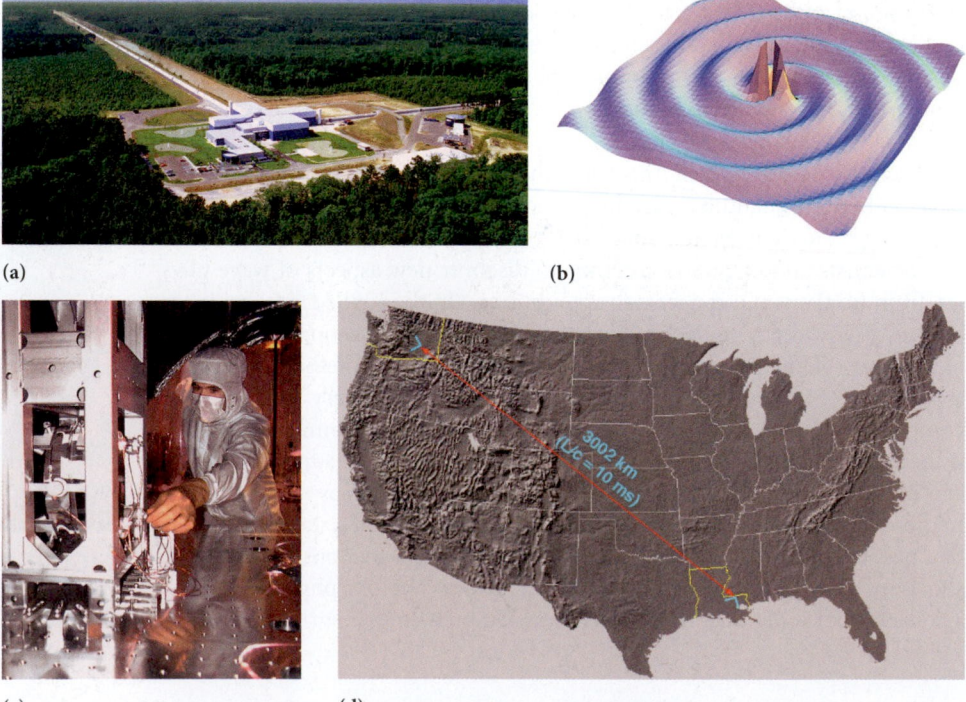

(a)

(b)

(c)

(d)

arranged in an L shape. These experiments push the envelope of what is technically possible and measurable. For example, the expected movement of the test masses due to a gravitational wave is on the order of only 10^{-18} m, which is smaller than the diameter of a hydrogen atom by a factor of 100 million, or 10^8. Measuring a deflection that small is a huge technological challenge.

Other laboratories in Europe (Virgo and GEO-600) and Japan (TAMA) are also trying to measure the effect of gravitational waves. Plans also exist for a space-based gravitational wave observatory, named LISA (Laser Interferometer Space Antenna), which will consist of three spacecraft arranged in an equilateral triangle with side length $5 \cdot 10^6$ km. Scientists have recently used supercomputers to simulate gravitational waves emitted from the collision of two massive black holes, as illustrated in Figure 15.33 in a calculation done by a group from NASA. Such waves might be intense enough to be detected on Earth.

FIGURE 15.33 Supercomputer simulation of the collision of two massive black holes, creating gravitational waves that may be observable.

WHAT WE HAVE LEARNED | EXAM STUDY GUIDE

- The wave equation describes the displacement $y(x,t)$ from the equilibrium position for any wave motion:
$$\frac{\partial^2}{\partial t^2} y(x,t) - v^2 \frac{\partial^2}{\partial x^2} y(x,t) = 0.$$

- For any wave, v is the wave velocity, λ is the wavelength, and f is the frequency, and they are related by $v = \lambda f$.

- The wave number is defined as $\kappa = 2\pi/\lambda$, just as the angular frequency is related to the period by $\omega = 2\pi/T$.

- Any function of the form $Y(\kappa x - \omega t + \phi_0)$ or $Y(\kappa x + \omega t + \phi_0)$ is a solution to the wave equation. The first function describes a wave traveling in the positive x-direction, and the second function describes a wave traveling in the negative x-direction.

- The speed of a transverse wave on a string is $v = \sqrt{F/\mu}$, where $\mu = m/L$ is the linear mass density of the string and F is the string tension.

- A one-dimensional sinusoidal wave is given by $y(x,t) = A\sin(\kappa x - \omega t + \phi_0)$, where the argument of the sine function is the phase of the wave and A is the amplitude of the wave.

- A three-dimensional spherical wave is described by $\psi(\vec{r},t) = \frac{A}{r}\sin(\kappa r - \omega t + \phi_0)$, and a plane wave is described by $\psi(\vec{r}, t) = A\sin(\kappa x - \omega t + \phi_0)$.

- The energy contained in a wave is $E = \frac{1}{2}\rho V\omega^2[A(r)]^2 = \frac{1}{2}\rho(A_\perp l)\omega^2[A(r)]^2 = \frac{1}{2}\rho(A_\perp vt)\omega^2[A(r)]^2$, where $A_\perp$ is the perpendicular cross-sectional area through which the wave passes.

- The power transmitted by a wave is $\bar{P} = \frac{E}{t} = \frac{1}{2}\rho A_\perp v\omega^2[A(r)]^2$, and its intensity is $I = \frac{1}{2}\rho v\omega^2[A(r)]^2$.

- For a spherical three-dimensional wave, the intensity is inversely proportional to the square of the distance to the source.

- The superposition principle holds for waves: Adding two solutions of the wave equation results in another valid solution.

- Waves can interfere in space and time, constructively or destructively, depending on their relative phases.

- Adding two traveling waves that are identical except for having velocity vectors that point in opposite directions yields a standing wave such as $y(x,t) = 2A\sin(\kappa x)\cos(\omega t)$, for which the dependence on space and time factorizes.

- The resonance frequencies (or harmonics) for waves on a string are given by $f_n = \frac{v}{\lambda_n} = n\frac{\sqrt{F}}{2L\sqrt{\mu}}$, for $n = 1, 2, 3, \dots$. The index n indicates the harmonic (for example, $n = 4$ corresponds to the fourth harmonic).

ANSWERS TO SELF-TEST OPPORTUNITIES

15.1 $f = 0.40$ Hz.

15.2 $A = 10$ cm, $\kappa = 0.31$ cm^{-1}, $\phi_0 = 0.30$.

15.3 The surface area of a sphere with radius r is $4\pi r^2$, and the total power is $P = IA_\perp$. With $r = 1.5 \cdot 10^{11}$ m and $I = 1.4$ kW/m^2, we get $P = 3.9 \cdot 10^{26}$ W.

15.4 Lowest resonance frequency: $f_1 = \frac{v}{\lambda_1} = \frac{v}{2L}$

Wave speed: $v = 2Lf_1 = 2(1.5 \cdot 10^{-6} \text{ m})(5.5 \cdot 10^7 \text{ s}^{-1}) = 165$ m/s.

PROBLEM-SOLVING GUIDELINES

1. If a wave fits the description of a one-dimensional traveling wave, you can write a mathematical equation describing it if you know the amplitude, A, of the motion and any two of these three quantities: velocity, v, wavelength, λ, and frequency, f. (Alternatively, you need to know two of the three quantities velocity, v, wave number, κ, and angular frequency, ω.)

2. If a wave is a one-dimensional standing wave on a string, its resonance frequencies are determined by the length, L, of the string and the speed, v, of the component waves. The speed of the wave, in turn, is determined by the tension, F, in the string and the linear mass density, μ, of the string.

MULTIPLE-CHOICE QUESTIONS

15.1 Fans at a local football stadium are so excited that their team is winning that they start "the wave" in celebration. Which of the following four statements is (are) true?

I. This wave is a traveling wave.

II. This wave is a transverse wave.

III. This wave is a longitudinal wave.

IV. This wave is a combination of a longitudinal wave and a transverse wave.

a) I and II   c) III only   e) I and III

b) II only   d) I and IV

15.2 You wish to decrease the speed of a wave traveling on a string to half its current value by changing the tension in the string. By what factor must you decrease the tension in the string?

a) 1   c) 2   e) none of the above

b) $\sqrt{2}$ d) 4

15.3 Suppose that the tension is doubled for a string on which a wave is propagated. How will the velocity of the wave change?

a) It will double.   c) It will be multiplied by $\sqrt{2}$.

b) It will quadruple.   d) It will be multiplied by $\frac{1}{2}$.

15.4 Which one of the following transverse waves has the greatest power?

a) a wave with velocity v, amplitude A, and frequency f

b) a wave with velocity v, amplitude $2A$, and frequency $f/2$

c) a wave with velocity $2v$, amplitude $A/2$, and frequency f

d) a wave with velocity $2v$, amplitude A, and frequency $f/2$

e) a wave with velocity $2v$, amplitude $A/2$, and frequency $2f$

15.5 The speed of light waves in air is greater than the speed of sound waves in air by *about* a factor of a million. Given a sound wave and a light wave of the same wavelength, both traveling through air, which statement about their frequencies is true?

a) The frequency of the sound wave will be about a million times greater than that of the light wave.

b) The frequency of the sound wave will be about a thousand times greater than that of the light wave.

c) The frequency of the light wave will be about a thousand times greater than that of the sound wave.

d) The frequency of the light wave will be about a million times greater than that of the sound wave.

e) There is insufficient information to determine the relationship between the two frequencies.

15.6 A string is made to oscillate, and a standing wave with three antinodes is created. If the tension in the string is increased by a factor of 4, the number of antinodes

a) increases.

b) remains the same.

c) decreases.

d) will equal the number of nodes.

15.7 The different colors of light we perceive are a result of the varying frequencies (and wavelengths) of the electromagnetic radiation. Infrared radiation has lower frequencies than does visible light, and ultraviolet radiation has higher frequencies than visible light does. The primary colors are red (R), yellow (Y), and blue (B). Order these colors by their wavelength, shortest to longest.

a) B, Y, R   b) B, R, Y   c) R, Y, B   d) R, B, Y

15.8 If transverse waves on a string travel with a velocity of 50 m/s when the string is under a tension of 20 N, what tension on the string is required for the waves to travel with a velocity of 30 m/s?

a) 7.2 N   c) 33 N   e) 45 N

b) 12 N   d) 40 N   f) 56 N

15.9 The intensity of electromagnetic waves from the Sun at the Earth is 1400 W/m^2. How much power does the Sun generate? The distance from the Earth to the Sun is $149.6 \cdot 10^6$ km.

a) $1.21 \cdot 10^{20}$ W

b) $2.43 \cdot 10^{24}$ W

c) $3.94 \cdot 10^{26}$ W

d) $2.11 \cdot 10^{28}$ W

e) $9.11 \cdot 10^{30}$ W

15.10 A guitar string is 0.750 m long and has a mass of 5.00 g. The string is tuned to E (660 Hz) when it vibrates at its fundamental frequency. What is the required tension on the string?

a) $2.90 \cdot 10^3$ N

b) $4.84 \cdot 10^3$ N

c) $6.53 \cdot 10^3$ N

d) $8.11 \cdot 10^3$ N

e) $1.23 \cdot 10^4$ N

CONCEPTUAL QUESTIONS

15.11 You and a friend are holding the two ends of a Slinky stretched out between you. How would you move your end of the Slinky to create (a) transverse waves or (b) longitudinal waves?

15.12 A steel cable consists of two sections with different cross-sectional areas, A_1 and A_2. A sinusoidal traveling wave is sent down this cable from the thin end of the cable. What happens to the wave on encountering the A_1/A_2 boundary? How do the speed, frequency and wavelength of the wave change?

15.13 Noise results from the superposition of a very large number of sound waves of various frequencies (usually in a continuous spectrum), amplitudes, and phases. Can interference arise with noise produced by two sources?

15.14 The $1/R^2$ dependency for intensity can be thought of as being due to the fact that the same power is being spread out over the surface of a larger and larger sphere. What happens to the intensity of a sound wave inside an enclosed space, such as a long hallway?

15.15 If two traveling waves have the same wavelength, frequency, and amplitude and are added appropriately, the result is a standing wave. Is it possible to combine two standing waves in some way to give a traveling wave?

15.16 A ping-pong ball is floating in the middle of a lake and waves begin to propagate on the surface. Can you think of a situation in which the ball remains stationary? Can you think of a situation involving a single wave on the lake in which the ball remains stationary?

15.17 Why do circular water waves on the surface of a pond decrease in amplitude as they travel away from the source?

15.18 Consider a wave on a string, with amplitude A and wavelength λ, traveling in one direction. Find the relationship between the maximum speed of any portion of string, v_{max}, and the wave speed, v.

15.19 Suppose a tightrope walker is standing in the middle of a rope a mile long. If one end of the rope is cut, how long will it be before the tightrope walker begins to fall? What would you have to know to answer this question?

15.20 Suppose a long line of cars waits at an intersection for the light to turn green. Once the light turns, the cars in the back cannot move right away, but have to wait until the cars ahead of them move. Explain this effect (called a *rarefaction* wave) in terms of the wave physics that we have studied in this chapter.

15.21 Derive the relationship for the speed of a wave on a string $v = \sqrt{F/\mu}$, where F is the tension force on the string and μ is the linear mass density of the string. Start with Derivation 15.1 and substitute the restoring force provided by the springs with a restoring force resulting from the tension of the string.

15.22 Derive the expression for the energy of a wave $E = \frac{1}{2}m\omega^2 A^2$, where m is the mass of the substance that is oscillating, ω is the angular velocity, and $A(r)$ is the amplitude. Start by expressing the kinetic energy and potential energy of the wave, and then add these to get the energy of the wave.

EXERCISES

A blue problem number indicates a worked-out solution is available in the Student Solutions Manual. One • and two •• indicate increasing level of problem difficulty.

Sections 15.1 through 15.3

15.23 One of the main things allowing humans to determine whether a sound is coming from the left or the right is the fact that the sound will reach one ear before the other. Given that the speed of sound in air is 343 m/s and that human ears are typically 20.0 cm apart, what is the maximum time resolution for human hearing that allows sounds coming from the left to be distinguished from sounds coming from the right? Why is it impossible for a diver to be able to tell from which direction the sound of a motorboat is coming? The speed of sound in water is $1.50 \cdot 10^3$ m/s.

15.24 Hiking in the mountains, you shout "hey," wait 2.00 s and shout again. What is the distance between the sound waves you cause? If you hear the first echo after 5.00 s, what is the distance between you and the point where your voice hit a mountain?

15.25 The displacement from equilibrium caused by a wave on a string is given by $y(x,t) = (-0.00200$ m$) \sin[(40.0$ m$^{-1})x - (800.$ s$^{-1})t]$. For this wave, what are the (a) amplitude, (b) number of waves in 1.00 m, (c) number of complete cycles in 1.00 s, (d) wavelength, and (e) speed?

•15.26 A traveling wave propagating on a string is described by the following equation:

$$y(x,t) = (5.00 \text{ mm})\sin\left((157.08 \text{ m}^{-1})x - (314.16 \text{ s}^{-1})t + 0.7854\right)$$

a) Determine the minimum separation, Δx_{min}, between two points on the string that oscillate in perfect opposition of phases (move in opposite directions *at all times*).

b) Determine the minimum separation, Δx_{AB}, between two points A and B on the string, if point B oscillates with a phase difference of 0.7854 rad compared to point A.

c) Find the number of crests of the wave that pass through point A in a time interval $\Delta t = 10.0$ s and the number of troughs that pass through point B in the same interval.

d) At what point along its trajectory should a linear driver connected to one end of the string at $x = 0$ start its oscillation to generate this sinusoidal traveling wave on the string?

••15.27 Consider a linear array of n masses, each equal to m, connected by $n + 1$ springs, all massless and having spring constant k, with the outer ends of the first and last springs fixed. The masses can move without friction in the linear dimension of the array.

a) Write the equations of motion for the masses.

b) Configurations of motion for which all parts of a system oscillate with the same angular frequency are called *normal modes* of the system; the corresponding angular frequencies are the system's *normal-mode angular frequencies*. Find the normal-mode angular frequencies of this array. Note that there are n normal modes.

Section 15.4

15.28 Show that the function $D = A \ln (x + vt)$ is a solution of the wave equation (equation 15.9).

15.29 A wave travels along a string in the positive x-direction at 30.0 m/s. The frequency of the wave is 50.0 Hz. At $x = 0$ and $t = 0$, the wave velocity is 2.50 m/s and the vertical displacement is $y = 4.00$ mm. Write the function $y(x,t)$ for the wave.

15.30 A wave on a string has a wave function given by

$$y(x,t) = (0.0200 \text{ m})\sin\left|\left(6.35 \text{ m}^{-1}\right)x + \left(2.63 \text{ s}^{-1}\right)t\right|.$$

a) What is the amplitude of the wave?

b) What is the period of the wave?

c) What is the wavelength of the wave?

d) What is the speed of the wave?

e) In which direction does the wave travel?

•**15.31** A sinusoidal wave traveling in the positive *x*-direction has a wavelength of 12.0 cm, a frequency of 10.0 Hz, and an amplitude of 10.0 cm. The part of the wave that is at the origin at *t* = 0 has a vertical displacement of 5.00 cm. For this wave, determine the

a) wave number,

d) speed,

b) period,

e) phase angle, and

c) angular frequency,

f) equation of motion.

•**15.32** A mass *m* hangs on a string that is connected to the ceiling. You pluck the string just above the mass, and a wave pulse travels up to the ceiling, reflects off the ceiling, and travels back to the mass. Compare the round-trip time for this wave pulse to that of a similar wave pulse on the same string if the attached mass is increased to 3.00*m*. (Assume that the string does not stretch in either case and the contribution of the mass of the string to the tension is negligible.)

•**15.33** Point *A* in the figure is a distance *d* below the ceiling. Determine how much longer it will take for a wave pulse to travel from point *A* to the ceiling along wire 1 than along wire 2.

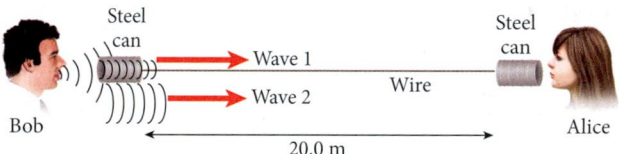

•**15.34** A particular steel guitar string has a mass per unit length of 1.93 g/m.

a) If the tension on this string is 62.2 N, what is the wave speed on the string?

b) For the wave speed to be increased by 1.00%, how much should the tension be changed?

•**15.35** Bob is talking to Alice using a tin can telephone, which consists of two steel cans connected by a 20.0-m-long taut steel wire (see the figure). The wire has a linear mass density of 6.13 g/m, and the tension on the wire is 25.0 N. The sound waves leave Bob's mouth, are collected by the can on the left, and then create vibrations in the wire, which travel to Alice's can and are transformed back into sound waves in air. Alice hears both the sound waves that have traveled through the wire (wave 1) and those that have traveled through the air (wave 2), bypassing the wire. Do these two kinds of waves reach her at the same time? If not, which wave arrives sooner and by how much? The speed of sound in air is 343 m/s. Assume that the waves on the string are transverse.

••**15.36** A wire of uniform linear mass density hangs from the ceiling. It takes 1.00 s for a wave pulse to travel the length of the wire. How long is the wire?

••**15.37** a) Starting from the general wave equation (equation 15.9), prove through direct derivation that the Gaussian wave packet described by the equation $y(x,t) = (5.00m)e^{-0.1(x-5t)^2}$ is indeed a traveling wave (that it satisfies the differential wave equation).

b) If *x* is specified in meters and *t* in seconds, determine the speed of this wave. On a single graph, plot this wave as a function of *x* at *t* = 0, *t* = 1.00 s, *t* = 2.00 s, and *t* = 3.00 s.

c) More generally, prove that any function *f*(*x*,*t*) that depends on *x* and *t* through a combined variable *x* ± *vt* is a solution of the wave equation, irrespective of the specific form of the function *f*.

Section 15.5

•**15.38** Suppose the slope of a beach underneath an ocean is 12.0 cm of dropoff for every 1.00 m of horizontal distance. A wave is moving inland, slowing down as it enters shallower water. What is its acceleration when it is 10.0 m from the shoreline? (Take the direction toward the shoreline as the positive *x*-direction.)

•**15.39** An earthquake generates three kinds of waves: surface waves (L waves), which are the slowest and weakest; shear (S) waves, which are transverse

waves and carry most of the energy; and pressure (P) waves, which are longitudinal waves and travel the fastest. The speed of P waves is approximately 7.0 km/s, and that of S waves is about 4.0 km/s. Animals seem to feel the P waves. If a dog senses the arrival of P waves and starts barking 30.0 s before an earthquake is felt by humans, approximately how far is the dog from the earthquake's epicenter?

Section 15.6

15.40 A string with a mass of 30.0 g and a length of 2.00 m is stretched under a tension of 70.0 N. How much power must be supplied to the string to generate a traveling wave that has a frequency of 50.0 Hz and an amplitude of 4.00 cm?

15.41 A string with a linear mass density of 0.100 kg/m is under a tension of 100. N. How much power must be supplied to the string to generate a sinusoidal wave of amplitude 2.00 cm and frequency 120. Hz?

•**15.42** A sinusoidal wave on a string is described by the equation *y* = (0.100 m) sin (0.750*x* − 40.0*t*), where *x* and *y* are in meters and *t* is in seconds. If the linear mass density of the string is 10.0 g/m, determine (a) the phase constant, (b) the phase of the wave at *x* = 2.00 cm and *t* = 0.100 s, (c) the speed of the wave, (d) the wavelength, (e) the frequency, and (f) the power transmitted by the wave.

Sections 15.7 and 15.8

15.43 In an acoustics experiment, a piano string with a mass of 5.00 g and a length of 70.0 cm is held under tension by running the string over a frictionless pulley and hanging a 250.-kg weight from it. The whole system is placed in an elevator.

a) What is the fundamental frequency of oscillation for the string when the elevator is at rest?

b) With what acceleration and in what direction (up or down) should the elevator move for the string to produce the proper frequency of 440. Hz, corresponding to middle A?

15.44 A string is 35.0 cm long and has a mass per unit length of $5.51 \cdot 10^{-4}$ kg/m. What tension must be applied to the string so that it vibrates at the fundamental frequency of 660. Hz?

15.45 A 2.00-m-long string of mass 10.0 g is clamped at both ends. The tension in the string is 150. N.

a) What is the speed of a wave on this string?

b) The string is plucked so that it oscillates. What is the wavelength and frequency of the resulting wave if it produces a standing wave with two antinodes?

•**15.46** Write the equation for a standing wave that has three antinodes of amplitude 2.00 cm on a 3.00-m-long string that is fixed at both ends and vibrates 15.0 times a second. The time *t* = 0 is chosen to be an instant when the string is flat. If a wave pulse were propagated along this string, how fast would it travel?

•**15.47** A 3.00-m-long string, fixed at both ends, has a mass of 6.00 g. If you want to set up a standing wave in this string having a frequency of 300. Hz and three antinodes, what tension should you put the string under?

•**15.48** A cowboy walks at a pace of about two steps per second, holding a glass of diameter 10.0 cm that contains milk. The milk sloshes higher and higher in the glass until it eventually starts to spill over the top. Determine the maximum speed of the waves in the milk.

•**15.49** Students in a lab produce standing waves on stretched strings connected to vibration generators. One such wave is described by the wave function $y(x,t) = (2.00 \text{ cm})\sin\left[(20.0 \text{ m}^{-1})x\right]\cos\left[(150. \text{ s}^{-1})t\right]$, where *y* is the transverse displacement of the string, *x* is the position along the string, and *t* is time. Rewrite this wave function in the form for a wave moving in the positive *x*-direction and a wave moving in the negative *x*-direction: $y(x,t) = f(x - vt) + g(x + vt)$; that is, find the functions *f* and *g* and the speed, *v*.

•**15.50** An array of wave emitters, as shown in the figure, emits a wave of wavelength λ that is to be detected at a distance *L* directly above the rightmost emitter. The distance between adjacent wave emitters is *d*.

a) Show that when $L \gg d$, the wave from the nth emitter (counting from right to left with $n = 0$ being the rightmost emitter) has to travel an extra distance of $\Delta s = n^2 (d^2/2L)$.

b) If $\lambda = d^2/2L$, will the interference at the detector be constructive or destructive?

c) If $\lambda = d^2/2L = 10^{-3}$ m and $L = 1.00 \cdot 10^3$ m, what is d, the distance between adjacent emitters?

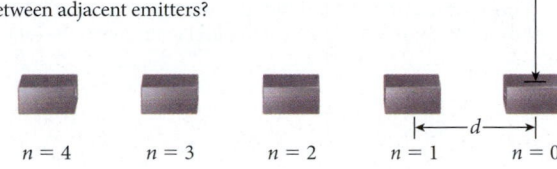

$n = 4 \qquad n = 3 \qquad n = 2 \qquad n = 1 \qquad n = 0$

••**15.51** A small ball floats in the center of a circular pool that has a radius of 5.00 m. Three wave generators are placed at the edge of the pool, separated by 120.°. The first wave generator operates at a frequency of 2.00 Hz. The second wave generator operates at a frequency of 3.00 Hz. The third wave generator operates at a frequency of 4.00 Hz. If the speed of each water wave is 5.00 m/s, and the amplitude of the waves is the same, sketch the height of the ball as a function of time from $t = 0$ to $t = 2.00$ s, assuming that the water surface is at zero height. Assume that all the wave generators impart a phase shift of zero. How would your answer change if one of the wave generators was moved to a different location at the edge of the pool?

••**15.52** A string with linear mass density $\mu = 0.0250$ kg/m under a tension of $F = 250.$ N is oriented in the x-direction. Two transverse waves of equal amplitude and with a phase angle of zero (at $t = 0$) but with different angular frequencies ($\omega = 3000.$ rad/s and $\omega/3 = 1000.$ rad/s) are created in the string by an oscillator located at $x = 0$. The resulting waves, which travel in the positive x-direction, are reflected at a distant point where both waves have a node, so there is a similar pair of waves traveling in the negative x-direction. Find the values of x at which the first two nodes in the standing wave are produced by these four waves.

••**15.53** The equation for a standing wave on a string with mass density μ is $y(x,t) = 2A \cos(\omega t) \sin(\kappa x)$. Show that the average kinetic energy and potential energy over time for this wave per unit length of string are given by $K_{ave}(x) = \mu\omega^2 A^2 \sin^2 \kappa x$ and $U_{ave}(x) = F(\kappa A)^2(\cos^2 \kappa x)$.

Additional Exercises

15.54 A sinusoidal wave traveling on a string is moving in the positive x-direction. The wave has a wavelength of 4 m, a frequency of 50.0 Hz, and an amplitude of 3.00 cm. What is the wave function for this wave?

15.55 A guitar string with a mass of 10.0 g is 1.00 m long and attached to the guitar at two points separated by 65.0 cm.

a) What is the frequency of the first harmonic of this string when it is placed under a tension of 81.0 N?

b) If the guitar string is replaced by a heavier one that has a mass of 16.0 g and is 1.00 m long, what is the frequency of the replacement string's first harmonic?

15.56 Write the equation for a sinusoidal wave propagating in the negative x-direction with a speed of 120. m/s, if a particle in the medium in which the wave is moving is observed to swing back and forth through a 6.00-cm range in 4.00 s. Assume that $t = 0$ is taken to be the instant when the particle is at $y = 0$ and that the particle moves in the positive y-direction immediately after $t = 0$.

15.57 Shown in the figure is a plot of the displacement, y, due to a sinusoidal wave traveling along a string as a function of time, t. What are the (a) period, (b) maximum speed, and (c) maximum acceleration of this wave perpendicular to the direction in which it is traveling?

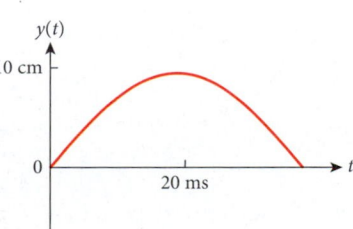

15.58 A 50.0-cm-long wire with a mass of 10.0 g is under a tension of 50.0 N. Both ends of the wire are held rigidly while it is plucked.

a) What is the speed of the waves on the wire?

b) What is the fundamental frequency of the standing wave?

c) What is the frequency of the third harmonic?

15.59 What is the wave speed along a brass wire with a radius of 0.500 mm stretched at a tension of 125 N? The density of brass is $8.60 \cdot 10^3$ kg/m^3.

15.60 Two steel wires are stretched under the same tension. The first wire has a diameter of 0.500 mm, and the second wire has a diameter of 1.00 mm. If the speed of waves traveling along the first wire is 50.0 m/s, what is the speed of waves traveling along the second wire?

15.61 The middle-C key (key 40) on a piano corresponds to a fundamental frequency of about 262 Hz, and the soprano-C key (key 64) corresponds to a fundamental frequency of 1046.5 Hz. If the strings used for both keys are identical in density and length, determine the ratio of the tensions in the two strings.

15.62 A sinusoidal wave travels along a stretched string. A point along the string has a maximum velocity of 1.00 m/s and a maximum displacement of 2.00 cm. What is the maximum acceleration at that point?

•**15.63** As shown in the figure, a sinusoidal wave travels to the right at a speed of v_1 along string 1, which has linear mass density μ_1. This wave has frequency f_1 and wavelength λ_1. Since string 1 is attached to string 2 (which has linear mass density $\mu_2 = 3\mu_1$), the first wave will excite a new wave in string 2, which will also move to the right. What is the frequency, f_2, of the wave produced in string 2? What is the speed, v_2, of the wave produced in string 2? What is the wavelength, λ_2, of the wave produced in string 2? Write all answers in terms of f_1, v_1, and λ_1.

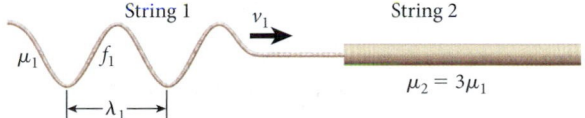

String 1 $\qquad v_1 \qquad$ String 2

$\mu_1 \qquad f_1$

λ_1

$\mu_2 = 3\mu_1$

•**15.64** The tension in a 2.70-m-long, 1.00-cm-diameter steel cable ($\rho = 7800.$ kg/m^3) is 840. N. What is the fundamental frequency of vibration of the cable?

•**15.65** A wave traveling on a string has the equation of motion $y(x,t) = 0.0200 \sin(5.00x - 8.00t)$, where x and y are in meters and t is in seconds.

a) Calculate the wavelength and the frequency of the wave.

b) Calculate its velocity.

c) If the linear mass density of the string is $\mu = 0.100$ kg/m, what is the tension on the string?

•**15.66** Calvin sloshes back and forth in his bathtub, producing a standing wave. What is the frequency of such a wave if the bathtub is 150. cm long and 80.0 cm wide and contains water that is 38.0 cm deep?

•**15.67** Consider a guitar string stretching 80.0 cm between its anchored ends. The string is tuned to play middle C, with a frequency of 261.6 Hz, when oscillating in its fundamental mode, that is, with one antinode between the ends. If the string is displaced 2.00 mm at its midpoint and released to produce this note, what are the wave speed, v, and the maximum speed, v_{max}, of the midpoint of the string?

15.68 Two waves traveling in opposite directions along a string fixed at both ends create a standing wave described by $y(x,t) = 1.00 \cdot 10^{-2} \sin(25.0x) \cos(1200.t)$. The string has a linear mass density of 0.0100 kg/m, and the tension in the string is supplied by a mass hanging from one end. If the string vibrates in its third harmonic, calculate (a) the length of the string, (b) the velocity of the waves, and (c) the mass of the hanging mass.

15.69 A sinusoidal transverse wave of wavelength 20.0 cm and frequency 500. Hz travels along a string in the positive z-direction. The wave oscillations take place in the xz-plane and have an amplitude of 3.00 cm. At time $t = 0$, the displacement of the string at $x = 0$ is $z = 3.00$ cm.

a) A photo of the wave is taken at $t = 0$. Make a simple sketch (including axes) of the string at this time.

b) Determine the speed of the wave.

c) Determine the wave's wave number.

d) If the linear mass density of the string is 30.0 g/m, what is the tension in the string?

e) Determine the function $D(z,t)$ that describes the displacement x that is produced in the string by this wave.

••**15.70** A heavy cable of total mass M and length $L = 5.00$ m has one end attached to a rigid support and the other end hanging free. A small transverse displacement is initiated at the bottom of the cable. How long does it take for the displacement to travel to the top of the cable?

MULTI-VERSION EXERCISES

15.71 A rubber band of mass 0.3491 g is stretched between two fingers, putting each side under a tension of 1.777 N. The overall stretched length of the band is 20.27 cm. One side of the band is plucked, setting up a vibration in 8.725 cm of the band's stretched length. What is the lowest-frequency vibration that can be set up on this part of the rubber band? Assume that the band stretches uniformly.

15.72 A rubber band of mass 0.4245 g is stretched between two fingers. The overall stretched length of the band is 20.91 cm. One side of the band is plucked, setting up a vibration in 8.117 cm of the band's stretched length. The lowest-frequency vibration that can be set up on this part of the rubber band is 184.2 Hz. What is the tension in each side of the rubber band? Assume that the band stretches uniformly.

15.73 A rubber band of mass 0.1701 g is stretched between two fingers, putting each side under a tension of 1.851 N. The overall stretched length of the band is 21.55 cm. One side of the band is plucked, setting up a vibration in the band's stretched length. The lowest-frequency vibration that can be set up on this part of the rubber band is 254.6 Hz. What is the length of the vibrating part of the band? Assume that the band stretches uniformly.

15.74 A satellite in a circular orbit around the Sun uses a square solar panel as a power source. The edges of the panel are 1.459 m long. The panel's efficiency is 16.57%. The satellite is $4.949 \cdot 10^7$ km from the Sun. How much power does the solar panel provide to the satellite? Assume that the total power output of the Sun is $3.937 \cdot 10^{26}$ W.

15.75 A satellite in a circular orbit around the Sun uses a square solar panel as a power source. The panel's efficiency is 16.87%. The satellite is $6.103 \cdot 10^7$ km from the Sun. The solar panel provides $5.215 \cdot 10^3$ W to the satellite. How long are the edges of the solar panel? Assume that the total power output of the Sun is $3.937 \cdot 10^{26}$ W.

15.76 A satellite in a circular orbit around the Sun uses a square solar panel as a power source. The edges of the panel are 2.375 m long. The satellite is $7.257 \cdot 10^7$ km from the Sun. The solar panel provides $5.768 \cdot 10^3$ W to the satellite. What is the efficiency of the solar panel? Assume that the total power output of the Sun is $3.937 \cdot 10^{26}$ W.

15.77 Consider a string with a linear mass density of 0.2833 g/cm and a length of 116.7 cm, which is oscillating as shown in the figure. The tension on the string is 18.25 N. At what frequency is the string vibrating?

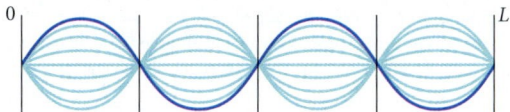

15.78 Consider a string with a linear mass density of 0.1291 g/cm and a length of 117.5 cm, which is oscillating as shown in the figure. The string is vibrating at a frequency of 93.63 Hz. What is the tension on the string?

15.79 Consider a string with a linear mass density of 0.1747 g/cm, which is oscillating as shown in the figure. The string is under a tension of 10.81 N and is vibrating at a frequency of 59.47 Hz. What is the length of the string?

16 Sound

FIGURE 16.1 U2 in the process of producing sound waves during their 2011 World Tour.

Recognizing sounds is one of the most important ways in which we learn about the world around us. As we'll see in this chapter, human hearing is very sensitive and can distinguish a wide range of frequencies and degrees of loudness. However, although interpreting sound probably developed in animals from earliest times, as it warned of predators or provided help in hunting prey, sound has also been a powerful source of cultural ritual and entertainment for humans (Figure 16.1). Music has been part of human life for as long as society has existed.

Sound is a type of wave, so this chapter's study of sound builds directly on the concepts about waves presented in Chapter 15. Some characteristics of sound waves also pertain to light waves and will be useful when we study those waves in later chapters. This chapter also examines sources of musical sounds and applications of sound in a wide range of other areas, from medicine to geography.

WHAT WE WILL LEARN

- Sound consists of longitudinal pressure waves and needs a medium in which to propagate.

- The speed of sound is usually higher in solids than in liquids and higher in liquids than in gases.

- The speed of sound in air depends on the temperature; it is approximately 343 m/s at normal atmospheric pressure and a temperature of 20 °C.

- The intensity of sound detectable by the human ear spans a large range and is usually expressed on a logarithmic scale in terms of decibels (dB).

- Sound waves from two or more sources can interfere in space and time, resulting in destructive or constructive interference.

- Interference of sound waves of slightly different frequencies results in oscillations in intensity known as beats. The beat frequency is the absolute value of the difference in the frequencies.

- The Doppler effect is the shift in the observed frequency of a sound due to the fact that the source moves (is approaching or receding) relative to the observer.

- If a sound source moves with a speed greater than that of sound, a shock wave, or Mach cone, develops.

- Standing resonant waves in open or closed pipes can be generated only at discrete wavelengths.

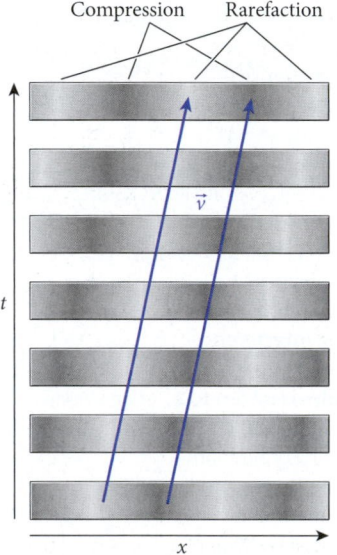

FIGURE 16.2 Propagation of a longitudinal pressure wave along an *x*-axis as a function of time.

FIGURE 16.3 You see fireworks before you hear their explosive sounds.

16.1 Longitudinal Pressure Waves

Sound is a pressure variation that propagates through some medium. In air, the pressure variation causes abnormal motion of air molecules in the direction of propagation; thus, a sound wave is longitudinal. When we hear a sound, our eardrums are set in vibration by the air next to them. If this air has a pressure variation that repeats with a certain frequency, the eardrums vibrate at this frequency.

Sound requires a medium for propagation. If the air is pumped out of a glass jar that contains a ringing bell, the sound ceases as the air is evacuated from the jar, even though the hammer clearly still hits the bell. Thus, sound waves originate from a source, or emitter, and need a medium to travel through. In spite of movie scenes that include a huge roar when a spaceship flies by or when a star or planet explodes, interstellar space is a silent place because it is a vacuum.

Figure 16.2 shows a continuous pressure wave composed of alternating variations of air pressure—a pressure excess (compression) followed by a pressure reduction (rarefaction). Plotting these variations along an *x*-axis for various times, as in Figure 16.2, allows us to deduce the wave speed.

Although sound has to propagate through a medium, that medium does not have to be air. When you are under water, you can still hear sounds, so you know that sound also propagates through liquids. In addition, you may have put your ear to a train track to obtain advance information on an approaching train (not recommended, by the way, because the train may be closer than you think). Thus, you know that sound propagates through solids. In fact, sound propagates faster and with less loss through metals than through air. If this were not so, putting your ear to the train track would be pointless.

Sound Velocity

When you watch fireworks on the Fourth of July, you see the explosions of the rockets before you hear the sounds (Figure 16.3). The reason is that the light emitted from the explosions reaches your eyes almost instantaneously, because the speed of light is approximately 300,000 km/s. However, the speed of sound is much slower, and so the sound waves from the explosions reach you some time after the light waves have already arrived.

How fast does sound propagate—in other words, what is the speed of sound? For a wave on a string, we saw in Chapter 15 that the wave speed is $v = \sqrt{F/\mu}$, where F is the string tension (a force) and μ is the linear mass density of the string. From this equation, you can think of v as the square root of the ratio of the restoring force to the inertial response. If the wave propagates through a three-dimensional medium instead of a one-dimensional string, the inertial response originates from the density of the medium, ρ. Chapter 13 discussed the elasticity of solids, and introduced the elastic modulus known as *Young's modulus, Y.* This modulus determines the fractional change in length of a thin rod as a function of the force per unit area applied to the rod. Analysis reveals that

Young's modulus is the appropriate force term for a wave propagating along a thin solid rod; so we obtain

$$v = \sqrt{\frac{Y}{\rho}} \qquad (16.1)$$

for the speed of sound in a thin solid rod.

The speed of sound in fluids—both liquids and gases—is similarly related to the bulk modulus, B, defined in Chapter 13, as determining a material's volume change in response to external pressure. Thus, the speed of sound in a gas or liquid is given by

$$v = \sqrt{\frac{B}{\rho}}. \qquad (16.2)$$

Although these dimensional arguments show that the speed of sound is proportional to the square root of the modulus divided by the appropriate density, a more fundamental analysis, such as that in the following derivation, shows that the proportionality constants have a value of exactly 1 in equations 16.1 and 16.2.

DERIVATION 16.1 / Speed of Sound

Suppose a fluid is contained in a cylinder, with a movable piston on one end (Figure 16.4). If this piston moves with speed v_p into the fluid, it will compress the fluid element in front of it. This fluid element, in turn, will move as a result of the pressure change; its leading edge will move with the speed of sound, v, which, by definition, is the speed of the pressure waves in the medium.

Pushing the piston into the fluid with a force, F, causes a pressure change, $p = F/A$ in the fluid (see Chapter 13), where A is the cross-sectional area of the cylinder and also of the piston. The force exerted on the fluid element of mass m causes an acceleration given by $\Delta v/\Delta t$:

$$F = m\frac{\Delta v}{\Delta t} = m\frac{v_p}{\Delta t} \Rightarrow p = \frac{F}{A} = \frac{m}{A}\frac{v_p}{\Delta t}.$$

Because the mass is the density of the fluid times its volume, $m = \rho V$, and the volume is that of the cylinder of base area A and length l, we can find the mass of the fluid element:

$$m = \rho V = \rho A l \Rightarrow p = \frac{m}{A}\frac{v_p}{\Delta t} = \frac{\rho A l}{A}\frac{v_p}{\Delta t} = \frac{\rho l v_p}{\Delta t}.$$

Since this fluid element responds to the compression by moving with speed v during the time interval Δt, the length of the fluid element that has experienced the pressure wave is $l = v\Delta t$, which finally yields the pressure difference:

$$p = \frac{\rho l v_p}{\Delta t} = \frac{\rho(v\Delta t)v_p}{\Delta t} = \rho v v_p. \qquad (i)$$

From the definition of the bulk modulus (see Chapter 13), $p = B\Delta V/V$. We can combine this expression for the pressure and that obtained in equation (i):

$$p = \rho v v_p = B\frac{\Delta V}{V}. \qquad (ii)$$

Again, the volume of the moving fluid, V, that has experienced the pressure wave is proportional to v since $V = Al = Av\Delta t$. In addition, the volume change in the fluid caused by pushing the piston into the cylinder is $\Delta V = Av_p\Delta t$. Thus, the ratio $\Delta V/V$ is equivalent to v_p/v, the ratio of the piston's speed to the speed of sound. Inserting this result into equation (ii) yields

$$\rho v v_p = B\frac{\Delta V}{V} = B\frac{v_p}{v} \Rightarrow$$

$$\rho v^2 = B \Rightarrow$$

$$v = \sqrt{\frac{B}{\rho}}.$$

As you can see, the speed with which the piston is pushed into the fluid cancels out. Therefore, it does not matter what the speed of the excitation is; the sound always propagates with the same speed in the medium.

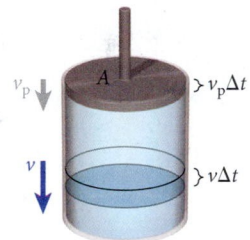

FIGURE 16.4 Piston compressing a fluid.

Concept Check 16.1

You are in the middle of a concert hall that is 120.0 m deep. What is the time difference between the arrival of the sound directly from the orchestra at your position and the arrival of the sound reflected from the back of the concert hall?

a) 0.010 s   d) 0.35 s

b) 0.056 s   e) 0.77 s

c) 0.11 s

Table 16.1	Speed of Sound in Some Common Substances	
	Substance	Speed of Sound (m/s)
Gases	Krypton	220
	Carbon dioxide	260
	Air	343
	Helium	960
	Hydrogen	1280
Liquids	Methanol	1143
	Mercury	1451
	Water	1480
	Seawater	1520
Solids	Lead	1160
	Concrete	3200
	Hardwood	4000
	Steel	5800
	Aluminum	6400
	Diamond	12,000

FIGURE 16.5 The 5-second rule for lightning strikes is due to the difference in speed between light and sound.

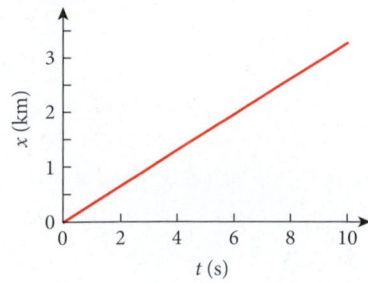

FIGURE 16.6 Distance x to the source (lightning strike) of the sound as a function of the time delay t in hearing the sound (thunder).

Equations 16.1 and 16.2 both state that the speed of sound in a given state of matter (gas, liquid, or solid) is inversely proportional to the square root of the density, which means that in two different gases, the speed of sound is higher in the one with the lower density. However, the values of Young's modulus for solids are much larger than the values of the bulk modulus for liquids, which are larger than those for gases. This difference is more important than the density dependence. From equations 16.1 and 16.2, it then follows that for the speeds of sound in solids, liquids, and gases, $v_{solid} > v_{liquid} > v_{gas}$. Representative values for the speed of sound in different materials under standard conditions of pressure (1 atm) and temperature (20 °C) are listed in Table 16.1.

We are most interested in the speed of sound in air, because this is the most important medium for sound propagation in everyday life. At normal atmospheric pressure and 20 °C, this speed of sound is

$$v_{air} = 343 \text{ m/s.} \tag{16.3}$$

Knowing the speed of sound in air explains the 5-second rule for thunderstorms: If 5 s or less pass between the instant you see a lightning strike (Figure 16.5) and the instant you hear the thunder, the lightning strike is 1 mi or less away. Since sound travels approximately 340 m/s, it travels 1700 m, or approximately 1 mi, in 5 s. The speed of light is approximately 1 million times greater than the speed of sound, so the visual perception of the lightning strike occurs with essentially no delay. Countries that use the metric system have a 3-second rule: A 3-s delay between lightning and thunder corresponds to 1 km of distance to the lightning strike (Figure 16.6).

The speed of sound in air depends (weakly) on the air temperature, T. The following linear dependence is obtained experimentally:

$$v(T) = (331 + 0.6 T/°C) \text{ m/s.} \tag{16.4}$$

EXAMPLE 16.1   Football Cheers

A student's apartment is located exactly 3.75 km from the football stadium. He is watching the game live on TV and sees the home team score a touchdown. According to his clock, 11.2 s pass after he hears the roar of the crowd on TV until he hears it again from outside.

PROBLEM

What is the temperature at game time?

SOLUTION

The TV signal moves with the speed of light and thus arrives at the student's TV at essentially the same instant as the action is happening in the stadium. Thus, the delay in the arrival of the roar at the student's apartment can be attributed entirely to the finite speed of sound. We find this speed of sound from the given data:

$$v = \frac{\Delta x}{\Delta t} = \frac{3750 \text{ m}}{11.2 \text{ s}} = 334.8 \text{ m/s.}$$

Using equation 16.4, we find the temperature at game time:

$$T = \frac{v(T) - 331}{0.6} °C = \frac{334.8 - 331}{0.6} °C = \frac{3.8}{0.6} °C = 6.4 °C.$$

DISCUSSION

A word of caution is in order: Several uncertainties are inherent in this result. First, if we redid the calculation for a time delay of 11.1 s, we would find the temperature to be 11.4 °C, or 5 °C higher than what we found using 11.2 s as the delay. Second, if 0.1 s makes such a big difference in the determined temperature, we need to ask if the TV signal really gets to the student's apartment instantaneously. The answer is no. If the student watches satellite TV, it takes about 0.2 s for the signal to make it up to the geostationary satellite and back down to the student's dish. In addition, short intentional delays are often inserted in TV broadcasts. Thus, our calculation is not very useful for practical purposes.

Sound Reflection

All waves, including sound waves, are at least partially reflected at the boundary between two different media. Sound waves are no exception. Sound waves that travel through air and then hit an object are partially reflected by the object. You can measure the distance to a distant large object by measuring the time between producing a short, loud sound and hearing that sound again after it has traveled to the object, reflected off the object, and returned to you. For example, if you were standing in Yosemite Valley and shouted in the direction of the flat face of Half Dome, 1.0 km away, the sound of your voice would carry across the valley, reflect off Half Dome, and return to you, making a round trip of 2.0 km. The speed of the sound of your voice is 343 m/s, so the time delay between shouting and hearing your reflected shout would be $t = (2000 \text{ m})/(343 \text{ m/s}) = 5.8$ s, a time interval that you could easily measure with a wristwatch.

This principle is used in ultrasound imaging for medical diagnostic purposes. Ultrasound waves have a frequency much higher than a human can hear, between 2 and 15 MHz. The frequency is chosen to provide detailed images and to penetrate deeply into human tissue. When the ultrasound waves encounter a change in density of the tissue, some of them are reflected back. Measuring how much time the ultrasound waves take to travel from the emitter to the receiver and recording how much of the emission is reflected as well as the direction of the original waves allows an image to be formed. The average value of the speed of ultrasound waves used for imaging human tissue is 1540 m/s. The time for these ultrasound waves to travel 2.5 cm and return is $t = (0.050 \text{ m})/(1540 \text{ m/s}) = 32$ µs, and the time to travel 10.0 cm and return is 130 µs. Thus, the ultrasound device needs to be able to measure accurately times in the range from 30 to 130 µs. A typical image of a fetus produced by ultrasound imaging is shown in Figure 16.7.

Bats and dolphins navigate using sound reflection (Figure 16.8). They emit sound waves in a frequency range from 14,000 Hz to well over 100,000 Hz in a specific direction and determine information about their surroundings from the reflected sound. This process of echolocation allows bats to navigate in darkness.

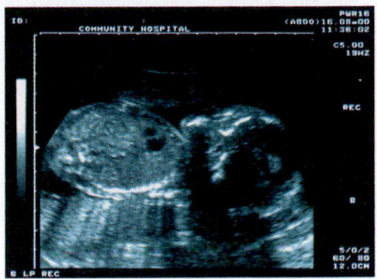

FIGURE 16.7 An image of a fetus produced by the reflection of ultrasound waves.

FIGURE 16.8 Bat flying in darkness, relying on echolocation for navigation.

16.2 Sound Intensity

In Chapter 15, the intensity of a wave was defined as the power per unit area. We saw that for spherical waves the intensity falls as the second power of the distance to the source, $I \propto r^{-2}$, giving the ratio

$$\frac{I(r_1)}{I(r_2)} = \left(\frac{r_2}{r_1}\right)^2 . \tag{16.5}$$

This relationship also holds for sound waves. Since the intensity is the power per unit area, its physical units are watts per square meter (W/m²).

Sound waves that can be detected by the human ear have a very large range of intensities, from whispers as low as 10^{-12} W/m² to the output of a jet engine or a rock band at close distance, which can reach 1 W/m². The oscillations in pressure for even the loudest sounds, at the pain threshold of 10 W/m², are on the order of only tens of micropascals (µPa). Normal atmospheric air pressure, by comparison, is 10^5 Pa. Thus, you can see that the air pressure varies by only 1 part in 10 billion, even for the loudest sounds. And the variation is several orders of magnitude less for the quietest sounds you can hear. This might give you a new appreciation for the capabilities of your ears.

Since human ears can register sounds over many orders of magnitude of intensity, a logarithmic scale is used to measure sound intensities. The unit of this scale is the bel (B), named for Alexander Graham Bell, but much more commonly used is the **decibel** (dB): 1 dB = 0.1 bel. The Greek letter β symbolizes the sound level measured on this decibel scale and is defined as

$$\beta = 10 \log \frac{I}{I_0} . \tag{16.6}$$

Here $I_0 = 10^{-12}$ W/m², which corresponds approximately to the minimum intensity that a human ear can hear. The notation "log" refers to the base-10 logarithm. (See Appendix A

Table 16.2	Levels of Sounds in Common Situations
Sound	**Sound Level (dB)**
Quietest sound heard	0
Background sound in library	30
Golf course	40–50
Street traffic	60–70
Train at railroad crossing	90
Dance club	110
Jackhammer	120
Jet taking off from aircraft carrier	130–150

for a refresher on logarithms.) Thus, a sound intensity 1000 times the reference intensity, I_0, gives $\beta = 10 \log 1000$ dB $= 10 \cdot 3$ dB $= 30$ dB. This sound level corresponds to a whisper not very far from your ear (Table 16.2).

Relative Intensity and Dynamic Range

The relative sound level, $\Delta\beta$, is the difference between two sound levels:

$$\Delta\beta = \beta_2 - \beta_1$$
$$= 10 \log \frac{I_2}{I_0} - 10 \log \frac{I_1}{I_0}$$
$$= 10 (\log I_2 - \log I_0) - 10 (\log I_1 - \log I_0)$$
$$= 10 \log I_2 - 10 \log I_1$$
$$\Delta\beta = 10 \log \frac{I_2}{I_1}. \tag{16.7}$$

The *dynamic range* is a measure of the relative sound levels of the loudest and the quietest sounds produced by a source. The dynamic range of a compact disc is approximately 90 dB, whereas the range was about 70 dB for vinyl records. Manufacturers list the dynamic range for all high-end loudspeakers and headphones as well. In general, the higher the dynamic range, the better the sound quality. However, the price usually increases with the dynamic range.

SOLVED PROBLEM 16.1 | Relative Sound Levels at a Rock Concert

Two friends attend a rock concert and bring along a sound meter. With this device, one of the friends measures a sound level of $\beta_1 = 105.0$ dB, whereas the other, who sits 4 rows (2.8 m) closer to the stage, measures $\beta_2 = 108.0$ dB.

PROBLEM
How far away are the two friends from the loudspeakers on stage?

SOLUTION

THINK We express the intensities at the two seats using equation 16.5 and the relative sound levels at the two seats using equation 16.7. We can combine these two equations to obtain an expression for the relative sound levels at the two seats. Knowing the distance between the two seats, we can then calculate the distance of the seats to the loudspeakers on the stage.

SKETCH Figure 16.9 shows the relative positions of the two friends.

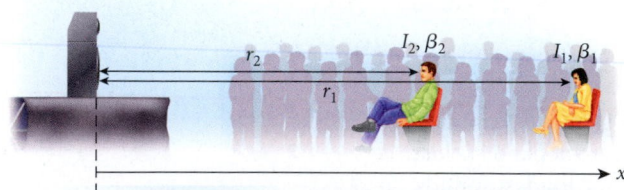

FIGURE 16.9 The relative distances, r_1 and r_2, of the two friends from the loudspeakers at a concert.

RESEARCH The distance of the first friend from the loudspeakers is r_1, and the distance of the second friend from the loudspeakers is r_2. The intensity of the sound at r_1 is I_1, and the intensity of the sound at r_2 is I_2. The sound level at r_1 is β_1, and the sound level at r_2 is β_2. We can express the two intensities in terms of the two distances using equation 16.5:

$$\frac{I_1}{I_2} = \left(\frac{r_2}{r_1}\right)^2. \tag{i}$$

We can then use equation 16.7 to relate the sound levels at r_1 and r_2 to the sound intensities:

$$\beta_2 - \beta_1 = 10 \log \frac{I_2}{I_1}. \tag{ii}$$

Combining equations (i) and (ii), we get

$$\beta_2 - \beta_1 = \Delta\beta = 10 \log \left(\frac{r_1}{r_2}\right)^2 = 20 \log \left(\frac{r_1}{r_2}\right). \tag{iii}$$

The relative sound level, $\Delta\beta$, is specified in the problem and we know that

$$r_2 = r_1 - 2.8 \text{ m}.$$

Thus, we can solve for the distance r_1 and then obtain r_2.

SIMPLIFY Substituting the relationship between the two distances into equation (iii) gives us

$$\Delta\beta = 20\log\left(\frac{r_1}{r_1 - 2.8\text{ m}}\right) \Rightarrow$$

$$10^{\Delta\beta/20} = \frac{r_1}{r_1 - 2.8\text{ m}} \Rightarrow$$

$$(r_1 - 2.8\text{ m})10^{\Delta\beta/20} = r_1 10^{\Delta\beta/20} - (2.8\text{ m})10^{\Delta\beta/20} = r_1 \Rightarrow$$

$$r_1 = \frac{(2.8\text{ m})10^{\Delta\beta/20}}{10^{\Delta\beta/20} - 1}.$$

CALCULATE Putting in the numerical values, we obtain the distance of the first friend from the loudspeakers:

$$r_1 = \frac{(2.8\text{ m})10^{(108.0\text{ dB} - 105.0\text{ dB})/20}}{10^{(108.0\text{ dB} - 105.0\text{ dB})/20} - 1} = 9.58726\text{ m}.$$

The distance of the second friend is

$$r_2 = (9.58726\text{ m}) - (2.8\text{ m}) = 6.78726\text{ m}.$$

ROUND We report our results to two significant figures:

$$r_1 = 9.6\text{ m} \quad\text{and}\quad r_2 = 6.8\text{ m}.$$

DOUBLE-CHECK To double-check, we substitute our results for the distances back into equation (iii) to verify that we get the specified sound level difference:

$$\Delta\beta = (108.0\text{ dB}) - (105.0\text{ dB}) = 3.0\text{ dB} = 20\log\left(\frac{9.6\text{ m}}{6.8\text{ m}}\right) = 3.0\text{ dB}.$$

Therefore, our answer seems reasonable, or at least consistent with what was originally stated in the problem.

Sound Attenuation

As sound travels through a medium, its intensity is reduced not only by the increasing distance from the source (the $1/r^2$ relationship discussed above) but also by scattering of the sound waves and absorption, which converts sound's energy into other forms of energy. This sound attenuation can be expressed as

$$I = I_0 e^{-\alpha x}, \tag{16.8}$$

where x is the path length and α is the absorption coefficient. The absorption coefficient is different for different materials and varies widely. Foams, for example, have very high absorption coefficients and are therefore used for sound insulation. The absorption coefficient also depends on the frequency of the sound; in general, it increases linearly with frequency.

Limits of Human Hearing

Our ears translate the pressure waves in air into what we perceive as sounds. The essential parts of the human ear are shown in a cutaway view in Figure 16.10. The ear canal opens to the outside and guides the sound waves to the eardrum, which vibrates when the sound waves hit it. Behind the eardrum is the inner ear, which is encased in some of the hardest bone in the human body. Inside the inner ear and connected to the eardrum are three small bones: hammer, anvil, and stirrup. The motions of the eardrum are transferred to the fluid-filled cochlea via these bones. In the cochlea, these motions are translated into nerve pulses, which are sent to the temporal lobe in the cerebral cortex.

Concept Check 16.2

At an intersection, you pull up next to a car whose windows are closed. You know that the people in the car are listening to loud music, but you cannot figure out what the song is. All you can hear is the rhythmic thumping of the bass because the materials of which the car is made

a) attenuate the high frequencies more than the low ones.

b) attenuate the low frequencies more than the high ones.

c) amplify the low frequencies.

d) amplify the high frequencies.

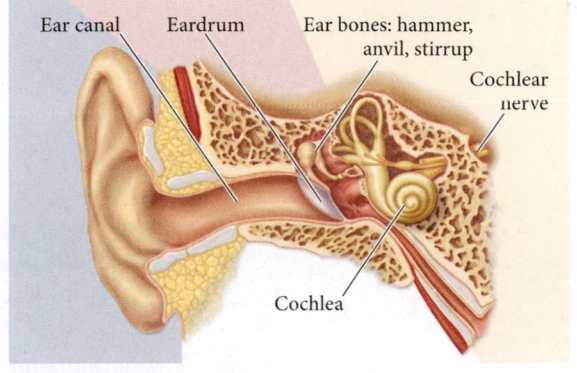

FIGURE 16.10 Parts of the human ear.

Human ears can detect sound waves with frequencies between approximately 20 Hz and 20,000 Hz. (Dogs, by the way, can detect even higher frequencies of sound.) As mentioned earlier, the threshold of human hearing was used as the anchor for the decibel scale. However, the ability of the human ear to detect sound depends strongly on the frequency of the sound, as well as the age of the human. Teenagers have no trouble hearing frequencies of 10,000 Hz, but most retirees cannot hear them at all. One example of such a high-pitched sound is the buzzing of cicadas in the summer, which many older people cannot hear anymore.

The sounds that we can hear best have frequencies around 1000 Hz. When you listen to music, the spectrum of sound typically extends over the entire range of human hearing. This occurs because musical instruments produce a characteristic mixture of the base frequencies of the notes played and several overtones at higher frequencies. If you play music on a stereo system and turn down the volume, you reduce the sound intensity in all frequency ranges approximately uniformly. Depending on how far down you turn the volume, you can approach sound intensities in the very high and very low frequencies that are very close to or below your hearing threshold. As a result, the music sounds flat. In order to correct for this perceived narrowing of the spectrum of the sound, many stereo systems have a loudness switch, which artificially enhances the power output of the very low and very high frequencies relative to the rest of the spectrum, giving the music a fuller sound at low volume settings.

A sound level above 130 dB will cause pain, and sound levels above 150 dB can rupture an eardrum. In addition, long-term exposure to sound levels above 120 dB causes loss of sensitivity in the human ear. For this reason, it is advisable to avoid listening to very loud music for long periods of time. In addition, where loud noises are part of the work environment, for example, on the deck of an aircraft carrier (Figure 16.11), it is necessary to wear ear protection to avoid ear damage.

FIGURE 16.11 A Navy launch officer ducks under the wing of an F/A-18F being launched off the USS *John C. Stennis*. Ear protection is required in this environment of high-intensity sound.

EXAMPLE 16.2 Wavelength Range of Human Hearing

The range of frequencies for sounds that the human ear can detect corresponds to a range of wavelengths.

PROBLEM

What is the range of wavelengths for the sounds to which the human ear is sensitive?

SOLUTION

The range of frequencies the ear can detect is 20–20,000 Hz. At room temperature, the speed of sound is 343 m/s. Wavelength, speed, and frequency are related by

$$v = \lambda f \Rightarrow \lambda = \frac{v}{f}.$$

For the lowest detectable frequency, we then obtain

$$\lambda_{max} = \frac{v}{f_{min}} = \frac{343 \text{ m/s}}{20 \text{ Hz}} = 17 \text{ m}.$$

Since the highest detectable frequency is 1000 times the lowest, the shortest wavelength of audible sound is 0.001 times the value we just calculated: $\lambda_{min} = 0.017$ m.

16.3   Sound Interference

Like all three-dimensional waves, sound waves from two or more sources can interfere in space and time. Let's first consider spatial interference of sound waves emitted by two **coherent sources**—that is, two sources that produce sound waves that have the same frequency and are in phase.

Figure 16.12 illustrates two loudspeakers emitting identical sinusoidal pressure fluctuations in phase. The circular arcs represent the maxima of the sound waves at some given instant in time. The horizontal line to the right of each speaker anchors a sine curve to show that an arc occurs wherever the sine has a maximum. The distance between neighboring arcs emanating from one source is exactly one wavelength, λ. You can clearly see that the arcs from the different loudspeakers intersect. Examples of such intersections are points A and C. If you count the maxima, you see that both A and C are exactly 8λ away from the lower loudspeaker, while A is 5λ away from the upper speaker and C is 6λ away. Thus, the path length difference, $\Delta r = r_2 - r_1$, is an integer multiple of the wavelength. This relationship is a general condition for constructive interference at a given spatial point:

$$\Delta r = n\lambda, \quad \text{for all } n = 0, \pm 1, \pm 2, \pm 3, \dots \quad \text{(constructive interference).} \tag{16.9}$$

Approximately halfway between points A and C in Figure 16.12 is point B. Just as for the other two points, the distance between B and the lower speaker is 8λ. However, this point is 5.5λ away from the upper speaker. As a result, at point B, the maximum of the sound wave from the lower speaker falls on a minimum of the sound wave from the upper speaker, and they cancel out. This means that destructive interference occurs at this point. You have probably already noticed that the path length difference is an odd number of half wavelengths. This is the general condition for destructive interference:

$$\Delta r = \left(n + \tfrac{1}{2}\right)\lambda, \quad \text{for all } n = 0, \pm 1, \pm 2, \pm 3, \dots \quad \text{(destructive interference).} \tag{16.10}$$

Wave interference can be visualized with the aid of a ripple tank, which is used in lecture demonstrations at many universities. Figure 16.13a shows such a tank, which is a clear rectangular plastic container filled with water, sitting on top of an overhead projector. A motor drives two pegs, which move up and down synchronously and create surface waves on the water. The resulting wave pattern is projected onto a screen. Figure 16.13b is a projection image that shows the wave interference clearly. Notice the lines revealing where the two waves enhance each other (interfere constructively) and where they cancel each other out (interfere destructively).

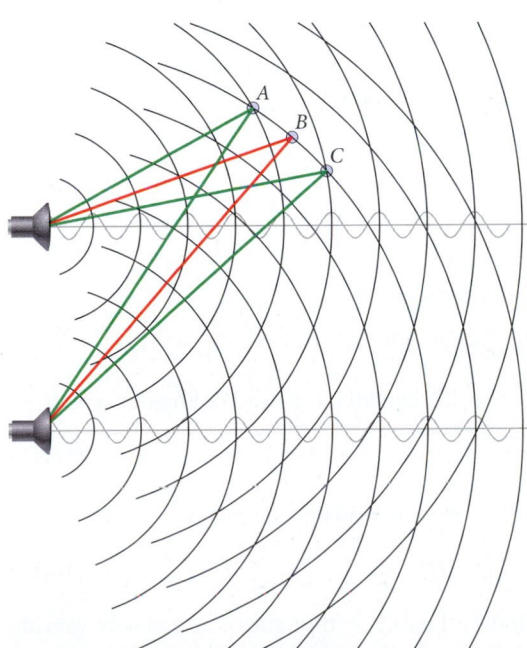

FIGURE 16.12 Interference of two identical sets of sinusoidal sound waves.

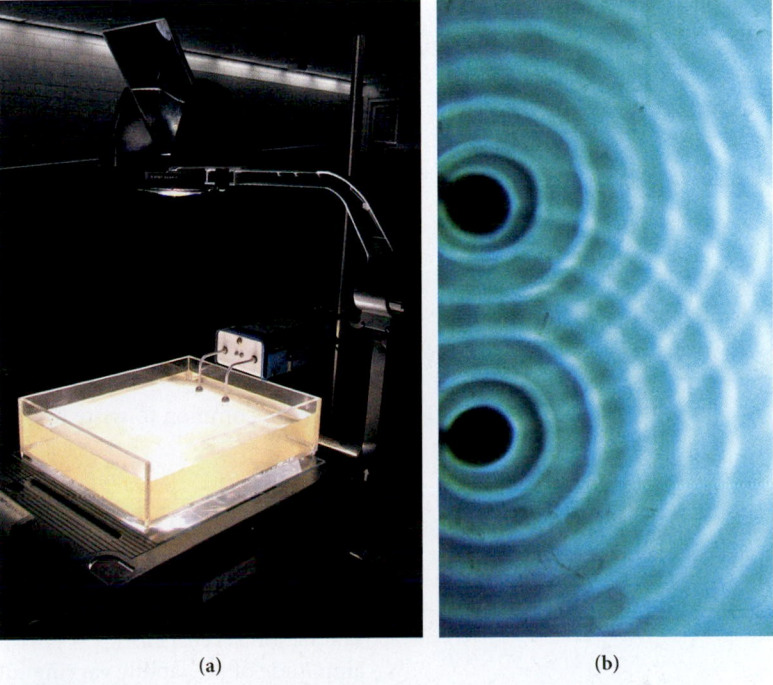

(a)   (b)

FIGURE 16.13 Setup for a lecture demonstration of wave phenomena. (a) The water-filled ripple tank, with a motor driving two pegs that generate surface waves on the water, sits on an overhead projector. (b) The projected image shows the interference of the water waves.

Why do we not detect the dead zones due to the lines of destructive interference in front of our stereo speakers when we listen to music at home? The answer is that these lines of interference depend on the wavelength and thus on the frequency. Sounds of different frequencies cancel each other out and add up maximally along different lines. Any note played by any instrument has a rich mixture of different frequencies; so, even if one particular frequency cancels, we can still hear all or most of the rest. Generally, we do not detect what is missing. In addition, all objects in a room scatter and reflect the sound emitted from stereo speakers to some degree, which makes detection of a dead zone even less likely. However, with pure sinusoidal tones, the existence of dead zones can be easily detected by ear. Nevertheless, spatial interference of sound waves does not play a huge role in daily life. In Chapter 34, where we discuss light waves, we will encounter many observable effects of wave interference.

Beats

It is also possible to have interference of two waves in time, which has much greater practical importance than interference in space does. To consider this effect, suppose an observer is located at some arbitrary point, x_0, in space, and two sound waves with slightly different frequencies and the same amplitude are emitted:

$$y_1(x_0,t) = A\sin(\kappa_1 x_0 + \omega_1 t + \phi_1) = A\sin(\omega_1 t + \tilde{\phi}_1)$$
$$y_2(x_0,t) = A\sin(\kappa_2 x_0 + \omega_2 t + \phi_2) = A\sin(\omega_2 t + \tilde{\phi}_2).$$

To obtain the right-hand side of each equation, we used the fact that the product of the wave number and the position is a constant for a given point in space and then just added that constant to the phase shift, which in this case is also a constant. We wrote these wave functions as one-dimensional waves, but the outcome is the same in three dimensions. For the next step, we simply set the two phase constants, $\tilde{\phi}_1$ and $\tilde{\phi}_2$, equal to zero, because they only cause a phase shift and have no other essential role. Therefore, the two time-dependent oscillations that form the starting waves are

$$y_1(x_0,t) = A\sin(\omega_1 t)$$
$$y_2(x_0,t) = A\sin(\omega_2 t).$$

Concept Check 16.3

If you detect a beat with two frequencies and then both frequencies are doubled, the resulting beat frequency will

a) stay the same.

b) double.

c) quadruple.

d) do any of the above.

e) do none of the above.

If we do this experiment using two angular frequencies, ω_1 and ω_2, that are close to each other, we can hear a sound that oscillates in volume as a function of time. Why does this happen?

In order to answer this question, we add the two sine waves. To add two sine functions, we can use an addition theorem for trigonometric functions:

$$\sin\alpha + \sin\beta = 2\cos\left[\tfrac{1}{2}(\alpha - \beta)\right]\sin\left[\tfrac{1}{2}(\alpha + \beta)\right].$$

Thus, we obtain for the two sine waves

$$\begin{aligned}y(x_0,t) &= y_1(x_0,t) + y_2(x_0,t)\\ &= A\sin(\omega_1 t) + A\sin(\omega_2 t)\\ &= 2A\cos\left[\tfrac{1}{2}(\omega_1 - \omega_2)t\right]\sin\left[\tfrac{1}{2}(\omega_1 + \omega_2)t\right].\end{aligned}$$

It is more common to write this result in terms of frequencies instead of angular speeds:

$$y(x_0,t) = 2A\cos\left[2\pi\tfrac{1}{2}(f_1 - f_2)t\right]\sin\left[2\pi\tfrac{1}{2}(f_1 + f_2)t\right]. \tag{16.11}$$

The term $\tfrac{1}{2}(f_1 + f_2)$ is simply the average of the two individual frequencies:

$$\overline{f} = \tfrac{1}{2}(f_1 + f_2). \tag{16.12}$$

The factor $2A\cos[2\pi\tfrac{1}{2}(f_1 - f_2)t]$ in equation 16.11 can be thought of as a slowly varying amplitude of the rapidly varying function $\sin[2\pi\tfrac{1}{2}(f_1 + f_2)t]$ when $|f_1 - f_2|$ is small.

When the cosine term has a maximum or minimum value (+1 or −1), a **beat** occurs; $|f_1 - f_2|$ is the beat frequency:

$$f_b = |f_1 - f_2|. \tag{16.13}$$

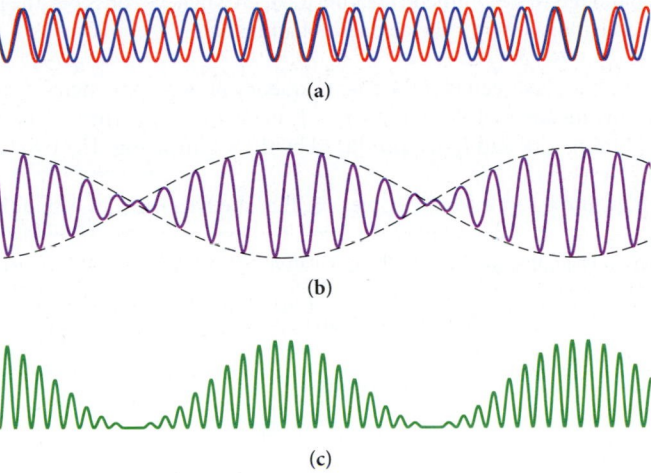

FIGURE 16.14 Beats of two sinusoidal tones. (a) The blue and red sine curves have frequencies that differ by 10%; (b) addition of the two curves; (c) square of the sum of the functions.

Two questions now arise: First, what happened to the factor $\frac{1}{2}$ in the cosine function of equation 16.11, and second, why is the beat frequency defined in terms of an absolute value? Use of the absolute value ensures that the beat frequency is positive, independent of which of the two frequencies is larger. For the cosine function, use of the absolute value does not matter at all, because $\cos|\alpha| = \cos\alpha = \cos(-\alpha)$. For the answer to why there is no factor $\frac{1}{2}$, take a look at Figure 16.14: Part (a) plots two sinusoidal tones as a function of time; part (b) shows the addition of the two functions, with the dashed envelope given by $\pm 2A\cos\left[2\pi\frac{1}{2}(f_1 - f_2)t\right]$. You can see that the amplitude of the addition of the two functions reaches a maximum twice for a given complete oscillation of the cosine function. This is perhaps more apparent from Figure 16.14c, the plot of the square of the function (equation 16.11), which is proportional to the sound intensity. Thus, the beat frequency, the frequency at which the volume oscillates up and down, is properly defined as shown in equation 16.13, omitting the factor $\frac{1}{2}$.

For example, striking two bars on a xylophone—one that emits at the proper frequency for middle A, 440 Hz, and another that emits at a frequency of 438 Hz—will produce two discernable volume oscillations per second. This is because the beat frequency is $f_b = \left|440\,\text{Hz} - 438\,\text{Hz}\right| = 2$ Hz, and the period of oscillation of the volume is therefore $T_b = 1/f_b = \frac{1}{2}$ s.

SOLVED PROBLEM 16.2 Tuning a Violin

Beats are a very effective tool for tuning instruments. For example, when an orchestra gets ready to perform, one member, usually the first violin chair, plays a reference tone, and the other members tune their instruments relative to that sound by listening for beats. Let's see how this is done.

PROBLEM
A violin string produces 3.00 beats per second when sounded along with a standard tuning fork of frequency 440. Hz.
a) What is the frequency of the violin's sound?
b) What should be done to tune the violin string to the same frequency as the tuning fork?

SOLUTION
THINK We know that two frequencies close to each other produce audible beats and that the beat frequency is the absolute value of the difference of these two frequencies. Thus, for part (a), we will obtain two values for the frequency of the violin's sound: one slightly above the tuning fork's reference frequency and one slightly below. For part (b), the obvious answer is that the tension in the string needs to be changed. In order to find out by how much, we need to apply concepts from Chapter 15, where we studied the relation between string tension and oscillation frequency.

– Continued

SKETCH Even though we recommend creating a sketch whenever possible, the present case is one of those few exceptions where a sketch does not add much.

RESEARCH We have just seen that the beat frequency of two frequencies, f_1 and f_2, is given by $f_b = |f_1 - f_2|$. This means that either $f_b = f_1 - f_2$ or $f_b = f_2 - f_1$, where f_1 is the frequency produced by the tuning fork and f_2 that produced by the violin string. The two values of f_b are the two solutions for part (a).

For part (b), we use a result from Chapter 15: that the frequency of a standing wave on a string is proportional to the square root of the string tension force, $f \propto \sqrt{F}$, which we can write as $f = a\sqrt{F}$, where a is a constant. Taking the derivative of both sides of this equation we get

$$df = ad\left(\sqrt{F}\right) = a\frac{1}{2}\frac{dF}{\sqrt{F}} = \frac{a}{2}\frac{dF}{\sqrt{F}}.$$

Dividing both sides of this equation by f gives us

$$\frac{df}{f} = \frac{\left(\dfrac{a}{2}\dfrac{dF}{\sqrt{F}}\right)}{a\sqrt{F}} = \frac{1}{2}\frac{dF}{F}.$$

So the required relative change in the tension force, dF/F, is twice the relative change in the frequency, df/f. In order to calculate by what fraction to change the string tension, we note we need to change the violin frequency f_2 by f_b to get to the desired frequency f_1. So $df = f_b$, the relative frequency change is $df/f = f_b/f_1$.

SIMPLIFY Part (a) has two solutions, $f_2 = f_1 - f_b$ and $f_2 = f_1 + f_b$. For part (b), we have

$$dF/F = 2\left(f_b/f_1\right).$$

CALCULATE The two answers for part (a) are $f_2 = 440.$ Hz – 3.00 Hz = 337. Hz and $f_2 = 440.$ Hz + 3.00 Hz = 443. Hz.

For part (b) we find that our relative frequency change is $f_b/f_1 = 3.00/440.$ and therefore $dF/F = 2(3.00/440.) = 0.013636$.

ROUND The two answers for part (a), 337. Hz and 443. Hz, need no further rounding. Rounding our answer for part (b) to three significant figures, we see that the string tension force should be increased or decreased by a factor of 0.0136 = 1.36%.

DOUBLE-CHECK Any time you hear a beat, you know that the two frequencies that are producing it have to be close to each other. Both of our answers for part (a) are close to the reference frequency, and so they satisfy this minimum requirement. Only a small adjustment in the string tension should be needed, and our result for part (b) confirms this. But should the string tension be increased or decreased to tune the string to the reference frequency? This question cannot be answered with the information given in the problem statement. In practice, one simply tightens the string. If the beats are then less frequent, one is on the right track and needs to tighten the string further. If the beats become more frequent with tightening, this means that the original frequency was already above the reference frequency, and the string tension should be decreased instead.

Sound Diffraction

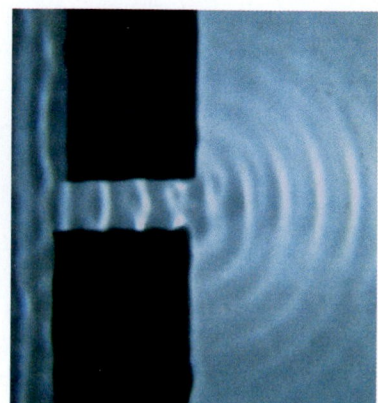

FIGURE 16.15 Ripple-tank projection image, in which a wave travels from left to right between two blocks and then spreads out spherically.

We can see an object only if the straight line between our eyes and the object is unobstructed. However, the situation is different for sound. We can hear somebody, for example, through an open doorway, even if we cannot see him or her. How is this possible? The reason is that the sound waves spread out spherically after they pass through the door opening and can reach any point in the room we are in. This effect is called *diffraction*. It is fairly easy to visualize diffraction using a ripple tank and projector. In the projected image in Figure 16.15, the white lines show the wave crests. You can see that the wavelength of the water surface wave (the distance between two neighboring crests) is comparable to the size of the opening between the two rectangular blocks (black) that were placed in the water. As we saw in Example 16.2, sound has wavelengths in the centimeter-to-meter range, making the ripple-tank demonstration a good analogy for what happens when sound passes through a door opening.

But light also consists of waves; so why don't we observe diffraction of light through doorways? The answer is that in order for pronounced diffraction effects to be observable, the wavelength has to be similar in size to or larger than the opening through which the wave propagates, and the wavelength range for light is very much smaller than a door opening. We will address the topic of light diffraction in much greater depth in Chapter 34.

Sound Localization

How do we know the direction from which a sound is coming? Since sound can undergo sizable diffraction, answering this question is not as easy as for light, for which we can simply point back to the source. Obviously, we have two ears, and if a sound's source is located on either our left or right side, the sound will travel different paths from the source to each ear. Our earlier discussion of the role of path differences in spatial sound interference may lead you to think that wave interference plays a role in sound localization. However, this is not the case.

One factor in sound localization is that the head "shadows" the ear that is farther away from the sound. This is the sound attenuation effect expressed by equation 16.8. Sound attenuation is frequency dependent, however, and plays the greatest role in localizing sounds in the higher frequency ranges. For frequencies below 800 Hz, the head causes very little attenuation, and the main clue used by the brain for sound localization is the interaural arrival-time difference, that is, the phase delay separating the arrivals of the sound at the two ears. For very low-frequency sounds, the wavelengths are large, on the order of several meters, which is much greater than the interaural distance. Localization of these sounds is very hard or even impossible. This is also the reason why your surround-sound system comes with only one subwoofer (which emits low-frequency sounds) but with at least four tweeters (which emit the high frequencies).

The field of psychoacoustics is devoted to topics like sound localization and has already made many breakthroughs, which, for example, leads to improvements in the design of concert halls and loudspeaker systems for touring rock bands, ensuring that there is not a bad seat in the concert venue.

Concept Check 16.4

Which of the following physical effects plays the smallest role in sound localization?

a) sound attenuation

b) interaural arrival-time difference

c) sound-intensity difference

d) phase difference

e) spatial interference

Active Noise Cancellation

We all know that noise can be muffled, or attenuated, and that some materials do this better than others. For example, if noise bothers you at night, you can use your pillow to cover your ears. Construction workers and other workers who have to spend extended periods in very loud environments wear ear-protecting headphones. Such ear protection reduces the amplitude of all sounds.

If you want to reduce background noise while listening to music, a technique to accomplish this is based on the principle of **active noise cancellation,** which relies on sound interference. An external sinusoidal sound wave arrives at the headphone and is recorded by a microphone (Figure 16.16). A processor inverts the phase of this sound wave and sends out the sound wave with the same frequency and amplitude, but opposite phase. The two sine waves add up (superposition principle), interfere destructively, and cancel out completely. At the same time, the speaker inside the headphones emits the music you want to hear, and the result is a listening experience that is free of background noise.

In practice, background noise never consists of a pure sinusoidal sound wave. Instead, many sounds of many different frequencies are mixed. In particular, the presence of high-frequency sounds constitutes a problem for active noise cancellation. However, this method works very well for periodic low-frequency sounds, such as the engine noise from airliners. Another application is in some luxury automobiles, where active noise cancellation techniques reduce wind and tire noise.

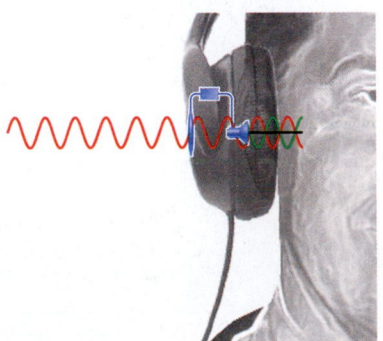

FIGURE 16.16 Active noise cancellation.

16.4 Doppler Effect

We've all had the experience where a train approaching a railroad crossing sounds its warning horn and then, as it moves past our position, the sound pitch changes from a higher frequency to a lower one. This change in frequency is called the **Doppler effect.**

In order to gain a qualitative understanding of the Doppler effect, consider Figure 16.17. In part (a), the leftmost column, a stationary sound source emits spherical waves that travel outward. The time evolution of these radial waves is depicted at six equally spaced points in time, from top to bottom. The maxima of the five sine waves radiate outward as circles, color coded in each frame from yellow to red.

When the source moves, the same sound waves are emitted from different points in space. However, from each of these points, the maxima again travel outward as circles, as shown in parts (b) and (c) in Figure 16.17. The only difference between these two columns is the speed with which the source of the sound waves moves from left to right. In each case, you can see a "crowding" of the sound waves in front of the emitter, on the right side of it in this context. This crowding is greater with higher source velocity and is the entire key to understanding the Doppler effect. An observer located to the right of the source—that is to say, with the source moving toward him or her—experiences more wave fronts per unit time and thus hears a higher frequency. An observer located to the left of the moving source, with the source moving away from him or her, experiences fewer wave fronts per unit time and thus hears a reduced frequency.

Quantitatively, the observed frequency, f_o, is given as

$$f_o = f\left(\frac{v_{\text{sound}}}{v_{\text{sound}} \pm v_{\text{source}}}\right), \qquad (16.14)$$

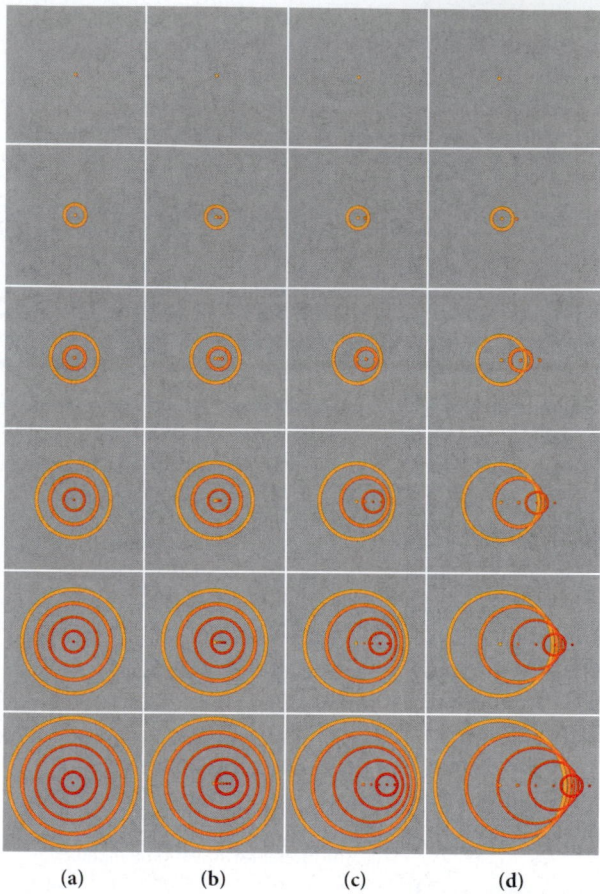

(a)   (b)   (c)   (d)

FIGURE 16.17 The Doppler effect for sound waves emitted at six equally spaced points in time: (a) stationary source; (b) source moving to the right; (c) source moving faster to the right; (d) source moving to the right faster than the speed of sound.

where f is the frequency of the sound as emitted by the source, and v_{sound} and v_{source} are the speeds of the sound and of the source, respectively. The upper sign (+) applies when the source moves away from the observer, and the lower sign (−) applies when the source moves toward the observer.

The Doppler effect also occurs if the source is stationary and the observer moves. In this case, the observed frequency is given by

$$f_o = f\left(\frac{v_{\text{sound}} \mp v_{\text{observer}}}{v_{\text{sound}}}\right) = f\left(1 \mp \frac{v_{\text{observer}}}{v_{\text{sound}}}\right). \qquad (16.15)$$

The upper sign (−) applies when the observer moves away from the source, and the lower sign (+) applies when the observer moves toward the source.

In each case, moving source or moving observer, the observed frequency is lower than the source frequency when observer and source are moving away from each other and is higher than the source frequency when they are moving toward each other.

DERIVATION 16.2 Doppler Shift

Let's begin with an observer at rest and a sound source moving toward the observer. If the sound source is at rest and emits a sound with frequency f, then the wavelength is $\lambda = v_{\text{sound}}/f$, which is also the distance between two successive wave crests. If the sound source moves toward the observer with speed v_{source}, then the distance between two successive crests as perceived by the observer is reduced to

$$\lambda_o = \frac{v_{\text{sound}} - v_{\text{source}}}{f}.$$

Using the relationship between wavelength, frequency, and the (fixed!) speed of sound, $v_{sound} = \lambda f$, we find the frequency of the sound as detected by the observer:

$$f_o = \frac{v_{sound}}{\lambda_o} = f\left(\frac{v_{sound}}{v_{sound} - v_{source}}\right).$$

This result proves equation 16.14 with the minus sign in the denominator. The plus sign is obtained by reversing the sign of the source velocity vector.

You may want to try the derivation for the case in which the observer is moving as an exercise.

Concept Check 16.5

Consider the case of a stationary observer and a source moving with constant velocity (less than the speed of sound) and emitting a sound of frequency f. If the source moves toward the observer, equation 16.14 predicts a higher observed frequency, f_+. If the source moves away from the observer, a lower observed frequency, f_-, results. Let's call the difference between the higher frequency and the original frequency Δ_+ and the difference between the original frequency and the lower frequency Δ_-: $\Delta_+ = f_+ - f$ and $\Delta_- = f - f_-$. Which statement about the frequency differences is correct?

a) $\Delta_+ > \Delta_-$

c) $\Delta_+ < \Delta_-$

e) It depends on the source velocity, v.

b) $\Delta_+ = \Delta_-$

d) It depends on the original frequency, f.

Finally, what happens if both the source and the observer move? The answer is that the expressions for the Doppler effect for moving source and moving observer simply combine, and the observed frequency is given by

$$f_o = f\left(\frac{v_{sound} \mp v_{observer}}{v_{sound}}\right)\left(\frac{v_{sound}}{v_{sound} \pm v_{source}}\right) = f\left(\frac{v_{sound} \mp v_{observer}}{v_{sound} \pm v_{source}}\right). \qquad (16.16)$$

Here the upper signs in the numerator and the denominator apply when observer or source is moving away from the other, and the lower signs apply when observer or source is moving toward the other. If the source and observer are moving in the same direction but with different speeds, the signs of the velocities must be chosen with respect to the fixed medium of the air.

SOLVED PROBLEM 16.3 | Doppler Shift of Reflected Ambulance Siren

PROBLEM

An ambulance has picked up an injured rock climber and is heading directly away from the canyon wall (where the climber was injured) at a speed of 31.3 m/s (70 mph). The ambulance's siren has a frequency of 400.0 Hz. After the ambulance turns off its siren, the injured rock climber can hear the reflected sound from the canyon wall for a few seconds. The speed of sound in air is $v_{sound} = 343$ m/s. What is the frequency of the reflected sound from the ambulance's siren as heard by the injured rock climber in the ambulance?

SOLUTION

THINK The ambulance can be considered to be a source moving with speed $v_{ambulance}$ and emitting a sound of frequency f. The frequency, f_1, of the siren at the fixed wall of the canyon can be calculated using the speed of sound, v_{sound}. The canyon wall reflects the sound of the siren back toward the moving ambulance. The climber in the ambulance can be thought of as a moving observer of the reflected sound of the siren. Combining the two Doppler shifts gives the frequency, f_2, of the reflected siren as heard on the moving ambulance.

SKETCH Figure 16.18 shows the ambulance moving away from the canyon wall with speed $v_{ambulance}$ while emitting a sound from its siren of frequency f.

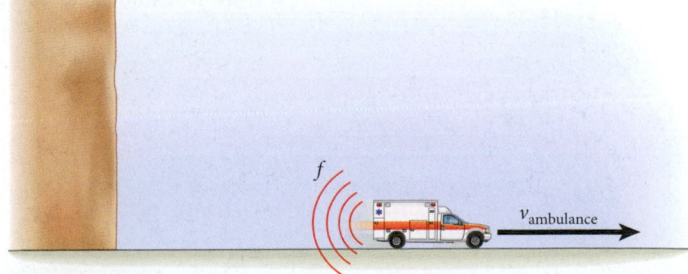

FIGURE 16.18 An ambulance moves away from a canyon wall with its siren operating.

– Continued

Concept Check 16.6

Now consider the case of a stationary source emitting a sound of frequency f and an observer moving with constant velocity (less than the speed of sound). If the observer moves toward the source, equation 16.15 predicts a higher observed frequency, f_+. If the observer moves away from the source, a lower observed frequency, f_-, results. Let's call the difference between the higher frequency and the original frequency Δ_+ and the difference between the original frequency and the lower frequency Δ_-: $\Delta_+ = f_+ - f$ and $\Delta_- = f - f_-$. Which statement about the frequency differences is correct?

a) $\Delta_+ > \Delta_-$

b) $\Delta_+ = \Delta_-$

c) $\Delta_+ < \Delta_-$

d) It depends on the original frequency, f.

e) It depends on the source velocity, v.

Concept Check 16.7

Suppose a source emitting a sound with frequency f moves to the right (in the positive x-direction) with a speed of 30 m/s. An observer is located to the right of the source and also moves to the right (in the positive x-direction) with a speed of 50 m/s. The observed frequency, f_o, will be _____ the original frequency, f.

a) lower than

c) higher than

b) the same as

RESEARCH The frequency, f_1, of the sound heard by a stationary observer standing at the canyon wall resulting from the sound of frequency f emitted by the ambulance's siren moving with speed $v_{ambulance}$ is given by equation 16.14:

$$f_1 = f\left(\frac{v_{sound}}{v_{sound} + v_{ambulance}}\right), \tag{i}$$

where the plus sign is applicable because the ambulance is moving away from the stationary observer. The sound is reflected from the canyon wall back toward the moving ambulance. Now the injured rock climber in the ambulance is a moving observer. We can calculate the frequency observed by the rock climber, f_2, using equation 16.15:

$$f_2 = f_1\left(1 - \frac{v_{ambulance}}{v_{sound}}\right), \tag{ii}$$

where the minus sign is used because the observer is moving away from the source.

SIMPLIFY We can combine equations (i) and (ii) to obtain

$$f_2 = f\left(\frac{v_{sound}}{v_{sound} + v_{ambulance}}\right)\left(1 - \frac{v_{ambulance}}{v_{sound}}\right).$$

You can see that this result is a special case of equation 16.16, where the speeds of observer and source are both the speed of the ambulance and both observer and source are moving away from each other.

CALCULATE Putting in the numerical values, we get

$$f_2 = (400 \text{ Hz})\left(\frac{343 \text{ m/s}}{343 \text{ m/s} + 31.3 \text{ m/s}}\right)\left(1 - \frac{31.3 \text{ m/s}}{343 \text{ m/s}}\right) = 333.1018 \text{ Hz}.$$

ROUND We report the result to three significant figures:

$$f_2 = 333 \text{ Hz}.$$

DOUBLE-CHECK The frequency of the reflected sound observed by the injured rock climber is approximately 20% lower than the original frequency of the ambulance's siren. The speed of the ambulance is about 10% of the speed of sound. The two Doppler shifts, one for the sound emitted from the moving ambulance and one for the reflected sound observed in the moving ambulance, should each shift the frequency by approximately 10%. So the observed frequency seems reasonable.

Doppler Effect in Two- and Three-Dimensional Space

So far, our discussion of the Doppler effect has assumed that the source (or observer) moves with constant velocity on a straight line, which passes right through the location of the observer (or source). However, if you are waiting at a railroad crossing, for example, a train does not pass directly through your location but has a perpendicular distance of closest approach, b, as shown in Figure 16.19. In this situation, what you hear is not an instantaneous switch from the higher frequency of the approaching train to the lower frequency of the receding train that equation 16.14 implies, but instead a gradual and smooth change from the higher to the lower frequency.

To obtain a quantitative expression of the Doppler effect for this case, we let $t = 0$ be the instant when the train is closest to the car, that is, as it crosses the road. For $t < 0$, the train is then moving toward the observer, and for $t > 0$, it is moving away from the observer. The origin of the spatial coordinate system is chosen as the location of the observer, the driver of the car. Then the distance of the train to the origin is given by

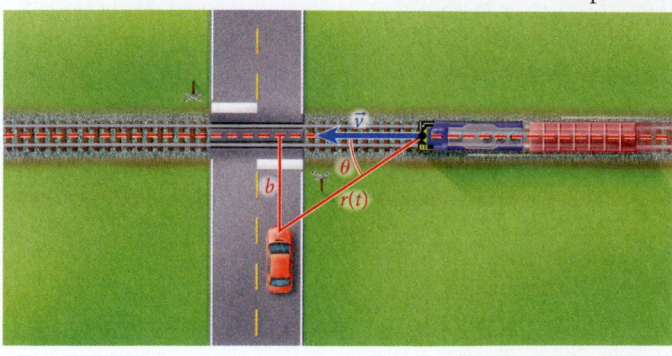

FIGURE 16.19 Doppler effect in a two-dimensional space—a car at a railroad crossing.

$r(t) = \sqrt{b^2 + v^2 t^2}$, where v is the (constant) speed of the train (see Figure 16.19). The angle θ between the train's velocity vector and the direction from which the sound is emitted relative to the car is given by

$$\cos\theta(t) = \frac{vt}{r(t)} = \frac{vt}{\sqrt{b^2 + v^2 t^2}}.$$

The velocity of the sound source as observed in the car as a function of time is the projection of the train's velocity vector onto the radial direction:

$$v_{source}(t) = v\cos\theta(t) = \frac{v^2 t}{\sqrt{b^2 + v^2 t^2}}.$$

We can now substitute for v_{source} in equation 16.14 and obtain

$$f_o(t) = f\frac{v_{sound}}{v_{sound} \pm v_{source}(t)} = f\frac{v_{sound}}{v_{sound} + \dfrac{v^2 t}{\sqrt{b^2 + v^2 t^2}}}. \tag{16.17}$$

In equation 16.17, the source velocity is negative for $t < 0$, corresponding to an approaching source, while the source velocity is positive for $t > 0$, corresponding to a receding source.

The formula of equation 16.17 may look complicated, but it is actually a smoothly varying function, as shown in Figure 16.20. Assume that the moving train emits a sound of frequency $f = 440$ Hz and that the sound's speed is 343 m/s. Figure 16.20a shows three plots of the frequency observed by a stationary observer in a car 10 m from the railroad crossing when a train moving at 20, 30, and 40 m/s passes through. Figure 16.20b shows plots of the frequencies observed by three stationary observers in cars parked at distances of 10, 40, and 70 m from the railroad crossing as a train passes at a speed of 40 m/s. These three curves converge toward the asymptotic frequencies of $f = 498$ Hz for the approaching train and $f = 394$ Hz for the receding train, which are the values predicted by equation 16.14. However, the closer the observer is to the railroad crossing, the more sudden the transition from the higher frequency to the lower one becomes.

It is fairly easy to set up a demonstration of the Doppler effect. Just drive down a road at constant speed, continuously honking your horn, and have a friend videotape you. Then you can analyze the video clip with a frequency analyzer and see how the frequencies of the car horn change as you pass by the camera. In Figure 16.21, the car drove along a country highway with a speed of 26.5 m/s and passed by the camera at a closest distance of 14.0 m. The six dominant frequencies of the car's horn, given by $f = n(441 \text{ Hz})$, $n = 1, 2, \ldots, 6$, are clearly visible in this figure as the narrow red bands representing high decibel values, and you can see the frequency shift as the car passes by. The superimposed dashed gray lines are the results of the calculations using equation 16.17 with a distance of closest approach $b = 14.0$ m and a source speed $v_{source} = 26.5$ m/s. You can see that they are in complete agreement with the experimental measurements.

Self-Test Opportunity 16.2

Show that the time derivative of the function determining the observed frequency (equation 16.17) at $t = 0$ is

given by $\left.\dfrac{df_o}{dt}\right|_{t=0} = -\dfrac{fv^2}{bv_{sound}}$.

Self-Test Opportunity 16.3

Derive a formula similar to equation 16.17 for an observer moving on a straight line and passing a stationary source at some nonzero distance of closest approach.

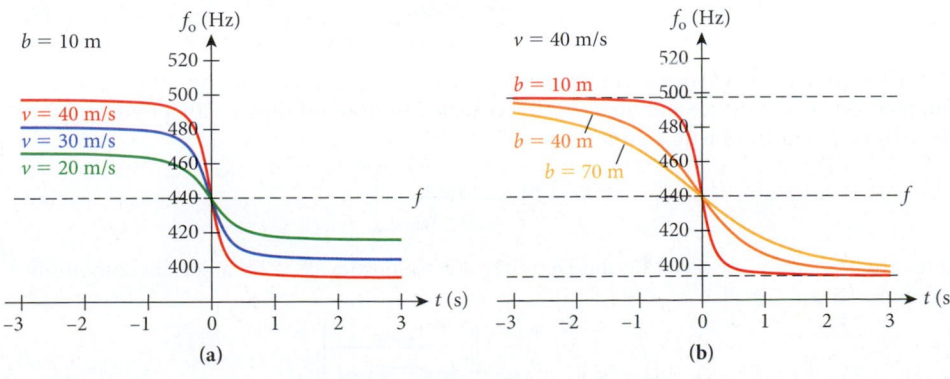

(a)   (b)

FIGURE 16.20 Time dependence of the frequency observed by a stationary observer: (a) source velocities of 20 m/s, 30 m/s, and 40 m/s and distance of closest approach of 10 m; (b) source velocity of 40 m/s and distances of closest approach of 10 m, 40 m, and 70 m.

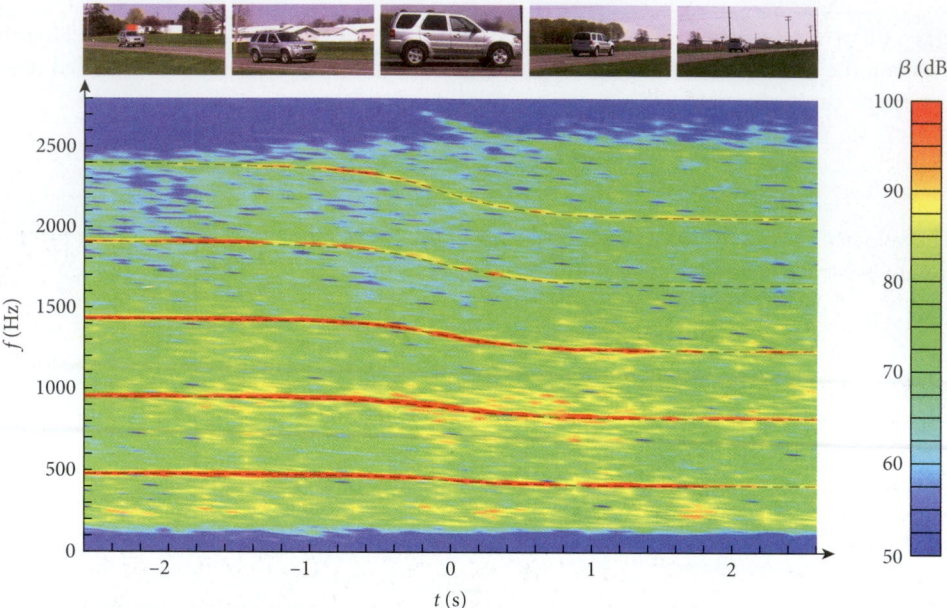

FIGURE 16.21 Demonstration of the Doppler effect.

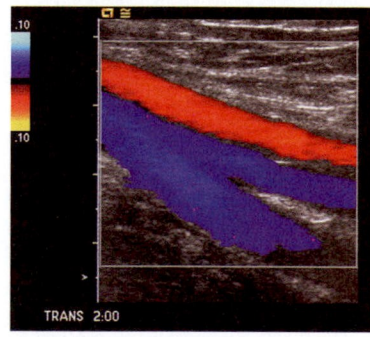

FIGURE 16.22 Doppler ultrasound image of blood flowing in the carotid artery. Red and blue indicate the speeds of the blood flow.

Applications of the Doppler Effect

In Section 16.1, we examined the application of ultrasound waves to the imaging of human tissues. The Doppler effect for ultrasound waves can be used to measure the speed of blood flowing in an artery, which can be an important diagnostic tool for cardiac disease. An example of the measurement of blood flow in the carotid artery is shown in Figure 16.22. This kind of measurement is implemented by transmitting ultrasound waves toward the flowing blood. The ultrasound waves reflect off the moving blood cells back toward the ultrasound device, where they are detected.

EXAMPLE 16.3   Doppler Ultrasound Measurement of Blood Flow

PROBLEM

What is the typical frequency change for ultrasound waves reflecting off blood flowing in an artery?

SOLUTION

Ultrasound waves have a typical frequency of $f = 2.0$ MHz. In an artery, blood flows with a speed of $v_{blood} = 1.0$ m/s. The speed of ultrasound waves in human tissue is $v_{sound} = 1540$ m/s. The blood cells can be thought of as moving observers for the ultrasound waves. Thus, if a blood cell is moving toward the source of the ultrasound, the frequency observed by the blood cell, f_1, is given by equation 16.15 with a plus sign

$$f_1 = f\left(1 + \frac{v_{blood}}{v_{sound}}\right). \tag{i}$$

The reflected ultrasound waves with frequency f_1 constitute a moving source. The stationary Doppler ultrasound device will then observe reflected ultrasound waves with a frequency, f_2, given by equation 16.14 with a minus sign

$$f_2 = f_1\left(\frac{v_{sound}}{v_{sound} - v_{blood}}\right). \tag{ii}$$

Thus, the frequency observed by the Doppler ultrasound device is obtained by combining equations (i) and (ii)

$$f_2 = f\left(1 + \frac{v_{blood}}{v_{sound}}\right)\left(\frac{v_{sound}}{v_{sound} - v_{blood}}\right).$$

Note that this equation is just a special case of equation 16.16, the expression for the Doppler effect with a moving source and a moving observer; here the speed of the observer and that of the source are both that of the blood cell. Putting in the numbers gives us

$$f_2 = (2.0 \text{ MHz})\left(1 + \frac{1.0 \text{ m/s}}{1540 \text{ m/s}}\right)\left(\frac{1540 \text{ m/s}}{1540 \text{ m/s} - 1.0 \text{ m/s}}\right) = 2.0026 \text{ MHz}.$$

The frequency difference, Δf, is then

$$\Delta f = f_2 - f = 2.6 \text{ kHz}.$$

This frequency change is the maximum observed when the blood is flowing as a result of a heart pulse. Between heart pulses, the blood slows down and almost stops. Combining the original ultrasound waves with frequency f and the reflected ultrasound waves with frequency f_2 yields a beat frequency, $f_{beat} = f_2 - f$, which varies from 2.6 kHz to zero as the heart pulses. These frequencies can be perceived by human ears, which means the beat frequency can simply be amplified and heard as a pulse monitor. This Doppler ultrasound effect is the basis of fetal heart probes.

Note that this example assumes that the blood is flowing directly toward the Doppler ultrasound device. In general, there is an angle between the direction of the transmitted ultrasound waves and the direction of the blood flow, which decreases the frequency change. Doppler ultrasound devices take this angle into account using information on the orientation of the artery.

The Doppler radar that you hear about on TV weather reports uses the Doppler effect for electromagnetic waves. Moving raindrops change the reflected radar waves to different frequencies and thus are detected by the Doppler radar. (However, the Doppler shift for electromagnetic waves such as radar waves is governed by a different formula than that used for sound. Electromagnetic waves will be discussed in Chapter 31.) Doppler radar is also used to detect rotation (and possible tornados) in thunderstorms by finding regions with motion both toward and away from the observer. Another common use of the Doppler effect is in the radar speed detectors employed by police around the world. And similar techniques are used by astronomers to measure red shifts of galaxies (more on this topic in Chapter 35 on relativity).

Mach Cone

We have yet to examine Figure 16.17d. In this column, the source speed exceeds the speed of sound. As you can see, in this case, the origin of a subsequent wave crest lies outside the circle formed by the previous wave crest.

Figure 16.23a shows the successive wave crests emitted by a source whose speed is greater than the speed of sound, called a **supersonic source**. You can see that all of these circles have a common tangent line that make an angle θ_M, called the **Mach angle**, with the direction of the velocity vector (in this case, the horizontal). This accumulation of wave fronts produces a large, abrupt, conical-shaped wave called a **shock wave**, or **Mach cone**. The Mach angle of the cone is given by

$$\theta_M = \sin^{-1}\left(\frac{v_{sound}}{v_{source}}\right). \tag{16.18}$$

This formula is valid only for source speeds that exceed the speed of sound, and in the limit where the source speed is equal to the speed of sound, θ_M approaches 90°. The greater the source speed, the smaller is the Mach angle.

DERIVATION 16.3 / Mach Angle

The derivation of equation 16.18 is quite straightforward if we examine Figure 16.23b, which shows the circle of a wave crest that was formed at some initial time, $t = 0$, but is now at a later time, Δt. The radius of this circle at time Δt is $v_{sound}\Delta t$ and is shown by the blue straight line. Note that the radius makes a 90° angle with the tangent line, indicated in black. However, during the

– Continued

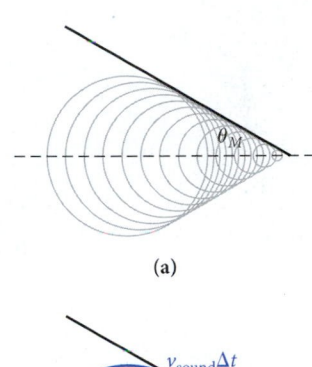

(a)

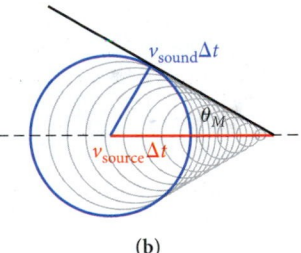

(b)

FIGURE 16.23 (a) The Mach cone, or shock wave. (b) Geometry for Derivation 16.3.

Concept Check 16.8

The Mach angle

a) increases with increasing speed of the source, as long as that speed remains below the speed of sound.

b) decreases with increasing speed of the source, as long as that speed remains below the speed of sound.

c) is independent of the speed of the source.

d) increases with increasing speed of the source, as long as that speed remains above the speed of sound.

e) decreases with increasing speed of the source, as long as that speed remains above the speed of sound.

same time, Δt, the source has moved a distance $v_{source}\Delta t$, which is indicated by the red line in the figure. The red, blue, and black lines in the figure form a right triangle. Then, by definition of the sine function, the sine of the Mach angle, θ_M, is

$$\sin\theta_M = \frac{v_{sound}\Delta t}{v_{source}\Delta t}.$$

Canceling out the common factor Δt and taking the inverse sine on both sides of this equation yields equation 16.18, as desired.

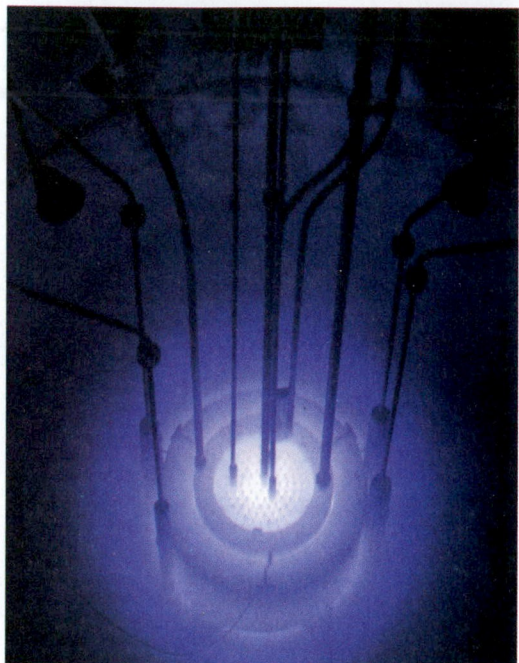

FIGURE 16.24 Cherenkov radiation (bluish glow) from the core of a nuclear reactor.

Interestingly, shock waves can be detected in all kinds of physical systems involving all kinds of waves. Motorboats can create shock waves relatively easily by moving across the water surface at high speeds. From these triangular bow waves, you can determine the speed of the surface waves on a lake, if you know the speed of the boat.

The speed of sound in atomic nuclei is approximately a third of the speed of light. Since particle accelerators can collide nuclei at speeds very close to the speed of light, shock waves should be created even in this environment. Some computer simulations have shown that this effect is present in collisions of atomic nuclei at very high energies, but definite experimental proof has been difficult to obtain. Several nuclear physics laboratories around the world are actively investigating this phenomenon.

Finally, it would seem unlikely that shock waves involving light could be observed, since the speed of light in vacuum is the maximum speed possible in nature, and thus the source velocity cannot exceed the speed of light in vacuum. When the Russian physicist Pavel Cherenkov (also sometimes spelled Čerenkov) proposed in the 1960s that this effect did occur, it was dismissed as impossible. However, so-called *Cherenkov radiation* (Figure 16.24) can be emitted when a source such as a high-energy proton or electron moving at close to the speed of light enters a medium, like water, where the speed of light is significantly lower. Then the source does move faster than the speed of light in that medium, and a Mach cone can form. Modern particle detectors make use of this Cherenkov radiation; measuring the emission angle allows the velocity of the particle that emitted the radiation to be calculated.

FIGURE 16.25 The supersonic Concorde jetliner taking off. Regular commercial service on the Concorde began in 1976 and ended in 2003.

EXAMPLE 16.4 | The Concorde

The speed of a supersonic aircraft is often given as a Mach number, M. A speed of Mach 1 ($M = 1$) means that the aircraft is traveling at the speed of sound. A speed of Mach 2 ($M = 2$) means that the aircraft is traveling at twice the speed of sound. The Concorde supersonic airliner (Figure 16.25) cruised at 60,000 feet, where the speed of sound is 295 m/s (660 mph). The maximum cruise speed of the Concorde was Mach 2.04 ($M = 2.04$).

PROBLEM
At this speed, what was the angle of the Mach cone produced by the Concorde?

SOLUTION
The angle of the Mach cone is given by equation 16.18:

$$\theta_M = \sin^{-1}\left(\frac{v_{sound}}{v_{source}}\right).$$

The speed of the source in this case is that of the Concorde, which is traveling at a speed of

$$v_{source} = Mv_{sound} = 2.04v_{sound}.$$

Thus, we can write the Mach angle as

$$\theta_M = \sin^{-1}\left(\frac{v_{\text{sound}}}{M v_{\text{sound}}}\right) = \sin^{-1}\left(\frac{1}{M}\right).$$

For the Concorde, we then have

$$\theta_M = \sin^{-1}\left(\frac{1}{2.04}\right) = 0.512 \text{ rad} = 29.4°.$$

16.5 Resonance and Music

Basically all musical instruments rely on the excitation of resonances to generate sound waves of discrete, reproducible, and predetermined frequencies. No discussion of sound could possibly be complete without a discussion of musical tones.

Tones

Which frequencies correspond to which musical tones? The answer to this question is not simple and has changed over time. The current twelve-tone equal temperament scale dates back several centuries. The publication of Johann Sebastian Bach's work *Das wohltemperierte Klavier (The Well-Tempered Clavier)* in 1722, comprised of 24 compositions in each of the possible major and minor scales, represented a landmark in the development of the modern tonal scale. Let's review these accepted values.

The tones associated with the white keys of the piano are called A, B, C, D, E, F, and G. The C closest to the left end of the keyboard is designated C1. The note below it, oddly enough, is B0. Thus, the sequence of tones for the white keys, from left to right, on a piano is A0, B0, C1, D1, E1, F1, G1, A1, B1, C2, D2, ... , A7, B7, C8. The scale is anchored with middle A (A4) at precisely 440 Hz. The next higher A (A5) is an octave higher, which means that its frequency is exactly twice that of middle A, or 880 Hz.

Between these two A's are 11 halftones: A-sharp/B-flat, B, C, C-sharp/D-flat, D, D-sharp/E-flat, E, F, F-sharp/G-flat, G, and G-sharp/A-flat. Ever since the work of J. S. Bach, these 11 halftones divide the octave into exactly 12 equal intervals, all of which differ by the same factor, a factor that will give the number 2 when multiplied with itself 12 times. This factor is $2^{1/12} = 1.0595$. The frequencies of all the tones can be found through successive multiplication with this factor. For example, D (D5) is 5 steps up from middle A: $(1.0595^5)(440 \text{ Hz}) = 587.3$ Hz. Frequencies of notes in other octaves are found by multiplication or division by a factor of 2. For example, the next higher D (D6) has a frequency of $2(587.3 \text{ Hz}) = 1174.7$ Hz.

Human voices have classifications based on the average ranges shown in Table 16.3. Thus, basically all songs sung by humans exhibit a range of frequencies of one order of magnitude, between 100 and 1000 Hz. No human voice and practically no musical instrument generates a pure single-frequency tone. Instead, musical tones have a rich admixture of overtones that gives each instrument and each human voice its own characteristics. For string instruments, the hollow wooden body of the instrument, which serves to amplify the sound of the strings, influences this admixture.

The touch-tone phones rely on sounds of preassigned frequencies. For each key that you press on the phone, two relatively pure tones are generated. Figure 16.26 shows the arrangement of these tones. Each key in a given row sets off the tone displayed to its right, and each key in a column triggers the tone displayed under it. If you press the 2 key, for example, the frequencies 697 Hz and 1336 Hz are produced.

Half-Open and Open Pipes

We discussed standing waves on strings in Chapter 15. These waves form the basis for the sounds produced by all string instruments. It is important to remember that standing waves can be excited on strings only at discrete

Table 16.3	Frequency Ranges for Classifications of Human Singers	
Classification	**Lowest Note**	**Highest Note**
Bass	E2 (82 Hz)	G4 (392 Hz)
Baritone	A2 (110 Hz)	A4 (440 Hz)
Tenor	D3 (147 Hz)	B4 (494 Hz)
Alto	G3 (196 Hz)	F5 (698 Hz)
Soprano	C4 (262 Hz)	C6 (1047 Hz)

FIGURE 16.26 The frequencies generated by the keys of a touch-tone phone.

resonance frequencies. Most percussion instruments, like drums, also work on the principle of exciting discrete resonances. However, the waveforms produced by these instruments are typically two-dimensional and are much more complicated than those produced on strings.

Wind instruments use half-open or open pipes to generate sounds. A *half-open pipe* is a pipe that is open at only one end and closed at the other end (such as a clarinet or trumpet); an open pipe is open on both ends (such as a flute). Figure 16.27a shows some of the possible standing waves in a half-open pipe. In the top pipe is the standing wave with the largest wavelength, for which the first antinode coincides with the length of the pipe. Since the distance between a node and the first antinode is a fourth of a wavelength, this resonance condition corresponds to $L = \frac{1}{4}\lambda$, where L is the length of the pipe. The middle and bottom pipes in Figure 16.27a have standing waves with smaller possible wavelengths, for which the second or third antinodes fall on the pipe opening, and $L = \frac{3}{4}\lambda$ and $L = \frac{5}{4}\lambda$, respectively. In general, the condition for a resonant standing wave in a half-open pipe is thus

$$L = \frac{2n-1}{4}\lambda, \quad \text{for} \quad n = 1, 2, 3, \dots .$$

Solving this equation for the possible wavelengths results in

$$\lambda_n = \frac{4L}{2n-1}, \quad \text{for} \quad n = 1, 2, 3, \dots . \tag{16.19}$$

Using $v = \lambda f$, we obtain the possible frequencies:

$$f_n = (2n-1)\frac{v}{4L}, \quad \text{for} \quad n = 1, 2, 3, \dots , \tag{16.20}$$

where n corresponds to the number of nodes.

Figure 16.27b displays the possible standing waves in an open pipe. In this case, there are antinodes at both ends of the pipe, and the condition for a resonant standing wave is

$$L = \frac{n}{2}\lambda, \quad n = 1, 2, 3, \dots .$$

This relationship results in wavelengths and frequencies of

$$\lambda_n = \frac{2L}{n}, \quad n = 1, 2, 3, \dots , \tag{16.21}$$

$$f_n = n\frac{v}{2L}, \quad n = 1, 2, 3, \dots , \tag{16.22}$$

where again n is the number of nodes.

For both half-open and open pipes, the fundamental frequency (first harmonic) is obtained with $n = 1$. For a half-open pipe, the first overtone (or second harmonic, for which $n = 2$) is three times as high as the fundamental frequency. However, for an open pipe, the first overtone ($n = 2$) is twice as high as the fundamental frequency, or one octave higher.

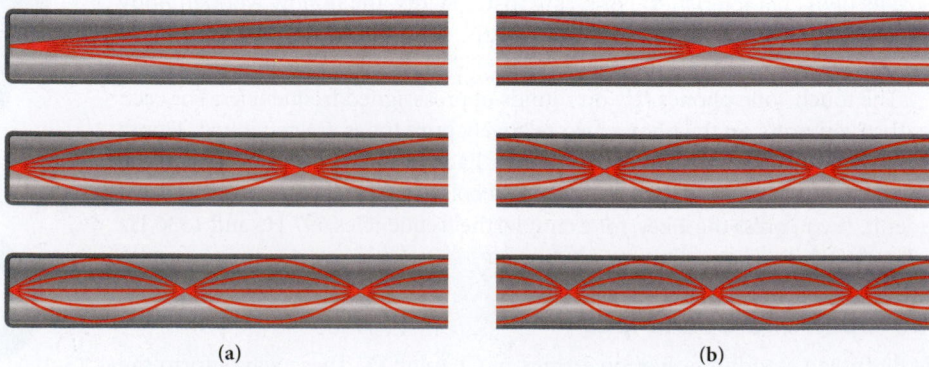

(a)   (b)

FIGURE 16.27 (a) Standing waves in a half-open pipe; (b) standing waves in an open pipe. The red lines represent the sound amplitude at various times.

EXAMPLE 16.5 A Pipe Organ

Church organs operate on the principle of creating standing waves in pipes. If you have ever been in an old cathedral, you have very likely seen an impressive collection of organ pipes (Figure 16.28).

PROBLEM

If you wanted to build an organ whose fundamental frequencies covered the frequency range of a piano, from A0 to C8, what range of lengths would you need to use for the pipes?

SOLUTION

The frequencies for A0 and C8 are $f_{A0} = 27.5$ Hz and $f_{C8} = 4186$ Hz. According to equations 16.20 and 16.22, the fundamental frequencies for a half-open pipe and an open pipe are

$$f_{1,\text{half-open}} = \frac{v}{4L} \quad \text{and} \quad f_{1,\text{open}} = \frac{v}{2L}.$$

FIGURE 16.28 Pipes of a church organ.

This means that the open pipes have to be twice as long as the corresponding half-open pipes for the same frequency. The lengths of the half-open pipes for the highest and lowest frequencies need to be

$$L_{\text{half-open,C8}} = \frac{v}{4 f_{C8}} = \frac{343 \text{ m/s}}{16{,}744 \text{ Hz}} = 0.0205 \text{ m}.$$

$$L_{\text{half-open,A0}} = \frac{v}{4 f_{A0}} = \frac{343 \text{ m/s}}{110 \text{ Hz}} = 3.118 \text{ m}.$$

COMMENT

Real organ pipes can indeed be as long as 10 ft. The part that produces the standing wave can be as short as 0.5 in; however, the base of a pipe is approximately 0.5 ft in length, so the shortest pipes typically observed in a church organ are at least that long.

SOLVED PROBLEM 16.4 Standing Wave in a Pipe

PROBLEM

A standing sound wave is set up in a pipe of length 0.410 m (Figure 16.29) containing air at normal pressure and temperature. What is the frequency of this sound?

FIGURE 16.29 A standing sound wave in a pipe.

SOLUTION

THINK First, we recognize that the standing sound wave shown in Figure 16.29 corresponds to one in a pipe that is closed on the left end and open on the right end. Thus, we are dealing with a standing wave in a half-open pipe. Four nodes are visible. Combining these observations and knowing the speed of sound, we can calculate the frequency of the standing wave.

SKETCH The standing wave with the pipe drawn in and the nodes marked is shown in Figure 16.30.

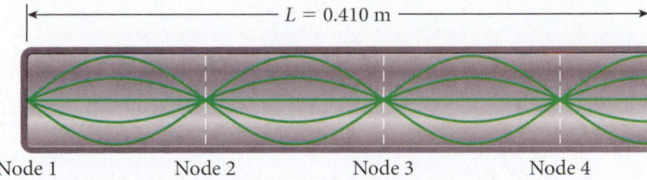

Node 1   Node 2   Node 3   Node 4

FIGURE 16.30 A standing sound wave with four nodes in a half-open pipe.

– Continued

RESEARCH   The frequency of this standing wave in a half-open pipe is given by

$$f_n = (2n-1)\frac{v}{4L},$$ (i)

where $n = 4$ is the number of nodes. The speed of sound in air is $v = 343$ m/s, and the length of the pipe is $L = 0.410$ m.

SIMPLIFY   For $n = 4$, we can write equation (i) for the frequency of the standing wave as

$$f_4 = \left[2(4)-1\right]\frac{v}{4L} = \frac{7v}{4L}.$$

CALCULATE   Putting in the numerical values, we get

$$f = \frac{7(343 \text{ m/s})}{4(0.410 \text{ m})} = 1464.02 \text{ Hz}.$$

ROUND   We report our result to three significant figures:

$$f = 1460 \text{ Hz}.$$

DOUBLE-CHECK   This frequency is well within the range of human hearing and just above the range of frequencies produced by the human voice. Therefore, our answer seems reasonable.

WHAT WE HAVE LEARNED | EXAM STUDY GUIDE

- Sound is a longitudinal pressure wave that requires a medium in which to propagate.

- The speed of sound in air at normal pressure and temperature (1 atm, 20 °C) is 343 m/s.

- In general, the speed of sound in a solid is given by $v = \sqrt{Y/\rho}$, and that in a liquid or a gas is given by $v = \sqrt{B/\rho}$.

- A logarithmic scale is used to measure sound intensities. The unit of this scale is the decibel (dB). The sound level, β, in this decibel scale is defined as $\beta = 10\log\frac{I}{I_0}$, where I is the intensity of the sound waves and $I_0 = 10^{-12}$ W/m², which corresponds approximately to the minimum intensity that a human ear can hear.

- Sound waves emitted by two coherent sources constructively interfere if the path length difference is $\Delta r = n\lambda$, for all $n = 0, \pm1, \pm2, \pm3, ...$, and destructively interfere if the path length difference is $\Delta r = (n + \frac{1}{2})\lambda$, for all $n = 0, \pm1, \pm2, \pm3, ...$.

- Two sine waves with equal amplitudes and slightly different frequencies produce beats with a beat frequency of $f_b = |f_1 - f_2|$.

- As a result of the Doppler effect, the frequency of sound from a moving source as perceived by a stationary observer

is $f_o = f\left(\dfrac{v_{\text{sound}}}{v_{\text{sound}} \pm v_{\text{source}}}\right)$, where f is the frequency of the sound as emitted by the source and v_{sound} and v_{source} are the speeds of sound and of the source, respectively. The upper sign (+) applies when the source moves away from the observer, and the lower sign (–) applies when the source moves toward the observer.

- The source of sound can be stationary while the observer moves. In this case, the observed frequency of the sound is given by $f_o = f\left(\dfrac{v_{\text{sound}} \mp v_{\text{observer}}}{v_{\text{sound}}}\right) = f\left(1 \mp \dfrac{v_{\text{observer}}}{v_{\text{sound}}}\right)$. Here the upper sign (–) applies when the observer moves away from the source, and the lower sign (+) applies when the observer moves toward the source.

- The Mach angle of the shock wave formed by a sound source moving at a supersonic speed is $\theta_M = \sin^{-1}\left(\dfrac{v_{\text{sound}}}{v_{\text{source}}}\right)$.

- The relationship between the length of a half-open pipe and the wavelengths of possible standing waves inside it is $L = \dfrac{2n-1}{4}\lambda$, for $n = 1, 2, ...$; for open pipes, the relationship is $L = \dfrac{n}{2}\lambda$, for $n = 1, 2, ...$.

ANSWERS TO SELF-TEST OPPORTUNITIES

16.1 $\beta = 10\log\dfrac{I}{I_0} \Rightarrow I = I_0\left(10^{\beta/10}\right)$

$P = IA = I\left(4\pi r^2\right) = I_0\left(10^{\beta/10}\right)\left(4\pi r^2\right)$

$P = 4\pi\left(10^{-12}\ \text{W/m}^2\right)10^{(80\ \text{dB})/10}\left(10.0\ \text{m}\right)^2 = 0.126\ \text{W}.$

16.2 $f_{\text{o}}(t) = f\dfrac{v_{\text{sound}}}{v_{\text{sound}} + \dfrac{v^2 t}{\sqrt{b^2 + v^2 t^2}}}$

$\dfrac{df_{\text{o}}(t)}{dt} = -fv_{\text{sound}}\dfrac{\left(-\dfrac{t^2 v^4}{\left(b^2 + v^2 t^2\right)^{3/2}} + \dfrac{v^2}{\sqrt{b^2 + v^2 t^2}}\right)}{\left(v_{\text{sound}} + \dfrac{v^2 t}{\sqrt{b^2 + v^2 t^2}}\right)^2}$

$\left.\dfrac{df_{\text{o}}(t)}{dt}\right|_{t=0} = -fv_{\text{sound}}\dfrac{\left(\dfrac{v^2}{\sqrt{b^2}}\right)}{\left(v_{\text{sound}}\right)^2} = -f\dfrac{v^2}{bv_{\text{sound}}}.$

16.3 $v_{\text{observer}}(t) = \dfrac{v^2 t}{\sqrt{b^2 + v^2 t^2}}$

$f_{\text{o}}(t) = f\dfrac{v_{\text{sound}} \mp v_{\text{observer}}(t)}{v_{\text{sound}}}$

$f_{\text{o}}(t) = f\dfrac{v_{\text{sound}} - \dfrac{v^2 t}{\sqrt{b^2 + v^2 t^2}}}{v_{\text{sound}}}.$

PROBLEM-SOLVING GUIDELINES

1. Be sure you're familiar with the properties of logarithms to solve problems involving the intensity of sound. In particular, remember that $\log(A/B) = \log A - \log B$ and that $\log x^n = n\log x$.

2. Situations involving the Doppler effect can be tricky. Sometimes you need to break the path of the sound wave into segments and treat each segment separately, depending on whether motion of the source or motion of the observer is occurring. Often a sketch is useful to identify which expression for the Doppler effect pertains to which part of the situation.

3. You don't need to memorize formulas for the overtone frequencies of open pipes or half-open pipes. Instead, remember that the closed end of a pipe is the location of a node and the open end of a pipe is the location of an antinode. Then you can reconstruct the standing waves from that point.

MULTIPLE-CHOICE QUESTIONS

16.1 You stand on the curb waiting to cross a street. Suddenly you hear the sound of a horn from a car approaching you at a constant speed. You hear a frequency of 80 Hz. After the car passes by, you hear a frequency of 72 Hz. At what speed was the car traveling?

a) 17 m/s

b) 18 m/s

c) 19 m/s

d) 20 m/s

16.2 A sound level of 50 decibels is

a) 2.5 times as intense as a sound of 20 decibels.

b) 6.25 times as intense as a sound of 20 decibels.

c) 10 times as intense as a sound of 20 decibels.

d) 100 times as intense as a sound of 20 decibels.

e) 1000 times as intense as a sound of 20 decibels.

16.3 A police car is moving in your direction, constantly accelerating, with its siren on. As it gets closer, the sound you hear will

a) stay at the same frequency.

b) drop in frequency.

c) increase in frequency.

d) More information is needed.

16.4 You are creating a sound wave by shaking a paddle in a liquid medium. How can you increase the speed of the resulting sound wave?

a) You shake the paddle harder to give the medium more kinetic energy.

b) You vibrate the paddle more rapidly to increase the frequency of the wave.

c) You create a resonance with a faster-moving wave in air.

d) All of these will work.

e) None of these will work.

f) Only (a) and (b) will work.

g) Only (a) and (c) will work.

h) Only (b) and (c) will work.

16.5 Standing on the sidewalk, you listen to the horn of a passing car. As the car passes, the frequency of the sound changes from high to low in a continuous manner; that is, there is no abrupt change in the perceived frequency. This occurs because

a) the pitch of the horn's sound changes continuously.

b) the intensity of the observed sound changes continuously.

c) you are not standing directly in the path of the moving car.

d) of all of the above reasons.

16.6 A sound meter placed 3 m from a speaker registers a sound level of 80 dB. If the volume on the speaker is then turned down so that the power is reduced by a factor of 25, what will the sound meter read?

a) 3.2 dB   c) 32 dB   e) 66 dB

b) 11 dB   d) 55 dB

16.7 What has the greatest effect on the speed of sound in air?

a) temperature of the air   c) wavelength of the sound

b) frequency of the sound   d) pressure of the atmosphere

16.8 A driver sits in her car at a railroad crossing. Three trains go by at different (constant) speeds, each emitting the same sound. Bored, the driver makes a recording of the sounds. Later, at home, she plots the frequencies as a function of time on her computer, resulting in the plots shown in the figure. Which of the three trains had the highest speed?

a) the one represented by the solid line

b) the one represented by the dashed line

c) the one represented by the dotted line

d) impossible to tell

16.9 Three physics faculty members sit in three different cars at a railroad crossing. Three trains go by at different (constant) speeds, each emitting the same sound. Each faculty member uses a cell phone to record the sound of a different train. On the next day, during a faculty meeting, they plot the frequencies as a function of time on a computer, resulting in the plots shown in the figure. Which of the three trains had the highest speed?

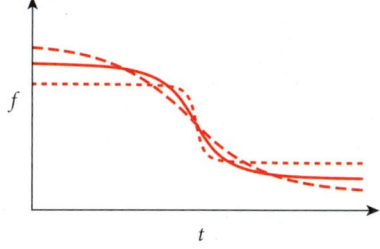

a) the one represented by the solid line

b) the one represented by the dashed line

c) the one represented by the dotted line

d) impossible to tell

16.10 In Question 16.9, which of the three faculty members was closest to the train tracks?

a) the one who recorded the sound represented by the solid line

b) the one who recorded the sound represented by the dashed line

c) the one who recorded the sound represented by the dotted line

d) impossible to tell

16.11 The beat frequency of two frequencies f_1 and f_2 is

a) the absolute value of the sum of the frequencies, $|f_1 + f_2|$.

b) the absolute value of the difference of the frequencies, $|f_1 - f_2|$.

c) the average of the two frequencies.

d) half of the sum of the two frequencies.

e) half of the absolute value of the difference of the two frequencies.

16.12 Suppose a source emitting a sound of frequency f moves toward the right (in the positive x-direction) with a speed of 30 m/s. An observer is located to the right of the source and also moves toward the right (in the positive x-direction) with a speed of 50 m/s. Which of the following is the correct expression for the frequency that the observer perceives, f_o?

a) $f_o = f \left| \dfrac{v_{sound} - 50 \text{ m/s}}{v_{sound} - 30 \text{ m/s}} \right|$

b) $f_o = f \left| \dfrac{v_{sound} + 50 \text{ m/s}}{v_{sound} - 30 \text{ m/s}} \right|$

c) $f_o = f \left| \dfrac{v_{sound} - 50 \text{ m/s}}{v_{sound} + 30 \text{ m/s}} \right|$

d) $f_o = f \left| \dfrac{v_{sound} + 50 \text{ m/s}}{v_{sound} + 30 \text{ m/s}} \right|$

16.13 Audible beats can be produced by two sine waves with frequencies f_1 and f_2, if

a) the two frequencies are close to each other.

b) f_1 is greater than f_2.

c) f_1 is smaller than f_2.

d) the two frequencies are identical.

16.14 If two loudspeakers at points A and B emit identical sine waves at the same frequency and constructive interference is observed at point C, then the

a) distance from A to C is the same as that from B to C.

b) points A, B, and C form an equilateral triangle.

c) difference between the distance from A to C and the distance from B to C is an integer multiple of the wavelength of the emitted waves.

d) difference between the distance from A to C and the distance from B to C is a half-integer multiple of the wavelength of the emitted waves.

CONCEPTUAL QUESTIONS

16.15 A (somewhat risky) way of telling if a train that cannot be seen or heard is approaching is by placing your ear on the rail. Explain why this works.

16.16 A classic demonstration of the physics of sound involves an alarm clock in a bell vacuum jar. The demonstration starts with air in the vacuum jar at normal atmospheric pressure and then the jar is evacuated to lower and lower pressures. Describe the expected outcome.

16.17 You are sitting near the back of a twin-engine commercial jet, which has the engines mounted on the fuselage near the tail. The nominal rotation speed for the turbofan in each engine is 5200 revolutions per minute, which is also the frequency of the dominant sound emitted by the engine. What audio clue would suggest that the engines are not perfectly synchronized and

that the turbofan in one of them is rotating approximately 1% faster than that in the other? How would you measure this effect if you had only your wristwatch (could measure time intervals only in seconds)? To hear this effect, go to http://vnatsci.ltu.edu/s_schneider/physlets/main/beats.shtml. This simulation works best with integer natural frequencies, so round your frequencies to the ones place.

16.18 You determine the direction from which a sound is coming by subconsciously judging the difference in time it takes to reach the right and left ears. Sound directly in front (or back) of you arrives at both ears at the same time; sound from your left arrives at your left ear before your right ear. What happens to this ability to determine the location of a sound if you are underwater? Will sounds appear to be located more in front or more to the side of where they actually are?

16.19 On a windy day, a child standing outside a school hears the school bell. If the wind is blowing toward the child from the direction of the bell, will it alter the frequency, the wavelength, or the velocity of the sound heard by the child?

16.20 A police siren contains two frequencies, producing a wavering sound (beats). Explain how the siren sound changes as a police car approaches, passes, and moves away from a pedestrian.

16.21 The Moon has no atmosphere. Is it possible to generate sound waves on the Moon?

16.22 When two pure tones with similar frequencies combine to produce beats, the result is a train of wave packets. That is, the sinusoidal waves are partially localized into packets. Suppose two sinusoidal waves of equal amplitude A, traveling in the same direction, have wave numbers κ and $\kappa + \Delta\kappa$ and angular frequencies ω and $\omega + \Delta\omega$, respectively. Let Δx be the length of a wave packet, that is, the distance between two nodes of the envelope of the combined sine functions. What is the value of the product $\Delta x \Delta \kappa$?

16.23 As you walk down the street, a convertible drives by on the other side of the street. You hear the deep bass speakers in the car thumping out a groovy but annoyingly loud beat. Estimate the intensity of the sound you perceive, and then calculate the minimum intensity that the poor ears of the convertible's driver and passengers are subject to.

16.24 If you blow air across the mouth of an empty soda bottle, you hear a tone. Why is it that if you put some water in the bottle, the pitch of the tone increases?

EXERCISES

Section 16.1

16.25 Standing on a sidewalk, you notice that you are halfway between a clock tower and a large building. When the clock strikes the hour, you hear an echo of the bell 0.500 s after hearing it directly from the tower. How far apart are the clock tower and the building?

16.26 Two farmers are standing on opposite sides of a very large empty field that is 510. m across. One farmer yells out some instructions, and 1.50 s pass until the sound reaches the other farmer. What is the temperature of the air?

16.27 The density of a sample of air is 1.205 kg/m^3, and the bulk modulus is $1.42 \cdot 10^5$ N/m^2.

a) Find the speed of sound in the air sample.

b) Find the temperature of the air sample.

•**16.28** You drop a stone down a well that is 9.50 m deep. How long is it before you hear the splash? The speed of sound in air is 343 m/s.

16.29 Electromagnetic radiation (light) consists of waves. More than a century ago, scientists thought that light, like other waves, required a medium (called the *ether*) to support its transmission. Glass, having a typical mass density of $\rho = 2500$ kg/m^3, also supports the transmission of light. What would the bulk modulus of glass have to be to support the transmission of light waves at a speed of $v = 2.0 \cdot 10^8$ m/s? Compare this to the actual bulk modulus of window glass, which is $5.0 \cdot 10^{10}$ N/m^2.

Section 16.2

16.30 Compare the intensity of sound at the pain level, 120 dB, with that at the whisper level, 20 dB.

16.31 The sound level in decibels is typically expressed as $\beta = 10 \log(I/I_0)$, but since sound is a pressure wave, the sound level can be expressed in terms of a pressure difference. Intensity depends on the amplitude squared, so the expression is $\beta = 20 \log(P/P_0)$, where P_0 is the smallest pressure difference noticeable by the ear: $P_0 = 2.00 \cdot 10^{-5}$ Pa. A loud rock concert has a sound level of 110. dB. Find the amplitude of the pressure wave generated by this concert.

16.32 At a state championship high school football game, the shout of a single person in the stands has an intensity level of about 50 dB at the center of the field. What would be the intensity level at the center of the field if all 10,000 fans at the game shouted from roughly the same distance away from that center point?

•**16.33** Two people are talking at a distance of 3.00 m from where you are, and you measure the sound intensity as $1.10 \cdot 10^{-7}$ W/m^2. Another student is 4.00 m away from the talkers. What sound intensity does the other student measure?

•**16.34** An excited child in a campground lets out a yell. Her parent, 1.20 m away, hears it as a 90.0-dB sound. A climber sits 850. m away from the child, on top of a nearby mountain. How loud does the yell sound to the climber?

•**16.35** While enjoyable, rock concerts can damage people's hearing. In the front row at a rock concert, 5.00 m from the speaker system, the sound intensity is 115.0 dB. How far back would you have to sit for the sound intensity to drop below the recommended safe level of 90.0 dB?

Section 16.3

16.36 Two sources, A and B, emit a sound of a certain wavelength. The sound emitted from both sources is detected at a point away from the sources. The sound from source A is a distance d from the observation point, whereas the sound from source B has to travel a distance of 3λ. What is the largest value of the wavelength, in terms of d, for the maximum sound intensity to be detected at the observation point? If $d = 10.0$ m and the speed of sound is 340. m/s, what is the frequency of the emitted sound?

16.37 A college student is at a concert and really wants to hear the music, so she sits between two in-phase loudspeakers, which point toward each other and are 50.0 m apart. The speakers emit sound at a frequency of 490. Hz. At the midpoint between the speakers, there will be constructive interference, and the music will be at its loudest. At what distance closest to the midpoint along the line between the two speakers could she also sit to experience the loudest sound?

16.38 A string of a violin produces 2 beats per second when sounded along with a standard fork of frequency 400. Hz. The beat frequency increases when the string is tightened.

a) What was the frequency of the violin at first?

b) What should be done to tune the violin?

•**16.39** You are standing against a wall opposite two speakers that are separated by 3.00 m, as shown in the figure. The two speakers begin emitting a 1372-Hz tone in phase. Where along the far wall should you stand so that the sound from the speakers is as soft as possible? Be specific; how far away from a spot centered between the speakers will you be? The far wall is 120. m from the wall that has the speakers. (Assume that the walls are good absorbers, and therefore, the contribution of reflections to the perceived sound is negligible.)

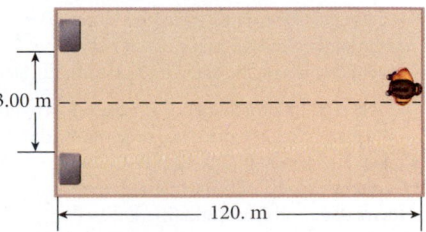

•**16.40** You are playing a note that has a fundamental frequency of 400. Hz on a guitar string of length 50.0 cm. At the same time, your friend plays a fundamental note on an open organ pipe, and 4 beats per second are heard. The mass per unit length of the string is 2.00 g/m. Assume that the speed of sound is 343 m/s.

a) What are the possible frequencies of the open organ pipe?

b) When the guitar string is tightened, the beat frequency decreases. Find the original tension in the string.

c) What is the length of the organ pipe?

•**16.41** Two 100.0-W speakers, A and B, are separated by a distance $D = 3.60$ m. The speakers emit in-phase sound waves at a frequency $f = 10,000.0$ Hz. Point P_1 is located at $x_1 = 4.50$ m and $y_1 = 0$ m; point P_2 is located at $x_2 = 4.50$ m and $y_2 = -\Delta y$.

a) Neglecting speaker B, what is the intensity, I_{A1} (in W/m^2), of the sound at point P_1 due to speaker A? Assume that the sound from the speaker is emitted uniformly in all directions.

b) What is this intensity in terms of decibels (sound level, β_{A1})?

c) When both speakers are turned on, there is a maximum in their combined intensities at P_1. As one moves toward P_2, this intensity reaches a single minimum and then becomes maximized again at P_2. How far is P_2 from P_1; that is, what is Δy? You may assume that $L \gg \Delta y$ and $D \gg \Delta y$, which will allow you to simplify the algebra by using $\sqrt{a \pm b} \approx a^{1/2} \pm \dfrac{b}{2a^{1/2}}$ when $a \gg b$.

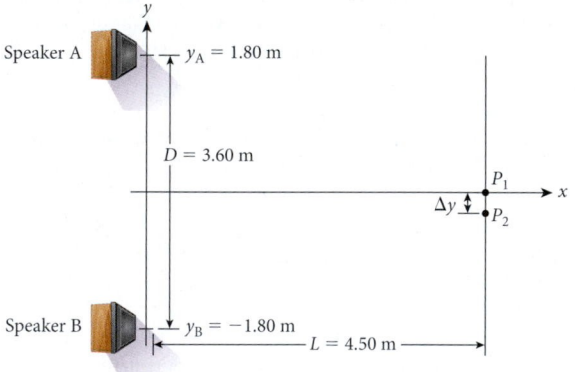

Section 16.4

16.42 A policeman with a very good ear and a good understanding of the Doppler effect stands on the shoulder of a freeway assisting a crew in a 40-mph work zone. He notices a car approaching that is honking its horn. As the car gets closer, the policeman hears the sound of the horn as a distinct B4 tone (494 Hz). The instant the car passes by, he hears the sound as a distinct A4 tone (440 Hz). He immediately jumps on his motorcycle, stops the car, and gives the motorist a speeding ticket. Explain his reasoning.

16.43 A meteorite hits the surface of the ocean at a speed of 8.80 km/s. What is the angle of the shock wave it produces (a) in the air just before hitting the ocean surface and (b) in the ocean just after entering? Assume that the speeds of sound in air and in water are 343 m/s and 1560 m/s, respectively.

16.44 A train whistle emits a sound at a frequency $f = 3000$. Hz when stationary. You are standing near the tracks when the train goes by at a speed of $v = 30.0$ m/s. What is the magnitude of the change in the frequency ($|\Delta f|$) of the whistle as the train passes? (Assume that the speed of sound is $v = 343$ m/s.)

•**16.45** You are driving along a highway at 30.0 m/s when you hear a siren. You look in the rear-view mirror and see a police car approaching you from behind with a constant speed. The frequency of the siren that you hear is 1300. Hz. Right after the police car passes you, the frequency of the siren that you hear is 1280. Hz.

a) How fast was the police car moving?

b) You are so nervous after the police car passes you that you pull off the road and stop. Then you hear another siren, this time from an ambulance

approaching from behind. The frequency of its siren that you hear is 1400. Hz. Once it passes, the frequency is 1200. Hz. What is the actual frequency of the ambulance's siren?

•**16.46** A bat flying toward a wall at a speed of 7.00 m/s emits an ultrasound wave with a frequency of 30.0 kHz. What frequency does the reflected wave have when it reaches the flying bat?

•**16.47** A plane flies at Mach 1.30, and its shock wave reaches a man on the ground 3.14 s after the plane passes directly overhead. Assume that the speed of sound is 343.0 m/s.

a) What is the Mach angle?

b) What is the altitude of the plane?

•**16.48** You are traveling in a car toward a hill at a speed of 40.0 mph. The car's horn emits sound waves of frequency 250. Hz, which move with a speed of 340. m/s.

a) Determine the frequency with which the waves strike the hill.

b) What is the frequency of the reflected sound waves you hear?

c) What is the beat frequency produced by the direct and the reflected sounds at your ears?

•**16.49** A car is parked at a railroad crossing. A train passes, and the driver of the car records the time dependence of the frequency of the sound emitted by the train, as shown in the figure.

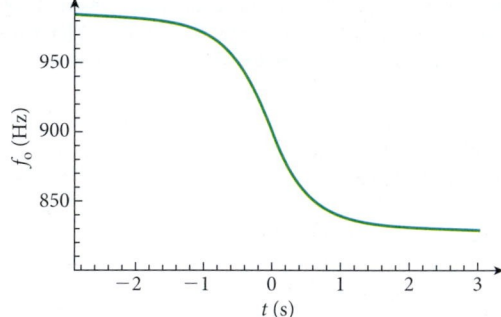

a) What is the frequency heard by someone riding on the train?

b) How fast is the train moving?

c) How far away from the train tracks is the driver of the car?

$$\left(Hint: \left.\dfrac{df_o}{dt}\right|_{t=0} = -\dfrac{fv^2}{b\,v_{sound}}.\right)$$

••**16.50** Many towns have tornado sirens, large elevated sirens used to warn locals of imminent tornados. In one small town, a siren is elevated 100. m off the ground. A car is being driven at 100. km/h directly away from this siren while it is emitting a 440.-Hz sound. What is the frequency of the sound heard by the driver as a function of the distance from the siren at which he starts? Plot this frequency as a function of the car's position up to 1000. m. Explain this plot in terms of the Doppler effect.

Section 16.5

16.51 A standing wave in a pipe with both ends open has a frequency of 440. Hz. The next higher overtone has a frequency of 660. Hz.

a) Determine the fundamental frequency.

b) How long is the pipe?

16.52 A bugle can be represented by a cylindrical pipe of length $L = 1.35$ m. The pipe is open at one end and closed at the other end (the end with the mouthpiece). Calculate the longest three wavelengths of standing waves inside the bugle. Also calculate the three lowest frequencies and the three longest wavelengths of the sound that is produced in the air around the bugle.

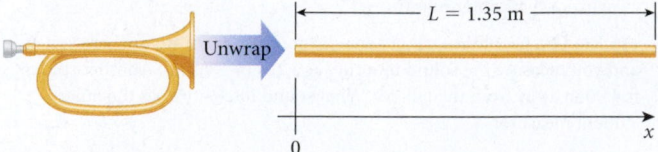

16.53 A soprano sings the note C6 (1047 Hz) across the mouth of a soda bottle. For a fundamental frequency equal to this note to be produced in the bottle, how far below the top of the bottle must the top of the liquid be?

16.54 A thin aluminum rod of length L = 2.00 m is clamped at its center. The speed of sound in aluminum is 5000. m/s. Find the lowest resonance frequency for vibrations in this rod.

•16.55 Find the resonance frequency of the ear canal. Treat it as a half-open pipe of diameter 8.0 mm and length 25 mm. Assume that the temperature inside the ear canal is body temperature (37 °C).

•16.56 A half-open pipe is constructed to produce a fundamental frequency of 262 Hz when the air temperature is 22.0 °C. It is used in an overheated building when the temperature is 35.0 °C. Neglecting thermal expansion in the pipe, what frequency will be heard?

Additional Exercises

16.57 A car horn emits a sound of frequency 400.0 Hz. The car is traveling at 20.0 m/s toward a stationary pedestrian when the driver sounds the horn. What frequency does the pedestrian hear?

16.58 An observer stands between two sound sources. Source A is moving away from the observer, and source B is moving toward the observer. Both sources emit sound of the same frequency. If both sources are moving with a speed $v_{sound}/2$, what is the ratio of the frequencies detected by the observer?

16.59 An F16 jet takes off from the deck of an aircraft carrier. A diver 1.00 km away from the ship is floating in the water with one ear below the surface and one ear above the surface. How much time elapses between the time he initially hears the jet's engines in one ear and the time he initially hears them in the other ear?

16.60 A train has a horn that produces a sound with a frequency of 311 Hz. Suppose you are standing next to a track when the train, horn blaring, approaches you at a speed of 22.3 m/s. By how much will the frequency of the sound you hear shift as the train passes you?

16.61 You have embarked on a project to build a five-pipe wind chime. The notes you have selected for your open pipes are presented in the table.

Note	Frequency (Hz)	Length (m)
G4	392	
A4	440	
B4	494	
F5	698	
C6	1047	

Calculate the length for each of the five pipes to achieve the desired frequency, and complete the table.

16.62 Two trains are traveling toward each other in still air at 25.0 m/s relative to the ground. One train is blowing a whistle at 300. Hz. Assume that the speed of sound is 343 m/s.

a) What frequency is heard by a man on the ground facing the whistle-blowing train?

b) What frequency is heard by a man on the other train?

•16.63 At a distance of 20.0 m from a sound source, the intensity of the sound is 60.0 dB. What is the intensity (in dB) at a point 2.00 m from the source? Assume that the sound radiates equally in all directions from the source.

•16.64 Two vehicles carrying speakers that produce a tone of frequency 1000.0 Hz are moving directly toward each other. Vehicle A is moving at 10.00 m/s and vehicle B is moving at 20.00 m/s. Assume that the speed of sound in air is 343.0 m/s, and find the frequencies that the driver of each vehicle hears.

•16.65 You are standing between two speakers that are separated by 80.0 m. Both speakers are playing a pure tone of 286 Hz. You begin running directly toward one of the speakers, and you measure a beat frequency of 10.0 Hz. How fast are you running?

•16.66 In a sound interference experiment, two identical loudspeakers are placed 4.00 m apart, facing in a direction perpendicular to the line that connects them. A microphone attached to a carrier sliding on a rail picks up the sound from the speakers at a distance of 400. m, as shown in the figure. The two speakers are driven in phase by the same signal generator at a frequency of 3400. Hz. Assume that the speed of sound in air is 340. m/s.

a) At what point(s) on the rail should the microphone be located for the sound reaching it to have maximum intensity?

b) At what point(s) should it be located for the sound reaching it to be zero?

c) What is the separation between two points of maximum intensity?

d) What is the separation between two points of zero intensity?

e) How would things change if the two loudspeakers produced sounds of the same frequency but different intensities?

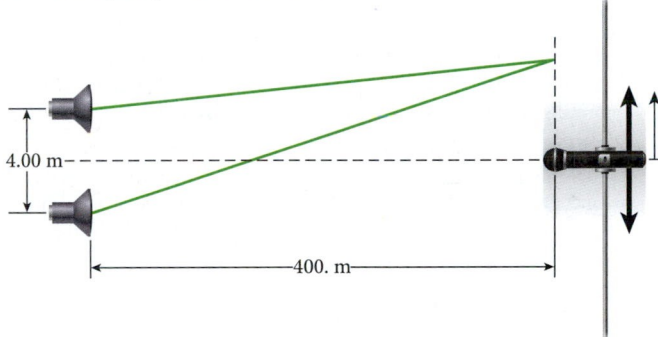

•16.67 A car traveling at 25.0 m/s honks its horn as it directly approaches the side of a large building. The horn produces a long sustained note of frequency f_0 = 230. Hz. The sound is reflected off the building back to the car's driver. The sound wave from the original note and that reflected off the building combine to create a beat frequency. What is the beat frequency that the driver hears (which tells him that he had better hit the brakes!)?

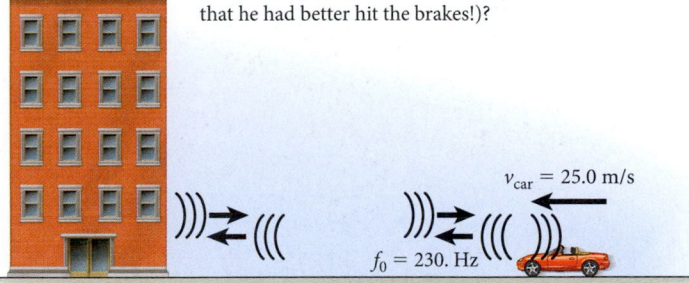

•16.68 Two identical half-open pipes each have a fundamental frequency of 500. Hz. What percentage change in the length of one of the pipes will cause a beat frequency of 10.0 Hz when they are sounded simultaneously?

•16.69 A source traveling to the right at a speed of 10.00 m/s emits a sound wave at a frequency of 100.0 Hz. The sound wave bounces off a reflector, which is traveling to the left at a speed of 5.00 m/s. What is the frequency of the reflected sound wave detected by a listener back at the source?

•16.70 In a suspense-thriller movie, two submarines, X and Y, approach each other, traveling at 10.0 m/s and 15.0 m/s, respectively. Submarine X "pings" submarine Y by sending a sonar wave of frequency 2000.0 Hz. Assume that the sound travels at 1500.0 m/s in the water.

a) Determine the frequency of the sonar wave detected by submarine Y.

b) What is the frequency detected by submarine X for the sonar wave reflected off submarine Y?

c) Suppose the submarines barely miss each other and begin to move away from each other. What frequency does submarine Y detect from the pings sent by X? How much is the Doppler shift?

••16.71 Consider a sound wave (that is, a longitudinal displacement wave) in an elastic medium with Young's modulus Y (solid) or bulk modulus B (fluid) and unperturbed density ρ_0. Suppose this wave is described by the wave function $\delta x(x,t)$, where δx denotes the displacement of a point in the medium from its equilibrium position, x is position along the path of the wave at equilibrium, and t is time. The wave can also be regarded as a pressure wave, described by wave function $\delta p(x,t)$, where δp denotes the change of pressure in the medium from its equilibrium value.

a) Find the relationship between $\delta p(x,t)$ and $\delta x(x,t)$, in general.

b) If the displacement wave is a pure sinusoidal function, with amplitude A, wave number κ, and angular frequency ω, given by $\delta x(x,t) = A\cos(\kappa x - \omega t)$, what is the corresponding pressure wave function, $\delta p(x,t)$? What is the amplitude of the pressure wave?

••16.72 Consider the sound wave of Problem 16.71.

a) Find the intensity, I, of the general wave, in terms of the wave functions $\delta x(x,t)$ and $\delta p(x,t)$.

b) Find the intensity of the sinusoidal wave of part (b), in terms of the displacement and amplitude of the pressure wave.

•16.73 Using the results of Problems 16.71 and 16.72, determine the displacement and amplitude of the pressure wave corresponding to a pure tone of frequency f = 1.000 kHz in air (density = 1.20 kg/m^3, speed of sound 343 m/s), at the threshold of hearing (β = 0.00 dB), and at the threshold of pain (β = 120. dB).

MULTI-VERSION EXERCISES

16.74 A person in a parked car sounds the horn. The frequency of the horn's sound is 489 Hz. A driver in an approaching car measures the frequency of the horn's sound as 509.4 Hz. What is the speed of the approaching car? (Use 343 m/s for the speed of sound.)

16.75 A person in a parked car sounds the horn. A driver in an approaching car, which is moving at a speed of 15.1 m/s, measures the frequency of the horn's sound as 579.4 Hz. What is the frequency of the sound that the person in the parked car hears? (Use 343 m/s for the speed of sound.)

16.76 A person in a parked car sounds the horn. The frequency of the horn's sound is 333 Hz. A driver in an approaching car is moving at a speed of 15.7 m/s. What is the frequency of the sound that she hears? (Use 343 m/s for the speed of sound.)

16.77 The tuba typically produces the lowest-frequency sounds in an orchestra. Consider a tuba to be a coiled tube of length 7.373 m. What is

the lowest-frequency sound a tuba can produce? Assume that the speed of sound is 343.0 m/s.

16.78 The tuba typically produces the lowest-frequency sounds in an orchestra. Consider a tuba to be a coiled tube. If the lowest frequency that the tuba can produce is 22.56 Hz, what is the length of the tube? Assume that the speed of sound is 343.0 m/s.

16.79 A metal bar has a Young's modulus of $266.3 \cdot 10^9$ N/m^2 and a mass density of 3497 kg/m^3. What is the speed of sound in this bar?

16.80 A metal bar has a Young's modulus of $112.1 \cdot 10^9$ N/m^2. The speed of sound in the bar is 5628 m/s. What is the mass density of this bar?

16.81 A metal bar has a mass density of 3579 kg/m^3. The speed of sound in this bar is 6642 m/s. What is Young's modulus for this bar?

17

Temperature

FIGURE 17.1 The star cluster known as the Pleiades, or Seven Sisters.

The Pleiades star cluster, shown in Figure 17.1, can be seen with the naked eye and was known to the ancient Greeks. What they could not know is that these stars exhibit some of the highest temperatures that occur in nature. Their surface temperatures range from 4,000 °C to 10,000 °C, depending on size and other factors. However, their interior temperatures can reach over 10 million °C, hot enough to vaporize any substance. On the other end of the temperature range, space itself, far from any stars, registers a temperature of roughly –270 °C.

This chapter begins our study of thermodynamics, including the concepts of temperature, heat, and entropy. In its widest sense, thermodynamics is the physics of energy and energy transfer—how energy is stored, how it is transformed from one kind to another, and how it can be made available to do work. We'll examine energy at the atomic and molecular level as well as at the macroscopic level of engines and machines.

This chapter looks at temperature—how it is defined and measured and how changes in temperature can affect objects. We'll consider various scales with which to quantify temperature as well as the ranges of temperatures observed in nature and in the lab.

In practical terms, it is almost impossible to describe temperature without also discussing heat, which is the subject of Chapter 18. Be sure you keep in mind the differences between these concepts as you study the next few chapters.

WHAT WE WILL LEARN

- Temperature is measured using any of several different physical properties of certain materials.

- The Fahrenheit temperature scale sets the temperature of the freezing point of water at 32 °F and that of the boiling point of water at 212 °F.

- The Celsius temperature scale sets the temperature of the freezing point of water at 0 °C and that of the boiling point of water at 100 °C.

- The Kelvin temperature scale is defined in terms of an absolute zero temperature, or the lowest temperature at which matter could theoretically exist. In the Kelvin temperature scale, the freezing point of water is 273.15 K, and the boiling point of water is 373.15 K.

- Heating a long, thin metal rod causes its length to increase linearly with the temperature, measured in K.

- Heating a liquid generally causes its volume to increase linearly with the temperature, measured in K.

- The Earth's average surface temperature in 2011 was approximately 14.46 °C, and it is increasing at a rate of about 0.2 °C per decade.

- Analysis of the cosmic microwave background radiation shows the temperature of "empty" intergalactic space to be 2.725 K.

17.1 Definition of Temperature

Temperature is a concept we all understand from experience. We hear weather forecasters tell us that the temperature will be 72 °F today. We hear doctors tell us that our body temperature is 98.6 °F. When we touch an object, we can tell whether it is hot or cold. If we put a hot object in contact with a cold object, the hot object will cool off and the cold object will warm up. If we measure the temperatures of the two objects after some time has passed, they will be equal. The two objects are then in **thermal equilibrium.**

Heat is the transfer between a system and its environment of **thermal energy**, which is energy in the form of random motion of the atoms and molecules making up the matter in question. Chapter 18 will quantify the concept of heat as thermal energy that is transferred because of a temperature difference.

The **temperature** of an object is related to the tendency of the object to transfer heat to or from its surroundings. Heat will transfer from the object to its surroundings if the object's temperature is higher than that of its surroundings. Heat will transfer to the object if its temperature is less than its surroundings. Note that cold is simply the absence of heat; there is no such thing as a flow of "coldness" between an object and its surroundings. If an object feels cold to your touch, this is simply a consequence of heat being transferred from your fingers to the object. (This is the macroscopic definition of temperature; we'll see in Chapter 19 that on a microscopic level, temperature is proportional to the kinetic energy of random motion of particles.)

Measuring temperature relies on the fact that if two objects are in thermal equilibrium with a third object, they are in thermal equilibrium with each other. This third object could be a **thermometer,** which measures temperature. This idea, often called the **Zeroth Law of Thermodynamics,** defines the concept of temperature and underlies the ability to measure temperature. That is, in order to find out if two objects have the same temperature, you do not need to bring them into thermal contact and monitor whether thermal energy transfers (which may be hard or even impossible in some instances). Instead, you can use a thermometer and measure each object's temperature separately; if your readings are the same, you know that the objects have the same temperature.

Temperature measurements can be taken using any of several common scales. Let's examine these.

Temperature Scales

Fahrenheit Scale

Several systems have been proposed and used to quantify temperature; the most widely used are the Fahrenheit, Celsius, and Kelvin scales. The **Fahrenheit temperature scale** was proposed in 1724 by Gabriel Fahrenheit, a German-born scientist living in Amsterdam. Fahrenheit also invented the mercury-expansion thermometer. The Fahrenheit scale went through several iterations. Fahrenheit finally defined the unit of the Fahrenheit scale (°F) by fixing 32 °F for the freezing point of water and 96 °F for the temperature of the human body as measured under the arm. Later, other scientists refined the scale by defining the freezing point of water as 32 °F and the boiling point of water as 212 °F. This temperature scale is used widely in the United States.

Celsius Scale

Anders Celsius, a Swedish astronomer, proposed the **Celsius temperature scale,** often called the *centigrade scale,* in 1742. Several iterations of this scale resulted in its unit (°C) being determined by setting the freezing point of water at 0 °C and the boiling point of water at 100 °C (at normal atmospheric pressure). This temperature scale is used worldwide, except in the United States.

Kelvin Scale

In 1848, William Thomson (Lord Kelvin), a British physicist, proposed another temperature scale, which is now called the **Kelvin temperature scale.** This scale is based on the existence of **absolute zero,** the minimum possible temperature.

The behavior of the pressure of gases at fixed volume as a function of temperature was studied, and the observed behavior was extrapolated to zero pressure to establish this absolute zero temperature. To see how this works, suppose we have a fixed volume of nitrogen gas in a hollow spherical aluminum container connected to a pressure gauge (Figure 17.2). There is enough nitrogen gas in the container that when the container is immersed in an ice-water bath, the pressure gauge reads 0.200 atm. We then place the container in boiling water. The pressure gauge now reads 0.273 atm. Therefore, the pressure has increased while the volume has stayed constant. Suppose we repeat this procedure with different starting pressures. Figure 17.3 summarizes the results, with four solid lines representing the different starting pressures. You can see that the pressure of the gas goes down as the temperature decreases. Conversely, lowering the pressure of the gas must lower its temperature. Theoretically, the lowest temperature of the gas can be determined by extrapolating the measured behavior until the pressure becomes zero. The dashed lines in Figure 17.3 show the extrapolations.

The relationship among pressure, volume, and temperature for a gas is the focus of Chapter 19. However, for now, you only need to understand the following observation: The data from the four sets of measurements starting at different initial pressures extrapolate to the same temperature at zero pressure. This temperature is called *absolute*

Ice water
0 °C

(a)

Boiling water
100 °C

(b)

FIGURE 17.2 A hollow aluminum sphere fitted with a pressure gauge and filled with nitrogen gas. (a) The container is held at 0 °C by placing it in ice water. (b) The container is held at 100 °C by placing it in boiling water.

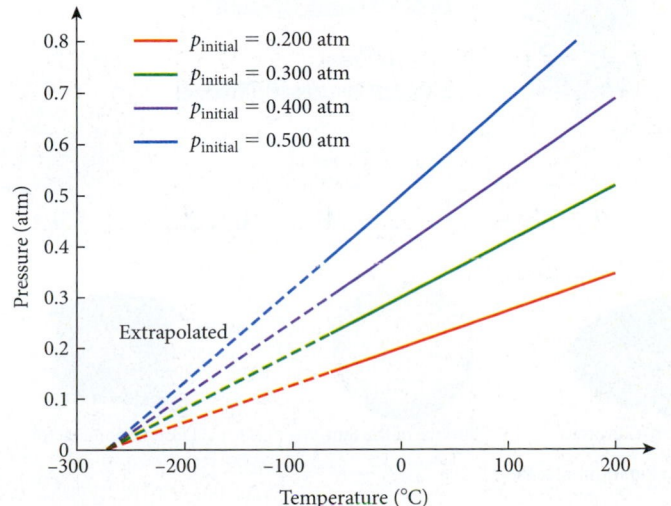

FIGURE 17.3 Solid lines: Pressures of a gas measured at fixed volume and different temperatures. Each line represents an experiment starting at a different initial pressure. Dashed lines: Extrapolation of the pressure of nitrogen gas in a fixed volume as the temperature is lowered.

zero and corresponds to –273.15 °C. When researchers attempt to lower the temperature of real nitrogen gas to very low temperatures, the linear relationship breaks down because the gaseous nitrogen liquefies, bringing subsequent interactions between the nitrogen molecules. In addition, different gases show slightly different behavior at very low temperatures. However, for low pressures and relatively high temperatures, this general result stands.

Absolute zero is the lowest temperature at which matter could theoretically exist. (Experimentally, it is impossible to reach absolute zero, just as it is impossible to build a perpetual motion machine.) We'll see in Chapter 19 that temperature corresponds to motion at the atomic and molecular level, so making an object reach absolute zero would imply that all motion in the atoms and molecules of the object ceases. However, in Chapters 36 and 37, we'll see that some motion of this type is required by quantum mechanics. This requirement, sometimes called the **Third Law of Thermodynamics,** means that absolute zero is never actually attainable.

Kelvin used the size of the Celsius degree (°C) as the size of the unit of his temperature scale, now called the **kelvin** (K). On the Kelvin scale, the freezing point of water is 273.15 K and the boiling point of water is 373.15 K. This temperature scale is used in many scientific calculations, as we'll see in the next few chapters. Because of these considerations, the kelvin is the standard SI unit for temperature. To achieve greater consistency, scientists have proposed defining the kelvin in terms of other fundamental constants, rather than in terms of the properties of water. These new definitions are scheduled to take effect by the year 2014.

17.2 Temperature Ranges

Temperature measurements span a huge range (shown in Figure 17.4, using a logarithmic scale), from the highest measured temperatures ($2 \cdot 10^{12}$ K), observed in relativistic heavy ion collisions (RHIC), to the lowest measured temperatures ($1 \cdot 10^{-10}$ K), observed in spin systems of rhodium atoms. The temperature in the center of the Sun is estimated to be $16 \cdot 10^6$ K, and at the surface of the Sun, it has been measured to be 5778 K. The lowest measured air temperature on the Earth's surface is 183.9 K (–89.2 °C) in Antarctica; the highest measured air temperature on the Earth's surface is 330.9 K (57.8 °C) in the

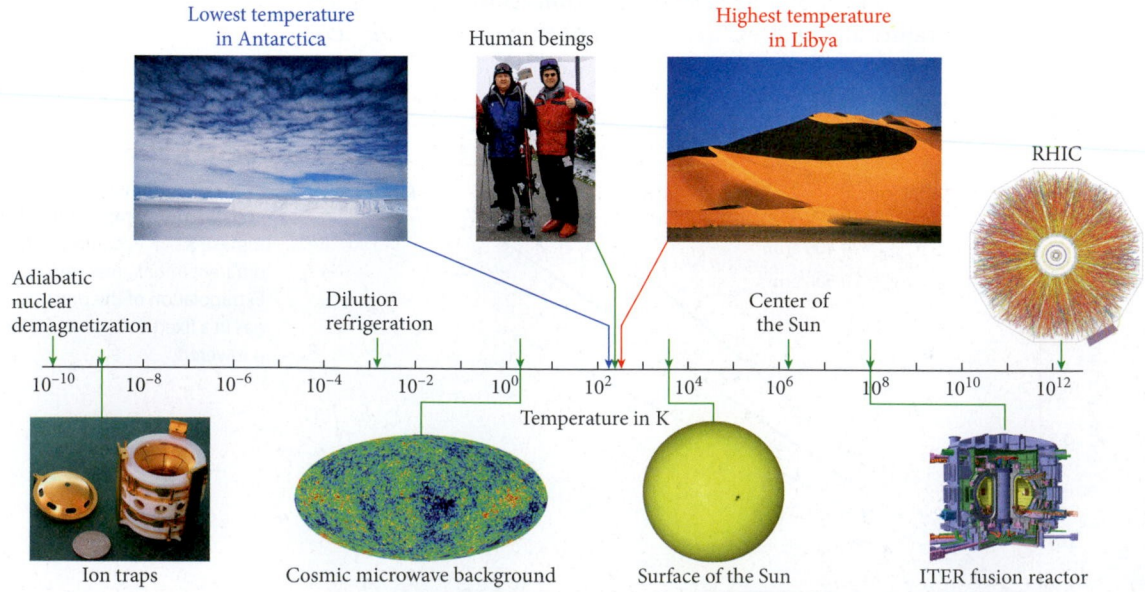

FIGURE 17.4 Range of observed temperatures, plotted on a logarithmic scale.

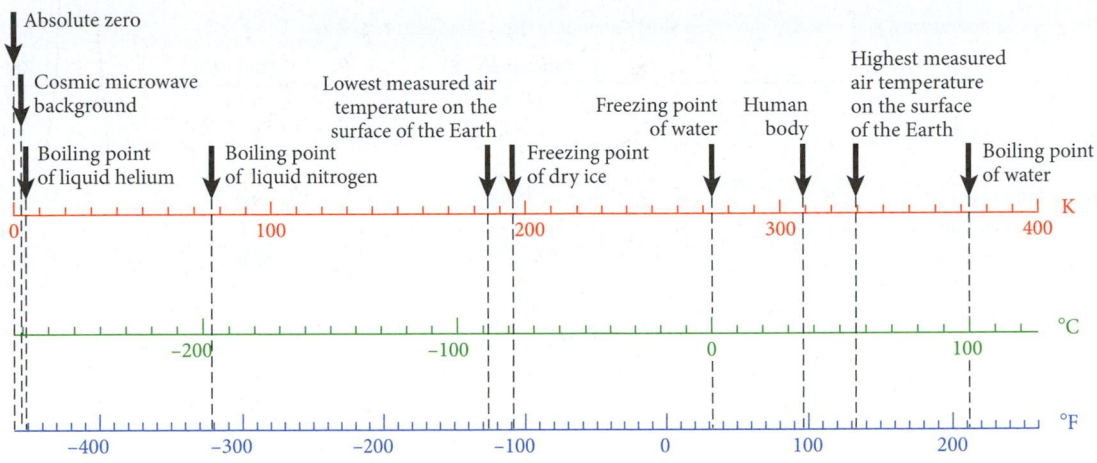

FIGURE 17.5 Representative temperatures, expressed in the three common temperature scales.

Sahara desert in Libya. You can see from Figure 17.4 that the observed temperature range on the surface of the Earth covers only a very small fraction of the range of observed temperatures. The cosmic microwave background radiation left over from the Big Bang 13.7 billion years ago has a temperature of 2.725 K (which is the temperature of "empty" intergalactic space), an observation that will be explained in more detail in later chapters. From Figure 17.4, you can see that temperatures attained in relativistic heavy ion collisions are 300 million times higher than the surface temperature of the Sun and that temperatures of atoms measured in ion traps are more than a billion times lower than the temperature of intergalactic space.

Figure 17.5 shows some representative temperatures above absolute zero and below 400 K, expressed in the Fahrenheit, Celsius, and Kelvin scales. This temperature range is shown on a linear scale.

In the following formulas for converting between the various temperature scales, the Fahrenheit temperature is T_F, measured in °F; the Celsius temperature is T_C, measured in °C; and the Kelvin temperature is T_K, measured in K.

Fahrenheit to Celsius:

$$T_C = \frac{5}{9}\left(T_F - 32 \ °F\right). \tag{17.1}$$

Celsius to Fahrenheit:

$$T_F = \frac{9}{5}T_C + 32 \ °C. \tag{17.2}$$

Celsius to Kelvin:

$$T_K = T_C + 273.15 \ °C. \tag{17.3}$$

Kelvin to Celsius:

$$T_C = T_K - 273.15 \ K. \tag{17.4}$$

The mean human body temperature taken orally is 36.8 °C, which corresponds to 98.2 °F. The often quoted mean oral temperature of 98.6 °F corresponds to a 19th-century measurement of 37 °C. This temperature is highlighted on the oral mercury thermometer in Figure 17.6. The difference between 98.6 °F and 98.2 °F thus corresponds to the error due to rounding the Celsius scale measurement of human body temperature to two digits. Table 17.1 lists human body temperature and other representative temperatures, expressed in the three temperature scales.

Self-Test Opportunity 17.1

At what temperature do the Fahrenheit and Celsius scales have the same numerical value?

FIGURE 17.6 An oral, mercury-expansion thermometer.

Table 17.1	Various Temperatures, Expressed in the Three Most Commonly Used Temperature Scales		
	Fahrenheit (°F)	Celsius (°C)	Kelvin (K)
Absolute zero	−459.67	−273.15	0
Water's freezing point	32	0	273.15
Water's boiling point	212	100	373.15
Typical human body temperature	98.2	36.8	310
Lowest measured air temperature	−129	−89.2	184
Highest measured air temperature	136	57.8	331
Lowest temperature ever measured in a lab	−459.67	−273.15	$1.0 \cdot 10^{-10}$
Highest temperature ever measured in a lab	$3.6 \cdot 10^{12}$	$2 \cdot 10^{12}$	$2 \cdot 10^{12}$
Cosmic background microwave radiation	−454.76	−270.42	2.73
Liquid nitrogen boiling point	−321	−196	77.3
Liquid helium boiling point	−452	−269	4.2
Temperature at the surface of the Sun	9,941	5,505	5,778
Temperature at the center of the Sun	$28 \cdot 10^{6}$	$16 \cdot 10^{6}$	$16 \cdot 10^{6}$
Average temperature of the surface of the Earth	59	15	288
Temperature at the center of the Earth	9,800	5,400	5,700

EXAMPLE 17.1 Room Temperature

Room temperature is often taken to be 72.0 °F.

PROBLEM
What is room temperature in the Celsius and Kelvin scales?

SOLUTION
Using equation 17.1, we can convert room temperature from degrees Fahrenheit to degrees Celsius:

$$T_C = \tfrac{5}{9}\left(72 \text{ °F} - 32 \text{ °F}\right) = 22.2 \text{ °C}.$$

Using equation 17.3, we can then express room temperature in kelvins as

$$T_K = 22 \text{ °C} + 273.15 \text{ °C} = 295. \text{ K}.$$

SOLVED PROBLEM 17.1 Temperature Conversion

PROBLEM
On a cold winter day, you notice that the temperature reading in Fahrenheit degrees is 10.0 degrees *higher* than that in Celsius degrees. By how many Celsius degrees does the temperature need to change for the Fahrenheit reading to be 10.0 degrees *lower* than the Celsius reading?

SOLUTION
THINK This appears to be a straightforward temperature conversion problem. The easiest way to solve this problem is to first figure out the temperature that meets the 10-degree-higher condition and then find the temperature that fulfills the 10-degree-lower condition. Our final answer is then the difference of those two temperatures.

SKETCH A sketch does not seem to help at first thought. However, as shown in Figure 17.7, we can graph the conversion line for Celsius and Fahrenheit temperatures and the two boundary conditions specified in the problem. This will allow us to solve the problem graphically.

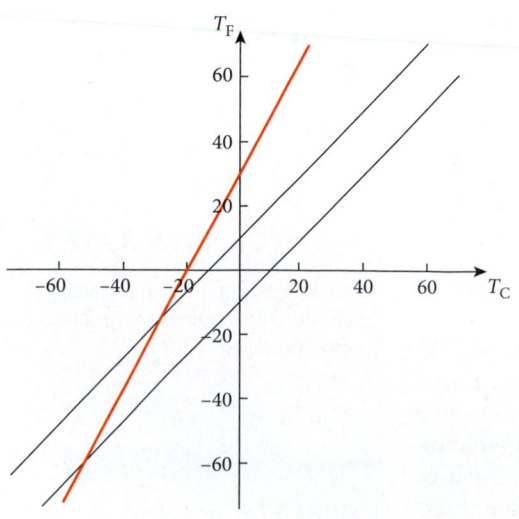

FIGURE 17.7 Conversion of Celsius temperatures to Fahrenheit (red line). The two black lines are the boundary conditions imposed in the problem.

RESEARCH The formula for converting from Celsius to Fahrenheit temperatures is $T_F = \tfrac{9}{5}T_C + 32$. For the first set of temperature readings, we have $T_F = T_C + 10$, and for the second set of readings, we have $T_F = T_C - 10$.

SIMPLIFY The temperature that meets the 10-degrees-higher condition can be obtained by setting the two expressions for T_F, $T_F = T_C + 10$ and $T_F = \frac{9}{5}T_C + 32$, equal, which results in

$$T_C + 10 = \tfrac{9}{5}T_C + 32 \Rightarrow -22 = \tfrac{4}{5}T_C \Rightarrow T_C = -27.5.$$

We find the temperature that meets the 10-degrees-lower condition in the same way:

$$T_C - 10 = \tfrac{9}{5}T_C + 32 \Rightarrow -42 = \tfrac{4}{5}T_C \Rightarrow T_C = -52.5.$$

CALCULATE Now, all that is left to do is take the difference between the two temperatures:

$$\Delta T_C = -52.5 - (-27.5) = -25.0.$$

ROUND The problem statement gave the temperatures with three significant digits, and so we express our final answer to that precision: $\Delta T_C = -25.0$. It would have to get colder by 25.0 °C for the Fahrenheit reading to be 10.0 degrees below the Celsius reading. (You might think that the number 32 in the temperature conversion equation has only two significant digits; however, this conversion constant is by definition exact, and so it does not degrade the precision of our final answer.)

DOUBLE-CHECK Our sketch provides an easy double-check: The two points at which the red line crosses the two parallel black lines appear to be at (or at least close to) the values we found algebraically.

Research at the Low-Temperature Frontier

How are very low temperatures attained in a lab? Researchers start by taking advantage of the properties of gases related to their pressure, volume, and temperature (again, Chapter 19 will go into these properties in much greater depth). In addition, phase changes (discussed in Chapter 18) from liquid to gas and back to liquid are crucial in producing low temperatures in the lab. For this discussion, we can simply note that when a liquid evaporates, the energy required to change the liquid to a gas is taken from the liquid—therefore, the liquid cools. For example, air is liquefied through a multistep process in which the air is compressed and cooled by removing the heat produced in the compression stage. Then, the fact that different gases have different temperatures at which they liquefy allows liquid nitrogen and liquid oxygen to be separated. At normal atmospheric pressure, liquid nitrogen has a boiling point of 77 K, while liquid oxygen has a boiling point of 90 K. Liquid nitrogen has many applications, not least of which is in classroom lecture demonstrations.

Even lower temperatures are achieved using the compression process, with liquid nitrogen extracting the heat as helium is compressed. Liquid helium boils at 4.2 K. Starting with liquid helium and reducing the pressure so that it evaporates, researchers can cool the remaining liquid down to about 1.2 K. For the isotope helium-3 (^{3}He), a temperature as low as 0.3 K can be reached with this evaporative-cooling technique.

To reach still lower temperatures, researchers use an apparatus called a *dilution refrigerator* (see Figure 17.8). A dilution refrigerator uses two isotopes of helium: ^{3}He (two protons and one neutron) and ^{4}He (two protons and two neutrons). (Isotopes are discussed in Chapter 40 on nuclear physics.) When the temperature of a mixture of liquid ^{3}He and liquid ^{4}He is reduced below 0.7 K, the two gases separate into a ^{3}He-poor phase and a ^{4}He-poor phase. Energy is required to move ^{3}He atoms into the ^{3}He-poor phase. If the atoms can be forced to cross the boundary between the two phases, the mixture can be cooled in a way similar to evaporative cooling. A dilution refrigerator can cool to temperatures around 10 mK, and the best commercial models can reach 2 mK.

Still lower temperatures can be reached with a gas of atoms confined to a trap. To attain these temperatures, researchers use techniques such as laser cooling. Laser cooling takes advantage of the electronic structure of certain atoms. (Atoms, transitions between their electron energy levels, and lasers will be covered in detail in Chapter 38.) Transitions between specific electron energy levels in an atom can cause photons with a wavelength close to visible light to be emitted. The inverse process can also occur, in which the atoms absorb photons. (Light absorption and emission are explained in Chapters 36 through 38.)

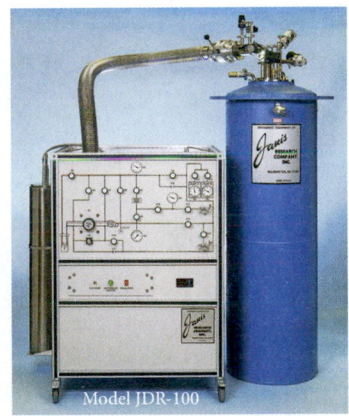

Model JDR-100

(a)

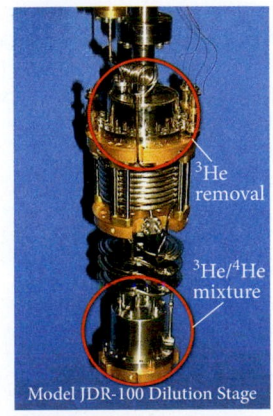

^{3}He removal

^{3}He/^{4}He mixture

Model JDR-100 Dilution Stage

(b)

FIGURE 17.8 (a) A commercial dilution refrigerator. (b) Interior of the dilution refrigerator with the ^{3}He/^{4}He mixture container (mixer) and the ^{3}He removal container (still).

FIGURE 17.9 A laser-cooling device for cooling trapped sodium atoms.

To cool atoms using laser cooling, researchers use a laser that emits light with a very specific wavelength that is longer than the wavelength of light emitted during an atomic transition. Thus, any atom moving toward the laser experiences a slightly shorter wavelength due to the Doppler effect (see Chapter 16 and Chapter 31), while any atom moving away from the laser experiences a longer wavelength. Atoms moving toward the laser absorb photons, but atoms moving away from the laser are not affected. Atoms that absorb photons re-emit them in random directions, effectively cooling the atoms. Temperatures approaching 10^{-9} K have been achieved using this method (Figure 17.9). Steven Chu, Claude Cohen-Tannoudji, and William D. Phillips were awarded the 1997 Nobel Prize in Physics for their work with laser cooling.

The coldest temperatures achieved in a lab have been accomplished using a technique called *adiabatic nuclear demagnetization*. The sample, typically a piece of rhodium metal, is first cooled using a dilution refrigerator. A strong magnetic field is then applied. (Magnetic fields will be discussed in Chapters 27 and 28.) A small amount of heat is produced in the rhodium metal, which is removed by the dilution refrigerator. Then the magnetic field is slowly turned off. As the magnetic field is reduced, the rhodium metal cools further, reaching temperatures as low as 10^{-10} K.

Research at the High-Temperature Frontier

How are high temperatures attained in the lab? The most common methods are by burning fuels or by creating explosions. For example, the yellow part of a candle flame has a temperature of 1400 K.

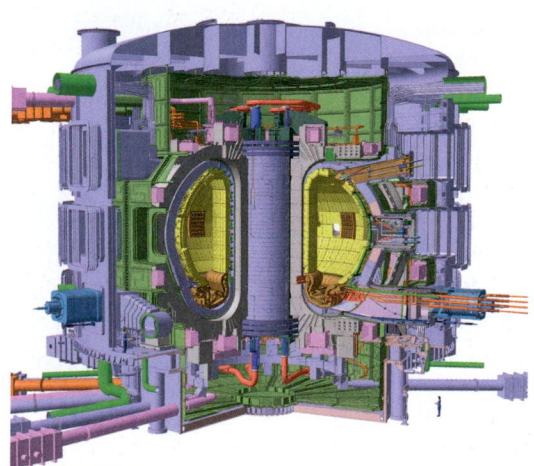

FIGURE 17.10 Cutaway drawing of the central core of ITER, the plasma fusion reactor to be built in France by the year 2019. For size comparison, a person is shown standing at the bottom.

The ITER fusion reactor to be constructed in Cadarache, France, by 2019 (Figure 17.10) is designed to fuse the hydrogen isotopes deuterium (^{2}H) and tritium (^{3}H) to yield helium (^{4}He) and a neutron, with a release of energy. This process of nuclear fusion (to be discussed in Chapter 40) is similar to the process that the Sun uses to fuse hydrogen into helium and thereby produce energy. In the Sun, the huge gravitational force compresses and heats the hydrogen nuclei to produce fusion. In ITER, magnetic confinement will be used to hold ionized hydrogen in the form of a plasma. A *plasma* is a state of matter in which electrons and nuclei move separately. A plasma cannot be contained in a physical container, because it is so hot that any contact with it would vaporize the container. In the fusion reactor, plasma is heated by flowing a current through it. In addition, the plasma is compressed and further heated by the applied magnetic field. At high temperatures, up to $9.9 \cdot 10^7$ K, and densities, ITER will produce usable energy from the fusion of hydrogen into helium.

It is possible to achieve even higher temperatures in particle accelerators. The highest temperature has been reached by colliding gold nuclei accelerated by the Relativistic Heavy Ion Collider at Brookhaven National Laboratory and by the Large Hadron Collider at the European CERN laboratory. When two large nuclei collide, a very hot system with a temperature of $2 \cdot 10^{12}$ K is created; this system is also very small ($\sim 10^{-14}$ m) and exists for very short periods of time ($\sim 10^{-22}$ s).

17.3   Measuring Temperature

How are temperatures measured? A device that measures temperature is called a *thermometer*. Any thermometer that can be calibrated directly by using a physical property is called a *primary thermometer*. A primary thermometer does not need calibration to external standard temperatures. An example of a primary thermometer is one based on the speed of sound in a gas. A *secondary thermometer* is one that requires external calibration to standard temperature references. Often, secondary thermometers are more sensitive than primary thermometers.

A common thermometer is the mercury-expansion thermometer, which is a secondary thermometer. This type of thermometer takes advantage of the thermal expansion of mercury (thermal expansion is discussed in Section 17.4). Other types of thermometers include bimetallic, thermocouple, chemoluminescence, and thermistor thermometers. It is also possible to measure the temperature of a material by studying the distribution of the speeds of the molecules inside that material.

To measure the temperature of an object or a system using a thermometer, the thermometer must be placed in thermal contact with the object or system. (*Thermal contact* is physical contact that allows relatively fast transfer of heat.) Heat will then transfer between the object or system and the thermometer, until they have the same temperature. A good thermometer should require as little thermal energy transfer as possible to reach thermal equilibrium, so that the temperature measurement does not significantly change the object's temperature. The thermometer should also be easily calibrated, so that anyone making the same measurement will get the same temperature.

Calibrating a thermometer requires reproducible conditions. It is difficult to reproduce the freezing point of water exactly, so scientists use a condition called the *triple point of water*. Solid ice, liquid water, and gaseous water vapor can coexist at only one temperature and pressure. By international agreement, the temperature of the triple point of water has been assigned the temperature 273.16 K (and a pressure of 611.73 Pa) for the calibration of thermometers.

Self-Test Opportunity 17.2

You have an uncalibrated thermometer, which is to be used to measure air temperatures. How would you calibrate the thermometer?

17.4   Thermal Expansion

Most of us are familiar in some way with **thermal expansion.** Perhaps you know that you can loosen a metal lid on a glass jar by heating the lid. You may have seen that bridge spans contain gaps in the roadway to allow for the expansion of sections of the bridge in warm weather. Or you may have observed that power lines sag in warm weather. And if you live in a colder climate, you may know that water can freeze in pipes in the winter and cause damage to the pipes as a result of its expansion.

Why do materials expand due to a change in temperature? In Chapter 13, we saw that all matter is made of atoms. The atoms vibrate, and the amplitude of their vibrations is a function of the temperature of the matter. The larger the amplitude of this vibrational motion, the greater the spaces between the atoms in solids and liquids. (Since gases occupy the entire space available to them, changes in vibrational amplitude have no effect on atomic spacing in gases. In Chapter 19, we will study the effect of temperature on gases.) As a very general rule, a higher temperature means larger vibrations and thus greater atomic spacing. Therefore, solids and liquids expand with increasing temperature. (The only exception to this general rule is at or close to the freezing or melting point, where some substances— most notably water—expand as they freeze and contract as they melt. This anomaly is due to the fact that the atoms assume places in an orderly crystal structure in the transition from liquid to solid phase.)

The thermal expansion of liquids and solids can be put to practical use. Bimetallic strips, which are often used in room thermostats, meat thermometers, and thermal protection devices in electrical equipment, take advantage of linear thermal expansion. (A bimetallic strip consists of two thin long strips of different metals, which are welded together.) A mercury thermometer uses volume expansion to provide precise measurements of temperature. Thermal expansion can occur as linear expansion, area expansion, or volume expansion; all three classifications describe the same phenomenon.

Linear Expansion

Let's consider a metal rod of length L (Figure 17.11). If we raise the temperature of the rod by $\Delta T = T_{final} - T_{initial}$, the length of the rod increases by the amount $\Delta L = L_{final} - L_{initial}$, given by

$$\Delta L = \alpha L \Delta T, \tag{17.5}$$

where α is the **linear expansion coefficient** of the metal from which the rod is constructed and the temperature difference is expressed in degrees Celsius or kelvins. The linear expansion coefficient is a constant for a given material within normal temperature ranges. Some typical linear expansion coefficients are listed in Table 17.2.

Table 17.2	Linear Expansion Coefficients of Some Common Materials
Material	$\alpha\ (10^{-6}\ {}^{\circ}C^{-1})$
Aluminum	22
Brass	19
Concrete	15
Copper	17
Diamond	1
Gold	14
Lead	29
Plate glass	9
Rubber	77
Steel	13
Tungsten	4.5

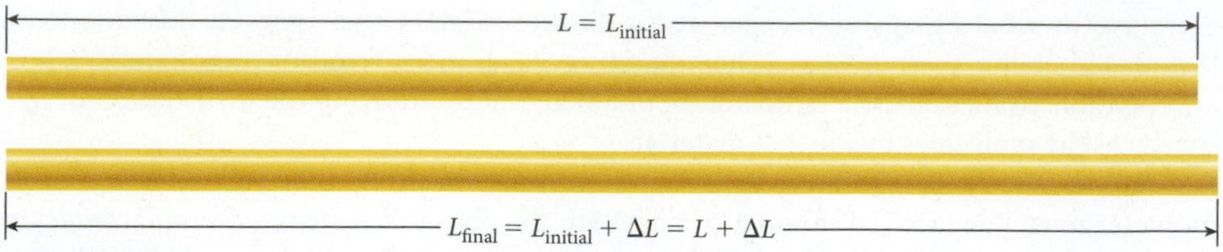

$$\overbrace{\qquad\qquad\qquad L = L_{\text{initial}} \qquad\qquad\qquad}$$

$$\underbrace{\qquad\qquad L_{\text{final}} = L_{\text{initial}} + \Delta L = L + \Delta L \qquad\qquad}$$

FIGURE 17.11 Thermal expansion of a rod with initial length L. (The thermally expanded rod at the bottom has been shifted so that the left edges coincide.)

FIGURE 17.12 The Mackinac Bridge across the Straits of Mackinac in Michigan is the third-longest suspension bridge in the United States.

EXAMPLE 17.2 | Thermal Expansion of the Mackinac Bridge

The center span of the Mackinac Bridge (Figure 17.12) has a length of 1158 m. The bridge is built of steel. Assume that the lowest possible temperature experienced by the bridge is –50 °C and the highest possible temperature is 50 °C.

PROBLEM

How much room must be made available for thermal expansion of the center span of the Mackinac Bridge?

SOLUTION

The linear expansion coefficient of steel is $\alpha = 13 \cdot 10^{-6}\ °\text{C}^{-1}$. Thus, the linear expansion of the center span of the bridge that must be allowed for is given by

$$\Delta L = \alpha L \Delta T = \left(13 \cdot 10^{-6}\ °\text{C}^{-1}\right)(1158\ \text{m})\left[50\ °\text{C} - \left(-50\ °\text{C}\right)\right] = 1.5\ \text{m}.$$

DISCUSSION

A change in length of 1.5 m is fairly large. How is this length change accommodated in practice? (Obviously, we cannot have gaps in the road surface.) The answer lies in expansion joints, which are metal connectors between bridge segments whose parts can move relative to each other. A popular type of expansion joint is the finger joint (see Figure 17.13). The Mackinac Bridge has two large finger joints at the towers to accommodate the expansion of the suspended parts of the roadway and eleven smaller finger joints and five sliding joints across the center span.

(a) (b)

FIGURE 17.13 Finger joints between road segments: (a) open and (b) closed.

From Table 17.2, you can see that various metals, such as brass and steel, have different linear expansion coefficients. This makes them useful in bimetallic strips. The following solved problem considers the result of heating a bimetallic strip.

SOLVED PROBLEM 17.2 | Bimetallic Strip

A straight bimetallic strip consists of a steel strip and a brass strip, each 1.25 cm wide and 30.5 cm long, welded together (see Figure 17.14a). Each strip is $t = 0.500$ mm thick. The bimetallic strip is heated uniformly along its length, as shown in Figure 17.14c. (It doesn't matter that the flame is on the right; the heating is uniform throughout the strip. If the flame were on the left, the strip would bend in the same direction!) The strip curves such that the radius of curvature is $R = 36.9$ cm.

PROBLEM

What is the temperature of the bimetallic strip after it is heated?

SOLUTION

THINK The bimetallic strip is constructed from two metals, steel and brass, that have different linear expansion coefficients, listed in Table 17.2. As the temperature of the

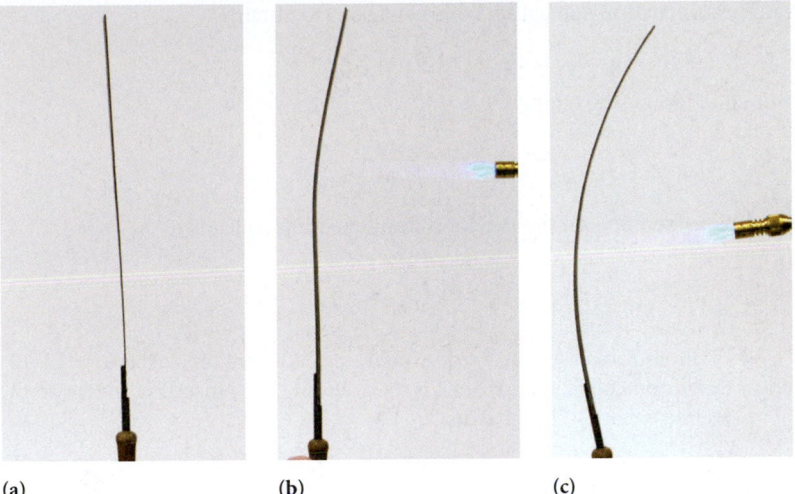

(a)   **(b)**   **(c)**

FIGURE 17.14 A bimetallic strip. (a) The bimetallic strip at room temperature. (b) The bimetallic strip as it begins to be heated by a gas torch (on the right edge of the frame). (c) The bimetallic strip heated to a uniform temperature over its full length.

bimetallic strip increases, the brass expands more than the steel does, so the strip curves toward the steel side. When the bimetallic strip is heated uniformly, both the steel and brass strips lie along the arc of a circle, with the brass strip on the outside and the steel strip on the inside. The ends of the bimetallic strip subtend the same angle, measured from the center of the circle. The arc length of each metal strip then equals the length of the bimetallic strip at room temperature plus the length due to linear thermal expansion. Equating the angle subtended by the steel strip to the angle subtended by the brass strip allows the temperature to be calculated.

SKETCH Figure 17.15 shows the bimetallic strip after it is heated. The angle subtended by the two ends of the strip is θ and the radius of the inner strip is r_1. The inner strip lies along a circle with radius $r_1 = 36.9$ cm.

RESEARCH The arc length, s_1, of the heated steel strip is $s_1 = r_1\theta$, where r_1 is the radius of the circle along which the steel strip lies, and θ is the angle subtended by the steel strip. Also, the arc length, s_2, of the heated brass strip is $s_2 = r_2\theta$, where r_2 is the radius of the circle along which the brass strip lies and θ is the *same angle* as that subtended by the steel strip. The two radii differ by the thickness, t, of the steel strip:

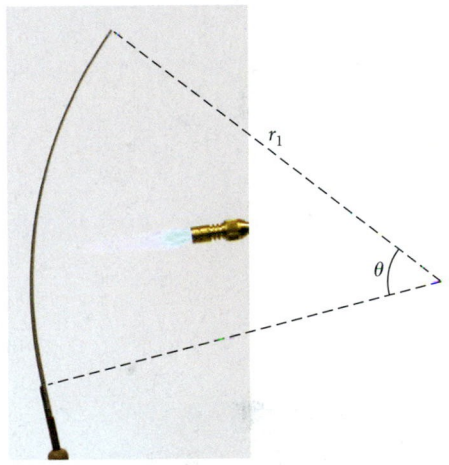

FIGURE 17.15 The bimetallic strip after it is heated, showing the angle subtended by the two ends of the strip.

$$r_2 = r_1 + t \Leftrightarrow t = r_2 - r_1. \qquad (i)$$

We can equate the expressions that are equal to the angle subtended by the two strips:

$$\theta = \frac{s_1}{r_1} = \frac{s_2}{r_2}. \qquad (ii)$$

The arc length, s_1, of the steel strip after it is heated is given by

$$s_1 = s + \Delta s_1 = s + \alpha_1 s \Delta T = s(1 + \alpha_1 \Delta T),$$

where s is the original length of the bimetallic strip. The factor α_1 is the linear expansion coefficient of steel, given in Table 17.2, and ΔT is the temperature difference between room temperature and the final temperature of the bimetallic strip. Correspondingly, the arc length, s_2, of the brass strip after being heated is given by

$$s_2 = s + \Delta s_2 = s + \alpha_2 s \Delta T = s(1 + \alpha_2 \Delta T),$$

where α_2 is the linear expansion coefficient of brass, given in Table 17.2.

SIMPLIFY We substitute the expressions for the arc lengths of the two strips after heating, s_1 and s_2, into equation (ii) to get

$$\frac{s(1 + \alpha_1 \Delta T)}{r_1} = \frac{s(1 + \alpha_2 \Delta T)}{r_2}.$$

Dividing both sides of this equation by the common factor s and multiplying by $r_1 r_2$ gives

$$r_2 + r_2 \alpha_1 \Delta T = r_1 + r_1 \alpha_2 \Delta T. \qquad \text{– Continued}$$

We rearrange this equation and gather common terms to obtain

$$r_2 - r_1 = r_1 \alpha_2 \Delta T - r_2 \alpha_1 \Delta T.$$

Solving this equation for the temperature difference gives

$$\Delta T = \frac{r_2 - r_1}{r_1 \alpha_2 - r_2 \alpha_1}.$$

Using the relationship between the two radii from equation (i) leads to

$$\Delta T = \frac{t}{r_1(\alpha_2 - \alpha_1) - t\alpha_1}. \qquad \text{(iii)}$$

CALCULATE From Table 17.2, the linear expansion coefficient for steel is $\alpha_1 = 13 \cdot 10^{-6} \, °\text{C}^{-1}$, and the linear expansion coefficient for brass is $\alpha_2 = 19 \cdot 10^{-6} \, °\text{C}^{-1}$. Inserting these and the other numerical values into equation (iii) gives us

$$\Delta T = \frac{0.500 \cdot 10^{-3} \, \text{m}}{(0.369 \, \text{m})(19 \cdot 10^{-6} \, °\text{C}^{-1} - 13 \cdot 10^{-6} \, °\text{C}^{-1}) - (0.500 \cdot 10^{-3} \, \text{m})(13 \cdot 10^{-6} \, °\text{C}^{-1})} = 226.5 \, °\text{C}.$$

ROUND Taking room temperature to be 20 °C and reporting our result to two significant figures gives us the temperature of the bimetallic strip after heating:

$$T = 20 \, °\text{C} + \Delta T = 250 \, °\text{C}.$$

DOUBLE-CHECK First, we check that the magnitude of the calculated temperature is reasonable. Our answer of 250 °C is well below the melting points of brass (900 °C) and steel (1450 °C), which is important because Figure 17.14c shows that the strip does not melt. Our answer is also significantly above room temperature, which is important because Figure 17.14a shows that the strip is straight at room temperature.

We can further check that the steel and brass strips do subtend the same angle. The angle subtended by the steel strip is

$$\theta_1 = \frac{s_1}{r_1} = \frac{30.5 \, \text{cm} + (30.5 \, \text{cm})(13 \cdot 10^{-6} \, °\text{C}^{-1})(226.5 \, °\text{C})}{36.9 \, \text{cm}} = 0.829 \, \text{rad} \equiv 47.5°.$$

The angle subtended by the brass strip is

$$\theta_2 = \frac{s_2}{r_2} = \frac{30.5 \, \text{cm} + (30.5 \, \text{cm})(19 \cdot 10^{-6} \, °\text{C}^{-1})(226.5 \, °\text{C})}{36.9 \, \text{cm} + 0.05 \, \text{cm}} = 0.829 \, \text{rad} \equiv 47.5°.$$

Note that because the thickness of the strips is small compared with the radius of the circle, we can rewrite equation (iii) as follows:

$$\Delta T \approx \frac{t}{r_1(\alpha_2 - \alpha_1)} = \frac{0.500 \cdot 10^{-3} \, \text{m}}{(0.369 \, \text{m})(19 \cdot 10^{-6} \, °\text{C}^{-1} - 13 \cdot 10^{-6} \, °\text{C}^{-1})} = 226 \, °\text{C}.$$

This agrees within rounding error with our calculated result. Thus, our answer seems reasonable.

Concept Check 17.4

Suppose the bimetallic strip in Figure 17.14 was made of aluminum on the right side and copper on the left side. Which way would the strip bend if it were heated in the same way as shown in the figure? (Consult Table 17.2 for the linear expansion coefficients of the two metals.)

a) It would bend to the right.

b) It would stay straight.

c) It would bend to the left.

Area Expansion

The effect of a change in temperature on the area of an object is analogous to using a copy machine to enlarge or reduce a picture. Each dimension of the object will change linearly with the change in temperature. Quantitatively, for a square object with side L (Figure 17.16), the area is given by $A = L^2$. Taking the differential of both sides of this equation, we get $dA = 2L \, dL$. If we make the approximations $\Delta A = dA$ and $\Delta L = dL$, we can write $\Delta A = 2L\Delta L$. Using equation 17.5, we then obtain

$$\Delta A = 2L(\alpha L \Delta T) = 2\alpha A \Delta T. \qquad \text{(17.6)}$$

Although a square was used to derive equation 17.6, it holds for a change in area of any shape.

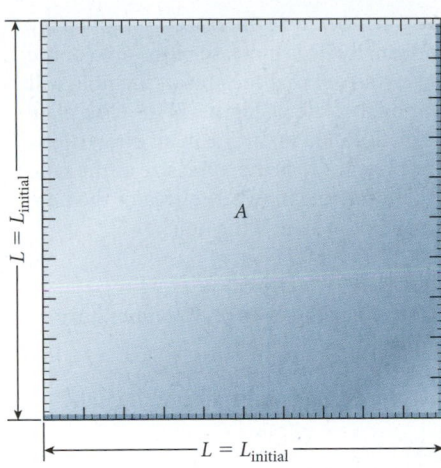 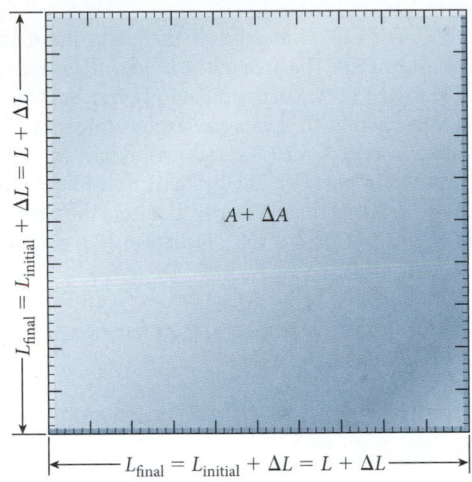

FIGURE 17.16 Thermal expansion of a square plate with side *L*.

SOLVED PROBLEM 17.3 | Expansion of a Plate with a Hole in It

A brass plate has a hole in it (Figure 17.17a); the hole has a diameter $d = 2.54$ cm. The hole is too small to allow a brass sphere to pass through (Figure 17.17b). However, when the plate is heated from 20.0 °C to 220.0 °C, the brass sphere passes through the hole in the plate.

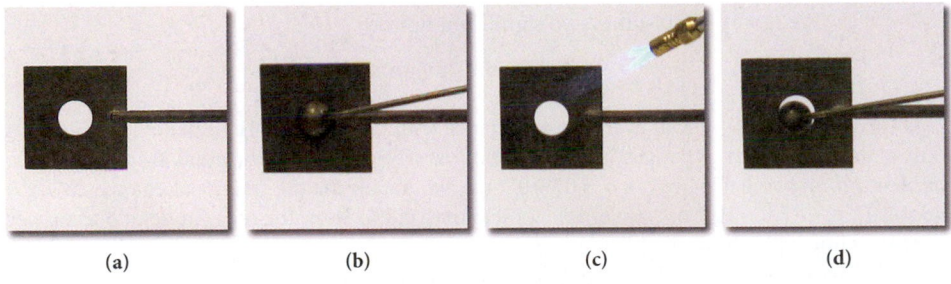

(a)   (b)   (c)   (d)

FIGURE 17.17 (a) The plate before it is heated. (b) A brass sphere will not pass through the hole in the unheated plate. (c) The plate is heated. (d) The same brass sphere passes through the hole in the heated brass plate.

PROBLEM
How much does the area of the hole in the brass plate increase as a result of heating?

SOLUTION
THINK The area of the brass plate increases as the temperature of the plate increases. Correspondingly, the area of the hole in the plate also increases. We can calculate the increase in the area of the hole using equation 17.6.

SKETCH Figure 17.18a shows the brass plate before it is heated, and Figure 17.18b shows the plate after it is heated.

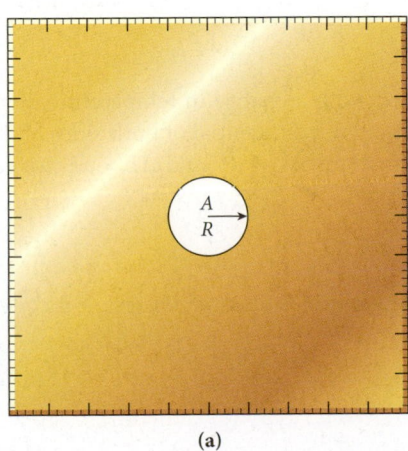 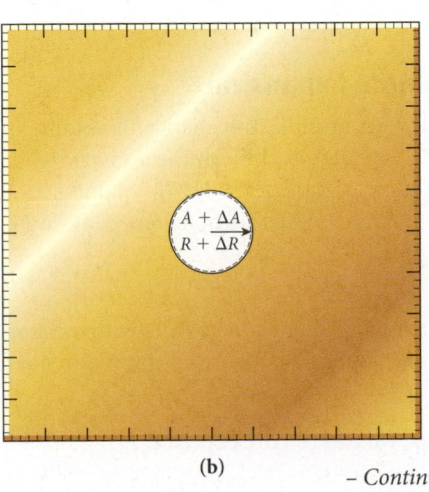

(a)   (b)

FIGURE 17.18 The thermal expansion of a plate with a hole in it: (a) before heating; (b) after heating.

– *Continued*

RESEARCH The area of the plate increases as the temperature increases, as given by equation 17.6. The area of the hole will increase proportionally. This increase in the area of the hole might seem surprising. However, you can convince yourself that the area of the hole will increase when the plate undergoes thermal expansion by looking at Figure 17.18. The plate with a hole at $T = 20$ °C is shown in part (a). The same plate scaled up by 5% in all dimensions is shown in part (b). The dashed circle in the hole in the plate in (b) is the same size as the hole in the original plate. Clearly, the hole in (b) is larger than the hole in (a). The area of the hole at $T = 20.0$ °C is $A = \pi R^2$. Equation 17.6 gives the change in the area of the hole:

$$\Delta A = 2\alpha A \Delta T,$$

where α is the linear expansion coefficient for brass and ΔT is the change in temperature of the brass plate.

SIMPLIFY Using $A = \pi R^2$ in equation 17.6, we have the change in the area of the hole:

$$\Delta A = 2\alpha\left(\pi R^2\right)\Delta T.$$

Remembering that $R = d/2$, we get

$$\Delta A = 2\alpha\pi\left(\tfrac{1}{2}d\right)^2\Delta T = \frac{\pi\alpha d^2\Delta T}{2}.$$

CALCULATE From Table 17.2, the linear expansion coefficient of brass is $\alpha = 19 \cdot 10^{-6}$ °C^{-1}. Thus, the change in the area of the hole is

$$\Delta A = \frac{\pi\left(19 \cdot 10^{-6}\ °\text{C}^{-1}\right)(0.0254\ \text{m})^2\left(220.0\ °\text{C} - 20.0\ °\text{C}\right)}{2} = 3.85098 \cdot 10^{-6}\ \text{m}^2.$$

ROUND We report our result to two significant figures:

$$\Delta A = 3.9 \cdot 10^{-6}\ \text{m}^2.$$

DOUBLE-CHECK From everyday experience with objects that are heated and cooled, we know that the relative changes in area are not very big. Since the original area of the hole is $A = \pi d^2/4 = \pi(0.0254\ \text{m})^2/4 = 5.07 \cdot 10^{-4}\ \text{m}^2$, we obtain for the fractional change $\Delta A/A = (3.9 \cdot 10^{-6}\ \text{m}^2)/(5.07 \cdot 10^{-4}\ \text{m}^2) = 7.6 \cdot 10^{-3}$, or less than 0.8%. Thus, the magnitude of our answer seems in line with physical intuition.

The change in the radius of the hole as the temperature increases is given by

$$\Delta R = \alpha R \Delta T = \left(19 \cdot 10^{-6}\ °\text{C}^{-1}\right)\left(\frac{0.0254\ \text{m}}{2}\right)(200\ °\text{C}) = 4.83 \cdot 10^{-5}\ \text{m}.$$

For that change in radius, the increase in the area of the hole is

$$\Delta A = \Delta\left(\pi R^2\right) = 2\pi R\Delta R = 2\pi\left(\frac{0.0254\ \text{m}}{2}\right)\left(4.8 \cdot 10^{-5}\ \text{m}\right) = 3.85 \cdot 10^{-6}\ \text{m}^2,$$

which agrees within rounding error with our result. Thus, our answer seems reasonable.

Volume Expansion

Now let's consider the change in volume of an object with a change in temperature. For a cube with side L, the volume is given by $V = L^3$. Taking the differential of both sides of this equation, we get $dV = 3L^2 dL$. Making the approximations $\Delta V = dV$ and $\Delta L = dL$, we can write $\Delta V = 3L^2\Delta L$. Then, using equation 17.5, we obtain

$$\Delta V = 3L^2\left(\alpha L\Delta T\right) = 3\alpha V\Delta T. \tag{17.7}$$

Because the change in volume with a change in temperature is often of interest, it is convenient to define the **volume expansion coefficient:**

$$\beta = 3\alpha. \tag{17.8}$$

Thus, we can rewrite equation 17.7 as

$$\Delta V = \beta V\Delta T. \tag{17.9}$$

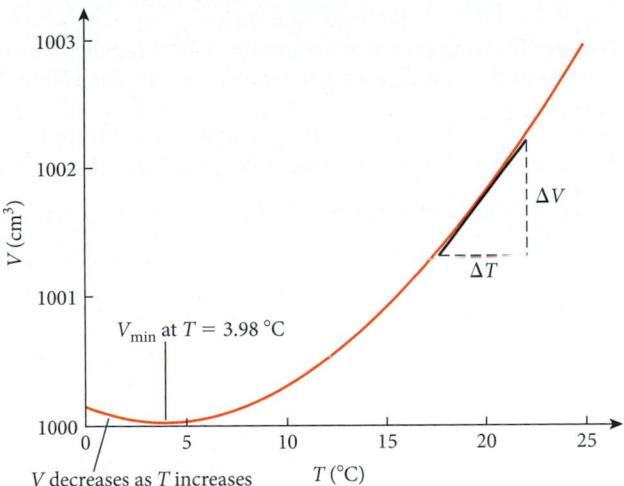

FIGURE 17.19 The dependence of the volume of 1 kg of water on temperature.

Although a cube was used to derive equation 17.9, the equation can be applied to a change in the volume of any shape. Some typical volume expansion coefficients are listed in Table 17.3.

Equation 17.9 applies to the thermal expansion of most solids and liquids. However, it does not describe the thermal expansion of water. Between 0 °C and about 4 °C, water contracts as the temperature increases (Figure 17.19). Water with a temperature of 4 °C is denser than water with a temperature just below 4 °C. This property of water has a dramatic effect on the way a lake freezes in the winter. As the air temperature drops from warm summer to cold winter temperatures, the water in a lake cools from the surface down. The cooler, denser water sinks to the bottom of the lake. However, as the temperature of the water on the surface falls below 4 °C, the downward motion ceases, and the cooler water remains at the surface of the lake, with the denser, warmer water below. The top layer eventually cools to 0 °C and then freezes. Ice is less dense than water, so the ice floats on the water. This newly formed layer of ice acts as insulation, which slows the freezing of the rest of the water in the lake. If water had the same thermal expansion properties as other common materials, instead of freezing from the top down, the lake would freeze from the bottom up, with warmer water remaining at the surface of the lake and cooler water sinking to the bottom. This would mean that lakes would freeze solid more often, and any forms of life in them that could not exist in ice would not survive the winter.

In addition, you can see from Figure 17.19 that the volume of a given amount of water never depends linearly on the temperature. However, the linear dependence of the volume of water on temperature can be approximated by considering a small temperature interval. The slope of the volume-temperature curve is $\Delta V/\Delta T$, so we can extract an effective volume expansion coefficient for small temperature changes. For example, the volume expansion coefficient for water at six different temperatures is given in Table 17.3; note that at 1 °C, $\beta = -47.8 \cdot 10^{-6}\,°C^{-1}$, which means that the volume of water will decrease as the temperature is increased.

Table 17.3	Volume Expansion Coefficients for Some Common Liquids
Material	$\beta\ (10^{-6}\,°C^{-1})$
Mercury	181
Gasoline	950
Kerosene	990
Ethyl alcohol	750
Water (1 °C)	−47.8
Water (4 °C)	0
Water (7 °C)	45.3
Water (10 °C)	87.5
Water (15 °C)	151
Water (20 °C)	207

EXAMPLE 17.3 | Thermal Expansion of Gasoline

You pull your car into a gas station on a hot summer day, when the air temperature is 40 °C. You fill your empty 55-L gas tank with gasoline that comes from an underground storage tank where the temperature is 12 °C. After paying for the gas, you decide to walk to the restaurant next door and have lunch. Two hours later, you come back to your car, realize that you left the cap off the gas tank, and discover that gasoline has spilled out of the gas tank onto the ground.

PROBLEM
How much gasoline has spilled?

– Continued

Concept Check 17.5

You have a metal cube, which you heat. After heating, the area of one of the cube's surfaces has increased by 0.02%. Which statement about the volume of the cube after heating is correct?

a) It has decreased by 0.02%.

b) It has increased by 0.02%.

c) It has increased by 0.01%.

d) It has increased by 0.03%.

e) Not enough information is given to determine the volume change.

SOLUTION

We know the following: The temperature of the gasoline you put in the tank starts out at 12 °C. The gasoline warms up to the outside air temperature of 40 °C. The volume expansion coefficient of gasoline is $950 \cdot 10^{-6}$ °C^{-1}.

While you were gone, the temperature of the gasoline changed from 12 °C to 40 °C. Using equation 17.9, we can find the change in volume of the gasoline as the temperature increases:

$$\Delta V = \beta V \Delta T = \left(950 \cdot 10^{-6} \text{ °C}^{-1}\right)\left(55 \text{ L}\right)\left(40 \text{ °C} - 12 \text{ °C}\right) = 1.5 \text{ L}.$$

Thus, the volume of the gasoline increases by 1.5 L as the temperature of the gasoline is raised from 12 °C to 40 °C. The gas tank was full when the temperature of the gasoline was 12 °C, so this excess volume spills out of the tank onto the ground.

17.5 Surface Temperature of the Earth

A report on daily surface temperatures is part of every weather report in newspapers and TV and radio newscasts. It is clear that it is usually colder at night than during the day, colder in the winter than in the summer, and hotter close to the Equator than near the poles. A current topic of intense discussion is whether the temperature of Earth is rising. A conclusive answer to this question requires data giving appropriate averages. The first average that is useful is over time. Figure 17.20 is a plot of the surface temperature of the Earth, time-averaged over one month (June 1992).

The time-averaged values of the temperature over the entire surface of Earth are obtained by systematically taking temperature measurements over the surface of the Earth, including the oceans. These measurements are then corrected for any systematic biases, such as the fact that many temperature-measuring stations are located in populated areas and many sparsely populated areas have few temperature measurements. After all of the corrections are taken into account, the result is the average surface temperature of Earth in a given year. The current year-round average surface temperature of Earth is approximately 287.7 K (14.6 °C). In Figure 17.21, this average global temperature is plotted for the years 1880 to 2011. You can see that since around 1900, the temperature has been increasing with time, indicating global warming. The blue horizontal line in the graph represents the average global temperature for the 20th century, 13.9 °C. The figure displays data compiled by an agency of the U.S. Department of Commerce. Agencies in other countries that have compiled similar data sets have arrived at slightly different average temperature values, mainly because of variations in averaging procedures. However, *all* estimates of the change in the average global temperature over time are virtually identical. For the past few decades, the Earth has been warming up at a rate of approximately 0.2 °C per decade.

Several models predict that the average global surface temperature of Earth will continue to increase. The magnitude of the increase in average global temperature over the past 155 years is around 1 °C, which does not seem like a large increase. However, combined with predicted future increases, it is sufficient to cause observable effects, such as the raising of ocean water levels, the disappearance of the Arctic ice cover in summers, climate shifts, and increased severity of storms and droughts around the world.

Figure 17.22 shows a record of the difference between the current average annual surface temperature in Antarctica and the average annual surface temperature over the last 420,000 years, which was determined from ice cores. Note that past temperatures were evaluated from measurements of carbon dioxide in the ice cores, that their values may be somewhat model-dependent, and that the resulting temperature differences may be significantly larger than the corresponding

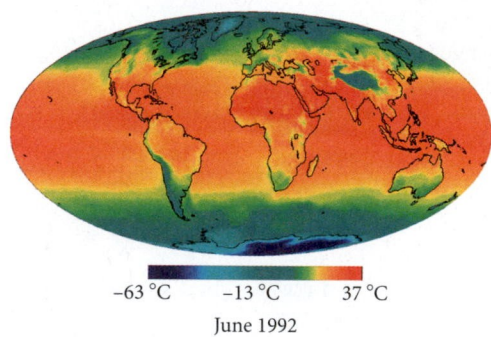

−63 °C   −13 °C   37 °C

June 1992

FIGURE 17.20 Time-averaged surface temperature of the Earth in June 1992. The colors represent a range of temperatures from −63 °C to +37 °C.

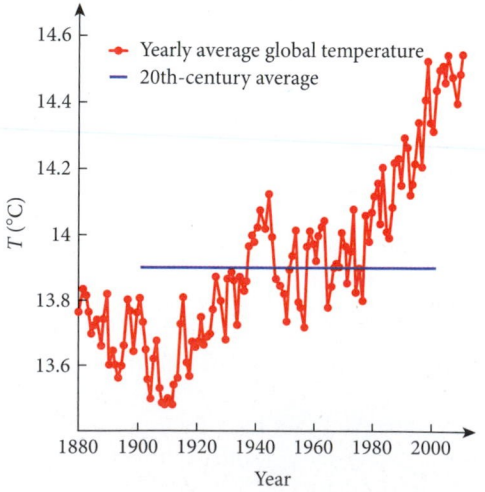

FIGURE 17.21 Annual average global surface temperature from 1880 to 2011 as measured by thermometers on land and in the ocean (red histogram). The blue horizontal line represents the average global temperature for the 20th century, 13.9 °C. (Source: Data compiled by the National Climatic Data Center, National Oceanic and Atmospheric Administration, U.S. Department of Commerce.)

global temperature differences. Several distinct periods are apparent in Figure 17.22. An interval of time when the temperature difference is around –7 °C corresponds to a period in which ice sheets covered parts of North America and Europe and is termed a *glacial period.* The last glacial period ended about 10,000 years ago. The warmer periods between glacial periods, called *interglacial periods,* correspond to temperature differences of around zero. In Figure 17.22, four glacial periods are visible, going back 400,000 years. An attempt to relate these temperature differences to differences in the heat received from the Sun due to variations in the Earth's orbit and the orientation of its rotational axis is known as the *Milankovitch Hypothesis.* However, these variations cannot account for all of the observed temperature differences.

The current warm, interglacial period began about 10,000 years ago and seems to be slightly cooler than previous interglacial periods. Previous interglacial periods have lasted from 10,000 to 25,000 years. However, human activities, such as the burning of fossil fuels and the resulting greenhouse effect (more on this in Chapter 18), are influencing the average global temperature. Models predict that the effect of these activities will be to warm the Earth, at least for the next several hundred years.

One effect of the warming of Earth's surface is a rise in sea level. Sea level has risen 120 m since the peak of the last glacial period, about 20,000 years ago, as a result of the melting of the glaciers that covered large areas of land. The melting of large amounts of ice resting on solid ground is the largest potential contributor to a further rise in sea level. For example, if all the ice in Antarctica melted, sea level would rise 61 m. If all the ice in Greenland melted, the rise in sea level would be 7 m. However, it would take several centuries for these large deposits of ice to melt completely, even if pessimistic predictions of climate models are correct. The rise in sea level due to thermal expansion is small compared with that due to the melting of large glaciers. The current rate of the rise in sea level is 3.2±0.4 mm/yr, as measured by radar altimeters on the TOPEX and Jason satellites. By the end of the 21st century, sea level is projected to rise by 1.0±0.3 m.

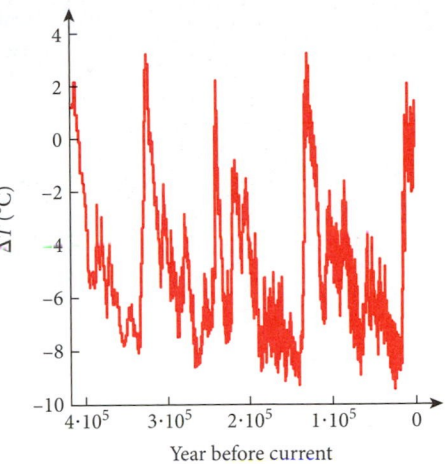

FIGURE 17.22 Average annual surface temperature of Antarctica in the past, extracted from carbon dioxide content of ice cores, relative to the present value.

Self-Test Opportunity 17.3

Identify the years corresponding to glacial and interglacial periods in Figure 17.22.

EXAMPLE 17.4 — Rise in Sea Level Due to Thermal Expansion of Water

The rise in the level of the Earth's oceans is of current concern. Oceans cover $3.6 \cdot 10^8$ km^2, slightly more than 70% of Earth's surface area. The average ocean depth is 3790 m. The surface ocean temperature varies widely, between 35 °C in the summer in the Persian Gulf and –2 °C in the Arctic and Antarctic regions. However, even if the ocean surface temperature exceeds 20 °C, the water temperature rapidly falls off as a function of depth and approaches 4 °C at a depth of 1000 m (Figure 17.23). The global average temperature of all seawater is approximately 3 °C. Table 17.3 lists a volume expansion coefficient of zero for water at a temperature of 4 °C. Thus, it is safe to assume that the volume of ocean water changes very little at a depth greater than 1000 m. For the top 1000 m of ocean water, let's assume a global average temperature of 10.0 °C and calculate the effect of thermal expansion.

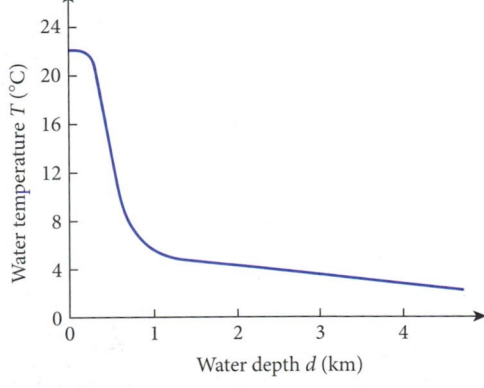

FIGURE 17.23 Average ocean water temperature as a function of depth below the surface.

PROBLEM

By how much would sea level change, solely as a result of the thermal expansion of water, if the water temperature of all the oceans increased by $\Delta T = 1.0$ °C?

SOLUTION

The volume expansion coefficient of water at 10.0 °C is $\beta = 87.5 \cdot 10^{-6}$ °C^{-1} (from Table 17.3), and the volume change of the oceans is given by equation 17.9, $\Delta V = \beta V \Delta T$, or

$$\frac{\Delta V}{V} = \beta \Delta T. \qquad \text{(i)}$$

We can express the total surface area of the oceans as $A = (0.7)4\pi R^2$, where R is the radius of Earth and the factor 0.7 reflects the fact that about 70% of the surface of the sphere is covered by water. We assume that the surface area of the oceans increases only minutely from the

– Continued

water moving up the shores and neglect the change in surface area due to this effect. Then, essentially all of the change of the oceans' volume will result from the change in the depth, and we can write

$$\frac{\Delta V}{V} = \frac{\Delta d \cdot A}{d \cdot A} = \frac{\Delta d}{d}. \tag{ii}$$

Combining equations (i) and (ii), we obtain an expression for the change in depth:

$$\frac{\Delta d}{d} = \beta \Delta T \Rightarrow \Delta d = \beta d \Delta T.$$

Inserting the numerical values, $d = 1000$ m, $\Delta T = 1.0\ ^\circ\text{C}$, and $\beta = 87.5 \cdot 10^{-6}\ ^\circ\text{C}^{-1}$, we obtain

$$\Delta d = (1000\ \text{m})(87.5 \cdot 10^{-6}\ ^\circ\text{C}^{-1})(1.0\ ^\circ\text{C}) = 9\ \text{cm}.$$

So, for every increase of the average ocean temperature by 1 °C, the sea level will rise by 9 cm (3.5 in). This rise is smaller than the anticipated rise due to the melting of the ice cover on Greenland or Antarctica but will contribute to the problem of coastal flooding.

17.6   Temperature of the Universe

In 1965, while working on an early radio telescope, Arno Penzias and Robert Wilson discovered the **cosmic microwave background radiation.** They detected "noise," or "static," that seemed to come from all directions in the sky. Penzias and Wilson figured out what was producing this noise (which earned them the 1978 Nobel Prize): It was electromagnetic radiation left over from the Big Bang, which occurred 13.7 billion years ago. (We'll discuss electromagnetic radiation in Chapter 31.) It is astonishing to realize that an "echo" of the Big Bang still reverberates in "empty" intergalactic space after such a long time. The wavelength of the cosmic background radiation is similar to the wavelength of the electromagnetic radiation used in a microwave oven. An analysis of the distribution of wavelengths of this radiation led to the deduction that the background temperature of the universe is 2.725 K (exactly how this analysis was done is described in Chapter 36 on quantum physics). George Gamov had already predicted a cosmic background temperature of 2.7 K in 1948, when it was not yet clear that the Big Bang is a scientifically established fact.

In 2001, the Wilkinson Microwave Anisotropy Probe (WMAP) satellite measured variations in the background temperature of the universe. This mission followed the successful Cosmic Background Explorer (COBE) satellite, launched in 1989, which resulted in the 2006 Nobel Prize for Physics being awarded to John Mather and George Smoot. The COBE and WMAP missions found that the cosmic microwave background radiation was very uniform in all directions, but small differences in the temperature were superimposed on the smooth background. The WMAP results for the background temperature in all directions are shown in Figure 17.24. The effects of the Milky Way galaxy have been subtracted. You can see that the variation in the background temperature in the universe is very small, since $\pm 200\ \mu\text{K}/2.725\ \text{K} = \pm 7.3 \cdot 10^{-5}$. From interpretation of these results and other observations, scientists deduced that the age of the universe is 13.7 billion years, with a margin

FIGURE 17.24 The temperature of the cosmic microwave background radiation everywhere in the universe. The colors represent a range of temperatures from 200 μK below to 200 μK above the average temperature of the cosmic microwave background radiation, which has an average temperature of 2.725 K.

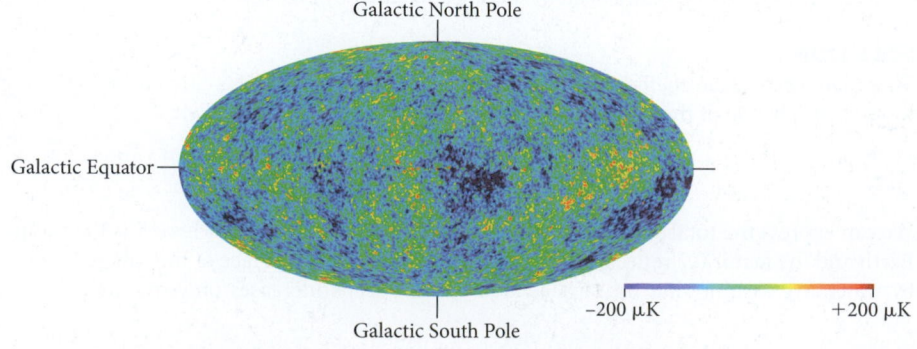

of error of less than 1%. In addition, scientists were able to deduce that the universe is composed of 4% ordinary matter, 23% dark matter, and 73% dark energy. Dark matter is matter that exerts an observable gravitational pull but seems to be invisible, as discussed in Chapter 12. Dark energy seems to be causing the expansion of the universe to speed up. (Chapter 39, on particle physics and cosmology, covers these phenomena in detail.) Both dark matter and dark energy are under intense current investigation, and understanding of them should improve in the next decade.

WHAT WE HAVE LEARNED | EXAM STUDY GUIDE

- The three commonly used temperature scales are the Fahrenheit scale, the Celsius scale, and the Kelvin scale.

- The Fahrenheit scale is used in the United States and defines the freezing point of water as 32 °F and the boiling point of water as 212 °F.

- The Celsius scale defines the freezing point of water as 0 °C and the boiling point of water as 100 °C.

- The Kelvin scale defines 0 K as absolute zero and the freezing point of water as 273.15 K. The size of the kelvin and the Celsius degree are the same.

- The conversion from °F to °C is given by $T_C = \frac{5}{9}(T_F - 32\ °F)$.

- The conversion from °C to °F is given by $T_F = \frac{9}{5}T_C + 32\ °C$.

- The conversion from °C to K is given by $T_K = T_C + 273.15\ °C$.

- The conversion from K to °C is given by $T_C = T_K - 273.15\ K$.

- The change in length, ΔL, of an object of length L as the temperature changes by ΔT is given by $\Delta L = \alpha L \Delta T$, where α is the linear expansion coefficient.

- The change in volume, ΔV, of an object with volume V as the temperature changes by ΔT is given by $\Delta V = \beta V \Delta T$, where β is the volume expansion coefficient.

ANSWERS TO SELF-TEST OPPORTUNITIES

17.1 Take $T = T_F = T_C$, and solve for T:

$$T = \frac{9}{5}T + 32$$

$$-\frac{4}{5}T = 32$$

$$T = -40\ °C\ or\ °F.$$

17.2 Use an ice-and-water mixture, which is at 0 °C, and boiling water, which is at 100 °C. Take the measurements and mark the corresponding places on the uncalibrated thermometer.

17.3 The glacial periods in years ago:

12,000 to 120,000

150,000 to 230,000

250,000 to 310,000

330,000 to 400,000.

The interglacial periods in years ago:

0 to 12,000

110,000 to 130,000

230,000 to 250,000

310,000 to 330,000

400,000 to 410,000.

PROBLEM-SOLVING GUIDELINES

1. Be consistent in your use of Celsius, Kelvin, or Fahrenheit temperatures. For calculating a temperature change, a Celsius degree is equal to a kelvin. If you need to find a temperature value, however, remember which temperature scale is asked for.

2. Thermal expansion has an effect similar to enlarging a photograph: All parts of the picture get expanded in the same way. Remember that a hole in an object expands with increasing temperature just as the object itself does. Sometimes you can simplify a problem involving volume expansion to one dimension and then use linear expansion coefficients. Be alert for these kinds of situations.

MULTIPLE-CHOICE QUESTIONS

17.1 Two mercury-expansion thermometers have identical reservoirs and cylindrical tubes made of the same glass but of different diameters. Which of the two thermometers can be calibrated to a better resolution?

a) The thermometer with the smaller-diameter tube will have better resolution.

b) The thermometer with the larger-diameter tube will have better resolution.

c) The diameter of the tube is irrelevant; it is only the volume expansion coefficient of mercury that matters.

d) Not enough information is given to tell.

17.2 For a class demonstration, your physics instructor uniformly heats a bimetallic strip that is held in a horizontal orientation. As a result, the bimetallic strip bends upward. This tells you that the coefficient of linear thermal expansion for metal T, on the top is _____ that of metal B, on the bottom.

a) smaller than   b) larger than   c) equal to

17.3 Two solid objects, A and B, are in contact. In which case will thermal energy transfer from A to B?

a) A is at 20 °C, and B is at 27 °C.

b) A is at 15 °C, and B is at 15 °C.

c) A is at 0 °C, and B is at –10 °C.

17.4 Which of the following bimetallic strips will exhibit the greatest sensitivity to temperature changes? That is, which one will bend the most as temperature increases?

a) copper and steel   d) aluminum and brass

b) steel and aluminum   e) copper and brass

c) copper and aluminum

17.5 The background temperature of the universe is

a) 6000 K.   c) 3 K.   e) 0 K.

b) 288 K.   d) 2.725 K.

17.6 Which air temperature feels coldest?

a) –40 °C   c) 233 K

b) –40 °F   d) All three are equal.

17.7 At what temperature do the Celsius and Fahrenheit temperature scales have the same numeric value?

a) –40 degrees   c) 40 degrees

b) 0 degrees   d) 100 degrees

17.8 The city of Yellowknife in the Northwest Territories of Canada is on the shore of Great Slave Lake. The average high temperature in July is 21 °C, and the average low in January is –31 °C. Great Slave Lake has a volume of 2090 km^3 and is the deepest lake in North America, with a depth of 614 m. What is the temperature of the water at the bottom of Great Slave Lake in January?

a) –31 °C   c) 0 °C   e) 32 °C

b) –10 °C   d) 4 °C

17.9 Which object has the higher temperature after being left outside for an entire winter night: a metal door knob or a rug?

a) The metal door knob has the higher temperature.

b) The rug has the higher temperature.

c) Both have the same temperature.

d) It depends on the outside temperature.

17.10 A concrete section of a bridge has a length of 10.0 m at 10.0 °C. By how much will the length of the section increase if its temperature increases to 40.0 °C?

a) 0.025 cm   c) 0.075 cm   e) 0.45 cm

b) 0.051 cm   d) 0.22 cm

17.11 Which of the following temperatures corresponds to the boiling point of water?

a) 0 °C   c) 0 K   e) 100 °F

b) 100 °C   d) 100 K

17.12 Which of the following temperatures corresponds to the freezing point of water?

a) 0 K   c) 100 °C   e) 100 °F

b) 0 °C   d) 100 K

17.13 The Zeroth Law of Thermodynamics tells us that

a) there is a temperature of absolute zero.

b) the freezing point of water is 0 °C.

c) two systems cannot be in thermal equilibrium with a third system.

d) thermal energy is conserved.

e) it is possible to construct a thermometer to measure the temperature of any system.

CONCEPTUAL QUESTIONS

17.14 A common way of opening a tight lid on a glass jar is to place it under warm water. The thermal expansion of the metal lid is greater than that of the glass jar; thus, the space between the two expands, making it easier to open the jar. Will this work for a metal lid on a container of the same kind of metal?

17.15 Would it be possible to have a temperature scale defined in such a way that the hotter an object or system got, the lower (less positive or more negative) its temperature was?

17.16 The solar corona has a temperature of about $1 \cdot 10^6$ K. However, a spaceship flying in the corona will not be burned up. Why is this?

17.17 Explain why it might be difficult to weld aluminum to steel or to weld any two unlike metals together.

17.18 Two solid objects are made of different materials. Their volumes and volume expansion coefficients are V_1 and V_2 and β_1 and β_2, respectively. It is observed that during a temperature change of ΔT, the volume of each object changes by the same amount. If $V_1 = 2V_2$, what is the ratio of the volume expansion coefficients?

17.19 Some textbooks use the unit K^{-1} rather than °C^{-1} for values of the linear expansion coefficient; see Table 17.2. How will the numerical values of the coefficient differ if expressed in K^{-1}?

17.20 You are outside on a hot day, with the air temperature at T_o. Your sports drink is at a temperature T_d in a sealed plastic bottle. There are a few remaining ice cubes in the sports drink, which are at a temperature T_i, but they are melting fast.

a) Write an inequality expressing the relationship among the three temperatures.

b) Give reasonable values for the three temperatures in degrees Celsius.

17.21 The Rankine temperature scale is an absolute temperature scale that uses Fahrenheit degrees; that is, temperatures are measured in Fahrenheit degrees, starting at absolute zero. Find the relationships between temperature values on the Rankine scale and those on the Fahrenheit, Kelvin, and Celsius scales.

17.22 The Zeroth Law of Thermodynamics forms the basis for the definition of temperature with regard to thermal energy. But the concept of temperature is used in other areas of physics. In a system with energy levels, such as electrons in an atom or protons in a magnetic field, the population of a level with energy E is proportional to the factor $e^{-E/k_B T}$, where T is the absolute temperature of the system and $k_B = 1.381 \cdot 10^{-23}$ J/K is Boltzmann's constant. In a two-level system with the levels' energies differing by ΔE, the ratio of the populations of the higher-energy and lower-energy levels is $p_{high}/p_{low} = e^{-\Delta E/k_B T}$. Such a system can have an infinite or even a negative absolute temperature. Explain the meaning of such temperatures.

17.23 Suppose a bimetallic strip is constructed of two strips of metals with linear expansion coefficients α_1 and α_2, where $\alpha_1 > \alpha_2$.

a) If the temperature of the bimetallic strip is reduced by ΔT, which way will the strip bend (toward the side made of metal 1 or the side made of metal 2)? Briefly explain.

b) If the temperature is increased by ΔT, which way will the strip bend?

17.24 For food storage, what is the advantage of placing a metal lid on a glass jar? (*Hint:* Why does running the metal lid under hot water for a minute help you open such a jar?)

17.25 A solid cylinder and a cylindrical shell, of identical outer radius and length and made of the same material, experience the same temperature increase, ΔT. Which of the two will expand to a larger outer radius?

EXERCISES

A blue problem number indicates a worked-out solution is available in the Student Solutions Manual. One • and two •• indicate increasing level of problem difficulty.

Sections 17.1 and 17.2

17.26 Express each of the following temperatures in degrees Celsius and in kelvins.

a) –19 °F   b) 98.6 °F   c) 52 °F

17.27 One thermometer is calibrated in degrees Celsius, and another in degrees Fahrenheit. At what temperature is the reading on the thermometer calibrated in degrees Celsius three times the reading on the other thermometer?

17.28 During the summer of 2007, temperatures as high as 47 °C were recorded in Southern Europe. The highest temperature ever recorded in the United States was 134 °F (at Death Valley, California, in 1913). What is the difference between these two temperatures, in degrees Celsius?

17.29 The lowest air temperature recorded on Earth is –129 °F in Antarctica. Convert this temperature to the Celsius scale.

17.30 What air temperature will feel twice as warm as 0 °F?

17.31 A piece of dry ice (solid carbon dioxide) sitting in a classroom has a temperature of approximately –79 °C.

a) What is this temperature in kelvins?

b) What is this temperature in degrees Fahrenheit?

17.32 In 1742, the Swedish astronomer Anders Celsius proposed a temperature scale on which water boils at 0.00 degrees and freezes at 100. degrees. In 1745, after the death of Celsius, Carolus Linnaeus (another Swedish scientist) reversed those standards, yielding the scale that is most commonly used today. Find room temperature (77.0 °F) on Celsius's original temperature scale.

17.33 At what temperature do the Kelvin and Fahrenheit scales have the same numeric value?

Section 17.4

17.34 How does the density of copper that is just above its melting temperature of 1356 K compare to that of copper at room temperature?

17.35 The density of steel is 7800. kg/m^3 at 20.0 °C. Find the density at 100.0 °C.

17.36 Two cubes with sides of length 100.0 mm fit in a space that is 201.0 mm wide, as shown in the figure. One cube is made of aluminum, and the other is made of brass. What temperature increase is required for the cubes to close the gap?

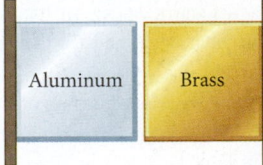

17.37 A brass piston ring is to be fitted onto a piston by first heating up the ring and then slipping it over the piston. The piston ring has an inner diameter of 10.00 cm and an outer diameter of 10.20 cm. The piston has an outer diameter of 10.10 cm, and a groove for the piston ring has an outer diameter of 10.00 cm. To what temperature must the piston ring be heated so that it will slip onto the piston?

17.38 An aluminum sphere of radius 10.0 cm is heated from 100.0 °F to 200.0 °F. Find (a) the volume change and (b) the radius change.

17.39 Steel rails for a train track are laid in a region subject to extremes of temperature. The distance from one juncture to the next is 5.2000 m, and the cross-sectional area of the rails is 60.0 cm^2. If the rails touch each other without buckling at the maximum temperature, 50.0 °C, how much space will there be between the rails at –10.0 °C?

17.40 Even though steel has a relatively low linear expansion coefficient ($\alpha_{steel} = 13 \cdot 10^{-6}$ °C^{-1}), the expansion of steel railroad tracks can potentially create significant problems on very hot summer days. To accommodate for the thermal expansion, a gap is left between consecutive sections of the track. If each section is 25.0 m long at 20.0 °C and the gap between sections is 10.0 mm wide, what is the highest temperature the tracks can take before the expansion creates compressive forces between sections?

17.41 A medical device used for handling tissue samples has two metal screws, one 20.0 cm long and made from brass ($\alpha_b = 18.9 \cdot 10^{-6}$ °C^{-1}) and the other 30.0 cm long and made from aluminum ($\alpha_a = 23.0 \cdot 10^{-6}$ °C^{-1}). A gap of 1.00 mm exists between the ends of the screws at 22.0 °C. At what temperature will the two screws touch?

•17.42 You are designing a precision mercury thermometer based on the thermal expansion of mercury ($\beta = 1.81 \cdot 10^{-4}$ °C^{-1}), which causes the mercury to expand up a thin capillary as the temperature increases. The equation for the change in volume of the mercury as a function of temperature is $\Delta V = \beta V_0 \Delta T$, where V_0 is the initial volume of the mercury and ΔV is the change in volume due to a change in temperature, ΔT. In response to a temperature change of 1.00 °C, the column of mercury in your precision thermometer should move a distance $D = 1.00$ cm up a cylindrical capillary of radius $r = 0.100$ mm. Determine the initial volume of mercury that allows this change. Then find the radius of a spherical bulb that contains this volume of mercury.

•17.43 On a hot summer day, a cubical swimming pool is filled to within 1.00 cm of the top with water at 21.0 °C. When the water warms to 37.0 °C, the pool is completely full. What is the depth of the pool?

•17.44 A steel rod of length 1.00 m is attached to the end of an aluminum rod of length 2.00 m (lengths measured at 22.0 °C). The combination rod is then heated to 200. °C. What is the change in length of the combination rod at 200. °C?

•17.45 A clock based on a simple pendulum is situated outdoors in Anchorage, Alaska. The pendulum consists of a mass of 1.00 kg that is hanging from a thin brass rod that is 2.000 m long. The clock is calibrated perfectly during a summer day with an average temperature of 25.0 °C. During the winter, the average temperature over one 24-h period is –20.0 °C. Find the time elapsed for that period according to the simple pendulum clock.

•17.46 In a thermometer manufacturing plant, a type of mercury thermometer is built at room temperature (20.0 °C) to measure temperatures in the 20.0 °C to 70.0 °C range, with a 1.00-cm³ spherical reservoir at the bottom and a 0.500-mm inner diameter expansion tube. The wall thickness of the reservoir and tube is negligible, and the 20.0 °C mark is at the junction between the spherical reservoir and the tube. The tubes and reservoirs are made of fused silica, a transparent *glass* form of SiO_2 that has a very low linear expansion coefficient ($\alpha = 0.400 \cdot 10^{-6}$ °C^{-1}). By mistake, the material used for one batch of thermometers was *quartz,* a transparent *crystalline* form of SiO_2 with a much higher linear expansion coefficient ($\alpha = 12.3 \cdot 10^{-6}$ °C^{-1}). Will the manufacturer have to scrap the batch, or will the thermometers work fine, within the expected uncertainty of 5% in reading the temperature? The volume expansion coefficient of mercury is $\beta = 181 \cdot 10^{-6}$ °C^{-1}.

•17.47 The ends of the two rods shown in the figure are separated by 5.00 mm at 25.0 °C. The left-hand rod is brass and 1.00 m long; the right-hand rod is steel and 1.00 m long. Assuming that the outside ends of both rods rest firmly against rigid supports, at what temperature will the inside ends of the rods just touch?

5.00 mm —▶| |◀—

•17.48 The figure shows a temperature-compensated pendulum in which lead and steel rods are arranged so that the pendulum's length is unaffected by changes in the temperature. Determine the length L of the two lead bars.

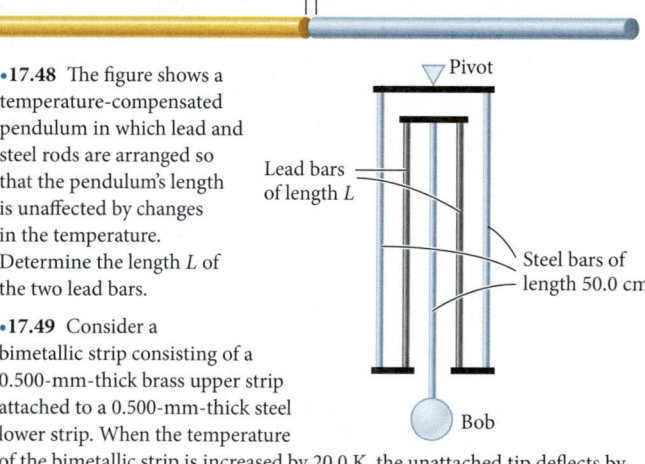

▽ Pivot

Lead bars of length L

Steel bars of length 50.0 cm

Bob

•17.49 Consider a bimetallic strip consisting of a 0.500-mm-thick brass upper strip attached to a 0.500-mm-thick steel lower strip. When the temperature of the bimetallic strip is increased by 20.0 K, the unattached tip deflects by 3.00 mm from its original straight position, as shown in the figure. What is the length of the strip at its original position?

3.00 mm

•17.50 Thermal expansion seems like a small effect, but it can engender tremendous, often damaging, forces. For example, steel has a linear expansion coefficient of $\alpha = 1.2 \cdot 10^{-5}$ °C^{-1} and a bulk modulus of $B = 160$ GPa. Calculate the pressure engendered in steel by a 1.0 °C temperature increase if no expansion is permitted.

•17.51 An iron horseshoe at room temperature is dunked into a cylindrical tank of water (radius of 10.0 cm) and causes the water level to rise by 0.250 cm. When the horseshoe is heated in the blacksmith's stove from room temperature to a temperature of $7.00 \cdot 10^2$ K, worked into its final shape, and then dunked back into the water, how much does the water level rise above the "no horseshoe" level (ignore any water that evaporates as the horseshoe enters the water)? *Note:* The linear expansion coefficient for iron is roughly that of steel: $11.0 \cdot 10^{-6}$ °C^{-1}.

•17.52 A clock has an aluminum pendulum with a period of 1.000 s at 20.0 °C. Suppose the clock is moved to a location where the average temperature is 30.0 °C. Determine (a) the new period of the clock's pendulum and (b) how much time the clock will lose or gain in 1 week.

•17.53 Using techniques similar to those that were originally developed for miniaturized semiconductor electronics, scientists and engineers create Micro-Electro-Mechanical Systems (MEMS). An example is an electrothermal actuator that is driven by heating its different parts using an electrical current. The device is used to position 125-μm-diameter optical fibers with submicron resolution and consists of thin and thick silicon arms connected in the shape of a U, as shown in the figure. The arms are not

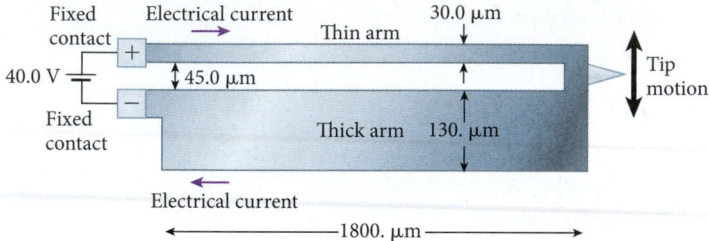

No current, both arms at 20.0 °C

Fixed contact

Electrical current →

Thin arm

30.0 μm

40.0 V

45.0 μm

Tip motion

Fixed contact

Thick arm   130. μm

Electrical current ←

◀— 1800. μm —▶

Human hair

~100 μm

attached to the substrate under the device but are free to move, whereas the electrical contacts (marked + and – in the figure) are attached to the substrate and cannot move. The thin arm is 30.0 μm wide, and the thick arm is 130. μm wide. Both arms are 1800. μm long. Electrical current flows through the arms, causing them to heat up. Although the same current flows through both arms, the thin arm has a greater electrical resistance than the thick arm and therefore dissipates more electrical power and gets substantially hotter. When current is made to flow through the arms, the thin arm reaches a temperature of 400. °C, and the thick arm reaches a temperature of 200. °C. Assume that the temperature in each arm is constant along the entire length of that arm (strictly speaking, this is not the case) and that the two arms remain parallel and bend only in the plane of the page at higher temperatures. How much and in which direction will the tip move? The linear expansion coefficient for this silicon is $3.20 \cdot 10^{-6}$ °C^{-1}.

•17.54 Another MEMS device, used for the same purpose as the one in Problem 17.53, has a different design. This electrothermal actuator consists of a thin V-shaped silicon beam, as shown in the figure. The beam is not attached to the substrate under the device but is free to move, whereas the electrical contacts (marked + and – in the figure) are attached to the substrate and cannot move. The beam spans a gap between the electrical contacts that is 1800 μm wide, and the two halves of the beam slant up from horizontal by 0.10 rad. Electrical current flows through the beam, causing it to heat up. When current is made to flow across the beam, the beam reaches a temperature of 500. °C. Assume that the temperature is constant along the entire length of the beam (strictly speaking, this is not the case). How much and in which direction will the tip move? The linear expansion coefficient for silicon is $3.2 \cdot 10^{-6}$ °C^{-1}.

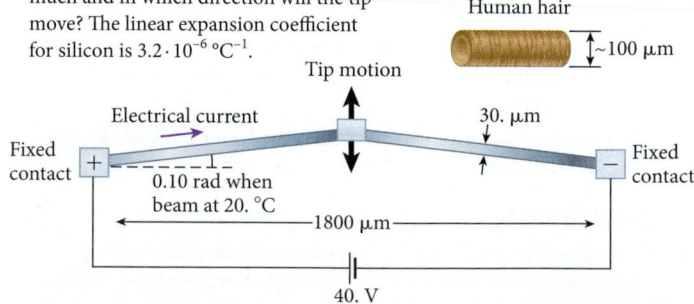

Human hair

~100 μm

Tip motion

Electrical current →

30. μm

Fixed contact   +

0.10 rad when beam at 20. °C

Fixed contact

◀— 1800 μm —▶

40. V

•17.55 The volume of 1.00 kg of liquid water over the temperature range from 0.00 °C to 50.0 °C fits reasonably well to the polynomial function $V = 1.00016 - (4.52 \cdot 10^{-5})T + (5.68 \cdot 10^{-6})T^2$, where the volume is measured in cubic meters and T is the temperature in degrees Celsius.

a) Use this information to calculate the volume expansion coefficient for liquid water as a function of temperature.

b) Evaluate your expression at 20.0 °C, and compare the value to that listed in Table 17.3.

••17.56 Suppose a bimetallic strip constructed of copper and steel strips of thickness 1.00 mm and length 25.0 mm has its temperature reduced by 5.00 K.

a) Determine the radius of curvature of the cooled strip (the radius of curvature of the interface between the two strips).

b) How far is the maximum deviation of the strip from its straight orientation?

Additional Exercises

17.57 A copper cube of side length 40. cm is heated from 20. °C to 120 °C. What is the change in the volume of the cube? The linear expansion coefficient of copper is $17 \cdot 10^{-6} \, °C^{-1}$.

17.58 When a 50.0-m-long metal pipe is heated from 10.0 °C to 40.0 °C, it lengthens by 2.85 cm.

a) Determine the linear expansion coefficient.

b) What type of metal is the pipe made of?

17.59 On a cool morning, with the temperature at 15.0 °C, a painter fills a 5.00-gal aluminum container to the brim with turpentine. When the temperature reaches 27.0 °C, how much fluid spills out of the container? The volume expansion coefficient for this brand of turpentine is $9.00 \cdot 10^{-4} \, °C^{-1}$.

17.60 A building having a steel infrastructure is $6.00 \cdot 10^2$ m high on a day when the temperature is 0.00 °C. How much taller is the building on a day when the temperature is 45.0 °C? The linear expansion coefficient of steel is $1.30 \cdot 10^{-5} \, °C^{-1}$.

17.61 In order to create a tight fit between two metal parts, machinists sometimes make the interior part larger than the hole into which it will fit and then either cool the interior part or heat the exterior part until they fit together. Suppose an aluminum rod with diameter D_1 (at 20. °C) is to be fit into a hole in a brass plate that has a diameter $D_2 = 10.000$ mm (at 20. °C). The machinists can cool the rod to 77.0 K by immersing it in liquid nitrogen. What is the largest possible diameter that the rod can have at 20. °C and just fit into the hole if the rod is cooled to 77.0 K and the brass plate is left at 20. °C? The linear expansion coefficients for aluminum and brass are $22 \cdot 10^{-6} \, °C^{-1}$ and $19 \cdot 10^{-6} \, °C^{-1}$, respectively.

D_1 $D_2 = 10.000$ mm

17.62 A military vehicle is fueled with gasoline in the United States, preparatory to being shipped overseas. Its fuel tank has a capacity of 213 L. When it is fueled, the temperature is 57 °F. At its destination, it may be required to operate in temperatures as high as 120. °F. What is the maximum volume of gasoline that should be put in its tank?

17.63 A mercury thermometer contains 8.00 mL of mercury. If the tube of the thermometer has a cross-sectional area of 1.00 mm², what should the spacing between the °C marks be?

17.64 A 14-gal container is filled with gasoline. Neglect the change in volume of the container, and find how many gallons are lost if the temperature increases by 27 °F. The volume expansion coefficient of gasoline is $9.6 \cdot 10^{-4} \, °C^{-1}$.

17.65 A highway of concrete slabs is to be built in the Libyan desert, where the highest air temperature recorded is 57.8 °C. The temperature is 20.0 °C during construction of the highway. The slabs are measured to be 12.0 m long at this temperature. How wide should the expansion cracks between the slabs be (at 20.0 °C) in order to prevent buckling at the highest temperatures?

17.66 An aluminum vessel with a volume capacity of 500. cm³ is filled with water to the brim at 20.0 °C. The vessel and contents are heated to 50.0 °C. As a result of the heating, will the water spill over the top, will there be room for more water to be added, or will the water level remain the same? Calculate the volume of water that will spill over or that could be added if either is the case.

17.67 By how much does the temperature of a given mass of kerosene need to change in order for its volume to increase by 1.00%?

17.68 A plastic-epoxy sheet has uniform holes of radius 1.99 cm. The holes are intended to allow solid ball bearings with an outer radius of 2.00 cm to just pass through. By how many degrees must the temperature of the plastic-epoxy sheet be raised so that the ball bearings will go through the holes? The linear expansion coefficient of plastic-epoxy is $1.30 \cdot 10^{-4} \, °C^{-1}$.

•17.69 A uniform brass disk of radius R and mass M with a moment of inertia I about its cylindrical axis of symmetry is at a temperature $T = 20.0$ °C. Determine the fractional change in its moment of inertia if it is heated to a temperature of 100. °C.

•17.70 A 25.01-mm-diameter brass ball sits at room temperature on a 25.00-mm-diameter hole made in an aluminum plate. The ball and plate are heated uniformly in a furnace, so both are at the same temperature at all times. At what temperature will the ball fall through the plate?

•17.71 In a pickup basketball game, your friend cracked one of his teeth in a collision with another player while attempting to make a basket. To correct the problem, his dentist placed a steel band of initial internal diameter 4.40 mm, and a cross-sectional area of width 3.50 mm, and thickness 0.450 mm on the tooth. Before placing the band on the tooth, he heated the band to 70.0 °C. What will be the tension in the band once it cools down to the temperature in your friend's mouth (36.8 °C)? The steel used for the band has a linear expansion coefficient of $\alpha = 13.0 \cdot 10^{-6} \, °C^{-1}$ and a Young's modulus of $Y = 200. \cdot 10^9 \, N/m^2$.

•17.72 Your physics teacher assigned a thermometer-building project. He gave you a glass tube with an inside diameter of 1.00 mm and a receptacle at one end. He also gave you 8.63 cm³ of mercury to pour into the tube, which filled the receptacle and some of the tube. You are to add marks indicating degrees Celsius on the glass tube. At what increments should the marks be put on? The volume expansion coefficient of mercury is $1.81 \cdot 10^{-4} \, °C^{-1}$.

•17.73 You are building a device for monitoring ultracold environments. Because the device will be used in environments where its temperature will change by 200. °C in 3.00 s, it must have the ability to withstand thermal shock (rapid temperature changes). The volume of the device is $5.00 \cdot 10^{-5} \, m^3$, and if the volume changes by $1.00 \cdot 10^{-7} \, m^3$ in a time interval of 5.00 s, the device will crack and be rendered useless. What is the maximum volume expansion coefficient that the material you use to build the device can have?

•17.74 A steel rod of length 1.0000 m and cross-sectional area $5.00 \cdot 10^{-4} \, m^2$ is placed snugly against two immobile end points. The rod is initially placed when the temperature is 0.00 °C. Find the stress in the rod when the temperature rises to 40.0 °C.

•17.75 A brass bugle can be thought of as a tube that is closed at one end and open at the other end. The brass from which the bugle is made has a linear expansion coefficient of $\alpha = 19.0 \cdot 10^{-6} \, °C^{-1}$. The overall length if the bugle is stretched out is 183.0 cm (at 20.0 °C). A bugle is played on a hot summer day (41.0 °C). Find the fundamental frequency if

a) only the change in air temperature is considered,

b) only the change in the length of the bugle is considered, and

c) both effects in parts (a) and (b) are taken into account.

MULTI-VERSION EXERCISES

17.76 At 26.45 °C, a steel bar is 268.67 cm long and a brass bar is 268.27 cm long. At what temperature will the two bars be the same length? Take the linear expansion coefficient of steel to be $13.00 \cdot 10^{-6}$ °C^{-1} and the linear expansion coefficient of brass to be $19.00 \cdot 10^{-6}$ °C^{-1}.

17.77 A steel bar and a brass bar are both at a temperature of 28.73 °C. The steel bar is 270.73 cm long. At a temperature of 214.07 °C, the two bars have the same length. What is the length of the brass bar at 28.73 °C? Take the linear expansion coefficient of steel to be $13.00 \cdot 10^{-6}$ °C^{-1} and the linear expansion coefficient of brass to be $19.00 \cdot 10^{-6}$ °C^{-1}.

17.78 A steel bar and a brass bar are both at a temperature of 31.03 °C. The brass bar is 272.47 cm long. At a temperature of 227.27 °C, the two bars have the same length. What is the length of the steel bar at 31.03 °C? Take the linear expansion coefficient of steel to be $13.00 \cdot 10^{-6}$ °C^{-1} and the linear expansion coefficient of brass to be $19.00 \cdot 10^{-6}$ °C^{-1}.

17.79 The main mirror of a telescope has a diameter of 5.093 m. The mirror is made of glass with a linear expansion coefficient of $3.749 \cdot 10^{-6}$ °C^{-1}. If the temperature of the mirror is raised by 33.37 °C, how much will the area of the mirror increase?

17.80 The main mirror of a telescope has a diameter of 4.553 m. When the temperature of the mirror is raised by 34.65 °C, the area of the mirror increases by $4.253 \cdot 10^{-3}$ m^2. What is the linear expansion coefficient of the glass of which the mirror is made?

17.81 The main mirror of a telescope has a diameter of 4.713 m. The mirror is made of glass with a linear expansion coefficient of $3.789 \cdot 10^{-6}$ °C^{-1}. By how much would the temperature of the mirror need to be raised to increase the area of the mirror by $4.750 \cdot 10^{-3}$ m^2?

17.82 A steel antenna for television broadcasting is 501.9 m tall on a summer day, when the outside temperature is 28.09 °C. The steel from which the antenna was constructed has a linear expansion coefficient of $13.89 \cdot 10^{-6}$ °C^{-1}. On a cold winter day, the temperature is –15.91 °C. What is the change in the height of the antenna?

17.83 A steel antenna for television broadcasting is 599.7 m tall on a summer day, when the outside temperature is 28.31 °C. On a cold winter day, when the temperature is –18.95 °C, the antenna is 0.4084 m shorter than on the summer day. What is the linear expansion coefficient of the steel from which the antenna was constructed?

17.84 A steel antenna for television broadcasting is 645.5 m tall on a summer day, when the outside temperature is 28.51 °C. The steel from which the antenna was constructed has a linear expansion coefficient of $14.93 \cdot 10^{-6}$ °C^{-1}. On a cold winter day, the antenna is 0.3903 m shorter than on the summer day. What is the temperature on the winter day?

18

Heat and the First Law of Thermodynamics

FIGURE 18.1 A thunderstorm.

arth's weather is driven by thermal energy in the atmosphere. The equatorial regions receive more of the Sun's radiation than the polar regions do; so warm air moves north and south from the Equator toward the poles to distribute the thermal energy more evenly. This transfer of thermal energy, called *convection*, sets up wind currents around the world, carrying clouds and rain as well as air. In extreme cases, rising warm air and sinking cooler air form dramatic storms, like the thunderstorm shown in Figure 18.1. Circular storms—tornados and hurricanes—are driven by the violent collision of warm air with cooler air.

This chapter examines the nature of heat and the mechanisms of thermal energy transfer. Heat is a form of energy that is transferred into or out of a system. Thus, heat is governed by a more general form of the law of conservation of energy, known as the First Law of Thermodynamics. We'll focus on this law in this chapter, along with some of its applications to thermodynamic processes and to changes in heat and temperature.

Heat is essential to life processes; no life could exist on Earth without heat from the Sun or from the Earth's interior. However, heat can also cause problems with the operation of electrical circuits, computers, motors, and other mechanical devices. Every branch of science and engineering has to deal with heat in one way or another, and thus the concepts in this chapter are important for all areas of research, design, and development.

WHAT WE WILL LEARN

- Heat is thermal energy that transfers between a system and its environment or between two systems, as a result of a temperature difference between them.

- The First Law of Thermodynamics states that the change in internal energy of a closed system is equal to the thermal energy absorbed by the system minus the work done by the system.

- Adding thermal energy to a substance in a given phase increases its temperature. This temperature increase

is inversely proportional to the heat capacity, C, of the substance.

- During a phase change from solid to liquid or from liquid to gas, the addition of thermal energy does not change the temperature of the mixed phase. The amount of thermal energy needed to complete a phase change is called its *latent heat*.

- The three main modes of transferring thermal energy are conduction, convection, and radiation.

18.1 Definition of Heat

Heat is one of the most common forms of energy in the universe, and we all experience it every day. Yet many people have misconceptions about heat that often cause confusion. For instance, a burning object such as a candle flame does not "have" heat that it emits when it becomes hot enough. Instead, the candle flame transmits energy, in the form of heat, to the air around it. To clarify these ideas, we need to start with clear and precise definitions of heat and the units used to measure it.

If you pour cold water into a glass and then put the glass on the kitchen table, the water will slowly warm until it reaches the temperature of the air in the room. Similarly, if you pour hot water into a glass and place it on the kitchen table, the water will slowly cool until it reaches the temperature of the air in the room. The warming or cooling takes place rapidly at first and then more slowly as the water comes into thermal equilibrium with the air in the kitchen. At thermal equilibrium, the water, the glass, and the air in the room are all at the same temperature.

The water in the glass is a **system** with temperature T_s, and the air in the kitchen is an **environment** with temperature T_e. If $T_s \neq T_e$, then the temperature of the system changes until it is equal to the temperature of the environment. A system can be simple or complicated; it is just any object or collection of objects we wish to examine. The difference between the environment and the system is that the environment is large compared to the system. The temperature of the system affects the environment, but we'll assume that the environment is so large that any changes in its temperature are imperceptible.

The change in temperature of the system is due to a transfer of energy between the system and its environment. This type of energy is **thermal energy,** an internal energy related to the motion of the atoms, molecules, and electrons that make up the system or the environment. Thermal energy in the process of being transferred from one body to another is called **heat** and is symbolized by Q. If thermal energy is transferred into a system, then $Q > 0$ (Figure 18.2a). That is, the system gains thermal energy when it receives heat from its environment. If thermal energy is transferred from the system to its environment, then $Q < 0$ (Figure 18.2c). If the system and its environment have the same temperature (Figure 18.2b), then $Q = 0$.

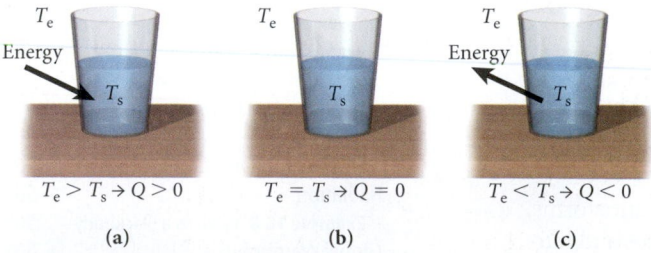

$T_e > T_s \rightarrow Q > 0$ $T_e = T_s \rightarrow Q = 0$ $T_e < T_s \rightarrow Q < 0$

(a)   (b)   (c)

FIGURE 18.2 (a) A system embedded in an environment that has a higher temperature. (b) A system embedded in an environment that has the same temperature. (c) A system embedded in an environment that has a lower temperature.

Definition

Heat, Q, is the energy transferred between a system and its environment (or between two systems) because of a temperature difference between them. When energy flows into the system, $Q > 0$; when energy flows out of the system, $Q < 0$.

18.2 Mechanical Equivalent of Heat

Recall from Chapter 5 that energy can also be transferred between a system and its environment as work done by a force acting on a system or by a system. The concepts of heat and work, discussed in Section 18.3, can be defined in terms of the transfer of energy to or from a system. We can refer to the internal energy of a system, but not to the heat contained in a system. If we observe hot water in a glass, we do not know whether thermal energy was transferred to the water or work was done on the water.

Heat is transferred energy and can be quantified using the SI unit of energy, the joule. Originally, heat was measured in terms of its ability to raise the temperature of water. The **calorie** (cal) was defined as the amount of heat required to raise the temperature of 1 gram of water by 1 °C. Another common measure of heat, still used in the United States, is the British thermal unit (BTU), defined as the amount of heat required to raise the temperature of 1 pound of water by 1 °F. However, the change in temperature of water as a function of the amount of thermal energy transferred to it depends on the original temperature of the water. Reproducible definitions of the calorie and the British thermal unit required that the measurements be made at a specified starting temperature.

In a classic experiment performed in 1843, the English physicist James Prescott Joule showed that the mechanical energy of an object could be converted into thermal energy. The apparatus Joule used consisted of a large mass supported by a rope that was run over a pulley and wound around an axle (Figure 18.3). As the mass descended, the unwinding rope turned a pair of large paddles in a container of water. Joule showed that the rise in temperature of the water was directly related to the mechanical work done by the falling object. Thus, Joule demonstrated that mechanical work could be turned into thermal energy, and he obtained a relationship between the calorie and the joule (the energy unit named in his honor).

The modern definition of the calorie is based on the joule. The calorie is defined to be exactly 4.186 J, with no reference to the change in temperature of water. The following are some conversion factors for energy units:

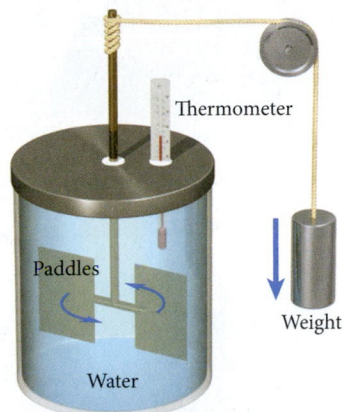

FIGURE 18.3 Apparatus for Joule's experiment to determine the mechanical equivalent of heat.

$$
\begin{aligned}
1\ \text{cal} &= 4.186\ \text{J} \\
1\ \text{BTU} &= 1055\ \text{J} \\
1\ \text{kWh} &= 3.60 \cdot 10^6\ \text{J} \\
1\ \text{kWh} &= 3412\ \text{BTU}.
\end{aligned}
\tag{18.1}
$$

The energy content of food is usually expressed in terms of calories. A food calorie, often called a *Calorie* or a *kilocalorie,* is not equal to the calorie just defined; a food calorie is equivalent to 1000 cal. The cost of electrical energy is usually stated in cents per kilowatt-hour (kWh).

EXAMPLE 18.1 | Candy Bar

PROBLEM
The label of a candy bar states that it has 275 Calories. What is the energy content of this candy bar in joules?

SOLUTION
Food calories are kilocalories and 1 kcal = 4186 J. The candy bar has 275 kcal, or

$$
275\ \text{kcal} \left(\frac{4186\ \text{J}}{1\ \text{kcal}} \right) = 1.15 \cdot 10^6\ \text{J}.
$$

DISCUSSION
Note that this energy is enough to raise a small truck with a weight of 22.2 kN (5000 lb) a distance of 52 m.

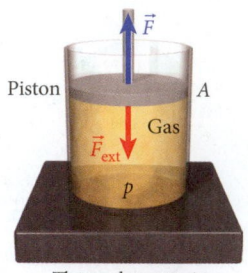

Piston

$\vec{F}$

A

Gas

$\vec{F}_{ext}$

p

Thermal reservoir

(a)

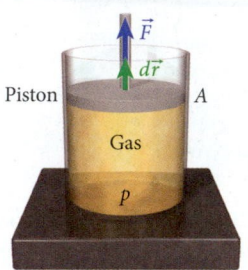

Piston

$\vec{F}$

$d\vec{r}$

A

Gas

p

Thermal reservoir

(b)

FIGURE 18.4 A gas-filled cylinder with a piston. The gas is in thermal contact with an infinite thermal reservoir. (a) An external force pushes on the piston, increasing the pressure in the gas. (b) The external force is removed, allowing the gas to push the piston out.

18.3 Heat and Work

Let's look at how energy can be transferred as heat or work between a system and its environment. We'll consider a system consisting of a gas-filled cylinder with a piston (Figure 18.4). The gas in the cylinder is described by a temperature T, a pressure p, and a volume V. We assume that the side wall of the cylinder does not allow heat to penetrate it. The gas is in thermal contact with an infinite thermal reservoir, which is an object so large that its temperature does not change even though it experiences thermal energy flow into it or out of it. (A thermal reservoir is an idealized construct that cannot actually exist. However, some large real-world systems that behave like thermal reservoirs include the ocean, the atmosphere, and the Earth itself.) This reservoir also has temperature T. In Figure 18.4a, an external force, $\vec{F}_{ext}$, pushes on the piston. The gas then pushes back with a force, $\vec{F}$, given by the pressure of the gas, p, times the area, A, of the piston: $F = pA$ (see Chapter 13).

To describe the behavior of this system, we consider the change from an initial state specified by pressure p_i, volume V_i, and temperature T_i and a final state specified by pressure p_f, volume V_f, and temperature T_f. Imagine changes made slowly so that the gas remains close to equilibrium, allowing measurements of p, V, and T. The progression from the initial state to the final state is a **thermodynamic process.** During this process, thermal energy may be transferred into the system (positive heat), or it may be transferred out of the system (negative heat).

When the external force is reduced (Figure 18.4b), the gas in the cylinder pushes the piston out a distance dr. The work done by the system in this process is

$$dW = \vec{F} \cdot d\vec{r} = (pA)(dr) = p(A\,dr) = p\,dV,$$

where $dV = A\,dr$ is the change in the volume of the system. Thus, the work done by the system in going from the initial configuration to the final configuration is given by

$$W = \int dW = \int_{V_i}^{V_f} p\,dV. \tag{18.2}$$

Note that during this change in volume, the pressure may also change. To evaluate this integral, we need to know the relationship between pressure and volume for the process. For example, if the pressure remains constant, we obtain

$$W = \int_{V_i}^{V_f} p\,dV = p\int_{V_i}^{V_f} dV = p(V_f - V_i) \quad \text{(for constant pressure).} \tag{18.3}$$

Equation 18.3 indicates that, for constant pressure, a negative change in the volume corresponds to negative work done by the system.

Figure 18.5 shows graphs of pressure versus volume, sometimes called pV**-diagrams.** The three parts of the figure show different paths, or ways to change the pressure and volume of a system from an initial condition to a final condition. Figure 18.5a illustrates a process that starts at an initial point i and proceeds to a final point f in such a way that the pressure decreases as the volume increases. The work done by the system is given by equation 18.2. The integral can be represented by the area under the curve, shown by the green shading in Figure 18.5a. In this case, the work done by the system is positive because the volume of the system increases.

Figure 18.5b illustrates a process that starts at an initial point i and proceeds to a final point f through an intermediate point m. The first step involves increasing the volume of the system while holding the pressure constant. One way to complete this step is to increase the temperature of the system as the volume increases, maintaining a constant pressure. The second step consists of decreasing the pressure while holding the volume constant. One way to accomplish this task is to lower the temperature by taking thermal energy away from the system. Again, the work done by the system can be represented by the area under the curve, shown by the green shading in Figure 18.5b. The work done by the system is positive because the volume of the system increases. However, the work done by the system originates only in the first step. In the second step, the system does no work because the volume does not change.

Figure 18.5c illustrates another process that starts at an initial point i and proceeds to a final point f through an intermediate point m. The first step involves decreasing the pressure of the system while holding the volume constant. The second step consists of increasing the volume while holding the pressure constant. Again, the work done by the system can be represented by the area under the curve, shown by the green shading in Figure 18.5c. The work done by the system is again positive because the volume of the system increases. However, the work done by the system originates only in the second step. In the first step, the system does no work because the volume does not change. The net work done in this process is less than the work done by the process in Figure 18.5b, as the green area is smaller. The thermal energy absorbed must also be less because the initial and final states are the same in these two processes, so the change in internal energy is the same. (This connection between work, heat, and the change in internal energy is discussed in more detail in Section 18.4.)

Thus, the work done by a system and the thermal energy transferred to the system depend on the manner in which the system moves from an initial point to a final point in a pV-diagram. This type of process is thus referred to as a **path-dependent process.**

The pV-diagrams in Figure 18.6 reverse the processes shown in Figure 18.5 by starting at the final points in Figure 18.5 and proceeding along the same path to the initial points. In each of these cases, the area under the curve represents the negative of the work done by the system in moving from an initial point i to a final point f. In all three cases, the work done by the system is negative because the volume of the system decreases.

Suppose a process starts at some initial point on a pV-diagram, follows some path, and returns to the original point. A path that returns to its starting point is called a **closed path.** Two examples of closed paths are illustrated in Figure 18.7. In Figure 18.7a, a process starts at an initial point i and proceeds to an intermediate point m_1; the volume is increased while the pressure is held constant. From m_1, the path goes to intermediate point m_2; the volume is held constant while the pressure decreases. From point m_2, the path goes to intermediate point m_3, with the pressure constant and the volume decreasing. Finally, the path goes from point m_3 to the final point, f, which is the same as the initial point, i. The area under the path from i to m_1 represents positive work done by the system (the volume increases), while the area under the path from m_2 to m_3 represents negative work done on the system (volume decreases). The addition of the positive and the negative work yields net positive work being done by the system, as depicted by the area of the green rectangle in Figure 18.7a. In this case, heat is positive.

Figure 18.7b shows the same path on a pV-diagram as in Figure 18.7a, but the process occurs in the opposite direction. The area under the path from intermediate point m_1 to intermediate point m_2 corresponds to positive work done by the system. The area under the path from intermediate point m_3 to final point f corresponds to negative work done on the system. The net work done by the system in this case is negative, and its magnitude is represented by the area of the orange rectangle in Figure 18.7b. In this case, the heat is negative.

The amount of work done by the system and the thermal energy absorbed by the system depend on the path taken on a pV-diagram, as well as on the direction in which the path is taken.

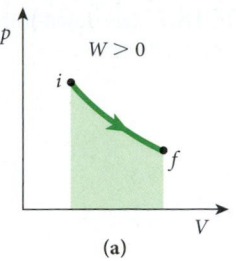

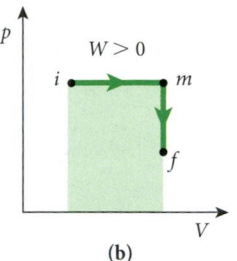

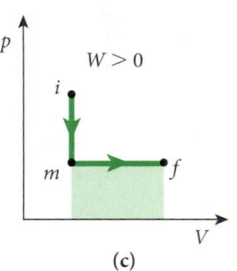

FIGURE 18.5 The paths of three different processes in pV-diagrams. (a) A process in which the pressure decreases as the volume increases. (b) A two-step process in which the first step consists of increasing the volume while keeping the pressure constant and the second step consists of decreasing the pressure while holding the volume constant. (c) Another two-step process, in which the first step consists of keeping the volume constant and decreasing the pressure and the second step consists of holding the pressure constant while increasing the volume. In all three cases, the green area below the curve represents the work done during the process.

Self-Test Opportunity 18.1

Consider the process shown in the pV-diagram. The path goes from point i to point f and back to point i. Is the work done by the system negative, zero, or positive?

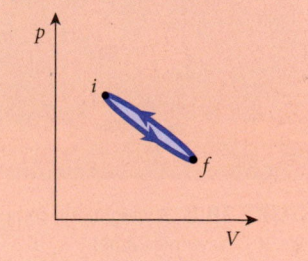

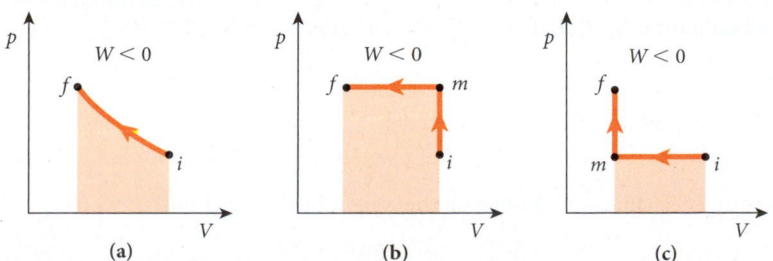

FIGURE 18.6 The paths of three processes in pV-diagrams. These are the reverse of the processes shown in Figure 18.5.

FIGURE 18.7 Two closed-path processes in *pV*-diagrams.

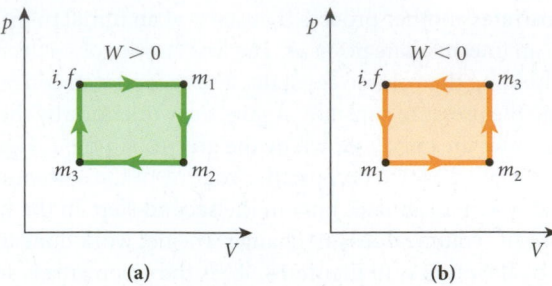

(a)   (b)

18.4 First Law of Thermodynamics

Thermodynamic systems are classified according to how they interact with their environment. An **open system** can exchange heat, energy, and mass with its surroundings. A **closed system** is a system into or out of which thermal energy can be transferred but from which no constituents can escape and to which no additional constituents are added. An **isolated system** is a system that has no interactions or exchanges of constituents with its environment.

Combining several of the concepts already covered in this chapter allows us to express the change in internal energy of a closed system in terms of heat and work as

$$\Delta E_{int} = E_{int,f} - E_{int,i} = Q - W. \tag{18.4}$$

This equation is known as the **First Law of Thermodynamics.** It can be stated as follows:

> The change in the internal energy of a closed system is equal to the heat acquired by the system minus the work done by the system.

In other words, energy is conserved. Heat and work can be transformed into internal energy, but no energy is lost. Note that here the work is done by the system, not done on the system. Essentially, the First Law of Thermodynamics extends the conservation of energy (first encountered in Chapter 6) beyond mechanical energy to include heat as well as work. (In other contexts, such as chemical reactions, work is defined as the work done on the system, which leads to a different sign convention for the work. The sign convention can be assigned in either way, as long as it is consistent.) Note also that the change in internal energy is path-independent, while changes in heat and work are path-dependent.

FIGURE 18.8 A weightlifter competes in the 2012 Olympic Games.

EXAMPLE 18.2 | Weightlifter

PROBLEM

A weightlifter snatches a barbell with mass $m = 180.0$ kg and moves it a distance $h = 1.25$ m vertically, as illustrated in Figure 18.8. If we consider the weightlifter to be a thermodynamic system, how much heat must he give off if his internal energy decreases by 4000. J?

SOLUTION

We start with the First Law of Thermodynamics (equation 18.4):

$$\Delta E_{int} = Q - W.$$

The work is the mechanical work done on the weight by the weightlifter:

$$W = mgh.$$

The heat is then given by

$$Q = \Delta E_{int} + W = \Delta E_{int} + mgh = -4000 \text{ J} + (180.0 \text{ kg})(9.81 \text{ m/s}^2)(1.25 \text{ m})$$

$$= -1790 \text{ J} = -0.428 \text{ kcal}.$$

The weightlifter cannot turn internal energy into useful work without giving off heat. Note that the decrease in internal energy of the weightlifter is less than 1 food calorie: (4000 J)/(4186 J/kcal) = 0.956 kcal. The heat output is only 0.428 kcal. Thus, rather small amounts of internal energy and heat are associated with the large effort required to lift the weight.

EXAMPLE 18.3 / Truck Sliding to a Stop

PROBLEM
The brakes of a moving truck with a mass $m = 3000.$ kg lock up. The truck skids to a stop on a horizontal road surface in a distance $L = 83.2$ m. The coefficient of kinetic friction between the truck tires and the surface of the road is $\mu_k = 0.600$. What happens to the internal energy of the truck?

SOLUTION
The First Law of Thermodynamics (equation 18.4) relates the internal energy, the heat, and the work done on the system:

$$\Delta E_{int} = Q - W.$$

In this case, no thermal energy is transferred to or from the truck because the process of sliding to a complete stop is sufficiently fast that there is no time to transfer thermal energy in appreciable quantity. Thus, $Q = 0$. The dissipation of energy from the friction force, W_f, is always negative (see Chapter 6). In this case, the energy dissipated is equal to original kinetic energy of the truck. The energy dissipated by the friction force is

$$W_f = -\mu_k mgL,$$

where mg is the normal force exerted on the road by the truck. Using the work-kinetic energy theorem, we have $\Delta K = W = W_f$ and

$$\Delta E_{int} = 0 - (-\mu_k mgL) = \mu_k mgL.$$

Putting in the numerical values gives

$$\Delta E_{int} = (0.600)(3000. \text{ kg})(9.81 \text{ m/s}^2)(83.2 \text{ m}) = 1.47 \text{ MJ}.$$

This increase in internal energy can warm the truck's tires. Chapter 6 discussed conservation of energy for conservative and nonconservative forces. Here we see that energy is conserved because mechanical work can be converted to internal energy.

DISCUSSION
Note that we have assumed that the truck is a closed system. However, when the truck begins to skid, the tires may leave skid marks on the road, removing matter and energy from the system. The assumption that no thermal energy is transferred to or from the truck in this process is also not exactly valid. In addition, the internal energy of the road surface will also increase as a result of the friction between the tires and the road, taking some of the total energy available. So, the 1.47 MJ added to the internal energy of the truck should be considered an upper limit. However, the basic lesson from this example is that the mechanical energy lost because of the action of nonconservative forces is transformed to internal energy of parts or all of the system, and the total energy is conserved.

18.5   First Law for Special Processes

The First Law of Thermodynamics—that is, the basic conservation of energy—holds for all kinds of processes involving a closed system, but energy can be transformed and thermal energy transported in some special ways in which only a single or a few of the variables that characterize the state of the system change. Some special processes, which occur often in physical situations, can be described using the First Law of Thermodynamics. These special processes are also usually the only ones for which we can compute numerical values for heat and work. For this reason, they will be discussed several times in the following chapters. Keep in mind that these processes are simplifications or idealizations, but they often approximate real-world situations fairly closely.

Adiabatic Processes

An **adiabatic process** is one in which no heat flows when the state of a system changes. This can happen, for example, if a process occurs quickly and there is not enough time for heat to be exchanged. Adiabatic processes are common because many physical processes do occur quickly enough that no thermal energy transfer takes place. For adiabatic processes, $Q = 0$ in equation 18.4, so

$$\Delta E_{int} = -W \quad \text{(for an adiabatic process)}. \tag{18.5}$$

Another situation in which an adiabatic process can occur is if the system is thermally insulated from its environment while pressure and volume changes occur. An example is compressing a gas in an insulated container or pumping air into a bicycle tire using a manual tire pump. The change in the internal energy of the gas is due only to the work done on the gas.

Constant-Volume Processes

Processes that occur at constant volume are called **isochoric processes.** For a process in which the volume is held constant, the system can do no work, so $W = 0$ in equation 18.4, giving

$$\Delta E_{int} = Q \quad \text{(for a process at constant volume)}. \tag{18.6}$$

An example of a constant-volume process is warming a gas in a rigid, closed container that is in contact with other bodies. No work can be done because the volume of the gas cannot change. The change in the internal energy of the gas occurs because of heat flowing into or out of the gas as a result of contact between the container and the other bodies. Cooking food in a pressure cooker is an isochoric process.

Closed-Path Processes

In a **closed-path process,** the system returns to the same state at which it started. Regardless of how the system reached that point, the internal energy must be the same as at the start, so $\Delta E_{int} = 0$ in equation 18.4. This gives

$$Q = W \quad \text{(for a closed-path process)}. \tag{18.7}$$

Thus, the net work done by a system during a closed-path process is equal to the thermal energy transferred into the system. Such cyclical processes form the basis of many kinds of heat engines (discussed in Chapter 20).

Free Expansion

If the thermally insulated (so $Q = 0$) container for a gas suddenly increases in size, the gas will expand to fill the new volume. During this **free expansion,** the system does no work and no heat is absorbed. That is, $W = 0$ and $Q = 0$, and equation 18.4 becomes

$$\Delta E_{int} = 0 \quad \text{(for free expansion of a gas)}. \tag{18.8}$$

To illustrate this situation, consider a box with a barrier in the center (Figure 18.9). A gas is confined in the left half of the box. When the barrier between the two halves is removed, the gas fills the total volume. However, the gas performs no work. This last statement requires a bit of explanation: In its free expansion, the gas does not move a piston or any other material device; thus, it does no work on anything. During the expansion, the gas particles move freely until they encounter the walls of the expanded container. The gas is not in equilibrium while it is expanding. For this system, we can plot the initial state and the final state on a pV-diagram, but not the intermediate state.

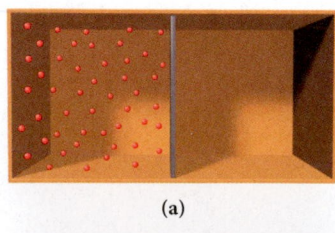
(a)

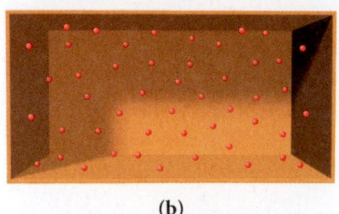
(b)

FIGURE 18.9 (a) A gas is confined to half the volume of a box. (b) The barrier separating the two halves is removed, and the gas expands to fill the volume.

Constant-Pressure Processes

Constant-pressure processes are called **isobaric processes.** Such processes are common in the study of the specific heat of gases. In an isobaric process, the volume can change, allowing the system to do work. Since the pressure is kept constant, $W = p(V_f - V_i) = p\Delta V$, according to equation 18.3. Thus, equation 18.4 can be written as follows:

$$\Delta E_{int} = Q - p\Delta V \quad \text{(for a process at constant pressure).} \tag{18.9}$$

An example of an isobaric process is the slow warming of a gas in a cylinder fitted with a frictionless piston, which can move to keep the pressure constant. The path of an isobaric process on a pV-diagram is a horizontal straight line. If the system moves in the positive-volume direction, the system is expanding. If the system moves in the negative-volume direction, the system is contracting. Cooking food in an open saucepan is another example of an isobaric process.

Constant-Temperature Processes

Constant-temperature processes are called **isothermal processes.** The temperature of the system is held constant through contact with an external thermal reservoir. Isothermal processes take place slowly enough that heat can be exchanged with the external reservoir to maintain a constant temperature. For example, heat can flow from a warm reservoir to the system, allowing the system to do work. The path of an isothermal process in a pV-diagram is called an **isotherm.** As we'll see in Chapter 19, the product of the pressure and the volume is constant for an ideal gas undergoing an isothermal process, giving the isotherm the form of a hyperbola. In addition, as we'll see in Chapter 20, isothermal processes play an important part in the analysis of devices that produce useful work from sources of heat.

18.6 Specific Heats of Solids and Fluids

Suppose a block of aluminum is at room temperature. If heat, Q, is then transferred to the block, the temperature of the block goes up proportionally to the amount of heat. The proportionality constant between the temperature difference and the heat is the **heat capacity,** C, of the block (or other object). Thus,

$$Q = C\Delta T, \tag{18.10}$$

where ΔT is the change in temperature.

The term *heat capacity* does not imply that an object contains a certain amount of heat. Rather, it tells how much heat is required to raise the temperature of the object by a given amount. The SI units for heat capacity are joules per kelvin (J/K).

The temperature change of an object due to heat can be described by the **specific heat,** c, which is defined as the heat capacity per unit mass, m:

$$c = \frac{C}{m}. \tag{18.11}$$

With this definition, the relationship between temperature change and heat can be written as

$$Q = cm\Delta T. \tag{18.12}$$

The SI units for the specific heat are J/(kg K). In practical applications, specific heat is also often expressed in cal/(g K). The units J/(kg K) and J/(kg °C) can be used interchangeably for specific heat since it is defined in terms of ΔT. The specific heats of various materials are given in Table 18.1.

Note that specific heat and heat capacity are usually measured in two ways. For most substances, they are measured under constant pressure (as in Table 18.1) and denoted by c_p and C_p. However, for fluids (gases and liquids), the specific heat and heat capacity can also be measured at constant volume and denoted by c_V and C_V. In general, measurements under constant pressure produce larger values, because mechanical work has to be performed in the process, and the difference is particularly large for gases. We'll discuss this in greater detail in Chapter 19.

Table 18.1	Specific Heats for Selected Materials	
	Specific Heat, c	
Material	**kJ/(kg K)**	**cal/(g K)**
Aluminum	0.900	0.215
Asphalt	0.92	0.22
Basalt	0.84	0.20
Copper	0.386	0.0922
Glass	0.840	0.20
Granite	0.790	0.189
Ice	2.06	0.500
Iron	0.450	0.107
Lead	0.129	0.0308
Limestone	0.217	0.908
Nickel	0.461	0.110
Sandstone	0.92	0.22
Steam	2.01	0.48
Steel	0.448	0.107
Water	4.19	1.00

The specific heat of a substance can also be defined in terms of the number of moles of the substance rather than its mass. This type of specific heat is called the *molar specific heat,* and is also discussed in Chapter 19.

The effect of differences in specific heats of substances can be observed readily at the seashore, where the sunlight transfers energy to the land and to the water roughly equally during the day. The specific heat of the land is about five times lower than the specific heat of the water. Thus, the land warms up more quickly than the water, and it warms the air above it more than the water warms the air above it. This temperature difference creates an onshore breeze during the day. The high specific heat of water helps to moderate climates around oceans and large lakes.

EXAMPLE 18.4 / Warming Water

PROBLEM

You have 2.00 L of water at a temperature of 20.0 °C. How much energy is required to raise the temperature of that water to 95.0 °C? Assuming that you use electricity to warm the water, how much will it cost at 10.0 cents per kilowatt-hour?

SOLUTION

The mass of 1.00 L of water is 1.00 kg. From Table 18.1, $c_{water} = 4.19$ kJ/(kg K). Therefore, the energy required to warm 2.00 kg of water from 20.0 °C to 95.0 °C is

$$Q = c_{water} m_{water} \Delta T = \left[4.19 \text{ kJ}/(\text{kg K}) \right] (2.00 \text{ kg})(95.0 \text{ °C} - 20.0 \text{ °C})$$

$$= 629{,}000 \text{ J}.$$

Using the conversion factor for converting joules to kilowatt-hours (equation 18.1), we calculate the cost of warming the water as

$$\text{Cost} = (629{,}000 \text{ J}) \left(\frac{10.0 \text{ cents}}{1 \text{ kWh}} \right) \left(\frac{1 \text{ kWh}}{3.60 \cdot 10^6 \text{ J}} \right) = 1.75 \text{ cents}.$$

Concept Check 18.1

Suppose you raise the temperature of copper block 1 from −10 °C to +10 °C, of copper block 2 from +20 °C to +40 °C, and of copper block 3 from +90 °C to +110 °C. Assuming that the blocks have the same mass, to which one did you add the most heat?

a) block 1

b) block 2

c) block 3

d) All of the blocks received the same amount of heat.

Calorimetry

A **calorimeter** is a device used to study internal energy changes by measuring thermal energy transfers. The thermal energy transfer and internal energy change may result from a chemical reaction, a physical change, or differences in temperature and specific heat. A simple calorimeter consists of an insulated container and a thermometer. For simple measurements, a Styrofoam cup and an ordinary alcohol thermometer will do. We'll assume that no heat is lost or gained in the calorimeter or the thermometer.

When two materials with different temperatures are placed in a calorimeter, heat will flow from the warmer substance to the cooler substance until the temperatures of the two substances are the same. No heat will flow into or out of the calorimeter. The heat lost by the warmer substance will be equal to the heat gained by the cooler substance.

A substance such as water with a high specific heat, $c = 4.19$ kJ/(kg K), requires more heat to raise its temperature by the same amount than does a substance with a low specific heat, like steel, with $c = 0.448$ kJ/(kg K). Solved Problem 18.1 illustrates the concepts of calorimetry.

SOLVED PROBLEM 18.1 / Water and Lead

PROBLEM

A metalsmith pours 3.00 kg of lead shot (which is the material used to fill shotgun shells) at a temperature of 94.7 °C into 1.00 kg of water at 27.5 °C in an insulated container, which acts as a calorimeter. What is the final temperature of the mixture?

SOLUTION

THINK The lead shot will give up heat and the water will absorb heat until both are at the same temperature. (Something that may not occur to you is that lead shot has a melting temperature, but it is significantly above 94.7 °C; so we do not have to deal with a phase change in this situation. The water will stay liquid, and the lead shot will stay solid. Section 18.7 addresses situations involving phase changes.)

SKETCH   Figure 18.10 shows the problem situation before and after the lead shot is added to the water.

RESEARCH   The heat lost by the lead shot, Q_{lead}, to its environment is given by $Q_{lead} = m_{lead}c_{lead}(T - T_{lead})$, where c_{lead} is the specific heat of lead, m_{lead} is the mass of the lead shot, T_{lead} is the original temperature of the lead shot, and T is the final equilibrium temperature.

The heat gained by the water, Q_{water}, is given by $Q_{water} = m_{water}c_{water}(T - T_{water})$, where c_{water} is the specific heat of water, m_{water} is the mass of the water, and T_{water} is the original temperature of the water.

The sum of the heat lost by the lead shot and the heat gained by the water is zero, because the process took place in an insulated container and because the total energy is conserved, a consequence of the First Law of Thermodynamics. So we can write

$$Q_{lead} + Q_{water} = 0 = m_{lead}c_{lead}(T - T_{lead}) + m_{water}c_{water}(T - T_{water}).$$

SIMPLIFY   We multiply through on both sides and reorder so that all of the terms containing the unknown temperature, T, are on the right side of the equation:

$$m_{lead}c_{lead}T_{lead} + m_{water}c_{water}T_{water} = m_{water}c_{water}T + m_{lead}c_{lead}T.$$

We can solve this equation for T by dividing both sides by $m_{lead}c_{lead} + m_{water}c_{water}$:

$$T = \frac{m_{lead}c_{lead}T_{lead} + m_{water}c_{water}T_{water}}{m_{lead}c_{lead} + m_{water}c_{water}}.$$

CALCULATE   Putting in the numerical values gives

$$T = \frac{(3.00 \text{ kg})[0.129 \text{ kJ/(kg K)}](94.7 \text{ °C}) + (1.00 \text{ kg})[4.19 \text{ kJ/(kg K)}](27.5 \text{ °C})}{(3.00 \text{ kg})[0.129 \text{ kJ/(kg K)}] + (1.00 \text{ kg})[4.19 \text{ kJ/(kg K)}]}$$

$$= 33.182 \text{ °C}.$$

ROUND   We report our result to three significant figures:

$$T = 33.2 \text{ °C}.$$

DOUBLE-CHECK   The final temperature of the mixture of lead shot and water is only 5.7 °C higher than the original temperature of the water. The masses of the lead shot and water differ by a factor of 3, but the specific heat of lead is much lower than the specific heat of water. Thus, it is reasonable that the lead has a large change in temperature and the water has a small change in temperature. To double-check, we calculate the heat lost by the lead,

$$Q_{lead} = m_{lead}c_{lead}(T - T_{lead}) = (3.00 \text{ kg})[0.129 \text{ kJ/(kg K)}](33.2 \text{ °C} - 94.7 \text{ °C})$$

$$= -23.8 \text{ kJ},$$

and compare the result to the heat gained by the water,

$$Q_{water} = m_{water}c_{water}(T - T_{water}) = (1.00 \text{ kg})[4.19 \text{ kJ/(kg K)}](33.2 \text{ °C} - 27.5 \text{ °C})$$

$$= 23.9 \text{ kJ}.$$

These results add up to zero within rounding error, as required.

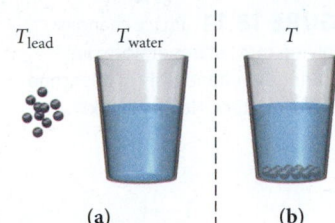

FIGURE 18.10   Lead shot and water (a) before and (b) after the lead shot is added to the water.

18.7   Latent Heat and Phase Transitions

As noted in Chapter 13, the three common **states of matter** (sometimes also called **phases**) are solid, liquid, and gas. We have been considering objects for which the change in temperature is proportional to the amount of heat that is added. This linear relationship between heat and temperature is, strictly speaking, an approximation, but it has high accuracy for solids and liquids. For a gas, adding heat will raise the temperature but may also change the pressure

FIGURE 18.11 Phase changes involving three states of water are occurring simultaneously in this scene from Yellowstone National Park.

or volume, depending on whether and how the gas is contained. Substances can have different specific heats depending on whether they are in a solid, liquid, or gaseous state.

If enough heat is added to a solid, it melts into a liquid. If enough heat is added to a liquid, it vaporizes into a gas. These are examples of **phase changes,** or **phase transitions** (Figure 18.11). During a phase change, the temperature of an object remains constant. The heat that is required to melt a solid, divided by its mass, is called the **latent heat of fusion,** L_{fusion}. Melting changes a substance from a solid to a liquid. The heat that is required to vaporize a liquid, divided by its mass, is called the **latent heat of vaporization,** $L_{\text{vaporization}}$. Vaporizing changes a substance from a liquid to a gas.

The temperature at which a solid melts to a liquid is the melting point, T_{melting}. The temperature at which a liquid vaporizes to a gas is the boiling point, T_{boiling}. The relationship between the mass of an object at its melting point and the heat needed to change the object from a solid to a liquid is given by

$$Q = mL_{\text{fusion}} \quad \left(\text{for } T = T_{\text{melting}} \right). \tag{18.13}$$

Similarly, the relationship between the mass of an object at its boiling point and the heat needed to change the object from a liquid to a gas is given by

$$Q = mL_{\text{vaporization}} \quad \left(\text{for } T = T_{\text{boiling}} \right). \tag{18.14}$$

The SI units for latent heats of fusion and vaporization are joules per kilogram (J/kg); units of calories per gram (cal/g) are also often used. The latent heat of fusion for a given substance is different from the latent heat of vaporization for that substance. Representative values for melting point, latent heat of fusion, boiling point, and latent heat of vaporization are listed in Table 18.2.

It is also possible for a substance to change directly from a solid to a gas. This process is called **sublimation.** For example, sublimation occurs when dry ice, which is solid (frozen) carbon dioxide, changes directly to gaseous carbon dioxide without passing through the liquid state. When a comet approaches the Sun, some of its frozen carbon dioxide sublimates, helping to produce the comet's visible tail.

If we continue to heat a gas, it will become ionized, which means that some or all of the electrons in the atoms of the gas are removed. An ionized gas and its free electrons form a state of matter called a **plasma.** Plasmas are very common in the universe; in fact, as much as 99% of the mass of the Solar System exists in the form of a plasma. The dusty cloud of gas known as the Omega/Swan Nebula (M17), shown in Figure 18.12, is also a plasma.

As mentioned in Chapter 13, there are other states of matter besides solid, liquid, gas, and plasma. For example, matter in a granular state has specific properties that differentiate it from the four major states of matter. Matter can also exist as a Bose-Einstein condensate in which individual atoms become indistinguishable, under very specific conditions at very

FIGURE 18.12 Image of the Omega/Swan Nebula (M17), taken by the Hubble Space Telescope, shows an immense region of gas ionized by radiation from young stars.

Table 18.2	Some Representative Melting Points, Boiling Points, Latent Heats of Fusion, and Latent Heats of Vaporization					
	Melting Point	**Latent Heat of Fusion, L_{fusion}**		**Boiling Point**	**Latent Heat of Vaporization, $L_{vaporization}$**	
Material	**(K)**	**(kJ/kg)**	**(cal/g)**	**(K)**	**(kJ/kg)**	**(cal/g)**
Hydrogen	13.8	58.6	14.0	20.3	452	108
Ethyl alcohol	156	104	24.9	351	858	205
Mercury	234	11.3	2.70	630	293	70.0
Water	273	334	79.7	373	2260	539
Aluminum	932	396	94.5	2740	10,500	2500
Copper	1359	205	49.0	2840	4730	1130

low temperatures, as we'll see in Chapter 36 on quantum physics. American physicists Eric Cornell and Carl Wieman, along with German physicist Wolfgang Ketterle, were awarded the Nobel Prize in Physics in 2001 for their studies of a Bose-Einstein condensate.

Cooling an object means reducing the internal energy of the object. As thermal energy is removed from a substance in the gaseous state, the temperature of the gas decreases, in relation to the specific heat of the gas, until the gas begins to condense into a liquid. This change takes place at a temperature called the *condensation point,* which is the same temperature as the boiling point of the substance. To convert all the gas to liquid requires the removal of an amount of heat corresponding to the latent heat of vaporization times the mass of the gas. If thermal energy continues to be removed, the temperature of the liquid will go down, as determined by the specific heat of the liquid, until the temperature reaches the freezing point, which is the same temperature as the melting point of the substance. To convert all of the liquid to a solid requires removing an amount of heat corresponding to the latent heat of fusion times the mass. If heat then continues to be removed from the solid, its temperature will decrease in relation to its specific heat.

Concept Check 18.2

Suppose you raise the temperature of 1 kg of H_2O from $-10\,°C$ to $+10\,°C$, of another 1 kg of H_2O from $+20\,°C$ to $+40\,°C$, and of another 1 kg of H_2O from $+90\,°C$ to $+110\,°C$. Which of the three changes in temperature requires the addition of the most heat?

a) first change

b) second change

c) third change

d) All three changes require the addition of the same amount of heat.

EXAMPLE 18.5 Warming Ice to Water and Water to Steam

PROBLEM
How much heat, Q, is required to convert 0.500 kg of ice (frozen water) at a temperature of $-30\,°C$ to steam at $140\,°C$?

SOLUTION
We solve this problem in steps, with each step corresponding to either a rise in temperature or a phase change. We first calculate how much heat is required to raise the temperature of the ice from $-30\,°C$ to $0\,°C$. The specific heat of ice is 2.06 kJ/(kg K), so the heat required is

$$Q_1 = cm\Delta T = \left[2.06 \text{ kJ/(kg K)}\right](0.500 \text{ kg})(30 \text{ K}) = 30.9 \text{ kJ}.$$

We continue to add heat to the ice until it melts. The temperature remains at $0\,°C$ until all the ice is melted. The latent heat of fusion for ice is 334 kJ/kg, so the heat required to melt all of the ice is

$$Q_2 = mL_{fusion} = (0.500 \text{ kg})(334 \text{ kJ/kg}) = 167 \text{ kJ}.$$

Once all the ice has melted to water, we continue to add heat until the water reaches the boiling point, at $100\,°C$. The required heat for this step is

$$Q_3 = cm\Delta T = \left[4.19 \text{ kJ/(kg K)}\right](0.500 \text{ kg})(100 \text{ K}) = 209.5 \text{ kJ}.$$

We continue to add heat to the water until it vaporizes. The heat required for vaporization is

$$Q_4 = mL_{vaporization} = (0.500 \text{ kg})(2260 \text{ kJ/kg}) = 1130 \text{ kJ}.$$

We now warm the steam and raise its temperature from $100\,°C$ to $140\,°C$. The heat necessary for this step is

$$Q_5 = cm\Delta T = \left[2.01 \text{ kJ/(kg K)}\right](0.500 \text{ kg})(40 \text{ K}) = 40.2 \text{ kJ}.$$

– Continued

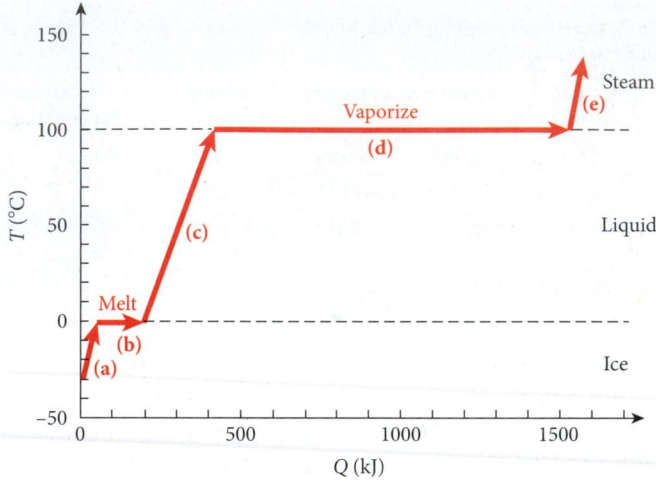

FIGURE 18.13   Plot of temperature versus heat added to change 0.500 kg of ice starting at a temperature of –30 °C to steam at 140 °C. The horizontal sections of the graph represent the latent heats of the two phase changes. The slopes of the other sections of the graph are the specific heats of water in its three phases (ice, liquid water, and steam).

Concept Check 18.3

You have a block of ice of mass m at a temperature of –3 °C in a thermally insulated container and you add the same mass m of liquid water at a temperature of 6 °C and let the mixture come to equilibrium. What is the final temperature of the mixture?

a) –3 °C   d) +4.5 °C

b) 0 °C   e) +6 °C

c) +3 °C

Thus, the total heat required is

$$Q = Q_1 + Q_2 + Q_3 + Q_4 + Q_5$$
$$= 30.9 \text{ kJ} + 167 \text{ kJ} + 209.5 \text{ kJ} + 1130 \text{ kJ} + 40.2 \text{ kJ} = 1580 \text{ kJ}.$$

Figure 18.13 shows a graph of the temperature of the water as a function of the heat added. In region (a), the temperature of the ice increases from –30 °C to 0 °C. In region (b), the ice melts to water while the temperature remains at 0 °C. In region (c), the water warms from 0 °C to 100 °C. In region (d), the water boils and changes to steam while the temperature remains at 100 °C. In region (e), the temperature of the steam increases from 100 °C to 140 °C.

Note that almost three quarters of the total heat in this entire process of warming the sample from –30 °C to 140 °C has to be spent to convert the liquid water to steam in the process of vaporization.

EXAMPLE 18.6   Work Done Vaporizing Water

Suppose 10.0 g of water at a temperature of 100.0 °C is in an insulated cylinder equipped with a piston that maintains a constant pressure of $p = 101.3$ kPa. Enough heat is added to the water to vaporize it to steam at a temperature of 100.0 °C. The volume of the water is $V_{\text{water}} = 10.0 \text{ cm}^3$, and the volume of the steam is $V_{\text{steam}} = 16,900 \text{ cm}^3$.

PROBLEM

What is the work done by the water as it vaporizes? What is the change in internal energy of the water?

SOLUTION

This process is carried out at a constant pressure. The work done by the vaporizing water is given by equation 18.3:

$$W = \int_{V_i}^{V_f} p \, dV = p \int_{V_i}^{V_f} dV = p(V_{\text{steam}} - V_{\text{water}}).$$

Putting in the numerical values gives the work done by the water as it increases in volume at a constant pressure:

$$W = (101.3 \cdot 10^3 \text{ Pa})(16,900 \cdot 10^{-6} \text{ m}^3 - 10.0 \cdot 10^{-6} \text{ m}^3) = 1710 \text{ J}.$$

The change in the internal energy of the water is given by the First Law of Thermodynamics (equation 18.4)

$$\Delta E_{\text{int}} = Q - W.$$

In this case, the thermal energy transferred is the heat required to vaporize the water. From Table 18.2, the latent heat of vaporization of water is $L_{\text{vaporization}} = 2260$ kJ/kg. Thus, the thermal energy transferred to the water is

$$Q = mL_{\text{vaporization}} = \left(10.0\cdot10^{-3}\text{ kg}\right)\left(2.260\cdot10^6\text{ J/kg}\right) = 22{,}600\text{ J}.$$

The change in the internal energy of the water is then

$$\Delta E_{\text{int}} = 22{,}600\text{ J} - 1710\text{ J} = 20{,}900\text{ J}.$$

Most of the added energy remains in the water as increased internal energy. This internal energy is related to the phase change of water to steam. The energy is used to overcome the attractive forces between the water molecules in the liquid state as it changes to the gaseous state.

18.8 Modes of Thermal Energy Transfer

The three main modes of thermal energy transfer are conduction, convection, and radiation. All three are illustrated by the campfire shown in Figure 18.14. **Radiation** is the transfer of thermal energy via electromagnetic waves. You can feel heat radiating from a campfire when you sit near it. **Convection** involves the physical movement of a substance (such as water or air) from thermal contact with one system to thermal contact with another system. The moving substance carries internal energy. You can see the thermal energy rising in the form of flames and hot air above the flames in the campfire. **Conduction** involves the transfer of thermal energy within an object (such as thermal energy transfer along a fire poker whose tip is hot) or the transfer of heat between two (or more) objects in thermal contact. Heat is conducted through a substance by the vibration of atoms and molecules and by motion of electrons. The water and food in the pots on the campfire are heated by conduction. The pots themselves are warmed by convection and radiation.

Conduction

Consider a bar of material with cross-sectional area A and length L (Figure 18.15a). This bar is put in physical contact with a thermal reservoir at a hotter temperature, T_h, and a thermal reservoir at a colder temperature, T_c, which means $T_h > T_c$ (Figure 18.15b). Heat then flows from the reservoir at the higher temperature to the reservoir at the lower temperature. The thermal energy transferred per unit time, P_{cond}, by the bar connecting the two heat reservoirs has been found to be given by

$$P_{\text{cond}} = \frac{Q}{t} = kA\frac{T_h - T_c}{L}, \tag{18.15}$$

where k is the **thermal conductivity** of the material of the bar. The SI units of thermal conductivity are W/(m K). Some typical values of thermal conductivity are listed in Table 18.3. Note the huge range of conductivities! The highest values are those for forms of pure carbon (diamond, carbon nanotubes, and especially graphene). The coinage metals (silver, gold, and copper) have very high thermal conductivities, but they are an order of magnitude lower than those of the forms of pure carbon. The materials at the bottom of the table (rubber, wood, paper, and polyurethane foam) are used for thermal insulation.

We can rearrange equation 18.15:

$$P_{\text{cond}} = \frac{Q}{t} = A\frac{T_h - T_c}{L/k} = A\frac{T_h - T_c}{R}, \tag{18.16}$$

where the **thermal resistance**, R, is defined to be

$$R = \frac{L}{k}. \tag{18.17}$$

The SI units of R are m^2 K/W. A higher R value means a lower rate of thermal energy transfer. A good insulator has a high R value. In the United States, thermal resistance is often specified by an R factor, which has the units ft^2 °F h/BTU. A common insulating material marked with an R factor of R-30 is shown in Figure 18.16. To convert an R factor from ft^2 °F h/BTU to m^2 K/W, divide the R factor by 5.678.

FIGURE 18.14 A campfire illustrates the three main modes of thermal energy transfer: conduction, convection, and radiation.

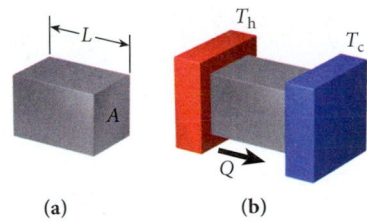

(a)   **(b)**

FIGURE 18.15 (a) A bar with cross-sectional area A and length L. (b) The bar is placed between two thermal reservoirs with temperatures T_h and T_c.

Table 18.3	Some Representative Thermal Conductivities
Material	**k [W/(m K)]**
Graphene	5000
Carbon nanotube	3500
Diamond	2000
Silver	420
Copper	390
Gold	320
Aluminum	220
Nickel	91
Iron	80
Lead	35
Stainless steel	20
Granite	3
Ice	2
Limestone	1.3
Concrete	0.8
Glass	0.8
Water	0.5
Rubber	0.16
Wood	0.16
Paper	0.05
Air	0.025
Polyurethane foam	0.02

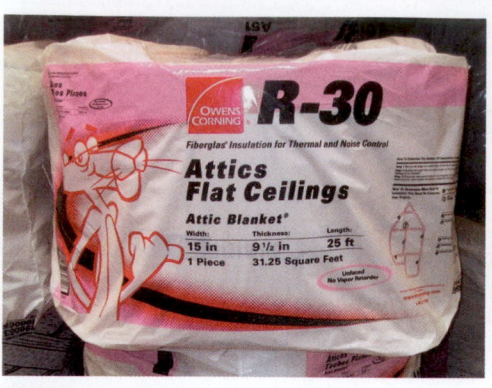
(a)

(b)

FIGURE 18.16 (a) A common insulating material with an *R* factor of R-30. (b) Insulation being installed in an attic.

Self-Test Opportunity 18.2

Show that the conversion factor for changing from ft^2 °F h/BTU to m^2 K/W is 5.678.

SOLVED PROBLEM 18.2 | **Thermal Energy Flow through a Copper/Aluminum Bar**

PROBLEM

A copper (Cu) bar of length $L = 90.0$ cm and cross-sectional area $A = 3.00$ cm^2 is in thermal contact at one end with a thermal reservoir at a temperature of 100. °C. The other end of the copper bar is in thermal contact with an aluminum (Al) bar of the same cross-sectional area and a length of 10.0 cm. The other end of the aluminum bar is in thermal contact with a thermal reservoir at a temperature of 1.00 °C. What is the thermal energy flow through the composite bar? The thermal conductivity of the copper is $k_{Cu} = 386$ W/(m K) and thermal conductivity of the aluminum is $k_{Al} = 220.$ W/(m K).

SOLUTION

THINK The thermal energy flow depends on the temperature difference between the two ends of the bar, the length and cross-sectional area of the bar, and the thermal conductivity of the materials. All the heat that flows from the high-temperature end must flow through both the copper and aluminum segments.

SKETCH Figure 18.17 is a sketch of the copper/aluminum bar.

RESEARCH We can use equation 18.15 to describe the thermal energy flow through a bar of length L and cross-sectional area A:

$$P_{cond} = kA\frac{T_h - T_c}{L}.$$

We have $T_h = 100.$ °C and $T_c = 1.00$ °C. We identify the temperature at the interface between the copper and aluminum segments as T. The thermal energy flow through the copper segment is then

$$P_{Cu} = k_{Cu} A\frac{T_h - T}{L_{Cu}}.$$

The thermal energy flow through the aluminum segment is

$$P_{Al} = k_{Al} A\frac{T - T_c}{L_{Al}}.$$

The thermal energy flow through the copper segment must equal the thermal energy flow through the aluminum segment, so we have

$$P_{Cu} = P_{Al}$$

$$= k_{Cu} A\frac{T_h - T}{L_{Cu}} = k_{Al} A\frac{T - T_c}{L_{Al}}. \qquad (i)$$

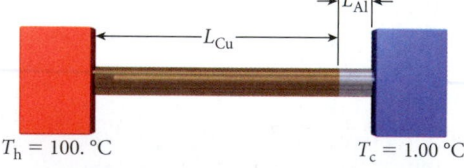

FIGURE 18.17 Copper/aluminum bar held at 100. °C at one end and 1.00 °C at the other end.

$T_h = 100.$ °C $T_c = 1.00$ °C

SIMPLIFY We can solve this equation for T. First, we divide through by A and then multiply both sides by $L_{Cu}L_{Al}$

$$k_{Cu}L_{Al}(T_h - T) = k_{Al}L_{Cu}(T - T_c).$$

Next, we multiply through on both sides and collect all terms with T on one side:

$$k_{Cu}L_{Al}T_h - k_{Cu}L_{Al}T = k_{Al}L_{Cu}T - k_{Al}L_{Cu}T_c$$

$$k_{Al}L_{Cu}T + k_{Cu}L_{Al}T = k_{Cu}L_{Al}T_h + k_{Al}L_{Cu}T_c$$

$$T(k_{Al}L_{Cu} + k_{Cu}L_{Al}) = k_{Cu}L_{Al}T_h + k_{Al}L_{Cu}T_c$$

$$T = \frac{k_{Cu}L_{Al}T_h + k_{Al}L_{Cu}T_c}{k_{Cu}L_{Al} + k_{Al}L_{Cu}}.$$

Substituting this expression for T into equation (i) gives us the thermal energy flow through the copper/aluminum bar.

CALCULATE Putting in the numerical values gives us

$$T = \frac{k_{Cu}L_{Al}T_h + k_{Al}L_{Cu}T_c}{k_{Cu}L_{Al} + k_{Al}L_{Cu}}$$

$$= \frac{[386 \text{ W/(m K)}](0.100 \text{ m})(373 \text{ K}) + [220. \text{ W/(m K)}](0.900 \text{ m})(274 \text{ K})}{[386 \text{ W/(m K)}](0.100 \text{ m}) + [220. \text{ W/(m K)}](0.900 \text{ m})}$$

$$= 290.1513 \text{ K}.$$

Putting this result for T into equation (i) gives us the thermal energy flow through the copper segment:

$$P_{Cu} = k_{Cu}A\frac{T_h - T}{L_{Cu}}$$

$$= [386 \text{ W/(m K)}](3.00 \cdot 10^{-4} \text{ m}^2)\frac{373 \text{ K} - 290.1513 \text{ K}}{0.900 \text{ m}}$$

$$= 10.6599 \text{ W}.$$

ROUND We report our result to three significant figures:

$$P_{Cu} = 10.7 \text{ W}.$$

DOUBLE-CHECK To double-check, let's calculate the thermal energy flow through the aluminum segment:

$$P_{Al} = k_{Al}A\frac{T - T_c}{L_{Al}} = [220 \text{ W/(m K)}](3.00 \cdot 10^{-4} \text{ m}^2)\frac{290.1513 \text{ K} - 274 \text{ K}}{0.100 \text{ m}} = 10.7 \text{ W}.$$

This agrees with our result for the copper segment.

EXAMPLE 18.7 | Roof Insulation

Suppose you insulate above the ceiling of a room with an insulation material having an R factor of R-30. The ceiling measures 5.00 m by 5.00 m. The temperature inside the room is 21.0 °C, and the temperature above the insulation is 40.0 °C.

PROBLEM
How much heat enters the room through the ceiling in a day if the room is maintained at a temperature of 21.0 °C?

SOLUTION
We solve equation 18.16 for the heat:

$$Q = At\frac{T_h - T_c}{R}.$$

Putting in the numerical values, we obtain (1 day has 86,400 s):

$$Q = (5.00 \text{ m} \cdot 5.00 \text{ m})(86,400 \text{ s})\frac{313 \text{ K} - 294 \text{ K}}{(30/5.678) \text{ m}^2 \text{ K/W}} = 7.77 \cdot 10^6 \text{ J}.$$

Concept Check 18.4

If you double the temperature (measured in kelvins) of an object, the thermal energy transferred away from it per unit time will

a) decrease by a factor of 2.

b) stay the same.

c) increase by a factor of 2.

d) increase by a factor of 4.

e) change by an amount that cannot be determined without knowing the temperature of the object's surroundings.

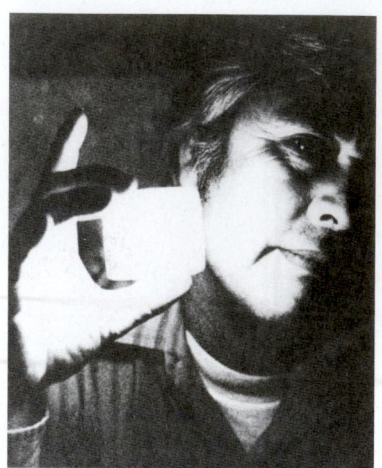

FIGURE 18.18 White-hot (1260 °C) ceramic tile used for the thermal protection system of the Space Shuttle is held by an unprotected hand.

Thermal insulation is a key component of spacecraft that have to reenter the Earth's atmosphere. The reentry process creates thermal energy from friction with air molecules. As a result of the high speed, a shock wave forms in front of the spacecraft, which deflects most of the thermal energy created in the process. However, excellent thermal insulation is still required to prevent this heat from being conducted to the frame of the spacecraft, which is basically made of aluminum and cannot stand temperatures significantly higher than 180 °C. The thermal protection system has to be of very low mass, like all parts of a spacecraft, and must be able to protect the spacecraft from very high temperatures, up to 1650 °C. In the spacecraft for Apollo missions, in which astronauts performed the Moon landings in the late 1960s and 1970s, and in planetary probes like the *Mars Rover,* the heat protection is simply an ablative heat shield, which burns off during entry into the atmosphere. For a reusable spacecraft like the Space Shuttle, such a shield is not an option, because it would require prohibitively expensive maintenance after each trip. The nose and wing edges of the Space Shuttle are instead covered with reinforced carbon, able to withstand the high reentry temperatures. The underside of the Space Shuttle is covered with passive heat protection, consisting of over 20,000 ceramic tiles made of 10% silica fibers and 90% empty space. They are such good thermal insulators that after being warmed to a temperature of 1260 °C (which is the maximum temperature the underside of the Space Shuttle encounters during reentry), they can be held by unprotected hands (see Figure 18.18). Thermal insulation to prevent heat loss is a very important consideration for homeowners. The following solved problem shows how such insulation can directly affect their pocketbooks.

SOLVED PROBLEM 18.3 / Cost of Warming a House in Winter

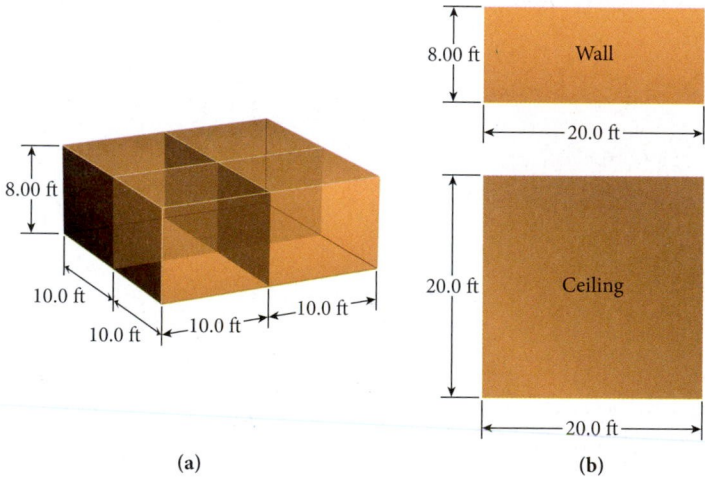

(a)

(b)

FIGURE 18.19 (a) An insulated four-room house. (b) One wall and the ceiling of the house.

You build a small house with four rooms (Figure 18.19a). Each room is 10.0 ft by 10.0 ft, and the ceiling is 8.00 ft high. The exterior walls are insulated with a material with an *R* factor of R-19, and the floor and ceiling are insulated with a material with an *R* factor of R-30. During the winter, the average temperature inside the house is 20.0 °C, and the average temperature outside the house is 0.00 °C. You warm the house for 6 months in winter using electricity that costs 9.5 cents per kilowatt-hour.

PROBLEM
How much do you pay to warm your house for the winter?

SOLUTION
THINK Using equation 18.16, we can calculate the total heat lost over 6 months through the walls (R-19) and through the floor and ceiling (R-30). We can then calculate how much it will cost to add this amount of heat to the house.

SKETCH The dimensions of one wall and the ceiling are shown in Figure 18.19b.

RESEARCH Each of the four exterior walls, as shown in Figure 18.19b, has an area of A_{wall}. The ceiling has an area of $A_{ceiling}$, shown in Figure 18.19b. The floor and ceiling have the same area. The thermal resistance of the walls is given by R-19; the thermal resistance of the floor and ceiling is given by R-30. Thus, the thermal resistance of the walls in SI units is

$$R_{wall} = \frac{19}{5.678} \, m^2 \, K/W = 3.346 \, m^2 \, K/W,$$

and the thermal resistance of the floor and ceiling is

$$R_{ceiling} = \frac{30}{5.678} \, m^2 \, K/W = 5.284 \, m^2 \, K/W.$$

SIMPLIFY We can calculate the heat lost per unit time using equation 18.16, taking the total area of the walls to be four times the area of one wall and the total area of the floor and ceiling as twice the area of the ceiling:

$$\frac{Q}{t} = 4A_{wall}\left(\frac{T_h - T_c}{R_{wall}}\right) + 2A_{ceiling}\left(\frac{T_h - T_c}{R_{ceiling}}\right) = 2(T_h - T_c)\left(\frac{2A_{wall}}{R_{wall}} + \frac{A_{ceiling}}{R_{ceiling}}\right).$$

CALCULATE The area of each exterior wall is $A_{wall} = (8.00 \text{ ft})(20.0 \text{ ft}) = 160.0 \text{ ft}^2 = 14.864 \text{ m}^2$. The area of the ceiling is $A_{ceiling} = (20.0 \text{ ft})(20.0 \text{ ft}) = 400.0 \text{ ft}^2 = 37.161 \text{ m}^2$. The number of seconds in 6 months is

$$t = (6 \text{ months})(30 \text{ days/month})(24 \text{ h/day})(3600 \text{ s/h}) = 1.5552 \cdot 10^7 \text{ s}.$$

The amount of heat lost in 6 months is then

$$Q = 2t(T_h - T_c)\left(\frac{2A_{wall}}{R_{wall}} + \frac{A_{ceiling}}{R_{ceiling}}\right)$$

$$= 2(1.5552 \cdot 10^7 \text{ s})(293 \text{ K} - 273 \text{ K})\left(\frac{2(14.864 \text{ m}^2)}{3.346 \text{ m}^2 \text{ K/W}} + \frac{37.161 \text{ m}^2}{5.284 \text{ m}^2 \text{ K/W}}\right)$$

$$= 9.902 \cdot 10^9 \text{ J}.$$

Calculating the total cost for 6 months of electricity for warming, we get

$$\text{Cost} = \left(\frac{\$0.095}{1 \text{ kWh}}\right)\left(\frac{1 \text{ kWh}}{3.60 \cdot 10^6 \text{ J}}\right)(9.902 \cdot 10^9 \text{ J}) = \$261.3041.$$

ROUND Since the input data were given to two significant figures, we report our result to two significant figures as well:

$$\text{Cost} = \$260.$$

(However, rounding down the amount on a bill is not acceptable to the utility company, and you would have to pay $261.30.)

DOUBLE-CHECK To double-check, let's use the R values the way a contractor in the United States might. The total area of the walls is $4(160 \text{ ft}^2)$, and the total area of the floor and ceiling is $2(400 \text{ ft}^2)$. The heat lost per hour through the walls is

$$\frac{Q}{t} = 4(160 \text{ ft}^2)\frac{20\left(\frac{9}{5}\right) \text{ °F}}{19 \text{ ft}^2 \text{ °F h/BTU}} = 1213 \text{ BTU/h}.$$

The heat lost through the ceiling and floor is

$$\frac{Q}{t} = 2(400 \text{ ft}^2)\frac{20\left(\frac{9}{5}\right) \text{ °F}}{30 \text{ ft}^2 \text{ °F h/BTU}} = 960 \text{ BTU/h}.$$

The total heat lost in 6 months (4320 h) is

$$Q = (4320 \text{ h})(1213 + 960) \text{ BTU/h}$$

$$= 9,387,360 \text{ BTU}.$$

The cost is then

$$\text{Cost} = \left(\frac{\$0.095}{1 \text{ kWh}}\right)\left(\frac{1 \text{ kWh}}{3412 \text{ BTU}}\right)(9,387,360 \text{ BTU}) = \$261.37.$$

This is close to our answer and seems reasonable.

Note that the cost of warming the small house for the winter seems low compared with real-world experience. The lowness of the cost is due to the fact that the house has no doors or windows and is well insulated.

Convection

If you hold your hand above a burning candle, you can feel the thermal energy transferred from the flame. The warmed air is less dense than the surrounding air and rises. The rising air carries thermal energy upward from the candle flame. This type of thermal energy transfer is called *convection*. Figure 18.20 shows a candle flame on Earth and on the orbiting Space Shuttle. You

(a)   (b)

FIGURE 18.20 Candle flames under (a) terrestrial conditions and (b) microgravity conditions.

FIGURE 18.21 A satellite image taken by the NASA TERRA satellite on May 5, 2001, uses false colors to show the temperature range of the water in the Atlantic Ocean off the east coast of the United States. The warm Gulf Stream is shown in red, and the east coast of the United States is shown in black.

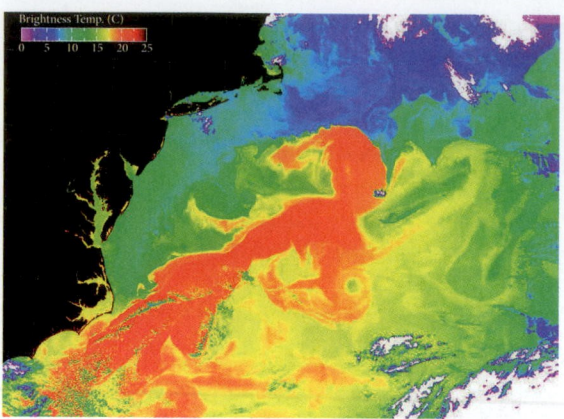

can see that the thermal energy travels upward from the candle burning on Earth but expands almost spherically from the candle aboard the Space Shuttle. The air in the Space Shuttle (actually in a small container on the Space Shuttle) has the same density in all directions, so the warmed air does not have a preferential direction in which to travel. The bottom of the flame on the candle aboard the Space Shuttle is extinguished because the wick carries thermal energy away. (In addition to convection involving a large-scale flow of mass, convection can occur via individual particles in a process called *diffusion,* which is not covered in this chapter.)

Most houses and office buildings in the United States have forced-air heating; that is, warm air is blown through heating ducts into rooms. This is an excellent example of thermal energy transfer via convection. The same air ducts are used in the summer for cooling the structures by blowing cooler air through the ducts and into the rooms, which is another example of convective thermal energy transfer. (But, of course, then the heat has the opposite sign, because the temperature in the room is lowered.)

Large amounts of energy are transferred by convection in the Earth's atmosphere and in the oceans. For example, the Gulf Stream carries warm water from the Gulf of Mexico northward through the Straits of Florida and up the east coast of the United States. The warmer temperature of the water in the Gulf Stream is revealed in the NASA satellite image in Figure 18.21. The Gulf Stream has a temperature around 20 °C as it flows with a speed of approximately 2 m/s up the east coast of the United States into the North Atlantic. The Gulf Stream then splits. One part continues to flow toward Britain and Western Europe, while the other part turns south along the African coast. The average temperature of Britain and Western Europe is approximately 5 °C higher than it would be without the thermal energy carried by the warmer waters of the Gulf Stream.

Some climate models predict that global warming may possibly threaten the Gulf Stream because of the melting of ice at the North Pole. The extra freshwater from the melting polar ice cap reduces the salinity of the water at the northern end of the Gulf Stream, which interferes with the mechanism that allows the cooled water from the Gulf Stream to sink and return to the south. Paradoxically, global warming could make Northern Europe colder.

SOLVED PROBLEM 18.4   Gulf Stream

Let's assume that a rectangular pipe of water 100 km wide and 500 m deep can approximate the Gulf Stream. The water in this pipe is moving with a speed of 2.0 m/s. The temperature of the water is 5.0 °C warmer than the surrounding water.

PROBLEM

Estimate how much power the Gulf Stream is carrying to the North Atlantic Ocean.

SOLUTION

THINK   We can calculate the volume flow rate by taking the product of the speed of the flow and the cross-sectional area of the pipe. Using the density of water, we can then calculate the mass flow rate. Using the specific heat of water and the temperature difference between the Gulf Stream water and the surrounding water, we can calculate the power carried by the Gulf Stream to the North Atlantic.

SKETCH Figure 18.22 shows the idealized Gulf Stream flowing to the northeast in the Atlantic Ocean.

RESEARCH We assume that the Gulf Stream has a rectangular cross section of width $w = 100$ km and depth $d = 500$ m. The area of the Gulf Stream flow is then

$$A = wd.$$

The speed of the flow of the Gulf Stream is assumed to be $v = 2.0$ m/s. The volume flow rate is given by

$$R_V = vA.$$

The density of seawater is $\rho = 1025$ kg/m^3. We can express the mass flow rate as

$$R_m = \rho R_V.$$

FIGURE 18.22 Idealized Gulf Stream flowing along the eastern coastline of the United States toward the North Atlantic.

The specific heat of water is $c = 4186$ J/(kg K). The heat required to raise the temperature of a mass m by ΔT is given by

$$Q = cm\Delta T.$$

The power carried by the Gulf Stream is equal to the heat per unit time:

$$\frac{Q}{t} = P = \frac{cm\Delta T}{t} = cR_m\Delta T.$$

(*Note:* The capital P symbolizes power; do not confuse this with the lowercase p used for pressure.)

SIMPLIFY The power carried by the Gulf Stream is given by

$$P = cR_m\Delta T = c\rho R_V \Delta T = c\rho vwd\Delta T.$$

CALCULATE The temperature difference is $\Delta T = 5\ °C = 5$ K. Putting in the numerical values gives

$$P = \left[4186\ \text{J}/\left(\text{kg K}\right)\right]\left(1025\ \text{kg/m}^3\right)\left(2.0\ \text{m/s}\right)\left(100\cdot10^3\ \text{m}\right)\left(500\ \text{m}\right)\left(5.0\ \text{K}\right)$$

$$= 2.1453\cdot10^{15}\ \text{W}.$$

ROUND We report our result to two significant figures:

$$P = 2.1\cdot10^{15}\ \text{W} = 2.1\ \text{PW} \quad (1\ \text{PW} = 1\ \text{petawatt} = 10^{15}\ \text{W}).$$

DOUBLE-CHECK To double-check this result, let's calculate how much power is incident on the Earth from the Sun. This total power is given by the cross-sectional area of Earth times the power incident on Earth per unit area:

$$P_{\text{total}} = \pi\left(6.4\cdot10^6\ \text{m}\right)^2\left(1400\ \text{W/m}^2\right) = 180\ \text{PW}.$$

Let's calculate how much of this power could be absorbed by the Gulf of Mexico. The intensity of sunlight at a distance from the Sun corresponding to the radius of Earth's orbit around the Sun is $S = 1400$ W/m^2. We can estimate the area of the Gulf of Mexico as $1.0\cdot10^6$ km^2 = $1.0\cdot10^{12}$ m^2. If the water of the Gulf of Mexico absorbed all the energy incident from the Sun for half of each day, then the power available from the Gulf would be

$$P = \frac{\left(1.0\cdot10^{12}\ \text{m}^2\right)\left(1400\ \text{W/m}^2\right)}{2} = 7.0\cdot10^{14}\ \text{W} = 0.7\ \text{PW},$$

which is less than our estimate of the power carried by the Gulf Stream. Thus, more of the Atlantic Ocean must be involved in providing energy for the Gulf Stream than just the Gulf of Mexico. In fact, the Gulf Stream gets its energy from a large fraction of the Atlantic Ocean. The Gulf Stream is part of a network of currents flowing in the Earth's oceans, induced by prevailing winds, temperature differences, and the Earth's topology and rotation.

Concept Check 18.5

The power carried by the Gulf Stream could be doubled by

a) decreasing its speed by a factor of 2.

b) decreasing its temperature by a factor of 2.

c) increasing its temperature by a factor of 2.

d) increasing its speed by a factor of 4.

e) none of the above.

Self-Test Opportunity 18.3

List some other convection phenomena encountered in everyday life.

Radiation

Radiation occurs via the transmission of electromagnetic waves. (In Chapter 31, we'll see how electromagnetic waves carry energy; for now, you can just accept this as an empirical fact.) Further, unlike mechanical and sound waves (covered in Chapters 15 and 16), these waves need no medium to support them. Thus, electromagnetic waves can carry energy from one site to another without any matter having to be present between the two sites. One example of the transport of energy via electromagnetic radiation occurs during a cell phone conversation; a cell phone acts as both a transmitter of radiation to the nearest cell tower and a receiver of radiation originating from that tower.

All objects emit electromagnetic radiation in the form of thermal radiation. The temperature of the object determines the radiated power of the object, $P_{radiated}$, which is given by the **Stefan-Boltzmann equation:**

$$P_{radiated} = \sigma \epsilon A T^4, \tag{18.18}$$

where $\sigma = 5.67 \cdot 10^{-8}$ W/K^{4}m^2 is called the **Stefan-Boltzmann constant,** ϵ is the **emissivity,** which has no units, and A is the surface (radiating) area. The temperature in equation 18.18 must be in kelvins, and it is assumed to be constant. The emissivity varies between 0 and 1, with 1 being the emissivity of an idealized object called a **blackbody.** A blackbody radiates 100% of its power, as given by equation 18.18, and absorbs 100% of any radiation incident on it. Although some real-world objects are close to being a blackbody, no perfect blackbodies exist; thus, the emissivity is always less than 1.

The earlier subsection on conduction discussed how the insulating ability of various materials is quantified by the R value. The heat loss of a house in winter or the heat gain in summer depends not only on conduction but also on radiation. New building techniques have aimed at increasing the efficiency of house insulation by using radiant barriers. A *radiant barrier* is a layer of material that effectively reflects electromagnetic waves, especially infrared radiation (which is the radiation we usually feel as warmth). The use of radiant barriers in house insulation is illustrated in Figure 18.23.

A radiant barrier is constructed of a reflective substance, usually aluminum. A typical commercial radiant barrier is shown in Figure 18.24. Its material is aluminum-coated polyolefin, which reflects 97% of infrared radiation. Tests by Oak Ridge National Laboratory of houses in Florida with and without radiant barriers have shown that summer heat gains of ceilings with R-19 insulation can be reduced by 16% to 42%, resulting in the reduction of air-conditioning costs by as much as 17%.

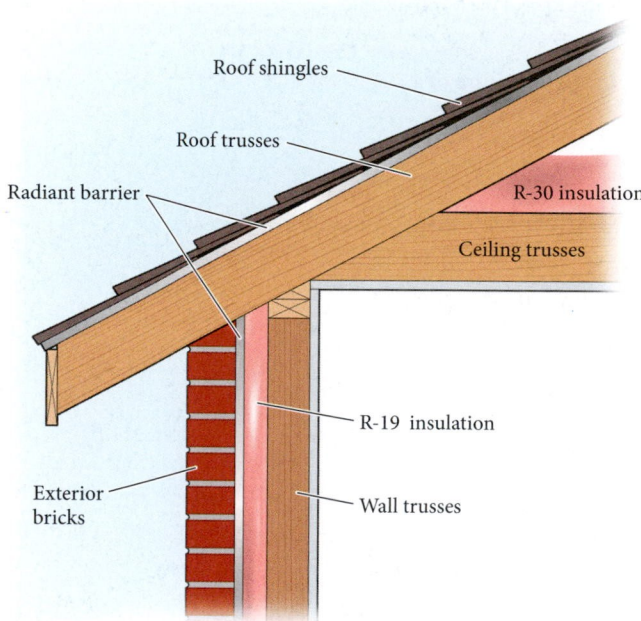

FIGURE 18.23 A schematic drawing of the corner of a house, showing part of the roof, part of the ceiling, and part of a wall. The roof consists of an outer layer of shingles, a radiant barrier, and trusses supporting the roof. The ceiling consists of ceiling trusses supporting insulation with an R factor of R-30. The wall consists of exterior bricks, a radiant barrier, insulation with an R factor of R-19, and supporting wall trusses.

The house illustrated in Figure 18.23 is designed to prevent heat from entering or leaving by conduction through its insulating layers, which have high R factors. The radiant barriers block heat from entering the house in the form of radiation. Unfortunately, this type of barrier also prevents the house from being warmed by the Sun in the winter. Heat gain or loss by convection is reduced by the dead space between the ceiling and the roof. Thus, the house is designed to reduce heat gain or loss by any of the three modes of thermal energy transfer: conduction, convection, or radiation. It is relatively straightforward to find out where a house is losing heat due to radiation by taking its picture with an infrared camera. Such images are called *thermograms*, and an example is shown in Figure 18.25. The red areas in the image show that this particular house loses the most heat through its roof, and thus an investment in better roof insulation might be called for.

FIGURE 18.24 One type of radiant barrier material, ARMA foil, produced by Energy Efficient Solutions.

FIGURE 18.25   Thermogram of a house.

EXAMPLE 18.8 / Earth as a Blackbody

Suppose that the Earth absorbed 100% of the incident energy from the Sun and then radiated all the energy back into space, just as a blackbody would.

PROBLEM
What would the temperature of the surface of Earth be?

SOLUTION
The intensity of sunlight arriving at the Earth is approximately $S = 1400$ W/m². The Earth absorbs energy as a disk with the radius of the Earth, R, whereas it radiates energy from its entire surface. At equilibrium, the energy absorbed equals the energy emitted:

$$(S)\left(\pi R^2\right) = (\sigma)(1)\left(4\pi R^2\right)T^4.$$

Solving for the temperature, we get

$$T = \sqrt[4]{\frac{S}{4\sigma}} = \sqrt[4]{\frac{1400 \text{ W/m}^2}{4\left(5.67 \cdot 10^{-8} \text{ W/K}^4\text{m}^2\right)}} = 280 \text{ K}.$$

This simple calculation gives a result close to the actual value of the average temperature of the surface of the Earth, which is about 288 K.

Concept Check 18.6

If the temperature (measured in kelvins) of an object is doubled, the heat it radiates per unit time will

a) decrease by a factor of 2.

b) stay the same.

c) increase by a factor of 2.

d) increase by a factor of 4.

e) increase by a factor of 16.

Global Warming

The difference between the temperature calculated for Earth as a blackbody in Example 18.8 and the actual temperature of the Earth's surface is partly due to the Earth's atmosphere, as illustrated in Figure 18.26.

Clouds in the Earth's atmosphere reflect 20% and absorb 19% of the Sun's energy. The atmosphere reflects 6% of the Sun's energy, and 4% is reflected by the surface of the Earth. The Earth's atmosphere transmits 51% of the energy from the Sun to the surface of the Earth. This solar energy is absorbed by the surface of the Earth and warms it, causing the surface to give off infrared radiation. Certain gases in the atmosphere—notably water vapor and carbon dioxide, in addition to other gases—absorb some of this infrared radiation, thereby trapping a fraction of the energy that would otherwise be radiated back into space. This trapping of thermal energy is called the **greenhouse effect**. The greenhouse effect keeps the Earth warmer than it would otherwise be and minimizes day-to-night temperature variations.

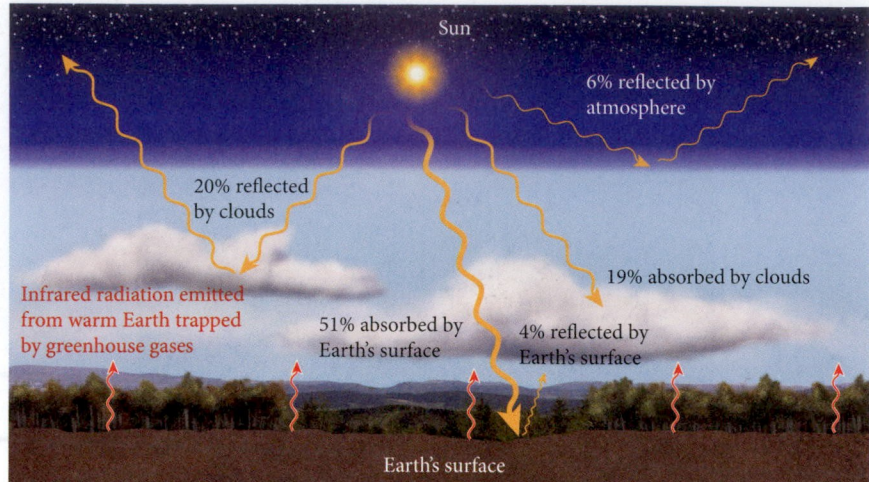

FIGURE 18.26 The Earth's atmosphere strongly affects the amount of energy absorbed by the Earth from the Sun.

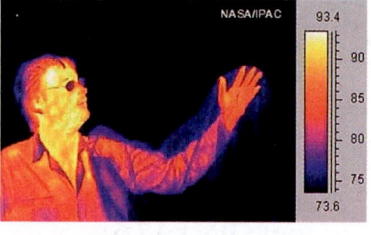

(a)

(b)

FIGURE 18.27 (a) A NASA scientist with his arm inside a black plastic bag, illuminated with visible light. (b) The same NASA scientist photographed with an infrared camera. The scientist radiated infrared radiation that passed through the black plastic bag.

The two photographs in Figure 18.27 illustrate how radiation of different wavelengths can penetrate materials differently. In Figure 18.27a, a NASA scientist has his arm inside a black plastic bag. The camera was sensitive to visible light, and his arm is not visible inside the bag. In Figure 18.27b, the same person was photographed with the lights off, using a camera sensitive to infrared radiation. The human body emits infrared radiation because its metabolism produces heat. This radiation passes through the black plastic that blocks visible light and the previously concealed arm is visible.

The burning of fossil fuels and other human activities have increased the amount of carbon dioxide in the Earth's atmosphere and increase the surface temperature by trapping infrared radiation that would otherwise be emitted into space.

Figure 18.28a shows the concentration of carbon dioxide in the air for the past 420,000 years. For this figure we have combined direct measurement from air samples at Mauna Loa station in Hawaii (green symbols) and the South Pole (orange symbols) with measurements from several ice core samples in Antarctica. Visible in this plot are glacial periods with relatively low carbon dioxide concentrations around 200 ppmv and interglacial periods with relatively high carbon dioxide concentrations around 275 ppmv. Part (b) of this figure shows the same data, but displayed only since the year 1000, which highlights

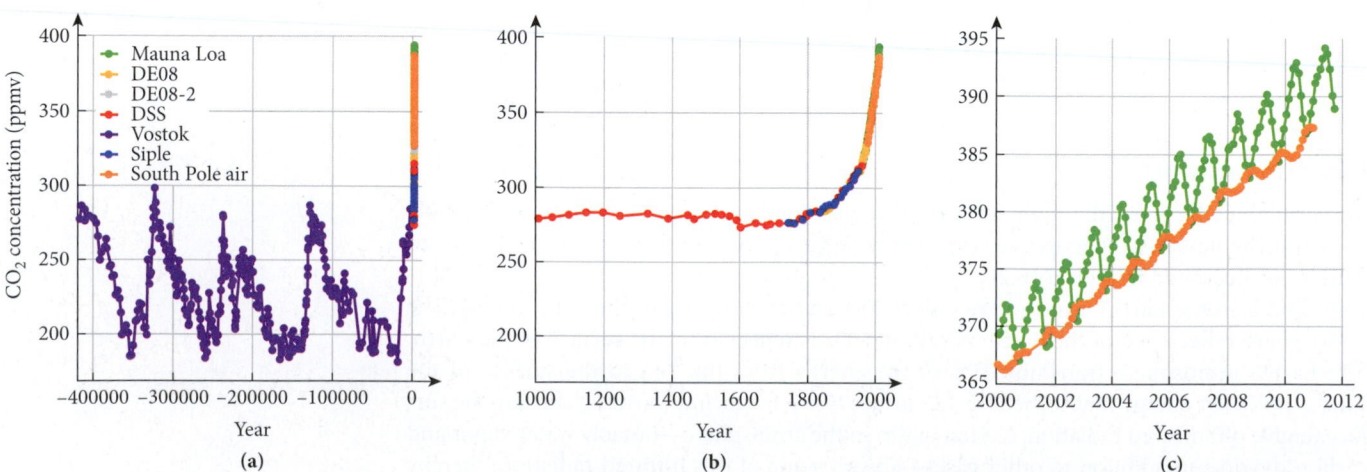

FIGURE 18.28 Concentration of carbon dioxide (CO_2) in the Earth's atmosphere in parts per million by volume (ppmv). (a) Concentration of carbon dioxide in the atmosphere during the last 420,000 years. The measurements shown are from air samples at Mauna Loa in Hawaii (green) and the South Pole (orange) and various ice core samples from Antarctica. (b) Display of the same data as in part (a), but only from 1000 AD to the present. (c) Display of the same data as in part (b), but only from 2000 to 2012.

the exponential growth of atmospheric CO_2 during the past two centuries. In part (c) of this figure we only display the data since the year 2000. At this resolution one can see the seasonal variation of the CO_2 concentration due to the growing seasons of the plants. The combination of direct measurements in the last 50 years and inferred concentrations from studies of ice cores indicates that the current concentration of carbon dioxide in the atmosphere is more than 30% higher than at any time in the 420,000 years before the Industrial Revolution (which occurred in the middle of the 18th century). Some researchers estimate that the current concentration of carbon dioxide is at its highest level in the past 20 million years. Models of the composition of the Earth's atmosphere based on current trends predict that the concentration of carbon dioxide will continue to increase in the next 100 years. This increase in the atmospheric concentration of carbon dioxide contributes to the observed global warming described in Chapter 17.

Worldwide, governments are reacting to these observations and predictions in many ways, including the Kyoto Protocol, which went into effect in 2005. In the Kyoto Protocol, signing nations agreed to make substantial cuts in their greenhouse gas emissions by the year 2012. However, several emerging nations were not required to reduce their greenhouse emissions. In 2012, international negotiations were underway to extend the Kyoto Protocol, but prospects were not bright. Canada had withdrawn from the treaty at the end of 2011 and the United States never ratified the previous agreement.

Heat in Computers

Thinking about thermal energy transfer may not bring your computer to mind right away. But cooling a computer is a major engineering problem. A typical desktop computer consumes between 100 and 150 W of electrical power, and a laptop uses between 25 and 70 W. As a general guideline, the higher the computer chips' frequency, the higher the power consumption. Most of this electrical power is converted into thermal energy and has to be removed from the computer. To accomplish this, computers have passive heat sinks, which consist of pieces of metal with large surface areas attached to the parts of the computer in need of cooling, mainly CPU and graphics chips (Figure 18.29). Passive heat sinks apply conduction to move the thermal energy away from the computer parts and then radiation to transfer the thermal energy to the surrounding air. An active heat sink includes a small fan to move more air past the metal surfaces to increase the thermal energy transfer. Of course, the fan also consumes electrical power and thus reduces the charge period of the battery in laptop computers.

While it may be only a slight annoyance to have a laptop computer warming your lap, cooling large computer clusters in data centers is very expensive. If a server farm contains 10,000 individual CPUs, its electrical power consumption is on the order of 1 MW, most of which is transformed to heat that has to be removed by very large air-conditioning systems. Thus, development of more efficient computer components, from power supplies to CPU chips, and more efficient cooling methods is currently the focus of much research.

Geothermal Power Resources

Earth has a solid inner core with a radius of approximately 1200 km, predominantly composed of iron (80%) and nickel. The inner core is encased by a liquid outer core with a radius of about 3400 km. Around the outer core is the mantle, which is mainly composed of oxygen, silicon, and magnesium and has a thickness of about 2800–2900 km. Earth's outermost layer is the crust, composed of solid tectonic plates with a thickness between 5 km and 70 km. Figure 18.30 shows these layers inside Earth.

Currently, the best estimate of the temperature at Earth's center is approximately 6000 K. This high temperature is mainly due to the radioactive decay of very long-lived isotopes of uranium, potassium, and thorium (radioactive decay is discussed in Chapter 40). The average surface temperature of Earth is approximately 290 K, so a very important question is how Earth's temperature

FIGURE 18.29 Active and passive heat sink technology mounted on a computer CPU.

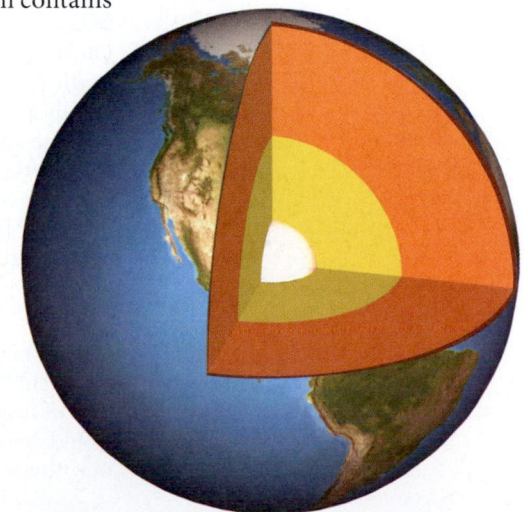

FIGURE 18.30 Structure of the Earth's interior (drawn to approximate scale), showing (from inside to outside) solid inner iron core, liquid outer core, mantle, and the comparatively thin crust.

FIGURE 18.31 Estimated temperatures at a depth of 6 km below ground level in the continental United States.

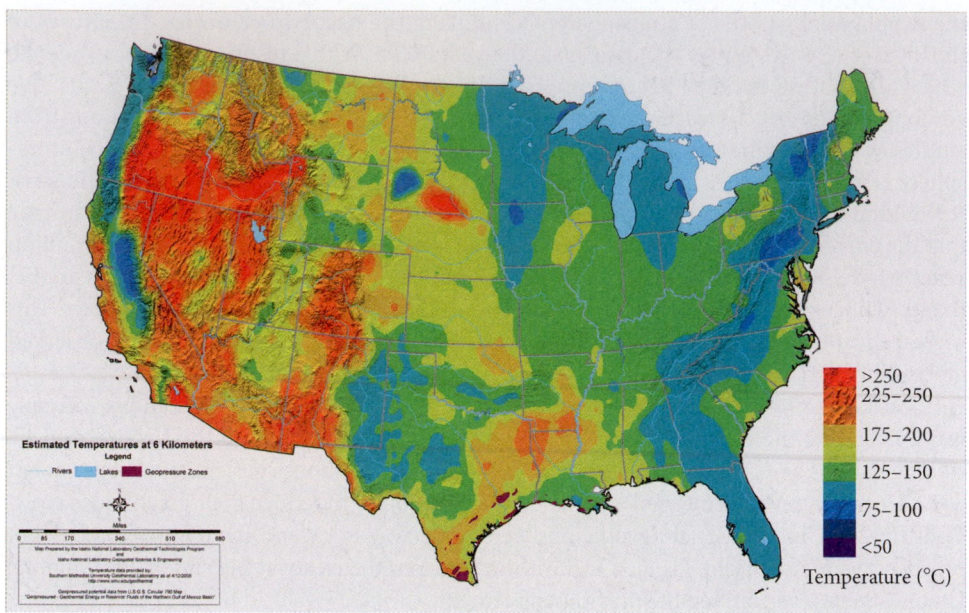

Estimated Temperatures at 6 Kilometers
Legend

Rivers   Lakes   Geopressure Zones

	Temperature (°C)
	>250
	225–250
	175–200
	125–150
	75–100
	<50

varies with depth. Near the edges of the tectonic plates, relatively high temperatures can be found fairly close to the surface; these are the source of volcanic events. Away from the plate edges, however, a simple rule of thumb is that the temperature in the crust increases by 25–30 K per kilometer of depth. This number is called the *geothermal gradient*. The temperature at the boundary between crust and mantle is between 800 K and 1200 K. Figure 18.31 shows the expected temperatures at a depth of 6 km below ground level in the continental United States.

EXAMPLE 18.9   Estimate of Earth's Internal Thermal Energy

Since Earth's core and mantle are at very high temperatures relative to its surface, there must be a lot of thermal energy available inside Earth.

PROBLEM
What is the thermal energy stored in Earth's interior?

SOLUTION
Obviously, we can make only a rough estimate, because the exact radial temperature profile of Earth is not known. Let's assume an average temperature of 3000 K, which is approximately half of the difference between the surface and core temperatures.

The specific heats (see Table 18.1) for the materials in the Earth's interior range from 0.45 kJ/(kg K) for iron to 0.92 kJ/(kg K) for rocks in the crust. In order to make our estimate, we will use an average value of 0.7 kJ/(kg K). The total mass of Earth is (see Table 12.1) $5.97 \cdot 10^{24}$ kg.

Inserting the numbers into equation 18.12, we find

$$Q_{\text{Earth}} = m_{\text{Earth}} c \Delta T = \left(6 \cdot 10^{24}\,\text{kg}\right)\left[0.7\,\text{kJ}/\left(\text{kg K}\right)\right]\left(3000\,\text{K}\right) = 10^{31}\,\text{J}.$$

Does it matter that some part of Earth's core is liquid and not solid? Should we account for the latent heat of fusion in our estimate? The answer is yes, in principle, but since the latent heat of fusion for metals is typically on the order of a few hundred kilojoules per kilogram, it would contribute only 10–20% of what the specific heat does in this case. For our order-of-magnitude estimate, we can safely neglect this contribution.

Since the interior of Earth is hotter than the surface, we expect heat to flow to the surface from the interior. How much heat flows per time unit cannot be determined solely from our estimate in Example 18.9. However, analysis of available data shows a global average of 87 mW/m² (65 mW/m² for the continents and 101 mW/m² for the oceans). Since the surface area of Earth is approximately $5.1 \cdot 10^{14}$ m², a total power of 44 TW emerges from below our feet. (For comparison, the entire power consumption of all humans is currently almost 16 TW. On the other hand, 44 TW represents only approximately 0.025% of the 175 PW of solar energy incident on Earth.)

Thus, tremendously large thermal resources are available in Earth's interior, and there is a huge flow of heat through the Earth's surface. The question is how to utilize these resources. In some countries, utilization of geothermal power is quite actively pursued; foremost among them is Iceland. Iceland has the advantage of being located where the Earth's crust is very thin; thus, the thermal resources of the Earth's interior can be tapped into comparatively easily. Five large geothermal power plants, like the one depicted in Figure 18.32, supply 25% of Iceland's electrical power and almost 90% of its hot water and home heating.

FIGURE 18.32 Nesjavellir Geothermal Power Station in Southwest Iceland, which generates 120 MW of electrical power.

SOLVED PROBLEM 18.5 | Enhanced Geothermal System

One way to tap into Earth's geothermal resources is by means of an *enhanced geothermal system* (EGS), in which two or more holes are drilled to a depth of several kilometers and then water is forced down one of the holes. If enough pressure is applied, the bedrock at the bottom of that borehole will develop small fissures that allow the water to percolate to the other borehole(s). This process is known as *hydraulic fracking*, or simply *fracking*. On the way through the bedrock, the water warms up to the temperature of that rock and emerges from the other borehole(s) as hot water or steam.

PROBLEM
How much thermal power does an enhanced geothermal system deliver, if it circulates 12.5 L of water per second and the water is heated from 35.1 °C to 126.5 °C?

SOLUTION
THINK We are asked to calculate the power, which is defined as energy per unit time. We are given the flow rate, which we can translate into mass per unit time. From this mass rate and the temperature difference, we can calculate the heat transported to the surface by the water per unit time, which is our answer.

SKETCH Figure 18.33 shows an enhanced geothermal system with three boreholes. The cold water is represented in blue, and the hot water or steam in red.

RESEARCH In Example 18.5, we found the heat required to raise the temperature of water through the phase change of vaporization. Following the approach of that example, we can calculate the heat required to raise the temperature of some mass, m, of water from T_1 to T_2:

$$Q = Q_1 + Q_2 + Q_3 = c_{water}m(100\ °C - T_1) + mL_{vaporization} + c_{steam}m(T_2 - 100\ °C).$$

The average power is given by $\bar{P} = \Delta Q / \Delta t$, where ΔQ is

$$\Delta Q = [c_{water}(100\ °C - T_1) + L_{vaporization} + c_{steam}(T_2 - 100\ °C)]\Delta m.$$

We know the volume flow rate, R_V, which is related to the mass flow rate, $R_m \equiv \Delta m/\Delta t$, via $R_m = R_V\rho$, where ρ is the mass density of water.

SIMPLIFY We can put all of the above together to obtain an expression for the average thermal power delivered by this system:

$$\bar{P} = \frac{\Delta Q}{\Delta t} = \left[c_{water}(100\ °C - T_1) + L_{vaporization} + c_{steam}(T_2 - 100\ °C)\right]\frac{\Delta m}{\Delta t}$$

$$= \left[c_{water}(100\ °C - T_1) + L_{vaporization} + c_{steam}(T_2 - 100\ °C)\right]R_V\rho.$$

– Continued

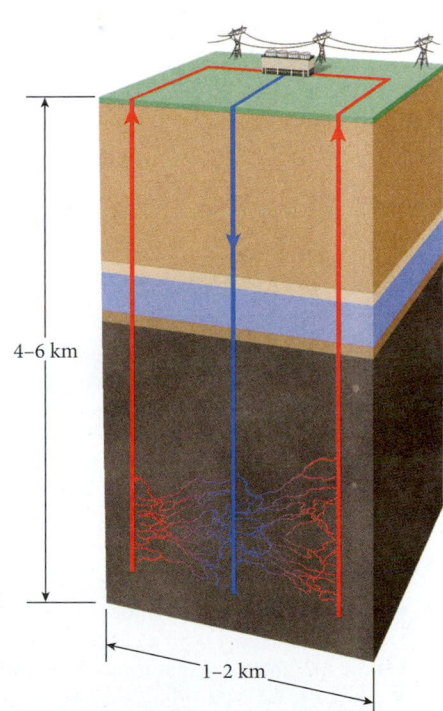

FIGURE 18.33 Diagram of an enhanced geothermal system.

4–6 km

1–2 km

CALCULATE

Now we insert the numerical values:

$$\overline{P} = \frac{\left(\left[4.19 \text{ kJ}/(\text{kg K})\right](100 \text{ °C} - 35.1 \text{ °C}) + (2260 \text{ kJ/kg}) + \left[2.01 \text{ kJ}/(\text{kg K})\right](126.5 \text{ °C} - 100 \text{ °C})\right)}{(12.5 \text{ L/s})(1.00 \text{ kg/L})}$$

$$= (272 \text{ kJ/kg} + 2260 \text{ kJ/kg} + 53.3 \text{ kJ/kg})(12.5 \text{ kg/s})$$

$$= 3.231625 \cdot 10^7 \text{ J/s}.$$

ROUND   We round to three significant digits and obtain $\overline{P} = 32.3$ MW.

DOUBLE-CHECK   Can we really expect to get 32.3 MW from this enhanced geothermal system? Not quite, because we have only determined the thermal power. Converting this power into electrical power leads to losses of more than 50%. In addition, the system's pumps and other equipment consume significant power. Thus, we can say that this system could deliver 5–10 MW, which is within the range of existing systems in operation around the world. Note that the main contribution to the power from this system arises from the latent heat of vaporization; this shows the importance of drilling into rock that is hot enough to generate steam, that is, at least 100 °C. But, as Figure 18.31 shows, this condition can be met in most of the continental United States, if the boreholes are drilled to a depth of 6 km.

As Solved Problem 18.5 shows, enhanced geothermal systems can make a significant contribution toward solving the global energy problem. To be sure, there are some risks involved in operating EGS power plants. For example, there have been instances of induced seismism (earthquake activity), and it is not entirely clear how much groundwater can be contaminated by the fracking process. However, recent experiments have used CO_2 as the pumping fluid instead of water, which has the advantages of contributing to carbon sequestration and avoiding the loss of water that is inherent in the fracking process.

WHAT WE HAVE LEARNED | EXAM STUDY GUIDE

- Heat is energy transferred between a system and its environment or between two systems because of a temperature difference between them.

- A calorie is defined in terms of the joule: 1 cal = 4.186 J.

- The work done by a system in going from an initial volume V_i to a final volume V_f is $W = \int dW = \int_{V_i}^{V_f} p \, dV$.

- The First Law of Thermodynamics states that the change in internal energy of a closed system is equal to the heat flowing into a system minus the work done by the system, or $\Delta E_{int} = Q - W$. The First Law of Thermodynamics states that energy in a closed system is conserved.

- An adiabatic process is one in which $Q = 0$.

- In a constant-volume process, $W = 0$.

- In a closed-loop process, $Q = W$.

- In an adiabatic free expansion, $Q = W = \Delta E_{int} = 0$.

- If heat, Q, is added to an object, its change in temperature, ΔT, is given by $\Delta T = \frac{Q}{C}$, where C is the heat capacity of the object.

- If heat, Q, is added to an object with mass m, its change in temperature, ΔT, is given by $\Delta T = \frac{Q}{cm}$, where c is the specific heat of the substance of which the object is made.

- The energy required to melt a solid to a liquid, divided by its mass, is the latent heat of fusion, L_{fusion}. During the melting process, the temperature of the system remains at the melting point, $T = T_{melting}$.

- The energy required to vaporize a liquid to a gas, divided by its mass, is the latent heat of vaporization, $L_{vaporization}$. During the vaporization process, the temperature of the system remains at the boiling point, $T = T_{boiling}$.

- If a bar of cross-sectional area A is placed between a thermal reservoir with temperature T_h and a thermal reservoir with temperature T_c, where $T_h > T_c$, the rate of heat through the bar is $P_{cond} = \frac{Q}{t} = A\frac{T_h - T_c}{R}$, where R is the thermal resistance of the material of the bar.

- The radiated power from an object with temperature T and surface area A is given by the Stefan-Boltzmann equation: $P_{radiated} = \sigma \epsilon A T^4$, where $\sigma = 5.67 \cdot 10^{-8}$ W/K^{4}m^2 is the Stefan-Boltzmann constant and ϵ is the emissivity.

ANSWERS TO SELF-TEST OPPORTUNITIES

18.1 The work is negative.

18.2

$$\left(\frac{1\,m^2\,K}{1\,W}\right)\left(\frac{(3.2808)^2\ ft^2}{m^2}\right)\left(\frac{\frac{9}{5}\ ^\circ F}{1\,K}\right)\left(\frac{1\,W\,s}{1\,J}\right)\left(\frac{1055\,J}{1\,BTU}\right)\left(\frac{1\,h}{3600\,s}\right)$$

$$=5.678\ \frac{ft^2\ ^\circ F\,h}{BTU}.$$

18.3 thunderstorms, jet stream, heating water in a pan, heating a house

PROBLEM-SOLVING GUIDELINES

1. When using the First Law of Thermodynamics, always check the signs of work and heat. In this book, work done on a system is positive, and work done by a system is negative; heat added to a system is positive, and heat given off to a system is negative. Some books define the signs of work and heat differently; be sure you know what sign convention applies for a particular problem.

2. Work and heat are path-dependent quantities, but a system's change in internal energy is path-independent. Thus, to calculate a change in internal energy, you can use any processes that start from the same initial position and end at the same final position in a pV-diagram. Work and heat may vary, depending on the path in the diagram, but their difference, $Q - W$, will remain the same. For these kinds of problems, be sure to clearly define the system and identify the initial and final conditions for each step of a process.

3. For calorimetry problems, conservation of energy demands that transfers of energy sum to zero. In other words, heat gained by one system must equal heat lost by the surroundings or some other system. This fact sets up the basic equation describing any transfer of heat between systems or objects.

4. For calculating amounts of heat and corresponding temperature changes, remember that specific heat refers to a heat change per unit mass of a substance, corresponding to a temperature change; for an object of known mass, you need to use the heat capacity. Also be aware of the possibility of a phase change. If a phase change is possible, break up the heat transfer process into steps, calculating heat corresponding to temperature changes and latent heat corresponding to phase changes. Remember that temperature changes are always final temperature minus initial temperature.

5. Be sure to check calorimetry calculations against reality. For example, if a final temperature is higher than an initial temperature, even though heat was extracted from a system, you may have overlooked a phase change.

MULTIPLE-CHOICE QUESTIONS

18.1 A 2.0-kg metal object with a temperature of 90 °C is submerged in 1.0 kg of water at 20 °C. The water-metal system reaches equilibrium at 32 °C. What is the specific heat of the metal?

a) 0.840 kJ/(kg K)

b) 0.129 kJ/(kg K)

c) 0.512 kJ/(kg K)

d) 0.433 kJ/(kg K)

18.2 A gas enclosed in a cylinder by a piston that can move without friction is warmed, and 1000 J of heat enters the gas. Assuming that the volume of the gas is constant, the change in the internal energy of the gas is

a) 0.

b) 1000 J.

c) –1000 J.

d) none of the above.

18.3 In the isothermal compression of a gas, the volume occupied by the gas is decreasing, but the temperature of the gas remains constant. In order for this to happen,

a) heat must enter the gas.

b) heat must exit the gas.

c) no heat exchange should take place between the gas and the surroundings.

18.4 Which surface should you set a pot on to keep it hotter for a longer time?

a) a smooth glass surface

b) a smooth steel surface

c) a smooth wood surface

d) a rough wood surface

18.5 Assuming that the severity of a burn increases as the amount of energy put into the skin increases, which of the following would cause the most severe burn (assume equal masses)?

a) water at 90 °C

b) copper at 110 °C

c) steam at 180 °C

d) aluminum at 100 °C

e) lead at 100 °C

18.6 In which type of process is no work done on a gas?

a) isothermal

b) isochoric

c) isobaric

d) none of the above

18.7 An aluminum block of mass $m_{Al} = 2.0$ kg and specific heat $c_{Al} = 910$ J/(kg K) is at an initial temperature of 1000 °C and is dropped into a bucket of water. The water has mass $m_{H_2O} = 12$ kg and specific heat $c_{H_2O} = 4190$ J/(kg K) and is at room temperature (25 °C). What is the approximate final temperature of the system when it reaches thermal equilibrium? (Neglect heat loss out of the system.)

a) 50 °C

b) 60 °C

c) 70 °C

d) 80 °C

18.8 A material has mass density ρ, volume V, and specific heat c. Which of the following is a correct expression for the heat exchange that occurs when the material's temperature changes by ΔT in degrees Celsius?

a) $(\rho c/V)\Delta T$

b) $(\rho cV)(\Delta T + 273.15)$

c) $(\rho cV)/\Delta T$

d) $\rho cV\Delta T$

18.9 Which of the following *does not* radiate heat?

a) ice cube

b) liquid nitrogen

c) liquid helium

d) a device at $T = 0.010$ K

e) all of the above

f) none of the above

18.10 Which of the following statements is (are) true?

a) When a system does work, its internal energy always decreases.

b) Work done on a system always decreases its internal energy.

c) When a system does work on its surroundings, the sign of the work is always positive.

d) Positive work done on a system is always equal to the system's gain in internal energy.

e) If you push on the piston of a gas-filled cylinder, the energy of the gas in the cylinder will increase.

18.11 How much thermal energy is needed to raise the temperature of a 3.0-kg copper block from 25.0 °C to 125 °C?

a) 116 kJ

b) 278 kJ

c) 421 kJ

d) 576 kJ

e) 761 kJ

18.12 How much thermal energy is needed to melt a 3.0-kg copper block that is initially at a temperature of 1359 K?

a) 101 kJ

b) 221 kJ

c) 390 kJ

d) 615 kJ

e) 792 kJ

18.13 Suppose you raise the temperature of 1 kg of water from −10 °C to +10 °C, then from +20 °C to +40 °C, and finally from +90 °C to +110 °C. Which of the three changes in temperature requires the addition of the *least* heat?

a) first change

b) second change

c) third change

d) All of the changes require the same amount of heat.

18.14 The seasonal variations in atmospheric CO_2 concentration shown in Figure 18.28 are caused by

a) the plant growth cycle, in which the plants in the Northern Hemisphere have the greater effect.

b) the plant growth cycle, in which the plants in the Southern Hemisphere have the greater effect.

c) the power usage of humans, which is much higher for the developed countries in the Northern Hemisphere.

d) the power usage of humans, which is much higher for the rapidly developing countries in the Southern Hemisphere.

CONCEPTUAL QUESTIONS

18.15 Estimate the power radiated by an average person. (Approximate the human body as a cylindrical blackbody.)

18.16 Several days after the end of a snowstorm, the roof of one house is still completely covered with snow, and another house's roof has no snow cover. Which house is most likely better insulated?

18.17 Why does tile feel so much colder to your feet after a bath than a bath rug? Why is this effect more striking when your feet are cold?

18.18 Can you think of a way to make a blackbody, a material that absorbs essentially all of the radiant energy falling on it, if you only have a material that reflects half the radiant energy that falls on it?

18.19 In 1883, the volcano on Krakatau Island in the Pacific erupted violently in the largest explosion in Earth's recorded history, destroying much of the island in the process. Global temperature measurements indicate that this explosion reduced the average temperature of Earth by about 1 °C during the next two decades. Why?

18.20 Fire walking is practiced in parts of the world for various reasons and is also a tourist attraction at some seaside resorts. How can a person walk across hot coals at a temperature well over 500 °F without burning his or her feet?

18.21 Why is a dry, fluffy coat a better insulator than the same coat when it is wet?

18.22 It has been proposed that global warming could be offset by dispersing large quantities of dust in the upper atmosphere. Why would this work, and how?

18.23 A thermos bottle fitted with a piston is filled with a gas. Since the thermos bottle is well insulated, no heat can enter or leave it. The piston is pushed in, compressing the gas.

a) What happens to the pressure of the gas? Does it increase, decrease, or stay the same?

b) What happens to the temperature of the gas? Does it increase, decrease, or stay the same?

c) Do any other properties of the gas change?

18.24 How would the rate of heat transfer between a thermal reservoir at a higher temperature and one at a lower temperature differ if the reservoirs were in contact with a 10-cm-long glass rod instead of a 10-m-long aluminum rod having an identical cross-sectional area?

18.25 Why might a hiker prefer a plastic bottle to an old-fashioned aluminum canteen for carrying his drinking water?

18.26 A girl has discovered a very old U.S. silver dollar and is holding it tightly in her little hands. Suppose that she put the silver dollar on the wooden (insulating) surface of a table, and then a friend came in from outside and placed on top of the silver dollar an equally old penny that she just found in the snow, where it had been left all night. Estimate the final equilibrium temperature of the system of the two coins in thermal contact.

EXERCISES

A blue problem number indicates a worked-out solution is available in the Student Solutions Manual. One • and two •• indicate increasing level of problem difficulty.

Sections 18.2 and 18.3

18.27 You are going to lift an elephant (mass = $5.0 \cdot 10^3$ kg) over your head (2.0 m vertical displacement).

a) Calculate the work required to do this. You will lift the elephant slowly (no tossing of the elephant allowed!). If you want, you can use a pulley system. (As you saw in Chapter 5, this does not change the energy required to lift the elephant, but it definitely reduces the force required to do so.)

b) How many doughnuts (250 food calories each) must you metabolize to supply the energy for this feat?

18.28 A gas has an initial volume of 2.00 m^3. It is expanded to three times its original volume through a process for which $P = \alpha V^3$, with $\alpha = 4.00$ N/m^{11}. How much work is done by the expanding gas?

18.29 How much work is done per cycle by a gas following the path shown on the pV-diagram?

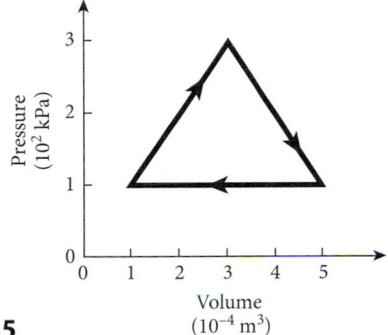

Sections 18.4 and 18.5

18.30 The internal energy of a gas is 500. J. The gas is compressed adiabatically, and its volume decreases by 100. cm^3. If the pressure applied on the gas during compression is 3.00 atm, what is the internal energy of the gas after the adiabatic compression?

Section 18.6

18.31 You have 1.00 cm^3 of each of the materials listed in Table 18.1 except ice and steam, all at room temperature, 22.0 °C. Which material has the highest temperature after 1.00 J of thermal energy is added to each sample? Which has the lowest temperature? What are these temperatures?

18.32 Suppose you mix 7.00 L of water at $2.00 \cdot 10^1$ °C with 3.00 L of water at 32.0 °C; the water is insulated so that no energy can flow into it or out of it. (You can achieve this, approximately, by mixing the two fluids in a foam cooler of the kind used to keep drinks cool for picnics.) The 10.0 L of water will come to some final temperature. What is this final temperature?

18.33 A 25.0-g piece of aluminum at 85.0 °C is dropped in 1.00 L of water at $1.00 \cdot 10^1$ °C, which is in an insulated beaker. Assuming that there is negligible heat loss to the surroundings, determine the equilibrium temperature of the system.

18.34 A 12.0-g lead bullet is shot with a speed of 250. m/s into a wooden wall. Assuming that 75.0% of the kinetic energy is absorbed by the bullet as heat (and 25.0% by the wall), what is the final temperature of the bullet?

•18.35 A 1.00-kg block of copper at 80.0 °C is dropped into a container with 2.00 L of water at 10.0 °C. Compare the magnitude of the change in energy of the copper to the magnitude of the change in energy of the water. Which value is larger?

•18.36 A 1.19-kg aluminum pot contains 2.31 L of water. Both pot and water are initially at 19.7 °C. How much heat must flow into the pot and the water to bring their temperature up to 95.0 °C? Assume that the effect of water evaporation during the heating process can be neglected and that the temperature remains uniform throughout the pot and the water.

•18.37 A metal brick found in an excavation was sent to a testing lab for nondestructive identification. The lab weighed the brick and found its mass to be 3.00 kg. The brick was heated to a temperature of 300. °C and dropped into an insulated copper calorimeter of mass 1.50 kg containing 2.00 kg of water at 20.0 °C. The final temperature at equilibrium was 31.7 °C. By calculating the specific heat from this data, can you identify the metal of which the brick is made?

•18.38 A $2.00 \cdot 10^2$ g piece of copper at a temperature of 450. K and a $1.00 \cdot 10^2$ g piece of aluminum at a temperature of $2.00 \cdot 10^2$ K are dropped into an insulated bucket containing $5.00 \cdot 10^2$ g of water at 280. K. What is the equilibrium temperature of the mixture?

••18.39 When an immersion glass thermometer is used to measure the temperature of a liquid, the temperature reading will be affected by an error due to heat transfer between the liquid and the thermometer. Suppose you want to measure the temperature of 6.00 mL of water in a Pyrex glass vial thermally insulated from the environment. The empty vial has a mass of 5.00 g. The thermometer you use is made of Pyrex glass as well and has a mass of 15.0 g, of which 4.00 g is the mercury inside the thermometer. The thermometer is initially at room temperature (20.0 °C). You place the thermometer in the water in the vial, and after a while, you read an equilibrium temperature of 29.0 °C. What was the actual temperature of the water in the vial *before* the temperature was measured? The specific heat of Pyrex glass around room temperature is 800. J/(kg K) and that of liquid mercury at room temperature is 140. J/(kg K).

Section 18.7

•18.40 Suppose 400. g of water at 30.0 °C is poured over a 60.0-g cube of ice with a temperature of −5.00 °C. If all the ice melts, what is the final temperature of the water? If all of the ice does not melt, how much ice remains when the water-ice mixture reaches equilibrium?

18.41 A person gave off 180. kcal of heat to evaporate water from her skin in a workout session. How much water did the person lose, assuming that the heat given off was used only to evaporate the water?

18.42 A 1.30-kg block of aluminum at 21.0 °C is to be melted and reshaped. How much heat must flow into the block in order to melt it?

18.43 The latent heat of vaporization of liquid nitrogen is about 200. kJ/kg. Suppose you have 1.00 kg of liquid nitrogen boiling at 77.0 K. If you supply heat at a constant rate of 10.0 W via an electric heater immersed in the liquid nitrogen, how long will it take to vaporize all of it? What is the time required to vaporize 1.00 kg of liquid helium, whose heat of vaporization is 20.9 kJ/kg?

•18.44 Suppose 0.0100 kg of steam (at 100.00 °C) is added to 0.100 kg of water (initially at 19.0 °C). The water is inside an aluminum cup of mass 35.0 g. The cup is inside a perfectly insulated calorimetry container that prevents heat exchange with the outside environment. Find the final temperature of the water after equilibrium is reached.

•18.45 Suppose $1.00 \cdot 10^2$ g of molten aluminum at 932 K is dropped into 1.00 L of water at room temperature, 22.0 °C.

a) How much water will boil away?

b) How much aluminum will solidify?

c) What will be the final temperature of the water-aluminum system?

d) Suppose the aluminum were initially at 1150 K. Could you still solve this problem using only the information given? What would be the result?

•18.46 In one of your rigorous workout sessions, you lost 150. g of water through evaporation. Assume that the amount of work done by your body was $1.80 \cdot 10^5$ J and that the heat required to evaporate the water came from your body.

a) Find the loss in internal energy of your body, assuming that the latent heat of vaporization is $2.42 \cdot 10^6$ J/kg.

b) Determine the minimum number of food calories that must be consumed to replace the internal energy lost (1 food calorie = 4186 J).

••18.47 Knife blades are often made of hardened carbon steel. In the hardening process, the blade is first heated to a temperature of 1346 °F and then cooled down rapidly by immersing it in a bath of water. To achieve the desired hardness, a blade needs to be brought from 1346 °F to a temperature below 500. °F. If a blade has a mass of 0.500 kg and the water is in an open copper container of mass 2.000 kg and sufficiently large volume, what is the minimum quantity of water that needs to be in the container for this hardening process to be successful? Assume that the water and copper container are initially at a temperature of 20.0 °C. Assume that the blade is not in direct mechanical (and thus thermal) contact with the container and that no water boils but all the water reaches 100. °C. Neglect cooling through radiation into the air. The heat capacity of copper around room temperature is $c_{copper} = 386$ J/(kg K). Use the data in the table below for the heat capacity of carbon steel.

Temperature Range (°C)	Heat Capacity (J/kg K)
150 to 200	519
200 to 250	536
250 to 350	553
350 to 450	595
450 to 550	662
550 to 650	754
650 to 750	846

••18.48 It has been postulated that ethanol (ethyl alcohol) "snow" falls near the poles of the planets Jupiter, Saturn, Uranus, and Neptune. If the polar regions of Uranus, defined to be north of latitude 75.0° N and south of latitude 75.0° S, experience 1.00 ft of ethanol snow, what is the minimum amount of energy lost to the atmosphere to produce this much snow from ethanol vapor? Assume that solid ethanol has a density of 1.00 g/cm³ and that ethanol snow—which is fluffy like Earth snow—is about 90.0% empty space. The specific heat capacities are 1.70 J/(g K) for ethanol vapor, 2.44 J/(g K) for ethanol liquid, and 2.42 J/(g K) for solid ethanol. How much power is dissipated if 1.00 ft of ethanol snow falls in one Earth day?

Section 18.8

18.49 A 100. mm by 100. mm by 5.00 mm slab of ice at 0.00 °C is placed on one of its larger faces on a 10.0-mm-thick metal disk that covers a pot of boiling water at normal atmospheric pressure. The time needed for the entire ice block to melt is measured to be 0.400 s. The density of ice is 920. kg/m³. Use the data in Table 18.3 to determine the metal the disk is most likely made of.

18.50 A copper sheet of thickness 2.00 mm is bonded to a steel sheet of thickness 1.00 mm. The outside surface of the copper sheet is held at a temperature of 100.0 °C and the steel sheet at 25.0 °C.

a) Determine the temperature of the copper-steel interface.

b) How much heat is conducted through 1.00 m² of the combined sheets per second?

18.51 The Sun is approximately a sphere of radius $6.963 \cdot 10^5$ km, at a mean distance $a = 1.496 \cdot 10^8$ km from the Earth. The *solar constant,* the intensity of solar radiation at the outer edge of Earth's atmosphere, is 1370. W/m². Assuming that the Sun radiates as a blackbody, calculate its surface temperature.

•18.52 An air-cooled motorcycle engine loses a significant amount of heat through thermal radiation according to the Stefan-Boltzmann equation. Assume that the ambient temperature is $T_0 = 27$ °C (300 K). Suppose the engine generates 15 hp (11 kW) of power and, because there are several deep surface fins, has a surface area of $A = 0.50$ m². A shiny engine has an emissivity $e = 0.050$, whereas an engine that is painted black has $e = 0.95$. Determine the equilibrium temperatures for the black engine and the shiny engine. (Assume that radiation is the only mode by which heat is dissipated from the engine.)

•18.53 One summer day, you decide to make a popsicle. You place a popsicle stick into an 8.00-oz glass of orange juice, which is at room temperature (71.0 °F). You then place the glass in the freezer, which is at –15.0 °F and has a cooling power of $4.00 \cdot 10^3$ BTU/h. How long does it take your popsicle to freeze?

•18.54 An ice cube at 0.00 °C measures 10.0 cm on a side. It sits on top of a copper block with a square cross section 10.0 cm on a side and a height of 20.0 cm. The bottom of the copper block is in thermal contact with a large pool of water at 90.0 °C. How long does it take the ice cube to melt? Assume that only the part in contact with the copper liquefies; that is, the cube gets shorter as it melts. The density of ice is 0.917 g/cm³.

•18.55 A single-pane window is a poor insulator. On a cold day, the temperature of the inside surface of the window is often much less than the room air temperature. Likewise, the outside surface of the window is likely to be much warmer than the outdoor air. The actual surface temperatures are strongly dependent on convection effects. For instance, suppose the air temperatures are 21.5 °C inside and –3.0 °C outside, the inner surface of the window is at 8.5 °C, and the outer surface is at 4.1 °C. At what rate will heat flow through the window? Take the thickness of the window to be 0.32 cm, the height to be 1.2 m, and the width to be 1.4 m.

•18.56 A cryogenic storage container holds liquid helium, which boils at 4.22 K. The container's outer shell has an effective area of 0.500 m² and is at $3.00 \cdot 10^2$ K. Suppose a student painted the outer shell of the container black, turning it into a pseudo-blackbody.

a) Determine the rate of heat loss due to radiation.

b) What is the rate at which the volume of the liquid helium in the container decreases as a result of boiling off? The latent heat of vaporization of liquid helium is 20.9 kJ/kg. The density of liquid helium is 0.125 kg/L.

•18.57 Mars is 1.52 times farther away from the Sun than the Earth is and has a diameter 0.532 times that of the Earth.

a) What is the intensity of solar radiation (in W/m²) on the surface of Mars?

b) Estimate the temperature on the surface of Mars.

••18.58 Two thermal reservoirs are connected by a solid copper bar. The bar is 2.00 m long, and the temperatures of the reservoirs are 80.0 °C and 20.0 °C.

a) Suppose the bar has a constant rectangular cross section, 10.0 cm on a side. What is the rate of heat flow through the bar?

b) Suppose the bar has a rectangular cross section that gradually widens from the colder reservoir to the warmer reservoir. The area A is determined by $A = (0.0100 \text{ m}^2)[1.00 + x/(2.00 \text{ m})]$, where x is the distance along the bar from the colder reservoir to the warmer one. Find the heat flow and the rate of change of temperature with distance at the colder end, at the warmer end, and at the middle of the bar.

••18.59 The radiation emitted by a blackbody at temperature T has a frequency distribution given by the Planck spectrum:

$$\epsilon_T(f) = \frac{2\pi h}{c^2}\left(\frac{f^3}{e^{hf/k_B T} - 1}\right),$$

where $\epsilon_T(f)$ is the energy density of the radiation per unit increment of frequency, v (for example, in watts per square meter per hertz), $h = 6.626 \cdot 10^{-34}$ J s is Planck's constant, $k_B = 1.38 \cdot 10^{-23}$ JK^{-1} is the Boltzmann constant, and c is the speed of light in vacuum. (We'll derive this distribution in Chapter 36 as a consequence of the quantum hypothesis of light, but here it can reveal something about radiation. Remarkably, the most accurately and precisely measured example of this energy distribution in nature is the cosmic microwave background radiation.) This distribution goes to zero in the limits $f \to 0$ and $f \to \infty$ with a single peak between those limits. As the temperature is increased, the energy density at each frequency value increases, and the peak shifts to a higher frequency value.

a) Find the frequency corresponding to the peak of the Planck spectrum, as a function of temperature.

b) Evaluate the peak frequency at $T = 6.00 \cdot 10^3$ K, approximately the temperature of the photosphere (surface) of the Sun.

c) Evaluate the peak frequency at $T = 2.735$ K, the temperature of the cosmic background microwave radiation.

d) Evaluate the peak frequency at $T = 300.$ K, which is approximately the surface temperature of Earth.

Additional Exercises

18.60 How much energy is required to warm 0.300 kg of aluminum from 20.0 °C to 100.0 °C?

18.61 The thermal conductivity of fiberglass batting, which is 4.00 in thick, is $8.00 \cdot 10^{-6}$ BTU/(ft °F s). What is the R value (in ft^2 °F h/BTU)?

18.62 Water is an excellent coolant as a result of its very high heat capacity. Calculate the amount of heat that is required to change the temperature of 10.0 kg of water by 10.0 K. Now calculate the kinetic energy of a car with $m = 1.00 \cdot 10^3$ kg moving at a speed of 27.0 m/s (60.4 mph). Compare the two quantities.

18.63 Approximately 95% of the energy developed by the filament in a spherical $1.0 \cdot 10^2$ W light bulb is dissipated through the glass bulb. If the thickness of the glass is 0.50 mm and the bulb's radius is 3.0 cm, calculate the temperature difference between the inner and outer surfaces of the glass. Take the thermal conductivity of the glass to be 0.80 W/(m K).

18.64 The label on a soft drink states that 12.0 fl. oz (355 g) provides 150. kcal. The drink is cooled to 10.0 °C before it is consumed. It then reaches body temperature of 37.0 °C. Find the net energy content of the drink. (*Hint:* You can treat the soft drink as being identical to water in terms of heat capacity.)

18.65 The human body transports heat from the interior tissues, at the temperature 37.0 °C, to the skin surface, at the temperature 27.0 °C, at a rate of 100. W. If the skin area is 1.50 m^2 and its thickness is 3.00 mm, what is the effective thermal conductivity, k, of skin?

•18.66 It has been said that sometimes lead bullets melt upon impact. Assume that a bullet receives 75.0% of the work done on it by a wall on impact as an increase in internal energy.

a) What is the minimum speed with which a 15.0-g lead bullet would have to hit a surface (assuming that the bullet stops completely and all the kinetic energy is absorbed by it) in order to begin melting?

b) What is the minimum impact speed required for the bullet to melt completely?

•18.67 Solar radiation at the Earth's surface has an intensity of about 1.4 kW/m^2. Assuming that Earth and Mars are blackbodies, calculate the intensity of sunlight at the surface of Mars.

•18.68 You become lost while hiking outside wearing only a bathing suit.

a) Calculate the power radiated from your body, assuming that your body's surface area is about 2.00 m^2 and your skin temperature is about 33.0 °C. Also, assume that your body has an emissivity of 1.00.

b) Calculate the net radiated power from your body when you are inside a shelter at 20.0 °C.

c) Calculate the net radiated power from your body when your skin temperature dropped to 27.0 °C.

•18.69 A 10.0-g ice cube at −10.0 °C is dropped into 40.0 g of water at 30.0 °C.

a) After enough time has passed to allow the ice cube and water to come to equilibrium, what is the temperature of the water?

b) If a second ice cube is added, what will the temperature be?

•18.70 Arthur Clarke wrote an interesting short story called "A Slight Case of Sunstroke." Disgruntled football fans came to the stadium one day equipped with mirrors and were ready to barbecue the referee if he favored one team over the other. Imagine the referee to be a cylinder filled with water of mass 60.0 kg at 35.0 °C. Also imagine that this cylinder absorbs all the light reflected on it from 50,000 mirrors. If the heat capacity of water is $4.20 \cdot 10^3$ J/(kg °C), how long will it take to raise the temperature of the water to 100. °C? Assume that the Sun gives out $1.00 \cdot 10^3$ W/m^2, the dimensions of each mirror are 25.0 cm by 25.0 cm, and the mirrors are held at an angle of 45.0°.

•18.71 If the average temperature of the North Atlantic is 12.0 °C and the Gulf Stream temperature averages 17.0 °C, estimate the net amount of heat the Gulf Stream radiates to the surrounding ocean. Use the details given in Solved Problem 18.4 (the length is about $8.00 \cdot 10^3$ km), and assume that $e = 0.930$. Don't forget to include the heat absorbed by the Gulf Stream.

•18.72 For a class demonstration, your physics instructor pours 1.00 kg of steam at 100.0 °C over 4.00 kg of ice at 0.00 °C and waits for the system to reach equilibrium, when he will measure the temperature. While the system reaches equilibrium, you are given the latent heats of ice and steam and the specific heat of water: $L_{ice} = 3.33 \cdot 10^5$ J/kg, $L_{steam} = 2.26 \cdot 10^6$ J/kg, $c_{water} = 4186$ J/(kg °C). You are asked to calculate the final equilibrium temperature of the system. What value do you find?

•18.73 Determine the ratio of the heat flow into a six-pack of aluminum soda cans to the heat flow into a 2.00-L plastic bottle of soda when both are taken out of the same refrigerator, that is, have the same initial temperature difference with the air in the room. Assume that each soda can has a diameter of 6.00 cm, a height of 12.0 cm, and a thickness of 0.100 cm. Use 205 W/(m K) as the thermal conductivity of aluminum. Assume that the 2.00-L bottle of soda has a diameter of 10.0 cm, a height of 25.0 cm, and a thickness of 0.100 cm. Use 0.100 W/(m K) as the thermal conductivity of plastic.

•18.74 The R factor for housing insulation gives the thermal resistance in units of ft^2 °F h/BTU. A good wall for harsh climates, corresponding to about 10.0 in of fiberglass, has $R = 40.0$ ft^2 °F h/BTU.

a) Determine the thermal resistance in SI units.

b) Find the heat flow per square meter through a wall that has insulation with an R factor of 40.0, when the outside temperature is −22.0 °C and the inside temperature is 23.0 °C.

•18.75 Suppose you have an attic room that measures 5.0 m by 5.0 m and is maintained at 21 °C when the outside temperature is 4.0 °C.

a) If you used R-19 insulation instead of R-30 insulation, how much more heat would exit this room in 1 day?

b) If electrical energy for heating the room costs 12 cents per kilowatt-hour, how much more would it cost you for heating for a period of 3 months at the same outside temperature with the R-19 insulation?

••18.76 A thermal window consists of two panes of glass separated by an air gap. Each pane of glass is 3.00 mm thick, and the air gap is 1.00 cm thick. Window glass has a thermal conductivity of 1.00 W/(m K), and air has a thermal conductivity of 0.0260 W/(m K). Suppose a thermal window separates a room at 20.00 °C from the outside at 0.00 °C.

a) What is the temperature at each of the four air-glass interfaces?

b) At what rate is heat lost from the room, per square meter of window?

c) Suppose the window had no air gap but consisted of a single layer of glass 6.00 mm thick. What would the rate of heat loss per square meter be then, under the same temperature conditions?

d) Heat conduction through the thermal window could be reduced essentially to zero by evacuating the space between the glass panes. Why is this not done?

MULTI-VERSION EXERCISES

18.77 Enhanced geothermal systems (EGS) consist of two or more boreholes that extend several kilometers below ground level into the hot bedrock. Since drilling these holes can cost millions, one concern is that the heat provided by the bedrock cannot pay back the initial investment. How long can 0.669 km^3 of granite deliver an average of 13.9 MW of power, if its initial temperature is 168.3 °C and its final temperature is 103.5 °C? [The density of granite is 2.75 times that of water, and its specific heat is 0.790 kJ/(kg °C).]

18.78 Enhanced geothermal systems (EGS) consist of two or more boreholes that extend several kilometers below ground level into the hot bedrock. Since drilling these holes can cost millions, one concern is that the heat provided by the bedrock cannot pay back the initial investment. Suppose 0.581 km^3 of granite can deliver an average of 15.7 MW of power for 164.6 yr. If, after that time, the temperature of the granite is 104.5 °C, what was its initial temperature (in degrees Celsius)? [The density of granite is 2.75 times that of water, and its specific heat is 0.790 kJ/(kg °C).]

18.79 Enhanced geothermal systems (EGS) consist of two or more boreholes that extend several kilometers below ground level into the hot bedrock. Since drilling these holes can cost millions, one concern is that

the heat provided by the bedrock cannot pay back the initial investment. Suppose 0.493 km^3 of granite is to be drawn on for 124.9 yr and in the process will cool from 169.9 °C to 105.5 °C. What is the average power the granite can deliver during that time? [The density of granite is 2.75 times that of water, and its specific heat is 0.790 kJ/(kg °C).]

18.80 An exterior wall is 5.183 m wide and 3.269 m tall. The wall is insulated with a material that has an R factor of 29. The outdoor temperature is 1.073 °C, and the indoor temperature is 23.37 °C. How much power is carried through the wall?

18.81 An exterior wall is 5.869 m wide and 3.289 m tall, and 69.71 W of power is carried through it. The outdoor temperature is 3.857 °C, and the indoor temperature is 24.21 °C. What is the R factor of the material with which the wall is insulated?

18.82 An exterior wall is 6.555 m wide and 3.311 m tall. The wall is insulated with a material that has an R factor of 34. When the outdoor temperature is 2.641 °C, 63.10 W of power is carried through the wall. What is the indoor temperature?

19

Ideal Gases

FIGURE 19.1 A scuba diver breathes compressed air under water.

Scuba diving (*scuba* was originally an acronym for "self-contained underwater breathing apparatus") is a popular activity in locations with warm seawater. However, it is also a great application of the physics of gases. The scuba diver shown in Figure 19.1 relies on a tank of compressed air for breathing underwater. The air is typically kept at a pressure of 200 atm (~3000 psi, or ~20 MPa). Before the diver can breathe this air, its pressure must be modified so that it is about the same as the pressure surrounding the diver's body in the water, which increases by about 1 atm for every 10 m below the surface (see Section 13.4).

In this chapter, we study the physics of gases. The results are based on an ideal gas, which doesn't actually exist, but many real gases behave approximately like an ideal gas in many situations. We first examine the properties of gases based on observations, including laws first stated by pioneers of hot-air balloon navigation, who had a very practical interest in the behavior of gases at high altitudes. Then we gain additional insights from the kinetic theory of an ideal gas, which applies mathematical analysis to gas particles under several assumed conditions.

The key properties of gases are the thermodynamic quantities of temperature, pressure, and volume, which is why this chapter is with the others on thermodynamics. However, the physics of gases has applications to many different areas of science, from astronomy to meteorology, from chemistry to biology. Many of these concepts will be important in later chapters.

WHAT WE WILL LEARN

- A gas consists of molecules that have enough spatial separation that the intermolecular bonding characteristic of liquids and solids is absent.

- The physical properties of a gas are pressure, volume, temperature, and number of molecules.

- An ideal gas is one in which the molecules of the gas are treated as point particles that do not interact with one another.

- The Ideal Gas Law gives the relationship among pressure, volume, temperature, and number of molecules of an ideal gas.

- The work done by an ideal gas is proportional to the change in volume of the gas if the pressure is constant.

- Dalton's Law says that the total pressure exerted by a mixture of gases is equal to the sum of the partial pressures exerted by each gas in the mixture.

- The specific heat of a gas can be calculated for both constant-volume and constant-pressure processes, and gases composed of monatomic, diatomic, or polyatomic molecules have different specific heats.

- Kinetic theory, describing the motion of the constituents of an ideal gas, accounts for its macroscopic properties, such as temperature and pressure.

- The temperature of a gas is proportional to the average kinetic energy of its molecules.

- The distribution of speeds of the molecules of a gas is described by the Maxwell speed distribution, and the distribution of kinetic energies of the molecules of a gas is described by the Maxwell kinetic energy distribution.

- The mean free path of a molecule in a gas is the average distance the molecule travels before encountering another molecule.

19.1 Empirical Gas Laws

Chapter 13 presented an overview of the states of matter. It introduced pressure as a physical quantity and stated that a gas is a special case of a fluid. Since the concept of temperature was introduced in Chapter 17, we can now look at how gases respond to changes in temperature, pressure, and volume. The gas laws introduced in this section provide the empirical evidence that will lead to the derivation of the Ideal Gas Law in the next section. We'll see that all of the empirical gas laws are special cases of the Ideal Gas Law.

A **gas** consists of molecules that are spatially separated such that the intermolecular bonding, which is characteristic of liquids and solids, is absent. A gas expands to fill the container in which it is placed. Thus, the volume of a gas is the volume of its container. This chapter uses the term *gas molecules* to refer to the constituents of a gas, although a gas may consist of atoms or molecules or may be a combination of atoms and molecules.

In Chapter 17, the temperature of a substance was defined in terms of its tendency to give off heat to its surroundings or to absorb heat from its surroundings. The pressure of a gas is defined as the force per unit area exerted by the gas molecules on the walls of a container. For many applications, standard temperature and pressure (STP) is defined as 0 °C (273.15 K) and 100.000 kPa.

Another characteristic of a gas, in addition to volume, temperature, and pressure, is the number of gas molecules in the volume of gas in a container. This number is expressed in terms of moles: 1 **mole** of a gas is defined to have $6.022 \cdot 10^{23}$ molecules. This number, known as *Avogadro's number,* was introduced in Chapter 13 and will be discussed further in connection with Avogadro's Law later in this section.

Several simple relationships exist among the four properties of a gas—pressure, volume, temperature, and number of molecules. This section covers four simple gas laws that relate these properties. All of these gas laws are named after their discoverers (Boyle, Charles, Gay-Lussac, Avogadro) and are empirical—that is, they were found from performing a series of measurements and not derived from some more fundamental theory. In the next section, we'll combine these four laws to form the Ideal Gas Law, relating all the macroscopic characteristics of gases. If you understand the Ideal Gas Law, all of the empirical gas laws follow immediately. Why not skip them, then? The answer is that these empirical gas laws form the basis from which the Ideal Gas Law is derived.

Boyle's Law

The first empirical gas law is Boyle's Law, which is also known in Europe as Mariotte's Law. English scientist Robert Boyle published this law in 1662; French scientist Edme Mariotte published a similar result in 1676. **Boyle's Law** states that the product of a gas's pressure, p, and its volume, V, at constant temperature is a constant (Figure 19.2). Mathematically, Boyle's Law is expressed as pV = constant (at constant temperature).

Another way to express Boyle's Law is to state that the product of the pressure, p_1, and volume, V_1, of a gas at time t_1 is equal to the product of the pressure, p_2, and the volume, V_2, of the same gas at the same temperature at another time, t_2:

$$p_1 V_1 = p_2 V_2 \quad \text{(at constant temperature).} \quad (19.1)$$

An everyday example of the application of Boyle's Law is breathing. When you take in a breath, your diaphragm expands, and the expansion produces a larger volume in your chest cavity. According to Boyle's Law, the air pressure in your lungs is reduced relative to the normal atmospheric pressure around you. The higher pressure outside your body then forces air into your lungs to equalize the pressure. To breathe out, your diaphragm contracts, reducing the volume of your chest cavity. This reduction in volume produces a higher pressure, which forces air out of your lungs.

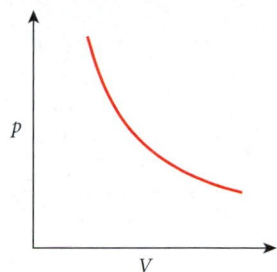

FIGURE 19.2 Relationship between pressure and volume as expressed by Boyle's Law.

Charles's Law

The second empirical gas law is **Charles's Law,** which states that for a gas kept at constant pressure, the volume of the gas, V, divided by its temperature, T, is constant (Figure 19.3). The French physicist (and pioneering high-altitude balloonist) Jacques Charles proposed this law in 1787. Mathematically, Charles's Law is V/T = constant (at constant pressure).

Another way to state Charles's Law is to state that the ratio of the temperature, T_1, and volume, V_1, of a gas at a given time, t_1, is equal to the ratio of the temperature, T_2, and the volume, V_2, of the same gas at the same pressure at another time, t_2:

$$\frac{V_1}{T_1} = \frac{V_2}{T_2} \Leftrightarrow \frac{V_1}{V_2} = \frac{T_1}{T_2} \quad \text{(at constant pressure).} \quad (19.2)$$

Note that the temperatures must be expressed in kelvins (K).

Because the density, ρ, of a given mass, m, of gas is $\rho = m/V$, Charles's Law can also be written as ρT = constant (at constant pressure). As a corollary to equation 19.2, we then have

$$\rho_1 T_1 = \rho_2 T_2 \Leftrightarrow \frac{\rho_1}{\rho_2} = \frac{T_2}{T_1} \quad \text{(at constant pressure).} \quad (19.3)$$

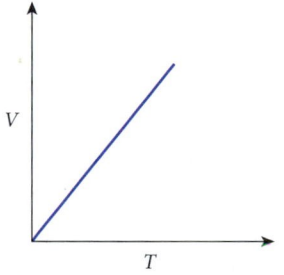

FIGURE 19.3 Relationship between volume and temperature as expressed by Charles's Law.

Gay-Lussac's Law

The third empirical gas law is **Gay-Lussac's Law,** which states that ratio of the pressure, p, of a gas to its temperature, T, at the same volume is constant (Figure 19.4). This law was presented in 1802 by the French chemist Joseph Louis Gay-Lussac (another avid high-altitude balloonist, by the way). Mathematically, Gay-Lussac's Law is expressed as p/T = constant (at constant volume).

Another way to express Gay-Lussac's Law is to state that the ratio of the pressure of a gas, p_1, and its temperature, T_1, at a given time, t_1, is equal to the ratio of the pressure, p_2, and the temperature, T_2, of the same gas at the same volume at another time, t_2:

$$\frac{p_1}{T_1} = \frac{p_2}{T_2} \quad \text{(at constant volume).} \quad (19.4)$$

Again, the temperatures must be in kelvins (K).

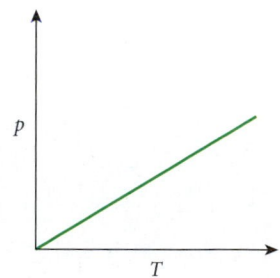

FIGURE 19.4 Relationship between pressure and temperature as expressed by Gay-Lussac's Law.

Avogadro's Law

The fourth empirical gas law deals with the quantity of a gas. **Avogadro's Law** states that the ratio of the volume of a gas, V, to the number of gas molecules, N, in that volume is constant if the pressure and temperature are held constant. This law was introduced in 1811 by the Italian physicist Amedeo Avogadro. Mathematically, Avogadro's Law is expressed as V/N = constant (at constant pressure and temperature).

Another way to express Avogadro's Law is to state that the ratio of the volume, V_1, and the number of molecules, N_1, of a gas at a given time, t_1, is equal to the ratio of the volume, V_2, and the number of molecules, N_2, of the same gas at the same pressure and temperature at another time, t_2:

$$\frac{V_1}{N_1} = \frac{V_2}{N_2} \quad \text{(at constant pressure and temperature).} \tag{19.5}$$

It has been found that a volume of 22.4 L ($1 \text{ L} = 10^{-3} \text{ m}^3$) of a gas at standard temperature and pressure contains $6.022 \cdot 10^{23}$ molecules. This number of molecules is called **Avogadro's number,** N_A. The currently accepted value for Avogadro's number is

$$N_A = (6.02214129 \pm 0.00000027) \cdot 10^{23}.$$

(As was pointed out in Chapter 13, Avogadro's number is defined to be the number of atoms in exactly 12 grams of the most common isotope of carbon, carbon-12.) One mole of any gas will have Avogadro's number of molecules. The number of moles is usually symbolized by n. Thus, the number of molecules, N, and the number of moles, n, of a gas are related by Avogadro's number:

$$N = nN_A. \tag{19.6}$$

Therefore, the mass in grams of one mole of a gas is numerically equal to the atomic mass or molecular mass of the constituent particles in atomic mass units. For example, nitrogen gas consists of molecules composed of two nitrogen atoms. Each nitrogen atom has an atomic mass number of 14. Thus, the nitrogen molecule has a molecular mass number of 28, and a 22.4-L volume of nitrogen gas at standard temperature and pressure has a mass of 28 g.

19.2 Ideal Gas Law

We can combine the empirical gas laws described in Section 19.1 to obtain a more general law relating the properties of gases, called the **Ideal Gas Law:**

$$pV = nRT, \tag{19.7}$$

where p, V, and T are the pressure, volume, and temperature, respectively, of n moles of a gas and R is the **universal gas constant.** The value of R is determined experimentally and is given by

$$R = (8.3144621 \pm 0.0000075) \text{ J/(mol K)}.$$

The constant R can also be expressed in other units, which changes its numerical value, for example, $R = 0.08205736$ L atm/(mol K).

The Ideal Gas Law was first stated by the French scientist Benoit Paul Émile Clapeyron in 1834. It is the main result of this chapter.

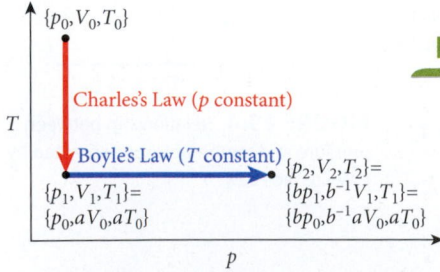

FIGURE 19.5 Path of the state of a gas in a plot of temperature versus pressure (Tp-diagram). The first part of the path takes place at a constant pressure and is described by Charles's Law. The second part of the path takes place at constant temperature and is described by Boyle's Law.

DERIVATION 19.1 | Ideal Gas Law

To derive the Ideal Gas Law, we start with one mole of an ideal gas, described by a pressure p_0, a volume V_0, and a temperature T_0, which can be indicated using the shorthand notation $\{p_0, V_0, T_0\}$. (Remember, an ideal gas has no interactions among its molecules.) This particular state of the gas is shown on a plot of temperature versus pressure (a Tp-diagram) in Figure 19.5.

Next, the state of the gas is changed by varying its volume and temperature while holding the pressure constant. In Figure 19.5, the gas's state goes from $\{p_0, V_0, T_0\}$ to $\{p_1, V_1, T_1\}$ on the Tp-diagram (red arrow). Using Charles's Law (equation 19.2), we can write

$$\frac{V_0}{T_0} = \frac{aV_0}{aT_0} = \frac{V_1}{T_1},$$

where we have multiplied the numerator and denominator of the ratio V_0/T_0 by a constant, a. We can describe the gas in this new state with

$$\{p_1, V_1, T_1\} = \{p_0, aV_0, aT_0\}. \tag{i}$$

From this point in the Tp-diagram, the pressure and volume of the gas change while the temperature is held constant. In Figure 19.5, the path goes from $\{p_1, V_1, T_1\}$ to $\{p_2, V_2, T_2\}$, as indicated by the blue arrow. Using Boyle's Law (equation 19.1), we can write

$$p_1V_1 = \frac{b}{b}p_1V_1 = (bp_1)(b^{-1}V_1) = p_2V_2,$$

where the product p_1V_1 is multiplied and divided by a constant, b. Now we can describe the state of the gas with

$$\{p_2, V_2, T_2\} = \{bp_1, b^{-1}V_1, T_1\}. \tag{ii}$$

Combining equations (i) and (ii) gives us

$$\{p_2, V_2, T_2\} = \{bp_0, b^{-1}aV_0, aT_0\}.$$

We can now write the ratio

$$\frac{p_2V_2}{T_2} = \frac{(bp_0)(b^{-1}aV_0)}{aT_0} = \frac{p_0V_0}{T_0}.$$

This implies that p_0V_0/T_0 is a constant, which we can call R, the universal gas constant.

This derivation was done for 1 mole of gas. For n moles of gas, Avogadro's Law says that with constant pressure and temperature, the volume of n moles of gas will be equal to the number of moles times the volume of one mole of gas. We can rearrange $p_0V_0/T_0 = R$ and multiply by n to get

$$np_0V_0 = nRT_0.$$

We now write the Ideal Gas Law as $pV/T = nR$, or, more commonly, as

$$pV = nRT,$$

where $p = p_0$, $V = nV_0$, and $T = T_0$ are the pressure, volume, and temperature, respectively, of n moles of gas.

The Ideal Gas Law can also be written in terms of the number of gas molecules instead of the number of moles of gas. This is sometimes useful for examining relationships at the molecular level of matter. In this form, the Ideal Gas Law is

$$pV = Nk_BT, \tag{19.8}$$

where N is the number of atoms or molecules and k_B is the **Boltzmann constant,** given by

$$k_B = \frac{R}{N_A}.$$

The currently accepted value of the Boltzmann constant is

$$k_B = (1.3806488 \pm 0.0000013) \cdot 10^{-23} \text{ J/K}.$$

The Boltzmann constant is an important fundamental physical constant that often arises in relationships based on atomic or molecular behavior. It will be used again later in this chapter and again in discussions of solid-state electronics and quantum mechanics. You may have noticed by now that the letter k is often used in mathematics and physics to denote a constant (the German word for *constant* is *Konstante,* which is where the use of k originated). However, the Boltzmann constant is of such basic significance that the subscript B is used to distinguish it from other physical constants denoted by k.

Another way to express the Ideal Gas Law for a constant number of moles of gas is

$$\frac{p_1V_1}{T_1} = \frac{p_2V_2}{T_2}, \tag{19.9}$$

Concept Check 19.1

The pressure inside a scuba tank is 205 atm at 22.0 °C. Suppose the tank is left out in the sun and the temperature of the compressed air inside the tank rises to 40.0 °C. What will the pressure inside the tank be?

a) 205 atm   d) 321 atm

b) 218 atm   e) 373 atm

c) 254 atm

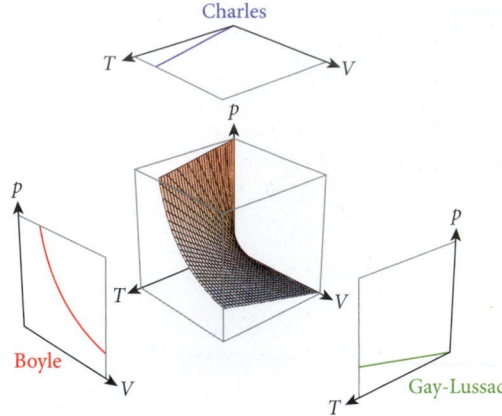

FIGURE 19.6 The relationship among the Ideal Gas Law and its special cases: Charles's Law (constant pressure), Boyle's Law (constant temperature), and Gay-Lussac's Law (constant volume).

where p_1, V_1, and T_1 are the pressure, volume, and temperature, respectively, at time 1 and p_2, V_2, and T_2 are the pressure, volume, and temperature, respectively, at time 2. The advantage of this formulation is that you do not have to know the numerical values of the universal gas constant or the Boltzmann constant to relate the pressures, volumes, and temperatures at two different times.

It is straightforward to confirm that Boyle's Law, Charles's Law, and Gay-Lussac's Law are embodied in the Ideal Gas Law. In Figure 19.6, a three-dimensional surface represents the relationship among the pressure, volume, and temperature of an ideal gas. At any fixed value of the pressure, the intersection of this surface with the temperature-volume plane is a line that reflects Charles's Law. The intersection of this surface with the pressure-temperature plane at any fixed value of the volume is a line that reflects Gay-Lussac's Law. And, finally, a line corresponding to Boyle's Law marks the intersection of this surface with the pressure-volume plane at any given temperature.

Note that the Ideal Gas Law has been obtained in this discussion by combining the empirical gas laws. The next section will analyze the behavior of ideal gas molecules in a container and thereby provide insight into the microscopic physical basis for the Ideal Gas Law. Let's first explore the consequences and predictive powers of the Ideal Gas Law with a few illustrative examples.

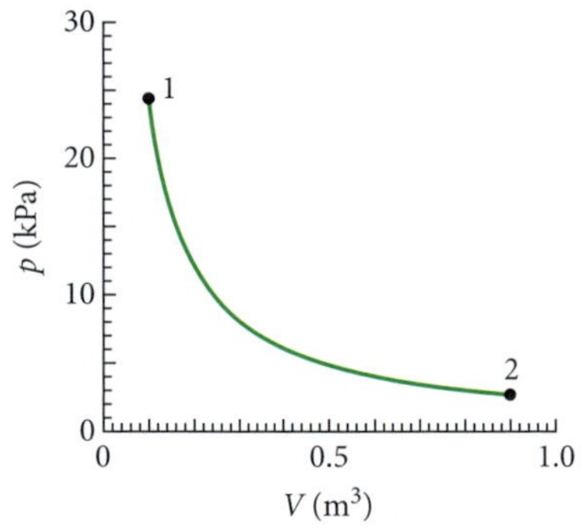

EXAMPLE 19.1 | Gas in a Cylinder

Consider a gas at a pressure of 24.9 kPa in a cylinder with a volume of 0.100 m³ and a piston. This pressure and volume corresponds to point 1 in Figure 19.7. The pressure of this gas has to be decreased to allow a manufacturing process to work efficiently. The piston is designed to increase the volume of the cylinder from 0.100 m³ to 0.900 m³ while keeping the temperature constant.

PROBLEM
What is the pressure of the gas at a volume of 0.900 m³?

SOLUTION
Point 2 in Figure 19.7 represents the pressure of the gas at a volume of 0.900 m³. You can see that the pressure decreases as the volume increases.

Since the temperature is constant ($T_1 = T_2$) in this situation, the Ideal Gas Law (in the form in equation 19.9), $p_1 V_1/T_1 = p_2 V_2/T_2$, reduces to $p_1 V_1 = p_2 V_2$. (Section 19.1 identified this special case of the Ideal Gas Law as Boyle's Law.) We can use this to calculate the new pressure:

$$p_2 = \frac{p_1 V_1}{V_2} = \frac{(24.9 \text{ kPa})(0.100 \text{ m}^3)}{0.900 \text{ m}^3} = 2.77 \text{ kPa}.$$

FIGURE 19.7 Plot of pressure versus volume for a gas described by Boyle's Law.

There are, of course, many applications where the temperature of a gas changes. Example 19.2 considers one.

EXAMPLE 19.2 | Cooling a Balloon

A balloon is blown up at room temperature and then placed in liquid nitrogen (Figure 19.8). The balloon is allowed to cool to liquid-nitrogen temperature and it shrinks dramatically. The cold balloon is then removed from the liquid nitrogen and allowed to warm back to room temperature. The balloon returns to its original volume.

PROBLEM
By what factor does the volume of the balloon decrease as the temperature of the air inside the balloon goes from room temperature to the temperature of liquid nitrogen?

FIGURE 19.8 The top row shows a balloon blown up at room temperature and then placed in liquid nitrogen. In the bottom row, the cold balloon is removed from the liquid nitrogen and allowed to warm to room temperature. The time between subsequent frames in each sequence is 20 s. Note that the liquid nitrogen causes a phase change in the air inside the balloon, causing it to contract more than the Ideal Gas Law alone predicts.

SOLUTION

To a good approximation, the gas inside the balloon is kept at constant pressure ($p_1 = p_2$), which you can verify from the fact that it is about equally hard to add air to a balloon of any size (for a reasonable range of balloon sizes). The Ideal Gas Law (equation 19.9), $p_1 V_1/T_1 = p_2 V_2/T_2$, in this case becomes $V_1/T_1 = V_2/T_2$. (In Section 19.1, this special case of the Ideal Gas Law was identified as Charles's Law.) Assume room temperature is 22 °C, or $T_1 = 295$ K, and the temperature of liquid nitrogen is $T_2 = 77.2$ K. We can calculate the fractional change in volume of the balloon:

$$\frac{V_2}{V_1} = \frac{T_2}{T_1} = \frac{77.2 \text{ K}}{295 \text{ K}} = 0.262.$$

DISCUSSION

The balloon in Figure 19.8 shrank by much more than the factor calculated from Charles's Law. (A ratio of cold volume to warm volume of 25% implies that the radius of the cold balloon should be 63% of the radius of the room-temperature balloon.) There are several reasons for the smaller than expected size of the cold balloon. First, the water vapor in the air in the balloon freezes out as ice particles. Second, some of the oxygen and some of the nitrogen in the air inside the balloon condense to liquids, to which the Ideal Gas Law does not apply.

Example 19.3 deals with the relationship of air temperature and air pressure.

EXAMPLE 19.3 Heat on the Golf Course

The 2007 PGA Championship was played in August in Oklahoma at the Southern Hills Country Club, at the highest average temperature of any major golf championship in history, reaching approximately 101 °F (38.3 °C = 311.5 K). Thus, the air density was lower than at cooler temperatures, causing the golf balls to fly farther. The players had to correct for this effect, just as they have to allow for longer flights when they play at higher elevations.

PROBLEM

The Southern Hills Country Club is at an elevation of 213 m (700 ft) above sea level. At what elevation would the Old Course at St. Andrews, Scotland, shown in Figure 19.9, have to be located, if the air temperature was 48 °F (8.9 °C = 282.0 K), for the golf balls to have the same increase in flight length as at Southern Hills?

SOLUTION

First, we calculate how the temperature influences the air density. We can use the relationship between the density and temperature for constant pressure, $\rho_1/\rho_2 = T_2/T_1$, equation 19.3. Putting in the known values (remember that we need to use kelvins for the air temperature), we obtain

$$\frac{\rho_1}{\rho_2} = \frac{T_2}{T_1} = \frac{282.0 \text{ K}}{311.5 \text{ K}} = 0.9053.$$

– *Continued*

FIGURE 19.9 Teeing off at the Old Course in St. Andrews, Scotland, at sea level and in late winter. At low elevations or low temperatures, the ball does not carry as far.

In other words, the air at 101 °F has only 90.53% of the density of air at 48 °F, at the same pressure. Now we have to compute the altitude that corresponds to the same density ratio. In Section 13.4, we derived a formula that relates the density to the altitude,

$$\rho(h) = \rho_0 e^{-h\rho_0 g/p_0},$$

where h is the height above sea level, ρ_0 is the density of air at sea level, and p_0 is the air pressure at sea level. The constants in the formula relating density to height can be written as

$$\frac{p_0}{\rho_0 g} = \frac{1.01 \cdot 10^5 \text{ Pa}}{\left(1.229 \text{ kg/m}^3\right)\left(9.81 \text{ m/s}^2\right)} = 8377 \text{ m}.$$

The elevation at Southern Hills is $h_1 = 213$ m, and the air density is

$$\rho(h_1) = \rho_0 e^{-h_1/(8377 \text{ m})}.$$

The air density at the Old Course at an elevation h_2 would be

$$\rho(h_2) = \rho_0 e^{-h_2/(8377 \text{ m})}.$$

Taking the ratio of these two densities gives

$$\frac{\rho(h_2)}{\rho(h_1)} = \frac{\rho_0 e^{-h_2/(8377 \text{ m})}}{\rho_0 e^{-h_1/(8377 \text{ m})}} = e^{-(h_2 - h_1)/(8377 \text{ m})}.$$

Setting this ratio equal to the ratio of the densities from the temperature difference gives

$$e^{-(h_2 - h_1)/(8377 \text{ m})} = 0.9053.$$

Taking the natural log of both sides of this equation leads to

$$-(h_2 - h_1)/(8377 \text{ m}) = \ln(0.9053),$$

which can be solved for h_2:

$$h_2 = h_1 - (8377 \text{ m})\ln(0.9053) = (213 \text{ m}) - (8377 \text{ m})\ln(0.9053) = 1046 \text{ m}.$$

Thus, at the temperature of 101 °F, the course at Southern Hills played as long as the Old Course would at an elevation of 1046 m (3432 ft) and a temperature of 48 °F.

DISCUSSION
Our solution neglected the effect of the high humidity at Southern Hills, which reduced the air density by an additional 2% relative to the density of dry air, causing the course to play like one with dry air and a temperature of 48 °F at an altitude over 4000 ft above sea level. (The effect of water vapor in the air on the air density is discussed later, in the subsection on Dalton's Law.)

SOLVED PROBLEM 19.1 | Pressure of a Planetary Nebula

PROBLEM
A planetary nebula is a cloud of mainly hydrogen gas with a density, ρ, of $1.0 \cdot 10^3$ molecules per cubic centimeter (cm^3) and a temperature, T, of $1.0 \cdot 10^4$ K. An example of a planetary nebula, the Ring Nebula, is shown in Figure 19.10. What is the pressure of the gas in the nebula?

SOLUTION
THINK   Because of the very low density of the gas in a planetary nebula, it is a very good approximation to treat that gas as an ideal gas and apply the Ideal Gas Law. We know the particle density, which is the number of gas molecules per unit volume, so we can replace the number of molecules in the Ideal Gas Law with the density times the volume of the nebula. The volume then cancels out, and we can solve for the pressure of the gas.

SKETCH   A rough sketch of a planetary nebula is shown in Figure 19.11.

RESEARCH   The Ideal Gas Law is given by

$$pV = Nk_B T, \tag{i}$$

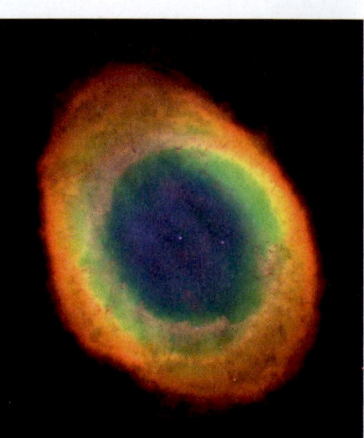

FIGURE 19.10 The Ring Nebula, as seen by the Hubble Space Telescope.

where p is the pressure, V is the volume, N is the number of molecules of gas, k_B is the Boltzmann constant, and T is the temperature. The number of molecules, N, is

$$N = \rho V, \qquad \text{(ii)}$$

where ρ is the number of molecules per unit volume in the nebula.

SIMPLIFY Substituting for N from equation (ii) into (i), we get

$$pV = N k_B T = \rho V k_B T,$$

which we can simplify to get an expression for the pressure of the gas in the nebula:

$$p = \rho k_B T.$$

CALCULATE Putting in the numerical values gives us

$$p = \rho k_B T = \left(\frac{1.0 \cdot 10^3 \ \text{molecules}}{1 \ \text{cm}^3} \right) \left(\frac{1 \ \text{cm}^3}{10^{-6} \ \text{m}^3} \right) \left(1.381 \cdot 10^{-23} \ \text{J/K} \right) \left(1.0 \cdot 10^4 \ \text{K} \right)$$

$$= 1.381 \cdot 10^{-10} \ \text{Pa}.$$

ROUND We report our result to two significant figures:

$$p = 1.4 \cdot 10^{-10} \ \text{Pa}.$$

DOUBLE-CHECK We can usually double-check our result by determining whether the unit(s) and the order of magnitude are appropriate. In this case, our answer has a unit of pressure (Pa), but we have no reference point for judging the order of magnitude. Obviously, the pressure of the gas in the nebula must be very low, much lower than the pressure of Earth's atmosphere, because the density of the gas molecules in the nebula is so small. In fact, the pressure we calculated is astonishingly low; as a comparison, the pressure of a good vacuum in a lab on Earth is $p_{\text{lab}} = 10^{-7}$ Pa, which is about 1000 times higher than the pressure in the nebula.

Note that although a gas generally expands to fill its container, in this case there is no container. The gas in a planetary nebula is held together by the gravitational interactions among its molecules.

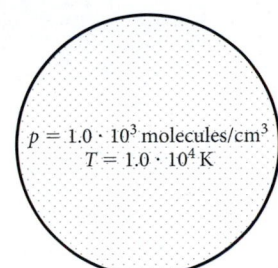

$p = 1.0 \cdot 10^3 \ \text{molecules/cm}^3$
$T = 1.0 \cdot 10^4 \ \text{K}$

FIGURE 19.11 An idealized planetary nebula.

Work Done by an Ideal Gas at Constant Temperature

Suppose we have an ideal gas at a constant temperature in a closed container whose volume can be changed, such as a cylinder with a piston. This setup allows us to perform an isothermal process, described in Chapter 18. For an isothermal process, the Ideal Gas Law says that the pressure is equal to a constant times the inverse of the volume: $p = nRT/V$. If the volume of the container is changed from an initial volume, V_i, to a final volume, V_f, the work done by the gas (see Chapter 18) is given by

$$W = \int_{V_i}^{V_f} p \, dV.$$

Substituting the expression for the pressure into this integral gives

$$W = \int_{V_i}^{V_f} p \, dV = (nRT) \int_{V_i}^{V_f} \frac{dV}{V} = nRT \left[\ln V \right]_{V_i}^{V_f},$$

which evaluates to

$$W = nRT \ln \left(\frac{V_f}{V_i} \right). \qquad \textbf{(19.10)}$$

Equation 19.10 indicates that the work done by the gas is positive if $V_f > V_i$ and is negative if $V_f < V_i$.

We can compare this result to those found in Chapter 18 for the work done by the gas under other assumptions. For example, if the volume is held constant rather than the temperature, the gas can do no work: $W = 0$. If the pressure is held constant, the work done by the gas is given by (see Chapter 18)

$$W = \int_{V_i}^{V_f} p\, dV = p \int_{V_i}^{V_f} dV = p\left(V_f - V_i\right) = p\Delta V.$$

Compressed Air Energy Storage

In a dentist's drill, the drill bit needs to rotate at a very high frequency of over 100,000 rpm. Putting an electric motor into the hand piece would add a lot of weight; therefore, the drill is driven by compressed air. Many other tools (including nail guns, air ratchets, and jackhammers) have the same constraints. Air compressors, which supply compressed air at pressures from 500 kPa to 1000 kPa, drive all of them. Citywide distribution systems for compressed air have been in use in cities such as Paris, France, since the end of the 19th century. Locomotives operating on compressed air have even been used in coal mines (where sparks from other types of engines could be fatal!).

Compressed air can be used to store energy temporarily, which is necessary in situations where power production is intermittent, for example, with wind farms and solar power plants. The use of large-scale storage facilities (see Figure 19.12) can smooth the time profile of the produced power. Energy storage and retrieval by these facilities are not as efficient as they are with pump storage facilities that use water, but in geographical locations with very little altitude variation, they may be the only way to accomplish energy storage. The first compressed air storage facility was built in 1978 in Huntorf, Germany, and a 110-MW facility, built in 1991 in McIntosh, Alabama, was the first in the United States.

FIGURE 19.12 Schematic diagram of a compressed air energy storage facility.

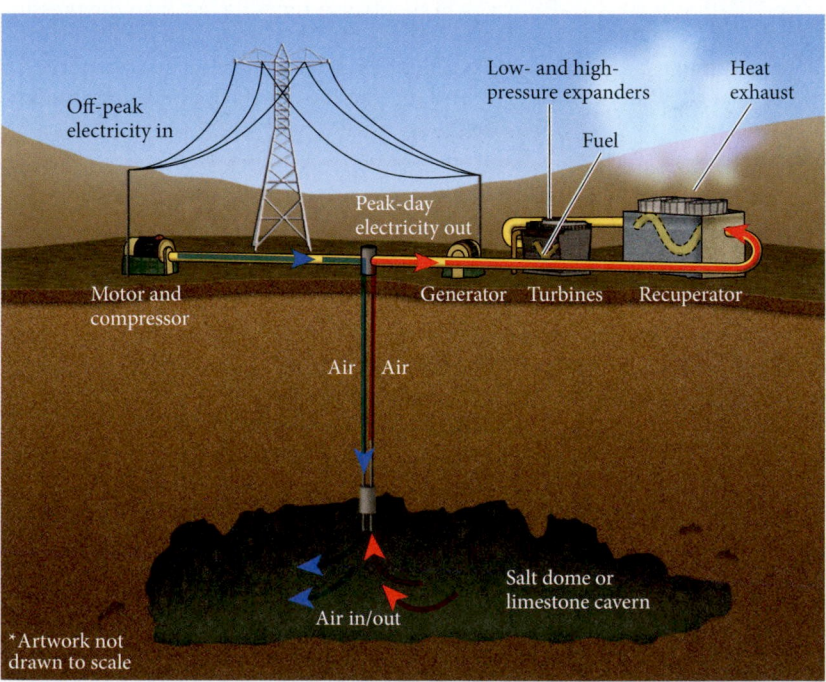

SOLVED PROBLEM 19.2 | Home Energy Storage

Homes located in areas with ample sunshine can harvest energy with solar panels. But in order to avoid buying electricity from the grid (or if a home is located off the grid, for example, in a rural area in a developing country), a homeowner needs some way to store the energy gathered by the solar cells for use during the times when the Sun isn't shining.

PROBLEM

Suppose a home's energy storage device is a 10.0-L tank and its solar panels deliver enough electricity to compress ambient air very slowly from atmospheric pressure to 23.1 MPa. What is the total energy stored in the air in the tank?

SOLUTION

THINK At first, it might seem that no energy can be stored by increasing the pressure of a gas inside a tank of fixed volume, because a constant volume implies that no work can be done. However, it is the volume of the gas that matters, not the volume of the container in the final state. If the air is compressed very slowly, then the assumption that it stays at constant temperature is fulfilled to a good approximation. Thus, we can use the concept of work done on an ideal gas at constant temperature, that is, work done by isothermal compression. (In Solved Problem 19.4, we will consider a case for which we cannot make the same assumption.)

SKETCH The pump used in the air compressor acts like a piston that changes the volume of the gas from its initial value to its final value (see Figure 19.13). The final value is the volume of the tank, and the initial value is the volume that the same amount of air would occupy under normal atmospheric pressure.

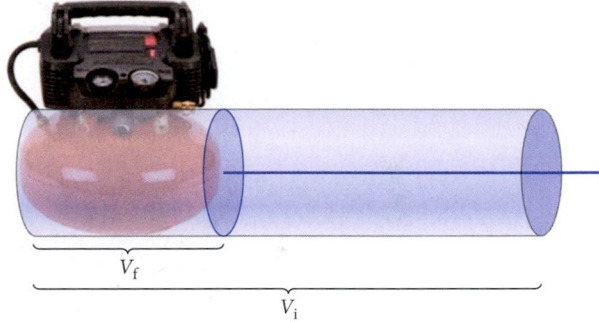

FIGURE 19.13 Compressing ambient air into a tank.

RESEARCH Since the process of isothermal compression of a gas applies here, we can use our result for the work done in such a process, $W = nRT \ln(V_f/V_i)$.

We are given the final volume of the air (10.0 L) but not the initial volume and both the final pressure (23.1 MPa) and the initial pressure (atmospheric pressure = 101 kPa). According to Boyle's Law (the Ideal Gas Law for constant temperature), the initial and final pressures and volumes are related by $p_i V_i = p_f V_f$.

SIMPLIFY Solving Boyle's Law for the ratio of the volumes yields

$$\frac{V_f}{V_i} = \frac{p_i}{p_f}.$$

Substituting for V_f/V_i from this version of Boyle's Law into our equation for the work done results in

$$W = nRT \ln\left(\frac{p_i}{p_f}\right).$$

Using the Ideal Gas Law, $pV = nRT$, we now substitute for nRT to obtain

$$W = p_f V_f \ln\left(\frac{p_i}{p_f}\right).$$

(Here we could have used the initial pressure and volume but chose to use the final pressure and volume because their values were given in the problem statement.)

CALCULATE Now we can insert the numerical values:

$$W = \left(23.1 \cdot 10^6 \text{ Pa}\right)\left(10.0 \cdot 10^{-3} \text{ m}^3\right) \ln(0.101/23.1)$$

$$= -1.254900 \cdot 10^6 \text{ J}.$$

ROUND We round our result to three significant figures: $W = -1.25 \cdot 10^6$ J. This result means that the energy stored in this compressed gas is $E = +1.25 \cdot 10^6$ J.

DOUBLE-CHECK Does the amount of stored energy we have calculated make sense? Since 1 kWh is 3.6 MJ, our result is approximately equal to $\frac{1}{3}$ kWh. This would be enough to power a couple of 40-W light bulbs for about 5 h, assuming that we could convert all of the stored energy into electricity, which is not possible. Thus, our result has to be taken as an upper limit. However, it seems plausible that a couple of light bulbs could be powered for a few hours with this relatively simple and cheap energy storage system.

Dalton's Law

How do ideal gas considerations change, if there is more than one type of gas in a given volume, as in the Earth's atmosphere, for example? **Dalton's Law** states that the pressure of a gas composed of a homogeneous mixture of different gases is equal to the sum of the partial pressures of the component gases. **Partial pressure** is defined as the pressure each gas would exert if the other gases were not present. Dalton's Law means that each gas is unaffected by the presence of other gases, as long as there is no interaction between the gas molecules. The law is named for John Dalton, a British chemist, who published it in 1801. Dalton's Law gives the total pressure, p_{total}, exerted by a mixture of m gases, each with partial pressure p_i:

$$p_{total} = p_1 + p_2 + p_3 + \cdots + p_m = \sum_{i=1}^{m} p_i. \qquad (19.11)$$

The total number of moles of gas, n_{total}, contained in a mixture of m gases is equal to the sum of the numbers of moles of each gas, n_i:

$$n_{total} = n_1 + n_2 + n_3 + \cdots + n_m = \sum_{i=1}^{m} n_i.$$

Then, the **mole fraction,** r_i, for each gas in a mixture is the number of moles of that gas divided by the total number of moles of gas:

$$r_i = \frac{n_i}{n_{total}}. \qquad (19.12)$$

The sum of the mole fractions is equal to 1: $\sum_{i=1}^{m} r_i = 1$. If the number of moles of gas increases, with the temperature and volume constant, the pressure must go up. We can then express each partial pressure as

$$p_i = r_i p_{total} = n_i \frac{p_{total}}{n_{total}}. \qquad (19.13)$$

This means that the partial pressures in a mixture of gases are simply proportional to the mole fractions of the gases present in the mixture.

Earth's Atmosphere

The atmosphere of the Earth has a mass of approximately $5.2 \cdot 10^{18}$ kg (~5 quadrillion metric tons; see Solved Problem 13.2). It is a mixture of several gases; see Table 19.1. The table lists only the components of dry air; in addition, air also contains water vapor (H_2O), which makes up approximately 0.25% of the entire atmosphere. (This sounds like a small number but equals approximately 10^{16} kg of water, which is about four times the volume of water in all the Great Lakes!) Near the Earth's surface, the water content of air ranges from less than 1% to over 3%, depending mainly on the air temperature and the availability of liquid water in the vicinity. Water vapor in the air *decreases* the average density of the atmosphere, because the molecular mass of water is 18, which is smaller than the molecular or atomic mass of almost all the other atmospheric gases.

Concept Check 19.3

What is the mass of 22.4 L of dry air at standard temperature and pressure?

a) 14.20 g   d) 32.22 g

b) 28.00 g   e) 60.00 g

c) 28.95 g

Concept Check 19.4

What is the oxygen pressure in the Earth's atmosphere at sea level?

a) 0   d) 4.8 atm

b) 0.21 atm   e) 20.9 atm

c) 1 atm

Table 19.1	Major Gases Making Up the Earth's Atmosphere		
Gas	**Percentage by Volume**	**Chemical Symbol**	**Molecular Mass**
Nitrogen	78.08	N_2	28.0
Oxygen	20.95	O_2	32.0
Argon	0.93	Ar	39.9
Carbon dioxide	0.039	CO_2	44.0
Neon	0.0018	Ne	20.2
Helium	0.0005	He	4.00
Methane	0.0002	CH_4	16.0
Krypton	0.0001	Kr	83.8

SOLVED PROBLEM 19.3 / Density of Air at STP

PROBLEM
What is the density of air at standard temperature and pressure (STP)?

SOLUTION
THINK We know that a mole of any gas at STP occupies a volume of 22.4 L. We know the relative fractions of the gases that comprise the atmosphere. We can combine Dalton's Law, Avogadro's Law, and the fractions of gases in air to obtain the mass of air contained in 22.4 L. The density is the mass divided by the volume.

SKETCH This is one of the rare occasions where a sketch is not helpful

RESEARCH We consider only the four gases in Table 19.1 that constitute a significant fraction of the atmosphere: nitrogen, oxygen, argon, and carbon dioxide. We know that 1 mole of any gas has a mass M (in grams) in a volume of 22.4 L at STP. From equation 19.12, we know that the mass of each gas in 22.4 L will be the fraction by volume times the molecular mass, so we have

$$m_{air} = r_{N_2} M_{N_2} + r_{O_2} M_{O_2} + r_{Ar} M_{Ar} + r_{CO_2} M_{CO_2}$$

and

$$\rho_{air} = \frac{m_{air}}{V_{air}} = \frac{m_{air}}{22.4 \text{ L}}.$$

SIMPLIFY The density of air is then

$$\rho_{air} = \frac{(0.7808 \cdot 28 \text{ g}) + (0.2095 \cdot 32 \text{ g}) + (0.0093 \cdot 40 \text{ g}) + (0.00039 \cdot 44 \text{ g})}{22.4 \text{ L}}.$$

CALCULATE Calculating the numerical result gives us

$$\rho_{air} = \frac{0.0218624 \text{ kg} + 0.006704 \text{ kg} + 0.000372 \text{ kg} + 0.000011716 \text{ kg}}{22.4 \cdot 10^{-3} \text{ m}^3}$$

$$= 1.29266 \text{ kg/m}^3.$$

ROUND We report our result to three significant figures:

$$\rho_{air} = 1.29 \text{ kg/m}^3.$$

DOUBLE-CHECK We can double-check our result by calculating the density of nitrogen gas alone at STP, since air is 78.08% nitrogen. The density of nitrogen at STP is

$$\rho_{N_2} = \frac{28.0 \text{ g}}{22.4 \text{ L}} = 1.25 \text{ g/L} = 1.25 \text{ kg/m}^3.$$

This result is close to and slightly lower than our result for the density of air; so our result is reasonable.

19.3 Equipartition Theorem

We have been discussing the macroscopic properties of gases, including volume, temperature, and pressure. To explain these properties in terms of the constituents of a gas, its molecules (or atoms), requires making several assumptions about the behavior of these molecules in a container. These assumptions, along with the results derived from them, are known as the **kinetic theory of an ideal gas.**

Suppose we have a gas in a container of volume V, and the gas fills this container evenly. We make the following assumptions:

- The number of molecules, N, is large, but the molecules themselves are small, so the average distance between molecules is large compared to their size. All molecules are identical, and each has mass m.

- The molecules are in constant random motion on straight-line trajectories. They do not interact with one another and can be considered point particles.
- The molecules have elastic collisions with the walls of the container.
- The volume, V, of the container is large compared to the size of the molecules. The container walls are rigid and stationary.

The kinetic theory explains how the microscopic properties of the gas molecules give rise to the macroscopic observables of pressure, volume, and temperature as related by the Ideal Gas Law. For now, we are mainly interested in what the average kinetic energy of the gas molecules is and how it relates to the temperature of the gas. This connection is called the **equipartition theorem**. In the next section, we use this theorem to derive expressions for the specific heats of gases. We'll then see that the equipartition theorem is intimately connected to the idea of degrees of freedom of the motion of the gas molecules. Finally, in Section 19.6, we'll return to the kinetic theory, considering the distribution of kinetic energies of the molecules (instead of just the average kinetic energy), and discuss the limitations of the concept of an ideal gas.

First, we obtain the average kinetic energy of the ideal gas by simply averaging the kinetic energies of the individual gas molecules:

$$K_{ave} = \frac{1}{N} \sum_{i=1}^{N} K_i = \frac{1}{N} \sum_{i=1}^{N} \tfrac{1}{2} m v_i^2 = \tfrac{1}{2} m \left(\frac{1}{N} \sum_{i=1}^{N} v_i^2 \right) = \tfrac{1}{2} m v_{rms}^2. \tag{19.14}$$

Thus, the **root-mean-square speed** of the gas molecules, v_{rms}, is defined as

$$v_{rms} = \sqrt{\frac{1}{N} \sum_{i=1}^{N} v_i^2}. \tag{19.15}$$

Note that the root-mean-square speed is *not the same* as the average speed of the gas molecules. But it can be considered a suitable average, because it immediately relates to the average kinetic energy, as seen in equation 19.14.

Derivation 19.2 shows how the average kinetic energy of the molecules in an ideal gas relates to the temperature of the gas. We'll see that the temperature of the gas is simply proportional to the average kinetic energy, with a proportionality constant $\tfrac{3}{2}$ times the Boltzmann constant:

$$K_{ave} = \tfrac{3}{2} k_B T. \tag{19.16}$$

This is the equipartition theorem, which states that gas molecules in thermal equilibrium have the same average kinetic energy, equal to $\tfrac{1}{2} k_B T$, associated with each of their three independent degrees of freedom. Thus, when we measure the temperature of a gas, we are determining the average kinetic energy of the molecules of the gas. This relationship is one of the key insights of kinetic theory and will be very useful in the rest of this chapter and in later chapters.

Let's see how to derive the equipartition theorem starting from Newtonian mechanics and using the ideal gas law.

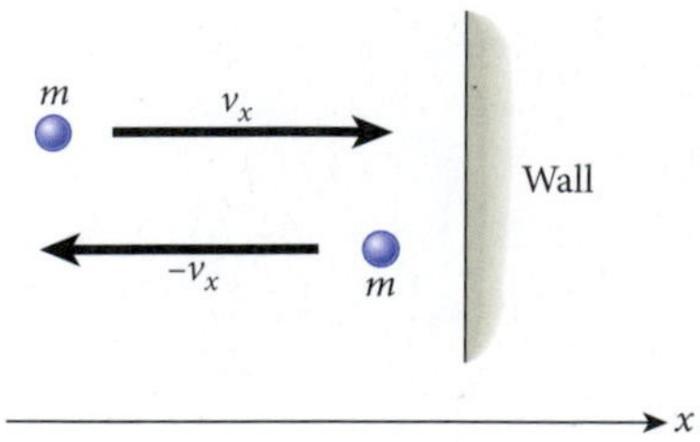

FIGURE 19.14 Gas molecule with mass m and velocity v_x bouncing off the wall of a container.

DERIVATION 19.2 / Average Kinetic Energy of Gas Molecules

The pressure of a gas in a container is determined by the interaction of the gas molecules with the walls of the container, that is, by the change in their momentum over a given time, which is a force. The pressure of the gas is a consequence of elastic collisions of the gas molecules with the walls of the container. When a gas molecule hits a wall, it bounces back with the same kinetic energy it had before the collision. We assume that the wall is stationary, and thus the component of the momentum of the gas molecule perpendicular to the surface of the wall is reversed in the collision (see Section 7.5).

Consider a gas molecule with mass m and an x-component of velocity, v_x, traveling perpendicular to the wall of a container (Figure 19.14). The gas molecule bounces off the wall and moves in the opposite direction with velocity $-v_x$. (Remember, we assumed elastic collisions between molecules and walls.) The change in the x-component of the momentum, Δp_x, of the gas molecule during the collision with the wall is

$$\Delta p_x = p_{f,x} - p_{i,x} = (m)(-v_x) - (m)(v_x) = -2mv_x.$$

To calculate the time-averaged force exerted by the gas on the wall of the container, we need to know not only how much the momentum changes during the collision but also how often a collision of a gas molecule with a wall occurs. To get an expression for the time between collisions, we assume the container is a cube with side length L (Figure 19.15). Then a collision with a particular wall, like that shown in Figure 19.14, occurs every time the gas molecule makes a complete trip from one side of the cube to the other and back again. This round trip covers a distance $2L$. We can express the time interval, Δt, between collisions with the wall as

$$\Delta t = \frac{2L}{v_x}.$$

We know from Newton's Second Law that the x-component of the force exerted by the wall on the gas molecule, $F_{wall,x}$, is given by

$$F_{wall,x} = \frac{\Delta p_x}{\Delta t} = \frac{-2mv_x}{(2L/v_x)} = -\frac{mv_x^2}{L},$$

where we have used $\Delta p_x = -2mv_x$ and $\Delta t = 2L/v_x$.

The x-component of the force exerted by the gas molecule on the wall, F_x, has the same magnitude as $F_{wall,x}$ but is in the opposite direction; that is, $F_x = -F_{wall,x}$, which is a direct consequence of Newton's Third Law (see Chapter 4). If there are N gas molecules, we can express the total force due to these gas molecules on the wall as

$$F_{tot,x} = \sum_{i=1}^{N} \frac{mv_{x,i}^2}{L} = \frac{m}{L} \sum_{i=1}^{N} v_{x,i}^2. \tag{i}$$

This result for the force on the wall located at $x = L$ is independent of the y- and z-components of the velocity vectors of the gas molecules. Since the molecules are in random motion, the average square of the x-components of the velocity is the same as the average square of the y-components or the z-components: $\sum_{i=1}^{N} v_{x,i}^2 = \sum_{i=1}^{N} v_{y,i}^2 = \sum_{i=1}^{N} v_{z,i}^2$. Since $v_i^2 = v_{x,i}^2 + v_{y,i}^2 + v_{z,i}^2$, we have $\sum_{i=1}^{N} v_{x,i}^2 = \frac{1}{3} \sum_{i=1}^{N} v_i^2$, or the average square of the x-components of the velocity is $\frac{1}{3}$ of the average square of the velocity. This is the reason for the name of the equipartition theorem: Each Cartesian velocity component has an equal part, $\frac{1}{3}$, of the overall kinetic energy. We can use this fact to rewrite equation (i) for the x-component of the force of the gas on the wall:

$$F_{tot,x} = \frac{m}{3L} \sum_{i=1}^{N} v_i^2.$$

Repeating this force analysis for the other five walls, or faces of the cube, we find that each experiences a force of the same magnitude, given by $F_{tot} = \frac{m}{3L} \sum_{i=1}^{N} v_i^2$. To find the pressure exerted by the gas molecules, we divide this force by the area of the wall, $A = L^2$:

$$p = \frac{F_{tot}}{A} = \frac{\frac{m}{3L} \sum_{i=1}^{N} v_i^2}{L^2} = \frac{m \sum_{i=1}^{N} v_i^2}{3L^3} = \frac{m}{3V} \sum_{i=1}^{N} v_i^2,$$

where we have used the cube's volume, $V = L^3$, in the last step. Using equation 19.15 for the root-mean-square speed of the gas molecules, we can express the pressure as

$$p = \frac{Nmv_{rms}^2}{3V}. \tag{ii}$$

This result holds for each face of the cube, and it can be applied to a volume of any shape. Multiplying both sides of equation (ii) by the volume leads to $pV = \frac{1}{3}Nmv_{rms}^2$, and since

– Continued

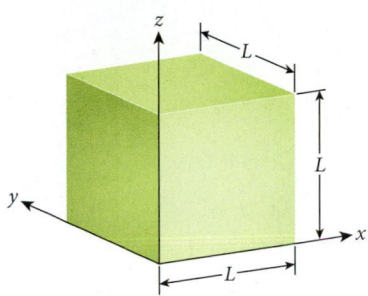

FIGURE 19.15 A cubical container with side length L. Three of the faces of the cube are in the xy-, xz-, and yz-planes.

the Ideal Gas Law (equation 19.8) states that $pV = Nk_BT$, we have $Nk_BT = \frac{1}{3}Nmv_{rms}^2$. Canceling out the factor of N on both sides leads to

$$3k_BT = mv_{rms}^2. \quad \text{(iii)}$$

Since $\frac{1}{2}mv_{rms}^2 = K_{ave}$, the average kinetic energy of a gas molecule, we have arrived at the desired result, equation 19.16:

$$K_{ave} = \frac{3}{2}k_BT.$$

Note that we can solve $3k_BT = mv_{rms}^2$, equation (iii) of Derivation 19.2, for the root-mean-square speed of the gas molecules:

$$v_{rms} = \sqrt{\frac{3k_BT}{m}}. \quad \text{(19.17)}$$

EXAMPLE 19.4   **Average Kinetic Energy of Air Molecules**

Suppose a roomful of air is at a temperature of 22.0 °C.

PROBLEM

What is the average kinetic energy of the molecules (and atoms) in the air?

SOLUTION

The average kinetic energy of the air molecules is given by equation 19.16:

$$K_{ave} = \frac{3}{2}k_BT = \frac{3}{2}\left(1.381 \cdot 10^{-23} \text{ J/K}\right)\left(273.15 \text{ K} + 22.0 \text{ K}\right) = 6.11 \cdot 10^{-21} \text{ J}.$$

Often, the average kinetic energy of air molecules is given in electron-volts (eV) rather than joules:

$$K_{ave} = \left(6.11 \cdot 10^{-21} \text{ J}\right)\frac{1 \text{ eV}}{1.602 \cdot 10^{-19} \text{ J}} = 0.0382 \text{ eV}.$$

A useful approximation to remember is that the average kinetic energy of air molecules at sea level and room temperature is about 0.04 eV.

Concept Check 19.5

The average kinetic energy of molecules of an ideal gas doubles if the _____ doubles.

a) temperature

b) pressure

c) mass of the gas molecules

d) volume of the container

e) none of the above

19.4   Specific Heat of an Ideal Gas

Chapter 18 addressed the specific heat of materials, which we now consider for gases. Gases can consist of atoms or molecules, and the internal energy of gases can be expressed in terms of their atomic and molecular properties. These relationships lead to the molar specific heats of ideal gases.

Let's begin with **monatomic gases,** which are gases in which atoms are not bound to other atoms. Monatomic gases include helium, neon, argon, krypton, and xenon (the noble gases). We assume that all the internal energy of a monatomic gas is in the form of translational kinetic energy. The average translational kinetic energy depends only on the temperature, as stated in equation 19.16.

The internal energy of a monatomic gas is the number of atoms, N, times the average translational kinetic energy of one atom in the gas:

$$E_{int} = NK_{ave} = N\left(\frac{3}{2}k_BT\right).$$

The number of atoms of the gas is the number of moles, n, times Avogadro's number, N_A: $N = nN_A$. Since $N_Ak_B = R$, we can express the internal energy as

$$E_{int} = \frac{3}{2}nRT. \quad \text{(19.18)}$$

Equation 19.18 shows that the internal energy of a monatomic gas depends only on the temperature of the gas.

Specific Heat at Constant Volume

Suppose an ideal monatomic gas at temperature T is held at constant volume. If heat, Q, is added to the gas, it has been shown empirically (see Chapter 18) that the temperature of the gas changes according to

$$Q = nC_V \Delta T \quad \text{(at constant volume)},$$

where C_V is the molar specific heat at constant volume. Because the volume of the gas is constant, it can do no work. Thus, we can use the First Law of Thermodynamics (see Chapter 18) to write

$$\Delta E_{int} = Q - W = nC_V \Delta T \quad \text{(at constant volume)}. \tag{19.19}$$

Remembering that the internal energy of a monatomic gas depends only on its temperature, we use equation 19.18 and obtain

$$\Delta E_{int} = \tfrac{3}{2} nR\Delta T.$$

Combining this result with $Q = nC_V\Delta T$ gives $nC_V\Delta T = \tfrac{3}{2}nR\Delta T$. Canceling the terms n and ΔT, which appear on both sides, we get an expression for the molar specific heat of an ideal monatomic gas at constant volume:

$$C_V = \tfrac{3}{2}R = 12.5 \text{ J/(mol K)}. \tag{19.20}$$

This value for the molar specific heat of an ideal monatomic gas agrees well with measured values for monatomic gases at standard temperature and pressure. These gases are primarily the noble gases. The molar specific heats for **diatomic gases** (gas molecules with two atoms) and **polyatomic gases** (gas molecules with more than two atoms) are higher than the molar specific heats for monatomic gases, as shown in Table 19.2. We'll discuss this difference later in this section.

We can substitute from equation 19.20 for $\tfrac{3}{2}R$ in equation 19.18 and obtain

$$E_{int} = nC_V T, \tag{19.21}$$

which means that the internal energy of an ideal gas depends only on n, C_V, and T.

The equation $\Delta E_{int} = nC_V\Delta T$ applies to all gases as long as the appropriate molar specific heat is used. According to this equation, the change in internal energy of an ideal gas depends only on n, C_V, and the change in temperature, ΔT. The change in internal energy does not depend on any corresponding pressure or volume change.

Specific Heat at Constant Pressure

Now let's consider the situation in which the temperature of an ideal gas is increased while the pressure of the gas is held constant. The added heat, Q, has been shown empirically to be related to the change in temperature as follows:

$$Q = nC_p\Delta T \quad \text{(for constant pressure)}, \tag{19.22}$$

Table 19.2	Some Typical Molar Specific Heats Obtained for Different Types of Gases		
Gas	**C_V [J/(mol K)]**	**C_p [J/(mol K)]**	**$\gamma = C_p/C_V$**
Helium (He)	12.5	20.8	1.66
Neon (Ne)	12.5	20.8	1.66
Argon (Ar)	12.5	20.8	1.66
Krypton (Kr)	12.5	20.8	1.66
Hydrogen (H_2)	20.4	28.8	1.41
Nitrogen (N_2)	20.7	29.1	1.41
Oxygen (O_2)	21.0	29.4	1.41
Carbon dioxide (CO_2)	28.2	36.6	1.29
Methane (CH_4)	27.5	35.9	1.30

where C_p is the molar specific heat at constant pressure. The molar specific heat at constant pressure is greater than the molar specific heat at constant volume because energy must be supplied to do work as well as to increase the temperature.

To relate the molar specific heat at constant volume to the molar specific heat at constant pressure, we start with the First Law of Thermodynamics: $\Delta E_{int} = Q - W$. In the previous section, we showed that $\Delta E_{int} = nC_V\Delta T$ applies for any pressure or volume change, including constant pressure. We substitute from equation 19.22 for Q. In Chapter 18, the work, W, done when the pressure remains constant was expressed as $W = p\Delta V$. With these substitutions into $\Delta E_{int} = Q - W$, we obtain

$$nC_V\Delta T = nC_p\Delta T - p\Delta V.$$

The Ideal Gas Law (equation 19.7) relates the change in volume to the change in temperature if the pressure is held constant: $p\Delta V = nR\Delta T$. Substituting this expression for the work, we finally have

$$nC_V\Delta T = nC_p\Delta T - nR\Delta T.$$

Every term in this equation contains the common factor $n\Delta T$. Dividing it out gives the very simple relation between the specific heats at constant volume and at constant pressure:

$$C_p = C_V + R. \tag{19.23}$$

Self-Test Opportunity 19.1

Using the data in Table 19.2, verify the relationship given in equation 19.23 between the molar specific heat at constant pressure and the molar specific heat at constant volume for an ideal gas.

Degrees of Freedom

As you can see in Table 19.2, the relationship $C_V = \frac{3}{2}R$ holds for monatomic gases but not for diatomic and polyatomic gases. This failure can be explained in terms of the possible degrees of freedom of motion for various types of molecules.

In general, a **degree of freedom** is a direction in which something can move. For a point particle in three-dimensional space, there are three orthogonal directions, which are independent of each other. For a collection of N point particles, there are $3N$ degrees of freedom for translational motion in three-dimensional space. But if the point particles are arranged in a solid object, they cannot move independently of each other, and the number of degrees of freedom is reduced to the translational degrees of freedom of the center of mass of the object (three), plus independent rotations of the object around the center of mass. In general, there are three possible independent rotations, giving a total of six degrees of freedom, three translational and three rotational.

Let's consider three different kinds of gases (Figure 19.16). The first is a monatomic gas, such as helium (He). The second is a diatomic gas, represented by nitrogen (N_2). The third is a polyatomic molecule, for example, methane (CH_4). The molecule of the polyatomic gas can rotate around all three coordinate axes and thus has three rotational degrees of freedom. The diatomic molecule shown in Figure 19.16 is aligned with the x-axis, and rotation about this axis will not yield a different configuration. Therefore, this molecule has only two rotational degrees of freedom, shown in the figure as being associated with rotations about the y- and z-axes. The monatomic gas consists of spherically symmetrical atoms, which have no rotational degrees of freedom. (Note that the argument presented here for rotational degrees of freedom does not really make sense for objects that obey classical Newtonian mechanics but does for those that follow the laws of quantum mechanics. We will discuss quantum physics in Chapters 36–40. For now, it is interesting to note that quantum mechanics has observable consequences in the macroscopic world, determining, for example, the specific heats of different gases.)

The various ways in which the internal energy of an ideal gas can be allocated are specified by the principle of **equipartition of energy,** which states that gas molecules in thermal equilibrium have the same average energy associated with each independent degree of freedom of their motion. The average energy per degree of freedom is given by $\frac{1}{2}k_BT$ for each gas molecule.

The equipartition theorem (equation 19.16) for gas molecules' average kinetic energy associated with temperature, $K_{ave} = \frac{3}{2}k_BT$, is consistent with the equipartition of the molecules' translational kinetic energy among three directions (or three degrees of freedom), x,

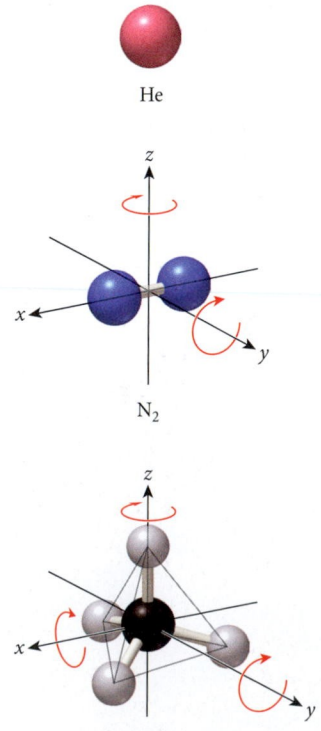

FIGURE 19.16 Rotational degrees of freedom for a helium atom (He), a nitrogen molecule (N_2), and a methane molecule (CH_4).

y, and z. The translational kinetic energy has three degrees of freedom, the average kinetic energy per translational degree of freedom is $\frac{1}{2}k_BT$, and the average kinetic energy of the gas molecules is given by $3\left(\frac{1}{2}k_BT\right)=\frac{3}{2}k_BT$.

Thus, the molar specific heat at constant volume predicted by the kinetic theory for monatomic gases by assuming three degrees of freedom for the translational motion is consistent with reality. However, the observed molar specific heats at constant volume are higher for diatomic gases than for monatomic gases. For polyatomic gases, the observed molar specific heats are even higher.

Specific Heat at Constant Volume for a Diatomic Gas

The nitrogen molecule in Figure 19.16 is lined up along the x-axis. If we look along the x-axis at the molecule, we see a point and thus are not able to discern any rotation around the x-axis. However, if we look along the y-axis or the z-axis, we can see that the nitrogen molecule can have appreciable rotational motion about either of those axes. Therefore, there are two degrees of freedom for rotational kinetic energy of the nitrogen molecule and, by extension, all diatomic molecules. Therefore, the average kinetic energy per diatomic molecule is $(3+2)\left(\frac{1}{2}k_BT\right)=\frac{5}{2}k_BT$.

This result implies that calculation of the molar specific heat at constant volume for a diatomic gas should be similar to that for a monatomic gas, but with two additional degrees of freedom for the average energy. Thus, we modify equation 19.20 to obtain the molar specific heat at constant volume for a diatomic gas:

$$C_V = \frac{3+2}{2}R = \frac{5}{2}R = \frac{5}{2}\left[8.31 \text{ J}/(\text{mol K})\right] = 20.8 \text{ J}/(\text{mol K}).$$

Comparing this value to the actual measured molar specific heats at constant volume given in Table 19.2 for the diatomic gases hydrogen, nitrogen, and oxygen, we see that this prediction from the kinetic theory of ideal gases agrees reasonably well with those measurements.

Specific Heat at Constant Volume for a Polyatomic Gas

Now let's consider the polyatomic gas, methane (CH_4), shown in Figure 19.16. The methane molecule is composed of four hydrogen atoms arranged in a tetrahedron, bonded to a carbon atom at the center. Looking along any of the three axes, we can discern rotation of this molecule. Thus, this particular molecule and all polyatomic molecules have three degrees of freedom related to rotational kinetic energy.

Thus, calculation of the molar specific heat at constant volume for a polyatomic gas should be similar to that for a monatomic gas, but with three additional degrees of freedom for the average energy. We modify equation 19.20 to obtain the molar specific heat at constant volume for a polyatomic gas:

$$C_V = \frac{3+3}{2}R = \frac{6}{2}R = 3\left[8.31 \text{ J}/(\text{mol K})\right] = 24.9 \text{ J}/(\text{mol K}).$$

Comparing this value to the actual measured molar specific heat at constant volume given in Table 19.2 for the polyatomic gas methane, we see that the value predicted by the kinetic theory of ideal gases is close to, but somewhat lower than, the measured value.

What accounts for the differences between predicted and actual values of the molar specific heats for diatomic and polyatomic molecules is the fact that these molecules can have internal degrees of freedom, in addition to translational and rotational degrees of freedom. The atoms of these molecules can oscillate with respect to one another. For example, imagine that the two oxygen atoms and one carbon atom of a carbon dioxide molecule are connected by springs. The three atoms could oscillate back and forth with respect to each other, which corresponds to extra degrees of freedom. The molar specific heat at constant volume is thus larger than the 24.9 J/(mol K) predicted for a polyatomic gas.

A certain minimum energy is required to excite the motion of internal oscillation in molecules. This fact cannot be understood from the viewpoint of classical mechanics; Chapters 36 through 38 will provide this understanding. For now, all you need to know is that a threshold effect is observed for vibrational degrees of freedom. In fact, the rotational degrees of freedom of diatomic and polyatomic gas molecules also exhibit this threshold effect. At low temperatures, the internal energy of the gas may be too low to enable the molecules to rotate. However, at standard temperature and pressure, most diatomic and polyatomic gases have a molar specific heat at constant volume that is consistent with their molecules' translational and rotational degrees of freedom.

Concept Check 19.6

The ratio of specific heat at constant pressure to specific heat at constant volume, C_p/C_V, for an ideal gas

a) is always equal to 1.

b) is always smaller than 1.

c) is always larger than 1.

d) can be smaller or larger than 1, depending on the degrees of freedom of the gas molecules.

Ratio of Specific Heats

Finally, it is convenient to take the ratio of the specific heat at constant pressure to that at constant volume. This ratio is conventionally denoted by the Greek letter γ:

$$\gamma \equiv \frac{C_p}{C_V}. \tag{19.24}$$

Inserting the predicted values for the specific heats of ideal gases with different degrees of freedom gives $\gamma = C_p/C_V = \frac{5}{2}R/\frac{3}{2}R = \frac{5}{3}$ for monatomic gases, $\gamma = C_p/C_V = \frac{7}{2}R/\frac{5}{2}R = \frac{7}{5}$ for diatomic gases, and $\gamma = C_p/C_V = \frac{8}{2}R/\frac{6}{2}R = \frac{4}{3}$ for polyatomic gases.

Table 19.2 also lists values for γ obtained from empirical data. Despite the fact that the empirical values of the specific heats for the gases other than the noble gases are not very close to the calculated values for an ideal gas, the values of the ratio γ are in very good agreement with our theoretical expectations.

19.5 Adiabatic Processes for an Ideal Gas

As we saw in Chapter 18, an adiabatic process is one in which the state of a system changes and no exchange of heat with the surroundings takes place during the change. An adiabatic process can occur when the system change occurs quickly. Thus, $Q = 0$ for an adiabatic process. From the First Law of Thermodynamics, we then have $\Delta E_{int} = -W$, as shown in Chapter 18.

Let's explore how the change in volume is related to the change in pressure for an ideal gas undergoing an adiabatic process, that is, a process in the absence of heat transfer. The relationship is given by

$$pV^\gamma = \text{constant} \quad \text{(for an adiabatic process)}, \tag{19.25}$$

where $\gamma = C_p/C_V$ is the ratio of the specific heat at constant pressure and the specific heat at constant volume for the ideal gas.

DERIVATION 19.3 | Pressure and Volume in an Adiabatic Process

In order to derive the relationship between pressure and volume for an adiabatic process, let's consider small differential changes in the internal energy during such a process: $dE_{int} = -dW$. In Chapter 18, we saw that $dW = p\,dV$, and therefore for the adiabatic process, we have

$$dE_{int} = -p\,dV. \tag{i}$$

However, equation 19.21 says that the internal energy is related in general to the temperature: $E_{int} = nC_VT$. Since we want to calculate dE_{int}, we take the derivative of this equation with respect to the temperature:

$$dE_{int} = nC_V\,dT. \tag{ii}$$

Combining equations (i) and (ii) for dE_{int} then yields

$$-p\,dV = nC_V\,dT. \tag{iii}$$

We can take the derivative of the Ideal Gas Law (equation 19.7), $pV = nRT$ (with n and R constant):

$$p\,dV + V\,dp = nR\,dT. \tag{iv}$$

Solving equation (iii) for dT and substituting that expression into equation (iv) gives us

$$p\,dV + V\,dp = nR\left(-\frac{p\,dV}{nC_V}\right).$$

Now we can collect the terms proportional to dV on the left-hand side, giving

$$\left(1 + \frac{R}{C_V}\right)p\,dV + V\,dp = 0. \tag{v}$$

Since $R = C_p - C_V$ from equation 19.23, we see that $1 + R/C_V = 1 + (C_p - C_V)/C_V = C_p/C_V$. Equation 19.24 represents this ratio of the two specific heats as γ. With this, equation (v) becomes $\gamma p\,dV + V\,dp = 0$. Dividing both sides by pV lets us finally write

$$\frac{dp}{p} + \gamma\frac{dV}{V} = 0.$$

Integration then yields

$$\ln p + \gamma \ln V = \text{constant}. \tag{vi}$$

Using the rules of logarithms, equation (vi) can also be written as $\ln (pV^\gamma) = \text{constant}$. We can exponentiate both sides of this equation to get the desired result:

$$pV^\gamma = \text{constant}.$$

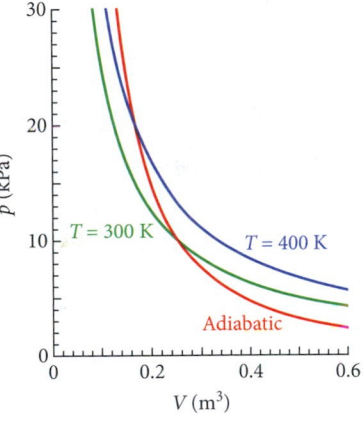

FIGURE 19.17   Plot of the behavior of a monatomic gas during three different processes on a pV-diagram. One process is adiabatic (red). The second process is isothermal with $T = 300$ K (green). The third process is also isothermal with $T = 400$ K (blue).

Another way of stating the relationship between pressure and volume for an adiabatic process involving an ideal gas is

$$p_f V_f^\gamma = p_i V_i^\gamma \quad \text{(for an adiabatic process)}, \tag{19.26}$$

where p_i and V_i are the initial pressure and volume, respectively, and p_f and V_f are the final pressure and volume, respectively.

Figure 19.17 plots the behavior of a gas during three different processes as a function of pressure and volume. The red curve represents an adiabatic process, in which pV^γ is constant, for a monatomic gas with $\gamma = \frac{5}{3}$. The other two curves represent isothermal (constant T) processes. One isothermal process occurs at $T = 300$ K, and the other occurs at $T = 400$ K. The adiabatic process shows a steeper decrease in pressure as a function of volume than the isothermal processes.

We can write an equation relating the temperature and volume of a gas in an adiabatic process by combining equation 19.25 and the Ideal Gas Law (equation 19.7):

$$pV^\gamma = \left(\frac{nRT}{V}\right)V^\gamma = (nR)TV^{\gamma-1} = \text{constant}.$$

Assuming that the gas is in a closed container, n is constant, and we can rewrite this relationship as $TV^{\gamma-1} = \text{constant}$ (for an adiabatic process), or

$$T_f V_f^{\gamma-1} = T_i V_i^{\gamma-1} \quad \text{(for an adiabatic process)}, \tag{19.27}$$

where T_i and V_i are the initial temperature and volume, respectively, and T_f and V_f are the final temperature and volume, respectively.

An adiabatic expansion as described by equation 19.27 occurs when you open a container that holds a cold, carbonated beverage. The carbon dioxide from the carbonation and the water vapor make up a gas with a pressure above atmospheric pressure. When this gas expands and some comes out of the container's opening, the system must do work. Because the process happens very quickly, no heat can be transferred to the gas; thus, the process is adiabatic, and the product of the temperature and the volume raised to the power $\gamma - 1$ remains constant. The volume of the gas increases, and the temperature must decrease. Thus, condensation occurs around the opening of the container.

Concept Check 19.7

A monatomic ideal gas occupies a volume V_i, which is then decreased to $\frac{1}{2}V_i$ via an adiabatic process. Which relationship is correct for the pressures of this gas?

a) $p_f = 2p_i$

b) $p_f = \frac{1}{2}p_i$

c) $p_f = 2^{5/3}p_i$

d) $p_f = \left(\frac{1}{2}\right)^{5/3}p_i$

Concept Check 19.8

A monatomic ideal gas occupies a volume V_i, which is then decreased to $\frac{1}{2}V_i$ via an adiabatic process. Which relationship is correct for the temperatures of this gas?

a) $T_f = 2^{2/3}T_i$

b) $T_f = 2^{5/3}T_i$

c) $T_f = \left(\frac{1}{2}\right)^{2/3}T_i$

d) $T_f = \left(\frac{1}{2}\right)^{5/3}T_i$

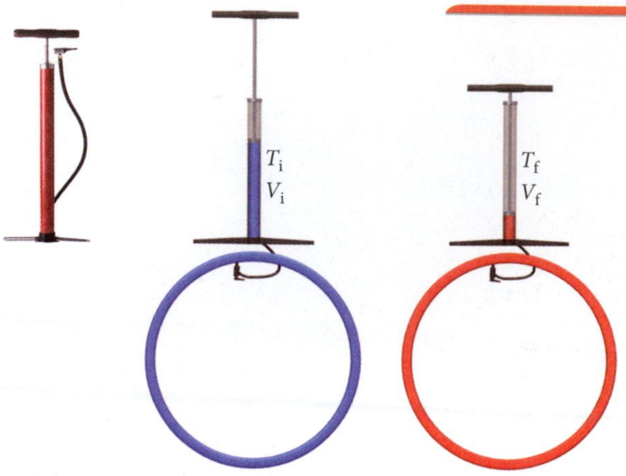

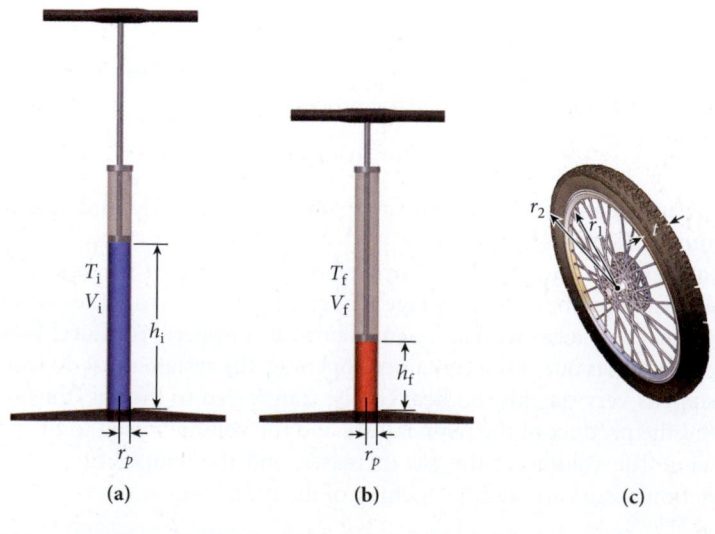

A hand-operated tire pump (Figure 19.18a) is used to inflate a bicycle tire. The tire pump consists of a cylinder with a piston. The tire pump is connected to the tire with a small hose. There is a valve between the pump and the tire that allows air to enter the tire, but not to exit. In addition, there is a valve that allows air to enter the pump after the compression stroke is complete and the piston is retracted to its original position. Figure 19.18b shows the pump and tire before the compression stroke. Figure 19.18c shows the pump and tire after the compression stroke.

The inner radius of the cylinder of the pump is $r_p = 1.00$ cm, the height of the active volume before the compression stroke is $h_i = 60.0$ cm, and the height of the active volume after the compression stroke is $h_f = 10.0$ cm. The tire can be considered to be a cylindrical ring with inner radius $r_1 = 66.0$ cm, outer radius $r_2 = 67.5$ cm, and thickness $t = 1.50$ cm. The temperature of the air in the pump and tire before the compression stroke is $T_i = 295$ K.

FIGURE 19.18 (a) A hand-operated tire pump used to inflate a bicycle tire. (b) The pump and tire before the compression stroke. (c) The pump and tire after the compression stroke.

PROBLEM

What is the temperature, T_f, of the air after one compression stroke?

SOLUTION

THINK The volume and temperature of the air before the compression stroke are V_i and T_i, respectively, and the volume and temperature of the air after the compression stroke are V_f and T_f, respectively. The compression stroke takes place quickly, so no heat can be transferred into or out of the system, and this process can be treated as an adiabatic process. The volume of air in the pump decreases when the pump handle is pushed down while the volume of air in the tire remains constant.

SKETCH The dimensions of the tire pump and the tire are shown in Figure 19.19.

RESEARCH The volume of air in the tire is

$$V_{tire} = t\left(\pi r_2^2 - \pi r_1^2\right).$$

The volume of air in the pump before the compression stroke is

$$V_{pump,i} = \pi r_p^2 h_i.$$

The volume of air in the pump after the compression stroke is

$$V_{pump,f} = \pi r_p^2 h_f.$$

Self-Test Opportunity 19.2

In Solved Problem 19.4, we assumed that the volume of the tire can be approximated as a cylindrical ring of thickness t. This implies a square cross section for the tire. Describe how our result would change if we approximated the cross section of the tire more realistically as a circle of diameter t.

FIGURE 19.19 (a) Dimensions of the tire pump before the compression stroke. (b) Dimensions of the tire pump after the compression stroke. (c) Dimensions of the bicycle tire.

The initial volume of air in the pump-tire system is $V_i = V_{pump,i} + V_{tire}$, and the final volume of air in the pump-tire system is $V_f = V_{pump,f} + V_{tire}$. We ignore the volume of air in the hose between the pump and the tire.

This process is adiabatic, so we can use equation 19.27 to describe the system before and after the compression stroke:

$$T_f V_f^{\gamma-1} = T_i V_i^{\gamma-1}. \tag{i}$$

SIMPLIFY We can solve equation (i) for the final temperature:

$$T_f = T_i \frac{V_i^{\gamma-1}}{V_f^{\gamma-1}} = T_i \left(\frac{V_i}{V_f}\right)^{\gamma-1}.$$

We can then insert the expressions for the initial volume and the final volume:

$$T_f = T_i \left(\frac{\pi r_p^2 h_i + t\left(\pi r_2^2 - \pi r_1^2\right)}{\pi r_p^2 h_f + t\left(\pi r_2^2 - \pi r_1^2\right)}\right)^{\gamma-1} = T_i \left(\frac{r_p^2 h_i + t\left(r_2^2 - r_1^2\right)}{r_p^2 h_f + t\left(r_2^2 - r_1^2\right)}\right)^{\gamma-1}.$$

CALCULATE Air is mostly diatomic gases (78% N_2 and 21% O_2, see Table 19.1), so we use $\gamma = \frac{7}{5}$, from Section 19.4. Putting in the numerical values then gives us

$$T_f = (295 \text{ K})\left(\frac{(0.0100 \text{ m})^2 (0.600 \text{ m}) + (0.0150 \text{ m})\left((0.675 \text{ m})^2 - (0.660 \text{ m})^2\right)}{(0.0100 \text{ m})^2 (0.100 \text{ m}) + (0.0150 \text{ m})\left((0.675 \text{ m})^2 - (0.660 \text{ m})^2\right)}\right)^{(7/5)-1}$$

$$= 313.162 \text{ K}.$$

ROUND We report our result to three significant figures:

$$T_f = 313 \text{ K}.$$

DOUBLE-CHECK The ratio of the initial volume of air in the pump-tire system to the final volume is

$$\frac{V_i}{V_f} = 1.16.$$

During the compression stroke, the temperature of the air increases from 295 K to 313 K, so the ratio of final temperature to initial temperature is

$$\frac{T_f}{T_i} = \frac{313 \text{ K}}{295 \text{ K}} = 1.06,$$

which seems reasonable because the final temperature should be equal to the initial temperature times the ratio of the initial and final volume raised to the power $\gamma - 1$: $(1.16)^{\gamma-1} = 1.16^{0.4} = 1.06$. If you depress the piston many times in a short period of time, each compression stroke will increase the temperature of the air in the tire by about 6%. Thus, the air in the tire and in the pump gets warmer as the tire is inflated.

Work Done by an Ideal Gas in an Adiabatic Process

We can also find the work done by a gas undergoing an adiabatic process from an initial state to a final state. In general, the work is given by

$$W = \int_i^f p\, dV.$$

For an adiabatic process, we can use equation 19.25 to write $p = cV^{-\gamma}$ (where c is a constant), and the integral determining the work becomes

$$W = \int_{V_i}^{V_f} cV^{-\gamma} dV = c\int_{V_i}^{V_f} V^{-\gamma} dV = c\left[\frac{V^{1-\gamma}}{1-\gamma}\right]_{V_i}^{V_f} = \frac{c}{1-\gamma}\left(V_f^{1-\gamma} - V_i^{1-\gamma}\right).$$

The constant c has the value $c = pV^\gamma$. Using the Ideal Gas Law in the form $p = nRT/V$, we have

$$c = \left(\frac{nRT}{V}\right)V^\gamma = nRTV^{\gamma-1}.$$

Now we insert this constant into the expression for the work and use equation 19.27. We finally find the work done by a gas undergoing an adiabatic process:

$$W = \frac{nR}{1-\gamma}\left(T_{\mathrm{f}} - T_{\mathrm{i}}\right). \tag{19.28}$$

19.6   Kinetic Theory of Gases

As promised earlier, this section examines some important results from the kinetic theory of gases beyond the equipartition theorem of Section 19.3.

Maxwell Speed Distribution

Equation 19.17 gives the root-mean-square speed of gas molecules at a given temperature. This is close to the average speed, but what is the distribution of speeds? That is, what is the probability that a gas molecule has some given speed between v and $v + dv$? The speed distribution, called the **Maxwell speed distribution,** or sometimes the *Maxwell-Boltzmann speed distribution,* is given by

$$f\left(v\right) = 4\pi\left(\frac{m}{2\pi k_{\mathrm{B}}T}\right)^{3/2} v^2 e^{-mv^2/2k_{\mathrm{B}}T}. \tag{19.29}$$

The units of the Maxwell speed distribution are $(\mathrm{m/s})^{-1}$. As required for a probability distribution, integration of this distribution with respect to dv yields 1: $\int_0^\infty f\left(v\right) dv = 1$.

A very important feature of probability distributions for observable quantities, including the one in equation 19.29 for the molecular speed, is that all kinds of physically meaningful averages can be calculated from them by simply multiplying the quantity to be averaged by the probability distribution and integrating over the entire range. For example, we can obtain the average speed for the Maxwell speed distribution by multiplying the speed, v, by the probability distribution, $f(v)$, and integrating this product, $vf(v)$, over all possible values of the speed from zero to infinity. The average speed is then given by

$$v_{\mathrm{ave}} = \int_0^\infty v f\left(v\right) dv = \sqrt{\frac{8 k_{\mathrm{B}} T}{\pi m}},$$

where we used the definite integral $\int_0^\infty x^3 e^{-x^2/a} dx = a^2/2$ in performing the integration.

To find the root-mean-square speed for the Maxwell speed distribution, we first find the average speed squared:

$$\left(v^2\right)_{\mathrm{ave}} = \int_0^\infty v^2 f\left(v\right) dv = \frac{3 k_{\mathrm{B}} T}{m},$$

where we employed the definite integral $\int_0^\infty x^4 e^{-x^2/a} dx = \frac{3}{8} a^{5/2} \sqrt{\pi}$. We then take the square root:

$$v_{\mathrm{rms}} = \sqrt{\left(v^2\right)_{\mathrm{ave}}} = \sqrt{\frac{3 k_{\mathrm{B}} T}{m}}.$$

As required, this agrees with equation 19.17 for the speed of gas molecules in an ideal gas.

The most probable speed for the Maxwell speed distribution—that is, the value at which $f(v)$ has a maximum—is calculated by taking the derivative of $f(v)$ with respect to v, setting that derivative equal to zero, and solving for v, which gives

$$v_{\mathrm{mp}} = \sqrt{\frac{2 k_{\mathrm{B}} T}{m}}.$$

Figure 19.20 shows the Maxwell speed distribution for nitrogen (N_2) molecules in air at a temperature of 295 K (22 °C), with the most probable speed, the average speed, and the root-mean-square speed. You can see in Figure 19.20 that the speeds of the nitrogen molecules are

Self-Test Opportunity 19.3

Show explicitly that integrating the Maxwell speed distribution from $v = 0$ to $v = \infty$ yields 1, as required for a probability distribution.

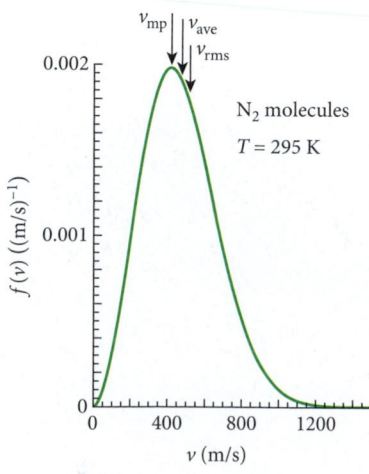

FIGURE 19.20 Maxwell speed distribution for nitrogen molecules with a temperature of 295 K plotted on a linear scale.

distributed around the average speed, v_{ave}. However, the distribution is not symmetrical around v_{rms}. A tail in the distribution extends to high velocities. You can also see that the most probable speed corresponds to the maximum of the distribution.

Figure 19.21 shows the Maxwell speed distribution for nitrogen molecules at four different temperatures. As the temperature is increased, the Maxwell speed distribution gets wider. Consequently, the maximum value of $f(v)$ has to get lower with increasing temperature, because the total area under the curve always equals 1, as required for a probability distribution.

In addition, at a given temperature, the Maxwell speed distribution for lighter atoms or molecules is wider and has more of a high-speed tail than the distribution for heavier atoms or molecules (Figure 19.22).

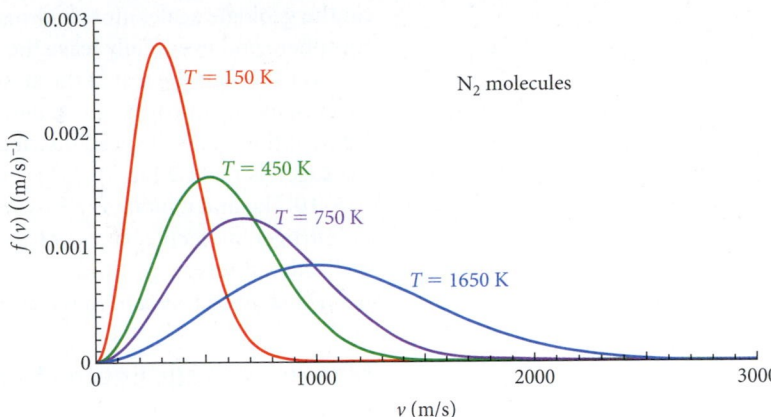

FIGURE 19.21   Maxwell speed distribution for N_2 molecules at four different temperatures.

Figure 19.22a shows the Maxwell speed distributions for hydrogen and nitrogen molecules plotted on a logarithmic scale at a temperature of 22 °C, which is characteristic of temperatures found at the surface of the Earth. How far the high-speed tails of these distributions extend is important to the composition of Earth's atmosphere. As we saw in Chapter 12, the escape speed, v_e, for any object near the surface of the Earth is 11.2 km/s. Thus, any air molecule that has at least this escape speed can leave Earth's atmosphere. No gas atoms or molecules are near the escape speed at this temperature. However, hydrogen gas is light compared to the other gases in the Earth's atmosphere and tends to move upward. The temperature of Earth's atmosphere depends on the altitude. At an altitude of 100 km above the surface, the Earth's atmosphere starts to become very warm and can reach temperatures of 1500 °C. This layer of the atmosphere is called the *thermosphere.*

Figure 19.22b shows the Maxwell speed distributions for hydrogen and nitrogen molecules plotted on a logarithmic scale at a temperature of 1500 °C. We can calculate the fraction of hydrogen molecules that will have a speed higher than the escape speed by evaluating the integral for the Maxwell speed distribution from the escape speed to infinity:

$$\text{Fraction of } H_2 \text{ escaping} = \int_{v=11.2 \text{ km/s}}^{\infty} f(v)dv = 7 \cdot 10^{-4} = 0.07\%.$$

Only a small fraction of the hydrogen molecules in the thermosphere have speeds above the escape speed. The hydrogen molecules with speeds above the escape speed will go out into space. The hydrogen molecules remaining behind have the same temperature as the rest of the thermosphere, and the speed distribution will reflect this temperature. Thus, over time

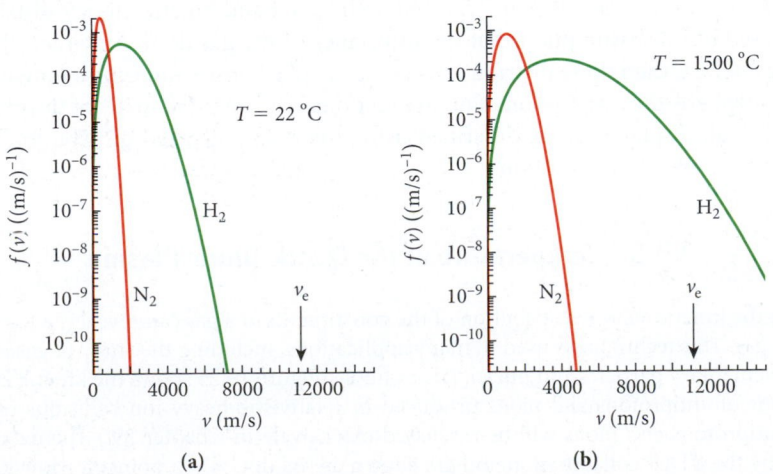

FIGURE 19.22   Maxwell speed distributions for hydrogen molecules and nitrogen molecules in the Earth's atmosphere, plotted on a logarithmic scale: (a) at a temperature of 22 °C; (b) at a temperature of 1500 °C.

on the geologic scale, most hydrogen molecules in the Earth's atmosphere rise to the thermosphere and eventually leave the Earth.

What about the rest of the atmosphere, which consists mainly of nitrogen and oxygen? Essentially no nitrogen molecules have speeds above the escape speed. The ratio of the value of the Maxwell speed distribution at the escape speed for nitrogen relative to that for hydrogen is $f_N(11.2 \text{ km/s})/f_H(11.2 \text{ km/s}) = 3 \cdot 10^{-48}$. Since the entire atmosphere has a mass of $5 \cdot 10^{18}$ kg and contains approximately 10^{44} nitrogen molecules, you can see that virtually no nitrogen molecule has a speed above the escape speed. Thus, there is very little hydrogen in Earth's atmosphere, while nitrogen and heavier gases (oxygen, argon, carbon dioxide, neon, and so on) remain in the atmosphere permanently.

Maxwell Kinetic Energy Distribution

Just as gas molecules have a distribution of speeds, they also have a distribution of kinetic energies. The **Maxwell kinetic energy distribution** (sometimes called the *Maxwell-Boltzmann kinetic energy distribution*) describes the energy spectra of gas molecules and is given by

$$g(K) = \frac{2}{\sqrt{\pi}}\left(\frac{1}{k_B T}\right)^{3/2} \sqrt{K}\, e^{-K/k_B T}. \tag{19.30}$$

The Maxwell kinetic energy distribution times dK gives the probability of observing a gas molecule with a kinetic energy between K and $K+dK$. Integration of this distribution with respect to dK yields 1, $\int_0^\infty g(K)\, dK = 1$, as required for any probability distribution. The unit for the Maxwell kinetic energy distribution is J^{-1}.

Figure 19.23 shows the Maxwell kinetic energy distribution for nitrogen molecules at a temperature of 295 K. The kinetic energies of the nitrogen molecules are distributed around the average kinetic energy. Like the Maxwell speed distribution, the Maxwell kinetic energy distribution is not symmetrical around K_{ave}, but has a significant tail at high kinetic energies. This high-kinetic-energy tail can carry information about the temperature of the gas.

Figure 19.24 shows the Maxwell kinetic energy distribution for nitrogen molecules at two temperatures, $T = 295$ K and $T = 1500$ K. The kinetic energy distribution at $T = 1500$ K is much flatter and extends to much higher kinetic energies than the kinetic energy distribution at $T = 295$ K. The high-kinetic-energy tails for the distributions in Figure 19.24 approach a simple exponential distribution of kinetic energy with a slope of $-1/k_B T$, as illustrated by the dashed lines for the two temperatures.

Note that the formulations given here for the Maxwell speed distribution and the Maxwell kinetic energy distribution do not take the effects of relativity into account. For example, the speed distribution extends up to infinite velocities, and no object can travel faster than the speed of light (to be discussed in detail in Chapter 35 on relativity). However, the relevant velocities for ideal gases are small compared with the speed of light. The highest speed shown in Figure 19.22 is $v = 1.5 \cdot 10^4$ m/s (for hydrogen molecules at $T = 1500$ K), which is only 0.005% of the speed of light. The Maxwell speed and kinetic energy distributions are also based on the assumption that the molecules of the gas do not interact. If the gas molecules interact, then these distributions do not apply. For gas molecules moving in the Earth's atmosphere, the assumption of no interactions is justified, so many of the properties of the Earth's atmosphere can be described using the Maxwell speed distribution and the Maxwell kinetic energy distribution.

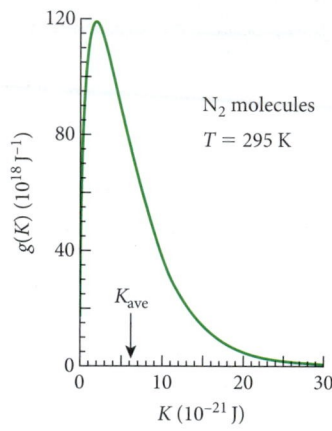

FIGURE 19.23 Maxwell kinetic energy distribution for nitrogen molecules at a temperature of 295 K, plotted on a linear scale.

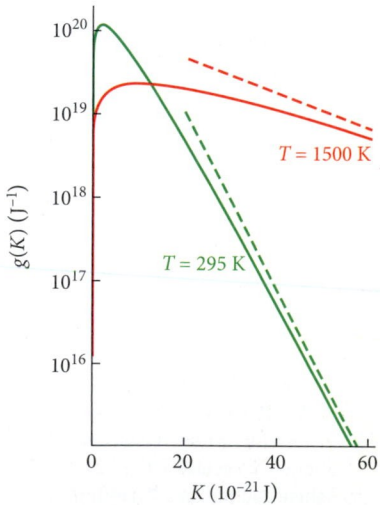

FIGURE 19.24 Maxwell kinetic energy distributions for nitrogen molecules at $T = 295$ K and $T = 1500$ K, plotted on a logarithmic scale.

EXAMPLE 19.5 Temperature of the Quark-Gluon Plasma

Measuring the kinetic energy distribution of the constituents of a gas can reveal the temperature of the gas. This technique is used in many applications, including the study of gases consisting of elementary particles and nuclei. For example, Figure 19.25 shows the kinetic energy distributions of antiprotons and pions produced in relativistic heavy ion collisions of gold nuclei. (Antiprotons and pions will be discussed extensively in Chapter 39.) The data were produced by the STAR collaboration and are shown as red dots. Data points for low kinetic

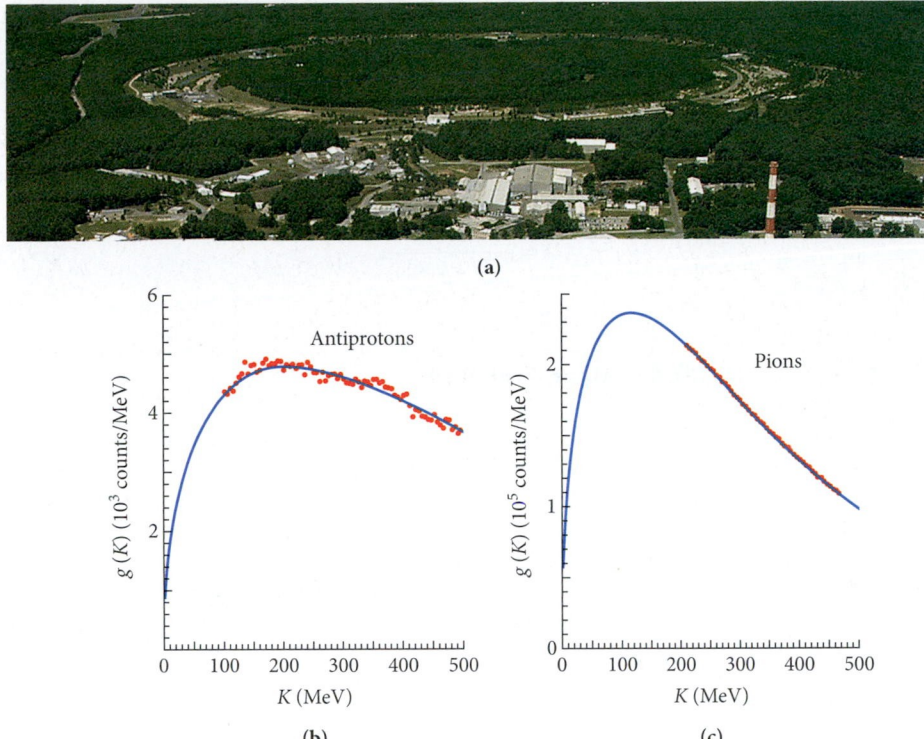

(a)

Antiprotons

Pions

(b)

(c)

FIGURE 19.25 (a) The 3.83-km-circumference ring of the Relativistic Heavy Ion Collider at Brookhaven National Laboratory. Experimental data on the kinetic energy distribution of (b) antiprotons and (c) pions produced in collisions of gold nuclei. These collisions become by far the hottest place in the Solar System for the short times they exist and produce a state of matter, the quark-gluon plasma, which existed for a fraction of a second after the Big Bang. The blue lines represent fits to the kinetic energy distributions using equation 19.30.

energies are missing, because the experimental apparatus cannot detect those energies. The blue lines represent fits to the data, using equation 19.30 with the best-fit value for the temperature, T. Values of $k_B T = 6.71 \cdot 10^{-11}$ J $= 419$ MeV for the antiprotons and $k_B T = 3.83 \cdot 10^{-11}$ J $= 239$ MeV for the pions were extracted, meaning that the temperatures of the antiprotons and pions are $4.86 \cdot 10^{12}$ K and $2.77 \cdot 10^{12}$ K, respectively. (These temperatures are several hundred million times higher than the temperature at the surface of the Sun!) These observations and others enable researchers to infer that a quark-gluon plasma with a temperature of $2 \cdot 10^{12}$ K is created for a very short time of approximately 10^{-22} s in these collisions.

Note: The connection between the temperature of the quark-gluon plasma and the temperature values extracted from the particle energy distributions is not entirely straightforward, because the collisions of the gold nuclei are explosive events. Also, an ideal gas is not the correct model. In addition, the kinetic energies in Figure 19.25 were determined with the relativistically correct formulation, which will be introduced in Chapter 35. Nevertheless, the degree to which the observed kinetic energy distributions follow the Maxwell distribution is astounding.

To end this subsection, let's consider nuclear fusion in the center of the Sun. This nuclear fusion is the source of energy that causes all of the Sun's radiation and thus makes life on Earth possible. The temperature at the center of the Sun is approximately $1.5 \cdot 10^7$ K. According to equation 19.16, the average kinetic energy of the hydrogen ions in the Sun's center is $\frac{3}{2} k_B T = 3.1 \cdot 10^{-16}$ J $= 1.9$ keV. However, for two hydrogen atoms to undergo nuclear fusion, they have to come close enough to each other to overcome their mutual electrostatic repulsion. (These topics will be discussed in detail in Chapters 21 on electrostatics and 40 on nuclear physics.) The minimum energy required to overcome the electrostatic repulsion is on the order of 1 MeV, or a few hundred times higher than the average kinetic energy of the hydrogen ions in the Sun's center. Thus, only hydrogen ions on the extreme tail of the Maxwell distribution have sufficient energy to participate in fusion reactions.

Mean Free Path

One of the assumptions about an ideal gas is that the gas molecules are point particles and do not interact with each other. But real gas molecules do have (very small, but nonzero) sizes and so have a chance of colliding with each other. Collisions of gas molecules with each other produce a scattering effect that causes the motion to be random and the distribution of their speeds to become the Maxwell distribution speed very quickly, independently of any initial speed distribution given to them.

How far do real gas molecules travel before they encounter another molecule? This distance is the **mean free path**, λ, of molecules in a gas. The mean free path is proportional to the temperature and inversely proportional to the pressure of the gas and the cross-sectional area of the molecule:

$$\lambda = \frac{k_B T}{\sqrt{2}\left(4\pi r^2\right)p}. \tag{19.31}$$

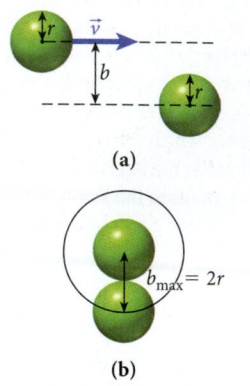

(a)

(b)

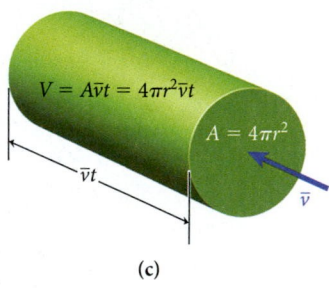

$V = A\bar{v}t = 4\pi r^2 \bar{v}t$

$A = 4\pi r^2$

$\bar{v}t$

$\bar{v}$

(c)

FIGURE 19.26 (a) A moving gas molecule approaches a stationary gas molecule with impact parameter b. (b) The cross section for scattering is represented by a circle with radius $2r$. (c) The volume swept out by molecules inside the cross-sectional area A in a time t when the molecules are traveling at speed $\bar{v}$.

DERIVATION 19.4 | Mean Free Path

To estimate the mean free path, we start with the geometric definition of the **cross section for scattering**. We assume that all the molecules in the gas are spheres with radius r. If two molecules touch, they will have an elastic collision, in which total momentum and total kinetic energy are conserved (see Chapter 7). Let's imagine that a moving gas molecule encounters another gas molecule at rest, as shown in Figure 19.26a.

The impact parameter b, as shown in Figure 19.26a, is the perpendicular distance between two lines connecting the centers of the two molecules and oriented parallel to the direction of the velocity of the moving molecule. The moving molecule will collide with the stationary molecule if $b \leq 2r$. Thus, the cross section for scattering is given by a circle with area

$$A = \pi\left(2r\right)^2 = 4\pi r^2,$$

as illustrated in Figure 19.26b. In a time period t, molecules with the average speed $\bar{v}$ will sweep out a volume

$$V = A\left(\bar{v}t\right) = \left(4\pi r^2\right)\left(\bar{v}t\right) = 4\pi r^2 \bar{v}t,$$

as shown in Figure 19.26c. Thus, the mean free path, λ, is the distance the molecule travels divided by the number of molecules it will collide with in that distance. The number of molecules that the molecule will encounter is just the volume swept out times the number of molecules per unit volume, n_V. The mean free path is then

$$\lambda = \frac{\bar{v}t}{V n_V} = \frac{\bar{v}t}{\left(4\pi r^2 \bar{v}t\right)n_V} = \frac{1}{\left(4\pi r^2\right)n_V}.$$

This result needs to be modified because we assumed that one moving molecule was incident on a stationary molecule. In an ideal gas, all the molecules are moving, so we must use the average relative velocity, v_{rel}, between the molecules rather than the average velocity of the molecules. It can be shown that $v_{rel} = \bar{v}\sqrt{2}$; so the volume swept out by the molecules changes to $V = 4\pi r^2 v_{rel} t = 4\pi r^2 \bar{v}\sqrt{2}t$. Thus, the expression for the mean free path becomes

$$\lambda = \frac{1}{\sqrt{2}\left(4\pi r^2\right)n_V}.$$

For diffuse gases, we can replace the number of molecules per unit volume with the ideal gas value, using equation 19.8:

$$n_V = \frac{N}{V} = \frac{p}{k_B T}.$$

The mean free path for a molecule in a real diffuse gas is then

$$\lambda = \frac{1}{\sqrt{2}\left(4\pi r^2\right)\left(\dfrac{p}{k_B T}\right)} = \frac{k_B T}{\sqrt{2}\left(4\pi r^2\right)p}.$$

Concept Check 19.9

The mean free path of a gas molecule doubles if the _____.

a) density of molecules goes down by a factor of 2

b) density of molecules doubles

c) diameter of the gas molecules doubles

d) diameter of the gas molecules goes down by a factor of 2

e) none of the above

Now that we have derived the formula for the mean free path, it is instructive to study a real case. Let's look at the most important gas, the Earth's atmosphere. An ideal gas has an "infinitely" long mean free path relative to the spacing between molecules in the gas. How closely a real gas approaches the ideal gas is determined by the ratio of the mean free path to the intermolecular spacing in the gas. Let's see how close Earth's atmosphere is to being an ideal gas.

EXAMPLE 19.6   Mean Free Path of Air Molecules

PROBLEM

What is the mean free path of a gas molecule in air at normal atmospheric pressure of 101.3 kPa and a temperature of 20.0 °C? Since air is 78.1% nitrogen, we'll assume for the sake of simplicity that air is 100% nitrogen. The radius of a nitrogen molecule is about 0.150 nm. Compare the mean free path to the radius of a nitrogen molecule and to the average spacing between the nitrogen molecules.

SOLUTION

We can use equation 19.31 to calculate the mean free path:

$$\lambda = \frac{\left(1.381 \cdot 10^{-23} \text{ J/K}\right)\left(293.15 \text{ K}\right)}{\sqrt{2}\left(4\pi\right)\left(0.150 \cdot 10^{-9} \text{ m}\right)^2\left(101.3 \cdot 10^3 \text{ Pa}\right)} = 9.99 \cdot 10^{-8} \text{ m}.$$

The ratio of the mean free path to the molecular radius is then

$$\frac{\lambda}{r} = \frac{9.99 \cdot 10^{-8} \text{ m}}{0.150 \cdot 10^{-9} \text{ m}} = 666.$$

The volume per particle in an ideal gas can be obtained using equation 19.8:

$$\frac{V}{N} = \frac{k_B T}{p}.$$

The average spacing between molecules is then the average volume to the $\frac{1}{3}$ power:

$$\left(\frac{V}{N}\right)^{1/3} = \left(\frac{k_B T}{p}\right)^{1/3} = \left[\frac{\left(1.381 \cdot 10^{-23} \text{ J/K}\right)\left(293.15 \text{ K}\right)}{\left(101.3 \cdot 10^3 \text{ Pa}\right)}\right]^{1/3} = 3.42 \cdot 10^{-9} \text{ m}.$$

Thus, the ratio between the mean free path and the average spacing between molecules is

$$\frac{9.99 \cdot 10^{-8} \text{ m}}{3.42 \cdot 10^{-9} \text{ m}} = 29.2.$$

The value of this ratio demonstrates that treating air as an ideal gas is a good approximation. Figure 19.27 compares the diameter of nitrogen molecules, the average distance between molecules, and the mean free path.

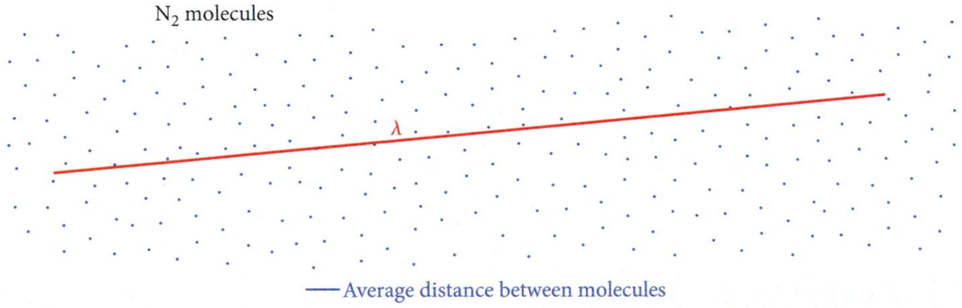

N₂ molecules

λ

—— Average distance between molecules

FIGURE 19.27 Scale drawing showing the size of nitrogen molecules, the mean free path of the molecules in the gas, and the average distance between molecules.

19.7   Real Gases

For an ideal gas, we assume that the molecules are point particles and have no interaction with each other. For real gases, we cannot make either of these assumptions. In Section 19.6, we saw that gas molecules can collide with one another because of their finite size. Since the molecules have a finite size, the volume that each molecule occupies is not available for other molecules and has to be excluded from the overall volume, V, of the container. In addition, the molecules can attract one another; since this attraction is a pairwise interaction, we expect it to be proportional to the square of the particle density, or inversely proportional to the square of the volume. Thus, the ideal gas equation of state, $pV = nRT$, has to be modified as follows:

$$\left(p + a\frac{n^2}{V^2}\right)(V - nb) = nRT, \tag{19.32}$$

where a is a parameter that measures the strength of the intermolecular attraction and b is the volume of 1 mole of the gas. Equation 19.32, called the *van der Waals equation*, is a more realistic equation of state.

Figure 19.28a is a plot of the ideal gas equation of state for 1 mole of an ideal gas, showing the pressure as a function of volume and temperature. This is basically the same plot presented in Figure 19.6, just from a different angle and over a smaller range of temperatures. Figure 19.28b shows how this plot changes as a result of the van der Waals corrections of equation 19.32. You can see that the pressure is lower, which is due to the attractive interactions between the molecules. But you can also see that the plot has a "dip" along all isotherms at lower temperatures. The reason for this dip is that at small volumes the molecules can form a liquid, whereas at large volumes they form a gas. The oscillations for a given isotherm indicate the region where liquid and gas coexist.

The oscillations in the coexistence region are remedied with Maxwell's construction (see Figure 19.28c). Removing the unphysical oscillations and connecting the isotherms in straight lines, as shown in Figure 19.28d, finally reveals the coexistence region between liquid and gas and the physical equation of state. The upper end point of the coexistence region is called the *critical point*. To the left of the coexistence region is the liquid phase, and to the right is the gas phase. Note how much more sharply the pressure rises along a given isotherm if the volume is reduced in the liquid region than in the gas region. This indicates how much more incompressible liquids are than gases.

The particular equation of state plotted in Figure 19.28 is for ethane, C_2H_6. For other gases, the plots are very similarly shaped, but different values of the parameters a and b in equation 19.32 lead to the coexistence region being located in a different part of the temperature-volume plane.

Self-Test Opportunity 19.4

Show that the ideal gas equation of state is a special case of the van der Waals equation in the limit of an infinitely large volume.

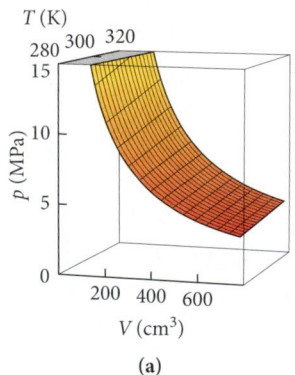

(a)

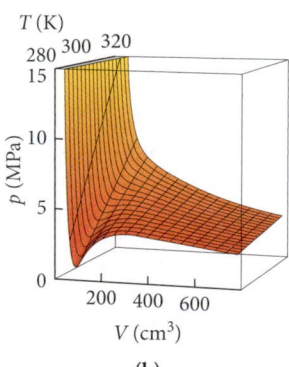

(b)

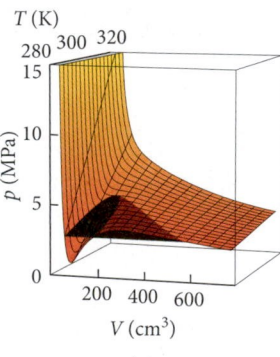

(c)

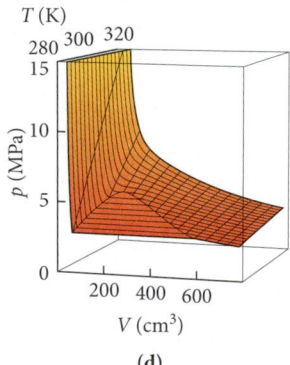

(d)

FIGURE 19.28 Pressure as a function of temperature and volume for 1 mole of ethane. (a) Ideal gas equation of state, (b) van der Waals equation, (c) Maxwell construction, (d) real equation of state for ethane.

WHAT WE HAVE LEARNED | EXAM STUDY GUIDE

- The Ideal Gas Law relates the pressure p, the volume V, the number of moles n, and the temperature T, of an ideal gas: $pV = nRT$, where $R = 8.314$ J/(mol K) is the universal gas constant. (Boyle's, Charles's, Gay-Lussac's, and Avogadro's Laws are all special cases of the Ideal Gas Law.)

- The Ideal Gas Law can also be expressed as $pV = Nk_BT$, where N is the number of molecules of gas and $k_B = 1.381 \cdot 10^{-23}$ J/K is the Boltzmann constant.

- Dalton's Law states that the total pressure exerted by a mixture of gases is equal to the sum of the partial pressures of the gases in the mixture:
$$p_{total} = p_1 + p_2 + p_3 + \cdots + p_n.$$

- The work done by an ideal gas at constant temperature in changing from an initial volume V_i to a final volume V_f is given by $W = nRT \ln(V_f/V_i)$.

- The equipartition theorem states that gas molecules in thermal equilibrium have the same average kinetic energy associated with each independent degree of freedom of their motion. The average kinetic energy per degree of freedom is $\frac{1}{2}k_BT$.

- The root-mean-square speed of the molecules of a gas is $v_{rms} = \sqrt{3k_BT/m}$, where m is the mass of each molecule.

- The molar specific heat of a gas at constant volume, C_V, is defined by $Q = nC_V\Delta T$, where Q is the heat, n is the number of moles of gas, and ΔT is the change in temperature of the gas.

- The molar specific heat of a gas at constant pressure, C_p, is defined by $Q = nC_p\Delta T$, where Q is the heat, n is the number of moles of gas, and ΔT is the change in temperature of the gas.

- The molar specific heat of a gas at constant pressure is related to the molar specific heat of a gas at constant volume by $C_p = C_V + R$.

- The molar specific heat at constant volume is $C_V = \frac{3}{2}R = 12.5$ J/(mol K) for a monatomic gas and $C_V = \frac{5}{2}R = 20.8$ J/(mol K) for a diatomic gas.

- For a gas undergoing an adiabatic process, $Q = 0$, and the following relationships hold: $pV^\gamma = $ constant and

$TV^{\gamma-1} = $ constant, where $\gamma = C_p/C_V$. For monatomic gases, $\gamma = \frac{5}{3}$; for diatomic gases, $\gamma = \frac{7}{5}$.

- The Maxwell speed distribution is given by
$$f(v) = 4\pi\left(\frac{m}{2\pi k_B T}\right)^{3/2} v^2 e^{-mv^2/2k_B T},$$ and $f(v)\,dv$ describes the probability that a molecule has a speed between v and $v + dv$.

- The Maxwell kinetic energy distribution is given by
$$g(K) = \frac{2}{\sqrt{\pi}}\left(\frac{1}{k_B T}\right)^{3/2} \sqrt{K}e^{-K/k_B T},$$ and $g(K)\,dK$ describes the probability that a molecule has a kinetic energy between K and $K + dK$.

- The mean free path, λ, of a molecule in an ideal gas is given by $\lambda = \dfrac{k_B T}{\sqrt{2}\left(4\pi r^2\right)p}$, where r is the radius of the molecule.

ANSWERS TO SELF-TEST OPPORTUNITIES

19.1 Compare C_p to $(C_V + R)$

Gas	C_V [J/(mol K)]	C_p [J/(mol K)]	$(C_V + R)$ [J/(mol K)]
Helium (He)	12.5	20.8	20.8
Neon (Ne)	12.5	20.8	20.8
Argon (Ar)	12.5	20.8	20.8
Krypton (Kr)	12.5	20.8	20.8
Hydrogen (H_2)	20.4	28.8	28.7
Nitrogen (N_2)	20.7	29.1	29.0
Oxygen (O_2)	21.0	29.4	29.1
Carbon dioxide (CO_2)	28.2	36.6	36.5
Methane (CH_4)	27.5	35.9	35.8

19.2 Approximate the tire's cross section as a circle of area $\pi(t/2)^2$. With this area, the volume of the tire is approximately $\pi(t/2)^2 2\pi(r_1+r_2)/2 = \pi^2 t^2(r_1+r_2)/4$. This

volume is $\pi/4 \approx 78.5\%$ of the volume used in Solved Problem 19.2. Thus, the change in the pump volume relative to the tire volume will be larger than in the solved problem, and so the final temperature should be slightly higher than that result. Following the same steps as in the solved problem gives a final temperature of 318 K.

19.3 Use the definite integral $\displaystyle\int_0^\infty x^2 e^{-x^2/a}\,dx = \frac{1}{4}a^{3/2}\sqrt{\pi}$.

19.4 For $V\to\infty$, we can neglect the excluded volume, nb, which is constant; thus, $\displaystyle\lim_{V\to\infty}(V - nb) = V$. Also, $\displaystyle\lim_{V\to\infty}(an^2/V^2) = 0$, because b and n are constant. Putting these together gives $\displaystyle\lim_{V\to\infty}\left((p + an^2/V^2)(V - nb)\right) = pV$, and pV is the left-hand side of the ideal gas equation of state.

PROBLEM-SOLVING GUIDELINES

1. It is often useful to list all the known and unknown quantities when applying the Ideal Gas Law, to clarify what you have and what you're looking for. This step includes identifying an initial state and a final state and listing the quantities for each state.

2. If you're dealing with moles of a gas, you need the form of the Ideal Gas Law that involves R. If you're dealing with numbers of molecules, you need the form involving k_B.

3. Pay close attention to units, which may be given in different systems. For temperatures, unit conversion does not just involve multiplicative constants; the different temperature scales are also offset by additive constants. To be on the safe side, temperatures should always be converted to kelvins. Also, be sure that the value of R or k_B you use is consistent with the units you're working with. In addition, remember that molar masses are usually given as grams; you might need to convert the masses to kilograms.

MULTIPLE-CHOICE QUESTIONS

19.1 A system can go from state i to state f as shown in the figure. Which relationship is true?

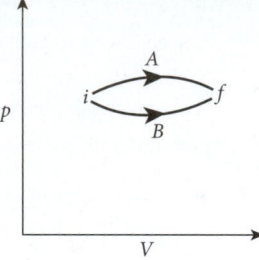

a) $Q_A > Q_B$

b) $Q_A = Q_B$

c) $Q_A < Q_B$

d) It is impossible to decide from the information given.

19.2 A tire on a car is inflated to a gauge pressure of 32 lb/in² at a temperature of 27 °C. After the car is driven for 30 mi, the pressure has increased to 34 lb/in². What is the temperature of the air inside the tire at this point?

a) 40. °C   b) 23 °C   c) 32 °C   d) 54 °C

19.3 Molar specific heat at constant pressure, C_p, is larger than molar specific heat at constant volume, C_V, for

a) a monoatomic ideal gas.

b) a diatomic atomic gas.

c) all of the above.

d) none of the above.

19.4 An ideal gas may expand from an initial pressure, p_i, and volume, V_i, to a final volume, V_f, isothermally, adiabatically, or isobarically. For which type of process is the heat that is added to the gas the largest? (Assume that p_i, V_i, and V_f are the same for each process.)

a) isothermal process

b) adiabatic process

c) isobaric process

d) All the processes have the same heat flow.

19.5 Which of the following gases has the highest root-mean-square speed?

a) nitrogen at 1 atm and 30 °C

b) argon at 1 atm and 30 °C

c) argon at 2 atm and 30 °C

d) oxygen at 2 atm and 30 °C

e) nitrogen at 2 atm and 15 °C

19.6 Two identical containers hold equal masses of gas, oxygen in one and nitrogen in the other. The gases are held at the same temperature. How does the pressure of the oxygen compare to that of the nitrogen?

a) $p_O > p_N$

b) $p_O = p_N$

c) $p_O < p_N$

d) cannot be determined from the information given

19.7 One mole of an ideal gas, at a temperature of 0 °C, is confined to a volume of 1.0 L. The pressure of this gas is

a) 1.0 atm.

b) 22.4 atm.

c) 1/22.4 atm.

d) 11.2 atm.

19.8 One hundred milliliters of liquid nitrogen with a mass of 80.7 g is sealed inside a 2-L container. After the liquid nitrogen heats up and turns into a gas, what is the pressure inside the container?

a) 0.05 atm

b) 0.08 atm

c) 0.09 atm

d) 9.1 atm

e) 18 atm

19.9 Consider a box filled with an ideal gas. The box undergoes a sudden free expansion from V_1 to V_2. Which of the following correctly describes this process?

a) Work done by the gas during the expansion is equal to $nRT \ln (V_2/V_1)$.

b) Heat is added to the box.

c) Final temperature equals initial temperature times (V_2/V_1).

d) The internal energy of the gas remains constant.

19.10 Compare the average kinetic energy at room temperature of a nitrogen molecule to that of an oxygen molecule. Which has the larger kinetic energy?

a) oxygen molecule

b) nitrogen molecule

c) They have the same kinetic energy.

d) It depends on the pressure.

19.11 Ten students take a 100-point exam. Their grades are 25, 97, 95, 100, 35, 32, 92, 75, 78, and 34 points. The average grade and the rms grade are, respectively,

a) 50.0 and 25.0.

b) 66.7 and 33.3.

c) 77.1 and 19.3.

d) 37.8 and 76.2.

e) 66.3 and 72.6.

19.12 A polyatomic ideal gas occupies an initial volume V_i, which is decreased to $\frac{1}{8}V_i$ via an adiabatic process. Which equation expresses the relationship between the initial and final pressures of this gas?

a) $p_f = 2p_i$

b) $p_f = 4p_i$

c) $p_f = 16p_i$

d) $p_f = \frac{17}{5}p_i$

e) $p_f = \frac{23}{7}p_i$

19.13 A polyatomic ideal gas occupies an initial volume V_i, which is decreased to $\frac{1}{8}V_i$ via an adiabatic process. Which equation expresses the relationship between the initial and final temperatures of this gas?

a) $T_f = \frac{2}{3}T_i$

b) $T_f = 2T_i$

c) $T_f = 16T_i$

d) $T_f = \frac{9}{5}T_i$

e) $T_f = 8T_i$

19.14 The reason there is almost no hydrogen in the Earth's atmosphere is that

a) hydrogen reacts with oxygen to form water.

b) sunlight destroys hydrogen.

c) the ozone layer absorbs hydrogen.

d) plants do not emit hydrogen.

e) hydrogen molecules in the thermosphere can have speeds that exceed Earth's escape velocity.

CONCEPTUAL QUESTIONS

19.15 Hot air is less dense than cold air and therefore experiences a net buoyant force and rises. Since hot air rises, the higher the elevation, the warmer the air should be. Therefore, the top of Mount Everest should be very warm. Explain why Mount Everest is colder than Death Valley.

19.16 The Maxwell speed distribution assumes that the gas is in equilibrium. Thus, if a gas, all of whose molecules were moving at the same speed, were given enough time, they would eventually come to satisfy the speed distribution. But the derivations of the kinetic theory presented in the text assumed that when a gas molecule hits the wall of a container, it bounces back with the same energy it had before the collision and that gas molecules exert no forces on each other. If gas molecules exchange energy

neither with the walls of their container nor with each other, how can they ever come to equilibrium? Is it not true that if they all had the same speed initially, some would have to slow down and others speed up, according to the Maxwell speed distribution?

19.17 When you blow hard on your hand, it feels cool, but when you breathe softly, it feels warm. Why?

19.18 Explain why the average velocity of air molecules in a closed auditorium is zero but their root-mean-square speed or average speed is not zero.

19.19 In a diesel engine, the fuel-air mixture is compressed rapidly. As a result, the temperature rises to the spontaneous combustion temperature for the fuel, causing the fuel to ignite. How is the rise in temperature possible, given the fact that the compression occurs so rapidly that there is not enough time for a significant amount of heat to flow into or out of the fuel-air mixture?

19.20 A cylinder with a sliding piston contains an ideal gas. The cylinder (with the gas inside) is heated by transferring the same amount of heat to the system in two different ways:

(1) The piston is blocked to prevent it from moving.

(2) The piston is allowed to slide frictionlessly to the end of the cylinder.

How does the final temperature of the gas under condition 1 compare to the final temperature of the gas under condition 2? Is it possible for the temperatures to be equal?

19.21 Show that the adiabatic bulk modulus, defined as $B = -V (dp/dV)$, for an ideal gas is equal to γp.

19.22 A monatomic ideal gas expands isothermally from $\{p_1, V_1, T_1\}$ to $\{p_2, V_2, T_1\}$. Then it undergoes an isochoric process, which takes it from $\{p_2, V_2, T_1\}$ to $\{p_1, V_2, T_2\}$. Finally the gas undergoes an isobaric compression, which takes it back to $\{p_1, V_1, T_1\}$.

a) Use the First Law of Thermodynamics to find Q for each of these processes.

b) Write an expression for total Q in terms of $p_1, p_2, V_1,$ and V_2.

19.23 Two atomic gases will react to form a diatomic gas: $A + B \rightarrow AB$. Suppose you perform this reaction with 1 mole each of A and B in a thermally isolated chamber, so no heat is exchanged with the environment. Will the temperature of the system increase or decrease?

19.24 A relationship that gives the pressure, p, of a substance as a function of its density, ρ, and temperature, T, is called an *equation of state*. For a gas with molar mass M, write the Ideal Gas Law as an equation of state.

19.25 The compression and rarefaction associated with a sound wave propagating in a gas are so much faster than the flow of heat in the gas that they can be treated as adiabatic processes.

a) Find the speed of sound, v_s, in an ideal gas of molar mass M.

b) In accord with Einstein's refinement of Newtonian mechanics, v_s cannot exceed the speed of light in vacuum, c. This fact implies a maximum temperature for an ideal gas. Find this temperature.

c) Evaluate the maximum temperature of part (b) for *monatomic* hydrogen gas (H).

d) What happens to the hydrogen at this maximum temperature?

19.26 The kinetic theory of an ideal gas takes into account not only translational motion of atoms or molecules but also, for diatomic and polyatomic gases, vibration and rotation. Will the temperature increase from a given amount of energy being supplied to a monatomic gas differ from the temperature increase due to the same amount of energy being supplied to a diatomic gas? Explain.

19.27 In the thermosphere, the layer of Earth's atmosphere extending from an altitude 100 km up to 700 km, the temperature can reach about 1500 °C. How would air in that layer feel to one's bare skin?

19.28 A glass of water at room temperature is left on the kitchen counter overnight. In the morning, the amount of water in the glass is smaller due to evaporation. The water in the glass is below the boiling point, so how is it possible for some of the liquid water to have turned into a gas?

EXERCISES

A blue problem number indicates a worked-out solution is available in the Student Solutions Manual. One • and two •• indicate increasing level of problem difficulty.

Section 19.1

19.29 A tire has a gauge pressure of 300. kPa at 15.0 °C. What is the gauge pressure at 45.0 °C? Assume that the change in volume of the tire is negligible.

19.30 A tank of compressed helium for inflating balloons is advertised as containing helium at a pressure of 2400. psi, which, when allowed to expand at atmospheric pressure, will occupy a volume of 244 ft³. Assuming no temperature change takes place during the expansion, what is the volume of the tank in cubic feet?

•19.31 Before embarking on a car trip from Michigan to Florida to escape the winter cold, you inflate the tires of your car to a manufacturer-suggested pressure of 33.0 psi while the outside temperature is 25.0 °F and then make sure your valve caps are airtight. You arrive in Florida two days later, and the temperature outside is a pleasant 72.0 °F.

a) What is the new pressure in your tires, in SI units?

b) If you let air out of the tires to bring the pressure back to the recommended 33.0 psi, what percentage of the original mass of air in the tires will you release?

•19.32 A 1.00-L volume of a gas undergoes an isochoric process in which its pressure doubles, followed by an isothermal process until the original pressure is reached. Determine the final volume of the gas.

•19.33 Suppose you have a pot filled with steam at 100.0 °C and at a pressure of 1.00 atm. The pot has a diameter of 15.0 cm and is 10.0 cm high. The mass of the lid is 0.500 kg. How hot do you need to heat the steam to lift the lid off the pot?

Section 19.2

•19.34 A quantity of liquid water comes into equilibrium with the air in a closed container, without completely evaporating, at a temperature of 25.0 °C. How many grams of water vapor does a liter of the air contain in this situation? The vapor pressure of water at 25.0 °C is 3.1690 kPa.

•19.35 Use the equation $\rho T = $ constant and information from Chapter 16 to estimate the speed of sound in air at 40.0 °C, given that the speed at 0.00 °C is 331 m/s.

19.36 Suppose 2.00 moles of an ideal gas is enclosed in a container of volume $1.00 \cdot 10^{-4}$ m³. The container is then placed in a furnace and raised to a temperature of 400. K. What is the final pressure of the gas?

19.37 Suppose 1.00 mole of an ideal gas is held at a constant volume of 2.00 L. Find the change in pressure if the temperature increases by 100. °C.

19.38 How much heat is added to the system when 2.00 kJ of work is performed by an ideal gas in an isothermal process? Give a reason for your answer.

•19.39 Liquid nitrogen, which is used in many physics research labs, can present a safety hazard if a large quantity evaporates in a confined space. The resulting nitrogen gas reduces the oxygen concentration, creating the risk of asphyxiation. Suppose 1.00 L of liquid nitrogen ($\rho = 808$ kg/m³) evaporates and comes into equilibrium with the air at 21.0 °C and 101 kPa. How much volume will it occupy?

•19.40 Liquid bromine (Br_2) is spilled in a laboratory accident and evaporates. Assuming that the vapor behaves as an ideal gas, with a temperature of 20.0 °C and a pressure of 101.0 kPa, find its density.

•19.41 Two containers contain the same gas at different temperatures and pressures, as indicated in the figure. The small container has a volume of 1.00 L, and the large container has a volume of 2.00 L. The two containers are then connected to each other using a thin tube, and the pressure and temperature in both containers are allowed to equalize. If the final temperature is 300. K, what is the final pressure? Assume that the connecting tube has negligible volume and mass.

Container 1:	Container 2:
2.00 L	1.00 L
600. K	200. K
$3.00 \cdot 10^5$ Pa	$2.00 \cdot 10^5$ Pa

•**19.42** A sample of gas for which $p = 1000.$ Pa, $V = 1.00$ L, and $T = 300.$ K is confined in a cylinder.

a) Find the new pressure if the volume is reduced to half of the original volume at the same temperature.

b) If the temperature is raised to 400. K in the process of part (a), what is the new pressure?

c) If the gas is then heated to 600. K from the initial value and the pressure of the gas becomes 3000. Pa, what is the new volume?

•**19.43** A cylinder of cross-sectional area 12.0 cm^2 is fitted with a piston, which is connected to a spring with a spring constant of 1000. N/m, as shown in the figure. The cylinder holds 0.005000 mole of an ideal gas at room temperature (23.0 °C). How far is the spring compressed if the temperature of the gas is raised to 150. °C?

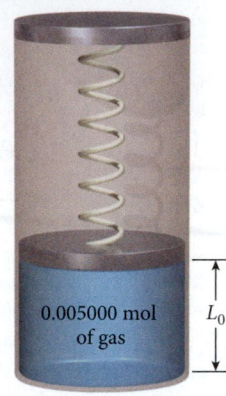

0.005000 mol of gas L_0

••**19.44** Air at 1.00 atm is inside a cylinder 20.0 cm in radius and 20.0 cm in length that sits on a table. The top of the cylinder is sealed with a movable piston. A 20.0-kg block is dropped onto the piston. From what height above the piston must the block be dropped to compress the piston by 1.00 mm? 2.00 mm? 1.00 cm?

Section 19.3

19.45 Interstellar space far from any stars is usually filled with atomic hydrogen (H) at a density of 1.00 atom/cm^3 and a very low temperature of 2.73 K.

a) Determine the pressure in interstellar space.

b) What is the root-mean-square speed of the atoms?

c) What would be the edge length of a cube that would contain atoms with a total of 1.00 J of energy?

19.46 a) What is the root-mean-square speed for a collection of helium-4 atoms at 300. K?

b) What is the root-mean-square speed for a collection of helium-3 atoms at 300. K?

19.47 Two isotopes of uranium, ^{235}U and ^{238}U, are separated by a gas diffusion process that involves combining them with fluorine to make the compound UF$_6$. Determine the ratio of the root-mean-square speeds of UF$_6$ molecules for the two isotopes. The masses of ^{235}UF$_6$ and ^{238}UF$_6$ are 349.03 amu and 352.04 amu, respectively.

19.48 The electrons that produce electric currents in a metal behave approximately like molecules of an ideal gas. The mass of an electron is $m_e = 9.109 \cdot 10^{-31}$ kg. If the temperature of the metal is 300.0 K, what is the root-mean-square speed of the electrons?

•**19.49** In a period of 6.00 s, $9.00 \cdot 10^{23}$ nitrogen molecules strike a section of a wall with an area of 2.00 cm^2. If the molecules move with a speed of 400.0 m/s and strike the wall head on in elastic collisions, what is the pressure exerted on the wall? (The mass of one N$_2$ molecule is $4.65 \cdot 10^{-26}$ kg.)

•**19.50** Assuming that the pressure remains constant, at what temperature is the root-mean-square speed of a helium atom equal to the root-mean-square speed of an air molecule at STP?

Section 19.4

19.51 Two copper cylinders, immersed in a water tank at 50.0 °C, contain helium and nitrogen, respectively. The helium-filled cylinder has a volume twice as large as the nitrogen-filled cylinder.

a) Calculate the average kinetic energy of a helium molecule and the average kinetic energy of a nitrogen molecule.

b) Determine the molar specific heat at constant volume (C_V) and at constant pressure (C_p) for the two gases.

c) Find γ for the two gases.

19.52 At room temperature, identical gas cylinders contain 10 moles of nitrogen gas and argon gas, respectively. Determine the ratio of energies stored in the two systems. Assume ideal gas behavior.

19.53 Calculate the change in internal energy of 1.00 mole of a diatomic ideal gas that starts at room temperature (293 K) when its temperature is increased by 2.00 K.

19.54 Treating air as an ideal gas of diatomic molecules, calculate how much heat is required to raise the temperature of the air in an 8.00 m by 10.0 m by 3.00 m room from 20.0 °C to 22.0 °C at 101 kPa. Neglect the change in the number of moles of air in the room.

•**19.55** What is the approximate energy required to raise the temperature of 1.00 L of air by 100. °C? The volume is held constant.

•**19.56** You are designing an experiment that requires a gas with $\gamma = 1.60$. However, from your physics lectures, you remember that no gas has such a γ value. However, you also remember that mixing monatomic and diatomic gases can yield a gas with such a γ value. Determine the fraction of diatomic molecules in a mixture of gases that has this value.

Section 19.5

19.57 Suppose 15.0 L of an ideal monatomic gas at a pressure of $1.50 \cdot 10^5$ kPa is expanded adiabatically (no heat transfer) until the volume is doubled.

a) What is the pressure of the gas at the new volume?

b) If the initial temperature of the gas was 300. K, what is its final temperature after the expansion?

19.58 A diesel engine works at a high compression ratio to compress air until it reaches a temperature high enough to ignite the diesel fuel. Suppose the compression ratio (ratio of volumes) of a specific diesel engine is 20.0 to 1.00. If air enters a cylinder at 1.00 atm and is compressed adiabatically, the compressed air reaches a pressure of 66.0 atm. Assuming that the air enters the engine at room temperature (25.0 °C) and that the air can be treated as an ideal gas, find the temperature of the compressed air.

19.59 Air in a diesel engine cylinder is quickly compressed from an initial temperature of 20.0 °C, an initial pressure of 1.00 atm, and an initial volume of 600. cm^3 to a final volume of 45.0 cm^3. Assuming the air to be an ideal diatomic gas, find the final temperature and pressure.

•**19.60** Suppose 6 L of a monatomic ideal gas, originally at 400. K and a pressure of 3.00 atm (called state 1), undergoes the following processes:

$1 \rightarrow 2$ isothermal expansion to $V_2 = 4V_1$

$2 \rightarrow 3$ isobaric compression

$3 \rightarrow 1$ adiabatic compression to its original state

Find the pressure, volume, and temperature of the gas in states 2 and 3. How many moles of the gas are there?

•**19.61** How much work is done by the gas during each of the three processes in Problem 19.60 and how much heat flows into the gas in each process?

••**19.62** Consider two geometrically identical cylinders of inner diameter 5.00 cm, one made of copper and the other of Teflon, each closed on the bottom and open on top. The two cylinders are immersed in a large-volume water tank at room temperature (20.0 °C), as shown in the figure. Note that the copper cylinder is an excellent conductor of heat and the Teflon cylinder is a good insulator. A frictionless piston with a rod and platter attached is placed in each cylinder. The mass of the piston-rod-platter assembly is 0.500 kg, and the cylinders are filled with helium gas so that initially both pistons are at equilibrium at 20.0 cm from the bottom of their respective cylinders.

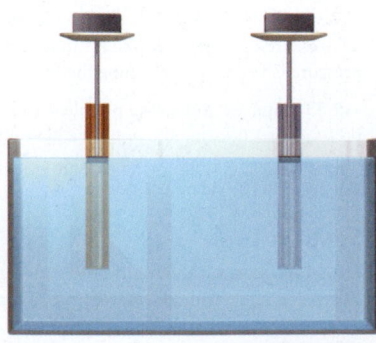

a) A 5.00-kg lead block is *slowly* placed on each platter, and the piston is *slowly* lowered until it reaches its final equilibrium state. Calculate the final height of each piston (as measured from the bottom of its cylinder).

b) If the two lead blocks are *dropped suddenly* on the platters, how will the final heights of the two pistons compare immediately after the lead blocks are dropped?

••**19.63** Chapter 13 examined the variation of pressure with altitude in the Earth's atmosphere, assuming constant temperature—a model known as the *isothermal atmosphere*. A better approximation is to treat the pressure variations with altitude as adiabatic. Assume that air can be treated as a diatomic ideal gas with effective molar mass $M_{air} = 28.97$ g/mol. Note that in reality many complexities of the Earth's atmosphere cannot be modeled as an adiabatic ideal gas, for example, the thermosphere.

a) Find the air pressure and temperature of the atmosphere as functions of altitude. Let the pressure at sea level be $p_0 = 101.0$ kPa and the temperature at sea level be 20.0 °C.

b) Determine the altitude at which the air pressure is half its sea level value and the temperature at this altitude. Also determine the altitude at which the air density is half its sea level value and the temperature at that altitude.

c) Compare these results with the isothermal model of Chapter 13.

Section 19.6

19.64 Consider nitrogen gas, N_2, at 20.0 °C. What is the root-mean-square speed of the nitrogen molecules? What is the most probable speed? What percentage of nitrogen molecules have a speed within 1.00 m/s of the most probable speed? (*Hint:* Assume that the probability of nitrogen atoms having speeds between 200.00 m/s and 202.00 m/s is constant.)

19.65 As noted in the text, the speed distribution of molecules in the Earth's atmosphere has a significant impact on its composition.

a) What is the average speed of a nitrogen molecule in the atmosphere, at a temperature of 18.0 °C and a (partial) pressure of 78.8 kPa?

b) What is the average speed of a hydrogen molecule at the same temperature and pressure?

19.66 A sealed container contains 1.00 mole of neon gas at STP. Estimate the number of neon atoms having speeds in the range from 200.00 m/s to 202.00 m/s. (*Hint:* Assume that the probability of neon atoms having speeds between 200.00 m/s and 202.00 m/s is constant.)

•**19.67** For a room at 0.00 °C and 1.00 atm, write an expression for the fraction of air molecules having speeds greater than the speed of sound, according to the Maxwell speed distribution. What is the average speed of each molecule? What is the root-mean-square speed? Assume that the air consists of uniform particles with a mass of 15.0 amu.

••**19.68** If a space shuttle is struck by a meteor, producing a circular hole 1.00 cm in diameter in the window, how long does it take before the pressure inside the cabin is reduced to half of its original value? Assume that the cabin of the shuttle is a cube with edges measuring 5.00 m and that the air inside is at a constant temperature of 21.0 °C at 1.00 atm and consists of uniform diatomic molecules each with a mass of 15 amu. (*Hint:* What fraction of the molecules are moving in the correct direction to exit through the hole, and what is their average speed? Derive an expression for the rate at which molecules exit the hole as a function of pressure, and thus of dp/dt.)

Additional Exercises

19.69 Suppose 1.00 mole of molecular nitrogen gas expands in volume very quickly, so no heat is exchanged with the environment during the process. If the volume increases from 1.00 L to 1.50 L, determine the work done on the environment if the gas's temperature drops from 22.0 °C to 18.0 °C. Assume ideal gas behavior.

19.70 Calculate the root-mean-square speed of air molecules at room temperature (22.0 °C) from the kinetic theory of an ideal gas.

19.71 At Party City, you purchase a helium-filled balloon with a diameter of 40.0 cm at 20.0 °C and at 1.00 atm.

a) How many helium atoms are inside the balloon?

b) What is the average kinetic energy of the atoms?

c) What is the root-mean-square speed of the atoms?

19.72 What is the total mass of all the oxygen molecules in a cubic meter of air at a temperature of 25.0 °C and a pressure of $1.01 \cdot 10^5$ Pa? Note that air is 20.9% (by volume) oxygen (molecular O_2), with the remainder being primarily nitrogen (molecular N_2).

19.73 The tires of a $3.00 \cdot 10^3$-lb car are filled to a gauge pressure (pressure added to atmospheric pressure) of 32.0 lb/in^2 while the car is on a lift at a repair shop. The car is then lowered to the ground.

a) Assuming that the internal volume of the tires does not change appreciably as a result of contact with the ground, give the absolute pressure in the tires in pascals when the car is on the ground.

b) What is the total contact area between the tires and the ground?

19.74 A gas expands at constant pressure from 3.00 L at 15.0 °C until the volume is 4.00 L. What is the final temperature of the gas?

19.75 An ideal gas has a density of 0.0899 g/L at 20.00 °C and 101.325 kPa. Identify the gas.

19.76 A tank of a gas consisting of 30.0% O_2 and 70.0% Ar has a volume of 1.00 m^3. It is moved from storage at 20.0 °C to the outside on a sunny day. Initially, the pressure gauge reads 1000. psi. After several hours, the pressure reading is 1500. psi. What is the temperature of the gas in the tank at this time?

19.77 Suppose 5.00 moles of an ideal monatomic gas expands at a constant temperature of 22.0 °C from an initial volume of 2.00 m^3 to 8.00 m^3.

a) How much work is done by the gas?

b) What is the final pressure of the gas?

•**19.78** A cylindrical chamber with a radius of 3.50 cm contains 0.120 mole of an ideal gas. It is fitted with a 450.-g piston that rests on top, enclosing a space of height 7.50 cm. As the cylinder is cooled by 15.0 °C, how far below the initial position does the piston end up?

•**19.79** Find the *most probable* kinetic energy for a molecule of a gas at temperature $T = 300.$ K. In what way does this value depend on the identity of the gas?

•**19.80** A closed auditorium of volume $2.50 \cdot 10^4$ m^3 is filled with 2000 people at the beginning of a show, and the air in the space is at a temperature of 293 K and a pressure of $1.013 \cdot 10^5$ Pa. If there were no ventilation, by how much would the temperature of the air rise during the 2.00-h show if each person metabolizes at a rate of 70.0 W?

•**19.81** At a temperature of 295. K, the vapor pressure of pentane (C_5H_{12}) is 60.7 kPa. Suppose 1.000 g of gaseous pentane is contained in a cylinder with diathermal (thermally conducting) walls and a piston to vary the volume. The initial volume is 1.000 L, and the piston is moved in slowly, keeping the temperature at 295 K. At what volume will the first drop of liquid pentane appear?

•**19.82** Helium gas at 22.0 °C and at 1.00 atm is contained in a well-insulated cylinder fitted with a piston. The piston is moved very slowly (adiabatic expansion) so that the volume increases to four times its original value.

a) What is the final pressure?

b) What is the final temperature of the gas?

MULTI-VERSION EXERCISES

19.83 Suppose 0.05839 mole of an ideal monatomic gas has a pressure of 1.000 atm at 273.15 K and is then cooled at constant pressure until its volume is reduced by 47.11%. What is the amount of work done by the gas?

19.84 Suppose an ideal monatomic gas has a pressure of 1.000 atm at 273.15 K and is then cooled at constant pressure until its volume is reduced by 47.53%. The amount of work done by the gas is $-7.540 \cdot 10^1$ J. How many moles of gas are there?

19.85 Suppose 0.03127 mole of an ideal monatomic gas with a pressure of 1.000 atm at 273.15 K is cooled at constant pressure and reduced in volume. The amount of work done by the gas is $-3.404 \cdot 10^1$ J. By what percentage was the volume of the gas reduced?

19.86 A 2.869-L bottle contains air plus 1.393 mole of sodium bicarbonate and 1.393 mole of acetic acid. These compounds react to produce 1.393 mole of carbon dioxide gas, along with water and sodium acetate. If the bottle is tightly sealed at atmospheric pressure ($1.013 \cdot 10^5$ Pa) before the reaction occurs, what is the pressure inside the bottle when the reaction is complete? Assume that the bottle is kept in a water bath that keeps the temperature in the bottle constant.

19.87 A bottle contains air plus 1.413 mole of sodium bicarbonate and 1.413 mole of acetic acid. These compounds react to produce 1.413 mole of carbon dioxide gas, along with water and sodium acetate. The bottle is tightly sealed at atmospheric pressure ($1.013 \cdot 10^5$ Pa) before the reaction occurs. The pressure inside the bottle when the reaction is complete is $1.064 \cdot 10^6$ Pa. What is the volume of the bottle? Assume that the bottle is kept in a water bath that keeps the temperature in the bottle constant.

19.88 A 3.787-L bottle contains air plus n moles of sodium bicarbonate and n moles of acetic acid. These compounds react to produce n moles of carbon dioxide gas, along with water and sodium acetate. The bottle is tightly sealed at atmospheric pressure ($1.013 \cdot 10^5$ Pa) before the reaction occurs. The pressure inside the bottle when the reaction is complete is $9.599 \cdot 10^5$ Pa. How many moles, n, of carbon dioxide gas are in the bottle? Assume that the bottle is kept in a water bath that keeps the temperature in the bottle constant.

20

The Second Law of Thermodynamics

FIGURE 20.1 (a) A steam locomotive. (b) The Thrust SSC.

(a)

(b)

The early studies of thermodynamics were motivated by the desire of scientists and engineers to discover the governing principles of engines and machines so that they could design more efficient and powerful ones. The steam engine could be used to power early locomotives and ships; however, overheated steam boilers could explode and cause damage and fatalities. Today, steam locomotives and ships are used primarily on scenic routes for tourists.

The principles that affect the efficiency of heat engines apply to all engines, from steam locomotives (Figure 20.1a) to jet engines. The Thrust SSC (supersonic car) shown in Figure 20.1b, used two powerful jet engines to set a land speed record of 763 mph in October 1997. However, for all its speed and power, it still produces more waste heat than usable energy, which is true of most engines.

In this chapter, we examine heat engines in theory and in practice. Their operation is governed by the Second Law of Thermodynamics—one of the most far-reaching and powerful statements in all of science. There are several different ways to express this law, including one involving a concept called *entropy*. The ideas discussed in this chapter have applications to practically all areas of science, including information processing, biology, and astronomy.

WHAT WE WILL LEARN

- Reversible thermodynamic processes are idealized processes in which a system can move from one thermodynamic state to another and back again, while remaining close to thermodynamic equilibrium.

- Most processes are irreversible; for example, the pieces of a broken coffee cup lying on the floor cannot spontaneously reassemble back into the cup. Practically all real-world thermodynamic processes are irreversible.

- Mechanical energy can be converted to thermal energy, and thermal energy can be converted to mechanical energy.

- A heat engine is a device that converts thermal energy into mechanical energy. This engine operates in a

thermodynamic cycle, returning to its original state periodically.

- No heat engine can have an efficiency of 100%. The Second Law of Thermodynamics places fundamental limitations on the efficiency of a heat engine.

- Refrigerators and air conditioners are heat engines that operate in reverse.

- The Second Law of Thermodynamics says that the entropy of an isolated system can never decrease.

- The entropy of a system can be related to the microscopic states of the constituents of the system.

20.1 Reversible and Irreversible Processes

As noted in Chapter 18, if you pour hot water into a glass and place that glass on a table, the water will slowly cool until it reaches the temperature of its surroundings. The air in the room will also warm, although imperceptibly. You would be astonished if, instead, the water got warmer and the air in the room cooled down slightly, while conserving energy. In fact, the First Law of Thermodynamics is satisfied for *both* of these scenarios. However, it is always the case that the water cools until it reaches the temperature of its surroundings. Thus, other physical principles are required to explain why the temperature changes one way and not the other.

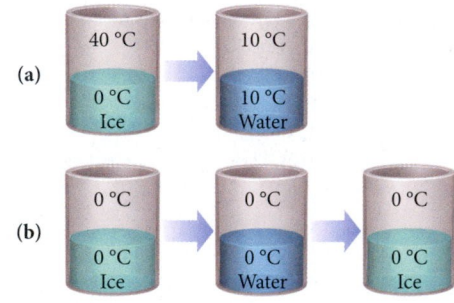

Practically all real-life thermodynamic processes are irreversible. For example, if a disk of ice at a temperature of 0 °C is placed in a metal can having a temperature of 40 °C, heat flows irreversibly from the can to the ice, as shown in Figure 20.2a. The ice will melt to water; the water will then warm and the can will cool, until the water and the can are at the same temperature (10 °C in this case). It is not possible to make small changes in any thermodynamic variable and return the system to the state corresponding to a warm can and frozen water.

However, we can imagine a class of idealized *reversible* processes. With a **reversible process,** a system is always close to being in thermodynamic equilibrium. In this case, making a small change in the state of the system can reverse any change in the thermodynamic variables of the system. For example, in Figure 20.2b, a disk of ice at a temperature of 0 °C is placed in a metal can that is also at a temperature of 0 °C. Raising the temperature of the can slightly will melt the ice to water. Then, lowering the temperature of the metal can will refreeze the water to ice, thus returning the system to its original state. (Technically speaking, the crystal structure of the ice would be different and thus the final state would be different from the initial state, but not in a way that concerns us here.)

FIGURE 20.2 Two thermodynamic processes: (a) An irreversible process in which a disk of ice at a temperature of 0 °C is placed inside a metal can at a temperature of 40 °C. (b) A reversible process in which a disk of ice at a temperature of 0 °C is placed inside a metal can at a temperature of 0 °C.

We can think of reversible processes as equilibrium processes in which the system always stays in thermal equilibrium or close to thermal equilibrium. If a system were actually in thermal equilibrium, no heat would flow and no work would be done by the system. Thus, a reversible process is an idealization. However, for a nearly reversible process, small temperature and pressure adjustments can keep a system close to thermal equilibrium.

On the other hand, if a process involves heat flow with a finite temperature difference, free expansion of a gas, or conversion of mechanical work to thermal energy, it is an **irreversible process.** It is not possible to make small changes to the temperature or pressure of the system and cause the process to proceed in the opposite direction. In addition, while an irreversible process is taking place, the system is not in thermal equilibrium.

The irreversibility of a process is related to the randomness or disorder of the system. For example, suppose you order a deck of cards by rank and suit and then take the first five cards, as illustrated in Figure 20.3a. You will observe a well-ordered system of cards, for example, all spades, starting with the ace and going down to the ten. Next, you put those five cards back in the deck, toss the deck of cards in the air, and let the cards fall on the floor. You pick up the cards one by one without looking at their rank or suit and then take the first five cards off the top of the deck. It is highly improbable that these five cards will be ordered by rank and suit. It is much more likely that you will see a result like the one shown in Figure 20.3b, that is, with cards in random order. The process of tossing and picking up the cards is irreversible.

Another example of an irreversible process is the free expansion of a gas, which we touched on in Chapter 18 and will discuss in greater detail in Section 20.7.

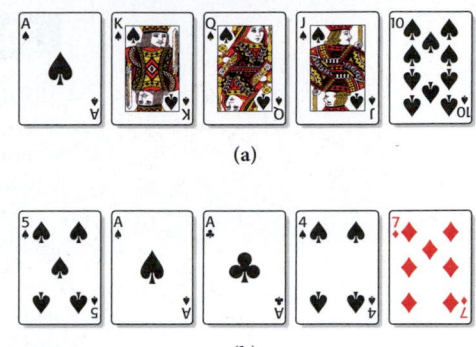

FIGURE 20.3 (a) The first five cards of a card deck ordered by rank and suit. (b) The first five cards of a deck that has been tossed in the air and picked up randomly.

Poincaré Recurrence Time

The common experience of déjà vu (French for "already seen") is the feeling of having experienced a certain situation sometime before. But do events really repeat? If processes leading from one event to another are truly irreversible, then an event cannot repeat exactly.

The French mathematician and physicist Henry Poincaré made an important contribution to this discussion in 1890, by famously stating his recurrence theorem. In it he postulates that certain isolated dynamical systems will return to a state arbitrarily close to their initial state after a sufficiently long time. This time is known as the *Poincaré recurrence time*, or simply the *Poincaré time*, of the system. This time can be calculated fairly straightforwardly for many systems that have only a finite number of different states, as the following example illustrates.

EXAMPLE 20.1 / Deck of Cards

Suppose you repeat the tossing and gathering of a deck of cards performed for Figure 20.3 again and again. Eventually, there is a chance that one of the random orders in which you pick up the cards will yield the sequence ace through ten of spades for the first five cards in the deck.

PROBLEM

If it takes 1 min on average to toss the cards into the air and then collect them into a stack again, how long can you expect it to take, on average, before you will find the sequence of ace through ten of spades as the first five cards in the deck?

SOLUTION

There are 52 cards in the deck. Each of them has the same $\frac{1}{52}$ probability of being on top. So the probability that the ace of spades ends up on top is $\frac{1}{52}$. If the ace is the top card, then there are 51 cards left. The probability that the king of spades is in the top position of the remaining 51 cards is $\frac{1}{51}$. Thus, the combined probability that the ace and king of spades are the top two cards in that order is $1/(52 \cdot 51) = 1/2652$. In the same way, the probability of the ordered sequence of ace through ten of spades occurring is $1/(52 \cdot 51 \cdot 50 \cdot 49 \cdot 48) = 1/311,875,200$.

So the average number of tries needed to obtain the desired sequence is 311,875,200. Since each of them takes 1 min, successful completion of this exercise would take approximately 593 yr.

DISCUSSION

We can almost immediately state the average number of tries it would take to get all 52 cards in the ordered sequence ace through two, for all four suits. It is $52 \cdot 51 \cdot 50 \cdot \ldots \cdot 3 \cdot 2 \cdot 1 = 52!$ If each of these attempts took a minute, it would require an average of $1.534 \cdot 10^{62}$ yr to obtain the desired order. But even after this time, there would be no guarantee that the ordered sequence had to appear. However, since there are 52! possible different sequences of the cards in the deck, at least one of the sequences would have appeared at least twice.

20.2 Engines and Refrigerators

The preceding discussion of the reversibility of thermodynamic processes might have seemed a bit abstract. However, now we are going to discuss engines, particularly those that operate in cycles, and the fine distinctions between reversible and irreversible processes are key to the operation of these engines.

A **heat engine** is a device that turns thermal energy into useful work. For example, an internal combustion engine or a jet engine extracts mechanical work from the thermal energy generated by burning a gasoline-air mixture (Figure 20.4a). For a heat engine to do work repeatedly, it must operate in a cycle. For example, work would be done if you put a firecracker under a soup can and exploded the firecracker. The thermal energy from the explosion would be turned into mechanical motion of the soup can. (However, the applications of this firecracker/soup-can engine are limited.)

An engine that operates in a cycle passes through various thermodynamic states and eventually returns to its original state. The cyclic processes that heat engines utilize always involve some kind of temperature change. We can think of the most basic heat engine as operating between two thermal reservoirs (Figure 20.4b). One thermal reservoir is at a high temperature, T_H, and the other thermal reservoir is at a low temperature, T_L. The engine takes heat, Q_H, from the high-temperature reservoir, turns some of that heat into mechanical work, W, and exhausts the remaining heat, Q_L (where $Q_L > 0$), to the low-temperature reservoir. According to the First Law of Thermodynamics (equivalent to the conservation of energy), $Q_H = W + Q_L$. Thus, to make an engine operate, it must be supplied energy in the form of Q_H, and then it will return useful work, W. The **efficiency, ϵ,** of an engine is defined as

$$\epsilon = \frac{W}{Q_H}.\qquad(20.1)$$

A **refrigerator,** like the one shown in Figure 20.5a, is a heat engine that operates in reverse. Instead of converting heat into work, the refrigerator uses work to move heat from a low-temperature thermal reservoir to a high-temperature thermal reservoir (Figure 20.5b). In a real refrigerator, an electric motor drives a compressor, which transfers thermal energy from the interior of the refrigerator to the air in the room. An air conditioner is also a refrigerator; it transfers thermal energy from the air in a room to the air outdoors.

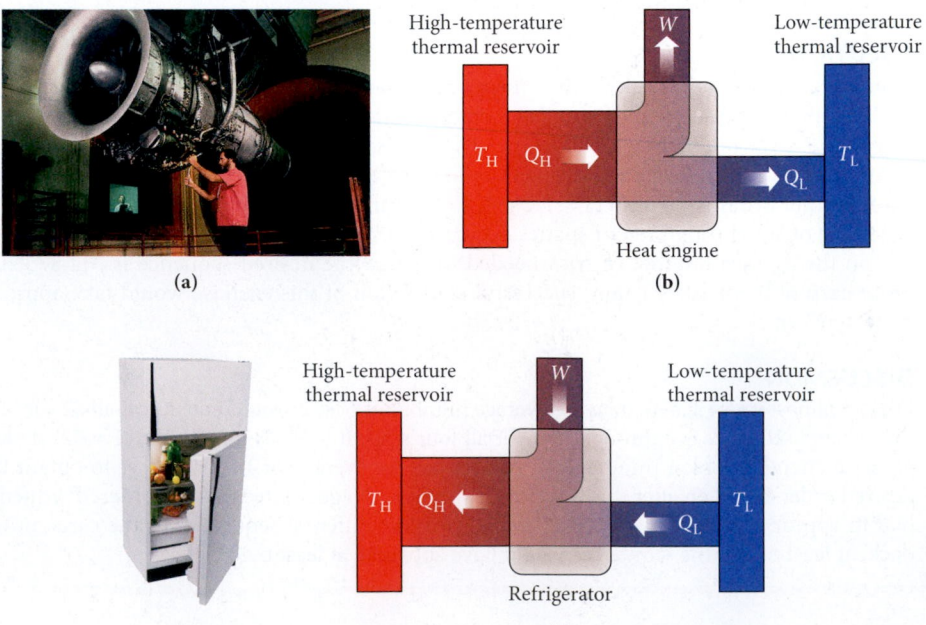

FIGURE 20.4 (a) A jet engine turns thermal energy into mechanical work. (b) Flow diagram of a heat engine operating between two thermal reservoirs. The engine takes heat from the high-temperature thermal reservoir, does some useful work, W, and exhausts the remaining heat to the low-temperature thermal reservoir.

FIGURE 20.5 (a) A household refrigerator. (b) Flow diagram of a refrigerator operating between two thermal reservoirs.

For a refrigerator, the First Law of Thermodynamics requires that $Q_L + W = Q_H$. It is desired that a refrigerator remove as much heat as possible from the cooler reservoir, Q_L, given the work, W, put into the refrigerator. The **coefficient of performance,** K, of a refrigerator is defined as

$$K = \frac{Q_L}{W}.$$ (20.2)

In the United States, refrigerators are usually rated by their annual energy usage, without quoting their actual efficiency. An efficient refrigerator is one whose energy use is comparable to the lowest energy use in its capacity class.

Air conditioners are often rated in terms of an energy efficiency rating (EER), defined as the heat removal capacity, H, in BTU/hour divided by the power P used in watts. The relationship between K and EER is

$$K = \frac{|Q_L|}{|W|} = \frac{Ht}{Pt} = \frac{H}{P} = \frac{\text{EER}}{3.41},$$

where t is a time interval. The factor 1/3.41 arises from the definition of a BTU per hour:

$$\frac{1\,\text{BTU/h}}{1\,\text{watt}} = \frac{(1055\,\text{J})/(3600\,\text{s})}{1\,\text{J/s}} = \frac{1}{3.41}.$$

Typical EER values for a room air conditioner range from 8 to 11, meaning that values of K range from 2.3 to 3.2. Thus, a typical room air conditioner can remove about three units of heat for every unit of energy used. Central air conditioners are often rated by a seasonally adjusted energy efficiency rating (SEER) that takes into account how long the air conditioner operates during a year.

A **heat pump** is a variation of a refrigerator that can be used to warm buildings. The heat pump warms the building by cooling the outside air. Just as for a refrigerator, $Q_L + W = Q_H$. However, for a heat pump, the quantity of interest is the heat put into the warmer reservoir, Q_H, rather than the heat removed from the cooler reservoir. Thus, the coefficient of performance of a heat pump is

$$K_{\text{heat pump}} = \frac{Q_H}{W},$$ (20.3)

where W is the work required to move the heat from the low-temperature reservoir to the high-temperature reservoir. The typical coefficient of performance of a commercial heat pump ranges from 3 to 4. Heat pumps do not work well when the outside temperature sinks below –18 °C (0 °F).

EXAMPLE 20.2 | Warming a House with a Heat Pump

A heat pump with a coefficient of performance of 3.500 is used to warm a house that loses 75,000. BTU/h of heat on a cold day. Assume that the cost of electricity is 10.00 cents per kilowatt-hour.

PROBLEM
How much does it cost to warm the house for the day?

SOLUTION
The coefficient of performance of a heat pump is given by equation 20.3:

$$K_{\text{heat pump}} = \frac{Q_H}{W}.$$

The work required to warm the house is then

$$W = \frac{Q_H}{K_{\text{heat pump}}}.$$

– Continued

The house is losing 75,000 BTU/h, which we convert to SI units:

$$\frac{75{,}000 \text{ BTU}}{1\text{ h}} \cdot \frac{1055\text{ J}}{1\text{ BTU}} \cdot \frac{1\text{ h}}{3600\text{ s}} = 21.98 \text{ kW}.$$

Therefore, the power required to warm the house is

$$P = \frac{W}{t} = \frac{Q_\text{H}/t}{K_\text{heat pump}} = \frac{21.98 \text{ kW}}{3.500} = 6.280 \text{ kW}.$$

The cost to warm the house for 24 h is then

$$\text{Cost} = \left(6.280 \text{ kW}\right)\left(24 \text{ h}\right)\frac{\$0.1000}{1 \text{ kWh}} = \$15.07.$$

20.3 Ideal Engines

An **ideal engine** is one that utilizes only reversible processes. Thus, the engine incorporates no "energy-wasting" effects, such as friction or viscosity. Remember, as defined in Section 20.2, a heat engine is a device that turns thermal energy into useful work. You might think that an *ideal* engine would be 100% efficient—would turn all the thermal energy supplied to it into useful mechanical work. However, we'll see that even the most efficient ideal engine cannot accomplish this. This fundamental inability to convert thermal energy totally into useful mechanical work lies at the very heart of the Second Law of Thermodynamics and of entropy, the two main topics of this chapter.

Carnot Cycle

An example of an ideal engine is the **Carnot engine,** which is the most efficient engine that operates between two temperature reservoirs and the only one to do so reversibly. The cycle of thermodynamic processes used by the Carnot engine is called the **Carnot cycle.**

In order to be reversible, a Carnot cycle consists of two isothermal processes and two adiabatic processes, as shown in Figure 20.6 in a pV-diagram (pressure-volume diagram). We can pick an arbitrary starting point for the cycle; let's say that it begins at point 1. The system first undergoes an isothermal process, during which the system expands and absorbs heat from a thermal reservoir at fixed temperature T_H (in Figure 20.6, $T_\text{H} = 400$ K). At point 2, the system begins an adiabatic expansion in which no heat is gained or lost. At point 3, the system begins another isothermal process, this time giving up heat to a second thermal reservoir at a lower temperature, T_L (in Figure 20.6, $T_\text{L} = 300$ K), while the system compresses. At point 4, the system begins a second adiabatic process in which no heat is gained or lost. The Carnot cycle is complete when the system returns to point 1.

The efficiency of the Carnot engine is not 100% but instead is given by

$$\epsilon = 1 - \frac{T_\text{L}}{T_\text{H}}. \tag{20.4}$$

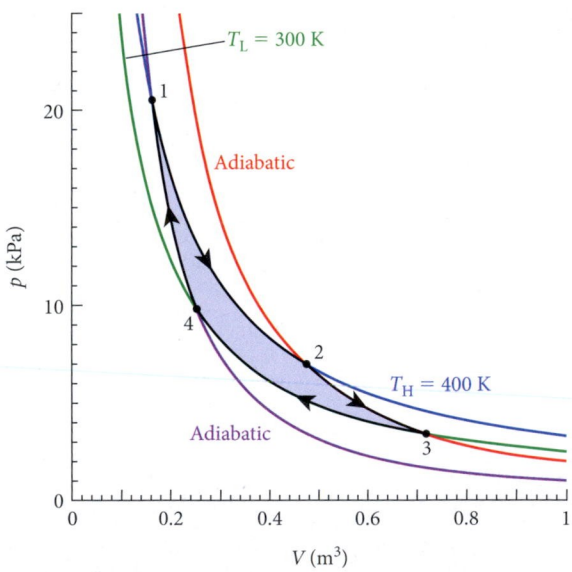

FIGURE 20.6 The Carnot cycle, consisting of two isothermal processes and two adiabatic processes.

Remarkably, the efficiency of a Carnot engine depends only on the temperature ratio of the two thermal reservoirs. For the engine shown in Figure 20.6, for example, the efficiency is $\epsilon = 1 - (300 \text{ K})/(400 \text{ K}) = 0.25$.

DERIVATION 20.1 / Efficiency of Carnot Engine

We can determine the work and heat gained and lost in a Carnot cycle. To do this, we assume that the system consists of an ideal gas in a container whose volume can change. This system is placed in contact with a thermal reservoir and originally has pressure p_1 and volume V_1, corresponding to point 1 in Figure 20.6. The thermal reservoir holds the temperature of the system constant at T_H. The volume of the system is then allowed to increase until the system reaches point 2 in Figure 20.6. At point 2, the pressure is p_2 and the volume is V_2. During this isothermal process, work, W_{12}, is done by the system, and heat, Q_H, is transferred to the system. Because this process is isothermal, the internal energy of the system does not change. Thus, the amount of work done by the system and the thermal energy transferred to the system are equal, given by (see the discussion in Section 19.2):

$$W_{12} = Q_H = nRT_H \ln\left(\frac{V_2}{V_1}\right). \tag{i}$$

The system is then removed from contact with the thermal reservoir whose temperature is T_H. The system continues to expand adiabatically to point 3 in Figure 20.6. At point 3, the expansion is stopped, and the system is put in contact with a second thermal reservoir that has temperature T_L. An adiabatic expansion means that $Q = 0$. In Chapter 19, we saw that the work done by an ideal gas undergoing an adiabatic process is $W = nR(T_f - T_i)/(1 - \gamma)$, where $\gamma = C_p/C_V$ and C_p is the specific heat of the gas at constant pressure and C_V is the specific heat of the gas at constant volume. Thus, the work done by the system in going from point 2 to point 3 is

$$W_{23} = \frac{nR}{1-\gamma}(T_L - T_H).$$

Remaining in contact with the thermal reservoir whose temperature is T_L, the system is compressed isothermally from point 3 to point 4. The work done by the system and the heat absorbed by the system are

$$W_{34} = -Q_L = nRT_L \ln\left(\frac{V_4}{V_3}\right). \tag{ii}$$

(In Chapter 18, this would have been written $W_{34} = Q_L$ because Q_L would be the heat added to the system and therefore $Q_L < 0$. However, in this chapter, we define Q_L as the heat delivered to the low-temperature reservoir, at $T = T_L$, and hence $Q_L > 0$.)

Finally, the system is removed from contact with the second thermal reservoir and compressed adiabatically from point 4 back to the original point, point 1. For this adiabatic process, $Q = 0$, and the work done by the system is

$$W_{41} = \frac{nR}{1-\gamma}(T_H - T_L).$$

Note that $W_{41} = -W_{23}$, since the work done in the adiabatic processes depends only on the initial and final temperatures. Therefore, $W_{41} + W_{23} = 0$, and for the complete cycle, the total work done by the system is

$$W = W_{12} + W_{23} + W_{34} + W_{41} = W_{12} + W_{34}.$$

Substituting the expressions for W_{12} and W_{34} from equations (i) and (ii) into the right side of the preceding equation, we have

$$W = nRT_H \ln\left(\frac{V_2}{V_1}\right) + nRT_L \ln\left(\frac{V_4}{V_3}\right).$$

We can now find the efficiency of the Carnot engine. The efficiency is defined in equation 20.1 as the ratio of the work done by the system, W, to the amount of heat supplied to the system, Q_H:

$$\epsilon = \frac{W}{Q_H} = \frac{nRT_H \ln\left(\frac{V_2}{V_1}\right) + nRT_L \ln\left(\frac{V_4}{V_3}\right)}{nRT_H \ln\left(\frac{V_2}{V_1}\right)} = 1 + \frac{T_L \ln\left(\frac{V_4}{V_3}\right)}{T_H \ln\left(\frac{V_2}{V_1}\right)}. \tag{iii}$$

– Continued

This expression for the efficiency of a Carnot engine contains the volumes at points 1, 2, 3, and 4. However, we can eliminate the volumes from this expression using the relation for adiabatic expansion of an ideal gas from Chapter 19, $TV^{\gamma-1} = $ constant.

For the adiabatic process from point 2 to point 3, we then have

$$T_H V_2^{\gamma-1} = T_L V_3^{\gamma-1}. \tag{iv}$$

Similarly, for the adiabatic process from point 4 to point 1, we have

$$T_L V_4^{\gamma-1} = T_H V_1^{\gamma-1}. \tag{v}$$

Forming ratios of both sides of equations (iv) and (v):

$$\frac{T_H V_2^{\gamma-1}}{T_H V_1^{\gamma-1}} = \frac{T_L V_3^{\gamma-1}}{T_L V_4^{\gamma-1}} \Rightarrow \frac{V_2}{V_1} = \frac{V_3}{V_4}.$$

Using this relationship between the volumes, we can now rewrite equation (iii) for the efficiency as

$$\epsilon = 1 + \frac{-T_L \ln\left(\dfrac{V_3}{V_4}\right)}{T_H \ln\left(\dfrac{V_2}{V_1}\right)} = 1 - \frac{T_L}{T_H},$$

which matches equation 20.4.

Can the efficiency of the Carnot engine reach 100%? To obtain 100% efficiency, T_H in equation 20.4 would have to be raised to infinity or T_L be lowered to absolute zero. Neither option is possible. Thus, the efficiency of a Carnot engine is always less than 100%.

Note that the total work done by the system during the Carnot cycle in Derivation 20.1 is $W = W_{12} + W_{34}$, and the two contributions to the total work due to the two isothermal processes are $W_{12} = Q_H$ and $W_{34} = -Q_L$. Therefore, as a direct consequence of the First Law of Thermodynamics, the total mechanical work from a Carnot cycle can also be written as

$$W = Q_H - Q_L. \tag{20.5}$$

Since the efficiency is defined in general as $\epsilon = W/Q_H$ (equation 20.1), the efficiency of a Carnot engine can thus also be expressed as

$$\epsilon = \frac{Q_H - Q_L}{Q_H}. \tag{20.6}$$

In this formulation, the efficiency of a Carnot engine is determined by the heat taken from the warmer reservoir minus the heat given back to the cooler reservoir. For this expression for the efficiency of a Carnot engine to yield an efficiency of 100%, the heat returned to the cooler reservoir would have to be zero. Conversely, if the efficiency of the Carnot engine is less than 100%, the engine cannot convert all the heat it takes in from the warmer reservoir to useful work.

French physicist Nicolas Leonard Sadi Carnot (1796–1832), who developed the Carnot cycle in the 19th century, proved the following statement, known as **Carnot's Theorem:**

> No heat engine operating between two thermal reservoirs can be more efficient than a Carnot engine operating between the two thermal reservoirs.

We won't present the proof of Carnot's Theorem. However, we'll note the idea behind the proof after discussing the Second Law of Thermodynamics in Section 20.5

We can imagine running a Carnot engine in reverse, creating a "Carnot refrigerator." The maximum coefficient of performance for such a refrigerator operating between two thermal reservoirs is

$$K_{max} = \frac{T_L}{T_H - T_L}. \tag{20.7}$$

Concept Check 20.1

What is the maximum (Carnot) coefficient of performance of a refrigerator in a room with a temperature of 22.0 °C ? The temperature inside the refrigerator is kept at 2.0 °C.

a) 0.10   d) 5.8

b) 0.44   e) 13.8

c) 3.0

Similarly, the maximum coefficient of performance for a heat pump operating between two thermal reservoirs is given by

$$K_{\text{heat pump max}} = \frac{T_H}{T_H - T_L}.$$ (20.8)

Note that the Carnot cycle constitutes the ideal thermodynamic process, which is the absolute upper limit on what is theoretically attainable. Real-world complications lower the efficiency drastically, as we'll see in Section 20.4. The typical coefficient of performance for a real refrigerator or heat pump is around 3.

Concept Check 20.2

What is the maximum (Carnot) coefficient of performance of a heat pump being used to warm a house to an interior temperature of 22.0 °C? The temperature outside the house is 2.0 °C.

a) 0.15   d) 6.5

b) 1.1   e) 14.8

c) 3.5

EXAMPLE 20.3 Work Done by a Carnot Engine

A Carnot engine takes 3000 J of heat from a thermal reservoir with a temperature $T_H = 500$ K and discards heat to a thermal reservoir with a temperature $T_L = 325$ K.

PROBLEM
How much work does the Carnot engine do in this process?

SOLUTION
We start with the definition of the efficiency, ϵ, of a heat engine (equation 20.1):

$$\epsilon = \frac{W}{Q_H},$$

where W is the work that the engine does and Q_H is the heat taken from the warmer thermal reservoir. We can then use equation 20.4 for the efficiency of a Carnot engine:

$$\epsilon = 1 - \frac{T_L}{T_H}.$$

Combining these two expressions for the efficiency gives us

$$\frac{W}{Q_H} = 1 - \frac{T_L}{T_H}.$$

Now we can express the work done by the Carnot engine:

$$W = Q_H\left(1 - \frac{T_L}{T_H}\right).$$

Putting in the numerical values gives the amount of work done by the Carnot engine:

$$W = (3000 \text{ J})\left(1 - \frac{325 \text{ K}}{500 \text{ K}}\right) = 1050 \text{ J}.$$

Self-Test Opportunity 20.1

What is the efficiency of the Carnot engine in Example 20.3?

EXAMPLE 20.4 Maximum Efficiency of an Electric Power Plant

In the United States, 85% of electricity is generated by burning fossil fuels to produce steam, which in turn drives alternators that produce electricity. Power plants can produce steam with a temperature as high as 600. °C by pressurizing the steam. The resulting waste heat must be exhausted into the environment, which is at a temperature of 20.0 °C.

PROBLEM
What is the maximum efficiency of such a power plant?

SOLUTION
The maximum efficiency of such a power plant is the efficiency of a Carnot heat engine (equation 20.4) operating between 20.0 °C and 600. °C:

$$\epsilon = 1 - \frac{T_L}{T_H} = 1 - \frac{293 \text{ K}}{873 \text{ K}} = 66.4\%.$$

– Continued

Real power plants achieve a lower efficiency of around 40%. However, many well-designed plants do not simply exhaust the waste heat into the environment. Instead, they employ *cogeneration* or *combined heat and power* (CHP). The heat that normally would be lost is used to heat nearby buildings or houses. This heat can even be used to operate air conditioners that cool nearby structures, a process called *trigeneration*. By employing CHP, modern power plants can make use of up to 90% of the energy used to generate electricity. For example, the power plant at Michigan State University (Figure 20.7) burns coal to generate electricity for the campus. The waste heat is used to heat campus buildings in the winter and air-condition them in the summer.

SOLVED PROBLEM 20.1 | Cost to Operate an Electric Power Plant

An electric power plant operates steam turbines at a temperature of 557 °C and uses cooling towers to keep the cooler thermal reservoir at a temperature of 38.3 °C. The operating cost of the plant for 1 yr is $52.0 million. The plant managers propose using the water from a nearby lake to lower the temperature of the cooler reservoir to 8.90 °C. Assume that the plant operates at maximum possible efficiency.

PROBLEM
How much will the operating cost of the plant be reduced in 1 yr as a result of the change in reservoir temperature? Assume that the plant generates the same amount of electricity.

SOLUTION
THINK We can calculate the efficiency of the power plant assuming that it operates at the theoretical Carnot efficiency, which depends only on the temperatures of the warmer thermal reservoir and the cooler thermal reservoir. We can calculate the amount of thermal energy put into the warmer reservoir and assume that the cost of operating the plant is proportional to that thermal energy. This thermal energy might come from the burning of fuel (coal, oil, or gas) or perhaps from a nuclear reactor. Lowering the temperature of the cooler reservoir will increase the efficiency of the power plant, and thus lower the required amount of thermal energy and the associated cost. The cost savings are then the operating cost of the original plant minus the operating cost of the improved plant.

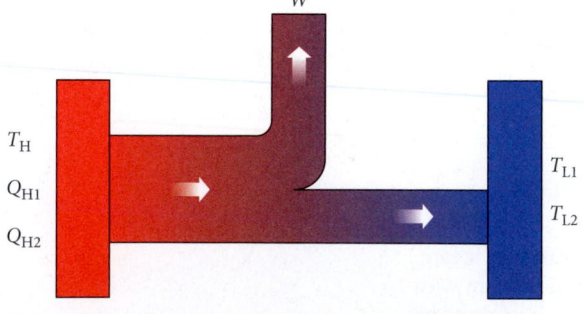

FIGURE 20.8 The thermodynamic quantities related to the operation of an electric power plant.

SKETCH Figure 20.8 is a sketch relating the thermodynamic quantities involved in the operation of an electric power plant. The temperature of the warmer reservoir is T_H, the thermal energy supplied to the original power plant is Q_{H1}, the thermal energy supplied to the improved power plant is Q_{H2}, the useful work done by the power plant is W, the original temperature of the cooler reservoir is T_{L1}, and the temperature of the improved cooler reservoir is T_{L2}.

RESEARCH The maximum efficiency of the power plant when operating with the cooling towers is given by the theoretical Carnot efficiency:

$$\epsilon_1 = \frac{T_H - T_{L1}}{T_H}. \tag{i}$$

The efficiency of the power plant under this condition can also be expressed as

$$\epsilon_1 = \frac{W}{Q_{H1}}, \tag{ii}$$

where Q_{H1} is the thermal energy supplied when the cooling towers are used. We can write similar expressions for the efficiency of the power plant when the lake is used to lower the temperature of the cooler reservoir:

$$\epsilon_2 = \frac{T_H - T_{L2}}{T_H} \tag{iii}$$

and

$$\epsilon_2 = \frac{W}{Q_{H2}}. \tag{iv}$$

The cost to operate the power plant is proportional to the thermal energy supplied to the warmer reservoir. We can thus equate the ratio of the original cost, c_1, to the lowered cost, c_2, with the ratio of the thermal energy originally required, Q_{H1}, to the thermal energy subsequently required, Q_{H2}:

$$\frac{c_1}{c_2} = \frac{Q_{H1}}{Q_{H2}}.$$

SIMPLIFY   We can express the new cost as

$$c_2 = c_1 \frac{Q_{H2}}{Q_{H1}}.$$

We can get an expression for the thermal energy originally required using equations (i) and (ii):

$$Q_{H1} = \frac{W}{\epsilon_1} = \frac{W}{\left(\dfrac{T_H - T_{L1}}{T_H}\right)} = \frac{T_H W}{T_H - T_{L1}}.$$

We can get a similar expression for the thermal energy required after the improvement using equations (iii) and (iv):

$$Q_{H2} = \frac{W}{\epsilon_2} = \frac{W}{\left(\dfrac{T_H - T_{L2}}{T_H}\right)} = \frac{T_H W}{T_H - T_{L2}}.$$

We can now express the new cost as

$$c_2 = c_1 \frac{Q_{H2}}{Q_{H1}} = c_1 \frac{\left(\dfrac{T_H W}{T_H - T_{L2}}\right)}{\left(\dfrac{T_H W}{T_H - T_{L1}}\right)} = c_1 \frac{T_H - T_{L1}}{T_H - T_{L2}}.$$

Note that the work W done by the power plant cancels out because the plant generates the same amount of electricity in both cases. The cost savings are then

$$c_1 - c_2 = c_1 - c_1 \frac{T_H - T_{L1}}{T_H - T_{L2}} = c_1 \left(1 - \frac{T_H - T_{L1}}{T_H - T_{L2}}\right) = c_1 \frac{T_{L1} - T_{L2}}{T_H - T_{L2}}.$$

CALCULATE   Putting in the numerical values, we get

$$c_1 - c_2 = c_1 \frac{T_{L1} - T_{L2}}{T_H - T_{L2}} = (\$52.0 \text{ million}) \frac{311.45 \text{ K} - 282.05 \text{ K}}{830.15 \text{ K} - 282.05 \text{ K}} = \$2.78927 \text{ million}.$$

ROUND   We report our result to three significant figures:

$$\text{Cost savings} = \$2.79 \text{ million}.$$

DOUBLE-CHECK   To double-check our result, we calculate the efficiency of the power plant using cooling towers:

$$\epsilon_1 = \frac{T_H - T_{L1}}{T_H} = \frac{830.15 \text{ K} - 311.45 \text{ K}}{830.15 \text{ K}} = 62.5\%.$$

The efficiency of the power plant using the lake to lower the temperature of the cooler reservoir is

$$\epsilon_2 = \frac{T_H - T_{L2}}{T_H} = \frac{830.15 \text{ K} - 282.05 \text{ K}}{830.15 \text{ K}} = 66.0\%.$$

These efficiencies seem reasonable. To check further, we verify that the ratio of the two efficiencies is equal to the inverse of the ratio of the two costs, because a higher efficiency means a lower cost. The ratio of the two efficiencies is

$$\frac{\epsilon_1}{\epsilon_2} = \frac{62.5\%}{66.0\%} = 0.947.$$

– Continued

The inverse ratio of the two costs is

$$\frac{c_2}{c_1} = \frac{\$52 \text{ million} - \$2.79 \text{ million}}{\$52 \text{ million}} = 0.946.$$

These ratios agree to within rounding error. Thus, our result seems reasonable.

20.4 Real Engines and Efficiency

A real heat engine based on the Carnot cycle is not practical. However, many practical heat engines that are in everyday use are designed to operate via cyclical thermodynamic processes. For an example of the operation of a real-world engine, let's examine the Otto cycle. Again, we assume an ideal gas as the working medium.

Otto Cycle

The **Otto cycle** is used in the modern internal combustion engines inside automobiles. This cycle consists of two adiabatic processes and two constant-volume processes (Figure 20.9) and is the default configuration for a four-cycle internal combustion engine. The piston-cylinder arrangement of a typical internal combustion engine is shown in Figure 20.10.

The thermal energy is provided by the ignition of a fuel-air mixture. The cycle starts with the piston at the top of the cylinder and proceeds through the following steps:

- *Intake stroke.* The piston moves down with the intake valve open, drawing in the fuel-air mixture (point 0 to point 1 in Figure 20.9 and Figure 20.10a), and the intake valve closes.

- *Compression stroke.* The piston moves upward, compressing the fuel-air mixture adiabatically (point 1 to point 2 in Figure 20.9 and Figure 20.10b).

- The spark plug ignites the fuel-air mixture, increasing the pressure at constant volume (point 2 to point 3 in Figure 20.9).

- *Power stroke.* Hot gases push the piston down adiabatically (point 3 to point 4 in Figure 20.9 and Figure 20.10c).

- When the piston is all the way down (point 4 in Figure 20.9), the exhaust valve opens. This reduces the pressure at constant volume, provides the heat rejection, and moves the system back to point 1.

- *Exhaust stroke.* The piston moves up, forcing out the burned gases (point 1 to point 0 in Figure 20.9 and Figure 20.10d), and the exhaust valve closes.

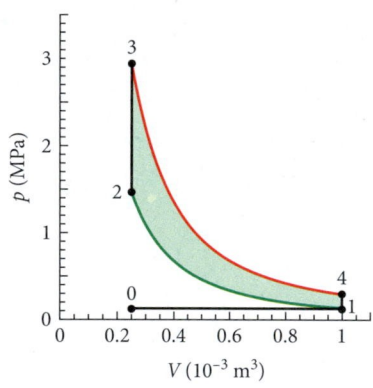

FIGURE 20.9 The Otto cycle, consisting of two adiabatic processes and two constant-volume processes.

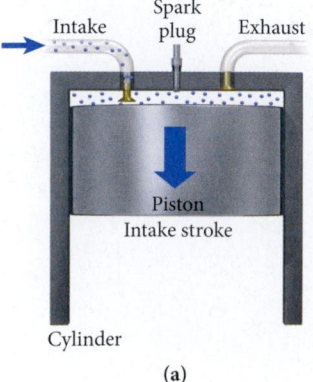

(a)

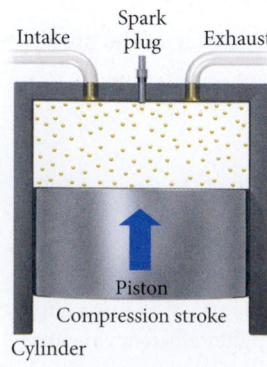

(b)

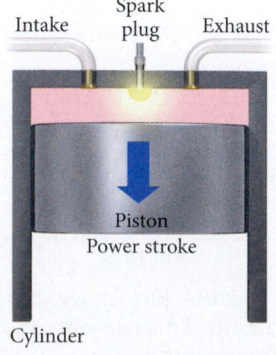

(c)

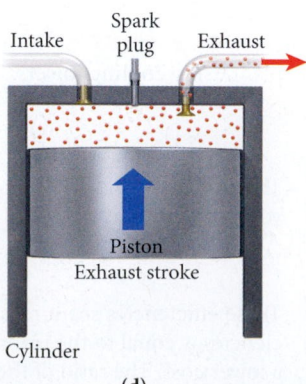

(d)

FIGURE 20.10 The four strokes of an internal combustion engine: (a) The intake stroke, during which the fuel-air mixture is drawn into the cylinder. (b) The compression stroke, in which the fuel-air mixture is compressed. (c) The power stroke, where the fuel-air mixture is ignited by the spark plug, releasing heat. (d) The exhaust stroke, during which the burned gases are pushed out of the cylinder.

The ratio of the expanded volume, V_1, to the compressed volume, V_2, is called the *compression ratio*:

$$r = \frac{V_1}{V_2}. \tag{20.9}$$

The efficiency, ϵ, of the Otto cycle can be expressed as a function of the compression ratio only:

$$\epsilon = 1 - r^{1-\gamma}. \tag{20.10}$$

(In contrast, for the Carnot cycle, equation 20.4 showed that the efficiency depends only on the ratio of two temperatures.)

DERIVATION 20.2 / Efficiency of the Otto Cycle

The Otto cycle begins at point 0 in Figure 20.9. The process from point 0 to point 1 simply draws the fuel-air mixture into the cylinder and plays no further role in the efficiency considerations. Neither does the inverse process from point 1 to point 0, the exhaust stroke, which removes the combustion products from the cylinder.

The process from point 1 to point 2 is adiabatic. The thermal energy transferred is zero, and the work done by the system (see Chapter 19) is given by

$$W_{12} = \frac{nR}{1-\gamma}(T_2 - T_1).$$

For the constant-volume process from point 2 to point 3, the work done by the system is zero, $W_{23} = 0$, and the thermal energy transferred to the system (see Chapter 18) is given by

$$Q_{23} = nC_V(T_3 - T_2).$$

The process from point 3 to point 4 is again adiabatic, and the thermal energy transferred is zero. The work done by the system is

$$W_{34} = \frac{nR}{1-\gamma}(T_4 - T_3).$$

The process from point 4 to point 1 takes place at constant volume, so the work done by the system is zero, $W_{41} = 0$, and the heat expelled from the system is

$$Q_{41} = nC_V(T_4 - T_1).$$

The total work done by the system is then simply the sum of the four contributions from the entire cycle (remember that $W_{23} = W_{41} = 0$)

$$W = W_{12} + W_{34} = \frac{nR}{1-\gamma}(T_2 - T_1) + \frac{nR}{1-\gamma}(T_4 - T_3) = \frac{nR}{1-\gamma}(T_2 - T_1 + T_4 - T_3).$$

The factor $R/(1-\gamma)$ can be expressed in terms of the specific heat capacity, C_V:

$$\frac{R}{1-\gamma} = \frac{C_p - C_V}{1 - \dfrac{C_p}{C_V}} = -C_V.$$

The efficiency of the Otto cycle is then (net work done divided by the heat gained by the system per cycle)

$$\epsilon = \frac{W}{Q_{23}} = \frac{-nC_V(T_2 - T_1 + T_4 - T_3)}{nC_V(T_3 - T_2)} = 1 - \frac{(T_4 - T_1)}{(T_3 - T_2)}. \tag{i}$$

We can use the fact that the process from point 1 to point 2 and the process from point 3 to point 4 are adiabatic to write (see Section 19.5)

$$T_1 V_1^{\gamma-1} = T_2 V_2^{\gamma-1} \Rightarrow T_2 = \left(\frac{V_1}{V_2}\right)^{\gamma-1} T_1 \tag{ii}$$

and

$$T_3 V_2^{\gamma-1} = T_4 V_1^{\gamma-1} \Rightarrow T_3 = \left(\frac{V_1}{V_2}\right)^{\gamma-1} T_4, \tag{iii}$$

– Continued

where V_1 is the volume at points 1 and 4 and V_2 is the volume at points 2 and 3. Substituting the expressions for T_2 and T_3 from equations (ii) and (iii) into equation (i) for the efficiency, we obtain equation 20.10:

$$\epsilon = 1 - \frac{(T_4 - T_1)}{\left(\dfrac{V_1}{V_2}\right)^{\gamma-1} T_4 - \left(\dfrac{V_1}{V_2}\right)^{\gamma-1} T_1} = 1 - \left(\frac{V_1}{V_2}\right)^{1-\gamma} = 1 - r^{1-\gamma}.$$

Self-Test Opportunity 20.2

How much would the theoretical efficiency of an internal combustion engine increase if the compression ratio were raised from 4 to 15?

Concept Check 20.3

Which of the four temperatures in the Otto cycle is the highest?

a) T_1

b) T_2

c) T_3

d) T_4

e) All four are identical.

As a numerical example, the compression ratio of the Otto cycle shown in Figure 20.9 is 4, and so, the efficiency of that Otto cycle is $\epsilon = 1 - 4^{1-7/5} = 1 - 4^{-0.4} = 0.426$. Thus, the theoretical efficiency of an engine operating on the Otto cycle with a compression ratio of 4 is 42.6%. Note, however, that this is the theoretical upper limit for the efficiency at this compression ratio. In principle, an internal combustion engine can be made more efficient by increasing the compression ratio, but practical factors prevent that approach. For example, if the compression ratio is too high, the fuel-air mixture will detonate before the compression is complete. Very high compression ratios put high stresses on the components of the engine. The actual compression ratios of gasoline-powered internal combustion engines range from 8 to 12.

Real Otto Engines

The actual efficiency of an internal combustion engine is about 20%. Why is this so much lower than the theoretical upper limit? There are many reasons. First, the "adiabatic" parts of the cycle do not really proceed without heat exchange between the gas in the piston and the engine block. This is obvious; we all know from experience that engines heat up while running, and this temperature rise is a direct consequence of heat leaking from the gas undergoing compression and ignition. Second, the gasoline-air mixture is not quite an ideal gas and thus has energy losses due to internal excitation. Third, during the ignition and heat rejection processes, the volume does not stay exactly constant, because both processes take some time, and the piston keeps moving continuously during that time. Fourth, during the intake and exhaust strokes, the pressure in the chamber is not exactly atmospheric pressure because of gas dynamics considerations. All of these effects, together with small friction losses, combine to reduce engine efficiency. All major car companies, as well as car racing crews, government and university labs, and even a few hobbyists, are continuously seeking ways to increase engine efficiency, because a higher efficiency means higher power output and/or better gas mileage. Since the oil crisis of the mid-1970s, the engine efficiency of U.S. automobiles has steadily increased (see Section 5.7), but even better miles-per-gallon performance is necessary for the United States to reduce its greenhouse gas emissions and its dependence on foreign oil.

SOLVED PROBLEM 20.2 | Efficiency of an Automobile Engine

A car with a gasoline-powered internal combustion engine travels with a speed of 26.8 m/s (60.0 mph) on a level road and uses gas at a rate of 6.92 L/100 km (34.0 mpg). The energy content of gasoline is 34.8 MJ/L.

PROBLEM

If the engine has an efficiency of 20.0%, how much power is delivered to keep the car moving at a constant speed?

SOLUTION

THINK We can calculate how much energy is being supplied to the engine by calculating the amount of fuel used and multiplying by the energy content of that fuel. The efficiency is the useful work divided by the energy being supplied; so, once we determine the energy supplied, we can find the useful work from the given efficiency of the engine. By dividing the work and the energy by an arbitrary time interval, we can determine the average power delivered.

SKETCH   Figure 20.11 shows a gasoline-powered automobile engine operating as a heat engine.

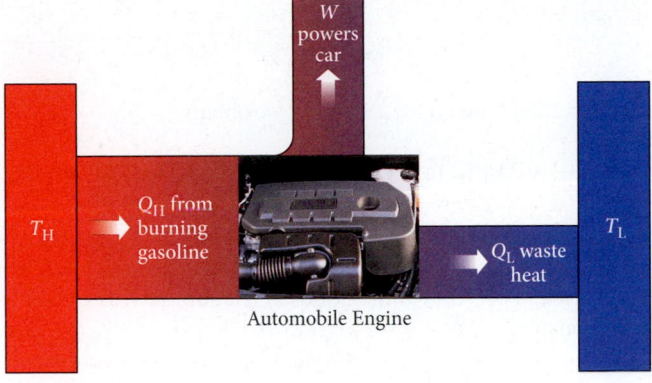

FIGURE 20.11   A gasoline-powered automobile engine operating as a heat engine to power a car traveling at constant speed v on a horizontal surface.

RESEARCH   The car is traveling at speed v. The rate, χ, at which the car burns gasoline, can be expressed in terms of the volume of gasoline burned per unit distance. We can calculate the volume of gasoline burned per unit time, V_t, by multiplying the speed of the car by the rate, χ, at which the car burns gasoline:

$$V_t = \chi v. \tag{i}$$

The energy per unit time supplied to the engine by consuming fuel is the power, P, given by the volume of gasoline used per unit time multiplied by the energy content of gasoline, E_g:

$$P = V_t E_g. \tag{ii}$$

The efficiency of the engine, ϵ, is given by equation 20.1:

$$\epsilon = \frac{W}{Q_H},$$

where W is the useful work and Q_H is the thermal energy supplied to the engine. If we divide both W and Q_H by a time interval, t, we get

$$\epsilon = \frac{W/t}{Q_H/t} = \frac{P_{\text{delivered}}}{P}, \tag{iii}$$

where $P_{\text{delivered}}$ is the power delivered by the engine of the car.

SIMPLIFY   We can combine equations (i) through (iii) to obtain

$$P_{\text{delivered}} = \epsilon P = \epsilon V_t E_g = \epsilon \chi v E_g.$$

CALCULATE   The power delivered is

$$P_{\text{delivered}} = \epsilon \chi v E_g = (0.200)\left(\frac{6.92 \text{ L}}{100 \cdot 10^3 \text{ m}}\right)(26.8 \text{ m/s})\left(34.8 \cdot 10^6 \text{ J/L}\right) = 12{,}907.7 \text{ W}.$$

ROUND   We report our result to three significant figures:

$$P_{\text{delivered}} = 12{,}900 \text{ W} = 12.9 \text{ kW} = 17.3 \text{ hp}.$$

DOUBLE-CHECK   To double-check our result, we calculate the power required to keep the car moving at 60.0 mph against air resistance. The power required, P_{air}, is equal to the product of the force of air resistance, F_{drag}, and the speed of the car:

$$P_{\text{air}} = F_{\text{drag}} v. \tag{iv}$$

The drag force created by air resistance is given by

$$F_{\text{drag}} = Kv^2. \tag{v}$$

The constant K has been found empirically to be

$$K = \tfrac{1}{2} c_d A \rho, \tag{vi}$$

– *Continued*

where c_d is the drag coefficient of the car, A is its front cross-sectional area, and ρ is the density of air. Combining equations (iv) through (vi) gives

$$P_{air} = F_{drag}v = \left(Kv^2\right)v = \tfrac{1}{2}c_d A\rho v^3.$$

Using $c_d = 0.33$, $A = 2.2 \text{ m}^2$, and $\rho = 1.29 \text{ kg/m}^3$, we obtain

$$P_{air} = \tfrac{1}{2}(0.33)\left(2.2 \text{ m}^2\right)\left(1.29 \text{ kg/m}^3\right)\left(26.8 \text{ m/s}\right)^3 = 9014 \text{ W} = 9.0 \text{ kW}.$$

This result for the power required to overcome air resistance ($P_{air} = 9.0 \text{ kW} = 12 \text{ hp}$) is about 70% of the calculated value for the power delivered ($P_{delivered} = 12.9 \text{ kW} = 17.3 \text{ hp}$). The remaining power is used to overcome other kinds of friction, such as rolling friction. Thus, our answer seems reasonable.

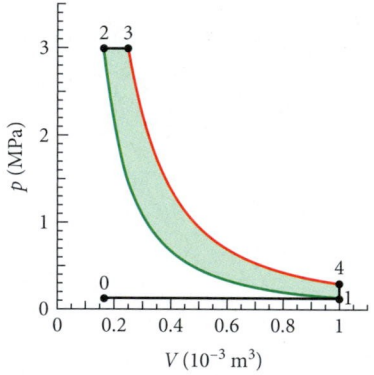

FIGURE 20.12 A pV-diagram for the Diesel cycle.

Diesel Cycle

Diesel engines and gasoline engines have somewhat different designs. Diesel engines do not compress a fuel-air mixture, but rather air only (path from point 1 to point 2 in Figure 20.12, green curve). The fuel is introduced (between point 2 and point 3) only after the air has been compressed. The thermal energy from the compression ignites the mixture (thus, no spark plug is required). This combustion process pushes the piston out at constant pressure. After combustion, the combustion products push the piston out farther in the same adiabatic manner as in the Otto cycle (path from point 3 to point 4, red curve). The process of heat rejection to the environment at constant volume (between points 4 and 1 in the diagram) also proceeds in the same way as in the Otto cycle, as do the intake stroke (path from 0 to 1) and exhaust stroke (path from 1 to 0). Diesel engines have a higher compression ratio and thus a higher efficiency than gasoline-powered four-stroke engines, although they have a slightly different thermodynamic cycle.

The efficiency of an ideal diesel engine is given by

$$\epsilon = 1 - r^{1-\gamma}\frac{\alpha^\gamma - 1}{\gamma(\alpha - 1)}, \tag{20.11}$$

where the compression ratio is again $r = V_1/V_2$ (just as for the Otto cycle), γ is again the ratio of the specific heats at constant pressure and constant volume, $\gamma = C_p/C_V$ (introduced in Chapter 19), and $\alpha = V_3/V_2$ is called the *cut-off ratio*, the ratio between the final and initial volumes of the combustion phase. The derivation of this formula proceeds similarly to Derivations 20.1 and 20.2 for the Carnot and Otto cycles but is omitted here.

Hybrid Cars

Hybrid cars combine a gasoline engine and an electric motor to achieve higher efficiency than a gasoline engine alone can achieve. The improvement in efficiency results from using a smaller gasoline engine than would normally be necessary and an electric motor, run off a battery charged by the gasoline engine, to supplement the gasoline engine when higher power is required. For example, the Ford Escape Hybrid (Figure 20.13) has a 114-kW (153-hp) gasoline engine coupled with a 70-kW (94-hp) electric motor, and the Ford Escape has a 128-kW (171-hp) gasoline engine. In addition, the gasoline engine used in the Ford Escape Hybrid, as well as in the hybrid vehicles from Toyota and Honda, applies a thermodynamic cycle different from the Otto cycle. This *Atkinson cycle* includes variable valve timing to increase the expansion phase of the process, allowing more useful work to be extracted from the energy consumed by the engine. An Atkinson-cycle engine produces less power per displacement than a comparable Otto-cycle engine but has higher efficiency.

In stop-and-go driving, hybrid cars have the advantage of using regenerative braking rather than normal brakes. Conventional brakes use friction to stop the car, which converts the kinetic energy of the car into wasted energy in the form of thermal energy (and wear on the brake pads). Regenerative brakes couple the wheels to an electric generator (which can be

Concept Check 20.4

In which part of the Diesel cycle in Figure 20.12 is heat added?

a) on the path from point 0 to point 1

b) on the path from point 1 to point 2

c) on the path from point 2 to point 3

d) on the path from point 3 to point 4

e) on the path from point 4 to point 1

Concept Check 20.5

During which part(s) of the Diesel cycle in Figure 20.12 is mechanical work done by the engine?

a) on the path from point 0 to point 1 and on the path from point 1 to point 0

b) on the path from point 1 to point 2

c) on the path from point 2 to point 3 and on the path from point 3 to point 4

d) on the path from point 4 to point 1

the electric motor driving the car), which converts the kinetic energy of the car into electrical energy. This energy is stored in the battery of the hybrid car to be used later. In addition, the gasoline engine of a hybrid car can be shut off when the car is stopped, so the car uses no energy while waiting at a stoplight.

In highway driving, hybrid cars take advantage of the fact that the gasoline engine can be run at a constant speed, corresponding to its most efficient operating speed, rather than having to run at a speed dictated by the speed of the car. Operation of the gasoline engine at a constant speed while the speed of the car changes is accomplished by a continuously variable transmission. The transmission couples both the gasoline engine and the electric motor to the drive wheels of the car, allowing the gasoline engine to run at its most efficient speed while taking power from the electric motor as needed.

FIGURE 20.13 The gasoline engine and electric motor of a hybrid car.

Efficiency and the Energy Crisis

The efficiencies of engines and refrigerators are not just theoretical constructs but have very important economic consequences, which are extremely relevant to solving the energy crisis. The efficiencies and performance coefficients calculated in this chapter with the aid of thermodynamic principles are theoretical upper limits. Real-world complications always reduce the actual efficiencies of engines and performance coefficients of refrigerators. But engineering research can overcome these real-world complications and provide better performance of real devices, approaching the ideal limits.

One impressive example of performance improvement has occurred with refrigerators sold in the United States, according to data compiled by Steve Chu (see Figure 20.14). Between 1975 and 2003, the average size of a refrigerator in U.S. kitchens increased by about 20%, but through a combination of tougher energy standards and research and development on refrigerator design and technology, the average power consumption fell by two-thirds, a total of 1200 kWh/yr, from 1800 kWh/yr in 1975 to 600 kWh/yr in 2003.

Since about 150 million new refrigerators and freezers are purchased each year in the United States, and each one saves approximately 1200 kWh/yr, a total energy savings of 180 billion kWh ($6.5 \cdot 10^{17}$ J = 0.65 EJ) is realized each year. In 2009, this savings was approximately twice as much as the combined energy produced through the use of wind power (30 billion kWh/yr), solar energy (2 billion kWh/yr), geothermal (14 billion kWh/yr), and biomass (54 billion kWh/yr).

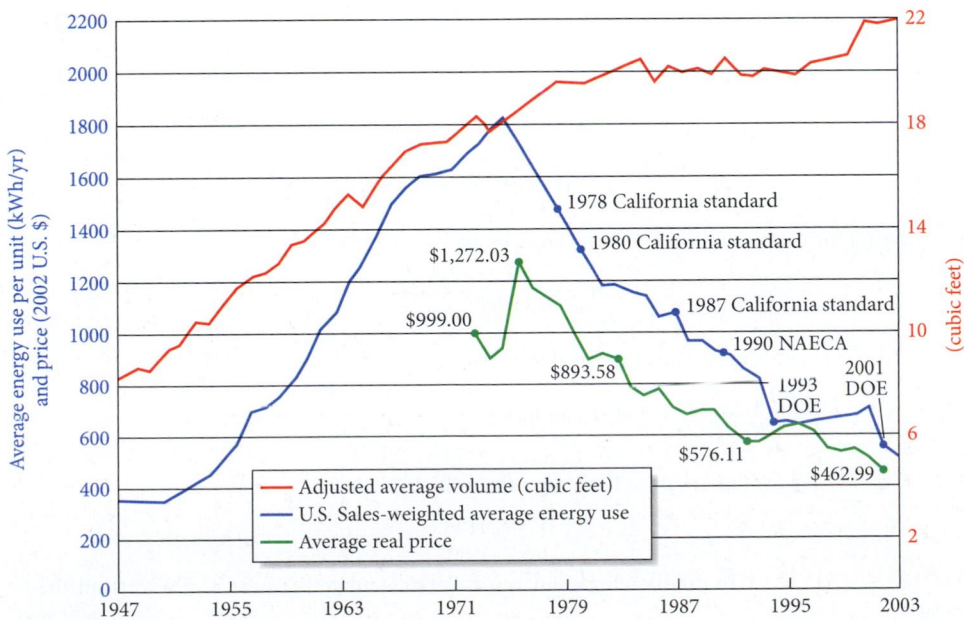

FIGURE 20.14 Average U.S. refrigerator volume (red line), price (green line), and energy use (blue line) from 1947 to 2003.

Energy consumption is directly proportional to energy cost, so the average U.S. household saves a total of almost $200 each year in electricity costs from using more efficient refrigerators relative to the 1975 standard. In addition, even though refrigerators have gotten much better, their price has fallen by more than half, also indicated in Figure 20.14. So a more energy-efficient refrigerator is not more expensive. And it keeps our food and favorite beverages just as cold as the 1975 model used to!

SOLVED PROBLEM 20.3   Freezing Water in a Refrigerator

Suppose we have 250 g of water at 0.00 °C. We want to freeze this water by putting it in a refrigerator operating in a room with a temperature of 22.0 °C. The temperature inside the refrigerator is maintained at –5.00 °C.

PROBLEM

What is the minimum amount of electrical energy that must be supplied to the refrigerator to freeze the water?

SOLUTION

THINK The amount of heat that must be removed depends on the latent heat of fusion and the given mass of water. The most efficient refrigerator possible is a Carnot refrigerator, so we will use the theoretical maximum coefficient of performance of such a refrigerator. Knowing the amount of heat to be removed from the low-temperature reservoir and the coefficient of performance, we can calculate the minimum energy that must be supplied.

SKETCH A heat flow diagram for the refrigerator is shown in Figure 20.15.

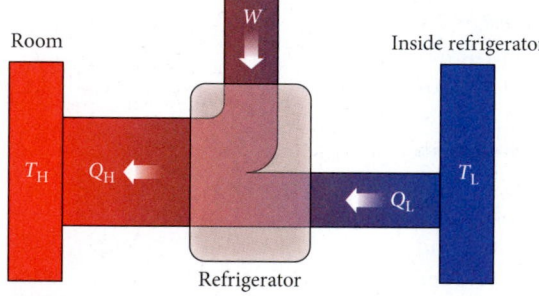

FIGURE 20.15 A heat flow diagram for a refrigerator that takes heat from its interior and exhausts it into the room using a source of electrical power.

RESEARCH The most efficient refrigerator possible is a Carnot refrigerator. The maximum coefficient of performance of a Carnot refrigerator is given by equations 20.2 and 20.7:

$$K_{max} = \frac{Q_L}{W} = \frac{T_L}{T_H - T_L}, \tag{i}$$

where Q_L is the heat removed from inside the refrigerator, W is the work (in terms of electrical energy) that must be supplied, T_L is the temperature inside the refrigerator, and T_H is the temperature of the room. The amount of heat that must be removed to freeze a mass m of water is given by (see Chapter 18)

$$Q_L = mL_{fusion}, \tag{ii}$$

where L_{fusion} = 334 kJ/kg is the latent heat of fusion of water (which can be found in Table 18.2).

SIMPLIFY We can solve equation (i) for the energy that must be supplied to the refrigerator:

$$W = Q_L \frac{T_H - T_L}{T_L}.$$

Substituting the expression for the heat removed from equation (ii), we get

$$W = \left(mL_{fusion}\right) \frac{T_H - T_L}{T_L}.$$

CALCULATE Putting in the numerical values gives

$$W = \frac{mL_{fusion}}{K} = \left(0.250 \text{ kg}\right)\left(334 \text{ kJ/kg}\right) \frac{295.15 \text{ K} - 268.15 \text{ K}}{268.15 \text{ K}} = 8.41231 \text{ kJ}.$$

ROUND We report our result to three significant figures:

$$W = 8.41 \text{ kJ}.$$

DOUBLE-CHECK To double-check our result, let's calculate the heat removed from the water:

$$Q_L = mL_{fusion} = \left(0.250 \text{ kg}\right)\left(334 \text{ kJ/kg}\right) = 83.5 \text{ kJ}.$$

Using our result for the energy required to freeze the water, we can calculate the coefficient of performance of the refrigerator:

$$K = \frac{Q_L}{W} = \frac{83.5 \text{ kJ}}{8.41 \text{ kJ}} = 9.93.$$

We can compare this result to the maximum coefficient of performance of a Carnot refrigerator:

$$K_{max} = \frac{T_L}{T_H - T_L} = \frac{268.15 \text{ K}}{295.15 \text{ K} - 268.15 \text{ K}} = 9.93.$$

Thus, our result seems reasonable.

We can see from this result that a very sizable amount of work has to be input to freeze a relatively small quantity of water, even in the idealized case of a Carnot refrigerator. The work calculated here represents a lower boundary for the work that needs to be done by our real refrigerators to make ice cubes. Thus, it very important to pay attention to the efficiency of real refrigerators.

20.5 The Second Law of Thermodynamics

We saw in Section 20.1 that the First Law of Thermodynamics is satisfied whether a glass of cold water warms on standing at room temperature or becomes colder. However, the Second Law of Thermodynamics is a general principle that puts constraints on the amount and direction of thermal energy transfer between systems and on the possible efficiencies of heat engines. This principle goes beyond the energy conservation of the First Law of Thermodynamics.

Chapter 18 discussed Joule's famous experiment that demonstrated that mechanical work could be converted completely into heat, as illustrated in Figure 20.16. In contrast, experiments show that it is impossible to build a heat engine that completely converts heat into work. This concept is illustrated in Figure 20.17.

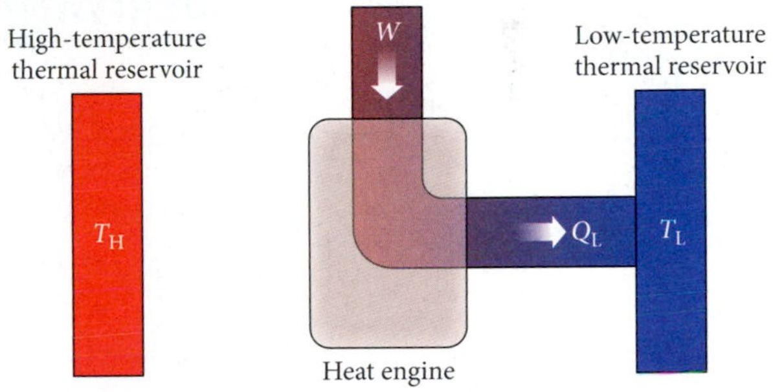

In other words, it is not possible to construct a 100% efficient heat engine. This fact forms the basis of the **Second Law of Thermodynamics:**

FIGURE 20.16 A heat flow diagram showing a system that converts work completely to heat.

It is impossible for a system to undergo a process in which it absorbs heat from a thermal reservoir at a given temperature and converts that heat completely to mechanical work without rejecting heat to a thermal reservoir at a lower temperature.

This formulation is often called the *Kelvin–Planck statement of the Second Law of Thermodynamics.* As an example, consider a book sliding on a table. The book slides to a stop, and the mechanical energy of motion is turned into thermal energy. This thermal energy takes the form of random motion of the molecules of the book, the air, and the table. It is impossible to convert this random motion back into organized motion of the book. It is, however, possible to convert *some* of the random motion related to thermal energy back into mechanical energy. Heat engines do that kind of conversion.

If the Second Law were not true, various impossible scenarios could occur. For example, an electric power plant could operate by taking heat from the surrounding air, and an ocean liner could power itself by taking heat from the seawater. These scenarios do not violate the First Law of Thermodynamics because energy is conserved. The fact that they cannot occur shows that the Second Law contains additional information about how nature works, beyond the principle of conservation of energy. The Second Law limits the ways in which energy can be used.

Another way to state the Second Law of Thermodynamics relates to refrigerators. We know that heat flows spontaneously from a warmer thermal reservoir to a cooler thermal reservoir. Heat

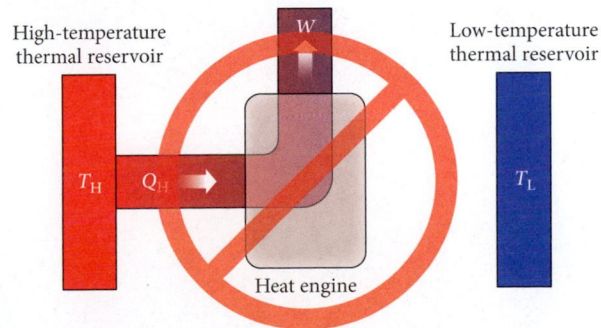

FIGURE 20.17 A heat flow diagram illustrating the impossible process of completely converting heat into useful work without rejecting heat into a low-temperature thermal reservoir.

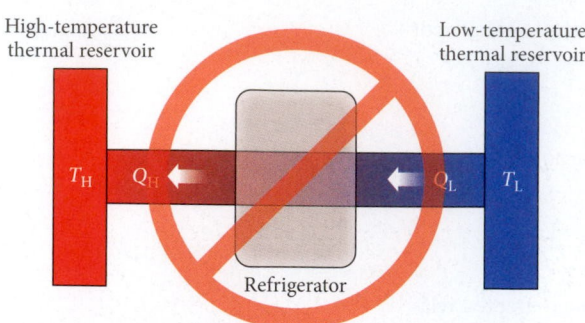

FIGURE 20.18 A heat flow diagram demonstrating the impossible process of moving heat from a low-temperature thermal reservoir to a high-temperature thermal reservoir without using any work.

never *spontaneously* flows from a cooler thermal reservoir to a warmer thermal reservoir. A refrigerator is a heat engine that moves heat from a cooler thermal reservoir to a warmer thermal reservoir; however, energy has to be supplied to the refrigerator for this transfer to take place, as shown in Figure 20.5b. The fact that it is impossible for a refrigerator to transfer thermal energy from a cooler reservoir to a warmer reservoir without using work is illustrated in Figure 20.18. This fact is the basis of another form of the Second Law of Thermodynamics:

> It is impossible for any process to transfer thermal energy from a cooler thermal reservoir to a warmer thermal reservoir without any work having been done to accomplish the transfer.

This equivalent formulation is often called the *Clausius statement of the Second Law of Thermodynamics.*

DERIVATION 20.3 / Carnot's Theorem

The following explains how to prove Carnot's Theorem, stated in Section 20.3. Suppose there are two thermal reservoirs, a heat engine working between these two reservoirs, and a Carnot engine operating in reverse as a refrigerator between the reservoirs, as shown in Figure 20.19.

Let's assume that the Carnot refrigerator operates with the theoretical efficiency given by

$$\epsilon_2 = \frac{W}{Q_{H2}}, \tag{i}$$

where W is the work required to put heat Q_{H2} into the high-temperature reservoir. We also assume that this required work is provided by a heat engine operating between the same two reservoirs. The efficiency of the heat engine can be expressed as

$$\epsilon_1 = \frac{W}{Q_{H1}}, \tag{ii}$$

where Q_{H1} is the heat removed from the high-temperature reservoir by the heat engine. Since equations (i) and (ii) both contain the same work, W, we can solve each of them for the work and then set the expressions equal to each other:

$$\epsilon_1 Q_{H1} = \epsilon_2 Q_{H2}.$$

FIGURE 20.19 A heat engine between two thermal reservoirs produces work. A Carnot engine operating in reverse as a refrigerator between two thermal reservoirs is being driven by the work produced by the heat engine.

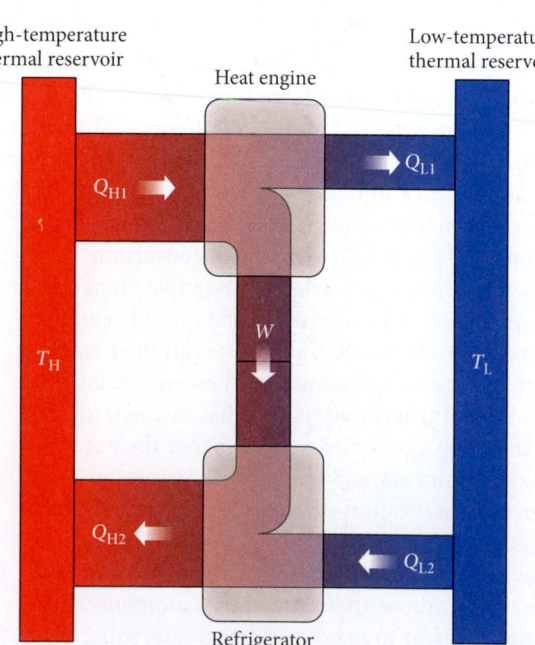

Thus, the ratio of the two efficiencies is

$$\frac{\epsilon_1}{\epsilon_2} = \frac{Q_{H2}}{Q_{H1}}.$$

If the efficiency of the heat engine is equal to the efficiency of the Carnot refrigerator ($\epsilon_1 = \epsilon_2$), then we have $Q_{H1} = Q_{H2}$, which means that the heat removed from the high-temperature reservoir is equal to the heat added to the high-temperature reservoir. If the efficiency of the heat engine is higher than the efficiency of the Carnot refrigerator ($\epsilon_1 > \epsilon_2$), then we have $Q_{H2} > Q_{H1}$, which means that more heat is gained than lost by the high-temperature reservoir. Applying the First Law of Thermodynamics to the overall system tells us that the heat engine working together with the Carnot refrigerator is then able to transfer thermal energy from the low-temperature reservoir to the high-temperature reservoir without any work being supplied. These two devices acting together would then violate the Clausius statement of the Second Law of Thermodynamics. Thus, no heat engine can be more efficient than a Carnot engine.

The conversion of mechanical work to thermal energy (such as by friction) and the flow of heat from a warm thermal reservoir to a cool one are irreversible processes. The Second Law of Thermodynamics says that these processes can only be partially reversed, thereby acknowledging their inherent one-way quality.

20.6 Entropy

The Second Law of Thermodynamics as stated in Section 20.5 is somewhat different from other laws presented in previous chapters, such as Newton's laws, because it is phrased in terms of impossibilities. However, the Second Law can be stated in a more direct manner using the concept of **entropy.**

Throughout the last three chapters, we have discussed the notion of thermal equilibrium. If two objects at different temperatures are brought into thermal contact, both of their temperatures will asymptotically approach a common equilibrium temperature. What drives this system to thermal equilibrium is entropy, and the state of thermal equilibrium is the state of maximum entropy.

The direction of thermal energy transfer is not determined by energy conservation but by the change in entropy of a system. The change in entropy of a system, ΔS, during a process that takes the system from an initial state to a final state is defined as

$$\Delta S = \int_i^f \frac{dQ}{T}, \tag{20.12}$$

where Q is the heat and T is the temperature in kelvins. The SI units for the change in entropy are joules per kelvin (J/K). It must be noted that equation 20.12 applies only to reversible processes; that is, the integration can only be carried out over a path representing a reversible process.

Note that entropy is defined in terms of its change from an initial to a final configuration. Entropy change is the physically meaningful quantity, not the absolute value of the entropy at any point. Another physical quantity for which only the change is important is the potential energy. The absolute value of the potential energy is always defined only relative to some arbitrary additive constant, but the change in potential energy between initial and final states is a precisely measurable physical quantity that gives rise to force(s). In a manner similar to the way connections between forces and potential energy changes were established in Chapter 6, this section shows how to calculate entropy changes for given temperature changes and amounts of heat and work in different systems. At thermal equilibrium, the entropy has an extremum (a maximum). At mechanically stable equilibrium, the net force is zero, and therefore the potential energy has an extremum (a minimum, in this case).

In an irreversible process in an isolated system, the entropy, S, of the system never decreases; it always increases or stays constant. In an isolated system, energy is always

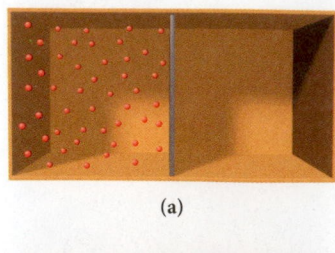

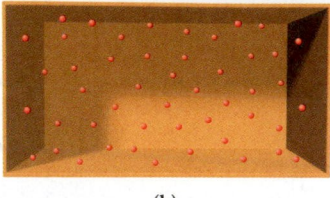

FIGURE 20.20　(a) A gas confined to half the volume of a box. (b) The barrier separating the two halves is removed, and the gas expands to fill the entire volume.

conserved, but entropy is not conserved. Thus, the change in entropy defines a direction for time; that is, time moves forward if the entropy of an isolated system is increasing.

The definition of entropy given by equation 20.12 rests on the macroscopic properties of a system, such as heat and temperature. Another definition of entropy, based on statistical descriptions of how the atoms and molecules of a system are arranged, is presented in the next section.

Since the integral in equation 20.12 can only be evaluated for a reversible process, how can we calculate the change in entropy for an irreversible process? The answer lies in the fact that entropy is a thermodynamic state variable, just like temperature, pressure, and volume. This means that we can calculate the entropy difference between a known initial state and a known final state even for an irreversible process if there is a reversible process (for which the integral in equation 20.12 can be evaluated!) that takes the system from the same initial to the same final state. Perhaps this is the subtlest point about thermodynamics conveyed in this entire chapter.

In order to illustrate this general method of computing the change in entropy for an irreversible process, let's return to a situation described in Chapter 18, the free expansion of a gas. Figure 20.20a shows a gas confined to the left half of a box. In Figure 20.20b, the barrier between the two halves has been removed, and the gas has expanded to fill the entire volume of the box. Clearly, once the gas has expanded to fill the entire volume of the box, the system will never spontaneously return to the state where all the gas molecules are located in the left half of the box. The state variables of the system before the barrier is removed are the initial temperature, T_i, the initial volume, V_i, and the initial entropy, S_i. After the barrier has been removed and the gas is again in equilibrium, the state of the system can be described in terms of the final temperature, T_f, the final volume, V_f, and the final entropy, S_f.

We cannot calculate the change in entropy of this system using equation 20.12 because the gas is not in equilibrium during the expansion phase. However, the change in the properties of the system depends only on the initial and final states, not on how the system got from one to the other. Therefore, we can choose a process that the system could have undergone for which we can evaluate the integral in equation 20.12.

In the free expansion of an ideal gas, the temperature remains constant; thus, it seems reasonable to use the isothermal expansion of an ideal gas. We can then evaluate the integral in equation 20.12 to calculate the change in entropy of the system undergoing an isothermal process:

$$\Delta S = \int_i^f \frac{dQ}{T} = \frac{1}{T} \int_i^f dQ = \frac{Q}{T}. \tag{20.13}$$

(We can move the factor $1/T$ outside the integral, because we are dealing with an isothermal process, for which the temperature is constant by definition.) As we saw in Chapter 19, the work done by an ideal gas in expanding from V_i to V_f at a constant temperature T is given by

$$W = nRT \ln\left(\frac{V_f}{V_i}\right).$$

For an isothermal process, the internal energy of the gas does not change; so $\Delta E_{int} = 0$. Thus, as shown in Chapter 18, we can use the First Law of Thermodynamics to write

$$\Delta E_{int} = W - Q = 0.$$

Consequently, for the isothermal process, the heat added to the system is

$$Q = W = nRT \ln\left(\frac{V_f}{V_i}\right).$$

The resulting entropy change for the isothermal process is then

$$\Delta S = \frac{Q}{T} = \frac{nRT \ln\left(\dfrac{V_f}{V_i}\right)}{T} = nR \ln\left(\frac{V_f}{V_i}\right). \tag{20.14}$$

The entropy change for the irreversible free expansion of a gas must be equal to the entropy change for the isothermal process because both processes have the same initial and final states and thus must have the same change in entropy.

For the irreversible free expansion of a gas, $V_f > V_i$, and so $\ln (V_f / V_i) > 0$. Thus, $\Delta S > 0$ because n and R are positive numbers. In fact, the change in entropy of any irreversible process is always positive.

Therefore, the Second Law of Thermodynamics can be stated in a third way.

> The entropy of an isolated system can never decrease.

EXAMPLE 20.5   Entropy Change for the Freezing of Water

Suppose we have 1.50 kg of water at a temperature of 0 °C. We put the water in a freezer, and enough heat is removed from the water to freeze it completely to ice at a temperature of 0 °C.

PROBLEM
How much does the entropy of the water-ice system change during the freezing process?

SOLUTION
The melting of ice is an isothermal process, so we can use equation 20.13 for the change in entropy:

$$\Delta S = \frac{Q}{T},$$

where Q is the heat that must be removed to change the water to ice at $T = 273.15$ K, the freezing point of water. The heat that must be removed to freeze the water is determined by the latent heat of fusion of water (ice), defined in Chapter 18. The heat that must be removed is

$$Q = mL_{fusion} = (1.50 \text{ kg})(334 \text{ kJ/kg}) = 501 \text{ kJ}.$$

Thus, the change in entropy of the water-ice system is

$$\Delta S = \frac{-501 \text{ kJ}}{273.15 \text{ K}} = -1830 \text{ J/K}.$$

Note that the entropy of the water-ice system of Example 20.5 decreased. How can the entropy of this system decrease? The Second Law of Thermodynamics states that the entropy of an isolated system can never decrease. However, the water-ice system is not an isolated system. The freezer used energy to remove heat from the water to freeze it and exhausted the heat into the local environment. Thus, the entropy of that environment increased more than the entropy of the water-ice system decreased. This is a very important distinction.

A similar analysis can be applied to the origins of complex life forms, which have much lower entropy than their surroundings. The development of life forms with low entropy is accompanied by an increase in the overall entropy of the Earth. In order for a living subsystem of Earth to reduce its own entropy at the expense of its environment, it needs a source of energy. This source of energy can be chemical bonds or other types of potential energy, which in the end arises from the energy provided to Earth by solar radiation.

Evolution toward more and more complex life forms is not in contradiction with the Second Law of Thermodynamics, because the evolving life forms do not form an isolated system. A contradiction between biological evolution and the Second Law of Thermodynamics is sometimes falsely claimed by opponents of evolution in the evolution/creationism debate. From a thermodynamics standpoint, this argument has to be rejected unequivocally.

EXAMPLE 20.6 | **Entropy Change for the Warming of Water**

Suppose we start with 2.00 kg of water at a temperature of 20.0 °C and warm the water until it reaches a temperature of 80.0 °C.

PROBLEM
What is the change in entropy of the water?

SOLUTION
We start with equation 20.12, relating the change in entropy to the integration of the differential flow of heat, dQ, with respect to temperature:

$$\Delta S = \int_i^f \frac{dQ}{T}. \tag{i}$$

The heat, Q, required to raise the temperature of a mass, m, of water is given by

$$Q = cm\Delta T, \tag{ii}$$

where $c = 4.19$ kJ/(kg K) is the specific heat of water. We can rewrite equation (ii) in terms of the differential change in heat, dQ, and the differential change in temperature, dT:

$$dQ = cm\, dT.$$

Then we can rewrite equation (i) as

$$\Delta S = \int_i^f \frac{dQ}{T} = \int_{T_i}^{T_f} \frac{cm\, dT}{T} = cm \int_{T_i}^{T_f} \frac{dT}{T} = cm \ln \frac{T_f}{T_i}.$$

With $T_i = 293.15$ K and $T_f = 353.15$ K, the change in entropy is

$$\Delta S = \left[4.19 \text{ kJ/}\left(\text{kg K}\right) \right]\left(2.00 \text{ kg}\right) \ln \frac{353.15 \text{ K}}{293.15 \text{ K}} = 1.56 \cdot 10^3 \text{ J/K}.$$

There is another important point regarding the macroscopic definition of entropy and the calculation of entropy with the aid of equation 20.12: The Second Law of Thermodynamics also implies that for all cyclical processes (such as the Carnot, Otto, and Diesel cycles)—that is, all processes for which the initial state is the same as the final state—the total entropy change over the entire cycle has to be greater than or equal to zero: $\Delta S \geq 0$. The entropy change is equal to zero for reversible processes, and greater than zero for irreversible ones. Note that the cyclic processes we have discussed here are all idealized reversible ones, but real processes are never quite completely reversible.

20.7 Microscopic Interpretation of Entropy

In Chapter 19, we saw that the internal energy of an ideal gas could be calculated by summing up the energies of the constituent particles of the gas. We can also determine the entropy of an ideal gas by studying the constituent particles. It turns out that this microscopic definition of entropy agrees with the macroscopic definition.

The ideas of order and disorder are intuitive. For example, a coffee cup is an ordered system. Smashing the cup by dropping it on the floor creates a system that is less ordered, or more disordered, than the original system. The disorder of a system can be described quantitatively using the concept of **microscopic states.** Another term for a microscopic state is a *degree of freedom*.

Suppose we toss n coins in the air, and half of them land heads up and half of them land tails up. The statement "half the coins are heads and half the coins are tails" is a description of the *macroscopic* state of the system of n coins. Each coin can have one of two *microscopic* states: heads or tails. Stating that half the coins are heads and half the coins are tails does not specify anything about the microscopic state of each coin, because there are many different possible ways to arrange the microscopic states of the coins without changing the macroscopic state.

However, if all the coins are heads or all the coins are tails, the microscopic state of each coin is known. The macroscopic state consisting of half heads and half tails is a *disordered* system because very little is known about the microscopic state of each coin. The macroscopic state with all heads or the macroscopic state with all tails is an *ordered* system because the microscopic state of each coin is known.

To quantify this concept, imagine tossing four coins in the air. There is only one way to get four heads with such a toss, four ways to get three heads and one tail, six ways to get two heads and two tails, four ways to get one head and three tails, and only one way to get four tails. Thus, there are five possible macroscopic states and sixteen possible microscopic states (see Figure 20.21, where heads are represented as red circles, and tails as blue circles).

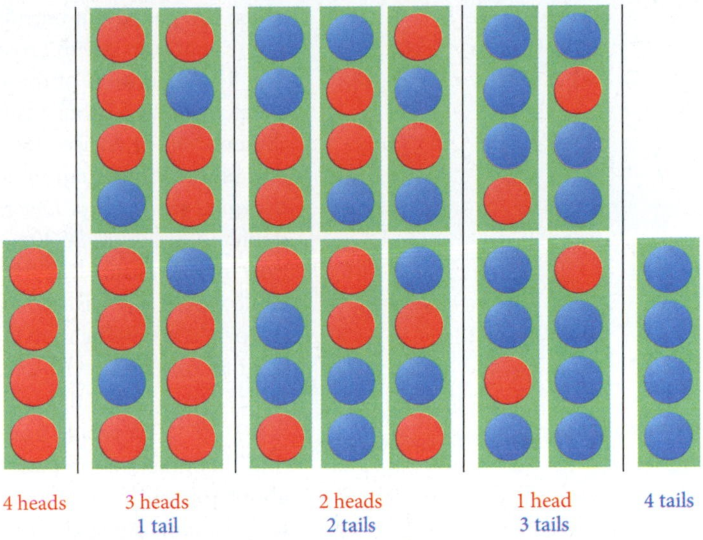

Now suppose we toss fifty coins in the air instead of four coins. There are $2^{50} = 1.13 \cdot 10^{15}$ possible microstates of this system of fifty tossed coins. The most probable macroscopic state consists of half heads and half tails. There are $1.26 \cdot 10^{14}$ possible microstates with half heads and half tails.

4 heads   3 heads   2 heads   1 head   4 tails
           1 tail   2 tails   3 tails

FIGURE 20.21 The sixteen possible microscopic states for four tossed coins, leading to five possible macroscopic states.

The probability that half the coins will be heads and half will be tails is 11.2%, while the probability of having all fifty coins land heads up is 1 in $1.13 \cdot 10^{15}$.

Let's apply these concepts to a real system of gas molecules: a mole of gas, or Avogadro's number of molecules, at pressure p, volume V, and temperature T. These three quantities describe the macroscopic state of the gas. The microscopic description of the system needs to specify the momentum and position of each molecule of the gas. Each molecule has three components of its momentum and three components of its position. Thus, at any given time, the gas can be in an extremely large number of microscopic states, depending on the positions and velocities of each of its $6.02 \cdot 10^{23}$ molecules. If the gas undergoes free expansion, the number of possible microscopic states increases and the system becomes more disordered. Because the entropy of a gas undergoing free expansion increases, the increase in disorder is related to the increase of entropy. This idea can be generalized as follows:

The most probable macroscopic state of a system is the state with the largest number of microscopic states, which is also the macroscopic state with the greatest disorder.

Let w be the number of possible microscopic states for a given macroscopic state. It can be shown that the entropy of the macroscopic state is given by

$$S = k_B \ln w, \tag{20.15}$$

where k_B is the Boltzmann constant. This equation was first written down by the Austrian physicist Ludwig Boltzmann and is his most significant accomplishment (it is chiseled into his tombstone). You can see from equation 20.15 that increasing the number of possible microscopic states increases the entropy.

The important aspect of a thermodynamic process is not the absolute entropy, but the change in entropy between an initial state and a final state. Taking equation 20.15 as the definition of entropy, the smallest number of microstates is one and the smallest entropy that can exist is then zero. According to this definition, entropy can never be negative. In practice, determining the number of possible microscopic states is difficult except for special systems. However, the change in the number of possible microscopic states can often be determined, thus allowing the change in entropy of the system to be found.

Consider a system that initially has w_i microstates and then undergoes a thermodynamic process to a macroscopic state with w_f microstates. The change in entropy is

$$\Delta S = S_f - S_i = k_B \ln w_f - k_B \ln w_i = k_B \ln \frac{w_f}{w_i}. \tag{20.16}$$

Thus, the change in entropy between two macroscopic states depends on the ratio of the number of possible microstates.

The definition of the entropy of a system in terms of the number of possible microstates leads to further insight into the Second Law of Thermodynamics, which states that the entropy of an isolated system can never decrease. This statement of the Second Law, combined with equation 20.15, means that *an isolated system can never undergo a thermodynamic process that lowers the number of possible microstates*. For example, if the process depicted in Figure 20.20 were to occur in reverse—that is, the gas underwent free contraction into a volume half its original size—the number of possible microstates for each molecule would decrease by a factor of 2. The probability of finding one gas molecule in half of the original volume then is $\frac{1}{2}$, and the probability of finding all the gas molecules in half of the original volume is $\left(\frac{1}{2}\right)^N$, where N is the number of molecules. If there are 100 gas molecules in the system, then the probability that all 100 molecules end up in half the original volume is $7.9 \cdot 10^{-31}$. We would have to check the system approximately $1/(7.9 \cdot 10^{-31}) \approx 10^{30}$ times to find the molecules in half the volume just once. Checking once per second, this would take about 10^{13} billion years, whereas the age of the universe is only 13.7 billion years. If the system contains Avogadro's number of gas molecules, then the probability that the molecules will all be in half of the volume is even smaller. Thus, although the probability that this process will happen is not zero, it is so small that we can treat it as zero. We can thus conclude that the Second Law of Thermodynamics, even if expressed in terms of probabilities, is never violated in any practical situation.

EXAMPLE 20.7 | Entropy Increase during Free Expansion of a Gas

Let's consider the free expansion of a gas like that shown in Figure 20.20. Initially 0.500 mole of nitrogen gas is confined to a volume of 0.500 m³. When the barrier is removed, the gas expands to fill the new volume of 1.00 m³.

PROBLEM
What is the change in entropy of the gas?

SOLUTION
We can use equation 20.14 to calculate the change in entropy of the system, assuming we can treat the system as an isothermal expansion of an ideal gas:

$$\Delta S = nR \ln\left(\frac{V_f}{V_i}\right) = nR \ln\left(\frac{1.00 \text{ m}^3}{0.500 \text{ m}^3}\right) = nR \ln 2 \qquad \text{(i)}$$

$$= (0.500 \text{ mole})\left[8.31 \text{ J/(mol K)}\right](\ln 2)$$

$$= 2.88 \text{ J/K}.$$

Another approach is to examine the number of microstates of the system before and after the expansion to calculate the change in entropy. In this system, the number of gas molecules is

$$N = nN_A,$$

where N_A is Avogadro's number. Before the expansion, there were w_i microstates for the gas molecules in the left half of the container. After the expansion, any of the molecules could be in the left half or the right half of the container. Therefore, the number of microstates after the expansion is

$$w_f = 2^N w_i.$$

Using equation 20.16 and remembering that $nR = Nk_B$, we can express the change in entropy of the system as

$$\Delta S = k_B \ln \frac{w_f}{w_i} = k_B \ln \frac{2^N w_i}{w_i} = Nk_B \ln 2 = nR \ln 2.$$

Thus, we get the same result for the change in entropy of a freely expanding gas by looking at the microscopic properties of the system as by using the macroscopic properties of the system in equation (i).

Concept Check 20.6

All reversible thermodynamic processes *always* proceed at

a) constant pressure.

b) constant temperature.

c) constant entropy.

d) constant volume.

e) none of the above.

Entropy Death

The universe is the ultimate isolated system. The vast majority of thermodynamic processes in the universe are irreversible, and thus the entropy of the universe as a whole is continuously increasing and asymptotically approaching its maximum. Thus, if the universe exists long enough, all energy will be evenly distributed throughout its volume. In addition, if the universe keeps expanding forever, then gravity—the only long-range force of importance—will not be able to pull objects together any more. This latter condition is the main difference from the early universe in the first moments after the Big Bang, when matter and energy were also distributed very evenly, as revealed by the analysis of the cosmic microwave background radiation (touched on in Chapter 17 and to be discussed in more depth in Chapter 39). But in the early universe, gravity was very strong, as a result of the concentration of matter in a very small space, and was able to develop minute fluctuations and contract matter into stars and galaxies. Thus, even though matter and energy were evenly distributed in the very early universe, the entropy was not near its maximum, and the entire universe was very far from thermal equilibrium.

In the long-term future of the universe, all stars, which currently represent sources of energy for other objects such as our planet, will eventually be extinct. Then, life will become impossible, because life needs an energy source to be able to lower entropy locally. As the universe asymptotically approaches its state of maximum entropy, every subsystem of the universe will reach thermodynamic equilibrium.

Sometimes this long-term fate of the universe is referred to as *heat death*. However, the temperature of the universe at that time will not be high, as this name may suggest, but very close to absolute zero and very nearly the same everywhere.

This is not something we have to worry about soon, because some estimates place the entropy death of the universe at about 10^{100} years into the future (with an uncertainty of many orders of magnitude). Obviously, the long-term future of the universe is very interesting and is currently an area of intense research. New discoveries about dark matter and dark energy may change our picture of the long-term future of the universe. But as it stands now, the universe will not go out with a bang, but with a whimper.

WHAT WE HAVE LEARNED | EXAM STUDY GUIDE

- In a *reversible* process, a system is always close to being in thermodynamic equilibrium. Making a small change in the state of the system can reverse any change in the thermodynamic variables of the system.

- An *irreversible* process involves heat flow with a finite temperature difference, free expansion of a gas, or conversion of mechanical work to thermal energy.

- A heat engine is a device that turns thermal energy into useful work.

- The efficiency, ϵ, of a heat engine is defined as $\epsilon = W/Q_H$, where W is the useful work extracted from the engine and Q_H is the energy provided to the engine in the form of heat.

- A refrigerator is a heat engine operating in reverse.

- The coefficient of performance, K, of a refrigerator is defined as $K = Q_L/W$, where Q_L is the heat extracted from the cooler thermal reservoir and W is the work required to extract that heat.

- An ideal heat engine is one in which all the processes involved are reversible.

- A Carnot engine uses the Carnot cycle, an ideal thermodynamic process consisting of two isothermal processes and two adiabatic processes. The Carnot engine is the most efficient engine that can operate between two thermal reservoirs.

- The efficiency of a Carnot engine is given by $\epsilon = (T_H - T_L)/T_H$, where T_H is the temperature of the warmer thermal reservoir and T_L is the temperature of the cooler thermal reservoir.

- The coefficient of performance of a Carnot refrigerator is $K_{max} = T_L/(T_H - T_L)$, where T_H is the temperature of the warmer thermal reservoir and T_L is the temperature of the cooler thermal reservoir.

- The Otto cycle describes the operation of internal combustion engines. It consists of two adiabatic processes and two constant-volume processes. The efficiency of an engine using the Otto cycle is given by $\epsilon = 1 - r^{1-\gamma}$, where $r = V_1/V_2$ is the compression ratio and $\gamma = C_p/C_V$.

- The Second Law of Thermodynamics can be stated as follows: It is impossible for a system to undergo a process in which it absorbs heat from a thermal reservoir at a given temperature and converts that heat completely to mechanical work without rejecting heat to a thermal reservoir at a lower temperature.

- The Second Law of Thermodynamics can also be stated as follows: It is impossible for any process to transfer thermal energy from a cooler thermal reservoir to a warmer thermal reservoir without any work having been done to accomplish the transfer.

- For reversible processes, the change in entropy of a system is defined as $\Delta S = \int_{i}^{f} \frac{dQ}{T}$, where dQ is the differential heat added to the system, T is the temperature, and the integration is carried out from an initial thermodynamic state to a final thermodynamic state.

- A third way to state the Second Law of Thermodynamics is as follows: The entropy of an isolated system can never decrease.

- The entropy of a macroscopic system can be defined in terms of the number of possible microscopic states, w, of the system: $S = k_B \ln w$, where k_B is the Boltzmann constant.

ANSWERS TO SELF-TEST OPPORTUNITIES

20.1 The efficiency is given by

$$\epsilon = 1 - \frac{T_L}{T_H} = 1 - \frac{325\ \text{K}}{500\ \text{K}} = 0.35 \quad \text{(or 35\%)}.$$

20.2 The ratio of the two efficiencies is

$$\frac{\epsilon_{15}}{\epsilon_{10}} = \frac{1 - 15^{-0.4}}{1 - 4^{-0.4}} = \frac{66.1\%}{42.6\%} = 1.55, \text{ which is a 55\% improvement.}$$

PROBLEM-SOLVING GUIDELINES

1. With problems in thermodynamics, you need to pay close attention to the signs of work (W) and heat (Q). Work done by a system is positive, and work done on a system is negative; heat that is transferred to a system is positive, and heat that is emitted by a system is negative.

2. Some problems involving heat engines deal with power and rate of thermal energy transfer. Power is work per unit time ($P = W/t$) and rate of thermal energy transfer is heat per unit time (Q/t). You can treat these quantities much like work and heat, but remember that they include a time dimension.

3. Entropy may seem a bit like energy, but these are very different concepts. The First Law of Thermodynamics is a conservation law—energy is conserved. The Second Law of Thermodynamics is not a conservation law—entropy is not conserved; it always either stays the same or increases and never decreases in an isolated system. You may need to identify initial and final states to calculate an entropy change, but remember that entropy can change.

4. Remember that the entropy of an isolated system always stays the same or increases (never decreases), but be sure a problem situation is really about an isolated system. Entropy of a system that is not isolated can decrease if offset by a larger entropy increase of the surroundings; don't assume that all entropy changes must be positive.

5. Do not apply the formula $\Delta S = \int_{i}^{f} \frac{dQ}{T}$ for entropy change without being sure you are dealing with a reversible process. If you need to calculate entropy change for an irreversible process, you first need to find an equivalent reversible process that connects the same initial and final states.

6. For problems involving heat engines, it is almost always useful to sketch a flow diagram showing the heat and the work, like the sketches in all the solved problems in this chapter.

MULTIPLE-CHOICE QUESTIONS

20.1 Which of the following processes always results in an increase in the energy of a system?

a) The system loses heat and does work on the surroundings.

b) The system gains heat and does work on the surroundings.

c) The system loses heat and has work done on it by the surroundings.

d) The system gains heat and has work done on it by the surroundings.

e) None of the above will increase the system's energy.

20.2 What is the magnitude of the change in entropy when 6.00 g of steam at 100 °C is condensed to water at 100 °C?

a) 46.6 J/K   c) 36.3 J/K

b) 52.4 J/K   d) 34.2 J/K

20.3 The change in entropy of a system can be calculated because

a) it depends only on the initial and final states.

b) any process is reversible.

c) entropy always increases.

d) None of the above applies.

20.4 An ideal gas undergoes an isothermal expansion. What will happen to its entropy?

a) It will increase.   c) It's impossible to determine.

b) It will decrease.   d) It will remain unchanged.

20.5 Which of the following processes (all constant-temperature expansions) produces the most work?

a) An ideal gas consisting of 1 mole of argon at 20 °C expands from 1 L to 2 L.

b) An ideal gas consisting of 1 mole of argon at 20 °C expands from 2 L to 4 L.

c) An ideal gas consisting of 2 moles of argon at 10 °C expands from 2 L to 4 L.

d) An ideal gas consisting of 1 mole of argon at 40 °C expands from 1 L to 2 L.

e) An ideal gas consisting of 1 mole of argon at 40 °C expands from 2 L to 4 L.

20.6 A heat engine operates with an efficiency of 0.5. What can the temperatures of the high-temperature and low-temperature reservoirs be?

a) $T_H = 600$ K and $T_L = 100$ K

b) $T_H = 600$ K and $T_L = 200$ K

c) $T_H = 500$ K and $T_L = 200$ K

d) $T_H = 500$ K and $T_L = 300$ K

e) $T_H = 600$ K and $T_L = 300$ K

20.7 The number of macrostates that can result from rolling a set of N six-sided dice is the number of different totals that can be obtained by adding the pips on the N faces that end up on top. The number of macrostates is

a) 6^N.   b) $6N$.   c) $6N - 1$.   d) $5N + 1$.

20.8 What capacity must a heat pump with a coefficient of performance of 3 have to heat a home that loses thermal energy at a rate of 12 kW on the coldest day of the year?

a) 3 kW   c) 10 kW   e) 40 kW

b) 4 kW   d) 30 kW

20.9 Which of the following statements about the Carnot cycle is (are) incorrect?

a) The maximum efficiency of a Carnot engine is 100% since the Carnot cycle is an ideal process.

b) The Carnot cycle consists of two isothermal processes and two adiabatic processes.

c) The Carnot cycle consists of two isothermal processes and two isentropic processes (constant entropy).

d) The efficiency of the Carnot cycle depends solely on the temperatures of the two thermal reservoirs.

20.10 Can a heat engine like the one shown in the figure operate?

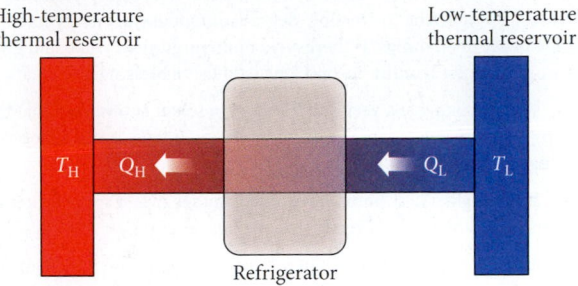

a) yes

b) no

c) need to know the specific cycle used by the engine to answer

d) yes, but only with a monatomic gas

e) yes, but only with a diatomic gas

20.11 An ideal heat engine is one that

a) uses only reversible processes.

b) uses only irreversible processes.

c) has an efficiency of 100%.

d) has an efficiency of 50%.

e) does no work.

20.12 When a coin is tossed, it can land heads up or tails up. You toss a coin 10 times, and it comes up heads every time. What is the probability that the coin will come up heads on the 11th toss?

a) 10%   c) 50%   e) 100%

b) 20%   d) 90%

20.13 In your rectangular living room, two robot vacuum cleaners move around randomly, cleaning the floor. What is the probability that both robot vacuum cleaners are in one half of the room at the same time?

a) 10%   c) 50%   e) 90%

b) 25%   d) 75%

20.14 A refrigerator operates by

a) doing work to move heat from a low-temperature thermal reservoir to a high-temperature thermal reservoir.

b) doing work to move heat from a high-temperature thermal reservoir to a low-temperature thermal reservoir.

c) using thermal energy to produce useful work.

d) moving heat from a low-temperature thermal reservoir to a high-temperature thermal reservoir without doing work.

e) moving heat from a high-temperature thermal reservoir to a low-temperature thermal reservoir without doing work.

CONCEPTUAL QUESTIONS

20.15 One of your friends begins to talk about how unfortunate the Second Law of Thermodynamics is, how sad it is that entropy must always increase, leading to the irreversible degradation of useful energy into heat and the decay of all things. Is there any counterargument you could give that would suggest that the Second Law is in fact a blessing?

20.16 While looking at a very small system, a scientist observes that the entropy of the system spontaneously decreases. If true, is this a Nobel-winning discovery or is it not that significant?

20.17 Why might a heat pump have an advantage over a space heater that converts electrical energy directly into thermal energy?

20.18 Imagine dividing a box into two equal parts, part A on the left and part B on the right. Four identical gas atoms, numbered 1 through 4, are placed in the box. What are most probable and second most probable distributions (for example, 3 atoms in A, 1 atom in B) of gas atoms in the box? Calculate the entropy, S, for these two distributions. Note that the configuration with 3 atoms in A and 1 atom in B and that with 1 atom in A and three atoms in B count as different configurations.

20.19 A key feature of thermodynamics is the fact that the internal energy, E_{int}, of a system and its entropy, S, are *state variables;* that is, they depend only on the thermodynamic state of the system and not on the processes by which it reached that state (unlike, for example, the heat content, Q). This means that the differentials $dE_{int} = T\,dS - p\,dV$ and $dS = T^{-1}dE_{int} + pT^{-1}dV$, where T is temperature (in kelvins), p is pressure, and V is volume, are exact differentials as defined in calculus. What relationships follow from this fact?

20.20 Other state variables useful for characterizing different classes of processes can be defined from E_{int}, S, p, and V. These include the enthalpy, $H \equiv E_{int} + pV$, the Helmholtz free energy, $A \equiv E_{int} - TS$, and the Gibbs free energy, $G \equiv E_{int} + pV - TS$.

a) Write the differential equations for dH, dA, and dG. Use the First Law to simplify.

b) All of these are also exact differentials. What relationships follow from this fact?

20.21 Prove that Boltzmann's microscopic definition of entropy, $S = k_B \ln w$, implies that entropy is an *additive* variable: Given two systems, A and B, in specified thermodynamic states, with entropies S_A and S_B, respectively, show that the corresponding entropy of the combined system is $S_A + S_B$.

20.22 Explain how it is possible for a heat pump like that in Example 20.2 to operate with a power of only 6.28 kW and heat a house that is losing thermal energy at a rate of 21.98 kW.

20.23 The temperature at the cloud tops of Saturn is approximately 150. K. The atmosphere of Saturn produces tremendous winds; wind speeds of 600. km/h have been inferred from spacecraft measurements. Can the wind chill factor on Saturn produce a temperature at (or below) absolute zero? How, or why not?

20.24 Is it a violation of the Second Law of Thermodynamics to capture all the exhaust heat from a steam engine and funnel it back into the system to do work? Why or why not?

20.25 You are given a beaker of water. What can you do to increase its entropy? What can you do to decrease its entropy?

EXERCISES

A blue problem number indicates a worked-out solution is available in the Student Solutions Manual. One • and two •• indicate increasing level of problem difficulty.

Section 20.2

20.26 With each cycle, a 2500.-W engine extracts 2100. J from a thermal reservoir at 90.0 °C and expels 1500. J into a thermal reservoir at 20.0 °C. What is the work done for each cycle? What is the engine's efficiency? How much time does each cycle take?

•**20.27** A refrigerator with a coefficient of performance of 3.80 is used to cool 2.00 L of mineral water from room temperature (25.0 °C) to 4.00 °C. If the refrigerator uses 480. W, how long will it take the water to reach 4.00 °C? Recall that the heat capacity of water is 4.19 kJ/(kg K) and the density of water is 1.00 g/cm^3. Assume that all other contents of the refrigerator are already at 4.00 °C.

•**20.28** The burning of fuel transfers $4.00 \cdot 10^5$ W of power into the engine of a 2000.-kg vehicle. If the engine's efficiency is 25.0%, determine the maximum speed the vehicle can achieve 5.00 s after starting from rest.

•**20.29** A heat engine consists of a heat source that causes a monatomic gas to expand, pushing against a piston, thereby doing work. The gas begins at a pressure of 300. kPa, a volume of 150. cm^3, and room temperature, 20.0 °C. On reaching a volume of 450. cm^3, the piston is locked in place, and the heat source is removed. At this point, the gas cools back to room temperature. Finally, the piston is unlocked and used to isothermally compress the gas back to its initial state.

a) Sketch the cycle on a pV-diagram.

b) Determine the work done on the gas and the heat flow out of the gas in each part of the cycle.

c) Using the results of part (b), determine the efficiency of the engine.

•**20.30** A heat engine cycle often used in refrigeration, is the *Brayton cycle,* which involves an adiabatic compression, followed by an isobaric expansion, an adiabatic expansion, and finally an isobaric compression. The system begins at temperature T_1 and transitions to temperatures T_2, T_3, and T_4 after respective parts of the cycle.

a) Sketch this cycle on a pV-diagram.

b) Show that the efficiency of the overall cycle is given by
$\epsilon = 1 - (T_4 - T_1)/(T_3 - T_2)$.

•**20.31** Suppose a Brayton engine (see Problem 20.30) is run as a refrigerator. In this case, the cycle begins at temperature T_1, and the gas is isobarically expanded until it reaches temperature T_4. Then the gas is adiabatically compressed, until its temperature is T_3. It is then isobarically compressed, and the temperature changes to T_2. Finally, it is adiabatically expanded until it returns to temperature T_1.

a) Sketch this cycle on a pV-diagram.

b) Show that the coefficient of performance of the engine is given by $K = (T_4 - T_1)/(T_3 - T_2 - T_4 + T_1)$.

Section 20.3

20.32 It is desired to build a heat pump that has an output temperature of 23.0 °C. Calculate the maximum coefficient of performance for the pump when the input source is (a) outdoor air on a cold winter day at –10.0 °C and (b) groundwater at 9.00 °C.

20.33 Consider a Carnot engine that works between thermal reservoirs with temperatures of 1000.0 K and 300.0 K. The average power of the engine is 1.00 kJ per cycle.

a) What is the efficiency of this engine?

b) How much energy is extracted from the warmer reservoir per cycle?

c) How much energy is delivered to the cooler reservoir?

20.34 A Carnot refrigerator is operating between thermal reservoirs with temperatures of 27.0 °C and 0.00 °C.

a) How much work will need to be input to extract 10.0 J of heat from the colder reservoir?

b) How much work will be needed if the colder reservoir is at –20.0 °C?

20.35 It has been suggested that the vast amount of thermal energy in the oceans could be put to use. The process would rely on the temperature difference between the top layer of the ocean and the bottom; the temperature of the water at the bottom is fairly constant, but the water temperature at the surface changes depending on the time of day, the season, and the weather. Suppose the temperature of the water at the top is 10.0 °C and the temperature of the water at the bottom is 4.00 °C. What is the maximum efficiency with which thermal energy could be extracted from the ocean under these conditions?

•**20.36** A Carnot engine takes an amount of heat $Q_H = 100.$ J from a high-temperature reservoir at temperature $T_H = 1000.$ °C, and exhausts the remaining heat into a low-temperature reservoir at $T_L = 10.0$ °C. Find the amount of work that is obtained from this process.

•**20.37** A Carnot engine operates between a warmer reservoir at a temperature T_1 and a cooler reservoir at a temperature T_2. It is found that increasing the temperature of the warmer reservoir by a factor of 2 while keeping the same temperature for the cooler reservoir increases the efficiency of the Carnot engine by a factor of 2 as well. Find the efficiency of the engine and the ratio of the temperatures of the two reservoirs in their original form.

•**20.38** A certain refrigerator is rated as being 32.0% as efficient as a Carnot refrigerator. To remove 100. J of heat from the interior at 0.00 °C and eject it to the outside at 22.0 °C, how much work must the refrigerator motor do?

Section 20.4

20.39 A refrigerator has a coefficient of performance of 5.00. If the refrigerator absorbs 40.0 cal of heat from the low-temperature reservoir in each cycle, what is the amount of heat expelled into the high-temperature reservoir?

20.40 A heat pump has a coefficient of performance of 5.00. If the heat pump absorbs 40.0 cal of heat from the cold outdoors in each cycle, what is the amount of heat expelled to the warm indoors?

20.41 An Otto engine has a maximum efficiency of 20.0%; find the compression ratio. Assume that the gas is diatomic.

•**20.42** An outboard motor for a boat is cooled by lake water at 15.0 °C and has a compression ratio of 10.0. Assume that the air is a diatomic gas.

a) Calculate the efficiency of the engine's Otto cycle.

b) Using your answer to part (a) and the fact that the efficiency of the Carnot cycle is greater than that of the Otto cycle, estimate the maximum temperature of the engine.

•**20.43** A heat engine uses 100. mg of helium gas and follows the cycle shown in the figure.

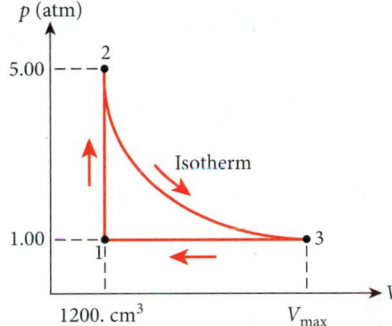

a) Determine the pressure, volume, and temperature of the gas at points 1, 2, and 3.

b) Determine the engine's efficiency.

c) What would be the maximum efficiency of the engine for it to be able to operate between the maximum and minimum temperatures?

••20.44 Gas turbine engines like those of jet aircraft operate on a thermodynamic cycle known as the *Brayton cycle*. The basic Brayton cycle, as shown in the figure, consists of two adiabatic processes—the compression and the expansion of gas through the turbine—and two isobaric processes. Heat is transferred to the gas during combustion in a constant-pressure process (path from point 2 to point 3) and removed from the gas in a heat exchanger during a constant-pressure process (path from point 4 to point 1). The key parameter for this cycle is the pressure ratio, defined as $r_p = p_{max}/p_{min} \equiv p_2/p_1$, and this is the only parameter needed to calculate the efficiency of a Brayton engine.

a) Determine an expression for the efficiency.

b) Calculate the efficiency of a Brayton engine that uses a diatomic gas and has a pressure ratio of 10.0.

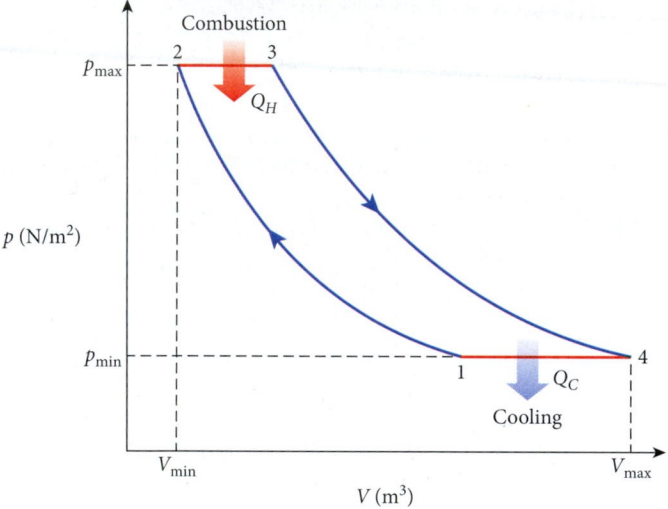

Sections 20.6 and 20.7

20.45 One end of a metal rod is in contact with a thermal reservoir at 700. K, and the other end is in contact with a thermal reservoir at 100. K. The rod and reservoirs make up an isolated system. If 8500. J are conducted from one end of the rod to the other uniformly (no change in temperature along the rod) what is the change in entropy of (a) each reservoir, (b) the rod, and (c) the system?

20.46 The entropy of a macroscopic state is given by $S = k_B \ln w$, where k_B is the Boltzmann constant and w is the number of possible microscopic states. Calculate the change in entropy when n moles of an ideal gas undergo free expansion to fill the entire volume of a box after a barrier between the two halves of the box is removed.

20.47 A proposal is submitted for a novel engine that will operate between 400. K and 300. K.

a) What is the theoretical maximum efficiency of the engine?

b) What is the total entropy change per cycle if the engine operates at maximum efficiency?

20.48 A 10.0-kg block initially slides at 10.0 m/s on a flat rough surface and eventually stops. If the block cools to the ambient temperature, which is 27.0 °C, what is the entropy change of the system?

20.49 Suppose an atom of volume V_A is inside a container of volume V. The atom can occupy any position within this volume. For this simple model, the number of states available to the atom is given by V/V_A. Now suppose the same atom is inside a container of volume $2V$. What will be the change in entropy?

•20.50 An ideal gas is enclosed in a cylinder with a movable piston at the top. The walls of the cylinder are insulated, so no heat can enter or exit. The gas initially occupies volume V_1 and has pressure p_1 and temperature T_1.

The piston is then moved very rapidly to a volume of $V_2 = 3V_1$. The process happens so rapidly that the enclosed gas does not do any work. Find p_2, T_2, and the change in entropy of the gas.

•20.51 Electrons have a property called *spin* that can be either up or down, analogous to the way a coin can be heads or tails. Consider five electrons. Calculate the entropy, S_{5up}, for the state where the spins of all five electrons are up. Calculate the entropy, S_{3up}, for the state where three spins are up and two are down.

•20.52 A 0.545-kg iron bar is taken from a forge at 1000.0 °C and dropped into 10.00 kg of water at 22.0 °C. Assuming that no energy is lost as heat to the surroundings as the water and bar reach their final temperature, determine the total entropy change of the water-bar system.

••20.53 If the Earth is treated as a spherical blackbody of radius 6371 km, absorbing heat from the Sun at a rate given by the solar constant (1370. W/m²) and immersed in space that has the approximate temperature $T_{sp} = 50.0$ K, it radiates heat back into space at an equilibrium temperature of 278.9 K. (This is a refinement of the model in used in Example 18.8.) Estimate the rate at which the Earth gains entropy in this model.

••20.54 Suppose a person metabolizes 2000. kcal/day.

a) With a core body temperature of 37.0 °C and an ambient temperature of 20.0 °C, what is the maximum (Carnot) efficiency with which the person can perform work?

b) If the person could work with that efficiency, at what rate, in watts, would he or she have to shed waste heat to the surroundings?

c) With a skin area of 1.50 m², a skin temperature of 27.0 °C, and an effective emissivity of $e = 0.600$, at what *net* rate does this person radiate heat to the 20.0 °C surroundings?

d) The rest of the waste heat must be removed by evaporating water, either as perspiration or from the lungs. At body temperature, the latent heat of vaporization of water is 575 cal/g. At what rate, in grams per hour, does this person lose water?

e) Estimate the rate at which the person gains entropy. Assume that all the required evaporation of water takes place in the lungs, at the core body temperature of 37.0 °C.

Additional Exercises

20.55 A nonpolluting source of energy is geothermal energy, from the heat inside the Earth. Estimate the maximum efficiency for a heat engine operating between the center of the Earth and the surface of the Earth (use Table 17.1).

20.56 In some of the thermodynamic cycles discussed in this chapter, one isotherm intersects one adiabatic curve. For an ideal gas, by what factor is the adiabatic curve steeper than the isotherm?

20.57 The internal combustion engines in today's cars operate on the Otto cycle. The efficiency of this cycle, $\epsilon_{Otto} = 1 - r^{1-\gamma}$, as derived in this chapter, depends on the compression ratio, $r = V_{max}/V_{min}$. Increasing the efficiency of an Otto engine can be done by increasing the compression ratio. This, in turn, requires fuel with a higher octane rating, to avoid self-ignition of the fuel-air mixture. The following table for a particular engine shows some octane ratings and the maximum compression ratio the engine must have before self-ignition (knocking) sets in.

Octane Rating of Fuel	Maximum Compression Ratio without Knocking
91	8.5
93	9.0
95	9.8
97	10.5

Calculate the maximum theoretical efficiency of an internal combustion engine running on each of these four types of gasoline and the percentage increase in efficiency between using fuel with an octane rating of 91 and using fuel with an octane rating of 97.

20.58 Consider a room air conditioner using a Carnot cycle at maximum theoretical efficiency and operating between the temperatures of 18.0 °C (indoors) and 35.0 °C (outdoors). For each 1.00 J of heat flowing out of the room into the air conditioner:

a) How much heat flows out of the air conditioner to the outdoors?

b) By approximately how much does the entropy of the room decrease?

c) By approximately how much does the entropy of the outdoor air increase?

20.59 Assume that it takes 0.0700 J of energy to heat a 1.00-g sample of mercury from 10.000 °C to 10.500 °C and that the heat capacity of mercury is constant, with a negligible change in volume as a function of temperature. Find the change in entropy if this sample is heated from 10.0 °C to 100. °C.

20.60 An inventor claims that he has created a water-driven engine with an efficiency of 0.200 that operates between thermal reservoirs at 4.0 °C and 20.0 °C. Is this claim valid?

20.61 Consider a system consisting of rolling a six-sided die. What happens to the entropy of the system if an additional die is added? Does it double? What happens to the entropy if the number of dice is three?

20.62 A 1200-kg car traveling at 30.0 m/s crashes into a wall on a hot day (27 °C). What is the total change in entropy?

20.63 If liquid nitrogen is boiled slowly—that is, reversibly—to transform it into nitrogen gas at a pressure $P = 100.0$ kPa, its entropy increases by $\Delta S = 72.1$ J/(mol K). The latent heat of vaporization of nitrogen at its boiling temperature at this pressure is $L_{vap} = 5.568$ kJ/mol. Using these data, calculate the boiling temperature of nitrogen at this pressure.

•**20.64** Find the net change in entropy when 100. g of water at 0 °C is added to 100. g of water at 100. °C.

•**20.65** A coal-burning power plant produces 3000. MW of thermal energy, which is used to boil water and produce supersaturated steam at 300. °C. This high-pressure steam turns a turbine producing 1000. MW of electrical power. At the end of the process, the steam is cooled to 30.0 °C and recycled.

a) What is the maximum possible efficiency of the plant?

b) What is the actual efficiency of the plant?

c) To cool the steam, river water runs through a condenser at a rate of $4.00 \cdot 10^7$ gal/h. If the water enters the condenser at 20.0 °C, what is its exit temperature?

•**20.66** Two equal-volume compartments of a box are joined by a thin wall as shown in the figure. The left compartment is filled with 0.0500 mole of helium gas at 500. K, and the right compartment contains 0.0250 mole of helium gas at 250. K. The right compartment also has a piston to which a force of 20.0 N is applied.

a) If the wall between the compartments is removed, what will the final temperature of the system be?

System 1
0.0500 mol He
at 500. K

System 2
0.0250 mol He
at 250. K

20.0 N

b) How much heat will be transferred from the left compartment to the right compartment?

c) What will be the displacement of the piston due to this transfer of heat?

d) What fraction of the heat will be converted into work?

•**20.67** A volume of 6.00 L of a monatomic ideal gas, originally at 400. K and a pressure of 3.00 atm (called state 1), undergoes the following processes, all done reversibly:

$1 \rightarrow 2$ isothermal expansion to $V_2 = 4V_1$

$2 \rightarrow 3$ isobaric compression

$3 \rightarrow 1$ adiabatic compression to its original state

Find the entropy change for each process.

•**20.68** The process shown in the pV-diagram is performed on 3.00 moles of a monatomic gas. Determine the amount of heat input for this process.

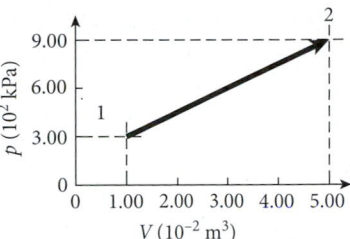

•**20.69** Suppose 1.00 mole of a monatomic ideal gas at a pressure of 4.00 atm and a volume of 30.0 L is isothermically expanded to a pressure of 1.00 atm and a volume of 120.0 L. Next, it is compressed at a constant pressure until its volume is 30.0 L, and then its pressure is increased at the constant volume of 30.0 L. What is the efficiency of this heat engine cycle?

••**20.70** Two cylinders, A and B, have equal inside diameters and pistons of negligible mass connected by a rigid rod. The pistons can move freely. The rod is a short tube with a valve. The valve is initially closed (see the figure).

Cylinder A and its piston are thermally insulated, and cylinder B is in thermal contact with a thermostat, which has temperature $T = 27.0$ °C. Initially, the piston of cylinder A is fixed, and inside the cylinder is a mass, $m = 0.320$ kg, of argon at a pressure higher than atmospheric pressure. Inside cylinder B, there is a mass of oxygen at normal atmospheric pressure. When the piston of cylinder A is freed, it moves very slowly, and at equilibrium, the volume of the argon in cylinder A is eight times higher, and the density of the oxygen has increased by a factor of 2. The thermostat receives heat $Q' = 7.479 \cdot 10^4$ J.

a) Based on the kinetic theory of an ideal gas, show that the thermodynamic process taking place in the cylinder A satisfies $TV^{2/3} = $ constant.

b) Calculate p, V, and T for the argon in the initial and final states. The molar mass of argon is $\mu = 39.95$ g/mol.

c) Calculate the final pressure of the mixture of the gases if the valve in the rod is opened.

MULTI-VERSION EXERCISES

20.71 A house's air conditioner has an energy efficiency rating of 10.47. The house absorbs thermal energy at a rate of 5.375 kJ/s. How much does it cost, in dollars, to run the air conditioner for a day if electricity costs 12.85¢/kWh?

20.72 A house's air conditioner has an energy efficiency rating of 10.71. The house absorbs thermal energy at a rate of 5.437 kJ/s. It costs $5.605 to run the air conditioner for a day. What is the electric company charging for electricity, in cents per kilowatt-hour?

20.73 An air conditioner is used to cool a house in the summer. The house absorbs thermal energy at a rate of 5.499 kJ/s. It costs $5.818 to run the air conditioner for a day when electricity costs 14.13¢/kWh. What is the energy efficiency rating of the air conditioner?

20.74 What is the minimum amount of work that must be done to extract 288.1 J of heat from a massive object at a temperature of 24.93 °C while releasing heat to a high-temperature reservoir at 195.3 °C?

20.75 The minimum amount of work that must be done to extract 425.5 J of heat from a massive object at a temperature of 25.05 °C while releasing

heat to a high-temperature reservoir is 118.5 J. What is the temperature of the high-temperature reservoir?

20.76 The minimum amount of work that must be done to extract 562.9 J of heat from a massive object at a low temperature while releasing heat to a high-temperature reservoir at 120.9 °C is 180.7 J. What is the temperature of the massive object?

20.77 A water-cooled engine produces 1833 W of power. Water enters the engine block at 11.25 °C and exits at 26.69 °C. The rate of water flow is 132.3 L/h. What is the engine's efficiency?

20.78 A water-cooled engine produces 1061 W of power and has an efficiency of 0.3591. Water enters the engine block at 11.35 °C and exits at 27.33 °C. What is the water flow rate, in liters per hour?

20.79 A water-cooled engine has an efficiency of 0.2815. Water enters the engine block at 11.45 °C and exits at 27.97 °C. The rate of water flow is 171.5 L/h. How much power does the engine produce?

21

Electrostatics

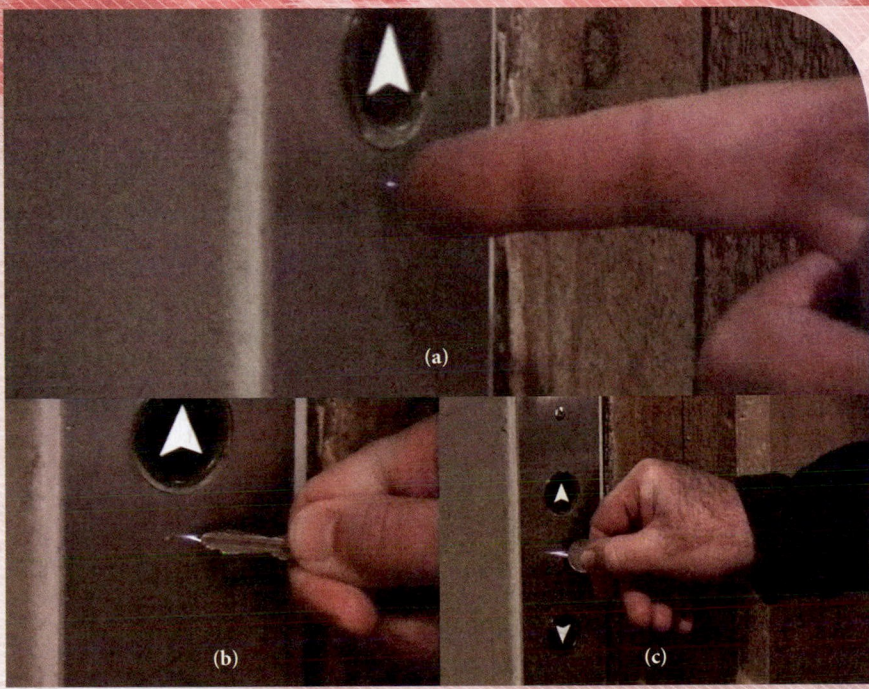

FIGURE 21.1 (a) A spark due to static electricity occurs between a person's finger and a metal surface near an elevator button. (b) and (c) Similar sparks are generated when the person holds a metal object like a car key or a coin, but are painless because the spark forms between the metal surface and the metal object.

Many people think of static electricity as the annoying spark that occurs when they reach for a metal object like a doorknob on a dry day, after they have been walking on a carpet (Figure 21.1). In fact, many electronics manufacturers place small metal plates on equipment so that users can discharge any spark on the plate and not damage the more sensitive parts of the equipment. However, static electricity is more than just an occasional annoyance; it is the starting point for any study of electricity and magnetism, forces that have changed human society as radically as anything since the discovery of fire or the wheel.

In this chapter, we examine the properties of electric charge. A moving electric charge gives rise to a separate phenomenon, called *magnetism,* which is covered in later chapters. Here we look at charged objects that are not moving—hence the term *electrostatics.* All objects have charge, since charged particles make up atoms and molecules. We often don't notice the effects of electrical charge because most objects are electrically neutral. The forces that hold atoms together and that keep objects separate even when they're in contact, are all electric in nature.

WHAT WE WILL LEARN

- Electric charge gives rise to a force between charged particles or objects.
- Electricity and magnetism together make up the electromagnetic force, one of the four fundamental forces of nature.
- There are two kinds of electric charge, positive and negative. Like charges repel, and unlike charges attract.
- Electric charge is quantized, meaning that it occurs only in integral multiples of a smallest elementary quantity. Electric charge is also conserved.
- Most materials around us are electrically neutral.
- The electron is an elementary particle, and its charge is the smallest observable quantity of electric charge.

- Insulators conduct electricity poorly or not at all. Conductors conduct electricity well but not perfectly—some energy losses occur.
- Semiconductors can be made to change between a conducting state and a nonconducting state.
- Superconductors conduct electricity perfectly.
- Objects can be charged directly by contact or indirectly by induction.
- The force that two stationary electric charges exert on each other is proportional to the product of the charges and varies as the inverse square of the distance between the two charges.
- Electrostatic forces between particles can be added as vectors by the process of superposition.

21.1 Electromagnetism

FIGURE 21.2 Lightning strikes over Seattle.

Perhaps no mystery puzzled ancient civilizations more than electricity, which they observed primarily in the form of lightning strikes (Figure 21.2). The destructive force inherent in lightning, which could set objects on fire and kill people and animals, seemed godlike. The ancient Greeks, for example, believed Zeus, father of the gods, had the ability to throw lightning bolts. The Germanic tribes ascribed this power to the god Thor and the Romans to the god Jupiter. Characteristically, the ability to cause lightning belonged to the god at the top (or near the top) of the hierarchy.

The ancient Greeks knew that if you rubbed a piece of amber with a piece of cloth, you could attract small, light objects with the amber. We now know that rubbing amber with a cloth transfers negatively charged particles called *electrons* from the cloth to the amber. (The words *electron* and *electricity* derive from the Greek word for amber.) Lightning also consists of a flow of electrons. The early Greeks and others also knew about naturally occurring magnetic objects called *lodestones,* which were found in deposits of magnetite, a mineral consisting of iron oxide. These objects were used to construct compasses as early as 300 BC.

The relationship between electricity and magnetism was not understood until the middle of the 19th century. The following chapters will reveal how electricity and magnetism can be unified into a common framework called *electromagnetism.* However, unification of forces does not stop there. During the early part of the 20th century, two more fundamental forces were discovered: the weak force, which operates in beta decay (in which an electron and a neutrino are spontaneously emitted from certain types of nuclei), and the strong force, which acts inside the atomic nucleus. We'll study these forces in more detail in Chapter 39 on particle physics. Currently, the electromagnetic and weak forces are viewed as two aspects of the electroweak force (Figure 21.3). For the phenomena discussed in this and the following chapters, this electroweak unification has no influence; it becomes important in the highest-energy particle collisions. Because the energy scale for the electroweak unification is so high, most textbooks continue to speak of four fundamental forces: gravitational, electromagnetic, weak, and strong.

Today, a large number of physicists believe that the electroweak force and the strong force can also be unified, that is, described in a common framework. Several theories propose ways to accomplish this, but so far experimental evidence is missing. Interestingly, the force that has been known longer than any of the other fundamental forces, gravity, seems to be hardest to shoehorn into a unified framework with the other fundamental forces. Quantum gravity, supersymmetry, and string theory are current foci of cutting-edge physics research in which theorists are attempting to construct this grand unification and dis-

Forces of Nature

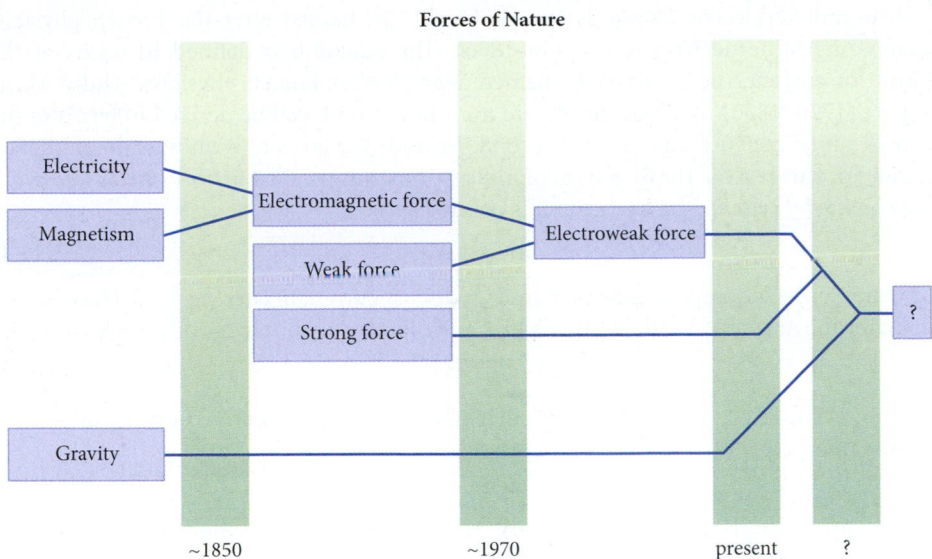

~1850   ~1970   present   ?

FIGURE 21.3 The history of the unification of fundamental forces.

cover the (hubristically named) Theory of Everything. They are mainly guided by symmetry principles and the conviction that nature must be elegant and simple.

We'll return to these considerations in Chapters 39 and 40. In this chapter, we consider electric charge, how materials react to electric charge, static electricity, and the forces resulting from electric charges. **Electrostatics** covers situations where charges stay in place and do not move.

21.2 Electric Charge

Let's look a little deeper into the cause of the electric sparks that you occasionally receive on a dry winter day if you walk across a carpet and then touch a metal doorknob. (Electrostatic sparks have even ignited gas fumes while someone is filling the tank at a gas station. This is not an urban legend; a few of these cases have been caught on gas station surveillance cameras.) The process that causes this sparking is called **charging.** Charging consists of the transfer of negatively charged particles, called **electrons,** from the atoms and molecules of the material of the carpet to the soles of your shoes. This charge can move relatively easily through your body, including your hands. The built-up electric charge discharges through the metal of the doorknob, creating a spark.

The two types of electric charge found in nature are **positive charge** and **negative charge.** Normally, objects around us do not seem to be charged; instead, they are electrically neutral. Neutral objects contain roughly equal numbers of positive and negative charges that largely cancel each other. Only when positive and negative charges are not balanced do we observe the effects of electric charge.

If you rub a glass rod with a cloth, the glass rod becomes charged and the cloth acquires a charge of the opposite sign. If you rub a plastic rod with fur, the rod and fur also become oppositely charged. If you bring two charged glass rods together, they repel each other. Similarly, if you bring two charged plastic rods together, they also repel each other. However, a charged glass rod and a charged plastic rod will attract each other. This difference arises because the glass rod and the plastic rod have opposite charge. These observations led to the following law:

Law of Electric Charges
Like charges repel and opposite charges attract.

The unit of electric charge is the **coulomb** (C), named after the French physicist Charles-Augustine de Coulomb (1736–1806). The coulomb is defined in terms of the SI unit for current, the ampere (A), named after another French physicist, André-Marie Ampère (1775–1836). Neither the ampere nor the coulomb can be derived in terms of the other SI units: meter, kilogram, and second. Instead, the ampere is another fundamental SI unit. For this reason, the SI system of units is sometimes called *MKSA (meter-kilogram-second-ampere) system*. The charge unit is defined as

$$1\text{ C} = 1\text{ A s}. \tag{21.1}$$

The definition of the ampere must wait until we discuss current in later chapters. However, we can define the magnitude of the coulomb by simply specifying the charge of a single electron:

$$q_e = -e, \tag{21.2}$$

where q_e is the charge and e has the (currently best accepted and experimentally measured) value

$$e = 1.602176565(35) \cdot 10^{-19}\text{ C}. \tag{21.3}$$

(Usually it is enough to carry only the first two to four significant digits of this mantissa. We will use a value of 1.602 in this chapter, but you should keep in mind that equation 21.3 gives the full accuracy to which this charge has been measured.)

The charge of the electron is an intrinsic property of the electron, just like its mass. The charge of the **proton,** another basic particle of atoms, is exactly the same magnitude as that of the electron, only the proton's charge is positive:

$$q_p = +e. \tag{21.4}$$

The choice of which charge is positive and which charge is negative is arbitrary. The conventional choice of $q_e < 0$ and $q_p > 0$ is due to the American statesman, scientist, and inventor Benjamin Franklin (1706–1790), who pioneered studies of electricity.

One coulomb is an extremely large unit of charge. We'll see later in this chapter just how big it is when we investigate the magnitude of the forces of charges on each other. Units of μC (microcoulombs, 10^{-6} C), nC (nanocoulombs, 10^{-9} C), and pC (picocoulombs, 10^{-12} C) are commonly used.

Benjamin Franklin also proposed that charge is conserved. Charge is not created or destroyed, simply moved from one object to another.

> **Law of Charge Conservation**
> The total electric charge of an isolated system is conserved.

This law is the fourth conservation law we have encountered so far, the first three being the conservation laws for total energy, momentum, and angular momentum. Conservation laws are a common thread that runs throughout all of physics and thus throughout this book as well.

It is important to note that there is a conservation law for charge, but *not* for mass. We'll see later in this book that mass and energy are not independent of each other. What is sometimes described in introductory chemistry as conservation of mass is not an exact conservation law, but only an approximation used to keep track of the number of atoms in chemical reactions. (It is a good approximation to a large number of significant figures but not an exact law, like charge conservation.) Conservation of charge applies to all systems, from the macroscopic system of plastic rod and fur down to systems of subatomic particles.

Elementary Charge

Electric charge occurs only in integral multiples of a minimum size. This is expressed by saying that charge is **quantized.** The smallest observable unit of electric charge is the charge of the electron, which is $-1.602 \cdot 10^{-19}$ C (as defined in equation 21.3).

Concept Check 21.1

How many electrons does it take to make 1.00 C of charge?

a) $1.60 \cdot 10^{19}$ d) $6.24 \cdot 10^{18}$

b) $6.60 \cdot 10^{19}$ e) $6.66 \cdot 10^{17}$

c) $3.20 \cdot 10^{16}$

The fact that electric charge is quantized was verified in an ingenious experiment carried out in 1910 by American physicist Robert A. Millikan (1868–1953) and known as the *Millikan oil drop experiment* (Figure 21.4). In this experiment, oil drops were sprayed into a chamber where electrons were knocked out of the drops by some form of radiation, usually X-rays. The resulting positively charged drops were allowed to fall between two electrically charged plates. Adjusting the charge of the plates caused the drops to stop falling and allowed their charge to be measured. What Millikan observed was that charge was quantized rather than continuous. (A quantitative analysis of this experiment will be presented in Chapter 23 on electric potential.) That is, this experiment and its subsequent refinements established that charge comes only in integer multiples of the charge of an electron. In everyday experiences with electricity, we do not notice that charge is quantized because most electrical phenomena involve huge numbers of electrons.

In Chapter 13, we discussed the facts that matter is composed of atoms and that an atom consists of a nucleus containing charged protons and neutral neutrons. A schematic drawing of a carbon atom is shown in Figure 21.5. A carbon atom has six protons and (usually) six neutrons in its nucleus. This nucleus is surrounded by six electrons. Note that this drawing is not to scale. In the actual atom, the distance of the electrons from the nucleus is much larger (by a factor on the order of 10,000) than the size of the nucleus. In addition, the electrons are shown in circular orbits, which is also not quite correct. In Chapter 38, we'll see that the locations of electrons in the atom can be characterized only by probability distributions.

As mentioned earlier, a proton has a positive charge with a magnitude that is *exactly* equal to the magnitude of the negative charge of an electron. In a neutral atom, the number of negatively charged electrons is equal to the number of positively charged protons. The mass of the electron is much smaller than the mass of the proton or the neutron. Therefore, most of the mass of an atom resides in the nucleus. Electrons can be removed from atoms relatively easily. For this reason, electrons are typically the carriers of electricity, rather than protons or atomic nuclei.

The electron is a fundamental particle and has no substructure: It is a point particle with zero radius (at least, according to current understanding). However, high-energy probes have been used to look inside the proton. A proton is composed of charged particles called *quarks,* held together by uncharged particles called *gluons.* Quarks have a charge of $\pm\frac{1}{3}$ or $\pm\frac{2}{3}$ times the charge of the electron. These fractionally charged particles cannot exist independently and have never been observed directly, despite numerous extensive searches. Just like the charge of an electron, the charges of quarks are intrinsic properties of these elementary particles.

A proton is composed of two *up quarks* (each with charge $+\frac{2}{3}e$) and one *down quark* (with charge $-\frac{1}{3}e$), giving the proton a charge of $q_p = (2)(+\frac{2}{3}e) + (1)(-\frac{1}{3}e) = +e$, as illustrated in Figure 21.6a. The electrically neutral neutron (hence the name!) is composed of an up quark and two down quarks, as shown in Figure 21.6b, so its charge is $q_n = (1)(+\frac{2}{3}e) + (2)(-\frac{1}{3}e) = 0$. In Chapter 39, we'll see that there are other, much more massive, quarks named *strange, charm, bottom,* and *top,* which have the same charges as the up and down quarks. There are also much more massive electron-like particles named *muon* and *tau.* But the basic fact remains that all of the matter in everyday experience is made up of electrons (with electrical charge $-e$), up and down quarks (with electrical charges $+\frac{2}{3}e$ and $-\frac{1}{3}e$, respectively), and gluons (zero charge).

It is remarkable that the charges of the quarks inside a proton add up to *exactly* the same magnitude as the charge of the electron. This fact is still a puzzle, pointing to some deep symmetry in nature that is not yet understood.

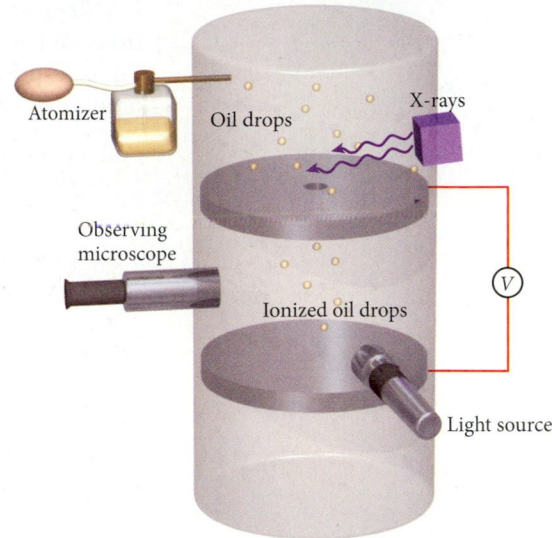

FIGURE 21.4 Schematic drawing of the Millikan oil drop experiment.

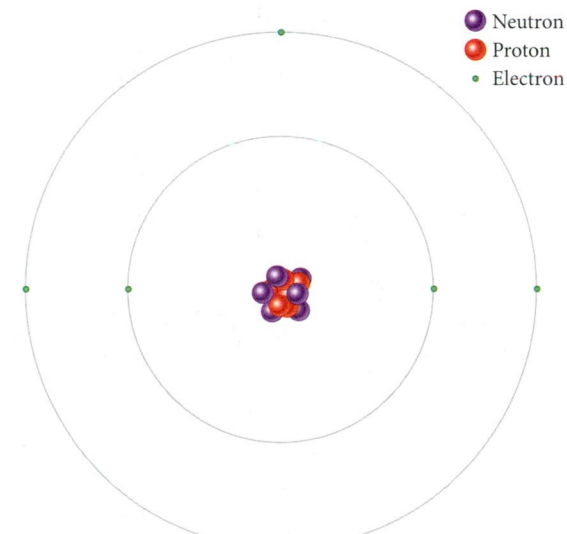

FIGURE 21.5 In a carbon atom, the nucleus contains six neutrons and six protons. The nucleus is surrounded by six electrons. Note that this drawing is schematic and not to scale.

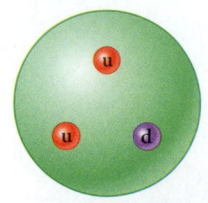

Proton

$$q_{\mathrm{p}} = +\tfrac{2}{3}e + \tfrac{2}{3}e - \tfrac{1}{3}e = +e$$

(a)

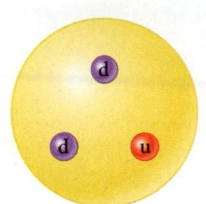

Neutron

$$q_{\mathrm{n}} = +\tfrac{2}{3}e - \tfrac{1}{3}e - \tfrac{1}{3}e = 0$$

(b)

FIGURE 21.6 (a) A proton contains two up quarks (u) and one down quark (d). (b) A neutron contains one up quark (u) and two down quarks (d).

Self-Test Opportunity 21.1

Give the charge of the following elementary particles or atoms in terms of the elementary charge $e = 1.602 \cdot 10^{-19}$ C.

a) proton

b) neutron

c) helium atom (two protons, two neutron, and two electrons)

d) hydrogen atom (one proton and one electron)

e) up quark

f) down quark

g) electron

h) alpha particle (two protons and two neutrons)

Because all macroscopic objects are made of atoms, which in turn are made of electrons and atomic nuclei consisting of protons and neutrons, the charge, q, of any object can be expressed in terms of the sum of the number of protons, N_{p}, minus the sum of the number of electrons, N_{e}, that make up the object:

$$q = e\left(N_{\mathrm{p}} - N_{\mathrm{e}}\right). \tag{21.5}$$

EXAMPLE 21.1 Net Charge

PROBLEM

If we wanted a block of iron of mass 3.25 kg to acquire a positive charge of 0.100 C, what fraction of the electrons would we have to remove?

SOLUTION

Iron has mass number 56. Therefore, the number of iron atoms in the 3.25-kg block is

$$N_{\mathrm{atom}} = \frac{(3.25 \text{ kg})(6.022 \cdot 10^{23}\,\text{atoms/mole})}{0.0560 \text{ kg/mole}} = 3.495 \cdot 10^{25} = 3.50 \cdot 10^{25} \text{ atoms.}$$

Note that we have used Avogadro's number, $6.022 \cdot 10^{23}$, and the definition of the mole, which specifies that the mass of 1 mole of a substance in grams is just the mass number of the substance—in this case, 56.

Because the atomic number of iron is 26, which equals the number of protons or electrons in an iron atom, the total number of electrons in the 3.25-kg block is

$$N_{\mathrm{e}} = 26 N_{\mathrm{atom}} = (26)(3.495 \cdot 10^{25}) = 9.09 \cdot 10^{26} \text{ electrons.}$$

We use equation 21.5 to find the number of electrons, $N_{\Delta\mathrm{e}}$, that we would have to remove. Because the number of electrons equals the number of protons in the original uncharged object, the difference in the number of protons and electrons is the number of removed electrons, $N_{\Delta\mathrm{e}}$:

$$q = e \cdot N_{\Delta\mathrm{e}} \Rightarrow N_{\Delta\mathrm{e}} = \frac{q}{e} = \frac{0.100 \text{ C}}{1.602 \cdot 10^{-19} \text{ C}} = 6.24 \cdot 10^{17}.$$

Finally, we obtain the fraction of electrons we would have to remove:

$$\frac{N_{\Delta\mathrm{e}}}{N_{\mathrm{e}}} = \frac{6.24 \cdot 10^{17}}{9.09 \cdot 10^{26}} = 6.87 \cdot 10^{-10}.$$

We would have to remove fewer than one in a billion electrons from the iron block in order to put the sizable positive charge of 0.100 C on it.

21.3 Insulators, Conductors, Semiconductors, and Superconductors

Materials that conduct electricity well are called **conductors.** Materials that do not conduct electricity are called **insulators.** (Of course, there are good and poor conductors and good and poor insulators, depending on the properties of the specific materials.)

The electronic structure of a material refers to the way in which electrons are bound to nuclei, as we'll discuss in later chapters. For now, we are interested in the relative propensity of the atoms of a material to either give up or acquire electrons. For insulators, no free movement of electrons occurs because the material has no loosely bound electrons that can escape from its atoms and thereby move freely throughout the material. Even when external charge is placed on an insulator, this external charge cannot move appreciably. Typical insulators are glass, plastic, and cloth.

On the other hand, materials that are conductors have an electronic structure that allows the free movement of some electrons. The positive charges of the atoms of a conducting material do not move, since they reside in the heavy nuclei. Typical solid conductors are metals. Copper, for example, is a very good conductor and is therefore used in electrical wiring.

Fluids and organic tissue can also serve as conductors. Pure distilled water is not a very good conductor. However, dissolving common table salt (NaCl), for example, in water improves its conductivity tremendously, because the positively charged sodium ions (Na^+) and negatively charged chlorine ions (Cl^-) can move within the water to conduct electricity. In liquids, unlike solids, positive as well as negative charge carriers are mobile. Organic tissue is not a very good conductor, but it conducts electricity well enough to make large currents dangerous to us. (We'll learn more about electric current in Chapters 25 and 26, where these terms, which are in everyday use, will be defined precisely.)

Semiconductors

A class of materials called **semiconductors** can change from being an insulator to being a conductor and back to an insulator again. Semiconductors were discovered only a little more than 50 years ago but are the backbone of the entire computer and consumer electronics industries. The first widespread use of semiconductors was in transistors (Figure 21.7a); modern computer chips (Figure 21.7b) perform the functions of millions of transistors. Computers and basically all modern consumer electronics products and devices (televisions, cameras, video game players, cell phones, etc.) would be impossible without semiconductors. Gordon Moore, cofounder of Intel, famously stated that due to advancing technology, the power of the average computer's CPU (central processing unit) doubles every 18 months, which is an empirical average over the last 5 decades. This doubling phenomenon is known as *Moore's Law.* Physicists have been and will undoubtedly continue to be the driving force behind this process of scientific discovery, invention, and improvement.

Semiconductors are of two kinds: intrinsic and extrinsic. Examples of *intrinsic semiconductors* are chemically pure crystals of gallium arsenide, germanium, or, especially, silicon. Engineers produce *extrinsic semiconductors* by *doping*, which is the addition of minute amounts (typically 1 part in 10^6) of other materials that can act as electron donors or electron receptors. Semiconductors doped with electron donors are called *n-type* (*n* stands for "negative charge"). If the doping substance acts as an electron receptor, the hole left behind by an electron that attaches to a receptor can also travel through the semiconductor and acts as an effective positive charge carrier. These semiconductors are consequently called *p-type* (*p* stand for "positive charge"). Thus, unlike normal solid conductors in which only negative charges move, semiconductors have movement of negative or positive charges (which are really electron holes, that is, missing electrons).

Superconductors

Superconductors are materials that have zero resistance to the conduction of electricity, as opposed to normal conductors, which conduct electricity well but with some losses. Materials are superconducting only at very low temperatures. A typical superconductor is a niobium-titanium alloy that must be kept near the temperature of liquid helium (4.2 K) to retain its superconducting properties. During the last 20 years, new materials called *high-T_c superconductors* (T_c stands for "critical temperature," which is the maximum temperature that allows superconductivity) have been developed. These are superconducting at the temperature at which nitrogen can exist as a liquid (77.3 K). Materials that are superconductors at room temperature (300 K) have not yet been found, but they would be extremely useful. Research directed at developing such materials and theoretically explaining what physical phenomena cause high-T_c superconductivity is currently in progress.

The topics of conductivity, superconductivity, and semiconductors will be discussed in more quantitative detail in Chapters 25 and 28.

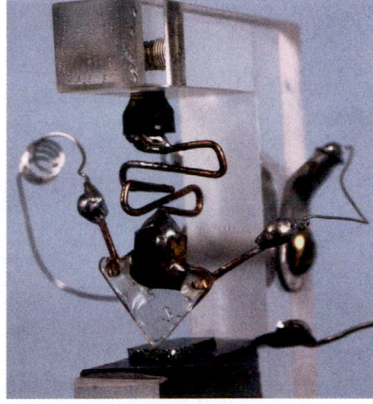

(a)

(b)

FIGURE 21.7 (a) Replica of the first transistor, invented in 1947 by John Bardeen, Walter H. Brattain, and William B. Shockley. (b) Modern computer chips made from silicon wafers contain many tens of millions of transistors.

21.4 Electrostatic Charging

Giving a static charge to an object is a process known as **electrostatic charging.** Electrostatic charging can be understood through a series of simple experiments. A power supply serves as a ready source of positive and negative charge. The battery in your car is a similar

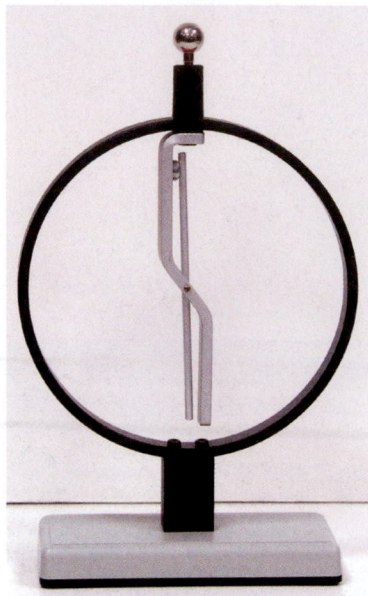

FIGURE 21.8 A typical electroscope used in lecture demonstrations.

power supply; it uses chemical reactions to create a separation between positive and negative charge. Several insulating paddles can be charged with positive or negative charge from the power supply. In addition, a conducting connection is made to the Earth. The Earth is a nearly infinite reservoir of charge, capable of effectively neutralizing electrically charged objects in contact with it. This taking away of charge is called **grounding,** and an electrical connection to the Earth is called a **ground.**

An **electroscope** is a device that gives an observable response when it is charged. You can build a relatively simple electroscope by using two strips of very thin metal foil that are attached at one end and are allowed to hang straight down adjacent to each other from an isolating frame. Kitchen aluminum foil is not suitable, because it is too thick, but hobby shops sell thinner metal foils. For the isolating frame, you can use a Styrofoam coffee cup turned sideways, for example.

The lecture-demonstration-quality electroscope shown in Figure 21.8 has two conductors that in their neutral position are touching and oriented in a vertical direction. One of the conductors is hinged at its midpoint so that it will move away from the fixed conductor if a charge appears on the electroscope. These two conductors are in contact with a conducting ball on top of the electroscope, which allows charge to be applied or removed easily.

An uncharged electroscope is shown in Figure 21.9a. The power supply is used to give a negative charge to one of the insulating paddles. When the paddle is brought near the ball of the electroscope, as shown in Figure 21.9b, the electrons in the conducting ball of the electroscope are repelled, which produces a net negative charge on the conductors of the electroscope. This negative charge causes the movable conductor to rotate because the stationary conductor also has negative charge and repels it. Because the paddle did not touch the ball, the charge on the movable conductors is **induced.** If the charged paddle is then taken away, as illustrated in Figure 21.9c, the induced charge reduces to zero, and the movable conductor returns to its original position, because the total charge on the electroscope did not change in the process.

If the same process is carried out with a positively charged paddle, the electrons in the conductors are attracted to the paddle and flow into the conducting ball. This leaves a net positive charge on the conductors, causing the movable conducting arm to rotate again. Note that the net charge of the electroscope is zero in both cases and that the motion of the conductor indicates only that the paddle is charged. When the positively charged paddle is removed, the movable conductor again returns to its original position. It is important to note that we cannot determine the sign of this charge!

On the other hand, if a negatively charged insulating paddle *touches* the ball of the electroscope, as shown in Figure 21.10b, electrons will flow from the paddle to the conductor, producing a net negative charge. When the paddle is removed, the charge remains and the movable arm remains rotated, as shown in Figure 21.10c. Similarly, if a positively charged insulating paddle touches the ball of the uncharged electroscope, the electroscope transfers electrons to the positively charged paddle and becomes positively charged. Again, both a

Concept Check 21.2

The hinged conductor moves away from the fixed conductor if a charge is applied to the electroscope, because

a) like charges repel each other.

b) like charges attract each other.

c) unlike charges attract each other.

d) unlike charges repel each other.

FIGURE 21.9 Inducing a charge: (a) An uncharged electroscope. (b) A negatively charged paddle is brought near the electroscope. (c) The negatively charged paddle is taken away.

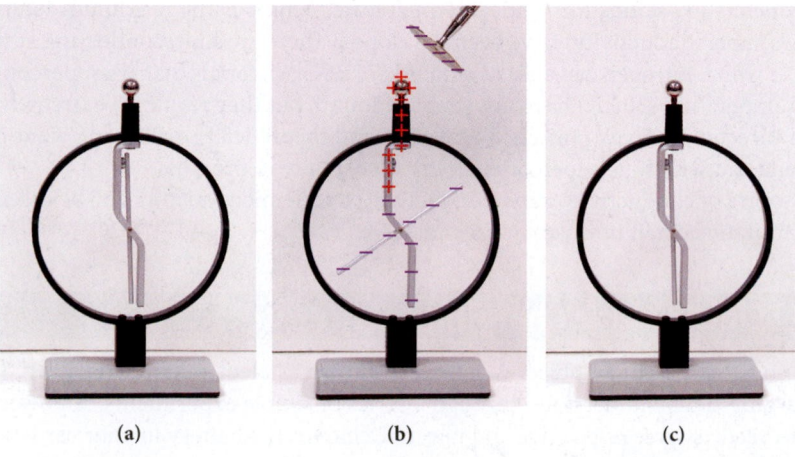

(a)   (b)   (c)

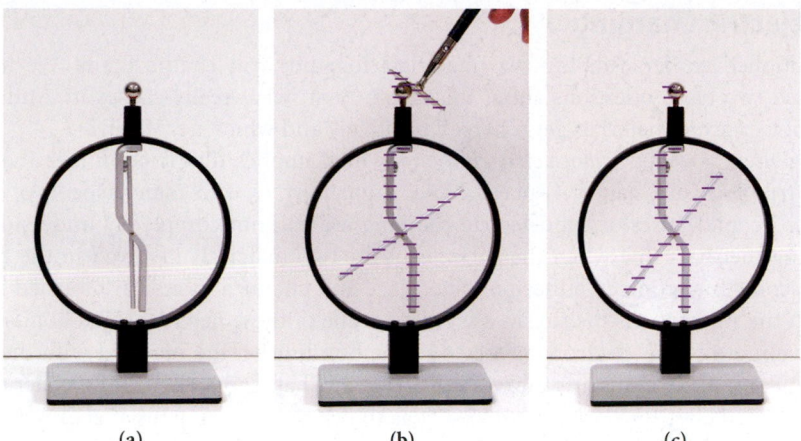

(a)　　　　　　　(b)　　　　　　　(c)

positively charged paddle and a negatively charged paddle have the same effect on the electroscope, and we have no way of determining whether the paddles are positively charged or negatively charged. This process is called **charging by contact.**

The two different kinds of charge can be demonstrated by first touching a negatively charged paddle to the electroscope, producing a rotation of the movable arm, as shown in Figure 21.10. If a positively charged paddle is then brought into contact with the electroscope, the movable arm returns to the uncharged position. The charge is neutralized (assuming that both paddles originally had the same absolute value of charge). Thus, there are two kinds of charge. However, because charges are manifestations of mobile electrons, a negative charge is an excess of electrons and a positive charge is a deficit of electrons.

The electroscope can be given a charge without touching it with the charged paddle, as shown in Figure 21.11. The uncharged electroscope is shown in Figure 21.11a. A negatively charged paddle is brought close to the ball of the electroscope but not touching it, as shown in Figure 21.11b. In Figure 21.11c, the electroscope is connected to a ground. Then, while the charged paddle is still close to but not touching the ball of the electroscope, the ground connection is removed in Figure 21.11d. Next, when the paddle is moved away from the electroscope in Figure 21.11e, the electroscope is still positively charged (but with a smaller deflection than in Figure 21.11b). The same process also works with a positively charged paddle. This process is called **charging by induction** and yields an electroscope charge that has the opposite sign from the charge on the paddle.

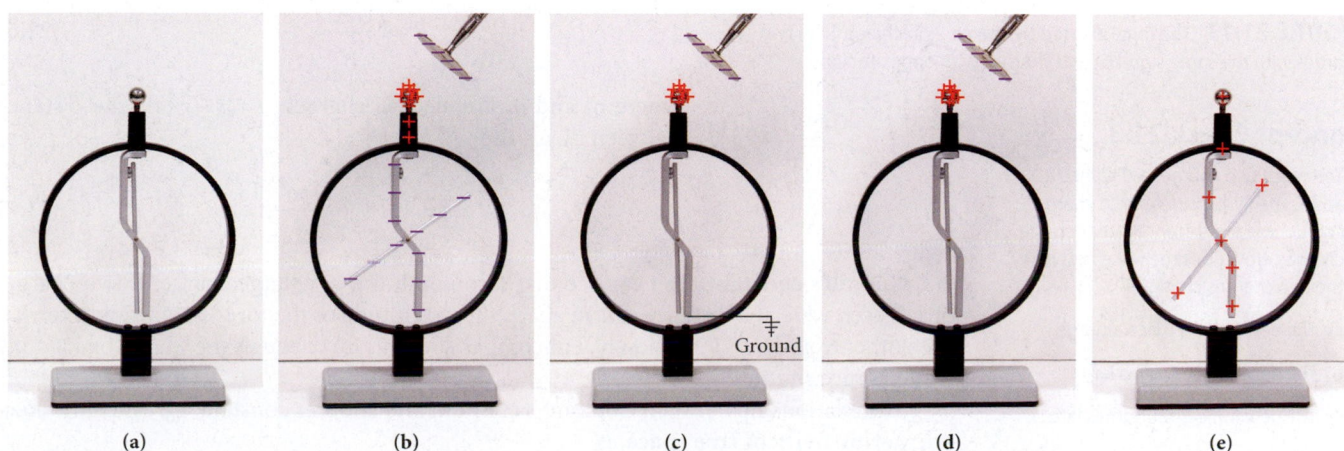

(a)　　　　　　(b)　　　　　　(c)　　　　　　(d)　　　　　　(e)

FIGURE **21.11** Charging by induction: (a) An uncharged electroscope. (b) A negatively charged paddle is brought close to the electroscope. (c) A ground is connected to the electroscope. (d) The connection to the ground is removed. (e) The negatively charged paddle is taken away, leaving the electroscope positively charged.

FIGURE 21.12 Triboelectric series for some common materials.

Triboelectric Charging

As mentioned earlier, rubbing two materials together will charge them. We have not addressed two basic questions about this effect: First, what really causes it? And second, which of two given materials gets charged positively and which negatively?

The effect is called *triboelectric charging*, which implies that friction (see the discussion of tribology in Chapter 4) plays a role. Amazingly, as with many aspects of friction, the microscopic causes of triboelectric charging are still not completely understood. The prevailing theory is that when the surfaces of the two materials involved in the charging process come into contact, adhesion takes place and chemical bonds are formed between atoms at the surfaces. As the surfaces separate, some of these newly formed bonds rupture and leave more of the electrons involved in the bonding on the material with the greater work function. We will discuss the work function in Chapter 36 when we consider the photoelectric effect, but for now we can view it as the electrostatic potential energy that binds electrons to atoms in a material.

Recent research results obtained by examining surfaces with atomic force microscopes suggest, however, that triboelectric charging can also occur when two pieces of the *same* material are rubbed against each other. In addition, it has been found that the charging process transfers not only electrons but sometimes also small specks (a few nanometers across) of material.

Which material gets charged positively and which negatively? This question has been answered by a long series of experiments whose results are summarized in Figure 21.12 in the form of a list of some common materials that may be rubbed together. If you rub two materials from this list against each other, the one nearer the top will receive a net positive charge and the other a net negative charge.

Finally, as a rule of thumb, more intense rubbing creates greater charge transfer. This occurs because more intense contact increases friction, which in turn creates more microscopic points of charge transfer on the surfaces of the materials.

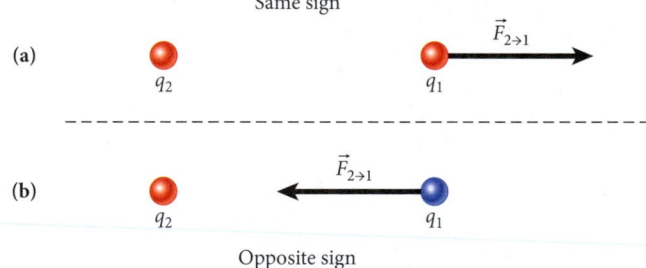

FIGURE 21.13 The force exerted by charge 2 on charge 1: (a) two charges with the same sign; (b) two charges with opposite signs.

21.5 Electrostatic Force—Coulomb's Law

The law of electric charges is evidence of a force between any two charges at rest. Experiments show that for the electrostatic force exerted by a charge q_2 on a charge q_1, $\vec{F}_{2\rightarrow1}$, the force on q_1 points toward q_2 if the charges have opposite signs and away from q_2 if the charges have like signs (Figure 21.13). This force on one charge due to another charge always acts along the line between the two charges. **Coulomb's Law** gives the magnitude of this force as

$$F = k\frac{|q_1 q_2|}{r^2},\tag{21.6}$$

where q_1 and q_2 are electric charges, $r = |\vec{r}_1 - \vec{r}_2|$ is the distance between them, and

$$k = 8.99 \cdot 10^9 \ \frac{\text{N m}^2}{\text{C}^2}\tag{21.7}$$

is **Coulomb's constant.** You can see that 1 coulomb is a *very* large charge. If two charges of 1 C each were at a distance of 1 m apart, the magnitude of the force they would exert on each other would be 8.99 billion N. For comparison, this force equals the weight of 450 fully loaded space shuttles!

The relationship between Coulomb's constant and another constant, ϵ_0, called the **electric permittivity of free space,** is

$$k = \frac{1}{4\pi\epsilon_0}.\tag{21.8}$$

Concept Check 21.3

You place two charges a distance r apart. Then you double each charge and double the distance between the charges. How does the force between the two charges change?

a) The new force is twice as large.

b) The new force is half as large.

c) The new force is four times larger.

d) The new force is four times smaller.

e) The new force is the same.

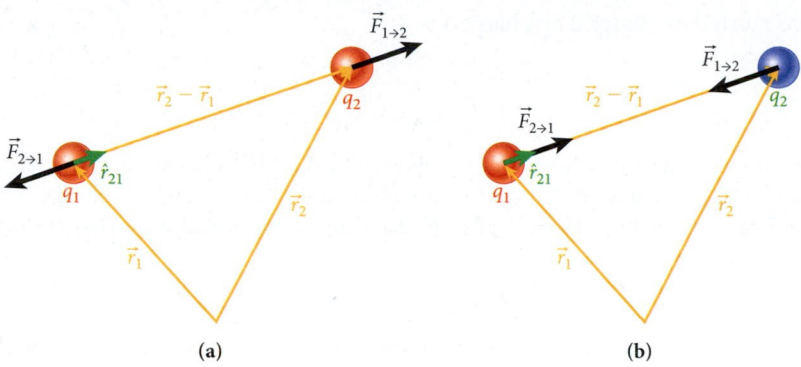

Consequently, the value of ϵ_0 is

$$\epsilon_0 = 8.85 \cdot 10^{-12} \frac{C^2}{N\,m^2}. \tag{21.9}$$

An alternative way of writing equation 21.6 is then

$$F = \frac{1}{4\pi\epsilon_0} \frac{|q_1 q_2|}{r^2}. \tag{21.10}$$

As you'll see in the next few chapters, some equations in electrostatics are more convenient to write with k, while others are more easily written in terms of $1/(4\pi\epsilon_0)$.

Note that the charges in equations 21.6 and 21.10 can be positive or negative, so the product of the charges can also be positive or negative. Since opposite charges attract and like charges repel, a negative value for the product $q_1 q_2$ signifies attraction and a positive value means repulsion.

Finally, Coulomb's Law for the force due to charge 2 on charge 1 can be written in vector form:

$$\vec{F}_{2\to1} = -k\frac{q_1 q_2}{r^3}(\vec{r}_2 - \vec{r}_1) = -k\frac{q_1 q_2}{r^2}\hat{r}_{21}. \tag{21.11}$$

In this equation, $\hat{r}_{21}$ is a unit vector pointing from q_2 to q_1 (see Figure 21.14). The negative sign indicates that the force is repulsive if both charges are positive or both charges are negative. In that case, $\vec{F}_{2\to1}$ points away from charge 2, as depicted in Figure 21.14a. On the other hand, if one of the charges is positive and the other negative, then $\vec{F}_{2\to1}$ points toward charge 2, as shown in Figure 21.14b.

If charge 2 exerts the force $\vec{F}_{2\to1}$ on charge 1, then the force $\vec{F}_{1\to2}$ that charge 1 exerts on charge 2 is simply obtained from Newton's Third Law (see Chapter 4): $\vec{F}_{1\to2} = -\vec{F}_{2\to1}$.

Superposition Principle

So far in this chapter, we have been dealing with two charges. Now let's consider three point charges, q_1, q_2, and q_3, at positions x_1, x_2, and x_3, respectively, as shown in Figure 21.15. The force exerted by charge 1 on charge 3, $\vec{F}_{1\to3}$, is given by

$$\vec{F}_{1\to3} = -\frac{kq_1 q_3}{\left(x_3 - x_1\right)^2}\hat{x}.$$

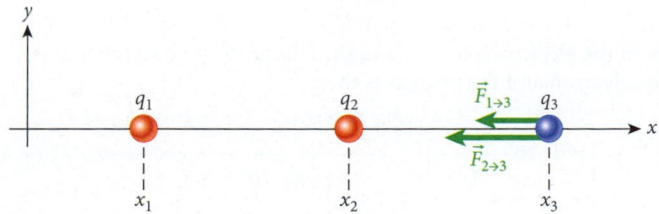

FIGURE 21.15 The forces exerted on charge 3 by charge 1 and charge 2.

The force exerted by charge 2 on charge 3 is

$$\vec{F}_{2\to3} = -\frac{kq_2q_3}{\left(x_3 - x_2\right)^2}\,\hat{x}.$$

The force that charge 1 exerts on charge 3 is not affected by the presence of charge 2. The force that charge 2 exerts on charge 3 is not affected by the presence of charge 1. In addition, the forces exerted by charge 1 and charge 2 on charge 3 add vectorially to produce a net force on charge 3:

$$\vec{F}_{net\to3} = \vec{F}_{1\to3} + \vec{F}_{2\to3}.$$

This superposition of forces is completely analogous to that described in Chapter 4 for the forces of gravity and friction.

In general, we can express the electrostatic force, $\vec{F}(\vec{r})$, acting on a charge q at position $\vec{r}$ due to a collection of charges, q_i, at positions $\vec{r}_i$ as

$$\vec{F}(\vec{r}) = kq\sum_{i=1}^{n} q_i\frac{\vec{r}_i - \vec{r}}{\left|\vec{r}_i - \vec{r}\right|^3}. \tag{21.12}$$

We obtain this result using superposition of the forces and equation 21.11 for each pairwise interaction.

EXAMPLE 21.2 Electrostatic Force inside the Atom

PROBLEM 1
What is the magnitude of the electrostatic force that the two protons inside the nucleus of a helium atom exert on each other?

SOLUTION 1
The two protons and two neutrons in the nucleus of the helium atom are held together by the strong force; the electrostatic force is pushing the protons apart. The charge of each proton is $q_p = +e$. A distance of approximately $r = 2\cdot10^{-15}$ m separates the two protons. Using Coulomb's Law, we can find the force:

$$F = k\frac{\left|q_p q_p\right|}{r^2} = \left(8.99\cdot10^9\;\frac{\text{N m}^2}{\text{C}^2}\right)\frac{\left(+1.6\cdot10^{-19}\;\text{C}\right)\left(+1.6\cdot10^{-19}\;\text{C}\right)}{\left(2\cdot10^{-15}\;\text{m}\right)^2} = 58\;\text{N}.$$

Therefore, the two protons in the atomic nucleus of a helium atom are being pushed apart with a force of 58 N (approximately the weight of a small dog). Considering the size of the nucleus, this is an astonishingly large force. Why do atomic nuclei not simply explode? The answer is that an even stronger force, the aptly named strong force, keeps them together.

PROBLEM 2
What is the magnitude of the electrostatic force between a gold nucleus and an electron of the gold atom in an orbit with radius $4.88\cdot10^{-12}$ m?

SOLUTION 2
The negatively charged electron and the positively charged gold nucleus attract each other with a force whose magnitude is

$$F = k\frac{\left|q_e q_N\right|}{r^2},$$

where the charge of the electron is $q_e = -e$ and the charge of the gold nucleus is $q_N = +79e$. The force between the electron and the nucleus is then

$$F = k\frac{\left|q_e q_N\right|}{r^2} = \left(8.99\cdot10^9\;\frac{\text{N m}^2}{\text{C}^2}\right)\frac{\left(1.60\cdot10^{-19}\;\text{C}\right)\left[(79)\left(1.60\cdot10^{-19}\;\text{C}\right)\right]}{\left(4.88\cdot10^{-12}\;\text{m}\right)^2} = 7.63\cdot10^{-4}\;\text{N}.$$

Thus, the magnitude of the electrostatic force exerted on an electron in a gold atom by the nucleus is about 100,000 times less than that between protons inside a nucleus.

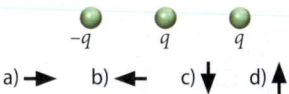

Note: The gold nucleus has a mass that is approximately 400,000 times that of the electron. But the force the gold nucleus exerts on the electron has exactly the same magnitude as the force that the electron exerts on the gold nucleus. You may say that this is obvious from Newton's Third Law (see Chapter 4), which is true. But it is worth emphasizing that this basic law holds for electrostatic forces as well.

EXAMPLE 21.3 | Equilibrium Position

PROBLEM

Two charged particles are placed as shown in Figure 21.16: $q_1 = 0.15$ μC is located at the origin, and $q_2 = 0.35$ μC is located on the positive x-axis at $x_2 = 0.40$ m. Where should a third charged particle, q_3, be placed to be at an equilibrium point (such that the forces on it sum to zero)?

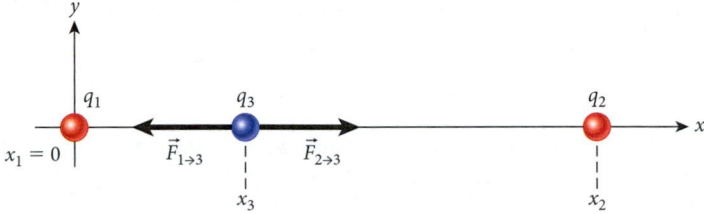

FIGURE 21.16 Placement of three charged particles. The third particle is shown as having a negative charge.

SOLUTION

Let's first determine where *not* to put the third charge. If the third charge is placed anywhere off the x-axis, there will always be a force component pointing toward or away from the x-axis. Thus, we can find an equilibrium point (a point where the forces sum to zero) only *on* the x-axis. The x-axis can be divided into three different segments: $x \le x_1 = 0$, $x_1 < x < x_2$, and $x_2 \le x$. For $x \le x_1 = 0$, the force vectors from both q_1 and q_2 acting on q_3 will point in the positive direction if the charge is negative and in the negative direction if the charge is positive. Because we are looking for a location where the two forces cancel, the segment $x \le x_1 = 0$ can be excluded. A similar argument excludes $x \ge x_2$.

In the remaining segment of the x-axis, $x_1 < x < x_2$, the forces from q_1 and q_2 on q_3 point in opposite directions. We look for the location, x_3, where the absolute magnitudes of both forces are equal and the forces thus sum to zero. We express the equality of the two forces as

$$\left| \vec{F}_{1 \to 3} \right| = \left| \vec{F}_{2 \to 3} \right|,$$

which we can rewrite as

$$k \frac{|q_1 q_3|}{(x_3 - x_1)^2} = k \frac{|q_3 q_2|}{(x_2 - x_3)^2}.$$

We now see that the magnitude and sign of the third charge do not matter because that charge cancels out, as does the constant k, giving us

$$\frac{q_1}{(x_3 - x_1)^2} = \frac{q_2}{(x_2 - x_3)^2}$$

or

$$q_1 (x_2 - x_3)^2 = q_2 (x_3 - x_1)^2. \tag{i}$$

Taking the square root of both sides and solving for x_3, we find

$$\sqrt{q_1} (x_2 - x_3) = \sqrt{q_2} (x_3 - x_1),$$

or

$$x_3 = \frac{\sqrt{q_1} x_2 + \sqrt{q_2} x_1}{\sqrt{q_1} + \sqrt{q_2}}.$$

We can take the square root of both sides of equation (i) because $x_1 < x_3 < x_2$, and so both of the roots, $x_2 - x_3$ and $x_3 - x_1$, are assured to be positive.

Inserting the numbers given in the problem statement, we obtain

$$x_3 = \frac{\sqrt{q_1} x_2 + \sqrt{q_2} x_1}{\sqrt{q_1} + \sqrt{q_2}} = \frac{\sqrt{0.15 \text{ μC}} (0.4 \text{ m})}{\sqrt{0.15 \text{ μC}} + \sqrt{0.35 \text{ μC}}} = 0.16 \text{ m}.$$

This result makes sense because we expect the equilibrium point to reside closer to the smaller charge.

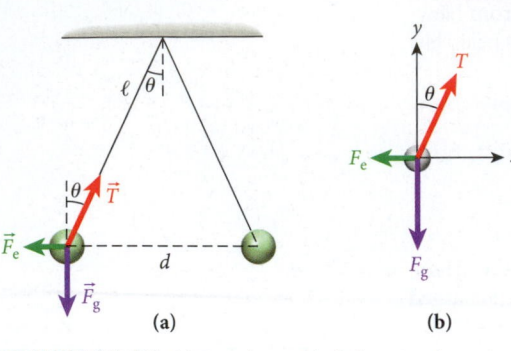

FIGURE 21.17 (a) Two charged balls hanging from the ceiling in their equilibrium position. (b) Free-body diagram for the left-hand charged ball.

SOLVED PROBLEM 21.1 | Charged Balls

PROBLEM

Two identical charged balls hang from the ceiling by insulated ropes of equal length, $\ell = 1.50$ m (Figure 21.17). A charge $q = 25.0$ μC is applied to each ball. Then the two balls hang at rest, and each supporting rope has an angle of 25.0° with respect to the vertical (Figure 21.17a). What is the mass of each ball?

SOLUTION

THINK Each charged ball has three forces acting on it: the force of gravity, the repulsive electrostatic force, and the tension in the supporting rope. Using the first condition for static equilibrium from Chapter 11, we know that the sum of all the forces on each ball must be zero. We can resolve the components of the three forces and set them equal to zero, allowing us to solve for the mass of the charged balls.

SKETCH A free-body diagram for the left-hand ball is shown in Figure 21.17b.

RESEARCH The condition for static equilibrium says that the sum of the x-components of the three forces acting on the ball must equal zero and the sum of y-components of these forces must equal zero. The sum of the x-components of the forces is

$$T \sin\theta - F_e = 0, \tag{i}$$

where T is the magnitude of the string tension, θ is the angle of the string relative to the vertical, and F_e is the magnitude of the electrostatic force. The sum of the y-components of the forces is

$$T \cos\theta - F_g = 0. \tag{ii}$$

The force of gravity, F_g, is just the weight of the charged ball:

$$F_g = mg, \tag{iii}$$

where m is the mass of the charged ball. The electrostatic force the two balls exert on each other is given by

$$F_e = k \frac{q^2}{d^2}, \tag{iv}$$

where d is the distance between the two balls. We can express the distance between the two balls in terms of the length of the string, ℓ, by looking at Figure 21.17a. We see that

$$\sin\theta = \frac{d/2}{\ell}.$$

We can then express the electrostatic force in terms of the angle with respect to the vertical, θ, and the length of the string, ℓ:

$$F_e = k \frac{q^2}{\left(2\ell \sin\theta\right)^2} = k \frac{q^2}{4\ell^2 \sin^2\theta}. \tag{v}$$

SIMPLIFY We divide equation (i) by equation (ii):

$$\frac{T \sin\theta}{T \cos\theta} = \frac{F_e}{F_g},$$

which, after the (unknown) string tension is canceled out, becomes

$$\tan\theta = \frac{F_e}{F_g}.$$

Substituting from equations (iii) and (v) for the force of gravity and the electrostatic force, we get

$$\tan\theta = \frac{k \dfrac{q^2}{4\ell^2 \sin^2\theta}}{mg} = \frac{kq^2}{4mg\ell^2 \sin^2\theta}.$$

Solving for the mass of the ball, we obtain

$$m = \frac{kq^2}{4g\ell^2 \sin^2\theta \tan\theta}.$$

CALCULATE Putting in the numerical values gives

$$m = \frac{\left(8.99\cdot10^9 \text{ N m}^2/\text{C}^2\right)\left(25.0 \text{ μC}\right)^2}{4\left(9.81 \text{ m/s}^2\right)\left(1.50 \text{ m}\right)^2 \left(\sin^2 25.0°\right)\left(\tan 25.0°\right)} = 0.764116 \text{ kg}.$$

ROUND We report our result to three significant figures:

$$m = 0.764 \text{ kg}.$$

DOUBLE-CHECK To double-check, we make the small-angle approximations that $\sin\theta \approx \tan\theta \approx \theta$ and $\cos\theta \approx 1$. The tension in the string then approaches mg, and we can express the x-components of the forces as

$$T\sin\theta \approx mg\theta = F_e = k\frac{q^2}{d^2} \approx k\frac{q^2}{(2\ell\theta)^2}.$$

Solving for the mass of the charged ball, we get

$$m = \frac{kq^2}{4g\ell^2\theta^3} = \frac{\left(8.99\cdot10^9 \text{ N m}^2/\text{C}^2\right)\left(25.0 \text{ μC}\right)^2}{4\left(9.81 \text{ m/s}^2\right)\left(1.50 \text{ m}\right)^2 \left(0.436 \text{ rad}\right)^3} = 0.768 \text{ kg},$$

which is close to our answer.

Electrostatic Precipitator

An application of electrostatic charging and electrostatic forces is the cleaning of emissions from coal-fired power plants. A device called an **electrostatic precipitator** (ESP) is used to remove ash and other particulates produced when coal is burned to generate electricity. The operation of this device is illustrated in Figure 21.18.

The ESP consists of wires and plates, with the wires held at a high negative voltage relative to the series of plates held at a positive voltage. (Here the term *voltage* is used colloquially; in Chapter 23, the concept will be defined in terms of electric potential difference.) In Figure 21.18, the exhaust gas from the coal-burning process enters the ESP from the left. Particulates passing near the wires pick up a negative charge. These particles are then attracted to one of the positive plates and stick there. The gas continues through the ESP, leaving the ash and other particulates behind. The accumulated material is then shaken off the plates to a hamper below. This waste can be used for many purposes, including construction materials and fertilizer. Figure 21.19 shows an example of a coal-fired power plant that incorporates an ESP.

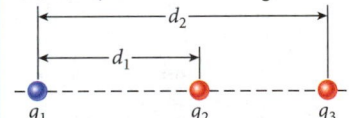
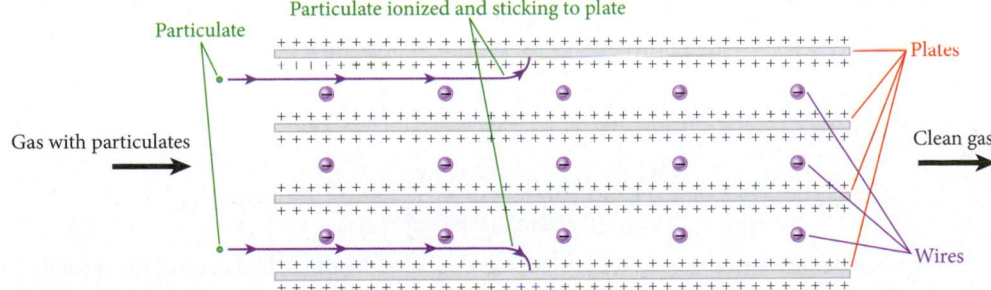

FIGURE 21.18 Operation of an electrostatic precipitator used to clean the exhaust gas of a coal-fired power plant. The view is from the top of the device.

FIGURE 21.19 A coal-fired power plant at Michigan State University that incorporates an electrostatic precipitator to remove particulates from its emissions.

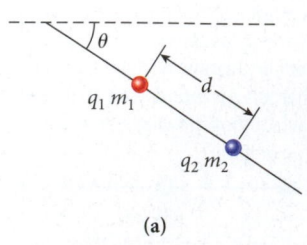

(a)

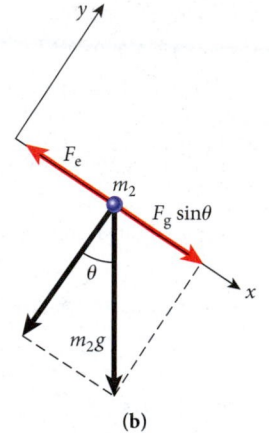

(b)

FIGURE 21.20 (a) Two charged beads on a wire. (b) Free-body diagram of the forces acting on the second bead.

SOLVED PROBLEM 21.2 | Bead on a Wire

PROBLEM

A bead with charge $q_1 = +1.28\ \mu\text{C}$ is fixed in place on an insulating wire that makes an angle of $\theta = 42.3°$ with respect to the horizontal (Figure 21.20a). A second bead with charge $q_2 = -5.06\ \mu\text{C}$ slides without friction on the wire. At a distance $d = 0.380$ m between the beads, the net force on the second bead is zero. What is the mass, m_2, of the second bead?

SOLUTION

THINK The force of gravity pulling the bead of mass m_2 down the wire is offset by the attractive electrostatic force between the positive charge on the first bead and the negative charge on the second bead. The second bead can be thought of as sliding on an inclined plane.

SKETCH Figure 21.20b shows a free-body diagram of the forces acting on the second bead. We have defined a coordinate system in which the positive x-direction is down the wire. The force exerted on m_2 by the wire can be omitted because this force has only a y-component, and we can solve the problem by analyzing just the x-components of the forces.

RESEARCH The attractive electrostatic force between the two beads balances the component of the force of gravity that acts on the second bead down the wire. The electrostatic force acts in the negative x-direction and its magnitude is given by

$$F_e = k\frac{|q_1 q_2|}{d^2}. \tag{i}$$

The x-component of the force of gravity acting on the second bead corresponds to the component of the weight of the second bead that is parallel to the wire. Figure 21.20b indicates that the component of the weight of the second bead down the wire is given by

$$F_g = m_2 g \sin\theta. \tag{ii}$$

SIMPLIFY For equilibrium, the electrostatic force and the gravitational force are equal: $F_e = F_g$. Substituting the expressions for these forces from equations (i) and (ii) yields

$$k\frac{|q_1 q_2|}{d^2} = m_2 g \sin\theta.$$

Solving this equation for the mass of the second bead gives us

$$m_2 = \frac{k|q_1 q_2|}{d^2 g \sin\theta}.$$

CALCULATE We put in the numerical values and get

$$m_2 = \frac{kq_1 q_2}{d^2 g \sin\theta} = \frac{\left(8.99 \cdot 10^9\ \text{N m}^2/\text{C}^2\right)(1.28\ \mu\text{C})(5.06\ \mu\text{C})}{(0.380\ \text{m})^2 \left(9.81\ \text{m/s}^2\right)(\sin 42.3°)} = 0.0610746\ \text{kg}.$$

ROUND We report our result to three significant figures:

$$m_2 = 0.0611\ \text{kg} = 61.1\ \text{g}.$$

DOUBLE-CHECK To double-check, let's calculate the mass of the second bead assuming that the wire is vertical, that is, $\theta = 90°$. We can then set the weight of the second bead equal to the electrostatic force between the two beads:

$$k\frac{|q_1 q_2|}{d^2} = m_2 g.$$

Solving for the mass of the second bead, we obtain

$$m_2 = \frac{kq_1 q_2}{d^2 g} = \frac{\left(8.99 \cdot 10^9\ \text{N m}^2/\text{C}^2\right)(1.28\ \mu\text{C})(5.06\ \mu\text{C})}{(0.380\ \text{m})^2 \left(9.81\ \text{m/s}^2\right)} = 0.0411\ \text{kg}.$$

As the angle of the wire relative to the horizontal decreases, the calculated mass of the second bead will increase. Our result of 0.0611 kg is somewhat higher than the mass that can be supported with a vertical wire, so it seems reasonable.

Laser Printer

Another example of a device that applies electro-static forces is the laser printer. The operation of such a printer is illustrated in Figure 21.21. The paper path follows the blue arrows. Paper is taken from the paper tray or fed manually through the alternate paper feed. The paper passes over a drum where the toner is placed on the surface of the paper and then passes through a fuser that melts the toner and permanently affixes it to the paper.

The drum consists of a metal cylinder coated with a special photosensitive material. The photosensitive surface is an insulator that retains charge in the absence of light, but dis-charges quickly if light is incident on the sur-face. The drum rotates so that its surface speed is the same as the speed of the moving paper. The basic principle of the operation of the drum is illustrated in Figure 21.22.

The drum is negatively charged with electrons using a wire held at high voltage. Then laser light is directed at the surface of the drum. Wherever the laser light strikes the surface of the drum, the surface at that point is discharged. A laser is used because its beam is narrow and remains focused. A line of the image being printed is written one pixel (picture element or dot) at a time using a laser beam directed by a moving mirror and a lens. A typical laser printer can write 600–1200 pixels per inch. The surface of the drum then passes by a roller that picks up toner from the toner cartridge. Toner consists of small, black, insulating particles composed of a plastic-like material. The toner roller is charged to the same negative voltage as the drum. Therefore, wher-ever the surface of the drum has been discharged, electrostatic forces deposit toner on the surface of the drum. Any portion of the drum surface that has not been exposed to the laser will not pick up toner.

As the drum rotates, it next comes in contact with the paper. The toner is then transferred from the surface of the drum to the paper. As the drum rotates, any remaining toner is scraped off and the surface is neutralized with an erase light or a rotating erase drum in preparation for printing the next image. The paper then continues on to the fuser, which melts the toner, producing a permanent image on the paper.

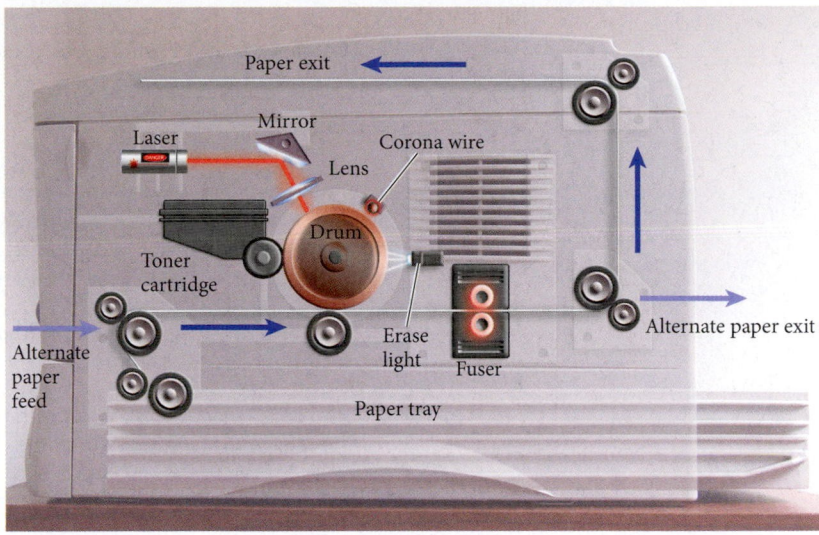

FIGURE 21.21 The operation of a typical laser printer.

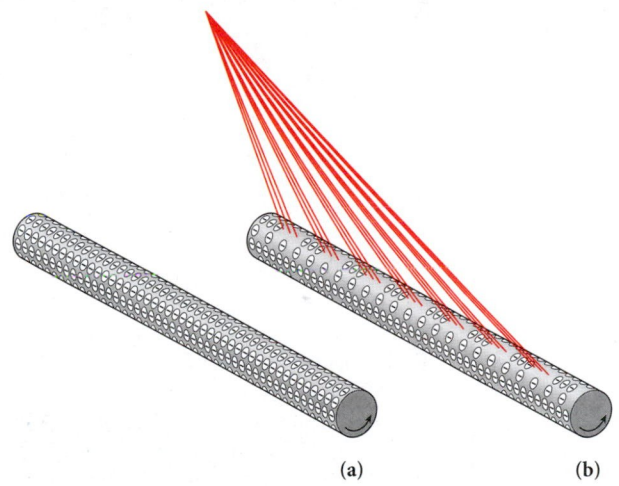

FIGURE 21.22 (a) The completely charged drum of a laser printer. This drum will produce a blank page. (b) A drum on which one line of information is being recorded by a laser. Wherever the laser strikes the charged drum, the negative charge is neutralized, and the discharged area will attract toner that will produce an image on the paper.

SOLVED PROBLEM 21.3 / Four Charged Objects

Consider four charges placed at the corners of a square with side length 1.25 m, as shown in Figure 21.23a.

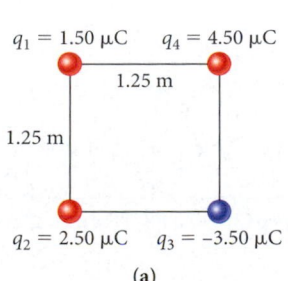

(a)

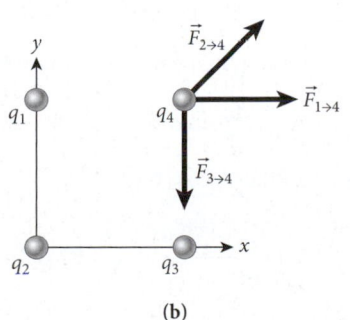

(b)

FIGURE 21.23 (a) Four charges placed at the corners of a square. (b) The forces exerted on q_4 by the other three charges.

– Continued

PROBLEM

What are the magnitude and the direction of the electrostatic force on q_4 resulting from the other three charges?

SOLUTION

THINK The electrostatic force on q_4 is the vector sum of the forces resulting from its interactions with the other three charges. Thus, it is important to avoid simply adding the individual force magnitudes algebraically. Instead we need to determine the individual force components in each spatial direction and add those to find the components of the net force vector. Then we need to calculate the length of that net force vector.

SKETCH Figure 21.23b shows the four charges in an xy-coordinate system with its origin at the location of q_2.

RESEARCH The net force on q_4 is the vector sum of the forces $\vec{F}_{1\to4}$, $\vec{F}_{2\to4}$, and $\vec{F}_{3\to4}$. The x-component of the summed forces is

$$F_x = k\frac{|q_1 q_4|}{d^2} + k\frac{|q_2 q_4|}{\left(\sqrt{2}d\right)^2}\cos 45° = \frac{kq_4}{d^2}\left(q_1 + \frac{q_2}{2}\cos 45°\right), \tag{i}$$

where d is the length of a side of the square and, as Figure 21.23b indicates, the x-component of $\vec{F}_{3\to4}$ is zero. The y-component of the summed forces is

$$F_y = k\frac{|q_2 q_4|}{\left(\sqrt{2}d\right)^2}\sin 45° - k\frac{|q_3 q_4|}{d^2} = \frac{kq_4}{d^2}\left(\frac{q_2}{2}\sin 45° + q_3\right), \tag{ii}$$

where, as Figure 21.23b indicates, the y-component of $\vec{F}_{1\to4}$ is zero.

The magnitude of the net force is given by

$$F = \sqrt{F_x^2 + F_y^2}, \tag{iii}$$

and the angle of the net force is given by

$$\tan\theta = \frac{F_y}{F_x}.$$

SIMPLIFY We substitute the expressions for F_x and F_y from equations (i) and (ii) into equation (iii):

$$F = \sqrt{\left[\frac{kq_4}{d^2}\left(q_1 + \frac{q_2}{2}\cos 45°\right)\right]^2 + \left[\frac{kq_4}{d^2}\left(\frac{q_2}{2}\sin 45° + q_3\right)\right]^2}.$$

We can rewrite this as

$$F = \frac{kq_4}{d^2}\sqrt{\left(q_1 + \frac{q_2}{2}\cos 45°\right)^2 + \left(\frac{q_2}{2}\sin 45° + q_3\right)^2}.$$

For the angle of the force, we get

$$\theta = \tan^{-1}\left(\frac{F_y}{F_x}\right) = \tan^{-1}\left(\frac{\frac{kq_4}{d^2}\left(\frac{q_2}{2}\sin 45° + q_3\right)}{\frac{kq_4}{d^2}\left(q_1 + \frac{q_2}{2}\cos 45°\right)}\right) = \tan^{-1}\left(\frac{\left(\frac{q_2}{2}\sin 45° + q_3\right)}{\left(q_1 + \frac{q_2}{2}\cos 45°\right)}\right).$$

CALCULATE Putting in the numerical values, we get

$$\frac{q_2}{2}\sin 45° = \frac{q_2}{2}\cos 45° = \frac{2.50\ \mu C}{2\sqrt{2}} = 0.883883\ \mu C.$$

The magnitude of the force is then

$$F = \frac{\left(8.99\cdot10^9\ \text{N m}^2/\text{C}^2\right)\left(4.50\ \mu C\right)}{\left(1.25\ \text{m}\right)^2}\sqrt{\left(1.50\ \mu C + 0.883883\ \mu C\right)^2 + \left(0.883883\ \mu C - 3.50\ \mu C\right)^2}$$

$$= 0.0916379\ \text{N}.$$

For the direction of the force, we obtain

$$\theta = \tan^{-1}\left(\frac{\left|\frac{q_2}{2}\sin 45° + q_3\right|}{\left|q_1 + \frac{q_2}{2}\cos 45°\right|}\right) = \tan^{-1}\left(\frac{(0.883883\ \mu C - 3.50\ \mu C)}{(1.50\ \mu C + 0.883883\ \mu C)}\right) = -47.6593°.$$

ROUND   We report our results to three significant figures:

$$F = 0.0916\ N$$

and

$$\theta = -47.7°.$$

DOUBLE-CHECK   To double-check our result, we calculate the magnitude of the three forces acting on q_4. For $\vec{F}_{1\to4}$, we get

$$F_{1\to4} = k\frac{|q_1 q_4|}{r_{14}^2} = \frac{\left(8.99\cdot10^9\ N\ m^2/C^2\right)(1.50\ \mu C)(4.50\ \mu C)}{(1.25\ m)^2} = 0.0388\ N.$$

For $\vec{F}_{2\to4}$, we get

$$F_{2\to4} = k\frac{|q_2 q_4|}{r_{24}^2} = \frac{\left(8.99\cdot10^9\ N\ m^2/C^2\right)(2.50\ \mu C)(4.50\ \mu C)}{\left[\sqrt{2}\,(1.25\ m)\right]^2} = 0.0324\ N.$$

For $F_{3\to4}$, we get

$$F_{3\to4} = k\frac{|q_3 q_4|}{r_{34}^2} = \frac{\left(8.99\cdot10^9\ N\ m^2/C^2\right)(3.50\ \mu C)(4.50\ \mu C)}{(1.25\ m)^2} = 0.0906\ N.$$

All three of the magnitudes of the individual forces are of the same order as our result for the net force. This gives us confidence that our answer is not off by a large factor.

The direction we obtained also seems reasonable, because it orients the resulting force downward and to the right, as could be expected from looking at Figure 21.23b.

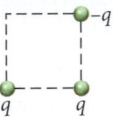
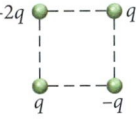
21.6   Coulomb's Law and Newton's Law of Gravitation

Coulomb's Law describing the electrostatic force between two electric charges, F_e, has a form similar to Newton's Law describing the gravitational force between two masses, F_g:

$$F_g = G\frac{m_1 m_2}{r^2} \quad \text{and} \quad F_e = k\frac{|q_1 q_2|}{r^2},$$

where m_1 and m_2 are the two masses, q_1 and q_2 are the two electric charges, and r is the distance of separation. Both forces vary with the inverse square of the distance. The electrostatic force can be attractive or repulsive because charges can have positive or negative signs. (See Figure 21.14a and b.) The gravitational force is always attractive because there is only one kind of mass. (For the gravitational force, only the case depicted in Figure 21.14b is possible.) The relative strengths of the forces are given by the proportionality constants k and G.

EXAMPLE 21.4   Forces between Electrons

Let's evaluate the relative strengths of the two interactions by calculating the ratio of the electrostatic force and the gravitational force that two electrons exert on each other. This ratio is given by

$$\frac{F_e}{F_g} = \frac{kq_e^2}{Gm_e^2}.$$

Because the dependence on distance is the same for both forces, there is no dependence on distance in the ratio of the two forces—it cancels out. The mass of an electron is $m_e = 9.109\cdot10^{-31}$ kg,

– Continued

Concept Check 21.11

The proton's mass is ~2000 times larger than the electron's mass. Therefore, the ratio F_e/F_g for two protons is _____ the value calculated in Example 21.4 for two electrons.

a) ~4 million times smaller than

b) ~2000 times smaller than

c) the same as

d) ~2000 times larger than

e) ~4 million times larger than

and its charge is $q_e = -1.602 \cdot 10^{-19}$ C. Using the value of Coulomb's constant given in equation 21.7, $k = 8.99 \cdot 10^9$ N m^2/C^2, and the value of the universal gravitational constant, $G = 6.67 \cdot 10^{-11}$ N m^2/kg^2, we find numerically

$$\frac{F_e}{F_g} = \frac{(8.99 \cdot 10^9 \text{ N m}^2/\text{C}^2)(1.602 \cdot 10^{-19} \text{ C})^2}{(6.67 \cdot 10^{-11} \text{ N m}^2/\text{kg}^2)(9.109 \cdot 10^{-31} \text{ kg})^2} = 4.17 \cdot 10^{42}.$$

Therefore, the electrostatic force between electrons is stronger than the gravitational force between them by more than 42 orders of magnitude.

Despite the relative weakness of the gravitational force, it is the only force that matters on the astronomical scale. The reason for this dominance is that all stars, planets, and other objects of astronomical relevance carry no net charge. Therefore, there is no net electrostatic interaction between them, and gravity dominates.

Coulomb's Law of electrostatics applies to macroscopic systems down to the atom, though subtle effects in atomic and subatomic systems require use of a more sophisticated approach called *quantum electrodynamics*. Newton's law of gravitation fails in subatomic systems and also must be modified for some phenomena in astronomical systems, such as the precessional motion of Mercury around the Sun. These fine details of the gravitational interaction are governed by Einstein's theory of general relativity.

The similarities between the gravitational and electrostatic interactions will be covered further in the next two chapters, which address electric fields and electric potential.

WHAT WE HAVE LEARNED | EXAM STUDY GUIDE

- There are two kinds of electric charge, positive and negative. Like charges repel, and unlike charges attract.

- The quantum (elementary quantity) of electric charge is $e = 1.602 \cdot 10^{-19}$ C.

- The electron has charge $q_e = -e$, and the proton has charge $q_p = +e$. The neutron has zero charge.

- The net charge of an object is given by e times the number of protons, N_p, minus e times the number of electrons, N_e, that make up the object: $q = e \cdot (N_p - N_e)$.

- The total charge in an isolated system is always conserved.

- Objects can be charged directly by contact or indirectly by induction.

- Coulomb's Law describes the force that two stationary charges exert on each other: $F = k\dfrac{|q_1 q_2|}{r^2} = \dfrac{1}{4\pi\epsilon_0}\dfrac{|q_1 q_2|}{r^2}.$

- The constant in Coulomb's Law is
$$k = \frac{1}{4\pi\epsilon_0} = 8.99 \cdot 10^9 \ \frac{\text{N m}^2}{\text{C}^2}.$$

- The electric permittivity of free space is
$$\epsilon_0 = 8.85 \cdot 10^{-12} \ \frac{\text{C}^2}{\text{N m}^2}.$$

ANSWERS TO SELF-TEST OPPORTUNITIES

21.1 a) +1 c) 0 e) $+\frac{2}{3}$ g) −1

 b) 0 d) 0 f) $-\frac{1}{3}$ h) +2

22.2 a) true c) false e) true

 b) false d) true

PROBLEM-SOLVING GUIDELINES

1. For problems involving Coulomb's Law, drawing a free-body diagram showing the electrostatic force vectors acting on a charged particle is often helpful. Pay careful attention to signs; a negative force between two particles indicates attraction, and a positive force indicates repulsion. Be sure that the directions of forces in the diagram match the signs of forces in the calculations.

2. Use symmetry to simplify your work. However, be careful to take account of charge magnitudes and signs as well

as distances. Two charges at equal distances from a third charge do not exert equal forces on that charge if they have different magnitudes or signs.

3. Units in electrostatics often have prefixes indicating powers of 10: Distances may be given in cm or mm; charges may be given in μC, nC, or pC; masses may be given in kg or g. Other units are also common. The best way to proceed is to convert all quantities to basic SI units, to be compatible with the value of k or $1/4\pi\epsilon_0$.

MULTIPLE-CHOICE QUESTIONS

21.1 When a metal plate is given a positive charge, which of the following is taking place?

a) Protons (positive charges) are transferred to the plate from another object.

b) Electrons (negative charges) are transferred from the plate to another object.

c) Electrons (negative charges) are transferred from the plate to another object, and protons (positive charges) are also transferred to the plate from another object.

d) It depends on whether the object conveying the charge is a conductor or an insulator.

21.2 The force between a charge of 25 μC and a charge of –10 μC is 8.0 N. What is the separation between the two charges?

a) 0.28 m   c) 0.45 m

b) 0.53 m   d) 0.15 m

21.3 A charge Q_1 is positioned on the x-axis at $x = a$. Where should a charge $Q_2 = -4Q_1$ be placed to produce a net electrostatic force of zero on a third charge, $Q_3 = Q_1$, located at the origin?

a) at the origin   c) at $x = -2a$

b) at $x = 2a$ d) at $x = -a$

21.4 Which one of these systems has the most negative charge?

a) 2 electrons

b) 3 electrons and 1 proton

c) 5 electrons and 5 protons

d) N electrons and $N - 3$ protons

e) 1 electron

21.5 Two point charges are fixed on the x-axis: $q_1 = 6.0$ μC is located at the origin, O, with $x_1 = 0.0$ cm, and $q_2 = -3.0$ μC is located at point A, with $x_2 = 8.0$ cm. Where should a third charge, q_3, be placed on the x-axis so that the total electrostatic force acting on it is zero?

a) 19 cm   c) 0.0 cm   e) –19 cm

b) 27 cm   d) 8.0 cm

21.6 Which of the following situations produces the largest net force on the charge Q?

a) Charge $Q = 1$ C is 1 m from a charge of –2 C.

b) Charge $Q = 1$ C is 0.5 m from a charge of –1 C.

c) Charge $Q = 1$ C is halfway between a charge of –1 C and a charge of 1 C that are 2 m apart.

d) Charge $Q = 1$ C is halfway between two charges of –2 C that are 2 m apart.

e) Charge $Q = 1$ C is a distance of 2 m from a charge of –4 C.

21.7 Two protons placed near one another with no other objects close by would

a) accelerate away from each other.

b) remain motionless.

c) accelerate toward each other.

d) be pulled together at constant speed.

e) move away from each other at constant speed.

21.8 Two lightweight metal spheres are suspended near each other from insulating threads. One sphere has a net charge; the other sphere has no net charge. The spheres will

a) attract each other.

b) exert no net electrostatic force on each other.

c) repel each other.

d) do any of these things depending on the sign of the net charge on the one sphere.

21.9 A metal plate is connected by a conductor to a ground through a switch. The switch is initially closed. A charge $+Q$ is brought close to the plate without touching it, and then the switch is opened. After the switch is opened, the charge $+Q$ is removed. What is the charge on the plate then?

a) The plate is uncharged.

b) The plate is positively charged.

c) The plate is negatively charged.

d) The plate could be either positively or negatively charged, depending on the charge it had before $+Q$ was brought near.

21.10 You bring a negatively charged rubber rod close to a grounded conductor without touching it. Then you disconnect the ground. What is the sign of the charge on the conductor after you remove the charged rod?

a) negative

b) positive

c) no charge

d) cannot be determined from the information given

21.11 When a rubber rod is rubbed with rabbit fur, the rod becomes

a) negatively charged.

b) positively charged.

c) neutral.

d) either negatively charged or positively charged, depending on whether the fur is always moved in the same direction or is moved back and forth.

21.12 When a glass rod is rubbed with a polyester scarf, the rod becomes

a) negatively charged.

b) positively charged.

c) neutral.

d) either negatively charged or positively charged, depending on whether the scarf is always moved in the same direction or is moved back and forth.

21.13 Consider an electron with mass m and charge $-e$ moving in a circular orbit with radius r around a fixed proton with mass M and charge $+e$. The electron is held in orbit by the electrostatic force between itself and the proton. Which one of the following expressions for the speed of the electron is correct?

a) $v = \sqrt{\dfrac{ke^2}{mr}}$ c) $v = \sqrt{\dfrac{2ke^2}{mr^2}}$ e) $v = \sqrt{\dfrac{ke^2}{2Mr}}$

b) $v = \sqrt{\dfrac{GM}{r}}$ d) $v = \sqrt{\dfrac{me^2}{kr}}$

21.14 Consider an electron with mass m and charge $-e$ located a distance r from a fixed proton with mass M and charge $+e$. The electron is released from rest. Which one of the following expressions for the magnitude of the initial acceleration of the electron is correct?

a) $a = \dfrac{2ke^2}{mMr}$ c) $a = \frac{1}{2}me^2k^2$ e) $a = \dfrac{ke^2}{mr^2}$

b) $a = \sqrt{\dfrac{2e^2}{mkr}}$ d) $a = \dfrac{2ke^2}{mr}$

CONCEPTUAL QUESTIONS

21.15 If two charged particles (the charge on each is Q) are separated by a distance d, there is a force F between them. What is the force if the magnitude of each charge is doubled and the distance between them changes to $2d$?

21.16 Suppose the Sun and the Earth were each given an equal amount of charge of the same sign, just sufficient to cancel their gravitational attraction. How many times the charge on an electron would that charge be? Is this number a large fraction of the number of charges of either sign in the Earth?

21.17 It is apparent that the electrostatic force is *extremely* strong, compared to gravity. In fact, the electrostatic force is the basic force governing phenomena in daily life—the tension in a string, the normal forces between surfaces, friction, chemical reactions, and so forth—except weight. Why then did it take so long for scientists to understand this force? Newton came up with his gravitational law long before electricity was even crudely understood.

21.18 Occasionally, people who gain static charge by shuffling their feet on the carpet will have their hair stand on end. Why does this happen?

21.19 Two positive charges, each equal to Q, are placed a distance $2d$ apart. A third charge, $-0.2Q$, is placed exactly halfway between the two positive charges and is displaced a distance $x \ll d$ (that is, x is much smaller than d) perpendicular to the line connecting the positive charges. What is the force on this charge? For $x \ll d$, how can you approximate the motion of the negative charge?

21.20 Why does a garment taken out of a clothes dryer sometimes cling to your body when you wear it?

21.21 Two charged spheres are initially a distance d apart. The magnitude of the force on each sphere is F. They are moved closer to each other such that the magnitude of the force on each of them is $9F$. By what factor has the distance between the two spheres changed?

21.22 How is it possible for one electrically neutral atom to exert an electrostatic force on another electrically neutral atom?

21.23 The scientists who first contributed to the understanding of the electrostatic force in the 18th century were well aware of Newton's law of gravitation. How could they deduce that the force they were studying was *not* a variant or some manifestation of the gravitational force?

21.24 Two charged particles move solely under the influence of the electrostatic forces between them. What shapes can their trajectories have?

21.25 Rubbing a balloon causes it to become negatively charged. The balloon then tends to cling to the wall of a room. For this to happen, must the wall be positively charged?

21.26 Two electric charges are placed on a line, as shown in the figure. Is it possible to place a charged particle (that is free to move) anywhere on the line between the two charges and have it not move?

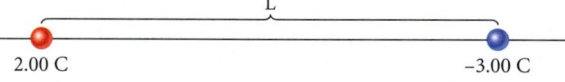

2.00 C   −3.00 C

21.27 Two electric charges are placed on a line as shown in the figure. Where on the line can a third charge be placed so that the force on that charge is zero? Does the sign or the magnitude of the third charge make any difference to the answer?

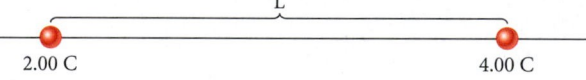

2.00 C   4.00 C

21.28 When a positively charged rod is brought close to a neutral conductor without touching it, will the rod experience an attractive force, a repulsive force, or no force at all? Explain.

21.29 When you exit a car and the humidity is low, you often experience a shock from static electricity created by sliding across the seat. How can you discharge yourself without experiencing a painful shock? Why is it dangerous to get back into your car while fueling your car?

EXERCISES

A blue problem number indicates a worked-out solution is available in the Student Solutions Manual. One • and two •• indicate increasing level of problem difficulty.

Section 21.2

21.30 How many electrons are required to yield a total charge of 1.00 C?

21.31 The *faraday* is a unit of charge frequently encountered in electrochemical applications and named for the British physicist and chemist Michael Faraday. It consists of exactly 1 mole of elementary charges. Calculate the number of coulombs in 1.000 faraday.

21.32 Another unit of charge is the *electrostatic unit* (esu). It is defined as follows: Two point charges, each of 1 esu and separated by 1 cm, exert a force of exactly 1 dyne on each other: 1 dyne $= 1$ g cm/s$^2 = 1 \cdot 10^{-5}$ N.

a) Determine the relationship between the esu and the coulomb.

b) Determine the relationship between the esu and the elementary charge.

21.33 A current of 5.00 mA is enough to make your muscles twitch. Calculate how many electrons flow through your skin if you are exposed to such a current for 10.0 s.

•**21.34** How many electrons does 1.00 kg of water contain?

•**21.35** The Earth is constantly being bombarded by cosmic rays, which consist mostly of protons. These protons are incident on the Earth's atmosphere from all directions at a rate of 1245. protons per square meter per second. Assuming that the depth of Earth's atmosphere is 120.0 km, what is the total charge incident on the atmosphere in 5.000 min? Assume that the radius of the surface of the Earth is 6378. km.

•**21.36** Performing an experiment similar to Millikan's oil drop experiment, a student measures these charge magnitudes:

$3.26 \cdot 10^{-19}$ C $5.09 \cdot 10^{-19}$ C $1.53 \cdot 10^{-19}$ C

$6.39 \cdot 10^{-19}$ C $4.66 \cdot 10^{-19}$ C

Find the charge on the electron using these measurements.

Section 21.3

•**21.37** A silicon sample is doped with phosphorus at 1 part per $1.00 \cdot 10^6$. Phosphorus acts as an electron donor, providing one free electron per atom. The density of silicon is 2.33 g/cm³, and its atomic mass is 28.09 g/mol.

a) Calculate the number of free (conduction) electrons per unit volume of the doped silicon.

b) Compare the result from part (a) with the number of conduction electrons per unit volume of copper wire, assuming that each copper atom produces one free (conduction) electron. The density of copper is 8.96 g/cm³, and its atomic mass is 63.54 g/mol.

Section 21.5

21.38 Two charged spheres are 8.00 cm apart. They are moved closer to each other by enough that the force on each of them increases four times. How far apart are they now?

21.39 Two identically charged particles separated by a distance of 1.00 m repel each other with a force of 1.00 N. What is the magnitude of the charges?

21.40 How far apart must two electrons be placed on the Earth's surface for there to be an electrostatic force between them equal to the weight of one of the electrons?

21.41 In solid sodium chloride (table salt), chloride ions have one more electron than they have protons, and sodium ions have one more proton than they have electrons. These ions are separated by about 0.28 nm. Calculate the electrostatic force between a sodium ion and a chloride ion.

21.42 In gaseous sodium chloride, chloride ions have one more electron than they have protons, and sodium ions have one more proton than they have electrons. These ions are separated by about 0.24 nm. Suppose a free electron is located 0.48 nm above the midpoint of the sodium chloride molecule. What are the magnitude and the direction of the electrostatic force the molecule exerts on it?

21.43 Calculate the magnitude of the electrostatic force the two up quarks inside a proton exert on each other if they are separated by a distance of 0.900 fm.

21.44 A $-4.00\text{-}\mu$C charge lies 20.0 cm to the right of a $2.00\text{-}\mu$C charge on the x-axis. What is the force on the $2.00\text{-}\mu$C charge?

•**21.45** Two initially uncharged identical metal spheres, 1 and 2, are connected by an insulating spring (unstretched length $L_0 = 1.00$ m, spring constant $k = 25.0$ N/m), as shown in the figure. Charges $+q$ and $-q$ are then placed on the spheres, and the spring contracts to length $L = 0.635$ m. Recall that the force exerted by a spring is $F_s = k\Delta x$, where Δx is the change in the spring's length from its equilibrium length. Determine the charge q. If the spring is coated with metal to make it conducting, what is the new length of the spring?

Before charging   After charging

$q_1 = +q$ $q_2 = -q$

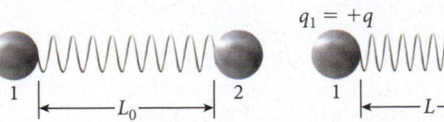

•**21.46** A point charge $+3q$ is located at the origin, and a point charge $-q$ is located on the x-axis at $D = 0.500$ m. At what location on the x-axis will a third charge, q_0, experience no net force from the other two charges?

•**21.47** Identical point charges Q are placed at each of the four corners of a rectangle measuring 2.00 m by 3.00 m. If $Q = 32.0$ μC, what is the magnitude of the electrostatic force on any one of the charges?

•**21.48** Charge $q_1 = 1.40 \cdot 10^{-8}$ C is placed at the origin. Charges $q_2 = -1.80 \cdot 10^{-8}$ C and $q_3 = 2.10 \cdot 10^{-8}$ C are placed at points (0.180 m,0.000 m) and (0.000 m,0.240 m), respectively, as shown in the figure. Determine the net electrostatic force (magnitude and direction) on charge q_3.

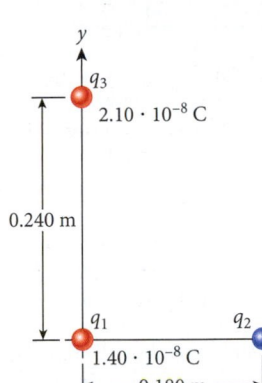

•**21.49** A positive charge Q is on the y-axis at a distance a from the origin, and another positive charge q is on the x-axis at a distance b from the origin.

a) For what value(s) of b is the x-component of the force on q a minimum?

b) For what value(s) of b is the x-component of the force on q a maximum?

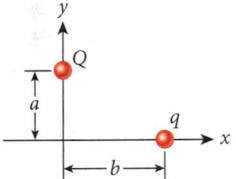

•**21.50** Find the magnitude and direction of the electrostatic force acting on the electron in the figure.

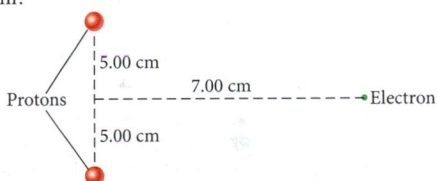

•**21.51** In a region of two-dimensional space, there are three fixed charges: +1.00 mC at (0,0), –2.00 mC at (17.0 mm,–5.00 mm), and +3.00 mC at (–2.00 mm,11.0 mm). What is the net force on the –2.00-mC charge?

•**21.52** Two cylindrical glass beads each of mass $m = 10.0$ mg are set on their flat ends on a horizontal insulating surface separated by a distance $d = 2.00$ cm. The coefficient of static friction between the beads and the surface is $\mu_s = 0.200$. The beads are then given identical charges (magnitude and sign). What is the minimum charge needed to start the beads moving?

•**21.53** A small ball with a mass of 30.0 g and a charge of –0.200 μC is suspended from the ceiling by a string. The ball hangs at a distance of 5.00 cm above an insulating floor. If a second small ball with a mass of 50.0 g and a charge of 0.400 μC is rolled directly beneath the first ball, will the second ball leave the floor? What is the tension in the string when the second ball is directly beneath the first ball?

•**21.54** A +3.00-mC charge and a –4.00-mC charge are fixed in position and separated by 5.00 m.

a) Where can a +7.00-mC charge be placed so that the net force on it is zero?

b) Where can a –7.00-mC charge be placed so that the net force on it is zero?

•**21.55** Four point charges, q, are fixed to the four corners of a square that is 10.0 cm on a side. An electron is suspended above a point at which its weight is balanced by the electrostatic force due to the four electrons, at a distance of 15.0 nm above the center of the square. What is the magnitude of the fixed charges? Express the charge both in coulombs and as a multiple of the electron's charge.

••**21.56** The figure shows a uniformly charged thin rod of length L that has total charge Q. Find an expression for the magnitude of the electrostatic force acting on an electron positioned on the axis of the rod at a distance d from the midpoint of the rod.

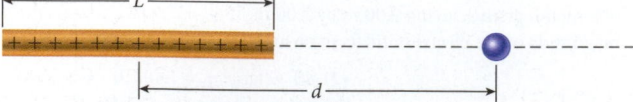

••**21.57** A negative charge, $-q$, is fixed at the coordinate (0,0). It is exerting an attractive force on a positive charge, $+q$, that is initially at coordinate $(x,0)$. As a result, the positive charge accelerates toward the negative charge. Use the binomial expansion $(1+x)^n \approx 1 + nx$, for $x \ll 1$, to show that when the positive charge moves a distance $\delta \ll x$ closer to the negative charge, the force that the negative charge exerts on it increases by $\Delta F = 2kq^2\delta/x^3$.

••**21.58** Two negative charges ($-q$ and $-q$) of equal magnitude are fixed at coordinates $(-d,0)$ and $(d,0)$. A positive charge of the same magnitude, q, and with mass m is placed at coordinate (0,0), midway between the two negative charges. If the positive charge is moved a distance $\delta \ll d$ in the positive y-direction and then released, the resulting motion will be that of a harmonic oscillator—the positive charge will oscillate between coordinates $(0,\delta)$ and $(0,-\delta)$. Find the net force acting on the positive charge when it moves to $(0,\delta)$ and use the binomial expansion $(1+x)^n \approx 1+nx$, for $x \ll 1$, to find an expression for the frequency of the resulting oscillation. (*Hint*: Keep only terms that are linear in δ.)

Section 21.6

21.59 Suppose the Earth and the Moon carried positive charges of equal magnitude. How large would the charge need to be to produce an electrostatic repulsion equal to 1.00% of the gravitational attraction between the two bodies?

21.60 The similarity of form of Newton's law of gravitation and Coulomb's Law caused some to speculate that the force of gravity is related to the electrostatic force. Suppose that gravitation is entirely electrical in nature—that an excess charge Q on the Earth and an equal and opposite excess charge $-Q$ on the Moon are responsible for the gravitational force that causes the observed orbital motion of the Moon about the Earth. What is the required size of Q to reproduce the observed magnitude of the gravitational force?

•**21.61** In the Bohr model of the hydrogen atom, the electron moves around the one-proton nucleus on circular orbits of well-determined radii, given by $r_n = n^2 a_B$, where $n = 1, 2, 3, \ldots$ is an integer that defines the orbit and $a_B = 5.29 \cdot 10^{-11}$m is the radius of the first (minimum) orbit, called the *Bohr radius*. Calculate the force of electrostatic interaction between the electron and the proton in the hydrogen atom for the first four orbits. Compare the strength of this interaction to the gravitational interaction between the proton and the electron.

•**21.62** Some of the earliest atomic models held that the orbital velocity of an electron in an atom could be correlated with the radius of the atom. If the radius of the hydrogen atom is $5.29 \cdot 10^{-11}$ m and the electrostatic force is responsible for the circular motion of the electron, what is the kinetic energy of this orbital electron?

21.63 For the atom described in Problem 21.62, what is the ratio of the gravitational force between electron and proton to the electrostatic force? How does this ratio change if the radius of the atom is doubled?

•**21.64** In general, astronomical objects are not exactly electrically neutral. Suppose the Earth and the Moon each carry a charge of $-1.00 \cdot 10^6$ C (this is approximately correct; a more precise value is identified in Chapter 22).

a) Compare the resulting electrostatic repulsion with the gravitational attraction between the Moon and the Earth. Look up any necessary data.

b) What effects does this electrostatic force have on the size, shape, and stability of the Moon's orbit around the Earth?

Additional Exercises

21.65 Eight 1.00-μC charges are arrayed along the y-axis located every 2.00 cm starting at $y = 0$ and extending to $y = 14.0$ cm. Find the force on the charge at $y = 4.00$ cm.

21.66 In a simplified Bohr model of the hydrogen atom, an electron is assumed to be traveling in a circular orbit of radius of about $5.29 \cdot 10^{-11}$ m around a proton. Calculate the speed of the electron in that orbit.

21.67 The nucleus of a carbon-14 atom (mass = 14 amu) has diameter of 3.01 fm. It has 6 protons and a charge of $+6e$.

a) What is the force on a proton located at 3.00 fm from the surface of this nucleus? Assume that the nucleus is a point charge.

b) What is the proton's acceleration?

21.68 Two charged objects experience a mutual repulsive force of 0.100 N. If the charge of one of the objects is reduced by half and the distance separating the objects is doubled, what is the new force?

21.69 A particle (charge = $+19.0$ μC) is located on the x-axis at $x = -10.0$ cm, and a second particle (charge = -57.0 μC) is placed on the x-axis at $x = +20.0$ cm. What is the magnitude of the total electrostatic force on a third particle (charge = -3.80 μC) placed at the origin $(x = 0)$?

21.70 Three point charges are positioned on the x-axis: $+64.0$ μC at $x = 0.00$ cm, $+80.0$ μC at $x = 25.0$ cm, and -160.0 μC at $x = 50.0$ cm. What is the magnitude of the electrostatic force acting on the $+64.0$-μC charge?

21.71 From collisions with cosmic rays and from the solar wind, the Earth has a net electric charge of approximately $-6.8 \cdot 10^5$ C. Find the charge that must be given to a 1.0-g object for it to be electrostatically levitated close to the Earth's surface.

21.72 Your sister wants to participate in the yearly science fair at her high school and asks you to suggest some exciting project. You suggest that she experiment with your recently created electron extractor to suspend her cat in the air. You tell her to buy a copper plate and bolt it to the ceiling in her room and then use your electron extractor to transfer electrons from the plate to the cat. If the cat weighs 7.00 kg and is suspended 2.00 m below the ceiling, how many electrons have to be extracted from the cat? Assume that the cat and the metal plate are point charges.

•**21.73** A 10.0-g mass is suspended 5.00 cm above a nonconducting flat plate, directly above an embedded charge of q (in coulombs). If the mass has the same charge, q, how much must q be so that the mass levitates (just floats, neither rising nor falling)? If the charge q is produced by adding electrons to the mass, by how much will the mass be changed?

•**21.74** Four point charges are placed at the following xy-coordinates:

$Q_1 = -1.00$ mC, at (-3.00 cm, 0.00 cm)

$Q_2 = -1.00$ mC, at ($+3.00$ cm, 0.00 cm)

$Q_3 = +1.024$ mC, at (0.00 cm, 0.00 cm)

$Q_4 = +2.00$ mC, at (0.00 cm, -4.00 cm)

Calculate the net force on charge Q_4 due to charges Q_1, Q_2, and Q_3.

•**21.75** Three 5.00-g Styrofoam balls of radius 2.00 cm are coated with carbon black to make them conducting and then are tied to 1.00-m-long threads and suspended freely from a common point. Each ball is given the same charge, q. At equilibrium, the balls form an equilateral triangle with sides of length 25.0 cm in the horizontal plane. Determine q.

•**21.76** Two point charges lie on the x-axis. If one point charge is 6.00 μC and lies at the origin and the other is -2.00 μC and lies at 20.0 cm, at what position must a third charge be placed to be in equilibrium?

•**21.77** Two beads with charges $q_1 = q_2 = +2.67$ μC are on an insulating string that hangs straight down from the ceiling as shown in the figure. The lower bead is fixed in place on the end of the string and has a mass $m_1 = 0.280$ kg. The second bead slides without friction on the string. At a distance $d = 0.360$ m between the

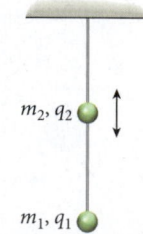

centers of the beads, the force of the Earth's gravity on m_2 is balanced by the electrostatic force between the two beads. What is the mass, m_2, of the second bead? (*Hint:* You can neglect the gravitational interaction between the two beads.)

•**21.78** Find the net force on a +2.00-C charge at the origin of an *xy*-coordinate system if there is a +5.00-C charge at (3.00 m,0.00) and a −3.00-C charge at (0.00,4.00 m).

•**21.79** Two spheres, each of mass $M = 2.33$ g, are attached by pieces of string of length $L = 45.0$ cm to a common point. The strings initially hang straight down, with the spheres just touching one another. An equal amount of charge, q, is placed on each sphere. The resulting forces on the spheres cause each string to hang at an angle of $\theta = 10.0°$ from the vertical. Determine q, the amount of charge on each sphere.

•**21.80** A point charge $q_1 = 100.$ nC is at the origin of an *xy*-coordinate system, a point charge $q_2 = −80.0$ nC is on the *x*-axis at $x = 2.00$ m, and a point charge $q_3 = −60.0$ nC is on the *y*-axis at $y = −2.00$ m. Determine the net force (magnitude and direction) on q_1.

•**21.81** A positive charge $q_1 = 1.00$ μC is fixed at the origin, and a second charge $q_2 = −2.00$ μC is fixed at $x = 10.0$ cm. Where along the *x*-axis should a third charge be positioned so that it experiences no force?

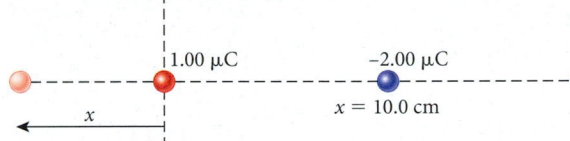

•**21.82** A bead with charge $q_1 = 1.27$ μC is fixed in place at the end of a wire that makes an angle of $\theta = 51.3°$ with the horizontal. A second bead with mass $m_2 = 3.77$ g and a charge of 6.79 μC slides without friction on the wire. What is the distance d at which the force of the Earth's gravity on m_2 is balanced by the electrostatic force between the two beads? Neglect the gravitational interaction between the two beads.

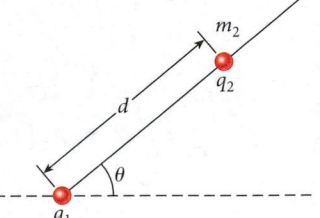

•**21.83** In the figure, the net electrostatic force on charge Q_A is zero. If $Q_A = +1.00$ nC, determine the magnitude of Q_0.

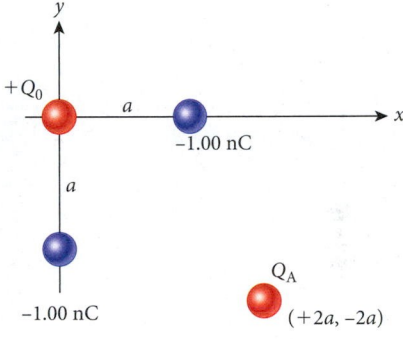

MULTI-VERSION EXERCISES

21.84 Two balls have the same mass, 0.9680 kg, and the same charge, 29.59 μC. They hang from the ceiling on strings of identical length, ℓ, as shown in the figure. If the angle of the strings with respect to the vertical is 29.79°, what is the length of the strings?

21.85 Two balls have the same mass and the same charge, 15.71 μC. They hang from the ceiling on strings of identical length, $\ell = 1.223$ m, as shown in the figure. The angle of the strings with respect to the vertical is 21.07°. What is the mass of each ball?

21.86 Two balls have the same mass, 0.9935 kg, and the same charge. They hang from the ceiling on strings of identical length, $\ell = 1.235$ m, as shown in the figure. The angle of the strings with respect to the vertical is 22.35°. What is the charge on each ball?

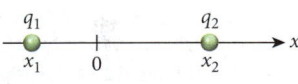

21.87 As shown in the figure, point charge q_1 is 3.979 μC and is located at $x_1 = −5.689$ m, and point charge q_2 is 8.669 μC and is located at $x_2 = 14.13$ m. What is the *x*-coordinate of the point at which the net force on a point charge of 5.000 μC will be zero?

21.88 As shown in the figure, point charge q_1 is 4.325 μC and is located at x_1, and point charge q_2 is 7.757 μC and is located at $x_2 = 14.33$ m. The *x*-coordinate of the point where the net force on a point charge of −3.000 μC is zero is 2.358 m. What is the value of x_1?

21.89 As shown in the figure, point charge q_1 is 4.671 μC and is located at $x_1 = −3.573$ m, and point charge q_2 is 6.845 μC and is located at x_2. The *x*-coordinate of the point where the net force on a point charge of −1.000 μC is zero is 4.625 m. What is the value of x_2?

Electric Fields and Gauss's Law

22

FIGURE 22.1 A great white shark can detect tiny electric fields generated by its prey.

The great white shark is one of the most feared predators on Earth (Figure 22.1). It has several senses that have evolved for hunting prey; for example, it can smell tiny amounts of blood from as far away as 5 km (3 mi). Perhaps more amazing, it has developed special organs (called the *ampullae of Lorenzini*) that can detect the tiny electric fields generated by the movement of muscles in an organism, whether a fish, a seal, or a human. However, just what are electric fields? In addition, how are they related to electric charges?

The concept of vector fields is one of the most useful and productive ideas in all of physics. This chapter explains what an electric field is and how it is connected to electrostatic charges and forces and then examines how to determine the electric field due to some given distribution of charge. This study leads us to one of the most important laws of electricity—Gauss's Law—which provides a relationship between electric fields and electrostatic charge. However, Gauss's Law has practical application only when the charge distribution has enough geometric symmetry to simplify the calculation, and even then, some other concepts related to electric fields are necessary in order to apply the equations. We'll examine another kind of field—magnetic fields—in Chapters 27 through 29. Then, Chapter 31 will show how Gauss's Law fits into a unified description of electric and magnetic fields—one of the finest achievements in physics, from both a practical and an esthetic point of view.

WHAT WE WILL LEARN

- An electric field represents the electric force at different points in space.
- Electric field lines represent the net force vectors exerted on a unit positive electric charge. They originate on positive charges and terminate on negative charges.
- The electric field of a point charge is radial, proportional to the charge, and inversely proportional to the square of the distance from the charge.
- An electric dipole consists of a positive charge and a negative charge of equal magnitude.
- The electric flux is the electric field component normal to an area times the area.
- Gauss's Law states that the electric flux through a closed surface is proportional to the net electric charge enclosed within the surface. This law provides simple ways to solve seemingly complicated electric field problems.
- The electric field inside a conductor is zero.
- The magnitude of the electric field due to a uniformly charged, infinitely long wire varies as the inverse of the perpendicular distance from the wire.
- The electric field due to an infinite sheet of charge does not depend on the distance from the sheet.
- The electric field outside a spherical distribution of charge is the same as the field of a point charge with the same total charge located at the sphere's center.

22.1 Definition of an Electric Field

In Chapter 21, we discussed the force between two or more point charges. When determining the net force exerted by other charges on a particular charge at some point in space, we obtain different directions for this force, depending on the sign of the charge that is the reference point. In addition, the net force is also proportional to the magnitude of the reference charge. The techniques used in Chapter 21 require us to redo the calculation for the net force each time we consider a different charge.

Dealing with this situation requires the concept of a **field,** which can be used to describe certain forces. An **electric field,** $E(r)$, is defined at any point in space, $\vec{r}$, as the net electric force on a charge, divided by that charge:

$$\vec{E}(\vec{r}) = \frac{\vec{F}(\vec{r})}{q}. \tag{22.1}$$

The units of the electric field are newtons per coulomb (N/C). This simple definition eliminates the cumbersome dependence of the electric force on the particular charge being used to measure the force. We can quickly determine the net force on any charge by using $\vec{F}(\vec{r}) = q\vec{E}(\vec{r})$, which is a trivial rearrangement of equation 22.1.

The electric force on a charge at a point is parallel (or antiparallel, depending on the sign of the charge in question) to the electric field at that point and proportional to the magnitude of the charge. The magnitude of the force is given by $F = |q|E$. The direction of the force on a positive charge is along $\vec{E}(\vec{r})$; the direction of the force on a negative charge is in the direction opposite to $\vec{E}(\vec{r})$.

If several sources of electric fields are present at the same time, such as several point charges, the electric field at any given point is determined by the superposition of the electric fields from all sources. This superposition follows directly from the superposition of forces introduced in our study of mechanics and discussed in Chapter 21 for electrostatic forces. The **superposition principle** for the total electric field, $\vec{E}_t$, at any point in space with coordinate $\vec{r}$, due to n electric field sources can be stated as

$$\vec{E}_t(\vec{r}) = \vec{E}_1(\vec{r}) + \vec{E}_2(\vec{r}) + \cdots + \vec{E}_n(\vec{r}). \tag{22.2}$$

22.2 Field Lines

An electric field can (and in most applications does) change as a function of the spatial coordinate. The changing direction and strength of the electric field can be visualized by means of **electric field lines.** These graphically represent the net vector force exerted on a

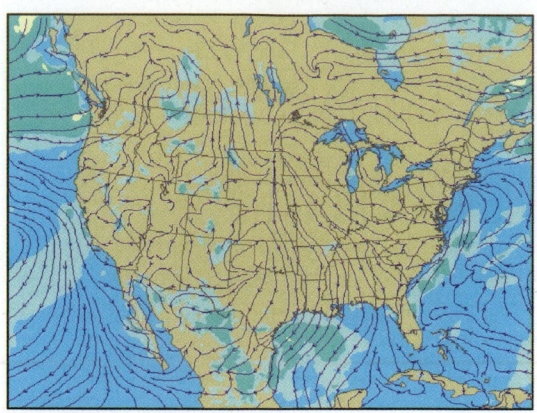

FIGURE 22.2 Streamlines of wind directions at the surface in the United States on March 23, 2008, from the National Weather Service.

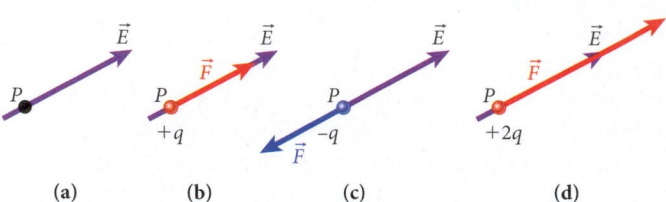

FIGURE 22.3 The force resulting from placing a charge in an electric field. (a) A point P on an electric field line. (b) A positive charge $+q$ placed at point P. (c) A negative charge $-q$ placed at point P. (d) A positive charge $+2q$ placed at point P.

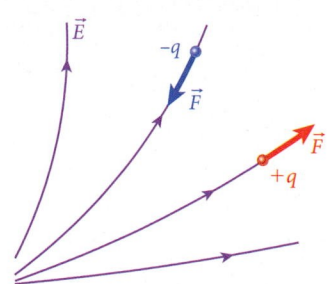

FIGURE 22.4 A nonuniform electric field. A positive charge $+q$ and a negative charge $-q$ placed in the field experience forces as shown. Each force is tangent to the electric field line.

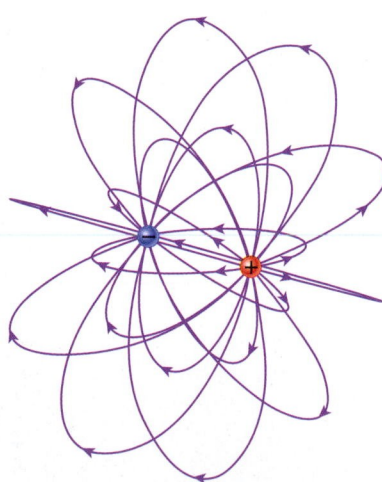

FIGURE 22.5 Three-dimensional representation of electric field lines from two point charges with opposite signs.

unit positive test charge. The representation applies separately for each point in space where the test charge might be placed. The direction of the field line at each point is the same as the direction of the force at that point, and the density of field lines is proportional to the magnitude of the force.

Electric field lines can be compared to the streamlines of wind directions, shown in Figure 22.2. These streamlines represent the force of the wind on objects at given locations, just as the electric field lines represent the electric force at specific points. A hot-air balloon can be used as a test particle for determining these wind streamlines. For example, a hot-air balloon launched in Dallas, Texas, would float from north to south in the situation depicted in Figure 22.2. Where the wind streamlines are close together, the speed of the wind is higher, so the balloon would move faster.

To draw an electric field line, we imagine placing a tiny positive charge at each point in the electric field. This charge is small enough that it does not affect the surrounding field. A small charge like this is sometimes called a **test charge.** We calculate the resultant force on the charge, and the direction of the force gives the direction of the field line. For example, Figure 22.3a shows a point in an electric field. In Figure 22.3b, a charge $+q$ is placed at point P, on an electric field line. The force on the charge is in the same direction as the electric field. In Figure 22.3c, a charge $-q$ is placed at point P, and the resulting force is in the direction opposite to the electric field. In Figure 22.3d, a charge $+2q$ is placed at point P, and the resulting force on the charge is in the direction of the electric field, with twice the magnitude of the force on the charge $+q$. We will follow the convention of depicting a positive charge as red and a negative charge as blue.

In a nonuniform electric field, the electric force at a given point is tangent to the electric field lines at that point, as illustrated in Figure 22.4. The force on a positive charge is in the direction of the electric field, and the force on a negative charge is in the direction opposite to the electric field.

Electric field lines point away from sources of positive charge and toward sources of negative charge. Each field line starts at a charge and ends at another charge. Electric field lines always originate on positive charges and terminate on negative charges.

Electric fields exist in three dimensions (Figure 22.5); however, this chapter usually presents two-dimensional depictions of electric fields for simplicity.

Point Charge

The electric field lines arising from an isolated point charge are shown in Figure 22.6. The field lines emanate in radial directions from the point charge. If the point charge is positive (Figure 22.6a), the field lines point outward, away from the charge; if the point charge is negative, the field lines point inward, toward the charge (Figure 22.6b). For an isolated positive point charge, the electric field lines originate at the charge and terminate on negative charges at infinity, and for a negative point charge, the electric field lines originate at positive charges at infinity and terminate at the charge. Note that the electric field lines are

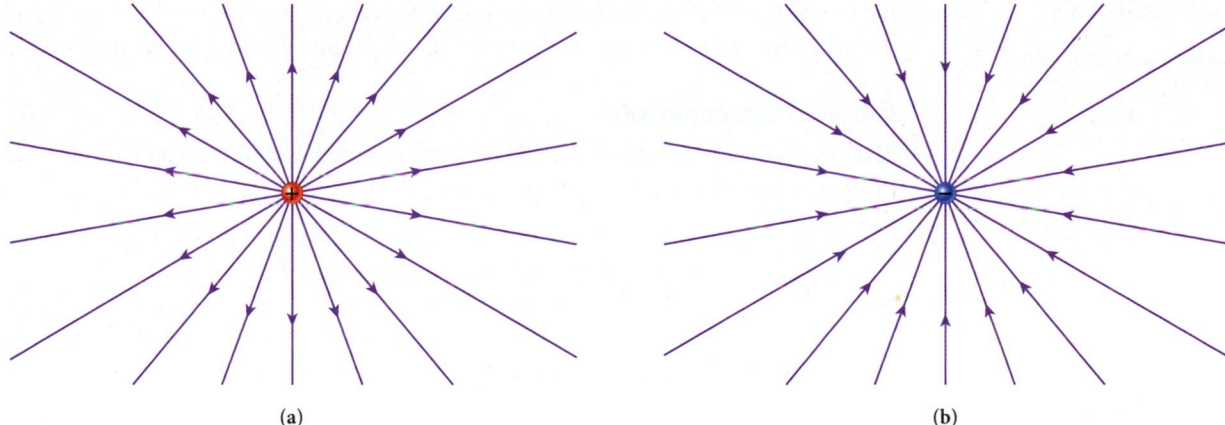

FIGURE 22.6 Electric field lines (a) from a single positive point charge and (b) to a single negative point charge.

closer together near the point charge and farther apart away from the point charge, indicating that the electric field becomes weaker with increasing distance from the charge. We'll examine the magnitude of the field quantitatively in Section 22.3.

Two Point Charges of Opposite Sign

We can use the superposition principle to determine the electric field from two point charges. Figure 22.7 shows the electric field lines for two oppositely charged point charges with the same magnitude. At each point in the plane, the electric field from the positive charge and the electric field from the negative charge add as vectors to give the magnitude and the direction of the resulting electric field. (Figure 22.5 shows the same field lines in three dimensions.)

As noted earlier, the electric field lines originate on the positive charge and terminate on the negative charge. At a point very close to either charge, the field lines are similar to those for a single point charge, since the effect of the more distant charge is small. Near the charges, the electric field lines are close together, indicating that the field is stronger in those regions. The fact that the field lines between the two charges connect indicates that an attractive force exists between the two charges.

Two Point Charges with the Same Sign

We can also apply the principle of superposition to two point charges with the same sign. Figure 22.8 shows the electric field lines for two point charges with the same sign and same magnitude. If both charges are positive (as in Figure 22.8), the electric field lines originate at the charges and terminate at infinity. If both charges are negative, the field lines originate at infinity and terminate at the charges. For two charges of the same sign, the field lines do not connect

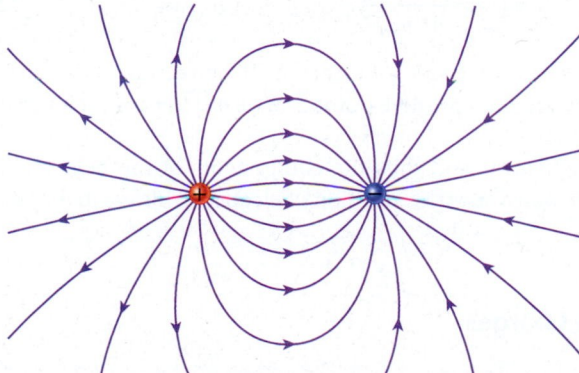

FIGURE 22.7 Electric field lines from two oppositely charged point charges. Each charge has the same magnitude.

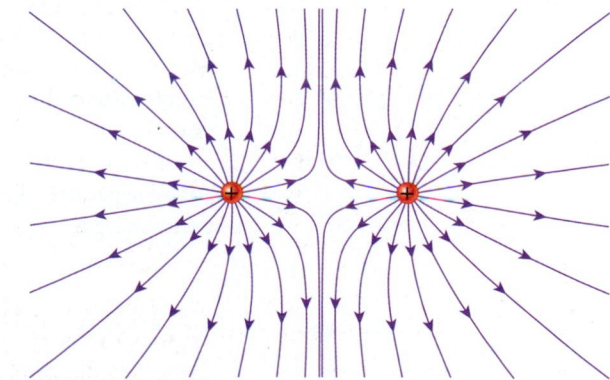

FIGURE 22.8 Electric field lines from two positive point charges with the same magnitude.

Concept Check 22.1

Which of the charges in the figure is (are) positive?

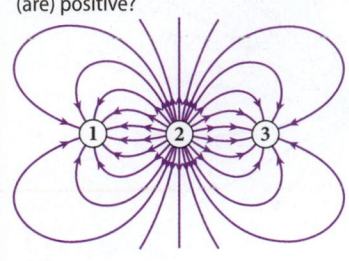

a) 1

b) 2

c) 3

d) 1 and 3

e) All three charges are positive.

the two charges. Rather, the field lines terminate on opposite charges at infinity. The fact that the field lines never terminate on the other charge signifies that the charges repel each other.

General Observations

The three simplest possible cases that we just examined lead to two general rules that apply to all field lines of all charge configurations:

1. *Field lines originate at positive charges and terminate at negative charges.*

2. *Field lines never cross.* This result is a consequence of the fact that the lines represent the electric field, which in turn is proportional to the net force that acts on a charge placed at a particular point. Field lines that crossed would imply that the net force points in two different directions at the same point, which is impossible.

Concept Check 22.2

Assuming that there are no charges in the four regions shown in the figure, which of the patterns could represent an electric field?

a) only 1

b) only 2

c) 2 and 3

d) 1 and 4

e) None of the patterns represent an electric field.

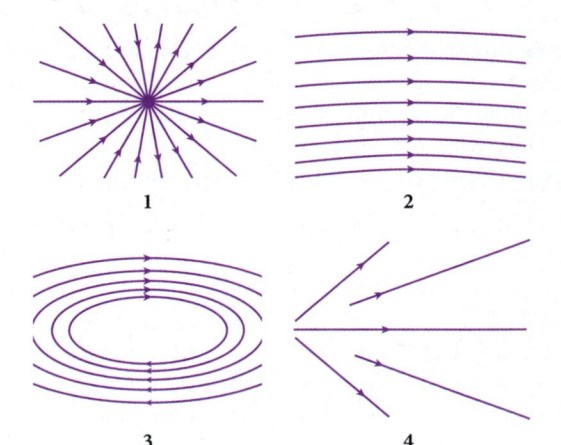

22.3 Electric Field due to Point Charges

The magnitude of the electric force on a point charge q_0 due to another point charge, q, is given by

$$F = \frac{1}{4\pi\epsilon_0} \frac{|qq_0|}{r^2}.$$ (22.3)

Taking q_0 to be a small test charge, we can express the magnitude of the electric field at the point where q_0 is and due to the point charge q as

$$E = \left|\frac{F}{q_0}\right| = \frac{1}{4\pi\epsilon_0} \frac{|q|}{r^2},$$ (22.4)

where r is the distance from the test charge to the point charge. The direction of this electric field is radial. The field points outward for a positive point charge and inward for a negative point charge.

An electric field is a vector quantity, and thus the components of the field must be added separately. Example 22.1 demonstrates the addition of electric fields created by three point charges.

FIGURE 22.9 Locations of three point charges.

EXAMPLE 22.1 | Three Charges

Figure 22.9 shows three fixed point charges: $q_1 = +1.50 \ \mu C$, $q_2 = +2.50 \ \mu C$, and $q_3 = -3.50 \ \mu C$. Charge q_1 is located at $(0,a)$, q_2 is located at $(0,0)$, and q_3 is located at $(b,0)$, where $a = 8.00$ m and $b = 6.00$ m.

PROBLEM
What electric field, $\vec{E}$, do these three charges produce at the point $P = (b,a)$?

SOLUTION
We must sum the electric fields from the three charges using equation 22.2. We proceed by summing component by component, starting with the field due to q_1:

$$\vec{E}_1 = E_{1,x}\hat{x} + E_{1,y}\hat{y}.$$

The field due to q_1 acts only in the x-direction at point (b,a), because q_1 has the same y-coordinate as P. Thus, $\vec{E}_1 = E_{1,x}\hat{x}$. We can determine $E_{1,x}$ using equation 22.4:

$$E_{1,x} = \frac{kq_1}{b^2}.$$

Note that the sign of $E_{1,x}$ is the same as the sign of q_1. Similarly, the field due to q_3 acts only in the y-direction at point (b,a). Thus, $\vec{E}_3 = E_{3,y}\hat{y}$, where

$$E_{3,y} = \frac{kq_3}{a^2}.$$

As shown in Figure 22.10, the electric field due to q_2 at P is given by

$$\vec{E}_2 = E_{2,x}\hat{x} + E_{2,y}\hat{y}.$$

Note that $\vec{E}_2$, the electric field due to q_2 at point P, points directly away from q_2, because $q_2 > 0$. (It would point directly toward q_2 if this charge were negative.) The magnitude of this electric field is given by

$$E_2 = \frac{k|q_2|}{a^2 + b^2}.$$

The component $E_{2,x}$ is given by $E_2 \cos\theta$, where $\theta = \tan^{-1}(a/b)$, and the component $E_{2,y}$ is given by $E_2 \sin\theta$.

Adding the components, the total electric field at point P is

$$\vec{E} = \left(E_{1,x} + E_{2,x}\right)\hat{x} + \left(E_{2,y} + E_{3,y}\right)\hat{y}$$

$$= \underbrace{\left(\frac{kq_1}{b^2} + \frac{kq_2\cos\theta}{a^2+b^2}\right)}_{E_x}\hat{x} + \underbrace{\left(\frac{kq_2\sin\theta}{a^2+b^2} + \frac{kq_3}{a^2}\right)}_{E_y}\hat{y}.$$

With the given values for a and b, we find $\theta = \tan^{-1}(8/6) = 53.1°$, and $a^2 + b^2 = (8.00\text{ m})^2 + (6.00\text{ m})^2 = 100\text{ m}^2$. We can then calculate the x-component of the total electric field as

$$E_x = \left(8.99\cdot10^9 \text{ N m}^2/\text{C}^2\right)\left|\frac{1.50\cdot10^{-6} \text{ C}}{(6.00\text{ m})^2} + \frac{\left(2.50\cdot10^{-6} \text{ C}\right)\left(\cos 53.1°\right)}{100\text{ m}^2}\right| = 509 \text{ N/C}.$$

The y-component is

$$E_y = \left(8.99\cdot10^9 \text{ N m}^2/\text{C}^2\right)\left|\frac{\left(2.50\cdot10^{-6} \text{ C}\right)\left(\sin 53.1°\right)}{100\text{ m}^2} + \frac{-3.50\cdot10^{-6} \text{ C}}{(8.00\text{ m})^2}\right| = -312 \text{ N/C}.$$

The magnitude of the field is

$$E = \sqrt{E_x^2 + E_y^2} = \sqrt{(509 \text{ N/C})^2 + (-312 \text{ N/C})^2} = 597 \text{ N/C}.$$

The direction of the field at point P is

$$\varphi = \tan^{-1}\left(\frac{E_y}{E_x}\right) = \tan^{-1}\left(\frac{-312 \text{ N/C}}{509 \text{ N/C}}\right) = -31.5°,$$

which means that the electric field points to the right and downward.

Note that even though the charges in this example are in microcoulombs and the distances are in meters, the electric fields are still large, showing that a microcoulomb is a large amount of charge.

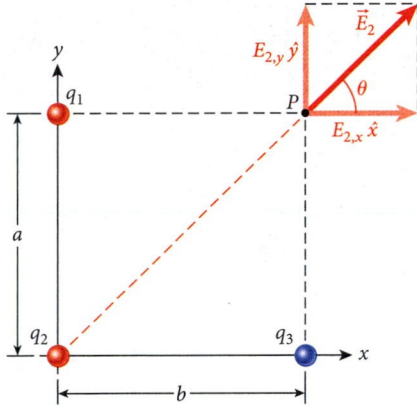

FIGURE 22.10 Electric field due to q_2 and its x- and y-components at point P.

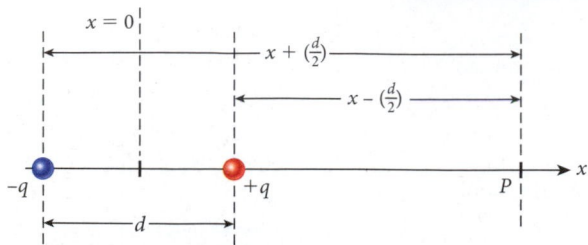

FIGURE 22.11 Calculation of the electric field from an electric dipole.

22.4 Electric Field due to a Dipole

A system of two equal (in magnitude) but oppositely charged point particles is called an **electric dipole**. The electric field from an electric dipole is given by the vector sum of the electric fields from the two charges. Figure 22.7 shows the electric field lines in two dimensions for an electric dipole.

The superposition principle allows us to determine the electric field due to two point charges through vector addition of the electric fields of the two charges. Let's consider the special case of the electric field due to a dipole along the axis of the dipole, defined as the line connecting the charges. This main symmetry axis of the dipole is assumed to be oriented along the x-axis (Figure 22.11).

The electric field, $\vec{E}$, at point P on the dipole axis is the sum of the field due to $+q$, denoted as $\vec{E}_+$, and the field due to $-q$, denoted as $\vec{E}_-$:

$$\vec{E} = \vec{E}_+ + \vec{E}_-.$$

Using equation 22.4, we can express the magnitude of the dipole's electric field along the x-axis, for $x > d/2$, as

$$E = \frac{1}{4\pi\epsilon_0}\frac{q}{r_+^2} + \frac{1}{4\pi\epsilon_0}\frac{-q}{r_-^2},$$

where r_+ is the distance between P and $+q$ and r_- is the distance between P and $-q$. Absolute value bars are not needed in this equation, because the first term on the right-hand side is positive and is greater than the second (negative) term. The electric field at all points on the x-axis (except at $x = \pm d/2$, where the two charges are located) is given by

$$\vec{E} = E_x\hat{x} = \frac{1}{4\pi\epsilon_0}\frac{q(x-d/2)}{r_+^3}\hat{x} + \frac{1}{4\pi\epsilon_0}\frac{-q(x+d/2)}{r_-^3}\hat{x}. \tag{22.5}$$

Now we examine the magnitude of $\vec{E}$ and restrict the value of x to $x > d/2$, where $E = E_x > 0$. Then we have

$$E = \frac{1}{4\pi\epsilon_0}\frac{q}{\left(x-\frac{1}{2}d\right)^2} - \frac{1}{4\pi\epsilon_0}\frac{q}{\left(x+\frac{1}{2}d\right)^2}.$$

With some rearrangement and keeping in mind that we want to obtain an expression that has the same form as the electric field from a point charge, we write the preceding equation as

$$E = \frac{q}{4\pi\epsilon_0 x^2}\left[\left(1-\frac{d}{2x}\right)^{-2} - \left(1+\frac{d}{2x}\right)^{-2}\right].$$

To find an expression for the electric field at a large distance from the dipole, we can make the approximation $x \gg d$ and use the binomial expansion. (Since $x \gg d$, we can drop terms containing the square of d/x and higher powers.) We obtain

$$E \approx \frac{q}{4\pi\epsilon_0 x^2}\left[\left(1+\frac{d}{x}-\cdots\right) - \left(1-\frac{d}{x}+\cdots\right)\right] = \frac{q}{4\pi\epsilon_0 x^2}\left(\frac{2d}{x}\right),$$

which can be rewritten as

$$E \approx \frac{qd}{2\pi\epsilon_0 x^3}. \tag{22.6}$$

Equation 22.6 can be simplified by defining a vector quantity called the **electric dipole moment**, $\vec{p}$. The direction of this dipole moment is from the negative charge to the positive charge, which is opposite to the direction of the electric field lines. The magnitude, p, of the electric dipole moment is given by

$$p = qd, \tag{22.7}$$

where q is the magnitude of either of the charges and d is the distance separating the two charges. With this definition, the expression for the magnitude of the electric field due to the dipole along the positive x-axis at a distance that is large compared with the separation between the two charges is

$$E = \frac{p}{2\pi\epsilon_0 |x|^3}. \tag{22.8}$$

Although not shown explicitly here, equation 22.8 is also valid for $x = \ll -d$. Also, an examination of equation 22.5 for $\vec{E}$ shows that $E_x > 0$ on either side of the dipole. In contrast to the field due to a point charge, which is inversely proportional to the square of the distance, the field due to a dipole is inversely proportional to the cube of the distance, according to equation 22.8.

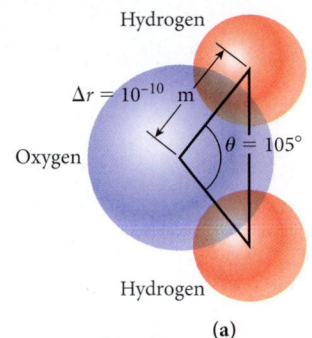

(a)

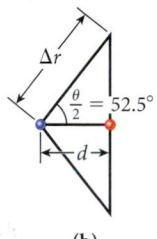

(b)

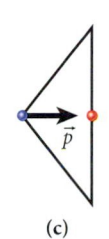

(c)

FIGURE 22.12 (a) Schematic drawing showing the geometry of a water molecule, H_2O, with atoms as spheres. (b) Diagram showing the effective positive (red dot on the right) and negative (blue dot on the left) charge centers. (c) Dipole moment assuming pointlike charges.

EXAMPLE 22.2 Water Molecule

The water molecule, H_2O, is arguably the most important one for life. It has a nonzero dipole moment, which is the basic reason why many organic molecules are able to bind to water. This dipole moment also allows water to be an excellent solvent for many inorganic and organic compounds.

Each water molecule consists of two atoms of hydrogen and one atom of oxygen, as shown in Figure 22.12a. The charge distribution of each of the individual atoms is approximately spherical. The oxygen atom tends to pull the negatively charged electrons toward itself, giving the hydrogen atoms slight positive charges. The three atoms are arranged so that the lines connecting the centers of the hydrogen atoms with the center of the oxygen atom have an angle of 105° between them (see Figure 22.12a).

PROBLEM

Suppose we approximate a water molecule by two positive charges at the locations of the two hydrogen nuclei (protons) and two negative charges at the location of the oxygen nucleus, with all charges of equal magnitude. What is the resulting electric dipole moment of water?

SOLUTION

The center of charge of the two positive charges, analogous to the center of mass of two masses, is located exactly halfway between the centers of the hydrogen atoms, as shown in Figure 22.12b. With the hydrogen-oxygen distance of $\Delta r = 10^{-10}$ m, as indicated in Figure 22.12a, the distance between the positive and negative charge centers is

$$d = \Delta r \cos\left(\frac{\theta}{2}\right) = \left(10^{-10}\ \text{m}\right)\left(\cos 52.5°\right) = 0.6 \cdot 10^{-10}\ \text{m}.$$

This distance times the transferred charge, $q = 2e$, is the magnitude of the dipole moment of water:

$$p = 2ed = \left(3.2 \cdot 10^{-19}\ \text{C}\right)\left(0.6 \cdot 10^{-10}\ \text{m}\right) = 2 \cdot 10^{-29}\ \text{C m}.$$

This result of an extremely oversimplified calculation actually comes close, within a factor of 3, to the measured value of $6.2 \cdot 10^{-30}$ C m. The fact that the real dipole moment of water is smaller than this calculated result is an indication that the two electrons of the hydrogen atoms are not pulled all the way to the oxygen but, on average, only one-third of the way.

Concept Check 22.4

An electrically neutral dipole is placed in an external electric field as shown in the figure in Concept Check 22.3. In which situation(s) is the net *torque* on the dipole zero?

a) 1 and 3

b) 2 and 4

c) 1 and 4

d) 2 and 3

e) 1 only

Concept Check 22.3

An electrically neutral dipole is placed in an external electric field as shown in the figure. In which situation(s) is the net *force* on the dipole zero?

a) 1 and 3

b) 2 and 4

c) 1 and 4

d) 2 and 3

e) 1 only

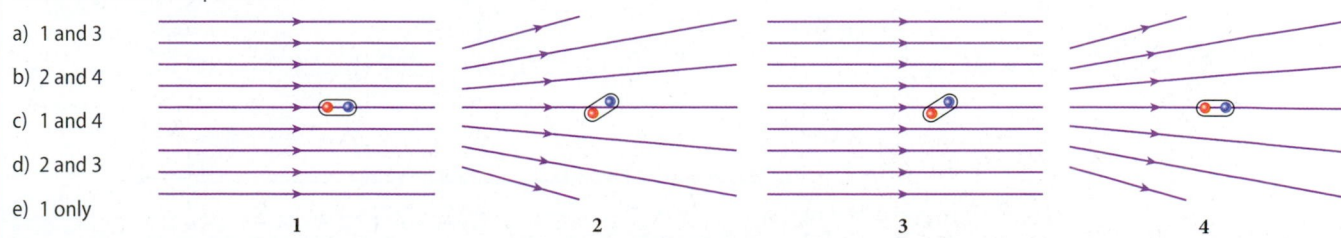

22.5   General Charge Distributions

We have determined the electric fields of a single point charge and of two point charges (an electric dipole). What if we want to determine the electric field due to many charges? Each individual charge creates an electric field, as described by equation 22.4, and because of the superposition principle, all of these electric fields can be added to find the net field at any point in space. But we have already seen in Example 22.1 that the addition of electric field vectors can be cumbersome for a collection of only three point charges. If we had to apply this method to, say, trillions of point charges, the task would be unmanageable even if we could use a supercomputer. Since real-world applications usually involve a very large number of charges, it is clear that we need a way to simplify the calculations. This can be accomplished by using an integral, if the large number of charges are arranged in space in some regular distribution. Of particular interest are two-dimensional distributions, where charges are located on the surface of a metallic object, and one-dimensional distributions, where charges are arranged along a wire. As we will see, integration can be a surprisingly simple way to solve problems involving such charge distributions, which would be very hard to analyze by the method of direct summation.

To prepare for the integration procedure, we divide the charge into differential elements of charge, dq, and find the electric field resulting from each differential charge element as if it were a point charge. If the charge is distributed along a one-dimensional object (a line), the differential charge may be expressed in terms of a charge per unit length times a differential length, or $\lambda\,dx$. If the charge is distributed over a surface (a two-dimensional object), dq is expressed in terms of a charge per unit area times a differential area, or $\sigma\,dA$. And, finally, if the charge is distributed over a three-dimensional volume, then dq is written as the product of a charge per unit volume and a differential volume, or $\rho\,dV$. That is,

$$\left.\begin{array}{l} dq = \lambda\,dx \\ dq = \sigma\,dA \\ dq = \rho\,dV \end{array}\right\} \text{ for a charge distribution } \left\{\begin{array}{l} \text{along a line;} \\ \text{over a surface;} \\ \text{throughout a volume.} \end{array}\right. \qquad (22.9)$$

The magnitude of the electric field resulting from the charge distribution is then obtained from the differential charge:

$$dE = k\frac{dq}{r^2}. \qquad (22.10)$$

In the following example, we find the electric field due to a finite line of charge.

EXAMPLE 22.3   Finite Line of Charge

To find the electric field along a line bisecting a finite length of wire with linear charge density λ, we integrate the contributions to the electric field from all the charge in the wire. We assume that the wire lies along the x-axis (Figure 22.13).

FIGURE 22.13 Calculating the electric field due to all the charge in a long wire by integrating the contributions to the electric field over the length of the wire.

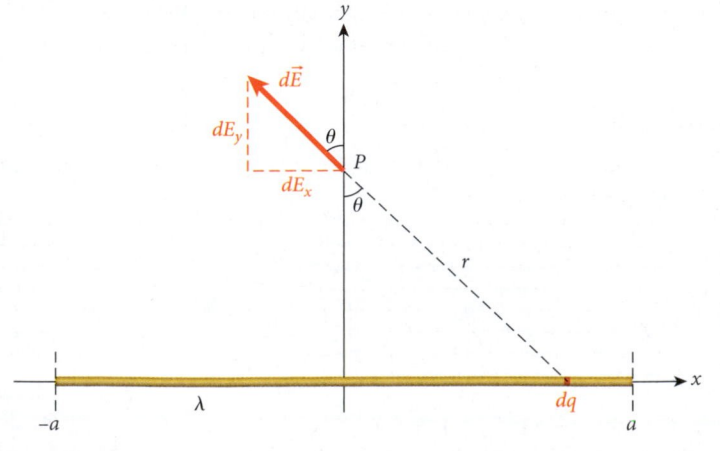

We also assume that the wire is positioned with its midpoint at $x = 0$, one end at $x = a$, and the other end at $x = -a$. The symmetry of the situation then allows us to conclude that there cannot be any electric force parallel to the wire (in the x-direction) along the line bisecting the wire. Along this line, the electric field can be only in the y-direction. We can then calculate the electric field due to all the charge for $x \geq 0$ and multiply the result by 2 to get the electric field for the whole wire.

We consider a differential charge, dq, on the x-axis, as shown in Figure 22.13. The magnitude of the electric field, dE, at a point $(0, y)$ due to this charge is given by equation 22.10,

$$dE = k\frac{dq}{r^2},$$

where $r = \sqrt{x^2 + y^2}$ is the distance from dq to point P. The component of the electric field perpendicular to the wire (in the y-direction) is then given by

$$dE_y = k\frac{dq}{r^2}\cos\theta,$$

where θ is the angle between the electric field produced by dq and the y-axis (see Figure 22.13). The angle θ is related to r and y because $\cos\theta = y/r$.

We can relate the differential charge to the differential distance along the x-axis through the linear charge density, λ: $dq = \lambda\,dx$. The electric field at a distance y from the long wire is then

$$E_y = 2\int_0^a dE_y = 2\int_0^a k\frac{dq}{r^2}\cos\theta = 2k\int_0^a \frac{\lambda\,dx}{r^2}\frac{y}{r} = 2k\lambda y\int_0^a \frac{dx}{\left(x^2 + y^2\right)^{3/2}}.$$

Evaluation of the integral on the right-hand side (with the aid of an integral table or a software package like Mathematica or Maple) gives us

$$\int_0^a \frac{dx}{\left(x^2 + y^2\right)^{3/2}} = \left[\frac{1}{y^2}\frac{x}{\sqrt{x^2 + y^2}}\right]_0^a = \frac{1}{y^2}\frac{a}{\sqrt{y^2 + a^2}}.$$

Thus, the electric field at a distance y along a line bisecting the wire is given by

$$E_y = 2k\lambda y\frac{1}{y^2}\frac{a}{\sqrt{y^2 + a^2}} = \frac{2k\lambda}{y}\frac{a}{\sqrt{y^2 + a^2}}.$$

Finally, when $a \to \infty$, that is, the wire becomes infinitely long, $a/\sqrt{y^2 + a^2} \to 1$, and we have for an infinitely long wire

$$E_y = \frac{2k\lambda}{y}.$$

In other words, the electric field decreases in inverse proportion to the distance from the wire.

Now let's tackle a problem with a slightly more complicated geometry, finding the electric field due to a ring of charge along the axis of the ring.

SOLVED PROBLEM 22.1 | Ring of Charge

PROBLEM
Consider a charged ring with radius $R = 0.250$ m (Figure 22.14). The ring has uniform linear charge density, and the total charge on the ring is $Q = +5.00\ \mu C$. What is the electric field at a distance $b = 0.500$ m along the axis of the ring?

SOLUTION
THINK The charge is evenly distributed around the ring. The electric field at position $x = b$ can be calculated by integrating the differential electric field due to a differential electric charge. By symmetry, the components of the electric field perpendicular to the axis of the ring integrate to zero, because the electric fields of charge elements on opposite sides of the axis cancel one another out. The resulting electric field is parallel to the axis of the circle.

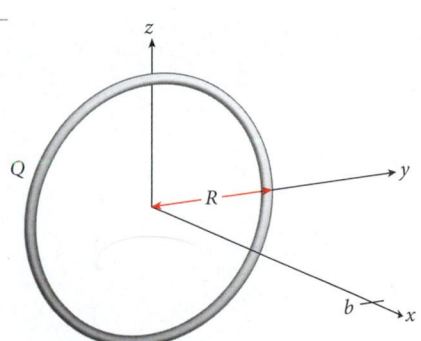

FIGURE 22.14 Charged ring with radius R and total charge Q.

– *Continued*

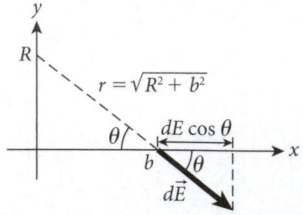

FIGURE 22.15 The geometry for the electric field along the axis of a ring of charge.

SKETCH Figure 22.15 shows the geometry for the electric field along the axis of the ring of charge.

RESEARCH The differential electric field, dE, at $x = b$ is due to a differential charge dq located at $y = R$ (see Figure 22.15). The distance from the point $(x = b, y = 0)$ to the point $(x = 0, y = R)$ is

$$r = \sqrt{R^2 + b^2}.$$

Again, the magnitude of $d\vec{E}$ is given by equation 22.10:

$$dE = k\frac{dq}{r^2}.$$

The magnitude of the component of $d\vec{E}$ parallel to the x-axis is given by

$$dE_x = dE\cos\theta = dE\frac{b}{r}.$$

SIMPLIFY We can find the total electric field by integrating its x-components over all the charge on the ring:

$$E_x = \int dE_x = \int_{\text{ring}} \frac{b}{r} k \frac{dq}{r^2}.$$

We need to integrate around the circumference of the ring of charge. We can relate the differential charge to the differential arc length, ds, as follows:

$$dq = \frac{Q}{2\pi R} ds.$$

We can then express the integral over the entire ring of charge as an integral around the arc length of a circle:

$$E_x = \int_0^{2\pi R} k\left(\frac{Q}{2\pi R}ds\right)\frac{b}{r^3} = \left(\frac{kQb}{2\pi R r^3}\right)\int_0^{2\pi R} ds = kQ\frac{b}{r^3} = \frac{kQb}{\left(R^2 + b^2\right)^{3/2}}.$$

CALCULATE Putting in the numerical values, we get

$$E_x = \frac{kQb}{\left(R^2 + b^2\right)^{3/2}} = \frac{\left(8.99 \cdot 10^9 \text{ N m}^2/\text{C}^2\right)\left(5.00 \cdot 10^{-6} \text{ C}\right)(0.500 \text{ m})}{\left[(0.250 \text{ m})^2 + (0.500 \text{ m})^2\right]^{3/2}} = 128,654 \text{ N/C}.$$

ROUND We report our result to three significant figures:

$$E_x = 1.29 \cdot 10^5 \text{ N/C}.$$

DOUBLE-CHECK We can check the validity of the formula we derived for the electric field by using a large distance from the ring of charge, such that $b \gg R$. In this case,

$$E_x = \frac{kQb}{\left(R^2 + b^2\right)^{3/2}} \stackrel{b \gg R}{\Rightarrow} E_x = \frac{kQb}{b^3} = k\frac{Q}{b^2},$$

which is the expression for the electric field due to a point charge Q at a distance b. We can also check the formula with $b = 0$:

$$E_x = \frac{kQb}{\left(R^2 + b^2\right)^{3/2}} \stackrel{b = 0}{\Rightarrow} E_x = 0,$$

which is what we would expect at the center of a ring of charge. Thus, our result seems reasonable.

22.6 Force due to an Electric Field

The force $\vec{F}$ exerted by an electric field $\vec{E}$ on a point charge q is given by $\vec{F} = q\vec{E}$, a simple restatement of the definition of the electric field in equation 22.1. Thus, the force exerted by the electric field on a positive charge acts in the same direction as the electric field. The force vector is always tangent to the electric field lines and points in the direction of the electric field if $q > 0$.

Concept Check 22.6

A small positively charged object could be placed in a uniform electric field at position A or position B in the figure. How do the electric forces on the object at the two positions compare?

a) The magnitude of the electric force on the object is greater at position A.

b) The magnitude of the electric force on the object is greater at position B.

c) There is no electric force on the object at either position A or position B.

d) The electric force on the object at position A has the same magnitude as the force on the object at position B but is in the opposite direction.

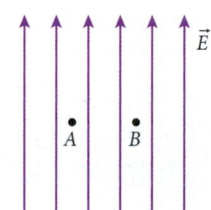

e) The electric force on the object at position A is the same nonzero electric force as that on the object at position B.

Concept Check 22.5

A small positively charged object is placed at rest in a uniform electric field as shown in the figure. When the object is released, it will

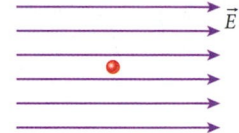

a) not move.

b) begin to move with a constant speed.

c) begin to move with a constant acceleration.

d) begin to move with an increasing acceleration.

e) move back and forth in simple harmonic motion.

The force at various locations on a positive charge due to the electric field in three dimensions is shown in Figure 22.16 for the case of two oppositely charged particles. (This is the same field as in Figure 22.5, but with some representative force vectors added.) You can see that the force on the positive charge is always tangent to the field lines and points in the same direction as the electric field. The force on the negative charge would point in the opposite direction.

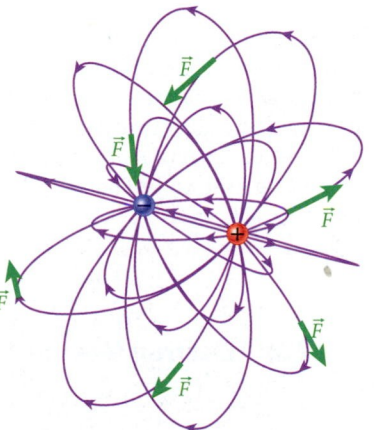

FIGURE 22.16 Direction of the force that an electric field produced by two opposite point charges exerts on a positive charge at various points in space.

Self-Test Opportunity 22.1

The figure shows a two-dimensional view of electric field lines due to two opposite charges. What is the direction of the electric field at the five points A, B, C, D, and E? At which of the five points is the magnitude of the electric field the largest?

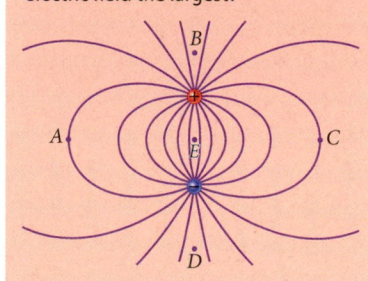

EXAMPLE 22.4 Time Projection Chamber

Nuclear physicists study new forms of matter by colliding gold nuclei at very high energies. In particle physics, new elementary particles are created and studied by colliding protons and antiprotons at the highest energies. These collisions create many particles that stream away from the interaction point at high speeds. A simple particle detector is not sufficient to identify these particles. A device that helps physicists study these collisions is a time projection chamber (TPC), found in most large particle detectors.

– Continued

One example of a TPC is the STAR TPC of the Relativistic Heavy Ion Collider at Brookhaven National Laboratory on Long Island, New York. The STAR TPC consists of a large cylinder filled with a gas (90% argon, 10% methane) that allows free electrons to move within it without being captured by the gas atoms or molecules.

Figure 22.17 shows the results of a collision of two gold nuclei that occurred in the STAR TPC. In such a collision, thousands of charged particles are created that pass through the gas inside the TPC. As these charged particles pass through the gas, they ionize the atoms of the gas, releasing free electrons. A constant electric field of magnitude 13,500 N/C is applied between the center of the TPC and the caps on the ends of the cylinder, and the field exerts an electric force on the freed electrons. Because the electrons have a negative charge, the electric field exerts a force in the direction opposite to the electric field. The electrons attempt to accelerate in the direction of the electric force, but they interact with the electrons of the molecules of the gas and begin to drift toward the caps with a constant speed of 5 cm/μs = $5 \cdot 10^4$ m/s $\approx$ 100,000 mph.

FIGURE 22.17 An event in the STAR TPC in which two gold nuclei have collided at very high energies at the point in the center of the image. Each colored line represents the track left behind by a subatomic particle produced in the collision.

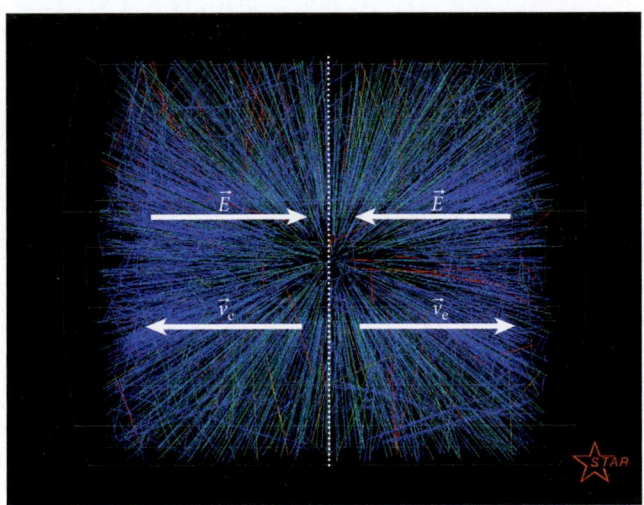

Each end cap of the cylinder has 68,304 detectors that can measure the charge as a function of the drift time of the electrons from the point where they were freed. Each detector has a specific (x,y) position. From measurements of the arrival time of the charge and the known drift speed of the electrons, the z-components of their positions can be calculated. Thus, the STAR TPC can produce a complete three-dimensional representation of the ionization track of each charged particle. These tracks are shown in Figure 22.17, where the colors represent the amount of ionization produced by each track.

SOLVED PROBLEM 22.2 | Electron Moving over a Charged Plate

PROBLEM

An electron with a kinetic energy of 2.00 keV (1 eV = $1.602 \cdot 10^{-19}$ J) is fired horizontally across a horizontally oriented charged conducting plate with a surface charge density of $+4.00 \cdot 10^{-6}$ C/m². Taking the positive direction to be upward (away from the plate), what is the vertical deflection of the electron after it has traveled a horizontal distance of 4.00 cm?

SOLUTION

THINK The initial velocity of the electron is horizontal. During its motion, the electron experiences a constant attractive force from the positively charged plate, which causes a constant acceleration downward. We can calculate the time it takes the electron to travel 4.00 cm in the horizontal direction and use this time to calculate the vertical deflection of the electron.

SKETCH Figure 22.18 shows the electron with initial velocity $\vec{v}_0$ in the horizontal direction. The initial position of the electron is taken to be at $x_0 = 0$ and $y = y_0$.

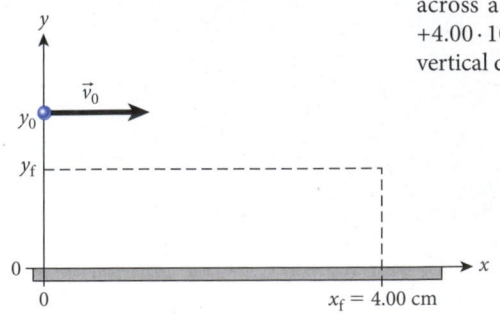

FIGURE 22.18 An electron moving to the right with initial velocity $\vec{v}_0$ over a charged conducting plate.

RESEARCH The time the electron takes to travel the given distance is

$$t = x_f/v_0, \tag{i}$$

where x_f is the final horizontal position and v_0 is the initial speed of the electron. While the electron is in motion, it experiences a force from the charged conducting plate. This force is directed downward (toward the plate) and has a magnitude given by

$$F = qE = e\frac{\sigma}{\epsilon_0}, \tag{ii}$$

where σ is the charge density on the conducting plate and e is the charge of an electron. This force causes a constant acceleration in the downward direction whose magnitude is given by $a = F/m$, where m is the mass of the electron. Using the expression for the force from equation (ii), we can express the magnitude of this acceleration as

$$a = \frac{F}{m} = \frac{e\sigma}{m\epsilon_0}. \tag{iii}$$

Note that this acceleration is constant. Thus, the vertical position of the electron as a function of time is given by

$$y_f = y_0 - \tfrac{1}{2}at^2 \Rightarrow y_f - y_0 = -\tfrac{1}{2}at^2. \tag{iv}$$

Finally, we can relate the electron's initial kinetic energy to its initial velocity through

$$K = \tfrac{1}{2}mv_0^2 \Rightarrow v_0^2 = \frac{2K}{m}. \tag{v}$$

SIMPLIFY We substitute the expressions for the time and the acceleration from equations (i) and (iii) into equation (iv) and obtain

$$y_f - y_0 = -\tfrac{1}{2}at^2 = -\frac{1}{2}\left(\frac{e\sigma}{m\epsilon_0}\right)\left(\frac{x_f}{v_0}\right)^2 = -\frac{e\sigma x_f^2}{2m\epsilon_0 v_0^2}. \tag{vi}$$

Now substituting the expression for the square of the initial speed from equation (v) into the right-hand side of equation (vi) gives us

$$y_f - y_0 = -\frac{e\sigma x_f^2}{2m\epsilon_0\left(\dfrac{2K}{m}\right)} = -\frac{e\sigma x_f^2}{4\epsilon_0 K}. \tag{vii}$$

CALCULATE We first convert the kinetic energy of the electron from electron-volts to joules:

$$K = (2.00\ \text{keV})\frac{1.602 \cdot 10^{-19}\ \text{J}}{1\ \text{eV}} = 3.204 \cdot 10^{-16}\ \text{J}.$$

Putting the numerical values into equation (vii), we get

$$y_f - y_0 = -\frac{e\sigma x_f^2}{4\epsilon_0 K} = -\frac{\left(1.602 \cdot 10^{-19}\ \text{C}\right)\left(4.00 \cdot 10^{-6}\ \text{C/m}^2\right)\left(0.0400\ \text{m}\right)^2}{4\left(8.85 \cdot 10^{-12}\ \text{C}^2/(\text{N m}^2)\right)\left(3.204 \cdot 10^{-16}\ \text{J}\right)} = -0.0903955\ \text{m}.$$

ROUND We report our result to three significant figures:

$$y_f - y_0 = -0.0904\ \text{m} = -9.04\ \text{cm}.$$

DOUBLE-CHECK The vertical deflection that we calculated is about twice the distance that the electron travels in the x-direction, which seems reasonable, at least in the sense of being of the same order of magnitude. Also, equation (vii) for the deflection has several features that should be present. First, the trajectory is parabolic, which we expect for a constant force and thus constant acceleration (see Chapter 3). Second, for zero surface charge density, we obtain zero deflection. Third, for very high kinetic energy, there is negligible deflection, which is also intuitively what we expect.

Dipole in an Electric Field

A point charge in an electric field experiences a force, given by equation 22.1. The electric force is always tangent to the electric field line passing through the point. The effect of an electric field on a dipole can be described in terms of the vector electric field, $\vec{E}$, and the

Concept Check 22.7

A negative charge $-q$ is placed in a nonuniform electric field as shown in the figure. What is the direction of the electric force on this negative charge?

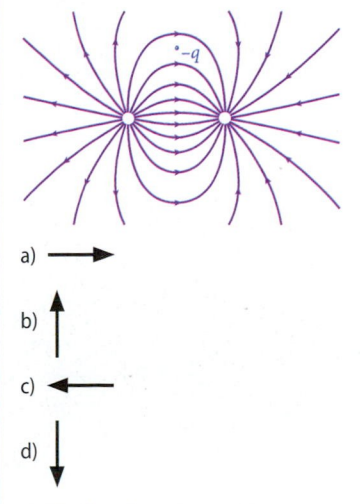

a) →

b) ↑

c) ←

d) ↓

e) The force is zero.

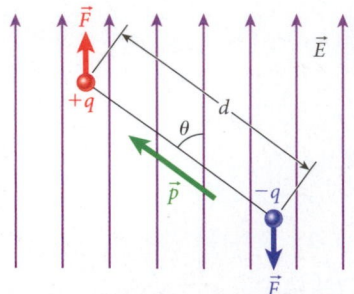

FIGURE 22.19 Electric dipole in an electric field.

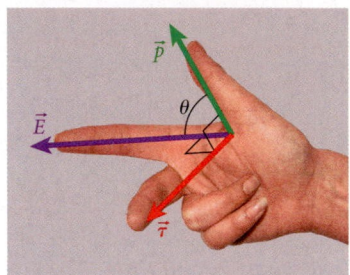

FIGURE 22.20 Right-hand rule for the vector product of the electric dipole moment and the electric field, producing the torque vector.

vector electric dipole moment, $\vec{p}$, without detailed knowledge of the charges making up the electric dipole.

To examine the behavior of an electric dipole, let's consider two charges, $+q$ and $-q$, separated by a distance d in a constant uniform electric field, $\vec{E}$ (Figure 22.19). (Note that we are now considering the forces acting on a dipole placed in an external field, as opposed to considering the field caused by the dipole, which we did in Section 22.4, and we also assume that the dipole field is small compared to $\vec{E}$ [so its effect on the uniform field can be ignored].) The electric field exerts an upward force on the positive charge and a downward force on the negative charge. Both forces have the magnitude qE. In Chapter 10, we saw that this situation gives rise to a torque, $\vec{\tau}$, given by $\vec{\tau} = \vec{r} \times \vec{F}$, where $\vec{r}$ is the moment arm and $\vec{F}$ is the force. The magnitude of the torque is $\tau = rF \sin\theta$.

As always, we can calculate the torque about any pivot point, so we can pick the location of the negative charge. Then, only the force on the positive charge contributes to the torque, and the length of the position vector is $r = d$, that is, the length of the dipole. Since, as already stated, $F = qE$, the expression for the torque on an electric dipole in an external electric field can be written as

$$\tau = qEd \sin\theta.$$

Remembering that the electric dipole moment is defined as $p = qd$, we obtain the magnitude of the torque:

$$\tau = pE \sin\theta. \tag{22.11}$$

Because the torque is a vector and must be perpendicular to both the electric dipole moment and the electric field, the relationship in equation 22.11 can be written as a vector product:

$$\vec{\tau} = \vec{p} \times \vec{E}. \tag{22.12}$$

As with all vector products, the direction of the torque is given by a right-hand rule. As shown in Figure 22.20, the thumb indicates the direction of the first term of the vector product, in this case $\vec{p}$, and the index finger indicates the direction of the second term, $\vec{E}$. The result of the vector product, $\vec{\tau}$, is then directed along the middle finger and is perpendicular to each of the two terms.

SOLVED PROBLEM 22.3 | Electric Dipole in an Electric Field

PROBLEM
An electric dipole with dipole moment of magnitude $p = 1.40 \cdot 10^{-12}$ C m is placed in a uniform electric field of magnitude $E = 498$ N/C (Figure 22.21a).

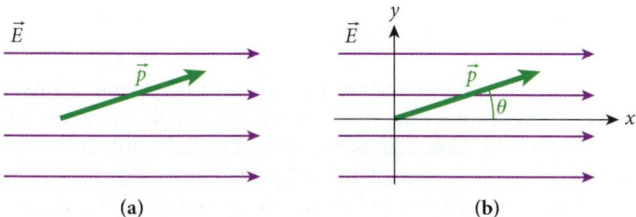

(a) (b)

FIGURE 22.21 (a) An electric dipole in a uniform electric field. (b) The electric field oriented in the x-direction and the dipole moment in the xy-plane.

At some instant (in time) the angle between the electric dipole moment and the electric field is $\theta = 14.5°$. What are the Cartesian components of the torque on the dipole?

SOLUTION
THINK The torque on the dipole is equal to the vector product of the electric field and the electric dipole moment.

SKETCH We assume that the electric field lines point in the x-direction and the electric dipole moment is in the xy-plane (Figure 22.21b). The z-direction is perpendicular to the plane of the page.

RESEARCH The torque on the electric dipole due to the electric field is given by

$$\vec{\tau} = \vec{p} \times \vec{E}.$$

Since the dipole is located in the xy-plane, the Cartesian components of the electric dipole moment are

$$\vec{p} = (p_x, p_y, 0).$$

Since the electric field is acting in the x-direction, its Cartesian components are

$$\vec{E} = (E_x, 0, 0) = (E, 0, 0).$$

SIMPLIFY From the definition of the vector product, we express the Cartesian components of the torque as

$$\vec{\tau} = (p_y E_z - p_z E_y)\hat{x} + (p_z E_x - p_x E_z)\hat{y} + (p_x E_y - p_y E_x)\hat{z}.$$

In this particular case, with E_y, E_z, and p_z all equal to zero, we have

$$\vec{\tau} = -p_y E_x \hat{z}.$$

The y-component of the dipole moment is $p_y = p\sin\theta$, and the x-component of the electric field is simply $E_x = E$. The magnitude of the torque is then

$$\tau = (p\sin\theta)E = pE\sin\theta,$$

and the direction of the torque is in the negative z-direction.

CALCULATE We insert the given numerical data and get

$$\tau = pE\sin\theta = (1.40\cdot10^{-12} \text{ C m})(498 \text{ N/C})(\sin14.5°) = 1.74565\cdot10^{-10} \text{ N m}.$$

ROUND We report our result to three significant figures:

$$\tau = 1.75\cdot10^{-10} \text{ N m}.$$

DOUBLE-CHECK From equation 22.11, we know that the magnitude of the torque is

$$\tau = pE\sin\theta,$$

which is the result we obtained using the explicit vector product. Applying the right-hand rule illustrated in Figure 22.20, we can determine the direction of the torque: With the right thumb representing the electric dipole moment and the right index finger representing the electric field, the right middle finger points into the page, which agrees with the result we found using the vector product. Thus, our result is correct.

Example 22.2 looked at the dipole moment of the water molecule. If water molecules are exposed to an external electric field, they experience a torque and thus begin to rotate. If the direction of the external electric field changes very rapidly, the water molecules perform rotational oscillations, which create heat. This is the principle of operation of a microwave oven. Microwave ovens use a frequency of 2.45 GHz for the oscillating electric field. (How an electric field is made to oscillate in time will be covered in Chapter 31 on electromagnetic waves.)

Electric fields also play a key role in human physiology, but these fields are time-varying and not static, like those studied in this chapter. (They will be covered in later chapters.) The evolution over time of electric fields in the human heart is measured by an electrocardiogram (ECG) (to be discussed, along with the functioning of pacemaker implants, in Chapter 26 on circuits). The human brain also generates continuously changing electrical fields through the activity of the neurons. These fields can be measured invasively by inserting electrodes through the skull and into the brain or by placing electrodes onto the surface of the exposed brain, usually during brain surgery. This technique is called electrocorticography (ECoG). An intense area of current research focuses on measuring and imaging brain electric fields noninvasively by attaching electrodes to the outside of the skull. However, since the skull itself dampens the electric fields, these techniques require great instrumental sensitivity and are still in their infancy. Perhaps the most exciting (or scary, depending on your point of view) research developments are in brain-computer interfaces. In this emerging field, electrical activity in the brain is used directly to control computers, and external stimuli are used to create electric fields inside the brain. Researchers in this area are motivated by the goal of helping people overcome physical disabilities, such as blindness or paralysis.

22.7 Electric Flux

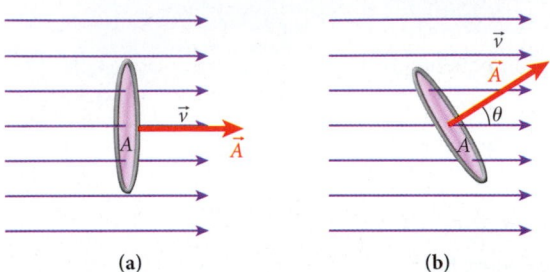

FIGURE 22.22 Water flowing with velocity of magnitude v through a ring of area A. (a) The area vector is parallel to the flow velocity. (b) The area vector is at an angle θ to the flow velocity.

Electric field calculations, like those in Example 22.3, can require quite a bit of work. However, in many common situations, particularly those with some geometric symmetry, a powerful technique for determining electric fields without having to explicitly calculate integrals can be used. This technique is based on *Gauss's Law,* one of the fundamental relations concerning electric fields. It will allow us to solve seemingly very complicated problems involving electric fields in an amazingly straightforward and simple fashion. However, to use Gauss's Law requires understanding of a concept called *electric flux.*

Imagine holding a ring with inside area A in a stream of water flowing with velocity $\vec{v}$, as shown in Figure 22.22. The area vector, $\vec{A}$, of the ring is defined as a vector with magnitude A pointing in a direction perpendicular to the plane of the ring. In Figure 22.22a, the area vector of the ring is parallel to the flow velocity, and the flow velocity is perpendicular to the plane of the ring. The product Av gives the amount of water passing through the ring per unit time (see Chapter 13), where v is the magnitude of the flow velocity. If the plane of the ring is tilted with respect to the direction of the flowing water (Figure 22.22b), the amount of water flowing through the ring is given by $Av\cos\theta$, where θ is the angle between the area vector of the ring and the direction of the velocity of the flowing water. The amount of water flowing through the ring is called the *flux,* $\Phi = Av\cos\theta = \vec{A}\cdot\vec{v}$. Since flux is a measure of volume per unit time, its units are cubic meters per second (m^3/s).

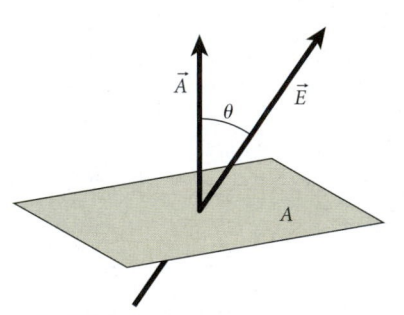

FIGURE 22.23 A uniform electric field $\vec{E}$ passing through an area $\vec{A}$.

An electric field is analogous to flowing water. Consider a uniform electric field of magnitude E passing through a given area A (Figure 22.23). Again, the area vector is $\vec{A}$, with a direction normal to the surface of the area and a magnitude A. The angle θ is the angle between the vector electric field and the area vector, as shown in Figure 22.23. The electric field passing through a given area A is called the **electric flux** and is given by

$$\Phi = EA\cos\theta. \tag{22.13}$$

In simple terms, the electric flux is proportional to the number of electric field lines passing through the area. We'll assume that the electric field is given by $\vec{E}(\vec{r})$ and that the area is a closed surface, rather than the open surface of a simple ring in flowing water. In this closed-surface case, the total, or net, electric flux is given by an integral of the electric field over the closed surface:

$$\Phi = \oiint \vec{E}\cdot d\vec{A}, \tag{22.14}$$

where $\vec{E}$ is the electric field at each differential area element $d\vec{A}$ of the closed surface. The direction of $d\vec{A}$ is outward from the closed surface. In equation 22.14, the loop on the integrals means that the integration is over a closed surface, and the two integral signs signify an integration over two variables. (*Note:* Some books use different notation for the integral over a closed surface, $\iint_S dA$ or just $\int_S dA$, but these refer to the same integration procedure as is represented in equation 22.14.) The differential area element $d\vec{A}$ must be described by two spatial variables, such as x and y in Cartesian coordinates or θ and ϕ in spherical coordinates.

Figure 22.24 shows a nonuniform electric field, $\vec{E}$, passing through a differential area element, $d\vec{A}$. A portion of the closed surface is also shown. The angle between the electric field and the differential area element is θ.

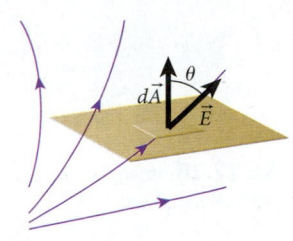

FIGURE 22.24 A nonuniform electric field, $\vec{E}$, passing through a differential area, $d\vec{A}$.

EXAMPLE 22.5 | Electric Flux through a Cube

Figure 22.25 shows a cube that has faces of area A in a uniform electric field, $\vec{E}$, that is perpendicular to the plane of one face of the cube.

PROBLEM
What is the net electric flux passing though the cube?

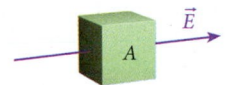

FIGURE 22.25 A cube with faces of area A in a uniform electric field, $\vec{E}$.

SOLUTION
The electric field in Figure 22.25 is perpendicular to the plane of one of the cube's six faces and therefore is also perpendicular to the opposite face. The area vectors of these two faces, $\vec{A}_1$ and $\vec{A}_2$, are shown in Figure 22.26a. The net electric flux passing through these two faces is

$$\Phi_{12} = \Phi_1 + \Phi_2 = \vec{E} \cdot \vec{A}_1 + \vec{E} \cdot \vec{A}_2 = -EA_1 + EA_2 = 0.$$

The negative sign arises for the flux through face 1 because the electric field and the area vector, $\vec{A}_1$, are in opposite directions. The area vectors of the remaining four faces are all perpendicular to the electric field, as shown in Figure 22.26b. The net electric flux passing through these four faces is

$$\Phi_{3456} = \Phi_3 + \Phi_4 + \Phi_5 + \Phi_6 = \vec{E} \cdot \vec{A}_3 + \vec{E} \cdot \vec{A}_4 + \vec{E} \cdot \vec{A}_5 + \vec{E} \cdot \vec{A}_6 = 0.$$

All the scalar products are zero because the area vectors of these four faces are perpendicular to the electric field. Thus, the net electric flux passing through the cube is

$$\Phi = \Phi_{12} + \Phi_{3456} = 0.$$

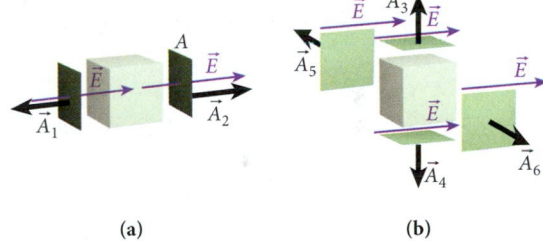

(a)   (b)

FIGURE 22.26 (a) The two faces of the cube that are perpendicular to the electric field. The area vectors are parallel and antiparallel to the electric field. (b) The four faces of the cube that are parallel to the electric field. The area vectors are perpendicular to the electric field.

22.8 Gauss's Law

To begin our discussion of Gauss's Law, let's imagine a box in the shape of a cube (Figure 22.27a), which is constructed of a material that does not affect electric fields. A positive test charge brought close to any surface of the box will experience no force. Now suppose a positive charge is inside the box and the positive test charge is brought close to the surface of the box (Figure 22.27b). The positive test charge experiences an outward force due to the positive charge inside the box. If the test charge is close to any surface of the box, it experiences the outward force. If twice as much positive charge is inside the box, the positive test charge experiences twice the outward force when brought close to any surface of the box.

Now suppose there is a negative charge inside the box (Figure 22.27c). When the positive test charge is brought close to one surface of the box, the charge experiences an inward force. If the positive test charge is close to any surface of the box, it experiences an inward force. Doubling the negative charge in the box doubles the inward force on the test charge when it is close to any surface of the box.

Self-Test Opportunity 22.3

The figure shows a cube with faces of area A and one face missing. This five-sided cubical object is in a uniform electric field, $\vec{E}$, perpendicular to one face. What is the net electric flux passing through the object?

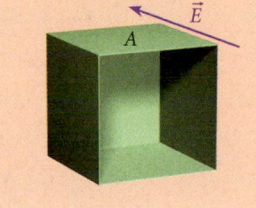

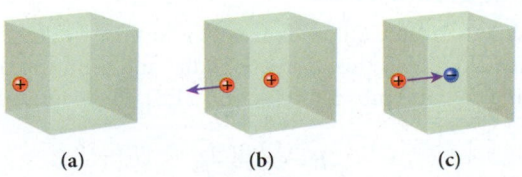

(a)   (b)   (c)

FIGURE 22.27 Three imaginary boxes constructed of material that does not affect electric fields. A positive test charge is brought up to the box from the left toward: (a) an empty box; (b) a box with a positive charge inside; (c) a box with a negative charge inside.

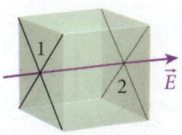

FIGURE 22.28 Imaginary empty box in a uniform electric field.

Concept Check 22.8

A cylinder made of an insulating material is placed in an electric field as shown in the figure. The net electric flux passing through the surface of the cylinder is

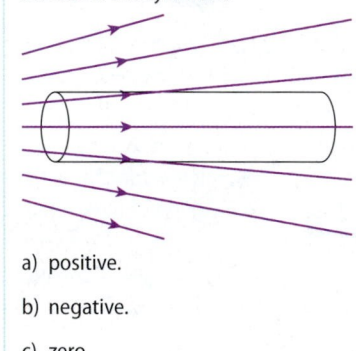

a) positive.

b) negative.

c) zero.

In analogy with flowing water, the electric field lines seem to be flowing out of the box containing positive charge and into the box containing negative charge.

Now let's imagine an empty box in a uniform electric field (Figure 22.28). If a positive test charge is brought close to side 1, it experiences an inward force. If the charge is close to side 2, it experiences an outward force. The electric field is parallel to the other four sides, so the positive test charge does not experience any inward or outward force when brought close to those sides. Thus, in analogy with flowing water, the net amount of electric field that seems to be flowing in and out of the box is zero.

Whenever a charge is inside the box, the electric field lines seem to be flowing in or out of the box. When there is no charge in the box, the net flow of electric field lines in or out of the box is zero. These observations and the definition of electric flux, which quantifies the concept of the flow of the electric field lines, lead to **Gauss's Law:**

$$\Phi = \frac{q}{\epsilon_0}. \tag{22.15}$$

Here q is the net charge inside a closed surface, called a **Gaussian surface.** The closed surface could be a box like that we have been discussing or any arbitrarily shaped closed surface. Usually, the shape of the Gaussian surface is chosen so as to reflect the symmetries of the problem situation.

Concept Check 22.9

The lines in the figure are electric field lines, and the circle is a Gaussian surface. For which case(s) is (are) the total electric flux nonzero?

a) 1 only

b) 2 only

c) 4, 5, and 6

d) 6 only

e) 1 and 2

An alternative formulation of Gauss's Law incorporates the definition of the electric flux (equation 22.14):

$$\oiint \vec{E} \cdot d\vec{A} = \frac{q}{\epsilon_0}. \tag{22.16}$$

According to equation 22.16, Gauss's Law states that the surface integral of the electric field components perpendicular to the area times the area is proportional to the net charge within the closed surface. This expression may look daunting, but it simplifies considerably in many cases and allows us to perform very quickly calculations that would otherwise be quite complicated.

Gauss's Law and Coulomb's Law

We can derive Gauss's Law from Coulomb's Law. To do this, we start with a positive point charge, q. The electric field due to this charge is radial and pointing outward, as we saw in Section 22.3. According to Coulomb's Law (Section 21.5), the magnitude of the electric field from this charge is

$$E = \frac{1}{4\pi\epsilon_0} \frac{q}{r^2}.$$

We now find the electric flux passing through a closed surface resulting from this point charge. For the Gaussian surface, we choose a spherical surface with radius r, with the charge at the center of the sphere, as shown in Figure 22.29. The electric field due to the positive point charge intersects each differential element of the surface of the Gaussian sphere perpendicularly. Therefore, at each point of this Gaussian surface, the electric field vector, $\vec{E}$, and the differential surface area vector, $d\vec{A}$, are parallel. The surface area vector will always point outward from the spherical Gaussian surface, but the electric field vector can point outward or inward depending on the sign of the charge. For a positive charge, the scalar product of the electric field and the surface area element is $\vec{E} \cdot d\vec{A} = E\, dA \cos 0° = E\, dA$. The electric flux in this case, according to equation 22.14, is

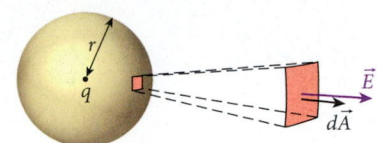

FIGURE 22.29 A spherical Gaussian surface with radius r surrounding a charge q. A closeup view of a differential surface element with area dA is shown.

$$\Phi = \oiint \vec{E} \cdot d\vec{A} = \oiint E\, dA.$$

Because the electric field has the same magnitude anywhere in space at a distance r from the point charge q, we can take E outside the integral:

$$\Phi = \oiint E\, dA = E \oiint dA.$$

Now what we have left to evaluate is the integral of the differential area over a spherical surface, which is given by $\oiint dA = 4\pi r^2$. Therefore, we have found from Coulomb's Law for the case of a point charge

$$\Phi = (E)\left(\oiint dA \right) = \left(\frac{1}{4\pi\epsilon_0} \frac{q}{r^2} \right)(4\pi r^2) = \frac{q}{\epsilon_0},$$

which is the same as the expression for Gauss's Law in equation 22.15. We have shown that Gauss's Law can be derived from Coulomb's Law for a positive point charge, but it can also be shown that Gauss's Law holds for any distribution of charge inside a closed surface.

Self-Test Opportunity 22.4

What changes in the preceding derivation of Gauss's Law if a negative point charge is used?

Shielding

Two important consequences of Gauss's Law are evident:

1. The electrostatic field inside any isolated conductor is always zero.
2. Cavities inside conductors are shielded from electric fields.

To examine these consequences, let's suppose a net electric field exists at some moment at some point inside an isolated conductor; see Figure 22.30a. But every conductor has free electrons inside it (blue circles in Figure 22.30b), which can move rapidly in response to any net external electric field, leaving behind positively charged ions (red circles in Figure 22.30b). The charges will move to the outer surface of the conductor, leaving no net accumulation of charge inside the volume of the conductor. These charges will in turn create an electric field inside the conductor (yellow arrows in Figure 22.30b), and they will move around until the electric field produced by them exactly cancels the external electric field. The net electric field thus becomes zero everywhere inside the conductor (Figure 22.30c).

If a cavity is scooped out of a conducting body, the net charge and thus the electric field inside this cavity is always zero, no matter how strongly the conductor is charged or how strong an external electric field acts on it. To prove this, we assume a closed Gaussian surface surrounds the cavity, completely inside the conductor. From the preceding discussion (see Figure 22.30), we know that at each point of this surface, the field is zero. Therefore, the net flux over this surface is also zero. By Gauss's Law, it then follows that this surface encloses zero net charge. If there were equal amounts of positive and negative charge on the cavity surface (and thus no net charge), this charge would not be stationary, as the positive and negative charges would be attracted to each other and would be free to move around the cavity surface to cancel each other. Therefore, any cavity inside a conductor is totally shielded from any external electric field. This effect is sometimes called **electrostatic shielding.**

A convincing demonstration of this shielding is provided by placing a plastic container filled with Styrofoam peanuts on top of a Van de Graaff generator, which serves as the source of strong electric field (Figure 22.31a). Charging the generator results in a large net charge accumulation on the dome, producing a strong electric field in the vicinity. Because

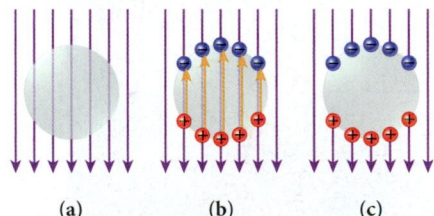

FIGURE 22.30 Shielding of an external electric field (purple vertical arrows) from the inside of a conductor.

Concept Check 22.10

A hollow, conducting sphere is initially given an evenly distributed negative charge. A positive charge $+q$ is brought near the sphere and placed at rest as shown in the figure. What is the direction of the electric field inside the hollow sphere?

a) →

b) ↑

c) ←

d) ↓

e) The field is zero.

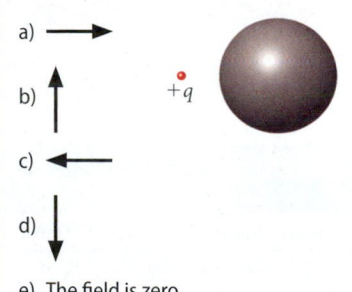

(a)   (b)

FIGURE 22.31 Styrofoam peanuts are put inside a container that is placed on top of a Van de Graaff generator, which is then charged. (a) The peanuts fly out of a nonconducting plastic container. (b) The peanuts remain within a metal can.

FIGURE 22.32 A person inside a Faraday cage is unharmed by a large voltage applied outside the cage, which produces a huge spark. This demonstration is performed several times daily at the Deutsches Museum in Munich, Germany.

of this field, the charges in the Styrofoam peanuts separate slightly, and the peanuts acquire small dipole moments. If the field were uniform, there would be no force on these dipoles. However, the nonuniform electric field does exert a force, even though the peanuts are electrically neutral. The peanuts thus fly out of the container. If the same Styrofoam peanuts are placed inside an open metal can, they do not fly out when the generator is charged (Figure 22.31b). The electric field easily penetrates the walls of the plastic container and reaches the Styrofoam peanuts, whereas, in accord with Gauss's Law, the conducting metal can provide shielding inside and prevents the Styrofoam peanuts from acquiring dipole moments.

The conductor surrounding the cavity does not have to be a solid piece of metal; even a wire mesh is sufficient to provide shielding. This can be demonstrated most impressively by seating a person inside a cage and then hitting the cage with a lightning-like electrical discharge (Figure 22.32). The person inside the cage is unhurt, even if he or she touches the metal of the cage *from the inside*. (It is important to realize that severe injuries can result if any body parts stick out of the cage, for example, if hands are wrapped around the bars of the cage!) This cage is called a *Faraday cage*, after British physicist Michael Faraday (1791–1867), who invented it.

A Faraday cage has important consequences, probably the most relevant of which is the fact that your car protects you from being hit by lightning while inside it—unless you drive a convertible. The sheet metal and steel frame that surround the passenger compartment provide the necessary shielding. (But as fiberglass, plastic, and carbon fiber begin to replace sheet metal in auto bodies, this shielding is not assured any more.)

Concept Check 22.11

A hollow, conducting sphere is initially uncharged. A positive charge, $+q_1$, is placed inside the sphere, as shown in the figure. Then, a second positive charge, $+q_2$, is placed near the sphere but outside it. Which of the following statements describes the net electric force on each charge?

a) There is a net electric force on $+q_2$ but not on $+q_1$.

b) There is a net electric force on $+q_1$ but not on $+q_2$.

c) Both charges are acted on by a net electric force with the same magnitude and in the same direction.

d) Both charges are acted on by a net electric force with the same magnitude but in opposite directions.

e) There is no net electric force on either charge.

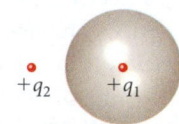

$+q_2$ $+q_1$

Table 22.1	Symbols for Charge Distributions	
Symbol	**Name**	**Unit**
λ	Charge per length	C/m
σ	Charge per area	C/m^2
ρ	Charge per volume	C/m^3

22.9   Special Symmetries

In this section we'll determine the electric field due to charged objects of different shapes. In Section 22.5, the charge distributions for different geometries were defined; see equation 22.9. Table 22.1 lists the symbols for these charge distributions and their units.

Cylindrical Symmetry

Using Gauss's Law, we can calculate the magnitude of the electric field due to a long straight conducting wire with uniform charge per unit length $\lambda > 0$. We first imagine a Gaussian surface in the form of a right cylinder with radius r and length L surrounding the wire so that the wire is along the axis of the cylinder (Figure 22.33). We can apply Gauss's Law to this Gaussian surface. From symmetry, we know that the electric field produced by the wire must be radial and perpendicular to the wire. What invoking symmetry means deserves further explanation because such arguments are very common.

First, we imagine rotating the wire about an axis along its length. This rotation would include all charges on the wire and their electric fields. However, the wire would still look the same after a rotation through any angle. The electric field created by the charge on the wire would therefore also be the same. From this argument, we conclude that the electric field cannot depend on the rotation angle around the wire. This conclusion is general: If an object has *rotational symmetry*, its electric field cannot depend on the rotation angle.

Second, if the wire is very long, it will look the same no matter where along its length it is viewed. If the wire is unchanged, its electric field is also unchanged. This observation means that there is no dependence on the coordinate along the wire. This symmetry is called *translational symmetry*. Since there is no preferred direction in space along the wire, there can be no electric field component parallel to the wire.

Returning to the Gaussian surface, we can see that the contribution to the integral in Gauss's Law (equation 22.16) from the ends of the cylinder is zero because the electric field is parallel to these surfaces and is thus perpendicular to the normal vectors from the surface. The electric field is perpendicular to the wall of the cylinder everywhere, so we have

$$\oiint \vec{E} \cdot d\vec{A} = EA = E\left(2\pi r L\right) = \frac{q}{\epsilon_0} = \frac{\lambda L}{\epsilon_0},$$

where $2\pi r L$ is the area of the wall of the cylinder. Solving this equation, we find the magnitude of the electric field due to a uniformly charged long straight wire:

$$E = \frac{\lambda}{2\pi\epsilon_0 r} = \frac{2k\lambda}{r}, \tag{22.17}$$

where r is the perpendicular distance to the wire. For $\lambda < 0$, equation 22.17 still applies, but the electric field points inward instead of outward. Note that this is the same result we obtained in Example 22.3 for the electric field due to a wire of infinite length—but attained here in a much simpler way!

You begin to see the great computational power contained in Gauss's Law, which can be used to calculate the electric field resulting from all kinds of charge distributions, both discrete and continuous. However, it is practical to use Gauss's Law only in situations where you can exploit some symmetry; otherwise, it is too difficult to calculate the flux.

It is instructive to compare the dependence of the electric field on the distance from a point charge and the distance from a long straight wire. For the point charge, the electric field falls off with the square of the distance, much faster than does the electric field due to the long wire, which decreases in inverse proportion to the distance.

Planar Symmetry

Assume a flat thin, infinite, nonconducting sheet of positive charge (Figure 22.34), with uniform charge per unit area $\sigma > 0$. Let's find the electric field a distance r from the surface of this infinite plane of charge.

To do this, we choose a Gaussian surface in the form of a closed right cylinder with cross-sectional area A and length $2r$, which cuts through the plane perpendicularly, as shown in Figure 22.34. Because the plane is infinite and the charge is positive, the electric field must be perpendicular to the ends of the cylinder and parallel to the cylinder wall. Using Gauss's Law, we obtain

$$\oiint \vec{E} \cdot d\vec{A} = \left(EA + EA\right) = \frac{q}{\epsilon_0} = \frac{\sigma A}{\epsilon_0},$$

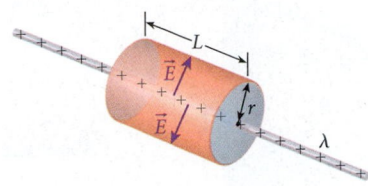

FIGURE 22.33 Long wire with charge per unit length λ surrounded by a Gaussian surface in the form of a right cylinder with radius r and length L. Representative electric field vectors are shown inside the cylinder.

Concept Check 22.12

A total of $1.45 \cdot 10^6$ excess electrons are placed on an initially electrically neutral wire of length 1.13 m. What is the magnitude of the electric field at a point at a perpendicular distance of 0.401 m away from the center of wire? (*Hint:* Assume that 1.13 m is close enough to "infinitely long.")

a) $9.21 \cdot 10^{-3}$ N/C

b) $2.92 \cdot 10^{-1}$ N/C

c) $6.77 \cdot 10^{1}$ N/C

d) $8.12 \cdot 10^{2}$ N/C

e) $3.31 \cdot 10^{3}$ N/C

Self-Test Opportunity 22.5

By how much does the answer to Concept Check 22.12 change if the assumption that the wire can be treated as being infinitely long is not made? (*Hint:* See Example 22.3.)

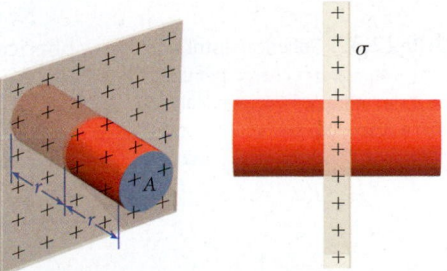

FIGURE 22.34 Infinite, flat, nonconducting sheet with charge density σ. Cutting through the plane perpendicularly is a Gaussian surface in the form of a right cylinder with cross-sectional area A parallel to the plane and height r above and below the plane.

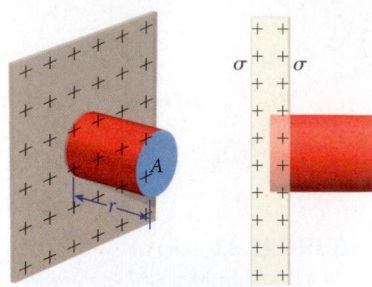

FIGURE 22.35 Infinite conducting plane with charge density σ on each surface and a Gaussian surface in the form of a right cylinder embedded in one side.

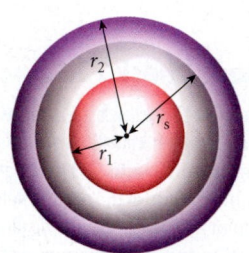

FIGURE 22.36 Spherical shell of charge with radius r_s along with a Gaussian surface with radius $r_2 > r_s$ and a second Gaussian surface with radius $r_1 < r_s$.

where σA is the charge enclosed in the cylinder. Thus, the magnitude of the electric field due to an infinite plane of charge is

$$E = \frac{\sigma}{2\epsilon_0}. \tag{22.18}$$

If $\sigma < 0$, then equation 22.18 still holds, but the electric field points toward the plane instead of away from it.

For an infinite conducting sheet with charge density $\sigma > 0$ on each surface, we can find the electric field by choosing a Gaussian surface in the form of a right cylinder. However, for this case, one end of the cylinder is embedded inside the conductor (Figure 22.35). The electric field inside the conductor is zero; therefore, there is no flux through the end of the cylinder enclosed in the conductor. The electric field outside the conductor must be perpendicular to the surface and therefore parallel to the wall of the cylinder and perpendicular to the end of the cylinder that is outside the conductor. Thus, the flux through the Gaussian surface is EA. The enclosed charge is given by σA, so Gauss's Law becomes

$$\oiint \vec{E} \cdot d\vec{A} = EA = \frac{\sigma A}{\epsilon_0}.$$

Thus, the magnitude of the electric field just outside the surface of a flat charged conductor is

$$E = \frac{\sigma}{\epsilon_0}. \tag{22.19}$$

Spherical Symmetry

To find the electric field due to a spherically symmetrical distribution of charge, we consider a thin spherical shell with charge $q > 0$ and radius r_s (Figure 22.36).

Here we use a spherical Gaussian surface with $r_2 > r_s$ that is concentric with the charged sphere. Applying Gauss's Law, we get

$$\oiint \vec{E} \cdot d\vec{A} = E\left(4\pi r_2^2\right) = \frac{q}{\epsilon_0}.$$

We can solve for the magnitude of the electric field, E, which is

$$E = \frac{1}{4\pi\epsilon_0} \frac{q}{r_2^2}.$$

If $q < 0$, the field points radially inward instead of radially outward from the spherical surfaces. For another spherical Gaussian surface, with $r_1 < r_s$, that is also concentric with the charged spherical shell, we obtain

$$\oiint \vec{E} \cdot d\vec{A} = E\left(4\pi r_1^2\right) = 0.$$

Thus, the electric field outside a spherical shell of charge behaves as if the charge were a point charge located at the center of the sphere, whereas the electric field is zero inside the spherical shell of charge.

Now let's find the electric field due to charge that is equally distributed throughout a spherical volume, with uniform charge density $\rho > 0$ (Figure 22.37). The radius of the sphere is r. We use a Gaussian surface in the form of a sphere with radius $r_1 < r$. From the symmetry of the charge distribution, we know that the electric field resulting from the charge is perpendicular to the Gaussian surface. Thus, we can write

$$\oiint \vec{E} \cdot d\vec{A} = E\left(4\pi r_1^2\right) = \frac{q}{\epsilon_0} = \frac{\rho}{\epsilon_0}\left(\tfrac{4}{3}\pi r_1^3\right),$$

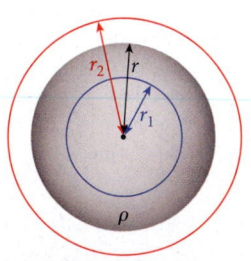

FIGURE 22.37 Spherical distribution of charge with uniform charge per unit volume ρ and radius r. Two spherical Gaussian surfaces are also shown, one with radius $r_1 < r$ and one with radius $r_2 > r$.

where $4\pi r_1^2$ is the area of the spherical Gaussian surface and $\tfrac{4}{3}\pi r_1^3$ is the volume enclosed by the Gaussian surface. From the preceding equation, we obtain the electric field at a radius r_1 inside a uniform distribution of charge:

$$E = \frac{\rho r_1}{3\epsilon_0}. \tag{22.20}$$

The total charge on the sphere can be called q_t, and it equals the total volume of the spherical charge distribution times the charge density:

$$q_t = \rho \tfrac{4}{3}\pi r^3.$$

The charge enclosed by the Gaussian surface then is

$$q = \frac{\text{volume inside } r_1}{\text{volume of charge distribution}} q_t = \frac{\tfrac{4}{3}\pi r_1^3}{\tfrac{4}{3}\pi r^3} q_t = \frac{r_1^3}{r^3} q_t.$$

With this the expression for the enclosed charge, we can rewrite Gauss's Law for this case as

$$\oiint \vec{E} \cdot d\vec{A} = E\left(4\pi r_1^2\right) = \frac{q_t}{\epsilon_0} \frac{r_1^3}{r^3},$$

which gives us

$$E = \frac{q_t r_1}{4\pi\epsilon_0 r^3} = \frac{kq_t r_1}{r^3}. \tag{22.21}$$

If we consider a Gaussian surface with a radius larger than the radius of the charge distribution, $r_2 > r$, we can apply Gauss's Law as follows:

$$\oiint \vec{E} \cdot d\vec{A} = E\left(4\pi r_2^2\right) = \frac{q_t}{\epsilon_0},$$

or

$$E = \frac{q_t}{4\pi\epsilon_0 r_2^2} = \frac{kq_t}{r_2^2}. \tag{22.22}$$

Thus, the electric field outside a uniform spherical distribution of charge is the same as the field due to a point charge of the same magnitude located at the center of the sphere.

Self-Test Opportunity 22.6

Consider a sphere of radius R with charge q uniformly distributed throughout the volume of the sphere. What is the magnitude of the electric field at a point $2R$ away from the center of the sphere?

SOLVED PROBLEM 22.4 | Nonuniform Spherical Charge Distribution

A spherically symmetrical but nonuniform charge distribution is given by

$$\rho(r) = \begin{cases} \rho_0\left(1 - \dfrac{r}{R}\right) & \text{for } r \leq R \\[2mm] 0 & \text{for } r > R, \end{cases}$$

where $\rho_0 = 10.0\ \mu\text{C/m}^3$ and $R = 0.250$ m.

PROBLEM
What is the electric field produced by this charge distribution at $r_1 = 0.125$ m and at $r_2 = 0.500$ m?

SOLUTION
THINK We can use Gauss's Law to determine the electric field as a function of radius if we employ a spherical Gaussian surface. The radius $r_1 = 0.125$ m is located inside the charge distribution. The charge enclosed inside the spherical surface at $r = r_1$ is given by an integral of the charge density from $r = 0$ to $r = r_1$. Outside the spherical charge distribution, the electric field is the same as that of a point charge whose magnitude is equal to the total charge of the spherical distribution.

SKETCH The charge density, ρ, as a function of radius, r, is plotted in Figure 22.38.

RESEARCH Gauss's Law (equation 22.16) tells us that $\oiint \vec{E} \cdot d\vec{A} = q/\epsilon_0$. Inside the nonuniform spherical charge distribution at a radius $r_1 < R$, Gauss's Law becomes

$$\epsilon_0 E\left(4\pi r_1^2\right) = \int_0^{V_1} \rho(r)\,dV = \int_0^{r_1} \rho_0\left(1 - \frac{r}{R}\right)\left(4\pi r^2\right)dr. \tag{i}$$

– Continued

FIGURE 22.38 Charge density as a function of radius for a nonuniform spherical charge distribution.

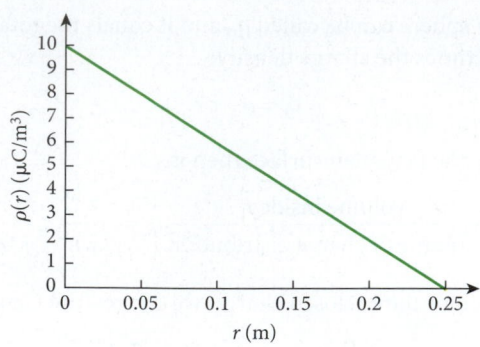

Carrying out the integral on the right-hand side of equation (i), we obtain

$$\int_0^{r_1} \rho_0\left(1 - \frac{r}{R}\right)\left(4\pi r^2\right) dr = 4\pi\rho_0 \int_0^{r_1}\left(r^2 - \frac{r^3}{R}\right) dr = 4\pi\rho_0\left(\frac{r_1^3}{3} - \frac{r_1^4}{4R}\right). \tag{ii}$$

SIMPLIFY   The electric field due to the charge inside $r_1 \le R$ is then given by

$$E = \frac{4\pi\rho_0\left(\dfrac{r_1^3}{3} - \dfrac{r_1^4}{4R}\right)}{\epsilon_0\left(4\pi r_1^2\right)} = \frac{\rho_0}{\epsilon_0}\left(\frac{r_1}{3} - \frac{r_1^2}{4R}\right). \tag{iii}$$

In order to calculate the electric field due to the charge inside $r_2 > R$, we need the total charge contained in the spherical charge distribution. We can obtain the total charge using equation (ii) with $r_1 = R$:

$$q_t = 4\pi\rho_0\left(\frac{R^3}{3} - \frac{R^4}{4R}\right) = 4\pi\rho_0\left(\frac{R^3}{3} - \frac{R^3}{4}\right) = 4\pi\rho_0\frac{R^3}{12} = \frac{\pi\rho_0 R^3}{3}.$$

The electric field outside the spherical charge distribution ($r_2 > R$) is then

$$E = \frac{1}{4\pi\epsilon_0}\frac{q_t}{r_2^2} = \frac{1}{4\pi\epsilon_0}\frac{\frac{1}{3}\pi\rho_0 R^3}{r_2^2} = \frac{\rho_0 R^3}{12\epsilon_0 r_2^2}. \tag{iv}$$

CALCULATE   The electric field at $r_1 = 0.125$ m is

$$E = \frac{\rho_0}{\epsilon_0}\left(\frac{r_1}{3} - \frac{r_1^2}{4R}\right) = \frac{10.0\ \mu C/m^3}{8.85\cdot 10^{-12}\ C^2/N\ m^2}\left(\frac{0.125\ m}{3} - \frac{(0.125\ m)^2}{4(0.250\ m)}\right) = 29{,}425.6\ N/C.$$

The electric field at $r_2 = 0.500$ m is

$$E = \frac{\rho_0 R^3}{12\epsilon_0 r_2^2} = \frac{\left(10.0\ \mu C/m^3\right)\left(0.250\ m\right)^3}{12\left(8.85\cdot 10^{-12}\ C^2/N\ m^2\right)\left(0.500\ m\right)^2} = 5885.12\ N/C.$$

ROUND   We report our results to three significant figures. The electric field at $r_1 = 0.125$ m is

$$E = 2.94\cdot 10^4\ N/C.$$

The electric field at $r_2 = 0.500$ m is

$$E = 5.89\cdot 10^3\ N/C.$$

DOUBLE-CHECK   The electric field at $r_1 = R$ can be calculated using equation (iii):

$$E = \frac{\rho_0}{\epsilon_0}\left(\frac{R}{3} - \frac{R^2}{4R}\right) = \frac{\rho_0 R}{12\epsilon_0} = \frac{\left(10.0\ \mu C/m^3\right)\left(0.250\ m\right)}{12\left(8.85\cdot 10^{-12}\ C^2/N\ m^2\right)} = 2.35\cdot 10^4\ N/C.$$

We can also use equation (iv) to find the electric field outside the spherical charge distribution but very close to the surface, where $r_2 \approx R$:

$$E = \frac{\rho_0 R^3}{12\epsilon_0 R^2} = \frac{\rho_0 R}{12\epsilon_0},$$

which is the same result we obtained using our result for $r_1 \leq R$. The calculated electric field at the surface of the charge distribution is lower than that at $r_1 = 0.125$, which may seem counterintuitive. An idea of the dependence of the magnitude of E on r is provided by the plot in Figure 22.39, which was created using equations (iii) and (iv).

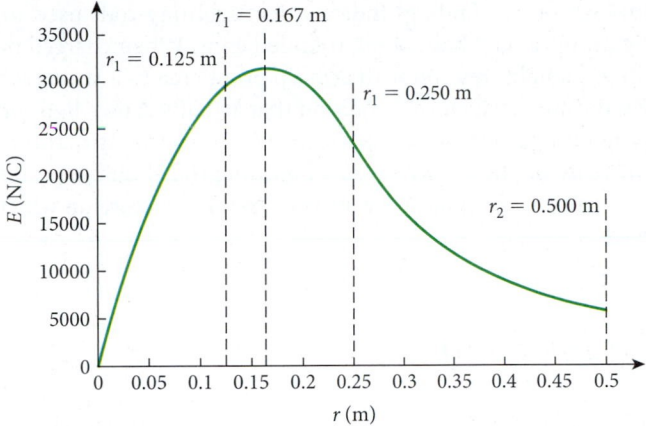

FIGURE 22.39 The electric field due to a nonuniform spherical distribution of charge as a function of the distance from the center of the sphere.

You can see that a maximum occurs in the electric field and that our result for $r_1 = 0.125$ m is less than this maximum value. We can calculate the radius at which the maximum occurs by differentiating equation (iii) with respect to r_1, setting the result equal to zero, and solving for r_1:

$$\frac{dE}{dr_1} = \frac{\rho_0}{\epsilon_0}\left(\frac{1}{3} - \frac{r_1}{2R}\right) = 0 \Rightarrow$$

$$\frac{1}{3} = \frac{r_1}{2R} \Rightarrow r_1 = \frac{2}{3}R.$$

Thus, we expect a maximum in the electric field at $r_1 = \frac{2}{3}R = 0.167$ m. The plot in Figure 22.39 does indeed show a maximum at that radius. It also shows the value of E at $r = 0.250$ to be smaller than that at $r = 0.125$ as we found in our calculation. Thus, our answers seem reasonable.

Concept Check 22.14

Suppose an uncharged hollow sphere made of a perfect insulator, for example a ping-pong ball, is resting on a perfect insulator. Some small amount of negative charge (say, a few hundred electrons) is placed at the north pole of the sphere. If you could check the distribution of the charge after a few seconds, what would you detect?

a) All of the added charge has vanished, and the sphere is again electrically neutral.

b) All of the added charge has moved to the center of the sphere.

c) All of the added charge is distributed uniformly over the surface of the sphere.

d) The added charge is still located at or very near the north pole of the sphere.

e) The added charge is performing a simple harmonic oscillation on a straight line between the south and north poles of the sphere.

Sharp Points and Lightning Rods

We have already seen that the electric field is perpendicular to the surface of a conductor. (To repeat, if there were a field component parallel to the conductor, then the charges inside the conductor would move until they reached equilibrium, which means no force or electric field component in the direction of motion, that is, along the surface of the conductor.) Figure 22.40a shows the distribution of charges on the surface of the end of a pointed conductor. Note that the charges are closer together at the sharp tip, where the curvature is largest. Near that sharp tip on the end of the conductor, the electric field looks much more like that due to a point charge, with the field lines spreading out radially (Figure 22.40b). Since the field lines are closer together near a sharp point on a conductor, the field is stronger near the sharp tip than on the flat part of the conductor.

Concept Check 22.13

Suppose an uncharged solid steel ball, for example, one of the steel balls used in an old-fashioned pinball machine, is resting on a perfect insulator. Some small amount of negative charge (say, a few hundred electrons) is placed at the north pole of the ball. If you could check the distribution of the charge after a few seconds, what would you detect?

a) All of the added charge has vanished, and the ball is again electrically neutral.

b) All of the added charge has moved to the center of the ball.

c) All of the added charge is distributed uniformly over the surface of the ball.

d) The added charge is still located at or very near the north pole of the ball.

e) The added charge is performing a simple harmonic oscillation on a straight line between the south and north poles of the ball.

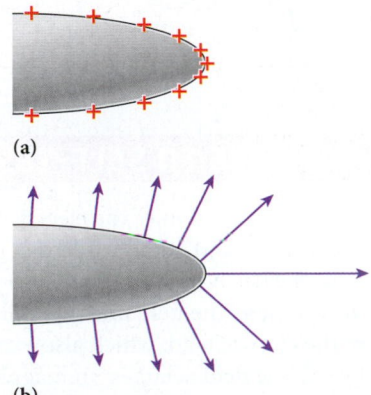

(a)

(b)

FIGURE 22.40 A sharp end of a conductor (with large curvature): (a) distribution of charges; (b) electric field at the surface of the conductor.

Benjamin Franklin proposed metal rods with sharp points as lightning rods (Figure 22.41). He reasoned that the sharp points would dissipate the electric charge built up in a storm, preventing the discharge of lightning. When Franklin installed such lightning rods, they were struck by lightning instead of the buildings to which they were attached. However, recent findings indicate that lightning rods used to protect structures from lightning should have blunt, rounded ends. When charged during thunderstorm conditions, a lightning rod with a sharp point creates a strong electric field that locally ionizes the air, producing a condition that actually causes lightning. Conversely, round-ended lightning rods are just as effective in protecting structures from lightning and do not increase lightning strikes. Any lightning rod should be carefully grounded to carry charge from a lightning strike away from the structure on which the lightning rod is mounted.

FIGURE 22.41 A lightning rod with a sharp point installed on the top of a building.

WHAT WE HAVE LEARNED | EXAM STUDY GUIDE

- The electric force, $\vec{F}(\vec{r})$, on a charge, q, due to an electric field, $\vec{E}(\vec{r})$, is given by $\vec{F}(\vec{r}) = q\vec{E}(\vec{r})$.

- The electric field at any point is equal to the sum of the electric fields from all sources: $\vec{E}_t(\vec{r}) = \vec{E}_1(\vec{r}) + \vec{E}_2(\vec{r}) + \cdots + \vec{E}_n(\vec{r})$.

- The magnitude of the electric field due to a point charge q at a distance r is given by $E(r) = \dfrac{1}{4\pi\epsilon_0}\dfrac{|q|}{r^2} = \dfrac{k|q|}{r^2}$. The electric field points radially away from a positive point charge and radially toward a negative charge.

- A system of two equal (in magnitude) but oppositely charged point particles is an electric dipole. The magnitude, p, of the electric dipole moment is given by $p = qd$, where q is the magnitude of either of the charges and d is the distance separating them. The electric dipole moment is a vector pointing from the negative toward the positive charge. On the dipole axis, the dipole produces an electric field of magnitude $E = \dfrac{p}{2\pi\epsilon_0|x|^3}$, where $|x| \gg d$.

- Gauss's Law states that the electric flux over an entire closed surface is equal to the enclosed charge divided by ϵ_0:
$$\oiint \vec{E}\cdot d\vec{A} = \dfrac{q}{\epsilon_0}.$$

- The differential electrical field is given by $dE = k\dfrac{dq}{r^2}$, and the differential charge is
$$\left.\begin{aligned} dq &= \lambda\,dx \\ dq &= \sigma\,dA \\ dq &= \rho\,dV \end{aligned}\right\} \text{for a charge distribution} \left\{\begin{aligned} &\text{along a line;} \\ &\text{over a surface;} \\ &\text{throughout a volume.} \end{aligned}\right.$$

- The magnitude of the electric field at a distance r from a long straight wire with uniform linear charge density $\lambda > 0$ is given by $E = \dfrac{\lambda}{2\pi\epsilon_0 r} = \dfrac{2k\lambda}{r}$.

- The magnitude of the electric field produced by an infinite nonconducting plane that has uniform charge density $\sigma > 0$ is $E = \frac{1}{2}\sigma/\epsilon_0$.

- The magnitude of the electric field produced by an infinite conducting plane that has uniform charge density $\sigma > 0$ on each side is $E = \sigma/\epsilon_0$.

- The electric field inside a closed conductor is zero.

- The electric field outside a charged spherical conductor is the same as that due to a point charge of the same magnitude located at the center of the sphere.

ANSWERS TO SELF-TEST OPPORTUNITIES

22.1 The direction of the electric field is downward at points A, C, and E and upward at points B and D. (There is an electric field at point E, even though there is no line drawn there; the field lines are only sample representations of the electric field, which also exists between the field lines.) The field is largest in magnitude at point E, which can be inferred from the fact that it is located where the field lines have the highest density.

22.2 The two forces acting on the two charges in the electric field create a torque on the electric dipole around its center of mass, given by

$$\tau = \big(\text{force}_+\big)\big(\text{moment arm}_+\big)\big(\sin\theta\big) + \big(\text{force}_-\big)\big(\text{moment arm}_-\big)\big(\sin\theta\big).$$

The length of the moment arm in both cases is $\frac{1}{2}d$, and the magnitude of the force is $F = qE$ for both charges. Thus, the torque on the electric dipole is

$$\tau = qE\left(\dfrac{d}{2}\sin\theta\right) + qE\left(\dfrac{d}{2}\sin\theta\right) = qEd\sin\theta.$$

22.3 The net electric flux passing though the object is EA. Remember, the object is not a closed surface; otherwise, the result would be zero.

22.4 The sign of the scalar product changes, because the electric field points radially inward: $\vec{E} \cdot d\vec{A} = E\,dA\cos 180° = -E\,dA$. But the magnitude of the electric field due to the negative charge is $E = \dfrac{1}{4\pi\epsilon_0} \dfrac{-q}{r^2}$. The two minus signs cancel, giving the same results for Coulomb's and Gauss's Laws for a point charge, independent of the sign of the charge.

22.5 For a wire of infinite length, $E_y = \dfrac{2k\lambda}{y}$; for a wire of finite length, $E_y = \dfrac{2k\lambda}{y}\dfrac{a}{\sqrt{y^2+a^2}}$. With the values given in Concept Check 22.12, $\dfrac{a}{\sqrt{y^2+a^2}} = \dfrac{0.565}{\sqrt{0.401^2+0.565^2}} = 0.815$.

Thus, the "infinitely long" approximation is off by ~18%.

22.6 The charged sphere acts like a point charge, so the electric field at $2R$ is

$$E = k\frac{q}{(2R)^2} = k\frac{q}{4R^2}.$$

PROBLEM-SOLVING GUIDELINES

1. Be sure to distinguish between the point where an electric field is being generated and the point where the electric field is being determined.

2. Some of the same guidelines for dealing with electrostatic charges and forces also apply to electric fields: Use symmetry to simplify your calculations; remember that the field is composed of vectors and thus you have to use vector operations instead of simple addition, multiplication, and so on; convert units to meters and coulombs for consistency with the given values of constants.

3. Remember to use the correct form of the charge density for field calculations: λ for linear charge density, σ for surface charge density, and ρ for volume charge density.

4. The key to using Gauss's Law is to choose the right shape for the Gaussian surface to exploit the symmetry of the problem situation. Cubical, cylindrical, and spherical Gaussian surfaces are typically useful.

5. Often, you can break a Gaussian surface into surface elements that are either perpendicular to or parallel to the electric field lines. If the field lines are perpendicular to the surface, the electric flux is simply the field strength times the area, EA, or $-EA$ if the field points inward instead of outward. If the field lines are parallel to the surface, the flux through that surface is zero. The total flux is the sum of the flux through each surface element of the Gaussian surface. Remember that zero flux through a Gaussian surface does not necessarily mean that the electric field is zero.

MULTIPLE-CHOICE QUESTIONS

22.1 In order to use Gauss's Law to calculate the electric field created by a known distribution of charge, which of the following *must* be true?

a) The charge distribution must be in a nonconducting medium.

b) The charge distribution must be in a conducting medium.

c) The charge distribution must have spherical or cylindrical symmetry.

d) The charge distribution must be uniform.

e) The charge distribution must have a high degree of symmetry that allows assumptions about the symmetry of its electric field to be made.

22.2 An electric dipole consists of two equal and opposite charges situated a small distance from each other. When the dipole is placed in a uniform electric field, which of the following statements is (are) true?

a) The dipole will not experience any net force from the electric field; since the charges are equal and have opposite signs, the individual effects will cancel out.

b) There will be no net force and no net torque acting on the dipole.

c) There will be a net force but no net torque acting on the dipole.

d) There will be no net force, but there will (in general) be a net torque acting on dipole.

22.3 A point charge, $+Q$, is located on the x-axis at $x = a$, and a second point charge, $-Q$, is located on the x-axis at $x = -a$. A Gaussian surface with radius $r = 2a$ is centered at the origin. The flux through this Gaussian surface is

a) zero.   c) less than zero.

b) greater than zero.   d) none of the above.

22.4 A charge of $+2q$ is placed at the center of an uncharged conducting shell. What will be the charges on the inner and outer surfaces of the shell, respectively?

a) $-2q$, $+2q$ c) $-2q$, $-2q$

b) $-q$, $+q$ d) $-2q$, $+4q$

22.5 Two infinite nonconducting plates are parallel to each other, with a distance $d = 10.0$ cm between them, as shown in the figure. Each plate carries a uniform charge distribution of $\sigma = 4.5$ μC/m². What is the electric field, $\vec{E}$, at point P (with $x_P = 20.0$ cm)?

a) 0 N/C

b) $2.54\hat{x}$ N/C

c) $(-5.08 \cdot 10^5)\hat{x}$ N/C

d) $(5.08 \cdot 10^5)\hat{x}$ N/C

e) $(-1.02 \cdot 10^6)\hat{x}$ N/C

f) $(1.02 \cdot 10^6)\hat{x}$ N/C

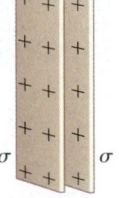 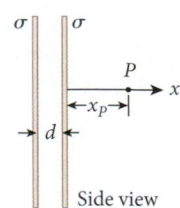

Side view

22.6 At which of the following locations is the electric field the strongest?

a) a point 1 m from a 1-C point charge

b) a point 1 m (perpendicular distance) from the center of a 1-m-long wire with 1 C of charge distributed on it

c) a point 1 m (perpendicular distance) from the center of a 1-m^2 sheet of charge with 1 C of charge distributed on it

d) a point 1 m from the surface of a charged spherical shell with a radius of 1 m

e) a point 1 m from the surface of a charged spherical shell with a radius of 0.5 m and a charge of 1 C

22.7 The electric flux through a spherical Gaussian surface of radius R centered on a charge Q is 1200 N/(C m^2). What is the electric flux through a cubic Gaussian surface of side R centered on the same charge Q?

a) less than 1200 N/(C m^2)

b) more than 1200 N/(C m^2)

c) equal to 1200 N/(C m^2)

d) cannot be determined from the information given

22.8 A single positive point charge, q, is at one corner of a cube with sides of length L, as shown in the figure. The net electric flux through the three adjacent sides is zero. The net electric flux through *each* of the other three sides is

a) $q/3\epsilon_0$.

b) $q/6\epsilon_0$.

c) $q/24\epsilon_0$.

d) $q/8\epsilon_0$.

22.9 Three –9-mC point charges are located at (0,0), (3 m,3 m), and (3 m,–3 m). What is the magnitude of the electric field at (3 m,0)?

a) $0.9 \cdot 10^7$ N/C

b) $1.2 \cdot 10^7$ N/C

c) $1.8 \cdot 10^7$ N/C

d) $2.4 \cdot 10^7$ N/C

e) $3.6 \cdot 10^7$ N/C

f) $5.4 \cdot 10^7$ N/C

g) $10.8 \cdot 10^7$ N/C

22.10 Which of the following statements is (are) true?

a) There will be *no* change in the charge on the inner surface of a hollow conducting sphere if additional charge is placed on the outer surface.

b) There will be some change in the charge on the inner surface of a hollow conducting sphere if additional charge is placed on the outer surface.

c) There will be *no* change in the charge on the inner surface of a hollow conducting sphere if additional charge is placed at the center of the sphere.

d) There will be some change in the charge on the inner surface of a hollow conducting sphere if additional charge is placed at the center of the sphere.

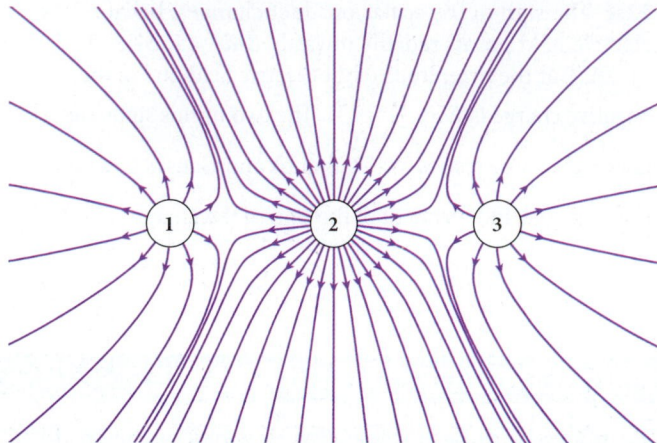

22.11 What are the signs of the charges in the configuration shown in the figure?

a) Charges 1, 2, and 3 are negative.

b) Charges 1, 2, and 3 are positive.

c) Charges 1 and 3 are positive, and 2 is negative.

d) Charges 1 and 3 are negative, and 2 is positive.

e) All that can be said is that the charges have the same sign.

22.12 Which of the following statements is (are) true?

a) Electric field lines point inward toward negative charges.

b) Electric field lines form circles around positive charges.

c) Electric field lines may cross.

d) Electric field lines point outward from positive charges.

e) A positive point charge released from rest will initially accelerate along a tangent to the electric field line at that point.

CONCEPTUAL QUESTIONS

22.13 Many people have been sitting in a car when it was struck by lightning. Why were they able to survive such an experience?

22.14 Why is it a bad idea to stand under a tree in a thunderstorm? What should one do instead to avoid getting struck by lightning?

22.15 Why do electric field lines never cross?

22.16 How is it possible that the flux through a closed surface does not depend on where inside the surface the charge is located (that is, the charge can be moved around inside the surface with no effect whatsoever on the flux)? If the charge is moved from just inside to just outside the surface, the flux changes discontinuously to zero, according to Gauss's Law. Does this really happen? Explain.

22.17 A solid conducting sphere of radius r_1 has a total charge of +3Q. It is placed inside (and concentric with) a conducting spherical shell of inner radius r_2 and outer radius r_3. Find the electric field in these regions: $r < r_1$, $r_1 < r < r_2$, $r_2 < r < r_3$, and $r > r_3$.

22.18 A thin rod has end points at $x = \pm 100$ cm. There is a total charge Q uniformly distributed along the rod.

a) What is the electric field very close to the midpoint of the rod?

b) What is the electric field a few centimeters (perpendicularly) from the midpoint of the rod?

c) What is the electric field very far (perpendicularly) from the midpoint of the rod?

22.19 A dipole is completely enclosed by a spherical surface. Describe how the total electric flux through this surface varies with the strength of the dipole.

22.20 Repeat Example 22.3, assuming that the charge distribution is $-\lambda$ for $-a < x < 0$ and $+\lambda$ for $0 < x < a$.

22.21 A negative charge is placed on a solid prolate spheroidal conductor (shown in cross section in the figure). Sketch the distribution of the charge on the conductor and the electric field lines due to the charge.

22.22 *Saint Elmo's fire* is an eerie glow that appears at the tips of masts and yardarms of sailing ships in stormy weather and at the tips and edges of the wings of aircraft in flight. St. Elmo's fire is an electrical phenomenon. Explain it, concisely.

22.23 A charge placed on a conductor of any shape forms a layer on the outer surface of the conductor. Mutual repulsion of the individual charge elements creates an outward pressure on this layer, called *electrostatic stress.* Treating the infinitesimal charge elements like tiles of a mosaic, calculate

the magnitude of this electrostatic stress in terms of the surface charge density, σ. Note that σ need not be uniform over the surface.

22.24 An electric dipole is placed in a uniform electric field as shown in the figure. What motion will the dipole have in the electric field? Which way will it move? Which way will it rotate?

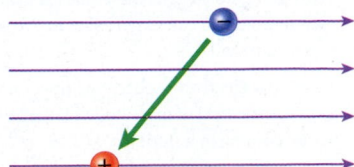

EXERCISES

A blue problem number indicates a worked-out solution is available in the Student Solutions Manual. One • and two •• indicate increasing level of problem difficulty.

Section 22.3

22.25 A point charge, $q = 4.00 \cdot 10^{-9}$ C, is placed on the x-axis at the origin. What is the electric field produced at $x = 25.0$ cm?

22.26 A +1.60-nC point charge is placed at one corner of a square (1.00 m on a side), and a –2.40-nC charge is placed on the corner diagonally opposite. What is the magnitude of the electric field at either of the other two corners?

22.27 A +48.00-nC point charge is placed on the x-axis at $x = 4.000$ m, and a –24.00-nC point charge is placed on the y-axis at $y = -6.000$ m. What is the direction of the electric field at the origin?

•22.28 Two point charges are placed at two of the corners of a triangle as shown in the figure. Find the magnitude and the direction of the electric field at the third corner of the triangle.

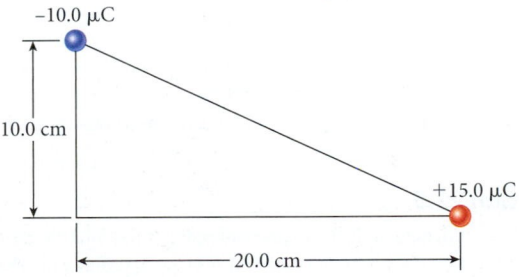

•22.29 A +5.00-C charge is located at the origin. A –3.00-C charge is placed at $x = 1.00$ m. At what finite distance(s) along the x-axis will the electric field be equal to zero?

•22.30 Three charges are on the y-axis. Two of the charges, each $-q$, are located $y = \pm d$, and the third charge, $+2q$, is located at $y = 0$. Derive an expression for the electric field at a point P on the x-axis.

Section 22.4

22.31 For the electric dipole shown in the figure, express the magnitude of the resulting electric field as a function of the perpendicular distance x from the center of the dipole axis. Comment on what the magnitude is when $x \gg d$.

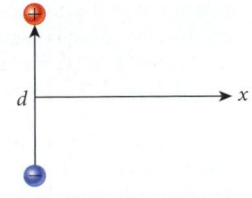

•22.32 Consider an electric dipole on the x-axis and centered at the origin. At a distance h along the positive x-axis, the magnitude of electric field due to the electric dipole is given by $k(2qd)/h^3$. Find a distance perpendicular to the x-axis and measured from the origin at which the magnitude of the electric field is the same.

Section 22.5

•22.33 A small metal ball with a mass of 4.00 g and a charge of 5.00 mC is located at a distance of 0.700 m above the ground in an electric field of 12.0 N/C directed to the east. The ball is then released from rest. What is the velocity of the ball after it has moved downward a vertical distance of 0.300 m?

•22.34 A charge per unit length $+\lambda$ is uniformly distributed along the positive y-axis from $y = 0$ to $y = +a$. A charge per unit length $-\lambda$ is uniformly distributed along the negative y-axis from $y = 0$ to $y = -a$. Write an expression for the electric field (magnitude and direction) at a point on the x-axis a distance x from the origin.

•22.35 A thin glass rod is bent into a semicircle of radius R. A charge $+Q$ is uniformly distributed along the upper half, and a charge $-Q$ is uniformly distributed along the lower half as shown in the figure. Find the magnitude and direction of the electric field $\vec{E}$ (in component form) at point P, the center of the semicircle.

•22.36 Two uniformly charged insulating rods are bent in a semicircular shape with radius $r = 10.0$ cm. If they are positioned so that they form a circle but do not touch and if they have opposite charges of +1.00 μC and –1.00 μC, find the magnitude and the direction of the electric field at the center of the composite circular charge configuration.

•22.37 A uniformly charged rod of length L with total charge Q lies along the y-axis, from $y = 0$ to $y = L$. Find an expression for the electric field at the point $(d,0)$ (that is, the point at $x = d$ on the x-axis).

••22.38 A charge Q is distributed evenly on a wire bent into an arc of radius R, as shown in the figure. What is the electric field at the center of the arc as a function of the angle θ? Sketch a graph of the electric field as a function of θ for $0 < \theta < 180°$.

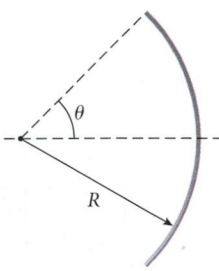

••22.39 A thin, flat washer is a disk with an outer diameter of 10.0 cm and a hole of diameter 4.00 cm in the center. The washer has a uniform charge distribution and a total charge of 7.00 nC. What is the electric field on the axis of the washer at a distance of 30.0 cm from the center of the washer?

Section 22.6

22.40 Research suggests that the electric fields in some thunderstorm clouds can be on the order of 10.0 kN/C. Calculate the magnitude of the electric force acting on a particle with two excess electrons in the presence of a 10.0-kN/C field.

22.41 An electric dipole has opposite charges of $5.00 \cdot 10^{-15}$ C separated by a distance of 0.400 mm. It is oriented at 60.0° with respect to a uniform electric field of magnitude $2.00 \cdot 10^3$ N/C. Determine the magnitude of the torque exerted on the dipole by the electric field.

22.42 Electric dipole moments of molecules are often measured in debyes (D), where 1 D = $3.34 \cdot 10^{-30}$ C m. For instance, the dipole moment of hydrogen chloride gas molecules is 1.05 D. Calculate the maximum torque such a molecule can experience in the presence of an electric field of magnitude 160.0 N/C.

22.43 An electron is observed traveling at a speed of $27.5 \cdot 10^6$ m/s parallel to an electric field of magnitude 11,400 N/C. How far will the electron travel before coming to a stop?

22.44 Two charges, $+e$ and $-e$, are a distance of 0.680 nm apart in an electric field, E, that has a magnitude of 4.40 kN/C and is directed at an angle of 45.0° with respect to the dipole axis. Calculate the dipole moment and thus the torque on the dipole in the electric field.

•22.45 A body of mass M, carrying charge Q, falls from rest from a height h (above the ground) near the surface of the Earth, where the gravitational acceleration is g and there is an electric field with a constant component E in the vertical direction.

a) Find an expression for the speed, v, of the body when it reaches the ground, in terms of M, Q, h, g, and E.

b) The expression from part (a) is not meaningful for certain values of M, g, Q, and E. Explain what happens in such cases.

•22.46 A water molecule, which is electrically neutral but has a dipole moment of magnitude $p = 6.20 \cdot 10^{-30}$ C m, is 1.00 cm away from a point charge $q = +1.00$ μC. The dipole will align with the electric field due to the charge. It will also experience a net force, since the field is not uniform.

a) Calculate the magnitude of the net force. (*Hint:* You do not need to know the precise size of the molecule, only that it is much smaller than 1 cm.)

b) Is the molecule attracted to or repelled by the point charge? Explain.

•22.47 A total of $3.05 \cdot 10^6$ electrons are placed on an initially uncharged wire of length 1.33 m.

a) What is the magnitude of the electric field a perpendicular distance of 0.401 m away from the midpoint of the wire?

b) What is the magnitude of the acceleration of a proton placed at that point in space?

c) In which direction does the electric field force point in this case?

Sections 22.7 and 22.8

22.48 Four charges are placed in three-dimensional space. The charges have magnitudes $+3q$, $-q$, $+2q$, and $-7q$. If a Gaussian surface encloses all the charges, what will be the electric flux through that surface?

22.49 The six faces of a cubical box each measure 20.0 cm by 20.0 cm, and the faces are numbered such that faces 1 and 6 are opposite to each other, as are faces 2 and 5, and faces 3 and 4. The flux through each face is given in the table. Find the net charge inside the cube.

Face	Flux (N m²/C)
1	−70.0
2	−300.0
3	−300.0
4	+300.0
5	−400.0
6	−500.0

22.50 A conducting solid sphere ($R = 0.15$ m, $q = 6.1 \cdot 10^{-6}$ C) is shown in the figure. Using Gauss's Law and two different Gaussian surfaces, determine the electric field (magnitude and direction) at point A, which is 0.0000010 m outside the conducting sphere. (*Hint:* One Gaussian surface is a sphere, and the other is a small right cylinder.)

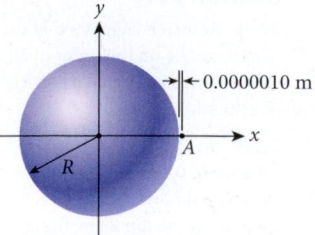

22.51 Electric fields of varying magnitudes are directed either inward or outward at right angles on the faces of a cube, as shown in the figure. What is the strength and direction of the field on the face F?

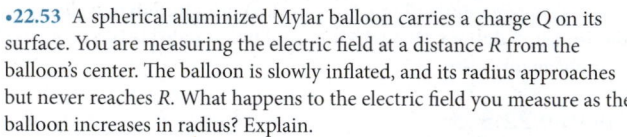

22.52 Consider a hollow spherical conductor with total charge $+5e$. The outer and inner radii are a and b, respectively. (a) Calculate the charge on the sphere's inner and outer surfaces if a charge of $-3e$ is placed at the center of the sphere. (b) What is the total net charge of the sphere?

•22.53 A spherical aluminized Mylar balloon carries a charge Q on its surface. You are measuring the electric field at a distance R from the balloon's center. The balloon is slowly inflated, and its radius approaches but never reaches R. What happens to the electric field you measure as the balloon increases in radius? Explain.

•22.54 A hollow conducting spherical shell has an inner radius of 8.00 cm and an outer radius of 10.0 cm. The electric field at the inner surface of the shell, E_i, has a magnitude of 80.0 N/C and points toward the center of the sphere, and the electric field at the outer surface, E_o, has a magnitude of 80.0 N/C and points away from the center of the sphere (see the figure). Determine the magnitude of the charge on the inner surface and on the outer surface of the spherical shell.

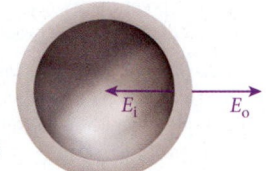

•22.55 A −6.00-nC point charge is located at the center of a conducting spherical shell. The shell has an inner radius of 2.00 m, an outer radius of 4.00 m, and a charge of +7.00 nC.

a) What is the electric field at $r = 1.00$ m?

b) What is the electric field at $r = 3.00$ m?

c) What is the electric field at $r = 5.00$ m?

d) What is the surface charge distribution, σ, on the outside surface of the shell?

Section 22.9

22.56 A solid, nonconducting sphere of radius a has total charge Q and a uniform charge distribution. Using Gauss's Law, determine the electric field (as a vector) in the regions $r < a$ and $r > a$ in terms of Q.

22.57 There is an electric field of magnitude 150.0 N/C, directed downward, near the surface of the Earth. What is the net electric charge on the Earth? You can treat the Earth as a spherical conductor of radius 6371 km.

22.58 A hollow metal sphere has inner and outer radii of 20.0 cm and 30.0 cm, respectively. As shown in the figure, a solid metal sphere of radius 10.0 cm is located at the center of the hollow sphere. The electric field at a point P, a distance of 15.0 cm from the center, is found to be $E_1 = 1.00 \cdot 10^4$ N/C, directed radially inward. At point Q, a distance of 35.0 cm from the center, the electric field is found to be $E_2 = 1.00 \cdot 10^4$ N/C, directed radially outward. Determine the total charge on (a) the surface of the inner sphere, (b) the inner surface of the hollow sphere, and (c) the outer surface of the hollow sphere.

22.59 Two parallel, infinite, nonconducting plates are 10.0 cm apart and have charge distributions of $+1.00$ μC/m² and -1.00 μC/m². What is the force on an electron in the space between the plates? What is the force on an electron located outside the two plates near the surface of one of the two plates?

22.60 An infinitely long charged wire produces an electric field of magnitude $1.23 \cdot 10^3$ N/C at a distance of 50.0 cm perpendicular to the wire. The direction of the electric field is toward the wire.

a) What is the charge distribution?

b) How many electrons per unit length are on the wire?

•22.61 A solid sphere of radius R has a nonuniform charge distribution $\rho = Ar^2$, where A is a constant. Determine the total charge, Q, within the volume of the sphere.

•22.62 Two parallel, uniformly charged, infinitely long wires are 6.00 cm apart and carry opposite charges with a linear charge density of $\lambda = 1.00$ µC/m. What are the magnitude and the direction of the electric field at a point midway between the two wires and 40.0 cm above the plane containing them?

•22.63 A sphere centered at the origin has a volume charge distribution of 120. nC/cm³ and a radius of 12.0 cm. The sphere is centered inside a conducting spherical shell with an inner radius of 30.0 cm and an outer radius of 50.0 cm. The charge on the spherical shell is –2.00 mC. What are the magnitude and the direction of the electric field at each of the following distances from the origin?

a) at $r = 10.0$ cm   c) at $r = 40.0$ cm

b) at $r = 20.0$ cm   d) at $r = 80.0$ cm

•22.64 A thin, hollow, metal cylinder of radius R has a surface charge distribution σ. A long, thin wire with a linear charge density $\lambda/2$ runs through the center of the cylinder. Find an expression for the electric field and determine the direction of the field at each of the following locations:

a) $r \le R$ b) $r \ge R$

•22.65 Two infinite sheets of charge are separated by 10.0 cm as shown in the figure. Sheet 1 has a surface charge distribution of $\sigma_1 = 3.00$ µC/m² and sheet 2 has a surface charge distribution of $\sigma_2 = -5.00$ µC/m². Find the total electric field (magnitude and direction) at each of the following locations:

a) at point P, 6.00 cm to the left of sheet 1

b) at point P', 6.00 cm to the right of sheet 1

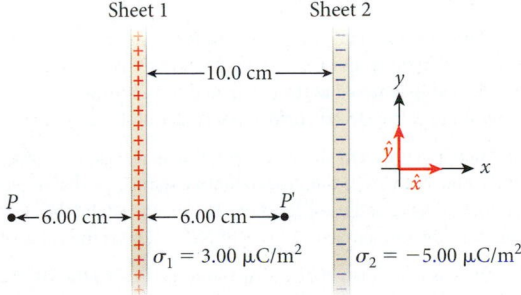

•22.66 A conducting solid sphere of radius 20.0 cm is located with its center at the origin of a three-dimensional coordinate system. A charge of 0.271 nC is placed on the sphere.

a) What is the magnitude of the electric field at point $(x,y,z) = (23.1 \text{ cm}, 1.10 \text{ cm}, 0.00 \text{ cm})$?

b) What is the angle of this electric field with the x-axis at this point?

c) What is the magnitude of the electric field at point $(x,y,z) = (4.10 \text{ cm}, 1.10 \text{ cm}, 0.00 \text{ cm})$?

••22.67 A solid nonconducting sphere of radius a has a total charge $+Q$ uniformly distributed throughout its volume. The surface of the sphere is coated with a very thin (negligible thickness) conducting layer of gold. A total charge of $-2Q$ is placed on this conducting layer. Use Gauss's Law to do the following.

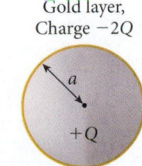

Gold layer,
Charge $-2Q$

a) Find the electric field $E(r)$ for $r < a$ (inside the sphere, up to and excluding the gold layer).

b) Find the electric field $E(r)$ for $r > a$ (outside the coated sphere, beyond the sphere and the gold layer).

c) Sketch the graph of $E(r)$ versus r. Comment on the continuity or discontinuity of the electric field, and relate this to the surface charge distribution on the gold layer.

••22.68 A solid nonconducting sphere has a volume charge distribution given by $\rho(r) = (\beta/r) \sin(\pi r/2R)$. Find the total charge contained in the spherical volume and the electric field in the regions $r < R$ and $r > R$. Show that the two expressions for the electric field equal each other at $r = R$.

••22.69 A very long cylindrical rod of nonconducting material with a 3.00-cm radius is given a uniformly distributed positive charge of 6.00 nC per centimeter of its length. Then a cylindrical cavity is drilled all the way through the rod, of radius 1 cm, with its axis located 1.50 cm from the axis of the rod. That is, if, at some cross section of the rod, x- and y-axes are placed so that the center of the rod is at $(x,y) = (0,0)$; then the center of the cylindrical cavity is at $(x,y) = (0,1.50)$. The creation of the cavity does not disturb the charge on the remainder of the rod that has not been drilled away; it just removes the charge from the region in the cavity. Find the electric field at the point $(x,y) = (2.00, 1.00)$.

••22.70 What is the electric field at a point P, which is at a distance $h = 20.0$ cm above an infinite sheet of charge that has a charge distribution of 1.30 C/m² and a hole of radius 5.00 cm whose center is directly below P, as shown in the figure? Plot the electric field as a function of h in terms of $\sigma/(2\epsilon_0)$.

P

$h = 20.0$ cm

5.00 cm

Additional Exercises

22.71 A cube has an edge length of 1.00 m. An electric field acting on the cube from outside has a constant magnitude of 150 N/C and its direction is also constant but unspecified (not necessarily along any edges of the cube). What is the total charge within the cube?

22.72 A carbon monoxide (CO) molecule has a dipole moment of approximately $8.0 \cdot 10^{-30}$ C m. If the carbon and oxygen atoms are separated by $1.2 \cdot 10^{-10}$ m, find the net charge on each atom and the maximum amount of torque the molecule would experience in an electric field of 500.0 N/C.

22.73 An infinitely long, solid cylinder of radius $R = 9.00$ cm, with a uniform charge per unit of volume of $\rho = 6.40 \cdot 10^{-8}$ C/m³, is centered about the y-axis. Find the magnitude of the electric field at a radius $r = 4.00$ cm from the center of this cylinder.

22.74 Find the magnitudes and the directions of the electric fields needed to counteract the weight of (a) an electron and (b) a proton at the Earth's surface.

22.75 A solid metal sphere of radius 8.00 cm, with a total charge of 10.0 µC, is surrounded by a metallic shell with a radius of 15.0 cm carrying a –5.00-µC charge. The sphere and the shell are both inside a larger metallic shell of inner radius 20.0 cm and outer radius 24.0 cm. The sphere and the two shells are concentric.

a) What is the charge on the inner wall of the larger shell?

b) If the electric field outside the larger shell is zero, what is the charge on the outer wall of the shell?

22.76 Two infinite, uniformly charged, flat, nonconducting surfaces are mutually perpendicular. One of the surfaces has a charge distribution of +30.0 pC/m², and the other has a charge distribution of –40.0 pC/m². What is the magnitude of the electric field at any point not on either surface?

22.77 There is an electric field of magnitude 150. N/C, directed vertically downward, near the surface of the Earth. Find the acceleration (magnitude and direction) of an electron released near the Earth's surface.

22.78 Suppose you have a large spherical balloon and you are able to measure the component E_n of the electric field normal to its surface. If you sum $E_n \, dA$ over the whole surface area of the balloon and obtain a magnitude of 10.0 N m²/C, what is the electric charge enclosed by the balloon?

22.79 A 30.0-cm-long uniformly charged rod is sealed in a container. The total electric flux leaving the container is $1.46 \cdot 10^6$ N m^2/C. Determine the linear charge distribution on the rod.

•**22.80** A long conducting wire with charge distribution λ and radius r produces an electric field of 2.73 N/C just outside its surface. What is the magnitude of the electric field just outside the surface of another wire with charge distribution 0.810λ and radius $6.50r$?

•**22.81** An object with mass $m = 1.00$ g and charge q is placed at point A, which is 0.0500 m above an infinitely large, uniformly charged, nonconducting sheet ($\sigma = -3.50 \cdot 10^{-5}$ C/m^2), as shown in the figure. Gravity is acting downward ($g = 9.81$ m/s^2). Determine the number, N, of electrons that must be added to or removed from the object for the object to remain motionless above the charged plane.

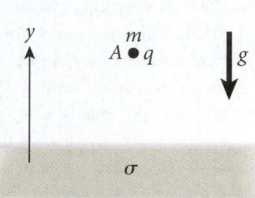

•**22.82** A proton enters the gap between a pair of metal plates (an electrostatic separator) that produce a uniform, vertical electric field between them. Ignore the effect of gravity on the proton.

a) Assuming that the length of the plates is 15.0 cm and that the proton approaches the plates with a speed of 15.0 km/s, what electric field strength should the plates be designed to provide so that the proton will be deflected vertically by $1.50 \cdot 10^{-3}$ rad?

b) What speed will the proton have after exiting the electric field?

c) Suppose the proton is one in a beam of protons that has been contaminated with positively charged kaons, particles whose mass is 494 MeV/c^2 ($8.81 \cdot 10^{-28}$ kg), while the mass of the proton is

938 MeV/c^2 ($1.67 \cdot 10^{-27}$ kg). The kaons have a charge of $+1e$, just like the protons. If the electrostatic separator is designed to give the protons a deflection of $1.20 \cdot 10^{-3}$ rad, what deflection will kaons with the same momentum as the protons experience?

•**22.83** Consider a uniform nonconducting sphere with a surface charge density $\rho = 3.57 \cdot 10^{-6}$ C/m^3 and a radius $R = 1.72$ m. What is the magnitude of the electric field 0.530 m from the center of the sphere?

••**22.84** A uniform sphere has a radius R and a total charge $+Q$, uniformly distributed throughout its volume. It is surrounded by a thick spherical shell carrying a total charge $-Q$, also uniformly distributed, and having an outer radius of $2R$. What is the electric field as a function of R?

••**22.85** If a charge is held in place above a large, flat, grounded, conducting slab, such as a floor, it will experience a downward force toward the floor. In fact, the electric field in the room above the floor will be exactly the same as that produced by the original charge plus a "mirror image" charge, equal in magnitude and opposite in sign, as far below the floor as the original charge is above it. Of course, there is no charge below the floor; the effect is produced by the surface charge distribution induced on the floor by the original charge.

a) Describe or sketch the electric field lines in the room above the floor.

b) If the original charge is 1.00 μC at a distance of 50.0 cm above the floor, calculate the downward force on this charge.

c) Find the electric field at (just above) the floor, as a function of the horizontal distance from the point on the floor directly under the original charge. Assume that the original charge is a point charge, $+q$, at a distance a above the floor. Ignore any effects of walls or ceiling.

d) Find the surface charge distribution $\sigma(\rho)$ induced on the floor.

e) Calculate the total surface charge induced on the floor.

MULTI-VERSION EXERCISES

22.86 A long, horizontal, conducting wire has the charge density $\lambda = 2.849 \cdot 10^{-12}$ C/m. A proton (mass = $1.673 \cdot 10^{-27}$ kg) is placed 0.6815 m above the wire and released. What is the magnitude of the initial acceleration of the proton?

22.87 A long, horizontal, conducting wire has the charge density λ. A proton (mass = $1.673 \cdot 10^{-27}$ kg) is placed 0.6897 m above the wire and released. The magnitude of the initial acceleration of the proton is $1.111 \cdot 10^7$ m/s^2. What is the charge density on the wire?

22.88 A long, horizontal, conducting wire has the charge density $\lambda = 6.055 \cdot 10^{-12}$ C/m. A proton (mass = $1.673 \cdot 10^{-27}$ kg) is placed a distance d above the wire and released. The magnitude of the initial acceleration of the proton is $1.494 \cdot 10^7$ m/s^2. What is the distance d?

22.89 There is a uniform charge distribution of $\lambda = 5.635 \cdot 10^{-8}$ C/m along a thin wire of length $L = 22.13$ cm. The wire is then curved into a semicircle that is centered at the origin and has a radius of $R = L/\pi$. Find the magnitude of the electric field at the center of the semicircle.

22.90 There is a uniform charge distribution λ along a thin wire of length $L = 10.55$ cm. The wire is then curved into a semicircle that is centered at the origin and has a radius of $R = L/\pi$. The magnitude of the electric field at the center of the semicircle is $3.117 \cdot 10^4$ N/C. What is the value of λ?

22.91 There is a uniform charge distribution of $\lambda = 6.005 \cdot 10^{-8}$ C/m along a thin wire of length L. The wire is then curved into a semicircle that is centered at the origin and has a radius of $R = L/\pi$. The magnitude of the electric field at the center of the semicircle is $2.425 \cdot 10^4$ N/C. What is the value of L?

23

Electric Potential

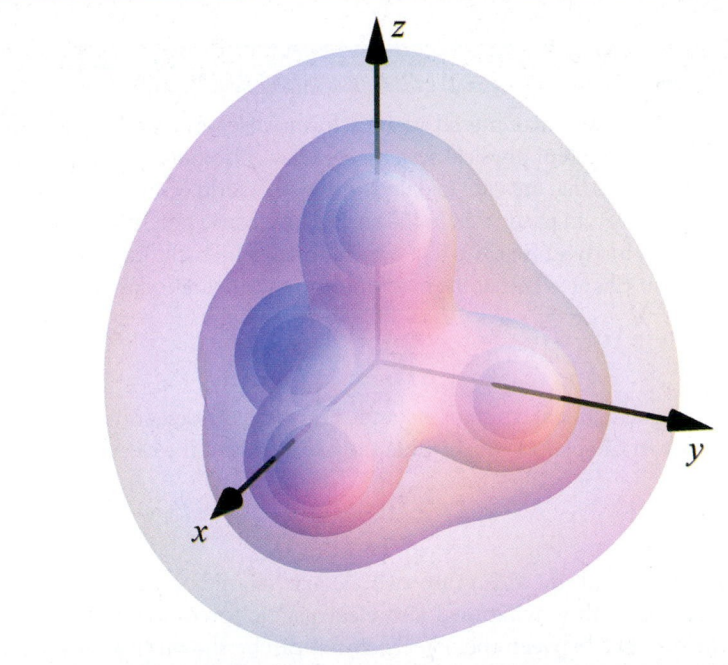

FIGURE 23.1 Five equipotential surfaces of a tetrahedral arrangement of four identical charges.

Energy is vital to all kinds of systems, from our own bodies to the world economy. A large fraction of the devices we rely on—including pacemakers, televisions, computers, cell phones, hair dryers, and locomotives—use electric energy. It is therefore important to understand how electrically powered devices process and store this energy. We start by looking at electric energy and electric potential in order to understand the relationship between them and see how to calculate both. Figure 23.1 shows the result of calculating the electric potential due to four identical charges arranged at the corners of a tetrahedron. The electric potential is visualized as five nested surfaces representing points at which its magnitude is constant. The highest values of the potential occur closest to the charges, and the equipotential surfaces approximate spheres around each charge. The equipotential surfaces with the lowest values are farthest away from the charges and approximate spheres around the center of mass of the combined charges.

WHAT WE WILL LEARN

- Electric potential energy is analogous to gravitational potential energy.
- The change in electric potential energy is proportional to the work done by the electric field on a charge.
- The electric potential at a given point in space is a scalar.
- The electric potential, V, of a point charge, q, is inversely proportional to the distance from that point charge.

- The electric potential can be derived from the electric field by integrating the electric field over a displacement.
- The electric potential at a given point in space due to a distribution of point charges equals the algebraic sum of the electric potentials due to the individual charges.
- The electric field can be derived from the electric potential by differentiating the electric potential with respect to displacement.

23.1   Electric Potential Energy

Throughout this book, we have encountered various forms of energy and have seen how energy conservation affects different physical systems. We have also noted the importance of energy conversion in processes that are vital to daily life and the world economy. Now we turn our attention to electric energy, in particular, to the storage of electric potential energy in batteries. An electric field has many similarities to a gravitational field, including its mathematical formulation. We saw in Chapter 12 that the magnitude of the gravitational force is given by

$$F_g = G\frac{m_1 m_2}{r^2},$$

where G is the universal gravitational constant, m_1 and m_2 are two masses, and r is the distance between the two masses. In Chapter 21, we saw that the magnitude of the electrostatic force is

$$F_e = k\frac{|q_1 q_2|}{r^2}, \tag{23.1}$$

where k is Coulomb's constant, q_1 and q_2 are two electric charges, and r is the distance between the two charges. Both gravitational and electrostatic forces depend only on the inverse square of the distance between the objects, and it can be shown that all such forces are conservative. Therefore, the electric potential energy, U, can be defined in analogy with the gravitational potential energy.

In Chapter 6, we saw that for any conservative force, the change in potential energy due to some spatial rearrangement of a system is equal to the negative of the work done by the conservative force during this spatial rearrangement. For a system of two or more particles, the work done by an electric force, W_e, when the system configuration changes from an initial state to a final state, is given in terms of the change in **electric potential energy,** ΔU:

$$\Delta U = U_f - U_i = -W_e, \tag{23.2}$$

where U_i is the initial electric potential energy and U_f is the final electric potential energy. Note that it does not matter *how* the system gets from the initial to the final state. The work is always the same, independent of the path taken. Chapter 6 noted that this path-independence of the work done by a force is a general feature of conservative forces.

As is the case for gravitational potential energy (see Chapter 12), a reference point for the electric potential energy must always be specified. It simplifies the equations and calculations if the zero point of the electric potential energy is assumed to be the configuration in which an infinitely large distance separates all the charges, which is exactly the same convention used for the gravitational potential energy. This assumption allows equation 23.2 for the change in electric potential energy to be rewritten as $\Delta U = U_f - 0 = U$, or

$$U = -W_{e,\infty}. \tag{23.3}$$

Even though the convention of zero potential energy at infinity is very useful and is universally accepted for a collection of point charges, in some physical situations there is a reason to select a reference potential energy at some point in space, which will not lead to a value of zero potential energy at infinite separation. Remember, all potential energies of conservative forces are fixed only within an arbitrary additive constant. So you need to pay attention to how this constant is chosen in a particular situation. One situation in which the potential energy at infinity is not set to zero is that involving a constant electric field.

Special Case: Charge in a Constant Electric Field

Let's consider a point charge, q, moving through a displacement, $\vec{d}$, in a constant electric field, $\vec{E}$ (Figure 23.2). The work done by a constant force $\vec{F}$ is $W = \vec{F} \cdot \vec{d}$. For this case, the constant force is created by a constant electric field, $\vec{F} = q\vec{E}$. Thus, the work done by the field on the charge is given by

$$W = q\vec{E} \cdot \vec{d} = qEd\cos\theta, \tag{23.4}$$

where θ is the angle between the electric force and the displacement. When the displacement is parallel to the electric field ($\theta = 0°$), the work done by the field on the charge is $W = qEd$. When the displacement is antiparallel to the electric field ($\theta = 180°$), the work done by the field is $W = -qEd$. Because the change in electric potential energy is related to the work done on the charge by $\Delta U = -W$, if $q > 0$, the charge loses potential energy when the displacement is in the same direction as the electric field and gains potential energy when the displacement is in the direction opposite to the electric field.

Figure 23.3a shows a mass, m, near the surface of the Earth, where it can be considered to be in a constant gravitational field, which points downward. From Chapter 6, we know that when the mass moves toward the surface of the Earth a distance h, the change in the gravitational potential energy of the mass is

$$\Delta U = -W = -\vec{F}_g \cdot \vec{d} = -mgh.$$

It is intuitive that the mass has less potential energy if it is closer to the surface of the Earth. Figure 23.3b shows a positive charge, q, in a constant electric field. If the charge moves a distance, d, in the same direction as the electric field, the change in the electric potential energy is

$$\Delta U = -W = -q\vec{E} \cdot \vec{d} = -qEd.$$

Thus, the electric potential energy of a charge in an electric field is analogous to the gravitational potential energy of a mass in Earth's gravitational field near the surface of Earth. (But, of course, the important difference between the two interactions is that masses come in only one variety, and exert gravitational attraction for one another, whereas charges can attract or repel each other. Thus, ΔU can change sign, depending on the signs of the charges.)

Special Case: Dipole in a Constant Electric Field

Now let's consider an electric dipole with dipole moment $\vec{p}$ moving through a constant electric field (see Figure 23.4). In Chapter 22, we saw that an electric dipole consists of a positive charge and a negative charge, which have equal magnitudes, meaning that the dipole has zero net charge. Since, according to equation 23.4, the work done in moving an object through a constant electric field is proportional to the charge on that object, the net work done in moving an electric dipole through a constant electric field is zero.

This fact may make it seem impossible to store potential energy in a system consisting of a dipole in a constant field. However, this is not the case. In Chapter 22, we saw that a dipole in a constant electric field experiences a torque, $\vec{\tau} = \vec{p} \times \vec{E}$, and so it is clear that the orientation of the dipole relative to the electric field is key. Let's see how the orientation of the dipole can result in the storage of potential energy.

In Chapter 10, we saw that the work done by a torque is given by $W = \int \vec{\tau}(\theta')d\vec{\theta}'$. If we apply an external torque opposing the torque that the dipole experiences from the electric field, we can express the work done by this external torque as follows:

$$W = \int_{\theta_0}^{\theta} \vec{\tau}(\theta')d\vec{\theta}' = \int_{\theta_0}^{\theta} -pE\sin\theta'd\theta' = -pE\int_{\theta_0}^{\theta} \sin\theta'd\theta' = pE(\cos\theta - \cos\theta_0).$$

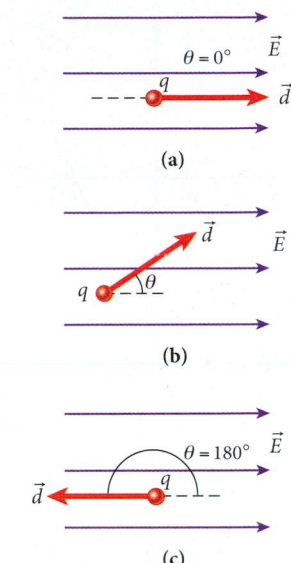

FIGURE 23.2 Work done by an electric field, $\vec{E}$, on a moving charge, q: (a) case where the displacement is in the same direction as the electric field, (b) general case, (c) case where the displacement is opposite to the direction of the electric field.

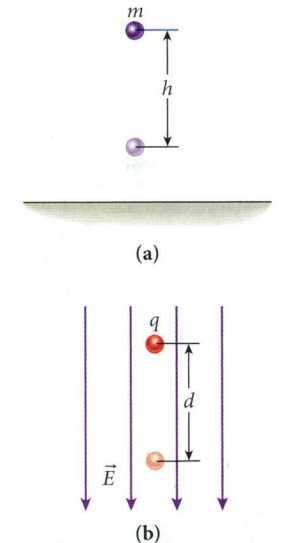

FIGURE 23.3 The analogy between gravitational potential energy and electric potential energy. (a) A mass falls in a gravitational field. (b) A positive charge moves in the same direction as an electric field.

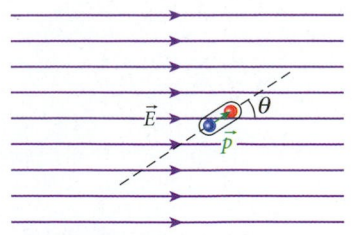

FIGURE 23.4 An electric dipole in a uniform electric field.

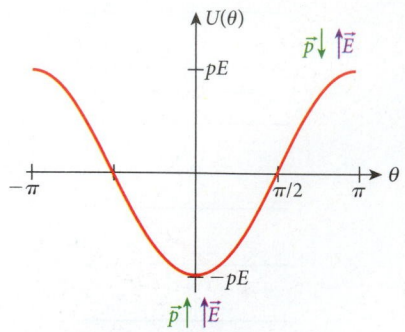

FIGURE 23.5 Potential energy as a function of the angle between an electric dipole and a constant external electric field.

With $W = -\Delta U = -(U - U_0) = U_0 - U$ (see Chapter 6), we then have the potential energy of an electric dipole in a constant electric field:

$$U = -pE\cos\theta = -\vec{p}\cdot\vec{E}, \qquad (23.5)$$

where we have chosen the integration constant U_0 so that the potential energy is zero for $\theta = \pi/2$.

Equation 23.5 indicates that the potential energy has a minimum at $\theta = 0$, where the dipole moment is parallel to the electric field; see Figure 23.5. Remember that electric field lines point from positive to negative charges and that the dipole moment is defined to point from the negative to the positive charge. When the dipole moment and electric field vectors are parallel, the negative charge of the dipole is closest to the positive charge that is generating the external electric field, and it makes physical sense that that configuration has the lowest energy.

23.2   Definition of Electric Potential

The potential energy of a charged particle, q, in an electric field depends on the magnitude of the charge as well as that of the electric field. A quantity that is independent of the charge on the particle is the **electric potential**, V, defined in terms of the electric potential energy as

$$V = \frac{U}{q}. \qquad (23.6)$$

Because U is proportional to q, V is independent of q, which makes it a useful variable. The electric potential, V, characterizes an electrical property of a point in space even when no charge, q, is placed at that point. In contrast to the electric field, which is a vector, the electric potential is a scalar. It has a value everywhere in space, but has no direction. In Chapter 6, we saw that we can always add an arbitrary constant to the potential energy without changing any observable consequence and that only differences in potential energy are physically meaningful. Since the electric potential is proportional to the potential energy, the same is true for it. Stating a difference in electric potentials is unambiguous, but stating a value for the electric potential itself always implies a normalization condition, which is usually that the potential is zero at an infinite distance.

The difference in electric potential, ΔV, between an initial point and final point, $V_f - V_i$, can be expressed in terms of the electric potential energy at each point:

$$\Delta V = V_f - V_i = \frac{U_f}{q} - \frac{U_i}{q} = \frac{\Delta U}{q}. \qquad (23.7)$$

Combining equations 23.2 and 23.7 yields a relationship between the change in electric potential and the work done by an electric field on a charge:

$$\Delta V = -\frac{W_e}{q}. \qquad (23.8)$$

Taking the electric potential energy to be zero at infinity, as in equation 23.3, gives the electric potential at a point as

$$V = -\frac{W_{e,\infty}}{q}, \qquad (23.9)$$

where $W_{e,\infty}$ is the work done by the electric field on the charge when it is brought in to the point from infinity. An electric potential can have a positive, a negative, or a zero value, but it has no direction.

The SI units for electric potential are joules/coulomb (J/C). This combination has been named the **volt** (V) for Italian physicist Alessandro Volta (1745–1827) (note the use of the roman V for the unit, whereas the italicized V is used for the physical quantity of electric potential):

$$1\,\text{V} \equiv \frac{1\,\text{J}}{1\,\text{C}}.$$

With this definition of the volt, the units for the magnitude of the electric field are

$$[E] = \frac{[F]}{[q]} = \frac{1\text{ N}}{1\text{ C}} = \left(\frac{1\text{ N}}{1\text{ C}}\right)\frac{1\text{ V}}{\left(\dfrac{1\text{ J}}{1\text{ C}}\right)}\left(\frac{1\text{ J}}{(1\text{ N})(1\text{ m})}\right) = \frac{1\text{ V}}{1\text{ m}}.$$

For the remainder of this book, the magnitude of an electric field will have units of V/m, which is the standard convention, instead of N/C. Note that an electric potential difference is often referred to as a "voltage," particularly in circuit analysis, because it is measured in volts.

EXAMPLE 23.1 Energy Gain of a Proton

A proton is placed between two parallel conducting plates in a vacuum (Figure 23.6). The difference in electric potential between the two plates is 450 V. The proton is released from rest close to the positive plate.

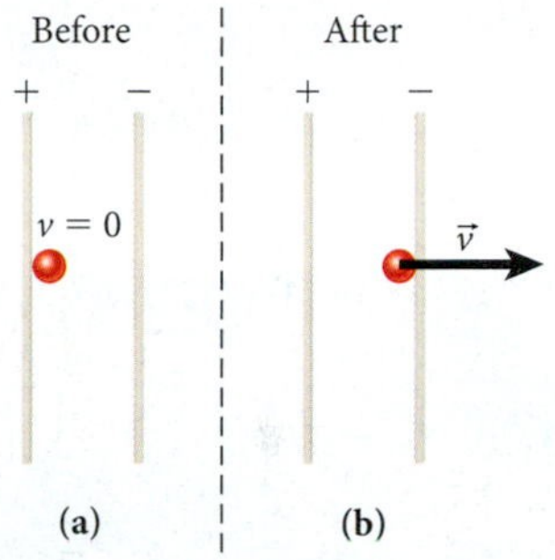

PROBLEM
What is the kinetic energy of the proton when it reaches the negative plate?

SOLUTION
The difference in electric potential, ΔV, between the two plates is 450 V. We can relate this potential difference across the two plates to the change in electric potential energy, ΔU, of the proton using equation 23.7:

$$\Delta V = \frac{\Delta U}{q}.$$

Because of the conservation of total energy, all the electric potential energy lost by the proton in crossing between the two plates is turned into kinetic energy due to the motion of the proton. We apply the law of conservation of energy, $\Delta K + \Delta U = 0$, where ΔU is the change in the proton's electric potential energy:

$$\Delta K = -\Delta U = -q\Delta V.$$

Because the proton started from rest, we can express its final kinetic energy as $K = -q\Delta V$. Therefore, the kinetic energy of the proton after crossing the gap between the two plates is

$$K = -\left(1.602\cdot10^{-19}\text{ C}\right)\left(-450\text{ V}\right) = 7.21\cdot10^{-17}\text{ J}.$$

FIGURE 23.6 A proton between two charged parallel conducting plates in a vacuum. (a) The proton is released from rest. (b) The proton has moved from the positive plate to the negative plate, gaining kinetic energy.

Concept Check 23.1

An electron is positioned and then released on the *x*-axis, where the electric potential has the value −20 V. Which of the following statements describes the subsequent motion of the electron?

a) The electron will move to the left (negative *x*-direction) because it is negatively charged.

b) The electron will move to the right (positive *x*-direction) because it is negatively charged.

c) The electron will move to the left (negative *x*-direction) because the electric potential is negative.

d) The electron will move to the right (positive *x*-direction) because the electric potential is negative.

e) Not enough information is given to predict the motion of the electron.

Because the acceleration of charged particles across a potential difference is often used in the measurement of physical quantities, a common unit for the kinetic energy of a singly charged particle, such as a proton or an electron, is the **electron-volt** (eV): 1 eV represents the energy gained by a proton ($q = 1.602\cdot10^{-19}$ C) accelerated across a potential difference of 1 V. The conversion between electron-volts and joules is

$$1\text{ eV} = 1.602\cdot10^{-19}\text{ J}.$$

The kinetic energy of the proton in Example 23.1 is then 450 eV, or 0.450 keV, which we could have obtained from the definition of the electron-volt without performing any calculations.

Batteries

A common means of creating electric potential is a battery. We'll see in Chapters 24 and 25 how a battery uses chemical reactions to provide a source of (nearly) constant potential difference between its two terminals. An assortment of batteries is shown in Figure 23.7.

At its simplest, a battery consists of two half-cells, filled with a conducting electrolyte (originally a liquid but now almost always a solid); see Figure 23.8. The electrolyte is separated into two equal portions by a barrier, which prevents the bulk of the electrolyte from passing through but allows charged ions to pass through. The negatively charged ions (anions) move toward the anode, and the positively charged ions move toward the cathode. This creates a potential difference between the two terminals of the battery. Thus, a battery is basically a device that converts chemical energy directly into electrical energy.

Research on battery technology is of current importance, because many mobile applications require a great deal of energy, from cell phones to laptop computers, from electrical cars to military gear. The weight of the batteries needs to be as small as possible, they need to be rapidly rechargeable for hundreds of cycles, they need to deliver as constant a potential difference as possible, and they need to be available at an affordable price. Thus, such research provides many scientific and engineering challenges.

One example of relatively recent battery technology is the lithium ion cell, which is often used in applications such as laptop computer batteries. A lithium ion battery has a much higher energy density (energy content per unit volume) than conventional batteries. A typical lithium ion cell, like the one in Figure 23.9, has a potential difference of 3.6 V. Lithium ion batteries have several other advantages over conventional batteries. They can be recharged hundreds of times. They have no "memory" effect and thus do not need to be conditioned to hold their charge. They hold their charge on the shelf. They also have some disadvantages. For example, if a lithium ion battery is completely discharged, it can no longer be recharged. The battery performs best if it is not charged to more than 80% of capacity and not discharged to less than 20% of its capacity. Heat degrades lithium ion batteries. If the batteries are discharged too quickly, the constituents can catch fire or explode. To deal with these problems, most commercial lithium ion battery packs have a small built-in electronic circuit that protects the battery pack. The circuit will not allow the battery to be overcharged or overly discharged; it will not allow charge to flow out of the battery so quickly that the battery will overheat. If the battery becomes too warm, the circuit disconnects the battery.

Currently, lithium ion batteries are being used in some electric-powered cars. The following example compares the energy carried by a battery-powered car and a gasoline-powered car.

(a)

(b)

FIGURE 23.7 (a) Some representative batteries (clockwise from upper left): rechargeable AA nickel metal hydride (NiMH) batteries in their charger, disposable 1.5-V AAA batteries, a 12-V lantern battery, a D-size battery, a lithium ion laptop battery, and a watch battery; (b) 330-V battery for a gas-electric hybrid SUV, filling the entire floor of the trunk.

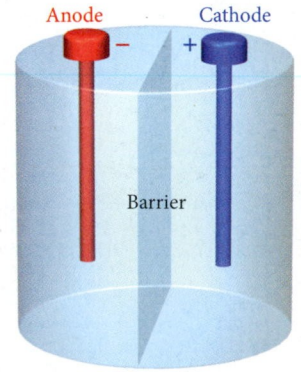

FIGURE 23.8 Schematic drawing of a battery.

FIGURE 23.9 A drawing showing the lithium ion battery pack of the Tesla electric-powered sports car. Also shown is one of the 6831 lithium ion cells that make up the battery pack.

EXAMPLE 23.2 | Battery-Powered Cars

Battery-powered cars produce no emissions and thus are an attractive alternative to gasoline-powered cars. Some of these cars, such as the Tesla sports car shown in Figure 23.10, are powered by batteries constructed of lithium ion cells.

The battery pack of the Tesla electric sports car (Figure 23.9) has the capacity to hold 53 kWh of energy. The battery pack is usually charged to 80% of its capacity and discharged to 20% of its capacity. A gasoline-powered car typically carries 50 L of gasoline, and gasoline has an energy content of 34.8 MJ/L.

FIGURE 23.10 The Tesla electric-powered sports car.

PROBLEM

How does the available energy in a lithium ion battery pack of an electric-powered car compare with the energy carried by a gasoline-powered car?

SOLUTION

Because not all of the energy can be extracted from a lithium ion battery without damaging it, the total usable energy is

$$E_{electric} = (80\% - 20\%)(53 \text{ kWh})\left(\frac{1000 \text{ W}}{1 \text{ kW}}\right)\left(\frac{3600 \text{ s}}{1 \text{ h}}\right) = 1.14 \cdot 10^8 \text{ J} = 114 \text{ MJ}.$$

A typical gasoline-powered car can carry 50 L of gasoline, which has an energy content of

$$E_{gasoline} = (50 \text{ L})(34.8 \text{ MJ/L}) = 1740 \text{ MJ}.$$

Thus, a typical gasoline-powered car carries 15 times as much energy as the Tesla electric-powered car. However, the efficiency of a gasoline-powered car is approximately 20%, while an electric-powered car can be close to 90% efficient. Thus, the usable energy of the electric-powered car is

$$E_{electric, usable} = 0.9(114 \text{ MJ}) = 103 \text{ MJ},$$

and the usable energy of the gasoline-powered car is

$$E_{gasoline, usable} = 0.2(1740 \text{ MJ}) = 348 \text{ MJ}.$$

The final numbers for the usable energy should be rounded to 1 significant digit in both cases. But the basic point is clear: You can see that electric-powered cars, even with lithium ion batteries, can carry less energy than gasoline-powered cars.

(a)

Van de Graaff Generator

One means of creating large electric potentials is a **Van de Graaff generator,** a device invented by the American physicist Robert J. Van de Graaff (1901–1967). Large Van de Graaff generators can produce electric potentials of millions of volts. More modest Van de Graaff generators, such as the one shown in Figure 23.11, can produce several hundred thousand volts and are often used in physics classrooms.

A Van de Graaff generator uses a *corona discharge* to apply a positive charge to a nonconducting moving belt. Putting a high positive voltage on a conductor with a sharp point creates the corona discharge. The electric field on the sharp point is much stronger than on the flat surface of the conductor (see Chapter 22). The air around the sharp point is ionized. The ionized air molecules have a net positive charge, which causes the ions to be repelled away from the sharp point and deposited on the rubber belt. The moving belt, driven by an electric motor, carries the charge up into a hollow metal sphere, where the charge is taken from the belt by a pointed contact connected to the metal sphere. The charge that builds up on the metal sphere distributes itself uniformly around the outside of the sphere. On the Van de Graaff generator shown in Figure 23.11, a voltage limiter is used to keep the generator from producing sparks larger than desired.

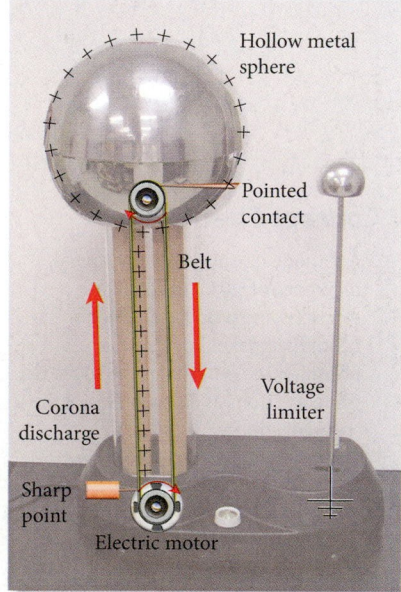

(b)

FIGURE 23.11 (a) A Van de Graaff generator used in physics classrooms. (b) The Van de Graaff generator can produce very high electric potentials by carrying charge from a corona discharge on a rubber belt up to a hollow metal sphere, where the charge is extracted from the belt by a sharp piece of metal attached to the inner surface of the sphere.

EXAMPLE 23.3 | Tandem Van de Graaff Accelerator

A Van de Graaff accelerator is a particle accelerator that uses high electric potentials for studying nuclear physics processes of astrophysical relevance. A tandem Van de Graaff accelerator with a terminal potential difference of 10.0 MV (10.0 million volts), is diagrammed in Figure 23.12. This terminal potential difference is created in the center of the accelerator by a larger, more sophisticated version of the classroom Van de Graaff generator. Negative ions are created in the ion source by attaching an electron to the atoms to be accelerated. The negative ions then accelerate toward the positively charged terminal. Inside the terminal, the ions pass through a thin foil that strips off electrons, producing positively charged ions that then are accelerated away from the terminal and out of the tandem accelerator.

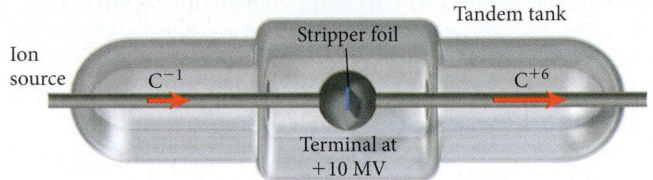

FIGURE 23.12   A tandem Van de Graaff accelerator.

PROBLEM 1
What is the highest kinetic energy that carbon nuclei can attain in this tandem accelerator?

SOLUTION 1
A tandem Van de Graaff accelerator has two stages of acceleration. In the first stage, each carbon ion has a net charge of $q_1 = -e$. After the stripper foil, the maximum charge any carbon ion can have is $q_2 = +6e$. The potential difference over which the ions are accelerated is $\Delta V = 10$ MV. The kinetic energy gained by each carbon ion is

$$\Delta K = |\Delta U| = |q_1 \Delta V| + |q_2 \Delta V| = K,$$

or

$$K = e\Delta V + 6e\Delta V = 7e\Delta V,$$

assuming that the initial speed of the ions is close to zero.

Putting in the numerical values, we get

$$K = 7\left(1.602 \cdot 10^{-19} \text{ C}\right)\left(10 \cdot 10^6 \text{ V}\right) = 1.12 \cdot 10^{-11} \text{ J}.$$

Nuclear physicists often use electron-volts instead of joules to express the kinetic energy of accelerated nuclei:

$$K = 7e\Delta V = 7e\left(10 \cdot 10^6 \text{ V}\right) = 7 \cdot 10^7 \text{ eV} = 70 \text{ MeV}.$$

PROBLEM 2
What is the highest speed that carbon nuclei can attain in this tandem accelerator?

SOLUTION 2
To determine the speed, we use the relationship between kinetic energy and speed:

$$K = \tfrac{1}{2}mv^2,$$

where $m = 1.99 \cdot 10^{-26}$ kg is the mass of the carbon nucleus. Solving this equation for the speed, we get

$$v = \sqrt{\frac{2K}{m}} = \sqrt{\frac{2\left(1.12 \cdot 10^{-11} \text{ J}\right)}{1.99 \cdot 10^{-26} \text{ kg}}} = 3.36 \cdot 10^7 \text{ m/s},$$

which is 11% of the speed of light.

Concept Check 23.2

A cathode ray tube uses a potential difference of 5.0 kV to accelerate electrons and produce an electron beam that makes images on a phosphor screen. What is the speed of these electrons as a percentage of the speed of light?

a) 0.025%   d) 4.5%

b) 0.22%   e) 14%

c) 1.3%

SOLVED PROBLEM 23.1 | Beam of Oxygen Ions

PROBLEM
Fully stripped (all electrons removed) oxygen (^{16}O) ions are accelerated from rest in a particle accelerator using a total potential difference of 10.0 MV = $1.00 \cdot 10^7$ V. The ^{16}O nucleus has 8 protons and 8 neutrons. The accelerator produces a beam of $3.13 \cdot 10^{12}$ ions per second. This

ion beam is completely stopped in a beam dump. What is the total power the beam dump has to absorb?

SOLUTION

THINK Power is energy per unit time. We can calculate the energy of each ion and then the total energy in the beam per unit time to obtain the power dissipated in the beam dump.

SKETCH Figure 23.13 illustrates a beam of fully stripped oxygen ions being stopped in a beam dump.

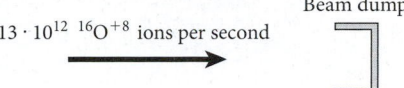

$3.13 \cdot 10^{12} \ {}^{16}O^{+8}$ ions per second

FIGURE 23.13 A beam of fully stripped oxygen ions stops in a beam dump.

RESEARCH The electric potential energy gained by each ion during the acceleration process is

$$U_{ion} = q\Delta V = ZeV,$$

where $Z = 8$ is the atomic number of oxygen, $e = 1.602 \cdot 10^{-19}$ C is the charge of a proton, and $V = 1.00 \cdot 10^7$ V is the electric potential across which the ions are accelerated.

SIMPLIFY The power of the beam, which is dissipated in the beam dump, is then

$$P = NU_{ion} = NZeV,$$

where $N = 3.13 \cdot 10^{12}$ ions/s is the number of ions per second stopped in the beam dump.

CALCULATE Putting in the numerical values, we get

$$P = NZeV = \left(3.13 \cdot 10^{12} \ \text{s}^{-1}\right)(8)\left(1.602 \cdot 10^{-19} \ \text{C}\right)\left(1.00 \cdot 10^7 \ \text{V}\right)$$
$$= 40.1141 \ \text{W}.$$

ROUND We report our result to three significant figures:

$$P = 40.1 \ \text{W}.$$

DOUBLE-CHECK We can relate the change in kinetic energy for each ion to the change in electric potential energy of each ion:

$$\Delta K = \Delta U = \tfrac{1}{2}mv^2 = U_{ion} = ZeV.$$

The mass of an oxygen nucleus is $2.66 \cdot 10^{-26}$ kg. The velocity of each ion is then

$$v = \sqrt{\frac{2ZeV}{m}} = \sqrt{\frac{2(8)\left(1.602 \cdot 10^{-19} \ \text{C}\right)\left(1.00 \cdot 10^7 \ \text{V}\right)}{2.66 \cdot 10^{-26} \ \text{kg}}} = 3.10 \cdot 10^7 \ \text{m/s},$$

which is about 10% of the speed of light, which seems reasonable for the velocity of the ions. Thus, our result seems reasonable.

23.3 Equipotential Surfaces and Lines

Imagine you had to map out a ski resort with three peaks, like the one shown in Figure 23.14a. In Figure 23.14b, lines of equal elevation have been superimposed on the peaks. You could walk along each of these lines, without ever going uphill or downhill, and would be guaranteed to reach the point from which you started. These lines are lines of constant gravitational potential, because the gravitational potential is a function of the elevation only, and the elevation remains constant on each of the lines. Figure 23.14c shows a top view of the contour lines of equal elevation, which mark the equipotential lines for the gravitational potential. If you have understood this figure, the following discussion of electric potential lines and surfaces should be easy to follow.

When an electric field is present, the electric potential has a value everywhere in space. Points that have the same electric potential form an **equipotential surface.** Charged particles can move along an equipotential surface without having any work done on them by the electric field. According to principles of electrostatics, the surface of a conductor must be an equipotential surface; otherwise, the free electrons on the conductor surface would accelerate. The discussion in Chapter 22 established that the electric field is zero everywhere inside the body of a conductor. This means that the entire volume of the conductor must be at the same potential; that is, the entire conductor is an equipotential.

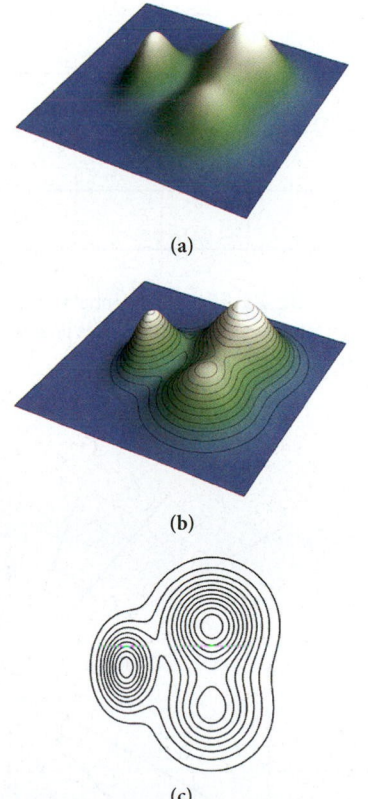

(a)

(b)

(c)

FIGURE 23.14 (a) Ski resort with three peaks; (b) the same peaks with lines of equal elevation superimposed; (c) the contour lines of equal elevation in a two-dimensional plot.

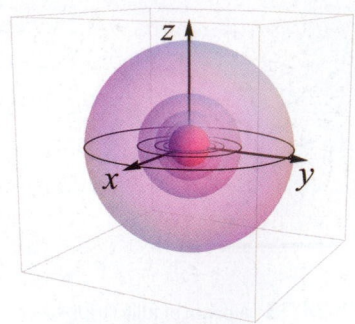

FIGURE 23.15 Concentric equipotential surfaces with values of 5 V, 4 V, 3 V, 2 V, and 1 V around a spherical conductor with a potential of 5 V centered at the origin of an *xyz*-coordinate system. The circles represent the intersections of the equipotential spheres with the *xy*-plane and are equipotential lines.

Equipotential surfaces exist in three dimensions (Figure 23.15); however, symmetries in the electric potential allow us to represent equipotential surfaces in two dimensions, as **equipotential lines** in the plane in which the charges reside. Before determining the shape and location of these equipotential surfaces, let's first look at some qualitative features of some of the simplest cases (for which the electric fields were determined in Chapter 22).

In drawing equipotential lines, we note that charges can move perpendicular to any electric field line without having any work done on them by the electric field, because according to equation 23.4, the scalar product of the electric field and the displacement is then zero. If the work done by the electric field is zero, the potential remains the same, by equation 23.8. Thus, *equipotential lines and planes are always perpendicular to the direction of the electric field.* (In Figure 23.14b, the elevation map of the ski resort, the equivalent of electric field lines would be the lines of steepest descent, which are, of course, always perpendicular to the lines of equal elevation.)

Before examining the particular equipotential surfaces resulting from different electric field configurations, let's note the two most important general observations of this section, which hold for all of the following cases:

1. The surface of any conductor forms an equipotential surface.

2. Equipotential surfaces are always perpendicular to the electric field lines at any point in space.

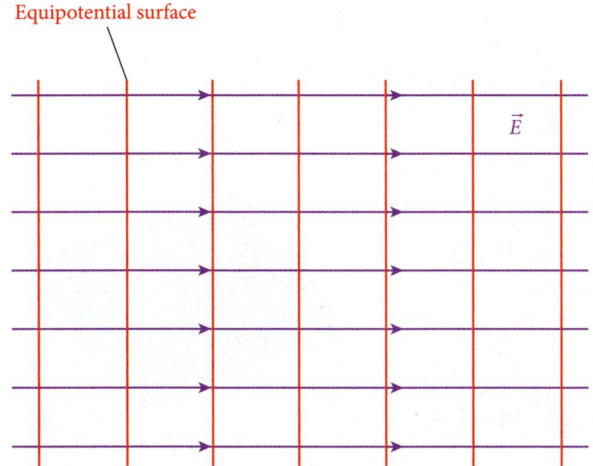

FIGURE 23.16 Equipotential surfaces (red lines) from a constant electric field. The purple lines with the arrowheads represent the electric field.

Constant Electric Field

A constant electric field has straight, equally spaced, and parallel field lines. Thus, such a field produces equipotential surfaces in the form of parallel planes, because of the condition that the equipotential surfaces or equipotential lines have to be perpendicular to the field lines. These planes are represented in two dimensions as equally spaced equipotential lines (Figure 23.16).

Single Point Charge

Figure 23.17 shows the electric field and corresponding equipotential lines due to a single point charge. The electric field lines extend radially from a positive point charge, as shown in Figure 23.17a. In this case, the field lines point away from the positive charge and terminate at infinity. For a negative charge, as shown in Figure 23.17b, the field lines originate at infinity and terminate at the negative charge. The equipotential lines are spheres centered on the point charge. (In the two-dimensional views shown in the figure, the circles represent the lines where the

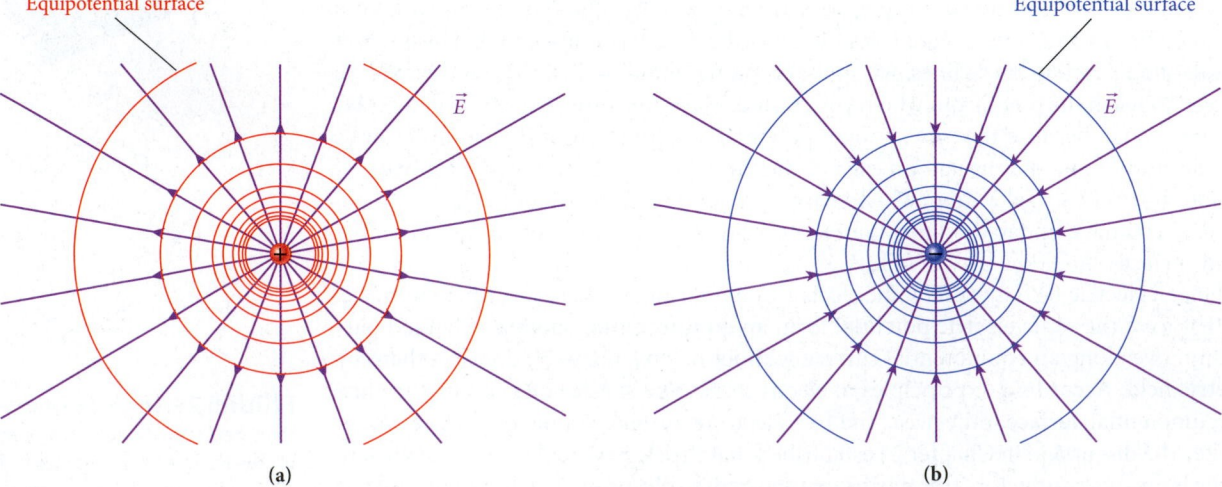

FIGURE 23.17 Equipotential surfaces and electric field lines from (a) a single positive point charge and (b) a single negative point charge.

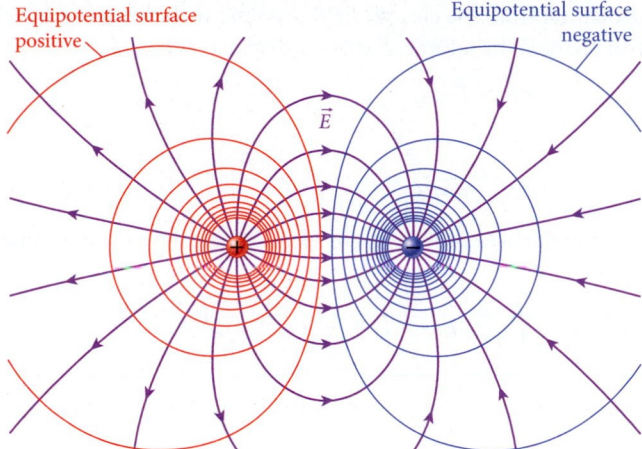

FIGURE 23.18 Equipotential surfaces created by point charges of the same magnitude but opposite sign. The red lines represent positive potential, and the blue lines represent negative potential. The purple lines with the arrowheads represent the electric field.

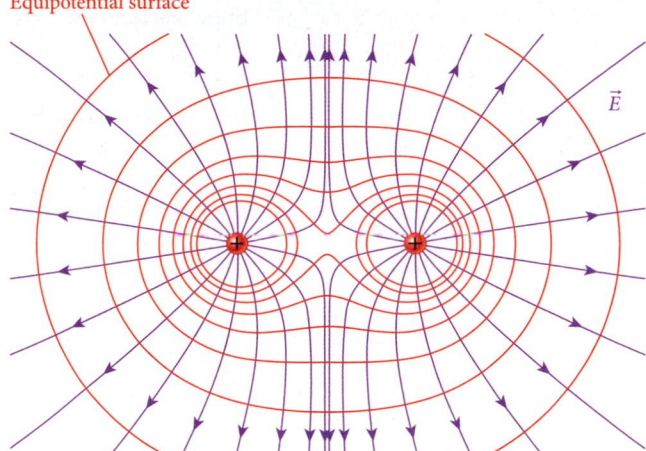

FIGURE 23.19 Equipotential surfaces (red lines) from two identical positive point charges. The purple lines with the arrowheads represent the electric field.

plane of the page cuts through equipotential spheres.) The values of the potential difference between neighboring equipotential lines are equal, producing equipotential lines that are close together near the charge and more widely spaced away from the charge. Note again that the equipotential lines are always perpendicular to the electric field lines. Equipotential surfaces do not have arrows like the field lines, because the potential is a scalar.

Two Oppositely Charged Point Charges

Figure 23.18 shows the electric field lines from two oppositely charged point charges, along with equipotential surfaces depicted as equipotential lines. An electrostatic force would attract these two point charges toward each other, but this discussion assumes that the charges are fixed in space and cannot move. The electric field lines originate at the positive charge and terminate on the negative charge. Again, the equipotential lines are always perpendicular to the electric field lines. The red lines in this figure represent positive equipotential surfaces, and the blue lines represent negative equipotential surfaces. Positive charges produce positive potential, and negative charges produce negative potential (relative to the value of the potential at infinity). Close to each charge, the resultant electric field lines and the resultant equipotential lines resemble those for a single point charge. Away from the vicinity of each charge, the electric field and the electric potential are the sums of the fields and potentials due to the two charges. The electric fields add as vectors, while the electric potentials add as scalars. Thus, the electric field is defined at all points in space in terms of a magnitude and a direction, while the electric potential is defined solely by its value at a given point in space and has no direction associated with it.

Two Identical Point Charges

Figure 23.19 shows electric field lines and equipotential surfaces resulting from two identical positive point charges. These two charges experience a repulsive electrostatic force. Because both charges are positive, the equipotential surfaces represent positive potentials. Again, the electric field and electric potential result from the sums of the fields and potentials, respectively, due to the two charges.

Self-Test Opportunity 23.1

Suppose the charges in Figure 23.18 were located at $(x,y) = (-10 \text{ cm}, 0)$ and $(x,y) = (+10 \text{ cm}, 0)$. What would the electric potential be along the y-axis ($x = 0$)?

Self-Test Opportunity 23.2

Suppose the charges in Figure 23.19 were located at $(x,y) = (-10 \text{ cm}, 0)$ and $(x,y) = (+10 \text{ cm}, 0)$. Would $(x,y) = (0,0)$ correspond to a maximum, a minimum, or a saddle point in the electric potential?

23.4 Electric Potential of Various Charge Distributions

The electric potential is defined as the work required to place a unit charge at a point, and work is a force acting over a distance. Also, the electric field can be defined as the force acting on a unit charge at a point. Therefore, it seems that the potential at a point should be related to the field strength at that point. In fact, electric potential and electric field are directly related; we can determine either one given an expression for the other.

To determine the electric potential from the electric field, we start with the definition of the work done on a particle with charge q by a force, $\vec{F}$, over a displacement, $d\vec{s}$:

$$dW = \vec{F} \cdot d\vec{s}.$$

In this case, the force is given by $\vec{F} = q\vec{E}$, so

$$dW = q\vec{E} \cdot d\vec{s}. \tag{23.10}$$

Integration of equation 23.10 as the particle moves in the electric field from some initial point to some final point gives

$$W = W_e = \int_i^f q\vec{E} \cdot d\vec{s} = q\int_i^f \vec{E} \cdot d\vec{s}.$$

Using equation 23.8 to relate the work done to the change in electric potential, we get

$$\Delta V = V_f - V_i = -\frac{W_e}{q} = -\int_i^f \vec{E} \cdot d\vec{s}.$$

As mentioned earlier, the usual convention is to set the electric potential to zero at infinity. With this convention, we can express the potential at some point $\vec{r}$ in space as

$$V(\vec{r}) - V(\infty) \equiv V(\vec{r}) = -\int_\infty^{\vec{r}} \vec{E} \cdot d\vec{s}. \tag{23.11}$$

Concept Check 23.3

In the figure, the lines represent equipotential lines. A charged object is moved from point P to point Q. How does the amount of work done on the object compare for these three cases?

a) All three cases involve the same work.

b) The most work is done in case 1.

c) The most work is done in case 2.

d) The most work is done in case 3.

e) Cases 1 and 3 involve the same amount of work, which is more than is involved in case 2.

Point Charge

Let's use equation 23.11 to determine the electric potential due to a point charge, q. The electric field due to a point charge, q (for now, taken as positive), at a distance r from the charge is given by

$$E = \frac{kq}{r^2}.$$

The direction of the electric field is radial from the point charge. Assume that the integration is carried out along a radial line from infinity to a point at a distance R from the point charge, such that $\vec{E} \cdot d\vec{s} = E\,dr$. Then we can use equation 23.11 to obtain

$$V(R) = -\int_\infty^R \vec{E} \cdot d\vec{s} = -\int_\infty^R \frac{kq}{r^2} dr = \left[\frac{kq}{r}\right]_\infty^R = \frac{kq}{R}.$$

Thus, the electric potential due to a point charge at a distance r from the charge is given by

$$V = \frac{kq}{r}. \tag{23.12}$$

Self-Test Opportunity 23.3

Obtaining equation 23.12 for the electric potential from a point charge involved integrating along a radial line from infinity to a point at a distance R from the point charge. How would the result change if the integration were carried out over a different path?

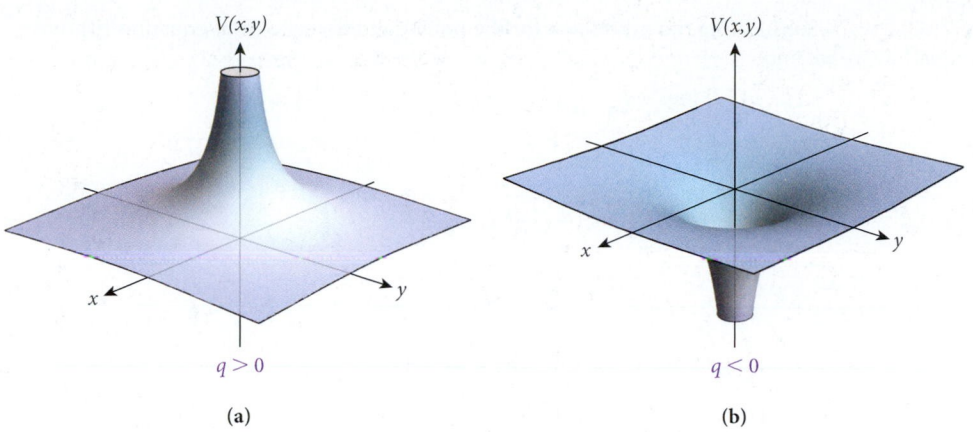

(a)   **(b)**

FIGURE 23.20 Electric potential due to (a) a positive point charge and (b) a negative point charge.

Concept Check 23.4

What is the electric potential 45.5 cm away from a point charge of 12.5 pC?

a) 0.247 V   d) 10.2 V

b) 1.45 V   e) 25.7 V

c) 4.22 V

Equation 23.12 also holds when $q < 0$. A positive charge produces a positive potential, and a negative charge produces a negative potential, as shown in Figure 23.20.

In Figure 23.20, the electric potential is calculated for all points in the xy-plane. The vertical axis represents the value of the potential at each point on the plane, $V(x,y)$, found using $r = \sqrt{x^2 + y^2}$. The potential is not calculated close to $r = 0$ because it becomes infinite there. You can see from Figure 23.20 how the circular equipotential lines shown in Figure 23.17 originate.

SOLVED PROBLEM 23.2 / Fixed and Moving Positive Charges

PROBLEM
A positive charge of 4.50 μC is fixed in place. A particle of mass 6.00 g and charge +3.00 μC is fired with an initial speed of 66.0 m/s directly toward the fixed charge from a distance of 4.20 cm away. How close does the moving charge get to the fixed charge before it comes to rest and starts moving away from the fixed charge?

SOLUTION
THINK The moving charge will gain electric potential energy as it nears the fixed charge. The negative of the change in potential energy of the moving charge is equal to the change in kinetic energy of the moving charge because $\Delta K + \Delta U = 0$.

SKETCH We set the location of the fixed charge at $x = 0$, as shown in Figure 23.21. The moving charge starts at $x = d_i$, moves with initial speed $v = v_0$, and comes to rest at $x = d_f$.

RESEARCH The moving charge gains electric potential energy as it approaches the fixed charge and loses kinetic energy until it stops. At that point, all the original kinetic energy of the moving charge has been converted to electric potential energy. Using energy conservation, we can write this relationship as

$$\Delta K + \Delta U = 0 \Rightarrow \Delta K = -\Delta U \Rightarrow$$
$$0 - \tfrac{1}{2} m v_0^2 = -q_{\text{moving}} \Delta V \Rightarrow$$
$$\tfrac{1}{2} m v_0^2 = q_{\text{moving}} \Delta V. \tag{i}$$

The electric potential experienced by the moving charge is due to the fixed charge, so we can write the change in potential as

$$\Delta V = V_f - V_i = k \frac{q_{\text{fixed}}}{d_f} - k \frac{q_{\text{fixed}}}{d_i} = k q_{\text{fixed}} \left(\frac{1}{d_f} - \frac{1}{d_i} \right). \tag{ii}$$

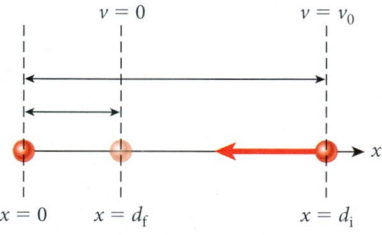

FIGURE 23.21 Two positive charges. One charge is fixed in place at $x = 0$, and the second charge begins moving with velocity $\vec{v}_0$ at $x = d_i$ and has zero velocity at $x = d_f$.

– Continued

SIMPLIFY Substituting the expression for the potential difference from equation (ii) into equation (i), we find

$$\frac{1}{2}mv_0^2 = q_{moving}\Delta V = kq_{moving}q_{fixed}\left(\frac{1}{d_f} - \frac{1}{d_i}\right) \Rightarrow$$

$$\frac{1}{d_f} - \frac{1}{d_i} = \frac{mv_0^2}{2kq_{moving}q_{fixed}} \Rightarrow$$

$$\frac{1}{d_f} = \frac{1}{d_i} + \frac{mv_0^2}{2kq_{moving}q_{fixed}}.$$

CALCULATE Putting in the numerical values, we get

$$\frac{1}{d_f} = \frac{1}{0.0420 \text{ m}} + \frac{(0.00600 \text{ kg})(66.0 \text{ m/s})^2}{2(8.99 \cdot 10^9 \text{ N m}^2/\text{C}^2)(3.00 \cdot 10^{-6}\text{C})(4.50 \cdot 10^{-6}\text{C})} = 131.485,$$

or

$$d_f = 0.00760545 \text{ m}.$$

ROUND We report our result to three significant figures:

$$d_f = 0.00761 \text{ m} = 0.761 \text{ cm}.$$

DOUBLE-CHECK The final distance of 0.761 cm is less than the initial distance of 4.20 cm. At the final distance, the electric potential energy of the moving charge is

$$U = q_{moving}V = q_{moving}\left(k\frac{q_{fixed}}{d_f}\right) = k\frac{q_{moving}q_{fixed}}{d_f}$$

$$= (8.99 \cdot 10^9 \text{ N m}^2/\text{C}^2)\frac{(3.00 \cdot 10^{-6} \text{ C})(4.50 \cdot 10^{-6} \text{ C})}{0.00761 \text{ m}} = 16.0 \text{ J}.$$

The electric potential energy at the initial distance is

$$U = q_{moving}V = q_{moving}\left(k\frac{q_{fixed}}{d_i}\right) = k\frac{q_{moving}q_{fixed}}{d_i}$$

$$= (8.99 \cdot 10^9 \text{ N m}^2/\text{C}^2)\frac{(3.00 \cdot 10^{-6} \text{ C})(4.50 \cdot 10^{-6} \text{ C})}{(0.0420 \text{ m})} = 2.9 \text{ J}.$$

The initial kinetic energy is

$$K = \tfrac{1}{2}mv^2 = \frac{(0.00600 \text{ kg})(66.0 \text{ m/s})^2}{2} = 13.1 \text{ J}.$$

We can see that the equation based on energy conservation, from which the solution process started, is satisfied:

$$\tfrac{1}{2}mv^2 = \Delta U$$

$$13.1 \text{ J} = 16.0 \text{ J} - 2.9 \text{ J} = 13.1 \text{ J}.$$

This gives us confidence that our result for the final distance is correct.

System of Point Charges

Assuming again that the electric potential is zero at an infinite distance from the origin, we calculate the electric potential due to a system of n point charges by adding the potentials due to all the charges:

$$V = \sum_{i=1}^{n} V_i = \sum_{i=1}^{n} \frac{kq_i}{r_i}. \tag{23.13}$$

Equation 23.13 can be proved by inserting the expression for the total electric field from n charges $\left(\vec{E}_t = \vec{E}_1 + \vec{E}_2 + \cdots + \vec{E}_n\right)$ into equation 23.11 and integrating term by term. The summation in equation 23.13 produces a potential at any point in space that has a value but no direction. Thus, calculating the potential due to a group of point charges is usually much simpler than calculating the electric field, which involves the addition of vectors.

EXAMPLE 23.4 / Superposition of Electric Potentials

Let's calculate the electric potential at a given point due to a system of point charges. Figure 23.22 shows three point charges: $q_1 = +1.50$ μC, $q_2 = +2.50$ μC, and $q_3 = -3.50$ μC. Charge q_1 is located at $(0,a)$, q_2 is located at $(0,0)$, and q_3 is located at $(b,0)$, where $a = 8.00$ m and $b = 6.00$ m.

The electric potential at point P is the sum of the potentials due to the three charges:

$$V = \sum_{i=1}^{3} \frac{kq_i}{r_i} = k\left(\frac{q_1}{r_1} + \frac{q_2}{r_2} + \frac{q_3}{r_3}\right) = k\left(\frac{q_1}{b} + \frac{q_2}{\sqrt{a^2+b^2}} + \frac{q_3}{a}\right)$$

$$= \left(8.99\cdot10^9 \text{ N m}^2/\text{C}^2\right)\left(\frac{1.50\cdot10^{-6}\text{ C}}{6.00\text{ m}} + \frac{2.50\cdot10^{-6}\text{ C}}{\sqrt{(8.00\text{ m})^2+(6.00\text{ m})^2}} + \frac{-3.50\cdot10^{-6}\text{ C}}{8.00\text{ m}}\right)$$

$$= 562 \text{ V}.$$

Note that the potential due to q_3 is negative at point P, but the sum of the potentials is positive.

This example is similar to Example 22.1, in which we calculated the electric field at point P due to three charges. Note that this calculation of the electric potential due to three charges is much simpler than that calculation.

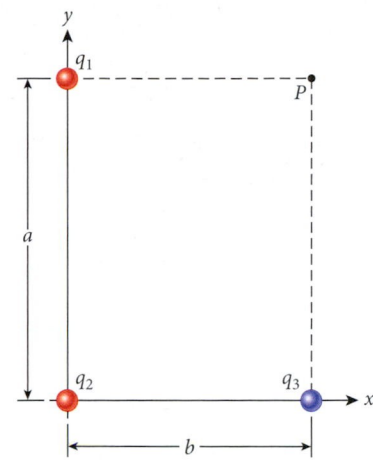

FIGURE 23.22 Electric potential at a point due to three point charges.

Concept Check 23.5

Two protons are located in space in the three ways shown in the figure. Rank the three cases from highest to lowest net electric potential, V, produced at point P.

a) $2 > 3 > 1$

b) All three potentials are the same.

c) $3 > 2 > 1$

d) The potentials are equal for cases 1 and 3, with the potential for case 2 lower.

e) $1 > 2 > 3$

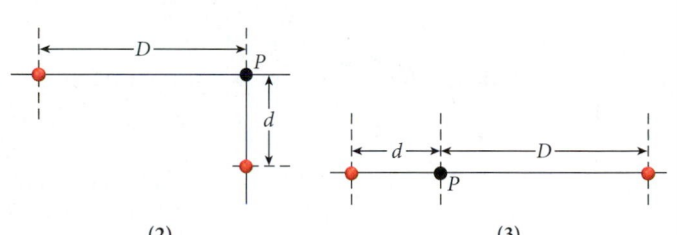

Concept Check 23.6

Three identical positive point charges are located at fixed points in space. Then charge q_2 is moved from its initial location to a final location as shown in the figure. Four different paths, marked (a) through (d), are shown. Path (a) follows the shortest line; path (b) takes q_2 around q_3; path (c) takes q_2 around q_3 and q_1; path (d) takes q_2 out to infinity and then to the final location. Which path requires the least work?

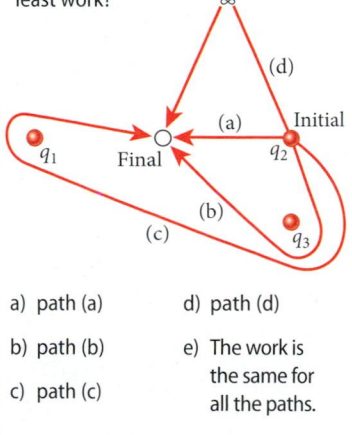

a) path (a)

b) path (b)

c) path (c)

d) path (d)

e) The work is the same for all the paths.

SOLVED PROBLEM 23.3 / Minimum Potential

PROBLEM
A charge of $q_1 = 0.829$ nC is placed at $r_1 = 0$ on the x-axis. Another charge of $q_2 = 0.275$ nC is placed at $r_2 = 11.9$ cm on the x-axis. At which point along the x-axis between the two charges does the electric potential resulting from both of them have a minimum?

SOLUTION

THINK We can express the electric potential due to the two charges as the sum of the electric potentials from the individual charges. To obtain the minimum potential, we take the derivative of the potential and set it equal to zero. We can then solve for the distance where the derivative is zero.

SKETCH Figure 23.23 shows the locations of the two charges.

– Continued

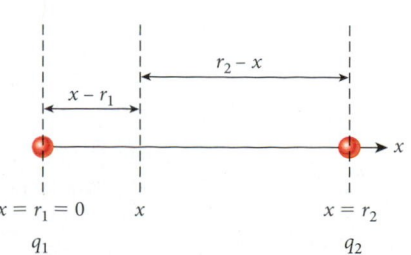

FIGURE 23.23 Two charges placed along the x-axis.

RESEARCH We can express the electric potential produced along the x-axis by the two charges as

$$V = V_1 + V_2 = k\frac{q_1}{x-r_1} + k\frac{q_2}{r_2-x} = k\frac{q_1}{x} + k\frac{q_2}{r_2-x}.$$

Note that the quantities x and $r_2 - x$ are always positive for $0 < x < r_2$. To find the minimum, we take the derivative of the electric potential:

$$\frac{dV}{dx} = -k\frac{q_1}{x^2} - k\frac{q_2}{(r_2-x)^2}(-1) = k\frac{q_2}{(r_2-x)^2} - k\frac{q_1}{x^2}.$$

SIMPLIFY Setting the derivative of the electric potential equal to zero and rearranging, we obtain

$$k\frac{q_2}{(r_2-x)^2} = k\frac{q_1}{x^2}.$$

Dividing out k and rearranging, we get

$$\frac{x^2}{(r_2-x)^2} = \frac{q_1}{q_2}.$$

Now we can take the square root and rearrange:

$$x = \pm(r_2-x)\sqrt{\frac{q_1}{q_2}}.$$

Because $x > 0$ and $(r_2 - x) > 0$, the sign must be positive. Solving for x, we get

$$x = \frac{r_2\sqrt{\dfrac{q_1}{q_2}}}{1+\sqrt{\dfrac{q_1}{q_2}}} = \frac{r_2}{\sqrt{\dfrac{q_2}{q_1}}+1}.$$

CALCULATE Putting in the numerical values results in

$$x = \frac{0.119\ \text{m}}{1+\sqrt{\dfrac{0.275\ \text{nC}}{0.829\ \text{nC}}}} = 0.0755097\ \text{m}.$$

ROUND We report our result to three significant figures:

$$x = 0.0755\ \text{m} = 7.55\ \text{cm}.$$

DOUBLE-CHECK We can double-check our result by plotting (for example, with a graphing calculator) the electric potential resulting from the two charges and graphically determining the minimum (Figure 23.24).

The minimum of the electric potential is located at $x = 7.55$ cm, which confirms our calculated result.

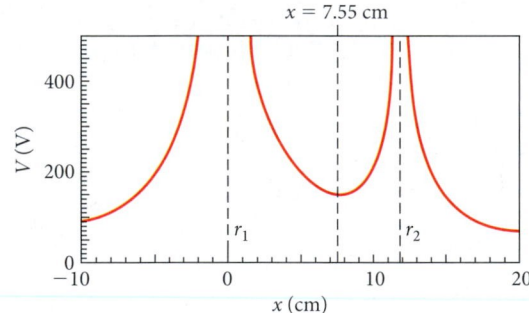

FIGURE 23.24 Graph of the electric potential resulting from two charges.

Continuous Charge Distribution

We can also determine the electric potential due to a continuous distribution of charge. To do this, we divide the charge into differential elements of charge, dq, and find the electric potential resulting from that differential charge as if it were a point charge. This is the way charge distributions were treated in determining electric fields in Chapter 22. The differential charge, dq, can be expressed in terms of a charge per unit length times a differential length, $\lambda\,dx$; in terms of a charge per unit area times a differential area, $\sigma\,dA$; or in terms of a charge per unit volume times a differential volume, $\rho\,dV$. The electric potential resulting from the charge distribution is obtained by integrating over the contributions from the differential charges. Let's consider an example involving the electric potential due to a one-dimensional charge distribution.

EXAMPLE 23.5 Finite Line of Charge

What is the electric potential at a distance d along the perpendicular bisector of a thin wire with length $2a$ and linear charge distribution λ (Figure 23.25)?

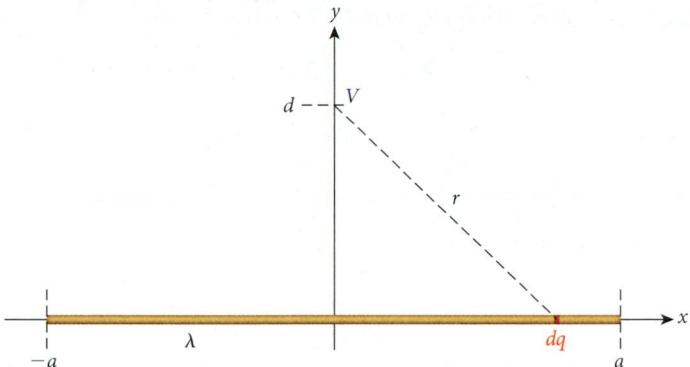

FIGURE 23.25 Calculating the electric potential due to a line of charge.

The differential electric potential, dV, at a distance d along the perpendicular bisector of the wire due to a differential charge, dq, is given by

$$dV = k\frac{dq}{r}.$$

The electric potential due to the whole wire is found by integrating dV along the length of the wire:

$$V = \int_{-a}^{a} dV = \int_{-a}^{a} k\,\frac{dq}{r}. \tag{i}$$

With $dq = \lambda\,dx$ and $r = \sqrt{x^2 + d^2}$, we can rewrite equation (i) as

$$V = \int_{-a}^{a} k\,\frac{\lambda\,dx}{\sqrt{x^2 + d^2}} = k\lambda \int_{-a}^{a} \frac{dx}{\sqrt{x^2 + d^2}}.$$

Finding this integral in a table or evaluating it with software gives

$$\int_{-a}^{a} \frac{dx}{\sqrt{x^2 + d^2}} = \left[\ln\!\left(x + \sqrt{x^2 + d^2}\right)\right]_{-a}^{a} = \ln\!\left(\frac{\sqrt{a^2 + d^2} + a}{\sqrt{a^2 + d^2} - a}\right).$$

Thus, the electric potential at a distance d along the perpendicular bisector of a finite line of charge is given by

$$V = k\lambda \ln\!\left(\frac{\sqrt{a^2 + d^2} + a}{\sqrt{a^2 + d^2} - a}\right).$$

SOLVED PROBLEM 23.4 Charged Disk

PROBLEM
A charge of 3.50 nC is uniformly applied to a disk of radius 1.00 cm. What is the electric potential at a distance of 4.50 mm from the disk along its symmetry axis, assuming, as usual, that the potential is zero at an infinite distance?

SOLUTION
THINK A point charge creates the electric potential $V(r) = kq/r$, but since the charge in this case is distributed over an area, we cannot use this relationship but have to perform an

– Continued

Self-Test Opportunity 23.4

Sketch the graph of the electric potential due to a hollow conducting charged sphere as a function of the radial coordinate r from zero to three times the sphere's radius, R.

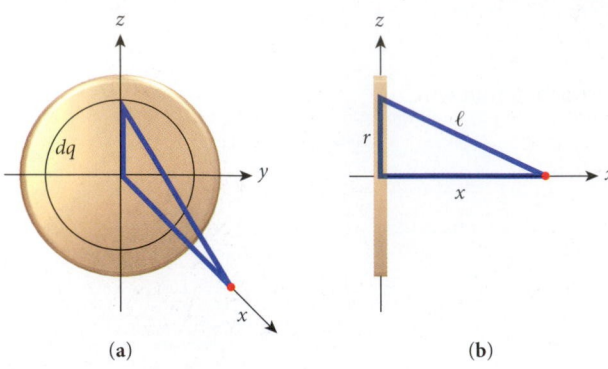

(a)   **(b)**

FIGURE 23.26 Electric potential on the symmetry axis of a disk: (a) front view, (b) side view.

integration. The general procedure for this integration is always the same: We divide the total charge into small increments, dq, compute the electric potential for each, and then integrate over all the charge increments. In this case, our task is to find the potential on a point along the symmetry axis of the disk, and so we should make use of this symmetry in our integration procedure.

SKETCH   A sketch of the problem situation is given in Figure 23.26.

RESEARCH   The surface charge density on the disk is $\sigma = q/A$, where A is the area of the disk, $A = \pi R^2$. Further, the charge is distributed symmetrically around the symmetry axis, the x-axis in Figure 23.26. This motivates the use of a thin ring of width dr for our differential charge unit: $dq = \sigma dA$, with $dA = 2\pi r dr$. You can see in Figure 23.26 that every point on the ring is an equal distance ℓ from the point (marked by the red dot) at which we want to evaluate the potential. The contribution of dq to the electric potential is then $dV = kdq/\ell$, and the total potential is $V = \int dV$. The last thing we need to do is to relate the distance ℓ to the distance x between the point and the center of the disk. From part (b) of Figure 23.26, you can see that this relationship is given by $\ell = \sqrt{r^2 + x^2}$.

SIMPLIFY   Putting the pieces together, we find that the potential on the symmetry axis of the disk as a function of the distance to the center is given by

$$V(x) = \int dV = \int \frac{k}{\ell} dq = \int \frac{k\sigma}{\ell} dA = \int \frac{k\sigma}{\ell} 2\pi r\, dr.$$

After inserting the expressions we found for the charge density, σ, and the distance, ℓ, we are ready to integrate over r from zero to the radius of the disk, R:

$$V(x) = \frac{2kq}{R^2} \int_0^R \frac{r}{\sqrt{x^2 + r^2}} dr = \frac{2kq}{R^2} \sqrt{x^2 + r^2} \Big|_0^R = \frac{2kq}{R^2} \left(\sqrt{x^2 + R^2} - x \right).$$

CALCULATE   Inserting the given numerical values, we find

$$V(4.5 \text{ mm}) = \frac{2(8.98755 \cdot 10^9 \text{ N m}^2/\text{C}^2)(3.5 \cdot 10^{-9} \text{C})}{(0.01 \text{ m})^2} \left(\sqrt{(4.5 \cdot 10^{-3} \text{ m})^2 + (0.01 \text{ m})^2} - 4.5 \cdot 10^{-3} \text{ m} \right)$$

$$= 4067.85 \text{ N m/C}.$$

ROUND   We round our final result to three significant figures: $V(4.5 \text{ mm}) = 4.07$ kV. (We have used 1 N m = 1 J and 1 J/1 C = 1 V to get the proper unit, the volt, for the potential.)

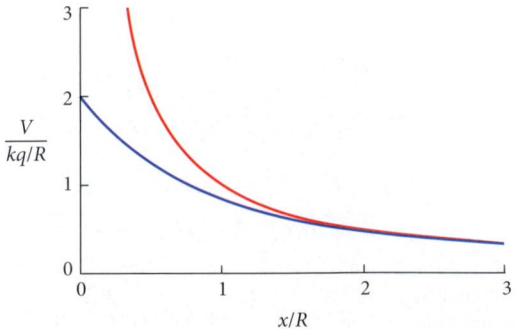

FIGURE 23.27 Comparison of the electric potential due to the uniformly charged disk with radius R (blue curve) and that due to a point charge (red curve).

DOUBLE-CHECK   We have already done the simple check of noting that the units of our answer are correct. As another check, we can look at the limiting case where the radius of the disk shrinks to zero, that is, where it becomes a point charge. The potential at a distance of 4.50 mm from a 3.50-nC point charge is

$$V_{\text{point}}(4.5 \text{ mm}) = \frac{(8.988 \cdot 10^9 \text{ N m}^2/\text{C}^2)(3.5 \cdot 10^{-9} \text{C})}{0.0045 \text{ m}} = 6.99 \text{ kV}.$$

This result is comforting because it is of the same order of magnitude as our calculated answer, just slightly larger, as we would expect. Figure 26.27 compares the graph of the electric potential due to the charged disk (blue curve) with the graph of the potential due to a point charge (red curve). As expected, the distribution of the charge does not matter at large distances, and our calculated potential approaches that for a point charge. For distances smaller than the radius of the disk, however, the difference becomes very pronounced. In particular, note that for $x \to 0$ the potential due to a uniformly charged disk does not diverge but has a maximum of $2kq/R$.

23.5 Finding the Electric Field from the Electric Potential

As we mentioned earlier, we can determine the electric field starting with the electric potential. This calculation uses equations 23.8 and 23.10:

$$-q\,dV = q\vec{E}\cdot d\vec{s},$$

where $d\vec{s}$ is a vector from an initial point to a final point located a small (infinitesimal) distance away. The component of the electric field, E_s, along the direction of $d\vec{s}$ is given by the partial derivative

$$E_s = -\frac{\partial V}{\partial s}. \tag{23.14}$$

(Chapter 15 on waves applied partial derivatives, and they were treated much like conventional derivatives, which we'll continue to do here.) Thus, we can find any component of the electric field by taking the partial derivative of the potential along the direction of that component. We can then write the components of the electric field in terms of partial derivatives of the potential:

$$E_x = -\frac{\partial V}{\partial x}; \quad E_y = -\frac{\partial V}{\partial y}; \quad E_z = -\frac{\partial V}{\partial z}. \tag{23.15}$$

The equivalent vector calculus formulation is $\vec{E} = -\vec{\nabla}V \equiv -(\partial V/\partial x,\ \partial V/\partial y,\ \partial V/\partial z)$, where the operator $\vec{\nabla}$ is called the **gradient.** Thus, the electric field can be determined either graphically, by measuring the negative of the change of the potential per unit distance perpendicular to an equipotential line, or analytically, by using equation 23.15.

Concept Check 23.7

Suppose an electric potential is described by $V(x, y, z) = -(5x^2 + y + z)$ in volts. Which of the following expressions describes the associated electric field, in units of volts per meter?

a) $\vec{E} = 5\hat{x} + 2\hat{y} + 2\hat{z}$

b) $\vec{E} = 10x\hat{x}$

c) $\vec{E} = 5x\hat{x} + 2\hat{y}$

d) $\vec{E} = 10x\hat{x} + \hat{y} + \hat{z}$

e) $\vec{E} = 0$

Concept Check 23.8

In the figure, the lines represent equipotential lines. How does the magnitude of the electric field, E, at point P compare for the three cases?

a) $E_1 = E_2 = E_3$

b) $E_1 > E_2 > E_3$

c) $E_1 < E_2 < E_3$

d) $E_3 > E_1 > E_2$

e) $E_3 < E_1 < E_2$

(1) 5 V, 10 V, 15 V, 20 V, P 25 V, 30 V

(2) 5 V, 10 V, 15 V, 20 V, P 25 V, 30 V

(3) 5 V, 10 V, 15 V, 20 V, P 25 V, 30 V

To visually reinforce the concepts of electric fields and potentials, the following example shows how a graphical technique can be used to find the field given the potential.

EXAMPLE 23.6 Graphical Extraction of the Electric Field

Let's consider a system of three point charges with values $q_1 = -6.00\ \mu C$, $q_2 = -3.00\ \mu C$, and $q_3 = +9.00\ \mu C$, located at positions $(x_1,y_1) = (1.5\ \text{cm}, 9.0\ \text{cm})$, $(x_2,y_2) = (6.0\ \text{cm}, 8.0\ \text{cm})$, and $(x_3,y_3) = (5.3\ \text{cm}, 2.0\ \text{cm})$. Figure 23.28 shows the electric potential, $V(x,y)$, resulting from these three charges, with equipotential lines calculated at potential values from -5000 V to 5000 V in 1000-V increments shown in Figure 23.29.

We can calculate the magnitude of the electric field at point P using equation 23.14 and graphical techniques. To perform this task, we use the green line in

– Continued

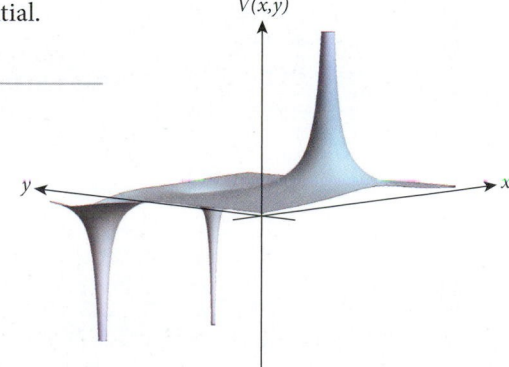

FIGURE 23.28 Electric potential due to three charges.

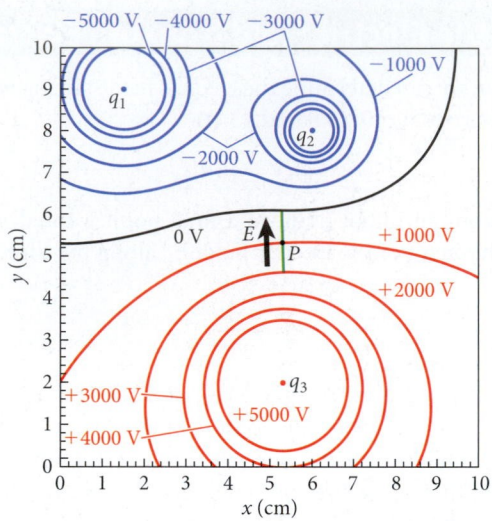

FIGURE 23.29 Equipotential lines for the electric potential due to three point charges.

Figure 23.29, which is drawn through point P perpendicular to the equipotential line because the electric field is always perpendicular to the equipotential lines, reaching from the equipotential line of 0 V to the line of 2000 V. As you can see from Figure 23.29, the length of the green line is 1.5 cm. Therefore, the magnitude of the electric field can be approximated as

$$|E_s| = \left| -\frac{\Delta V}{\Delta s} \right| = \left| \frac{(+2000 \text{ V}) - (0 \text{ V})}{1.5 \text{ cm}} \right| = 1.3 \cdot 10^5 \text{ V/m,}$$

where Δs is the length of the line through point P. The negative sign in equation 23.14 indicates that the direction of the electric field between neighboring equipotential lines points from the 2000-V equipotential line to the zero potential line.

Concept Check 23.9

In the figure, the lines represent equipotential lines. A positive charge is placed at point P, and then another positive charge is placed at point Q. Which set of vectors best represents the relative magnitudes and directions of the electric field forces exerted on the positive charges at P and Q?

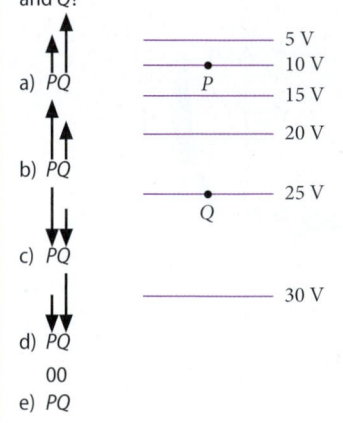

a) PQ

b) PQ

c) PQ

d) PQ

00

e) PQ

In Chapter 22, we derived an expression for the electric field along the perpendicular bisector of a finite line of charge:

$$E_y = \frac{2k\lambda}{y} \frac{a}{\sqrt{y^2 + a^2}}.$$

In Example 23.5, we found an expression for the electric potential along the perpendicular bisector of a finite line of charge; here we replace the coordinate d used in that example with the distance in the y-direction:

$$V = k\lambda \ln\left(\frac{\sqrt{y^2 + a^2} + a}{\sqrt{y^2 + a^2} - a} \right), \tag{23.16}$$

We can find the y-component of the electric field from the potential using equation 23.15:

$$E_y = -\frac{\partial V}{\partial y}$$

$$= -\frac{\partial \left(k\lambda \ln\left(\frac{\sqrt{y^2 + a^2} + a}{\sqrt{y^2 + a^2} - a} \right) \right)}{\partial y}$$

$$= -k\lambda \left(\frac{\partial \left(\ln\left(\sqrt{y^2 + a^2} + a \right) \right)}{\partial y} - \frac{\partial \left(\ln\left(\sqrt{y^2 + a^2} - a \right) \right)}{\partial y} \right).$$

Taking the partial derivative (remember that we can treat the partial derivative like a regular derivative), we obtain for the first term

$$\frac{\partial}{\partial y}\left(\ln\left(\sqrt{y^2+a^2}+a\right)\right)=\left(\underbrace{\frac{1}{\sqrt{y^2+a^2}+a}}_{\text{derivative of ln}}\right)\left(\underbrace{\frac{1}{2}\frac{1}{\sqrt{y^2+a^2}}}_{\text{derivative of }\sqrt{y^2+a^2}}\right)\underbrace{(2y)}_{\text{derivative of }y^2}=\frac{y}{y^2+a^2+a\sqrt{y^2+a^2}},$$

where the fact that the derivative of the natural log function is $d(\ln x)/dx = 1/x$ and the chain rule of differentiation have been used. (The outer and inner derivatives are indicated under the terms that they generate.) A similar expression can be found for the second term. Using the values of the derivatives, we find the component of the electric field:

$$E_y = -k\lambda\left(\frac{y}{y^2+a^2+a\sqrt{y^2+a^2}}-\frac{y}{y^2+a^2-a\sqrt{y^2+a^2}}\right)=\frac{2k\lambda}{y}\frac{a}{\sqrt{y^2+a^2}}.$$

This result is the same as that for the electric field in the y-direction derived in Chapter 22 by integrating over a finite line of charge.

23.6 Electric Potential Energy of a System of Point Charges

Section 23.1 discussed the electric potential energy of a point charge in a given external electric field, and Section 23.4 described how to calculate the electric potential due to a system of point charges. This section combines these two pieces of information to find the electric potential energy of a system of point charges. Consider a system of charges that are infinitely far apart. To bring these charges into proximity with each other, work must be done on the charges, which changes the electric potential energy of the system. The electric potential energy of a system of point charges is defined as the work required to bring the charges together from being infinitely far apart.

As an example, let's find the electric potential energy of a system of two point charges (Figure 23.30). Assume that the two charges start at an infinite separation. We then bring point charge q_1 into the system. Because the system without charges has no electric field and no corresponding electric force, this action does not require that any work be done on the charge. Keeping this charge stationary, we bring the second point charge, q_2, from infinity to a distance r from q_1. Using equation 23.6, we can write the electric potential energy of the system as

$$U = q_2V, \tag{23.17}$$

where

$$V = \frac{kq_1}{r}. \tag{23.18}$$

Thus, the electric potential energy of this system of two point charges is

$$U = \frac{kq_1q_2}{r}. \tag{23.19}$$

From the work-energy theorem, the work, W, that must be done on the particles to bring them together and keep them stationary is equal to U. If the two charges have the same sign, $W = U > 0$, positive work must be done to bring them together from infinity and keep them motionless. If the two charges have opposite signs, negative work must be done to bring them together from infinity and hold them motionless. To determine U for more than two point charges, we assemble them from infinity one charge at a time, in any order.

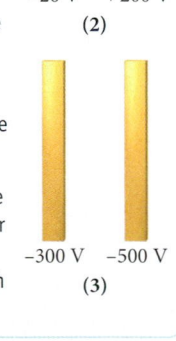
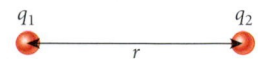

q_1 q_2
r

FIGURE 23.30 Two point charges separated by a distance r.

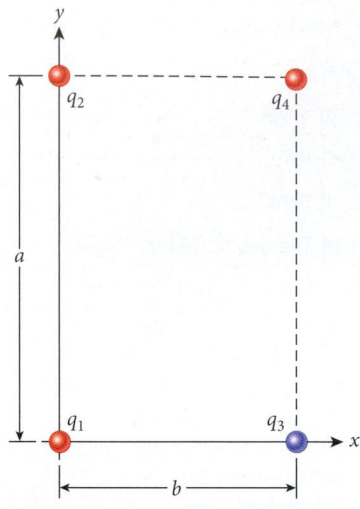

FIGURE 23.31 Calculating the potential energy of a system of four point charges.

EXAMPLE 23.7 | Four Point Charges

Let's calculate the electric potential energy of a system of four point charges, shown in Figure 23.31. The four point charges have the values $q_1 = +1.0\ \mu C$, $q_2 = +2.0\ \mu C$, $q_3 = -3.0\ \mu C$, and $q_4 = +4.0\ \mu C$. The charges are placed with $a = 6.0$ m and $b = 4.0$ m.

PROBLEM
What is the electric potential energy of this system of four point charges?

SOLUTION
We begin the calculation with the four charges infinitely far apart and assume that the electric potential energy is zero in that configuration. We bring in q_1 and position that charge at $(0,0)$. This action does not change the electric potential energy of the system. Now we bring in q_2 and place that charge at $(0,a)$. The electric potential energy of the system is now

$$U = \frac{kq_1q_2}{a}.$$

Bringing q_3 in from an infinite distance and placing it at $(b,0)$ changes the potential energy of the system through the interaction of q_3 with q_1 and the interaction of q_3 with q_2. The new potential energy is

$$U = \frac{kq_1q_2}{a} + \frac{kq_1q_3}{b} + \frac{kq_2q_3}{\sqrt{a^2+b^2}}.$$

Finally, bringing in q_4 and placing it at (b,a) changes the potential energy of the system through interactions with q_1, q_2, and q_3, bringing the total electric potential energy of the system to

$$U = \frac{kq_1q_2}{a} + \frac{kq_1q_3}{b} + \frac{kq_2q_3}{\sqrt{a^2+b^2}} + \frac{kq_1q_4}{\sqrt{a^2+b^2}} + \frac{kq_2q_4}{b} + \frac{kq_3q_4}{a}.$$

Note that the order in which the charges are brought from infinity will not change this result. (You can try a different order to verify this statement.) Putting in the numerical values, we obtain

$$U = \left(3.0 \cdot 10^{-3}\ J\right) + \left(-6.7 \cdot 10^{-3}\ J\right) + \left(-7.5 \cdot 10^{-3}\ J\right) +$$
$$\left(5.0 \cdot 10^{-3}\ J\right) + \left(1.8 \cdot 10^{-2}\ J\right) + \left(-1.8 \cdot 10^{-2}\ J\right) = -6.2 \cdot 10^{-3}\ J.$$

From the calculation in Example 23.7, we extrapolate the result to obtain a formula for the electric potential energy of a collection of point charges:

$$U = k \sum_{ij\,(pairings)} \frac{q_iq_j}{r_{ij}}, \qquad (23.20)$$

where i and j label each pair of charges, the summation is over each pair ij (for all $i \neq j$), and r_{ij} is the distance between the charges in each pair. An alternative way to write this double sum is

$$U = \frac{1}{2}k \sum_{j=1}^{n} \sum_{i=1,i\neq j}^{n} \frac{q_iq_j}{|\vec{r}_i - \vec{r}_j|},$$

which is more explicit than the equivalent formulation of equation 23.20.

WHAT WE HAVE LEARNED | EXAM STUDY GUIDE

- The change in the electric potential energy, ΔU, of a point charge moving in an electric field is equal to the negative of the work done on the point charge by the electric field W_e:
  $\Delta U = U_f - U_i = -W_e$.

- The change in electric potential energy, ΔU, is equal to the charge, q, times the change in electric potential, ΔV: $\Delta U = q\Delta V$.

- Equipotential surfaces and equipotential lines represent locations in space that have the same electric potential. Equipotential surfaces are always perpendicular to the electric field lines.

- A surface of a conductor is an equipotential surface.

- The change in electric potential can be determined from the electric field by integrating over the field:

  $\Delta V = -\int_i^f \vec{E} \cdot d\vec{s}$. Setting the potential equal to zero at infinity gives $V = \int_i^\infty \vec{E} \cdot d\vec{s}$.

- The electric potential due to a point charge, q, at a distance r from the charge is given by $V = \dfrac{kq}{r}$.

- The electric potential due to a system of n point charges can be expressed as an algebraic sum of the individual potentials: $V = \sum_{i=1}^n V_i$.

- The electric field can be determined from gradients of the electric potential in each component direction:
  $E_x = -\dfrac{\partial V}{\partial x}, \; E_y = -\dfrac{\partial V}{\partial y}, \; E_z = -\dfrac{\partial V}{\partial z}$.

- The electric potential energy of a system of two point charges is given by $U = \dfrac{kq_1 q_2}{r}$.

ANSWERS TO SELF-TEST OPPORTUNITIES

23.1 The electric potential along the y-axis is zero.

23.2 $(x,y) = (0,0)$ corresponds to a saddle point.

23.3 Nothing would change. The electrostatic force is conservative, and for a conservative force, the work is path-independent.

23.4

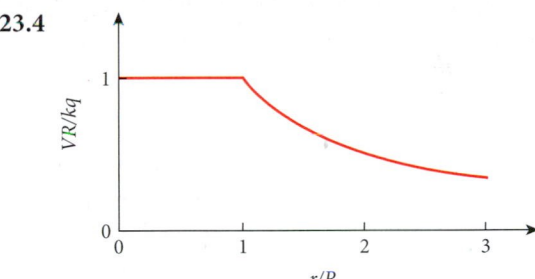

PROBLEM-SOLVING GUIDELINES

1. A common source of error in calculations is confusing the electric field, $\vec{E}$, the electric potential energy, U, and the electric potential, V. Remember that an electric field is a vector quantity produced by a charge distribution; electric potential energy is a property of the charge distribution; and electric potential is a property of the field. Be sure you know what it is that you're calculating.

2. Be sure to identify the point with respect to which you are calculating the potential energy or potential. Like calculations involving electric fields, calculations involving potentials can use a linear charge distribution (λ), a planar charge distribution (σ), or a volume charge distribution (ρ).

3. Since potential is a scalar, the total potential due to a system of point charges is calculated by simply adding the individual potentials due to all the charges. For a continuous charge distribution, you need to calculate the potential by integrating over the differential charge. Assume that the potential produced by the differential charge is the same as the potential from a point charge!

MULTIPLE-CHOICE QUESTIONS

23.1 A positive charge is released and moves along an electric field line. This charge moves to a position of

a) lower potential and lower potential energy.

b) lower potential and higher potential energy.

c) higher potential and lower potential energy.

d) higher potential and higher potential energy.

23.2 A proton is placed midway between points A and B. The potential at point A is -20 V, and the potential at point B $+20$ V. The potential at the midpoint is 0 V. The proton will

a) remain at rest.

b) move toward point B with constant velocity.

c) accelerate toward point A.

d) accelerate toward point B.

e) move toward point A with constant velocity.

23.3 What would be the consequence of setting the potential at +100 V at infinity, rather than taking it to be zero there?

a) Nothing; the field and the potential would have the same values at every finite point.

b) The electric potential would become infinite at every finite point, and the electric field could not be defined.

c) The electric potential everywhere would be 100 V higher, and the electric field would be the same.

d) It would depend on the situation. For example, the potential due to a positive point charge would drop off more slowly with distance, so the magnitude of the electric field would be less.

23.4 In which situation is the electric potential the highest?

a) at a point 1 m from a point charge of 1 C

b) at a point 1 m from the center of a uniformly charged spherical shell with a radius of 0.5 m and a total charge of 1 C

c) at a point 1 m from the center of a uniformly charged rod with a length of 1 m and a total charge of 1 C

d) at a point 2 m from a point charge of 2 C

e) at a point 0.5 m from a point charge of 0.5 C

23.5 The amount of work done to move a positive point charge q on an equipotential surface of 1000 V relative to that done to move the charge on an equipotential surface of 10 V is

a) the same.   c) more.

b) less.   d) dependent on the distance the charge moves.

23.6 A solid conducting sphere of radius R is centered about the origin of an xyz-coordinate system. A total charge Q is distributed uniformly on the surface of the sphere. Assuming, as usual, that the electric potential is zero at an infinite distance, what is the electric potential at the center of the conducting sphere?

a) zero   c) $Q/2\pi\epsilon_0 R$

b) $Q/\epsilon_0 R$ d) $Q/4\pi\epsilon_0 R$

23.7 Which of the following angles between an electric dipole moment and an applied electric field will result in the most stable state?

a) 0 rad   d) The electric dipole moment is not

b) $\pi/2$ rad   stable under any condition in an

c) π rad   applied electric field.

23.8 A positive point charge is to be moved from point A to point B in the vicinity of an electric dipole. Which of the three paths shown in the figure will result in the most work being done by the dipole's electric field on the point charge?

a) path 1

b) path 2

c) path 3

d) The work is the same on all three paths.

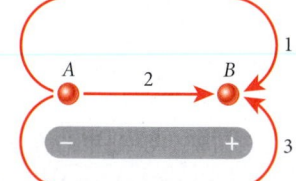

23.9 Each of the following pairs of charges are separated by a distance d. Which pair has the highest potential energy?

a) +5 C and +3 C   d) All pairs have the same

b) +5 C and –3 C   potential energy.

c) –5 C and +3 C

23.10 A negatively charged particle revolves in a clockwise direction around a positively charged sphere. The work done on the negatively charged particle by the electric field of the sphere is

a) positive.   b) negative.   c) zero.

23.11 A hollow conducting sphere of radius R is centered about the origin of an xyz-coordinate system. A total charge Q is distributed uniformly over the surface of the sphere. Assuming, as usual, that the electric potential is zero at an infinite distance, what is the electric potential at the center of the sphere?

a) zero

b) $2kQ/R$

c) kQ/R

d) $kQ/2R$

e) $kQ/4R$

23.12 A solid conducting sphere of radius R has a charge Q evenly distributed over its surface, producing an electric potential V_0 at the surface. How much charge must be added to the sphere to increase the potential at the surface to $2V_0$?

a) $Q/2$

b) Q

c) $2Q$

d) Q^2

e) $2Q^2$

23.13 Which one of the following statements is not true?

a) Equipotential lines are parallel to the electric field lines.

b) Equipotential lines for a point charge are circular.

c) Equipotential surfaces exist for any charge distribution.

d) When a charge moves on an equipotential surface, the work done on the charge is zero.

23.14 If a proton and an alpha particle (composed of two protons and two neutrons) are each accelerated from rest through the same potential difference, how do their resulting speeds compare?

a) The proton has twice the speed of the alpha particle.

b) The proton has the same speed as the alpha particle.

c) The proton has half the speed of the alpha particle.

d) The speed of the proton is $\sqrt{2}$ times the speed of the alpha particle.

e) The speed of the alpha particle is $\sqrt{2}$ times the speed of the proton.

CONCEPTUAL QUESTIONS

23.15 High-voltage power lines are used to transport electricity cross country. These wires are favored resting places for birds. Why don't the birds die when they touch the wires?

23.16 You have heard that it is dangerous to stand under trees in electrical storms. Why?

23.17 Can two equipotential lines cross? Why or why not?

23.18 Why is it important, when soldering connectors onto a piece of electronic circuitry, to leave no pointy protrusions from the solder joints?

23.19 Using Gauss's Law and the relation between electric potential and electric field, show that the potential outside a uniformly charged sphere

is identical to the potential of a point charge placed at the center of the sphere and equal to the total charge of the sphere. What is the potential at the surface of the sphere? How does the potential change if the charge distribution is not uniform but has spherical (radial) symmetry?

23.20 A metal ring has a total charge q and a radius R, as shown in the figure. Without performing any calculations, predict the value of the electric potential and the electric field at the center of the circle.

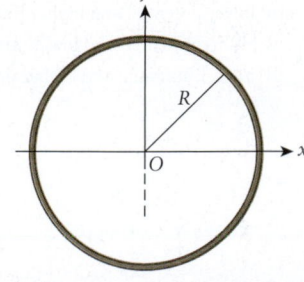

23.21 Find an integral expression for the electric potential at a point on the z-axis a distance H from a half-disk of radius R (see the figure). The half-disk has uniformly distributed charge over its surface, with charge distribution σ.

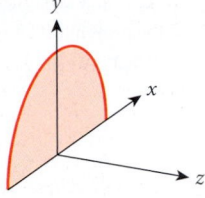

23.22 An electron moves away from a proton. Describe how the potential it encounters changes. Describe how its potential energy is changing.

23.23 The electric potential energy of a continuous charge distribution can be found in a way similar to that used for systems of point charges in Section 23.6, by breaking the distribution up into suitable pieces. Find the electric potential energy of an *arbitrary* spherically symmetrical charge distribution, $\rho(r)$. Do **not** assume that $\rho(r)$ represents a point charge, that it is constant, that it is piecewise-constant, or that it does or does not end at any finite radius, r. Your expression must cover all possibilities. Your expression may include an integral or integrals that cannot be evaluated without knowing the specific form of $\rho(r)$. (*Hint:* A spherical pearl is built up of thin layers of nacre added one by one.)

EXERCISES

A blue problem number indicates a worked-out solution is available in the Student Solutions Manual. One • and two •• indicate increasing level of problem difficulty.

Section 23.1

23.24 In molecules of gaseous sodium chloride, the chloride ion has one more electron than proton, and the sodium ion has one more proton than electron. These ions are separated by about 0.236 nm. How much work would be required to increase the distance between these ions to 1.00 cm?

•**23.25** A metal ball with a mass of $3.00 \cdot 10^{-6}$ kg and a charge of $+5.00$ mC has a kinetic energy of $6.00 \cdot 10^8$ J. It is traveling directly at an infinite plane of charge with a charge distribution of $+4.00$ C/m². If it is currently 1.00 m away from the plane of charge, how close will it come to the plane before stopping?

Section 23.2

23.26 An electron is accelerated from rest through a potential difference of 370. V. What is its final speed?

23.27 How much work would be done by an electric field in moving a proton from a point at a potential of $+180$. V to a point at a potential of -60.0 V?

23.28 What potential difference is needed to give an alpha particle (composed of 2 protons and 2 neutrons) 200. keV of kinetic energy?

23.29 A proton, initially at rest, is accelerated through a potential difference of 500. V. What is its final velocity?

23.30 A 10.0-V battery is connected to two parallel metal plates placed in a vacuum. An electron is accelerated from rest from the negative plate toward the positive plate.

a) What kinetic energy does the electron have just as it reaches the positive plate?

b) What is the speed of the electron just as it reaches the positive plate?

•**23.31** A proton gun fires a proton from midway between two plates, A and B, which are separated by a distance of 10.0 cm; the proton initially moves at a speed of 150.0 km/s toward plate B. Plate A is kept at zero potential, and plate B at a potential of 400.0 V.

a) Will the proton reach plate B?

b) If not, where will it turn around?

c) With what speed will it hit plate A?

•**23.32** Fully stripped (all electrons removed) sulfur (^{32}S) ions are accelerated from rest in an accelerator that uses a total voltage of $1.00 \cdot 10^9$ V. ^{32}S has 16 protons and 16 neutrons. The accelerator produces a beam consisting of $6.61 \cdot 10^{12}$ ions per second. This beam of ions is completely stopped in a beam dump. What is the total power the beam dump has to absorb?

Section 23.4

23.33 Two point charges are located at two corners of a rectangle, as shown in the figure.

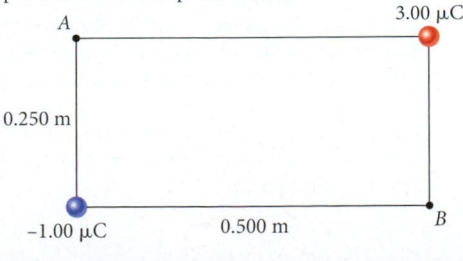

a) What is the electric potential at point A?

b) What is the potential difference between points A and B?

23.34 Four identical point charges ($+1.61$ nC) are placed at the corners of a rectangle, which measures 3.00 m by 5.00 m. If the electric potential is taken to be zero at infinity, what is the potential at the geometric center of this rectangle?

23.35 If a Van de Graaff generator has an electric potential of $1.00 \cdot 10^5$ V and a diameter of 20.0 cm, find how many more protons than electrons are on its surface.

23.36 One issue encountered during the exploration of Mars has been the accumulation of static charge on land-roving vehicles, resulting in a potential of 100. V or more. Calculate how much charge must be placed on the surface of a sphere of radius 1.00 m for the electric potential just above the surface to be 100. V. Assume that the charge is uniformly distributed.

23.37 A charge $Q = +5.60$ µC is uniformly distributed on a thin cylindrical plastic shell. The radius, R, of the shell is 4.50 cm. Calculate the electric potential at the origin of the xy-coordinate system shown in the figure. Assume that the electric potential is zero at points infinitely far away from the origin.

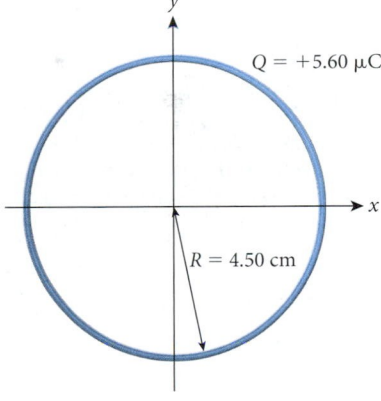

23.38 A hollow spherical conductor with a 5.00-cm radius has a surface charge of 8.00 nC.

a) What is the potential 8.00 cm from the center of the sphere?

b) What is the potential 3.00 cm from the center of the sphere?

c) What is the potential at the center of the sphere?

23.39 Find the potential at the center of curvature of the (thin) wire shown in the figure. It has a (uniformly distributed) charge per unit length of $\lambda = 3.00 \cdot 10^{-8}$ C/m and a radius of curvature of $R = 8.00$ cm.

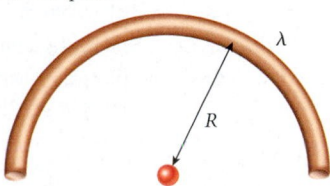

•**23.40** Consider a dipole with charge q and separation d. What is the potential a distance x from the center of this dipole at an angle θ with respect to the dipole axis, as shown in the figure?

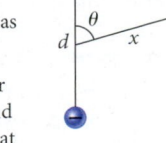

•**23.41** A spherical water drop 50.0 µm in diameter has a uniformly distributed charge of $+20.0$ pC. Find (a) the potential at its surface and (b) the potential at its center.

•**23.42** Consider an electron in the ground state of the hydrogen atom, separated from the proton by a distance of 0.0529 nm.

a) Viewing the electron as a satellite orbiting the proton in the electric potential, calculate the speed of the electron in its orbit.

b) Calculate an effective escape speed for the electron.

c) Calculate the energy of an electron having this speed, and from it determine the energy that must be given to the electron to ionize the hydrogen atom.

•**23.43** Four point charges are arranged in a square with side length $2a$, where $a = 2.70$ cm. The charges have the same magnitude, 1.50 nC; three of them are positive and one is negative as shown in the figure. What is the value of the electric potential generated by these four point charges at point $P = (0,0,c)$, where $c = 4.10$ cm?

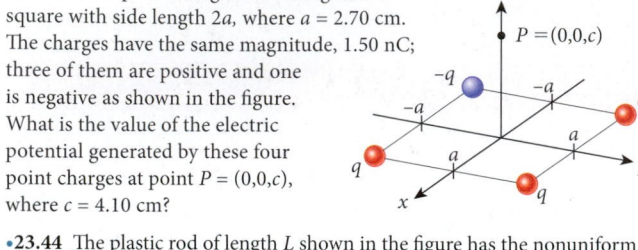

•**23.44** The plastic rod of length L shown in the figure has the nonuniform linear charge distribution $\lambda = cx$, where c is a positive constant. Find an expression for the electric potential at point P on the y-axis, a distance y from one end of the rod.

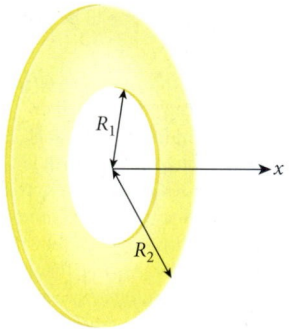

••**23.45** An electric field varies in space according to this equation: $\vec{E} = E_0 x e^{-x} \hat{x}$.

a) For what value of x does the electric field have its largest value, x_{max}?

b) What is the potential difference between the points at $x = 0$ and $x = x_{max}$?

••**23.46** Derive an expression for electric potential along the axis (the x-axis) of a disk with a hole in the center, as shown in the figure, where R_1 and R_2 are the inner and outer radii of the disk. What would the potential be if $R_1 = 0$?

Section 23.5

23.47 An electric field is established in a nonuniform rod. A voltmeter is used to measure the potential difference between the left end of the rod and a point a distance x from the left end. The process is repeated, and it is found that the data are described by the relationship $\Delta V = 270.x^2$, where ΔV has the units V/m². What is the x-component of the electric field at a point 13.0 cm from the left end?

23.48 Two parallel plates are held at potentials of +200.0 V and –100.0 V. The plates are separated by 1.00 cm.

a) Find the electric field between the plates.

b) An electron is initially placed halfway between the plates. Find its kinetic energy when it hits the positive plate.

23.49 A 2.50-mg dust particle with a charge of 1.00 μC falls at a point $x = 2.00$ m in a region where the electric potential varies according to $V(x) = (2.00 \text{ V/m}^2)x^2 - (3.00 \text{ V/m}^3)x^3$. With what acceleration will the particle start moving after it touches down?

23.50 The electric potential in a volume of space is given by $V(x,y,z) = x^2 + xy^2 + yz$. Determine the electric field in this region at the coordinate (3,4,5).

•**23.51** The electric potential inside a 10.0-m-long linear particle accelerator is given by $V = (3000 - 5x^2/\text{m}^2)$ V, where x is the distance from the left plate along the accelerator tube, as shown in the figure.

a) Determine an expression for the electric field along the accelerator tube.

b) A proton is released (from rest) at $x = 4.00$ m. Calculate the acceleration of the proton just after it is released.

c) What is the impact speed of the proton when (and if) it collides with the plate?

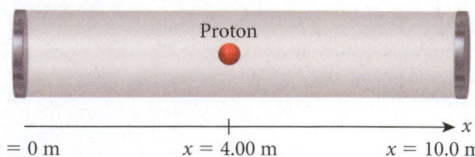

$x = 0$ m $x = 4.00$ m $x = 10.0$ m

•**23.52** An infinite plane of charge has a uniform charge distribution of +4.00 nC/m² and is located in the yz-plane at $x = 0$. A +11.0-nC fixed point charge is located at $x = +2.00$ m.

a) Find the electric potential $V(x)$ on the x-axis from $0 < x < +2.00$ m.

b) At what position(s) on the x-axis between $x = 0$ and $x = +2.00$ m is the electric potential a minimum?

c) Where on the x-axis between $x = 0$ m and $x = +2.00$ m could a positive point charge be placed and not move?

•**23.53** Use $V = \dfrac{kq}{r}$, $E_x = -\dfrac{\partial V}{\partial x}$, $E_y = -\dfrac{\partial V}{\partial y}$, and $E_z = -\dfrac{\partial V}{\partial z}$ to derive the expression for the electric field of a point charge, q.

•**23.54** Show that an electron in a one-dimensional electrical potential, $V(x) = Ax^2$, where the constant A is a positive real number, will execute simple harmonic motion about the origin. What is the period of that motion?

••**23.55** The electric field, $\vec{E}(\vec{r})$, and the electric potential, $V(\vec{r})$, are calculated from the charge distribution, $\rho(\vec{r})$, by integrating Coulomb's Law and then the electric field. In the other direction, the field and the charge distribution are determined from the potential by suitably differentiating. Suppose the electric potential in a large region of space is given by $V(r) = V_0 \exp(-r^2/a^2)$, where V_0 and a are constants and $r = \sqrt{x^2 + y^2 + z^2}$ is the distance from the origin.

a) Find the electric field $\vec{E}(\vec{r})$ in this region.

b) Determine the charge density, $\rho(\vec{r})$, in this region, which gives rise to the potential and field.

c) Find the total charge in this region.

d) Roughly sketch the charge distribution that could give rise to such an electric field.

••**23.56** The electron beam emitted by an electron gun is controlled (steered) with two sets of parallel conducting plates: a horizontal set to control the vertical motion of the beam, and a vertical set to control the horizontal motion of the beam. The beam is emitted with an initial velocity of $2.00 \cdot 10^7$ m/s. The width of the plates is $d = 5.00$ cm, the separation between the plates is $D = 4.00$ cm, and the distance between the edge of the plates and a target screen is $L = 40.0$ cm. In the absence of any applied voltage, the electron beam hits the origin of the xy-coordinate system on the observation screen. What voltages need to be applied to the two sets of plates for the electron beam to hit a target placed on the observation screen at coordinates $(x,y) = (0 \text{ cm}, 8.00 \text{ cm})$?

Section 23.6

23.57 Nuclear fusion reactions require that positively charged nuclei be brought into close proximity, against the electrostatic repulsion. As a simple example, suppose a proton is fired at a second, stationary proton from a large distance away. What kinetic energy must be given to the moving proton to get it to come within $1.00 \cdot 10^{-15}$ m of the target? Assume that there is a head-on collision and that the target is fixed in place.

23.58 Fission of a uranium nucleus (containing 92 protons) produces a barium nucleus (56 protons) and a krypton nucleus (36 protons). The fragments fly apart as a result of electrostatic repulsion; they ultimately emerge with a total of 200. MeV of kinetic energy. Use this information to estimate the size of the uranium nucleus; that is, treat the barium and krypton nuclei as point charges and calculate the separation between them at the start of the process.

23.59 A deuterium ion and a tritium ion each have charge $+e$. What work has to be done on the deuterium ion in order to bring it within $1.00 \cdot 10^{-14}$ m of the tritium ion? This is the distance within which the two ions can fuse, as a result of strong nuclear interactions that overcome electrostatic repulsion, to produce a helium-5 nucleus. Express the work in electron-volts.

•**23.60** Three charges, q_1, q_2, and q_3, are located at the corners of an equilateral triangle with side length of 1.20 m. Find the work done in each of the following cases:

a) to bring the first particle, $q_1 = 1.00$ pC, to P from infinity

b) to bring the second particle, $q_2 = 2.00$ pC, to Q from infinity

c) to bring the last particle, $q_3 = 3.00$ pC, to R from infinity

d) Find the total potential energy stored in the final configuration of q_1, q_2, and q_3.

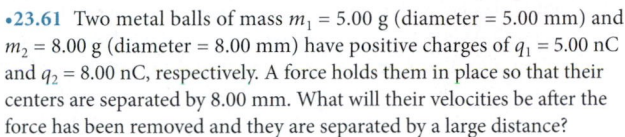

•**23.61** Two metal balls of mass $m_1 = 5.00$ g (diameter = 5.00 mm) and $m_2 = 8.00$ g (diameter = 8.00 mm) have positive charges of $q_1 = 5.00$ nC and $q_2 = 8.00$ nC, respectively. A force holds them in place so that their centers are separated by 8.00 mm. What will their velocities be after the force has been removed and they are separated by a large distance?

Additional Exercises

23.62 Two protons at rest and separated by 1.00 mm are released simultaneously. What is the speed of either at the instant when the two are 10.0 mm apart?

23.63 A 12-V battery is connected between a hollow metal sphere with a radius of 1 m and a ground, as shown in the figure. What are the electric field and the electric potential inside the hollow metal sphere?

12 V

23.64 A solid metal ball with a radius of 3.00 m has a charge of 4.00 mC. If the electric potential is zero far away from the ball, what is the electric potential at each of the following positions?

a) at $r = 0$ m, the center of the ball

b) at $r = 3.00$ m, on the surface of the ball

c) at $r = 5.00$ m

23.65 An insulating sheet in the xz-plane is uniformly charged with a charge distribution $\sigma = 3.50 \cdot 10^{-6}$ C/m^2. What is the change in potential when a charge of $Q = 1.25$ μC is moved from position A to position B in the figure?

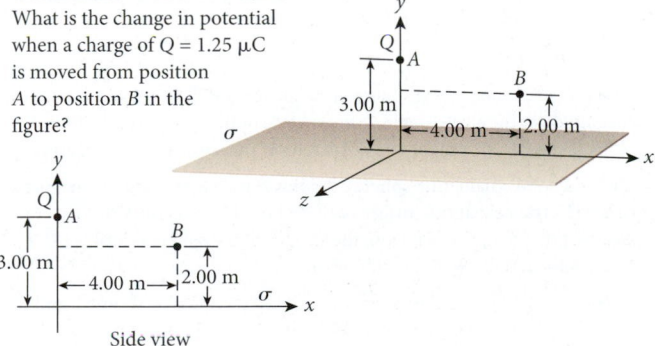

Side view

23.66 Suppose that an electron inside a cathode ray tube starts from rest and is accelerated by the tube's voltage of 21.9 kV. What is the speed (in km/s) with which the electron (mass $= 9.11 \cdot 10^{-31}$ kg) hits the screen of the tube?

23.67 A conducting solid sphere (radius of $R = 18.0$ cm, charge of $q = 6.10 \cdot 10^{-6}$ C) is shown in the figure. Calculate the electric potential at a point 24.0 cm from the center (point A), a point on the surface (point B), and at the center of the sphere (point C). Assume

that the electric potential is zero at points infinitely far away from the origin of the coordinate system.

23.68 A classroom Van de Graaff generator accumulates a charge of $1.00 \cdot 10^{-6}$ C on its spherical conductor, which has a radius of 10.0 cm and stands on an insulating column. Neglecting the effects of the generator base or any other objects or fields, find the potential at the surface of the sphere. Assume that the potential is zero at infinity.

23.69 A Van de Graaff generator has a spherical conductor with a radius of 25.0 cm. It can produce a maximum electric field of $2.00 \cdot 10^6$ V/m. What are the maximum voltage and charge that it can hold?

23.70 A proton with a speed of $1.23 \cdot 10^4$ m/s is moving from infinity directly toward a second proton. Assuming that the second proton is fixed in place, find the position where the moving proton stops momentarily before turning around.

23.71 Two metal spheres of radii $r_1 = 10.0$ cm and $r_2 = 20.0$ cm, respectively, have been positively charged so that each has a total charge of 100. μC.

a) What is the ratio of their surface charge distributions?

b) If the two spheres are connected by a copper wire, how much charge flows through the wire before the system reaches equilibrium?

23.72 The solid metal sphere of radius $a = 0.200$ m shown in the figure has a surface charge distribution of σ. The potential difference between the surface of the sphere and a point P at a distance $r_P = 0.500$ m from the center of the sphere is $\Delta V = V_{surface} - V_P = +4\pi V = +12.566$ V. Determine the value of σ.

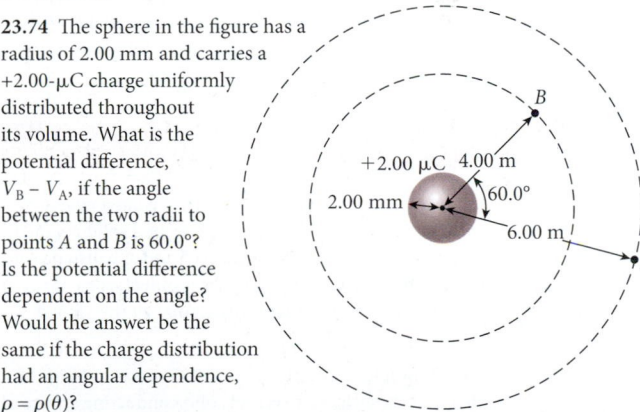

23.73 A particle with a charge of $+5.00$ μC is released from rest at a point on the x-axis, where $x = 0.100$ m. It begins to move as a result of the presence of a $+9.00$-μC charge that remains fixed at the origin. What is the kinetic energy of the particle at the instant it passes the point $x = 0.200$ m?

23.74 The sphere in the figure has a radius of 2.00 mm and carries a $+2.00$-μC charge uniformly distributed throughout its volume. What is the potential difference, $V_B - V_A$, if the angle between the two radii to points A and B is 60.0°? Is the potential difference dependent on the angle? Would the answer be the same if the charge distribution had an angular dependence, $\rho = \rho(\theta)$?

•**23.75** Two metallic spheres have radii of 10.0 cm and 5.00 cm, respectively. The magnitude of the electric field on the surface of each sphere is 3600. V/m. The two spheres are then connected by a long, thin metal wire. Determine the magnitude of the electric field on the surface of each sphere when they are connected.

•**23.76** A ring with charge Q and radius R is in the yz-plane and centered on the origin. What is the electric potential a distance x above the center of the ring? Derive the electric field from this relationship.

•**23.77** A charge of 0.681 nC is placed at $x = 0$. Another charge of 0.167 nC is placed at $x_1 = 10.9$ cm on the x-axis.

a) What is the combined electric potential of these two charges at $x = 20.1$ cm, also on the x-axis?

b) At which point(s) on the x-axis does this potential have a minimum?

•**23.78** A point charge of +2.00 µC is located at (2.50 m, 3.20 m). A second point charge of –3.10 µC is located at (–2.10 m, 1.00 m).

a) What is the electric potential at the origin?

b) Along a line passing through both point charges, at what point(s) is (are) the electric potential(s) equal to zero?

•**23.79** A total charge of $Q = 4.20 \cdot 10^{-6}$ C is placed on a conducting sphere (sphere 1) of radius $R = 0.400$ m.

a) What is the electric potential, V_1, at the surface of sphere 1 assuming that the potential infinitely far away from it is zero? (*Hint:* What is the change in potential if a charge is brought from infinitely far away, where $V(\infty) = 0$, to the surface of the sphere?)

b) A second conducting sphere (sphere 2) of radius $r = 0.100$ m with an initial net charge of zero ($q = 0$) is connected to sphere 1 using a long thin metal wire. How much charge flows from sphere 1 to sphere 2 to bring them into equilibrium? What are the electric fields at the surfaces of the two spheres at equilibrium?

•**23.80** A thin line of charge is aligned along the positive y-axis from $0 \le y \le L$, with $L = 4.0$ cm. The charge is not uniformly distributed but has a charge per unit length of $\lambda = Ay$, with $A = 8.00 \cdot 10^{-7}$ C/m^2. Assuming that the electric potential is zero at infinite distance, find the electric potential at a point on the x-axis as a function of x. Give the value of the electric potential at $x = 3.00$ cm.

•**23.81** Two fixed point charges are on the x-axis. A charge of –3.00 mC is located at $x = +2.00$ m and a charge of +5.00 mC is located at $x = –4.00$ m.

a) Find the electric potential, $V(x)$, for an arbitrary point on the x-axis.

b) At what position(s) on the x-axis is $V(x) = 0$?

c) Find $E(x)$ for an arbitrary point on the x-axis.

•**23.82** One of the greatest physics experiments in history measured the charge-to-mass ratio of an electron, q/m. If a uniform potential difference is created between two plates, atomized particles—each with an integral

amount of charge—can be suspended in space. The assumption is that the particles of unknown mass, M, contain a net number, n, of electrons of mass m and charge q. For a plate separation of d, what is the potential difference necessary to suspend a particle of mass M containing n net electrons? What is the acceleration of the particle if the voltage is cut in half? What is the acceleration of the particle if the voltage is doubled?

•**23.83** A uniform linear charge distribution of total positive charge Q has the shape of a half-circle of radius R, as shown in the figure.

a) Without performing any calculations, predict the electric potential produced by this linear charge distribution at point O.

b) Confirm, through direct calculations, your prediction of part (a).

c) Make a similar prediction for the electric field.

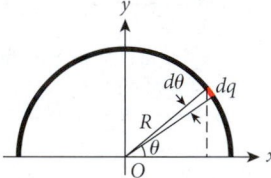

••**23.84** A point charge Q is placed a distance R from the center of a conducting sphere of radius a, with $R > a$ (the point charge is outside the sphere). The sphere is grounded, that is, connected to a distant, unlimited source and/or sink of charge at zero potential. (Neither the distant ground nor the connection directly affects the electric field in the vicinity of the charge and sphere.) As a result, the sphere acquires a charge opposite in sign to Q, and the point charge experiences an attractive force toward the sphere.

a) Remarkably, the electric field *outside the sphere* is the same as would be produced by the point charge Q plus an imaginary *mirror-image* point charge q, with magnitude and location that make the set of points corresponding to the surface of the sphere an equipotential of potential zero. That is, the imaginary point charge produces the same field contribution outside the sphere as the actual surface charge on the sphere. Calculate the value and location of q. (*Hint:* By symmetry, q must lie somewhere on the axis that passes through the center of the sphere and the location of Q.)

b) Calculate the force exerted on point charge Q and directed toward the sphere, in terms of the original quantities Q, R, and a.

c) Determine the actual nonuniform surface charge distribution on the conducting sphere.

MULTI-VERSION EXERCISES

23.85 A solid conducting sphere of radius $R_1 = 1.206$ m has a charge of $Q = 1.953$ µC evenly distributed over its surface. A second solid conducting sphere of radius $R_2 = 0.6115$ m is initially uncharged and at a distance of 10.00 m from the first sphere. The two spheres are momentarily connected with a wire, which is then removed. What is the charge on the second sphere?

23.86 A solid conducting sphere of radius $R_1 = 1.435$ m has a charge of Q evenly distributed over its surface. A second solid conducting sphere of radius $R_2 = 0.6177$ m is initially uncharged and at a distance of 10.00 m from the first sphere. The two spheres are momentarily connected with a wire, which is then removed. The resulting charge on the second sphere is 0.9356 µC. What was the original charge, Q, on the first sphere?

23.87 A solid conducting sphere of radius R_1 has a charge of $Q = 4.263$ µC evenly distributed over its surface. A second solid conducting sphere of radius $R_2 = 0.6239$ m is initially uncharged and at a distance of 10.00 m from the first sphere. The two spheres are momentarily connected with a

wire, which is then removed. The resulting charge on the second sphere is 1.162 µC. What is the radius of the first sphere?

23.88 A solid conducting sphere of radius $R = 1.895$ m is charged, and the magnitude of the electric field at the surface of the sphere is $3.165 \cdot 10^5$ V/m. What is the electric potential 29.81 cm from the surface of the sphere?

23.89 A solid conducting sphere of radius R is charged, and the magnitude of the electric field at the surface of the sphere is $3.269 \cdot 10^5$ V/m. The electric potential 32.37 cm from the surface of the sphere is $2.843 \cdot 10^5$ V. What is the radius, R, of the sphere?

23.90 A solid conducting sphere of radius $R = 1.351$ m is charged, and the magnitude of the electric potential 34.95 cm from the surface of the sphere is $3.618 \cdot 10^5$ V. What is the magnitude of the electric field at the surface of the sphere?

24

Capacitors

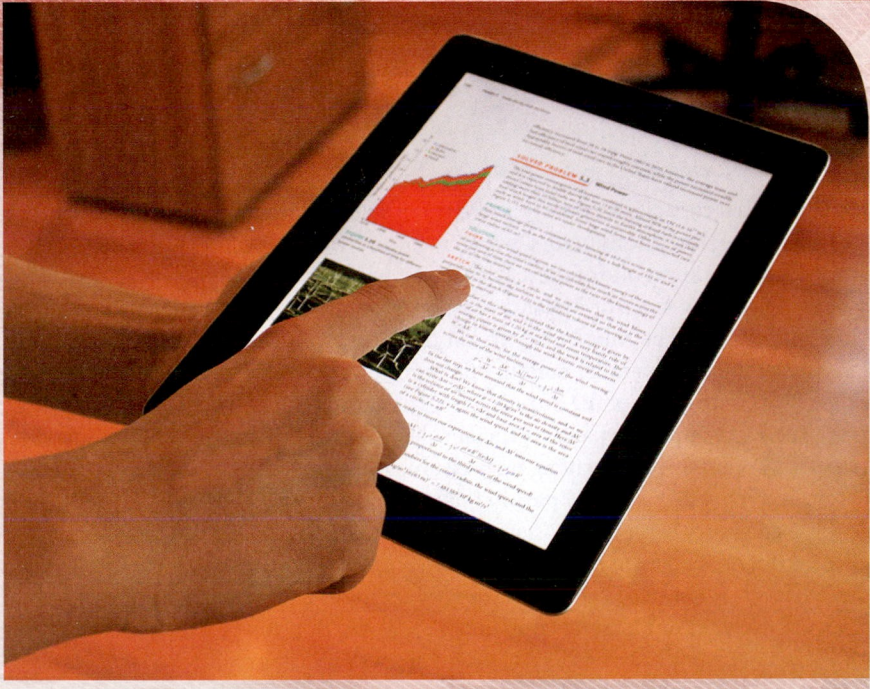

FIGURE 24.1 Interacting with the touch screen of an iPad.

Touch screens, such as the one shown in Figure 24.1, have become very common, found on everything from computer screens to cell phones to voting machines. They work in several ways, one of which involves using a property of conductors called *capacitance,* which we'll study in this chapter. Capacitance appears whenever two conductors—any two conductors—are separated by a small distance. The contact of a finger with a touch screen causes a change in capacitance that can be detected.

Capacitors have the very useful capability of storing electric charge and then releasing it very quickly. Thus, they are useful in camera flash attachments, cardiac defibrillators, and even experimental fusion reactors—anything that needs a large electric charge delivered quickly. Most circuits of any kind contain at least one capacitor. However, capacitance has a downside. It can also appear where it's not wanted, for example, between neighboring conductors in a tiny electronics circuit, where it can create "cross-talk"—unwanted interference between circuit components.

Because capacitors are one of the basic elements of electric circuits, this chapter examines how they function in simple circuits. The next two chapters will cover additional basic circuit elements and their uses.

WHAT WE WILL LEARN

- Capacitance is the ability to store electric charge.
- Capacitors usually consist of two separated conductors or conducting plates.
- A capacitor can store charge on one plate, and there is typically an equal and opposite charge on the other plate.
- The capacitance of a capacitor is the charge stored on the plates divided by the resulting electric potential difference.
- A capacitor can store electric potential energy.
- A common type of capacitor is the parallel plate capacitor, consisting of two flat parallel conducting plates.

- The capacitance of a given capacitor depends on its geometry.
- In a circuit, capacitors wired in parallel or in series can be replaced by an equivalent capacitance.
- The capacitance of a given capacitor is increased when a dielectric material is placed between the plates.
- A dielectric material reduces the electric field between the plates of a capacitor as a result of the alignment of molecular dipole moments in the dielectric material.

24.1 Capacitance

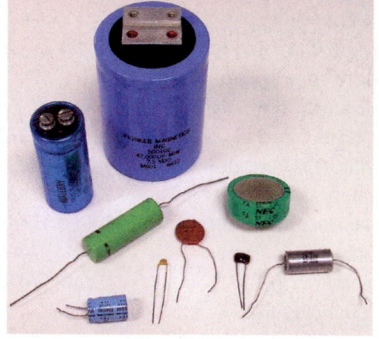

FIGURE 24.2 Some representative types of capacitors.

FIGURE 24.3 Two sheets of metal foil separated by an insulating layer.

FIGURE 24.4 The metal foil and Mylar sandwich shown in Figure 24.3 can be rolled up with an insulating layer to produce a capacitor with a compact geometry.

Figure 24.2, shows that capacitors come in a variety of sizes and shapes. In general, a **capacitor** consists of two separated conductors, which are usually called *plates* even if they are not simple planes. If we take apart one of these capacitors, we might find two sheets of metal foil separated by an insulating layer of Mylar, as shown in Figure 24.3. The sandwiched layers of metal foil and Mylar can be rolled up with another insulating layer into a compact form that does not resemble two parallel conductors, as shown in Figure 24.4. This technique produces capacitors with some of the physical formats shown in Figure 24.2. The insulating layer between the two metal foils plays a crucial role in the characteristics of the capacitor.

To study the properties of capacitors, we'll assume a convenient geometry and then generalize the results. Figure 24.5 shows a **parallel plate capacitor,** which consists of two parallel conducting plates, each with area A, separated by a distance, d, and assumed to be in a vacuum. The capacitor is charged by placing a charge of $+q$ on one plate and a charge of $-q$ on the other plate. (It is not necessary to put exactly opposite charges onto the two plates of the capacitor to charge it; any difference in charge will do. But, for practical purposes, the overall device should remain neutral, and this requires charges of equal magnitude and opposite sign on the two plates.) Because the plates are conductors, they are equipotential surfaces; thus, the electrons on the plates will distribute themselves uniformly over the surfaces.

Let's apply the results obtained in Chapter 23 to determine the electric potential and electric field for the parallel plate capacitor. (In principle, we could do this by calculating the electric potential and electric field for continuous charge distributions. However, for this physical configuration we would need to use a computer to provide the solution.) Let's place the origin of the coordinate system in the middle between the two plates, with the x-axis aligned with the two plates. Figure 24.6 shows a three-dimensional plot of the electric potential, $V(x,y)$, in the xy-plane, similar to the plots in Chapter 23.

The potential in Figure 24.6 has a very steep (and approximately linear) drop between the two plates and a more gradual drop outside the plates. This means that the electric field can be expected to be strongest between the plates and weaker outside. Figure 24.7a presents a contour plot of the electric potential shown in Figure 24.6 for the two parallel plates. Negative potential values are shaded in green, and positive values in pink. The equipotential lines, which are the lines where the three-dimensional equipotential surfaces intersect the xy-plane, displayed in Figure 24.6 are also shown in this plot, as are representations of the two plates. Note that the equipotential lines between the two plates are all parallel to each other and equally spaced.

In Figure 24.7b, the electric field lines have been added to the contour plot. The electric field is determined using $\vec{E}(\vec{r}) = -\vec{\nabla}V(\vec{r})$, introduced in Chapter 23. Far away from the two plates, the electric field looks very similar to that generated by a dipole composed of two point charges. It is easy to see that the electric field lines are perpendicular to the potential contour lines (which represent the equipotential surfaces!) everywhere in space.

But the electric field lines in Figure 24.7b do not convey adequate information about the magnitude of the electric field. Another representation of the electric field, in Figure 24.7c, displays the electric field vectors at regularly spaced grid points in the *xy*-plane. (The contour shading of the potential has been removed to reduce visual clutter.) In this plot, the field strength at each point of the grid is proportional to the size of the arrow at that point. You can clearly see that the electric field between the two plates is perpendicular to the plates and much larger in magnitude than the field outside the plates. The field in the space outside the plates is called the *fringe field*. If the plates are moved closer together, the electric field between the plates remains the same, while the fringe field is reduced.

The potential difference, ΔV, between the two parallel plates of the capacitor is proportional to the amount of charge on the plates. The proportionality constant is the **capacitance,** C, of the device, defined as

$$C = \left| \frac{q}{\Delta V} \right|. \tag{24.1}$$

The capacitance of a device depends on the area of the plates and the distance between them but not on the charge or the potential difference. (This will be shown for this and other geometries in the following sections.) By definition, the capacitance is a positive number. It tells how much charge is required to produce a given potential difference between the plates. The larger the capacitance, the more charge is required to produce a given potential difference. (Note that it is a common practice to use V, not ΔV, to represent potential difference. Be sure you understand when V is being used for potential and when it is being used for potential difference.)

Equation 24.1, the definition of capacitance, can be rewritten in this commonly used form:

$$q = C\Delta V.$$

Equation 24.1 indicates that the units of capacitance are the units of charge divided by the units of potential, or coulombs per volt. A new unit was assigned to capacitance, named after British physicist Michael Faraday (1791–1867). This unit is called the **farad** (F):

$$1\,\text{F} = \frac{1\,\text{C}}{1\,\text{V}}. \tag{24.2}$$

One farad represents a very large capacitance. Typically, capacitors have a capacitance in the range from $1\ \mu\text{F} = 1 \cdot 10^{-6}$ F to 1 pF $= 1 \cdot 10^{-12}$ F.

With the definition of the farad, we can write the electric permittivity of free space, ϵ_0 (introduced in Chapter 21), as $8.85 \cdot 10^{-12}$ F/m.

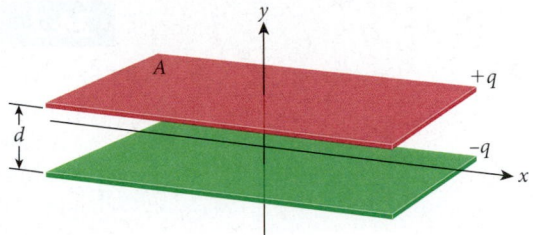

FIGURE 24.5 Parallel plate capacitor consisting of two conducting plates, each having area *A*, separated by a distance *d*.

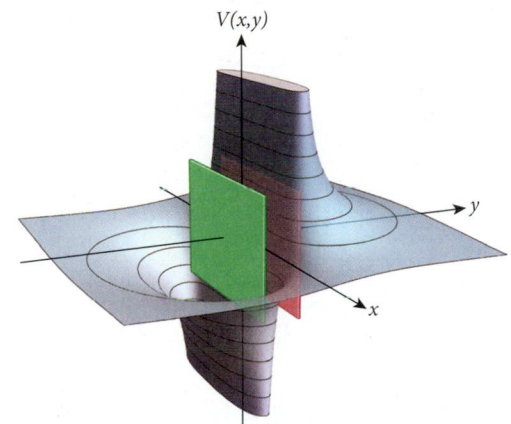

FIGURE 24.6 Electric potential in the *xy*-plane for the two oppositely charged parallel plates (superimposed) of Figure 24.5.

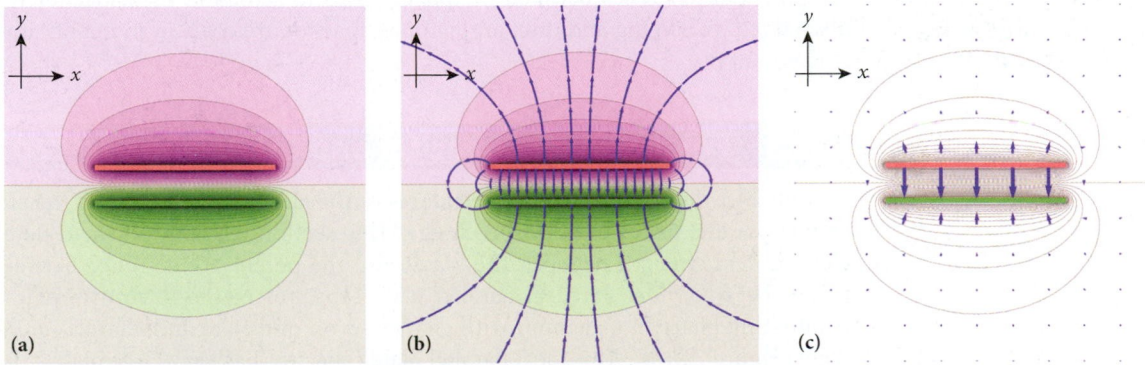

FIGURE 24.7 (a) Two-dimensional contour plot of the same potential as in Figure 24.6. (b) Contour plot with electric field lines superimposed. (c) Electric field strength at regularly spaced points in the *xy*-plane represented by the sizes of the arrows.

24.2 Circuits

The next few chapters will introduce more and more complex and interesting circuits. So let's look at what a circuit is, in general.

An **electric circuit** consists of simple wires or other conducting paths that connect circuit elements. These circuit elements can be capacitors, which we'll examine in depth in this chapter. Other important circuit elements are resistors and galvanometers (introduced in Chapter 25), the voltmeter and the ammeter (introduced in Chapter 26), transistors (which can be used as on-off switches or as amplifiers; also covered in Chapter 26), and inductors (discussed in Chapter 29).

Circuits usually need some kind of power, which can be provided either by a battery or by an AC (alternating current) power source. The concept of a battery, a device that maintains a potential difference across its terminals through chemical reactions, was introduced in Chapter 23; for the purpose of a circuit, it can be viewed simply as an external source of electrostatic potential difference, something that delivers a fixed potential difference (which is commonly called *voltage*). An AC power source can produce the same result with a specially designed circuit that maintains a fixed potential difference. Chapter 29 on induction and Chapter 30 on electromagnetic oscillations and current will discuss AC power sources in more detail. Figure 24.8 shows the symbols for circuit elements, which are used throughout this and the following chapters.

————	Wire	—Ⓖ—	Galvanometer
⊣⊢	Capacitor	—Ⓥ—	Voltmeter
—ⱴⱴⱴ—	Resistor	—Ⓐ—	Ammeter
—ⓑⓑⓑ—	Inductor	⊣⊦	Battery
—•⟋—	Switch	—Ⓐ⌄—	AC source

FIGURE 24.8 Commonly used symbols for circuit elements.

Charging and Discharging a Capacitor

A capacitor is charged by connecting it to a battery or to a constant-voltage power supply to create a circuit. Charge flows to the capacitor from the battery or power supply until the potential difference across the capacitor is the same as the supplied voltage. If the capacitor is disconnected, it retains its charge and potential difference. A real capacitor is subject to charge leaking away over time. However, in this chapter, we'll assume that an isolated capacitor retains its charge and potential difference indefinitely.

Figure 24.9 illustrates this charging process with a circuit diagram. In this diagram, the lines represent conducting wires. The battery (power supply) is represented by the symbol ⊤̵, which is labeled with plus and minus signs indicating the potential assignments of the terminals and with the potential difference, V. The capacitor is represented by the symbol ⊥⊤, which is labeled C. This circuit also contains a switch. When the switch is between positions a and b, the battery is not connected and the circuit is open. When the switch is at position a, the circuit is closed; the battery is connected across the capacitor, and the capacitor charges. When the switch is at position b, the circuit is closed in a different manner. The battery is removed from the circuit, the two plates of the capacitor are connected to each other, and charge can flow from one plate to the other through the wire, which now forms a physical connection between the plates. When the charge has dissipated on the two plates, the potential difference between the plates drops to zero, and the capacitor is said to be discharged. (Charging and discharging of a capacitor are covered in quantitative detail in Chapter 26.)

Concept Check 24.1

The figure shows a charged capacitor. What is the net charge on the capacitor?

a) $(+q) + (-q) = 0$

b) $|+q| + |-q| = 0$

c) $|+q| + |-q| = 2q$

d) $(+q) + (-q) = 2q$

e) q

$+q$
$-q$

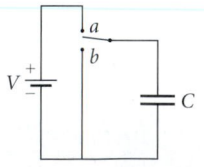

V ⊣⊦

FIGURE 24.9 Simple circuit used for charging and discharging a capacitor.

24.3 Parallel Plate Capacitor and Other Types of Capacitors

Section 24.1 discussed the general features of the electric potential and the electric field of two parallel plates of opposite charge. This section examines how to determine the electric field strength between the plates and the potential difference between the two plates. Let's consider an ideal parallel plate capacitor in the form of a pair of parallel conducting plates in a vacuum with charge $+q$ on one plate and charge $-q$ on the other plate (Figure 24.10). (This ideal parallel plate capacitor has very large plates, that are very close together, much closer than shown in Figure 24.10. This configuration allows us to neglect the fringe field, the small electric field outside the space between the plates, shown in Figure 24.7c.) When the plates are charged, the upper plate has charge $+q$ and the lower plate has charge $-q$. The electric field between the two plates points from the positively charged plate downward toward the negatively charged plate. The field near the ends of the plates, the fringe field (compare Figure 24.7), can be neglected; that is, we can assume that the electric field is constant, with magnitude E, everywhere between the plates and zero elsewhere. The electric field is always perpendicular to the surface of the two parallel plates.

The electric field can be found using Gauss's Law:

$$\oiint \vec{E} \cdot d\vec{A} = \frac{q}{\epsilon_0}. \tag{24.3}$$

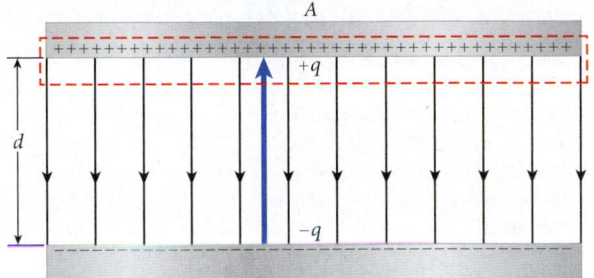

FIGURE 24.10 Side view of a parallel plate capacitor consisting of two plates with the same surface area, A, and separated by a small distance, d. The red dashed line is a Gaussian surface. The black arrows pointing downward represent the electric field. The blue arrow indicates an integration path.

How do we evaluate the integral over the Gaussian surface (whose cross section is outlined by a red dashed line in Figure 24.10)? We add the contributions from the top, the bottom, and the sides. The sides of the Gaussian surface are very small, so we can ignore the contributions from the fringe field. The top surface passes through the conductor, where the electric field is zero (remember shielding; see Chapter 22). This leaves only the bottom part of the Gaussian surface. The electric field vectors point straight down and are perpendicular to the conductor surfaces. The vector normal to the surface, $d\vec{A}$, points in the same direction and is thus parallel to $\vec{E}$. Therefore, the scalar product is $\vec{E} \cdot d\vec{A} = E \, dA \cos 0° = E \, dA$. For the integral over the Gaussian surface, we then have

$$\oiint \vec{E} \cdot d\vec{A} = \underset{\text{bottom}}{\iint} E \, dA = E \underset{\text{bottom}}{\iint} dA = EA,$$

where A is the area of the plate. In other words, for the parallel plate capacitor, Gauss's Law yields

$$EA = \frac{q}{\epsilon_0}, \tag{24.4}$$

where A is the surface area of the positively charged plate and q is the magnitude of the charge on the positively charged plate. The charge on each plate resides entirely on the inside surface because of the presence of opposite charge on the other plate.

The electric potential difference across the two plates in terms of the electric field is

$$\Delta V = -\int_{i}^{f} \vec{E} \cdot d\vec{s}. \tag{24.5}$$

The path of integration is chosen to be from the negatively charged plate to the positively charged plate, along the blue arrow in Figure 24.10. Since the electric field is antiparallel to this integration path (see Figure 24.10), the scalar product is $\vec{E} \cdot d\vec{s} = E \, ds \cos 180° = -E \, ds$. Thus, the integral in equation 24.5 reduces to

$$\Delta V = Ed = \frac{qd}{\epsilon_0 A},$$

where we used equation 24.4 to relate the electric field to the charge. Combining this expression for the potential difference and the definition of capacitance (equation 24.1) gives an expression for the capacitance of a parallel plate capacitor:

$$C = \left| \frac{q}{\Delta V} \right| = \frac{\epsilon_0 A}{d}. \tag{24.6}$$

Note that the capacitance of a parallel plate capacitor depends only on the area of the plates and the distance between the plates. In other words, only the geometry of a capacitor affects its capacitance. The amount of charge on the capacitor or the potential difference between its plates does not affect its capacitance.

Concept Check 24.2

Suppose you charge a parallel plate capacitor using a battery and then remove the battery, isolating the capacitor and leaving it charged. You then move the plates of the capacitor farther apart. The potential difference between the plates will

a) increase.

b) decrease.

c) stay the same.

d) not be determinable.

Self-Test Opportunity 24.1

You charge a parallel plate capacitor using a battery. You then remove the battery and isolate the capacitor. If you decrease the distance between the plates of the capacitor, what will happen to the electric field between the plates?

Concept Check 24.3

Suppose you have a parallel plate capacitor with area A and plate separation d, but space constraints on a circuit board force you to reduce the area of the capacitor by a factor of 2. What do you have to do to compensate and retain the same value of the capacitance?

a) reduce d by a factor of 2

b) increase d by a factor of 2

c) reduce d by a factor of 4

d) increase d by a factor of 4

EXAMPLE 24.1   Area of a Parallel Plate Capacitor

A parallel plate capacitor has plates that are separated by 1.00 mm (Figure 24.11).

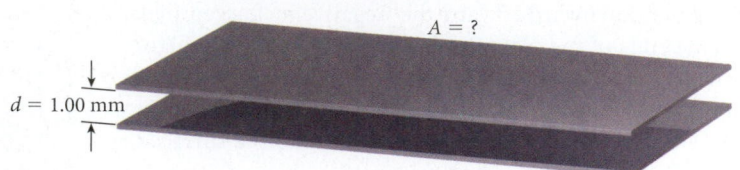

$A = ?$

$d = 1.00$ mm

FIGURE 24.11   A parallel plate capacitor with plates separated by 1.00 mm.

PROBLEM

What is the area required to give this capacitor a capacitance of 1.00 F?

SOLUTION

The capacitance is given by

$$C = \frac{\epsilon_0 A}{d}. \tag{i}$$

Solving equation (i) for the area and putting in $d = 1.00 \cdot 10^{-3}$ m and $C = 1.00$ F, we get

$$A = \frac{dC}{\epsilon_0} = \frac{\left(1.00 \cdot 10^{-3}\ \text{m}\right)\left(1.00\ \text{F}\right)}{\left(8.85 \cdot 10^{-12}\ \text{F/m}\right)} = 1.13 \cdot 10^8\ \text{m}^2.$$

If these plates were square, each one would be 10.6 km by 10.6 km (6.59 mi by 6.59 mi)! This result emphasizes that a farad is an extremely large amount of capacitance.

Cylindrical Capacitor

Consider a capacitor constructed of two collinear conducting cylinders with vacuum between them (Figure 24.12). The inner cylinder has radius r_1, and the outer cylinder has radius r_2. The inner cylinder has charge $-q$, and the outer cylinder has charge $+q$. The electric field between the two cylinders is then directed radially inward and perpendicular to the surfaces of both cylinders. As for a parallel plate capacitor, we assume that the cylinders are long and that there is essentially no fringe field near their ends.

We can apply Gauss's Law to find the electric field between the two cylinders, using a Gaussian surface in the form of a cylinder with radius r and length L that is collinear with the two cylinders of the capacitor, as shown in Figure 24.12. The enclosed charge is then $-q$, because only the negatively charged surface of the capacitor is inside the Gaussian surface. The normal vector to the Gaussian surface, $d\vec{A}$, points radially outward and is thus antiparallel to the electric field. This means that $\vec{E} \cdot d\vec{A} = E\, dA \cos 180° = -E\, dA$. Applying Gauss's Law and using the fact that the surface of the cylinder has area $A = 2\pi r L$ results in

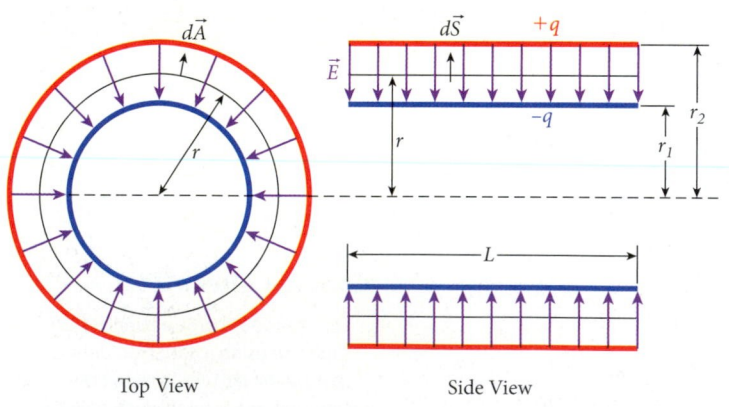

Top View   Side View

FIGURE 24.12   Cylindrical capacitor consisting of two long collinear conducting cylinders. The black circle represents a Gaussian surface. The purple arrows represent the electric field.

$$\oiint \vec{E} \cdot d\vec{A} = -E \oiint dA = -E\, 2\pi r L = \frac{-q}{\epsilon_0}. \tag{24.7}$$

Equation 24.7 can be rearranged to give an expression for the magnitude of the electric field:

$$E = \frac{q}{\epsilon_0\, 2\pi r L}, \quad \text{for } r_1 < r < r_2.$$

The potential difference between the two cylindrical capacitor plates is obtained by integrating over the electric field, $\Delta V = -\int_i^f \vec{E} \cdot d\vec{s}$. For the integration path in the radial direction from the negatively charged cylinder at r_1 to the positively charged cylinder at r_2, the electric field is antiparallel to the path. Thus, $\vec{E} \cdot d\vec{s}$ in equation 24.5 becomes $-E\,dr$. Therefore,

$$\Delta V = -\int_i^f \vec{E} \cdot d\vec{s} = \int_{r_1}^{r_2} E\,dr = \int_{r_1}^{r_2} \frac{q}{\epsilon_0 2\pi r L}\,dr = \frac{q}{\epsilon_0 2\pi L}\ln\!\left(\frac{r_2}{r_1}\right).$$

This expression for the potential difference and equation 24.1 yield an expression for the capacitance:

$$C = \left|\frac{q}{\Delta V}\right| = \frac{q}{\dfrac{q}{\epsilon_0 2\pi L}\ln\!\left(r_2/r_1\right)} = \frac{2\pi\epsilon_0 L}{\ln\!\left(r_2/r_1\right)}. \tag{24.8}$$

Just as for a parallel plate capacitor, the capacitance of a cylindrical capacitor depends only on its geometry.

Spherical Capacitor

Now let's consider a spherical capacitor formed by two concentric conducting spheres with radii r_1 and r_2 and vacuum between them (Figure 24.13). The inner sphere has charge $+q$, and the outer sphere has charge $-q$. The electric field is perpendicular to the surfaces of both spheres and points radially from the inner, positively charged sphere to the outer, negatively charged sphere, as shown by the purple arrows in Figure 24.13. (Previously, for the parallel plate and cylindrical capacitors, the integration was from the negative to the positive charge. Here we'll see what happens when the direction is reversed.) To determine the magnitude of the electric field, we employ Gauss's Law, using a Gaussian surface consisting of a sphere concentric with the two spherical conductors and having a radius r such that $r_1 < r < r_2$. The electric field is also perpendicular to the Gaussian surface everywhere, so we have

$$\oiint \vec{E} \cdot d\vec{A} = EA = E\left(4\pi r^2\right) = \frac{q}{\epsilon_0}. \tag{24.9}$$

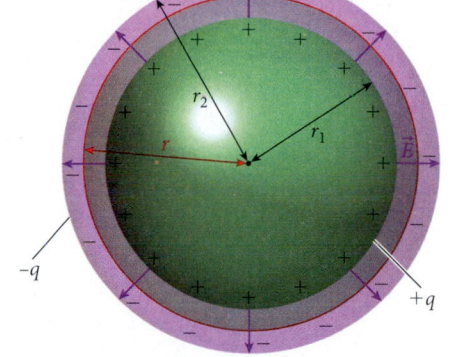

FIGURE 24.13 Spherical capacitor consisting of two concentric conducting spheres. The Gaussian surface is represented by the red circle of radius r.

Solving equation 24.9 for E gives

$$E = \frac{q}{4\pi\epsilon_0 r^2}, \qquad \text{for } r_1 < r < r_2.$$

For the potential difference, we proceed in a fashion similar to that used for the cylindrical capacitor and obtain

$$\Delta V = -\int_i^f \vec{E} \cdot d\vec{s} = -\int_{r_1}^{r_2} E\,dr = -\int_{r_1}^{r_2} \frac{q}{4\pi\epsilon_0 r^2}\,dr = -\frac{q}{4\pi\epsilon_0}\left(\frac{1}{r_1} - \frac{1}{r_2}\right).$$

In this case, $\Delta V < 0$. Why? Because we integrated from the positive charge to the negative charge! The positive charge is at a higher potential than the negative one, resulting in a negative potential difference. Equation 24.1 gives the capacitance of a spherical capacitor as the absolute value of the charge divided by the absolute value of the potential difference:

$$C = \left|\frac{q}{\Delta V}\right| = \frac{q}{\dfrac{q}{4\pi\epsilon_0}\left(\dfrac{1}{r_1} - \dfrac{1}{r_2}\right)} = \frac{4\pi\epsilon_0}{\left(\dfrac{1}{r_1} - \dfrac{1}{r_2}\right)}.$$

This can be rewritten in a more convenient form:

$$C = 4\pi\epsilon_0 \frac{r_1 r_2}{r_2 - r_1}. \tag{24.10}$$

Note that again the capacitance depends only on the geometry of the device.

Concept Check 24.4

If the inner and outer radii of a spherical capacitor are increased by a factor of 2, what happens to the capacitance?

a) It is reduced by a factor of 4.

b) It is reduced by a factor of 2.

c) It stays the same.

d) It is increased by a factor of 2.

e) It is increased by a factor of 4.

We can obtain the capacitance of a single conducting sphere from equation 24.10 by assuming that the outer spherical conductor is infinitely far away. With $r_2 = \infty$ and $r_1 = R$, the capacitance of an isolated spherical conductor is given by

$$C = 4\pi\epsilon_0 R. \tag{24.11}$$

24.4 Capacitors in Circuits

As stated earlier, a circuit is a set of electrical devices connected by conducting wires. Capacitors can be wired in circuits in different ways, but the two most fundamental ones are parallel connection and series connection.

Capacitors in Parallel

Figure 24.14 shows a circuit with three capacitors in **parallel connection**. Each of the three capacitors has one plate wired directly to the positive terminal of a battery with potential difference V and one plate wired directly to the negative terminal of that battery. The same circuit appears in the upper part of Figure 24.15, and the lower part of this figure shows the value of the potential at each part of the circuit in a three-dimensional plot. This illustrates that all capacitor plates connected to the positive terminal of the battery are at the same potential. The other plates of the capacitors are all at the potential of the negative terminal of the battery (set to zero). (The negative and positive terminals of the battery are joined with a light blue sheet to show that these two terminals are part of the same device and to provide a better visual representation of the potential difference between the two terminals. The plates of each capacitor are joined by a light gray band.)

The key insight provided by Figure 24.15 is that the potential difference across each of the three capacitors is the same, ΔV. Thus, for the three capacitors in this circuit, we have

$$q_1 = C_1 \Delta V$$
$$q_2 = C_2 \Delta V$$
$$q_3 = C_3 \Delta V.$$

In general, the charge on each capacitor can have a different value. The three capacitors can be viewed as one equivalent capacitor that holds a total charge q, given by

$$q = q_1 + q_2 + q_3 = C_1 \Delta V + C_2 \Delta V + C_3 \Delta V = \left(C_1 + C_2 + C_3\right)\Delta V.$$

Thus, the equivalent capacitance for this capacitor is

$$C_{eq} = C_1 + C_2 + C_3.$$

This result can be extended to any number, n, of capacitors connected in parallel:

$$C_{eq} = \sum_{i=1}^{n} C_i. \tag{24.12}$$

In other words, the equivalent capacitance of a system of capacitors in parallel is just the sum of the capacitances. Thus, several capacitors in parallel in a circuit can be replaced with an equivalent capacitance given by equation 24.12, as shown in Figure 24.16.

Capacitors in Series

Figure 24.17 shows a circuit with three capacitors in **series connection**. In this configuration, the battery produces an equal charge of $+q$ on the right plate of each capacitor and an equal charge of $-q$ on the left plate of each capacitor. This fact can be made clear by starting when the capacitors are uncharged. The battery is then connected to the series arrangement of the three capacitors. The positive plate of C_3 is connected to the positive terminal of the battery and begins to collect positive charge supplied by the battery. This positive charge induces a negative charge of equal magnitude onto the other plate of C_3.

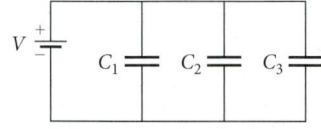

FIGURE 24.14 Simple circuit with a battery and three capacitors in parallel.

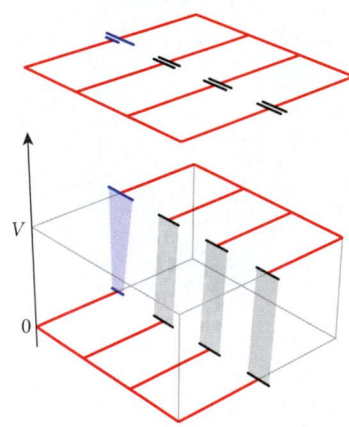

FIGURE 24.15 The potential in different parts of the circuit of Figure 24.14.

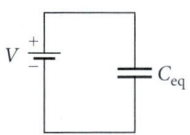

FIGURE 24.16 The three capacitors in Figure 24.14 can be replaced with an equivalent capacitance.

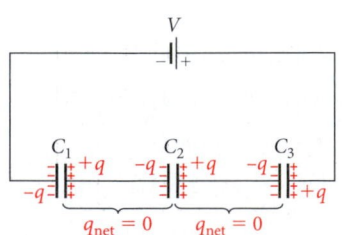

FIGURE 24.17 Simple circuit with three capacitors in series.

The negatively charged plate of C_3 is connected to the right plate of C_2, which then becomes positively charged because no net charge can accumulate on the isolated section consisting of the left plate of C_3 and the right plate of C_2. The positively charged plate of C_2 induces a negative charge of equal magnitude onto the other plate of C_2. In turn, the negatively charged plate of C_2 leaves a positive charge on the plate of C_1 connected to it, which induces a negative charge onto the left plate of C_1. The negatively charged plate of C_1 is connected to the negative terminal of the battery. Thus, charge flows from the battery, charging the positive plate of C_3 to a charge of value $+q$, and inducing a corresponding charge of $-q$ on the negatively charged plate of C_1. Therefore, each capacitor does indeed end up with the same charge.

When the three capacitors in the circuit in Figure 24.17 are charged, the sum of the potential drops across all three must equal the potential difference supplied by the battery. This is illustrated in Figure 24.18, a three-dimensional representation of the potential in the circuit with the three capacitors in series, similar to that in Figure 24.15. (Note that the potential drops at the three capacitors in series are not equal; this is true in general for a series connection.)

As you can see from Figure 24.18, the potential drops across the three capacitors must add up to the total potential difference, ΔV, supplied by the battery. Because each capacitor has the same charge, we have

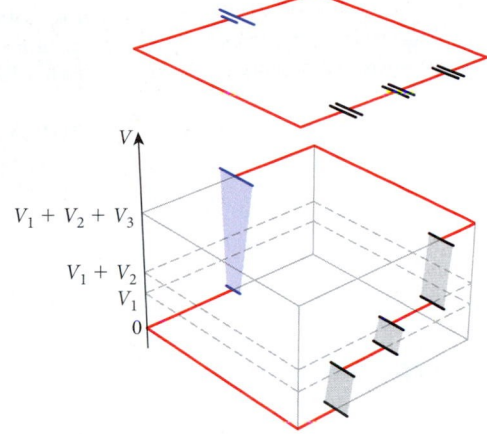

FIGURE 24.18 The potential in a circuit with three capacitors in series.

$$\Delta V = \Delta V_1 + \Delta V_2 + \Delta V_3 = \frac{q}{C_1} + \frac{q}{C_2} + \frac{q}{C_3} = q\left(\frac{1}{C_1} + \frac{1}{C_2} + \frac{1}{C_3}\right).$$

The equivalent capacitance can be written as

$$\Delta V = \frac{q}{C_{eq}},$$

where

$$\frac{1}{C_{eq}} = \frac{1}{C_1} + \frac{1}{C_2} + \frac{1}{C_3}. \tag{24.13}$$

Thus, the three capacitors in series in the circuit shown in Figure 24.17 can be replaced with an equivalent capacitance given by equation 24.13, yielding the same circuit diagram as that in Figure 24.16.

Concept Check 24.5

For a circuit with three capacitors in series, the equivalent capacitance must always be

a) equal to the largest of the three individual capacitances.

b) equal to the smallest of the three individual capacitances.

c) larger than the largest of the three individual capacitances.

d) smaller than the smallest of the three individual capacitances.

Concept Check 24.6

The potential drop for a circuit with three capacitors of different individual capacitances in series connection is

a) the same across each capacitor and has the same value as the potential difference supplied by the battery.

b) the same across each capacitor and has $\frac{1}{3}$ of the value of the potential difference supplied by the battery.

c) largest across the capacitor with the smallest capacitance.

d) largest across the capacitor with the largest capacitance.

For a system of n capacitors, equation 24.13 generalizes to

$$\frac{1}{C_{eq}} = \sum_{i=1}^{n} \frac{1}{C_i}. \tag{24.14}$$

Thus, the capacitance of a system of capacitors in series is always less than the smallest capacitance in the system.

Finding equivalent capacitances for capacitors in series and in parallel allows problems involving complicated circuits to be solved, as the following example illustrates.

Self-Test Opportunity 24.2

What is the equivalent capacitance for four 10.0-μF capacitors connected in series? What is the equivalent capacitance for four 10.0-μF capacitors connected in parallel?

Concept Check 24.7

Three capacitors, each with capacitance C, are connected as shown in the figure. What is the equivalent capacitance for this arrangement of capacitors?

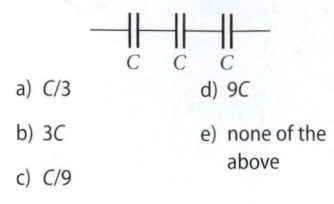

C C C

a) $C/3$

b) $3C$

c) $C/9$

d) $9C$

e) none of the above

Concept Check 24.8

Three capacitors, each with capacitance C, are connected as shown in the figure. What is the equivalent capacitance for this arrangement of capacitors?

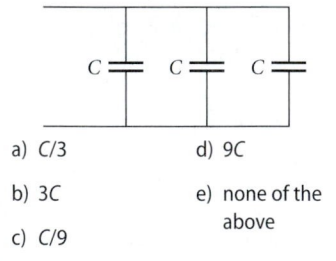

C C C

a) $C/3$

b) $3C$

c) $C/9$

d) $9C$

e) none of the above

EXAMPLE 24.2   System of Capacitors

PROBLEM

Consider the circuit shown in Figure 24.19a, a complicated-looking arrangement of five capacitors with a battery. What is the combined capacitance of this set of five capacitors? If each capacitor has a capacitance of 5.0 nF, what is the equivalent capacitance of the arrangement? If the potential difference of the battery is 12 V, what is the charge on each capacitor?

SOLUTION

This problem may look complicated at first, but it can be simplified into sequential steps, using the rules for equivalent capacitances of capacitors in series and in parallel. We begin with the innermost circuit structures and work outward.

STEP 1

Looking at capacitors 1 and 2 in Figure 24.19a, we see right away that they are in parallel. Because capacitor 3 is some distance away, it is less obvious that it is also in parallel with 1 and 2. However, the upper plates of all three of these capacitors are connected by wires and are thus at the same potential. The same goes for their lower plates, so all three are indeed in parallel. According to equation 24.12, the equivalent capacity for these three capacitors is

$$C_{123} = \sum_{i=1}^{3} C_i = C_1 + C_2 + C_3.$$

This replacement is shown in Figure 24.19b.

STEP 2

In Figure 24.19b, C_{123} and C_4 are in series. Thus, their equivalent capacitance is, according to equation 24.14,

$$\frac{1}{C_{1234}} = \frac{1}{C_{123}} + \frac{1}{C_4} \Rightarrow C_{1234} = \frac{C_{123}C_4}{C_{123}+C_4}.$$

This replacement is shown in Figure 24.19c.

STEP 3

Finally, C_{1234} and C_5 are in parallel in Figure 24.19c. Therefore, we can repeat the calculation for two capacitors in parallel and find the equivalent capacitance of all five capacitors:

$$C_{12345} = C_{1234} + C_5 = \frac{C_{123}C_4}{C_{123}+C_4} + C_5 = \frac{(C_1+C_2+C_3)C_4}{C_1+C_2+C_3+C_4} + C_5.$$

This result gives us the simple circuit shown in Figure 24.19d.

FIGURE 24.19 System of capacitors: (a) original circuit configuration; (b) reducing parallel capacitors to their equivalent; (c) reducing series capacitors to their equivalent; (d) equivalent capacitance for the entire set of capacitors.

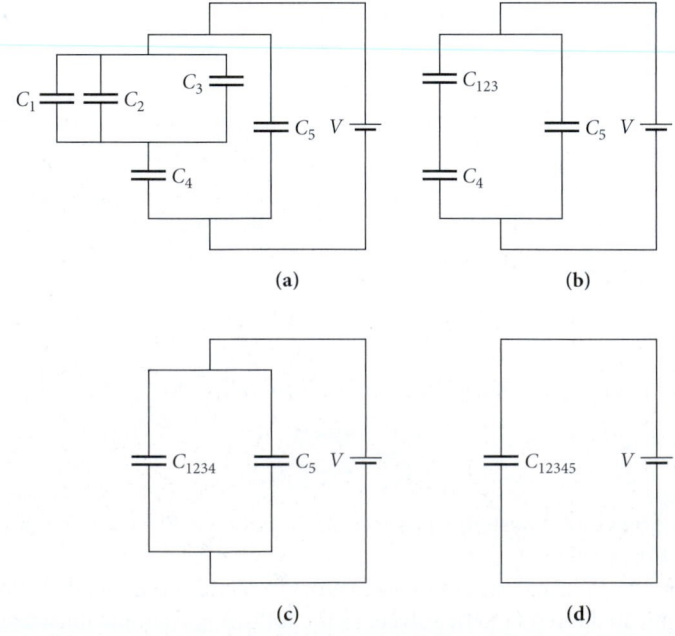

STEP 4: INSERT THE NUMBERS FOR THE CAPACITORS

We can now find the equivalent capacitance if all the capacitors have identical 5.0-nF capacitances:

$$\left(\frac{(5.0+5.0+5.0)5.0}{5.0+5.0+5.0+5.0}+5.0\right)nF = 8.8 \ nF.$$

As you can see, more than half of the total capacitance of this arrangement is provided by capacitor 5 alone. This result shows that you need to be extremely careful about how you arrange capacitors in circuits.

STEP 5: CALCULATE THE CHARGES ON THE CAPACITORS

C_{1234} and C_5 are in parallel. Thus, they have the same potential difference across them, 12 V. The charge on C_5 is then

$$q_5 = C_5 \Delta V = (5.0 \ nF)(12 \ V) = 60. \ nC.$$

C_{1234} is composed of C_{123} and C_4 in series. Thus, C_{123} and C_4 must have the same charge q_4, so

$$\Delta V = \Delta V_{123} + \Delta V_4 = \frac{q_4}{C_{123}} + \frac{q_4}{C_4} = q_4 \left(\frac{1}{C_{123}} + \frac{1}{C_4}\right).$$

The charge on C_4 is then

$$q_4 = \Delta V \frac{C_{123}C_4}{C_{123}+C_4} = \Delta V \frac{(C_1+C_2+C_3)C_4}{C_1+C_2+C_3+C_4} = (12 \ V)\frac{(15 \ nF)(5.0 \ nF)}{20.0 \ nF} = 45 \ nC.$$

C_{123} is equivalent to three capacitors in parallel, and it also has the same charge as C_4, or 45 nC. The three capacitors C_1, C_2, and C_3 have the same capacitance, the same potential difference across them as they are in parallel, and the sum of the charge on these three capacitors must equal 45 nC. Therefore, we can calculate the charge on C_1, C_2, and C_3:

$$q_1 = q_2 = q_3 = \frac{45 \ nC}{3} = 15 \ nC.$$

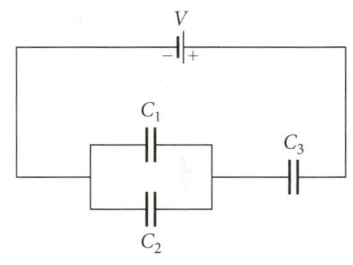
24.5 Energy Stored in Capacitors

Capacitors are extremely useful for storing electric potential energy. They are much more useful than batteries if the potential energy has to be converted into other energy forms very quickly. One application of capacitors for the storage and rapid release of electric potential energy is described in Example 24.4, about the high-powered laser at the National Ignition Facility. Let's examine how much energy can be stored in a capacitor.

A battery must do work to charge a capacitor. This work can be conceptualized in terms of changing the electric potential energy of the capacitor. In order to accomplish the charging process, charge has to be moved against the potential between the two capacitor plates. As noted earlier in this chapter, the larger the charge of the capacitor is, the larger the potential difference between the plates. This means that the more charge already on the capacitor, the harder it becomes to add a differential amount of charge to the capacitor. The differential work, dW, done by a battery with potential difference ΔV to put a differential charge, dq, on a capacitor with capacitance C is

$$dW = \Delta V' dq' = \frac{q'}{C} dq',$$

where $\Delta V'$ and q' are the instantaneous (increasing) potential difference and charge, respectively, on the capacitor during the charging process. The total work, W_t, required to bring the capacitor to its full charge, q, is given by

$$W_t = \int dW = \int_0^q \frac{q'}{C} dq' = \frac{1}{2}\frac{q^2}{C}.$$

This work is stored as electric potential energy:

$$U = \frac{1}{2}\frac{q^2}{C} = \frac{1}{2}C(\Delta V)^2 = \frac{1}{2}q\Delta V. \tag{24.15}$$

All three of the formulations for the stored electric potential energy in equation 24.15 are equally valid. Each can be transformed to one of the others by using $q = C\Delta V$ and eliminating one of the three quantities in favor of the other two.

The **electric energy density,** u, is defined as the electric potential energy per unit volume:

$$u = \frac{U}{\text{volume}}.$$

(*Note:* V is not used to represent the volume here, because in this context it is reserved for the potential.)

For the special case of a parallel plate capacitor that has no fringe field, it is easy to calculate the volume enclosed between two plates of area A separated by a perpendicular distance d. It is the area of each plate times the distance between the plates, or Ad. Using equation 24.15 for the electric potential energy, we obtain

$$u = \frac{U}{Ad} = \frac{\frac{1}{2}C(\Delta V)^2}{Ad} = \frac{C(\Delta V)^2}{2Ad}.$$

Using equation 24.6 for the capacitance of a parallel plate capacitor with vacuum between the plates, we get

$$u = \frac{(\epsilon_0 A/d)(\Delta V)^2}{2Ad} = \frac{1}{2}\epsilon_0\left(\frac{\Delta V}{d}\right)^2.$$

Recognizing that $\Delta V/d$ is the magnitude of the electric field, E, we obtain an expression for the electric energy density for a parallel plate capacitor:

$$u = \frac{1}{2}\epsilon_0 E^2. \tag{24.16}$$

This result, although derived for a parallel plate capacitor, is in fact much more general. The electric potential energy stored in any electric field per unit volume occupied by that field can be described using equation 24.16.

Concept Check 24.10

How much energy is stored in the 180-μF capacitor of a camera flash unit charged to 300.0 V?

a) 1.22 J   d) 115 J

b) 8.10 J   e) 300 J

c) 45.0 J

EXAMPLE 24.3 | Thundercloud

Suppose a thundercloud with a width of 2.0 km and a length of 3.0 km hovers at an altitude of 0.50 km over a flat area. The cloud carries a charge of 160 C, and the ground has no charge.

PROBLEM 1
What is the potential difference between the cloud and the ground?

SOLUTION 1
We can approximate the cloud-ground system as a parallel plate capacitor. Its capacitance is, according to equation 24.6,

$$C = \frac{\epsilon_0 A}{d} = \frac{(8.85 \cdot 10^{-12}\ \text{F/m})(2.0\ \text{km})(3.0\ \text{km})}{0.50\ \text{km}} = 0.11\ \mu\text{F}.$$

Because we know the charge carried by the cloud, 160 C, it is tempting to insert this value into the relationship among charge, capacitance, and potential difference (equation 24.1) to find the desired answer. However, a parallel plate capacitor with a charge of $+q$ on one plate and $-q$ on the other has a charge difference of $2q$ between the plates. For the cloud-ground system, $2q = 160$ C, or $q = 80.$ C. Alternatively, we can think of the cloud as a charged insulator and use the result from Section 22.9, that the field due to a plane sheet of charge is $E = \sigma/2\epsilon_0$, to justify the factor of $\frac{1}{2}$. Now we can use equation 24.1 and obtain

$$\Delta V = \frac{q}{C} = \frac{80.\ \text{C}}{0.11\ \mu\text{F}} = 7.3 \cdot 10^8\ \text{V}.$$

The potential difference is more than 700 million volts!

PROBLEM 2
Lightning strikes require electric field strengths of approximately 2.5 MV/m. Are the conditions described in the problem statement sufficient for a lightning strike?

SOLUTION 2

We use the potential difference between cloud and ground and the given distance between them to calculate the magnitude of the electric field:

$$E = \frac{\Delta V}{d} = \frac{7.3 \cdot 10^8 \text{ V}}{0.50 \text{ km}} = 1.5 \text{ MV/m}.$$

From this result, we can conclude that no lightning will develop in these conditions. However, if the cloud drifted over a radio tower, the electric field strength would likely increase and lead to a lightning discharge.

PROBLEM 3

What is the total electric potential energy contained in the field between this thundercloud and the ground?

SOLUTION 3

From equation 24.15, the total electric potential energy stored in the cloud-ground system is

$$U = \tfrac{1}{2} q \Delta V = 0.5(80. \text{ C})(7.3 \cdot 10^8 \text{ V}) = 2.9 \cdot 10^{10} \text{ J}.$$

For comparison, this energy is sufficient to run a typical 1500-W hair dryer for more than 5000 hours.

SOLVED PROBLEM 24.1 | Energy Stored in Capacitors

PROBLEM

Suppose many capacitors, each with $C = 90.0 \text{ μF}$, are connected in parallel across a battery with a potential difference of $\Delta V = 160.0$ V. How many capacitors are needed to store 95.6 J of energy?

SOLUTION

THINK The equivalent capacitance of many capacitors connected in parallel is given by the sum of the capacitances of all the capacitors. We can calculate the energy stored from the equivalent capacitance of the capacitors in parallel and the potential difference of the battery.

SKETCH Figure 24.20 shows a circuit with n capacitors connected in parallel across a battery.

RESEARCH The equivalent capacitance, C_{eq}, of n capacitors, each with capacitance C, connected in parallel is

$$C_{eq} = C_1 + C_2 + \cdots + C_n = nC.$$

The energy stored in the capacitors is then given by

$$U = \tfrac{1}{2} C_{eq} (\Delta V)^2 = \tfrac{1}{2} nC(\Delta V)^2. \tag{i}$$

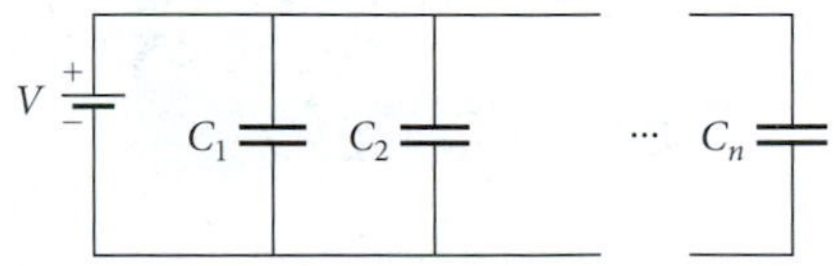

FIGURE 24.20 A circuit with n capacitors connected in parallel across a battery.

SIMPLIFY Solving equation (i) for the required number of capacitors gives us

$$n = \frac{2U}{C(\Delta V)^2}.$$

CALCULATE Putting in the numerical values, we get

$$n = \frac{2(95.6 \text{ J})}{(90.0 \cdot 10^{-6} \text{ C})(160.0 \text{ V})^2} = 82.986.$$

ROUND We report our result as an integer number of capacitors:

$$n = 83 \text{ capacitors.}$$

DOUBLE-CHECK The capacitance of 83 capacitors with $C = 90.0$ μF is

$$C_{eq} = 83(90.0 \text{ μF}) = 0.00747 \text{ F.}$$

– Continued

Charging this capacitor with a 160-V battery produces a stored energy of

$$U = \tfrac{1}{2} C_{eq} (\Delta V)^2 = \tfrac{1}{2}(0.00747 \text{ F})(160.0 \text{ V})^2 = 95.6 \text{ J}.$$

Thus, our answer for the number of capacitors is consistent.

FIGURE 24.21 Forty-eight power-conditioning modules in one of four bays at the National Ignition Facility at Lawrence Livermore National Laboratory.

EXAMPLE 24.4 The National Ignition Facility

The National Ignition Facility (NIF) is a high-powered laser designed to produce fusion reactions similar to those that occur in the Sun. The laser uses a short, high-energy pulse of light to heat and compress a small pellet containing isotopes of hydrogen. The laser is powered by 192 power-conditioning modules (Figure 24.21), each containing twenty 300.-μF capacitors connected in parallel and charged to 24.0 kV. The capacitors are charged over a period of 90.0 s. The laser is then fired by discharging all the energy stored in the capacitors in 400. μs.

PROBLEM 1
How much energy is stored in the capacitors of NIF?

SOLUTION 1
The capacitors are connected in parallel. Thus, the equivalent capacitance of each power-conditioning module is

$$C_{eq} = 20(300. \ \mu F) = 6.00 \text{ mF}.$$

The energy stored in each power-conditioning module is

$$U = \tfrac{1}{2} C_{eq} (\Delta V)^2 = \tfrac{1}{2}\left(6.00 \cdot 10^{-3} \text{ F}\right)\left(24.0 \cdot 10^3 \text{ V}\right)^2 = 1.73 \text{ MJ}.$$

Thus, the total energy stored in all the capacitors of NIF is

$$U_{total} = 192(1.73 \text{ MJ}) = 332 \text{ MJ}.$$

PROBLEM 2
What is the average power released by the power-conditioning modules during the laser pulse?

SOLUTION 2
The power is the energy per unit time, which is given by

$$P = \frac{\Delta U}{\Delta t} = \frac{332 \text{ MJ}}{400. \ \mu s} = \frac{332 \cdot 10^6 \text{ J}}{400. \cdot 10^{-6} \text{ s}} = 8.30 \cdot 10^{11} \text{ W} = 0.830 \text{ TW}.$$

In comparison, the average electrical power generated in the United States in 2010 was 0.47 TW. Of course, the 0.830 TW of power delivered to the NIF laser is maintained for only a small fraction of a second.

(a)

FIGURE 24.22 (a) An automatic external defibrillator (AED) in its holder on a wall. (b) Schematic diagram showing where to place the hands-free electrodes.

Defibrillator

An important application of capacitors is the portable automatic external defibrillator (AED), a device designed to shock the heart of a person who is in ventricular fibrillation. A typical AED is shown in Figure 24.22a.

Being in ventricular fibrillation means that the heart is not beating in a regular pattern. Instead, the signals that control the beating of the heart are erratic, preventing the heart from performing its function of maintaining regular blood circulation throughout the body. This condition must be treated within a few minutes to avoid permanent damage or death. Having many AED devices located in accessible public places allows quick treatment of this condition.

An AED provides a pulse of electrical current intended to stimulate the heart to beat regularly. Typically, an AED is designed to analyze a person's heartbeat automatically, determine if the person is in ventricular fibrillation, and administer the electrical pulse if required. The operator of the AED must attach the electrodes of the AED to the chest of the person experiencing the problem and push the start button. The AED will

(b)

do nothing if the person is not in ventricular defibrillation. If the AED determines that the person is in ventricular fibrillation, the AED will instruct the operator to press the button to initiate the electrical pulse. Note that an AED is not designed to restart a heart that is not beating. Rather it is designed to restore a regular heartbeat when the heart is beating erratically.

Typically, an AED delivers 150 J of electric energy to the patient, administered through a pair of electrodes that are attached to the chest area (see Figure 24.22b). This energy is stored by charging a capacitor through a special circuit from a low-voltage battery. This capacitor typically has a capacitance of 100. μF and is charged in 10. s. The power used during charging is

$$P = \frac{E}{t} = \frac{150 \text{ J}}{10. \text{ s}} = 15 \text{ W},$$

which is within the capability of a simple battery. The energy of the capacitor is then discharged in 10. ms. The instantaneous power during the discharge is

$$P = \frac{E}{t} = \frac{150 \text{ J}}{10. \text{ ms}} = 15 \text{ kW},$$

which is beyond the capability of a small, portable battery but well within the capabilities of a well-designed capacitor.

The energy stored in the capacitor is $U = \frac{1}{2}C(\Delta V)^2$. When the capacitor is charged, its potential difference is

$$\Delta V = \sqrt{\frac{2U}{C}} = \sqrt{\frac{2(150 \text{ J})}{100. \cdot 10^{-6} \text{ F}}} = 1.7 \text{ kV}.$$

When the AED delivers an electrical current, the capacitor is charged from a battery contained in the AED. The capacitor is then discharged through the person to stimulate the heart to beat in a regular manner. Most AEDs can deliver the electrical current many times without recharging the battery.

24.6 Capacitors with Dielectrics

The capacitors we have been discussing have air or vacuum between the plates. However, capacitors for just about any commercial application have an insulating material, called a **dielectric,** between the two plates. This dielectric serves several purposes: First, it maintains the separation between the plates. Second, the dielectric insulates the two plates from each other electrically. Third, the dielectric allows the capacitor to maintain a higher potential difference than it could with only air between the plates. Lastly, a dielectric increases the capacitance of the capacitor. We'll see that this ability to increase the capacitance is due to the molecular structure of the dielectric.

Filling the space between the plates of a capacitor completely with a dielectric increases its capacitance by a numerical factor called the **dielectric constant,** κ. We'll assume that the dielectric fills the entire volume between the capacitor plates, unless explicitly stated otherwise. Solved Problem 24.2 considers an example where the filling is only partial.

The capacitance, C, of a capacitor containing a dielectric with dielectric constant κ between its plates is given by

$$C = \kappa C_{\text{air}}, \tag{24.17}$$

where C_{air} is the capacitance of the capacitor without the dielectric.

Placing a dielectric between the plates of a capacitor has the effect of lowering the electric field between the plates (see Section 24.7) and allowing more charge to be stored in the capacitor. For example, the electric field between the plates of a parallel plate capacitor, given by equation 24.4, is modified for a parallel plate capacitor with a dielectric to

$$E = \frac{E_{\text{air}}}{\kappa} = \frac{q}{\kappa \epsilon_0 A} = \frac{q}{\epsilon A}. \tag{24.18}$$

The constant ϵ_0 is the electric permittivity of free space, previously encountered in Coulomb's Law. The right-hand side of equation 24.18 was obtained by replacing the factor $\kappa \epsilon_0$

Table 24.1	Dielectric Constants and Dielectric Strengths for Some Representative Materials	
Material	**Dielectric Constant, κ**	**Dielectric Strength (kV/mm)**
Vacuum	1	
Air (1 atm)	1.00059	2.5
Liquid nitrogen	1.454	
Teflon	2.1	60
Polyethylene	2.25	50
Benzene	2.28	
Polystyrene	2.6	24
Lexan	2.96	16
Mica	3–6	150–220
Paper	3	16
Mylar	3.1	280
Plexiglas	3.4	30
Polyvinyl chloride (PVC)	3.4	29
Glass	5	14
Neoprene	16	12
Germanium	16	
Glycerin	42.5	
Water	80.4	65
Strontium titanate	310	8

Note that these values are approximate and are for room temperature.

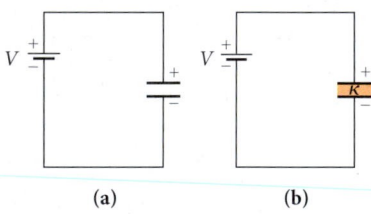

FIGURE 24.23 Parallel plate capacitor connected to a battery: (a) with no dielectric; (b) with a dielectric inserted between the plates.

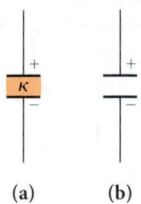

FIGURE 24.24 Isolated capacitor: (a) with dielectric and (b) with dielectric removed.

with ϵ, the **electric permittivity** of the dielectric. In other words, the electric permittivity of a dielectric is the product of the electric permittivity of free space (the vacuum) and the dielectric constant of the dielectric:

$$\epsilon = \kappa\epsilon_0. \tag{24.19}$$

Note that this replacement of ϵ_0 by ϵ is all that is needed to generalize expressions for the capacitance, such as equations 24.6, 24.8, and 24.10, from the values applicable for a capacitor with vacuum between its plates to those appropriate when the capacitor is completely filled with a dielectric. We can now see how the capacitance is increased by adding a dielectric between the plates. The potential difference across a parallel plate capacitor is

$$\Delta V = Ed = \frac{qd}{\kappa\epsilon_0 A}.$$

Therefore, we can write the capacitance as

$$C = \frac{q}{\Delta V} = \frac{\kappa\epsilon_0 A}{d} = \kappa C_{air}.$$

The **dielectric strength** of a material is a measure of its ability to withstand potential difference. If the electric field strength in the dielectric exceeds the dielectric strength, the dielectric breaks down and begins to conduct charge between the plates via a spark, which usually destroys the capacitor. Thus, a useful capacitor must contain a dielectric that not only provides a given capacitance but also enables the device to hold the required potential difference without breaking down. Capacitors are normally specified by the value of their capacitance and by the maximum potential difference that they are designed to handle.

The dielectric constant of vacuum is defined to be 1, and the dielectric constant of air is close to 1.0. The dielectric constants and dielectric strengths of air and of other common materials used as dielectrics are listed in Table 24.1.

EXAMPLE 24.5 | Parallel Plate Capacitor with a Dielectric

PROBLEM 1
Consider a parallel plate capacitor without a dielectric and with capacitance $C = 2.00$ μF connected to a battery with potential difference $\Delta V = 12.0$ V (Figure 24.23a). What is the charge stored in the capacitor?

SOLUTION 1
Using the definition of capacitance (equation 24.1), we have

$$q = C\Delta V = \left(2.00 \cdot 10^{-6} \text{ F}\right)\left(12.0 \text{ V}\right) = 2.40 \cdot 10^{-5} \text{ C}.$$

PROBLEM 2
In Figure 24.23b, a dielectric with $\kappa = 2.50$ has been inserted between the plates of the capacitor, completely filling the space between them. Now what is the charge on the capacitor?

SOLUTION 2
The capacitance of the capacitor is increased by the dielectric:

$$C = \kappa C_{air}.$$

The charge is

$$q = \kappa C_{air}\Delta V = \left(2.50\right)\left(2.00 \cdot 10^{-6} \text{ F}\right)\left(12.0 \text{ V}\right) = 6.00 \cdot 10^{-5} \text{ C}.$$

The charge on the capacitor increases when the capacitance increases because the battery maintains a constant potential difference across the capacitor. The battery provides the additional charge until the capacitor is fully charged.

PROBLEM 3
Now suppose the capacitor is disconnected from the battery (Figure 24.24a). The capacitor, which is now isolated, maintains its charge of $q = 6.00 \cdot 10^{-5}$ C and its potential difference of

$\Delta V = 12.0$ V. What happens to the charge and potential difference if the dielectric is removed, keeping the capacitor isolated (Figure 24.24b)?

SOLUTION 3

The charge on the isolated capacitor cannot change when the dielectric is removed because there is nowhere for the charge to flow. Thus, the potential difference on the capacitor is

$$\Delta V = \frac{q}{C} = \frac{6.00 \cdot 10^{-5} \text{ C}}{2.00 \cdot 10^{-6} \text{ F}} = 30.0 \text{ V}.$$

The potential difference increases because removing the dielectric increases the electric field and the resulting potential difference between the plates.

PROBLEM 4

Does removing the dielectric change the energy stored in the capacitor?

SOLUTION 4

The energy stored in a capacitor is given by equation 24.15. Before the dielectric was removed, the energy in the capacitor was

$$U = \tfrac{1}{2}C(\Delta V)^2 = \tfrac{1}{2}\kappa C_{air}(\Delta V)^2 = \tfrac{1}{2}(2.50)(2.00 \cdot 10^{-6} \text{ F})(12.0 \text{ V})^2 = 3.60 \cdot 10^{-4} \text{ J}.$$

After the dielectric is removed, the energy is

$$U = \tfrac{1}{2}C_{air}(\Delta V)^2 = \tfrac{1}{2}(2.00 \cdot 10^{-6} \text{ F})(30.0 \text{ V})^2 = 9.00 \cdot 10^{-4} \text{ J}.$$

The increase in energy from $3.60 \cdot 10^{-4}$ J to $9.00 \cdot 10^{-4}$ J when the dielectric is removed is due to the work done on the dielectric in pulling it out of the electric field between the plates.

Concept Check 24.12

What would happen if the dielectric in the capacitor of Example 24.5 were pulled halfway out and then released?

a) The dielectric would be pulled back into the capacitor.

b) The dielectric would heat up rapidly.

c) The dielectric would be pushed out of the capacitor.

d) The capacitor plates would heat up rapidly.

e) The dielectric would remain in the halfway position, and no heating would be observed.

SOLVED PROBLEM 24.2 | Capacitor Partially Filled with a Dielectric

PROBLEM

A parallel plate capacitor is constructed of two square conducting plates with side length $L = 10.0$ cm (Figure 24.25a). The distance between the plates is $d = 0.250$ cm. A dielectric with dielectric constant $\kappa = 15.0$ and thickness 0.250 cm is inserted between the plates. The dielectric is $L = 10.0$ cm wide and $L/2 = 5.00$ cm long, as shown in Figure 24.25a. What is the capacitance of this capacitor?

SOLUTION

THINK We have a parallel plate capacitor that is partially filled with a dielectric. We can treat this capacitor as two capacitors in parallel. One capacitor is a parallel plate capacitor with plate area $A = L(L/2)$ and air between the plates; the second capacitor is a parallel plate capacitor with plate area $A = L(L/2)$ and a dielectric between the plates.

SKETCH Figure 24.25b shows a representation of the partially filled capacitor as two capacitors in parallel: one filled with a dielectric and the other filled with air.

RESEARCH The capacitance, C_1, of a parallel plate capacitor is given by equation 24.6:

$$C_1 = \frac{\epsilon_0 A}{d},$$

where A is the area of the plates and d is the separation between them. If a dielectric is placed between the plates, the capacitance becomes

$$C_2 = \kappa \frac{\epsilon_0 A}{d},$$

where κ is the dielectric constant. For two capacitors, C_1 and C_2, in parallel, the effective capacitance, C_{12}, is given by

$$C_{12} = C_1 + C_2.$$

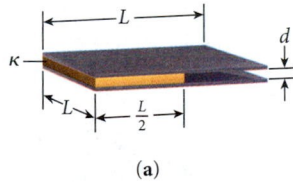

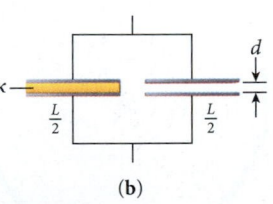

FIGURE 24.25 (a) A parallel plate capacitor with square plates of side length L separated by distance d with a dielectric that is L wide and $L/2$ long and has a dielectric constant κ inserted between the plates. (b) The partially filled capacitor represented as two capacitors in parallel.

– Continued

SIMPLIFY Substituting the expressions for the two individual capacitances into the sum, we get

$$C_{12} = \frac{\epsilon_0 A}{d} + \kappa \frac{\epsilon_0 A}{d} = (\kappa + 1)\frac{\epsilon_0 A}{d}. \tag{i}$$

The area of the plates for each capacitor is

$$A = (L)(L/2) = L^2/2.$$

Inserting the expression for the area into equation (i) gives the capacitance of the partially filled capacitor:

$$C_{12} = (\kappa + 1)\frac{\epsilon_0 (L^2/2)}{d} = \frac{(\kappa + 1)\epsilon_0 L^2}{2d}.$$

CALCULATE Putting in the numerical values, we get

$$C_{12} = \frac{(15.0 + 1)(8.85 \cdot 10^{-12} \text{ F/m})(0.100 \text{ m})^2}{2(0.00250 \text{ m})} = 2.832 \cdot 10^{-10} \text{ F}.$$

ROUND We report our result to three significant figures:

$$C_{12} = 2.83 \cdot 10^{-10} \text{ F} = 283 \text{ pF}.$$

DOUBLE-CHECK To double-check our answer, we calculate the capacitance of the capacitor without any dielectric:

$$C = \frac{\epsilon_0 A}{d} = \frac{(8.85 \cdot 10^{-12} \text{ F/m})(0.100 \text{ m})^2}{0.0025 \text{ m}} = 3.54 \cdot 10^{-11} \text{ F} = 35.4 \text{ pF}.$$

We then calculate the capacitance of the capacitor if completely filled with dielectric:

$$C = \kappa \frac{\epsilon_0 A}{d} = (15.0)(35.4 \text{ pF}) = 5.31 \cdot 10^{-10} \text{ F} = 531 \text{ pF}.$$

Our result for the partially filled capacitor is half of the sum of these two results, so it seems reasonable.

EXAMPLE 24.6 Capacitance of a Coaxial Cable

Coaxial cables are used to transport signals, for example, TV signals, between devices with minimum interference from the surroundings. A 20.0-m-long coaxial cable is composed of a conductor and a coaxial conducting shield around the conductor. The space between the conductor and the shield is filled with polystyrene. The radius of the conductor is 0.250 mm, and the radius of the shield is 2.00 mm (Figure 24.26).

PROBLEM
What is the capacitance of the coaxial cable?

SOLUTION
We can think of the conductor of the coaxial cable as a cylinder because all the charge on the conductor resides on its surface. From Table 24.1, the dielectric constant for polystyrene is 2.6. We can treat the coaxial cable as a cylindrical capacitor with $r_1 = 0.250$ mm and $r_2 = 2.00$ mm, filled with a dielectric with $\kappa = 2.6$. Then, we can use equation 24.8 to find the capacitance of the coaxial cable:

$$C = \kappa \frac{2\pi\epsilon_0 L}{\ln(r_2/r_1)} = \frac{2.6(2\pi)(8.85 \cdot 10^{-12} \text{ F/m})(20.0 \text{ m})}{\ln\left[(2.00 \cdot 10^{-3} \text{ m})/(2.50 \cdot 10^{-4} \text{ m})\right]} = 1.4 \cdot 10^{-9} \text{ F} = 1.4 \text{ nF}.$$

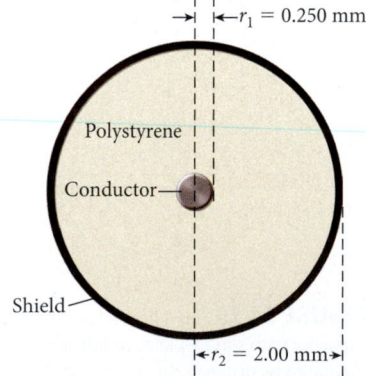

FIGURE 24.26 Cross section of a coaxial cable.

$r_1 = 0.250$ mm

Polystyrene

Conductor

Shield

$r_2 = 2.00$ mm

One interesting application of capacitance and the dielectric constant is in measuring liquid nitrogen levels in cryostats (containers insulated to maintain cold temperatures). It is often difficult to conduct a visual exam to determine how much liquid nitrogen is left in a cryostat. However, if one determines the capacitance, C, of the empty cryostat, then, when completely filled with liquid nitrogen, the cryostat should have a capacitance of $\kappa C = 1.454C$, because liquid nitrogen has a dielectric constant of 1.454. The capacitance varies smoothly as a function of the fullness between the maximum value $\kappa C = 1.454C$ for the completely full cryostat and the value C for the empty cryostat, giving an easy way to determine how full the cryostat is.

SOLVED PROBLEM 24.3 | Charge on a Cylindrical Capacitor

PROBLEM
Consider a cylindrical capacitor with inner radius r_1 = 10.0 cm, outer radius r_2 = 12.0 cm, and length L = 50.0 cm (Figure 24.27a). A dielectric with dielectric constant κ =12.5 fills the volume between the two cylinders (Figure 24.27b). The capacitor is connected to a 100.0-V battery and charged completely. What is the charge on the capacitor?

SOLUTION
THINK We have a cylindrical capacitor filled with a dielectric. When the capacitor is connected to the battery, charge will accumulate on the capacitor until the capacitor is fully charged. We can calculate the amount of charge on the capacitor.

SKETCH A circuit diagram with the cylindrical capacitor connected to a battery is shown in Figure 24.28.

RESEARCH The capacitance, C, of a cylindrical capacitor is given by equation 24.8:

$$C = \frac{2\pi\epsilon_0 L}{\ln(r_2/r_1)},$$

where r_1 is the inner radius of the capacitor, r_2 is the outer radius of the capacitor, and L is the length of the capacitor. With a dielectric between the plates, the capacitance becomes

$$C = \kappa \frac{2\pi\epsilon_0 L}{\ln(r_2/r_1)}, \tag{i}$$

where κ is the dielectric constant. For a capacitor with capacitance C charged to a potential difference ΔV, the charge q is given by equation 24.1:

$$q = C\Delta V. \tag{ii}$$

SIMPLIFY Combining equations (i) and (ii) gives

$$q = C\Delta V = \left(\kappa \frac{2\pi\epsilon_0 L}{\ln(r_2/r_1)}\right)\Delta V = \frac{2\kappa\pi\epsilon_0 L\Delta V}{\ln(r_2/r_1)}.$$

CALCULATE Putting in the numerical values, we get

$$q = \frac{2\kappa\pi\epsilon_0 L\Delta V}{\ln(r_2/r_1)} = \frac{2(12.5)\pi(8.85\cdot10^{-12}\ \text{F/m})(0.500\ \text{m})(100.0\ \text{V})}{\ln[(0.120\ \text{m})/(0.100\ \text{m})]} = 19.0618\cdot10^{-8}\ \text{C}.$$

ROUND We report our result to three significant figures:

$$q = 19.1\cdot10^{-8}\ \text{C} = 191\ \text{nC}.$$

DOUBLE-CHECK Our answer is a very small fraction of a coulomb of charge, so it seems reasonable.

24.7 Microscopic Perspective on Dielectrics

Let's consider what happens at the atomic and molecular level when a dielectric is placed in an electric field. There are two types of dielectric materials: polar dielectrics and nonpolar dielectrics.

A **polar dielectric** is a material composed of molecules that have a permanent electric dipole moment due to their structure. A common example of such a molecule is

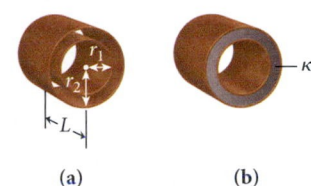

FIGURE 24.27 (a) A cylindrical capacitor with inner radius r_1, outer radius r_2, and length L. (b) A dielectric with dielectric constant κ is inserted between the cylinders.

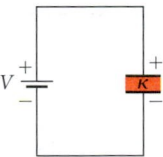

FIGURE 24.28 A cylindrical capacitor connected to a battery.

FIGURE 24.29 Polar molecules:
(a) randomly distributed and (b) oriented
by an external electric field.

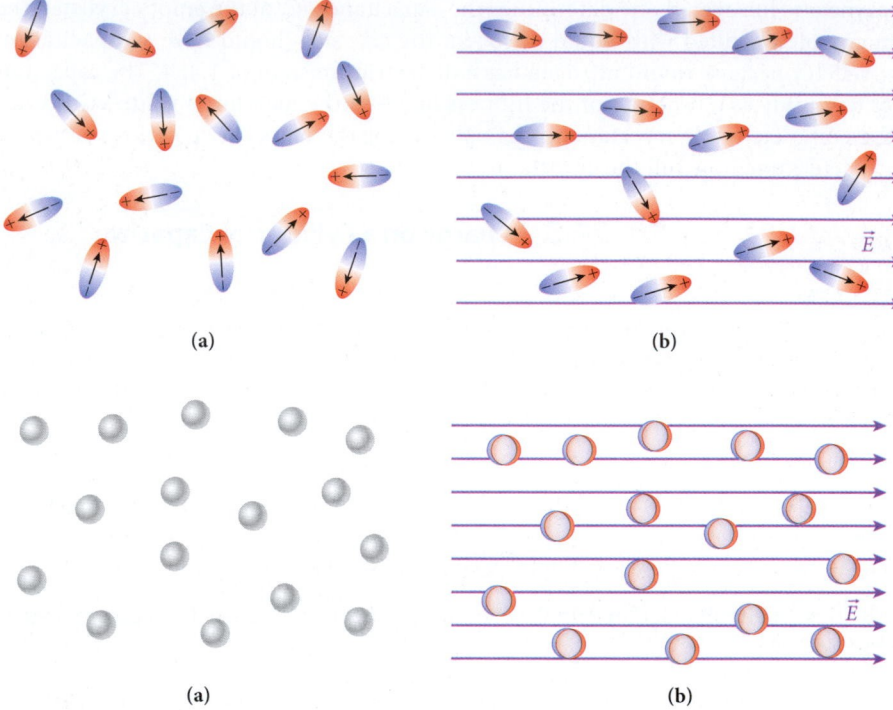

(a)   (b)

FIGURE 24.30 Nonpolar molecules:
(a) with no electric dipole moment and
(b) with an electric dipole moment
induced by an external electric field.

(a)   (b)

water. Normally, the directions of electric dipoles are randomly distributed (Figure 24.29a). However, when an electric field is applied to these polar molecules, they tend to align with the field (Figure 24.29b).

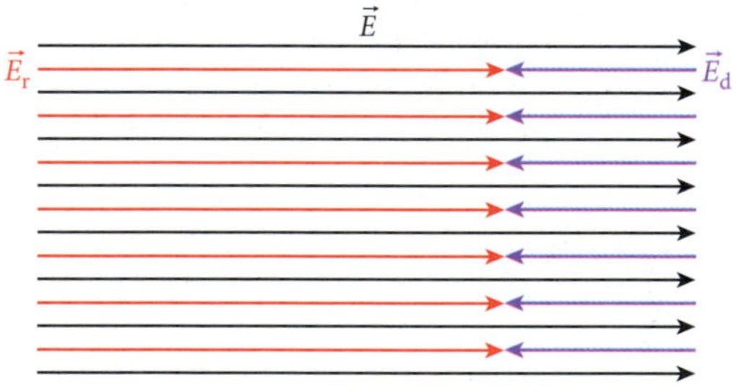

A **nonpolar dielectric** is a material composed of atoms or molecules that have no inherent electric dipole moment (Figure 24.30a). These atoms or molecules can be induced to have a dipole moment under the influence of an external electric field (Figure 24.30b). The opposite directions of the electric forces acting on the negative and positive charges in the atom or molecule displace these two charge distributions and produce an induced electric dipole moment.

In both polar and nonpolar dielectrics, the fields resulting from the aligned electric dipole moments tend to partially cancel the original external electric field (Figure 24.31). For an electric field, $\vec{E}$, applied across a capacitor with a dielectric between the plates, the resulting electric field, $\vec{E}_r$, inside the capacitor is just the original field plus the electric field induced in the dielectric material, $\vec{E}_d$:

$$\vec{E}_r = \vec{E} + \vec{E}_d,$$

or

$$E_r = E - E_d.$$

FIGURE 24.31 Partial cancellation of the applied electric field across a parallel plate capacitor by the electric dipoles of a dielectric.

Note that the resulting electric field points in the same direction as the original field but is smaller in magnitude. The dielectric constant is given by $\kappa = E/E_r$.

Electrolytic Capacitors

Instead of consisting of two conducting plates with the gap filled with a dielectric, a capacitor can have one of the plates replaced by an ion-conducting liquid, or electrolyte, that is, a liquid that contains ions that can move freely through it. Most commonly, these *electrolytic capacitors* are constructed of two pieces of aluminum foil, of which one is coated with an insulating oxide layer. The two foil pieces are separated by a paper spacer, which is saturated with the electrolyte. The oxide layer typically has a dielectric constant of ~10 and a dielectric strength of 20–30 kV/mm. This layer can therefore be very thin, and this type of electrolytic capacitor has a relatively high charge capacity.

The main drawback of an electrolytic capacitor is that it is polarized and one electrode must always be maintained at a positive potential relative to the other. A reverse potential difference as low as 1–2 V will destroy the oxide layer and lead to short-circuiting and destruction of the capacitor.

Supercapacitors

As we've seen in this chapter, 1 F is a huge amount of capacitance. Even the National Ignition Facility (NIF), which needs the highest possible energy storage, uses only 300-μF capacitors. However, it is possible to create *supercapacitors* (also called *ultracapacitors*) with vastly greater capacitance. This is accomplished by using a material with a very large surface area between the capacitor plates. Activated charcoal is one possibility, as it has a very large surface area because of its foamlike structure at a nanoscale level. Two layers of activated charcoal are given charges of opposite polarity and are separated by an insulating material (represented by the red line in Figure 24.32b). This allows each side of the supercapacitor to store oppositely charged free ions from the electrolyte. The separation between the electrolyte ions and the charges on the activated char-

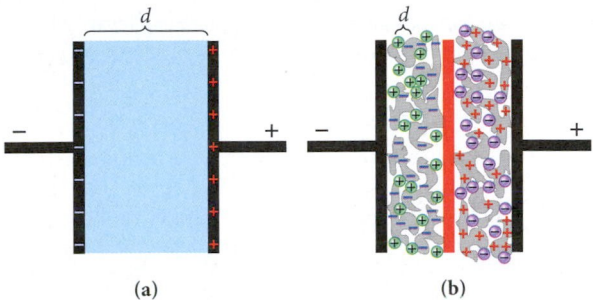

FIGURE 24.32 Comparison of (a) a conventional parallel plate capacitor and (b) a supercapacitor filled with activated charcoal.

coal is typically on the order of nanometers (nm), that is, millions of times smaller than in conventional capacitors. The activated charcoal provides a surface area many orders of magnitude larger than in conventional capacitors. Since, as noted in Section 24.3, the capacitance is proportional to the surface area and inversely proportional to the plate separation, this technology has resulted in commercially available capacitors with capacitances on the order of kilofarads (kF), that is, millions of times larger than those used in the NIF.

Why is the NIF not using supercapacitors? The answer is that these supercapacitors can only function with potential differences of up to 2–3 V. The highest-capacity commercially available supercapacitors have capacitance values of up to 5 kF. Using $U = \frac{1}{2}C(\Delta V)^2$ and $\Delta V = 2$ V shows that a supercapacitor can hold 10 kJ. The 300-μF capacitors used at NIF, when charged to 24 kV, can hold 86.4 kJ. In addition, they can also be discharged much more rapidly, which is crucial for fulfilling the high power requirement of the NIF's laser.

However, supercapacitors can reach energy storage capabilities that rival those of conventional batteries. In addition, supercapacitors can be charged and discharged millions of times, compared to perhaps thousands of times for rechargeable batteries. This, along with their very short charging time, makes them potentially suitable for many applications. For example, there is intense research into using these supercapacitors for electric vehicles. A bus based on this energy storage technology, named *capabus*, is currently in use in Shanghai, China (see Figure 24.33).

A promising line of research on improving the potential difference that supercapacitors can employ is investigating the use of carbon nanotubes and graphene instead of activated charcoal. The first laboratory prototypes are very promising, and commercial products based on this approach could be in use within a few years. Research has also succeeded in improving the energy storage capabilities of supercapacitors while lowering the prices. A 5-kF supercapacitor that cost about $5,000 in 2000 can now be purchased for 1% of that price.

FIGURE 24.33 Supercapacitor-powered bus recharging at a bus stop in Shanghai, China.

WHAT WE HAVE LEARNED | EXAM STUDY GUIDE

- The capacitance of a capacitor—its ability to store charge—is defined in terms of the charge, q, that can be stored on the capacitor and the potential difference, V, across the plates: $q = C\Delta V$.

- The farad is the unit of capacitance: $1\text{ F} = \dfrac{1\text{ C}}{1\text{ V}}$.

- The capacitance of a parallel plate capacitor with plates of area A with a vacuum (or air) between the plates separated by distance d is given by $C = \dfrac{\epsilon_0 A}{d}$.

- The capacitance of a cylindrical capacitor of length L consisting of two collinear cylinders with a vacuum (or air) between the cylinders with inner radius r_1 and outer radius r_2 is given by $C = \dfrac{2\pi\epsilon_0 L}{\ln(r_2/r_1)}$.

- The capacitance of a spherical capacitor consisting of two concentric spheres with vacuum (or air) between the spheres with inner radius r_1 and outer radius r_2 is given by $C = 4\pi\epsilon_0 \dfrac{r_1 r_2}{r_2 - r_1}$.

- The electric energy density, u, between the plates of a parallel plate capacitor with vacuum (or air) between the plates is given by $u = \frac{1}{2}\epsilon_0 E^2$.

- A system of n capacitors connected in parallel in a circuit can be replaced by an equivalent capacitance given by the sum of the capacitances of the capacitors: $C_{eq} = \displaystyle\sum_{i=1}^{n} C_i$.

- A system of n capacitors connected in series in a circuit can be replaced by an equivalent capacitance given by the reciprocal of the sum of the reciprocal capacitances of the capacitors: $\dfrac{1}{C_{eq}} = \displaystyle\sum_{i=1}^{n} \dfrac{1}{C_i}$.

- When the space between the plates of a capacitor is filled with a dielectric whose dielectric constant is κ, the capacitance increases relative to the capacitance in air: $C = \kappa C_{air}$.

- The electric potential energy stored in a capacitor is $U = \frac{1}{2}q^2/C = \frac{1}{2}C(\Delta V)^2 = \frac{1}{2}q\Delta V$.

ANSWERS TO SELF-TEST OPPORTUNITIES

24.1 The electric field remains constant.

24.2 Series: $\dfrac{1}{C_{eq}} = \dfrac{4}{C} \Rightarrow C_{eq} = \frac{1}{4}C = 2.50\ \mu\text{F}.$

Parallel: $C_{eq} = 4C = 40.0\ \mu\text{F}.$

24.3 $100\text{ V} = d(2500\text{ V/mm}) \Rightarrow d = 0.04\text{ mm}.$

PROBLEM-SOLVING GUIDELINES

1. Remember that saying that a capacitor has charge q means that one plate has charge $+q$ and the other plate has charge $-q$. Be sure you understand how a charge applied to a capacitor is distributed between the two conducting plates; review Example 24.3 if you're uncertain about this.

2. It is always a good idea to draw a circuit diagram when solving a problem involving a circuit, if one is not supplied. Identifying series and parallel connections can take some practice, but is usually an important first step in reducing a complicated-looking circuit to an equivalent circuit that is straightforward to deal with. Remember that capacitors connected in series all have the same charge, and capacitors connected in parallel all have the same potential difference.

3. You can remember most of the important results for a capacitor with a dielectric if you remember that a dielectric increases the capacitance. (This is what makes a dielectric useful.) If your calculations show a reduced capacitance with a dielectric, recheck your work.

4. You can calculate the energy stored in a capacitor if you know two out of these three quantities: the charge on a plate, the capacitance of the capacitor, and the potential difference between the plates. Make sure you take advantage of equation 24.15 in the appropriate form.

MULTIPLE-CHOICE QUESTIONS

24.1 In the circuit shown in the figure, the capacitance for each capacitor is C. The equivalent capacitance for these three capacitors is

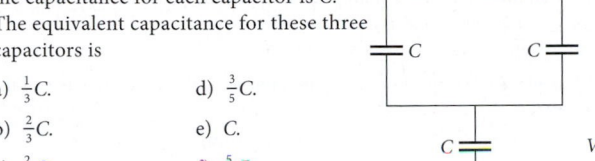

a) $\frac{1}{3}C$.

b) $\frac{2}{3}C$.

c) $\frac{2}{5}C$.

d) $\frac{3}{5}C$.

e) C.

f) $\frac{5}{3}C$.

24.2 A parallel plate capacitor of capacitance C has plates of area A with distance d between them. When the capacitor is connected to a battery supplying potential difference V, it has a charge of magnitude Q on its plates. While the capacitor is connected to the battery, the distance between the plates is decreased by a factor of 3. The magnitude of the charge on the plates and the capacitance will then be

a) $\frac{1}{3}Q$ and $\frac{1}{3}C$.

b) $\frac{1}{3}Q$ and $3C$.

c) $3Q$ and $3C$.

d) $3Q$ and $\frac{1}{3}C$.

24.3 The distance between the plates of a parallel plate capacitor is reduced by half and the area of the plates is doubled. What happens to the capacitance?

a) It remains unchanged.

b) It doubles.

c) It quadruples.

d) It is reduced by half.

24.4 Which of the following capacitors has the largest charge?

a) a parallel plate capacitor with an area of 10 cm^2 and a plate separation of 2 mm connected to a 10-V battery

b) a parallel plate capacitor with an area of 5 cm^2 and a plate separation of 1 mm connected to a 10-V battery

c) a parallel plate capacitor with an area of 10 cm^2 and a plate separation of 4 mm connected to a 5-V battery

d) a parallel plate capacitor with an area of 20 cm^2 and a plate separation of 2 mm connected to a 20-V battery

e) All of the capacitors have the same charge.

24.5 Two identical parallel plate capacitors are connected in a circuit as shown in the figure. Initially the space between the plates of each capacitor is filled with air. Which of the following changes will double the total amount of charge stored on both capacitors with the same applied potential difference?

a) Fill the space between the plates of C_1 with glass (dielectric constant of 4) and leave C_2 as is.

b) Fill the space between the plates of C_1 with Teflon (dielectric constant of 2) and leave C_2 as is.

c) Fill the space between the plates of both C_1 and C_2 with Teflon (dielectric constant of 2).

d) Fill the space between the plates of both C_1 and C_2 with glass (dielectric constant of 4).

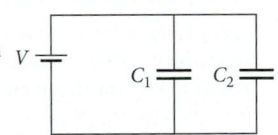

24.6 The space between the plates of an isolated parallel plate capacitor is filled with a slab of dielectric material. The magnitude of the charge Q on each plate is kept constant. If the dielectric material is removed the energy stored in the capacitor

a) increases.

b) stays the same.

c) decreases.

d) may increase or decrease.

24.7 Which of the following is (are) proportional to the capacitance of a parallel plate capacitor?

a) the charge stored on each conducting plate

b) the potential difference between the two plates

c) the separation distance between the two plates

d) the area of each plate

e) all of the above

f) none of the above

24.8 A dielectric with the dielectric constant $\kappa = 4$ is inserted into a parallel plate capacitor, filling $\frac{1}{3}$ of the volume, as shown in the figure. If the capacitance of the capacitor without the dielectric is C, what is the capacitance of the capacitor with the dielectric?

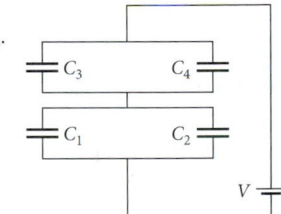

a) 0.75C

b) C

c) 2C

d) 4C

e) 6C

24.9 A parallel plate capacitor is connected to a battery for charging. After some time, while the battery is still connected to the capacitor, the distance between the capacitor plates is doubled. Which of the following is (are) true?

a) The electric field between the plates is halved.

b) The potential difference of the battery is halved.

c) The capacitance doubles.

d) The potential difference across the plates does not change.

e) The charge on the plates does not change.

24.10 Referring to the figure, decide which of the following equations is (are) true. Assume that all of the capacitors have different capacitances. The potential difference across capacitor C_1 is V_1. The potential difference across capacitor C_2 is V_2. The potential difference across capacitor C_3 is V_3. The potential difference across capacitor C_4 is V_4. The charge stored in capacitor C_1 is q_1. The charge stored in capacitor C_2 is q_2. The charge stored in capacitor C_3 is q_3. The charge stored in capacitor C_4 is q_4.

a) $q_1 = q_3$

b) $V_1 + V_2 = V$

c) $q_1 + q_2 = q_3 + q_4$

d) $V_1 + V_2 = V_3 + V_4$

e) $V_1 + V_3 = V$

24.11 You have N identical capacitors, each with capacitance C, connected in series. The equivalent capacitance of this system of capacitors is

a) NC.

b) C/N.

c) N^2C.

d) C/N^2.

e) C.

24.12 You have N identical capacitors, each with capacitance C, connected in parallel. The equivalent capacitance of this system of capacitors is

a) NC.

b) C/N.

c) N^2C.

d) C/N^2.

e) C.

24.13 When a dielectric is placed between the plates of a charged, isolated capacitor, the electric field inside the capacitor

a) increases.

b) decreases.

c) stays the same.

d) increases if the charge on the plates is positive.

e) decreases if the charge on the plates is positive.

24.14 A parallel plate capacitor with a dielectric filling the volume between its plates is charged. The charge is

a) stored on the plates.

b) stored on the dielectric.

c) stored both on the plates and in the dielectric.

CONCEPTUAL QUESTIONS

24.15 Must a capacitor's plates be made of conducting material? What would happen if two insulating plates were used instead of conducting plates?

24.16 Does it take more work to separate the plates of a charged parallel plate capacitor while it remains connected to the charging battery or after it has been disconnected from the charging battery?

24.17 When working on a piece of equipment, electricians and electronics technicians sometimes attach a grounding wire to the equipment even after turning the device off and unplugging it. Why would they do this?

24.18 Table 24.1 does not list a value of the dielectric constant for any good conductor. What value would you assign to it?

24.19 A parallel plate capacitor is charged with a battery and then disconnected from the battery, leaving a certain amount of energy stored in the capacitor. The separation between the plates is then increased. What happens to the energy stored in the capacitor? Discuss your answer in terms of energy conservation.

24.20 You have an electric device containing a 10.0-μF capacitor, but an application requires an 18.0-μF capacitor. What modification can you make to your device to increase its capacitance to 18.0-μF?

24.21 Two capacitors with capacitances C_1 and C_2 are connected in series. Show that, no matter what the values of C_1 and C_2 are, the equivalent capacitance is always less than the smaller of the two capacitances.

24.22 Two capacitors, with capacitances C_1 and C_2, are connected in series. A potential difference, V_0, is applied across the combination of capacitors. Find the potential differences V_1 and V_2 across the individual capacitors, in terms of V_0, C_1, and C_2.

24.23 An isolated solid spherical conductor of radius 5.00 cm is surrounded by dry air. It is given a charge and acquires potential V, with the potential at infinity assumed to be zero.

a) Calculate the maximum magnitude V can have.

b) Explain clearly and concisely *why* there is a maximum.

24.24 A parallel plate capacitor with capacitance C is connected to a power supply that maintains a constant potential difference, V. A slab of dielectric, with dielectric constant κ, is then inserted into and completely fills the previously empty space between the plates.

a) What was the energy stored on the capacitor before the insertion of the dielectric?

b) What was the energy stored after the insertion of the dielectric?

c) Was the dielectric pulled into the space between the plates, or did it have to be pushed in? Explain.

24.25 A parallel plate capacitor with square plates of edge length L separated by a distance d is given a charge Q, then disconnected from its power source. A close-fitting square slab of dielectric, with dielectric constant κ, is then inserted into the previously empty space between the plates. Calculate the force with which the slab is pulled into the capacitor during the insertion process.

24.26 A cylindrical capacitor has an outer radius R and a separation d between the cylinders. Determine what the capacitance approaches in the limit where $d \ll R$. (*Hint:* Express the capacitance in terms of the ratio d/R and then examine what happens as that ratio becomes very small compared to 1.) Explain why the limit on the capacitance makes sense.

24.27 A parallel plate capacitor is constructed from two plates of different areas. If this capacitor is initially uncharged and then connected to a battery, how will the amount of charge on the big plate compare to the amount of charge on the small plate?

24.28 A parallel plate capacitor is connected to a battery. As the plates are moved farther apart, what happens to each of the following?

a) the potential difference across the plates

b) the charge on the plates

c) the electric field between the plates

EXERCISES

A blue problem number indicates a worked-out solution is available in the Student Solutions Manual. One • and two •• indicate increasing level of problem difficulty.

Section 24.3

24.29 Supercapacitors, with capacitances of 1.00 F or more, are made with plates that have a spongelike structure with a very large surface area. Determine the surface area of a supercapacitor that has a capacitance of 1.00 F and an effective separation between the plates of $d = 1.00$ mm.

24.30 A potential difference of 100. V is applied across the two collinear conducting cylinders shown in the figure. The radius of the outer cylinder is 15.0 cm, the radius of the inner cylinder is 10.0 cm, and the length of the two cylinders is 40.0 cm. How much charge is applied to each of the cylinders? What is the magnitude of the electric field between the two cylinders?

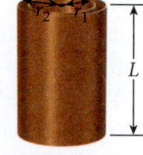

24.31 What is the radius of an isolated spherical conductor that has a capacitance of 1.00 F?

24.32 A spherical capacitor is made from two thin concentric conducting shells. The inner shell has radius r_1, and the outer shell has radius r_2. What is the fractional difference in the capacitances of this spherical capacitor and a parallel plate capacitor made from plates that have the same area as the inner sphere and the same separation $d = r_2 - r_1$ between them?

24.33 Calculate the capacitance of the Earth. Treat the Earth as an isolated spherical conductor of radius 6371 km.

24.34 Two concentric metal spheres are found to have a potential difference of 900. V when a charge of $6.726 \cdot 10^{-8}$ C is applied to them. The radius of the outer sphere is 0.210 m. What is the radius of the inner sphere?

•24.35 A capacitor consists of two parallel plates, but one of them can move relative to the other as shown in the figure. Air fills the space between the plates, and the capacitance is 32.0 pF when the separation between plates is $d = 0.500$ cm.

a) A battery supplying a potential difference $V = 9.00$ V is connected to the plates. What is the charge distribution, σ, on the left plate? What are the capacitance, C', and the charge distribution, σ', when d is changed to 0.250 cm?

b) With $d = 0.500$ cm, the battery is disconnected from the plates. The plates are then moved so that $d = 0.250$ cm. What is the potential difference V', between the plates?

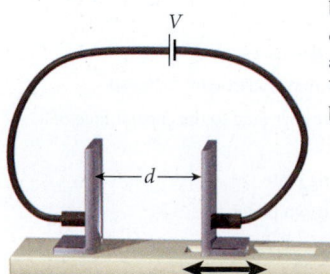

Section 24.4

24.36 Determine all the values of equivalent capacitance you can create using any combination of three identical capacitors with capacitance C.

24.37 A large parallel plate capacitor with plates that are square with side length 1.00 cm and are separated by a distance of 1.00 mm is dropped and damaged. Half of the areas of the two plates are pushed closer together to a distance of 0.500 mm. What is the capacitance of the damaged capacitor?

24.38 Three capacitors with capacitances $C_1 = 3.10$ nF, $C_2 = 1.30$ nF, and $C_3 = 3.70$ nF are wired to a battery with $V = 14.9$ V, as shown in the figure. What is the potential drop across capacitor C_2?

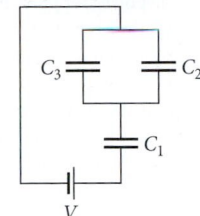

24.39 Four capacitors with capacitances $C_1 = 3.50$ nF, $C_2 = 2.10$ nF, $C_3 = 1.30$ nF, and $C_4 = 4.90$ nF are wired to a battery with $V = 10.3$ V, as shown in the figure. What is the equivalent capacitance of this set of capacitors?

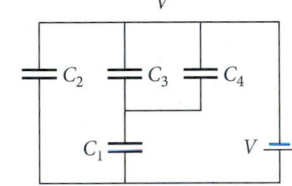

24.40 The capacitors in the circuit shown in the figure have capacitances $C_1 = 18.0$ μF, $C_2 = 11.3$ μF, $C_3 = 33.0$ μF, and $C_4 = 44.0$ μF. The potential difference is $V = 10.0$ V. What is the total charge the power source must supply to charge this arrangement of capacitors?

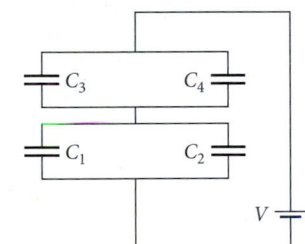

24.41 Six capacitors are connected as shown in the figure.

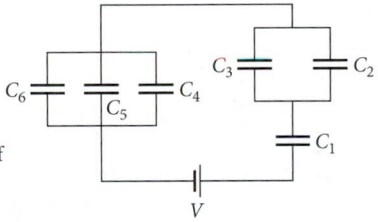

a) If $C_3 = 2.300$ nF, what does C_2 have to be to yield an equivalent capacitance of 5.000 nF for the combination of the two capacitors?

b) For the same values of C_2 and C_3 as in part (a), what is the value of C_1 that will give an equivalent capacitance of 1.914 nF for the combination of the three capacitors?

c) For the same values of C_1, C_2, and C_3 as in part (b), what is the equivalent capacitance of the whole set of capacitors if the values of the other capacitances are $C_4 = 1.300$ nF, $C_5 = 1.700$ nF, and $C_6 = 4.700$ nF?

d) If a battery with a potential difference of 11.70 V is connected to the capacitors as shown in the figure, what is the total charge on the six capacitors?

e) What is the potential drop across C_5 in this case?

•24.42 A potential difference of $V = 80.0$ V is applied across a circuit with capacitances $C_1 = 15.0$ nF, $C_2 = 7.00$ nF, and $C_3 = 20.0$ nF, as shown in the figure. What is the magnitude and sign of q_{3l}, the charge on the left plate of C_3 (marked by point A)? What is the electric potential, V_3, across C_3? What is the magnitude and sign of the charge q_{2r}, on the right plate of C_2 (marked by point B)?

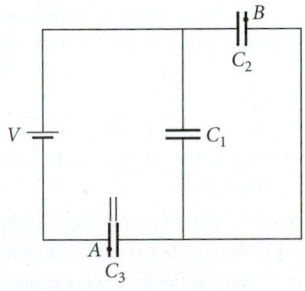

•24.43 Fifty parallel plate capacitors are connected in series. The distance between the plates is d for the first capacitor, $2d$ for the second capacitor, $3d$

for the third capacitor, and so on. The area of the plates is the same for all the capacitors. Express the equivalent capacitance of the whole set in terms of C_1 (the capacitance of the first capacitor).

•24.44 A 5.00-nF capacitor charged to 60.0 V and a 7.00-nF capacitor charged to 40.0 V are connected negative plate to negative plate. What is the final charge on the 7.00-nF capacitor?

Section 24.5

24.45 When a capacitor has a charge of magnitude 60.0 μC on each plate, the potential difference across the plates is 12.0 V. How much energy is stored in this capacitor when the potential difference across its plates is 120. V?

24.46 The capacitor in an automatic external defibrillator is charged to 7.50 kV and stores 2400. J of energy. What is its capacitance?

24.47 The Earth has an electric field of 150. V/m near its surface. Find the electrical energy contained in each cubic meter of air near the surface.

•24.48 The potential difference across two capacitors in series is 120. V. The capacitances are $C_1 = 1.00 \cdot 10^3$ μF and $C_2 = 1.50 \cdot 10^3$ μF.

a) What is the total capacitance of this pair of capacitors?

b) What is the charge on each capacitor?

c) What is the potential difference across each capacitor?

d) What is the total energy stored by the capacitors?

•24.49 Neutron stars are thought to have electric dipole ($\vec{p}$) layers at their surfaces. If a neutron star with a 10.0-km radius has a dipole layer 1.00 cm thick with charge distributions of $+1.00$ μC/cm^2 and -1.00 μC/cm^2 on the surface, as indicated in the figure, what is the capacitance of this star? What is the electric potential energy stored in the neutron star's dipole layer?

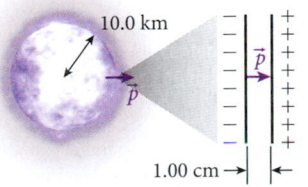

•24.50 A $4.00 \cdot 10^3$-nF parallel plate capacitor is connected to a 12.0-V battery and charged.

a) What is the charge Q on the positive plate of the capacitor?

b) What is the electric potential energy stored in the capacitor?

The $4.00 \cdot 10^3$-nF capacitor is then disconnected from the 12.0-V battery and used to charge three uncharged capacitors, a 100.-nF capacitor, a 200.-nF capacitor, and a 300.-nF capacitor, connected in series.

c) After charging, what is the potential difference across each of the four capacitors?

d) How much of the electrical energy stored in the $4.00 \cdot 10^3$-nF capacitor was transferred to the other three capacitors?

•24.51 The figure shows a circuit with $V = 12.0$ V, $C_1 = 500.$ pF, and $C_2 = 500.$ pF. The switch is closed, to A, and the capacitor C_1 is fully charged. Find (a) the energy delivered by the battery and (b) the energy stored in C_1. Then the switch is thrown to B and the circuit is allowed to reach equilibrium. Find (c) the total energy stored at C_1 and C_2. (d) Explain the energy loss, if there is any.

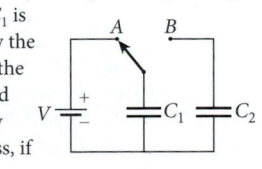

••24.52 The Earth is held together by its own gravity. But it is also a charge-bearing conductor.

a) The Earth can be regarded as a conducting sphere of radius 6371 km, with electric field $\vec{E} = (-150. \text{ V/m})\hat{r}$ at its surface, where $\hat{r}$ is a unit vector directed radially outward. Calculate the total electric potential energy associated with the Earth's electric charge and field.

b) The Earth has gravitational potential energy, akin to the electric potential energy. Calculate this energy, treating the Earth as a uniform solid sphere. (*Hint: $dU = -(Gm/r)dm$.*)

c) Use the results of parts (a) and (b) to address this question: To what extent do electrostatic forces affect the structure of the Earth?

Section 24.6

24.53 Two parallel plate capacitors have identical plate areas and identical plate separations. The maximum energy each can store is determined by the maximum potential difference that can be applied before dielectric breakdown occurs. One capacitor has air between its plates, and the other has Mylar. Find the ratio of the maximum energy the Mylar capacitor can store to the maximum energy the air capacitor can store.

24.54 A capacitor has parallel plates, with half of the space between the plates filled with a dielectric material of constant κ and the other half filled with air as shown in the figure. Assume that the plates are square, with sides of length L, and that the separation between the plates is s. Determine the capacitance as a function of L.

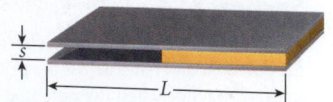

24.55 Calculate the maximum surface charge distribution that can be maintained on any surface surrounded by dry air.

24.56 Thermocoax is a type of coaxial cable used for high-frequency filtering in cryogenic quantum computing experiments. Its stainless steel shield has an inner diameter of 0.350 mm, and its Nichrome conductor has a diameter of 0.170 mm. Nichrome is used because its resistance doesn't change much in going from room temperature to near absolute zero. The insulating dielectric is magnesium oxide (MgO), which has a dielectric constant of 9.70. Calculate the capacitance per meter of Thermocoax.

24.57 A parallel plate capacitor has square plates of side $L = 10.0$ cm and a distance $d = 1.00$ cm between the plates. Of the space between the plates, $\frac{1}{5}$ is filled with a dielectric with dielectric constant $\kappa_1 = 20.0$. The remaining $\frac{4}{5}$ of the space is filled with a different dielectric, with $\kappa_2 = 5.00$. Find the capacitance of the capacitor.

•24.58 A 4.0-nF parallel plate capacitor with a sheet of Mylar ($\kappa = 3.1$) filling the space between the plates is charged to a potential difference of 120 V and is then disconnected.

a) How much work is required to completely remove the sheet of Mylar from the space between the two plates?

b) What is the potential difference between the plates of the capacitor once the Mylar is completely removed?

•24.59 The volume between the two cylinders of a cylindrical capacitor is half filled with a dielectric whose dielectric constant is κ and is connected to a battery with a potential difference ΔV. What is the charge placed on the capacitor? What is the ratio of this charge to the charge placed on an identical capacitor with no dielectric connected in the same way across the same potential drop?

•24.60 A dielectric slab with thickness d and dielectric constant $\kappa = 2.31$ is inserted in a parallel place capacitor that has been charged by a 110.-V battery and has area $A = 100.$ cm^2 and separation distance $d = 2.50$ cm.

a) Find the capacitance, C, the potential difference, V, the electric field, E, the total charge stored on the capacitor Q, and electric potential energy stored in the capacitor, U, before the dielectric material is inserted.

b) Find C, V, E, Q, and U when the dielectric slab has been inserted and the battery is still connected.

c) Find C, V, E, Q, and U when the dielectric slab is in place and the battery is disconnected.

•24.61 A parallel plate capacitor has a capacitance of 120. pF and a plate area of 100. cm^2. The space between the plates is filled with mica whose dielectric constant is 5.40. The plates of the capacitor are kept at 50.0 V.

a) What is the strength of the electric field in the mica?

b) What is the amount of free charge on the plates?

c) What is the amount of charge induced on the mica?

••24.62 Design a parallel plate capacitor with a capacitance of 47.0 pF and a capacity of 7.50 nC. You have available conducting plates, which can be cut to any size, and Plexiglas sheets, which can be cut to any size and machined to any thickness. Plexiglas has a dielectric constant of 3.40 and a dielectric strength of $3.00 \cdot 10^7$ V/m. You must make your capacitor as compact as possible. Specify all relevant dimensions. Ignore any fringe field at the edges of the capacitor plates.

••24.63 A parallel plate capacitor consisting of a pair of rectangular plates, each measuring 1.00 cm by 10.0 cm, with a separation between the plates of 0.100 mm, is charged by a power supply at a potential difference of $1.00 \cdot 10^3$ V. The power supply is then removed, and without being discharged, the capacitor is placed in a vertical position over a container holding de-ionized water, with the short sides of the plates in contact with the water, as shown in the figure. Using energy considerations, show that the water will rise between the plates. Neglecting other effects, determine the system of equations that can be used to calculate the height to which the water rises between the plates. You do not have to solve the system.

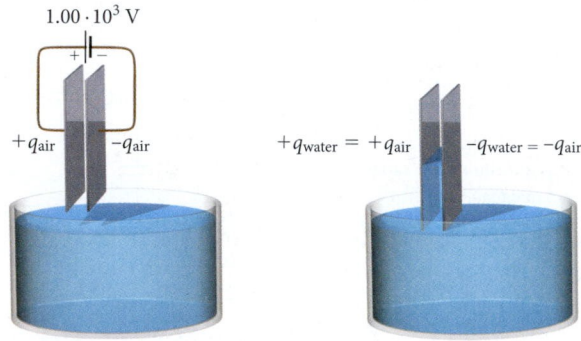

$1.00 \cdot 10^3$ V

$+q_{air}$ $-q_{air}$ $+q_{water} = +q_{air}$ $-q_{water} = -q_{air}$

Additional Exercises

24.64 Two circular metal plates of radius 0.610 m and thickness 7.10 mm are used in a parallel plate capacitor. A gap of 2.10 mm is left between the plates, and half of the space (a semicircle) between them is filled with a dielectric for which $\kappa = 11.1$ and the other half is filled with air. What is the capacitance of this capacitor?

24.65 Considering the dielectric strength of air, what is the maximum amount of charge that can be stored on the plates of a capacitor that are a distance of 15 mm apart and have an area of 25 cm^2?

24.66 The figure shows three capacitors in a circuit: $C_1 = 2.00$ nF and $C_2 = C_3 = 4.00$ nF. Find the charge on each capacitor when the potential difference applied is $V = 1.50$ V.

24.67 A capacitor with a vacuum between its plates is connected to a battery and then the gap is filled with Mylar. By what percentage is its energy-storing capacity increased?

24.68 A parallel plate capacitor with a plate area of 12.0 cm^2 and air in the space between the plates, which are separated by 1.50 mm, is connected to a 9.00-V battery. If the plates are pulled back so that the separation increases to 2.75 mm, how much work is done?

24.69 Suppose you want to make a 1.00-F capacitor using two square sheets of aluminum foil. If the sheets of foil are separated by a single piece of paper (thickness of about 0.100 mm and $\kappa = 5.00$), find the size of the sheets of foil (the length of each edge).

24.70 A 4.00-pF parallel plate capacitor has a potential difference of 10.0 V across it. The plates are 3.00 mm apart, and the space between them contains air.

a) What is the charge on the capacitor?

b) How much energy is stored in the capacitor?

c) What is the area of the plates?

d) What would the capacitance of this capacitor be if the space between the plates were filled with polystyrene?

24.71 A four-capacitor circuit is charged by a battery, as shown in the figure. The capacitances are $C_1 = 1.00$ mF, $C_2 = 2.00$ mF, $C_3 = 3.00$ mF, and $C_4 = 4.00$ mF, and the battery potential is $V_B = 1.00$ V. When the circuit is at equilibrium, point D has potential $V_D = 0.00$ V. What is the potential, V_A, at point A?

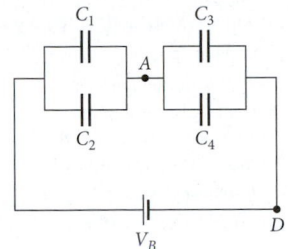

24.72 How much energy can be stored in a capacitor with two parallel plates, each with an area of 64.0 cm^2 and separated by a gap of 1.30 mm, filled with porcelain whose dielectric constant is 7.00, and holding equal and opposite charges of magnitude 420. μC?

24.73 A quantum mechanical device known as the *Josephson junction* consists of two overlapping layers of superconducting metal (for example, aluminum at 1.00 K) separated by 20.0 nm of aluminum oxide, which has a dielectric constant of 9.10. If this device has an area of 100. μm^2 and a parallel plate configuration, estimate its capacitance.

24.74 Three capacitors with capacitances $C_1 = 6.00$ μF, $C_2 = 3.00$ μF, and $C_3 = 5.00$ μF are connected in a circuit as shown in the figure, with an applied potential of V. After the charges on the capacitors have reached their equilibrium values, the charge Q_2 on the second capacitor is found to be 40.0 μC.

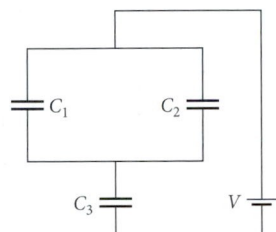

a) What is the charge, Q_1, on capacitor C_1?

b) What is the charge, Q_3, on capacitor C_3?

c) How much voltage, V, was applied across the capacitors?

24.75 For a science project, a fourth-grader cuts the tops and bottoms off two soup cans of equal height, 7.24 cm, and with radii of 3.02 cm and 4.16 cm, puts the smaller one inside the larger, and hot-glues them both on a sheet of plastic, as shown in the figure. Then she fills the gap between the cans with a special "soup" (dielectric constant of 63.0). What is the capacitance of this arrangement?

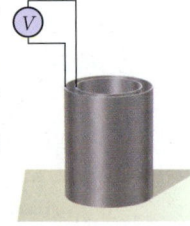

24.76 The Earth can be thought of as a spherical capacitor. If the net charge on the Earth is $-7.80 \cdot 10^5$ C, find (a) the capacitance of the Earth and (b) the electric potential energy stored on the Earth's surface.

•24.77 A parallel plate capacitor with air in the gap between the plates is connected to a 6.00-V battery. After charging, the energy stored in the capacitor is 72.0 nJ. Without disconnecting the capacitor from the battery, a dielectric is inserted into the gap and an additional 317 nJ of energy flows from the battery to the capacitor.

a) What is the dielectric constant of the dielectric?

b) If each of the plates has an area of 50.0 cm^2, what is the charge on the positive plate of the capacitor after the dielectric has been inserted?

c) What is the magnitude of the electric field between the plates before the dielectric is inserted?

d) What is the magnitude of the electric field between the plates after the dielectric is inserted?

•24.78 An 8.00-μF capacitor is fully charged by a 240.-V battery, which is then disconnected. Next, the capacitor is connected to an initially uncharged capacitor of capacitance C, and the potential difference across it is found to be 80.0 V. What is C? How much energy ends up being stored in the second capacitor?

•24.79 A parallel plate capacitor consists of square plates of edge length 2.00 cm separated by a distance of 1.00 mm. The capacitor is charged with

a 15.0-V battery, and the battery is then removed. A 1.00-mm-thick sheet of nylon (dielectric constant of 3.50) is slid between the plates. What is the average force (magnitude and direction) on the nylon sheet as it is inserted into the capacitor?

•24.80 A proton traveling along the x-axis at a speed of $1.00 \cdot 10^6$ m/s enters the gap between the plates of a 2.00-cm-wide parallel plate capacitor. The surface charge distributions on the plates are given by $\sigma = \pm 1.00 \cdot 10^{-6}$ C/m^2. How far has the proton been deflected sideways (Δy) when it reaches the far edge of the capacitor? Assume that the electric field is uniform inside the capacitor and zero outside.

•24.81 A parallel plate capacitor has square plates of side $L = 10.0$ cm separated by a distance $d = 2.50$ mm, as shown in the figure. The capacitor is charged by a battery with potential difference $V_0 = 75.0$ V; the battery is then disconnected.

a) Determine the capacitance, C_0, and the electric potential energy, U_0, stored in the capacitor at this point.

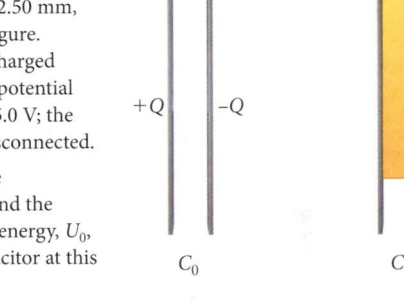

Charged capacitor   Charged capacitor with Plexiglas slab inserted

b) A slab made of Plexiglas ($\kappa = 3.40$) is then inserted so that it fills $\frac{2}{3}$ of the volume between the plates, as shown in the figure. Determine the new capacitance, C', the new potential difference between the plates, V', and the new electric potential energy, U', stored in the capacitor.

c) Neglecting gravity, did the inserter of the dielectric slab have to do work or not?

•24.82 A typical AAA battery has stored energy of about 3400 J. (Battery capacity is typically listed as 625 mA h, meaning that much charge can be delivered at approximately 1.5 V.) Suppose you want to build a parallel plate capacitor to store this amount of energy, using a plate separation of 1.0 mm and with air filling the space between the plates.

a) Assuming that the potential difference across the capacitor is 1.50 V, what must the area of each plate be?

b) Assuming that the potential difference across the capacitor is the maximum that can be applied without dielectric breakdown occurring, what must the area of each plate be?

c) Is either capacitor a practical replacement for the AAA battery?

•24.83 Two parallel plate capacitors, C_1 and C_2, are connected in series to a 96.0-V battery. Both capacitors have plates with an area of 1.00 cm^2 and a separation of 0.100 mm; C_1 has air between its plates, and C_2 has that space filled with porcelain (dielectric constant of 7.00 and dielectric strength of 5.70 kV/mm).

a) After charging, what are the charges on each capacitor?

b) What is the total energy stored in the two capacitors?

c) What is the electric field between the plates of C_2?

•24.84 The plates of parallel plate capacitor A consist of two metal discs of identical radius, $R_1 = 4.00$ cm, separated by a distance $d = 2.00$ mm, as shown in the figure.

a) Calculate the capacitance of this parallel plate capacitor with the space between the plates filled with air.

b) A dielectric in the shape of a thick-walled cylinder of outer radius $R_1 = 4.00$ cm, inner radius $R_2 = 2.00$ cm, thickness $d = 2.00$ mm, and dielectric constant $\kappa = 2.00$ is placed between the plates, coaxial with them, as shown in the figure. Calculate the capacitance of capacitor B, with this dielectric.

c) The dielectric cylinder is removed, and instead a solid disc of radius R_1 made of the same dielectric is placed between the plates to form capacitor C, as shown in the figure. What is the new capacitance?

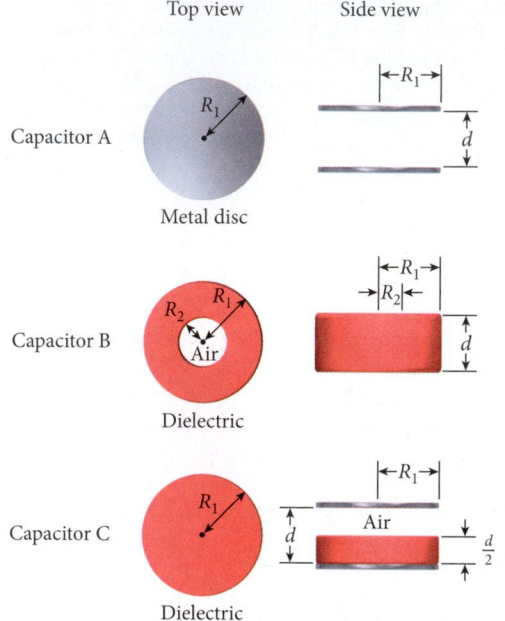

Capacitor A

Top view   Side view

Metal disc

Capacitor B

Dielectric

Capacitor C

Dielectric

•**24.85** A 1.00-μF capacitor charged to 50.0 V and a 2.00-μF capacitor charged to 20.0 V are connected, with the positive plate of each connected to the negative plate of the other. What is the final charge on the 1.00-μF capacitor?

•**24.86** The capacitance of a spherical capacitor consisting of two concentric conducting spheres with radii r_1 and r_2 ($r_2 > r_1$) is given by $C = 4\pi\epsilon_0 r_1 r_2/(r_2 - r_1)$. Suppose that the space between the spheres, from r_1 up to a radius R ($r_1 < R < r_2$) is filled with a dielectric for which $\epsilon = 10\epsilon_0$.

Find an expression for the capacitance, and check the limits when $R = r_1$ and $R = r_2$.

•**24.87** In the figure, a parallel plate capacitor is connected to a 300.-V battery. With the capacitor connected, a proton is fired with a speed of $2.00 \cdot 10^5$ m/s from (through) the negative plate of the capacitor at an angle θ with the normal to the plate.

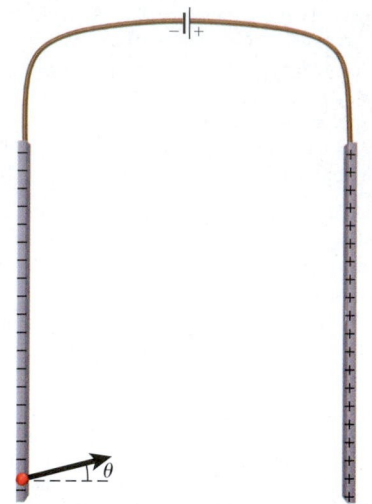

a) Show that the proton cannot reach the positive plate of the capacitor, regardless of what the angle θ is.

b) Sketch the trajectory of the proton between the plates.

c) Assuming that $V = 0$ at the negative plate, calculate the potential at the point between the plates where the proton reverses its motion in the x-direction.

d) Assuming that the plates are long enough for the proton to stay between them throughout its motion, calculate the speed (magnitude only) of the proton as it collides with the negative plate.

••**24.88** For the parallel plate capacitor with dielectric shown in the figure, prove that for a given thickness of the dielectric slab, the capacitance does not depend on the position of the slab relative to the two conducting plates (that is, it does not depend on the values of d_1 and d_3).

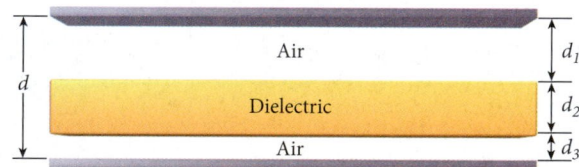

MULTI-VERSION EXERCISES

24.89 The battery of an electric car stores 53.63 MJ of energy. How many supercapacitors, each with capacitance $C = 3.361$ kF at a potential difference of 2.121 V, are required to supply this amount of energy?

24.90 The battery of an electric car stores 60.51 MJ of energy. If 6990 supercapacitors, each with capacitance $C = 3.423$ kF, are required to supply this amount of energy, what is the potential difference across each supercapacitor?

24.91 The battery of an electric car stores 67.39 MJ of energy. If 6845 supercapacitors, each with capacitance C and charged to a potential difference of 2.377 V, can supply this amount of energy, what is the value of C for each supercapacitor?

24.92 A parallel plate capacitor with vacuum between the plates has a capacitance of 3.547 μF. A dielectric material with $\kappa = 4.617$ is placed between the plates, completely filling the volume between them. The

capacitor is then connected to a battery that maintains a potential difference of 10.03 V across the plates. How much work is required to pull the dielectric material out of the capacitor?

24.93 A parallel plate capacitor with vacuum between the plates has a capacitance of 3.607 μF. A dielectric material is placed between the plates, completely filling the volume between them. The capacitor is then connected to a battery that maintains a potential difference of 11.33 V across the plates. The dielectric material is pulled out of the capacitor, which requires $4.804 \cdot 10^{-4}$ J of work. What is the dielectric constant of the material?

24.94 A parallel plate capacitor with vacuum between the plates has a capacitance of 3.669 μF. A dielectric material with $\kappa = 3.533$ is placed between the plates, completely filling the volume between them. The capacitor is then connected to a battery that maintains a potential difference V across the plates. The dielectric material is pulled out of the capacitor, which requires $7.389 \cdot 10^{-4}$ J of work. What is the potential difference, V?

25

Current and Resistance

FIGURE 25.1 A current flowing through a wire makes this light bulb shine.

Electric lighting is so commonplace that you don't even think about it. You walk into a dark room and simply flick on a switch, bringing the room to nearly daytime brightness (Figure 25.1). However, what happens when the switch is flicked depends ultimately on principles of physics and engineering devices that took decades to develop and refine.

This chapter is the first to focus on electric charges in motion. It presents some of the fundamental concepts that we'll use in Chapter 26 to analyze basic electric circuits, which are integral to all electronics applications. These chapters concentrate on the electrical effects of moving charges, but you should be aware that moving charges also give rise to other effects, which we'll start to examine in Chapter 27 on magnetism.

WHAT WE WILL LEARN

- Electric current at a point in a circuit is the rate at which net charge moves past that point.

- Direct current is current flowing in a direction that does not change with time. The direction of current is defined as the direction in which positive charge would be moving.

- The current density passing a given point in a conductor is the current per cross-sectional area.

- The conductivity of a material characterizes the ability of that material to conduct current. Its inverse is resistivity.

- The resistance of a device depends on its geometry and on the material of which it is made.

- The resistivity of a conductor increases approximately linearly with temperature.

- Electromotive force (usually referred to as emf) is a potential difference in an electric circuit.

- Ohm's Law states that the potential drop across a device is equal to the current flowing through the device times the resistance of the device.

- A simple circuit consists of a source of emf and resistors connected in series or in parallel.

- In a circuit diagram, an equivalent resistance can replace resistors connected in series or in parallel.

- The power in a circuit is the product of the current and the voltage drop.

- A diode conducts current in one direction but not in the opposite direction.

25.1 Electric Current

Up to this point, our study of electricity has focused on electrostatics, which deals with the properties of stationary electric charges and fields. Electric circuits were introduced in the discussion of capacitors in Chapter 24, but it covered only situations involving fully charged capacitors, where the charge is at rest. If electrostatics were all there was to electricity, it would not be nearly as important to modern society as it is. The world-changing impact of electricity is due to the properties of charges in motion, or electric current. All electrical devices rely on some kind of current for their operation.

Let's start by looking at a few very simple experiments. When you were a kid, you very likely had some toys that ran on batteries, and probably some of them contained small light bulbs. Consider a very simple circuit consisting only of a battery, a switch, and a light bulb (see Figure 25.2). If the switch is open, as in Figure 25.2a, the light bulb does not shine. If the switch is closed, as in Figure 25.2b, the light bulb turns on. We all know why this happens—because a current flows through the closed circuit. In Chapter 24, we saw that the battery provides a potential difference for the circuit. In this chapter, we'll examine what it means for the current to flow, what the physical basis of current is, and how it is related to the potential difference provided by the battery. We'll see that the light bulb acts as a resistor in the circuit, and we'll examine how resistors behave.

Let's begin by considering the simple experiment performed in Figure 25.2c, where the battery's orientation is the reverse of what it is in Figure 25.2b. The light bulb shines just

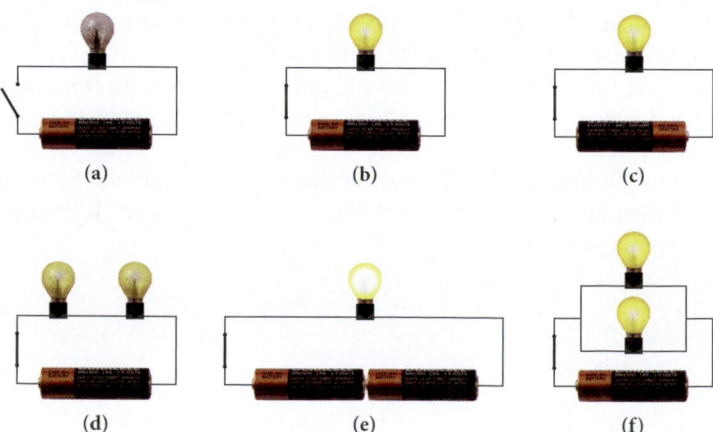

FIGURE 25.2 Experiments with batteries and light bulbs.

the same, despite the fact that the sign of the potential difference provided by the battery is reversed. (The positive terminal of the battery is on the end with the copper color.) In Figure 25.2d, two light bulbs are in the circuit, one behind the other. (Section 25.5 will focus on this arrangement of resistors, which is a connection *in series*.) Each of the two light bulbs shines with significantly less intensity than the single bulb in Figure 25.2c, so the current in the circuit may be smaller than before. On the other hand, two batteries in series, as in Figure 25.2e, double the potential difference in the circuit, and the bulb shines significantly brighter. Finally, if separate wires are used to connect the two light bulbs to a single battery, as shown in Figure 25.2f, the light bulbs shine with about the same intensity as in Figure 25.2b or Figure 25.2c. This way of wiring resistors in a circuit is called a *parallel connection* and will be explored in Section 25.6.

Quantitatively, the **electric current,** *i*, is the net charge passing a given point in a given time, divided by that time. The random motion of electrons in a conductor is not a current, in spite of the fact that large amounts of charge are moving past a given point, because no *net* charge flows. If net charge *dq* passes a point during time *dt*, the current at that point is, by definition,

$$i = \frac{dq}{dt}. \tag{25.1}$$

The net amount of charge passing a given point in time *t* is the integral of the current with respect to time:

$$q = \int dq = \int_0^t i \, dt'. \tag{25.2}$$

Total charge is conserved, implying that charge flowing in a conductor is never lost. Therefore, the same amount of charge flows into one end of a conductor as emerges from the other end.

The unit of current, coulombs per second, was given the name **ampere** (abbreviated A, or sometimes amp), after the French physicist André Ampère (1775–1836):

$$1 \, A = \frac{1 \, C}{1 \, s}.$$

Some typical currents are 1 A for a light bulb, 200 A for the starter in your car, $1 \, mA = 1 \cdot 10^{-3} A$ to power your MP3 player, 1 nA for the currents in your brain's neurons and synaptic connections, and $10,000 \, A = 10^4 \, A$ in a lightning strike (for a short time). The smallest currents that can be measured are those from individual electrons tunneling in scanning tunneling microscopes and are on the order of 10 pA. The largest current in the Solar System is the solar wind, which is in the GA range. Other examples of the broad range of currents are shown in Figure 25.3.

You should remember this handy safety rule related to the orders of magnitudes of currents: 1-10-100. That is, 1 mA of current flowing through a human body can be felt (as

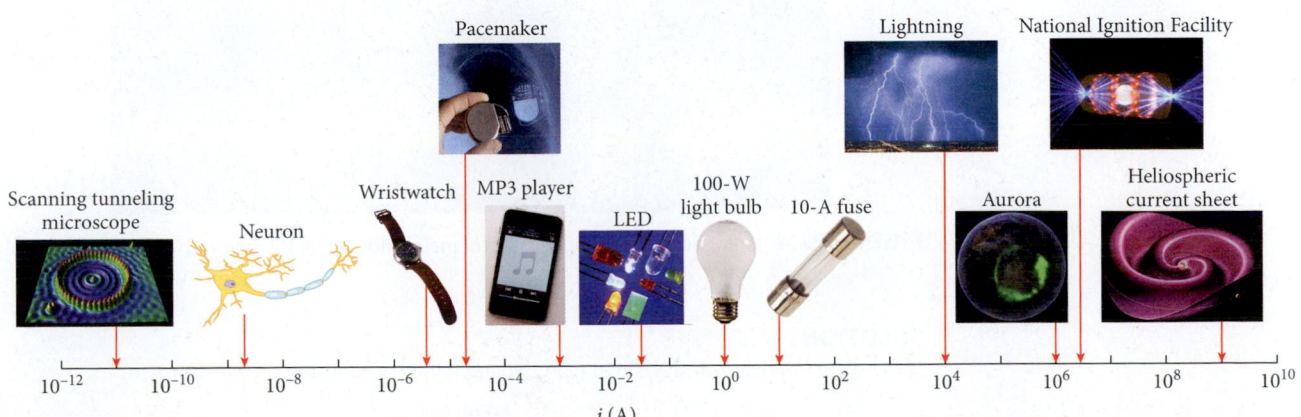

FIGURE 25.3 Examples of electrical currents ranging from 1 pA to 10 GA.

a tingle, usually), 10 mA of current makes muscles contract to the point that the person cannot let go of the wire carrying the current, and 100 mA is sufficient to stop the heart.

A current that flows in only one direction, which does not change with time, is called **direct current.** (Current that flows first in one direction and then in the opposite direction is called *alternating current* and will be discussed in Chapter 30.) In this chapter, the direction of the current flowing in a conductor is indicated by an arrow.

Physically, the charge carriers in a conductor are electrons, which are negatively charged. However, by convention, positive current is defined as flowing from the positive to the negative terminal. The reason for this counterintuitive definition of current direction is that the definition originated in the second half of the 19th century, when it was not known that electrons are the charge carriers responsible for current. And so the current direction was simply defined as the direction in which the positive charges would flow.

Self-Test Opportunity 25.1

A typical rechargeable AA battery is rated at 700 mAh. How long can this battery provide a current of 100 μA ?

EXAMPLE 25.1 Iontophoresis

There are three ways to administer anti-inflammatory medication. The painless way is the oral one—simply swallowing the drug. However, this method typically leads to a small amount of the drug in the affected tissue, on the order of 1 μg. The second way is to have the drug injected locally with a needle. This hurts but can deposit on the order of 10 mg of the drug in the affected tissue—four orders of magnitude more than with the oral method. However, since the 1990s, a third method has been available, which is also painless and can deposit on the order of 100 μg of a drug in the area where it is needed. This method, called *iontophoresis*, uses (very weak) electrical currents that are sent through the patient's tissue (Figure 25.4). The iontophoresis device consists of a battery and two electrodes (plus other electronic circuitry that allows the nurse to control the strength of the current applied). The anti-inflammatory drug, usually dexamethasone, is applied to the underside of the negatively charged electrode. A current flows through the patient's skin and deposits the drug in the tissue to a depth of up to 1.7 cm.

PROBLEM

A nurse wants to administer 80 μg of dexamethasone to the heel of an injured soccer player. If she uses an iontophoresis device that applies a current of 0.14 mA, as shown in Figure 25.4, how long does the administration of the dose take? Assume that the instrument has an application rate of 650 μg/C and that the current flows at a constant rate.

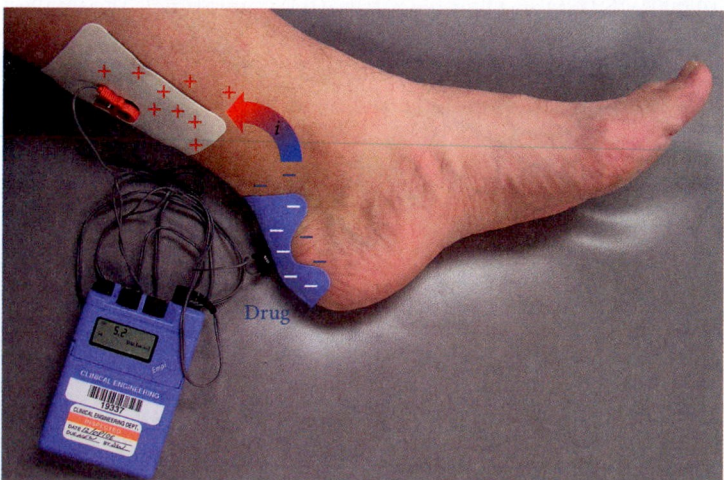

FIGURE 25.4 Iontophoresis is the application of medication under the skin with the aid of electrical current.

SOLUTION

If the drug application rate is 650 μg/C, to apply 80 μg requires a total charge of

$$q = \frac{80\ \mu g}{650\ \mu g/C} = 0.123\ C.$$

The current flows at a constant rate, so the integral in equation 25.2 is simply

$$q = \int_0^t i\, dt' = it.$$

Solving for t and inserting the numbers, we find

$$q = it \Rightarrow t = \frac{q}{i} = \frac{0.123 \text{ C}}{0.14 \cdot 10^{-3} \text{ A}} = 880 \text{ s}.$$

The iontophoresis treatment of the athlete will take approximately 15 min.

25.2 Current Density

Consider a current flowing in a conductor. For a perpendicular plane through the conductor, the current per unit area flowing through the conductor at that point (the cross-sectional area A in Figure 25.5) is the **current density, $\vec{J}$**. The direction of $\vec{J}$ is defined as the direction of the velocity of the positive charges (or opposite to the direction of negative charges) crossing the plane. The current flowing through the plane is

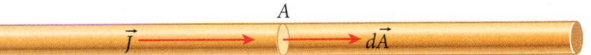

FIGURE 25.5 Segment of a conductor (wire), with a perpendicular plane intersecting it and forming a cross-sectional area A.

$$i = \int \vec{J} \cdot d\vec{A}, \tag{25.3}$$

where $d\vec{A}$ is the differential area element of the perpendicular plane, as indicated in Figure 25.5. If the current is uniform and perpendicular to the plane, then $i = JA$, and the magnitude of the current density can be expressed as

$$J = \frac{i}{A}. \tag{25.4}$$

In a conductor that is not carrying current, the conduction electrons move randomly. When current flows through the conductor, electrons still move randomly but also have an additional **drift velocity, $\vec{v}_d$**, in the direction opposite to that of the electric field driving the current. The magnitude of the velocity of random motion is on the order of 10^6 m/s, while the magnitude of the drift velocity is on the order of 10^{-4} m/s or even less. With such a slow drift velocity, you might wonder why a light comes on almost immediately after you turn on a switch. The answer is that the switch establishes an electric field almost immediately throughout the circuit (with a speed on the order of $3 \cdot 10^8$ m/s), causing the free electrons in the entire circuit (including in the light bulb) to move almost instantly.

The current density is related to the drift velocity of the moving electrons. Consider a conductor with cross-sectional area A and electric field $\vec{E}$ applied to it. Suppose the conductor has n conduction electrons per unit volume, and assume that all the electrons have the same drift velocity and that the current density is uniform. The negatively charged electrons will drift in a direction opposite to that of the electric field. In a time interval dt, each electron moves a net distance $v_d\, dt$. The volume of electrons passing a cross section of the conductor in time dt is then $Av_d\, dt$, and the number of electrons in this volume is $nAv_d\, dt$. Each electron has charge $-e$, so the charge dq that flows through the area in time dt is

$$dq = -nev_d A\, dt. \tag{25.5}$$

Therefore, the current is

$$i = \frac{dq}{dt} = -nev_d A. \tag{25.6}$$

The resulting current density is

$$J = \frac{i}{A} = -nev_d. \tag{25.7}$$

Equation 25.7 was derived in one spatial dimension, as is appropriate for a wire. However, it can be readily generalized to arbitrary directions in three-dimensional space:

$$\vec{J} = -(ne)\vec{v}_d.$$

You can see that the drift velocity vector is antiparallel to the current density vector, as stated before.

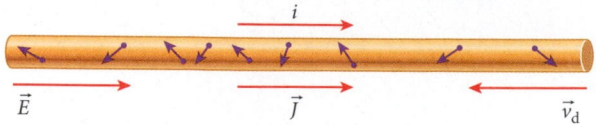

FIGURE 25.6 Electrons moving in a wire from right to left, causing a current in the direction from left to right.

Figure 25.6 shows a schematic drawing of a wire carrying a current. The physical current carriers are negatively charged electrons. In Figure 25.6, these electrons are moving to the left with drift velocity $\vec{v}_d$. However, the electric field, the current density, and the current are all directed to the right because of the convention that these quantities refer to positive charges. You may find this convention somewhat confusing, but you'll need to keep it in mind.

SOLVED PROBLEM 25.1 | **Drift Velocity of Electrons in a Copper Wire**

PROBLEM

You are playing "Galactic Destroyer" on your video game console. Your game controller operates at 12 V and is connected to the main box with an 18-gauge copper wire of length 1.5 m. As you fly your spaceship into battle, you hold the joystick in the forward position for 5.3 s, sending a current of 0.78 mA to the console. How far have the electrons in the wire moved during those few seconds, while on the screen your spaceship crossed half of a star system?

SOLUTION

THINK To find out how far electrons in a wire move during a given time interval, we need to calculate their drift velocity. To determine the drift velocity for electrons in a copper wire carrying a current, we need to find the density of charge-carrying electrons in copper. Then, we can apply the definition of the charge density to calculate the drift velocity.

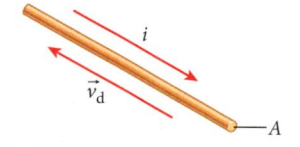

FIGURE 25.7 A copper wire with cross-sectional area A carrying a current, i.

SKETCH A copper wire with cross-sectional area A carrying a current, i, is shown in Figure 25.7, which also shows that, by convention, the electrons drift in the direction opposite to the direction of the current.

RESEARCH We obtain the distance x traveled by the electrons during time t from

$$x = v_d t,$$

where v_d is the magnitude of the drift velocity of the electrons. The drift velocity is related to the current density via equation 25.7:

$$\frac{i}{A} = -nev_d \qquad (i)$$

where i is the current, A is the cross-sectional area (0.823 mm^2 for an 18-gauge wire), n is the density of electrons, and $-e$ is the charge of an electron. The density of electrons is defined as

$$n = \frac{\text{number of conduction electrons}}{\text{volume}}.$$

We can calculate the density of electrons by assuming there is one conduction electron per copper atom. The density of copper is

$$\rho_{Cu} = 8.96 \text{ g/cm}^3 = 8960 \text{ kg/m}^3.$$

One mole of copper has a mass of 63.5 g and contains $6.02 \cdot 10^{23}$ atoms. Thus, the density of electrons is

$$n = \left(\frac{1 \text{ electron}}{1 \text{ atom}}\right)\left(\frac{6.02 \cdot 10^{23} \text{ atoms}}{63.5 \text{ g}}\right)\left(\frac{8.96 \text{ g}}{1 \text{ cm}^3}\right)\left(\frac{10^6 \text{ cm}^3}{1 \text{ m}^3}\right) = 8.49 \cdot 10^{28} \frac{\text{electrons}}{\text{m}^3}.$$

SIMPLIFY We solve equation (i) for the magnitude of the drift velocity:

$$v_d = \frac{i}{neA}.$$

Thus, the distance traveled by the electrons is

$$x = v_d t = \frac{i}{neA}t.$$

CALCULATE Putting in the numerical values, we get

$$x = v_d t = \frac{it}{neA} = \frac{\left(0.78 \cdot 10^{-3} \text{ A}\right)(5.3 \text{ s})}{\left(8.49 \cdot 10^{28} \text{ m}^{-3}\right)\left(1.602 \cdot 10^{-19} \text{ C}\right)\left(0.823 \text{ mm}^2\right)}$$

$$= \left(6.96826 \cdot 10^{-8} \text{ m/s}\right)(5.3 \text{ s})$$

$$= 3.69318 \cdot 10^{-7} \text{ m}.$$

ROUND We report our results to two significant figures:

$$v_d = 7.0 \cdot 10^{-8} \text{ m/s},$$

and

$$x = 3.7 \cdot 10^{-7} \text{ m} = 0.37 \text{ μm}.$$

DOUBLE-CHECK Our result for the magnitude of the drift velocity turns out to be a stunningly small number. Earlier, it was stated that typical drift velocities are on the order of 10^{-4} m/s or smaller. Since the current is proportional to the drift velocity, a relatively small current implies a relatively small drift velocity. An 18-gauge wire can carry a current of several amperes, so the current specified in the problem statement is less than 1% of the maximum current. Therefore, the fact that our calculated drift velocity is less than 1% of 10^{-4} m/s, a typical drift velocity for high currents, is reasonable.

The distance we calculated for the movement of the electrons is less than 0.001 of the thickness of a fingernail, a very small distance compared to the length of the wire. This result provides a valuable reminder that the electromagnetic field moves with nearly the speed of light (in vacuum) inside a conductor and causes all conduction electrons to drift basically at the same time. Therefore, the signal from your game controller arrives almost instantaneously at the console, despite the incredibly slow pace of the individual electrons.

25.3 Resistivity and Resistance

Some materials conduct electricity better than others. Applying a given potential difference across a good conductor results in a relatively large current; applying the same potential difference across an insulator produces little current. The **resistivity**, ρ, is a measure of how strongly a material opposes the flow of electric current. The **resistance**, R, is a material's opposition to the flow of electric current.

If a known electric potential difference, ΔV, is applied across a conductor (some physical device or material that conducts current) and the resulting current, i, is measured, the resistance of that conductor is given by

$$R = \frac{\Delta V}{i}. \tag{25.8}$$

The units of resistance are volts per ampere, a combination that was given the name **ohm** and the symbol Ω (the capital Greek letter omega), in honor of the German physicist Georg Simon Ohm (1789–1854):

$$1 \, \Omega = \frac{1 \text{ V}}{1 \text{ A}}.$$

Rearrangement of equation 25.8 results in

$$i = \frac{\Delta V}{R}, \tag{25.9}$$

which states that for a given potential difference, ΔV, the current, i, is inversely proportional to the resistance, R. This equation is commonly referred to as **Ohm's Law.** A rearrangement of equation 25.9, $\Delta V = iR$, is sometimes also referred to as Ohm's Law.

Sometimes devices are described in terms of the **conductance**, G, defined as

$$G = \frac{i}{\Delta V} = \frac{1}{R}.$$

Conductance has the SI derived unit of siemens (S), in honor of German inventor and industrialist Ernst Werner von Siemens (1816–1892):

$$1\,S = \frac{1\,A}{1\,V} = \frac{1}{1\,\Omega}.$$

In some conductors, the resistivity depends on the direction in which the current is flowing. This chapter assumes that the resistivity of a material is uniform for all directions of the current.

The resistance of a device depends on the material of which the device is made as well as its geometry. As stated earlier, the resistivity of a material characterizes how much it opposes the flow of current. The resistivity is defined in terms of the magnitude of the applied electric field, E, and the magnitude of the resulting current density, J:

$$\rho = \frac{E}{J}. \tag{25.10}$$

The units of resistivity are

$$[\rho] = \frac{[E]}{[J]} = \frac{V/m}{A/m^2} = \frac{V\,m}{A} = \Omega\,m.$$

Table 25.1	The Resistivity and the Temperature Coefficient of the Resistivity for Some Representative Conductors	
Material	**Resistivity, ρ, at 20 °C ($10^{-8}\,\Omega$ m)**	**Temperature Coefficient, α ($10^{-3}\,K^{-1}$)**
Silver	1.62	3.8
Copper	1.72	3.9
Gold	2.44	3.4
Aluminum	2.82	3.9
Brass	3.9	2
Tungsten	5.51	4.5
Nickel	7	5.9
Iron	9.7	5
Steel	11	5
Tantalum	13	3.1
Lead	22	4.3
Constantan	49	0.01
Stainless steel	70	1
Mercury	95.8	0.89
Nichrome	108	0.4

The values for steel and stainless steel depend strongly on the type of steel.

Table 25.1 lists the resistivities of some representative conductors at 20 °C. As you can see, typical values for the resistivity of metal conductors used in wires are on the order of $10^{-8}\,\Omega$ m. For example, copper has a resistivity of about $2 \cdot 10^{-8}\,\Omega$ m. Several metal alloys listed in Table 25.1 have useful properties. For example, wire made from Nichrome (80% nickel and 20% chromium) is often used as a heating element in devices such as toasters. The next time you are toasting an English muffin, look inside the toaster. The glowing elements are probably Nichrome wires. The resistivity of Nichrome ($108 \cdot 10^{-8}\,\Omega$ m) is about 50 times that of copper. Thus, when current is run through the Nichrome wires of the toaster, the wires dissipate power and heat up until they glow with a dull red color, while the copper wires of the electrical cord that connects the toaster to the wall remain cool.

Sometimes materials are specified in terms of their **conductivity**, σ, rather than their resistivity, ρ; conductivity is defined as

$$\sigma = \frac{1}{\rho}.$$

The units of conductivity are $(\Omega\,m)^{-1}$.

The resistance of a conductor can be found from its resistivity and its geometry. For a homogeneous conductor of length L and constant cross-sectional area A, the equation $\Delta V = -\int \vec{E} \cdot d\vec{s}$ from Chapter 23 can be used to relate the electric field, E, and the electric potential difference, ΔV, across the conductor:

$$E = \frac{\Delta V}{L}.$$

Note that in contrast to electrostatics, where the surface of any conductor is an equipotential surface and has no electric field inside and no current flowing through it, the conductor in this situation has $\Delta V \neq 0$ and $\vec{E} \neq 0$, causing a current to flow. The magnitude of the current density is the current divided by the cross-sectional area:

$$J = \frac{i}{A}.$$

From the definition of resistivity (equation 25.10) and using $J = i/A$ and Ohm's Law (equation 25.8), we obtain

$$\rho = \frac{E}{J} = \frac{\Delta V/L}{i/A} = \frac{\Delta V}{i} \frac{A}{L} = \frac{iR}{i} \frac{A}{L} = R \frac{A}{L}.$$

Rearranging terms yields an expression for the resistance of a conductor in terms of the resistivity of its constituent material, the length, and the cross-sectional area:

$$R = \rho \frac{L}{A}. \tag{25.11}$$

Note that in the limit $A \to \infty$, equation 25.11 implies that $R \to 0$. Therefore, the equation is only an approximation, because the conduction electrons will still encounter some resistance in traveling through some length L of a material of density ρ. However, for all wires in actual use, equation 25.11 is an excellent approximation.

Size Convention for Wires

The American Wire Gauge (AWG) size convention for wires specifies diameters and thus cross-sectional areas on a logarithmic scale. The AWG size convention is shown in Table 25.2. The wire gauge is related to the diameter: The higher the gauge number, the thinner the wire. For large-diameter wires, gauge numbers consist of one or more zeros, as shown in Table 25.2. A 00-gauge wire is equivalent to a -1-gauge, a 000-gauge wire is equivalent to a -2-gauge, and so on. By definition, a 36-gauge wire has a diameter of exactly 0.005 in, and a 0000-gauge wire has a diameter of exactly 0.46 in. (These sizes appear in red in Table 25.2.) There are 39 gauge values from 0000-gauge to 36-gauge, and the gauge number is a logarithmic representation of the wire diameter. Therefore, the formula to convert from the AWG gauge to the wire diameter, in inches, is $d = (0.005)92^{(36-n)/39}$, where n is the gauge number. Typical residential wiring uses 12-gauge to 10-gauge wires. An important rule of thumb is that a reduction by 3 gauges doubles the cross-sectional area of the wire. Examining equation 25.11, you can see that to cut the resistance of a given length of wire in half, you have to reduce the gauge number by 3.

Table 25.2	Wire Diameters and Cross-Sectional Areas as Defined by the American Wire Gauge Convention		
AWG	d (in)	d (mm)	A (mm²)
000000	0.5800	14.733	170.49
00000	0.5165	13.120	135.20
0000	0.46	11.684	107.22
000	0.4096	10.405	85.029
00	0.3648	9.2658	67.431
0	0.3249	8.2515	53.475
1	0.2893	7.3481	42.408
⋯			
8	0.1285	3.2636	8.3656
9	0.1144	2.9064	6.6342
10	0.1019	2.5882	5.2612
11	0.0907	2.3048	4.1723
12	0.0808	2.0525	3.3088
13	0.0720	1.8278	2.6240
14	0.0641	1.6277	2.0809
15	0.0571	1.4495	1.6502
16	0.0508	1.2908	1.3087
17	0.0453	1.1495	1.0378
18	0.0403	1.0237	0.8230
⋯			
35	0.0056	0.1426	0.0160
36	0.005	0.1270	0.0127
37	0.0045	0.1131	0.0100
⋯			

EXAMPLE 25.2   Resistance of a Copper Wire

Standard wires that electricians put into residential housing have fairly low resistance.

PROBLEM
What is the resistance of the 100.0-m standard 12-gauge copper wire that is typically used in wiring household electrical outlets?

SOLUTION
A 12-gauge copper wire has a diameter of 2.053 mm (see Table 25.2). Its cross-sectional area is then

$$A = 3.31 \text{ mm}^2.$$

Using the value for the resistivity of copper from Table 25.1 and equation 25.11, we find

$$R = \rho \frac{L}{A} = (1.72 \cdot 10^{-8} \ \Omega \ \text{m}) \frac{100.0 \text{ m}}{3.31 \cdot 10^{-6} \text{ m}^2} = 0.520 \ \Omega.$$

Concept Check 25.1

If the diameter of the wire in Example 25.2 is doubled, its resistance will

a) increase by a factor of 4.

b) increase by a factor of 2.

c) stay the same.

d) decrease by a factor of 2.

e) decrease by a factor of 4.

Resistor Codes

In many applications, circuit design calls for a range of resistances in various parts of a circuit. Commercially available resistors, such as those shown in Figure 25.8a, have a wide range of resistances. Resistors are commonly made of carbon enclosed in a plastic cover that looks like a medicine capsule, with wires sticking out at the ends for electrical connection. The value of the resistance is indicated by three or four color bands on the plastic covering. The first two bands indicate numbers for the mantissa, the third represents a power of 10, and the fourth indicates a tolerance for the range of values. For the mantissa and power of 10, the numbers associated with the colors are black = 0, brown = 1, red = 2, orange = 3, yellow = 4, green = 5,

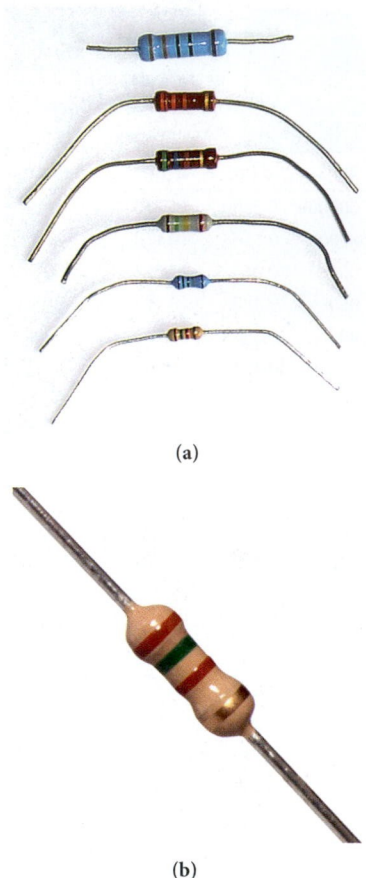

(a)

(b)

FIGURE 25.8 (a) Selection of resistors with various resistances. (b) Color-coding of a 150-Ω resistor.

Concept Check 25.2

If the temperature of a copper wire with a resistance of 100 Ω is increased by 25 K, the resistance will

a) increase by approximately 10 Ω.

b) increase by approximately 4 mΩ.

c) decrease by approximately 4 mΩ.

d) decrease by approximately 10 Ω.

e) stay the same.

blue = 6, purple = 7, gray = 8, and white = 9. For the tolerance, brown means 1%, red means 2%, gold means 5%, silver means 10%, and no band at all means 20%. For example, the single resistor shown in Figure 25.8b has the colors (left to right) brown, green, brown, and gold. From the code, the resistance of this resistor is $15 \cdot 10^1 \ \Omega = 150 \ \Omega$, with a tolerance of 5%.

Temperature Dependence and Superconductivity

The values of resistivity and resistance vary with temperature. For metals, this dependence on temperature is linear over a broad range of temperatures. An empirical relationship for the temperature dependence of the resistivity of a metal is

$$\rho - \rho_0 = \rho_0 \alpha \left(T - T_0 \right), \tag{25.12}$$

where ρ is the resistivity at temperature T, ρ_0 is the resistivity at temperature T_0, and α is the **temperature coefficient of electric resistivity** for the particular conductor.

In everyday applications, the temperature dependence of the resistance is often important. Equation 25.11 states that the resistance of a device depends on its length and cross-sectional area. These quantities depend on temperature, as we saw in Chapter 17; however, the temperature dependence of linear expansion is much smaller than the temperature dependence of resistivity for a particular conductor. Thus, the temperature dependence of the resistance of a conductor can be approximated as

$$R - R_0 = R_0 \alpha \left(T - T_0 \right). \tag{25.13}$$

Note that equations 25.12 and 25.13 deal with temperature differences, so the temperatures can be expressed in degrees Celsius or kelvins (but not in degrees Fahrenheit!). Also note that T_0 does not necessarily have to be room temperature.

Values of α for representative conductors are listed in Table 25.1. Note that common metal conductors such as copper have a temperature coefficient of electric resistivity on the order of $4 \cdot 10^{-3} \ K^{-1}$. However, one metal alloy, constantan (60% copper and 40% nickel), has the special characteristic that its temperature coefficient of electric resistivity is very small: $\alpha = 1 \cdot 10^{-5} \ K^{-1}$. The name of this alloy comes from shortening the phrase "constant resistance." The small temperature coefficient of constantan combined with its relatively high resistivity of $4.9 \cdot 10^{-7} \ \Omega$ m makes it useful for precision resistors whose resistances have little dependence on temperature. Note also that Nichrome has a relative small temperature coefficient, $4 \cdot 10^{-4} \ K^{-1}$, which makes it suitable for the construction of heating elements, as noted earlier.

According to equation 25.12, most materials have a resistivity that varies linearly with temperature under ordinary circumstances. However, some materials do not follow this rule at low temperatures. At very low temperatures, the resistivity of some materials goes to exactly zero. These materials are called **superconductors.** Superconductors have applications in the construction of magnets for devices such as magnetic resonance imagers (MRI). Magnets constructed with superconductors use less power and can produce stronger magnetic fields than magnets constructed with conventional resistive conductors. A more extensive discussion of superconductivity will be presented in Chapter 28.

The resistance of some semiconducting materials actually decreases as the temperature increases, which implies a negative temperature coefficient of electric resistivity. These materials are often employed in high-resolution detectors for optical measurements or in particle detectors. Such devices must be kept cold to keep their resistance high, which is accomplished with refrigerators or liquid nitrogen.

A *thermistor* is a semiconductor whose resistance depends strongly on temperature. Thermistors are used to measure temperature. The temperature dependence of the resistance of a typical thermistor is shown in Figure 25.9a. Here you can see that the resistance of a thermistor falls with increasing temperature. This drop is in contrast to the increase in resistance of a copper wire over the same temperature range, shown in Figure 25.9b.

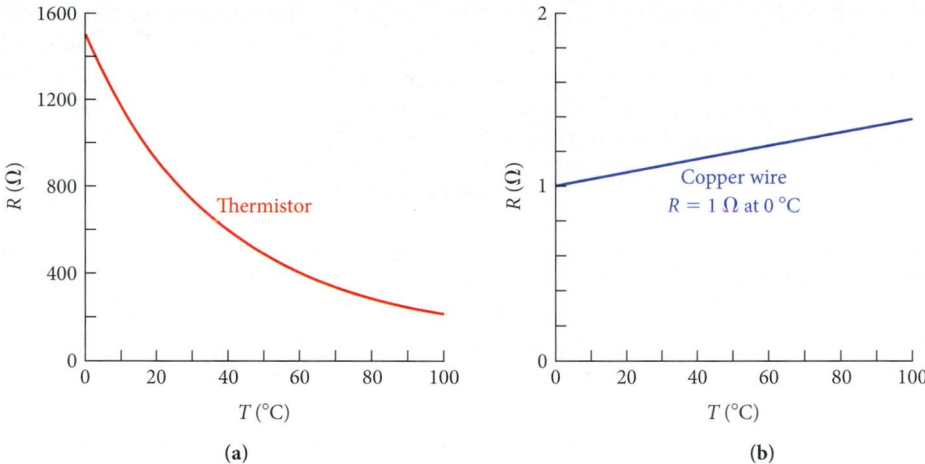

FIGURE 25.9 (a) The temperature dependence of the resistance of a thermistor. (b) The temperature dependence of the resistance of a copper wire that has a resistance of 1 Ω at $T = 0$ °C.

Microscopic Basis of Conduction in Solids

Conduction of current through solids results from the motion of electrons. In a metal conductor such as copper, the atoms of the metal form a regular array called a *crystal lattice*. The outermost electrons of each atom are essentially free to move randomly in this lattice. When an electric field is applied, the electrons drift in the direction opposite to that of the electric field. Resistance to drift occurs when electrons interact with the metal atoms in the lattice. When the temperature of the metal is increased, the motion of the atoms in the lattice increases. This, in turn, increases the probability that electrons will interact with the atoms, effectively increasing the resistance of the metal. A detailed discussion of the interaction between the lattice atoms and the conduction electrons is beyond the scope of this book. The approximate relations given by equations 25.12 and 25.13, which result from this interaction, are sufficient for our purposes.

The atoms of a semiconductor are also arranged in a crystal lattice. However, the outermost electrons of the atoms of the semiconductor are not free to move about within the lattice. To move about, the electrons must be given enough energy to attain an energy state where they can move freely. Thus, a typical semiconductor has a higher resistance than a metal conductor because it has many fewer conduction electrons. In addition, when a semiconductor is heated, many more electrons gain enough energy to move freely; thus, the resistance of the semiconductor decreases as its temperature increases.

25.4 Electromotive Force and Ohm's Law

For current to flow through a resistor, a potential difference must be established across the resistor. This potential difference, supplied by a battery or other device, is termed an **electromotive force,** abbreviated **emf.** (Electromotive force is not a force at all, but rather a potential difference. The term is still in widespread use but mainly in the form of its abbreviation, pronounced "ee-em-eff.") A device that maintains a potential difference is called an *emf device* and does work on the charge carriers. The potential difference created by the emf device is represented as V_{emf}. This text assumes that emf devices have terminals to which a circuit can be connected. The emf device is assumed to maintain a constant potential difference, V_{emf}, between these terminals.

Examples of emf devices are batteries, electric generators, and solar cells. **Batteries,** discussed in Chapters 23 and 24, produce emf through chemical reactions. Electric generators create emf from mechanical motion. Solar cells convert light energy from the Sun to electric energy. If you examine a battery, you will find its potential difference (sometimes colloquially called "voltage") written on it. This "voltage" is the potential difference (emf) that the battery can provide to a circuit. (Note that a battery is a source of constant emf, it does *not* supply constant current to a circuit.) Rechargeable batteries also display a rating in mAh (milliampere-hour), which provides information on the total charge the battery can deliver when fully charged. The mAh is another unit of charge:

$$1 \text{ mAh} = (10^{-3} \text{ A})(3600 \text{ s}) = 3.6 \text{ As} = 3.6 \text{ C}.$$

Electrical components in a circuit can be sources of emf, capacitors, resistors, or other electrical devices. These components are connected with conducting wires. At least one component must be a source of emf because the potential difference created by the emf device is what drives the current through the circuit. You can think of an emf device as the pump in a water pipeline; without the pump, the water sits in the pipe and doesn't move. Once the pump is turned on, the water moves through the pipe in a continuous flow.

An electric circuit starts and ends at an emf device. Since the emf device maintains a constant potential difference, V_{emf}, between its terminals, positive current leaves the device at the higher potential of its positive terminal and enters its negative terminal at a lower potential. This lower potential is conventionally set to zero.

Consider a simple circuit of the form shown in Figure 25.10, where a source of emf provides a potential difference, V_{emf}, across a resistor with resistance R. Note an important convention for circuit diagrams: A resistor is always symbolized by a zigzag line, and it is assumed that all of the resistance, R, is concentrated there. The wires connecting the different circuit elements are represented by straight lines; it is implied that they do not have a resistance. Physical wires do, of course, have some resistance, but it is assumed to be negligible for the purpose of the diagram.

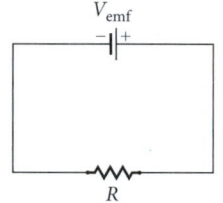

V_{emf}

R

FIGURE 25.10 Simple circuit containing a source of emf and a resistor.

For a circuit like the one shown in Figure 25.10, the emf device provides the potential difference that creates the current flowing through the resistor. Therefore, in this case, Ohm's Law (equation 25.9) can be written in terms of the external emf as

$$V_{emf} = iR. \tag{25.14}$$

Note that, unlike Newton's law of gravitation or the law of conservation of energy, Ohm's Law is not a law of nature. It is not even obeyed by all resistors. For many resistors, called *ohmic resistors,* the current is directly proportional to the potential difference across the resistor over a wide range of temperatures and a wide range of applied potential differences. For other resistors, called *non-ohmic resistors,* current and potential difference are not directly proportional at all. Non-ohmic resistors include many kinds of transistors, which means that many modern electronic devices do not obey Ohm's Law. We'll take a closer look at one of these devices, the diode, in Section 25.8. Nevertheless, a large class of materials and devices (such as conventional wires, for example) do obey Ohm's Law, and thus it is worth devoting attention to its consequences. The remainder of this chapter (with the exception of Section 25.8) treats resistors as ohmic devices; that is, devices that obey Ohm's Law.

The current, i, that flows through the resistor in Figure 25.10 also flows through the source of emf and the wires connecting the components. Because the wires are assumed to have zero resistance (as noted above), the change in potential of the current must occur in the resistor, according to Ohm's Law. This change is referred to as the **potential drop** across the resistor. Thus, the circuit shown in Figure 25.10 can be represented in a different way, making it clearer where the potential drop happens and showing which parts of the circuit are at which potential. Figure 25.11a shows the circuit in Figure 25.10. Figure 25.11b shows the same circuit but with the vertical dimension representing the value of the electric potential at different points around the circuit. The potential difference is supplied by the source of emf, and the entire potential drop occurs across the single resistor. (Remember, the convention is that the lines connecting the circuit elements in a circuit diagram represent wires with no resistance. Therefore, these connecting wires are represented in Figure 25.11b by horizontal lines, signifying that the entire wire is at exactly the same potential.) Ohm's Law

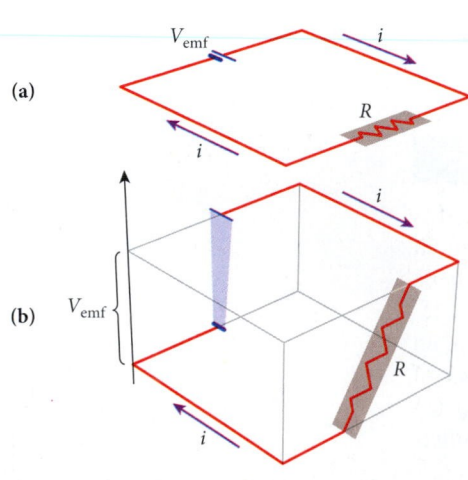

(a)

V_{emf}

i

R

i

(b)

V_{emf}

i

R

i

FIGURE 25.11 (a) Conventional representation of a simple circuit with a resistor and a source of emf. (b) Three-dimensional representation of the same circuit, displaying the potential at each point in the circuit. The current in the circuit is shown for both views.

applies for the potential drop across the resistor, and the current in the circuit can be calculated using equation 25.9.

Figure 25.11 illustrates an important point about circuits. Sources of emf add potential difference to a circuit, and potential drops through resistors reduce potential in the circuit. However, the total potential difference on any closed path around the complete circuit must be zero. This is a straightforward consequence of the law of conservation of energy. An analogy with gravity may help: You can gain and lose potential energy by moving up and down in a gravitational field (for example, by climbing up and down hills), but if you arrive back at the same point from which you started, the net energy gained or lost is exactly zero. The same holds for current flowing in a circuit: It does not matter how many potential drops or sources of emf are encountered on any closed loop; a given point always has the same value of the electric potential. The current can flow through the loop in either direction with the same result.

Resistance of the Human Body

This short introduction of resistance and Ohm's Law leads to a point about electrical safety. It was mentioned earlier that currents above 100 mA can be deadly if they flow through human heart muscle. Ohm's Law makes it clear that the resistance of the human body determines whether a given potential difference—say, from a car battery—can be dangerous. Since we usually handle tools with our hands, the most relevant measure for the human body's resistance, R_{body}, is the resistance along a path from the fingertips of one hand to the fingertips of the other hand. (Note that the heart is pretty much in the middle of this path!) For most people this resistance is in the range 500 kΩ < R_{body} < 2 MΩ. Most of this resistance comes from the skin, in particular, the layers of dead skin on the outside. However, if the skin is wet, its conductivity is drastically increased, and consequently, the body's resistance is drastically lowered. For a given potential difference, Ohm's Law implies that the current then drastically increases. Handling electrical devices in wet environments or touching them with your tongue is thus a very bad idea.

Wires in a circuit can have sharp points where they are cut. If these points penetrate the skin at the fingertips, the resistance of the skin is eliminated, and the fingertip-to-fingertip resistance is very drastically lowered. If a wire penetrates a blood vessel, the human body's resistance decreases even further, because blood has a high salinity and is thus a good conductor. In this case, even relatively small potential differences from batteries can have a deadly effect.

25.5 Resistors in Series

A circuit can contain more than one resistor and/or more than one source of emf. The analysis of circuits with multiple resistors requires different techniques. Let's first examine resistors connected in series.

Two resistors, R_1 and R_2, are connected in series with one source of emf with potential difference V_{emf} in the circuit shown in Figure 25.12. The potential drop across resistor R_1 is denoted by ΔV_1, and the potential drop across resistor R_2 by ΔV_2. The two potential drops must sum to the potential difference supplied by the source of emf:

$$V_{emf} = \Delta V_1 + \Delta V_2.$$

The crucial insight is that the same current must flow through all the elements of the circuit. How do we know this? Remember, at the beginning of this chapter, current was defined as the rate of change of the charge in time: $i = dq/dt$. Current has to be the same everywhere along a wire, and also in a resistor, because charge is conserved everywhere. No charge is lost or gained along the wire, and so the current is the same everywhere around the loop in Figure 25.12.

To clear up a common misconception, note that *there is no such thing as current getting "used up" in a resistor*. No matter how many resistors are connected in series, the current that flows into the first one is the same current that flows out of the last one. An analogy to water flowing in a pipe may help: No matter how long the pipe is and how many bends it may have, all water that flows into one end has to come out the other end.

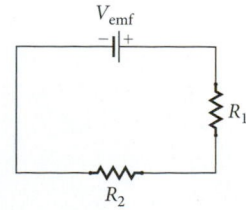

FIGURE 25.12 Circuit with two resistors in series with one source of emf.

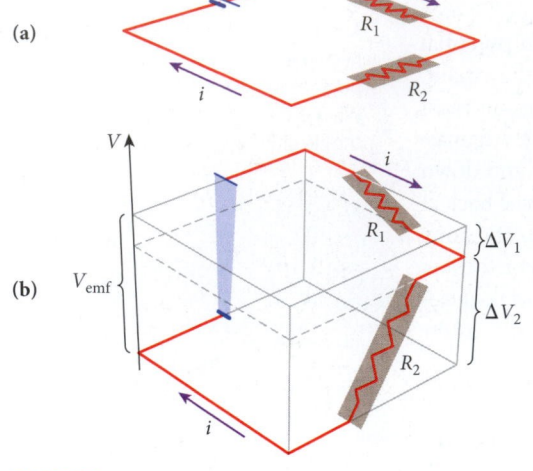

FIGURE 25.13 (a) Conventional representation of a simple circuit with two resistors in series and a source of emf. (b) Three-dimensional representation of the same circuit, displaying the potential at each point in the circuit. The current in the circuit is shown for both views.

Thus, the current flowing through each resistor in Figure 25.12 is the same. For each resistor, we can apply Ohm's Law and get

$$V_{emf} = iR_1 + iR_2.$$

An equivalent resistance, R_{eq}, can replace the two individual resistances:

$$V_{emf} = iR_1 + iR_2 = iR_{eq},$$

where

$$R_{eq} = R_1 + R_2.$$

Thus, two resistors in series can be replaced with an equivalent resistance equal to the sum of the two resistances. Figure 25.13 illustrates the potential drops in the series circuit of Figure 25.12, using a three-dimensional view.

The expression for the equivalent resistance of two resistors in series can be generalized to a circuit with n resistors in series:

$$R_{eq} = \sum_{i=1}^{n} R_i \quad \text{(for resistors in series).} \tag{25.15}$$

That is, if resistors are connected in a single path so that the same current flows through all of them, their total resistance is just the sum of their individual resistances.

Concept Check 25.3

What are the relative magnitudes of the two resistances in Figure 25.13?

a) $R_1 < R_2$

b) $R_1 = R_2$

c) $R_1 > R_2$

d) Not enough information is given in the figure to compare the resistances.

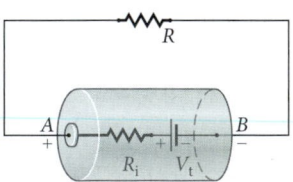

FIGURE 25.14 Battery (gray cylinder) with internal resistance R_i connected to an external resistor, R.

EXAMPLE 25.3 | Internal Resistance of a Battery

When a battery is not connected in a circuit, the potential difference across its terminals is V_t. When the battery is connected in series with a resistor with resistance R, current i flows through the circuit. When the current is flowing, the potential difference, V_{emf}, across the terminals of the battery is less than V_t. This drop occurs because the battery has an internal resistance, R_i, which can be thought of as being in series with the external resistor (Figure 25.14). That is,

$$V_t = iR_{eq} = i(R + R_i).$$

The battery is depicted by the gray cylinder in Figure 25.14. The terminals of the battery are represented by points A and B.

PROBLEM

Consider a battery that has $V_t = 12.0$ V when it is not connected to a circuit. When a 10.0-Ω resistor is connected with the battery, the potential difference across the battery's terminals drops to 10.9 V. What is the internal resistance of the battery?

SOLUTION

The current flowing through the external resistor is given by

$$i = \frac{\Delta V}{R} = \frac{10.9 \text{ V}}{10.0 \text{ }\Omega} = 1.09 \text{ A}.$$

The current flowing in the complete circuit, including the battery, must be the same as the current flowing in the external resistor. Thus, we have

$$V_t = iR_{eq} = i(R + R_i)$$

$$(R + R_i) = \frac{V_t}{i}$$

$$R_i = \frac{V_t}{i} - R = \frac{12.0 \text{ V}}{1.09 \text{ A}} - 10.0 \text{ }\Omega = 1.00 \text{ }\Omega.$$

The internal resistance of the battery is 1.00 Ω. Batteries with internal resistance are said to be *nonideal*. Unless otherwise specified, batteries in circuits will be assumed to have zero internal resistance. Such batteries are said to be *ideal*. An ideal battery maintains a constant potential difference between its terminals, independent of the current flowing.

Whether a battery can still provide energy cannot be determined by simply measuring the potential difference across the terminals. Instead, you must place a resistance on the battery and then measure the potential difference. If the battery is no longer functional, it may still provide its rated potential difference when not connected, but its potential difference may drop to zero when connected to an external resistance. Some brands of batteries have built-in devices to measure the functioning potential difference simply by pressing on a particular spot on the battery and observing an indicator.

Resistor with a Nonconstant Cross Section

Up to now the discussion has assumed that a resistor has the same cross-sectional area, A, and the same resistivity, ρ, everywhere along its length (this was the implicit assumption in the derivation leading to equation 25.11). This is, of course, not always the case. How do we handle the analysis of a resistor whose cross-sectional area is a function of the position x along the resistor, $A(x)$, and/or whose resistivity can change as a function of position, $\rho(x)$? We simply divide the resistor into many very short pieces of length Δx and sum over all of them, since equation 25.15 says that the total resistance is the sum of all of the resistances of the individual short pieces; then we take the limit $\Delta x \to 0$. If this sounds like an integration to you, you are right. The general formula for computing the resistance of a resistor of length L with a nonuniform cross-sectional area, $A(x)$, is

$$R = \int_0^L \frac{\rho(x)}{A(x)} dx. \tag{25.16}$$

A concrete example will help clarify this equation.

SOLVED PROBLEM 25.2 / Brain Probe

Chapter 22 mentioned the technique of electrocorticography (ECoG), which researchers use to measure the electric field generated by neurons in the brain. Some of these measurements can only be done by inserting very thin wires into the brain in order to probe directly into neurons. These wires are insulated, with only a very short tip exposed, which is pulled into a very fine conical tip. ECoG is being used to treat an epilepsy patient in Figure 25.15.

PROBLEM

If the wire used for ECoG is made of tungsten and has a diameter of 0.74 mm and the tip has a length of 2.0 mm and is sharpened to a diameter of 2.4 μm at the end, what is the resistance of the tip? (The resistivity of tungsten is listed in Table 25.1 as $5.51 \cdot 10^{-8}$ Ω m.)

SOLUTION

THINK First, why might one want to know the resistance? To measure electrical fields or potential differences in neurons, probes with a large resistance, say, on the order of kilo-ohms, cannot be used because the fields or differences will not be detectable. However, since resistance is inversely proportional to the cross-sectional area, a very small area means a relatively large resistance. The problem statement says that the probe has a very fine tip, much pointier than any sewing needle. Hence the need to find out the resistance of the probe before inserting it into the brain!

Clearly, we are dealing with a case of nonconstant cross-sectional area, and so we will need to perform the integration of equation 25.16. However, since the tip is entirely made of tungsten, the resistivity is constant throughout its volume, which will simplify the task.

SKETCH Figure 25.16a shows a three-dimensional view of the tip, and Figure 25.16b presents a cut through its symmetry plane and the integration path.

RESEARCH The research part is quite simple for this problem, because we already know which equation we need to use. However, equation 25.16 needs to be altered to reflect the fact that the resistivity is constant throughout the tip:

$$R = \rho \int_0^L \frac{1}{A(x)} dx, \tag{i}$$

– Continued

– Continued

Concept Check 25.4

Three identical resistors, R_1, R_2, and R_3, are wired together as shown in the figure. An electric current is flowing through the three resistors. The current through R_2

$R_1 \quad R_2 \quad R_3$

a) is the same as the current through R_1 and R_3.

b) is a third of the current through R_1 and R_3.

c) is twice the sum of the current through R_1 and R_3.

d) is three times the current through R_1 and R_3.

e) cannot be determined.

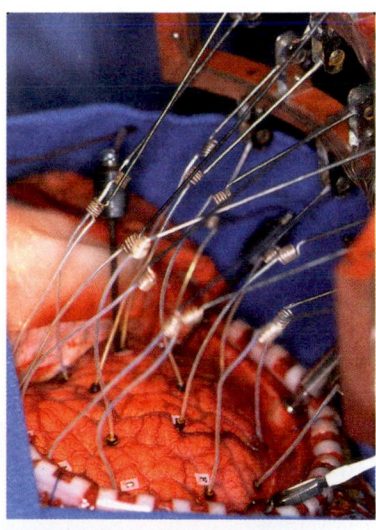

FIGURE 25.15 Electrocorticography performed with electrode grids on the cerebral cortex of an epilepsy patient.

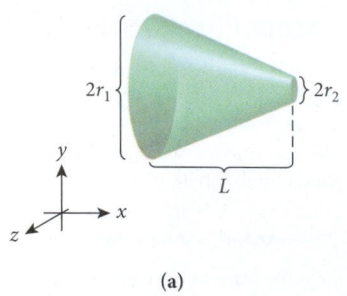

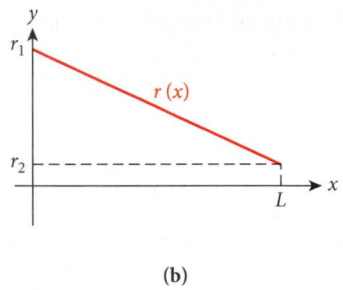

FIGURE 25.16 (a) Shape of the tip of the probe. (b) Coordinate system for the integration.

where $A(x)$ is the area of a circle, $A(x) = \pi[r(x)]^2$. The radius of the circle falls linearly from r_1 to r_2 (see Figure 25.16b):

$$r(x) = r_1 + \frac{(r_2 - r_1)x}{L}. \qquad \text{(ii)}$$

SIMPLIFY   We substitute the expression for the radius from equation (ii) into the formula for the area and then substitute the resulting expression for $A(x)$ into equation (i). We arrive at

$$R = \rho \int_0^L \frac{1}{\pi\left(r_1 + (r_2 - r_1)x/L\right)^2} dx.$$

This integral may look daunting at first sight, but except for x all the other quantities are constants. We consult an integration table or software and find

$$R = -\frac{\rho L}{\pi(r_2 - r_1)\left(r_1 + (r_2 - r_1)x/L\right)}\Bigg|_0^L = \frac{\rho L}{\pi r_1 r_2}.$$

CALCULATE   Putting in the numerical values, we get

$$R = \frac{\left(5.51 \cdot 10^{-8} \ \Omega \ \text{m}\right)\left(2.0 \cdot 10^{-3} \ \text{m}\right)}{\pi(0.37 \cdot 10^{-3} \ \text{m})(1.2 \cdot 10^{-6} \ \text{m})} = 7.90039 \cdot 10^{-2} \ \Omega.$$

ROUND   We report our result to the two significant figures to which the geometric properties of the tip were given:

$$R = 7.9 \cdot 10^{-2} \ \Omega.$$

DOUBLE-CHECK   The value of 79 mΩ seems a very small resistance to a current that has to pass through a tip that is sharpened to a diameter of 2.4 μm. On the other hand, the tip has a very small length, which argues for a small resistance. An additional confidence builder is the fact that the units worked out properly.

However, there are also a few tests we can perform to convince ourselves that at least the asymptotic limits of the solution to $R = \rho L/(\pi r_1 r_2)$ are reasonable. First, as the length approaches zero, so does the resistance, as expected. Second, as the radius of either end of the tip approaches zero, the formula predicts an infinite resistance, which is also expected.

25.6   Resistors in Parallel

Instead of being connected in series, which causes all the current to pass through both resistors, two resistors can be connected in parallel, which divides the current between them, as shown in Figure 25.17. Again, to better illustrate the potential drops, Figure 25.18 shows the same circuit in a three-dimensional view.

In this case, the potential drop across each resistor is equal to the potential difference provided by the source of emf. Using Ohm's Law (equation 25.14) for the current i_1 in R_1 and the current i_2 in R_2, we have

$$i_1 = \frac{V_{\text{emf}}}{R_1}$$

and

$$i_2 = \frac{V_{\text{emf}}}{R_2}.$$

The total current from the source of emf, i, must be

$$i = i_1 + i_2.$$

Inserting the expressions for i_1 and i_2, we obtain

$$i = i_1 + i_2 = \frac{V_{\text{emf}}}{R_1} + \frac{V_{\text{emf}}}{R_2} = V_{\text{emf}}\left(\frac{1}{R_1} + \frac{1}{R_2}\right).$$

FIGURE 25.17 Circuit with two resistors connected in parallel and a single source of emf.

Ohm's Law (equation 25.14) can be rewritten as

$$i = V_{emf}\left(\frac{1}{R_{eq}}\right).$$ (a)

Thus, two resistors connected in parallel can be replaced with an equivalent resistance given by

$$\frac{1}{R_{eq}} = \frac{1}{R_1} + \frac{1}{R_2}.$$

In general, the equivalent resistance for n resistors connected in parallel is given by

$$\frac{1}{R_{eq}} = \sum_{i=1}^{n}\frac{1}{R_i} \quad \text{(resistors in parallel).}$$ (25.17)

Clearly, combining resistors in series and in parallel to form equivalent resistances allows circuits with various combinations of resistors to be analyzed in a way analogous to the analysis of combinations of capacitors performed in Chapter 24.

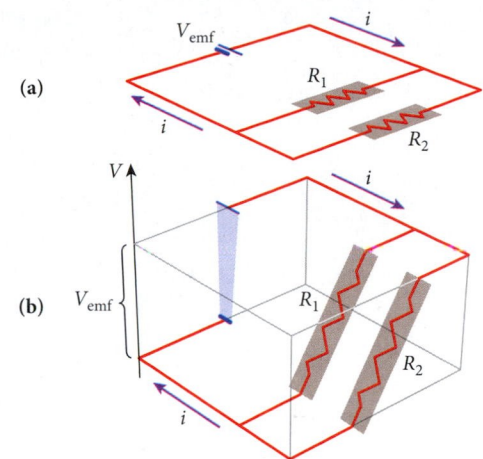

FIGURE 25.18 (a) Conventional representation of a simple circuit with two resistors in parallel and a source of emf. (b) Three-dimensional representation of the same circuit, displaying the potential at each point in the circuit.

Concept Check 25.5

Three identical resistors, R_1, R_2, and R_3, are wired together as shown in the figure. An electric current is flowing from point A to point B. The current flowing through R_2

a) is the same as the current through R_1 and R_3.

b) is a third of the current through R_1 and R_3.

c) is twice the sum of the current through R_1 and R_3.

d) is three times the current through R_1 and R_3.

e) cannot be determined.

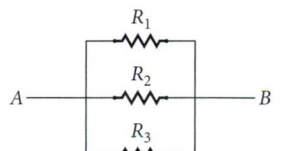

Concept Check 25.6

Which combination of resistors has the highest equivalent resistance?

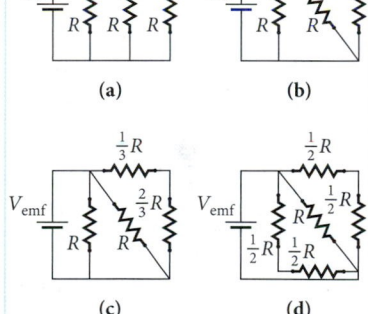

a) combination (a)

b) combination (b)

c) combination (c)

d) combination (d)

e) The equivalent resistance is the same for all four.

EXAMPLE 25.4 Equivalent Resistance in a Circuit with Six Resistors

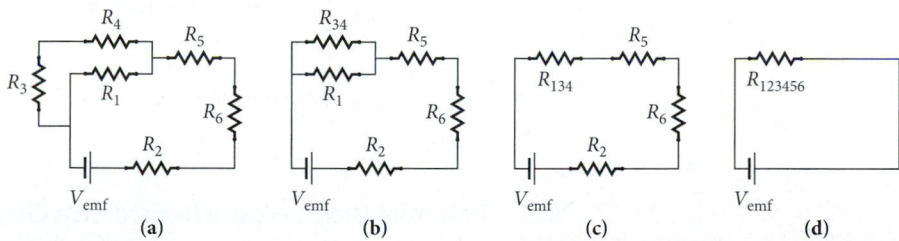

FIGURE 25.19 (a) Circuit with six resistors. (b)–(d) Steps in combining these resistors to determine the equivalent resistance.

PROBLEM

Figure 25.19a shows a circuit with six resistors, R_1 through R_6. What is the current flowing through resistors R_2 and R_3 in terms of V_{emf} and R_1 through R_6?

SOLUTION

We begin by identifying parts of the circuit that are clearly wired in parallel or in series. The current flowing through R_2 is the current flowing from the source of emf. We note that R_3 and R_4 are in series. Thus, we can write

$$R_{34} = R_3 + R_4.$$ (i)

– Continued

This substitution is made in Figure 25.19b. This figure shows us that R_{34} and R_1 are in parallel. We can then write

$$\frac{1}{R_{134}} = \frac{1}{R_1} + \frac{1}{R_{34}},$$

or

$$R_{134} = \frac{R_1 R_{34}}{R_1 + R_{34}}.$$ (ii)

This substitution is depicted in Figure 25.19c. From this figure, we can see that R_2, R_5, R_6, and R_{134} are in series. Thus, we can write

$$R_{123456} = R_2 + R_5 + R_6 + R_{134}.$$ (iii)

This substitution is shown in Figure 25.19d. We substitute for R_{34} and R_{134} from equations (i) and (ii) into equation (iii):

$$R_{123456} = R_2 + R_5 + R_6 + \frac{R_1 R_{34}}{R_1 + R_{34}} = R_2 + R_5 + R_6 + \frac{R_1(R_3 + R_4)}{R_1 + R_3 + R_4}.$$

Thus, i_2, the current flowing through R_2, is given by

$$i_2 = \frac{V_{emf}}{R_{123456}}.$$

Now we turn to the determination of the current flowing through R_3. Current i_2 is also flowing through the equivalent resistance R_{134} that contains R_3 (see Figure 25.19c). Thus, we can write

$$V_{134} = i_2 R_{134},$$

where V_{134} is the potential drop across the equivalent resistance R_{134}. The resistor R_1 and the equivalent resistance R_{34} are in parallel. Thus, V_{34}, the potential drop across R_{34}, is the same as the potential drop across R_{134} which is V_{134}. The resistors R_3 and R_4 are in series, and thus, i_3, the current flowing through R_3, is the same as i_{34}, the current flowing through R_{34}. We can thus write

$$V_{34} = V_{134} = i_{34} R_{34} = i_3 R_{34}.$$

Now we can express i_3 in terms of V and R_1 through R_6:

$$i_3 = \frac{V_{134}}{R_{34}} = \frac{i_2 R_{134}}{R_{34}} = \frac{\left(\dfrac{V_{emf}}{R_{123456}}\right) R_{134}}{R_{34}} = \frac{V_{emf} R_{134}}{R_{34} R_{123456}} = \frac{V_{emf}\left(\dfrac{R_1 R_{34}}{R_1 + R_{34}}\right)}{R_{34} R_{123456}} = \frac{V_{emf} R_1}{R_{123456}(R_1 + R_{34})}$$

or

$$i_3 = \frac{V_{emf} R_1}{\left(R_2 + R_5 + R_6 + \dfrac{R_1(R_3 + R_4)}{R_1 + R_3 + R_4}\right)(R_1 + R_3 + R_4)} = \frac{V_{emf} R_1}{(R_2 + R_5 + R_6)(R_1 + R_3 + R_4) + R_1(R_3 + R_4)}.$$

SOLVED PROBLEM 25.3   **Potential Drop across a Resistor in a Circuit**

PROBLEM

The circuit shown in Figure 25.20a has four resistors and a battery with $V_{emf} = 149$ V. The values of the four resistors are $R_1 = 17.0\ \Omega$, $R_2 = 51.0\ \Omega$, $R_3 = 114.0\ \Omega$, and $R_4 = 55.0\ \Omega$. What is the magnitude of the potential drop across R_2?

SOLUTION

THINK The resistors R_2 and R_3 are in parallel and can be replaced with an equivalent resistance, R_{23}. The resistors R_1 and R_4 are in series with R_{23}. The current flowing through R_1, R_4, and R_{23} is the same because they are in series. We can obtain the current in the circuit by calculating the equivalent resistance for R_1, R_4, and R_{23} and using Ohm's Law. The potential drop across R_{23} is equal to the current flowing in the circuit times R_{23}. The potential drop across R_2 is the same as the potential drop across R_{23} because R_2 and R_3 are in parallel.

SKETCH The potential drop across resistor R_2 is illustrated in Figure 25.20b.

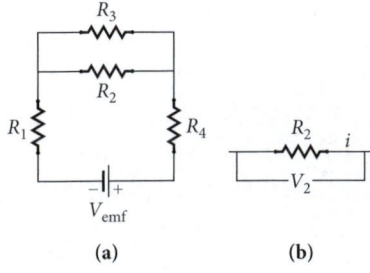

FIGURE 25.20 (a) A circuit with four resistors and a battery. (b) Potential drop across resistor R_2.

RESEARCH The equivalent resistance for R_2 and R_3 can be calculated using equation 25.17:

$$\frac{1}{R_{23}} = \frac{1}{R_2} + \frac{1}{R_3}. \qquad (i)$$

The equivalent resistance of the three resistors in series can be found using equation 25.15:

$$R_{eq} = \sum_{i=1}^{n} R_i = R_1 + R_{23} + R_4.$$

Finally, we obtain the current in the circuit using Ohm's Law:

$$V_{emf} = iR_{eq} = i(R_1 + R_{23} + R_4).$$

SIMPLIFY The potential drop, V_2, across R_2 is equal to the potential drop, V_{23}, across the equivalent resistance R_{23}:

$$V_2 = V_{23} = iR_{23} = \frac{V_{emf}}{R_1 + R_{23} + R_4} R_{23} = \frac{R_{23}V_{emf}}{R_1 + R_{23} + R_4}. \qquad (ii)$$

We can solve equation (i) for R_{23} to obtain

$$R_{23} = \frac{R_2 R_3}{R_2 + R_3}.$$

We can then use equation (ii) to determine the potential drop V_2 as

$$V_2 = \frac{\left(\dfrac{R_2 R_3}{R_2 + R_3}\right)V_{emf}}{R_1 + \left(\dfrac{R_2 R_3}{R_2 + R_3}\right) + R_4} = \frac{R_2 R_3 V_{emf}}{R_1(R_2 + R_3) + R_2 R_3 + R_4(R_2 + R_3)},$$

which we can rewrite as

$$V_2 = \frac{R_2 R_3 V_{emf}}{(R_1 + R_4)(R_2 + R_3) + R_2 R_3}.$$

CALCULATE Putting in the numerical values, we get

$$V_2 = \frac{R_2 R_3 V_{emf}}{(R_1 + R_4)(R_2 + R_3) + R_2 R_3}$$

$$= \frac{(51.0\ \Omega)(114.0\ \Omega)(149\ V)}{(17.0\ \Omega + 55.0\ \Omega)(51.0\ \Omega + 114.0\ \Omega) + (51.0\ \Omega)(114.0\ \Omega)}$$

$$= 48.9593\ V.$$

ROUND We report our result to three significant figures:

$$V = 49.0\ V.$$

DOUBLE-CHECK You may be tempted to avoid completing the analytic solution as we've done here. Instead, you may want to insert numbers earlier, for example, into the expression for R_{23}. So, to double-check our result, let's calculate the current in the circuit explicitly and then calculate the potential drop across R_{23} using that current. The equivalent resistance for R_2 and R_3 in parallel is

$$R_{23} = \frac{R_2 R_3}{R_2 + R_3} = \frac{(51.0\ \Omega)(114.0\ \Omega)}{51.0\ \Omega + 114.0\ \Omega} = 35.2\ \Omega.$$

The current in the circuit is then

$$i = \frac{V_{emf}}{R_1 + R_{23} + R_4} = \frac{149.0\ V}{17.0\ \Omega + 35.2\ \Omega + 55.0\ \Omega} = 1.39\ A.$$

The potential drop across R_2 is then

$$V_2 = iR_{23} = (1.39\ A)(35.2\ \Omega) = 48.9\ V,$$

which agrees with our result within rounding error. It is reassuring that both methods lead to the same answer.

We can also check that the potential drops across R_1, R_{23}, and R_4 sum to V_{emf}, as they should since R_1, R_{23}, and R_4 are in series. The potential drop across R_1 is $V_1 = iR_1 = (1.39\ A)(17.0\ \Omega) = 23.6\ V$. The potential drop across R_4 is $V_4 = iR_4 = (1.39\ A)(55.0\ \Omega) = 76.5\ V$. So the total potential drop is $V_{total} = V_1 + V_{23} + V_4 = (23.6\ V) + (48.9\ V) + (76.5\ V) = 149\ V$, which is equal to V_{emf}. Thus, our answer is consistent.

Concept Check 25.7

As more identical resistors, R, are added to the circuit shown in the figure, the resistance between points A and B will

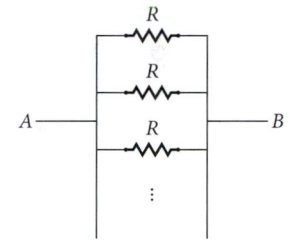

a) increase.

b) stay the same.

c) decrease.

d) change in an unpredictable manner.

Concept Check 25.8

Three light bulbs are connected in series with a battery that delivers a constant potential difference, V_{emf}. When a wire is connected across light bulb 2 as shown in the figure, light bulbs 1 and 3

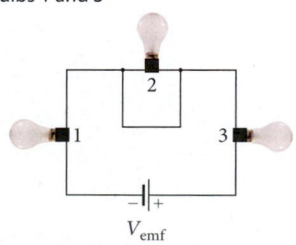

a) burn just as brightly as they did before the wire was connected.

b) burn more brightly than they did before the wire was connected.

c) burn less brightly than they did before the wire was connected.

d) go out.

25.7 Energy and Power in Electric Circuits

Consider a simple circuit in which a source of emf with potential difference ΔV causes a current, i, to flow. The work required from the emf device to move a differential amount of charge, dq, from the negative terminal to the positive terminal (within the emf device) is equal to the increase in electric potential energy of that charge, dU:

$$dU = dq\,\Delta V.$$

Remembering that current is defined as $i = dq/dt$, we can rewrite the differential electric potential energy as

$$dU = i\,dt\,\Delta V.$$

Using the definition of power, $P = dU/dt$, and substituting into it the expression for the differential potential energy, we obtain

$$P = \frac{dU}{dt} = \frac{i\,dt\,\Delta V}{dt} = i\Delta V.$$

Thus, the product of the current times the potential difference gives the power supplied by the source of emf. By conservation of energy, this power is equal to the power dissipated in a circuit containing one resistor. In a more complicated circuit, each resistor will dissipate power at the rate given by this equation, where i and ΔV refer to the current through and potential difference across that resistor. Ohm's Law (equation 25.9) leads to different formulations of the power:

$$P = i\Delta V = i^2 R = \frac{\left(\Delta V\right)^2}{R}. \tag{25.18}$$

The unit of power (as noted in Chapter 5) is the watt (W). Electrical devices, such as light bulbs, are rated in terms of how much power they consume. Your electric bill depends on how much electrical energy your appliances consume, and this energy is measured in kilowatt-hours (kWh).

Where does this energy go? This question will be addressed quantitatively in Chapter 30 when alternating currents are discussed. Qualitatively, much or most of the energy dissipated in resistors is converted into heat. This phenomenon is employed in incandescent lighting, where heating a metal filament to a very high temperature causes it to emit light. The heat dissipated in electrical circuits is a huge problem for large-scale computer systems and server farms for the biggest Internet databases. These computer systems use thousands of processors for computing applications that can be parallelized. All of these processors emit heat, and very expensive cooling has to be provided to offset it. It turns out that the cost of cooling is one of the most stringent boundary conditions limiting the maximum size of these supercomputers.

Some of the power dissipated in circuits can be converted into mechanical energy by motors. The functioning of electric motors requires an understanding of magnetism and will be covered later.

EXAMPLE 25.5 | Temperature Dependence of a Light Bulb's Resistance

A 100-W light bulb is connected in series to a source of emf with $V_{\text{emf}} = 100$ V. When the light bulb is lit, the temperature of its tungsten filament is 2520 °C.

PROBLEM
What is the resistance of the light bulb's tungsten filament at room temperature (20 °C)?

SOLUTION
The resistance of the filament when the light bulb is lit can be obtained using equation 25.18:

$$P = \frac{V_{\text{emf}}^2}{R}.$$

We rearrange this equation and substitute the numerical values to get the resistance of the filament:

$$R = \frac{V_{emf}^2}{P} = \frac{(100 \text{ V})^2}{100 \text{ W}} = 100 \ \Omega.$$

The temperature dependence of the filament's resistance is given by equation 25.13:

$$R - R_0 = R_0 \alpha (T - T_0).$$

We solve for the resistance at room temperature, R_0:

$$R = R_0 + R_0 \alpha (T - T_0) = R_0 \left[1 + \alpha (T - T_0) \right]$$

$$R_0 = \frac{R}{1 + \alpha (T - T_0)}.$$

Using the temperature coefficient of resistivity for tungsten from Table 25.1, we get

$$R_0 = \frac{R}{1 + \alpha (T - T_0)} = \frac{100 \ \Omega}{1 + \left(4.5 \cdot 10^{-3} \ °C^{-1} \right)\left(2520 \ °C - 20 \ °C \right)} = 8.2 \ \Omega.$$

High-Voltage Direct Current Power Transmission

The transmission of electrical power from power-generating stations to users of electricity is of great practical interest. Often, electrical power-generating stations are located in remote areas, and thus the power must be transmitted long distances. This is particularly true for clean power sources, such as hydroelectric dams and large solar farms in deserts.

The power, P, transmitted to users is the product of the current, i, and the potential difference, ΔV, in the power line: $P = i\Delta V$. Thus, the current required for a given power is $i = P/\Delta V$, and a higher potential difference means a lower current in the power line. Equation 25.18 indicates that the power dissipated in an electrical power transmission line, P_{loss}, is given by $P_{loss} = i^2 R$. The resistance, R, of the power line is fixed; thus, decreasing the power lost during transmission means reducing the current carried in the transmission line. This reduction is accomplished by transmitting the power using a very high potential difference and a very low current. Looking at equation 25.18, you might argue that we could also write $P_{loss} = (\Delta V)^2/R$ and that a high potential difference means a large power loss instead of a small power loss. However, ΔV in this equation is the potential drop across the power line, not the potential difference at which the electrical power is being transmitted. The potential drop across the power line is $V_{drop} = iR$, which is much lower than the high potential difference used to transmit the electrical power. The expressions for the transmitted power and the dissipated power can be combined to give $P_{loss} = (P/\Delta V)^2 R = P^2 R/(\Delta V)^2$, which means that for a given amount of power, the dissipated power decreases as the square of the potential difference used to transmit the power.

Normally, electrical power generation and transmission use alternating currents. As we'll see in Chapter 30, alternating currents have the advantage that it is easy to raise or lower the potential difference via transformers. However, alternating currents have the inherent disadvantage of high power losses. High-voltage direct current (HVDC) transmission lines do not have this problem and suffer only power losses due to the resistance of the power line. However, HVDC transmission lines have the extra requirement that alternating current must be converted to direct current for transmission and the direct current must be converted back to alternating current at the destination.

Chapter 5 mentioned the electrical energy produced by the Itaipú Dam on the Paraná River in Brazil and Paraguay. Part of the power produced by this hydroelectric plant is transmitted via the world's largest HVDC transmission line a distance of about 800 km from the Itaipú Dam to São Paulo, Brazil, one of the ten largest metropolitan areas in the world. The transmission line carries 6300 MW of electrical power using direct current with a potential difference of ± 600 kV. The station at the Itaipú Dam that converts the alternating current to direct current is shown in Figure 25.21a. The station in São Paulo that converts the transmitted direct current back to alternating current is shown in Figure 25.21b.

(a)

(b)

FIGURE 25.21 (a) The station that converts alternating current to direct current at the Itaipú Dam on the Paraná River in Brazil and Paraguay. (b) The station that converts the transmitted direct current back to alternating current in São Paulo, Brazil.

Self-Test Opportunity 25.3

Consider a battery with internal resistance R_i. What external resistance, R, will undergo the maximum heating when connected to this battery?

Future applications of HVDC power transmission include the transmission of power from solar power stations located in remote areas in the southwest United States to densely populated areas, such as large cities in California and Texas. Some of the most promising research on power transmission is focusing on the use of superconducting wires for this purpose. As mentioned earlier, once a material becomes superconducting, its resistance approaches zero. Therefore, it can transmit electrical power without significant losses. The most promising materials for superconducting power transmission wires are *high-temperature superconductors*, discovered only in 1986. These wires have to be cooled with liquid nitrogen, but we are approaching the point where their widespread use in long-distance power transmission can become a reality.

SOLVED PROBLEM 25.4 | Size of Wire for a Power Line

PROBLEM

Imagine you are designing the HVDC power line from the Itaipú Dam on the Paraná River in Brazil and Paraguay to the city of São Paulo in Brazil. The power line is 800 km long and transmits 6300 MW of power at a potential difference of 1.20 MV. (Figure 25.22 shows an HVDC line.)

The electric company requires that no more than 25% of the power be lost in transmission. If the line consists of one wire made out of copper and having a circular cross section, what is the minimum diameter of the wire?

SOLUTION

THINK Knowing the power transmitted and the potential difference with which it is transmitted, we can calculate the current carried in the line. We can then express the power lost in terms of the resistance of the transmission line. With the current and the resistance of the wire, we can write an expression for the power lost during transmission. The resistance of the wire is a function of the diameter of the wire, the length of the wire, and the resistivity of copper. We can then solve for the diameter of the wire that will keep the power loss within the specified limit.

FIGURE 25.22 HVDC power transmission line.

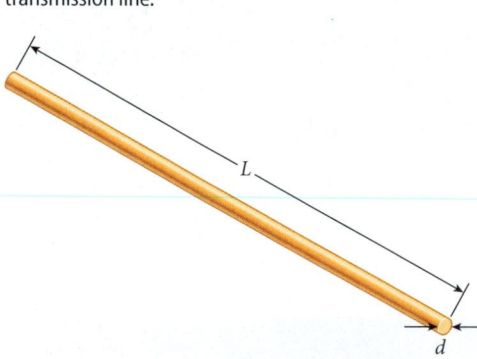

FIGURE 25.23 An HVDC transmission line consisting of a copper conductor (not to scale).

SKETCH A sketch of a copper wire of length L and diameter d is shown in Figure 25.23.

RESEARCH The power, P, carried in the line is related to the current, i, and the potential difference, ΔV: $P = i\Delta V$. The power lost in transmission, P_{lost}, can be related (see equation 25.18) to the current in the wire and the resistance, R, of the wire:

$$P_{lost} = i^2 R. \tag{i}$$

The resistance of the wire is given by equation 25.11:

$$R = \rho_{Cu} \frac{L}{A}, \tag{ii}$$

where ρ_{Cu} is the resistivity of copper, L is the length of the wire, and A is the cross-sectional area of the wire.

The cross-sectional area of the wire is the area of a circle:

$$A = \pi \left(\frac{d}{2}\right)^2 = \frac{\pi d^2}{4},$$

where d is the diameter of the wire. Thus, with the area of a circle substituted for A, equation (ii) becomes

$$R = \rho_{Cu} \frac{L}{\pi d^2 / 4}. \tag{iii}$$

SIMPLIFY We can solve $P = i\Delta V$ for the current in the wire:

$$i = \frac{P}{\Delta V}.$$

Substituting this expression for the current and that for the resistance from equation (iii) into equation (i) for the lost power gives

$$P_{\text{lost}} = \left(\frac{P}{\Delta V}\right)^2 \left(\rho_{\text{Cu}} \frac{L}{\pi d^2/4}\right) = \frac{4P^2 \rho_{\text{Cu}} L}{\pi (\Delta V)^2 d^2}.$$

The fraction of lost power relative to total power, f, is

$$\frac{P_{\text{lost}}}{P} = \frac{\left(\dfrac{4P^2 \rho_{\text{Cu}} L}{\pi (\Delta V)^2 d^2}\right)}{P} = \frac{4P \rho_{\text{Cu}} L}{\pi (\Delta V)^2 d^2} = f.$$

Solving this equation for the diameter of the wire gives

$$d = \sqrt{\frac{4P \rho_{\text{Cu}} L}{f \pi (\Delta V)^2}}.$$

CALCULATE   Putting in the numerical values gives us

$$d = \sqrt{\frac{4\left(6300 \cdot 10^6 \text{ W}\right)\left(1.72 \cdot 10^{-8} \text{ } \Omega \text{ m}\right)\left(800 \cdot 10^3 \text{ m}\right)}{(0.25)\pi\left(1.20 \cdot 10^6 \text{ V}\right)^2}} = 0.0175099 \text{ m}.$$

ROUND   Rounding to three significant figures gives us the minimum diameter of the copper wire:

$$d = 1.75 \text{ cm}.$$

DOUBLE-CHECK   To double-check our result, let's calculate the resistance of this transmission line. Using our calculated value for the diameter, we can find the cross-sectional area and then, using equation 25.11, obtain

$$R = \rho_{\text{Cu}} \frac{L}{\pi d^2/4} = \frac{4\rho_{\text{Cu}} L}{\pi d^2} = \frac{4\left(1.72 \cdot 10^{-8} \text{ } \Omega \text{ m}\right)\left(800 \cdot 10^3 \text{ m}\right)}{\pi \left(1.75 \cdot 10^{-2} \text{ m}\right)^2} = 57.2 \text{ } \Omega.$$

The current transmitted is

$$i = \frac{P}{V} = \frac{6300 \cdot 10^6 \text{ W}}{1.20 \cdot 10^6 \text{ V}} = 5250 \text{ A}.$$

The power lost is then

$$P = i^2 R = (5250 \text{ A})^2 (57.2 \text{ } \Omega) = 1580 \text{ MW},$$

which is close (within rounding error) to 25% of the total power of 6300 MW. Thus, our result seems reasonable.

FIGURE 25.24   (a) The circuit of Figure 25.2c, but with a diode included. (b) Reversing the potential difference from the battery causes the current to stop flowing and the light bulb to stop shining.

25.8   Diodes: One-Way Streets in Circuits

Section 25.4 stated that many resistors obey Ohm's Law. It was noted, however, that there are also non-ohmic resistors that do not obey Ohm's Law. A very common and extremely useful example is a diode. A *diode* is an electronic device that is designed to conduct current in one direction and not in the other direction. Remember that Figure 25.2c showed that a light bulb was still shining with the same intensity when the battery it was connected to was reversed. If a diode (represented by the symbol ▶⊢) is added to the same circuit, the diode prevents the current from flowing when the potential difference delivered by the battery is reversed; see Figure 25.24. The diode acts like a one-way street for the current.

Figure 25.25 shows current versus potential difference for a 3-Ω ohmic resistor and a silicon diode. The resistor obeys Ohm's Law, with the current flowing in the opposite direction when the potential difference is negative. The plot of current versus potential difference for the resistor is a straight line with a slope of $\frac{1}{3}$ Ω. The silicon diode is wired so that it will not conduct

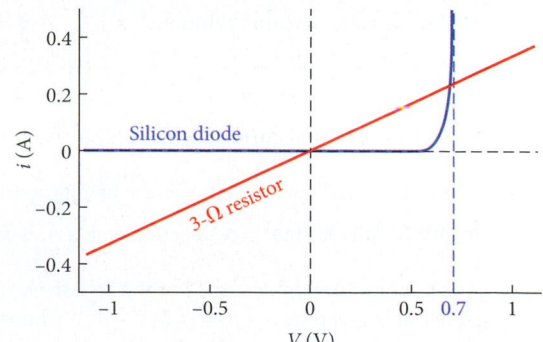

FIGURE 25.25   Current as a function of potential difference for a resistor (red) and a diode (blue).

FIGURE 25.26 Giant LED screen used during the opening ceremony of the 2008 Olympic Games in Beijing.

Self-Test Opportunity 25.4

Suppose the battery in Figure 25.24 has a potential difference of 1.5 V across its terminals and the diode is a silicon diode like the one in Figure 25.25. What are the potential drops across the diode and the light bulb in parts (a) and (b) of Figure 25.24?

any current when there is a negative potential difference. This silicon diode, like most, will conduct current if the potential difference is above 0.7 V. For potential differences above this threshold, the diode is essentially a conductor; below this threshold, the diode will not conduct current. The turn-on of the diode above the threshold potential difference increases exponentially; it can be close to instantaneous, as is visible in Figure 25.25.

Diodes are very useful for converting alternating current to direct current, as we'll see in Chapter 30. The fundamental physics principles that underlie the functioning of diodes require an understanding of quantum mechanics.

One particularly useful kind of diode is the light-emitting diode (LED), which not only regulates current in a circuit but also emits light of a single wavelength in a very controlled way. LEDs that emit light of many different wavelengths have been manufactured, and they emit light much more efficiently than conventional incandescent bulbs do. Light intensity is measured in lumens (lm). Light sources can be compared in terms of how many lumens they produce per watt of electrical power. During the last decade, intensive research into LED technology has resulted in huge increases in LED output efficiency, which has reached values of 130 to 170 lm/W. This compares very favorably with conventional incandescent lights (which are in the range from 5 to 20 lm/W), halogen lights (20 to 30 lm/W), and even fluorescent high-efficiency lights (30 to 95 lm/W). Prices for LEDs (in particular, "white" LEDs) are still comparatively high but are expected to decrease significantly. The United States uses over 100 billion kWh of electrical energy for lighting alone each year, which is approximately 10% of the total U.S. energy consumption. Universal use of LED lighting could save 70% to 90% of those 100 billion kWh, approximately the annual energy output of 10 nuclear power plants (~1 GW power each).

LEDs are also used in large display screens, where high light output is desirable. Perhaps the most impressive of these was showcased during the opening ceremony of the 2008 Beijing Olympics (Figure 25.26). It used 44,000 individual LEDs and measured an astounding 147 m by 22 m.

WHAT WE HAVE LEARNED | EXAM STUDY GUIDE

- Current, i, is defined as the rate at which charge, q, flows past a particular point: $i = \dfrac{dq}{dt}$.

- The magnitude of the average current density, J, at a given cross-sectional area, A, in a conductor is given by $J = \dfrac{i}{A}$.

- The magnitude of the current density, J, is related to the magnitude of the drift velocity, v_d, of the current-carrying charges, $-e$, by $J = \dfrac{i}{A} = -nev_d$, where n is the number of charge carriers per unit volume.

- The resistivity, ρ, of a material is defined in terms of the magnitudes of the electric field applied across the material, E, and the resulting current density, J: $\rho = \dfrac{E}{J}$.

- The resistance, R, of a specific device having resistivity ρ, length L, and constant cross-sectional area A, is $R = \rho \dfrac{L}{A}$.

- The temperature dependence of the resistivity of a material is given by $\rho - \rho_0 = \rho_0 \alpha(T - T_0)$, where ρ is the final resistivity, ρ_0 is the initial resistivity, α is the

temperature coefficient of electric resistivity, T is the final temperature, and T_0 is the initial temperature.

- The electromotive force, or emf, is a potential difference created by a device that drives current through a circuit.

- Ohm's Law states that when a potential difference, ΔV, appears across a resistor, R, the current, i, flowing through the resistor is $i = \dfrac{\Delta V}{R}$.

- Resistors connected in series can be replaced with an equivalent resistance, R_{eq}, given by the sum of the resistances of the resistors: $R_{eq} = \displaystyle\sum_{i=1}^{n} R_i$.

- Resistors connected in parallel can be replaced with an equivalent resistance, R_{eq}, given by $\dfrac{1}{R_{eq}} = \displaystyle\sum_{i=1}^{n} \dfrac{1}{R_i}$.

- The power, P, dissipated by a resistor, R, through which a current, i, flows is given by $P = i\Delta V = i^2 R = \dfrac{(\Delta V)^2}{R}$, where ΔV is the potential drop across the resistor.

ANSWERS TO SELF-TEST OPPORTUNITIES

25.1 $\dfrac{700 \text{ mAh}}{0.1 \text{ mA}} = 7000 \text{ h} \approx 292 \text{ days.}$

25.2 $\Delta V = iR \Rightarrow i = \dfrac{\Delta V}{R} = \dfrac{1.50 \text{ V}}{10.0 \ \Omega} = 0.150 \text{ A.}$

25.3 The maximum heating of the external resistance occurs when the external resistance is equal to the internal resistance.

$$V_t = V_{emf} + iR_i = i\left(R + R_i\right)$$

$$P_{heat} = i^2 R = \dfrac{V_t^2 R}{\left(R + R_i\right)^2}$$

$$\dfrac{dP_{heat}}{dR} = -\dfrac{2V_t^2 R}{\left(R + R_i\right)^3} + \dfrac{V_t^2}{\left(R + R_i\right)^2} = 0 \text{ at extremum.}$$

From this, it follows that $R = R_i$.

You can check that this extremum is a maximum by taking the second derivative at $R = R_i$:

$$\left.\dfrac{d^2 P_{heat}}{dR^2}\right|_{R=R_i} = V_t^2\left(\dfrac{6}{16R^3} - \dfrac{5}{8R^3}\right) = -\dfrac{V_t^2}{R^3}\left(\dfrac{1}{2}\right) < 0.$$

25.4 The sum of the two potential drops, across the diode and across the light bulb, has to equal 1.5 V, the potential difference supplied by the battery. In part (b), the diode prevents any current from flowing; therefore, the potential drop across the diode is 1.5 V and that across the light bulb is zero. In part (a), the potential drop across the diode is 0.7 V (see Figure 25.25), and therefore that across the light bulb is 1.5 V – 0.7 V = 0.8 V.

PROBLEM-SOLVING GUIDELINES

1. If a circuit diagram is not given as part of the problem statement, draw one yourself and label all the given values and unknown components. Indicate the direction of the current, starting from the emf source. (Don't worry about getting the direction of the current wrong in your diagram; if you guess wrong, your final answer for the current will be a negative number.)

2. Sources of emf supply potential to a circuit, and resistors reduce potential in the circuit. However, be careful to check the direction of the potential of the emf source relative to

that of the current; a current flowing in the direction opposite to the potential of an emf device picks up a negative potential difference.

3. The sum of the potential drops across resistors in a circuit equals the net amount of emf supplied to the circuit. (This is a consequence of the law of conservation of energy.)

4. In any given wire segment, the current is the same everywhere. (This is a consequence of the law of conservation of charge.)

MULTIPLE-CHOICE QUESTIONS

25.1 If the current through a resistor is increased by a factor of 2, how does this affect the power that is dissipated?

a) It decreases by a factor of 4.

b) It increases by a factor of 2.

c) It decreases by a factor of 8.

d) It increases by a factor of 4.

25.2 You make a parallel connection between two resistors, resistor A having a very large resistance and resistor B having a very small resistance. The equivalent resistance for this combination will be

a) slightly greater than the resistance of resistor A.

b) slightly less than the resistance of resistor A.

c) slightly greater than the resistance of resistor B.

d) slightly less than the resistance of resistor B.

25.3 Two cylindrical wires, 1 and 2, made of the same material, have the same resistance. If the length of wire 2 is twice that of wire 1, what is the ratio of their cross-sectional areas, A_1 and A_2?

a) $A_1/A_2 = 2$

b) $A_1/A_2 = 4$

c) $A_1/A_2 = 0.5$

d) $A_1/A_2 = 0.25$

25.4 All three light bulbs in the circuit shown in the figure are identical. Which of the three shines the brightest?

a) A

b) B

c) C

d) A and B

e) All three are equally bright.

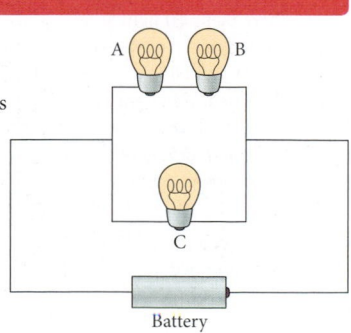

25.5 All of the six light bulbs in the circuit shown in the figure are identical. Which ordering correctly expresses the relative brightness of the bulbs? (*Hint:* The more current flowing through a light bulb, the brighter it is!)

a) $A = B > C = D > E = F$

b) $A = B = E = F > C = D$

c) $C = D > A = B = E = F$

d) $A = B = C = D = E = F$

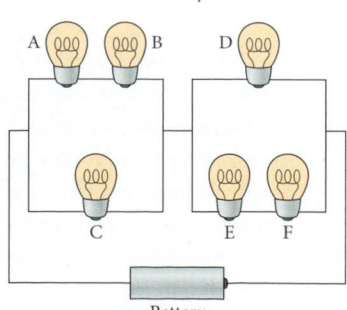

25.6 Which of the arrangements of three identical light bulbs shown in the figure draws the most current from the battery?

a) A   d) All three draw equal current.

b) B   e) A and C are tied for drawing the most current.

c) C

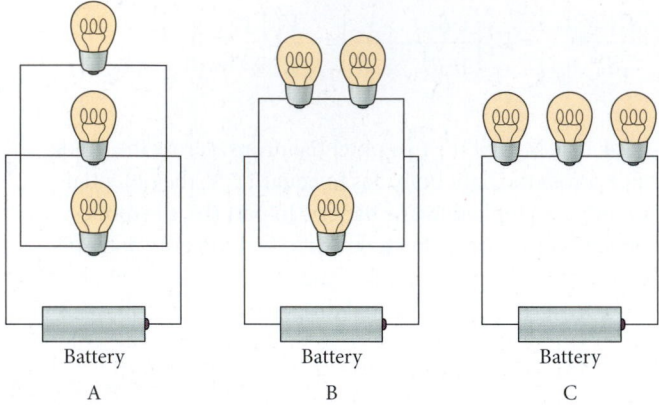

25.7 Which of the arrangements of three identical light bulbs shown in the figure has the highest resistance?

a) A   d) All three have equal resistance.

b) B   e) A and C are tied for having the highest resistance.

c) C

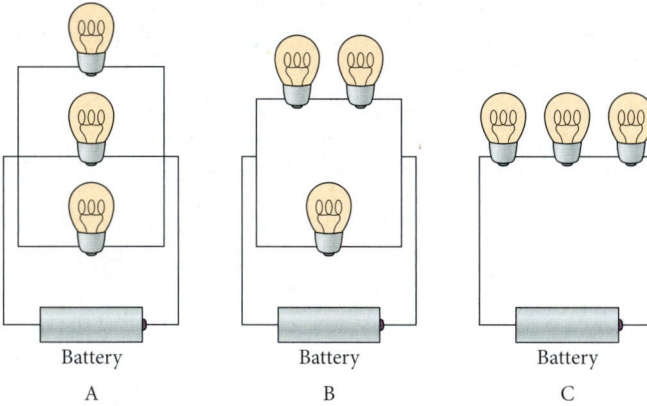

25.8 Three identical light bulbs are connected as shown in the figure. Initially the switch is closed. When the switch is opened (as shown in the figure), bulb C goes off. What happens to bulbs A and B?

a) Bulb A gets brighter, and bulb B gets dimmer.

b) Both bulbs A and B get brighter.

c) Both bulbs A and B get dimmer.

d) Bulb A gets dimmer, and bulb B gets brighter.

25.9 Which of the following wires has the largest current flowing through it?

a) a 1-m-long copper wire of diameter 1 mm connected to a 10-V battery

b) a 0.5-m-long copper wire of diameter 0.5 mm connected to a 5-V battery

c) a 2-m-long copper wire of diameter 2 mm connected to a 20-V battery

d) a 1-m-long copper wire of diameter 0.5 mm connected to a 5-V battery

e) All of the wires have the same current flowing through them.

25.10 Ohm's Law states that the potential difference across a device is equal to

a) the current flowing through the device times the resistance of the device.

b) the current flowing through the device divided by the resistance of the device.

c) the resistance of the device divided by the current flowing through the device.

d) the current flowing through the device times the cross-sectional area of the device.

e) the current flowing through the device times the length of the device.

25.11 A constant electric field is maintained inside a semiconductor. As the temperature is lowered, the magnitude of the current density inside the semiconductor

a) increases.   c) decreases.

b) stays the same.   d) may increase or decrease.

25.12 Which of the following is an incorrect statement?

a) The currents through electronic devices connected in series are equal.

b) The potential drops across electronic devices connected in parallel are equal.

c) More current flows across the smaller resistance when two resistors are connected in parallel.

d) More current flows across the smaller resistance when two resistors are connected in series.

25.13 Identical batteries are connected in three different arrangements to the same light bulb as shown in the figure. Assume that the batteries have no internal resistance. In which arrangement will the light bulb shine the brightest?

a) A   d) The bulb will have the same brightness in all three arrangements.

b) B

c) C   e) The bulb will not light in any of the arrangements.

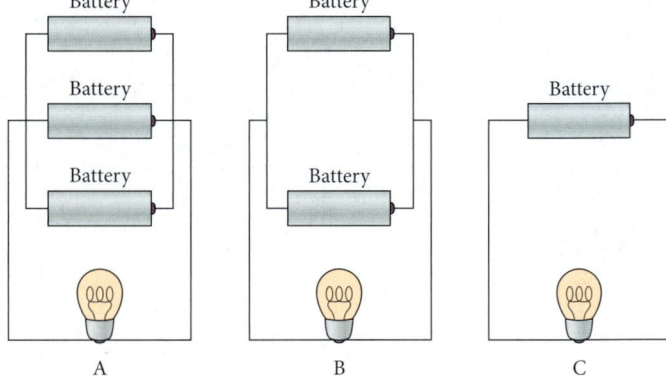

25.14 Identical batteries are connected in three different arrangements to the same light bulb as shown in the figure. Assume that the batteries have no internal resistance. In which arrangement will the light bulb shine the brightest?

a) A   d) The bulb will have the same brightness in all three arrangements.

b) B

c) C   e) The bulb will not light in any of the arrangements.

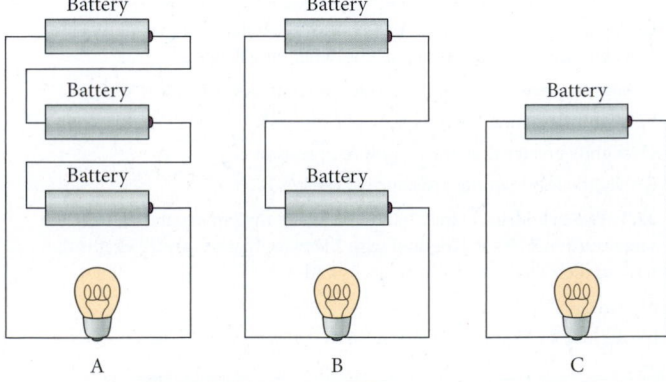

CONCEPTUAL QUESTIONS

25.15 What would happen to the drift velocity of electrons in a wire if the resistance due to collisions between the electrons and the atoms in the crystal lattice of the metal disappeared?

25.16 Why do light bulbs typically burn out just as they are turned on rather than while they are lit?

25.17 Two identical light bulbs are connected to a battery. Will the light bulbs be brighter if they are connected in series or in parallel?

25.18 Two resistors with resistances R_1 and R_2 are connected in parallel. Demonstrate that, no matter what the actual values of R_1 and R_2 are, the equivalent resistance is always less than the smaller of the two resistances.

25.19 Show that for resistors connected in series, it is always the highest resistance that dissipates the most power, while for resistors connected in parallel, it is always the lowest resistance that dissipates the most power.

25.20 For the connections shown in the figure, determine the current i_1 in terms of the total current, i, and R_1 and R_2.

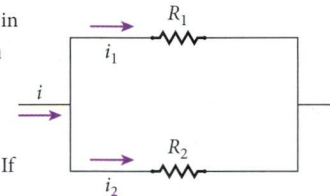

25.21 An infinite number of resistors are connected in parallel. If $R_1 = 10 \ \Omega$, $R_2 = 10^2 \ \Omega$, $R_3 = 10^3 \ \Omega$, and so on, show that $R_{eq} = 9 \ \Omega$.

25.22 You are given two identical batteries and two pieces of wire. The red wire has a higher resistance than the black wire. You place the red wire across the terminals of one battery and the black wire across the terminals of the other battery. Which wire gets hotter?

25.23 Should light bulbs (ordinary incandescent bulbs with tungsten filaments) be considered ohmic resistors? Why or why not? How would this be determined experimentally?

25.24 A charged-particle beam is used to inject a charge, Q_0, into a small, irregularly shaped region (not a cavity, just some region within the solid block) in the interior of a block of ohmic material with conductivity σ and permittivity ϵ at time $t = 0$. Eventually, all the injected charge will move to the outer surface of the block, but how quickly?

a) Derive a differential equation for the charge, $Q(t)$, in the injection region as a function of time.

b) Solve the equation from part (a) to find $Q(t)$ for all $t \geq 0$.

c) For copper, a good conductor, and for quartz (crystalline SiO_2), an insulator, calculate the time for the charge in the injection region to decrease by half. Look up the necessary values. Assume that the effective "dielectric constant" of copper is 1.00000.

25.25 Show that the drift speed of free electrons in a wire does not depend on the cross-sectional area of the wire.

25.26 Rank the brightness of the six identical light bulbs in the circuit in the figure. Each light bulb may be treated as an identical resistor with resistance R.

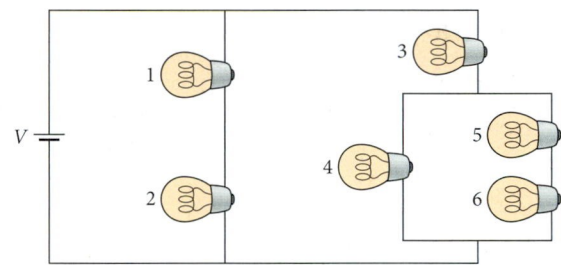

25.27 Two conductors of the same length and radius are connected to the same emf device. If the resistance of one is twice that of the other, to which conductor is more power delivered?

EXERCISES

A blue problem number indicates a worked-out solution is available in the Student Solutions Manual. One • and two •• indicate increasing level of problem difficulty.

Sections 25.1 and 25.2

25.28 How many protons are in the beam traveling close to the speed of light in the Tevatron at Fermilab, which is carrying 11 mA of current around the 6.3-km circumference of the main Tevatron ring?

25.29 What is the current density in an aluminum wire having a radius of 1.00 mm and carrying a current of 1.00 mA? What is the drift speed of the electrons carrying this current? The density of aluminum is $2.70 \cdot 10^3 \ kg/m^3$, and 1 mole of aluminum has a mass of 26.98 g. There is one conduction electron per atom in aluminum.

•**25.30** A copper wire has a diameter $d_{Cu} = 0.0500$ cm, is 3.00 m long, and has a charge-carrier density of $8.50 \cdot 10^{28}$ electrons/m^3. As shown in the figure, the copper wire is attached to an equal length of aluminum wire with a diameter $d_{Al} = 0.0100$ cm and a charge-carrier density of $6.02 \cdot 10^{28}$ electrons/m^3. A current of 0.400 A flows through the copper wire.

a) What is the ratio of the current densities in the two wires, J_{Cu}/J_{Al}?

b) What is the ratio of the drift velocities in the two wires, v_{d-Cu}/v_{d-Al}?

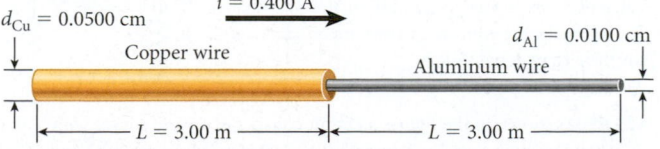

•**25.31** A current of 0.123 mA flows in a silver wire whose cross-sectional area is 0.923 mm^2.

a) Find the density of electrons in the wire, assuming that there is one conduction electron per silver atom.

b) Find the current density in the wire assuming that the current is uniform.

c) Find the electrons' drift speed.

Section 25.3

25.32 What is the resistance of a copper wire of length $l = 10.9$ m and diameter $d = 1.30$ mm? The resistivity of copper is $1.72 \cdot 10^{-8} \ \Omega$ m.

25.33 Two conductors are made of the same material and have the same length L. Conductor A is a hollow tube with inside diameter 2.00 mm and outside diameter 3.00 mm; conductor B is a solid wire with radius R_B. What value of R_B is required for the two conductors to have the same resistance measured between their ends?

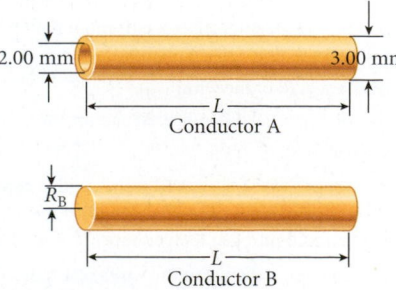

25.34 A copper coil has a resistance of 0.100 Ω at room temperature (20.0 °C). What is its resistance when it is cooled to –100. °C?

25.35 What gauge of aluminum wire will have the same resistance per unit length as 12-gauge copper wire?

25.36 A rectangular wafer of pure silicon, with resistivity $\rho = 2300\ \Omega$ m, measures 2.00 cm by 3.00 cm by 0.0100 cm. Find the maximum resistance of this rectangular wafer between any two faces.

•**25.37** A copper wire that is 1.00 m long and has a radius of 0.500 mm is stretched to a length of 2.00 m. What is the fractional change in resistance, $\Delta R/R$, as the wire is stretched? What is $\Delta R/R$ for a wire of the same initial dimensions made out of aluminum?

•**25.38** The most common material used for sandpaper, silicon carbide, is also widely used in electrical applications. One common device is a tubular resistor made of a special grade of silicon carbide called *carborundum*. A particular carborundum resistor (see the figure) consists of a thick-walled cylindrical shell (a pipe) of inner radius $a = 1.50$ cm, outer radius $b = 2.50$ cm, and length $L = 60.0$ cm. The resistance of this carborundum resistor at 20.0 °C is 1.00 Ω.

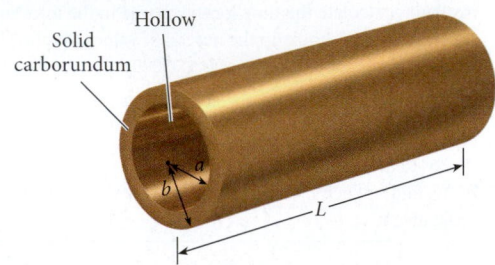

a) Calculate the resistivity of carborundum at room temperature. Compare this to the resistivities of the most commonly used conductors (copper, aluminum, and silver).

b) Carborundum has a high temperature coefficient of resistivity: $\alpha = 2.14 \cdot 10^{-3}\ \text{K}^{-1}$. If, in a particular application, the carborundum resistor heats up to 300. °C, what is the percentage change in its resistance between room temperature (20.0 °C) and this operating temperature?

••**25.39** As illustrated in the figure, a current, i, flows through the junction of two materials with the same cross-sectional area and with conductivities σ_1 and σ_2. Show that the total amount of charge at the junction is $\epsilon_0 i(1/\sigma_2 - 1/\sigma_1)$.

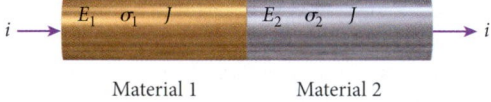

Material 1   Material 2

Section 25.4

25.40 A potential difference of 12.0 V is applied across a wire of cross-sectional area 4.50 mm^2 and length 1000. km. The current passing through the wire is $3.20 \cdot 10^{-3}$ A.

a) What is the resistance of the wire?

b) What type of wire is this?

25.41 One brand of 12.0-V automotive battery used to be advertised as providing "600 cold-cranking amps." Assuming that this is the current the battery supplies if its terminals are shorted, that is, connected to negligible resistance, determine the internal resistance of the battery. (*IMPORTANT:* **Do not** attempt such a connection as it could be lethal!)

25.42 A copper wire has radius $r = 0.0250$ cm, is 3.00 m long, has resistivity $\rho = 1.72 \cdot 10^{-8}\ \Omega$ m, and carries a current of 0.400 A. The wire has a charge-carrier density of $8.50 \cdot 10^{28}$ electrons/m^3.

a) What is the resistance, R, of the wire?

b) What is the electric potential difference, ΔV, across the wire?

c) What is the electric field, E, in the wire?

•**25.43** A 34-gauge copper wire ($A = 0.0201$ mm^2), with a constant potential difference of 0.100 V applied across its 1.00 m length at room temperature (20.0 °C), is cooled to liquid nitrogen temperature (77 K = –196 °C).

a) Determine the percentage change in the wire's resistance during the drop in temperature.

b) Determine the percentage change in current flowing in the wire.

c) Compare the drift speeds of the electrons at the two temperatures.

Section 25.5

25.44 A resistor of unknown resistance and a 35.0-Ω resistor are connected across a 120.-V emf device in such a way that an 11.0-A current flows. What is the value of the unknown resistance?

25.45 A battery has a potential difference of 14.50 V when it is not connected in a circuit. When a 17.91-Ω resistor is connected across the battery, the potential difference of the battery drops to 12.68 V. What is the internal resistance of the battery?

25.46 When a battery is connected to a 100.-Ω resistor, the current is 4.00 A. When the same battery is connected to a 400.-Ω resistor, the current is 1.01 A. Find the emf supplied by the battery and the internal resistance of the battery.

•**25.47** A light bulb is connected to a source of emf. There is a 6.20 V drop across the light bulb and a current of 4.10 A flowing through the light bulb.

a) What is the resistance of the light bulb?

b) A second light bulb, identical to the first, is connected in series with the first bulb. The potential drop across the bulbs is now 6.29 V, and the current through the bulbs is 2.90 A. Calculate the resistance of each light bulb.

c) Why are your answers to parts (a) and (b) not the same?

Section 25.6

25.48 What is the current in the 10.0-Ω resistor in the circuit in the figure?

25.49 What is the equivalent resistance of the five resistors in the circuit in the figure?

•**25.50** What is the current in the circuit shown in the figure when the switch is (a) open and (b) closed?

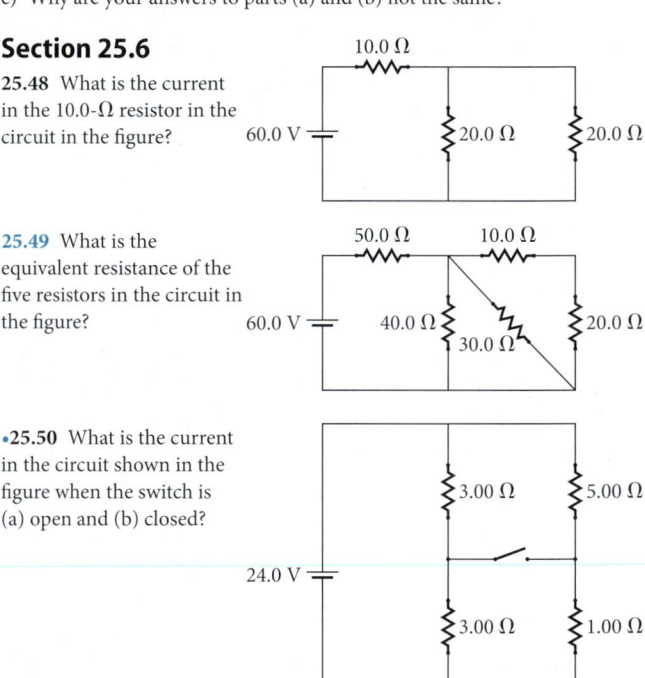

25.51 For the circuit shown in the figure, $R_1 = 6.00\ \Omega$, $R_2 = 6.00\ \Omega$, $R_3 = 2.00\ \Omega$, $R_4 = 4.00\ \Omega$, $R_5 = 3.00\ \Omega$, and the potential difference is 12.0 V.

a) What is the equivalent resistance for the circuit?

b) What is the current through R_5?

c) What is the potential drop across R_3?

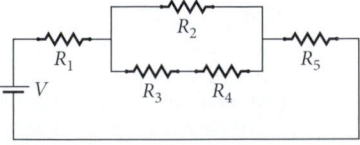

25.52 Four resistors are connected in a circuit as shown in the figure. What value of R_1, expressed as a multiple of R_0, will make the equivalent resistance for the circuit equal to R_0?

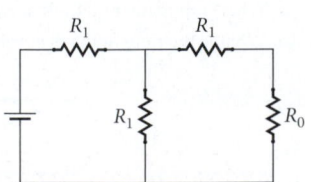

•**25.53** As shown in the figure, a circuit consists of an emf source with $V = 20.0$ V and six resistors. Resistors $R_1 = 5.00\ \Omega$ and $R_2 = 10.00\ \Omega$ are connected in series. Resistors $R_3 = 5.00\ \Omega$ and $R_4 = 5.00\ \Omega$ are connected in parallel and are in series with R_1 and R_2. Resistors $R_5 = 2.00\ \Omega$ and $R_6 = 2.00\ \Omega$ are connected in parallel and are also in series with R_1 and R_2.

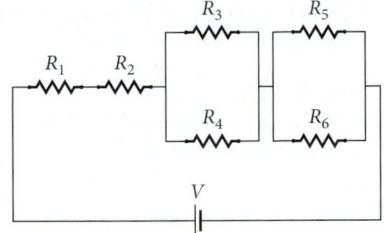

a) What is the potential drop across each resistor?

b) How much current flows through each resistor?

•**25.54** When a 40.0-V emf device is placed across two resistors in series, a current of 10.0 A flows through each of the resistors. When the same emf device is placed across the same two resistors in parallel, the current through the emf device is 50.0 A. What is the magnitude of the larger of the two resistances?

Section 25.7

25.55 A voltage spike causes the line voltage in a home to jump rapidly from 110. V to 150. V. What is the percentage increase in the power output of a 100.-W tungsten-filament incandescent light bulb during this spike, assuming that the bulb's resistance remains constant?

25.56 A thundercloud similar to the one described in Example 24.3 produces a lightning bolt that strikes a radio tower. If the lightning bolt transfers 5.00 C of charge in about 0.100 ms and the potential remains constant at 70.0 MV, find (a) the average current, (b) the average power, (c) the total energy, and (d) the effective resistance of the air during the lightning strike.

25.57 A hair dryer consumes 1600. W of power and operates at 110. V. (Assume that the current is DC. In fact, these are root-mean-square values of AC quantities, but the calculation is not affected. Chapter 30 covers AC circuits in detail.)

a) Will the hair dryer trip a circuit breaker designed to interrupt the circuit if the current exceeds 15.0 A?

b) What is the resistance of the hair dryer when it is operating?

25.58 How much money will a homeowner owe an electric company if he turns on a 100.00-W incandescent light bulb and leaves it on for an entire year? (Assume that the cost of electricity is $0.12000/kWh and that the light bulb lasts that long.) The same amount of light can be provided by a 26.000-W compact fluorescent light bulb. What would it cost the home-owner to leave one of those on for a year?

25.59 Three resistors are connected across a battery as shown in the figure.

a) How much power is dissipated across the three resistors?

b) Determine the potential drop across each resistor.

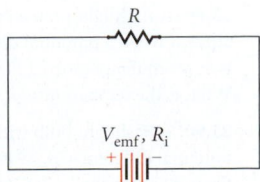

25.60 Suppose an AAA battery is able to supply 625 mAh before its potential drops below 1.50 V. How long will it be able to supply power to a 5.00-W bulb before the potential drops below 1.50 V?

•**25.61** Show that the power supplied to the circuit in the figure by the battery with internal resistance R_i is maximum when the resistance of the resistor in the circuit, R, is equal to R_i. Determine the power supplied to R. For practice, calculate the power dissipated by a 12.0-V battery with an internal resistance of 2.00 Ω when $R = 1.00\ \Omega$, $R = 2.00\ \Omega$, and $R = 3.00\ \Omega$.

•**25.62** A water heater consisting of a metal coil that is connected across the terminals of a 15.0-V power supply is able to heat 250 mL of water from room temperature to boiling point in 45.0 s. What is the resistance of the coil?

•**25.63** A potential difference of $V = 0.500$ V is applied across a block of silicon with resistivity $8.70 \cdot 10^{-4}\ \Omega$ m. As indicated in the figure, the dimensions of the silicon block are width $a = 2.00$ mm and length $L = 15.0$ cm. The resistance of the silicon block is 50.0 Ω, and the density of charge carriers is $1.23 \cdot 10^{23}\ \text{m}^{-3}$. Assume that the current density in the block is uniform and that current flows in silicon according to Ohm's Law. The total length of 0.500-mm-diameter copper wire in the circuit is 75.0 cm, and the resistivity of copper is $1.69 \cdot 10^{-8}\ \Omega$ m.

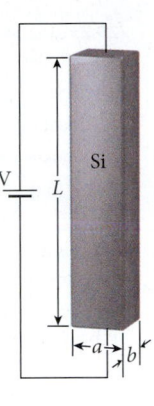

a) What is the resistance, R_w, of the copper wire?

b) What are the direction and the magnitude of the electric current, i, in the block?

c) What is the thickness, b, of the block?

d) On average, how long does it take an electron to pass from one end of the block to the other?

e) How much power, P, is dissipated by the block?

f) As what form of energy does this dissipated power appear?

Additional Exercises

25.64 In an emergency, you need to run a radio that uses 30.0 W of power when attached to a 10.0-V power supply. The only power supply you have access to provides 25.0 kV, but you do have a large number of 25.0-Ω resistors. If you want the power to the radio to be as close as possible to 30.0 W, how many resistors should you use, and how should they be connected (in series or in parallel)?

25.65 A certain brand of hot dog cooker applies a potential difference of 120. V to opposite ends of the hot dog and cooks it by means of the heat produced. If 48.0 kJ is needed to cook each hot dog, what current is needed to cook three hot dogs simultaneously in 2.00 min? Assume a parallel connection.

25.66 A circuit consists of a copper wire of length 10.0 m and radius 1.00 mm connected to a 10.0-V battery. An aluminum wire of length 5.00 m is connected to the same battery and dissipates the same amount of power. What is the radius of the aluminum wire?

25.67 The resistivity of a conductor is $\rho = 1.00 \cdot 10^{-5}\ \Omega$ m. If a cylindrical wire is made of this conductor, with a cross-sectional area of $1.00 \cdot 10^{-6}\ \text{m}^2$, what should the length of the wire be for its resistance to be 10.0 Ω?

25.68 Two cylindrical wires of identical length are made of copper and aluminum. If they carry the same current and have the same potential difference across their length, what is the ratio of their radii?

25.69 Two resistors with resistances 200. Ω and 400. Ω are connected (a) in series and (b) in parallel with an ideal 9.00-V battery. Compare the power delivered to the 200.-Ω resistor.

25.70 What are (a) the conductance and (b) the radius of a 3.50-m-long iron heating element for a 110.-V, 1500.-W heater?

25.71 A 100.-W, 240.-V European light bulb is used in an American household, where the electricity is delivered at 120. V. What power will it consume?

25.72 A modern house is wired for 115 V, and the current is limited by circuit breakers to a maximum of 200. A. (For the purpose of this problem, treat these as DC quantities.)

a) Calculate the minimum total resistance the circuitry in the house can have at any time.

b) Calculate the maximum electrical power the house can consume.

•**25.73** A 12.0-V battery with an internal resistance $R_i = 4.00\ \Omega$ is attached across an external resistor of resistance R. Find the maximum power that can be delivered to the resistor.

•**25.74** A multiclad wire consists of a zinc core of radius 1.00 mm surrounded by a copper sheath of thickness 1.00 mm. The resistivity of zinc is $\rho = 5.964 \cdot 10^{-8}\ \Omega$ m. What is the resistance of a 10.0-m-long strand of this wire?

•**25.75** The Stanford Linear Accelerator accelerated a beam consisting of $2.0 \cdot 10^{14}$ electrons per second through a potential difference of $2.0 \cdot 10^{10}$ V.

a) Calculate the current in the beam.

b) Calculate the power of the beam.

c) Calculate the effective ohmic resistance of the accelerator.

•**25.76** In the circuit shown in the figure, $R_1 = 3.00 \ \Omega$, $R_2 = 6.00 \ \Omega$, $R_3 = 20.0 \ \Omega$, and $V_{emf} = 12.0$ V.

a) Determine a value for the equivalent resistance.

b) Calculate the magnitude of the current flowing through R_3 on the top branch of the circuit (marked with a vertical arrow).

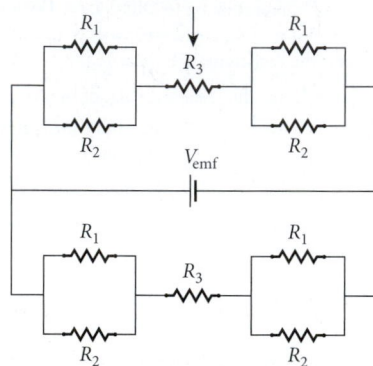

•**25.77** Three resistors are connected to a power supply with $V = 110.$ V as shown in the figure.

a) Find the potential drop across R_3.

b) Find the current in R_1.

c) Find the rate at which thermal energy is dissipated from R_2.

$R_1 = 2.00 \ \Omega$

$V = 110.$ V

$R_2 = 3.00 \ \Omega$ $R_3 = 6.00 \ \Omega$

•**25.78** A battery with $V = 1.500$ V is connected to three resistors as shown in the figure.

a) Find the potential drop across each resistor.

b) Find the current in each resistor.

$R_2 = 4.00 \ \Omega$

$R_1 = 2.00 \ \Omega$

$V = 1.500$ V

$R_3 = 6.00 \ \Omega$

•**25.79** A 2.50-m-long copper cable is connected across the terminals of a 12.0-V car battery. Assuming that it is completely insulated from its environment, how long after the connection is made will the copper start to melt? (Useful information: copper has a mass density of 8960 kg/m³, a melting point of 1359 K, and a specific heat of 386 J/kg/K.)

•**25.80** A piece of copper wire is used to form a circular loop of radius 10.0 cm. The wire has a cross-sectional area of 10.0 mm². Points A and B are 90.0° apart, as shown in the figure. Find the resistance between points A and B.

•**25.81** Two conducting wires have identical lengths $L_1 = L_2 = L = 10.0$ km and identical circular cross sections of radius $r_1 = r_2 = r = 1.00$ mm. One wire is made of steel (with resistivity $\rho_{steel} = 40.0 \cdot 10^{-8} \ \Omega$ m); the other is made of copper (with resistivity $\rho_{copper} = 1.68 \cdot 10^{-8} \ \Omega$ m).

a) Calculate the ratio of the power dissipated by the two wires, P_{copper}/P_{steel}, when they are connected in parallel and a potential difference of $V = 100.$ V is applied to them.

b) Based on this result, how do you explain the fact that conductors for power transmission are made of copper and not steel?

•**25.82** Before bendable tungsten filaments were developed, Thomas Edison used carbon filaments in his light bulbs. Though carbon has a very high melting temperature (3599 °C), its sublimation rate is high at high temperatures. So carbon-filament bulbs were kept at lower temperatures, thereby rendering them dimmer than later tungsten-based bulbs. A typical carbon-filament bulb requires an average power of 40. W, when 110 volts is applied across it, and has a filament temperature of 1800 °C. Carbon, unlike copper, has a negative temperature coefficient of resistivity: $\alpha = -0.00050$ °C⁻¹. Calculate the resistance at room temperature (20. °C) of this carbon filament.

••**25.83** A material is said to be *ohmic* if an electric field, $\vec{E}$, in the material gives rise to current density $\vec{J} = \sigma \vec{E}$, where the conductivity, σ, is a constant independent of $\vec{E}$ or $\vec{J}$. (This is the precise form of Ohm's Law.) Suppose in some material an electric field, $\vec{E}$, produces current density, $\vec{J}$, not necessarily related by Ohm's Law; that is, the material may or may not be ohmic.

a) Calculate the rate of energy dissipation (sometimes called *ohmic heating* or *joule heating*) per unit volume in this material, in terms of $\vec{E}$ and $\vec{J}$.

b) Express the result of part (a) in terms of $\vec{E}$ alone and $\vec{J}$ alone, for $\vec{E}$ and $\vec{J}$ related via Ohm's Law, that is, in an ohmic material with conductivity σ or resistivity ρ.

MULTI-VERSION EXERCISES

25.84 A high-voltage direct current (HVDC) transmission line carries electrical power a distance of 643.1 km. The line transmits 7935 MW of power at a potential difference of 1.177 MV. If the HVDC line consists of one copper wire of diameter 2.353 cm, what fraction of the power is lost in transmission?

25.85 A high-voltage direct current (HVDC) transmission line carries 5319 MW of electrical power a distance of 411.7 km. The HVDC line consists of one copper wire of diameter 2.125 cm. The fraction of the power lost in transmission is $7.538 \cdot 10^{-2}$. What is the potential difference in the line?

25.86 A high-voltage direct current (HVDC) transmission line carries 5703 MW of electrical power at a potential difference of 1.197 MV. The HVDC line consists of one copper wire of diameter 1.895 cm. The fraction of the power lost in transmission is $1.166 \cdot 10^{-1}$. How long is the line?

25.87 The reserve capacity (RC) of a car battery is defined as the number of minutes the battery can provide 25.0 A of current at a potential difference of 10.5 V. Thus, the RC indicates how long the battery can power a car whose charging system has failed. How much energy is stored in a car battery with an RC of 110.0?

25.88 The reserve capacity (RC) of a car battery is defined as the number of minutes the battery can provide 25.0 A of current at a potential difference of 10.5 V. Thus, the RC indicates how long the battery can power a car whose charging system has failed. If a car battery stores $1.843 \cdot 10^6$ J of energy, what is its RC?

25.89 A flashlight bulb with a tungsten filament draws 374.3 mA of current when a potential difference of 3.907 V is applied. When the bulb is at room temperature (20.00 °C) and is not lit, its resistance is 1.347 Ω. What is the temperature of the tungsten filament when the bulb is lit?

25.90 A flashlight bulb with a tungsten filament draws 420.1 mA when a potential difference of 3.949 V is applied. The temperature of the tungsten filament when the bulb is lit is 1291 °C. What is the resistance of the bulb when it is at room temperature (20.00 °C) and is not lit?

25.91 When a flashlight bulb with a tungsten filament is lit, the applied potential difference is 3.991 V and the temperature of the tungsten filament is $1.110 \cdot 10^3$ °C. The resistance of the bulb when it is at room temperature (20.00 °C) and is not lit is 1.451 Ω. What current does the bulb draw when it is lit?

26

Direct Current Circuits

FIGURE 26.1 A circuit board can have hundreds of circuit components connected by metallic conducting paths.

The electric circuit, such as the one shown in Figure 26.1, undoubtedly changed the world. Modern electronics continues to change human society, at a faster and faster pace. It took 38 years for radio to reach 50 million users in the United States. However, it took only 13 years for television to reach that number of users, 10 years for cable TV, 5 years for the Internet, and 3 years for cell phones.

This chapter examines the techniques used to analyze circuits that cannot be broken down into simple series and parallel connections. Modern electronics design depends on millions of different circuits, each with its own purpose and configuration. However, regardless of how complicated a circuit becomes, the basic rules for analyzing it are the ones presented in this chapter.

Some of the circuits analyzed in this chapter contain not only resistors and emf devices but also capacitors. In these circuits, the current is not steady, but changes with time. Time-varying currents will be covered more thoroughly in later chapters, which introduce additional circuit components.

WHAT WE WILL LEARN

- Some circuits cannot be reduced to a single loop; complex circuits can be analyzed using Kirchhoff's rules.

- Kirchhoff's Junction Rule states that the algebraic sum of the currents at any junction in a circuit must be zero.

- Kirchhoff's Loop Rule states that the algebraic sum of the potential changes around any closed loop in a circuit must be zero.

- Single-loop circuits can be analyzed using Kirchhoff's Loop Rule.

- Multiloop circuits must be analyzed using both Kirchhoff's Junction Rule and Kirchhoff's Loop Rule.

- The current in a circuit that contains a resistor and a capacitor varies exponentially with time, with a characteristic time constant given by the product of the resistance and the capacitance.

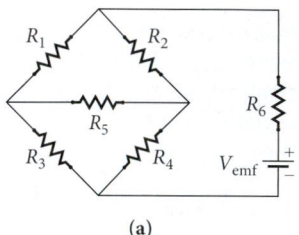

(a)

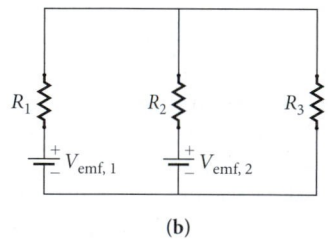

(b)

FIGURE 26.2 Two examples of circuits that cannot be reduced to simple combinations of parallel and series resistors.

26.1 Kirchhoff's Rules

In Chapter 25, we considered several kinds of direct current (DC) circuits, each containing one emf device along with resistors connected in series or in parallel. Some seemingly complicated circuits contain multiple resistors in series or in parallel that can be replaced with an equivalent resistance. However, we did not consider circuits containing multiple sources of emf. In addition, there are single-loop and multiloop circuits with emf devices and resistors that cannot be reduced to simple circuits containing parallel or series connections. Figure 26.2 shows two examples of such circuits. This chapter explains how to analyze these kinds of circuits using **Kirchhoff's rules.**

Kirchhoff's Junction Rule

A **junction** is a place in a circuit where three or more wires are connected to each other. Each connection between two junctions in a circuit is called a **branch.** A branch can contain any number of different circuit elements and the wires between them. Each branch can have a current flowing, and this current is the same everywhere in the branch. This fact leads to **Kirchhoff's Junction Rule:**

The sum of the currents entering a junction must equal the sum of the currents leaving the junction.

With a positive sign assigned (arbitrarily) to currents entering the junction and a negative sign to those exiting the junction, Kirchhoff's Junction Rule is expressed mathematically as

$$\text{Junction: } \sum_{k=1}^{n} i_k = 0. \tag{26.1}$$

How do you know which currents enter a junction and which exit the junction when you make a drawing like the one shown in Figure 26.3? You don't; you simply assign a direction for each current along a given wire. If an assigned direction turns out to be wrong, you will obtain a negative number for that particular current in your final solution.

Kirchhoff's Junction Rule is a direct consequence of the conservation of electric charge. Junctions do not have the capability of storing charge. Thus, charge conservation requires that all charges streaming into a junction also leave the junction, which is exactly what Kirchhoff's Junction Rule states.

According to Kirchhoff's Junction Rule, at each junction in a multiloop circuit, the current flowing into the junction must equal the current flowing out of the junction. For example, Figure 26.3 shows a single junction, a, with a current, i_1, entering the junction and two currents, i_2 and i_3, leaving the junction. According to Kirchhoff's Junction Rule, in this case,

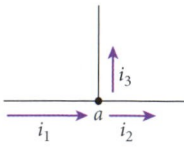

FIGURE 26.3 A single junction from a multiloop circuit.

$$\sum_{k=1}^{3} i_k = i_1 - i_2 - i_3 = 0 \Rightarrow i_1 = i_2 + i_3.$$

Kirchhoff's Loop Rule

A **loop** in a circuit is any set of connected wires and circuit elements forming a closed path. If you follow a loop, eventually you will get to the same point from which you started. For example, in the circuit diagram shown in Figure 26.2b, three possible loops can be identified. These three loops are shown in different colors (red, green, and blue) in Figure 26.4. The blue loop includes resistors 1 and 2, emf sources 1 and 2, and their connecting wires. The red loop includes resistors 2 and 3, emf source 2, and their connecting wires. Finally, the green loop includes resistors 1 and 3, emf source 1, and their connecting wires. Note that any given wire or circuit element can be and usually is part of more than one loop.

You can move through any loop in a circuit in either a clockwise or a counterclockwise direction. Figure 26.4 shows a clockwise path through each of the loops, as indicated by the arrows. But the direction of the path taken around the loop is irrelevant as long as your choice is followed consistently all the way around the loop.

Summing the potential differences from all circuit elements encountered along any given loop yields the total potential difference of the complete path along the loop. **Kirchhoff's Loop Rule** then states:

> The potential difference around a complete circuit loop must sum to zero.

Kirchhoff's Loop Rule is a direct consequence of the fact that electric potential is single-valued. This means that the electric potential energy of a conduction electron at a point in the circuit has one specific value. Suppose this rule were not valid. Then we could analyze the potential changes of a conduction electron in going around a loop and find that the electron had a different potential energy when it returned to its starting point. The potential energy of this electron would change at a point in the circuit, in obvious contradiction of energy conservation. In other words, Kirchhoff's Loop Rule is simply a consequence of the law of conservation of energy.

Application of Kirchhoff's Loop Rule requires conventions for determining the potential drop across each element of the circuit. This depends on the assumed direction of the current and the direction of the analysis. For emf sources, the rules are straightforward, since minus and plus signs (as well as short and long lines) indicate which side of the emf source is at the higher potential. The potential drop for an emf source is in the direction from minus to plus or from short line to long line. As noted earlier, the assignment of the current directions and the choice of a clockwise or counterclockwise path around a loop are arbitrary. Any direction will give the same information, as long as it is applied consistently around a loop. The conventions used to analyze circuit elements in a loop are summarized in Table 26.1 and Figure 26.5, where the magnitude of the current through the circuit element is i. (The labels in the rightmost column of Table 26.1 correspond to the parts of Figure 26.5.)

If we move around a loop in a circuit in the same direction as the current, the potential changes across resistors will be negative. If we move around the loop in the opposite direction from the current, the potential changes across the resistors will be positive. If

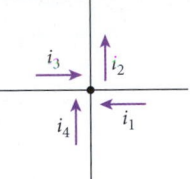
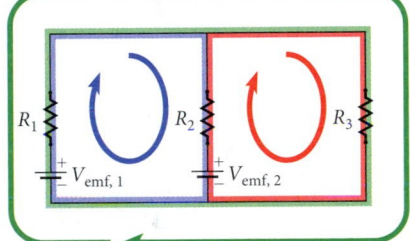

FIGURE 26.4 The three possible loops (indicated in red, green, and blue) for the circuit diagram shown in Figure 26.2b.

Table 26.1	Conventions Used to Determine the Sign of Potential Changes Around a Single-Loop Circuit Containing Several Resistors and Sources of emf	
Element	**Direction of Analysis**	**Potential Change**
R	Same as current	$-iR$	(a)
R	Opposite to current	$+iR$	(b)
V_{emf}	Same as emf	$+V_{emf}$	(c)
V_{emf}	Opposite to emf	$-V_{emf}$	(d)

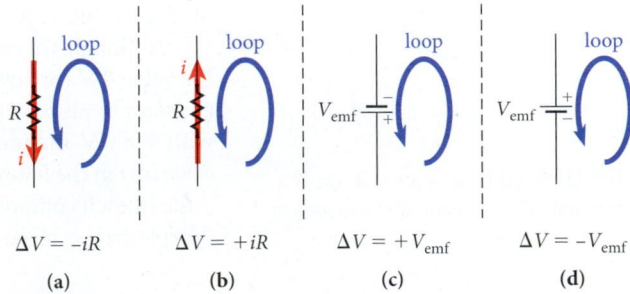

FIGURE 26.5 Sign convention for potential changes in analyzing loops.

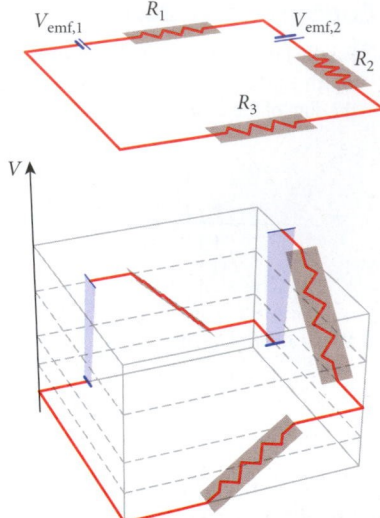

FIGURE 26.6 Loop with multiple sources of emf and multiple resistors.

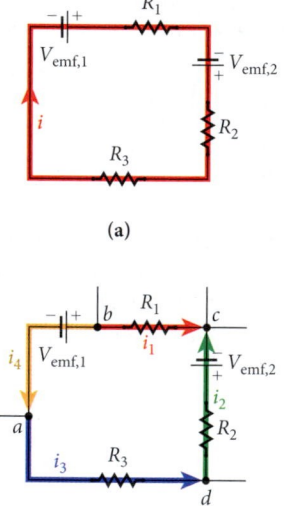

FIGURE 26.7 The same circuit loop as in Figure 26.6: (a) as an isolated single loop; (b) as a loop connected to other circuit branches.

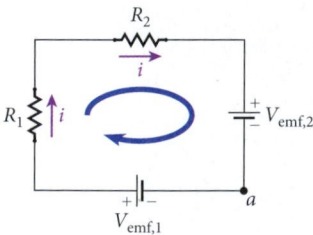

FIGURE 26.8 A single-loop circuit containing two resistors and two sources of emf in series.

we move around a loop in such a way that we pass through an emf source from the negative to the positive terminal, this component contributes a positive potential difference. If we pass through an emf source from the positive to the negative terminal, that component contributes a negative potential difference.

With these conventions, Kirchhoff's Loop Rule is written in mathematical form as

$$\text{Closed loop:} \quad \sum_{j=1}^{m} V_{\text{emf},j} - \sum_{k=1}^{n} i_k R_k = 0. \tag{26.2}$$

To clarify Kirchhoff's Loop Rule, Figure 26.6 shows a loop with two sources of emf and three resistors, using the three-dimensional display employed in Chapters 24 and 25, where the value of the electric potential, V, is represented in the vertical dimension. The most important point visualized by Figure 26.6 is that one complete turn around the loop always ends up at the same value of the potential as at the starting point. This is exactly what Kirchhoff's Loop Rule (equation 26.2) claims. An analogy with downhill skiing may help. When you ski, you are moving around in the gravitational potential, up and down the mountain. A ski lift corresponds to a source of emf, lifting you to a higher value of the gravitational potential. A downhill ski run corresponds to a resistor. (The annoying horizontal traverses between runs correspond to the wires in a circuit—both the wires and the horizontal traverses are at constant potential.) Thus, starting at $V_{\text{emf},1}$ and proceeding clockwise around the loop in Figure 26.6 is analogous to a ski outing in which you take two different lifts and ski down three different runs. And the important point, which is obvious in skiing, is that you return to the same altitude (the same value of the gravitational potential) from which you started, once you have completed the round trip.

A final point about loops is illustrated by Figure 26.7. Figure 26.7a reproduces the circuit shown in Figure 26.6 as a single isolated loop. Since this loop does not have junctions and thus has only one branch (which is the entire loop), the same current i flows everywhere in the loop. In Figure 26.7b, this loop is connected at four junctions (labeled a, b, c, and d) to other parts of a more extended circuit. Now this loop has four branches, each of which can have a different current flowing through it, as illustrated by the different-colored arrows in the figure. The point is this: In both parts of the figure, Kirchhoff's Loop Rule holds for the loop shown. The relative values of the electric potential between any two circuit elements in Figure 26.7b are the same as those shown in Figure 26.6, independent of the currents that are forced through the different branches of the loop by the rest of the circuit. (In our downhill skiing analogy, the currents correspond to different numbers of skiers on the lifts and the downhill runs. Obviously the number of skiers on the hill does not have any influence on the steepness of the hill.)

26.2 Single-Loop Circuits

Let's begin analyzing general circuits by considering a circuit containing two sources of emf, $V_{\text{emf},1}$ and $V_{\text{emf},2}$, and two resistors, R_1 and R_2, connected in series in a single loop, as shown in Figure 26.8. Note that $V_{\text{emf},1}$ and $V_{\text{emf},2}$ have opposite polarity. In this single-loop circuit, there are no junctions, and so the entire circuit consists of a single branch. The current is the same everywhere in the loop. To illustrate the potential changes across the components of this circuit, Figure 26.9 shows a three-dimensional view.

Although we could arbitrarily pick any point in the circuit of Figure 26.8 and assign it the value 0 V (or any other value of the potential, because we can always add a global additive constant to all potential values without changing the physical outcome), we start at point a with $V = 0$ V and proceed around the circuit in a clockwise direction (indicated by blue elliptical arrow in the figure). Because the components of the circuit are in series, the current, i, is the same in each component, and we assume that the current is flowing in the clockwise direction (purple arrows in the figure). The first circuit component along the clockwise path from point a

is the source of emf, $V_{emf,1}$, which produces a positive potential gain of $V_{emf,1}$. Next is resistor R_1, which produces a potential drop given by $\Delta V_1 = iR_1$. Continuing around the loop, the next component is resistor R_2, which produces a potential drop given by $\Delta V_2 = iR_2$. Next, we encounter a second source of emf, $V_{emf,2}$. This source of emf is wired into the circuit with its polarity opposite that of $V_{emf,1}$. Thus, this component produces a potential *drop* with magnitude $V_{emf,2}$, rather than a potential *gain*. We have now completed the loop and are back at $V = 0$ V. Using equation 26.2, we sum the potential changes of this loop as follows:

$$V_{emf,1} - \Delta V_1 - \Delta V_2 - V_{emf,2} = V_{emf,1} - iR_1 - iR_2 - V_{emf,2} = 0.$$

To show that the direction in which we move through a loop, clockwise or counterclockwise, is arbitrary, let's analyze the same circuit in the counterclockwise direction, starting at point a (see Figure 26.10). The first circuit element is $V_{emf,2}$, which produces a positive potential gain. The next element is R_2. Because we have assumed that the current is in the clockwise direction and we are analyzing the loop in the counterclockwise direction, the potential change for R_2 is $+iR_2$, according to the conventions listed in Table 26.1. Proceeding to the next element in the loop, R_1, we use a similar argument to designate the potential change for this resistor as $+iR_1$. The final element in the circuit is $V_{emf,1}$, which is aligned in a direction opposite to that of our analysis, so the potential change across this element is $-V_{emf,1}$. Kirchhoff's Loop Rule then gives us

$$+V_{emf,2} + iR_2 + iR_1 - V_{emf,1} = 0.$$

You can see that the clockwise and counterclockwise loop directions give the same information, which means that the direction in which we choose to analyze the circuit does not matter.

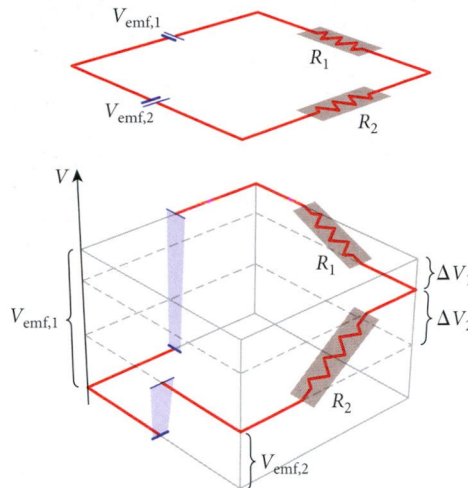

FIGURE 26.9 Three-dimensional representation of the single-loop circuit in Figure 26.8, containing two resistors and two sources of emf in series.

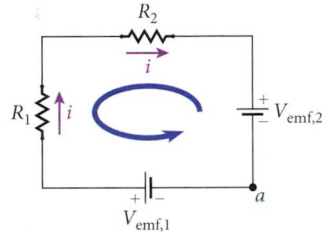

FIGURE 26.10 The same loop as in Figure 26.8, but analyzed in the counterclockwise direction.

SOLVED PROBLEM 26.1 | Charging a Battery

A 12.0-V battery with internal resistance $R_i = 0.200\ \Omega$ is being charged by a battery charger that is capable of delivering a current of magnitude $i = 6.00$ A.

PROBLEM
What is the minimum emf the battery charger must supply to be able to charge the battery?

SOLUTION
THINK The battery charger, which is an external source of emf, must have enough potential difference to overcome the potential difference of the battery and the potential drop across the battery's internal resistance. The battery charger must be hooked up so that its positive terminal is connected to the positive terminal of the battery to be charged. We can think of the battery's internal resistance as a resistor in a single-loop circuit that also contains two sources of emf with opposite polarities.

SKETCH Figure 26.11 shows a diagram of the circuit, consisting of a battery with potential difference V_t and internal resistance R_i connected to an external source of emf, V_e. The yellow shaded area represents the battery's physical dimensions. Note that the positive terminal of the battery charger is connected to the positive terminal of the battery.

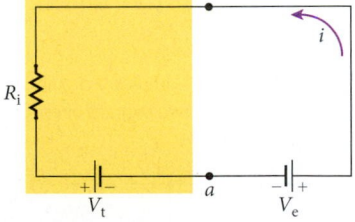

FIGURE 26.11 Circuit consisting of a battery with internal resistance connected to an external source of emf.

RESEARCH We can apply Kirchhoff's Loop Rule to this circuit. We assume a current flowing counterclockwise around the circuit, as shown in Figure 26.11. The potential changes around the circuit must sum to zero. We sum the potential changes starting at point b and moving in a counterclockwise direction:

$$-iR_i - V_t + V_e = 0.$$

– Continued

SIMPLIFY   We can solve this equation for the required potential difference of the charger:

$$V_e = iR_i + V_t,$$

where i is the current that the charger supplies.

CALCULATE   Putting in the numerical values gives us

$$V_e = iR_i + V_t = (6.00 \text{ A})(0.200 \text{ } \Omega) + 12.0 \text{ V} = 13.20 \text{ V}.$$

ROUND   We report our result to three significant figures:

$$V_e = 13.2 \text{ V}.$$

DOUBLE-CHECK   Our result indicates that the battery charger has to have a higher potential difference than the specified potential difference of the battery, which is reasonable. A typical charger for a 12-V battery has a potential difference of around 14 V.

26.3   Multiloop Circuits

Analyzing multiloop circuits requires both Kirchhoff's Loop Rule and Kirchhoff's Junction Rule. The procedure for analyzing a multiloop circuit consists of identifying complete loops and junction points in the circuit and applying Kirchhoff's rules to these parts of the circuit separately. Analyzing the single loops in a multiloop circuit with Kirchhoff's Loop Rule and the junctions with Kirchhoff's Junction Rule results in a system of coupled equations in several unknown variables. These equations can be solved for the quantities of interest using various techniques, including direct substitution. Example 26.1 illustrates the analysis of a multiloop circuit.

EXAMPLE 26.1   Multiloop Circuit

Consider the circuit shown in Figure 26.12. This circuit has three resistors, R_1, R_2, and R_3, and two sources of emf, $V_{emf,1}$ and $V_{emf,2}$. The red arrows show the direction of potential drop across the emf sources. This circuit cannot be resolved into simple series or parallel connections. To analyze this circuit, we need to assign directions to the currents flowing through the resistors. We can choose these directions arbitrarily (knowing that if we choose the wrong direction, the resulting current value will be negative). Figure 26.13 shows the circuit with the assigned directions for the currents shown by the purple arrows.

Let's consider junction b first. The current entering the junction must equal the current leaving it, so we can write

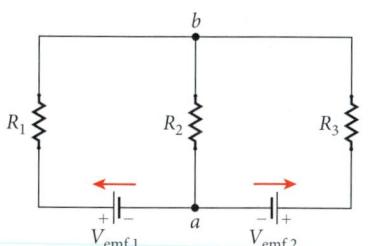

FIGURE 26.12   Multiloop circuit with three resistors and two sources of emf.

$$i_2 = i_1 + i_3. \tag{i}$$

Looking at junction a, we again equate the incoming current and the outgoing current to get

$$i_1 + i_3 = i_2,$$

which provides the same information obtained for junction b. Note that this is a typical result: If a circuit has n junctions, it is possible to obtain at most $n - 1$ independent equations from application of Kirchhoff's Junction Rule. (In this case, $n = 2$, so we can get only one independent equation.)

At this point we cannot determine the currents in the circuit because we have three unknown values and only one equation. Therefore, we need two more independent equations. To get these equations, we apply Kirchhoff's Loop Rule. We can identify three loops in the circuit shown in Figure 26.13:

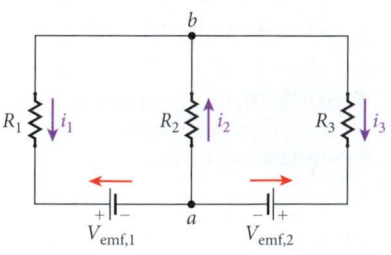

FIGURE 26.13   Multiloop circuit with the assumed direction of the current through the resistors indicated.

1. the left half of the circuit, including the elements R_1, R_2, and $V_{emf,1}$;

2. the right half of the circuit, including the elements R_2, R_3, and $V_{emf,2}$; and

3. the outer loop, including the elements R_1, R_3, $V_{emf,1}$, and $V_{emf,2}$.

Applying Kirchhoff's Loop Rule to the left half of the circuit, using the assumed directions for the currents and analyzing the loop in a counterclockwise direction starting at junction b, we obtain

$$-i_1 R_1 - V_{emf,1} - i_2 R_2 = 0,$$

or

$$i_1 R_1 + V_{\text{emf},1} + i_2 R_2 = 0. \qquad \text{(ii)}$$

Applying the Loop Rule to the right half of the circuit, again starting at junction b and analyzing the loop in a clockwise direction, we get

$$-i_3 R_3 - V_{\text{emf},2} - i_2 R_2 = 0,$$

or

$$i_3 R_3 + V_{\text{emf},2} + i_2 R_2 = 0. \qquad \text{(iii)}$$

Applying the Loop Rule to the outer loop, starting at junction b and working in a clockwise direction, gives us

$$-i_3 R_3 - V_{\text{emf},2} + V_{\text{emf},1} + i_1 R_1 = 0.$$

This equation provides no new information because we can also obtain it by subtracting equation (iii) from equation (ii). For all three loops, we obtain equivalent information if we analyze them in either a counterclockwise direction or a clockwise direction or if we start at any other point and move around the loops from there.

With three equations, (i), (ii), and (iii), and three unknowns, i_1, i_2, and i_3, we can solve for the unknown currents in several ways. For example, we could put the three equations in matrix format and then solve them using Kramer's method on a calculator. This is a recommended method for complicated circuits with many equations and many unknowns. However, for this example, we can proceed by substituting from equation (i) into the other two, thus eliminating i_2. We then solve one of the two resulting equations for i_1 and substitute from that into the other to obtain an expression for i_3. Substituting back then gives solutions for i_2 and i_1:

$$i_1 = -\frac{(R_2 + R_3)V_{\text{emf},1} - R_2 V_{\text{emf},2}}{R_1 R_2 + R_1 R_3 + R_2 R_3}$$

$$i_2 = -\frac{R_3 V_{\text{emf},1} + R_1 V_{\text{emf},2}}{R_1 R_2 + R_1 R_3 + R_2 R_3}$$

$$i_3 = -\frac{-R_2 V_{\text{emf},1} + (R_1 + R_2)V_{\text{emf},2}}{R_1 R_2 + R_1 R_3 + R_2 R_3}.$$

Note: You do not need to remember this particular solution or the linear algebra used to get there. However, the general method of applying Kirchhoff's rules for loops and junctions, and assigning currents in arbitrary directions is the central idea of circuit analysis.

Concept Check 26.2

In the circuit in the figure, there are three identical resistors. The switch, S, is initially open. When the switch is closed, what happens to the current flowing in R_1?

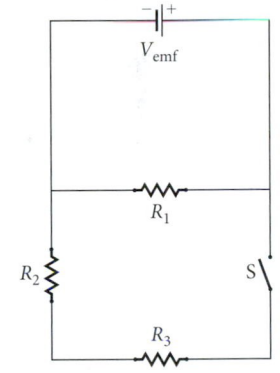

a) The current in R_1 decreases.

b) The current in R_1 increases.

c) The current in R_1 stays the same.

SOLVED PROBLEM 26.2 | The Wheatstone Bridge

The Wheatstone bridge is a particular circuit used to measure unknown resistances. The circuit diagram of a Wheatstone bridge is shown in Figure 26.14. This circuit consists of three known resistances, R_1, R_3, and a variable resistor, R_v, as well as an unknown resistance, R_u. A source of emf, V, is connected across junctions a and c. A sensitive ammeter (a device used to measure current, discussed in Section 26.4) is connected between junctions b and d. The Wheatstone bridge is used to determine R_u by varying R_v until the ammeter between b and d shows no current flowing. When the ammeter reads zero, the bridge is said to be balanced.

PROBLEM
Determine the unknown resistance, R_u, in the Wheatstone bridge shown in Figure 26.14. The known resistances are $R_1 = 100.0 \ \Omega$ and $R_3 = 110.0 \ \Omega$, and $R_v = 15.63 \ \Omega$ when the current through the ammeter is zero, indicating that the bridge is balanced.

SOLUTION
THINK The circuit has four resistors and an ammeter, and each component can have a current flowing through it. However, in this case, with $R_v = 15.63 \ \Omega$, there is no current flowing through the ammeter. Setting this current to zero leaves four unknown currents through the four resistors, and we therefore need four equations. We can use Kirchhoff's rules to analyze two loops, adb and cbd, and two junctions, b and d.

SKETCH Figure 26.15 shows the Wheatstone bridge with the assumed directions for currents i_1, i_3, i_u, i_v, and i_A.

– Continued

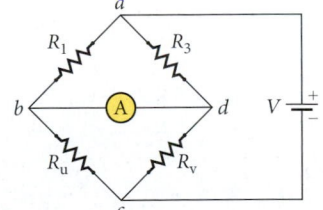

FIGURE 26.14 Circuit diagram of a Wheatstone bridge.

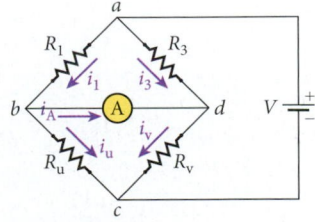

FIGURE 26.15 The Wheatstone bridge with the assumed current directions indicated.

RESEARCH   We first apply Kirchhoff's Loop Rule to loop *adb*, starting at *a* and going clockwise, to obtain

$$-i_3 R_3 + i_A R_A + i_1 R_1 = 0, \qquad\qquad\qquad (i)$$

where R_A is the resistance of the ammeter. We apply Kirchhoff's Loop Rule again to loop *cbd*, starting at *c* and going clockwise, to get

$$+i_u R_u - i_A R_A - i_v R_v = 0. \qquad\qquad\qquad (ii)$$

Now we can use Kirchhoff's Junction Rule at junction *b* to obtain

$$i_1 = i_A + i_u. \qquad\qquad\qquad (iii)$$

Another application of Kirchhoff's Junction Rule, at junction *d*, gives

$$i_3 + i_A = i_v. \qquad\qquad\qquad (iv)$$

SIMPLIFY   When the current through the ammeter is zero ($i_A = 0$), we can rewrite equations (i) through (iv) as follows:

$$i_1 R_1 = i_3 R_3 \qquad\qquad\qquad (v)$$

$$i_u R_u = i_v R_v \qquad\qquad\qquad (vi)$$

$$i_1 = i_u \qquad\qquad\qquad (vii)$$

and

$$i_3 = i_v. \qquad\qquad\qquad (viii)$$

Dividing equation (vi) by equation (v) gives us

$$\frac{i_u R_u}{i_1 R_1} = \frac{i_v R_v}{i_3 R_3},$$

which we can rewrite using equations (vii) and (viii):

$$R_u = \frac{R_1}{R_3} R_v.$$

Self-Test Opportunity 26.1

Show that when the current through the ammeter is zero, the equivalent resistance for the resistors in the Wheatstone bridge in Figure 26.14 is given by

$$R_{eq} = \frac{(R_1 + R_u)(R_3 + R_v)}{R_1 + R_u + R_3 + R_v}.$$

CALCULATE   Putting in the numerical values, we get

$$R_u = \frac{R_1}{R_3} R_v = \frac{100.0 \ \Omega}{110.0 \ \Omega} 15.63 \ \Omega = 14.20901 \ \Omega.$$

ROUND   We report our result to four significant figures:

$$R_u = 14.21 \ \Omega.$$

DOUBLE-CHECK   Our result for the resistance of the unknown resistor is similar to the value for the variable resistor. Thus, our answer seems reasonable, because the other two resistors in the circuit also have resistances that are approximately equal.

General Observations on Circuit Networks

An important observation about solving circuit problems is that, in general, the complete analysis of a circuit requires knowing the current flowing in each branch of the circuit. We use Kirchhoff's Junction Rule and Loop Rule to establish equations relating the currents, and we need as many linearly independent equations as there are branches to guarantee that we can obtain a solution to the system.

Let's consider the abstract example shown in Figure 26.16, where all circuit components except the wires have been omitted. This circuit has four junctions, shown in blue in Figure 26.16a. Six branches connect these junctions, shown in Figure 26.16b. We therefore need six linearly independent equations relating the currents in these branches. It was noted earlier that not all equations obtained by applying Kirchhoff's rules to a circuit are linearly independent. This fact is worth repeating: If a circuit has *n* junctions, it is possible to obtain at most *n* − 1 independent equations from application of the Junction Rule. (For the circuit in Figure 26.16, *n* = 4, so we can get only three independent equations.)

The Junction Rule alone is not enough for the complete analysis of any circuit. It is generally best to write as many equations as possible for junctions and then augment them with equations obtained from loops. Figure 26.16c shows that there are six possible loops

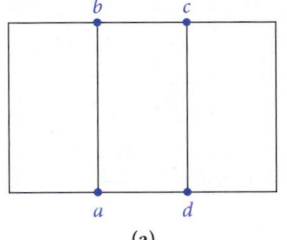
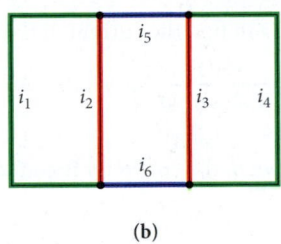
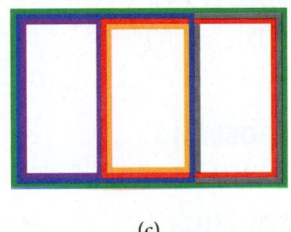

(a)   **(b)**   **(c)**

FIGURE 26.16 Circuit network consisting of (a) four junctions, (b) six branches, (c) six possible loops.

in this network, which are marked in different colors. Clearly, there are more loops than we need to analyze to obtain three equations. This is again a general observation: The system of equations that can be set up by considering all possible loops is overdetermined. Thus, you'll always have the freedom to select particular loops to augment the equations obtained from analyzing the junctions. As a general rule of thumb, it is best to choose loops with fewer circuit elements, which often makes the subsequent linear algebra considerably simpler. In particular, if you are asked to find the current in a particular branch of a network, choosing the appropriate loop may allow you to avoid setting up a lengthy set of equations and let you solve the problem with just one equation. So it pays to devote some attention to the selection of loops! Do not write down more equations than you need to solve for the unknowns in any particular problem. This will only complicate the algebra. However, once you have the solution, you can use one or more of the unused loops to check your values.

26.4 Ammeters and Voltmeters

A device used to measure current is called an **ammeter.** A device used to measure potential difference is called a **voltmeter.** To measure the current, an ammeter must be wired in a circuit in *series*. Figure 26.17 shows an ammeter connected in a circuit in a way that allows it to measure the current i. To measure the potential difference, a voltmeter must be wired in *parallel* with the component across which the potential difference is to be measured. Figure 26.17 shows a voltmeter placed in the circuit to measure the potential drop across resistor R_1.

It is important to realize that these instruments must be able to make measurements while disturbing the circuit as little as possible. Thus, ammeters are designed to have as low a resistance as possible, usually on the order of 1 Ω, so they do not have an appreciable effect on the currents they measure. Voltmeters are designed to have as high a resistance as possible, usually on the order of 10 MΩ (10^7 Ω), so they have a negligible effect on the potential differences they are measuring.

In practice, measurements of current and potential difference are made with a digital multimeter that can switch between functioning as an ammeter and functioning as a voltmeter. It displays the results with a numerical digital display, which includes the sign of the potential difference or current. Most digital multimeters can also measure the resistance of a circuit component; that is, they can function as an **ohmmeter.** The digital multimeter performs this task by applying a known potential difference and measuring the resulting current. This test is useful for determining circuit continuity and the status of fuses, as well as measuring the resistance of resistors.

EXAMPLE 26.2   Voltmeter in a Simple Circuit

Consider a simple circuit consisting of a source of emf with voltage $V_{emf} = 150.$ V and a resistor with resistance $R = 100.$ kΩ (Figure 26.18). A voltmeter with resistance $R_V = 10.0$ MΩ is connected across the resistor.

PROBLEM 1

What is the current in the circuit before the voltmeter is connected?   *– Continued*

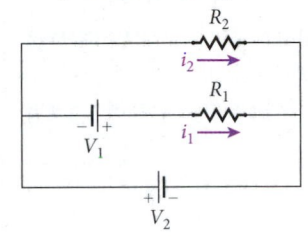
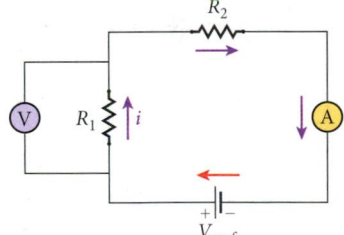

FIGURE 26.17 Placement of an ammeter and a voltmeter in a simple circuit.

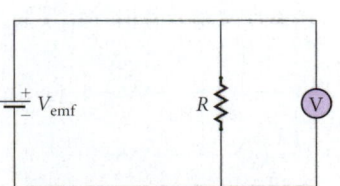

FIGURE 26.18 A simple circuit with a voltmeter connected in parallel across a resistor.

SOLUTION 1

Ohm's Law says that $V = iR$, so we can find the current in the circuit:

$$i = \frac{V_{emf}}{R} = \frac{150. \text{ V}}{100. \cdot 10^3 \text{ } \Omega} = 1.50 \cdot 10^{-3} \text{ A} = 1.50 \text{ mA}.$$

PROBLEM 2

What is the current in the circuit when the voltmeter is connected across the resistor?

SOLUTION 2

The equivalent resistance of the resistor and the voltmeter connected in parallel is given by

$$\frac{1}{R_{eq}} = \frac{1}{R} + \frac{1}{R_V}.$$

Solving for the equivalent resistance and putting in the numerical values, we get

$$R_{eq} = \frac{RR_V}{R + R_V} = \frac{\left(100. \cdot 10^3 \text{ } \Omega\right)\left(10.0 \cdot 10^6 \text{ } \Omega\right)}{100. \cdot 10^3 \text{ } \Omega + 10.0 \cdot 10^6 \text{ } \Omega} = 9.90 \cdot 10^4 \text{ } \Omega = 99.0 \text{ k}\Omega.$$

The current is then

$$i = \frac{V_{emf}}{R_{eq}} = \frac{150. \text{ V}}{9.90 \cdot 10^4 \text{ } \Omega} = 1.52 \cdot 10^{-3} \text{ A} = 1.52 \text{ mA}.$$

The current in the circuit increases by 0.02 mA when the voltmeter is connected because the parallel combination of the resistor and the voltmeter has a lower resistance than that of the resistor alone. However, the effect is small, even with this relatively large resistance ($R = 100.$ kΩ).

Concept Check 26.4

Which of the circuits shown in the figure will not function properly?

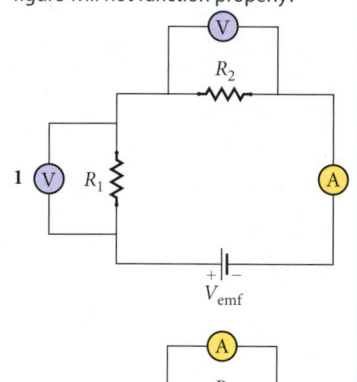

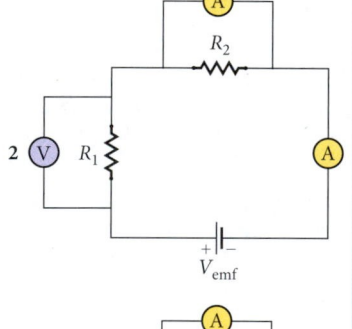

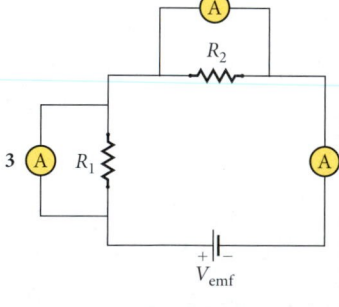

a) 1

b) 2

c) 3

d) 1 and 2

e) 2 and 3

SOLVED PROBLEM 26.3 | **Increasing the Range of an Ammeter**

PROBLEM

An ammeter can be used to measure different ranges of current by adding a current divider in the form of a shunt resistor connected in parallel with the ammeter. A *shunt resistor* is simply a resistor with a very small resistance. Its name arises from the fact that when it is connected in parallel with the ammeter, whose resistance is larger, most of the current is shunted through it, bypassing the meter. The sensitivity of the ammeter is therefore decreased allowing it to measure larger currents. Suppose an ammeter produces a full-scale reading when a current of $i_{int} = 5.10$ mA passes through it. The ammeter has an internal resistance of $R_i = 16.8$ Ω. To use this ammeter to measure a maximum current of $i_{max} = 20.2$ A, what should be the resistance of the shunt resistor, R_s, connected in parallel with the ammeter?

SOLUTION

THINK The shunt resistor connected in parallel with the ammeter needs to have a substantially lower resistance than the internal resistance of the ammeter. Most of the current will then flow through the shunt resistor rather than the ammeter.

SKETCH Figure 26.19 shows a shunt resistor, R_s, connected in parallel with an ammeter.

RESEARCH The two resistors are connected in parallel, so the potential difference across each resistor is the same. The potential difference that gives a full-scale reading on the ammeter is

$$\Delta V_{fs} = i_{int} R_i. \tag{i}$$

From Chapter 25, we know that the equivalent resistance of the two resistors in parallel is given by

$$\frac{1}{R_{eq}} = \frac{1}{R_i} + \frac{1}{R_s}. \tag{ii}$$

The voltage drop across the equivalent resistance must equal the voltage drop across the ammeter that gives a full-scale reading when current i_{max} is flowing through the circuit. Therefore, we can write

$$\Delta V_{fs} = i_{max} R_{eq}. \tag{iii}$$

SIMPLIFY Combining equations (i) and (iii) for the potential difference gives

$$\Delta V_{fs} = i_{int} R_i = i_{max} R_{eq}. \tag{iv}$$

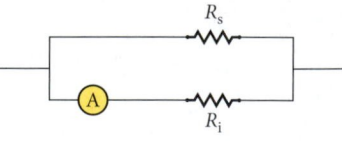

FIGURE 26.19 An ammeter with a shunt resistor connected across it in parallel.

We can rearrange equation (iv) and substitute for R_{eq} in equation (ii):

$$\frac{i_{max}}{i_{int}R_i} = \frac{1}{R_{eq}} = \frac{1}{R_i} + \frac{1}{R_s}. \tag{v}$$

Solving equation (v) for the shunt resistance gives us

$$\frac{1}{R_s} = \frac{i_{max}}{i_{int}R_i} - \frac{1}{R_i} = \frac{1}{R_i}\left(\frac{i_{max}}{i_{int}} - 1\right) = \frac{1}{R_i}\left(\frac{i_{max} - i_{int}}{i_{int}}\right),$$

or

$$R_s = R_i \frac{i_{int}}{i_{max} - i_{int}}.$$

CALCULATE   Putting in the numerical values, we get

$$R_s = R_i \frac{i_{int}}{i_{max} - i_{int}} = \left(16.8\ \Omega\right)\frac{5.10\cdot10^{-3}\ A}{20.2\ A - 5.10\cdot10^{-3}\ A}$$
$$= 0.00424266\ \Omega.$$

ROUND   We report our result to three significant figures:

$$R_s = 0.00424\ \Omega.$$

DOUBLE-CHECK   The equivalent resistance of the ammeter and the shunt resistor connected in parallel is given by equation (ii). Solving that equation for the equivalent resistance and putting in the numbers gives

$$R_{eq} = \frac{R_i R_s}{R_i + R_s} = \frac{\left(16.8\ \Omega\right)\left(0.00424\ \Omega\right)}{16.8\ \Omega + 0.00424\ \Omega} = 0.00424\ \Omega.$$

Thus, the equivalent resistance of the ammeter and the shunt resistor connected in parallel is approximately equal to the resistance of the shunt resistor. This low equivalent resistance is necessary for a current-measuring instrument, which must be placed in series in a circuit. If the current-measuring device has a high resistance, its presence will disturb the measurement of the current.

26.5 RC Circuits

So far in this chapter, we have dealt with circuits containing sources of emf and resistors. The currents in these circuits do not vary in time. Now we consider circuits that contain capacitors (see Chapter 24), as well as sources of emf and resistors. Called **RC circuits**, these circuits have currents that *do* vary with time. The simplest circuit operations that involve time-dependent currents are the charging and discharging of a capacitor. Understanding these time-dependent processes involves the solution of some simple differential equations. After magnetism and magnetic phenomena are introduced in Chapters 27 through 29, time-dependent currents will be discussed again in Chapter 30, which will build on the techniques introduced here.

Charging a Capacitor

Consider a circuit with a source of emf, V_{emf}, a resistor, R, and a capacitor, C (Figure 26.20). Initially, the switch is open and the capacitor is uncharged, as shown in Figure 26.20a. When the switch is closed (Figure 26.20b), current begins to flow in the circuit, building up opposite charges on the plates of the capacitor and thus creating a potential difference, ΔV, across the capacitor. Current flows because of the source of emf, which maintains a constant voltage. When the capacitor is fully charged (Figure 26.20c), no more current flows in the circuit. The potential difference across the plates is then equal to the voltage provided by the source of emf, and the magnitude of the total charge, q_{tot}, on each plate of the capacitor is $q_{tot} = CV_{emf}$.

While the capacitor is charging, we can analyze the current, i, flowing in the circuit (assumed to flow from the negative to the positive terminal inside the voltage source) by applying Kirchhoff's Loop Rule to the loop in Figure 26.20b in the counterclockwise direction:

$$V_{emf} - V_R - V_C = V_{emf} - i(t)R - q(t)/C = 0,$$

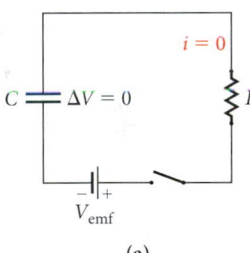

(a)

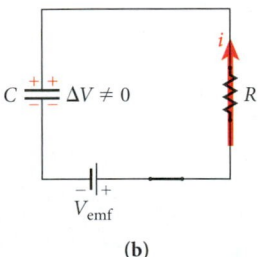

(b)

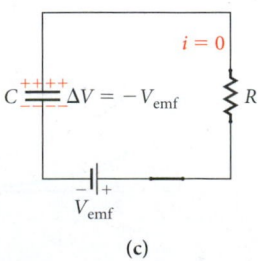

(c)

FIGURE 26.20 A basic RC circuit, containing a source of emf, a resistor, and a capacitor: (a) with the switch open; (b) a short time after the switch is closed; (c) a long time after the switch is closed.

where V_C is the potential drop across the capacitor and $q(t)$ is the charge on the capacitor at a given time t. The change of the charge on the capacitor plates due to the current is $i(t) = dq(t)/dt$, and we can rewrite the preceding equation as

$$R\frac{dq(t)}{dt} + \frac{q(t)}{C} = V_{\text{emf}},$$

or

$$\frac{dq(t)}{dt} + \frac{q(t)}{RC} = \frac{V_{\text{emf}}}{R}. \tag{26.3}$$

This differential equation relates the charge to its time derivative. The discussion of damped oscillations in Chapter 14 involved similar differential equations. It seems appropriate to try an exponential form for the solution of equation 26.3 because an exponential function is the only type that has the property of having a derivative that is identical to itself. Because equation 26.3 also has a constant term, the trial solution needs to have a constant term. We therefore try a solution with a constant and an exponential and for which $q(0) = 0$:

$$q(t) = q_{\text{max}}\left(1 - e^{-t/\tau}\right),$$

where the constants q_{max} and τ are to be determined. Substituting this trial solution back into equation 26.3, we obtain

$$q_{\text{max}}\frac{1}{\tau}e^{-t/\tau} + \frac{1}{RC}q_{\text{max}}\left(1 - e^{-t/\tau}\right) = \frac{V_{\text{emf}}}{R}.$$

Now we collect the time-dependent terms on the left-hand side and the time-independent terms on the right-hand side:

$$q_{\text{max}}e^{-t/\tau}\left(\frac{1}{\tau} - \frac{1}{RC}\right) = \frac{V_{\text{emf}}}{R} - \frac{1}{RC}q_{\text{max}}.$$

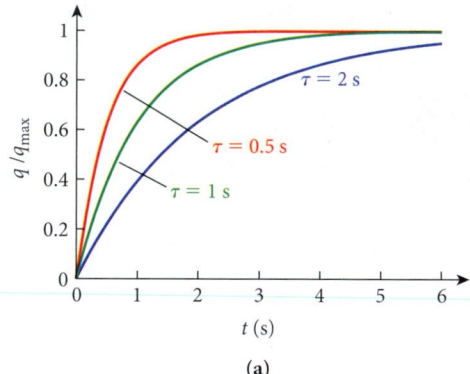

This equation can only be true for all times if both sides are equal to zero. From the left-hand side, we then find

$$\tau = RC. \tag{26.4}$$

Thus, the constant τ (called the **time constant**) is simply the product of the capacitance and the resistance. From the right-hand side, we find an expression for the constant q_{max}:

$$q_{\text{max}} = CV_{\text{emf}}.$$

Thus, the differential equation for charging the capacitor (equation 26.3) has the solution

$$q(t) = CV_{\text{emf}}\left(1 - e^{-t/RC}\right). \tag{26.5}$$

Note that at $t = 0$, $q = 0$, which is the initial condition before the circuit components were connected. At $t = \infty$, $q = q_{\text{max}} = CV_{\text{emf}}$, which is the steady-state condition in which the capacitor is fully charged. The time dependence of the charge on the capacitor is shown in Figure 26.21a for three different values of the time constant τ.

The current flowing in the circuit is obtained by differentiating equation 26.5 with respect to time:

$$i = \frac{dq}{dt} = \left(\frac{V_{\text{emf}}}{R}\right)e^{-t/RC}. \tag{26.6}$$

From equation 26.6, at $t = 0$, the current in the circuit is V_{emf}/R, and at $t = \infty$, the current is zero, as shown in Figure 26.21b.

How do we know that the solution we have found for equation 26.3 is the only solution? This is not obvious from the preceding discussion, but the solution is unique. (A proof is generally provided in a course on differential equations.)

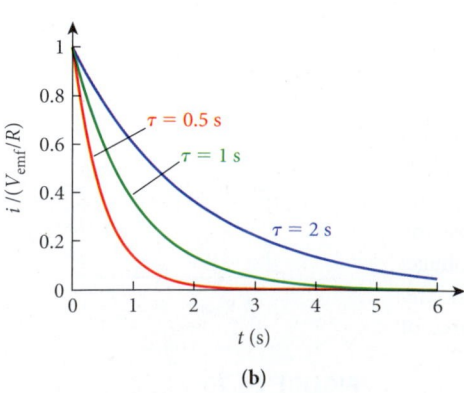

FIGURE 26.21 Charging a capacitor: (a) charge on the capacitor as a function of time; (b) current flowing through the resistor as a function of time.

Discharging a Capacitor

Now let's consider a circuit containing only a resistor, R, and a fully charged capacitor, C, obtained by moving the switch in Figure 26.22 from position 1 to position 2. The charge on the capacitor before the switch is moved is q_{max}. In this case, current will flow in the circuit until the capacitor is completely discharged. While the capacitor is discharging, we can apply Kirchhoff's Loop Rule around the loop in the clockwise direction and obtain

$$-i(t)R - V_C = -i(t)R - \frac{q(t)}{C} = 0.$$

We can rewrite this equation using the definition of current:

$$\frac{R\,dq(t)}{dt} + \frac{q(t)}{C} = 0. \qquad (26.7)$$

The solution to equation 26.7 is obtained using the same method as for equation 26.3, except that equation 26.7 has no constant term and $q(0) > 0$. Thus, we try a solution of the form $q(t) = q_{max}e^{-t/\tau}$, which leads to

$$q(t) = q_{max}e^{-t/RC}. \qquad (26.8)$$

At $t = 0$, the charge on the capacitor is q_{max}. At $t = \infty$, the charge on the capacitor is zero.

We can obtain the current by differentiating equation 26.8 as a function of time:

$$i(t) = \frac{dq}{dt} = -\left(\frac{q_{max}}{RC}\right)e^{-t/RC}. \qquad (26.9)$$

At $t = 0$, the current in the circuit is $-q_{max}/RC$. At $t = \infty$, the current in the circuit is zero. Plotting the time dependence of the charge on the capacitor and the current flowing through the resistor for the discharging process would result in exponentially decreasing curves like those in Figure 26.21b.

The equations describing the time dependence of the charging and the discharging of a capacitor all involve the exponential factor $e^{-t/RC}$. Again, the product of the resistance and the capacitance is defined as the time constant of an RC circuit: $\tau = RC$. According to equation 26.5, after an amount of time equal to the time constant, the capacitor will have been charged to 63% of its maximum value. Thus, an RC circuit can be characterized by specifying the time constant. A large time constant means that it takes a long time to charge the capacitor; a small time constant means that it takes a short time to charge the capacitor.

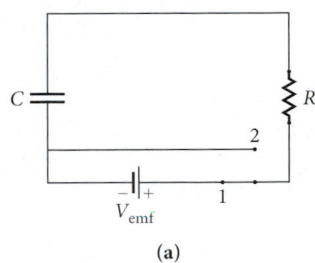

(a)

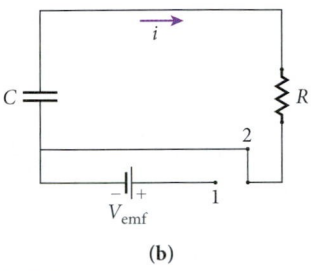

(b)

FIGURE 26.22 RC circuit containing an emf source, a resistor, a capacitor, and a switch. The capacitor is (a) charged with the switch in position 1 and (b) discharged with the switch in position 2.

Concept Check 26.5

To discharge a capacitor in an RC circuit very quickly, what should the values of the resistance and the capacitance be?

a) Both should be as large as possible.

b) Resistance should be as large as possible, and capacitance as small as possible.

c) Resistance should be as small as possible, and capacitance as large as possible.

d) Both should be as small as possible.

EXAMPLE 26.3 | Time Required to Charge a Capacitor

Consider a circuit consisting of a 12.0-V battery, a 50.0-Ω resistor, and a 100.0-μF capacitor wired in series. The capacitor is initially completely discharged.

PROBLEM

How long after the circuit is closed will it take to charge the capacitor to 90% of its maximum charge?

SOLUTION

The charge on the capacitor as a function of time is given by

$$q(t) = q_{max}\left(1 - e^{-t/RC}\right),$$

where q_{max} is the maximum charge on the capacitor. We want to know the time until $q(t)/q_{max} = 0.90$, which can be obtained from

$$\left(1 - e^{-t/RC}\right) = \frac{q(t)}{q_{max}} = 0.90,$$

or

$$0.10 = e^{-t/RC}. \qquad (i)$$

– Continued

Taking the natural log of both sides of equation (i), we get

$$\ln 0.10 = -\frac{t}{RC},$$

or

$$t = -RC\ln 0.10 = -(50.0\ \Omega)(100\cdot10^{-6}\ \text{F})(-2.30) = 0.0115\ \text{s} = 11.5\ \text{ms}.$$

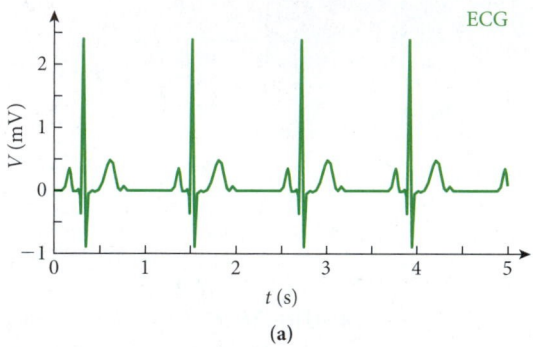

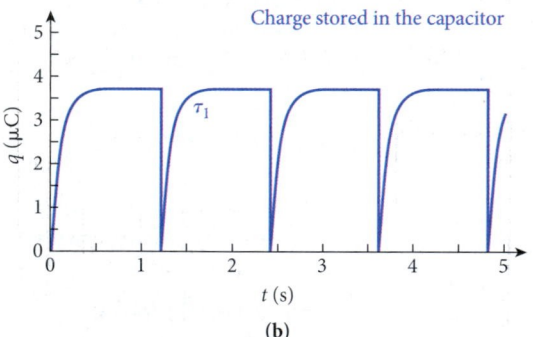

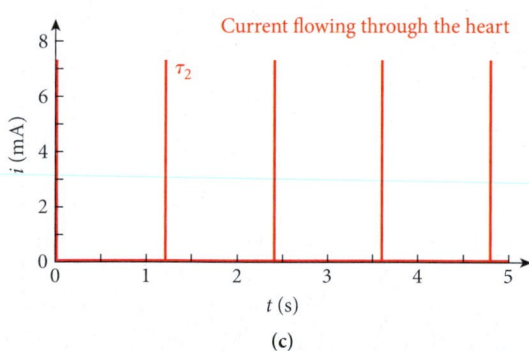

FIGURE 26.23 (a) An electrocardiogram (ECG) showing four regular heartbeats. (b) The charge stored in the pacemaker's capacitor as a function of time. (c) The current flowing through the heart due to the discharging of the pacemaker's capacitor.

Pacemaker

A normal human heart beats at regular intervals, sending blood through the body. The heart's own electrical signals regulate its beating. These electrical signals can be measured through the skin using an electrocardiograph. This device produces a graph of potential difference versus time, which is called an *electrocardiogram*, or *ECG* (sometimes *EKG*, from the German word *Electrokardiograph*). Figure 26.23a shows how an ECG would represent four regular heartbeats occurring at a rate of 72 beats per minute. Doctors and medical personnel can use an ECG to diagnose the health of the heart.

Sometimes the heart does not beat regularly and needs help to maintain its proper rhythm. This help can be provided by a pacemaker, an electrical circuit that sends electrical pulses to the heart at regular intervals, replacing the heart's usual electrical signals and stimulating the heart to beat at prescribed intervals. A pacemaker is implanted in the patient and connected directly to the heart, as illustrated in Figure 26.24.

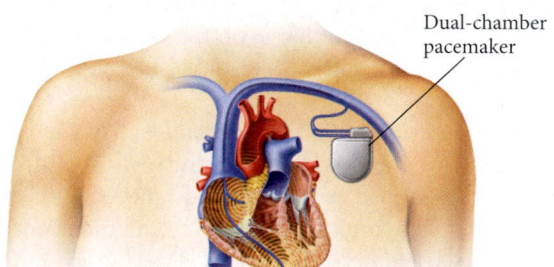

FIGURE 26.24 A modern pacemaker implanted in a patient. The pacemaker sends electrical pulses to two chambers of the heart to help keep it beating regularly.

Concept Check 26.6

In the circuit shown in the figure, the capacitor, *C*, is initially uncharged. Immediately after the switch is closed,

a) the current flowing through R_1 is zero.

b) the current flowing through R_1 is larger than that through R_2.

c) the current flowing through R_2 is larger than that through R_1.

d) the current flowing through R_1 is the same as that through R_2.

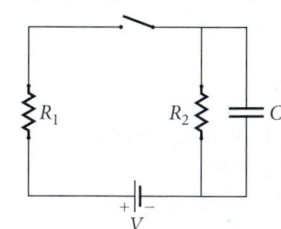

Concept Check 26.7

In the circuit shown in the figure, the switch is closed. After a long time,

a) the current through R_1 is zero.

b) the current flowing through R_1 is larger than that through R_2.

c) the current flowing through R_2 is larger than that through R_1.

d) the current flowing through R_1 is the same as that through R_2.

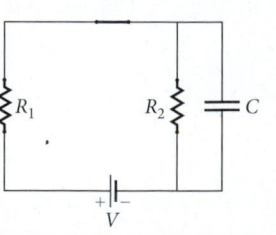

EXAMPLE 26.4 Circuit Elements of a Pacemaker

Let's analyze the circuit shown in Figure 26.25, which simulates the function of a pacemaker. This pacemaker circuit operates by charging a capacitor, C, for some time using a battery with voltage V_{emf} and a resistor R_1, as illustrated in Figure 26.25a, where the switch is open. Closing the switch, as in Figure 26.25b, shorts the capacitor across the heart, and the capacitor discharges through the heart in a short time to stimulate the heart to beat. Thus, this circuit operates as a pacemaker by keeping the switch open for the time between heartbeats, closing the switch for a short time to stimulate a heartbeat, and then opening the switch again.

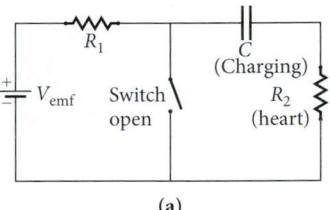

(a)

PROBLEM

What values of the capacitance, C, and the resistance, R_1, should be used in a pacemaker?

SOLUTION

We assume that the heart acts as a resistor with a value $R_2 = 500 \ \Omega$ and that the source of emf is a lithium ion battery (discussed in Chapter 23). Such a battery has a very high energy density and a voltage of 3.7 V. A normal heart rate is in the range from 60 to 100 beats per minute. However, the pacemaker might need to stimulate the heart to beat faster, so it should be capable of running at 180 beats per minute, which means that the capacitor may be charged up to 180 times a minute. Thus, the minimum time between discharges, t_{min}, is

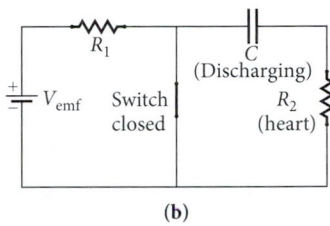

(b)

FIGURE 26.25 (a) A simplified pacemaker circuit in charging mode. (b) The pacemaker circuit in discharging mode.

$$t_{min} = \frac{1}{180 \text{ beats/min}} = \left(\frac{1 \text{ min}}{180}\right)\left(\frac{60 \text{ s}}{1 \text{ min}}\right) = 0.333 \text{ s}.$$

Equation 26.5 gives the charge, q, as a function of time, t, for the maximum charge, q_{max}, for a given time constant, $\tau_1 = R_1C$:

$$q = q_{max}\left(1 - e^{-t/\tau_1}\right).$$

Rearranging this equation gives us

$$\frac{q}{q_{max}} = f = 1 - e^{-t/\tau_1},$$

where f is the fraction of the charge capacity of the capacitor. Let's assume the capacitor must be charged to 95% of its maximum charge in time t_{min}. Solving for the time constant gives us

$$\tau_1 = R_1C = -\frac{t_{min}}{\ln(1-f)} = -\frac{0.333 \text{ s}}{\ln(1-0.95)} = 0.111 \text{ s}.$$

Thus, the time constant for the charging should be $\tau_1 = R_1C = 100$ ms. The time constant for discharging, τ_2, needs to be small to produce short pulses of a high current to stimulate the heart. Letting $\tau_2 = 0.500$ ms, which is on the order of the narrow electrical pulse in the ECG, we have

$$\tau_2 = R_2C = 0.500 \text{ ms}.$$

We can solve for the required capacitance and substitute the value 500 Ω for R_2:

$$C = \frac{0.000500 \text{ s}}{500 \ \Omega} = 1.00 \ \mu\text{F}.$$

The resistance required for the charging circuit can now be related to the time constant for the charging and the value of the capacitance we just calculated:

$$\tau_1 = R_1C = 0.100 \text{ s} = R_1(1.00 \ \mu\text{F}),$$

which gives us

$$R_1 = \frac{0.100 \text{ s}}{1.00 \cdot 10^{-6} \text{ F}} = 100 \text{ k}\Omega.$$

Figure 26.23b shows the charge in the capacitor of the simulated pacemaker as a function of time for a heart rate of 72 heartbeats per minute. You can see that the capacitor charges to nearly full capacity before it is discharged. Figure 26.23c shows the current that flows through the heart when the capacitor discharges. The current pulse is narrow, lasting less than a millisecond. This pulse stimulates the heart to beat, as illustrated by the ECG in Figure 26.23a. The rate at which the heart beats is controlled by the rate at which the pacemaker's switch is closed and opened, which is controlled by a microprocessor.

FIGURE 26.26 The main components of a neuron.

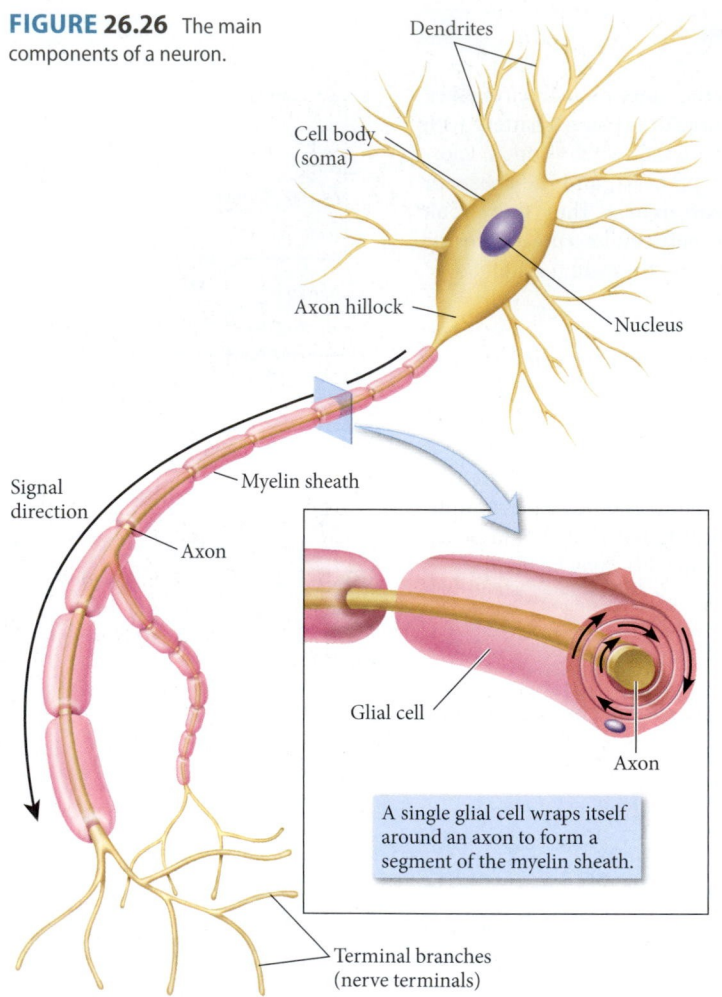

FIGURE 26.26 The main components of a neuron.

Dendrites

Cell body (soma)

Axon hillock

Nucleus

Signal direction

Myelin sheath

Axon

Glial cell

Axon

A single glial cell wraps itself around an axon to form a segment of the myelin sheath.

Terminal branches (nerve terminals)

Neuron

The type of cell responsible for transmitting and processing signals in the nervous systems and brains of humans and other animals is a neuron (Figure 26.26). The neuron conducts the necessary currents by electrochemical means, via the movement of ions (mainly Na^+, K^+, and Cl^-). Neurons receive signals from other neurons through dendrites and send signals to other neurons through an axon. The axon can be quite long (for example, in the spinal cord) and is covered with an insulating myelin sheath. The signals are received from and sent to other neurons or cells in the sense organs and other tissues. All of this, and much more, can be learned in an introductory biology or physiology course. Here we'll take a look at a neuron as a basic circuit that processes signals.

An input signal has to be strong enough to get a neuron to fire, that is, to send an output signal down the axon. It is a crude but reasonable approximation to represent the main cell body of a neuron, the soma, as a basic RC circuit that processes these signals. A diagram of this RC circuit is shown in Figure 26.27. A capacitor and a resistor are connected in parallel to an input and an output potential. The typical potential values for neurons are on the order of ±50 mV relative to the background of the surrounding tissue. If $\Delta V = V_{in} - V_{out} \neq 0$, a current flows through the circuit. Part of this current flows through the resistor, but part of the current simultaneously charges the capacitor until it reaches the potential difference between input and output potentials. The potential difference between the capacitor plates rises exponentially, according to $V_C(t) = (V_{in} - V_{out})(1 - e^{-t/RC})$, just as for the process of charging a capacitor in an RC circuit. (The time constant, $\tau = RC$, is on the order of 10 ms, assuming a capacitance of 1 nF and a resistance of 10 MΩ.) If the external source of potential difference is then removed from the circuit, the capacitor discharges with the same time constant, and the potential across the capacitor decays exponentially, according to $V_C(t) = V_0 e^{-t/RC}$. This simple time dependence captures the basic response of a neuron. Figure 26.28 shows the potential difference across the capacitor of this model neuron, while it is charged for 30 ms and then discharged.

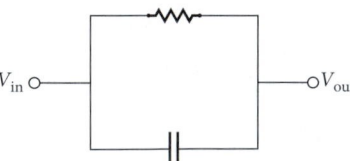

FIGURE 26.27 Simplified model of a neuron as an RC circuit.

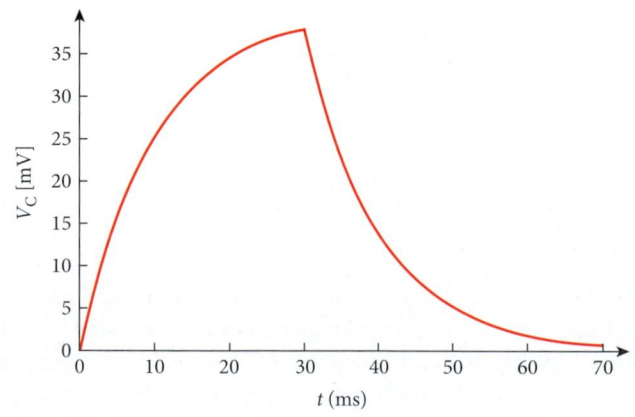

FIGURE 26.28 Potential difference on the capacitor in a model neuron.

SOLVED PROBLEM 26.4 | Rate of Energy Storage in a Capacitor

A resistor with $R = 2.50$ MΩ and a capacitor with $C = 1.25$ μF are connected in series with a battery for which $V_{emf} = 12.0$ V. At $t = 2.50$ s after the circuit is closed, what is the rate at which energy is being stored in the capacitor?

THINK When the circuit is closed, the capacitor begins to charge. The rate at which energy is stored in the capacitor is given by the time derivative of the amount of energy stored in the capacitor, which is a function of the charge on the capacitor.

SKETCH Figure 26.29 shows a diagram of the series circuit containing a battery, a resistor, and a capacitor.

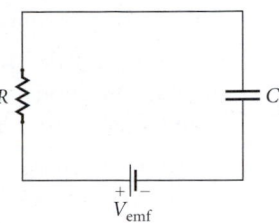

FIGURE 26.29 Series circuit containing a battery, a resistor, and a capacitor.

RESEARCH The charge on the capacitor as a function of time is given by equation 26.5:

$$q(t) = CV_{emf}\left(1 - e^{-t/RC}\right).$$

The energy stored in a capacitor that has charge q is given by (see Chapter 24)

$$U = \tfrac{1}{2}\frac{q^2}{C}. \tag{i}$$

The time derivative of the energy stored in the capacitor is then

$$\frac{dU}{dt} = \frac{d}{dt}\left(\tfrac{1}{2}\frac{q^2(t)}{C}\right) = \frac{q(t)}{C}\frac{dq(t)}{dt}. \tag{ii}$$

The time derivative of the charge is the current, i. Thus, we can replace dq/dt with the expression given by equation 26.6:

$$i(t) = \frac{dq(t)}{dt} = \left(\frac{V_{emf}}{R}\right)e^{-t/RC}. \tag{iii}$$

SIMPLIFY We can express the rate of change of the energy stored in the capacitor by combining equations (i) through (iii):

$$\frac{dU}{dt} = \frac{q(t)}{C}i(t) = \frac{CV_{emf}\left(1 - e^{-t/RC}\right)}{C}\left(\frac{V_{emf}}{R}\right)e^{-t/RC} = \frac{V_{emf}^2}{R}e^{-t/RC}\left(1 - e^{-t/RC}\right).$$

CALCULATE We first calculate the value of the time constant, $\tau = RC$:

$$RC = \left(2.50 \cdot 10^6 \ \Omega\right)\left(1.25 \cdot 10^{-6} \ F\right) = 3.125 \ s.$$

We can then calculate the rate of change of the energy stored in the capacitor:

$$\frac{dU}{dt} = \frac{(12.0 \ V)^2}{2.50 \cdot 10^6 \ \Omega}e^{-(2.50 \ s)/(3.125 \ s)}\left(1 - e^{-(2.50 \ s)/(3.125 \ s)}\right) = 1.42521 \cdot 10^{-5} \ W.$$

ROUND We report our result to three significant figures:

$$\frac{dU}{dt} = 1.43 \cdot 10^{-5} \ W.$$

DOUBLE-CHECK The current at $t = 2.50$ s is

$$i(2.50 \ s) = \left(\frac{12.0 \ V}{2.50 \ M\Omega}\right)e^{-(2.50 \ s)/(3.125 \ s)} = 2.16 \cdot 10^{-6} \ A.$$

The rate of energy dissipation in the resistor at this time is

$$P = \frac{dU}{dt} = i^2 R = \left(2.16 \cdot 10^{-6} \ A\right)^2 \left(2.50 \cdot 10^6 \ \Omega\right) = 1.16 \cdot 10^{-5} \ W.$$

The rate at which the battery delivers energy to the circuit at this time is given by

$$P = \frac{dU}{dt} = iV_{emf} = \left(2.16 \cdot 10^{-6} \ A\right)(12.0 \ V) = 2.59 \cdot 10^{-5} \ W.$$

Energy conservation dictates that at any time the energy supplied by the battery is either dissipated as heat in the resistor or stored in the capacitor. In this case, the power supplied by the battery, $2.59 \cdot 10^{-5}$ W, is equal to the power dissipated as heat in the resistor, $1.16 \cdot 10^{-5}$ W, plus the rate at which energy is stored in the capacitor, $1.43 \cdot 10^{-5}$ W. Thus, our answer is consistent.

WHAT WE HAVE LEARNED | EXAM STUDY GUIDE

- Kirchhoff's rules for analyzing circuits are as follows:

  - Kirchhoff's Junction Rule: The sum of the currents entering a junction must equal the sum of the currents leaving a junction.

  - Kirchhoff's Loop Rule: The potential difference around a complete loop must sum to zero.

- In applying Kirchhoff's Loop Rule, the sign of the potential change for each circuit element is determined by the direction of the current and the direction of analysis. The conventions are

  - Sources of emf in the same direction as the direction of analysis provide potential gains, while sources opposite to the analysis direction give potential drops.

  - For resistors, the magnitude of the potential change is $|iR|$, where i is the assumed current and R is the resistance. The sign of the potential change depends on the (known or assumed) direction of the current as well as the direction of analysis. If these directions are the same, the resistor produces a potential drop. If the directions are opposite, the resistor produces a potential gain.

- An RC circuit contains a resistor of resistance R and a capacitor of capacitance C. The time constant, τ, is given by $\tau = RC$.

- In an RC circuit, the charge, q, as a function of time for a charging capacitor with capacitance C is given by $q(t) = CV_{emf}(1 - e^{-t/RC})$, where V_{emf} is the voltage supplied by the source of emf and R is the resistance of the resistor.

- In an RC circuit, the charge, q, as a function of time for a discharging capacitor with capacitance C is given by $q(t) = q_{max}e^{-t/RC}$, where q_{max} is the size of the charge on the capacitor plates at $t = 0$ and R is the resistance of the resistor.

ANSWERS TO SELF-TEST OPPORTUNITIES

26.1 The resistors R_1 and R_u are in series and have an equivalent resistance of $R_{1u} = R_1 + R_u$. The resistors R_3 and R_v are in series and have an equivalent resistance of $R_{3v} = R_3 + R_v$. The equivalent resistances R_{1u} and R_{3v} are in parallel. Thus, we can write

$$\frac{1}{R_{eq}} = \frac{1}{R_{1u}} + \frac{1}{R_{3v}},$$

or

$$R_{eq} = \frac{R_{1u}R_{3v}}{R_{1u} + R_{3v}} = \frac{(R_1 + R_u)(R_3 + R_v)}{R_1 + R_u + R_3 + R_v}.$$

26.2 When the headlights are on, the battery is supplying a modest amount of current to the lights, and the potential drop across the internal resistance of the battery is small. The starter motor is wired in parallel with the lights. When the starter motor engages, it draws a large current, producing a noticeable drop in potential across the internal resistance of the battery and causing less current to flow to the headlights.

26.3 $q = q_{max}e^{-t/RC}$

$$\frac{q}{q_{max}} = 0.01 = e^{-t/RC} \Rightarrow \ln 0.01 = -\frac{t}{RC}$$

$$t = -RC\ln 0.01 = -(100 \ \Omega)(1.00 \cdot 10^{-3} \ \text{F})(\ln 0.01) = 0.461 \ \text{s}.$$

PROBLEM-SOLVING GUIDELINES

1. It is always helpful to label everything in a circuit diagram, including all the given information and all the unknowns, as well as pertinent currents, branches, and junctions. Redraw the diagram at a larger scale if you need more space for clarity.

2. Remember that the directions you choose for currents and for the path around a circuit loop are arbitrary. If your choice turns out to be incorrect, a negative value for the current will result.

3. Review the signs for potential changes given in Table 26.1. Moving through a circuit loop in the same direction as the assumed current means that an emf device produces a positive potential change in the direction from negative to positive within the device and that the potential change

across a resistor is negative. Sign errors are common, and it pays to stick to the conventions to avoid such errors.

4. Sources of emf or resistors may be parts of two separate loops. Count each circuit component as a part of each loop it is in, according to the sign conventions you've adopted for that loop. A resistor may yield a potential drop in one loop and a potential gain in the other loop.

5. It is always possible to use Kirchhoff's rules to write more equations than you need to solve for unknown currents in a circuit's branches. Write as many equations as possible for junctions, and then augment them with equations representing loops. But not all loops are created equal; you need to select them carefully. As a rule of thumb, choose loops with fewer circuit elements.

MULTIPLE-CHOICE QUESTIONS

26.1 A resistor and a capacitor are connected in series. If a second identical capacitor is connected in series in the same circuit, the time constant for the circuit will

a) decrease.   b) increase.   c) stay the same.

26.2 A resistor and a capacitor are connected in series. If a second identical resistor is connected in series in the same circuit, the time constant for the circuit will

a) decrease.   b) increase.   c) stay the same.

26.3 A circuit consists of a source of emf, a resistor, and a capacitor, all connected in series. The capacitor is fully charged. How much current is flowing through it?

a) $i = V/R$ b) zero   c) neither (a) nor (b)

26.4 Which of the following will reduce the time constant in an RC circuit?

a) increasing the dielectric constant of the capacitor

b) adding an additional 20 m of wire between the capacitor and the resistor

c) increasing the voltage of the battery

d) adding an additional resistor in parallel with the first resistor

e) none of the above

26.5 Kirchhoff's Junction Rule states that

a) the algebraic sum of the currents at any junction in a circuit must be zero.

b) the algebraic sum of the potential changes around any closed loop in a circuit must be zero.

c) the current in a circuit with a resistor and a capacitor varies exponentially with time.

d) the current at a junction is given by the product of the resistance and the capacitance.

e) the time for the current development at a junction is given by the product of the resistance and the capacitance.

26.6 How long will it take, as a multiple of the time constant, τ, for the capacitor in an RC circuit to be 98% charged?

a) 9τ c) 90τ e) 0.98τ

b) 0.9τ d) 4τ

26.7 A capacitor C is initially uncharged. At time $t = 0$, the capacitor is attached through a resistor R to a battery. The energy stored in the capacitor increases, eventually reaching a value U as $t \to \infty$. After a time equal to the time constant $\tau = RC$, the energy stored in the capacitor is given by

a) U/e.   c) $U(1 - 1/e)^2$.

b) U/e^2.   d) $U(1 - 1/e)$.

26.8 Which of the following has the same unit as the electromotive force (emf)?

a) current

b) electric potential

c) electric field

d) electric power

e) none of the above

26.9 The capacitor in each circuit in the figure is first charged by a 10-V battery with no internal resistance. Then, the switch is flipped from position A to position B, and the capacitor is discharged through various resistors. For which circuit is the total energy dissipated by the resistor the largest?

26.10 Two resistors, $R_1 = 3.00 \ \Omega$ and $R_2 = 5.00 \ \Omega$, are connected in series with a battery and an ammeter, as shown in the figure. The battery supplies $V_{emf} = 8.00$ V, and the ammeter has the resistance $R_A = 1.00 \ \Omega$. What is the current measured by the ammeter?

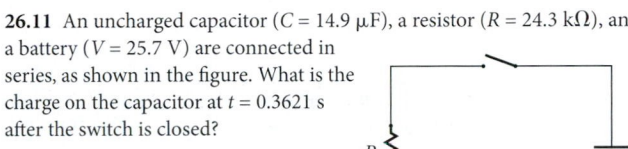

a) 0.500 A

b) 0.750 A

c) 0.889 A

d) 1.00 A

e) 1.50 A

26.11 An uncharged capacitor ($C = 14.9 \ \mu$F), a resistor ($R = 24.3 \ k\Omega$), and a battery ($V = 25.7$ V) are connected in series, as shown in the figure. What is the charge on the capacitor at $t = 0.3621$ s after the switch is closed?

a) $5.48 \cdot 10^{-5}$ C

b) $7.94 \cdot 10^{-5}$ C

c) $1.15 \cdot 10^{-5}$ C

d) $1.66 \cdot 10^{-4}$ C

e) $2.42 \cdot 10^{-4}$ C

26.12 Kirchhoff's Loop Rule states that

a) the algebraic sum of the currents around a complete circuit loop must be zero.

b) the resistances around a complete circuit loop must sum to zero.

c) the sources of emf around a complete circuit loop must sum to zero.

d) the sum of the potential differences around a complete circuit loop must be greater than zero.

e) the potential differences around a complete circuit loop must sum to zero.

26.13 Which of the following statements are true?

1. An ideal ammeter should have infinite resistance.

2. An ideal ammeter should have zero resistance.

3. An ideal voltmeter should have infinite resistance.

4. An ideal voltmeter should have zero resistance.

a) 1 and 3

b) 2 and 4

c) 2 and 3

d) 1 and 4

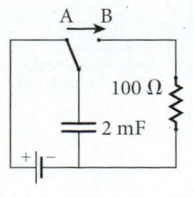

(a)

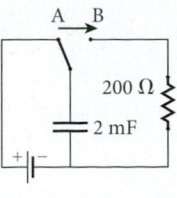

(b)

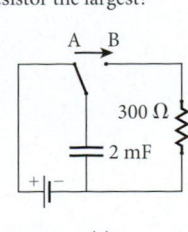

(c)

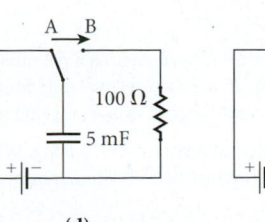

(d)

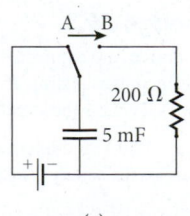

(e)

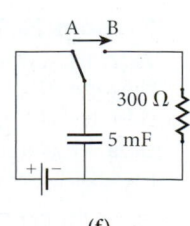

(f)

CONCEPTUAL QUESTIONS

26.14 You want to measure simultaneously the potential difference across and the current through a resistor, *R*. As the circuit diagrams show, there are two ways to connect the two instruments—ammeter and voltmeter—in the circuit. Comment on the result of the measurement using each configuration.

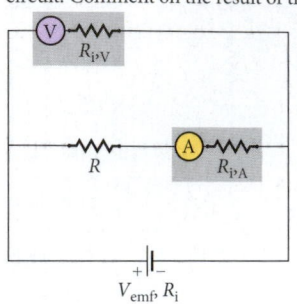 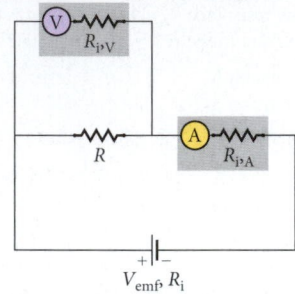

26.15 If the capacitor in an RC circuit is replaced with two identical capacitors connected in series, what happens to the time constant for the circuit?

26.16 You want to accurately measure the resistance, R_{device}, of a new device. The figure shows two ways to accomplish this task. On the left, an ohmmeter produces a current through the device and measures that current, *i*, and the potential difference, ΔV, across the device. This potential difference includes the potential drops across the wires leading to and from the device and across the contacts that connect the wires to the device. These extra resistances cannot always be neglected, especially if the device has low resistance. This technique is called *two-probe measurement* since two probe wires are connected to the device. The resulting current, *i*, is measured with an ammeter. The total resistance is then determined by dividing ΔV by *i*. For this configuration, what is the resistance that the ohmmeter measures? In the alternative configuration, shown on the right, a similar current source is used to produce and measure the current through the device, but the potential difference, ΔV, is measured directly across the device with a nearly ideal voltmeter with extremely large internal resistance. This technique is called a *four-probe measurement* since four probe wires are connected to the device. What resistance is being measured

in this four-probe configuration? Is it different from that being assessed by the two-probe measurement? Why or why not? (*Hint*: Four-probe measurements are used extensively by scientists and engineers and are especially useful for accurate measurements of the resistance of materials or devices with low resistance.)

26.17 Explain why the time constant for an RC circuit increases with *R* and with *C*. (The answer "That's what the formula says" is not sufficient.)

26.18 A battery, a resistor, and a capacitor are connected in series in an RC circuit. What happens to the current through a resistor after a long time? Explain using Kirchhoff's rules.

26.19 How can you light a 1.0-W, 1.5-V bulb with your 12.0-V car battery?

26.20 A multiloop circuit contains a number of resistors and batteries. If the emf values of all the batteries are doubled, what happens to the currents in all the components of the circuit?

26.21 A multiloop circuit of resistors, capacitors, and batteries is switched on at $t = 0$, at which time all the capacitors are uncharged. The initial distribution of currents and potential differences in the circuit can be analyzed by treating the capacitors as if they were connecting wires or closed switches. The final distribution of currents and potential differences, which occurs after a long time has passed, can be analyzed by treating the capacitors as open segments or open switches. Explain why these tricks work.

26.22 Voltmeters are always connected in parallel with a circuit component, and ammeters are always connected in series. Explain why.

26.23 You wish to measure both the current through and the potential difference across some component of a circuit. It is not possible to do this simultaneously and accurately with ordinary voltmeters and ammeters. Explain why not.

26.24 Two light bulbs for use at 110 V are rated at 60 W and 100 W, respectively. Which has the filament with lower resistance?

26.25 Two capacitors in series are charged through a resistor. Identical capacitors are instead connected in parallel and charged through the same resistor. How do the times required to fully charge the two sets of capacitors compare?

26.26 The figure shows a circuit consisting of a battery connected to a resistor and a capacitor, which is fully discharged initially, in series with a switch.

a) What is the current in the circuit at any time *t*?

b) Calculate the total energy provided by the battery from $t = 0$ to $t = \infty$.

c) Calculate the total energy dissipated from the resistor over the same time period.

d) Is energy conserved in this circuit?

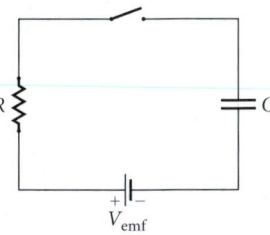

EXERCISES

A blue problem number indicates a worked-out solution is available in the Student Solutions Manual. One • and two •• indicate increasing level of problem difficulty.

Sections 26.1 through 26.3

26.27 Two resistors, R_1 and R_2, are connected in series across a potential difference, ΔV_0. Express the potential drop across each resistor individually, in terms of these quantities. What is the significance of this arrangement?

26.28 A battery has $V_{emf} = 12.0$ V and internal resistance $r = 1.00\ \Omega$. What resistance, *R*, can be put across the battery to extract 10.0 W of power from it?

26.29 Three resistors are connected across a battery as shown in the figure. What values of *R* and V_{emf} will produce the indicated currents?

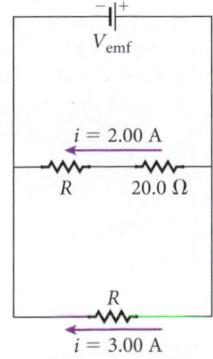

•**26.30** Find the equivalent resistance for the circuit in the figure.

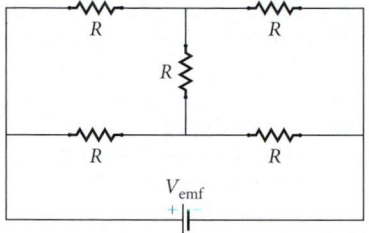

•**26.31** The dead battery of your car provides a potential difference of 9.950 V and has an internal resistance of 1.100 Ω. You charge it by connecting it with jumper cables to the live battery of another car. The live battery provides a potential difference of 12.00 V and has an internal resistance of 0.01000 Ω, and the starter resistance is 0.07000 Ω.

a) Draw the circuit diagram for the connected batteries.

b) Determine the current in the live battery, in the dead battery, and in the starter immediately after you closed the circuit.

•**26.32** In the circuit shown in the figure, $V_1 = 1.50$ V, $V_2 = 2.50$ V, $R_1 = 4.00$ Ω, and $R_2 = 5.00$ Ω. What is the magnitude of the current, i_1, flowing through resistor R_1?

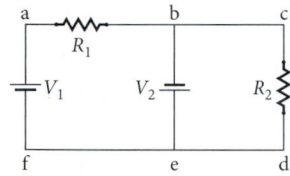

•**26.33** The circuit shown in the figure consists of two batteries supplying voltages V_A and V_B and three light bulbs with resistances R_1, R_2, and R_3. Calculate the magnitudes of the currents i_1, i_2, and i_3 flowing through the bulbs. Indicate the correct directions of current flow on the diagram. Calculate the power, P_A and P_B, supplied by battery A and by battery B.

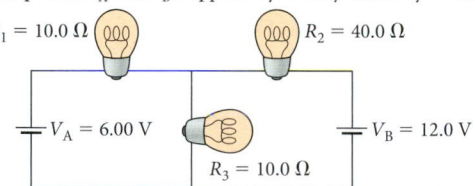

•**26.34** In the circuit shown in the figure, $R_1 = 5.00$ Ω, $R_2 = 10.0$ Ω, $R_3 = 15.0$ Ω, $V_{emf,1} = 10.0$ V, and $V_{emf,2} = 15.0$ V. Using Kirchhoff's Loop and Junction Rules, determine the currents i_1, i_2, and i_3 flowing through R_1, R_2, and R_3, respectively, in the direction indicated in the figure.

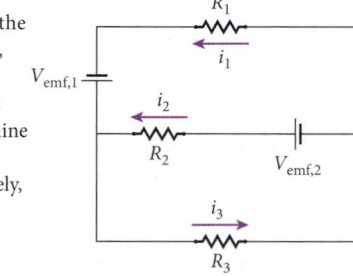

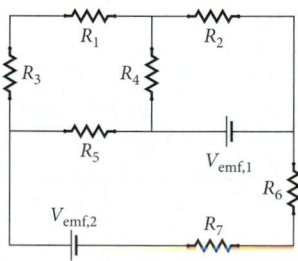

•**26.35** For the circuit shown in the figure, find the magnitude and the direction of the current through each resistor and the power supplied by each battery, using the following values: $R_1 = 4.00$ Ω, $R_2 = 6.00$ Ω, $R_3 = 8.00$ Ω, $R_4 = 6.00$ Ω, $R_5 = 5.00$ Ω, $R_6 = 10.0$ Ω, $R_7 = 3.00$ Ω, $V_{emf,1} = 6.00$ V, and $V_{emf,2} = 12.0$ V.

•**26.36** A Wheatstone bridge is constructed using a 1.00-m-long Nichrome wire (the purple line in the figure) with a conducting contact that can slide along the wire. A resistor, $R_1 = 100.$ Ω, is placed on one side of the bridge, and another resistor, R, of unknown resistance, is placed on the other side. The contact is moved along the Nichrome wire, and it is found that the ammeter reading

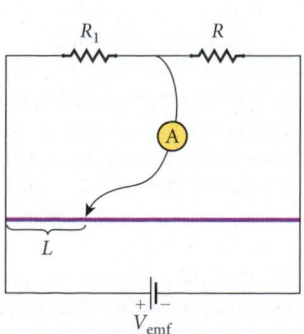

is zero for $L = 25.0$ cm. Knowing that the wire has a uniform cross section throughout its length, determine the unknown resistance.

••**26.37** A "resistive ladder" is constructed with identical resistors, R, making up its legs and rungs, as shown in the figure. The ladder has "infinite" height; that is, it extends very far in one direction. Find the equivalent resistance of the ladder, measured between its "feet" (points A and B), in terms of R.

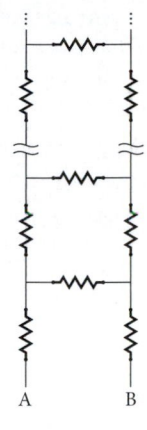

••**26.38** Consider an "infinite," that is, very large, two-dimensional square grid of identical resistors, R, as shown in the figure. Find the equivalent resistance of the grid, as measured across any individual resistor. (*Hint:* Symmetry and superposition are very helpful in solving this problem.)

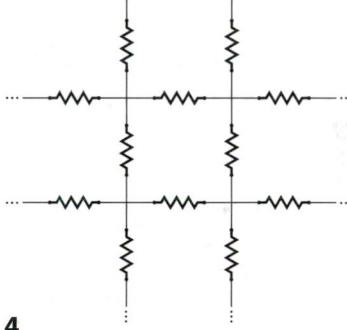

Section 26.4

26.39 To extend the useful range of an ammeter, a shunt resistor, R_{shunt}, is placed in parallel with the ammeter as shown in the figure. If the internal resistance of the ammeter is $R_{i,A}$, determine the resistance that the shunt resistor has to have to extend the useful range of the ammeter by a factor N. Then, calculate the resistance the shunt resistor has to have to allow an ammeter with an internal resistance of 1.00 Ω and a maximum range of 1.00 A to measure currents up to 100. A. What fraction of the total 100.-A current flows through the ammeter, and what fraction flows through the shunt resistor?

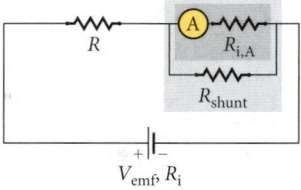

26.40 To extend the useful range of a voltmeter, an additional resistor, R_{series}, is placed in series with the voltmeter as shown in the figure. If the internal resistance of the voltmeter is $R_{i,V}$, determine the resistance that the added series resistor has to have to extend the useful range of the voltmeter by a factor N. Then, calculate the resistance the series resistor has to have to allow a voltmeter with an internal resistance of 1.00 MΩ (10^6 Ω) and a maximum range of 1.00 V to measure potential differences up to 100. V. What fraction of the total 100.-V potential drop occurs across the voltmeter, and what fraction of that drop occurs across the added series resistor?

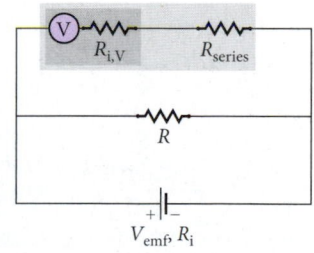

26.41 As shown in the figure, a 6.0000-V battery is used to produce a current through two identical resistors, R, each having a resistance of

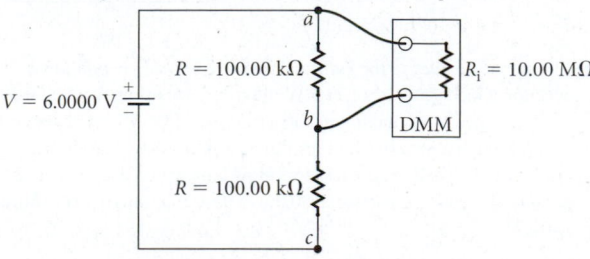

100.00 kΩ. A digital multimeter (DMM) is used to measure the potential difference across the first resistor. DMMs typically have an internal resistance of 10.00 MΩ. Determine the potential differences V_{ab} (the potential difference between points a and b, which is the difference the DMM measures) and V_{bc} (the potential difference between points b and c, which is the difference across the second resistor). Nominally, $V_{ab} = V_{bc}$, but this may not be the case here. How can this measurement error be reduced?

26.42 You want to make an ohmmeter to measure the resistance of unknown resistors. You have a battery with voltage $V_{emf} = 9.00$ V, a variable resistor, R, and an ammeter that measures current on a linear scale from 0 to 10.0 mA.

a) What resistance should the variable resistor have so that the ammeter gives its full-scale (maximum) reading when the ohmmeter is shorted?

b) Using the resistance from part (a), what is the unknown resistance if the ammeter reads $\frac{1}{4}$ of its full scale?

•**26.43** A circuit consists of two 1.00-kΩ resistors in series with an ideal 12.0-V battery.

a) Calculate the current flowing through each resistor.

b) A student trying to measure the current flowing through one of the resistors inadvertently connects an ammeter in parallel with that resistor rather than in series with it. How much current will flow through the ammeter, assuming that it has an internal resistance of 1.00 Ω?

•**26.44** A circuit consists of two 100.-kΩ resistors in series with an ideal 12.0-V battery.

a) Calculate the potential drop across one of the resistors.

b) A voltmeter with internal resistance 10.0 MΩ is connected in parallel with one of the two resistors in order to measure the potential drop across the resistor. By what percentage will the voltmeter reading deviate from the value you determined in part (a)? (*Hint:* The difference is rather small so it is helpful to solve algebraically first to avoid a rounding error.)

Section 26.5

26.45 Initially, switches S_1 and S_2 in the circuit shown in the figure are open and the capacitor has a charge of 100. mC. About how long will it take after switch S_1 is closed for the charge on the capacitor to drop to 5.00 mC?

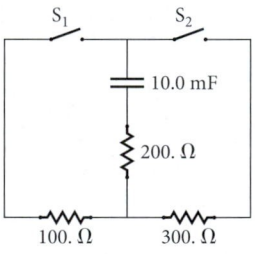

26.46 What is the time constant for the discharging of the capacitors in the circuit shown in the figure? If the 2.00-µF capacitor initially has a potential difference of 10.0 V across its plates, how much charge is left on it after the switch has been closed for a time equal to half of the time constant?

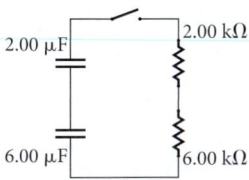

26.47 The circuit shown in the figure has a switch, S, two resistors, $R_1 = 1.00$ Ω and $R_2 = 2.00$ Ω, a 12.0-V battery, and a capacitor with $C = 20.0$ µF. After the switch is closed, what will the maximum charge on the capacitor be? How long after the switch has been closed will the capacitor have 50.0% of this maximum charge?

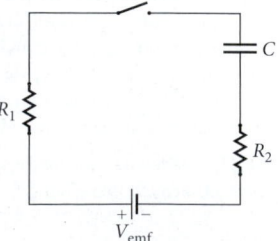

26.48 In the movie *Back to the Future*, time travel is made possible by a flux capacitor, which generates 1.21 GW of power. Assuming that a 1.00-F capacitor is charged to its maximum capacity with a 12.0-V car battery and is discharged through a resistor, what resistance is necessary to produce a peak power output of 1.21 GW in the resistor? How long would it take for a 12.0-V car battery to charge the capacitor to 90.0% of its maximum capacity through this resistor?

26.49 During a physics demonstration, a fully charged 90.0-µF capacitor is discharged through a 60.0-Ω resistor. How long will it take for the capacitor to lose 80.0% of its initial energy?

•**26.50** Two parallel plate capacitors, C_1 and C_2, are connected in series with a 60.0-V battery and a 300.-kΩ resistor, as shown in the figure. Both capacitors have plates with an area of 2.00 cm² and a separation of 0.100 mm. Capacitor C_1 has air between its plates, and capacitor C_2 has the gap filled with a certain porcelain (dielectric constant of 7.00 and dielectric strength of 5.70 kV/mm). The switch is closed, and a long time passes.

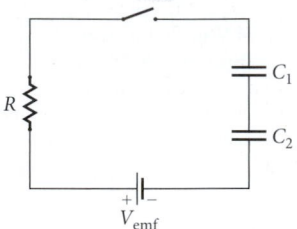

a) What is the charge on capacitor C_1?

b) What is the charge on capacitor C_2?

c) What is the total energy stored in the two capacitors?

d) What is the electric field inside capacitor C_2?

•**26.51** A parallel plate capacitor with $C = 0.0500$ µF has a separation between its plates of $d = 50.0$ µm. The dielectric that fills the space between the plates has dielectric constant $\kappa = 2.50$ and resistivity $\rho = 4.00 \cdot 10^{12}$ Ω m. What is the time constant for this capacitor? (*Hint:* First calculate the area of the plates for the given C and κ, and then determine the resistance of the dielectric between the plates.)

•**26.52** A 12.0-V battery is attached to a 2.00-mF capacitor and a 100.-Ω resistor. Once the capacitor is fully charged, what is the energy stored in it? What is the energy dissipated as heat by the resistor as the capacitor is charging?

•**26.53** A capacitor bank is designed to discharge 5.00 J of energy through a 10.0-kΩ resistor array in under 2.00 ms. To what potential difference must the bank be charged, and what must the capacitance of the bank be?

•**26.54** The circuit in the figure has a capacitor connected to a battery, two switches, and three resistors. Initially, the capacitor is uncharged and both of the switches are open.

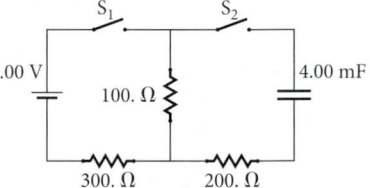

a) Switch S_1 is closed. What is the current flowing out of the battery immediately after switch S_1 is closed?

b) After about 10.0 min, switch S_2 is closed. What is the current flowing out of the battery immediately after switch S_2 is closed?

c) What is the current flowing out of the battery about 10.0 min after switch S_2 has been closed?

d) After another 10.0 min, switch S_1 is opened. How long will it take until the current in the 200.-Ω resistor is below 1.00 mA?

•**26.55** In the circuit shown in the figure, $R_1 = 10.0$ Ω, $R_2 = 4.00$ Ω, and $R_3 = 10.0$ Ω, and the capacitor has capacitance $C = 2.00$ µF.

a) Determine the potential difference, ΔV_C, across the capacitor after switch S has been closed for a long time.

b) Determine the energy stored in the capacitor when switch S has been closed for a long time.

c) After switch S is opened, how much energy is dissipated through R_3?

••**26.56** A cube of gold that is 2.50 mm on a side is connected across the terminals of a 15.0-µF capacitor that initially has a potential difference of 100.0 V between its plates.

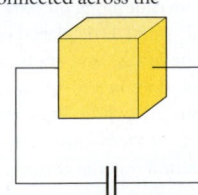

a) What time is required to fully discharge the capacitor?

b) When the capacitor is fully discharged, what is the temperature of the gold cube?

••26.57 A "capacitive ladder" is constructed with identical capacitors, C, making up its legs and rungs, as shown in the figure. The ladder has "infinite" height; that is, it extends very far in one direction. Calculate the equivalent capacitance of the ladder, measured between its "feet" (points A and B), in terms of C.

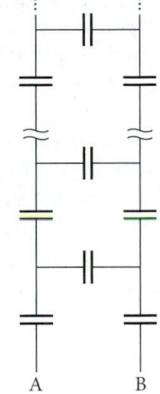

Additional Exercises

26.58 In the circuit in the figure, the capacitors are completely uncharged. The switch is then closed for a long time.

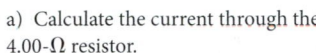

a) Calculate the current through the 4.00-Ω resistor.

b) Find the potential difference across the 4.00-Ω, 6.00-Ω, and 8.00-Ω resistors.

c) Find the potential difference across the 1.00-μF capacitor.

26.59 The ammeter your physics instructor uses for in-class demonstrations has internal resistance $R_i = 75.0 \, \Omega$ and measures a maximum current of 1.50 mA. The same ammeter can be used to measure currents of much greater magnitudes by wiring a shunt resistor of relatively small resistance, R_{shunt}, in parallel with the ammeter. (a) Sketch the circuit diagram, and explain why the shunt resistor connected in parallel with the ammeter allows it to measure larger currents. (b) Calculate the resistance the shunt resistor has to have to allow the ammeter to measure a maximum current of 15.0 A.

26.60 Many electronics devices can be dangerous even after they are shut off. Consider an RC circuit with a 150.-μF capacitor and a 1.00-MΩ resistor connected to a 200.-V power source for a long time and then disconnected and shorted, as shown in the figure. How long will it be until the potential difference across the capacitor drops to below 50.0 V?

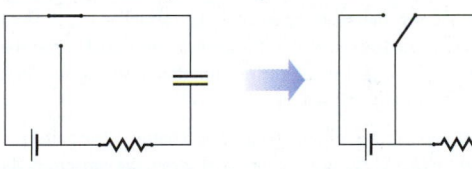

26.61 Design a circuit like that shown in the figure to operate a strobe light. The capacitor discharges power through the light bulb filament (resistance of 2.50 kΩ) in 0.200 ms and charges through a resistor R, with a repeat cycle of 1.00 kHz. What capacitor and resistor should be used?

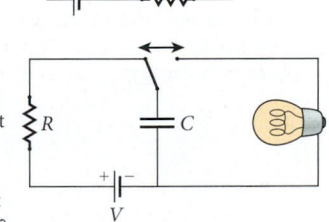

26.62 An ammeter with an internal resistance of 53.0 Ω measures a current of 5.25 mA in a circuit containing a battery and a total resistance of 1130 Ω. The insertion of the ammeter alters the resistance of the circuit, and thus the measurement does not give the actual value of the current in the circuit without the ammeter. Determine the actual value of the current.

•26.63 In the circuit shown in the figure, a 10.0-μF capacitor is charged by a 9.00-V battery with the two-way switch kept in position X for a long time. Then the switch is suddenly flicked to position Y. What current flows through the 40.0-Ω resistor

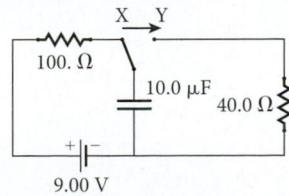

a) immediately after the switch moves to position Y?

b) 1.00 ms after the switch moves to position Y?

•26.64 How long will it take for the current in a circuit to drop from its initial value to 1.50 mA if the circuit contains two 3.80-μF capacitors that are initially uncharged, two 2.20-kΩ resistors, and a 12.0-V battery all connected in series?

26.65 An RC circuit has a time constant of 3.10 s. At $t = 0$, the process of charging the capacitor begins. At what time will the energy stored in the capacitor reach half of its maximum value?

•26.66 For the circuit shown in the figure, determine the charge on each capacitor when (a) switch S has been closed for a long time and (b) switch S has been open for a long time.

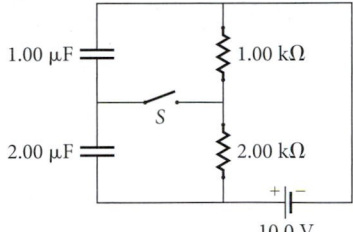

•26.67 Three resistors, $R_1 = 10.0 \, \Omega$, $R_2 = 20.0 \, \Omega$, and $R_3 = 30.0 \, \Omega$, are connected in a multiloop circuit, as shown in the figure. Determine the amount of power dissipated in the three resistors.

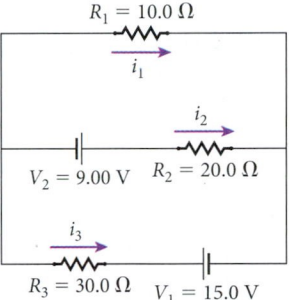

•26.68 The figure shows a circuit containing two batteries and three resistors. The batteries provide $V_{emf,1} = 12.0$ V and $V_{emf,2} = 16.0$ V and have no internal resistance. The resistors have resistances of $R_1 = 30.0 \, \Omega$, $R_2 = 40.0 \, \Omega$, and $R_3 = 20.0 \, \Omega$. Find the magnitude of the potential drop across R_2.

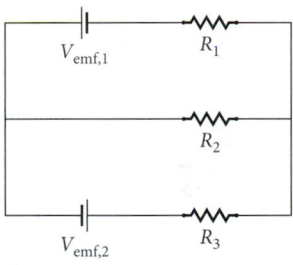

26.69 The figure shows a spherical capacitor. The inner sphere has radius $a = 1.00$ cm, and the outer sphere has radius $b = 1.10$ cm. The battery supplies $V_{emf} = 10.0$ V, and the resistor has a value of $R = 10.0$ MΩ.

a) Determine the time constant of the RC circuit.

b) Determine how much charge has accumulated on the capacitor after switch S has been closed for 0.1 ms.

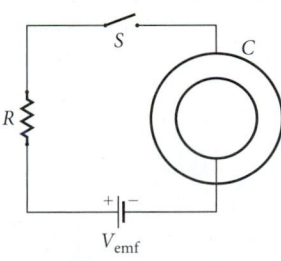

•26.70 Write the set of equations that determines the three currents in the circuit shown in the figure. (Assume that the capacitor is initially uncharged.)

•26.71 Consider a series RC circuit with $R = 10.0 \, \Omega$, $C = 10.0 \, \mu$F and $V = 10.0$ V.

a) How much time, expressed as a multiple of the time constant, does it take for the capacitor to be charged to half of its maximum value?

b) At this instant, what is the ratio of the energy stored in the capacitor to its maximum possible value?

c) Now suppose the capacitor is fully charged. At time $t = 0$, the original circuit is opened and the capacitor is allowed to discharge across another

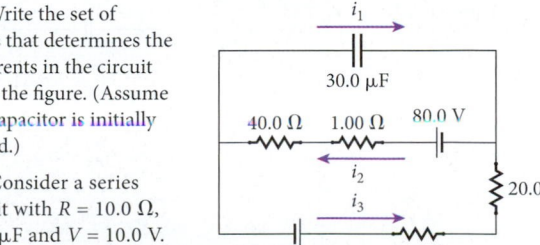

resistor, $R' = 1.00\ \Omega$, that is connected across the capacitor. What is the time constant for the discharging of the capacitor?

d) How many seconds does it take for the capacitor to discharge half of its maximum stored charge, Q?

•**26.72** a) What is the current in the 5.00-Ω resistor in the circuit shown in the figure?

b) What is the power dissipated in the 5.00-Ω resistor?

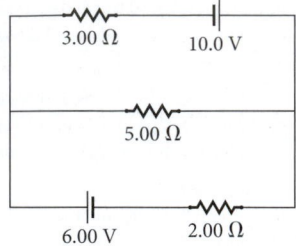

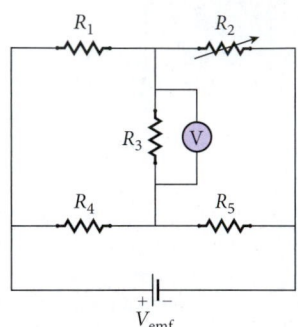

•**26.73** In the Wheatstone bridge shown in the figure, the known resistances are $R_1 = 8.00\ \Omega$, $R_4 = 2.00\ \Omega$, and $R_5 = 6.00\ \Omega$, and the battery supplies $V_{emf} = 15.0$ V. The variable resistance R_2 is adjusted until the potential difference across R_3 is zero ($V = 0$). Find i_2 (the current through resistor R_2) at this time.

••**26.74** Consider the circuit with five resistors and two batteries (with no internal resistance) shown in the figure.

a) Write a set of equations that will allow you to solve for the current in each of the resistors.

b) Solve the equations from part (a) for the current in the 4.00-Ω resistor.

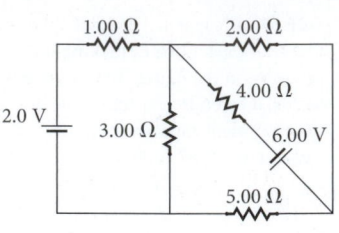

••**26.75** Consider an "infinite," that is, very large, two-dimensional square grid of identical capacitors, C, shown in the figure. Find the effective capacitance of the grid, as measured across any individual capacitor.

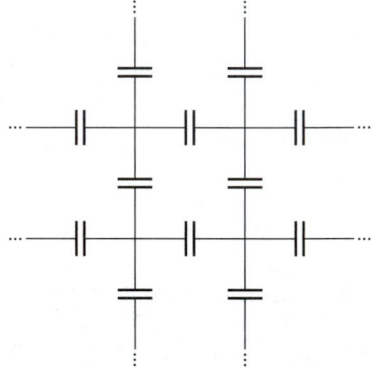

MULTI-VERSION EXERCISES

26.76 A 11.45-V battery with internal resistance $R_i = 0.1373\ \Omega$ is to be charged by a battery charger that is capable of delivering a current $i = 9.759$ A. What is the minimum emf the battery charger must be able to supply in order to charge the battery?

26.77 A battery with internal resistance $R_i = 0.1415\ \Omega$ is being charged by a battery charger that delivers a current $i = 5.399$ A. The battery charger supplies an emf of 14.51 V. What is the potential difference across the terminals of the battery?

26.78 A 16.05-V battery with internal resistance R_i is being charged by a battery charger that is capable of delivering a current $i = 6.041$ A. The battery charger supplies an emf of 16.93 V. What is the internal resistance, R_i, of the battery?

26.79 The single-loop circuit shown in the figure has $V_{emf,1} = 21.01$ V, $V_{emf,2} = 10.75$ V, $R_1 = 23.37\ \Omega$, and $R_2 = 11.61\ \Omega$. What is the current flowing in the circuit?

26.80 The single-loop circuit shown in the figure has $V_{emf,1} = 16.37$ V,

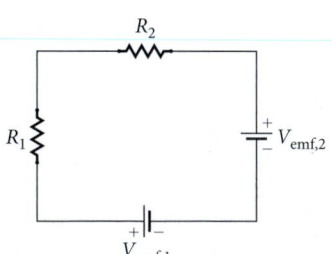

$V_{emf,2} = 10.81$ V, and $R_1 = 24.65\ \Omega$. The current flowing in the circuit is 0.1600 A. What is the resistance R_2?

26.81 The single-loop circuit shown in the figure has $V_{emf,1} = 17.75$ V, $R_1 = 25.95\ \Omega$, and $R_2 = 13.59\ \Omega$. The current flowing in the circuit is 0.1740 A. What is $V_{emf,2}$?

26.82 A 15.19-mF capacitor is fully charged using a battery that supplies $V_{emf} = 131.1$ V. The battery is disconnected, and a 616.5-Ω resistor is connected across the capacitor. What current will be flowing through the resistor after 3.871 s?

26.83 A capacitor is fully charged using a battery that supplies $V_{emf} = 133.1$ V. The battery is disconnected, and a 655.1-Ω resistor is connected across the capacitor. The current flowing through the resistor after 1.743 s is 0.1745 A. What is the capacitance of the capacitor?

26.84 A 19.79-mF capacitor is fully charged using a battery. The battery is disconnected, and a 693.5-Ω resistor is connected across the capacitor. The current flowing through the resistor after 6.615 s is 0.1203 A. What is the emf supplied by the battery?

27

Magnetism

FIGURE 27.1 The Shanghai Maglev Train in the Pudong Airport station. The inset is a display inside the train showing the maximum speed of 430 km/h (267 mph) attained during the 7-minute 20-second trip from the airport to downtown Shanghai.

W hile much of our knowledge of magnetism has been known for two centuries and some of it since ancient times, new and exciting technical applications of this phenomenon continue to be developed today. Many of the devices we use every day make use of magnetism. Cars, computers, power generators, almost anything that uses an electric motor, and almost any information storage technology are just a few examples. One particularly impressive example is Shanghai's Maglev Train, shown in Figure 27.1, which floats on artificially generated magnetic fields and is thus able to reach very high speeds. In order to appreciate these innovations, however, we need to start with the basics of magnetism.

This chapter is the first to consider magnetism, describing magnetic fields and magnetic forces and their effects on charged particles and currents. We'll continue to study magnetism in the next few chapters, describing the causes of magnetic fields and their connection to electric fields. You will see that electricity and magnetism are really parts of the same universal force, called the *electromagnetic force*; their connection is one of the most spectacular successes in physical theory.

WHAT WE WILL LEARN

- Permanent magnets exist in nature. A magnet always has a north pole and a south pole. A single magnetic north pole or south pole cannot be isolated—magnetic poles always come in pairs.

- Opposite magnetic poles attract, and like poles repel.

- Breaking a bar magnet in half results in two new magnets, each with a north and a south pole.

- A magnetic field exerts a force on a moving charged particle.

- The force exerted on a charged particle moving in a magnetic field is perpendicular to both the magnetic field and the velocity of the particle.

- The torque on a current-carrying loop can be expressed in terms of the vector product of the magnetic dipole moment of the loop and the magnetic field.

- The Hall effect can be used to measure magnetic fields.

27.1   Permanent Magnets

In the region of Magnesia (in central Greece), the ancient Greeks found several types of naturally occurring minerals that attract and repel each other and attract certain kinds of metal, such as iron. They also, if floating freely, line up with the North and South Poles of the Earth. These minerals are various forms of iron oxide and are called **permanent magnets.** Other examples of permanent magnets include refrigerator magnets and magnetic door latches, which are made of compounds of iron, nickel, or cobalt. If you touch an iron bar to a piece of the mineral lodestone (magnetic magnetite), the iron bar will be magnetized. If you float this iron bar in water, it will align with the Earth's magnetic poles. The end of the magnet that points north is called the **north magnetic pole,** and the other end is called the **south magnetic pole.**

If two permanent magnets are brought close together with the two north poles or two south poles almost touching, the magnets repel each other (Figure 27.2a). If a north pole and a south pole are brought close together, the magnets attract each other (Figure 27.2b). What is called the North Pole of Earth is actually a magnetic south pole, which is why it attracts the north pole of permanent magnets.

Breaking a permanent magnet in half does not yield one north pole and one south pole. Instead, two new magnets, each with its own north and south pole result (Figure 27.3). Unlike electric charge, which exists as separate positive (proton) and negative (electron) charges, no separate magnetic monopoles (isolated north and south poles) exist. Scientists have carried out extensive searches for magnetic monopoles, and none has been found. The discussion of the source of magnetism in this chapter will help you understand why there are no magnetic monopoles.

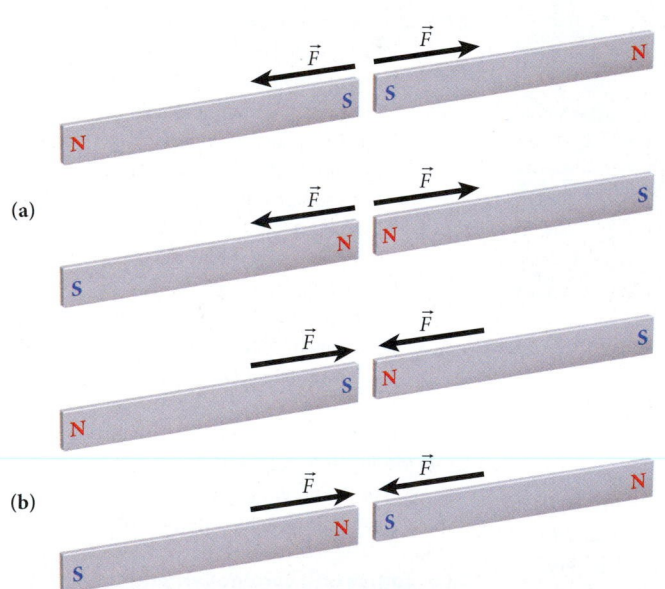

FIGURE 27.2   (a) Like magnetic poles repel; (b) unlike magnetic poles attract.

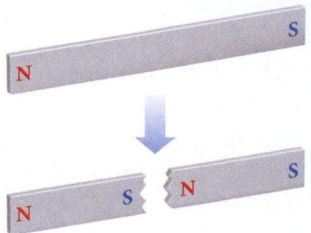

FIGURE 27.3 Breaking a bar magnet in half yields two magnets, each with its own north and south pole.

Magnetic Field Lines

Permanent magnets interact with each other at some distance, without touching. In analogy with the gravitational field and the electric field, the concept of a **magnetic field** is used to describe the magnetic force. The vector $\vec{B}(\vec{r})$ denotes the magnetic field vector at any given point in space.

Like an electric field, a magnetic field is represented using field lines. The magnetic field vector is always tangent to the **magnetic field lines.** The magnetic field lines from a permanent bar magnet are shown in Figure 27.4a. As with electric field lines, closer spacing between lines indicates higher field strength. In an electric field, the electric force on a positive test charge points in the same direction as the electric field vector.

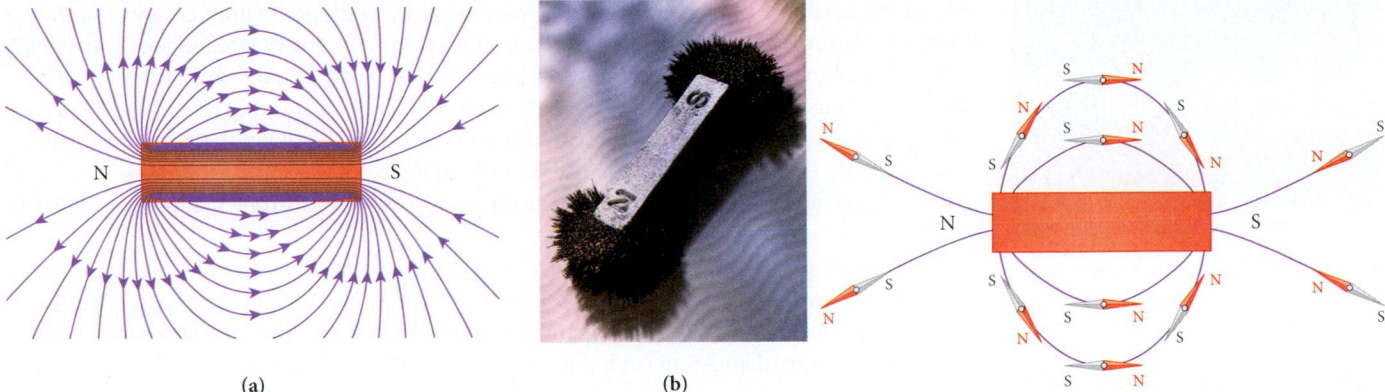

FIGURE 27.4 (a) Computer-generated magnetic field lines from a permanent bar magnet. (b) Iron filings align themselves with the magnetic field lines and make them visible.

FIGURE 27.5 Using a compass needle to determine the direction of the magnetic field from a bar magnet.

However, because no magnetic monopole exists, the magnetic force cannot be described in an analogous way.

The direction of the magnetic field is established in terms of the direction in which a compass needle points. A compass needle, with a north pole and a south pole, will orient itself so that its north pole points in the direction of the magnetic field. Thus, the direction of the field can be determined at any point by noting the direction in which a compass needle placed at that point points, as illustrated in Figure 27.5 for a bar magnet.

Externally, magnetic field lines appear to originate on north poles and terminate on south poles, but these field lines are actually closed loops that penetrate the magnet itself. This formation of loops is an important difference between electric and magnetic field lines (for static fields—this statement does not apply to time-dependent fields, as we'll see in subsequent chapters). Recall that electric field lines start at positive charges and end on negative charges. However, because no magnetic monopoles exist, magnetic field lines cannot start or stop at particular points. Instead, they form closed loops that do not start or stop anywhere. We'll see later that this difference is important in describing the interaction of electric and magnetic fields. If you see a field pattern and don't know at first if it is an electric field or a magnetic field, check for closed loops. If you find some, it is a magnetic field; if the field lines do not form loops, it is an electric field.

Earth's Magnetic Field

Earth itself is a magnet, with a magnetic field similar to the magnetic field of a bar magnet (Figure 27.4). This magnetic field is important because it protects us from high-energy radiation from space called *cosmic rays*. These cosmic rays consist mostly of charged particles that are deflected away from Earth's surface by its magnetic field. The poles of Earth's magnetic field do not coincide with the geographic poles, defined as the points where Earth's rotation axis intersects its surface.

Figure 27.6 shows a cross section of Earth's magnetic field lines. The field lines are close together, forming a surface that wraps around Earth like a doughnut. Earth's magnetic field is distorted by the solar wind, a flow of ionized particles, mainly protons, emitted by the Sun and moving outward from the Sun at approximately 400 km/s. Two bands of charged particles captured from the solar wind circle the Earth. These are called the **Van Allen radiation belts** (Figure 27.6), after James A. Van Allen (1914–2006),

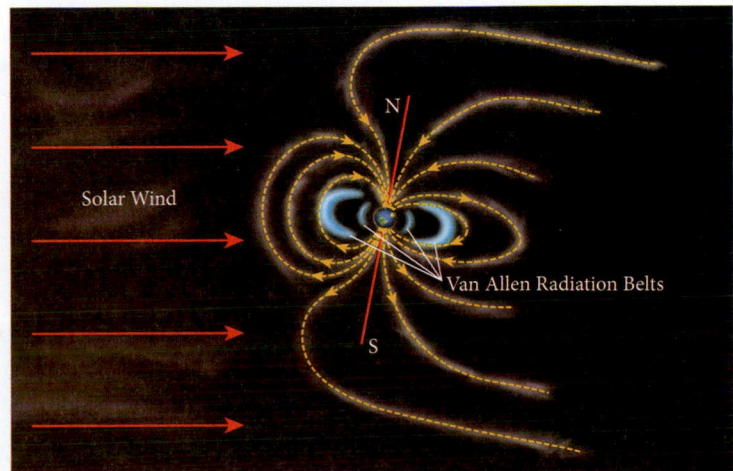

FIGURE 27.6 Cross section through Earth's magnetic field. The dashed lines represent the magnetic field lines. The axis defined by the north and south magnetic poles (red line) currently forms an angle of approximately 11° with the rotation axis.

FIGURE 27.7 Aurora borealis over Finland, photographed from the International Space Station.

FIGURE 27.8 Aurora on Saturn, photographed by the Hubble Space Telescope.

who discovered them in the early days of space flight by putting radiation counters on satellites. The Van Allen radiation belts come closest to Earth near the north and south magnetic poles, where the charged particles trapped within the belts often collide with atoms in the atmosphere, exciting them. These excited atoms emit light of different colors as they collide and lose energy; the results are the fabulous **aurora borealis** (northern lights) in the northern latitudes (see Figure 27.7) and the **aurora australis** (southern lights) in the southern latitudes. Aurorae are not unique to Earth; they have been seen on other planets with strong magnetic fields, such as Jupiter and Saturn (shown in Figure 27.8).

Earth's magnetic poles move at a current rate of up to 40 km in a single year. Right now, the magnetic north pole is located approximately 2800 km away from the geographic South Pole, at the edge of Antarctica, and is moving toward Australia. The magnetic south pole is located in the Canadian Arctic and, if its present rate of motion continues, will reach Siberia in 2050. Earth's magnetic field has decreased steadily at a rate of about 7% per century since it was first measured accurately around 1840. At that rate, Earth's magnetic field will disappear in a few thousand years. However, some geological evidence indicates that the magnetic field of Earth has reversed itself approximately 170 times in the past 100 million years. The last reversal occurred about 770,000 years ago. Thus, rather than disappear, Earth's magnetic field may reverse its direction. What is the cause of Earth's magnetic field? Surprisingly, the answer to this question is not known exactly and is under intense current research. Most likely, it is caused by strong electrical currents inside the Earth, caused by the spinning liquid iron-nickel core. This spinning is often referred to as the *dynamo effect*. (We'll see how currents create magnetic fields in Chapter 28.)

Because the geographic North Pole and the magnetic north pole are not in the same location, a compass needle generally does not point exactly to the geographic North Pole. This difference is called the **magnetic declination.** The magnetic declination is taken to be positive when magnetic north is east of true north and negative when magnetic north is west of true north. The magnetic north pole currently lies on a line that passes through southeastern Missouri, western Illinois, eastern Iowa, and western Wisconsin. Along this line, the magnetic declination is zero. West of this line, the magnetic declination is positive and reaches 18° in Seattle. East of this line, the declination is negative, up to –18° in Maine. A map showing the magnetic declinations in the United States as of 2004 is presented in Figure 27.9.

Because the positions of Earth's magnetic poles move with time, the magnetic declinations for all locations on Earth's surface also change with time. For example, Figure 27.10 shows the estimated magnetic declination for Lansing, Michigan, for the period 1900–2004. A similar graph can be drawn for any location on Earth.

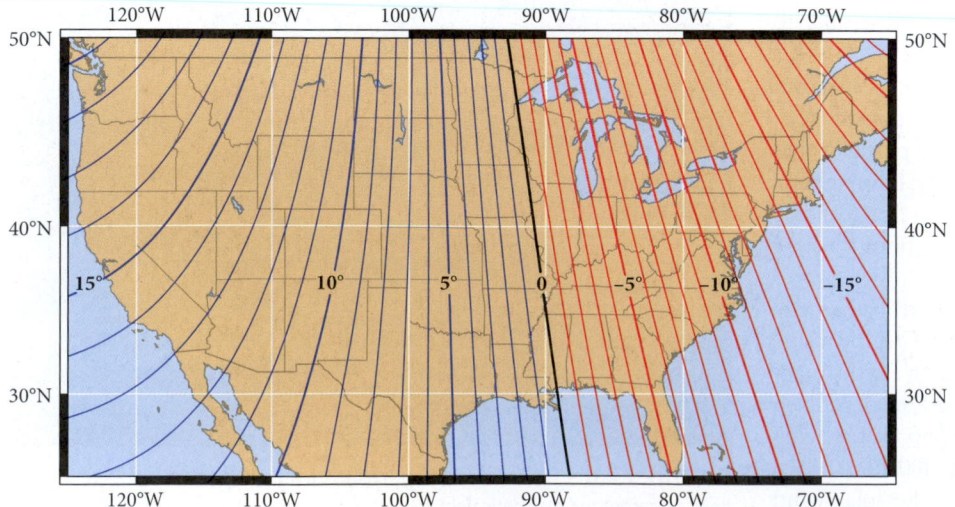

FIGURE 27.9 Magnetic declinations in the United States in 2004. Red lines represent negative magnetic declinations, and blue lines signify positive magnetic declinations. Lines of magnetic declination are separated by 1 degree.

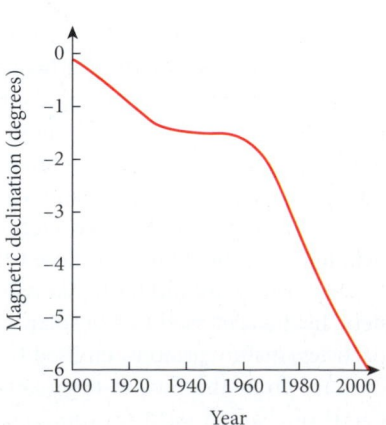

FIGURE 27.10 The magnetic declination at Lansing, Michigan, from 1900 to 2004.

Superposition of Magnetic Fields

If several sources of magnetic field, such as several permanent magnets, are close together, the magnetic field at any given point in space is given by the superposition of the magnetic fields from all the sources. This superposition of fields follows directly from the superposition of forces introduced in Chapter 4. The superposition principle for the total magnetic field, $\vec{B}_{\text{total}}(\vec{r})$, due to n magnetic field sources can be stated as

$$\vec{B}_{\text{total}}(\vec{r}) = \vec{B}_1(\vec{r}) + \vec{B}_2(\vec{r}) + \cdots + \vec{B}_n(\vec{r}). \tag{27.1}$$

This superposition principle for magnetic fields is exactly analogous to the superposition principle for electric fields, presented in Chapter 22.

27.2 Magnetic Force

The qualitative discussion in the preceding section pointed out that a magnetic field has a direction, along the magnetic field lines. The magnitude of a magnetic field is determined by examining its effect on a moving charged particle. We'll start with a constant magnetic field and study its effect on a single charge. As a reminder, we saw in Chapter 22 that the electric field exerts a force on a charge given by $\vec{F}_E = q\vec{E}$. Experiments such as the one shown in Figure 27.11 show that a magnetic field does not exert a force on a charge at rest but only on a *moving* charge.

A magnetic field is defined in terms of the force exerted by the field on a moving charged particle. The magnetic force exerted by a magnetic field on a moving charged particle with charge q moving with velocity $\vec{v}$ is given by

$$\vec{F}_B = q\vec{v} \times \vec{B}. \tag{27.2}$$

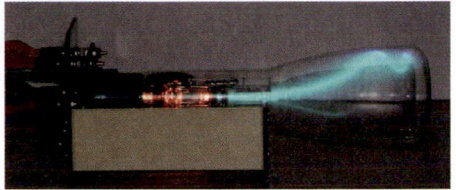

FIGURE 27.11 A beam of electrons (bluish green), made visible by a small amount of gas in an evacuated tube, is bent by a magnet (at the right edge of picture).

The direction of the force is perpendicular to both the velocity of the moving charged particle and the magnetic field (Figure 27.12). This statement is right-hand rule 1. The right-hand rule gives the force direction on a positive charge given the known velocity and field directions. However, for a negative charge the force will be in the opposite direction.

The magnitude of the magnetic force on a moving charged particle is

$$F_B = |q|vB\sin\theta, \tag{27.3}$$

where θ is the angle between the velocity of the charged particle and the magnetic field. (The angle θ is always between 0° and 180°, and therefore, $\sin\theta \geq 0$.) You can see that no magnetic force acts on a charged particle moving parallel to a magnetic field because in that case $\theta = 0°$. If a charged particle is moving perpendicularly to the magnetic field, $\theta = 90°$ and (for fixed values of v and B) the magnitude of the magnetic force has its maximum value of

$$F_B = |q|vB \quad \left(\text{for } \vec{v} \perp \vec{B}\right). \tag{27.4}$$

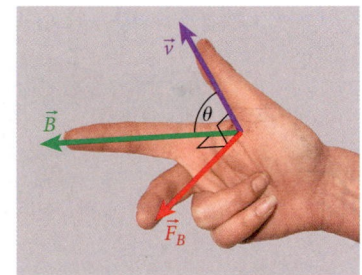

FIGURE 27.12 Right-hand rule 1 for the force exerted by a magnetic field, $\vec{B}$, on a particle with charge q moving with velocity $\vec{v}$. To find the direction of the magnetic force, point your thumb in the direction of the velocity of the moving charged particle and your index finger in the direction of the magnetic field, and your middle finger then gives you the direction of the magnetic force.

Magnetic Force and Work

Equation 27.2 established that the magnetic force is the vector product of the velocity vector and magnetic field vector and thus is perpendicular to both vectors. This implies that $\vec{F}_B \cdot \vec{v} = 0$ and, since the force is the product of mass and acceleration, also that $\vec{a} \cdot \vec{v} = 0$. In Chapter 9 on circular motion, we saw that this condition means that the direction of the velocity vector can change but the magnitude of the velocity vector, the speed, remains the same. Therefore, the kinetic energy, $\frac{1}{2}mv^2$, remains constant for a particle subjected to a magnetic force, and the magnetic force *does no work* on the moving particle.

This is a profound result: A constant magnetic field cannot be used to do work on a particle. The kinetic energy of a particle moving in a constant magnetic field remains constant, even though the direction of the particle's velocity vector can change as a function of time while the particle is moving through the magnetic field. An electric field, on the other hand, can easily be used to do work on a particle.

Units of Magnetic Field Strength

To discuss the motion of charges in magnetic fields, we need to know what units are used to measure the magnetic field strength. Solving equation 27.4 for the field strength and inserting the units of the other quantities gives

$$[F_B]=[q][v][B] \Rightarrow [B]=\frac{[F_B]}{[q][v]}=\frac{\text{N s}}{\text{C m}}.$$

Because the ampere (A) is defined as 1 C/s, (N s)/(C m) = N/(A m). The unit of magnetic field strength has been named the **tesla** (T), in honor of Croatian-born American physicist and inventor Nikola Tesla (1856–1943):

$$1\,\text{T} = 1\,\frac{\text{N s}}{\text{C m}} = 1\,\frac{\text{N}}{\text{A m}}.$$

A tesla is a rather large amount of magnetic field strength. Sometimes magnetic field strength is given in gauss (G), which is not an SI unit:

$$1\,\text{G} = 10^{-4}\,\text{T}.$$

For example, the strength of Earth's magnetic field at Earth's surface is on the order of 0.5 G ($5 \cdot 10^{-5}$ T). It varies with location from 0.2 G to 0.6 G, as illustrated in Figure 27.13.

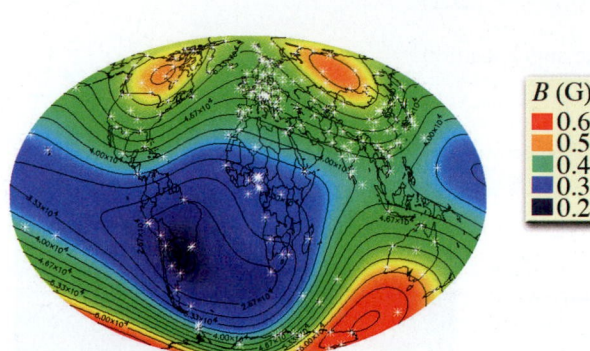

FIGURE 27.13 Global map of the strength of Earth's magnetic field.

B (G)
■ 0.6
■ 0.5
■ 0.4
■ 0.3
■ 0.2

SOLVED PROBLEM 27.1   **Cathode Ray Tube**

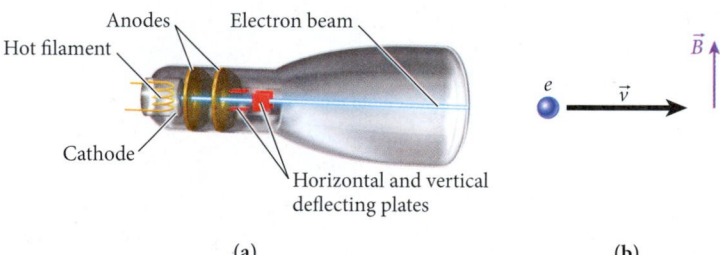

FIGURE 27.14   (a) A cathode ray tube. (b) Electrons moving with velocity $\vec{v}$ enter a constant magnetic field.

PROBLEM

Consider a cathode ray tube similar to the one shown in Figure 27.11. In this tube, a potential difference of $\Delta V = 111$ V accelerates electrons horizontally (starting essentially from rest) in an electron gun, as shown in Figure 27.14a. The electron gun has a specially coated filament that emits electrons when heated. A negatively charged cathode controls the number of electrons emitted. Positively charged anodes focus and accelerate the electrons into a beam. Downstream from the anodes are horizontal and vertical deflecting plates. Beyond the electron gun is a constant magnetic field with magnitude $B = 3.40 \cdot 10^{-4}$ T. The direction of the magnetic field is upward, perpendicular to the initial velocity of the electrons. What is the magnitude of the acceleration of the electrons due to the magnetic field? (The mass of an electron is $9.11 \cdot 10^{-31}$ kg.)

SOLUTION

THINK   The electrons gain kinetic energy in the electron gun of the cathode ray tube. The gain in kinetic energy of each electron is equal to the charge of the electron times the potential difference. The speed of the electrons can be found from the definition of kinetic energy. The magnetic force on an electron can be found from the electron charge, the electron velocity, and the strength of the magnetic field, and it is equal to the mass of the electron times its acceleration.

SKETCH   Figure 27.14b shows an electron, moving with velocity $\vec{v}$, entering a constant magnetic field that is perpendicular to the electron path.

RESEARCH   The change in kinetic energy, ΔK, of the electrons plus the change in potential energy of the electrons is equal to zero:

$$\Delta K + \Delta U = \tfrac{1}{2}mv^2 + q\Delta V = 0.$$

Since, in this case, $q = -e$, we see that

$$e\Delta V = \tfrac{1}{2}mv^2, \qquad\qquad\qquad (i)$$

where ΔV is the magnitude of the potential difference across which the electrons were accelerated and m is the mass of an electron. We can solve equation (i) for the speed of the electrons:

$$v = \sqrt{\frac{2e\Delta V}{m}}. \tag{ii}$$

The magnitude of the force exerted by the magnetic field on the electrons is given by equation 27.3:

$$F_B = evB\sin 90° = evB,$$

where $-e$ is the charge of an electron and B is the magnitude of the magnetic field. According to Newton's Second Law, $F_{net} = ma$. Since the only force present is the magnetic one, we have

$$F_B = ma = evB, \tag{iii}$$

where a is the magnitude of the acceleration of the electrons.

SIMPLIFY We can rearrange equation (iii) and substitute the expression for the speed of the electrons from equation (ii) to obtain the acceleration of the electrons:

$$a = \frac{evB}{m} = \frac{eB\sqrt{\frac{2e\Delta V}{m}}}{m} = B\sqrt{2\Delta V\frac{e^3}{m^3}}.$$

CALCULATE Putting in the numerical values gives us

$$a = (3.40\cdot10^{-4}\text{ T})\sqrt{2(111\text{ V})\frac{\left(1.602\cdot10^{-19}\text{ C}\right)^3}{\left(9.11\cdot10^{-31}\text{ kg}\right)^3}} = 3.7357\cdot10^{14}\text{ m/s}^2.$$

ROUND We report our result to three significant figures:

$$a = 3.74\cdot10^{14}\text{ m/s}^2.$$

DOUBLE-CHECK The calculated acceleration is tremendously large, almost 40 trillion times the Earth's gravitational acceleration. So we certainly want to double-check. We first calculate the speed of the electrons:

$$v = \sqrt{\frac{2e\Delta V}{m}} = \sqrt{\frac{2\left(1.602\cdot10^{-19}\text{ C}\right)(111\text{ V})}{9.11\cdot10^{-31}\text{ kg}}} = 6.25\cdot10^6\text{ m/s}.$$

A speed of 6250 km/s may seem large, but it is reasonable for electrons because it is only 2% of the speed of light. The magnetic force on each electron is then

$$F_B = evB = \left(1.602\cdot10^{-19}\text{ C}\right)\left(6.25\cdot10^6\text{ m/s}\right)\left(3.40\cdot10^{-4}\text{ T}\right) = 3.40\cdot10^{-16}\text{ N}.$$

The acceleration is very large because the mass of an electron is very small.

Concept Check 27.1

In what direction will the electron in Figure 27.14b be deflected as it enters the constant magnetic field?

a) into the page

b) out of the page

c) upward

d) downward

e) no deflection

Self-Test Opportunity 27.1

Three particles, each with charge $q = 6.15$ μC and speed $v = 465$ m/s, enter a uniform magnetic field with magnitude $B = 0.165$ T (see the figure). What is the magnitude of the magnetic force on each of the particles?

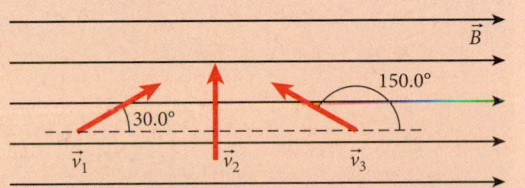

27.3 Motion of Charged Particles in a Magnetic Field

The fact that the force due to a magnetic field acting on a moving charged particle is perpendicular to both the field and the particle's velocity makes this force different from any we've considered so far. However, the tools we use to analyze this force—Newton's laws and the laws of conservation of energy, momentum, and angular momentum—are the same.

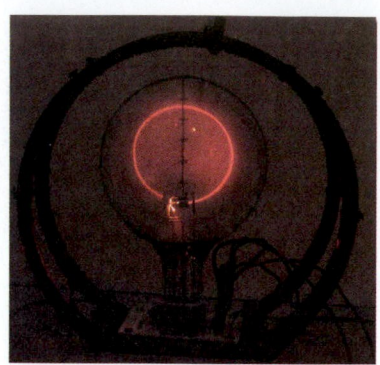

FIGURE 27.15 Electron beam bent into a circular path by the magnetic field generated by two coils.

Paths of Moving Charged Particles in a Constant Magnetic Field

Suppose you drive your car at constant speed around a circular track. The friction between the tires and the road provides the centripetal force that keeps the car moving in a circle. This force always points toward the center of the circle and creates a centripetal acceleration (discussed in Chapter 9). A similar physical situation occurs when a particle with charge q and mass m moves with velocity $\vec{v}$ perpendicular to a uniform magnetic field, $\vec{B}$, as illustrated in Figure 27.15.

In this situation, the particle moves in a circle with constant speed v and the magnetic force of magnitude $F_B = |q|vB$ supplies the centripetal force that keeps the particle moving in a circle. Particles with opposite charges and the same mass will orbit in opposite directions at the same orbital radius. For example, electrons and positrons are elementary particles with the same mass; the electron has a negative charge, and the positron has a positive charge. Figure 27.16 is a bubble chamber photograph showing two electron-positron pairs. A bubble chamber is a device that can track charged particles moving in a constant magnetic field. (The pairs of electrons and positrons were created by interactions of elementary particles, which will be covered in detail in Chapter 39.) Pair 1 has an electron and a positron that have the same relatively low speed. The particles initially travel in a circle. However, as they move through the bubble chamber, they slow down. (This slowdown is not due to the magnetic force but to collisions of the particles with the molecules of the gas in the bubble chamber.) Thus, the radius of the circle gets smaller and smaller, creating a spiral. The electron and positron in pair 2 have a much higher speed. Their tracks are curved but do not form a complete circle before the particles exit the bubble chamber.

If the velocity of a charged particle is parallel (or antiparallel) to the magnetic field, the particle experiences no magnetic force and continues to travel in a straight line.

For motion perpendicular to a magnetic field, as in Figure 27.15, the force required to keep a particle moving with speed v in a circle with radius r is the centripetal force:

$$F = \frac{mv^2}{r}.$$

Setting this expression for the centripetal force equal to that for the magnetic force, we obtain

$$vB|q| = \frac{mv^2}{r}.$$

Rearranging gives an expression for the radius of the circle in which the particle is traveling:

$$r = \frac{mv}{|q|B}. \tag{27.5}$$

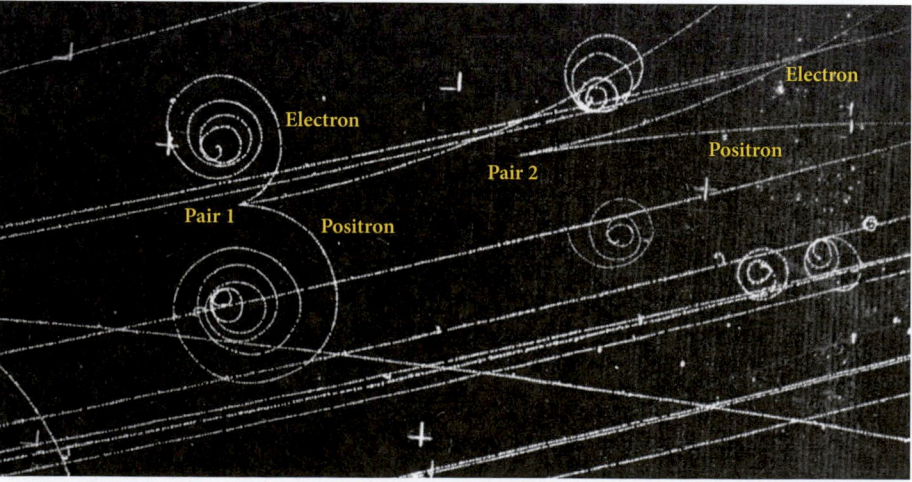

FIGURE 27.16 Bubble chamber photograph showing two electron-positron pairs. The bubble chamber is located in a constant magnetic field pointing directly out of the page.

A common way to express this relationship is in terms of the magnitude of the momentum of the particle:

$$Br = \frac{p}{|q|}.$$ (27.6)

If the velocity $\vec{v}$ is not perpendicular to $\vec{B}$, then the velocity component perpendicular to $\vec{B}$ causes circular motion while the parallel velocity component is unaffected by $\vec{B}$ and drags this orbit into a helical shape.

Time Projection Chamber

Particle physicists create new elementary particles by colliding larger particles at the highest energies. In these collisions, many particles stream away from the interaction point at high speeds. A simple particle detector is not sufficient to identify these particles. A device that can help physicists study these collisions is a time projection chamber (TPC). The STAR TPC was described in Example 22.4.

Figure 27.17 shows collisions of two protons and two gold nuclei that occurred in the center of the STAR TPC. The proton-proton collision creates dozens of particles; the gold-gold collision creates thousands of particles. Each charged particle leaves a track in the TPC. The color assigned by a computer to the track represents the ionization density of the track as particles pass through the gas of the TPC. As they pass through the gas, the particles ionize the atoms of the gas, releasing free electrons. The gas allows the free electrons to drift without recombining with positive ions. Electric fields applied between the center of the TPC and the end caps of the cylinder exert an electric force on the free electrons, making them drift toward the end caps, where they are recorded electronically. Using the drift time and the recording positions, computer software reconstructs the trajectories that the particles took through the TPC. The particles produced in the collisions have a velocity component that is perpendicular to the TPC's magnetic field and thus have circular trajectories.

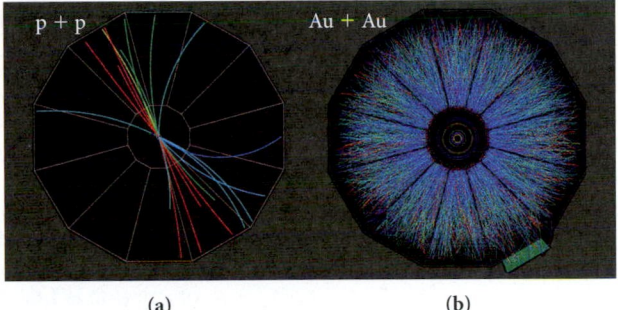

(a)   (b)

FIGURE 27.17 Curved tracks left by the motion of charged particles produced in collisions of (a) two protons, each with kinetic energy of 100 GeV, and (b) two gold nuclei, each with kinetic energy of 100 GeV.

EXAMPLE 27.1 | Transverse Momentum of a Particle in the TPC

One track of a moving charged particle from Figure 27.17a is shown in Figure 27.18. The radius of the circular trajectory this particle is following is $r = 2.300$ m. The magnitude of the magnetic field in the TPC is $B = 0.5000$ T. We can assume that the particle has charge $|q| = 1.602177 \cdot 10^{-19}$ C.

PROBLEM

What is the component of the particle's momentum that is perpendicular to the magnetic field?

SOLUTION

We'll call this component the transverse momentum of the particle, p_t. We use equation 27.6, replacing p with p_t, because the magnetic force depends only on p_t, and not on the component of the momentum that is parallel to $\vec{B}$:

$$Br = \frac{p_t}{|q|}.$$

We can express the magnitude of the transverse momentum of the particle in terms of the magnitude of the TPC's magnetic field and the absolute value of the charge of the particle:

$$p_t = |q|Br = (1.602177 \cdot 10^{-19} \text{ C})(0.5000 \text{ T})(2.300 \text{ m}) = 1.843 \cdot 10^{-19} \text{ kg m/s}.$$

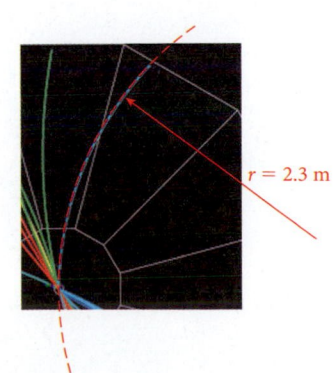

$r = 2.3$ m

FIGURE 27.18 Circle fitted to the trajectory of one of the charged particles produced from the proton-proton collision in the STAR TPC shown in Figure 27.17a.

– *Continued*

Instead of these SI units for momentum, particle physicists often use MeV/c (recall Example 7.5). Since 1 MeV $=1.602 \cdot 10^{-13}$ J, the conversion between kg m/s and MeV/c is

$$1 \text{ MeV}/c = \frac{1.60218 \cdot 10^{-13} \text{ J}}{2.99792 \cdot 10^8 \text{ m/s}} = 5.3443 \cdot 10^{-22} \text{ kg m/s} \Leftrightarrow 1 \text{ kg m/s} = 1.87115 \cdot 10^{21} \text{ MeV}/c.$$

Therefore, our result for the transverse momentum is

$$p_t = 344.8 \text{ MeV}/c.$$

The analysis of a particle's transverse momentum carried out here can be done by automated computer algorithms on up to approximately 5000 charged particles created in a single collision of two gold nuclei. This very complex task takes about 30 seconds for a computer (3-GHz processor) to finish. In contrast, the TPC can record up to 1000 events per second, which corresponds to 1 millisecond per event.

EXAMPLE 27.2 | The Solar Wind and Earth's Magnetic Field

Section 27.1 discussed the Van Allen radiation belts that trap particles emitted from the Sun. The Sun throws approximately 1 million tons of matter into space every second. This matter is mostly protons traveling at a speed of around 400. km/s.

PROBLEM

If protons from the Sun are incident perpendicular to Earth's magnetic field (which has a magnitude of 50.0 μT at the Equator), what is the radius of the orbit of the protons? The mass of a proton is $1.67 \cdot 10^{-27}$ kg.

SOLUTION

Equation 27.5 relates the magnitude of the magnetic field, B, the radius of a circular orbit, r, and the speed, v, of a particle with mass m and charge q traveling perpendicular to a magnetic field:

$$r = \frac{mv}{|q|B}.$$

Putting in the numerical values, we get

$$r = \frac{\left(1.67 \cdot 10^{-27} \text{ kg}\right)\left(400. \cdot 10^3 \text{ m/s}\right)}{\left(1.602 \cdot 10^{-19} \text{ C}\right)\left(50.0 \cdot 10^{-6} \text{ T}\right)} = 83.5 \text{ m}.$$

Thus, the protons of the solar wind orbit around the Earth's magnetic field lines at the Equator in circles of radius 83.5 m. Protons that are incident on the Earth's magnetic field away from the Equator are not traveling perpendicular to the magnetic field, so their orbital radius is larger. However, the magnetic field lines are closer together, meaning that the field is stronger toward the poles. The protons thus spiral along the field lines as they approach the poles. The shape of the Earth's magnetic field forces these protons traveling toward the poles to reverse and travel back toward the Equator, trapping the protons in the Van Allen radiation belts. Thus, the Earth's magnetic field completely blocks the solar wind from reaching the Earth's surface. This is vital, because the blocked cosmic radiation would otherwise make it impossible for higher organisms to live on Earth by ionizing (removing electrons from) atoms and destroying large molecules, for example, DNA.

Cyclotron Frequency

If a particle performs a complete circular orbit inside a uniform magnetic field—for example, like the electrons in the beam shown in Figure 27.15—then the period of revolution, T, of the particle is the circumference of the circle divided by the speed:

$$T = \frac{2\pi r}{v} = \frac{2\pi m}{|q|B}. \tag{27.7}$$

The frequency, f, of the motion of the charged particle is the inverse of the period:

$$f = \frac{1}{T} = \frac{|q|B}{2\pi m}.$$ (27.8)

The angular speed, ω, of the motion is

$$\omega = 2\pi f = \frac{|q|B}{m}.$$ (27.9)

Thus, the frequency and the angular speed of the particle's motion are independent of the particle's speed and thus independent of the particle's kinetic energy. This fact is used in cyclotrons, which is why ω as given by equation 27.9 is referred to as the **cyclotron frequency.** In a cyclotron, particles are accelerated to higher and higher kinetic energies, and the fact that the cyclotron frequency is independent of the kinetic energy makes designing a cyclotron much easier.

EXAMPLE 27.3 / Energy of a Cyclotron

A cyclotron is a particle accelerator (Figure 27.19). The golden horn-shaped pieces of metal shown in the figure (historically called *dees*) have alternating electric potentials applied to them, so a positively charged particle always has a negatively charged dee ahead when it emerges from under any dee, which is now positively charged. The resulting electric field accelerates the particle. Because the cyclotron sits in a strong magnetic field, the particle's trajectory is curved. The radius of the trajectory is proportional to the magnitude of the particle's momentum, according to equation 27.6, so the accelerated particle spirals outward until it reaches the edge of the magnetic field (where its path is no longer bent by the field) and is extracted. According to equation 27.9, the angular frequency is independent of the particle's momentum or energy, so the frequency with which the polarity of the dees is changed does not have to be adjusted as the particle is accelerated. (This holds true only as long as the speed of the accelerated particles does not approach a sizable fraction of the speed of light, as we'll see in Chapter 35 on relativity. To compensate for relativistic effects, the magnetic field of a cyclotron increases with the orbital radius of the accelerated particles.)

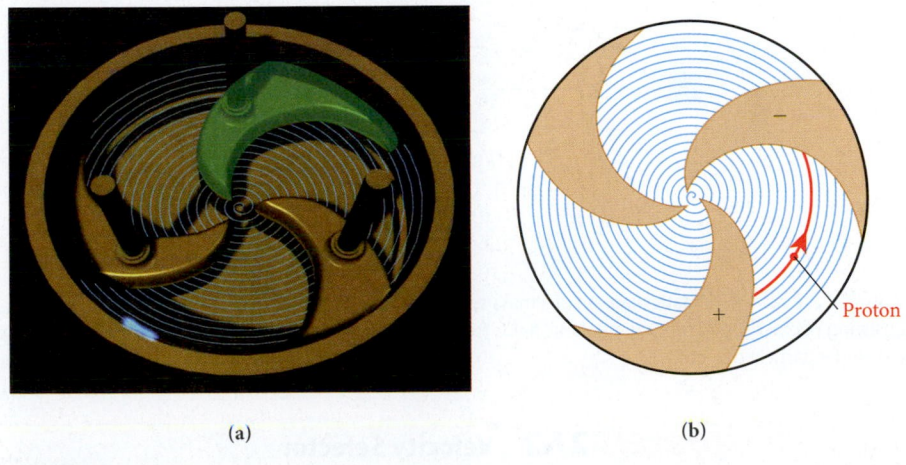

(a)   (b)

FIGURE 27.19 (a) Computer-generated drawing of the central section of the K500 superconducting cyclotron at the National Superconducting Cyclotron Laboratory at Michigan State University, with the spiral trajectory of an accelerated particle superimposed. One of the three dees of the cyclotron is highlighted in green. (b) Top view of the K500, showing a proton being accelerated between two dees.

PROBLEM

What is the kinetic energy, in mega-electron-volts (MeV), of a proton extracted from a cyclotron with radius $r = 1.81$ m, if the magnetic field of the cyclotron is uniform and has magnitude $B = 0.851$ T? The mass of a proton is $1.67 \cdot 10^{-27}$ kg.

SOLUTION

We can solve equation 27.5 for the speed, v, of the proton:

$$v = \frac{r|q|B}{m}.$$

– Continued

Self-Test Opportunity 27.2

A uniform magnetic field is directed out of the page (the standard notation of a dot within a circle represents the tip of the arrowhead of a field line). A charged particle is traveling in the plane of the page, as shown by the arrows in the figure.

a) Is the charge of the particle positive or negative?

b) Is the particle slowing down, speeding up, or moving at constant speed?

c) Is the magnetic field doing work on the particle?

Concept Check 27.2

Cosmic rays would continuously bombard Earth's surface if most of them were not deflected by Earth's magnetic field. Given that Earth is approximately a magnetic dipole (see Figure 27.6), the intensity of cosmic rays incident on its surface is greatest at the

a) North and South Poles.

b) Equator.

c) middle latitudes.

We substitute this expression for v into the equation for kinetic energy:

$$K = \tfrac{1}{2}mv^2 = \tfrac{1}{2}m\left(\frac{r|q|B}{m}\right)^2 = \frac{r^2q^2B^2}{2m}.$$

Putting in the given numbers, we get the kinetic energy in joules:

$$K = \frac{(1.81\ \text{m})^2\,(1.602\cdot10^{-19}\ \text{C})^2\,(0.851\ \text{T})^2}{2(1.67\cdot10^{-27}\ \text{kg})} = 1.82\cdot10^{-11}\ \text{J}.$$

Since 1 eV = $1.602\cdot10^{-19}$ J and 1 MeV = 10^6 eV, we have

$$K = 1.82\cdot10^{-11}\ \text{J}\left(\frac{1\ \text{eV}}{1.602\cdot10^{-19}\ \text{J}}\right)\left(\frac{1\ \text{MeV}}{10^6\ \text{eV}}\right) = 114\ \text{MeV}.$$

Mass Spectrometer

One application of the motion of charged particles in a magnetic field is a **mass spectrometer,** which allows precise determination of atomic and molecular masses and can be useful for carbon dating and the analysis of unknown chemical compounds. A mass spectrometer operates by ionizing the atoms or molecules to be studied and accelerating them through an electric potential. The ions are then passed through a velocity selector (described further in Solved Problem 27.2), which allows only ions with a given velocity to pass through and blocks the remaining ions. The ions then enter a region of constant magnetic field. In the magnetic field, the radius of curvature of the orbit of each ion is given by equation 27.5: $r = mv/|q|B$. Assuming that all the atoms or molecules are singly ionized (have a charge of +1 or –1), the radius of curvature is proportional to the mass of the ion. A schematic diagram of a mass spectrometer is shown in Figure 27.20.

Ions with different masses will have orbits with different radii in the constant magnetic field. For example, in Figure 27.20, ions with orbital radius r_1 have a smaller mass than ions with orbital radius r_2. The particle detector measures the distances from the entrance point, d_1 and d_2, which can be related to the orbital radii and thus the mass of the ions.

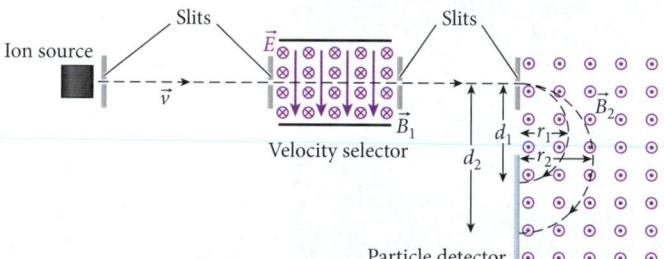

FIGURE 27.20 Schematic diagram of a mass spectrometer showing an ion source, a velocity selector consisting of crossed electric and magnetic fields (see Solved Problem 27.2), a region of constant magnetic field, and a particle detector.

SOLVED PROBLEM 27.2 | Velocity Selector

Protons are accelerated from rest through an electric potential difference of $\Delta V = 14.0$ kV. The protons enter a velocity selector, consisting of a parallel plate capacitor in a constant magnetic field, directed perpendicularly into the plane of the page in Figure 27.21a. The electric field between the plates of the parallel plate capacitor is $\vec{E} = 4.30\cdot10^5$ V/m, directed along the plane of the page and downward in Figure 27.21a. This arrangement of perpendicular electric and magnetic fields is referred to as *crossed fields*.

PROBLEM

What magnetic field is required for the protons to move through the velocity selector without being deflected?

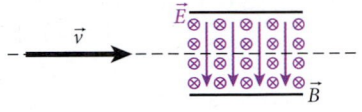

(a)

SOLUTION

THINK   For a proton to move on a straight line without deflection requires that the net force on the proton be zero. Since the proton has a certain velocity and since the magnetic force depends on the velocity, it is plausible that this condition of zero net force cannot be realized for arbitrary speeds of the proton—hence the name *velocity selector*.

SKETCH   Figure 27.21b shows the electric and magnetic forces on the protons as they pass through the velocity selector. Note that the two forces point in opposite directions.

RESEARCH   The change in kinetic energy of the protons plus the change in electric potential energy is equal to zero, which can be expressed as

$$\Delta K = -\Delta U = \tfrac{1}{2}mv^2 = e\Delta V,$$

where m is the mass of a proton, v is the speed of the proton after acceleration, e is the charge of the proton, and ΔV is the electric potential difference across which the protons were accelerated. The speed of the protons after acceleration is

$$v = \sqrt{\frac{2e\Delta V}{m}}. \tag{i}$$

When the protons enter the velocity selector, the direction of the electric force is in the direction of the electric field, which is downward (negative y-direction). The magnitude of the electric force is

$$F_E = eE, \tag{ii}$$

where E is the magnitude of the electric field in the velocity selector. Right-hand rule 1 gives the direction of the magnetic force: With your thumb in the direction of the velocity of the protons (positive x-direction) and your index finger in the direction of the magnetic field (into the page), your middle finger points up (positive y-direction). Thus, the direction of the magnetic force on the protons is upward. The magnitude of the magnetic force is given by

$$F_B = evB, \tag{iii}$$

where B is the magnitude of the magnetic field in the velocity selector.

SIMPLIFY   The condition that allows the protons to pass though the velocity selector without being deflected is that the electric force balances the magnetic force, or $F_E = F_B$. Using equations (ii) and (iii), we can express this condition as

$$eE = evB.$$

Solving for the magnetic field, B, and substituting for v from equation (i), we obtain

$$B = \frac{E}{\sqrt{\dfrac{2e\Delta V}{m}}} = E\sqrt{\frac{m}{2e\Delta V}}.$$

CALCULATE   Putting in the numerical values gives us

$$B = \left(4.30 \cdot 10^5 \text{ V/m}\right)\sqrt{\frac{1.67 \cdot 10^{-27} \text{ kg}}{2\left(1.602 \cdot 10^{-19} \text{ C}\right)\left(14.0 \cdot 10^3 \text{ V}\right)}} = 0.262371 \text{ T}.$$

ROUND   We report our result to three significant figures:

$$B = 0.262 \text{ T}.$$

DOUBLE-CHECK   We verify that the electric force is equal to the magnetic force. The electric force is

$$F_E = eE = \left(1.602 \cdot 10^{-19} \text{ C}\right)\left(4.30 \cdot 10^5 \text{ V/m}\right) = 6.89 \cdot 10^{-14} \text{ N}.$$

To calculate the magnitude of the magnetic force, we need to find the speed of the protons:

$$v = \sqrt{\frac{2e\Delta V}{m}} = \sqrt{\frac{2\left(1.602 \cdot 10^{-19} \text{ C}\right)\left(14.0 \cdot 10^3 \text{ V}\right)}{1.67 \cdot 10^{-27} \text{ kg}}} = 1.64 \cdot 10^6 \text{ m/s}.$$

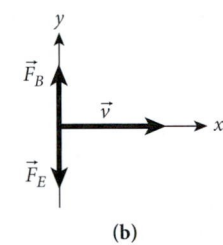

(b)

FIGURE 27.21   (a) A proton entering a velocity selector, consisting of crossed electric and magnetic fields. (b) Electric and magnetic forces on a proton passing through the combined fields.

– Continued

This speed is 0.55% of the speed of light, which is not totally impossible. The magnitude of the magnetic force is then

$$F_B = evB = \left(1.602 \cdot 10^{-19}\ \text{C}\right)\left(1.64 \cdot 10^6\ \text{m/s}\right)(0.262\ \text{T}) = 6.88 \cdot 10^{-14}\ \text{N},$$

which agrees with the value of the electric force within rounding error. Thus, our result seems reasonable.

Magnetic Levitation

An interesting application of magnetic force is **magnetic levitation,** a situation in which an upward magnetic force on an object balances the downward gravitational force, achieving static equilibrium with no need for direct contact of surfaces. But if you try to balance a magnet over another magnet by orienting the north poles (or south poles) toward each other, you'll see right away that this is not possible. Instead, one of the magnets will simply flip over, and then the opposite poles will point toward each other, and the attractive force between them will cause the two magnets to snap together. As we saw in Chapter 11, a stable equilibrium requires a local minimum of the potential energy, which does not exist for the pure repulsive interaction of two like magnetic poles.

Figure 27.22 shows a toy called the Levitron demonstrating the principle of magnetic levitation. The magnetic top is spun on a plate and then lifted to the proper height and released. The top can remain suspended for several minutes. How does this toy work, considering the requirement for stable equilibrium just mentioned? The answer is that the rapid rotation of the top provides a sufficiently large angular momentum and creates a potential energy barrier that prevents the magnet from flipping over.

Of course, there are other ways to create stable magnetic levitation systems, all of them involving multiple magnets attached rigidly to each other. Magnetic levitation has real-world applications in magnetic levitation (maglev) trains. These trains have several advantages over trains that run on normal steel rails: There are no moving parts to wear out, there is less vibration, and reduced friction means that high speeds are possible. Several maglev trains are already in service around the world and more are being planned. One example is the Shanghai Maglev Train (Figure 27.1), which operates between the Shanghai Pudong Airport and downtown Shanghai and reaches speeds of up to 430 km/h (267 mph).

The Shanghai Maglev Train operates using magnets attached to the cars (Figure 27.23). These are normal, non-superconducting magnetic coils with electronic feedback to produce stable levitation and guidance. The train cars are held 15 cm above the guideway to allow clearance of any objects that may be on the guideway. The levitation and guidance magnets are held at a distance of 10 mm from the guideway, which is constructed of a magnetic material. The propulsion of the train is provided by magnetic fields built into the guideway. The train propulsion system operates like an electric motor (see Section 27.5) whose circular loops have been unwrapped to a linear configuration.

Maglev trains that use superconducting magnets have been tested, but some technical problems have yet to be resolved, including the maintenance of the superconducting coils and the exposure of the passengers to high magnetic fields.

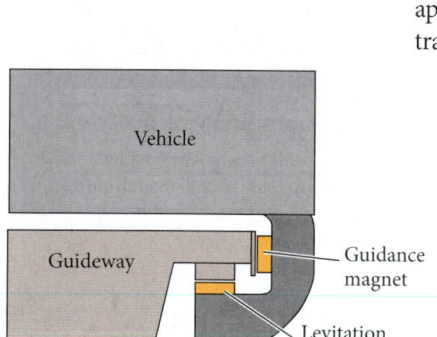

FIGURE 27.22 The Levitron, a toy demonstrating the magnetic levitation of a spinning magnet above a base magnet.

FIGURE 27.23 Cross section of one side of a train car of the Shanghai Maglev Train. The levitation magnets lift the cars 15 cm off the guideway, and the guidance magnets keep the cars centered on the guideway. The magnets are all mounted on the moving vehicle.

27.4   Magnetic Force on a Current-Carrying Wire

Consider a wire carrying a current, i, in a constant magnetic field, $\vec{B}$ (Figure 27.24a). The magnetic field exerts a force on the moving charges in the wire. The charge, q, flowing past a point in the wire in a given time, t, is $q = ti$. During this time, the charge occupies a length, L, of wire given by $L = v_{\text{d}}t$, where v_{d} is the drift speed (the magnitude of the drift velocity) of the charge carriers in the wire. Thus, we obtain

$$q = ti = \frac{L}{v_{\text{d}}}i. \tag{27.10}$$

The magnitude of the magnetic force is then

$$F_B = qv_d B \sin\theta = \left(\frac{L}{v_d}i\right)v_d B \sin\theta = iLB\sin\theta, \qquad (27.11)$$

where θ is the angle between the direction of the current flow and the direction of the magnetic field. The direction of the force is perpendicular to both the current and the magnetic field and is given by a variant of right-hand rule 1, with the current in the direction of the velocity of a charged particle, as illustrated in Figure 27.24b. This variant of right-hand rule 1 takes advantage of the fact that current can be thought of as charges in motion.

Equation 27.11 can be expressed as a vector product:

$$\vec{F}_B = i\vec{L} \times \vec{B}, \qquad (27.12)$$

where the notation $i\vec{L}$ represents the current in a length of wire. Equation 27.12 is simply a reformulation of equation 27.2 for the case in which the moving charges make up a current flowing in a wire. Since physical situations involving currents are far more common than those involving the motion of an isolated charged particle, equation 27.12 is the most useful form for determining the magnetic force in practical applications.

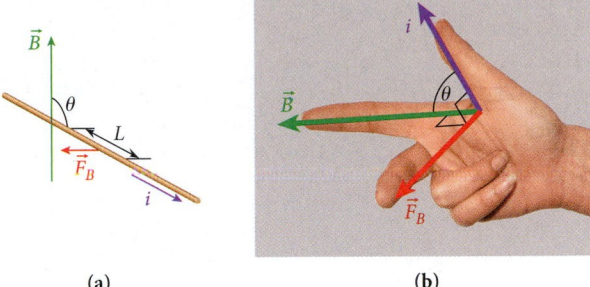

FIGURE 27.24 (a) Magnetic force on a current-carrying wire. (b) A variant of right-hand rule 1 giving the direction of the magnetic force on a current-carrying wire. To determine the direction of the force on a current-carrying wire using your right hand, point your thumb in the direction of the current and your index finger in the direction of the magnetic field; then your middle finger will point in the direction of the force.

EXAMPLE 27.4 | Force on the Voice Coil of a Loudspeaker

A loudspeaker produces sound by exerting a magnetic force on a voice coil in a magnetic field, as shown in Figure 27.25. The movable voice coil is connected to a speaker cone that actually produces the sounds. The magnetic field is produced by the two permanent magnets as shown. The magnitude of the magnetic field is $B = 1.50$ T. The voice coil is composed of $n = 100$ turns of wire carrying a current, $i = 1.00$ mA. The diameter of the voice coil is $d = 2.50$ cm.

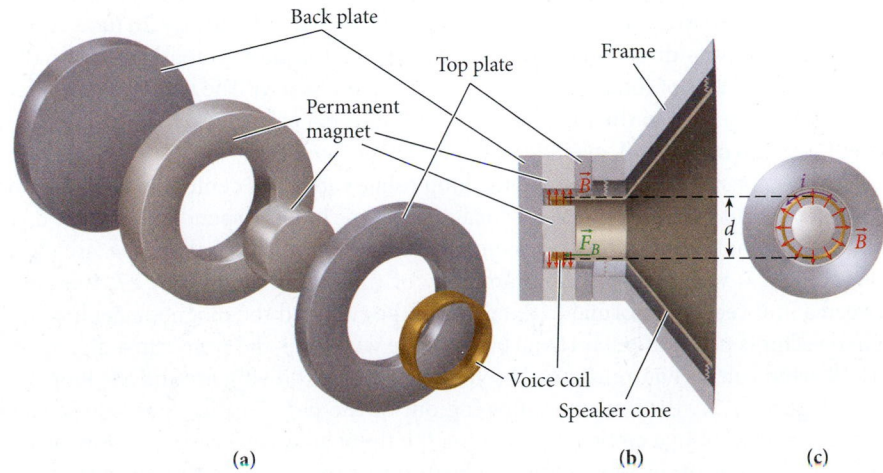

FIGURE 27.25 Schematic diagram of a loudspeaker: (a) an exploded three-dimensional view of the driver end of the loudspeaker; (b) a cross-sectional side view of the loudspeaker; (c) a front view of the driver end of the loudspeaker.

PROBLEM
What is the magnetic force exerted by the magnetic field on the loudspeaker's voice coil?

SOLUTION
The magnitude of the magnetic force on the voice coil is given by equation 27.11:

$$F = iLB\sin\theta,$$

– Continued

Concept Check 27.3

The figure shows a wire lying along the x-axis with a current, i, flowing in the negative x-direction. The wire is in a uniform magnetic field. The magnetic force, $\vec{F}_B$, acts on the wire in the positive z-direction. The magnetic field is oriented so that the force is maximum. What is the direction of the magnetic field?

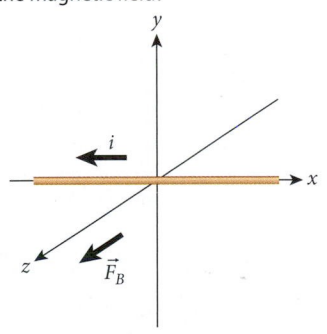

a) positive y-direction

b) negative x-direction

c) negative y-direction

d) positive z-direction

e) negative z-direction

where L is the length of wire carrying current i in the magnetic field with magnitude B. The wire makes an angle θ with the magnetic field. In this case, the wire is always perpendicular to the magnetic field, so $\theta = 90°$. The length of wire in the voice coil is given by the number of turns, n, times the circumference, πd, of each turn

$$L = n\pi d.$$

Thus, the force on the voice coil is

$$F = i(n\pi d)B(\sin 90°) = n\pi idB.$$

Putting in the numerical values, we get

$$F = n\pi idB = (100)(\pi)(1.00 \cdot 10^{-3}\ \text{A})(2.50 \cdot 10^{-2}\ \text{m})(1.50\ \text{T}) = 0.0118\ \text{N}.$$

From the right-hand rule 1 illustrated in Figure 27.24b, the direction of the force exerted by the magnetic field on the voice coil is toward the left in Figure 27.25b and perpendicularly into the page in Figure 27.25c.

If the current in the voice coil is reversed, the force will be in the opposite direction. If the current is proportional to the amplitude of a sound wave, sound waves can be reproduced in the cone of the loudspeaker. This basic idea is used in most speakers and headphones.

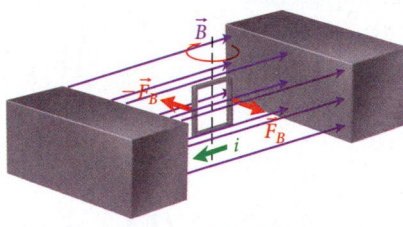

FIGURE 27.26 A primitive element of an electric motor consisting of a current-carrying loop in a magnetic field.

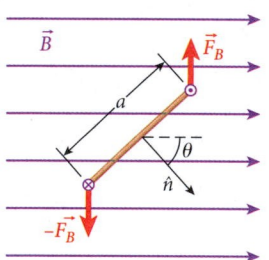

FIGURE 27.27 Bottom view of a current-carrying loop in a magnetic field, showing the forces acting on the loop.

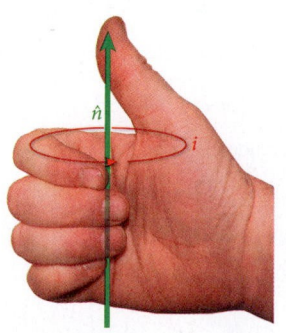

FIGURE 27.28 Right-hand rule 2 gives the direction of the unit normal vector for a current-carrying loop. According to the rule, if you curl the fingers of your right hand in the direction of the current in the loop, your thumb points in the direction of the unit normal vector.

27.5   Torque on a Current-Carrying Loop

Electric motors rely on the magnetic force exerted on a current-carrying wire. This force is used to create a torque that turns a shaft. Let's consider a simple electric motor, consisting of a single square loop carrying a current, i, in a constant magnetic field, $\vec{B}$. The loop is oriented so that its horizontal sections are parallel to the magnetic field and its vertical sections are perpendicular to the magnetic field, as shown in Figure 27.26. The magnitude of the magnetic force on the two vertical sections of the loop is given by equation 27.11 with $\theta = 90°$:

$$F = iLB.$$

The direction of the magnetic force is given by the variant of right-hand rule 1 illustrated in Figure 27.24b. The two magnetic forces, $\vec{F}_B$ and $-\vec{F}_B$, shown in Figure 27.26 have equal magnitudes and opposite directions. These forces create a torque that tends to rotate the loop around a vertical axis of rotation. These two forces sum to zero. The two horizontal sections of the loop are parallel to the magnetic field and thus experience no magnetic force. Thus, no net force acts on the coil, even though a torque is produced.

Now we consider the case where the loop rotates about its center. As the loop turns in the magnetic field, the forces on the vertical sides of the loop, perpendicular to the direction of the field, do not change. The forces on the square loop, with side length a, are illustrated in Figure 27.27, which shows a bottom view of the coil. In Figure 27.27, θ is the angle between a unit vector, $\hat{n}$, normal to the plane of the coil, and the magnetic field, $\vec{B}$. The unit normal vector is perpendicular to the plane of the wire loop and points in a direction given by right-hand rule 2 (Figure 27.28), based on the current flowing around the loop.

In Figure 27.27, the current is flowing out of the page in the right side of the loop, indicated by the dot in a circle (representing the tip of an arrowhead) and flowing into the page in the left side of the loop, indicated by the cross in a circle (representing the tail of an arrow). The magnitude of the force on each of these vertical segments is

$$F = iaB.$$

The forces on the two horizontal segments of the loop are parallel or antiparallel to the axis of rotation and do not cause a torque, and these two forces sum to zero. Therefore, there is no net force on the loop.

The sum of the torques on the two vertical segments of the square loop gives the net torque exerted on the loop about its center:

$$\tau_1 = (iaB)\left(\frac{a}{2}\right)\sin\theta + (iaB)\left(\frac{a}{2}\right)\sin\theta = ia^2 B\sin\theta = iAB\sin\theta, \qquad (27.13)$$

where the index 1 on τ_1 indicates that it is the torque on a single loop and $A = a^2$ is the area of the loop. The reason that the loop continues to rotate and doesn't stop at $\theta = 0°$ is that it is connected to a device called a **commutator,** which causes the current to change directions as the coil rotates. This commutator consists of a split ring, with one end of the loop connected to each half of the loop, as shown in Figure 27.29. The current in the loop switches direction two times for every complete rotation of the loop.

If the single loop is replaced with a coil consisting of many loops wound closely together, the torque on the coil is found by multiplying the torque on a loop, τ_1 from equation 27.13, by the number of windings (loops in the coil), N:

$$\tau = N\tau_1 = NiAB\sin\theta. \tag{27.14}$$

Does this expression for the torque hold for other shapes with area A, other than squares? The answer is yes, though we will not prove it here.

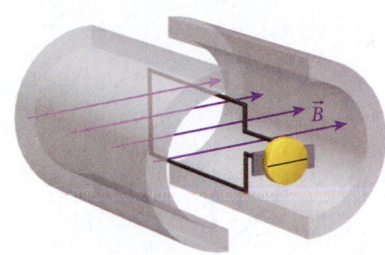

FIGURE 27.29 A wire loop connected to a source of current through a commutator ring.

Concept Check 27.4

The top view of a current-carrying loop in a constant magnetic field is shown in the figure. The torque on the loop will cause it to rotate

a) clockwise.

b) counterclockwise.

c) not at all.

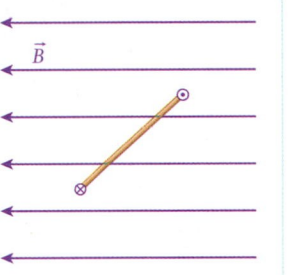

27.6 Magnetic Dipole Moment

A current-carrying coil can be described with one parameter, which contains information about a key characteristic of the coil in a magnetic field. The magnitude of the **magnetic dipole moment,** $\vec{\mu}$, of a current-carrying coil of wire is defined to be

$$\mu = NiA, \tag{27.15}$$

where N is the number of windings, i is the current through the wire, and A is the area of the loops. The direction of the magnetic dipole moment is given by right-hand rule 2 and is the direction of the unit normal vector, $\hat{n}$. Using equation 27.15, we can rewrite equation 27.14 as

$$\tau = \left(NiA\right)B\sin\theta = \mu B\sin\theta. \tag{27.16}$$

The torque on a magnetic dipole is given by

$$\vec{\tau} = \vec{\mu}\times\vec{B}. \tag{27.17}$$

That is, the torque on a current-carrying coil is the vector product of the magnetic dipole moment of the coil and the magnetic field.

SOLVED PROBLEM 27.3 Torque on a Rectangular Current-Carrying Loop

A rectangular loop with height $h = 6.50$ cm and width $w = 4.50$ cm is in a uniform magnetic field of magnitude $B = 0.250$ T, which points in the negative y-direction (Figure 27.30a). The loop makes an angle of $\theta = 33.0°$ with the y-axis, as shown in the figure. The loop carries a current of magnitude $i = 9.00$ A in the direction indicated by the arrows.

PROBLEM
What is the magnitude of the torque on the loop around the z-axis?

– Continued

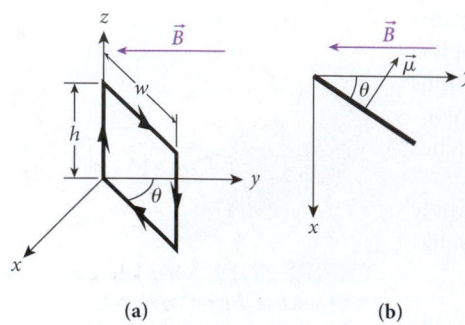

FIGURE 27.30 (a) A rectangular loop carrying a current in a magnetic field. (b) View of the rectangular loop looking down on the *xy*-plane. The magnetic dipole moment is perpendicular to the plane of the loop, with a direction determined by right-hand rule 2.

SOLUTION

THINK The torque on the loop is equal to the vector cross product of the magnetic dipole moment and the magnetic field. The magnetic dipole moment is perpendicular to the plane of the loop, with the direction given by right-hand rule 2.

SKETCH Figure 27.30b is a view of the loop looking down on the *xy*-plane.

RESEARCH The magnitude of the magnetic dipole moment of the loop is

$$\mu = NiA = iwh. \tag{i}$$

The magnitude of the torque on the loop is

$$\tau = \mu B \sin\theta_{\mu B}, \tag{ii}$$

where $\theta_{\mu B}$ is the angle between the magnetic dipole moment and the magnetic field. From Figure 27.30b, we can see that

$$\theta_{\mu B} = \theta + 90°. \tag{iii}$$

SIMPLIFY We can combine equations (i), (ii), and (iii) to obtain

$$\tau = iwhB \sin(\theta + 90°).$$

CALCULATE Putting in the numerical values, we get

$$\tau = (9.00 \text{ A})(4.50 \cdot 10^{-2} \text{ m})(6.50 \cdot 10^{-2} \text{ m})(0.250 \text{ T})\left[\sin(33.0° + 90°)\right]$$

$$= 0.0055195 \text{ N m}.$$

ROUND We report our result to three significant figures:

$$\tau = 5.52 \cdot 10^{-3} \text{ N m}.$$

DOUBLE-CHECK The magnitude of the force on each of the vertical segments of the loop is

$$F_B = ihB = (9.00 \text{ A})(6.50 \cdot 10^{-2} \text{ m})(0.250 \text{ T}) = 0.146 \text{ N}.$$

The magnitude of the torque is then the magnitude of the force on the vertical segment that is not along the *z*-axis times the moment arm (which is *w*) times the sine of the angle between the force and the moment arm:

$$\tau = Fw \sin(33.0° + 90°) = 0.146 \text{ N}(4.50 \cdot 10^{-2} \text{ m})\left[\sin(33.0° + 90°)\right] = 5.52 \cdot 10^{-3} \text{ N m}.$$

This is the same as the result calculated above.

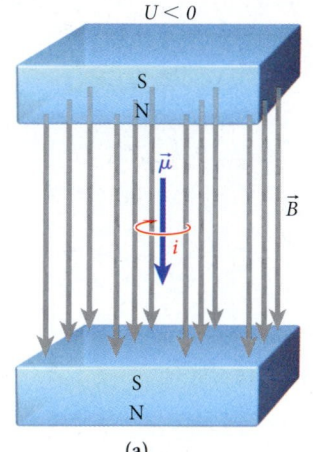

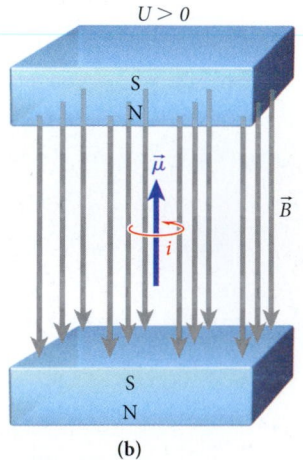

FIGURE 27.31 Magnetic dipole moment vector in an external magnetic field: (a) magnetic dipole and external magnetic field are parallel, resulting in a negative potential energy; (b) magnetic dipole and external magnetic field are antiparallel, resulting in a positive potential energy.

A magnetic dipole has potential energy in an external magnetic field. If the dipole moment is aligned with the magnetic field, the dipole has its minimum potential energy. If the dipole moment is oriented in a direction opposite to the external field, the dipole has its maximum potential energy. From Chapter 10, the work done by a torque is

$$W = \int_{\theta_0}^{\theta} \tau(\theta')d\theta'. \tag{27.18}$$

Using the work-energy theorem and equation 27.16 and setting $\theta_0 = 90°$, we can express the magnetic potential energy, *U*, of a magnetic dipole in an external magnetic field, $\vec{B}$, as

$$W = \int_{\theta_0}^{\theta} \tau(\theta')d\theta' = \int_{\theta_0}^{\theta} \mu B \sin\theta' d\theta' = -\mu B \cos\theta \Big|_{\theta_0}^{\theta} = U(\theta) - U(90°),$$

or

$$U(\theta) = -\mu B \cos\theta = -\vec{\mu} \cdot \vec{B}, \tag{27.19}$$

where θ is the angle between the magnetic dipole moment and the external magnetic field.

The lowest value, $-\mu B$, of the potential energy of a magnetic dipole in an external magnetic field is achieved when the dipole's magnetic moment vector is parallel to the external magnetic field vector, and the highest value, $+\mu B$, results when the two vectors are antiparallel (see Figure 27.31). This dependence of potential energy on orientation occurs in diverse physical situations, for which magnetic dipoles in external magnetic fields are a simple model. So far, the only magnetic dipoles we have discussed are current-carrying loops. However, other types of magnetic dipoles exist, including bar magnets and even the Earth. In addition, elementary charged particles such as protons have intrinsic magnetic dipole moments.

27.7 Hall Effect

Consider a conductor carrying a current, i, flowing in a direction perpendicular to a magnetic field, $\vec{B}$ (Figure 27.32a). The electrons in the conductor are moving with velocity $\vec{v}_d$ in the direction opposite to the current. The moving electrons experience a force perpendicular to their velocity, causing them to move toward one edge of the conductor. After some time, many electrons have moved to one edge of the conductor, creating a net negative charge on that edge and leaving a net positive charge on the opposite edge of the conductor. This charge distribution creates an electric field, $\vec{E}$, which exerts a force on the electrons in a direction opposite to that exerted by the magnetic field. When the magnitude of the force exerted on the electrons by the electric field is equal to the magnitude of the force exerted on them by the magnetic field, the net number of electrons on the edges of the conductor no longer changes with time. This result is called the **Hall effect.** The potential difference, ΔV_H, between the edges of the conductor when equilibrium is reached is termed the **Hall potential difference,** given by

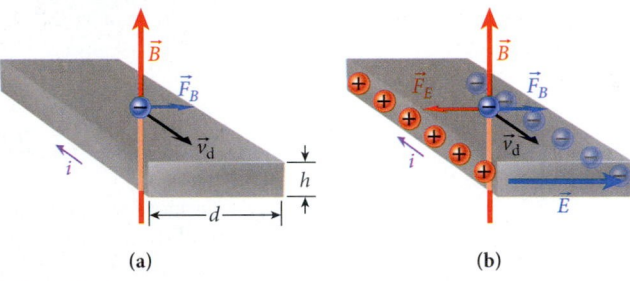

FIGURE 27.32 (a) A conductor carrying a current in a magnetic field. The charge carriers are electrons. (b) The electrons have drifted to one side of the conductor, leaving a net positive charge on the opposite side. This distribution of charges creates an electric field. The potential difference across the conductor is the Hall potential difference.

$$\Delta V_H = Ed, \tag{27.20}$$

where d is the width of the conductor and E is the magnitude of the created electric field. (See Chapter 23 for the relationship between the electric potential difference and the constant electric field.)

The Hall effect can be used to demonstrate that the charge carriers in metals are negatively charged. If the charge carriers in a metal were positive and moving in the direction of the current shown in Figure 27.32a, those positive charges would collect on the same edge of the conductor as the electrons in Figure 27.32b, giving an electric field with the opposite sign. Thus, the charge carriers in conductors are negatively charged and must be electrons. The Hall effect also establishes that in some semiconductors the charge carriers are electron holes (missing electrons), which appear to be positively charged carriers.

The Hall effect can also be used to determine a magnetic field by measuring the current flowing through the conductor and the resulting potential difference across the conductor. To obtain the formula for the magnetic field, we start with the equilibrium condition of the Hall effect, that the magnitudes of the magnetic and electric forces are equal:

$$F_E = F_B \Rightarrow eE = v_d Be \Rightarrow B = \frac{E}{v_d} = \frac{\Delta V_H}{v_d d}, \tag{27.21}$$

where substitution for E from equation 27.20 is used in the last step. In Chapter 25, we saw that the drift speed, v_d, of an electron in a conductor can be related to the magnitude of the current density, J, in the conductor:

$$J = \frac{i}{A} = nev_d,$$

where A is the cross-sectional area of the conductor and n is the number of electrons per unit volume in the conductor. As shown in Figure 27.32a, the cross-sectional area is given by $A = dh$, where d is the width and h is the height of the conductor. Solving $i/A = nev_d$ for the drift speed and substituting hd for A gives

$$v_d = \frac{i}{Ane} = \frac{i}{hdne}.$$

Substituting this expression for v_d into equation 27.21, we have

$$B = \frac{\Delta V_H}{v_d d} = \frac{\Delta V_H dhne}{id} = \frac{\Delta V_H hne}{i}. \tag{27.22}$$

Thus, equation 27.22 gives the magnetic field strength (magnitude) from a measured value of the Hall potential difference, ΔV_H, and the known height, h, and density of charge carriers, n, of the conductor. Equivalently, a rearranged form of equation 27.22 can be used to find the Hall voltage if the magnetic field strength is known:

$$\Delta V_H = \frac{iB}{neh}. \tag{27.23}$$

EXAMPLE 27.5 | Hall Effect

Suppose we use a Hall probe to measure the magnitude of a constant magnetic field. The Hall probe is a strip of copper with a height, h, of 2.00 mm. We measure a voltage of 0.250 μV across the probe when we run a current of 1.25 A through it.

PROBLEM

What is the magnitude of the magnetic field?

SOLUTION

The magnetic field is given by equation 27.22:

$$B = \frac{\Delta V_H hne}{i}.$$

We have been given the values of V_H, h, and i, and we know e. The density of the electrons, n, is defined as the number of electrons per unit volume:

$$n = \frac{\text{number of electrons}}{\text{volume}}.$$

The density of copper is $\rho_{Cu} = 8.96 \text{ g/cm}^3 = 8960 \text{ kg/m}^3$, and 1 mole of copper has a mass of 63.5 g and $6.02 \cdot 10^{23}$ atoms. Each copper atom has one conduction electron. Thus, the density of the electrons is

$$n = \left(\frac{1 \text{ electron}}{1 \text{ atom}}\right)\left(\frac{6.02 \cdot 10^{23} \text{ atoms}}{63.5 \text{ g}}\right)\left(\frac{8.96 \text{ g}}{1 \text{ cm}^3}\right)\left(\frac{1.0 \cdot 10^6 \text{ cm}^3}{1 \text{ m}^3}\right) = 8.49 \cdot 10^{28} \frac{\text{electrons}}{\text{m}^3}.$$

We can now calculate the magnitude of the magnetic field:

$$B = \frac{\left(0.250 \cdot 10^{-6} \text{ V}\right)\left(0.002 \text{ m}\right)\left(8.49 \cdot 10^{28} \frac{\text{electrons}}{\text{m}^3}\right)\left(1.602 \cdot 10^{-19} \text{ C}\right)}{1.25 \text{ A}} = 5.44 \text{ T}.$$

Concept Check 27.5

A conductor is carrying a current in a constant magnetic field, as shown in the figure. The resulting electric field due to the Hall effect is

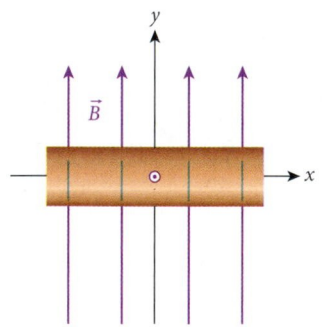

a) in the positive x-direction.

b) in the negative x-direction.

c) in the positive y-direction.

d) in the negative y-direction.

WHAT WE HAVE LEARNED | EXAM STUDY GUIDE

- Magnetic field lines indicate the direction of a magnetic field in space. Magnetic field lines do not end on magnetic poles, but form closed loops instead.

- The magnetic force on a particle with charge q moving with velocity $\vec{v}$ in a magnetic field, $\vec{B}$, is given by $\vec{F} = q\vec{v} \times \vec{B}$. Right-hand rule 1 gives the direction of the force.

- For a particle with charge q moving with speed v perpendicular to a magnetic field of magnitude B, the magnitude of the magnetic force on the moving charged particle is $F = |q| vB$.

- The unit of magnetic field is the tesla, abbreviated T.

- The average magnitude of the Earth's magnetic field at the surface is approximately $0.5 \cdot 10^{-4}$ T.

- A particle with mass m and charge q moving with speed v perpendicular to a magnetic field with magnitude B has a trajectory that is a circle with radius $r = mv / |q| B$.

- The cyclotron frequency, ω, of a particle with charge q and mass m moving in a circular orbit in a constant magnetic field of magnitude B is given by $\omega = |q| B/m$.

- The force exerted by a magnetic field, $\vec{B}$, on a length of wire, $\vec{L}$, carrying a current, i, is given by $\vec{F} = i\vec{L} \times \vec{B}$. The magnitude of this force is $F = iLB \sin\theta$, where θ is the angle between the direction of the current and the direction of the magnetic field.

- The magnitude of the torque on a loop carrying a current, i, in a magnetic field with magnitude B is $\tau = iAB \sin\theta$, where A is the area of the loop and θ is the angle between a unit vector normal to the loop and the direction of the magnetic field. Right-hand rule 2 gives the direction of the unit normal vector to the loop.

- The magnitude of the magnetic dipole moment of a coil carrying a current, i, is given by $\mu = NiA$, where N is the number of loops (windings) and A is the area of a loop. The direction of the dipole moment is given by right-hand rule 2 and is the direction in which the unit normal vector points.

- The Hall effect results when a current, i, flowing through a conductor with height h in a magnetic field of magnitude B produces a potential difference across the conductor (the Hall potential difference), given by $\Delta V_H = iB/neh$, where n is the number of electrons per unit volume and e is the magnitude of an electron's charge.

ANSWERS TO SELF-TEST OPPORTUNITIES

27.1 Particle 1: $F_B = qvB \sin\theta =$
$(6.15 \cdot 10^{-6} \text{ C})(465 \text{ m/s})(0.165 \text{ T})(\sin 30.0°) = 2.36 \cdot 10^{-4} \text{ N}$.

Particle 2: $F_B = qvB \sin\theta =$
$(6.15 \cdot 10^{-6} \text{ C})(465 \text{ m/s})(0.165 \text{ T})(\sin 90.0°) = 4.72 \cdot 10^{-4} \text{ N}$.

Particle 3: $F_B = qvB \sin\theta =$
$(6.15 \cdot 10^{-6} \text{ C})(465 \text{ m/s})(0.165 \text{ T})(\sin 150.0°) = 2.36 \cdot 10^{-4} \text{ N}$.

27.2 a) positive

b) slowing down

c) no (Therefore, another force must be acting on the particle to slow it down.)

27.3 $\Delta U = U_{max} - U_{min} = 2\mu B = 2iAB =$
$2(2.00 \text{ A})(0.100 \text{ m}^2)(0.500 \text{ T}) = 0.200 \text{ J}$.

PROBLEM-SOLVING GUIDELINES

1. When working with magnetic fields and forces, you need to sketch a clear diagram of the problem situation in three dimensions. Often, a separate sketch of the velocity and magnetic field vectors (or of the length of wire and the field vectors) is useful to visualize the plane in which they lie, since the magnetic force will be perpendicular to that plane.

2. Remember that the right-hand rules apply for positive charges and currents. If a charge or a current is negative, you can use the right-hand rule, but the force will then be in the opposite direction.

3. A particle in both an electric and a magnetic field experiences an electric force, $\vec{F}_E = q\vec{E}$, and a magnetic force, $\vec{F}_B = q\vec{v} \times \vec{B}$. Be sure you take the vector sum of the individual forces.

MULTIPLE-CHOICE QUESTIONS

27.1 A magnetic field is oriented in a certain direction in a horizontal plane. An electron moves in a certain direction in the horizontal plane. For this situation, there

a) is one possible direction for the magnetic force on the electron.

b) are two possible directions for the magnetic force on the electron.

c) are infinite possible directions for the magnetic force on the electron.

27.2 A particle with charge q is at rest when a magnetic field is suddenly turned on. The field points in the z-direction. What is the direction of the net force acting on the charged particle?

a) in the x-direction   c) The net force is zero.

b) in the y-direction   d) in the z-direction

27.3 Which of the following has the largest cyclotron frequency?

a) an electron with speed v in a magnetic field with magnitude B

b) an electron with speed $2v$ in a magnetic field with magnitude B

c) an electron with speed $v/2$ in a magnetic field with magnitude B

d) an electron with speed $2v$ in a magnetic field with magnitude $B/2$

e) an electron with speed $v/2$ in a magnetic field with magnitude $2B$

27.4 In the Hall effect, a potential difference produced across a conductor of finite thickness in a magnetic field by a current flowing through the conductor is given by

a) the product of the density of electrons, the charge of an electron, and the conductor's thickness divided by the product of the magnitudes of the current and the magnetic field.

b) the reciprocal of the expression described in part (a).

c) the product of the charge on an electron and the conductor's thickness divided by the product of the density of electrons and the magnitudes of the current and the magnetic field.

d) the reciprocal of the expression described in (c).

e) none of the above.

27.5 An electron (with charge $-e$ and mass m_e) moving in the positive x-direction enters a velocity selector. The velocity selector consists of crossed electric and magnetic fields: $\vec{E}$ is directed in the positive y-direction, and $\vec{B}$ is directed in the positive z-direction. For a velocity v (in the positive x-direction), the net force on the electron is zero, and the electron moves straight through the velocity selector. With what velocity will a proton (with charge $+e$ and mass $m_p = 1836\, m_e$) move straight through the velocity selector?

a) v

b) $-v$

c) $v/1836$

d) $-v/1836$

27.6 In which direction does a magnetic force act on an electron that is moving in the positive x-direction in a magnetic field pointing in the positive z-direction?

a) the positive y-direction

b) the negative y-direction

c) the negative x-direction

d) any direction in the xy-plane

27.7 A charged particle is moving in a constant magnetic field. Which of the following statements concerning the magnetic force exerted on the particle is (are) true? (Assume that the magnetic field is not parallel or antiparallel to the velocity.)

a) It does no work on the particle.

b) It may increase the speed of the particle.

c) It may change the velocity of the particle.

d) It can act only on the particle while the particle is in motion.

e) It does not change the kinetic energy of the particle.

27.8 An electron moves in a circular trajectory with radius r_i in a constant magnetic field. What is the final radius of the trajectory when the magnetic field is doubled?

a) $\dfrac{r_i}{4}$

b) $\dfrac{r_i}{2}$

c) r_i

d) $2r_i$

e) $4r_i$

27.9 Protons in the solar wind reach Earth's magnetic field with a speed of 400 km/s. If the magnitude of this field is $5.0 \cdot 10^{-5}$ T and the velocity of the protons is perpendicular to it, what is the cyclotron frequency of the protons after entering the field?

a) 122 Hz

b) 233 Hz

c) 321 Hz

d) 432 Hz

e) 763 Hz

27.10 An isolated segment of wire of length $L = 4.50$ m carries a current of magnitude $i = 35.0$ A at an angle $\theta = 50.3°$ with respect to a constant magnetic field of magnitude $B = 6.70 \cdot 10^{-2}$ T (see the figure). What is the magnitude of the magnetic force on the wire?

a) 2.66 N

b) 3.86 N

c) 5.60 N

d) 8.12 N

e) 11.8 N

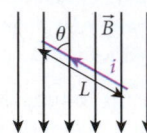

27.11 A coil is composed of circular loops of radius $r = 5.13$ cm and has $N = 47$ windings. A current, $i = 1.27$ A, flows through the coil, which is inside a homogeneous magnetic field of magnitude 0.911 T. What is the maximum torque on the coil due to the magnetic field?

a) 0.148 N m

b) 0.211 N m

c) 0.350 N m

d) 0.450 N m

e) 0.622 N m

27.12 A copper conductor has a current, $i = 1.41$ A, flowing through it, perpendicular to a constant magnetic field, $B = 4.94$ T, as shown in the figure. The conductor is $d = 0.100$ m wide and $h = 2.00$ mm high. What is the electric potential difference between points 1 and 2?

a) $2.56 \cdot 10^{-9}$ V

b) $5.12 \cdot 10^{-9}$ V

c) $7.50 \cdot 10^{-8}$ V

d) $2.56 \cdot 10^{-7}$ V

e) $9.66 \cdot 10^{-7}$ V

CONCEPTUAL QUESTIONS

27.13 Draw on the xyz-coordinate system and specify (in terms of the unit vectors $\hat{x}$, $\hat{y}$, and $\hat{z}$) the direction of the magnetic force on each of the moving particles shown in the figures. *Note:* The positive y-axis is toward the right, the positive z-axis is toward the top of the page, and the positive x-axis is directed out of the page.

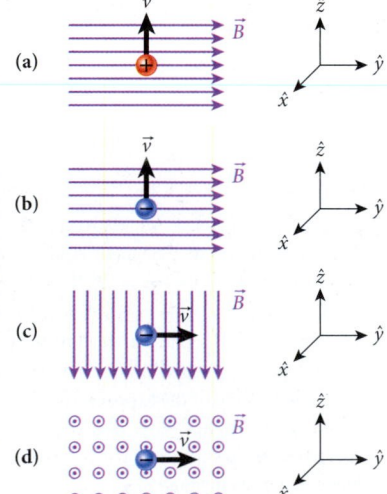

27.14 A particle with mass m, charge q, and velocity v enters a magnetic field of magnitude B and with direction perpendicular to the initial velocity of the particle. What is the work done by the magnetic field on the particle? How does this affect the particle's motion?

27.15 An electron is moving with a constant velocity. When it enters an electric field that is perpendicular to its velocity, the electron will follow a _____ trajectory. When the electron enters a magnetic field that is perpendicular to its velocity, it will follow a _____ trajectory.

27.16 A proton, moving in negative y-direction in a magnetic field, experiences a force of magnitude F, acting in the negative x-direction.

a) What is the direction of the magnetic field producing this force?

b) Does your answer change if the word "proton" in the statement is replaced by "electron"?

27.17 It would be mathematically possible, for a region with zero current density, to define a scalar magnetic potential analogous to the electrostatic potential: $V_B(\vec{r}) = -\int_{\vec{r}_0}^{\vec{r}} \vec{B} \cdot d\vec{s}$, or $\vec{B}(\vec{r}) = -\nabla V_B(\vec{r})$. However, this has not been done. Explain why not.

27.18 A current-carrying wire is positioned within a large, uniform magnetic field, $\vec{B}$. However, the wire experiences no force. Explain how this might be possible.

27.19 A charged particle moves under the influence of an electric field only. Is it possible for the particle to move with a constant speed? What if the electric field is replaced with a magnetic field?

27.20 A charged particle travels with speed v, at an angle θ with respect to the z-axis. It enters at time $t = 0$ a region of space where there is a magnetic field of magnitude B in the positive z-direction. When does it emerge from this region of space?

27.21 An electron is traveling horizontally from the northwest toward the southeast in a region of space where the Earth's magnetic field is directed horizontally toward the north. What is the direction of the magnetic force on the electron?

27.22 At the Earth's surface, there is an electric field that points approximately straight down and has magnitude 150 V/m. Suppose you had a tuneable electron gun (you can release electrons with whatever kinetic energy you like) and a detector to determine the direction of motion of the electrons when they leave the gun. Explain how you could use the gun to find the direction toward the north magnetic pole.

Specifically, what kinetic energy would the electrons need to have? (*Hint:* It might be easier to think about finding which direction is east or west.)

27.23 The work done by the magnetic field on a charged particle in motion in a cyclotron is zero. How, then, can a cyclotron be used as a particle accelerator, and what essential feature of the particle's motion makes it possible?

EXERCISES

A blue problem number indicates a worked-out solution is available in the Student Solutions Manual. One • and two •• indicate increasing level of problem difficulty.

Section 27.2

27.24 A proton moving with a speed of $4.00 \cdot 10^5$ m/s in the positive y-direction enters a uniform magnetic field of 0.400 T pointing in the positive x-direction. Calculate the magnitude of the force on the proton.

27.25 The magnitude of the magnetic force on a particle with charge $-2e$ moving with speed $v = 1.00 \cdot 10^5$ m/s is $3.00 \cdot 10^{-18}$ N. What is the magnitude of the magnetic field component perpendicular to the direction of motion of the particle?

•27.26 A particle with a charge of $+10.0$ μC is moving at 300. m/s in the positive z-direction.

a) Find the minimum magnetic field required to keep it moving in a straight line at constant speed if there is a uniform electric field of magnitude 100. V/m pointing in the positive y-direction.

b) Find the minimum magnetic field required to keep the particle moving in a straight line at constant speed if there is a uniform electric field of magnitude 100. V/m pointing in the positive z-direction.

•27.27 A particle with a charge of 20.0 μC moves along the x-axis with a speed of 50.0 m/s. It enters a magnetic field given by $\vec{B} = 0.300\,\hat{y} + 0.700\,\hat{z}$, in teslas. Determine the magnitude and the direction of the magnetic force on the particle.

••27.28 The magnetic field in a region in space (where $x > 0$ and $y > 0$) is given by $\vec{B} = (x - az)\hat{y} + (xy - b)\hat{z}$, where a and b are positive constants. An electron moving with a constant velocity, $\vec{v} = v_0\hat{x}$, enters this region. What are the coordinates of the points at which the net force acting on the electron is zero?

Section 27.3

27.29 A proton is accelerated from rest by a potential difference of 400. V. The proton enters a uniform magnetic field and follows a circular path of radius 20.0 cm. Determine the magnitude of the magnetic field.

27.30 An electron with a speed of $4.00 \cdot 10^5$ m/s enters a uniform magnetic field of magnitude 0.0400 T at an angle of 35.0° to the magnetic field lines. The electron will follow a helical path.

a) Determine the radius of the helical path.

b) How far forward will the electron have moved after completing one circle?

27.31 A particle with mass m and charge q is moving within both an electric field and a magnetic field, $\vec{E}$ and $\vec{B}$. The particle has velocity $\vec{v}$, momentum $\vec{p}$, and kinetic energy K. Find general expressions for $d\vec{p}/dt$ and dK/dt, in terms of these seven quantities.

27.32 The Earth is showered with particles from space known as *muons*. They have a charge identical to that of an electron but are many times heavier ($m = 1.88 \cdot 10^{-28}$ kg). Suppose a strong magnetic field is established in a lab ($B = 0.500$ T) and a muon enters this field with a velocity of $3.00 \cdot 10^6$ m/s at a right angle to the field. What will be the radius of the muon's resulting orbit?

27.33 An electron in a magnetic field moves counterclockwise on a circle in the xy-plane, with a cyclotron frequency of $\omega = 1.20 \cdot 10^{12}$ Hz. What is the magnetic field, $\vec{B}$?

27.34 An electron with energy equal to $4.00 \cdot 10^2$ eV and an electron with energy equal to $2.00 \cdot 10^2$ eV are trapped in a uniform magnetic field and move in circular paths in a plane perpendicular to the magnetic field. What is the ratio of the radii of their orbits?

•27.35 A proton with an initial velocity given by $(1.00\hat{x} + 2.00\hat{y} + 3.00\hat{z})$ $(10^5$ m/s) enters a magnetic field given by $(0.500$ T)$\hat{z}$. Describe the motion of the proton.

•27.36 Initially at rest, a small copper sphere with a mass of $3.00 \cdot 10^{-6}$ kg and a charge of $5.00 \cdot 10^{-4}$ C is accelerated through a 7000.-V potential difference before entering a magnetic field of magnitude 4.00 T, directed perpendicular to its velocity. What is the radius of curvature of the sphere's motion in the magnetic field?

•27.37 Two particles with masses m_1 and m_2 and charges q and $2q$ travel with the same velocity, v, and enter a magnetic field of strength B at the same point, as shown in the figure. In the magnetic field, they move in semicircles with radii R and $2R$. What is the ratio of their masses? Is it possible to apply an electric field that would cause the particles to move in a straight line in the magnetic field? If yes, what would be the magnitude and the direction of the field?

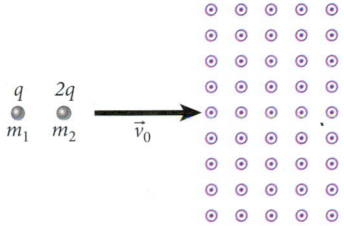

•27.38 The figure shows a schematic diagram of a simple mass spectrometer, consisting of a velocity selector and a particle detector and being used to separate singly ionized atoms ($q = +e = 1.602 \cdot 10^{-19}$ C) of gold (Au) and molybdenum (Mo). The electric field inside the velocity selector has magnitude $E = 1.789 \cdot 10^4$ V/m and points toward the top of the page, and the magnetic field has magnitude $B_1 = 1.000$ T and points out of the page.

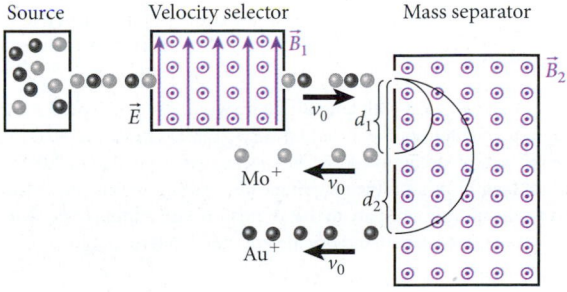

a) Draw the electric force vector, $\vec{F}_E$, and the magnetic force vector, $\vec{F}_B$, acting on the ions inside the velocity selector.

b) Calculate the velocity, v_0, of the ions that make it through the velocity selector (those that travel in a straight line). Does v_0 depend on the type of ion (gold versus molybdenum), or is it the same for both types of ions?

c) Write the equation for the radius of the semicircular path of an ion in the particle detector: $R = R(m, v_0, q, B_2)$.

d) The gold ions (represented by the dark gray spheres) exit the particle detector at a distance $d_2 = 40.00$ cm from the entrance slit, while the molybdenum ions (represented by the light gray spheres) exit the particle detector at a distance $d_1 = 19.81$ cm from the entrance slit. The mass of a gold ion is $m_{gold} = 3.271 \cdot 10^{-25}$ kg. Calculate the mass of a molybdenum ion.

••27.39 A small particle accelerator for accelerating ^{3}He$^+$ ions is shown in the figure. The ^{3}He$^+$ ions exit the ion source with a kinetic energy of 4.00 keV. Regions 1 and 2 contain magnetic fields directed into the page, and region 3 contains an electric field directed from left to right. The ^{3}He$^+$ ion beam exits the accelerator from a hole on the right that is 7.00 cm below the ion source, as shown in the figure.

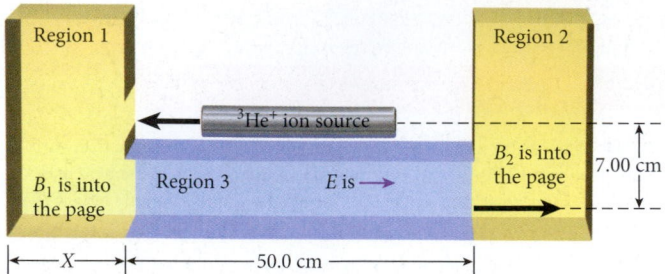

a) If $B_1 = 1.00$ T and region 3 is 50.0 cm long with $E = 60.0$ kV/m, what value should B_2 have to cause the ions to move straight through the exit hole after being accelerated twice in region 3?

b) What minimum width X should region 1 have?

c) What is the velocity of the ions when they leave the accelerator?

Section 27.4

27.40 A straight wire of length 2.00 m carries a current of 24.0 A. It is placed on a horizontal tabletop in a uniform horizontal magnetic field. The wire makes an angle of 30.0° with the magnetic field lines. If the magnitude of the force on the wire is 0.500 N, what is the magnitude of the magnetic field?

27.41 As shown in the figure, a straight conductor parallel to the x-axis can slide without friction on top of two horizontal conducting rails that are parallel to the y-axis and a distance of $L = 0.200$ m apart, in a vertical magnetic field of 1.00 T. A 20.0-A current is maintained through the conductor. If a string is connected exactly at the center of the conductor and passes over a frictionless pulley, what mass m suspended from the string allows the conductor to be at rest?

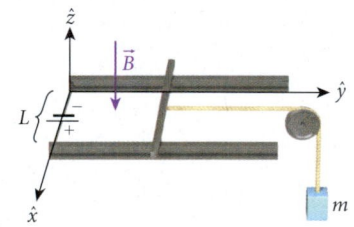

27.42 A copper wire of radius 0.500 mm is carrying a current at the Earth's Equator. Assuming that the magnetic field of the Earth has magnitude 0.500 G at the Equator and is parallel to the surface of the Earth and that the current in the wire flows toward the east, what current is required to allow the wire to levitate?

•27.43 A copper sheet with length 1.00 m, width 0.500 m, and thickness 1.00 mm is oriented so that its largest surface area is perpendicular to a magnetic field of strength 5.00 T. The sheet carries a current of 3.00 A across its length. What is the magnitude of the force on this sheet? How does this magnitude compare to that of the force on a thin copper wire carrying the same current and oriented perpendicularly to the same magnetic field?

•27.44 A conducting rod of length L slides freely down an inclined plane, as shown in the figure. The plane is inclined at an angle θ from the horizontal. A uniform magnetic field of strength B acts in the positive y-direction. Determine the magnitude and the direction of the current that would have to be passed through the rod to hold it in position on the inclined plane.

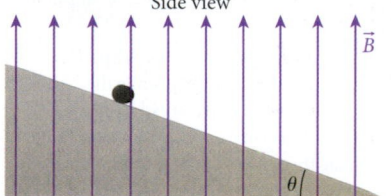
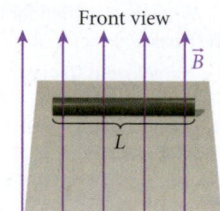

•27.45 A square loop of wire, with side length $d = 8.00$ cm, carries a current of magnitude $i = 0.150$ A and is free to rotate. It is placed between the poles of an electromagnet that produce a uniform magnetic field of 1.00 T. The loop is initially placed so that its normal vector, $\hat{n}$, is at a 35.0° angle relative to the direction of the magnetic field vector, with the angle θ defined as shown in the figure. The wire is copper (with a density of $\rho = 8960$ kg/m^3), and its diameter is 0.500 mm. What is the magnitude of the initial angular acceleration of the loop when it is released?

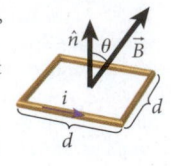

•27.46 A rail gun accelerates a projectile from rest by using the magnetic force on a current-carrying wire. The wire has radius $r = 5.10 \cdot 10^{-4}$ m and is made of copper having a density of $\rho = 8960$ kg/m^3. The gun consists of rails of length $L = 1.00$ m in a constant magnetic field of magnitude $B = 2.00$ T, oriented perpendicular to the plane defined by the rails. The wire forms an electrical connection across the rails at one end of the rails. When triggered, a current of $1.00 \cdot 10^4$ A flows through the wire, which accelerates the wire along the rails. Calculate the final speed of the wire as it leaves the rails. (Neglect friction.)

Sections 27.5 and 27.6

•27.47 A square loop of wire of side length ℓ lies in the xy-plane, with its center at the origin and its sides parallel to the x- and y-axes. It carries a current, i, flowing in the counterclockwise direction, as viewed looking down the z-axis from the positive direction. The loop is in a magnetic field given by $\vec{B} = (B_0/a)(z\hat{x} + x\hat{z})$, where B_0 is a constant field strength, a is a constant with the dimension of length, and $\hat{x}$ and $\hat{z}$ are unit vectors in the positive x-direction and positive z-direction. Calculate the net force on the loop.

27.48 A rectangular coil with 20 windings carries a current of 2.00 mA flowing in the counterclockwise direction. It has two sides that are parallel to the y-axis and have length 8.00 cm and two sides that are parallel to the x-axis and have length 6.00 cm. A uniform magnetic field of 50.0 μT acts in the positive x-direction. What torque must be applied to the loop to hold it steady?

27.49 A coil consists of 120. circular loops of wire of radius 4.80 cm. A current of 0.490 A runs through the coil, which is oriented vertically and is free to rotate about a vertical axis (parallel to the z-axis). It experiences a uniform horizontal magnetic field in the positive x-direction. When the coil is oriented parallel to the x-axis, a force of 1.20 N applied to the edge of the coil in the positive y-direction can keep it from rotating. Calculate the strength of the magnetic field.

27.50 Twenty loops of wire are tightly wound around a round pencil that has a diameter of 6.00 mm. The pencil is then placed in a uniform 5.00-T magnetic field, as shown in the figure. If a 3.00-A current is present in the coil of wire, what is the magnitude of the torque on the pencil?

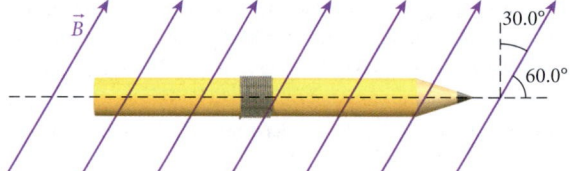

•27.51 A copper wire with density $\rho = 8960$ kg/m^3 is formed into a circular loop of radius 50.0 cm. The cross-sectional area of the wire is $1.00 \cdot 10^{-5}$ m^2, and a potential difference of 0.0120 V is applied to the wire. What is the maximum angular acceleration of the loop when it is placed in a magnetic field of magnitude 0.250 T? The loop rotates about an axis in the plane of the loop that corresponds to a diameter.

27.52 A simple galvanometer is made from a coil that consists of N loops of wire of area A. The coil is attached to a mass, M, by a light rigid rod of length L. With no current in the coil, the mass hangs straight down, and the coil lies in a horizontal plane. The coil is in a uniform magnetic field of magnitude B that is oriented horizontally. Calculate the angle from the vertical of the rigid rod as a function of the current, i, in the coil.

27.53 Show that the magnetic dipole moment of an electron orbiting in a hydrogen atom is proportional to its angular momentum, L: $\mu = -eL/(2m)$, where $-e$ is the charge of the electron and m is its mass.

•27.54 The figure shows a top view of a current-carrying ring of wire having a diameter $d = 8.00$ cm, which is suspended from the ceiling via a thin string. A 1.00-A current flows in the ring in the direction indicated in the figure. The ring is connected to one end of a spring with a spring constant of 100. N/m. When the ring is in the position shown in the figure, the spring is at its equilibrium length, ℓ. Determine the extension of the spring when a magnetic field of magnitude $B = 2.00$ T is applied parallel to the plane of the ring, as shown in the figure.

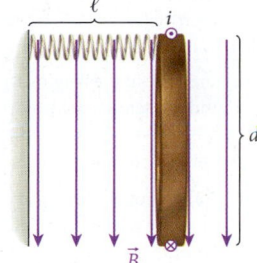

•27.55 A coil of wire consisting of 40 rectangular loops, with width 16.0 cm and height 30.0 cm, is placed in a constant magnetic field given by $\vec{B} = 0.0650T\hat{x} + 0.250T\hat{z}$. The coil is hinged to a fixed thin rod along the y-axis (along segment da in the figure) and is originally located in the xy-plane. A current of 0.200 A runs through the wire.

a) What are the magnitude and the direction of the force, $\vec{F}_{ab}$, that $\vec{B}$ exerts on segment ab of the coil?

b) What are the magnitude and the direction of force, $\vec{F}_{bc}$, that $\vec{B}$ exerts on segment bc of the coil?

c) What is the magnitude of the net force, F_{net}, that $\vec{B}$ exerts on the coil?

d) What are the magnitude and the direction of the torque, $\vec{\tau}$, that $\vec{B}$ exerts on the coil?

e) In what direction, if any, will the coil rotate about the y-axis (viewed from above and looking down that axis)?

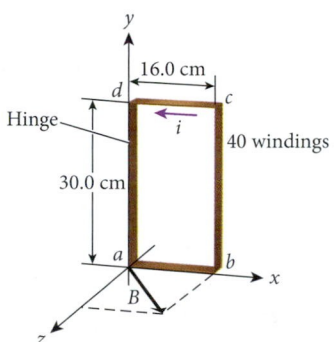

Section 27.7

27.56 A high electron mobility transistor (HEMT) controls large currents by applying a small voltage to a thin sheet of electrons. The density and mobility of the electrons in the sheet are critical for the operation of the HEMT. HEMTs consisting of AlGaN/GaN/Si are being studied because they promise better performance at higher currents, temperatures, and frequencies than conventional silicon HEMTs can achieve. In one study, the Hall effect was used to measure the density of electrons in one of these new HEMTs. When a current of 10.0 μA flows through the length of the electron sheet, which is 1.00 mm long, 0.300 mm wide, and 10.0 nm thick, a magnetic field of 1.00 T perpendicular to the sheet produces a voltage of 0.680 mV across the width of the sheet. What is the density of electrons in the sheet?

•27.57 The figure shows schematically a setup for a Hall effect measurement using a thin film of zinc oxide of thickness 1.50 μm. The current, i, across the thin film is 12.3 mA, and the Hall potential, V_H, is −20.1 mV when a magnetic field of magnitude $B = 0.900$ T is applied perpendicular to the current flow.

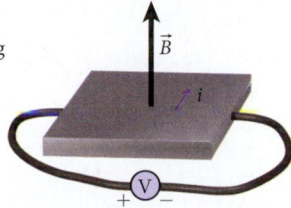

a) What are the charge carriers in the thin film? [*Hint:* They can be either electrons with charge $-e$ or electron holes (missing electrons) with charge $+e$.]

b) Calculate the density of charge carriers in the thin film.

Additional Exercises

27.58 A cyclotron in a magnetic field of 9.00 T is used to accelerate protons to 50.0% of the speed of light. What is the cyclotron frequency of

these protons? What is the radius of their trajectory in the cyclotron? What are the cyclotron frequency and the trajectory radius of the same protons in the Earth's magnetic field? Assume that the Earth's magnetic field is 0.500 G.

27.59 A straight wire carrying a current of 3.41 A is placed at an angle of 10.0° to the horizontal between the pole tips of a magnet producing a field of 0.220 T upward. The poles' tips each have a 10.0 cm diameter. The magnetic force causes the wire to move out of the space between the poles. What is the magnitude of that force?

27.60 An electron is moving at $v = 6.00 \cdot 10^7$ m/s perpendicular to the Earth's magnetic field. If the field strength is $0.500 \cdot 10^{-4}$ T, what is the radius of the electron's circular path?

27.61 A straight wire with a constant current running through it is in Earth's magnetic field, at a location where the magnitude is 0.430 G. What is the minimum current that must flow through the wire for a 10.0-cm length of it to experience a force of 1.00 N?

27.62 A circular coil with a radius of 10.0 cm has 100 turns of wire and carries a current, $i = 100.$ mA. It is free to rotate in a region with a constant horizontal magnetic field given by $\vec{B} = (0.0100$ T$)\hat{x}$. If the unit normal vector to the plane of the coil makes an angle of 30.0° with the horizontal, what is the magnitude of the net torque acting on the coil?

27.63 At $t = 0$ an electron crosses the positive y-axis (so $x = 0$) at 60.0 cm from the origin with velocity $2.00 \cdot 10^5$ m/s in the positive x-direction. It is in a uniform magnetic field.

a) Find the magnitude and the direction of the magnetic field that will cause the electron to cross the x-axis at $x = 60.0$ cm.

b) What work is done on the electron during this motion?

c) How long will the trip from y-axis to x-axis take?

•27.64 A 12.0-V battery is connected to a 3.00-Ω resistor in a rigid rectangular loop of wire measuring 3.00 m by 1.00 m. As shown in the figure, a length $\ell = 1.00$ m of wire at the end of the loop extends into a 2.00 m by 2.00 m region with a magnetic field of magnitude 5.00 T, directed into the page. What is the net force on the loop?

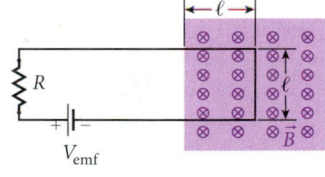

27.65 An alpha particle ($m = 6.64 \cdot 10^{-27}$ kg, $q = +2e$) is accelerated by a potential difference of 2700. V and moves in a plane perpendicular to a constant magnetic field of magnitude 0.340 T, which curves the trajectory of the alpha particle. Determine the radius of curvature and the period of revolution.

•27.66 In a certain area, the electric field near the surface of the Earth is given by $\vec{E} = (-150.$ N/C$)\hat{z}$, and the Earth's magnetic field is given by $\vec{B} = (50.0 \, \mu$T$)\hat{r}_N - (20.0 \, \mu$T$)\hat{z}$, where $\hat{z}$ is a unit vector pointing vertically upward and $\hat{r}_N$ is a horizontal unit vector pointing due north. What velocity, $\vec{v}$, will allow an electron in this region to move in a straight line at constant speed?

•27.67 A helium leak detector uses a mass spectrometer to detect tiny leaks in a vacuum chamber. The chamber is evacuated with a vacuum pump and then sprayed with helium gas on the outside. If there is any leak, the helium molecules pass through the leak and into the chamber, whose volume is sampled by the leak detector. In the spectrometer, helium ions are accelerated and released into a tube, where their motion is perpendicular to an applied magnetic field, $\vec{B}$, and they follow a circular path of radius r and then hit a detector. Estimate the velocity required if the radius of the ions' circular path is to be no more than 5.00 cm, the magnetic field is 0.150 T, and the mass of a helium-4 atom is about $6.64 \cdot 10^{-27}$ kg. Assume that each ion is singly ionized (has one electron less than the neutral atom). By what factor does the required velocity change if helium-3 atoms, which have about $\frac{3}{4}$ as much mass as helium-4 atoms, are used?

•27.68 In your laboratory, you set up an experiment with an electron gun that emits electrons with energy of 7.50 keV toward an atomic target. What deflection (magnitude and direction) will Earth's magnetic field (0.300 G) produce in the beam of electrons if the beam is initially directed due east and covers a distance of 1.00 m from the gun to the target? (*Hint:* First calculate the radius of curvature, and then determine how far away from a straight line the electron beam has deviated after 1.00 m.)

•**27.69** A proton enters the region between the two plates shown in the figure moving in the x-direction with a speed $v = 1.35 \cdot 10^6$ m/s. The potential of the top plate is 200. V, and the potential of the bottom plate is 0 V. What are the direction and the magnitude of the magnetic field, $\vec{B}$, that is required between the plates for the proton to continue traveling in a straight line in the x-direction?

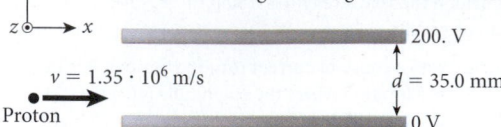

•**27.70** An electron moving at a constant velocity, $\vec{v} = v_0 \hat{x}$, enters a region in space where a magnetic field is present. The magnetic field, $\vec{B}$, is constant and points in the z-direction. What are the magnitude and the direction of the magnetic force acting on the electron? If the width of the region where the magnetic field is present is d, what is the minimum velocity the electron must have in order to escape this region?

•**27.71** A 30-turn square coil with a mass of 0.250 kg and a side length of 0.200 m is hinged along a horizontal side and carries a 5.00-A current. It is placed in a magnetic field pointing vertically downward and having a magnitude of 0.00500 T. Determine the angle that the plane of the coil makes with the vertical when the coil is in equilibrium. Use $g = 9.81$ m/s^2.

•**27.72** A semicircular loop of wire of radius R is in the xy-plane, centered about the origin. The wire carries a current, i, counterclockwise around the semicircle, from $x = -R$ to $x = +R$ on the x-axis. A magnetic field, $\vec{B}$, is pointing out of the plane, in the positive z-direction. Calculate the net force on the semicircular loop.

•**27.73** A proton moving at speed $v = 1.00 \cdot 10^6$ m/s enters a region in space where a magnetic field given by $\vec{B} = (-0.500 \text{ T})\hat{z}$ exists. The velocity vector of the proton is at an angle $\theta = 60.0°$ with respect to the positive z-axis.

a) Analyze the motion of the proton and describe its trajectory (in qualitative terms only).

b) Calculate the radius, r, of the trajectory projected onto a plane perpendicular to the magnetic field (in the xy-plane).

c) Calculate the period, T, and frequency, f, of the motion in that plane.

d) Calculate the pitch of the motion (the distance traveled by the proton in the direction of the magnetic field in 1 period).

MULTI-VERSION EXERCISES

27.74 A small aluminum ball with a mass of 5.063 g and a charge of 11.03 C is moving northward at 3079 m/s. You want the ball to travel in a horizontal circle with a radius of 2.137 m and in a clockwise direction when viewed from above. Ignoring gravity, what is the magnitude of the magnetic field that must be applied to the aluminum ball to cause it to move in this way?

27.75 A small aluminum ball with a charge of 11.17 C is moving northward at 3131 m/s. You want the ball to travel in a horizontal circle with a radius of 2.015 m and in a clockwise direction when viewed from above. Ignoring gravity, the magnitude of the magnetic field that must be applied to the aluminum ball to cause it to move in this way is $B = 0.8000$ T. What is the mass of the ball?

27.76 A small charged aluminum ball with a mass of 3.435 g is moving northward at 3183 m/s. You want the ball to travel in a horizontal circle with a radius of 1.893 m and in a clockwise direction when viewed from above. Ignoring gravity, the magnitude of the magnetic field that must be applied to the aluminum ball to cause it to move in this way is $B = 0.5107$ T. What is the charge on the ball?

27.77 A velocity selector is used in a mass spectrometer to produce a beam of charged particles with uniform velocity. Suppose the fields in a selector are given by $\vec{E} = (1.749 \cdot 10^4 \text{ V/m})\hat{x}$ and $\vec{B} = (46.23 \text{ mT})\hat{y}$. Find the speed with which a charged particle can travel through the selector in the z-direction without being deflected.

27.78 A velocity selector is used in a mass spectrometer to produce a beam of charged particles with uniform velocity. Suppose the fields in a selector are given by $\vec{E} = (2.207 \cdot 10^4 \text{ V/m})\hat{x}$ and $\vec{B} = B_y \hat{y}$. The speed with which a charged particle can travel through the selector in the z-direction without being deflected is $4.713 \cdot 10^5$ m/s. What is the value of B_y?

27.79 A velocity selector is used in a mass spectrometer to produce a beam of charged particles of uniform velocity. Suppose the fields in a selector are given by $\vec{E} = E_x \hat{x}$ and $\vec{B} = (47.45 \text{ mT})\hat{y}$. The speed with which a charged particle can travel through the selector in the z-direction without being deflected is $5.616 \cdot 10^5$ m/s. What is the value of E_x?

28

Magnetic Fields of Moving Charges

FIGURE 28.1 This array of 96 iron atoms holds 1 byte (8 bits) of information in the magnetic fields of the atoms—an impressive feat of data storage.

We saw in Chapter 27 that a magnetic field can affect the path of a charged particle or the flow of a current. In this chapter, we consider magnetic fields *caused* by electric currents. Any charged particle generates a magnetic field, if it is moving. Various magnetic fields are caused by different distributions of current. The powerful electromagnets used in industry, in physics research, in medical diagnostics, and in other applications are primarily solenoids—coils of current-carrying wire with hundreds or thousands of loops. We'll see in this chapter why solenoids generate particularly useful magnetic fields. We will also see that an individual atom can have a permanent magnetic field. Figure 28.1 shows how the magnetic fields of iron atoms (shown in red and blue) can be used to store information. This technology is still in its infancy (the image was produced in 2012). However, if it can be scaled up, it will allow 10 terabytes of data to be stored on a square centimeter of iron surface, a density that is higher than what is currently achievable by a factor of more than 1000.

WHAT WE WILL LEARN

- Moving charges (currents) create magnetic fields.
- The magnetic field created by a current flowing in a long, straight wire varies inversely with the distance from the wire.
- Two parallel wires carrying current in the same direction attract each other. Two parallel wires carrying current in opposite directions repel each other.
- Ampere's Law is used to calculate the magnetic field caused by certain symmetrical current distributions, just as Gauss's Law is useful in calculating electric fields in situations having spatial charge symmetry.

- The magnetic field inside a long, straight wire varies linearly with the distance from the center of the wire.
- A solenoid is an electromagnet that can be used to produce a constant magnetic field with a large volume.
- Some atoms can be thought of as small magnets created by motion of electrons in the atom.
- Materials can exhibit three kinds of intrinsic magnetism: diamagnetism, paramagnetism, and ferromagnetism.
- Superconducting magnets can be used to produce very strong magnetic fields.

28.1 Biot-Savart Law

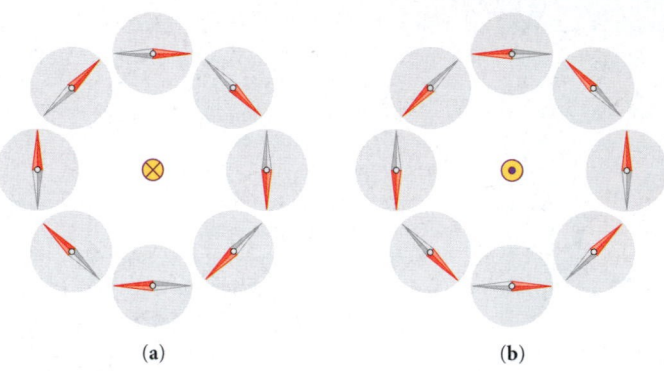

(a)   (b)

FIGURE 28.2 Wire (yellow circle) with current running through it: (a) into the page (indicated by the cross); (b) out of the page (indicated by the dot). The orientation of a compass needle placed close to the wire is shown at different locations around the wire.

Chapter 27 introduced magnetic fields and field lines by showing how a compass needle orients itself near a permanent magnet. In a similar demonstration, a strong current is run through a (very long, straight) wire (with or without insulation). If a compass needle is then brought close to the wire, the compass needle orients itself relative to the wire in the way shown in Figure 28.2. This observation was first made by the Danish physicist Hans Oersted (1777–1851) in 1819 while conducting a demonstration for students during a lecture. We conclude that the current in the wire produces a magnetic field. Since the direction of the compass needle indicates the direction of the magnetic field, we further conclude that the magnetic field lines form circles around this current-carrying wire. Note the difference between parts (a) and (b) of Figure 28.2: When the direction of the current is reversed, the orientation of the compass needle is also reversed. Although the figure doesn't show this, if the compass needle is moved farther and farther away from the wire, eventually it again orients itself in the direction of the magnetic field of the Earth. This indicates that the magnetic field produced by the wire gets weaker as a function of increasing distance from the wire. The next question we need to answer is this: How can we determine the magnetic field produced by a moving charge?

To describe the electric field in terms of the electric charge, we showed that (see Chapter 22):

$$dE = \frac{1}{4\pi\epsilon_0} \frac{|dq|}{r^2},$$

where dq is a charge element. The electric field points in a radial direction (inward toward or outward from the electric charge, depending on the sign of the charge), so

$$d\vec{E} = \frac{1}{4\pi\epsilon_0} \frac{dq}{r^3} \vec{r} = \frac{1}{4\pi\epsilon_0} \frac{dq}{r^2} \hat{r}.$$

The situation is slightly more complicated for a magnetic field because a current element, $i\,d\vec{s}$, that produces a magnetic field has a direction, as opposed to a nondirectional point charge that produces an electric field. As a result of a long series of experiments involving tests similar to that depicted in Figure 28.2 and conducted in the early 19th century, the French scientists Jean-Baptiste Biot (1774–1862) and Felix Savart (1791–1841) established that the magnetic field produced by a current element, $i\,d\vec{s}$, is given by

$$d\vec{B} = \frac{\mu_0}{4\pi} \frac{i\,d\vec{s} \times \vec{r}}{r^3} = \frac{\mu_0}{4\pi} \frac{i\,d\vec{s} \times \hat{r}}{r^2}. \tag{28.1}$$

Here $d\vec{s}$ is a vector of differential length ds pointing in the direction in which the current flows along the conductor and $\vec{r}$ is the position vector measured *from* the current element *to* the point at which the field is to be determined. Figure 28.3 depicts the physical situation described by this formula, which is called the **Biot-Savart Law.**

The constant μ_0 in equation 28.1 is called the **magnetic permeability of free space** and has the value

$$\mu_0 = 4\pi \cdot 10^{-7} \frac{\text{T m}}{\text{A}}. \tag{28.2}$$

From equation 28.1 and Figure 28.3, you can see that the direction of the magnetic field produced by the current element $i\,d\vec{s}$ is perpendicular to both the position vector and the current element. The magnitude of the magnetic field is given by

$$dB = \frac{\mu_0}{4\pi} \frac{i\,ds\sin\theta}{r^2}, \tag{28.3}$$

where θ (with possible values between 0° and 180°) is the angle between the direction of the position vector and the current element. The direction of the magnetic field is given by a variant of right-hand rule 1, introduced in Chapter 27. To determine the direction of the magnetic field using your right hand, point your thumb in the direction of the differential current element and your index finger in the direction of the position vector, and your middle finger will point in the direction of the differential magnetic field.

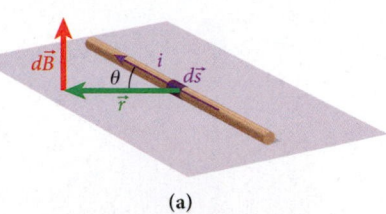

(a)

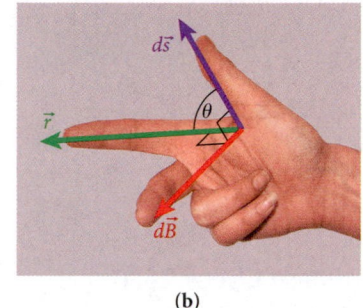

(b)

FIGURE 28.3 (a) Three-dimensional depiction of the Biot-Savart Law. The differential magnetic field is perpendicular to both the differential current element and the position vector. (b) Right-hand rule 1 applied to the quantities involved in the Biot-Savart Law.

28.2 Magnetic Fields due to Current Distributions

Chapter 27 addressed the superposition principle for magnetic fields. Using this super-position principle, we can compute the magnetic field at any point in space as the sum of the differential magnetic fields described by the Biot-Savart Law. This section examines the magnetic fields generated by the most common configurations of current-carrying wires.

Magnetic Field from a Long, Straight Wire

Let's first examine the magnetic field from an infinitely long, straight wire carrying a current, i. We consider the magnetic field, $d\vec{B}$, at a point P that is a perpendicular distance $r_\perp$ from the wire (Figure 28.4). The magnitude of the field, dB, at that point due to the current element $i\,ds$ is given by equation 28.3; the direction of the field is given by $d\vec{s} \times \vec{r}$ and is out of the page. We find the magnetic field from the right half of the wire and multiply by 2 to get the magnetic field from the whole wire. Thus, the magnitude of the magnetic field at a perpendicular distance $r_\perp$ from the wire is given by

FIGURE 28.4 Magnetic field from a long, straight current-carrying wire.

$$B = 2\int_0^\infty dB = 2\int_0^\infty \frac{\mu_0}{4\pi} \frac{i\,ds\sin\theta}{r^2} = \frac{\mu_0 i}{2\pi} \int_0^\infty \frac{ds\sin\theta}{r^2}.$$

We can relate r and θ to $r_\perp$ and s (r, s, and θ) by $r = \sqrt{s^2 + r_\perp^2}$ and $\sin\theta = \sin(\pi - \theta) = r_\perp / \sqrt{s^2 + r_\perp^2}$ (see Figure 28.4). Substituting for r and $\sin\theta$ in the preceding expression for B gives

$$B = \frac{\mu_0 i}{2\pi} \int_0^\infty \frac{r_\perp \, ds}{(s^2 + r_\perp^2)^{3/2}}.$$

Evaluating this definite integral, we find

$$B = \frac{\mu_0 i}{2\pi} \left[\frac{1}{r_\perp^2} \frac{r_\perp s}{(s^2 + r_\perp^2)^{1/2}} \right]_0^\infty = \frac{\mu_0 i}{2\pi r_\perp} \left[\frac{s}{(s^2 + r_\perp^2)^{1/2}} \right]_{s\to\infty} - 0.$$

FIGURE 28.5 Right-hand rule 3 for the magnetic field from a current-carrying wire.

For $s \gg r_\perp$, the term in the brackets approaches the value 1. Therefore, the magnitude of the magnetic field at a perpendicular distance $r_\perp$ from a long, straight wire carrying a current, i, is

$$B = \frac{\mu_0 i}{2\pi r_\perp}. \tag{28.4}$$

The direction of the magnetic field at any point is found by applying right-hand rule 1 to the current element and position vectors shown in Figure 28.4. This results in a new right-hand rule, called *right-hand rule 3*, which can be used to determine the direction of the magnetic field due to a current-carrying wire. If you grab the wire with your right hand so that your thumb points in the direction of the current, your fingers will curl in the direction of the magnetic field (Figure 28.5).

Looking along a current-carrying wire would reveal that the magnetic field lines form concentric circles (Figure 28.6). Notice from the distance between the field lines that the field is strongest near the wire and drops off in proportion to $1/r_\perp$, as indicated by equation 28.4.

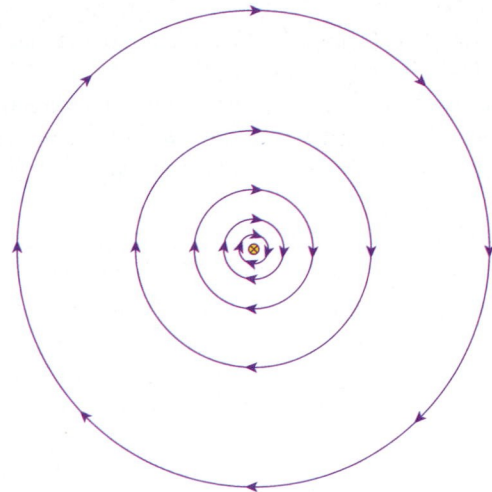

FIGURE 28.6 Magnetic field lines around a long, straight wire (yellow circle at the center) that is perpendicular to the page and carries a current flowing into the page, signified by the cross.

Two Parallel Wires

Let's examine the case in which two parallel wires are carrying current. The two wires exert magnetic forces on each other because the magnetic field of one wire exerts a force on the moving charges in the second wire. The magnitude of the magnetic field created by a current-carrying wire is given by equation 28.4. This magnetic field is always perpendicular to the wire with a direction given by right-hand rule 3 (Figure 28.5).

Let's first consider wire 1 carrying a current, i_1, toward the right, as shown in Figure 28.7a. The magnitude of the magnetic field a perpendicular distance d from wire 1 is

$$B_1 = \frac{\mu_0 i_1}{2\pi d}. \tag{28.5}$$

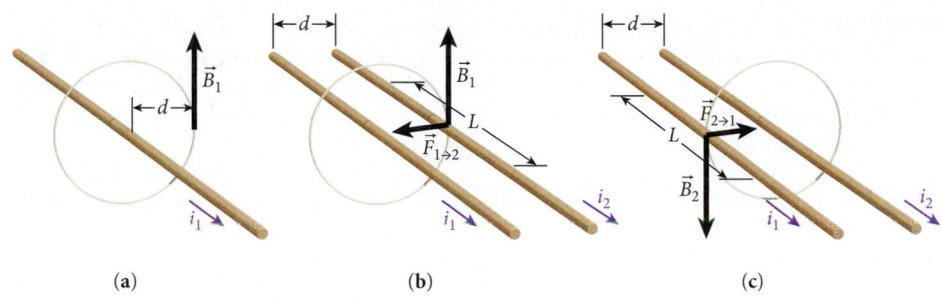

(a)   (b)   (c)

The direction of $\vec{B}_1$ is given by right-hand rule 3 and is shown for a particular point in Figure 28.7a.

Now consider wire 2 carrying a current, i_2, in the same direction as i_1 and placed parallel to wire 1 at a distance d from it (Figure 28.7b). The magnetic field due to wire 1 exerts a magnetic force on the moving charges in the current flowing in wire 2. In Chapter 27, we saw that the magnetic force on a current-carrying wire is given by

$$\vec{F} = i\vec{L} \times \vec{B}.$$

The magnitude of the magnetic force on a length, L, of wire 2 is then

$$F = iLB\sin\theta = i_2 LB_1, \tag{28.6}$$

because $\vec{B}_1$ is perpendicular to wire 2 and thus $\theta = 90°$. Substituting for B_1 from equation 28.5 into equation 28.6, we find the magnitude of the force exerted by wire 1 on a length L of wire 2:

$$F_{1\to 2} = i_2 L\left(\frac{\mu_0 i_1}{2\pi d}\right) = \frac{\mu_0 i_1 i_2 L}{2\pi d}. \tag{28.7}$$

According to right-hand rule 1, $\vec{F}_{1\to 2}$ points toward wire 1 and is perpendicular to both wires. An analogous calculation allows us to deduce that the force from wire 2 on a length, L, of wire 1 has the same magnitude and opposite direction: $\vec{F}_{2\to 1} = -\vec{F}_{1\to 2}$. This result is shown in Figure 28.7c and is a simple consequence of Newton's Third Law.

Concept Check 28.4

Two parallel wires are near each other, as shown in the figure. Wire 1 carries a current i, and wire 2 carries a current $2i$. Which statement about the magnetic forces that the two wires exert on each other is correct?

a) The two wires exert no forces on each other.

b) The two wires exert attractive forces of the same magnitude on each other.

c) The two wires exert repulsive forces of the same magnitude on each other.

d) Wire 1 exerts a stronger force on wire 2 than wire 2 exerts on wire 1.

e) Wire 2 exerts a stronger force on wire 1 than wire 1 exerts on wire 2.

Wire 1   Wire 2

SOLVED PROBLEM 28.1 / Magnetic Field from Four Wires

Four wires are each carrying a current of magnitude $i = 1.00$ A. The wires are located at the four corners of a square with side $a = 3.70$ cm. Two of the wires are carrying current into the page, and the other two are carrying current out of the page (Figure 28.8).

PROBLEM
What is the y-component of the magnetic field at the center of the square?

– Continued

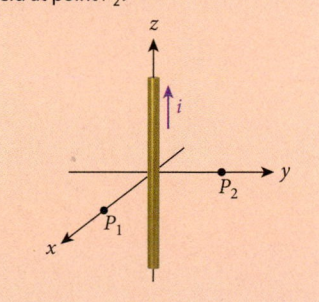

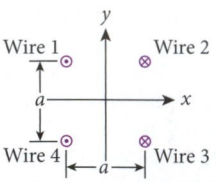

FIGURE 28.8 Four wires located at the corners of a square. Two of the wires are carrying current into the page, and the other two are carrying current out of the page.

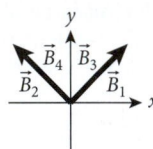

FIGURE 28.9 The magnetic fields from the four current-carrying wires.

SOLUTION

THINK The magnetic field at the center of the square is the vector sum of the magnetic fields from the four current-carrying wires. The magnitude of the magnetic field due to each of the four wires is the same. The direction of the magnetic field from each wire is determined using right-hand rule 3.

SKETCH Figure 28.9 shows the magnetic fields from the four wires: $\vec{B}_1$ is the magnetic field from wire 1, $\vec{B}_2$ is the magnetic field from wire 2, $\vec{B}_3$ is the magnetic field from wire 3, and $\vec{B}_4$ is the magnetic field from wire 4. Note that $\vec{B}_2$ and $\vec{B}_4$ are equal and $\vec{B}_1$ and $\vec{B}_3$ are equal.

RESEARCH The magnitude of the magnetic field from each of the four wires is given by

$$B = \frac{\mu_0 i}{2\pi r} = \frac{\mu_0 i}{2\pi \left(a/\sqrt{2}\right)},$$

where $a/\sqrt{2}$ is the distance from each wire to the center of the square.

Right-hand rule 3 gives us the directions of the magnetic fields, which are shown in Figure 28.9. The y-component of each of the magnetic fields is given by

$$B_y = B \sin 45°.$$

SIMPLIFY The sum of the y-components of the four magnetic fields is

$$B_{y,\text{sum}} = 4B_y = 4B\sin 45° = 4\frac{\mu_0 i}{2\pi \left(a/\sqrt{2}\right)}\left(\frac{1}{\sqrt{2}}\right) = \frac{2\mu_0 i}{\pi a},$$

where we have used $\sin 45° = 1/\sqrt{2}$.

CALCULATE Putting in the numerical values gives us

$$B_{y,\text{sum}} = \frac{2\mu_0 i}{\pi a} = \frac{2\left(4\pi \cdot 10^{-7}\ \text{T m/A}\right)(1.00\ \text{A})}{\pi\left(3.70 \cdot 10^{-2}\ \text{m}\right)} = 2.16216 \cdot 10^{-5}\ \text{T}.$$

ROUND We report our result to three significant figures:

$$B_{y,\text{sum}} = 2.16 \cdot 10^{-5}\ \text{T}.$$

DOUBLE-CHECK To double-check our result, we calculate the magnitude of the magnetic field at the center of the square due to one of the wires:

$$B = \frac{\mu_0 i}{2\pi \left(a/\sqrt{2}\right)} = \frac{\left(4\pi \cdot 10^{-7}\ \text{T m/A}\right)(1.00\ \text{A})\sqrt{2}}{2\pi\left(3.70 \cdot 10^{-2}\ \text{m}\right)} = 7.64 \cdot 10^{-6}\ \text{T}.$$

The sum of the y-components is then

$$B_{y,\text{sum}} = \frac{4\left(7.64 \cdot 10^{-6}\ \text{T}\right)}{\sqrt{2}} = 2.16 \cdot 10^{-5}\ \text{T},$$

which agrees with our result.

Definition of the Ampere

The force $F_{1\to 2}$ described by equation 28.7 is used in the SI definition of the ampere: An *ampere* (A) is the constant current that, if maintained in two straight, parallel conductors of infinite length and negligible circular cross section, placed 1 m apart in vacuum, would produce between these conductors a force of $2 \cdot 10^{-7}$ N per meter of length. This physical situation is described by equation 28.7 for the force between two parallel current-carrying wires with $i_1 = i_2 = $ exactly 1 A, $d = $ exactly 1 m, and $F_{1\to 2} = $ exactly $2 \cdot 10^{-7}$ N. We can solve equation 28.7 for μ_0:

$$\mu_0 = \frac{\left(2\pi d\right)F_{1\to 2}}{i_1 i_2 L} = \frac{2\pi\left(1\ \text{m}\right)\left(2 \cdot 10^{-7}\ \text{N}\right)}{\left(1\ \text{A}\right)\left(1\ \text{A}\right)\left(1\ \text{m}\right)} = \text{exactly}\ 4\pi \cdot 10^{-7}\ \frac{\text{T m}}{\text{A}},$$

which indicates that the magnetic permeability of free space is *defined* to be exactly $\mu_0 = 4\pi \cdot 10^{-7}$ T m/A (see equation 28.2).

In Chapter 21, when Coulomb's Law was introduced, the value of the electric permittivity of free space, ϵ_0, was given: $\epsilon_0 = 8.85 \cdot 10^{-12}$ C²/(N m²). Since 1 A = 1 C/s and 1 T = 1 (N s)/(C m) (see Chapter 27), the product of the two constants, ϵ_0 and μ_0, is

$$\mu_0\epsilon_0 = \left(4\pi \cdot 10^{-7}\ \frac{\text{T m}}{\text{A}}\right)\left(8.85 \cdot 10^{-12}\ \frac{\text{C}^2}{\text{N m}^2}\right) = 1.11 \cdot 10^{-17}\ \frac{\text{s}^2}{\text{m}^2},$$

which has the units of the inverse of the square of a speed. Thus, $1/\sqrt{\mu_0\epsilon_0}$ gives the value of this speed as that of the speed of light, $c = 3.00 \cdot 10^8$ m/s. This is by no means a coincidence, as we'll see in later chapters. For now, it is sufficient to state the empirical finding:

$$c = \frac{1}{\sqrt{\mu_0\epsilon_0}}.$$

Since the magnetic permeability of free space is defined to be exactly $\mu_0 = 4\pi \cdot 10^{-7}$ T m/A and the speed of light is defined as exactly $c = 299{,}792{,}458$ m/s (see the discussion in Chapter 1), the expression $c = 1/\sqrt{\mu_0\epsilon_0}$ also fixes the value of the electric permittivity of free space.

EXAMPLE 28.1 | Force on a Loop

A long, straight wire is carrying a current of magnitude $i_1 = 5.00$ A toward the right (Figure 28.10). A square loop with sides of length $a = 0.250$ m is placed with its sides parallel and perpendicular to the wire at a distance $d = 0.100$ m from the wire. The square loop carries a current of magnitude $i_2 = 2.20$ A in the counterclockwise direction.

PROBLEM
What is the net magnetic force on the square loop?

SOLUTION
The force on the square loop is due to the magnetic field created by the current flowing in the straight wire. Right-hand rule 3 tells us that the magnetic field from the current flowing in the wire is directed into the page in the region where the loop is located (see Figure 28.10). The right-hand rule and equation 28.4 tell us that the resulting force on the left side of the loop is toward the right, and the force on the right side of the loop is toward the left. In Figure 28.10, these two forces are represented by green arrows. The two forces are equal in magnitude and opposite in direction, so they sum to zero. The force on the top side of the loop is downward (red arrow in Figure 28.10, pointing in the negative y-direction), and its magnitude is given by equation 28.7:

$$F_{\text{down}} = \frac{\mu_0 i_1 i_2 a}{2\pi d},$$

where a is the length of the top side of the loop. The force on the bottom side of the loop is upward (the other red arrow in Figure 28.10, pointing in the positive y-direction), and its magnitude is given by

$$F_{\text{up}} = \frac{\mu_0 i_1 i_2 a}{2\pi(d+a)}.$$

Thus, we can express the net magnetic force on the loop as

$$\vec{F} = (F_{\text{up}} - F_{\text{down}})\hat{y}.$$

Putting in the numbers results in

$$\vec{F} = \frac{\left(4\pi \cdot 10^{-7}\ \text{T m/A}\right)(5.00\ \text{A})(2.20\ \text{A})(0.250\ \text{m})}{2\pi}\left(\frac{1}{0.350\ \text{m}} - \frac{1}{0.100\ \text{m}}\right)\hat{y} = (-3.93 \cdot 10^{-6})\,\hat{y}\ \text{N}.$$

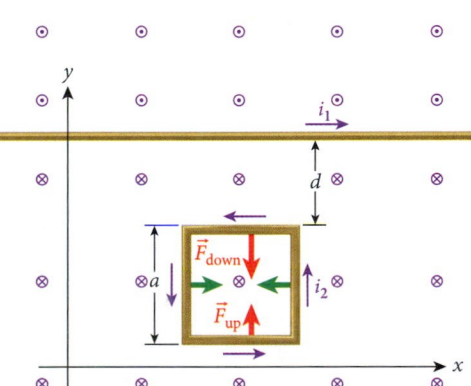

FIGURE 28.10 A current-carrying wire and a square loop.

Concept Check 28.5

A wire is carrying a current, i, in the positive y-direction, as shown in the figure. The wire is located in a uniform magnetic field, $\vec{B}$, oriented in such a way that the magnetic force on the wire is maximized. The magnetic force acting on the wire, $\vec{F}_B$, is in the negative x-direction. What is the direction of the magnetic field?

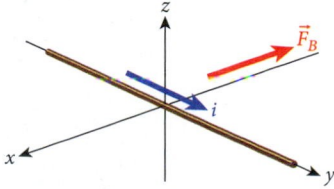

a) the positive x-direction

b) the negative x-direction

c) the negative y-direction

d) the positive z-direction

e) the negative z-direction

SOLVED PROBLEM 28.2 | Electromagnetic Rail Accelerator

Electromagnetic rail accelerators are being developed for accelerating fuel pellets in fusion experiments and for launching spacecraft into orbit. The U.S. Navy is experimenting with electromagnetic rail accelerators that can launch projectiles at very high speeds, such as the

– Continued

FIGURE 28.11   U.S. Navy rail gun.

rail gun shown in Figure 28.11. In this rail gun, currents are run through two parallel conducting rails that are connected by a movable conductor oriented perpendicular to the rails. The projectile is attached to the movable conductor. For this example, we'll assume that the rail gun consists of two parallel rails of cross-sectional radius $r = 5.00$ cm, whose centers are separated by a distance $d = 25.0$ cm and which have a length $L = 5.00$ m, and that the rail gun accelerates the projectile to a kinetic energy of $K = 32.0$ MJ. The projectile also functions as the movable conductor.

PROBLEM

How much current is required to accelerate the projectile?

SOLUTION

THINK   The currents in the two rails are in opposite directions. The current flowing through the movable conductor is perpendicular to the two currents in the rails. The magnetic fields from the two rails are in the same direction and exert forces on the movable conductor in the same direction. The force from the magnetic field of each rail depends on the distance from the rail. Thus, we must integrate the force along the distance between the two rails to obtain the total force. The total force on the movable conductor is twice the force from the magnetic field from one rail. The kinetic energy gained by the projectile is the total force exerted by the magnetic fields of the two rails times the distance over which the force acts.

SKETCH   Figure 28.12 shows top and cross-sectional views of the rails and the movable conductor.

RESEARCH   The current-carrying movable conductor, which completes the circuit between the two rails, is also the projectile and is accelerated by the magnetic forces produced by the two rails. The force exerted on the projectile depends on the distance, x, from the center of a rail, as illustrated in Figure 28.12b. Thus, to calculate the total force on the projectile, we must integrate over the length of the projectile. We use equation 28.4 to find the magnitude of the magnetic field, B_1, from current i flowing in rail 1 at a distance x from the center of the rail:

$$B_1 = \frac{\mu_0 i}{2\pi x}.$$

According to equation 28.6, the magnitude of the differential force, dF_1, exerted on a differential length, dx, of the projectile by the magnetic field from rail 1 is

$$dF_1 = i(dx)B_1 = i(dx)\left(\frac{\mu_0 i}{2\pi x}\right).$$

The direction of the force is given by right-hand rule 3, which tells us that the force is upward in the plane of the page in Figure 28.12a and into the page in Figure 28.12b. The magnitude of the force on the projectile is given by integrating dF_1 over the length of the projectile:

$$F_1 = \int_r^{d-r} dF_1 = \int_r^{d-r} \frac{\mu_0 i^2}{2\pi} \frac{dx}{x} = \frac{\mu_0 i^2}{2\pi} [\ln x]_r^{d-r} = \frac{\mu_0 i^2}{2\pi}\left(\ln(d-r) - \ln r\right) = \frac{\mu_0 i^2}{2\pi}\ln\left(\frac{d-r}{r}\right). \tag{i}$$

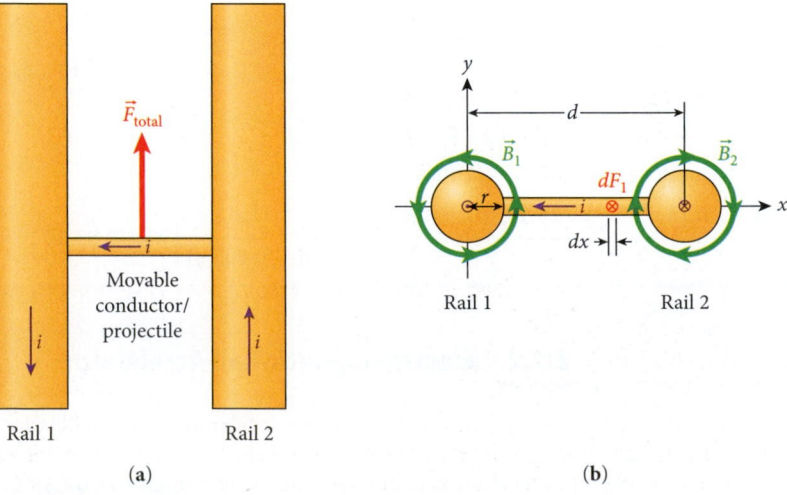

(a)   (b)

FIGURE 28.12   Schematic diagram of a rail gun: (a) top view; (b) cross-sectional view.

Because the magnetic field from rail 2 is in the same direction as the magnetic field from rail 1, the force exerted by the magnetic field from rail 2 on the projectile is the same as that from rail 1. Thus, the magnitude of the total force exerted on the projectile is

$$F_{\text{total}} = 2F_1. \tag{ii}$$

The kinetic energy gained by the projectile is equal (by the work-energy theorem introduced in Chapter 6) to the magnitude of the force exerted times the distance over which the force acts

$$K = F_{\text{total}} L. \tag{iii}$$

SIMPLIFY We can combine equations (i), (ii), and (iii) to obtain

$$K = 2F_1 L = 2\left[\frac{\mu_0 i^2}{2\pi} \ln\left(\frac{d-r}{r}\right)\right] L = \frac{\mu_0 L i^2}{\pi} \ln\left(\frac{d-r}{r}\right).$$

Solving this equation for the current gives us

$$i = \sqrt{\frac{K\pi}{\mu_0 L \ln\left(\dfrac{d-r}{r}\right)}}.$$

CALCULATE Putting in the numerical values, we get

$$i = \sqrt{\frac{K\pi}{\mu_0 L \ln\left(\dfrac{d-r}{r}\right)}} = \sqrt{\frac{\left(32.0 \cdot 10^6 \text{ J}\right)\pi}{\left(4\pi \cdot 10^{-7} \dfrac{\text{T m}}{\text{A}}\right)(5.00 \text{ m}) \ln\left(\dfrac{25.0 \text{ cm} - 5.00 \text{ cm}}{5.00 \text{ cm}}\right)}} = 3.397287 \cdot 10^6 \text{ A}.$$

ROUND We report our result to three significant figures:

$$i = 3.40 \cdot 10^6 \text{ A} = 3.40 \text{ MA}.$$

DOUBLE-CHECK Let's double-check our result by looking at the force exerted on the projectile. Putting our result for the current into equation (i) for the magnitude of the force exerted by the magnetic field of rail 1 gives

$$F_1 = \frac{\mu_0 i^2}{2\pi} \ln\left(\frac{d-r}{r}\right) = \frac{\left(4\pi \cdot 10^{-7} \dfrac{\text{T m}}{\text{A}}\right)\left(3.40 \cdot 10^6 \text{ A}\right)^2}{2\pi} \ln\left(\frac{25.0 \text{ cm} - 5.00 \text{ cm}}{5.00 \text{ cm}}\right) = 3.20 \cdot 10^6 \text{ N}.$$

We can approximate the magnitude of this force by assuming that the force is constant along the conductor and is equal to the value at $x = d/2$:

$$F_1 = i(d-2r)B = i\left(\frac{\mu_0 i(d-2r)}{2\pi\left(\dfrac{d}{2}\right)}\right) = \frac{\mu_0 i^2}{\pi} \frac{d-2r}{d}$$

$$= \frac{\left(4\pi \cdot 10^{-7} \dfrac{\text{T m}}{\text{A}}\right)\left(3.40 \cdot 10^6 \text{ A}\right)^2}{\pi} \frac{[25.0 \text{ cm} - 2(5.00 \text{ cm})]}{25.0 \text{ cm}} = 2.77 \cdot 10^6 \text{ N}.$$

This value is within a factor of 2 of our calculated value for the force, which seems reasonable. However, just to be sure, let's calculate the kinetic energy of the projectile using the calculated value of the force:

$$K = F_{\text{total}} L = 2F_1 L = 2\left(3.20 \cdot 10^6 \text{ N}\right)(5.00 \text{ m}) = 32.0 \cdot 10^6 \text{ J}.$$

This result agrees with the 32.0 MJ assumed in the problem situation.

Note that if a projectile of mass $m = 5.00$ kg were given a kinetic energy of 32.0 MJ by this rail gun, its speed would be

$$v = \sqrt{\frac{2K}{m}} = \sqrt{\frac{2\left(32.0 \cdot 10^6 \text{ J}\right)}{5.00 \text{ kg}}} = 3.58 \text{ km/s}.$$

The rail gun would be capable of launching a projectile at 10 times the speed of sound, which is much larger than the typical speed of a bullet, about 3 times the speed of sound.

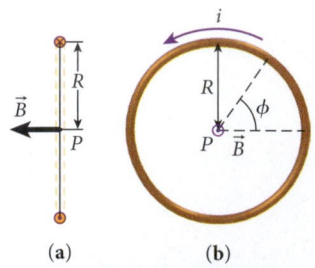

(a)   **(b)**

FIGURE 28.13 A circular loop of radius R carrying a current, i: (a) side view; (b) front view. The cross on the upper yellow circle in part (a) signifies that the current at the top of the loop is into the page, and the dot on the lower yellow circle in part (a) signifies that the current at the bottom of the loop is out of the page. Point P is located at the center of the loop.

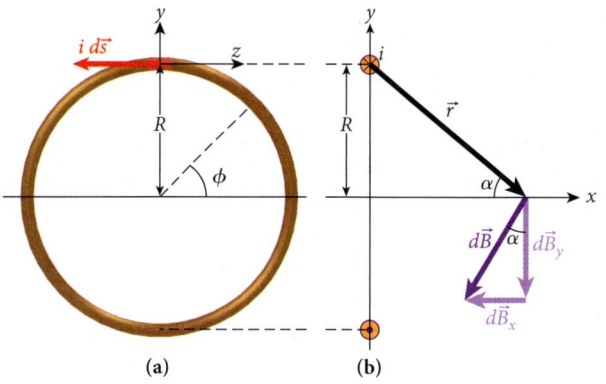

(a)   **(b)**

FIGURE 28.14 Geometry for calculating the magnetic field along the axis of a current-carrying loop: (a) front view, (b) side view.

Magnetic Field due to a Wire Loop

Now let's find the magnetic field at the center of a circular current-carrying loop of wire. Figure 28.13a shows a cross section of a circular loop with radius R carrying a current, i. Applying equation 28.3, $dB = \mu_0 i\, ds \sin\theta/(4\pi r^2)$, to this case, we can see that $r = R$ and $\theta = 90°$ for every current element $i\, ds$ along the loop. For the magnitude of the magnetic field at the center of the loop from each current element, we get

$$dB = \frac{\mu_0}{4\pi}\frac{i\, ds \sin 90°}{R^2} = \frac{\mu_0}{4\pi}\frac{i\, ds}{R^2}.$$

Going around the loop in Figure 28.13b, we can relate the angle ϕ to the current element by $ds = R\, d\phi$, allowing us to calculate the magnitude of the magnetic field at the center of the loop:

$$B = \int dB = \int_0^{2\pi} \frac{\mu_0}{4\pi}\frac{iR\, d\phi}{R^2} = \frac{\mu_0 i}{2R}. \qquad (28.8)$$

Keep in mind that equation 28.8 only gives the magnitude of the magnetic field at the center of the loop, where the magnitude is $B(r = 0) = \frac{1}{2}\mu_0 i/R$. To determine the direction of the magnetic field, we again use a variant of right-hand rule 1. Using your right hand, point your thumb in the direction of the current element (into the page for the upper circle in Figure 28.13a marked with a cross), and your index finger in the direction of the radial vector from the current element (down); your middle finger then points to the left. Using right-hand rule 3 (Figure 28.5), we also find that the current shown in Figure 28.13 produces a magnetic field $\vec{B}$ directed toward the left.

Now let's find the magnetic field from the loop along the axis of the loop rather than at the center (Figure 28.14). We set up a coordinate system such that the axis of the loop lies along the x-axis and the center of the loop is located at $x = 0$, $y = 0$, and $z = 0$. The radial vector $\vec{r}$ is the displacement to any point along the x-axis from a current element, $i\, d\vec{s}$, along the loop. The current element shown in Figure 28.14 lies in the negative z-direction. The radial vector $\vec{r}$ lies in the xy-plane and so is perpendicular to the current element. This situation is the same for any current element around the loop. Therefore, we can employ equation 28.3 with $\theta = 90°$ and obtain an expression for the magnitude of the differential magnetic field at any point along the x-axis:

$$dB = \frac{\mu_0}{4\pi}\frac{i\, ds \sin 90°}{r^2} = \frac{\mu_0}{4\pi}\frac{i\, ds}{r^2}.$$

A variant of right-hand rule 1 gives the direction of the differential magnetic field: Using your right hand, point your thumb in the direction of the differential current element (negative z-direction) and your index finger in the direction of the radial vector (positive x-direction and negative y-direction); the direction of the differential magnetic field will be given by your middle finger (negative x-direction and negative y-direction). The differential magnetic field is shown in Figure 28.14. To obtain the complete magnetic field, we need to integrate over the differential current element. From the symmetry of the situation, we can see that the y-component of the differential magnetic field, dB_y, will integrate to zero. The x-component of the differential magnetic field, dB_x, is given by

$$dB_x = dB \sin\alpha = \frac{\mu_0}{4\pi}\frac{i\, ds}{r^2}\sin\alpha,$$

where α is the angle between $\vec{r}$ and the x-axis (see Figure 28.14b). We can express the magnitude of $\vec{r}$ in terms of x and R as $r = \sqrt{x^2 + R^2}$ and express $\sin\alpha$ in terms of x and the radius of the loop R as $\sin\alpha = R/\sqrt{x^2 + R^2}$. We can then rewrite the expression for the x-component of the differential magnetic field as

$$dB_x = \frac{\mu_0}{4\pi}\frac{i\, ds}{x^2 + R^2}\frac{R}{\sqrt{x^2 + R^2}} = \frac{\mu_0 i\, ds}{4\pi}\frac{R}{\left(x^2 + R^2\right)^{3/2}}.$$

This expression for B_x is independent of the location of the current element, and so the integral to find the magnitude of the total field can be simplified to

$$B_x = \int dB_x = \frac{\mu_0 iR}{4\pi \left(x^2 + R^2 \right)^{3/2}} \int ds.$$

Going around the loop, we can relate the angle ϕ to the current element by $ds = R\, d\phi$ (see Figure 28.14a), allowing us to calculate the magnetic field along the axis of the loop:

$$B_x = \frac{\mu_0 iR}{4\pi \left(x^2 + R^2 \right)^{3/2}} \int_0^{2\pi} R\, d\phi = \frac{\mu_0 i 2\pi}{4\pi} \frac{R^2}{\left(x^2 + R^2 \right)^{3/2}},$$

or

$$B_x = \frac{\mu_0 i}{2} \frac{R^2}{\left(x^2 + R^2 \right)^{3/2}}. \qquad (28.9)$$

From our earlier application of the variant of right-hand rule 1, we know that the magnetic field along the axis of the loop is in the negative x-direction, as shown in Figure 28.14. We can also apply right-hand rule 3 to obtain the direction of the magnetic field: At any point on the loop, point your thumb tangent to the loop in the direction of the current and your fingers will curl in a direction that indicates that the field inside the loop is in the negative x-direction.

Using more advanced techniques and with the aid of a computer, we can determine the magnetic field produced by a current-carrying loop at other points in space. The magnetic field lines from a wire loop are shown in Figure 28.15. The value for the magnetic field given by equation 28.8 is valid only at the center point of Figure 28.15. The value for the magnetic field given by equation 28.9 is valid only along the axis of the loop.

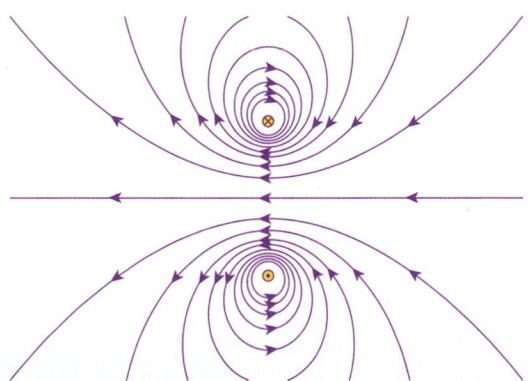

FIGURE 28.15 Magnetic field lines from a loop of wire carrying a current, looking at the loop edge-on. The upper yellow circle with a cross indicates current directed into the page, and the lower yellow circle with a dot indicates current directed out of the page.

SOLVED PROBLEM 28.3 | Field from a Wire Containing a Loop

A loop with radius $r = 8.30$ mm is formed in the middle of a long, straight insulated wire carrying a current of magnitude $i = 26.5$ mA (Figure 28.16a).

PROBLEM
What is the magnitude of the magnetic field at the center of the loop?

SOLUTION
THINK The magnetic field at the center of the loop is equal to the vector sum of the magnetic fields from the long, straight wire and from the loop.

SKETCH The magnetic field from the long, straight wire, $\vec{B}_{\text{wire}}$, and the magnetic field from the loop, $\vec{B}_{\text{loop}}$, are shown in Figure 28.16b.

– Continued

Self-Test Opportunity 28.3

Show that equation 28.9 for the magnitude of the magnetic field along the axis of a current-carrying loop reduces to equation 28.8 for the magnitude of the magnetic field at the center of a current-carrying loop.

Concept Check 28.6

Two identical wire loops carry the same current, i, as shown in the figure. What is the direction of the magnetic field at point P?

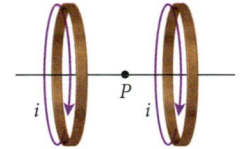

a) upward (toward the top of the page)

b) toward the right

c) downward

d) toward the left

e) The magnetic field at point P is zero.

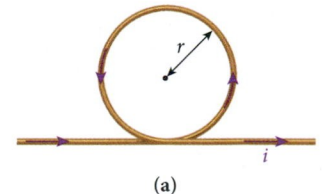

(a)

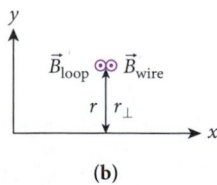

(b)

FIGURE 28.16 (a) A loop with radius r in a long, straight insulated wire carrying a current, i. (b) The magnetic field from the wire and the magnetic field from the loop, displaced slightly for clarity.

RESEARCH Using right-hand rule 3, we find that both magnetic fields point out of the page at the center of the loop, as illustrated in Figure 28.16b. Thus, we can add the magnitudes of the magnetic field produced by the wire and the magnetic field produced by the loop. The magnetic field produced by the wire at the center of the loop has a magnitude given by equation 28.4:

$$B_{\text{wire}} = \frac{\mu_0 i}{2\pi r_\perp},$$

where $r_\perp$ is the perpendicular distance from the wire, which is equal to r, the radius of the loop. The magnitude of the magnetic field produced by the loop at its center is given by equation 28.8:

$$B_{\text{loop}} = \frac{\mu_0 i}{2r}.$$

SIMPLIFY We add the magnitudes of the two magnetic fields since the vectors are in the same direction:

$$B = B_{\text{wire}} + B_{\text{loop}} = \frac{\mu_0 i}{2\pi r} + \frac{\mu_0 i}{2r} = \frac{\mu_0 i}{2r}\left(\frac{1}{\pi} + 1\right).$$

CALCULATE Putting in the numerical values, we get

$$B = \frac{\mu_0 i}{2r}\left(\frac{1}{\pi} + 1\right) = \frac{\left(4\pi \cdot 10^{-7} \text{ T m/A}\right)\left(26.5 \cdot 10^{-3} \text{ A}\right)}{2\left(8.30 \cdot 10^{-3} \text{ m}\right)}\left(\frac{1}{\pi} + 1\right) = 2.64463 \cdot 10^{-6} \text{ T}.$$

ROUND We report our result to three significant figures and note the direction of the field:

$$\vec{B} = 2.64 \cdot 10^{-6} \text{ T, out of the page.}$$

DOUBLE-CHECK To double-check our result, we calculate the magnitudes of the magnetic fields from the wire and from the loop separately. The magnitude of the magnetic field from the wire is

$$B_{\text{wire}} = \frac{\mu_0 i}{2\pi r} = \frac{\left(4\pi \cdot 10^{-7} \text{ T m/A}\right)\left(26.5 \cdot 10^{-3} \text{ A}\right)}{2\pi\left(8.30 \cdot 10^{-3} \text{ m}\right)} = 6.385 \cdot 10^{-7} \text{ T}.$$

The magnitude of the magnetic field from the loop is

$$B_{\text{loop}} = \frac{\mu_0 i}{2r} = \frac{\left(4\pi \cdot 10^{-7} \text{ T m/A}\right)\left(26.5 \cdot 10^{-3} \text{ A}\right)}{2\left(8.30 \cdot 10^{-3} \text{ m}\right)} = 2.006 \cdot 10^{-6} \text{ T}.$$

The sum of these two magnitudes matches our result:

$$6.385 \cdot 10^{-7} \text{ T} + 2.006 \cdot 10^{-6} \text{ T} = 2.64 \cdot 10^{-6} \text{ T}.$$

28.3 Ampere's Law

Recall from Chapter 22 that calculating the electric field resulting from a distribution of electric charge can require evaluating a difficult integral. However, if the charge distribution has cylindrical, spherical, or planar symmetry, we can apply Gauss's Law and obtain the electric field in an elegant manner. Similarly, calculating the magnetic field due to an arbitrary distribution of current elements using the Biot-Savart Law (equation 28.1) may involve the evaluation of a difficult integral. Alternatively, we can avoid using the Biot-Savart Law and instead apply **Ampere's Law** to calculate the magnetic field from a distribution of current elements when the distribution has cylindrical or other symmetry. Often, problems can be solved with much less effort in this way than by using a direct integration. The mathematical statement of Ampere's Law is

$$\oint \vec{B} \cdot d\vec{s} = \mu_0 i_{\text{enc}}. \tag{28.10}$$

The symbol $\oint$ means that the integrand, $\vec{B} \cdot d\vec{s}$, is integrated over a closed loop, called an **Amperian loop**. This loop is chosen so that the integral in equation 28.10 is not difficult to evaluate, a procedure similar to that used in applying Gauss's Law. The total current enclosed in this loop is i_{enc}, which is also similar to Gauss's Law, where the chosen closed surface encloses a total net charge.

As an example of how Ampere's Law is used, consider the five currents shown in Figure 28.17, which are all perpendicular to the plane. An Amperian loop, represented by the red line, encloses currents i_1, i_2, and i_3 and excludes currents i_4 and i_5. By Ampere's Law, the closed-loop integral over the magnetic field resulting from these three currents is given by

$$\oint \vec{B} \cdot d\vec{s} = \oint B \cos\theta \, ds = \mu_0 \left(i_1 - i_2 + i_3 \right),$$

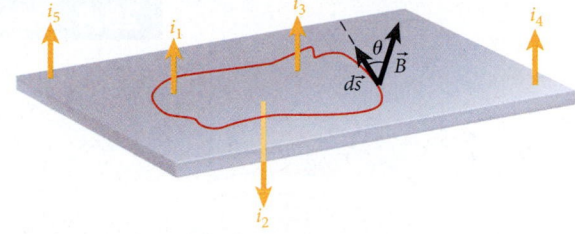

FIGURE 28.17 Five currents and an Amperian loop.

where θ is the angle between the direction of the magnetic field, $\vec{B}$, and the direction of the element of length, $d\vec{s}$, at each point along the Amperian loop. The integration over the Amperian loop can be done in either direction. Figure 28.17 indicates a direction of integration from the direction of $d\vec{s}$, along with the resulting magnetic field. The sign of the contributing currents can be determined using a right-hand rule: Curl your fingers in the direction of integration, and then currents in the same direction as your thumb are positive. Two of the three currents in the Amperian loop are positive, and one is negative. Adding the three currents is simple, but the integral $\oint B \cos\theta \, ds$ cannot be easily evaluated. However, let's examine some special situations in which Amperian loops contain symmetrical distributions of current that can be exploited to carry out the integration.

Magnetic Field inside a Long, Straight Wire

Figure 28.18 shows a current, i, flowing out of the page in a wire with a circular cross section of radius R. This current is uniformly distributed over the cross-sectional area of the wire. To find the magnetic field due to this current, we use an Amperian loop with radius $r_\perp$, represented by the red circle. If $\vec{B}$ had an outward (or inward) component, by symmetry, it would have an outward (or inward) component at all points around the loop, and the corresponding magnetic field line could never be closed. Therefore, $\vec{B}$ must be tangential to the Amperian loop. Thus, we can rewrite the integral of Ampere's Law as

$$\oint \vec{B} \cdot d\vec{s} = B \oint d\vec{s} = B 2\pi r_\perp.$$

We can calculate the enclosed current from the ratio of the area of the Amperian loop to the cross-sectional area of the wire:

$$i_{enc} = i \frac{A_{loop}}{A_{wire}} = i \frac{\pi r_\perp^2}{\pi R^2}.$$

Thus, we obtain

$$2\pi B r_\perp = \mu_0 i \frac{\pi r_\perp^2}{\pi R^2},$$

or

$$B = \left(\frac{\mu_0 i}{2\pi R^2} \right) r_\perp. \tag{28.11}$$

Let's compare the expressions for the magnitudes of the magnetic field outside and inside the wire—equations 28.4 and 28.11. First, substituting R for $r_\perp$ in both expressions, we obtain the same result for the magnetic field magnitude at the surface of the wire in both cases: $B(R) = \mu_0 i / 2\pi R$. Both equations provide the same solution at the wire's surface. Inside the wire, we find that the magnetic field magnitude rises linearly with $r_\perp$ up to the value of $B(R) = \mu_0 i / (2\pi R)$ and from there falls off with the inverse of $r_\perp$. Figure 28.19 shows this dependence in the graph at the bottom. The upper part of the figure depicts the cross section through the wire (golden area), the magnetic field lines (black circles, spaced to indicate the strength of the magnetic field), and the magnetic field vectors at selected points in space (red arrows).

Concept Check 28.7

Three wires are carrying currents of the same magnitude, i, in the directions shown in the figure. Four Amperian loops (a), (b), (c), and (d) are shown. For which Amperian loop is the magnitude of $\oint \vec{B} \cdot d\vec{s}$ the greatest?

a) loop a
b) loop b
c) loop c
d) loop d

e) All four loops yield the same value of $\oint \vec{B} \cdot d\vec{s}$.

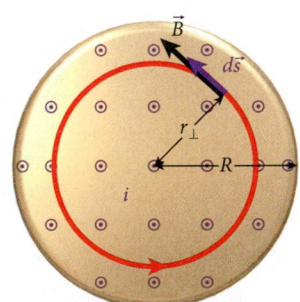

FIGURE 28.18 Using Ampere's Law to find the magnetic field produced inside a long, straight wire.

28.4 Magnetic Fields of Solenoids and Toroids

We have seen that current flowing through a single loop of wire produces a magnetic field that is not uniform, as illustrated in Figure 28.15. However, real-world applications often require a uniform magnetic field. A device commonly used to produce a uniform magnetic field is the **Helmholtz coil** (Figure 28.20a). A Helmholtz coil consists of two coaxial wire loops. Each coaxial loop consists of multiple loops (windings or turns) of a single wire, and therefore acts magnetically like a single loop.

The magnetic field lines from a Helmholtz coil are shown in Figure 28.20b. You can see that there is a region of uniform magnetic field (characterized by horizontal parallel segments of the field lines) in the center between the loops, in contrast to the field from a single loop shown in Figure 28.15. Again, these field lines were calculated with the aid of a computer to provide a qualitative understanding of the geometry of magnetic fields.

Taking multiple loops a step further, Figure 28.21 shows the magnetic field lines from four coaxial wire loops. The region of uniform magnetic field in the center of the loops is expanded, but note that the field is not uniform near the wires and near the two ends.

A strong uniform magnetic field is produced by a **solenoid**, consisting of many loops of a wire wound close together. Figure 28.22 shows the magnetic field lines from a solenoid with 600 turns, or loops. You can see that the magnetic field lines are very close together on the inside of the solenoid and far apart on the outside. Like that inside the Helmholtz coil (Figure 28.20b), the magnetic field is uniform inside the solenoid coil. The spacing of the field lines is a measure of the strength of the magnetic field, and you can see that the magnetic field is much stronger inside the solenoid than outside the solenoid.

An *ideal* solenoid has a magnetic field of zero outside and of a uniform constant finite value inside. To determine the magnitude of the magnetic field inside an ideal solenoid, we can apply Ampere's Law (equation 28.10) to a section of a solenoid far from its ends (Figure 28.23). To do so, we first choose an Amperian loop over which to carry out the integration. A judicious choice, shown by the red rectangle in Figure 28.23, encloses some current and

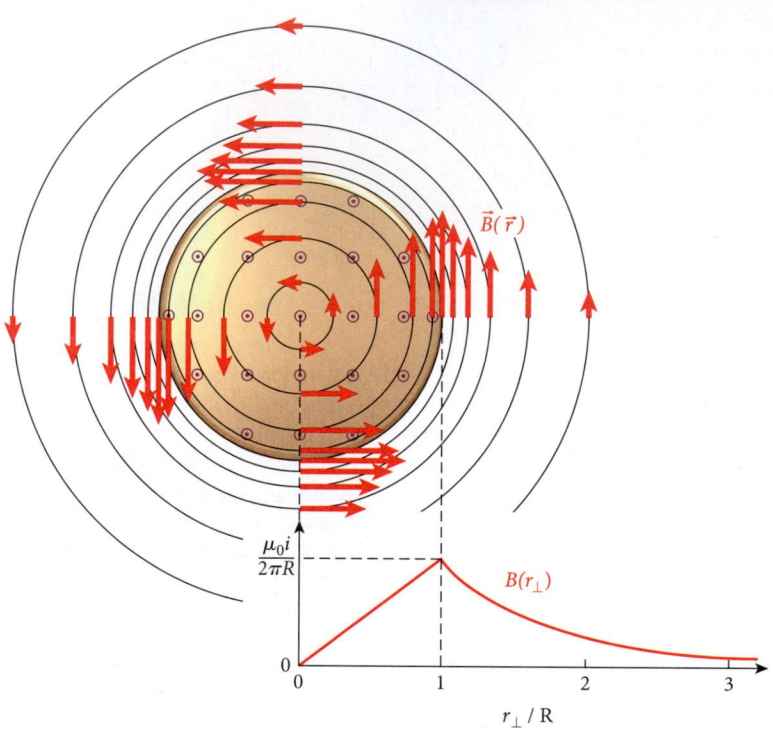

FIGURE 28.19 Radial dependence of the magnetic field for a wire with current flowing out of the page.

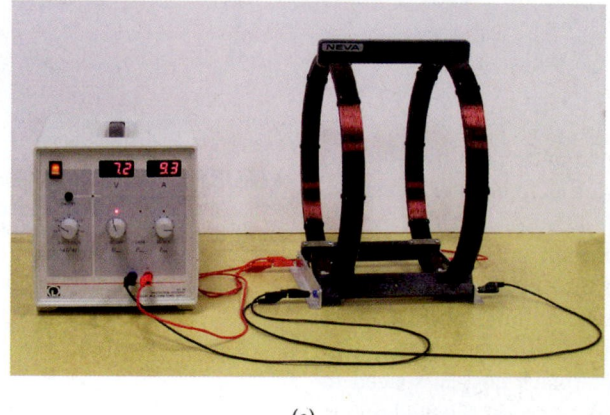

(a)

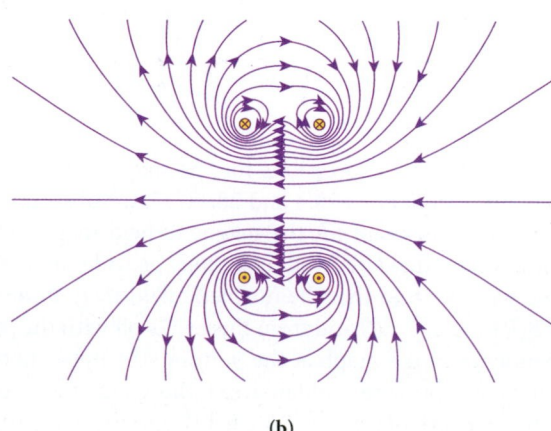

(b)

FIGURE 28.20 (a) A typical Helmholtz coil used in physics labs generates a nearly uniform magnetic field in its interior. (b) Magnetic field lines for a Helmholtz coil.

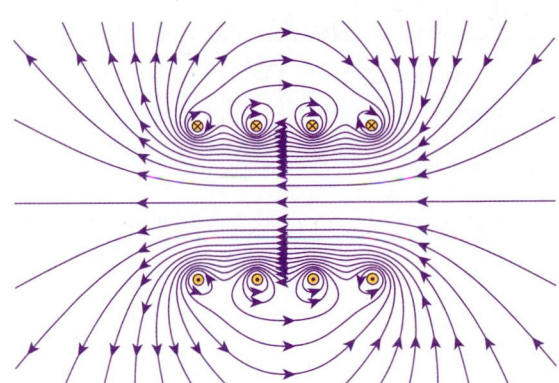

FIGURE 28.21 Magnetic field lines resulting from four coaxial wire loops with many windings.

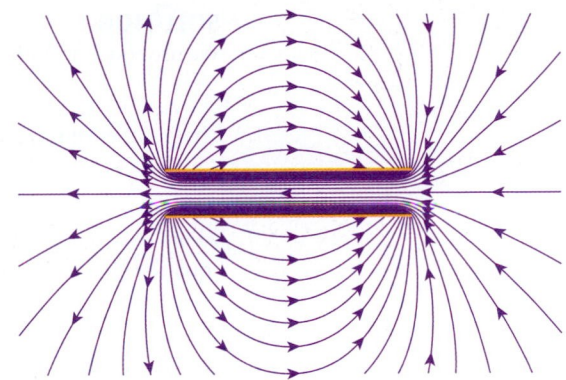

FIGURE 28.22 Magnetic field lines for a solenoid with 600 turns. The current along the top of the solenoid is flowing into the page, and the current along the bottom of the solenoid is flowing out of the page.

exploits the symmetry of the solenoid as well as simplifying the evaluation of the integral:

$$\oint \vec{B} \cdot d\vec{s} = \int_a^b \vec{B} \cdot d\vec{s} + \int_b^c \vec{B} \cdot d\vec{s} + \int_c^d \vec{B} \cdot d\vec{s} + \int_d^a \vec{B} \cdot d\vec{s}.$$

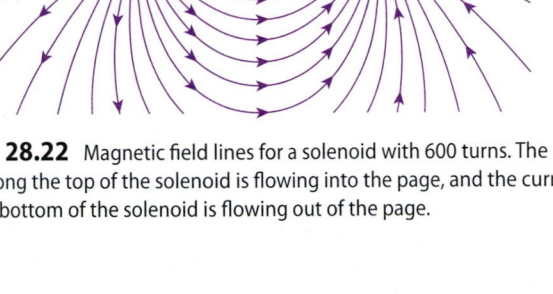

The value of the third integral on the right-hand side, between points c and d in the interior of the solenoid, is Bh. The values of the second and fourth integrals are zero because the magnetic field is perpendicular to the direction of integration. The first integral, between the points a and b in the exterior of the ideal solenoid, is zero because the magnetic field outside of an ideal solenoid is zero. Thus, the value of the integral over the entire Amperian loop is Bh.

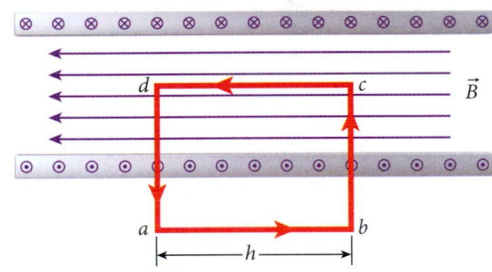

FIGURE 28.23 Amperian loop for determining the magnitude of the magnetic field of an ideal solenoid.

The enclosed current is the current in the turns of the solenoid that are within the Amperian loop. The current is the same in each turn because the solenoid is made from one wire and the same current flows through each turn. Thus, the enclosed current is just the number of turns times the current:

$$i_{enc} = nhi,$$

where n is the number of turns per unit length. Therefore, according to Ampere's Law, we have

$$Bh = \mu_0 nhi.$$

Thus, the magnitude of the magnetic field inside an ideal solenoid is

$$B = \mu_0 ni. \qquad (28.12)$$

Equation 28.12 is valid only away from the ends of the solenoid. Note that B does not depend on position inside the solenoid: An ideal solenoid creates a constant and uniform magnetic field inside itself and no field outside itself. A real-world solenoid, like the one shown in Figure 28.22, has fringe fields near its ends but can still produce a high-quality uniform magnetic field.

EXAMPLE 28.2 Solenoid

The solenoid of the STAR detector at the Brookhaven National Laboratory, New York, discussed in Chapter 27, has a magnetic field of magnitude 0.50 T when carrying a current of 0.40 kA. The solenoid is 8.0 m long.

PROBLEM

What is the number of turns in this solenoid, assuming that it is an ideal solenoid?

– Continued

SOLUTION

We use equation 28.12 to calculate the magnitude of the magnetic field of an ideal solenoid:

$$B = \mu_0 n i. \tag{i}$$

The number of turns per unit length is given by

$$n = \frac{N}{L}, \tag{ii}$$

where N is the number of turns and L is the length of the solenoid. Substituting for n from equation (ii) into equation (i), we get

$$B = \mu_0 i \frac{N}{L}. \tag{iii}$$

Solving equation (iii) for the number of turns, we obtain

$$N = \frac{BL}{\mu_0 i} = \frac{(0.50 \text{ T})(8.0 \text{ m})}{\left(4\pi \cdot 10^{-7} \ \dfrac{\text{T m}}{\text{A}}\right)(0.40 \cdot 10^3 \text{ A})} = 8.0 \cdot 10^3 \text{ turns.}$$

Concept Check 28.8

You have a solenoid with a fixed number of turns connected to a power supply that can supply a fixed amount of current. To double the field inside the solenoid, you can

a) double the radius of the solenoid.

b) halve the radius of the solenoid.

c) double the length of the solenoid.

d) halve the length of the solenoid.

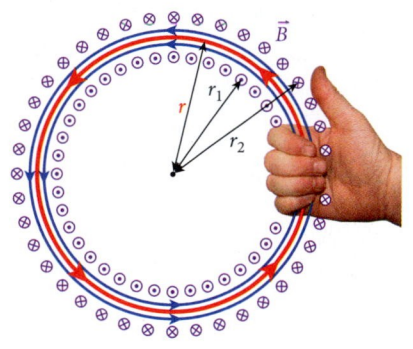

FIGURE 28.24 Toroidal magnet with Amperian loop (red) in the form of a circle with radius r. Right-hand rule 4 states that if you place the fingers of your right hand in the direction of the current flow, your thumb shows the direction of the magnetic field inside the toroid.

If a solenoid is bent so that the two ends meet (Figure 28.24), it acquires a doughnut shape (a torus), with the wire forming a series of loops, each with the same current flowing through it. This device is called a *toroidal magnet*, or **toroid**. Just as for an ideal solenoid, the magnetic field outside the coils of an ideal toroidal magnet is zero. The magnitude of the magnetic field inside the toroid coil can be calculated by using Ampere's Law and choosing an Amperian loop in the form of a circle with radius r, such that $r_1 < r < r_2$, where r_1 and r_2 are the inner and outer radii of the toroid. The magnetic field is always tangential to the Amperian loop, so we have

$$\oint \vec{B} \cdot d\vec{s} = 2\pi r B.$$

The enclosed current is the number of loops (or turns), N, in the toroid times the current, i, in the wire (in each loop); so, Ampere's law gives us

$$2\pi r B = \mu_0 N i.$$

Therefore, the magnitude of the magnetic field inside of the toroid is given by

$$B = \frac{\mu_0 N i}{2\pi r}. \tag{28.13}$$

Note that, unlike the magnetic field inside a solenoid, the magnitude of the magnetic field inside a toroid does depend on the radius. As the radius increases, the magnitude of the magnetic field decreases. The direction of the magnetic field can be obtained using *right-hand rule 4*: If you wrap the fingers of your right hand around the toroid in the direction of the current, as shown in Figure 28.24, your thumb points in the direction of the magnetic field inside the toroid.

SOLVED PROBLEM 28.4 | Field of a Toroidal Magnet

A toroidal magnet is made from 202 m of copper wire that is capable of carrying a current of magnitude $i = 2.40$ A. The toroid has an average radius $R = 15.0$ cm and a cross-sectional diameter $d = 1.60$ cm (Figure 28.25a).

PROBLEM

What is the largest magnetic field that can be produced at the average toroidal radius, R?

SOLUTION

THINK The number of turns in the toroidal magnet is given by the length of the wire divided by the circumference of the cross-sectional area of the coil. With these parameters, the magnetic field of the toroidal magnet at $r = R$ can be calculated.

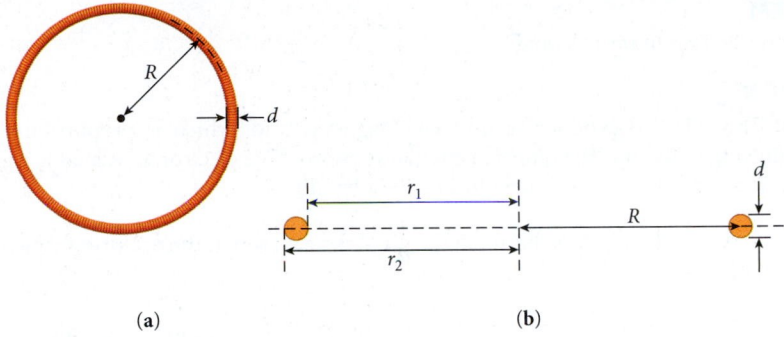

FIGURE 28.25 (a) A toroidal magnet. (b) Cross section of the toroidal magnet.

SKETCH Figure 28.25b shows a cross-sectional cut of the toroidal magnet.

RESEARCH The magnitude of the magnetic field of a toroidal magnet is given by equation 28.13:

$$B = \frac{\mu_0 N i}{2\pi R}, \tag{i}$$

where N is the number of turns and R is the radius at which the magnetic field is measured. The number of turns, N, is given by the length, L, of the wire divided by the circumference of the cross-sectional area:

$$N = \frac{L}{\pi d}, \tag{ii}$$

where d is the diameter of the cross-sectional area of the toroid.

SIMPLIFY We can combine equations (i) and (ii) to obtain an expression for B:

$$B = \frac{\mu_0 (L/\pi d) i}{2\pi R} = \frac{\mu_0 L i}{2\pi^2 R d}.$$

CALCULATE Putting in the numerical values gives us

$$B = \frac{\mu_0 L i}{2\pi^2 R d} = \frac{\left(4\pi \cdot 10^{-7} \text{ T m/A}\right)(202 \text{ m})(2.40 \text{ A})}{2\pi^2 \left(15.0 \cdot 10^{-2} \text{ m}\right)\left(1.60 \cdot 10^{-2} \text{ m}\right)} = 0.0128597 \text{ T}.$$

ROUND We report our result to three significant figures:

$$B = 1.29 \cdot 10^{-2} \text{ T}.$$

DOUBLE-CHECK As a double-check, we calculate the magnitude of the field inside a solenoid that has the same length as the circumference of the toroidal magnet. The number of turns per unit length is

$$n = \frac{L/\pi d}{2\pi R} = \frac{L}{2\pi^2 R d} = \frac{(202 \text{ m})}{2\pi^2 \left(15.0 \cdot 10^{-2} \text{ m}\right)\left(1.60 \cdot 10^{-2} \text{ m}\right)} = 4264 \text{ turns/m}.$$

The magnitude of the magnetic field inside a solenoid with that number of turns per unit length is

$$B = \mu_0 n i = \left(4\pi \cdot 10^{-7} \text{ T m/A}\right)\left(4264 \text{ m}^{-1}\right)(2.40 \text{ A}) = 1.29 \cdot 10^{-2} \text{ T}.$$

Thus, our answer for the magnitude of the field inside the toroid seems reasonable.

SOLVED PROBLEM 28.5 | Electron Motion in a Solenoid

An ideal solenoid has 200. turns/cm. An electron inside the coil of the solenoid moves in a circle with radius $r = 3.00$ cm perpendicular to the solenoid's axis. The electron moves with a speed of $v = 0.0500c$, where c is the speed of light.

– Continued

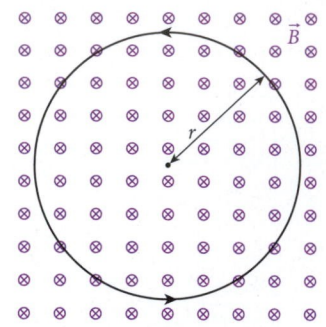

FIGURE 28.26 Electron traveling in a circular path inside a solenoid.

PROBLEM

What is the current in the solenoid?

SOLUTION

THINK The solenoid produces a uniform magnetic field, which is proportional to the current flowing in its coil. The radius of circular motion of the electron is related to the speed of the electron and the magnetic field inside the solenoid.

SKETCH Figure 28.26 shows the circular path of the electron in the uniform magnetic field of the solenoid.

RESEARCH The magnitude of the magnetic field inside the solenoid is given by

$$B = \mu_0 n i, \tag{i}$$

where i is the current in the solenoid and n is the number of turns per unit length. The magnetic force provides the centripetal force needed for the electron to move in a circle and so the radius of the electron's path can be related to B:

$$r = \frac{mv}{eB}, \tag{ii}$$

where m is the electron's mass, v is its speed, and e is the magnitude of the electron's charge.

SIMPLIFY Combining equations (i) and (ii), we have

$$r = \frac{mv}{e(\mu_0 n i)}.$$

Solving this equation for the current in the solenoid, we obtain

$$i = \frac{mv}{er\mu_0 n}. \tag{iii}$$

CALCULATE The speed of the electron was specified in terms of the speed of light:

$$v = 0.0500c = 0.0500\left(3.00 \cdot 10^8 \ \text{m/s}\right) = 1.50 \cdot 10^7 \ \text{m/s}.$$

Putting this and the other numerical values into equation (iii), we get

$$i = \frac{mv}{er\mu_0 n} = \frac{\left(9.11 \cdot 10^{-31} \ \text{kg}\right)\left(1.50 \cdot 10^7 \ \text{m/s}\right)}{\left(1.602 \cdot 10^{-19} \ \text{C}\right)\left(3.00 \cdot 10^{-2} \ \text{m}\right)\left(4\pi \cdot 10^{-7} \ \text{T m/A}\right)\left(200 \cdot 10^2 \ \text{m}^{-1}\right)}$$

$$= 0.113132 \ \text{A}.$$

ROUND We report our result to three significant figures:

$$i = 0.113 \ \text{A}.$$

DOUBLE-CHECK To double-check our result, we use it to calculate the magnitude of the magnetic field inside the solenoid:

$$B = \mu_0 n i = \left(4\pi \cdot 10^{-7} \ \text{T m/A}\right)\left(200 \cdot 10^2 \ \text{m}^{-1}\right)(0.113 \ \text{A}) = 0.00284 \ \text{T}.$$

This magnitude of magnetic field seems reasonable. Thus, our calculated value for the current in the solenoid seems reasonable.

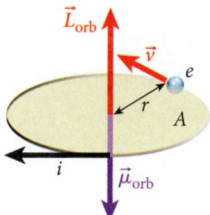

FIGURE 28.27 An electron moving with constant speed in a circular orbit in an atom.

28.5 Atoms as Magnets

The atoms that make up all matter contain moving electrons, which form current loops that produce magnetic fields. In most materials, these current loops are randomly oriented and produce no net magnetic field. Some materials have a fraction of these current loops aligned. These materials, called *magnetic materials* (Section 28.6), do produce a net magnetic field. Other materials can have their current loops aligned by an external magnetic field and become magnetized.

Let's consider a highly simplified model of the atom: An electron moving at constant speed v in a circular orbit with radius r (Figure 28.27). We can think of the moving charge of the electron as a current, i. Current is defined as the charge passing a particular point per unit of time. For this case, the charge is the charge of the electron, with magnitude e, and the time is related to the period, T, of the electron's orbit. Thus, the magnitude of the current is given by

$$i = \frac{e}{T} = \frac{e}{2\pi r/v} = \frac{ve}{2\pi r}.$$

The magnitude of the magnetic dipole moment of the orbiting electron is given by

$$\mu_{orb} = iA = \frac{ve}{2\pi r}\left(\pi r^2\right) = \frac{ver}{2}. \tag{28.14}$$

The magnitude of the orbital angular momentum of the electron is

$$L_{orb} = rp = rmv,$$

where m is the mass of the electron. Solving equation 28.14 for v and substituting that expression into the expression for the orbital angular momentum gives us

$$L_{orb} = rm\left(\frac{2\mu_{orb}}{er}\right) = \frac{2m\mu_{orb}}{e}.$$

Because the magnetic dipole moment and the angular momentum are vector quantities, we can write

$$\vec{\mu}_{orb} = -\frac{e}{2m}\vec{L}_{orb}, \tag{28.15}$$

where the negative sign is needed because the current is defined in terms of the direction of the flow of positive charge.

EXAMPLE 28.3 | Orbital Magnetic Moment of the Hydrogen Atom

Assume that the hydrogen atom consists of an electron moving with speed v in a circular orbit with radius r around a stationary proton. Also assume that the centripetal force keeping the electron moving in a circle is the electrostatic force between the proton and the electron. The radius of the orbit of the electron is $r = 5.29 \cdot 10^{-11}$ m. (This radius is derived using concepts discussed in Chapter 38 on atomic physics.)

PROBLEM
What is the magnitude of the orbital magnetic moment of the hydrogen atom?

SOLUTION
The magnitude of the orbital magnetic moment is

$$\left|\mu_{orb}\right| = \frac{e}{2m}L_{orb} = \frac{e}{2m}(rmv) = \frac{erv}{2}. \tag{i}$$

Equating the magnitude of the centripetal force keeping the electron moving in a circle with that of the electrostatic force between the electron and the proton gives us

$$\frac{mv^2}{r} = k\frac{e^2}{r^2},$$

where k is the Coulomb constant. We can solve this equation for the speed of the electron:

$$v = e\sqrt{\frac{k}{mr}}. \tag{ii}$$

Substituting for v from equation (ii) into equation (i) gives us

$$\left|\mu_{orb}\right| = \frac{er}{2}\left(e\sqrt{\frac{k}{mr}}\right) = \frac{e^2}{2}\sqrt{\frac{kr}{m}}.$$

When we put in the various numerical values, we get

$$\left|\mu_{orb}\right| = \frac{\left(1.602\cdot10^{-19}\,\text{C}\right)^2}{2}\sqrt{\frac{\left(8.99\cdot10^9\,\text{N m}^2/\text{C}^2\right)\left(5.29\cdot10^{-11}\,\text{m}\right)}{9.11\cdot10^{-31}\,\text{kg}}} = 9.27\cdot10^{-24}\,\text{A m}^2.$$

– Continued

This result agrees with experimental measurements of the orbital magnetic moment of the hydrogen atom. However, other predictions about the properties of hydrogen and other atoms that are based on the idea that electrons in atoms have circular orbits disagree with experimental observations. Thus, detailed description of the magnetic properties of atoms must incorporate phenomena described by quantum physics, which will be covered in Chapter 36.

Spin

The magnetic dipole moment from the orbital motion of electrons is not the only contribution to the magnetic moment of atoms. Electrons and other elementary particles have their own, intrinsic magnetic moments, due to their spin. The phenomenon of spin will be covered thoroughly in the discussion of quantum physics in Chapters 36 through 39, but some facts about spin and its connection to a particle's intrinsic angular momentum have been discovered experimentally and do not require an understanding of quantum mechanics. Electrons, protons, and neutrons all have a spin of magnitude $s = \frac{1}{2}$. The magnitude of the angular momentum of these particles is $S = \hbar\sqrt{s(s+1)}$, and the z-component of the angular momentum can have a value of either $S_z = -\frac{1}{2}\hbar$ or $S_z = +\frac{1}{2}\hbar$, where $\hbar$ is Planck's constant divided by 2π. This spin cannot be explained by orbital motion of some substructure in the particles. Electrons, for example, are apparently true point particles. Thus, spin is an intrinsic property, similar to mass or electric charge.

The magnetic character of bulk matter is determined largely by electron spin magnetic moments. The magnetic moment of a particle with spin $\vec{\mu}_s$ is related to its spin angular momentum, $\vec{S}$, via

$$\vec{\mu}_s = g\frac{q}{2m}\vec{S}, \tag{28.16}$$

where q is the charge of the particle and m is its mass. The quantity g is dimensionless and is called the *g-factor*. For the electron, its numerical value is $g = -2.0023193043622(15)$, one of the most precisely measured quantities in nature. If you compare this equation to equation 28.15 for the magnetic dipole moment due to the orbital angular momentum, you see that they are very similar.

28.6   Magnetic Properties of Matter

In Chapter 27, we saw that magnetic dipoles do not experience a net force in a homogeneous external magnetic field, but do experience a torque. This torque drives a single free dipole to an orientation in which it is antiparallel to the external field, because this is the state with the lowest magnetic potential energy. We've just seen in Section 28.5 that atoms can have magnetic dipoles. What happens when matter (which is composed of atoms) is exposed to an external magnetic field?

The dipole moments of the atoms in a material can point in different directions or in the same direction. The **magnetization, $\vec{M}$,** of a material is defined as the net dipole moment created by the dipole moments of the atoms in the material per unit volume. The magnetic field, $\vec{B}$, inside the material then depends on the external magnetic field, $\vec{B}_0$, and the magnetization, $\vec{M}$:

$$\vec{B} = \vec{B}_0 + \mu_0\vec{M}, \tag{28.17}$$

where μ_0 is again the magnetic permeability of free space. Instead of including the external magnetic field, $\vec{B}_0$, it is customary to use the **magnetic field strength, $\vec{H}$:**

$$\vec{H} = \frac{\vec{B}_0}{\mu_0}. \tag{28.18}$$

With this definition of the magnetic field strength, equation 28.17 can be written

$$\vec{B} = \mu_0(\vec{H} + \vec{M}). \tag{28.19}$$

Since the unit of magnetic field is $[B] = $ T and the unit of the magnetic permeability is $[\mu_0] = $ T m/A, the units of magnetization and magnetic field strength are $[M] = [H] = $ A/m.

Diamagnetism and Paramagnetism

The question not yet answered is how the magnetization depends on the external magnetic field, $\vec{B}_0$, or, equivalently, on the magnetic field strength, $\vec{H}$. For most materials (not all!) this relationship is linear:

$$\vec{M} = \chi_m \vec{H}, \tag{28.20}$$

where the proportionality constant χ_m is called the **magnetic susceptibility** of the material (Table 28.1). But there are materials that do not obey the simple linear relationship of equation 28.20, and the most prominent among those are ferromagnets, which we'll discuss in the next subsection. Let's first examine diamagnetic and paramagnetic materials, for which equation 28.20 holds.

If $\chi_m < 0$, the dipoles inside the material tend to arrange themselves to oppose an external magnetic field, just like free dipoles. In this case, the magnetization vector points in the direction opposite to the magnetic field strength vector. Materials with $\chi_m < 0$ are said to be *diamagnetic*. Most materials exhibit **diamagnetism.** In diamagnetic materials, a weak magnetic dipole moment is induced by an external magnetic field in a direction opposite to the direction of the external field. The induced magnetic field disappears when the external field is removed. If the external field is nonuniform, interaction of the induced dipole moment of the diamagnetic material with the external field creates a force directed from a region of greater magnetic field strength to a region of lower magnetic field strength.

An example of biological material exhibiting diamagnetism is shown in Figure 28.28. Diamagnetic forces induced by a nonuniform external magnetic field of 16 T are levitating a live frog. (This experience apparently did not bother the frog.) The normally negligible diamagnetic force is large enough in this case to overcome gravity.

If the magnetic susceptibility in equation 28.20 is greater than zero, $\chi_m > 0$, the magnetization of the material points in the same direction as the magnetic field strength. Note that χ_m for a vacuum is 0. This property is **paramagnetism,** and materials that exhibit it are said to be *paramagnetic*. Materials containing certain transition elements (including actinides and rare earths) exhibit paramagnetism. Each atom of these elements has a permanent magnetic dipole, but normally these dipole moments are randomly oriented and produce no net magnetic field. However, in the presence of an external magnetic field, some of these magnetic dipole moments align in the same direction as the external field. When the external field is removed, the induced magnetic dipole moment disappears. If the external field is nonuniform, this induced magnetic dipole moment interacts with the external field to produce a force directed from a region of lower magnetic field strength to a region of higher magnetic field strength—just the opposite of the effect of diamagnetism.

Substituting the expression for $\vec{M}$ from equation 28.20 into equation 28.19 for the magnetic field, $\vec{B}$, inside a material gives

$$\vec{B} = \mu_0(\vec{H} + \vec{M}) = \mu_0(\vec{H} + \chi_m \vec{H}) = \mu_0(1 + \chi_m)\vec{H}. \tag{28.21}$$

In analogy to the relative electric permittivity introduced in Chapter 24, the **relative magnetic permeability,** κ_m, is commonly defined as

$$\kappa_m = 1 + \chi_m. \tag{28.22}$$

Then, the magnetic permeability, μ, of a material can be expressed as

$$\mu = (1 + \chi_m)\mu_0 = \kappa_m \mu_0. \tag{28.23}$$

Replacing μ_0 with μ in the Biot-Savart Law (equation 28.1) and Ampere's Law (equation 28.10) enables us to use these laws for calculating the magnetic field in a particular material.

Finally for paramagnetic materials, the magnitude of the magnetization is temperature dependent. Conventionally, this temperature dependence is expressed via *Curie's Law*:

$$M = \frac{cB}{T}, \tag{28.24}$$

where c is Curie's constant, B is the magnitude of the magnetic field, and T is the temperature in kelvins.

Table 28.1	Values of Magnetic Susceptibility for Some Common Diamagnetic and Paramagnetic Materials
Material	**Magnetic Susceptibility, χ_m**
Aluminum	$+2.2 \cdot 10^{-5}$
Bismuth	$-1.66 \cdot 10^{-4}$
Diamond (carbon)	$-2.1 \cdot 10^{-5}$
Graphite (carbon)	$-1.6 \cdot 10^{-5}$
Hydrogen	$-2.2 \cdot 10^{-9}$
Lead	$-1.8 \cdot 10^{-5}$
Lithium	$+1.4 \cdot 10^{-5}$
Mercury	$-2.9 \cdot 10^{-5}$
Oxygen	$+1.9 \cdot 10^{-6}$
Platinum	$+2.65 \cdot 10^{-4}$
Silicon	$-3.7 \cdot 10^{-6}$
Sodium	$+7.2 \cdot 10^{-6}$
Sodium chloride (NaCl)	$-1.4 \cdot 10^{-5}$
Tungsten	$+6.8 \cdot 10^{-5}$
Uranium	$+4.0 \cdot 10^{-4}$
Vacuum	0
Water	$-9 \cdot 10^{-6}$

FIGURE 28.28 A live frog being levitated by a strong magnetic field at the High Field Magnet Laboratory, Radboud University Nijmegen, The Netherlands.

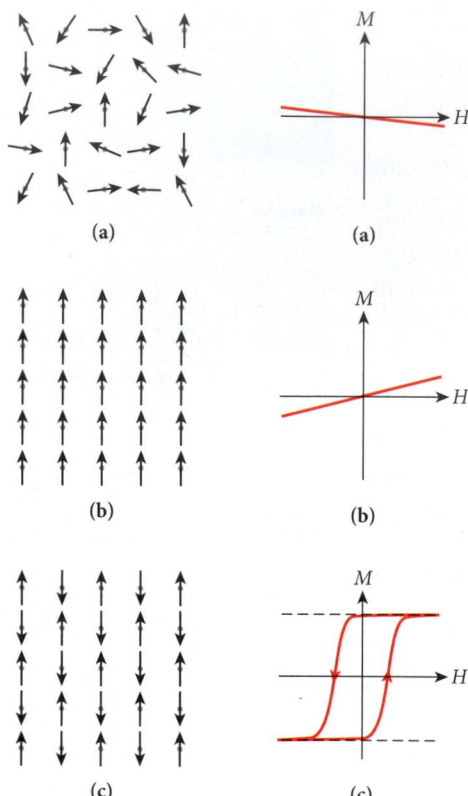

FIGURE 28.29 Magnetic domains: (a) randomly oriented; (b) perfect ferromagnetic order; (c) perfect antiferromagnetic order. (Note that this illustration only shows the directions of individual magnetic domains in an idealized scenario. The magnetic domains also can, and usually have, very different magnitudes.)

FIGURE 28.30 Magnetization as a function of the magnetic field strength: (a) for diamagnetic materials; (b) for paramagnetic materials; (c) the hysteresis loop for ferromagnetic materials.

Ferromagnetism

The elements iron, nickel, cobalt, gadolinium, and dysprosium—and alloys containing these elements—exhibit **ferromagnetism**. A ferromagnetic material shows long-range ordering at the atomic level, which causes the dipole moments of atoms to line up with each other in a limited region called a **domain.** Within a domain, the magnetic field can be strong. However, in bulk samples of the material, domains are randomly oriented, leaving no net magnetic field. Figure 28.29a shows randomly oriented magnetic dipole moments in a domain, and Figure 28.29b shows perfect ferromagnetic order. Figure 28.29c illustrates the interesting case of perfect antiferromagnetic order, in which the interaction between neighboring magnetic dipole moments causes them to be oriented in opposite directions. This ordering can be realized only at very low temperatures.

An external magnetic field can align domains as shown in Figure 28.29b, as a result of the interaction between the magnetic dipole moments of the domain and the external field. As a result, a ferromagnetic material retains all or some of its induced magnetism when the external magnetic field is removed, since the domains stay aligned. In addition, the magnetic field produced by a current in a solenoid or a toroid will be larger if a ferromagnetic material is present in the device. But in contrast to diamagnetic and paramagnetic materials, ferromagnetic materials do not obey the simple linear relationship given in equation 28.20. The domains retain their orientations, and thus the material exhibits a nonzero magnetization even in the absence of an external magnetic field. (This is why permanent magnets exist.)

Figure 28.30 illustrates the dependence of the magnetization on the magnetic field strength for the three types of materials we've discussed. Figure 28.30a and Figure 28.30b show the linear dependence according to equation 28.20 for diamagnetic and paramagnetic materials, respectively. Figure 28.30c shows the typical hysteresis loop obtained for ferromagnetic materials. The arrows on the red curve show the direction in which the magnetization process develops, and the dashed lines represent the maximum magnetization (positive and negative) possible. For any point on this hysteresis loop, the magnetization can be expressed in terms of an effective value of the magnetic permeability, μ, of the ferromagnetic material, similar to what is given in equation 28.23; however, this permeability is *not* a constant but depends on the applied magnetic field strength and even on the path by which that value of the field strength was attained. Regardless, the values of the effective permeability, μ, for ferromagnetic materials can be much larger than those for paramagnetic materials (greater by a factor of up to 10^4).

Ferromagnetism exhibits temperature dependence. At a certain temperature, called the *Curie temperature,* ferromagnetic materials cease to exhibit ferromagnetism. At this point, the ferromagnetic order due to the interaction of the dipole moments in these materials is overwhelmed by the thermal motion. For iron, the Curie temperature is 768 °C. Figure 28.31 shows a simple demonstration in which heating a permanent ferromagnet (Figure

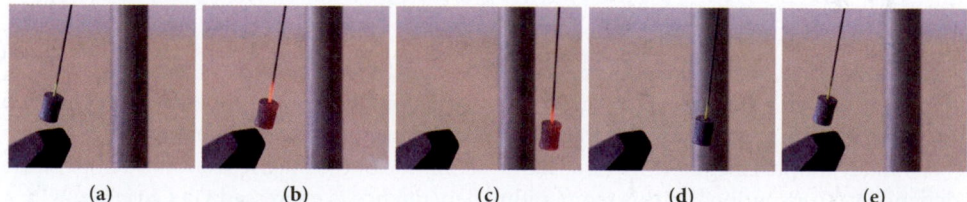

FIGURE 28.31 Demonstration of the temperature dependence of ferromagnetism: (a) A permanent magnet forms the bob of a pendulum and is deflected from the vertical by another magnet (lower left corner of each frame); (b) the magnet is heated and its temperature begins to increase; (c) the temperature of the magnet is high enough that its magnetic field is diminished and it hangs at a much smaller angle; (d) as the magnet cools, it begins to recover its original magnetic field; and (e) it returns to its original position.

28.31b) diminishes the attraction between it and another magnet (Figure 28.31c). As the magnet subsequently cools (Figure 28.31d), it again becomes a permanent magnet (Figure 28.31e).

28.7 Magnetism and Superconductivity

Magnets for industrial applications and scientific research can be constructed using ordinary resistive wire with current flowing through it. A typical magnet of this type is a large solenoid. The current flowing through the wire of the magnet produces resistive heating, and the heat is usually removed by low-conductivity water flowing through hollow conductors. (Low-conductivity water has been purified so that it does not conduct electricity.) These room-temperature magnets typically produce magnetic fields with strengths up to 1.5 T and are usually relatively inexpensive to construct but expensive to operate because of the high cost of electricity.

Some applications, such as magnetic resonance imaging (MRI), require magnetic fields of the highest possible magnitude to ensure the best signal-to-noise ratio in the measurements. To achieve these fields, magnets are constructed using superconducting coils rather than resistive coils. Such a magnet can produce a stronger field than a room-temperature magnet, with a magnitude of 10 T or higher. Materials such as mercury and lead exhibit superconductivity at liquid helium temperatures, but some metals that are good conductors at room temperature, such as copper and gold, never become superconducting. The disadvantage of a superconducting magnet is that the conductor must be kept at the temperature of liquid helium, which is approximately 4 K (although recent discoveries described later in this section are easing this limitation). Thus, the magnet must be enclosed in a cryostat filled with liquid helium to keep it cold. An advantage of a superconducting magnet is the fact that once the current is established in the coil of the magnet, it will continue to flow until it is removed by external means. However, the energy saving realized by having no resistive loss in the coil is at least partially offset by the expenditure of energy required to keep the superconducting coil cold.

When current flows through superconducting mercury or lead, the magnetic field inside the material becomes zero. The reduction to zero of the magnetic field inside a material cooled sufficiently that it becomes a superconductor is called the *Meissner effect*. Above the critical temperature, T_c, for the transition to superconductivity, the Meissner effect disappears, and the material becomes a normal conductor (Figure 28.32).

Figure 28.33 shows an impressive demonstration of the Meissner effect: A piece of a superconductor (cooled to a temperature below its critical temperature) causes a permanent magnet to float above it by expelling the magnet's intrinsic magnetic field. It achieves this because superconducting currents on its surface produce a magnetic field opposed to the applied field, which yields a net field of zero inside the superconductor and repulsion between the fields above the superconductor.

The conductor used in a superconducting magnet is specially designed to overcome the Meissner effect. Modern superconductors are constructed from filaments of niobium-titanium alloy embedded in solid copper. The niobium-titanium filaments have microscopic domains in which a magnetic field can exist without being excluded. The copper serves as a mechanical support and can take over the current load should the superconductor become normally conducting. This type of superconductor can produce magnetic fields with magnitudes as high as 15 T.

During the last two decades, physicists and engineers have discovered new materials that are superconducting at temperatures well above 4 K. Critical temperatures of up to 160 K have been reported for these *high-temperature superconductors,* which means that they can be made superconducting by cooling them with liquid nitrogen. Many researchers around the world are looking for materials that are superconducting at room temperature. These materials would revolutionize many areas of industry, in particular, transportation and the power grid. Superconducting power transmission lines are already used in some pilot projects, such as the 574-megawatt transmission line used by the Long Island Power Authority (Figure 28.34), which utilizes American

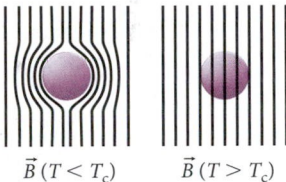

FIGURE 28.32 The Meissner effect, in which a material excludes external magnetic fields from its interior when it is cooled below the critical temperature at which it becomes superconducting.

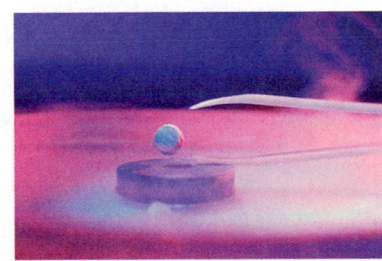

FIGURE 28.33 Through the Meissner effect, a superconductor expels the magnetic field of a permanent magnet, which thus hovers above it.

FIGURE 28.34 The 574-megawatt high-temperature superconducting wire power transmission line used by the Long Island Power Authority. Photo courtesy of AMSC.

Superconductor's high-temperature superconducting wires and a liquid air cooling system. One major future project using superconducting wires for high-current power transmission of up to 5 gigawatts in each cable is the Tres Amigas SuperStation in Clovis, New Mexico. It will connect the two main North American power grids, the Eastern Interconnection and the Western Interconnection, with the Texas Interconnection.

WHAT WE HAVE LEARNED | EXAM STUDY GUIDE

- The magnetic permeability of free space, μ_0, is given by $4\pi \cdot 10^{-7}$ T m/A.

- The Biot-Savart Law, $d\vec{B} = \dfrac{\mu_0}{4\pi} \dfrac{i\, d\vec{s} \times \hat{r}}{r^2}$, describes the differential magnetic field, $d\vec{B}$, caused by a current element, $i\, d\vec{s}$, at a position $\vec{r}$ relative to the current element.

- The magnitude of the magnetic field at a distance $r_\perp$ from a long, straight wire carrying a current i is $B = \mu_0 i / 2\pi r_\perp$.

- The magnetic field magnitude at the center of a loop with radius R carrying a current i is $B = \mu_0 i / 2R$.

- Ampere's Law is given by $\oint \vec{B} \cdot d\vec{s} = \mu_0 i_{enc}$, where $d\vec{s}$ is the integration path and i_{enc} is the current enclosed in a chosen Amperian loop.

- The magnitude of the magnetic field inside a solenoid carrying a current i and having n turns per unit length is $B = \mu_0 n i$.

- The magnitude of the magnetic field inside a toroid having N turns and a radius r and carrying a current is given by $B = \mu_0 N i / 2\pi r$.

- For an electron with charge $-e$ and mass m moving in a circular orbit, the magnetic dipole moment can be related to the orbital angular momentum through $\vec{\mu}_{orb} = -\dfrac{e}{2m} \vec{L}_{orb}$.

- For diamagnetic and paramagnetic materials, the magnetization is proportional to the magnetic field strength: $\vec{M} = \chi_m \vec{H}$. Ferromagnetic materials follow a hysteresis loop and thus deviate from this linear relationship.

- The magnetic field inside a diamagnetic or paramagnetic material is due to the external magnetic field strength and the magnetization:

$$\vec{B} = \mu_0(\vec{H} + \vec{M}) = \mu_0(\vec{H} + \chi_m \vec{H}) = \mu_0(1 + \chi_m)\vec{H} = \mu_0 \kappa_m \vec{H} = \mu \vec{H},$$

where κ_m is the relative magnetic permeability.

- The four right-hand rules related to magnetic fields are shown in Figure 28.35. Right-hand rule 1 gives the direction of the magnetic force on a charged particle moving in a magnetic field. Right-hand rule 2 gives the direction of the unit normal vector for a current-carrying loop. Right-hand rule 3 gives the direction of the magnetic field from a current-carrying wire. Right-hand rule 4 gives the direction of the magnetic field inside a toroidal magnet.

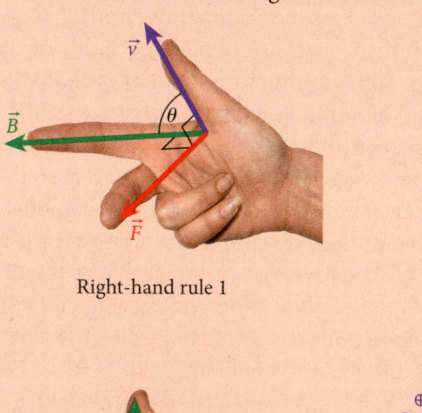

Right-hand rule 1

Right-hand rule 2

Right-hand rule 3

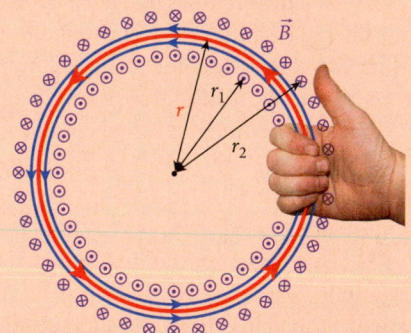

Right-hand rule 4

FIGURE 28.35 Four right-hand rules related to magnetic fields.

ANSWERS TO SELF-TEST OPPORTUNITIES

28.1 The magnetic field at point P_1 points in the positive y-direction. The magnetic field at point P_2 points in the negative x-direction.

28.2 Two parallel wires carrying current in the same direction attract each other. Two parallel wires carrying current in opposite directions repel each other.

28.3 $B_x = \dfrac{\mu_0 i}{2} \dfrac{R^2}{\left(0^2 + R^2\right)^{3/2}} = \dfrac{\mu_0 i}{2} \dfrac{R^2}{R^3} = \dfrac{\mu_0 i}{2R}.$

PROBLEM-SOLVING GUIDELINES

1. When using the Biot-Savart Law, you should always draw a diagram of the situation, with the current element highlighted. Check for simplifying symmetries before proceeding with calculations; you can save yourself a significant amount of work.

2. When applying Ampere's Law, choose an Amperian loop that has some geometrical symmetry, in order to simplify the evaluation of the integral. Often, you can use right-hand rule 3 to choose the direction of integration along the loop: Point your thumb in the direction of the net current through the loop and your fingers curl in the direction of integration. This method will also remind you to sum the currents through the Amperian loop to determine the enclosed current.

3. Remember the superposition principle for magnetic fields: The net magnetic field at any point in space is the vector sum of the individual magnetic fields generated by different objects. Make sure you do not simply add the magnitudes. Instead, you generally need to add the spatial components of the different sources of magnetic field separately.

4. All of the principles governing the motion of charged particles in magnetic fields and all of the problem-solving guidelines presented in Chapter 27 still apply. It does not matter if the magnetic field is due to a permanent magnet or an electromagnet.

5. In order to calculate the magnetic field in a material, you can use the formulas derived from Ampere's Law and Biot-Savart's Law, but you have to replace μ_0 with $\mu = (1 + \chi_m)\mu_0 = \kappa_m\mu_0$.

MULTIPLE-CHOICE QUESTIONS

28.1 Two long, straight wires are parallel to each other. The wires carry currents of different magnitudes. If the amount of current flowing in each wire is doubled, the magnitude of the force between the wires will be

a) twice the magnitude of the original force.

b) four times the magnitude of the original force.

c) the same as the magnitude of the original force.

d) half of the magnitude of the original force.

28.2 A current element produces a magnetic field in the region surrounding it. At any point in space, the magnetic field produced by this current element points in a direction that is

a) radial from the current element to the point in space.

b) parallel to the current element.

c) perpendicular to the current element and to the radial direction.

28.3 The number of turns in a solenoid is doubled, and its length is halved. How does its magnetic field change?

a) It doubles. c) It quadruples.

b) It is halved. d) It remains unchanged.

28.4 Consider two parallel current-carrying wires. The magnetic fields cause attractive forces between the wires, so it appears that the magnetic field due to one wire is doing work on the other wire. How is this explained?

a) The magnetic force can do no work on isolated charges; this says nothing about the work it can do on charges confined in a conductor.

b) Since only an electric field can do work on charges, it is actually the electric fields doing the work here.

c) This apparent work is due to another type of force.

28.5 In a solenoid in which the wires are wound such that each loop touches the adjacent ones, which of the following will increase the magnetic field inside the magnet?

a) making the radius of the loops smaller

b) increasing the radius of the wire

c) increasing the radius of the solenoid

d) decreasing the radius of the wire

e) immersing the solenoid in gasoline

28.6 Two insulated wires cross at a 90° angle. Currents are sent through the two wires. Which one of the figures best represents the configuration of the wires if the current in the horizontal wire flows in the positive x-direction and the current in the vertical wire flows in the positive y-direction?

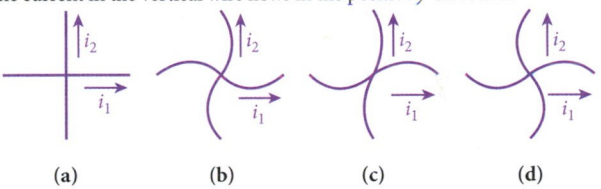

(a)   (b)   (c)   (d)

28.7 What is a good rule of thumb for designing a simple magnetic coil? Specifically, given a circular coil of radius ~1 cm, what is the approximate magnitude of the magnetic field, in gausses per amp per turn? (*Note:* 1 G = 0.0001 T.)

a) 0.0001 G/(A-turn) c) 1 G/(A-turn)

b) 0.01 G/(A-turn) d) 100 G/(A-turn)

28.8 A solid cylinder carries a current that is uniform over its cross section. Where is the magnitude of the magnetic field the greatest?

a) at the center of the cylinder's cross section

b) in the middle of the cylinder

c) at the surface

d) none of the above

28.9 Two long, straight wires have currents flowing in them in the same direction, as shown in the figure. The force between the wires is

a) attractive. b) repulsive. c) zero.

28.10 In a magneto-optic experiment, a liquid sample in a 10-mL spherical vial is placed in a highly uniform magnetic field, and a laser beam is directed through the sample. Which of the following should be used to create the uniform magnetic field required by the experiment?

a) a 5-cm-diameter flat coil consisting of one turn of 4-gauge wire

b) a 10-cm-diameter, 20-turn, single-layer, tightly wound coil made of 18-gauge wire

c) a 2-cm-diameter, 10-cm-long, tightly wound solenoid made of 18-gauge wire

d) a set of two coaxial 10-cm-diameter coils at a distance of 5 cm apart, each consisting of one turn of 4-gauge wire

28.11 Assume that a lightning bolt can be modeled as a long, straight line of current. If 15.0 C of charge passes by a point in $1.50 \cdot 10^{-3}$ s, what is the magnitude of the magnetic field at a distance of 26.0 m from the lightning bolt?

a) $7.69 \cdot 10^{-5}$ T
c) $4.21 \cdot 10^{-2}$ T
e) $2.22 \cdot 10^{2}$ T

b) $9.22 \cdot 10^{-3}$ T
d) $1.11 \cdot 10^{-1}$ T

28.12 Two solenoids have the same length, but solenoid 1 has 15 times more turns and $\frac{1}{9}$ as large a radius and carries 7 times as much current as solenoid 2. Calculate the ratio of the magnitude of the magnetic field inside solenoid 1 to that of the magnetic field inside solenoid 2.

a) 105
c) 144
e) 197

b) 123
d) 168

28.13 The wire in the figure carries a current i and contains a circular arc of radius R and angle $\pi/2$ and two straight sections that are mutually perpendicular and, if extended, would intersect the center, C, of the arc. What is the magnetic field at point C due to the wire?

a) $B = \dfrac{\mu_0 i}{2R}$

b) $B = \dfrac{\mu_0 i}{4R}$

c) $B = \dfrac{\mu_0 i}{6R}$

d) $B = \dfrac{\mu_0 i}{8R}$

e) $B = \dfrac{\mu_0 i}{12R}$

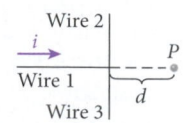

28.14 Wire 1 carries a current i_1, and wire 2 carries a current i_2 in the opposite direction, as shown in the figure. What is the direction of the force exerted by wire 1 on a length L of wire 2?

a) toward wire 1

b) away from wire 1

c) Wire 1 does not exert a force on wire 2 in this situation.

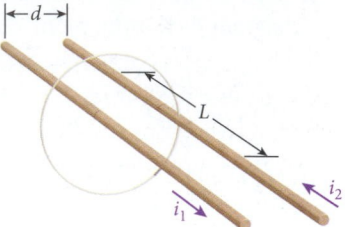

CONCEPTUAL QUESTIONS

28.15 Many electrical applications use twisted-pair cables in which the ground and signal wires spiral about each other. Why?

28.16 Discuss how the accuracy of a compass needle in showing the true direction of north can be affected by the magnetic field due to currents in wires and appliances in a residential building.

28.17 Can an ideal solenoid, one with no magnetic field outside the solenoid, exist? If not, does that invalidate the derivation of the magnetic field inside the solenoid (Section 28.4)?

28.18 Conservative forces tend to act on objects in such a way as to minimize a system's potential energy. Use this principle to explain the direction of the force on the current-carrying loop described in Example 28.1.

28.19 Two particles, each with charge q and mass m, are traveling in a vacuum on parallel trajectories a distance d apart and at a speed v (much less than the speed of light). Calculate the ratio of the magnitude of the magnetic force that each exerts on the other to the magnitude of the electric force that each exerts on the other: $F_{\mathrm{m}}/F_{\mathrm{e}}$.

28.20 A long, straight, cylindrical tube of inner radius a and outer radius b carries a total current i uniformly across its cross section. Determine the magnitude of the magnetic field due to the tube at the midpoint between the inner and outer radii.

28.21 Three identical straight wires are connected in a T, as shown in the figure. If current i flows into the junction, what is the magnetic field at point P, a distance d from the junction?

28.22 In a certain region, there is a constant and uniform magnetic field, $\vec{B}$. Any electric field in the region is also unchanging in time. Find the current density, $\vec{J}$, in this region.

28.23 The magnetic character of bulk matter is determined largely by electron spin magnetic moments, rather than by orbital dipole moments. (Nuclear contributions are negligible, as the proton's spin magnetic moment is about 658 times smaller than that of the electron.) If the atoms or molecules of a substance have unpaired electron spins, the associated magnetic moments give rise to paramagnetism or to ferromagnetism if the

interactions between atoms or molecules are strong enough to align them in domains. If the atoms or molecules have no net unpaired spins, then magnetic perturbations of electrons' orbits give rise to diamagnetism.

a) Molecular hydrogen gas (H_2) is weakly diamagnetic. What does this imply about the spins of the two electrons in the hydrogen molecule?

b) What would you expect the magnetic behavior of atomic hydrogen gas (H) to be?

28.24 Exposed to sufficiently high magnetic fields, materials *saturate*, or approach a maximum magnetization. Would you expect the saturation (maximum) magnetization of paramagnetic materials to be much less than, roughly the same as, or much greater than that of ferromagnetic materials? Explain why.

28.25 A long, straight wire carries a current, as shown in the figure. A single electron is shot directly toward the wire from above. The trajectory of the electron and the wire are in the same plane. Will the electron be deflected from its initial path, and if so, in which direction?

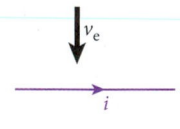

28.26 A square loop, with sides of length L, carries current i. Find the magnitude of the magnetic field from the loop at the center of the loop, as a function of i and L.

28.27 A current of constant density, J_0, flows through a very long cylindrical conducting shell with inner radius a and outer radius b. What is the magnetic field in the regions $r < a$, $a < r < b$, and $r > b$? Does $B_{a<r<b} = B_{r>b}$ for $r = b$?

28.28 Parallel wires, a distance D apart, carry a current, i, in opposite directions as shown in the figure. A circular loop, of radius $R = D/2$, has the same current flowing in a counterclockwise direction. Determine the magnitude and the direction of the magnetic field from the loop and the parallel wires at the center of the loop, as a function of i and R.

28.29 The current density in a cylindrical conductor of radius R varies as $J(r) = J_0 e^{-r/R}$ (in the region from zero to R). Express the magnitude of the magnetic field in the regions $r < R$ and $r > R$. Produce a sketch of the radial dependence, $B(r)$.

EXERCISES

A blue problem number indicates a worked-out solution is available in the Student Solutions Manual. One • and two •• indicate increasing level of problem difficulty.

Sections 28.1 and 28.2

28.30 Two long parallel wires are separated by 3.0 mm. The current flowing in one of the wires is twice that in the other wire. If the magnitude of the force on a 1.0-m length of one of the wires is 7.0 μN, what are the magnitudes of the two currents?

28.31 An electron is shot from an electron gun with a speed of $4.0 \cdot 10^5$ m/s and moves parallel to and at a distance of 5.0 cm above a long, straight wire carrying a current of 15 A. Determine the magnitude and the direction of the acceleration of the electron the instant it leaves the electron gun.

28.32 An electron moves in a straight line at a speed of $5.00 \cdot 10^6$ m/s. What are the magnitude and the direction of the magnetic field created by the moving electron at a distance $d = 5.00$ m ahead of it on its line of motion? How does the answer change if the moving particle is a proton?

28.33 Suppose that the magnetic field of the Earth were due to a single current moving in a circle of radius $2.00 \cdot 10^3$ km through the Earth's molten core. The strength of the Earth's magnetic field on the surface near a magnetic pole is about $6.00 \cdot 10^{-5}$ T. About how large a current would be required to produce such a field?

28.34 A square ammeter has sides of length 3.00 cm. The sides of the ammeter are capable of measuring the magnetic field they are subject to. When the ammeter is clamped around a wire carrying a direct current, as shown in the figure, the average value of the magnetic field measured in the sides is 3.00 G. What is the current in the wire?

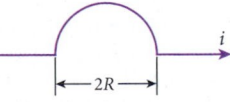

3.00 cm

•28.35 A long, straight wire carrying a 2.00-A current lies along the x-axis. A particle with charge $q = -3.00$ μC moves parallel to the y-axis through the point $(x,y,z) = (0,2,0)$. Where in the xy-plane should another long, straight wire be placed so that there is no magnetic force on the particle at the point where it crosses the plane?

•28.36 Find the magnetic field in the center of a wire semicircle like that shown in the figure, with radius $R = 10.0$ cm, if the current in the wire is $i = 12.0$ A.

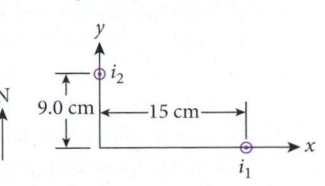

$\leftarrow 2R \rightarrow$

•28.37 Two very long wires run parallel to the z-axis, as shown in the figure. They each carry a current, $i_1 = i_2 = 25.0$ A, flowing in the direction of the positive z-axis. The magnetic field of the Earth is given by $\vec{B} = (2.60 \cdot 10^{-5})\hat{y}$ T (in the xy-plane and pointing due north). A magnetic compass needle is placed at the origin. Determine the angle θ between the compass needle and the x-axis. (*Hint:* The compass needle will align its axis along the direction of the net magnetic field.)

N

9.0 cm ⊢―15 cm―⊢

$\circ i_2$

$\circ i_1$

→ x

•28.38 Two identical coaxial coils of wire of radius 20.0 cm are directly on top of each other, separated by a 2.00-mm gap. The lower coil is on a flat table and has a current i in the clockwise direction; the upper coil carries an identical current and has a mass of 0.0500 kg. Determine the magnitude and the direction that the current in the upper coil has to have to keep it levitated at the distance 2.00 mm above the lower coil.

•28.39 A long, straight wire lying along the x-axis carries a current, i, flowing in the positive x-direction. A second long, straight wire lies along

the y-axis and has a current i in the positive y-direction. What are the magnitude and the direction of the magnetic field at point $z = b$ on the z-axis?

•28.40 A square loop of wire with a side length of 10.0 cm carries a current of 0.300 A. What is the magnetic field in the center of the square loop?

•28.41 The figure shows the cross section through three long wires with a linear mass distribution of 100. g/m. They carry currents i_1, i_2, and i_3 in the directions shown. Wires 2 and 3 are 10.0 cm apart and are attached to a vertical surface, and each carries a current of 600. A. What current, i_1, will allow wire 1 to "float" at a perpendicular distance $d = 10.0$ cm from the vertical surface? (Neglect the thickness of the wires.)

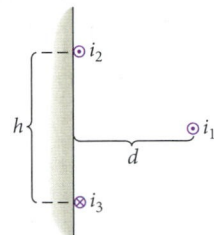

$\odot i_2$

h ⟨

$\odot i_1$

d

$\otimes i_3$

•28.42 A hairpin configuration is formed of two semi-infinite straight wires that are 2.00 cm apart and joined by a semicircular piece of wire (whose radius must be 1.00 cm and whose center is at the origin of xyz-coordinates). The top wire lies along the line $y = 1.00$ cm, and the bottom wire lies along the line $y = -1.00$ cm; these two wires are in the left side ($x < 0$) of the xy-plane. The current in the hairpin is 3.00 A, and it is directed toward the right in the top wire, clockwise around the semicircle, and to the left in the bottom wire. Find the magnetic field at the origin of the coordinate system.

•28.43 A long, straight wire is located along the x-axis ($y = 0$ and $z = 0$). The wire carries a current of 7.00 A in the positive x-direction. What are the magnitude and the direction of the force on a particle with a charge of 9.00 C located at $(+1.00$ m$,+2.00$ m$,0)$, when it has a velocity of 3000. m/s in each of the following directions?

a) the positive x-direction   c) the negative z-direction

b) the positive y-direction

•28.44 A long, straight wire has a 10.0-A current flowing in the positive x-direction, as shown in the figure. Close to the wire is a square loop of copper wire that carries a 2.00-A current in the direction shown. The near side of the loop is $d = 0.500$ m away from the wire. The length of each side of the square is $a = 1.00$ m.

a) Find the net force between the two current-carrying objects.

b) Find the net torque on the loop.

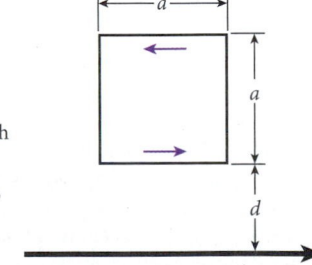

$\leftarrow a \rightarrow$

a

d

28.45 A square box with sides of length 1.00 m has one corner at the origin of a coordinate system, as shown in the figure. Two coils are attached to the outside of the box. One coil is on the box face that is in the xz-plane at $y = 0$, and the second is on the box face in the yz-plane at $x = 1.00$ m. Each of the coils has a diameter of 1.00 m and contains 30.0 turns of wire carrying a current of 5.00 A in each turn. The current in each coil is clockwise when the coil is viewed from outside the box. What are the magnitude and the direction of the magnetic field at the center of the box?

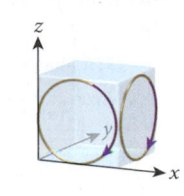

z

x

Section 28.3

28.46 The current density in a cylindrical conductor of radius R varies as $J(r) = J_0 r/R$ (in the region from zero to R). Express the magnitude of the magnetic field in the regions $r < R$ and $r > R$. Produce a sketch of the radial dependence, $B(r)$.

28.47 The figure shows a cross section across the diameter of a long, solid, cylindrical conductor. The radius of the cylinder is $R = 10.0$ cm. A current of 1.35 A is uniformly distributed throughout the conductor and is flowing out of the page. Calculate the direction and the magnitude of the magnetic field at positions $r_a = 0.0$ cm, $r_b = 4.00$ cm, $r_c = 10.0$ cm, and $r_d = 16.0$ cm.

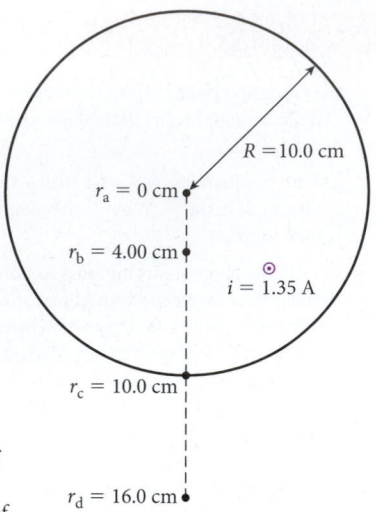

$R = 10.0$ cm
$r_a = 0$ cm
$r_b = 4.00$ cm
$i = 1.35$ A
$r_c = 10.0$ cm
$r_d = 16.0$ cm

••28.48 A coaxial wire consists of a copper core of radius 1.00 mm surrounded by a copper sheath of inside radius 1.50 mm and outside radius 2.00 mm. A current, i, flows in one direction in the core and in the opposite direction in the sheath. Graph the magnitude of the magnetic field as a function of the distance from the center of the wire.

•28.49 A very large sheet of conducting material located in the xy-plane, as shown in the figure, has a uniform current flowing in the y-direction. The current density is 1.5 A/cm. Use Ampere's Law to calculate the direction and the magnitude of the magnetic field just above the center of the sheet (not close to any edges).

Current density = 1.5 A/cm
i
z y
x

Section 28.4

28.50 A current of 2.00 A is flowing through a 1000-turn solenoid of length $L = 40.0$ cm. What is the magnitude of the magnetic field inside the solenoid?

28.51 Solenoid A has twice the diameter, three times the length, and four times the number of turns of solenoid B. The two solenoids have currents of equal magnitudes flowing through them. Find the ratio of the magnitude of the magnetic field in the interior of solenoid A to that of solenoid B.

28.52 A long solenoid (diameter of 6.00 cm) is wound with 1000 turns per meter of thin wire through which a current of 0.250 A is maintained. A wire carrying a current of 10.0 A is inserted along the axis of the solenoid. What is the magnitude of the magnetic field at a point 1.00 cm from the axis?

28.53 A long, straight wire carries a current of 2.5 A.

a) What is the strength of the magnetic field at a distance of 3.9 cm from the wire?

b) If the wire still carries 2.5 A, but is used to form a long solenoid with 32 turns per centimeter and a radius of 3.9 cm, what is the strength of the magnetic field inside the solenoid?

28.54 Figure 28.20a shows a Helmholtz coil used to generate uniform magnetic fields. Suppose the Helmholtz coil consists of two sets of coaxial wire loops with 15 turns of radius $R = 75.0$ cm, which are separated by R, and each coil carries a current of 0.123 A flowing in the same direction. Calculate the magnitude and the direction of the magnetic field in the center between the coils.

•28.55 A particle detector utilizes a solenoid that has 550 turns of wire per centimeter. The wire carries a current of 22 A. A cylindrical detector that lies within the solenoid has an inner radius of 0.80 m. Electron and positron beams are directed into the solenoid parallel to its axis. What is the minimum momentum perpendicular to the solenoid axis that a particle can have if it is to be able to enter the detector?

Sections 28.5 through 28.7

28.56 An electron has a spin magnetic moment of magnitude $\mu = 9.285 \cdot 10^{-24}$ A m^2. Consequently, it has energy associated with its orientation in a magnetic field. If the difference between the energy of an electron that is "spin up" in a magnetic field of magnitude B and the energy of one that is "spin down" in the same magnetic field (where "up" and "down" refer to the direction of the magnetic field) is $9.460 \cdot 10^{-25}$ J, what is the field magnitude, B?

28.57 When a magnetic dipole is placed in a magnetic field, it has a natural tendency to minimize its potential energy by aligning itself with the field. If there is sufficient thermal energy present, however, the dipole may rotate so that it is no longer aligned with the field. Using $k_B T$ as a measure of the thermal energy, where k_B is Boltzmann's constant and T is the temperature in kelvins, determine the temperature at which there is sufficient thermal energy to rotate the magnetic dipole associated with a hydrogen atom from an orientation parallel to an applied magnetic field to one that is antiparallel to the applied field. Assume that the strength of the field is 0.15 T.

28.58 Aluminum becomes superconducting at a temperature around 1.0 K if exposed to a magnetic field of magnitude less than 0.0105 T. Determine the maximum current that can flow in an aluminum superconducting wire with radius $R = 1.0$ mm.

28.59 If you want to construct an electromagnet by running a current of 3.00 A through a solenoid with 500. windings and length 3.50 cm and you want the magnetic field inside the solenoid to have the magnitude $B = 2.96$ T, you can insert a ferrite core into the solenoid. What value of the relative magnetic permeability should this ferrite core have in order to make this work?

28.60 What is the magnitude of the magnetic field inside a long, straight tungsten wire of circular cross section with diameter 2.4 mm and carrying a current of 3.5 A, at a distance of 0.60 mm from its central axis?

•28.61 You charge up a small rubber ball of mass 200. g by rubbing it over your hair. The ball acquires a charge of 2.00 μC. You then tie a 1.00-m-long string to it and swing it in a horizontal circle, providing a centripetal force of 25.0 N. What is the magnetic moment of the system?

•28.62 Consider a model of the hydrogen atom in which an electron orbits a proton in the plane perpendicular to the proton's spin angular momentum (and magnetic dipole moment) at a distance equal to the Bohr radius, $a_0 = 5.292 \cdot 10^{-11}$ m. (This is an oversimplified classical model.) The spin of the electron is allowed to be either parallel to the proton's spin or antiparallel to it; the orbit is the same in either case. But since the proton produces a magnetic field at the electron's location, and the electron has its own intrinsic magnetic dipole moment, the energy of the electron differs depending on its spin. The magnetic field produced by the proton's spin may be modeled as a dipole field, like the electric field due to an electric dipole discussed in Chapter 22. Calculate the energy difference between the two electron-spin configurations. Consider only the interaction between the magnetic dipole moment associated with the electron's spin and the field produced by the proton's spin.

••28.63 Consider an electron to be a uniformly dense sphere of charge, with a total charge of $-e = -1.602 \cdot 10^{-19}$ C, spinning at an angular frequency, ω.

a) Write an expression for its classical angular momentum of rotation, L.

b) Write an expression for its magnetic dipole moment, μ.

c) Find the ratio, $\gamma_e = \mu/L$, known as the *gyromagnetic ratio*.

Additional Exercises

28.64 Two 50-turn coils, each with a diameter of 4.00 m, are placed 1.00 m apart, as shown in the figure. A current of 7.00 A is flowing in the wires of both coils; the direction of the current is clockwise for both coils when viewed from the left. What is the magnitude of the magnetic field in the center between the two coils?

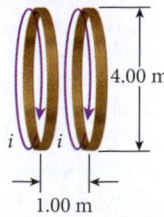

4.00 m
i i
1.00 m

28.65 The wires in the figure are separated by a vertical distance d. Point B is at the midpoint between the two wires; point A is a distance $d/2$ from the lower wire. The horizontal distance between A and B is much larger than d. Both wires carry the same current, i. The strength of the magnetic field at point A is 2.00 mT. What is the strength of the field at point B?

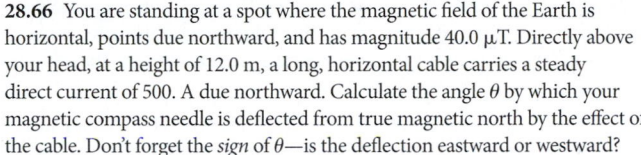

28.66 You are standing at a spot where the magnetic field of the Earth is horizontal, points due northward, and has magnitude 40.0 μT. Directly above your head, at a height of 12.0 m, a long, horizontal cable carries a steady direct current of 500. A due northward. Calculate the angle θ by which your magnetic compass needle is deflected from true magnetic north by the effect of the cable. Don't forget the *sign* of θ—is the deflection eastward or westward?

28.67 The magnetic dipole moment of the Earth is approximately $8.0 \cdot 10^{22}$ A m². The source of the Earth's magnetic field is not known; one possibility might be the circulation of ions in the Earth's molten outer core. Assume that the circulating ions move a circular loop of radius 2500 km. What "current" must they produce to yield the observed field?

28.68 A circular wire loop has a radius $R = 0.12$ m and carries a current $i = 0.10$ A. The loop is placed in the xy-plane in a uniform magnetic field given by $\vec{B} = -1.5\hat{z}$ T, as shown in the figure. Determine the direction and the magnitude of the loop's magnetic moment and calculate the potential energy of the loop in the position shown. If the wire loop can move freely, how will it orient itself to minimize its potential energy, and what is the value of the lowest potential energy?

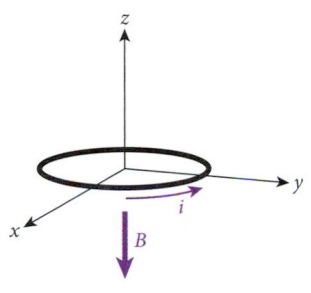

28.69 A 0.90-m-long solenoid has a radius of 5.0 mm. When the wire carries a 0.20-A current, the magnetic field inside the solenoid is 5.0 mT. How many turns of wire are there in the solenoid?

28.70 In a coaxial cable, the solid core carries a current i. The sheath also carries a current i but in the opposite direction, and it has an inner radius a and an outer radius b. The current density is equally distributed over each conductor. Find an expression for the magnetic field at a distance $a < r < b$ from the center of the core.

•28.71 A 50-turn rectangular coil of wire with dimensions 10.0 cm by 20.0 cm lies in a horizontal plane, as shown in the figure. The axis of rotation of the coil is aligned north and south. It carries a current $i = 1.00$ A and is in a magnetic field pointing from west to east. A mass of 50.0 g hangs from one side of the coil. Determine the strength the magnetic field has to have to keep the coil in the horizontal orientation.

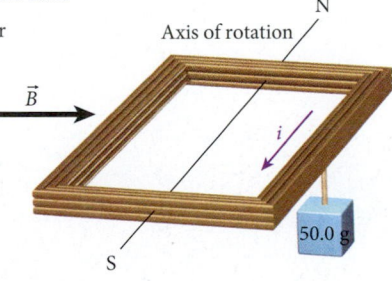

•28.72 Two long, straight parallel wires are separated by a distance of 20.0 cm. Each wire carries a current of 10.0 A in the same direction. What is the magnitude of the resulting magnetic field at a point that is 12.0 cm from each wire?

•28.73 A particle with a mass of 1.00 mg and a charge q is moving at a speed of 1000. m/s along a horizontal path 10.0 cm below and parallel to a straight current-carrying wire. Determine q if the magnitude of the current in the wire is 10.0 A.

•28.74 A conducting coil consisting of n turns of wire is placed in a uniform magnetic field given by $\vec{B} = 2.00\,\hat{y}$ T, as shown in the figure. The radius of the

coil is $R = 5.00$ cm, and the angle between the magnetic field vector and the unit normal vector to the coil is $\theta = 60.0°$. The current through the coil is $i = 5.00$ A.

a) Specify the direction of the current in the coil, given the direction of the magnetic dipole moment, $\vec{\mu}$, shown in the figure.

b) Calculate the number of turns, n, the coil must have for the torque on the loop to be 3.40 N m.

c) If the radius of the loop is decreased to $R = 2.50$ cm, what should the number of turns, N, be for the torque to remain unchanged? Assume that i, B, and θ stay the same.

Side view

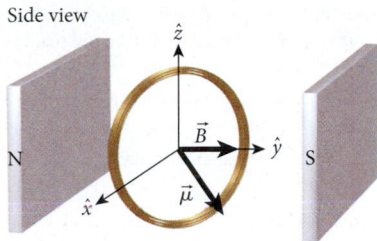

Top view

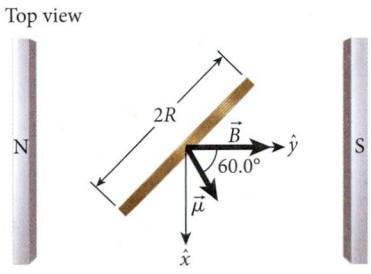

•28.75 A loop of wire of radius $R = 25.0$ cm has a smaller loop of radius $r = 0.900$ cm at its center, with the planes of the two loops perpendicular to each other. When a current of 14.0 A is passed through both loops, the smaller loop experiences a torque due to the magnetic field produced by the larger loop. Determine this torque, assuming that the smaller loop is sufficiently small that the magnetic field due to the larger loop is the same across its entire surface.

•28.76 Two wires, each 25.0 cm long, are connected to two separate 9.00-V batteries, as shown in the figure. The resistance of the first wire is 5.00 Ω, and that of the other wire is unknown (R). If the separation between the wires is 4.00 mm, what value of R will produce a force of magnitude $4.00 \cdot 10^{-5}$ N between them? Is the force attractive or repulsive?

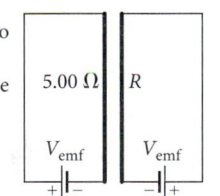

•28.77 A proton is moving under the combined influence of an electric field ($E = 1000.$ V/m) and a magnetic field ($B = 1.20$ T), as shown in the figure.

a) What is the acceleration of the proton at the instant it enters the crossed fields?

b) What would the acceleration be if the direction of the proton's motion were reversed?

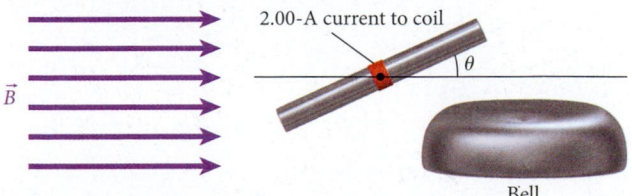

•28.78 A toy airplane of mass 0.175 kg, with a charge of 36 mC, is flying with a speed of 2.8 m/s at a height of 17.2 cm above and parallel to a wire, which is carrying a 25-A current. The airplane experiences some acceleration. Determine this acceleration.

•28.79 An electromagnetic doorbell has been constructed by wrapping 70 turns of wire around a long, thin rod, as shown in the figure. The rod has a mass of 30.0 g, a length of 8.00 cm, and a cross-sectional area of 0.200 cm². The rod is free to pivot about an axis through its center, which is also the center of the coil. Initially, the rod makes an angle of $\theta = 25.0°$ with the horizontal. When $\theta = 0.00°$, the rod strikes a bell. A uniform magnetic field of 900. G is directed at an angle $\theta = 0.00°$.

a) If a current of 2.00 A is flowing in the coil, what is the torque on the rod when $\theta = 25.0°$?

b) What is the angular velocity of the rod when it strikes the bell?

•28.80 Two long, parallel wires separated by a distance d carry currents in opposite directions. If the left-hand wire carries a current $i/2$ and the right-hand wire carries a current i, determine where the magnetic field is zero.

•28.81 A horizontally oriented coil of wire of radius 5.00 cm that carries a current, i, is being levitated by the south pole of a vertically oriented bar magnet suspended above its center, as shown in the figure. If the magnetic field on all parts of the coil makes an angle $\theta = 45.0°$ with the vertical, determine the magnitude and the direction of the current needed to keep the coil floating in midair. The magnitude of the magnetic field is $B = 0.0100$ T, the number of turns in the coil is $N = 10.0$, and the total coil mass is 10.0 g.

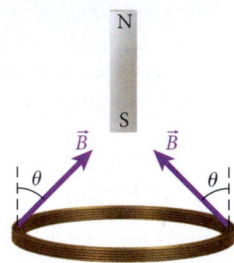

•28.82 As shown in the figure, a long, hollow, conducting cylinder of inner radius a and outer radius b carries a current that is flowing out of the page. Suppose that $a = 5.00$ cm, $b = 7.00$ cm, and the current $i = 100.$ mA, uniformly distributed over the cylinder wall (between a and b). Find the magnitude of the magnetic field at each of the following distances r from the center of the cylinder:

a) $r = 4.00$ cm

b) $r = 6.50$ cm

c) $r = 9.00$ cm

•28.83 A wire of radius R carries a current i. The current density is given by $J = J_0(1 - r/R)$, where r is measured from the center of the wire and J_0 is a constant. Use Ampere's Law to find the magnetic field inside the wire at a distance $r < R$ from the central axis.

•28.84 A circular wire of radius 5.0 cm has a current of 3.0 A flowing in it. The wire is placed in a uniform magnetic field of 5.0 mT.

a) Determine the maximum torque on the wire.

b) Determine the range of the magnetic potential energy of the wire.

MULTI-VERSION EXERCISES

28.85 The loop shown in the figure is carrying a current of 3.857 A, and the distance $r = 1.411$ m. What is the magnitude of the magnetic field at point P inside the loop?

28.86 The loop shown in the figure is carrying a current of 3.961 A. The magnitude of the magnetic field at point P inside the loop is $7.213 \cdot 10^{-7}$ T. What is the value of r?

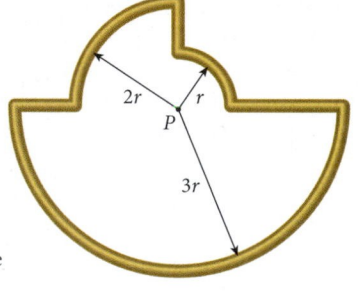

28.87 In the current-carrying loop shown in the figure, the distance $r = 2.329$ m. The magnitude of the magnetic field at point P inside the loop is $5.937 \cdot 10^{-7}$ T. What is the current in the loop?

28.88 A toroidal magnet has an inner radius of 1.895 m and an outer radius of 2.075 m. When the wire carries a 33.45-A current, the magnetic field at a distance of 1.985 m from the center of the toroid is 66.78 mT. How many turns of wire are there in the toroid?

28.89 A toroidal magnet has an inner radius of 1.121 m and an outer radius of 1.311 m. The magnetic field at a distance of 1.216 m from the center of the toroid is 78.30 mT. There are 22,381 turns of wire in the toroid. What is the current in the toroid?

28.90 A toroidal magnet has an inner radius of 1.351 m and an outer radius of 1.541 m. The wire carries a 49.13-A current, and there are 24,945 turns in the toroid. What is the magnetic field at a distance of 1.446 m from the center of the toroid?

29

Electromagnetic Induction

FIGURE 29.1 The Grand Coulee Dam on the Columbia River in the state of Washington is the largest single producer of electricity in the United States. Shown here are the giant generators, which apply the physical principle of induction to produce electricity.

M ost of us take electric power for granted—we flip a switch and we have power for lighting, heating, and entertainment. But the vast network that supplies this power—called *the grid*—depends on large generators that convert mechanical energy into electrical energy (Figure 29.1). The physical principles that enable this conversion are the subject of this chapter.

In Chapter 27, we saw that a magnetic field can affect the path of charged particles, or electric currents, and in Chapter 28, we saw that an electric current generates a magnetic field. In this chapter, we'll see that a changing magnetic field generates an electric current, and thus an electric field. Note the word "changing" here; just as a magnetic field is generated only when electrical charges are in motion, an electric field is generated only when a magnetic field is in motion (relative to a conductor) or otherwise changes as a function of time. This symmetry will turn out to be a key part of the unified description of electricity and magnetism presented in Chapter 31.

WHAT WE WILL LEARN

- A changing magnetic field inside a conducting loop induces a current in the loop.

- A changing current in a loop induces a current in a nearby loop.

- Faraday's Law of Induction states that a potential difference is induced in a loop when there is a change in the magnetic flux through the loop.

- Magnetic flux is the product of the magnitude of the average magnetic field and the perpendicular area that it penetrates.

- Lenz's Law states that the current induced in a loop by a changing magnetic flux produces a magnetic field that opposes this change in magnetic flux.

- A changing magnetic field induces an electric field.

- The inductance of a device is a measure of its opposition to changes in current flowing through it.

- Electric motors and electric generators are everyday applications of magnetic induction.

- A simple single-loop circuit with an inductor and a resistor has a characteristic time constant given by the inductance divided by the resistance.

- Energy is stored in a magnetic field.

29.1 Faraday's Experiments

Some of the great discoveries about electricity and magnetism took place in the late 18th and early 19th centuries. In 1750, the American Benjamin Franklin showed that lightning is a form of electricity, with his famous kite-flying experiment. (Perhaps the most amazing aspect of that experiment was that he was not killed by the lightning strike.) In 1799, the Italian Alessandro Volta constructed the first battery, called a *voltaic pile* at the time. In 1820, the Danish physicist Hans Christian Oersted demonstrated that an electric current could produce a magnetic field strong enough to deflect a compass needle. (He performed his experiment during a lecture to his students, making it one of the most productive lecture demonstrations in the history of science.)

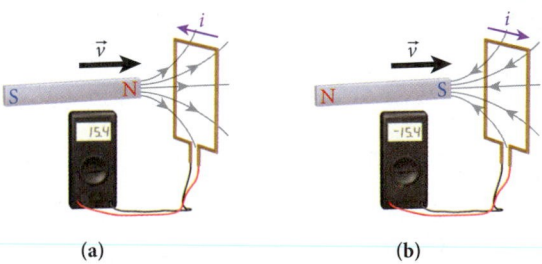

FIGURE 29.2 Moving a magnet toward a wire loop induces a current to flow in the loop. (a) With the north pole of the magnet pointing toward the loop, a positive current results. (b) With the south pole of the magnet pointing toward the loop, a negative current results.

However, the experiments that are most relevant to this chapter were performed in the 1830s by the British chemist and physicist Michael Faraday and independently by the American physicist Joseph Henry. Their work demonstrated that a changing magnetic field could generate a potential difference in a conductor, strong enough to produce an electric current. This discovery is of basic importance to all the electrical and magnetic devices we use every day, from computers to cell phones, from televisions to credit cards, from the tiniest batteries to the largest electrical power grids. Both Faraday and Henry had fundamental electrical units named after them, and justifiably so.

To understand Faraday's experiments, consider a wire loop connected to an ammeter. A bar magnet is some distance from the loop with its north pole pointing toward the loop. While the magnet is stationary, no current flows in the loop. However, if the magnet is moved toward the loop (Figure 29.2a), a counterclockwise current flows in the loop as indicated by the positive current in the ammeter. If the magnet is moved toward the loop faster, a larger current is induced in the loop. If the magnet is reversed, so the south pole points toward the loop (Figure 29.2b), and moved toward the loop, current flows in the loop in the opposite direction. If the north pole of the magnet points toward the loop, and the magnet is then moved *away* from the loop (Figure 29.3a), a negative clockwise current, as indicated on the meter in Figure 29.3a, is induced in the loop. If the south pole of the magnet points toward the loop, and the magnet is moved away from the loop (Figure 29.3b), a positive current is induced.

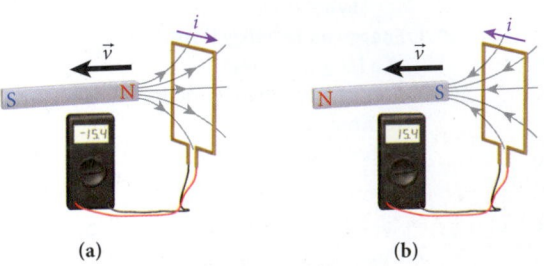

FIGURE 29.3 Moving a magnet away from a wire loop also induces a current to flow in the loop. (a) With the north pole of the magnet pointing toward the loop, a negative current results. (b) With the south pole of the magnet pointing toward the loop, a positive current results.

The four results illustrated in Figures 29.2 and 29.3 can be replicated by holding the magnets stationary and moving the loops. For example, with the arrangement shown in Figure 29.2a, if the loop is moved toward the stationary magnet, a positive current flows in the loop.

Similar effects can be observed using two conducting loops (Figure 29.4). If a constant current is flowing through loop 1, no current is induced in loop 2. If the current in loop 1 is increased, a current is induced in loop 2 in the opposite direction. Thus, not only does the increasing current in the first loop induce a current in the second loop, but the induced current is in the opposite direction. Furthermore, if current is flowing in loop 1 in the same direction as before and is then *decreased* (Figure 29.5), the current induced in loop 2 flows in the *same* direction as the current in loop 1.

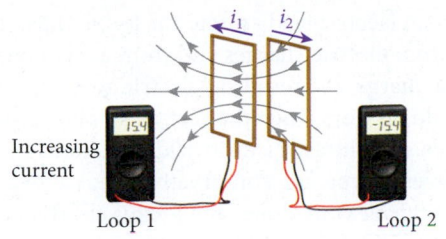

FIGURE 29.4 An increasing current in loop 1 induces a current in the opposite direction in loop 2. (The magnetic field lines shown are those produced by the current 1 flowing through loop 1.)

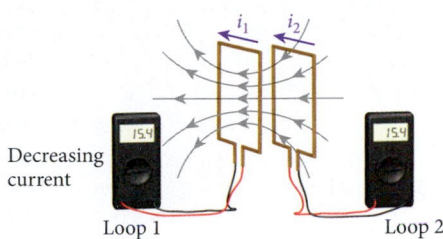

FIGURE 29.5 A decreasing current in loop 1 induces a current in the same direction in loop 2.

All of the phenomena illustrated in these four figures can be explained by Faraday's Law of Induction, discussed in Section 29.2, and by Lenz's Law, discussed in Section 29.3.

Concept Check 29.1

The four figures show a bar magnet and a low-voltage light bulb connected to the ends of a conducting loop. The plane of the loop is perpendicular to the dashed line. In case 1, the loop is stationary, and the magnet is moving away from the loop. In case 2, the magnet is stationary, and the loop is moving toward the magnet. In case 3, both the magnet and the loop are stationary, but the area of the loop is increasing. In case 4, the magnet is stationary, and the loop is rotating about its center. In which of these situations will the bulb light up?

Case 1

Case 2

Case 3

Case 4

a) case 1

b) cases 1 and 2

c) cases 1, 2, and 3

d) cases 1, 2, and 4

e) all four cases

29.2 Faraday's Law of Induction

From the observations in the preceding section, we see that a changing magnetic field through a loop induces a current in the loop. We can visualize the change in magnetic field as a change in the number of magnetic field lines passing through the loop. **Faraday's Law of Induction** in its qualitative form states:

> A potential difference is induced in a loop when the number of magnetic field lines passing through the loop changes with time.

The rate of change of the magnetic field lines determines the induced potential difference. The existence of this potential difference means that the changing magnetic field actually creates an electric field around the loop! Thus, there are two ways of producing an electric field: from electric charges and from a changing magnetic field. If the electric field arises from a charge, the resulting electric force on a test charge is conservative. Conservative forces do no work when they act on an object whose path starts and ends at the same point in space. In contrast, electric fields generated by changing magnetic fields give rise to electric forces that are *not* conservative. Thus, a test particle that moves around a circular loop once will have work done on it by this electric field. In fact, the amount of the work done is the induced potential difference times the charge of the test particle.

The magnetic field lines are quantified by the magnetic flux, in analogy with the electric flux. Chapter 22 introduced Gauss's Law for electric fields and defined the electric flux as the surface integral of an electric field passing through a differential element of area, dA. Mathematically, $\Phi_E = \iint \vec{E} \cdot d\vec{A}$, where $d\vec{A}$ is a vector of magnitude dA that is perpendicular to the differential area. By analogy, for a magnetic field, **magnetic flux** is defined as the surface integral of the magnetic field passing through a differential element of area:

$$\Phi_B = \iint \vec{B} \cdot d\vec{A}, \tag{29.1}$$

where $\vec{B}$ is the magnetic field at each differential area element, $d\vec{A}$, of a closed surface. The loop in the symbol for a surface integral means that the integration is over a closed surface. The two integrals signify integration over two variables. The differential area element, $d\vec{A}$, must be described by two spatial variables, such as x and y in Cartesian coordinates or θ and ϕ in spherical coordinates. With a closed surface, the differential area vector, $d\vec{A}$, always points out of the enclosed volume and is perpendicular to the surface everywhere.

Integration of the electric flux over a closed surface (see Chapter 22) yields Gauss's Law: $\oiint \vec{E} \cdot d\vec{A} = q/\epsilon_0$. That is, the integral of the electric flux over a closed surface is equal to the enclosed electric charge, q, divided by the electric permittivity of free space, ϵ_0. Integration of the magnetic flux over a *closed* surface yields zero:

$$\oiint \vec{B} \cdot d\vec{A} = 0. \tag{29.2}$$

This result is often termed **Gauss's Law for Magnetic Fields.** You might think that the integral of the magnetic flux over a closed surface would be equal to the enclosed "magnetic charge" divided by the magnetic permeability of free space. However, there are no free magnetic charges, no magnetic monopoles, no separate north poles or separate south poles. Magnetic poles are always found in pairs. Thus, Gauss's Law for Magnetic Fields is another way of stating that magnetic monopoles do not exist. (Extensive searches for magnetic monopoles have been conducted since the 1980s, but not successfully. However, several string theories and grand unified theories—more on those in Chapter 39—predict that magnetic monopoles exist.) Another way to state Gauss's Law for Magnetic Fields is that magnetic field lines have no beginning or end but form a continuous loop.

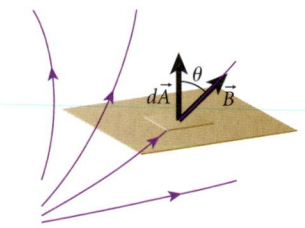

FIGURE 29.6 A nonuniform magnetic field $\vec{B}$ passing through a differential area, $d\vec{A}$.

Figure 29.6 shows a nonuniform magnetic field, $\vec{B}$, passing through a differential area element, $d\vec{A}$. A portion of the closed surface is also shown. The angle between the magnetic field and the differential area vector is θ.

Consider the special case of a flat loop of area A in a constant magnetic field, as illustrated in Figure 29.7. For this case, we can rewrite equation 29.1 as

$$\Phi_B = BA\cos\theta, \tag{29.3}$$

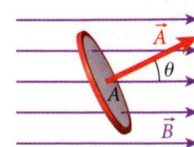

FIGURE 29.7 A flat loop of area A in a constant magnetic field, $\vec{B}$. The magnetic field makes an angle θ with respect to the surface normal vector of the loop.

where B is the magnitude of the constant magnetic field, A is the area of the loop, and θ is the angle between the surface normal vector to the plane of the loop and the magnetic field lines. Thus, if the magnetic field is perpendicular to the plane of the loop, $\theta = 0°$ and $\Phi_B = BA$. If the magnetic field is parallel to the plane of the loop, $\theta = 90°$ and $\Phi_B = 0$.

The unit of magnetic flux is $[\Phi_B] = [B][A] = \text{T m}^2$. This unit has received a special name, the **weber** (Wb):

$$1\text{ Wb} = 1\text{ T m}^2. \tag{29.4}$$

Faraday's Law of Induction is stated quantitatively, in terms of the magnetic flux, as follows:

The magnitude of the potential difference, ΔV_{ind}, induced in a conducting loop is equal to the time rate of change of the magnetic flux through the loop.

Faraday's Law of Induction is thus expressed by the equation

$$\Delta V_{\text{ind}} = -\frac{d\Phi_B}{dt}. \tag{29.5}$$

The negative sign in equation 29.5 is necessary because the induced potential difference establishes an induced current whose magnetic field tends to oppose the flux change. Section 29.3 on Lenz's Law discusses this phenomenon in detail.

The magnetic flux can be changed in several ways, including changing the magnitude of the magnetic field, changing the area of the loop, or changing the angle the loop makes with respect to the magnetic field. In all situations that involve some form of motion of a conductor relative to the source of a magnetic field, the induced potential difference is called a **motional emf.**

Induction in a Flat Loop inside a Magnetic Field

Let's apply equation 29.5 to a flat wire loop inside a uniform magnetic field, where *uniform* means that the field has the same value (same magnitude and same direction) at all points in space at a given time but can vary in time. This arrangement is the simplest case we can address. According to equation 29.3, the magnetic flux in this case is given by $\Phi_B = BA\cos\theta$. According to equation 29.5, the induced potential difference is then

$$\Delta V_{\text{ind}} = -\frac{d\Phi_B}{dt} = -\frac{d}{dt}(BA\cos\theta). \tag{29.6}$$

We can use the product rule from calculus to expand this derivative:

$$\Delta V_{\text{ind}} = -A\cos\theta\frac{dB}{dt} - B\cos\theta\frac{dA}{dt} + AB\sin\theta\frac{d\theta}{dt}. \tag{29.7}$$

Because the time derivative of the angular displacement is the angular velocity, $d\theta/dt = \omega$, the induced potential difference in a flat loop inside a uniform magnetic field is

$$\Delta V_{\text{ind}} = -A\cos\theta\frac{dB}{dt} - B\cos\theta\frac{dA}{dt} + \omega AB\sin\theta. \tag{29.8}$$

Holding two of the three variables in equation 29.8 (A, B, and θ) constant results in the following *three special* cases:

1. Holding the area of the loop and its orientation relative to the magnetic field constant but varying the magnetic field in time yields

$$A \text{ and } \theta \text{ constant: } \Delta V_{\text{ind}} = -A\cos\theta\frac{dB}{dt}. \tag{29.9}$$

2. Holding the magnetic field as well as the orientation of the loop relative to the magnetic field constant but changing the area of the loop that is exposed to the magnetic field yields

$$B \text{ and } \theta \text{ constant: } \Delta V_{\text{ind}} = -B\cos\theta\frac{dA}{dt}. \tag{29.10}$$

3. Holding the magnetic field constant and keeping the area of the loop fixed but allowing the angle between the two to change as a function of time yields

$$A \text{ and } B \text{ constant: } \Delta V_{\text{ind}} = \omega AB\sin\theta. \tag{29.11}$$

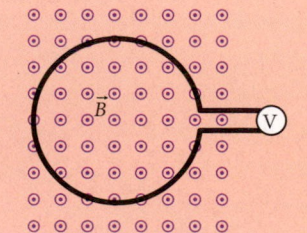

The following examples illustrate the first two cases. Section 29.4 addresses the third case, which has the most useful technical applications, leading directly to electric motors and generators.

EXAMPLE 29.1 | **Potential Difference Induced by a Changing Magnetic Field**

A current of 600 mA is flowing in an ideal solenoid, resulting in a magnetic field of 0.025 T inside the solenoid. Then the current increases with time, t, according to

$$i(t) = i_0 \left[1 + (2.4 \text{ s}^{-2}) t^2 \right].$$

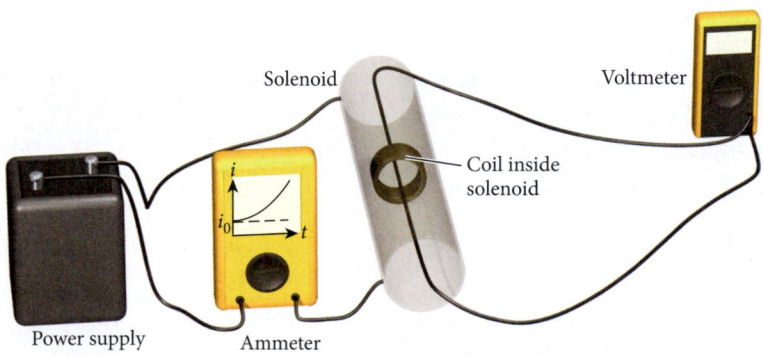

FIGURE 29.8 A time-varying current applied to a solenoid induces a potential difference in a coil.

PROBLEM

If a circular coil of radius 3.4 cm with $N = 200$ windings is located inside the solenoid with its normal vector parallel to the magnetic field (Figure 29.8), what is the induced potential difference in the coil at $t = 2.0$ s?

SOLUTION

First, we compute the area of the coil. Since it is circular, its area is πR^2. However, it has N windings, and thus the area is that of N loops of area πR^2. The net effect is that the number of windings acts as a simple multiplier for the loop area, and the total effective area of the coil is

$$A = N\pi R^2 = 200\pi (0.034 \text{ m})^2 = 0.73 \text{ m}^2. \quad \text{(i)}$$

The magnetic field inside an ideal solenoid is $B = \mu_0 n i$, where n is the number of windings per unit length, and i is the current (see Chapter 28). Because the magnetic field is proportional to the current, we immediately obtain for the time dependence of the magnetic field in this case

$$B(t) = B_0 \left[1 + (2.4 \text{ s}^{-2}) t^2 \right],$$

with $B_0 = \mu_0 n i_0 = 0.025$ T, according to the problem statement.

Further, in this case, the area of the coil and the angle between each loop and the magnetic field (which is zero) are constant. Therefore, equation 29.9 applies. We then find for the induced potential difference, where the area A already accounts for the number of windings, as shown in equation (i):

$$\Delta V_{\text{ind}} = -A\cos\theta \frac{dB}{dt}$$

$$= -A\cos\theta \frac{d}{dt} \left[B_0 \left(1 + \left(2.4 \text{ s}^{-2} \right) t^2 \right) \right]$$

$$= -AB_0 \cos\theta \left(2 \left(2.4 \text{ s}^{-2} \right) t \right)$$

$$= -(0.73 \text{ m}^2)(0.025 \text{ T})(\cos 0°)\left(4.8 \text{ s}^{-2} \right) t$$

$$= (-0.088 \text{ V/s}) t.$$

At time $t = 2.0$ s, the induced potential difference in the coil is $\Delta V_{\text{ind}} = -0.18$ V.

An important general point is made by Example 29.1: The potential difference induced in a coil with N windings and area A is simply N times the potential difference induced in a single loop of area A. Equations 29.8 through 29.11 are valid for multiloop coils, and the only way that the number of windings enters the calculations is as a multiplier in determining the effective area of the coil.

Concept Check 29.2

A power supply is connected to loop 1 and an ammeter as shown in the figure. Loop 2 is close to loop 1 and is connected to a voltmeter. A graph of the current i through loop 1 as a function of time, t, is also shown in the figure. Which graph best describes the induced potential difference, ΔV_{ind}, in loop 2 as a function of time, t?

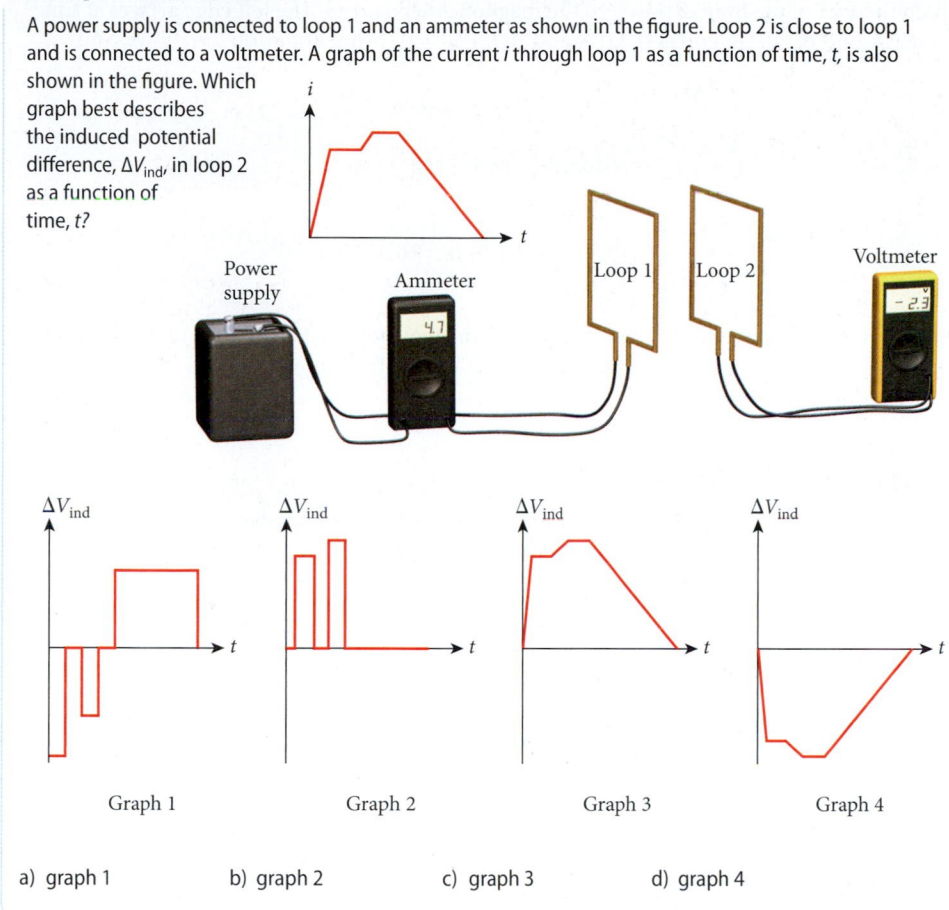

Graph 1   Graph 2   Graph 3   Graph 4

a) graph 1   b) graph 2   c) graph 3   d) graph 4

EXAMPLE 29.2   Potential Difference Induced by a Moving Loop

A rectangular wire loop of width $w = 3.1$ cm and depth $d_0 = 4.8$ cm is pulled out of the gap between two permanent magnets. A magnetic field of magnitude $B = 0.073$ T is present throughout the gap (Figure 29.9).

PROBLEM
If the loop is removed at a constant speed of 1.6 cm/s, what is the induced voltage in the loop as a function of time?

SOLUTION
This situation corresponds to the special case of induction due to an area change, governed by equation 29.10. The magnetic field and the orientation of the loop relative to the field remain constant. We assume that the angle between the magnetic field vector and the area vector is zero. What changes is the area of the loop that is exposed to the magnetic field. With a gap as narrow as that shown in Figure 29.9, very little field occurs outside the gap, so the effective area of the loop exposed to the field is $A(t) = (w)(d(t))$, where $d(t) = d_0 - vt$ is the depth of the part of the loop inside the magnetic field at time t. While the entire loop is still inside the gap, no voltage is produced. Letting the time of arrival of the right edge of the loop at the right end of the gap be $t = 0$, we have

$$A(t) = (w)(d(t)) = w(d_0 - vt).$$

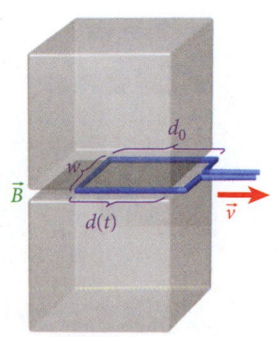

FIGURE 29.9 A wire loop (blue) is pulled out of a gap between two magnets.

– Continued

This formula holds until the left edge of the loop reaches the right end of the gap, after which the area of the loop exposed to the magnetic field is zero. The left edge arrives at time $t_f = d/v = (4.8 \text{ cm})/(1.6 \text{ cm/s}) = 3.0 \text{ s}$, and $A(t > t_f) = 0$. From equation 29.10, we find

$$\Delta V_{ind} = -B\cos\theta \frac{dA}{dt}$$

$$= -B\cos\theta \frac{d}{dt}\left[w(d_0 - vt)\right]$$

$$= wvB\cos\theta$$

$$= (0.031 \text{ m})(0.016 \text{ m/s})(0.073 \text{ T})\cos0°$$

$$= 3.6 \cdot 10^{-5} \text{ V}.$$

During the time interval between 0 and 3 s, a constant potential difference of 36 μV is induced, and no potential difference is induced outside this time interval.

Concept Check 29.3

A long wire carries a current, i, as shown in the figure. A square loop moves in the same plane as the wire, as indicated. In which cases will the loop have an induced current?

a) cases 1 and 2

b) cases 1 and 3

c) cases 2 and 3

d) None of the loops will have an induced current.

e) All of the loops will have an induced current.

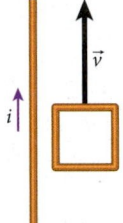

Case 1

Case 2

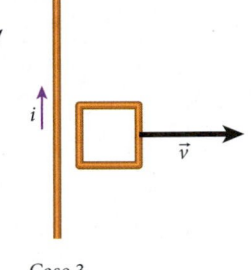

Case 3

29.3   Lenz's Law

Lenz's Law provides a rule for determining the direction of an induced current in a loop. An induced current will have a direction such that the magnetic field *due to* the induced current *opposes* the change in the magnetic flux that *induces* the current. The direction of the induced current can be used to determine the locations of higher and lower potential.

Let's apply Lenz's Law to the situations described in Section 29.1. The physical situation shown in Figure 29.2a involves moving a magnet toward a loop with the north pole pointed toward the loop. In this case, the magnetic field lines point away from the north pole of the magnet. As the magnet moves toward the loop, the magnitude of the magnetic field within the loop, in the direction pointing toward the loop, increases as depicted in Figure 29.10a. Lenz's Law states that the current induced in the loop tends to oppose the change in magnetic flux. The induced magnetic field, $\vec{B}_{ind}$, then points in the opposite direction from that of the field due to the magnet.

In Figure 29.2b, a magnet is moved toward a loop with the south pole pointed toward the loop. In this case, the magnetic field lines point toward the south pole of the magnet. As the magnet moves toward the loop, the magnitude of the field in the direction pointing toward the south pole increases, as depicted in Figure 29.10b. Lenz's Law states that the induced current creates a magnetic field that tends to oppose the increase in magnetic flux. This induced field points in the opposite direction from that of the field lines due to the magnet.

Similarly, Figure 29.10c and Figure 29.10d represent the physical situations depicted in Figure 29.3a and Figure 29.3b, respectively. In these two cases, the magnitude of the magnetic flux is decreasing, and a current is induced that produces a magnetic field opposing this decrease. In both cases, a current is induced in the loop that creates a magnetic field pointing in the same direction as the magnetic field from the magnet.

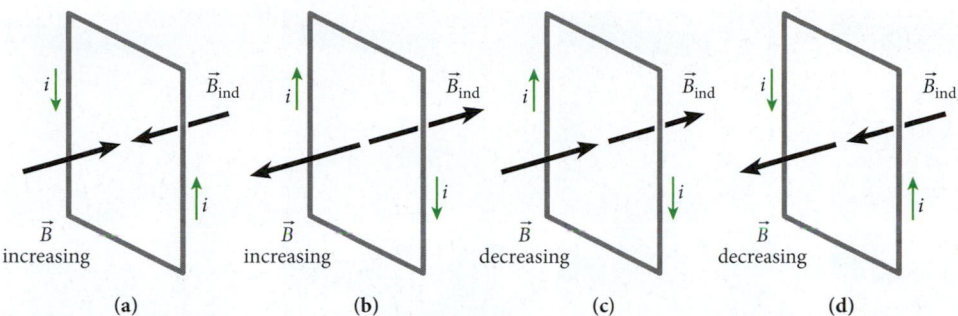

FIGURE 29.10 Relationship of an external magnetic field, $\vec{B}$, the induced current, i, and the magnetic field, $\vec{B}_{ind}$, resulting from that induced current: (a) An increasing magnetic field pointing to the right induces a current that creates a magnetic field pointing to the left. (b) An increasing magnetic field pointing to the left induces a current that creates a magnetic field pointing to the right. (c) A decreasing magnetic field pointing to the right induces a current that creates a magnetic field pointing to the right. (d) A decreasing magnetic field pointing to the left induces a current that creates a magnetic field pointing to the left.

For two loops with one having a changing current, Lenz's Law is applied in the same way. The increasing current in loop 1 in Figure 29.4 induces a current in loop 2 that creates a magnetic field opposing the increase in magnetic flux, as depicted in Figure 29.10b. The decreasing current in loop 1 in Figure 29.5 induces a current in loop 2 that creates a magnetic field opposing the decrease in magnetic flux, as depicted in Figure 29.10d.

Self-Test Opportunity 29.2

A square conducting loop with very small resistance is moved at constant speed from a region with no magnetic field through a region of constant magnetic field and then into a region with no magnetic field, as shown in the figure. As the loop enters the magnetic field, what is the direction of the induced current? As the loop leaves the magnetic field, what is the direction of the induced current?

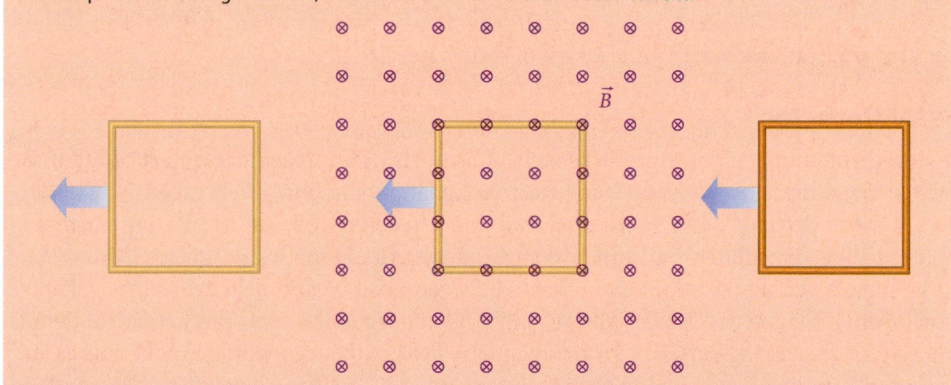

Self-Test Opportunity 29.3

Suppose Lenz's Law instead stated that the induced magnetic field augments the magnetic flux, meaning that Faraday's Law of Induction would be written as $\Delta V_{ind} = +d\Phi_B/dt$, that is, with a positive instead of a negative sign. What would be the consequences? Can you explain why this would lead to a contradiction?

Eddy Currents

Let's consider two pendulums, each with a nonmagnetic conducting metal plate at the end that is designed to pass through the gap between strong permanent magnets (Figure 29.11). One metal plate is solid, and the other has slots cut in it. The pendulums are pulled to one side and released. The pendulum with the solid metal plate stops in the gap, while the slotted plate passes through the magnetic field, only slowing slightly. This demonstration illustrates the very important phenomenon of induced **eddy currents.** As the pendulum with the solid plate enters the magnetic field between the magnets, Lenz's Law says that the changing magnetic flux induces currents that tend to oppose the change in flux. These currents produce induced magnetic fields opposing the external field that created the currents. These induced magnetic fields interact with the external magnetic field (via their spatial gradients) to stop the pendulum. Larger induced currents produce larger induced magnetic fields and thus lead to more rapid deceleration of the pendulum. In the slotted plate, the

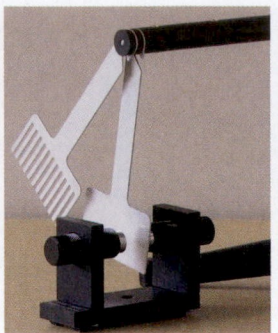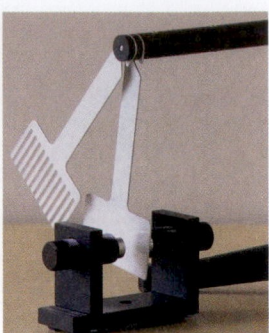

FIGURE 29.11 Two pendulums, one consisting of an arm and a solid metal plate and a second consisting of an arm and a slotted metal plate. The five frames are in time sequence from left to right, with the two pendulums starting their motion together in the second frame from the left. The pendulum with the solid plate stops in the gap, while the pendulum with the slotted plate passes through the gap.

induced eddy currents are broken up by the slots, and the slotted plate passes through the magnetic field, only slowing slightly. Eddy currents are not like the undirected and uniform current induced in the loop in Example 29.2 but are instead swirling eddies like those seen in turbulent flowing water.

Where does the energy contained in the motion of the pendulum with the solid plate in Figure 29.11 go—in other words, how do the eddy currents stop the pendulum? The answer is that the eddy currents disperse heat in the metal because of its finite resistance, as discussed in Chapter 25. The stronger the induced eddy currents are, the more rapidly energy is converted from the pendulum's motion into heat. This is why the slotted plate, with its much smaller induced eddy currents, is only slowed slightly as it passes through the gap between the magnets (although the slowing will stop it eventually).

Eddy currents are often undesirable, forcing equipment designers to minimize them by segmenting or laminating electrical devices that must operate in an environment of changing magnetic fields. However, eddy currents can also be useful and are employed in certain practical applications, such as the brakes of train cars.

Metal Detector

Passing through metal detectors, especially at airports, is an unavoidable part of life these days. A metal detector works by using electromagnetic induction, often called *pulse induction*. A metal detector has a transmitter coil and a receiver coil. An alternating current is applied to the transmitter coil, which then produces an alternating magnetic field. (*Alternating* means varying as a function of time between positive and negative values. Chapter 30 will provide more precise definitions, physical consequences, and mathematical details concerning alternating current.) As the magnetic field of the transmitter coil increases and decreases, it induces a current in the receiving coil that tends to counteract the change in the magnetic flux produced by the transmitter coil. The induced current in the receiver coil is measured when nothing but air is between the coils. If a conductor in the form of a metal object passes between transmitter and receiver coils, a current will be induced in the metal object in the form of eddy currents. These eddy currents will act to counter the increases and decreases of the changing magnetic field produced by the transmitter coil, which in turn induces a current in the receiver coil that tends to counter the increase in current in the metal. The measured current in the receiver coil will be less when any metal object is present between the two coils.

A schematic diagram of an airport metal detector is shown in Figure 29.12. A transmitter coil and a receiver coil are located on opposite sides of an entry door. The person or object to be scanned passes through the door between the two coils. Suppose that the current in the transmitter coil is flowing in the direction shown and increasing. A current will be induced in the metal plate in the opposite direction and will tend to oppose the increase in the current in the transmitter coil. The increasing current in the metal plate will induce a current in the receiver coil that is in the opposite direction and

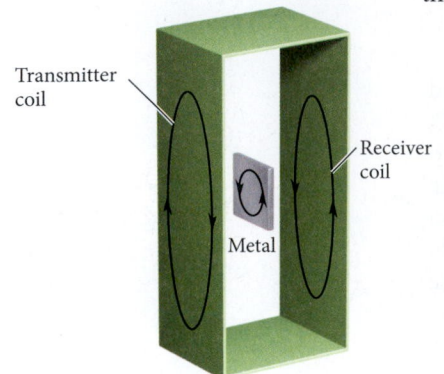

FIGURE 29.12 Schematic diagram of an airport metal detector.

tends to oppose the increase in the current in the metal plate (not shown in the diagram). Thus, the metal plate induces a current in the receiver coil that flows in the same direction as the current in the transmitter coil. Without the metal plate, the increasing current in the transmitter coil induces a current in the opposite direction in the receiver coil that tends to oppose the increase in the current in the transmitter coil (as shown in the diagram). Thus, the overall effect of the metal plate in the metal detector is to decrease the observed current in the receiver coil. The metal object does not have to be a flat plate; any piece of metal, provided it is large enough, will have currents induced in it that can be detected by measuring the induced current in the receiver coil.

Metal detectors are also used to control traffic lights. In this application, a rectangular wire loop, which serves as both transmitter and receiver coil, is embedded in the road surface. A pulse of current is passed through the loop, which induces eddy currents in any metal near the loop. The current in the loop is measured after the current pulse is completed. When a car moves onto the road surface above the loop, eddy currents induced in the metal of the car cause a different current to be measured between pulses, which then triggers the traffic light to switch to green (after an appropriate delay to allow other vehicles to clear the intersection, of course!). On older road surfaces, you can often see rectangular-shaped scars in the asphalt resulting from retrofitting intersections with these induction loops.

Induced Potential Difference on a Wire Moving in a Magnetic Field

Consider a conducting wire of length ℓ moving with constant velocity $\vec{v}$ perpendicular to a constant magnetic field, $\vec{B}$, directed into the page (Figure 29.13). The wire is oriented so that it is perpendicular to the velocity and to the magnetic field. The magnetic field exerts a force, $\vec{F}_B$, on the conduction electrons in the wire, causing them to move downward. This motion of the electrons produces a net negative charge at the bottom end of the wire and a net positive charge at the top end of the wire. This charge separation produces an electric field, $\vec{E}$, which exerts on the conduction electrons a force, $\vec{F}_E$, that tends to cancel the magnetic force. After some time, the two forces become equal in magnitude (but opposite in direction) producing a zero net force:

$$F_B = evB = F_E = eE. \qquad (29.12)$$

Thus, the induced electric field can be expressed by

$$E = vB. \qquad (29.13)$$

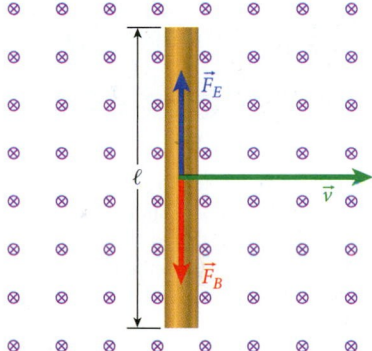

FIGURE 29.13 A moving conductor in a constant magnetic field. The magnetic and electric forces on the conduction electrons are shown.

Concept Check 29.4

A metal bar is moving with constant velocity $\vec{v}$ through a uniform magnetic field pointing into the page, as shown in the figure.

Which of the following most accurately represents the charge distribution on the surface of the metal bar?

a) distribution 1

b) distribution 2

c) distribution 3

d) distribution 4

e) distribution 5

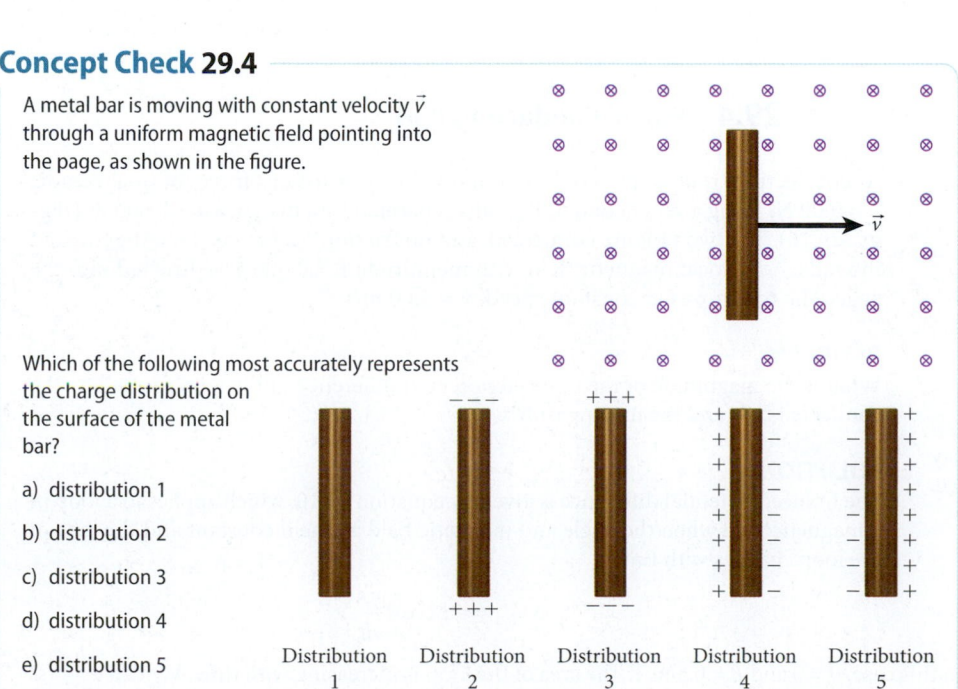

Distribution 1   Distribution 2   Distribution 3   Distribution 4   Distribution 5

Because the electric field is constant in the wire, it produces a potential difference between the two ends of the wire given by

$$E = \frac{\Delta V_{ind}}{\ell} = vB. \qquad (29.14)$$

The induced potential difference between the ends of the wire is then

$$\Delta V_{ind} = v\ell B. \qquad (29.15)$$

This is one form of motional emf, mentioned in Section 29.2.

EXAMPLE 29.3 | Satellite Tethered to a Space Shuttle

In 1996, the Space Shuttle *Columbia* deployed a tethered satellite on a wire out to a distance of 20. km (Figure 29.14). The wire was oriented perpendicular to the Earth's magnetic field at that point, and the magnitude of the field was $B = 5.1 \cdot 10^{-5}$ T. *Columbia* was traveling at a speed of 7.6 km/s.

PROBLEM
What was the potential difference induced between the ends of the wire?

SOLUTION
We can use equation 29.15 to determine the induced potential difference between the ends of the wire. The length of the wire is $L = 20.0$ km, and the speed of the wire through the magnetic field of the Earth ($B = 5.1 \cdot 10^{-5}$ T) is the same as the speed of the Space Shuttle, which is $v = 7.6$ km/s. Thus, we have

$$\Delta V_{ind} = vLB = (7.6 \cdot 10^3 \text{ m/s})(20.0 \cdot 10^3 \text{ m})(5.1 \cdot 10^{-5} \text{ T}) = 7.8 \text{ kV}.$$

The astronauts on the Space Shuttle measured a current of about 0.5 A at a voltage of 3.5 kV. The circuit consisted of the deployed wire and ionized atoms in space as the return path for the current. The wire broke just as the deployment length reached 20 km, but the generation of electric current from the motion of a spacecraft had been demonstrated.

FIGURE 29.14 (a) An artist's conception of the Space Shuttle *Columbia* and the tethered satellite. (b) A photograph of the tethered satellite being deployed from *Columbia*.

Solved Problem 28.2 concerned an electromagnetic rail accelerator. The next example focuses on the phenomenon of induction in a similar system.

EXAMPLE 29.4 | Pulled Conducting Rod

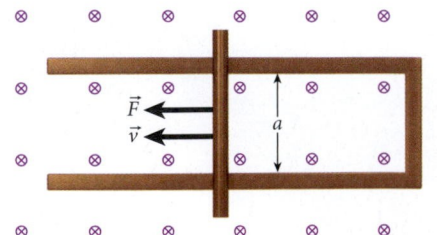

FIGURE 29.15 A conducting rod is pulled along two conducting rails with a constant velocity in a constant magnetic field directed into the page.

A conducting rod is pulled horizontally by a constant force of magnitude, $F = 5.00$ N, along a set of conducting rails separated by a distance $a = 0.500$ m (Figure 29.15). The two rails are connected, and no friction occurs between the rod and the rails. A uniform magnetic field with magnitude $B = 0.500$ T is directed into the page. The rod moves at constant speed, $v = 5.00$ m/s.

PROBLEM
What is the magnitude of the induced potential difference in the loop formed by the connected rails and the moving rod?

SOLUTION
The induced potential difference is given by equation 29.10, which applies to a loop in a magnetic field when the angle and magnetic field are held constant and the area of the loop changes with time:

$$\Delta V_{ind} = -B\cos\theta \frac{dA}{dt}.$$

In this case, $\theta = 0$ and $B = 0.500$ T. The area of the loop is increasing with time. We can express the area of the loop in terms of A_0, the area before the rod started moving, and an additional area given by the product of the speed of the loop and the time for which the loop has been moving times the distance, a, between the rails:

$$A = A_0 + a(vt) = A_0 + vta.$$

The change of the loop's area as a function of time is then

$$\frac{dA}{dt} = \frac{d}{dt}(A_0 + vta) = va.$$

Thus, the magnitude of the induced potential difference is

$$\Delta V_{\text{ind}} = \left| -B\cos\theta \frac{dA}{dt} \right| = vaB. \tag{i}$$

Inserting the numerical values, we obtain

$$\Delta V_{\text{ind}} = (5.00 \text{ m/s})(0.500 \text{ m})(0.500 \text{ T}) = 1.25 \text{ V}.$$

Note that equation (i), $\Delta V_{\text{ind}} = vaB$, which we derived from Faraday's Law of Induction, has the same form as equation 29.15 for the potential difference induced in a wire moving in a magnetic field, which was derived using the magnetic force on moving charges.

Concept Check 29.5

A wire loop is placed in a uniform magnetic field. Over a period of 2 s, the loop is shrunk. Which statement about the induced potential difference is correct?

a) There will be some induced potential difference.

b) There will be no induced potential difference because the loop changes size along one axis and not the other.

c) There will be no induced potential difference because the loop is not closed.

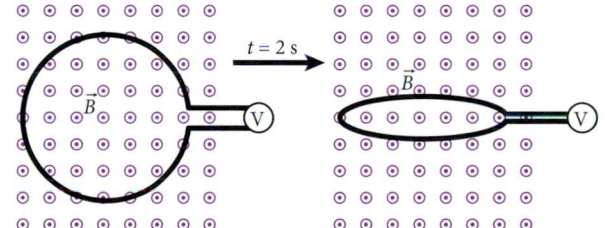

d) There will be no induced potential difference because the loop is shrinking.

SOLVED PROBLEM 29.1 | Power from a Rotating Rod

A conducting rod with length $\ell = 8.17$ cm rotates around one of its ends in a uniform magnetic field that has a magnitude $B = 1.53$ T and is directed parallel to the rotation axis of the rod (Figure 29.16). The other end of the rod slides on a frictionless conducting ring. The rod makes 6.00 revolutions per second. A resistor, $R = 1.63$ mΩ, is connected between the rotating rod and the conducting ring.

PROBLEM
What is the power dissipated in the resistor due to magnetic induction?

SOLUTION

THINK We can calculate the potential difference induced in a conductor of length ℓ moving with speed v perpendicular to a magnetic field of magnitude B. However, the rotating rod has different speeds at different radii, $v(r)$. Therefore, we must calculate the potential difference induced in the rod by integrating $Bv(r)$ over the length of the rod. From the induced potential difference, we can calculate the power dissipated in the resistor.

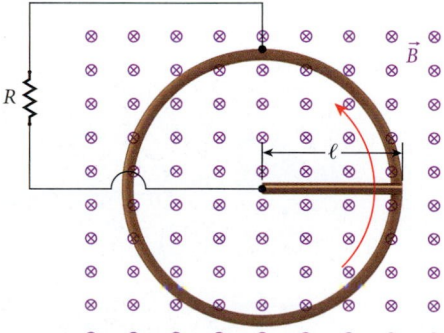

FIGURE 29.16 Conducting rod rotating in a constant magnetic field directed into the page.

SKETCH Figure 29.17 shows the speed as a function of radius for the conducting rod.

RESEARCH The potential difference, ΔV_{ind}, induced on a conductor of length ℓ moving with speed v perpendicular to a magnetic field of magnitude B is given by equation 29.15:

$$\Delta V_{\text{ind}} = v\ell B. \qquad\qquad - \textit{Continued}$$

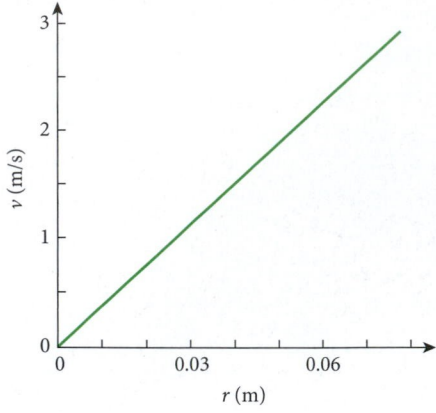

FIGURE 29.17 Speed as a function of radius for the conducting rod.

However, in this case, different parts of the conducting rod are moving at different speeds. We can express the speed of the different parts of the rod as a function of distance r from the axis of rotation:

$$v(r) = \frac{2\pi r}{T},$$

where $v(r)$ is the speed of the rod at distance r and T is the period of the rotation. We can then find the potential difference induced in the rotating rod over its length, ℓ:

$$\Delta V_{\text{ind}} = \int_0^\ell v(r)B\,dr. \tag{i}$$

The power dissipated in the resistor is given by

$$P = \frac{\Delta V_{\text{ind}}^2}{R}. \tag{ii}$$

SIMPLIFY Evaluating the definite integral in equation (i) gives

$$\Delta V_{\text{ind}} = \int_0^\ell \left(\frac{2\pi r}{T}\right) B\,dr = \frac{2\pi B}{T}\frac{\ell^2}{2} = \frac{\pi B \ell^2}{T}. \tag{iii}$$

Substituting the expression for ΔV_{ind} from equation (iii) into equation (ii) leads to an expression for the power dissipated in the resistor:

$$P = \frac{\left(\pi B \ell^2 / T\right)^2}{R} = \frac{\pi^2 B^2 \ell^4}{RT^2}.$$

CALCULATE The period is the inverse of the frequency. The frequency is $f = 6.00$ Hz, so the period is

$$T = \frac{1}{f} = \frac{1}{6.00}\text{ s}.$$

Putting in the numerical values gives us

$$P = \frac{\pi^2 B^2 \ell^4}{RT^2} = \frac{\pi^2 (1.53\text{ T})^2 (0.0817\text{ m})^4}{\left(1.63\cdot 10^{-3}\ \Omega\right)\left(\dfrac{1}{6.00}\text{ s}\right)^2} = 22.7345\text{ W}.$$

ROUND We report our result to three significant figures:

$$P = 22.7\text{ W}.$$

DOUBLE-CHECK To double-check our result, we consider a conducting rod of the same length moving perpendicularly to the same magnetic field with a speed equal to the speed of the center of the rotating rod, which is

$$v(\ell/2) = \frac{2\pi(\ell/2)}{T} = \frac{2\pi L}{2T} = \frac{2\pi(0.0817\text{ m})}{2\left(\dfrac{1}{6.00}\text{ s}\right)} = 1.54\text{ m/s}.$$

The induced potential difference across the conducting rod moving perpendicularly would be

$$\Delta V_{\text{ind}} = v\ell B = (1.54\text{ m/s})(0.0817\text{ m})(1.53\text{ T}) = 0.193\text{ V}.$$

The power dissipated in the resistor would then be

$$P = \frac{\Delta V_{\text{ind}}^2}{R} = \frac{(0.193\text{ V})^2}{1.63\cdot 10^{-3}\ \Omega} = 22.9\text{ W},$$

which is close to our result within rounding error. Thus, our result seems reasonable.

Finally, note that there is a possible additional source of potential difference between the two ends of the rod. Our solution assumed that the potential difference between the two ends is due exclusively to the magnetic induction. However, all charge carriers inside the rod are forced on a circular path due to the rotation. This requires a centripetal force, which should in principle reduce the potential difference between the two ends of the rod. However, for the small angular velocity of the rod in this problem, this effect is negligible.

29.4 Generators and Motors

The third special case of the basic induction process described in Section 29.2 is by far the most interesting technologically. In this case, the angle between the conducting loop and the magnetic field is varied over time, while keeping the area of the loop as well as the magnetic field strength constant. In this situation, equation 29.11 can be used to apply Faraday's Law of Induction to the generation and application of electric current. A device that produces electric current from mechanical motion is called an **electric generator**. A device that produces mechanical motion from electric current is called an **electric motor**. Figure 29.18 shows a very simple electric motor.

A simple generator consists of a loop forced to rotate in a fixed magnetic field. The force that causes the loop to rotate can be supplied by hot steam running over a turbine, as occurs in nuclear and coal-fired power plants. (Power plants actually use multiple loops in order to increase the power output.) On the other hand, the loop can be made to rotate by flowing water or wind to generate electricity in a pollution-free way.

Figure 29.19 shows two types of simple generators. In a direct-current generator, the rotating loop is connected to an external circuit through a split commutator ring, as illustrated in Figure 29.19a. As the loop turns, the connection is reversed twice per revolution, so the induced potential difference always has the same sign. Figure 29.19b shows a similar arrangement used to produce an alternating current. An alternating current is a current that varies in time between positive and negative values, with the variation often showing a sinusoidal form. Each end of the loop is connected to the external circuit through its own solid slip ring. Thus, this generator produces an induced potential difference that varies from positive to negative and back. A generator that produces alternating voltages and the resulting alternating current is also called an **alternator**. Figure 29.20 shows the induced potential difference as a function of time for each type of generator.

The devices in Figure 29.19 could also be used as motors by supplying current to the loop and using its resulting motion to do work.

FIGURE 29.18 A very simple electric motor used for lecture demonstrations. It consists of a pair of permanent magnets on the outside and two solenoids, through which current is sent, on the inside.

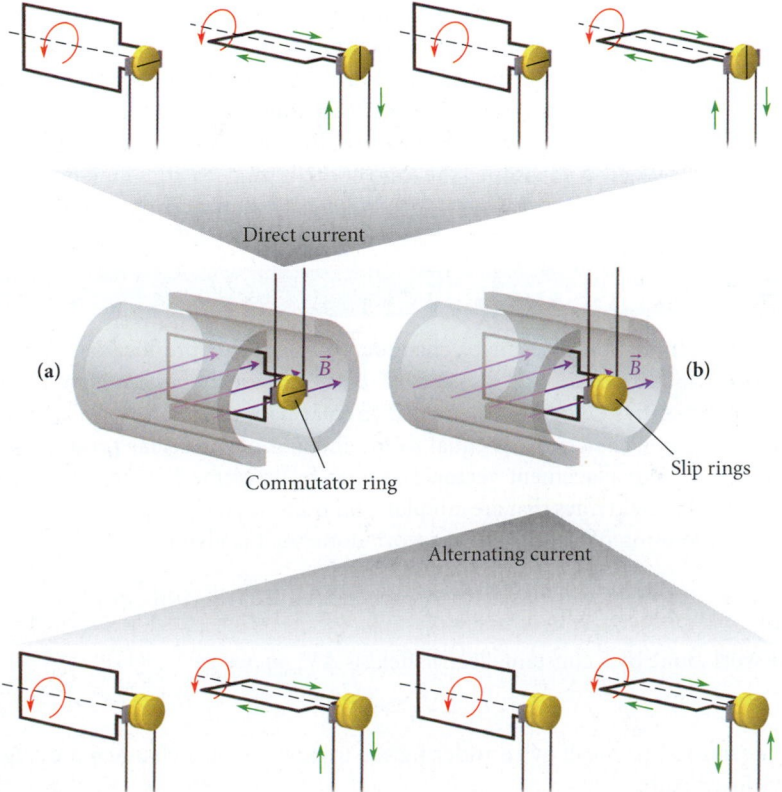

FIGURE 29.19 (a) A simple direct-current (DC) generator/motor. (b) A simple alternating-current (AC) generator/motor.

Self-Test Opportunity 29.4

A generator is operated by rotating a coil of N turns in a constant magnetic field of magnitude B at a frequency f. The resistance of the coil is R, and the cross-sectional area of the coil is A. Decide whether each of the following statements is true or false.

a) The average induced potential difference doubles if the frequency, f, is doubled.

b) The average induced potential difference doubles if the resistance, R, is doubled.

c) The average induced potential difference doubles if the magnetic field's magnitude, B, is doubled.

d) The average induced potential difference doubles if the area, A, is doubled.

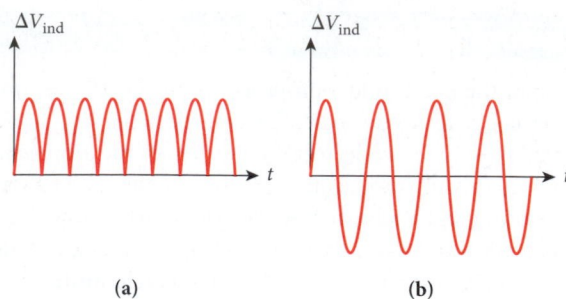

FIGURE 29.20 Induced potential difference as a function of time for (a) a simple direct-current generator; (b) a simple alternating-current generator.

Real-world generators and motors are considerably more complex than the simple examples in Figure 29.19. For example, instead of permanent magnets, current flowing in coils creates the required magnetic field. Many closely spaced loops are employed to make more efficient use of the rotational motion. Multiple loops also resolve the problem that a simple, one-loop motor could stop in a position where the current through the loop produces no torque. The magnetic field may also change with time in phase with the rotating loop. In some generators and motors, the loops (coils) are fixed and the magnet rotates.

Regenerative Braking

Hybrid cars are propelled by a combination of gasoline power and electrical power. One attractive feature of a hybrid vehicle is that it is capable of **regenerative braking.** When the brakes are used to slow or stop a nonhybrid vehicle, the kinetic energy of the vehicle is turned into heat in the brake pads. This heat dissipates into the environment, and energy is lost. In a hybrid car, the brakes are connected to an electric motor (Figure 29.21), which functions as a generator, charging the car's battery. Thus, the kinetic energy of the car is partially recovered during braking, and this energy can later be used to propel the car, contributing to its efficiency and greatly increasing its gas mileage in stop-and-go driving.

FIGURE 29.21 Regenerative brakes on the rear wheels of a hybrid vehicle are used to slow the vehicle and convert the vehicle's kinetic energy into electrical energy that is stored in a lithium ion battery.

29.5 Induced Electric Field

Faraday's Law of Induction states that a changing magnetic flux produces an induced potential difference, which can lead to an induced current. What are the consequences of this effect?

Consider a positive charge q moving in a circular path with radius r in an electric field, $\vec{E}$. The work done on the charge is equal to the integral of the scalar product of the force and the differential displacement vector. For now, let's assume that the electric field $\vec{E}$ is constant, that it has field lines that are circular, and that the charge moves along one of these lines. In one revolution of the charge, the work done on it is given by

$$\oint \vec{F} \cdot d\vec{s} = \oint q\vec{E} \cdot d\vec{s} = \oint q\cos 0° E\, ds = qE \oint ds = qE(2\pi r).$$

Since the work done by a constant electric field is $\Delta V_{\text{ind}} q$, we get

$$\Delta V_{\text{ind}} = 2\pi r E.$$

We can generalize this result by considering the work done on a charge q moving along an arbitrary closed path:

$$W = \oint \vec{F} \cdot d\vec{s} = q \oint \vec{E} \cdot d\vec{s}.$$

Again substituting $\Delta V_{ind} q$ for the work done, we obtain

$$\Delta V_{ind} = \oint \vec{E} \cdot d\vec{s}. \qquad (29.16)$$

Now we can express the induced potential difference in a different way by combining equation 29.5 with equation 29.16:

$$\oint \vec{E} \cdot d\vec{s} = -\frac{d\Phi_B}{dt}. \qquad (29.17)$$

Equation 27.17 states that a changing magnetic flux induces an electric field. This equation can be applied to any closed path in a changing magnetic field, even if no conductor exists in the path.

29.6   Inductance of a Solenoid

Consider a long solenoid with N turns carrying a current, i. This current creates a magnetic field in the center of the solenoid, resulting in a magnetic flux, Φ_B. The same magnetic flux goes through each of the N windings of the solenoid. It is customary to define the **flux linkage** as the product of the number of windings and the magnetic flux, or $N\Phi_B$. Equation 29.1 defined the magnetic flux as $\Phi_B = \iint \vec{B} \cdot d\vec{A}$. Inside the solenoid, the magnetic field vector and the surface normal vector, $d\vec{A}$, are parallel. And we saw in Chapter 28 that the magnitude of the magnetic field inside the solenoid is $B = \mu_0 n i$, where $\mu_0 = 4\pi \cdot 10^{-7}$ T m/A is the magnetic permeability of free space, i is the current, and n is the number of windings per unit length ($n = N/\ell$). Therefore, the magnetic flux in the interior of a solenoid is proportional to the current flowing through the solenoid, which trivially means that the flux linkage is also proportional to the current. We can express this proportionality as

$$N\Phi_B = Li, \qquad (29.18)$$

using a proportionality constant, L, called the **inductance.** (*Note:* Use of the letter L to represent the inductance is the convention. Although L is also used for the physical quantity of length and the physical quantity of angular momentum, inductance is not connected to either of these in any way.)

Inductance is a measure of the flux linkage produced by a solenoid per unit of current. The unit of inductance is the **henry** (H), named after American physicist Joseph Henry (1797–1878) and given by

$$[L] = \frac{[\Phi_B]}{[i]} \Rightarrow 1\,\text{H} = \frac{1\,\text{T m}^2}{1\,\text{A}}. \qquad (29.19)$$

The definition of the henry given in equation 29.19 means that the magnetic permeability of free space can also be given as $\mu_0 = 4\pi \cdot 10^{-7}$ H/m.

Now let's use equation 29.18 to find the inductance of a solenoid with cross-sectional area A and length ℓ. The flux linkage for this solenoid is

$$N\Phi_B = (n\ell)(BA), \qquad (29.20)$$

where n is the number of turns per unit length and B is the magnitude of the magnetic field inside the solenoid. Thus, the inductance is given by

$$L = \frac{N\Phi_B}{i} = \frac{(n\ell)(\mu_0 n i)(A)}{i} = \mu_0 n^2 \ell A. \qquad (29.21)$$

This expression for the inductance of a solenoid is good for long solenoids because fringe field effects at the ends of such a solenoid are small. You can see from equation 29.21 that the inductance of a solenoid depends only on the geometry (length, area, and number of turns) of the device. This dependence of inductance on geometry alone holds for all coils and solenoids, just as the capacitance of any capacitor depends only on its geometry.

Any solenoid has an inductance, and when a solenoid is used in an electric circuit, it is called an *inductor,* simply because its inductance is its most important property as far as the current flow is concerned.

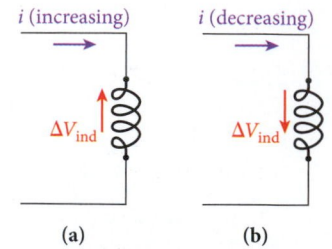

FIGURE 29.22 (a) Self-induced potential difference in an inductor when the current is increasing. (b) Self-induced potential difference in an inductor when the current is decreasing.

29.7 Self-Induction and Mutual Induction

Consider the situation in which two coils, or inductors, are close to each other, and a changing current in the first coil produces a magnetic flux in the second coil. However, the changing current in the first coil also induces a potential difference in that coil, and thus the magnetic field from that coil also changes. This phenomenon is called **self-induction**. The resulting potential difference is termed the *self-induced potential difference*. Changing the current in the first coil also induces a potential difference in the second coil. This phenomenon is called **mutual induction**.

According to Faraday's Law of Induction, the self-induced potential difference for any inductor is given by

$$\Delta V_{\text{ind},L} = -\frac{d(N\Phi_B)}{dt} = -\frac{d(Li)}{dt} = -L\frac{di}{dt}, \tag{29.22}$$

where equation 29.18 allows us to substitute Li for $N\Phi_B$. Thus, in any inductor, a self-induced potential difference appears when the current changes with time. This self-induced potential difference depends on the time rate of change of the current and the inductance of the device.

Lenz's Law provides the direction of the self-induced potential difference. The negative sign in equation 29.22 provides the clue that the induced potential difference always opposes any change in current. For example, Figure 29.22a shows current flowing through an inductor and increasing with time. Thus, the self-induced potential difference will oppose the increase in current. In Figure 29.22b, the current flowing through an inductor is decreasing with time. Thus, a self-induced potential difference will oppose the decrease in current. We have assumed that these inductors are ideal inductors; that is, they have no resistance. All induced potential differences manifest themselves across the connections of the inductor. Inductors with resistance are treated in Section 29.8.

Now let's consider two adjacent coils with their central axes aligned (Figure 29.23). Coil 1 has N_1 turns, and coil 2 has N_2 turns. The current in coil 1 produces a magnetic field, $\vec{B}_1$. The flux linkage in coil 2 resulting from the magnetic field in coil 1 is $N_2\Phi_{1\to2}$. The mutual inductance, $M_{1\to2}$, of coil 2 due to coil 1 is defined as

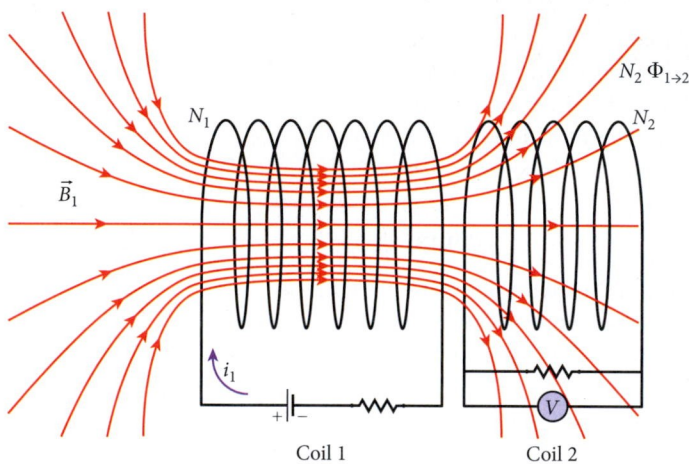

FIGURE 29.23 Coil 1 has current i_1. Coil 2 has a voltmeter capable of measuring small, induced potential differences.

$$M_{1\to2} = \frac{N_2\Phi_{1\to2}}{i_1}. \tag{29.23}$$

Multiplying both sides of equation 29.23 by i_1 yields

$$i_1 M_{1\to2} = N_2\Phi_{1\to2}.$$

If the current in coil 1 changes with time, we can write

$$M_{1\to2}\frac{di_1}{dt} = N_2\frac{d\Phi_{1\to2}}{dt}.$$

The right-hand side of this equation is similar to the right-hand side of Faraday's Law of Induction (equation 29.5). Thus, we can write

$$\Delta V_{\text{ind},2} = -M_{1\to2}\frac{di_1}{dt}. \tag{29.24}$$

Now let's reverse the roles of the two coils (Figure 29.24). The current, i_2, in coil 2 produces a magnetic field, $\vec{B}_2$. The flux linkage in coil 1 resulting from the magnetic field in coil 2 is $N_1\Phi_{2\to1}$. Using the same analysis we applied to determine the mutual inductance of coil 2 due to coil 1, we find

$$\Delta V_{\text{ind},1} = -M_{2\to1}\frac{di_2}{dt}, \tag{29.25}$$

where $M_{2\to1}$ is the mutual inductance of coil 1 due to coil 2.

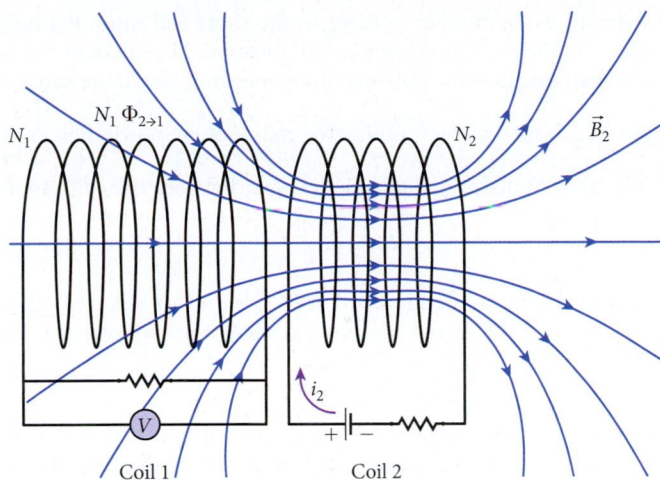

FIGURE 29.24 Coil 2 has current i_2. Coil 1 has a voltmeter capable of measuring small, induced potential differences.

We see that the potential difference induced in one coil is proportional to the change of current in the other coil. The proportionality constant is the mutual induction. If we switched the indexes 1 and 2 and repeated the entire analysis of the coils' effects on each other, we could show that

$$M_{1\to 2} = M_{2\to 1} = M.$$

We can then rewrite equations 29.24 and 29.25 as

$$\Delta V_{\text{ind},2} = - M \frac{di_1}{dt} \qquad (29.26)$$

and

$$\Delta V_{\text{ind},1} = - M \frac{di_2}{dt}, \qquad (29.27)$$

where M is the **mutual inductance** between the two coils. The SI unit of mutual inductance is the henry. One major application of mutual inductance is in transformers, which are discussed in Chapter 30.

SOLVED PROBLEM 29.2 | Mutual Induction of a Solenoid and a Coil

A long solenoid with a circular cross section of radius $r_1 = 2.80$ cm and $n = 290$ turns/cm is inside and coaxial with a short coil that has a circular cross section of radius $r_2 = 4.90$ cm and $N = 31$ turns (Figure 29.25a). The current in the solenoid is increased at a constant rate from zero to $i = 2.20$ A over a time interval of 48.0 ms.

PROBLEM
What is the potential difference induced in the short coil while the current is changing?

SOLUTION
THINK The potential difference induced in the short coil is due to the changing current flowing in the solenoid. According to equation 29.23, the mutual inductance of the short

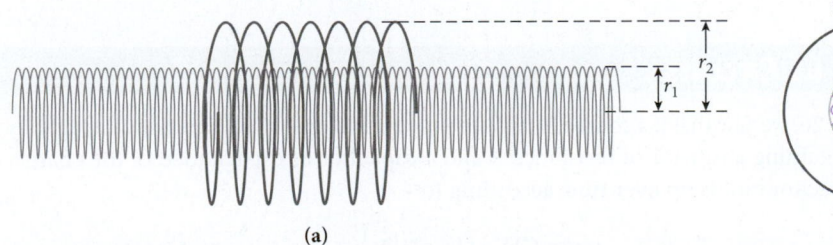

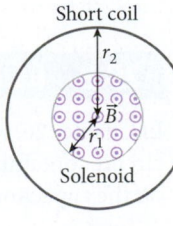

Short coil

FIGURE 29.25 (a) A long solenoid of radius r_1 inside a short coil of radius r_2. (b) View of the two coils looking down the central axis.

Solenoid

(a)

(b)

– Continued

coil due to the solenoid is the number of turns in the short coil times the magnetic flux of the solenoid, divided by the current flowing in the solenoid. Having determined the mutual inductance, we can then calculate the potential difference induced in the short coil.

SKETCH Figure 29.25b shows a view of the two coils looking down their central axis.

RESEARCH We can formulate the mutual inductance between the coil and the solenoid as

$$M = \frac{N\Phi_{s\to c}}{i}, \tag{i}$$

where N is the number of turns in the short coil, $N\Phi_{s\to c}$ is the flux linkage in the coil resulting from the magnetic field of the solenoid, and i is the current in the solenoid. The flux can be expressed as

$$\Phi_{s\to c} = BA, \tag{ii}$$

where B is the magnitude of the magnetic field inside the solenoid and A is its cross-sectional area. Recall from Chapter 28 that for a solenoid, the magnitude of the magnetic field is

$$B = \mu_0 n i,$$

where n is the number of turns per unit length. The cross-sectional area of the solenoid is

$$A = \pi r_1^2. \tag{iii}$$

The potential difference induced in the short coil is then

$$\Delta V_{ind} = -M \frac{di}{dt}.$$

SIMPLIFY We can combine equations (i), (ii), and (iii) to obtain the mutual inductance between the two coils:

$$M = \frac{NBA}{i} = \frac{N(\mu_0 n i)(\pi r_1^2)}{i} = N\pi\mu_0 n r_1^2.$$

Then the potential difference induced in the short coil is

$$\Delta V_{ind} = -\left(N\pi\mu_0 n r_1^2\right)\frac{di}{dt}.$$

CALCULATE The change in the current is constant, so

$$\frac{di}{dt} = \frac{2.20 \text{ A}}{48.0 \cdot 10^{-3} \text{ s}} = 45.8333 \text{ A/s}.$$

The mutual inductance between the two coils is

$$M = (31)\pi\left(4\pi \cdot 10^{-7} \text{ T m/A}\right)\left(290 \cdot 10^2 \text{ m}^{-1}\right)\left(2.80 \cdot 10^{-2} \text{ m}\right)^2 = 0.0027825 \text{ H}.$$

The potential difference induced in the short coil is then

$$\Delta V_{ind} = -(0.0027825 \text{ H})(45.8333 \text{ A/s}) = -0.127531 \text{ V}.$$

ROUND We report our result to three significant figures:

$$\Delta V_{ind} = -0.128 \text{ V}.$$

DOUBLE-CHECK The magnitude of the potential difference induced in the short outer coil is 128 mV, which is a magnitude that could be attained by moving a strong bar magnet in and out of a coil. Thus, our result seems reasonable.

Concept Check 29.6

Two identical coils are shown in the figure. Coil 1 has a current i flowing in the direction shown. When the switch in the circuit containing coil 1 is opened, what happens in coil 2?

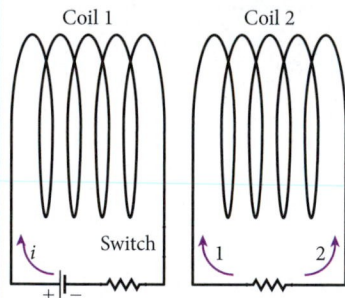

a) A current is induced in coil 2 that flows in direction 1.

b) A current is induced in coil 2 that flows in direction 2.

c) No current is induced in coil 2.

29.8 RL Circuits

In Chapter 26, we saw that if a source of emf supplying a voltage, V_{emf}, is put into a single-loop circuit containing a resistor of resistance R and a capacitor of capacitance C, the charge, q, on the capacitor builds up over time according to

$$q = CV_{emf}\left(1 - e^{-t/\tau_{RC}}\right),$$

where the time constant of the circuit, $\tau_{RC} = RC$, is the product of the resistance and the capacitance. The same time constant governs the decrease of the initial charge, q_0, on the capacitor if the source of emf is suddenly removed and the circuit is short-circuited:

$$q = q_0 e^{-t/\tau_{RC}}.$$

If a source of emf is placed in a single-loop circuit containing a resistor with resistance R and an inductor with inductance L, called an **RL circuit**, a similar phenomenon occurs. Figure 29.26 shows a circuit in which a source of emf is connected to a resistor and an inductor in series. If the circuit included only the resistor and not the inductor, the current would increase almost instantaneously to the value given by Ohm's Law, $i = V_{emf}/R$, as soon as the switch was closed. However, in the circuit with both the resistor and the inductor, the increasing current flowing through the inductor creates a self-induced potential difference that tends to oppose the increase in current. As time passes, the change in current decreases, and the opposing self-induced potential difference also decreases. After a long time, the current becomes steady at the value V_{emf}/R.

We can use Kirchhoff's Loop Rule to analyze this circuit, assuming that the current, i, at any given time is flowing through the circuit in a counterclockwise direction. With current flowing counterclockwise around the circuit, the source of emf provides a gain in potential, $+V_{emf}$, and the resistor causes a drop in potential, $-iR$. The self-inductance of the inductor produces a drop in potential because it is opposing the increase in current. The drop in potential due to the inductor is proportional to the time rate of change of the current, as given by equation 29.22. Thus, we can write the sum of the potential drops around the circuit as

$$V_{emf} - iR - L\frac{di}{dt} = 0.$$

We can rewrite this equation as

$$L\frac{di}{dt} + iR = V_{emf}. \tag{29.28}$$

The solution to this differential equation is obtained in exactly the same way as the solution to the differential equation for the RC circuit was obtained in Chapter 26. The solution, which can be checked by substituting it into equation 29.28 is

$$i(t) = \frac{V_{emf}}{R}\left(1 - e^{-t/(L/R)}\right). \tag{29.29}$$

The quantity L/R is the time constant of the RL circuit:

$$\tau_{RL} = \frac{L}{R}. \tag{29.30}$$

This time dependence of the current in an RL circuit is shown in Figure 29.27a for three different values of the time constant.

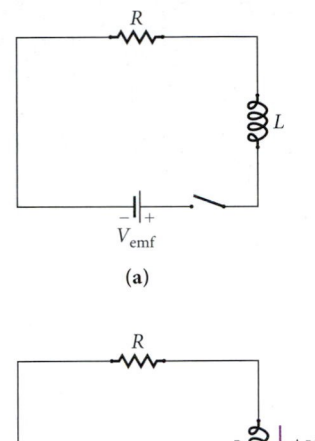

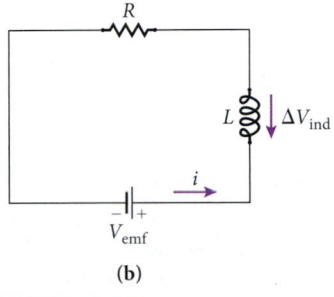

FIGURE 29.26 Single-loop circuit with a source of emf, a resistor, and an inductor: (a) switch open; (b) switch closed. When the switch is closed, the current flowing in the direction shown increases. A potential difference is induced across the inductor in the opposite direction, as shown.

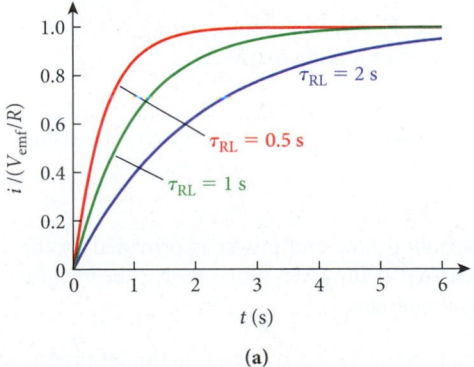

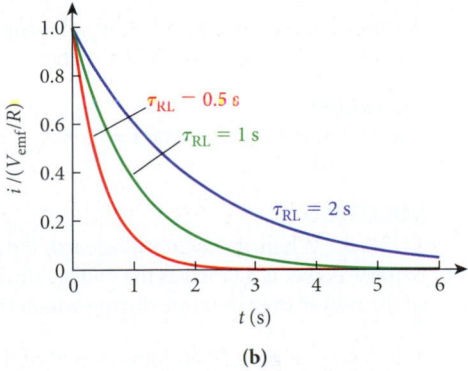

FIGURE 29.27 Time dependence of the current flowing through an RL circuit. (a) Current as a function of time when a resistor, an inductor, and a source of emf are connected in series. (b) Current as a function of time when the source of emf is suddenly removed from an RL circuit that has been connected for a long time.

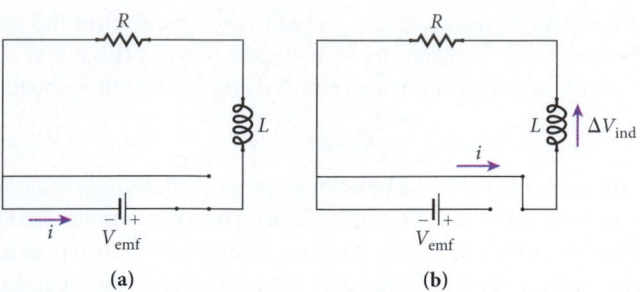

FIGURE 29.28 Single-loop circuit with a source of emf, a resistor, and an inductor. (a) The circuit with the source of emf connected. Current is flowing in the direction shown. (b) The source of emf is removed, and the resistor and the inductor are connected. Current flows in the same direction as before but is decreasing. A potential difference is induced across the inductor in the same direction as the current, as shown.

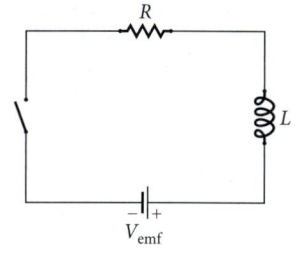
Looking at equation 29.29, you can see that for $t = 0$, the current is zero. For $t \to \infty$, the current is given by $i = V_{emf}/R$, which is as expected.

Now consider the circuit depicted in Figure 29.28, in which a source of emf was first connected and then suddenly removed. We can use equation 29.28 with $V_{emf} = 0$ to describe the time dependence of this circuit:

$$L\frac{di}{dt} + iR = 0. \tag{29.31}$$

The resistor causes a potential drop, and the inductor has a self-induced potential difference that tends to oppose the decrease in current. The solution of equation 29.31 is

$$i(t) = i_0 e^{-t/\tau_{RL}}. \tag{29.32}$$

The initial conditions when the source of emf is connected can be used to determine the initial current: $i_0 = V_{emf}/R$. Equation 29.32 describes a single-loop circuit with a resistor and an inductor that initially has a current i_0. The current drops exponentially with time with a time constant $\tau_{RL} = L/R$, and after a long time, the current in the circuit is zero. The current in this RL circuit as a function of time for three different values of the time constant is plotted in Figure 29.27b.

RL circuits can be used as timers to turn on devices at particular intervals and can also be used to filter out noise. However, these applications are usually handled with similar RC circuits because small capacitors are available in a wider range of capacitances than are inductors. The real value of inductors becomes apparent in circuits with all three components, resistors, capacitors, and inductors, which are covered in Chapter 30.

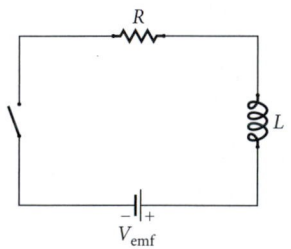

FIGURE 29.29 An RL circuit with a switch.

SOLVED PROBLEM 29.3 | Work Done by a Battery

A series circuit contains a battery that supplies $V_{emf} = 40.0$ V, an inductor with $L = 2.20$ H, a resistor with $R = 160.0\ \Omega$, and a switch, connected as shown in Figure 29.29.

PROBLEM
The switch is closed at time $t = 0$. How much work is done by the battery between $t = 0$ and $t = 1.65 \cdot 10^{-2}$ s?

SOLUTION
THINK   When the switch is closed, current begins to flow and power is provided by the battery. Power is defined as the voltage times the current at any given time. Work is the integral of the power over the time during which the circuit operates.

SKETCH   Figure 29.30 shows a plot of the current in the RL circuit as a function of time.

RESEARCH   The power in the circuit at any time t after the switch is closed is given by

$$P(t) = V_{emf}\, i(t), \tag{i}$$

where $i(t)$ is the current in the circuit. The current as a function of time for this circuit is given by equation 29.29,

$$i(t) = \frac{V_{emf}}{R}\left(1 - e^{-t/\tau_{RL}}\right), \tag{ii}$$

where $\tau_{RL} = L/R$. The work done by the battery is the integral of the power over the time in which the circuit has been in operation:

$$W = \int_0^T P(t)dt, \tag{iii}$$

where T is the time after the switch is closed.

SIMPLIFY We can combine equations (i), (ii), and (iii) to obtain

$$W = \int_0^T \frac{V_{emf}^2}{R}\left(1 - e^{-t/\tau_{RL}}\right)dt.$$

Evaluating the definite integral gives us

$$W = \frac{V_{emf}^2}{R}\left[t + \tau_{RL}e^{-t/\tau_{RL}}\right]_0^T = \frac{V_{emf}^2}{R}\left[T + \tau_{RL}\left(e^{-T/\tau_{RL}} - 1\right)\right]. \tag{iv}$$

CALCULATE First, we calculate the value of the time constant:

$$\tau_{RL} = \frac{L}{R} = \frac{2.20 \text{ H}}{160.0 \ \Omega} = 1.375 \cdot 10^{-2} \text{ s}.$$

Putting all the numerical values into equation (iv) then gives

$$W = \frac{(40.0 \text{ V})^2}{160.0 \ \Omega}\left[\left(1.65 \cdot 10^{-2} \text{ s}\right) + \left(1.375 \cdot 10^{-2} \text{ s}\right)\left(e^{-\left(1.65 \cdot 10^{-2} \text{ s}\right)/\left(1.375 \cdot 10^{-2} \text{ s}\right)} - 1\right)\right] = 0.0689142 \text{ J}.$$

ROUND We report our result to three significant figures:

$$W = 6.89 \cdot 10^{-2} \text{ J}.$$

DOUBLE-CHECK To double-check our result, we assume that the current in the circuit is constant in time and equal to half of the final current:

$$i_{ave} = \frac{i(T)}{2} = \frac{\left(V_{emf}/R\right)\left(1 - e^{-T/\tau_{RL}}\right)}{2} = \frac{\left(40.0 \text{ V}/160.0 \ \Omega\right)\left(1 - e^{-\left(1.65 \cdot 10^{-2} \text{ s}\right)/\left(1.375 \cdot 10^{-2} \text{ s}\right)}\right)}{2} = 0.0874 \text{ A}.$$

This current would correspond to the average current if the current increased linearly with time. The work done would then be

$$W = PT = i_{ave}V_{emf}T = (0.0874 \text{ A})(40.0 \text{ V})\left(1.65 \cdot 10^{-2} \text{ s}\right) = 5.77 \cdot 10^{-2} \text{ J}.$$

This value is less than, but close to, our calculated result. Thus, our result seems reasonable.

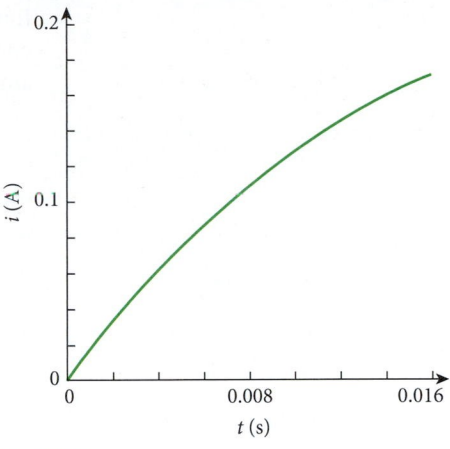

FIGURE 29.30 Current in the RL circuit as a function of time.

29.9 Energy and Energy Density of a Magnetic Field

We can think of an inductor as a device that can store energy in a magnetic field, similar to the way a capacitor can store energy in an electric field. The energy stored in the electric field of a capacitor is given by

$$U_E = \tfrac{1}{2}\frac{q^2}{C}.$$

Consider an inductor connected to a source of emf. Current begins to flow through the inductor, producing a self-induced potential difference opposing the increase in current. The instantaneous power provided by the source of emf is the product of the current and the voltage of the emf source, V_{emf}. Using equation 29.28 with $R = 0$, we can write

$$P = V_{emf}i = \left(L\frac{di}{dt}\right)i. \tag{29.33}$$

Integrating this power over the time it takes to reach a final current i in the circuit yields the energy provided by the source of emf. Since there are no resistive losses in this circuit, this amount of energy must be stored in the magnetic field of the inductor. Therefore,

$$U_B = \int_0^t P\,dt = \int_0^i Li'\,di' = \tfrac{1}{2}Li^2. \tag{29.34}$$

Equation 29.34 has a form similar to the analogous equation for a capacitor's electric field, with q replaced by i and $1/C$ replaced by L.

Now let's consider an ideal solenoid with length ℓ, cross-sectional area A, and n turns per unit length, carrying current i. The energy stored in the magnetic field of the solenoid according to equation 29.21 is

$$U_B = \tfrac{1}{2}Li^2 = \tfrac{1}{2}\mu_0 n^2 \ell A i^2.$$

The magnetic field occupies the volume enclosed by the solenoid, which is given by ℓA. Thus, the energy density, u_B, of the magnetic field of the solenoid is

$$u_B = \frac{\tfrac{1}{2}\mu_0 n^2 \ell A i^2}{\ell A} = \tfrac{1}{2}\mu_0 n^2 i^2.$$

Since $B = \mu_0 n i$ for a solenoid, the energy density of the magnetic field of a solenoid can be expressed as

$$u_B = \frac{1}{2\mu_0} B^2. \tag{29.35}$$

Although we derived this expression for the special case of a solenoid, it applies to magnetic fields in general.

Concept Check 29.8

A long solenoid has a circular cross section of radius $r = 8.10$ cm, a length $\ell = 0.540$ m, and $n = 2.00 \cdot 10^4$ turns/m. The solenoid stores 42.5 mJ of energy when it carries a current i. If the current is doubled, to $2i$, the energy stored in the solenoid

a) decreases by a factor of 4.

b) decreases by a factor of 2.

c) remains the same.

d) increases by a factor of 2.

e) increases by a factor of 4.

29.10 Applications to Information Technology

Computers and many consumer electronics devices use magnetization and induction to store and retrieve information. Examples are computer hard drives, videotapes, audiotapes, and the magnetic strips on credit cards. During the last decade, the use of storage media based on other technologies, such as the optical storage of information on CDs and DVDs and the flash memory cards in digital cameras, has increased; however, magnetic storage devices are still a technological mainstay and the basis of a multibillion dollar industry.

Computer Hard Drive

One device that stores information using magnetization and induction is the computer hard drive. The hard drive stores information in the form of *bits,* the binary code consisting of zeros and ones. Eight bits make a *byte,* which can represent a number or an alphanumeric character. A modern hard drive can hold up to 2 terabytes ($2 \cdot 10^{12}$ bytes) of information. A hard drive consists of one or more rotating platters with a ferromagnetic coating accessed by a movable read/write head, as shown in Figure 29.31.

The read/write head can be positioned to access any one of many tracks on the rotating platter. The operation of a read/write head in a conventional hard drive is illustrated in Figure 29.32a. As the coated platter moves below the read/write head, a pulse of current in one direction magnetizes the surface of the platter to represent a binary one, or a pulse of current in the opposite direction magnetizes the surface to represent a binary zero. In Figure 29.32a, a binary one is shown as a red arrow pointing to the right, and a binary zero is shown as a green arrow pointing to the left. In read mode, when the magnetized areas of the platter pass beneath the read sensor, a positive or negative current is induced, and the electronics of the hard drive can tell if the information is a zero or a one. The method used to encode and read back data shown in Figure 29.32a is called *longitudinal encoding* because the magnetic fields of the magnetized areas of the platter are parallel or antiparallel to the motion of the platter. The data storage capacity of hard drives was increased by

Platter

Read/write head

FIGURE 29.31 The read/write head and spinning platter inside a computer hard drive.

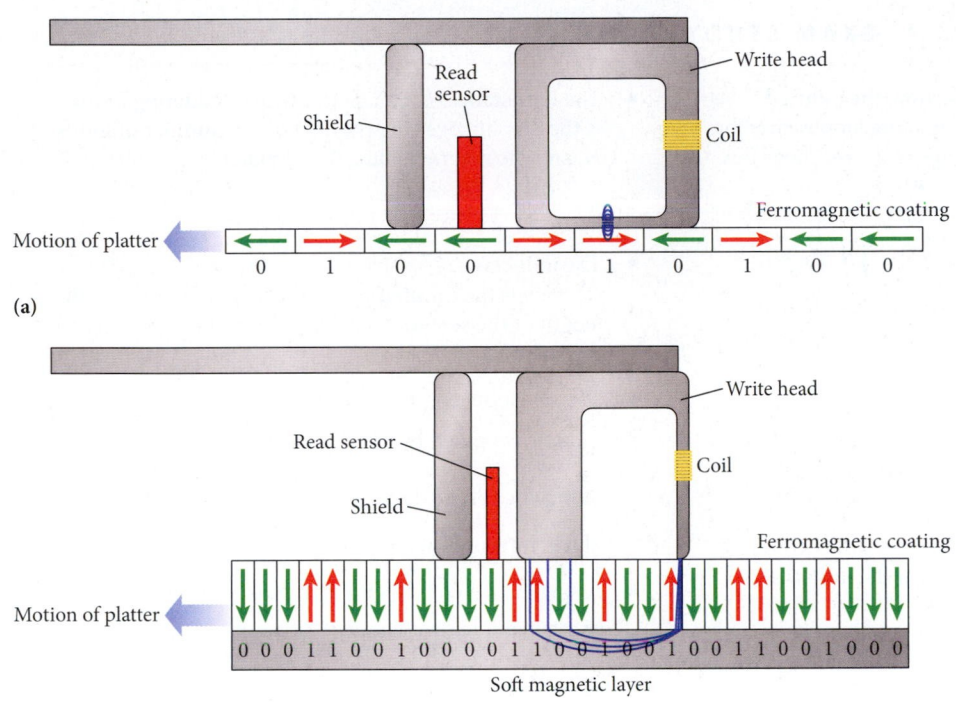

FIGURE 29.32 The read/write head of a computer hard drive. (a) Longitudinal encoding of information on the spinning platter. (b) Perpendicular encoding of information on the spinning platter.

making the magnetized areas smaller and by adding more platters and read/write heads. However, manufacturers found it difficult to construct hard drives holding more than 250 gigabytes ($250 \cdot 10^9$ bytes) using this technique. As the manufacturers made the bits smaller, the bits interfered with each other, causing some bits to flip randomly and introduce errors in the stored data.

Recently, the technique of *perpendicular encoding* of data has been developed, illustrated in Figure 29.32b. Again, a read/write head is used above a spinning platter coated with a ferromagnetic substance. However, in this case, the magnetic fields are perpendicular to the surface of the platter, which allows a tighter packing of the bits and increases the capacity of the hard drive. The platter is constructed with a thicker ferromagnetic coating and a soft magnetic material on the bottom that acts to contain the magnetic field lines. Note that the magnetic field lines at the pointed end of the write head are very close together, whereas the magnetic field lines returning to the blunt end of the write head are widely spaced. Thus, the ferromagnetic coating of the platter is strongly magnetized in the up or down direction, depending on the direction of the current pulse through the coil of the write head, while the bits closer to the read sensor are not affected.

Hard drives using perpendicular encoding also incorporate the phenomenon called *giant magnetoresistance (GMR),* which allows the construction of a very small and sensitive read sensor. The French physicist Albert Fert and the German physicist Peter Grünberg received the Nobel Prize in Physics in 2007 for the discovery of this effect. Hard drives with information storage capacities of up to 2 terabytes or more that use perpendicular encoding and GMR read sensors are widely available. iPods with a storage capacity above 64 GB are an example of a device using this technology (iPod Touch and iPhones use a different storage technology that has no moving parts). The fact that you can watch full-length movies on your iPod and carry many thousands of songs in it as well is a direct result of physics research performed during the last two decades. And as research in nanoscience and nanotechnology continues to produce exciting results for technological applications, the amazing growth in the capabilities of consumer electronics devices will continue for the foreseeable future.

WHAT WE HAVE LEARNED | EXAM STUDY GUIDE

- According to Faraday's Law of Induction, the induced potential difference, ΔV_{ind}, in a conducting loop is given by the negative of the time rate of change of the magnetic flux passing through the loop: $\Delta V_{ind} = -\dfrac{d\Phi_B}{dt}$.

- The magnetic flux, Φ_B, is given by $\Phi_B = \displaystyle\iint \vec{B} \cdot d\vec{A}$, where $\vec{B}$ is the magnetic field and $d\vec{A}$ is the differential area element defined by a vector normal to the surface through which the magnetic field passes.

- For a constant magnetic field, $\vec{B}$, the magnetic flux, Φ_B, passing through an area, A, is given by $\Phi_B = BA\cos\theta$, where θ is the angle between the magnetic field vector and a normal vector to the area.

- Lenz's Law states that a changing magnetic flux through a conducting loop induces a current in the loop that opposes the change in magnetic flux.

- A magnetic field that is changing in time induces an electric field given by $\displaystyle\oint \vec{E} \cdot d\vec{s} = -\dfrac{d\Phi_B}{dt}$, where the integration is done over any closed path in the magnetic field.

- The inductance, L, of a device with conducting loops is the flux linkage (the product of the number of loops, N, and the magnetic flux, Φ_B) divided by the current, i:
$$L = \frac{N\Phi_B}{i}.$$

- The inductance of a solenoid is given by $L = \mu_0 n^2 \ell A$, where n is the number of turns per unit length, ℓ is the length of the solenoid, and A is the cross-sectional area of the solenoid.

- The self-induced potential difference, $\Delta V_{ind,L}$, for any inductor is given by $\Delta V_{ind,L} = -L\dfrac{di}{dt}$, where L is the inductance of the device and $\dfrac{di}{dt}$ is the time rate of change of the current flowing through the inductor.

- A single-loop circuit with an inductance L and a resistance R has a characteristic time constant of $\tau_{RL} = \dfrac{L}{R}$.

- The energy, U_B, stored in the magnetic field of an inductor with an inductance L and carrying a current i is given by $U_B = \frac{1}{2}Li^2$.

ANSWERS TO SELF-TEST OPPORTUNITIES

29.1 $\dfrac{dB}{dt} = -\dfrac{0.500 \text{ T}}{0.250 \text{ s}} = -2.00 \text{ T/s}$

$\Delta V_{ind} = -\dfrac{d\Phi_B}{dt} = -\dfrac{d(BA)}{dt} = -\pi r^2 \dfrac{dB}{dt}$

$r = \sqrt{\dfrac{|\Delta V_{ind}|}{\pi |dB/dt|}} = \sqrt{\dfrac{1.24 \text{ V}}{\pi (2.00 \text{ T/s})}} = 0.444 \text{ m.}$

29.2 As the loop enters the magnetic field, the magnetic flux is increasing. The current induced in the loop will be in the counterclockwise direction to oppose the increase in the flux. As the loop exits the magnetic field, the magnetic flux

is decreasing. The current induced in the loop will be in a clockwise direction to oppose the decrease in the flux.

29.3 If the induced potential difference were equal to the change in the magnetic flux, then any increase in the flux going through a coil (perhaps from just a minute random fluctuation in the ambient magnetic field in the room) would lead to an induced potential difference, which would produce a current in the coil, which would act to increase the flux, which would lead to a larger induced potential difference, a larger current, and an even larger increase in flux. In other words, a runaway situation would result, which clearly violates energy conservation.

29.4 a) true   b) false   c) true   d) true

PROBLEM-SOLVING GUIDELINES

1. To solve a problem involving electromagnetic induction, first ask: What is making the magnetic flux change? If the magnetic field is changing, you need to use equation 29.9; if the area through which the flux passes is changing, you need to use equation 29.10; and if the orientation between the magnetic field and the area is changing, you need to use equation 29.11. You don't need to memorize these equations, as long as you remember Faraday's Law (equation 29.5) and the definition of magnetic flux (equation 29.1).

2. Once you know which elements of the problem situation are constant and which are changing, use Lenz's Law to determine the direction of the induced current and the locations of higher and lower potential. Then, you can pick a direction for the differential area vector, $d\vec{A}$, of the flux and calculate the unknown quantities.

MULTIPLE-CHOICE QUESTIONS

29.1 A solenoid with 200 turns and a cross-sectional area of 60 cm^2 has a magnetic field of 0.60 T along its axis. If the field is confined within the solenoid and changes at a rate of 0.20 T/s, the magnitude of the induced potential difference in the solenoid will be

a) 0.0020 V.   b) 0.02 V.   c) 0.001 V.   d) 0.24 V.

29.2 The rectangular loop of wire in Figure 29.9 is pulled with a constant acceleration from a region of zero magnetic field into a region of a uniform magnetic field. During this process, the current induced in the loop

a) will be zero.

b) will be some constant value that is not zero.

c) will increase linearly with time.

d) will increase exponentially with time.

e) will increase linearly with the square of the time.

29.3 Which of the following will induce a current in a loop of wire in a uniform magnetic field?

a) decreasing the strength of the field

b) rotating the loop about an axis parallel to the field

c) moving the loop within the field

d) all of the above

e) none of the above

29.4 Faraday's Law of Induction states that

a) a potential difference is induced in a loop when there is a change in the magnetic flux through the loop.

b) the current induced in a loop by a changing magnetic field produces a magnetic field that opposes this change in magnetic field.

c) a changing magnetic field induces an electric field.

d) the inductance of a device is a measure of its opposition to changes in current flowing through it.

e) magnetic flux is the product of the average magnetic field and the area perpendicular to it that it penetrates.

29.5 A conducting ring is moving from left to right through a uniform magnetic field, as shown in the figure. In which region(s) is there an induced current in the ring?

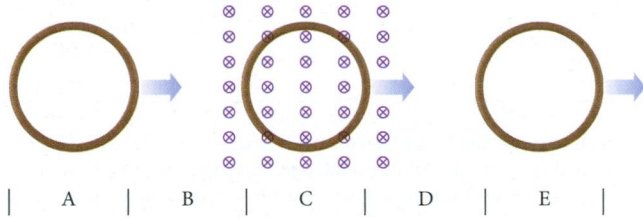

a) regions B and D

b) regions B, C, and D

c) region C

d) regions A through E

29.6 A circular loop of wire moving in the *xy*-plane with a constant velocity in the negative *x*-direction enters a uniform magnetic field, which covers the region in which $x < 0$, as shown in the figure. The surface normal vector of the loop points in the direction of the magnetic field. Which of the following statements is correct?

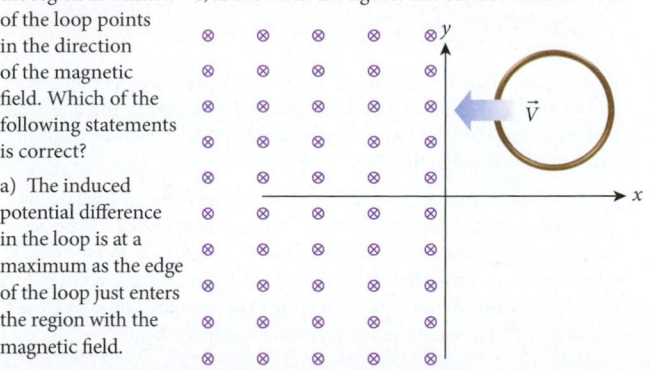

a) The induced potential difference in the loop is at a maximum as the edge of the loop just enters the region with the magnetic field.

b) The induced potential difference in the loop is at a maximum when one fourth of the loop is in the region with the magnetic field.

c) The induced potential difference in the loop is at a maximum when the loop is halfway into the region with the magnetic field.

d) The induced potential difference in the loop is constant from the instant the loop starts to enter the region with the magnetic field.

29.7 Which of the following statements regarding self-induction is correct?

a) Self-induction occurs only when a direct current is flowing through a circuit.

b) Self-induction occurs only when an alternating current is flowing through a circuit.

c) Self-induction occurs when either a direct current or an alternating current is flowing through a circuit.

d) Self-induction occurs when either a direct current or an alternating current is flowing through a circuit as long as the current is varying.

29.8 You have a light bulb, a bar magnet, a spool of wire that you can cut into as many pieces as you want, and nothing else. How can you get the bulb to light up?

a) You can't. The bulb needs electricity to light it, not magnetism.

b) You cut a length of wire, connect the light bulb to the two ends of the wire, and pass the magnet through the loop that is formed.

c) You cut two lengths of wire and connect the magnet and the bulb in series.

29.9 Calculate the potential difference induced between the tips of the wings of a Boeing 747-400 with a wingspan of 64.67 m when it is in level flight at a speed of 913 km/h. Assume that the magnitude of the downward component of the Earth's magnetic field is $B = 5.00 \cdot 10^{-5}$ T.

a) 0.820 V   c) 10.4 V   e) 225 V

b) 2.95 V   d) 30.1 V

29.10 A long solenoid with a circular cross section of radius $r_1 = 2.80$ cm and $n = 290$ turns/cm is inside of and coaxial with a short coil that has a circular cross section of radius $r_2 = 4.90$ cm and $N = 31$ turns. Suppose the current in the short coil is increased steadily from zero to $i = 2.80$ A in 18.0 ms. What is the magnitude of the potential difference induced in the solenoid while the current in the short coil is changing?

a) 0.0991 V   c) 0.233 V   e) 0.750 V

b) 0.128 V   d) 0.433 V

29.11 A long solenoid has a circular cross section of radius $r = 8.10$ cm, a length $\ell = 0.540$ m, and $n = 2.00 \cdot 10^4$ turns/m. The solenoid is carrying a current of magnitude $i = 4.04 \cdot 10^{-3}$ A. How much energy is stored in the magnetic field of the solenoid?

a) $2.11 \cdot 10^{-7}$ J   c) $4.57 \cdot 10^{-5}$ J   e) $4.55 \cdot 10^{-1}$ J

b) $8.91 \cdot 10^{-6}$ J   d) $6.66 \cdot 10^{-3}$ J

29.12 Suppose the current in the short coil in Solved Problem 29.2 is increased steadily from zero to $i = 2.80$ A in 18.0 ms. What is the magnitude of the potential difference induced in the solenoid while the current in the short coil is changing?

a) 0.0991 V   c) 0.233 V   e) 0.750 V

b) 0.128 V   d) 0.433 V

29.13 Suppose the length of the rotating rod in Solved Problem 29.1 is increased by a factor of 2. By what factor does the power dissipated in the resistor change?

a) $\frac{1}{2}$ c) 4   e) 16

b) 2   d) 8

29.14 Suppose the resistance of the resistor in Solved Problem 29.1 is increased by a factor of 2. By what factor does the power dissipated in the resistor change?

a) $\frac{1}{2}$ c) 4   e) 16

b) 2   d) 8

CONCEPTUAL QUESTIONS

29.15 When you plug a refrigerator into a wall socket, on occasion, a spark appears between the prongs. What causes this?

29.16 People with pacemakers or other mechanical devices as implants are often warned to stay away from large machinery or motors. Why?

29.17 Chapter 14 discussed damped harmonic oscillators, in which the damping force is velocity dependent and always opposes the motion of the oscillator. One way of producing this type of force is to use a piece of metal, such as aluminum, that moves through a nonuniform magnetic field. Explain why this technique is capable of producing a damping force.

29.18 In a popular lecture demonstration, a cylindrical permanent magnet is dropped down a long aluminum tube as shown in the figure. Neglecting friction of the magnet against the inner walls of the tube and assuming that the tube is very long compared to the size of the magnet, will the magnet accelerate downward with an acceleration equal to g (free fall)? If not, describe the eventual motion of the magnet. Does it matter if the north pole or south pole of the magnet is on the lower side?

29.19 A popular demonstration of eddy currents involves dropping a magnet down a long metal tube and a long glass or plastic tube. As the magnet falls through a tube, the magnetic flux changes as the magnet moves toward or away from each part of the tube.

a) Which tube has the larger voltage induced in it?

b) Which tube has the larger eddy currents induced in it?

29.20 The current in a very long, tightly wound solenoid with radius a and n turns per unit length varies over time according to the equation $i(t) = Ct^2$, where the current i is in amps and the time t is in seconds, and C is a constant with appropriate units. Concentric with the solenoid is a conducting ring of radius r, as shown in the figure.

a) Write an expression for the potential difference induced in the ring.

b) Write an expression for the magnitude of the electric field induced at an arbitrary point on the ring.

c) Is the ring necessary for the induced electric field to exist?

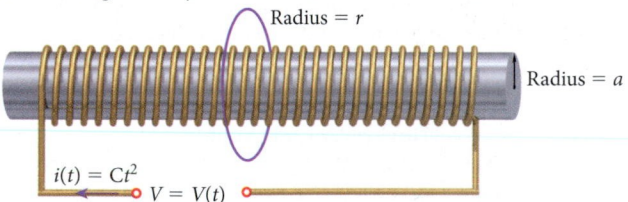

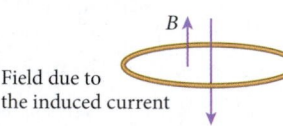

29.21 A circular wire ring experiences an increasing magnetic field in the upward direction, as shown in the figure. What is the direction of the induced current in the ring?

29.22 A square conducting loop with sides of length L is rotating at a constant angular speed, ω, in a uniform magnetic field of magnitude B. At time $t = 0$, the loop is oriented so that the direction normal to the loop is aligned with the magnetic field. Find an expression for the potential difference induced in the loop as a function of time.

29.23 A solid metal disk of radius R is rotating around its center axis at a constant angular speed of ω. The disk is in a uniform magnetic field of magnitude B that is oriented normal to the surface of the disk. Calculate the magnitude of the potential difference between the center of the disk and the outside edge.

29.24 Large electric fields are certainly a hazard to the human body, as they can produce dangerous currents, but what about large magnetic fields? A man 1.80 m tall walks at 2.00 m/s perpendicular to a horizontal magnetic field of 5.0 T; that is, he walks between the pole faces of a very big magnet. (Such a magnet can, for example, be found in the National Superconducting Cyclotron Laboratory at Michigan State University.) Given that his body is full of conducting fluids, estimate the potential difference induced between his head and feet.

29.25 At Los Alamos National Laboratories, one means of producing very large magnetic fields is the *EPFCG (explosively-pumped flux compression generator)*, which is used to study the effects of a high-power electromagnetic pulse (EMP) in electronic warfare. Explosives are packed and detonated in the space between a solenoid and a small copper cylinder coaxial with and inside the solenoid, as shown in the figure. The explosion occurs in a very short time and collapses the cylinder rapidly. This rapid change creates inductive currents that keep the magnetic flux constant while the cylinder's radius shrinks by a factor of r_i/r_f. Estimate the magnetic field produced, assuming that the radius is compressed by a factor of 14 and the initial magnitude of the magnetic field, B_i, is 1.0 T.

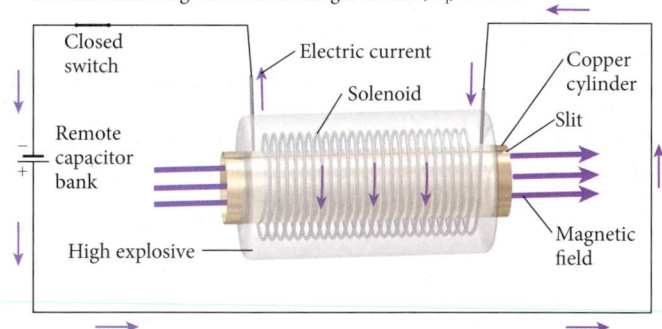

29.26 A metal hoop is laid flat on the ground. A magnetic field that is directed upward, out of the ground, is increasing in magnitude. As you look down on the hoop from above, what is the direction of the induced current in the hoop?

29.27 The wire of a tightly wound solenoid is unwound and then rewound to form another solenoid with double the diameter of the first solenoid. By what factor will the inductance change?

EXERCISES

A blue problem number indicates a worked-out solution is available in the Student Solutions Manual. One • and two •• indicate increasing level of problem difficulty.

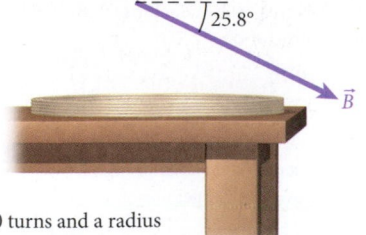

Sections 29.1 and 29.2

29.28 A circular coil of wire with 20 turns and a radius of 40.0 cm is laying flat on a horizontal tabletop as shown

in the figure. There is a uniform magnetic field extending over the entire table with a magnitude of 5.00 T and directed to the north and downward, making an angle of 25.8° with the horizontal. What is the magnitude of the magnetic flux through the coil?

29.29 When a magnet in an MRI is abruptly shut down, the magnet is said to be *quenched*. Quenching can occur in as little as 20.0 s. Suppose a magnet with an initial field of 1.20 T is quenched in 20.0 s, and the final field is approximately zero. Under these conditions, what is the average induced potential difference around a conducting loop of radius 1.00 cm (about the size of a wedding ring) oriented perpendicular to the field?

29.30 An 8-turn coil has square loops measuring 0.200 m along a side and a resistance of 3.00 Ω. It is placed in a magnetic field that makes an angle of 40.0° with the plane of each loop. The magnitude of this field varies with time according to $B = 1.50t^3$, where t is measured in seconds and B in teslas. What is the induced current in the coil at $t = 2.00$ s?

29.31 A metal loop has an area of 0.100 m² and is placed flat on the ground. There is a uniform magnetic field pointing due west, as shown in the figure. This magnetic field initially has a magnitude of 0.123 T, which decreases steadily to 0.075 T during a period of 0.579 s. Find the potential difference induced in the loop during this time.

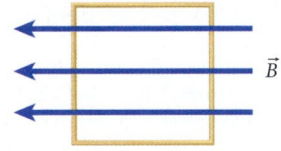

•29.32 A respiration monitor has a flexible loop of copper wire, which wraps about the chest. As the wearer breathes, the radius of the loop of wire increases and decreases. When a person in the Earth's magnetic field (assume $0.426 \cdot 10^{-4}$ T) inhales, what is the average current in the loop, assuming that it has a resistance of 30.0 Ω and increases in radius from 20.0 cm to 25.0 cm over 1.00 s? Assume that the magnetic field is perpendicular to the plane of the loop.

•29.33 A circular conducting loop with radius a and resistance R_2 is concentric with a circular conducting loop with radius $b \gg a$ (b much greater than a) and resistance R_1. A time-dependent voltage is applied to the larger loop; its slow sinusoidal variation in time is given by $V(t) = V_0 \sin\omega t$, where V_0 and ω are constants with dimensions of voltage and inverse time, respectively. Assuming that the magnetic field throughout the inner loop is uniform (constant in space) and equal to the field at the center of the loop, derive expressions for the potential difference induced in the inner loop and the current i through that loop.

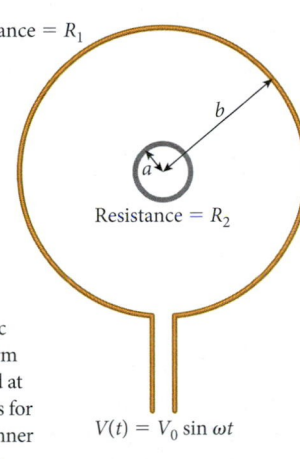

Resistance = R_1

Resistance = R_2

$V(t) = V_0 \sin \omega t$

••29.34 A long solenoid with cross-sectional area A_1 surrounds another long solenoid with cross-sectional area $A_2 < A_1$ and resistance R. Both solenoids have the same length and the same number of turns. A current given by $i = i_0 \cos\omega t$ is flowing through the outer solenoid. Find an expression for the magnetic field in the inner solenoid due to the induced current.

Section 29.3

29.35 The conducting loop in the shape of a quarter-circle shown in the figure has a radius of 10.0 cm and a resistance of 0.200 Ω. The magnetic field strength within the dotted circle of radius 3.00 cm is initially 2.00 T. The magnetic field strength then decreases from 2.00 T to 1.00 T in 2.00 s. Find (a) the magnitude and (b) the direction of the induced current in the loop.

29.36 A supersonic aircraft with a wingspan of 10.0 m is flying over the north magnetic pole (in a magnetic field of magnitude 0.500 G oriented perpendicular to the ground) at a speed of three times the speed of sound (Mach 3). What is the potential difference between the tips of the wings? Assume that the wings are made of aluminum.

•29.37 A helicopter hovers above the north magnetic pole in a magnetic field of magnitude 0.426 G and oriented perpendicular to the ground. The helicopter rotors are 10.0 m long, are made of aluminum, and rotate about the hub with a rotational speed of $1.00 \cdot 10^4$ rpm. What is the potential difference from the hub to the end of a rotor?

•29.38 An elastic circular conducting loop expands at a constant rate over time such that its radius is given by $r(t) = r_0 + vt$, where $r_0 = 0.100$ m and

$v = 0.0150$ m/s. The loop has a constant resistance of $R = 12.0 \Omega$ and is placed in a uniform magnetic field of magnitude $B_0 = 0.750$ T, perpendicular to the plane of the loop, as shown in the figure. Calculate the direction and the magnitude of the induced current, i, at $t = 5.00$ s.

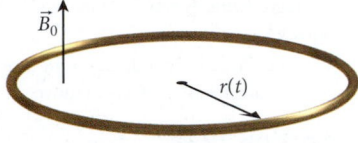

•29.39 A rectangular frame of conducting wire has negligible resistance and width w and is held vertically in a magnetic field of magnitude B, as shown in the figure. A metal bar with mass m and resistance R is placed across the frame, maintaining contact with the frame. Derive an expression for the terminal velocity of the bar if it is allowed to fall freely along this frame starting from rest. Neglect friction between the wires and the metal bar.

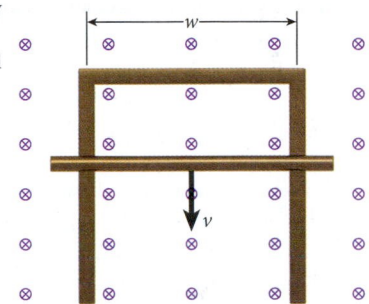

•29.40 Two parallel conducting rails with negligible resistance are connected at one end by a resistor of resistance R, as shown in the figure. The rails are placed in a magnetic field $\vec{B}_{ext}$, which is perpendicular to the plane of the rails. This magnetic field is uniform and time independent. The distance between the rails is ℓ. A conducting rod slides without friction on top of the two rails at constant velocity $\vec{v}$.

Three-dimensional view

a) Using Faraday's Law of Induction, calculate the magnitude of the potential difference induced in the moving rod.

b) Calculate the magnitude of the induced current in the rod, i_{ind}.

c) Show that for the rod to move at a constant velocity as shown, it must be pulled with an external force, $\vec{F}_{ext}$, and calculate the magnitude of this force.

d) Calculate the work done, W_{ext}, and the power generated, P_{ext}, by the external force in moving the rod.

e) Calculate the power used (dissipated) by the resistor, P_R. Explain the correlation between this result and those of part (d).

Top view
$\vec{B}_{ext}$ into the page

$\vec{B}_{ind}$ out of page

•29.41 A long, straight wire runs along the y-axis. The wire carries a current in the positive y-direction that is changing as a function of time according to $i = 2.00$ A + $(0.300$ A/s$)t$. A loop of wire is located in the xy-plane near the y-axis, as shown in the figure. The loop has dimensions 7.00 m by 5.00 m and is 1.00 m away from the wire. What is the induced potential difference in the wire loop at $t = 10.0$ s?

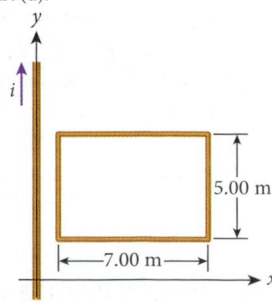

5.00 m

7.00 m

••29.42 The long, straight wire in the figure has a current $i = 1.00$ A flowing in it. A square loop with 10.0-cm sides and a resistance of 0.0200 Ω is positioned 10.0 cm away from the wire. The loop is then moved in the positive x-direction with a speed $v = 10.0$ cm/s.

a) Find the direction of the induced current in the loop.

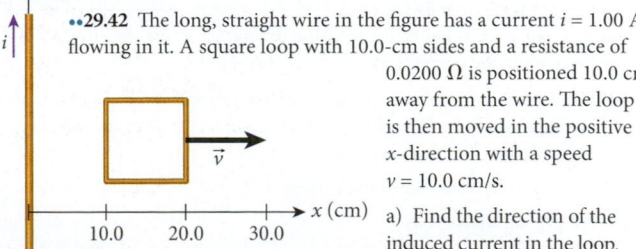

b) Identify the directions of the magnetic forces acting on all sides of the square loop.

c) Calculate the direction and the magnitude of the net force acting on the loop at the instant it starts to move.

Section 29.4

29.43 A simple generator consists of a loop rotating inside a constant magnetic field (see Figure 29.19). If the loop is rotating with frequency f, the magnetic flux is given by $\Phi(t) = BA\cos(2\pi ft)$. If $B = 1.00$ T and $A = 1.00$ m^2, what must the value of f be for the maximum induced potential difference to be 110. V?

•29.44 A motor has a single loop inside a magnetic field of magnitude 0.870 T. If the area of the loop is 300. cm^2, find the maximum angular speed possible for this motor when connected to a source of emf providing 170. V.

•29.45 Your friend decides to produce electrical power by turning a coil of $1.00 \cdot 10^5$ circular loops of wire around an axis perpendicular to the Earth's magnetic field, which has a local magnitude of 0.300 G. The loops have a radius of 25.0 cm.

a) If your friend turns the coil at a frequency of 150. Hz, what peak current will flow in a resistor, $R = 1.50$ kΩ, connected to the coil?

b) The average current flowing in the coil will be 0.7071 times the peak current. What will be the average power obtained from this device?

Sections 29.6 and 29.7

29.46 Find the mutual inductance of the solenoid and the coil described in Example 29.1 and the induced potential difference in the coil at $t = 2.0$ s using the techniques described in Section 29.7. How do the results for the induced potential difference compare?

29.47 The figure shows the current through a 10.0-mH inductor over a time interval of 8.00 ms. Draw a graph showing the self-induced potential difference, $\Delta V_{ind,L}$, for the inductor over the same interval.

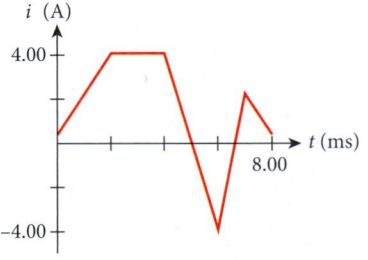

•29.48 A short coil with radius $R = 10.0$ cm contains $N = 30.0$ turns and surrounds a long solenoid with radius $r = 8.00$ cm containing $n = 60$ turns per centimeter. The current in the short coil is increased at a constant rate from zero to $i = 2.00$ A in a time of $t = 12.0$ s. Calculate the induced potential difference in the long solenoid while the current is increasing in the short coil.

Section 29.8

29.49 Consider an RL circuit with resistance $R = 1.00$ MΩ and inductance $L = 1.00$ H, which is powered by a 10.0-V battery.

a) What is the time constant of the circuit?

b) If the switch is closed at time $t = 0$, what is the current just after that time? After 2.00 μs? When a long time has passed?

29.50 In the circuit in the figure, $R = 120.$ Ω, $L = 3.00$ H, and $V_{emf} = 40.0$ V. After the switch is closed, how long will it take the current in the inductor to reach 300. mA?

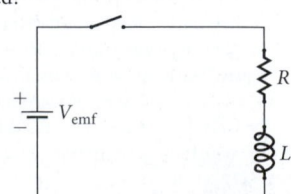

29.51 The current is increasing at a rate of 3.60 A/s in an RL circuit with $R = 3.25$ Ω and $L = 440.$ mH. What is the potential difference across the circuit at the moment when the current in the circuit is 3.00 A?

•29.52 In the circuit in the figure, a battery supplies $V_{emf} = 18.0$ V and $R_1 = 6.00$ Ω, $R_2 = 6.00$ Ω, and $L = 5.00$ H. Calculate each of the following *immediately* after the switch is closed:

a) the current flowing out of the battery

b) the current through R_1

c) the current through R_2

d) the potential difference across R_1

e) the potential difference across R_2

f) the potential difference across L

g) the rate of current change across R_1

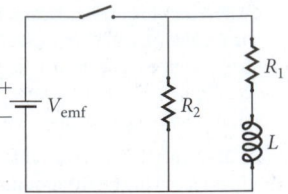

•29.53 In the circuit in the figure, a battery supplies $V_{emf} = 18.0$ V and $R_1 = 6.00$ Ω, $R_2 = 6.00$ Ω, and $L = 5.00$ H. Calculate each of the following a *long time* after the switch is closed:

a) the current flowing out of the battery

b) the current through R_1

c) the current through R_2

d) the potential difference across R_1

e) the potential difference across R_2

f) the potential difference across L

g) the rate of current change across R_1

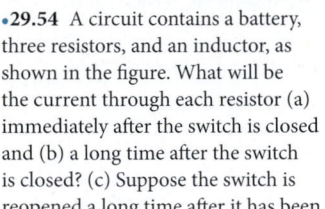

•29.54 A circuit contains a battery, three resistors, and an inductor, as shown in the figure. What will be the current through each resistor (a) immediately after the switch is closed and (b) a long time after the switch is closed? (c) Suppose the switch is reopened a long time after it has been closed. What is the current in each resistor? After a long time?

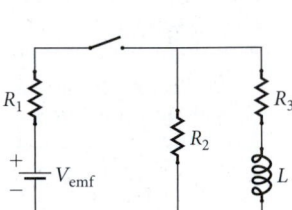

Section 29.9

29.55 Having just learned that there is energy associated with magnetic fields, an inventor sets out to tap the energy associated with the Earth's magnetic field. What volume of space near Earth's surface contains 1.00 J of energy, assuming the strength of the magnetic field to be $5.00 \cdot 10^{-5}$ T?

29.56 A clinical MRI (magnetic resonance imaging) superconducting magnet can be approximated as a solenoid with a diameter of 1.00 m, a length of 1.50 m, and a uniform magnetic field of 3.00 T. Determine (a) the energy density of the magnetic field and (b) the total energy in the solenoid.

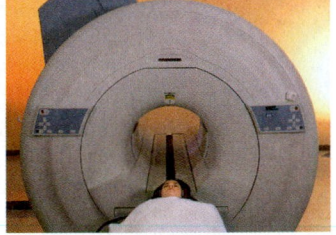

29.57 A *magnetar* (magnetic neutron star) has a magnetic field near its surface of magnitude $4.00 \cdot 10^{10}$ T.

a) Calculate the energy density of this magnetic field.

b) The Special Theory of Relativity associates energy with any mass m at rest according to $E_0 = mc^2$ (more on this in Chapter 35). Find the rest mass density associated with the energy density of part (a).

•29.58 An emf of 20.0 V is applied to a coil with an inductance of 40.0 mH and a resistance of 0.500 Ω.

a) Determine the energy stored in the magnetic field when the current reaches $\frac{1}{4}$ of its maximum value.

b) How long does it take for the current to reach this value?

•29.59 A student wearing a 15.0-g gold band with radius 0.750 cm (and with a resistance of 61.9 $\mu\Omega$ and a specific heat capacity of $c = 129$ J/kg °C) on her finger moves her finger from a region having a magnetic field of 0.0800 T, pointing along her finger, to a region with zero magnetic field in 40.0 ms. As a result of this action, thermal energy is added to the band due to the induced current, which raises the temperature of the band. Calculate the temperature rise in the band, assuming that all the energy produced is used in raising the temperature.

•29.60 A coil with N turns and area A, carrying a constant current, i, flips in an external magnetic field, $\vec{B}_{ext}$, so that its dipole moment switches from opposition to the field to alignment with the field. During this process, induction produces a potential difference that tends to reduce the current in the coil. Calculate the work done by the coil's power supply to maintain the constant current.

••29.61 An electromagnetic wave propagating in vacuum has electric and magnetic fields given by $\vec{E}(\vec{x},t) = \vec{E}_0\cos(\vec{k} \cdot \vec{x} - \omega t)$ and $\vec{B}(\vec{x},t) = \vec{B}_0\cos(\vec{k} \cdot \vec{x} - \omega t)$, where $\vec{B}_0$ is given by $\vec{B}_0 = \vec{k} \times \vec{E}_0/\omega$ and the wave vector $\vec{k}$ is perpendicular to both $\vec{E}_0$ and $\vec{B}_0$. The magnitude of $\vec{k}$ and the angular frequency ω satisfy the dispersion relation, $\omega/|\vec{k}| = (\mu_0\epsilon_0)^{-1/2}$, where μ_0 and ϵ_0 are the permeability and permittivity of free space, respectively. Such a wave transports energy in both its electric and magnetic fields. Calculate the ratio of the energy densities of the magnetic and electric fields, u_B/u_E, in this wave. Simplify your final answer as much as possible.

Additional Exercises

29.62 A wire of length $\ell = 10.0$ cm is moving with constant velocity in the xy-plane; the wire is parallel to the y-axis and moving along the x-axis. If a magnetic field of magnitude 1.00 T is pointing along the positive z-axis, what must the velocity of the wire be in order for a potential difference of 2.00 V to be induced across it?

29.63 The magnetic field inside the solenoid in the figure changes at the rate of 1.50 T/s. A conducting coil with 2000 turns surrounds the solenoid, as shown. The radius of the solenoid is 4.00 cm, and the radius of the coil is 7.00 cm. What is the potential difference induced in the coil?

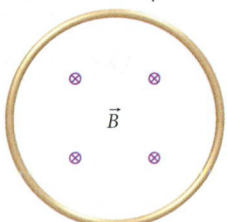

29.64 An ideal battery (with no internal resistance) supplies V_{emf} and is connected to a superconducting (no resistance!) coil of inductance L at time $t = 0$. Find the current in the coil as a function of time, $i(t)$. Assume that all connections also have zero resistance.

29.65 A 100-turn solenoid of length 8.00 cm and radius 6.00 mm carries a current of 0.400 A from right to left. The current is then reversed so that it flows from left to right. By how much does the energy stored in the magnetic field inside the solenoid change?

29.66 The electric field near the Earth's surface has a magnitude of 150. N/C, and the magnitude of the Earth's magnetic field near the surface is typically 50.0 μT. Calculate and compare the energy densities associated with these two fields. Assume that the electric and magnetic properties of air are essentially those of a vacuum.

29.67 What is the inductance in a series RL circuit in which $R = 3.00$ kΩ if the current increases to $\frac{1}{2}$ of its final value in 20.0 μs?

29.68 A 100.-V battery is connected in series with a 500.-Ω resistor. According to Faraday's Law of Induction, current can never change instantaneously, so there is always some "stray" inductance. Suppose the stray inductance is 0.200 μH. How long will it take the current to build up to within 0.500% of its final value of 0.200 A after the resistor is connected to the battery?

29.69 A single loop of wire with an area of 5.00 m² is located in the plane of the page, as shown in the figure. A time-varying magnetic field in the region of the loop is directed into the page, and its magnitude is given by $B = 3.00$ T + (2.00 T/s)t. At $t = 2.00$ s, what are the induced potential difference in the loop and the direction of the induced current?

29.70 A 9.00-V battery is connected through a switch to two identical resistors and an ideal inductor, as shown in the figure. Each of the resistors has a resistance of 100. Ω, and the inductor has an inductance of 3.00 H. The switch is initially open.

a) Immediately after the switch is closed, what is the current in resistor R_1 and in resistor R_2?

b) At 50.0 ms after the switch is closed, what is the current in resistor R_1 and in resistor R_2?

c) At 500. ms after the switch is closed, what is the current in resistor R_1 and in resistor R_2?

d) After a long time (> 10.0 s), the switch is opened again. Immediately after the switch is opened, what is the current in resistor R_1 and in resistor R_2?

e) At 50.0 ms after the switch is opened, what is the current in resistor R_1 and in resistor R_2?

f) At 500. ms after the switch is opened, what is the current in resistor R_1 and in resistor R_2?

•29.71 A long solenoid with length 3.00 m and $n = 290$. turns/m carries a current of 3.00 A. It stores 2.80 J of energy. What is the cross-sectional area of the solenoid?

•29.72 A rectangular conducting loop with dimensions a and b and resistance R is placed in the xy-plane. A magnetic field of magnitude B passes through the loop. The magnetic field is in the positive z-direction and varies in time according to $B = B_0(1 + c_1t^3)$, where c_1 is a constant with units of $1/s^3$. What is the direction of the current induced in the loop, and what is its value at $t = 1$ s (in terms of a, b, R, B_0, and c_1)?

•29.73 A circuit contains a 12.0-V battery, a switch, and a light bulb connected in series. When the light bulb has a current of 0.100 A flowing in it, it just starts to glow. This bulb draws 2.00 W when the switch has been closed for a long time. The switch is opened, and an inductor is added to the circuit, in series with the bulb. If the light bulb begins to glow 3.50 ms after the switch is closed again, what is the magnitude of the inductance? Ignore any time needed to heat the filament, and assume that you are able to observe a glow as soon as the current in the filament reaches the 0.100-A threshold.

•29.74 A circular loop of area A is placed perpendicular to a time-varying magnetic field of magnitude $B(t) = B_0 + at + bt^2$, where B_0, a, and b are constants.

a) What is the magnetic flux through the loop at $t = 0$?

b) Derive an equation for the induced potential difference in the loop as a function of time.

c) What are the magnitude and the direction of the induced current if the resistance of the loop is R?

•29.75 A conducting rod of length 50.0 cm slides over two parallel metal bars placed in a magnetic field with a magnitude of 1.00 kG, as shown in the figure. The ends of the rods are connected by two resistors, $R_1 = 100.$ Ω and $R_2 = 200.$ Ω. The conducting rod moves with a constant speed of 8.00 m/s.

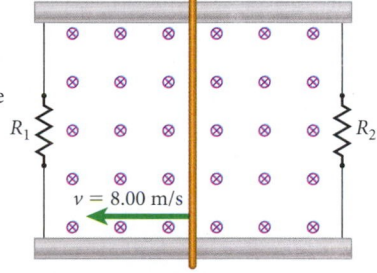

a) What are the currents flowing through the two resistors?

b) What power is delivered to the resistors?

c) What force is needed to keep the rod moving with constant velocity?

•29.76 A rectangular wire loop (dimensions of $h = 15.0$ cm and $w = 8.00$ cm) with resistance $R = 5.00$ Ω is mounted on a door, as shown in the figure. The Earth's magnetic field, $B_E = 2.60 \cdot 10^{-5}$ T, is uniform and perpendicular to the surface of the closed door (the surface is in the xz-plane). At time $t = 0$, the door is opened (right edge moves toward the y-axis) at a constant rate, with an opening angle of $\theta(t) = \omega t$, where $\omega = 3.50$ rad/s. Calculate the direction and the magnitude of the current induced in the loop, $i(t = 0.200$ s).

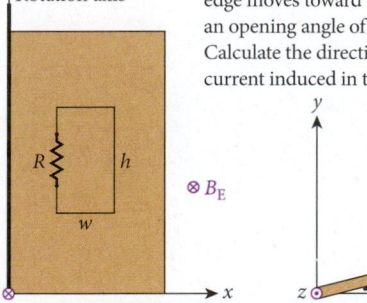

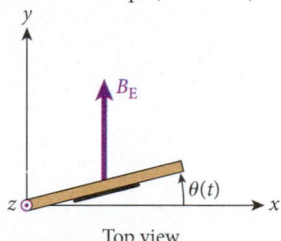

Front view   Top view

•**29.77** A steel cylinder with radius 2.50 cm and length 10.0 cm rolls without slipping down a ramp that is inclined at 15.0° above the horizontal and has a length (along the ramp) of 3.00 m. What is the induced potential difference between the ends of the cylinder as the cylinder leaves the bottom of the ramp, if the downward slope of the ramp points in the direction of the Earth's magnetic field at that location? (Use 0.426 G for the local strength of the Earth's magnetic field.)

•**29.78** The figure shows a circuit in which a battery is connected to a resistor and an inductor in series.

a) What is the current in the circuit at any time t after the switch is closed?

b) Calculate the total energy provided by the battery from $t = 0$ to $t = L/R$.

c) Calculate the total energy dissipated in the resistor over the same time period.

d) Is energy conserved in this circuit?

•**29.79** As shown in the figure, a rectangular (60.0 cm long by 15.0 cm wide) circuit loop with resistance 35.0 Ω is held parallel to the xy-plane with one half inside a uniform magnetic field. A magnetic field given by $\vec{B} = 2.00\hat{z}$ T is directed along the positive z-axis to the right of the dashed line; there is no external magnetic field to the left of the dashed line.

a) Calculate the magnitude of the force required to move the loop to the left at a constant speed of 10.0 cm/s while the right end of the loop is still in the magnetic field.

b) What power is expended to pull the loop out of the magnetic field at this speed?

c) What is the power dissipated by the resistor?

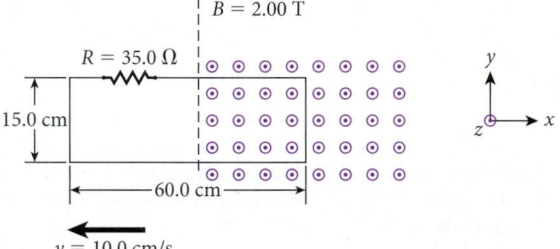

MULTI-VERSION EXERCISES

29.80 What is the resistance in an RL circuit with $L = 33.03$ mH if the time required for the current to reach 75% of its maximum value is 3.350 ms?

29.81 What is the inductance in an RL circuit with $R = 17.88$ Ω if the time required for the current to reach 75% of its maximum value is 3.450 ms?

29.82 For an RL circuit with $R = 21.84$ Ω and $L = 55.93$ mH, how long does it take the current to reach 75% of its maximum value?

29.83 A wedding ring (of diameter 1.95 cm) is tossed into the air and given a spin, resulting in an angular velocity of 13.3 rev/s. The rotation axis is a diameter of the ring. If the magnitude of the Earth's magnetic field at the ring's location is $4.77 \cdot 10^{-5}$ T, what is the maximum induced potential difference in the ring?

29.84 A wedding ring is tossed into the air and given a spin, resulting in an angular velocity of 13.5 rev/s. The rotation axis is a diameter of the ring. The magnitude of the Earth's magnetic field is $4.97 \cdot 10^{-5}$ T at the ring's location. If the maximum induced voltage in the ring is $1.446 \cdot 10^{-6}$ V, what is the diameter of the ring?

29.85 A wedding ring of diameter 1.63 cm is tossed into the air and given a spin, resulting in an angular velocity of 13.7 rev/s. The rotation axis is a diameter of the ring. If the maximum induced voltage in the ring is $6.556 \cdot 10^{-7}$ V, what is the magnitude of the Earth's magnetic field at this location?

30 Alternating Current Circuits

FIGURE 30.1 Almost all of the devices typically found on a student's desk are driven by alternating current.

Alternating Current Circuits

Almost all the devices in our homes operate on electricity. For example, Figure 30.1 shows a small sample of the consumer electronics devices (a notebook computer, a laser printer, a desk lamp, a smartphone, and headphones) that make our lives easier and more enjoyable by turning electricity into light and sound and by allowing us to communicate with the world. Refrigerators, televisions, washing machines, microwave ovens, air conditioners, hair dryers, and many other devices on which we rely also operate on electricity.

Worldwide, over 2 terawatts (TW) of electric power are consumed, with one-quarter of the total used in North America. World electric power use has doubled in the last quarter-century, driven mostly by very strong growth in East Asia. Almost all of this electricity is delivered to consumers via alternating current, which is the main topic of this chapter.

In Chapters 24 through 26 and 29, we examined direct current (DC) circuits, in which current flows in one direction. In this chapter, we study alternating current (AC) circuits, which use time-varying emf. AC circuits contain the same circuit elements (resistors, capacitors, and inductors) as DC circuits but show interesting physical phenomena of tremendous importance for technological applications. In this chapter, we will learn what produces oscillations in currents and voltages, how the effective resistance in a circuit depends on the oscillation frequency, how this dependence is used in filters, and how alternating voltages and currents are transformed and rectified.

WHAT WE WILL LEARN

- Voltages and currents in single-loop circuits containing an inductor and a capacitor oscillate with a characteristic frequency.

- Voltages and currents in single-loop circuits containing a resistor, an inductor, and a capacitor also oscillate with a characteristic frequency, but these oscillations are damped over time.

- A single-loop circuit containing a time-varying source of emf and a resistor has currents and voltages that are in phase and time-varying.

- A single-loop circuit containing a time-varying source of emf and a capacitor has a time-varying current and voltage that are out of phase by $+\pi/2$ rad ($+90°$), with the current leading the voltage. The voltage and the current in such a circuit are related by the capacitive reactance.

- A single-loop circuit containing a time-varying source of emf and an inductor has a time-varying current and voltage that are out of phase by $-\pi/2$ rad ($-90°$), with the voltage leading the current. The voltage and the current in such a circuit are related by the inductive reactance.

- A single-loop circuit containing a time-varying source of emf and a resistor, a capacitor, and an inductor has a time-varying current and voltage. The phase difference between the current and voltage depends on the values of the resistance, the capacitance, and the inductance and on the frequency of the emf source.

- A single-loop circuit containing a time-varying source of emf and a resistor, a capacitor, and an inductor has a resonant frequency determined by the values of the inductance and the capacitance.

- The impedance of an alternating-current circuit is similar to the resistance of a direct-current circuit, but the impedance depends on the frequency of the time-varying source of emf.

- Transformers can raise (or lower) alternating voltages while lowering (or raising) alternating currents.

- Rectifiers convert alternating current to direct current.

30.1   LC Circuits

Previous chapters introduced three circuit elements: capacitors, resistors, and inductors. We have examined simple single-loop circuits containing resistors and capacitors (RC circuits) or resistors and inductors (RL circuits). Now we'll consider simple single-loop circuits containing inductors and capacitors: **LC circuits.** We'll see that LC circuits have currents and voltages that vary sinusoidally with time, rather than increasing or decreasing exponentially with time, like currents and voltages in RC and RL circuits. These variations of voltage and current in LC circuits are called **electromagnetic oscillations.**

To understand electromagnetic oscillations, consider a simple single-loop circuit containing an inductor and a capacitor (Figure 30.2). Recall that the energy stored in the electric field of a capacitor with capacitance C is given by (see Chapter 24)

$$U_E = \tfrac{1}{2}\frac{q^2}{C},$$

where q is the magnitude of the charge on the capacitor plates. The energy stored in the magnetic field of an inductor with inductance L is given by (see Chapter 29)

$$U_B = \tfrac{1}{2}Li^2,$$

where i is the current flowing through the inductor. Figure 30.2 shows how these energies vary with time in this LC circuit. This figure may remind you of Figure 14.16, which showed potential energy and kinetic energy as a function of time for a mass oscillating on a spring. The resemblance is not coincidental! Much of this chapter's mathematical description of electromagnetic oscillations follows closely that developed in Chapter 14 for mechanical oscillations.

In Figure 30.2a, the capacitor is initially fully charged (with the positive charge on the bottom plate) and then connected to the circuit. At that time, the energy in the circuit is contained entirely in the electric field of the capacitor. The capacitor begins to discharge through the inductor in Figure 30.2b. At this point, current is flowing through the inductor, which generates a magnetic field. (A green arrow or label below each circuit diagram indicates the direction and the magnitude of the instantaneous current, i.) Now part of the

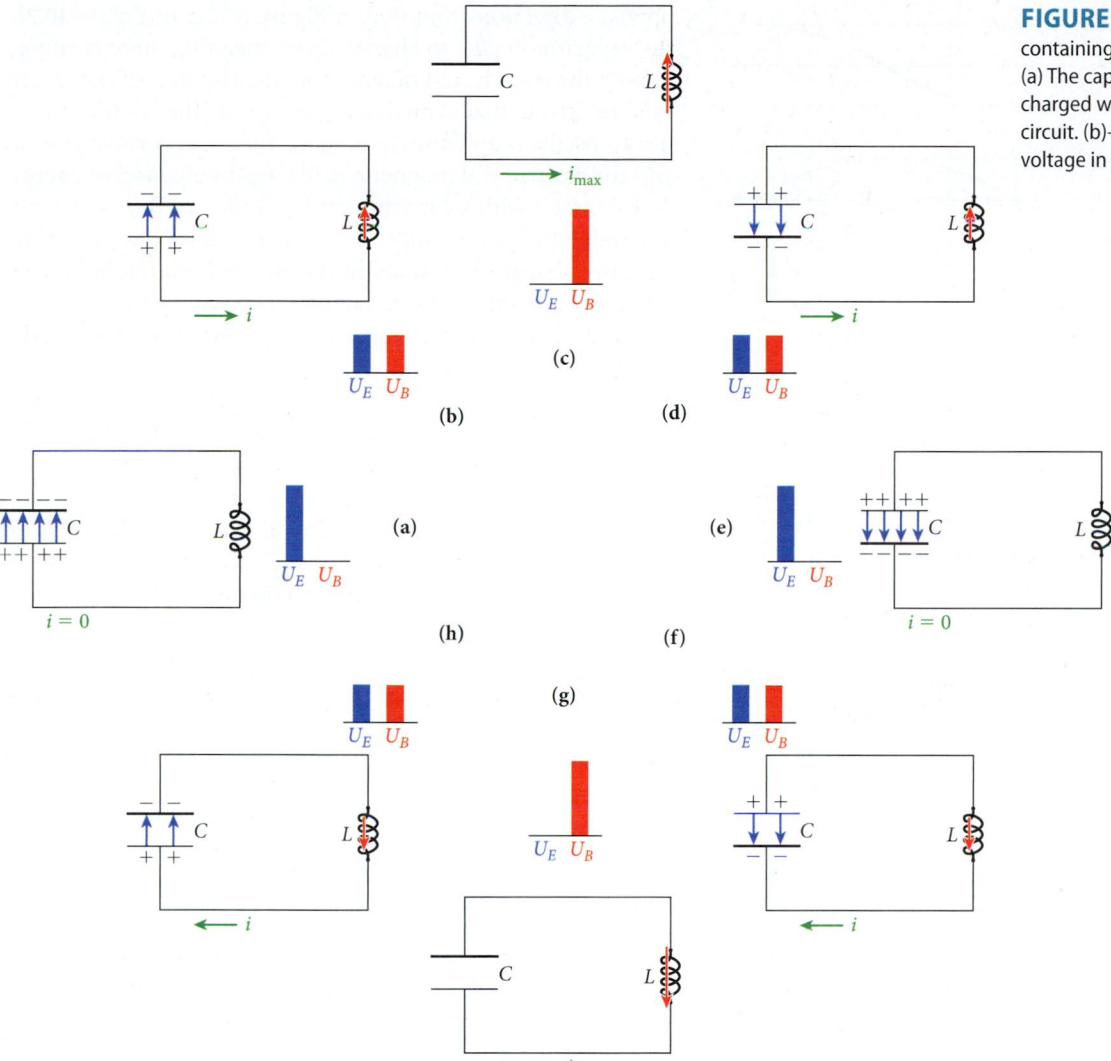

FIGURE 30.2 A single-loop circuit containing a capacitor and an inductor. (a) The capacitor is initially completely charged when it is connected to the circuit. (b)–(h) The current and the voltage in the circuit oscillate over time.

energy of the circuit is stored in the electric field of the capacitor and part in the magnetic field of the inductor. The current begins to level off as the inductor's increasing magnetic field induces an emf that opposes the current. In Figure 30.2c, the capacitor is completely discharged, and maximum current is flowing through the inductor. (When the magnitude of i has its maximum value, it is designated as i_{max} in the figure.) All the energy of the circuit is now stored in the magnetic field of the inductor. However, the current continues to flow, decreasing from its maximum value, which causes the magnetic field in the inductor to decrease. In Figure 30.2d, the capacitor begins to charge with the opposite polarity (positive charge on the top plate). Energy is again stored in the electric field of the capacitor, as well as in the magnetic field of the inductor.

In Figure 30.2e, the energy in the circuit is again entirely contained in the electric field of the capacitor. Note that the electric field now points in the opposite direction from the original field in Figure 30.2a. The current is zero, as is the magnetic field in the inductor. In Figure 30.2f, the capacitor begins to discharge again, producing a current flowing in the direction opposite to that in parts (b) through (d) of the figure; this current in turn creates a magnetic field in the opposite direction in the inductor. Again, part of the energy is stored in the electric field and part in the magnetic field. In Figure 30.2g, the energy is all stored in the magnetic field of the inductor, but with the magnetic field in the opposite direction from that in Figure 30.2c and with the maximum current in the

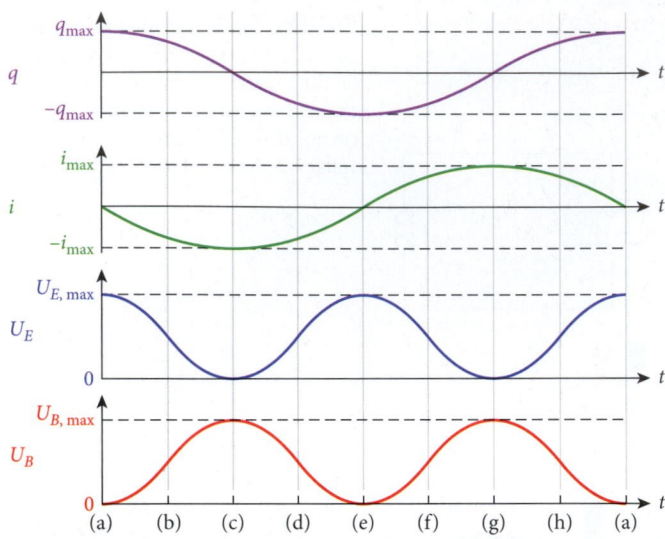

FIGURE 30.3 Variation of the charge, current, electric energy, and magnetic energy as a function of time for a simple, single-loop LC circuit. The letters along the bottom refer to the parts of Figure 30.2.

opposite direction from that in Figure 30.2c. In Figure 30.2h, the capacitor begins to charge again, meaning there is energy in both the electric and magnetic fields. The state of the circuit then returns to that shown in Figure 30.2a. The circuit continues to oscillate indefinitely because there is no resistor in it, and the electric and magnetic fields together conserve energy. A real circuit with a capacitor and an inductor does not oscillate indefinitely; instead, the oscillations die away with time because of small resistances in the circuit (covered in Section 30.3) or electromagnetic radiation (covered in Chapter 31).

The charge on either capacitor plate and the current in the LC circuit vary sinusoidally, as shown in Figure 30.3, where q_{max} refers to the maximum charge on the capacitor plate that is initially positively charged (the lower plate in Figure 30.2a). The energy in the electric field depends on the square of the charge on the capacitor, and the energy in the magnetic field depends on the square of the current in the inductor. Thus, the electric energy, U_E, and the magnetic energy, U_B, vary between zero and their respective maximum values as a function of time.

Concept Check 30.1

Figure 30.2a shows that the charge on the capacitor in an LC circuit is largest when the current is zero. What about the potential difference across the capacitor?

a) The potential difference across the capacitor is largest when the current is the largest.

b) The potential difference across the capacitor is largest when the charge is the largest.

c) The potential difference across the capacitor does not change.

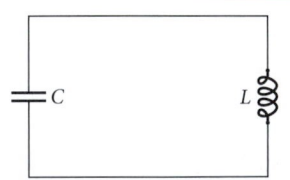

FIGURE 30.4 A single-loop LC circuit containing an inductor and a capacitor.

30.2 Analysis of LC Oscillations

Now, let's quantify the phenomena described in the preceding section. Consider a single-loop circuit containing a capacitor of capacitance C and an inductor of inductance L, but no resistor and no resistive losses in the circuit wire, as illustrated in Figure 30.4. We can write the total energy in the circuit, U, as the sum of the electric energy in the capacitor and the magnetic energy in the inductor:

$$U = U_E + U_B.$$

Using the expressions for the electric energy and the magnetic energy in terms of the charge and the current, $U_E = \frac{1}{2}(q^2/C)$ and $U_B = \frac{1}{2}Li^2$, we obtain

$$U = U_E + U_B = \frac{1}{2}\frac{q^2}{C} + \frac{1}{2}Li^2.$$

Because we have assumed zero resistance, no energy can be lost to heat, and the energy in the circuit will remain constant, because the electric field and the magnetic field together conserve energy. Thus, the derivative with respect to time of the energy in the circuit is zero:

$$\frac{dU}{dt} = \frac{d}{dt}\left(\frac{1}{2}\frac{q^2}{C} + \frac{1}{2}Li^2\right) = \frac{q}{C}\frac{dq}{dt} + Li\frac{di}{dt} = 0.$$

By definition, the current is the time derivative of the charge, $i = dq/dt$, and therefore, the time derivative of the current is the second derivative of the charge:

$$\frac{di}{dt} = \frac{d}{dt}\left(\frac{dq}{dt}\right) = \frac{d^2q}{dt^2}.$$

With this expression for di/dt, the preceding equation for the time derivative of the total energy, dU/dt, becomes

$$\frac{q}{C}\frac{dq}{dt} + L\frac{dq}{dt}\frac{d^2q}{dt^2} = \frac{dq}{dt}\left(\frac{q}{C} + L\frac{d^2q}{dt^2}\right) = 0.$$

We can rewrite this equation as

$$\frac{d^2q}{dt^2} + \frac{q}{LC} = 0. \tag{30.1}$$

(We discard the solution $dq/dt = 0$ because it corresponds to the situation where there is initially no charge on the capacitor.) This differential equation has the same form as that for simple harmonic motion, which describes the position, x, of an object with mass m connected to a spring with spring constant k:

$$\frac{d^2x}{dt^2} + \frac{k}{m}x = 0.$$

We saw in Chapter 14 that the solution of this differential equation for the position as a function of time was a sinusoidal function: $x = x_{max} \cos(\omega_0 t + \phi)$, where ϕ is a phase constant and the angular frequency, ω_0, is given by $\omega_0 = \sqrt{k/m}$.

Simply substituting q for x and $1/LC$ for k/m in the differential equation for simple harmonic motion leads to the analogous solution for equation 30.1. Thus, the charge as a function of time in an LC circuit is given by

$$q = q_{max} \cos(\omega_0 t - \phi), \tag{30.2}$$

where q_{max} is the magnitude of the maximum charge in the circuit and ϕ is the phase constant, which is determined by the initial conditions for a given situation. (Note that the convention for electromagnetic oscillations is to use a negative sign in front of ϕ.) The angular frequency is given by

$$\omega_0 = \sqrt{\frac{1}{LC}} = \frac{1}{\sqrt{LC}}. \tag{30.3}$$

The current is given by the time derivative of equation 30.2:

$$i = \frac{dq}{dt} = \frac{d}{dt}\left(q_{max} \cos(\omega_0 t - \phi)\right) = -\omega_0 q_{max} \sin(\omega_0 t - \phi).$$

Since the maximum current in the circuit is $i_{max} = \omega_0 q_{max}$, we get

$$i = -i_{max} \sin(\omega_0 t - \phi). \tag{30.4}$$

Equations 30.2 and 30.4 correspond to the top two curves in Figure 30.3 with $\phi = 0$. We can write expressions for the electric energy and the magnetic energy as functions of time:

$$U_E = \frac{1}{2}\frac{q^2}{C} = \frac{1}{2}\frac{\left[q_{max}\cos(\omega_0 t - \phi)\right]^2}{C} = \frac{q_{max}^2}{2C}\cos^2(\omega_0 t - \phi),$$

and

$$U_B = \frac{1}{2}Li^2 = \frac{L}{2}\left[-i_{max}\sin(\omega_0 t - \phi)\right]^2 = \frac{L}{2}i_{max}^2 \sin^2(\omega_0 t - \phi).$$

Since $i_{max} = \omega_0 q_{max}$ and $\omega_0 = 1/\sqrt{LC}$, we can write

$$\frac{L}{2}i_{max}^2 = \frac{L}{2}\omega_0^2 q_{max}^2 = \frac{q_{max}^2}{2C}.$$

Thus, we can express the magnetic energy as a function of time as follows:

$$U_B = \frac{q_{max}^2}{2C}\sin^2(\omega_0 t - \phi).$$

Note that both the electric energy and the magnetic energy have a maximum value equal to $q_{max}^2/(2C)$ and a minimum of zero.

Concept Check 30.2

In Figure 30.3, suppose $t = 0$ at point (c). What is the phase constant in this case? (Define a clockwise current as positive.)

a) 0

b) $\pi/2$

c) π

d) $3\pi/2$

e) none of the above

We can obtain an expression for the total energy in the circuit, U, by summing the electric and magnetic energies and then using the trigonometric identity $\sin^2\theta + \cos^2\theta = 1$:

$$U = U_E + U_B = \frac{q_{max}^2}{2C}\cos^2(\omega_0 t - \phi) + \frac{q_{max}^2}{2C}\sin^2(\omega_0 t - \phi)$$

$$= \frac{q_{max}^2}{2C}\left[\sin^2(\omega_0 t - \phi) + \cos^2(\omega_0 t - \phi)\right]$$

$$= \frac{q_{max}^2}{2C} = \frac{L}{2}i_{max}^2.$$

Thus, the total energy in the circuit remains constant with time and is proportional to the square of the original charge put on the capacitor.

EXAMPLE 30.1 Characteristics of an LC Circuit

A circuit contains a capacitor, with $C = 1.50$ µF, and an inductor, with $L = 3.50$ mH (Figure 30.4). The capacitor is fully charged using a 12.0-V battery and is then connected to the circuit.

PROBLEMS

What is the angular frequency of the circuit? What is the total energy in the circuit? What is the charge on the capacitor after $t = 2.50$ s?

SOLUTIONS

The angular frequency of the circuit is given by

$$\omega_0 = \frac{1}{\sqrt{LC}} = \frac{1}{\sqrt{(3.50 \cdot 10^{-3}\text{ H})(1.50 \cdot 10^{-6}\text{ F})}} = 1.38 \cdot 10^4 \text{ rad/s}.$$

The total energy in the circuit is

$$U = \frac{q_{max}^2}{2C}.$$

The maximum charge on the capacitor is

$$q_{max} = CV_{emf} = (1.50 \cdot 10^{-6}\text{ F})(12.0\text{ V})$$

$$= 1.80 \cdot 10^{-5}\text{ C}.$$

Thus, we can calculate the initial energy stored in the electric field of the capacitor, which is the same as the total energy in the circuit:

$$U = \frac{q_{max}^2}{2C} = \frac{(1.80 \cdot 10^{-5}\text{ C})^2}{2(1.50 \cdot 10^{-6}\text{ F})} = 1.08 \cdot 10^{-4} \text{ J}.$$

The charge on the capacitor as a function of time is given by

$$q = q_{max}\cos(\omega_0 t - \phi).$$

To determine the constant ϕ, we remember that $q = q_{max}$ at $t = 0$, so

$$q(0) = q_{max} = q_{max}\cos[(\omega_0)(0) - \phi] = q_{max}\cos(-\phi) = q_{max}\cos\phi.$$

Thus, we see that $\phi = 0$, and we can write the charge as a function of time as follows:

$$q = q_{max}\cos\omega_0 t.$$

Putting in the values $q_{max} = 1.80 \cdot 10^{-5}$ C, $\omega_0 = 1.38 \cdot 10^4$ rad/s, and $t = 2.50$ s, we get

$$q = (1.80 \cdot 10^{-5}\text{ C})\cos[(1.38 \cdot 10^4\text{ rad/s})(2.50\text{ s})] = 1.02 \cdot 10^{-5}\text{ C}.$$

Self-Test Opportunity 30.1

The frequency of oscillation of an LC circuit is 200.0 kHz. At $t = 0$, the capacitor has its maximum positive charge on the lower plate. Decide whether each of the following statements is true or false.

a) At $t = 2.50$ µs, the charge on the lower plate has its maximum negative value.

b) At $t = 5.00$ µs, the current in the circuit is at its maximum value.

c) At $t = 2.50$ µs, the energy in the circuit is stored completely in the inductor.

d) At $t = 1.25$ µs, half the energy in the circuit is stored in the capacitor and half the energy is stored in the inductor.

30.3 Damped Oscillations in an RLC Circuit

Now let's consider a single-loop circuit that has a capacitor and an inductor but also a resistor—an **RLC circuit,** as shown in Figure 30.5. We saw in the preceding section that the energy of a circuit with a capacitor and an inductor remains constant and that the energy is transformed from electric to magnetic and back again with no losses. However, if a resistor is present in the circuit, the current flow produces ohmic losses, which show up as thermal energy. Thus, the energy of the circuit decreases because of these losses. The rate of energy loss is given by

$$\frac{dU}{dt} = -i^2 R.$$

We can rewrite the change in the energy in the circuit as a function of time:

$$\frac{dU}{dt} = \frac{q}{C}\frac{dq}{dt} + Li\frac{di}{dt} = -i^2 R.$$

Again, since $i = dq/dt$ and $di/dt = d^2q/dt^2$, we can write

$$\frac{q}{C}\frac{dq}{dt} + Li\frac{di}{dt} + i^2 R = \frac{q}{C}\frac{dq}{dt} + L\frac{dq}{dt}\frac{d^2q}{dt^2} + \left(\frac{dq}{dt}\right)^2 R = 0,$$

or

$$\frac{d^2q}{dt^2} + \frac{R}{L}\frac{dq}{dt} + \frac{1}{LC}q = 0. \qquad (30.5)$$

The solution of this differential equation (for small damping, meaning sufficiently small values of the resistance) is

$$q = q_{max}e^{-Rt/2L}\cos\omega t, \qquad (30.6)$$

where

$$\omega = \sqrt{\omega_0^2 - \left(\frac{R}{2L}\right)^2} \qquad (30.7)$$

and $\omega_0 = 1/\sqrt{LC}$.

The calculus used in arriving at the solution in equation 30.6 is not shown. You can verify that the solution satisfies equation 30.5 by straightforward substitution from equations 30.6 and 30.7 into 30.5. You can also refer back to Chapter 14, where it was shown that the equation of motion for a weakly damped (or underdamped) mechanical oscillator has a similar solution. Chapter 14 also covered overdamped and critically damped oscillations.

If the capacitor in the single-loop RLC circuit of Figure 30.5 is charged and then connected in the circuit, the charge on the capacitor will vary sinusoidally with time while decreasing in amplitude (Figure 30.6). Taking the derivative of equation 30.6 shows that the current, $i = dq/dt$, has an amplitude that is damped at the same rate that the charge is damped and that this amplitude also varies sinusoidally with time. After some time, no current remains in the circuit.

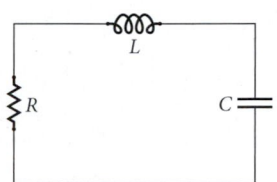

FIGURE 30.5 A single-loop RLC circuit containing a resistor, an inductor, and a capacitor.

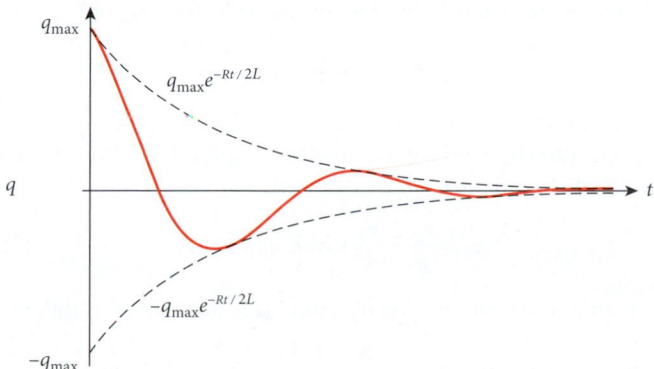

FIGURE 30.6 Graph of the charge on the capacitor as a function of time in a circuit containing a capacitor, an inductor, and a resistor.

We can analyze the energy in the circuit as a function of time by calculating the energy stored in the electric field of the capacitor:

$$U_E = \frac{1}{2}\frac{q^2}{C} = \frac{q_{max}^2}{2C}e^{-Rt/L}\cos^2\omega t.$$

Thus, U_E and U_B both decrease exponentially in time, and therefore so does the total energy in the circuit, $U_E + U_B$.

30.4 Driven AC Circuits

So far, we have been studying circuits that contain a source of constant emf or that start with a constant charge and contain energy that then oscillates between electric and magnetic fields. However, many interesting effects occur in a circuit in which the current oscillates continuously. Let's investigate some of these effects, starting with a time-varying source of emf and then considering in turn a resistor, a capacitor, and an inductor connected to this source.

Alternating Driving emf

A source of emf can be capable of producing a time-varying voltage, as opposed to the sources of constant emf considered in previous chapters. We'll assume that the source of time-varying emf provides a sinusoidal voltage as a function of time, the *driving emf,* given by

$$V_{emf} = V_{max}\sin\omega t, \tag{30.8}$$

where ω is the angular frequency of the emf and V_{max} is the maximum amplitude or value of the emf.

The current induced in a circuit containing a source of time-varying emf will also vary sinusoidally with time. This time-varying current is called **alternating current (AC).** However, the alternating current may not always remain in phase with the time-varying emf. The current, i, as a function of time is given by

$$i = I\sin(\omega t - \phi), \tag{30.9}$$

where I is the amplitude of the current and the angular frequency of the time-varying current is the same as that of the driving emf, but the phase constant ϕ is not zero. Note that, as is the convention, the phase constant is preceded by a negative sign.

Circuit with a Resistor

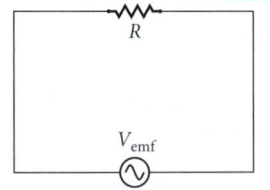

FIGURE 30.7 Single-loop circuit with a resistor and a source of time-varying emf.

Let's begin our analysis of RLC circuits with alternating current by considering a circuit containing only a resistor and a source of time-varying emf (Figure 30.7). Applying Kirchhoff's Loop Rule to this circuit, we obtain

$$V_{emf} - v_R = 0,$$

where v_R is the voltage drop across the resistor. Substituting v_R for V_{emf} in equation 30.8, we get

$$v_R = V_{max}\sin\omega t = V_R\sin\omega t,$$

where V_R is the maximum voltage drop across the resistor. Note that the voltage as a function of time is represented by a lowercase v and the amplitude of the voltage with uppercase V. According to Ohm's Law, $V = iR$, so we can write

$$i_R = \frac{v_R}{R} = \frac{V_R}{R}\sin\omega t = I_R\sin\omega t. \tag{30.10}$$

Thus, the current amplitude and the voltage amplitude are related as follows:

$$V_R = I_R R. \tag{30.11}$$

Figure 30.8a shows the voltage across the resistor and the current through it as functions of time. The time-varying current can be represented by a phasor, $\vec{I}_R$, and the

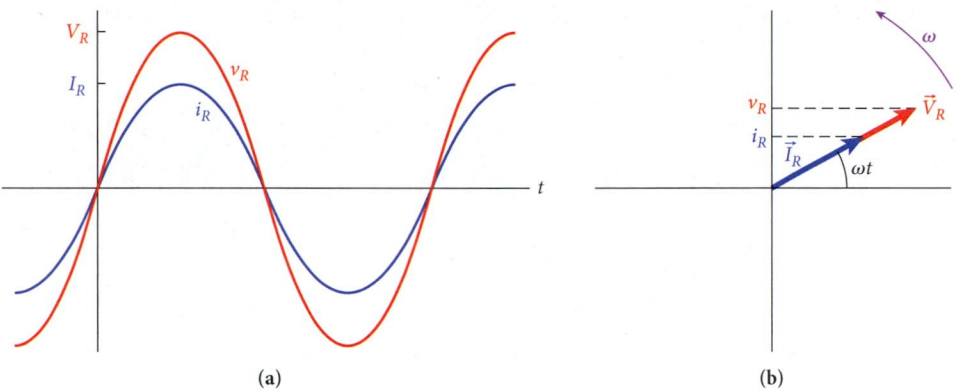

FIGURE 30.8 Alternating voltage and current for a single-loop circuit containing a source of time-varying emf and a resistor: (a) voltage and current as functions of time; (b) phasors representing voltage and current, showing that they are in phase.

time-varying voltage by a phasor, $\vec{V}_R$ (Figure 30.8b). A **phasor** is a counterclockwise-rotating vector (with its tail fixed at the origin) whose projection on the vertical axis represents the sinusoidal variation of the particular quantity in time. The angular velocity of the phasors in Figure 30.8b is ω of equation 30.10. The current flowing through the resistor and the voltage across the resistor are in phase, which means that the phase difference between the current and the voltage is zero.

Circuit with a Capacitor

Now let's examine a circuit that contains a capacitor and a source of time-varying emf (Figure 30.9). The voltage across the capacitor is given by Kirchhoff's Loop Rule,

$$V_{\text{emf}} - v_C = 0,$$

where v_C is the voltage drop across the capacitor. Thus, we have

$$v_C = V_{\text{max}} \sin \omega t = V_C \sin \omega t,$$

where V_C is the maximum voltage across the capacitor. Since $q = CV$ for a capacitor, we can write

$$q = Cv_C = CV_C \sin \omega t.$$

However, we want an expression for the current (rather than the charge) as a function of time. Therefore, we take the derivative with respect to time of the preceding equation:

$$i_C = \frac{dq}{dt} = \frac{d(CV_C \sin \omega t)}{dt} = \omega CV_C \cos \omega t.$$

This equation can be written in a form comparable to that of equation 30.10 by defining a quantity that is similar to resistance, called the **capacitive reactance,** X_C:

$$X_C = \frac{1}{\omega C}. \tag{30.12}$$

This definition allows us to express the current, i_C, as

$$i_C = \frac{V_C}{X_C} \cos \omega t,$$

or, with $I_C = V_C/X_C$, as

$$i_C = I_C \cos \omega t.$$

We can use $\cos \theta = \sin(\theta + \pi/2)$ to express this result in a form analogous to that of equation 30.10:

$$i_C = I_C \sin(\omega t + \pi/2). \tag{30.13}$$

This expression for the current flowing in a circuit with only a capacitor is similar to the expression for the current flowing in a circuit with only a resistor, except that current and voltage are out of phase by $\pi/2$ rad (90°). Figure 30.10a shows the voltage and the current as functions of time.

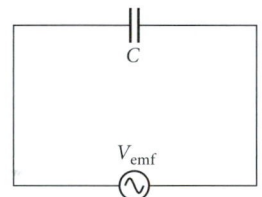

FIGURE 30.9 Single-loop circuit with a capacitor and a source of time-varying emf.

Concept Check 30.4

A circuit containing a capacitor (Figure 30.9) has a source of time-varying emf that provides a voltage given by $v_C = V_C \sin \omega t$. What is the current, i_C, through the capacitor when the potential difference across it is largest ($v_C = V_{\text{max}}$)?

a) $i_C = 0$

b) $i_C = +I_{\text{max}}$

c) $i_C = -I_{\text{max}}$

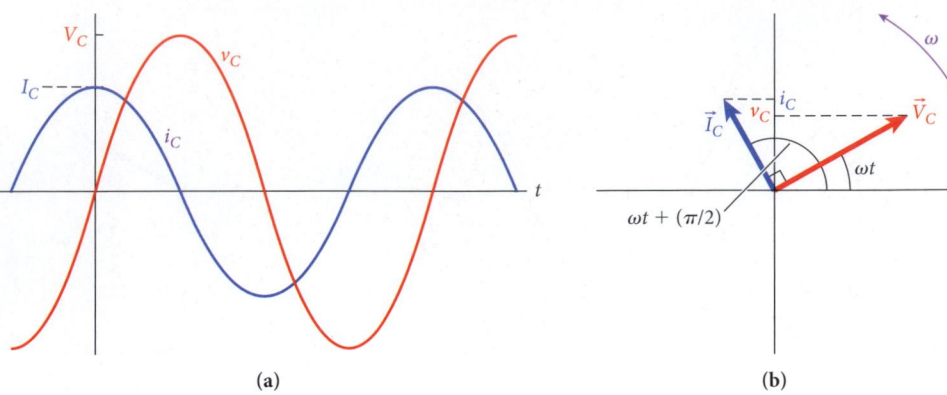

FIGURE 30.10 Alternating voltage and current for a single-loop circuit containing a source of time-varying emf and a capacitor: (a) voltage and current as functions of time; (b) phasors representing voltage and current, showing that they are out of phase by $\pi/2$ rad (90°).

The corresponding phasors, $\vec{I}_C$ and $\vec{V}_C$, shown in Figure 30.10b, indicate that *for a circuit with only a capacitor, the current leads the voltage.* The amplitude of the voltage across the capacitor and the amplitude of the current through the capacitor are related by

$$V_C = I_C X_C. \tag{30.14}$$

This equation resembles Ohm's Law with the capacitive reactance replacing the resistance. One major difference between the capacitive reactance and the resistance is that the capacitive reactance depends on the angular frequency of the time-varying emf.

Circuit with an Inductor

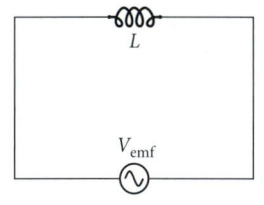

FIGURE 30.11 Single-loop circuit with an inductor and a source of time-varying emf.

Now let's consider a circuit with a source of time-varying emf and an inductor (Figure 30.11). We again apply Kirchhoff's Loop Rule to this circuit to obtain the voltage across the inductor:

$$v_L = V_{\max} \sin \omega t = V_L \sin \omega t,$$

where V_L is the maximum voltage across the inductor. A changing current in an inductor induces an emf given by

$$v_L = L \frac{di_L}{dt}.$$

Note that for positive di/dt the voltage drop across the inductor is positive because the direction of current is the direction of decreasing potential. Thus, we can write

$$L \frac{di_L}{dt} = V_L \sin \omega t,$$

or

$$\frac{di_L}{dt} = \frac{V_L}{L} \sin \omega t.$$

We are interested in the current rather than its time derivative, so we integrate the preceding equation:

$$i_L = \int \frac{di_L}{dt} dt = \int \frac{V_L}{L} \sin \omega t \, dt = -\frac{V_L}{\omega L} \cos \omega t.$$

Here we set the constant of integration to zero because we are not interested in solutions that contain both an oscillating and a constant current. The **inductive reactance,** which, like the capacitive reactance, is similar to resistance, is defined as

$$X_L = \omega L. \tag{30.15}$$

Using the inductive reactance, we can express i_L as

$$i_L = -\frac{V_L}{X_L} \cos \omega t = -I_L \cos \omega t,$$

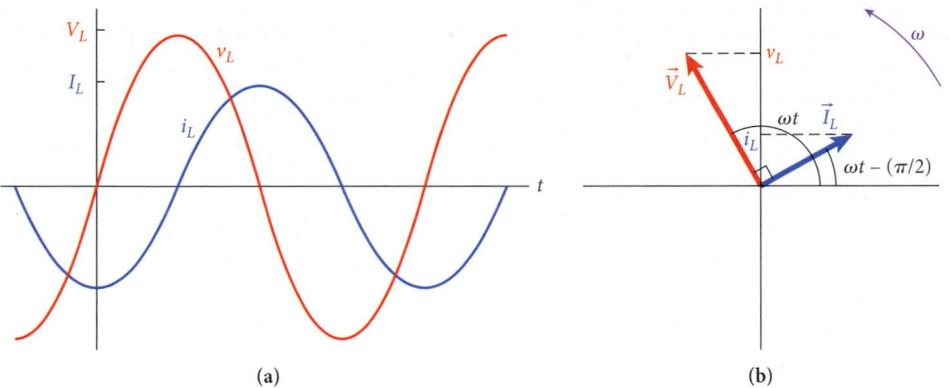

(a)   (b)

FIGURE 30.12 Alternating voltage and current for a single-loop circuit containing a source of time-varying emf and an inductor: (a) voltage and current as functions of time; (b) phasors representing voltage and current, showing that they are out of phase by $-\pi/2$ rad ($-90°$).

where I_L is the maximum current. Thus,

$$V_L = I_L X_L,$$

which again resembles Ohm's Law except that the inductive reactance depends on the angular frequency of the time-varying emf.

Because $-\cos\theta = \sin(\theta - \pi/2)$, we can rewrite $i_L = -I_L \cos\omega t$ as follows:

$$i_L = I_L \sin\left(\omega t - \pi/2\right). \tag{30.16}$$

Thus, the current flowing in a circuit with an inductor and a source of time-varying emf is out of phase with the voltage by $-\pi/2$ rad. Figure 30.12a shows voltage and current as functions of time. The corresponding phasors, $\vec{I}_L$ and $\vec{V}_L$, are shown in Figure 30.12b, which shows that *for a circuit with an inductor, the voltage leads the current.*

30.5 Series RLC Circuit

Now we're ready to consider a single-loop circuit that has all three circuit elements, along with a source of time-varying emf (Figure 30.13). We will not present a full mathematical analysis of this RLC circuit but will use phasors to analyze the important aspects.

The time-varying current in the simple RLC circuit can be described by a phasor, $\vec{I}_m$ (Figure 30.14). The projection of $\vec{I}_m$ on the vertical axis represents the current i flowing in the circuit as a function of time, t, where the angle of the phasor is given by $\omega t - \phi$ such that

$$i = I_m \sin\left(\omega t - \phi\right).$$

The current i and the voltages across the circuit components have different phases with respect to the time-varying emf, as we saw in the previous section:

- *For the resistor,* the voltage v_R and the current i are in phase with each other, and the voltage phasor, $\vec{V}_R$, is in phase with $\vec{I}_m$.
- *For the capacitor,* the current i leads the voltage v_C by $\pi/2$ rad ($90°$), so the voltage phasor, $\vec{V}_C$, has an angle that is $\pi/2$ rad ($90°$) less than the angles of $\vec{I}_m$ and $\vec{V}_R$.
- *For the inductor,* the current i lags behind the voltage v_L by $\pi/2$ rad ($90°$), so the voltage phasor, $\vec{V}_L$, has an angle that is $\pi/2$ rad ($90°$) greater than the angles of $\vec{I}_m$ and $\vec{V}_R$.

The voltage phasors for the RLC circuit are shown in Figure 30.15. The instantaneous voltage across each component is represented by the projection of the respective phasor on the vertical axis.

The total voltage drop across all components, V, is given by

$$V = v_R + v_C + v_L. \tag{30.17}$$

The total voltage, V, can be thought of as the projection on the vertical axis of the phasor $\vec{V}_m$, representing the time-varying emf in the circuit (Figure 30.16). The phasors in Figure 30.15 rotate together, so equation 30.17 holds at any time. The voltage phasors must sum as vectors to

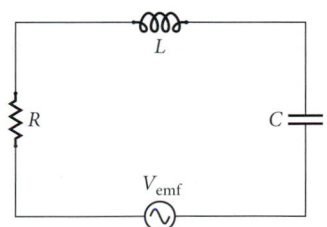

FIGURE 30.13 A single-loop circuit containing a source of time-varying emf, a resistor, an inductor, and a capacitor.

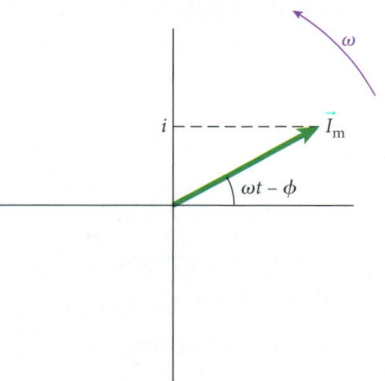

FIGURE 30.14 Phasor $\vec{I}_m$ representing the current i flowing in an RLC circuit.

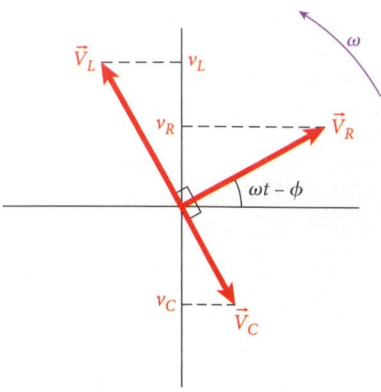

FIGURE 30.15 Voltage phasors for an RLC series circuit. The phasor $\vec{V}_R$ is in phase with the phasor $\vec{I}_m$ representing the current in the circuit.

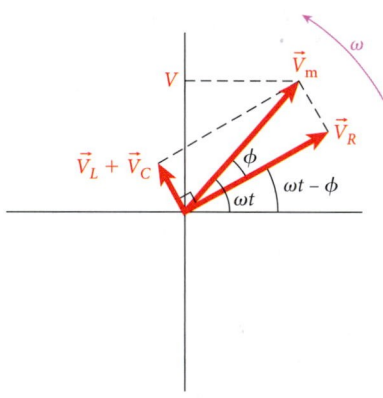

FIGURE 30.16 Sum of the voltage phasors in an RLC series circuit.

Concept Check 30.5

A circuit like that shown in Figure 30.13 containing a capacitor, an inductor, and a resistor connected in series with a source of time-varying emf has $V_{emf} = V_m \sin \omega t$. At a point in time when V_{emf} is increasing, how is the current in the circuit behaving?

a) The current is increasing.

b) The current is decreasing.

c) The current is not changing.

d) The current may be increasing or decreasing.

match $\vec{V}_m$ in order to satisfy equation 30.17 at all times. This vector sum is shown in Figure 30.16. In this figure, the sum of the two phasors $\vec{V}_L$ and $\vec{V}_C$ has been replaced with $\vec{V}_L + \vec{V}_C$. The vector sum of $\vec{V}_L + \vec{V}_C$ and $\vec{V}_R$ must equal $\vec{V}_m$. Thus, we can write

$$V_m^2 = V_R^2 + \left(V_L - V_C\right)^2, \tag{30.18}$$

because the vectors $\vec{V}_L$ and $\vec{V}_C$ always point in opposite directions, and $\vec{V}_m$ is perpendicular to both. Now we can substitute our previously derived expressions for V_R, V_L, and V_C into equation 30.18, taking the amplitude of the current in all three components to be I_m because they are connected in series:

$$V_m^2 = \left(I_m R\right)^2 + \left(I_m X_L - I_m X_C\right)^2.$$

We can then solve for the amplitude of the current in the circuit:

$$I_m = \frac{V_m}{\sqrt{R^2 + \left(X_L - X_C\right)^2}}.$$

The denominator of the term on the right-hand side is called the **impedance, Z**:

$$Z = \sqrt{R^2 + \left(X_L - X_C\right)^2}. \tag{30.19}$$

The impedance of a circuit depends on the frequency of the time-varying emf. This time dependence is expressed explicitly when substitutions are made for the capacitive reactance, X_C, and the inductive reactance, X_L:

$$Z = \sqrt{R^2 + \left(\omega L - \frac{1}{\omega C}\right)^2}. \tag{30.20}$$

The impedance of an AC circuit has the unit ohm (Ω), just like the resistance in a DC circuit. We can then write

$$I_m = \frac{V_m}{\sqrt{R^2 + \left(\omega L - \frac{1}{\omega C}\right)^2}} = \frac{V_m}{Z}. \tag{30.21}$$

The current flowing in an AC circuit depends on the difference between the inductive reactance and the capacitive reactance and is called the *total reactance*. The phase constant, ϕ, can be expressed in terms of this difference. The phase constant is defined as the phase difference between the voltage phasors $\vec{V}_R$ and $\vec{V}_m$ depicted in Figure 30.16. Thus, we can express the phase constant as

$$\phi = \tan^{-1}\left(\frac{V_L - V_C}{V_R}\right).$$

Since $V_L = X_L I_m$, $V_C = X_C I_m$, and $V_R = R I_m$, this can be rewritten as follows:

$$\phi = \tan^{-1}\left(\frac{X_L - X_C}{R}\right).$$

Using $X_C = 1/\omega C$ and $X_L = \omega L$, we can then obtain the frequency dependence of the phase constant:

$$\phi = \tan^{-1}\left(\frac{\omega L - (\omega C)^{-1}}{R}\right). \tag{30.22}$$

The current in the RLC circuit can now be written as

$$i = I_m \sin\left(\omega t - \phi\right), \tag{30.23}$$

where I_m is the magnitude of the phasor $\vec{I}_m$. The voltage across all the components in the circuit is given by the time-varying source of emf:

$$V = V_{emf}(t) = V_m \sin \omega t, \tag{30.24}$$

where V_m is the magnitude of the phasor $\vec{V}_m$.

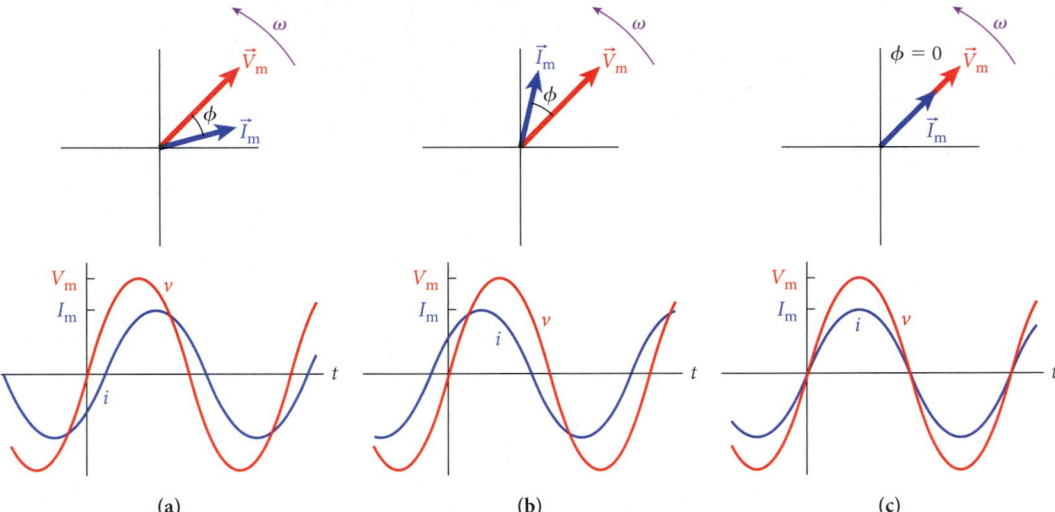

FIGURE 30.17 Current and voltage as functions of time for an RLC circuit with: (a) $X_L > X_C$; (b) $X_L < X_C$; (c) $X_L = X_C$.

Thus, three conditions are possible for an AC circuit containing a resistor, a capacitor, and an inductor connected in series:

- For $X_L > X_C$, ϕ is positive, and the current in the circuit lags behind the voltage in the circuit. This circuit is similar to a circuit with only an inductor, except that the phase constant is not necessarily $\pi/2$ rad (90°), as illustrated in Figure 30.17a.

- For $X_L < X_C$, ϕ is negative, and the current in the circuit leads the voltage in the circuit. This circuit is similar to a circuit with only a capacitor, except that the phase constant is not necessarily $-\pi/2$ rad (−90°), as illustrated in Figure 30.17b.

- For $X_L = X_C$, ϕ is zero, and the current in the circuit is in phase with the voltage in the circuit. This circuit is similar to a circuit with only a resistance, as illustrated in Figure 30.17c. When $\phi = 0$, the circuit is said to be in **resonance.**

The current amplitude, I_m, in the series RLC circuit depends on the frequency of the time-varying emf, as well as on L and C. Inspection of equation 30.21 shows that the maximum current occurs when

$$\omega L - \frac{1}{\omega C} = 0,$$

which corresponds to $\phi = 0$ and $X_L = X_C$. The angular frequency, ω_0, at which the maximum current occurs, called the **resonant angular frequency,** is

$$\omega_0 = \frac{1}{\sqrt{LC}}.$$

A Practical Example

Now let's look at a real circuit (Figure 30.18). The diagram for this circuit is shown in Figure 30.13. The circuit has a source of time-varying emf with $V_m = 7.5$ V and also has $L = 8.2$ mH, $C = 100$ µF, and $R = 10$ Ω. The maximum current, I_m, was measured as a function of the ratio of the angular frequency of the time-varying emf to the resonant angular frequency, ω/ω_0 (Figure 30.19). Red circles indicate the results of the measurements. The maximum value of the current occurs, as expected, at the resonant angular frequency. However, with $R = 10$ Ω, the relationship between I_m and ω/ω_0, given by equation 30.21, results in the green curve, which does not reproduce the measured results. To better describe the current, we must remember that in a real circuit, the inductor has a resistance, even at the resonant frequency. The black curve in Figure 30.19 corresponds to equation 30.21 with $R = 15.4$ Ω.

The resonant behavior of an RLC circuit resembles the response of a damped mechanical oscillator (Chapter 14). Figure 30.20 shows the calculated maximum current, I_m, as

Self-Test Opportunity 30.3

Consider a series RLC circuit like the one shown in Figure 30.13. The circuit is driven at an angular frequency ω by the time-varying emf. The resonant angular frequency is ω_0. Decide whether each of the following statements is true or false.

a) If $\omega = \omega_0$, the voltage and the current are in phase.

b) If $\omega < \omega_0$, the voltage lags behind the current.

c) If $\omega > \omega_0$, then $X_C > X_L$.

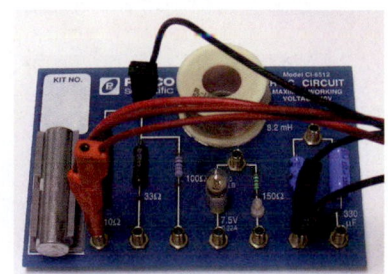

FIGURE 30.18 Real series circuit containing an 8.2-mH inductor, a 10-Ω resistor, and a 100-µF capacitor.

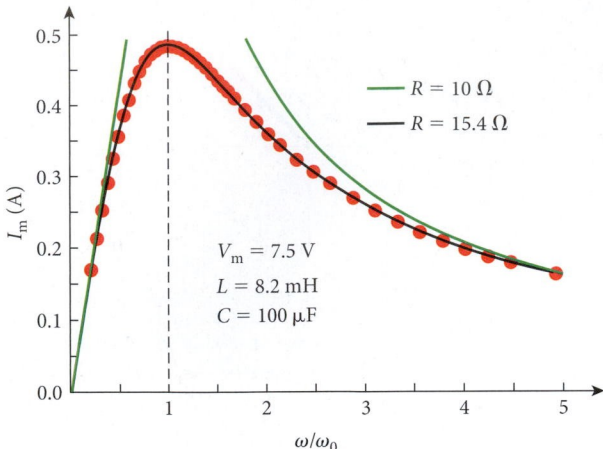

FIGURE 30.19 Graph of the maximum current, I_m, versus the ratio of the angular frequency, ω, of the time-varying emf to the resonance frequency, ω_0, for an RLC circuit. Red dots represent measurements. The text explains the green and black curves.

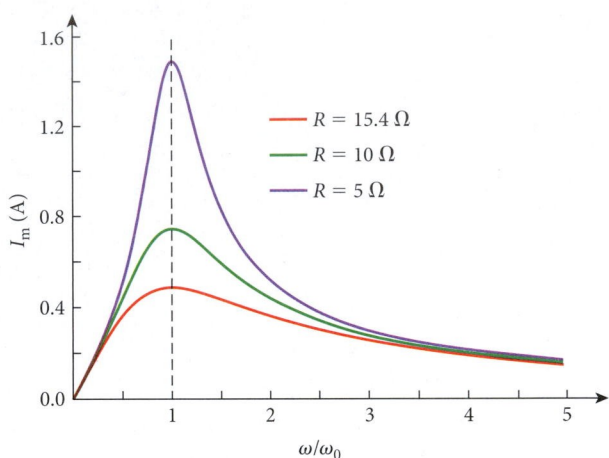

FIGURE 30.20 Graph of the maximum current, I_m, versus the ratio of the angular frequency, ω, of the time-varying emf to the resonance frequency, ω_0, for three series RLC circuits, with $L = 8.2$ mH, $C = 100$ μF, and three different resistances.

a function of the ratio of the angular frequency of the time-varying emf to the resonant angular frequency, ω/ω_0, for a series RLC circuit with $V_m = 7.5$ V, $L = 8.2$ mH, $C = 100$ μF, and three different resistances. You can see that as the resistance is lowered, the maximum current at the resonant angular frequency increases, producing a more pronounced peak.

Self-Test Opportunity 30.4

Consider a series RLC circuit like the one shown in Figure 30.13. Decide whether each of the following statements is true or false.

a) The current through the resistor is the same as the current through the inductor at all times.

b) In an ideal scenario, energy is dissipated in the resistor but not in the capacitor or in the inductor.

c) The voltage drop across the resistor is the same as the voltage drop across the inductor at all times.

EXAMPLE 30.2 Characterizing an RLC Circuit

Suppose an RLC series circuit like the one shown in Figure 30.13 has $R = 91.0$ Ω, $C = 6.00$ μF, and $L = 60.0$ mH. The source of time-varying emf has an angular frequency of $\omega = 64.0$ rad/s.

PROBLEM
What is the impedance of this circuit?

SOLUTION
Normally, we would solve this problem by obtaining an expression for the impedance in terms of the quantities provided. However, instead we'll calculate several intermediate numerical answers to gain insight into the characteristics of this circuit.

The impedance is given by $Z = \sqrt{R^2 + (X_L - X_C)^2}$. To see which of the quantities on the right-hand side has the greatest impact on the impedance, we calculate the quantities individually. The inductive reactance is

$$X_L = \omega L = (64.0 \text{ rad/s})(60.0 \cdot 10^{-3} \text{ H}) = 3.84 \ \Omega.$$

The capacitive reactance is

$$X_C = \frac{1}{\omega C} = \frac{1}{(64.0 \text{ rad/s})(6.00 \cdot 10^{-6} \text{ F})} = 2.60 \text{ k}\Omega.$$

We see that the impedance of this circuit is dominated by the capacitive reactance at the given value of the angular frequency. This type of circuit is called *capacitive*.

Putting in our results for the capacitive and inductive reactances, we calculate the impedance:

$$Z = \sqrt{R^2 + (X_L - X_C)^2} = \sqrt{(91.0 \ \Omega)^2 + (3.84 \ \Omega - 2.60 \cdot 10^3 \ \Omega)^2} = 2.60 \text{ k}\Omega.$$

That is, the inductive reactance and the resistance are completely negligible within rounding error. For comparison, the impedance of this circuit when it is in resonance is

$$Z = \sqrt{R^2 + (X_L - X_C)^2} = \sqrt{(91.0 \ \Omega)^2 + 0} = 91.0 \ \Omega.$$

This result implies that the circuit as described in the problem statement is far from resonance, which is consistent with the very different values we obtained for the capacitive and inductive reactances. (Remember that at resonance these two reactances have the same value!)

Frequency Filters

We have been analyzing circuits that have a time-varying emf with a single frequency. However, many applications involve time-varying emfs that reflect a superposition of many frequencies. In some situations, certain frequencies need to be filtered out of this kind of circuit. (Series RLC circuits can be used as frequency filters.) One example of such a circuit can be found in DSL (digital subscriber line) filters for making connections to the Internet over a household telephone line. A typical DSL filter is shown in Figure 30.21.

A DSL Internet connection operates at high frequencies and is connected to a household's regular phone line. The high operating frequency of the DSL connection causes noise on the phones in the house. Therefore, a **band-pass filter** is normally installed on all the phones in the house to filter out the high-frequency noise created by the DSL Internet connection. Frequency filters can be designed to pass low frequencies and block high frequencies (*low-pass filter*) or to pass high frequencies and block low frequencies (*high-pass filter*). A low-pass filter can be combined with a high-pass filter to allow a range of frequencies to pass (*band-pass filter*) and block the frequencies outside that range.

Figure 30.22 shows two examples of a low-pass filter, where V_{in} is a time-varying emf with many frequencies. A low-pass filter is essentially a voltage divider. Part of the original voltage passes through the circuit, while part goes to ground. For the RC version of the low-pass filter, shown in Figure 30.22a, low frequencies essentially have an open circuit, while high frequencies are preferentially sent to ground. Thus, only signals with low frequencies will pass through the filter. This behavior makes sense because current going to ground must pass through a capacitor that essentially blocks the flow of current for low frequencies because the capacitor plates are charging, while rapidly changing current does not allow charge to build up on the plates of the capacitor, allowing current to flow. For the RL version, shown in Figure 30.22b, low frequencies easily pass through the inductor while high frequencies are blocked. This effect arises because the self-induced emf in an inductor opposes rapid changes in current, effectively blocking current through the inductor at high frequencies, while a slow change in current produces a much smaller opposing emf, allowing current to flow.

To quantify the performance of the low-pass filter in Figure 30.22a, we define the input section of the circuit to be the resistor and the capacitor. The impedance of this section is $Z_{in} = \sqrt{R^2 + X_C^2}$. The impedance of the output section is just $Z_{out} = X_C$. The ratio of the emf into the filter and the emf emerging from the filter is

$$\frac{V_{out}}{V_{in}} = \frac{Z_{out}}{Z_{in}}. \tag{30.25}$$

The ratio of the emfs can then be written as

$$\frac{V_{out}}{V_{in}} = \frac{X_C}{\sqrt{R^2 + X_C^2}} = \frac{1}{\sqrt{\left(\dfrac{R}{X_C}\right)^2 + 1}} = \frac{1}{\sqrt{1 + \omega^2 R^2 C^2}}. \tag{30.26}$$

For the RL version of the low-pass filter, shown in Figure 30.22b, $Z_{in} = \sqrt{R^2 + X_L^2}$ and $Z_{out} = R$, allowing us to write

$$\frac{V_{out}}{V_{in}} = \frac{R}{\sqrt{R^2 + X_L^2}} = \frac{1}{\sqrt{1 + \left(\omega^2 L^2 / R^2\right)}}. \tag{30.27}$$

The *breakpoint frequency*, ω_B, between the responses to low and high frequencies is the frequency at which the ratio V_{out}/V_{in} is $1/\sqrt{2} = 0.707$. At that frequency for the RC version, we have

$$\frac{1}{\sqrt{1 + \omega_B^2 R^2 C^2}} = \frac{1}{\sqrt{2}},$$

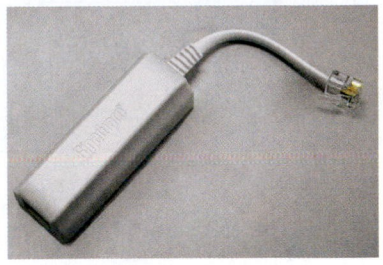

FIGURE 30.21 A typical band-pass filter for phones connected to a household circuit that has a DSL Internet connection.

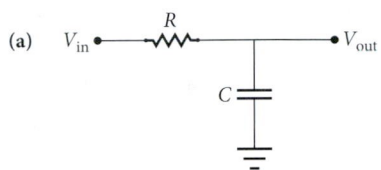

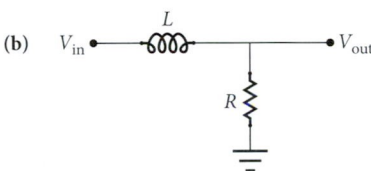

FIGURE 30.22 Two low-pass filters: (a) RC version; (b) RL version.

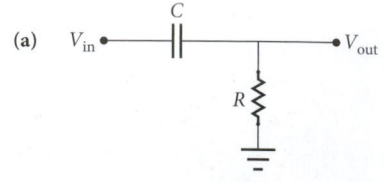

(a)

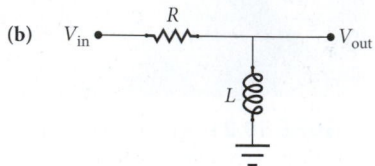

(b)

FIGURE 30.23 Two high-pass filters:
(a) RC version; (b) RL version.

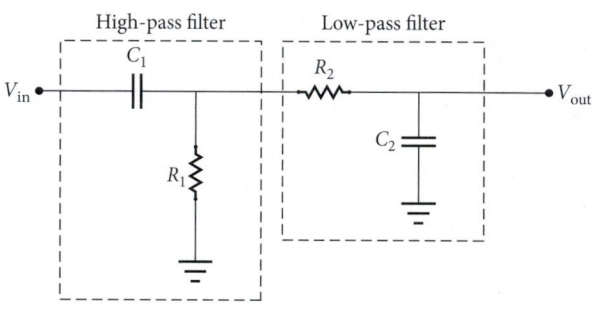

FIGURE 30.24 A band-pass filter consisting of a high-pass filter connected in series with a low-pass filter.

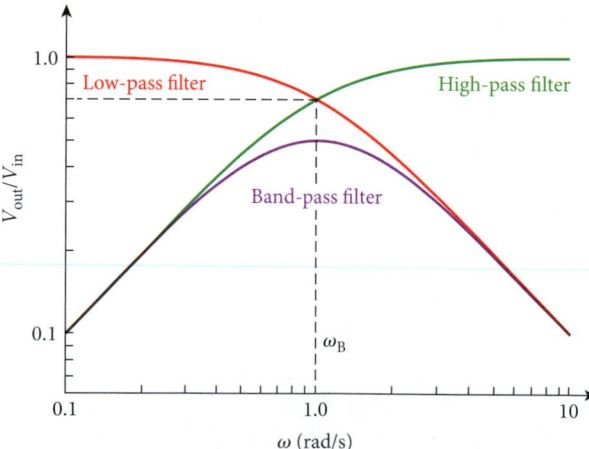

FIGURE 30.25 The frequency response of a low-pass filter, a high-pass filter, and a band-pass filter.

Concept Check 30.6

The frequency response for a band-pass filter plotted in Figure 30.25 is the _____ of the frequency responses for the low-pass and high-pass filters.

a) sum   d) ratio

b) product   e) It is none of the above.

c) difference

from which we can solve for the breakpoint frequency:

$$\omega_B = \frac{1}{RC}. \qquad (30.28)$$

For the RL version of the low-pass filter, the breakpoint frequency is obtained from equation 30.27:

$$\omega_B = \frac{R}{L}. \qquad (30.29)$$

Figure 30.23 shows two examples of a high-pass filter. A high-pass filter is also a voltage divider. For the RC version of the high-pass filter, shown in Figure 30.23a, signals with low frequencies cannot pass the capacitor while signals with high frequencies pass through easily. This behavior makes sense because the signal must pass through a capacitor that essentially blocks the flow of current for low frequencies because the capacitor plates are charging, while rapidly changing current does not allow charge to build up on the plates of the capacitor, allowing current to flow. For the RL version, shown in Figure 30.23b, signals with low frequency have essentially an open circuit to ground, while signals with high frequencies are blocked from reaching ground. Thus, only signals with high frequencies will be passed through the filter. This effect arises because the self-induced emf in an inductor opposes rapid changes in current, effectively blocking current through the inductor at high frequencies, while a slow change in current produces a much smaller opposing emf, allowing current to flow.

For the RC version of the high-pass filter in Figure 30.23a, the impedance of the input section is $Z_{in} = \sqrt{R^2 + X_C^2}$, while the impedance of the output section is $Z_{out} = R$. The ratio of the output emf to the input emf is then

$$\frac{V_{out}}{V_{in}} = \frac{R}{\sqrt{R^2 + X_C^2}} = \frac{1}{\sqrt{1 + X_C^2/R^2}} = \frac{1}{\sqrt{1 + \frac{1}{\omega^2 R^2 C^2}}}. \qquad (30.30)$$

For the RL version of the high-pass filter shown in Figure 30.23b, the ratio of the output emf to the input emf is

$$\frac{V_{out}}{V_{in}} = \frac{X_L}{\sqrt{R^2 + X_L^2}} = \frac{1}{\sqrt{\frac{R^2}{X_L^2} + 1}} = \frac{1}{\sqrt{1 + \frac{R^2}{\omega^2 L^2}}}. \qquad (30.31)$$

For these high-pass filters, as the frequency increases, the ratio of output emf to input emf approaches 1, while for low frequencies, the ratio of output emf to input emf goes to zero. The breakpoint frequencies for the high-pass filters are the same as for the low-pass filters: $\omega_B = 1/(RC)$ for the RC version and $\omega_B = R/L$ for the RL version.

An example of a band-pass filter is shown in Figure 30.24. The band-pass filter consists of a high-pass filter in series with a low-pass filter. Thus, both high and low frequencies are suppressed, and a narrow band of frequencies is allowed to pass through the filter.

Figure 30.25 shows the frequency response of a low-pass filter and a high-pass filter with $R = 50.0 \ \Omega$ and $C = 20.0$ mF. For this combination of resistance and capacitance, the breakpoint frequency is

$$\omega_B = \frac{1}{RC} = \frac{1}{(50.0 \ \Omega)(20.0 \cdot 10^{-3} \ \text{F})} = 1.00 \ \text{rad/s}.$$

Also shown in Figure 30.25 is the frequency response of a band-pass filter with $R_1 = R_2 = 50.0 \ \Omega$ and $C_1 = C_2 = 20.0$ mF.

EXAMPLE 30.3 | **Crossover Circuit for Audio Speakers**

One way to improve the performance of an audio system is to send high frequencies to a small speaker called a *tweeter* and low frequencies to a large speaker called a *woofer*. Figure 30.26 shows a simple crossover circuit that preferentially passes high frequencies to a tweeter and low frequencies to a woofer. The crossover circuit consists of an RC high-pass filter and an RL low-pass filter connected in parallel to the output of the audio amplifier. The speakers act as resistors, as shown in Figure 30.26. The capacitance and the resistance of this crossover circuit are $C = 10.0\ \mu F$ and $L = 10.0$ mH. The speakers each have a resistance of $R = 8.00\ \Omega$.

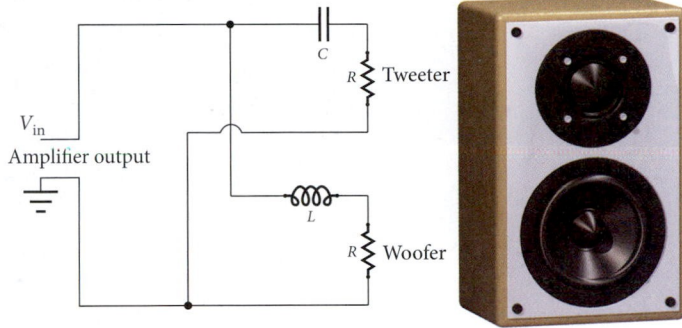

FIGURE 30.26 Crossover circuit for audio speakers.

PROBLEM

What is the crossover frequency for this crossover circuit?

SOLUTION

We can use equation 30.27 for the response of the RL low-pass filter and equation 30.30 for the response of the RC high-pass filter and equate the two responses:

$$\frac{V_{out}}{V_{in}} = \frac{R}{\sqrt{R^2 + X_L^2}} = \frac{R}{\sqrt{R^2 + X_C^2}}.$$

Thus, the responses of the low-pass filter and the high-pass filter are the same when

$$X_L = X_C. \qquad (i)$$

We can rewrite equation (i) as

$$\omega_{crossover} L = \frac{1}{\omega_{crossover} C},$$

where $\omega_{crossover}$ is the crossover angular frequency. Thus, the crossover angular frequency is

$$\omega_{crossover} = \frac{1}{\sqrt{LC}}.$$

We want to determine the crossover frequency, and since $f = \omega/2\pi$, we have

$$f_{crossover} = \frac{\omega_{crossover}}{2\pi} = \frac{1}{2\pi\sqrt{LC}}.$$

Putting in the numerical values, we get

$$f_{crossover} = \frac{1}{2\pi\sqrt{LC}} = \frac{1}{2\pi\sqrt{(10.0\ \text{mH})(10.0\ \mu F)}} = 503.\ \text{Hz}.$$

Figure 30.27 shows the response of the crossover circuit as a function of frequency. The low-pass response and the high-pass response cross at $f_{crossover} = 503.$ Hz, sending higher frequencies predominantly to the tweeter and lower frequencies predominantly to the woofer.

This simple crossover circuit would not produce ideal audio performance over a broad range of frequencies and speaker designs. More sophisticated crossover circuits that have better performance deal with mid-range frequencies as well.

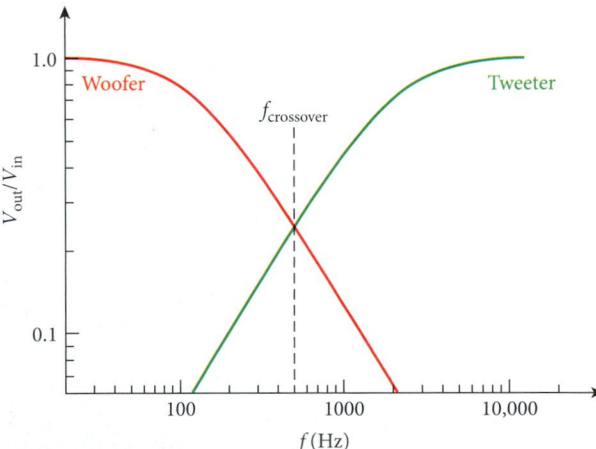

FIGURE 30.27 The response of the crossover circuit as a function of frequency.

30.6 Energy and Power in AC Circuits

When an RLC circuit is in operation, some of the energy in the circuit is stored in the electric field of the capacitor, some of the energy is stored in the magnetic field of the inductor, and some energy is dissipated in the form of heat in the resistor. In most applications, we are interested in the steady-state behavior of the circuit, behavior that occurs after initial (transient) effects die out. (A full mathematical analysis would also account for the transient effects, which die out exponentially in a way similar to that described by equation 30.6 for a single-loop RLC circuit with no emf source.) The sum of the energy stored in the capacitor and the inductor does not change in the steady state, as we saw in Section 30.2. Therefore, the energy transferred from the source of emf to the circuit is transferred to the resistor.

The rate at which energy is dissipated in the resistor is the power, P, given by

$$P = i^2 R = \left[I \sin(\omega t - \phi) \right]^2 R = I^2 R \sin^2(\omega t - \phi), \tag{30.32}$$

where the alternating current, i, is given by equation 30.9. We can express the average power, $\langle P \rangle$, using the fact that the average value of $\sin^2(\omega t - \phi)$ over a full oscillation is $\frac{1}{2}$:

$$\langle P \rangle = \tfrac{1}{2} I^2 R = \left(\frac{I}{\sqrt{2}} \right)^2 R.$$

In calculations of power and energy, it is common to use the **root-mean-square (rms) current,** I_{rms}. (In general, *root-mean-square*, or *rms*, means the square root of the mean of the square of the specific quantity.) From equation 30.32, we have $i^2 = \left[I \sin(\omega t - \phi) \right]^2$, and the mean (or average) of i^2 is $I^2/2$. Thus, $I_{rms} = I/\sqrt{2}$. We can then write the average power as

$$\langle P \rangle = I_{rms}^2 R. \tag{30.33}$$

In a similar way, we can define the root-mean-square values of other time-varying quantities, such as the voltage:

$$V_{rms} = \frac{V_m}{\sqrt{2}}. \tag{30.34}$$

The current and voltage values normally quoted for alternating currents and measured by AC ammeters and voltmeters are I_{rms} and V_{rms}. For example, wall sockets in the United States provide $V_{rms} = 110$ V, which corresponds to a maximum voltage of $\sqrt{2}(110 \text{ V}) \approx 156$ V.

We can rewrite equation 30.21 in terms of root-mean-square values by multiplying both sides of the equation by $1/\sqrt{2}$:

$$I_{rms} = \frac{V_{rms}}{Z} = \frac{V_{rms}}{\sqrt{R^2 + \left(\omega L - \dfrac{1}{\omega C} \right)^2}}. \tag{30.35}$$

This form is used most often to describe the characteristics of AC circuits.

We can describe the average power dissipated in an AC circuit in a different way by starting with equation 30.33:

$$\langle P \rangle = I_{rms}^2 R = \frac{V_{rms}}{Z} I_{rms} R = I_{rms} V_{rms} \frac{R}{Z}. \tag{30.36}$$

From Figure 30.16, we see that the cosine of the phase constant is equal to the ratio of the maximum value of the voltage across the resistor to the maximum value of the time-varying emf:

$$\cos \phi = \frac{V_R}{V_m} = \frac{IR}{IZ} = \frac{R}{Z}. \tag{30.37}$$

We can thus rewrite equation 30.36 as follows:

$$\langle P \rangle = I_{rms} V_{rms} \cos \phi. \tag{30.38}$$

This expression gives the average power dissipated in an AC circuit, where the term $\cos \phi$ is called the **power factor.** You can see that for $\phi = 0$, maximum power is dissipated in the circuit; that is, the maximum power is dissipated in an AC circuit when the frequency of the time-varying emf matches the resonant frequency of the circuit.

We can combine equations 30.19, 30.35, and 30.36 to obtain an expression for the average power as a function of the angular frequency, the inductance, the resistance, and the resonant frequency:

$$\langle P \rangle = I_{rms} V_{rms} \frac{R}{Z} = \frac{V_{rms}}{\sqrt{R^2 + \left(\omega L - \dfrac{1}{\omega C} \right)^2}} V_{rms} \frac{R}{\sqrt{R^2 + \left(\omega L - \dfrac{1}{\omega C} \right)^2}},$$

or simply

$$\langle P \rangle = \frac{V_{rms}^2 R}{R^2 + \left(\omega L - \dfrac{1}{\omega C} \right)^2}.$$

Since $\omega_0 = 1/\sqrt{LC}$, we can write $C = 1/(L\omega_0^2)$, and thus, we find for the average power for a series RLC circuit in terms of the angular frequency:

$$\langle P \rangle = \frac{V_{rms}^2 R}{R^2 + \left(\omega L - \dfrac{L\omega_0^2}{\omega} \right)^2} = \frac{V_{rms}^2 R \omega^2}{R^2 \omega^2 + L^2 \left(\omega^2 - \omega_0^2 \right)^2}. \tag{30.39}$$

Typically all voltages, currents, and powers in AC circuits are specified as root-mean-square values. For example, the typical 110-V AC wall circuit has $V_{rms} = 110$ V, and the ubiquitous 1000-W hair dryer uses $P_{rms} = 1000$ W.

SOLVED PROBLEM 30.1 Voltage Drop across an Inductor

A series RLC circuit has a source of time-varying emf that supplies $V_{rms} = 170.0$ V, a resistance $R = 820.0 \ \Omega$, an inductance $L = 30.0$ mH, and a capacitance $C = 0.290$ mF. The circuit is operating at its resonant frequency.

PROBLEM
What is the root-mean-square voltage drop across the inductor?

SOLUTION
THINK   At the resonant frequency, the impedance of the circuit is equal to the resistance. We can calculate the root-mean-square current in the circuit. The voltage drop across the inductor is then the product of the root-mean-square current in the circuit and the inductive reactance.

SKETCH   A diagram of the series RLC circuit is shown in Figure 30.28.

RESEARCH   At resonance, the impedance of the circuit is

$$Z = \sqrt{R^2 + \left(X_L - X_C \right)^2} = R.$$

At resonance, the root-mean-square current, I_{rms}, in the circuit is given by

$$V_{rms} = I_{rms} R.$$

The root-mean-square voltage drop across the inductor, V_L, at resonance is

$$V_L = I_{rms} X_L,$$

where the inductive reactance X_L is defined as

$$X_L = \omega L$$

and ω is the angular frequency at which the circuit is operating. The resonant angular frequency, ω_0, of the circuit is

$$\omega_0 = \frac{1}{\sqrt{LC}}.$$

SIMPLIFY   Combining all these equations gives us the voltage drop across the inductor at resonance:

$$V_L = \left(\frac{V_{rms}}{R} \right)(\omega_0 L) = \frac{L V_{rms}}{R} \frac{1}{\sqrt{LC}} = \frac{V_{rms}}{R} \sqrt{\frac{L}{C}}.$$

CALCULATE   Putting in the numerical values gives us

$$V_L = \frac{170.0 \text{ V}}{820.0 \ \Omega} \sqrt{\frac{30.0 \cdot 10^{-3} \text{ H}}{0.290 \cdot 10^{-3} \text{ F}}} = 2.10861 \text{ V}.$$

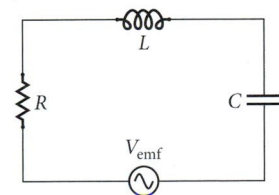

FIGURE 30.28   A series RLC circuit.

– *Continued*

ROUND We report our result to three significant figures:

$$V_L = 2.11 \text{ V}.$$

DOUBLE-CHECK The root-mean-square voltage drop across the capacitor is

$$V_C = \left(\frac{V_{rms}}{R}\right)\left(\frac{1}{\omega_0 C}\right) = \frac{V_{rms}}{RC}\sqrt{LC} = \left(\frac{V_{rms}}{R}\right)\sqrt{\frac{L}{C}},$$

which is the same as the root-mean-square voltage drop across the inductor. At resonance, the instantaneous voltage drop across the inductor is the negative of the voltage drop across the capacitor. Thus, the rms voltage across the capacitor should be the same as the rms voltage across the inductor. Thus, our result seems reasonable.

SOLVED PROBLEM 30.2 | Power Dissipated in an RLC Circuit

A series RLC circuit has a source of emf providing $V_{rms} = 120.0$ V at a frequency $f = 50.0$ Hz, as well as an inductor, $L = 0.500$ H, a capacitor, $C = 3.30$ μF, and a resistor, $R = 276.$ Ω.

PROBLEM
What is the average power dissipated in the circuit?

SOLUTION
THINK The average power dissipated in the circuit is the root-mean-square current times the root-mean-square voltage, but it depends on the angular frequency of the source of emf. The current in the circuit can be found using the impedance.

SKETCH A diagram of a series RLC circuit is shown in Figure 30.28.

RESEARCH The angular frequency, ω, of the source of emf is

$$\omega = 2\pi f.$$

The impedance, Z, of the circuit is

$$Z = \sqrt{R^2 + (X_L - X_C)^2},$$

where the inductive reactance is given by

$$X_L = \omega L$$

and the capacitive reactance is given by

$$X_C = \frac{1}{\omega C}.$$

We can find the root-mean-square current, I_{rms}, in the circuit using the relationship

$$V_{rms} = I_{rms} Z.$$

The average power dissipated in the circuit, $\langle P \rangle$, is given by

$$\langle P \rangle = I_{rms} V_{rms} \cos\phi,$$

where ϕ is the phase constant between the voltage and the current in the circuit:

$$\phi = \tan^{-1}\left(\frac{X_L - X_C}{R}\right).$$

SIMPLIFY We can combine all these equations to obtain an expression for the average power dissipated in the circuit:

$$\langle P \rangle = \frac{V_{rms}}{Z} V_{rms} \cos\phi = \frac{V_{rms}^2}{\sqrt{R^2 + (X_L - X_C)^2}} \cos\phi.$$

CALCULATE First, we calculate the inductive reactance:

$$X_L = \omega L = 2\pi f L = 2\pi (50.0 \text{ Hz})(0.500 \text{ H}) = 157.1 \text{ Ω}.$$

Next, we calculate the capacitive reactance:

$$X_C = \frac{1}{\omega C} = \frac{1}{2\pi f C} = \frac{1}{2\pi (50.0 \text{ Hz})(3.30 \cdot 10^{-6} \text{ F})} = 964.6 \ \Omega.$$

The phase constant is then

$$\phi = \tan^{-1}\left(\frac{X_L - X_C}{R}\right) = \tan^{-1}\left(\frac{157.1 \ \Omega - 964.6 \ \Omega}{276 \ \Omega}\right) = -1.241 \text{ rad} = -71.13°.$$

We now calculate the average power dissipated in the circuit:

$$\langle P \rangle = \frac{(120.0 \text{ V})^2}{\sqrt{(276 \ \Omega)^2 + (157.1 \ \Omega - 964.6 \ \Omega)^2}} \cos(-1.241 \text{ rad}) = 5.46477 \text{ W}.$$

ROUND We report our result to three significant figures:

$$\langle P \rangle = 5.46 \text{ W}.$$

DOUBLE-CHECK To double-check our result, we can calculate the power that would be dissipated in the circuit if it were operating at the resonant frequency. At the resonant frequency, the maximum power is dissipated in the circuit, and the impedance of the circuit is equal to the resistance of the resistor. Thus, the maximum average power is

$$\langle P \rangle_{\text{max}} = \frac{V_{\text{rms}}^2}{R} = \frac{(120.0 \text{ V})^2}{276 \ \Omega} = 52.2 \text{ W}.$$

Our result for the power dissipated at $f = 50.0$ Hz is lower than the maximum average power, so it seems plausible.

Quality Factor

The **quality factor,** Q, of a series RLC circuit is defined as

$$Q = \frac{\omega_0 L}{R} = \frac{1}{R}\sqrt{\frac{L}{C}}. \tag{30.40}$$

The quality factor is the ratio of total energy stored in the system divided by the energy dissipated per cycle of the oscillation. This is the same definition used for mechanical oscillators in Chapter 14. For a series RLC circuit, the quality factor characterizes the selectivity of the circuit. The higher the value of Q, the more selective the circuit, that is, the more precisely a given frequency can be isolated (as in an AM radio receiver, discussed next). The lower the value of Q, the less selective the circuit becomes.

AM Radio Receiver

Let's look at a typical example of a selective series RLC circuit, an AM radio receiver. An AM radio receiver can be constructed using a series RLC circuit in which the time-varying emf is supplied by an antenna that picks up transmissions from a distant radio station broadcasting at a given frequency and converts those transmissions to voltage, as illustrated in Figure 30.29.

Figure 30.30 is a plot of the average power as a function of the frequency of the signal received on the antenna for the circuit shown in Figure 30.29, assuming that $R = 0.09111 \ \Omega$, $L = 5.000 \ \mu$H, $C = 6.693$ nF, and $V_{\text{rms}} = 3.500$ mV. The resonant angular frequency for this circuit is

$$\omega_0 = \frac{1}{\sqrt{LC}} = \frac{1}{\sqrt{(5.000 \cdot 10^{-6} \text{ H})(6.693 \cdot 10^{-9} \text{ F})}} = 5.466 \cdot 10^6 \text{ rad/s},$$

which corresponds to a resonant frequency of

$$f_0 = \frac{\omega_0}{2\pi} = \frac{5.466 \cdot 10^6 \text{ rad/s}}{2\pi} = 870.0 \text{ kHz}.$$

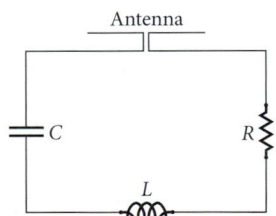

FIGURE 30.29 A series RLC circuit with the source of time-varying emf replaced by an antenna. This circuit can function as an AM radio receiver.

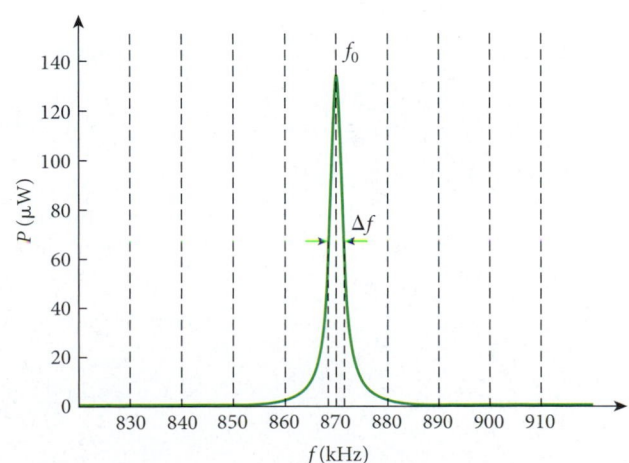

FIGURE 30.30 Power response of a series RLC circuit functioning as an AM radio receiver. The label Δf indicates the *full width at half maximum*, or the difference between the frequencies where the power has half the value that it has at its maximum value, at frequency f_0.

The quality factor for this circuit is

$$Q = \frac{\omega_0 L}{R} = \frac{\left(5.466 \cdot 10^6 \text{ rad/s}\right)\left(5.000 \cdot 10^{-6} \text{ H}\right)}{0.09111 \ \Omega} = 300.0.$$

A method for determining the approximate quality factor of a series RLC circuit uses the formula

$$Q = \frac{\omega_0}{\Delta\omega} = \frac{f_0}{\Delta f},$$

where $\Delta\omega$ and Δf are the full widths at half maximum for the angular frequency and the frequency, respectively, on the power response curve. The higher the value of Q, the narrower the power response to frequency. In Figure 30.30, $\Delta f = 2.9$ kHz, which gives a quality factor of

$$Q = \frac{f_0}{\Delta f} = \frac{870.0 \text{ kHz}}{2.9 \text{ kHz}} = 300.0.$$

This is the same result obtained using the formula that defines the quality factor in equation 30.40. Note that these two formulas for the quality factor have the same results only for high Q!

The alternative formula for the quality factor of a series RLC circuit is similar to the expression given in Chapter 14 for the quality of a weakly damped mechanical oscillator, $Q \approx \dfrac{\omega_0}{2\omega_\gamma}$, where ω_0 is the resonant angular frequency and ω_γ is the damping angular frequency.

In Figure 30.30, the frequencies of adjacent channels in the AM radio band are indicated by the vertical dashed lines located 10 kHz apart. The response of the series RLC circuit allows the AM receiver to tune in one station and exclude the adjacent channels.

SOLVED PROBLEM 30.3 / Unknown Inductance in an RL Circuit

Consider a series RL circuit with a source of time-varying emf. In this circuit, $V_{\text{rms}} = 33.0$ V with $f = 7.10$ kHz and $R = 83.0 \ \Omega$. A current $I_{\text{rms}} = 0.158$ A flows in the circuit.

PROBLEM
What is the magnitude of the inductance, L?

SOLUTION
THINK   The specified voltage and current are implicitly root-mean-square values. We can relate the voltage and current through the impedance of the circuit. The impedance of this circuit depends on the resistance and the inductance, as well as the frequency of the source of emf.

SKETCH   A diagram of the circuit is shown in Figure 30.31.

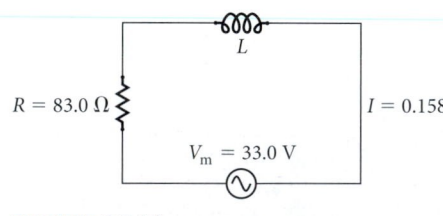

$R = 83.0 \ \Omega$ $I = 0.158$ A

$V_m = 33.0$ V

FIGURE 30.31   A series RL circuit.

RESEARCH   We can relate the time-varying emf, V_m, and the impedance Z in the circuit:

$$V_m = IZ. \tag{i}$$

The impedance is given by

$$Z = \sqrt{R^2 + (X_L - X_C)^2} = \sqrt{R^2 + X_L^2}, \tag{ii}$$

where R is the resistance, X_L is the inductive reactance, and X_C is the capacitive reactance, which is zero. The angular frequency, ω, of the circuit is given by

$$\omega = 2\pi f,$$

where f is the frequency. We can express the inductive reactance as

$$X_L = \omega L. \tag{iii}$$

SIMPLIFY   We can combine equations (i), (ii), and (iii) to obtain

$$Z^2 = R^2 + X_L^2 = \left(\frac{V_m}{I}\right)^2 = R^2 + (\omega L)^2. \tag{iv}$$

Rearranging equation (iv) gives

$$\omega L = \sqrt{\frac{V_m^2}{I^2} - R^2}.$$

And, finally, we find the unknown inductance:

$$L = \frac{1}{2\pi f}\sqrt{\frac{V_m^2}{I^2} - R^2}.$$

CALCULATE Putting in the numerical values gives us

$$L = \frac{1}{2\pi\left(7.10\cdot10^3\ \text{s}^{-1}\right)}\sqrt{\frac{\left(33.0\ \text{V}\right)^2}{\left(0.158\ \text{A}\right)^2} - \left(83.0\ \Omega\right)^2} = 0.0042963\ \text{H}.$$

ROUND We report our result to three significant figures:

$$L = 4.30\cdot10^{-3}\ \text{H} = 4.30\ \text{mH}.$$

DOUBLE-CHECK To double-check our result for the inductance, we first calculate the inductive reactance:

$$X_L = 2\pi fL = 2\pi\left(7.10\cdot10^3\ \text{s}^{-1}\right)\left(4.30\cdot10^{-3}\ \text{H}\right) = 192.\ \Omega.$$

The impedance of the circuit is then

$$Z = \sqrt{R^2 + X_L^2} = \sqrt{\left(83.0\ \Omega\right)^2 + \left(192\ \Omega\right)^2} = 209.\ \Omega.$$

We use this value of Z to calculate a value of V_m:

$$V_m = IZ = \left(0.158\ \text{A}\right)\left(209.\ \Omega\right) = 33.0\ \text{V},$$

which agrees with the value specified in the problem statement. Thus, our result is consistent.

30.7 Transformers

This section discusses the root-mean-square values of currents and voltages, rather than the maximum or instantaneous values. The result obtained for the power is always the average power when root-mean-square values are used. This practice is the convention normally followed by scientists, engineers, and electricians dealing with AC circuits.

In an AC circuit that has only a resistor, the phase constant is zero. Thus, we can express the power as

$$P = IV. \tag{30.41}$$

For a given power delivered to a circuit, the application dictates the choice of high current or high voltage. For example, to provide enough power to operate a computer or a vacuum cleaner, using a high voltage might be dangerous. The design of electric generators is complicated by the use of high voltages. Therefore, in these devices, lower voltages and higher currents are advantageous.

However, the transmission of electric power requires the opposite condition. The power dissipated in a transmission line is given by $P = I^2R$. Thus, the power lost in a line, like those in Figure 30.32a is proportional to the square of the current in the line. As an example, consider a power plant that produces 500. MW of power. If the power is transmitted at 350. kV, the current in the power lines will be

$$I = \frac{P}{V} = \frac{500.\ \text{MW}}{350.\ \text{kV}} = \frac{5.00\cdot10^8\ \text{W}}{3.50\cdot10^5\ \text{V}} = 1.43\ \text{kA}.$$

If the total resistance of the power lines is 50. Ω, the power lost in the transmission lines is

$$P = I^2R = \left(1.43\ \text{kA}\right)^2\left(50.0\ \Omega\right) = 102.\ \text{MW},$$

or about 20% of the generated power. A similar calculation would show that transmitting the power at 200. kV instead of 350. kV would increase the power loss by a factor of 3.1. Thus, approximately 60% of the power generated would be lost in transmission. This is why the transmission of electric power is always done at the highest possible voltage.

The ability to change voltage allows electric power to be generated and used at low, safe voltages but transmitted at the highest practical voltage. Alternating currents and voltages are transformed from high to low values by a device called, appropriately, a **transformer.** A transformer

(a)

(b)

FIGURE 30.32 (a) High-voltage power lines; (b) transformers for residential power lines.

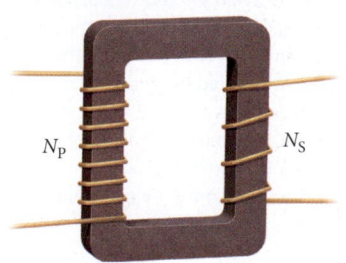

FIGURE 30.34 Transformer with N_P primary windings and N_S secondary windings.

that takes voltages from lower to higher values is called a *step-up transformer;* a transformer that takes voltages from higher to lower values is called a *step-down transformer.* Transformers are the main components of, for example, your cell phone charger (Figure 30.33) and the power supply for your MP3 player, your laptop, and pretty much every other consumer electronics device. Most of these devices require voltages of 12 V or less, but the grid delivers 110 V to outlets in the United States, necessitating the transformers that clutter your desk drawers.

A transformer consists of two sets of coils wrapped around an iron core (Figure 30.34). The primary coil, with N_P turns, is connected to a source of emf described by

$$V_{emf} = V_{max} \sin \omega t.$$

We'll assume that the primary coil acts as an inductor. The primary circuit has the current and voltage out of phase by $\pi/2$ rad (90°), so the power factor, $\cos \phi$, is zero. Thus, the source of emf is not delivering any power to the transformer if only the primary coil is connected. In other words, if the secondary coil is not connected to a closed circuit, the transformer does not draw any power. For example, your cell phone charger does not draw power if it is plugged into the wall socket but the other end is not connected to the cell phone. (This statement is not absolutely true; there is finite resistance in the wires in the primary coil, which this description is neglecting.)

The secondary coil of a transformer has N_S turns. The time-varying emf in the primary coil induces a time-varying magnetic field in the iron core. This core passes through the secondary coil. Thus, a time-varying voltage is induced in the secondary coil, as described by Faraday's Law of Induction:

$$V_{emf} = -N \frac{d\Phi_B}{dt},$$

where N is the number of turns and Φ_B is the magnetic flux. Because of the iron core, both the primary and secondary coils experience the same changing magnetic flux. Thus,

$$V_S = -N_S \frac{d\Phi_B}{dt}$$

and

$$V_P = -N_P \frac{d\Phi_B}{dt},$$

where V_S and V_P are the voltages across the secondary and primary windings, respectively. Dividing the first of these two equations by the other and rearranging gives

$$\frac{V_P}{N_P} = \frac{V_S}{N_S},$$

or

$$V_S = V_P \frac{N_S}{N_P}. \tag{30.42}$$

The transformer changes the voltage of the primary circuit to a secondary voltage, given by the ratio of the number turns in the secondary coil divided by the number of turns in the primary coil.

If a resistor, R, is connected across the secondary windings, a current, I_S, will begin to flow through the secondary coil. The power in the secondary circuit is then $P_S = I_S V_S$. This current induces a time-varying magnetic field that induces a voltage in the primary coil, so that the emf source then produces enough current, I_P, to maintain the original voltage. This current, I_P, is in phase with the voltage because of the resistor, so power can be transmitted to the transformer. Energy conservation requires that the power delivered to the primary coil be transferred to the secondary coil, so we can write

$$P_P = I_P V_P = P_S = I_S V_S.$$

Using equation 30.42, we can express the current in the secondary circuit as

$$I_S = I_P \frac{V_P}{V_S} = I_P \frac{N_P}{N_S}. \tag{30.43}$$

The current in the secondary circuit is equal to the current in the primary circuit multiplied by the ratio of the number of primary turns divided by the number of secondary turns.

When the secondary circuit begins to draw current, current must be supplied to the primary circuit. Since $V_S = I_S R$ in the secondary circuit, we can use equations 30.42 and 30.43 to write

$$I_P = \frac{N_S}{N_P} I_S = \frac{N_S}{N_P} \frac{V_S}{R} = \frac{N_S}{N_P} \left(V_P \frac{N_S}{N_P} \right) \frac{1}{R} = \left(\frac{N_S}{N_P} \right)^2 \frac{V_P}{R}. \qquad (30.44)$$

The effective resistance of the primary circuit can be expressed in terms of $V_P = I_P R_P$, so that the effective resistance is

$$R_P = \frac{V_P}{I_P} = V_P \left(\frac{N_P}{N_S} \right)^2 \frac{R}{V_P} = \left(\frac{N_P}{N_S} \right)^2 R. \qquad (30.45)$$

Note that we have assumed that there are no losses in the transformer, that the primary coil is only an inductor, that there are no losses in magnetic flux between the primary and secondary coils, and that the secondary circuit has the only resistance. Real transformers do have some losses. Part of these losses result from the fact that the alternating magnetic fields from the coils induce eddy currents in the iron core of the transformer. To counter this effect, transformer cores are constructed by laminating layers of metal to inhibit the formation of eddy currents. Modern transformers can transform voltages with very little loss.

Another application of transformers is **impedance matching.** The power transfer between a source of emf and a device that uses power is at a maximum when the impedance is the same in both. Often, the source of emf and the intended device do not have the same impedance. A common example is a stereo amplifier and its speakers. Usually, the amplifier has high impedance and the speakers have low impedance. A transformer placed between the amplifier and the speakers can help match the impedances of the devices, producing a more efficient power transfer.

30.8 Rectifiers

Many electronic devices require direct current rather than alternating current. However, many common sources of electrical power provide alternating current. Therefore, this current must be converted to direct current to operate electronic equipment. A **rectifier** is a device that converts alternating current to direct current. Most rectifiers use an electronic component that was described in Section 25.8—the diode. A diode is designed to allow current to flow in one direction and not in the other direction. The symbol for a diode is ▸|, and the direction of the arrowhead signifies the direction in which the diode will conduct current.

Let's start with a simple circuit containing a source of time-varying emf, a resistor, and a diode, as shown in Figure 30.35b. The voltage provided by the source of emf is alternately positive and negative, as shown in Figure 30.35a. Note that both ends of the source of emf are connected simultaneously so that when one end produces a positive voltage, the other end produces a negative voltage. The circuit in Figure 30.35b produces current in the resistor that indeed flows in only one direction. However, the circuit blocks half the current, as illustrated in Figure 30.35c. Thus, this type of circuit is often termed a **halfwave rectifier.**

To allow all the current to flow in one direction, the type of circuit shown in Figure 30.36 is employed. Again the voltage alternates between positive and negative, as shown in Figure 30.36a. Two equivalent circuit diagrams are shown in Figure 30.36b and Figure 30.36c. All the current in the resistor flows in one direction, as illustrated in Figure 30.36d. This type of circuit is called a **fullwave rectifier.**

To illustrate how the fullwave rectifier works, Figure 30.37 shows instantaneous views of the circuit with positive and negative voltage. In Figure 30.37a, the voltage from the

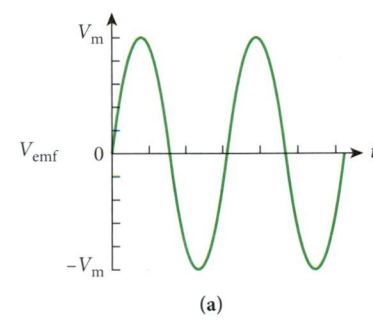

(a)

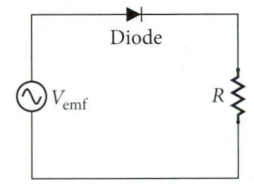

(b)

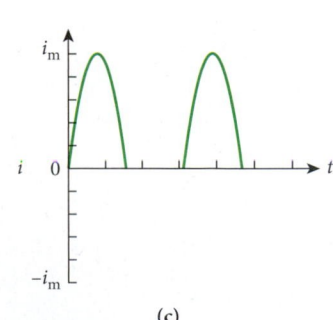

(c)

FIGURE 30.35 Circuit containing a source of time-varying emf, a resistor, and a diode, forming a halfwave rectifier: (a) the emf as a function of time; (b) the circuit diagram; (c) the current flowing through the circuit as a function of time.

FIGURE 30.36 Circuit containing a source of time-varying emf, a resistor, and four diodes, forming a fullwave rectifier: (a) the emf as a function of time; (b) the circuit diagram; (c) alternative way of drawing the circuit diagram; (d) the current flowing through the circuit as a function of time.

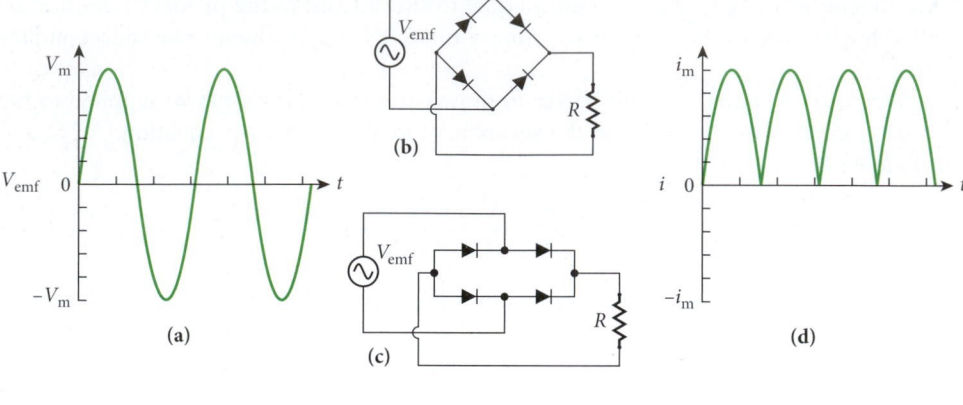

FIGURE 30.37 A fullwave rectifier with the diodes that are not conducting current at the given instant in gray. The current in the resistor always flows in the same direction. (a) Positive voltage. (b) Negative voltage.

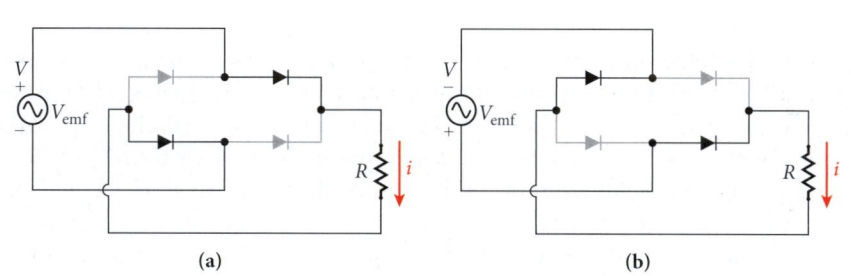

Self-Test Opportunity 30.6

A typical alternator found in automobiles produces three-phase alternating current. Each phase is shifted by 120° from the next phase. Draw the circuit diagram for the fullwave rectifier for this alternator, based on the circuit diagram in Figure 30.36c but incorporating six diodes instead of four.

source of emf is positive. The black diodes are conducting current, while the gray diodes are not. The voltage is reversed in Figure 30.37b, and the current flows through the other pair of diodes; the current in the resistor is still in the same direction.

Although the fullwave rectifier does indeed convert alternating current to direct current, the resulting direct current varies with time. This variance, often called *ripple*, can be smoothed out by adding a capacitor to the output of the rectifier, creating an RC circuit with a time constant governed by the choice of R and C, as shown in Figure 30.38. In Figure 30.38a, the time-varying emf is shown, and the circuit diagram is presented in Figure 30.38b. The direct current is filtered by the added capacitor. The resulting current as a function of time is shown in Figure 30.38c. The current still varies with time but much less so than the current flowing out of the fullwave rectifier without a capacitor.

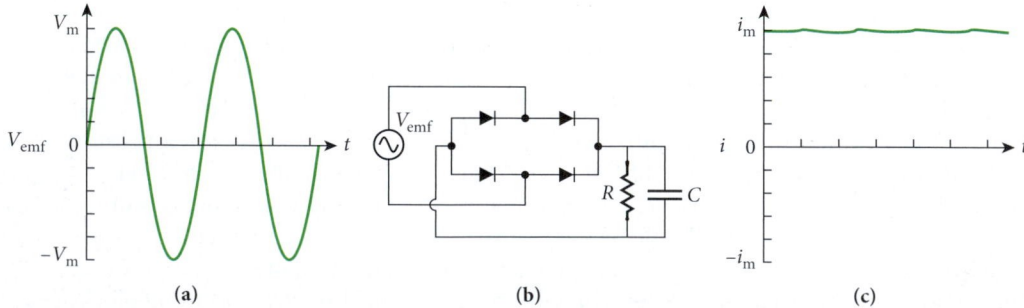

FIGURE 30.38 Circuit containing a source of time-varying emf, a resistor, a capacitor, and four diodes, forming a filtered fullwave rectifier: (a) the emf as a function of time; (b) the circuit diagram; (c) the current flowing through the circuit as a function of time.

WHAT WE HAVE LEARNED | EXAM STUDY GUIDE

- The energy stored in the electric field of a capacitor with capacitance C and charge q is given by $U_E = \frac{1}{2}(q^2/C)$; the energy stored in the magnetic field of an inductor with inductance L that is carrying current i is given by $U_B = \frac{1}{2}Li^2$.

- The current in a single-loop circuit containing an inductor and a capacitor (an LC circuit) oscillates with a frequency given by $\omega_0 = 1/\sqrt{LC}$.

- The current in a single-loop circuit containing a resistor, an inductor, and a capacitor (an RLC circuit) oscillates with a frequency given by $\omega = \sqrt{\omega_0^2 - (R/2L)^2}$, where $\omega_0 = 1/\sqrt{LC}$.

- The charge, q, on a capacitor in a single-loop RLC circuit oscillates and decreases exponentially with time according to $q = q_{max}\, e^{-Rt/2L} \cos(\omega t)$, where q_{max} is the original charge on the capacitor.

- For a single-loop circuit containing a source of time-varying emf and a resistor, R, $V_R = I_R R$, where V_R and I_R are the voltage and the current, respectively.

- For a single-loop circuit containing a source of time-varying emf that has frequency ω and a capacitor, $V_C = I_C X_C$, where V_C and I_C are the voltage and the current, respectively, and $X_C = 1/\omega C$ is the capacitive reactance.

- For a single-loop circuit containing a source of time-varying emf that has frequency ω and an inductor,

- $V_L = I_L X_L$, where V_L and I_L are the voltage and the current, respectively, and $X_L = \omega L$ is the inductive reactance.

- For a single-loop RLC circuit containing a source of time-varying emf that has frequency ω, $V = IZ$, where V and I are the voltage and the current, respectively, and $Z = \sqrt{R^2 + (X_L - X_C)^2}$ is the impedance.

- The phase constant, ϕ, between the current and the voltage in a single-loop RLC circuit containing a source of time-varying emf that has frequency ω is given by $\phi = \tan^{-1}\left(\dfrac{X_L - X_C}{R}\right)$.

- The average power in a single-loop RLC circuit containing a source of time-varying emf that has frequency ω is given by $\langle P \rangle = I_{rms} V_{rms} \cos\phi$, where $I_{rms} = I_m/\sqrt{2}$ and $V_{rms} = V_m/\sqrt{2}$.

- All currents, voltages, and powers quoted for alternating-current (AC) circuits are typically root-mean-square values.

- A transformer with N_P windings in the primary coil and N_S windings in the secondary coil can convert a primary alternating voltage, V_P, to a secondary alternating voltage, V_S, given by $V_S = V_P \dfrac{N_S}{N_P}$, and a primary alternating current, I_P, to a secondary alternating current, I_S, given by $I_S = I_P \dfrac{N_P}{N_S}$.

ANSWERS TO SELF-TEST OPPORTUNITIES

30.1 For this circuit, $\omega_0 = 2\pi f = 2\pi(200 \text{ kHz})$ rad/s = $4\pi \cdot 10^5$ rad/s; $\phi = 0$ since $q(0) = q_{max}$

a) true $(\cos\omega_0 t = -1)$

b) false $(\sin\omega_0 t = 0)$

c) false $(\cos\omega_0 t = -1$ and $\sin\omega_0 t = 0)$

d) false $(\cos\omega_0 t = 0$ and $\sin\omega_0 t = 1)$

30.2 It can be seen that k/m corresponds to $1/LC$, and $b/2m$ corresponds to $R/2L$. Thus, the inductance, L, plays the role of the mass, m, the capacitance, C, corresponds to the inverse spring constant, $1/k$, and the resistance, R, has the function of the damping constant, b.

30.3 a) true   b) true   c) false

30.4 a) true   b) true   c) false

30.5 The resistance is inversely proportional to the inverse of the area of the wire and thus inversely proportional to the square of the radius. Therefore, a wire that was 10 times thicker would have a resistance that was 100 times lower.

30.6

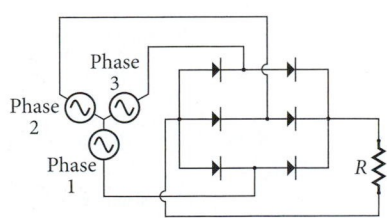

Three-phase bridge rectifier

PROBLEM-SOLVING GUIDELINES

1. Most problems concerning AC circuits require you to calculate resistance, capacitive reactance, inductive reactance, or impedance. Be sure that you understand what each of these quantities is and how to use them in calculating currents and voltages.

2. You will often have to distinguish between the instantaneous current or voltage in a circuit and the root-mean-square or maximum value of current or voltage. The common convention is to use lowercase i and v for instantaneous values and uppercase I and V for constant values (with subscripts as necessary). Be sure you use notation that is clear so that you won't become confused during the calculations.

3. Remember the phase relations for AC circuits: For a resistor, current and voltage are in phase; for a capacitor, current leads voltage; for an inductor, current lags voltage.

4. Phasors add by vector operations, not by simple scalar arithmetic. Whenever you use phasors to determine current or voltage, check the results by checking the phase relationships given in the preceding guideline.

5. It is usually easier to work with angular frequency (ω) than with frequency (f) in analyzing AC circuits. Most often, you will be given an angular frequency in the problem statement, but if you are given a frequency, convert it to an angular frequency by multiplying it by 2π.

MULTIPLE-CHOICE QUESTIONS

30.1 A 200-Ω resistor, a 40.0-mH inductor and a 3.0-μF capacitor are connected in series with a source of time-varying emf that provides 10.0 V at a frequency of 1000 Hz. What is the impedance of the circuit?

a) 200 Ω

b) 228 Ω

c) 342 Ω

d) 282 Ω

30.2 For which values of f is $X_L > X_C$?

a) $f > 2\pi(LC)^{1/2}$

b) $f > (2\pi LC)^{-1}$

c) $f > (2\pi(LC)^{1/2})^{-1}$

d) $f > 2\pi LC$

30.3 Which statement about the phase relation between the electric and magnetic fields in an LC circuit is correct?

a) When one field is at its maximum, the other is also, and the same for the minimum values.

b) When one field is at maximum strength, the other is at minimum (zero) strength.

c) The phase relation, in general, depends on the values of L and C.

30.4 For the band-pass filter shown in Figure 30.25, how can the width of the frequency response be increased?

a) increase R_1

b) decrease C_1

c) increase R_2

d) increase C_2

e) do any of the above

30.5 The phase constant, ϕ, between the voltage and the current in an AC circuit depends on the _____.

a) inductive reactance

b) capacitive reactance

c) resistance

d) all of the above

30.6 The AM radio band covers the frequency range from 520 kHz to 1610 kHz. Assuming a fixed inductance in a simple LC circuit, what ratio of capacitance is necessary to cover this frequency range? That is, what is the value of C_h/C_l, where C_h is the capacitance for the highest frequency and C_l is the capacitance for the lowest frequency?

a) 9.59

b) 0.104

c) 0.568

d) 1.76

30.7 In the RLC circuit in the figure, $R = 60\ \Omega$, $L = 3$ mH, $C = 4$ mF, and the source of time-varying emf has a peak voltage of 120 V. What should the angular frequency, ω, be to produce the largest current in the resistor?

a) 4.2 rad/s

b) 8.3 rad/s

c) 204 rad/s

d) 289 rad/s

e) 5000 rad/s

f) 20,000 rad/s

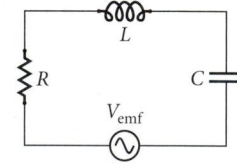

30.8 A standard North American wall socket plug is labeled 110 V. This label indicates the _____ value of the voltage.

a) average

b) maximum

c) root-mean-square (rms)

d) instantaneous

30.9 A circuit contains a source of time-varying emf, which is given by $V_{emf} = 120.0 \sin[(377\ \text{rad/s})t]$ V, and a capacitor with capacitance $C = 5.00\ \mu$F. What is the current in the circuit at $t = 1.00$ s?

a) 0.226 A

b) 0.451 A

c) 0.555 A

d) 0.750 A

e) 1.25 A

30.10 A source of time-varying emf supplies $V_{max} = 115.0$ V at $f = 60.0$ Hz in a series RLC circuit in which $R = 374\ \Omega$, $L = 0.310$ H, and $C = 5.50\ \mu$F. What is the impedance of this circuit?

a) 321 Ω

b) 523 Ω

c) 622 Ω

d) 831 Ω

e) 975 Ω

CONCEPTUAL QUESTIONS

30.11 What is the impedance of a series RLC circuit when the frequency of time-varying emf is set to the resonant frequency of the circuit?

30.12 Estimate the total energy stored in the 5.00 km of space above Earth's surface if the average magnitude of the magnetic field at Earth's surface is about $0.500 \cdot 10^{-4}$ T.

30.13 In a DC circuit containing a capacitor, a current will flow through the circuit for only a very short time, while the capacitor is being charged or discharged. On the other hand, a steady alternating current will flow in a circuit containing the same capacitor but powered by a source of time-varying emf. Does it mean that charges are crossing the gap (dielectric) of the capacitor?

30.14 In an RL circuit with alternating current, the current lags behind the voltage. What does this mean, and how can it be explained qualitatively, based on the phenomenon of electromagnetic induction?

30.15 In Solved Problem 30.3, the voltage supplied by the source of time-varying emf is 33.0 V, the voltage across the resistor is $V_R = IR = 13.1$ V, and the voltage across the inductor is $V_L = IX_L = 30.3$ V. Does this circuit obey Kirchhoff's rules?

30.16 Why is rms power specified for an AC circuit, not average power?

30.17 Why can't we use a universal charger that plugs into a household outlet to charge all our electrical devices—cell phone, toy dog, can opener, and so on—rather than using a separate charger with its own transformer for each device?

30.18 If you use a parallel plate capacitor with air in the gap between the plates as part of a series RLC circuit in a generator, you can measure current flowing through the generator. Why is it that the air gap in the capacitor does not act like an open switch, blocking all current flow in the circuit?

30.19 A common configuration of wires has twisted pairs as opposed to straight, parallel wires. What is the technical advantage of using twisted pairs of wires versus straight, parallel pairs?

30.20 In a classroom demonstration, an iron core is inserted into a large solenoid connected to an AC power source. The effect of the core is to magnify the magnetic field in the solenoid by the relative magnetic permeability, κ_m, of the core (where κ_m is a dimensionless constant, substantially greater than unity for a ferromagnetic material, introduced in Chapter 28) or, equivalently, to replace the magnetic permeability of free space, μ_0, with the magnetic permeability of the core, $\mu = \kappa_m \mu_0$.

a) The measured root-mean-square current drops from approximately 10 A to less than 1 A and remains at the lower value. Explain why.

b) What would happen if the power source were DC?

30.21 Along Capitol Drive in Milwaukee, Wisconsin, there are a large number of radio broadcasting towers. Contrary to expectation, radio reception there is terrible; unwanted stations often interfere with the one tuned in. Given that a car radio tuner is a resonant oscillator—its resonant frequency is adjusted to that of the desired station—explain this crosstalk phenomenon.

30.22 A series RLC circuit is in resonance when driven by a sinusoidal voltage at its resonant frequency, $\omega_0 = (LC)^{-1/2}$. But if the same circuit is driven by a *square-wave voltage* (which is alternately on and off for equal time intervals), it will exhibit resonance at its resonant frequency and at $\frac{1}{3}$, $\frac{1}{5}$, $\frac{1}{7}$, ... , of this frequency. Explain why.

30.23 Is it possible for the voltage amplitude across the inductor in a series RLC circuit to exceed the voltage amplitude of the voltage supply? Why or why not?

30.24 Why can't a transformer be used to step up or step down the voltage in a DC circuit?

30.25 The figure shows a circuit with a source of constant emf connected in series to a resistor, an inductor, and a capacitor. What is the steady-state current flow through the circuit?

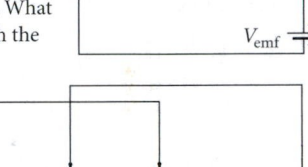

•30.26 An RLC circuit has a capacitor, a resistor, and an inductor connected in parallel, as shown in the figure, and a source of time-varying emf providing V_{rms} at a frequency f. Find an expression for I_{rms} in terms of V_{rms}, f, L, C, and R.

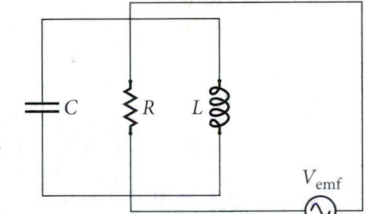

EXERCISES

A blue problem number indicates a worked-out solution is available in the Student Solutions Manual. One • and two •• indicate increasing level of problem difficulty.

Sections 30.1 and 30.2

30.27 For the LC circuit in the figure, $L = 32.0$ mH and $C = 45.0$ μF. The capacitor is charged to $q_0 = 10.0$ μC, and at $t = 0$, the switch is closed. At what time is the energy stored in the capacitor first equal to the energy stored in the inductor?

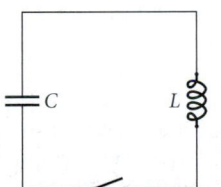

30.28 A 2.00-μF capacitor is fully charged by being connected to a 12.0-V battery. The fully charged capacitor is then connected to a 0.250-H inductor. Calculate
(a) the maximum current in the inductor and (b) the frequency of oscillation of the LC circuit.

30.29 An LC circuit consists of a 1.00-mH inductor and a fully charged capacitor. After 2.10 ms, the energy stored in the capacitor is half of its original value. What is the capacitance?

30.30 The time-varying current in an LC circuit where $C = 10.0$ μF is given by $i(t) = (1.00\text{A}) \sin (1200.t)$, where t is in seconds.

a) At what time after $t = 0$ does the current reach its maximum value?

b) What is the total energy of the circuit?

c) What is the inductance, L?

•30.31 A 10.0-μF capacitor is fully charged by a 12.0-V battery and is then disconnected from the battery and allowed to discharge through a 0.200-H inductor. Find the first three times when the charge on the capacitor is 80.0-μC, taking $t = 0$ as the instant when the capacitor is connected to the inductor.

•30.32 A 4.00-mF capacitor is connected in series with a 7.00-mH inductor. The peak current in the wires between the capacitor and the inductor is 3.00 A.

a) What is the total electric energy in this circuit?

b) Write an expression for the charge on the capacitor as a function of time, assuming the capacitor is fully charged at $t = 0$ s.

Section 30.3

30.33 A circuit contains a 4.50-nF capacitor and a 4.00-mH inductor. If some charge is placed initially on the capacitor, an oscillating current with angular frequency ω_0 is produced. By what factor does this angular frequency change if a 1.00-kΩ resistor is connected in series with the capacitor and the inductor?

•30.34 An RLC oscillator circuit contains a 50.0-Ω resistor and a 1.00-mH inductor. What capacitance is necessary for the time constant of the circuit (the 1/e value) to be equal to the oscillation period? Plot the voltage across the resistor as a function of time.

•30.35 A 2.00-μF capacitor was fully charged by being connected to a 12.0-V battery. The fully charged capacitor is then connected in series with a resistor and an inductor: $R = 50.0$ Ω and $L = 0.200$ H. Calculate the damped frequency of the resulting circuit.

•30.36 An LC circuit consists of a capacitor, $C = 2.50$ μF, and an inductor, $L = 4.00$ mH. The capacitor is fully charged using a battery and then connected to the inductor. An oscilloscope is used to measure the frequency of the oscillations in the circuit. Next, the circuit is opened, and a resistor, R, is inserted in series with the inductor and the capacitor. The capacitor is again fully charged using the same battery and then connected to the circuit. The angular frequency of the damped oscillations in the

RLC circuit is found to be 20.0% less than the angular frequency of the oscillations in the LC circuit.

a) Determine the resistance of the resistor.

b) How long after the capacitor is reconnected in the circuit will the amplitude of the damped current through the circuit be 50.0% of the initial amplitude?

c) How many complete damped oscillations will have occurred in that time?

Section 30.4

30.37 At what frequency will a 10.0-μF capacitor have the reactance $X_C = 200. \; \Omega$?

30.38 A capacitor with capacitance $C = 5.00 \cdot 10^{-6}$ F is connected to an AC power source having a peak value of 10.0 V and $f = 100.$ Hz. Find the reactance of the capacitor and the maximum current in the circuit.

Section 30.5

30.39 A series circuit contains a 100.0-Ω resistor, a 0.500-H inductor, a 0.400-μF capacitor, and a source of time-varying emf providing 40.0 V.

a) What is the resonant angular frequency of the circuit?

b) What current will flow through the circuit at the resonant frequency?

30.40 A variable capacitor used in an RLC circuit produces a resonant frequency of 5.0 MHz when its capacitance is set to 15 pF. What will the resonant frequency be when the capacitance is increased to 380 pF?

30.41 Determine the phase constant and the impedance of the RLC circuit shown in the figure when the frequency of the time-varying emf is 1.00 kHz, $C = 100. \; \mu$F, $L = 10.0$ mH, and $R = 100. \; \Omega$.

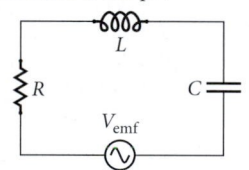

30.42 What is the resonant frequency of the series RLC circuit of Problem 30.41 if $C = 4.00 \; \mu$F, $L = 5.00$ mH, and $R = 1.00$ kΩ? What is the maximum current in the circuit if $V_m = 10.0$ V at the resonant frequency?

•30.43 In a series RLC circuit, $V = (12.0 \; \text{V})(\sin\omega t)$, $R = 10.0 \; \Omega$, $L = 2.00$ H, and $C = 10.0 \; \mu$F. At resonance, determine the amplitude of the voltage across the inductor. Is the result reasonable, considering that the voltage supplied to the entire circuit has an amplitude of 12.0 V?

•30.44 An AC power source with $V_m = 220.$ V and $f = 60.0$ Hz is connected in a series RLC circuit. The resistance, R, inductance, L, and capacitance, C, of this circuit are, respectively, 50.0 Ω, 0.200 H, and 0.0400 mF. Find each of the following quantities:

a) the inductive reactance

b) the capacitive reactance

c) the impedance of the circuit

d) the maximum current through the circuit at this frequency

e) the maximum potential difference across each circuit element

•30.45 The series RLC circuit shown in the figure has $R = 2.20 \; \Omega$, $L = 9.30$ mH, $C = 2.27$ mF, $V_m = 110.$ V, and $\omega = 377$ rad/s.

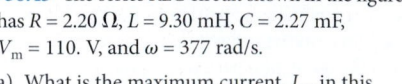
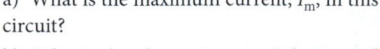

a) What is the maximum current, I_m, in this circuit?

b) What is the phase constant, ϕ, between the voltage and the current?

c) The capacitance, C, can be varied. What value of C will allow the largest current amplitude oscillations to occur, and what are the magnitudes of this current, I'_m, and the phase angle, ϕ', between the current and the voltage?

•30.46 Design an RC high-pass filter that passes a signal with frequency 5.00 kHz, has a ratio $V_{out}/V_{in} = 0.500$, and has an impedance of 1.00 kΩ at very high frequencies.

a) What components will you use?

b) What is the phase of V_{out} relative to V_{in} at the frequency of 5.00 kHz?

••30.47 Design an RC high-pass filter that rejects 60.0-Hz line noise from a circuit used in a detector. Your criteria are reduction of the amplitude of the line noise by a factor of 1000. and total impedance at high frequencies of 2.00 kΩ.

a) What components will you use?

b) What is the frequency range of the signals that will be passed with at least 90.0% of their amplitude?

Section 30.6

30.48 What is the maximum value of the AC voltage whose root-mean-square value is (a) 110 V or (b) 220 V?

30.49 The quality factor, Q, of a circuit can be defined by $Q = \omega_0(U_E + U_B)/P$. Express the quality factor of a series RLC circuit in terms of its resistance R, inductance L, and capacitance C.

30.50 A label on a hair dryer reads "110V 1250W." What is the peak current in the hair dryer, assuming that it behaves like a resistor?

30.51 A radio tuner has a resistance of 1.00 $\mu\Omega$, a capacitance of 25.0 nF, and an inductance of 3.00 mH.

a) Find the resonant frequency of this tuner.

b) Calculate the power in the circuit if a signal at the resonant frequency produces an emf across the antenna of $V_{rms} = 1.50$ mV.

•30.52 A circuit contains a 100.-Ω resistor, a 0.0500-H inductor, a 0.400-μF capacitor, and a source of time-varying emf connected in series. The time-varying emf corresponds to $V_{rms} = 50.0$ V at a frequency of 2000. Hz.

a) Determine the current in the circuit.

b) Determine the voltage drop across each component of the circuit.

c) How much power is drawn from the source of emf?

•30.53 The figure shows a simple FM antenna circuit in which $L = 8.22 \; \mu$H and C is variable (the capacitor can be tuned to receive a specific station). The radio signal from your favorite FM station produces a sinusoidal time-varying emf with an amplitude of 12.9 μV and a frequency of 88.7 MHz in the antenna.

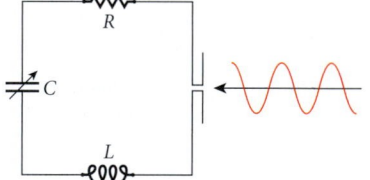

a) To what value, C_0, should you tune the capacitor in order to best receive this station?

b) Another radio station's signal produces a sinusoidal time-varying emf with the same amplitude, 12.9 μV, but with a frequency of 88.5 MHz in the antenna. With the circuit tuned to optimize reception at 88.7 MHz, what should the value, R_0, of the resistance be in order to reduce by a factor of 2 (compared to the current if the circuit were optimized for 88.5 MHz) the current produced by the signal from this station?

Section 30.7

30.54 The transmission of electric power occurs at the highest possible voltage to reduce losses. By how much could the power loss be reduced by raising the voltage by a factor of 10.0?

30.55 Treat the solenoid and coil of Solved Problem 29.2 as a transformer.

a) Find the root-mean-square voltage in the coil if the solenoid has a root-mean-square voltage of 120 V and a frequency of 60. Hz. The length of the solenoid is 12.0 cm.

b) What is the voltage in the coil if the frequency is 0 Hz (DC current)?

30.56 A transformer has 800 turns in the primary coil and 40 turns in the secondary coil.

a) What happens if an AC voltage of 100. V is across the primary coil?

b) If the initial AC current is 5.00 A, what is the output current?

c) What happens if a DC current at 100. V flows into the primary coil?

d) If the initial DC current is 5.00 A, what is the output current?

30.57 A transformer contains a primary coil with 200 turns and a secondary coil with 120 turns. The secondary coil drives a current I through a 1.00-kΩ resistor. If an input voltage of $V_{rms} = 75.0$ V is applied across the primary coil, what is the power dissipated in the resistor?

Section 30.8

30.58 Consider the filtered fullwave rectifier shown in the figure. If the frequency of the source of time-varying emf is 60. Hz, what is the frequency of the resulting current?

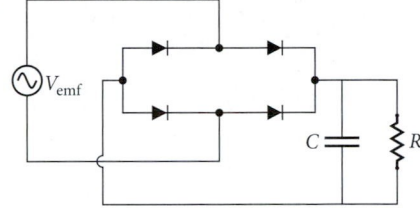

30.59 A voltage $V_{rms} = 110$ V at a frequency of 60. Hz is applied to the primary coil of a transformer. The transformer has a ratio $N_P/N_S = 11$. The secondary coil is used as the source of V_{emf} for the filtered fullwave rectifier of Problem 30.58.

a) What is the maximum voltage in the secondary coil of the transformer?

b) What is the DC voltage provided to the resistor?

Additional Exercises

30.60 A vacuum cleaner motor can be viewed as an inductor with an inductance of 100. mH. For a 60.0-Hz AC voltage of $V_{rms} = 115$. V, what capacitance must be in series with the motor to maximize the power output of the vacuum cleaner?

30.61 When you turn the dial on a radio to tune it, you are adjusting a variable capacitor in an LC circuit. Suppose you tune to an AM station broadcasting at a frequency of 1000. kHz, and there is a 10.0-mH inductor in the tuning circuit. When you have tuned in the station, what is the capacitance of the capacitor?

30.62 A series RLC circuit has a source of time-varying emf providing 12.0 V at a frequency f_0, with $L = 7.00$ mH, $R = 100.$ Ω, and $C = 0.0500$ mF.

a) What is the resonant frequency of this circuit?

b) What is the average power dissipated in the resistor at this resonant frequency?

30.63 What are the maximum values of (a) current and (b) voltage when an incandescent 60-W light bulb (at 110 V) is connected to a wall plug labeled 110 V?

30.64 A 360-Hz source of emf is connected in a circuit consisting of a capacitor, a 25-mH inductor, and an 0.80-Ω resistor. For the current and the voltage to be in phase, what should the value of C be?

30.65 What is the resistance in an RLC circuit with $L = 65.0$ mH and $C = 1.00$ μF if the circuit loses 3.50% of its total energy as thermal energy in each cycle?

30.66 A transformer with 400. turns in its primary coil and 20. turns in its secondary coil is designed to deliver an average power of 1200. W with a maximum voltage of 60.0 V. What is the maximum current in the primary coil?

30.67 A 5.00-μF capacitor in series with a 4.00-Ω resistor is charged with a 9.00-V battery for a long time by closing the switch (position a in the figure). The capacitor is then discharged through an inductor ($L = 40.0$ mH) by closing the switch (position b) at $t = 0$.

a) Determine the maximum current through the inductor.

b) What is the first time at which the current is at its maximum?

•30.68 In the RC high-pass filter shown in the figure, $R = 10.0$ kΩ and $C = 0.0470$ μF. What is the 3.00-dB frequency of this circuit (where dB means basically the same for electric current as it did for sound in Chapter 16)? That is, at what frequency does the ratio of output voltage to input voltage satisfy $20 \log (V_{out}/V_{in}) = -3.00$?

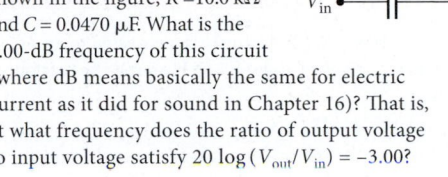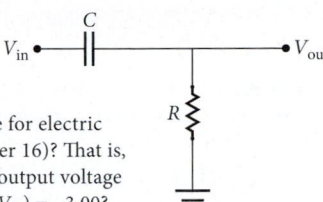

•30.69 The discussion of RL, RC, and RLC circuits in this chapter has assumed a purely resistive resistor, one whose inductance and capacitance are exactly zero. While the capacitance of a resistor can generally be neglected, inductance is an intrinsic part of the resistor. Indeed, one of the most widely used resistors, the wire-wound resistor, is nothing but a solenoid made of highly resistive wire. Suppose a wire-wound resistor of unknown resistance is connected to a DC power supply. At a voltage of $V = 10.0$ V across the resistor, the current through the resistor is 1.00 A. Next, the same resistor is connected to an AC power source providing $V_{rms} = 10.0$ V at a variable frequency. When the frequency is 20.0 kHz, a current, $I_{rms} = 0.800$ A, is measured through the resistor.

a) Calculate the resistance of the resistor.

b) Calculate the inductive reactance of the resistor.

c) Calculate the inductance of the resistor.

d) Calculate the frequency of the AC power source at which the inductive reactance of the resistor exceeds its resistance.

•30.70 In a certain RLC circuit, a 20.0-Ω resistor, a 10.0-mH inductor, and a 5.00-μF capacitor are connected in series with an AC power source for which $V_{rms} = 10.0$ V and $f = 100.$ Hz. Calculate

a) the amplitude of the current,

b) the phase between the current and the voltage, and

c) the maximum voltage across each component.

•30.71 a) A loop of wire 5.00 cm in diameter is carrying a current of 2.00 A. What is the energy density of the magnetic field at its center?

b) What current has to flow in a straight wire to produce the same energy density at a point 4.00 cm from the wire?

•30.72 A 75,000-W light bulb (yes, there are such things!) operates at $I_{rms} = 200.$ A and $V_{rms} = 440.$ V in a 60.0-Hz AC circuit. Find the resistance, R, and self-inductance, L, of this bulb. Its capacitive reactance is negligible.

•30.73 Show that the power dissipated in a resistor connected to an AC power source with a frequency ω oscillates with a frequency 2ω.

•30.74 A 300.-Ω resistor is connected in series with a 4.00-μF capacitor and a source of time-varying emf providing $V_{rms} = 40.0$ V.

a) At what frequency will the potential drop across the capacitor equal that across the resistor?

b) What is the rms current through the circuit when this occurs?

•30.75 An electromagnet consists of 200 loops and has a length of 10.0 cm and a cross-sectional area of 5.00 cm^2. Find the resonant frequency of this electromagnet when it is attached to the Earth (treat the Earth as a spherical capacitor).

•30.76 Laboratory experiments with series RLC circuits require some care, as these circuits can produce large voltages at resonance. Suppose you have a 1.00-H inductor (not difficult to obtain) and a variety of resistors and capacitors. Design a series RLC circuit that will resonate at a frequency (not an angular frequency) of 60.0 Hz and will produce at resonance a magnification of the voltage across the capacitor or the inductor by a factor of 20.0 times the input voltage or the voltage across the resistor.

•30.77 A particular RC low-pass filter has a breakpoint frequency of 200. Hz. At what frequency will the output voltage divided by the input voltage be 0.100?

MULTI-VERSION EXERCISES

30.78 An inductor with inductance $L = 42.1$ mH is connected to an AC power source that supplies $V_{emf} = 19.1$ V at $f = 605$ Hz. Find the reactance of the inductor.

30.79 An inductor with inductance $L = 52.5$ mH is connected to an AC power source that supplies $V_{emf} = 19.9$ V at $f = 669$ Hz. Find the maximum current in the circuit.

30.80 An inductor with inductance L is connected to an AC power source that supplies $V_{emf} = 20.7$ V at $f = 733$ Hz. If the reactance of the inductor is to be 81.52 Ω, what should the value of L be?

30.81 An inductor with inductance L is connected to an AC power source that supplies $V_{emf} = 21.5$ V at $f = 797$ Hz. If the maximum current in the circuit is to be 0.1528 A, what should the value of L be?

31

Electromagnetic Waves

FIGURE 31.1 The solar power plant Solúcar PS10, located near Seville, Spain, concentrates the power from the Sun's electromagnetic radiation to produce electricity.

In the last few chapters we have studied electric and magnetic fields and seen how they can change over time. The present chapter concludes our study of electromagnetism and focuses on how the interaction between electric and magnetic fields gives rise to electromagnetic waves.

Literally every place on Earth is bathed in electromagnetic waves, of which the most obvious type is probably visible light. Other types of electromagnetic waves that are familiar to you range from TV and radio waves to microwaves and X-rays. Our cell phones, wireless Internet connections, and the GPS system all utilize electromagnetic waves, as do tanning booths and laser pointers. In this chapter, we'll examine the nature of electromagnetic waves, including their similarities and differences and their transmission of energy and pressure to other objects. Solar power plants, like the one shown in Figure 31.1, are impressive demonstrations of how the energy contained in electromagnetic waves emitted by the Sun can be harnessed.

Most of the results in this chapter apply only to electromagnetic waves propagating through vacuum. For practical purposes, however, electromagnetic waves propagating through Earth's atmosphere can be treated in the same way. There are some significant differences for electromagnetic waves propagating through other media, but these are not addressed in this chapter.

The next chapter is the first to focus on *optics*—the properties and behavior of light—and many of those properties apply to other electromagnetic waves as well.

WHAT WE WILL LEARN

- Changing electric fields induce magnetic fields, and changing magnetic fields induce electric fields.

- Maxwell's equations describe electromagnetic phenomena.

- Electromagnetic waves have both electric and magnetic fields.

- Solutions of Maxwell's equations can be expressed in terms of sinusoidally varying traveling waves.

- For an electromagnetic wave, the electric field is perpendicular to the magnetic field and both fields are perpendicular to the direction in which the wave is traveling.

- The speed of light can be expressed in terms of constants related to electric and magnetic fields.

- Light is an electromagnetic wave.

- Electromagnetic waves can transport energy and momentum.

- The intensity of an electromagnetic wave is proportional to the square of the root-mean-square magnitude of the electric field of the wave.

- The direction of the electric field of a traveling electromagnetic wave is called the *polarization direction*.

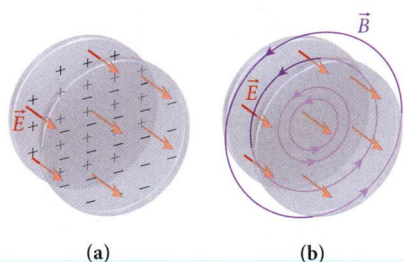

FIGURE 31.2 (a) A charged circular capacitor. The red arrows represent the electric field between the plates. (b) A capacitor with charge increasing with time. The red arrows represent the electric field, and the purple loops represent the induced magnetic field.

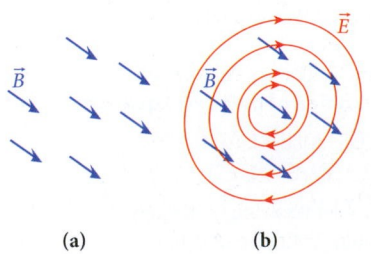

FIGURE 31.3 (a) A constant uniform magnetic field. (b) A uniform magnetic field increasing with time, which induces an electric field represented by the red loops.

31.1   Maxwell's Law of Induction for Induced Magnetic Fields

In Chapter 29, we saw that a changing magnetic field induces an electric field. According to Faraday's Law of Induction,

$$\oint \vec{E} \cdot d\vec{s} = -\frac{d\Phi_B}{dt},$$ (31.1)

where $\vec{E}$ is the electric field induced around a closed loop by the changing magnetic flux, Φ_B, through that loop. In a similar way, a changing electric field induces a magnetic field. **Maxwell's Law of Induction** (named for British physicist James Clerk Maxwell, 1831–1879) describes this phenomenon as follows:

$$\oint \vec{B} \cdot d\vec{s} = \mu_0 \epsilon_0 \frac{d\Phi_E}{dt},$$ (31.2)

where $\vec{B}$ is the magnetic field induced around a closed loop by a changing electric flux, Φ_E, through that loop. This equation is similar to equation 31.1 except for the constant $\mu_0 \epsilon_0$ and the lack of a negative sign. The constant is a consequence of the SI units used for magnetic fields. The fact that the right-hand side of equation 31.2 does not have a negative sign implies that the induced magnetic field has the opposite sign from that of an induced electric field when both are induced under similar conditions, as we'll see shortly.

First, note that it is *not at all obvious* that equation 31.2 can be written in analogy to equation 31.1. When Maxwell first wrote this equation, it represented a major step forward in the unification of electricity and magnetism. Faraday discovered his law in 1831, but it took a quarter of a century for Maxwell to come up with the counterpart. What is stated here as simple fact is in reality the first great conceptual leap toward the unification of all physical forces of nature. This unification began with Maxwell's work one and a half centuries ago and continues in modern physics research today.

A circular capacitor can be used to illustrate an induced magnetic field (Figure 31.2). For the capacitor shown in Figure 31.2a, the charge is constant, and a constant electric field appears between the plates. There is no magnetic field. For the capacitor shown in Figure 31.2b, the charge is increasing with time. Thus, the electric flux between the plates is increasing with time. A magnetic field, $\vec{B}$, is induced, represented by the purple loops, which also indicate the direction of $\vec{B}$. Along each loop, the magnetic field vector has the same magnitude and is directed tangentially to the loop. When the charge stops increasing, the electric flux remains constant and the magnetic field disappears.

Next, consider a uniform magnetic field that is also constant in time, as in Figure 31.3a. In Figure 31.3b, the magnetic field is still uniform in space but is increasing with time, which induces an electric field, shown by the red loops. The electric field vector has constant magnitude along each loop and is directed tangentially to the loops as shown. Note that this induced electric field points in the opposite direction from the induced magnetic field caused by an increasing electric field (Figure 31.2b).

Now recall Ampere's Law:

$$\oint \vec{B} \cdot d\vec{s} = \mu_0 i_{\text{enc}}, \tag{31.3}$$

which relates the integral around a loop of the dot product of the magnetic field and the differential displacement along the loop, $\vec{B} \cdot d\vec{s}$, to the current flowing through the loop. However, Maxwell realized that this equation is incomplete because it does not account for contributions to the magnetic field caused by changing electric fields. Equations 31.2 and 31.3 can be combined to produce a description of magnetic fields created by moving charges and by changing electric fields:

$$\oint \vec{B} \cdot d\vec{s} = \mu_0 \epsilon_0 \frac{d\Phi_E}{dt} + \mu_0 i_{\text{enc}}. \tag{31.4}$$

Equation 31.4 is called the **Maxwell-Ampere Law.** You can see that for the case of constant current, such as current flowing in a conductor, this equation reduces to Ampere's Law. For the case of a changing electric field without current flowing, such as the electric field between the plates of a capacitor, this equation reduces to Maxwell's Law of Induction. It is important to realize that the Maxwell-Ampere Law describes two different sources of magnetic field: the conventional current (as discussed in Chapter 28) and the time-varying electric flux (examined in more detail in the following subsection).

Displacement Current

Looking at the Maxwell-Ampere Law (equation 31.4), you can see that the expression $\epsilon_0 d\Phi_E/dt$ on the right-hand side of the equation must have the units of current. Although no actual "current" is "displaced," this term is called the **displacement current,** i_{d}:

$$i_{\text{d}} = \epsilon_0 \frac{d\Phi_E}{dt}. \tag{31.5}$$

With this definition, we can rewrite equation 31.4 as

$$\oint \vec{B} \cdot d\vec{s} = \mu_0 (i_{\text{d}} + i_{\text{enc}}).$$

Again, let's consider a parallel plate capacitor with circular plates, now placed in a circuit in which a current, i, is flowing while the capacitor is charging (Figure 31.4). For a parallel plate capacitor, the charge, q, is related to the electric field between the plates, E, as follows (see Chapter 24)

$$q = \epsilon_0 A E$$

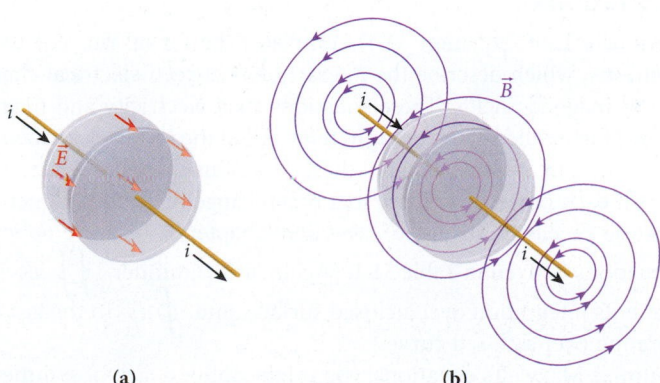

(a)   (b)

FIGURE 31.4 A parallel plate capacitor in a circuit being charged by a current, i: (a) the electric field between the plates at a given instant; (b) the magnetic field around the wires and between the plates of the capacitor.

Concept Check 31.1

The displacement current, i_d, for the charging circular capacitor with radius R shown in the figure is equal to the conduction current, i, in the wires. Points 1 and 3 are located a perpendicular distance r from the wires, and point 2 is located the same perpendicular distance r from the center of the capacitor, such that $r > R$. Rank the magnetic fields at points 1, 2, and 3, from largest magnitude to smallest.

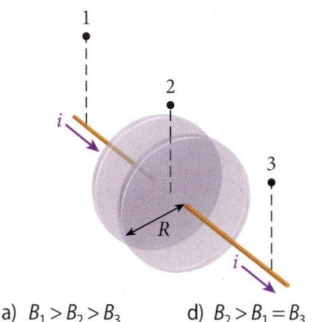

a) $B_1 > B_2 > B_3$

b) $B_3 > B_2 > B_1$

c) $B_1 = B_3 > B_2$

d) $B_2 > B_1 = B_3$

e) $B_1 = B_2 = B_3$

where A is the area of the plates. The current, i, in the circuit can be obtained by taking the time derivative of this equation:

$$i = \frac{dq}{dt} = \epsilon_0 A \frac{dE}{dt}. \tag{31.6}$$

Assuming that the electric field between the plates of the capacitor is uniform, we can obtain an expression for the displacement current:

$$i_d = \epsilon_0 \frac{d\Phi_E}{dt} = \epsilon_0 \frac{d(AE)}{dt} = \epsilon_0 A \frac{dE}{dt}. \tag{31.7}$$

Thus, the current in the circuit, i, given by equation 31.6, is equal to the displacement current, i_d, given by equation 31.7. Although no actual current is flowing between the plates of the capacitor, in the sense that no actual charges move across the capacitor gap from one plate to the other, the displacement current can be used to calculate the induced magnetic field.

To calculate the magnetic field between the two circular plates of the capacitor, we assume that the volume between the two plates can be replaced with a conductor of radius R carrying current i_d. In Chapter 28, we saw that the magnetic field at a perpendicular distance r from the center of the capacitor is given by

$$B = \left(\frac{\mu_0 i_d}{2\pi R^2} \right) r \qquad \text{(for } r < R\text{)}.$$

The system outside the capacitor can be treated as a current-carrying wire; so the magnetic field at a perpendicular distance r from the wire is

$$B = \frac{\mu_0 i_d}{2\pi r} \qquad \text{(for } r > R\text{)}.$$

Concept Check 31.2

The displacement current, i_d, for the charging circular capacitor with radius R shown in the figure is equal to the conduction current, i, in the wires. Points 1 and 3 are located a perpendicular distance r from the wires, and point 2 is located the same perpendicular distance r from the center of the capacitor, such that $r < R$. Rank the magnetic fields at points 1, 2, and 3, from largest magnitude to smallest.

a) $B_1 > B_2 > B_3$

b) $B_3 > B_2 > B_1$

c) $B_1 = B_3 > B_2$

d) $B_2 > B_1 = B_3$

e) $B_1 = B_2 = B_3$

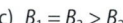

Maxwell's Equations

The Maxwell-Ampere Law (equation 31.4) completes the set of four equations known as **Maxwell's equations,** which describe the interactions between electrical charges, currents, electric fields, and magnetic fields. These equations treat electricity and magnetism as two aspects of a unified force called **electromagnetism.** All of the results described previously for electricity and magnetism are still valid, but these equations show how electric and magnetic fields interact with each other, giving rise to a broad range of electromagnetic phenomena. This chapter focuses on electromagnetic waves and Chapter 34 on wave optics. A summary of Maxwell's equations is given in Table 31.1. (Again, as a reminder, $\oiint d\vec{A}$ in the first two equations represents integration over a closed surface, and $\oint d\vec{s}$ in the last two equations indicates integration over a closed curve.)

If you scrutinize Maxwell's equations, you might notice a lack of symmetry between $\vec{E}$ and $\vec{B}$. This difference arises from the fact that electric charges exist in isolation and a corresponding current appears when charges move, but apparently no isolated, stationary magnetic charges occur in nature. Particles that hypothetically have a single magnetic charge (a north pole or a south pole, but not both) are called *magnetic monopoles,* but empiri-

Table 31.1	Maxwell's Equations Describing Electromagnetic Phenomena	
Name	**Equation**	**Description**
Gauss's Law for Electric Fields	$\oiint \vec{E} \cdot d\vec{A} = \dfrac{q_{enc}}{\epsilon_0}$	The net electric flux through a closed surface is proportional to the net enclosed electric charge.
Gauss's Law for Magnetic Fields	$\oiint \vec{B} \cdot d\vec{A} = 0$	The net magnetic flux through a closed surface is zero (no magnetic monopoles exist).
Faraday's Law of Induction	$\oint \vec{E} \cdot d\vec{s} = -\dfrac{d\Phi_B}{dt}$	An electric field is induced by a changing magnetic flux.
Maxwell-Ampere Law	$\oint \vec{B} \cdot d\vec{s} = \mu_0 \epsilon_0 \dfrac{d\Phi_E}{dt} + \mu_0 i_{enc}$	A magnetic field is induced by a changing electric flux or by a current.

cally it has been found that magnetic poles always come in pairs, a north pole together with a south pole. There is no fundamental reason for the absence of magnetic monopoles, and many experiments have searched unsuccessfully for them. The most sensitive of these experiments was MACRO, which used a massive detector that operated for many years in a laboratory located deep under the Gran Sasso mountain in Italy. MACRO searched for magnetic monopoles in cosmic rays, without success.

We'll now begin our study of **electromagnetic waves.** Electromagnetic waves consist of electric and magnetic fields, can travel through vacuum without any supporting medium, and do not involve moving charges or currents. The existence of electromagnetic waves was first demonstrated in 1888 by the German physicist Heinrich Hertz (1857–1894). Hertz used an RLC circuit that induced a current in an inductor that drove a spark gap. A spark gap consists of two electrodes that, when a potential difference is applied across them, produce a spark by exciting the gas between the electrodes. Hertz placed a loop and a small spark gap several meters apart. He observed that sparks were induced in the remote loop in a pattern that correlated with the electromagnetic oscillations in the primary RLC circuit. Thus, electromagnetic waves were able to travel through space without any medium to support them. For this contribution and others, the basic unit of oscillation, cycles per second, was named the hertz (Hz) in his honor.

31.2 Wave Solutions to Maxwell's Equations

As Section 31.7 will show, it is possible to use advanced calculus to derive a general wave equation from Maxwell's equations starting with those equations in differential form. However, we'll first assume that electromagnetic waves propagating in vacuum (no moving charges or currents) have the form of a traveling wave and show that this form satisfies Maxwell's equations.

Proposed Solution

We assume the following equations express the electric and magnetic fields in a particular electromagnetic wave that happens to be traveling in the positive x-direction:

$$\vec{E}(\vec{r},t) = E_{max} \sin(\kappa x - \omega t)\hat{y}$$

and

$$\vec{B}(\vec{r},t) = B_{max} \sin(\kappa x - \omega t)\hat{z}, \tag{31.8}$$

where $\kappa = 2\pi/\lambda$ is the wave number and $\omega = 2\pi f$ is the angular frequency of a wave with wavelength λ and frequency f. Note that the magnitudes of both fields have no dependence on the y- or z-coordinates, only on the x-coordinate and time. This type of wave, in which the electric and magnetic field vectors lie in a plane, is called a **plane wave.** Equation

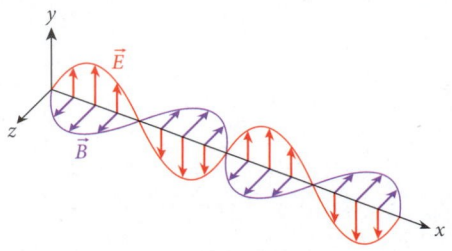

FIGURE 31.5 Representation of an electromagnetic wave traveling in the positive x-direction at a given instant.

Concept Check 31.3

An electromagnetic plane wave is traveling through vacuum. The electric field of the wave is given by $\vec{E} = E_{max} \cos(\kappa x - \omega t)\hat{y}$. Which of the following equations describes the magnetic field of the wave?

a) $\vec{B} = B_{max} \cos(\kappa x - \omega t)\hat{x}$

b) $\vec{B} = B_{max} \cos(\kappa y - \omega t)\hat{y}$

c) $\vec{B} = B_{max} \cos(\kappa z - \omega t)\hat{z}$

d) $\vec{B} = B_{max} \cos(\kappa y - \omega t)\hat{z}$

e) $\vec{B} = B_{max} \cos(\kappa x - \omega t)\hat{z}$

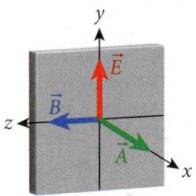

FIGURE 31.6 Gaussian surface (gray box) around a portion of the vector representation of an electromagnetic wave traveling in the positive x-direction. The area vector is shown for the front face of the Gaussian surface.

31.8 indicates that this particular electromagnetic wave is traveling in the positive x-direction because, as the time t increases, the coordinate x has to increase to maintain the same value for the fields. The wave described by equation 31.8 is shown in Figure 31.5.

In the particular case illustrated in Figure 31.5, the electric field is completely in the y-direction and the magnetic field is completely in the z-direction; that is, both fields are perpendicular to the direction of wave propagation. It turns out that the electric field is always perpendicular to the direction the wave is traveling and is always perpendicular to the magnetic field. However, in general, for an electromagnetic wave that is propagating along the x-axis, the electric field can point anywhere in the yz-plane.

The representation of the wave in Figure 31.5 is an instantaneous abstraction. The vectors shown represent the magnitude and the direction for the electric and magnetic fields; however, you should realize that these fields are not solid objects. Nothing made of matter actually moves left and right or up and down as the wave travels. The vectors pointing left and right and up and down represent the electric and magnetic fields.

Showing that the traveling wave described by equation 31.8 satisfies all of Maxwell's equations involves quite a bit of vector calculus but also uses many of the concepts that have been developed in the preceding chapters. The following subsections work through this process in detail, one equation at a time.

Gauss's Law for Electric Fields

Let's start with Gauss's Law for Electric Fields. For an electromagnetic wave in vacuum, there is no enclosed charge anywhere ($q_{enc} = 0$); thus, we must show that the proposed solution of equation 31.8 satisfies

$$\oiint \vec{E} \cdot d\vec{A} = 0. \tag{31.9}$$

We choose a rectangular box as a Gaussian surface enclosing a portion of the vector representation of the wave (Figure 31.6). For the faces of the box in the yz-plane, $\vec{E} \cdot d\vec{A}$ is zero because the vectors $\vec{E}$ and $d\vec{A}$ are perpendicular to each other. The same is true for the faces in the xy-plane. The faces in the xz-plane contribute $+EA_1$ and $-EA_1$, where A_1 is the area of the top face and the bottom face. Thus, the integral is zero, and Gauss's Law for Electric Fields is satisfied.

If we analyzed the vector representation at different times, we would get a different electric field. However, because the electric field is always in the y-direction, the integral will always be zero.

Gauss's Law for Magnetic Fields

For Gauss's Law for Magnetic Fields, we must show that

$$\oiint \vec{B} \cdot d\vec{A} = 0. \tag{31.10}$$

We again use the closed surface in Figure 31.6 for the integration. For the faces in the yz-plane and for those in the xz-plane, $\vec{B} \cdot d\vec{A}$ is zero because the vectors $\vec{B}$ and $d\vec{A}$ are perpendicular to each other. The faces in the xy-plane contribute $+BA_2$ and $-BA_2$, where A_2 is the area of each of the two faces in the xy-plane. Thus, the integral is zero, and Gauss's Law for Magnetic Fields is satisfied.

Faraday's Law of Induction

Now let's address Faraday's Law of Induction:

$$\oint \vec{E} \cdot d\vec{s} = -\frac{d\Phi_B}{dt}. \tag{31.11}$$

To evaluate the integral on the left-hand side of this equation, we assume a closed loop in the xy-plane that has width dx and height h and goes from a to b to c to d and back to a, as depicted by the gray rectangle in Figure 31.7. The differential area $d\vec{A} = \hat{n}dA = \hat{n}hdx$ of this rectangle has its unit surface normal vector, $\hat{n}$, pointing in the positive z-direction. Note that the electric and magnetic fields change with distance along the x-axis. Thus, going from point x to point $x + dx$, the electric field changes from $\vec{E}(x)$ to $\vec{E}(x+dx) = \vec{E}(x) + d\vec{E}$.

To evaluate the integral of equation 31.11 over the closed loop, we split the loop into four pieces, integrating counterclockwise from a to b, b to c, c to d, and d to a. The contributions to the integral that are parallel to the x-axis, from integrating from b to c and from d to a, are zero because the electric field is always perpendicular to the direction of integration. For the integrations in the y-direction, a to b and c to d, the electric field is parallel or antiparallel to the direction of integration; therefore, the scalar product reduces to a conventional product. Because the electric field is independent of the y-coordinate, it can be taken out of the integral. Thus, the integral along each of the segments in the y-direction is a simple product of the integrand (the magnitude of the electric field at the corresponding x-coordinate) and the length of the integration interval (h), times -1 for the integration in the negative y-direction because $\vec{E}$ is antiparallel to the direction of integration. Thus, the integral of equation 31.11 evaluates to

$$\oint \vec{E} \cdot d\vec{s} = E \int_a^b ds - E \int_c^d ds = (E + dE)(h) - Eh = (dE)(h).$$

The right-hand side of equation 31.11 is given by

$$-\frac{d\Phi_B}{dt} = -A\frac{dB}{dt} = -(h)(dx)\frac{dB}{dt},$$

where dt is the time during which the wave travels a distance dx. Thus, we have

$$(h)(dE) = -(h)(dx)\frac{dB}{dt},$$

or

$$\frac{dE}{dx} = -\frac{dB}{dt}. \tag{31.12}$$

The derivatives dE/dx and dB/dt are each taken with respect to a single variable, although both E and B depend on both x and t. Thus, we can more appropriately write equation 31.12 using partial derivatives:

$$\frac{\partial E}{\partial x} = -\frac{\partial B}{\partial t}. \tag{31.13}$$

Using the assumed forms for the electric and magnetic fields (equation 31.8), we can expand the partial derivatives:

$$\frac{\partial E}{\partial x} = \frac{\partial}{\partial x}\left(E_{max}\sin(\kappa x - \omega t)\right) = \kappa E_{max}\cos(\kappa x - \omega t),$$

and

$$\frac{\partial B}{\partial t} = \frac{\partial}{\partial t}\left(B_{max}\sin(\kappa x - \omega t)\right) = -\omega B_{max}\cos(\kappa x - \omega t).$$

Substituting these expressions into equation 31.13 gives

$$\kappa E_{max}\cos(\kappa x - \omega t) = -\left[-\omega B_{max}\cos(\kappa x - \omega t)\right].$$

The angular frequency and wave number are related via

$$\frac{\omega}{\kappa} = \frac{2\pi f}{(2\pi/\lambda)} = f\lambda = c, \tag{31.14}$$

where c is the speed of the wave. (In general, we could use v for the speed of this wave. However, we choose to use c, because, as we'll see, all electromagnetic waves propagate in vacuum with a characteristic speed, the speed of light, which is conventionally represented by c.) Thus, we have

$$\frac{E_{max}}{B_{max}} = \frac{\omega}{\kappa} = c.$$

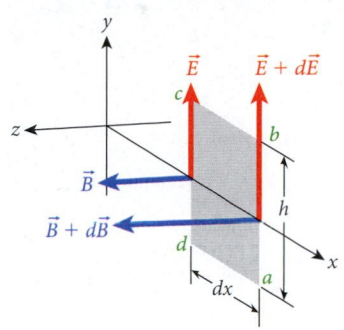

FIGURE 31.7 Two snapshots of the electric and magnetic fields in an electromagnetic wave. The gray area represents an integration loop for Faraday's Law.

We can use equation 31.8 to rewrite this equation in terms of the ratio of the magnitudes of the fields at a fixed place and time as

$$\frac{E}{B} = \frac{\left|\vec{E}(\vec{r},t)\right|}{\left|\vec{B}(\vec{r},t)\right|} = \frac{E_{\max}\left|\sin(\kappa x - \omega t)\right|}{B_{\max}\left|\sin(\kappa x - \omega t)\right|} = c. \tag{31.15}$$

Thus, equation 31.8 satisfies Faraday's Law of Induction if the ratio of the electric and magnetic field magnitudes is c.

Maxwell-Ampere Law

Finally, we address the Maxwell-Ampere Law. For electromagnetic waves, in which no current flows, we can write

$$\oint \vec{B} \cdot d\vec{s} = \mu_0 \epsilon_0 \frac{d\Phi_E}{dt}. \tag{31.16}$$

To evaluate the integral on the left-hand side of this equation, we assume a closed loop in the xz-plane that has width dx and height h, represented by the gray rectangle in Figure 31.8. The differential area of this rectangle is oriented in the positive y-direction.

The integral around the loop in the counterclockwise direction (a to b to c to d to a) is given by

$$\oint \vec{B} \cdot d\vec{s} = Bh - (B + dB)(h) = -(dB)(h). \tag{31.17}$$

As before, the parts of the loop parallel to the x-axis do not contribute to the integral. The right-hand side of equation 31.16 can be written as

$$\mu_0 \epsilon_0 \frac{d\Phi_E}{dt} = \mu_0 \epsilon_0 A \frac{dE}{dt} = \mu_0 \epsilon_0 (h)(dx) \frac{dE}{dt}, \tag{31.18}$$

where dt is the time during which the wave travels a distance dx. Substituting from equations 31.17 and 31.18 into equation 31.16, we get

$$-(dB)(h) = \mu_0 \epsilon_0 (h)(dx) \frac{dE}{dt}.$$

Expressing this equation in terms of partial derivatives, as we did for equation 31.12, we get

$$-\frac{\partial B}{\partial x} = \mu_0 \epsilon_0 \frac{\partial E}{\partial t}.$$

Now, using equation 31.8, we have

$$-\left[\kappa B_{\max} \cos(\kappa x - \omega t)\right] = -\mu_0 \epsilon_0 \omega E_{\max} \cos(\kappa x - \omega t),$$

or

$$\frac{E_{\max}}{B_{\max}} = \frac{\kappa}{\mu_0 \epsilon_0 \omega} = \frac{1}{\mu_0 \epsilon_0 c}.$$

We can express this equation in terms of the electric and magnetic field magnitudes as before:

$$\frac{E}{B} = \frac{1}{\mu_0 \epsilon_0 c}. \tag{31.19}$$

Equation 31.8 satisfies the Maxwell-Ampere Law if the ratio of the electric and magnetic field magnitudes is given by $1/\mu_0 \epsilon_0 c$.

The Speed of Light

From equations 31.15 and 31.19, we can conclude that

$$\frac{E}{B} = \frac{1}{\mu_0 \epsilon_0 c} = c,$$

which leads to

$$c = \frac{1}{\sqrt{\mu_0 \epsilon_0}}. \tag{31.20}$$

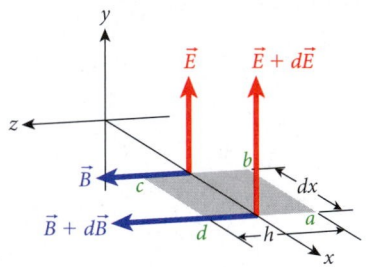

FIGURE 31.8 Representations of the electric and magnetic fields in an electromagnetic wave at a given instant. The gray area represents an integration loop for the Maxwell-Ampere Law.

Thus, the speed of an electromagnetic wave can be expressed in terms of the two fundamental constants related to electric and magnetic fields: the magnetic permeability and the electric permittivity of free space (vacuum). Putting the accepted values of these constants into equation 31.20 gives

$$c = \frac{1}{\sqrt{\left(4\pi \cdot 10^{-7}\ \text{H/m}\right)\left(8.85 \cdot 10^{-12}\ \text{F/m}\right)}} = 3.00 \cdot 10^{8}\ \text{m/s}.$$

This calculated speed is equal to the measured speed of light. This equality means that all electromagnetic waves travel (in vacuum) at the speed of light and suggests that light is an electromagnetic wave.

Equation 31.15 states $E/B = c$. Even though c is a very large number, equation 31.15 does not mean that the electric field magnitude is much larger than the magnetic field magnitude. In fact, electric and magnetic fields are measured in different units, so a direct comparison is not possible.

The speed of light plays an important role in the theory of special relativity, which we'll examine in Chapter 35. The speed of light is always the same in any reference frame. Thus, if you send an electromagnetic wave out in a specific direction, any observer, regardless of whether that observer is moving toward you or away from you or in another direction, will see that wave moving at the speed of light. This amazing result, along with the plausible postulate that the laws of physics are the same for all inertial observers, leads to the theory of special relativity.

The speed of light can be measured extremely precisely, much more precisely than the meter could be determined from the original reference standard. Therefore, the speed of light is now defined to be precisely

$$c = 299{,}792{,}458\ \text{m/s}. \tag{31.21}$$

The definition of the meter is now simply the distance that light can traverse in vacuum in a time interval of $1/299{,}792{,}458$ s.

31.3   The Electromagnetic Spectrum

All electromagnetic waves travel at the speed of light. However, the wavelength and the frequency of electromagnetic waves vary dramatically. The speed of light, c, the wavelength, λ, and the frequency, f, are related by

$$c = \lambda f. \tag{31.22}$$

Examples of electromagnetic waves are light, radio waves, microwaves, X-rays, and gamma rays. Three applications of electromagnetic waves are shown in Figure 31.9.

The **electromagnetic spectrum** is illustrated in Figure 31.10, which includes electromagnetic waves with wavelengths ranging from 1000 m and longer to less than 10^{-12} m, with corresponding frequencies ranging from 10^{5} to 10^{20} Hz. Electromagnetic waves with wavelengths (and frequencies) in certain ranges are identified by characteristic names:

- **Visible light** refers to electromagnetic waves that we can see with our eyes, with wavelengths from 400 nm (blue) to 700 nm (red). The response of the human eye peaks at around 550 nm (green) and drops off quickly away from that wavelength.

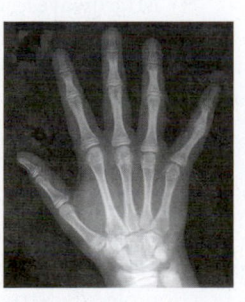

(a)   (b)   (c)

FIGURE 31.9 (a) Very Large Array radio telescope. (b) False color radar image of the surface of Venus. (c) X-ray image of a hand.

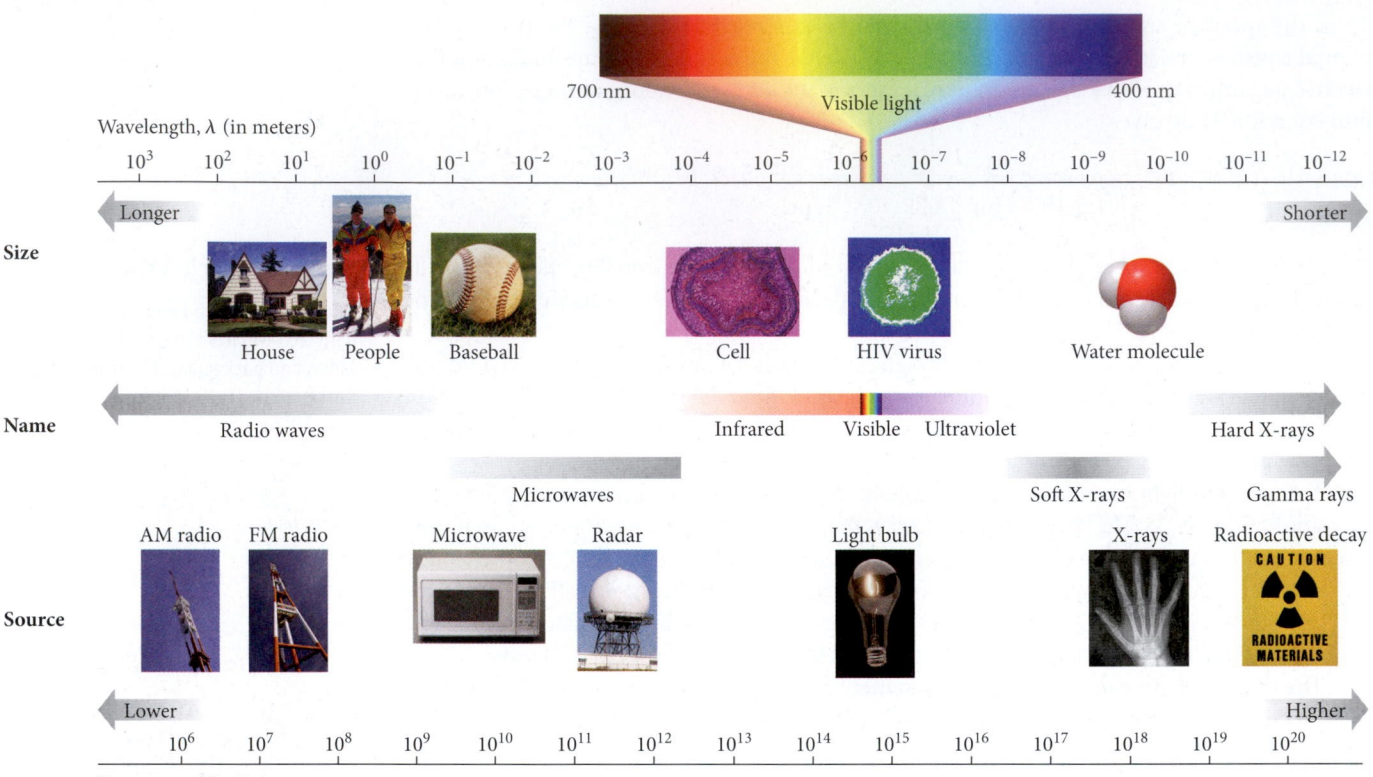

FIGURE 31.10 The electromagnetic spectrum.

Other wavelengths of electromagnetic waves are invisible to the human eye. However, we can detect them by other means.

- **Infrared waves**, with wavelengths just longer than visible light up to around 10^{-4} m, are felt as warmth. Detectors of infrared waves can be used to measure heat leaks in homes and offices, as well as locate brewing volcanoes. Many animals have developed the ability to see infrared waves, so they can see in the dark. Infrared beams are also used in automatic faucets in public restrooms and in remote control units for TV and DVD players.

- **Ultraviolet rays**, with wavelengths just shorter than visible light down to a few nanometers (10^{-9} m), can damage skin and cause sunburn. Fortunately, Earth's atmosphere, particularly its ozone layer, prevents most of the Sun's ultraviolet rays from reaching Earth's surface. Ultraviolet rays are used in hospitals to sterilize equipment and also produce optical properties such as fluorescence.

- **Radio waves** have frequencies ranging from several hundred kHz (AM radio) to 100 MHz (FM radio). They are also widely used in astronomy because they can pass through clouds of dust and gas that block visible light; the Very Large Array shown in Figure 31.9a is a collection of telescopes that utilize radio waves.

- **Microwaves,** used to pop popcorn in microwave ovens and transmit phone messages through relay towers or satellites, have frequencies around 10 GHz. Radar uses waves with wavelengths between those of radio waves and microwaves, which enable them to travel easily through the atmosphere and reflect off objects from the size of a baseball to the size of a storm cloud. Figure 31.9b shows a radar image of the surface of Venus, which is always obscured by clouds that block visible light.

- **X-rays** used to produce medical images, such the one shown in Figure 31.9c, have wavelengths on the order of 10^{-10} m. This length is about the same as the distance between atoms in a solid crystal, so X-rays are used to determine the detailed molecular structure of any material that can be crystallized.

Self-Test Opportunity 31.2

An FM radio station broadcasts at 90.5 MHz, and an AM radio station broadcasts at 870 kHz. What are the wavelengths of these electromagnetic waves?

■ **Gamma rays** emitted in the decay of radioactive nuclei have very short wavelengths, on the order of 10^{-12} m, and can cause damage to human cells. They are often used in medicine to destroy cancer cells or other malignant tissues that are hard to reach.

Communication Frequency Bands

The frequency ranges assigned to radio and television broadcasts are shown in Figure 31.11. The range of frequencies assigned to AM (amplitude modulation) radio stations is from 535 kHz to 1705 kHz. FM (frequency modulation) radio stations use the frequencies between 88.0 MHz and 108.0 MHz. VHF (very high frequency) television operates in two ranges: 54.0 MHz to 88.0 MHz for channels 2 through 6, and 174.0 MHz to 216.0 MHz for channels 7 through 13. UHF (ultra-high frequency) television channels 14 through 51 broadcast in the range from 470.0 MHz to 698.0 MHz. Most high-definition television (HDTV) broadcasts use the UHF band and channels 14 through 51.

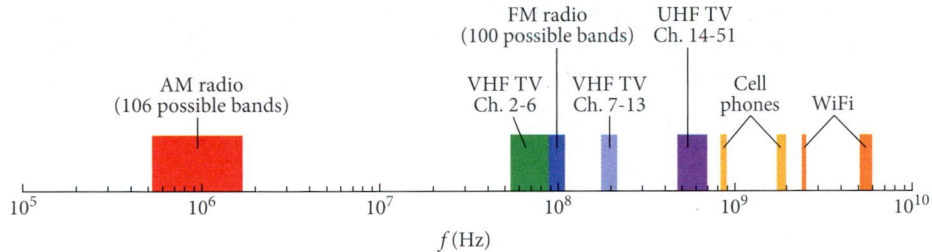

FIGURE 31.11 The frequency bands assigned to radio broadcasts, television broadcasts, cell phones, and WiFi computer connections in the United States.

A radio or television station transmits a carrier signal on a given frequency. The carrier signal is a sine wave with a frequency equal to the frequency of the broadcasting station. In the case of AM broadcasts, the amplitude of the carrier wave is modified by the information being transmitted, as illustrated in Figure 31.12a. The modulation of the amplitude of the carrier signal carries the transmitted message. Figure 31.12a shows a simple sine wave, indicating that a simple tone is being transmitted. The signal is received by a tuned RLC circuit whose resonant frequency is equal to the frequency of the carrier signal. The current induced in the circuit is proportional to the message being transmitted. AM transmission is vulnerable to noise and signal loss because the message is proportional to the amplitude of the signal, which can change if conditions vary.

For FM transmission, the frequency of the carrier signal is modified by the message to produce a modulated signal, as shown in Figure 31.12b. This type of transmission is much less affected by noise and signal loss because the message is extracted from frequency shifts of the carrier signal, rather than from changes in the amplitude of the carrier signal. FM radio receivers commonly use a *Foster-Seeley discriminator* to demodulate the FM signal.

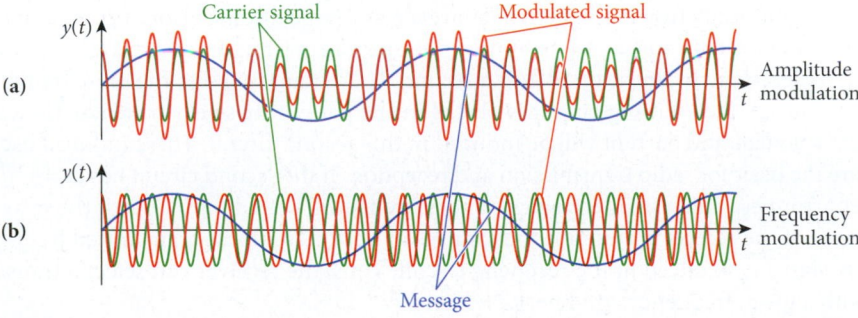

FIGURE 31.12 (a) Amplitude modulation. (b) Frequency modulation. For both cases, the green curve represents the carrier signal, the red curve represents the modulated signal, and the blue curve signifies the information being transmitted.

A Foster-Seeley discriminator uses an RLC circuit tuned to the frequency of the carrier signal and connected to two diodes, resembling the fullwave rectifier discussed in Chapter 30. If the input equals the carrier frequency, the two halves of the tuned circuit produce the same rectified voltage and the output is zero. As the frequency of the carrier signal changes, the balance between the two halves of the rectified circuit changes, resulting in a voltage proportional to the frequency deviation of the carrier signal. A Foster-Seeley discriminator is sensitive to amplitude variations and is usually coupled with a limiter amplifier stage to desensitize it to variations in the strength of the carrier wave by allowing lower power signals to pass through it unaffected while removing the peaks of the signals that exceed a certain power level.

HDTV transmitters broadcast information digitally in the form of zeros and ones. One byte of information contains eight bits, where a bit is a zero or a one. The screen is subdivided into picture elements (pixels) with digital representations of the red, green, and blue color of each pixel. Currently, the highest resolution for HDTV is 1080i, which has 1920 pixels in the horizontal direction and 1080 pixels in the vertical direction. Half the picture (every other horizontal line) is updated 60 times every second, and the two halves of the image are interlaced to form the complete image. (See the subsection "Applications of Polarization" in Section 31.6 for more information on video formats.) HDTV is broadcast using a compression-decompression (codec) technique, typically the standard known as MPEG-2, to reduce the amount of data that must be transmitted. A typical HDTV station broadcasts about 17 megabytes of information per second.

Cell phone transmissions occur in the frequency bands from 824 to 894 MHz and 1.85 to 1.99 GHz. WiFi data connections for computers operate in the ranges from 2.401 to 2.484 GHz (for the international standard; the U.S. standard band has an upper limit of 2.473 GHz) and from 5.15 to 5.85 GHz. These frequencies are in the microwave range, and some people worry about prolonged exposure to electromagnetic waves emitted by cell phones and WiFi. However, the relatively low power of these devices combined with the fact that the energy of these waves is much lower than that of other waves that are commonly encountered, such as visible light, argue that such exposure presents little danger. Chapter 37 on quantum mechanics will discuss the energy of the photons associated with electromagnetic waves.

Traveling Electromagnetic Waves

Subatomic processes can produce electromagnetic waves such as gamma rays, X-rays, and light. Electromagnetic waves can also be produced by an RLC circuit connected to an antenna (Figure 31.13). The connection between the RLC circuit and the antenna occurs through a transformer. A dipole antenna is used to approximate an electric dipole. The voltage and current in the antenna vary sinusoidally with time and cause the flow of charge in the antenna to oscillate with the frequency, ω_0, of the RLC circuit. The accelerating charges create **traveling electromagnetic waves.** These waves travel from the antenna at speed c and with frequency $f = \omega_0/(2\pi)$.

The traveling electromagnetic waves propagate as wave fronts spreading out spherically from the antenna. However, at a large distance from the antenna, the wave fronts appear to be almost flat, or planar. Thus, such a traveling wave is described by equation 31.8. If a second RLC circuit tuned to the same frequency, ω_0, as the emitting circuit is placed in the path of these electromagnetic waves, voltage and current will be induced in this second circuit. These induced oscillations are the basis for radio transmission and reception. If the second circuit has $\omega = 1/\sqrt{LC}$, different from ω_0, much smaller voltages and currents will be induced. Only if the resonant frequency of the receiving circuit is the same as or very close to the transmitted frequency will any signal be induced in the receiving circuit. Thus, the receiver can select a transmission with a given frequency and reject all others.

The principle of transmission of electromagnetic waves was discovered by Heinrich Hertz in 1888, as described in Section 31.1, and was used by the Italian physicist Guglielmo Marconi (1874–1937) to transmit wireless signals.

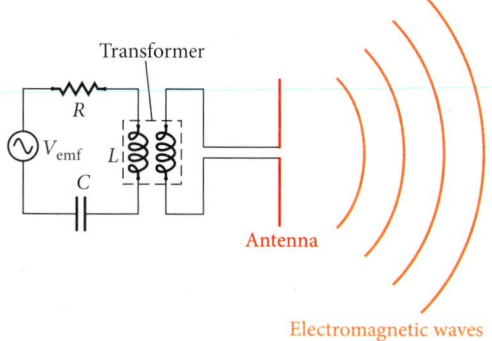

FIGURE 31.13 An RLC circuit coupled to an antenna that emits traveling electromagnetic waves.

31.4 Poynting Vector and Energy Transport

When you walk out into the sunlight, you feel warmth. If you stay out too long in the bright sunshine, you become sunburned. These phenomena are caused by electromagnetic waves emitted from the Sun. These electromagnetic waves carry energy that was generated in the nuclear reactions in the core of the Sun.

The rate at which energy is transported by an electromagnetic wave is usually defined in terms of a vector, $\vec{S}$, given by

$$\vec{S} = \frac{1}{\mu_0} \vec{E} \times \vec{B}. \qquad (31.23)$$

This quantity is called the **Poynting vector,** after British physicist John Poynting (1852–1914), who first discussed its properties. The magnitude of $\vec{S}$ is related to the instantaneous rate at which energy is transported by an electromagnetic wave over a given area, or more simply, the instantaneous power per unit area:

$$S = |\vec{S}| = \left(\frac{\text{power}}{\text{area}}\right)_{\text{instantaneous}}. \qquad (31.24)$$

The units of the Poynting vector are thus watts per square meter (W/m^2).

For an electromagnetic wave, where $\vec{E}$ is perpendicular to $\vec{B}$, equation 31.23 yields

$$S = \frac{1}{\mu_0} EB.$$

According to equation 31.15, the magnitudes of the electric field and the magnetic field are directly related via $E/B = c$. Thus, we can express the instantaneous power per unit area of an electromagnetic wave in terms of the magnitude of the electric field or that of the magnetic field. Since it is usually easier to measure an electric field than a magnetic field, the instantaneous power per unit area is given by

$$S = \frac{1}{c\mu_0} E^2. \qquad (31.25)$$

We can now substitute a sinusoidal form for the electric field, $E = E_{\text{max}} \sin(\kappa x - \omega t)$, and obtain an expression for the transmitted power per unit area. However, the usual means of describing the power per unit area in an electromagnetic wave is the intensity, I, of the wave, given by

$$I = S_{\text{ave}} = \left(\frac{\text{power}}{\text{area}}\right)_{\text{ave}} = \frac{1}{c\mu_0} \left[E_{\text{max}}^2 \sin^2\left(\kappa x - \omega t\right) \right]_{\text{ave}}.$$

The units of intensity are the same as the units of the Poynting vector, W/m^2. The time-averaged value of $\sin^2(\kappa x - \omega t)$ is $\frac{1}{2}$, so we can express the intensity as

$$I = \frac{1}{c\mu_0} E_{\text{rms}}^2, \qquad (31.26)$$

where $E_{\text{rms}} = E_{\text{max}}/\sqrt{2}$.

Because the magnitudes of the electric and magnetic fields of an electromagnetic wave are related by $E = cB$ and c is such a large number, you might conclude that the energy transported by the electric field is much larger than the energy transported by the magnetic field. Actually these energies are *the same.* To see this, recall from Chapters 24 and 29 that the energy density of an electric field is given by

$$u_E = \tfrac{1}{2}\epsilon_0 E^2,$$

and the energy density of a magnetic field is given by

$$u_B = \frac{1}{2\mu_0} B^2.$$

If we substitute $E = cB$ and $c = 1/\sqrt{\mu_0\epsilon_0}$ into the expression for the energy density of the electric field, we get

$$u_E = \tfrac{1}{2}\epsilon_0(cB)^2 = \tfrac{1}{2}\epsilon_0\left(\frac{B}{\sqrt{\mu_0\epsilon_0}}\right)^2 = \frac{1}{2\mu_0}B^2 = u_B. \tag{31.27}$$

Thus, the energy density of the electric field is the same as the energy density of the magnetic field everywhere in the electromagnetic wave.

(a)

(b)

FIGURE 31.14 (a) Photovoltaic solar panels mounted on the roof of a house. (b) A plug-in electric car capable of driving 37 miles on a single charge.

EXAMPLE 31.1 | Using Solar Panels to Charge an Electric Car

Suppose that photovoltaic (solar power to electric power) solar panels (Figure 31.14a) can be mounted on the roof of your house at a cost per area of $\eta = \$355/m^2$. You have an electric car (Figure 31.14b) that requires a charge corresponding to an energy of $U = 10.0$ kWh for a day of local driving. The solar panels convert solar power to electricity with an efficiency $\epsilon = 14.1\%$ and have an area A. Suppose that sunlight is incident on your solar panels for $\Delta t = 4.00$ h a day with an average intensity of $S_{ave} = 600.$ W/m^2.

PROBLEM
How much do you need to spend on solar panels to give your electric car its daily charge?

SOLUTION
We equate the total amount of energy produced by the solar panels to the energy required to charge the car:

$$U_{produced} = P\Delta t = U.$$

The amount of power produced by the solar panels is the average intensity of the sunlight times the area of the solar panels times the efficiency of the solar panels:

$$P = \epsilon A S_{ave}.$$

Thus, the total area required is

$$A = \frac{P}{\epsilon S_{ave}} = \frac{(U/\Delta t)}{\epsilon S_{ave}} = \frac{U}{\epsilon S_{ave}\Delta t}.$$

The total cost will then be

$$\text{Cost} = \eta A = \frac{\eta U}{\epsilon S_{ave}\Delta t}.$$

Putting in the numerical values gives us

$$\text{Cost} = \frac{\eta U}{\epsilon S_{ave}\Delta t} = \frac{(\$355/m^2)(10.0\text{ kWh})}{(0.141)(0.600\text{ kW/m}^2)(4.00\text{ h})} = \$10,500.$$

If you drove your electric car 37 miles per day every day for 10. years, that would work out to 7.8 cents per mile. In contrast, the cost would be 20. cents per mile for a gasoline-powered car with a 20. mpg rating and gas costing \$4.00 per gallon.

   This is not a large cost savings over the 10-year period (\$10,500 for solar versus \$27,000 for gasoline power) considering an electric car may cost more than a gasoline-powered car. However, your solar-powered electric car would be a completely carbon-neutral, zero-emission mode of transportation (the same as riding your bicycle, without the exercise benefits). Material scientists are working intensely to increase the efficiency of commercially available solar cells, and mass production has dramatically lowered the cost of solar panels in recent years (more than 70% since the first edition of this textbook was published) and is expected to continue to do so in the future. So solar-powered electric cars could soon be a viable and attractive alternative to gasoline-powered cars.

EXAMPLE 31.2 The Root-Mean-Square Electric and Magnetic Fields from Sunlight

The average intensity of sunlight at the Earth's surface is approximately 1400 W/m^2, if the Sun is directly overhead.

PROBLEM

What are the root-mean-square electric and magnetic fields of these electromagnetic waves?

SOLUTION

The intensity of sunlight can be related to the root-mean-square electric field using equation 31.26:

$$I = \frac{1}{c\mu_0} E_{rms}^2.$$

Solving for the root-mean-square electric field gives us

$$E_{rms} = \sqrt{Ic\mu_0} = \sqrt{\left(1400 \text{ W/m}^2\right)\left(3.00 \cdot 10^8 \text{ m/s}\right)\left(4\pi \cdot 10^{-7} \text{ T m/A}\right)}$$

$$= 730 \text{ V/m}.$$

In comparison, the root-mean-square electric field in a typical home is 5–10 V/m. Standing directly under an electric power transmission line, one would experience a root-mean-square electric field of 200–10,000 V/m depending on the conditions.

The root-mean-square magnetic field of sunlight is

$$B_{rms} = \frac{E_{rms}}{c} = \frac{730 \text{ V/m}}{3.00 \cdot 10^8 \text{ m/s}} = 2.4 \text{ μT}.$$

In comparison, the root-mean-square value of the Earth's magnetic field is 50 μT, the root-mean-square magnetic field found in a typical home is 0.5 μT, and the root-mean-square magnetic field under a power transmission line is 2 μT.

31.5 Radiation Pressure

When you walk out into the sunlight, you feel warmth, but you do not feel any force from the sunlight. Sunlight is exerting a pressure on you, but that pressure is so small that you cannot notice it. Because the electromagnetic waves making up sunlight are radiated from the Sun and travel to the Earth, they are referred to as **radiation.** Chapter 18 discussed radiation as one way to transfer heat. As we'll see in Chapter 40 on nuclear physics, this type of radiation is not necessarily the same as radioactive radiation resulting from the decay of unstable nuclei. However, radio waves, infrared waves, visible light, and X-rays are all fundamentally the same electromagnetic radiation. (Which is not to say, however, that all kinds of electromagnetic radiation have the same effect on the human body. For example, UV light can give you sunburn and even trigger skin cancer, whereas there is no credible evidence that radiation emitted from cell phones can cause cancer.)

Let's calculate the magnitude of the pressure exerted by these radiated electromagnetic waves. Electromagnetic waves carry energy, U, as shown in Section 31.4. Electromagnetic waves also have linear momentum, $\vec{p}$. This concept is subtle because electromagnetic waves have no mass, and we saw in Chapter 7 that momentum is equal to mass multiplied by velocity. Maxwell showed that if a plane wave of radiation is totally absorbed by a surface (perpendicular to the direction of the plane wave) for a time interval, Δt, and an amount of energy, ΔU, is absorbed by the surface in that process, then the magnitude of the momentum transferred to that surface by the wave in that time interval is

$$\Delta p = \frac{\Delta U}{c}.$$

Chapter 35 on relativity will show that this relationship between energy and momentum holds for massless objects; for now, it is stated as a fact, without proof.

The magnitude of the force on the surface is then $F = \Delta p/\Delta t$ (Newton's Second Law). The total energy, ΔU, absorbed by an area A of the surface during the time interval Δt is equal to the product of the area, the time interval, and the radiation intensity, I (introduced in Section 31.4): $\Delta U = IA\Delta t$. Therefore, the magnitude of the force exerted by the electromagnetic wave on this area is

$$F = \frac{\Delta p}{\Delta t} = \frac{\Delta U}{c\Delta t} = \frac{IA\Delta t}{c\Delta t} = \frac{IA}{c}.$$

Since pressure is defined as force (magnitude) per unit area, the radiation pressure, p_r, is

$$p_r = \frac{F}{A},$$

and, consequently,

$$p_r = \frac{I}{c} \quad \text{(for total absorption).} \tag{31.28}$$

Equation 31.28 states that the radiation pressure due to electromagnetic waves is simply the intensity divided by the speed of light, but only for the case of total absorption of the radiation by the surface.

The other limiting case is total reflection of the electromagnetic waves. In that case, the momentum transfer is *twice* as big as for total absorption, just like the momentum transfer from a ball to a wall is twice as big in a perfectly elastic collision as in a perfectly inelastic collision. In the perfectly elastic collision, the ball's initial momentum is reversed and $\Delta p = p_i - (-p_i) = 2p_i$, whereas for the totally inelastic collision, $\Delta p = p_i - 0 = p_i$, as explained in Chapter 7. So, the radiation pressure for the case of perfect reflection of the electromagnetic waves off a surface is

$$p_r = \frac{2I}{c} \quad \text{(for perfect reflection).} \tag{31.29}$$

The radiation pressure from sunlight is comparatively small. The intensity of sunlight at the Earth's surface is at most 1400 W/m², when the Sun is directly overhead and there are no clouds in the sky. (This can happen only between the Tropics of Cancer and Capricorn, located at ±23° latitude relative to the Equator.) Thus, the maximum radiation pressure for sunlight that is totally absorbed is

$$p_r = \frac{I}{c} = \frac{1400 \text{ W/m}^2}{3 \cdot 10^8 \text{ m/s}} = 4.67 \cdot 10^{-6} \text{ N/m}^2 = 4.67 \text{ }\mu\text{Pa}.$$

For comparison, atmospheric pressure is 101 kPa (see Chapter 13), which is greater than the sunlight's radiation pressure on the surface of Earth by more than a factor of 20 billion. Another useful comparison is the lowest pressure difference that human hearing can detect, which is generally quoted as approximately 20 μPa for sounds in the 1-kHz frequency range, where the human ear is most sensitive (see Chapter 16).

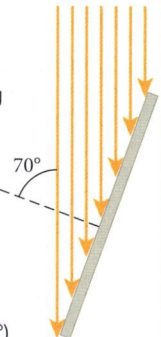

EXAMPLE 31.3 | **Radiation Pressure from a Laser Pointer**

A green laser pointer has a power of 1.00 mW. You shine the laser pointer perpendicularly on a mirror, which reflects the light. The spot of light on the mirror is 2.00 mm in diameter.

PROBLEM

What force does the light from the laser pointer exert on the mirror?

SOLUTION

The intensity of the light is given by

$$I = \frac{\text{power}}{\text{area}} = \frac{1.00 \cdot 10^{-3} \text{ W}}{\pi \left(1.00 \cdot 10^{-3} \text{ m}\right)^2} = 318 \text{ W/m}^2.$$

The radiation pressure for a perfectly reflecting surface is given by equation 31.29 and also is equal to the force exerted by the light divided by the area over which it acts:

$$p_r = \frac{\text{force}}{\text{area}} = \frac{2I}{c}.$$

Thus, the force exerted on the mirror is

$$\text{Force} = (\text{area})\left(\frac{2I}{c}\right) = \pi\left(1.0 \cdot 10^{-3} \text{ m}\right)^2 \frac{2\left(318 \text{ W/m}^2\right)}{3.00 \cdot 10^8 \text{ m/s}} = 6.66 \cdot 10^{-12} \text{ N}.$$

Self-Test Opportunity 31.3

Suppose you have a satellite in orbit around the Sun, as shown in the figure. The orbit is in the counterclockwise direction looking down on the north pole of the Sun. You want to deploy a solar sail consisting of a large, totally reflecting mirror, which can be oriented so that it is perpendicular to the light coming from the Sun or at an angle with respect to the light coming from the Sun. Describe the effect on the orbit of your satellite for the three deployment angles shown in the figure.

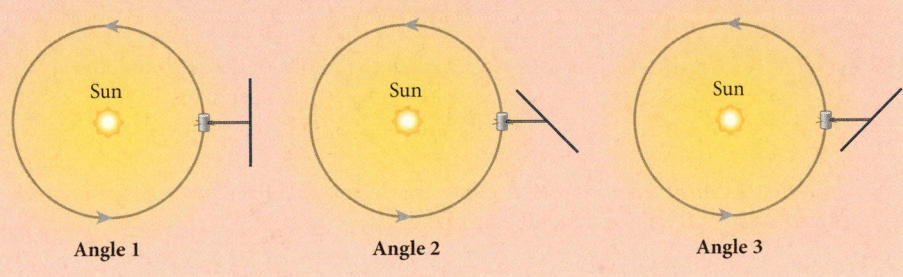

Angle 1   Angle 2   Angle 3

SOLVED PROBLEM 31.1 | Solar Stationary Satellite

Suppose researchers want to place a satellite that will remain stationary above the north pole of the Sun in order to study the Sun's long-term rotational characteristics. The satellite will have a totally reflecting solar sail and be located at a distance of $1.50 \cdot 10^{11}$ m from the center of the Sun. The intensity of sunlight at that distance is 1400 W/m^2. The plane of the solar sail is perpendicular to a line connecting the satellite and the center of the Sun. The mass of the satellite and sail is 100.0 kg.

PROBLEM
What is the required area of the solar sail?

SOLUTION
THINK In an equilibrium position for the satellite, the area of the solar sail times the radiation pressure from the Sun produces a force that is balanced by the gravitational force between the satellite and the Sun. We can equate these two forces and solve for the area of the solar sail.

SKETCH Figure 31.15 is a diagram of the satellite with a solar sail near the Sun.

RESEARCH The satellite will be stationary if the force of gravity, F_g, is balanced by the force from the radiation pressure of sunlight, F_{rp}:

$$F_g = F_{rp}.$$

The force corresponding to the radiation pressure from sunlight is equal to the radiation pressure, p_r, times the area of the solar sail, A:

$$F_{rp} = p_r A.$$

The radiation pressure can be expressed in terms of the intensity of the sunlight, I, incident on the totally reflecting solar sail:

$$p_r = \frac{2I}{c}.$$

The force of gravity between the satellite and the Sun is given by

$$F_g = G\frac{m m_{\text{Sun}}}{d^2},$$

– Continued

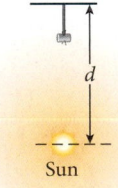

FIGURE 31.15 A satellite with a solar sail near the Sun.

where G is the universal gravitational constant, m is the mass of the satellite and sail, m_{Sun} is the mass of the Sun, and d is the distance between the satellite and the Sun.

SIMPLIFY We can combine all these equations to obtain

$$\left(\frac{2I}{c}\right)A = G\frac{mm_{Sun}}{d^2}.$$

Solving for the area of the solar sail gives us

$$A = G\frac{cmm_{Sun}}{2Id^2}.$$

CALCULATE Putting in the numerical values, we get

$$A = \left(6.67\cdot10^{-11}\ \text{m}^3\ \text{kg}^{-1}\ \text{s}^{-2}\right)\frac{\left(3.00\cdot10^8\ \text{m/s}\right)\left(100.0\ \text{kg}\right)\left(1.99\cdot10^{30}\ \text{kg}\right)}{2\left(1400\ \text{W/m}^2\right)\left(1.50\cdot10^{11}\ \text{m}\right)^2} = 63{,}206.2\ \text{m}^2.$$

ROUND We report our result to three significant figures:

$$A = 6.32\cdot10^4\ \text{m}^2.$$

DOUBLE-CHECK If the solar sail were circular, the radius of the sail would be

$$R = \sqrt{\frac{A}{\pi}} = \sqrt{\frac{6.32\cdot10^4\ \text{m}^2}{\pi}} = 142\ \text{m}.$$

We can relate the thickness of the sail, t, times the density, ρ, of the material from which the sail is constructed to the mass per unit area of the sail:

$$t\rho = \frac{m}{A}.$$

If the sail were composed of a sturdy material such as kapton ($\rho = 1420\ \text{kg/m}^3$) and had a mass of 75 kg, the thickness of the sail would be

$$t = \frac{75\ \text{kg}}{\left(1420\ \text{kg/m}^3\right)\left(6.32\cdot10^4\ \text{m}^2\right)} = 8.36\cdot10^{-7}\ \text{m} = 0.836\ \mu\text{m}.$$

Kapton is a polyimide film developed to remain stable in the wide range of temperatures found in space; from near absolute zero to over 600 K. Current production techniques cannot produce kapton this thin. However, the required areal mass density may be realizable using other materials in the future.

SOLVED PROBLEM 31.2 | Laser-Powered Sailing

One idea for propelling long-range spacecraft involves using a high-powered laser beam, rather than sunlight, focused on a large totally reflecting sail. The spacecraft could then be propelled from Earth. Suppose a 10.0-GW laser could be focused at long distances. The spacecraft has a mass of 200.0 kg, and its reflecting sail is large enough to intercept all of the light emitted by the laser.

PROBLEM
Neglecting gravity, how long would it take the spacecraft to reach a speed of 30.0% of the speed of light, starting from rest?

SOLUTION
THINK The radiation pressure from the laser produces a constant force on the spacecraft's sail, resulting in a constant acceleration. Using the constant acceleration, we can calculate the time to reach the final speed starting from rest.

SKETCH Figure 31.16 is a diagram of a laser focusing light on the spacecraft with a totally reflecting sail.

FIGURE 31.16 A laser focusing its light on a spacecraft with a totally reflecting sail.

RESEARCH The radiation pressure, p_r, from the light with intensity I produced by the laser is

$$p_r = \frac{2I}{c}.$$

Pressure is defined as force, F, per unit area, A, of the spot the beam produces on the sail, so we can write

$$\frac{2I}{c} = \frac{F}{A}.$$

The intensity of the laser is given by the power, P, of the laser divided by the spot's area, A. Assuming that the sail of the spacecraft can intercept the entire laser beam, we can write

$$\frac{F}{A} = \frac{2(P/A)}{c}.$$

Solving for the force exerted by the laser beam on the sail and using Newton's Second Law, we can write

$$F = \frac{2P}{c} = ma. \tag{i}$$

SIMPLIFY We can solve equation (i) for the acceleration:

$$a = \frac{2P}{mc}.$$

Assuming that all the power of the laser remains focused on the sail of the spacecraft, the spacecraft will experience a constant acceleration. Then, the final speed, v, of the spacecraft can be related to the time it takes to reach that speed through

$$v = at = 0.300c.$$

Solving for the time gives us

$$t = \frac{0.300c}{2P/mc} = \frac{0.300mc^2}{2P}.$$

CALCULATE Putting in the numerical values gives us

$$t = \frac{0.300mc^2}{2P} = \frac{0.300(200.0 \text{ kg})(3.00 \cdot 10^8 \text{ m/s})^2}{2(10.0 \cdot 10^9 \text{ W})} = 270{,}000{,}000 \text{ s}.$$

ROUND We report our result to three significant figures:

$$t = 270{,}000{,}000 \text{ s} = 8.56 \text{ yr}.$$

DOUBLE-CHECK To double-check our result, we calculate the acceleration of the spacecraft:

$$a = \frac{2P}{mc} = \frac{2(10.0 \cdot 10^9 \text{ W})}{(200 \text{ kg})(3.00 \cdot 10^8 \text{ m/s})} = 0.333 \text{ m/s}^2.$$

This acceleration is 3% of the acceleration due to gravity at the surface of the Earth. This acceleration is produced by a laser with 10 times the power of a typical power station, which must run continuously for 8.56 yr. The distance the spacecraft will travel during that time is

$$x = \tfrac{1}{2}at^2 = \tfrac{1}{2}(0.333 \text{ m/s}^2)(2.70 \cdot 10^8 \text{ s})^2 = 1.21 \cdot 10^{16} \text{ m} = 1.28 \text{ light-years},$$

which is slightly more than $1\frac{1}{4}$ times the distance light travels in a year. The laser must remain focused on the spacecraft at this distance. Thus, although our calculations seem reasonable, the requirements for a laser-driven spacecraft with a reflecting sail seem difficult to achieve. In Chapter 35, we'll see that we must modify this calculation because the speed involved is a significant fraction of the speed of light.

31.6 Polarization

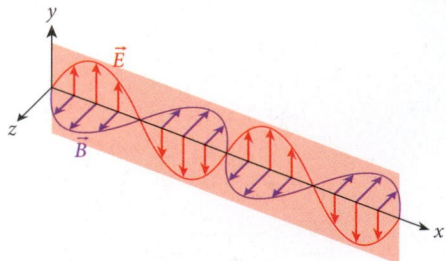

FIGURE 31.17 An electromagnetic wave with the plane of oscillation of the electric field shown in pink.

For the electromagnetic wave represented in Figure 31.5, the electric field always points along the y-axis. The direction in which the wave is traveling is the positive x-direction, so the electric field of the electromagnetic wave lies within a plane of oscillation (Figure 31.17).

We can visualize the polarization of an electromagnetic wave by looking at the electric field vector of the wave in the yz-plane, which is perpendicular to the direction in which the wave is traveling (Figure 31.18a). The electric field vector changes from the positive y-direction to the negative y-direction and back again as the wave travels. The electric field of the wave oscillates in the y-direction only, never changing its orientation. This type of wave is called a **plane-polarized wave** in the y-direction.

The electromagnetic waves making up the light emitted by most common light sources, such as the Sun or an incandescent light bulb, have random polarizations. Each wave has its electric field vector oscillating in a different plane. Such light is called **unpolarized light.** Light from an unpolarized source can be represented by many vectors like the ones shown in Figure 31.18a, but with random orientations (Figure 31.18b). Unpolarized light can also be represented by summing the y-components and the z-components separately to produce the net y- and z-components. Unpolarized light has equal components in the y- and z-directions (Figure 31.19a). If there is less net polarization in the y-direction than in the z-direction, then the light is said to be partially polarized in the z-direction (Figure 31.19b).

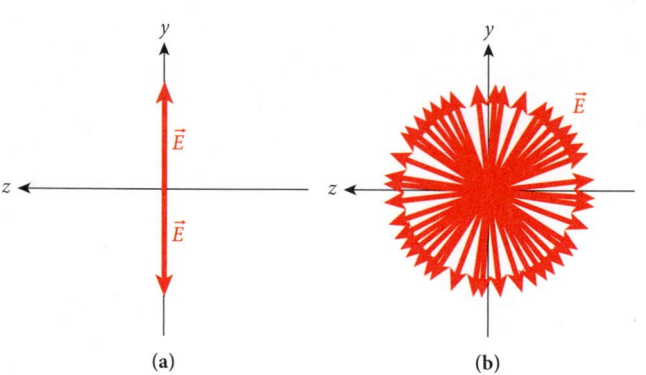

FIGURE 31.18 (a) Electric field vectors in the yz-plane, defining the plane of polarization to be the xy-plane. (b) Electric field vectors oriented at random angles.

Unpolarized light can be transformed to polarized light by passing the unpolarized light through a polarizer. A **polarizer** allows only one component of the electric field vectors of the light waves to pass through. One way to make a polarizer is to produce a material that consists of long parallel chains of molecules. This discussion will not go into the details of the molecular structure but will simply characterize each polarizer by a polarizing direction. Unpolarized light passing through a polarizer emerges polarized in the polarizing direction (Figure 31.20). The components of the light that have the same direction as the polarizer are transmitted, but the components of the light that are perpendicular to the polarizer are absorbed.

Now let's consider the intensity of the light that passes through a polarizer. Unpolarized light with intensity I_0 has equal components in the y- and z-directions. After passing through

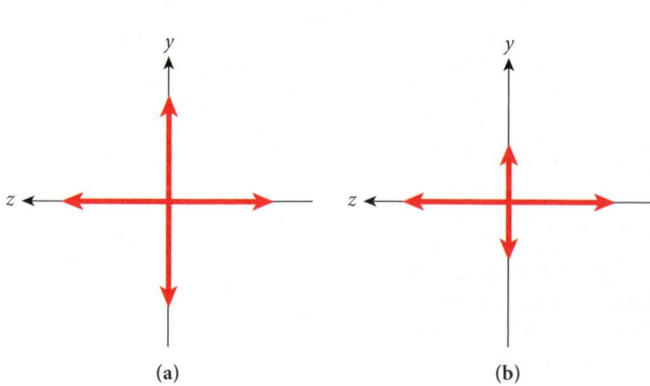

FIGURE 31.19 (a) Net components of the electric field for unpolarized light. (b) Net components of the electric field for partially polarized light.

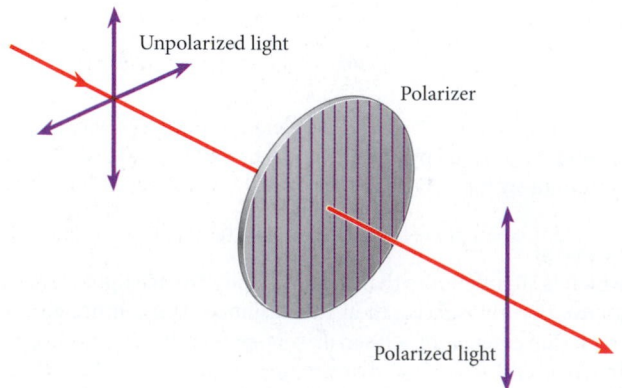

FIGURE 31.20 Unpolarized light passing through a vertical polarizer. After the light passes through the polarizer, it is vertically polarized.

a vertical polarizer, only the y-component (or vertical component) remains. The intensity, I, of the light that passes through the polarizer is given by

$$I = \tfrac{1}{2} I_0, \tag{31.30}$$

because the unpolarized light had equal contributions from the y- and z-components and only the y-components are transmitted by the polarizer. The factor $\tfrac{1}{2}$ applies only to the case of unpolarized light passing through a polarizer.

Let's consider polarized light passing through a polarizer (Figure 31.21). If the polarizer axis is parallel to the polarization direction of the incident polarized light, all of the light will be transmitted with the original polarization (Figure 31.21a). If the polarizing angle of the polarizer is perpendicular to the polarization of polarized light, no light will be transmitted (Figure 31.21b).

What happens when polarized light is incident on a polarizer and the polarization of the light is neither parallel nor perpendicular to the polarizing angle of the polarizer (Figure 31.22)? Let's assume that the angle between the incident polarized light and the polarizing angle is θ. The magnitude of the transmitted electric field, E, is given by

$$E = E_0 \cos\theta,$$

where E_0 is the magnitude of the electric field of the incident polarized light. From equation 31.26, we can see that the intensity of the light before passing through the polarizer, I_0, is given by

$$I_0 = \frac{1}{c\mu_0} E_{rms}^2 = \frac{1}{2c\mu_0} E_0^2.$$

After the light passes through the polarizer, the intensity, I, is given by

$$I = \frac{1}{2c\mu_0} E^2.$$

We can express the transmitted intensity in terms of the initial intensity as follows:

$$I = \frac{1}{2c\mu_0} E^2 = \frac{1}{2c\mu_0}\left(E_0 \cos\theta\right)^2 = I_0 \cos^2\theta. \tag{31.31}$$

This equation is called the **Law of Malus.** It applies only to polarized light incident on a polarizer.

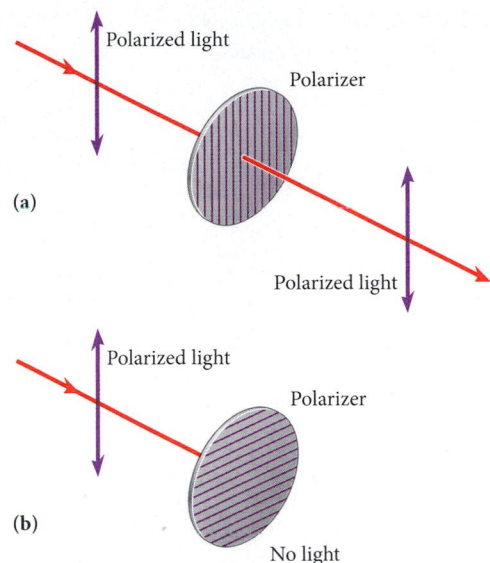

FIGURE 31.21 (a) Vertically polarized light incident on a vertical polarizer. (b) Vertically polarized light incident on a horizontal polarizer.

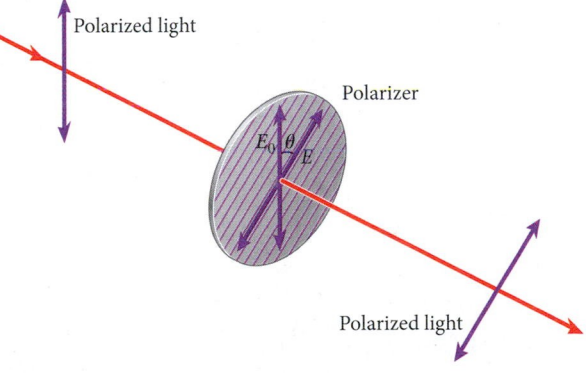

FIGURE 31.22 Polarized light passing through a polarizer whose polarizing angle is neither parallel nor perpendicular to the polarization of the incident light.

EXAMPLE 31.4   Three Polarizers

Suppose that unpolarized light with intensity I_0 is initially incident on the first of three polarizers in a line. The first polarizer has a polarizing direction that is vertical. The second polarizer has a polarizing angle of 45.0° with respect to the vertical. The third polarizer has a polarizing angle of 90.0° with respect to the vertical.

PROBLEM
What is the intensity of the light after passing through all three polarizers, in terms of the initial intensity?

SOLUTION
Figure 31.23 illustrates the light passing through the three polarizers. The intensity of the unpolarized light is I_0. The intensity of the light after passing through the first polarizer is

$$I_1 = \tfrac{1}{2} I_0.$$

The intensity of the light after passing through the second polarizer is

$$I_2 = I_1 \cos^2\left(45° - 0°\right) = I_1 \cos^2 45° = \tfrac{1}{2} I_0 \cos^2 45°.$$

– *Continued*

FIGURE 31.23 Unpolarized light passing through three polarizers.

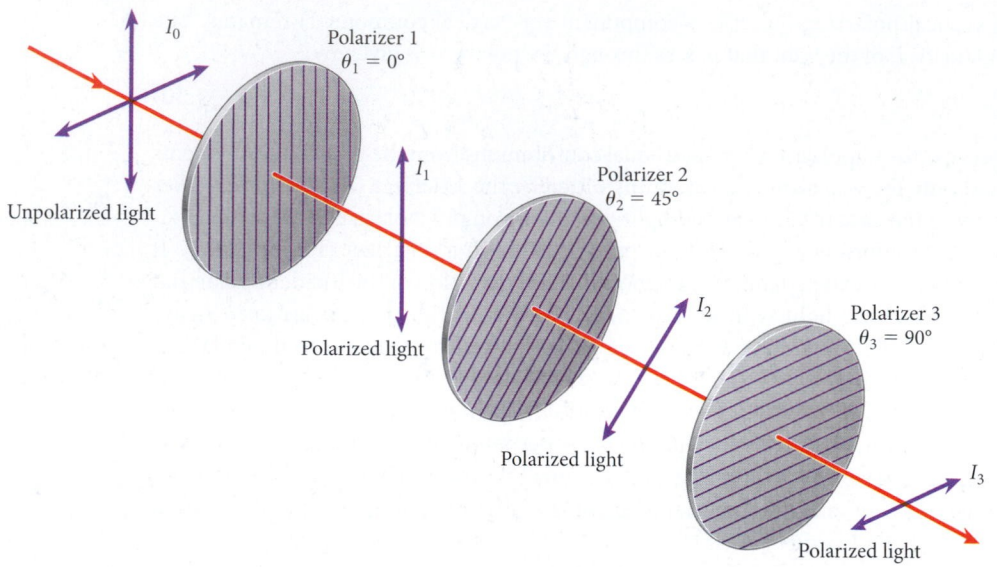

The intensity of the light after passing through the third polarizer is

$$I_3 = I_2 \cos^2\left(90° - 45°\right) = I_2 \cos^2 45° = \tfrac{1}{2} I_0 \cos^4 45°,$$

or $I_3 = I_0/8$.

The fact that $\tfrac{1}{8}$ of the light's initial intensity is transmitted is somewhat surprising because polarizers 1 and 3 have polarizing angles that are perpendicular to each other. Acting by themselves, polarizers 1 and 3 would block all of the light. Yet by adding an additional obstacle (polarizer 2) between these two polarizers, $\tfrac{1}{8}$ of the original intensity gets through. A series of polarizers with small differences in their polarizing angles can thus be used to rotate the polarization direction of light with only modest losses in intensity.

Concept Check 31.7

The figure shows unpolarized light incident on polarizer 1 with polarizing angle $\theta_1 = 0°$ and then on polarizer 2 with polarizing angle $\theta_2 = 90°$, which results in no light passing through. If polarizer 3 with polarizing angle $\theta_3 = 50°$ is placed between polarizers 1 and 2, which of the following statements is true?

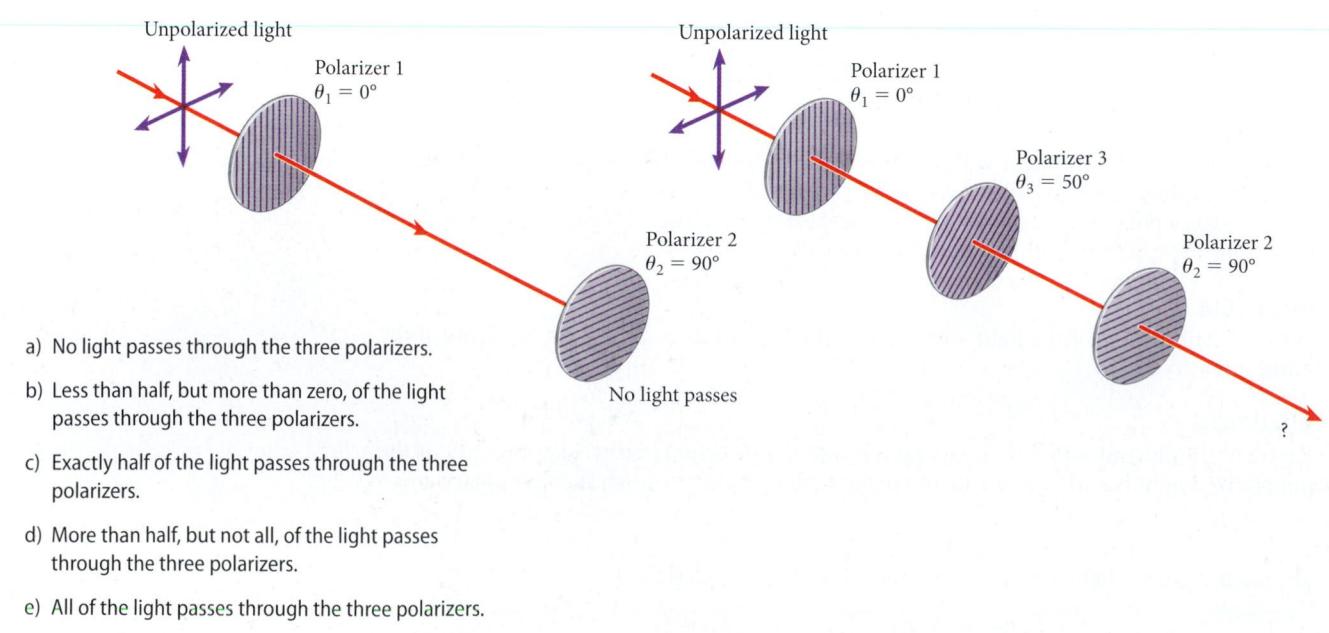

a) No light passes through the three polarizers.

b) Less than half, but more than zero, of the light passes through the three polarizers.

c) Exactly half of the light passes through the three polarizers.

d) More than half, but not all, of the light passes through the three polarizers.

e) All of the light passes through the three polarizers.

SOLVED PROBLEM 31.3 | Multiple Polarizers

Suppose you have light polarized in the vertical direction and want to rotate the polarization to the horizontal direction ($\theta = 90.0°$). If you pass the vertically polarized light through a polarizer whose polarizing angle is horizontal, all the light will be blocked. If, instead, you use a series of ten polarizers, each of which has a polarizing angle θ that is 9.00° more than that of the preceding one, with the first polarizer having $\theta = 9.00°$, you can rotate the polarization by 90.0° and still have light passing through.

PROBLEM
What fraction of the intensity of the incident light is transmitted through the ten polarizers?

SOLUTION
THINK Each polarizer is rotated 9.00° from the preceding polarizer. Thus, each polarizer transmits a fraction of the intensity equal to $f = \cos^2 9°$. The fraction transmitted is then f^{10}.

SKETCH Figure 31.24 shows the direction of the initial polarization and the orientations of the ten polarizers.

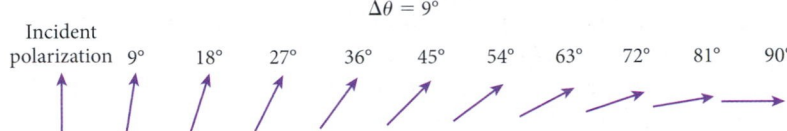

FIGURE 31.24 The direction of the polarization of the incident light and the direction of the polarizing angles of ten polarizers.

RESEARCH The intensity, I, of polarized light passing through a polarizer whose polarizing direction makes an angle θ with the polarization of the incident light is given by

$$I = I_0 \cos^2\theta,$$

where I_0 is the intensity of the incident polarized light. In this case, the polarizing direction of each successive polarizer is rotated $\Delta\theta = 9°$ relative to the polarizing direction of the preceding polarizer. Thus, each polarizer reduces the intensity by the factor

$$\frac{I}{I_0} = \cos^2\Delta\theta.$$

SIMPLIFY The reduction in intensity after the light has passed through ten polarizers, each with its polarization direction differing from that of the preceding polarizer by $\Delta\theta$ is

$$\frac{I_{10}}{I_0} = \left(\cos^2\Delta\theta\right)^{10}.$$

CALCULATE Putting in the numerical values, we get

$$\frac{I_{10}}{I_0} = \left(\cos^2 9°\right)^{10} = 0.780546.$$

ROUND We report our result to three significant figures:

$$\frac{I_{10}}{I_0} = 0.781 = 78.1\%.$$

DOUBLE-CHECK Using ten polarizers, each rotated by 9° more than the preceding one, the polarization of the incident polarized light was rotated by 90° and 78.1% of the light was transmitted, whereas using one polarizer rotated by 90° would have blocked all of the incident light. To see if our answer is reasonable, let's assume that instead of ten polarizers, we use n polarizers, each rotated by an angle $\Delta\theta = \theta_{max}/n$, where $\theta_{max} = 90°$. For each polarizer, the angle between its polarization direction and that of the preceding polarizer is small, so we can use the small-angle approximation for $\cos\Delta\theta$ to write

$$\frac{I_1}{I_0} \approx \left(1 - \frac{(\Delta\theta)^2}{2}\right)^2.$$

– Continued

The intensity of the light that passes through the n polarizers is then

$$\frac{I_n}{I_0} \approx \left(1 - \frac{(\theta_{max}/n)^2}{2}\right)^{2n} = \left(1 - \frac{\theta_{max}^2}{2n^2}\right)^{2n}.$$

For large n,

$$\frac{I_n}{I_0} \approx 1 = 100\%.$$

Using ten polarizers to rotate the polarization of the incident polarized light allowed 78.1 % of the light to pass. Using more polarizers with smaller changes in the polarization direction would allow the transmission to approach 100%. Thus, our result seems reasonable.

Applications of Polarization

Polarization has many practical applications. Sunglasses often have a polarized coating that blocks reflected light, which is usually polarized. A computer's or television's liquid crystal display (LCD) has an array of liquid crystals sandwiched between two polarizers whose polarizing angles are rotated 90° with respect to each other. Normally, the liquid crystal rotates the polarization of the light between the two polarizers so that light passes through. An array of addressable electrodes applies a varying voltage across each of the liquid crystals, causing the liquid crystals to rotate the polarization less, darkening the area covered by the electrode. The television or computer monitor screen can then display a large number of picture elements, or *pixels*, that produce a high-resolution image.

Figure 31.25a shows a top view of the layers of an LCD screen. Unpolarized light is emitted by a backlight. This light passes through a vertical polarizer. The polarized light then passes through a transparent layer of conducting pixel pads. These pads are designed to put varying amounts of voltage across the next layer, which is composed of liquid crystals, with respect to the transparent common electrode. If no voltage is applied across the liquid crystals, they rotate the polarization of the incident light by 90°. This light with rotated polarization can then pass through the transparent common electrode, the color filter, the horizontal polarizer, and the screen cover. When voltage of varying magnitude is applied to the pixel pad, the liquid crystals rotate the polarization of the incident light by a varying amount. When the full voltage is applied to the pixel pad, the polarization of the incident light is not rotated, and the horizontal polarizer blocks any light transmitted through the transparent common electrode and the color filter.

Figure 31.25b shows a front view of a small segment of the LCD screen, illustrating how the screen produces an image. The image is created by an array of pixels. Each pixel is subdivided into three subpixels: one red, one green, and one blue. By varying the voltage across each subpixel, a superposition of red, green, and blue light is created, producing a color on that pixel. It is difficult to connect a single wire to each subpixel, however. A high-definition 1080p LCD screen has 1080 times 1920 times 3 subpixels,

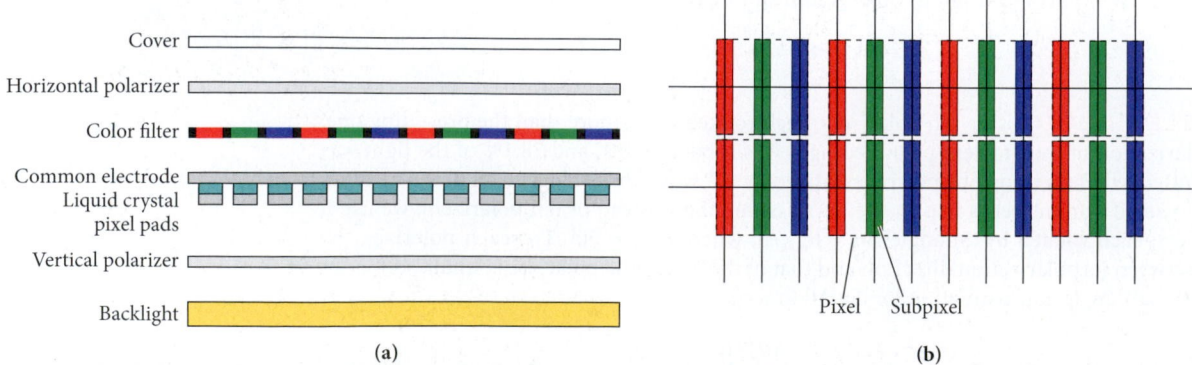

FIGURE 31.25 (a) Top view of the layers making up an LCD screen. (b) Front view of a subset of pixels and subpixels on an LCD screen.

or 6,220,800 subpixels. These subpixels are connected in columns and rows, as shown in Figure 31.25b. To turn on a subpixel, a voltage from both a column and a row must be applied. Thus, the subpixels are turned on one row at a time. With the voltage for one row on, the voltages for the subpixels in the desired columns are turned on. A small capacitor holds the voltage until the row is turned on again.

A high-definition 1080p LCD screen is scanned 60 times a second, producing a complete image in each scan. On a high-definition 1080i screen, every other row of the image is scanned 60 times a second and then the two images are interlaced. Another high-definition standard is 720p, which scans 720 rows 60 times a second with a horizontal resolution of 1280 pixels. The 720p and 1080i standards are in common use in television broadcasting. The standard resolution for a television image is 480i, with every other row being updated 60 times a second and producing 640 columns of pixels.

Viewing a 3D movie also involves the use of polarization filters. Moviegoers are equipped with glasses that have different polarization filters built into each lens. The projection equipment generates two different images with different polarizations on the screen. These images are also slightly offset from each other, and the viewer's brain constructs the 3D illusion by combining the two offset images. Older projection systems use linear polarization filters with polarization directions perpendicular to each other. However, this type of system yields diminished 3D effects when viewers tilt their heads sideways.

Modern 3D movie projection equipment uses circular polarization filters. Circular polarized light comes in two orthogonal varieties, left-circular and right-circular. Circular polarization filters work just like linear polarization filters, subject to a corresponding version of the Law of Malus. The 3D viewing glasses used with this modern equipment have one lens that passes left-circular and one that passes right-circular polarized light. This type of glasses produces 3D effects that are not affected when viewers tilt their heads.

There is one important difference between linearly polarized and circular polarized light: Linear polarized light remains in the same state of polarization after being reflected by a mirror, whereas circular polarized light changes its state from left-circular polarized to right-circular polarized, or vice versa. Figure 31.26 shows an interesting experiment you can do yourself: Hold a pair of 3D viewing glasses with circular polarization in front of a mirror and take a picture through one of the lenses. You can see that the light passing through the left lens, reflecting off the mirror, and then passing through the same lens again gets totally attenuated. In contrast, the light passing through the left lens, reflecting off the mirror, and then passing through the right lens is transmitted without being attenuated. If 3D glasses with linear polarization filters were used for this experiment instead, the light would pass through the lenses at half of its original intensity (equation 31.30), and the entire image of the camera phone could be seen in the mirror.

FIGURE 31.26 A camera phone takes a picture of its reflection in a mirror through 3D viewing glasses with circular polarization filters. One lens of the glasses polarizes light in a left-circular fashion, and the other polarizes it in a right-circular fashion.

31.7 Derivation of the Wave Equation

Table 31.1 lists the four equations known as Maxwell's equations in integral form. There are also equivalent differential versions of these equations, which is the way they are usually printed on T-shirts and posters:

$$\vec{\nabla}\cdot\vec{E} = \frac{\rho}{\epsilon_0},$$

$$\vec{\nabla}\times\vec{E} = -\frac{\partial}{\partial t}\vec{B},$$

$$\vec{\nabla}\cdot\vec{B} = 0,$$

and

$$\vec{\nabla}\times\vec{B} = \epsilon_0\mu_0\frac{\partial}{\partial t}\vec{E} + \mu_0\vec{j},$$

where ρ is the charge density (charge q per unit volume) and $\vec{j}$ is the current density. In vacuum and in the absence of charges, both are zero; $\rho = 0$ and $\vec{j} = 0$. The symbol $\vec{\nabla}$ is the gradient operator, which represents the vector with the partial derivatives in each spatial direction. In Cartesian coordinates, it is $\vec{\nabla} = (\partial/\partial x, \partial/\partial y, \partial/\partial z)$.

In Section 31.2, we saw that electromagnetic waves as described by equation 31.8 are valid solutions to all the Maxwell equations in vacuum. However, strictly speaking, we have not yet written the wave equation that the electric field and the magnetic field obey. Now, with the aid of the differential form of Maxwell's equations, we can derive the wave equation for the electric field, which is

$$\frac{\partial^2}{\partial t^2}\vec{E} - c^2\nabla^2\vec{E} = 0. \qquad (31.32)$$

The wave equation for the magnetic field is

$$\frac{\partial^2}{\partial t^2}\vec{B} - c^2\nabla^2\vec{B} = 0. \qquad (31.33)$$

DERIVATION 31.1 | Wave Equation for the Electric Field in Vacuum

To derive the wave equation for the electric field in vacuum, we take the vector product of the second Maxwell equation and the gradient operator, $\vec{\nabla}$:

$$\vec{\nabla}\times\vec{\nabla}\times\vec{E} = -\vec{\nabla}\times\frac{\partial}{\partial t}\vec{B}. \qquad (i)$$

On the right-hand side of equation (i), we can interchange the order of the time derivative and the spatial derivative:

$$-\vec{\nabla}\times\frac{\partial}{\partial t}\vec{B} = -\frac{\partial}{\partial t}(\vec{\nabla}\times\vec{B}) = -\frac{\partial}{\partial t}\left(\epsilon_0\mu_0\frac{\partial\vec{E}}{\partial t}\right) = -\epsilon_0\mu_0\frac{\partial^2}{\partial t^2}\vec{E}. \qquad (ii)$$

The second step of this transformation makes use of the third Maxwell equation with $\vec{j} = 0$, which is appropriate in vacuum. The left-hand side of equation (i) is a double vector product. Chapter 1 introduced the BAC-CAB rule for double vector products: $\vec{A}\times(\vec{B}\times\vec{C}) = \vec{B}(\vec{A}\cdot\vec{C}) - \vec{C}(\vec{A}\cdot\vec{B})$. Applying this rule, we find

$$\vec{\nabla}\times\vec{\nabla}\times\vec{E} = \vec{\nabla}(\vec{\nabla}\cdot\vec{E}) - \nabla^2\vec{E} = -\nabla^2\vec{E}, \qquad (iii)$$

where the first Maxwell equation $\left(\text{in vacuum: } \vec{\nabla}\cdot\vec{E} = 0\right)$ is used in the second step. The symbol ∇^2 is the scalar product of the gradient operator with itself: $\nabla^2 = \partial^2/\partial x^2 + \partial^2/\partial y^2 + \partial^2/\partial z^2$. If we substitute from equations (ii) and (iii) into equation (i) and use the fact that the speed of light is $c = 1/\sqrt{\mu_0\epsilon_0}$ (equation 31.20) we obtain the desired wave equation:

$$\frac{\partial^2}{\partial t^2}\vec{E} - \frac{1}{\epsilon_0\mu_0}\nabla^2\vec{E} = \frac{\partial^2}{\partial t^2}\vec{E} - c^2\nabla^2\vec{E} = 0.$$

This implies that electromagnetic waves moving at the speed of light are indeed a solution of Maxwell's equations, as discussed (but not exactly proven) in Section 31.2.

Self-Test Opportunity 31.4

Derive the wave equation for the magnetic field (equation 31.33) in the same way as Derivation 31.1 handles the wave equation for the electric field.

Self-Test Opportunity 31.5

Show that $\vec{E}(\vec{r},t) = E_{max}\sin(\kappa x - \omega t)\hat{y}$ and $\vec{B}(\vec{r},t) = B_{max}\sin(\kappa x - \omega t)\hat{z}$ are indeed solutions of the wave equation for the electric and magnetic fields.

WHAT WE HAVE LEARNED | EXAM STUDY GUIDE

- When a capacitor is being charged, a displacement current can be visualized between the plates, given by $i_d = \epsilon_0 d\Phi_E/dt$, where Φ_E is the electric flux.

- Maxwell's equations describe how electrical charges, currents, electric fields, and magnetic fields affect each other, forming a unified theory of electromagnetism.

  - Gauss's Law for Electric Fields, $\oiint \vec{E} \cdot d\vec{A} = q_{enc}/\epsilon_0$, relates the net electric flux through a closed surface to the net enclosed electric charge.

  - Gauss's Law for Magnetic Fields, $\oiint \vec{B} \cdot d\vec{A} = 0$, states that the net magnetic flux through any closed surface is zero.

  - Faraday's Law of Induction, $\oint \vec{E} \cdot d\vec{s} = -d\Phi_B/dt$, relates the induced electric field to the changing magnetic flux.

  - The Maxwell-Ampere Law, $\oint \vec{B} \cdot d\vec{s} = \mu_0\epsilon_0 d\Phi_E/dt + \mu_0 i_{enc}$, relates the induced magnetic field to the changing electric flux and to the current.

- For an electromagnetic wave traveling in the positive x-direction, the electric and magnetic fields can be described by $\vec{E}(\vec{r},t) = E_{max} \sin(\kappa x - \omega t)\hat{y}$ and $\vec{B}(\vec{r},t) = B_{max} \sin(\kappa x - \omega t)\hat{z}$, where $\kappa = 2\pi/\lambda$ is the wave number and $\omega = 2\pi f$ is the angular frequency.

- The magnitudes of the electric and magnetic fields of an electromagnetic wave at any fixed time and place are related by the speed of light, $E = cB$.

- The speed of light can be related to the two basic electromagnetic constants: $c = 1/\sqrt{\mu_0\epsilon_0}$.

- The instantaneous power per unit area carried by an electromagnetic wave is the magnitude of the Poynting vector, $S = [1/(c\mu_0)]E^2$, where E is the magnitude of the electric field.

- The intensity of an electromagnetic wave is defined as the average power per unit area carried by the wave, $I = S_{ave} = [1/(c\mu_0)]E_{rms}^2$, where E_{rms} is the root-mean-square magnitude of the electric field.

- For an electromagnetic wave, the energy density carried by the electric field is $u_E = \frac{1}{2}\epsilon_0 E^2$, and the energy density carried by the magnetic field is $u_B = [1/(2\mu_0)]B^2$. For any such wave, $u_E = u_B$.

- The radiation pressure exerted by electromagnetic waves of intensity I is given by $p_r = I/c$ if the electromagnetic waves are totally absorbed or by $p_r = 2I/c$ if the waves are perfectly reflected.

- The polarization of an electromagnetic wave is given by the direction of the electric field vector.

- The intensity of unpolarized light that has passed through a polarizer is $I = I_0/2$, where I_0 is the intensity of the unpolarized light incident on the polarizer.

- The intensity of polarized light that has passed through a polarizer is $I = I_0 \cos^2\theta$, where I_0 is the intensity of the polarized light incident on the polarizer and θ is the angle between the polarization of the incident polarized light and the polarizing angle of the polarizer.

ANSWERS TO SELF-TEST OPPORTUNITIES

31.1 $t = \dfrac{d}{c} = \dfrac{8.30 \cdot 10^{16} \text{ m}}{3.00 \cdot 10^8 \text{ m/s}} = 2.77 \cdot 10^8 \text{ s} = 8.77 \text{ yr.}$

31.2 $c = \lambda f \Rightarrow \lambda = \dfrac{c}{f}$

$\lambda_{FM} = \dfrac{3.00 \cdot 10^8 \text{ m}}{90.5 \cdot 10^6 \text{ Hz}} = 3.31 \text{ m}$

$\lambda_{AM} = \dfrac{3.00 \cdot 10^8 \text{ m}}{870 \cdot 10^3 \text{ Hz}} = 345 \text{ m.}$

31.3 Deployment angle 1 will produce an elliptical orbit with the Sun at one focus. The force from radiation pressure depends on the inverse square of the distance, just as the force of gravity does. Thus, the orbit will become an ellipse, just as if the mass of the Sun or the

mass of the object were suddenly reduced slightly. Because the force is perpendicular to the velocity of the satellite, the energy of the satellite is not affected.

Deployment angle 2 will result in a growing orbit. The resulting force from the reflected light produces a component of force that is in the same direction as the velocity of the spacecraft. Thus, the spacecraft gains energy, and the radius of the orbit increases. Note that the speed of the spacecraft decreases but its total energy increases.

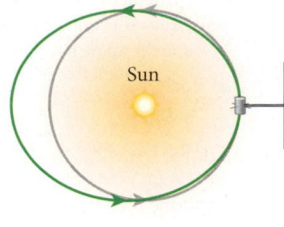

Angle 1

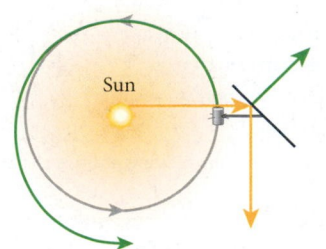

Angle 2

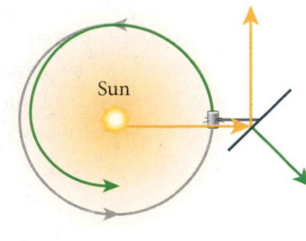

Angle 3

Deployment angle 3 will result in a shrinking orbit. The resulting force from the reflected light produces a component of force that is in the opposite direction from the velocity of the spacecraft. Thus, the spacecraft loses energy, and the radius of the orbit decreases. Note that the speed of the spacecraft increases but its total energy decreases.

31.4 Take the vector product of the gradient operator $\vec{\nabla}$ and the fourth Maxwell equation: $\vec{\nabla} \times \vec{\nabla} \times \vec{B} = \epsilon_0 \mu_0 \vec{\nabla} \times \dfrac{\partial}{\partial t} \vec{E}.$

The right-hand side of this equation is

$$\epsilon_0 \mu_0 \vec{\nabla} \times \frac{\partial}{\partial t} \vec{E} = \epsilon_0 \mu_0 \frac{\partial}{\partial t} (\vec{\nabla} \times \vec{E}) = \epsilon_0 \mu_0 \frac{\partial}{\partial t} \left(-\frac{\partial \vec{B}}{\partial t} \right) = -\epsilon_0 \mu_0 \frac{\partial^2}{\partial t^2} \vec{B}.$$

The left-hand side is $\vec{\nabla} \times \vec{\nabla} \times \vec{B} = \vec{\nabla}(\vec{\nabla} \cdot \vec{B}) - \nabla^2 \vec{B} = -\nabla^2 \vec{B}.$

31.5 $\dfrac{\partial^2}{\partial t^2} \sin(\kappa x - \omega t) = -\omega^2 \sin(\kappa x - \omega t)$

and

$\dfrac{\partial^2}{\partial x^2} \sin(\kappa x - \omega t) = -\kappa^2 \sin(\kappa x - \omega t).$

Thus, this function is a solution for $c = \omega/\kappa.$

PROBLEM-SOLVING GUIDELINES

1. The same basic relationships that characterize any waves apply to electromagnetic waves. Remember that $c = \lambda f$ and $\omega = c\kappa$, where c is the speed of an electromagnetic wave. Review Chapter 15 if necessary.

2. It is often helpful to draw a diagram showing the direction of the wave motion and the orientation of both the electric and magnetic fields. Remember the relationships between $\vec{E}$ and $\vec{B}$ for both magnitude and direction, including $E/B = (\mu_0 \epsilon_0)^{-1/2} = c$ for electromagnetic waves.

MULTIPLE-CHOICE QUESTIONS

31.1 Which of the following phenomena can be observed for electromagnetic waves but not for sound waves?

a) interference   c) polarization   e) scattering

b) diffraction   d) absorption

31.2 Which of the following statements concerning electromagnetic waves are incorrect? (Select all that apply.)

a) Electromagnetic waves in vacuum travel at the speed of light.

b) The magnitudes of the electric field and the magnetic field are equal.

c) Only the electric field vector is perpendicular to the direction of the wave's propagation.

d) Both the electric field vector and the magnetic field vector are perpendicular to the direction of propagation.

e) An electromagnetic wave carries energy only when $E = B$.

31.3 The international radio station Voice of Slobbovia announces that it is "transmitting to North America on the 49-meter band." Which frequency is the station transmitting on?

a) 820 kHz

b) 6.12 MHz

c) 91.7 MHz

d) The information given tells nothing about the frequency.

31.4 Which of the following exerts the largest amount of radiation pressure?

a) a 1-mW laser pointer on a 2-mm-diameter spot 1 m away

b) a 200-W light bulb on a 4-mm-diameter spot 10 m away

c) a 100-W light bulb on a 2-mm-diameter spot 4 m away

d) a 200-W light bulb on a 2-mm-diameter spot 5 m away

e) All of the above exert the same pressure.

31.5 What is the direction of the net force on the moving positive charge in the figure?

a) into the page   c) out of the page

b) toward the right   d) toward the left

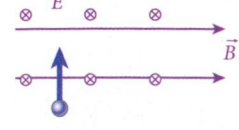

31.6 A proton moves perpendicularly to crossed electric and magnetic fields as shown in the figure. What is the direction of the net force on the proton?

a) toward the left

b) toward the right

c) into the page

d) out of the page

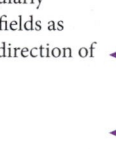

31.7 It is speculated that isolated magnetic "charges" (magnetic monopoles) may exist somewhere in the universe. Which of Maxwell's equations, (1) Gauss's Law for Electric Fields, (2) Gauss's Law for Magnetic Fields, (3) Faraday's Law of Induction, and/or (4) the Maxwell-Ampere Law, would be altered by the existence of magnetic monopoles?

a) only (2)   c) (2) and (3)

b) (1) and (2)   d) only (3)

31.8 According to Gauss's Law for Magnetic Fields, all magnetic field lines form a complete loop. Therefore, the direction of the magnetic field $\vec{B}$ points from _____ pole to _____ pole outside of an ordinary bar magnet and from _____ pole to _____ pole inside the magnet.

a) north, south, north, south   c) south, north, south, north

b) north, south, south, north   d) south, north, north, south

31.9 Unpolarized light with intensity $I_{\text{in}} = 1.87 \text{ W/m}^2$ passes through two polarizers. The emerging polarized light has intensity $I_{\text{out}} = 0.383 \text{ W/m}^2$. What is the angle between the two polarizers?

a) 23.9°   c) 50.2°   e) 88.9°

b) 34.6°   d) 72.7°

31.10 The average intensity of sunlight at the Earth's surface is approximately 1400 W/m^2, if the Sun is directly overhead. The average distance between the Earth and the Sun is $1.50 \cdot 10^{11}$ m. What is the average power emitted by the Sun?

a) $99.9 \cdot 10^{25}$ W   c) $6.3 \cdot 10^{27}$ W   e) $5.9 \cdot 10^{29}$ W

b) $4.0 \cdot 10^{26}$ W   d) $4.3 \cdot 10^{28}$ W

CONCEPTUAL QUESTIONS

31.11 In a polarized light experiment, a setup similar to the one in Figure 31.23 is used. Unpolarized light with intensity I_0 is incident on polarizer 1. Polarizers 1 and 3 are crossed (at a 90° angle), and their orientations are fixed during the experiment. Initially, polarizer 2 has its polarizing angle at 45°. Then, at time $t = 0$, polarizer 2 starts to rotate with angular velocity ω about the direction of propagation of light in a clockwise direction as viewed by an observer looking toward the light source. A photodiode is used to monitor the intensity of the light emerging from polarizer 3.

a) Determine an expression for this intensity as a function of time.

b) How would the expression from part (a) change if polarizer 2 were rotated about an axis parallel to the direction of propagation of the light but displaced by a distance $d < R$, where R is the radius of the polarizer?

31.12 A dipole antenna is located at the origin with its axis along the z-axis. As electric current oscillates up and down the antenna, polarized electromagnetic radiation travels away from the antenna along the positive y-axis. What are the possible directions of electric and magnetic fields at point A on the y-axis? Explain.

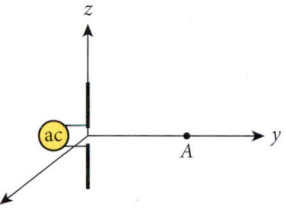

31.13 Does the information in Section 31.6 affect the answer to Example 31.2 regarding the root-mean-square magnitude of electric field at the Earth's surface from the Sun?

31.14 Maxwell's equations predict that there are no magnetic monopoles. If these monopoles existed, how would the motion of charged particles change as they approached such a monopole?

31.15 If two communication signals were sent at the same time to the Moon, one via radio waves and one via visible light, which one would arrive at the Moon first?

31.16 Show that Ampere's Law is not necessarily consistent if the surface through which the flux is to be calculated is a closed surface, but that the Maxwell-Ampere Law always is. (Hence, Maxwell's introduction of his law of induction and the displacement current are not optional; they are logically necessary.) Show also that Faraday's Law of Induction does not suffer from this consistency problem.

31.17 Maxwell's equations and Newton's laws of motion are mutually inconsistent; the great edifice of classical physics is fatally flawed. Explain why.

31.18 Practically everyone who has studied the electromagnetic spectrum has wondered how the world would appear if we could see over a range of frequencies comparable to the ten octaves over which we can hear rather than the less than one octave over which we can see. (An octave refers to a factor of 2 in frequency.) But this is practically impossible. Why?

31.19 Isotropic electromagnetic waves expand uniformly outward in all directions in three dimensions. Electromagnetic waves from a small, isotropic source are not plane waves, which have constant maximum amplitudes.

a) How does the maximum amplitude of the electric field of radiation from a small, isotropic source vary with distance from the source?

b) Compare this with the electrostatic field of a point charge.

31.20 A pair of sunglasses is held in front of a flat-panel computer monitor (which is on) so that the lenses are always parallel to the display. As the lenses are rotated, it is noticed that the intensity of light coming from the display and passing through the lenses is varying. Why?

31.21 Two polarizing filters are crossed at 90°, so when light is shined from behind the pair of filters, no light passes through. A third filter is inserted between the two, initially aligned with one of them. Describe what happens as the intermediate filter is rotated through an angle of 360°.

EXERCISES

A blue problem number indicates a worked-out solution is available in the Student Solutions Manual. One • and two •• indicate increasing level of problem difficulty.

Section 31.1

31.22 An electric field of magnitude 200.0 V/m is directed perpendicular to a circular planar surface with radius 6.00 cm. If the electric field increases at a rate of 10.0 V/(m s), determine the magnitude and the direction of the magnetic field at a radial distance 10.0 cm away from the center of the circular area.

•31.23 A wire of radius 1.00 mm carries a current of 20.0 A. The wire is connected to a parallel plate capacitor with circular plates of radius $R = 4.00$ cm and a separation between the plates of $s = 2.00$ mm. What is the magnitude of the magnetic field due to the changing electric field at a point that is a radial distance of $r = 1.00$ cm from the center of the parallel plates? Neglect edge effects.

•31.24 The current flowing in a solenoid that is 20.0 cm long and has a radius of 2.00 cm and 500. turns decreases from 3.00 A to 1.00 A in 0.100 s. Determine the magnitude of the induced electric field inside the solenoid 1.00 cm from its center.

31.25 A parallel plate capacitor has air between disk-shaped plates of radius 4.00 mm that are coaxial and 1.00 mm apart. Charge is being accumulated on the plates of the capacitor. What is the displacement current between the plates at an instant when the rate of charge accumulation on the plates is 10.0 μC/s?

•31.26 A parallel plate capacitor has circular plates of radius 10.0 cm that are separated by a distance of 5.00 mm. The potential across the capacitor

is increased at a constant rate of 1.20 kV/s. Determine the magnitude of the magnetic field between the plates at a distance $r = 4.00$ cm from the center.

•31.27 The voltage across a cylindrical conductor of radius r, length L, and resistance R varies with time. The time-varying voltage causes a time-varying current, i, to flow in the cylinder. Show that the displacement current equals $\epsilon_0 \rho di/dt$, where ρ is the resistivity of the conductor.

Section 31.2

31.28 The amplitude of the electric field of an electromagnetic wave is 250. V/m. What is the amplitude of the magnetic field of the electromagnetic wave?

31.29 Determine the distance in feet that light can travel in vacuum during 1.00 ns.

31.30 How long does it take light to travel from the Moon to the Earth? From the Sun to the Earth? From Jupiter to the Earth?

31.31 Alice made a telephone call from her home telephone in New York to her fiancé stationed in Baghdad, about 10,000 km away, and the signal was carried on a telephone cable. The following day, Alice called her fiancé again from work using her cell phone, and the signal was transmitted via a satellite 36,000 km above the Earth's surface, halfway between New York and Baghdad. Estimate the time taken for the signals sent by (a) the telephone cable and (b) via the satellite to reach Baghdad, assuming that the signal speed in both cases is the same as the speed of light, c. Would there be a noticeable delay in either case?

•31.32 Electric and magnetic fields in many materials can be analyzed using the relationships for these fields in vacuum, but substituting relative values of the permittivity and the permeability, $\epsilon = \kappa \epsilon_0$ and $\mu = \kappa_m \mu_0$, for their vacuum values, where κ is the dielectric constant and κ_m the relative permeability of the material. Calculate the ratio of the speed of electromagnetic waves in vacuum to their speed in such a material.

Section 31.3

31.33 The wavelength range for visible light is 400 nm to 700 nm (see Figure 31.10) in air. What is the frequency range of visible light?

31.34 The antenna of a cell phone is a straight rod 8.0 cm long. Calculate the operating frequency of the signal from this phone, assuming that the antenna length is $\frac{1}{4}$ of the wavelength of the signal.

•31.35 Suppose an RLC circuit in resonance is used to produce a radio wave of wavelength 150 m. If the circuit has a 2.0-pF capacitor, what size inductor is used?

•31.36 Three FM radio stations covering the same geographical area broadcast at frequencies 91.1, 91.3, and 91.5 MHz, respectively. What is the maximum allowable wavelength width of the band-pass filter in a radio receiver such that the FM station 91.3 can be played free of interference from FM 91.1 or FM 91.5? Use $c = 3.00 \cdot 10^8$ m/s, and calculate the wavelength to an uncertainty of 1 mm.

Section 31.4

31.37 A monochromatic point source of light emits 1.5 W of electromagnetic power uniformly in all directions. Find the Poynting vector at a point situated at each of the following locations:

a) 0.30 m from the source

b) 0.32 m from the source

c) 1.00 m from the source

31.38 Consider an electron in a hydrogen atom, which is 0.050 nm from the proton in the nucleus.

a) What electric field does the electron experience?

b) In order to produce an electric field whose root-mean-square magnitude is the same as that of the field in part (a), what intensity must a laser light have?

31.39 A 3.00-kW carbon dioxide laser is used in laser welding. If the beam is 1.00 mm in diameter, what is the amplitude of the electric field in the beam?

31.40 Suppose that charges on a dipole antenna oscillate slowly at a rate of 1.00 cycle/s, and the antenna radiates electromagnetic waves in a region of space. If someone measured the time-varying magnetic field in the region and found its maximum to be 1.00 mT, what would be the maximum electric field, E, in the region, in units of volts per meter? What is the period of the charge oscillation? What is the magnitude of the Poynting vector?

31.41 Calculate the average value of the Poynting vector, S_{ave}, for an electromagnetic wave having an electric field of amplitude 100. V/m.

a) What is the average energy density of this wave in J/m³?

b) How large is the amplitude of the magnetic field?

•31.42 The most intense beam of light that can propagate through dry air must have an electric field whose maximum amplitude is no greater than the breakdown value for air: $E_{max}^{air} = 3.0 \cdot 10^6$ V/m, assuming that this value is unaffected by the frequency of the wave.

a) Calculate the maximum amplitude the magnetic field of this wave can have.

b) Calculate the intensity of this wave.

c) What happens to a wave more intense than this?

••31.43 A continuous-wave (cw) argon-ion laser beam has an average power of 10.0 W and a beam diameter of 1.00 mm. Assume that the intensity of the beam is the same throughout the cross section of the beam (which is not true, as the actual distribution of intensity is a Gaussian function).

a) Calculate the intensity of the laser beam. Compare this with the average intensity of sunlight at Earth's surface (1400. W/m²).

b) Find the root-mean-square electric field in the laser beam.

c) Find the average value of the Poynting vector over time.

d) If the wavelength of the laser beam is 514.5 nm in vacuum, write an expression for the instantaneous Poynting vector, where the instantaneous Poynting vector is zero at $t = 0$ and $x = 0$.

e) Calculate the root-mean-square value of the magnetic field in the laser beam.

••31.44 A voltage, V, is applied across a cylindrical conductor of radius r, length L, and resistance R. As a result, a current, i, is flowing through the conductor, which gives rise to a magnetic field, B. The conductor is placed along the y-axis, and the current is flowing in the positive y-direction. Assume that the electric field is uniform throughout the conductor.

a) Find the magnitude and the direction of the Poynting vector at the surface of the conductor of the static electric and magnetic fields.

b) Show that $\int \vec{S} \cdot d\vec{A} = i^2 R$.

Section 31.5

31.45 Radiation from the Sun reaches the Earth at a rate of 1.40 kW/m² above the atmosphere and at a rate of 1.00 kW/m² on an ocean beach.

a) Calculate the maximum values of E and B above the atmosphere.

b) Find the pressure and the force exerted by the radiation on a person lying flat on the beach who has an area of 0.750 m² exposed to the Sun.

31.46 Scientists have proposed using the radiation pressure of sunlight for travel to other planets in the Solar System. If the intensity of the electromagnetic radiation produced by the Sun is about 1.40 kW/m² near the Earth, what size would a sail have to be to accelerate a spaceship with a mass of 10.0 metric tons at 1.00 m/s²?

a) Assume that the sail absorbs all the incident radiation.

b) Assume that the sail perfectly reflects all the incident radiation.

31.47 A solar sail is a giant circle (with a radius $R = 10.0$ km) made of a material that is perfectly reflecting on one side and totally absorbing on the other side. In deep space, away from other sources of light, the cosmic microwave background will provide the primary source of radiation incident on the sail. Assuming that this radiation is that of an ideal black body at $T = 2.725$ K, calculate the net force on the sail due to its reflection and absorption. Also assume that any heat transferred to the sail will be conducted away, and that the photons are incident perpendicular to the surface of the sail.

•31.48 Two astronauts are at rest in outer space, one 20.0 m from the Space Shuttle and the other 40.0 m from the shuttle. Using a 100.0-W laser, the astronaut located 40.0 m away from the shuttle decides to propel the other astronaut toward the Space Shuttle. He focuses the laser on a piece of totally reflecting fabric on her space suit. If her total mass with equipment is 100.0 kg, how long will it take her to reach the Space Shuttle?

•31.49 A laser that produces a spot of light that is 1.00 mm in diameter is shone perpendicularly on the center of a thin, perfectly reflecting circular (2.00 mm in diameter) aluminum plate mounted vertically on a flat piece of cork that floats on the surface of the water in a large beaker. The mass of this "sailboat" is 0.100 g, and it travels 2.00 mm in 63.0 s. Assuming that the laser power is constant in the region where the sailboat is located during its motion, what is the power of the laser? (Neglect air resistance and the viscosity of water.)

•31.50 A tiny particle of density 2000. kg/m³ is at the same distance from the Sun as the Earth is ($1.50 \cdot 10^{11}$ m). Assume that the particle is spherical and perfectly reflecting. What would its radius have to be for the outward radiation pressure on it to be 1.00% of the inward gravitational attraction of the Sun? (Take the Sun's mass to be $2.00 \cdot 10^{30}$ kg.)

•31.51 Silica aerogel, an extremely porous, thermally insulating material made of silica, has a density of 1.00 mg/cm³. A thin circular slice of aerogel has a diameter of 2.00 mm and a thickness of 0.10 mm.

a) What is the weight of the aerogel slice (in newtons)?

b) What are the intensity and the radiation pressure of a 5.00-mW laser beam of diameter 2.00 mm on the sample?

c) How many 5.00-mW lasers with a beam diameter of 2.00 mm would be needed to make the slice float in the Earth's gravitational field? Use $g = 9.81$ m/s².

Section 31.6

31.52 Two polarizers are out of alignment by 30.0°. If light of intensity 1.00 W/m² and initially polarized halfway between the polarizing angles of the two filters passes through both filters, what is the intensity of the transmitted light?

31.53 A 10.0-mW vertically polarized laser beam passes through a polarizer whose polarizing angle is 30.0° from the horizontal. What is the power of the laser beam when it emerges from the polarizer?

•31.54 Unpolarized light of intensity I_0 is incident on a series of five polarizers, with the polarization direction of each rotated 10.0° from that of the preceding one. What fraction of the incident light will pass through the series?

•31.55 A laser produces light that is polarized in the vertical direction. The light travels in the positive y-direction and passes through two polarizers, which have polarizing angles of 35.0° and 55.0° from the vertical, as shown in the figure. The laser beam is collimated (neither converging nor expanding), has a circular cross section with a diameter of 1.00 mm, and has an average power of 15.0 mW at point A. At point C, what are the amplitudes of the electric and magnetic fields, and what is the intensity of the laser light?

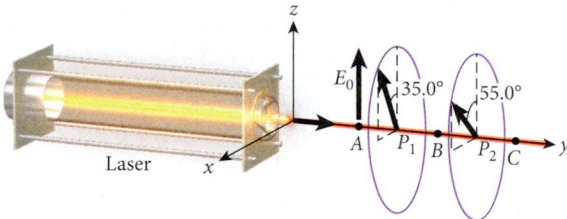

Additional Exercises

31.56 A laser beam takes 50.0 ms to be reflected back from a totally reflecting sail on a spacecraft. How far away is the sail?

31.57 A house with a south-facing roof has photovoltaic panels on the roof. The photovoltaic panels have an efficiency of 10.0% and occupy an area with dimensions 3.00 m by 8.00 m. The average solar radiation incident on the panels is 300. W/m², averaged over all conditions for a year. How many kilowatt-hours of electricity will the solar panels generate in a 30-day month?

31.58 What is the radiation pressure due to Betelgeuse (which has a luminosity, or power output, 10,000 times that of the Sun) at a distance from it equal to that of Uranus's orbit from the Sun?

31.59 A 200.-W laser produces a beam with a cross-sectional area of 1.00 mm² and a wavelength of 628 m. What is the amplitude of the electric field in the beam?

31.60 What is the wavelength of the electromagnetic waves used for cell phone communications at 848.97 MHz?

31.61 As shown in the figure, sunlight is coming straight down (negative z-direction) on a solar panel (of length $L = 1.40$ m and width $W = 0.900$ m) on the Mars rover *Spirit*. The amplitude of the electric field in the solar radiation is 673 V/m and is uniform (the radiation has the same amplitude everywhere). If the solar panel converts solar radiation to electrical power with an efficiency of 18.0%, how much average power can the panel generate?

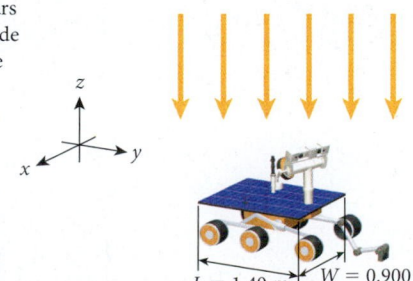

31.62 A 14.9-μF capacitor, a 24.3-kΩ resistor, a switch, and a 25.0-V battery are connected in series. What is the rate of change of the electric field between the plates of the capacitor at $t = 0.3621$ s after the switch is closed? The area of the plates is 1.00 cm².

31.63 A focused 300-W spotlight delivers 40% of its light within a circular area with a diameter of 2 m. What is the root-mean-square electric field in this illuminated area?

31.64 What is the electric field amplitude of an electromagnetic wave whose magnetic field amplitude is $5.00 \cdot 10^{-3}$ T?

31.65 What is the distance between successive heating antinodes in a microwave oven's cavity? A microwave oven typically operates at a frequency of 2.4 GHz.

31.66 The solar constant measured by Earth satellites is roughly 1400 W/m².

a) Find the maximum electric field of the electromagnetic radiation from the Sun.

b) Find the maximum magnetic field of these electromagnetic waves.

•31.67 The peak electric field at a distance of 2.25 m from a light bulb is 21.2 V/m.

a) What is the peak magnetic field there?

b) What is the power output of the bulb?

•31.68 If the peak electric field due to a star whose radius is twice that of the Sun is 44.0 V/m at a distance of 15 AU, what is its temperature? Treat the star as a blackbody.

•31.69 A 5.00-mW laser pointer has a beam diameter of 2.00 mm.

a) What is the root-mean-square value of the electric field in this laser beam?

b) Calculate the total electromagnetic energy in 1.00 m of this laser beam.

•31.70 At the surface of the Earth, the Sun delivers an estimated 1.00 kW/m² of energy. Suppose sunlight hits a 10.0 m by 30.0 m roof at an angle of 90.0°.

a) Estimate the total power incident on the roof.

b) Find the radiation pressure on the roof.

•31.71 The National Ignition Facility has the most powerful laser in the world; it uses 192 beams to aim 500. TW of power at a spherical pellet of diameter 2.00 mm. How fast would a pellet of density 2.00 g/cm³ accelerate if only one of the laser beams hits it for 1.00 ns and 2.00% of the light is absorbed?

•31.72 A resistor consists of a solid cylinder of radius r and length L. The resistor has resistance R and is carrying current i. Use the Poynting vector to calculate the power radiated out of the surface of the resistor.

•31.73 A radio tower is transmitting 30.0 kW of power equally in all directions. Assume that the radio waves that hit the Earth are reflected.

a) What is the magnitude of the Poynting vector at a distance of 12.0 km from the tower?

b) What is the root-mean-square value of the electric force on an electron at this location?

•31.74 Quantum theory says that electromagnetic waves actually consist of discrete packets—photons—each with energy $E = \hbar\omega$, where $\hbar = 1.054572 \cdot 10^{-34}$ J s is Planck's reduced constant and ω is the angular frequency of the wave.

a) Find the momentum of a photon.

b) Find the magnitude of angular momentum of a photon. Photons are *circularly polarized;* that is, they are described by a superposition of two plane-polarized waves with equal field amplitudes, equal frequencies, and perpendicular polarizations, one-quarter of a cycle (90° or $\pi/2$ rad) out of phase, so the electric and magnetic field vectors at any fixed point rotate in a circle with the angular frequency of the waves. It can be shown that a circularly polarized wave of energy U and angular frequency ω has an angular momentum of magnitude $L = U/\omega$. (The direction of the angular momentum is given by the thumb of the right hand, when the fingers are curled in the direction in which the field vectors circulate.)

c) The ratio of the angular momentum of a particle to $\hbar$ is its spin quantum number. Determine the spin quantum number of the photon.

•31.75 A microwave operates at 250. W. Assuming that the waves emerge from a point source emitter on one side of the oven, how long does it take to melt an ice cube 2.00 cm on a side that is 10.0 cm away from the emitter if 10.0% of the photons that strike the cube are absorbed by it? How many photons of wavelength 10.0 cm hit the ice cube per second? Assume a cube density of 0.960 g/cm^3.

•**31.76** An industrial carbon dioxide laser produces a beam of radiation with average power of 6.00 kW at a wavelength of 10.6 μm. Such a laser can be used to cut steel up to 25 mm thick. The laser light is polarized in the x-direction, travels in the positive z-direction, and is collimated (neither diverging or converging) at a *constant* diameter of 100.0 μm. Write the

equations for the laser light's electric and magnetic fields as a function of time and of position z along the beam. Recall that $\vec{E}$ and $\vec{B}$ are vectors. Leave the overall phase unspecified, but be sure to check the relative phase between $\vec{E}$ and $\vec{B}$.

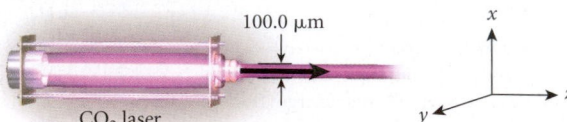

MULTI-VERSION EXERCISES

31.77 During the testing of a new light bulb, a sensor is placed 31.9 cm from the bulb. It records an intensity of 182.9 W/m^2 for the radiation emitted by the bulb. What is the root-mean-square value of the electric field at the sensor's location?

31.78 During the testing of a new light bulb, a sensor is placed 42.1 cm from the bulb. It records an intensity of 191.4 W/m^2 for the radiation emitted by the bulb. What is the root-mean-square value of the magnetic field at the sensor's location?

31.79 During the testing of a new light bulb, a sensor is placed 52.5 cm from the bulb. It records a root-mean-square value of $9.142 \cdot 10^{-7}$ T for the magnetic field of the radiation emitted by the bulb. What is the intensity of that radiation at the sensor's location?

31.80 During the testing of a new light bulb, a sensor is placed 17.7 cm from the bulb. It records a root-mean-square value of 279.9 V/m for the electric field of the radiation emitted from the bulb. What is the intensity of that radiation at the sensor's location?

31.81 To visually examine sunspots through a telescope, astronomers have to reduce the intensity of the sunlight to avoid harming their retinas. They

accomplish this intensity reduction by mounting two polarizers on the telescope. The first polarizer has a polarizing angle of 28.1° relative to the horizontal, and the second has a polarizing angle of 88.6°. By what fraction is the intensity of the incident sunlight reduced by the polarizers?

31.82 To visually examine sunspots through a telescope, astronomers have to reduce the intensity of the sunlight to avoid harming their retinas. They accomplish this intensity reduction by mounting two polarizers on the telescope. The first polarizer has a polarizing angle of 38.3° relative to the horizontal. If the astronomers want to reduce the intensity of the sunlight by a factor of 0.7584, what polarizing angle should the second polarizer have with the horizontal? Assume that this angle is greater than that of the first polarizer.

31.83 To visually examine sunspots through a telescope, astronomers have to reduce the intensity of the sunlight to avoid harming their retinas. They accomplish this intensity reduction by mounting two linear polarizers on the telescope. The second polarizer has a polarizing angle of 110.6° relative to the horizontal. If the astronomers want to reduce the intensity of the sunlight by a factor of 0.7645, what polarizing angle should the first polarizer have with the horizontal? Assume that this angle is smaller than that of the second polarizer.

32

Geometric Optics

FIGURE 32.1 Primary and secondary rainbows formed by refraction and reflection of light in raindrops over Ka'anapali Beach on Maui, Hawaii.

The study of light is divided into three fields: geometric optics, wave optics, and quantum optics. In Chapter 31, we learned that light is an electromagnetic wave, and in Chapter 34 we will deal with the wave properties of light. This chapter discusses **geometric optics,** in which light is characterized as rays. Quantum optics makes use of the fact that light is quantized, its energy localized in point particles called *photons* (Chapters 36 and 37).

Rainbows (Figure 32.1) can be seen only when there are water droplets in the air and the Sun is behind the observer. Why is that? The reason has to do with how raindrops reflect and refract light—the two optical processes that are the main subjects of this chapter.

We have seen that light is a wave, but in this chapter we will examine systems in which its wavelength is small compared to other physical dimensions of the system. Then we can ignore the wave character of light and consider only how light travels through air—or glass, or water, or any other transparent medium. It turns out that thinking about light in this way is enough to explain the optics of mirrors, lenses, and other optical devices, including prisms and even rainbows. Later, in Chapter 34, we will examine systems in which the wavelength is not negligibly small and see how that characteristic gives rise to other optical effects.

This chapter primarily considers visible light, but keep in mind that the laws of reflection, refraction, and image formation also apply to other kinds of electromagnetic waves. For instance, many useful properties of radio waves are based on reflection and refraction.

WHAT WE WILL LEARN

- In cases where the wavelength is small compared to other length scales in a physical system, light waves can be modeled by light rays, moving on straight-line trajectories and representing the direction of a propagating light wave.

- The law of reflection states that the angle of incidence is equal to the angle of reflection.

- Mirrors can focus light and produce images governed by the mirror equation, which states that the inverse of the object distance plus the inverse of the image distance equals the inverse of the focal length of the mirror.

- Light is refracted (changes direction) when it is incident on a boundary between two optically transparent media.

- Snell's Law governs refraction at the boundary between different transparent materials. It relates the angles of incident and refracted light rays and the indices of refraction of the two materials.

- When light crosses the boundary between two media, and the index of refraction of the second medium is less than the index of refraction of the first medium, there is a critical angle of incidence above which refraction cannot take place. Instead, the light is totally reflected.

- We will learn to use the laws of geometrical optics for a wide variety of applications, from simple bathroom mirrors to optical fibers in endoscopes to rainbows.

32.1 Light Rays and Shadows

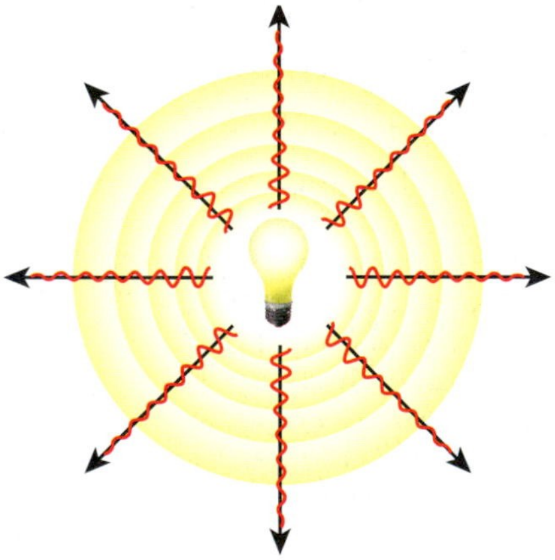

In Chapter 31, we saw that electromagnetic waves spread spherically from a point source. The concentric yellow spheres in Figure 32.2 represent the spreading of spherical **wave fronts** of the light emitted from a light bulb. (A *front* is a locus of points that have the same instantaneous value for the electric field.) The black arrows are the **light rays,** which are perpendicular to the wave fronts at every point in space and point in the direction of propagation of the light. The undulating red lines represent the oscillating electric field.

Light waves far away from their source can be treated as plane waves whose wave fronts are traveling in a straight line (Figure 32.3). These traveling plane waves can be represented by parallel arrows perpendicular to the surface of the planes. In this chapter, we will treat light as a ray traveling in a straight line while in a homogeneous medium. Viewing light as rays will enable us to analyze and solve a broad range of practical problems, both geometrically and by means of various constructions.

Everyday experience tells us that light travels in a straight line. We cannot easily see that light has a wave structure or a quantum structure. The reason for this is simply that the wavelengths of visible light (400–700 nm) are small compared to the structures that form our everyday experiences. (The few notable exceptions, including soap films, will be discussed in Chapter 34.)

A person standing outside in the bright sunshine sees shadows cast by objects in the sunlight. The edges of the shadows appear reasonably sharp, so the person theorizes that light travels in straight lines and the objects block the light rays from the Sun that strike them. A shadow is created where light is intercepted, while bright areas are created where the unintercepted light rays continue in straight lines and strike the ground or other surface. The shadow is not completely black, because light scatters from other sources and partially illuminates the shadowed area. And the edge of the shadow is also not completely sharp, because the Sun has a diameter that is not negligible. But still, the creation of shadows lends credibility to the hypothesis of straight-line motion of light.

This observation can be made in a more controlled manner by placing a piece of cardboard containing a small hole in front of a bright projector light bulb (Figure 32.4). This produces a round bright spot on the screen (the image). A smaller hole produces a smaller image on the screen. A larger hole produces a larger image.

If the size of the light source is small enough (pointlike), the similar triangles in Figure 32.4 can be used to find the relationship between the size of the hole (r), the size of the

FIGURE 32.2 Light spreading from a source. Yellow: spherical wave fronts; red: oscillating electric field; black: rays.

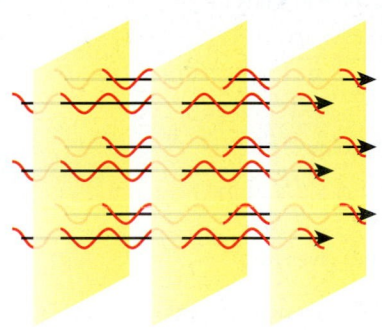

FIGURE 32.3 Planes representing wave fronts of a traveling light wave. The red sinusoidal oscillations represent the oscillating electric or magnetic field. The black arrows are the corresponding light rays, which are always perpendicular to the wave front.

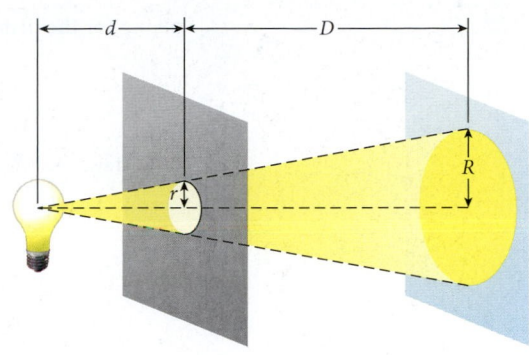

FIGURE 32.4 Light shining through an opening onto a screen.

image (R), the distance between light source and the hole (d), and the distance between the hole and the screen (D):

$$\frac{r}{d} = \frac{R}{d+D}. \qquad (32.1)$$

This equation is sometimes referred to as the **law of rays.**

SOLVED PROBLEM 32.1 | Shadow of a Ball

The light from a small light bulb creates a shadow of a ball on a wall. The diameter of the ball is 14.3 cm, and the diameter of the shadow of the ball on the wall is 27.5 cm. The ball is 1.99 m away from the wall.

PROBLEM
How far is the light bulb from the wall?

SOLUTION
THINK The light bulb is small, which means that we can neglect its size and treat it as a point source. The triangle formed by the light bulb and the ball is similar to the triangle formed by the light bulb and the shadow. The distance from the ball to the wall is given, so we can solve for the distance from the light bulb to the ball, add that distance to the distance of the ball from the wall, and obtain the distance of the light bulb from the wall.

SKETCH Figure 32.5 shows a sketch of a light bulb casting a shadow of a ball on a wall.

RESEARCH The triangle formed by the light bulb and the ball is similar to the triangle formed by the light bulb and the shadow. Using equation 32.1 we can write

$$\frac{r}{d} = \frac{R}{d+D},$$

where r is the radius of the ball, R is the radius of the shadow cast on the wall, d is the distance of the light bulb from the ball, and D is the distance of the ball from the wall.

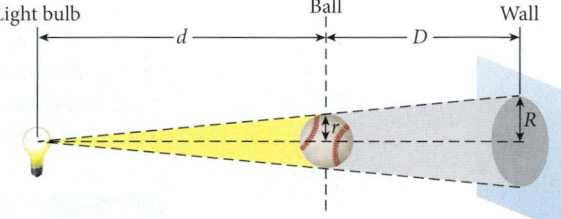

FIGURE 32.5 A light bulb casts a shadow of a ball on a wall.

SIMPLIFY We can rearrange the law of rays to obtain

$$d+D = \frac{Rd}{r}.$$

Collecting the terms involving the distance from the light bulb to the ball on the left side gives us

$$d\left(1 - \frac{R}{r}\right) = -D.$$

– Continued

Solving for the distance from the light bulb to the ball leads to

$$d = \frac{D}{\dfrac{R}{r} - 1} = \frac{rD}{R - r}.$$

The distance of the light bulb from the wall, $d_{\text{light bulb}}$, is then

$$d_{\text{light bulb}} = d + D = \frac{rD}{R - r} + D.$$

CALCULATE   Putting in the numerical values gives us

$$d_{\text{light bulb}} = \frac{rD}{R - r} + D = \frac{(1.99 \text{ m})(0.143 \text{ m})}{(0.275 \text{ m}) - (0.143 \text{ m})} + 1.99 \text{ m} = 2.15583 + 1.99 = 4.14583 \text{ m}.$$

ROUND   We report our result to three significant figures:

$$d_{\text{light bulb}} = 4.15 \text{ m}.$$

DOUBLE-CHECK   To double-check our result, we calculate the ratio r/d and compare that result with the ratio $R/(d + D)$, which must have the same value according to the law of rays (equation 32.1). We note from above that the value of d, the distance from the bulb to the ball, is 2.16 m. The first ratio is

$$\frac{r}{d} = \frac{0.143 \text{ m}}{2.16 \text{ m}} = 0.0662.$$

The second ratio is

$$\frac{R}{d + D} = \frac{0.275 \text{ m}}{4.15 \text{ m}} = 0.0663.$$

These values agree within rounding errors, so our answer seems reasonable.

As a second double-check, we can examine limiting cases and see if they support our result. What happens to our result in the limiting case where the distance d from the light bulb to the ball is large compared to the distance D from the ball to the wall? This is a case similar to sunlight hitting the ball and throwing a shadow. We know that the shadow is approximately the same size as the ball in this case. And the limiting case of our formula $r/d = R/(d + D)$ bears this out, because for $d \gg D$ we see that $r/d = R/(d + D) \approx R/d$, and so $r \approx R$.

What happens in the reverse case, where $d \ll D$, that is, the light bulb is very close to the ball, as compared to the distance between ball and wall? The size of the shadow diverges. Our formula also shows this limiting case, because in this case $R = (d + D)r/d \approx rD/d \gg r$, that is, the radius of the shadow becomes very large relative to the radius of the ball.

Concept Check 32.1

In Solved Problem 32.1, we neglected the size of the light bulb and treated it as a point source. What happens if the light bulb is too large to make this assumption, for example, if it is larger than the baseball?

a) The shadow of the ball will not be sharp but will be fuzzy around the edge.

b) No shadow will be formed.

c) Multiple shadows will be formed.

d) It is impossible to make a general statement about this situation.

Let us finish this introductory discussion of light rays and ray tracing with a very important observation: The direction of the rays is reversible. In the following discussions of reflection off surfaces and refraction at boundaries between different media, light rays will be drawn with an implied direction, emerging from objects and then scattering or encountering a boundary between materials and refracting. But all of the drawings are equally valid if the direction of the rays is reversed. Keep this in mind as we proceed through the following examples and derivations.

32.2   Reflection and Plane Mirrors

Some objects (light bulbs, fires, the Sun, etc.) emit light and are thus primary light sources. Objects that are not primary light sources can be seen because they reflect light. There are two different kinds of reflection, diffuse and specular. In diffuse reflection, light waves hitting the object's surface are scattered randomly. In specular reflection, they are all reflected in the same way. Most objects show diffuse reflection, where the color of the reflected light is a property not just of the wavelength of the incoming light before reflection, but also of the surface properties of the object reflecting the light. The difference between diffuse and specular reflection lies in the roughness of the surface on the scale of the wavelength of the light. For a surface showing specular reflection, the local surface normal vectors (red arrows in Figure 32.6b) are aligned, whereas they are not for a surface showing diffuse reflection (Figure 32.6a).

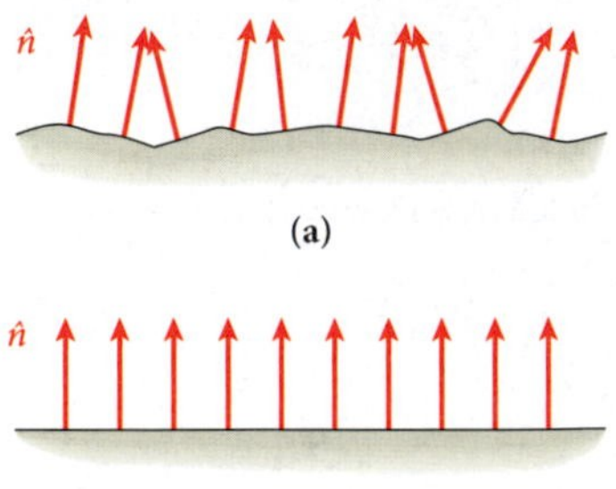

FIGURE 32.6 Orientation of surface normal vectors for a surface showing (a) diffuse reflection and one showing (b) specular reflection.

A **mirror** is a surface that reflects light in specular fashion. Here we will deal only with perfect mirrors. A *perfect mirror* is a mirror that does not absorb any light and that reflects 100% of incident light, independent of the intensity of that light. In Chapter 31 we saw that light is a type of electromagnetic wave. Visible light consists of electromagnetic waves with wavelengths of approximately 400 nm to 700 nm. For a mirror to be considered perfect, it must reflect 100% of the incoming light in at least this range of wavelengths. Mirror perfection is not easily achieved.

Conventional bathroom mirrors consist of a piece of glass with a metal coating on the back side. They are usually sufficient for our purposes, but they are not perfect mirrors. You can see this with the aid of a medicine cabinet with three mirrored doors. Fold out the left and right doors to the point where they are facing each other and look parallel, and then stick your head between them. You will see a huge number of your own reflections, with the light emitted from your head bouncing back and forth many times between the two mirrors until it reaches your eyes. The images formed from light that has undergone multiple reflections are noticeably dimmer (Figure 32.7). The reason is that each reflection absorbs a fraction of the intensity of the incoming light, perhaps on the order of 1%.

In 1998 Yoel Fink, then a graduate student at MIT, invented the *omnidirectional dielectric mirror*, which can be considered perfect in the sense defined above, with absorption losses of less than 0.0001%. He did this by using many alternating, approximately 1 μm thick, layers of a polymer and a semiconducting glass. While initially developed with a grant from the Defense Advanced Research Projects Agency (DARPA), this technology has been used to create vastly improved laser surgery tools. This is yet another example of how modern advanced physics research continues to lead to stunning technological breakthroughs, even in a field as long-established as geometric optics.

Let's start with flat, plane mirrors. For reflection of light rays incident on the surface of a plane mirror, there is a simple rule, known as the **law of reflection:** The angle of incidence, θ_i, is equal to the angle of reflection, θ_r. These angles are always measured from the surface normal, which is defined to be a line perpendicular to the surface of the plane mirror. In addition, the incident ray, the normal, and the reflected ray all lie in the same plane (Figure 32.8).

Mathematically, the law of reflection is given by

$$\theta_r = \theta_i. \tag{32.2}$$

Parallel rays incident on a plane mirror are reflected in such a way that the reflected rays are also parallel (Figure 32.9), because every normal to the surface (dashed line in Figure 32.8) is parallel to other normals.

Image Formed by a Plane Mirror

An **image** can be formed by light reflected from a plane mirror. For example, when you stand in front of a mirror, you see an image of yourself that appears to be behind the mirror. This perception occurs because the brain assumes that light rays reaching the eyes have traveled in straight lines with no change in direction. Thus, if light rays appear to originate at a point (say, behind the mirror), the eye (or a camera) sees a source of light at that point, whether or not an actual light source is there. This type of image, from which the light does not emanate, is referred to as a **virtual image.** By their nature, virtual images cannot be displayed on a screen. In contrast, **real images** are formed at a location at which you could physically put an object, like a screen or a charge-coupled device (CCD) from a camera.

An example will clarify the notion of a virtual image. In Figure 32.10, every point on the surface of the candle flame emits light, with the light rays moving out in radial directions. An observer can locate the candle flame in space by tracing the light rays back to the point where they intersect. Most of these light rays are drawn in gray. The rays that hit the mirror are highlighted in black. Each of them gets reflected according to the law of reflection (equation 32.2), with the angle of reflection relative to the surface normal being equal to the angle of incidence. The reflected rays also have a point at which they intersect. This point is found by continuing the reflected rays behind the mirror (dashed black lines). Therefore, for the observer, the light seems to come from a point behind the mirror. This point is the location of the image formed by the mirror.

FIGURE 32.7 Light reflection between two almost exactly parallel plane mirrors.

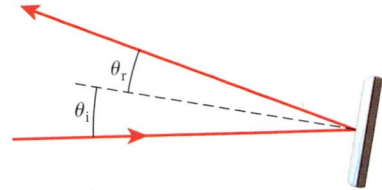

FIGURE 32.8 The angle of incidence equals the angle of reflection for reflection of light off a plane mirror. The dashed line is normal (perpendicular) to the mirror.

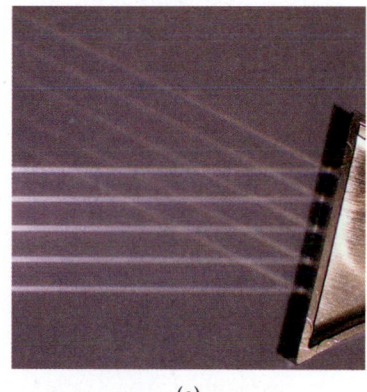

(a)

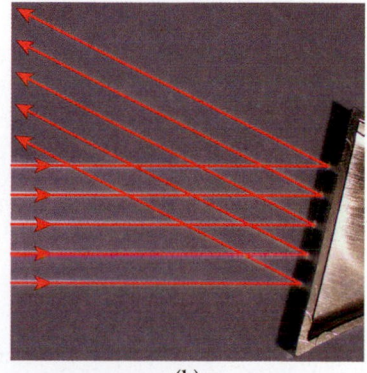

(b)

FIGURE 32.9 (a) Parallel light rays reflect off a plane mirror. (b) Arrows superimposed on the light rays.

FIGURE 32.10 Image of a candle, as
seen in a mirror (blue line).

FIGURE 32.10 Image of a candle, as
seen in a mirror (blue line).

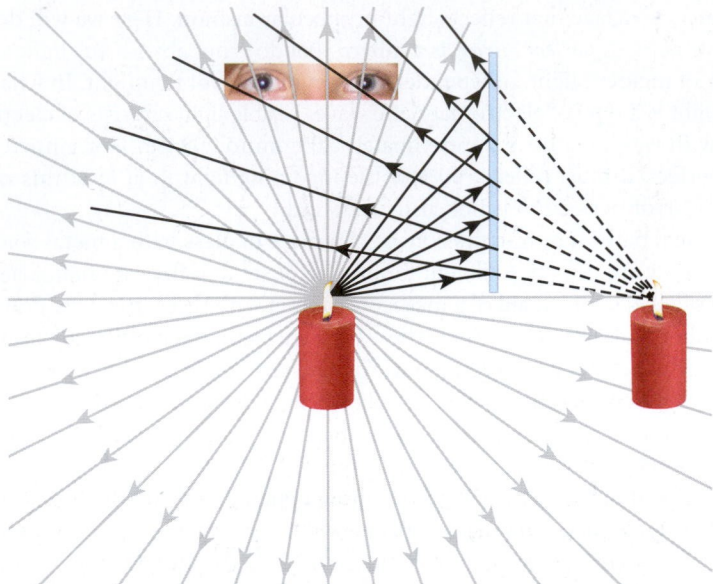

Images formed by plane mirrors are left-right reversed because the light rays incident on the surface of the mirror are reflected back on the other side of the normal. Thus, we have the term *mirror image*. We will explore this point a bit more below, with the aid of Figure 32.12 and Figure 32.13.

But first let's consider quantitatively the process of forming images with a plane mirror (Figure 32.11) by using the technique of ray tracing. It turns out that a few rays are sufficient to construct an image, and we are free to select the most convenient ones.

For this image construction, we choose the case where an object with height h_o is placed a distance d_o from a plane mirror. Following the common convention, the object is represented by an arrow, which indicates the object's height and orientation. The object is oriented so that the tail of the arrow is on the **optic axis,** which is defined as a normal to the plane of the mirror. (For a plane mirror, the optic axis can be shown as going through any point on the mirror, but later when we examine curved mirrors, we will find that there is a unique location for the optic axis.) Three light rays determine where the image is formed.

1. The first light ray emanates from the bottom of the arrow along the optic axis. This ray is reflected directly back on itself. The extrapolation of this reflected ray along the optic axis to the right of the mirror indicates that the bottom of the image is located on the optic axis.

2. The second light ray starts from the top of the arrow parallel to the optic axis and is reflected directly back on itself. The extrapolation of this ray past the mirror is shown in Figure 32.11 as a dashed line.

3. The third ray starts from the top of the arrow, strikes the mirror where the optic axis intersects the mirror, and is reflected with an angle equal to its angle of incidence. The extrapolation of the reflected ray is shown as a dashed line in Figure 32.11.

FIGURE 32.11 Ray diagram for the
image formed by a plane mirror.

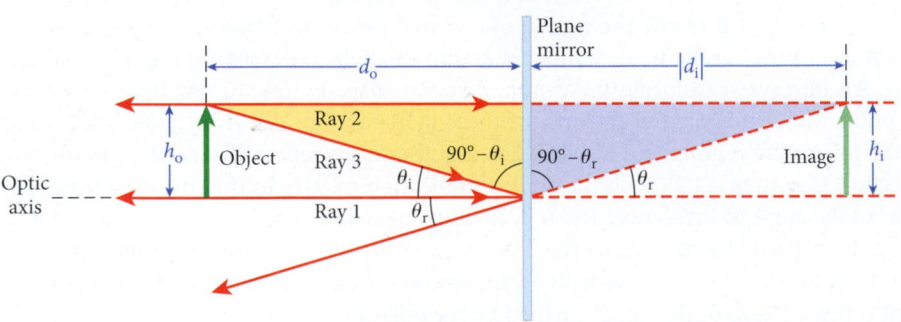

The extrapolations of these last two rays intersect at the point where the top of the image forms (Figure 32.11). It turns out that extrapolations of all rays from the arrow tip that hit the mirror—not just the two rays shown here—intersect at the top of the image. Thus, the mirror produces a virtual image on the opposite side of the mirror. This image has a height h_i and is located a distance d_i to the right of the mirror. In Figure 32.11, the yellow triangle is congruent with the blue triangle. Thus,

$$h_i = h_o \tag{32.3}$$

and

$$|d_i| = |d_o|. \tag{32.4}$$

By convention, the sign of the image distance for a virtual image produced by a mirror is negative. Thus, we use the absolute value of the image distance in equation 32.4. The image produced by a plane mirror appears to be the same distance behind the mirror as the object is in front of the mirror. In addition, the image appears upright and the same size as the object.

Why are mirror images left-right reversed but not up-down reversed? A person standing in front of a plane mirror sees an image of herself standing at the same distance behind the mirror as she is standing in front of the mirror (Figure 32.12). The image seen by the person can be constructed with two light rays as shown, but of course light rays are coming from every visible point of the person. The image is upright (it has the same orientation as the object, not upside-down) and virtual (the image is formed behind the surface of the mirror).

What about the left-right reversal? In Figure 32.13, the virtual image is again constructed with two rays, but all other rays behave the same way. The figure shows that the mirror actually does a front-to-back reversal, not a left-right or top-bottom reversal. If you hold an arrow pointing to the right and look in the mirror, the arrow still points to the right! You can see that the real person in Figure 32.13 has his watch on his right arm. He perceives that his virtual self has the watch on the left arm only because his brain imagines that the image is formed by a 180° rotation through a vertical axis, and not by a front-to-back reversal. If you find this too complicated, perhaps the following visualization will help you: Suppose you paint the letters representing your university's name on your forehead to get yourself ready for the big game. Then you put a length of clear packaging tape over the letters. As you pull the tape off, some of the paint sticks to it. If you pull the tape straight away from you, you see the back side of the tape that contacted the paint, and the letters are on it mirror-reversed. Looking at your image in the mirror is just like looking at the back side of the tape.

If you look at an image of yourself using a webcam on your computer, you see yourself as other people see you—the image from the webcam is not reversed front-to-back. This image is confusing after your long experience with seeing yourself in a mirror. Every movement you make seems to be made by your image in the opposite direction from what you expect. For this reason, some videoconferencing software now presents images that incorporate the front-to-back reversal inherent in mirror images, which makes for a more natural experience.

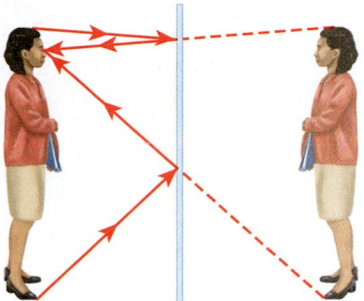

FIGURE 32.12 A person standing in front of a mirror sees a virtual image of herself.

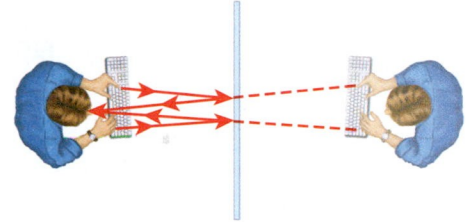

FIGURE 32.13 A person sitting in front of a mirror sees a mirror image of himself.

Concept Check 32.2

An observer (represented by the eye in the figure) is looking at a light bulb reflected in a plane mirror. To the observer, where does the image of the light bulb appear to be?

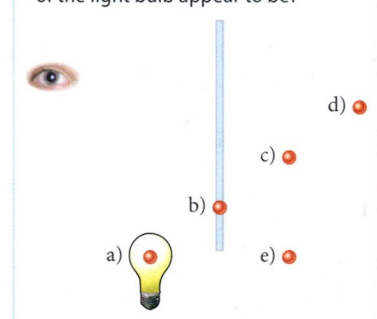

EXAMPLE 32.1 Full-Length Mirror

PROBLEM
A person who is 184 cm (6 ft $\frac{1}{2}$ in) tall wants to buy a mirror in which he can see his entire body. His eyes are 8 cm from the top of his head. What is the minimum height of the mirror that is needed?

SOLUTION
For simplicity, we represent the person as a pole 184 cm tall with "eyes" 8 cm from the top (this number does not matter, as the discussion below will show), as shown in Figure 32.14.

– Continued

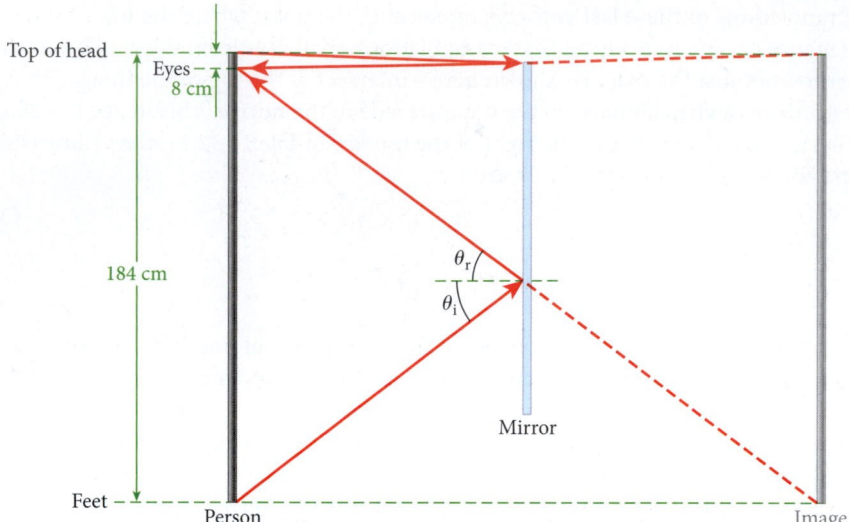

FIGURE 32.14 Distances and angles for a person standing in front of a mirror.

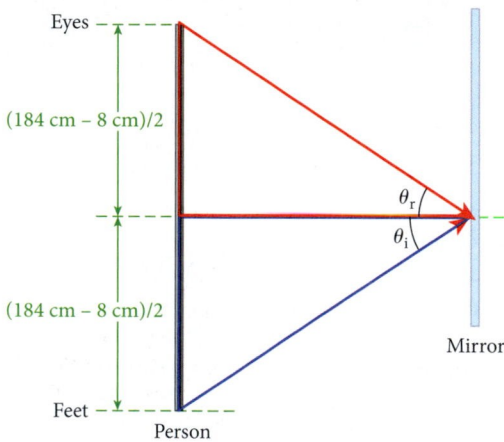

FIGURE 32.15 Two congruent triangles formed by the light from the person's feet.

First, consider where the light from the person's feet must travel to reach his eyes. The light leaving the feet is represented by a red arrow in Figure 32.14. The angle of incidence on the mirror, θ_i, is equal to the angle of reflection from the mirror, θ_r. We can draw two triangles that include these angles (Figure 32.15).

These two triangles are congruent because $\theta_i = \theta_r$, they each have a right angle, and they share a common side. Thus, the vertical sides of each triangle must have the same length. The sum of these two sides is equal to the height of the person minus the distance from the top of the person's head to his eyes. Therefore, the vertical side of each triangle has the length (184 cm − 8 cm)/2, as indicated in Figure 32.15.

We can now see that the bottom edge of the mirror needs to be at a height of (184 cm − 8 cm)/2 = 88 cm above the floor. A similar analysis of two congruent triangles gives us the location of the top edge of the mirror, which needs to be (8 cm)/2 = 4 cm below the top of the person's head. Therefore, the minimum height of the mirror is 184 cm − 4 cm − 88 cm = 92 cm. This height is exactly half the height of the person. Thus, a mirror that is half a person's height affords the person a full-length view. This result does not depend on the distance of the eyes from the top of the person's head or on how close to the mirror the person stands. However, it does depend on how the mirror is hung: The top of the mirror must be halfway between the eyes and the top of the head.

32.3   Curved Mirrors

When light is reflected from the surface of a curved mirror, the light rays follow the law of reflection at each point on the surface. The light rays that are parallel before they strike the mirror are reflected in different directions, depending on the part of the mirror that they strike. Depending on whether the mirror is concave or convex, the light rays can converge or diverge. What kind of curved mirrors are most useful? Parabolic mirrors work best in precise optical devices, as we will see later. However, spherical mirrors are used in many practical applications. Since geometrical constructions are somewhat simpler with spherical mirrors, we start with them. We will see that geometric optics for these mirrors is valid only for small angle scattering and that spherical aberration arises with larger angles.

Focusing Spherical Mirrors

Consider a spherical mirror with a reflecting surface on its inside. This is a **concave mirror.** Figure 32.16 represents this spherical reflecting surface as a segment of a circle. The optic

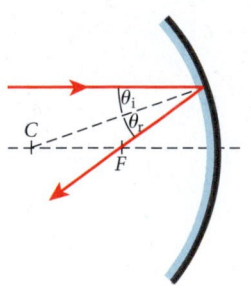

FIGURE 32.16 A horizontal light ray is reflected through the focal point of a concave mirror.

axis of the mirror, represented in the figure by a horizontal dashed line, is a line through the center of the sphere, which is labeled *C*. Imagine that a horizontal light ray above and parallel to the optic axis is incident on the surface of the mirror. The law of reflection applies at the point where the light ray strikes the mirror. The normal to the surface at this point, a dashed line in Figure 32.16, is a radial line that goes through the center of the sphere. For the isosceles triangle in Figure 32.16, you can see that each short side is about half the length of the long side, provided that θ_r is small. Thus, the reflected ray crosses the optic axis approximately halfway between the mirror and point *C*.

Now suppose there are many horizontal light rays incident on this spherical mirror (Figure 32.17). Each light ray obeys the law of reflection at the point where it strikes the mirror. Thus, each ray will cross the optic axis halfway between the mirror and point *C*. This crossing point, *F*, is called the **focal point**. Note that only horizontal rays incident on the mirror close to the optic axis will be reflected through the focal point, because only for such rays is the condition of a small reflection angle satisfied. (The distance of the incident ray from the optic axis in Figure 32.16 is probably the upper limit for satisfying that condition.) Unless otherwise specified, all horizontal rays are assumed to be close enough to the optic axis to pass through *F* on reflection.

Point *F* is halfway between point *C* and the surface of the mirror. Point *C* is located at the center of the sphere, so the distance of *C* from the surface of the mirror is just the radius of the mirror, *R*. The **focal length** is the distance along the optical axis between the mirror's surface and the focal point. Therefore the focal length *f* of a converging spherical mirror is

$$f = \frac{R}{2} \quad \text{(converging mirror)}. \qquad (32.5)$$

Light rays incident on an actual converging mirror are shown in Figure 32.18.

Now let's consider forming images with a converging mirror, such as the one in Figure 32.19. An object with height h_o is placed a distance d_o from the mirror, where $d_o > f$. The object is represented by an arrow, which indicates the height and orientation of the object. The object is oriented so that the tail of the arrow is on the optic axis, which, as before, is a normal to the surface of the spherical mirror along a line passing through the center *C* of the sphere. Four light rays determine where the image is formed.

1. The first light ray emanates from the bottom of the arrow along the optic axis and is usually not shown. This ray merely indicates that the bottom of the image is located on the optic axis.

2. The second light ray starts from the top of the arrow, is parallel to the optic axis, and is reflected through the focal point of the mirror.

3. The third ray starts from the top of the arrow, passes through the center of the sphere, *C*, and is reflected back on itself.

4. The fourth ray starts from the top of the arrow, passes through the focal point, *F*, and is reflected back parallel to the optic axis.

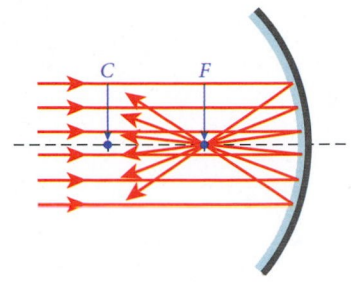

FIGURE 32.17 Many parallel light rays reflected through the focal point of a concave mirror.

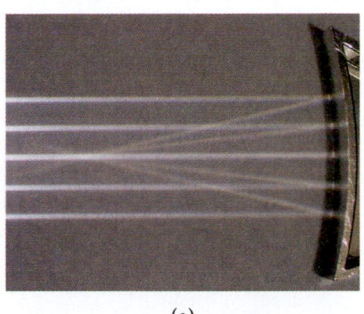

(a)

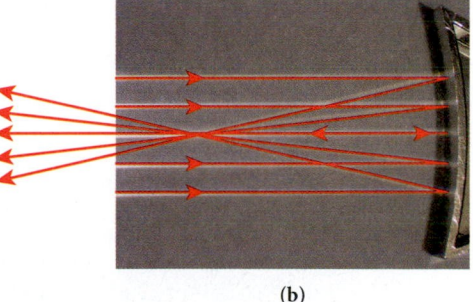

(b)

FIGURE 32.18 (a) Parallel light rays reflected to the focal point by a converging mirror. (b) Same image with arrows superimposed.

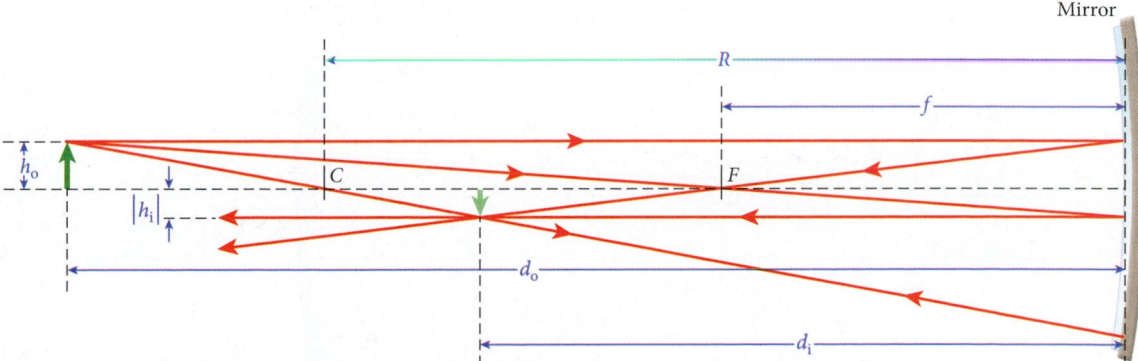

FIGURE 32.19 Image produced by a converging mirror when the object distance is greater than the focal length of the mirror.

The last three reflected rays intersect at the point where the image of the top of the object is formed (Figure 32.19). It turns out that all rays from the arrow tip that strike the mirror—not just the three shown here—intersect at the top of the image when they are reflected. Thus, we say that the mirror focuses the rays to form the image.

The reconstruction of the special case shown in Figure 32.19 shows a real image (on the same side of the mirror as the object, not behind the mirror) with height h_i at a distance d_i from the surface of the mirror. The image height, h_i, is assigned a negative value to denote that the image is inverted. By convention, the image distance, d_i, is defined as positive because the image is on the same side of the mirror as the object. The image is inverted and in this case is reduced in size relative to the object. An image is called real when a screen placed at the image location can display a sharp projection of the image. For a virtual image, the light rays do not go through the image and thus no light reaches a screen placed at the image location.

Now let's reconstruct another case for a converging mirror, where $d_o < f$ (Figure 32.20). The object stands on the optic axis, and three light rays determine where the image is formed.

1. The first ray merely indicates that the tail of the image lies on the optic axis and is usually not shown.

2. The second ray starts from the top of the object, is parallel to the optic axis, and is reflected through the focal point.

3. The third ray leaves the top of the object along a radius and is reflected back on itself through the center of the sphere.

The reflected rays are clearly diverging. To determine the location of the image, we must extrapolate the reflected rays to the other side of the mirror. These extrapolations intersect at a distance d_i from the surface of the mirror, producing an image with height h_i.

In this case, the image is formed on the opposite side of the mirror from the object, a virtual image. (By convention, the distance d_i is assigned a negative value to denote that the image is virtual.) To an observer, the image appears to be behind the mirror, as with a plane mirror. The image is upright and larger than the object. These results for $d_o < f$ are quite different from those for $d_o > f$. Mirrors used for shaving or applying cosmetics are usually converging mirrors; the user puts his or her face closer than the focal length, producing a large, upright image.

Before treating diverging mirrors, let's formalize the sign conventions for distances and heights.

1. All distances on the same side of the mirror as the object are defined to be positive, and all distances on the opposite side of the mirror from the object are defined to be negative. Thus, f and d_o are positive for converging mirrors.

2. For real images, d_i is positive. For virtual images, d_i is negative.

3. If the image is upright, then h_i is positive; if the image is inverted, h_i is negative.

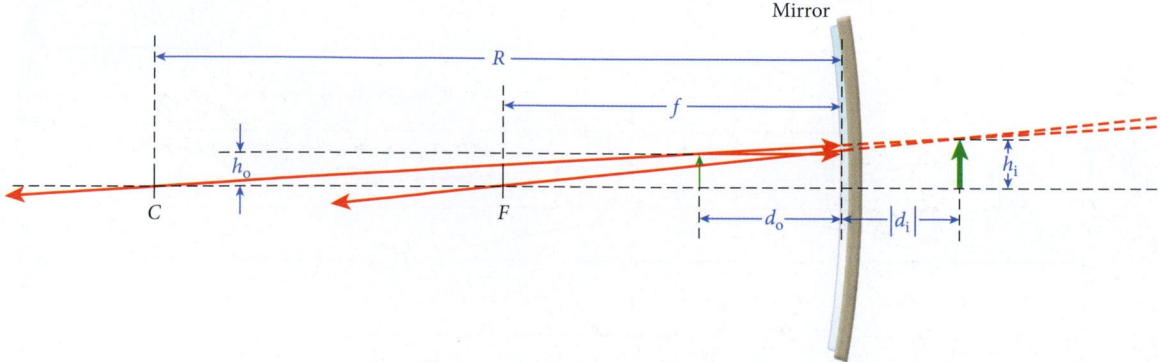

FIGURE 32.20 Image produced by a converging mirror when the object distance is less than the focal length of the mirror.

We can derive the **mirror equation** (equation 32.6), which relates the object distance, d_o, the image distance, d_i, and the focal length of a mirror, f:

$$\frac{1}{d_o} + \frac{1}{d_i} = \frac{1}{f}.$$ (32.6)

The signs of the terms in this equation are based on the conventions just defined.

DERIVATION 32.1 / Spherical-Mirror Equation

The spherical-mirror equation can be derived by starting with a converging mirror that has a focal length f. An object h_o tall stands at a distance d_o from the surface of the mirror on the optic axis (Figure 32.21).

Trace a ray from the top of the object parallel to the optic axis and reflect it through the focal point. Trace a second ray from the top of the object through the focal point and reflect it parallel to the optic axis. The image is formed with height h_i at distance d_i from the mirror. Form a right triangle with height h_o and base d_o (green triangle 1 in Figure 32.21). Form a second right triangle with $-h_i > 0$ as the height and d_i as the base (green triangle 2 in Figure 32.21). From the law of reflection for a ray striking the mirror on the optic axis, the indicated angles in the two triangles are the same, so the two triangles are similar. Thus,

$$\frac{h_o}{d_o} = \frac{-h_i}{d_i} \quad \text{or} \quad \frac{-h_i}{h_o} = \frac{d_i}{d_o}.$$ (i)

Now let us look at the geometry for two different triangles. One right triangle (yellow triangle 4 in Figure 32.22) is defined by the height $-h_i$ and the base length $d_i - f$. The second triangle (yellow triangle 3 in Figure 32.22) is defined by the height h_o and the base f. The two triangles are similar, so

$$\frac{h_o}{f} = \frac{-h_i}{d_i - f} \quad \text{or} \quad \frac{-h_i}{h_o} = \frac{d_i - f}{f}.$$

Substituting from equation (i) for the ratio of the heights gives

$$\frac{d_i}{d_o} = \frac{d_i - f}{f}.$$

Multiplying both sides of this equation by f yields

$$\frac{fd_i}{d_o} = d_i - f \Rightarrow \frac{fd_i}{d_o} + f = d_i.$$

Finally, dividing both sides of this equation by the product fd_i leads to the spherical-mirror equation:

$$\frac{1}{d_o} + \frac{1}{d_i} = \frac{1}{f}.$$

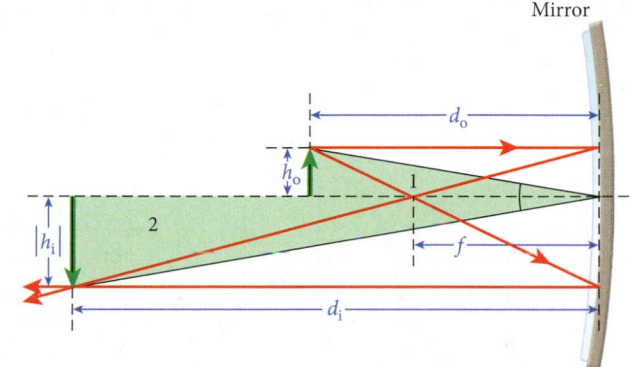

FIGURE 32.21 The two similar triangles 1 and 2 in the derivation of the spherical-mirror equation.

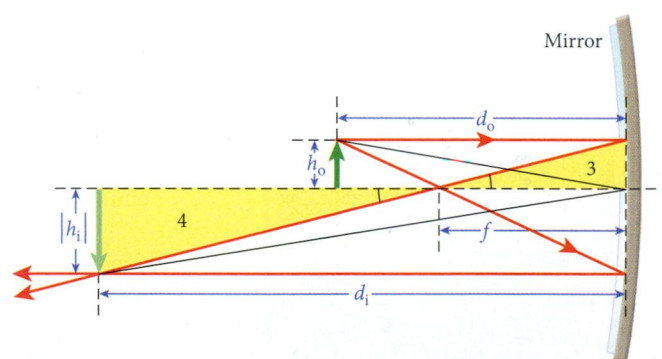

FIGURE 32.22 The two similar triangles 3 and 4 in the derivation of the spherical-mirror equation.

Concept Check 32.3

The mirror equation tells us that

a) moving an object farther from a converging mirror means that the image will be formed farther from the mirror.

b) moving an object farther from a converging mirror means that the image will be formed closer to the mirror.

c) either of the above could occur, depending on the focal length of the mirror.

The **magnification**, m, of a mirror is defined as the ratio of the size of the image to the size of the object, $m = h_i/h_o$. In the Derivation 32.1, we saw that this ratio is the negative of the ratio of the image distance to the object distance. Thus, we have

$$m = \frac{h_i}{h_o} = -\frac{d_i}{d_o}.$$ (32.7)

Note that m is negative for the situation used in the derivation. Algebraically, this occurs because $h_i < 0$. The significance of $m < 0$ is that it tells us that the image is inverted. Table 32.1 summarizes the characteristics of images formed by a converging mirror for five different classes of object distances.

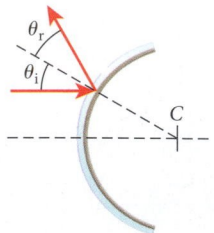

FIGURE 32.23 Reflection of a horizontal light ray from a diverging spherical mirror.

Table 32.1	**Image Characteristics for Converging Mirrors**		
Case	**Image Type**	**Image Orientation**	**Magnification**
$d_o < f$	Virtual	Upright	Enlarged
$d_o = f$	Real	Upright	Image at infinity
$f < d_o < 2f$	Real	Inverted	Enlarged
$d_o = 2f$	Real	Inverted	Same size
$d_o > 2f$	Real	Inverted	Reduced

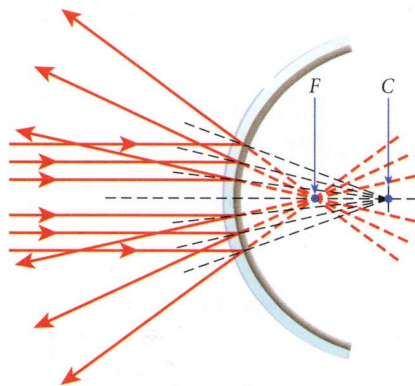

FIGURE 32.24 The reflection of parallel light rays from the surface of a diverging mirror.

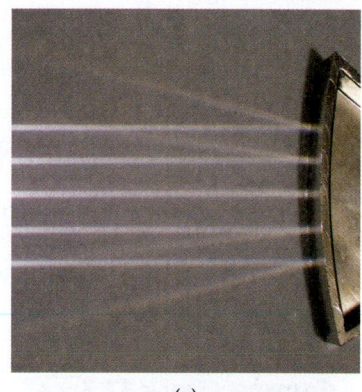

(a)

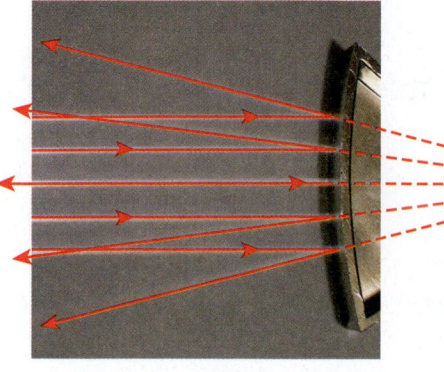

(b)

FIGURE 32.25 (a) Parallel light rays reflected from a diverging spherical mirror. (b) Arrows corresponding to the light rays, with the dashed lines representing the extrapolated reflected rays.

Diverging Spherical Mirrors

Suppose we have a spherical mirror with the reflecting surface on the outside of the sphere (Figure 32.23). This is a **convex mirror,** and the reflected rays diverge. In Figure 32.23, this spherical reflecting surface is indicated by a semicircle. The optic axis of the mirror is a line through the center of the sphere, represented by the horizontal dashed line. Imagine that a horizontal light ray above the optic axis is incident on the surface of the mirror. At the point where the light ray strikes the mirror, the law of reflection applies.

In contrast to the normal for a converging mirror, the normal to the surface of a diverging mirror points *away* from the center of the sphere. When we extrapolate the normal through the surface of the sphere, it intersects the optic axis of the sphere at the sphere's center, marked C in Figure 32.23. When we observe the reflected ray, it seems to be coming from inside the sphere.

Suppose many horizontal light rays are incident on this spherical mirror (Figure 32.24). Each light ray obeys the law of reflection. The rays diverge and do not seem to form any kind of image. However, if the reflected rays are extrapolated through the surface of the mirror, they all intersect the optic axis at one point. This point is the focal point of this diverging mirror.

Figure 32.25a shows five parallel light rays incident on an actual diverging spherical mirror. The dashed lines in Figure 32.25b represent the extrapolation of the reflected rays to show the focal point.

Now let's discuss images formed by diverging mirrors (Figure 32.26). Again we use three rays.

1. The first ray establishes that the tail of the arrow lies on the optic axis and is usually not shown.

2. The second ray starts from the top of the object, travels parallel to the optic axis, and is reflected from the surface of the mirror in such a way that its extrapolation crosses the optic axis a distance from the surface of the mirror equal to the focal length of the mirror.

3. The third ray begins at the top of the object and is directed so that its extrapolation intersects the center of the sphere. This ray is reflected back on itself.

The reflected rays diverge, but the extrapolated rays converge a distance d_i from the surface of the mirror on the side of the mirror opposite the object. The rays converge at a distance h_i above the optic axis, forming an upright, reduced image on the side of the mirror opposite the object. This image is virtual because the light rays do not actually go through it. These characteristics (upright, reduced, virtual image) are valid for all object distances for a diverging mirror. Such mirrors are often used in stores to give clerks at the front of the store a wide view down the aisles and as passenger side rear-view mirrors on cars.

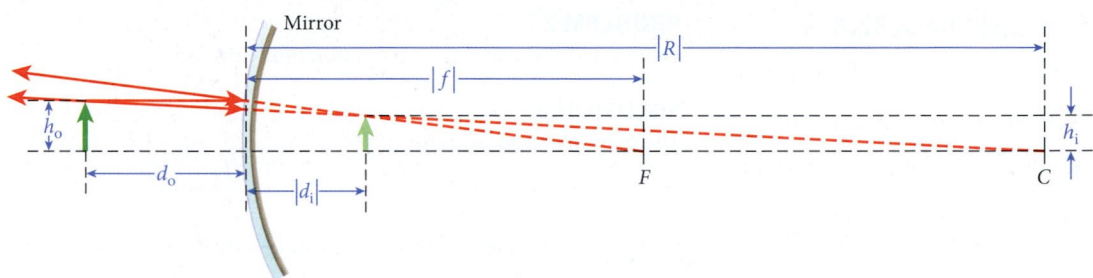

FIGURE 32.26 Image formed by an object placed in front of a diverging spherical mirror.

In the case of a diverging mirror, the focal length, f, is negative because the focal point of the mirror is on the side opposite the object. A negative value is also assigned to the radius of a diverging mirror. Thus,

$$f = \frac{R}{2} \quad \text{(with } R < 0 \text{ for a diverging mirror).}$$

The object distance, d_o, is always taken to be positive. Rearranging the mirror equation (equation 32.6), which is also valid for a diverging mirror, gives

$$d_i = \frac{d_o f}{d_o - f}. \tag{32.8}$$

If d_o is always positive and f is always negative, d_i is always negative. Applying equation 32.7 for the magnification, we find that m is always positive. Looking at Figure 32.26 will also convince you that the image is always reduced in size. Thus, a diverging mirror (even if $d_o > |f|$) always produces a virtual, upright, and reduced image.

Self-Test Opportunity 32.1

You are standing 2.50 m from a diverging mirror with focal length $f = -0.500$ m. What do you see in the mirror? (The answer "myself" is not good enough!)

EXAMPLE 32.2 | Image Formed by a Converging Mirror

Consider an object 5.00 cm tall placed 55.0 cm from a converging mirror with focal length 20.0 cm (Figure 32.27).

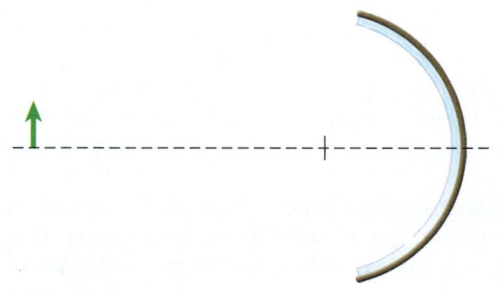

FIGURE 32.27 An object (green arrow) whose image is to be formed by a converging mirror.

PROBLEM 1
Where is the image produced?

SOLUTION 1
We can use the mirror equation to find the image distance, d_i, in terms of the object distance, d_o, and the focal length of the mirror, f:

$$\frac{1}{d_o} + \frac{1}{d_i} = \frac{1}{f}.$$

The mirror is converging, so the focal length is positive. The image distance is

$$d_i = \frac{d_o f}{d_o - f} = \frac{(55.0 \text{ cm})(20.0 \text{ cm})}{55.0 \text{ cm} - 20.0 \text{ cm}} = 31.4 \text{ cm.}$$

– Continued

Concept Check 32.4

Moving the object in Example 32.2 10% closer to the mirror would produce an image that was

a) closer to the mirror and larger than before.

b) closer to the mirror and smaller than before.

c) farther from the mirror and larger than before.

d) farther from the mirror and smaller than before.

e) It is not possible to predict the image's characteristics.

PROBLEM 2

What are the size and the orientation of the image?

SOLUTION 2

The magnification, m, is given by

$$m = \frac{h_i}{h_o} = -\frac{d_i}{d_o},$$

where h_o is the height of the object and h_i is the height of the image. The image height is then

$$h_i = -h_o \frac{d_i}{d_o} = -(5.00 \text{ cm})\frac{31.4 \text{ cm}}{55.0 \text{ cm}} = -2.85 \text{ cm}.$$

The magnification is

$$m = \frac{h_i}{h_o} = \frac{-2.85 \text{ cm}}{5.00 \text{ cm}} = -0.570.$$

Thus, the image is inverted and reduced.

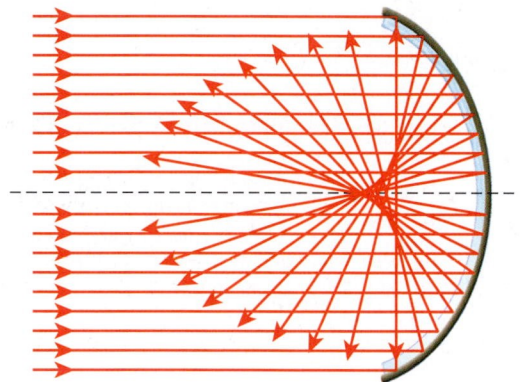

FIGURE 32.28 Parallel light rays incident on a spherical converging mirror, demonstrating spherical aberration.

Spherical Aberration

The equations we have derived for spherical mirrors $\left(\frac{1}{d_o} + \frac{1}{d_i} = \frac{1}{f} \right.$ and $m = \frac{h_i}{h_o} = -\frac{d_i}{d_o} \left.\right)$ apply only to light rays that are close to the optic axis. If the light rays are far from the optic axis, they will not be focused through the focal point of the mirror, leading to a distorted image. Strictly speaking, there is no precise focal point in this situation. This condition is called **spherical aberration.**

Figure 32.28 shows several light rays not close to the optic axis incident on a spherical converging mirror. The rays farther from the optic axis are reflected in such a way that they cross the optic axis closer to the mirror than do rays that strike the mirror closer to the axis. As the rays approach the optic axis, they are reflected through points closer and closer to the focal point.

DERIVATION 32.2 / Spherical Aberration for Converging Mirrors

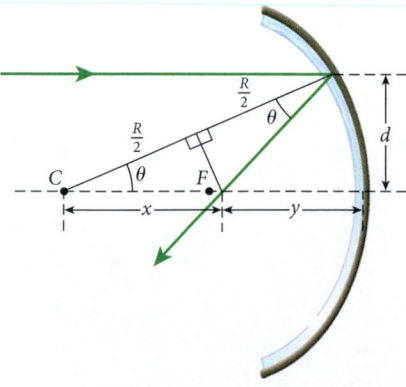

FIGURE 32.29 Geometry for the spherical aberration of a converging spherical mirror.

Up until now we have assumed that light rays parallel to the optic axis and close to it are reflected in such a way that they cross the optic axis at the focal point of the mirror. The focal point is at half the radius of curvature of the mirror. However, light rays that are far from the optic axis are not reflected through the focal point. To derive an expression for the point at which rays parallel to but far from the optic axis cross the axis, we use the geometry shown in Figure 32.29.

Start with a light ray parallel to the optic axis and at a distance d from it. The ray reflects and crosses the axis a distance y from the mirror. The ray makes an angle θ with respect to the normal to the mirror's surface, which is a radius R of the spherical mirror. The law of reflection tells us that the angle of incidence of the ray on the surface of the mirror is equal to the angle of reflection.

Define the distance from the center of the spherical mirror to the point at which the reflected ray crosses the optic axis as x. Draw a perpendicular line from the point where the ray crosses the axis to the normal. This line forms two congruent right triangles with angle θ, hypotenuse x, and adjacent side $R/2$, such that

$$\cos\theta = \frac{R/2}{x} \Rightarrow x = \frac{R}{2\cos\theta}.$$

We can also express θ in terms of d as

$$\sin\theta = \frac{d}{R} \Rightarrow \theta = \sin^{-1}\left(\frac{d}{R}\right).$$

The distance y is given by

$$y = R - x = R - \frac{R}{2\cos\theta} = R\left(1 - \frac{1}{2\cos\theta}\right) = R\left(1 - \frac{1}{2\cos\left(\sin^{-1}\left(\frac{d}{R}\right)\right)}\right).$$

Remembering the trigonometric identity $\cos(\sin^{-1}(x)) = \sqrt{1-x^2}$, we can rewrite our result for y as

$$y = R\left(1 - \frac{1}{2\sqrt{1-\left(\frac{d}{R}\right)^2}}\right) = \frac{R}{2}\left(2 - \frac{1}{\sqrt{1-\left(\frac{d}{R}\right)^2}}\right) = \frac{R}{2}\left(2 - \left[1-\left(\frac{d}{R}\right)^2\right]^{-1/2}\right).$$

We can see that for $d \ll R$, $1 - (d/R)^2 \approx 1$ and $y \approx R/2$, which agrees with equation 32.5. We can make a better approximation by writing the Maclaurin series expansion

$$\left(1-x^2\right)^{-1/2} \approx 1 + \frac{x^2}{2} \quad \text{(for } x \ll 1\text{)}.$$

Taking $x = (d/R) \ll 1$, we can make the approximation

$$y = \frac{R}{2}\left(2 - \left[1-\left(\frac{d}{R}\right)^2\right]^{-1/2}\right) \approx \frac{R}{2}\left(2 - \left[1 + \frac{1}{2}\left(\frac{d}{R}\right)^2\right]\right) = \frac{R}{2}\left(1 - \frac{d^2}{2R^2}\right).$$

Thus, the amount of spherical aberration is given approximately by

$$\frac{R}{2} - y \approx \frac{d^2}{4R} \quad \left(\text{for } \frac{d}{R} \ll 1\right).$$

Self-Test Opportunity 32.2

For a converging spherical mirror with $R = 7.20$ cm, assume that the incident light rays are not close to the optic axis of the mirror but are parallel to the axis. Calculate the position at which the reflected light ray intersects the optic axis if the incident ray is

a) 0.720 cm away from the optical axis.

b) 0.800 cm away from the optical axis.

c) 1.80 cm away from the optical axis.

d) 3.60 cm away from the optical axis.

Parabolic Mirrors

Parabolic mirrors have a surface that reflects light from a distant source to the focal point from anywhere on the mirror. Thus, the full size of the mirror can be used to collect light and form images that do not suffer from spherical aberrations. Figure 32.30 shows vertical light rays incident on a parabolic mirror. All rays are reflected through the focal point of the mirror.

If the surface of a parabolic mirror is described by the equation $y(x) = ax^2$, then its focal point is located at the point $(x = 0, y = 1/(4a))$. Its focal length is therefore

$$f = \frac{1}{4a}. \tag{32.9}$$

Parabolic mirrors are more difficult to produce than spherical mirrors and are accordingly more expensive. Most large reflecting telescopes use parabolic mirrors in order to avoid spherical aberration. Many automobile headlights use parabolic reflectors for the same reason, but these send light in the opposite direction: The light source is placed at the focal point, and the reflector sends the light out in a strong beam parallel to the optic axis, as illustrated in Figure 32.31. (However, incandescent headlights are likely to be replaced with LED headlights, which will use small lenses to accomplish the same thing.)

Figure 32.31a shows the low-beam and high-beam headlights for a car. The parabolic reflectors for headlights are faceted. A facet is a flat area on the reflector. These facets help produce the light distribution desired for the headlight. Figure 32.31b shows the light bulb for the high-beam headlight at the focus of a parabolic mirror that functions as a reflector for the light produced by the light bulb, producing a focused parallel beam of light.

The satellite TV antennas ("dishes") found on many rooftops are also parabolic in shape. These satellite dishes are not mirrors in the sense that they do not reflect visible light in specular fashion. But they are parabolic reflectors for radiation in the wavelength range used for satellite TV transmission.

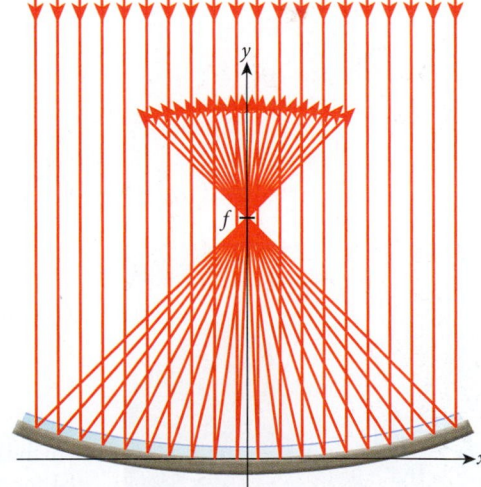

FIGURE 32.30 Light rays reflected by a parabolic mirror.

FIGURE 32.31 (a) Headlight assembly for a car. The left light is the low-beam headlight, and the right light is the high-beam headlight. (b) Diagram of the high-beam headlight showing the light bulb at the focus of the parabolic reflector.

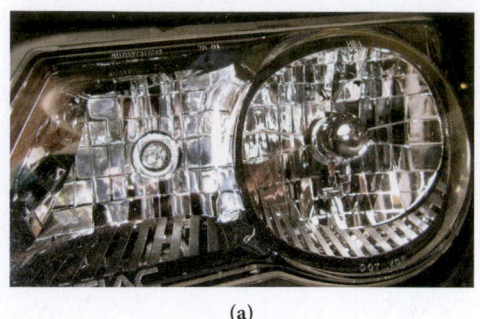

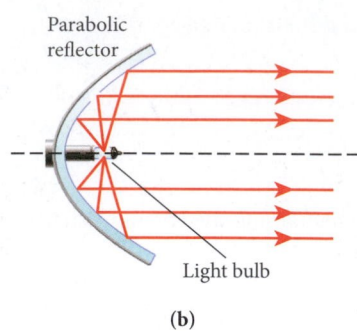

Parabolic reflector

Light bulb

(a)   (b)

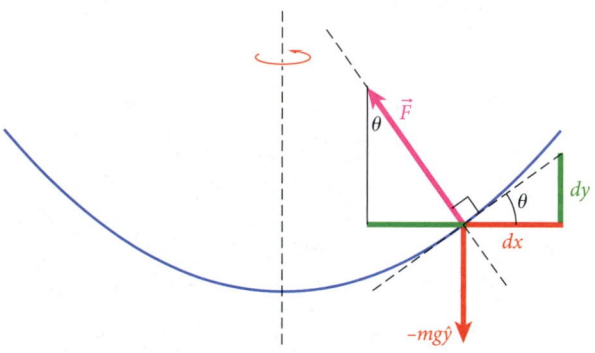

FIGURE 32.32 Geometry of a rotational parabola and free-body diagram for a fluid surface element. The force $\vec{F}$ exerted by the fluid on a surface element is shown in magenta. It needs to balance the force of gravity (red) and provide the centripetal force (green) necessary to keep the surface element moving in a circular path.

Self-Test Opportunity 32.3

Show that the derivation of $y(x) = (\omega^2/2g)x^2$ for the surface of a rotating liquid can be accomplished using energy considerations.

FIGURE 32.33 University of British Columbia liquid-mirror telescope.

Rotation Parabolas

For precise optical applications, clearly it is best to use parabolic mirrors. For example, in Chapter 33, we will see that huge parabolic mirrors are used in telescopes. One very interesting way to create a parabolic mirror is to put a liquid into rotational motion. At every point on the surface of the liquid, the surface will be perpendicular to the force from the fluid acting on that surface element. This force, $\vec{F}$, adds to the force of gravity acting on the surface element, $-mg\hat{y}$, to provide the net centripetal force, which is needed to keep the surface element on a circular path (Figure 32.32). In the xy-coordinate system chosen here, the centripetal force is $-m\omega^2 x\hat{x}$ (refer to Chapter 9 on circular motion).

The angle θ of the surface element with respect to the horizontal is given by $\tan\theta = dy/dx$ (see Figure 32.32). The same angle also can be used to express the components of the force $\vec{F}$. The vertical component of $\vec{F}$ has to balance the force of gravity, and the horizontal component has to provide the net centripetal force:

$$F\cos\theta = mg$$

$$F\sin\theta = m\omega^2 x.$$

Dividing these two equations by each other leads to $\tan\theta = (\omega^2/g)x$. Previously we showed that $\tan\theta = dy/dx$, so

$$\frac{dy}{dx} = \frac{\omega^2}{g}x.$$

Integration then results in the desired shape of the surface

$$y(x) = \frac{\omega^2}{2g}x^2,$$

which is a parabola. Because the focal length of a parabola of the form $y = ax^2$ is $f = 1/(4a)$, the focal length of this parabolic mirror made of a rotating liquid is

$$f = \frac{g}{2\omega^2}.$$

Rotation of liquid surfaces has been successfully used to construct large telescope mirrors. One such mirror is shown in Figure 32.33. A very large version of such a telescope has been designed to be installed on the Moon. While this may sound like science fiction at the present, such a telescope would be much cheaper to construct than one with a solid mirror. Since the Moon has no atmosphere, the telescope would not suffer from the atmospheric distortions that all ground-based telescopes suffer from on Earth. And it could be constructed on a much larger scale than what is possible with satellite-based telescopes like the Hubble Space Telescope. (We will cover telescopes in much greater detail in Chapter 33.)

Concentrating Solar Power

The energy contained in solar radiation can be harnessed by solar power plants, which accomplish this in two different ways. One involves photovoltaic processes, which we will discuss in Chapter 36. The other method focuses the Sun's rays onto a focal point, where their energy can be used for heating water and/or generating steam, which in turn can be used for electricity production.

One ray-focusing design uses parabolic troughs (see Figure 32.34). The cross section of one of these troughs looks like Figure 32.30. At the focal point of the parabola is a liquid-filled heat transfer tube, which captures the solar energy for processing into electricity. As the Sun moves across the sky during the day, the parabolic troughs have to be rotated so that their optical axes point in the direction of the Sun.

Another design for a solar power plant uses a large array of flat mirrors to focus the Sun's rays on a focal point at the top of a tower (see Figure 32.35). At that focal point, temperatures of a few thousand degrees can be reached, allowing for very efficient conversion of solar energy into electricity.

FIGURE 32.34 Large-scale solar farm utilizing parabolic troughs.

FIGURE 32.35 The world's first commercial tower solar farm, an 11-MW installation in Sanlúcar la Mayor, Spain.

32.4 Refraction and Snell's Law

Light travels at different speeds in different optically transparent materials. The ratio of the speed of light in vacuum to the speed of light in a material is called the **index of refraction** of the material. The index of refraction, n, is therefore defined as

$$n = \frac{c}{v}, \tag{32.10}$$

where c is the speed of light in vacuum and v is the speed of light in the material. The speed of light in a physical medium such as glass is always less than the speed of light

Table 32.2	Indices of Refraction for Some Common Materials*
Material	**Index of Refraction**
Gases	
Air	1.000271
Helium	1.000036
Carbon dioxide	1.00045
Liquids	
Water	1.333
Methyl alcohol	1.329
Ethyl alcohol	1.362
Glycerine	1.473
Benzene	1.501
Typical oil	1.5
Solids	
Ice	1.310
Calcium fluoride	1.434
Fused quartz	1.46
Salt	1.544
Polystyrene	1.49
Typical Lucite	1.5
Typical glass	1.5
Quartz	1.544
Diamond	2.417

*For light with wavelength 589.3 nm

in vacuum. Thus, the index of refraction of a material is always greater than or equal to 1, and by definition the index of refraction of vacuum is 1. Table 32.2 lists the indices of refraction for some common materials. In general, the index of refraction is a function of the wavelength of the light, but the table gives typical average values for visible light. We will return to the wavelength dependence of the index of refraction in a later subsection on chromatic dispersion.

When light crosses the boundary between two transparent materials, some of it is reflected, but usually some of it crosses the boundary into the other material and in the process is refracted. **Refraction** means that the light rays do not travel in a straight line across the boundary, but change direction. When light crosses a boundary from a medium with a lower index of refraction, n_1, to a medium with a higher index of refraction, n_2, the light rays change their direction and bend *toward* the normal to the boundary (Figure 32.36). The change in direction toward the normal means that the angle of refraction, θ_2, is less than the angle of incidence, θ_1.

Figure 32.37 shows actual light rays in air incident on the boundary between air and glass. The light rays are refracted toward the normal. (The reflected light is also observable.)

When light crosses a boundary from a medium with a higher index of refraction n_1 to a medium with a lower index of refraction n_2, the light rays are bent *away* from the normal (Figure 32.38). Bending away from the normal means that $\theta_2 > \theta_1$.

Figure 32.39 shows actual light rays in glass incident on the boundary between glass and air. The light rays are refracted away from the normal.

From measurements of the angles of incident and refracted rays in media with different indices of refraction, we can construct a law of refraction. This law of refraction based on empirical observations can be expressed as

$$n_1 \sin \theta_1 = n_2 \sin \theta_2, \tag{32.11}$$

where n_1 and θ_1 are the index of refraction and angle of incidence (relative to the surface normal) in the first medium, and n_2 and θ_2 are the index of refraction and angle of the transmitted ray (relative to the surface normal) in the second medium. This law of refraction is also called **Snell's Law.** It cannot be proven using ray optics alone; however, in Chapter 34 this law is derived using wave optics.

The index of refraction of air is very close to 1, as shown in Table 32.2, and in this book we will always assume that the index of refraction of air is $n_{air} = 1$. It is very common

(a)

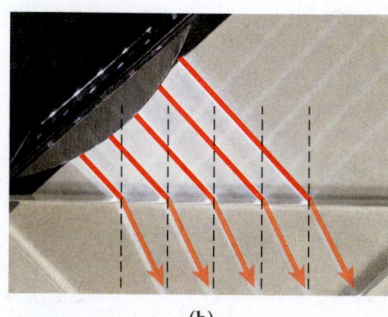

(b)

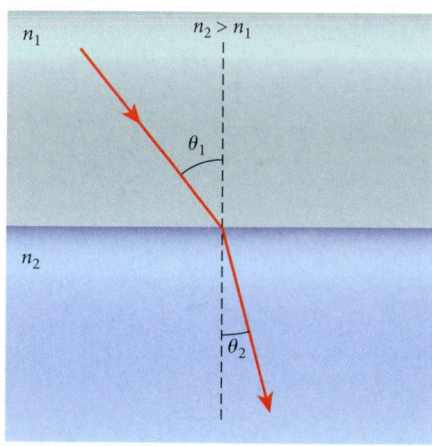

FIGURE 32.36 Light rays refracted at the boundary between two optical media with $n_1 < n_2$. (The reflected light ray is not shown.)

FIGURE 32.37 (a) Light rays refracted when crossing the boundary between air and glass. (b) Arrows superimposed on the light rays. The dashed lines are normal to the boundary.

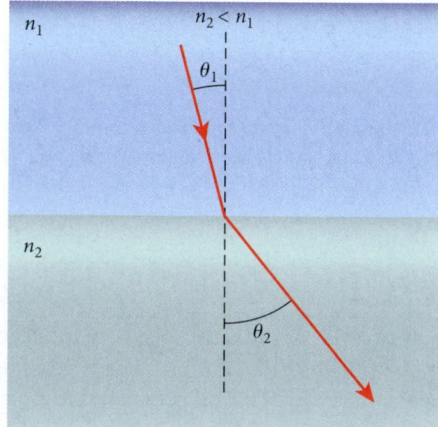

FIGURE 32.38 Light rays refracted at the boundary between two optical media with $n_1 > n_2$.

for light to be incident on various media from air, so we can write the formula for refraction of light incident on a surface of a medium with index of refraction n_{medium} as

$$n_{medium} = \frac{\sin \theta_{air}}{\sin \theta_{medium}}, \qquad (32.12)$$

where θ_{air} is the angle of incidence (relative to the surface normal) for light in air and θ_{medium} is the angle of the refracted light (relative to the surface normal) in the medium. Note that θ_{air} is always greater than θ_{medium}!

Also note that for light coming out of a medium into air, the refraction formula is the same (equation 32.12)! This is a special case of the general statement advanced earlier: All ray-tracing figures remain valid if the direction of the light rays is reversed. For example, changing the direction of the arrows on the red line in Figure 32.38 still yields a physically valid path for a light ray crossing the boundary between the two media.

Fermat's Principle

We have just stated that Snell's Law cannot be proven using ray optics alone. But there is a qualifier to this statement: If you accept Fermat's Principle, then Snell's Law emerges automatically. **Fermat's Principle**, or the **principle of least time**, as expressed by Fermat in 1657, states that the path taken by a ray of light between two points in space is the path that takes the least time.

How can we prove Snell's Law this way? The key is the connection between the speed of light in a medium, v, and the medium's index of refraction, n, which was established in equation 32.10. If light moves at different speeds in two media, then Fermat's Principle is analogous to the logic used in determining who won the aquathlon race, which we did in Example 2.6 and the following text. In that problem, we calculated the angle at which a contestant must swim to shore in a race that consists of a swim to shore (with one given speed) and a run along the shore (with another speed). Winning the race means taking the minimum time, which is exactly what Fermat's Principle postulates.

If you accept Fermat's Principle and you want to prove Snell's Law, then all you have to do is to revisit the solution to Example 2.6. But of course you can ask where Fermat's Principle comes from. Using the minimum time to win a race seems obvious. But what makes the light rays understand that they should take the minimum time to get from point A to point B? The answer to this can be found in Huygens's Principle. However, to gain an understanding of the physics underlying that principle, you need to wait until Chapter 34, when we revisit the model of light as a wave.

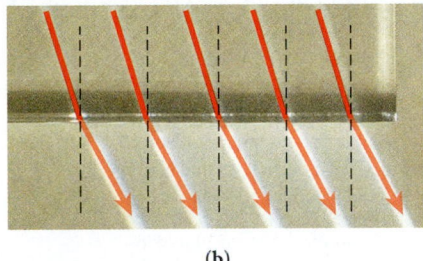

(a)

(b)

FIGURE 32.39 (a) Light rays refracted when crossing the boundary between glass and air. (b) Arrows superimposed on the light rays. The dashed lines are normal to the boundary.

EXAMPLE 32.3 | Apparent Depth

You are standing on the edge of a pond and looking at the water at an angle of $\theta_2 = 45.0°$ to the vertical, as shown in Figure 32.40. You see a fish in the pond. The actual depth of the fish is $d_{actual} = 1.50$ m.

PROBLEM
What is the apparent depth of the fish from your vantage point?

SOLUTION
Light rays from the fish are refracted at the surface of the water into your eyes at a 45.0° angle as shown in Figure 32.40. In the figure we show a central ray as representative of a collection of rays emitted from one point on the body of the fish (semitransparent cone). Where the eyes see these rays intersect is where you locate the fish.

The apparent depth of the fish is located along the upper dashed horizontal line in Figure 32.40, extrapolated from the direction of the light rays as they enter your eye. This extrapolated line makes an angle $\theta_2 = 45.0°$ with the normal to the surface of the water.

– Continued

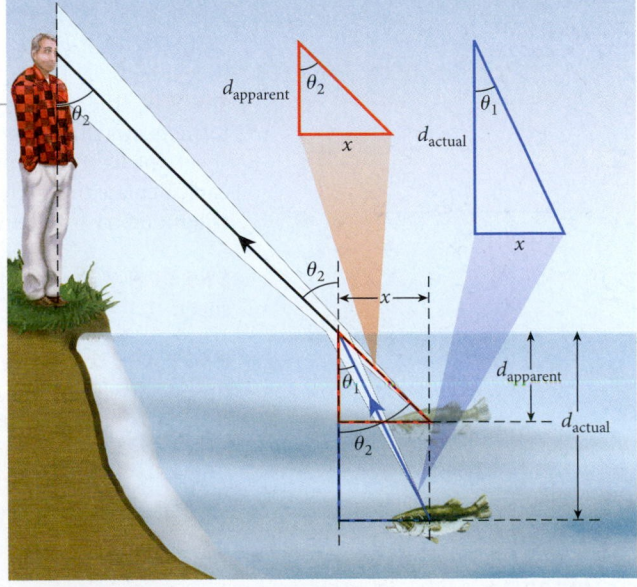

FIGURE 32.40 You are looking at the surface of a pond at an angle of $\theta_2 = 45.0°$, and you see a fish under the water. The rays that are emitted from one point on the body of the fish are shown as a semitransparent cone, with the center line of the cone represented by a solid ray.

Using Snell's Law, we can calculate the angle θ_1 that the light rays from the fish must make at the surface of the water:

$$n_{air}\sin\theta_2 = n_{water}\sin\theta_1.$$

Taking $n_{air}=1.00$ and $n_{water}=1.333$ from Table 32.2, we can calculate θ_1:

$$\theta_1 = \sin^{-1}\left(\frac{\sin\theta_2}{n_{water}}\right) = \sin^{-1}\left(\frac{\sin 45.0°}{1.333}\right) = 32.0°.$$

We define the horizontal distance between the point where the light rays intersect the surface of the water and a point on the back of the fish as x, as shown in Figure 32.40. We can then define the red triangle and the blue triangle. From the red triangle, we get

$$\tan\theta_2 = \frac{x}{d_{apparent}},$$

and from the blue triangle, we obtain

$$\tan\theta_1 = \frac{x}{d_{actual}}.$$

Solving each of these two equations for x and setting the resulting expressions equal to each other gives us

$$d_{actual}\tan\theta_1 = d_{apparent}\tan\theta_2,$$

which we can solve for the apparent depth of the fish:

$$d_{apparent} = \frac{d_{actual}\tan\theta_1}{\tan\theta_2} = \frac{(1.50\ \text{m})\tan(32.0°)}{\tan(45.0°)} = 0.937\ \text{m}.$$

Thus, the fish appears to be closer to the water surface than it actually is.

Concept Check 32.7

If the same fish swimming at the same depth is seen by a different observer whose line of sight is at an angle larger than that given in Example 32.3, then, to that observer, the fish's apparent depth will be

a) larger than that calculated in Example 32.3.

b) the same as that calculated in Example 32.3.

c) smaller than that calculated in Example 32.3.

SOLVED PROBLEM 32.2 | Displacement of Light Rays in Transparent Material

A light ray in air is incident on a sheet of transparent material with thickness $t = 5.90$ cm and with index of refraction $n = 1.50$. The angle of incidence is $\theta_{air} = 38.5°$ (Figure 32.41a).

PROBLEM
What is the distance d that the light ray is displaced after it has passed through the sheet?

SOLUTION

THINK The light ray refracts toward the normal as it enters the transparent sheet and refracts away from the normal as it exits the transparent sheet. After passing through the sheet, the light ray is parallel to the incident light ray, but is displaced. Using Snell's Law, we can calculate the angle of refraction in the transparent sheet. Having that angle, we can use trigonometry to calculate the displacement of the light ray after passing through the sheet.

SKETCH Figure 32.41b shows the geometry for the light ray passing through the transparent sheet.

RESEARCH The light ray is incident on the sheet with an angle θ_{air}. Snell's Law relates the incident angle to the refracted angle, θ_{medium}, through the index of refraction of the transparent material, n:

$$n = \frac{\sin\theta_{air}}{\sin\theta_{medium}}.$$

We can solve this equation for the refraction angle:

$$\theta_{medium} = \sin^{-1}\left(\frac{\sin\theta_{air}}{n}\right). \tag{i}$$

The distance the light ray travels in the transparent sheet is L. Using Figure 32.41b, we can relate the thickness of the sheet, t, to the refraction angle:

$$\cos\theta_{medium} = \frac{t}{L}. \tag{ii}$$

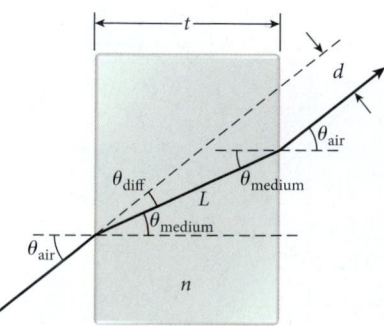

(a)

(b)

FIGURE 32.41 (a) Light ray incident on a sheet of transparent material with thickness t and index of refraction n. (b) Angles of incidence and refraction as the ray enters and exits the sheet.

If θ_{diff} is the difference between the incident angle and the refraction angle, as shown in Figure 32.41b, then

$$\theta_{diff} = \theta_{air} - \theta_{medium}. \qquad (iii)$$

We can see in Figure 32.41b that the displacement of the light ray, d, can be related to the distance the light ray travels in the sheet and the difference between the incident angle and the refracted angle:

$$\sin\theta_{diff} = \frac{d}{L}. \qquad (iv)$$

SIMPLIFY We can combine equations (ii), (iii), and (iv) to obtain

$$d = L\sin\theta_{diff} = \frac{t}{\cos\theta_{medium}}\sin(\theta_{air} - \theta_{medium}), \qquad (v)$$

where, from equation (i),

$$\theta_{medium} = \sin^{-1}\left(\frac{\sin\theta_{air}}{n}\right).$$

CALCULATE We first calculate the angle of refraction, using equation (i):

$$\theta_{medium} = \sin^{-1}\left(\frac{\sin(38.5°)}{1.50}\right) = 24.5199°.$$

Then we calculate the displacement of the light ray, using equation (v):

$$d = \frac{5.90 \text{ cm}}{\cos(24.5199°)}\sin(38.5° - 24.5199°) = 1.56663 \text{ cm}.$$

ROUND We report our result to three significant figures:

$$d = 1.57 \text{ cm}.$$

DOUBLE-CHECK If the light ray were incident perpendicularly on the sheet ($\theta_{air} = 0°$), the displacement of the ray would be zero. To calculate what the displacement would be in the limit as the incident angle approaches $\theta_{air} = 90°$, we can rewrite equation (v) as

$$d = \frac{t\sin(\theta_{air} - \theta_{medium})}{\cos\theta_{medium}} = t\frac{\sin\theta_{air}\cos\theta_{medium} - \cos\theta_{air}\sin\theta_{medium}}{\cos\theta_{medium}},$$

using the trigonometric identity $\sin(\alpha - \beta) = \sin\alpha\cos\beta - \cos\alpha\sin\beta$. Taking $\theta_{air} = 90°$, we get

$$d = t\frac{1(\cos\theta_{medium}) - 0(\sin\theta_{medium})}{\cos\theta_{medium}} = t.$$

Thus, our result must be between zero and the thickness of the sheet. Our result is $d = 1.57$ cm, while $d = 5.90$ cm for $\theta_{air} = 90°$, so our result seems reasonable.

> **Concept Check 32.8**
>
> In Solved Problem 32.2, the sheet of transparent material was surrounded by air. If the sheet were surrounded by water instead, what would happen to the displacement of the light ray with the same angle of incidence? (Assume that the index of refraction of the transparent material is still larger than that of water.)
>
> a) It would be larger.
>
> b) It would be the same.
>
> c) It would be smaller.

Total Internal Reflection

Now let's consider light traveling in an optical medium with index of refraction n_1 and crossing a boundary into another optical medium with a lower index of refraction n_2 such that $n_2 < n_1$. In this case, the light is bent away from the normal. As we increase the angle of incidence, θ_1, the angle of the transmitted light, θ_2, can approach 90°. When θ_1 exceeds the angle for which $\theta_2 = 90°$, **total internal reflection** takes place instead of refraction; all of the light is reflected internally. The critical angle, θ_c, at which total internal reflection takes place is given by

$$\frac{n_2}{n_1} = \frac{\sin\theta_1}{\sin\theta_2} = \frac{\sin\theta_c}{\sin 90°}$$

or, because $\sin 90° = 1$,

$$\sin\theta_c = \frac{n_2}{n_1} \quad (\text{for } n_2 \leq n_1). \qquad (32.13)$$

You can see from this equation that total internal reflection can occur only for light traveling from a medium with a higher index of refraction into a medium with a lower index of refraction, because the sine of an angle cannot be greater than 1. At angles less than θ_c, some light is reflected and some is transmitted. At angles greater than θ_c, all the light is reflected and none is transmitted.

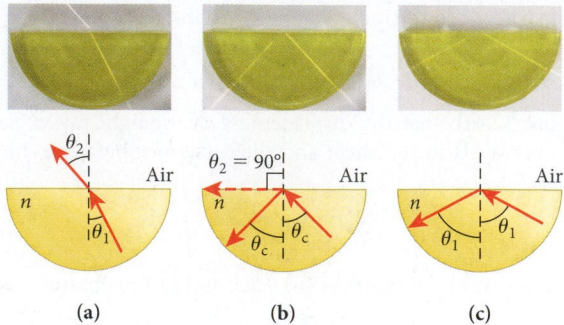

FIGURE 32.42 Light rays traveling through glass with index of refraction n from the lower right to the upper left are incident on the boundary between glass and air. The top panels are photographs and the lower panels are drawings illustrating the paths of the light rays. (a) Light is refracted at the boundary. (A small amount of light is also reflected by the boundary, but is too weak to show up in the photograph.) (b) Light is incident on the boundary at the critical angle for total internal reflection. (c) Light is totally internally reflected at the boundary.

If the second medium is air, then we can take $n_2 = 1$ and obtain an expression for the critical angle for total internal reflection for light leaving a medium with index of refraction n and entering air:

$$\sin \theta_c = \frac{1}{n}. \qquad (32.14)$$

Total internal reflection is illustrated in Figure 32.42. In this figure, light rays are traveling in glass with index of refraction n from lower right to upper left. In Figure 32.42a, light rays are incident on the boundary between glass and air and are refracted according to Snell's Law. In Figure 32.42b, light rays are incident on the boundary at the critical angle of total internal reflection, θ_c. At this angle, the refracted rays would have an angle of $\theta_2 = 90°$, as depicted by the dashed arrow. Light rays in Figure 32.42c are incident on the boundary with an angle greater than the critical angle of total internal reflection, $\theta_1 > \theta_c$, so all the light is reflected.

Optical Fibers

An important application of total internal reflection is the transmission of light in **optical fibers.** Light is shone into a fiber so that the angle of incidence at the end surface of the fiber is greater than the critical angle for total internal reflection. The light is then transported the length of the fiber as it bounces repeatedly from the inner surface. Thus, optical fibers can be used to transport light from a source to a destination. Figure 32.43 shows a bundle of optical fibers with their ends open to the camera. The other end of each fiber is optically connected to a light source. Note that optical fibers can transport light in directions other than a straight line. The fiber may be bent, as long as the radius of curvature of the bend is not small enough to allow the light traveling in the fiber to have angles of incidence θ_i less than θ_c. If $\theta_i < \theta_c$, the light will be absorbed by the cladding around the core of the fiber. These arguments are applicable to optical fibers with core diameters greater than 10 μm, that is, large compared to the wavelength of light.

One type of optical fiber used for digital communications consists of a glass core surrounded by cladding, which is composed of glass with a lower index of refraction than the core. The cladding is coated to prevent damage (Figure 32.44).

FIGURE 32.43 Light transported from a light source by optical fibers.

FIGURE 32.44 The structure of an optical fiber incorporating total internal reflection.

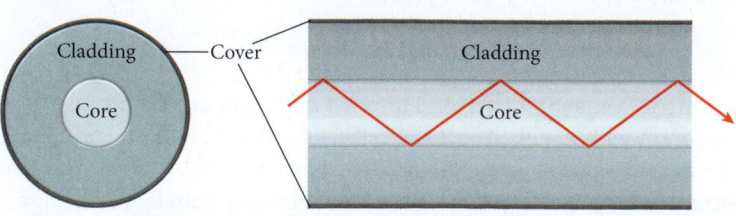

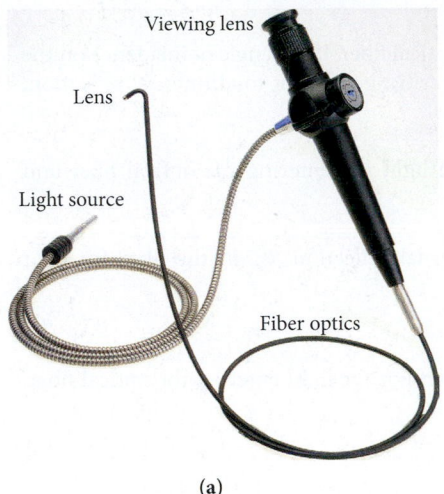

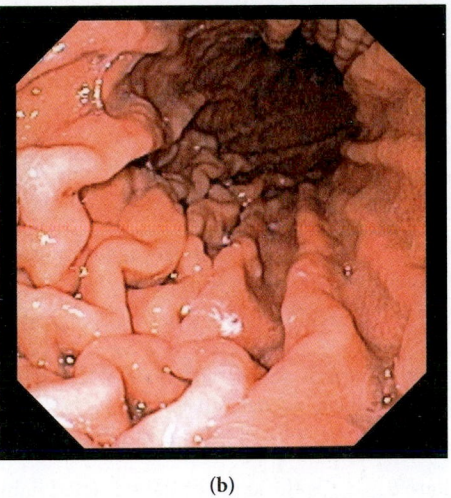

FIGURE 32.45 (a) An endoscope. (b) An image inside the stomach using an endoscope.

Viewing lens

Lens

Light source

Fiber optics

(a)

(b)

For a typical commercial optical fiber, the core material is silicon dioxide (SiO_2) doped with germanium (Ge) to increase its index of refraction. The typical commercial fiber can transmit light 500 m with small losses. The light is generated using light-emitting diodes (LEDs) that produce light with a long wavelength. The light is generated as short pulses. Small light losses do not affect digital signals because digital signals are transmitted as binary bits rather than as analog signals. As long as the bits are registered correctly, the information is transmitted flawlessly, as opposed to analog signals, which would be directly degraded by any signal loss. Every 500 m the signals are received, amplified, and retransmitted. Thus, data immune to interference can be transmitted at very high rates over long distances. These advantages explain why fiber optics forms the backbone of the modern Internet.

Optical fibers are also used for analog signal transmission, if the distance over which the signal has to be transported is not too long, up to a few meters. Applications in which optical fibers are used for analog signal transmission include the endoscope and related devices such as the borescope. An endoscope is used to look inside the human body without performing surgery, while a borescope is used to view hard-to-reach places in machinery. A typical endoscope is shown in Figure 32.45a. An endoscope uses a small lens to produce an image of the area of interest. The image produced by the lens is focused on one end of a bundle of thousands of optical fibers. The fiber-optics bundle is small enough in diameter and flexible enough to be inserted into the human body through an orifice such as the esophagus. Each fiber transports one pixel of the image to the other end of the bundle of fibers, where a second lens reproduces the image for viewing by the health professional. An endoscope may have 7,000 to 25,000 image-producing fibers. In addition, another set of optical fibers carries light to illuminate the region of interest. Often the lens end of the endoscope can be moved (articulated) remotely to view the desired area.

SOLVED PROBLEM 32.3 / Optical Fiber

Consider a long optical fiber with index of refraction $n = 1.265$ that is surrounded by air. (There is no cladding.) The end of the fiber is polished so that it is flat and perpendicular to the length of the fiber. A light ray from a laser is incident from air onto the center of the circular face of the end of the fiber.

PROBLEM

What is the maximum angle of incidence for this light ray such that it will be confined and transported by the optical fiber? (Neglect any reflection as the light ray enters the fiber.)

– Continued

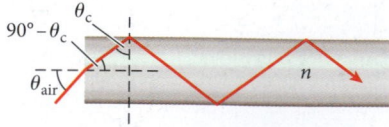

FIGURE 32.46 Light entering an optical fiber and undergoing total internal reflection.

SOLUTION

THINK The light ray will refract as it enters the optical fiber. If the angle of incidence on the end surface of the optical fiber is greater than the critical angle for total internal reflection, then the light is transmitted in the fiber without loss.

SKETCH Figure 32.46 shows a sketch of the light ray entering the optical fiber and reflecting off its inner surface.

RESEARCH The critical angle for total internal reflection, θ_c, in the fiber for light entering from air is given by

$$\sin\theta_c = \frac{1}{n},\tag{i}$$

where n is the index of refraction of the optical fiber. For the light entering the optical fiber, Snell's Law tells us that, since $n_{air} = 1$,

$$\sin\theta_{air} = n\sin\theta_{medium},\tag{ii}$$

where θ_{medium} is the angle of the refracted light ray in the fiber. From Figure 32.46 we can see that

$$\theta_{medium} = 90° - \theta_c.\tag{iii}$$

SIMPLIFY We can solve equation (ii) to obtain the maximum angle of incidence,

$$\theta_{air} = \sin^{-1}\left(n\sin\theta_{medium}\right).$$

Using equation (iii) we can write

$$\theta_{air} = \sin^{-1}\left(n\sin\left(90° - \theta_c\right)\right) = \sin^{-1}\left(n\cos\theta_c\right),$$

where we have used the trigonometric identity $\sin(90° - \alpha) = \cos\alpha$. We can then use equation (i) to obtain

$$\theta_{air} = \sin^{-1}\left(n\cos\left(\sin^{-1}\left(\frac{1}{n}\right)\right)\right).$$

CALCULATE Putting in the numerical values, we get

$$\theta_{air} = \sin^{-1}\left((1.265)\cos\left(\sin^{-1}\left(\frac{1}{1.265}\right)\right)\right) = 50.7816°.$$

ROUND We report our result to four significant figures, because the index of refraction was given with this accuracy:

$$\theta_{air} = 50.78°.$$

DOUBLE-CHECK The critical angle of total internal reflection for the optical fiber is

$$\theta_c = \sin^{-1}\left(\frac{1}{1.265}\right) = 52.23°.$$

Applying Snell's Law at the end surface of the optical fiber gives us

$$\theta_{medium} = \sin^{-1}\left(\frac{\sin(50.78°)}{1.265}\right) = 37.77°.$$

Thus, $\theta_{medium} = 37.77° = 90° - \theta_c = 90° - 52.23° = 37.77°$, and our result appears reasonable.

Mirages

A mirage is often associated with traveling in the desert. You seem to see an oasis with water in the distance. As you approach the welcome sight, it disappears. However, you don't have to be traveling in the desert to observe this phenomenon. You can often see a mirage when traveling on a long, straight highway on a hot day. An example of such a mirage is shown in Figure 32.47a.

The mirage is caused by refraction in the air near the surface of the hot road. The air near the surface of the road is warmer than the air farther from the surface. As shown in

(a)

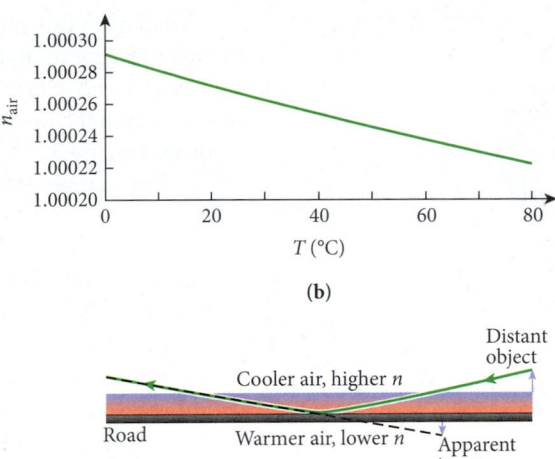

(b)

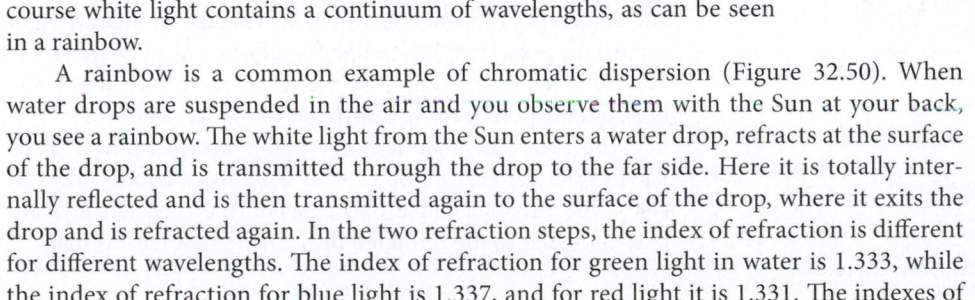

(c)

Figure 32.47b, the index of refraction of air goes down as its temperature goes up. Thus, the air near the road has a lower index of refraction than the cooler air above it. As light from distant objects passes through the warmer air layer, it is refracted upward, as illustrated in Figure 32.47c. The appearance of water is created by light refracted from the sky. You see what seems to be light reflected off the surface of water, as shown in Figure 32.47a. Other objects are also visible in the mirage shown in Figure 32.47a, such as trees and car headlights. As you keep driving, you soon find that there is no water standing on the road.

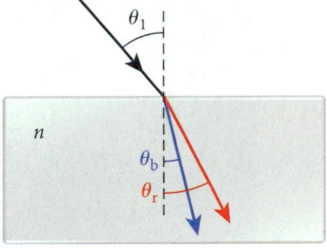

FIGURE 32.48 Chromatic dispersion for light refracted across the boundary of two optical media.

Chromatic Dispersion

The index of refraction of an optical medium depends on the wavelength of the light traveling in that medium. This dependence of the index of refraction on the wavelength of the light means that light of different colors is refracted differently at the boundary between two optical media. This effect is called **chromatic dispersion.**

In general, the index of refraction for a given optical medium is greater for shorter wavelengths than for longer wavelengths. Therefore, blue light is refracted more than red light. We can see that $\theta_b < \theta_r$ in Figure 32.48.

White light consists of a superposition of all visible wavelengths. When a beam of white light is projected onto a glass prism (Figure 32.49), the incident white light separates into the different visible wavelengths, because each wavelength is refracted at a different angle. Figure 32.49 shows three refracted light rays, for red, green, and blue light, but of course white light contains a continuum of wavelengths, as can be seen in a rainbow.

A rainbow is a common example of chromatic dispersion (Figure 32.50). When water drops are suspended in the air and you observe them with the Sun at your back, you see a rainbow. The white light from the Sun enters a water drop, refracts at the surface of the drop, and is transmitted through the drop to the far side. Here it is totally internally reflected and is then transmitted again to the surface of the drop, where it exits the drop and is refracted again. In the two refraction steps, the index of refraction is different for different wavelengths. The index of refraction for green light in water is 1.333, while the index of refraction for blue light is 1.337, and for red light it is 1.331. The indexes of refraction for all the colors form a continuum of values.

A typical rainbow is shown in Figure 32.1. You can see the blue light along the inner curve of the rainbow and the red light on the outer curve. The arc of the rainbow represents an average 42° angle from the direction of the sunlight. Another feature of rainbows

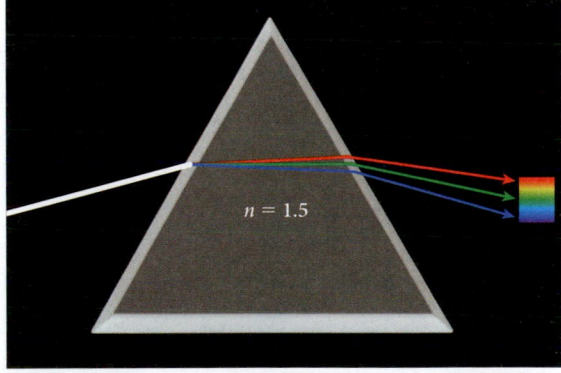

FIGURE 32.49 White light incident on a glass prism is separated into its component colors by chromatic dispersion.

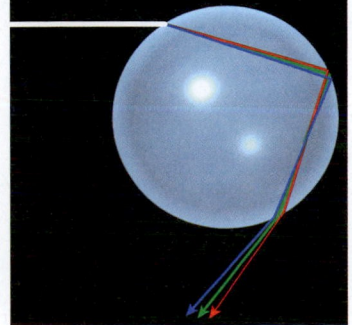

FIGURE 32.50 Chromatic dispersion in a spherical drop of water produces a rainbow.

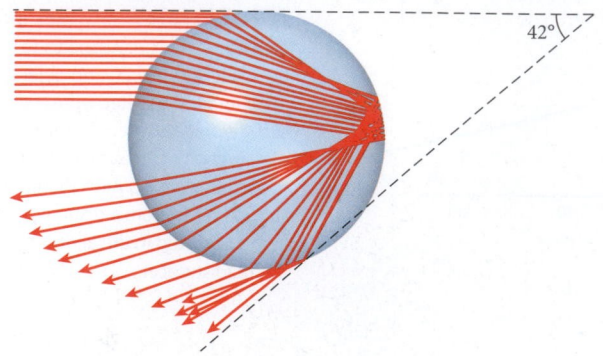

FIGURE 32.51 The paths taken by parallel light rays in a spherical drop of water.

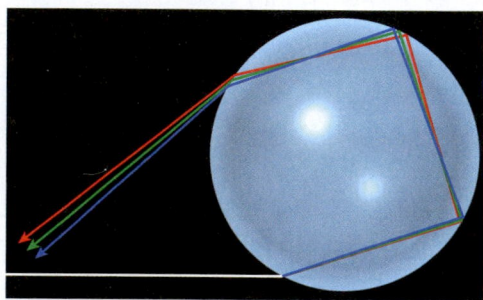

FIGURE 32.52 Chromatic dispersion and double reflection in a raindrop to produce a secondary rainbow.

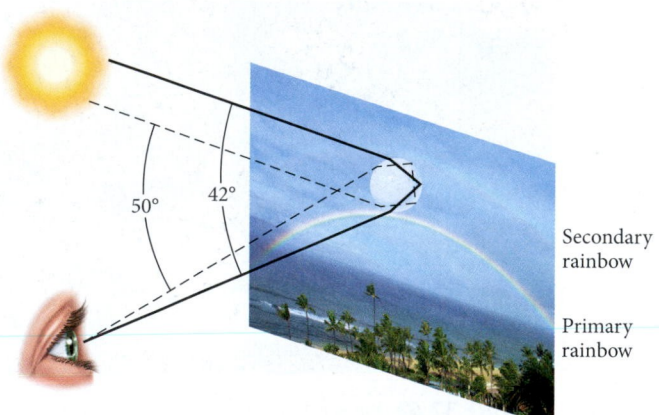

FIGURE 32.53 Geometry of the scattering of light in forming primary and secondary rainbows.

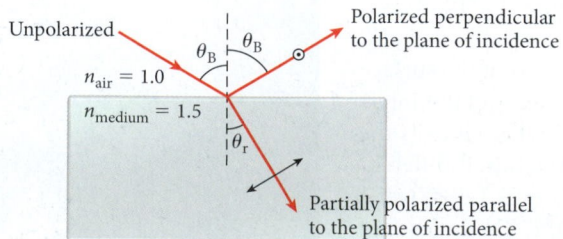

FIGURE 32.54 Unpolarized light is incident from air on a surface. When the incident angle is equal to the Brewster angle, θ_B, the reflected light is 100% polarized perpendicular to the plane of incidence. The refracted light is partially polarized parallel to the plane of incidence.

is evident in this photograph: The region inside the arc of the rainbow appears to be brighter than the region outside the arc. You can understand this phenomenon by looking at the path light rays take as they are refracted and reflected in the raindrops (Figure 32.51).

You can see from this diagram that most of the light that is refracted and reflected back to the observer has an angle less than 42° from the direction of the sunlight. At angles smaller than 42°, most of the light is refracted and reflected back to the observer, although no separation of the different wavelengths is observed because dispersed colors from one ray merge with those from another ray and form white light. At larger angles, no light is sent back to an observer—thus, outside the arc of the rainbow, much less light appears. However, some light still appears there because it is scattered from other sources.

Figure 32.1 also contains a secondary rainbow. The secondary rainbow appears at a larger angle than the primary rainbow, and in it the order of the colors is reversed. The secondary rainbow is created by light that reflects twice inside the raindrop (Figure 32.52). In contrast to the situation shown in Figure 32.50, when the ray in Figure 32.52 strikes the drop surface the third time (after one refraction and one reflection), not all of the light gets refracted and leaves the drop. Instead some of it gets internally reflected again and then refracted out of the drop, forming the secondary rainbow. In Figure 32.52, you can see that the angles of the emerging blue and red light are reversed compared to the blue and red light shown in Figure 32.50 for the primary rainbow. (For the secondary rainbow there is also a preferred angle, in this case approximately 50°, at which parallel rays entering the drop scatter for the secondary rainbow.)

Finally, combining our observations for the primary and secondary rainbows yields a quantitative understanding of the physical phenomena underlying Figure 32.1. The observed angles of the primary and secondary rainbows with respect to the sunlight are sketched in Figure 32.53.

Polarization by Reflection

The phenomenon of polarization was described in Chapter 31. Here we deal with a special way to create polarized, or at least partially polarized, light. When light is incident from air on an optical medium such as glass or water, some light is reflected and some light is refracted. The light reflected in this situation is partially polarized. When light is reflected at a certain angle, called the **Brewster angle**, θ_B, the reflected light is completely horizontally polarized, as illustrated in Figure 32.54.

The Brewster angle is named after Scottish physicist Sir David Brewster (1781–1868), who demonstrated this effect in 1815. The Brewster angle occurs when the reflected rays are perpendicular to the refracted rays. Snell's Law tells us that

$$n_{air} \sin\theta_B = n_{medium} \sin\theta_r,$$

where θ_r is the angle of the refracted ray. Figure 32.54 shows that

$$180° = \theta_B + \theta_r + 90°$$

because the reflected rays and the refracted rays are perpendicular to each other. This relationship between the angles can be rewritten as

$$\theta_r = 90° - \theta_B.$$

Substituting this result into Snell's Law gives

$$n_{air} \sin\theta_B = n_{medium} \sin(90° - \theta_B) = n_{medium} \cos\theta_B,$$

which can be rearranged to finally obtain

$$\frac{n_{medium}}{n_{air}} = \tan\theta_B, \qquad (32.15)$$

where we have made use of the trigonometric identity $\tan\theta = \sin\theta/\cos\theta$. Equation 32.15 is called **Brewster's Law.**

The fact that light in air reflected off the surface of water is partially horizontally polarized means that the glare of sunlight off surfaces of water can be blocked by sunglasses that are covered with a polarization filter that allows only vertically polarized light to pass through.

WHAT WE HAVE LEARNED | EXAM STUDY GUIDE

- The angle of incidence equals the angle of reflection, $\theta_i = \theta_r$.

- The focal length of a spherical mirror is equal to half its radius of curvature, $f = \frac{1}{2}R$. The radius, R, is positive for converging (concave) mirrors and negative for diverging (convex) mirrors.

- For images formed by spherical mirrors, the object distance, the image distance, and the focal length of the mirror are related by the mirror equation, $\frac{1}{d_o} + \frac{1}{d_i} = \frac{1}{f}$. Here d_o is always positive, while d_i is positive if the image is on the same side of the mirror as the object and negative if the image is on the other side. The focal length, f, is positive for converging (concave) mirrors and negative for diverging (convex) mirrors.

- The index of refraction of a medium is defined as the ratio of the speed of light in vacuum to the speed of light in the medium, $n = c/v$.

- Light is refracted (changes direction) as it crosses the boundary between two media with different indices of refraction. This refraction is governed by Snell's Law: $n_1 \sin\theta_1 = n_2 \sin\theta_2$.

- The critical angle for total internal reflection at the boundary between two media with different indices of refraction is given by $\sin\theta_c = \frac{n_2}{n_1}$ (for $n_2 \le n_1$).

- The Brewster angle, θ_B, is the angle of incidence of light from air to a medium with a higher index of refraction at which the reflected light is completely horizontally polarized. The angle is given by Brewster's Law: $\tan\theta_B = n_{medium}/n_{air}$.

ANSWERS TO SELF-TEST OPPORTUNITIES

32.1 $d_i = \dfrac{d_o f}{d_o - f} = \dfrac{(2.50 \text{ m})(-0.500 \text{ m})}{(2.50 \text{ m}) - (-0.500 \text{ m})} = -0.417 \text{ m},$

$$m = -\frac{d_i}{d_o} = -\frac{-0.417 \text{ m}}{2.50 \text{ m}} = 0.167.$$

Thus, you see an upright, smaller version of yourself.

32.2 We can use this result from Derivation 32.2:

$$y = R\left(1 - \frac{1}{2\cos\left(\sin^{-1}(d/R)\right)}\right).$$

As the position gets closer to the optic axis, the distance for the crossing point gets closer to the focal length, $f = R/2 = 3.60$ cm.

Distance from Axis (cm)	Distance from Mirror (cm)
0.720	3.58
0.800	3.58
1.80	3.48
3.60	3.04

32.3 Kinetic energy of rotation is $K = \frac{1}{2}m\omega^2 x^2$; potential energy is $U = mgy$. Set them equal and find $y = x^2\omega^2/2g$.

32.4 Depending on the time of day and of the year, the Sun moves across the sky on a path that depends on the azimuthal and polar angles. So, in general, one axis of rotation is not sufficient to keep the Sun's incident rays always normal to a surface. However, rotation about the long axis is sufficient to keep the Sun in the plane created by the optical axis of the parabola and the rotation axis. The variation in the radiation intensity is then a function of the cosine of the angle of the Sun relative to the optical axis in that plane. However, rotating the parabolic troughs about an axis perpendicular to their long axis would be very expensive. So the loss of the intensity of sunlight has to be accepted for economic reasons.

PROBLEM-SOLVING GUIDELINES

1. Drawing a clear, well-labeled diagram should be the first step in solving almost any optics problem. Include all the information you know and all the information you need to find. Remember that angles are measured from the normal to a surface, not to the surface itself.

2. You normally need to draw only two principal rays to locate an image formed by mirrors, but you should draw a third ray as a check that the first two are drawn correctly. Even if you need to solve a problem using the mirror equation, an accurate drawing can help you approximate the answers as a check on your calculations.

3. When you calculate distances for mirrors, remember the sign conventions and be careful to use them correctly. If a diagram indicates an inverted image, say, but your result does not have a negative sign, go back to the starting equation and check the signs of all distances and focal lengths.

4. Remember that when a ray diagram for refraction shows a light ray crossing the boundary from a medium with a lower index of refraction into a medium with a higher index of refraction, the refracted ray will be closer to the normal than the incident ray.

MULTIPLE-CHOICE QUESTIONS

32.1 Legend says that Archimedes set the Roman fleet on fire as it was invading Syracuse. Archimedes created a huge _____ mirror and used it to focus the Sun's rays on the Roman vessels.

a) plane

b) parabolic diverging

c) parabolic converging

32.2 Which of the following interface combinations has the smallest critical angle?

a) light traveling from ice to diamond

b) light traveling from quartz to Lucite

c) light traveling from diamond to glass

d) light traveling from Lucite to diamond

e) light traveling from Lucite to quartz

32.3 For specular reflection of a light ray, the angle of incidence

a) must be equal to the angle of reflection.

b) is always less than the angle of reflection.

c) is always greater than the angle of reflection.

d) is equal to 90°—the angle of reflection.

e) may be greater than, less than, or equal to the angle of reflection.

32.4 Standing by a pool filled with water, under what condition will you see a reflection of the scenery on the opposite side through total internal reflection of the light from the scenery?

a) Your eyes are level with the water.

b) You observe the pool at an angle of 41.8°.

c) There is no condition under which this can occur.

d) You observe the pool at an angle of 48.2°.

32.5 You are using a mirror and a camera to make a self-portrait. You focus the camera on yourself through the mirror. The mirror is a distance D away from you. To what distance should you set the range of focus on the camera?

a) D b) $2D$ c) $D/2$ d) $4D$

32.6 What is the magnification for a plane mirror?

a) +1

b) −1

c) greater than +1

d) not defined for a plane mirror

32.7 A small object is placed in front of a converging mirror with radius $R = 7.50$ cm so that the image distance equals the object distance. How far is this object from the mirror?

a) 2.50 cm

b) 5.00 cm

c) 7.50 cm

d) 10.0 cm

e) 15.0 cm

32.8 Sunlight strikes a piece of glass at an angle of incidence of $\theta_i = 33.4°$. What is the difference between the angle of refraction of a red light ray ($\lambda = 660.0$ nm) and that of a violet light ray ($\lambda = 410.0$ nm)? The glass's index of refraction is $n = 1.520$ for red light and $n = 1.538$ for violet light.

a) 0.03°   c) 0.19°   e) 0.82°

b) 0.12°   d) 0.26°

32.9 The image of yourself that you see in a bathroom mirror is

a) left-right inverted.

b) front-to-back inverted.

c) bottom-to-top inverted.

d) not inverted.

32.10 Where do you have to place an object in front of a concave mirror with focal length f for the image to be the same size as the object?

a) at $d_o = 0.5f$ c) at $d_o = 2f$ e) none of the above

b) at $d_o = f$ d) at $d_o = 2.5f$

32.11 You are looking down into a swimming pool at an angle of 20° relative to the vertical, and you see a coin at the bottom of the pool. This coin appears to you to be at

a) a lesser depth than it really is.

b) the same depth as it really is.

c) a greater depth than it really is.

32.12 You are looking straight down into a swimming pool, that is, at an angle of 0° relative to the vertical, and you see a coin at the bottom of the pool. This coin appears to you to be at

a) a lesser depth than it really is.

b) the same depth as it really is.

c) a greater depth than it really is.

CONCEPTUAL QUESTIONS

32.13 The figure shows the difference between the refractive index profile of a *step index fiber* and the refractive index profile of a *graded index fiber.* Applying geometric optics principles, comment on the path followed by a light ray entering each of the optic fibers.

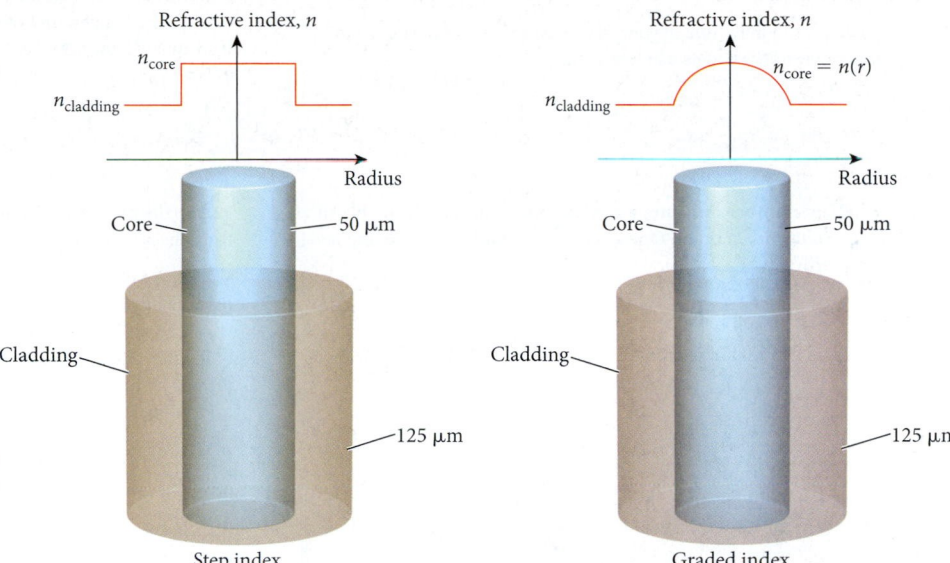

Step index

Graded index

32.14 A slab of Plexiglas 2.00 cm thick with index of refraction 1.51 is placed over a physics textbook. The slab has parallel sides. The text is at height $y = 0$. Consider two rays of light, A and B, that originate at the letter "t" in the text under the Plexiglas and travel toward an observer who is above the Plexiglas, looking down. **Draw** on the figure the apparent *y*-position of the letter under the Plexiglas as seen by the observer. (*Hint:* From where in the Plexiglas do the two rays appear to originate for the observer?) You can easily do this experiment yourself. If you do not have a block of glass or Plexiglas, you can try placing a flat-bottomed drinking glass over the text.

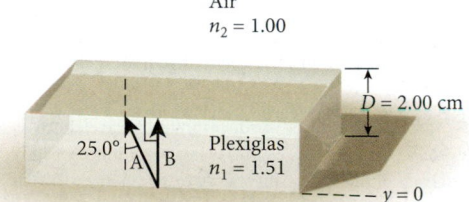

32.15 A concave spherical mirror is used to create an image of an object 5.00 cm tall that is located at position $x = 0$ cm, which is 20.0 cm from point C, the center of curvature of the mirror, as shown in the figure. The magnitude of the radius of curvature of the mirror is 10.0 cm. Without changing the mirror, how can you reduce the spherical aberration produced by it? Are there any disadvantages to your approach to reducing the spherical aberration?

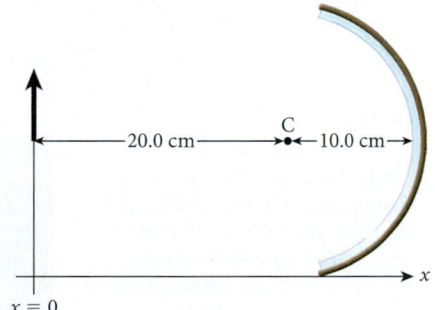

32.16 If you look at an object at the bottom of a pool, the pool looks less deep than it actually is.

a) From what you have learned, calculate how deep a pool seems to be if it is actually 4 feet deep and you look directly down on it. The index of refraction of water is 1.33.

b) Would the pool look more or less deep than it actually is if you looked at it from an angle other than vertical? Answer this qualitatively, without using an equation.

32.17 Why does refraction occur? That is, what is the physical reason a wave moves with a different velocity when it passes from one medium into another?

32.18 Many optical fibers have minimum specified bending radii. Why?

32.19 A physics student is eying a steel drum, the top part of which has the approximate shape of a concave spherical surface. The surface is sufficiently polished that the student can just make out the reflection of her finger when she places it above the drum. As she slowly moves her finger toward the surface and then away from it, you ask her what she is doing. She replies that she is estimating the radius of curvature of the drum. How can she do that?

32.20 State whether the following is true or false and explain your answer: The wavelength of He-Ne laser light in water is less than its wavelength in air. (The index of refraction of water is 1.33.)

32.21 Among the instruments Apollo astronauts left on the Moon were reflectors for bouncing laser beams back to Earth. These made it possible to measure the distance from the Earth to the Moon with unprecedented precision (uncertainties of a few centimeters in 384,000 km), for the study of both celestial mechanics and Earth's plate tectonics. The reflectors consist not of ordinary mirrors, but of arrays of *corner cubes*, each constructed of three square plane mirrors fixed perpendicular to each other, as adjacent faces of a cube. Why? Explain the functioning and advantages of this design.

32.22 A 45°-45°-90° triangular prism can be used to reverse a light beam: The light enters perpendicular to the hypotenuse of the prism, reflects off each leg, and emerges perpendicular to the hypotenuse again. The surfaces of the prism are *not* silvered. If the prism is made of glass with index of refraction $n_{glass} = 1.520$ and is surrounded by air, the light beam will be reflected with a minimum loss of intensity (there are reflection losses as the light enters and leaves the prism).

a) Will this work if the prism is *under water,* which has the index of refraction $n_{H_2O} = 1.333$?

b) Such prisms are used, in preference to mirrors, to bend the optical path in quality binoculars. Why?

32.23 An object is imaged by a converging spherical mirror as shown in the figure. Suppose a black cloth is put between the object and the mirror so that it covers everything above the optic axis. How will the image be affected?

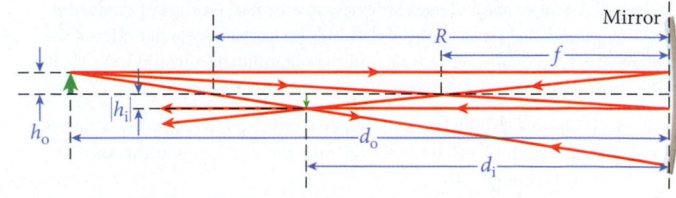

32.24 You are under water in a pond and look up through the smooth surface of the water at the Sun in the sky. Is the Sun in fact higher in the sky than it appears to you, or is it lower?

32.25 Holding a spoon in front of your face, convex side toward you, estimate the location of the image and its magnification.

32.26 A *solar furnace* uses a large parabolic mirror (such mirrors can be several stories high) to focus the light of the Sun to heat a target. A large solar furnace can melt metals. Is it possible to attain temperatures exceeding 6000 K (the temperature of the photosphere of the Sun) in a solar furnace? If so, how? If not, why not?

EXERCISES

A blue problem number indicates a worked-out solution is available in the Student Solutions Manual. One • and two •• indicate increasing level of problem difficulty.

Section 32.2

32.27 A person sits 1.00 m in front of a plane mirror. What is the location of the image?

32.28 A periscope consists of two flat mirrors and is used for viewing objects when an obstacle impedes direct viewing. Suppose Curious George is looking through a periscope, in which the two flat mirrors are separated by a distance $L = 0.400$ m, at the Man in the Yellow Hat, whose hat is at $d_o = 3.00$ m from the upper mirror. What is the distance D of the final image of the yellow hat from the lower mirror?

32.29 A person stands at a point P relative to two plane mirrors oriented at 90.0°, as shown in the figure. How far away from each other do the person's images appear to be?

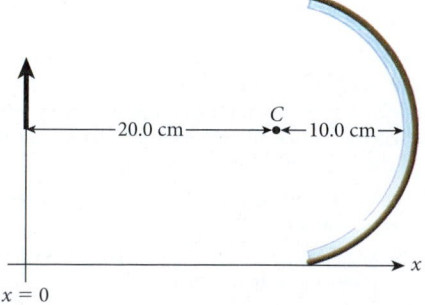

•32.30 Even the best mirrors absorb or transmit some of the light incident on them. The highest-quality mirrors might reflect 99.997% of the incident light. Suppose a cubical "room," 3.00 m on an edge, were constructed with such mirrors for the walls, floor, and ceiling. How slowly would such a room get dark? Estimate the time required for the intensity of light in such a room to fall to 1.00% of its initial value after the only light source in the room was switched off.

Section 32.3

32.31 The radius of curvature of a convex mirror is –25.0 cm. What is its focal length?

32.32 A concave spherical mirror is used to create an image of an object 5.00 cm tall that is located at position $x = 0$ cm, which is 20.0 cm from point C, the center of curvature of the mirror, as shown in the figure. The magnitude of the radius of curvature of the mirror is 10.0 cm. Calculate the position x_i where the image is formed. Use the coordinate system given in the drawing. What is the height h_i of the image? Is the image upright (pointing up) or inverted (pointing down)? Is it real or virtual?

32.33 Convex mirrors are often used as sideview mirrors on cars. Many such mirrors display the warning "Objects in mirror are closer than they appear." Assume that a convex sideview mirror has a radius of curvature of 14.0 m and that a car is 11.0 m behind the mirror. For a flat mirror, the image distance would be 11.0 m, and the magnification would be 1. Find the image distance and the magnification for this convex mirror.

32.34 A 5.00-cm-tall object is placed 30.0 cm away from a convex mirror with a focal length of –10.0 cm. Determine the size, orientation, and position of the image.

32.35 The magnification of a convex mirror is 0.60× for an object 2.0 m from the mirror. What is the focal length of this mirror?

•32.36 An object is located at a distance of 100. cm from a concave mirror of focal length 20.0 cm. Another concave mirror of focal length 5.00 cm is located 20.0 cm in front of the first concave mirror. The reflecting sides of the two mirrors face each other. What is the location of the final image formed by the two mirrors and the total magnification produced by them in combination?

••32.37 The shape of an *elliptical* mirror is described by the curve $\frac{x^2}{a^2} + \frac{y^2}{b^2} = 1$, with semimajor axis a and semiminor axis b. The foci of this ellipse are at the points $(c,0)$ and $(-c,0)$, with $c = (a^2 - b^2)^{1/2}$. Show that any light ray in the xy-plane that passes through one focus is reflected through the other. "Whispering galleries" make use of this phenomenon for reflecting sound waves.

Section 32.4

32.38 What is the speed of light in crown glass, whose index of refraction is 1.52?

32.39 An optical fiber with an index of refraction of 1.50 is used to transport light of wavelength 400. nm. What is the critical angle for transporting light through this fiber without loss? If the fiber is immersed in water? In oil?

32.40 A helium-neon laser produces light of wavelength $\lambda_{vac} = 632.8$ nm in vacuum. If this light passes into water, with index of refraction $n = 1.333$, what will each of the following characteristics be?

a) speed
b) frequency
c) wavelength
d) color

32.41 A light ray is incident from water, whose index of refraction is 1.33, on a plate of glass whose index of refraction is 1.73. What angle of incidence will result in fully polarized reflected light?

•32.42 Suppose you are standing at the bottom of a swimming pool whose surface is perfectly calm. Looking up, you see a circular window to the "outer world." If your eyes are approximately 2.00 m beneath the surface, what is the diameter of this circular window?

•32.43 A ray of light of a particular wavelength is incident on an equilateral triangular prism whose index of refraction for light of this wavelength is 1.23. The ray is parallel to the base of the prism as it approaches and enters the prism at the midpoint of one of the sides, as shown in the figure. What is the direction of the ray when it emerges from the prism?

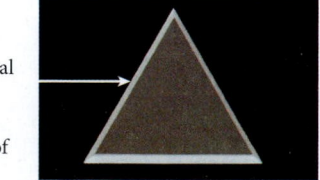

•32.44 A collimated laser beam strikes the left side (A) of a glass block at an angle of 20.0° with respect to the horizontal, as shown in the figure. The block has an index of refraction of 1.55 and is surrounded by air, with an index of

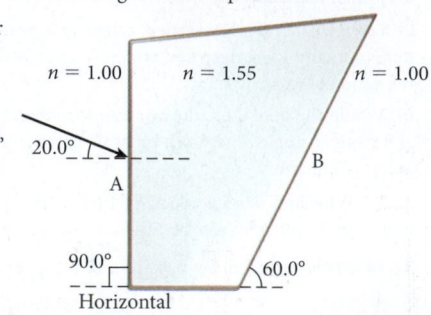

refraction of 1.00. The left side of the glass block is vertical (90.0° from horizontal) while the right side (B) is at an angle of 60.0° from the horizontal. Determine the angle θ_{BT} **with respect to the horizontal** at which the light exits surface B.

•**32.45** In a step index fiber, the index of refraction undergoes a discontinuity (jump) at the core-cladding boundary, as shown in the figure. Infrared light with wavelength 1550 nm propagates through such a fiber by total internal reflection at the core-cladding boundary. The index of refraction of the core for the infrared light is $n_{core} = 1.48$. If the maximum angle, α_{max}, at which light can enter the fiber with no light lost into the cladding is $\alpha_{max} = 14.033°$, calculate the percent difference between the index of refraction of the core and the index of refraction of the cladding.

••**32.46** Refer to Figure 32.51 and prove that the arc of the primary rainbow represents the 42° angle from the direction of the sunlight.

••**32.47** Use Fermat's Principle to derive the law of reflection.

••**32.48** Fermat's Principle, from which geometric optics can be derived, states that light travels by a path that minimizes the time of travel between points. Consider a light beam that travels a horizontal distance D and a vertical distance h, through two large flat slabs of material that have a vertical interface between them. One slab has a thickness $D/2$ and an index of refraction n_1, and the other has a thickness $D/2$ and an index of refraction n_2. Write the equation relating the indices of refraction and the angles from the horizontal that the light beam makes at the interface (θ_1 and θ_2) which minimize the time for this travel.

Additional Exercises

32.49 Suppose your height is 2.00 m and you are standing 50.0 cm in front of a plane mirror.

a) What is the image distance?

b) What is the image height?

c) Is the image inverted or upright?

d) Is the image real or virtual?

32.50 A light ray of wavelength 700. nm traveling in air ($n_1 = 1.00$) is incident on a boundary with a liquid ($n_2 = 1.63$).

a) What is the frequency of the refracted ray?

b) What is the speed of the refracted ray?

c) What is the wavelength of the refracted ray?

32.51 You have a spherical mirror with a radius of curvature of +20.0 cm (so it is concave facing you). You are looking at an object whose image size you want to double so you can see it better. At what locations could you put the object? Where will the images be for these object locations, and will they be real or virtual?

32.52 You are submerged in a swimming pool. What is the maximum angle at which you can see light coming from above the pool surface? That is, what is the angle for total internal reflection from water into air?

32.53 Light hits the surface of water at an incident angle of 30.0° with respect to the normal to the surface. What is the angle between the reflected ray and the refracted ray?

32.54 A spherical metallic Christmas tree ornament has a diameter of 8.00 cm. If Saint Nicholas is by the fireplace, 1.56 m away, where will he see his reflection in the ornament? Is the image real or virtual?

32.55 One of the factors that cause a diamond to sparkle is its relatively small critical angle. Compare the critical angle for diamond in air with that for diamond in water.

32.56 What kind of image, virtual or real, is formed by a converging mirror when the object is placed a distance away from the mirror that is

a) beyond the center of curvature of the mirror,

b) between the center of curvature and half of the distance to the center of curvature, and

c) closer than half of the distance to the center of curvature?

32.57 At what angle θ shown in the figure must a beam of light enter the water for the reflected beam to make an angle of 40.0° with respect to the normal of the water's surface?

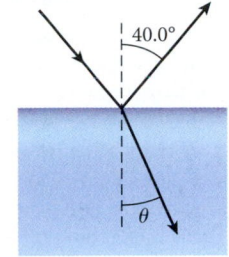

•**32.58** A concave mirror forms a real image twice as large as the object. The object is then moved, and the new real image is three times the size of the object. If the image is 75 cm from its initial position, how far was the object moved and what is the focal length of the mirror?

•**32.59** How deep does a point in the middle of a 3.00-m-deep pool appear to be to a person standing beside the pool at a position 2.00 m horizontally from the point? Take the index of refraction for the liquid in the pool to be 1.30 and that for air to be 1.00.

•**32.60** What is the smallest incident angle θ_i at which the light beam shown in the figure will undergo total internal reflection in the prism, if the index of refraction of the prism for light of this wavelength is 1.500?

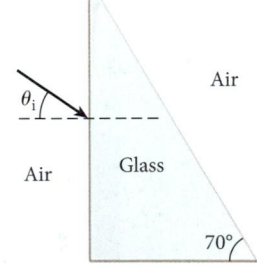

32.61 Reflection and refraction, like all classical features of light and other electromagnetic waves, are governed by the Maxwell equations. The Maxwell equations are *time-reversal invariant*, which means that any solution of the equations that is reversed in time is also a solution.

a) Suppose some configuration of electric charge density, ρ, current density, $\vec{j}$, electric field, $\vec{E}$, and magnetic field, $\vec{B}$, is a solution of the Maxwell equations. What is the corresponding time-reversed solution?

b) How, then, do one-way mirrors work?

••**32.62** Refer to Example 32.3 and use the numbers provided there. Further, assume that your eyes are at a height of 1.70 m above the water.

a) Calculate the time it takes for light to travel on the path from the fish to your eyes.

b) Calculate the time light would take on a straight-line path from the fish to your eyes.

c) Calculate the time light would take on a path from the fish vertically upward to the water surface and then straight to your eyes.

d) Calculate the time light would take on the straight-line path from the apparent location of the fish to your eyes.

e) What can you say about Fermat's Principle from the results of parts (a) through (d)?

32.63 If you want to construct a liquid mirror of focal length 2.50 m, with what angular velocity do you have to rotate your liquid?

••**32.64** One proposal for a space-based telescope is to put a large rotating liquid mirror on the Moon. Suppose you want to use a liquid mirror that is 100.0 m in diameter and has a focal length of 347.5 m. The gravitational acceleration on the Moon is 1.62 m/s².

a) What angular velocity does your mirror have?

b) What is the linear speed of a point on the perimeter of the mirror?

c) How high above the center is the perimeter of the mirror?

MULTI-VERSION EXERCISES

32.65 The origin of a coordinate system is placed at the center of curvature of a spherical mirror with radius of curvature $R = 58.1$ cm (see the figure). An object is placed at $x_o = 19.7$ cm. What is the x-coordinate of the image?

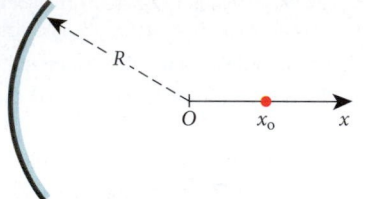

32.66 The origin of a coordinate system is placed at the center of curvature of a spherical mirror with radius of curvature $R = 59.3$ cm (see the figure). An object is placed at $x_o = 39.5$ cm. What is the mirror's magnification?

32.67 The origin of a coordinate system is placed at the center of a spherical mirror with radius of curvature R (see the figure). An object is placed at $x_o = 10.1$ cm. The image is formed at $x = -7.405$ cm. What is the value of R?

32.68 The origin of a coordinate system is placed at the center of curvature of a spherical mirror with radius of curvature R (see the figure). An object is placed at $x_o = 10.3$ cm. The image is formed at $x = -7.548$ cm. What is the focal length of the mirror?

─────────

32.69 A layer of methyl alcohol, with index of refraction 1.329, rests on a block of ice, with index of refraction 1.310. A ray of light passes through the methyl alcohol at an angle of $\varphi_1 = 61.07°$ relative to the alcohol-ice

boundary. What is the angle φ_2 relative to the boundary at which the ray passes through the ice?

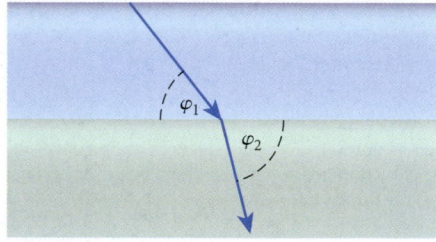

32.70 A layer of a transparent material rests on a block of fused quartz, whose index of refraction is 1.460. A ray of light passes through the unknown material at an angle of $\varphi_1 = 63.65°$ relative to the boundary between the materials and is refracted at an angle of $\varphi_2 = 70.26°$ relative to the boundary. What is the index of refraction of the unknown material?

32.71 A layer of carbon dioxide, with index of refraction 1.00045, rests on a block of ice, with index of refraction 1.310. A ray of light passes through the carbon dioxide at an angle of φ_1 relative to the boundary between the materials and then passes through the ice at an angle of $\varphi_2 = 72.06°$ relative to the boundary. What is the value of φ_1?

32.72 A layer of water, with index of refraction 1.333, rests on a block of an unknown transparent material. A ray of light passes through the water at an angle of $\varphi_1 = 68.77°$ relative to the boundary between the materials and then passes through the unknown material at an angle of $\varphi_2 = 72.98°$ relative to the boundary. What is the speed of light in the unknown material?

33

Lenses and Optical Instruments

FIGURE 33.1 (a) The Whirlpool Galaxy. (b) Marine diatoms living in Antarctica.

The two images in Figure 33.1 are some of the largest and smallest objects we can optically observe with visible light. The image in Figure 33.1a is the Whirlpool Galaxy (M51), with a diameter of about 76,000 light-years $(7 \cdot 10^{20}$ m), and a distance from Earth of about 23 million light-years $(2 \cdot 10^{23}$ m). The image in Figure 33.1b shows assorted diatoms found living between crystals of annual sea ice in Antarctica. Diatoms are on the order of 20 microns $(2 \cdot 10^{-5}$ m) in length. Over the years, the ability to form images of these kinds of objects has completely changed our understanding of biology, astronomy, geology, engineering—in fact, just about every branch of science and technology.

All optical image-forming instruments work from a combination of lenses or mirrors. In a sense, the ideas in this chapter are simply applications of the principles discussed in Chapter 32. However, understanding how an image is formed is essential to interpreting the subject of the image. In this chapter we will examine a variety of image-forming instruments, including the human eye. The same principles of geometric optics govern image formation using other kinds of radiation; we will see some examples of this as well.

WHAT WE WILL LEARN

- Lenses can focus light and produce images governed by the Thin-Lens Equation, which states that the inverse of the object distance plus the inverse of the image distance equals the inverse of the focal length of the lens.

- Optical instruments are combinations of mirrors and lenses.

- A single converging lens can be used as a magnifier.

- Two-lens systems are often used in optical instruments.

- Placing a converging lens close to a diverging lens can produce a zoom lens.

- A camera is an optical instrument that records real images produced by a lens.

- The human eye is an optical instrument governed by the Thin-Lens Equation. Various types of lenses are used to correct defects in vision.

- Microscopes are systems of lenses designed to magnify the image of close but very small objects.

- Telescopes are systems of lenses or mirrors designed to magnify the image of distant but very large objects.

33.1 Lenses

When light is refracted while crossing a curved boundary between two different media, the light rays obey the law of refraction at each point on the boundary. The angle at which the light rays cross the boundary (with respect to the local normal to the boundary) is different along the curve, so the refracted angle is different at different points along the curve. A spherical curved boundary between two optically transparent media forms a spherical surface. If light enters a medium through one spherical surface and then returns to the original medium through another spherical surface, the device that has the spherical surfaces is called a **lens.** Light rays that are initially parallel before they strike the lens are refracted in different directions, depending on the part of the lens they strike. Depending on the shape of the lens, the light rays can be focused or can diverge.

Converging Lenses

Figure 33.2a shows an example of a converging lens. A red light ray hits the left surface. Part of the ray gets reflected, and part of it enters the lens and gets refracted at the surface. It travels in a straight line through the lens until it hits the right surface, where part of it gets reflected and another part of it gets refracted as it leaves the lens. In Figure 33.2b we highlight the path of the twice-refracted ray in yellow. We also draw the two local tangent lines to the curved surfaces of the lens (short-dashed black lines) and the normal lines (long-dashed black lines). It is apparent that the law of reflection and Snell's Law of refraction, which we developed in the previous chapter, also hold here.

In Figure 33.3a we show a picture of five parallel light rays hitting the lens and being focused into the same point. In Figure 33.3b we represent the lens by a single dashed black line through its optical center and idealize the double refraction of the real rays with the single refraction of the yellow lines. This is called the *thin-lens approximation.* And from Figure 33.3b it is apparent that this approximation is justified in this case. Notice also that the center ray in Figure 33.3 does not get refracted. This means that it forms the optic axis.

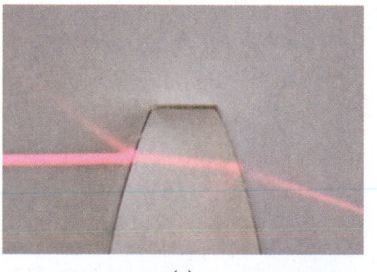

(a)

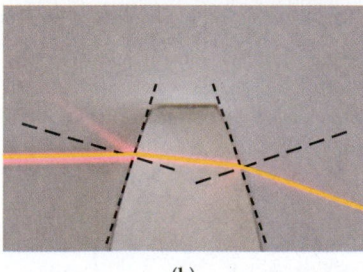

(b)

FIGURE 33.2 (a) A red light ray enters a converging lens from the right and gets refracted at both surfaces. (b) The local tangent lines (short-dashed black lines) and their normal lines (long-dashed) at the locations where the ray hits the surfaces are superimposed.

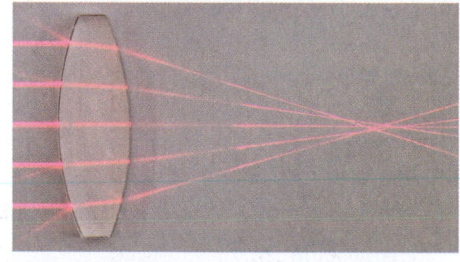

(a)

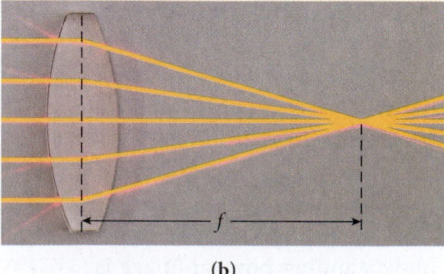

(b)

FIGURE 33.3 (a) Five parallel red light rays hit a converging lens from the left and focused at a single point. (b) Representation of the lens by a single dashed line through its center, and refraction of the idealized rays (yellow lines) through the same focal point.

Concept Check 33.1

If you shift all parallel rays in Figure 33.3 a bit up from their present location, the focal point will

a) move a bit down from its present location.

b) move a bit up from its present location.

c) remain unchanged.

d) move a bit right from its present location.

e) move a bit left from its present location.

Just like for the case of the curved mirrors in Chapter 32, the distance between the optical center and the focal point is called the focal length, f. It is easily verified experimentally that sending parallel rays into the lens from the right makes them converge at a focal point to the left of the lens, which is the same distance from the optic center f as the focal point on the right. The inverse of the focal length is called the **lens power**, P,

$$P = \frac{1}{f}. \tag{33.1}$$

The power of the lens is measured in **diopters**, D, with $1\,\text{D} = 1\,\text{m}^{-1}$. Eyeglasses and contact lenses are typically characterized in terms of their diopter value.

Lenses can be used to form images, again following the same principles of ray tracing that we developed in the previous chapter. Figure 33.4 shows the geometric construction of the formation of an image using a converging lens. For convenience we use a simple arrow as our object and place its tail on the optic axis. This placement ensures that the tail of the arrow image is also on the optic axis, and we only have to worry about imaging the tip of the arrow. Our object has a height h_o and is located a distance d_o from the center of the lens.

Three special rays are useful for constructing images:

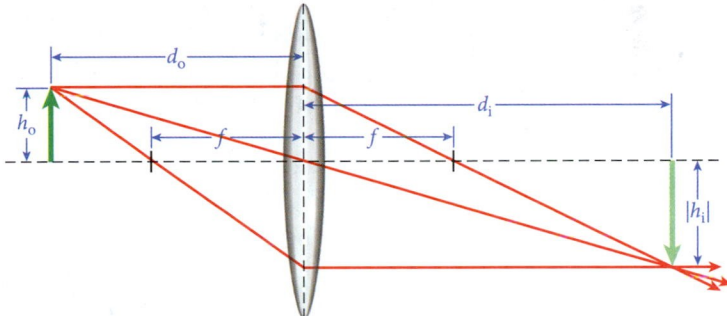

FIGURE 33.4 Real image produced by a converging lens for the case that the object distance is larger than the focal length.

1. The first ray again is drawn from the tip of the arrow parallel to the optic axis and is refracted through the focal point on the opposite side of the lens.

2. A second ray is drawn from the tip of the arrow straight through the center of the lens.

3. A third ray is drawn from the tip of the arrow through the focal point on the same side of the lens and is then refracted parallel to the optic axis.

These three rays all intersect at the same point, which marks the tip of the arrow image. Obviously, if you have drawn any two of the three rays and found their intersection, then you are done, but the third ray can provide a useful double-check. Perhaps equally obvious, but still worth mentioning is the fact that all rays emerging from the object and hitting the lens are focused in the same spot of the image. The only thing that makes the three rays highlighted by us special is that they are easy to construct.

Figure 33.4 shows the case in which $d_o > f$. In this case we find an image that is upside down (inverted). The image is formed on the opposite side of the lens from the object, and

Concept Check 33.2

If you move the object in Figure 33.4 to the left of its present location, the image will

a) move to the left and become larger.

b) move to the left and become smaller.

c) remain unchanged.

d) move to the right and become larger.

e) move to the right and become smaller.

Concept Check 33.3

If you cover the top half of the lens in Figure 33.4 with a material that does not allow light to pass through, then

a) the top half of the image will vanish.

b) the bottom half of the image will vanish.

c) the entire image will vanish.

d) the entire image will still be visible, but will be fainter.

e) nothing will change.

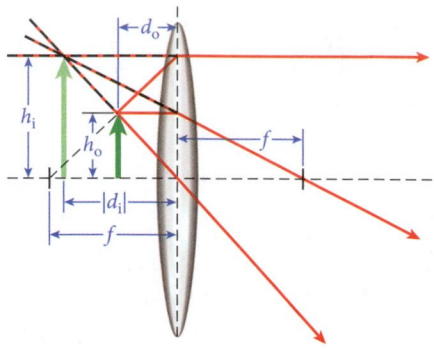

FIGURE 33.5 Imaginary image produced by a converging lens for the case that the object distance is smaller than the focal length.

by convention it is called a real image. In Figure 33.5 we look at the case $d_o < f$. You can see that in this case the three rays have to be extended backwards to intersect. This means that the image is formed on the same side of the lens as the object, and by convention it is called a virtual image. You can see from Figure 33.5 that the image is upright.

Diverging Lenses

The discussion of diverging lenses follows very much the same ideas as those for converging lenses. In Figure 33.6 we show the path of a light ray through a diverging lens, in the same way that we have done it in Figure 33.2 for a converging lens.

It is clear from Figure 33.6 that the ray is refracted away from the optic axis, whereas in the case of the converging lens the ray is refracted toward the optic axis. This becomes even more apparent in Figure 33.7, where we examine five parallel rays entering the diverging lens. As the name suggests, the rays diverge from each other. However, as you can see in Figure 33.7b, if one extends the refracted rays (yellow-black dashed lines) on the opposite side of the lens, then they also intersect at one point on the optic axis, the focal point.

Diverging lenses can also be used to form images. Figure 33.8 shows the geometric construction for the formation of an image using a diverging lens. As you can see, the same three special rays used for converging lenses can also be used for diverging lenses, but you have to be aware that you need to extrapolate the refracted rays on the other side of the lens.

Lens Equations

For curved mirrors we went from ray tracing and image construction to quantitative calculation of image distances and heights by deriving the mirror equation, which provides a relationship between focal length f, object distance d_o, and image distance d_i. It turns out that the same formal relationship between these physical quantities,

$$\frac{1}{d_o} + \frac{1}{d_i} = \frac{1}{f},$$ (33.2)

also holds for lenses and is called the **Thin-Lens Equation**. However, in order for this relationship to work, certain sign conventions have to be followed, which are summarized in Table 33.1.

The magnification is again defined as the ratio of image size to object size, and just like for mirrors, it is

$$m = \frac{h_i}{h_o} = -\frac{d_i}{d_o}.$$ (33.3)

The derivation of equations 33.2 and 33.3 is in complete analogy to the derivation 32.1 in the previous chapter for the spherical-mirror equation.

For curved mirrors we had seen that the focal length was simply half of the radius of curvature. For lenses, we can have a different curvature on opposite sides of the lens. In addition, we also expect the index of refraction of the lens to play a role. What we find is the **Lens-Maker's Formula**,

$$\frac{1}{f} = \frac{n_{lens} - n_{medium}}{n_{medium}} \left(\frac{1}{R_1} - \frac{1}{R_2} \right),$$ (33.4)

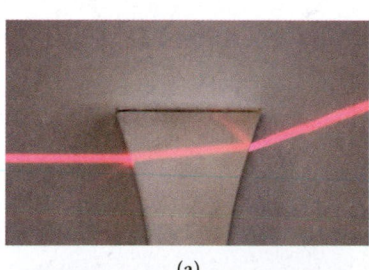

(a)

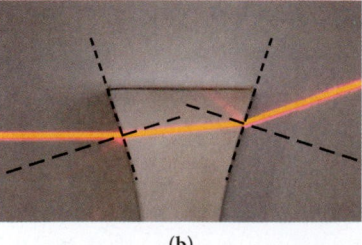

(b)

FIGURE 33.6 (a) A red light ray enters a diverging lens from the right and gets refracted at both surfaces. (b) The local tangent lines (short-dashed black lines) and their normal lines (long-dashed) at the locations where the ray hits the surfaces are superimposed.

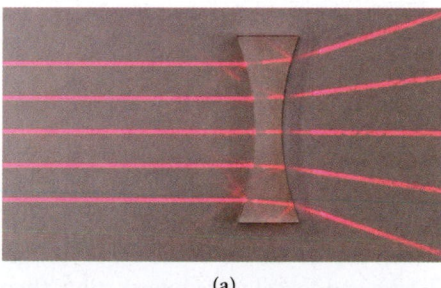

(a)

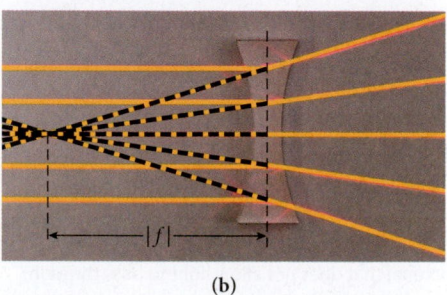

(b)

FIGURE 33.7 (a) Five parallel red light rays hit a converging lens from the left and focused at a single point. (b) Representation of the lens by a single dashed line through its center, and refraction of the idealized rays (yellow lines) through the same focal point.

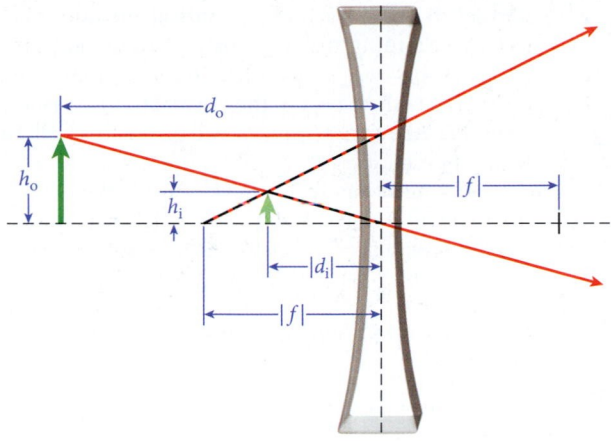

FIGURE 33.8 Image produced by a diverging lens of object placed a distance from the lens larger than the focal length of the lens.

Concept Check 33.4

If you move the object in Figure 33.8 to the left of its present location, the image will

a) move to the left and become larger.

b) move to the left and become smaller.

c) remain unchanged.

d) move to the right and become larger.

e) move to the right and become smaller.

Table 33.1	Sign Conventions for Image Construction with a Single Lens	
	Converging Lens	**Diverging Lens**
Object distance	$d_o > 0$	$d_o > 0$
Focal length	$f_{conv.} > 0$	$f_{div.} < 0$
Image distance	Same side as object: $d_i < 0$ Opposite side: $d_i > 0$	Same side as object: $d_i < 0$ Opposite side: $d_i > 0$

where n_{lens} and n_{medium} are the indices of refraction of the lens and the medium surrounding it, respectively, and R_1 and R_2 are the radii of curvature of the two lens surfaces. The radii of curvature can have positive and negative signs, and the sign convention is that the radius of curvature is positive if the ray is hitting the circle segment from the outside, and it is negative if the ray is hitting the circle segment from the inside (see Figure 33.9).

Note the role of the medium surrounding the lens in equation 33.4: if the index of refraction of the medium is smaller than that of the lens, a converging lens might look like the one shown in Figure 33.3; however, if the same lens is put into a medium with $n_{medium} > n_{lens}$, the same lens acts as a diverging lens! In most cases the medium surrounding the lens is air, with an index of refraction of 1, in which case equation 33.4 simplifies to

$$\frac{1}{f} = (n_{lens} - 1)\left(\frac{1}{R_1} - \frac{1}{R_2}\right). \qquad (33.5)$$

Even though equation 33.5 is only a special case of equation 33.4, it is far more useful in practical applications, because most lenses in optical instruments are surrounded by air as the optical medium.

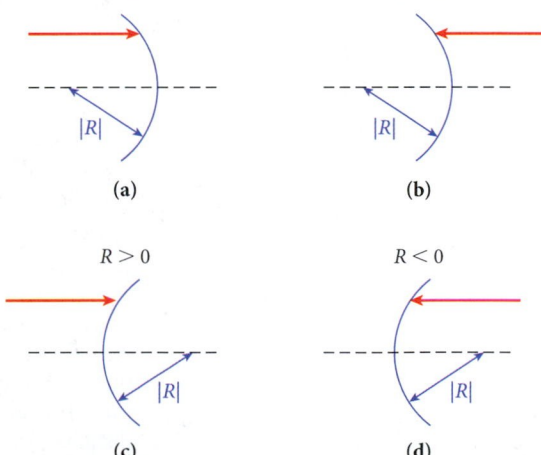

FIGURE 33.9 Sign convention for the radius of curvature of a lens surface (blue curve).

DERIVATION 33.1 / Lens-Maker's Formula

We start the derivation of the Lens-Maker's Formula by assuming that we have light traveling in a medium with index of refraction n_{medium} incident on a spherical surface of an optical medium with index of refraction n_{lens} and radius of curvature R (Figure 33.10). We draw the optic axis as a line perpendicular to the spherical face of the optical medium, through the center C of the spherical face. We assume a light ray originating from a point source at a distance d_o from the lens at an angle α with respect to the optic axis. This ray makes an angle θ_1 with respect to a

– *Continued*

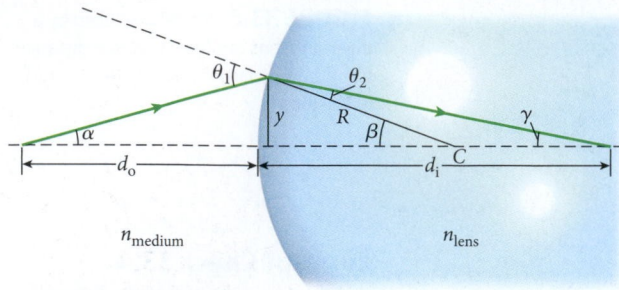

FIGURE 33.10 Light traveling in a medium incident on a spherical surface of a lens.

normal to the surface of the optical medium and then gets refracted according to Snell's Law of refraction (see Chapter 32), $n_{\text{medium}}\sin\theta_1 = n_{\text{lens}}\sin\theta_2$, where θ_2 is the angle the refracted light ray makes with respect to the normal to the surface. If α is a small angle, then the angles θ_1 and θ_2 will be small as well, and we can write $n_{\text{medium}}\theta_1 = n_{\text{lens}}\theta_2$.

Looking at Figure 33.10, we can see the relationship between the angles $\theta_1 = \alpha + \beta$ and $\theta_2 = \beta - \gamma$. Inserting these into our small-angle approximation of Snell's Law results in

$$n_{\text{medium}}(\alpha + \beta) = n_{\text{lens}}(\beta - \gamma).$$

From Figure 33.3 and again using the small-angle approximation, we can see

$$\alpha \approx \tan\alpha \approx \frac{y}{d_{\text{o}}}, \quad \beta \approx \tan\beta \approx \frac{y}{R}, \quad \gamma \approx \tan\gamma \approx \frac{y}{d_{\text{i}}}.$$

Inserting these relationships for the three angles into the previous equation yields

$$n_{\text{medium}}\left(\frac{y}{d_{\text{o}}} + \frac{y}{R}\right) = n_{\text{lens}}\left(\frac{y}{R} - \frac{y}{d_{\text{i}}}\right).$$

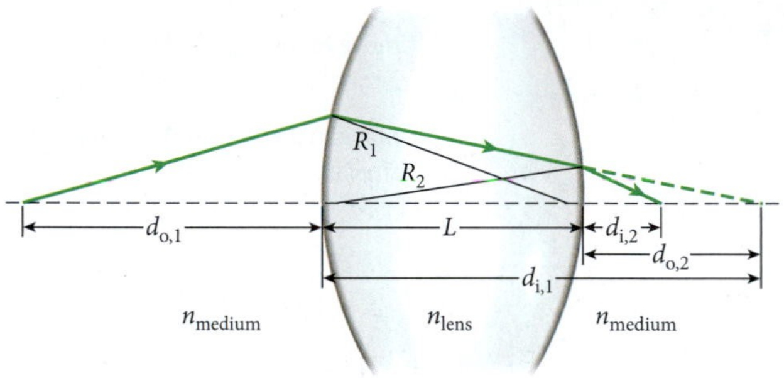

Finally, we cancel out the factor y on both sides of this equation and rearrange, and find

$$\frac{n_{\text{medium}}}{d_{\text{o}}} + \frac{n_{\text{lens}}}{d_{\text{i}}} = \frac{n_{\text{lens}} - n_{\text{medium}}}{R}.$$

The quantity $P_{\text{s}} \equiv (n_{\text{lens}} - n_{\text{medium}})/R$ is called the surface power of the lens surface. What have we accomplished so far? We have derived an equation for the image distance for a given object distance and for a given radius of curvature of the lens, for refraction of light from one surface of the lens. But our lens has a front and a back surface (see Figure 33.11). Therefore our equation is applicable for both surfaces. Adding indices "1" and "2" we then have two equations,

$$\frac{n_{\text{medium}}}{d_{\text{o,1}}} + \frac{n_{\text{lens}}}{d_{\text{i,1}}} = \frac{n_{\text{lens}} - n_{\text{medium}}}{R_1} \quad \text{and} \quad \frac{n_{\text{lens}}}{d_{\text{o,2}}} + \frac{n_{\text{medium}}}{d_{\text{i,2}}} = \frac{n_{\text{medium}} - n_{\text{lens}}}{R_2}.$$

FIGURE 33.11 A lens of thickness L. The left face of the lens has radius R_1 and the right face of the lens has radius R_2.

If the rays come into the lens from the right, as indicated in Figure 33.11, they are refracted on surface 1 and form an image at $d_{\text{i,1}}$. Since the image is real, according to our sign convention in Table 33.1, $d_{\text{i,1}} > 0$. That image acts as the object for the refraction of rays on surface 2. You can see from Figure 33.11 that we measure the distance for refraction off surface 2 relative to that surface, and that $|d_{\text{i,1}}| = L + |d_{\text{o,2}}|$, where L is the lens thickness.

We now make the additional assumption that our lens is thin enough that we can neglect its thickness L relative to the image and object distances in the problem. In this approximation we then have $|d_{\text{i,1}}| \approx |d_{\text{o,2}}|$. Our rays travel from left to right, and positive object distances imply that the object is to the left of the lens. However, as you can see from Figure 33.11, the object for the refraction off the second surface is to the right of the lens, which means a negative object distance. Therefore $d_{\text{o,2}} = -d_{\text{i,1}}$.

With this result we can combine the two equations for the surface powers of the two lenses and obtain

$$\left.\begin{array}{l} \dfrac{n_{\text{medium}}}{d_{\text{o,1}}} + \dfrac{n_{\text{lens}}}{d_{\text{i,1}}} = \dfrac{n_{\text{lens}} - n_{\text{medium}}}{R_1} \Rightarrow \dfrac{n_{\text{lens}}}{d_{\text{i,1}}} = \dfrac{n_{\text{lens}} - n_{\text{medium}}}{R_1} - \dfrac{n_{\text{medium}}}{d_{\text{o,1}}} \\[2ex] \dfrac{n_{\text{lens}}}{d_{\text{o,2}}} + \dfrac{n_{\text{medium}}}{d_{\text{i,2}}} = \dfrac{n_{\text{medium}} - n_{\text{lens}}}{R_2} \Rightarrow \dfrac{n_{\text{lens}}}{d_{\text{o,2}}} = \dfrac{n_{\text{medium}} - n_{\text{lens}}}{R_2} - \dfrac{n_{\text{medium}}}{d_{\text{i,2}}} = -\dfrac{n_{\text{lens}}}{d_{\text{i,1}}} \end{array}\right\} \Rightarrow$$

$$\frac{n_{\text{lens}} - n_{\text{medium}}}{R_1} - \frac{n_{\text{medium}}}{d_{\text{o,1}}} = -\frac{n_{\text{medium}} - n_{\text{lens}}}{R_2} + \frac{n_{\text{medium}}}{d_{\text{i,2}}}.$$

Rearranging now results in

$$\frac{n_{\text{medium}}}{d_{\text{o,1}}} + \frac{n_{\text{medium}}}{d_{\text{i,2}}} = \frac{n_{\text{lens}} - n_{\text{medium}}}{R_1} - \frac{n_{\text{medium}} - n_{\text{lens}}}{R_2} \Rightarrow$$

$$\frac{1}{d_{\text{o,1}}} + \frac{1}{d_{\text{i,2}}} = \frac{n_{\text{lens}} - n_{\text{medium}}}{n_{\text{medium}}}\left(\frac{1}{R_1} - \frac{1}{R_2}\right).$$

The combination of the two refracting surfaces thus maps an object located at $d_{o,1}$ into the image located at $d_{i,2}$. Using the Thin-Lens Equation 33.2, we see that the left side of our result is simply $1/f$, and we arrive at our desired result for the Lens-Maker's Formula

$$\frac{1}{f} = \frac{n_{lens} - n_{medium}}{n_{medium}} \left(\frac{1}{R_1} - \frac{1}{R_2} \right).$$

Lenses do not necessarily have the same curvature on the entrance and exit surfaces. Lenses can be formed in several different ways, as illustrated in Figure 33.12 for several different lenses in air. For example, light incident from the left on the converging meniscus lens shown in Figure 33.12 would first encounter a convex surface ($R_1 > 0$) and then a second convex surface ($R_2 > 0$). For the converging meniscus lens, $R_1 < R_2$, therefore $1/R_1 > 1/R_2$, so the focal length of the lens given by the Lens-Maker's Formula in air (equation 33.5) is positive, producing a converging lens. For the diverging meniscus lens shown in Figure 33.12, the first surface is convex ($R_1 > 0$) and the second surface is also convex ($R_2 > 0$). However, for the diverging meniscus lens, $R_1 > R_2$ and $1/R_1 < 1/R_2$, so the focal length is negative, producing a diverging lens. Meniscus lenses are commonly used in corrective eyeglasses.

For a converging lens, we find that for $d_o > f$ we always get a real, inverted image, i.e., an upside-down image formed on the opposite side of the lens. If $d_o = f$, then $1/d_i = 0$, and the image is located at infinity. For a converging lens and $d_o < f$ we always get a virtual, upright, and enlarged image on the same side of the lens as the object. The results for all values of the object distance are summarized in Table 33.2 for quick reference. Diverging lenses always produce an image that is virtual, upright, and reduced in size.

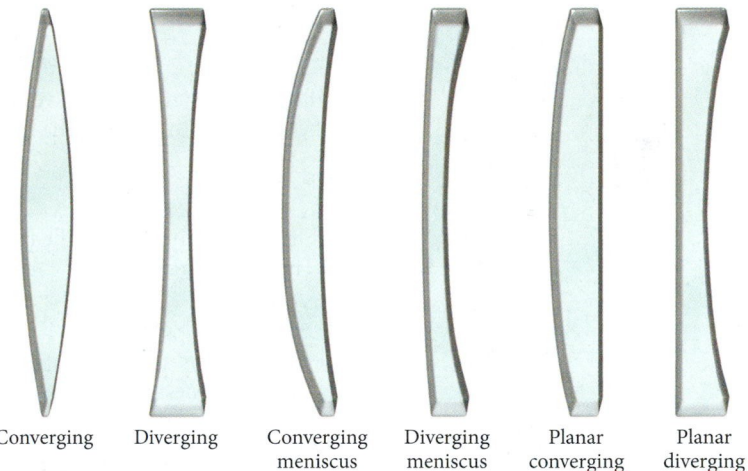

Converging   Diverging   Converging meniscus   Diverging meniscus   Planar converging   Planar diverging

FIGURE 33.12 Different types of lenses created by different radii of curvature on the entrance and exit surfaces.

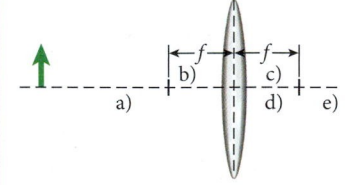
Table 33.2	Image Characteristics for Converging Lenses		
Case	**Image Type**	**Image Orientation**	**Magnification**
$f < d_o < 2f$	Real	Inverted	Enlarged
$d_o = 2f$	Real	Inverted	Same size
$d_o > 2f$	Real	Inverted	Reduced
$d_o = f$			Infinity
$d_o < f$	Virtual	Upright	Enlarged

Self-Test Opportunity 33.2

In the four diagrams shown in the figure below, the solid arrow represents the object and the dashed arrow represents the image. The dashed rectangle represents a single optical element. The possible optical elements include a plane mirror, a converging mirror, a diverging mirror, a diverging lens, and a converging lens. Match each diagram with the corresponding optical element.

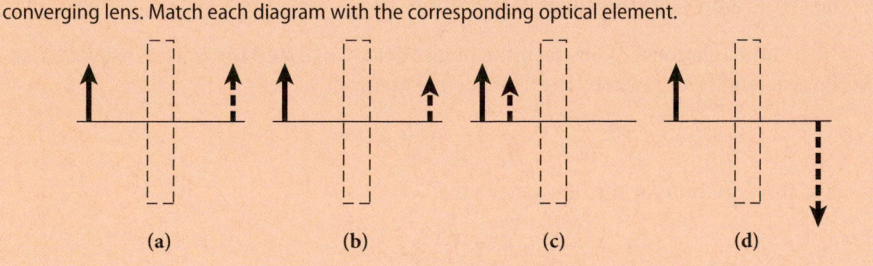

(a)   (b)   (c)   (d)

EXAMPLE 33.1 | Image Formed by a Thin Lens

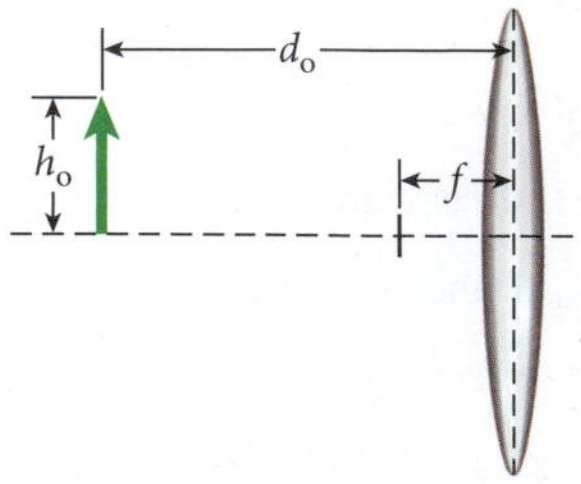

We place an object with height $h_o = 5.00$ cm at a distance $d_o = 16.0$ cm from a thin converging lens with focal length $f = 4.00$ cm (Figure 33.13).

PROBLEM
What is the image distance? What is the linear magnification of the image? What is the image height?

SOLUTION
The image distance can be calculated using the Thin-Lens Equation (equation 33.2),

$$\frac{1}{d_o} + \frac{1}{d_i} = \frac{1}{f}.$$

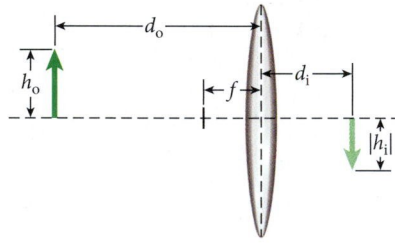

Solving for the image distance, we get

$$d_i = \frac{d_o f}{d_o - f} = \frac{(16.0 \text{ cm})(4.00 \text{ cm})}{16.0 \text{ cm} - 4.00 \text{ cm}} = 5.33 \text{ cm}.$$

The image distance is positive. Therefore, the image is real and appears on the opposite side of the lens from the object. The magnification formula for lenses is given by

$$m = \frac{h_i}{h_o} = -\frac{d_i}{d_o}.$$

Thus, we can calculate the magnification using the given object distance and the calculated image distance:

$$m = -\frac{d_i}{d_o} = -\frac{5.33 \text{ cm}}{16.0 \text{ cm}} = -0.333.$$

The magnification is negative, so the image is inverted. The magnitude of the magnification is less than one, so the image is reduced. We can now calculate the image height:

$$h_i = mh_o = (-0.333)(5.00 \text{ cm}) = -1.67 \text{ cm}.$$

The image height is negative, so the image is inverted, as we expected from the negative magnification.

The resulting image is illustrated in Figure 33.14.

FIGURE 33.13 An object placed in front of a thin converging lens.

FIGURE 33.14 Image formed by a converging thin lens.

SOLVED PROBLEM 33.1 | Image of the Moon

PROBLEM
An image of the Moon is focused onto a screen, using a converging lens of focal length $f = 50.0$ cm. The radius of the Moon is $R = 1.737 \cdot 10^6$ m, and the mean distance between Earth and the Moon is $d = 3.844 \cdot 10^8$ m. What is the radius of the Moon in the image on the screen?

SOLUTION
THINK The linear magnification of the image produced by the lens can be expressed in terms of the ratio of the image size to the object size and in terms of the ratio of image distance to the object distance.

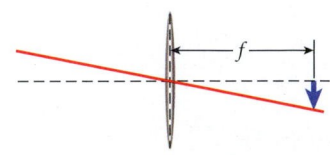

SKETCH Figure 33.15 shows a sketch of the image of the Moon produced by a converging lens. A blue arrow represents the image of the Moon.

FIGURE 33.15 Image of the Moon produced by a converging lens.

RESEARCH The relationship between the object distance d_o and the image distance d_i for a lens with focal length f is given by the Thin-Lens Equation,

$$\frac{1}{d_o} + \frac{1}{d_i} = \frac{1}{f}.$$

We can solve this equation for the image distance,

$$d_i = \frac{d_o f}{d_o - f}.$$

In this case, the object distance is the distance from Earth to the Moon. Because the object distance is much greater than the focal length of the lens, we can write

$$d_i \approx \frac{d_o f}{d_o} = f.$$

The linear magnitude of the magnification m for a lens can be written as

$$m = -\frac{d_i}{d_o} = \frac{h_i}{h_o}$$

where h_i is the height (radius) of the image and h_o is the height of the object, which in this case is the radius of the Moon.

SIMPLIFY We can solve the previous equation for the image height,

$$h_i = -h_o \frac{d_i}{d_o}.$$

Taking the object height to be the radius of the Moon R, the object distance to be the distance from Earth to the Moon d, and the image distance to be the focal length of the lens f, we can write

$$h_i = -R\frac{f}{d}.$$

CALCULATE Putting in the numerical values, we have

$$h_i = -R\frac{f}{d} = \left(1.737 \cdot 10^6 \text{ m}\right)\frac{0.500 \text{ m}}{3.844 \cdot 10^8 \text{ m}} = -0.00225937 \text{ m}.$$

ROUND We report our result for the radius of the image of the Moon on the screen to three significant figures,

$$h_i = -2.26 \cdot 10^{-3} \text{ m} = -2.26 \text{ mm}.$$

DOUBLE-CHECK The Moon is relatively close to Earth and we can see the Moon as a disk easily with the naked eye. Thus, it seems reasonable that we could produce an image of the Moon with radius 2.26 mm on a screen with a lens of focal length 50.0 cm. The negative sign for the image height means that our image of the Moon is inverted.

SOLVED PROBLEM 33.2 | Two Positions of a Converging Lens

A light bulb is at a distance $d = 1.45$ m away from a screen. A converging lens with focal length $f = 15.3$ cm forms an image of the light bulb on the screen for two lens positions.

PROBLEM
What is the distance between these two positions?

SOLUTION
THINK The sum of the object distance of the lens plus the image distance of the lens is equal to the distance of the light bulb from the screen. Using the lens equation, we can solve for the possible image distances using the quadratic equation. The distance between the two image distances is the distance between the two lens positions.

SKETCH Figure 33.16 shows a sketch of the lens placed between the light bulb and the screen.

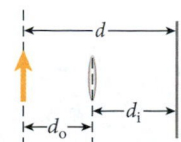

FIGURE 33.16 A lens is placed between a light bulb and a screen.

RESEARCH The Thin-Lens Equation is

$$\frac{1}{d_o} + \frac{1}{d_i} = \frac{1}{f}, \tag{i}$$

where d_o is the object distance, d_i is the image distance, and f is the focal length. We can express the object distance in terms of the image distance and the distance of the light bulb from the screen d,

$$d_o = d - d_i. \tag{ii}$$

– Continued

Substituting equation (ii) into equation (i) gives us

$$\frac{1}{d-d_i}+\frac{1}{d_i}=\frac{1}{f}.$$

We can rearrange this equation to get

$$d_i+(d-d_i)=\frac{d_i(d-d_i)}{f}.$$

Collecting terms and multiplying by f gives us

$$df = d_i d - d_i^2. \tag{iii}$$

SIMPLIFY We can rewrite equation (iii) so that we can recognize it as a quadratic equation, for which we can find the solutions:

$$d_i^2 - d_i d + df = 0.$$

The solutions of this equation are

$$d_i = \frac{d \pm \sqrt{d^2 - 4df}}{2},$$

where one solution corresponds to the + sign and the other solution corresponds to the – sign.

CALCULATE The solution corresponding to the + sign is

$$d_{i+} = \frac{(1.45 \text{ m}) + \sqrt{(1.45 \text{ m})^2 - 4(1.45 \text{ m})(0.153 \text{ m})}}{2} = 1.27616 \text{ m}.$$

The solution corresponding to the – sign is

$$d_{i-} = \frac{(1.45 \text{ m}) - \sqrt{(1.45 \text{ m})^2 - 4(1.45 \text{ m})(0.153 \text{ m})}}{2} = 0.173842 \text{ m}.$$

The difference between the two positions is $\Delta d_i = 1.10232$ m.

ROUND We report our result to three significant figures,

$$\Delta d_i = 1.10 \text{ m}.$$

(*Note:* Even though we have found two positions for the lens that allow an image to be formed, it is not correct to assume that the image has the same size in both cases.)

DOUBLE-CHECK To double-check our answer, we substitute our solutions for the image distance into the Thin-Lens Equation and show that they work. For the first solution, $d_i = 1.276$ m, so the corresponding object distance is $d_o = d - d_i = 0.174$ m. The lens equation then tells us

$$\frac{1}{0.174}+\frac{1}{1.276}=\frac{1}{0.153 \text{ m}},$$

which agrees within round-off errors. For the second solution, we simply reverse the role of the image distance and object distance in the Thin-Lens Equation, and we get the same answer. Thus our result seems reasonable.

FIGURE 33.17 A typical magnifying glass.

33.2 Magnifier

One way to make an object appear larger is to bring it closer. However, if the object is brought too close to the eye, the object will appear fuzzy. The position closest to the eye that an object can be placed at and remain in focus is called the "near point," as discussed in detail below. Another way to make an object appear larger is to use a **magnifier** or magnifying glass (Figure 33.17).

A magnifier is nothing more than a converging lens that produces an enlarged virtual image of an object. This image appears at a distance that is at or beyond the near point of the eye, so an observer can clearly see the image. The **angular magnification** m_θ of the magnifier is defined as the ratio of the apparent angle subtended by the image to the angle subtended by the object when located at the near point.

Figure 33.18 shows the geometry of a magnifier. Assume an object with height h_o. Without a magnifier, the largest angle θ_1 that you can attain and still see the object clearly occurs when you place the object at the near point d_{near} (Figure 33.18a). You can get a magnified image of the object by placing the object just inside the focal length of a converging lens (Figure 33.18b). You then look through the lens at the image, which is intentionally located at least as far away as the near point. Therefore you can see the enlarged, upright, virtual image. The angle subtended by the image is θ_2.

The angular magnification of the magnifier is defined as

$$m_\theta = \frac{\theta_2}{\theta_1}. \tag{33.6}$$

Figure 33.18a shows that the angle subtended by the object without the magnifier is given by

$$\tan\theta_1 = \frac{h_o}{d_{near}}.$$

Figure 33.18b shows that the angle subtended by the image of the object is given by

$$\tan\theta_2 = \frac{h_o}{f},$$

where f is the focal length of the lens. We assume that the object is placed at the focal length of the lens, so the image is at minus infinity. (Recall from Section 33.1 that a ray through the center of the lens is not bent. It is this ray that forms the hypotenuse of the right triangle used for this equation.) We then make a small-angle approximation to get $\tan\theta_1 \approx \theta_1$ and $\tan\theta_2 \approx \theta_2$. Thus, the angular magnification of a magnifier can be written as

$$m_\theta = \frac{\theta_2}{\theta_1} \approx \frac{h_o/f}{h_o/d_{near}} = \frac{d_{near}}{f}.$$

Assuming a typical value for the near point of 25 cm (which is the value for a middle-aged person; for a 20-year-old the near point may be closer to 10 cm), the angular magnification can be written as

$$m_\theta \approx \frac{d_{near}}{f} \approx \frac{0.25 \text{ m}}{f}. \tag{33.7}$$

The magnification obtained in equation 33.7 is not the maximum magnification that is achievable with a magnifier. That maximum can be reached if the final image can be placed at the near point. Using the lens equation with $d_i = -d_{near}$ to find d_o and then using $\theta_2 = h_o/d_o$, we find $m_\theta = (0.25 \text{ m}/f) + 1$, if $d_{near} = 0.25$ m. However, it can be quite delicate to obtain a sharp image of the object in this configuration; this is why most people prefer to use the magnifying glass as we described above. Henceforth, unless otherwise stated, we will assume that the image is at infinity and will use equation 33.7.

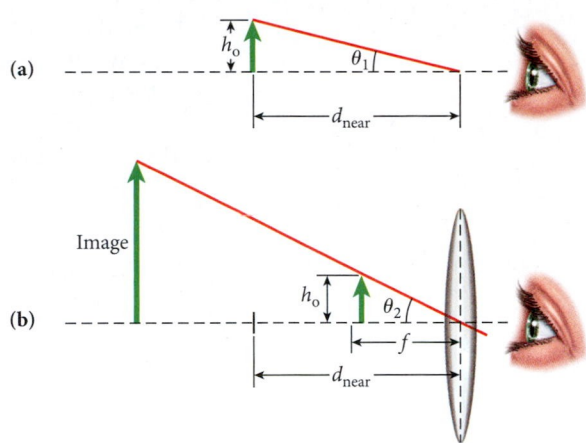

FIGURE 33.18 The geometry of a magnifier. (a) Viewing an object at the near point. (b) Viewing a magnified image of the object.

Concept Check 33.7

What is the focal length (in meters) of a magnifying glass that gives an angular magnification of 6?

a) 0.010 m   d) 0.042 m

b) 0.021 m   e) 0.055 m

c) 0.035 m

33.3 Systems of Two or More Optical Elements

We have seen that a lens or mirror can produce an image of an object. This image can then, in turn, be used as the object for a second lens or mirror. The recurring theme for all two-lens systems is that the image of the first lens becomes the object of the second lens. Let's start our study of multi-lens systems by considering a two-lens system consisting of two converging lenses, placed as shown in Figure 33.19.

An object with height $h_{o,1}$ is placed a distance $d_{o,1}$ from the first lens, which has a focal length f_1. An image is produced at a distance $d_{i,1}$ given by the Thin-Lens Equation (33.2),

$$\frac{1}{d_{o,1}} + \frac{1}{d_{i,1}} = \frac{1}{f_1}.$$

Given the value of $d_{o,1}$ that we have chosen, this image is real and inverted. This image then becomes the object for the second lens. The height of this object is the same as the height of the image produced by the first lens. The object is located a distance $d_{o,2}$ from the second lens, which has a focal length f_2. An image is formed at a distance $d_{i,2}$ again governed by the Thin-Lens Equation,

$$\frac{1}{d_{o,2}} + \frac{1}{d_{i,2}} = \frac{1}{f_2}.$$

For the parameters of the system depicted in Figure 33.19, the final image of the second lens is real and inverted, compared to the second object (first image). The linear magnification of the first lens is given by $m_1 = h_{i,1}/h_{o,1}$ and the linear magnification of the second lens is given by $m_2 = h_{i,2}/h_{o,2}$. The product of the linear magnifications of the two lenses gives the linear magnification of the two-lens system:

$$m_{12} = m_1 m_2 = \left(\frac{h_{i,1}}{h_{o,1}}\right)\left(\frac{h_{i,2}}{h_{o,2}}\right) = \frac{h_{i,2}}{h_{o,1}}. \qquad (33.8)$$

From equation 33.8 we can see that the image produced by this two-lens system is real and upright. Thus, this system of lenses can be used to produce real images that are not inverted. This kind of image formation cannot be done with a single converging lens.

Now let's consider a two-lens system with one converging lens and one diverging lens (Figure 33.20). An object with height $h_{o,1}$ is placed a distance $d_{o,1}$ from the first lens, which has a focal length f_1. Again, an image is produced at a distance $d_{i,1}$ given by the Thin-Lens Equation (equation 33.2). This particular image is real, inverted, and enlarged. Just like in the previous case, this image generated by the first lens then becomes the object for the second lens. This second image is virtual, upright with respect to the first image, and reduced. The combined linear magnification of the two lenses is again the product of the linear magnifications of the individual lenses and also given by $m_{12} = m_1 m_2 = h_{i,2}/h_{o,1}$, just like in the case discussed previously. Figure 33.20 shows that the final image produced by this two-lens system is virtual and inverted with respect to the original object.

Now let's take this same two-lens system of a converging lens followed by a diverging lens and put the two lenses close together so that their separation x is less than their individual focal lengths (Figure 33.21). This two-lens system acts as a converging lens. By varying the distance x we can vary the effective focal length of the converging lens system. This arrangement is called a **zoom lens.**

FIGURE 33.19 A system of two converging lenses.

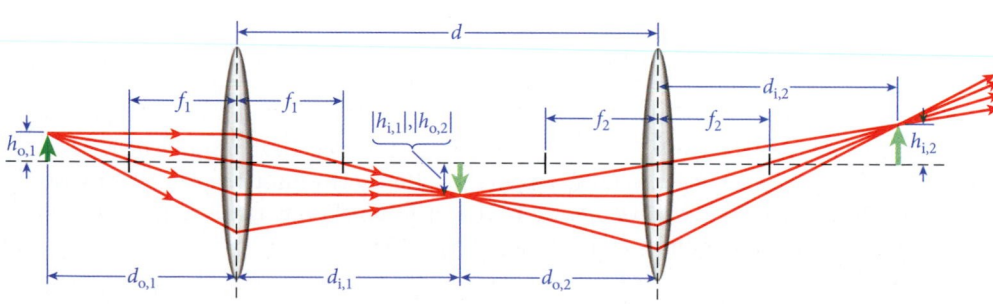

FIGURE 33.20 A two-lens system consisting of a converging lens and a diverging lens.

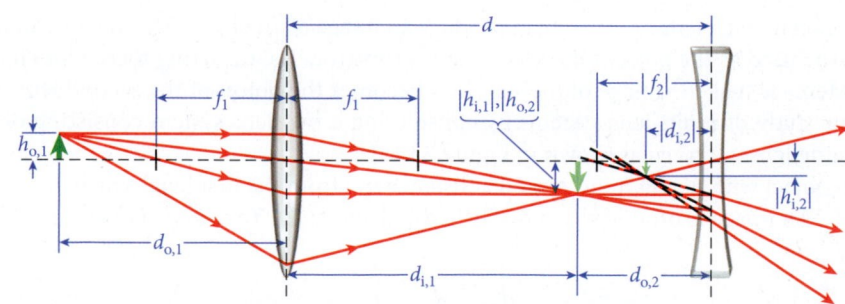

The effective focal length of this two-lens system is defined as the distance from the center of the first lens to the position of the final image for an object originally located at infinity. The first lens has a focal length f_1, which means that objects placed at a large distance will produce an image at an image distance of $d_{i,1} = f_1$. We can understand this result using the Thin-Lens Equation,

$$\frac{1}{d_{o,1}} + \frac{1}{d_{i,1}} = \frac{1}{\infty} + \frac{1}{d_{i,1}} = 0 + \frac{1}{d_{i,1}} = \frac{1}{d_{i,1}} = \frac{1}{f_1}.$$

Assuming $f_1 > x$, the image produced by the first lens is generated on the right side of the second lens. This means that the object distance for the second lens must be negative in this case, because we have defined positive object distances to be to the left of a lens. The object distance for the second lens is

$$d_{o,2} = -\left(d_{i,1} - x\right) = x - d_{i,1} = x - f_1.$$

Using this image as the object for the second lens, we get

$$\frac{1}{d_{o,2}} + \frac{1}{d_{i,2}} = \frac{1}{x - f_1} + \frac{1}{d_{i,2}} = \frac{1}{f_2}.$$

This equation can be solved for $d_{i,2}$ to get the effective focal length of the zoom lens system,

$$f_{\text{eff}} = x + d_{i,2} = x + \frac{f_2\left(x - f_1\right)}{x - \left(f_2 + f_1\right)}. \tag{33.9}$$

Equation 33.9 and Figure 33.21 show that when the lenses are close together, the effective focal length is longer; when the lenses are farther apart, the effective focal length is shorter. Thus, by adjusting the distance between the converging lens and the diverging lens, we can produce an effective lens with varying focal length, as is done in zoom lenses of cameras. Note that for cameras, this result is useful only for $f_{\text{eff}} > x$ because objects at infinity must produce a real image on the digital image recorder (for a digital camera) or on the film (for an old-fashioned conventional camera).

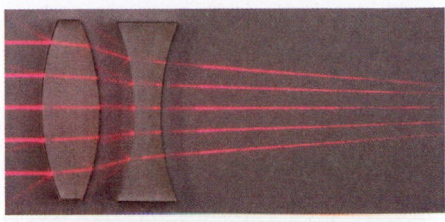
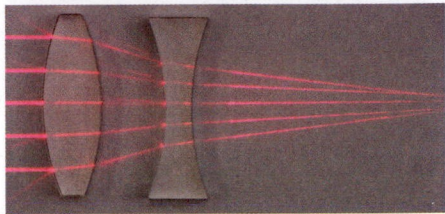
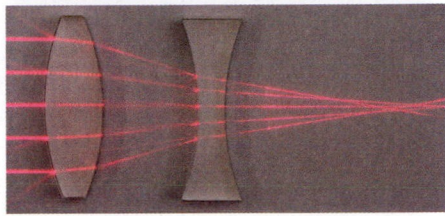

FIGURE 33.21 A zoom lens system consisting of a converging lens followed by a diverging lens. Three different distances between the two lenses are shown.

Concept Check 33.8

In Figure 33.21 the lenses are arranged so that the light first hits the converging lens and then the diverging one. What happens if you switch the two lenses so the the light first hits the diverging one?

a) You still have a zoom lens with the approximately same effective focal length.

b) You have a diverging zoom lens.

c) You cannot form an image.

d) You still have a zoom lens, but with a much longer effective focal length.

Self-Test Opportunity 33.3

Use equation 33.9 to show that the effective focal length f_{eff} of two thin lenses with focal lengths f_1 and f_2 close together is given by

$$\frac{1}{f_{\text{eff}}} = \frac{1}{f_1} + \frac{1}{f_2}.$$

SOLVED PROBLEM 33.3 | Image Produced by Two Lenses

Consider a system of two lenses. The first lens is a converging lens with a focal length $f_1 = 21.4$ cm. The second lens is a diverging lens with focal length $f_2 = -34.4$ cm. The center of the second lens is $d = 80.0$ cm to the right of the center of the first lens. An object is placed a distance $d_{o,1} = 63.5$ cm to the left of the first lens. These lenses produce an image of the object.

PROBLEM

Where is the image produced by the second lens located with respect to the center of the second lens? What are the characteristics of the image? What is the linear magnification of the final image with respect to the original object?

SOLUTION

THINK The Thin-Lens Equation (which we will here call simply the lens equation) tells us where the first lens produces an image of the object placed to the left of the first lens. The image

Self-Test Opportunity 33.4

A normal 35-mm camera has a lens with a focal length of 50 mm. Suppose you replace the normal lens with a zoom lens whose focal length can be varied from 50 mm to 200 mm and use the camera to photograph an object at a very large distance. Compared to a 50-mm lens, what magnification of the image would be achieved using the 200-mm focal length?

– Continued

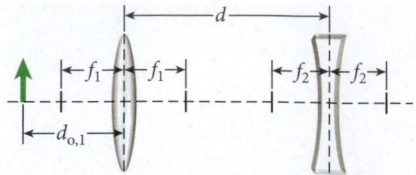

FIGURE 33.22 A system of two lenses with focal lengths and object distance marked.

produced by the first lens becomes the object for the second lens. We again use the lens equation to locate the image produced by the second lens.

SKETCH Figure 33.22 shows a sketch of the object, the first lens, and the second lens.

RESEARCH The lens equation applied to the first lens gives

$$\frac{1}{d_{o,1}} + \frac{1}{d_{i,1}} = \frac{1}{f_1},$$

where $d_{i,1}$ is the image distance for the first lens. Solving for the image distance of the first lens gives

$$d_{i,1} = \frac{d_{o,1} f_1}{d_{o,1} - f_1}. \tag{i}$$

The lens equation applied to the second lens gives

$$\frac{1}{d_{o,2}} + \frac{1}{d_{i,2}} = \frac{1}{f_2},$$

where $d_{o,2}$ is the object distance for the second lens and $d_{i,2}$ is the image distance for the second lens. We can solve for the image distance for the second lens,

$$d_{i,2} = \frac{d_{o,2} f_2}{d_{o,2} - f_2}. \tag{ii}$$

The object for the second lens is the image produced by the first lens. Thus, we can relate the object distance for the second lens to the image distance of the first lens and the distance between the two lenses:

$$d_{o,2} = d - d_{i,1}. \tag{iii}$$

SIMPLIFY We can substitute equation (iii) into equation (ii) to obtain

$$d_{i,2} = \frac{\left(d - d_{i,1}\right) f_2}{\left(d - d_{i,1}\right) - f_2}. \tag{iv}$$

We can then substitute equation (i) into equation (iv) to obtain an expression for the image distance of the second lens in terms of the quantities given in the problem:

$$d_{i,2} = \frac{\left(d - \left(\frac{d_{o,1} f_1}{d_{o,1} - f_1}\right)\right) f_2}{\left(d - \left(\frac{d_{o,1} f_1}{d_{o,1} - f_1}\right)\right) - f_2}.$$

CALCULATE Putting in the numerical values gives us

$$d_{i,2} = \frac{\left((80.0\ \text{cm}) - \left(\frac{(63.5\ \text{cm})(21.4\ \text{cm})}{(63.5\ \text{cm}) - (21.4\ \text{cm})}\right)\right)(-34.4\ \text{cm})}{\left((80.0\ \text{cm}) - \left(\frac{(63.5\ \text{cm})(21.4\ \text{cm})}{(63.5\ \text{cm}) - (21.4\ \text{cm})}\right)\right) - (-34.4\ \text{cm})} = -19.9902\ \text{cm}.$$

ROUND We report our result for the location of the image to three significant figures,

$$d_{i,2} = -20.0\ \text{cm}.$$

DOUBLE-CHECK To double-check our result and calculate the quantities needed to answer the remaining parts of the problem, we calculate the position of the image produced by the first lens:

$$d_{i,1} = \frac{d_{o,1} f_1}{d_{o,1} - f_1} = \left(\frac{(63.5\ \text{cm})(21.4\ \text{cm})}{(63.5\ \text{cm}) - (21.4\ \text{cm})}\right) = 32.3\ \text{cm}.$$

The image distance for the first lens is positive, so the image is real and formed 32.3 cm to the right of the first lens. The object distance for the second lens is then

$$d_{o,2} = d - d_{i,1} = 47.7\ \text{cm}.$$

The image formed by the first lens is 47.7 cm to the left of the second lens, which seems reasonable. We can then calculate the image distance for the second lens:

$$d_{i,2} = \frac{d_{o,2}f_2}{d_{o,2}-f_2} = \left| \frac{(47.7\text{ cm})(-34.4\text{ cm})}{(47.7\text{ cm})-(-34.4\text{ cm})} \right| = -20.0\text{ cm},$$

which agrees with our result. Thus, our answer for the distance of the image from the center of lens 2 seems reasonable.

The final image is virtual because it is on the same side of lens 2 as the object for lens 2, which we know because the image distance for lens 2 is negative. The linear magnification for the final image compared with the original object is

$$m = m_1 m_2 = \left(-\frac{d_{i,1}}{d_{o,1}}\right)\left(-\frac{d_{i,2}}{d_{o,2}}\right) = \left(-\frac{32.3\text{ cm}}{63.5\text{ cm}}\right)\left(-\frac{-20.0\text{ cm}}{47.7\text{ cm}}\right) = -0.213.$$

The image is reduced because $|m| < 1$. The image is inverted because $m < 0$.

SOLVED PROBLEM 33.4 Image Produced by a Lens and a Mirror

An object is placed a distance $d_{o,1} = 25.6$ cm to the left of a converging lens with focal length $f_1 = 20.6$ cm. A converging mirror with focal length $f_2 = 10.3$ cm is placed a distance $d = 120.77$ cm to the right of the lens.

PROBLEM
What is the magnification of the image produced by the lens-and-mirror combination?

SOLUTION
THINK The lens will produce a real, inverted image of the object. This image becomes the object for the converging mirror. The object distance for the mirror is the distance between the lens and the mirror minus the image distance of the lens. The overall magnification is the magnification of the lens times the magnification of the mirror.

SKETCH Figure 33.23 shows a sketch of the object, the lens, and the mirror.

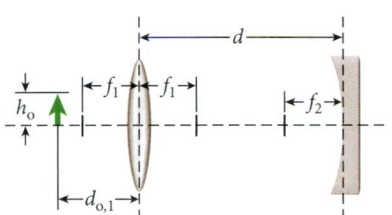

FIGURE 33.23 An object being imaged by a lens-and-mirror combination.

RESEARCH The Thin-Lens Equation tells us that the image distance $d_{i,1}$ for the lens is given by

$$d_{i,1} = \frac{d_{o,1}f_1}{d_{o,1}-f_1}. \tag{i}$$

The mirror equation tells us that the image distance for the mirror is

$$d_{i,2} = \frac{d_{o,2}f_2}{d_{o,2}-f_2}, \tag{ii}$$

where the object distance for the mirror is

$$d_{o,2} = d - d_{i,1}. \tag{iii}$$

The magnification m of the lens-and-mirror system is given by

$$m = m_1 m_2, \tag{iv}$$

where m_1 is the magnification of the lens and m_2 is the magnification of the mirror.

SIMPLIFY The magnification of the lens using equation (i) is

$$m_1 = -\frac{d_{i,1}}{d_{o,1}} = -\frac{\left(\dfrac{d_{o,1}f_1}{d_{o,1}-f_1}\right)}{d_{o,1}} = -\frac{f_1}{d_{o,1}-f_1} = \frac{f_1}{f_1-d_{o,1}}. \tag{v}$$

Similarly, the magnification of the mirror using equation (ii) is

$$m_2 = -\frac{d_{i,2}}{d_{o,2}} = -\frac{\left(\dfrac{d_{o,2}f_2}{d_{o,2}-f_2}\right)}{d_{o,2}} = -\frac{f_2}{d_{o,2}-f_2} = \frac{f_2}{f_2-d_{o,2}}. \tag{vi}$$

– *Continued*

We can then write for the overall magnification, using equations (iii), (v), and (vi) in equation (iv)

$$m = \left(\frac{f_1}{f_1 - d_{o,1}}\right)\left(\frac{f_2}{f_2 - d_{o,2}}\right) = \left(\frac{f_1}{f_1 - d_{o,1}}\right)\left(\frac{f_2}{f_2 - (d - d_{i,1})}\right).$$

Finally, substituting from equation (i) for the image distance for the lens gives us

$$m = \left(\frac{f_1}{f_1 - d_{o,1}}\right)\frac{f_2}{f_2 - \left(d - \dfrac{d_{o,1}f_1}{d_{o,1} - f_1}\right)}.$$

CALCULATE   Putting in the numerical values gives us

$$m = \left(\frac{(20.6\text{ cm})}{(20.6\text{ cm}) - (25.6\text{ cm})}\right)\frac{(10.3\text{ cm})}{(10.3\text{ cm}) - \left((120.77\text{ cm}) - \dfrac{(25.6\text{ cm})(20.6\text{ cm})}{(25.6\text{ cm}) - (20.6\text{ cm})}\right)}$$

$$m = 8.490596.$$

ROUND   We report our result to three significant figures,

$$m = 8.49.$$

DOUBLE-CHECK   To double-check our result, we first calculate the image distance for the lens,

$$d_{i,1} = \frac{d_{o,1}f_1}{d_{o,1} - f_1} = \frac{(25.6\text{ cm})(20.6\text{ cm})}{(25.6\text{ cm}) - (20.6\text{ cm})} = 105.47\text{ cm}.$$

The object distance for the mirror is then

$$d_{o,2} = d - d_{i,1} = 120.77\text{ cm} - 105.47\text{ cm} = 15.3\text{ cm}.$$

The image distance for the mirror is then

$$d_{i,2} = \frac{d_{o,2}f_2}{d_{o,2} - f_2} = \frac{(15.3\text{ cm})(10.3\text{ cm})}{(15.3\text{ cm}) - (10.3\text{ cm})} = 31.52\text{ cm}.$$

The magnification of the image is then

$$m = \frac{d_{i,1}}{d_{o,1}}\frac{d_{i,2}}{d_{o,2}} = \frac{105.5\text{ cm}}{25.6\text{ cm}}\frac{31.52\text{ cm}}{15.3\text{ cm}} = 8.49,$$

which agrees with our result.

In general it is always a good idea to keep with our seven-step method for problem solving and only insert numerical values at the very end. However, for problems involving two or more optical elements, it may be reasonable to make an exception to this general procedure. You can see from the preceding two solved problems that the formulas can get quite cumbersome. Many students find it simpler to compute intermediate image distances and heights numerically first and then use these values to continue further. Arguably, this leads to fewer errors in calculations, in particular under exam pressure. In the last two solved problems we have used this approach as our double-checks. Feel free to employ either technique in solving multi-element optics problems.

33.4   Human Eye

The human eye can be considered an optical instrument. The eye is nearly spherical in shape, about 2.5 cm in diameter (Figure 33.24). The front part of the eye is more sharply curved than the rest of the eye and is covered with the *cornea*. Behind the cornea is a fluid called the *aqueous humor*. Behind that is the *lens,* composed of a fibrous jelly. The lens is held in place by ligaments that connect it to the *ciliary muscles,* which allow the lens to change shape and thus change its focus. Behind the lens is the *vitreous humor*.

The index of refraction of the two fluids in the eye is 1.34, close to that of water (1.33). The index of refraction of the material making up the lens is about 1.40. Thus, most refraction of light rays occurs at the air/cornea boundary, which has the largest difference in indices of refraction.

Refraction at the cornea and lens surfaces produces a real, inverted image on the *retina* of the eye. Cells in the retina called rods and cones convert the image from light to electrical impulses. These impulses are then sent to the brain through the *optic nerve*. The brain interprets the inverted image so that we see it upright. In front of the lens is the *iris,* the colored part of the eye, which partially opens or closes to regulate the amount of light incident on the retina.

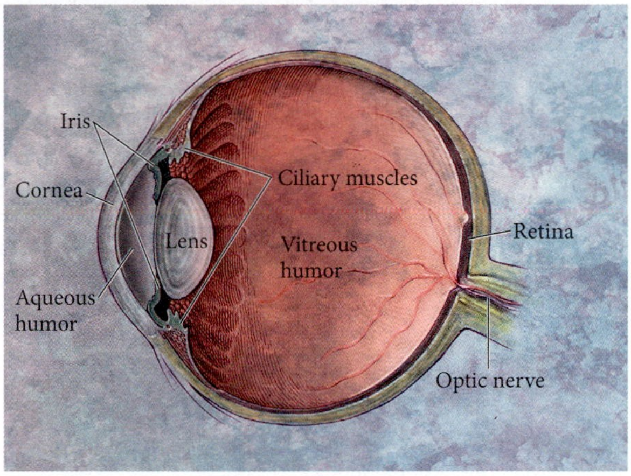

For an object to be seen clearly, the image must be formed at the location of the retina (Figure 33.25a). The shape of the eye cannot be changed; so changing the shape of the lens must control the distance of the image. For a distant object, relaxing the lens focuses the image at the retina. For close objects, the ciliary muscle increases the curvature of the lens to again focus the image on the retina. This process is called **accommodation.**

The extremes over which distinct vision is possible are called the far point and near point. The **far point** of a normal eye is infinity. The **near point** of a normal eye depends on the ability of the eye to focus. This ability changes with age. A young child can focus on objects as near as 7 cm. As a person ages, the near point increases. Typically, a 50-year-old person has a near point of 40 cm.

Several common vision defects result from incorrect focal distances. In the case of **myopia** (nearsightedness), the image is produced in front of the retina (Figure 33.25b). In the case of **hyperopia** (farsightedness), the image would be produced behind the retina (Figure 33.25c) if it could, but the retina absorbs the light before an image can be formed. Myopia can be corrected using diverging lenses, while hyperopia can be corrected using converging lenses.

FIGURE 33.24 Drawing of the human eye, showing its major features.

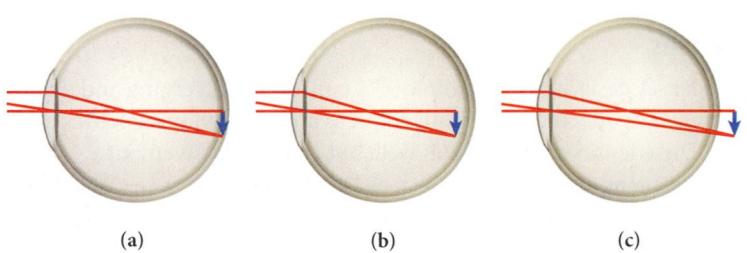

(a) (b) (c)

FIGURE 33.25 (a) Image produced by the lens of a normal-sighted person; (b) image produced by the lens of a nearsighted person; (c) image produced by a the lens of farsighted person.

EXAMPLE 33.2 Corrective Lenses

PROBLEM 1
Optometrists often quote the power, P, of a corrective lens rather than its focal length, f (see equation 33.1). What is the power of the corrective lens for a myopic (nearsighted) person whose uncorrected far point is 15 cm?

SOLUTION 1
The corrective lens must form a virtual, upright image of an object located at infinity, with the image located 15 cm in front of the lens (Figure 33.26).

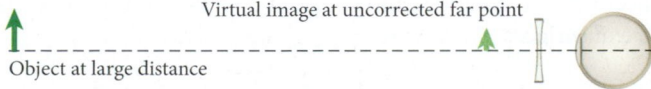

Virtual image at uncorrected far point

Object at large distance

FIGURE 33.26 Geometry of a corrective lens for nearsightedness.

Thus, the object distance d_o is ∞ and the image distance d_i is -15 cm. From the lens equation

$$\frac{1}{d_o} + \frac{1}{d_i} = \frac{1}{f},$$

– Continued

we have

$$\frac{1}{\infty} + \frac{1}{-0.15 \text{ m}} = \frac{1}{f} = -6.7 \text{ diopter.}$$

The required lens is a diverging lens with a power of –6.7 diopter and a focal length of –0.15 m.

PROBLEM 2

A hyperopic (farsighted) person whose uncorrected near point is 75 cm wishes to read a newspaper at a distance of 25 cm. What is the power of the corrective lens needed for this person to read the paper?

SOLUTION 2

The corrective lens must produce a virtual, upright image of the newspaper at the uncorrected near point of the person's vision (Figure 33.27).

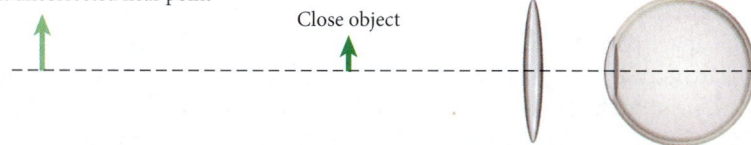

FIGURE 33.27 Geometry of a corrective lens for farsightedness.

The object and image are on the same side of the lens, so the image distance is negative. Thus, the object distance is 25 cm and the image distance is –75 cm:

$$\frac{1}{0.25 \text{ m}} + \frac{1}{-0.75 \text{ m}} = \frac{1}{f} = +2.7 \text{ diopter.}$$

The required lens is a converging lens with a power of +2.7 diopter, corresponding to a focal length of +0.37 m.

Concept Check 33.9

A hyperopic (farsighted) person can read a newspaper at a distance of $d = 125$ cm, but no closer. What power lens should this person use for reading if they wish to hold the newspaper at a distance of 25 cm from their eyes?

a) –3.5 diopter

b) –1.25 diopter

c) +0.50 diopter

d) +2.5 diopter

e) +3.2 diopter

Contact Lenses

The corrective lenses described in Example 33.2 consist of converging and diverging lenses. The converging lenses are convex on both surfaces and the diverging lenses are concave on both surfaces. These lenses correct vision well, but can be inconvenient.

A more convenient type of corrective lens is the **contact lens.** A contact lens is placed directly on the cornea of the eye, relieving the person from having to wear external glasses. These lenses are convex on the entrance surface and concave on the exit surface, similar to the meniscus lenses discussed in Section 33.1 (Figure 33.12). The concave exit surface is placed directly on the eye, and of course its curvature is determined by the curvature of the eyeball. It is possible to produce contact lenses that are converging or diverging (Figure 33.28) by varying the curvature of the entrance surface.

The contact lens shown in Figure 33.28a has $R_1 > R_2$ and thus has a negative focal length, corresponding to a diverging lens. The contact lens shown in Figure 33.28b has $R_1 < R_2$, which gives a positive focal length and a converging lens. Note, however, that the index of refraction of typical materials used for contact lenses is in the range between 1.4 and 1.8 and thus higher than the index of refraction of the cornea, which is slightly less than 1.4. In order to find the power of the contact lens, we cannot simply use the Lens-Maker's Formula 33.4, because the index of refraction of the medium next to the entrance and exit surfaces is different. Instead, we have to examine the change in the surface power of the cornea due to the presence of the contact lens and compare it to the sum of the surface powers of the entrance and exit surfaces of the lens.

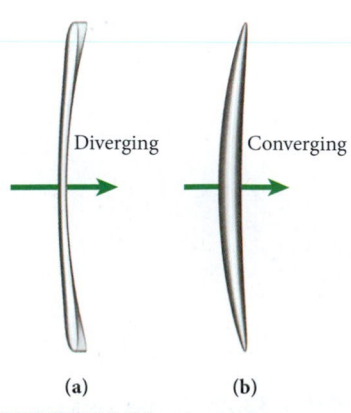

FIGURE 33.28 (a) A diverging contact lens. (b) A converging contact lens. The green arrows represent the direction of the light traveling through the lenses.

LASIK Surgery

An alternative to corrective lenses has been developed in which the cornea is altered to produce the desired optical response of the human eye. One such method, laser-assisted in situ keratomileusis **(LASIK) surgery,** uses a laser to modify the curvature of the cornea.

An example of LASIK surgery used to correct myopia is shown in Figure 33.29. Part (a) shows a myopic human eye, with the image produced in front of the retina. The effective

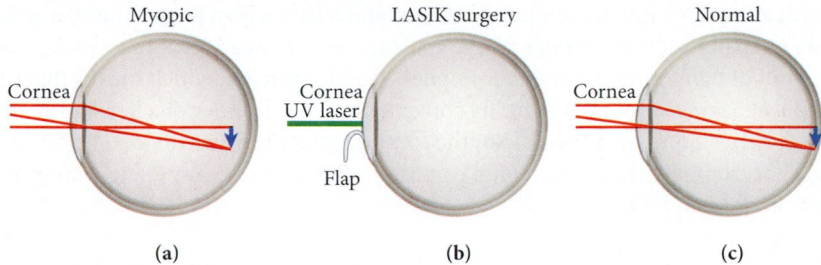

Myopic   LASIK surgery   Normal

Cornea   Cornea   Cornea
                UV laser

                Flap

(a)   (b)   (c)

FIGURE 33.29 (a) A myopic (nearsighted) human eye, where the image is formed in front of the retina. (b) In LASIK surgery, a flap is cut from the surface of the cornea and part of the stroma is removed using a UV laser. The flap is then folded back and the cornea heals. (c) Normal vision is produced by focusing the image on the retina.

focal length of this eye is too short. The LASIK procedure begins with the cutting of a flap off the surface of the cornea and folding it back (Figure 33.29b), exposing the inner part of the cornea, called the stroma. An ultraviolet (UV) laser (wavelength 193 nm) is used to remove material in the stroma with very short laser pulses of duration on the order of 10 ns and energy of the order of 1 mJ, producing a flatter surface. This flatter surface corresponds to a larger radius of curvature for the cornea. (The laser operates with ultraviolet light that breaks down the molecular structure of the cornea without heating the surface.) The flap is then folded back and the cornea heals. The Lens-Maker's Formula (equation 33.4) tells us that increasing the front radius R_1 will increase the effective focal length f of the eye. Thus, the surgery allows the image to be produced on the retina (Figure 33.29c).

The technique described here is specific for myopic patients. The LASIK procedure does not work as well for hyperopic patients or for patients with astigmatism (irregular curvature of the cornea). To treat hyperopic patients, the laser must remove material in the stroma around the center of the cornea to increase the curvature of the surface.

33.5 Camera

The **camera** is an optical instrument consisting of a body that excludes light and contains a system of lenses that focuses an image of an object on a recording medium such as photographic film or a digital sensor. A camera with a digital sensor is often called a digital camera, which is our main interest in the present section. A drawing of a digital camera is shown in Figure 33.30.

Normally we refer to the system of lenses in the camera as just the lens of the camera, neglecting the sophisticated multiple elements required to produce a high-quality image. The lens of the camera produces a real, inverted image of an object on the digital sensor. The camera lens is designed so that it can be moved to produce a sharp, focused image on the digital image recorder, depending on the image distance and the focal length of the lens. Many digital cameras have lenses that can be adjusted to have different focal lengths. A small digital camera, such as the one in Figure 33.30, with a 6× zoom lens can vary the focal length of the lens from 5.8 mm to 34.8 mm.

The lens has an opening area. An iris can be used to limit this opening area. The opening area is termed the **aperture** of the lens. The aperture is important because the amount of light that the camera can collect is proportional to the aperture. Because the aperture is usually a circle, we usually characterize the aperture by its diameter.

The size of the lens is often referred to as the **f-number** of the lens. The f-number of a lens is defined as the focal length of the lens divided by the diameter of the aperture of the lens:

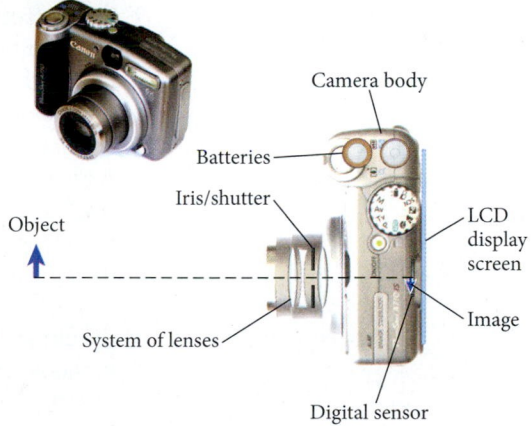

Camera body

Batteries

Iris/shutter

Object   LCD
                               display
                               screen

                               Image

System of lenses

Digital sensor

FIGURE 33.30 A typical digital camera produces an image on a digital sensor.

$$f\text{-number} = \frac{\text{focal length}}{\text{aperture diameter}} = \frac{f}{D}. \qquad (33.10)$$

A lens with a small f-number is called a fast lens, and a lens with a large f-number is called a slow lens. A small f-number implies a large aperture, which means that the lens can collect a large amount of light. A large f-number implies a small aperture, which means that the lens cannot collect a large amount of light. By convention, the f-number of a lens is usually written as $f/\#$, where # represents the ratio of the focal length of the lens divided by the aperture diameter. For example, the small digital camera in Figure 33.30 has a lens with f-number ranging from $f/2.8$ to $f/4.8$.

The aperture of the camera can be controlled by a variable iris, which limits the amount of light incident on the digital sensor. Conventionally, the f-number of a lens was set by twisting a ring on the lens. The ring had multiple indents placed at stops to help the photographer select the required amount of light. These **f-stops** were placed at intervals that changed the amount of light accepted by a factor of two. Because the aperture depends on the square of the diameter, the f-stops were spaced a factor of $\sqrt{2}$ apart. Some of the standard f-stops are $f/2$, $f/2.8$, $f/4$, $f/5.6$, $f/8$, $f/11$, and $f/16$. Modern digital cameras usually set the required f-number automatically.

The digital image sensor requires a specific amount of light to form a good image. The iris can function as a shutter to control the amount of time the sensor sees and the total amount of light that is incident on the sensor. Unlike a film camera, the shutter of a digital camera is usually open to allow the photographer to see exactly what the camera sees on the LCD screen on the back of the camera. The total amount of light energy is the product of the time the light is incident on the digital sensor times the aperture times the light intensity. A digital sensor can also be programmed to accept light only over a given time interval without the use of a mechanical shutter. A typical time interval for an exposure varies from $1/60^{th}$ of a second to $1/1000^{th}$ of a second. Short exposure times allow the camera to capture a moving image without blurring. To prevent unwanted light interfering with a recorded image, the shutter closes while the image is being read out.

In addition to controlling the amount of light incident on the digital sensor, the iris can also affect the image by changing the **depth of field** of the image, which is the range in object distance where the image is in focus. If the iris is small (larger f-number), the range of angles of incident light is restricted, and a wider range of object distances will still produce a focused image on the digital sensor. If the iris is wide open (smaller f-number), a larger range of angles are admitted into the camera, meaning that only a narrow range of object distances will produce a focused image on the digital sensor. Thus, the photographer can increase the depth of field of an image by closing down the aperture and increasing the exposure time, or decrease the depth of field by opening the aperture and decreasing the exposure time.

The standard photographic film used for many years was 36 mm wide and 24 mm tall and was usually called 35-mm film. Most digital sensors are much smaller than 35-mm film. A typical small digital camera has a digital sensor that is 5.76 mm wide and 4.29 mm tall. This small size affects the magnification of the image produced by the camera, depending on the focal length of the lens. The **angle of view** of the camera, α, is more relevant to photography than the magnification. The angle of view can be defined in terms of the horizontal angle, the vertical angle, or the diagonal angle, where horizontal and vertical refer to the geometry of the film or digital sensor. We will consider the horizontal angle of view here, but the other two versions can be easily calculated from our results for the horizontal angle of view.

We can derive an expression for the horizontal angle of view, α, starting with our definition of linear magnification, m,

$$m = \frac{h_i}{h_o} = -\frac{d_i}{d_o},$$

where h_i is the image height, h_o is the object height, d_i is the image distance, and d_o is the object distance. Because photographers are not concerned about whether the image is upright or inverted, we will use the absolute value of the heights and distances in this calculation. In Figure 33.31 we show a schematic drawing containing an object, its corresponding image on film or a digital sensor, and the lens.

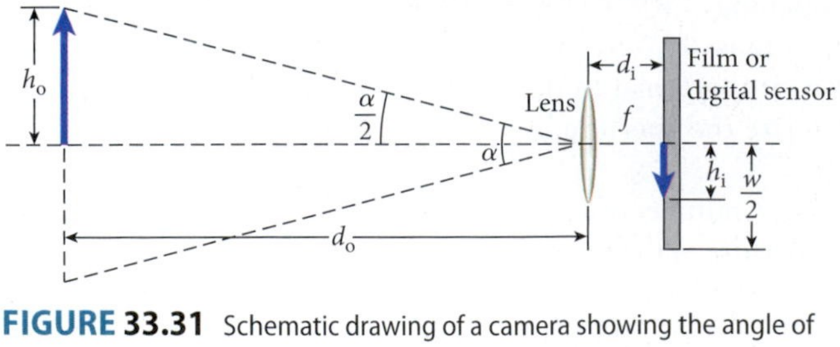

FIGURE 33.31 Schematic drawing of a camera showing the angle of view.

The horizontal width of the sensor is w. The object as shown in Figure 33.31 subtends an angle of $\alpha/2$ and the horizontal angle of view is α. The maximum image size that the film or digital sensor can detect occurs when the image height is equal to half the width of the film or digital sensor, $h_i = w/2$. If the object is not too close to the camera, the image distance is approximately equal to the focal length f of the lens, so we can use the definition of the linear magnification to write

$$\frac{w/2}{h_o} = \frac{d_i}{d_o} \approx \frac{f}{d_o} \Rightarrow \frac{h_o}{d_o} \approx \frac{w}{2f}.$$

From Figure 33.31, we can see that the horizontal angle of view can be related to the object height and object distance as

$$\frac{h_o}{d_o} = \tan(\alpha/2).$$

Combining these two equations for the ratio h_o/d_o gives us

$$\frac{h_o}{d_o} = \frac{w}{2f} = \tan(\alpha/2).$$

Thus, the horizontal angle of view is

$$\alpha = 2\tan^{-1}\left(\frac{w}{2f}\right). \tag{33.11}$$

As an example of the effect of changing the focal length on a picture taken with a camera, we show three photos of a statue taken with different zoom settings in Figure 33.32.

The 1× zoom setting corresponds to a focal length of 5.8 mm, the 3× setting corresponds to a focal length of 17.4 mm, and the 6× setting corresponds to a focal length of 34.8 mm. The horizontal angle of view is shown for the each of the three zoom settings in Figure 33.33.

The horizontal angle of view for the 1× zoom setting of $\alpha = 52.8°$ is close to the horizontal angle of view of normal human vision of $\alpha \sim 50°$–$60°$. (Our peripheral vision extends to approximately 180°, but the inner 50°–60° dominate our perception.) The 6× zoom setting has a narrow horizontal angle of view of $\alpha = 9.5°$, providing a magnified image of the statue. For the camera to have a wider horizontal angle of view (wide-angle view), it would need a lens with a focal length less than 5.8 mm.

The digital sensor is composed of millions of electronic picture elements (pixels). For example, the small digital camera in Figure 33.30 uses a charge-coupled device (CCD) digital sensor with an array of pixels 3072 pixels wide and 2304 pixels tall for a total of 7,077,888 pixels (7.1 megapixels). Each pixel responds to light by liberating electrons. An analog-to-digital

(a) **(b)** **(c)**

FIGURE 33.32 Photos of a statue using three different zooms: (a) 1× zoom; (b) 3× zoom; (c) 6× zoom.

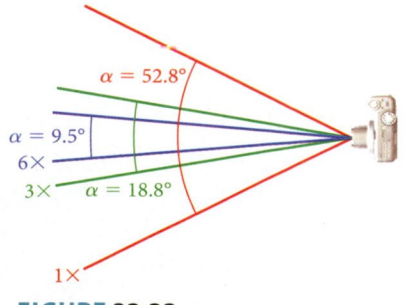

FIGURE 33.33 The horizontal angle of view for a small digital camera with 1×, 3×, and 6× zooms.

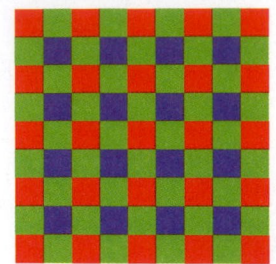

FIGURE 33.34 A portion of a Bayer filter used to produce color images with digital sensors.

converter (ADC) reads out the pixels by digitizing the number of liberated electrons. The pixels in each row are digitized in order. Then the next row is shifted down and digitized until all the pixels are digitized. The question now remains: How does the CCD produce a color picture? The CCD responds only to the intensity of light, not the wavelength. The answer is that light from the image focused on the CCD first passes through a Bayer filter as illustrated in Figure 33.34.

The Bayer filter consists of rows of alternating pixels of red and green, followed by a row of alternating pixels of blue and green. There are more green pixels than red or blue pixels because the eye responds better to green than to red or blue. A small microprocessor in the camera calculates the best color for each pixel, based on the intensity of the red, green, and blue light passed by the Bayer filter.

A computer projector works much like a digital camera, except that the object is at the focus of the lens, which projects an image on a distant screen. The object for the projector can be a small light-transmitting LCD (liquid crystal display, see Chapter 31) screen that displays the image to be projected. A bright light shines through the LCD to produce the object for the lens. Another type of projector is the DLP (digital light processor), which uses small mirrors etched on a computer chip to produce the image. Each mirror represents a pixel. Colors are produced by alternately bathing the chip in red, green, and blue light. The image produced by the mirrors becomes the object for a lens, which projects the image on a screen. Movie theaters commonly use digital projectors based on DLP.

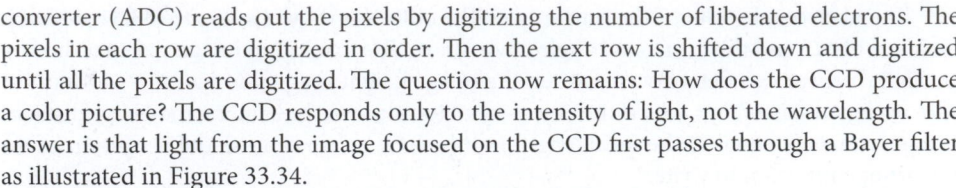

EXAMPLE 33.3 / Focal Length of a Simple Point-and-Shoot Camera

For many occasions, a simple point-and-shoot camera is sufficient. One example is an inexpensive commercial camera that uses 35-mm film (36 mm wide and 24 mm tall) like the one shown in Figure 33.35. You could envision another version that uses a digital sensor that is 4.0 mm wide and 3.0 mm tall.

PROBLEM

What are the required focal lengths of the lenses for these cameras if they are to have a horizontal angle of view of $\alpha = 46°$?

SOLUTION

The expression for the horizontal angle of view is given by equation 33.11:

$$\alpha = 2\tan^{-1}\left(\frac{w}{2f}\right).$$

We can rearrange this equation to obtain the focal length

$$f = \frac{w}{2\tan(\alpha/2)}.$$

The focal length for the 35-mm film version is

$$f = \frac{w}{2\tan(\alpha/2)} = \frac{36 \text{ mm}}{2\tan(46°/2)} = 42 \text{ mm}.$$

The focal length for the digital sensor version is

$$f = \frac{w}{2\tan(\alpha/2)} = \frac{4.0 \text{ mm}}{2\tan(46°/2)} = 4.7 \text{ mm}.$$

The digital sensor version would be considerably smaller than the 35-mm film version.

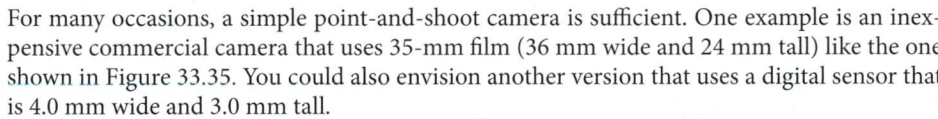

FIGURE 33.35 A simple point-and-shoot camera with fixed focal length.

33.6 Microscope

The simplest **microscope** is a system of two lenses. For example, Figure 33.36 shows a microscope constructed of two thin lenses. The first lens is a converging lens of short focal length, f_o, called the **objective lens.** The second lens is another converging lens of greater focal length, f_e, called the **eyepiece.**

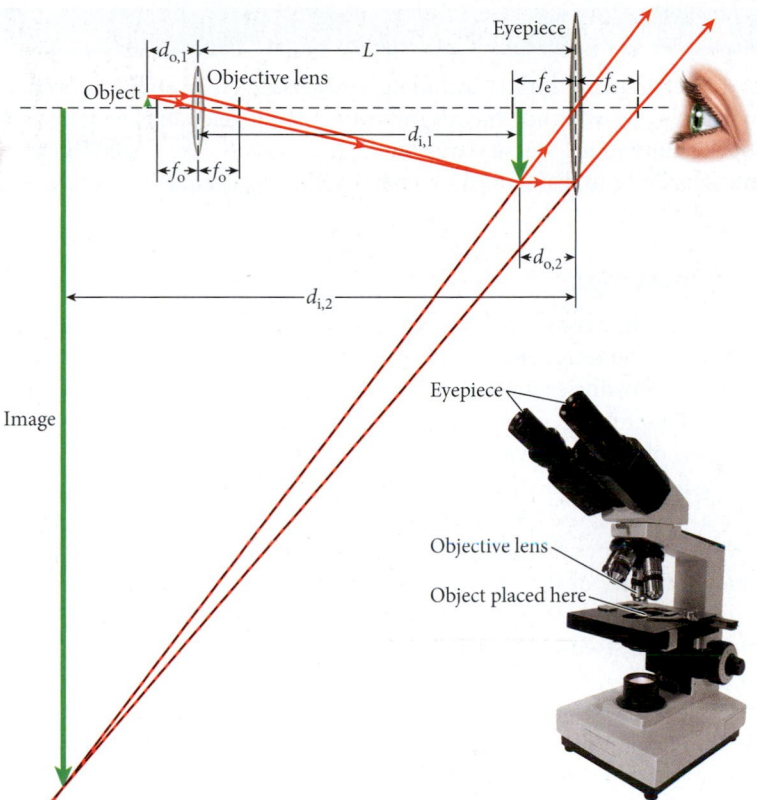

FIGURE 33.36 Geometry of the image formed by a microscope. The objective lens, eyepiece, and object for a real microscope are also shown.

The object to be observed is placed just outside the focal length f_o of the objective lens, so that $d_{o,1} \approx f_o$. This arrangement allows the objective lens to form a real, inverted, and enlarged image of the object some distance from the objective lens. This image then becomes the object for the eyepiece lens. This intermediate image is placed just inside the focal length f_e of the eyepiece lens, so that the eyepiece lens produces a virtual, upright, and enlarged image of the intermediate image and $d_{o,2} \approx f_e$. The resulting magnification of the microscope is the product of the magnification from each lens.

Let L be the distance between the two lenses, and assume $L \gg f_o, f_e$. Using the notation of Section 33.3, this means that $d_{i,1} \approx L$. Thus, the linear magnification of the microscope is

$$m = \frac{d_{i,1}d_{i,2}}{d_{o,1}d_{o,2}} = -\frac{(0.25 \text{ m})L}{f_o f_e}, \tag{33.12}$$

where we have used $d_{i,2} = -0.25$ m because we assume that the final image is produced at a comfortable viewing distance of 0.25 m.

EXAMPLE 33.4 Magnification of a Microscope

Consider a microscope that consists of an objective lens and an eyepiece lens separated by 15 cm. The focal length of the objective lens is 2 mm, and the focal length of the eyepiece lens is 20 mm. Assume the final image is produced at a distance of 25 cm.

PROBLEM

What is the magnitude of the linear magnification of this microscope?

SOLUTION

Taking our expression for the linear magnification of a microscope (equation 33.12), we have

$$|m| = \frac{(0.25 \text{ m})L}{f_o f_e} = \frac{(0.25 \text{ m})(0.15 \text{ m})}{(0.002 \text{ m})(0.020 \text{ m})} = 940.$$

Concept Check 33.10

A microscope is intended to have a linear magnification whose magnitude is 330. If the objective lens has a power of 350 diopter and the eyepiece lens has a power of 10.0 diopter, how long must the microscope be? Assuming final image is produced at a distance of 25 cm.

a) 37.7 cm   d) 65.0 cm

b) 40.0 cm   e) 75.0 cm

c) 51.3 cm

33.7 | Telescope

Telescopes come in many forms, including **refracting telescopes** and **reflecting telescopes.** In this chapter we study the magnification of the telescope, which is a measure of the telescope's ability to help us see large but distant objects. The resolution of a telescope, which is the capability to distinguish two nearby objects, is equally important and will be covered in Chapter 34.

Refracting Telescope

The refracting telescope consists of two converging lenses—the objective and the eyepiece. A modern commercial refracting telescope used by amateur astronomers is shown in Figure 33.37. In the following examples, we represent a telescope using two thin lenses. However, an actual refracting telescope uses more sophisticated lenses.

Because the object to be viewed is at a large distance, the incoming light rays can be considered to be parallel. Thus, the objective lens forms a real image of the distant object at distance f_o (Figure 33.38). The eyepiece is placed so that the image formed by the objective is a distance f_e from the eyepiece. Thus, the eyepiece forms a virtual, magnified image at infinity of the image formed by the objective, again producing parallel rays. In Figure 33.38, the parallel light rays from the distant object are shown incident on the objective lens. Red-and-black dashed lines depict the parallel light rays forming the virtual image.

Because the telescope deals with objects at large distances, it is not helpful to determine the magnification of the telescope using the formula for linear magnification which involves distances found from the lens equation. Therefore, we define the angular magnification of the telescope as –1 times the angle observed in the eyepiece θ_e divided by the angle subtended by the object being viewed θ_o (Figure 33.38),

$$m_\theta = -\frac{\theta_e}{\theta_o} = -\frac{f_o}{f_e}. \tag{33.13}$$

FIGURE 33.37 A modern telescope used by amateur astronomers. This telescope has an 80-mm diameter objective lens with a 400-mm focal length.

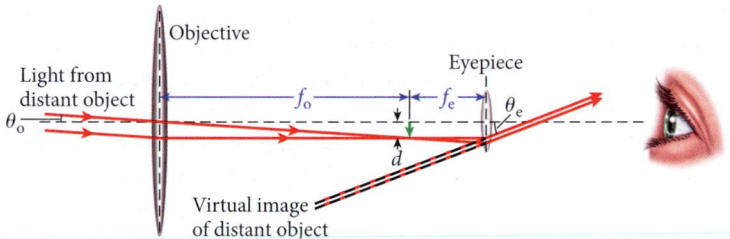

FIGURE 33.38 Geometry of an image formed by a refracting telescope.

DERIVATION 33.2 | Angular Magnification

The angle θ_o is the angle subtended by a distant object. The height of the image produced by the objective lens is d (Figure 33.38). This image is produced at the focal length of the objective lens, f_o. Because the focal length of the objective lens is large compared with the image size,

$$\theta_o \approx \tan\theta_o = \frac{d}{f_o}.$$

The image is placed at the focal point of the eyepiece lens. The apparent angle in the eyepiece, θ_e, can be written as

$$\theta_e \approx \tan\theta_e = \frac{d}{f_e},$$

again assuming that the image size is small compared with the focal length of the eyepiece. The ratio of the apparent eyepiece angle to the objective angle gives the angular magnification:

$$\frac{\theta_e}{\theta_o} = \frac{d/f_e}{d/f_o} = \frac{f_o}{f_e}.$$

Thus the angular magnification of a refracting telescope is

$$m_\theta = -\frac{f_o}{f_e},$$

where the minus signs means that the image is inverted.

For example, the refracting telescope shown in Figure 33.37 has an objective lens with focal length $f_o = 400$ mm and an eyepiece with focal length $f_e = 9.70$ mm, which gives the telescope a magnification of

$$m = -\frac{f_o}{f_e} = -\frac{400 \text{ mm}}{9.7 \text{ mm}} = -41.$$

EXAMPLE 33.5 | Magnification of a Refracting Telescope

The world's largest refracting telescope at that time was completed in 1897 and is still in use and housed at the Yerkes Observatory in Williams Bay, Wisconsin (between Chicago and Milwaukee). It has an objective lens of diameter 1.0 m (40 in) with a focal length of 19 m (62 ft).

PROBLEM
What should the focal length of the eyepiece be to give a magnification of magnitude 250?

SOLUTION
The focal length of the objective lens is given as $f_o = 19$ m. The absolute value of the magnification is to be $|m| = 250$. The magnification is given by $|m| = f_o/f_e$. This gives us the focal length of the eyepiece lens as

$$f_e = \frac{f_o}{|m|} = \frac{19 \text{ m}}{250} = 0.076 \text{ m} = 7.6 \text{ cm}.$$

Reflecting Telescope

Most large astronomical telescopes are reflecting telescopes, with the objective lens replaced with a concave mirror. (However, the eyepiece of a reflecting telescope is still a lens.) An example of such a telescope is the Southern Astrophysical Research (SOAR) telescope shown in Figure 33.39, which has a 4.1-m diameter primary mirror. Large mirrors are necessary to gather as much light as possible to produce high-quality images from distant, faint astronomical objects. The light-gathering capability of a telescope is in practice much more important than the resulting magnification of the telescope.

Large mirrors are easier to fabricate and maintain in position than large lenses. Also, mirrors don't have chromatic aberration, so they work for a larger range of wavelengths. Finally, the size of refracting telescopes is limited because large lenses, which can be supported only around their edges, tend to sag due to their own weight. By contrast, modern reflecting telescopes use a large number (on the order of 100) individual

FIGURE 33.39 The SOAR telescope with a 4.1-m diameter primary mirror.

Self-Test Opportunity 33.5

A refracting telescope using two converging lenses is often called a Keplerian telescope. The eyepiece lens is placed a distance $d = f_o + f_e$ from the objective lens. Another type of refracting telescope uses a converging lens for the objective lens and a diverging lens for the eyepiece lens.

The eyepiece lens in this telescope is placed at a distance $d = f_o + f_e$, where $f_e < 0$. This type of telescope is often called a Galilean telescope. Discuss some advantages of a Galilean telescope versus a Keplerian telescope.

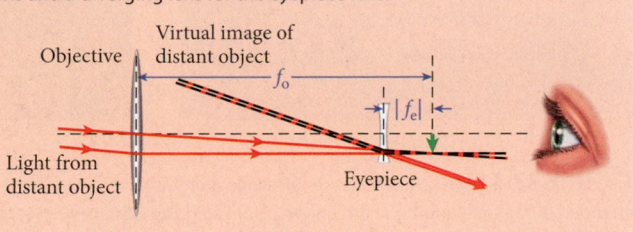

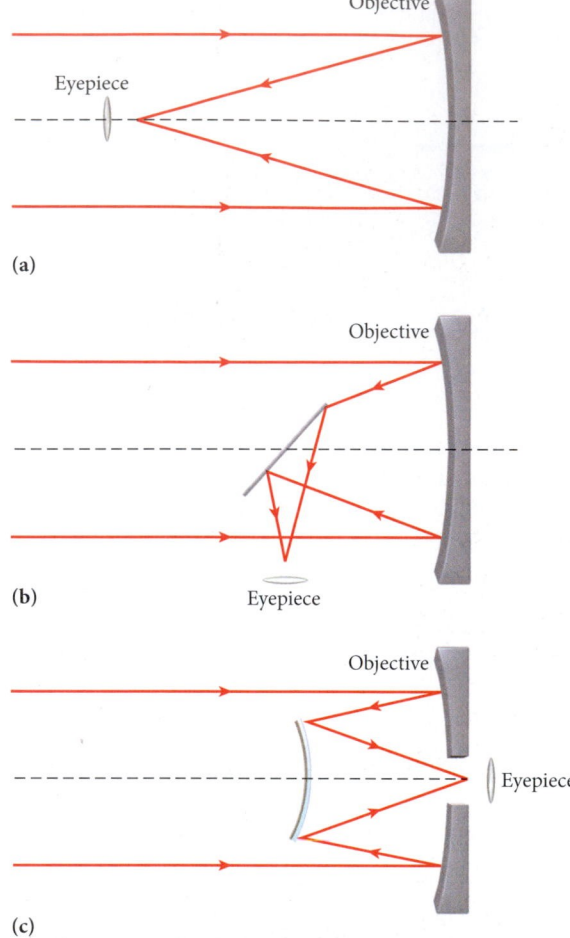

(a)

(b)

(c)

FIGURE 33.40 (a) Geometry of a standard reflecting telescope. (b) Geometry of a Newtonian reflecting telescope, with the eyepiece to the side. (c) Geometry of a Cassegrain reflecting telescope, with a secondary mirror and an eyepiece in the rear. (The relative sizes of the secondary mirror and eyepieces are not to scale with the primary mirror and are exaggerated.)

computer-controlled force actuators on the backside of the mirror to keep it as closely as possible in the ideal parabolic shape required for optimum image quality. These force actuators typically can each exert pulling or pushing forces of the order of 100 N on the mirror. They enable the mirror surface to stay within a few tens of nanometers of the ideal parabolic shape, even as the entire telescope is gradually rotated and tilted to stay locked on the astronomical object that it is tracking through the night sky during long-exposure imaging.

Various types of reflecting telescopes have been developed. Figure 33.40 shows three simple examples. The simplest form of reflecting telescope incorporates a parabolic mirror and an eyepiece (Figure 33.40a). This geometry is impractical because the observer must be placed in the direction from which the light comes. In the Newtonian solution to a reflecting telescope, a plane mirror at an angle of 45° reflects the light outside the structure of the telescope to an eyepiece (Figure 33.40b). In the Cassegrain geometry, a secondary convex hyperboloid mirror, placed perpendicular to the optic axis of the mirror, reflects the light through a hole in the center of the mirror (Figure 33.40c). In the last two cases, the secondary mirror is small enough that it absorbs a small (but not insignificant!) fraction of the incoming light. In all three cases, an eyepiece is used to magnify the image produced by the objective mirror. The SOAR telescope shown in Figure 33.39 uses the geometry shown in Figure 33.40b.

Hubble Space Telescope

The Hubble Space Telescope (HST), named for American astronomer Edwin Hubble (1889–1953), was deployed April 25, 1990, from the Space Shuttle (Figure 33.41). The HST orbits Earth 590 km above its surface, far above the atmosphere that disturbs the images gathered by ground-based telescopes. The HST is a reflecting telescope of Ritchey-Chrétian design arranged in Cassegrain geometry, but using a concave hyperbolic objective mirror, rather than a paraboloid mirror and a convex hyperbolic secondary mirror as is used in the traditional Cassegrain design. This design gives the HST a wide field of view and eliminates spherical aberration. The objective mirror is 2.40 m in diameter and has an effective focal length of 57.6 m. The secondary mirror

Concept Check 3.11

Suppose a reflecting telescope consists of a concave spherical mirror with a radius of curvature $R = 17.0$ m and an eyepiece lens of focal length $f_e = 29.0$ cm. What is the magnitude of the magnification of this telescope?

a) 29.3   d) 66.1

b) 45.0   e) 78.9

c) 58.6

(a)

(b)

FIGURE 33.41 The Hubble Space Telescope is arguably the world's most famous optical instrument and has changed our understanding of the universe. (a) The Hubble Space Telescope in orbit. (b) Hubble photo of the gas pillars in the Eagle Nebula (M16): Pillars of Creation in a star-forming region.

Concept Check 33.12

Reflecting telescopes with large mirror diameters are better than those with smaller diameters. Why?

a) Because they have a shorter focal length.

c) Because they have a greater magnification.

b) Because they have a longer focal length.

d) Because they can collect more light.

(a)

is 0.267 m in diameter and is located 4.91 m from the objective mirror. The secondary mirror can be moved under ground control to produce the best focus. The eyepiece is replaced by a set of electronic instruments specialized for various astronomical tasks.

The original HST objective mirror was produced with a flaw caused by a defective testing instrument. The mirror had been polished very precisely, but unfortunately it was polished very precisely to a wrong shape. The maximum deviation from the perfect shape of the mirror was only 2.3 μm, but the resulting spherical aberrations were catastrophic (see Figure 33.42a). In December 1993, a Space Shuttle Servicing Mission deployed the Corrective Optics Space Telescope Axial Replacement (COSTAR) package, which corrected the flaw in the objective mirror and allowed the HST to begin revolutionizing our understanding of the universe. The two images of the galaxy M100 shown in Figure 33.42 demonstrate the image quality of the HST before and after the installation of COSTAR. In February 1997, two new instrument packages were installed in the HST on Shuttle Servicing Mission 2. These instruments contained their own optical corrections. In December 1997, the Shuttle Servicing Mission 3A was launched to correct a serious problem with the HST's stabilizing gyroscopes. In March 2002, the Shuttle Servicing Mission 3B added several new instruments. Shuttle Servicing Mission 4 in May 2009 replaced two failed instruments (spectrograph and advanced camera) and installed two new instruments (spectrograph and wide field camera) as well as new batteries and gyroscopes.

(b)

FIGURE 33.42 Two images of the galaxy M100. (a) An image produced by the Hubble Space Telescope just after it was launched. (b) The same object photographed after the optics were corrected.

James Webb Space Telescope

The planned replacement for the HST is the James Webb Space Telescope (JWST), named for NASA's second administrator, James E. Webb (1906–1992). This project is planned for launch in the year 2014. The objective mirror for the JWST will be 6.5 m in diameter and will be composed of 18 mirror segments. An artist's conception of the JWST is shown in Figure 33.43.

The JWST will use infrared light to study the universe. Infrared light can penetrate the dust clouds that occur in our galaxy and elsewhere and block visible light. The JWST will orbit the Earth at a distance of 1.5 million km (about four times the Earth–Moon distance), in such a way that Earth will always be between the Sun and the JWST. Having the Sun always blocked by Earth will allow continuous viewing and shield the cryogenic infrared detectors aboard the JWST from changes in sunlight. The JWST is also equipped with a multiple-layer sunshield as shown in Figure 33.43.

CHANDRA X-Ray Observatory

The JWST will carry out observations using infrared light, which reflects well from mirrors. However, other types of electromagnetic waves do not reflect from the surfaces of mirrors. For example, if X-rays are incident on the surface of a mirror perpendicularly, they will penetrate the surface of the mirror rather than reflect. However, if the X-rays are incident on the surface of the mirror at a large angle, the X-rays will be reflected. The CHANDRA X-ray Observatory (CXO) produces images with X-rays using the mirrors shown in Figure 33.44a.

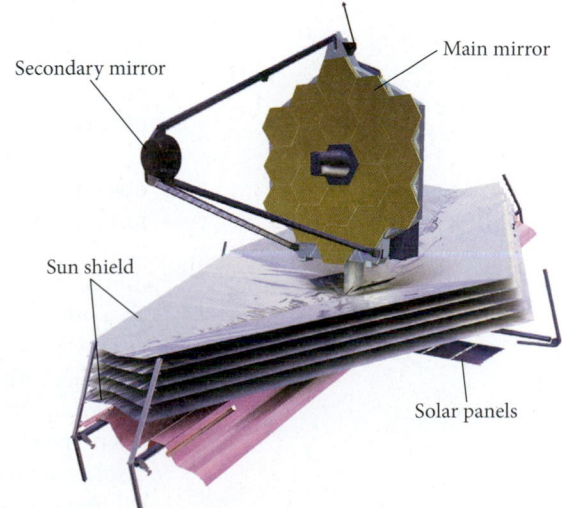

Secondary mirror

Main mirror

Sun shield

Solar panels

FIGURE 33.43 The planned James Webb Space Telescope. (Drawing from the Space Telescope Science Institute)

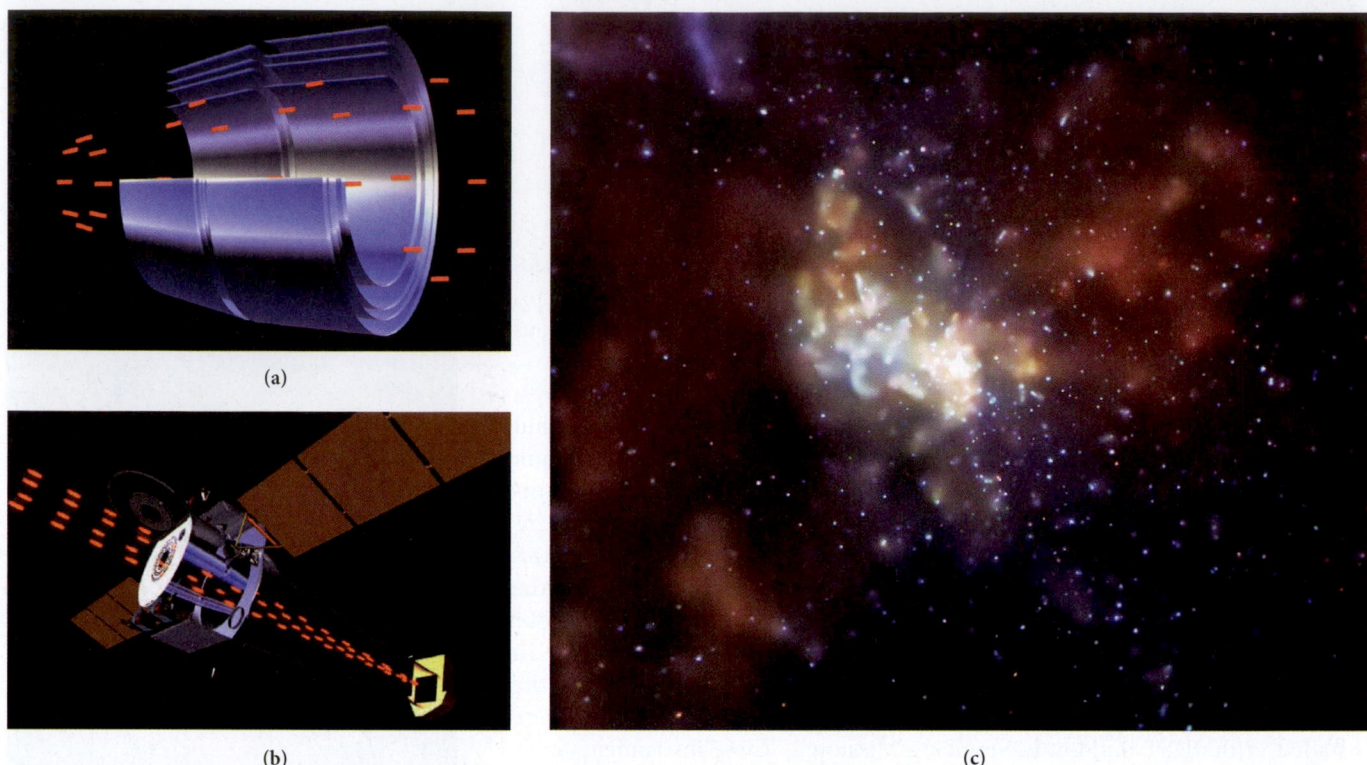

(a)

(b)

(c)

FIGURE 33.44 The CHANDRA X-ray Observatory. (a) The cylindrical mirrors used to focus X-rays. (b) The path of X-rays through CHANDRA. c) An X-ray image produced by CHANDRA of the center of the Milky Way Galaxy.

The mirrors of the CXO form images just as lenses and reflecting mirrors do (Figure 33.44b), although the geometry of the X-ray lenses is different. These images formed using X-rays can give us information about high-energy regions of the universe such as the very active galactic center of the Milky Way Galaxy, shown in Figure 33.44c.

33.8 Laser Tweezers

After discussing the largest optical instruments with which we can explore the largest structures of the universe, we want to devote the last section of this chapter to an optical instrument with which we can manipulate some of the smallest structures.

As shown in Example 31.3, the force exerted on an object by light is on the order of 10^{-12} N (= 1 piconewton). This force is too small to affect macroscopic objects. However, physicists using very intense lasers focused to a small area can exert forces sufficient to manipulate objects as small as a single atom. These devices are called optical traps or **laser tweezers.**

Laser tweezers are constructed by focusing an intense laser beam to a point using the objective lens of a microscope. The force exerted by a laser pointer on a piece of paper produces a force in the direction of the original laser beam. In the physical realization of laser tweezers, the laser beam is focused in such a way that the light is more intense in the middle of the light distribution. In addition, the focusing produces light rays that converge on a point.

Let's consider the effect of the focused laser light on a spherical, optically transparent object. This object could be a small plastic sphere or a living cell that is approximately spherical. Define the original direction of the laser light as the z-direction, so the xy-plane lies perpendicular to the incident direction. Figure 33.45a shows the object in the yz-plane, displaced slightly in the negative y-direction. Light coming from the center of the distribution of light is more intense and also (as we will see in Chapter 34) is refracted (or bent) downward. Light coming from the edge of the distribution is less intense and is refracted upward. The resultant change in momentum of the incident light rays in the y-direction is downward. To conserve momentum, the object must recoil in the positive y-direction. Thus, the intense focused light of the laser produces a restoring force on the object if it is not positioned at $y = 0$, and the object is trapped in the y-direction.

In the z-direction, which is the direction of the incident laser light, trapping also occurs because of the converging rays produced by the focusing of the light by the lens (Figure 33.45b). In this case, the object is just to the right of the focus point. The incident rays are refracted such that they are more parallel to the incident direction than before. Thus the component of the momentum of light in the z-direction has been increased and the transparent object must recoil in the opposite direction to conserve momentum. If the object were to the left of the focus point, it would experience a force to the right. Thus the object is trapped in the direction parallel to the incident laser light as well as the direction perpendicular to the light.

This technique relies on the transmission of the incident light. If this technique is applied to an object that is not transparent, the incident light will be reflected and will produce a force pushing the object in the general direction of the incident light.

Laser tweezers have been used to trap cells, bacteria, viruses, and small polystyrene beads. They have also been used to manipulate DNA strands and to study molecular-size motors.

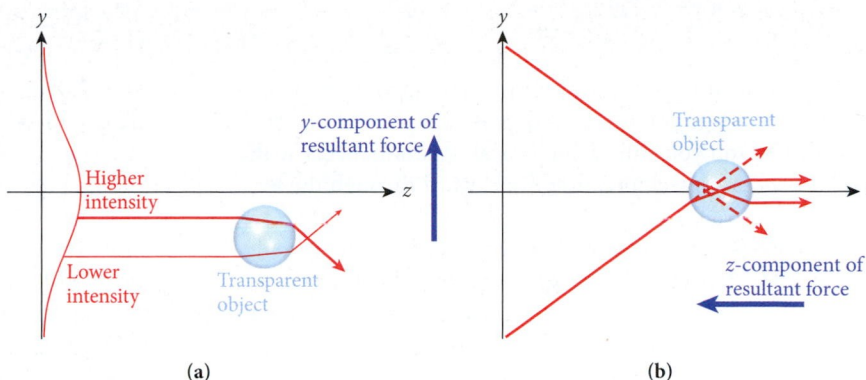

FIGURE 33.45 The effect of focused laser light on a small, spherical, optically transparent object. (a) The curved red line on the left represents the varying intensity of the wave. In addition, the restoring force in the y-direction is shown. (b) A restoring force also occurs in the z-direction.

WHAT WE HAVE LEARNED | EXAM STUDY GUIDE

- For images formed by lenses, the object distance, the image distance, and the focal length of the lens are related by the Thin-Lens Equation, $\dfrac{1}{d_o} + \dfrac{1}{d_i} = \dfrac{1}{f}$. Here d_o is always positive, whereas d_i is positive if the image is on the opposite side of the lens as the object and negative if on the same side. The focal length f is positive for converging lenses and negative for diverging lenses.

- The Lens-Maker's Formula relates the curvature of both sides of a lens and its index of refraction to its focal length, $\dfrac{1}{f} = (n-1)\left(\dfrac{1}{R_1} - \dfrac{1}{R_2}\right)$.

- The diopter is a dimensionless number defined as the inverse of the focal length in meters.

- The angular magnification of a simple magnifier with an image at infinity is given by $m_\theta \approx \dfrac{0.25\ \text{m}}{f}$. (Assumed here is a typical value of 0.25 m for the near point of the average middle-aged person.)

- The linear magnification of a microscope is given by $m = -\dfrac{(0.25\ \text{m})L}{f_o f_e}$, where L is the distance between the two lenses, f_o is the focal length of the objective lens, and f_e is the focal length of the eyepiece lens. (Final image assumed formed at distance of 0.25 m.)

- The angular magnification of a telescope is given by $m_\theta = -\dfrac{\theta_e}{\theta_o} = -\dfrac{f_o}{f_e}$, where f_o is the focal length of the objective lens or mirror and f_e is the focal length of the eyepiece lens.

- The f-number of a camera lens is defined as the focal length of the lens f divided by the diameter of the aperture of the lens D, $f\text{-number} = \dfrac{\text{focal length}}{\text{aperture diameter}} = \dfrac{f}{D}$.

- The horizontal angle of view of a camera is $\alpha = 2\tan^{-1}\left(\dfrac{w}{2f}\right)$, where w is the width of the film or digital image sensor and f is the focal length of the camera lens.

ANSWERS TO SELF-TEST OPPORTUNITIES

33.1 The answer is different for real and imaginary images. For a real image you can exchange the object and image locations, and you still get a physical situation, because the light rays can travel forward or backward through the lens. In other words, if you place an object in Figure 33.4 where the image is now, you would get a reduced-size image. In Figure 33.5 however, the image is virtual and not obtained where the rays physically cross, but where the extrapolations of the rays cross. In this case you cannot simply interchange image and object. For an object placed inside the focal length of a converging lens you *always* obtain an enlarged image.

33.2 a) plane mirror   c) diverging lens

     b) diverging mirror   d) converging lens

33.3 Taking $x = 0$ in equation 33.9 gives us $f_{eff} = f_2 f_1 / (f_2 + f_1)$. We can invert this equation to get

$$\frac{1}{f_{eff}} = \frac{f_2 + f_1}{f_2 f_1} = \frac{1}{f_1} + \frac{1}{f_2}.$$

33.4 The magnitude of the magnification is given by $|m| = d_i / d_o$. If the object distance is infinity, then $d_i = f$. Thus the 200-mm focal length lens will produce a magnification of four times that of the normal 50-mm lens.

33.5 Comparing with Figure 33.38, we can see that a Galilean telescope produces an upright image, which is useful for terrestrial telescopes, opera glasses, and binoculars. The Keplerian telescope produces an inverted image. A Galilean telescope can be shorter than a Keplerian telescope, which is also useful for optical devices such as opera glasses or binoculars.

PROBLEM-SOLVING GUIDELINES

Basically the same guidelines as in the previous chapter also apply here, because for lenses and optical instruments we simply use the same ray-tracing laws as for mirrors.

1. Drawing a large, clear, well-labeled diagram should be the first step in solving almost any optics problem. Include all the information you know and all the information you need to find. Remember that angles are measured from the normal to a surface, not to the surface itself.

2. You need to draw only two principal rays to locate an image formed by mirrors or lenses, but you should draw a third ray as a check that they are drawn correctly. Even if you need to solve a problem using the mirror equation or Thin-Lens Equation, an accurate drawing can help you approximate the answers as a check on your calculations. But keep in mind that in some lens configurations with large magnifica-

tion factors, even minute drawing inaccuracies can translate into large errors in image size or location. So it is also not a good idea to rely strongly on your drawing alone.

3. When you calculate distances for mirrors or lenses, remember the sign conventions and be careful to use them correctly. If a diagram indicates an inverted image, say, but your calculation does not have a negative sign, go back to the starting equation and check the signs of all distances and focal lengths.

4. In problems involving two or more optical elements it is usually required to calculate the intermediate image position after each element. It may be advisable to compute this position numerically before proceeding to the next optical element in the calculation. Often this prevents the problem from getting too complex.

MULTIPLE-CHOICE QUESTIONS

33.1 For a microscope to work as intended, the separation between the objective lens and the eyepiece must be such that the intermediate image produced by the objective lens will occur at a distance (as measured from the optical center of the eyepiece)

a) slightly larger than the focal length.

b) slightly smaller than the focal length.

c) equal to the focal length.

d) The position of the intermediate image is irrelevant.

33.2 Which one of the following is not a characteristic of a simple two-lens astronomical refracting telescope?

a) The final image is virtual.

b) The objective forms a virtual image.

c) The final image is inverted.

33.3 A converging lens will be used as a magnifying glass. In order for this to work, the object must be placed at a distance

a) $d_o > f.$ c) $d_o < f.$

b) $d_o = f.$ d) None of the above.

33.4 An object is moved from a distance of 30 cm to a distance of 10 cm in front of a converging lens of focal length 20 cm. What happens to the image?

a) Image goes from real and upright to real and inverted.

b) Image goes from virtual and upright to real and inverted.

c) Image goes from virtual and inverted to real and upright.

d) Image goes from real and inverted to virtual and upright.

e) None of the above.

33.5 What type of lens is a magnifying glass?

a) converging
c) spherical
e) plain

b) diverging
d) cylindrical

33.6 LASIK surgery uses a laser to modify the

a) curvature of the retina.

b) index of refraction of the aqueous humor.

c) curvature of the lens.

d) curvature of the cornea.

33.7 What is the focal length of a flat sheet of transparent glass?

a) zero
c) thickness of the glass

b) infinity
d) undefined

33.8 Where is the image formed if an object is placed 25 cm from the eye of a nearsighted person. What kind of a corrective lens should the person wear?

a) Behind the retina. Converging lenses.

b) Behind the retina. Diverging lenses.

c) In front of the retina. Converging lenses.

d) In front of the retina. Diverging lenses.

33.9 An object is placed on the left of a converging lens at a distance that is less than the focal length of the lens. The image produced will be

a) real and inverted.
c) virtual and inverted.

b) virtual and upright.
d) real and upright.

33.10 What would you expect to happen to the magnitude of the power of a lens when it is placed in water ($n = 1.33$)?

a) It would increase.

b) It would decrease.

c) It would stay the same.

d) It would depend if the lens was converging or diverging.

33.11 An unknown lens forms an image of an object that is 24 cm away from the lens, inverted, and a factor of 4 larger in size than the object. Where is the object located?

a) 6 cm from the lens on the same side of the lens

b) 6 cm from the lens on the other side of the lens

c) 96 cm from the lens on the same side of the lens

d) 96 cm from the lens on the other side of the lens

e) No object could have formed this image.

33.12 A single lens with two convex surfaces made of sapphire with index of refraction $n = 1.77$ has surfaces with radii of curvature $R_1 = 27.0$ cm and $R_2 = -27.0$ cm. What is the focal length of this lens in air?

a) 17.5 cm
c) 30.7 cm
e) 54.0 cm

b) 27.0 cm
d) 40.8 cm

33.13 A single lens with two convex surfaces made of sapphire with index of refraction $n = 1.77$ has surfaces with radii of curvature $R_1 = 27.0$ cm and $R_2 = -27.0$ cm. What is the focal length of this lens in water ($n = 1.33$)?

a) 17.5 cm
d) 40.8 cm

b) 27.0 cm
e) 54.0 cm

c) 30.7 cm

33.14 An object is placed 15.0 cm to the left of a converging lens with focal length $f = 5.00$ cm, as shown in the figure. Where is the image formed?

a) 1.25 cm to the right of the lens

b) 2.75 cm to the left of the lens

c) 3.75 cm to the right of the lens

d) 5.00 cm to the left of the lens

e) 7.50 cm to the right of the lens

33.15 The eyepiece for the Yerkes telescope has a focal length of approximately 10 cm. The Hubble Space Telescope can reach an effective angular magnification factor of 8000. What would the focal length of the objective lens at Yerkes need to be in order to reach the same factor of 8000?

a) 8 m
d) 0.00125 m

b) 80 m
e) 800 m

c) 0.125 m

33.16 Where is the image of the objective lens of a microscope located and what does it look like?

a) real and upright
c) virtual and upright

b) real and inverted
d) virtual and inverted

33.17 A pea sits at the focal point of a converging lens, 3 cm to the left of that lens on the optical axis. A second identical lens is at a distance of 12 cm to the right from the first one. Where is the image of the pea located, and is it real or virtual?

a) A virtual image 3 cm to the left of lens 2

b) A real image 3 cm to the left of lens 2

c) A virtual image 3 cm to the right of lens 2

d) A real image 3 cm to the right of lens 2

e) No image can be formed, because the 2 lenses are more than 2 focal lengths apart.

CONCEPTUAL QUESTIONS

33.18 Several small drops of paint (less than 1 mm in diameter) splatter on a painter's eyeglasses, which are approximately 2 cm in front of the painter's eyes. Do the dots appear in what the painter sees? How do the dots affect what the painter sees?

33.19 When a diver with 20/20 vision removes her mask underwater, her vision becomes blurry. Why is this the case? Does the diver become nearsighted (eye lens focuses in front of retina) or farsighted (eye lens focuses behind retina)? As the index of refraction of the medium approaches that of the lens, where does the object get imaged? Typically, the index of refraction for water is 1.33, while the index of refraction for the lens in a human eye is 1.40.

33.20 In H.G. Wells's classic story *The Invisible Man,* a man manages to change the index of refraction of his body to 1.0; thus, light would not bend as it enters his body (assuming he is in air and not swimming). If the index of refraction of his eyes were equal to one, would he be able to see? If so, how would things appear?

33.21 Astronomers sometimes place filters in the path of light as it passes through their telescopes and optical equipment. The filters allow only a single color to pass through. What are the advantages of this? What are the disadvantages?

33.22 Is it possible to start a fire by focusing the light of the Sun with ordinary eyeglasses? How, or why not?

33.23 Will the magnification produced by a simple magnifier increase, decrease, or stay the same when the object and the lens are both moved from air into water?

33.24 Is it possible to design a system that will form an image without lenses or mirrors? If so, how? and what drawbacks, if any, would it have?

33.25 A lens system widely used in machine vision for dimensional measurement is the so-called telecentric lens system. In its basic configuration, it consists of two thin lenses of focal lengths f_1 and f_2, respectively, placed a distance $d = f_1 + f_2$ apart, and a small circular aperture called a *stop* aperture placed at the common focal point between the two lenses. The purpose of such a system is to provide a magnification that is independent of the distance between the object and the lens system, within a specified range of distances that define the so-called *depth of field* of the system.

a) Draw a ray diagram through the system.

b) Determine the magnification of such a system.

c) Determine the requirements that must be met by the two lenses in order for a telecentric system to be able to image with maximum resolution an object with a diameter of 50.0 mm on a so-called $\frac{1}{2}$-inch CCD camera head (detector dimensions are ~ 6.50 mm × 5.00 mm).

33.26 Lenses or lens systems for photography are rated by both focal length and "speed." The "speed" of a lens is measured by its *f*-number or *f*-stop, the ratio of its focal length to its aperture diameter. For many commercially available lenses this number is approximately a power of $\sqrt{2}$, e.g., 1.4, 2.0, 2.8, 4.0, etc.; the lens is provided with a mechanical iris diaphragm that can be set to different apertures or *f*-stops. The smaller the *f*-number, the "faster" the lens. "Faster" lenses are more expensive, as a wide aperture requires a higher-quality lens.

a) Explain the connection between *f*-number, (i.e., aperture diameter) and "speed."

b) Look up the necessary data and calculate the *f*-numbers of (the primary mirrors of) the Keck 10-meter telescope, the Hubble Space Telescope, and the Arecibo radio telescope.

33.27 The *f*-number of a photographic system determines not just its speed but also its "depth of field," the range of distances over which objects remain in acceptable focus. Low *f*-numbers correspond to small depth of field, high *f*-numbers to large depth of field. Explain this.

33.28 Mirrors for astronomical instruments are invariably *first-surface* mirrors: The reflective coating is applied to the surface exposed to the incoming light. Household mirrors, on the other hand, are *second-surface* mirrors: The coating is applied to the back of the glass or plastic material of the mirror. (You can tell the difference by bringing the tip of an object

close to the surface of the mirror. Object and image will nearly touch with a first-surface mirror; a gap will remain between them with a second-surface mirror.) Explain the reasons for these design differences.

33.29 When sharing binoculars with a friend, you notice that you have to readjust the focus when he has been using it (he wears glasses, but removes them to use the binoculars). Why?

33.30 Using ray tracing in the diagram below, find the image of the object (upright arrow) in the following system having a diverging (double-concave) lens. Is the image real or virtual? Is the image height less than or greater than the object height?

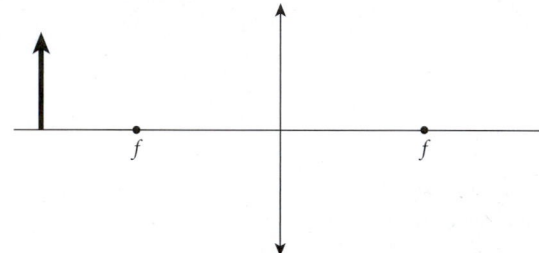

33.31 You have made a simple telescope from two convex lenses. The objective lens is the one of the two lenses that is closer to the object being observed. What kind of image is produced by the eyepiece lens if the eyepiece is closer to the objective lens than the image produced by the objective lens?

33.32 What kind of lens is used in eyeglasses to correct the vision of someone who is

a) nearsighted?

b) farsighted?

33.33 A physics student epoxies two converging lenses to the opposite ends of a $2.0 \cdot 10^1$-cm-long tube. One lens has a focal length of $f_1 = 6.0$ cm and the other has a focal length of $f_2 = 3.0$ cm. She wants to use this device as a microscope. Which end should she look through to obtain the highest magnification of an object?

EXERCISES

A blue problem number indicates a worked-out solution is available in the Student Solutions Manual. One • and two •• indicate increasing level of problem difficulty.

Section 33.1

33.34 An object of height h is placed at a distance d_o on the left side of a converging lens of focal length f ($f < d_o$).

a) What must d_o be in order for the image to form at a distance $3f$ on the right side of the lens?

b) What will be the magnification?

33.35 An object is 6.0 cm from a converging thin lens along the axis of the lens. If the lens has a focal length of 9.0 cm, determine the image magnification.

33.36 A biconvex *ice lens* is made so that the radius of curvature of the front surface is 15.0 cm and that for the back is 20.0 cm. Determine how far you would put dry twigs if you wish to start a fire with your ice lens.

33.37 As a high-power laser engineer you need to focus a 1.06-mm diameter laser beam to a 10.0-μm diameter spot 20.0 cm behind the lens. What focal length lens would you use?

33.38 A plastic cylinder of length $3.0 \cdot 10^1$ cm has its ends ground to convex (from the rod outward) spherical surfaces, each having radius of curvature $1.0 \cdot 10^1$ cm. A small object is placed $1.0 \cdot 10^1$ cm from the left end. How far will the image of the object lie from the right end, if the index of refraction of the plastic is 1.5?

•**33.39** The object (upright arrow) in the following system has a height of 2.5 cm and is placed 5.0 cm away from a converging (convex) lens with a focal length of 3.0 cm. What is the magnification of the image? Is the image upright or inverted? Confirm your answers by ray tracing.

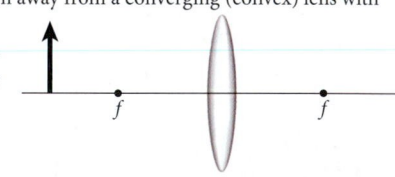

•**33.40** Demonstrate that the minimum distance possible between a real object and its real image through a thin convex lens is $4f$, where f is the focal length of the lens.

•**33.41** An air-filled cavity bound by two spherical surfaces is created inside a glass block. The two spherical surfaces have radii of 30.0 cm and 20.0 cm, respectively, and the thickness of the cavity is 40.0 cm (see diagram below). A light-emitting diode (LED) is embedded inside the block a distance of 60.0 cm in front of the cavity. Given $n_{\text{glass}} = 1.50$ and $n_{\text{air}} = 1.00$, and using only paraxial light rays (i.e., in the paraxial approximation):

a) Calculate the final position of the image of the LED through the air-filled cavity.

b) Draw a ray diagram showing how the image is formed.

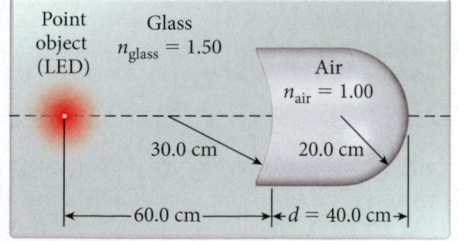

Section 33.2

33.42 To study a tissue sample better, a pathologist holds a 5.00-cm focal length magnifying glass 3.00 cm from the sample. How much magnification can he get from the lens?

33.43 Suppose a magnifying glass has a focal length of 5.0 cm. Find the magnifying power of this glass when the object is placed at the near point.

33.44 What is the focal length of a magnifying glass if a 1.0-mm object appears to be 10. mm?

•**33.45** A person with a near-point distance of 24.0 cm finds that a magnifying glass gives an angular magnification that is 1.25 times larger when the image of the magnifier is at the near point than when the image is at infinity. What is the focal length of the magnifying glass?

Section 33.3

33.46 A beam of parallel light, 1.00 mm in diameter passes through a lens with a focal length of 10.0 cm. Another lens, this one of focal length 20.0 cm, is located behind the first lens so that the light traveling out from it is again parallel.

a) What is the distance between the two lenses?

b) How wide is the outgoing beam?

33.47 How large does a 5.0-mm insect appear when viewed with a system of two identical lenses of focal length 5.0 cm separated by a distance 12.0 cm if the insect is 10.0 cm from the first lens? Is the image real or virtual? Inverted or upright?

•**33.48** Three converging lenses of focal length 5.0 cm are arranged with a spacing of $2.0 \cdot 10^1$ cm between them, and are used to image an insect $2.0 \cdot 10^1$ cm away.

a) Where is the image?

b) Is it real or virtual?

c) Is it upright or inverted?

•**33.49** Two identical thin convex lenses, each of focal length f, are separated by a distance $d = 2.5f$. An object is placed in front of the first lens at a distance $d_{o,1} = 2f$.

a) Calculate the position of the final image of the object through the system of lenses.

b) Calculate the total transverse magnification of the system.

c) Draw the ray diagram for this system and show the final image.

d) Describe the final image (real or virtual, erect or inverted, larger or smaller) in relation to the initial object.

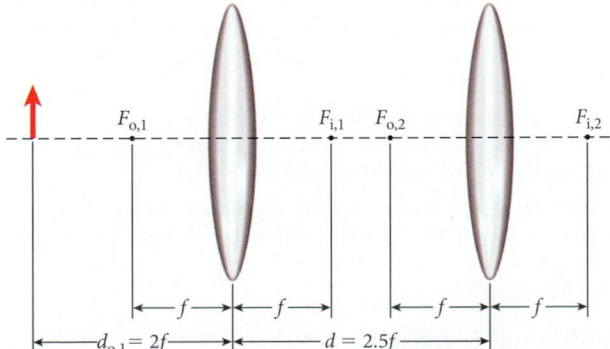

•**33.50** Two converging lenses with focal lengths 5.00 cm and 10.0 cm, respectively, are placed 30.0 cm apart. An object of height $h = 5.00$ cm is placed 10.0 cm to the left of the 5.00-cm lens. What will be the position and height of the final image produced by this lens system?

33.51 Two lenses are used to create an image for a source 10.0 cm tall that is located 30.0 cm to the left of the first lens, as shown in the figure. Lens L_1 is a biconcave lens made of crown glass (index of refraction $n = 1.55$) and has a radius of curvature of 20.0 cm for both surface 1 and 2. Lens L_2 is 40.0 cm to the right of the first lens L_1. Lens L_2 is a converging lens

with a focal length of 30.0 cm. At what distance relative to the object does the image get formed? Determine this position by sketching rays and calculating algebraically.

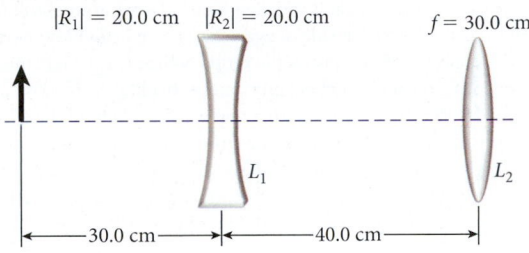

••**33.52** Sophisticated optical systems can be analyzed and designed with the aid of techniques from linear algebra. Light rays (or particle beams) are described at any point along the axis of the system by a two-component column vector containing y, the distance of the ray from the optic axis, and dy/dx, the slope of the ray. Components of the system are described by 2×2 matrices which incorporate their effects on the ray; combinations of components are described by products of these matrices.

a) A thin lens does not alter the position of a ray, but increases (diverging) or decreases (converging) its slope an amount proportional to the distance of the ray from the axis. Construct the matrix for a thin lens of focal length f.

b) A space between components does not alter the slope of a ray; the distance of the ray from the axis changes by the slope of the ray times the length of the space. Write the matrix for a space of length x.

c) Write the matrix for the two-lens "zoom lens" system described in Section 33.3.

Section 33.4

33.53 The typical length of a human eyeball is 2.50 cm.

a) What is the effective focal length of the two-lens system made from a normal person's cornea and lens when viewing objects far away?

b) What is the effective focal length for viewing objects at the near point?

33.54 Using your answers from the previous question, and given that the cornea in a typical human eye has a fixed focal length of 2.33 cm, what range of focal lengths does the lens in a typical eye have?

33.55 Jane has a near point of 125 cm and wishes to read from a computer screen 40. cm from her eye.

a) What is the object distance?

b) What is the image distance?

c) What is the focal length?

d) What is the power of the corrective lens needed?

e) Is the corrective lens diverging or converging?

33.56 Bill has a far point of 125 cm and wishes to see distant objects clearly.

a) What is the object distance?

b) What is the image distance?

c) What is the focal length?

d) What is the power of the corrective lens needed?

e) Is the corrective lens diverging or converging?

33.57 A person wearing bifocal glasses is reading a newspaper a distance of 25 cm. The lower part of the lens is converging for reading and has a focal length of 70. cm. The upper part of the lens is diverging for seeing at distances far away and has a focal length of 50. cm. What are the uncorrected near and far points for the person?

33.58 The radius of curvature for the outer part of the cornea is 8.0 mm, the inner portion is relatively flat. If the index of refraction of the cornea and the aqueous humor is 1.34:

a) Find the power of the cornea.

b) If the combination of the lens and the cornea has a power of 50. diopter, find the power of the lens (assume the two are touching).

•33.59 As objects are moved closer to the human eye, the ciliary muscle causes the lens at the front of the eye to become more curved, thereby decreasing the focal length of the lens. The shortest focal length f_{min} of this lens is typically 2.3 cm. What is the closest one can bring an object to a normal human eye (see left side of figure below) and still have the image of the object projected sharply on the retina, which is 2.5 cm behind the lens? Now consider a nearsighted human eye that has the same f_{min} but is stretched horizontally, with a retina that is 3.0 cm behind the lens (see right side of figure below). What is the closest one can bring an object to this nearsighted human eye and still have the image of the object projected sharply on the retina. Compare the maximum angular magnification produced by this nearsighted eye with that of the normal eye.

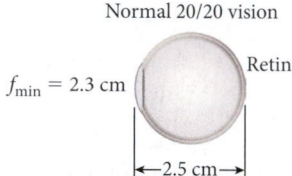
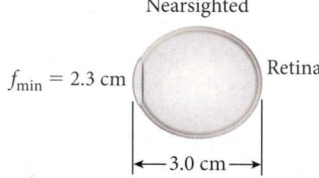

Normal 20/20 vision   Nearsighted

Retina   Retina

f_{min} = 2.3 cm f_{min} = 2.3 cm

←2.5 cm→   ←3.0 cm→

•33.60 A person is nearsighted (myopic). The power of his eyeglass lenses is −5.75 diopters, and he wears the lenses 1.00 cm in front of his corneas. What is the prescribed power of his contact lenses?

Section 33.5

33.61 An amateur photographer attempts to build a custom zoom lens using a converging lens followed by a diverging lens. The two lenses are separated by a distance x = 50. mm as shown below. If the focal length of the first lens is $2.0 \cdot 10^2$ mm and the focal length of the second lens is $-3.0 \cdot 10^2$ mm, what will the effective focal length of this compound lens be? What will f_{eff} be if the lens separation is changed to $1.0 \cdot 10^2$ mm?

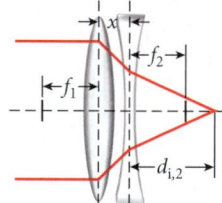
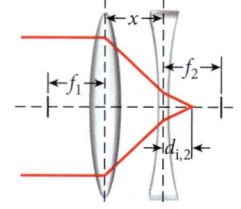

33.62 For a certain camera, the distance between the lens and the film is 10.0 cm. You observe that objects that are very far away appear properly focused. How far from the film would you have to move the lens in order to properly focus an object 100. cm away?

33.63 A camera has a lens with a focal length of 60. mm. Suppose you replace the normal lens with a zoom lens whose focal length can be varied from 35. mm to 250. mm and use the camera to photograph an object very far away. Compared to a 60.-mm lens, what magnification of the image would be achieved using the 240.-mm focal length?

•33.64 A camera lens usually consists of a combination of two or more lenses to produce a good-quality image. Suppose a camera lens has two lenses—a diverging lens of focal length 10.0 cm and a converging lens of focal length 5.00 cm. The two lenses are held 7.00 cm apart. A flower of length 10.0 cm, to be pictured, is held upright at a distance 50.0 cm in front of the diverging lens; the converging lens is placed behind the diverging lens. What are the location, orientation, size, and the magnification of the final image?

Section 33.6

33.65 A student finds a tube 20. cm long with a lens attached to one end. The attached lens has a focal length of 0.70 cm. The student wants to use the tube and lens to make a microscope with a magnification of $3.0 \cdot 10^2 \times$. What focal length lens should the student attach to the other end of the tube?

33.66 The objective lens in a laboratory microscope has a focal length of 3.00 cm and provides an overall magnification of $1.00 \cdot 10^2$. What is the focal length of the eyepiece if the distance between the two lenses is 30.0 cm?

33.67 You have found in the lab an old microscope, which has lost its eyepiece. It still has its objective lens, and markings indicate that its focal length is 7.00 mm. You can put in a new eyepiece, which goes in 20.0 cm from the objective. You need a magnification of about 200. Assume you want the comfortable viewing distance for the final image to be 25.0 cm. You find in a drawer eyepieces marked 2.00-, 4.00-, and 8.00-cm focal length. Which is your best choice?

33.68 A microscope has a 2.00-cm focal length eyepiece and a 0.800-cm objective lens. For a relaxed normal eye, calculate the position of the object if the distance between the lenses is 16.2 cm.

•33.69 Suppose you want to design a microscope so that for a fixed distance between the two lenses, the magnitude of the magnification will vary from 150 to 450 as you substitute eyepieces of various focal lengths.

a) If the longest focal length for an eyepiece is $6.0 \cdot 10^1$ mm, what will the shortest focal length be?

b) If you want the distance between the eyepiece and objective to be 35 cm, what should the focal length of the objective be?

Section 33.7

33.70 A refracting telescope has the objective lens of focal length 10.0 m. Assume it is used with an eyepiece of focal length 2.00 cm. What is the magnification of this telescope?

33.71 What is the magnification of a telescope with $f_o = 1.00 \cdot 10^2$ cm and $f_e = 5.00$ cm?

33.72 A simple telescope is formed using an eyepiece of focal length 25.0 mm and an objective of focal length 80.0 mm. Calculate the angle subtended by the image of the Moon when viewed through this telescope from the Earth.

33.73 Galileo discovered the moons of Jupiter in the fall of 1609. He used a telescope of his own design that had an objective lens with a focal length of $f_o = 40.0$ inches and an eyepiece lens with a focal length of $f_e = 2.00$ inches. Calculate the magnifying power of Galileo's telescope.

33.74 Two distant stars are separated by an angle of 35 arcseconds. If you have a refracting telescope whose objective lens focal length is 3.5 m, what focal length eyepiece do you need in order to observe the stars as though they were separated by 35 arcminutes?

•33.75 A 180× astronomical telescope is adjusted for a relaxed eye when the two lenses are 1.30 m apart. What is the focal length of each lens?

•33.76 Two refracting telescopes are used to look at craters on the Moon. The objective focal length of both telescopes is 95.0 cm and the eyepiece focal length of both telescopes is 3.80 cm. The telescopes are identical except for the diameter of the lenses. Telescope A has an objective diameter of 10.0 cm while the lenses of telescope B are scaled up by a factor of two, so that its objective diameter is 20.0 cm.

a) What are the angular magnifications of telescopes A and B?

b) Do the images produced by the telescopes have the same brightness? If not, which is brighter and by how much?

••33.77 Some reflecting telescope mirrors utilize a rotating tub of mercury to produce a large parabolic surface. If the tub is rotating on its axis with an angular frequency ω, show that the focal length of the resulting mirror is: $f = g/2\omega^2$.

Additional Exercises

33.78 An object 4.0 cm high is projected onto a screen using a converging lens with a focal length of 35 cm. The image on the screen is 56 cm. Where are the lens and the object located relative to the screen?

33.79 A person with normal vision picks up a nearsighted friend's eyeglasses and attempts to focus on objects around her while wearing them. She is able to focus only on very distant objects while wearing the eyeglasses, and is completely unable to focus on objects that are near to her. Estimate the prescription of her friend's eyeglasses in diopters.

33.80 Suppose the near point of your eye is $2.0 \cdot 10^1$ cm and the far point is infinity. If you put on -0.20 diopter spectacles, what will be the range over which you will be able to see objects distinctly?

33.81 A fish is swimming in an aquarium at an apparent depth $d = 1.00 \cdot 10^1$ cm. At what depth should you grab in order to catch it?

33.82 A classmate claims that by using a 40.0-cm focal length mirror, he can project onto a screen a 10.0-cm tall bird located 100. m away. He claims that the image will be no less than 1.00 cm tall and inverted. Will he make good on his claim?

33.83 An object is 6.0 cm from a thin lens along the axis of the lens. If the lens has a focal length of 9.0 cm, determine the image distance.

33.84 A thin spherical lens is fabricated from glass so that it bulges outward in the middle on both sides. The glass lens has been ground so that the surfaces are part of a sphere with a radius of 25.0 cm on one side and a radius of 30.0 cm on the other. What is the power of this lens in diopters?

33.85 A diverging lens is fabricated from glass such that one surface of the lens is convex and the other surface is concave. The glass lens has been ground so that the convex surface is part of a sphere with a radius of 45 cm and the concave surface is part of a sphere with a radius of $2.0 \cdot 10^1$ cm. What is the power of this lens in diopters?

33.86 A person who is farsighted can see clearly an object that is at least 2.5 m away. To be able to read a book $2.0 \cdot 10^1$ cm away, what kind of corrective glasses should he purchase?

33.87 You are experimenting with a magnifying glass (consisting of a single converging lens) at a table. You discover that by holding the magnifying glass 92.0 mm above your desk, you can form a real image of a light that is directly overhead. If the distance between the light and the table is 2.35 m, what is the focal length of the lens?

33.88 A girl forgets to put on her glasses and finds that she needs to hold a book 15 cm from her eyes in order to clearly see the print.

a) If she were to hold the book 25 cm away, what type of corrective lens would she need to use to clearly see the print?

b) What is the focal length of the lens?

33.89 The focal length of the lens of a camera is 38.0 mm. How far must the lens be moved to change focus from a person 30.0 m away to one that is 5.00 m away?

33.90 A telescope is advertised as providing a magnification of magnitude 41 using an eyepiece of focal length $4.0 \cdot 10^1$ mm. What is the focal length of the objective?

•33.91 Determine the position and size of the final image formed by a system of elements consisting of an object 2.0 cm high located at $x = 0$ m, a converging lens with focal length $5.0 \cdot 10^1$ cm located at $x = 3.0 \cdot 10^1$ cm and a plane mirror located at $x = 7.0 \cdot 10^1$ cm.

33.92 The distance from the lens (actually a combination of the cornea and the crystalline lens) to the retina at the back of the eye is 2.0 cm. If light is to focus on the retina,

a) what is the focal length of the lens when viewing a distant object?

b) what is the focal length of the lens when viewing an object 25 cm away from the front of the eye?

33.93 You visit your eye doctor and discover that you require lenses having a diopter value of -8.4. Are you nearsighted or farsighted? With uncorrected vision, how far away from your eyes must you hold a book to read clearly?

33.94 Jack has a near point of 32 cm and uses a magnifier of 25 diopter.

a) What is the magnification if the final image is at infinity?

b) What is the magnification if the final image is at the near point?

33.95 Where is the image and what is the magnification if you hold a 2.0-inch-diameter clear glass marble 1.0 ft in front of you?

•33.96 A diverging lens with $f = -30.0$ cm is placed 15.0 cm behind a converging lens with $f = 20.0$ cm. Where will an object at infinity in front of the converging lens be focused?

•33.97 An instructor wants to use a lens to project a real image of a light bulb onto a screen 1.71 m from the bulb. In order to get the image to be twice as large as the bulb, what focal length lens will be needed?

•33.98 An old refracting telescope placed in a museum as an exhibit is 55 cm long and has magnification of 45. What are the focal lengths of its objective and eyepiece lenses?

•33.99 A converging lens of focal length $f = 50.0$ cm is placed 175 cm to the left of a metallic sphere of radius $R = 100.$ cm. An object of height $h = 20.0$ cm is placed 30.0 cm to the left of the lens. What is the height of the image formed by the metallic sphere?

•33.100 When performing optical spectroscopy (for example, photoluminescence or Raman spectroscopy), a laser beam is focused on the sample to be investigated by means of a lens having a focal distance f. Assume that the laser beam exits a pupil D_0 in diameter that is located at a distance d_0 from the focusing lens. For the case when the image of the exit pupil forms on the sample, calculate

a) at what distance d_i from the lens is the sample located.

b) the diameter D_i of the laser spot (image of the exit pupil) on the sample.

c) the numerical results for: $f = 10.0$ cm, $D_0 = 2.00$ mm, $d_0 = 1.50$ m.

•33.101 For a person whose near point is 115 cm, so that he can read a computer monitor at 55 cm, what power of reading glasses should his optician prescribe, keeping the lens–eye distance of 2.0 cm for his spectacles?

•33.102 A telescope has been properly focused on the Sun. You want to observe the Sun visually, but to protect your sight you don't want to look through the eyepiece; rather, you want to project an image of the Sun on a screen 1.5 m behind (the original position of) the eyepiece, and observe that. If the focal length of the eyepiece is 8.0 cm, how must you move the eyepiece?

MULTI-VERSION EXERCISES

33.103 A lens of a pair of eyeglasses with index of refraction 1.723 has a power of 4.29 D in air. What is the power of this lens if it is put in water with $n = 1.333$?

33.104 A lens of a pair of eyeglasses has a power of 5.55 D in air. The power of this same lens is 2.262 if it is put in water with $n = 1.333$. What is the index of refraction of this lens?

33.105 A lens of a pair of eyeglasses with index of refraction 1.735 has a power of 2.794 D in water, with $n = 1.333$. What is the power of this lens if it is put in air?

33.106 A telescope, consisting of two lenses, has an angular magnification of absolute value 81.4 and an objective lens with refractive power 0.234 D. What is the refractive power of the eyepiece of the telescope?

33.107 A telescope, consisting of two lenses, has an objective lens with refractive power 0.186 D and an eyepiece with focal length of 5.33 cm. What is the absolute value of the angular magnification of this telescope?

33.108 A telescope, consisting of two lenses, has an objective lens with focal length 646.7 cm and an eyepiece with focal length 5.41 cm. What is the absolute value of its angular magnification?

Wave Optics

34

FIGURE 34.1 Soap bubble showing colors from interference phenomena.

Soap bubbles consist of clear water with a bit of dish soap mixed in. Why, then, do we see the colors of the rainbow when we look at soap bubbles in the air (Figure 34.1)? It turns out that the optics of soap bubbles are not the same as the reflections and refractions that give rise to rainbows (Chapter 32). Instead, the colors of bubbles are due to interference of light of various wavelengths—a wave effect, not explained by geometric optics.

We have studied image formation by assuming that light travels as straight rays, without regard to the nature of light: for example, whether it consists of particles or waves. In this chapter, we look at optical effects best explained by the wave nature of light—a study sometimes called *physical optics,* as distinct from geometric optics. You may want to review material on waves in Chapter 15, especially the concept of interference of waves. We will also study a wave property called diffraction, which we did not discuss in Chapter 15 but which becomes important when dealing with very small wavelengths such as those of light. The two properties, interference and *diffraction,* explain not only the colors of soap bubbles, but also such issues as how well we can see distant objects as separate and not blurred together.

The material in this chapter does not completely answer the questions of what light is and how it behaves. We will see in following chapters that wondering about the nature of light led to some of the greatest developments in physics in the 20th century, ideas that are often grouped together under the name *modern physics.* The ideas in this chapter will be pivotal in our examination of the modern understanding of light, matter, time, and space.

WHAT WE WILL LEARN

- The wave nature of light leads to phenomena that cannot be explained using geometric optics.

- Superposed light waves that are in phase interfere constructively; superposed light waves that are 180° out of phase interfere destructively.

- Superposed light waves that have traveled different distances can interfere constructively or destructively, depending on the path length difference.

- Light waves spread out after passing through a narrow slit or after encountering an obstacle. This spreading out is called *diffraction*.

- Light waves that have the same phase and frequency are called *coherent light waves*.

- Coherent light incident on one narrow slit produces a diffraction pattern; coherent light incident on two narrow slits produces an interference pattern.

- Interference can occur when light is is partially reflected from each of the two surfaces (front and back) of a thin film.

- An interferometer is a device designed to measure lengths or changes in length using the interference of light.

- Diffraction can limit the ability of a telescope or a camera to resolve distant objects.

- A diffraction grating has many narrow slits or rulings and can be used with a single-wavelength light source to produce an interference pattern that consists of narrow bright fringes separated by wide dark areas.

- X-ray diffraction can be used to study the atomic structure of materials.

34.1 Light Waves

We learned in Chapter 31 that light is an electromagnetic wave. However, normally we do not think of light as a wave, because its wavelength is so short that we usually do not notice its wave behavior. Thus, in Chapters 32 and 33, we discussed light as rays, a description that is appropriate for all physical situations where we can neglect the wavelength of light in comparison to all other physical dimensions. These rays travel in straight lines except when they are reflected off a mirror or refracted at the boundary between two optical media.

Now we want to address physical situations where we cannot approximate the wavelength of light as negligibly small any more. In this chapter, we discuss the wave nature of light, and we apply to light many of the wave concepts (superposition, coherence, interference, boundary reflection, and others) we developed in Chapter 15. This will sometimes lead to rather surprising results. But all of these results can be verified experimentally.

One way to reconcile the wave nature of light with the geometric optical properties of light is to use **Huygens's Principle,** developed by Dutch physicist Christiaan Huygens (1629–1695). Huygens proposed a wave theory of light in 1678, long before Maxwell developed his theories of light. Huygens's Principle states that every point on a propagating wave front serves as a source of spherical secondary wavelets. Later, the envelope of these secondary wavelets becomes a new wave front. If the original wave has frequency f and speed v, the secondary wavelets have the same f and v. Diagrams of phenomena based on Huygens's Principle are called *Huygens constructions*.

Figure 34.2 shows a Huygens construction for a plane wave. We start with a plane wave traveling at the speed of light, c. We assume point sources of spherical wavelets along the wave front, as shown. These wavelets also travel at c, so at a time Δt the wavelets have traveled a distance $c\Delta t$. Assuming many point sources along the wave front, Figure 34.2 shows that the

FIGURE 34.2 Huygens construction for a plane wave traveling vertically upward.

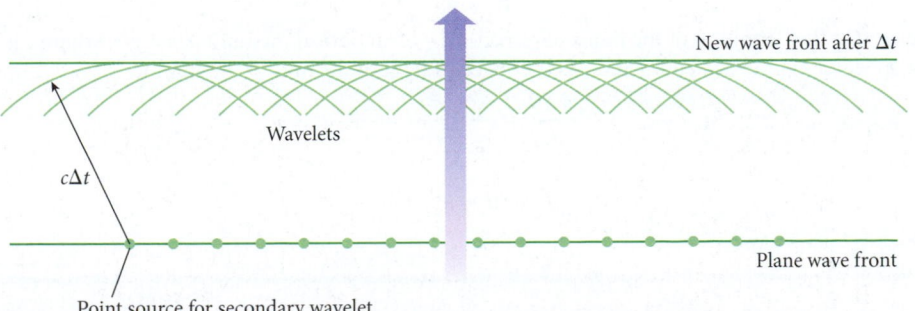

New wave front after Δt

Wavelets

$c\Delta t$

Plane wave front

Point source for secondary wavelet

envelope of these wavelets forms a new wave front parallel to the original wave front. Thus, the wave continues to travel in a straight line with the original frequency and speed.

In Chapter 32, the index of refraction n in a medium was defined as the ratio of the speed of light in vacuum, c, to the speed of light in that medium, v:

$$n = \frac{c}{v}. \tag{34.1}$$

Using this definition, Snell's Law can be expressed as

$$n_1 \sin\theta_1 = n_2 \sin\theta_2. \tag{34.2}$$

Chapter 32 showed that this law correctly describes refraction phenomena at the boundaries between different media, but the derivation of Snell's Law was deferred to the present chapter. With Huygens's Principle in hand, we can now perform this derivation.

DERIVATION 34.1 / Snell's Law

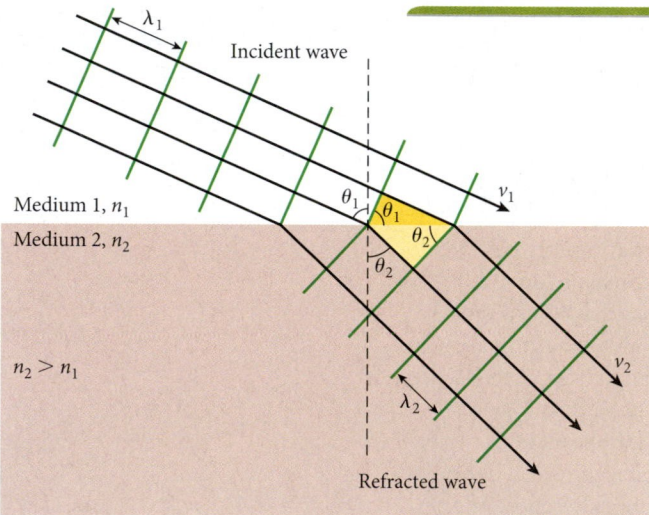

FIGURE 34.3 A Huygens construction of a wave traveling in an optical medium and incident on the boundary with a second optical medium. The green lines represent wave fronts; the black arrows denote rays.

A Huygens construction can be used to derive Snell's Law for refraction between two optical media with different indexes of refraction. Assume a wave with wave fronts separated by a wavelength λ_1 traveling with speed v_1 in an optically clear medium and incident on the boundary with a second optically clear medium (Figure 34.3).

The angle of the incident wave front (green line in medium 1) with respect to the boundary is θ_1, which is also the angle that the light ray (black arrow) makes with respect to a normal to the boundary. When the wave enters the second medium, it travels with speed v_2. According to Huygens's Principle, the wave fronts are the result of wavelet propagation at the speed of the original wave, so the separation of the wave fronts in the second medium can be written in terms of the wavelength in the second medium, λ_2. Thus, the time interval between wave fronts for the first medium is λ_1/v_1, and the time interval for the second medium is λ_2/v_2. The essential point is that this time interval is the same for waves on either side of the boundary. (If these intervals were not equal, fronts would be appearing or disappearing mysteriously!) So, we have:

$$\frac{\lambda_1}{v_1} = \frac{\lambda_2}{v_2} \Leftrightarrow \frac{\lambda_1}{\lambda_2} = \frac{v_1}{v_2}.$$

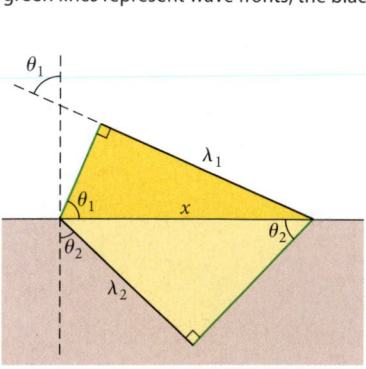

FIGURE 34.4 Expanded section of Figure 34.3, showing the incident wave front and its direction as well as the transmitted wave front and its direction.

Thus, the wavelengths of the light in the two media are proportional to the speeds of the light in those media.

We can get a relation between the angle of the incident wave fronts with the boundary, θ_1, and the angle of the transmitted wave fronts with the boundary, θ_2, by analyzing the yellow shaded region of Figure 34.3, shown in more detail in Figure 34.4.

From Figure 34.4, and using trigonometry, it follows that

$$\sin\theta_1 = \frac{\lambda_1}{x} \quad \text{and} \quad \sin\theta_2 = \frac{\lambda_2}{x}.$$

Dividing the first of these two equations by the other gives

$$\frac{\sin\theta_1}{\sin\theta_2} = \frac{\lambda_1}{\lambda_2} = \frac{v_1}{v_2}.$$

Given that the definition of the index of refraction n of an optical medium is $n = c/v$, where c is the speed of light in a vacuum and v is the speed of light in the optical medium,

$$\frac{\sin\theta_1}{\sin\theta_2} = \frac{v_1}{v_2} = \frac{c/n_1}{c/n_2} = \frac{n_2}{n_1},$$

or

$$n_1 \sin\theta_1 = n_2 \sin\theta_2,$$

which is Snell's Law.

We have seen that the wavelength of light changes when going from vacuum to an optical medium with index of refraction greater than 1. Taking the result $\lambda_1/v_1 = \lambda_2/v_2$ from Derivation 34.1, with medium 1 being a vacuum while medium 2 has an index of refraction n, we can write

$$\lambda_n = \lambda \frac{v}{c} = \frac{\lambda}{n}.$$

Thus, the wavelength of light is shorter in a medium with an index of refraction greater than 1 than it is in a vacuum. The frequency f of this light can be calculated from $v = \lambda f$. The frequency f_n of light traveling in the medium is then given by

$$f_n = \frac{v}{\lambda_n} = \frac{c/n}{\lambda/n} = \frac{c}{\lambda} = f. \tag{34.3}$$

Thus, the frequency of light traveling in an optical medium with $n > 1$ is the same as the frequency of that light traveling in a vacuum. (Equation 34.3 is equivalent to the statement made earlier that the time interval between wave fronts for the first medium equals the time interval between wave fronts for the second medium.)

Concept Check 34.2

A light ray passes through a stack of transparent plastic blocks, as shown in the figure. Which of the blocks has the highest index of refraction?

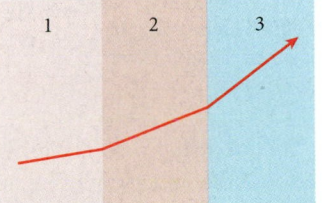

a) 1

b) 2

c) 3

d) They all have the same index of refraction.

e) The answer cannot be determined from the information given.

Definition

Coherent light is light made up of waves with the same wavelength that are in phase with each other. A useful source of coherent light is a laser. Light from light bulbs or sunlight, on the other hand, is **incoherent,** meaning that the waves may have different wavelengths and phase relationships with each other.

Self-Test Opportunity 34.1

Water has an index of refraction of 1.33, which means that the wavelengths of light emitted from objects are different when propagating in water than when propagating in air. Does this mean that you see objects in different colors when looking at them underwater? (You can do this experiment in your bathtub, if you want.)

Concept Check 34.1

Blocks of two different transparent materials are sitting in air and have identical light rays of the same wavelength incident on them at the same angle. Examining the figure, what can you say about the speed of light in these two blocks?

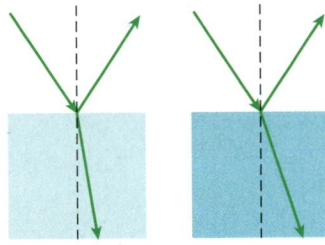

a) The speed of light is the same in both blocks.

b) The speed of light is greater in the block on the left.

c) The speed of light is greater in the block on the right.

d) The relative speed of light in the two blocks cannot be determined from the information given.

34.2 Interference

Sunlight is composed of light with a broad range of frequencies and corresponding wavelengths. We often see different colors separated out of sunlight after it refracts and reflects in raindrops, forming a rainbow. We also sometimes see various colors from sunlight due to constructive and destructive **interference** phenomena in thin transparent layers of materials, such as soap bubbles or thin films of oil floating on water. In contrast to a rainbow, this thin-film effect is due to interference.

The geometric optics of Chapters 32 and 33 cannot explain interference. Interference phenomena can be understood only by taking into account the wave nature of light. This section considers light waves that have the same wavelength in vacuum. Interference takes place when such light waves are superposed. If the light waves are in phase, they interfere constructively, as illustrated in Figure 34.5.

In Figure 34.5a and Figure 34.5b, the electric field component in the y-direction is plotted for two electromagnetic waves traveling in the x-direction. The two waves are in phase. Saying that the two waves are in phase is the same as saying that the phase difference between the two waves is zero. A phase difference of 2π radians (360°), corresponding to starting with two waves in phase and then displacing one of the waves by one full wavelength, will also

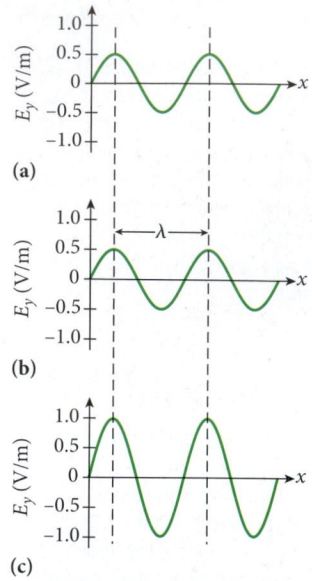

(a)

(b)

(c)

FIGURE 34.5 (a, b) Light waves in phase, with the same amplitude and wavelength, λ; (c) in-phase superposition of the two light waves demonstrates constructive interference, producing a wave with twice the amplitude.

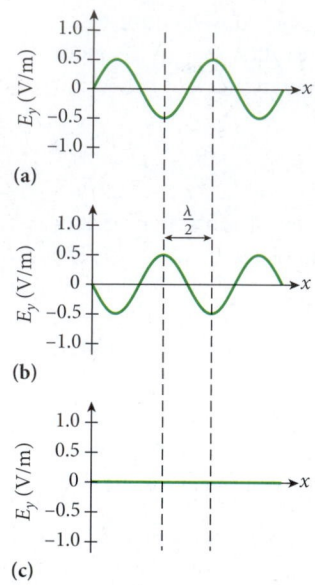

(a)

(b)

(c)

FIGURE 34.6 (a, b) Light waves out of phase by $\lambda/2$, with the same amplitude and wavelength; (c) superposition of the two light waves demonstrates destructive interference.

produce two waves that are in phase. If each light wave is traveling from its own point of origin, a phase difference will occur where they meet, related to the path difference between the two waves, even though they start in phase. The criterion for **constructive interference** is characterized by a path length difference, Δx, given by

$$\Delta x = m\lambda \quad (\text{for } m = 0, \pm 1, \pm 2, \ldots). \tag{34.4}$$

Figure 34.5c shows the two waves constructively interfering. The amplitudes of the two waves add, giving a wave with the same frequency but twice the amplitude of the two original waves.

If the two identical light waves are out of phase by π radians (180°) (Figure 34.6a and Figure 34.6b), the amplitudes of the waves will sum to zero everywhere when they meet. This is the condition for **destructive interference** (Figure 34.6c).

Figure 34.6a and Figure 34.6b show that this situation is equivalent to starting with two waves in phase and then displacing one of the waves by half of one wavelength ($\lambda/2$). Again, if we think of the two light waves as being emitted from different sources, the phase difference can be related to the path difference. Destructive interference takes place if the path difference is a half wavelength plus an integer times the wavelength:

$$\Delta x = \left(m + \tfrac{1}{2}\right)\lambda \quad (\text{for } m = 0, \pm 1, \pm 2, \ldots). \tag{34.5}$$

The following subsections will discuss interference phenomena caused by light waves with the same wavelength that are initially in phase but travel different distances or travel with different speeds (in different media) to reach the same point. At this point, the light waves are superposed and can interfere. Whether the interference is constructive or destructive for coherent light depends only on the path length difference, which is a recurring theme that will dominate all of the discussions that follow.

The different interference phenomena caused by various path differences of coherent light are summarized in Figure 34.7. Most of this chapter will explore the different conditions that lead to the interference patterns shown on the screens in the three parts of this figure. The following subsection explores pure double-slit interference (Figure 34.7a), which assumes that two very narrow slits of width comparable to the wavelength of the light are separated by a distance that is large compared to their individual widths. Section 34.3 will explore the case of coherent light hitting a single slit with a width a few times the

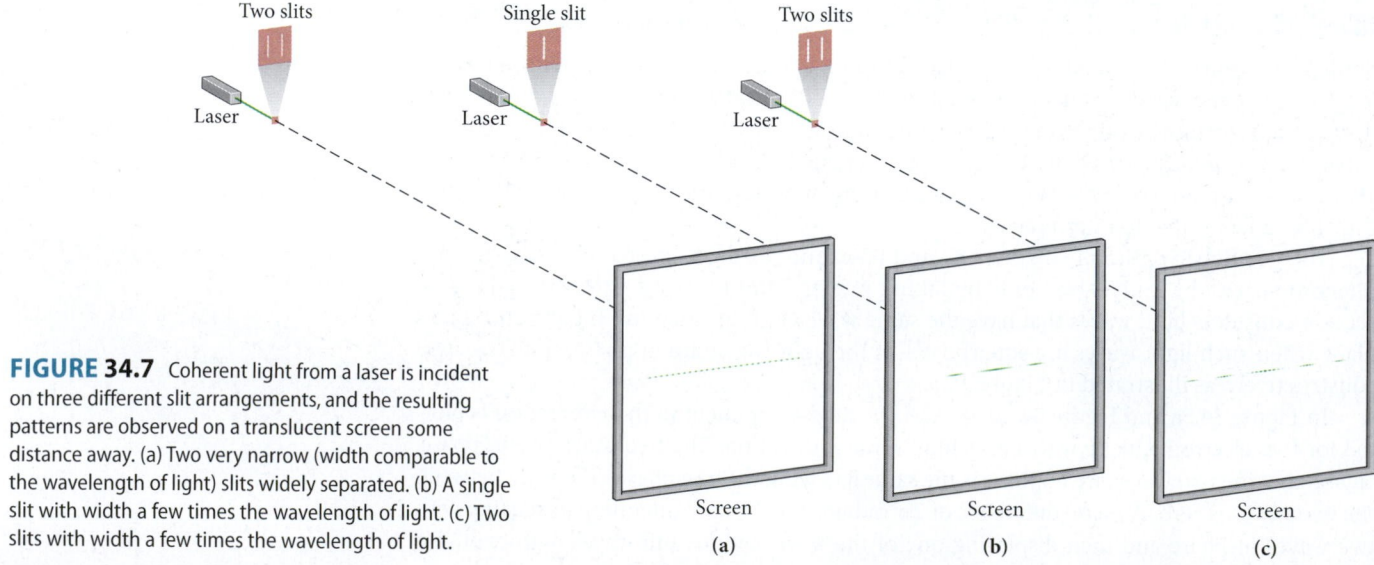

FIGURE 34.7 Coherent light from a laser is incident on three different slit arrangements, and the resulting patterns are observed on a translucent screen some distance away. (a) Two very narrow (width comparable to the wavelength of light) slits widely separated. (b) A single slit with width a few times the wavelength of light. (c) Two slits with width a few times the wavelength of light.

wavelength of the coherent light hitting it (Figure 34.7b). That section will also combine the effects of both cases and discuss double-slit interference with diffraction (Figure 34.7c). Detailed discussion of these cases will clarify the idea that path length differences between coherent light waves are responsible for all of them.

Double-Slit Interference

Our first example of the interference of light is **Young's double-slit experiment,** named for English physicist Thomas Young (1773–1829), who carried out this investigation in 1801. Assume that monochromatic coherent light such as that from a laser is incident on a pair of slits of width comparable to or even smaller than the wavelength of light, as shown in Figure 34.7a. (Strictly speaking, it is not necessary to use monochromatic coherent light from a laser to carry out this experiment, but doing so simplifies the following explanation of double-slit interference.)

For each slit, we use a Huygens construction, assuming that all the light passing through the slit is due to wavelets emitted from a single point at the center of the slit (Figure 34.8a). In this figure, spherical wavelets are emitted from this point. Assume that the slit is much narrower than the wavelength of light (the wavelength of green laser light is approximately 532 nm) so that the source of the wavelets can be represented with one point. Note that the slit opening is barely visible to the naked eye in this case. Even the two very narrow lines representing the magnified slits in Figure 34.7a are much too wide in comparison with the limiting case of extremely narrow slits that we are studying here.

Figure 34.8b shows two slits like the one in Figure 34.8a. A distance d separates the slits. Again, monochromatic coherent light is incident from the left and a source of spherical wavelets is at the center of each slit. The dashed lines represent lines along which constructive interference occurs. Placing a screen to the right of the slits will produce an alternating pattern of bright lines and dark lines, corresponding to constructive and destructive interference between the light waves emitted from the two slits.

To quantify these lines of constructive interference, we expand and simplify Figure 34.8b in Figure 34.9. In this figure, the two lines r_1 and r_2 represent the distances from the centers of slit S_1 and slit S_2, respectively, to a point P on a screen placed a distance L away from the slits. A line drawn from a point midway between the two slits to point P on the screen makes an angle θ with respect to a line drawn from the slits perpendicular to the screen. The point P on the screen is a distance y above this centerline.

To further quantify the two-slit geometry, we expand and simplify Figure 34.9 in Figure 34.10. In this figure, we assume that the screen has been placed a large distance L away from the slits, so the lines S_1P and S_2P are essentially parallel to each other and to the line drawn from the center of the two slits to point P. Another line is drawn from S_1 perpendicular to S_1P and S_2P, making a triangle with sides d, b, and Δx. The quantity Δx is the path length difference, $r_2 - r_1$.

This path length difference, $\Delta x = r_2 - r_1$, produces a phase difference for light originating from the two slits and illuminating the screen at point P. The path length difference can be expressed in terms of the distance d between the slits and the angle θ at which the light is observed:

$$\sin\theta = \frac{\Delta x}{d},$$

or

$$\Delta x = d\sin\theta.$$

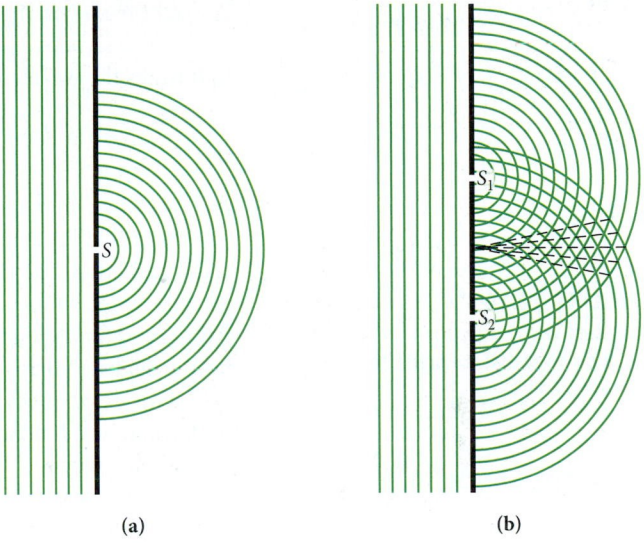

FIGURE 34.8 Huygens construction for a coherent light wave (green laser beam) incident from the left on (a) a single slit and (b) two slits, S_1 and S_2. The dashed lines indicate where constructive interference occurs.

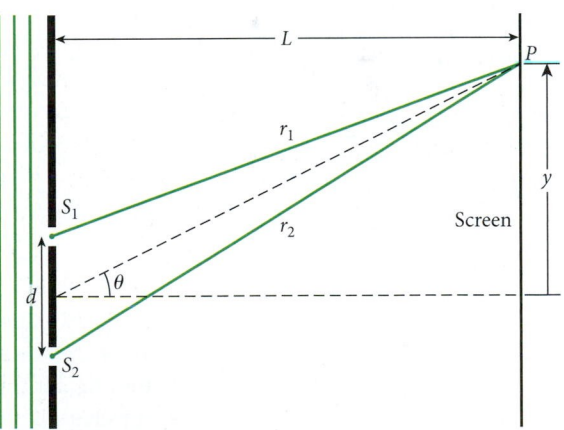

FIGURE 34.9 Expanded view of coherent light incident on two slits. The green lines to the right of the slits represent the distances light must travel from S_1 and S_2 to a point P on the screen.

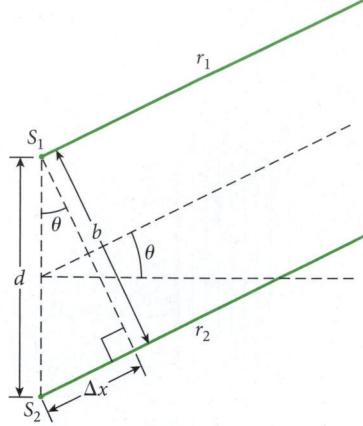

FIGURE 34.10 Expanded view of two slits, where the screen is placed far enough away that the green lines S_1P and S_2P are parallel.

Concept Check 34.3

In a Young's double-slit experiment where $L = 1.0$ m and $d = 0.10$ cm, light with a wavelength of 500. nm is used. There is a dark fringe at 0.75 mm to the left of the center of the diffraction pattern on the screen. What will happen to the intensity of the light at this location if one of the slits is covered up?

a) The intensity will decrease.

b) The intensity will increase.

c) Nothing will happen to the intensity.

d) It depends on which of the two slits is covered up.

For constructive interference, this path length difference must be an integer multiple of the wavelength of the incident light:

$$\Delta x = d\sin\theta = m\lambda \quad \left(\text{for } m = 0, \pm 1, \pm 2, \ldots\right) \quad \begin{pmatrix}\text{bright fringe,} \\ \text{constructive interference}\end{pmatrix}. \quad (34.6)$$

A bright fringe on the screen signals constructive interference.

For destructive interference, the path length difference must be an integer plus one-half times the wavelength:

$$\Delta x = d\sin\theta = \left(m + \tfrac{1}{2}\right)\lambda \quad \left(\text{for } m = 0, \pm 1, \pm 2, \ldots\right) \quad \begin{pmatrix}\text{dark fringe,} \\ \text{destructive interference}\end{pmatrix}. \quad (34.7)$$

A dark fringe on the screen signals destructive interference.

Note that for constructive interference and $m = 0$, we obtain $\theta = 0°$, which means that $\Delta x = 0$ and there is a bright fringe at 0°. This bright fringe is called the *central maximum*. The integer m determines the **order** of the fringe. The order has a different meaning for bright fringes and for dark fringes. For example, using equation 34.6 with $m = 1$ gives the angle of the first-order bright fringe, $m = 2$ gives the second-order bright fringe, and so on. Using equation 34.7 with $m = 0$ gives the angle of the first-order dark fringe, $m = 1$ gives the second-order dark fringe, etc. For both bright and dark fringes, the first-order fringe is the one closest to the central maximum.

For the bright fringes close to the central maximum, the angle θ is small and can be approximated by $\sin\theta \approx \tan\theta = y/L$ (see Figure 34.9). Thus, equation 34.6 can be expressed as

$$d\sin\theta = d\frac{y}{L} = m\lambda \quad \left(\text{for } m = 0, \pm 1, \pm 2, \ldots\right),$$

or

$$y = \frac{m\lambda L}{d} \quad \left(\text{for } m = 0, \pm 1, \pm 2, \ldots\right), \quad (34.8)$$

which gives the distances along the screen of the bright fringes from the central maximum.

Similarly, the distances along the screen of the dark fringes from the central maximum can be expressed as

$$y = \frac{\left(m + \tfrac{1}{2}\right)\lambda L}{d} \quad \left(\text{for } m = 0, \pm 1, \pm 2, \ldots\right). \quad (34.9)$$

The positions of the centers of the bright and dark fringes are described by equations 34.8 and 34.9. But the intensity of the light at any point on the screen can also be calculated. We begin by assuming that the light emitted at each slit is in phase. The electric field of the light waves can be described by $\vec{E}_m = E_{\text{max}}\sin\omega t$, where E_{max} is the amplitude of the wave and ω is the angular frequency. When the light waves arrive at the screen from the two slits, they have traveled different distances, and so can have different phases. The electric field of the light arriving from S_1 and hitting a given point on the screen can be expressed as

$$\vec{E}_{m1} = E_{\text{max}}\sin\left(\omega t\right),$$

and the electric field of the light arriving at the same point from S_2 can be expressed as

$$\vec{E}_{m2} = E_{\text{max}}\sin\left(\omega t + \phi\right),$$

where ϕ is the phase constant of $\vec{E}_{m2}$ with respect to $\vec{E}_{m1}$. Figure 34.11a shows the two phasors $\vec{E}_{m1}$ and $\vec{E}_{m2}$.

The sum of the two phasors $\vec{E}_{m1}$ and $\vec{E}_{m2}$ is shown in Figure 34.11b, which also shows that the magnitude E of the sum of the two phasors is

$$E = 2E_{\text{max}}\cos\left(\phi/2\right).$$

Chapter 31 showed that the intensity of an electromagnetic wave is proportional to the square of the amplitude of the electric field; so we have

$$\frac{I}{I_{\text{max}}} = \frac{E^2}{E_{\text{max}}^2}.$$

Using the result for the magnitude of the electric field just obtained, we get

$$I = 4I_{max} \cos^2 (\phi/2)$$

for the intensity of the total wave at point P as a function of the phase difference between the two light waves.

Now the phase difference must be related to the path length difference. Figure 34.10 shows that the path length difference Δx causes a phase shift given by

$$\phi = \frac{\Delta x}{\lambda}(2\pi),$$

because when $\Delta x = \lambda$, the phase shift is $\phi = 2\pi$. Noting that $\Delta x = d \sin\theta$, the phase constant can be expressed as

$$\phi = \frac{2\pi d}{\lambda} \sin\theta.$$

Thus, we can write an equation for the intensity of the light produced by the interference from two slits:

$$I = 4I_{max} \cos^2 \left(\frac{\pi d}{\lambda} \sin\theta \right).$$

Finally, we can obtain an expression for the intensity pattern on a screen resulting from coherent light incident on two narrow widely separated slits using the approximation discussed before, which is valid near the central maximum and θ is small enough that $\sin\theta \approx \tan\theta = y/L$:

$$I = 4I_{max} \cos^2 \left(\frac{\pi d y}{\lambda L} \right). \tag{34.10}$$

For example, if the screen is $L = 2.0$ m away from the slits, the slits are separated by $d = 1.0 \cdot 10^{-5}$ m, and the wavelength of the incident light is $\lambda = 550$ nm, we get the intensity pattern shown in Figure 34.12. In this figure, the intensity varies from $4I_{max}$ to zero. Covering one slit produces an intensity of I_{max} at all values of y. Illuminating both slits with light that has random phases produces an intensity of $2I_{max}$ at all values of y. Only when both slits are illuminated with coherent light do we observe the oscillatory pattern with respect to y that is characteristic of two-slit interference.

Thin-Film Interference and Newton's Rings

Another way of producing interference phenomena is by using light that is partially reflected from the front and back surfaces of a thin film. A **thin film** is an optically clear material with a thickness on the order of a few wavelengths of light. Examples of thin films include the walls of soap bubbles and thin layers of oil floating on water. When the light reflected off the front surface interferes with the light reflected off the back surface of the thin film, we see the color corresponding to the particular wavelength of light that is interfering constructively.

Consider light traveling in an optical medium with index of refraction n_1 that strikes a second optical medium with index of refraction n_2, as illustrated in Figure 34.13.

One possibility is that the light can be transmitted through the boundary, as illustrated in Figure 34.13a and Figure 34.13b. In these cases, the phase of the light does not change, independent of whether $n_1 < n_2$ or $n_1 > n_2$. A second possibility is that the light is reflected. In this case, the phase of the light can change, depending on the index of refraction of the two optical media. If $n_1 < n_2$, the phase of the reflected wave is changed by

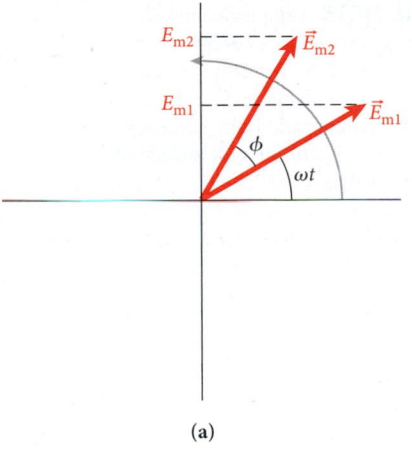

(a)

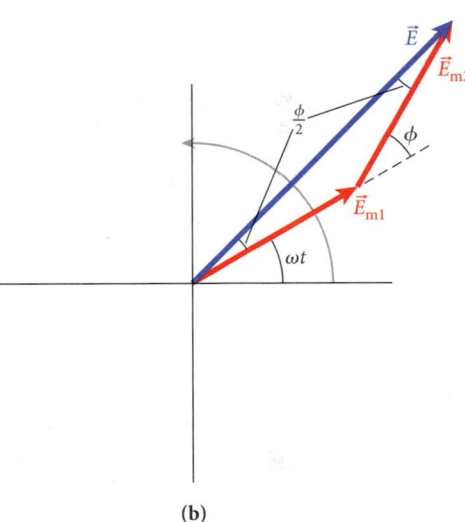

(b)

FIGURE 34.11 (a) Two phasors $\vec{E}_{m1}$ and $\vec{E}_{m2}$ separated by a phase ϕ. (b) Sum of the two phasors $\vec{E}_{m1}$ and $\vec{E}_{m2}$.

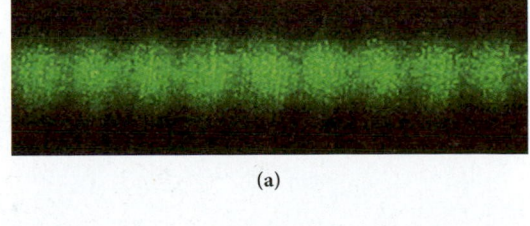

(a)

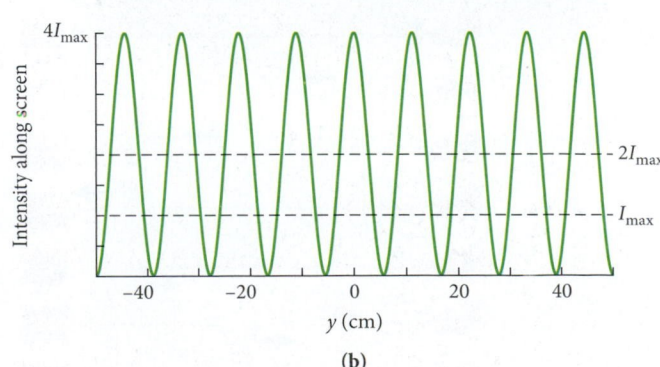

(b)

FIGURE 34.12 Intensity pattern for two-slit interference using light with wavelength 550 nm incident on two narrow slits separated by 10^{-5} m and at a distance of 2 m from the screen. (a) Photograph of light intensity on the screen. (b) Graph of the light intensity along the screen.

FIGURE 34.13 Light traveling in an optical medium with index of refraction n_1 is incident on a boundary with a second optical medium with index of refraction n_2. (a) Light is transmitted with no phase change for $n_1 < n_2$. (b) Light is transmitted with no phase change for $n_1 > n_2$. (c) Light is reflected with a 180° phase change for $n_1 < n_2$. (d) Light is reflected with no phase change for $n_1 > n_2$.

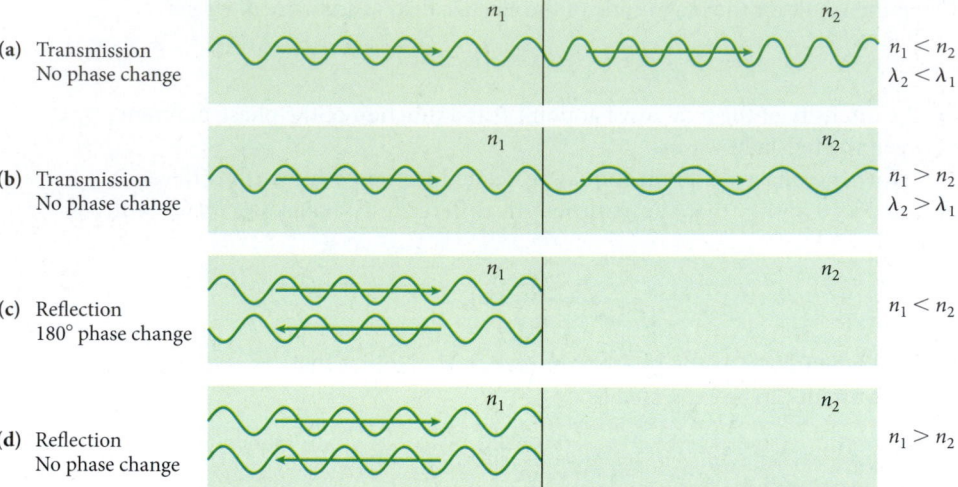

180° (corresponding to half a wavelength), as shown in Figure 34.13c. If $n_2 > n_1$, then no phase change occurs, as depicted in Figure 34.13d. (The reflected waves have been displaced vertically for clarity's sake!)

The reason for the phase change upon reflection follows from the theory of electromagnetic waves, and the derivation is beyond the scope of this book. However, in Section 15.4, we studied the reflection of waves on a string from a boundary, and this mechanical analog is sufficient to understand the basic reason for the 180° phase change in one case and no phase change in the other. Figure 34.14 shows a rigid (left) and a flexible (right) connection of a string to a wall. The reflection of light at the boundary between the optically less dense (lower value of n) medium and the optically more dense (higher value of n) medium corresponds to the situation on the left, and the reverse case to the situation on the right.

Let's begin our analysis of thin films by studying a thin film of thickness t with index of refraction n and with air on both sides (Figure 34.15). Assume that monochromatic light is incident perpendicular to the surface of the film. For clarity and simplicity, we show only the rays we are interested in, with a small nonzero angle of incidence so they can be distinguished. When the light wave reaches the boundary between air and the film, part of the wave is reflected and part of the wave is transmitted. The reflected wave undergoes a phase shift of half a wavelength when it is reflected because $n_{air} < n$. The light that is transmitted has no phase shift and continues to the back surface of the film. At the back surface, again part of the wave is transmitted and part of the wave is reflected. The transmitted light passes through the film completely. The reflected light has no phase shift because $n > n_{air}$, and it travels back to the front surface of the film. At the front surface, some of the light reflected from the back

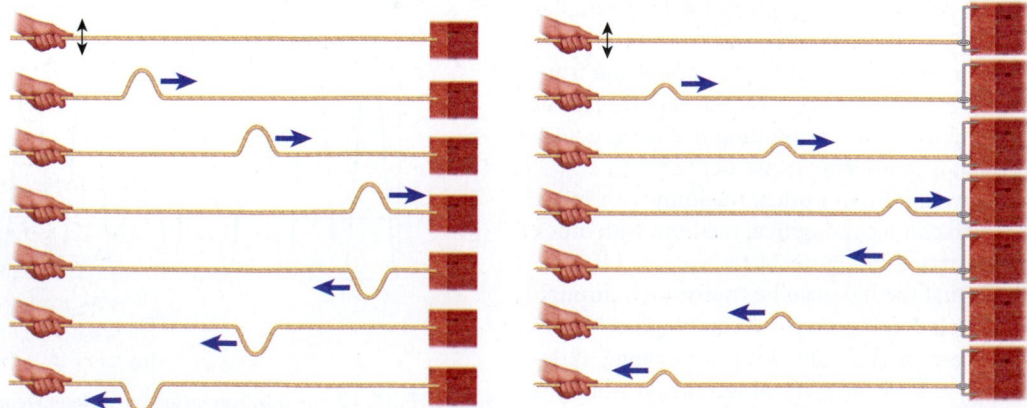

FIGURE 34.14 Sketch of the time sequence (top to bottom) of the reflection of a wave pulse on a string at a wall. Left: Rigid connection to the wall results in a 180° phase change in the reflected wave. Right: A movable connection to the wall results in no phase change.

surface is transmitted and some is reflected. The transmitted light has no phase shift, and it emerges from the film and interferes with the light that was reflected when the original wave first struck the film. The transmitted and then reflected light has traveled a longer distance than the originally reflected light and has a phase shift determined by the path length difference. This difference is twice the thickness t of the film.

The fact that the originally reflected light has undergone a phase shift of half a wavelength but the transmitted and then reflected light has not means that the criterion for *constructive* interference is given by $\Delta x = (m + \frac{1}{2})\lambda = 2t$ (for $m = 0,1,2,...$). The wavelength λ refers to the wavelength of the light traveling in the thin film, which has index of refraction n. The wavelength of the light traveling in air is related to the wavelength of the light traveling in the film by $\lambda = \lambda_{air}/n$. We can then write

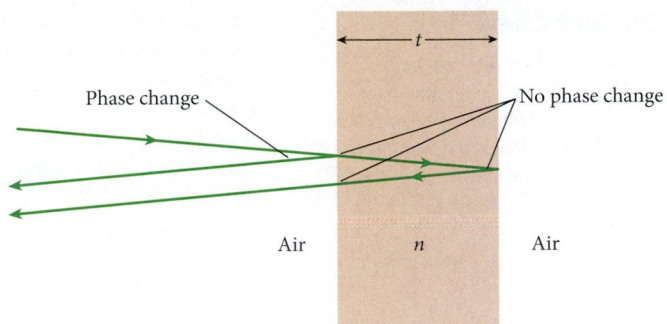

FIGURE 34.15 Light waves in air incident on a thin film with index of refraction n and thickness t.

$$\left(m + \tfrac{1}{2}\right)\frac{\lambda_{air}}{n} = 2t \quad \left(\text{for } m = 0,1,2,...\right) \begin{pmatrix} \text{thin film,} \\ \text{constructive interference} \end{pmatrix}. \qquad (34.11)$$

The minimum thickness, t_{min}, that will produce constructive interference corresponds to $m = 0$:

$$t_{min} = \frac{\lambda_{air}}{4n}. \qquad (34.12)$$

An oil slick or the wall of a soap bubble has varying thicknesses and thus affects different wavelengths differently, often creating a rainbow effect.

EXAMPLE 34.1   Lens Coating

Many high-quality lenses are coated to prevent reflections. The coating is designed to set up destructive interference for light reflected from the surface of the lens. Assume that the coating is magnesium fluoride, which has $n_{coating} = 1.38$, and that the lens is glass with $n_{lens} = 1.51$.

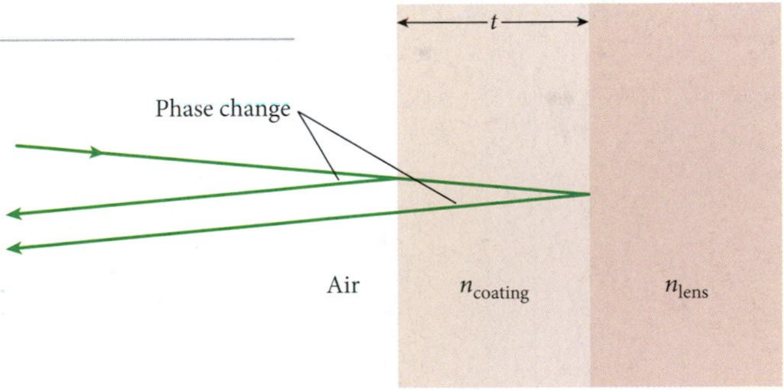

PROBLEM

What is the minimum thickness of the coating that will produce destructive interference for light with a wavelength in air of 550 nm?

FIGURE 34.16 Light incident on a coated lens.

SOLUTION

Assume that the light is incident perpendicularly (or at least close to perpendicularly) on the surface of the coated lens (Figure 34.16). Light reflected at the surface of the coating undergoes a phase change of half a wavelength, because $n_{air} < n_{coating}$. The light transmitted through the coating has no phase change. Light reflected at the boundary between the coating and the lens also undergoes a phase change of half a wavelength, because $n_{coating} < n_{lens}$. This reflected light travels back through the coating and exits with no other phase change. Thus, both the light reflected from the coating and the light reflected from the lens have undergone a phase change of half a wavelength. Therefore, the criterion for destructive interference is

$$\left(m + \tfrac{1}{2}\right)\frac{\lambda_{air}}{n_{coating}} = 2t \quad \left(\text{for } m = 0,1,2,...\right).$$

The minimum thickness of the coating that will produce destructive interference corresponds to $m = 0$:

$$t_{min} = \frac{\lambda_{air}}{4n_{coating}} = \frac{550 \cdot 10^{-9} \text{ m}}{4(1.38)} = 9.96 \cdot 10^{-8} \text{ m} = 99.6 \text{ nm}.$$

– Continued

Concept Check 34.4

If the same coating material of the same thickness as in Example 34.1 were used on a different lens with an index of refraction less than that of the coating, the result would be

a) reduced reflection and thus *more* light entering the lens than would occur without the coating.

b) enhanced reflection and thus *less* light entering the lens than would occur without the coating.

c) the same amount of light entering the lens as would do so without the coating.

d) no light at all (or close to none) entering the lens.

FIGURE 34.17 Newton's rings produced with white light.

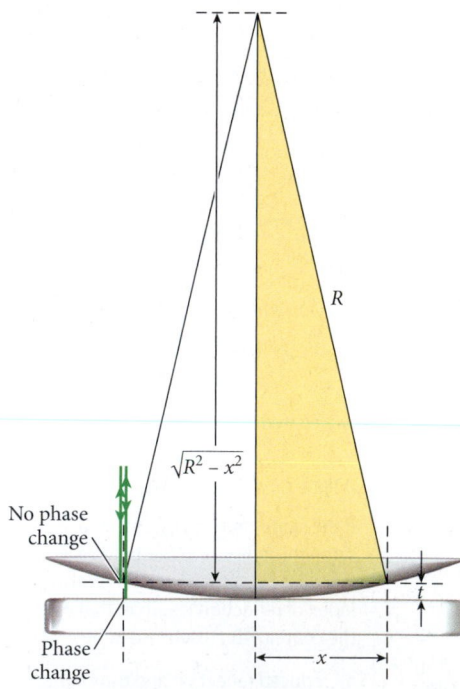

FIGURE 34.18 The geometry of Newton's rings. The bottom of the top piece of glass is a spherical surface, while the bottom plate is planar.

Lens coatings that fulfill this destructive interference condition are referred to as *quarter-lambda coatings*. Because this type of coating causes destructive interference of reflected light, it prevents light from reflecting itself and thus allows more light to enter the lens than would do so if it were not present. Expensive camera lenses have quarter-lambda coatings for wavelengths in the middle of the optical wavelength band, around 500 nm. But remember that a quarter-lambda coating can work only if the index of refraction of the coating is less than that of the lens.

A phenomenon similar to thin-film interference is **Newton's rings.** This effect is an interference pattern caused by the reflection of light between two glass surfaces, a spherical surface and an adjacent flat surface, as shown in Figure 34.17.

Newton's rings are caused by interference between light reflected from the bottom of the spherically curved glass and light reflected from the top of the flat plate of glass, as shown in Figure 34.18. (Note that the curvature of the spherical surface is exaggerated in this diagram. In real situations, R is much larger than x.)

Newton's rings are concentric circles, alternately dark and bright. The dark circles correspond to destructive interference, and the bright circles correspond to constructive interference. The example shown in Figure 34.17 was produced with white light to show different colored rings. For the calculation of Newton's rings, we assume monochromatic, coherent light with wavelength λ in air. The distance between the curved surface and the flat surface, t, is a function of the horizontal distance from the point where the spherical surface touches the flat surface, as shown in Figure 34.18. The Pythagorean Theorem applied to the yellow triangle in Figure 34.18 gives the relationship

$$t = R - \sqrt{R^2 - x^2} = R - R\sqrt{1 - \left(\frac{x}{R}\right)^2},$$

where R is the radius of curvature of the spherical glass surface and x is the horizontal distance from the point where the curved glass touches the flat glass to the point where the light enters the glass.

As mentioned above, in Figure 34.18, the curvature of the spherical surface has been exaggerated for clarity. The actual glass surface used to produce Newton's rings has a large radius of curvature compared with the distance from the touching point, so that $x/R \ll 1$. Thus, we can approximate

$$\sqrt{1 - \left(\frac{x}{R}\right)^2} \approx 1 - \frac{1}{2}\left(\frac{x}{R}\right)^2 \quad \left(\text{for } \frac{x}{R} \ll 1\right).$$

Now we can write an approximate expression for the distance between the surfaces:

$$t \approx R - R\left(1 - \frac{1}{2}\left(\frac{x}{R}\right)^2\right) = \frac{1}{2}\frac{x^2}{R}.$$

The path length difference for the light reflecting off the bottom of the curved surface versus the light reflecting off the top of the flat surface is $2t$, which can be written

$$2t = 2\left(\frac{1}{2}\frac{x^2}{R}\right) = \frac{x^2}{R}.$$

The light reflecting off the bottom of the curved surface has no phase change because the index of refraction of glass is higher than the index of refraction of air. The light reflecting off the top of the flat glass plate suffers a phase change of half a wavelength because the index of refraction of glass is larger than the index of refraction of air. Thus, the criterion for constructive interference is

$$\left(m + \tfrac{1}{2}\right)\lambda = 2t \quad (\text{for } m = 0,1,2,\ldots).$$

Combining this constructive interference criterion with the result for the thickness, and defining x_m as the radius of the bright circle for a given m, gives

$$\left(m+\tfrac{1}{2}\right)\lambda = \frac{x_m^2}{R} \quad \left(\text{for } m=0,1,2,...\right),$$

which can be solved for the radius of the bright circles:

$$x_m = \sqrt{R\left(m+\tfrac{1}{2}\right)\lambda} \quad \left(\text{for } m=0,1,2,..., \text{ and } \frac{x_m}{R}\ll 1\right). \tag{34.13}$$

SOLVED PROBLEM 34.1 Air Wedge

Light of wavelength λ = 516 nm is incident perpendicularly on two glass plates. The glass plates are spaced at one end by a thin piece of kapton film. Due to the wedge of air created by this film, 29 bright interference fringes are observed across the top plate, with a dark fringe at the end by the film.

PROBLEM
How thick is the film?

SOLUTION

THINK Light passes through the top plate, reflects from the top surface of the bottom plate, and then interferes with light reflected from the bottom surface of the top plate. A phase change occurs when the light is reflected from the bottom plate, so the criterion for constructive interference is that the path length is equal to an integer plus one-half times the wavelength. The criterion for destructive interference is that the path difference is an integer times the wavelength.

SKETCH Figure 34.19 is a sketch showing the two glass plates with a thin piece of film separating them at one end. Light is incident from the top.

RESEARCH At any point along the plates, the criterion for constructive interference is given by

$$2t = \left(m+\tfrac{1}{2}\right)\lambda,$$

where t is the separation between the plates, m is an integer, and λ is the wavelength of the incident light. There are 29 bright fringes visible. The first bright fringe corresponds to $m = 0$, and the 29th bright fringe corresponds to $m = 28$. Immediately past the 29th bright fringe is a dark fringe where the piece of film is located. The criterion for destructive interference is $2t = n\lambda$, where n is an integer.

SIMPLIFY The dark fringe located at the end of the glass plate, where the film is located, is given by

$$2t = \left(28+\tfrac{1}{2}+\tfrac{1}{2}\right)\lambda = 29\lambda,$$

where the factor $28 + \tfrac{1}{2}$ produces the constructive interference and the extra $\tfrac{1}{2}$ produces the dark fringe. Thus, we can solve for the separation of the plates, which corresponds to the thickness of the film:

$$t = \frac{29}{2}\lambda.$$

CALCULATE Putting in the numerical values gives

$$t = \frac{29}{2}\left(516\cdot10^{-9}\text{ m}\right) = 0.00000748\text{ m}.$$

ROUND We report our result to three significant figures:

$$t = 7.48\cdot10^{-6}\text{ m}.$$

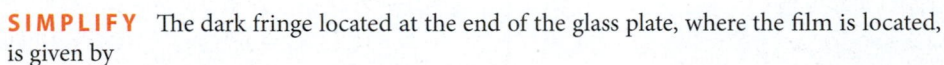

FIGURE 34.19 Two glass plates with a thin film separating the plates at one end. The green arrows represent the incident light from the top. The angle of the light is exaggerated for clarity. The interfering light is reflected off the bottom of the top plate and the top of the bottom plate.

– Continued

DOUBLE-CHECK Occasionally throughout this book we need to add reminders that checking our results just for the right units and expected order of magnitude can do a lot to prevent simple errors. The meter is certainly the right unit for a physical length, in this case, the film thickness. At first glance, you may think that $\sim 10^{-5}$ m, on the order of 1/100th of the thickness of a fingernail, may be impossibly thin for a solid film. However, an Internet search on *kapton film* will show that 7.5 μm is within the range of thicknesses in which kapton film is produced. Thus, our answer seems plausible.

Interferometer

An **interferometer** is a device designed to measure lengths or small changes in length using the fringes produced by the interference of light. An interferometer can measure changes in lengths to an accuracy of a fraction of the wavelength of the light it is using. The interferometer described here is similar to, but simpler than, the one constructed by Albert Michelson at the Case Institute in Cleveland, Ohio, in 1887.

A photograph and a diagram of a commercial interferometer used in physics labs are shown in Figure 34.20. A more detailed drawing of this same interferometer is shown in Figure 34.21. This particular interferometer uses a laser light source that emits coherent light with wavelength $\lambda = 632.8$ nm. The light passes through a defocusing lens that spreads out the very narrowly focused laser beam. The light then passes through a half-silvered mirror, m_1. Part of the light is reflected toward the adjustable mirror, m_3, and part of the light is transmitted to the movable mirror, m_2. The distance between m_1 and m_2 is x_2, and the distance between m_1 and m_3 is x_3. The transmitted light is totally reflected from m_2 back toward m_1. The reflected light is also totally reflected from m_3 back toward m_1. Part of the light reflected from m_2 is then reflected by m_1 toward the viewing screen; the rest of the light is transmitted and not considered in this analysis. Part of the light from m_3 is transmitted through m_1; the rest is reflected and not considered. The alignment of the light is such that the two beams that strike the viewing screen are collinear; the separation shown in Figure 34.21 is simply for clarity.

The light beams from mirrors m_2 and m_3 that strike the viewing screen interfere, based on their path length differ-

(a)

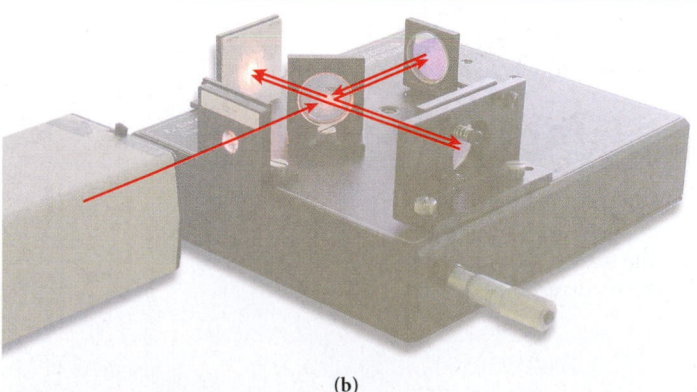

(b)

FIGURE 34.20 A Michelson-type interferometer used in an introductory physics laboratory. (a) Photograph; (b) schematic drawing of the light paths. The laser beam is split by the central half-silvered mirror. One part of the light continues in the same direction, is reflected, returns, and is reflected by the half-silvered mirror to the screen. A second part of the light is reflected by the half-silvered mirror, and is then reflected back through the half-silvered mirror to the screen.

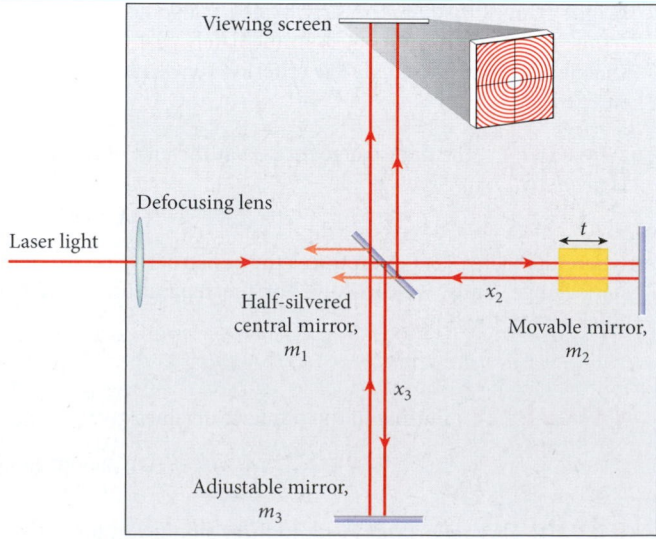

FIGURE 34.21 Top view of a Michelson-type interferometer, showing the detailed light path.

ence. Both beams undergo two reflections, each resulting in a phase change of 180°, so the condition for constructive interference is

$$\Delta x = m\lambda \quad \left(\text{for } m = 0, \pm 1, \pm 2, \ldots \right).$$

The two different paths have a path length difference of

$$\Delta x = 2x_2 - 2x_3 = 2\left(x_2 - x_3\right).$$

The viewing screen displays concentric circles or linear fringes corresponding to constructive and destructive interference, depending on the type of defocusing lens used and the tilt of the mirrors. If the movable mirror, m_2, is moved a distance $\lambda/2$, the fringes shift by one fringe. Thus, this type of interferometer can be used to measure changes in distance on the order of a fraction of a wavelength of light, depending on how well the shift of the interference fringes can be measured.

Another type of measurement can be made with this interferometer by placing a material with index of refraction n and thickness t in the path of the light traveling to the movable mirror m_2, as depicted in Figure 34.21. The path length difference in terms of the number of wavelengths will change because the wavelength of light in the material, λ_n, is different from the wavelength in air, λ. The wavelength in the material is related to the wavelength in air by

$$\lambda_n = \frac{\lambda}{n}.$$

Thus, the number of wavelengths in the material is

$$N_{\text{material}} = \frac{2t}{\lambda_n} = \frac{2tn}{\lambda}.$$

The number of wavelengths that would have been there if the light traveled through only air is

$$N_{\text{air}} = \frac{2t}{\lambda}.$$

Thus, the difference in the number of wavelengths is

$$N_{\text{material}} - N_{\text{air}} = \frac{2tn}{\lambda} - \frac{2t}{\lambda} = \frac{2t}{\lambda}\left(n - 1\right). \tag{34.14}$$

When the material is placed between m_1 and m_2, an observer will see a shift of one fringe for every wavelength shift in the path length difference. Thus, we can substitute the number of fringes shifted for $N_{\text{material}} - N_{\text{air}}$ in equation 34.14, and, given the index of refraction, we can obtain the thickness of the material. Alternatively, we can insert material with a well-known thickness and determine the index of refraction. This can be done by rotating the material slowly and counting the fringes.

Self-Test Opportunity 34.2

The wavelength of a monochromatic light source is measured using a Michelson interferometer. When the movable mirror is moved a distance $d = 0.250$ mm, $N = 1200$ fringes move by the screen. What is the wavelength of the light?

34.3 Diffraction

Any wave passing through an opening experiences diffraction. **Diffraction** means that the wave spreads out on the other side of the opening rather than the opening casting a sharp shadow. Diffraction is most noticeable when the opening is about the same size as the wavelength of the wave. With water surface waves, this is easily visualized: Figure 34.22 shows a satellite picture of a Caribbean resort that has a seawall to protect the beach from erosion; the diffraction of the ocean waves through the opening in the seawall is evident. The same effect applies to light waves. If light passes through a narrow slit, it produces a characteristic pattern of light and dark areas called a *diffraction pattern*. Light passing a sharp edge also exhibits a diffraction pattern.

Huygens's Principle describes this spreading out, and a Huygens construction can be used to quantify the diffraction phenomenon. For example, Figure 34.23a shows coherent light incident on an opening, which has a width comparable to the wavelength of the light. Rather than continuing in a narrow beam, the light spreads out on the other side of the opening. We can describe this spreading out by using a Huygens construction and assuming

FIGURE 34.22 Satellite image of diffraction of ocean waves passing through an opening in a seawall at a Caribbean resort.

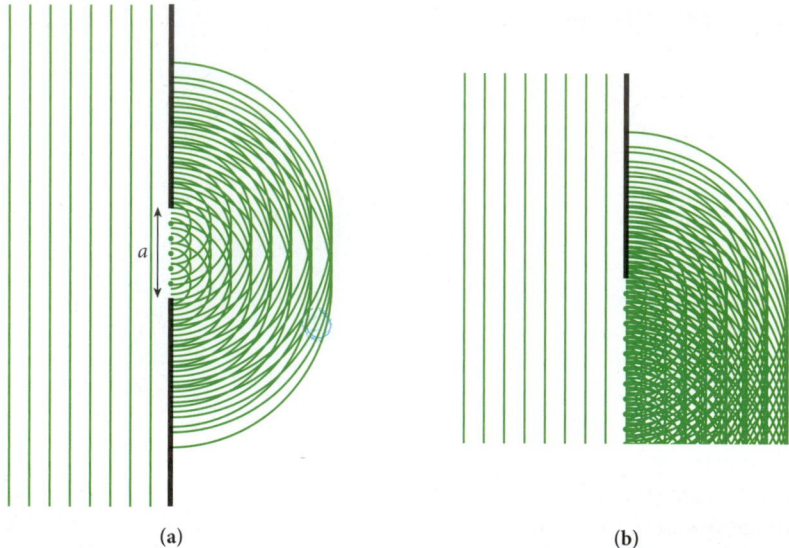

FIGURE 34.23 (a) Coherent light incident on an opening with a width comparable to the wavelength of the light. The dots represent sources of spherical wavelets in a Huygens construction. (b) Coherent light incident on a barrier. The light diffracts around the barrier.

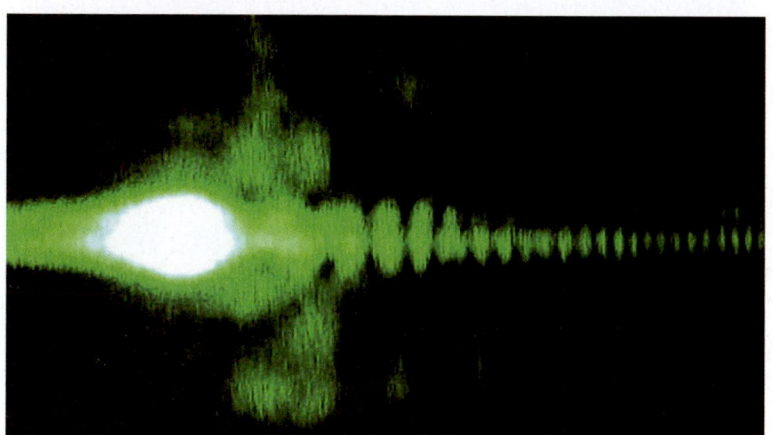

FIGURE 34.24 Light from a green laser incident on the vertical edge of a razor blade as viewed on a distant screen.

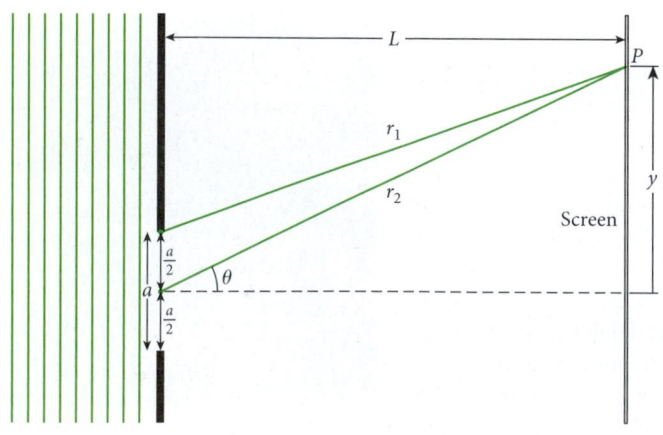

FIGURE 34.25 Geometry for determining the location of the first dark fringe from a single slit, using two rays from the slit.

that spherical wavelets are emitted at several points inside the opening. The resulting light waves on the right side of the opening undergo interference and produce a characteristic diffraction pattern.

Light waves can also go around the edges of barriers, as shown in Figure 34.23b. In this case, the light far from the edge of the barrier continues to travel like the light waves shown in Figure 34.2. The light near the edge of the barrier seems to bend around the barrier, in a way that can be represented using point sources of wavelets near the edge.

Figure 34.24 shows a picture of an experiment, where light from a green laser is blocked by the vertical edge of a razor blade, and the resulting diffraction pattern is viewed on a distant screen. The bright spot is the central maximum and the less bright lines are higher-order diffraction maxima.

Diffraction phenomena cannot be explained by geometric optics. As we will see, diffraction effects rather than geometric effects often limit the resolution of an optical instrument. The following subsections will present a quantitative examination of single-slit diffraction, diffraction by a circular opening, and double-slit diffraction; Section 34.4 will consider diffraction by a grating. In all cases, similar relationships for the location of the diffraction maxima and minima result, modified only by the individual geometries.

Single-Slit Diffraction

Consider the diffraction of light through a single slit of width a, which is comparable to the wavelength of light passing through the slit, as illustrated in Figure 34.7b. The calculation is aided by a Huygens construction, as shown in Figure 34.23a.

Assume that the light passing through the single slit is described by spherical wavelets emitted from a distribution of points located in the slit. The light emitted from these points will superpose and interfere, based on the path length for each wavelet at each position. At a distant screen, we observe an intensity pattern characteristic of diffraction, consisting of bright and dark fringes. For the case of two-slit interference, we were able to work out the equations for the bright fringes based on constructive interference. For diffraction, we will restrict ourselves to analyzing the dark fringes of destructive interference.

To study this interference, we redraw and simplify Figure 34.23a as shown in Figure 34.25. Assume that coherent light with wavelength λ is incident on a slit with width a, producing an interference pattern on a screen a distance L away. We employ a simple method of analyzing pairs of light waves emitted from points in the slit. We start with light emitted from the top edge of the slit and from the center of the slit, as shown in Figure 34.25. To analyze the path difference, we show an expanded version of Figure 34.25 in Figure 34.26.

For this quantitative analysis, we assume that the screen distance L is very large compared to the size of the slit opening, a.

Then, the two green lines representing paths of lengths r_1 and r_2 in Figure 34.26 are parallel and both make an angle θ with the line running perpendicularly from the center of the slit to the screen. Therefore, the path length difference for these two rays is given by

$$\sin\theta = \frac{\Delta x}{a/2} \Leftrightarrow \Delta x = \frac{1}{2}a\sin\theta.$$

The criterion for the first dark fringe is

$$\Delta x = \frac{a\sin\theta}{2} = \frac{\lambda}{2} \Rightarrow a\sin\theta = \lambda.$$

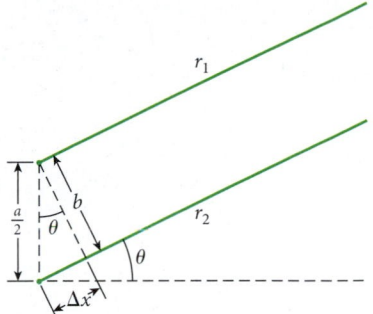

FIGURE 34.26 Expanded version of the geometry for determining the location of the first dark fringe from diffraction by a single slit.

Although we chose one ray originating from the top edge of the slit and one from the middle of the slit to locate the first dark fringe, we could have used any two rays that originated at a distance $a/2$ apart inside the slit. That is, all wavelets can be paired up so that the separation within a pair is $a/2$. (Thus, we take into account the entire slit.) Each pair destructively interferes, so an overall dark fringe appears on the screen.

Now consider four rays instead of two (Figure 34.27). Here we choose a ray from the top edge of the slit and three more rays originating from points spaced $a/4$ apart. The geometry in Figure 34.27 can be expanded to represent the case with the screen far away, as shown in Figure 34.28. Clearly, the path length difference between the pairs of rays labeled r_1 and r_2, r_2 and r_3, and r_3 and r_4 is given by

$$\sin\theta = \frac{\Delta x}{a/4} \Leftrightarrow \Delta x = \frac{1}{4}a\sin\theta.$$

The criterion for a dark fringe arising from these three pairs of rays is

$$\frac{a\sin\theta}{4} = \frac{\lambda}{2} \Rightarrow a\sin\theta = 2\lambda.$$

All wavelets could be grouped into four rays similar to those shown in Figure 34.28. Each group destructively interferes, so an overall dark fringe is visible on the screen. Thus, the condition $a\sin\theta = 2\lambda$ describes the second dark fringe.

At this point, we can easily generalize and take groups of six and eight rays and describe the third and fourth dark fringes, and so forth. The result is that the dark fringes from single-slit diffraction can be described by

$$a\sin\theta = m\lambda \quad (\text{for } m = \pm 1, \pm 2, \pm 3, \dots). \tag{34.15}$$

If the screen is placed a sufficiently large distance from the slits, the angle θ is small and can be approximated by $\sin\theta \approx \tan\theta = y/L$. Thus, the position of the dark fringes on the screen can be expressed as

$$\frac{ay}{L} = m\lambda \quad (\text{for } m = \pm 1, \pm 2, \pm 3, \dots),$$

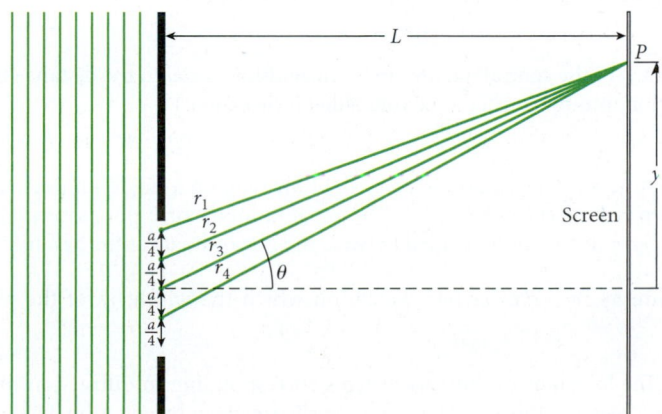

FIGURE 34.27 Geometry for determining the location of the second dark fringe from a single slit, using four rays from the slit.

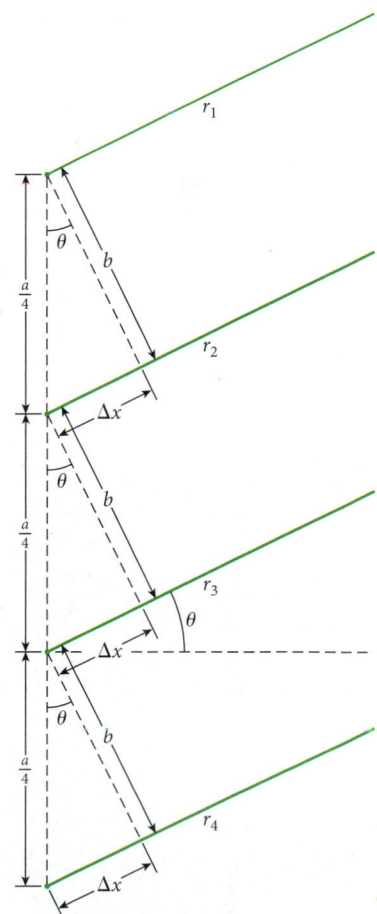

FIGURE 34.28 Expanded version of the geometry for determining the location of the second dark fringe from diffraction by a single slit.

or

$$y = \frac{m\lambda L}{a} \quad (\text{for } m = \pm 1, \pm 2, \pm 3, ...) \quad \begin{pmatrix} \text{positions of} \\ \text{dark fringes} \end{pmatrix}. \tag{34.16}$$

Detailed analysis of the spherical wavelets shows that the intensity I relative to I_0, the intensity at $\theta = 0°$, is

$$I = I_0 \left(\frac{\sin \alpha}{\alpha} \right)^2, \tag{34.17}$$

where

$$\alpha = \frac{\pi a}{\lambda} \sin \theta. \tag{34.18}$$

Equation 34.17 shows that this expression for the intensity I is zero for $\sin \alpha = 0$ (unless $\alpha = 0°$), which means that $\alpha = m\pi$ for $m = 1,2,3,...$. For the special case of $\alpha = 0°$ corresponding to $\theta = 0°$, we use

$$\lim_{\alpha \to 0} \left(\frac{\sin \alpha}{\alpha} \right) = 1.$$

Therefore,

$$m\pi = \frac{\pi a}{\lambda} \sin \theta \quad (\text{for } m = \pm 1, \pm 2, \pm 3, ...),$$

or

$$a \sin \theta = m\lambda \quad (\text{for } m = \pm 1, \pm 2, \pm 3, ...),$$

which gives the same result for the diffraction minima as equation 34.15.

For dark fringes close to the central maximum, the angle θ is small and can be approximated by $\sin \theta \approx \tan \theta = y/L$. Thus, equation 34.18 can be expressed as

$$\alpha = \frac{\pi a y}{\lambda L}. \tag{34.19}$$

For example, if the screen is $L = 2.0$ m away from the slit, the slit has a width of $a = 5.0 \cdot 10^{-6}$ m, and the wavelength of the incident light is $\lambda = 550$ nm, we get the intensity pattern shown in Figure 34.29. An intensity distribution like that shown in Figure 34.29 is called a *Fraunhofer diffraction pattern*.

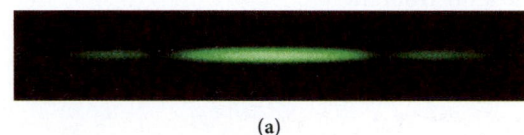

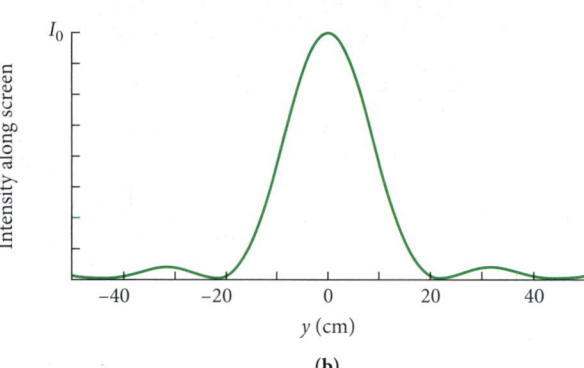

FIGURE 34.29 Intensity pattern for diffraction through a single slit. (a) Photograph of the light on the screen. (b) Graph of the intensity along the screen.

SOLVED PROBLEM 34.2 | Width of Central Maximum

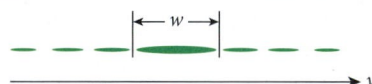

FIGURE 34.30 Single-slit diffraction pattern observed on a screen.

The single-slit diffraction pattern shown in Figure 34.30 was produced with light of wavelength $\lambda = 510.0$ nm. The screen on which the pattern was projected was located a distance $L = 1.40$ m from the slit. The slit had a width of $a = 0.100$ mm.

PROBLEM

What is the width w of the central maximum? (The width is equal to the distance between the two first diffraction minima located on either side of the center.)

SOLUTION

THINK The diffraction pattern produced by light incident on a single slit peaks at $y = 0$ and falls to a minimum on both sides of the peak. The first minimum corresponds to $m = 1$. Therefore, the width of the central maximum is equal to twice the y-coordinate of the first minimum.

SKETCH Figure 34.30 serves as our sketch, on which the width, w, of the maximum is indicated.

RESEARCH The locations of the dark fringes corresponding to diffraction minima on a screen when the screen is sufficiently far from the slit are given by equation 34.16:

$$y = \frac{m\lambda L}{a} \quad (\text{for } m = \pm 1, \pm 2, \pm 3, ...),$$

where λ is the wavelength of the light incident on the single slit, m is the order of the minimum, L is the distance from the slit to the screen, and a is the width of the slit. For single-slit diffraction, the position $y = 0$ corresponds to the position of the central maximum. The first diffraction minimum corresponds to $m = 1$. Therefore, the distance Δy from the central minimum to the position of the first minimum is given by

$$\Delta y = \frac{\lambda L}{a}.$$

SIMPLIFY The width of the central maximum, w, is then

$$w = 2\Delta y = \frac{2\lambda L}{a}.$$

CALCULATE Putting in the numerical values gives us

$$w = \frac{2\left(510.0\cdot 10^{-9}\text{ m}\right)\left(1.40\text{ m}\right)}{\left(1.00\cdot 10^{-4}\text{ m}\right)} = 0.01428\text{ m}.$$

ROUND We report our result to three significant figures:

$$w = 0.01428\text{ m} = 14.3\text{ cm}.$$

DOUBLE-CHECK The width of the central maximum projected on the screen is relatively small. A slit that is large compared with the wavelength of light shows little diffraction, while a slit that has a width comparable to or smaller than the wavelength of light produces a broad diffraction pattern. The ratio of the wavelength of the light to the slit width is

$$\frac{\lambda}{a} = 510.0\text{ nm}/1.00\cdot 10^{-4}\text{ m} = 0.00510.$$

Thus, our finding of a narrow central maximum appears reasonable.

Diffraction by a Circular Opening

So far, we have considered interference through two slits and diffraction through a single slit. Now we consider diffraction of light through a circular opening. Diffraction through a circular opening applies to observing objects with telescopes having circular mirrors/lenses or with cameras having circular lenses. The **resolution** of a telescope or a camera (that is, its ability to distinguish two point objects as being separate) is limited by diffraction.

The first diffraction minimum for light with wavelength λ passing through a circular opening with diameter d is given by

$$\sin\theta = 1.22\frac{\lambda}{d}, \tag{34.20}$$

where θ is the angle from the perpendicular line through the center of the opening to the first diffraction minimum. This result is similar to the result from a single slit except for the factor 1.22. We will not derive this expression here. Figure 34.31 shows a diffraction pattern for red laser light with wavelength $\lambda = 633$ nm passing through a circular opening with diameter 0.040 mm and projected on a screen located 1.00 m from the opening. The diameter of the innermost dark circle is measured to be 3.9 cm. The first diffraction minimum is clearly visible in Figure 34.31.

Figure 34.32 depicts three different situations for the observation of two distant point objects using a lens. In Figure 34.32a, the angular separation is clearly large enough to resolve the objects. In Figure 34.32b, the angular separation is large enough to just barely resolve the objects. In Figure 34.32c, the angular separation is too small to resolve the objects.

If a circular lens is used to observe two distant point objects, whose angular separation is small (such as two stars), diffraction limits the ability of the lens to distinguish these two objects. The criterion for being able to separate two point objects is based on the idea that if the central maximum of the first object is located at the first diffraction minimum of the

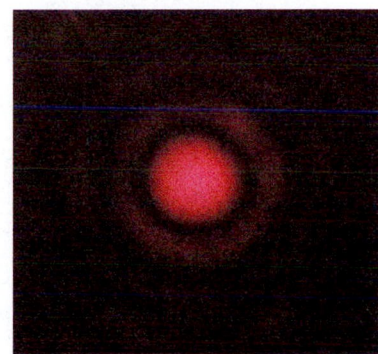

FIGURE 34.31 The diffraction pattern for red laser light with $\lambda = 633$ nm passing through a circular opening with diameter 0.040 mm and projected on a distant screen.

FIGURE 34.32 Diffraction through a circular opening: representing the image you might see with a lens observing two distant point objects. The top row presents a two-dimensional representation of each intensity pattern, and the bottom row shows a three-dimensional representation. (a) The angular separation is large enough to clearly resolve the two objects; (b) the angular separation is marginally large enough to resolve the two objects; (c) the angular separation of the two objects is too small to allow them to be resolved.

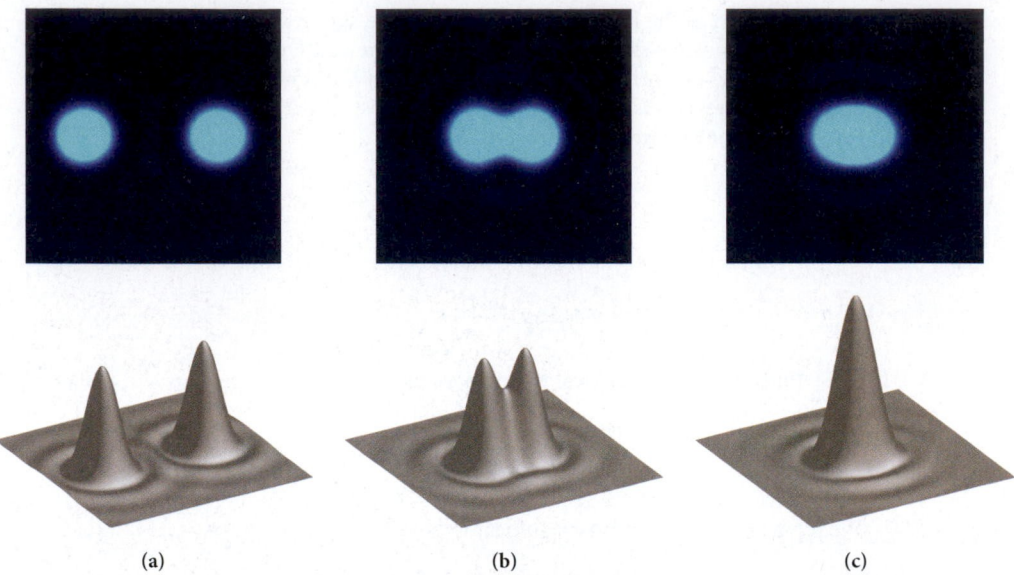

(a)   (b)   (c)

second object, the objects are just resolved. This criterion, called **Rayleigh's Criterion,** is expressed as

$$\theta_R = \sin^{-1}\left(\frac{1.22\lambda}{d}\right), \qquad (34.21)$$

where θ_R is the minimum resolvable angular separation in radians, λ is the wavelength of the light used to observe the objects, and d is the diameter of the lens or mirror.

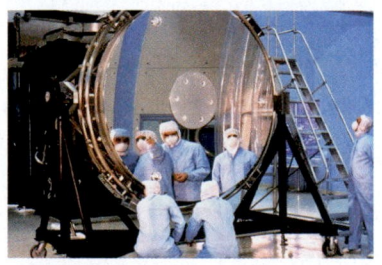

FIGURE 34.33 The main mirror of the Hubble Space Telescope has a diameter of 2.4 m. The hyperbolic shape of the mirror is precise to 32 nm, which means that if the mirror were the size of Earth, any bumps in the glass would be 17 cm high or less.

EXAMPLE 34.2 | Rayleigh's Criterion for the Hubble Space Telescope

PROBLEM

The diameter of the main mirror in the Hubble Space Telescope (Figure 34.33) is 2.4 m. What is the minimum angular resolution of the Hubble Space Telescope for green light?

SOLUTION

Using Rayleigh's Criterion, with green light of wavelength $\lambda = 550$ nm, we get

$$\theta_R = 1.22\frac{550 \cdot 10^{-9}\,\text{m}}{2.4\,\text{m}} = 2.8 \cdot 10^{-7}\,\text{rad} = 1.6 \cdot 10^{-5}\,\text{degrees},$$

which corresponds to the angle subtended by a dime (diameter of 17.9 mm) located 64 km away.

Self-Test Opportunity 34.3

Can individual watchtowers on the Great Wall of China be seen by astronauts on the Space Shuttle orbiting the Earth at an altitude of 190 km? Assume that the Great Wall is 10.0 m wide.

Concept Check 34.8

You are driving your car and listening to music on the radio. Your car is equipped with AM radio ($f \approx 1$ MHz), FM radio ($f \approx 100$ MHz), and XM satellite radio ($f = 2.3$ GHz). You enter a tunnel with a circular opening of diameter 10 m. Which kind of radio signal will you be able to receive the longest as you continue to move into the tunnel?

a) AM   b) FM   c) XM

SOLVED PROBLEM 34.3 | Spy Satellite

You have been assigned the job of designing a camera lens for a spy satellite. This satellite will orbit the Earth at an altitude of 201 km. The camera is sensitive to light with a wavelength of 607 nm. The camera must be able to resolve objects on the ground that are 0.490 m apart.

PROBLEM
What is the minimum diameter of the lens?

SOLUTION
THINK The resolution of this camera lens will be limited by diffraction. We can apply Rayleigh's Criterion to calculate the minimum diameter of the lens, given the angle subtended by two objects on the ground as viewed from the spy satellite orbiting above.

SKETCH Figure 34.34 shows a sketch of the spy satellite observing two objects on the ground.

RESEARCH The Rayleigh Criterion for resolving two objects separated by an angle θ_R using light with wavelength λ is given by

$$\theta_R = \sin^{-1}\left(\frac{1.22\lambda}{d}\right),$$

where d is the diameter of the circular camera lens in the spy satellite.
  The angle required for the performance requirement of the spy satellite is given by

$$\theta_s = \tan^{-1}\left(\frac{\Delta x}{h}\right),$$

where Δx is the distance between the two objects on the ground and h is the height of the spy satellite above the ground.

SIMPLIFY We can equate θ_R and θ_s to get

$$\sin^{-1}\left(\frac{1.22\lambda}{d}\right) = \tan^{-1}\left(\frac{\Delta x}{h}\right).$$

Solving for the diameter of the camera lens gives us

$$d = \frac{1.22\lambda}{\sin\left[\tan^{-1}\left(\dfrac{\Delta x}{h}\right)\right]}.$$

CALCULATE Putting in the numerical values gives us

$$d = \frac{1.22\left(607\cdot 10^{-9}\text{ m}\right)}{\sin\left[\tan^{-1}\left(\dfrac{0.490\text{ m}}{201\cdot 10^{3}\text{ m}}\right)\right]} = 0.30377\text{ m}.$$

ROUND We report our result to three significant figures:

$$d = 0.304\text{ m}.$$

DOUBLE-CHECK To double-check our result, we make the small-angle approximation for the Rayleigh Criterion and the angle subtended by the two objects. For the case where the wavelength of light is much smaller than the aperture of the camera, we can write

$$\sin(\theta_R) = \frac{1.22\lambda}{d} \approx \theta_R.$$

For the angle at the camera in the spy satellite, we can write

$$\tan(\theta_s) = \frac{\Delta x}{h} \approx \theta_s.$$

Thus,

$$\frac{1.22\lambda}{d} = \frac{\Delta x}{h},$$

which can be solved for the minimum diameter of the camera lens:

$$d = \frac{1.22\lambda h}{\Delta x} = \frac{1.22\left(607\cdot 10^{-9}\text{ m}\right)\left(201\cdot 10^{3}\text{ m}\right)}{0.490\text{ m}} = 0.304\text{ m}.$$

This agrees with our result within round-off error. Thus, our result seems reasonable.

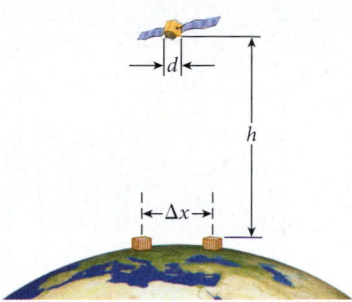

FIGURE 34.34 A spy satellite at height h observing two objects on the ground a distance Δx apart.

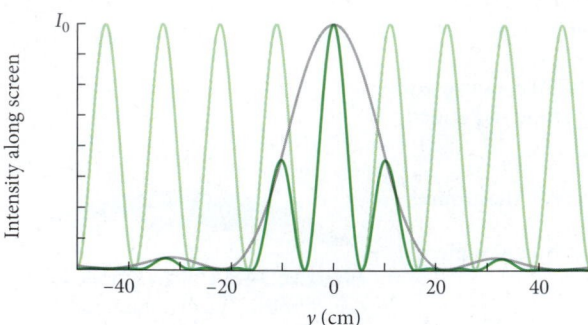

FIGURE 34.35 Intensity pattern from a double slit. The dark green line is the observed intensity pattern. The faint gray line is the intensity for a single slit with the same width as the two slits. The faint green line is the intensity distribution for two very narrow slits separated by the same distance as the double slits that produced the dark green line.

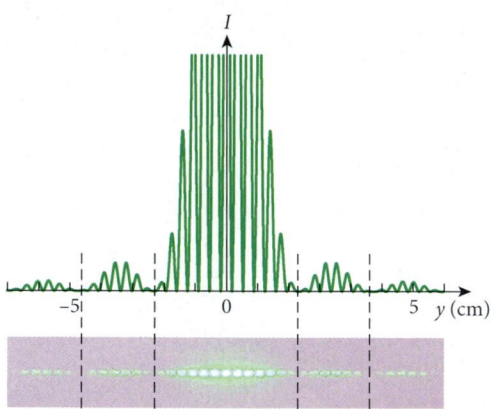

FIGURE 34.36 Photograph of the intensity pattern produced by a double slit illuminated by green light from a laser is below the graph of the intensity of the light. The central lines extends above the plotted scale in the vertical direction but have been cut off to allow the lower-intensity lines to be visible. The predicted intensity pattern is shown for $a = 0.0452$ mm, $d = 0.300$ mm, and $\lambda = 532$ nm. The dashed lines mark the diffraction minima.

Double-Slit Diffraction

Section 34.2 discussed the interference pattern produced by two slits. That analysis assumed that the slits themselves were very narrow compared with the wavelength of light, $a \ll \lambda$. For these narrow slits, the diffraction maximum is very wide, with peaks in the intensity that have the same value at all angles (see Figure 34.12).

However, with double slits for which the condition $a \ll \lambda$ is not met, as illustrated in Figure 34.7c, not all the interference fringes have the same intensity. For diffraction effects, the intensity of the interference pattern from double slits can be shown to be given by

$$I = I_0 \cos^2 \beta \left(\frac{\sin \alpha}{\alpha} \right)^2, \tag{34.22}$$

where $\alpha = \pi a \sin \theta / \lambda$, $\beta = \pi d \sin \theta / \lambda$, and d is the distance between the slits. If the screen is placed a sufficiently large distance from the slits, the angle θ is small and can be approximated by $\sin \theta \approx \tan \theta = y/L$. Then α and β can be approximated as $\alpha = \pi a y / \lambda L$ and $\beta = \pi d y / \lambda L$.

If the screen is $L = 2.0$ m away from the slits, each slit has a width of $a = 5.0 \cdot 10^{-6}$ m, the slits are $d = 1.0 \cdot 10^{-5}$ m apart, and the wavelength of the incident light is $\lambda = 550$ nm, we get the intensity pattern shown by the dark green line in Figure 34.35.

Figure 34.35 shows that the positions of the maxima on the screen are not changed from those for two very narrow slits. However, the maximum intensity is modulated by the diffraction intensity distribution, shown in faint gray. The diffraction pattern in Figure 34.35 forms an envelope for the interference intensity distribution. If one of the two slits were covered, only the diffraction pattern would be seen.

Figure 34.36 shows a photograph of an interference/diffraction pattern from a double slit projected on a screen that is 2 m away using light of wavelength $\lambda = 532$ nm from a green laser. The slits have a width of $a = 4.52 \cdot 10^{-5}$ m and are $d = 3.00 \cdot 10^{-4}$ m apart. The central maxima, which consist of all the two-slit maxima located within the single-slit central maximum envelope, are intentionally overexposed in the photograph to allow the secondary maxima to be observed. Figure 34.36 also shows the predicted intensity pattern, effectively overexposed by choosing the maximum value of the plotted intensity to be 38% of the maximum predicted intensity, which allows the secondary maxima to be visible. The first two single-slit diffraction minima on each side of the central maxima are marked with dashed lines.

34.4 Gratings

We have discussed diffraction and interference for a single slit and for two slits. How do diffraction and interference apply to a system of many slits? Putting many slits together forms a device called a **diffraction grating**. A diffraction grating has a large number of slits, or rulings, placed very close together. A diffraction grating can also be constructed using an opaque material containing grooves rather than slits; this type is called a *reflection* grating. A diffraction grating produces an intensity pattern consisting of narrow bright fringes separated by wide dark areas. This characteristic pattern results because having many slits means there can be destructive interference a small distance away from the maxima.

A portion of a diffraction grating is shown in Figure 34.37. This drawing shows coherent light with wavelength λ incident on a series of narrow slits, each separated by a distance d. A diffraction pattern is produced on a screen a long distance L away.

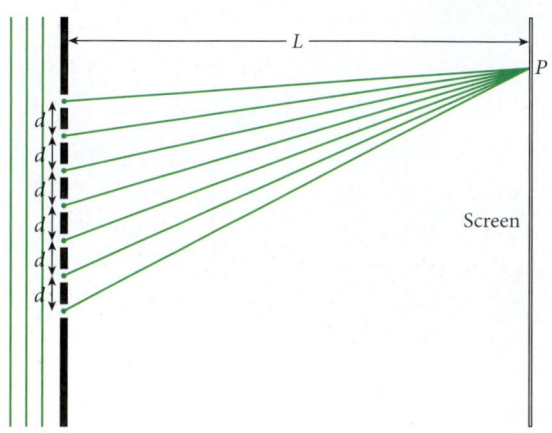

FIGURE 34.37 A portion of a diffraction grating.

Figure 34.37 can be expanded, as we did for the single-slit and double-slit cases (again using the limit that $L \gg d$ so that all of the rays drawn are parallel), to enable us to analyze the path length difference for the light from each of the slits to the screen (Figure 34.38). We assume that L is so large that the light rays are approximately parallel to one another.

The distance d is called the *grating spacing*. If the grating's width is W, the number N of slits (or rulings) is $N = W/d$. Diffraction gratings are often specified in terms of the number n_l of slits (or rulings) per unit length. We can obtain d from the specified n_l using $d = 1/n_l$.

Let's calculate the path length differences for the paths shown in Figure 34.38. Using an adjacent pair of rays, the path length difference is $\Delta x = d \sin \theta$. To produce bright lines, or constructive interference, this path length difference must be an integer multiple of the wavelength, so

$$d \sin \theta = m\lambda \quad \left(\text{for } m = 0, \pm 1, \pm 2, \ldots\right). \tag{34.23}$$

The values of m correspond to different bright lines. For $m = 0$, we have the central maximum at $\theta = 0$. For $m = 1$, we have the first-order maximum. For $m = 2$, we have the second-order maximum, and so on. Typically, diffraction gratings are designed to produce large angular separations between the maxima, so we do not make the small-angle approximation when discussing diffraction gratings.

Because diffraction gratings produce widely spaced narrow maxima, they can be used to determine the wavelength of monochromatic light by rearranging equation 34.23,

$$\lambda = \frac{d}{m} \sin \theta \quad \left(\text{for } m = 0, \pm 1, \pm 2, \ldots\right). \tag{34.24}$$

Monochromatic light incident on a diffraction grating produces lines on a screen at widely separated angles. For example, when a focused laser beam strikes a diffraction grating, a set of widely spaced spots is created, as illustrated in Figure 34.39.

In addition, diffraction gratings can be used to separate out different wavelengths of light from a spectrum of wavelengths. For example, sunlight is dispersed into multiple sets of rainbow-like colors as a function of θ. If the light is composed of several discrete wavelengths, the light is separated into sets of lines corresponding to each of those wavelengths.

The quality of a diffraction grating can be quantified in terms of its dispersion. The **dispersion** is the measure of the ability of a diffraction grating to spread apart the various wavelengths in a given order. Dispersion is defined by $D = \Delta\theta/\Delta\lambda$, where $\Delta\theta$ is the angular separation between two lines with wavelength difference $\Delta\lambda$. We can get an expression for the dispersion by differentiating equation 34.24 with respect to λ:

$$\frac{d\theta}{d\lambda} = \frac{d}{d\lambda}\left(\sin^{-1}\left(\frac{m\lambda}{d}\right)\right) = \frac{1}{\sqrt{1 - \left(\frac{m\lambda}{d}\right)^2}} \frac{m}{d} = \frac{m}{\sqrt{d^2 - (m\lambda)^2}}.$$

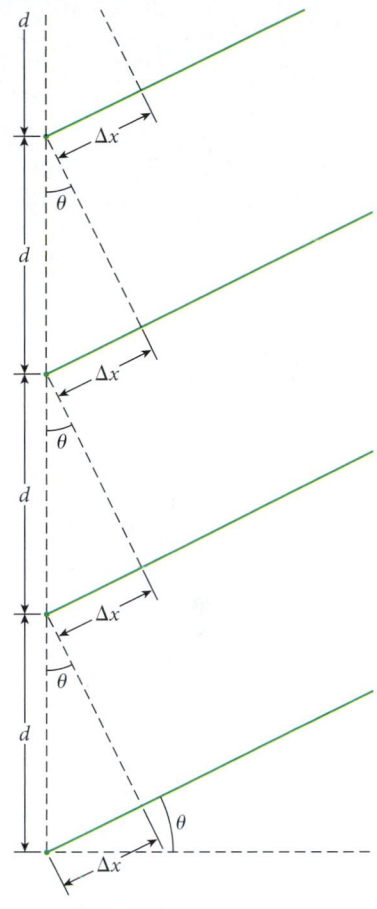

FIGURE 34.38 Expanded drawing of the geometry for a diffraction grating, assuming that the screen is far away compared to the spacing between the slits of the grating.

FIGURE 34.39 The diffraction pattern produced on a screen by green laser light incident on a diffraction grating with grating spacing $n_l = 787$ slits/cm. The spots corresponding to the central maximum, to $m = \pm 1$, and to $m = \pm 2$ are shown.

Concept Check 34.9

Figure 34.39 shows that the diffraction pattern from a grating has bright spots that are widely separated. By contrast, the double-slit pattern in Figure 34.36 has broad maxima closely spaced. Why?

a) because the condition for constructive interference is fulfilled for only a very few angles for a grating

b) because the condition for destructive interference is almost always fulfilled for a grating

c) (a) and (b)

d) because the slits in the grating are so small

e) none of the above

Because $m\lambda/d = \sin\theta$, we can express $d\theta/d\lambda$ as

$$\frac{d\theta}{d\lambda} = \frac{1}{\sqrt{1-\sin^2\theta}}\frac{m}{d} = \frac{m}{d\cos\theta},$$

where we have used the identity $\sin^2\theta + \cos^2\theta = 1$. Taking intervals of θ and λ that are not too large, we can write an expression for the dispersion of a diffraction grating:

$$D = \frac{\Delta\theta}{\Delta\lambda} = \frac{m}{d\cos\theta} \quad (\text{for } m = \pm 1, \pm 2, \pm 3, \ldots). \tag{34.25}$$

The dispersion of a diffraction grating increases as the distance d between slits gets smaller, and as the order m gets higher. Note that the dispersion does not depend on the number of slits, N.

The **resolving power**, R, of a diffraction grating is a measure of its ability to resolve closely spaced maxima. This ability depends on the width of each maximum. Consider a diffraction grating used to resolve light of two wavelengths, λ_1 and λ_2, with $\lambda_{\text{ave}} = (\lambda_1 + \lambda_2)/2$ and $\Delta\lambda = |\lambda_2 - \lambda_1|$. We define the resolving power of the grating as

$$R = \frac{\lambda_{\text{ave}}}{\Delta\lambda_{\text{min}}}, \tag{34.26}$$

where $\Delta\lambda_{\text{min}}$ is the minimum value of $\Delta\lambda$ for which the wavelengths are resolved.

In order to discuss the resolving power, we need an expression for the width of each maximum. The width of each maximum is determined by the position of the first minimum on one side of the central maximum. We define the angular half-width, θ_{hw}, of the maximum as the angle between the maximum and this first minimum (Figure 34.40).

Equation 34.25 tells us the angular spread for a given $\Delta\lambda$. The two wavelengths can barely be resolved if this spread $\Delta\theta = \theta_{\text{hw}}$. To determine the position of the first minimum, we do a single-slit diffraction analysis using the whole grating as the single slit (Figure 34.41). This approach is justified if the number of slits N is large. In any case, it provides a useful and meaningful approximation, while a precise mathematical calculation would tend to obscure the physics involved.

The angle of the first minimum for single-slit diffraction can be obtained from the condition $a\sin\theta = \lambda$, where Nd is substituted for the slit width a: $Nd\sin\theta_{\text{hw}} = \lambda$. Because θ_{hw} is small, we can write $\sin\theta_{\text{hw}} \approx \theta_{\text{hw}}$, or

$$\theta_{\text{hw}} = \frac{\lambda}{Nd}.$$

We can show (though we don't do so here) that the width of the maxima for other orders can be written as

$$\theta_{\text{hw}} = \frac{\lambda}{Nd\cos\theta},$$

where θ is the angle corresponding to the maximum intensity for the particular order. Substituting θ_{hw} for $\Delta\theta$ in equation 34.25, we get

$$\frac{\Delta\theta}{\Delta\lambda} = \frac{\lambda}{Nd\cos\theta\Delta\lambda} = \frac{m}{d\cos\theta},$$

or

$$R = \frac{\lambda}{\Delta\lambda} = Nm, \tag{34.27}$$

where we have taken $\lambda \approx (\lambda + (\lambda + \Delta\lambda))/2$. Note that the resolving power of a diffraction grating depends only on the total number of slits and the order.

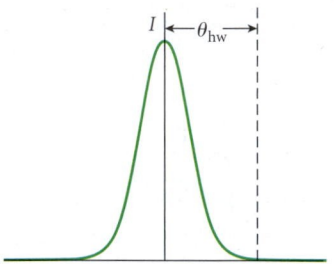

FIGURE 34.40 The angular half-width of the central maximum for a diffraction grating.

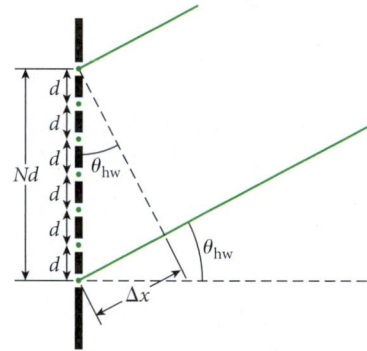

FIGURE 34.41 Calculation of the half-width of the central maximum for a diffraction grating, treating the entire grating as a single slit.

FIGURE 34.42 Different colors resulting from constructive interference of sunlight striking a blank DVD.

EXAMPLE 34.3   CD or DVD as Diffraction Grating

Diffraction gratings can take the form of a series of narrow slits that light passes through or a series of closely spaced grooves that reflect light. The resulting diffraction pattern is the same. We can therefore think of the spiral grooves of a CD or a DVD as a diffraction grating. In Figure 34.42, a DVD is shown reflecting various colors from sunlight.

If we shine a green laser pointer with wavelength 532 nm perpendicular to the surface of a CD, we observe a diffraction pattern in the form of bright spots on a screen placed perpendicular to the CD at a perpendicular distance $L = 1.6$ cm away from the point where the laser beam hits the CD (Figure 34.43 and Figure 34.44). The spacing between the grooves in the CD is $d = 1.60 \cdot 10^{-6}$ m $= 1.6$ μm.

PROBLEM

What is the horizontal distance from the surface of the CD edge to the spots observed on the screen?

SOLUTION

We start with our expression for the angle of constructive interference from a diffraction grating,

$$d \sin \phi_m = m\lambda \quad (\text{for } m = 0, \pm 1, \pm 2, \pm 3, \dots),$$

where d is the distance between adjacent grooves, λ is the wavelength of the light, m is the order, and ϕ_m is the angle of the mth order maximum.

The wavelength that we need for this calculation is the wavelength of light in the polycarbonate plastic from which the CD is made, which has an index of refraction $n = 1.55$; this wavelength is: $\lambda = \lambda_{air}/n$. Thus, the condition for the diffraction maxima is given by

$$d \sin \phi_m = \frac{m\lambda_{air}}{n} \Leftrightarrow n \sin \phi_m = \frac{m\lambda_{air}}{d}.$$

This light ray must next pass through the surface of the CD, where it is refracted. Applying Snell's Law at the surface and taking $n_{air} = 1$, we get

$$n \sin \phi_m = 1 \cdot \sin \theta_m = \sin \theta_m = \frac{m\lambda_{air}}{d}.$$

We can rewrite this equation as $d \sin \theta_m = m\lambda_{air}$, which is exactly what we would have gotten if we had treated the CD as a diffraction grating in air.

Starting with $m = 0$, we find that the central maximum produces a spot at $\theta_0 = 0°$, which means that the spot is produced in front of the laser pointer. You can see this diffraction maximum in the reflection of the laser pointer from the surface of the CD in Figure 34.43.

Moving to $m = 1$, we find the angle of the first-order diffraction maximum, θ_1:

$$\theta_1 = \sin^{-1} \left(\frac{\lambda_{air}}{d} \right) = \sin^{-1} \left(\frac{532 \cdot 10^{-9} \text{ m}}{1.60 \cdot 10^{-6} \text{ m}} \right) = 19.4°.$$

Looking at Figure 34.44a, we can see that $\tan \theta_1 = L/y_1$, so we calculate the distance of the first-order spot along the screen as $y_1 = (1.6 \text{ cm})/(\tan 19.4°) = 4.54$ cm. The

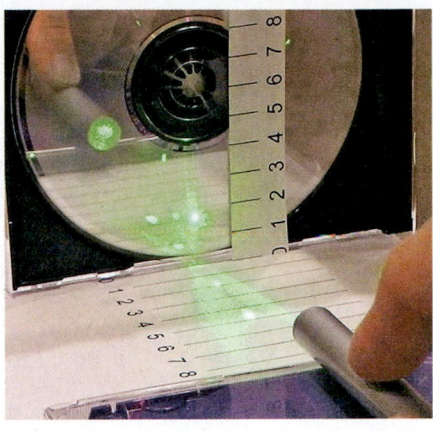

(a)

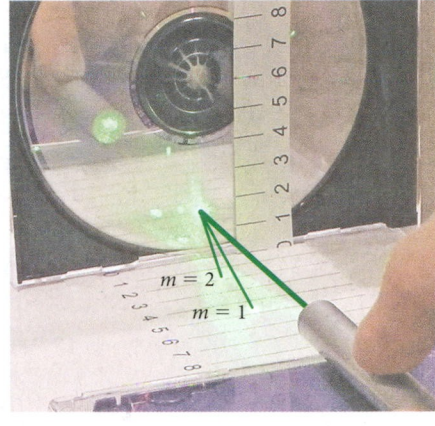

(b)

FIGURE 34.43 Using a CD as a diffraction grating. (a) A green laser pointer shining perpendicularly on the bottom surface of a CD. (b) Green lines illustrate the light from the laser pointer, and the order of each diffraction spot is labeled. The vertical and horizontal rulers have markings every centimeter.

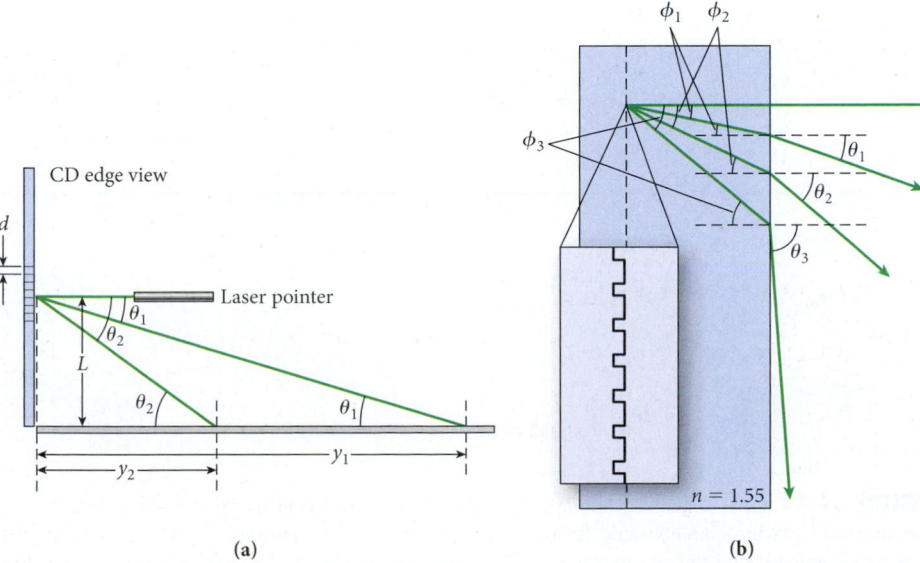

(a)   (b)

FIGURE 34.44 Geometry of using a CD as a diffraction grating. (a) Side view of the geometry of the measurement shown in Figure 34.43. A horizontal white screen is used to locate the diffraction maxima. (b) Expanded drawing of the diffraction grating inside the polycarbonate CD, illustrating the diffraction and refraction that occur.

– Continued

photograph in Figure 34.43 shows that this calculation is in good agreement with what we observe for $m = 1$.

The angle of the second-order maximum is given by

$$\theta_2 = \sin^{-1}\left(\frac{2\lambda_{air}}{d}\right) = \sin^{-1}\left(\frac{2\cdot532\cdot10^{-9}\ m}{1.60\cdot10^{-6}\ m}\right) = 41.7°.$$

Thus, the position of the second-order spot is $y_2 = (1.6\ cm)/(\tan 41.7°) = 1.80\ cm$, which is also in good agreement with what we observe for $m = 2$ in Figure 34.43.

The angle of the third-order maximum is

$$\theta_3 = \sin^{-1}\left(\frac{3\lambda_{air}}{d}\right) = \sin^{-1}\left(\frac{3\cdot532\cdot10^{-9}\ m}{1.60\cdot10^{-6}\ m}\right) = 85.9°.$$

The position of the third-order spot is $y_3 = (1.6\ cm)/(\tan 85.9°) = 0.11\ cm$. The third-order spot is not clearly visible in Figure 34.43 because the third-order maximum is dimmer than the first and second maxima and because the angle at which the spot must be observed is very close to 90°.

What about a fourth-order maximum? For the angle of this maximum, we would have $\theta_4 = \sin^{-1}(4\lambda_{air}/d)$. However, for this order, $4\lambda_{air}/d = 4(532\ nm)/(1.6\ \mu m) = 1.33$, which cannot occur because $|\sin\theta| \leq 1$. Therefore, only three spots can appear on the screen, only two of which are easily visible.

If we carried out the same experiment using a red laser with $\lambda_{air} = 633\ nm$, we would get only two maxima, occurring at $\theta_1 = 23.3°$ and $\theta_2 = 52.3°$.

Self-Test Opportunity 34.4

What maxima would you observe if you repeated the experiment of Example 34.3 with a DVD and a green laser pointer? (The separation between grooves on a DVD is 740 nm compared with 1600 nm for a CD.)

Blu-Ray Discs

Earlier, in Example 9.2, we calculated the length of a CD track and also showed microscopic images of CD surfaces. Now we want to explore how Blu-ray discs and other optical discs store digital information, and how computers and consumer electronic devices read them. CDs, DVDs, and Blu-ray optical discs all operate on similar principles. Figure 34.45c and Figure 34.45d show schematic cross sections of a Blu-ray disc.

A Blu-ray disc, like a CD or DVD, stores digital information in terms of ones and zeros. These ones and zeros are encoded in the location of the edges of the high (land) areas and low (pit) areas in the aluminum layer shown in Figure 34.45. The high and low areas rotate with the Blu-ray disc and pass over a blue solid-state laser that emits light with a wavelength of $\lambda = 405\ nm$ in air. A schematic diagram of the blue laser assembly is shown in Figure 34.46.

The Blu-ray player incorporates several concepts of wave optics presented in this chapter, including a diffraction grating and destructive interference, as well as polarization, which was discussed in Chapter 31. A solid-state laser produces light with wavelength

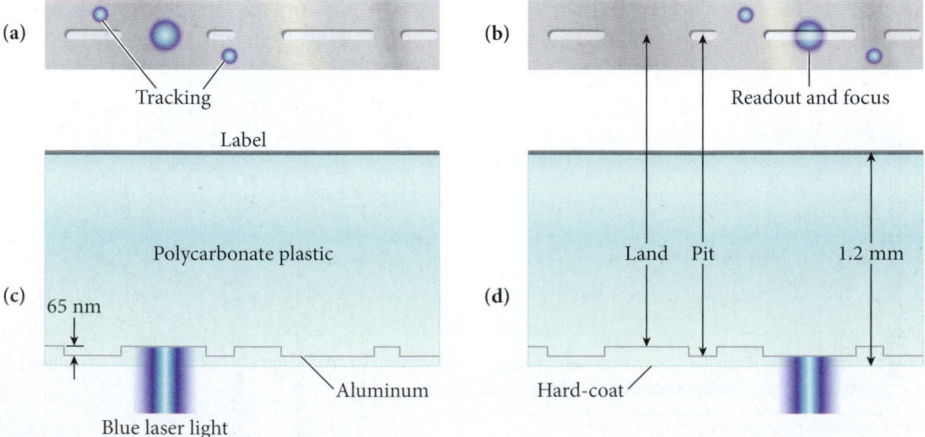

FIGURE 34.45 Cross section of a Blu-ray disc. (a) View from the bottom showing the laser data beam reflecting from only the aluminum land. (b) View from the bottom showing the laser data beam reflecting from both aluminum land and pit simultaneously. (c) Edge view showing the laser data beam focused on the aluminum land. (d) Edge view showing the laser beam focused on the aluminum pit (and land).

$\lambda = 405$ nm in air. This wavelength is shorter than the wavelength of the laser light used to read CDs by a factor of almost 2. The shorter wavelength allows the land and pit areas to be smaller, which allows more data to be stored. This light is passed through a diffraction grating. The central maximum and the two first-order maxima are shown in Figure 34.46. The higher-order maxima are not used. The light in the central maximum is used to read the data from the disc and to maintain the focus of the beam. The light from the two first-order maxima is used for tracking the data on the disc.

After passing through the diffraction grating, the light passes through a polarizer and then proceeds to a polarizing beam splitter. The light is directed upward with a turning mirror. The light is then focused on the Blu-ray disc after passing through a quarter-wave plate that rotates the polarization. The effect of the polarizing elements is to separate the light reflected from the Blu-ray disc surface and direct it into the photodiode array while minimizing the direct light from the laser traveling into the photodiode array.

When the light from the central maximum is shining completely on the land, as illustrated in Figure 34.45a and Figure 34.45c, all the light is reflected into the photodiode. When the light from the laser is shining on a pit, as shown in Figure 34.45b and Figure 34.45d, light is reflected from both the pit and the land. In this case, the light that reflects from the land area travels farther than the light reflected from the pit area. In both cases, the light undergoes a phase change when it is reflected. Thus, the analysis of destructive interference due to a coating on a lens can be applied to this case. We denote the difference in height between the land and pit areas as t, so the path length difference for the light from the high and low areas is $2t$. The criterion for destructive interference is

$$\left(m+\frac{1}{2}\right)\frac{\lambda_{air}}{n_{polycarbonate}} = 2t \quad \left(\text{for } m = 0, \pm 1, \pm 2, ...\right),$$

where $n_{polycarbonate}$ is the index of refraction of the polycarbonate of which the Blu-ray disc is made and λ_{air} is the wavelength of the light emitted by the laser. For $m = 0$,

$$t = \frac{\lambda_{air}}{4n_{polycarbonate}}.$$

For the Blu-ray laser, $\lambda_{air} = 405$ nm and the refractive index of the polycarbonate is 1.58; therefore, the thickness required for destructive interference is

$$t = \frac{\lambda_{air}}{4n_{polycarbonate}} = \frac{405 \text{ nm}}{4(1.58)} = 64.1 \text{ nm},$$

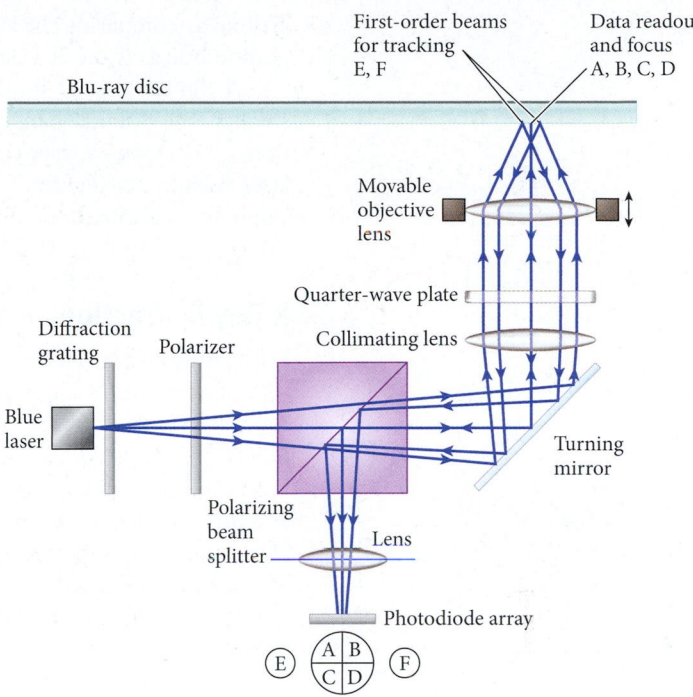

FIGURE 34.46 A schematic diagram of the blue laser assembly of a Blu-ray disc reader.

which is close to the difference between the high and low areas shown in Figure 34.45.

As the disc spins, the land and pit areas pass over the laser. When the land areas are over the laser, all of the light is reflected and the photodiode registers a given voltage. When the pit areas pass over the laser, part of the light is lost to destructive interference and the photodiode reads a lower voltage. Whenever the voltage changes—from high to low, or low to high—the Blu-ray player records a one. Otherwise, the player records a zero. This pattern of zeros and ones is translated into the normal digital code used in computers by the 8-14 method, which converts the 14 bits encoded on the disc into 8 bits of digital information. In addition, 3 bits are added to each set of 14 bits to allow the reader to maintain its tracking. The light from the central maximum is projected onto a photodiode that is segmented into four parts, A, B, C, and D, as shown in Figure 34.46. The balance of these four signals is used to adjust the distance of the movable objective lens from the surface of the disc. The tracking is done by comparing the signals from the two first-order maxima that are registered in the photodiode as E and F, as shown in Figure 34.46.

A Blu-ray disc is similar to a CD or a DVD except that the high and low areas are smaller. The distance between grooves is 1.6 μm on a CD and 0.74 μm on a DVD. In addition, a DVD uses a laser that emits light with a wavelength of 650 nm, while a CD uses a laser with a wavelength of 780 nm. A Blu-ray disc uses a groove spacing of 0.32 μm and can hold 25 gigabytes of digital information. A CD can hold up to 700 megabytes of digital information, while a DVD can hold 4.7 gigabytes.

X-Ray Diffraction and Crystal Structure

Wilhelm Röntgen (1845–1923) discovered X-rays in 1895. His experiments suggested that X-rays were electromagnetic waves with a wavelength of about 10^{-10} m. At about the same time, the study of crystalline solids suggested that their atoms were arranged in a regular repeating pattern with a spacing of about 10^{-10} m between the atoms. Putting these two ideas together, Max von Laue (1879–1960) proposed in the early 1900s that a crystal could serve as a three-dimensional diffraction grating for X-rays. In 1912, von Laue, Walter Friederich (1883–1963), and Paul Knipping (1883–1935) did the first **X-ray diffraction** experiment, which showed diffraction of X-rays by a crystal. Soon afterward, Sir William Henry Bragg (1862–1942) and his son William Lawrence Bragg (1890–1971) derived Bragg's Law (given below) and carried out a series of experiments involving X-ray diffraction by crystals.

Let's assume we have a cubic crystal, with each atom in the lattice a distance a away from its neighboring atoms in all three directions (Figure 34.47). We can imagine various planes of atoms in this crystal. For example, the horizontal planes are composed of atoms spaced a distance a apart, with the planes themselves spaced a distance a from one another. If X-rays are incident on these planes, the rows of atoms in the crystalline lattice can act like a diffraction grating for the X-rays. The X-rays can be thought of as scattering from the atoms (Figure 34.48).

Interference effects are caused by path length differences. When X-rays scatter off one plane, all the waves remain in phase as long as the incident angle equals the reflected angle. However, for two adjacent planes, Figure 34.49 shows that the path length difference for the scattered X-rays from the two planes is

$$\Delta x = \Delta x_1 + \Delta x_2 = 2a \sin\theta, \tag{34.28}$$

where θ is the angle between the incoming X-rays and the plane of atoms. (Note that unfortunately this convention is different from that used for all the other cases we have considered, where the angle is always measured relative to the surface normal!) Thus, the criterion for constructive interference due to Bragg scattering is given by

$$2a \sin\theta = m\lambda \quad \left(\text{for } m = 0, \pm 1, \pm 2, \ldots\right). \tag{34.29}$$

This equation is known as **Bragg's Law.**

When X-rays are incident on a crystal, several different planes can function as diffraction gratings. Some examples are illustrated in Figure 34.50. These planes do not have the spacing a between them.

To study the atomic structure of a substance using X-ray diffraction, X-rays can be scattered nearly parallel to the surface of a sample (Figure 34.51a). Alternatively, the X-rays can be transmitted through the sample and detected on the opposite side of the sample (Figure 34.51b). For the parallel scattering method, the angle of incidence θ should equal the angle of observation. For the transmission method, the observed angle is twice the

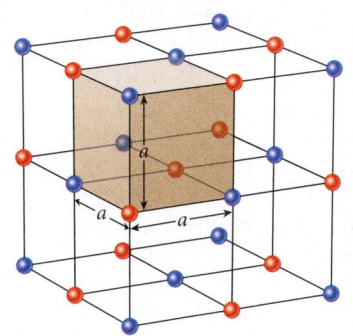

FIGURE 34.47 A cubic crystal lattice with interatomic spacing a.

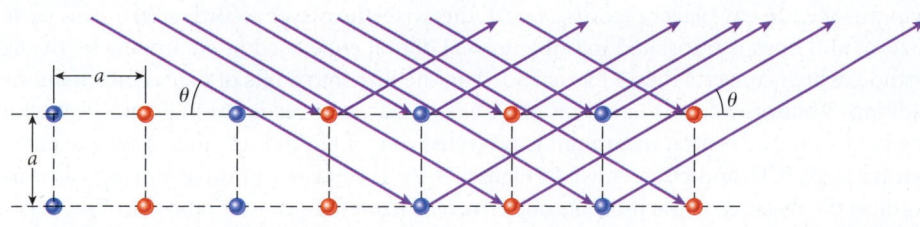

FIGURE 34.48 Schematic diagram of X-rays scattering off planes of atoms in a crystal.

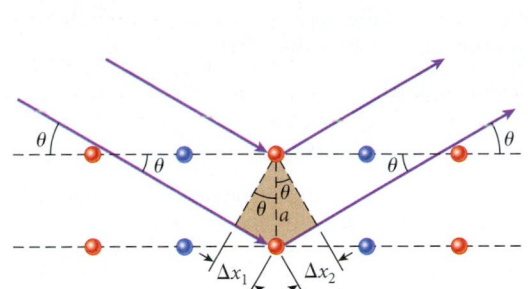

FIGURE 34.49 Path length difference for X-rays scattered from two adjacent planes.

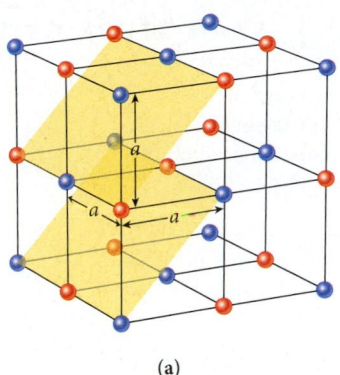

FIGURE 34.50 Examples of planes that could function as diffraction gratings for X-rays in a cubic crystalline lattice.

FIGURE 34.51 Two geometries for studying the atomic structure of a sample using X-ray diffraction. (a) X-rays scattered nearly parallel to the surface; (b) X-rays transmitted through the sample.

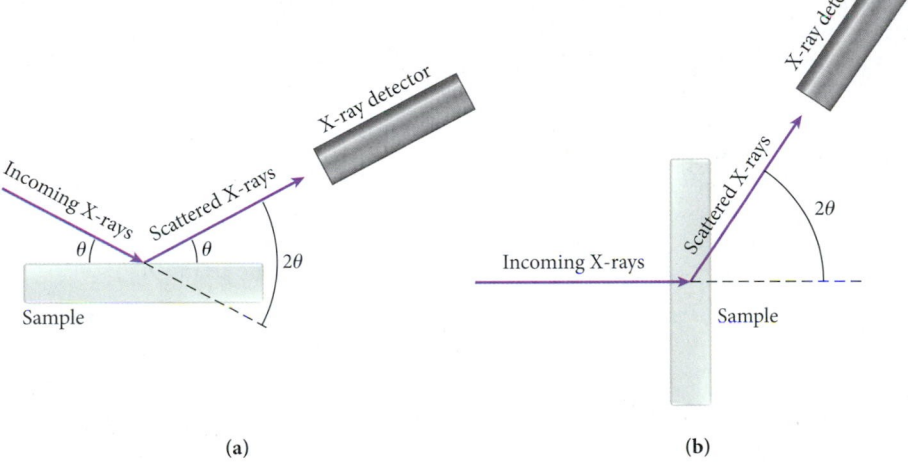

Bragg angle θ. By measuring the intensity of the X-rays as a function of θ, we can determine details of the structure of the material being studied.

Figure 34.52 shows a sample image obtained from the scattering of X-rays by a protein, known as "3Clpro," which was fixed in a crystalline structure. The small gray and black dots in the image constitute the diffraction pattern of the protein. The concentric rings are a result of X-rays scattering off water molecules, which surround the protein and are oriented in random directions. The white horizontal bar with the small white square in the middle of the image is the beam stop, which prevents the X-rays that did not scatter from reaching the detector and damaging it. From the diffraction patterns obtained using many different orientations of the sample relative to the X-ray beam, computer analysis can reconstruct the three-dimensional spatial structure.

Modern particle accelerators—such as the National Synchrotron Light Source at Brookhaven National Laboratory, the Advanced Light Source at Lawrence Berkeley National Laboratory, or the Advanced Photon Source at Argonne National Laboratory (Figure 34.53), and many others around the world—are used to produce high-quality, intense beams of X-rays to carry out research in condensed matter and materials science. Similar scattering information can be gathered by bombarding crystalline structures with intense neutron beams. (To understand how this works, you have to wait until we examine quantum mechanics in Chapter 37.) Intense neutron beams for materials science research are used at the Spallation Neutron Source at Oak Ridge National Laboratory. These huge X-ray and neutron scattering facilities cost hundreds of millions of dollars, but are absolutely essential tools for investigating the nanoscale structure of materials. They are the basic research tools needed to achieve modern and future nanotechnology advances.

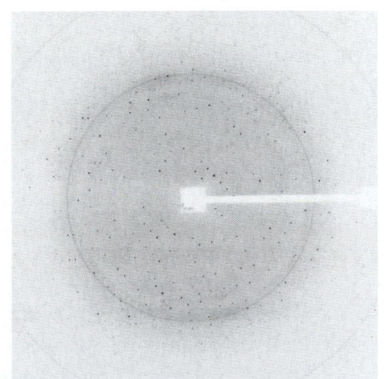

FIGURE 34.52 X-ray diffraction image of a protein.

FIGURE 34.53 Advanced Photon Source.

WHAT WE HAVE LEARNED | EXAM STUDY GUIDE

- Huygens's Principle states that every point on a propagating wave front serves as a source of spherical secondary wavelets. A geometric analysis based on this principle is called a Huygens construction.

- The condition for two coherent waves with wavelength λ to interfere constructively is $\Delta x = m\lambda$ (for $m = 0, \pm 1, \pm 2, \ldots$), where Δx is the path length difference between the two waves.

- The condition for two coherent waves with wavelength λ to interfere destructively is $\Delta x = (m + \frac{1}{2})\lambda$ (for $m = 0, \pm 1, \pm 2, \ldots$), where Δx is the path length difference between the two waves.

- The angle of a bright fringe from two narrow slits spaced a distance d apart and illuminated by coherent light with wavelength λ is given by $d \sin\theta = m\lambda$ (for $m = 0, \pm 1, \pm 2, \ldots$). On a screen a distance L away, the position y of the bright fringe from the central maximum along the screen is given by $y = \dfrac{m\lambda L}{d}$ (for $m = 0, \pm 1, \pm 2, \ldots$).

- The angle of a dark fringe from two narrow slits spaced a distance d apart and illuminated by coherent light with wavelength λ is given by $d \sin\theta = (m + \frac{1}{2})\lambda$ (for $m = 0, \pm 1, \pm 2, \ldots$). On a screen a distance L away, the position y of the dark fringe from the central maximum along the screen is given by $y = \dfrac{\left(m + \frac{1}{2}\right)\lambda L}{d}$ (for $m = 0, \pm 1, \pm 2, \ldots$).

- The condition for constructive interference for light with wavelength λ_{air} in a thin film of thickness t and index of refraction n in air is $\left(m + \frac{1}{2}\right)\dfrac{\lambda_{air}}{n} = 2t$ (for $m = 0, 1, 2, \ldots$).

- The radii of the bright circles in Newton's rings are given by $x_m = \sqrt{R\left(m + \frac{1}{2}\right)\lambda}$ (for $m = 0, 1, 2, \ldots$), where R is the radius of curvature of the upper curved glass surface and λ is the wavelength of the incident light.

- The angle of a dark fringe from a single slit of width a illuminated by light with wavelength λ is given by $a \sin\theta = m\lambda$ (for $m = \pm 1, \pm 2, \pm 3, \ldots$).

- The angle θ of the first minimum from a circular aperture with diameter d illuminated with light of wavelength λ is $\sin\theta = 1.22\dfrac{\lambda}{d}$. This equation also supplies Rayleigh's Criterion. The angle in the equation expresses the minimum resolvable angle between two distant objects for a telescope primary lens or mirror or a camera lens with diameter d.

- The angle θ of the maxima from a diffraction grating illuminated with light of wavelength λ is given by $\theta = \sin^{-1}\left(\dfrac{m\lambda}{d}\right)$ (for $m = 0, \pm 1, \pm 2, \ldots$), where d is the distance between the slits (or rulings) of the grating.

- The dispersion of a diffraction grating is given by $D = \dfrac{m}{d\cos\theta}$ (for $m = \pm 1, \pm 2, \pm 3, \ldots$), where d is the distance between the slits (or rulings) of the grating.

- The resolving power of a diffraction grating is given by $R = Nm$ (for $m = 1, 2, 3, \ldots$), where N is the number of slits (or rulings) in the grating.

- For X-rays scattering off planes of atoms separated by a distance a, the condition for constructive interference is $2a \sin\theta = m\lambda$ (for $m = 0, 1, 2, \ldots$). The angle θ is the angle between incoming X-rays and the plane of atoms and the angle of observation of the X-rays.

ANSWERS TO SELF-TEST OPPORTUNITIES

34.1 There is no change in colors. After all, the light has to propagate through your eyeball before it reaches your retina, and the index of refraction of your eyeball does not change when your head is underwater.

34.2 $\lambda = \dfrac{2d}{N} = \dfrac{2\left(0.25 \cdot 10^{-3} \text{ m}\right)}{1200} = 4.17 \cdot 10^{-7} \text{ m} = 417 \text{ nm}.$

34.3 Assume that $\lambda = 550$ nm and that the diameter of the human pupil is $d = 7.00$ mm.

$$\theta_R = \sin^{-1}\left(\frac{1.22\lambda}{d}\right) = 6.71 \cdot 10^{-5} \text{ rad}$$

$$\theta_{wall} = \frac{y}{L} = \frac{10.0 \text{ m}}{190 \cdot 10^3 \text{ m}} = 5.26 \cdot 10^{-5} \text{ rad}$$

Individual watchtowers on the Great Wall of China are difficult to see from the orbiting Space Shuttle.

34.4 The first maximum would be $\theta_1 = \sin^{-1}(532 \text{ nm}/740 \text{ nm}) = 46.0°$. No other maxima are possible.

PROBLEM-SOLVING GUIDELINES

1. A sketch of the optical situation is almost always helpful. It is simplest to use rays, not wave fronts, but remember that you are dealing with wave effects. Use the diagram to clearly identify any path differences involved in the problem situation.

2. The basic idea of wave optics is that constructive interference occurs when the path difference is an integer number of wavelengths, while destructive interference occurs when

the path difference is an odd-integer number of half wavelengths. Always start with this concept and take into account any additional phase changes due to reflection.

3. Remember that a phase change occurs when light traveling in a less dense medium reflects from a more dense medium. If light reflects from a less dense medium, no phase change occurs.

MULTIPLE-CHOICE QUESTIONS

34.1 Suppose the distance between the slits in a double-slit experiment is $2.00 \cdot 10^{-5}$ m. A beam of light with a wavelength of 750 nm is shone on the slits. What is the angular separation between the central maximum and an adjacent maximum?

a) $5.00 \cdot 10^{-2}$ rad

b) $4.50 \cdot 10^{-2}$ rad

c) $3.75 \cdot 10^{-2}$ rad

d) $2.50 \cdot 10^{-2}$ rad

34.2 When two light waves, both with wavelength λ and amplitude A, interfere constructively, they produce a light wave of the same wavelength but with amplitude $2A$. What is the intensity of this light wave?

a) same intensity as the original waves

b) double the intensity of the original waves

c) quadruple the intensity of the original waves

d) Not enough information is given.

34.3 A laser beam with wavelength 633 nm is split into two beams by a beam splitter. One beam goes to mirror 1, a distance L from the beam splitter, and returns to the beam splitter, while the other beam goes to mirror 2, a distance $L + \Delta x$ from the beam splitter, and returns to the beam splitter. The beams then recombine and travel to a detector together. If $L = 1.00000$ m and $\Delta x = 1.00$ mm, which best describes the kind of interference observed at the detector? (*Hint*: To double-check your answer, you may need to use a formula that was originally intended for combining two beams in a different geometry.)

a) purely constructive

b) purely destructive

c) mostly constructive

d) mostly destructive

e) neither constructive nor destructive

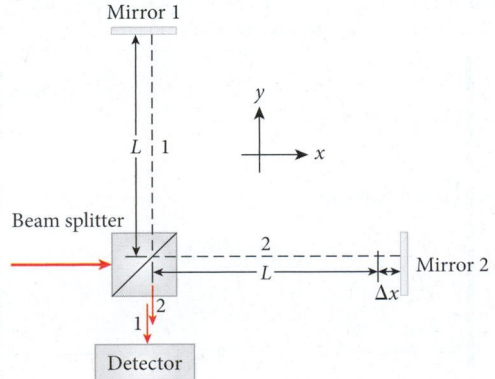

34.4 Which type of light incident on a grating with 1000 rulings with a spacing of 2.00 μm would produce the largest number of maxima on a screen 5.00 m away?

a) blue light of wavelength 450 nm

b) green light of wavelength 550 nm

c) yellow light of wavelength 575 nm

d) red light of wavelength 625 nm

e) need more information

34.5 If the wavelength of light illuminating a double slit is halved, the fringe spacing is

a) halved.

b) doubled.

c) not changed.

d) changed by a factor of $1/\sqrt{2}$.

34.6 A red laser pointer shines light with a wavelength of 635 nm on a diffraction grating with 300 slits/mm. A screen is placed a distance of 2.0 m behind the diffraction grating to observe the diffraction pattern. How far away from the central maximum will the next bright spot be on the screen?

a) 39 cm

b) 76 cm

c) 94 cm

d) 4.2 m

e) 9.5 m

34.7 Newton's rings are interference patterns caused by the reflection of light between two glass surfaces. What color is the center of Newton's rings produced with white light?

a) white

b) black

c) red

d) violet

34.8 In Young's double-slit experiment, both slits were illuminated by a laser beam and the interference pattern was observed on a screen. If the viewing screen is moved farther from the slits, what happens to the interference pattern?

a) The pattern gets brighter.

b) The pattern gets brighter, and the maxima are closer together.

c) The pattern gets less bright, and the maxima are farther apart.

d) There is no change in the pattern.

e) The pattern becomes unfocused.

f) The pattern disappears.

34.9 Light of wavelength = 560.0 nm enters a block of clear plastic from air at an incident angle of $\theta_i = 36.1°$ with respect to the normal. The angle of refraction is $\theta_r = 21.7°$. What is the speed of the light inside the plastic?

a) $1.16 \cdot 10^8$ m/s

b) $1.31 \cdot 10^8$ m/s

c) $1.67 \cdot 10^8$ m/s

d) $1.88 \cdot 10^8$ m/s

e) $3.00 \cdot 10^8$ m/s

34.10 Huygens's Principle says that each point of a wave front in a slit is a point source of light emitting a spherical wavelet. A Huygens construction applies

a) to any point anywhere in the path of the wave front.

b) to any point in the path of the wave front where matter is present.

c) only in slits.

34.11 If Huygens's Principle holds everywhere, why does a laser beam not spread out?

a) All the light waves that spread in the perpendicular direction from the beam interfere destructively.

b) It does spread out, but the spread is so small that we don't notice it.

c) Huygens's Principle isn't true in general; it only applies to slits, edges, and other obstacles.

d) Lasers employ additional special beams to keep the main beam from spreading.

34.12 A pair of thin slits is separated by a distance $d = 1.40$ mm and is illuminated with light of wavelength 460.0 nm. What is the separation between adjacent interference maxima on a screen a distance $L = 2.90$ m away?

a) 0.00332 mm

b) 0.556 mm

c) 0.953 mm

d) 1.45 mm

e) 3.23 mm

CONCEPTUAL QUESTIONS

34.13 What happens to a double-slit interference pattern if

a) the wavelength is increased?

b) the separation between the slits is increased?

c) the experimental apparatus is placed in water?

34.14 Estimate the frequency of an ultrasonic (sound) wave for which diffraction effects would be as small as they are for visible light.

34.15 Why are radio telescopes so much larger than optical telescopes? Does an X-ray telescope also have to be larger than an optical telescope?

34.16 Can light pass through a single slit narrower than its wavelength? If not, why not? If so, describe the distribution of the light beyond the slit.

34.17 One type of hologram consists of bright and dark fringes produced on photographic film by interfering laser beams. If the hologram is illuminated with white light, the image will be reproduced multiple times, in different pure colors at different sizes.

a) Explain why.

b) Which colors correspond to the largest and smallest images, and why?

34.18 A double slit is positioned in front of an incandescent light bulb. Will an interference pattern be produced?

34.19 Many astronomical observatories, especially radio observatories, are coupling several telescopes together. What are the advantages of this?

34.20 In a single-slit diffraction pattern, there is a bright central maximum surrounded by successively dimmer higher-order maxima. Farther away from the central maximum, eventually no more maxima are observed. Is this because the remaining maxima are too dim? Or is there an upper limit to the number of maxima that can be observed, no matter how good the observer's eyes, for a given slit and light source?

34.21 Which close pair of stars will be more easily resolvable with a telescope: two red stars or two blue ones? Assume the binary star systems are the same distance from Earth and are separated by the same angle.

34.22 A red laser pointer shines on a diffraction grating, producing a diffraction pattern on a screen behind the grating. If the red laser pointer is replaced with a green laser pointer, will the green bright spots on the screen be closer together or farther apart than the red bright spots were?

EXERCISES

A blue problem number indicates a worked-out solution is available in the Student Solutions Manual. One • and two •• indicate increasing level of problem difficulty.

Section 34.1

34.23 A helium-neon (He-Ne) laser has a wavelength of 632.8 nm.

a) What is the wavelength of this light as it passes through Lucite with index of refraction $n = 1.500$?

b) What is the speed of the light in the Lucite?

34.24 It is common knowledge that the visible light spectrum extends approximately from 400 nm to 700 nm. Roughly, 400 nm to 500 nm corresponds to blue light, 500 nm to 550 nm corresponds to green, 550 nm to 600 nm to yellow-orange, and above 600 nm to red. In an experiment, red light with a wavelength of 632.8 nm from a He-Ne laser is refracted into a fish tank filled with water (with index of refraction 1.333). What is the wavelength of the laser light in water, and what color will it have in water?

Section 34.2

34.25 What minimum path length difference is needed to cause a phase shift of $\pi/4$ in light of wavelength 700. nm?

34.26 Coherent, monochromatic light of wavelength 450.0 nm is emitted from two locations and detected at another location. The path length difference between the two routes taken by the light is 20.25 cm. Will the two light waves interfere destructively or constructively at the detection point?

34.27 A Young's double-slit experiment is performed with monochromatic green light ($\lambda = 540$ nm). The separation between the slits is 0.100 mm, and

the interference pattern on a screen has the first side maximum 5.40 mm from the center of the pattern. How far away from the slits is the screen?

34.28 For a double-slit experiment, two 1.50-mm-wide slits are separated by a distance of 1.00 mm. The slits are illuminated by laser light with wavelength 633 nm. If a screen is placed 5.00 m away from the slits, determine the separation of the bright fringes on the screen.

•**34.29** Coherent monochromatic light with wavelength $\lambda = 514$ nm is incident on two thin slits that are separated by a distance $d = 0.500$ mm. The intensity of the radiation at a screen 2.50 m away is 180.0 W/cm². Determine the position $y_{1/3}$ at which the intensity at the central peak (at $y = 0$) drops to $I_0/3$ (where I_0 is the intensity at $\theta = 0°$).

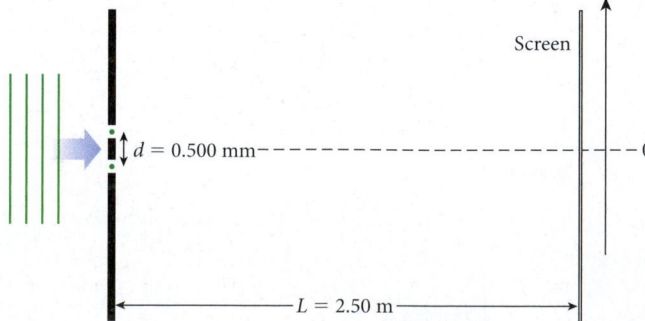

•**34.30** In a double-slit experiment, He-Ne laser light of wavelength 633 nm produced an interference pattern on a screen placed some distance from the slits. When one of the slits was covered with a thin glass slide of thickness 12.0 μm, the central bright fringe shifted to the point occupied earlier by the 10th dark fringe (see the figure). What is the index of refraction of the glass slide?

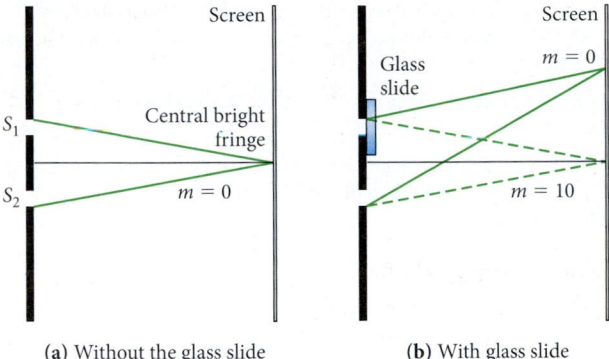

(a) Without the glass slide   **(b)** With glass slide

34.31 Suppose the thickness of a thin soap film ($n = 1.32$) surrounded by air is nonuniform and gradually tapers. Monochromatic light of wavelength 550 nm illuminates the film. At the thinnest end, a dark fringe is observed. How thick is the film at the two dark fringes closest to that fringe?

34.32 White light (400. nm $< \lambda <$ 700. nm) shines onto a puddle of water ($n = 1.33$). There is a thin (100.0 nm thick) layer of oil ($n = 1.47$) on top of the water. What wavelengths of light would you see reflected?

34.33 Some mirrors for infrared lasers are constructed with alternating layers of hafnia and silica. Suppose you want to produce constructive interference from a thin film of hafnia ($n = 1.90$) on BK-7 glass ($n = 1.51$) using infrared radiation of wavelength 1.06 μm. What is the smallest film thickness that would be appropriate, assuming that the laser beam is oriented at right angles to the film?

34.34 Sometimes thin films are used as filters to prevent certain colors from entering a lens. Suppose an infrared filter is to be designed to prevent 800.0-nm light from entering a lens. Find the minimum thickness for a film of MgF$_2$ ($n = 1.38$) that will prevent this light from entering the lens.

•34.35 White light shines on a sheet of mica that has a uniform thickness of 1.30 μm. When the reflected light is viewed using a spectrometer, it is noted that light with wavelengths of 433.3 nm, 487.5 nm, 557.1 nm, 650.0 nm, and 780.0 nm is not present in the reflected light. What is the index of refraction of the mica?

•34.36 A single beam of coherent light ($\lambda = 633 \cdot 10^{-9}$ m) is incident on two glass slides, which are touching at one end and are separated by a 0.0200-mm-thick sheet of paper on the other end, as shown in the figure. Beam 1 reflects off the bottom surface of the top slide, and beam 2 reflects off the top surface of the bottom slide. Assume that all the beams are perfectly vertical and are perpendicular to both slides, that is, the slides are nearly parallel (the angle is exaggerated in the figure); the beams are shown at angles in the figure so that they are easier to identify. Beams 1 and 2 recombine at the location of the eye. The slides are 8.00 cm long. Starting from the left end ($x = 0$), at what positions, x_{bright}, do bright bands appear to the observer above the slides? How many bright bands are observed?

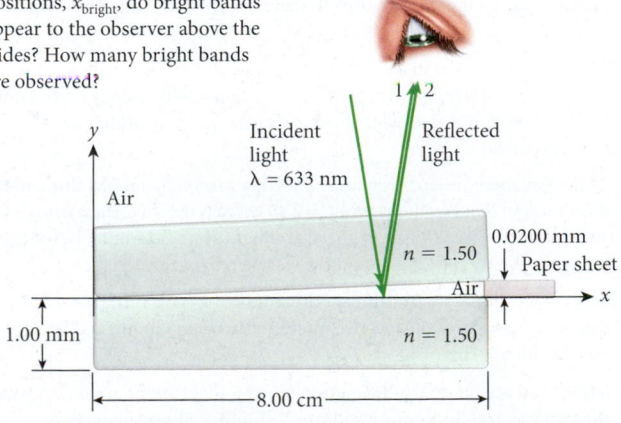

•34.37 A common interference setup for seeing Newton's rings consists of a plano-convex lens placed on a plane mirror and illuminated from above at normal incidence with monochromatic light. In an experiment using a plano-convex lens with focal length $f = 80.00$ cm and index of refraction $n_l = 1.500$, the radius of the third bright circle is found to be 0.8487 mm. Determine the wavelength of the monochromatic light.

34.38 A Michelson interferometer is used in a class of commercially available optical instruments called *wavelength meters*. In a wavelength meter, the interferometer is illuminated simultaneously with parallel beams from a reference laser of known wavelength and an unknown laser. The movable mirror of the interferometer is then displaced by a distance Δd, and the number of fringes produced by each laser and shifting past a reference point (a photo detector) is counted. In a given wavelength meter, a red He-Ne laser ($\lambda_{Red} = 632.8$ nm) is used as a reference laser. When the movable mirror of the interferometer is displaced by a distance Δd, the shifting of $\Delta N_{Red} = 6.000 \cdot 10^4$ red fringes and $\Delta N_{unknown} = 7.780 \cdot 10^4$ fringes is observed by the photo detector.

a) Calculate the wavelength of the unknown laser.

b) Calculate the displacement, Δd, of the movable mirror.

34.39 Monochromatic blue light ($\lambda = 449$ nm) is beamed into a Michelson interferometer. How many fringes shift on the screen when the movable mirror is moved a distance $d = 0.381$ mm?

•34.40 At the Long-baseline Interferometer Gravitational-wave Observatory (LIGO) facilities in Hanford, Washington, and Livingston, Louisiana, laser beams of wavelength 550.0 nm travel along perpendicular paths 4.000 km long. Each beam is reflected along its path and back 100 times before the beams are combined and compared. If a gravitational wave increases the length of one path and decreases the other, each by 1.000 part in 10^{21}, what is the resulting phase difference between the two beams?

Section 34.3

34.41 Light of wavelength 653 nm illuminates a single slit. If the angle between the first dark fringes on either side of the central maximum is 32.0°, what is the width of the slit?

34.42 An instructor uses light of wavelength 633 nm to create a diffraction pattern with a slit of width 0.135 mm. How far away from the slit must the instructor place the screen in order for the full width of the central maximum to be 5.00 cm?

34.43 What is the largest slit width for which there are no minima when the wavelength of the incident light on the single slit is 600. nm?

34.44 Plane microwaves are incident on a single slit of width 2.00 cm. The second minimum is observed at an angle of 43.0°. What is the wavelength of the microwaves?

34.45 The Large Binocular Telescope (LBT), on Mount Graham near Tucson, Arizona, has two 8.4-m-diameter primary mirrors. The mirrors are centered a distance of 14.4 m apart, thus producing a mirror with an effective diameter of 14.4 m. What is the minimum angular resolution of the LBT for green light ($\lambda = 550$ nm)?

34.46 A canvas tent has a single, tiny hole in its side. On the opposite wall of the tent, 2.0 m away, you observe a dot (due to sunlight incident upon the hole) of width 2.0 mm, with a faint ring around it. What is the size of the hole in the tent? (Assume a wave length of 570 nm for the sunlight.)

34.47 Calculate and compare the angular resolutions of the Hubble Space Telescope (aperture diameter, 2.40 m; wavelength, 450. nm), the Keck Telescope (aperture diameter, 10.0 m; wavelength, 450. nm), and the Arecibo radio telescope (aperture diameter, 305 m; wavelength, 0.210 m). Assume that the resolution of each instrument is limited by diffraction.

34.48 The Hubble Space Telescope is capable of resolving optical images to an angular resolution of $2.80 \cdot 10^{-7}$ rad with its 2.40-m mirror (see Figure 34.33). How large would a radio telescope have to be in order to image an object with the same resolution, assuming that the wavelength of the radio waves emitted by the object is 10.0 cm?

34.49 Think of the pupil of your eye as a circular aperture 5.00 mm in diameter. Assume you are viewing light of wavelength 550. nm, to which your eyes are maximally sensitive.

a) What is the minimum angular separation at which you can distinguish two stars?

b) What is the maximum distance at which you can distinguish the two headlights of a car mounted 1.50 m apart?

34.50 A red laser pointer shines light with a wavelength of 635 nm on a double slit, producing a diffraction pattern on a screen that is 1.60 m behind the double slit. The central maximum of the diffraction pattern has a width of 4.20 cm, and the fourth bright spot is missing on both sides. What is the width of the individual slits, and what is the separation between them?

•**34.51** A double slit is opposite the center of a 1.8-m-wide screen that is 2.0 m away. The slit separation is 24 μm, and the width of each slit is 7.2 μm. How many bright fringes are visible on the screen, including the central maximum, if the slit is illuminated by 600.-nm light?

•**34.52** A two-slit apparatus is covered with a red (670-nm) filter. When white light shines on the filter, nine interference maxima appear on a screen, within the 4.50-cm-wide central diffraction maximum. When a blue (450-nm) filter replaces the red filter, how many interference maxima will there be in the central diffraction maximum, and how wide will that diffraction maximum be?

34.53 The intensity pattern observed in a two-slit experiment is presented in the figure. The red line represents the actual intensity measured as a function of angle, while the green line represents the envelope of the single-slit interference pattern.

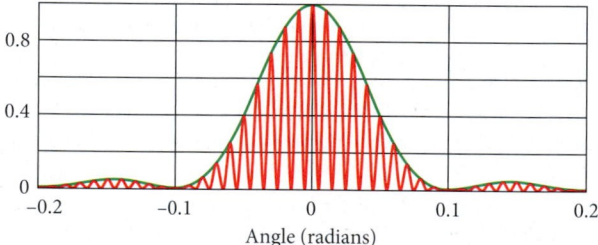

a) Determine the slit width a in terms of the wavelength λ of the light used in the experiment.

b) Determine the center-to-center slit separation d in terms of the wavelength λ.

c) Using the information in the graph, determine the ratio of slit width a to the center-to-center separation between the slits, d.

d) Can you calculate the wavelength of light, actual slit separation, and slit width?

Section 34.4

34.54 Two different wavelengths of light are incident on a diffraction grating. One wavelength is 600. nm, and the other is unknown. If the third-order bright fringe of the unknown wavelength appears at the same position as the second-order bright fringe of the 600.-nm light, what is the value of the unknown wavelength?

34.55 Light from an argon laser strikes a diffraction grating that has 7020 slits per centimeter. The central and first-order maxima are separated by 0.332 m on a wall 1.00 m from the grating. Determine the wavelength of the laser light.

•**34.56** A 5.000-cm-wide diffraction grating with 200 slits is used to resolve two closely spaced lines (a doublet) in a spectrum. The doublet consists of two wavelengths, $\lambda_a = 629.8$ nm and $\lambda_b = 630.2$ nm. The light illuminates the entire grating at normal incidence.

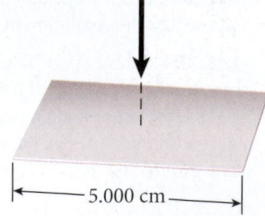

Calculate to *four significant digits* the angles θ_{1a} and θ_{1b} with respect to the normal at which the first-order diffracted beams for the two wavelengths, λ_a and λ_b, respectively, will be reflected from the grating. Note that this is not 0°! What order of diffraction is required to resolve these two lines using this grating?

•**34.57** A diffraction grating has $4.00 \cdot 10^3$ slits/cm and has white light (400.–700. nm) incident on it. What wavelength(s) will be visible at 45.0°?

34.58 What is the wavelength of the X-rays if first-order Bragg diffraction is observed at 23.0° relative to the crystal surface, with an interatomic distance of 0.256 nm?

Additional Exercises

34.59 How many slits per centimeter must a grating have if there are to be no second-order maxima or minima for any visible wavelength (400.–700. nm)?

34.60 Many times, radio antennas occur in pairs. They then produce constructive interference in one direction while producing destructive interference in another direction—acting as a directional antenna—so that their emissions don't overlap with nearby stations. How far apart at a minimum should a local radio station, operating at 88.1 MHz, place its pair of antennas operating in phase so that no emission occurs along a line 45.0° from the line joining the antennas?

34.61 A laser produces a coherent beam of light that does not spread (diffract) as much in comparison to light from other sources, like an incandescent bulb. Lasers therefore have been used for very accurate measurements of large distances, such as the distance between the Moon and the Earth. In one such experiment, a laser pulse (wavelength of 633 nm) is fired at the Moon. What should be the size of the circular aperture of the laser source in order to produce a central maximum of 1.00-km diameter on the surface of the Moon? The distance between the Moon and the Earth is $3.84 \cdot 10^5$ km.

34.62 A diffraction grating with exactly 1000 slits per centimeter is illuminated by a He-Ne laser of wavelength 633 nm.

a) What is the highest order of diffraction that could be observed with this grating?

b) What would be the highest order if there were exactly 10,000 slits per centimeter?

34.63 The thermal stability of a Michelson interferometer can be improved by submerging it in water. Consider an interferometer that is submerged in water, measuring light from a monochromatic source that is in air. If the movable mirror moves a distance $d = 0.200$ mm, exactly $N = 800$ fringes are shifted on the screen. What is the original wavelength (in air) of the monochromatic light?

34.64 A Blu-ray player uses a blue laser that produces light with a wavelength in air of 405 nm. If the disc is protected with polycarbonate ($n = 1.58$), determine the minimum thickness of the disc for destructive interference. Compare this value to that for CDs illuminated by infrared light.

34.65 An airplane is made invisible to radar by coating it with a 5.00-mm-thick layer of an antireflective polymer with index of refraction $n = 1.50$. What is the wavelength of the radar waves for which the plane is made invisible?

34.66 Coherent monochromatic light passes through parallel slits and then onto a screen that is at a distance $L = 2.40$ m from the slits. The narrow slits are a distance $d = 2.00 \cdot 10^{-5}$ m apart. If the minimum spacing between bright spots is $y = 6.00$ cm, find the wavelength of the light.

34.67 Determine the minimum thickness of a soap film ($n = 1.32$) that would produce constructive interference when illuminated by light of wavelength 550. nm.

34.68 You are making a diffraction grating that is to be used to resolve the two spectral lines in the sodium D doublet, at wavelengths of

588.9950 nm and 589.5924 nm, by at least 2.00 mm on a screen that is 80.0 cm from the grating. The rulings are to cover a distance of 1.50 cm on the grating. What is the minimum number of rulings you should have on the grating?

34.69 A Michelson interferometer is illuminated with a 600.-nm light source. How many fringes are observed to shift if one of the mirrors of the interferometer is moved a distance of 200. μm?

34.70 What is the smallest object separation you can resolve with your naked eye? Assume that the diameter of your pupil is 3.5 mm and that your eye has a near point of 25 cm and a far point of infinity.

34.71 With a telescope with an objective of diameter 12.0 cm, how close can two features on the Moon be and still be resolved? Take the wavelength of the light to be 550. nm, near the center of the visible spectrum.

34.72 There is air on both sides of a soap film. What is the smallest thickness that the soap film ($n = 1.420$) can have and appear dark if illuminated with 500.-nm light?

34.73 X-rays with a wavelength of 1.00 nm are scattered off two small tumors in a human body. If the two tumors are a distance of 10.0 cm away from the

X-ray detector, which has an entrance aperture of diameter 1.00 mm, what is the minimum separation between the two tumors that will allow the X-ray detector to determine that there are two tumors instead of one?

•**34.74** Glass with an index of refraction of 1.50 is inserted into one arm of a Michelson interferometer that uses a 600.-nm light source. This causes the fringe pattern to shift by exactly 1000 fringes. How thick is the glass?

•**34.75** White light is shone on a very thin layer of mica ($n = 1.57$), and above the mica layer, interference maxima for light of two wavelengths (and no other in between) are seen: 516.9 nm and 610.9 nm. What is the thickness of the mica layer?

•**34.76** A Newton's ring apparatus consists of a convex lens with a large radius of curvature R placed on a flat glass disc. (a) Show that the horizontal distance x from the center, the thickness d of the air gap, and the radius of curvature R are related by $x^2 = 2Rd$. (b) Show that the radius of the nth constructive interference ring is given by $x_n = [(n + \frac{1}{2})\lambda R]^{1/2}$. (c) How many bright rings may be seen if the apparatus is illuminated by red light of wavelength 700. nm with $R = 10.0$ m and the plane glass disc diameter 5.00 cm?

MULTI-VERSION EXERCISES

34.77 In a double-slit experiment, the slits are $2.49 \cdot 10^{-5}$ m apart. If light of wavelength 477 nm passes through the slits, what will be the distance between the third-order and fourth-order bright fringes on a screen 1.23 m away?

34.78 In a double-slit experiment, the slits are $3.41 \cdot 10^{-5}$ m apart. The distance between the second-order and third-order bright fringes on a screen 1.63 m away is 2.30 cm. What is the wavelength of the light?

34.79 A double-slit experiment using light of wavelength 485 nm produces an interference pattern on a screen that is 2.01 m away from the slits. The distance between the first-order and third-order maxima on the screen is measured to be 4.50 cm. What is the separation between the slits?

34.80 A double-slit experiment using light of wavelength 489 nm produces an interference pattern on a screen. The slit separation is $1.25 \cdot 10^{-2}$ mm. The distance between the first-order and fourth-order maxima on the screen is measured to be 28.05 cm. What is the distance between the slits and the screen?

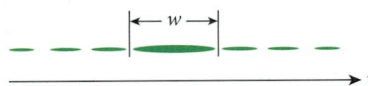

34.81 The single-slit diffraction pattern shown in the figure was produced with light of wavelength 495 nm. The screen on which the pattern was projected is located a distance of 2.77 m from the slit. The slit has a width of 2.73 mm. What is the width w of the central maximum?

34.82 The single-slit diffraction pattern shown in the figure was produced with light from a laser. The screen on which the pattern was projected is located a distance of 3.17 m from the slit. The slit has a width of 0.555 mm. The width of the central maximum is $w = 5.81$ mm. What is the wavelength of the laser light?

34.83 The single-slit diffraction pattern shown in the figure was produced with light of wavelength 503 nm. The screen on which the pattern was projected is located a distance of 3.55 m from the slit. The width of the central maximum is $w = 5.71$ mm. What is the width of the slit?

34.84 The single-slit diffraction pattern shown in the figure was produced with light of wavelength 507 nm incident on a slit of width 0.693 mm. The width of the central maximum is $w = 5.75$ mm. How far is the screen on which the pattern was projected from the slit?

35

Relativity

FIGURE 35.1 A photograph of Einstein's Cross, which is formed by gravitational lensing of a distant quasar by the galaxy in the center.

The picture in Figure 35.1 is not a cluster of stars, as you might assume at first glance. Instead, the central bright spot is a galaxy at a great distance from Earth, and the four spots around it are all images of an even more distant object behind it. The path traveled by the light from the more distant object (called a *quasar*, or *quasi-stellar object*) is curved by the gravitational pull of the intermediate galaxy to form four surrounding images. The fact that light can be bent by gravitation was not predicted by Newton's laws but was deduced from Einstein's theory of relativity. The image in the photograph, called *Einstein's Cross*, is one of many observations that confirm Einstein's theory of relativity.

The theory of relativity changes our understanding of space and time in very basic ways. And some of its consequences seem contradictory to our everyday experiences. The idea that a meter stick in a train station may not be the same length for a person standing in the station as for a person passing by on a train seems preposterous. Some people believe that relativity is impossible and claim that it only describes human perceptions or is "only a theory." However, relativity has been tested as much as any idea in science and has been confirmed every time. Time and space are not independent of the reference frame, and it is not just a wild idea or an optical illusion—it is how our universe works.

WHAT WE WILL LEARN

- Light always moves at the same speed in vacuum, independent of the velocity of the source or the observer.

- The two postulates of special relativity are (1) all physical laws are the same in all inertial reference frames and (2) the speed of light in vacuum is invariant. Inertial frames are reference frames that move with constant velocity.

- From the postulates of special relativity, it follows that measurements of time and space intervals are different for different observers who move relative to each other.

- For high speeds, the Lorentz transformation must be used between reference frames instead of the Galilean transformation.

- Velocities do not add linearly. Velocity addition must be done using a Lorentz transformation so that it conforms to the postulate that the speed of light cannot be exceeded.

- Kinetic energy and momentum and their relationship need new definitions.

- All of Newtonian mechanics derives from relativistic mechanics as a limiting case for speeds small compared to the speed of light.

This chapter focuses primarily on the special theory of relativity, which is called *special* because it deals with the special case of motion with constant velocity, that is, with zero acceleration. Brief mention will be made of some ideas from the general theory of relativity, which does deal with accelerated motion. Although the concepts may seem strange at first, the mathematics are not particularly difficult. Newtonian mechanics can be thought of as a special case of Einstein's work, giving the same results in ordinary situations. But special relativity arrives at quite different results for motion at a significant fraction of the speed of light. These results play a major role in the physics of the very small (high-energy particle physics and quantum mechanics) and the physics of the very large (astronomy and cosmology).

35.1 Space, Time, and the Speed of Light

Chapters 15 and 16 demonstrated that mechanical waves and sound waves need a medium in which to propagate. Chapter 31 showed that light is an electromagnetic wave, and that all electromagnetic waves can propagate through vacuum. However, this knowledge is relatively new in science, only a little more than 100 years old. Up until 1887, scientists believed that light needed a medium in which to propagate, and they called this medium the *luminiferous aether,* or simply the **aether.** (There is another spelling, *ether,* for this medium, but since it is also used as the name for a chemical compound, we avoid it here.) This idea of the aether brought up a question: What exactly is this medium? Light from very distant stars and galaxies reaches our eyes, so it is evidently able to propagate outside of Earth's atmosphere. This observation implies that all of space must be filled with this medium. How could this medium be detected?

If all of space were filled with aether, then Earth would have to move relative to this aether on its path around the Sun. Chapter 3 demonstrated that the motion of the medium makes a clear difference to a trajectory. For example, an airplane moving through wind has a different ground speed if it moves perpendicularly to the wind than if it moves with or against the wind. Chapter 16 showed that this same basic principle applies to the propagation of sound waves though a medium.

In the 19th century, huge efforts were made to measure the speed of light. From the point of view of science history, this quest is a fascinating story in itself. However, detecting the aether does not require knowing the precise speed of light. It only requires knowing that the motion of Earth relative to the aether would imply different speeds of light measured in the lab, depending on the direction of the light's velocity with respect to the aether. This effect is exactly what Albert Michelson and Edward Morley at the Case Institute in Cleveland, Ohio, set out to measure in 1887. They used an ingenious device called an *interferometer* (see Chapter 34, which includes a picture of a Michelson-type interferometer).

What they found stunned the world of physics: a null result! There was no measurable difference! Light moves with exactly the same speed in every direction, and no change in motion relative to the aether could be detected.

Physicists struggled to explain this astounding result. Two leading theorists, Hendrik Lorentz (1853–1928) and George Fitzgerald (1851–1922), came up with the idea that objects moving through the aether become length-contracted just enough to offset the change in the speed of light with direction. It took the genius of Albert Einstein (1879–1955), however, to make the conceptual leap required for a new insight and its astounding consequences: The aether does not exist. Thinking through the fact that the speed of light is constant for all observers, independent of the observer's motion, led Einstein to the formulation of the theory of special relativity, the subject of this chapter.

Einstein's Postulates and Reference Frames

In 1905, a 26-year-old Swiss patent clerk, fresh out of a rather undistinguished university physics career, wrote three scientific articles that shook the scientific world. And what's more amazing, he did this in his spare time! These three papers were

1. A paper explaining that the *photoelectric effect* is due to the quantum nature of light. This explanation earned Einstein the 1921 Nobel Prize in Physics. We will consider this effect in Chapter 36.

2. A paper explaining that *Brownian motion*, which is the motion of very small particles in water or other solutions, is due to collisions with molecules and atoms. This result provided compelling arguments that atoms really exist. (This fact was not at all clear before Einstein's work.)

3. Finally, and most important for this chapter, a paper presenting the theory of special relativity.

Einstein stated two postulates, from which all of special relativity followed. To understand this, we first need a definition: An **inertial reference frame** is a reference frame in which an object accelerates only when a net external force is acting on it. An inertial reference frame moves with constant velocity with respect to any other inertial reference frame. A noninertial frame is a frame for which the point of origin experiences an acceleration. For example, a diver who jumps off a diving board and is in free fall is not in an inertial reference frame, because she is experiencing a net acceleration. Unless stated otherwise, the phrase *reference frame* in this chapter refers to an *inertial* reference frame. With this definition, we can state Einstein's postulates:

Postulate 1: The laws of physics are the same in every inertial reference frame, independent of the motion of the reference frame.

Postulate 2: The speed of light, c, is the same in every inertial reference frame.

The value of the speed of light is

$$c = 299{,}792{,}458 \text{ m/s}. \tag{35.1}$$

As was noted in Chapter 1, this value for the speed of light is the accepted exact value, because it serves as the basis for the definition of the SI unit of the meter. Handy approximate values for the speed of light are $c \approx 186{,}000$ mi/s, or $c \approx 1$ ft/ns in British units, and the very commonly used $c \approx 3 \cdot 10^8$ m/s.

The first of Einstein's two postulates should not raise any objections. In its motion around the Sun, the Earth moves through space with a speed of more than 29 km/s $\approx$ 65,000 mph. Because the Earth orbits very nearly in a circle, the Earth's velocity vector relative to the Sun continually changes direction. However, we still expect that a physics measurement follows laws of nature that are independent of the season in which the measurement is made (neglecting minor effects due to the small accelerations of the Earth and of the Sun relative to the Milky Way).

The second postulate explains the null result that Michelson and Morley measured. However, this idea is not quite so easy to digest. Let's conduct a thought experiment: Suppose you are flying in a rocket through space with a speed of $c/2$, directly toward Earth. You shine a laser in the forward direction. The light of the laser has a speed c. Naively, and with what we know about velocity addition so far, we would expect the light of the laser to have a speed of $c + (c/2) = 1.5c$ when observed on Earth. This result is what we would predict based on the discussion of relative motion in Chapter 3.

This velocity addition, however, only works for speeds that are small relative to the speed of light. Einstein's second postulate says that the speed of light as seen on Earth is still c. Later in this chapter, we will state a rule for velocity addition that is correct for all speeds. For now, note that c is the maximum possible speed that any object can have in any reference frame. This statement is astounding and leads to all kinds of interesting and seemingly counterintuitive consequences—all of which have nonetheless been experimentally verified. Therefore, we now know that Einstein's theory is almost certainly correct. We may never think of space and time the same way again!

Beta and Gamma Factors

Because the speed of light plays such an important role in relativity, we introduce two commonly used dimensionless quantities, beta and gamma, that depend only on the (constant) speed of light, c, and the velocity, $\vec{v}$, of an object:

$$\vec{\beta} = \frac{\vec{v}}{c} \tag{35.2}$$

and

$$\gamma = \frac{1}{\sqrt{1-\beta^2}} = \frac{1}{\sqrt{1-(v/c)^2}}. \tag{35.3}$$

We will use the notation $\beta \equiv |\vec{\beta}|$. Note that for $v \equiv |\vec{v}| \leq c$, equations 35.2 and 35.3 mean that $\beta \leq 1$ and $\gamma \geq 1$.

It is instructive to plot γ as a function of β (Figure 35.2). For speeds that are small compared to the speed of light, β is very small, approximately equal to zero. In that case, γ is very close to 1. However, as β approaches 1, γ *diverges*—that is, γ grows larger and larger and eventually becomes infinite when $\beta = 1$.

There is a useful approximation that is valid for low speeds. In such cases, $|\vec{v}|$ is small compared to c, and therefore β is small compared to 1. By using the mathematical series expansion $(1-x)^{-1/2} = 1 + \frac{1}{2}x^2 + \frac{3}{8}x^4 + \cdots$, we can approximate γ as

$$\gamma \approx 1 + \tfrac{1}{2}\beta^2 = 1 + \tfrac{1}{2}\left(\frac{v}{c}\right)^2 \quad \text{(for } \beta \text{ small compared to 1).} \tag{35.4}$$

Self-Test Opportunity 35.1

A light-year is the distance light travels in a year. Calculate that distance in meters.

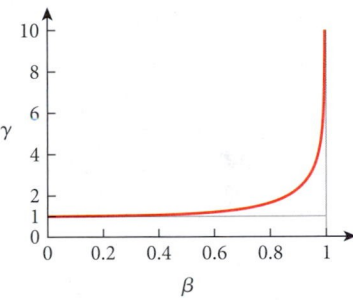

FIGURE 35.2 Dependence of γ on β.

Self-Test Opportunity 35.2

At what fraction of the speed of light would the relativistic effect (deviation between the correct relativistic expression and its nonrelativistic approximation) be 5%? At what fraction of the speed of light would the relativistic effect be 50%?

EXAMPLE 35.1   *Apollo* Spacecraft

On its way to the Moon, the *Apollo* spacecraft reached speeds of $4.0 \cdot 10^4$ km/h relative to Earth, almost an order of magnitude (10 times) faster than any jet aircraft.

PROBLEM
What are the values of the relativistic factors β and γ in this case?

SOLUTION
First, convert the speed to SI units:

$$v = 4.0 \cdot 10^4 \text{ km/h} = 4.0 \cdot 10^4 \text{ km/h} \cdot (1000 \text{ m/km})/(3600 \text{ s/h}) \approx 1.1 \cdot 10^4 \text{ m/s}$$

To compute β, simply divide the spacecraft's speed by the speed of light:

$$\beta = \frac{v}{c} = \frac{1.1 \cdot 10^4 \text{ m/s}}{3.0 \cdot 10^8 \text{ m/s}} = 3.7 \cdot 10^{-5}.$$

Now substitute this result into the formula for γ and obtain

$$\gamma = \frac{1}{\sqrt{1-\beta^2}} = \frac{1}{\sqrt{1-(3.7 \cdot 10^{-5})^2}} = 1 + 6.9 \cdot 10^{-10} = 1.00000000069.$$

From Example 35.1, you can see that almost all motion of macroscopic objects involves values of β that are very close to zero and values of γ that are very close to 1. In all such cases, as we will see throughout this chapter, it is safe to neglect the effects of relativity and calculate the nonrelativistic approximation. However, in many interesting situations, relativity must be kept in mind. This difference is the subject of this chapter.

Light Cones

As a corollary of the two Einstein postulates, we find that nothing can propagate with a speed greater than the speed of light in vacuum. (Strictly speaking, there is the possibility that there exist particles for which the speed of light is not the *upper* but the *lower* limit to their speed. These hypothetical particles, called *tachyons*, will not be discussed any further in this book.) If nothing can propagate with a speed greater than the speed of light in vacuum, then limits exist on how events can influence each other. Two people cannot exchange signals with each other at speeds exceeding the speed of light. Therefore, instantaneous effects of events originating at one point in space on another point in space are impossible. It simply takes time for a signal or a causative influence to propagate through space.

To investigate this result, let's consider the one-dimensional case shown in Figure 35.3. An event occurs at time $t = 0$ at location $x = x_0$ (red dot). A signal that announces this event in space and time can propagate no faster than the speed of light, that is, with a velocity in the interval between $v = -c$ and $v = +c$. The tan-colored triangle in Figure 35.3 shows the region in space and time in which the event can be observed. This region is called the posi-tive **light cone** of the event $(0,x_0)$. The blue point located at time t_1 and position x_1 is able to receive a signal that the original event has happened; however, the green point located at time t_2 and position x_2 is not able to receive such a signal. This implies that the event rep-resented by the red dot cannot possibly have caused the event represented by the green dot in Figure 35.3. The two events, red and green, cannot be causally connected—one cannot have caused the other.

Conversely, there is a region in the past that is able to influence the event at $(0,x_0)$. This region is the negative light cone. Figure 35.4a shows both light cones in the conventional representation with a vertical time axis. The event for which the light cone is displayed is conveniently moved to the origin of the coordinate system. Why is the light cone called a "cone" and not a "triangle"? The answer is shown in Figure 35.4b. For two space coordinates x and y, the condition $v = \sqrt{v_x^2 + v_y^2} = c$ sweeps out a cone in the (2+1)-dimensional x,y,t space. Only events inside the negative light cone can influence events at the origin, at the apex of the light cone. Conversely, only events inside the positive light cone can be influ-enced by the event located at the origin.

Often the axes of light-cone diagrams are scaled so that they have the same units. One way to scale the axes is to multiply the time axis by the speed of light, so that both axes have the units of length. Another method involves dividing the x-axis by the speed of light, so that both axes have the units of time. In this way, the speed of light becomes a diagonal line in the light-cone diagram, which is very useful for making quantitative arguments, as can be seen in the following example.

In a light-cone diagram, we can draw world lines. A **world line** is the trajectory of an object in space and time. This kind of plot is often referred to as a *graph in space-time*, reflecting how these two dimensions are intertwined in relativity theory. A typical plot con-taining several world lines is shown in Figure 35.5. This type of plot represents motion in only one space dimension, x, along with time. Let's imagine we have an object initially located at $x = 0$ and $t = 0$ in this plot. If the object is not moving in the x-direction, the object traces out a vertical line in this plot (arrow 1). If the object is moving in the positive x-direction with a constant speed, its trajectory is represented by a path pointing up and to the right (arrow 2). If the object is moving with a constant speed in the negative x-direction, its trajectory is depicted by a path pointing up and to the left (arrow 3). An object traveling with the speed of light in the positive x-direction is represented by a trajectory with a 45° angle with respect to the vertical axis (arrow 4). An object traveling with the speed of light in the negative x-direction is represented by a trajectory with a −45° angle with respect to the vertical axis (arrow 5).

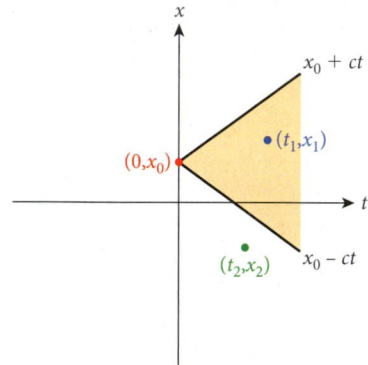

FIGURE 35.3 Light cone of a point in space.

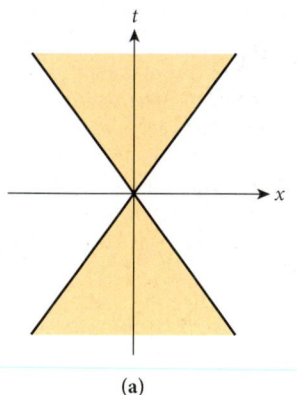

(a)

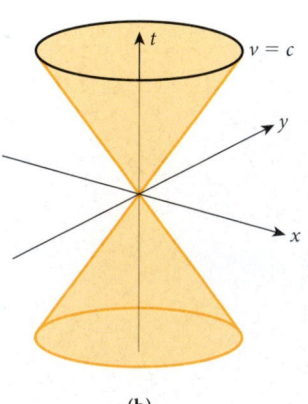

(b)

FIGURE 35.4 Conventional representation of the positive and negative light cones, with the time axis pointing vertically up. (a) (1+1)-dimensional space-time, (b) (2+1)-dimensional space-time.

Space-Time Intervals

In Newtonian mechanics, we can easily determine the distance between two points: $\Delta r = |\vec{r_1} - \vec{r_2}| = \sqrt{(x_1 - x_2)^2 + (y_1 - y_2)^2 + (z_1 - z_2)^2}$. If two events take place at different times, then the time difference between them is $\Delta t = t_2 - t_1$. In view of the discussion of light cones and causality, we introduce the space-time interval s between two events 1 and 2. We define s as follows:

$$s^2 = c^2(\Delta t)^2 - (\Delta r)^2. \tag{35.5}$$

Depending on the sign of s^2, we can distinguish three types of space-time intervals—timelike intervals, lightlike intervals, and spacelike intervals:

$$\left.\begin{array}{lll} s^2 > 0 & \Rightarrow & c^2\Delta t^2 > \Delta r^2 \quad \text{timelike} \\ s^2 = 0 & \Rightarrow & c^2\Delta t^2 = \Delta r^2 \quad \text{lightlike} \\ s^2 < 0 & \Rightarrow & c^2\Delta t^2 < \Delta r^2 \quad \text{spacelike} \end{array}\right\} \text{space-time intervals.} \tag{35.6}$$

Lightlike space-time intervals are on the surface of the light cone, timelike space-time intervals are in the interior of the light cone, and spacelike space-time intervals are on the exterior. In a timelike interval, we can define a proper time interval, $\Delta\tau$, which is the time between two events measured by an observer traveling with a clock in an inertial frame between these events, with the observer's path intersecting the world line of each event as that event occurs. This **proper time** is

$$\Delta\tau = \sqrt{\Delta t^2 - \Delta r^2/c^2}. \tag{35.7}$$

Note that with the timelike condition that $s^2 > 0$, the proper time is a real number (with a time unit). The existence of timelike (or lightlike) intervals between two events means that the two events can be causally connected.

If two events are separated by a spacelike interval, then they cannot be causally connected—that is, neither of the two events can possibly trigger the other one. We can define a proper distance, $\Delta\sigma$, between two events separated by a spacelike interval,

$$\Delta\sigma = \sqrt{\Delta r^2 - c^2\Delta t^2}, \tag{35.8}$$

which is a real number (with a length unit).

35.2 Time Dilation and Length Contraction

In our everyday experience, we consider time and space as being absolute, without restriction or qualification. By "absolute," we mean that observers in all inertial reference frames measure the same value for the length of any object or for the duration of any event. However, if we follow Einstein's postulates to their logical conclusions, this is not the case. The concepts of nonabsolute time and space lead to conclusions that sound like science fiction, but they have been verified experimentally.

Time Dilation

One of the most remarkable consequences of the theory of relativity is that time measurement is not independent of the reference frame. Instead, time that elapses between two events in a moving reference frame where events occur at different locations is dilated (made longer) when compared to the time interval in the rest frame (where the events occur at the same location):

$$\Delta t = \gamma\Delta t_0 = \frac{\Delta t_0}{\sqrt{1 - (v/c)^2}}. \tag{35.9}$$

This **time dilation** means that if a clock advances Δt_0 while at rest, an observer in motion relative to the clock sees the clock advancing by $\Delta t > \Delta t_0$. That time interval Δt depends on the speed v with which the observer is moving relative to the clock! This result is truly revolutionary. However, it has experimentally verifiable consequences, as we will see in Example 35.2.

Concept Check 35.1

An event 0 takes place at some point in space-time, as shown in the figure. Which of the five other events in space-time—A, B, C, D, and/or E—can be influenced by the event 0?

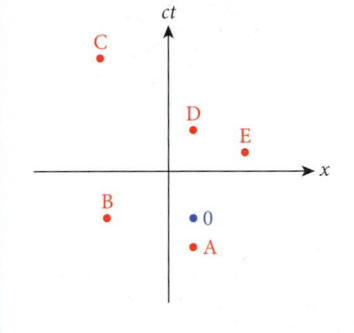

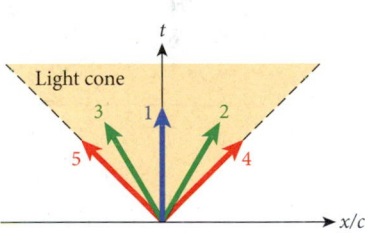

FIGURE 35.5 A graph in space-time showing a positive light cone. Object at rest: blue arrow; objects moving with the speed of light: red arrows; objects moving with a constant speed less than c: green arrows.

Concept Check 35.2

An event 0 takes place at some point in space-time, as shown in the figure. Which of the five other events in spacetime—A, B, C, D, and/or E—form a timelike space-time interval with the event 0?

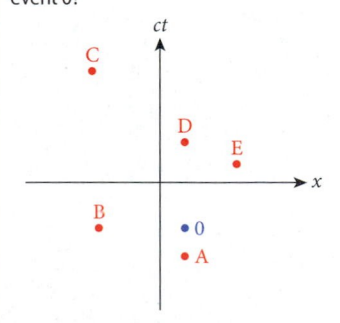

DERIVATION 35.1 | Time Dilation

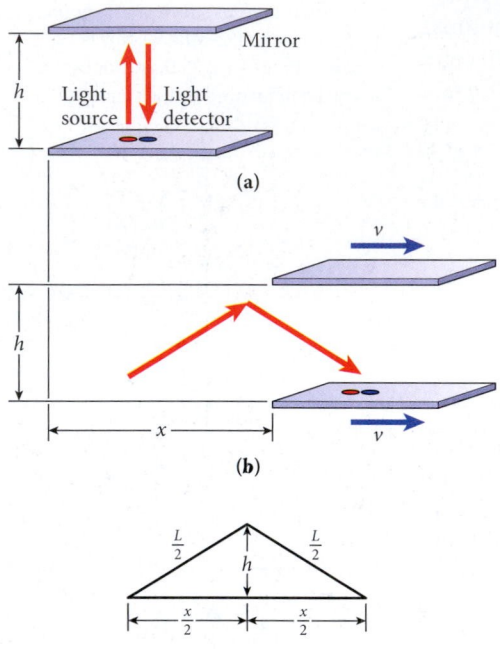

FIGURE 35.6 Measuring time in two different reference frames: (a) The mirror is at rest in this reference frame; (b) the apparatus is moving with velocity v in this reference frame. (c) Triangle formed by the reflected light in the moving reference frame.

How can we derive time dilation? We know that relativistic effects involve the speed of light, so let's construct a clock that keeps time by bouncing a vertical light beam off a mirror and detecting the reflection (Figure 35.6a).

If we know the distance h between the mirror and the light source, then the time for the beam to go up and down is

$$\Delta t_0 = \frac{2h}{c},$$

where the subscript zero refers to the fact that the observer of this time interval is not moving relative to the measurement apparatus. (It takes the same time, h/c, for the light beam to go up as for it to go down.)

Now let's have an observer move to the left with a speed v in the horizontal direction. We call this the negative x-direction, so this observer sees the clock moving in the positive x-direction with a speed v (Figure 35.6b). You can see from Figure 35.6c that in this case the observer sees the light beam as having a path length of

$$L = 2\sqrt{h^2 + (x/2)^2}, \tag{i}$$

where x is the distance that the clock moves while the light beam is passing through the air. This result is simply a consequence of the Pythagorean Theorem. We have tacitly assumed that h is the same for observers in the two reference frames depicted in Figure 35.6. If h were not the same, then we could distinguish one reference frame from the other, in violation of Einstein's first postulate.

The moving observer would say that $x = v\Delta t$, where Δt refers now to the time interval for the light to be emitted and then detected in this observer's system. We can also use the above equation relating h and Δt_0, $h = c\Delta t_0/2$, to eliminate h.

At this point, the second postulate becomes essential. The light beams can travel only with c, the constant speed of light, which is independent of the velocity of the observer! Thus, we have the relation

$$L = c\Delta t. \tag{ii}$$

We substitute for L from equation (ii) into equation (i) and then substitute for h and x to get

$$c\Delta t = 2\sqrt{(c\Delta t_0/2) + (v\Delta t/2)} = \sqrt{(c\Delta t_0)^2 + (v\Delta t)^2}.$$

Solving this equation for Δt, we obtain

$$\Delta t = \frac{\Delta t_0}{\sqrt{1 - (v/c)^2}} = \gamma \Delta t_0.$$

In the clock's rest frame—Figure 35.6a—the time interval is Δt_0; this quantity could represent the time between two clicks of the clock. In the frame in which the clock is moving—Figure 35.6b—the time interval between these two clicks is $\Delta t = \gamma \Delta t_0 > \Delta t_0$. Thus, we say that moving clocks run slow, meaning that the time interval measured is longer when measured in any reference frame in which a clock is moving.

Because the velocity of light is independent of the speed of the observer, we had to admit that the time measured by the two observers is different. This step is remarkably bold.

Just to be clear, there is *nothing* wrong with the clocks, and there's nothing special about a moving clock. Nothing special happens to the clocks due to the high velocities. What is happening is that time itself is slowed down, according to an observer in a different reference frame. This means that everything is slowed down according to this observer, including our motion and our body clocks. We would appear to move in slow motion, and would age slower, compared to the observer in the other reference frame. It is also interesting to note that this effect is reciprocal. That is, observers in two inertial reference frames moving with respect to each other each see the other's clock running slow! But this illustrates the central idea of special relativity—there is no absolute motion; all inertial frames are equally valid for making measurements.

Finally, a remark on notation: In Section 35.1, we introduced the concept of proper time (see equation 35.7). Sometimes you will see the time measured in the rest frame of a clock referred to as *proper time*. The proper time interval used in Derivation 35.1 is Δt_0.

It is one thing to postulate and then mathematically prove the effect of time dilation. It is something else entirely to observe this effect in the laboratory or in nature. Amazingly, this has been done, as discussed in Example 35.2.

EXAMPLE 35.2 ⟋ Muon Decay

A *muon* is an unstable subatomic particle with a mean lifetime, τ_0, of only 2.2 μs. This lifetime can be observed easily when muons decay at rest in the lab. However, when muons are produced in flight at very high speeds, their mean lifetime becomes time dilated. In 1977, an experiment carried out at the European CERN particle accelerator produced muons with a speed of $v = 0.9994c \Rightarrow \beta = 0.9994$. In this case, we have for γ:

$$\gamma = \frac{1}{\sqrt{1-\beta^2}} = \frac{1}{\sqrt{1-0.9994^2}} = 29.$$

Therefore, the mean lifetime, τ, of the muons is expected to be longer by a factor of $\gamma = 29$ at this speed than if they were at rest:

$$\tau = \gamma\tau_0 = 29\,(2.2\ \mu s) = 64\ \mu s.$$

It is quite straightforward to measure this effect by measuring the time between the production and the decay of the muons. Alternatively, you could move some distance away from the production site and see if the muons still reach you. Without the effect of time dilation, during its lifetime of 2.2 μs before it decayed, a muon with this speed could move a distance of only

$$x = v\tau_0 = (0.9994c)\,(2.2\ \mu s) = 660\ m.$$

However, with the effect of time dilation, the distance traveled becomes

$$x = v\tau = v\gamma\tau_0 = (0.9994c)\,29\,(2.2\ \mu s) = 19\ km.$$

To do this experiment, you just need to see how far away from the muon's production site you can detect the decay products. The CERN experiment verified the relativity prediction of time dilation. This measurement shows that time dilation is indeed a real effect. Particles live longer the faster they move. (However, they live the usual time according to clocks in their rest frame.) In fact, giant particle accelerators still in development will collide muons. For these accelerators, muons need to be transported over distances of many kilometers. The fact that these muon colliders are possible at all is due to the effect of time dilation.

Do relativistic effects relate only to subatomic particles and have no relevance for macroscopic objects? No. In 1971, scientists flew four extremely precise atomic clocks around the Earth, once in each direction. They observed that the clocks flying eastward lost 59±10 ns, while the clocks flying westward gained 275±21 ns, compared to a ground-based atomic clock. Thus, the effect of time dilation was confirmed by this experiment with macroscopic clocks. Of course, the effect was incredibly small because the speed of an airliner is small compared with the speed of light. The clocks lost 59 ns and gained 275 ns in 3 days (259,200 s), a few parts per trillion. The quantitative explanation of this time-dilation experiment also involves general relativity. But this is not the point we want to make here. The main point is that this experiment produced a measurable effect, and not a zero effect, as we would expect in the absence of time dilation. The basic fact that this experiment proves is that time can no longer be considered an absolute quantity.

Length Contraction

Relativity implies not only that time is variant and dilated as a function of speed, but also that length is not invariant. The length L of an object moving with speed v shows a **length contraction** relative to its length in its own rest frame, called its **proper length**, L_0. We find that

$$L = \frac{L_0}{\gamma} = L_0\sqrt{1-(v/c)^2}. \tag{35.10}$$

Concept Check 35.3

A clock on a spaceship shows that a time interval of 1.00 s has expired. If this clock is moving with a speed of 0.860c relative to a stationary observer, what time period has expired in the reference frame of the stationary observer?

a) 0.860 s   d) 1.77 s

b) 1.00 s   e) 1.96 s

c) 1.25 s

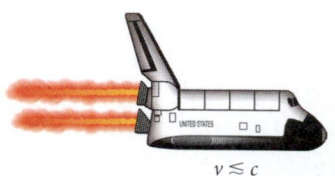

$v \lesssim c$

At rest

FIGURE 35.7 Illustration of length contraction (not to scale!).

DERIVATION 35.2 / Length Contraction

Imagine that you want to measure the length of a space shuttle (Figure 35.7) that has a speed v in your reference frame and a proper length L_0. The way to measure the length without using a meter stick is simple: You can hook up a laser beam to a clock fixed in the laboratory. When the tip of the shuttle breaks the laser beam, it starts the clock, and when the end of the shuttle passes that same point and the laser beam is not blocked anymore, the clock stops. This time interval is Δt_0, or the proper time, and thus $L = v\Delta t_0$.

Inside the shuttle, using a clock fixed to the inside, the measured time during which the laser beam is blocked by the shuttle is $\Delta t = \gamma \Delta t_0$ because of time dilation. Thus, an astronaut inside the shuttle observes that a moving clock with speed v emitted light when passing the shuttle's tip and once again (Δt later) when passing the tail and deduces that $L_0 = v\Delta t = v\gamma \Delta t_0$. Thus, we have

$$\frac{L}{L_0} = \frac{v\Delta t_0}{v\gamma \Delta t_0} \quad \text{or} \quad L = \frac{L_0}{\gamma}.$$

Note that for this thought experiment, it is essential that the length measured is *along the direction of motion*. All lengths perpendicular to the direction of motion remain the same (see Figure 35.7).

FIGURE 35.8 A NASCAR race car.

As you can see from Derivations 35.1 and 35.2, the phenomena of time dilation and length contraction are quite intimately related. The speed of light being constant for all observers implies time dilation, which has been experimentally confirmed many times. Time dilation, in turn, implies length contraction.

The essential fact to remember about length contraction is that moving objects are shorter. They don't just appear shorter—they *are* shorter as measured by an observer in the frame in which the object is moving. This contraction is another mind-bending consequence of the postulates of special relativity.

EXAMPLE 35.3 / Length Contraction of a NASCAR Race Car

You see a NASCAR race car (Figure 35.8) go by at a constant speed of $v = 89.4$ m/s (200 mph). When stopped in the pits, the race car has a length of 5.232 m.

PROBLEM

What is the change in length of the NASCAR race car from your reference frame in the grandstands? Assume that the car is moving perpendicular to your line of sight.

SOLUTION

The length of the race car will be contracted because of its motion. The proper length of the race car is $L_0 = 5.232$ m. The length in your reference frame is given by equation 35.10:

$$L = \frac{L_0}{\gamma} = L_0\sqrt{1 - (v/c)^2} \approx L_0\left(1 - \frac{1}{2}\left(\frac{v}{c}\right)^2\right) = L_0 - \Delta L,$$

where the change in the length of the race car is

$$\Delta L = L_0 \frac{1}{2}\left(\frac{v}{c}\right)^2.$$

Here we have applied a series expansion, $(1 - x^2)^{1/2} = 1 - \frac{1}{2}x^2 + \cdots$, as we did in equation 35.4. The car's speed is small compared with the speed of light, so $v/c \ll 1$ and our expansion is well justified. Thus, the race car is shorter by

$$\Delta L = L_0 \frac{1}{2}\left(\frac{v}{c}\right)^2 = \frac{5.232 \text{ m}}{2}\left(\frac{89.4 \text{ m/s}}{3.00 \cdot 10^8 \text{ m/s}}\right)^2 = 2.32 \cdot 10^{-13} \text{ m}.$$

The car's change in length is smaller than the diameter of a typical atom. So, the length contraction of objects moving at everyday speeds is not noticeable.

Twin Paradox

We have seen that a time interval (say, between clock ticks) depends on the speed of the object (say, a clock) in the frame of an observer, $\Delta t = \gamma \Delta t_0$. Let's perform a little thought experiment:

Astronaut Alice has a twin brother, Bob. At the age of 20, Alice boards a spaceship that flies to a space station 3.25 light-years away from Earth and then returns. The spaceship is a good one and can fly with a speed of 65.0% of the speed of light, resulting in a gamma factor of $\gamma = 1.32$. In Bob's reference frame, the total distance traveled by Alice is 2(3.25 light-years) = 6.50 light-years.

In Alice's reference frame, she travels a distance of $d = 6.50$ light-years/$\gamma = 4.92$ light-years, because the distance between Earth and the space station is length-contracted in her reference frame. Thus, the time it takes Alice to complete the trip is

$$t = d/v = (4.92 c \cdot \text{years})/0.650 c = 7.57 \text{ years}.$$

While the entire trip back and forth takes Alice 7.6 years in Alice's reference frame, time dilation means that $7.6\gamma = 10.0$ years pass in Bob's reference frame. Therefore, when Alice steps out of the spaceship after her trip, she will be 27.6 years old, whereas Bob will be 30.0 years old.

Now we can put ourselves into the reference frame of Alice: In Alice's frame, she was at rest and Bob was moving at 65.0% of the speed of light. Therefore, Alice should have aged 1.32 times more than Bob aged. Because Alice knows that she has aged 7.57 years, she expects her brother Bob to be only 20 + 7.57/1.32 = 25.8 years old when they meet again. Both siblings cannot be younger than each other. This apparent inconsistency is called the *twin paradox*. Which of these two views is right?

The apparent paradox is resolved when we realize that although Bob remains in an inertial reference frame at rest on the Earth for the duration, astronaut Alice lives in two different inertial frames during her round-trip. During the outbound leg, she is moving away from Earth and toward the distant space station. When she reaches the space station, she turns around and travels with a constant speed back from the space station to Earth. Thus, the symmetry is broken between the two twins.

We can analyze the paths of the two twins in space-time using the techniques of light cones and world lines, plotting time in an inertial rest frame versus the positions of both twins in one direction, the x-direction. We analyze the problem from both the point of view of Bob and the point of view of Alice. We start by analyzing the trip in the rest frame of stay-at-home twin Bob, as shown in Figure 35.9, where the axes are scaled so that the units for both are years.

In Figure 35.9, Bob's speed is always zero and he remains at $x = 0$. A red vertical line represents Bob's trajectory. In contrast, Alice is initially moving with a speed of 65.0% of the speed of light ($v = 0.650c$) away from Earth. A blue line labeled $v = 0.650c$ depicts Alice's outbound trajectory. We define the positive x-direction as pointing from the Earth to the distant space station. Each twin wants to keep in touch with the other. Thus, each twin sends an electronic birthday card to the other twin on their birthday in each one's reference frame. These messages travel with the speed of light. Bob sends his electronic message directly toward the space station, and Alice sends her electronic greeting directly back toward Earth. Bob's messages are shown as red arrows pointing up and to the right. Alice's messages are shown as blue arrows pointing up and to the left. When the message arrows cross the trajectory of one of the twins, that twin receives and enjoys an electronic birthday card.

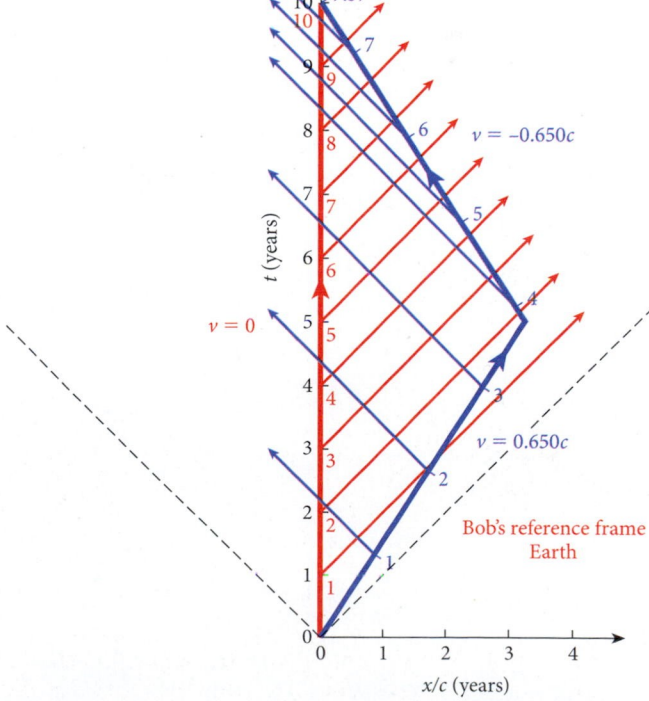

FIGURE 35.9 Plot showing the velocity of the two twins in Bob's reference frame, which is at rest on Earth. The thick, vertical red line represents Bob's trajectory. The two thick blue lines depict Alice's trajectory. Thin red lines labeled by red numbers (corresponding to the years since Alice left) represent Bob's birthday messages. Thin blue lines labeled by blue numbers (also corresponding to the years since Alice left) depict Alice's birthday messages. The dashed lines show the light cone at $t = 0$.

After 5 years pass in Bob's frame and 3.79 years pass in Alice's frame, Alice reaches the space station and turns for home. Bob is getting a little worried by now because he has received only two birthday cards in 5 years. Alice is not feeling much better, since she has received only one message in 3.79 years. After Alice turns around, she receives eight electronic birthday cards in the next 3.79 years. Bob receives five more greetings in the remaining 5 years. When Alice arrives back home on Earth, she gets a firsthand 30th birthday greeting from Bob, but Alice is not ready to celebrate her 28th birthday yet. Alice's age is 27.6 years. Alice received a total of ten birthday greetings while Bob received only seven.

Now let's analyze the same trip from the inertial rest frame corresponding to Alice's outbound leg (Figure 35.10) to show that we can get the same answer from Alice's point of view. In this reference frame, Bob and Earth are both traveling in the negative x-direction with a velocity $v = -0.650c$. During the outbound portion of the trip, Alice's velocity is zero in her reference frame. The space station is traveling toward Alice with a speed of 65.0% of the speed of light, so the distance of 3.25 light-years is covered in 3.79 years because of length contraction. Note that this part of the two diagrams in Figure 35.9 and Figure 35.10 is completely symmetric.

When the space station reaches Alice, the symmetry in the two representations of Figures 35.9 and 35.10 is broken. Alice begins to travel with a speed fast enough to catch up with Earth in the negative x-direction. To establish a relative speed of 65.0% of the speed of light with respect to Earth, Alice must travel with a speed of 91.4% of the speed of light. (This relativistic addition of velocities is discussed in Section 35.3.)

FIGURE 35.10 Plot showing the velocity of the two twins in the reference frame of Alice's outbound leg. The thick red line represents Bob's trajectory. The two thick blue lines depict Alice's trajectory. Thin red lines labeled by red numbers (corresponding to the years since Alice left) represent Bob's birthday messages. Thin blue lines labeled by blue numbers (corresponding to the years since Alice left) depict Alice's birthday messages. The dashed lines represent the light cone at $t = 0$.

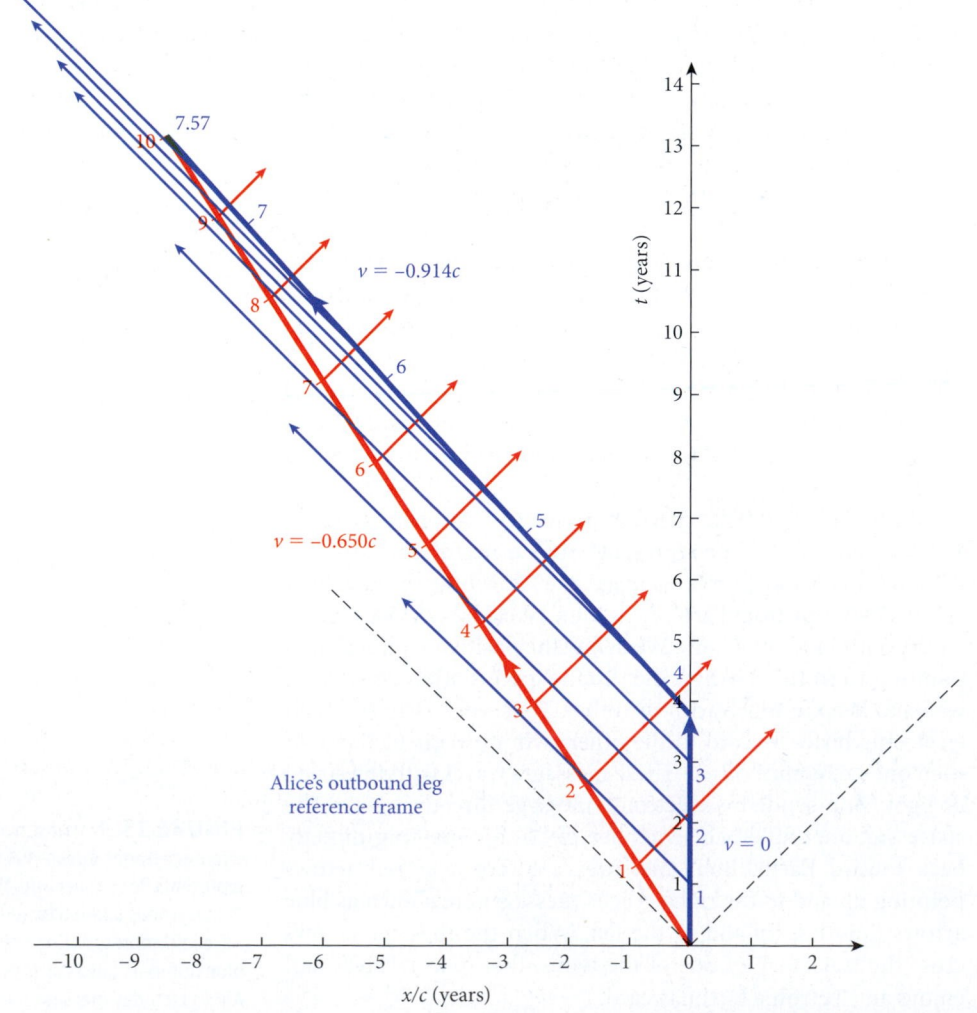

Again Bob receives two birthday greetings in the first 5 years in his reference frame, while Alice again receives one birthday greeting before she starts moving toward Earth. As Alice streaks for Earth at a speed of 0.914c, she receives eight birthday greetings, while Bob again receives five. When the twins are reunited on Earth, Bob is 30.0 years old and Alice is 27.6 years old. This result using Alice's outbound reference frame is the same as the one we obtained using Bob's reference frame. Thus, the twin paradox is resolved.

Note that we analyzed the twin paradox in terms of special relativity only. You might worry about the parts of Alice's journey that involved acceleration. Alice had to accelerate to 65.0% of the speed of light to begin her journey to the space station, and then she had to slow down, stop, and reaccelerate to a speed of 65.0% of the speed of light to return to Earth. Even at a constant acceleration of three times the acceleration of gravity, it would take almost 3 months to reach a speed of 65.0% of the speed of light from rest and the same amount of time to stop, starting at a speed of 65.0% of the speed of light. However, we could postulate various scenarios to remove these objections or at least minimize their importance. For example, we could simply make the trip longer, making the acceleration phase negligible. Acceleration is necessary to explain the twin paradox because Alice must reverse her course to return to Earth, changing her inertial reference frame. However, the effects of general relativity are not needed to explain the twin paradox.

Relativistic Frequency Shift

If time is not invariant, then quantities that depend explicitly on time may also be expected to change as a function of an object's velocity. The frequency f is one such quantity.

In Chapter 16, we discussed the Doppler effect for sound waves—the change in sound's frequency due to the relative motion between observer, source, and medium. We have seen, however, that light does not need a medium in which to propagate. Therefore, the **relativistic frequency shift** is only remotely connected to the Doppler effect. (However, it is often informally referred to as *the relativistic Doppler effect* or the *Doppler effect for electromagnetic radiation*.) If v is the relative velocity between source and observer and the relative motion occurs in a radial direction (source and observer are moving either directly toward each other or directly away from each other), then the formula for the observed frequency, f, of light with a given frequency f_0 is

$$f = f_0 \sqrt{\frac{c \mp v}{c \pm v}}, \tag{35.11}$$

where the upper signs (– in the numerator and + in the denominator) are used for motion away from each other and the lower signs (+ in the numerator and – in the denominator) for motion toward each other. (A transverse Doppler shift also occurs, purely due to time-dilation effects, but we will not consider this effect here.)

Because the relationship $c = \lambda f$ between speed, wavelength, and frequency is still valid, we get for the wavelength:

$$\lambda = \lambda_0 \sqrt{\frac{c \pm v}{c \mp v}}. \tag{35.12}$$

When looking out through our telescopes into the universe, we find that just about all galaxies send light toward us that is **red-shifted,** meaning $\lambda > \lambda_0$ (red is the visible color with the longest wavelength). This means that these galaxies have $v > 0$; that is, they are moving away from us. (If an object is moving toward us, then $\lambda < \lambda_0$ and the light is said to be **blue-shifted.**) In addition, astoundingly, the farther away galaxies are from us, the more they are red-shifted. This observation is clear evidence for an ever-expanding universe.

Often astronomers quote a **red-shift parameter,** also referred to as just the *red-shift.* This quantity is defined as the ratio of the shift in the wavelength of light divided by the wavelength of that light as observed when the source is at rest:

$$z = \frac{\Delta \lambda}{\lambda_0} = \frac{\lambda - \lambda_0}{\lambda_0} = \sqrt{\frac{c \pm v}{c \mp v}} - 1. \tag{35.13}$$

SOLVED PROBLEM 35.1 / Galactic Red-Shift

During a deep-space survey, a galaxy is observed to have a red-shift of $z = 0.450$.

PROBLEM
With what speed is that galaxy moving away from us?

SOLUTION

THINK The red-shift is a function of the speed v with which the galaxy moves away from us. We can solve equation 35.13 for the speed in terms of the red-shift.

SKETCH Figure 35.11 shows a sketch of the galaxy moving away from Earth.

FIGURE 35.11 A galaxy moving away from Earth (not to scale).

RESEARCH We start with equation 35.13, relating the red-shift, z, to the speed v with which the galaxy is moving away from us:

$$z = \sqrt{\frac{c+v}{c-v}} - 1.$$

We can rearrange this equation to get

$$\left(z+1\right)^2 = \frac{c+v}{c-v}$$

$$c\left(z+1\right)^2 - v\left(z+1\right)^2 = c+v.$$

Gathering the multiplicative factors of v on one side and those of c on the other side, we get

$$v + v\left(z+1\right)^2 = c\left(z+1\right)^2 - c.$$

SIMPLIFY We can now write an expression for the velocity of the galaxy moving away from us in terms of the speed of light:

$$\frac{v}{c} = \frac{\left(z+1\right)^2 - 1}{\left(z+1\right)^2 + 1}.$$

CALCULATE Putting in the numerical values gives us

$$\frac{v}{c} = \frac{\left(0.450+1\right)^2 - 1}{\left(0.450+1\right)^2 + 1} = 0.355359.$$

ROUND We report our result to three significant figures:

$$\frac{v}{c} = 0.355.$$

DOUBLE-CHECK A galaxy moving away from us with a velocity of 35.5% of the speed of light seems reasonable (at least it does not exceed the speed of light).

35.3 Lorentz Transformation

In studying time dilation and length contraction, we have seen how time and space intervals can be transformed from one inertial reference frame to another. These transformations of time and space can be combined in going from one reference frame into another that moves with speed v relative to the first one.

To be specific, suppose we have two reference frames, F (with coordinate system x, y, z and time t) and F' (with coordinate system x', y', z' and time t'), which are at the same point $x = x'$, $y = y'$, $z = z'$ (this arrangement can always be accomplished by a simple shift in the location of the origin of the coordinate systems) at the same time $t = t' = 0$, and moving with velocity v relative to each other along the x-axis. We want to know how to describe an event in frame F' (which is some occurrence, like a firecracker explosion) that has coordinates x, y, z and is happening at time t as observed in frame F, using coordinates x', y', z' and time t' (Figure 35.12).

Nonrelativistically, this transformation is given by

$$x' = x - vt$$
$$y' = y$$
$$z' = z$$
$$t' = t.$$

(35.14)

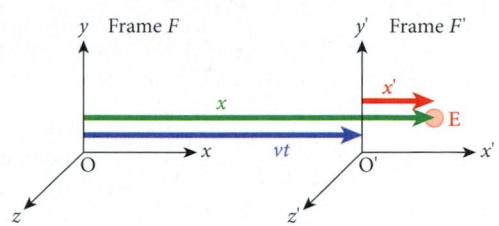

FIGURE 35.12 The Lorentz transformation connects two reference frames moving at a constant velocity v relative to one another.

This transformation is known as the **Galilean transformation,** which we encountered in Chapter 3 when discussing motion in two and three dimensions. These equations are correct as long as v is small compared to c. Before we considered time dilation, the last one of these equations would have seemed utterly trivial. Now this is no longer the case.

The transformation that is valid for all velocities of magnitude v is called the **Lorentz transformation,** first given by the Dutch physicist Hendrik Lorentz (1853–1928):

$$x' = \gamma(x - vt)$$
$$y' = y$$
$$z' = z$$
$$t' = \gamma(t - vx/c^2).$$

(35.15)

We can also construct the inverse transformation:

$$x = \gamma(x' + vt')$$
$$y = y'$$
$$z = z'$$
$$t = \gamma(t' + vx'/c^2).$$

(35.16)

In the limit of small speeds ($\gamma = 1$, $\beta = v/c = 0$), the Galilean transformation follows from the Lorentz transformation as a special case, as can be easily seen. However, for speeds that are not small compared to the speed of light, the Lorentz transformation includes the effects of both time dilation and length contraction.

DERIVATION 35.3 / Lorentz Transformation

There is an ongoing discussion on what constitutes the "simplest" derivation of the Lorentz transformation. Here we follow the paper by J.-M. Lévy, who published this derivation in the *American Journal of Physics,* Vol. 75, No. 7 (2007), pp. 615–618. In this derivation, we make use of the fact that we have already derived the effect of length contraction (Derivation 35.2).

Let's look at Figure 35.12. Assume that $y = y' = 0$, $z = z' = 0$ and consider the event E in frame F (with coordinate system x, y, z and time t). The green arrow of length x marks the position vector of the event E relative to the origin O of frame F. We can use simple vector addition and find that the green arrow, $\overline{OE}$, is the vector sum

$$\overline{OE} = \overline{OO'} + \overline{O'E},$$

(i)

where $\overline{OO'}$ is the displacement vector of the origin of frame F' relative to the origin of F, and $\overline{O'E}$ is the displacement vector from the origin of F' to event E. This vector addition equation (i) is valid independent of the frame in which we are evaluating our results, and now all we need to do is to express equation (i) in both frames.

First, let's evaluate equation (i) in frame F, where the length of the vector $\overline{OE}$ is simply x and the length of the vector $\overline{OO'}$ is vt. Because x' is measured in frame F', expressing it in frame F results in length contraction; that is, in frame F, the length of the vector $\overline{O'E}$ can be expressed as x'/γ. This means that in frame F we obtain the following for equation (i):

$$x = vt + x'/\gamma.$$

(ii)

Now let's perform the same task in frame F' (with coordinate system x', y', z' and time t'). In this frame, we find that the length of $\overline{O'E}$ is x' and the length of $\overline{OO'}$ is vt'. But frame F is

– *Continued*

moving relative to the rest frame, F', and therefore, in frame F', x is length-contracted. Thus, the length of $\overline{OE}$ is x/γ in frame F'. Consequently, in F', we evaluate equation (i) as

$$x/\gamma = vt' + x'. \tag{iii}$$

Now all that is left are some algebraic manipulations. First, we subtract vt from both sides of equation (ii) and then multiply both sides by γ, resulting in

$$\gamma(x - vt) = x', \tag{iv}$$

which is the first equation of the Lorentz transformation (equation 35.15). Multiplying both sides of equation (iii) by γ gives us the first equation of the inverse Lorentz transformation (equation 35.16),

$$x = \gamma(x' + vt'). \tag{v}$$

If we insert the expression for x' from equation (iv) into (v), we find

$$x = \gamma(\gamma(x - vt) + vt') = \gamma^2(x - vt) + \gamma vt' \Rightarrow$$
$$x/\gamma^2 = x - vt + vt'/\gamma \Rightarrow$$
$$-(1 - \gamma^{-2})x = -vt + vt'/\gamma.$$

Self-Test Opportunity 35.3

By inserting the expression for x from equation (v) in Derivation 35.3 into equation (iv), one is able to derive the equation $t = \gamma(t' + vx'/c^2)$. Can you show this?

Since $\gamma = 1/\sqrt{1 - v^2/c^2}$, it follows that $\gamma^2 = 1/(1 - v^2/c^2)$ and therefore $1 - \gamma^{-2} = v^2/c^2$. This leads to

$$-xv^2/c^2 = -vt + vt'/\gamma$$
$$-xv/c^2 = -t + t'/\gamma$$
$$\gamma(t - xv/c^2) = t',$$

which we recognize as the last equation in the Lorentz transformation (equation 35.15).

Concept Check 35.7

Which of the following are invariants under the nonrelativistic Galilean transformation of equation 35.14?

a) x c) z

b) y d) t

Invariants

Under a Lorentz transformation between inertial frames, the coordinate along the direction of relative motion of the two inertial frames (chosen to be the x-coordinate here) changes, as does the time. The two coordinates perpendicular to the velocity vector of the relative motion (chosen to be the y- and z-coordinates here) remain the same in both frames. They are **invariants** under the Lorentz transformation.

What other, if any, kinematical invariants can one construct? At this point, we will not attempt to be exhaustive but will focus on one, which was introduced earlier. This invariant is the space-time interval of equation 35.5, $s^2 = c^2 \Delta t^2 - \Delta r^2$.

DERIVATION 35.4 **Lorentz-Invariance of Space-Time Intervals**

We need to show that the space-time interval is the same in different inertial frames, that is, that $s^2 = s'^2$. Since we have an expression for s^2 in terms of the coordinates and we know how the coordinates transform under a Lorentz transformation, we can construct our proof:

$$s'^2 = c^2 \Delta t'^2 - \Delta r'^2$$
$$= c^2 \Delta t'^2 - \Delta x'^2 - \Delta y'^2 - \Delta z'^2$$
$$= c^2 \gamma^2 (\Delta t - v\Delta x/c^2)^2 - \gamma^2(\Delta x - v\Delta t)^2 - \Delta y^2 - \Delta z^2$$
$$= c^2 \gamma^2 (\Delta t^2 - 2v\Delta x \Delta t/c^2 + v^2 \Delta x^2/c^4) - \gamma^2(\Delta x^2 - 2v\Delta x \Delta t + v^2 \Delta t^2) - \Delta y^2 - \Delta z^2.$$

In the second step, we simply replaced $\Delta r'^2$ by the sum of the squares of the components. In the third step, we applied the Lorentz transformation to each coordinate interval, and in the fourth step, we evaluated the squared terms. Now we can carry out the multiplications and then cancel the underlined terms:

$$s'^2 = c^2 \gamma^2 \Delta t^2 \underline{-2v\gamma^2 \Delta x \Delta t} + v^2 \gamma^2 \Delta x^2/c^2 - \gamma^2 \Delta x^2 \underline{+2v\gamma^2 \Delta x \Delta t} - v^2 \gamma^2 \Delta t^2 - \Delta y^2 - \Delta z^2$$
$$= c^2 \gamma^2 \Delta t^2 + v^2 \gamma^2 \Delta x^2/c^2 - \gamma^2 \Delta x^2 - v^2 \gamma^2 \Delta t^2 - \Delta y^2 - \Delta z^2 \tag{i}$$
$$= c^2 \Delta t^2 (\gamma^2 - v^2 \gamma^2/c^2) - \Delta x^2 (\gamma^2 - v^2 \gamma^2/c^2) - \Delta y^2 - \Delta z^2.$$

Now we need to evaluate the expression in parentheses in the last line of equation (i):

$$\gamma^2 - v^2\gamma^2/c^2 = \frac{1}{1-v^2/c^2} - \frac{v^2/c^2}{1-v^2/c^2} = \frac{1-v^2/c^2}{1-v^2/c^2} = 1.$$

Using this identity, we find for the space-time interval:

$$s'^2 = c^2\Delta t^2 \underbrace{(\gamma^2 - v^2\gamma^2/c^2)}_{1} - \Delta x^2 \underbrace{(\gamma^2 - v^2\gamma^2/c^2)}_{1} - \Delta y^2 - \Delta z^2$$

$$= c^2\Delta t^2 - \Delta x^2 - \Delta y^2 - \Delta z^2$$

$$= c^2\Delta t^2 - \Delta r^2$$

$$= s^2.$$

We find indeed that $s^2 = s'^2$, meaning that the space-time interval is invariant under Lorentz transformation.

Other invariants are introduced below.

Relativistic Velocity Transformation

Let's return to the problem we touched on in Section 35.1. Suppose you are flying in a rocket through space with a speed of $c/2$ directly toward Earth, and you shine a laser in the forward direction. Naively, you might expect that the light of the laser would have a speed of $c + (c/2) = 1.5c$ when observed on Earth. This observation would follow from the velocity-addition formula in Chapter 3, $\vec{v}_{pg} = \vec{v}_{ps} + \vec{v}_{sg}$, where the subscripts p, g, and s refer to the projectile (light, in this case), the ground, and the spaceship. However, according to Einstein, the light of the laser has a speed c in either the rocket's or the Earth's reference frame.

Because the result using the nonrelativistic velocity-addition formula contradicts the postulate that no speed is greater than c, we clearly need a new formula to add velocities correctly. We will restrict the use of **relativistic velocity addition** to one-dimensional motion of the two frames relative to each other in the positive or negative x-directions. If the velocity of an object has a value of $\vec{u} = (u_x, u_y, u_z)$ in frame F, and if frame F' moves with velocity $\vec{v} = (v, 0, 0)$ relative to frame F, then the velocity $\vec{u}' = (u'_x, u'_y, u'_z)$ of that object in frame F' has the value given by

$$u'_x = \frac{u_x - v}{1 - \dfrac{vu_x}{c^2}}$$

$$u'_y = \frac{u_y/\gamma}{1 - \dfrac{vu_x}{c^2}} \qquad (35.17)$$

$$u'_z = \frac{u_z/\gamma}{1 - \dfrac{vu_x}{c^2}}.$$

The inverse transformation from frame F' to frame F is given by

$$u_x = \frac{u'_x + v}{1 + \dfrac{vu'_x}{c^2}}$$

$$u_y = \frac{u'_y/\gamma}{1 + \dfrac{vu'_x}{c^2}} \qquad (35.18)$$

$$u_z = \frac{u'_z/\gamma}{1 + \dfrac{vu'_x}{c^2}}.$$

We have restricted all motion to the positive and negative x-directions, just as was done to arrive at the formula for the Lorentz transformation. However, $\vec{u}$, $\vec{u}'$, and $\vec{v}$ are all vectors and can have positive and negative values, depending on the particular situation to which these transformation equations are applied. In addition, these formulas can also be used to add and subtract two velocities appropriately.

DERIVATION 35.5 | Velocity Transformation

To see how a velocity transforms from one frame to another, we can start with the Lorentz transformation, which tells us how spatial and time coordinates transform. Then we can take the time derivative of the x-coordinate, which gives us the velocity component in the direction of the relative motion of the two frames with respect to each other (we call this velocity u, because we have already used the symbol v for the velocity of the frames relative to each other):

$$u'_x = \frac{dx'}{dt'} = \frac{\gamma(dx - vdt)}{\gamma(dt - vdx/c^2)}.$$

All we have done up to this point is to express the differentials dx' and dt' in the frame F, with the aid of the Lorentz transformation of equations 35.15. Now we cancel out the factor γ in the numerator and denominator and find

$$u'_x = \frac{dx - vdt}{dt - vdx/c^2} = \frac{\dfrac{dx}{dt} - v}{1 - \dfrac{v}{c^2}\dfrac{dx}{dt}}.$$

Substituting from $u_x = dx/dt$, we then get the velocity transformation for the longitudinal component (the component along the direction of the relative motion between the two frames):

$$u'_x = \frac{u_x - v}{1 - \dfrac{vu_x}{c^2}}.$$

Now we can conduct a similar exercise for the transverse components, that is, the components of the velocity vector perpendicular to the direction of relative motion between the two frames. We show this explicitly for the y-direction, but the z-direction is found using the same arguments. Again we take the derivative and Lorentz-transform the differentials, recognizing that $dy' = dy$:

$$u'_y = \frac{dy'}{dt'} = \frac{dy}{\gamma(dt - vdx/c^2)}.$$

Dividing the numerator and denominator by dt then leads to the desired result:

$$u'_y = \frac{\dfrac{dy}{dt}}{\gamma\left(1 - \dfrac{v}{c^2}\dfrac{dx}{dt}\right)} = \frac{u_y/\gamma}{1 - \dfrac{vu_x}{c^2}}.$$

Self-Test Opportunity 35.4

Start with $u_x = dx/dt$, use the inverse Lorentz transformation, and show that you arrive at

$$u_x = \frac{u'_x + v}{1 + \dfrac{vu'_x}{c^2}}.$$

Note that in the limit when the speeds are small compared to the speed of light, $|v| \ll c$ and $|u| \ll c$, the term vu_x/c^2 in the denominator of the velocity transformation equation becomes small compared to 1, and γ approaches 1, so equations 35.17 and 35.18 approach the nonrelativistic limit of $u'_x = u_x - v$ and $u_x = u'_x + v$, $u'_y = u_y$, and $u'_z = u_z$.

Back to our example: Now you can see that $u' = c$ "plus" $v = c/2$ does not yield $1.5c$, but instead

$$u = \frac{u' + c/2}{1 + \dfrac{u'c/2}{c^2}} = \frac{c + c/2}{1 + \dfrac{cc/2}{c^2}} = c.$$

Therefore this analysis works out just right: The speed of light is the same in every reference frame! Thus, equations 35.17 and 35.18 have the proper nonrelativistic limit and at the same time satisfy the postulates of special relativity.

SOLVED PROBLEM 35.2 / Particles in an Accelerator

Suppose you have an electron and an alpha particle (nucleus of a helium atom) moving through a piece of beam pipe inside a particle accelerator. The particles are traveling in the same direction. The speed of the electron is $0.830c$, and the speed of the alpha particle is $0.750c$, both measured by a stationary observer in the lab.

PROBLEM

What is the speed of the alpha particle as observed from the electron, in terms of the speed of light?

SOLUTION

THINK The electron is overtaking the alpha particle, because it is faster, as measured by the stationary observer in the lab. One could be tempted to make constructions of the kind $\vec{v}_1 - \vec{v}_2$. However, since we now have the relativistic velocity transformation, it is far easier to cast this problem in this way. Assume that all given velocities occur in the positive x-direction. All we need to do is to be careful which velocities we identify as v, u_x, and u'_x. We can then use equation 35.17 or equation 35.18 to find our answer.

SKETCH Figure 35.13 is a sketch of the electron passing the alpha particle.

RESEARCH We have two reference frames, our original frame F with origin on Earth and frame F' with origin on the electron, moving with constant velocity v relative to the lab.

Finding the velocity u'_x of the alpha particle in the frame of the electron involves the relativistic transformation of velocities, where the velocity u_x of the alpha particle in the lab's frame is already known. We can then use equation 35.17 to accomplish our goal of finding

$$u'_x = \frac{u_x - v}{1 - \dfrac{vu_x}{c^2}}.$$

SIMPLIFY There is not much to simplify in this case. Most of the work here was in defining which quantity is measured in which frame and identifying the correct relative velocity of the frames.

CALCULATE Putting in the numerical values gives us

$$u'_x = \frac{u_x - v}{1 - \dfrac{vu_x}{c^2}} = \frac{0.750c - 0.830c}{1 - \dfrac{(0.830c)(0.750c)}{c^2}} = -0.2119205c.$$

ROUND We report the result to three significant figures:

$$u'_x = -0.212c.$$

DOUBLE-CHECK The negative sign for the velocity of the alpha particle as seen by the electron makes sense because the alpha particle would be seen to be approaching the electron in the negative x-direction. If we calculate the relative velocity between the electron and the alpha particle nonrelativistically, we obtain

$$v_{rel} = v_{alpha} - v_{electron} = 0.750c - 0.830c = -0.080c.$$

Looking at our solution, we can see that the nonrelativistic velocity difference is modified by a factor of $|1/(1 - u_x v/c^2)|$. As the velocities of the electron and the alpha particle become a significant fraction of the speed of light, this factor will become larger than 1, and the magnitude of the relativistic velocity difference will be larger than the magnitude of the nonrelativistic difference. Thus, our result seems reasonable.

SOLVED PROBLEM 35.3 / Same Velocity

For some purposes, particle physicists need to have beams of electrons that have exactly the same velocity as beams of protons. For a proton, $m_p c^2 = 938$ MeV. For an electron, $m_e c^2 = 0.511$ MeV.

PROBLEM

If you have a proton beam with kinetic energy 2.50 GeV, what is the required kinetic energy for a beam of electrons with the same velocity?

– Continued

Concept Check 35.8

We have just seen that the real sum of two positive velocities consistent with relativity is less than the sum resulting from nonrelativistic velocity addition. What about velocity differences? Assume that two velocity vectors, $\vec{v}_1$ and $\vec{v}_2$, both point in the positive x-direction. Call the relative velocity between the two $\vec{v}_r$. Which of the following is correct?

a) $\left|\vec{v}_r\right|_{nonrelativistic} < \left|\vec{v}_r\right|_{relativistic}$

b) $\left|\vec{v}_r\right|_{nonrelativistic} = \left|\vec{v}_r\right|_{relativistic}$

c) $\left|\vec{v}_r\right|_{nonrelativistic} > \left|\vec{v}_r\right|_{relativistic}$

d) Either (a) or (c) is correct, depending on which of the two velocities, $\vec{v}_1$ or $\vec{v}_2$, is larger.

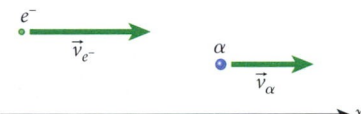

FIGURE 35.13 Electron moving past an alpha particle.

SOLUTION

THINK Particles with the same velocity v have the same value of γ. We can write expressions for the electron and the proton relating γ, the total energy, and the rest mass, solve them for γ, and equate them. The total energy of a particle is equal to the rest mass plus the kinetic energy. We can then solve for the required kinetic energy of the electron.

SKETCH This is one of the rare times when a sketch is not needed.

RESEARCH We can relate the energy E of a particle and γ through

$$E = \gamma m c^2.$$

We can therefore write the energy for an electron as $E_e = \gamma m_e c^2$ and the energy for a proton as $E_p = \gamma m_p c^2$. We can then equate γ for the electron and the proton:

$$\gamma = \frac{E_e}{m_e c^2} = \frac{E_p}{m_p c^2}.$$

The energy of the electron can be written as the sum of kinetic energy and rest energy, $E_e = K_e + m_e c^2$. In the same way, we write for the proton $E_p = K_p + m_p c^2$.

SIMPLIFY

We can combine the preceding equations to get

$$\frac{K_e + m_e c^2}{m_e c^2} = \frac{K_p + m_p c^2}{m_p c^2}.$$

Solving this for the kinetic energy of the electron, we obtain

$$K_e = m_e c^2 \left(\frac{K_p + m_p c^2}{m_p c^2} \right) - m_e c^2.$$

CALCULATE Putting in the numerical values gives us

$$K_e = (0.511 \text{ MeV}) \left(\frac{2500 \text{ MeV} + 938 \text{ MeV}}{938 \text{ MeV}} \right) - (0.511 \text{ MeV}) = 1.36194 \text{ MeV}.$$

ROUND We report our result with three significant figures:

$$K_e = 1.36 \text{ MeV}.$$

DOUBLE-CHECK We can double-check our result by calculating γ for the electron and the proton. For the electron, we have

$$\gamma = \frac{E_e}{m_e c^2} = \frac{1.36 \text{ MeV} + 0.511 \text{ MeV}}{0.511 \text{ MeV}} = 3.66.$$

For the proton,

$$\gamma = \frac{E_p}{m_p c^2} = \frac{2.50 \text{ GeV} + 938 \text{ MeV}}{938 \text{ MeV}} = 3.67,$$

which agrees with our calculated value for the electron within rounding error. Thus, our result seems reasonable.

35.4 Relativistic Momentum and Energy

Length, time, and velocity are not the only concepts that need revision within the theory of special relativity. Energy and momentum need revising as well. Again we are guided by the principle that physical laws should be invariant under the transformation from one frame to another and the realization that for speeds small compared to the speed of light we should again recover the nonrelativistic relationships derived earlier in the mechanics chapters.

Momentum

In Newtonian mechanics (see Chapter 7) momentum is defined as the product of velocity and mass: $\vec{p} = m\vec{v}$. Because we have now seen that an object's speed cannot exceed c, either the definition of momentum has to be changed or a maximum momentum is possible for a given particle. It turns out that the first possibility is the correct one.

The definition of momentum that is consistent with the theory of special relativity is

$$\vec{p} = \gamma\, m\vec{u},$$
(35.19)

where m is the **rest mass** of the particle (its mass as measured in a frame where it is at rest), $\vec{u}$ is the velocity of the particle in some frame, $\gamma = \dfrac{1}{\sqrt{1 - u^2/c^2}}$, and $\vec{p}$ is the momentum of the particle in that frame. Equation 35.19 is valid only for particles having mass $m > 0$. We discuss massless particles later in this section. (Unfortunately, many older textbooks use the notion of *relativistic mass*, γm. This notion has been almost universally rejected. Mass is now considered invariant—that is, independent of the speed at which an object moves in a particular reference frame.)

In Figure 35.14, the two formulas for the momentum are compared. Blue shows the correct formula, and red the nonrelativistic approximation. The velocity is shown in terms of c, and the momentum in terms of mc. Up to speeds of approximately $c/2$, the two formulas give pretty much the same result, but then the correct relativistic momentum goes to infinity as v approaches c.

Newton's Second Law does not have to be modified in the relativistic limit. It remains

$$\vec{F}_{\text{net}} = \frac{d}{dt}\vec{p},$$
(35.20)

where $\vec{p}$ is the relativistic momentum.

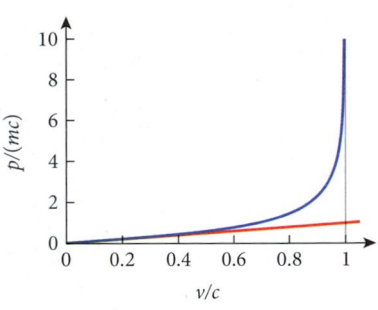

FIGURE 35.14 Momentum as a function of speed. The blue line plots the exact formula; the red line, the nonrelativistic approximation.

Energy

If momentum needs a change of definition, energy can also be expected to be in need of revision.

In our nonrelativistic considerations, we found that the total energy of a particle in the absence of an external potential is just its kinetic energy, $K = \frac{1}{2}mv^2$. In the relativistic case, we find that we have to consider the contribution of the mass to the energy of a particle. Einstein found that the energy of a particle with mass m at rest is

$$E_0 = mc^2.$$
(35.21)

(Arguably the most famous formula in all of science!)

If the particle is in motion, then the energy increases by the same factor γ by which time is dilated for a moving particle. The general case for the energy is then

$$E = \gamma E_0 = \gamma mc^2.$$
(35.22)

The correct formula for the kinetic energy is obtained by subtracting the "rest energy" from the total energy:

$$K = E - E_0 = (\gamma - 1)E_0 = (\gamma - 1)mc^2.$$
(35.23)

The nonrelativistic formula for kinetic energy works at speeds that are small compared to the speed of light. So, how can we recover the nonrelativistic approximation $K = \frac{1}{2}mv^2$ from the general relativistic formula?

This result is actually quite straightforward, because for small speeds we already found that $\gamma = 1 + \frac{1}{2}\beta^2 = 1 + \frac{1}{2}v^2/c^2$. We can insert this expression for γ into the kinetic energy formula (equation 35.23) and get

$$K_{\text{small } v} = (\gamma - 1)mc^2 \approx (1 + \tfrac{1}{2}v^2/c^2 - 1)mc^2 = \tfrac{1}{2}mv^2.$$
(35.24)

DERIVATION 35.6 | Energy

Let's go back to the work–kinetic energy theorem (see Chapter 5) and calculate the consequences of using the Lorentz transformation. For simplicity we use a one-dimensional case with motion along the x-direction only. By definition, work is the integral of the force:

$$W = \int_{x_0}^{x} F\, dx = \int_{x_0}^{x} \frac{dp}{dt}\, dx.$$

– Continued

Here we want to assume that x_0 is the coordinate at $t = 0$ and that the particle has zero speed and therefore zero kinetic energy at $t = 0$. We can calculate dp/dt by using equation 35.19 and taking the time derivative:

$$\frac{dp}{dt} = \frac{d}{dt}(\gamma mv) = \frac{d}{dt}\left(\frac{mv}{\sqrt{1-v^2/c^2}}\right) = \frac{m}{\sqrt{1-v^2/c^2}}\frac{dv}{dt} + (-\tfrac{1}{2})(-2v/c^2)\frac{mv}{(1-v^2/c^2)^{3/2}}\frac{dv}{dt}$$

$$= \left(\frac{m(1-v^2/c^2)}{(1-v^2/c^2)^{3/2}} + \frac{mv^2/c^2}{(1-v^2/c^2)^{3/2}}\right)\frac{dv}{dt} = \frac{m}{(1-v^2/c^2)^{3/2}}\frac{dv}{dt},$$

where we have used the product and chain rules of differentiation. We can insert this result into the above integral for the work. By using the substitution of variables $dx = vdt$, we then find

$$W = \int_0^t \frac{m}{(1-v^2/c^2)^{3/2}}\frac{dv}{dt}v\,dt = \int_0^v \frac{mv\,dv}{(1-v^2/c^2)^{3/2}}.$$

Evaluating this integral, we arrive at

$$W = \left.\frac{mc^2}{\sqrt{1-v^2/c^2}}\right|_0^v = \frac{mc^2}{\sqrt{1-v^2/c^2}} - mc^2 = (\gamma-1)mc^2.$$

Since the work–kinetic energy theorem states that $W = \Delta K$, and since in this case we started with a kinetic energy of zero, this leads us to

$$K = (\gamma-1)mc^2,$$

which is equation 35.23. We also see from the evaluation of the integral that the term mc^2 arises from the contribution of $v = 0$ and can thus be identified as the rest energy of equation 35.21.

Momentum-Energy Relationship

We found that the nonrelativistic kinetic energy and momentum of an object are related via $K = p^2/2m$. Therefore, it is appropriate to ask what the correct general relationship between energy and momentum is. With $E = \gamma mc^2$ and $\vec{p} = \gamma m\vec{v}$ as our starting point, we find

$$E^2 = p^2c^2 + m^2c^4. \tag{35.25}$$

DERIVATION 35.7 | Energy-Momentum Relation

We start with our equations for momentum and energy and square each of them:

$$\vec{p} = \gamma m\vec{v} \Rightarrow p^2 = \gamma^2 m^2 v^2,$$
$$E = \gamma mc^2 \Rightarrow E^2 = \gamma^2 m^2 c^4.$$

The square of the relativistic factor γ appears in both equations. Expressed in terms of v and c, it is

$$\gamma^2 = \frac{1}{1-\beta^2} = \frac{1}{1-v^2/c^2} = \frac{c^2}{c^2-v^2}.$$

Now we can evaluate the expression for the square of the energy, obtaining

$$E^2 = \frac{c^2}{c^2-v^2}m^2c^4 = \frac{c^2-v^2+v^2}{c^2-v^2}m^2c^4 = m^2c^4 + \frac{v^2}{c^2-v^2}m^2c^4$$

$$= m^2c^4 + \frac{c^2}{c^2-v^2}m^2v^2c^2 = m^2c^4 + \gamma^2 m^2 v^2 c^2$$

$$= m^2c^4 + p^2c^2.$$

In the second step of this computation, we simply added and subtracted v^2 in the numerator, then realized that we could split the numerator in such a way that the factor $c^2 - v^2$ would cancel the denominator. In the next step, we rearranged the factors in the second term of the sum so that we could factor out γ again, and finally recognized that $\gamma^2 m^2 v^2 = p^2$, which gives us our desired result.

The energy-momentum relationship of equation 35.25 is often presented in the form of its square root,

$$E = \sqrt{p^2 c^2 + m^2 c^4}.$$ (35.26)

Note that equation 35.25 can also be rewritten in the form

$$E^2 - p^2 c^2 = m^2 c^4.$$ (35.27)

The mass m of a particle is a scalar invariant, and the speed c in vacuum is also invariant, so it follows from this simple rearrangement that $E^2 - p^2 c^2$ is an invariant, just like the space-time interval $s^2 = c^2 \Delta t^2 - \Delta r^2$ is an invariant.

A special case of equation 35.27 is relevant for particles with zero mass. (Photons, the particle representation for all radiation, including visible light, are examples of massless particles.) If a particle has $m = 0$, then the energy-momentum relationship simplifies considerably, and the momentum becomes proportional to the energy:

$$E = pc \quad (\text{for } m = 0).$$ (35.28)

Speed, Energy, and Momentum

For particles with $m > 0$, dividing the absolute value of the momentum, $p = \gamma m v$, by the energy, $E = \gamma m c^2$, gives

$$\frac{p}{E} = \frac{\gamma m v}{\gamma m c^2} = \frac{v}{c^2} \Rightarrow v = \frac{pc^2}{E},$$

or, equivalently,

$$\beta = \frac{v}{c} = \frac{pc}{E}.$$ (35.29)

This formula is often very useful for determining the relativistic factor β from the known energy and momentum of a particle. However, it also provides another energy-momentum relationship that is useful in practice:

$$\beta = \frac{pc}{E} \Rightarrow p = \frac{\beta E}{c} \quad \text{or} \quad E = \frac{pc}{\beta}.$$ (35.30)

EXAMPLE 35.4 | Electron at 0.99c

We have seen that in some applications it is advantageous to use the energy unit electron-volt (eV), $1 \text{ eV} = 1.602 \cdot 10^{-19}$ J. The rest energy of an electron is $E_0 = 5.11 \cdot 10^5 \text{ eV} = 0.511 \text{ MeV}$, and its rest mass is $m = 0.511 \text{ MeV}/c^2$.

PROBLEM

If an electron has a speed that is 99.0% of that of light, what are its total energy, kinetic energy, and momentum?

SOLUTION

First, we calculate γ for this case:

$$\gamma = \frac{1}{\sqrt{1 - \beta^2}} = \frac{1}{\sqrt{1 - (0.990)^2}} = 7.09.$$

The total energy of the electron is given by

$$E = \gamma E_0 = 7.09(0.511 \text{ MeV}) = 3.62 \text{ MeV}.$$

The kinetic energy is therefore

$$K = (\gamma - 1)E_0 = 6.09(0.511 \text{ MeV}) = 3.11 \text{ MeV}.$$

The momentum of this electron is then

$$p = \frac{\beta E}{c} = \frac{(0.990)(3.62 \text{ MeV})}{c} = 3.58 \text{ MeV}/c.$$

Interestingly, we can accelerate electrons to 99% of the speed of light with quite small accelerators, which can fit in labs of typical physics buildings. However, going above $0.99c$ becomes very expensive. To get electrons to 99.9999999% of the speed of light requires a giant particle accelerator. The SLAC linear accelerator at Stanford, over 3 km long, is one such machine.

The equation $E = mc^2$ implies that energy and mass are related and can be interconverted. In chemistry class, you may have heard the expression "conservation of mass," which means that in chemical reactions between different molecules, the number of atoms of each kind has to remain the same. However, this is merely an accounting device, and not a strict conservation law. The conservation laws of energy and momentum still hold in relativistic kinematics, but the total mass in a reaction is *not* conserved exactly. This fact is particularly evident in decays of particles, and Example 35.5 examines one such case.

EXAMPLE 35.5 | Kaon Decay

The neutral kaon is a particle that can decay into a positive pion and a negative pion:

$$K^0 \rightarrow \pi^+ + \pi^-.$$

Particles will be discussed in much greater detail in Chapter 39, but the kinematics of particle decays can be understood just from the principles of relativity.

PROBLEM
What is the kinetic energy of each of the two pions after the decay of the kaon? The mass of the neutral kaon is 497.65 MeV/c^2, and the masses of the positive and negative pions are each 139.57 MeV/c^2. Assume that the neutral kaon is at rest before the decay occurs.

SOLUTION
Momentum and energy are conserved in this decay. The momentum before and after the decay of the kaon is zero. Thus,

$$\vec{p}_{\pi^+} + \vec{p}_{\pi^-} = 0.$$

We see that the magnitudes of the momenta of the two pions are the same. Because the masses of the two pions are also the same, their kinetic energies after the decay must be the same:

$$K_{\pi^+} = K_{\pi^-} = K,$$

where K is the kinetic energy of each pion after the decay. The total energy of the kaon before the decay is

$$E = m_{K^0}c^2,$$

because the kaon is at rest. After the decay, the total energy is

$$E = E_{\pi^+} + E_{\pi^-} = \left(m_{\pi^+}c^2 + K_{\pi^+}\right) + \left(m_{\pi^-}c^2 + K_{\pi^-}\right).$$

Energy conservation means that the total energy before and after the decay is the same:

$$m_{K^0}c^2 = \left(m_{\pi^+}c^2 + K_{\pi^+}\right) + \left(m_{\pi^-}c^2 + K_{\pi^-}\right).$$

We can rearrange this equation to get

$$m_{K^0}c^2 - m_{\pi^+}c^2 - m_{\pi^-}c^2 = K_{\pi^+} + K_{\pi^-} = 2K,$$

remembering that the kinetic energies of the π^+ and the π^- are the same. Solving for the kinetic energy of each pion gives us

$$K = \frac{m_{K^0}c^2 - m_{\pi^+}c^2 - m_{\pi^-}c^2}{2}.$$

Putting in the values of the masses given in the problem statement leads us to the result:

$$K = \tfrac{1}{2}(497.65 \text{ MeV} - 139.57 \text{ MeV} - 139.57 \text{ MeV}) = 109.26 \text{ MeV}.$$

Note that the entire kinetic energy of the two pions, which are the decay products from the kaon at rest, comes from the mass difference between the mass of the kaon and the sum of the masses of the two pions.

The conversion of mass into energy is used in nuclear fission reactions, where heavy atomic nuclei are split into smaller parts, liberating a large amount of kinetic energy in the process, which in turn is converted into heat and ultimately electrical energy. This is the basis for the nuclear power industry and will be covered in much greater detail in Chapter

40 on nuclear physics. There we will also discuss nuclear fusion, which is ultimately based on the same mass-energy relation, and which powers most stars and will hopefully soon be used in commercially available fusion reactors for electrical power generation.

Lorentz Transformation of Momentum and Energy

Derivation 35.3 showed that the position vector $\vec{r}$ and the time t in some frame F can be expressed in some other frame F' as the position vector $\vec{r}'$ and the time t' via a Lorentz transformation. What about momentum and energy?

In equation 35.19, the momentum vector is defined as $\vec{p} = \gamma\, m\vec{u}$. From this followed equation 35.22 for the energy, $E = \gamma mc^2$. Velocities can be transformed from one frame to another using equation 35.17 via the relationship $u_x' = (u_x - v)/(1 - vu_x/c^2)$. Therefore, the Lorentz transformation of momentum and energy can be written as

$$
\begin{aligned}
p_x' &= \gamma(p_x - vE/c^2) \\
p_y' &= p_y \\
p_z' &= p_z \\
E' &= \gamma(E - vp_x).
\end{aligned}
\tag{35.31}
$$

We leave the proof of this relationship to an end-of-chapter exercise. With the aid of this Lorentz transformation, one can show that $E^2 - p^2c^2 = E'^2 - p'^2c^2$, that is, that $E^2 - p^2c^2$ is a Lorentz invariant. This proof is analogous to Derivation 35.4 and is also left as an end-of-chapter exercise.

Two-Body Collisions

Lorentz transformations and Lorentz invariants help in dealing with very common problems in relativistic kinematics, those of two-body collisions. We have just stated that $E^2 - p^2c^2$ is a Lorentz invariant. This definition can be extended to the energies and momentum vectors of two particles, as observed in a certain frame (Figure 35.15). (We use subscripts 1 and 2 to identify the individual particles.) It is common to introduce the quantity S (pronounced simply as *ess*) as

$$
S \equiv (E_1 + E_2)^2 - (\vec{p}_1 + \vec{p}_2)^2 c^2,
\tag{35.32}
$$

where E_1 and $\vec{p}_1$ are the total energy and momentum of particle 1, and E_2 and $\vec{p}_2$ are the total energy and momentum of particle 2. Since $E^2 - p^2c^2$ is a Lorentz invariant, so is S. Thus, we are free to transform our observed quantities into a frame in which they are easiest to evaluate. A convenient frame is the center-of-mass frame. We denote the quantities in the center-of-mass frame with a superscript cm. In Chapter 8, we found that in this frame, the two momentum vectors of the two particles are exactly equal in magnitude and opposite in direction: $\vec{p}_1^{\,cm} = -\vec{p}_2^{\,cm}$. Evaluating S in this frame, we see that

$$
S = (E_1^{cm} + E_2^{cm})^2 - (\underbrace{\vec{p}_1^{\,cm} + \vec{p}_2^{\,cm}}_{0})^2 c^2 = (E_1^{cm} + E_2^{cm})^2.
$$

This shows the physical meaning of S: The square root of S is equal to the sum of the energies of the two particles. That is, it is the total available energy in the center-of-mass frame:

$$
\sqrt{S} = E_1^{cm} + E_2^{cm}.
\tag{35.33}
$$

Thus, $\sqrt{S}$ provides information on the maximum energy that can be used for physical processes in two-body collisions. Because the energy of each of the two particles can also be written as the sum of kinetic energy plus mass energy, equation 35.33 can also be written as

$$
\sqrt{S} = K_1^{cm} + m_1c^2 + K_2^{cm} + m_2c^2.
$$

For practical considerations, the other of the two most interesting inertial frames, besides the center-of-mass frame, is the laboratory frame. The following example shows a useful relationship between the laboratory and the center-of-mass frames.

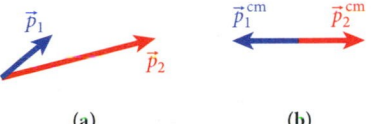

FIGURE 35.15 Momentum vectors of two particles in two different inertial frames. (a) Arbitrary frame; (b) center-of-mass frame.

EXAMPLE 35.6 / Colliders vs. Fixed-Target Accelerators

Suppose you want to create collisions between two protons in order to produce new particles. Two kinds of accelerators can accomplish this. In one, you can have one proton at rest and shoot the other one at it with some kinetic energy. This is the method used in a fixed-target accelerator. A more complicated way to generate collisions is to accelerate the two protons to the same kinetic energy and run them into each other from opposite directions. This technique requires much greater precision, but has a huge payoff if large center-of-mass energies are needed.

PROBLEM

The Relativistic Heavy Ion Collider (RHIC) at Brookhaven National Laboratory can produce proton beams with a kinetic energy of 250 GeV, traveling in opposite directions. What kinetic energy would a proton beam produced by a fixed-target accelerator need to have to reach the same center-of-mass energy?

SOLUTION

The rest mass of the proton is $m_p = 0.938$ GeV/c^2. It is small compared to the kinetic energies in this problem, but if we do not neglect it, we will arrive at a formula that is applicable for all beam energies.

In the center-of-mass frame, the total available energy is ($m_1 = m_2 = m_p$ and $K_1^{cm} = K_2^{cm} \equiv K^{cm}$):

$$S = (2K^{cm} + 2m_p c^2)^2. \tag{i}$$

In the lab frame, which is used by the fixed-target accelerator, we can start with equation 35.32, which is of course valid in all inertial frames, and evaluate

$$S = (E_1^{lab} + E_2^{lab})^2 - (\vec{p}_1^{lab} + \vec{p}_2^{lab})^2 c^2$$

$$= (E_1^{lab})^2 + 2E_1^{lab} E_2^{lab} + (E_2^{lab})^2 - (\vec{p}_1^{lab})^2 c^2 - 2(\vec{p}_1^{lab} \cdot \vec{p}_2^{lab})c^2 - (\vec{p}_2^{lab})^2 c^2$$

$$= \left[(E_1^{lab})^2 - (\vec{p}_1^{lab})^2 c^2\right] + \left[(E_2^{lab})^2 - (\vec{p}_2^{lab})^2 c^2\right] + 2E_1^{lab} E_2^{lab} - 2(\vec{p}_1^{lab} \cdot \vec{p}_2^{lab})c^2.$$

The terms in the square brackets are the squared invariant mass energies of the proton in each case; substituting for the terms in the square brackets from equation 35.27 gives

$$S = 2m_p^2 c^4 + 2E_1^{lab} E_2^{lab} - 2(\vec{p}_1^{lab} \cdot \vec{p}_2^{lab})c^2. \tag{ii}$$

One of the protons, say proton 1, is at rest ($\vec{p}_1^{lab} = 0$, $E_1^{cm} = m_p c^2$). This eliminates the scalar product in equation (ii). For the other proton, we can write the total energy as the sum of kinetic energy and mass energy and then obtain

$$S = 2m_p^2 c^4 + 2m_p c^2 (K^{lab} + m_p c^2) = 4m_p^2 c^4 + 2m_p c^2 K^{lab}. \tag{iii}$$

Because S is invariant, we can set equations (i) and (iii) equal and find

$$4m_p^2 c^4 + 2m_p c^2 K^{lab} = (2K^{cm} + 2m_p c^2)^2$$

$$= 4(K^{cm})^2 + 8K^{cm} m_p c^2 + 4m_p^2 c^4.$$

We solve this for the kinetic energy in the lab frame and find

$$K^{lab} = 4K^{cm} + \frac{2(K^{cm})^2}{m_p c^2}.$$

Clearly, for beam kinetic energies that are small compared to the proton's mass energy, the linear term in this expression dominates. (The linear term is exactly what a nonrelativistic calculation would have found; see Chapter 8.) But for large kinetic energies, the quadratic term dominates, as shown in Figure 35.16. Knowing that the cost of an accelerator increases with the kinetic energy, we can clearly see that only colliders can be used to attain the very high center-of-mass energies needed for cutting-edge particle and nuclear physics experiments.

Finally, here is the numerical answer: If we insert 250 GeV (the value for proton beams at RHIC) as the center-of-mass kinetic energy for each of the two beams, we find that a fixed-target accelerator would need to attain a beam energy of over 134,000 GeV (134 TeV).

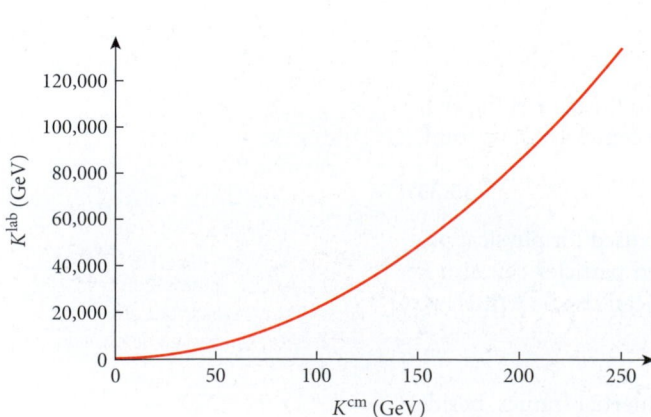

FIGURE 35.16 Equivalent fixed-target beam kinetic energy as a function of the center-of-mass kinetic energy in proton-proton collisions.

35.5 General Relativity

So far, we have talked about only special relativity. The theory of general relativity encompasses all of special relativity but, in addition, provides a theory of gravity. We have seen one theory of gravity that has been used rather successfully since the time of Newton. In the Newtonian theory of gravity, the gravitational force acting on a mass m due to another mass M is given as

$$F_g = \frac{GmM}{r^2} = ma, \tag{35.34}$$

where we have used Newton's Second Law to relate the force to m and the acceleration a. The mass m appears twice in this equation, but there is one very fine difference. The mass that appears on the right side is called the *inertial mass*. This mass is the mass that undergoes the acceleration. However, the quantity that has a role in the force of gravity is also a mass, the *gravitational mass*. Even in our most sensitive experimental tests, we find that inertial mass and gravitational mass are exactly equal, within experimental uncertainty.

The Newtonian theory of gravity served us extremely well. It was in almost complete agreement with all experimental observations. However, some small problems occurred in precision observations, such as a discrepancy in the exact orbit of the planet Mercury, that could not be explained with the Newtonian theory of gravity.

Albert Einstein again had the decisive insight, in 1907: *"If a person falls freely, he will not feel his own weight"* (Figure 35.17). This observation means that you cannot distinguish whether you are in an accelerating reference frame or subject to a gravitational force. This idea led to the famous **Equivalence Principle:**

FIGURE 35.17 Einstein's observation: A free-falling object becomes weightless.

> All local freely falling nonrotating laboratories are equivalent for the performance of all physical experiments.

From this principle, we can prove that space and time are locally curved due to the presence of masses and, in turn, that *curved space-time* affects the motion of masses, determining how they move.

This concept may seem difficult to visualize, but an example might help. Consider a flat rubber sheet that symbolizes space in two dimensions. If you put a bowling ball onto this rubber sheet, it will deform the sheet in the way shown in Figure 35.18a. Any mass that now comes rolling along the surface of the rubber sheet experiences the curvature of the sheet and thus moves as if attracted to the bowling ball. From the viewpoint of general relativity, however, this scenario is a free motion along the shortest path in curved space-time. In Figure 35.18b, a second mass comes along, which causes its own deformation of space that results in a mutual attraction, but the principle remains the same.

One of the most striking predictions of general relativity concerns the motion of light. Because light does not have mass, the Newtonian law of gravity would suggest that gravity cannot affect the motion of light. The theory of general relativity, however, holds that light

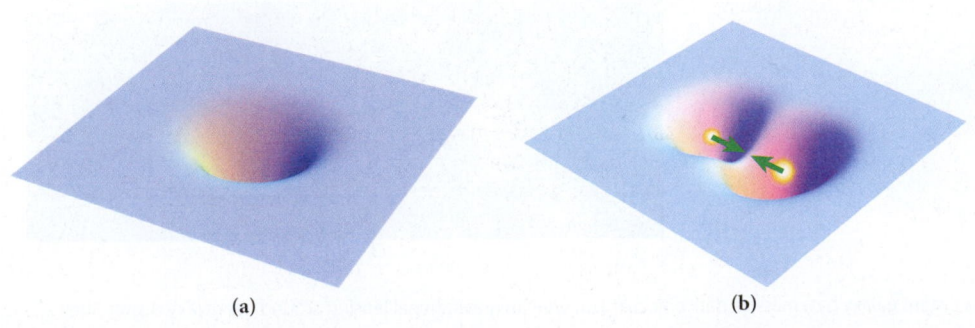

FIGURE 35.18 The deformation of space due to the presence of a massive object. (a) One object deforms space around it; (b) two objects attract each other as a result of their deformation of space.

(a)   (b)

FIGURE 35.19 Photograph taken by Eddington's team of a solar eclipse on May 29, 1919, on the island of Principe, near Africa. The horizontal bars mark the apparent positions of the distant star.

moves through curved space-time along the shortest path and thus should be deflected by the presence of large masses.

Observations by British astronomer Sir Arthur Eddington (1882–1944) during a solar eclipse in 1919 confirmed this spectacular prediction. Figure 35.19 shows the photograph taken by Eddington's team during the solar eclipse.

To visualize what was happening, we again use our rubber sheet analogy. Light from a distant star that passes near the Sun is deflected, as shown in Figure 35.20. The mass of the Sun acts like a lens. The consequence is that an observer sees two stars, or even arcs, instead of just one star. Eddington measured the undeflected direction of the distant star 6 months later by photographing stars at night when the Sun was not near the path of their light. The observed displacement was about $\frac{1}{5}$ the diameter of the image of the distant star. General relativity predicted that the angular separation of the observed image and the actual direction of the star would be 1.74", in good agreement with the observed angular separation. These measurements could be carried out only during a solar eclipse; otherwise, the light from the Sun would simply overpower the light from the stars. The systematic accuracy of these results has been criticized, but subsequent measurements in several solar eclipses have verified them.

Recent observations with the Hubble Space Telescope have demonstrated gravitational lensing by massive dark objects (Figure 35.21). Figure 35.21a shows the light from a distant galaxy following a curved path around an unseen, massive dark object. This effect can produce two or more images of the distant galaxy, as illustrated, or it can produce arcs of light originating from the distant galaxy. Several arcs resulting from the gravitational lensing by a massive dark object are visible in Figure 35.21b.

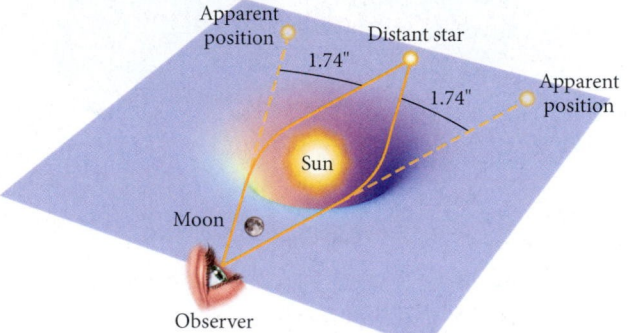

FIGURE 35.20 Light from a distant star is bent around the Sun. The angle is greatly exaggerated for clarity. Such apparent stars are visible close to the Sun only during a solar eclipse.

Black Holes

When light comes close enough to an object of sufficient mass, the light cannot escape from the curvature in space-time generated by that object. What we mean by "close enough" is defined by the **Schwarzschild radius:**

$$R_S = \frac{2GM}{c^2}. \tag{35.35}$$

Every mass has a Schwarzschild radius that can be easily calculated. This formula can be arrived at by setting the escape speed found in Chapter 12, $v_{esc} = \sqrt{\dfrac{2GM}{R}}$, equal to the speed of light and rearranging. If this radius lies within the interior of the object, nothing

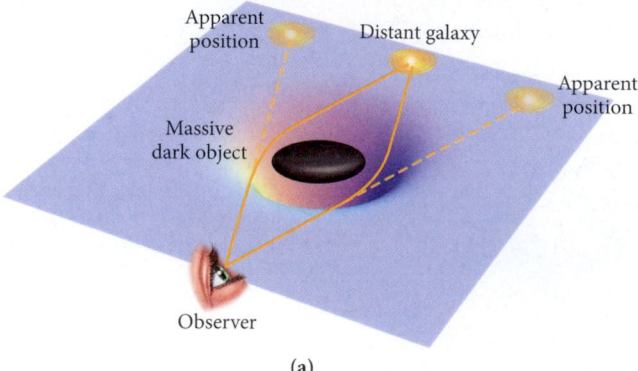

(a)   (b)

FIGURE 35.21 (a) Gravitational lensing of a distant galaxy by a massive dark object; (b) arcs due to gravitational lensing around the galaxy cluster Abell 2218.

remarkable happens. For example, the Schwarzschild radius of Earth (representing the mass of Earth with a point mass of the same magnitude) is

$$R_{S,E} = \frac{2GM_E}{c^2} = \frac{2(6.67 \cdot 10^{-11}\ \text{Nm}^2/\text{kg}^2)(6.0 \cdot 10^{24}\ \text{kg})}{(3.0 \cdot 10^8\ \text{m/s})^2} = 8.9\ \text{mm}.$$

This radius is less than $\frac{1}{2}$ inch and obviously much less than the radius of Earth. The Schwarzschild radius has physical meaning only if the mass M for which it is entirely contained inside the radius.

However, at the end of their lives, stars that had original masses larger than about 15 solar masses collapse to such high densities that their radii are less than their Schwarzschild radii. No information on such an object can escape from it to the outside. We call such an object a **black hole.** (If we could compact the entire Earth to a sphere with less than $\frac{1}{2}$ inch radius, it would turn into a black hole.) Supermassive black holes with masses equal to millions of solar masses are at the centers of many galaxies, including our own. Example 12.4 pointed out the existence of a very massive black hole in the center of the Milky Way and showed that the mass of this black hole is approximately $3.7 \cdot 10^6$ times the mass of our Sun.

Gravitational Waves

One particularly intriguing prediction of general relativity is the existence of gravitational waves. These waves can be created by large masses in strongly accelerated motion. Chapter 15 already touched on the efforts to detect gravitational waves with modern gravitational wave detectors like LIGO (Laser Interferometer Gravitational-Wave Observatory). LIGO consists of two observatories, one in Hanford, Washington, and the other in Livingston, Louisiana, with a distance of 3002 km between them. An even more sensitive gravitational wave observatory is on the drawing board. Tentatively named LISA (Laser Interferometer Space Antenna), it will consist of three spacecraft arranged in an equilateral triangle of side length 5 million km (Figure 35.22).

Perhaps the best chance for observing a source of gravitational wave emission is provided by *binary pulsars*, which consist of two neutron stars (extremely small dense stars with diameters of about 20 km) orbiting each other at a very close distance, with one of the pair being a pulsar (emitting pulses of electromagnetic radiation at very precise time intervals, usually in the range of milliseconds to seconds). These binary pulsars emit very intense gravitational waves. Since the total energy of the system is conserved, a binary pulsar's orbital period decreases over time. This effect was first observed in 1974 by Russell Hulse and his thesis advisor, Joseph Taylor, who shared the 1993 Nobel Prize in Physics for their discovery. They were also able to show that the loss in pulsar period length is consistent with the predictions of general relativity.

Concept Check 35.9

What is the Schwarzschild radius of a black hole with a mass of 14.6 solar masses ($m_{Sun} = 1.99 \cdot 10^{30}$ kg)?

a) 43.2 cm

b) 55.1 m

c) 1.55 km

d) 43.1 km

e) $4.55 \cdot 10^4$ km

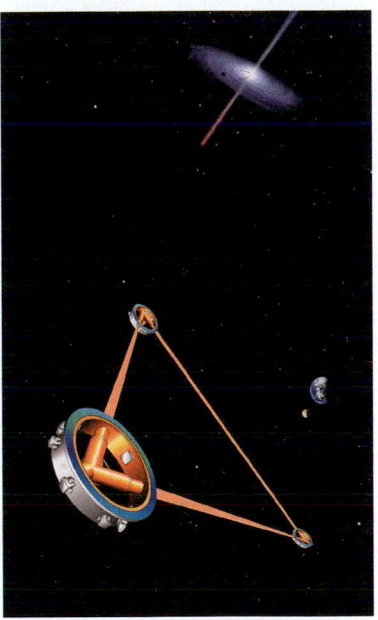

FIGURE 35.22 Artist's concept of LISA (Laser Interferometer Space Antenna), which has been proposed as the next generation of gravitational wave detector.

35.6 Relativity in Our Daily Lives: GPS

While you may argue that most relativistic effects are important only in outer space, at impossibly large speeds, or at the beginning of our universe, relativity touches all of our lives through one technology. This application is the **Global Positioning System (GPS),** which consists of 24 satellites that orbit the Earth (Figure 35.23), at an altitude of about 20,000 km above ground, with a period of one half of a sidereal day [1 sidereal day = the time for the Earth to complete one full rotation so that the stars have the same position again in the night sky = 86,164.09074 s $\approx$ (1 − 1/365.2425)(86,400 s); see also Example 9.3].

GPS operates by having clocks on all satellites synchronized to a very high precision. The modern atomic clocks on board these satellites have fractional time stability to typically 1 part in 10^{13}. By sending synchronized timing signals, the satellites enable users with a GPS receiver to determine their own location in space and time. Typically, a receiver can detect signals from at least four satellites simultaneously. For each satellite, the receiver's position in space, $\vec{r}_r$, and time, t_r, relative to that of satellite i can be determined from the equation

$$\left| \vec{r}_r - \vec{r}_i \right| = c \left| t_r - t_i \right|. \tag{35.36}$$

FIGURE 35.23 GPS satellite in orbit. (Image courtesy of Lockheed Martin Corporation)

Because the positions in time and space of each satellite are known, equation 35.36 is an equation with four unknown quantities: the three spatial coordinates and the time on the clock of the receiver. Detecting signals from four satellites gives the receiver four equations with four unknown quantities. This system of equations provides solutions with an astonishing accuracy of about 1 m.

For these equations to hold, we must rely on the second postulate of special relativity. However, other relativistic effects also need to be included. The satellites move with speeds of approximately 4 km/s relative to Earth, and time-dilation effects cause frequency shifts in the clocks of 1 part in 10^{10} (satellite atomic clocks slow by about 7 μs/day compared to ground-based clocks), which is about 1000 times too large to be ignored. In addition, gravitational corrections due to general relativity are at least of the same magnitude. Therefore, correct application of the theory of relativity is essential for the proper functioning of the Global Positioning System.

WHAT WE HAVE LEARNED | EXAM STUDY GUIDE

- Light always moves at the same speed, independent of the velocity of the source and the observer: $c = 2.99792458 \cdot 10^8$ m/s.

- The two postulates of special relativity are

  (1) The laws of physics are the same in every inertial reference frame, independent of the motion of the reference frame.

  (2) The speed of light, c, is the same in every inertial reference frame.

- The dimensionless quantities $\beta \leq 1$ and $\gamma \geq 1$ are defined as follows:
  $$\beta = |\vec{v}/c| \text{ and } \gamma = \frac{1}{\sqrt{1-\beta^2}} = \frac{1}{\sqrt{1-(v/c)^2}}.$$

- Time is dilated as a function of the speed of the observer, according to
  $$\Delta t = \gamma \Delta t_0 = \frac{\Delta t_0}{\sqrt{1-(v/c)^2}},$$
  where Δt_0 is the time measured in the rest frame.

- Length in the direction of motion is contracted as a function of the speed of the observer, according to
  $$L = \frac{L_0}{\gamma} = L_0 \sqrt{1-(v/c)^2},$$
  where L_0 is the proper length, measured in its rest frame.

- The relativistic frequency shift for light from a moving source is
  $$f = f_0 \sqrt{\frac{c \mp v}{c \pm v}},$$
  and the corresponding wavelength correction is
  $$\lambda = \lambda_0 \sqrt{\frac{c \pm v}{c \mp v}}.$$

  The red-shift parameter is defined as
  $$z = \frac{\Delta \lambda}{\lambda} = \sqrt{\frac{c \pm v}{c \mp v}} - 1.$$

- The Lorentz transformation for space and time coordinates is
  $$x' = \gamma(x - vt); \ y' = y; \ z' = z; \ t' = \gamma(t - vx/c^2).$$

- The relativistic velocity transformation is
  $$u' = \frac{u-v}{1-vu/c^2},$$
  and the inverse transformation is
  $$u = \frac{u'+v}{1+vu'/c^2},$$
  where all velocities are in the same direction.

- The rest energy of a particle is $E_0 = mc^2$.

- The relativistic expression for momentum is $\vec{p} = \gamma m\vec{v}$, and that for energy is $E = \gamma E_0 = \gamma mc^2$.

- The relationship between energy and momentum is given by $E^2 = p^2c^2 + m^2c^4$.

- The Lorentz transformation for momentum and energy is
  $$p'_x = \gamma(p_x - vE/c^2); \ p'_y = p_y; \ p'_z = p_z; \ E' = \gamma(E - vp_x).$$

- The Lorentz invariant,
  $$S \equiv (E_1 + E_2)^2 - (\vec{p}_1 + \vec{p}_2)^2 c^2,$$
  is the square of the total available energy in the center-of-mass frame.

- The Equivalence Principle of the theory of general relativity states that all local freely falling nonrotating laboratories are equivalent for the performance of all physical experiments.

- The Schwarzschild radius of a massive object is given by
  $$R_S = \frac{2GM}{c^2}.$$

ANSWERS TO SELF-TEST OPPORTUNITIES

35.1 1 light-year = $(3.00 \cdot 10^8 \text{ m/s})(365.25 \text{ days/year})$ $(24 \text{ hours/day})(3600 \text{ s/hour}) = 9.47 \cdot 10^{15}$ m

35.2

5% effect:

Exact: $\beta = \sqrt{1 - \gamma^{-2}} = \sqrt{1 - (1.05)^{-2}} = 0.31 = 31\%$ of c

Approximation: $\beta \approx \sqrt{2(\gamma - 1)} = \sqrt{2(1.05 - 1)} = 0.32 = 33\%$ of c

50% effect:

Exact: $\beta = \sqrt{1 - \gamma^{-2}} = \sqrt{1 - (1.50)^{-2}} = 0.75 = 75\%$ of c.

35.3

$\gamma\{\gamma(x' + vt') - vt\} = \gamma^2(x' + vt') - \gamma vt = x' \Rightarrow (\text{divide by } \gamma^2)$

$x' + vt' - vt/\gamma = x'/\gamma^2 \Rightarrow (\text{subtract } x')$

$vt' - vt/\gamma = x'(\gamma^{-2} - 1) = -x'v^2/c^2 \Rightarrow (\text{divide by } v)$

$t' - t/\gamma = -x'v/c^2 \Rightarrow (\text{rearrange})$

$t' + x'v/c^2 = t/\gamma \Rightarrow (\text{multiply with } \gamma)$

$t = \gamma(t' + x'v/c^2).$

35.4 $u_x = \dfrac{dx}{dt} = \dfrac{\gamma(dx' + vdt')}{\gamma(dt' + vdx'/c^2)}, u' = \dfrac{\dfrac{dx'}{dt'} + v}{1 + \dfrac{v}{c^2}\dfrac{dx'}{dt'}} = \dfrac{u'_x + v}{1 + \dfrac{vu'_x}{c^2}}.$

PROBLEM-SOLVING GUIDELINES

1. The first step in attacking any relativity problem is to identify the information. What are the events involved in the problem? What are the reference frames? What is the proper time, and what is the proper length? Review Section 35.2 and be sure you know how to recognize the proper time and proper length.

2. Given the subtleties of relativity theory, checking answers is very important. Check that time has dilated and length

has contracted for a reference frame that is moving relative to a stationary frame.

3. For the Lorentz transformation equations, be sure to identify the two reference frames, as well as the direction of the velocity between them and the velocity of any motion within each frame.

MULTIPLE-CHOICE QUESTIONS

35.1 The most important fact we learned about the aether is that

a) no experimental evidence of its effects was ever found.

b) its existence was proven experimentally.

c) it transmits light in all directions equally.

d) it transmits light faster in the longitudinal direction.

e) it transmits light slower in the longitudinal direction.

35.2 If spaceship A is traveling at 70% of the speed of light relative to an observer at rest, and spaceship B is traveling at 90% of the speed of light relative to an observer at rest, which of the following have the greatest velocity as measured by an observer in spaceship B?

a) a cannonball shot from A to B at 50% of the speed of light as measured in A's reference frame

b) a ball thrown from B to A at 50% of the speed of light as measured in B's reference frame

c) a particle beam shot from a stationary observer to B at 70% of the speed of light as measured in the stationary reference frame

d) a beam of light shot from A to B, traveling at the speed of light in A's reference frame

e) All of the above have the same velocity as measured in B's reference frame.

35.3 A particle of rest mass m_0 travels at a speed $v = 0.20c$. How fast must the particle travel in order for its momentum to increase to twice its original value?

a) 0.40c

b) 0.10c

c) 0.38c

d) 0.42c

e) 0.99c

35.4 Which quantity is invariant—that is, has the same value—in all reference frames?

a) time interval, Δt

b) space interval, Δx

c) velocity, v

d) space-time interval, $c^2(\Delta t)^2 - (\Delta x)^2$

35.5 Two twins, A and B, are in deep space on similar rockets traveling in opposite directions with a relative speed of $c/4$. After a while, twin A turns around and travels back toward twin B again, so that their relative speed is $c/4$. When they meet again, is one twin younger, and if so, which twin is younger?

a) Twin A is younger.

b) Twin B is younger.

c) The twins are the same age.

d) Each twin thinks the other is younger.

35.6 A proton has a momentum of 3.0 GeV/c. With what velocity is it moving relative to a stationary observer?

a) 0.31c

b) 0.33c

c) 0.91c

d) 0.95c

e) 3.2c

35.7 A square of area 100 m² that is at rest in a reference frame is moving with a speed $(\sqrt{3}/2)c$. Which of the following statements is incorrect?

a) $\beta = \sqrt{3}/2$

b) $\gamma = 2$

c) To an observer at rest, it looks like a square with an area less than 100 m².

d) The length along the direction of motion is contracted by a factor of $\frac{1}{2}$.

35.8 Consider a particle moving with a speed less than 0.5*c*. If the speed of the particle is doubled, by what factor will the momentum increase?

a) less than 2   b) 2   c) greater than 2

35.9 Three groups of experimenters measure the average decay time for a specific type of radioactive particle. Group 1 accelerates the particles to 0.5*c*, moving from left to right, and then measures the decay time in the beam, obtaining a result of 20 ms. Group 2 accelerates the particles to −0.5*c*, from right to left, and then measures the decay time in the beam. Group 3 keeps the particles at rest in a container and measures their decay time. Which of the following is true about these measurements?

a) Group 2 measures a decay time of 20 ms.

b) Group 2 measures a decay time less than 20 ms.

c) Group 3 measures a decay time of 20 ms.

d) Both (a) and (c) are true.

e) Both (b) and (c) are true.

35.10 A person in a spaceship holds a meter stick parallel to the motion of the ship as it passes by the Earth with $\gamma = 2$. What length does an observer at rest on the Earth measure for the meter stick?

a) 2 m   c) 0.5 m   e) none of the above

b) 1 m   d) 0.0 m

35.11 A person in a spaceship holds a meter stick as the ship moves parallel to the Earth's surface with $\gamma = 2$. What does the *person in the ship* notice if she rotates the stick from parallel to perpendicular to the ship's motion?

a) The stick becomes shorter.

b) The stick becomes longer.

c) The length of the stick stays the same.

35.12 A person in a spaceship holds a meter stick as the ship moves parallel to the Earth's surface with $\gamma = 2$. What does an *observer on the Earth* notice as the person in the spaceship rotates the stick from parallel to perpendicular to the ship's motion?

a) The stick becomes shorter.

b) The stick becomes longer.

c) The length of the stick stays the same.

35.13 As the velocity of an object increases, so does its energy. What does this imply?

a) At very low velocities, the object's energy is equal to its mass times c^2.

b) In order for an object with mass *m* to reach the speed of light, infinite energy is required.

c) Only objects with $m = 0$ can travel at the speed of light.

d) all of the above

e) none of the above

35.14 In the LHC at CERN in Geneva, Switzerland, two beams of protons are accelerated to $E = 7$ TeV. The proton's rest mass is approximately $m = 1$ GeV/c^2. When the two proton beams collide, what is the rest mass of the heaviest particle that can possibly be created?

a) 1 GeV/c^2

b) 2 GeV/c^2

c) 7 TeV/c^2

d) 14 TeV/c^2

CONCEPTUAL QUESTIONS

35.15 In mechanics, one often uses the model of a perfectly rigid body to determine the motion of physical objects (see, for example, Chapter 10 on rotation). Explain how this model contradicts Einstein's special theory of relativity.

35.16 Use light cones and world lines to help solve the following problem. Eddie and Martin are throwing water balloons very fast at a target. At $t = -13$ μs, the target is at $x = 0$, Eddie is at $x = -2$ km, and Martin is at $x = 5$ km; all three remain in these positions for all time. The target is hit at $t = 0$. Who made the successful shot? Prove this using the light cone for the target. When the target is hit, it sends out a radio signal. When does Martin know the target has been hit? When does Eddie know the target has been hit? Use the world lines to show this. Before starting to draw the diagrams, consider this: If your *x* position is measured in kilometers and you are plotting *t* versus *x*/*c*, what unit must *t* be in?

35.17 A gravitational lens should produce a halo effect and not arcs. Given that the light travels not only to the right and left of the intervening massive object but also to the top and bottom, why do we typically see only arcs?

35.18 Suppose you are explaining the theory of relativity to a friend, and you tell him that nothing can go faster than 300,000 km/s. He says that is obviously false: Suppose a spaceship traveling past you at 200,000 km/s, which is perfectly possible according to what you are saying, fires a torpedo straight ahead whose speed is 200,000 km/s relative to the spaceship, which is also perfectly possible; then, he says, the torpedo's speed is 400,000 km/s. How would you answer him?

35.19 Consider a positively charged particle moving at constant speed parallel to a current-carrying wire, in the direction of the current. As you know (after studying Chapters 27 and 28), the particle is attracted to the wire by the magnetic force due to the current. Now suppose another observer moves along with the particle, so according to him the particle is at rest. Of course, a particle at rest feels no magnetic force. Does that observer see the particle attracted to the wire or not? How can that be? (Either answer seems

to lead to a contradiction: If the particle is attracted, it must be by an electric force because there is no magnetic force, but there is no electric field from a neutral wire; if the particle is not attracted, the observer sees that the particle is, in fact, moving toward the wire.)

35.20 At rest, a rocket has an overall length of *L*. A garage at rest (built for the rocket by the lowest bidder) is only *L*/2 in length. Luckily, the garage has both a front door and a back door, so that when the rocket flies at a speed of $v = 0.866c$, it fits entirely into the garage. However, according to the rocket pilot, the rocket has length *L* and the garage has length *L*/4. How does the rocket pilot observe that the rocket does not fit into the garage?

35.21 A rod at rest on Earth makes an angle of 10° with the *x*-axis. If the rod is moved along the *x*-axis, what happens to this angle, as viewed by an observer on the ground?

35.22 An astronaut in a spaceship flying toward Earth's Equator at half the speed of light observes Earth to be an oblong solid, wider and taller than it appears deep, rotating around its long axis. A second astronaut flying toward Earth's North Pole at half the speed of light observes Earth to be a similar shape but rotating about its short axis. Why does this not present a contradiction?

35.23 Consider two clocks carried by observers in a reference frame moving at speed *v* in the positive *x*-direction relative to Earth's rest frame. Assume that the two reference frames have parallel axes and that their origins coincide when clocks at that point in both frames read zero. Suppose the clocks are separated by a distance *l* in the *x*'-direction in their own reference frame; for instance, $x' = 0$ for one clock and $x' = l$ for the other, with $y' = z' = 0$ for both. Determine the readings t' on both clocks as functions of the time coordinate *t* in Earth's reference frame.

35.24 Prove that in all cases, adding two sub-light-speed velocities relativistically will always yield a sub-light-speed velocity. Consider motion in one spatial dimension only.

35.25 A famous result in Newtonian dynamics is that if a particle in motion collides elastically with an identical particle at rest, the two particles emerge from the collision on perpendicular trajectories. Does the same hold in the special theory of relativity? Suppose a particle of rest mass m and total energy E collides with an identical particle at rest, and the two particles emerge from the collision with new velocities. Are those velocities necessarily perpendicular? Explain.

35.26 Suppose you are watching a spaceship orbiting Earth at 80% of the speed of light. What is the length of the ship as viewed from the center of the orbit?

EXERCISES

A blue problem number indicates a worked-out solution is available in the Student Solutions Manual. One • and two •• indicate increasing level of problem difficulty.

Section 35.1

35.27 Find the speed of light in *feet per nanosecond,* to three significant figures.

35.28 Find the value of g, the gravitational acceleration at Earth's surface, in light-years per year per year, to three significant figures.

35.29 Michelson and Morley used an interferometer to show that the speed of light is constant, regardless of Earth's motion through any purported luminiferous aether. An analogy can be made with the different times it takes a rowboat to travel two different round-trip paths in a river that flows at a constant velocity (u) downstream. Let one path be for a distance D directly across the river, then back again; and let the other path be the same distance D directly upstream, then back again. Assume that the rowboat travels at a constant speed v (with respect to the water) for both trips. Neglect the time it takes for the rowboat to turn around. Find the ratio of the cross-stream time to the upstream-downstream time, as a function of the given constants.

35.30 What is the value of γ for a particle moving at a speed of $0.800c$?

Section 35.2

35.31 An astronaut on a spaceship traveling at a speed of $0.50c$ is holding a meter stick parallel to the direction of motion.

a) What is the length of the meter stick as measured by another astronaut on the spaceship?

b) If an observer on Earth could observe the meter stick, what would be the length of the meter stick as measured by that observer?

35.32 A spacecraft travels along a straight line from Earth to the Moon, a distance of $3.84 \cdot 10^8$ m. Its speed measured on Earth is $0.50c$.

a) How long does the trip take, according to a clock on Earth?

b) How long does the trip take, according to a clock on the spacecraft?

c) Determine the distance between Earth and the Moon if it were measured by a person on the spacecraft.

35.33 A 30.-year-old says goodbye to her 10.-year-old son and leaves on an interstellar trip. When she returns to Earth, both she and her son are 40. years old. What was the speed of the spaceship?

35.34 If a muon is moving at 90.0% of the speed of light, how does its measured lifetime compare to its lifetime of $2.2 \cdot 10^{-6}$ s when it is in the rest frame of a laboratory?

35.35 A fire truck 10.0 m long needs to fit into a garage 8.00 m long (at least temporarily). How fast must the fire truck be going to fit entirely inside the garage, at least temporarily? How long does it take for the truck to get inside the garage, from

a) the garage's point of view?

b) the fire truck's point of view?

35.36 In Jules Verne's classic *Around the World in Eighty Days,* Phileas Fogg travels around the world in, according to his calculation, 81 days. Because he crossed the International Date Line, he actually made it only 80 days. How fast would he have to go in order to have time dilation make 80 days seem like 81? (Of course, at this speed, it would take a lot less than 1 day to get around the world. . . .)

•**35.37** Suppose NASA discovers a planet just like Earth orbiting a star just like the Sun. This planet is 35 light-years away from our Solar System. NASA quickly plans to send astronauts to this planet, but with the condition that the astronauts not age more than 25 years during the journey.

a) At what speed must the spaceship travel, in Earth's reference frame, so that the astronauts age 25 years during their journey?

b) According to the astronauts, what will be the distance of their trip?

•**35.38** Consider a meter stick at rest in a reference frame F. It lies in the x,y-plane and makes an angle of 37° with the x-axis. The reference frame F then moves with a constant velocity v parallel to the x-axis of another reference frame F'.

a) What is the velocity of the meter stick measured in F' at an angle 45° to the x-axis?

b) What is the length of the meter stick in F' under these conditions?

•**35.39** A spaceship shaped like an isosceles triangle has a width of 20.0 m and a length of 50.0 m. What is the angle between the base of the ship and the side of the ship as measured by a stationary observer if the ship is moving past the observer at a speed of $0.400c$? Plot this angle as a function of the speed of the ship.

35.40 How fast must you be traveling relative to a blue light (480 nm) for it to appear red (660 nm)?

35.41 In your physics class, you have just learned about the relativistic frequency shift, and you decide to amaze your friends at a party. You tell them that once you drove through a stoplight and when you were pulled over, you did not get ticketed because you explained to the police officer that the relativistic Doppler shift made the red light of wavelength 650 nm appear green to you, with a wavelength of 520 nm. If your story were true, how fast would you have been traveling?

35.42 A meteor made of pure kryptonite (yes, we know that there really isn't such a thing as kryptonite . . .) is moving toward Earth. If the meteor eventually hits Earth, the impact will cause severe damage, threatening life as we know it. If a laser hits the meteor with light of wavelength 560 nm, the meteor will blow up. The only laser on Earth powerful enough to hit the meteor produces light with a 532-nm wavelength. Scientists decide to launch the laser in a spacecraft and use special relativity to get the right wavelength. The meteor is moving very slowly, so there is no correction for relative velocities. At what speed does the spaceship need to move so that the laser light will have the right wavelength, and should it travel toward or away from the meteor?

35.43 Radar-based speed detection works by sending an electromagnetic wave out from a source and examining the Doppler shift of the reflected wave. Suppose a wave of frequency 10.6 GHz is sent toward a car moving away from the source at a speed of 32.0 km/h. What is the difference between the frequency of the wave emitted by the source and the frequency of the wave an observer in the car would detect?

•**35.44** A He-Ne laser onboard a spaceship moving toward a remote space station emits a beam of red light directed toward the space station. The wavelength of the light in the beam, as measured by a wavelength meter on the spaceship, is 632.8 nm. If the astronauts on the space station see the beam as a blue beam of light with a measured wavelength of 514.5 nm, what is the relative speed of the spaceship with respect to the space station? What is the red-shift parameter, z, in this case?

Section 35.3

35.45 Sam sees two events as simultaneous:

(i) Event A occurs at the point (0,0,0) at the instant 0:00:00 universal time.

(ii) Event B occurs at the point (500. m,0,0) at the same moment.

Tim, moving past Sam with a velocity of $0.999c\hat{x}$, also observes the two events.

a) Which event occurs first in Tim's reference frame?

b) How long after the first event does the second event happen in Tim's reference frame?

35.46 Use relativistic velocity addition to reconfirm that the speed of light with respect to any inertial reference frame is c. Assume one-dimensional motion along a common x-axis.

35.47 You are driving down a straight highway at a speed of $v = 50.0$ m/s relative to the ground. An oncoming car travels with the same speed in the opposite direction. What relative speed do you observe for the oncoming car?

35.48 A rocket ship approaching Earth at $0.90c$ fires a missile toward Earth with a speed of $0.50c$, relative to the rocket ship. As viewed from Earth, how fast is the missile approaching Earth?

35.49 In the twin paradox example (in Section 35.2), Alice boards a spaceship that flies to a space station 3.25 light-years away and then returns with a speed of $0.65c$.

a) Calculate the total distance Alice traveled during the trip, as measured by Alice.

b) Using the total distance from part (a), calculate the total time duration for the trip, as measured by Alice.

•35.50 In the twin paradox example, Alice boards a spaceship that flies to a space station 3.25 light-years away and then returns with a speed of $0.650c$. View the trip in terms of Alice's reference frame.

a) Show that Alice must travel with a speed of $0.914c$ to establish a relative speed of $0.650c$ with respect to Earth when she is returning to Earth.

b) Calculate the time duration for Alice's return flight to Earth at the speed of $0.914c$.

•35.51 Robert, standing at the rear end of a railroad car of length 100. m, shoots an arrow toward the front end of the car. He measures the velocity of the arrow as $0.300c$. Jenny, who was standing on the platform, saw all of this as the train passed her with a velocity of $0.750c$. Determine the following as observed by Jenny:

a) the length of the car

b) the velocity of the arrow

c) the time taken by arrow to cover the length of the car

d) the distance covered by the arrow

••35.52 Consider motion in one spatial dimension. For any velocity v, define the parameter θ via the relation $v = c\tanh\theta$, where c is the speed of light in vacuum. This quantity is called the *velocity parameter* or the *rapidity* corresponding to velocity v.

a) Prove that for two velocities, which add via a Lorentz transformation, the corresponding velocity parameters simply add algebraically, that is, like Galilean velocities.

b) Consider two reference frames in motion at speed v in the x-direction relative to one another, with axes parallel and origins coinciding when clocks at the origin in both frames read zero. Write the Lorentz transformation between the two coordinate systems entirely in terms of the velocity parameter corresponding to v and the coordinates.

Section 35.4

35.53 What is the speed of a particle whose momentum is $p = mc$?

35.54 An electron's rest mass is 0.511 MeV/c^2.

a) How fast must an electron be moving if its energy is to be 10 times its rest energy?

b) What is the momentum of the electron at this speed?

35.55 The Relativistic Heavy Ion Collider (RHIC) can produce colliding beams of gold nuclei with beam kinetic energy of $A \cdot 100$. GeV in the center-of-mass frame, where A is the number of nucleons in a gold nucleus (197). You can approximate the mass energy of a nucleon as 1.00 GeV. What is the equivalent beam kinetic energy for a fixed-target accelerator? (See Example 35.6.)

35.56 How much work is required to accelerate a proton from rest up to a speed of $0.997c$?

35.57 In proton accelerators used to treat cancer patients, protons are accelerated to $0.61c$. Determine the energy of each proton, expressing your answer in mega-electron-volts (MeV).

•35.58 In some proton accelerators, proton beams are directed toward each other to produce head-on collisions. Suppose that in such an accelerator, protons move with a speed relative to the lab reference frame of $0.9972c$.

a) Calculate the speed of approach of one proton with respect to another one with which it is about to collide head on. Express your answer as a multiple of c, using six significant figures.

b) What is the kinetic energy of each proton (in units of MeV) in the laboratory reference frame?

c) What is the kinetic energy of one of the colliding protons (in units of MeV) in the rest frame of the other proton?

•35.59 The hot filament of the electron gun in a cathode ray tube releases electrons with nearly zero kinetic energy. The electrons are next accelerated under a potential difference of 5.00 kV, before being steered toward the phosphor on the screen of the tube.

a) Calculate the kinetic energy acquired by an electron under this accelerating potential difference.

b) Is the electron moving at relativistic speed?

c) What are the electron's total energy and momentum? (Give both values, relativistic and nonrelativistic, for both quantities.)

35.60 Consider a one-dimensional collision at relativistic speeds between two particles with masses m_1 and m_2. Particle 1 is initially moving with a speed of $0.700c$ and collides with particle 2, which is initially at rest. After the collision, particle 1 recoils with a speed of $0.500c$, while particle 2 starts moving with a speed of $0.200c$. What is the ratio m_2/m_1?

•35.61 In an elementary-particle experiment, a particle of mass m is fired, with momentum mc, at a target particle of mass $2\sqrt{2}m$. The two particles form a single new particle (in a completely inelastic collision). Find the following:

a) the speed of the projectile before the collision

b) the mass of the new particle

c) the speed of the new particle after the collision

•35.62 Show that momentum and energy transform from one inertial frame to another as $p'_x = \gamma(p_x - vE/c^2)$; $p'_y = p_y$; $p'_z = p_z$; $E' = \gamma(E - vp_x)$. (*Hint:* Look at Derivation 35.4 for the space-time Lorentz transformation.)

•35.63 Show that $E^2 - p^2c^2 = E'^2 - p'^2c^2$, that is, that $E^2 - p^2c^2$ is a Lorentz invariant. (*Hint:* Look at Derivation 35.4, which shows that the space-time interval is a Lorentz invariant.)

Sections 35.5 and 35.6

35.64 The deviation of the space-time geometry near the Earth from the flat space-time of the special theory of relativity can be gauged by the ratio Φ/c^2, where Φ is the Newtonian gravitational potential at the Earth's surface. Find the value of this quantity.

35.65 Calculate the Schwarzschild radius of a black hole with the mass of

a) the Sun.

b) a proton. How does this result compare with the size scale of 10^{-15} m usually associated with a proton?

35.66 Assuming that the speed of GPS satellites is approximately 4.00 km/s relative to Earth, calculate how much slower per day the atomic clocks on the satellites run, compared to stationary atomic clocks on Earth.

35.67 What is the Schwarzschild radius of the black hole at the center of our Milky Way? (*Hint:* The mass of this black hole was determined in Example 12.4.)

Additional Exercises

35.68 In order to fit a 50.0-foot-long stretch limousine into a 35.0-foot-long garage, how fast would the limousine driver have to be moving, in the garage's reference frame? Comment on what happens to the garage in the limousine's reference frame.

35.69 Using relativistic expressions, compare the momentum of two electrons, one moving at $2.00 \cdot 10^8$ m/s and the other moving at $2.00 \cdot 10^3$ m/s. What is the percent difference between nonrelativistic momentum values and these values?

35.70 Rocket A passes Earth at a speed of $0.75c$. At the same time, rocket B passes Earth moving with a speed of $0.95c$ relative to Earth in the same direction. How fast is B moving relative to A when it passes A?

35.71 Determine the difference in kinetic energy of an electron traveling at $0.9900c$ and one traveling at $0.9999c$, first using standard Newtonian mechanics and then using special relativity.

35.72 Right before take-off, a passenger on a plane flying from town A to town B synchronizes his clock with the clock of his friend who is waiting for him in town B. The plane flies with a constant velocity of 240 m/s. The moment the plane touches the ground, the two friends check their clocks simultaneously. The clock of the passenger on the plane shows that it took exactly 3.00 h to travel from A to B. Ignoring any effects of acceleration:

a) Will the clock of the friend waiting in B show a shorter or a longer time interval?

b) What is the difference between the readings of the two clocks?

35.73 The explosive yield of the atomic bomb dropped on Hiroshima near the end of World War II was approximately 15.0 kilotons of TNT. One kiloton corresponds to about $4.18 \cdot 10^{12}$ J of energy. Find the amount of mass that was converted into energy in this bomb.

35.74 At what speed will the length of a meter stick appear to be 90.0 cm?

35.75 What is the relative speed between two objects approaching each other head on, if each is traveling at speed of $0.600c$ as measured by an observer on Earth?

35.76 An old song contains these lines: "While driving in my Cadillac, what to my surprise; a little Nash Rambler was following me, about one-third my size." The singer of that song assumes that the Nash Rambler is driving at a similar velocity. Suppose, though, rather than actually being one-third the Cadillac's size, the proper length of the Rambler is the same as that of the Cadillac. What would be the velocity of the Rambler relative to the Cadillac for the song's observation to be accurate?

35.77 You shouldn't invoke time dilation due to your relative motion with respect to the rest of the world as an excuse for being late to class. While it is true that relative to those at rest in the classroom, your time while you are in motion runs more slowly, the difference is negligible. Suppose over a weekend you drove from your college in the Midwest to New York City and back, a round-trip of 2200. mi, driving for 20.0 h in each direction. By what amount, at most, would your watch differ from your professor's watch?

35.78 A spaceship is traveling at two-thirds of the speed of light directly toward a stationary asteroid. If the spaceship turns on its headlights, what will be the speed of the light traveling from the spaceship to the asteroid as observed by

a) someone on the spaceship?

b) someone on the asteroid?

35.79 Two stationary space stations are separated by a distance of 100. light-years, as measured by someone on one of the space stations. A spaceship traveling at $0.950c$ relative to the space stations passes by one of them heading directly toward the other one. How long will it take to reach the other space station, as measured by someone on the spaceship? How much time will have passed for a traveler on the spaceship as it travels from one space station to the other, as measured by someone on one of the space stations? Round the answers to the nearest year.

35.80 An electron is accelerated from rest through a potential of $1.0 \cdot 10^6$ V. What is its final speed?

35.81 In the age of interstellar travel, an expedition is mounted to an interesting star 2000.0 light-years from Earth. To make it possible to get volunteers for the expedition, the planners guarantee that the round-trip to the star will take no more than 10.000% of a normal human lifetime. (At the time, the normal human lifetime is 400.00 years.) What is the minimum speed with which the ship carrying the expedition must travel?

•**35.82** What is the energy of a particle with speed of $0.800c$ and a momentum of $1.00 \cdot 10^{-20}$ N s?

•**35.83** In a high-speed football game, a running back traveling at 55.0% of the speed of light relative to the field throws the ball to a receiver running in the same direction at 65.0% of the speed of light relative to the field. The speed of the ball relative to the running back is 80.0% of the speed of light.

a) How fast does the receiver perceive the speed of the ball to be?

b) If the running back shines a flashlight at the receiver, how fast will the photons appear to be traveling to the receiver?

•**35.84** You have been presented with a source of electrons, ^{14}C, having kinetic energy equal to 0.305 times the rest energy. Suppose you have a pair of detectors that can detect the passage of the electrons without disturbing them. You wish to show that the relativistic expression for momentum is correct and the nonrelativistic expression is incorrect. If a 2.0-m-long baseline between your detectors is used, what timing accuracy is needed to show that the relativistic expression for momentum is correct?

•**35.85** A spacecraft travels a distance of $1.00 \cdot 10^{-3}$ light-years in 20.0 h, as measured by an observer stationed on Earth. How long does the journey take as measured by the captain of the spacecraft?

•**35.86** More significant than the kinematic features of the special theory of relativity are the dynamical processes it describes that Newtonian dynamics does not. Suppose a hypothetical particle with rest mass 1.000 GeV/c^2 and kinetic energy 1.000 GeV collides with an identical particle at rest. Amazingly, the two particles fuse to form a single new particle. Total energy and momentum are both conserved in the collision.

a) Find the momentum and speed of the first particle.

b) Find the rest mass and speed of the new particle.

••**35.87** Although it deals with inertial reference frames, the special theory of relativity describes accelerating objects without difficulty. Of course, uniform acceleration no longer means $dv/dt = g$, where g is a constant, since that would have v exceeding c in a finite time. Rather, it means that the acceleration *experienced* by the moving body is constant: In each increment of the body's own proper time, $d\tau$, the body experiences a velocity increment $dv = g \, d\tau$ as measured in the inertial frame in which the body is momentarily at rest. (As it accelerates, the body encounters a sequence of such frames, each moving with respect to the others.) Given this interpretation:

a) Write a differential equation for the velocity v of the body, moving in one spatial dimension, as measured in the inertial frame in which the body was initially at rest (the "ground frame"). You can simplify your equation by remembering that squares and higher powers of differentials can be neglected.

b) Solve the equation from part (a) for $v(t)$, where both v and t are measured in the ground frame.

c) Verify that $v(t)$ behaves appropriately for small and large values of t.

d) Calculate the position of the body, $x(t)$, as measured in the ground frame. For convenience, assume that the body is at rest at ground-frame time $t = 0$ and at ground-frame position $x = c^2/g$.

e) Identify the trajectory of the body on a space-time diagram (a *Minkowski diagram,* for Hermann Minkowski) with coordinates x and ct, as measured in the ground frame.

f) For $g = 9.81 \text{ m/s}^2$, calculate how much time it takes the body to accelerate from rest to 70.7% of c, measured in the ground frame, and how much ground-frame distance the body covers in this time.

MULTI-VERSION EXERCISES

35.88 A gold nucleus of rest mass 183.473 GeV/c^2 is accelerated from $0.5785c$ to $0.8433c$. How much work is done on the gold nucleus in this process?

35.89 A gold nucleus of rest mass 183.473 GeV/c^2 is accelerated from $0.4243c$ to some final speed. In this process, 140.779 GeV of work is done on the gold nucleus. What is the final speed of the gold nucleus as a fraction of c?

35.90 A gold nucleus of rest mass 183.473 GeV/c^2 is accelerated from some initial speed to a final speed of $0.8475c$. In this process, 137.782 GeV of work is done on the gold nucleus. What was the initial speed of the gold nucleus as a fraction of c?

35.91 Two identical nuclei, each with rest mass 50.30 GeV/c^2, are accelerated in a collider to a kinetic energy of 503.01 GeV and made to

collide head on. If one of the two nuclei were instead kept at rest, what would the kinetic energy of the other nucleus have to be for the collision to achieve the same center-of-mass energy?

35.92 Two identical nuclei are accelerated in a collider to a kinetic energy of 621.38 GeV and made to collide head on. If one of the two nuclei were instead kept at rest, the kinetic energy of the other nucleus would have to be 15,161.70 GeV for the collision to achieve the same center-of-mass energy. What is the rest mass of each of the nuclei?

35.93 A nucleus with rest mass 23.94 GeV/c^2 is at rest in the lab. An identical nucleus is accelerated to a kinetic energy of 10,868.96 GeV and made to collide with the first nucleus. If instead the two nuclei were made to collide head on in a collider, what would the kinetic energy of each nucleus have to be for the collision to achieve the same center-of-mass energy?

36 Quantum Physics

FIGURE 36.1 The Long Island Solar Farm, built on the grounds of Brookhaven National Laboratory on Long Island, New York, in 2011, uses photovoltaic panels to generate a peak electric power of 32 MW.

Photovoltaic cells convert light into electricity. Solar farms like that shown in Figure 36.1 use these cells, and so do the sensors in elevator and garage doors. In order to understand how these devices convert light into moving electrons, we need to look at another aspect of light, which will lead us into the quantum world. When light interacts with material objects, it often reveals a particle-like nature: Tiny wave packets called *photons* interact with individual atoms or molecules or biological cells. The photoelectric effect is impressive physical proof that light can have particle characteristics, and we will study it in the present chapter. Einstein provided the explanation for the photoelectric effect in 1905; a century later, we apply this effect to harness the power of solar radiation and help satisfy humanity's ever-increasing need for electrical power.

Does the existence of photons mean that light is not a wave after all? We will see in this chapter that the difference between a wave and a particle is not clear-cut. At very small scales, waves can act like particles, and particles can act like waves. This discovery led to revolutionary changes in our understanding of physics, as far-reaching as the changes in space and time described by the theory of relativity. The remaining chapters of this book are devoted to the new ideas that arose from the discovery of photons—known as quantum physics.

WHAT WE WILL LEARN

- On the basis of the quantum hypothesis, it is possible to derive Planck's Radiation Law for the power radiated in a given frequency interval. The quantum approach avoids the ultraviolet catastrophe that makes the classical explanation unphysical at short wavelengths (high frequencies), and it contains the classical radiation laws as limits.

- What we normally think of as a wave, such as light, has particle characteristics. The photoelectric effect is explained with the quantum hypothesis by saying that light consists of elementary quanta called *photons*. The energy of a photon is equal to its frequency times Planck's constant.

- The Compton effect is the scattering of a high-energy photon (X-ray) off an electron. The observations for the scattering of X-rays are explained by kinematics, which assumes that the photon has particle characteristics.

- What we normally think of as matter also has wave characteristics. The de Broglie wavelength of a particle is defined as Planck's constant divided by the magnitude of the particle's momentum; it is the fundamental wavelength associated with a matter wave.

- The Heisenberg Uncertainty Relation stipulates that the product of the uncertainty in momentum and the uncertainty in position, measured simultaneously, has an absolute lower bound. An energy-time uncertainty relation has the same lower bound as that for momentum and position.

- Elementary quantum particles have an intrinsic property called *spin*, which has the dimension of angular momentum. Spin is quantized. Particles are divided into two categories: fermions with spins that are half-integer multiples of Planck's constant divided by 2π, and bosons with spins that are integer multiples of Planck's constant divided by 2π.

- The Pauli Exclusion Principle states that no two fermions can occupy the same quantum state at the same time. This means that in any given atom, no two fermions can have exactly identical quantum numbers, which are numbers that characterize the quantum state of the particle. Bosons can condense at a low temperature in such a way that most of them occupy the same quantum state.

36.1 The Nature of Matter, Space, and Time

By now you have probably accepted the notion that matter consists of constituents called *atoms*. Originally atoms were thought to be indivisible, hence the name *atom,* which derives from the Greek word $\alpha\tau o\mu o\varsigma$ ("individual, indivisible"). We will see that atoms actually do have substructure. Each atom consists of a "cloud" of electrons surrounding a nucleus, which in turn consists of neutrons and protons. It is currently thought that electrons have no substructure, but protons and neutrons are known to be composed of three quarks, held together by gluons (see Section 21.2). These quarks and gluons are also thought to be elementary—that is, they are thought to lack substructure. How physicists have arrived at these deductions and conclusions is the subject of Chapters 37–40. For now, however, it suffices to point out that matter is granular—it consists of extremely small, indivisible pieces.

What about time and space—are they granular, too? In Chapter 35 on relativity, we learned some rather surprising results about the connection between time and space. However, we have not yet considered the question of whether time or space can be subdivided into infinitesimally small quantities. Calculus assumes that time is continuous, not granular, because limits of $\Delta t \rightarrow 0$ are used to arrive at the definitions of velocity and acceleration. Energy and momentum are related to each other in ways similar to the ways space and time are related. This immediately raises the question of whether energy and momentum are continuous quantities, or if some smallest granule of energy and some elementary quantity of momentum exist. For example, a spinning ball has rotational kinetic energy. By increasing its angular speed, we also increase its kinetic energy. But can we make the increment infinitesimally small, or is there some smallest quantum of energy that we are forced to add?

Let's start this investigation by taking another look at light. Light can be considered an electromagnetic wave, as shown in Chapter 31. In our studies of light, we first explored geometric optics in Chapters 32 and 33, examining image formation with mirrors, lenses, and other optical instruments. In those chapters, we assumed that light consists of rays that move along straight lines. When we looked in Chapter 34 at interference and diffraction, we were forced to invoke the wave character of light. In that chapter, we found that the wave character of light comes into play only when we explore spatial dimensions on the order of the wavelength of light and that geometric optics provides very good approximations for spatial dimensions that are very large compared to the wavelength.

Does our previous description of light as an electromagnetic wave suffice to describe all the phenomena we can observe? The answer is no, as we will see in this chapter.

36.2 Blackbody Radiation

When we talked about thermal radiation in Chapter 18, we introduced the idealized concept of a blackbody. This idealization can be realized to good accuracy by looking at the radiation coming from a small hole in a large cavity kept at a temperature T. If we look at the visible light that emerges from such a hole at room temperature, the hole appears black because all the light that enters the cavity through the hole is scattered and ultimately absorbed by the walls. However, at much higher temperatures, the hole begins to glow in the visible part of the electromagnetic spectrum. Everyday examples of visible light approximating blackbody radiation include the dull red color of the cooking elements of electric stoves, the bright light from the filament of an incandescent light bulb, the light from the Sun, and the red glow of volcanic lava (Figure 36.2).

Let's first briefly review what was known about blackbody radiation from classical wave physics and from empirical observations. The *Stefan-Boltzmann Radiation Law* for the total intensity I (energy radiated per unit time and unit area) of this blackbody radiation is

$$I = \int_0^\infty \epsilon(\lambda)d\lambda = \sigma T^4. \tag{36.1}$$

FIGURE 36.2 Volcanic lava emits light and is a very good approximation of a blackbody radiator.

Here $\epsilon(\lambda)$ is the **spectral emittance** (often called the *spectral radiance*) as a function of wavelength. It is the power radiated per unit area and wavelength and has the SI units of W m^{-3}. The integral extends over all possible wavelengths from zero to infinity, and σ is the Stefan-Boltzmann constant,

$$\sigma = 5.670400(40)\cdot10^{-8} \text{ W m}^{-2} \text{ K}^{-4}.$$

The most important feature of the Stefan-Boltzmann Radiation Law (equation 36.1) is that the total intensity of the radiation increases with the fourth power of the temperature.

In 1896, German physicist Wilhelm Wien (1864–1928) empirically derived **Wien's Law** to describe the spectral emittance of a blackbody:

$$\epsilon_{\text{Wien}}(\lambda) = \frac{a}{\lambda^5}e^{-b/\lambda T} \quad \text{(approximation for small } \lambda\text{)}, \tag{36.2}$$

where a and b are constants. Wien's Law succeeded in describing the spectral emittance of blackbodies for short wavelengths but was less successful in describing the spectral emittance for long wavelengths. The **Wien Displacement Law** summarizes another important experimental finding about the spectral emittance. It states that the spectral emittance peaks at a certain wavelength, λ_m, and that this wavelength depends on the temperature:

$$\lambda_m T = \text{constant} = 2.90\cdot10^{-3} \text{ K m}. \tag{36.3}$$

Using the representation of light as an electromagnetic wave, the English physicists Lord Rayleigh and Sir James Jeans managed to derive an expression for the spectral emittance of blackbody radiation:

$$\epsilon_{\text{RJ}}(\lambda) = \frac{2\pi ck_B T}{\lambda^4} \quad \text{(approximation for large } \lambda\text{)}. \tag{36.4}$$

Here c is the speed of light and k_B is Boltzmann's constant,

$$k_B = 1.3806488(13)\cdot10^{-23} \text{ J/K}.$$

This solution, however, had one glaring fault: As $\lambda \to 0$, the expression on the right side of equation 36.4 diverges. This problem later became known as the *ultraviolet catastrophe* (recall that ultraviolet radiation has small wavelengths). If equation 36.4 were correct for all wavelengths, then the integral in equation 36.1 would diverge and the intensity radiated by a black-body would become infinite at any temperature. Clearly, this is impossible! However, for large wavelengths, the result obtained by Rayleigh and Jeans fits the experimental observations.

In order to provide a formula for the spectral emittance that fits the observations at all wavelengths, the German physicist Max Planck took a radical step in 1900. He proposed that the energy contained in light and all other electromagnetic radiation interacts with solid objects in discrete bundles. He hypothesized that the energy of a bundle is proportional to the frequency of the radiation:

$$E = hf,$$

(36.5)

where h is **Planck's constant** and has the value

$$h = 6.62606957(29) \cdot 10^{-34} \text{ J s}.$$

(36.6)

We have already introduced the energy unit the electron-volt, $1 \text{ eV} = 1.602177 \cdot 10^{-19}$ J, and Planck's constant can also be expressed in terms of that unit: $h = 4.13567 \cdot 10^{-15}$ eV s.

As you will see later in this chapter, many formulas in quantum physics involve Planck's constant divided by 2π. Because this is a relatively common occurrence, it is customary to use the notation $\hbar$ to denote this ratio:

$$\hbar \equiv \frac{h}{2\pi} = 1.05457 \cdot 10^{-34} \text{ J s} = 6.5821 \cdot 10^{-16} \text{ eV s}.$$

(36.7)

The wavelength and frequency of light are still related to the speed via $c = \lambda f$, so we can also write, instead of equation 36.5,

$$E = hf = \frac{hc}{\lambda}.$$

(36.8)

Planck's Radiation Law, which Planck based on the quantized-energy hypothesis, is

$$I_T(f) = \frac{2h}{c^2} \frac{f^3}{e^{hf/k_B T} - 1}.$$

(36.9)

Here $I_T(f)df$ is the power (amount of energy per unit time) radiated in the frequency range between f and $f + df$ by a blackbody at temperature T per unit surface area of the blackbody opening and per unit solid angle. The name for $I_T(f)$ is the *specific intensity*, or the **spectral brightness,** and its SI units are W m^{-2} sr^{-1} Hz^{-1}. We will revisit Planck's result later in this chapter in order to understand the dependence of the spectral brightness on frequency and temperature.

Note that the spectral brightness does not depend on direction. The spectral brightness, $I_T(f)$, can be integrated over the entire hemisphere of all possible directions to give the spectral emittance, $\epsilon_T(f)$:

$$\epsilon_T(f) = \int_\Omega I_T(f) \cos\theta \, d\Omega = \int_0^{\pi/2} \left(\int_0^{2\pi} I_T(f) d\phi \right) \sin\theta \cos\theta \, d\theta$$

$$= I_T(f) \int_0^{2\pi} d\phi \int_0^{\pi/2} \sin\theta \cos\theta \, d\theta$$

$$= I_T(f) 2\pi \tfrac{1}{2}$$

$$= \pi I_T(f).$$

Thus, we see that the spectral emittance is larger than the spectral brightness by exactly the factor π:

$$\epsilon_T(f) = \pi I_T(f).$$

(36.10)

The factor $\cos\theta$ in the first integral is the projection of the normal vector to the hole, $\hat{n}$, in the direction of the radiation, so $\cos\theta$ represents the effective reduction of the unit emission area as a function of the polar angle. The integral over the solid angle of the hemisphere is a double integral over the angle θ from 0 to $\pi/2$ and the angle ϕ from 0 to 2π (Figure 36.3).

These integrals are easily evaluated, as shown, because $I_T(f)$ does not depend on the angles. Combining equation 36.10 with equation 36.9, we obtain

$$\epsilon_T(f) = \frac{2\pi h}{c^2} \frac{f^3}{e^{hf/k_B T} - 1}. \tag{36.11}$$

The spectral emittance, $\epsilon_T(f)$, has the SI units of W m^{-2} Hz^{-1}.

We can also write the spectral brightness and the spectral emittance as functions of the wavelength instead of the frequency. To do this, we use $c = \lambda f$:

$$I_T(\lambda) = I_T(f)\left|\frac{df}{d\lambda}\right| = I_T(f)\left|\frac{d}{d\lambda}\left(\frac{c}{\lambda}\right)\right| = I_T(f)\frac{c}{\lambda^2}.$$

Therefore, the spectral brightness as a function of wavelength is given as

$$I_T(\lambda) = \frac{2hc^2}{\lambda^5 \left(e^{hc/\lambda k_B T} - 1\right)}. \tag{36.12}$$

Its SI units are W m^{-3} sr^{-1}. The spectral emittance can be obtained as a function of wavelength by integrating the spectral brightness over all emission angles, resulting in a multiplicative factor of π and giving

$$\epsilon(\lambda) = \frac{2\pi hc^2}{\lambda^5 \left(e^{hc/\lambda k_B T} - 1\right)}, \tag{36.13}$$

which has SI units of W m^{-3}. All four versions of Planck's Law (equations 36.9, 36.11, 36.12, and 36.13) are equally valid. Care must be taken, however, when talking about the spectral brightness or the spectral emittance. They differ by a factor of π, as shown, because the spectral emittance is the integral of the spectral brightness over all emission angles in the hemisphere.

Figure 36.4 displays plots of the Planck Radiation Law for the spectral emittance as a function of wavelength for three temperatures. The top curve is calculated for a temperature of 5800 K, approximately the surface temperature of the Sun. The spectral emittance has a maximum near a wavelength of 500 nm (green-blue), in accordance with the result of equation 36.3, the Wien Displacement Law. The other two curves are for 5400 K, peaking at 540 nm (yellow-green), and 5000 K, peaking at 580 nm (orange). Overlaid on the graph are the colors of the visible spectrum.

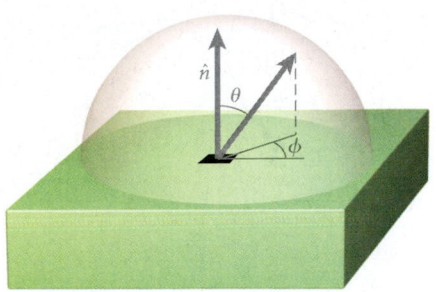

FIGURE 36.3 Blackbody radiating through the small black hole in all directions of the hemisphere.

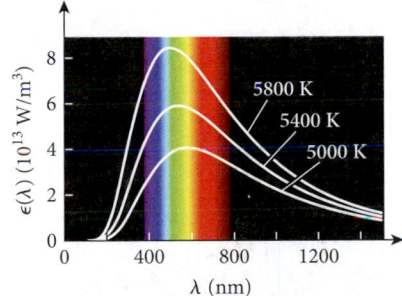

FIGURE 36.4 Planck spectral emittance as a function of wavelength for three temperatures: 5800 K, 5400 K, and 5000 K, from top to bottom.

DERIVATION 36.1 | Radiation Laws

Let's show that Wien's Law, the Rayleigh-Jeans Law, the Stefan-Boltzmann Radiation Law, and the Wien Displacement Law can be derived from Planck's Radiation Law (equation 36.13).

Wien's Law: For small values of λ, the argument of the exponential function in Planck's Radiation Law becomes large, which allows us to write

$$\frac{1}{e^{hc/\lambda k_B T} - 1} \approx e^{-hc/\lambda k_B T}.$$

We then obtain an expression for Wien's Law as a limiting case of Planck's Law:

$$\epsilon(\lambda) = \frac{2\pi hc^2}{\lambda^5 \left(e^{hc/\lambda k_B T} - 1\right)} \approx \frac{2\pi hc^2}{\lambda^5} e^{-hc/\lambda k_B T},$$

which has the same dependence on wavelength as equation 36.2, with the constants now defined as $a = 2\pi hc^2$ and $b = hc/k_B$.

Rayleigh-Jeans Law: For large values of λ, the argument of the exponential function in Planck's Radiation Law becomes small. We can expand the exponential function for small arguments as $e^x \approx 1 + x$. In this case, we then have $e^{hc/\lambda k_B T} - 1 \approx hc/\lambda k_B T$, and we find for large values of the wavelength:

$$\epsilon(\lambda) = \frac{2\pi hc^2}{\lambda^5 \left(e^{hc/\lambda k_B T} - 1\right)} \approx \frac{2\pi hc^2}{\lambda^5 \left(hc/\lambda k_B T\right)} = \frac{2\pi c k_B T}{\lambda^4},$$

which is the Rayleigh-Jeans Law as given in equation 36.4.

– *Continued*

Wien Displacement Law: For this law, we need to find the wavelength for which $\epsilon(\lambda)$ reaches a maximum; that is, we need to take the derivative with respect to the wavelength and find the root. The derivative is

$$\frac{d\epsilon(\lambda)}{d\lambda} = \frac{d}{d\lambda}\left(\frac{2\pi hc^2}{\lambda^5\left(e^{hc/\lambda k_B T}-1\right)}\right)$$

$$= -\frac{10\pi hc^2}{\lambda^6\left(e^{hc/\lambda k_B T}-1\right)} + \frac{2\pi h^2 c^3 e^{hc/\lambda k_B T}}{\lambda^7\left(e^{hc/\lambda k_B T}-1\right)^2 k_B T}$$

$$= \frac{2\pi hc^2}{\lambda^7\left(e^{hc/\lambda k_B T}-1\right)^2 k_B T}\left(-5\lambda k_B T\left(e^{hc/\lambda k_B T}-1\right)+hce^{hc/\lambda k_B T}\right).$$

Except for the uninteresting case where $T \to \infty$, this expression can be zero only if the numerator is zero. Therefore, we need to solve

$$-5\lambda_m k_B T\left(e^{hc/\lambda_m k_B T}-1\right)+hce^{hc/\lambda_m k_B T} = 0,$$

where λ_m is the value of the wavelength for which the Planck Radiation Law has its maximum. If we substitute from $u = hc/\lambda_m k_B T$, then this equation reduces to

$$5\left(e^u - 1\right) = ue^u \Rightarrow 5 - 5e^{-u} = u.$$

This equation can be solved by simple iteration, by using a spreadsheet, for example. The trivial root is $u = 0$, but we are not interested in this solution, which corresponds to an infinite wavelength. Therefore, we start our iteration at the value 1 and find $5 - 5e^{-1} = 3.1606$. Then we insert this new value, finding $5 - 5e^{-3.1606} = 4.7880$, and so on. We reach convergence very quickly and obtain

$$u = \frac{hc}{\lambda_m k_B T} = 4.9651 \Rightarrow \lambda_m T = \frac{hc}{k_B 4.9651}.$$

Using the values of the constants h, c, and k_B, we then find

$$\lambda_m T = 2.898 \cdot 10^{-3} \text{ K m},$$

which is in complete agreement with the experimentally found value given by equation 36.3.

Stefan-Boltzmann Law: To obtain the total intensity radiated, I, we need to integrate Planck's Radiation Law over the wavelength, from zero to infinity. We find

$$I = \int_0^\infty \epsilon(\lambda)d\lambda = \int_0^\infty \frac{2\pi hc^2}{\lambda^5\left(e^{hc/\lambda k_B T}-1\right)}d\lambda = \frac{2k_B^4 \pi^5}{15h^3 c^2}T^4 = \sigma T^4.$$

Inserting the values for Planck's constant, Boltzmann's constant, and the speed of light, we can verify that the Stefan-Boltzmann constant is indeed given by

$$\sigma = \frac{2k_B^4 \pi^5}{15h^3 c^2} = 5.6704 \cdot 10^{-8} \text{ W m}^{-2}\text{ K}^{-4}.$$

Thus, we see that the radiation law derived by Planck contains the previously known radiation laws as special cases, as illustrated in Figure 36.5.

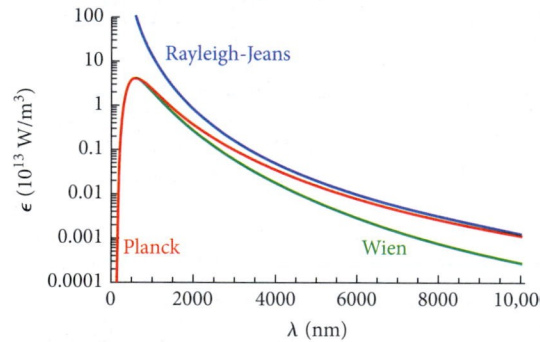

FIGURE 36.5 Comparison of Planck's Radiation Law, the Rayleigh-Jeans Law, and Wien's Law (using the constants from Derivation 36.1) for $T = 5000$ K.

Concept Check 36.1

The visible spectrum of light extends from approximately 380 nm (violet-blue) to 780 nm (red). What is the corresponding range of photon energies in units of electron-volts?

a) 1.59 eV to 3.26 eV

b) $2.54 \cdot 10^{-19}$ eV to $5.23 \cdot 10^{-19}$ eV

c) $0.38 \cdot 10^{15}$ eV to $0.78 \cdot 10^{15}$ eV

d) 190 eV to 390 eV

Planck's Law is consistent with Wien's Law for short wavelengths and agrees with the Rayleigh-Jeans Law at long wavelengths. This success earned immediate acceptance for Planck's Radiation Law for blackbodies, even though it was based on the radical assumption of quantized energy states. In particular, one can see from Derivation 36.1 how the quantum hypothesis resolves and avoids the classical ultraviolet catastrophe discussed earlier (see equation 36.4): For a given frequency f, the energy hf is needed to create a photon. As the frequency increases, it becomes less and less likely that the system can supply the energy needed for the creation of a photon. This leads to a cutoff at high frequencies and thus at low wavelengths, in agreement with observations. Thus, the observed avoidance of the ultraviolet catastrophe is a direct consequence of the quantum nature of light.

The most stunning example of a blackbody spectrum is obtained from the cosmic background radiation. This radiation is a remnant of the Big Bang and is astonishingly uniform in the entire universe. The COBE satellite in 1990 and more recently the WMAP satellite have substantiated this in amazing detail. As indicated in Figure 36.6, the COBE mission found that the cosmic background radiation is that of a perfect blackbody at a temperature of 2.725 ± 0.001 K; that is, the entire universe is a perfect blackbody radiator. George Smoot and John Mather, the leaders of the COBE team, received the 2006 Nobel Prize in Physics for the achievements of that mission. (A more detailed discussion of the cosmic background radiation is presented in Chapter 39.)

Just as blackbody radiation was used to measure the temperature of the universe, it can be used to measure the temperature of objects without physically touching them. If an object is hot enough, it will radiate photons in the visible range, as mentioned earlier in this section. For example, the temperature of molten iron in a steel factory can be measured by analyzing the photons radiated from the red-hot iron. Objects close to room temperature radiate photons mainly in the infrared range. Modern infrared thermometers can measure a person's temperature by observing the infrared radiation from the person's eardrum. Infrared thermometers are also used to measure the temperatures of food and electrical components.

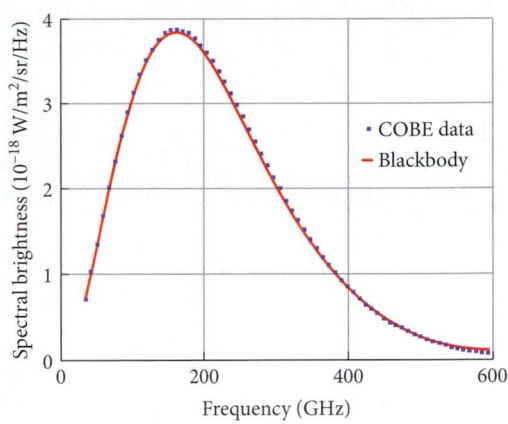

FIGURE 36.6 The spectral brightness of the cosmic background radiation as a function of frequency. The blue squares indicate data obtained by the COBE satellite, while the red curve is a best fit of a Planck spectrum at a temperature of 2.725 K.

36.3 Photoelectric Effect

Planck's hypothesis about the existence of some smallest possible discrete energy quanta was originally viewed as a computational construction, not as a real revolution in physics. This view changed, however, in 1905, with Einstein's explanation of the photoelectric effect. Einstein solved the puzzle of the photoelectric effect by proposing that light behaves as if it consists of localized bundles, or *quanta*, of energy-light. The photoelectric effect was originally discovered by Heinrich Hertz in 1886 and definitively demonstrated in 1916 by Robert A. Millikan, who quantitatively verified all of Einstein's predictions. Einstein's explanation of the photoelectric effect earned him the 1921 Nobel Prize in Physics. The quantized nature of light was conclusively demonstrated in 1923 by Arthur Holly Compton, as we will see in Section 36.4. The American chemist Gilbert Lewis (1875–1946) coined the term **photon** in 1926 to refer to these quanta of energy-light. For the remainder of this chapter, photons will be considered to be the quanta of light and all electromagnetic radiation.

In the **photoelectric effect,** light knocks electrons out of the surface of a suitable metal, creating an electric current. To see a practical application of the photoelectric effect, we have to look no further than the door of an elevator. How does the door sense that someone is standing in the opening? The answer is a photosensor, which usually consists of a light source and a light receptor utilizing the photoelectric effect. If an object, such as a person, is located between the light source and the receptor, the receptor no longer receives light and triggers electric switches to keep the elevator door open. Similarly, automatic garage-door openers are required to use a photosensor to avoid crushing a person on the way down.

A series of experiments can be performed to examine the photoelectric effect. Figure 36.7 shows the basic setup. On the left side is a light source, which could be a light bulb as shown or a light-emitting diode (which is utilized in many photocircuits) or just plain sunlight. On the right side is a photosensor, consisting of a piece of metal (photocathode, rectangular shape) and a metal plate (anode, black line) housed in an evacuated glass container. A metal often used for the photocathode is cesium. The photosensor is part of a circuit with a voltage source and an ammeter. Between the light source and the photosensor

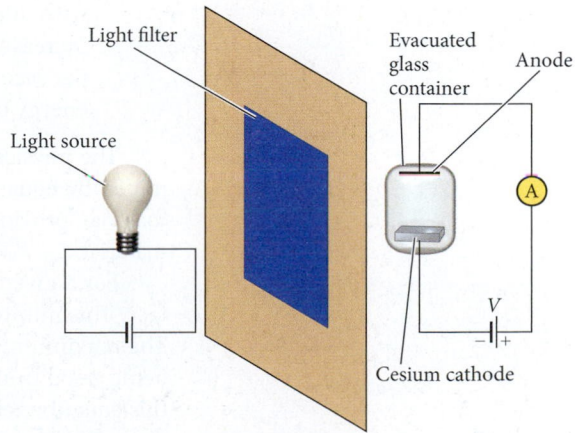

FIGURE 36.7 Schematic circuit diagram for the photoelectric effect.

is a filter that passes only one color of light (blue in this case). Experiments yield the following observations:

- With voltage set at $V = 0$ and a blue filter, a current is detectable in the ammeter. This indicates that electrons are crossing the gap between the photocathode and the anode. If the intensity of the light is increased, the measured current rises as well, indicating that more electrons move across the gap.

- With a red filter, no current is detectable in the ammeter. This finding does not change as a function of the light intensity.

- With a blue filter and a positive value of V, the current flowing through the ammeter rises. As voltage is changed to increasingly more negative values, the current measured in the ammeter is gradually reduced and then stops at some threshold value of the voltage.

From these experiments, one can conclude that electrons must get released from the surface of the metal by the light striking it. These electrons must have a kinetic energy, and the maximum of this kinetic energy can be measured by applying a negative voltage to the anode. The electrons need to overcome this voltage to cross the gap from the photocathode to the anode. If the electrons (charge $q = -e$) start with a maximum kinetic energy of K_{max} from the surface of the cathode and just reach the anode with zero kinetic energy after overcoming a potential of $V = -V_0$, then, from the work-energy theorem, we find

$$W = \Delta K + \Delta U = (0 - K_{max}) + [(-e)(-V_0) - 0] = -K_{max} + eV_0 = 0 \Rightarrow$$

$$eV_0 = K_{max} = \tfrac{1}{2}mv_{max}^2. \tag{36.14}$$

(The nonrelativistic approximation $\tfrac{1}{2}mv^2$ for the kinetic energy of the electrons can be used here, because the electrons are moving slowly in this case.) The potential V_0 is called the *stopping potential,* and for a given material it depends on the color of the light, which implies that it has a frequency dependence. Careful measurements reveal a linear dependence of V_0 on the frequency f. In addition, below a certain frequency, the stopping potential becomes zero. Light with a lower frequency is not able to give the electrons in the photocathode enough energy to escape from its surface.

The key conceptual problems of the photoelectric effect from the viewpoint of classical wave physics can be summarized as follows:

- Classically, a beam of light of any frequency can eject electrons from a metal, as long as the light has enough intensity. However, observations show that the incident light beam must have a frequency greater than the minimum value, f_{min}, regardless of its intensity. Einstein's explanation was that the energy of the energy-light quantum (the photon) is proportional to the frequency, $E = hf$.

- Classically, the maximum kinetic energy of the ejected electrons should increase with increasing intensity of the light beam. However, observations show that increasing the intensity of the light beam increases the number of electrons ejected per second, not their energy; only increasing the frequency of the light increases the energy of the ejected electrons.

The physical picture of the photoelectric effect is that a photon with an energy determined by equation 36.5 hits an electron at the surface of the metal and ejects it from the material, provided that the electron gains sufficient energy to overcome its attraction to the material.

For a given material, a minimum energy is required to free an electron from the surface. This minimum energy, called the **work function,** ϕ, is a constant for a given material. The maximum kinetic energy that an electron can have after colliding with a photon and being freed from the metal surface is then $K_{max} = hf - \phi$. Because K_{max} cannot be negative, this equation tells us that there is a minimum (threshold) light frequency necessary for the photoelectric effect to occur:

$$f_{min} = \phi / h, \tag{36.15}$$

Element	ϕ (eV)	f_{min} (10^{15} Hz)	λ_{max} (nm)	Element	ϕ (eV)	f_{min} (10^{15} Hz)	λ_{max} (nm)
Aluminum	4.1	0.99	302	Magnesium	3.7	0.89	335
Beryllium	5	1.21	248	Mercury	4.5	1.09	276
Cadmium	4.1	0.99	302	Nickel	5	1.21	248
Calcium	2.9	0.70	428	Niobium	4.3	1.04	288
Carbon	4.8	1.16	258	Potassium	2.3	0.56	539
Cesium	2.1	0.51	590	Platinum	6.3	1.52	197
Cobalt	5	1.21	248	Selenium	5.1	1.23	243
Copper	4.7	1.14	264	Silver	4.7	1.14	264
Gold	5.1	1.23	243	Sodium	2.3	0.56	539
Iron	4.5	1.09	276	Uranium	3.6	0.87	344
Lead	4.1	0.99	302	Zinc	4.3	1.04	288

Table 36.1 Work Functions and Corresponding Minimum Frequencies and Maximum Wavelengths for Common Elements

which is consistent with the experimental results. Table 36.1 shows work functions and corresponding threshold frequencies and cutoff wavelengths for various metals. Using the connection between the maximum kinetic energy and the stopping potential in equation 36.14, we then find for the frequency dependence of the stopping potential in the photoelectric effect:

$$eV_0 = hf - \phi. \qquad (36.16)$$

EXAMPLE 36.1 / Work Function

Suppose you are using a circuit like that in Figure 36.7 and you have a photosensor with an unknown material for the photocathode. When using light of wavelength 250 nm (ultraviolet), you find that you have to apply a stopping potential of 2.86 V to eliminate the current. When using light of wavelength 392 nm (blue-violet), you measure a value of 1.00 V for the stopping potential, and using a wavelength of 540 nm (orange), you measure a stopping potential of 0.170 V.

PROBLEM
What is the value of the work function for this material?

SOLUTION
It is perhaps easiest to solve this problem in graphical form. Equation 36.16 shows that the stopping potential, the work function, and the frequency have a linear relationship, and because linear relationships can be drawn as straight lines, it is best to convert the given wavelengths into corresponding frequencies by using $f = c/\lambda$. We then plot the stopping potential, V_0, as a function of frequency for the three given data points (Figure 36.8). When we fit a straight line through these data points, the line gives us a value of -2.1 V at $f = 0$. Using equation 36.16, we then find for the work function:

$$eV_0(f = 0) = -\phi \Rightarrow \phi = -eV_0(f = 0) = -e(-2.1 \text{ V}) = 2.1 \text{ eV}.$$

Looking at Table 36.1, we can see that the material used in this photosensor is probably cesium, which has the lowest work function of all the metals listed.

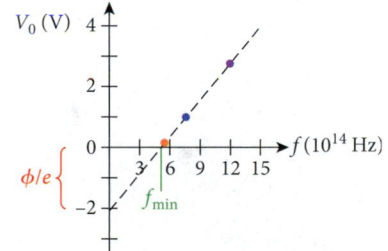

FIGURE 36.8 Applied stopping potential as a function of the frequency of light for a particular photosensor material.

Self-Test Opportunity 36.1

Assuming that the work function in Example 36.1 indeed has a value of 2.1 eV, find the maximum kinetic energy that emitted electrons can have for the three given cases of light of different wavelengths incident on the photosensor.

Chapter 35, on relativity, showed that the momentum and the energy of any particle are related by $E^2 = p^2c^2 + m^2c^4$. A photon has zero mass, so its energy and momentum are related by $E = pc$. (When the symbol p is used with no arrow, it stands for the magnitude of the momentum; however, we may refer to p simply as the momentum.) For the momentum of a photon, we can then write, from equation 36.8:

$$p = \frac{E}{c} = \frac{hf}{c} = \frac{h}{\lambda}. \qquad (36.17)$$

Thus, we see that a photon's momentum and energy are both proportional to the frequency and inversely proportional to the wavelength of the corresponding electromagnetic radiation. A photon acts like a particle, even though it is described by a frequency. Light acts like a wave even though it consists of particles called photons. This **wave-particle duality** of light is conceptually hard to grasp, and it kept physicists and philosophers busy for the first part of the 20th century.

Several practical problems are associated with detecting single photons of visible light. First, as you found in Concept Check 36.1, each photon has energy only in the range between 1.6 and 3.3 eV. Expressed in SI units, this is in the range between $2.6 \cdot 10^{-19}$ and $5.3 \cdot 10^{-19}$ J—a very small amount of energy. Even the best photocathodes have a quantum efficiency of only 30% or less for a photon in this energy range, meaning that at most only 30% of the photons hitting the photocathode actually manage to knock out an electron. Second, one liberated electron (or photoelectron) represents only a very tiny charge and thus only a very tiny current. To register an easily measurable current, many electrons need to be generated from each photoelectron. Many practical devices accomplish this with photomultiplier tubes.

A **photomultiplier tube** makes use of the fact that an electron that hits a metal surface with a kinetic energy on the order of 100 eV usually knocks out several electrons in the process. Therefore, a photocathode and an anode are combined in an evacuated glass tube with several intermediate plates, called *dynodes*. Each dynode is kept at a potential difference of several hundred volts relative to its neighbors (Figure 36.9). Commercially available photomultiplier tubes have chains of up to $n = 14$ dynodes, and each dynode produces a statistical average of η electrons for each electron that hits it, where η can have values of up to 3.5. The total gain of such a photomultiplier tube is then η^n. For $n = 14$ and $\eta = 3.4$, for example, this gain factor is $2.76 \cdot 10^7$—that is, almost 28 million electrons are produced at the anode for each photoelectron knocked out of the photocathode by a single photon.

Another application of the detection of photons is night-vision devices. A schematic diagram of a night-vision device is shown in Figure 36.10b. In a night-vision device, an entire image of an object is focused on the photocathode by the objective lens. The incident photons

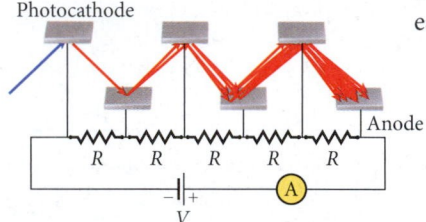

FIGURE 36.9 Schematic diagram of a photomultiplier tube. The blue arrow represents a single photon, and the red arrows represent electrons.

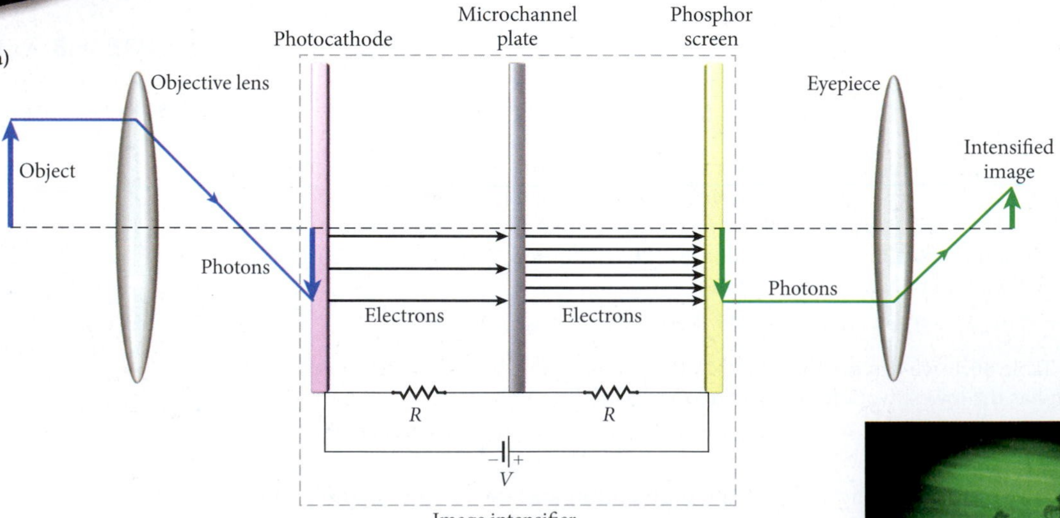

FIGURE 36.10 A night-vision device produces an intensified image of an object in low-light conditions. (a) Photo of a night-vision device; (b) schematic diagram; (c) image produced with a night-vision device.

cause the photocathode to release electrons. A potential difference of around 1000 V accelerates these electrons into a microchannel plate that contains an array of millions of channels that function like miniature photomultiplier tubes, multiplying the number of electrons by a factor of around 10^4. A second potential difference of around 1000 V between the microchannel plate and a phosphor screen accelerates the multiplied electrons so that they hit the screen and cause the emission of green light where they strike the screen, forming an intensified image. This light is then focused by the eyepiece to produce an image like the one shown in Figure 36.10c. This type of night-vision device works by amplifying low light levels. There is another type of night-vision device that uses infrared light emitted by warm objects to observe those objects in the dark.

Other methods are also used to detect single photons and convert them into an electric signal. Most notable among these are charge-coupled devices (CCDs) and complementary metal oxide semiconductor devices (CMOS), which form the basis of every digital camera and video recorder. However, an understanding of the physics underlying a CCD or a CMOS requires knowledge of semiconductors.

EXAMPLE 36.2   Photons from a Laser Pointer

Any object that you see emits photons that travel from the object to your retinas, where the photons trigger electric signals that are sent to your brain. Let's look at a light source to begin to understand the numbers of photons involved.

PROBLEM
What is the approximate number of photons emitted per second by a 5.00-mW green laser pointer?

SOLUTION
Green laser pointers usually operate at a wavelength of 532 nm. A wavelength of 532 nm corresponds to a frequency of

$$f = \frac{c}{\lambda} = \frac{2.998 \cdot 10^8 \text{ m/s}}{5.32 \cdot 10^{-7} \text{ m}} = 5.635 \cdot 10^{14} \text{ Hz}.$$

With Planck's hypothesis, $E = hf$, we can calculate the energy contained in a single photon emitted by the green laser pointer:

$$E = hf = (6.626 \cdot 10^{-34} \text{ J s})(5.635 \cdot 10^{14} \text{ s}^{-1}) = 3.73 \cdot 10^{-19} \text{ J}.$$

Because the laser pointer is rated at 5.00 mW, it emits 5.00 mJ of energy each second. The number of photons emitted in each second is therefore

$$n = \frac{5.00 \cdot 10^{-3} \text{ J}}{3.73 \cdot 10^{-19} \text{ J}} = 1.34 \cdot 10^{16}.$$

In other words, this little handheld laser pointer emits over 13 million billion photons each second!

DISCUSSION
The energy of a single photon emitted from this green laser pointer can be expressed in units of eV s as well:
$$E = hf = (4.13567 \cdot 10^{-15} \text{ eV s})(5.635 \cdot 10^{14} \text{ s}^{-1}) = 2.33 \text{ eV}.$$

Now you may begin to grasp the usefulness of the electron-volt as the energy unit for dealing with atomic and quantum phenomena. The typical energy scale for processes in this realm is the electron-volt.

Concept Check 36.2

You have a source of light with a given intensity and wavelength. You reduce the wavelength while leaving the intensity the same. Which of the following statements is true?

a) More photons per second will be emitted from the light source.

b) Fewer photons per second will be emitted from the light source.

c) The number of photons emitted per second will remain the same, but the energy of each one will be reduced.

d) The number of photons emitted per second will remain the same, but the energy of each one will be increased.

e) The number of photons emitted per second will remain the same, but each one will move slower.

Self-Test Opportunity 36.2

Calculate the number of photons in the visible spectrum emitted each second by the Sun. To solve this problem, you need to know that the Sun's radiation has an intensity of 1370 W/m^2 on Earth, which is at a distance of 148 million km from the Sun. From this information, you can calculate the total power output of the Sun. Then, by examining Figure 36.4, you can see that approximately $\frac{1}{4}$ of the photons in the Sun's radiation are in the visible spectrum.

36.4   Compton Scattering

The discussion of the electromagnetic spectrum in Chapter 31 characterized X-rays as those electromagnetic waves with frequencies approximately 100 to 100,000 times higher than visible light. Viewing the electromagnetic spectrum as consisting of photons, we see that X-ray photons have energies of hundreds to hundreds of thousands of electron-volts. X-rays can be produced by accelerating electrons until they have several thousand electron-volts (several keV) of kinetic energy and then shooting them into a metal foil. The deceleration of the electrons in the foil creates the X-rays. These X-rays are called **Bremsstrahlung,** which is the German word for "deceleration radiation." (Another process also occurs in which atoms become excited and then emit X-rays of certain energies; this process is discussed in Chapter 38.) Classical electromagnetic theory can make certain predictions for electromagnetic radiation produced with accelerated charges, but a complete understanding really requires a theory called **quantum electrodynamics,** developed in the 1950s by American physicists Julian Schwinger (1918–1994) and Richard Feynman (1918–1988) and Japanese physicist Sin-Itiro Tomonaga (1906–1979), for which they shared the 1965 Nobel Prize in Physics.

For now, consider what happens when an X-ray scatters off an electron. First, what does the wave picture of light predict? If a wave hits a stationary small object like an electron, Huygens's Principle tells us that a spherical wave originates from the object, scattering (reflecting) the incoming wave. The scattered wave has the same frequency and wavelength as the incoming wave. However, in 1923, American physicist Arthur Holly Compton (1892–1962) discovered that X-rays scattered off electrons at rest produce X-rays with longer wavelengths than the original X-rays. A larger wavelength implies a smaller frequency, and thus lower energy and momentum for the X-ray photons, according to equation 36.8.

If one accepts that photons have particle-like properties of momentum and energy, then the interaction of an X-ray and an electron can be analyzed just like the scattering of one billiard ball off another. Because photons move with the speed of light and because the energies of X-rays are not negligible compared to the mass of the electron, we have to employ relativistic dynamics (Chapter 35) and cannot simply use the formulas developed in Chapter 7 on momentum and collisions. However, conservation laws of energy and momentum are still applied to arrive at the desired result.

Let's call the energy of the X-ray photon before the collision E and after the collision E'. Then the magnitudes of the corresponding momenta of the photon before and after the collision are $p = E/c$ and $p' = E'/c$. The electron has no momentum before the collision because we assume that it is at rest. During the collision, it receives a momentum $\vec{p}_e$. A diagram of the scattering process is shown in Figure 36.11. The energy of the electron before the collision is simply its rest energy, $m_e c^2$, and its energy after the collision is

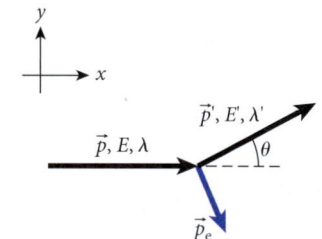

FIGURE 36.11 Momentum conservation in Compton scattering.

$$E_e = \sqrt{\left(p_e c\right)^2 + \left(m_e c^2\right)^2}.$$

Energy and momentum conservation during the collision then imply

$$\vec{p}' + \vec{p}_e = \vec{p} \tag{36.18}$$

and

$$E' + E_e = E + m_e c^2. \tag{36.19}$$

Derivation 36.2 shows that the final wavelength of the X-ray can be expressed as

$$\lambda' = \lambda + \frac{h}{m_e c}(1 - \cos\theta), \tag{36.20}$$

where θ is the angle between the incoming and outgoing photon. This is the formula for **Compton scattering,** linking the wavelength of the photon after scattering to the wavelength of the incoming photon.

DERIVATION 36.2 Compton Scattering

To derive equation 36.20, we isolate $\vec{p}_e$ in equation 36.18 and E_e in equation 36.19. For the first equation, this results in $\vec{p}_e = \vec{p} - \vec{p}'$. We square both sides, to obtain

$$p_e^2 = \left(\vec{p} - \vec{p}'\right)^2 = p^2 + p'^2 - 2pp'\cos\theta. \tag{i}$$

Rearranging and squaring each side of equation 36.19 yields

$$E_e = E - E' + m_e c^2 \Rightarrow$$

$$E_e^2 = \left(E - E' + m_e c^2\right)^2 \Rightarrow$$

$$p_e^2 c^2 + m_e^2 c^4 = \left(E - E'\right)^2 + m_e^2 c^4 + 2\left(E - E'\right)m_e c^2,$$

where we have made use of the relativistic energy-momentum relation, $E_e^2 = p_e^2 c^2 + m_e^2 c^4$ (see Section 35.4) on the left-hand side of the last line of this calculation. We can also use the relation between momentum and energy, $E = pc$, for the photon and obtain

$$p_e^2 c^2 + m_e^2 c^4 = \left(p - p'\right)^2 c^2 + m_e^2 c^4 + 2\left(p - p'\right)m_e c^3.$$

Subtracting the term $m_e^2 c^4$ from both sides and then dividing by a common factor c^2 gives

$$p_e^2 = \left(p - p'\right)^2 + 2\left(p - p'\right)m_e c. \tag{ii}$$

Equations (i) and (ii) have the same left-hand side, and so their right-hand sides must be equal:

$$p^2 + p'^2 - 2pp'\cos\theta = \left(p - p'\right)^2 + 2\left(p - p'\right)m_e c.$$

Using $(p - p')^2 = p^2 + p'^2 - 2pp'$, we then get

$$p^2 + p'^2 - 2pp'\cos\theta = p^2 + p'^2 - 2pp' + 2\left(p - p'\right)m_e c \Rightarrow$$

$$2pp'(1 - \cos\theta) = 2\left(p - p'\right)m_e c.$$

Now we use the relationship in equation 36.17 between a photon's momentum and wavelength, $p = h/\lambda$, and find

$$2\frac{h}{\lambda}\frac{h}{\lambda'}(1 - \cos\theta) = 2\left(\frac{h}{\lambda} - \frac{h}{\lambda'}\right)m_e c \Rightarrow \frac{h}{m_e c}(1 - \cos\theta) = \lambda' - \lambda,$$

which is the result stated in equation 36.20.

The ratio of the constants $h/m_e c$ has the dimension of length, as can be seen from equation 36.20. This characteristic of an electron is called the **Compton wavelength** of the electron and has the value

$$\lambda_e = \frac{h}{m_e c} = \frac{6.626 \cdot 10^{-34} \text{ J s}}{(9.109 \cdot 10^{-31} \text{ kg})(2.998 \cdot 10^8 \text{ m/s})} = 2.426 \cdot 10^{-12} \text{ m}. \tag{36.21}$$

Self-Test Opportunity 36.3

The electrons inside a metal are not quite at rest, but have kinetic energies of a few electron-volts. Why is it permissible to assume that the electron is at rest in the derivation of the Compton scattering formula?

SOLVED PROBLEM 36.1 | Compton Scattering

PROBLEM

An X-ray photon with a frequency of $3.3530 \cdot 10^{19}$ strikes an electron in a metal foil, and the scattered photon is detected at an angle of $32.300°$ relative to the direction of the incoming photon.

a) What are the energies of the incoming and scattered photons, in electron-volts?
b) What is the kinetic energy of the electron after the collision?
c) What is the magnitude of the momentum of the electron after the collision?

SOLUTION

THINK Let's begin with the incoming photon. Converting its frequency into energy is a straightforward application of $E = hf$. Since we want the answer to have units of electron-volts, we need to use Planck's constant expressed in eV s: $h = 4.13567 \cdot 10^{-15}$ eV s. Now, for the electron, we know that energy is conserved in the scattering of a photon off an electron. (This comes as no surprise, since energy conservation was one of the starting principles in our derivation of the Compton scattering formula.) Thus, the kinetic energy that the electron receives in this scattering process is simply equal to the energy lost by the photon.

SKETCH If we assume that the incoming photon travels along the positive x-axis, that the scattering event takes place in the xy-plane, and that the photon is deflected to positive values of y, we can use Figure 36.11 for our sketch.

– Continued

RESEARCH   a) We only need to insert the given value for the incoming photon's frequency into $E = hf$ to obtain that photon's energy. To find the energy of the scattered photon, we need to use the Compton scattering formula. For this equation, we need the wavelength of the incoming photon, which we can obtain from the frequency via $\lambda = c/f$. For the Compton scattering process, we have just derived that $\lambda' = \lambda + h(1 - \cos\theta)/(m_e c)$. Once we have the wavelength of the scattered photon, we can convert that value to an energy value using $E' = hf' = hc/\lambda'$.

b) As stated above, total energy is conserved. The electron is assumed to be stationary before the scattering event; that is, it has zero kinetic energy. The final kinetic energy of the electron is thus simply equal to the energy lost by the photon: $K_e = E - E'$. Finally, we can calculate the absolute value of the momentum of the electron from the energy-momentum relation derived in Chapter 35:

$$p_e = \frac{1}{c}\sqrt{\left(K_e + m_e c^2\right)^2 - m_e^2 c^4} \, .$$

SIMPLIFY   We can evaluate the expression for the energy of the scattered photon as follows:

$$E' = \frac{hc}{\lambda'} = \frac{hc}{\lambda + \dfrac{h}{m_e c}(1 - \cos\theta)} = \frac{hc}{\dfrac{hc}{E} + \dfrac{h}{m_e c}(1 - \cos\theta)} = \frac{1}{\dfrac{1}{E} + \dfrac{1}{m_e c^2}(1 - \cos\theta)} \, .$$

For the electron's momentum, we find

$$p_e = \frac{1}{c}\sqrt{\left(K_e + m_e c^2\right)^2 - m_e^2 c^4} = \frac{1}{c}\sqrt{K_e^2 + 2 K_e m_e c^2} \, .$$

CALCULATE   a) The energy of the incoming X-ray photon is

$$E = hf = (4.13567 \cdot 10^{-15} \text{ eV s})(3.3530 \cdot 10^{19} \text{ s}^{-1}) = 138.67 \text{ keV}.$$

The energy of the scattered X-ray photon is

$$E' = \frac{1}{\dfrac{1}{138.67 \text{ keV}} + \dfrac{1}{511.00 \text{ keV}}(1 - \cos 32.300°)} = 133.08 \text{ keV}.$$

b) The kinetic energy of the electron after the collision is

$$K_e = E - E' = 138.67 \text{ keV} - 133.08 \text{ keV} = 5.59 \text{ keV}.$$

c) The electron's momentum is

$$p_e = \frac{1}{c}\sqrt{(5.59 \text{ keV})^2 + 2(5.59 \text{ keV})(511.00 \text{ keV})} = 75.8 \text{ keV}/c.$$

ROUND   The computed values have been rounded to the proper number of significant figures. But why do the values for the electron's kinetic energy and momentum have only three digits? After all, the input values were given to five significant figures. The answer is that we were able to compute the initial and final photon energies only to the nearest 0.01 keV, and since the electron's kinetic energy is the difference between these two energies, its accuracy is limited to the nearest 0.01 keV.

DOUBLE-CHECK   In order to obtain the electron's momentum, we could use momentum conservation and find

$$\vec{p}_e = \vec{p} - \vec{p}\,'.$$

Since we have calculated the energies of the photons before and after the collision, we can obtain the magnitudes of the momentum vectors for the incoming and scattered photons in units of keV/c:

$$p = \frac{E}{c} = 138.67 \text{ keV}/c,$$

$$p' = \frac{E}{c} = 133.08 \text{ keV}/c.$$

The momentum of the incoming photon, $\vec{p}$, has only an x-component, because we assumed that it traveled along the positive x-axis. Thus,

$$p_x = 138.67 \text{ keV}/c \quad \text{and} \quad p_y = 0.$$

The angle of the scattered photon was specified in the problem statement as 32.300°. We then obtain for the Cartesian components of the momentum vector for this photon:

$$p'_x = (133.08 \text{ keV}/c)\cos(32.300°) = 112.49 \text{ keV}/c$$

and

$$p'_y = (133.08 \text{ keV}/c)\sin(32.300°) = 71.112 \text{ keV}/c.$$

Thus, the components of the electron's momentum are

$$p_{e,x} = p_x - p'_x = 138.67 \text{ keV}/c - 112.49 \text{ keV}/c = 26.18 \text{ keV}/c$$

and

$$p_{e,y} = p_y - p'_y = -71.11 \text{ keV}/c.$$

Now we can obtain the magnitude of the electron's momentum vector by the usual procedure of taking the square root of the sum of the squares of the components:

$$p_e = \sqrt{p_{e,x}^2 + p_{e,y}^2} = 75.8 \text{ keV}/c.$$

This is the same result as we obtained above, showing that our two methods of solution are consistent with each other.

Self-Test Opportunity 36.4

Why do you have to use X-rays to observe the Compton effect? Can you explain why Compton scattering is not observable for visible light?

36.5 Matter Waves

So far, we have established that photons are the quantum particles of light and of all other electromagnetic radiation. However, everything we have said about the wave nature of light is still true; for example, we can demonstrate interference and diffraction, which are typical wave phenomena. Looking at light as quantum particles does not invalidate the wave picture of light, just as the ray picture of light can be seen as a special limiting case of the more general wave description of light.

Given the quantum character of light and the particle character of electromagnetic waves, do things that we ordinarily think of as particles, like electrons and atoms, also have wave properties? This is exactly what Prince Louis de Broglie (1892–1987), a French graduate student at the time, proposed in 1923. His original publication, all of four pages long, contained this hypothesis, and he later was awarded the Nobel Prize in Physics in 1929.

If particles have wave character, then what is the appropriate wavelength? For light, we found (equation 36.17) that the momentum of a photon is $p = h/\lambda$. De Broglie used the same relationship for particles and proposed as the relevant wavelength for **matter waves:**

$$\lambda = \frac{h}{p} = \frac{h}{mv\gamma} = \frac{h}{mv}\sqrt{1 - \frac{v^2}{c^2}}. \qquad (36.22)$$

This wavelength, the de Broglie wavelength, depends on the mass m and speed v of a particle. Equation 36.22 used the relativistic form of the momentum, $p = mv\gamma$, but the literature often presents this nonrelativistic approximation for the **de Broglie wavelength:**

$$\lambda = \frac{h}{mv} \quad \text{(nonrelativistic approximation)}. \qquad (36.23)$$

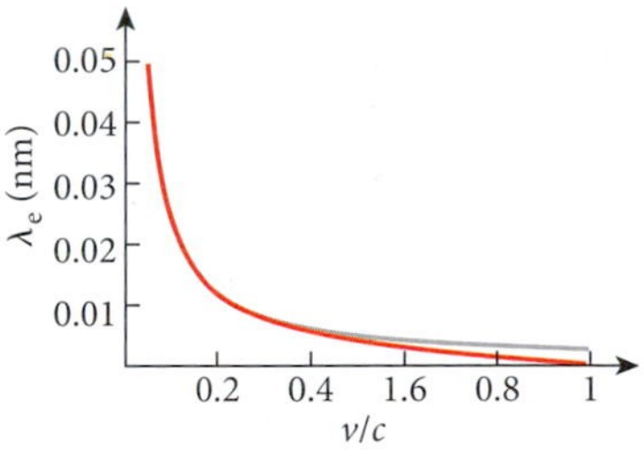

As Figure 36.12 shows for the case of an electron, at speeds of up to 40% of the speed of light, the nonrelativistic approximation is very close to the exact result given by equation 36.22.

As you can see from Figure 36.12, the de Broglie wavelength of an electron, even one moving at 10% of the speed of light, is on the order of one-tenth of a nanometer. What are typical de Broglie wavelengths for macroscopic objects? Example 36.3 shows a calculation.

FIGURE 36.12 De Broglie wavelength of an electron as a function of its speed. Red curve is the exact result; gray curve is the nonrelativistic approximation.

Concept Check 36.3

Which of the following statements is true?

a) More massive and faster objects have bigger de Broglie wavelengths than less massive and slower ones.

b) Less massive and faster objects have bigger de Broglie wavelengths than more massive and slower ones.

c) More massive and slower objects have bigger de Broglie wavelengths than less massive and faster ones.

d) Less massive and slower objects have bigger de Broglie wavelengths than more massive and faster ones.

EXAMPLE 36.3 | De Broglie Wavelength of a Raindrop

Raindrop diameters vary from approximately 0.50 mm to 5.0 mm. Drops at the lower end of this size range fall with speeds of 2 m/s; those at the upper end, with speeds up to 9 m/s.

PROBLEM

What is the range of de Broglie wavelengths for raindrops?

SOLUTION

The mass and diameter of a raindrop are related via

$$m = V\rho = \tfrac{4}{3}\pi r^3 \rho = \tfrac{1}{6}\pi d^3 \rho,$$

– Continued

where ρ is the density, r is the radius, and d is the diameter of a drop. The density of water is $\rho = 1000 \text{ kg/m}^3$, so the mass of a drop with $d = 0.50$ mm is $6.5 \cdot 10^{-8}$ kg, and the mass of a drop with $d = 5.0$ mm is $6.5 \cdot 10^{-5}$ kg.

For the extremely small speeds under consideration in this example, we are fully justified in using the nonrelativistic approximation, $\lambda = h/mv$, for the de Broglie wavelength. We then obtain for the smallest raindrops

$$\lambda = \frac{h}{mv} = \frac{6.626 \cdot 10^{-34} \text{ J s}}{(6.5 \cdot 10^{-8} \text{ kg})(2 \text{ m/s})} = 5 \cdot 10^{-27} \text{ m},$$

and for the largest raindrops, $\lambda = 1 \cdot 10^{-30}$ m.

DISCUSSION

Even the smallest and slowest-moving raindrops have de Broglie wavelengths that are many orders of magnitude smaller than diameters of individual atoms, which are approximately 10^{-10} m. Thus, we can safely ignore all implications of matter waves for macroscopic objects. Any object that is big enough for us to see it with the unaided eye and that moves fast enough that we can discern some kind of motion at all has a de Broglie wavelength so small that any observation of quantum wave phenomena for the object is ruled out.

Self-Test Opportunity 36.5

Plot the de Broglie wavelength of an electron as a function of its kinetic energy, for kinetic energies between 1 and 1000 eV. Is there a visible difference in this range of energies if you use the nonrelativistic approximation, $p = \sqrt{2mK}$?

As we have seen in Solved Problem 36.1, the kinetic energy–momentum relationship for a particle is given by

$$p = \frac{1}{c}\sqrt{E^2 - m^2 c^4} = \frac{1}{c}\sqrt{(K + mc^2)^2 - m^2 c^4} = \frac{1}{c}\sqrt{K^2 + 2Kmc^2}.$$

Therefore, the de Broglie wavelength can also be written as a function of the kinetic energy of a particle:

$$\lambda = \frac{h}{p} = \frac{hc}{\sqrt{K^2 + 2Kmc^2}}.$$

So far we have only discussed the theoretical possibility of matter waves, as described by de Broglie's hypothesis. Where is the experimental proof? Before we look at this proof, it is helpful to revisit material from Chapter 34 on wave optics and see what makes a wave a wave.

Double-Slit Experiment with Particles

To provide experimental evidence for the existence of matter waves, it is necessary to show that objects such as electrons, neutrons, protons, or whole atoms, which have mass and are normally thought of as particles, can exhibit wavelike behavior. Chapter 34 showed that the two main physical effects that characterize waves are interference and diffraction. What sort of experiment could show interference and/or diffraction of matter?

Young was able to demonstrate the wave character of light by shining light through two narrow slits, separated by a distance d. The interference pattern produced in this way is shown in Figure 36.13.

In the double-slit experiment, bright interference fringes appear on the screen a distance L away from the slits. Provided that the line from the center of the two slits to a bright fringe makes a small angle with the perpendicular to the screen, the distance from this fringe to a neighboring fringe is given by

$$\Delta x = \frac{\lambda L}{d}. \tag{36.24}$$

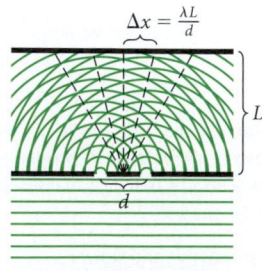

FIGURE 36.13 Double-slit interference for light.

However, in the derivation in Chapter 34 of this formula for the interference maxima, it was necessary to use the condition that the width of an individual slit is on the order of the wavelength of the light wave. For moderately fast-moving electrons, the de Broglie wavelength is on the order of one-tenth of a nanometer or less, and thus is more than three orders of magnitude smaller than the wavelength of visible light. Producing a double slit with sufficiently small slit separation d and individual slit width a for carrying out a double-slit experiment using electrons presents a sizable technical challenge. Also, if we want to use heavier particles, such as protons or neutrons, then the technical challenges become even bigger because the de Broglie wavelength is inversely proportional to the mass of a particle moving with a given speed.

For this reason, a double-slit experiment to verify the wave character of electrons was not carried out right after de Broglie proposed his revolutionary idea. Instead, an experiment in 1927 by American physicists Clinton Davisson (1881–1958) and Lester H. Germer (1896–1971), working at the Bell Telephone Laboratories in New Jersey, provided proof for the physical existence of matter waves. Davisson and Germer continued their earlier work on Bragg scattering of X-rays off crystals and scattered a beam of electrons off a nickel crystal, observing interference patterns similar to those produced by X-rays. Today, beams of neutrons can also be scattered off crystals; American physicist Clifford G. Shull (1915–2001) and Canadian physicist Bertram Brockhouse (1918–2003) received the 1994 Nobel Prize in Physics for pioneering such techniques. X-rays, electrons, and neutrons all provide equivalent Bragg-scattering patterns, proof that matter waves are real.

Only since the early 1960s has there existed sufficiently precise technology to conduct double-slit experiments with electrons. Let's think about what should happen in such an experiment and then compare our prediction with the outcome of the experiment. The basic setup of the experiment is shown in Figure 36.14. An electron is emitted from a plate in an electron gun (the plate is heated in order to do this, but the heater is not shown here) and then accelerated by a voltage V. It then passes through a double slit on its way to the screen.

If electrons behave like particles, they will travel in straight lines from the electron gun through one of the slits and on to the screen. We then predict that we will see the electrons as two lines on the screen—the images of the two slits. Since the electrons might get slightly deflected as they pass though a slit, the distribution of the electrons passing through each slit will be somewhat smeared out. Because the slit separation d is very small, the distributions of the electrons passing through the two slits will overlap on the screen. This classical particle-like expectation leads to a distribution of electrons hitting the screen as sketched in Figure 36.15a. Here the distribution from the electrons passing through the left slit is shown in blue, and that of the electrons passing through the right slit in red. Shown in green is the sum of the two, the total intensity distribution that should be recorded if our particle-like expectation bears out.

On the other hand, if electrons show wavelike characteristics, then their intensity distribution on the screen should be governed by the combined effects of interference and diffraction, just as we observed for light in Chapter 34 on wave optics. In this case (compare Section 34.3), the intensity as a function of the coordinate x along the screen should be

$$I(x) = I_{max} \cos^2\left(\frac{\pi d}{\lambda L} x\right) \left(\frac{\lambda L}{\pi a x} \sin\left(\frac{\pi a}{\lambda L} x\right)\right)^2. \tag{36.25}$$

Here d is the slit separation, a is the width of the individual slits, and L is the distance between the double slit and the screen. This is the same terminology used in Young's double-slit experiment with light, but now λ is the de Broglie wavelength of the electron (equation 36.22). The function $I(x)$ of equation 36.25 is graphed in Figure 36.15b.

Figure 36.16 shows what the pattern of electrons striking the screen (yellow spots) might look like for an intensity distribution as given by equation 36.25, with a de Broglie wavelength of 12.2 pm, a slit separation of 3.0 nm, a slit width of 1.0 nm, and a distance to the screen of 1.0 m. The lower part of Figure 36.16 is the counting histogram—that is, the number of electron hits in a given interval of the coordinate x along the screen. Overlaid in red over the blue counting histogram is the intensity distribution of equation 36.25.

What is the outcome of the actual experiment? Electrons can be shot through the slits and at the screen individually, so the pattern can be observed to emerge as a function of time. This experiment was performed by P. G. Merli and colleagues in 1976; the outcome is shown in Figure 36.17. The lowest panel shows the final outcome of this experiment, with the interference fringes clearly visible. The experiment clearly demonstrates that the electrons show wavelike interference phenomena.

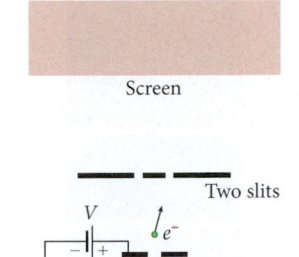

FIGURE 36.14 Experimental setup for an electron double-slit experiment.

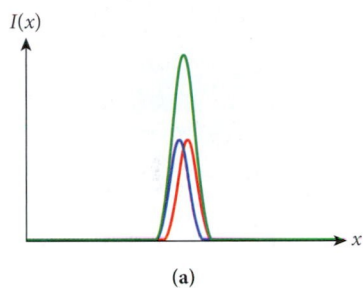

(a)

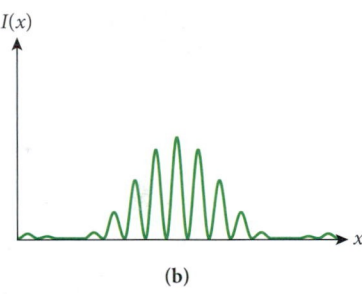

(b)

FIGURE 36.15 Intensity distribution (number of electrons hitting the screen per unit length) along the screen in the electron double-slit experiment. (a) Classical particle-like expectation; (b) expected distribution if electrons show matter-wave characteristics.

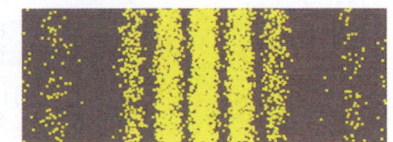

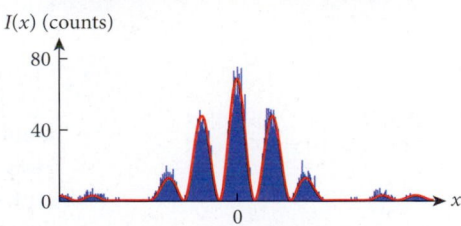

FIGURE 36.16 Connection between hits of individual electrons on the screen (yellow spots in upper part) and the count histogram (blue, lower part), with the predicted intensity distribution (red curve).

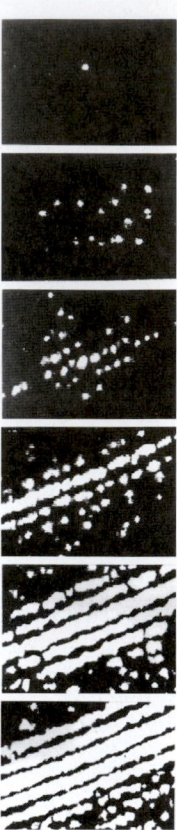

FIGURE 36.17 Experimental double-slit electron interference pattern as it develops over time, from top to bottom.

However, Figure 36.17 shows more: You can see that each individual electron leaves a mark by hitting a certain very localized area of the screen. Therefore, the idea that each electron somehow gets distributed over the entire screen, proportional to the overall intensity distribution, is not true either.

Because photons have particle properties, it might be expected that photons striking the double slits would exhibit the granularity shown in Figure 36.17. When the double-slit experiment is performed using one photon at a time, a pattern similar to that shown for electrons in Figure 36.17 is indeed observed.

So, what makes an individual electron follow the path that it does? This is the central question of quantum physics. Finding the answer will tell us most of what is essential about the atomic quantum world.

Are both slits essential for this interference pattern to emerge? In other words, does the electron somehow move through both slits at the same time? If one slit is closed, the interference pattern on the screen disappears, and only one maximum—the image of that slit—is produced. The result corresponds to either the red or the blue curve in Figure 36.15a, depending on which slit is covered.

What if, instead, it were possible to determine which slit the electron moves through without closing the other slit? Perhaps this could be accomplished by taking advantage of the fact that the electron has a charge and thus represents a current as it moves through a slit. It might be possible to measure this current.

Concept Check 36.4

Can you guess the outcome of measuring the current to determine which slit the electron passes through? If we measure the current representing the electron passing through the slits, which of the following will be the result?

a) Exactly one-half of each individual electron passes through each of the two slits.

b) Each individual electron passes through just one slit. The electrons passing through the left slit cause the left part of the interference pattern, and the electrons passing through the right slit cause the right part.

c) Each individual electron passes through just one slit. The electrons passing through the

left slit cause the right part of the interference pattern, and the electrons passing through the right slit cause the left part.

d) Each individual electron passes through just one slit, but as we conduct measurements to try to determine this, the interference pattern on the screen is destroyed, and we observe only the central maximum on the screen.

We find that any attempt to associate an electron with a particular slit destroys the interference pattern. This outcome should become more acceptable when we further explore the wave character of electrons and other objects that we normally think of as particles.

36.6 Uncertainty Relation

How precisely is it possible to measure physical properties such as location, momentum, energy, and time? In addition, to what precision can two properties be measured simultaneously? This question is never considered in classical mechanics, where it is assumed that all dynamic quantities can be measured with arbitrary precision with improved instrumentation.

However, in the quantum realm, where particles exhibit wave character (de Broglie matter waves) and where waves behave like particles (photons), the answer is not so simple. How can the precise location of a wave be specified, for example? Perhaps more important, does the process of measuring a physical property of a quantum object influence the outcome of that measurement, as well as all future measurements? For instance, when we measure the position of an object, we typically record the light waves emitted from that object. However, these emitted light waves also carry momentum, as we have seen in this chapter. Thus, we can anticipate that the process of measuring a particle's position and its momentum cannot be done simultaneously with arbitrary precision.

Let the uncertainty in a measurement of a particle's position be Δx and the uncertainty in the measurement of its momentum be Δp_x, in the same sense as is done in statistics. In statistics, the outcome of a series of independent measurements of a quantity is quoted in terms of the mean value, which is the average of the measurements, plus or minus the standard deviation, which is a measure of the width of the distribution of the measurements. When a physical measurement is performed in the lab, the result also must be expressed in terms of the average value plus or minus the uncertainty in the measurement. (This uncertainty can be of statistical or systematic origin, but we are not concerned with this distinction.)

The astonishing statement arising from quantum physics is that the momentum and the position of an object cannot be measured simultaneously with arbitrary precision. The more precisely we attempt to measure an object's momentum, the less precise the information on its position has to become, and vice versa. This physical statement is cast in mathematical terms in the form of the **Heisenberg Uncertainty Relation:**

$$\Delta x \cdot \Delta p_x \geq \tfrac{1}{2}\hbar. \tag{36.26}$$

This relation was discovered in 1927 by the German physicist Werner Heisenberg (1901–1976) and caused a revolutionary change in our understanding of the measurement process, as well as of our fundamental ability to know the physical world. Chapter 37 will return to this uncertainty relation and use it for calculations. For now, to motivate this relation, we will use the same considerations suggested by Heisenberg in his original paper. This heuristic derivation uses the so-called gamma-ray microscope.

DERIVATION 36.3 | Gamma-Ray Microscope and the Uncertainty Relation

If you want to see something in a microscope, you have to bounce light off that object and catch the reflected light in the lens. The minimum size, Δx, of the object that you can resolve with the microscope is limited by diffraction (see Chapter 34) and is given by

$$\Delta x = \frac{\lambda}{2\sin\alpha} \approx \frac{\lambda\ell}{d}. \tag{i}$$

Here λ is the wavelength of the light and α is the opening angle (Figure 36.18). In this figure, d is the diameter of the lens opening of the microscope, and ℓ is the distance between the object and the lens, which we assume to be large compared to the lens opening; then $2\sin\alpha \approx d/\ell$. To resolve small sizes, we need to employ light with a short wavelength, that is, gamma rays. These gamma-ray photons carry a momentum $p = h/\lambda$ (see equation 36.17).

When the object (an electron) is hit with gamma rays (yellow circle, indicating a beam pointing into the page), the photons bounce off the object and get deflected into the lens opening via Compton scattering. In one extreme case, a photon can bounce off the electron and get deflected to the right edge of the opening. This photon has a momentum $\vec{p}_{\gamma 1}$, as shown in the figure. Its momentum component in the x-direction is $p_{\gamma 1,x} = p_{\gamma 1}\sin\alpha = h\sin\alpha/\lambda$. The electron receives the opposite x-component of momentum, $p_{e1,x}$, pointing to the left, due to recoil. At the other extreme, a photon receives momentum $\vec{p}_{\gamma 2}$ after the collision with the electron and gets scattered to the left edge of the lens opening, which causes a recoil of the electron with the same magnitude as before, but in the opposite direction.

We can detect that a photon has entered the lens opening but cannot detect where along the distance d it has done so. This means that it is undetermined which recoil the electron received. Thus, the momentum of the electron has an uncertainty of

$$\Delta p_x = 2|p_{\gamma 1,x}| = \frac{2h\sin\alpha}{\lambda}. \tag{ii}$$

Combining equations (i) and (ii), we find

$$\Delta p_x = \frac{2h\sin\alpha}{\lambda} = \frac{h}{\Delta x}.$$

Thus, the product of the minimum uncertainty in the size of the electron and the uncertainty in the momentum of the electron is Planck's constant, h.

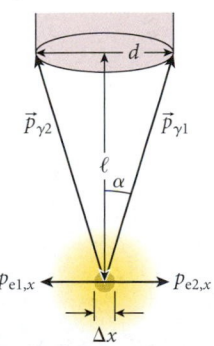

FIGURE 36.18 Geometry of a gamma-ray microscope and the momenta of a photon-electron interaction.

– Continued

DISCUSSION

You may feel that this argument involves hand waving, and to a certain degree, it does. Our result is larger by a factor of 4π higher than the exact answer for the minimum product of momentum and coordinate uncertainties given in equation 36.26. However, an exact numerical relation is not the point of this analysis using a gamma-ray microscope. Instead, it shows a certain lower limit of the uncertainties in coordinate and corresponding momentum that can be observed simultaneously. This in itself is an astounding fact and a consequence of quantum physics.

Heisenberg also stated another uncertainty relation, for the uncertainties in the measurement of the energy of an object, ΔE, and in the measurement of the time, Δt. It is formally similar to the coordinate-momentum uncertainty relation of equation 36.26:

$$\Delta E \cdot \Delta t \geq \tfrac{1}{2}\hbar. \qquad (36.27)$$

DERIVATION 36.4 / **Energy-Time Uncertainty Relation**

The coordinate-momentum uncertainty relation straightforwardly motivates the time-energy relation for a nonrelativistic free particle. For such a particle, the energy is all kinetic energy, and thus the uncertainty in the energy is

$$\Delta E = \Delta\left(\frac{p^2}{2m}\right) = \frac{\Delta(p^2)}{2m} = \frac{2p\Delta p}{2m} = v\Delta p.$$

The uncertainty in the time is given by

$$\Delta t = \frac{\Delta x}{v}.$$

Multiplying these two results, we find

$$\Delta E \cdot \Delta t = (v\Delta p)\cdot(\Delta x/v) = \Delta x \cdot \Delta p \geq \tfrac{1}{2}\hbar,$$

where we have used equation 36.26 in the last step.

The energy-time uncertainty relation stipulates that classical energy conservation can be violated by some amount of energy for some time interval, because the quantum state may not have a sharp energy value. The larger the "violation" of energy conservation is, however, the shorter the time interval for this "violation" will be.

Do these uncertainty relations conflict with our everyday experience? In other words, how important are the fundamental limitations imposed by the uncertainty relations? Let us examine an example.

Concept Check 36.5

In Example 36.4, if all other parameters remain the same but the mass of the car is doubled, the resulting uncertainty in the speed will be

a) the same.

b) half as large.

c) one-quarter as large.

d) twice as large.

e) four times as large.

EXAMPLE 36.4 / **Trying to Get Out of a Speeding Ticket**

PROBLEM

On the German autobahn, some sections actually do have a speed limit. Occasionally the German police set up speed traps in these sections. They measure the speed of a vehicle and take a picture of the driver at the same time, as proof that they really have the right person committing the offense. A German physics student receives such a picture of herself and her car (a BMW 318Ci of mass 1462 kg, including the driver and the gas in the tank), with a notification that she was driving 132 km/h in a zone with a 100-km/h speed limit. She notices that the picture the police took is very sharp and fixes her position to an uncertainty of 1 mm. She argues that the uncertainty relation prevents the police from determining her speed precisely, and she therefore should not get a speeding ticket, or at least not get one for driving more than 30 km/h above the speed limit, which would cause her to lose her license. Will this strategy be successful?

SOLUTION

If the judge knows physics, the student will not be successful. Here is why: If the mass of the car is known precisely enough, then the uncertainty in the speed is found from the uncertainty in the momentum as

$$\Delta v = \frac{1}{m}\Delta p.$$

Using the uncertainty relation, we find the uncertainty in the momentum to be $\Delta p \geq \frac{1}{2}\hbar/\Delta x$, leading us to a value of the uncertainty in the speed of

$$\Delta v = \frac{1}{m}\Delta p \geq \frac{\hbar}{2m\Delta x}.$$

Using the constant (see equation 36.7) $\hbar = 1.05457 \cdot 10^{-34}$ J s and the values of $\Delta x = 10^{-3}$ m and $m = 1462$ kg given in the problem statement, we find

$$\Delta v \geq \frac{1.05457 \cdot 10^{-34} \text{ J s}}{2(1462 \text{ kg})(10^{-3} \text{ m})} = 3.6 \cdot 10^{-35} \text{ m/s}.$$

The restriction on the minimum measurement uncertainty in the speed due to the uncertainty relation is 35 orders of magnitude too small to be useful for the student's defense!

Self-Test Opportunity 36.6

How precisely would the position of the car in Example 36.4 have to be fixed to validate the student's claim that she should not have to surrender her driver's license because the uncertainty in her speed had to be larger than 2.00 km/h?

The uncertainty relation is perhaps the most important result discussed in this chapter and has far-reaching consequences. It sets a fundamental limit on how precisely we can measure and therefore know anything about our world. For macroscopic objects, the effect of the uncertainty relation can safely be neglected in almost all applications. However, an insurmountable limit exists to the precision of simultaneous measurements of pairs of variables, such as momentum and position, or energy and time. The analysis using the gamma-ray microscope seems to imply that the attempt to measure the location somehow imparts a simple recoil onto the object being measured. But in the quantum world, the objects normally characterized as particles have wave character, just as what we normally understand as a wave has particle character. At its core, the uncertainty relation derives from this particle-wave duality. This will be explored in detail in Chapter 37, where we will learn how to calculate properties with the tools of quantum mechanics.

36.7 Spin

Stern-Gerlach Experiment

In 1920, two German physicists, Otto Stern (1888–1969) and Walther Gerlach (1889–1979), performed one of the most influential experiments in the history of quantum physics. In this experiment, they tried to distinguish between a classical and a quantum description of atoms.

The basic setup of the Stern-Gerlach experiment is shown in Figure 36.19. An oven produces a gas of (electrically neutral) silver atoms. These atoms are allowed to escape from the oven through a hole and form a "beam" of silver atoms, which move along a straight line (green dots in the figure). The silver atoms then enter a strong inhomogeneous magnetic field and continue on to a screen.

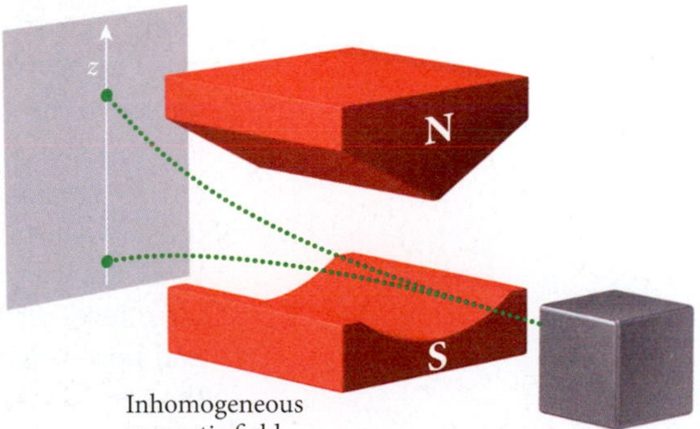

Section 28.5 mentioned that atoms can have a magnetic moment,

$$\vec{\mu} = -\frac{e}{2m}\vec{L},$$

Inhomogeneous magnetic field

Furnace emitting silver atoms

FIGURE 36.19 Basic setup of the Stern-Gerlach experiment.

where $\vec{L}$ is the angular momentum. Section 28.5 considered only the magnetic moment due to a charge in a circular orbit, and the angular momentum was the orbital angular momentum. Section 27.6 showed that the potential energy of a magnetic dipole in a magnetic field is $U = -\vec{\mu}\cdot\vec{B}$. The force is then the negative value of the gradient of the potential energy. In a magnetic field that changes only as a function of the z-coordinate, the force is

$$F_z = -\frac{\partial}{\partial z}\left(-\vec{\mu}\cdot\vec{B}\right) = \mu_z\frac{\partial B}{\partial z}.$$

Stern and Gerlach tried to determine whether the orbital angular momentum was quantized. Sending a beam of atoms that have a magnetic moment through an inhomogeneous magnetic field causes a deflection of the beam, due to the interaction of the field gradient, $\partial B/\partial z$, with the z-component of the magnetic moment of each atom, μ_z. Classically, the magnetic moment can have any value along the z-axis in the interval between $-|\vec{\mu}|$ and $+|\vec{\mu}|$, which should produce a line on the screen that corresponds to all possible deflections. However, in the quantum realm, only discrete values of the z-component of the angular momentum are possible, which leads to discrete spots on the screen, separated by empty areas, as illustrated in Figure 36.19.

We know now that the orbital angular momentum of a silver atom in its ground state is zero—this will be shown in Chapter 38, where states of atoms and their orbital angular momenta are calculated. Thus, if the Stern-Gerlach experiment had measured the splitting of the beam of silver atoms due only to the orbital angular momentum, the experiment would have been a failure. However, the total angular momentum of the atom is *not* just the orbital angular momentum. In addition, each elementary particle in the atom has an intrinsic angular momentum, which has no classical equivalent. The intrinsic angular momentum is called **spin.**

Spin of Elementary Particles and the Pauli Exclusion Principle

All elementary particles have a characteristic intrinsic angular momentum, or spin. There are two fundamentally different groups of elementary particles: those that have integer values (in multiples of Planck's constant $\hbar$) of spin, and those that have half-integer values. Some elementary particles have spin 0, but for our purposes, we count this as an integer spin.

Elementary particles with half-integer spins are called **fermions,** in honor of the Italian-American physicist Enrico Fermi (1901–1954). Fermions include the electron, the proton, and the neutron—in other words, all the building blocks of the matter we see around us. A fermion with spin $\frac{1}{2}\hbar$ can have one of two different spin quantum states, indicated by $+\frac{1}{2}\hbar$ and $-\frac{1}{2}\hbar$, which by convention represent the projection of the spin onto the z-coordinate axis. We will encounter other quantum numbers in the following chapters, but for now all you need to know is that spin $\frac{1}{2}\hbar$ particles come in two varieties, which are usually called *spin-up* and *spin-down*.

Elementary particles with integer values of spin are called **bosons,** in honor of the Indian physicist Satyendra Nath Bose (1894–1974). Photons are bosons and have spin $1\hbar$.

In Chapters 39 and 40 on elementary particles and nuclear physics, we will return to spin and give other examples of particles with half-integer and integer-valued spins and try to find a more general organizing principle. For now, we can use the two fundamentally different classes of particles—bosons and fermions—as working concepts.

One extremely important rule for fermions is the **Pauli Exclusion Principle:** No two fermions can occupy the same quantum state at the same time at the same location. In any given atom, no two fermions can have exactly identical quantum numbers. Because energy is quantized, each energy quantum state in a given system can be occupied by at most two fermions: one spin-up and the other spin-down. In the remaining chapters of this book, we will return repeatedly to the consequences of this exclusion principle. Chapter 37 introduces wave functions and then shows that two-particle wave functions for fermions and bosons have fundamentally different symmetries. Chapter 38 will show the effects of the Pauli Exclusion Principle in the construction of multi-electron atoms and the resulting periodic table of the elements. Chapters 39 and 40 will show that this principle gives rise to the Fermi energy inside the nucleus.

36.8 Spin and Statistics

Chapter 19 presented the probability distribution function for *classical* identical particles. For these identical particles, the distribution function is called the **Maxwell-Boltzmann distribution:**

$$g(E) = \frac{2}{\sqrt{\pi}} \left(\frac{1}{k_B T} \right)^{3/2} \sqrt{E} e^{-E/k_B T},$$

where E is the energy, T is the temperature, and k_B is Boltzmann's constant. (In deriving the Maxwell-Boltzmann distribution, we examined only the kinetic energy distribution of molecules in a gas, but this function is valid for all energy distributions of classical particles.) In this derivation, we neglected all quantum effects. Now we want to examine how quantum effects change this distribution.

The Maxwell-Boltzmann distribution can be written in a slightly different way for particles in discrete energy states, E_i. For a total of N particles in a system, the expected number of particles with energy E_i is N_i, and the expected fraction of particles in energy state E_i is $n_i = N_i/N$, where

$$n_i = \frac{N_i}{N} = \frac{g_i e^{-E_i/k_B T}}{Z} = \frac{g_i}{e^{(E_i - \mu)/k_B T}}. \tag{36.28}$$

Here g_i is the degeneracy of energy state E_i, that is, the number of different states in the system that have the same energy E_i. The quantity μ is called the *chemical potential* and has the same units as energy. The chemical potential is the amount by which the energy of the system would change if an additional particle were added, while the remaining properties of the system remained fixed. The quantity Z is called the **partition function:**

$$Z = \sum_i g_i e^{-E_i/k_B T}. \tag{36.29}$$

The partition function encodes the thermodynamic properties of the system, and taking the appropriate derivatives of the partition function can reproduce many of the properties of the system.

The derivation of the Maxwell-Boltzmann distribution assumed that each of the particles in the ensemble is a classical particle. What we mean by *classical* is that each particle is distinguishable from the others. However, quantum particles are indistinguishable from like particles. (For example, one proton cannot be distinguished from another proton.) Thus, the distribution function has to be modified appropriately.

To see how this can work in practice, let's consider the simplest way of distributing two particles into two different states. Label these states a and b (Figure 36.20). First, consider the case of distinguishable particles, labeled 1 and 2. This is the classical Maxwell-Boltzmann distribution. In this case, there are four different configurations of the system. We can have both particles in state a, we can have particle 1 in state a and particle 2 in state b, we can have particle 1 in state b and particle 2 in state a, and finally we can have both particles in state b. This means that this two-particle system can have four different states (configurations) that it can be in, if the two particles are distinguishable. Next, consider two spin-0 bosons, indistinguishable quantum particles. This is known as the **Bose-Einstein distribution.** In this case, both particles can be in state a or both in state b, just as in the case of distinguishable particles. However, if one of them is in state a and the other is in state b, then it does not matter which one is which, because they are indistinguishable. Therefore, the two-particle system has only three different states for bosons, and not four as in the case of distinguishable particles.

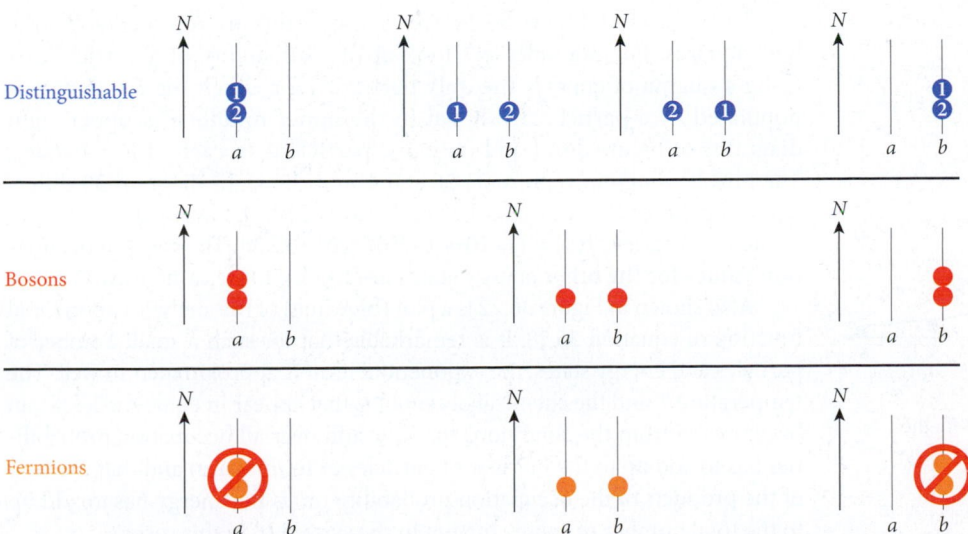

FIGURE 36.20 Distributing two particles over two different states.

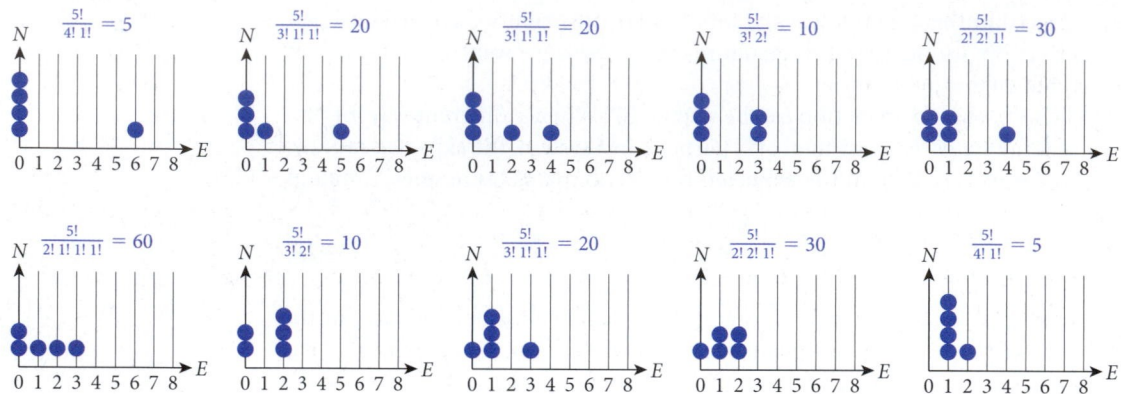

FIGURE 36.21 All possible partitions of 6 quanta of energy among 5 distinguishable particles.

Finally, the easiest case is obtained for two identical spin-$\frac{1}{2}$ fermions (for example, two electrons with spin-up). This case is called the **Fermi-Dirac distribution.** The Pauli Exclusion Principle states that these two identical fermions cannot occupy the same quantum state. (In the lowest panel of Figure 36.20, this fact is indicated by the "forbidden" sign, a red circle with a diagonal bar, over the two configurations in which both fermions would reside in the same state.) In this case, the two-particle system of identical fermions can be in only one configuration, where the quantum states a and b are each occupied by one fermion.

Now let us turn to the more complicated case of the energy distribution for the case of a system of 5 particles that share a total of 6 quanta of energy in a one-dimensional model. For classical distinguishable particles, the 6 energy quanta can be shared by one particle carrying all 6 energy quanta and the other 4 particles each with zero energy. This sharing of the total energy of 6 quanta can be written as $6 = 6+0+0+0+0$. Since the particles are distinguishable, we have to keep track of which one of the 5 carries 6 quanta; this energy partition can happen in 5 different ways. The energy can also be distributed as $6 = 5+1+0+0+0$. Each of the 5 particles can be the carrier of 5 quanta, and then each of the remaining 4 particles can carry the remaining 1 quantum. Thus, this particular energy partition can happen in $g_i = 5 \cdot 4 = 20$ different ways. Figure 36.21 shows all possible partitions of 6 energy quanta over 5 particles. The blue numbers above the diagrams show the number of different ways the same configuration can happen by exchanging all possible particles. The total number of different energy partitions among the 5 particles is the sum of all of the numbers listed above the diagrams and is 210.

Now we can calculate the probability of each energy state being occupied by counting how many partitions each energy state occurs in, multiplied by how often this partition occurs, and then dividing the result by the total number of partitions.

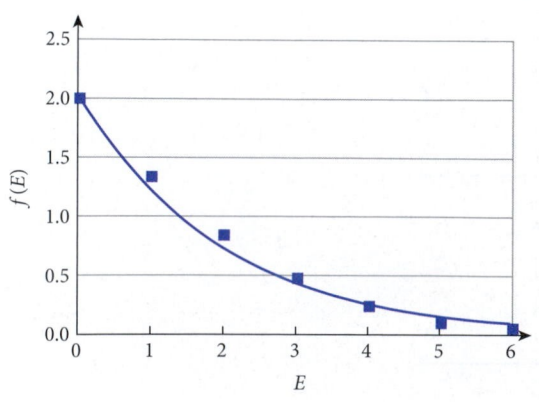

FIGURE 36.22 Average occupation values for the energy states, corresponding to the partitions of Figure 36.21.

This is done in Figure 36.22, where the results are represented by the blue squares. For example, let's look at the probability of a particle carrying 4 quanta of energy. The only partitions for which the $E = 4$ state is populated by a particle are shown in the upper middle and upper right diagrams of Figure 36.21 and only one particle is in each of $E = 4$ states. The middle diagram represents 20 states and the right diagram 30 different states. Since the total number of states is 210, the average occupation of the $E = 4$ state is $f(4) = (1 \cdot 20 + 1 \cdot 30)/210 = 0.238$. The average occupation values for the other energy states are found in the same way.

Also shown in Figure 36.22 is a plot (blue line) of the analytic exponential function of equation 36.28. It is remarkable that for such a small number of particles and energy states, the exponential limit is approximated so well. The temperature T and the chemical potential μ that appear in equation 36.28 can be extracted from the conditions that the sum over all occupation probabilities has to add up to the number of particles (5 in this case) and that the sum of the products of the occupation probability times the energy has to add up to the total number of energy quanta in the system (6 in this case).

Now we can ask how these considerations have to change if we are dealing with indistinguishable particles. For bosons, we have done most of the work already, because we can arrange the bosons as in Figure 36.21. However, since the bosons are indistinguishable, permutations of the different particles do not yield new states. Therefore, the weight factor that counts the number of states for a given arrangement of particles among energy states is always 1 for each partition diagram in Figure 36.21. Thus, bosons have only 10 energy partitions, instead of the 210 we found for distinguishable particles. If we now calculate the average occupation of the $E = 4$ state, we find it is $f(4) = (1+1)/10 = 0.2$.

The resulting distribution for identical bosons is displayed in Figure 36.23 (red dots) and compared to the classical case (blue line) that we just discussed. The distributions are remarkably similar, but important small differences appear, which are not just numerical artifacts. For instance, the occupation of the $E = 0$ state is slightly enhanced in the boson case relative to the classical distinguishable particles. As increasingly more particles are added to the system, this effect will become increasingly pronounced. It is another manifestation of the fact that bosons often occupy the same states as other bosons.

For the distribution of 6 energy quanta over 5 fermions, it might appear that the fermions can be distributed into the energy states as in Figure 36.21. However, the Pauli Exclusion Principle does not allow this. Each energy state can be occupied by at most 1 spin-up fermion and 1 spin-down fermion. Thus, the occupation of each level cannot exceed 2. This eliminates 7 partitions from Figure 36.21, because these have 3 or more particles in the same energy state. The only 3 partitions left are those shown in Figure 36.24. To illustrate, suppose the system has 3 spin-up and 2 spin-down fermions. For the left and right partitions in Figure 36.24, only 1 fermion is in a singly occupied state, which therefore has to be the unpaired spin-up. The middle partition diagram shows 3 fermions in singly occupied states, and 2 of these have to be spin-up and 1 spin-down. Since each of the 3 energy states can hold the spin-up fermion, this middle diagram represents a total of 3 states. This means that a total of 5 states are available to this system. Calculating the occupation of the $E = 4$ state now results in $f(4) = 1/5 = 0.2$ (which is, by chance, the same value as for the bosons). The orange line in Figure 36.23 shows the average occupations for all energy states. The occupation for the $E = 0$ state is suppressed in the case of fermions relative to classical particles.

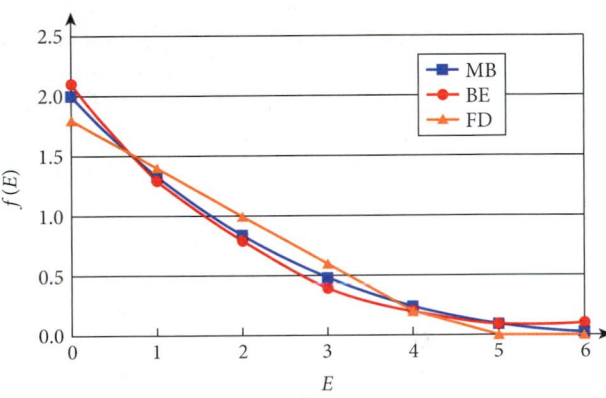

FIGURE 36.23 Comparison of the Maxwell-Boltzmann (MB), Bose-Einstein (BE), and Fermi-Dirac (FD) distributions for 6 energy quanta distributed among 5 particles.

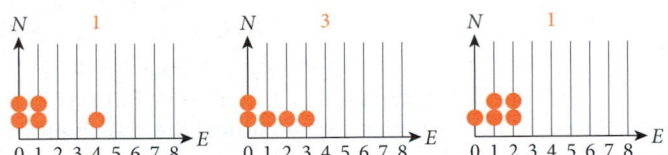

FIGURE 36.24 Possible ways to distribute 6 energy quanta among 5 fermions.

Concept Check 36.6

Which of the following distributions are possible ways to distribute 4 energy quanta over 5 fermions with spin $\frac{1}{2}\hbar$?

a)   b)   c)   d)   e)

Self-Test Opportunity 36.7

Plot the average occupation values for the distribution of 8 energy quanta over 5 particles. How many total states are available for each of these types of particles: fermions, bosons, and classical particles?

Let's now write the distributions for the much larger sets of particles present in physical systems. For bosons, the limiting Bose-Einstein distribution of the number of particles in a given energy state, E_i, is

$$N_i = \frac{1}{e^{(E_i - \mu)/k_B T} - 1}.$$ (36.30)

For fermions, the Fermi-Dirac distribution is

$$N_i = \frac{1}{e^{(E_i - \mu)/k_B T} + 1}.$$ (36.31)

The number of fermions with energy E_i can be found by multiplying the Fermi-Dirac distribution (equation 36.31) by the degeneracy g_i. (In our fermion example, this number is $g_i = 2$ for each state, because the Pauli Exclusion Principle says that each state can only be occupied by one spin-up and one spin-down fermion.)

Sometimes these distributions are expressed by using the definition of the absolute activity,

$$z = e^{\mu/k_B T}.$$

The Bose-Einstein distribution is then written as

$$N_i = \frac{1}{e^{E_i/k_B T}/z - 1},$$

and the Fermi-Dirac distribution is

$$N_i = \frac{1}{e^{E_i/k_B T}/z + 1}.$$

Both distribution functions approach the Maxwell-Boltzmann distribution for negligibly small absolute activity, $z \ll 1$.

For energy states that are sufficiently close to one another, the discrete values of the energy states, E_i, can be replaced by a continuous energy variable, E. Then the probability of finding a particle at energy E in the Maxwell-Boltzmann distribution can be written as

$$f_{MB}(E) = \frac{1}{ae^{E/k_B T}},$$ (36.32)

where a is a normalization constant that is a function of the chemical potential and the degeneracy. In the same way, the probability of finding a particle at energy E in the Fermi-Dirac distribution can be written as

$$f_{FD}(E) = \frac{1}{ae^{E/k_B T} + 1}.$$ (36.33)

The Fermi-Dirac distribution is plotted in Figure 36.25 as a function of the ratio of the energy divided by the chemical potential. The function is shown for three different temperatures. For $k_B T \ll \mu$, this function approaches a step function that falls from a value of 1 to 0 at $E = \mu$. For higher temperatures, the transition from high occupancy to low occupancy becomes increasingly smooth.

For the Bose-Einstein case,

$$f_{BE}(E) = \frac{1}{ae^{E/k_B T} - 1}.$$ (36.34)

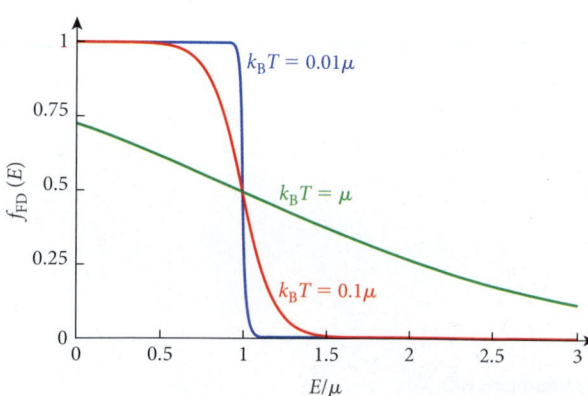

FIGURE 36.25 Fermi-Dirac distribution for three different temperatures as a function of the ratio of energy to chemical potential.

As we stated, photons are bosons and thus are subject to Bose-Einstein statistics. For the special case of photons, the chemical potential is zero, and thus the normalization constant in equation 36.34 has the value $a = 1$. If the photon energy approaches zero, then $e^{E/k_B T}$ approaches 1, and the denominator vanishes in equation 36.34. This means that the occupation of states with very low energy can increase without limit for photons.

Let's take another look at Planck's radiation formula (equation 36.9) for the spectral brightness as a function of the wavelength: $I_T(f) = 2hc^{-2} f^3 / \left(e^{hf/k_B T} - 1 \right)$.

Since $E = hf$ for photons, the denominator contains the factor $e^{E/k_B T} - 1$, which is also the case for the Bose-Einstein distribution for photons. Thus, the spectral brightness can be expressed as a function of photon energy:

$$I_T(E) = \frac{2E^3}{h^2 c^2 \left(e^{E/k_B T} - 1\right)} = \frac{2E^3}{h^2 c^2} f_{BE}(E).$$

(36.35)

Thus, the radiation formula obtained by Planck says that the spectral brightness as a function of photon energy is proportional to the third power of the energy times the Bose-Einstein probability of finding a photon at that energy.

Bose-Einstein Condensate

In his original paper in 1924, Bose discussed the blackbody spectrum for photons. Einstein extended this work in the same year to atoms with integer spin. Einstein noticed that at very low temperatures, a large fraction of the atoms go into the lowest energy quantum state: "One part condenses, the rest remains a 'saturated ideal gas.'" For this to happen, the atoms must be close enough to their neighbors that their de Broglie waves (see equation 36.23) overlap with one another. It can be shown that this condition implies, for a density ρ of atoms per unit volume, that $\rho \cdot \lambda^3 > 2.61$, where λ is the de Broglie wavelength.

How can one obtain a *Bose-Einstein condensate (BEC)* in the laboratory? Carl Wieman and Eric Cornell used a magnetic trap to collect rubidium atoms, which form spin-1 bosons at low temperatures and in a strong magnetic field. They used laser cooling, but this was still not sufficient to reach the low temperature required for transition to a BEC. By gradually reducing the depth of the trap, they allowed the atoms with higher energies to escape from the trap, leaving only the lower-energy atoms behind. With this method of evaporative cooling, they managed to cool the rubidium atoms to temperatures of a few nanokelvins (nK). They then switched off the trap and allowed the trapped collection of atoms to expand. A simple ideal gas will expand due to thermal motion of the atoms, but the BEC expands only at the minimal rate required by the Heisenberg Uncertainty Relation. A short while later, Wieman and Cornell imaged the expanding cloud, which by this time had reached a spatial extension on the order of 0.2 mm, and observed the distribution shown in Figure 36.26, which shows the results at three different temperatures. Clearly, the picture at 200 nK is qualitatively different from that at 400 nK. At 200 nK, the central peak indicates the presence of the BEC. The right panel shows the situation at a temperature of 50 nK, where essentially all the rubidium atoms are part of the BEC.

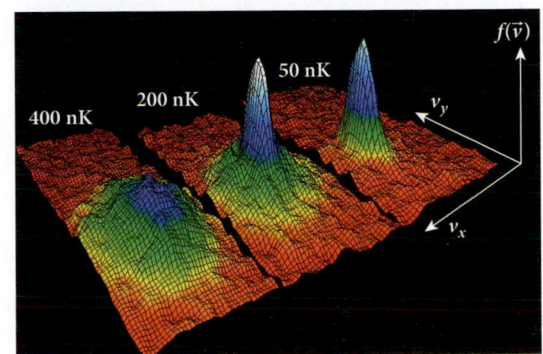

FIGURE 36.26 Bose-Einstein condensation of rubidium atoms in a magnetic trap. This picture was produced by the Wieman-Cornell group in 1995 and is included in their Nobel lecture. (The axes were added by this book's authors.) The middle image shows the appearance of the Bose-Einstein condensate. In the right image, almost all atoms are in the condensate.

Only six years after this discovery, their work on the BEC resulted in the 2001 Nobel Prize in Physics for Wieman and Cornell, which they shared with Wolfgang Ketterle, whose group had performed similar work on BECs for other systems at the same time. The study of BECs now flourishes in many laboratories around the world, revealing many astounding facts about atoms, condensates, and quantum mechanics.

SOLVED PROBLEM 36.2 | Rubidium Bose-Einstein Condensate

PROBLEM

What is the minimum density of rubidium atoms in a magnetic trap at a temperature of 200. nK that can produce the onset of Bose-Einstein condensation? (*Hint:* The mass of a rubidium-87 atom is $1.5 \cdot 10^{-25}$ kg.)

SOLUTION

THINK The text stated that the rubidium atoms must be close enough to one another that they can overlap in a quantum mechanical sense. This criterion was expressed as $\rho \cdot \lambda^3 > 2.61$. Thus, our task amounts to finding the de Broglie wavelength of rubidium atoms in thermal motion at a temperature of 200 nK.

– Continued

SKETCH A sketch is not really helpful in this case.

RESEARCH We have already stated that the criterion for the density is

$$\rho \cdot \lambda^3 > 2.61. \tag{i}$$

The de Broglie wavelength, λ, was given in equation 36.22:

$$\lambda = \frac{h}{p}. \tag{ii}$$

To find the momentum p of a rubidium atom, we use the classical energy equipartition theorem, introduced in Chapter 19, which states that the kinetic energy of an atom in a gas is given by the thermal energy at the given temperature:

$$E = \frac{p^2}{2m} = \tfrac{3}{2} k_B T \Rightarrow p = \sqrt{3 m k_B T}. \tag{iii}$$

SIMPLIFY Substituting from equation (iii) into equation (ii), we find for the de Broglie wavelength in this case:

$$\lambda = \frac{h}{p} = \frac{h}{\sqrt{3 m k_B T}}. \tag{iv}$$

Solving equation (i) for the density and then substituting for the de Broglie wavelength from equation (iv) results in

$$\rho > \frac{2.61}{\lambda^3} = \frac{2.61 (3 m k_B T)^{3/2}}{h^3}.$$

CALCULATE Inserting the values of the constants ($h = 6.63 \cdot 10^{-34}$ J s, $k_B = 1.38 \cdot 10^{-23}$ J/K) and the given temperature ($T = 200$ nK $= 2 \cdot 10^{-7}$ K), we obtain for the density

$$\rho > \frac{2.61 \left(3 \left(1.5 \cdot 10^{-25} \text{ kg} \right) \left(1.38 \cdot 10^{-23} \text{ J/k} \right) \left(2.00 \cdot 10^{-7} \text{ K} \right) \right)^{3/2}}{\left(6.63 \cdot 10^{-34} \text{ J s} \right)^3} = 1.2396 \cdot 10^{19} \text{ m}^{-3}.$$

ROUND For the minimum density that could result in the formation of a BEC, our appropriately rounded result is $\rho > 1.2 \cdot 10^{19}$ m^{-3}.

DOUBLE-CHECK The de Broglie wavelength of the rubidium atoms is

$$\lambda = \frac{6.63 \cdot 10^{-34} \text{ J s}}{\sqrt{3 \left(1.5 \cdot 10^{-25} \text{ kg} \right) \left(1.38 \cdot 10^{-23} \text{ J/K} \right) \left(2.00 \cdot 10^{-7} \text{ K} \right)}} = 5.9 \cdot 10^{-7} \text{ m}.$$

Note that this is a factor of 1000 larger than the atomic diameter of rubidium, which is $5.0 \cdot 10^{-10}$ m.

Is a density of 10^{19} atoms/m^3 a large or a small value? Let us compare it to the density of water molecules and air molecules. One cubic meter of liquid water contains $3 \cdot 10^{28}$ water molecules, and one cubic meter of air contains $3 \cdot 10^{25}$ nitrogen and oxygen molecules. Thus, the gas density of rubidium atoms in the trap is approximately a million times lower than the density of air at normal atmospheric conditions.

If we have a million atoms in a trap at a density of 10^{19} atoms/m^3 in an approximately spherical configuration, what is the radius of this sphere? The total volume occupied by the million atoms is

$$V = \frac{N}{\rho} = \frac{10^6}{10^{19} \text{ m}^{-3}} = 10^{-13} \text{ m}^3.$$

Because the volume of a sphere is $V = \tfrac{4}{3} \pi r^3$, we find that the radius of the sphere containing a million atoms at this density is

$$r = \left(\frac{3}{4\pi} V \right)^{1/3} = 30 \ \mu\text{m}.$$

This makes it clear that the Bose-Einstein condensate could not be imaged directly in the experiment conducted by Cornell and Wieman. This is why they needed to turn off the trap and let its contents expand by at least a factor of 10 in each direction before they were able to produce the images shown in Figure 36.26.

WHAT WE HAVE LEARNED | EXAM STUDY GUIDE

- Planck's constant is $h = 6.62606957(29) \cdot 10^{-34}$ J s, and the energy of a photon is $E = hf$.

- Planck's Radiation Law for the power radiated in the frequency interval between f and $f + df$, or the spectral brightness as a function of frequency, is

$$I_T(f) = \frac{2h}{c^2} \frac{f^3}{e^{hf/k_B T} - 1}.$$

  It avoids the ultraviolet divergence that made its classical predecessors unphysical in the high-frequency region, and it contains the classical radiation laws as limiting cases.

- The spectral brightness as a function of wavelength is

$$I_T(\lambda) = \frac{2hc^2}{\lambda^5 \left(e^{hc/\lambda k_B T} - 1 \right)}.$$

- The spectral emittance is related to the spectral brightness by a simple multiplicative factor of π: $\epsilon_T(f) = \pi I_T(f)$.

- Integration of the spectral emittance over all frequencies (or wavelengths) yields the radiated intensity, $I = \int\limits_0^\infty$,

  where $\epsilon(\lambda) d\lambda = \sigma T^4$ the Stefan-Boltzmann constant is

$$\sigma = \frac{2k_B^4 \pi^5}{15 h^3 c^2} = 5.6704 \cdot 10^{-8} \text{ W m}^{-2}\text{ K}^{-4}.$$

- What we normally think of as a wave has particle character. The photoelectric effect is explained with the quantum hypothesis by assigning particle properties to the photon, the elementary quantum of light. This quantum hypothesis gives the correct frequency dependence of the stopping potential, $eV_0 = hf - \phi$, where ϕ is the work function. The work function is a constant that depends on the material used.

- The Compton effect describes the photon wavelength λ' after a photon of wavelength λ scatters off an electron:

$$\lambda' = \lambda + \frac{h}{m_e c}(1 - \cos\theta).$$

- The Compton wavelength of the electron is

$$\lambda_e = \frac{h}{m_e c} = 2.426 \cdot 10^{-12} \text{ m}.$$

- What we normally think of as matter also has wave characteristics. The de Broglie wavelength is defined as

$$\lambda = \frac{h}{p}.$$

  Matter waves can be demonstrated by performing double-slit experiments with electrons, which yield the same kind of interference patterns as for photons.

- The Heisenberg Uncertainty Relation stipulates that the product of the uncertainty in momentum and the uncertainty in position has an absolute lower bound, $\Delta x \cdot \Delta p_x \geq \frac{1}{2}\hbar$. The energy-time uncertainty relation is $\Delta E \cdot \Delta t \geq \frac{1}{2}\hbar$.

- Elementary quantum particles have the intrinsic property of spin, which has the dimensions of angular momentum. The spin is quantized, dividing particles into two categories: fermions with spins that are half-integer multiples of $\hbar$, and bosons with spins that are integer multiples of $\hbar$.

- The Pauli Exclusion Principle states that no two fermions can occupy the same quantum state at the same time and at the same location. In any given atom, no two fermions can have exactly identical quantum numbers.

- Bosons can condense at very low temperatures in such a way that a very significant fraction of them occupy the same quantum state.

ANSWERS TO SELF-TEST OPPORTUNITIES

36.1 $K_{max} = hf - \phi$: For $\lambda = 250$, 392, and 540 nm and $\phi = 2.1$ eV, $K_{max} = 2.9$, 1.1, and 0.20 eV.

36.2 148 million km $= 148 \cdot 10^6 \cdot 10^3$ m $= 1.48 \cdot 10^{11}$ m.

$P = \left(4\pi r^2\right) I = 4\pi \left(1.48 \cdot 10^{11} \text{ m}\right)^2 \left(1370 \text{ W/m}^2\right) = 3.77 \cdot 10^{26}$ W.

In the visible spectrum, $P_{visible} = P/4 = 9.43 \cdot 10^{25}$ W.

In 1 s, $9.43 \cdot 10^{25}$ J of energy is emitted by the Sun.

Assume that the average wavelength of the visible photons is $\lambda = 550$ nm. The energy of each photon is then

$$E = hf = h\frac{c}{\lambda} = \left(6.626 \cdot 10^{-34} \text{ J s}\right)\frac{3.00 \cdot 10^8 \text{ m/s}}{550 \cdot 10^{-9} \text{ m}} = 3.61 \cdot 10^{-19} \text{ J}.$$

$$N = P_{visible}/E = \left(9.43 \cdot 10^{25} \text{ W}\right)/\left(3.61 \cdot 10^{-19} \text{ J}\right)$$

$$= 2.6 \cdot 10^{44} \text{ visible photons per second.}$$

36.3 Momentum of 1.00-eV electron: $m_e c^2 = 511$ keV.

$$E^2 = \left(m_e c^2\right)^2 + \left(pc\right)^2 = \left(m_e c^2 + K\right)^2$$

$$pc = \sqrt{\left(m_e c^2 + K\right)^2 - \left(m_e c^2\right)^2}$$

$$= \sqrt{\left(511 \text{ keV} + 1.00 \cdot 10^{-3} \text{ keV}\right)^2 - \left(511 \text{ keV}\right)^2} = 1.01 \text{ keV}.$$

Momentum of 100-keV X-ray:

$$\left(m_e c^2\right)^2 + \left(pc\right)^2 = \left(m_e c^2 + K\right)^2$$

$$pc = K = 100 \text{ keV}.$$

Momentum of electron is 1% of that of X-ray, or negligible.

36.4 The maximum change in wavelength of a photon undergoing Compton scattering occurs when $\theta = 180°$, so

$$\Delta\lambda_{max} = \frac{h}{m_e c}(1 - \cos 180°) = \frac{2h}{m_e c}$$

$$= \frac{2(6.63 \cdot 10^{-34} \text{ J s})}{(9.11 \cdot 10^{-31} \text{ kg})(3.00 \cdot 10^8 \text{ m/s})} = 4.85 \cdot 10^{-12} \text{ m}$$

$$= 0.00485 \text{ nm}.$$

This maximum change in wavelength is much smaller than the wavelength of visible light, 400 – 700 nm, so it would not be observable.

36.5

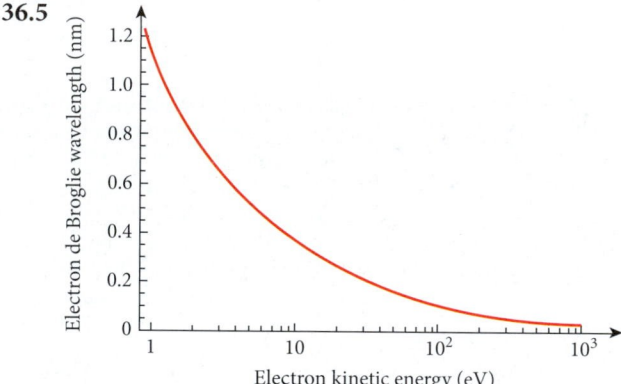

The de Broglie wavelength for an electron with a kinetic energy of 1000 eV is
$\lambda = 0.0387879$ nm.
Calculating the momentum nonrelativistically gives us
$\lambda = 0.0388068$ nm.
The difference is small.

36.6 $\Delta v = 2.00$ km/h $= 0.556$ m/s.

$$\Delta x \cdot \Delta p \geq \tfrac{1}{2}\hbar$$

$$\Delta x \geq \frac{\hbar}{2\Delta p} = \frac{\hbar}{2m\Delta v} = \frac{1.05457 \cdot 10^{-34} \text{ J s}}{2(1462 \text{ kg})(0.556 \text{ m/s})} = 6.49 \cdot 10^{-38} \text{ m}.$$

36.7 Numbers of states are 16 for fermions, 18 for bosons, and 495 for classical particles.

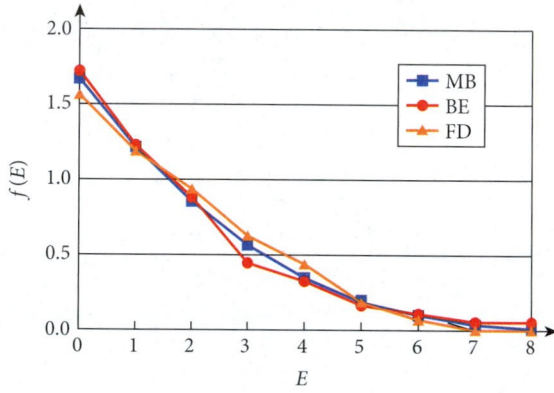

PROBLEM-SOLVING GUIDELINES

1. The starting point for most calculations involving photons or matter waves is to relate the particle properties energy, E, and momentum, p, to the wave properties wavelength, λ, and frequency, f. The key relations are $E = hf$, $p = E/c = hf/c = h/\lambda$, and $\lambda = h/p$.

2. Be careful and consistent in applying units. Often, converting all units to meters and kilograms will help you keep track of unit exponents. Use of electron-volts will of-

ten simplify your calculations, but be sure to use Planck's constant with the appropriate units: $h = 6.626 \cdot 10^{-34}$ J s or $h = 4.136 \cdot 10^{-15}$ eV s.

3. For checking your work, it is helpful to keep in mind some rough orders of magnitude: the size of an atom is 10^{-10} m; the mass of an electron is 10^{-30} kg; the charge of a proton or electron is 10^{-19} C; at room temperature, $k_B T = \frac{1}{40}$ eV.

MULTIPLE-CHOICE QUESTIONS

36.1 Ultraviolet light of wavelength 350 nm is incident on a material with a stopping potential of 0.25 V. The work function of the material is

a) 4.0 eV.

b) 3.3 eV.

c) 2.3 eV.

d) 5.2 eV.

36.2 The existence of a cutoff frequency in the photoelectric effect

a) cannot be explained using classical physics.

b) shows that the model provided by classical physics is not correct in this case.

c) shows that a photon model of light should be used in this case.

d) shows that the energy of the photon is proportional to its frequency.

e) All of the above are true.

36.3 To have a larger photocurrent, which of the following should occur? (Select all the changes that apply.)

a) brighter light

b) dimmer light

c) higher frequency

d) lower frequency

36.4 Which of the following has the smallest de Broglie wavelength?

a) an electron traveling at 80% of the speed of light

b) a proton traveling at 20% of the speed of light

c) a carbon nucleus traveling at 70% of the speed of light

d) a helium nucleus traveling at 80% of the speed of light

e) a lithium nucleus traveling at 50% of the speed of light

36.5 A blackbody is an ideal system that

a) absorbs 100% of the light incident upon it, but cannot emit light of its own.

b) radiates 100% of the power it generates, but cannot absorb radiation of its own.

c) either absorbs 100% of the light incident upon it, or radiates 100% of the power it generates.

d) absorbs 50% of the light incident upon it, and emits 50% of the radiation it generates.

e) blackens completely any body that comes in contact with it.

36.6 Which one of the following statements is true if the intensity of a light beam is increased while its frequency is kept the same?

a) The photons gain higher speeds.

b) The energy of the photons is increased.

c) The number of photons per unit time is increased.

d) The wavelength of the light is increased.

36.7 Which of the following has the highest temperature?

a) a white-hot object   c) a blue-hot object

b) a red-hot object

36.8 Electrons having a narrow range of kinetic energies are impinging on a double slit, with separation D between the slits. The electrons form an interference pattern on a phosphorescent screen, with a separation Δx between the fringes. If the spacing between the slits is reduced to $D/2$, the separation between the fringes will be

a) Δx.   c) $\Delta x/2$.

b) $2\Delta x$.   d) none of these.

CONCEPTUAL QUESTIONS

36.9 Why is a white-hot object hotter than a red-hot object?

36.10 Answer in your own words the question of whether an electron is a particle or a wave.

36.11 If I look in a mirror while wearing a blue shirt, my reflection is wearing a blue shirt, not a red shirt. But according to the Compton effect, the photons that bounce off the mirror should have a lower energy and therefore a longer wavelength. Explain why my reflection shows the same color shirt as I am wearing.

36.12 Vacuum in deep space is not empty, but a boiling sea of particles and antiparticles that are constantly forming and annihilating each other. Determine the maximum lifetime for a proton-antiproton pair to form there without violating Heisenberg's Uncertainty Relation.

36.13 Consider a universe where Planck's constant is 5 J s. How would a game of tennis change? Consider the interactions of individual players with the ball and the interaction of the ball with the net.

36.14 In classical mechanics, what information is needed in order to predict where a particle with no net force on it will be at some later time? Why is this prediction not possible in quantum mechanics?

36.15 What would a classical physicist expect to be the result of shining an increasingly brighter UV lamp on a metal surface, in terms of the energy of emitted electrons? How does this differ from what the theory of the photoelectric effect predicts?

36.16 Which is more damaging to human tissue, a 60-W source of visible light or a 2-mW source of X-rays? Explain your choice.

36.17 Neutrons are spin-$\frac{1}{2}$ fermions. An unpolarized beam of neutrons has equal numbers of neutrons in the $+\frac{1}{2}$ and $-\frac{1}{2}$ spin states. When a beam of unpolarized neutrons is passed through a gas of unpolarized ^{3}He atoms, the neutrons can be absorbed by the ^{3}He atoms to create ^{4}He atoms. If the ^{3}He atoms are polarized, so that the spins of the neutrons in their nuclei are all aligned, will the same number of neutrons from the unpolarized neutron beam be absorbed by the polarized ^{3}He atoms? How well are neutrons in each of the two spin states of the unpolarized neutron beam absorbed by the ^{3}He?

36.18 You are performing a photoelectric effect experiment. Using a photocathode made of cesium, you first illuminate it with a green laser beam ($\lambda = 514.5$ nm) of power 100 mW. Next, you double the power of the laser beam, to 200 mW. How will the energies per electron of the electrons emitted by the photocathode compare for the two cases?

EXERCISES

A blue problem number indicates a worked-out solution is available in the Student Solutions Manual. One • and two •• indicate increasing level of problem difficulty.

Section 36.2

36.19 Calculate the peak wavelengths of

a) the solar light received by Earth, and

b) the light emitted by the Earth.

Assume that the surface temperatures of the Sun and the Earth are 5800. K and 300. K, respectively.

36.20 Calculate the range of temperatures for which the peak emission of blackbody radiation from a hot filament occurs within the visible range of the electromagnetic spectrum. Take the visible spectrum as extending from 380 nm to 780 nm. What is the total intensity of the radiation from the filament at the two temperatures at the ends of the range?

36.21 Ultra-high-energy gamma rays with energies up to $3.5 \cdot 10^{12}$ eV, come from the equator of the Milky Way. What is the wavelength of this radiation? How does its energy compare to the rest energy of a proton?

36.22 Consider the radiation emitted by an object at room temperature (20. °C). For radiation at the peak of the spectral energy density, calculate

a) the wavelength,   c) the energy of one photon.

b) the frequency, and

•**36.23** The temperature of your skin is approximately 35.0 °C.

a) Assuming that your skin is a blackbody, what is the peak wavelength of the radiation it emits?

b) Assuming a total surface area of 2.00 m^2, what is the total power emitted by your skin?

c) Given your answer to part (b), why don't you glow as brightly as a light bulb?

•**36.24** A pure, defect-free semiconductor material will absorb the electromagnetic radiation incident on it only if the energy of the individual photons in the incident beam is larger than a threshold value known as the *band-gap* of the semiconductor. Otherwise, the material will be transparent to the photons. The known room-temperature band-gaps for germanium, silicon, and gallium-arsenide, three widely used semiconductors, are 0.66 eV, 1.12 eV, and 1.42 eV, respectively.

a) Determine the room-temperature transparency range of each semiconductor.

b) Compare these with the transparency range of ZnSe, a semiconductor with a band-gap of 2.67 eV, and explain the yellow color observed experimentally for the ZnSe crystals.

c) Which of these materials could be used to detect the 1550-nm optical communications wavelength?

•**36.25** The mass of a dime is 2.268 g, its diameter is 17.91 mm, and its thickness is 1.350 mm. Determine

a) the radiant energy emitted by a dime per second at room temperature,

b) the number of photons leaving the dime per second (assuming that all photons have the wavelength of the peak of the distribution for this estimate), and

c) the volume of air that has an energy equal to that of 1 s of radiation from the dime.

Section 36.3

36.26 The work function of a certain material is 5.8 eV. What is the photoelectric threshold frequency for this material?

36.27 What is the maximum kinetic energy of the electrons ejected from a sodium surface by light of wavelength 470 nm?

36.28 The threshold wavelength for the photoelectric effect for a specific alloy is 400. nm. What is the work function in electron-volts?

36.29 In a photoelectric effect experiment, a laser beam of unknown wavelength is shone on a cesium photocathode (work function of $\phi = 2.100$ eV). It is found that a stopping potential of 0.310 V is needed to eliminate the current. Next, the same laser is shone on a photocathode made of an unknown material, and a stopping potential of 0.110 V is needed to eliminate the current.

a) What is the work function for the unknown material?

b) What metal is a possible candidate for the unknown material?

36.30 You illuminate a zinc surface with 550-nm light. How high do you have to turn up the stopping potential to eliminate the photoelectric current completely?

36.31 White light, $\lambda = 400.$ to 750. nm, falls on barium ($\phi = 2.48$ eV).

a) What is the maximum kinetic energy of electrons ejected from the metal?

b) Would the longest-wavelength light eject electrons?

c) What wavelength of light would eject electrons with zero kinetic energy?

•**36.32** To determine the work function of the material of a photodiode, you measured a maximum kinetic energy of 1.50 eV corresponding to a certain wavelength. Later, you decreased the wavelength by 50.0% and found the maximum kinetic energy of the photoelectrons to be 3.80 eV. From this information, determine

a) the work function of the material, and

b) the original wavelength.

Section 36.4

36.33 X-rays of wavelength $\lambda = 0.120$ nm are scattered from carbon. What is the Compton wavelength shift for photons deflected at a 90.0° angle relative to the incident beam?

36.34 A 2.0-MeV X-ray photon is scattered off a free electron at rest and deflected by an angle of 53°. What is the wavelength of the scattered photon?

36.35 A photon with a wavelength of 0.30 nm collides with an electron that is initially at rest. If the photon rebounds at an angle of 160°, how much energy did it lose in the collision?

•**36.36** X-rays having an energy of 400.0 keV undergo Compton scattering from a target electron. The scattered rays are detected at an angle of 25.0° relative to the incident rays. Find

a) the kinetic energy of the scattered X-rays, and

b) the kinetic energy of the recoiling electron.

•**36.37** Consider the case in which photons scatter off a free proton.

a) If 140.-keV X-rays deflect off a proton at 90.0°, what is their fractional change in energy, $(E_0 - E)/E_0$?

b) What initial energy would a photon have to have to experience a 1.00% change in energy at 90.0° scattering?

•**36.38** An X-ray photon with an energy of 50.0 keV strikes an electron that is initially at rest inside a metal. The photon is scattered at an angle of 45°. What are the kinetic energy and the momentum (magnitude and direction) of the electron after the collision? You may use the nonrelativistic relationship between the kinetic energy and the momentum of the electron.

Section 36.5

36.39 Calculate the wavelength of

a) a 2.00-eV photon, and

b) an electron with a kinetic energy of 2.00 eV.

36.40 What is the de Broglie wavelength of a $2.000 \cdot 10^3$-kg car moving at a speed of 100.0 km/h?

36.41 A nitrogen molecule of mass $m = 4.648 \cdot 10^{-26}$ kg has a speed of 300.0 m/s.

a) Determine its de Broglie wavelength.

b) How far apart are the double slits if a beam of nitrogen molecules creates an interference pattern with fringes 0.30 cm apart on a screen 70.0 cm in front of the slits?

36.42 Alpha particles are accelerated through a potential difference of 20.0 kV. What is their de Broglie wavelength?

36.43 Consider an electron whose de Broglie wavelength is equal to the wavelength of green light (550 nm).

a) Treating the electron nonrelativistically, what is its speed?

b) Does your calculation confirm that a nonrelativistic treatment is sufficient?

c) Calculate the kinetic energy of the electron in electron-volts.

•**36.44** After you tell your 60.0-kg roommate about de Broglie's hypothesis that particles of momentum p have wave characteristics with wavelength $\lambda = h/p$, he starts thinking of his fate as a wave and asks you if he could be diffracted when passing through the 90.0-cm-wide doorway of your dorm room.

a) What is the maximum speed at which your roommate could pass through the doorway and be significantly diffracted?

b) If it takes one step to pass through the doorway, how long should it take your roommate to make that step (assuming that the length of his step is 0.75 m) for him to be diffracted?

c) What is the answer to your roommate's question? (*Hint:* Assume that significant diffraction occurs when the width of the diffraction aperture is less than 10.0 times the wavelength of the wave being diffracted.)

••**36.45** Consider de Broglie waves for a Newtonian particle of mass m, momentum $p = mv$, and energy $E = p^2/(2m)$, that is, waves with wavelength $\lambda = h/p$ and frequency $f = E/h$.

a) For these waves, calculate the dispersion relation, $\omega = \omega(\kappa)$, where κ is the wave number.

b) Calculate the phase velocity ($v_p = \omega/\kappa$) and the group velocity ($v_g = \partial\omega/\partial\kappa$) of these waves.

••36.46 Now consider de Broglie waves for a (relativistic) particle of mass m, momentum $p = mv\gamma$, and total energy $E = mc^2\gamma$, with $\gamma = [1 - (v/c)^2]^{-1/2}$. The waves have wavelength $\lambda = h/p$ and frequency $f = E/h$ as in Problem 36.45, but with the relativistic momentum and energy values.

a) For these waves, calculate the dispersion relation, $\omega = \omega(\kappa)$, where κ is the wave number.

b) Calculate the phase velocity ($v_p = \omega/\kappa$) and the group velocity ($v_g = \partial\omega/\partial\kappa$) of these waves.

c) Which corresponds to the classical velocity of the particle?

Section 36.6

36.47 A 50.0-kg particle has a de Broglie wavelength of 20.0 cm.

a) How fast is the particle moving?

b) What is the smallest uncertainty in the particle's speed if its position uncertainty is 20.0 cm?

36.48 During the period of time required for light to pass through a hydrogen atom ($r = 0.53 \cdot 10^{-10}$ m), what is the least uncertainty in the atom's energy? Express your answer in electron-volts.

36.49 A free neutron ($m = 1.67 \cdot 10^{-27}$ kg) has a mean lifetime of 882 s. What is the minimum uncertainty in its mass (in kg)?

36.50 Suppose that Fuzzy, a quantum-mechanical duck, lives in a world in which Planck's constant is $\hbar = 1.00$ J s. Fuzzy has a mass of 0.500 kg and initially is known to be within a 0.750-m-wide pond. What is the minimum uncertainty in Fuzzy's speed? Assuming that this uncertainty prevails for 5.00 s, how far away could Fuzzy be from the pond after 5.00 s?

36.51 An electron is confined to a box with a dimension of 20.0 μm. What is the minimum speed the electron can have?

•36.52 A dust particle of mass $1.00 \cdot 10^{-16}$ kg and diameter 5.00 μm is confined to a box of length 15.0 μm.

a) How will you know whether the particle is at rest?

b) If the particle is not at rest, what will be the range of its velocity?

c) At the lower limit of the velocity range, how long will it take to move a distance of 1.00 μm?

Section 36.8

••36.53 A quantum state of energy E can be occupied by any number n of bosons, including $n = 0$. At the absolute temperature T, the probability of finding n particles in the state is given by $P_n = N \exp(-nE/k_B T)$, where k_B is Boltzmann's constant and N is the normalization factor determined by the requirement that all the probabilities sum to unity. Calculate the mean value of n, that is, the average occupation, of this state, given this probability distribution.

•36.54 For the same quantum state as in Problem 36.53, with a probability distribution of the same form but with fermions, the only possible occupation values are $n = 0$ and $n = 1$. Calculate the mean occupation $\langle n \rangle$ of the state in this case.

••36.55 Consider a system made up of N particles. The energy per particle is given by $\langle E \rangle = (\Sigma E_i e^{-E_i/k_B T})/Z$, where Z is the partition function defined in equation 36.29. If this is a two-state system with $E_1 = 0$ and $E_2 = E$ and $g_1 = g_2 = 1$, calculate the heat capacity of the system, defined as $N(d\langle E \rangle/dT)$, and approximate its behavior at very high and very low temperatures (that is, $k_B T \gg 1$ and $k_B T \ll 1$).

Additional Exercises

36.56 Given that the work function of tungsten is 4.55 eV, what is the stopping potential in an experiment using tungsten cathodes at 360 nm?

36.57 Find the ratio of the de Broglie wavelengths of a 100.-MeV proton and a 100.-MeV electron.

36.58 An *einstein* (E) is a unit of measurement equal to Avogadro's number ($6.02 \cdot 10^{23}$) of photons. How much energy is contained in 1 E of violet light ($\lambda = 400.$ nm)?

36.59 In baseball, a 100.-g ball can travel as fast as 100. mph. The *Voyager* spacecraft, with a mass of about 250. kg, is currently traveling at 125,000 km/h.

a) What is the de Broglie wavelength of this ball?

b) What is the de Broglie wavelength of *Voyager*?

36.60 What is the minimum uncertainty in the velocity of a 1.0-ng particle that is at rest on the head of a 1.0-mm-wide pin?

36.61 A photovoltaic device uses monochromatic light of wavelength 700. nm that is incident normally on a surface of area 10.0 cm². Calculate the photon flux rate if the light intensity is 0.300 W/cm².

36.62 The solar constant measured by Earth satellites is roughly 1400. W/m². Though the Sun emits light of different wavelengths, the peak of the wavelength spectrum is at 500. nm.

a) Find the corresponding photon frequency.

b) Find the corresponding photon energy.

c) Find the number flux of photons (number of photons per unit area per unit time) arriving at Earth, assuming that all light emitted by the Sun has the same peak wavelength.

36.63 Two silver plates in vacuum are separated by 1.0 cm and have a potential difference of 5.0 V between them that opposes electron flow. What is the largest wavelength of light that can be incident on the cathode and produce a current in the anode?

36.64 How many photons per second must strike a surface of area 10.0 m² to produce a force of 0.100 N on the surface, if the wavelength of the photons is 600. nm? Assume that the photons are absorbed.

36.65 Suppose the wave function describing an electron predicts a statistical spread of $1.00 \cdot 10^{-4}$ m/s in the electron's velocity. What is the corresponding statistical spread in its position?

36.66 What is the temperature of a blackbody whose peak emitted wavelength is in the X-ray portion of the spectrum?

36.67 A nocturnal bird's eye can detect monochromatic light of frequency $5.8 \cdot 10^{14}$ Hz with a power as small as $2.333 \cdot 10^{-17}$ W. What is the corresponding number of photons per second that the nocturnal bird's eye can detect?

36.68 A particular ultraviolet laser produces radiation of wavelength 355 nm. Suppose this is used in a photoelectric experiment with a calcium sample. What will the stopping potential be?

36.69 What is the wavelength of an electron that is accelerated from rest through a potential difference of $1.00 \cdot 10^{-5}$ V?

•36.70 Compton used photons of wavelength 0.0711 nm.

a) What is the wavelength of the photons scattered at $\theta = 180.°$?

b) What is energy of these photons?

c) If the target were a proton and not an electron, how would your answer to part (a) change?

•36.71 Calculate the number of photons originating at the Sun that are received in the Earth's upper atmosphere per year.

•36.72 Scintillation detectors for gamma rays transfer the energy of a gamma-ray photon to an electron within a crystal, via the photoelectric effect or Compton scattering. The electron transfers its energy to atoms in the crystal, which re-emit it as a light flash detected by a photomultiplier tube. The charge pulse produced by the photomultiplier tube is proportional to the original energy in the crystal; this can be measured so that an energy spectrum can be displayed. Gamma rays absorbed in the photoelectric effect are recorded as a *photopeak* in the spectrum, at the full energy of the rays. The Compton-scattered electrons are also recorded, at a range of lower energies known as the *Compton plateau*. The highest-energy electrons form the *Compton edge* of the plateau. Gamma-ray photons scattered by 180.° appear as a *backscatter peak* in the spectrum. For gamma-ray photons of energy 511 KeV, calculate the energies of the Compton edge and the backscatter peak in the spectrum.

MULTI-VERSION EXERCISES

36.73 A stationary free electron in a gas is struck by a 6.37-nm X-ray, which experiences an increase in wavelength of 1.13 pm. How fast is the electron moving after the interaction with the X-ray?

36.74 A stationary free electron in a gas is struck by an X-ray with an energy of 175.37 eV, which experiences a decrease in energy of 28.52 meV. How fast is the electron moving after the interaction with the X-ray?

36.75 A stationary free electron in a gas is struck by an X-ray with an energy of 159.98 eV. After the collision, the speed of the electron is measured to be 92.17 km/s. By how much did the energy of the X-ray decrease?

36.76 A stationary free electron in a gas is struck by an X-ray with a wavelength of 5.43 nm. After the collision, the speed of the electron is measured to be 132.7 km/s. By how much did the wavelength of the X-ray increase?

36.77 A stationary free electron in a gas is struck by an X-ray with a wavelength of 6.13 nm. After the collision, the speed of the electron is measured to be 118.5 km/s. By how much did the frequency of the X-ray change?

36.78 An accelerator boosts a proton's kinetic energy so that its de Broglie wavelength is $3.63 \cdot 10^{-15}$ m. What is the total energy of the proton?

36.79 An accelerator boosts a proton's kinetic energy so that its de Broglie wavelength is $4.43 \cdot 10^{-15}$ m. What is the kinetic energy of the proton?

36.80 An accelerator boosts a proton's kinetic energy so that its de Broglie wavelength is $1.71 \cdot 10^{-15}$ m. What is the speed of the proton?

37

Quantum Mechanics

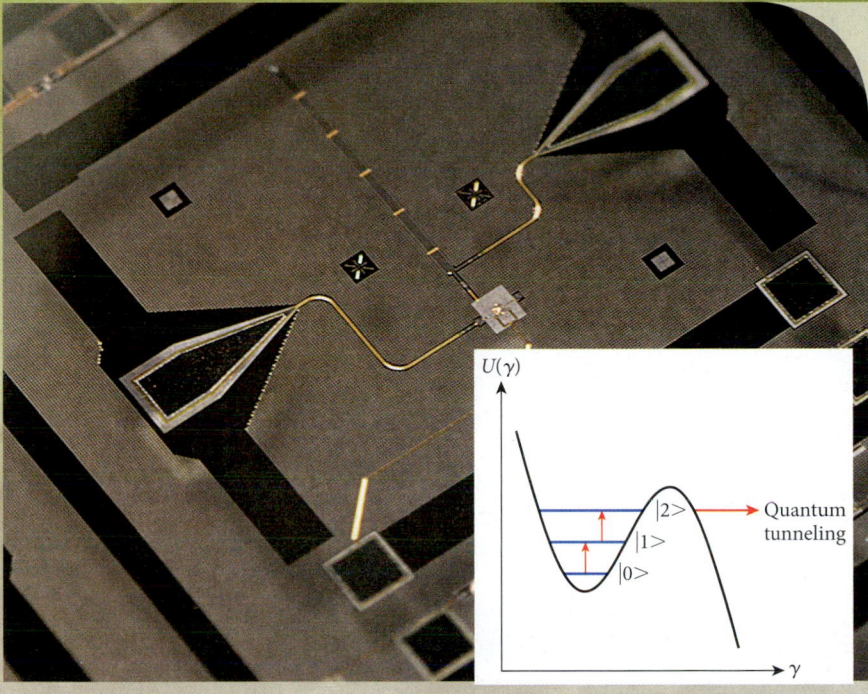

FIGURE 37.1 One of the first experimental devices used for quantum computing. Inset (lower right): schematic representation of a potential for the quantum states used in this process.

Chapter 36 introduced some of the basic ideas of quantum physics, including photons, matter waves, and the uncertainty principle. In this chapter, we extend these ideas to examine particle dynamics on the atomic scale. We emphasize physical concepts rather than mathematical details, which can become pretty involved. However, the ideas of quantum mechanics have produced practical discoveries that could not be imagined using classical physics alone.

Many early researchers in modern physics had serious doubts about what wave functions really are and why probability distributions take a central role in predicting results. For example, the Danish physicist Niels Bohr (1885–1962) once said, "Anyone who has not been shocked by quantum physics has not understood it." And the great American physicist Richard Feynman (1918–1988) wrote, "I think I can safely say that nobody understands quantum mechanics . . . Do not keep saying to yourself, if you can possibly avoid it, 'But how can it be like that? . . .' Nobody knows how it can be like that." Yet now, in the 21st century, quantum mechanics has been established as one of the most accurate and comprehensive areas in all of physics.

When we talked about light waves and sound waves, we discussed the concepts of spherical and planar waves. These waves are specific examples of wave functions that solve particular wave equations. We can now ask what the wave function of the electron is, or what the wave function of any other object that we conventionally think of as a particle is. This question will lead us to consider mechanics in the

WHAT WE WILL LEARN

- A particle is described by its complex wave function. The absolute square of the complex wave function is the probability of finding the particle at some position. The integral of the square of the wave function over all of space equals 1.

- The time-independent Schrödinger equation is the nonrelativistic wave equation for a particle in a potential energy distribution, $U(x)$.

- Wave functions that are solutions to the problem of a particle confined to an infinite potential well are sine functions. The energy value of a solution is proportional to the square of the quantum number of the solution. The solutions to the problem of a potential of finite height have exponential tails reaching into the classically forbidden regions.

- If a particle encounters a potential barrier of finite height and width, it can tunnel through this barrier, even if it has energy less than the height of the barrier. The probability of the particle tunneling through the barrier depends exponentially on the width of the barrier.

- The wave function solutions for the harmonic oscillator potential are Hermite polynomials. The corresponding energy eigenvalues are equally spaced. Oscillator wave functions with $n = 0$ have the minimum product of momentum and position uncertainties allowed by the uncertainty principle.

- The correspondence principle states that if the energy difference, ΔE, between neighboring energy states becomes small relative to the total energy, E, then the quantum solution approaches its classical limit.

- The Hamiltonian operator (or simply the Hamiltonian, for short) is the operator that yields the total energy. It is linear. Thus, a linear combination of two solutions to the Schrödinger equation, which is based on the Hamiltonian, is also a solution.

- The two-particle wave function for (noninteracting) bosons is the symmetrized product of one-particle wave functions, and the two-particle wave function for (noninteracting) fermions is the antisymmetrized product.

quantum world, conventionally called *quantum mechanics,* which explores the observable consequences of the wavelike behavior of particles. Recent years have witnessed extensive attempts to go beyond simply understanding quantum systems and to manipulate them and use them for purposes like quantum computing (see Figure 37.1), which carry the promise of revolutionizing nanoscience and nanotechnology.

37.1 Wave Function

Let's briefly review what we have accomplished so far in our investigation of the quantum world. In Chapter 36, we saw that the photoelectric effect and Compton scattering can be explained only if light has particle properties. However, in Chapter 31 on electromagnetic waves we saw that light is also an electromagnetic wave of the form $E = E_{max} \sin(\kappa x - \omega t)$, where κ is the wave number and ω is the angular frequency. We also saw that the intensity of the wave is proportional to the square of the electric field. We used this relationship again in Chapter 34 on wave optics to calculate the intensity of the interference pattern of two light waves. Because the oscillation of the electric field in space and time represents the wave function of light, clearly, for light, the intensity is proportional to the square of the wave function.

What is the corresponding wave function of an electron? Or more generally: What is the quantum wave function of any particle?

In common notation, $\psi(\vec{r},t)$ is used to represent the **wave function** to denote that it depends on the spatial coordinate $\vec{r}$ and the time t. It is common to use the lowercase Greek letter psi, ψ, for the wave function. We used $y(\vec{r},t)$ to describe the wave function in Chapter 15 when we examined general wave phenomena. However, in the early 20th century, the pioneers of quantum physics used $\psi(\vec{r},t)$, which became the standard notation. In addition, representing the wave function, which is not a coordinate or a distance, by y could be confused with the common usage of y.

A lot of physical insight can be gained by studying mechanics problems in one spatial dimension. For the one-dimensional wave function, we use the notation $\psi(x,t)$. To start with the simplest case, we'll look at the wave function in coordinate space, returning to the time dependence later. If we are interested in only the spatial distribution of the wave function, we use the notation $\psi(x)$.

What is the wave function $\psi(x)$ of the electrons hitting the screen in the double-slit experiment of Section 36.5? Just as for photons, the wave function is proportional to the square root of the intensity, with the intensity given by equation 36.25. The intensity is shown in Figure 37.2a. Figure 37.2b and Figure 37.2c show two possible wave functions that correspond to this intensity distribution, which differ from each other by a multiplicative factor of –1. Note that the wave function can have positive and negative values, whereas the intensity has a minimum value of zero (that is, it is never negative).

Wave Function and Probability

What is the physical meaning of the quantum wave function of a particle? Figure 37.3 shows the connection between the positions at which the electrons hit the screen and the intensity distribution resulting from double-slit interference.

Clearly, the number of counts per bin of some unit length (yellow histogram), and the intensity (blue line) are proportional to each other. The number of counts per bin divided by the total number of electrons is the probability that an electron hits in the interval marked by the bin. We denote the bin width by dx, because we want to use infinitesimally small bins. Thus, the probability of an electron striking in the interval between x and $x+dx$ is proportional to $I(x)dx$. Finally, because we have just established that the intensity is proportional to the square of the wave function, we can express the probability $\Pi(x)$ of the electron hitting in the interval between x and $x+dx$ as

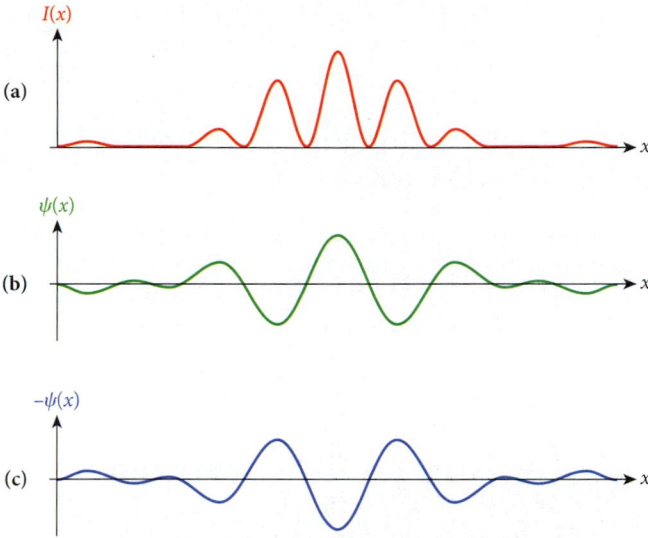

FIGURE 37.2 (a) Intensity distribution for electrons hitting the screen in the electron double-slit experiment; (b,c) two possible wave functions corresponding to this intensity distribution.

$$\Pi(x)dx = |\psi(x)|^2 dx. \tag{37.1}$$

This is equivalent to saying that $|\psi(x)|^2$ is the probability density of finding the particle at position x. The probability is a dimensionless number between 0 and 1, and dx has the dimension of length, so the probability density must have the physical dimension of inverse length.

Why is the square of the absolute value of the wave function used in equation 37.1, instead of just the square of the wave function? It turns out that wave functions can have a complex amplitude. Thus, a positive real number is ensured only by taking the absolute square of this amplitude.

The probability of finding a particular electron at any position in space must be 1, because it must be somewhere in space. This gives the normalization condition for the wave function,

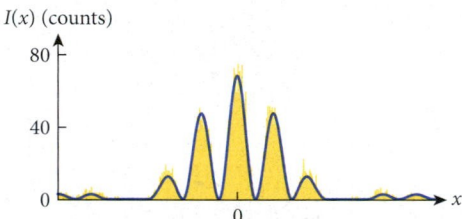

FIGURE 37.3 Intensity distribution (number of electrons hitting the screen per unit length) along the screen in the electron double-slit experiment. The yellow histogram shows the number of electrons hitting the screen in a coordinate interval.

$$\int_{-\infty}^{\infty} |\psi(x)|^2 dx = \int_{-\infty}^{\infty} \psi^*(x)\psi(x)dx = 1, \tag{37.2}$$

where $\psi^*(x)$ is the complex conjugate of $\psi(x)$. This normalization condition will allow us to determine the amplitude of the wave function. Equation 37.2 says that the integral of the absolute square of the wave function over all of space has the value 1, because if we look everywhere in space, then the integrated probability of finding the electron somewhere is 1.

With the aid of complex numbers (see Appendix A for a refresher on complex numbers), we can write a convenient form for the coordinate space dependence of the wave function of a freely moving particle. This wave function must represent a traveling wave, since the particle is traveling freely; thus, it can be expected that the wave function will be sinusoidal, like the waveform for an electromagnetic wave. This wave function depends on space and time. Time dependence will be treated later. For now, we focus on the dependence on the x-coordinate alone, setting $t = 0$. The particle's wave function is written as a linear combination of sine and cosine oscillations with wave number $\kappa = 2\pi/\lambda$, where λ is the de Broglie wavelength introduced in Chapter 36:

$$\psi(x) = C\cos(\kappa x) + D\sin(\kappa x).$$

Using complex number notation, the same wave function can be written as

$$\psi(x) = Ae^{i\kappa x} + Be^{-i\kappa x}. \tag{37.3}$$

The coefficients C and D and A and B are, in general, complex numbers. The coefficients A and B can be chosen to suit the particular physical situation, under the condition that the overall wave function is normalized, according to equation 37.2.

Why is it more convenient to use the complex numbers? An answer to this question is given in the next subsection in relation to the momentum that corresponds to a wave function.

Momentum

Suppose we have a freely moving particle with a wave function of the form given by equation 37.3. For now, we set $B = 0$. This will simplify our work, and we will soon learn the significance of the coefficient B. Therefore, the particle's wave function is

$$\psi(x) = Ae^{i\kappa x}. \tag{37.4}$$

What is the momentum associated with this wave function? According to the de Broglie relation, $\lambda = h/p$, the wave number can be expressed as

$$\kappa = \frac{2\pi}{\lambda} = \frac{2\pi}{h/p} = p\frac{2\pi}{h} = \frac{p}{\hbar},$$

where $\hbar = h/2\pi$. Consequently,

$$p = \hbar\kappa. \tag{37.5}$$

What operation applied to the wave function will result in a product of the wave function times the momentum? Let's try this *Ansatz*:

$$\hat{\mathbf{p}}\psi(x) = -i\hbar\frac{d}{dx}\psi(x). \tag{37.6}$$

Here we have used the notation

$$\hat{\mathbf{p}} = -i\hbar\frac{d}{dx}$$

to denote the **momentum operator,** that is, the operator that results in the product of the momentum and the wave function when applied to the wave function. This way of writing the momentum operator is motivated by noticing that taking the derivative of $\psi(x) = Ae^{i\kappa x}$ with respect to x is equivalent to multiplying the wave function by the factor $i\kappa$. However, this momentum operator turns out to be more general and works for all wave functions, not just that of equation 37.4.

We can take the derivative and convince ourselves that our *Ansatz* makes sense for this wave function:

$$-i\hbar\frac{d}{dx}\psi(x) = -i\hbar\frac{d}{dx}Ae^{i\kappa x} = -i\hbar A\frac{d}{dx}e^{i\kappa x} = -i\hbar A(i\kappa)e^{i\kappa x} = \hbar\kappa Ae^{i\kappa x} = \hbar\kappa\psi(x) = p\psi(x).$$

Thus, applying momentum operator to the wave function $\psi(x) = Ae^{i\kappa x}$ indeed gives us the product of the wave function with its momentum $p = \hbar\kappa$. Therefore, it is appropriate to interpret the wave function (equation 37.4) as that of a free particle moving with positive momentum $p = \hbar\kappa$. Conversely, the wave function $\psi(x) = Be^{-i\kappa x}$ describes a free particle moving with a negative x-component of momentum, $-p$, that is, along the negative x-axis. That is the significance of the coefficient B. The more general wave function (equation 37.3) therefore describes a superposition of left- and right-moving waves.

Kinetic Energy

We have found that we can apply a momentum operator to the quantum wave function of a particle in order to find the particle's momentum. Are there other operators that can be applied to wave functions to find the equivalent classical quantities? One such classical quantity is the kinetic energy of a particle. Classically, the kinetic energy can be written as $K = p^2/2m$, so, following the idea of equation 37.6, we introduce an operator for the kinetic energy:

$$\hat{\mathbf{K}}\psi(x) = \frac{1}{2m}\hat{\mathbf{p}}^2\psi(x) = \frac{1}{2m}\left(-i\hbar\frac{d}{dx}\right)^2\psi(x) = -\frac{\hbar^2}{2m}\frac{d^2}{dx^2}\psi(x). \tag{37.7}$$

This equation can be taken to be the general definition of the kinetic energy operator. What happens in the special case where this operator is applied to the wave function of a freely moving particle with momentum $p = \hbar\kappa$, one that has a wave function $\psi(x) = Ae^{i\kappa x}$? In this case, we obtain

$$-\frac{\hbar^2}{2m}\frac{d^2}{dx^2}\psi(x) = -\frac{\hbar^2}{2m}\frac{d^2}{dx^2}Ae^{i\kappa x} = -i\kappa\frac{\hbar^2}{2m}\frac{d}{dx}Ae^{i\kappa x} = \kappa^2\frac{\hbar^2}{2m}Ae^{i\kappa x} = \frac{p^2}{2m}Ae^{i\kappa x} = KAe^{i\kappa x}.$$

Indeed, this operator works as advertised. Applying the kinetic energy operator to the wave function of a free particle yields the particle's kinetic energy times the wave function. Note that inserting the wave function $\psi(x) = Be^{-i\kappa x}$ would have resulted in the same value for the kinetic energy, because $(-\kappa)^2 = \kappa^2$. Thus, for the superposition of a left- and a right-moving wave with wave number κ, applying the kinetic energy operator to the wave function of equation 37.3 results in

$$\hat{\mathbf{K}}(Ae^{i\kappa x} + Be^{-i\kappa x}) \equiv -\frac{\hbar^2}{2m}\frac{d^2}{dx^2}(Ae^{i\kappa x} + Be^{-i\kappa x}) = \frac{\hbar^2\kappa^2}{2m}(Ae^{i\kappa x} + Be^{-i\kappa x}). \tag{37.8}$$

37.2 Time-Independent Schrödinger Equation

Given that an electron can be represented by a wave function, what is the equation of motion that describes how this wave function depends on space and time? Such an equation would give solutions for the wave function that are consistent with the observations discussed so far in this chapter. In particular, the solutions should have de Broglie wavelengths of the kind found in Chapter 36, and the absolute squares of the solutions should reproduce the results of the double-slit experiment. The Austrian physicist Erwin Schrödinger (1887–1961) found such an equation in 1925, and it now carries his name. It is the foundation for all of non-relativistic quantum mechanics. By "nonrelativistic," we mean all physical cases where the speeds of the objects are small compared to the speed of light, allowing the kinetic energy to be expressed as $p^2/2m$. The following discussion applies the time-independent Schrödinger equation to electrons, but this equation also holds for any other object conventionally thought of as a particle.

For now, we look only at one-dimensional problems and investigate their static solutions, that is, their solutions independent of time. The **time-independent Schrödinger equation** for this case is

$$-\frac{\hbar^2}{2m}\frac{d^2\psi(x)}{dx^2} + U(x)\psi(x) = E\psi(x). \tag{37.9}$$

In this equation, $U(x)$ represents the potential energy, which can be different for different positions, and E is the total mechanical energy of the wave. We have already introduced the first term in this equation as the kinetic energy times the wave function, so we may think of the time-independent Schrödinger equation as an expression of the law of energy conservation for the wave function:

$$\left(K(x) + U(x)\right)\psi(x) = E\psi(x).$$

If the potential energy is zero everywhere, then $U(x) = 0$ and the total energy is equal to the kinetic energy. For this case, we have already found the solution because the time-independent Schrödinger equation then simply reduces to equation 37.8. In the absence of potential energy, the solution to the equation is thus $\psi(x) = Ae^{i\kappa x} + Be^{-i\kappa x}$.

What is the solution for a very large potential energy? Physicists like to use the limit of an *infinite* potential energy in order to obtain a simple solution. For an infinite potential energy at some point in space, the time-independent Schrödinger equation demands that either the energy E is infinite or the wave function has the value zero at that position. We are not interested in the unphysical, infinite-energy case, so the wave function has to be zero at that position. Therefore, it is physically impossible for a particle to be in a region with infinite potential energy. We call this region a *forbidden region*.

Starting from the simple limiting cases of zero potential energy and infinite potential energy, we can construct more complicated wave function solutions to potential energy distributions. To find these solutions, we have to keep in mind that

- the wave function must be continuous in space (that is, it must not make any "jumps," which would create positions where the derivatives of the wave function are undefined),

- the wave function must be zero in regions with infinite potential energy, as just discussed, and

- the wave function is normalized (that is, it fulfills the normalization condition of equation 37.2).

37.3 Infinite Potential Well

Our first example of a potential energy distribution in space is that of an **infinite potential well.** This is the simplest case mathematically, and it provides insight into interesting physical situations. For an infinite potential well, we define the potential energy as a function of the spatial coordinate x as follows:

$$U(x) = \begin{cases} \infty & \text{for } x < 0 \\ 0 & \text{for } 0 \leq x \leq a \\ \infty & \text{for } x > a. \end{cases} \tag{37.10}$$

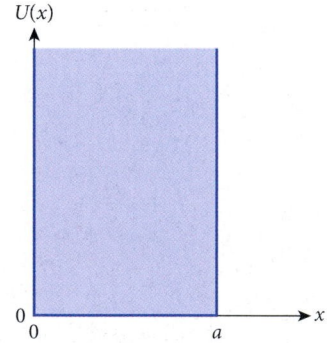

FIGURE 37.4 An infinite potential well. The allowed region with noninfinite values of the potential is shaded in blue.

For this case, the wave function is zero for all values outside of the coordinate space interval between 0 and a (Figure 37.4), which implies a zero value at the interval boundaries. Inside this coordinate space interval, the wave function is of the kind given by equation 37.3, $\psi(x) = Ae^{i\kappa x} + Be^{-i\kappa x}$. As stated before, this solution is mathematically equivalent to $\psi(x) = C \cos(\kappa x) + D \sin(\kappa x)$. This way of writing the wave function is advantageous in this case, because we know that $\psi(0) = 0$. Substituting $x = 0$ into the wave function gives

$$\psi(0) = C\cos(\kappa \cdot 0) + D\sin(\kappa \cdot 0) = C = 0.$$

Thus, the coefficient C must be zero, and the cosine term cannot contribute to the solution, giving the solution $\psi(x) = D \sin(\kappa x)$ in the interval between 0 and a. This makes physical sense, because the superposition of the left-traveling wave $Be^{-i\kappa x}$ and the right-traveling wave $Ae^{i\kappa x}$ results in a standing wave, which is the appropriate form for a wave trapped between two infinitely high potential barriers.

Thus, our preliminary result is that the wave function can be written as

$$\psi(x) = \begin{cases} 0 & \text{for } x < 0 \\ D \sin(\kappa x) & \text{for } 0 \leq x \leq a \\ 0 & \text{for } x > a. \end{cases}$$

Because the wave function must be continuous, there are constraints on the possible solutions inside the interval between 0 and a, since only those solutions that vanish at the boundaries can be used. At $x = 0$, this vanishing happens automatically because the sine function is always zero at this point. However, at $x = a$, the boundary condition is

$$\psi(x = a) = D\sin(\kappa a) = 0.$$

This implies that not all values of the wavelength $\lambda = 2\pi/\kappa$ are possible, but only those for which

$$\kappa a = \frac{2\pi a}{\lambda} = n\pi \quad \text{(for all } n = 1, 2, 3, ...)$$

is fulfilled, because the sine function is zero whenever its argument is an integer multiple of π. (Technically, $n = 0$ is also permitted, but for any finite value of a this implies an unphysical infinite wavelength. Negative integers would also be permitted, but because $\sin(-x) = -\sin(x)$, these do not give us additional solutions beyond those already contained in the positive integers.) Thus, the only possible wavelengths in this potential well are those given by

$$\lambda_n = \frac{2a}{n} \quad \text{(for all } n = 1, 2, 3, ...). \tag{37.11}$$

Then the only possible solutions of the time-independent Schrödinger equation for the case of a particle in an infinite potential well are

$$\psi(x) = \begin{cases} 0 & \text{for } x < 0 \\ D\sin\left(\dfrac{n\pi x}{a}\right) \text{ with } n = 1, 2, 3, ... & \text{for } 0 \le x \le a \\ 0 & \text{for } x > a. \end{cases}$$

We are almost done, but we still have to determine the amplitude D of the wave function. In order to obtain it, we use the normalization condition (equation 37.2):

$$1 = \int_{-\infty}^{\infty} |\psi(x)|^2\, dx = \int_{-\infty}^{0} |\psi(x)|^2\, dx + \int_{0}^{a} |\psi(x)|^2\, dx + \int_{a}^{\infty} |\psi(x)|^2\, dx$$

$$= 0 + \int_{0}^{a} \left| D\sin\left(\frac{n\pi x}{a}\right) \right|^2 dx + 0$$

$$= |D|^2 \int_{0}^{a} \sin^2\left(\frac{n\pi x}{a}\right) dx$$

$$= |D|^2 \frac{a}{2}.$$

Therefore, $|D|^2 = 2/a$. Any complex number z can be written as $z = re^{i\theta}$, and $|z| = r$ for any phase angle θ. The magnitude of D is $\sqrt{2/a}$, so in general we can write $D = \sqrt{2/a}\, e^{i\theta}$. For simplicity, we select a phase angle of 0, and thus we have the complete solution for the wave function of a particle confined to a potential well with infinitely high walls:

$$\psi(x) = \begin{cases} 0 & \text{for } x < 0 \\ \sqrt{\dfrac{2}{a}}\sin\left(\dfrac{n\pi x}{a}\right) \text{ with } n = 1, 2, 3, ... & \text{for } 0 \le x \le a \\ 0 & \text{for } x > a. \end{cases} \tag{37.12}$$

In equation 37.12, each value of n corresponds to a different possible wave function for a particle in an infinite well. To distinguish these wave functions, each of them is labeled with an index that indicates the number n, the **principal quantum number**. Thus, $\psi_1(x)$ denotes the solution for quantum number 1, and so on. Figure 37.5a shows the wave function solutions for the four lowest quantum numbers.

FIGURE 37.5 (a) Wave functions corresponding to the four lowest quantum numbers for a particle in an infinite potential well; (b) probability of finding the particle with the wave function of part (a) at any particular position in the well.

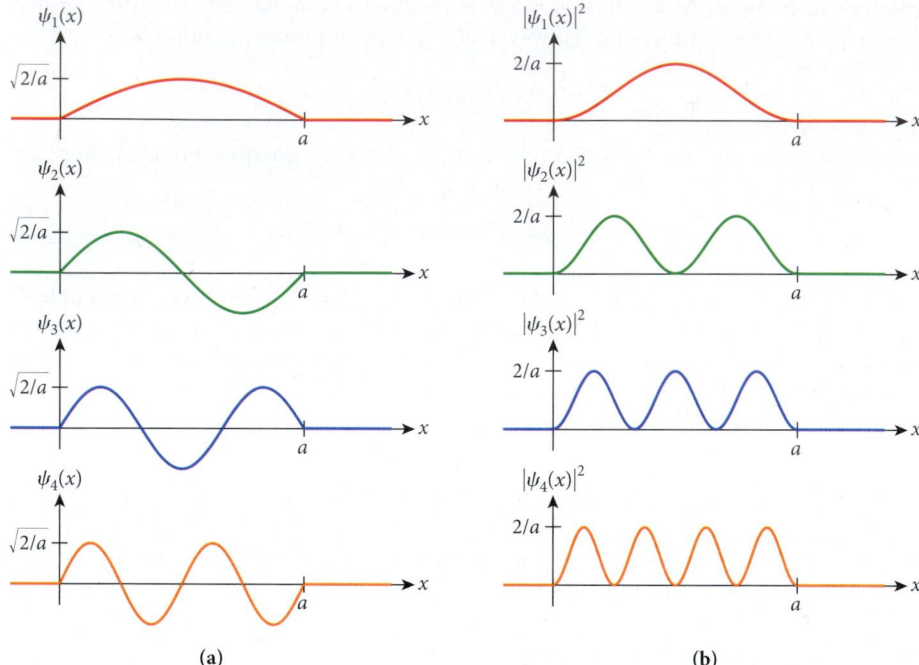

(a)   (b)

Earlier, we saw that the probability of finding a particle at a given position is proportional to the absolute square of the wave function. Figure 37.5b shows the absolute squares of the wave functions that correspond to the four lowest quantum numbers. Each plot shows the relative likelihood of finding the particle at particular positions in the well when performing a position measurement.

The emergence of simple integer numbers—that is, quantum numbers—is a typical feature of quantum mechanical systems. Boundary conditions on the wave functions force the existence of solutions that are quantized. The discussions of atoms, atomic nuclei, and elementary particles in Chapters 38 and 39 will show that there are different quantum numbers in many situations.

Chapter 15's solution for a vibrating string has many similarities to the solution for a particle in an infinite potential well. Because the string is clamped at both ends, nodes of the wave function appear at the ends, allowing only the fundamental oscillation and its harmonics. The infinitely high potential outside the interval [0,a] accomplishes the same effect of "clamping down" the wave function of a particle at the boundaries of the potential well. Therefore, the standing wave solutions for the classical vibrating string and the quantum particle in an infinite potential well are mathematically identical.

Finally, let's compare the quantum probability distribution to the classical one for the case we have just considered. In the classical case, the particle remains trapped between 0 and a. Between these two points, it moves with constant speed, $v = \sqrt{2E/m}$.

The probability of finding a particle at a given point in space is proportional to the fraction, dt, of the total time the particle spends in the vicinity, dx, of the point x. In turn, this time interval dt is inversely proportional to the speed, $v(x)$, the particle has at this point, $dt = dx/v(x)$. Thus, the classical probability of finding a particle in the interval between x and $x + dx$ is

$$\Pi_c(x)dx \propto \frac{1}{v(x)}. \tag{37.13}$$

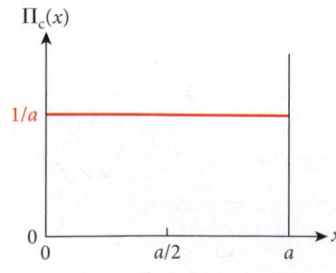

FIGURE 37.6 Classical probability distribution for finding a particle at any particular position trapped in an infinite potential well.

If v is independent of position, then the classical probability of finding a particle must be a constant as a function of the position. The resulting probability distribution is shown in Figure 37.6.

Comparing the distribution in Figure 37.6 with the quantum mechanical probability distribution in Figure 37.5b, we see clear differences. Each quantum mechanical probability distribution oscillates between 0 and a maximum value of $2/a$, whereas the classical probability value is the average of this oscillation, $1/a$. The relationship between classical and quantum mechanical probability distributions is discussed in more detail later in this chapter.

Energy of a Particle

The time-independent Schrödinger equation (37.9) depends on the total energy of the particle. Now that we have the complete wave function solution for a particle in an infinite potential well, let's find the total energy that corresponds to this solution. Outside the interval between $x = 0$ and $x = a$, the wave function is zero, so the particle has zero probability of residing in that region. Thus, the outside region does not contribute to the energy. Inside the interval between 0 and a, there is no potential energy. Therefore, the total energy is equal to the kinetic energy of the particle. Equation 37.7 allows us to calculate this kinetic energy:

$$\hat{\mathbf{K}}\psi(x) = -\frac{\hbar^2}{2m}\frac{d^2\psi(x)}{dx^2}.$$

For each quantum number n, however, a different result will be obtained for the kinetic energy and therefore the total energy. We label this total energy corresponding to the quantum number n as E_n. For the case of the particle in the infinite potential well,

$$E_n\psi_n(x) = -\frac{\hbar^2}{2m}\frac{d^2\psi_n(x)}{dx^2}$$

$$= -\frac{\hbar^2}{2m}\frac{d^2}{dx^2}\left(\sqrt{\frac{2}{a}}\sin\left(\frac{n\pi x}{a}\right)\right)$$

$$= -\frac{\hbar^2}{2m}\left(\frac{n\pi}{a}\right)\frac{d}{dx}\left(\sqrt{\frac{2}{a}}\cos\left(\frac{n\pi x}{a}\right)\right)$$

$$= \frac{\hbar^2}{2m}\left(\frac{n\pi}{a}\right)^2\left(\sqrt{\frac{2}{a}}\sin\left(\frac{n\pi x}{a}\right)\right)$$

$$= \frac{\hbar^2}{2m}\left(\frac{n\pi}{a}\right)^2\psi_n(x).$$

The total energy E_n corresponding to the quantum number n is thus proportional to the square of n:

$$E_n = \frac{\hbar^2\pi^2}{2ma^2}n^2 \quad \text{(for } n = 1, 2, 3, \ldots\text{).} \tag{37.14}$$

The total energy also depends inversely on the square of the width of the potential well, a, and inversely on the mass of the particle, m. The total energies can be represented as horizontal lines in a plot of energy versus position. Such a plot, called an **energy-level diagram**, is shown in Figure 37.7. Because the energies of these states are lower than the (infinite) height of the potential well, the states are referred to as **bound states**. The infinite potential well has an infinite number of bound states.

This solution for an infinite potential well is sometimes called a "particle in a box" and it allows us to simply model an electron bound to an atom or a proton bound in an atomic nucleus, as the following example shows.

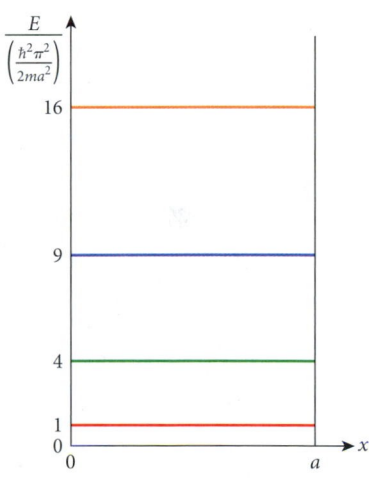

FIGURE 37.7 The four lowest energy levels in an infinite potential well. The colors of the energy levels correspond to the colors of the corresponding wave functions in Figure 37.5.

Concept Check 37.1

If the width of the potential well is reduced to half its original value, the energy of the $n = 3$ wave function will

a) stay the same.

b) be reduced by a factor of 2.

c) be reduced by a factor of 4.

d) be increased by a factor of 2.

e) be increased by a factor of 4.

EXAMPLE 37.1 Electron in a Box

PROBLEM

What is the kinetic energy of the wave function with the lowest quantum number for an electron confined to a box of width $2.00\ \text{Å} \equiv 2.000 \cdot 10^{-10}$ m?

– Continued

SOLUTION

The mass of the electron is $m = 9.109 \cdot 10^{-31}$ kg. Substituting numerical values into equation 37.14 gives

$$E_1 = \frac{\hbar^2 \pi^2}{2ma^2} = \frac{(1.0546 \cdot 10^{-34} \text{ J s})^2 \pi^2}{2(9.109 \cdot 10^{-31} \text{ kg})(2.000 \cdot 10^{-10} \text{ m})^2} = 1.506 \cdot 10^{-18} \text{ J}.$$

Alternatively, this energy value can be expressed in units of electron-volts:

$$E_1 = (1.506 \cdot 10^{-18} \text{ J})/(1.602 \cdot 10^{-19} \text{ J/eV}) = 9.401 \text{ eV}.$$

DISCUSSION

Chapter 38 will show that atoms have typical diameters of 10^{-10} m, so this result allows us to estimate that the typical energy scale for electrons in atoms has to be on the order of 10 eV. Real atoms are much more complicated than a simple model of a box with infinitely high walls, but this model allows us to make educated guesses about the energy scales involved in a physical problem.

Performing the same exercise for a proton (with a mass approximately 1800 times that of the electron) confined to a box of width $4.0 \cdot 10^{-15}$ m, which is the typical dimension of an atomic nucleus, yields the answer of $13 \cdot 10^6$ eV. We conclude that typical nuclear energy scales are on the order of 13 MeV, a remarkably good first guess.

Self-Test Opportunity 37.2

How do the solutions for the wave functions and energies of a particle in an infinite potential well change if the potential energy function of equation 37.10 is replaced with with one that is still infinite outside the interval between 0 and a, but has a constant value of $c \neq 0$ inside this interval?

Multidimensional Wells

The idea of a one-dimensional infinite potential well can be extended to a two-dimensional well in which a particle is confined in the xy-plane, or even to a rectangular box in three-dimensional space. We will not derive the full solutions for these situations, but instead will stress the general features.

First, let's think about what will be different as our calculations are extended from one to two spatial dimensions. The potential energy in this case can be a function of both variables, x and y, and so we write it as $U(x,y)$. This means that the wave function must also be written as a function of these two variables, $\psi(x,y)$. In our classical considerations of kinetic energy, we saw that it can be expressed as

$$K = \frac{p^2}{2m} = \frac{p_x^2}{2m} + \frac{p_y^2}{2m}.$$

In analogy with what we derived for the one-dimensional case (see equation 37.6), the x- and y-components of the momentum operator can be written as

$$\hat{p}_x \psi(x,y) = -i\hbar \frac{\partial}{\partial x} \psi(x,y)$$

$$\hat{p}_y \psi(x,y) = -i\hbar \frac{\partial}{\partial y} \psi(x,y).$$

The only change relative to the one-dimensional case is that partial derivatives must be used, as is appropriate for multivariable calculus. (A partial derivative with respect to one variable treats the other variables as constants.) Then the kinetic energy operator for the quantum wave function can be written as

$$\hat{K}\psi(x,y) = \frac{1}{2m}\hat{p}_x^2 \psi(x,y) + \frac{1}{2m}\hat{p}_y^2 \psi(x,y)$$

$$= -\frac{\hbar^2}{2m}\frac{\partial^2}{\partial x^2}\psi(x,y) - \frac{\hbar^2}{2m}\frac{\partial^2}{\partial y^2}\psi(x,y).$$

Finally, the two-dimensional time-independent Schrödinger equation is

$$-\frac{\hbar^2}{2m}\frac{\partial^2 \psi(x,y)}{\partial x^2} - \frac{\hbar^2}{2m}\frac{\partial^2 \psi(x,y)}{\partial y^2} + U(x,y)\psi(x,y) = E\psi(x,y).$$

To proceed, we need to specify the shape of the potential energy well. If the potential energy can be written as a sum of a function that depends only on x and another that depends only on y—that is to say, $U(x,y) = U_1(x) + U_2(y)$—then the problem becomes **separable.** This means that the wave function is a product of two functions and that each one depends on only one variable: $\psi(x,y) = \psi_1(x) \cdot \psi_2(y)$.

Further, if we use the simplifying case of a two-dimensional rectangular infinite potential well,

$$U(x,y) = U_1(x) + U_2(y),$$

$$\text{with } U_1(x) = \begin{cases} \infty & \text{for } x < 0 \\ 0 & \text{for } 0 \le x \le a, \\ \infty & \text{for } x > a \end{cases} \quad U_2(y) = \begin{cases} \infty & \text{for } y < 0 \\ 0 & \text{for } 0 \le y \le b \\ \infty & \text{for } y > b, \end{cases}$$

then the solutions we obtain are products of the wave functions given by equation 37.12 in the x- and y-directions. Explicitly, these wave functions are

$$\psi(x,y) = \psi_1(x) \cdot \psi_2(y) \tag{37.15}$$

$$\psi_1(x) = \begin{cases} 0 & \text{for } x < 0 \\ \sqrt{\dfrac{2}{a}} \sin\left(\dfrac{n_x \pi x}{a}\right) & \text{with } n_x = 1,2,3,\ldots \text{ for } 0 \le x \le a \\ 0 & \text{for } x > a \end{cases}$$

$$\psi_2(y) = \begin{cases} 0 & \text{for } y < 0 \\ \sqrt{\dfrac{2}{b}} \sin\left(\dfrac{n_y \pi y}{b}\right) & \text{with } n_y = 1,2,3,\ldots \text{ for } 0 \le y \le b \\ 0 & \text{for } y > b. \end{cases}$$

These solutions are displayed in Figure 37.8 for the lowest values of the quantum numbers n_x and n_y.

In the same way that we arrived at the solution for the energy values corresponding to the one-dimensional quantum numbers n in equation 37.14, we obtain for the energies corresponding to these two-dimensional wave functions:

$$E_{n_x,n_y} = \frac{\hbar^2 \pi^2}{2ma^2} n_x^2 + \frac{\hbar^2 \pi^2}{2mb^2} n_y^2. \tag{37.16}$$

Self-Test Opportunity 37.3

Using the same considerations, write the wave functions and energies for a three-dimensional rectangular infinite potential well.

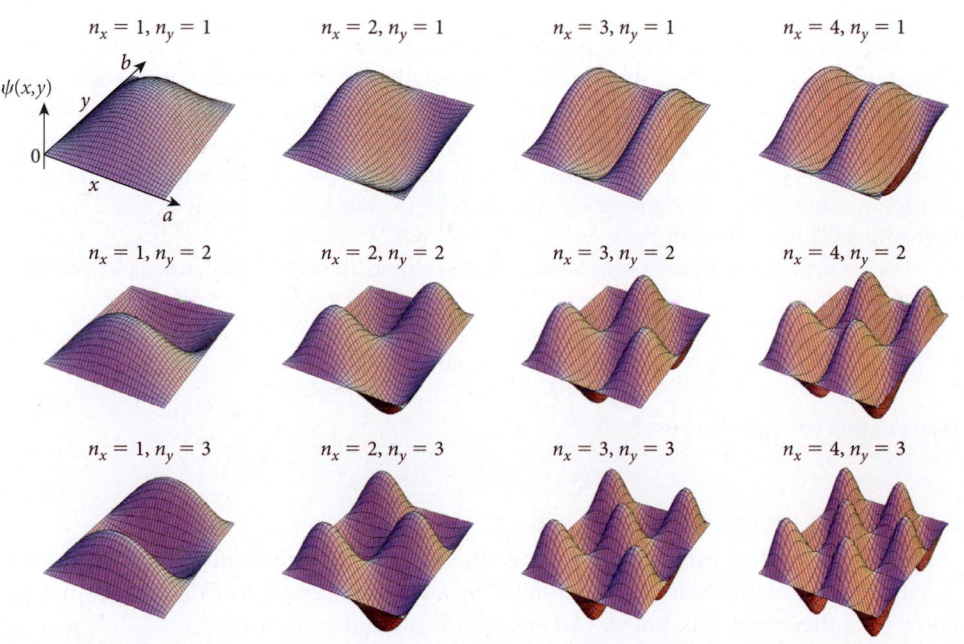

$n_x = 1, n_y = 1$ $n_x = 2, n_y = 1$ $n_x = 3, n_y = 1$ $n_x = 4, n_y = 1$

$n_x = 1, n_y = 2$ $n_x = 2, n_y = 2$ $n_x = 3, n_y = 2$ $n_x = 4, n_y = 2$

$n_x = 1, n_y = 3$ $n_x = 2, n_y = 3$ $n_x = 3, n_y = 3$ $n_x = 4, n_y = 3$

FIGURE 37.8 Wave functions corresponding to the lowest quantum numbers for a particle in a two-dimensional rectangular infinite potential well.

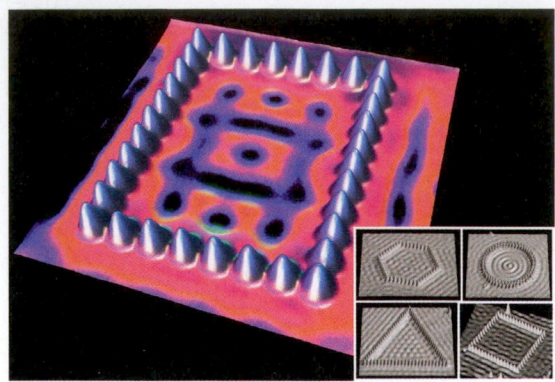

FIGURE 37.9 Two-dimensional rectangular quantum corral made of individual iron atoms, arranged on a copper surface. The color coding inside the corral shows the electron wave probability density. The gray-scale insets show several corrals of different shapes, and the ripples indicate the electron wave probability density. These arrangements were created using a scanning tunneling microscope.

If, for example, $a = b$ (a square potential well), then the same energy value can usually be obtained in more than one way. For instance, the states with quantum numbers ($n_x = 1$, $n_y = 2$) and ($n_x = 2$, $n_y = 1$) have the same energy. This situation is referred to as *degeneracy*.

If the potential well is not rectangular, then the solution usually cannot be written in a simple form. However, in many cases the problem can still be solved numerically. Experimentally, very high (not quite infinite) two-dimensional potential wells for electrons can be generated by arranging several atoms on a flat surface in the shape of a corral. Figure 37.9 shows one example, where the color coding (large image) or the ripples (gray-scale inset) represent the probability distributions for the electron wave function inside corrals consisting of iron atoms arranged in different shapes on a flat copper surface. The instrument used to arrange the atoms in the ways shown and to generate these experimental images is a scanning tunneling microscope, which will be discussed in more detail later in this chapter.

To conclude this discussion of the infinite potential well, let's stress the main point. Even though the infinite potential well is a fairly unrealistic abstraction, it exhibits some characteristic features that are common to all quantum systems: discreteness of energy levels, nonzero ground-state energy, increased separation of energy levels under tightened constraints, a quantum number for each dimension, and so on. These features, combined with the relative mathematical simplicity, make the infinite potential well worth studying. It introduces the qualitatively new aspects of quantum physics without clouding them with too many mathematical complications.

37.4 Finite Potential Wells

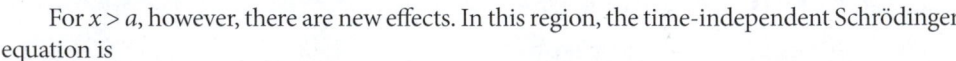

Let's return to the one-dimensional case and solve a problem that is a bit more complicated than that of a potential well with infinitely high potential outside it. Here we consider the case where the wall of the well is not infinitely high, but instead has a finite height.

The shape of the **finite potential well** we want to study is shown in Figure 37.10: The potential energy $U(x)$ is zero inside the interval from 0 to a, is infinite for all values of $x < 0$, and has a finite constant value of $U_1 > 0$ for $x > a$:

$$U(x) = \begin{cases} \infty & \text{for } x < 0 \\ 0 & \text{for } 0 \leq x \leq a \\ U_1 & \text{for } x > a. \end{cases}$$

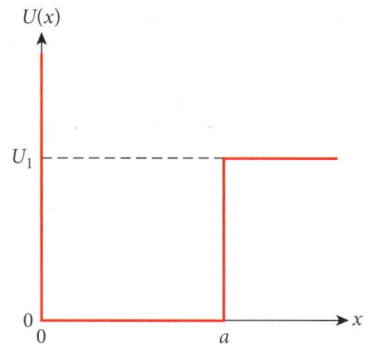

FIGURE 37.10 Finite potential energy well.

Just as in the case of the infinite potential well, we construct a solution for each of the three regions separately and then match these solutions to each other at the boundaries. The solution remains the same as before for $x < 0$, where the wave function must have a constant value of zero. In the interval from 0 to a, the wave function again must have the general form $\psi(x) = C \cos(\kappa x) + D \sin(\kappa x)$. And again, because the wave function has to be continuous and $\psi(0) = 0$, the coefficient C must be 0, giving the solution $\psi(x) = D \sin(\kappa x)$ in the interval between 0 and a.

For $x > a$, however, there are new effects. In this region, the time-independent Schrödinger equation is

$$-\frac{\hbar^2}{2m} \frac{d^2 \psi(x)}{dx^2} + U_1 \psi(x) = E \psi(x).$$

Rearranging this equation leads to

$$\frac{d^2 \psi(x)}{dx^2} = \frac{2m(U_1 - E)}{\hbar^2} \psi(x). \tag{37.17}$$

Let's look at the form of this equation: The second derivative of the wave function is equal to the wave function itself, multiplied by a constant, $2m(U_1 - E)/\hbar^2$. We do not yet know what the energy E is, but we can distinguish two different cases.

Case 1: Energy Larger Than the Well Depth

For $E > U_1$, we have $2m(U_1 - E)/\hbar^2 < 0$, and we obtain oscillatory solutions of the same kind as those we found inside the interval between 0 and a, but with a different wavelength and thus a different wave number:

$$\psi(x) = F\cos(\kappa'x) + G\sin(\kappa'x) \quad (\text{for } x > a \text{ and } E > U_1).$$

How are the wave numbers κ' and κ related to each other? Remember that inside the interval $[0,a]$ there is no potential energy, the total energy is all kinetic energy, and we found that $E = p^2/2m = \hbar^2\kappa^2/2m$. For $x > a$ and $E > U_1$, the kinetic energy is then simply $E - U_1$, and thus $E - U_1 = \hbar^2\kappa'^2/2m$. Therefore,

$$\kappa' = \sqrt{\kappa^2 - \frac{2mU_1}{\hbar^2}}. \tag{37.18}$$

Thus, the relationship of the wave numbers is $\kappa' < \kappa$, and so the wavelength of the spatial oscillation is larger in the region where $U(x) = U_1 > 0$ than in the region where $U(x) = 0$.

The wave function for this case when the total energy is larger than the height of the potential well, $E > U_1$, is then

$$\psi(x) = \begin{cases} 0 & \text{for } x < 0 \\ D\sin(\kappa x) & \text{for } 0 \leq x \leq a \\ F\cos(\kappa'x) + G\sin(\kappa'x) & \text{for } x > a. \end{cases}$$

How do we determine the complete solution? We must find the values of the amplitude coefficients D, F, and G. These three numbers can be determined from the following three conditions:

- The wave function must be continuous at the boundary $x = a$ (as explained earlier).
- The derivative of the wave function must be continuous at $x = a$ (as explained below).
- The absolute square of the wave function must be normalized to 1.

The rationale for the second condition is that the momentum is related to the derivative of the wave function, so if the derivative of the wave function were discontinuous, then the momentum would be undefined at the point of discontinuity. Note that the wave numbers κ and κ' are not further constrained by these conditions, and neither are the possible values of the energy E. All energies larger than the value U_1 turn out to be possible.

We will not work through all of the algebra; this is usually done during an upper-level course on quantum mechanics. From the structure we have worked out so far, however, we can make a sketch (Figure 37.11) of the type of wave function that we can expect. This wave function is continuous and has a continuous derivative everywhere. Also, in general, the oscillations in the region with $U > 0$ have larger wavelength (because of equation 37.18) and amplitude (because of the larger wavelength and the first two conditions given above) than the oscillations in the region with $U = 0$. We emphasize once more that an infinite continuum of wavelengths, not just discrete wavelengths, is allowed in this case.

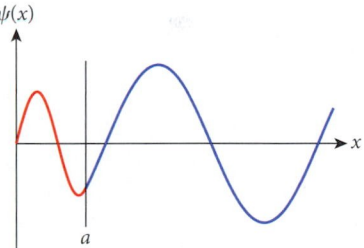

FIGURE 37.11 Wave function for the finite potential well, with energy larger than the height of the potential well.

Case 2: Energy Smaller Than the Well Depth, Bound States

In the second case, the energy of the particle is smaller than the depth of the potential well, so the wave functions that fulfill the condition $E < U_1$ form bound states. Part of our task is to find out if there are any bound states at all. Because even the lowest possible energy of a wave function in the infinite well has a finite nonzero value, we expect that for very shallow potential wells, no bound state occurs.

For $E < U_1$, the constant $2m(U_1 - E)/\hbar^2$ has a value larger than zero. Thus, instead of oscillatory solutions, we obtain exponential solutions to the differential equation 37.17. To show this in quantitative detail, we introduce a constant γ:

$$\gamma = \sqrt{\frac{2m(U_1 - E)}{\hbar^2}}. \tag{37.19}$$

Then the differential equation 37.17 that we have to solve for the region $x > a$ becomes

$$\frac{d^2\psi(x)}{dx^2} = \gamma^2 \psi(x).$$

It has this solution:

$$\psi(x) = Fe^{-\gamma x} + Ge^{\gamma x} \quad \text{(for } x > a \text{ and } E < U_1 \text{)}.$$

(That this is a solution can be verified by taking its second derivative and inserting that back into the differential equation.) We discard the exponentially rising term $e^{\gamma x}$ because it becomes infinite as $x \to \infty$. Then the wave function would not be normalizable; that is, the integral of the absolute square of the wave function could not have the value 1 according to equation 37.2. Thus, $G = 0$, and our solution becomes

$$\psi(x) = Fe^{-\gamma x} \quad \text{(for } x > a \text{ and } E < U_1 \text{)}.$$

Again, the wave function for all three regions can be written as follows:

$$\psi(x) = \begin{cases} 0 & \text{for } x < 0 \\ D\sin(\kappa x) & \text{for } 0 \leq x \leq a \\ Fe^{-\gamma x} & \text{for } x > a. \end{cases} \tag{37.20}$$

The conditions for a normalized wave function that is continuous and has a continuous derivative at $x = a$ provide three equations for the two unknown amplitudes D and F, as well as the wave number κ. The other constant that appears in this solution, γ, is not independent of κ. This can be seen by using the defining equation (37.19) for γ and the fact that when the potential energy is zero in the region $0 \leq x \leq a$, the total energy E, which is constant, is simply $E = \hbar^2 \kappa^2 / 2m$. This results in

$$\gamma^2 = \frac{2m(U_1 - E)}{\hbar^2} = \frac{2mU_1}{\hbar^2} - \frac{2mE}{\hbar^2} = \frac{2mU_1}{\hbar^2} - \kappa^2. \tag{37.21}$$

Before we examine a representative case in greater detail, let's first think about the general features of the solution that we expect to obtain. Just as in the case of an infinite potential well, the wave function in the case of a finite potential well oscillates in sine form in the region where the potential energy vanishes. However, the wave function in the latter case is allowed to "leak" through the wall, although it has an exponentially decreasing value as x increases beyond a. This leakage allows the wavelength for $x < a$ to be somewhat longer than the wavelength in the infinite well because now the function spills over into $x > a$. A longer wavelength implies a smaller wave number. Since the total energy is proportional to the square of the wave number, the energy values corresponding to the wave functions in the finite potential well can be expected to be lower than their counterparts in the infinite well. Another way of stating this result is to say that the wave functions in the infinite well are more *localized*, and thus have larger kinetic energies, than their corresponding counterparts in the well of finite depth.

SOLVED PROBLEM 37.1 / Finite Potential Well

PROBLEM

Suppose an electron is to have at least two bound states in a well of the shape indicated in Figure 37.10. If the width of the well is $a = 1.30$ nm, how high does the potential step U_1 (the depth of the well) need to be for the wavelength of the $n = 2$ state to be 20.0% greater than it would be in an infinite potential well of the same width?

SOLUTION

THINK This problem must be approached in steps. First, equation 37.11 gave the result for the infinite well, $\lambda_n = 2a/n$, for all $n = 1,2,3,\ldots$ Thus, we obtain for $n = 2$ the wavelength $\lambda_2 = 2a/2 = a$. (As usual, we do not insert the value for a at this point, but leave the result general and insert the numbers only at the end.)

A wavelength 20.0% larger than that for the infinite well in this case is $\lambda_2' = 1.20\lambda_2 = 1.20a$. The corresponding value of the wave number is thus $\kappa_2' = 2\pi/\lambda_2' = 2\pi/(1.20a)$. Since the wave number is inversely proportional to the wavelength, and the wavelength needs to increase by 20.0%, the wave number needs to be reduced by the same factor: $\kappa_2' = \kappa_2/1.20$. (Again, we postpone inserting numerical values for the wave numbers until the end.)

SKETCH We can use Figure 37.10 as our sketch.

RESEARCH The general wave function for this problem is given by equation 37.20:

$$\psi_2(x) = \begin{cases} 0 & \text{for } x < 0 \\ D\sin(\kappa_2' x) & \text{for } 0 \le x \le a \\ Fe^{-\gamma x} & \text{for } x > a. \end{cases}$$

Let's look at the boundary conditions at $x = a$. Demanding continuity of the wave function requires that

$$D\sin(\kappa_2' a) = Fe^{-\gamma a} \Rightarrow \sin(\kappa_2' a) = \frac{F}{D}e^{-\gamma a}. \tag{i}$$

We take the derivative of the wave function, and then look at the continuity condition for that derivative. The derivative is

$$\frac{d}{dx}\psi_2(x) = \begin{cases} 0 & \text{for } x < 0 \\ \kappa_2' D\cos(\kappa_2' x) & \text{for } 0 \le x \le a \\ -\gamma Fe^{-\gamma x} & \text{for } x > a. \end{cases}$$

A continuous first derivative at $x = a$ thus requires that

$$\kappa_2' D\cos(\kappa_2' a) = -\gamma Fe^{-\gamma a} \Rightarrow -\frac{\kappa_2'}{\gamma}\cos(\kappa_2' a) = \frac{F}{D}e^{-\gamma a}. \tag{ii}$$

SIMPLIFY Equations (i) and (ii) have the same right-hand sides. Therefore, their left-hand sides must also be equal, and we obtain

$$-\frac{\kappa_2'}{\gamma}\cos(\kappa_2' a) = \sin(\kappa_2' a) \Rightarrow$$

$$\gamma = -\kappa_2' \cot(\kappa_2' a). \tag{37.22}$$

Taking the square of equation 37.22 results in

$$\gamma^2 = \kappa_2'^2 \cot^2(\kappa_2' a). \tag{iii}$$

However, we also have the following expression for γ^2 from equation 37.21:

$$\gamma^2 = \frac{2mU_1}{\hbar^2} - \kappa_2'^2. \tag{iv}$$

Combining equations (iii) and (iv) gives

$$\kappa_2'^2 \cot^2(\kappa_2' a) = \frac{2mU_1}{\hbar^2} - \kappa_2'^2 \Rightarrow$$

$$\frac{2mU_1}{\hbar^2} = \kappa_2'^2\left(1 + \cot^2(\kappa_2' a)\right) \Rightarrow$$

$$U_1 = \frac{\hbar^2\kappa_2'^2}{2m}\left(1 + \cot^2(\kappa_2' a)\right).$$

Formally, this is our answer.

CALCULATE Now we can insert the numbers. We have already noted that $\kappa_2' = 2\pi/\lambda_2' = 2\pi/(1.20a)$; therefore, $\kappa_2' a = 2\pi/1.20$. We can then calculate the factor $\cot^2(\kappa_2' a)$:

$$\cot^2(\kappa_2' a) = \cot^2(2\pi/1.20) = 0.3333 = \tfrac{1}{3}.$$

Therefore, the potential step has to have a height of $1 + \tfrac{1}{3} = \tfrac{4}{3}$ times the energy E_2 of the wave function ψ_2. This energy is

$$E_2 = \frac{\hbar^2\kappa_2'^2}{2m} = \frac{\hbar^2(2\pi/1.20a)^2}{2m} = \frac{h^2}{2.88ma^2}$$

$$= \frac{(6.626\cdot10^{-34}\text{ J s})^2}{2.88(9.109\cdot10^{-31}\text{ kg})(1.30\cdot10^{-9}\text{ m})^2}$$

$$= 9.90\cdot10^{-20}\text{ J} = 0.618\text{ eV}.$$

– Continued

ROUND To three significant figures, our answer is $U_1 = \frac{4}{3}E_2 = 0.824$ eV.

DOUBLE-CHECK Note that $\kappa_2' = \kappa_2/1.20$ as demanded by the problem statement. Since $E \propto \kappa^2$ for the potential well, the energy E_2 is lower than that in the infinite well by a factor of $(1/1.20)^2 = 0.694$.

This problem is fairly typical. It shows how to use the conditions of continuity of the wave function and its derivatives at boundaries to set up equations that provide the undetermined constants in the generic solution. In this case, the generic solution was of the same type as equation 37.20, with constants U_1, D, F, κ_2', and γ, which were to be determined. The continuity of the wave function was expressed in equation (i), and the continuity of the derivative in equation (ii). A third equation was obtained from the relationship between γ and κ_2' expressed in equation 37.21. Because κ_2' was specified by the problem statement, we would need four equations to solve for the remaining four unknown quantities. The continuity of ψ and its derivative at the boundary and the relationship between γ and κ_2' would be enough to solve for U_1. To find D and F, we would need to use the normalization condition (equation 37.2).

Solved Problem 37.1 presented a condition that predetermined the shape of the wave function by specifying the desired value of the wave number. A much more conventional problem is figuring out what wave functions fit into a potential of given depth. We will do this in the following example, simplifying the task by using the same numbers as in Solved Problem 37.1.

EXAMPLE 37.2 | Bound States

PROBLEM
An electron is trapped in a finite potential well of the kind shown in Figure 37.10 with a depth of $U_1 = 0.824$ eV and a width of $a = 1.30$ nm. What wave numbers correspond to the possible bound states in this well?

SOLUTION
The well depth is the same as in Solved Problem 37.1, so one solution for the wave number κ should correspond to the value $2\pi/(1.20a)$, the starting point in that problem. We also found two conditions for the exponential decay constant γ. Equation 37.22 gave $\gamma = -\kappa\cot(\kappa a)$, and equation 37.21 gave $\gamma = \sqrt{2mU_1/\hbar^2 - \kappa^2}$. Combining these results gives

$$\sqrt{\frac{2mU_1}{\hbar^2} - \kappa^2} = -\kappa\cot(\kappa a) \Rightarrow$$

$$\sqrt{\frac{2a^2mU_1}{\hbar^2} - (\kappa a)^2} = -(\kappa a)\cot(\kappa a). \qquad (i)$$

We omitted the index 2 for the wave number κ from equation 37.22, because we want to search for all possible values of κ that satisfy equation (i). This equation usually does not have an algebraic solution, but we can solve it numerically quite straightforwardly. The numerical constants in this case are $a = 1.30$ nm and $2a^2mU_1/\hbar^2 = 36.5$. Thus, we have to solve the equation

$$\sqrt{36.5 - y^2} = -y\cot y, \text{ with } y = \kappa a.$$

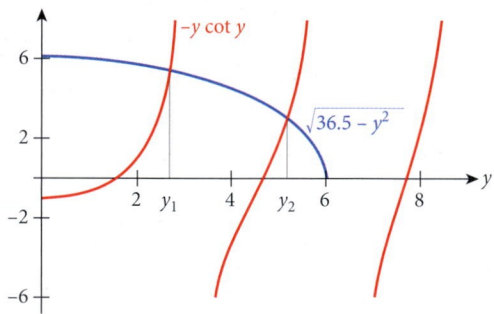

FIGURE 37.12 Finding the wave numbers for the bound states.

Figure 37.12 plots the two functions $\sqrt{36.5 - y^2}$ (in blue) and $-y\cot y$ (in red) and shows the positions where they intersect.

As you can see, there are only two positions where the two functions have the same value. Therefore, the potential well of depth $U_1 = 0.824$ eV and width $a = 1.30$ nm has only two bound states. Numerically, we find $y_1 = \kappa_1 a = 2.68$ and $y_2 = \kappa_2 a = 5.24$. This second value is nothing other than $2\pi/1.20$, the value in Solved Problem 37.1. However, the value of $\kappa_1 = 2.68/a$ is new information. Note that, unlike the case of the infinite potential well, $\kappa_2 \neq 2\kappa_1$ for the finite potential well.

In conclusion, Figure 37.13 shows the energies that correspond to the two bound states, overlaid on the shape of the potential energy function used for the potential well.

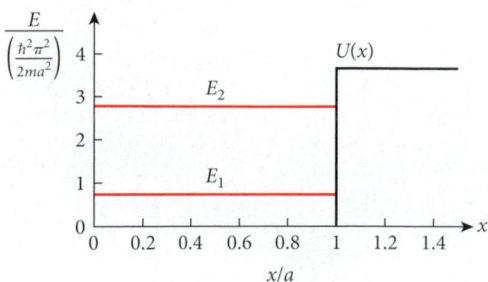

FIGURE 37.13 Energies of the lowest two possible wave functions in the finite potential well.

Figure 37.14a shows the wave functions that correspond to the two values found in Example 37.2 for the wave number, as functions of the spatial coordinate x/a. The upper graph is for the wave function times $\sqrt{a}$, corresponding to wave number κ_1 and energy E_1, and the lower graph is for κ_2 and E_2. The red part of the curve corresponds to the sinusoidal part of the wave function in the potential-free region. The blue part is the exponential penetration of the wave function into the region where the potential energy is greater than the energy of the electron; this is the classically forbidden region. It is apparent from the figure that the two parts of each wave function are properly matched, since each wave function is continuous and has a continuous derivative at the boundary $x/a = 1$. In order to obtain each of these wave functions, we would have to solve a system of equations of the kind used in Solved Problem 37.1.

The fact that an electron can penetrate the potential boundary and move into the classically forbidden region where $x > a$ is a phenomenon that is unique to the quantum world. Classically, the electron would simply bounce off the wall and could only be found at some point between $x = 0$ and $x = a$. In the quantum world, this is not the case any more. According to equation 37.1, the probability of the electron being in the spatial interval dx can be calculated as $\Pi(x) = |\psi(x)|^2 \, dx$.

Figure 37.14b plots $a|\psi(x)|^2$ for both of the bound-state wave functions. Integrating the area under the curve in the top graph gives the result that, for the bound state with wave number κ_1, the probability of finding the electron between $x = 0$ and $x = a$ is 96.9% (red area), and the probability of finding the electron in the classically forbidden region, $x > a$, is 3.1% (blue area). If the electron resides in the bound state with wave number κ_2, it has a probability of 18.6% of being found in the classically forbidden region and a probability of 81.4% of being found in $0 < x < a$. Since the energy of the second state is higher than that of the first state and close to the depth of the potential well, the electron in the second state can penetrate farther into the classically forbidden region and consequently has a greater probability of being found there.

Tunneling

If the wave function can reach into the classically forbidden region, then what happens if the potential step of finite height shown in Figure 37.10 has only a finite width? This situation is shown in Figure 37.15a, which shows a plot of the potential energy function

$$U(x) = \begin{cases} \infty & \text{for } x < 0 \\ 0 & \text{for } 0 \le x \le a \\ U_1 & \text{for } a < x < b \\ 0 & \text{for } x \ge b. \end{cases}$$

Concept Check 37.2

If the width of the potential well in Example 37.2 is doubled, to 2.6 nm, and the depth remains the same, the number of bound states will

a) stay the same.

b) increase.

c) decrease.

Self-Test Opportunity 37.4

Can you use the conditions of continuity of the wave function and of its derivative at the boundary $x = b$ to derive the values of the phase shift ϕ and the amplitude G for the wave function for the tunneling of a particle through a potential energy step?

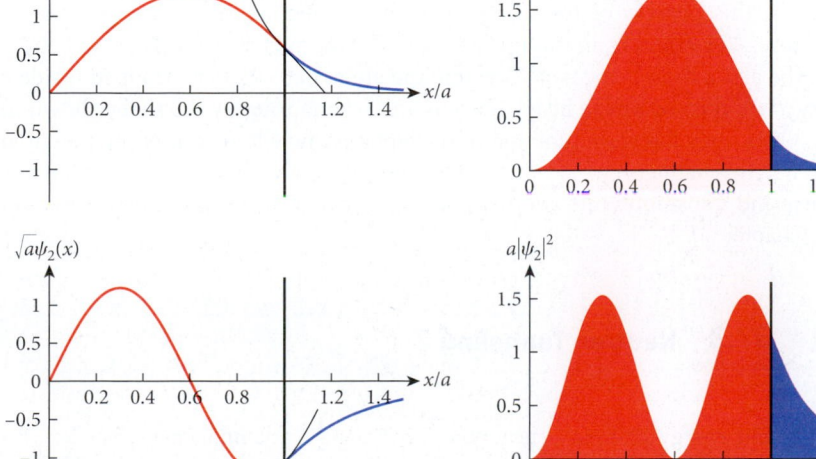

(a)   (b)

FIGURE 37.14 (a) The two wave functions with lowest values of the wave number for a finite potential step at $x = a$ as a function of x/a. The red line shows the sine function for the zero-potential region, and the blue line shows the exponential function in the classically forbidden region. (b) Corresponding probability distributions for finding the electron at a given coordinate. The blue area indicates the probability of finding the electron in the classically forbidden region. All four plots have scale factors such that the boundary at $x = a$ appears at the value of 1 on the abscissa and the areas in part (b) are the probabilities for finding the particle within the corresponding abscissa region.

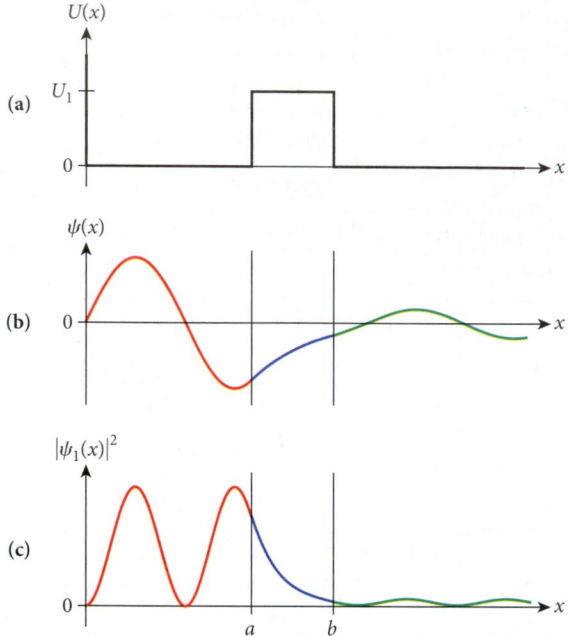

FIGURE 37.15 (a) Potential energy step of finite height and finite width; (b) wave function tunneling through the potential energy barrier; (c) probability density distribution for finding the particle at a position x.

Again, the wave function in the different regions can be written as follows:

$$\psi(x) = \begin{cases} 0 & \text{for } x < 0 \\ D\sin(\kappa x) & \text{for } 0 \le x \le a \\ Fe^{-\gamma x} & \text{for } a < x < b \\ G\sin(\kappa x + \phi) & \text{for } x \ge b. \end{cases}$$

Note that the same value can be used for the wave number κ on both sides of the barrier in this case because the potential is zero in both of these regions. In the first part of this section we solved for the wave function in the region $0 \le x \le b$. The wave function has a sinusoidal part (red part of the curve in Figure 37.15b) in the region between 0 and a and an exponentially decreasing part (blue part of the curve in part b) between a and b. What is different here is that the wave function does not continue to decay exponentially beyond $x = b$, but it oscillates in sinusoidal fashion for $x > b$ with the same wave number κ as it has between 0 and a. This process is called **tunneling**. The exact shape of the wave function (green part of the curve in part b) for $x > b$ is again determined by matching the wave function and its first derivative at the boundary $x = b$, just as we did before at $x = a$.

Figure 37.15c shows the plot of the probability density for this wave function. Clearly, there is some nonzero probability that the wave function will tunnel through the potential energy barrier and emerge on the other side. Classically, by contrast, a particle located at $0 < x < a$ with the same total energy would not have enough energy to escape and would remain trapped forever in the region $0 \le x \le a$.

The **transmission coefficient**, T, is defined as the ratio of the absolute square of the wave amplitude at the exit from the barrier to the absolute square of the wave amplitude at the entrance to the barrier. For the wave function $Fe^{-\gamma x}$ in the region of the barrier, the transmission coefficient is found to be

$$T = \frac{|\psi(b)|^2}{|\psi(a)|^2} = \frac{|Fe^{-\gamma b}|^2}{|Fe^{-\gamma a}|^2} = |e^{-\gamma(b-a)}|^2 = e^{-2\gamma(b-a)}. \tag{37.23}$$

Thus, the transmission coefficient depends exponentially on the width of the barrier, $b - a$. Note that the probability that the particle is to the left of the barrier ($x \le a$) is proportional to $|\psi(a)|^2$. The probability that the particle is to the right of the barrier ($x \ge b$) is proportional to $|\psi(b)|^2$. The transmission coefficient T therefore measures the probability that a particle hitting the barrier on the left will emerge on the right.

Fascinatingly, this process of tunneling through a potential energy barrier can be observed in nature—for example, in the alpha decay of heavy nuclei. It is a crude, but effective, model of the alpha decay process to imagine the alpha particle to be trapped inside a potential energy well formed by the heavy nucleus and having roughly the shape shown in Figure 37.15a. Over time, the wave function of the alpha particle leaks out of the potential energy well. When this happens, we say that the nucleus *alpha decays*, meaning that it emits an alpha particle and transmutes into another nucleus. The details of this decay process will be covered in Chapter 40.

EXAMPLE 37.3 | Neutron Tunneling

PROBLEM

If a neutron of kinetic energy 22.4 MeV encounters a rectangular potential energy barrier of height 36.2 MeV and width 8.40 fm, what is the probability that the neutron will be able to tunnel through this barrier?

SOLUTION

The tunneling probability was given in equation 37.23 as $\Pi_t = T = e^{-2\gamma(b-a)}$. Knowing that the width of the barrier is $b - a = 8.40$ fm, all we have to do to solve this problem is to figure out the value of the decay constant γ. Equation 37.19 applies in this case, and we can write

$$\gamma = \sqrt{\frac{2m(U_1 - E)}{\hbar^2}}.$$

In order to put in numbers, we can express $\hbar$ in units that are very useful in nuclear and particle physics: $\hbar c = 197.33$ MeV fm $\Leftrightarrow \hbar = 197.33$ MeV fm/c. The mass of the neutron has the value $m_n = 1.6749 \cdot 10^{-27}$ kg $= 939.57$ MeV/c^2. Inserting the numerical values, we find for the decay constant:

$$\gamma = \sqrt{\frac{2(939.57 \text{ MeV}/c^2)(36.2 \text{ MeV} - 22.4 \text{ MeV})}{(197.33 \text{ MeV fm}/c)^2}} = 0.816 \text{ fm}^{-1}.$$

Therefore, we obtain for the tunneling probability:

$$\Pi_t = e^{-2\gamma(b-a)} = e^{-2(0.816)(8.40)} = 1.11 \cdot 10^{-6}.$$

Scanning Tunneling Microscope

In 1981, the Swiss physicist Heinrich Rohrer (1933–) and the German physicist Gerd Binnig (1947–) discovered that the tunneling effect can be used to image surfaces of materials, and, for the first time, they obtained images of atoms. They were awarded half of the 1986 Nobel Prize in Physics for this discovery. Their work was a conceptual leap and a great technical achievement at the time, but the underlying basic physics is actually relatively straightforward to understand applying the concepts we have developed so far.

Figure 37.16a plots the potential of an electron in an atom in black. Chapter 23 showed that this potential is the Coulomb potential, $U(r) \propto 1/r$. This potential is represented as a function of one spatial coordinate, and the atom sits at x_0. The blue curve is a qualitative sketch of the absolute square of the electron wave function in this potential. (Chapter 38 will show exact calculations for the hydrogen atom.) In Figure 37.16b–d, a second atom, located at x_1, is moved closer and closer to the atom located at x_0. As we move the second atom, we monitor the resulting potential distribution, as well as the corresponding wave function. The potential barrier that prevents the electron from moving from the atom at x_0 to the atom at x_1 becomes narrower and at the same time lower as the atoms are moved closer to each other. Thus, the electron wave function is able to tunnel through the barrier. Quantitative calculations show that the tunnel current rises very steeply if the distance between the two atoms decreases below a certain point.

The technique used by the **scanning tunneling microscope (STM)** is to very carefully move a tip the size of a single atom closer and closer to the surface of the material that we want to probe and to record the current due to tunneling electrons (Figure 37.17). How is it possible to produce a tip the size of a single atom? To start with, a very thin wire is cut at an angle. This produces a very fine tip, but not the size of a single atom. However, one of the

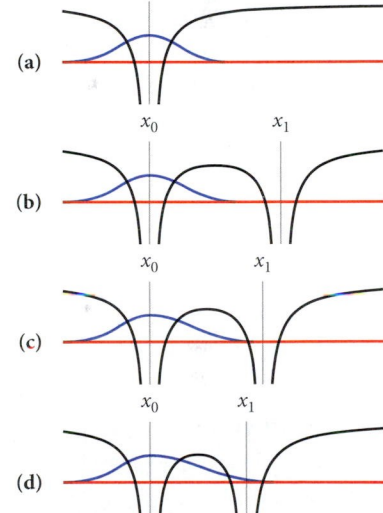

FIGURE 37.16 Black curves: potential of an electron in an atom as a function of the spatial coordinate; blue curve: sketch of the corresponding electron probability distribution. (a) Single isolated atom; (b, c, d) two atoms at various relative distances.

FIGURE 37.17 (a) Scanning the surface with a single-atom tip; (b) resulting image of the surface of a platinum sample.

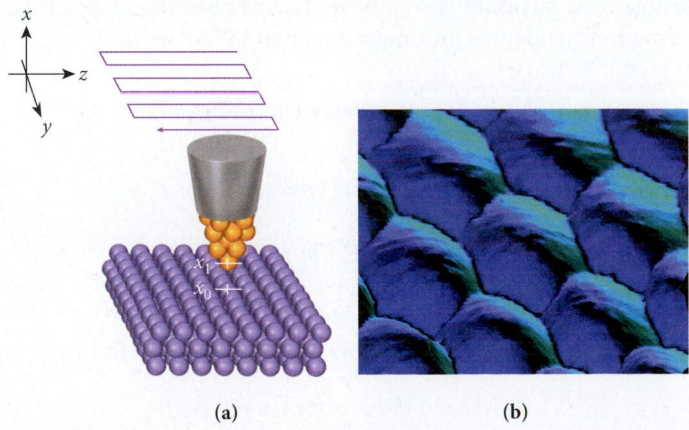

(a)   (b)

atoms at the end will usually stick out just a tiny bit farther than those surrounding it. This tiny bit is sufficient, because the tunnel current depends sensitively on the distance. Thus, this tip acts like a single atom.

The tip is moved up and down in a feedback loop that attempts to keep the measured tunnel current at a constant value. Then the tip is guided over the surface in a scanning path as shown by the purple line in Figure 37.17a. This process produces images of surfaces at an atomic resolution. For example, the surface of a piece of platinum is shown in Figure 37.17b.

37.5 Harmonic Oscillator

The three most commonly used potentials in quantum mechanical calculations are the potential well, the central $1/r$ potential, and the oscillator potential. Chapter 38 will address the central $1/r$ potential, and Sections 37.3 and 37.4 extensively investigated the potential well. This leaves only the oscillator potential.

Classical Harmonic Oscillator

Chapter 14, which covered harmonic oscillations, showed that oscillators occur in many physical situations. Now the question arises: What is the quantum representation of a harmonic oscillator? That is, we want to know the possible wave functions and energies that correspond to the following potential energy function for a harmonic oscillator:

$$U(x) = \tfrac{1}{2}kx^2 = \tfrac{1}{2}m\omega_0^2 x^2.$$

This potential energy function is plotted in Figure 37.18. Here k is the *spring constant* and is measured in units of N/m, while ω_0 is the angular frequency, $\omega_0 = \sqrt{k/m}$. (*Note:* In this chapter, we use several similar symbols: k is the symbol for the spring constant, κ is used for the wave number, K is the value of the kinetic energy, and $\hat{\mathbf{K}}$ is the kinetic energy operator. It is easy to confuse them, so be extra careful!)

Before we go on, let's first review the situation for a classical particle that is a harmonic oscillator in a potential well—for example, a mass on a spring or a pendulum. For a given total energy E, the kinetic energy of this particle can be calculated by taking the difference between the total energy and the potential energy: $K(x) = E - U(x)$. Keeping in mind that kinetic energy cannot assume negative values, we then obtain a region in coordinate space in which a particle with a given total energy is allowed to be, labeled "classically allowed" in Figure 37.19.

The points at which the kinetic energy (blue curve in Figure 37.19) reaches the value zero form the boundary of this classically allowed region. These points are called *classical turning points* because a classical oscillator "turns around" here. Outside the turning points lies the classically forbidden region (shaded region in Figure 37.19), into which a classical particle with total mechanical energy E is never able to penetrate.

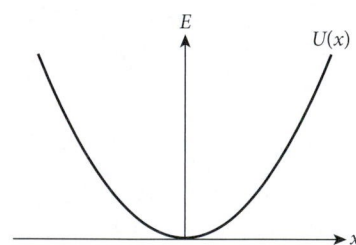

FIGURE 37.18 Potential energy as a function of position for a harmonic oscillator.

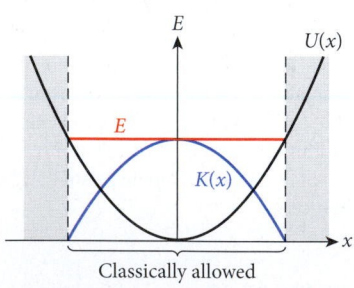

FIGURE 37.19 Classically allowed region for a particle in an oscillator potential.

Quantum Harmonic Oscillator

Now let's investigate the quantum harmonic oscillator. Inserting its potential energy function into the time-independent Schrödinger equation (37.9), we find

$$-\frac{\hbar^2}{2m}\frac{d^2\psi(x)}{dx^2} + \tfrac{1}{2}m\omega_0^2 x^2 \psi(x) = E\psi(x) \Rightarrow$$

$$\frac{d^2\psi(x)}{dx^2} = -\frac{2m}{\hbar^2}\left(E - \tfrac{1}{2}m\omega_0^2 x^2\right)\psi(x) \Rightarrow \tag{37.24}$$

$$\frac{d^2\psi(x)}{dx^2} = -\left(\frac{2mE}{\hbar^2} - \frac{m^2\omega_0^2}{\hbar^2}x^2\right)\psi(x).$$

Again we find that the energy can assume only discrete values. In this case, they are

$$E_n = \left(n + \tfrac{1}{2}\right)\hbar\omega_0 \quad (\text{for } n = 0,1,2,\dots). \tag{37.25}$$

The wave functions have the general form of Gaussians $\left(e^{-ax^2}\right)$ multiplied by special polynomials, called *Hermite polynomials*. The wave functions corresponding to the lowest values of the energy quantum number, n, are

$$\psi_0(x) = \frac{1}{\sqrt{\sigma}\,\pi^{1/4}} e^{-x^2/2\sigma^2}$$

$$\psi_1(x) = \frac{1}{\sqrt{\sigma}\,\pi^{1/4}} \frac{1}{\sqrt{2}} \left(2\frac{x}{\sigma}\right) e^{-x^2/2\sigma^2}$$

$$\psi_2(x) = \frac{1}{\sqrt{\sigma}\,\pi^{1/4}} \frac{1}{\sqrt{8}} \left(4\frac{x^2}{\sigma^2} - 2\right) e^{-x^2/2\sigma^2} \qquad (37.26)$$

$$\psi_3(x) = \frac{1}{\sqrt{\sigma}\,\pi^{1/4}} \frac{1}{\sqrt{48}} \left(8\frac{x^3}{\sigma^3} - 12\frac{x}{\sigma}\right) e^{-x^2/2\sigma^2}$$

$$\psi_4(x) = \frac{1}{\sqrt{\sigma}\,\pi^{1/4}} \frac{1}{\sqrt{384}} \left(16\frac{x^4}{\sigma^4} - 48\frac{x^2}{\sigma^2} + 12\right) e^{-x^2/2\sigma^2},$$

where the constant σ is defined as

$$\sigma = \sqrt{\frac{\hbar}{m\omega_0}}. \qquad (37.27)$$

This constant has the physical dimension of length and is the half-width of the Gaussian. The general form of the wave function for the harmonic oscillator can be written as

$$\psi_n(x) = \frac{1}{\sqrt{\sigma}\,\pi^{1/4}} \frac{1}{\sqrt{n!\,2^n}} H_n(x/\sigma) e^{-x^2/2\sigma^2}. \qquad (37.28)$$

The Hermite polynomial $H_n(x)$ is a polynomial of rank n; that is, the highest power of x is x^n. This polynomial can be defined in terms of the derivatives of the Gaussian function:

$$H_n(x) = (-1)^n e^{x^2} \frac{d^n}{dx^n} e^{-x^2}.$$

DERIVATION 37.1 | Oscillator Wave Function and Energy

The entire derivation of the complete solution (equation 37.28) is somewhat lengthy and will not be shown here. Still, it is instructive to verify that one of the solutions given in equation 37.26 satisfies equation 37.24. We show as an example that $\psi_1(x)$ is indeed a solution to the time-independent Schrödinger equation (37.24) with energy $E_1 = \left(\frac{1}{2} + 1\right)\hbar\omega_0 = \frac{3}{2}\hbar\omega_0$.

We start by taking the second derivative of $\psi_1(x)$ with respect to x. To do this more efficiently, we rewrite the wave function as

$$\psi_1(x) = \frac{1}{\sqrt{\sigma}\,\pi^{1/4}} \frac{1}{\sqrt{2}} \left(2\frac{x}{\sigma}\right) e^{-x^2/2\sigma^2} = Axe^{-x^2/2\sigma^2}.$$

Here A is just a normalization constant, so the first and second derivatives of this function are

$$\frac{d}{dx}\psi_1(x) = Ae^{-x^2/2\sigma^2} - A\frac{x^2}{\sigma^2} e^{-x^2/2\sigma^2}$$

$$\frac{d^2}{dx^2}\psi_1(x) = -A\frac{3x}{\sigma^2} e^{-x^2/2\sigma^2} + A\frac{x^3}{\sigma^4} e^{-x^2/2\sigma^2}.$$

Inserting this second derivative into the time-independent Schrödinger equation results in

$$\frac{d^2\psi(x)}{dx^2} = -\left(\frac{2mE}{\hbar^2} - \frac{m^2\omega_0^2}{\hbar^2}x^2\right)\psi(x) \Rightarrow$$

$$-A\frac{3x}{\sigma^2} e^{-x^2/2\sigma^2} + A\frac{x^3}{\sigma^4} e^{-x^2/2\sigma^2} = -\left(\frac{2mE}{\hbar^2} - \frac{m^2\omega_0^2}{\hbar^2}x^2\right)Axe^{-x^2/2\sigma^2}$$

$$= -\left(\frac{2mE}{\hbar^2}x - \frac{m^2\omega_0^2}{\hbar^2}x^3\right)Ae^{-x^2/2\sigma^2}.$$

– Continued

Now we factor out the common term $Ae^{-x^2/2\sigma^2}$ and sort the terms by powers of x:

$$Ae^{-x^2/2\sigma^2}\left(\left(\frac{2mE}{\hbar^2}-\frac{3}{\sigma^2}\right)x+\left(\frac{1}{\sigma^4}-\frac{m^2\omega_0^2}{\hbar^2}\right)x^3\right)=0. \tag{i}$$

Because $Ae^{-x^2/2\sigma^2}\neq0$, equation (i) can be fulfilled for all x only if the coefficients of x and x^3 are both equal to 0. For the coefficient of x^3, we find

$$\frac{1}{\sigma^4}-\frac{m^2\omega_0^2}{\hbar^2}=0\Rightarrow\sigma=\sqrt{\frac{\hbar}{m\omega_0}}.$$

This confirms equation 37.27. We further find for the coefficient of x in equation (i):

$$\frac{2mE}{\hbar^2}-\frac{3}{\sigma^2}=0\Rightarrow E=\tfrac{3}{2}\frac{\hbar^2}{m\sigma^2}.$$

Using the result of equation 37.27, which we just derived, we then find

$$E=\tfrac{3}{2}\frac{\hbar^2}{m\sigma^2}=\tfrac{3}{2}\frac{\hbar^2}{m(\hbar/m\omega_0)}=\tfrac{3}{2}\hbar\omega_0.$$

Thus, we have shown that $\psi_1(x)$ as defined in equation 37.26 is indeed a solution to the time-independent Schrödinger equation with energy E_1 as given in equation 37.25.

It is important to note that the possible energy values for a one-dimensional harmonic oscillator are evenly spaced, with a constant energy difference of $\hbar\omega_0$ between neighboring energy levels. A wide variety of condensed matter systems and other systems studied in atomic and nuclear physics exhibit discrete energy spectra with constant spacing between the levels.

The wave functions of the quantum harmonic oscillator given in equation 37.26 are plotted in Figure 37.20. To avoid possible confusion, notice that two different plots using different vertical axes are superimposed on each other, with a common horizontal x-axis. The black parabola shows the potential energy as a function of x, and the red horizontal lines show the possible total energies that correspond to solutions of the time-independent Schrödinger equation. The vertical axes for both plots are in units of energy. The intersections of the red total-energy lines with the potential-energy parabola mark the classical turning points and are indicated by the short vertical green lines. However, the blue lines show the wave functions and do not have units of energy, and thus correspond to a different vertical scale. For each wave function, the $\psi=0$ line is adjusted to fall on the horizontal line that marks the function's lowest energy value. This provides an elegant way of displaying the wave functions, their classical turning points, and the corresponding energy values, all in the same plot. We can quantitatively show how each wave function "leaks" into the classically forbidden region.

Figure 37.21 shows the probability distributions that correspond to the wave functions. Again, they are simply the absolute squares of the wave functions. Note that the higher the

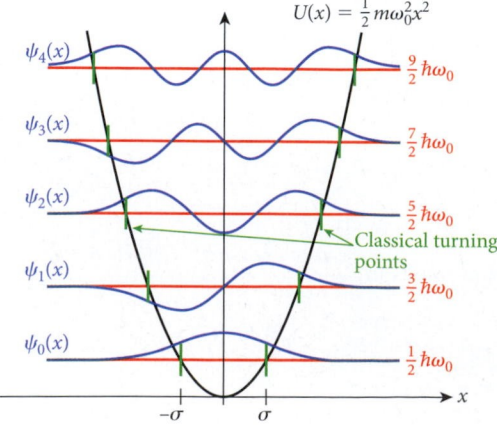

FIGURE 37.20 Wave functions (blue) corresponding to the five lowest energy values (red) of the harmonic oscillator potential (black). The classical turning points for each wave function are indicated by short vertical green lines.

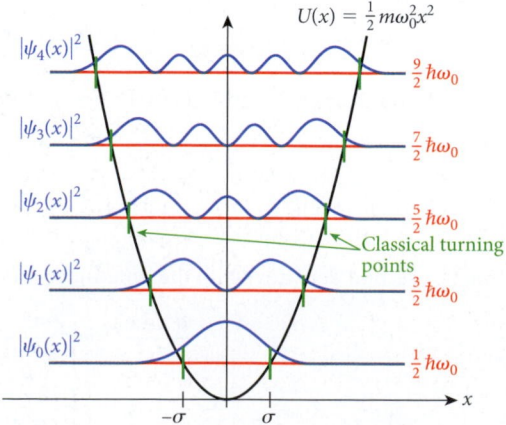

FIGURE 37.21 Probability distributions for finding an electron in a harmonic oscillator potential with a particular energy at a particular point in space.

quantum number n, the smaller the penetration range of the corresponding wave function beyond the classical turning point. This result can be understood by looking at the shape of the potential $U(x)$: For higher values of the energy E, the potential becomes steeper at the positions of the classical turning points for that energy.

SOLVED PROBLEM 37.2 | Half-Oscillator

PROBLEM
Suppose you know that an electron is confined to the following potential, where a is a constant:

$$U(x) = \begin{cases} a^2 x^2 & \text{for } x > 0 \\ \infty & \text{for } x \leq 0. \end{cases}$$

What is the energy of the first excited state, if the ground-state energy of the electron in this potential has a value of 3.5 eV?

SOLUTION

THINK The potential here is a mixture of an infinite potential well, where $U(x) = \infty$ outside certain regions in space, and a harmonic oscillator potential, $U(x) = \frac{1}{2}m\omega_0^2 x^2$, with the constant $a = \sqrt{\frac{1}{2}m\omega_0^2}$. The key to solving this problem is that the only constraint that an infinite potential wall imposes on a wave function is that it has to be zero at the interface between the infinite and the finite part of the potential function. In this case, this requirement can be written as $\psi(x = 0) = 0$.

SKETCH Figure 37.22 shows the shape of the potential function given in the problem statement.

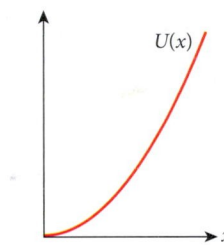

FIGURE 37.22 Half-oscillator potential.

RESEARCH We have already established that the wave functions on the left side ($x \leq 0$) have to be 0. On the right side, the wave functions need to be those of a harmonic oscillator as given in equation 37.26, where the lowest-energy wave functions in the harmonic oscillator potential are

$$\psi_0(x) = \frac{1}{\sqrt{\sigma}\pi^{1/4}} e^{-x^2/2\sigma^2}$$

$$\psi_1(x) = \frac{1}{\sqrt{\sigma}\pi^{1/4}} \frac{1}{\sqrt{2}} \left(2\frac{x}{\sigma}\right) e^{-x^2/2\sigma^2}$$

$$\psi_2(x) = \frac{1}{\sqrt{\sigma}\pi^{1/4}} \frac{1}{\sqrt{8}} \left(4\frac{x^2}{\sigma^2} - 2\right) e^{-x^2/2\sigma^2}$$

$$\psi_3(x) = \frac{1}{\sqrt{\sigma}\pi^{1/4}} \frac{1}{\sqrt{48}} \left(8\frac{x^3}{\sigma^3} - 12\frac{x}{\sigma}\right) e^{-x^2/2\sigma^2}$$

$$\psi_4(x) = \frac{1}{\sqrt{\sigma}\pi^{1/4}} \frac{1}{\sqrt{384}} \left(16\frac{x^4}{\sigma^4} - 48\frac{x^2}{\sigma^2} + 12\right) e^{-x^2/2\sigma^2}$$

with the constant σ defined as $\sigma = \sqrt{\hbar/m\omega_0}$, and the energies corresponding to these wave functions are

$$E_n = \left(n + \tfrac{1}{2}\right)\hbar\omega_0 \quad \text{(for } n = 0,1,2,...\text{)}.$$

SIMPLIFY Clearly, the wave functions $\psi_0(x)$, $\psi_2(x)$, $\psi_4(x)$, and all others with even n have nonzero values for $x = 0$ and thus do not qualify as solutions for the half-oscillator. These are the even-parity wave functions with the property $\psi(-x) = \psi(x)$. However, the odd-parity wave functions, $\psi_1(x)$, $\psi_3(x)$, and all others with odd n, have the required property that $\psi(x = 0) = 0$. In general, odd-parity wave functions have the property $\psi(-x) = -\psi(x)$ and thus must vanish at the origin in order to be continuous.

This means that the ground state for this electron must have the wave function

$$\psi_{gs}(x) = \frac{1}{\sqrt{\sigma}\pi^{1/4}} \frac{1}{\sqrt{2}} \left(2\frac{x}{\sigma}\right) e^{-x^2/2\sigma^2}$$

with energy

$$E_{gs} = (1 + \tfrac{1}{2})\hbar\omega_0 = \tfrac{3}{2}\hbar\omega_0, \qquad \text{– Continued}$$

and the first excited state must have the wave function

$$\psi_{ex}(x) = \frac{1}{\sqrt{\sigma}\pi^{1/4}} \frac{1}{\sqrt{48}}\left(8\frac{x^3}{\sigma^3} - 12\frac{x}{\sigma}\right)e^{-x^2/2\sigma^2}$$

with energy

$$E_{ex} = (3+\tfrac{1}{2})\hbar\omega_0 = \tfrac{7}{2}\hbar\omega_0.$$

Thus, the ratio of the energy of the first excited state to that of the ground state is

$$\frac{E_{ex}}{E_{gs}} = \frac{\tfrac{7}{2}\hbar\omega_0}{\tfrac{3}{2}\hbar\omega_0} = \frac{7}{3}.$$

CALCULATE The energy of the ground state was given in the problem statement as 3.5 eV; so we find that

$$E_{ex} = \tfrac{7}{3}E_{gs} = \tfrac{7}{3}\cdot 3.5 \text{ eV} = 8.1666666 \text{ eV}.$$

ROUND We round our answer to two significant digits:

$$E_{ex} = 8.2 \text{ eV}.$$

DOUBLE-CHECK For oscillator wave functions, it is characteristic that the energy difference between neighboring energy levels is constant. This is also the case for this half-oscillator, for which only the odd-parity wave functions are valid solutions. If we calculated the energy of the second excited state, we would find a value of $\tfrac{11}{3}E_{gs} = 12.8$ eV.

37.6 Wave Functions and Measurements

In Sections 37.3, 37.4, and 37.5, we calculated the wave functions for particular quantum systems. While we were able to obtain analytic solutions in the cases we considered, numerical calculations performed with computers are necessary to obtain the wave functions for almost all potentials representing actual physical systems. However, the common features we have identified so far characterize all single-particle wave functions: a discrete number of nodes, associations with particular values of energy, and the fact that the probability of finding a particle somewhere is proportional to the absolute square of the wave function.

In experimental situations, the probability distribution can usually be measured. However, a complete reconstruction of the wave function is generally impossible, because the phase is usually unknown and not knowable. This fact was illustrated by our selection of the value of 0 for the phase angle when finding the wave function of a particle confined to an infinite potential well (equation 37.12). Reconstructing quantum wave functions with the aid of additional constraints is a research topic of great current interest. As an example of a result of such studies, Figure 37.23 displays the quantum wave function for a particular experimental system obtained recently by Jeff Lundeen and colleagues at the Institute for National Measurement Standards, Ottawa, Canada.

Suppose we obtain a quantum wave function for a particular situation. How can this wave function be used to gain information about the physical properties of the object with which it is associated? How can the object's position, velocity, momentum, kinetic energy, and so on be calculated?

Equation 37.2 introduced the normalization condition for the wave function:

$$\int_{-\infty}^{\infty} |\psi(x)|^2 \, dx = \int_{-\infty}^{\infty} \psi^*(x)\psi(x)\,dx = 1.$$

To measure the average value of any observable quantity, we apply an operator that corresponds to this operation to the wave function ψ (for example, $\hat{\mathbf{p}}\psi = p\psi$). Then we multiply the result by ψ^* (to get the probability $\psi^*\psi$ in the integrand) and integrate over all of space. For example, to figure out the average momentum associated with a wave function, we apply the momentum operator (equation 37.6) to the wave function and then integrate:

$$\langle p \rangle = \int_{-\infty}^{\infty} \psi^*(x)\hat{\mathbf{p}}\psi(x)dx = -i\hbar \int_{-\infty}^{\infty} \psi^*(x)\frac{d}{dx}\psi(x)dx. \qquad (37.29)$$

FIGURE 37.23 The wave function obtained in a quantum optics experiment artistically superimposed on the optical bench elements used in the experiment. (*Source:* Jeff Lundeen and Charles Bamber, *Nature* 2011.)

In a similar way, any quantity that is a function of the momentum can be calculated using integrals that involve the derivative of the wave function. For example, the average kinetic energy can be obtained as follows:

$$\langle K \rangle = \frac{1}{2m}\langle p^2 \rangle = \frac{1}{2m}\int_{-\infty}^{\infty}\psi^*(x)\hat{\mathbf{p}}^2\psi(x)dx = -\frac{\hbar^2}{2m}\int_{-\infty}^{\infty}\psi^*(x)\frac{d^2}{dx^2}\psi(x)dx. \qquad (37.30)$$

It is somewhat more straightforward to find the average values for quantities that depend on the position x. To find the average position, we can evaluate this integral:

$$\langle x \rangle = \int_{-\infty}^{\infty}\psi^*(x)x\psi(x)dx. \qquad (37.31)$$

The average values denoted with angle brackets $\langle\ \rangle$ are often referred to as **expectation values,** that is, the expected outcomes of measurements. This description of the formal process of finding average values is rather abstract, so let's look at an example to see how actual numbers are obtained.

EXAMPLE 37.4   Position and Energy

PROBLEM
What is the average position of an electron (mass $m = 9.109 \cdot 10^{-31}$ kg $= 511$ keV/c^2) in the $n = 0$ wave function of the harmonic oscillator potential with an angular frequency constant of $\omega_0 = 1.00 \cdot 10^{16}$ s^{-1}? What is the electron's average kinetic energy?

SOLUTION
According to equation 37.26, the solution to the time-independent Schrödinger equation for this harmonic oscillator potential is

$$\psi_0(x) = Ae^{-x^2/2\sigma^2} = \frac{1}{\sqrt{\sigma}\pi^{1/4}}e^{-x^2/2\sigma^2}.$$

The electron's mass and the oscillator's angular frequency enter into the width of the potential, $\sigma = \sqrt{\hbar/m\omega_0}$. Here we again use the normalization constant $A = 1/(\sqrt{\sigma}\pi^{1/4})$ to simplify the writing of the integrals.

First, we calculate the average position using equation 37.31:

$$\langle x \rangle = \int_{-\infty}^{\infty}\psi^*(x)x\psi(x)dx$$

$$= \int_{-\infty}^{\infty}Ae^{-x^2/2\sigma^2}xAe^{-x^2/2\sigma^2}dx \qquad (i)$$

$$= A^2\int_{-\infty}^{\infty}xe^{-x^2/\sigma^2}dx$$

$$= 0.$$

These steps deserve an explanation. First, when we inserted the wave function, we used the fact that it is real, with no imaginary part, so in this case $\psi^*(x) = \psi(x)$. In the next step, we removed the constant factor A^2 from the integral and multiplied the two Gaussians by adding their arguments. Finally, we arrived at a value of 0 for the integral by realizing that x is an odd function and e^{-x^2/σ^2} is an even function of x; therefore, their product must be an odd function of x, resulting in an integral of zero value when the integration limits are symmetrical.

In fact, we could have looked at Figure 37.21 and immediately realized this answer from the fact that the probability distribution $|\psi_0(x)|^2$ is symmetric with respect to 0. However, because the integral in equation (i) is the first of several similar ones, it is instructive to go through all the steps explicitly to confirm that we indeed find the expected solution.

– Continued

Now, let's address the more complicated problem of finding the electron's average kinetic energy. According to equation 37.30, we have to solve

$$\langle K \rangle = -\frac{\hbar^2}{2m} \int_{-\infty}^{\infty} \psi^*(x) \frac{d^2}{dx^2} \psi(x) dx$$

(ii)

$$= -\frac{\hbar^2}{2m} \int_{-\infty}^{\infty} A e^{-x^2/2\sigma^2} \frac{d^2}{dx^2} A e^{-x^2/2\sigma^2} dx.$$

Taking the first and second derivatives of the oscillator wave function for $n = 0$ yields

$$\frac{d}{dx} A e^{-x^2/2\sigma^2} = -A \frac{x}{\sigma^2} e^{-x^2/2\sigma^2}$$

$$\frac{d^2}{dx^2} A e^{-x^2/2\sigma^2} = A \left(\frac{x^2}{\sigma^4} - \frac{1}{\sigma^2} \right) e^{-x^2/2\sigma^2}.$$

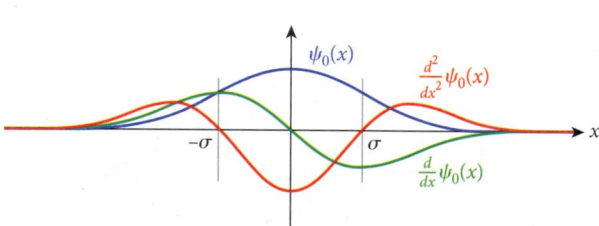

These derivatives are plotted in Figure 37.24, where the classical turning points are also shown. For the quantum wave function, these are the points where the second derivative of the wave function has a value of 0. (We can see from equation 37.9 that if $E = U$, then $\frac{d^2}{dx^2} \psi(x) = 0$.)

FIGURE 37.24 The $n = 0$ oscillator wave function (blue) and its first (green) and second (red) derivatives as a function of position. The gray vertical lines mark the classical turning point.

Now we can proceed with the integration in equation (ii):

$$\langle K \rangle = -\frac{A^2 \hbar^2}{2m} \int_{-\infty}^{\infty} e^{-x^2/2\sigma^2} \frac{d^2}{dx^2} e^{-x^2/2\sigma^2} dx$$

$$= -\frac{A^2 \hbar^2}{2m} \int_{-\infty}^{\infty} e^{-x^2/2\sigma^2} \left(\frac{x^2}{\sigma^4} - \frac{1}{\sigma^2} \right) e^{-x^2/2\sigma^2} dx$$

$$= -\frac{A^2 \hbar^2}{2m} \int_{-\infty}^{\infty} e^{-x^2/\sigma^2} \left(\frac{x^2}{\sigma^4} - \frac{1}{\sigma^2} \right) dx$$

$$= -\frac{A^2 \hbar^2}{2m\sigma^4} \int_{-\infty}^{\infty} x^2 e^{-x^2/\sigma^2} dx + \frac{A^2 \hbar^2}{2m\sigma^2} \int_{-\infty}^{\infty} e^{-x^2/\sigma^2} dx$$

$$= -\frac{A^2 \hbar^2}{2m\sigma^4} \frac{1}{2} \sqrt{\pi} \sigma^3 + \frac{A^2 \hbar^2}{2m\sigma^2} \sqrt{\pi} \sigma$$

$$= \frac{\hbar^2}{4m\sigma^2}.$$

In the last step, we used $A^2 = 1/(\sigma \sqrt{\pi})$, which follows from the definition of A. We are almost done. Using $\sigma = \sqrt{\hbar/m\omega_0}$ (see equation 37.27), we find

$$\langle K \rangle = \frac{\hbar^2}{4m(\hbar/m\omega_0)} = \frac{1}{4} \hbar \omega_0.$$

Finally, inserting the given value for ω_0 gives us our numerical answer:

$$\langle K \rangle = \frac{1}{4} \hbar \omega_0 = \frac{1}{4}(1.055 \cdot 10^{-34} \text{ J s})(1.00 \cdot 10^{16} \text{ s}^{-1})$$

$$= 2.64 \cdot 10^{-19} \text{ J}$$

$$= 1.65 \text{ eV}.$$

DISCUSSION

Our result amounts to exactly half of the total energy of the oscillator in the $n = 0$ state! Also, note that this answer depends only on the value of the angular frequency, ω_0, and not on the mass, m. Another particle with a different mass, trapped in the same oscillator potential, would have the same average kinetic energy in the $n = 0$ state.

Self-Test Opportunity 37.5

Calculate the average value of the electron's momentum in the $n = 0$ state, or simply state the result as a result of symmetry arguments.

Self-Test Opportunity 37.6

Calculate the average value of the electron's kinetic energy in the $n = 0,1,2,3$ states of the harmonic oscillator. Is there a simple shortcut, or do you have to evaluate the integral for a new wave function each time?

Uncertainty Relationship for Oscillator Wave Functions

Chapter 36 explored the fundamental importance of the Heisenberg Uncertainty Relation, $\Delta x \cdot \Delta p \geq \frac{1}{2}\hbar$. There we were able to motivate the relationship only by examining the gamma-ray microscope that Heisenberg envisioned as a *Gedankenexperiment* (thought experiment) in his original 1927 paper on the uncertainty relation. The same paper also contained a much more general mathematical proof. We will not reconstruct this proof here. However, we can calculate the uncertainties in momentum and position for the oscillator wave functions that we have found. Let's do this for the $n = 0$ state.

The square of the uncertainty in position is $(\Delta x)^2 = \left\langle \left(x - \langle x \rangle \right)^2 \right\rangle$. We just found for the $n = 0$ state that $\langle x \rangle = 0$. In this case (and only when $\langle x \rangle = 0$), we find $(\Delta x)^2 = \langle x^2 \rangle$. Thus, to find the uncertainty in position, we have to evaluate the following integral:

$$(\Delta x)^2 = \langle x^2 \rangle = \int_{-\infty}^{\infty} A e^{-x^2/2\sigma^2} x^2 A e^{-x^2/2\sigma^2} \, dx$$

$$= A^2 \int_{-\infty}^{\infty} x^2 e^{-x^2/\sigma^2} \, dx$$

$$= \frac{1}{2} A^2 \sqrt{\pi} \sigma^3 = \frac{1}{2}\sigma^2 \Rightarrow$$

$$\Delta x = \frac{\sigma}{\sqrt{2}}.$$

For this wave function, the uncertainty in momentum is $(\Delta p)^2 = \left\langle \left(p - \langle p \rangle \right)^2 \right\rangle = \langle p^2 \rangle$, because the average momentum is also zero. Furthermore, the average kinetic energy for this wave function, $\langle K \rangle = \langle p^2 \rangle / 2m$, was calculated in Example 37.4. Thus, we find that

$$(\Delta p)^2 = \langle p^2 \rangle = 2m \langle K \rangle = 2m \frac{\hbar^2}{4m\sigma^2} = \frac{\hbar^2}{2\sigma^2} \Rightarrow$$

$$\Delta p = \frac{\hbar}{\sqrt{2}\sigma}.$$

Multiplying the uncertainties in position and momentum for the harmonic oscillator in the $n = 0$ state gives

$$\Delta x \cdot \Delta p = \frac{\sigma}{\sqrt{2}} \frac{\hbar}{\sqrt{2}\sigma} = \frac{1}{2}\hbar.$$

This means that the $n = 0$ state of the harmonic oscillator is a state with the physically minimum possible uncertainty. Also, note that this result is independent of the width σ and thus does not depend on the angular frequency or the mass. It can also be shown that the product of the uncertainties in position and momentum for an oscillator wave function with quantum number n is given by $\Delta x \cdot \Delta p = (\frac{1}{2} + n)\hbar$.

37.7 Correspondence Principle

When we investigated relativistic mechanics, we found a smooth transition to nonrelativistic (Newtonian) mechanics as an object's speed becomes small compared to the speed of light. In the limit $v \ll c$, the classical mechanical case is recovered from the relativistic description. In a similar way, we can ask if we can recover the classical mechanical case from quantum mechanics, and in what limit. To address this question, we examine the probability distribution for finding an electron in an oscillator potential at a particular point in coordinate space; then we compare the results for the classical and quantum oscillators.

Just as we did for the classical particle in a box, we can calculate the classical probability distribution for finding a particle with total energy E in an oscillator potential.

Equation 37.13 states that the classical probability of finding a particle at some point in coordinate space is inversely proportional to the particle's speed at that point. The speed can be determined as a function of position for the harmonic oscillator potential using energy conservation:

$$E = K + U = \frac{1}{2}mv^2 + \frac{1}{2}m\omega_0^2 x^2.$$

Solving this equation for the speed gives

$$v = \sqrt{\frac{2E}{m} - x^2\omega_0^2}.$$

Thus, the classical probability distribution of a particle in a harmonic oscillator potential can be written as follows:

$$\Pi_c(x) = \frac{\omega_0}{\pi\sqrt{\dfrac{2E}{m} - x^2\omega_0^2}}.$$

This function is defined only between the two classical turning points located at

$$x_t = \pm\sqrt{\frac{2E}{m\omega_0^2}}.$$

(The factors of ω_0 in the numerator and π in the denominator of the probability distribution function ensure that its integral equals 1.) Figure 37.25 shows the resulting probability distribution, with the classical turning points marked by the two green vertical lines.

The quantum probability distributions in Figure 37.21, however, do not seem to agree at all with the classical probability distribution in Figure 37.25. In the classical case, the probability of finding the particle peaks near the classical turning points. In the quantum case shown in Figure 37.21, on the other hand, the probability distribution seems flat or even peaked in the middle. From our experience with a particle in a box, we are perhaps not surprised that the quantum wave function penetrates partially beyond the classical turning points. However, it should cause some concern that the gross features of the classical and quantum probability distributions do seem not to correspond to each other at all.

This observation changes when we examine a quantum oscillator wave function for a large value of the quantum number n, as shown in Figure 37.26 for $n = 20$. For higher quantum numbers, the quantum oscillator probability distribution still oscillates and still penetrates slightly beyond the classical turning points. However, it starts to oscillate around an average value that corresponds to the classical limit.

Recall that the energy difference between neighboring allowed quantum oscillator energies is a constant, $\Delta E = \hbar\omega_0$. If this energy difference, ΔE, becomes small relative to the total energy, E, that is, if $\Delta E/E \ll 1$—which is equivalent to saying that the quantum number is large—then the quantum mechanical solution approaches the classical limit. This is a general feature of quantum mechanics, conventionally called the **correspondence principle.**

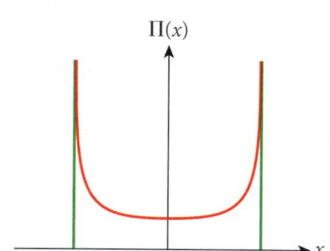

FIGURE 37.25 Classical probability distribution for finding a particle at a particular position in an oscillator potential (red curve); classical turning points (green vertical lines).

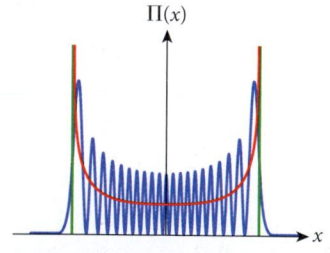

FIGURE 37.26 The classical probability distribution of Figure 37.25 with the probability distribution for the $n = 20$ quantum oscillator wave function superimposed.

37.8 Time-Dependent Schrödinger Equation

Up to now, we have concerned ourselves only with time-independent problems—that is, with probability distributions that do not change in time. However, electrons and other particles move around. How can the time dependence of the corresponding matter wave be described mathematically? Besides describing the dependence on the spatial coordinate x, the wave function also has to account for the dependence on the time t. We will still restrict the particle to motion in one spatial direction, and we will assume that the potential energy is constant in time; that is, it is a function of the coordinate x only. The time-dependent Schrödinger equation is then

$$-\frac{\hbar^2}{2m}\frac{\partial^2}{\partial x^2}\Psi(x,t)+U(x)\Psi(x,t)=i\hbar\frac{\partial}{\partial t}\Psi(x,t). \tag{37.32}$$

Here we use $\Psi(x,t)$ to denote the quantum wave function that is dependent on space and time. At this point, we are not interested in the general solutions to this partial differential equation or in deriving it from some general principles. However, we would like to see if we can find special solutions that allow us to write the general wave function as a product of two functions that each depend on only one variable:

$$\Psi(x,t)=\psi(x)\chi(t).$$

This method of solving a partial differential equation is commonly called *separation of variables*. Let's see what happens if we insert this *Ansatz* into the time-dependent Schrödinger equation (37.32):

$$-\frac{\hbar^2}{2m}\frac{\partial^2}{\partial x^2}\left(\psi(x)\chi(t)\right)+U(x)\left(\psi(x)\chi(t)\right)=i\hbar\frac{\partial}{\partial t}\left(\psi(x)\chi(t)\right)\Rightarrow$$

$$-\chi(t)\frac{\hbar^2}{2m}\frac{\partial^2}{\partial x^2}\psi(x)+\chi(t)U(x)\psi(x)=i\hbar\psi(x)\frac{\partial}{\partial t}\chi(t).$$

We divide both sides of this equation by the product $\psi(x)\chi(t)$ and find

$$\left(-\frac{\hbar^2}{2m}\frac{\partial^2}{\partial x^2}\psi(x)+U(x)\psi(x)\right)\frac{1}{\psi(x)}=\left(i\hbar\frac{\partial}{\partial t}\chi(t)\right)\frac{1}{\chi(t)}.$$

Now the left-hand side is a function of x only, and the right-hand side is a function of t only. This equality can hold for all x and t only if each side is equal to the same constant. Motivated by the time-independent Schrödinger equation (37.9), we call this constant E, the energy that entered into that equation. The left-hand side of the above equation then leads to

$$\left(-\frac{\hbar^2}{2m}\frac{d^2}{dx^2}\psi(x)+U(x)\psi(x)\right)\frac{1}{\psi(x)}=E\Rightarrow$$

$$-\frac{\hbar^2}{2m}\frac{d^2}{dx^2}\psi(x)+U(x)\psi(x)=E\psi(x),$$

which we recognize as the time-independent Schrödinger equation. With the same argument, we find for the right-hand side of the above equation

$$E=\left(i\hbar\frac{d}{dt}\chi(t)\right)\frac{1}{\chi(t)}\Rightarrow$$

$$\frac{d}{dt}\chi(t)=-\frac{i}{\hbar}E\chi(t)\Rightarrow$$

$$\chi(t)=Ae^{-iEt/\hbar}=Ae^{-i\omega t},$$

where we have again introduced the angular frequency ω via $E=\hbar\omega$. We can set the normalization constant A equal to 1, because $\left|e^{-i\omega t}\right|^2=1$, and thus the normalization condition for the wave function is fulfilled. Our overall solution is therefore

$$\Psi(x,t)=\psi(x)e^{-i\omega t}, \tag{37.33}$$

with $\psi(x)$ being a normalized solution to the time-independent Schrödinger equation with energy $E=\hbar\omega$. A wave function of the form of equation 37.33 is called a *stationary solution*, or a **stationary state,** of the time-dependent Schrödinger equation (37.32). The minimum condition for the existence of such stationary states is that the potential energy be time independent, that is, constant in time. A stationary state is a state with an exact energy if the potential energy is constant in time. This is just as in Newtonian mechanics, where the total mechanical energy E is conserved. Keep in mind that there can be many other solutions to equation 37.32, but the special class of solutions with well-defined energy can be written in the separable form of equation 37.33.

Eigenfunctions and Eigenvalues

Just as we defined the operator $\hat{\mathbf{K}}$ for the kinetic energy in equation 37.7, we can also introduce an operator $\hat{\mathbf{H}}$, which, applied to the wave function, yields the product of the total energy and the wave function:

$$\hat{\mathbf{H}}\psi(x) = E\psi(x). \tag{37.34}$$

If you compare this to the time-independent Schrödinger equation, you see that this operator can be formally written as

$$\hat{\mathbf{H}} = -\frac{\hbar^2}{2m}\frac{d^2}{dx^2} + U(x) = \hat{\mathbf{K}} + \hat{\mathbf{U}}. \tag{37.35}$$

The operator $\hat{\mathbf{H}}$ is called the **Hamiltonian operator.** In linear algebra, if it is possible to apply an operator to a function and obtain a constant times that same function, then the function is called an **eigenfunction** and the constant an **eigenvalue.** Thus, a stationary state is an eigenfunction of the Hamiltonian operator $\hat{\mathbf{H}}$ with eigenvalue E.

The Hamiltonian operator is linear: For any functions $\xi_1(x)$ and $\xi_2(x)$ and constants a_1 and a_2,

$$\hat{\mathbf{H}}\big(a_1\xi_1(x) + a_2\xi_2(x)\big) = a_1\hat{\mathbf{H}}\xi_1(x) + a_2\hat{\mathbf{H}}\xi_2(x). \tag{37.36}$$

Further, if two functions $\psi_1(x)$ and $\psi_2(x)$ are solutions with eigenvalues E_1 and E_2, then applying the Hamiltonian operator to the linear combination $\psi(x) = a_1\psi_1(x) + a_2\psi_2(x)$ yields

$$\begin{aligned}
\hat{\mathbf{H}}\psi(x) &= \hat{\mathbf{H}}\big(a_1\psi_1(x) + a_2\psi_2(x)\big) \\
&= a_1\hat{\mathbf{H}}\psi_1(x) + a_2\hat{\mathbf{H}}\psi_2(x) \\
&= a_1 E_1\psi_1(x) + a_2 E_2\psi_2(x).
\end{aligned}$$

Note that $\psi(x) = a_1\psi_1(x) + a_2\psi_2(x)$ is *not* an eigenfunction of $\hat{\mathbf{H}}$ in the general case where $E_1 \neq E_2$.

37.9 Many-Particle Wave Function

So far, we have discussed quantum mechanics only for cases in which a single particle is present. To proceed further, we have to discuss the general features of the wave function when two or more particles are present.

Two-Particle Wave Function

Let's start with two particles and assume that we know the wave function for the case where only one of these particles is present. Further, let's again restrict ourselves to the static (time-independent) case. Then the Schrödinger equation for the one-particle wave function, $\psi(x)$, is given by equation 37.9. If two particles are present in the same potential, we need to determine how each of the particles interacts with the external potential and how they interact with each other. The general case of the two particles interacting with each other is outside the scope of this book. However, considering the case in which the two particles do not interact with each other lets us gain important physical insight and is helpful in many physical situations.

First, let's think about the notation. We want to characterize the coordinate of particle 1 as x_1 and that of particle 2 as x_2. Then the single-particle wave function of particle 1 in state a is denoted as $\psi_a(x_1)$ and that of particle 2 in state b, as $\psi_b(x_2)$. For the two-particle wave function as a function of both coordinates x_1 and x_2, we use the notation $\Psi(x_1, x_2)$.

Recall the discussion of spin and statistics in Section 36.8. We now need to discuss three different cases:

Distinguishable Particles—An example is an electron and a neutron in an infinite potential well. This is the most straightforward case, because the wave function for the two-particle state is simply the product of the wave functions for the two single-particle states:

$$\Psi(x_1, x_2) = \psi_a(x_1) \cdot \psi_b(x_2).$$ (37.37)

Identical Bosons—An example is two pions (discussed in Chapters 39 and 40), one of them in state a and the other in state b. Since bosons of one species (identical bosons) are indistinguishable particles, we cannot say definitely that boson 1 is in state a and boson 2 is in state b. It is equally possible that boson 2 is in state a and boson 1 is in state b. The two-particle wave function has to be symmetric (stay the same) under exchange of the indexes of the two particles. Instead of just writing the two-particle wave function as the product of the single-particle wave functions, we have to add an exchange term, as in the following:

$$\Psi^B(x_1, x_2) = \frac{1}{\sqrt{2}}\left(\psi_a(x_1) \cdot \psi_b(x_2) + \psi_a(x_2) \cdot \psi_b(x_1)\right).$$ (37.38)

Here we use the superscript B to remind us that this is the symmetrized two-particle wave function for bosons. The factor $1/\sqrt{2}$ in front of the sum ensures that the two-particle wave function is normalized to 1.

Identical Fermions—An example is two electrons in an atom, one of them in state a and the other in state b. Fermions of one species are also indistinguishable, but they cannot occupy the same state simultaneously. The two-particle wave function thus has to be antisymmetric (change its sign) under the exchange of the indexes of the two particles, so $\Psi = 0$ when $a = b$, as we discuss below. Mathematically, this is expressed as

$$\Psi^F(x_1, x_2) = \frac{1}{\sqrt{2}}\left(\psi_a(x_1) \cdot \psi_b(x_2) - \psi_a(x_2) \cdot \psi_b(x_1)\right).$$ (37.39)

Again we use a superscript, in this case F, to remind us that this is the antisymmetrized two-particle wave function for identical fermions.

Concept Check 37.4

In equation 37.39, the term $\psi_a(x_2) \cdot \psi_b(x_1)$ has a negative sign, and the term $\psi_a(x_1) \cdot \psi_b(x_2)$ has a positive sign. How important is this sign convention?

a) These are the only signs possible for the two terms, because the second term is the exchange term and thus must have a negative sign.

b) It does not matter which term gets which sign; all that matters is that they have opposite signs. If you multiply a wave function by an overall factor of -1, you still obtain a valid wave function.

c) The sign of the exchange term is arbitrary; it could be positive or negative. But the first term must always be positive.

d) Both terms can have either sign, and each of the four sign combinations $(++,--,-+,+-)$ leads to a valid wave function.

You can see that the expressions for the two-particle wave function of bosons (equation 37.38) and fermions (equation 37.39) differ only in the sign of the exchange term, positive for bosons and negative for fermions. Note that the wave function in equation 37.39 fulfills the basic requirement of the Pauli Exclusion Principle, which states that two fermions cannot occupy the same quantum state at the same time. If we set $b = a$ in equation 37.39, we obtain

$$\Psi^F(x_1, x_2) = \frac{1}{\sqrt{2}}\left(\psi_a(x_1) \cdot \psi_a(x_2) - \psi_a(x_2) \cdot \psi_a(x_1)\right) = 0.$$

Further, if we exchange the indexes of the two fermions, we find

$$\Psi^F(x_2, x_1) = \frac{1}{\sqrt{2}}\left(\psi_a(x_2) \cdot \psi_b(x_1) - \psi_a(x_1) \cdot \psi_b(x_2)\right)$$

$$= -\frac{1}{\sqrt{2}}\left(\psi_a(x_1) \cdot \psi_b(x_2) - \psi_a(x_2) \cdot \psi_b(x_1)\right)$$

$$= -\Psi^F(x_1, x_2).$$

Thus, if we exchange the positions of the two fermions, the wave function undergoes a sign change.

This change in sign is central to several current research topics in physics: The sign change in the two-fermion wave function severely restricts the use of computer modeling for physical systems in which fermions can trade places as part of the time evolution of the system. The problem is that this sign change causes many quantities that must be calculated for the computer simulations to average out to zero, and then two quantities that must be divided by each other are often each very close to zero, which results in huge uncertainties in the numerical results. This *fermion sign problem* is universal in many-body theoretical physics, and physicists have suggested many clever approaches to try to overcome the computational limitations that it imposes, so far without success.

Can we also have a symmetric wave function in coordinate space for a two-fermion system? The answer is yes! The requirement of antisymmetrization of the two-fermion wave function means that the overall wave function has to satisfy equation 37.39. However, the wave function is the product of the spin wave function and the coordinate-space wave function. Thus, if the spin wave function is symmetric under exchange (two fermions with identical spin projection), then the coordinate-space wave function has to be antisymmetric. Also, if the spin wave function is antisymmetric (two fermions with opposite spin projection), then the coordinate-space wave function has to be symmetric under exchange, as Example 37.5 shows.

EXAMPLE 37.5 | Hydrogen Molecule

One of the most impressive examples of the importance of exchange symmetry in a wave function is covalent bonding in the hydrogen molecule. A hydrogen molecule consists of two hydrogen atoms, each consisting of one electron and one proton. In an isolated hydrogen atom, the electron is bound to the proton by the Coulomb potential, which we saw in Chapter 23 is proportional to $1/r$. Chapter 38 will show how to calculate the wave function of the electron in the hydrogen atom. For now, we need to know only that this wave function has its largest value at the origin, the position of the nucleus, and falls off exponentially in the radial direction.

When two hydrogen atoms are in close proximity to each other, they can interact by sharing both of their electrons equally. This process of sharing electrons in forming a chemical bond is called *covalent bonding*, as opposed to *ionic bonding*, where one or more electrons get pulled off one atom and then are predominantly held by the other atom.

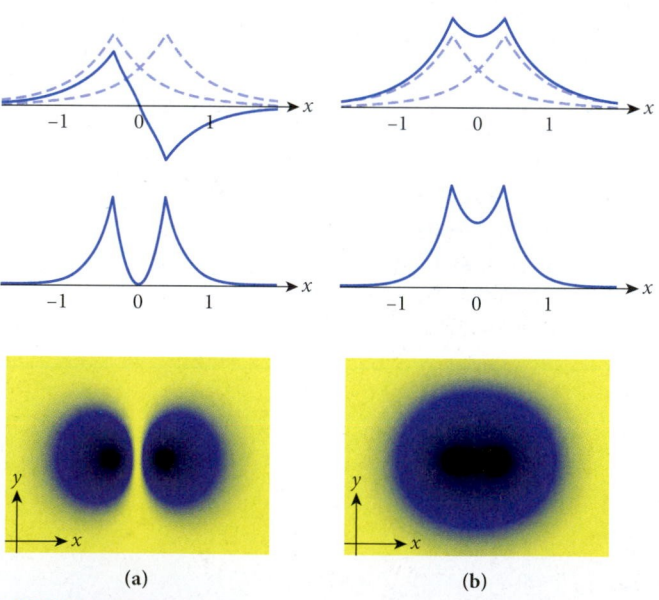

Figure 37.27a shows the antisymmetric coordinate-space two-electron wave function for the hydrogen molecule, and Figure 37.27b shows the symmetric case. The top row shows the wave functions along the axis through both nuclei, which you can compare to the two single-electron wave functions centered at ±0.37 Å, the equilibrium separation of the two nuclei in the hydrogen molecule. The middle row shows the absolute square of each wave function, which is the probability density of finding an electron at a given value of the x-coordinate. The bottom row shows the probability density of finding the electron in the xy-plane, where yellow corresponds to the lowest values and black to the highest.

For the symmetric coordinate-space wave function (Figure 37.27b), a significant portion of the probability of finding an electron is shifted to the region between the two nuclei. This means that both nuclei experience an attractive potential in the direction of the other nucleus, due to the Coulomb interaction of the nuclei with the electrons. The depth of this potential due to the exchange interaction is 4.52 eV at the equilibrium separation of 0.74 Å, leading to a net attractive force between the two atoms that produces the covalent bonding in the H_2 molecule. The condition for this bonding is that the two electrons must have opposite spin projections—one *spin-up* and one *spin-down*. If both have the same spin projection, then the coordinate-space wave function needs to be antisymmetric, as shown in Figure 37.27a. The resulting net depletion of the electron density in the region between the two nuclei leads to a repulsive potential at all separation distances and thus no bonding.

FIGURE 37.27 Coordinate-space electron wave functions in the hydrogen molecule: (a) antisymmetric wave function; (b) symmetric wave function. Shown are the two-particle wave functions (top row), probability densities (middle row) along the major axis, and density distributions in the two-dimensional plane (bottom row).

Many-Fermion Wave Function

Equation 37.39 is equivalent to the determinant of a two-by-two matrix of the single-particle wave functions:

$$\Psi^{F}(x_1, x_2) = \frac{1}{\sqrt{2}} \begin{vmatrix} \psi_a(x_1) & \psi_a(x_2) \\ \psi_b(x_1) & \psi_b(x_2) \end{vmatrix}.$$

This determinant is commonly called a **Slater determinant.** The Slater determinant can be generalized to a system consisting of many fermions. If there are n fermions, then the n-fermion wave function is the determinant of an $n \times n$ matrix, where each row is a given single-particle state and each column is a particle coordinate:

$$\Psi^F(x_1, x_2, ..., x_n) = \frac{1}{\sqrt{n!}} \begin{vmatrix} \psi_1(x_1) & \psi_1(x_2) & \cdots & \psi_1(x_n) \\ \psi_2(x_1) & \psi_2(x_2) & \cdots & \psi_2(x_n) \\ \vdots & \vdots & \ddots & \vdots \\ \psi_n(x_1) & \psi_n(x_2) & \cdots & \psi_n(x_n) \end{vmatrix}. \tag{37.40}$$

This Slater determinant contains all possible permutations of the single-particle wave functions over all particle coordinates. Again, the factor $1/\sqrt{n!}$ ensures that the many-fermion wave function is normalized—that is, that the integral over all space of the absolute square of the wave function is equal to 1 when the single-particle wave functions are properly normalized.

Quantum Computing

Currently, one of the most active research areas in quantum mechanics is the field of quantum computing. A large number of physicists around the world are using different physical systems to try to implement quantum computers. Among the many systems considered are chains of trapped atoms, quantum dots in semiconductors, electrons on the surface of liquid helium, and collections of buckyballs (C_{60} molecules shaped like soccer balls). New ideas regarding which systems to use are generated on a regular basis, and many groups are working on refining them. What all proposed quantum computers have in common is that they work with many-particle wave functions.

To understand the potential power of quantum computing, let's first look at the workings of a classical computer. A classical computer is based on algorithms that are applied to digitized information, which is stored in *bits*. A bit can be in one of two states, either on or off. Conventionally, the numbers 1 and 0 represent these two states. Note that nothing appears between 0 and 1 in today's conventional computers; it's all or nothing. A *byte* is a collection of 8 bits. Thus, a possible representation for one particular value of a byte could be 01100010, in which the 8 bits are in their on or off states as indicated by the numbers, one digit for each bit.

Now let's consider a quantum two-state system, in which transitions between states can be induced through some interaction with an outside force but do not happen spontaneously through emission or absorption of a single photon. Left alone, a particle in one of the two states should stay in that state for a long time.

Conventionally, the two quantum states are denoted as $|0\rangle$ and $|1\rangle$ to make an analogy with a classical bit that can have a value of either 0 or 1. However, a quantum mechanical system can be in a superposition of two states, as noted earlier in this chapter. Therefore, the quantum equivalent of the bit, the **qubit,** is defined as a wave function that is a superposition of the two states $|0\rangle$ and $|1\rangle$:

$$|\psi\rangle = c_1 |1\rangle + c_0 |0\rangle,$$

where c_1 and c_0 are complex numbers and represent the probabilities that the system is in state $|1\rangle$ or in state $|0\rangle$, respectively (Figure 37.28). The numbers c_1 and c_0 are not completely independent, because the requirement that the sum of the probabilities that the particle is in state $|0\rangle$ or $|1\rangle$ equals 1 necessitates $|c_1|^2 + |c_0|^2 = 1$.

Conceptually, it is straightforward to generalize from one qubit to n qubits. Assume a collection of individual two-state systems that are not independent of one another. Then the state of the overall system can be written as the sum of the combinations of the individual states. For a system of two qubits, the wave function can be written as

$$|\psi\rangle = c_{11} |11\rangle + c_{10} |10\rangle + c_{01} |01\rangle + c_{00} |00\rangle,$$

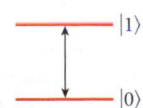

FIGURE 37.28 Simple pictorial representation of a single qubit.

where c_{00}, c_{01}, c_{10}, and c_{11} are again complex numbers, with the normalization condition that $|c_{11}|^2 + |c_{10}|^2 + |c_{01}|^2 + |c_{00}|^2 = 1$. For three qubits, we can write

$$|\psi\rangle = c_{111}|111\rangle + c_{110}|110\rangle + c_{101}|101\rangle + c_{100}|100\rangle$$
$$+ c_{011}|011\rangle + c_{010}|010\rangle + c_{001}|001\rangle + c_{000}|000\rangle.$$

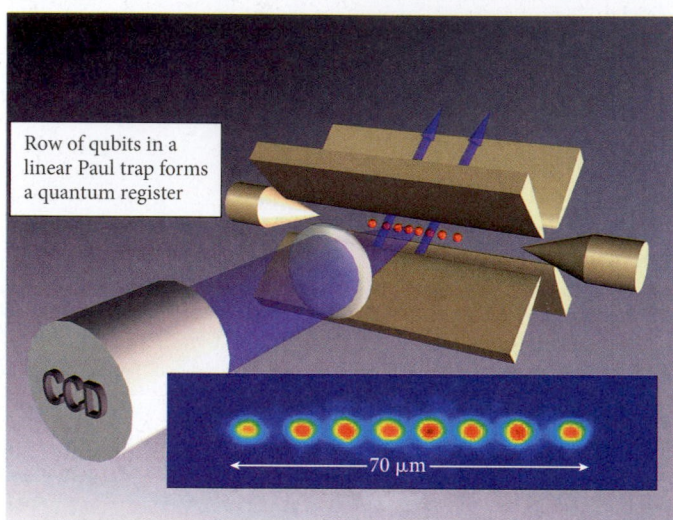

FIGURE 37.29 A schematic diagram of a possible implementation of a quantum computer as a chain of eight calcium ions held inside a Paul trap, which is a device that uses oscillating electric fields to trap ions. The inset is a picture of the induced fluorescence of the trapped ions.

We can easily generalize this equation to hold for larger numbers of qubits. The number of terms in the wave function for 1 qubit is 2, for 2 qubits is 4, and for 3 qubits is 8—the number of terms in the wave function for n qubits is 2^n. All of the promise of quantum computing power arises from this number, 2^n. If a classical computer operates on a register of n bits, then it executes n operations simultaneously. However, if a quantum computer operates on a register of n qubits, then it executes 2^n operations simultaneously. A diagram of one proposed way to implement a quantum computer is shown in Figure 37.29.

Many cautionary statements are in order here. First, probing the wave function of the qubit states necessarily changes the information stored in the qubits. This requires developing special error-correction algorithms, which correct an error without destroying the quantum state. Second, an actual quantum computer that can do more than execute just a few simple demonstration operations has yet to be implemented. Furthermore, quantum computing is by its very construction probabilistic in nature. A calculation must be repeated several times to obtain a reliable result. Only for a small number of algorithms has it been proven that a quantum computer can take full advantage of the theoretically possible 2^n simultaneous operations. One of these algorithms is the factorization of large integers into prime numbers. This problem has attracted a huge amount of interest and also funding, because the encryption schemes most commonly used rely on the fact that present-day computers cannot factorize large integers in a human lifetime. By comparison, a working quantum computer with approximately 100 qubits could do this job in seconds.

37.10 Antimatter

In Section 37.8, we discussed the Hamiltonian operator, which represents the total energy of a particle. However, we used the classical formula, $K = p^2/2m$, for the kinetic energy, which is appropriate only for speeds that are small compared to the speed of light. In Chapter 35 on relativity, we noted that the more general relationship between momentum and energy is

$$E^2 = p^2c^2 + m^2c^4.$$

This raises a question: Is it possible to construct a framework of quantum physics that is also correct for speeds that are *not* small compared to the speed of light? In other words, can we construct a relativistic quantum theory? The first person who figured out how to do this was the British physicist Paul Adrien Maurice Dirac (1902–1984). He published his famous **Dirac equation** in 1928 and for this accomplishment shared the 1933 Nobel Prize in Physics with Erwin Schrödinger.

We will not attempt to derive or even motivate the Dirac equation. Most university physics departments never even mention this equation in general introductory courses. However, we can look at the equation and its notation, compare it to the equivalent nonrelativistic Schrödinger equation, and explore some of its physical consequences. (You can treat the following couple of pages as a resource you may want to return to in the course of future studies. Or perhaps seeing this equation will motivate you to explore this topic further at the present time.)

The Dirac equation for a "free" particle—that is, a particle that is not subjected to any external potential or force—is

$$\left(\gamma^0 mc^2 + \sum_{i=1}^{3} \gamma^i \hat{p}_i c \right) \Psi(\vec{r},t) = i\hbar \frac{\partial}{\partial t} \Psi(\vec{r},t).$$ (37.41)

In this equation, c is the speed of light, m is the rest mass of the particle, and $\hat{p}_i$ is the momentum operator in one of the three orthogonal directions (x, y, z), which we introduced in Section 37.1. The symbols $\gamma^0, \gamma^1, \gamma^2, \gamma^3$ refer to the so-called Dirac matrices:

$$\gamma^0 = \begin{pmatrix} 1 & 0 & 0 & 0 \\ 0 & 1 & 0 & 0 \\ 0 & 0 & -1 & 0 \\ 0 & 0 & 0 & -1 \end{pmatrix}, \quad \gamma^1 = \begin{pmatrix} 0 & 0 & 0 & 1 \\ 0 & 0 & 1 & 0 \\ 0 & -1 & 0 & 0 \\ -1 & 0 & 0 & 0 \end{pmatrix}, \quad \gamma^2 = \begin{pmatrix} 0 & 0 & 0 & -i \\ 0 & 0 & i & 0 \\ 0 & i & 0 & 0 \\ -i & 0 & 0 & 0 \end{pmatrix}, \quad \gamma^3 = \begin{pmatrix} 0 & 0 & 1 & 0 \\ 0 & 0 & 0 & -1 \\ -1 & 0 & 0 & 0 \\ 0 & 1 & 0 & 0 \end{pmatrix}.$$

The wave function $\Psi(\vec{r},t)$ in equation 37.41 is a four-component *spinor*, instead of a one-component scalar, as is the case for the quantum mechanical wave functions that we have discussed so far:

$$\Psi(\vec{r},t) = \begin{pmatrix} \Psi_1(\vec{r},t) \\ \Psi_2(\vec{r},t) \\ \Psi_3(\vec{r},t) \\ \Psi_4(\vec{r},t) \end{pmatrix}.$$

The Dirac equation (37.41) is the relativistically correct extension of the time-dependent Schrödinger equation (37.32) for the three-dimensional potential-free case:

$$\left(-\frac{\hbar^2}{2m}\frac{\partial^2}{\partial x^2} - \frac{\hbar^2}{2m}\frac{\partial^2}{\partial y^2} - \frac{\hbar^2}{2m}\frac{\partial^2}{\partial z^2} \right) \Psi(\vec{r},t) = i\hbar \frac{\partial}{\partial t} \Psi(\vec{r},t).$$

We can find the solutions of the Dirac equation by proceeding in a manner similar to what we did for the Schrödinger equation. For example, we can express the time dependence of the solutions for the free particle as

$$\Psi(\vec{r},t) = \Psi_0(\vec{r})e^{-iEt/\hbar}.$$ (37.42)

The static wave function $\Psi_0(\vec{r})$ satisfies the time-independent Dirac equation with the energy eigenvalue E:

$$\left(\gamma^0 mc^2 + \sum_{i=1}^{3} \gamma^i p_i c \right) \Psi_0(\vec{r}) = E\Psi_0(\vec{r}).$$ (37.43)

We look for solutions of the form of a plane wave moving along one particular coordinate axis, conventionally chosen to be the z-axis:

$$\Psi_0(\vec{r}) = we^{ipz/\hbar}.$$ (37.44)

Here p is the momentum of the particle. (Because it is moving along the z-axis, its momentum component along that axis is also the absolute value of its momentum, that is, the length of its momentum vector.) Also, w is a constant four-component spinor. Inserting this *Ansatz* for $\Psi_0(\vec{r})$ into the static Dirac equation (37.43) results in the matrix equation

$$\begin{pmatrix} mc^2 & 0 & pc & 0 \\ 0 & mc^2 & 0 & -pc \\ pc & 0 & -mc^2 & 0 \\ 0 & -pc & 0 & -mc^2 \end{pmatrix} \begin{pmatrix} w_1 \\ w_2 \\ w_3 \\ w_4 \end{pmatrix} = E \begin{pmatrix} w_1 \\ w_2 \\ w_3 \\ w_4 \end{pmatrix}.$$ (37.45)

The solutions for the possible energies are

$$E(p) = \pm\sqrt{m^2 c^4 + p^2 c^2}.$$ (37.46)

This is satisfying, because it matches the energy-momentum relationship that we arrived at in Chapter 35 using relativistic mechanics. The upper two components of the spinor w, or w_1 and w_2, represent the solutions for positive energies: one for spin-up and one for spin-down. The lower two components, w_3 and w_4, are the spin-up and spin-down solutions for negative energies.

The importance of the Dirac equation lies in the fact that it unifies the two extensions of classical mechanics into one consistent description. Chapter 35 showed that a relativistic description is needed for object speeds that are comparable to the speed of light and that relativistic mechanics contains classical mechanics as a special case for speeds that are small compared to the speed of light. In Chapter 36 and this chapter, we have found that we need a quantum description for very small systems and that the correspondence principle also establishes classical mechanics as a special case of quantum mechanics. Dirac's equation contains relativistic quantum mechanics and has nonrelativistic quantum mechanics as a special case; it thus provides a very satisfying and aesthetically appealing general framework.

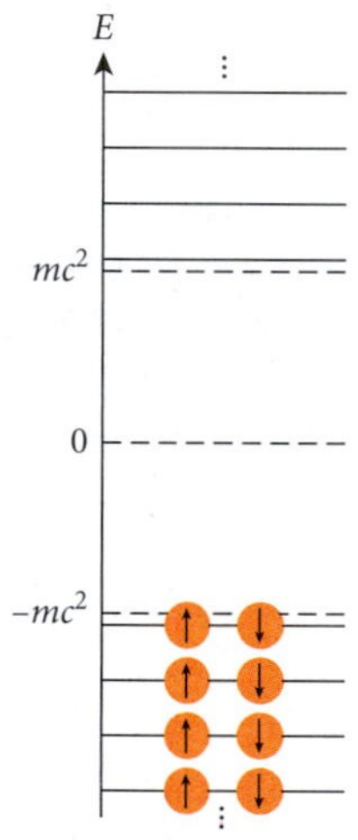

What about the negative energy solutions in equation 37.46? How can these be interpreted? So far, in our discussion of quantum mechanics, we have assumed that the energy of a free particle is always positive. This enables us to think of the concept of the ground state as the state with lowest energy. However, if states with negative energies are available, what prevents an electron placed into a positive energy state from decaying into a negative energy state and then on through a sequence of successively lower energy states by emitting a photon each time?

To overcome this problem of endless radiation, Dirac proposed the solution illustrated in Figure 37.30: All possible negative energy states are already occupied completely by one spin-up and one spin-down particle. This picture of what we perceive to be empty space as a "sea" of occupied negative-energy electron states is commonly called the **Dirac sea.** What are these postulated particles in negative energy states? Do they have the same mass as the electron? Are they real or just a mental construct to help a flawed model survive consistency checks? In particular, is it possible to remove one of these negative energy electrons from the Dirac sea and create a hole in its place?

The answer to this last question is yes. A hole in the Dirac sea acts just like a spin-$\frac{1}{2}$ particle with the same mass as an electron, but with an opposite charge and spin projection. To create such a hole in the Dirac sea, one of the electrons in a negative energy state must be lifted to a positive energy state. Lifting the electron across the forbidden gap from $-mc^2$ to mc^2 requires depositing into the vacuum an energy of at least twice the mass of the electron times the speed of light squared (Figure 37.31). Because the mass-energy of the electron is 511 keV, at least 1.022 MeV of energy is needed to lift an electron into a positive energy state and at the same time create a hole in the Dirac sea.

This positively charged hole in the Dirac sea with the same mass as the electron is not just a convenient algebraic trick, but a real particle. The American physicist Carl Anderson (1905–1991) discovered this particle, named the *positron*, in 1932 using photographic plates exposed to cosmic radiation. This experiment provided a triumphant confirmation of Dirac's theory. Despite this success, the idea of the Dirac sea with an infinite number of particles filling all available energy states does not seem very elegant. In particular, the notion that the vacuum has infinite energy seems counterintuitive. More modern quantum field theories have replaced some of these concepts, but the basic difficulties of infinite quantities are still present, leading many practitioners to believe that further breakthroughs are needed.

The **positron** is the **antiparticle** to the electron, an example of **antimatter.** Every fermion obeys the Dirac equation, and so the concept of a filled Dirac sea applies to the vacuum

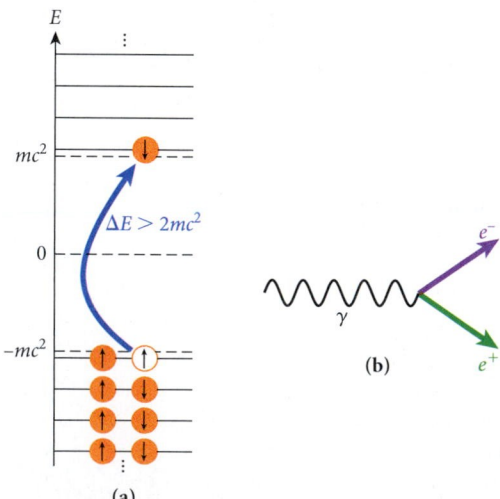

FIGURE 37.30 The Dirac vacuum, where all negative energy states are completely occupied.

FIGURE 37.31 Creation of a hole in the Dirac sea: electron-positron pair creation. (a) Energy diagram; (b) sketch in coordinate space.

ground state. Thus, each fermion has an antiparticle. The Italian physicist Emilio Segrè (1905–1989) and American Owen Chamberlain (1920–2006) discovered the antiparticle to the proton, the antiproton, in 1955. In 1995, physicists detected the first atoms of antihydrogen at CERN. Since 2000, CERN has had an "antimatter factory," in which ultracold antihydrogen atoms are produced and studied. One of the many questions to be answered is whether antimatter responds to gravity in the same way that regular matter does.

   An antiparticle can also annihilate if it meets its particle partner. In this **annihilation** process, sketched in Figure 37.32, the entire mass of both particles is converted into energy. Thus, from the combination of quantum theory and relativity emerges a new picture of the relationship between mass and energy: Mass and energy can be interconverted, and mass is just one more form of energy.

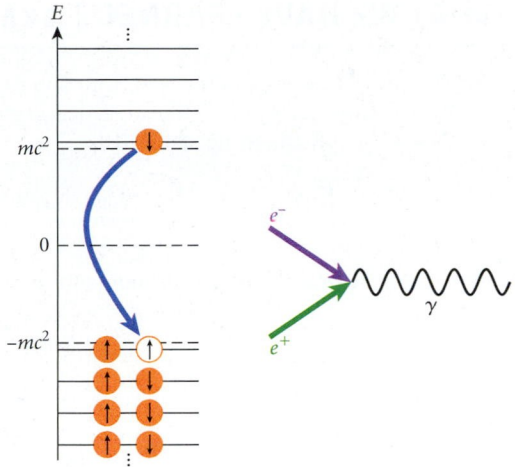

FIGURE 37.32 Electron-positron annihilation.

EXAMPLE 37.6 | Matter Annihilation

Matter-antimatter annihilation is a popular energy source in science fiction books and movies. This idea conveniently ignores the fact that there is no source of antimatter anywhere on Earth or in our Solar System, or, most likely, anywhere in the entire universe. (We will investigate why this is so in Chapters 39 and 40 on subatomic physics.) All antimatter that could be used as an energy source would have to be created first, and at least as much energy would be required to do this as could be gained later from the annihilation. Nevertheless, it is interesting to figure out how much energy would be released in the annihilation process.

PROBLEM 1
How much energy would be released if you could let a soft drink can full of antiwater annihilate with regular water?

SOLUTION 1
A soft drink can holds 0.330 L of liquid. If this liquid is water, then its mass is 0.330 kg, because water has a density of 1000 kg/m^3 and so 1 L of water has a mass of 1 kg. Since the mass of an antiparticle (which makes up antimatter) is the same as the mass of the corresponding particle, the mass of the antiwater is also 0.330 kg. Therefore, we need to annihilate a total of 2 × 0.330 kg of mass (0.330 kg of antiwater and 0.330 kg of water).

   The energy contained in this mass is obtained by multiplying the total mass by the square of the speed of light:

$$E = mc^2 = (0.660 \text{ kg})(3.00 \cdot 10^8 \text{ m/s})^2 = 5.94 \cdot 10^{16} \text{ J}.$$

To see how much energy this is, let's consider a follow-up question.

PROBLEM 2
The largest nuclear power plants produce 1200 MW. How long would such a nuclear power plant have to operate in order to produce the same amount of energy as the soft drink can of antiwater?

SOLUTION 2
A 1200-MW power plant produces $1.2 \cdot 10^9$ J of usable energy each second. Thus, the number of seconds it takes to produce $E = 5.94 \cdot 10^{16}$ J is

$$\frac{5.94 \cdot 10^{16} \text{ J}}{1.2 \cdot 10^9 \text{ J/s}} = 4.95 \cdot 10^7 \text{ s}.$$

This is more than 1.5 years!

WHAT WE HAVE LEARNED | EXAM STUDY GUIDE

- The absolute square of the wave function $\psi(x)$ is the probability density of finding a particle at some position. The wave function is normalized as

$$\int_{-\infty}^{\infty} |\psi(x)|^2 \, dx = \int_{-\infty}^{\infty} \psi^*(x)\psi(x) \, dx = 1.$$

- The operator for the particle's momentum is

$$\hat{\mathbf{p}}\psi(x) = -i\hbar \frac{d}{dx}\psi(x).$$

- The operator for the particle's kinetic energy is

$$\hat{\mathbf{K}}\psi(x) = \frac{1}{2m}\hat{\mathbf{p}}^2\psi(x) = \frac{1}{2m}\left(-i\hbar\frac{d}{dx}\right)^2\psi(x) = -\frac{\hbar^2}{2m}\frac{d^2}{dx^2}\psi(x).$$

- The time-independent Schrödinger equation for a particle in a potential $U(x)$ is given by

$$-\frac{\hbar^2}{2m}\frac{d^2\psi(x)}{dx^2} + U(x)\psi(x) = E\psi(x).$$

- The wave function solution for a particle confined to an infinite potential well with walls at $x = 0$ and $x = a$ is

$$\psi(x) = \begin{cases} 0 & \text{for } x < 0 \\ \sqrt{\dfrac{2}{a}}\sin\left(\dfrac{n\pi x}{a}\right) \text{ with } n = 1,2,3,\ldots & \text{for } 0 \le x \le a \\ 0 & \text{for } x > a. \end{cases}$$

The corresponding energy eigenvalues are

$$E_n = \frac{\hbar^2\pi^2}{2ma^2}n^2 \quad (\text{for } n = 1,2,3,\ldots).$$

- The solution for a finite potential well, with $E > U_1$,

$$U(x) = \begin{cases} \infty & \text{for } x < 0 \\ 0 & \text{for } 0 \le x \le a \\ U_1 & \text{for } x > a, \end{cases}$$

is

$$\psi(x) = \begin{cases} 0 & \text{for } x < 0 \\ D\sin(\kappa x) & \text{for } 0 \le x \le a \\ F\cos(\kappa' x) + G\sin(\kappa' x) & \text{for } x > a, \end{cases}$$

with $\kappa' = \sqrt{\kappa^2 - \dfrac{2mU_1}{\hbar^2}}$.

- The solution for the same finite potential well, with $E < U_1$, is

$$\psi(x) = \begin{cases} 0 & \text{for } x < 0 \\ D\sin(\kappa x) & \text{for } 0 \le x \le a, \\ Fe^{-\gamma x} & \text{for } x > a, \end{cases}$$

with $\gamma^2 = \dfrac{2m(U_1 - E)}{\hbar^2} = \dfrac{2mU_1}{\hbar^2} - \kappa^2$.

- For a potential step of the form

$$U(x) = \begin{cases} \infty & \text{for } x < 0 \\ 0 & \text{for } 0 \le x \le a \\ U_1 & \text{for } a < x < b \\ 0 & \text{for } x \ge b, \end{cases}$$

the wave function is

$$\psi(x) = \begin{cases} 0 & \text{for } x < 0 \\ D\sin(\kappa x) & \text{for } 0 \le x \le a \\ Fe^{-\gamma x} & \text{for } a < x < b \\ G\sin(\kappa x + \varphi) & \text{for } x \ge b. \end{cases}$$

The probability that a particle with energy less than the barrier height will tunnel through the barrier is

$$\Pi_t = \frac{|\psi(b)|^2}{|\psi(a)|^2} = \frac{\left|Fe^{-\gamma b}\right|^2}{\left|Fe^{-\gamma a}\right|^2} = \left|e^{-\gamma(b-a)}\right|^2 = e^{-2\gamma(b-a)},$$

which depends exponentially on the barrier width.

- The wave function solutions for the harmonic oscillator potential $U(x) = \frac{1}{2}kx^2 = \frac{1}{2}m\omega_0^2 x^2$ are

$$\psi_n(x) = \frac{1}{\sqrt{\sigma}\pi^{1/4}}\frac{1}{\sqrt{n!\,2^n}}H_n(x/\sigma)e^{-x^2/2\sigma^2},$$

where H_n is a Hermite polynomial and

$$\sigma = \sqrt{\frac{\hbar}{m\omega_0}}.$$

The corresponding energy eigenvalues are equally spaced:

$$E_n = \left(n + \tfrac{1}{2}\right)\hbar\omega_0 \quad (\text{for } n = 0,1,2,\ldots).$$

- The quantum mechanical expectation value of a quantity is found by integrating, over the entire space, ψ^* times the result obtained when the corresponding operator acts on ψ. The position expectation value is

$$\langle x \rangle = \int_{-\infty}^{\infty} \psi^*(x)x\psi(x)\,dx,$$

and the momentum expectation value is

$$\langle p \rangle = \int_{-\infty}^{\infty} \psi^*(x)\hat{\mathbf{p}}\psi(x)\,dx = -i\hbar\int_{-\infty}^{\infty}\psi^*(x)\frac{d}{dx}\psi(x)\,dx.$$

- The ground-state ($n = 0$) oscillator wave function has the minimum product of the uncertainties in position and momentum allowed by the uncertainty principle.

- The correspondence principle states that if the energy difference, ΔE, between neighboring quantum energy states becomes small relative to the total energy, E, that is, if $\Delta E \ll E$, the quantum solution approaches its classical limit.

- The time-dependent Schrödinger equation is

$$-\frac{\hbar^2}{2m}\frac{\partial^2}{\partial x^2}\Psi(x,t) + U(x)\Psi(x,t) = i\hbar\frac{\partial}{\partial t}\Psi(x,t)$$

with the solution $\Psi(x,t) = \psi(x)e^{-i\omega t}$.

- The Hamiltonian operator is

$$\hat{\mathbf{H}} = -\frac{\hbar^2}{2m}\frac{d^2}{dx^2} + U(x) = \hat{\mathbf{K}} + U(x).$$

  It is linear, so a linear combination of two solutions to the Schrödinger equation is also a solution.

- The two-particle wave function for (noninteracting) bosons is the symmetrized product of the one-particle wave functions:

$$\Psi^{\mathrm{B}}(x_1, x_2) = \frac{1}{\sqrt{2}}\big(\psi_a(x_1)\cdot\psi_b(x_2) + \psi_a(x_2)\cdot\psi_b(x_1)\big).$$

- The two-particle wave function for (noninteracting) fermions is the antisymmetrized product of the one-particle wave functions:

$$\Psi^{\mathrm{F}}(x_1, x_2) = \frac{1}{\sqrt{2}}\big(\psi_a(x_1)\cdot\psi_b(x_2) - \psi_a(x_2)\cdot\psi_b(x_1)\big).$$

- Quantum computing defines a qubit as

$$|\psi\rangle = c_1|1\rangle + c_0|0\rangle.$$

  The superposition of qubits gives a quantum computer the (still only theoretical!) ability to evaluate many calculations simultaneously, providing a great advantage over conventional digital computing.

- The Dirac equation,

$$\left(\gamma^0 mc^2 + \sum_{i=1}^{3}\gamma^i \hat{p}_i c\right)\Psi(\vec{r},t) = i\hbar\frac{\partial}{\partial t}\Psi(\vec{r},t),$$

  is the generalization of the time-dependent Schrödinger equation for the relativistically correct relationship between energy and momentum. It leads to negative energy states and antiparticles.

ANSWERS TO SELF-TEST OPPORTUNITIES

37.1 Use $e^{\pm i\theta} = \cos\theta \pm i\sin\theta$ (Euler).

$$\begin{aligned}
\psi(x) &= Ae^{i\kappa x} + Be^{-i\kappa x} \\
&= A\big(\cos(\kappa x) + i\sin(\kappa x)\big) + B\big(\cos(\kappa x) - i\sin(\kappa x)\big) \\
&= (A+B)\cos(\kappa x) + i(A-B)\sin(\kappa x) \\
&= C\cos(\kappa x) + D\sin(\kappa x), \quad \text{where}
\end{aligned}$$

$C = A+B$ and $D = i(A-B)$.

37.2 Adding a constant energy, c, everywhere in space leaves the wave function solutions unchanged. However, all energies are shifted up by the constant c.

37.3 The wave function is the product of three one-dimensional wave functions, $\psi(x,y,z) = \psi_1(x)\cdot\psi_2(y)\cdot\psi_3(z)$:

$$\psi_1(x) = \begin{cases} 0 & \text{for } x < 0 \\ \sqrt{2/a}\,\sin(n_x\pi x/a) \text{ with } n_x = 1,2,3,\dots & \text{for } 0 \le x \le a \\ 0 & \text{for } x > a, \end{cases}$$

$$\psi_2(y) = \begin{cases} 0 & \text{for } y < 0 \\ \sqrt{2/b}\,\sin(n_y\pi y/b) \text{ with } n_y = 1,2,3,\dots & \text{for } 0 \le y \le b \\ 0 & \text{for } y > b, \end{cases}$$

$$\psi_3(z) = \begin{cases} 0 & \text{for } z < 0 \\ \sqrt{2/c}\,\sin(n_z\pi z/c) \text{ with } n_z = 1,2,3,\dots & \text{for } 0 \le z \le c \\ 0 & \text{for } z > c. \end{cases}$$

$$E_{n_x,n_y,n_z} = \frac{\hbar^2\pi^2}{2ma^2}n_x^2 + \frac{\hbar^2\pi^2}{2mb^2}n_y^2 + \frac{\hbar^2\pi^2}{2mc^2}n_z^2.$$

37.4 A continuous wave function at point b requires that

$$Fe^{-\gamma b} = G\sin(\kappa b + \phi), \tag{i}$$

and a continuous derivative of the wave function at the same point requires that

$$-\gamma Fe^{-\gamma b} = \kappa G\cos(\kappa b + \phi). \tag{ii}$$

Divide equation (i) by equation (ii) to obtain

$\tan(\kappa b + \phi) = \dfrac{\kappa}{-\gamma}$. Solve this for the phase shift, ϕ:

$$\phi = \tan^{-1}\left(\frac{\kappa}{-\gamma}\right) - \kappa b.$$

To get the amplitude, G, it is easiest to divide equation (ii) by κ, then square both sides of equation (i) and equation (ii). This results in

$$F^2 e^{-2\gamma b} = G^2\sin^2(\kappa b + \phi)$$
$$(\gamma/\kappa)^2 F^2 e^{-2\gamma b} = G^2\cos^2(\kappa b + \phi).$$

Now add these two equations, which yields

$$(1 + (\gamma/\kappa)^2)F^2 e^{-2\gamma b} = G^2.$$

Taking the square root gives us the expression for G:

$$G = \sqrt{1 + (\gamma/\kappa)^2}\, Fe^{-\gamma b}.$$

37.5 Start with the definition and insert the wave function:

$$\langle p \rangle = \int_{-\infty}^{\infty} \psi^*(x)\hat{p}\psi(x)dx = -i\hbar \int_{-\infty}^{\infty} \psi^*(x)\frac{d}{dx}\psi(x)dx$$

$$\psi_0(x) = Ae^{-x^2/2\sigma^2}$$

$$\frac{d}{dx}\psi_0(x) = \frac{Ax}{\sigma^2}e^{-x^2/2\sigma^2}$$

$$= \int_{-\infty}^{\infty}\left(Ae^{-x^2/2\sigma^2}\right)\left(\frac{Ax}{\sigma^2}e^{-x^2/2\sigma^2}\right)dx = \frac{A}{\sigma^2}\int_{-\infty}^{\infty} xe^{-x^2/\sigma^2}\,dx = 0.$$

37.6 For the harmonic oscillator, we find

$$E_n = \left(n + \tfrac{1}{2}\right)\hbar\omega_0 \quad (\text{for } n = 0,1,2,...).$$

The average kinetic energies are

$$\langle K_0 \rangle = E_0 / 2 = \frac{\hbar\omega_0}{4}$$

$$\langle K_1 \rangle = E_1 / 2 = \frac{3\hbar\omega_0}{4}$$

$$\langle K_2 \rangle = E_2 / 2 = \frac{5\hbar\omega_0}{4}$$

$$\langle K_3 \rangle = E_3 / 2 = \frac{7\hbar\omega_0}{4}.$$

PROBLEM-SOLVING GUIDELINES

1. Quantum mechanics dominates all of modern physics, and many different systems have been and continue to be investigated. However, most of these systems do not have straightforward analytic solutions and require the use of computers. On the introductory level, most problems can be mapped to (in)finite potential wells or harmonic oscillator potentials.

2. Wave functions are often determined by boundary conditions and symmetries. Be sure to exploit these in your solution attempts.

3. Quantum mechanics and classical mechanics are not totally different from each other. Classical mechanics is the limit of quantum mechanics when the fact that $\hbar$ is nonzero can be neglected. Be sure that your quantum solutions do not violate classical mechanics in that limit.

4. Typically, you will find discrete energies for quantum solutions. Fixed energy differences between neighboring states point to oscillator potentials, whereas linearly increasing energy differences point to infinite potential wells.

MULTIPLE-CHOICE QUESTIONS

37.1 The wavelength of an electron in an infinite potential well is $\alpha/2$, where α is the width of the well. Which state is the electron in?

a) $n = 3$

c) $n = 4$

b) $n = 6$

d) $n = 2$

37.2 For which of the following states will the particle never be found in the exact center of a square infinite potential well?

a) the ground state

b) the first excited state

c) the second excited state

d) any of the above

e) none of the above

37.3 The probability of finding an electron at a particular location in a hydrogen atom is directly proportional to

a) its energy.

b) its momentum.

c) its wave function.

d) the square of its wave function.

e) the product of the position coordinate and the square of the wave function.

f) none of the above.

37.4 Is the superposition of two wave functions, which are solutions to the time-independent Schrödinger equation for the same potential energy, also a solution to the Schrödinger equation?

a) no

b) yes

c) depends on the value of the potential energy

d) only if $\dfrac{d^2\psi(x)}{dx^2} = 0$

37.5 Let κ be the wave number of a particle moving in one dimension with velocity v. If the velocity of the particle is doubled, to $2v$, then the magnitude of the wave number is

a) κ.

c) $\kappa/2$.

b) 2κ.

d) none of these.

37.6 An electron is in a square infinite potential well of width a: $U(x) = \infty$, for $x < 0$ and $x > a$. If the electron is in the first excited state, $\psi(x) = A\sin(2\pi x/a)$, at what position(s) is the probability function a maximum?

a) 0

c) $a/2$

e) both $a/4$ and $3a/4$

b) $a/4$

d) $3a/4$

37.7 Which of the following statements is (are) true?

a) The energy of electrons is always discrete.

b) The energy of a bound electron is continuous.

c) The energy of a free electron is discrete.

d) The energy of an electron is discrete when it is bound to an ion.

37.8 Which of the following statements is (are) true?

a) In a one-dimensional quantum harmonic oscillator, the energy levels are evenly spaced.

b) In a one-dimensional infinite potential well, the energy levels are evenly spaced.

c) The minimum total energy possible for a classical harmonic oscillator is zero.

d) The correspondence principle states that because the minimum possible total energy for the classical simple harmonic oscillator is zero, the expected value for the fundamental state ($n = 0$) of the one-dimensional quantum harmonic oscillator should also be zero.

e) The $n = 0$ state of a one-dimensional quantum harmonic oscillator is the state with the minimum possible uncertainty, $\Delta x \Delta p$.

37.9 Simple harmonic oscillation occurs when the potential energy function is equal to $\frac{1}{2}kx^2$, where k is a constant. What happens to the ground-state energy level if k is increased?

a) It increases.

b) It remain the same.

c) It decreases.

37.10 A particle with energy $E = 5$ eV approaches a barrier of height $U = 8$ eV. Quantum mechanically there is a nonzero probability that the particle will tunnel through the barrier. If the barrier height is slowly decreased, the probability that the particle will deflect off the barrier will

a) decrease.   b) increase.   c) not change.

CONCEPTUAL QUESTIONS

37.11 Is the following statement true or false? The larger the amplitude of a Schrödinger wave function, the larger its kinetic energy. Explain your answer.

37.12 For a particle trapped in a square infinite potential well of length L, what happens to the probability that the particle is found between 0 and $L/2$ as the particle's energy increases?

37.13 Think about what happens to wave functions for a particle in a square infinite potential well as the quantum number n approaches infinity. Does the probability distribution in that limit obey the correspondence principle? Explain.

37.14 Show by symmetry arguments that the expectation value of the momentum for an even-n state of a one-dimensional harmonic oscillator is zero.

37.15 Is it possible for the expectation value of the position of an electron to correspond to a position where the electron's probability function, $\Pi(x)$, is zero? If it is possible, give a specific example.

37.16 Sketch the two lowest-energy wave functions for an electron in an infinite potential well that is 20 nm wide and a finite potential well that is 1 eV deep and is also 20 nm wide. Using your sketches, can you determine whether the energy levels in the finite potential well are lower, the same, or higher than in the infinite potential well?

37.17 In the cores of white dwarf stars, carbon nuclei are thought to be locked into very ordered lattices because the temperature is very low, $\sim 10^4$ K. Consider a one-dimensional lattice in which the atoms are separated by 20 fm (1 fm = $1 \cdot 10^{-15}$ m). Approximate the Coulomb potentials of the two outside atoms as following a quadratic relationship, and assume that the atoms' vibrational motions are small. What energy state would the central carbon atom be in at this temperature? (Use $E = \frac{3}{2}k_BT$.)

37.18 For a square finite potential well, you have seen solutions for particle energies greater than and less than the well depth. Show that these solutions are equal outside the potential well if the particle energy is equal to the well depth. Explain your answer and the possible difficulty with it.

37.19 Consider the energies allowed for bound states of a half-harmonic oscillator, having the potential

$$U(x) = \begin{cases} \frac{1}{2}m\omega_0^2 x^2 & \text{for } x > 0 \\ \infty & \text{for } x \leq 0. \end{cases}$$

Using simple arguments based on the characteristics of normalized wave functions, what are the energies allowed for bound states in this potential?

37.20 Suppose $\psi(x)$ is a properly normalized wave function describing the state of an electron. Consider a second wave function, $\psi_{new}(x) = e^{i\phi}\psi(x)$, for some real number ϕ. How does the probability density associated with ψ_{new} compare to that associated with ψ?

37.21 A particle of mass m is in the potential $U(x) = U_0 \cosh(x/a)$, where U_0 and a are constants. Show that the ground-state energy of the particle can be *estimated* as

$$E_0 \cong U_0 + \frac{1}{2}\hbar\left(\frac{U_0}{ma^2}\right)^{1/2}.$$

37.22 The time-independent Schrödinger equation for a nonrelativistic free particle of mass m is obtained from the energy relationship $E = p^2/(2m)$ by replacing E and p with appropriate derivative operators, as suggested by the de Broglie relations. Using this procedure, derive a quantum wave equation for a *relativistic* particle of mass m, for which the energy relation is $E^2 - p^2c^2 = m^2c^4$, *without* taking any square root of this relation.

EXERCISES

A blue problem number indicates a worked-out solution is available in the Student Solutions Manual. One • and two •• indicate increasing level of problem difficulty.

Section 37.1

37.23 A neutron has a kinetic energy of 10.0 MeV. What size object would have to be used to observe neutron diffraction effects? Is there anything in nature of this size that could serve as a target to demonstrate the wave nature of 10.0-MeV neutrons?

37.24 Given the complex function $f(x) = (8 + 3i) + (7 - 2i)x$ of the real variable x, what is $|f(x)|^2$?

Section 37.3

37.25 Determine the two lowest energies of the wave function of an electron in a box of width $2.0 \cdot 10^{-9}$ m.

37.26 Determine the three lowest energies of the wave function of a proton in a box of width $1.0 \cdot 10^{-10}$ m.

37.27 What is the ratio of the energy difference between the ground state and the first excited state for a square infinite potential well of length L and that for a square infinite potential well of length $2L$? That is, find $(E_2 - E_1)_L/(E_2 - E_1)_{2L}$.

•**37.28** An electron is confined in a one-dimensional infinite potential well of width 1.0 nm. Calculate

a) the energy difference between the second excited state and the ground state, and

b) the wavelength of light emitted when an electron makes this transition.

•**37.29** Find the wave function for a particle in an infinite square well centered at the origin, with the walls at $\pm a/2$.

•**37.30** In Example 37.1, we calculated the energy of the wave function with the lowest quantum number for an electron confined to a one-dimensional box of width 2.00 Å. However, atoms are *three-dimensional* entities with a typical diameter of 1.00 Å = 10^{-10} m. It would seem then that the better approximation would be an electron trapped in a three-dimensional infinite potential well (a potential cube with sides of 1.00 Å).

a) Derive an expression for the wave function and the corresponding energies of an electron in a three-dimensional rectangular infinite potential well.

b) Calculate the lowest energy allowed for the electron in this case.

Section 37.4

37.31 An electron is confined to a potential well shaped as shown in Figure 37.10. The width of the well is $1.0 \cdot 10^{-9}$ m, and $U_1 = 2.0$ eV. Is the $n = 3$ state bound in this well?

37.32 If a proton of kinetic energy 18.0 MeV encounters a rectangular potential energy barrier of height 29.8 MeV and width $1.00 \cdot 10^{-15}$ m, what is the probability that the proton will tunnel through the barrier?

•**37.33** Suppose the kinetic energy of the neutron described in Example 37.3 is increased by 15%. By what factor does the neutron's probability of tunneling through the barrier increase?

•**37.34** A beam of electrons moving in the positive x-direction encounters a potential barrier that is 2.51 eV high and 1.00 nm wide. Each electron has a kinetic energy of 2.50 eV, and the electrons arrive at the barrier at a rate of 1000. electrons/s. What is the rate I_T (in electrons/s) at which electrons pass through the barrier, on average? What is the rate I_R (in electrons/s) at which electrons bounce back from the barrier, on average? Determine and compare the wavelengths of the electrons before and after they pass through the barrier.

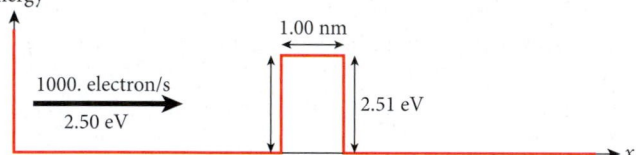

•**37.35** Consider an electron approaching a potential barrier 2.00 nm wide and 7.00 eV high. What is the energy of the electron if it has a 10.0% probability of tunneling through this barrier?

•**37.36** Consider an electron in a three-dimensional box—with infinite potential walls—of dimensions 1.00 nm × 2.00 nm × 3.00 nm. Find the quantum numbers n_x, n_y, and n_z and the energies (in eV) of the six lowest energy levels. Are any of these levels degenerate, that is, do any distinct quantum states have identical energies?

•**37.37** In a scanning tunneling microscope, the probability that an electron from the probe will tunnel through a 0.100-nm gap is 0.100%. Calculate the work function of the probe.

••**37.38** Consider a square potential, $U(x) = 0$ for $x < -\alpha$, $U(x) = -U_0$ for $-\alpha \leq x \leq \alpha$, where U_0 is a positive constant, and $U(x) = 0$ for $x > \alpha$. For $E > 0$, the solutions of the time-independent Schrödinger equation in the three regions will be the following:

For $x < -\alpha$, $\psi(x) = e^{i\kappa x} + Re^{-i\kappa x}$, where $\kappa^2 = 2mE/\hbar^2$ and R is the amplitude of a reflected wave.

For $-\alpha \leq x \leq \alpha$, $\psi(x) = Ae^{i\kappa' x} + Be^{-i\kappa' x}$, and $(\kappa')^2 = 2m(E + U_0)/\hbar^2$.

For $x > \alpha$, $\psi(x) = Te^{i\kappa x}$ where T is the amplitude of the transmitted wave.

Match $\psi(x)$ and $d\psi(x)/dx$ at $-\alpha$ and α and find an expression for R. What is the condition for which $R = 0$ (that is, there is no reflected wave)?

••**37.39** a) Determine the wave function and the energy levels for the bound states of an electron in the symmetrical one-dimensional potential well of finite depth shown in the figure.

b) If the penetration distance η into the classically forbidden region is defined as the distance at which the wave function decreases to $1/e$ of its value at the edge of the well, determine an expression for this penetration distance.

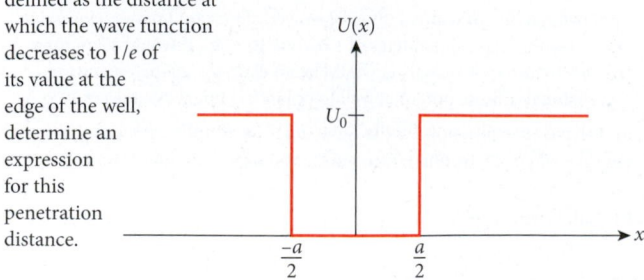

c) The electrons in a typical GaAs-GaAlAs quantum-well laser diode are confined within a one-dimensional potential well like the one shown in the figure, of width 1 nm and depth 0.300 eV. Numerical solutions to the Schrödinger equation show that there is only one possible bound state for the electrons in this case, with energy of 0.125 eV. Calculate the penetration distance for these electrons.

Section 37.5

37.40 An oxygen molecule has a vibrational mode that behaves approximately like a simple harmonic oscillator with frequency $2.99 \cdot 10^{14}$ rad/s. Calculate the energy of the ground state and the first two excited states.

37.41 An electron in a harmonic oscillator potential emits a photon with a wavelength of 360 nm as it undergoes a $3 \rightarrow 1$ quantum jump. What is the wavelength of the photon emitted in a $3 \rightarrow 2$ quantum jump? (*Hint:* The energy of the photon is equal to the energy difference between the initial and the final state of the electron.)

37.42 An experimental measurement of the energy levels of a hydrogen molecule, H_2, shows that they are evenly spaced and separated by about $9 \cdot 10^{-20}$ J. A reasonable model of one of the hydrogen atoms would then seem to be that of a particle in a simple harmonic oscillator potential. Assuming that the hydrogen atom is attached by a spring with a spring constant k to the other atom in the molecule, what is the spring constant k?

•**37.43** Calculate the ground-state energy for an electron confined to a cubic potential well with sides equal to twice the Bohr radius ($R = 0.0529$ nm). Determine the spring constant that would give this same ground-state energy for a harmonic oscillator.

•**37.44** A particle in a harmonic oscillator potential has the initial wave function $\Psi(x,0) = A [\psi_0(x) + \psi_1(x)]$. Normalize $\Psi(x,0)$.

•**37.45** The ground-state wave function for a harmonic oscillator is given by $\psi_0(x) = A_2 e^{-x^2/2b^2}$.

a) Determine the normalization constant, A_2.

b) Determine the probability that a quantum harmonic oscillator in the $n = 0$ state will be found in the classically forbidden region.

Section 37.6

37.46 A particle is in a square infinite potential well of width L and is in the $n = 3$ state. What is the probability that, when observed, the particle is found to be in the rightmost 10.0% of the well?

37.47 An electron is confined between $x = 0$ and $x = L$. The wave function of the electron is $\psi(x) = A\sin(2\pi x/L)$. The wave function is zero for the regions $x < 0$ and $x > L$.

a) Determine the normalization constant, A.

b) What is the probability of finding the electron in the region $0 \leq x \leq L/3$?

37.48 Find the probability of finding an electron trapped in a one-dimensional infinite well of width 2.00 nm in the $n = 2$ state between 0.800 and 0.900 nm (assume that the left edge of the well is at $x = 0$ and the right edge is at $x = 2.00$ nm).

•**37.49** An electron is trapped in a one-dimensional infinite potential well that is $L = 300$. pm wide. What is the probability that the electron in the first excited state will be detected in an interval between $x = 0.500 L$ and $x = 0.750 L$?

Section 37.8

•**37.50** Find the uncertainty of x for the wave function $\Psi(x,t) = Ae^{-\lambda x^2} e^{-i\omega t}$.

•**37.51** Write a wave function $\Psi(\vec{r},t)$ for a nonrelativistic free particle of mass m moving in three dimensions with momentum $\vec{p}$, including the correct time dependence as required by the Schrödinger equation. What is the probability density associated with this wave?

•**37.52** Suppose a quantum particle is in a stationary state with a wave function $\Psi(x,t)$. The calculation of $\langle x \rangle$, the expectation value of the particle's position, is shown in the text. Calculate $d\langle x \rangle/dt$ (**not** $\langle dx/dt \rangle$).

••37.53 Although quantum systems are frequently characterized by their stationary states, a quantum particle is not required to be in such a state unless its energy has been measured. The actual state of the particle is determined by its initial conditions. Suppose a particle of mass m in a one-dimensional potential well with infinite walls (a box) of width a is actually in a state with the wave function

$$\Psi(x,t) = \frac{1}{\sqrt{2}}[\Psi_1(x,t) + \Psi_2(x,t)],$$

where Ψ_1 denotes the stationary state with quantum number $n = 1$ and Ψ_2 denotes the state with $n = 2$. Calculate the probability density distribution for the position x of the particle in this state.

Section 37.10

37.54 In Chapter 40, you will see that a nuclear-fusion reaction between two protons (creating a deuteron, a positron, and a neutrino) releases 0.42 MeV of energy. Nuclear fusion is what causes the stars to shine, and if we can harness it, we can solve the world's energy problems. Compare the energy released in this fusion reaction with what would be released by the annihilation of a proton and an antiproton.

37.55 Particle-antiparticle pairs are occasionally created out of empty space. Considering energy-time uncertainty, how long would such pairs be expected to exist at most if they consist of

a) an electron and a positron?

b) a proton and an antiproton?

37.56 A positron and an electron annihilate, producing two 2.0-MeV gamma rays moving in opposite directions. Calculate the kinetic energy of the electron when the kinetic energy of the positron is twice that of the electron.

Additional Exercises

37.57 Calculate the energy of the first excited state of a proton in a one-dimensional infinite potential well of width $\alpha = 1.00$ nm.

37.58 Electrons in a scanning tunneling microscope encounter a potential barrier that has a height of $U = 4.0$ eV above their total energy. By what factor does the tunneling current change if the tip moves a net distance of 0.10 nm farther from the surface?

37.59 An electron is confined in a three-dimensional cubic space of L^3 with infinite potentials.

a) Write down the normalized solution of the wave function for the ground state.

b) How many energy states are available between the ground state and the second excited state? (Take the electron's spin into account.)

37.60 A mass-and-spring harmonic oscillator used for classroom demonstrations has the angular frequency $\omega_0 = 4.45$ s^{-1}. If this oscillator has a total (kinetic plus potential) energy $E = 1.00$ J, what is its corresponding quantum number, n?

37.61 The neutrons in a parallel beam, each having kinetic energy $\frac{1}{40}$ eV (which approximately corresponds to room temperature), are directed through two slits 0.50 mm apart. How far apart will the peaks of the interference pattern be on a screen 1.5 m away?

37.62 Find the ground-state energy (in eV) of an electron in a one-dimensional box, if the box is of length $L = 0.100$ nm.

37.63 An approximately one-dimensional potential well can be formed by surrounding a layer of GaAs with layers of Al$_x$Ga$_{1-x}$As. The GaAs layers can be fabricated in thicknesses that are integral multiples of the single-layer thickness, 0.28 nm. Some electrons in the GaAs layer behave as if they were trapped in a box. For simplicity, treat the box as a one-dimensional infinite potential well and ignore the interactions between the electrons and the Ga and As atoms (such interactions are often accounted for by replacing the actual electron mass with an effective electron mass). Calculate the energy of the ground state in this well for these cases:

a) 2 GaAs layers

b) 5 GaAs layers

37.64 Consider a water molecule in the vapor state in a room 4.00 m × 10.0 m × 10.0 m.

a) What is the ground-state energy of this molecule, treating it as a simple particle in a box?

b) Compare this energy to the average thermal energy of such a molecule, taking the temperature to be 300. K.

c) What can you conclude from the two numbers you just calculated?

37.65 A neutron moves between rigid walls 8.4 fm apart. What is the energy of its $n = 1$ state?

•37.66 A surface is examined using a scanning tunneling microscope (STM). For the range of the gap, L, between the tip of the probe and the sample surface, assume that the electron wave function for the atoms under investigation falls off exponentially as $|\psi| = e^{-(10.0 \text{ nm}^{-1})a}$. The tunneling current through the tip of the probe is proportional to the tunneling probability. In this situation, what is the ratio of the current when the tip is 0.400 nm above a surface feature to the current when the tip is 0.420 nm above the surface?

•37.67 An electron is trapped in a one-dimensional infinite well of width 2.00 nm. It starts in the $n = 4$ state, and then goes into the $n = 2$ state, emitting radiation with energy corresponding to the energy difference between the two states. What is the wavelength of the radiation?

•37.68 Two long, straight wires that lie along the same line have a separation at their tips of 2.00 nm. The potential energy of an electron in this gap is about 1.00 eV higher than it is in the conduction band of the two wires. Conduction-band electrons have enough energy to contribute to the current flowing in the wire. What is the probability that a conduction electron from one wire will be found in the other wire after arriving at the gap?

•37.69 Consider an electron that is confined to a one-dimensional infinite potential well of width $a = 0.10$ nm, and another electron that is confined to a three-dimensional (cubic) infinite potential well with sides of length $a = 0.10$ nm. Let the electron confined to the cube be in its ground state. Determine the difference in energy ground state of the two electrons and the excited state of the one-dimensional electron that minimizes the difference in energy with the three-dimensional electron.

•37.70 An electron with an energy of 129 KeV is trapped in a potential well defined by an infinite potential at $x < 0$ and a potential barrier of finite height U_1 extending from $x = 529.2$ fm to $x = 2116.8$ fm (1 fm = $1 \cdot 10^{-15}$ m), as shown in the figure. It is found that the electron can be detected beyond the barrier with a probability of 10%. Calculate the height of the potential barrier.

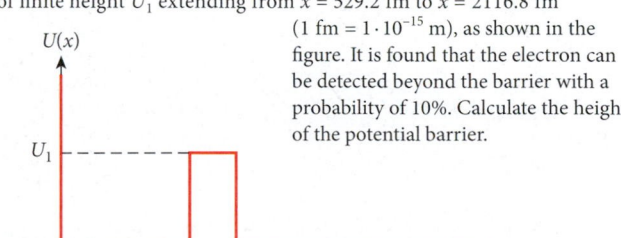

•37.71 Consider an electron that is confined to the xy-plane by a two-dimensional rectangular infinite potential well. The width of the well is w in the x-direction and $2w$ in the y-direction. What is the lowest energy that is shared by more than one distinct state, that is, two different states having the same energy?

MULTI-VERSION EXERCISES

37.72 A 5.15-MeV alpha particle (mass = 3.7274 GeV/c^2) inside a heavy nucleus encounters a barrier whose average height is 15.5 MeV and whose width is 11.7 fm (1 fm = $1 \cdot 10^{-15}$ m). What is the probability that the alpha particle will tunnel through the barrier? (*Hint:* A potentially useful value is $\hbar c$ = 197.327 MeV fm.)

37.73 A 6.31-MeV alpha particle (mass = 3.7274 GeV/c^2) inside a heavy nucleus encounters a barrier whose average height is 15.7 MeV and whose width is 13.7 fm (1 fm = $1 \cdot 10^{-15}$ m). What is the value of the decay constant, γ (in fm^{-1})? (*Hint:* A potentially useful value is $\hbar c$ = 197.327 MeV fm.)

37.74 An alpha particle (mass = 3.7274 GeV/c^2) inside a heavy nucleus encounters a barrier whose average height is 15.7 MeV and whose width is 15.5 fm (1 fm = $1 \cdot 10^{-15}$ m). The decay constant, γ, is measured to be 1.257 fm^{-1}. What is the kinetic energy of the alpha particle? (*Hint:* A potentially useful value is $\hbar c$ = 197.327 MeV fm.)

37.75 A 8.59-MeV alpha particle (mass = 3.7274 GeV/c^2) inside a heavy nucleus encounters a barrier whose average height is 15.9 MeV. The tunneling probability is measured to be $1.042 \cdot 10^{-18}$. What is the width of the barrier? (*Hint:* A potentially useful value is $\hbar c$ = 197.327 MeV fm.)

37.76 An electron with a mass of $9.109 \cdot 10^{-31}$ kg is trapped inside a one-dimensional infinite potential well of width 13.5 nm. What is the energy difference between the $n = 5$ and the $n = 1$ states?

37.77 A proton with a mass of $1.673 \cdot 10^{-27}$ kg is trapped inside a one-dimensional infinite potential well of width 23.9 nm. What is the quantum number, n, of the state that has an energy difference of $1.08 \cdot 10^{-3}$ meV with the $n = 2$ state?

37.78 A particle is trapped inside a one-dimensional infinite potential well of width 19.3 nm. The energy difference between the $n = 2$ and the $n = 1$ states is $2.639 \cdot 10^{-25}$ J. What is the mass of the particle?

38

Atomic Physics

FIGURE 38.1 A sodium laser (nearly vertical yellow line) is fired into the night sky to provide a laser guide for the adaptive optics of the telescopes of the Keck Observatory in Mauna Kea, Hawaii. The long exposure of the photograph make the stars appear as (very slightly) curved lines.

Lasers were first developed less than 50 years ago, but they have become common in many areas of daily life. They are used in Blu-ray, DVD, and CD players, in bar-code readers in stores, even in light shows at rock concerts and other events. They also have important scientific applications, for example, in astronomy. Observatories fire lasers into the sky (Figure 38.1) and see how the light is distorted by atmospheric disturbance. Computers then feed this information back to adaptive mirrors that have movable sections that correct for atmospheric motion, producing steady, sharp images.

Lasers work by producing photons of light as electrons move between energy levels in atoms, a topic we will study in this chapter. The concept of atomic energy levels was part of the Bohr model of the atom, which was one of the earliest attempts at applying quantum ideas to atomic structure. However, despite some outstanding successes, the Bohr model did not fully explain how particles function together within atoms; that explanation requires application of the quantum wave mechanics of Chapter 37. We will present an overview of this application, skipping over many of the mathematical details, and then show how the results underlie the operation of X-ray tubes and lasers.

Chapters 36 and 37 surveyed the phenomena that can be explained by the quantization hypothesis, and we solved some simple systems by following the laws imposed by quantum mechanics. With this introductory understanding of quantum mechanics, we have the tools to understand atoms.

WHAT WE WILL LEARN

- The narrow lines observed in the spectra of some atoms result from the transitions of electrons from one state to another.

- The Bohr model of the atom, which is based on a quantization condition for the angular momentum, was an early success in explaining the observed line spectra of hydrogen. The electrons' orbital radii and the atom's energy levels obtained with the Bohr model are physically correct. However, the Bohr model does not obtain the correct values of the quantum numbers and is thus intrinsically flawed.

- To go beyond the Bohr model and find for the correct electron wave functions of the hydrogen atom, we have to solve the Schrödinger equation for the electrons in the Coulomb potential due to the nucleus.

- The complete solution of the Schrödinger equation for the hydrogen atom is the product of the angular part and the radial part. The quantum numbers from these solutions are the radial quantum number, $n = 1,2,3, ...$; the orbital angular momentum quantum number, $\ell = 0,1, ..., n - 1$; and the magnetic quantum number, $m = -\ell, ..., \ell$.

- The wave functions for other atoms can be determined in a manner similar to that used for the hydrogen wave functions. The Pauli Exclusion Principle requires that each possible state be occupied by at most two electrons, one with spin down and one with spin up.

- Lasers work on the basis of population inversion, in which electrons are lifted into a metastable state and subsequently stimulated to emit photons when they make transitions from the metastable state to the ground state.

FIGURE 38.2 A prism decomposes white light into different colors (wavelengths).

38.1   Spectral Lines

Chapter 32 showed that the index of refraction of light in a prism depends on the wavelength. Figure 38.2 shows how white light (coming in from the left) is split into different wavelength components by a prism, as a result of the dependence of the index of refraction on wavelength.

A **spectrometer** can be constructed using a prism. This instrument displays the spectral components of light emitted by all kinds of objects. Figure 38.3 shows a kind of spectrometer often used in introductory physics laboratories. One arm of the spectrometer points toward a light source and provides a narrowly focused thin beam of light, which strikes the central prism and is then spectrally decomposed. The other spectrometer arm contains the ocular lenses and is used for observations. It rotates around a central pivot axis located under the center of the prism. With the spectrometer, we can measure the angle between the two arms and thus determine the angle of deflection of each color. The index of refraction of the prism is known, and thus Snell's Law can be applied to precisely determine the wavelengths of the light.

In the late 19th century, spectrometers were used in many laboratories to study gases. Researchers trapped a gas in an electric discharge tube and then sent a current through the tube, causing the gas to emit light. (The black box in the right part of Figure 38.3 holds the purple-glowing electric discharge tube with its power supply.) Incidentally, this way of generating light is the basis for neon lights.

A spectrometer pointed at the Sun will spectrally decompose light into a spectrum like the one shown in Figure 38.2—that is, an array of different colors. Measuring the intensity, or spectral emittance, of the different colors of the Sun's light makes it possible to measure the temperature of the Sun's surface, on the assumption that the Sun emits a blackbody spectrum. An expensive spectrometer is not needed to see that the Sun emits a broad spectrum of all colors; a child's toy prism can do the job.

However, if you pointed a spectrometer at a discharge tube filled with pure hydrogen and looked at the spectrum emitted by this gas, you might be very surprised. Instead of a broad continuous spectrum, you would see a few thin lines of particular colors. At most there are four lines: red (wavelength 656 nm, also historically called H-alpha), teal blue (wavelength 486 nm, H-beta), deep blue (434 nm, H-gamma), and finally violet (410 nm, H-delta). Other discrete lines appear in the hydrogen spectrum, but they lie outside the range of

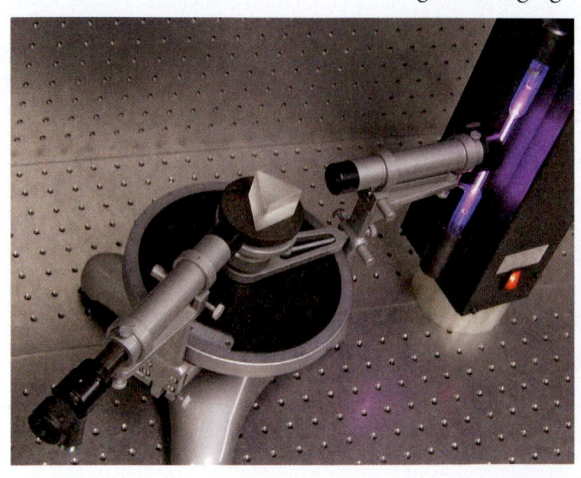

FIGURE 38.3 Spectrometer with prism.

wavelengths accessible by the unaided human eye. Still, it is possible to determine the wavelengths of these other spectral lines, and this was done to very good precision in the 19th century. The lower part of Figure 38.4 shows where the hydrogen lines are located as a function of their wavelengths, λ. Shown are the four visible lines in their characteristic colors. In addition, the invisible lines are indicated with thin white lines. These lines fall into distinct groups, named after their discoverers: the Lyman, Balmer, Paschen, and Brackett series. (Other series, not shown, were discovered at wavelengths above 2000 nm.)

Purely from studying patterns in the values for the wavelengths, the Swiss mathematician and hobbyist-physicist Johann Balmer (1825–1898) found in 1885 an empirical formula that predicted the wavelengths in the Balmer group of the hydrogen spectrum:

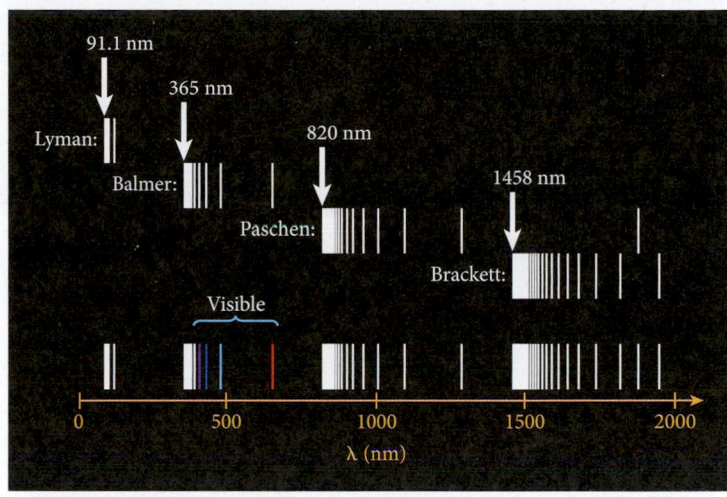

FIGURE 38.4 Calculated spectral lines of hydrogen.

$$\lambda = (364.56 \text{ nm}) \frac{n^2}{n^2 - 4} \quad \text{(for } n = 3, 4, 5, \dots\text{)}. \tag{38.1}$$

Three years later, the Swedish physicist Johannes Rydberg (1854–1919) was able to generalize the Balmer formula to include all other line series (Lyman, Paschen, Brackett, etc.) in the hydrogen spectrum. The **Rydberg formula** for the wavelengths in the hydrogen spectrum is

$$\frac{1}{\lambda} = R_H \left(\frac{1}{n_1^2} - \frac{1}{n_2^2} \right), \quad \text{with } n_1 < n_2. \tag{38.2}$$

The numbers n_1 and n_2 are simple integers; R_H is the **Rydberg constant** for hydrogen and has the dimension of inverse length. The value of R_H is

$$R_H = 1.097373 \cdot 10^7 \text{ m}^{-1}.$$

For $n_1 = 1$ and $n_2 = 2, 3, 4, \dots, \infty$, we obtain the wavelengths of the Lyman series; for $n_1 = 2$ and $n_2 = 3, 4, 5, \dots, \infty$, we obtain the Balmer series; and so on.

EXAMPLE 38.1 Spectral Lines

PROBLEM
Figure 38.4 gives the wavelength of the leftmost line in each of the first four series in the hydrogen spectrum. What are the wavelengths for the rightmost lines in these series?

SOLUTION
The wavelengths of the lines in the hydrogen spectrum are described by the Rydberg formula (equation 38.2):

$$\frac{1}{\lambda} = R_H \left(\frac{1}{n_1^2} - \frac{1}{n_2^2} \right).$$

Because $n_1 < n_2$, the term $1/n_1^2$ is always greater than $1/n_2^2$. For a given fixed value of n_1, an increasing value of n_2 means that a smaller number, $1/n_2^2$, is subtracted from $1/n_1^2$ and thus $1/\lambda$ increases. Therefore, the wavelength decreases monotonically with increasing n_2.

Now that we understand this dependence, we can see that the leftmost and therefore smallest wavelength in each series must correspond to the largest value of n_2, which is $n_2 \approx \infty$. This is also the reason why the lines become so densely spaced at this end of the series: There are an infinite number of values of n_2 with almost exactly the same wavelength.

Conversely, the rightmost and therefore largest value of the wavelength in each series must correspond to the smallest possible value of n_2 in that series. Since we have the condition

– Continued

$n_1 < n_2$, this smallest possible value is $n_{2,\min} = n_1 + 1$. Using this relation, we find for the maximum wavelength in each series:

Lyman series: $n_1 = 1,\ n_{2,\min} = 1 + 1 = 2 \Rightarrow \dfrac{1}{\lambda_{\max}} = R_H\left(1 - \dfrac{1}{4}\right) \Rightarrow \lambda_{\max} = 121.5 \text{ nm}$

Balmer series: $n_1 = 2,\ n_{2,\min} = 2 + 1 = 3 \Rightarrow \dfrac{1}{\lambda_{\max}} = R_H\left(\dfrac{1}{4} - \dfrac{1}{9}\right) \Rightarrow \lambda_{\max} = 656 \text{ nm}$

Paschen series: $n_1 = 3,\ n_{2,\min} = 3 + 1 = 4 \Rightarrow \dfrac{1}{\lambda_{\max}} = R_H\left(\dfrac{1}{9} - \dfrac{1}{16}\right) \Rightarrow \lambda_{\max} = 1875 \text{ nm}$

Brackett series: $n_1 = 4,\ n_{2,\min} = 4 + 1 = 5 \Rightarrow \dfrac{1}{\lambda_{\max}} = R_H\left(\dfrac{1}{16} - \dfrac{1}{25}\right) \Rightarrow \lambda_{\max} = 4050 \text{ nm}.$

DISCUSSION

The maximum wavelength in the Balmer series corresponds to the red H-alpha line in the visible spectrum and is thus probably the most prominent line in the entire hydrogen spectrum. You can see that the highest values of the wavelengths in each of the first two series are lower than the lowest value in the next higher series. However, this changes with the Paschen series. Its longest wavelength line has a wavelength of 1875 nm, which is greater than all but a few wavelengths in the Brackett series, which has its minimum wavelength at a value of 1458 nm. The lines with the highest wavelengths in the Lyman, Balmer, and Paschen series are shown in Figure 38.4, whereas the highest wavelength of the Brackett series lies far outside the range of wavelengths displayed.

Before we go on, let's emphasize the main point: Hydrogen gas can emit light when excited, but the light cannot have any wavelength but only the well-defined wavelengths described by the Rydberg formula. Chapters 36 and 37, which discussed all kinds of quantized phenomena, lead us to look for an explanation for the discreteness of the hydrogen spectrum within quantum mechanics. The next section covers Bohr's quantum description, which is valuable for its insight, its historical significance, and its limited success. The rigorous quantum mechanical analysis of the hydrogen atom will be presented in Section 38.3.

Other elements in their gaseous state also show similar **line spectra** with discrete lines. In particular, some elements (such as lithium and sodium) have hydrogen-like spectra in their gaseous state. In general, however, the line spectra of other atoms are more complicated than the simplest case of the hydrogen atom. The characteristic line spectra often serve as atomic "fingerprints" for detecting the presence of specific chemical elements. Astronomers use this technique when they want to determine the elemental composition of stars, for example.

38.2 Bohr's Model of the Atom

An atom consists of a nucleus, which contains positively charged protons and uncharged neutrons, surrounded by negatively charged electrons that are bound to it by Coulomb interactions. In an electrically neutral atom, the number of protons is equal to the number of electrons. The ionization of an atom—that is, the removal of one or more electrons— causes the remaining ion to be positively charged. The typical size of an atom is on the order of $d \approx 10^{-10}$ m, and the typical size of the atomic nucleus is smaller than that by a factor of 10,000. Hydrogen—the most abundant element in the universe—is also the simplest atom in the universe, consisting of only one proton in the central nucleus with one electron orbiting it.

It is not unusual nowadays for children to learn most of these basic facts about atoms in elementary school. However, at the beginning of the 20th century, the atom was unknown territory, and it was not at all clear what its basic structure was. For example, one possible model of an atom was the "plum pudding" model. In this model, the entire atom was filled with positive charge, and the electrons were equally distributed over the entire atomic volume like raisins in a plum pudding. However, the 1909 scattering experiments of Ernest Rutherford, Hans Geiger, and Ernest Marsden brought clarity and led to the currently accepted model. The physics of the atomic nucleus will be discussed in Chapters 39 and 40.

After Rutherford's experiments, physicists believed that electrons orbit the nucleus, not unlike the way in which the Earth and the other planets orbit the Sun. Let's quantify what this means for the simplest atom, hydrogen. Chapter 9 showed that motion in a circular orbit requires a constant centripetal acceleration. Because the gravitational force acting on the electron due to the nucleus is so weak, the centripetal force is provided by the Coulomb force (see Chapter 21). This leads us to

$$k\frac{e^2}{r^2} = \mu\frac{v^2}{r}, \tag{38.3}$$

where $k = 8.98755 \cdot 10^9$ N m^2/C^2 is the Coulomb constant. This equation simply restates Newton's Second Law, with the force given by the Coulomb force and the acceleration as the centripetal acceleration.

A remark on the notation μ for the mass is in order. Up to now, we have been using m for the mass of the electron. However, the electron in a hydrogen atom is not moving around the proton; instead, both are moving around a common center of mass. We can incorporate this effect by introducing the **reduced mass, μ:**

$$\mu = \frac{mM}{m+M}, \tag{38.4}$$

where $m = 9.10938291(40) \cdot 10^{-31}$ kg is the electron mass and $M = 1.672621777(74) \cdot 10^{-27}$ kg is the proton mass. This reduced mass has the numerical value $\mu = 9.10442 \cdot 10^{-31}$ kg. Because the proton mass is 1836 times the electron mass, the term $M/(M + m)$ is very close to 1, and thus $\mu \approx m$ to 1 part in 2000. Thus, we could still use the notation m for the mass in this case, and it would be close enough for most purposes. However, we will use the reduced mass μ from now on. An advantage to using μ in our formulations is that historically, a quantum number m was introduced in the solution to the hydrogen problem, and we want to avoid confusion.

This classical picture of a miniature planetary system of electrons in orbit around a central nucleus, which plays the role that the Sun does in the Solar System, has immense appeal and gives rise to a symmetry between the macro-cosmos and the micro-cosmos that just rings true. Right away, though, it runs into a fatal conceptual flaw: An accelerated charge radiates away energy. An electron moving on a classical circular orbit and experiencing constant centripetal acceleration would quickly lose energy and spiral into the nucleus, thus destroying the atom. Niels Bohr addressed this flaw in 1913.

Quantization of Orbital Angular Momentum

In 1913, the Danish physicist Niels Bohr (1885–1962) made an *ad hoc* assumption that the angular momentum of the electron in its orbit around the nucleus can assume only discrete values. Chapter 10 introduced the angular momentum vector of a point particle as, $\vec{L} = \vec{r} \times \vec{p}$. Bohr's idea was to require that the only possible stable electron orbits should be those whose angular momentum is an integer multiple of $\hbar$ (Planck's constant divided by 2π):

$$L = |\vec{r} \times \vec{p}| = r\mu v = n\hbar \quad \text{(for } n = 1,2,3,...\text{)}. \tag{38.5}$$

Why impose this condition? Dimensionally it works out, because both angular momentum and Planck's constant have units of kg m^2 s^{-1} = J s. However, of course, this fact alone is not sufficient. In the following discussion, we show that the quantization of angular momentum also leads to the quantization of energy. In addition, if the electron's energy is quantized, then the electron cannot radiate arbitrarily small amounts of energy and spiral into the nucleus.

Let's first work out the consequences and predictions of the Bohr postulate and then discuss its meaning and the validity of the model. To begin the calculations for the **Bohr model of the hydrogen atom,** we start with equation 38.3 for the centripetal force and multiply both sides by the factor μr^3. This leads to

$$k\frac{e^2}{r^2} = \mu\frac{v^2}{r} \Rightarrow r\mu ke^2 = \mu^2 v^2 r^2.$$

On the right-hand side, we can see that $\mu^2 v^2 r^2$ is the square of the angular momentum. From Bohr's quantization condition (equation 38.5), this is equal to $n^2 \hbar^2$. Therefore,

$$r \mu k e^2 = \mu^2 v^2 r^2 = n^2 \hbar^2.$$

Solving this for the orbital radius, r, results in

$$r = \frac{\hbar^2}{\mu k e^2} n^2 \equiv a_0 n^2. \tag{38.6}$$

This equation gives the allowed radii in the Bohr model of the hydrogen atom. They are proportional to the square of the quantum number n, and the proportionality constant, a_0, is called the **Bohr radius:**

$$a_0 = \frac{\hbar^2}{\mu k e^2} = \frac{\left(1.05457 \cdot 10^{-34} \text{ J s}\right)^2}{\left(9.10442 \cdot 10^{-31} \text{ kg}\right)\left(8.98755 \cdot 10^9 \text{ N m}^2/\text{C}^2\right)\left(1.60218 \cdot 10^{-19} \text{ C}\right)^2}$$

$$a_0 = 5.295 \cdot 10^{-11} \text{ m} = 0.05295 \text{ nm} = 0.5295 \text{ Å}.$$

The value of the Bohr radius is given here in nanometers (nm) as well as angstroms (Å), with $1 \text{ Å} = 10^{-10} \text{ m} = 0.1 \text{ nm}$. Inserting this result for the Bohr radius back into equation 38.3, we find for the speed v:

$$v = \sqrt{\frac{k e^2}{\mu a_0 n^2}} = \frac{1}{n} \sqrt{\frac{\left(8.988 \cdot 10^9 \text{ N m}^2/\text{C}^2\right)\left(1.602 \cdot 10^{-19} \text{ C}\right)^2}{\left(9.104 \cdot 10^{-31} \text{ kg}\right)\left(5.295 \cdot 10^{-11} \text{ m}\right)}} = \frac{1}{n} 2.188 \cdot 10^6 \text{ m/s} = \frac{1}{n} 0.007297 c.$$

This speed is 0.73% of the speed of light for $n = 1$ and decreases monotonically for the higher orbits. Therefore, the nonrelativistic approximation that we are using is justified. The total energy of the electron in orbit is the sum of its potential and kinetic energies:

$$E = \tfrac{1}{2} \mu v^2 - k \frac{e^2}{r} = -\tfrac{1}{2} k \frac{e^2}{r} = -\tfrac{1}{2} k \frac{e^2}{a_0 n^2} = -E_0 \frac{1}{n^2}. \tag{38.7}$$

The second step of this equation uses the general result for satellite motion (see Chapter 12) that the kinetic energy has exactly half of the magnitude of the potential energy, so the total energy is half of the potential energy. (This relation is often referred to as the *virial theorem*.) The constant E_0 can be calculated by inserting the values of the constants:

$$E_0 = \frac{k e^2}{2 a_0} = \frac{\left(8.988 \cdot 10^9 \text{ N m}^2/\text{C}^2\right)\left(1.602 \cdot 10^{-19} \text{ C}\right)^2}{2\left(5.295 \cdot 10^{-11} \text{ m}\right)} = 2.18 \cdot 10^{-18} \text{ J} = 13.6 \text{ eV}.$$

Thus, the allowed energies for electrons in orbit around the nucleus of the hydrogen atom are $E(n = 1) = -13.6$ eV, $E(n = 2) = -3.40$ eV, $E(n = 3) = -1.51$ eV, and so on, approaching zero energy from below as $n \rightarrow \infty$. Combining equations 38.6 and 38.7 gives $E = -E_0 a_0/r$. Figure 38.5 shows E plotted versus r and indicates that only certain values of E and r, corresponding to n being an integer in equations 38.6 and 38.7, are allowed.

In many textbooks, the Bohr radius is defined in terms of the mass of the electron, m_e, rather than the reduced mass, μ, of the electron in the hydrogen atom:

$$a_{0,m_e} = \frac{\hbar^2}{m_e k e^2}.$$

The Bohr radius has the value $a_{0,m_e} = 5.292 \cdot 10^{-11}$ m using this definition.

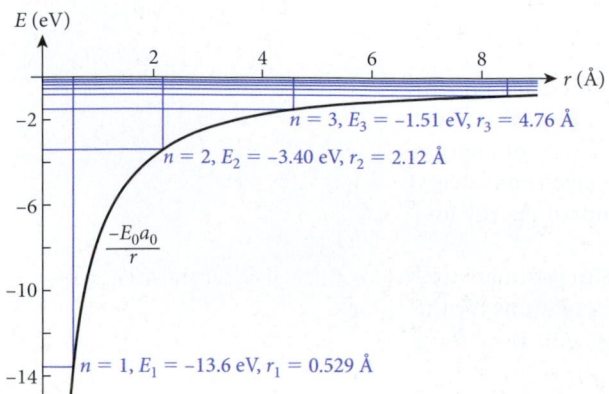

E (eV)

$n = 3, E_3 = -1.51$ eV, $r_3 = 4.76$ Å

$n = 2, E_2 = -3.40$ eV, $r_2 = 2.12$ Å

$-\dfrac{E_0 a_0}{r}$

$n = 1, E_1 = -13.6$ eV, $r_1 = 0.529$ Å

FIGURE 38.5 Energies and radii for allowed electron orbits in the Bohr model of hydrogen.

Spectral Lines in the Bohr Model

In the Bohr model, an electron cannot radiate away small amounts of energy and thus spiral into the nucleus, because of the quantization of angular momentum. However, transitions between states are still allowed. Bohr postulated that an electron in a higher energy state with quantum number n_2 could "jump" to a lower energy state with quantum number n_1, and the atom would emit a photon with energy equal to the difference in the energies between the two states. The photon energy is related to its frequency via Planck's relation, $E = hf$: $E_{n_2} = E_{n_1} + hf$. For light, the frequency is related to the wavelength via $f = c/\lambda$, where c is the speed of light, giving

$$E_{n_2} = E_{n_1} + \frac{hc}{\lambda}. \tag{38.8}$$

The formula for the energies (equation 38.7) then gives

$$-\frac{ke^2}{2a_0}\frac{1}{n_2^2} = -\frac{ke^2}{2a_0}\frac{1}{n_1^2} + \frac{hc}{\lambda} \Rightarrow$$

$$\frac{1}{\lambda} = \frac{ke^2}{2hca_0}\left(\frac{1}{n_1^2} - \frac{1}{n_2^2}\right).$$

This equation is structurally identical to equation 38.2. Furthermore, the product of the constants in this equation evaluates to

$$\frac{ke^2}{2hca_0} = \frac{\left(8.988 \cdot 10^9 \text{ N m}^2/\text{C}^2\right)\left(1.602 \cdot 10^{-19} \text{ C}\right)^2}{2\left(6.626 \cdot 10^{-34} \text{ J s}\right)\left(2.998 \cdot 10^8 \text{ m/s}\right)\left(5.295 \cdot 10^{-11} \text{ m}\right)} = 1.097 \cdot 10^7 \text{ m}^{-1}.$$

This agrees with the value of the Rydberg constant, which was determined from the experimental data, to four significant figures. That is, we can write

$$R_\text{H} = \frac{ke^2}{2hca_0}.$$

Given $a_0 = \hbar^2/\mu ke^2$, the Rydberg constant can also be written as

$$R_\text{H} = \frac{ke^2}{2hc(\hbar^2/\mu ke^2)} = \frac{\mu k^2 e^4}{4\pi c\hbar^3}.$$

In other words, the Bohr model is able to explain the structure of the hydrogen line spectrum and can be used to derive the experimentally found value of the Rydberg constant from the fundamental constants of the electron mass m (contained in the reduced mass, μ), charge quantum e, Planck's constant h (equal to $2\pi\hbar$), the speed of light c, and Coulomb's constant k. This was properly celebrated as an astounding success and lent considerable credence to the seemingly *ad hoc* assumption of quantized angular momentum of the electron in its orbit around the central nucleus. Transitions between energy levels can be depicted in an **energy-level diagram,** as is done in Figure 38.6 for the Balmer series.

However, at its heart, the Bohr model assumes that the electron is a classical point particle. The model was refined several times, but after Louis de Broglie postulated matter waves in 1923 (see Chapter 36) and Schrödinger published his wave equation in 1926 (see Chapter 37), it became clear that the Bohr model had to be replaced by a quantum mechanical theory of the hydrogen atom. In addition to the conceptual problems associated with a classical point particle, the Bohr model was also flawed in that it postulated an orbital angular momentum of $\hbar$ for the lowest energy

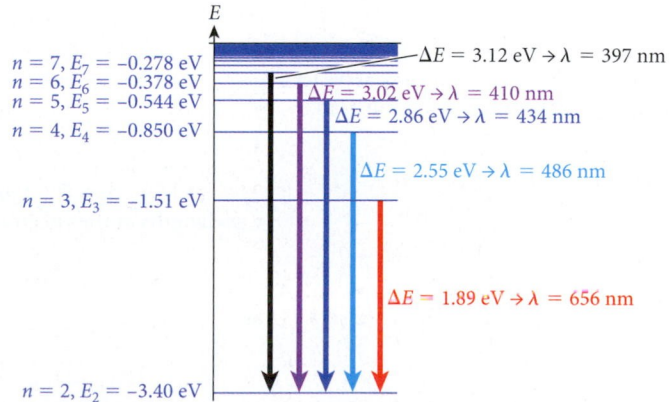

FIGURE 38.6 The first few electron transitions corresponding to the Balmer series in hydrogen.

state—the ground state—of hydrogen, in contradiction to experimental evidence that points to an orbital angular momentum of zero for this state. Nevertheless, the Bohr model was a very instructive first attempt at solving the hydrogen problem, and its impressive success in explaining line spectra hints that the real solution to this problem must somehow be close to what the Bohr model postulates.

SOLVED PROBLEM 38.1 | Paschen Series for Doubly Ionized Lithium

We learned that the Lyman, Balmer, Paschen, Brackett, and other series in the hydrogen spectrum correspond to transitions in the hydrogen atom, in which the atom's (single) electron "jumps" between two discrete energy states. We can observe the equivalent line spectra for one-electron ions of other atoms. Doubly ionized lithium is such an ion.

PROBLEM
What is the wavelength of the second line of the Paschen series of doubly ionized lithium?

SOLUTION
THINK Lithium has an atomic number of $Z = 3$. A doubly ionized lithium ion has two fewer electrons than the neutral atom. This leaves one electron in the Coulomb potential due to the nucleus, and we obtain hydrogen-like wave functions. The only difference from the case of the hydrogen atom is that we have $-kZe^2/r = -3ke^2/r$ for the Coulomb potential in this case. Thus, the formulas for the line spectra of the hydrogen atom still hold, but we need to find the appropriate Rydberg constant for this case.

SKETCH Though the Bohr model of the hydrogen atom has fatal shortcomings, it is still useful to sketch the Bohr radii (Figure 38.7) to help clarify which transition is responsible for the spectral line.

RESEARCH The wavelengths for hydrogen-like line spectra are given by equation 38.2:

$$\frac{1}{\lambda} = R_H \left(\frac{1}{n_1^2} - \frac{1}{n_2^2} \right), \quad \text{with } n_1 < n_2.$$

For the Rydberg constant, R_H, we had correctly obtained from the Bohr model

$$R_H = \frac{\mu k^2 e^4}{4\pi c \hbar^3},$$

where μ is the reduced mass of the electron (which for this lithium ion is even closer to the exact electron mass than in the hydrogen atom, where it deviated by only 1 part in 2000 from the real value).

In this case, we need to replace ke^2 by $3ke^2$ to obtain the Rydberg constant for the doubly ionized lithium ion. It is 9 times bigger ($(3ke^2)^2 = 9(ke^2)^2 = 9k^2e^4$) than the value of $R_H = 1.097373 \cdot 10^7 \text{ m}^{-1}$ for the hydrogen atom.

Finally, from Figure 38.7, we see that $n_1 = 3$ for the Paschen series and that the second line in that series must have $n_2 = 5$.

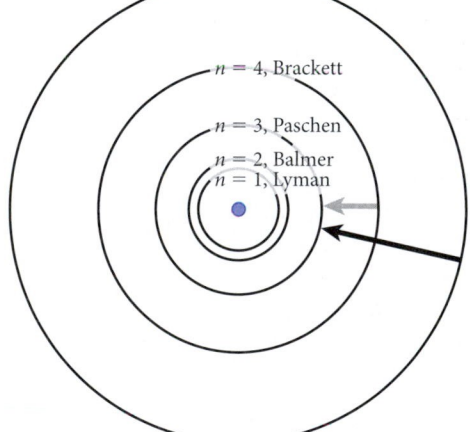

FIGURE 38.7 Transition responsible for the second line of the Paschen series (in black).

SIMPLIFY Inserting $9R_H$ instead of R_H and $n_1 = 3$ and $n_2 = 5$ into the formula for the wavelengths of the spectral lines, we obtain

$$\frac{1}{\lambda} = 9R_H \left(\frac{1}{9} - \frac{1}{25} \right)$$

or

$$\lambda = \left(9R_H \left(\frac{1}{9} - \frac{1}{25} \right) \right)^{-1}.$$

CALCULATE Now it is time to put in the numbers:

$$\lambda = \left(9(1.097373 \cdot 10^7 \text{ m}^{-1}) \left(\frac{1}{9} - \frac{1}{25} \right) \right)^{-1} = 1.423855 \cdot 10^{-7} \text{ m}.$$

ROUND Within the assumptions stated, our result is exact in the sense that we used only the measured value of the Rydberg constant and integer values for the quantum numbers. However, the reduced mass of the electron is not exactly the same for the doubly ionized lithium ion as it is for the hydrogen atom, so we have a deviation on the order of 1 part in 2000, and it is prudent to round our result to three or at most four digits. Our final answer is thus

$$\lambda = 1.424 \cdot 10^{-7}\ \text{m} = 142.4\ \text{nm}.$$

DOUBLE-CHECK   Figure 38.4 shows that the Paschen lines in the hydrogen spectrum start at 820 nm. The second line in that series thus must have a wavelength greater than 820 nm. From our research above, it is clear that the wavelength for the second line in the Paschen series for the lithium ion should be smaller by a factor of 9 than this value, which it is. So we have additional confidence that our result is realistic.

38.3 Hydrogen Electron Wave Function

To improve on the Bohr model of the hydrogen atom, we have to return to what we learned about quantum mechanics in Chapter 37 and solve the Schrödinger equation for the electron in a Coulomb potential. To find the bound states of the electron, we need to solve the time-independent Schrödinger equation and find the energy eigenvalues, E_n. We hope our solution will resemble equation 38.7, $E_n = -13.6/n^2$ eV, because this prediction of the Bohr model was found to be in agreement with experimental data.

A word of caution before we start on this task: Part of the mathematics involved in solving this problem is tedious and perhaps as advanced as any we will present. This is the bad news. The good news is that this problem is very instructive, providing a glimpse of many deep features of quantum mechanics and of ways to illuminate them. The math will not be worked out in all its details, and sometimes the outcome of a calculation will simply be stated. Do not become discouraged as we study this essential quantum system. It will not only help you understand the hydrogen atom, but also give insight into the entire periodic table of the elements.

The potential in the Schrödinger equation is the Coulomb potential, $U(\vec{r}) = -ke^2/r$. Note that this potential depends only on the radial distance to the origin (where the nucleus is located), and not on the angular direction of the electron relative to the nucleus. The hydrogen atom is an object in the real world and thus exists in three-dimensional space. According to Chapter 37, the Schrödinger equation in three-dimensional space must be written as

$$-\frac{\hbar^2}{2\mu}\frac{\partial^2 \psi}{\partial x^2} - \frac{\hbar^2}{2\mu}\frac{\partial^2 \psi}{\partial y^2} - \frac{\hbar^2}{2\mu}\frac{\partial^2 \psi}{\partial z^2} + U\psi = -\frac{\hbar^2}{2\mu}\nabla^2 \psi + U\psi = E\psi.$$

The Laplacian operator, ∇^2, in Cartesian coordinates is $\nabla^2 = \partial^2/\partial x^2 + \partial^2/\partial y^2 + \partial^2/\partial z^2$. This operator appeared in Chapter 37 in the section on antimatter. It is a seemingly straightforward generalization of the second derivative d^2/dx^2, which appears in the one-dimensional Schrödinger equation.

Now we need to choose a coordinate system. The potential energy term U depends only on the radial coordinate r, so it is advantageous to use spherical coordinates, r, θ, ϕ, as shown in Figure 38.8. Then the Schrödinger equation reads

$$-\frac{\hbar^2}{2\mu}\nabla^2 \psi(r,\theta,\phi) - k\frac{e^2}{r}\psi(r,\theta,\phi) = E\psi(r,\theta,\phi), \qquad (38.9)$$

where the Coulomb potential is used for the potential U.

With spherical coordinates, however, there is a price to pay: The Laplacian operator looks much more complicated:

$$\nabla^2 \psi = \frac{1}{r}\frac{\partial^2}{\partial r^2}(r\psi) + \frac{1}{r^2 \sin\theta}\frac{\partial}{\partial\theta}\left(\sin\theta\frac{\partial}{\partial\theta}\psi\right) + \frac{1}{r^2 \sin^2\theta}\frac{\partial^2}{\partial\phi^2}\psi. \qquad (38.10)$$

Is this worth the effort? The answer is certainly yes if the wave function does not depend on the angular coordinates but only on the radial one. Therefore, we try this method first and look for spherically symmetric solutions.

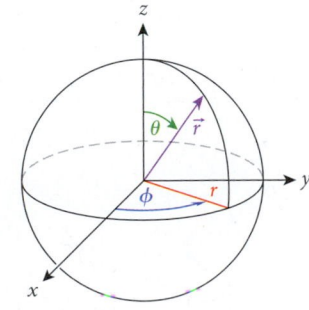

FIGURE 38.8   Definition of the spherical coordinates.

Spherically Symmetric Solutions

Motivated by the Bohr model's success, we first investigate whether there are spherically symmetric solutions. If these exist, they can only be a function of the radial coordinate, r. Let's work with this assumption and see where it leads us. The Laplacian operator simplifies greatly because the derivatives with respect to θ and ϕ vanish, leaving us with

$$\nabla^2 \psi(r) = \frac{1}{r} \frac{\partial^2}{\partial r^2} \big(r\psi(r)\big) = \frac{1}{r} \frac{d^2}{dr^2} \big(r\psi(r)\big).$$

Notice that we replaced the partial derivative by the conventional derivative, because the wave function in this case depends on only one variable, r. Thus, for the Schrödinger equation, we now have

$$-\frac{\hbar^2}{2\mu} \frac{1}{r} \frac{d^2}{dr^2} \big(r\psi(r)\big) - k\frac{e^2}{r} \psi(r) = E\psi(r). \tag{38.11}$$

The solution of this differential equation is tedious, but the result is very interesting. We find that, similar to the particle in a box or the harmonic oscillator, the energy can assume only certain values of E_n, given by

$$E_n = -\frac{\mu k^2 e^4}{2\hbar^2} \frac{1}{n^2} \quad \text{(for } n = 1, 2, 3, \dots\text{)}.$$

This is exactly the form of the energy eigenvalues we obtained for the Bohr model (equation 38.7) with $E_0 = \mu k^2 e^4 / 2\hbar^2 = 13.6$ eV, an energy unit sometimes called a *rydberg*.

The wave functions that correspond to the lowest values of the quantum number n are

$$\psi_1(r) = A_1 e^{-r/a_0}$$

$$\psi_2(r) = A_2 \left(1 - \frac{r}{2a_0}\right) e^{-r/2a_0} \tag{38.12}$$

$$\psi_3(r) = A_3 \left(1 - \frac{2r}{3a_0} + \frac{2r^2}{27a_0^2}\right) e^{-r/3a_0},$$

where the subscript on ψ represents the corresponding value of n. Here A_1, A_2, and A_3 are normalization constants that can be determined from the condition that the integral of the absolute square of the wave function is 1:

$$\int |\psi_n(r)|^2 \, d^3r = \int_0^\infty 4\pi r^2 |\psi_n(r)|^2 \, dr = 1.$$

Figure 38.9a shows these solutions, and Figure 38.9b shows their probability densities, weighted with the factor r^2 to account for the effect of the volume element increasing with rising radius. The solutions for the three lowest values of the radial quantum number n are

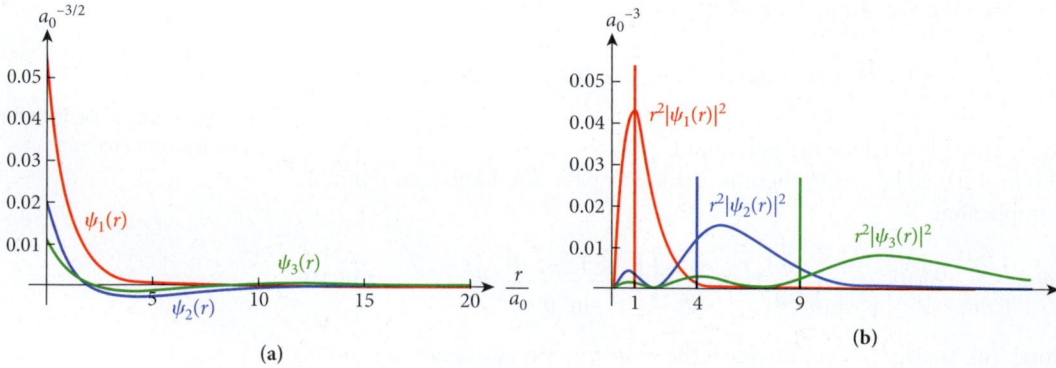

FIGURE 38.9 Spherically symmetric solutions to the Schrödinger equation for the three lowest values of the radial quantum number for the hydrogen atom. (a) Wave functions; (b) weighted probability densities. The vertical lines in (b) represent the Bohr model's predictions.

shown as a function of the radial coordinate divided by the Bohr radius, a_0. Vertical lines indicate the locations of the semiclassical Bohr orbits that correspond to these values of n. For $n = 1$, the maximum of the probability density for the wave function corresponds to the Bohr orbit, whereas this is not the case for $n > 1$.

EXAMPLE 38.2 Normalization of the Hydrogen Wave Function

PROBLEM

The hydrogen wave function corresponding to $n = 1$ has the form given in equation 38.12: $\psi_1(r) = A_1 e^{-r/a_0}$. Using the condition that the integral of the absolute square of the wave function is 1, determine A_1.

SOLUTION

The condition that the integral of the absolute square of the wave function is 1 gives us

$$\int |\psi_n(r)|^2 \, d^3r = \int_0^\infty 4\pi r^2 |\psi_n(r)|^2 \, dr = 1.$$

Putting in the given wave function leads to

$$\int_0^\infty 4\pi r^2 \left| A_1 e^{-r/a_0} \right|^2 dr = 4\pi A_1^2 \int_0^\infty r^2 e^{-2r/a_0} \, dr.$$

The definite integral can be expressed as $\int_0^\infty x^2 e^{-cx} \, dx = 2/c^3$. Taking $c = 2/a_0$, we get

$$4\pi A_1^2 \int_0^\infty r^2 e^{-2r/a_0} \, dr = 4\pi A_1^2 \left[2\left(\frac{a_0}{2}\right)^3 \right] = \pi a_0^3 A_1^2 = 1.$$

We can solve this equation for A_1:

$$A_1 = \frac{1}{\sqrt{\pi a_0^3}} = \frac{1}{\sqrt{\pi} \, a_0^{3/2}}.$$

Thus, the hydrogen wave function corresponding to $n = 1$ is

$$\psi_1(r) = \frac{1}{\sqrt{\pi} \, a_0^{3/2}} e^{-r/a_0}.$$

Angular Momentum

Chapter 37 introduced quantum operators for the momentum and kinetic energy, where the classical value of the momentum is replaced by a constant times the derivative in coordinate space, $-i\hbar d/dx$. In three-dimensional space, the momentum operator is a vector, for which each individual component is a partial derivative in its respective direction. In Cartesian coordinates, this operator is written as

$$\hat{\mathbf{p}}\psi(x,y,z) = -i\hbar \left(\frac{\partial}{\partial x}, \frac{\partial}{\partial y}, \frac{\partial}{\partial z} \right) \psi(x,y,z) \equiv -i\hbar \vec{\nabla} \psi(x,y,z).$$

This derivative operator is called the *gradient,* for which we use the common shorthand notation, $\vec{\nabla}$. In spherical coordinates, the gradient is given by

$$\vec{\nabla}\psi(r,\theta,\phi) = \left(\frac{\partial}{\partial r}, \frac{1}{r}\frac{\partial}{\partial \theta}, \frac{1}{r\sin\theta}\frac{\partial}{\partial \phi} \right) \psi(r,\theta,\phi).$$

Then the equation for the momentum operator becomes

$$\hat{\mathbf{p}}\psi(x,y,z) = -i\hbar\vec{\nabla}\psi(r,\theta,\phi) = -i\hbar \left(\frac{\partial}{\partial r}, \frac{1}{r}\frac{\partial}{\partial \theta}, \frac{1}{r\sin\theta}\frac{\partial}{\partial \phi} \right) \psi(r,\theta,\phi).$$

Now we can write the angular momentum operator by simply replacing the momentum vector by the gradient operator:

$$\hat{\mathbf{L}}\psi = -i\hbar\vec{r}\times\vec{\nabla}\psi.$$ (38.13)

(Remember, classically the angular momentum is $\vec{L}=\vec{r}\times\vec{p}$.) According to the rules for quantum measurements, we can now calculate the expectation value of the angular momentum operator:

$$\langle L\rangle = \int \psi^*\hat{\mathbf{L}}\psi d^3 r = -i\hbar\int\psi^*\vec{r}\times\vec{\nabla}\psi d^3 r.$$ (38.14)

It is particularly interesting to look at the outcome of this measurement of the angular momentum for the case of spherically symmetric wave functions, that is, wave functions that depend only on the radial coordinate r, and not on any angular coordinate, θ or ϕ. Then we obtain

$$\langle L\rangle = \int \psi^*(r)\hat{\mathbf{L}}\psi(r)d^3 r = -i\hbar\int\psi^*(r)\vec{r}\times\vec{\nabla}\psi(r)d^3 r = 0.$$ (38.15)

DERIVATION 38.1 | Angular Momentum Expectation Value

To show that equation 38.15 is true, we use the gradient operator in spherical coordinates. Applying it to the spherically symmetric wave function results in

$$\vec{\nabla}\psi(r)=\left(\frac{\partial}{\partial r},\frac{1}{r}\frac{\partial}{\partial\theta},\frac{1}{r\sin\theta}\frac{\partial}{\partial\phi}\right)\psi(r)=\hat{r}\frac{\partial}{\partial r}\psi(r),$$

because the partial derivatives in the θ and ϕ directions are both zero when the wave function does not depend on either of these angular coordinates. We have again used the notation $\hat{r}$ for the unit vector in the radial direction. In terms of this unit vector, we can write the vector $\vec{r}$ as $\vec{r}=r\hat{r}$. Now the vector product in the integral of equation 38.15 simplifies to

$$-i\hbar\int\psi^*(r)\vec{r}\times\vec{\nabla}\psi(r)d^3 r = -i\hbar\int\psi^*(r)r\hat{r}\times\hat{r}\frac{\partial}{\partial r}\psi(r)d^3 r.$$

For any vector, the vector product with itself has the value zero. Thus, $\hat{r}\times\hat{r}=0$, and the entire integral has the value zero.

This is a very general result that holds for any spherically symmetric wave function: A wave function that does not depend on the angular coordinate θ or ϕ must have an expectation value of zero for the angular momentum.

The spherically symmetric solutions for the hydrogen atom, $\psi_n(r)$, obtained in the previous subsection, depend only on the radius, so all have angular momentum zero. It is not permissible to interpret the quantum number n as one of angular momentum, as the Bohr model did, despite the fact that the energy eigenvalues of the quantum mechanical spherically symmetric solutions reproduce exactly the discrete energy values of the Bohr model. This fact implies that *the Bohr model is fatally flawed*: We cannot start with a classical particle in orbit around a central nucleus, demand quantization of angular momentum, and then obtain consistent results. Despite its astounding success in explaining the hydrogen line spectra in great detail and to great precision, the Bohr model is wrong, and its success was accidental.

Full Solution

Does the Schrödinger equation for hydrogen have solutions that are not spherically symmetric and that have nonzero angular momentum? The answer is yes. We will only sketch out how to get to these solutions. Typically, the mathematics needed to arrive at them is lengthy and complicated. However, it is also instructive and something to look forward to in a more advanced course, where this topic will be covered in greater detail.

In the following discussion, we introduce many different functions. Keep in mind that these are not meant to be memorized. The goal here is for you to understand the general features of the solutions, not their exact form. Even the most seasoned professionals in the field do not know all the functions by heart and look them up in tables when they are needed.

Separation of Variables

We start with a technique we used successfully in finding solutions for the time-dependent Schrödinger equation in one spatial dimension: the separation of variables. To achieve this, we assume that the full wave function can be written as a product of three functions, each of which is a function of only one variable:

$$\psi(r,\theta,\phi) = f(r)g(\theta)h(\phi). \tag{38.16}$$

We insert this trial solution into the Schrödinger equation (38.9) to find

$$-\frac{\hbar^2}{2\mu}\nabla^2\Big(f(r)g(\theta)h(\phi)\Big) - k\frac{e^2}{r}\Big(f(r)g(\theta)h(\phi)\Big) = E\Big(f(r)g(\theta)h(\phi)\Big).$$

Now we evaluate the action of the Laplacian operator (in spherical coordinates, equation 38.10) on this product of functions:

$$\nabla^2\Big(f(r)g(\theta)h(\phi)\Big) = \frac{1}{r}\frac{\partial^2}{\partial r^2}\Big(r\Big(f(r)g(\theta)h(\phi)\Big)\Big)$$

$$+ \frac{1}{r^2\sin\theta}\frac{\partial}{\partial\theta}\left(\sin\theta\frac{\partial}{\partial\theta}\Big(f(r)g(\theta)h(\phi)\Big)\right)$$

$$+ \frac{1}{r^2\sin^2\theta}\frac{\partial^2}{\partial\phi^2}\Big(f(r)g(\theta)h(\phi)\Big)$$

$$= g(\theta)h(\phi)\frac{1}{r}\frac{\partial^2}{\partial r^2}\Big(rf(r)\Big)$$

$$+ \frac{f(r)h(\phi)}{r^2\sin\theta}\frac{\partial}{\partial\theta}\left(\sin\theta\frac{\partial}{\partial\theta}g(\theta)\right)$$

$$+ \frac{f(r)g(\theta)}{r^2\sin^2\theta}\frac{\partial^2}{\partial\phi^2}h(\phi).$$

Inserting this result into the Schrödinger equation and multiplying both sides of the equation by the product $-2\mu r^2/\hbar^2 f(r)g(\theta)h(\phi)$, we arrive at

$$\frac{r}{f(r)}\frac{\partial^2}{\partial r^2}\Big(rf(r)\Big) + \frac{1}{g(\theta)\sin\theta}\frac{\partial}{\partial\theta}\left(\sin\theta\frac{\partial}{\partial\theta}g(\theta)\right)$$

$$+ \frac{1}{h(\phi)\sin^2\theta}\frac{\partial^2}{\partial\phi^2}h(\phi) + \frac{2\mu r^2}{\hbar^2}\left(E + k\frac{e^2}{r}\right) = 0.$$

Now we can see that some terms depend only on the radial variable r, and not on the angular variables θ and ϕ, while other terms depend on θ and/or ϕ, but not on r. We rearrange the preceding equation to get all r-dependent terms on the left-hand side and all others on the right-hand side:

$$\frac{r}{f(r)}\frac{\partial^2}{\partial r^2}\Big(rf(r)\Big) + \frac{2\mu r^2}{\hbar^2}\left(E + k\frac{e^2}{r}\right) =$$

$$-\frac{1}{g(\theta)\sin\theta}\frac{\partial}{\partial\theta}\left(\sin\theta\frac{\partial}{\partial\theta}g(\theta)\right) - \frac{1}{h(\phi)\sin^2\theta}\frac{\partial^2}{\partial\phi^2}h(\phi). \tag{38.17}$$

This equation can be true only if each side is equal to the same constant. We make a choice that may look strange at first and call this constant $\ell(\ell + 1)$.

Radial Part

Using the integration constant $\ell(\ell + 1)$ gives us the radial equation (left-hand side of equation 38.17):

$$\frac{r}{f(r)}\frac{d^2}{dr^2}\left(rf(r)\right)+\frac{2\mu r^2}{\hbar^2}\left(E+k\frac{e^2}{r}\right)=\ell(\ell+1).$$

(Again, we replaced the partial derivatives with conventional ones because the function f depends on only one variable.) If we multiply this equation by $-f(r)\hbar^2/2\mu r^2$, it has the form

$$-\frac{\hbar^2}{2\mu}\frac{1}{r}\frac{d^2}{dr^2}\left(rf(r)\right)-k\frac{e^2}{r}f(r)+\frac{\ell(\ell+1)\hbar^2}{2\mu r^2}f(r)=Ef(r). \tag{38.18}$$

Comparing this equation with the Schrödinger equation for the spherically symmetric solution (equation 38.11), we see that the only difference is the term $\ell(\ell + 1)\hbar^2/2\mu r^2$, which plays the role of an additional potential, in addition to the Coulomb potential. (For $\ell = 0$, we recover the spherically symmetric solution.) What is this additional potential term?

An argument from the orbital motion of a classical particle in a central potential may help. To work out our classical argument, let's start with energy conservation, $\frac{1}{2}\mu v^2 + U(r) = E$. (Remember, this is the classical equivalent of the Schrödinger equation.) Now we can split the velocity into radial and tangential parts and get $\frac{1}{2}\mu v^2 = \frac{1}{2}\mu v_r^2 + \frac{1}{2}\mu v_t^2 = \frac{1}{2}\mu v_r^2 + \frac{1}{2}\mu(\omega r)^2$. Further note that the angular momentum for the case of a point particle in circular motion is $L = \mu r v_t = \mu r^2 \omega$. This definition also holds for all motion in a central potential, not just circular motion, and the angular momentum is a conserved quantity. Thus, we can write $\frac{1}{2}\mu(\omega r)^2 = \frac{1}{2}L^2/\mu r^2$. Therefore, our classical equation for energy conservation in this case is $\frac{1}{2}\mu v_r^2 + U(r) + L^2/2\mu r^2 = E$.

The $\ell(\ell + 1)\hbar^2/2\mu r^2$ term now seems plausible as a term originating from conservation of angular momentum, if we are allowed to substitute the quantum mechanical $\ell(\ell + 1)\hbar^2$ for the classical L^2. Later we will see that these solution functions are also the eigenfunctions for the quantum mechanical operator corresponding to L^2 with eigenvalue $\ell(\ell + 1)\hbar^2$; so this substitution works out just as advertised. Now you can begin to understand the seemingly arbitrary choice for the integration constant, $\ell(\ell + 1)$. The $\ell(\ell + 1)\hbar^2/2\mu r^2$ term represents the quantum mechanical angular momentum barrier in the rotating system, and we will soon see that ℓ is an integer. We say "barrier" because this term is positive and increases as r decreases, just like a potential energy barrier. Figure 38.10 shows the effective potential from equation 38.18 for various values of ℓ.

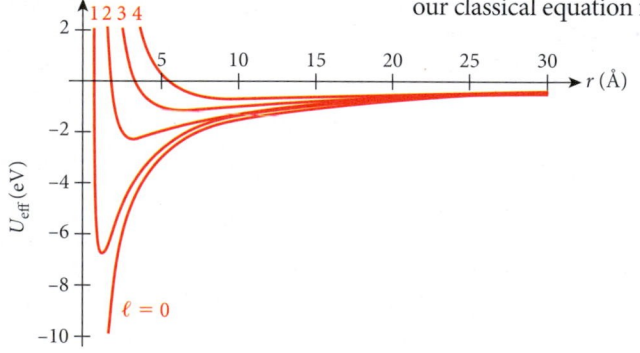

FIGURE 38.10 Effective potential for (from bottom to top) angular momentum quantum numbers $\ell = 0, 1, 2, 3$, and 4.

What are the solutions to the differential equation for the radial function, equation 38.18? In general, they are products of exponential functions and polynomials. For very large r, the exponential function dominates the asymptotic behavior of the solutions. The solutions depend on two quantum numbers. First is the quantum number n, which was present in the spherically symmetric solutions presented earlier. Second is the quantum number ℓ, just introduced for the separation of variables, which we motivated as the quantum number of angular momentum. (The next section will give further evidence for this assignment.) For the sake of completeness, we write out the solutions for the radial functions, but you are not expected to memorize them:

$$f_{n\ell}(r)=\sqrt{\frac{4(n-\ell-1)!}{a_0^3 n^4 (n+\ell)!}}\,e^{-r/na_0}\left(\frac{2r}{na_0}\right)^{\ell}L_{n-\ell-1}^{2\ell+1}\left(\frac{2r}{na_0}\right). \tag{38.19}$$

These solutions exist only for integers $n > \ell$ and $\ell \geq 0$. Therefore, for a given value of n, the quantum number ℓ can have the values

$$0 \leq \ell \leq n-1.$$

The factor $L_{n-\ell-1}^{2\ell+1}\left(2r/na_0\right)$ in equation 38.19 is an associated Laguerre polynomial. These polynomials have been tabulated in reference books, where you can look them up. Their exact form is not important here, but we note the following general considerations. The product $(2r/na_0)^{\ell}L_{n-\ell-1}^{2\ell+1}\left(2r/na_0\right)$ is a polynomial of rank $n - 1$ for any value of ℓ. This polynomial has $n - \ell - 1$ positive roots and ℓ roots at $r = 0$. An example for $n = 3$ is shown in Figure 38.11, with the positive roots of the three possible polynomials indicated by the colored arrows.

Instead of examining equation 38.19 further in abstract terms, it is perhaps more instructive to write the radial solutions for the lowest values of n and ℓ:

$$f_{10}(r) = 2a_0^{-3/2} e^{-r/a_0}$$

$$f_{20}(r) = \frac{1}{\sqrt{2}} a_0^{-3/2} \left(1 - \frac{r}{2a_0} \right) e^{-r/2a_0}$$

$$f_{21}(r) = \frac{1}{2\sqrt{6}} a_0^{-3/2} \left(\frac{r}{a_0} \right) e^{-r/2a_0}$$

$$f_{30}(r) = \frac{2}{3\sqrt{3}} a_0^{-3/2} \left(1 - \frac{2r}{3a_0} + \frac{2r^2}{27a_0^2} \right) e^{-r/3a_0}$$

$$f_{31}(r) = \frac{8}{27\sqrt{6}} a_0^{-3/2} \left(\frac{r}{a_0} - \frac{r^2}{6a_0^2} \right) e^{-r/3a_0}$$

$$f_{32}(r) = \frac{4}{81\sqrt{30}} a_0^{-3/2} \left(\frac{r^2}{a_0^2} \right) e^{-r/3a_0}.$$

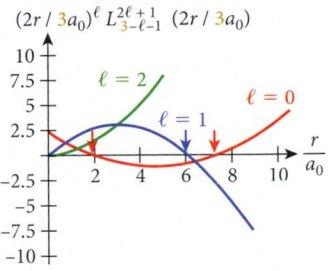

FIGURE 38.11 Polynomial terms in the radial wave function for $n = 3$.

Comparing these solutions to the spherically symmetric ones (equation 38.12) shows that for each value of n, the solutions with $\ell = 0$ correspond to the spherically symmetric solutions.

Because the lowest values of n correspond to the lowest values of the energy, the radial functions listed here are overwhelmingly the most important for the hydrogen atom.

If we plot the radial functions for the lowest allowed values of the quantum numbers n and ℓ, an interesting picture emerges. Figure 38.12a shows the six wave functions listed above. It is important to note that the solutions for a radial quantum number n are polynomials of rank $n - 1$, multiplied by an exponentially decreasing term that assures that the wave functions approach zero for very large values of the radius r. All wave functions with $\ell > 0$ have a value of 0 at $r = 0$; only those with $\ell = 0$ assume a finite nonzero value at $r = 0$. (In the figure, it may appear that the $\ell = 0$ solutions diverge as they approach $r = 0$, but this is misleading. If the scale were expanded, we would see them turn over, because the exponential function has a value of 1 at $r = 0$.)

Figure 38.12b shows the probability of finding an electron in a state with quantum numbers n and ℓ at a distance r from the nucleus of the hydrogen atom. The wave function f_{10}—the only wave function for $n = 1$—is red, and the two possible wave functions for $n = 2$ (f_{20} and f_{21}) are blue. The three possible wave functions for $n = 3$, which are f_{30}, f_{31}, and f_{32}, are shown in green. For the group of wave functions with a given value of n, all weighted probability densities plotted in Figure 38.12b peak at a similar distance from the origin. They form a **shell**. This appearance of shells is a universal phenomenon in atomic and nuclear physics and is a unique quantum effect. We will return to the concept of shells repeatedly in our discussions of these subjects.

An interesting side note: The radial wave functions $f_{n,n-1}$ all peak at values of r that correspond to the radii predicted by the Bohr model for the corresponding quantum number n.

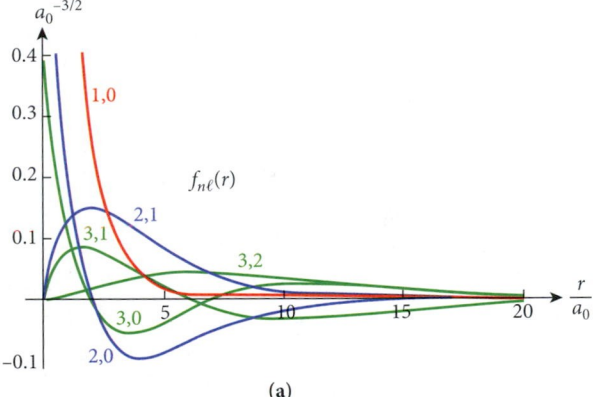

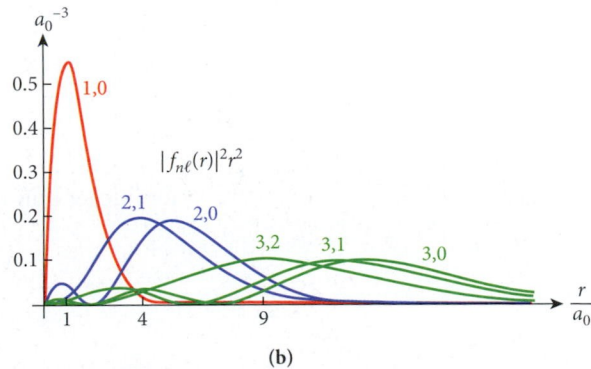

FIGURE 38.12 (a) Radial wave functions for the lowest values of the quantum numbers n and ℓ; (b) corresponding weighted probability densities.

Angular Part

Let's return to the Schrödinger equation in spherical coordinates (equation 38.17). If the left-hand side is equal to $\ell(\ell + 1)$, so is the right-hand side:

$$\ell(\ell+1) = -\frac{1}{g(\theta)\sin\theta} \frac{\partial}{\partial\theta} \left(\sin\theta \frac{\partial}{\partial\theta} g(\theta) \right) - \frac{1}{h(\phi)\sin^2\theta} \frac{\partial^2}{\partial\phi^2} h(\phi). \tag{38.20}$$

This equation is still a function of the two variables θ and ϕ. We can rearrange the terms and obtain an equation in which all θ-dependent terms are on the left-hand side and all ϕ-dependent terms are on the right-hand side. This is accomplished by multiplying both sides of equation 38.20 by $\sin^2\theta$ and then moving the first term on the right to the left:

$$\ell(\ell+1)\sin^2\theta + \frac{\sin\theta}{g(\theta)}\frac{\partial}{\partial\theta}\left(\sin\theta\frac{\partial}{\partial\theta}g(\theta)\right) = -\frac{1}{h(\phi)}\frac{\partial^2}{\partial\phi^2}h(\phi).$$

Again, we have separated the variables, and for this equation to be true, both sides need to be equal to the same constant. Let us call this constant m^2. (Remember, we are using μ for the mass of the electron, not m.)

The two resulting ordinary differential equations for the angular variables are

$$\frac{d^2}{d\phi^2}h(\phi) = -m^2 h(\phi) \tag{38.21}$$

$$\ell(\ell+1)\sin^2\theta + \frac{\sin\theta}{g(\theta)}\frac{d}{d\theta}\left(\sin\theta\frac{d}{d\theta}g(\theta)\right) = m^2. \tag{38.22}$$

Note that the form of differential equation 38.21 is exactly the same as that for a simple harmonic oscillator, which we studied extensively in Chapter 14. Perhaps you can therefore guess the solution to equation 38.21 right away: It is a linear combination of $\sin(m\phi)$ and $\cos(m\phi)$. Alternatively, using the conventional complex notation, it is

$$h(\phi) = Ae^{im\phi}. \tag{38.23}$$

Here A is a normalization constant, which we can deal with later when we consider the overall normalization of the wave function. What is more important is that the same wave function must be obtained when 2π is added to the angle ϕ, because this addition corresponds to one complete 360° rotation in the xy-plane. This implies that

$$e^{im\phi} = e^{im(\phi+2\pi)} \Rightarrow e^{2\pi im} = 1 \Rightarrow m = 0, \pm 1, \pm 2, \ldots.$$

Thus, we have found that our integration constant m must be an integer.

The solution to equation 38.22 for θ is not so easy to obtain, so we simply state it here:

$$g(\theta) = BP_\ell^m(\cos\theta) = B(-1)^{|m|}\sin^{|m|}\theta\left(\frac{d}{d(\cos\theta)}\right)^{|m|}P_\ell(\cos\theta). \tag{38.24}$$

Here B is another normalization constant. Just like the constant A, we ignore it for now. The functions P_ℓ^m are called *associated Legendre functions*, and the functions P_ℓ are *Legendre polynomials*:

$$P_\ell(x) \equiv \frac{1}{2^\ell \ell!}\left(\frac{d}{dx}\right)^\ell (x^2-1)^\ell \quad \text{and} \quad P_\ell^m(x) = (-1)^m\left(1-x^2\right)^{m/2}\left(\frac{d}{dx}\right)^m P_\ell(x),$$

for integer non-negative values of ℓ. The first few Legendre polynomials are

$$P_0(x) = 1$$

$$P_1(x) = \frac{1}{2}\frac{d}{dx}\left(x^2-1\right) = x$$

$$P_2(x) = \frac{1}{8}\frac{d^2}{dx^2}\left(x^2-1\right)^2 = \frac{1}{2}\left(3x^2-1\right)$$

$$P_3(x) = \frac{1}{48}\frac{d^3}{dx^3}\left(x^2-1\right)^3 = \frac{1}{2}\left(5x^3-3x\right)$$

$$P_4(x) = \frac{1}{384}\frac{d^4}{dx^4}\left(x^2-1\right)^4 = \frac{1}{8}\left(35x^4-30x^2+3\right)$$

$$P_5(x) = \frac{1}{3840}\frac{d^5}{dx^5}\left(x^2-1\right)^5 = \frac{1}{8}\left(63x^5-70x^3+15x\right).$$

These are shown in Figure 38.13.

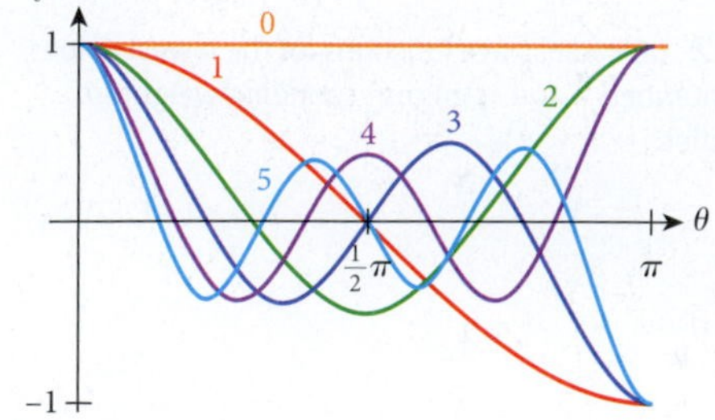

FIGURE 38.13 Legendre polynomials for $\ell = 0, 1, 2, 3, 4$, and 5.

You can see that a Legendre polynomial P_ℓ is a polynomial of rank ℓ. If you take the nth-derivative of a polynomial of rank ℓ, with $n > \ell$, you obtain zero. This means that an upper limit exists for the integer $|m|$:

$$|m| \le \ell \Rightarrow -\ell \le m \le \ell.$$

Therefore, for any value of ℓ, there are $2\ell + 1$ possible values of m. The associated Legendre functions $P_\ell^m(\cos\theta)$ with the lowest values of ℓ are

	$m = 0$	$m = 1$	$m = 2$	$m = 3$
$\ell = 0$	1			
$\ell = 1$	$\cos\theta$	$-\sin\theta$		
$\ell = 2$	$\frac{1}{2}(3\cos^2\theta - 1)$	$-3\cos\theta\sin\theta$	$3\sin^2\theta$	
$\ell = 3$	$\frac{1}{2}(5\cos^3\theta - 3\cos\theta)$	$-\frac{3}{2}\sin\theta(5\cos^2\theta - 1)$	$15\cos\theta\sin^2\theta$	$-15\sin^3\theta.$

The expressions for negative values of m do not have to be written separately, because only the absolute value of the quantum number m appears in the defining equation 38.24.

The products of the functions, $g(\theta)h(\phi)$, properly normalized, are called the **spherical harmonics**, Y_ℓ^m:

$$Y_\ell^m(\theta,\phi) = \sqrt{\frac{(\ell - |m|)!}{(\ell + |m|)!} \frac{(2\ell + 1)}{4\pi}} P_\ell^m(\cos\theta)e^{im\phi}. \tag{38.25}$$

The spherical harmonics describe the angular dependence of the electron wave functions for the hydrogen atom. They have a complex phase factor $e^{im\phi}$, but are otherwise real-valued. Figure 38.14 plots the absolute values of the spherical harmonics. These absolute values are symmetric under rotation about the z-axis, but in general, for $m \ne 0$, this is not the case for the real part or for the imaginary part.

What else is special about the spherical harmonics? It can be shown (we will not do this here) that they are also eigenfunctions of the operator for the square of the angular momentum, $\hat{L}^2$, with eigenvalues $\ell(\ell + 1)\hbar^2$, and of the operator for the projection of the angular momentum on the z-axis, $\hat{L}_z$, with eigenvalue $m\hbar$:

$$\hat{L}^2 Y_\ell^m(\theta,\phi) = \ell(\ell + 1)\hbar^2 Y_\ell^m(\theta,\phi) \tag{38.26}$$

$$\hat{L}_z Y_\ell^m(\theta,\phi) = m\hbar Y_\ell^m(\theta,\phi). \tag{38.27}$$

Thus, the quantum number ℓ is a measure of the absolute value of the total orbital angular momentum—that is, the length of the angular momentum vector—and the quantum number m measures the length of the projection of the angular momentum vector along the z-axis.

You might now be able to guess what the operators $\hat{L}^2$ and $\hat{L}_z$ look like in spherical coordinates. They are

$$\hat{L}^2 = -\hbar^2 \frac{1}{\sin\theta} \frac{\partial}{\partial\theta}\left(\sin\theta \frac{\partial}{\partial\theta}\right) - \hbar^2 \frac{1}{\sin^2\theta} \frac{\partial^2}{\partial\phi^2}$$

and

$$\hat{L}_z = -i\hbar \frac{\partial}{\partial\phi}.$$

Comparing these two expressions to equations 38.20 and 38.21 shows why equations 38.26 and 38.27 must be true. The construction of the spherical harmonics as solutions to the angular part of a rotationally invariant Schrödinger equation means that they must be eigenfunctions to the operators for the square of the angular momentum and for the projection of the angular momentum on the symmetry axis.

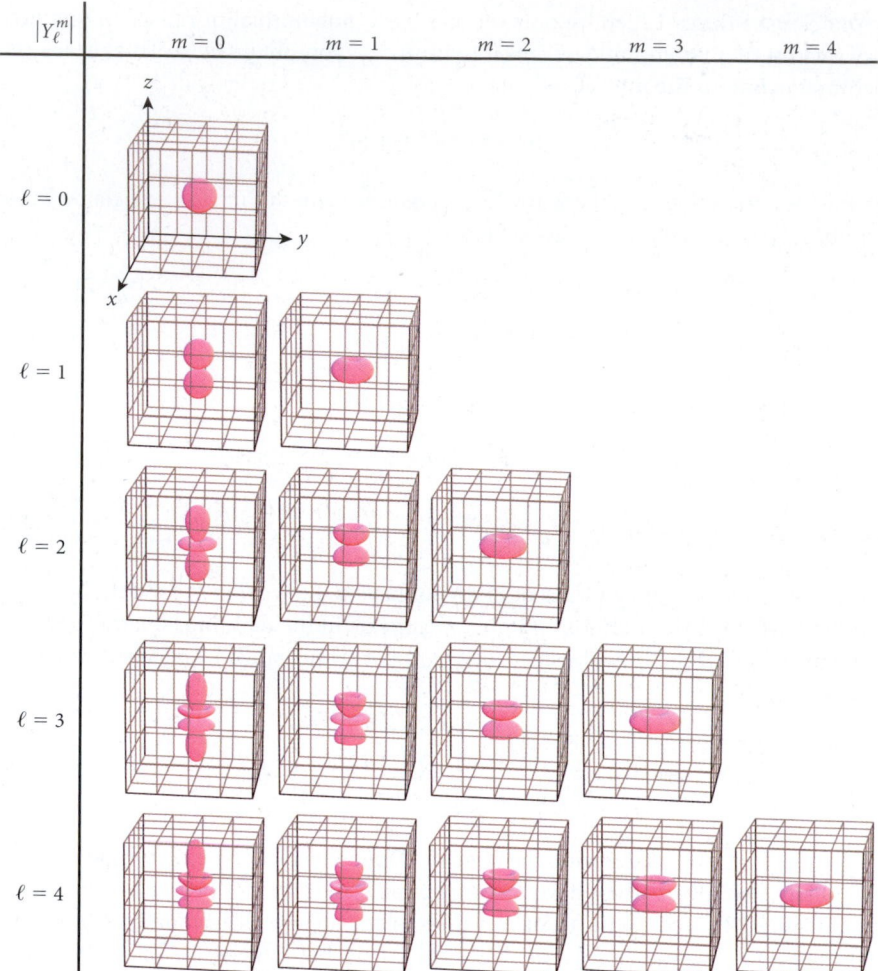

FIGURE 38.14 Absolute value of the spherical harmonics for the lowest values of the angular momentum quantum numbers. Each box extends from −1 to 1 in each Cartesian direction.

Complete Solution

Let's collect the parts of our complete solution. We have separated the variables and constructed the solutions as products of the radial part of the wave function (equation 38.19) and the angular part (equation 38.25):

$$\psi_{n\ell m}(r,\theta,\phi) = f_{n\ell}(r)Y_{\ell m}(\theta,\phi). \tag{38.28}$$

The quantum numbers of these solutions are the **radial quantum number,** $n = 1,2,3, \dots$; the **orbital angular momentum quantum number,** $\ell = 0,1, \dots ,n-1$; and the **magnetic quantum number,** $m = -\ell, \dots ,\ell$. The energy eigenvalue corresponding to a particular eigenfunction that is a solution depends only on the radial quantum number:

$$E_n = -\frac{1}{n^2}E_0 = -\frac{1}{n^2}\frac{\mu k^2 e^4}{2\hbar^2}. \tag{38.29}$$

The solutions for a given value of the radial quantum number n all peak at a similar distance from the center, approximately given by the radius of the corresponding semiclassical Bohr orbit, $r_n = n^2 a_0 = n^2\hbar^2/\mu ke^2$. (For the wave functions with the maximum possible angular momentum quantum number, $\ell = n-1$, this is an exact statement.) Thus, the wave functions for a given radial quantum number n form a shell. The shells are sometimes (especially in chemistry) labeled with capital letters according to their radial quantum number:

$$K \quad L \quad M \quad N \quad O \quad P \quad Q \quad \dots$$
$$n = \quad 1 \quad 2 \quad 3 \quad 4 \quad 5 \quad 6 \quad 7 \quad \dots.$$

Therefore, the lowest energy shell is the *K*-shell, and the shells with successively higher radial quantum number *n* follow alphabetically.

Within a given shell, all wave functions have the same energy. Traditionally, they are labeled with letters according to their orbital angular momentum quantum number:

$$s \quad p \quad d \quad f \quad g \quad h \quad i \quad k \quad ...$$
$$\ell = \quad 0 \quad 1 \quad 2 \quad 3 \quad 4 \quad 5 \quad 6 \quad 7 \quad ....$$

Here the first two angular momentum states receive the letters *s* (monopole) and *p* (dipole), and from $\ell = 2$ (quadrupole) on, the angular momentum states follow the alphabet successively, starting with the letter *d*, but skipping *e* and *j*. (Why this ordering? This is how the nomenclature was developed historically.) Thus, a 4*d* state implies that the radial quantum number is 4 and the orbital angular momentum quantum number is 2. The energies and level assignments are shown in Figure 38.15.

We can also plot the solution wave functions. We have the choice of displaying the wave function, its real or imaginary part, or its absolute value. All wave functions with $m = 0$ are real, and we can easily plot these wave functions—for example, in slices through space. Figure 38.16 shows the wave functions $\psi_{n\ell m}$ for the lowest values of the quantum numbers *n* and ℓ with $m = 0$ in a slice through coordinate space, in the *xz*-plane with $y = 0$. Blue shades imply positive values of the wave function, and red shades indicate negative values. Yellow represents values close to zero; a scale is given on the right side. In each case, a region between $\pm 30 a_0$ is shown in

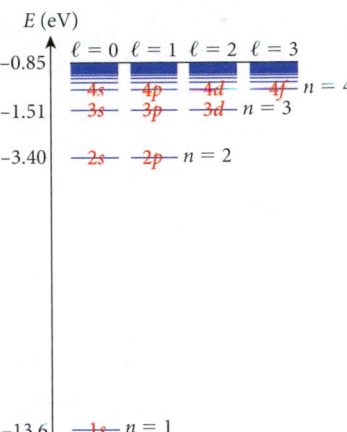

FIGURE 38.15 Wave function quantum numbers, energies, and label assignments.

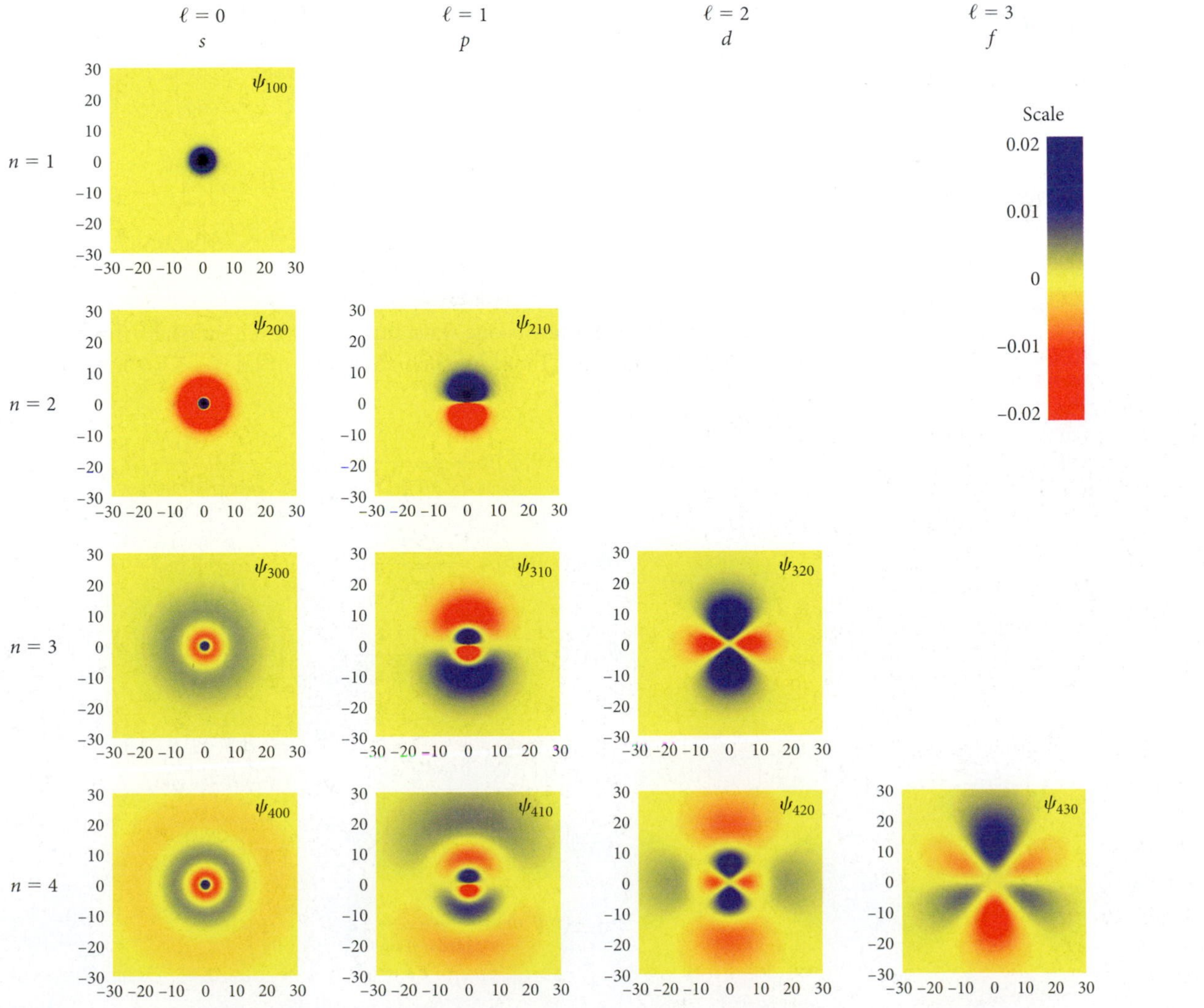

FIGURE 38.16 Hydrogen atom electron wave functions in the *xz*-plane for $y = 0$, $\psi_{n\ell m}(x,0,z)$. The coordinates in the *xz*-plane are displayed in multiples of the classical Bohr radius, a_0. The scale on the right indicates how the colors in the plots represent the value of $\psi_{n\ell m}$.

each direction, x and z. As you can see, all s-states ($\ell = 0$ states) appear as concentric circles in the contour plots, because they are all spherically symmetric. The states with nonzero angular momentum show beautiful patterns of alternating regions of positive and negative values.

We can see regularities in the patterns that emerge. As n increases, an outer shell becomes populated, while the structure of the inner shell wave function with the same angular momentum quantum number gets preserved and squeezed down, with a spherical node (yellow rings) separating it from the neighboring shells. All p-waves show twofold symmetries, as expected from dipolar shapes. In the same way, the d-waves exhibit fourfold quadrupole structure, and the f-waves show a characteristic sixfold symmetry.

Finally, we list the wave functions for the innermost shells—those with the lowest values for the radial quantum numbers and thus the lowest energies. In the $n = 1$ shell, there is only one possible wave function, and it is in an s-state:

$$\psi_{100}(r,\theta,\phi) = \frac{1}{\sqrt{\pi} a_0^{3/2}} e^{-r/a_0}. \tag{38.30}$$

Self-Test Opportunity 38.3

Referring to the explicit form of the solution, sketch a plot of $\psi_{211}(r,\theta,\phi)$ similar to those in Figure 38.16.

Concept Check 38.2

How many wave functions are possible in the $n = 5$ shell?

a) 9   d) 21

b) 14   e) 25

c) 16

Self-Test Opportunity 38.4

Verify that equation 38.30 is indeed the ground-state hydrogen wave function by inserting it into the Schrödinger equation. For added difficulty, try the same exercise for one of the wave functions with $n = 2$.

Note that the value of the energy eigenvalue that corresponds to this wave function is the lowest possible, $E_1 = -13.6$ eV. An electron that resides in this state cannot radiate away a photon to go to a lower energy state. Therefore, the state with this energy is the ground state of the electron in the hydrogen atom, and equation 38.30 is the ground-state wave function. You can compare equation 38.30 with the results of Example 38.2.

In the $n = 2$ shell, one wave function in the s-state is possible, plus three p-waves, for a total of four wave functions, all of them corresponding to an energy eigenvalue of $E_2 = -3.40$ eV:

$$\psi_{200}(r,\theta,\phi) = \frac{1}{2\sqrt{2\pi} a_0^{3/2}} e^{-r/2a_0} \left(1 - \frac{r}{2a_0}\right)$$

$$\psi_{210}(r,\theta,\phi) = \frac{1}{2\sqrt{2\pi} a_0^{3/2}} e^{-r/2a_0} \left(\frac{r}{2a_0}\right) \cos\theta$$

$$\psi_{21\pm1}(r,\theta,\phi) = \mp \frac{1}{4\sqrt{\pi} a_0^{3/2}} e^{-r/2a_0} \left(\frac{r}{2a_0}\right) \sin\theta e^{\pm i\phi}.$$

In the $n = 3$ shell, there is one wave function in the s-subshell, three wave functions in the p-subshell, and five in the d-subshell, for a total of $1 + 3 + 5 = 9$ wave functions, all with energy $E_3 = -1.51$ eV:

$$\psi_{300}(r,\theta,\phi) = \frac{1}{3\sqrt{3\pi} a_0^{3/2}} e^{-r/3a_0} \left(1 - \frac{2r}{3a_0} + \frac{2r^2}{27a_0^2}\right)$$

$$\psi_{310}(r,\theta,\phi) = \frac{2\sqrt{2}}{27\sqrt{\pi} a_0^{3/2}} e^{-r/3a_0} \left(\frac{r}{a_0} - \frac{r^2}{6a_0^2}\right) \cos\theta$$

$$\psi_{31\pm1}(r,\theta,\phi) = \mp \frac{2}{27\sqrt{\pi} a_0^{3/2}} e^{-r/3a_0} \left(\frac{r}{a_0} - \frac{r^2}{6a_0^2}\right) \sin\theta e^{\pm i\phi}$$

$$\psi_{320}(r,\theta,\phi) = \frac{1}{81\sqrt{6\pi} a_0^{3/2}} e^{-r/3a_0} \left(\frac{r^2}{a_0^2}\right) \left(3\cos^2\theta - 1\right)$$

$$\psi_{32\pm1}(r,\theta,\phi) = \mp \frac{1}{81\sqrt{\pi} a_0^{3/2}} e^{-r/3a_0} \left(\frac{r^2}{a_0^2}\right) \cos\theta \sin\theta e^{\pm i\phi}$$

$$\psi_{32\pm2}(r,\theta,\phi) = \frac{1}{162\sqrt{\pi} a_0^{3/2}} e^{-r/3a_0} \left(\frac{r^2}{a_0^2}\right) \sin^2\theta e^{\pm 2i\phi}.$$

In general, we can have $2\ell+1$ wave functions with different m values for a given orbital angular momentum ℓ. In addition, since we can have $n - 1$ different orbital angular momentum states for a given radial quantum number n, the total number $N(n)$ of possible wave functions in a given shell with radial quantum number n is

$$N(n) = \sum_{\ell=0}^{n-1} (2\ell+1) = n^2. \qquad (38.31)$$

Does the procedure of separation of variables ensure that we have found *all* solutions? We will not prove this, but the answer is yes, because the set of solutions that we have found forms a complete basis for all functions in this space. We will not elaborate on this statement, which is covered in an advanced course on quantum physics.

We add a postscript to the discussion of the hydrogen atom. This system is surely the best-studied system in all of quantum physics, and one of the few for which an exact solution is possible. The differential equations that result, as well as their solutions, show an incredible amount of structure. It is perhaps surprising that this arguably simplest real quantum system exhibits such richness, and perhaps even more surprising that we can construct a mathematical description that captures this complexity—which is visible, for example, in the plots of Figure 38.16. The fact that the natural world arranges itself in this way and then can be captured by our mathematical constructions seems absolutely miraculous. We hope that the somewhat tedious mathematical derivations did not obscure the beauty that underlies this physical system and the theory that describes it.

38.4   Other Atoms

Now that we have a good understanding of the wave functions, energies, and possible transitions between levels for the hydrogen atom, we can ask if we can explain other atoms as well. What do we need to change in our formalism to explain other atoms? First, the charge Z of the nucleus changes. For hydrogen, $Z = 1$, but for higher values of Z, the potential needs to change to

$$U(r) = k\frac{Ze^2}{r}.$$

To be electrically neutral, an atom with Z protons must have Z electrons. These electrons will reside in the lowest energy states available to them. It is easy to see into which state the first electron can be put: It is the $1s$ state. We can generalize equation 38.29 for the hydrogen atom to see that the energy of this first electron is

$$E_1 = -\frac{\mu k^2 Z^2 e^4}{2\hbar^2}. \qquad (38.32)$$

We arrive at this result by simply replacing e^2 by Ze^2 in the Schrödinger equation. It is a useful exercise to examine the mathematics for the hydrogen atom and convince yourself that equation 38.32 indeed holds. (In equation 38.32, the value of μ is the reduced mass, calculated using equation 38.4 with M the mass of the nucleus.)

Where do we place the second electron? The answer to this is not so straightforward, because any given electron will interact with all other electrons of the same atom, which partially shield the charge Z of the nucleus. This problem cannot be solved with the exact analytical methods developed earlier. It needs approximation methods that are beyond the scope of this book. However, we can gain some qualitative understanding with the tools at our disposal.

The general idea of radial shells and subshells with fixed angular momentum quantum numbers, which we developed for the hydrogen atom, remains valid for other atoms. The partial screening of the Coulomb potential of the nucleus acts differently on different levels. The states with $\ell = 0$ have probability distributions that are peaked at the origin. For these states, there is a low probability that another electron lies closer to the origin, partially shielding the nuclear charge. For states with larger values of the angular momentum quantum number, however, shielding plays a more important role. This leads to the distribution of energy levels sketched in Figure 38.17. At first glance, this figure looks similar to Figure 38.15, which showed

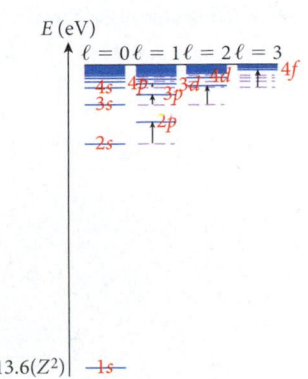

FIGURE 38.17 Energy levels in atoms with several electrons.

the plot for the hydrogen atom. Here, however, the energy scale is different by a factor of Z^2. The locations of energy levels in the absence of shielding are shown by the dashed blue lines, while the proper locations of the levels are indicated by the solid blue lines. For a few levels, small black arrows show the shifts. What is the main difference between Figure 38.17 and Figure 38.15? In Figure 38.17, different angular momentum quantum numbers ℓ for the same values of the main quantum number n result in different energy values, whereas they all have the same value for the hydrogen atom. This is due to the other electrons partially shielding the nuclear charge in atoms with more than one electron. If we sort the levels by increasing energy, starting with the lowest one, we find that the progression is 1s, 2s, 2p, 3s, 3p, 4s, 3d, 4p, 5s, 4d, 5p, 6s, ...

Electrons have a spin quantum number $\frac{1}{2}$. Thus, $2(\frac{1}{2}) + 1 = 2$ electrons can occupy each wave function with its set of fixed quantum numbers—one electron with spin up and the other with spin down. This fact is due to the Pauli Exclusion Principle. From that principle and Figure 38.17, we can determine in which order the electrons are distributed over the energy levels, as we fill each level in Figure 38.17 with two electrons successively.

Appendix D contains the ground-state electron configurations for all elements. These are written in the conventional shorthand notation with the number of electrons in a given level as a superscript right after the notation for the level. For example, fluorine ($Z = 9$) is listed with the ground-state electron configuration of $1s^2 2s^2 2p^5$. This means that two electrons occupy each of the 1s and 2s levels, and the remaining five electrons are in the 2p state. The electron configuration of aluminum ($Z = 13$) is listed as $[\text{Ne}]3s^2 3p$. This means that the occupation of the 1s, 2s, and 2p levels in aluminum is the same as in neon (that is, they are fully occupied) and aluminum has two additional electrons in the 3s state, plus one additional electron in the 3p state.

When all electrons of a given atom are in the lowest possible energy states available to them, the atom is in its ground state. Note that as Z increases, electrons sometimes fill an s state in a higher n shell rather than a state with lower n and higher angular momentum. (For example, see $Z = 19$, 37, and 55.) This occurs because the angular momentum barrier has made the lower n state correspond to a higher electron energy.

Just as in the hydrogen atom, electrons in other atoms can be excited into higher energy states by absorbing photons. However, the resulting line spectra are generally much more complicated than that of the hydrogen atom. It is considerably easier to answer the question of how much energy it takes to remove the least strongly bound electron from an atom. The process of removing an electron from an atom is called **ionization,** so we want to know the single-electron **ionization energy.** To lift the least strongly bound electron from its state with energy lower than zero to an energy just above zero, the ionization energy is the same as the magnitude of the energy of the level occupied by the least strongly bound electron. Figure 38.18 presents this ionization energy for every known atom (except for astatine, $Z = 85$, for which the ionization energy has not been measured yet). Can we explain the regularities of the ionization energies shown in this figure?

Concept Check 38.3

Element 111 was discovered in 1994 and named roentgenium in 2004. It exists for only a few seconds before it decays. The electron configuration of roentgenium has not been determined yet. What would you predict it to be?

a) $[\text{Xe}]4f^{14}5d^{10}6s^2 6p^6$

b) $[\text{Rn}]4f^{14}5d^{10}6s^2 6p^6$

c) $[\text{Rn}]5f^{14}6d^9 7s^2$

d) $[\text{Xe}]5f^{14}6d^9 7s^2$

FIGURE 38.18 Single-electron ionization energies for all elements.

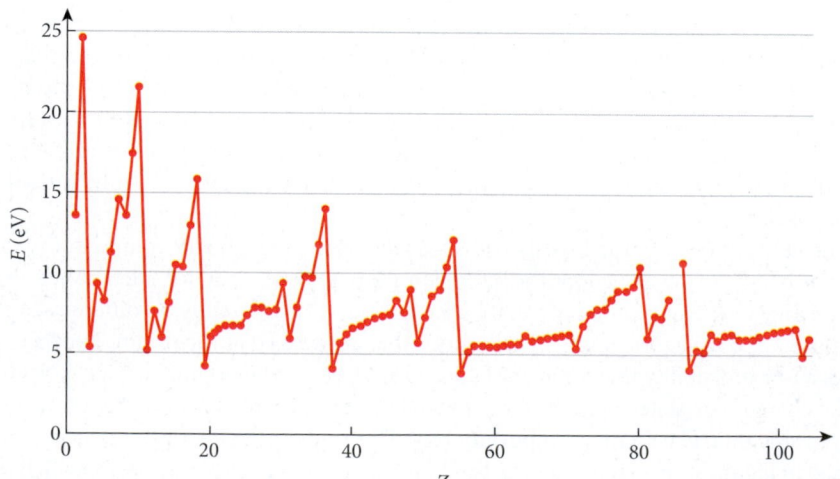

Let's start with hydrogen. The single electron of hydrogen resides in the 1s state in the ground state. Its energy is −13.6 eV, so it takes +13.6 eV to liberate this single electron.

Helium: Helium has two electrons, and they can both be accommodated in the 1s state. With $Z = 2$, the nuclear Coulomb potential is twice as strong as that of the hydrogen atom. The presence of the other electron partially screens this potential, but the second electron in the helium atom is still bound more strongly than the single electron in hydrogen. Experimentally, we find a very large value of 24.6 eV for the ionization energy of helium, the largest value for any element. What does this mean for the chemical properties of helium? It is very difficult to remove an electron from helium and add it to any other atom to form a chemical bond. Conversely, it is also not possible to add another electron to the $n = 1$ shell in helium, because it is already completely filled with two electrons. An additional electron would have to be added to the $n = 2$ shell and thus be much more weakly bound to helium. Therefore, helium is chemically completely inert and is called a **noble gas.**

Lithium: Because $Z = 3$, lithium atoms have three electrons. The first two reside in the 1s state and completely fill the $n = 1$ level, just as in helium. The third electron needs to go in the next higher state, the 2s state. From Figure 38.12b, we see that the radial wave function of the 2s state is localized farther away from the center of the atom. We then expect the third electron to be less strongly bound, and indeed, we find an ionization energy of only 5.39 eV for lithium. Consequently, lithium acts as an excellent electron donor in chemical reactions.

Beryllium: This element has four protons and four electrons. The fourth electron fits into the 2s state, filling this angular momentum subshell completely. However, remember that for $n = 2$, we have angular momentum quantum numbers of 0 and 1. Thus, beryllium is not a closed-shell atom like helium. The ionization energy of beryllium is 9.32 eV, much higher than that of lithium, but also much lower than that of helium.

Boron through neon: Starting with boron, electrons populate the 2p subshell, which can hold $2(2\ell+1) = 2(2 \cdot 1+1) = 6$ electrons. The first electron put into the 2p subshell is alone in this subshell. Again examining Figure 38.12b for guidance, we see that the 2s and 2p states have similar average distances of the radial wave functions from the center. Therefore, the ionization energy of boron ($Z = 5$) is slightly lower than that of beryllium, but not nearly as low as that of lithium. Adding more electrons to the 2p subshell increases the ionization energy successively. Thus, it is increasingly difficult to remove an electron from the elements carbon, nitrogen, oxygen, and fluorine, and they are not good electron donors in chemical reactions. However, they become better electron receptors as more electrons are added to the 2p subshell. In particular, fluorine, with only one electron missing from an otherwise closed $n = 2$ shell, is chemically very aggressive. Neon ($Z = 10$) completes the 2p subshell as well as the $n = 2$ shell. Neon thus has chemical properties very similar to those of helium—it is also chemically inert and is a noble gas. Because the radial wave functions in the $n = 2$ shell are on average farther from the atom's center than those in the $n = 1$ shell, the electrons in this shell should experience less attraction to the central nucleus and thus have lower ionization energies. Comparing the values for helium (24.6 eV) and neon (21.6 eV) confirms this expectation.

Some systematic trends in the progression of ionization energies are becoming evident. What looks like a wild and irregular zigzag curve in Figure 38.18 begins to exhibit structure. It is this structure that underlies the **periodic table of chemical elements** (Figure 38.19). In the periodic table, elements that are filling the $n = 1$ shell are in the first row, and those that are filling the $n = 2$ shell are in the second row. When an electron is added to a new shell, this corresponds to a new row in the periodic table. In each row, the electrons are added one by one, going from left to right. The elements with completely filled subshells, meaning that any additional electron would have to be placed into the next higher major shell, end up in the rightmost column; those that start a new shell are in the leftmost column.

The first two rows in the periodic table are identical to the first two major shells, but in general, a given row in the periodic table is not identical to a major shell. In the third row, the 3d electrons are missing. Because their energy levels are higher than those of the 4s electrons, the 4s electrons are added before them. Because the 4s electrons start a new row,

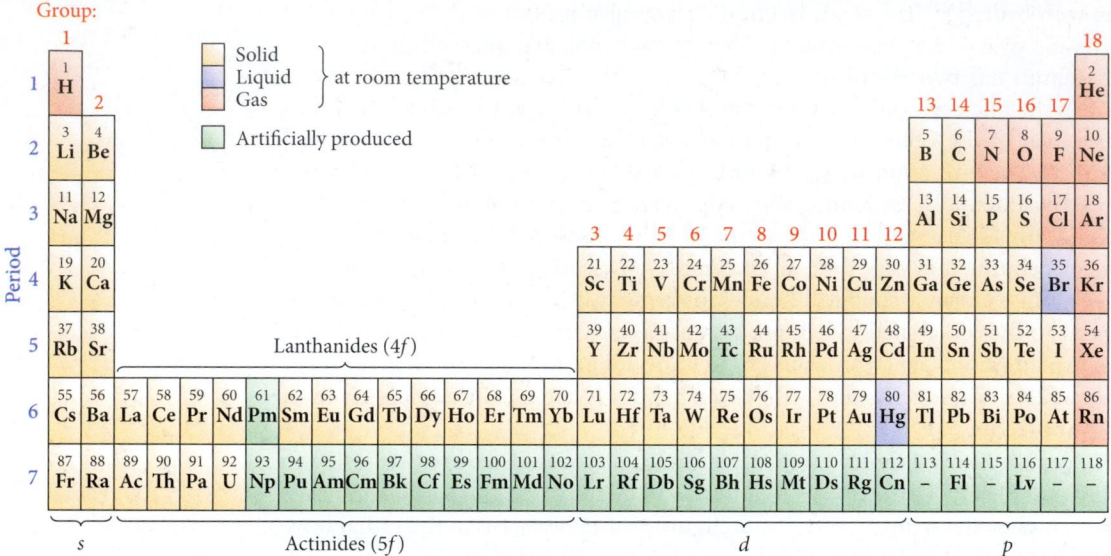

FIGURE 38.19 Periodic table of the elements.

the elements scandium through zinc, in which electrons fill the 3*d* subshell, are members of the fourth row of the periodic table, not the third.

Elements in a given column have a similar electron configuration in their outermost shell. The completely filled inner shells have very little influence on an atom's interactions with other atoms or photons. The electrons in the outermost shells of atoms, called the **valence electrons,** primarily determine chemical behavior. Therefore, elements in a given column are chemically similar in many ways.

Note that the lowest row in Figure 38.19 consists mostly of artificially produced atoms. Atoms with atomic numbers larger than 92 (uranium) are not stable and thus not found in nature. They can be created in the lab and exist for some time before they decay. As a general rule of thumb, the higher the atomic number is, the shorter the time the artificially produced atom exists before it decays. Elements 113, 115, 116, 117, and 118 were discovered only in the first decade of the 21st century, and each exists for only a few milliseconds to seconds before decay. (Chapter 40 investigates the reasons for the short lifetimes of atoms with high atomic numbers.)

Using the arrangement of the elements in the periodic table, we can display the ionization energies for all elements up to $Z = 104$ in a three-dimensional representation (Figure 38.20). In general, the ionization energy decreases as n increases in the radial shells. Also, the ionization energy increases from left to right in the periodic table, as more electrons are added to a given angular momentum subshell. This trend indicates how we can gain great insights into chemistry from an understanding of the quantum mechanical wave functions of electrons in a central Coulomb potential. It seems miraculous that a general understanding of atoms with many electrons can be gained from the solution for the comparatively simple hydrogen atom with only one electron. Nevertheless, judging from experimental data on the ionization energies, the general principle works.

Note that the Pauli Exclusion Principle, which we discussed in Chapter 37, is absolutely crucial for understanding multi-electron atoms. If it were not true, all atoms would look a lot like hydrogen, and our world as we know it could not exist.

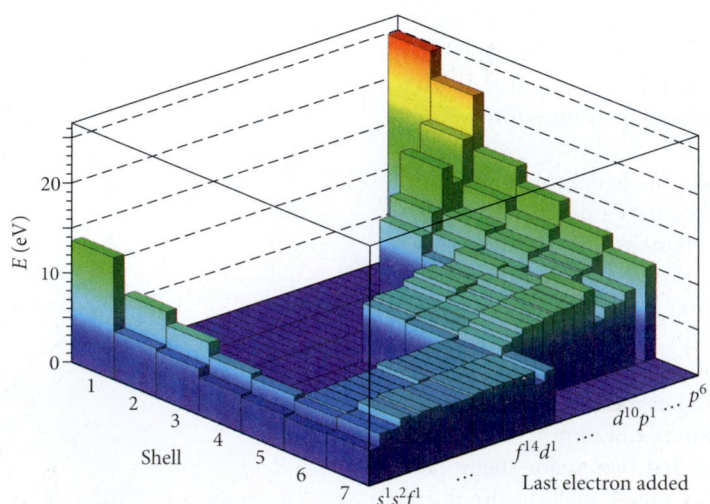

FIGURE 38.20 Ionization energies for the elements up to $Z = 104$ (rutherfordium) in the periodic table.

EXAMPLE 38.3 Ionization Energy of the Helium Atom

As mentioned in the text, the ionization energy of helium has been measured to be 24.6 eV. A helium atom ($Z = 2$) has two electrons in the $n = 1$ shell. These two electrons are in the ground state but interact with each other.

PROBLEM

Estimate the ionization energy of helium.

SOLUTION

The ground-state energy of electrons, if they do not interact with each other, is given by equation 38.32:

$$E_1 = -\frac{\mu k^2 Z^2 e^4}{2\hbar^2}.$$

The reduced mass is closer to the mass of the electron because the mass of the helium nucleus is approximately four times more than the mass of a proton. The ground-state energy for one of the electrons in the helium atom is

$$E_1 = -\frac{\mu k^2 Z^2 e^4}{2\hbar^2} = -Z^2 \frac{\mu k^2 e^4}{2\hbar^2} = -Z^2 (13.6 \text{ eV}) = -4(13.6 \text{ eV}) = -54.4 \text{ eV}.$$

We can estimate the energy of the interaction between the two electrons by remembering that the two electrons repel each other because they are both negatively charged. Thus, the electrons will try to be as far apart as possible. The Bohr model assumes that the electrons are in an orbit with a fixed radius, so they will be on opposite sides of the orbit, which means they are separated by its diameter. The radius of the orbit is given by equation 38.6 with e^2 replaced by Ze^2:

$$r = \frac{\hbar^2}{\mu k Z e^2} = \frac{a_0}{Z} = a_0/2.$$

The electric potential energy of the two electrons separated by a distance d is given in Chapter 23 as $U = ke^2/d$. This energy is positive, which means it lowers the energy required to ionize the helium atom. The electric potential energy is

$$U = \frac{ke^2}{d} = \frac{ke^2}{2(a_0/2)} = \frac{\left(8.99 \cdot 10^9 \text{ N m}^2/\text{C}^2\right)\left(1.602 \cdot 10^{-19} \text{ C}\right)^2}{\left(5.295 \cdot 10^{-11} \text{ m}\right)} = 4.36 \cdot 10^{-18} \text{ J} = 27.2 \text{ eV}.$$

The difference between the ground-state energy and the electric potential energy between the two electrons is 27.2 eV, which is close to the measured ionization energy of 24.6 eV.

DISCUSSION

This analysis is oversimplified. The actual interaction between the two electrons is more complicated. However, our simple analysis gives some insight into the magnitude of the energies involved.

Concept Check 38.4

Element 118 was discovered in 2006 and exists for approximately 1 millisecond before it decays. Only a few atoms of this element have been produced. The ionization energy for element 118 has not been measured yet. What would you predict it to be?

a) approximately 0 eV (the atom is unstable)

b) approximately 2 eV

c) approximately 5 eV

d) approximately 10 eV

e) approximately 20 eV

X-Ray Production

Arguably the discovery of X-rays by Wilhelm Conrad Röntgen (1845–1923) in 1895 was one of the events that kicked off the modern physics era. X-rays are penetrating radiation consisting of high-energy photons. Typical photon energies for diagnostic medical X-rays are between 25 keV and 140 keV, with mammograms using a typical maximum energy of 25 keV and dental X-rays using a typical maximum energy of 60 keV. X-rays for baggage screening at airports have energies up to 160 keV. All of these X-rays can be produced by X-ray tubes (Figure 38.21).

How does an X-ray tube work? The physical principle is straightforward: A metal filament is heated and emits electrons from its surface. These electrons are then accelerated across an

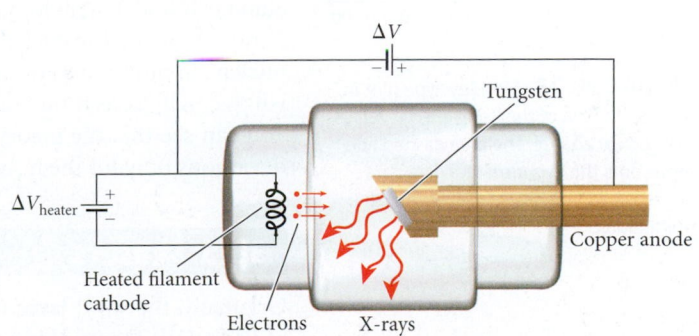

FIGURE 38.21 Schematic diagram of an X-ray tube.

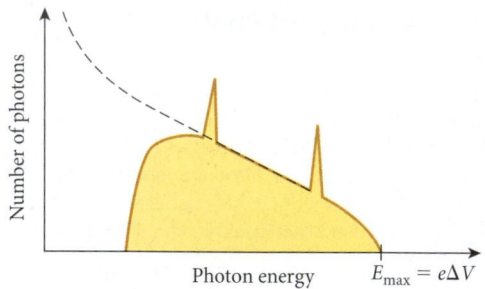

FIGURE 38.22 Sketch of an X-ray spectrum with a *bremsstrahlung* continuum plus sharp peaks due to electron transitions between atomic shells.

Concept Check 38.5

In the X-ray spectrum in Figure 38.22, suppose one of the peaks corresponds to a *K*-shell transition and the other to an *L*-shell transition. Which of the following is true?

a) The peak with the higher photon energy corresponds to the *K*-shell transition.

b) The peak with the higher photon energy corresponds to the *L*-shell transition.

c) Either peak could correspond to either transition, depending on the electrostatic potential difference applied in the X-ray tube.

d) The peaks cannot be identified, without knowing what material was used for the anode.

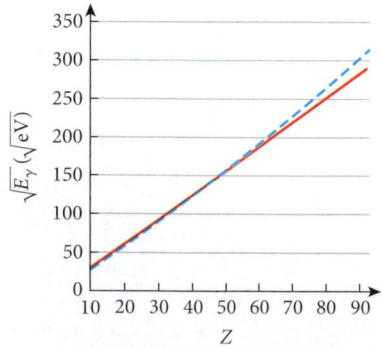

FIGURE 38.23 Moseley-type plot of the square root of the X-ray energy versus the nuclear charge. The blue dashed line represents the experimental data, and the red solid line gives the theoretical prediction.

electrostatic potential difference, ΔV. They hit the surface of the anode, which is typically made of metal, conventionally tungsten. As the electrons hit the surface of the metal, they can produce X-rays in two different ways. One is *bremsstrahlung* (German for "braking radiation"). As the electrons penetrate the surface of the metal, they experience a strong deceleration, which in turn causes the emission of photons. This process generates a continuous distribution of photons sketched by the dashed line in Figure 38.22. Note that this distribution terminates at a maximum energy, which is the total kinetic energy gained by the electron from the process of acceleration through the electrostatic potential, $E_{max} = e\Delta V$, and is converted to a single photon of energy $hf_{max} = hc/\lambda_{min}$. The low-energy X-rays cannot penetrate through the walls of the X-ray tube. Therefore, the X-ray continuum is cut off at lower energies as well, as indicated in the figure.

The other process by which X-rays are produced occurs when the accelerated electrons collide with electrons in the atoms of the anode and knock them out of their atomic shells. If a strongly bound electron from one of the inner shells gets knocked out, then an electron from the outer shells can make a transition into that inner shell and emit a photon of a fixed energy. This process is exactly the same as the one that leads to the spectral lines of hydrogen, which was discussed extensively in Section 38.2. However, the most important difference is that atoms with $Z > 1$ have electrons that are bound more strongly, and the corresponding photon energies are much larger than in the hydrogen line spectrum. The resulting discrete lines appear superimposed as spikes above the *bremsstrahlung* continuum, as sketched for two transitions in Figure 38.22. Since the photon energy is given as the difference between the energies of the initial and final states of the electron, these discrete peaks are characteristic of the anode material used.

These peaks are named according to the shell (*K* shell for $n = 1$, *L* shell for $n = 2$, ...) that the electron transitions into, with an index (α, β, ...) corresponding to the shell above, from which the electron transitions. A K_α X-ray corresponds to the transition from the $n = 2$ to the $n = 1$ shell.

We can even make predictions about the energies of the X-rays resulting from these transitions of electrons between atomic shells. According to equation 38.32, the energy of the innermost electron is $E_1 = -\mu k^2 Z^2 e^4/2\hbar^2 = -(13.6 \text{ eV})Z^2$. The transition energies should then be a factor of Z^2 bigger than those in the hydrogen atom. We can also take into account that for the outer shells the inner electrons partially screen the nuclear charge, which should result in an enhancement over the hydrogen case by a factor of $(Z - Z_{screen})^2$. The screening charge, Z_{screen}, depends only on the electron shell under consideration, not on the charge of the nucleus, Z. Therefore, we can predict that the X-ray energy corresponding to a particular transition in the atom should depend on the square of the nuclear charge. This means that a plot of the square root of the energy of the X-ray resulting from a particular transition should show a linear dependence on the nuclear charge. This type of plot was first produced by Henry Moseley (1887–1915) in 1913, and it contributed tremendously to convincing the scientific community of the validity of atomic models.

Figure 38.23 shows a plot of the type introduced by Moseley for the square root of the energy of the X-rays produced by of the K_α transition for all atoms from neon to uranium. The blue dashed line represents the experimental data, collected by the National Institute of Standards and Technology. The red line is based on the theory that the dependence of the X-ray energy on the nuclear charge is given by $(Z - Z_{screen})^2$, with Z_{screen} independent of the nucleus in question. (For this plot, $Z_{screen} = 1$, because the other electron still in the *K* shell will partially screen the nuclear charge so that it seems to be 1 less than its actual value.) You can see that the theory is a good approximation to the experimental data, with small deviations only for the heaviest atoms.

38.5 Lasers

Originally, the word **laser** was an acronym for *Light Amplification by Stimulated Emission of Radiation*. The word is now so commonly used that it is not written in all capital letters, as is the case for an acronym. Lasers, invented in 1960, are used today in all kinds of practical applications, such as Blu-ray, DVD and CD players and recorders, laser surgery, laser pointers, guidance systems, and precision measurement and survey systems. Lasers come in a large variety of sizes, power outputs, and beam colors and use many different materials. However, a few characteristics are common to practically all lasers. First, all lasers must have a medium (a gas, liquid, or solid) in which atoms can be excited to a higher energy state. Also, it must be possible to create a *population inversion,* in which more atoms are in the excited state than in the ground state. Second, a laser has to have a resonator cavity, usually simply a pair of parallel mirrors. Third, a laser has to have a means of pumping energy into the lasing medium.

We will examine the basic physics involved using a helium-neon (He-Ne) gas laser as an example. The lowest energy levels of helium and neon are sketched in Figure 38.24. The two valence electrons of helium reside in the lowest possible level, the 1s subshell, as discussed in Section 38.4. The ten electrons of neon fill the 1s, 2s, and 2p subshells. The valence electrons of neon in its ground state reside in the 2p subshell.

Figure 38.24 shows the ground states of the two atoms (blue for helium, green for neon) as horizontal lines at zero energy. The lowest possible excited state for a single electron in helium is the 2s state, which lies at an energy of 20.61 eV above the ground state. Photons carry angular momentum $\ell = 1\hbar$, and the 1s and 2s states both have zero angular momentum, so it is not possible for a photon to cause a transition from the 1s to the 2s state or from the 2s to the 1s state.

Figure 38.24b shows the lowest energy levels of a single electron in neon atoms. (The other nine electrons remain in their respective shells: two each in the 1s and 2s angular momentum states, and five in the 2p angular momentum state.) The states of highest interest for the present purpose (2p, 3p, and 5s; solid green lines) are in a separate column from those that are of lesser interest (3s, 4s, and 4p; dashed green lines). The energy difference between the 2p ground state and the 5s excited state in neon is 20.66 eV, very close to the 20.61 eV by which the 2s state lies above the 1s state in helium. The difference between these two energies is only 0.05 eV, which is very close to the average kinetic energy of gas molecules at room temperature (see Example 19.4). The energy of the 3p state in neon is 18.70 eV above the ground state. A transition from the 3p (angular momentum $\ell = 1\hbar$) to the 5s state (angular momentum 0) is thus possible by absorption of a photon of wavelength λ (see equation 38.8):

$$E_{5s} = E_{3p} + \frac{hc}{\lambda} \Rightarrow$$

$$\lambda = \frac{hc}{E_{5s} - E_{3p}} = \frac{1240 \text{ eV nm}}{(20.66 - 18.70) \text{ eV}} = \frac{1240 \text{ eV nm}}{1.96 \text{ eV}} = 633 \text{ nm}.$$

Conversely, the transition from the 5s to the 3p state proceeds via the emission of a photon of wavelength 633 nm. This process causes the emission of red light, which is characteristic for this type of laser.

You may ask why the 5s state cannot emit a photon via a transition to the 4p state. In fact, this transition is also observed in practice and corresponds to a photon wavelength of 3390 nm. Another possible transition is between the 4s and 3p states, which causes the emission of a photon with wavelength of 1150 nm. In total, more than 10 laser transitions are known in this system, but for practical purposes the most useful one is the 5s-to-3p transition, because it is in the visible region.

Stimulated Emission and Population Inversion

Figure 38.25 illustrates the possible interaction processes of photons with atoms. The discussion of spectral lines in Sections 38.1 and 38.2 examined the connection between the transition between states of different energy in atoms and the emission and absorption of photons.

Figure 38.25a depicts the emission of a photon from an atom in an excited state and the absorption of a photon with the appropriate energy to put the atom into the excited state. Figure 38.25b shows what happens if a large number of atoms are present in a system, a few of which are in the excited state. The excited atoms decay by emitting photons of fixed energy, but the direction of emission is random. Finally, Figure 38.25c shows the conditions

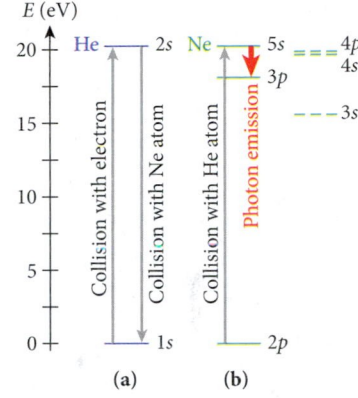

FIGURE 38.24 Energy levels relative to their respective ground states for (a) helium and (b) neon atoms.

FIGURE 38.25 Interaction of photons (red sine waves) with atoms (blue dots for ground state, red dots for excited state). (a) Emission and absorption processes for an isolated atom. (b) Decay of excited atoms in a mixture of atoms in the ground and excited states. (c) Stimulated emission of photons from a population-inverted mixture of atoms in the excited and ground states.

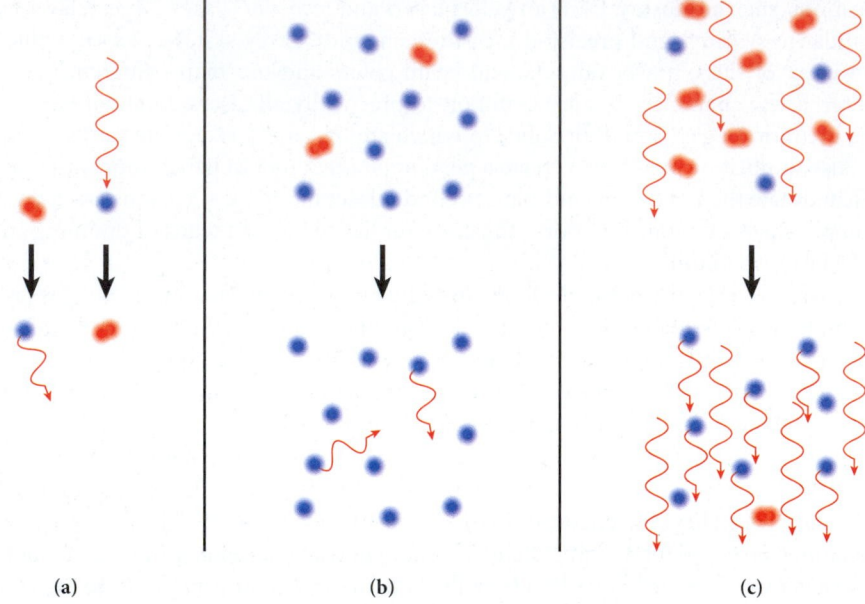

(a)   (b)   (c)

for **stimulated emission:** A number of coherent photons, all with the same energy and moving in the same spatial direction, are sent into a system of atoms, in which the population of atoms in the state with higher energy is greater than the population with lower energy. Photons are bosons, so they prefer to occupy quantum states that are already occupied by other photons. Thus, the presence of a large number of coherent photons causes the preferred emission of photons into the same quantum state (same energy and same direction of motion) already occupied by the coherent photons.

For stimulated emission to occur, the coherent photons sent into the system must have the same energy as the emitted photons, which is the energy difference between the two states in the atom. Then each photon of coherent light sent into the system of atoms can also be absorbed by an atom in the lower energy state, lifting it into the higher energy state. If more atoms are in the lower energy state than in the higher energy state, the net effect is an overall reduction in the number of photons. Thus, light amplification through stimulated emission can only be successful if the population of atoms in higher energy states is greater than the population in lower energy states.

Chapter 19 showed that the probability of atoms (or molecules) in a gas having an energy E at a given temperature T is proportional to the exponential factor $e^{-E/k_B T}$. Thus, the ratio of the population of the higher energy state to the population of the lower energy state is

$$\frac{n_{\text{higher}}}{n_{\text{lower}}} = e^{-\Delta E/k_B T},$$

where ΔE is the energy difference between the higher and the lower energy states. Let's put in some typical numbers. Previously, we found that the energy difference between the $5s$ and $3p$ states in neon is $\Delta E = 1.96$ eV. At room temperature ($T = 300$ K), this means that for every atom found in the $3p$ state, there are only $e^{-1.96/(300 \cdot 8.617 \cdot 10^{-5})} = 1.2 \cdot 10^{-33}$ atoms in the $5s$ state. (Here the Boltzmann constant is expressed in the convenient units of electron-volts per kelvin: $k_B = 8.617 \cdot 10^{-5}$ eV/K.)

Since a positive energy difference ΔE always implies that $e^{-\Delta E/k_B T} < 1$, the population of the state with higher energy is always less than that of the state with lower energy, if the system of atoms is in thermal equilibrium.

Achieving the population inversion required for light amplification through stimulated emission requires systems for which **metastable states** exist—that is, states that have long lifetimes because a simple decay is forbidden. The energy-level diagrams of helium and neon in Figure 38.24 show that the $2s$ state in helium is just such a metastable state, because it cannot decay via emission of a single photon. This state can be populated through collisions of helium

Concept Check 38.6

What is the population ratio of the $5s$ state relative to the $3p$ state at a temperature of 5000 K?

a) $1.2 \cdot 10^{-33}$ d) 0.51

b) $4.3 \cdot 10^{-23}$ e) 2.4

c) 0.011

atoms with electrons, thereby "pumping" the excited state. This part of the laser works similarly to a simple neon sign, where gas atoms are excited via an electron discharge. The helium atoms can then transfer this energy to the neon atoms in the 5s state, which has approximately the same energy relative to neon's ground state, through collisions between excited helium atoms with neon atoms in the ground state. If enough energy is pumped into the system, population inversion between the 5s and 3p states in neon can be achieved by populating the 5s state faster than it can decay into the 3p state. The utilization of a metastable state to achieve population inversion is very common in all kinds of laser systems.

Figure 38.26 shows a demonstration model of a helium-neon laser. The discharge for pumping the lasing medium takes place in the gas cell. It is enclosed between a pair of parallel mirrors, which bounce the coherent photons back and forth many times to maximize the stimulated emission. The mirror on the right side is only 99% reflective; this allows the laser beam to exit from the cell.

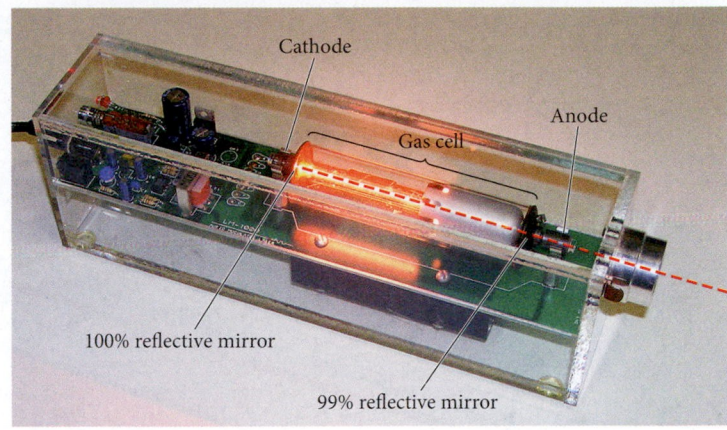

FIGURE 38.26 Demonstration model of a laser. The lasing medium, a mixture of helium and neon gas, is enclosed between two parallel mirrors. The laser beam itself is not visible but is indicated by the red dashed line.

EXAMPLE 38.4 | Number of Photons from a Pulsed Ruby Laser

The lasing medium in a ruby laser consists mostly of alumina (Al_2O_3) and a small amount of chromium, which is responsible for the red color of the ruby. This laser emits a pulse of light ($\lambda = 694$ nm) with a power of 3.00 kW over a period of 10.0 ns.

PROBLEM
How many chromium atoms undergo stimulated emission to produce this pulse?

SOLUTION
Each chromium atom that undergoes stimulated emission gives off one photon. The total energy of the pulse is equal to the number of photons, which is equal to the number of excited chromium atoms, times the energy of each photon. The total energy of the light contained in the pulse, E_{total}, is equal to the power P of the pulse times the time duration t of the pulse:

$$E_{total} = Pt = (3.00 \text{ kW})(10.0 \text{ ns}) = 3.0 \cdot 10^{-5} \text{ J}.$$

The energy of each photon is

$$E = hf = h\frac{c}{\lambda} = (6.626 \cdot 10^{-34} \text{ J s})\frac{3.00 \cdot 10^8 \text{ m/s}}{694 \cdot 10^{-9} \text{ m}} = 2.86 \cdot 10^{-19} \text{ J}.$$

Thus, the number of chromium atoms, N, undergoing stimulated emission during the pulse is

$$N = \frac{E_{total}}{E} = \frac{3.00 \cdot 10^{-5} \text{ J}}{2.86 \cdot 10^{-19} \text{ J}} = 1.05 \cdot 10^{14}.$$

In comparison, the total number of atoms in the ruby is around 10^{23}.

Research and Lasers

Lasers come in a huge variety. Laser pointers emit light with a power of up to 5 mW and typically use semiconductor devices called *diodes*. Carbon-dioxide lasers, another type of gas laser, emit laser beams with wavelengths of 9.6 and 10.6 μm at a power of up to 100 kW. These lasers are used in industrial applications for cutting and welding. Other devices include chemical lasers, dye lasers, and solid-state lasers.

Current research on advanced lasers focuses on the generation of extremely short laser pulses of duration less than 1 picosecond (10^{-12} s). At this timing resolution, it begins to become possible to obtain information on the establishment of chemical bonds in real time.

FIGURE 38.27 Cutaway diagram of the National Ignition Facility (NIF) laser system. The lasers are focused on the fusion pellets in the target chamber.

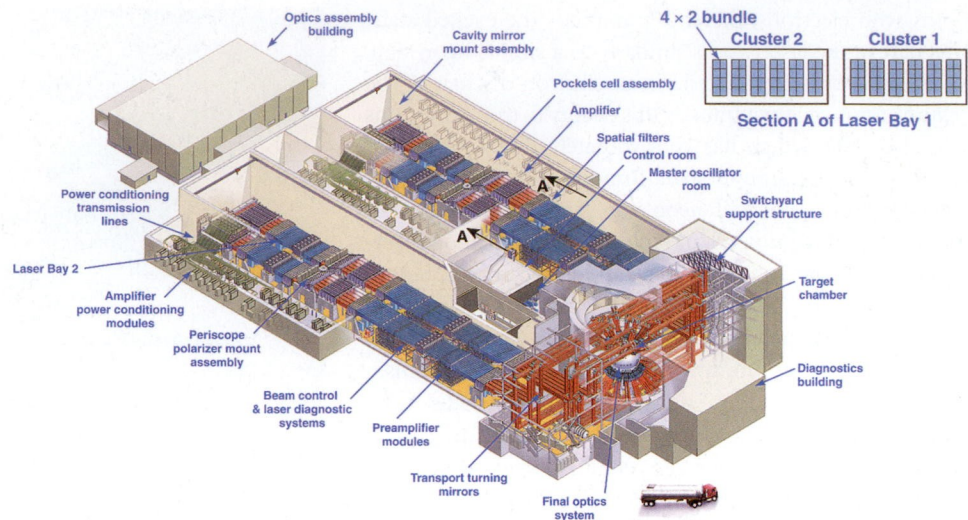

FIGURE 38.28 The Cluster 3 beam path in Laser Bay 2 of the NIF, which contains 6 bundles of 8 beams for a total of 48 beam lines. It was completed in October 2006, and all 6 bundles were operational by December 2006. Laser beams travel over 1000 feet before they reach the target chamber.

Another focus of current research interest is free-electron lasers, in which the lasing medium is a relativistic electron beam. The electron beam passes through an alternating periodic magnetic field, which causes transverse "wiggles" in the motion of the electrons, generating coherent electromagnetic radiation. A laser beam generated in this way is tunable over a wide range of wavelengths, from millimeters to wavelengths in the visible range.

By using ions instead of atoms, it is possible to produce X-ray lasers, which have much higher-energy beams because each photon has a higher energy than in the case of optical lasers. This is also a very active area of research, in particular because of its potential medical and military applications.

The largest laser in the world was completed in 2009 and is housed at the National Ignition Facility (NIF; see Figures 38.27 and 38.28) at the Lawrence Livermore National Laboratory in California. Used to study nuclear fusion, it is a pulsed laser that produces 192 separate beams, consisting of short-lived laser pulses of total energy 1.8 MJ and duration 4 ns. Thus, the peak power of this laser is

$$P_{\max} = E / \Delta t = (1.8 \cdot 10^6 \text{ J}) / (4 \cdot 10^{-9} \text{ s}) = 5 \cdot 10^{14} \text{ W} = 500 \text{ TW}.$$

WHAT WE HAVE LEARNED | EXAM STUDY GUIDE

- The narrow lines observed in the spectra of atoms are due to the transitions of electrons from one state to another.

- The wavelengths of the photons emitted by an excited hydrogen atom are given by the Rydberg formula:

$$\frac{1}{\lambda} = R_H \left(\frac{1}{n_1^2} - \frac{1}{n_2^2} \right), \quad \text{with } n_1 < n_2,$$

  where $R_H = 1.097373 \cdot 10^7 \text{ m}^{-1}$ is the Rydberg constant.

- The Bohr model of the atom was an early success in explaining the observed line spectra. It is based on the quantization condition for the angular momentum

$$L = |\vec{r} \times \vec{p}| = r\mu v = n\hbar \quad (\text{for } n = 1, 2, 3, \ldots).$$

- The orbital radii in the Bohr model are given by

$$r = \frac{\hbar^2}{\mu k e^2} n^2 = a_0 n^2.$$

  The Bohr radius is $a_0 = 5.295 \cdot 10^{-11} \text{ m} = 0.05295 \text{ nm} = 0.5295 \text{ Å}.$

- To obtain the correct electron wave functions for the hydrogen atom, we need to solve the Schrödinger equation for the electron in the Coulomb potential of the nucleus:

$$-\frac{\hbar^2}{2\mu} \nabla^2 \psi(r,\theta,\phi) - k\frac{e^2}{r}\psi(r,\theta,\phi) = E\psi(r,\theta,\phi).$$

- The full solution of this Schrödinger equation can be obtained by separation of variables: $\psi(r, \theta, \phi) = f(r)g(\theta)h(\phi)$.

- The angular parts of the solution are spherical harmonics, which are simultaneously eigenfunctions of the squared angular momentum operator, $\hat{L}^2 Y_\ell^m(\theta,\phi) = \ell(\ell+1)\hbar^2 Y_\ell^m(\theta,\phi)$, and the operator for the z-projection of the angular momentum vector, $\hat{L}_z Y_\ell^m(\theta,\phi) = m\hbar Y_\ell^m(\theta,\phi)$.

- The complete solution for the hydrogen eigenfunction is the product of the angular part and the radial part: $\psi_{n\ell m}(r,\theta,\phi) = f_{n\ell}(r)Y_{\ell m}(\theta,\phi)$.

- The quantum numbers of these solutions are the radial quantum number, $n = 1,2,3, ...$; the orbital angular momentum quantum number, $\ell = 0,1, ... ,n - 1$; and the magnetic quantum number, $m = -\ell, ... , \ell$. The energy eigenvalue corresponding to a particular eigenfunction that is a solution depends only on the radial quantum number and is the same as for the Bohr model:

$$E_n = -\frac{1}{n^2} E_0 = -\frac{1}{n^2} \frac{\mu k^2 e^4}{2\hbar^2}.$$

- The hydrogen ground-state wave function is

$$\psi_{100}(r,\theta,\phi) = \frac{1}{\sqrt{\pi} a_0^{3/2}} e^{-r/a_0}.$$

- The total number, $N(n)$, of possible wave functions in a given shell with radial quantum number n is

$$N(n) = \sum_{\ell=0}^{n-1} (2\ell+1) = n^2.$$

- The wave functions for other atoms can be determined in a manner similar to the hydrogen wave functions. The Pauli Exclusion Principle states that each possible state is occupied by at most two electrons, one with spin down and one with spin up.

- The order in which the angular momentum states are filled follows their energies and is 1s, 2s, 2p, 3s, 3p, 4s, 3d, 4p, 5s, 4d, 5p, 6s, ...

- Lasers work on the basis of population inversion. Electrons are lifted into a metastable state, followed by stimulated emission of photons with energy corresponding to the transition between the metastable state and the ground state.

PROBLEM-SOLVING GUIDELINES

1. For photons emitted or absorbed in atomic transitions, you can always calculate the energy as the difference between the final and initial electron energies. The Bohr model of the atom may be fundamentally flawed, but it is still useful in many calculations involving electron energies in hydrogen-like atoms.

2. The hydrogen atom is a quantum system, so all of the problem-solving guidelines presented in Chapters 36 and 37 apply to problems involving this system. In particular, the hydrogen wave functions for a given quantum state can be used to obtain expectation values of quantum operators via integration.

3. The Pauli Exclusion Principle is of fundamental importance in understanding how subshells get filled in many-electron atoms. Making sure that no two electrons occupy the same quantum state in the same atom at the same time avoids many simple errors.

ANSWERS TO SELF-TEST OPPORTUNITIES

38.1 We need to take the limit as $n_2 \to \infty$ and find

$\frac{1}{\lambda} = R_H \left(\frac{1}{n_1^2} - \frac{1}{n_2^2} \right) \xrightarrow[n_2 \to \infty]{} \frac{R_H}{n_1^2}$. With $R_H = 1.097373 \cdot 10^7 \text{ m}^{-1}$, we find for the Lyman ($n_1 = 1$), Balmer (2), Paschen (3), and Brackett (4) series: $\lambda_{\min} = 91.1$ nm, 365 nm, 820 nm, 1458 nm.

38.2 For Balmer, $n_1 = 2, n_2 = n$: $\frac{1}{\lambda} = R_H \left(\frac{1}{4} - \frac{1}{n^2} \right) = R_H \left(\frac{n^2 - 4}{4n^2} \right)$.

Therefore, the wavelengths in this series are $\lambda = \frac{4}{R_H} \frac{n^2}{n^2 - 4} = (364.5 \text{ nm}) \frac{n^2}{n^2 - 4}$.

38.3 $\psi_{211}(x,0,z)$ has a shape like $\psi_{210}(x,0,z)$, but rotated 90° counterclockwise in the xz-plane.

38.4 Insert the wave function, $\psi_{100}(r,\theta,\phi) = \frac{1}{\sqrt{\pi} a_0^{3/2}} e^{-r/a_0}$, into the Schrödinger equation:

$-\frac{\hbar^2}{2\mu} \frac{1}{r} \frac{d^2}{dr^2} \left(r\psi(r) \right) - k\frac{e^2}{r} \psi(r) = E\psi(r)$. First, take the

derivative: $\frac{d}{dr} \left(r \frac{1}{\sqrt{\pi} a_0^{3/2}} e^{-r/a_0} \right) = \frac{1}{\sqrt{\pi} a_0^{3/2}} e^{-r/a_0} \left(1 - \frac{r}{a_0} \right)$;

then take the second derivative:

$\frac{d}{dr} \left(\frac{1}{\sqrt{\pi} a_0^{3/2}} e^{-r/a_0} \left(1 - \frac{r}{a_0} \right) \right) = -\frac{e^{-r/a_0}}{a_0 \sqrt{\pi} a_0^{3/2}} \left(2 - \frac{r}{a_0} \right)$. Insert back

into the Schrödinger equation:

$\left(-\frac{\hbar^2}{2\mu} \frac{1}{r} \right) \left[-\frac{e^{-r/a_0}}{a_0 \sqrt{\pi} a_0^{3/2}} \left(2 - \frac{r}{a_0} \right) \right] - k\frac{e^2}{r} \left(\frac{1}{\sqrt{\pi} a_0^{3/2}} e^{-r/a_0} \right) =$

$E \left(\frac{1}{\sqrt{\pi} a_0^{3/2}} e^{-r/a_0} \right)$. Now cancel common factors and find

$E = -\frac{\mu k^2 e^4}{2\hbar^2}$, which means that $\psi_{100}(r,\theta,\phi)$ is indeed a

solution to the Schrödinger equation with the correct value for the energy.

MULTIPLE-CHOICE QUESTIONS

38.1 The wavelength of the fourth line in the Lyman series of the hydrogen spectrum is

a) 80.0 nm.   b) 85.0 nm.   c) 90.2 nm.   d) 94.9 nm.

38.2 The electron in a certain hydrogen atom is in the $n = 5$ state. Which of the following could be the ℓ and m values for the electron?

a) 5, –3   b) 4, –5   c) 3, –2   d) 4, –6

38.3 A muon has the same charge as an electron but a mass that is 207 times greater. The negatively charged muon can bind to a proton to form a new type of hydrogen atom. How does the binding energy, $E_{B\mu}$, of the muon in the ground state of a muonic hydrogen atom compare with the binding energy, E_{Be}, of an electron in the ground state of a conventional hydrogen atom?

a) $|E_{B\mu}| \approx |E_{Be}|$

b) $|E_{B\mu}| \approx 100 |E_{Be}|$

c) $|E_{B\mu}| \approx |E_{Be}|/100$

d) $|E_{B\mu}| \approx 200 |E_{Be}|$

e) $|E_{B\mu}| \approx |E_{Be}|/200$

38.4 Which of the following can be used to explain why you can't walk through walls?

a) Coulomb repulsive force

b) the strong nuclear force

c) gravity

d) Pauli Exclusion Principle

e) none of the above

38.5 How many antinodes are there in a system with the quantum number $n = 6$?

a) 12   b) 6   c) 3   d) 5

38.6 Transition metals can be defined as the elements in which the d subshells go from empty to full. How many transition metals are there in each period?

a) 2   b) 6   c) 10   d) 14

38.7 What is the shortest wavelength photon that can be emitted by singly ionized helium (He^+)?

a) 0.00 nm   d) 91.0 nm

b) 22.8 nm   e) 365 nm

c) 46.0 nm

38.8 An electron made a transition between allowed states, emitting a photon. Which of the following physical constants is (are) needed to calculate the energy of the photon from the measured wavelength?

a) the Plank constant, h

b) the basic electric charge, e

c) the speed of light in vacuum, c

d) the Stefan-Boltzmann constant, σ

CONCEPTUAL QUESTIONS

38.9 The common depiction of an atom, with electrons tracing elliptical orbits centered on the nucleus, is an icon of the Atomic Age. Given what you know of the physics of atoms, what's wrong with this picture?

38.10 Given that the hydrogen atom has an infinite number of energy levels, why can't a hydrogen atom in the ground state absorb all possible wavelengths of light?

38.11 Compare the difference in energy levels for increasing n for an infinite square well, a quantum harmonic oscillator, and the hydrogen atom.

38.12 What would happen to the energy levels of a hydrogen atom if the Coulomb force doubled in strength? What would happen to the sizes of atoms?

38.13 Which model of the hydrogen atom—the Bohr model or the quantum mechanical model—predicts that the electron spends more time near the nucleus?

38.14 For $\ell < 4$, which values of ℓ and m correspond to wave functions that have their maximum probability in the xy-plane?

38.15 *Hund's rule*, a component of the *Aufbauprinzip* (construction principle), states that as one moves across the periodic table, with increasing atomic number, the available subshells are filled successively with one electron in each orbital, their spins all parallel; only when all orbitals in a subshell contain one electron are second electrons, with spins opposite to the first, placed in the orbitals. Explain why the ground-state electron configurations of successive elements should follow this pattern.

38.16 The energy-level diagram for a four-level laser is presented in the figure, which includes information about the relative times required for the various transitions involved in the operation of such a laser. Explain why the nonradiative transitions have to be fast (higher rate of transitions per second) compared to the slower laser transition.

38.17 The hydrogen atom wave function ψ_{200} is zero when $r = 2a_0$. Does this mean that the electron in that state can never be observed at a distance of $2a_0$ from the nucleus or that the electron can never be observed passing through the spherical surface defined by $r = 2a_0$? Is there a difference between those two descriptions?

38.18 A 10-eV electron collides with (but is not captured by) a hydrogen atom in its ground state. Calculate the wavelengths of all photons that might be emitted.

EXERCISES

A blue problem number indicates a worked-out solution is available in the Student Solutions Manual. One • and two •• indicate increasing level of problem difficulty.

Section 38.1

38.19 What is the shortest wavelength of light that a hydrogen atom will emit?

38.20 Determine the wavelength of the second line in the Paschen series.

38.21 The Pfund series results from emission/absorption of photons due to transitions of electrons in a hydrogen atom to/from the $n = 5$ energy level from/to higher energy levels. What are the shortest and longest wavelengths of lines in the Pfund series? Are any of these in the visible portion of the electromagnetic spectrum?

38.22 An electron in the second excited state of a hydrogen atom jumps to the ground state. What are the possible colors and wavelengths of the light emitted as a result of the jump?

Section 38.2

38.23 Calculate the energy of the fifth excited state of a hydrogen atom.

38.24 Hydrogen atoms are bombarded with 13.1-eV electrons. Determine the shortest wavelength of light the atoms will emit.

38.25 The Rydberg constant with a finite mass of the nucleus is given by $R_{modified} = R_H/(1+m/M)$, where m and M are the masses of an electron and a nucleus, respectively. Calculate the modified value of the Rydberg constant for

a) a hydrogen atom, and

b) a positronium (in a positronium, the "nucleus" is a positron, which has the same mass as an electron).

38.26 A muon is a particle very similar to an electron. It has the same charge but its mass is $1.88 \cdot 10^{-28}$ kg.

a) Calculate the reduced mass for a hydrogen-like muonic atom consisting of a single proton and a muon.

b) Calculate the ionization energy for such an atom, assuming that the muon is initially in its ground state.

•**38.27** An excited hydrogen atom emits a photon with an energy of 1.133 eV. What were the initial and final states of the hydrogen atom before and after emitting the photon?

•**38.28** An 8.00-eV photon is absorbed by an electron in the $n = 2$ state of a hydrogen atom. Calculate the final speed of the electron.

•**38.29** Assume that Bohr's quantized energy levels apply to planetary orbits. Derive an equation similar to equation 38.6 that gives the allowed radii of those orbits, and estimate the principal quantum number for the Earth's orbit.

•**38.30** Prove that the period of an electron in the nth Bohr orbit is given by $T = n^3/(2cR_H)$, with $n = 1,2,3, \ldots$.

Section 38.3

38.31 What are the largest and smallest possible values for the angular momentum L of an electron in the $n = 5$ shell?

38.32 Electrons with the same value of the quantum number n are said to occupy the same electron shell, $K, L, M, N,$ or higher. Calculate the maximum allowed number of electrons for the

a) K shell,   b) L shell, and   c) M shell.

•**38.33** What is the angle between the total angular momentum vector and the z-axis for a hydrogen atom in the stationary state $(3,2,1)$?

•**38.34** A hydrogen atom is in its fifth excited state, with principal quantum number $n = 6$. The atom emits a photon with a wavelength of 410 nm. Determine the maximum possible orbital angular momentum of the electron after emission.

•**38.35** The radial wave function for hydrogen in the 1s state is given by $R_{1s} = A_1 e^{-r/a_0}$, where the normalization constant, A_1, was found in Example 38.2.

a) Calculate the probability density at $r = a_0/2$.

b) The 1s wave function has a maximum at $r = 0$ but the 1s radial probability density peaks at $r = a_0$. Explain this difference.

•**38.36** For the wave function $\psi_{100}(r)$ in equation 38.30, find the value of r for which the function $P(r) = 4\pi r^2 |\psi_{100}(r)|^2$ is a maximum.

••**38.37** An electron in a hydrogen atom is in the 2s state. Calculate the probability of finding the electron within a Bohr radius ($a_0 = 0.05295$ nm) of the proton. The 2s wave function for hydrogen is

$$\psi_{2s}(r) = \frac{1}{4\sqrt{2\pi a_0^3}}\left(2 - \frac{r}{a_0}\right)e^{-r/2a_0}.$$

Evaluating the integral is a bit tedious, so you may want to consider using a program such as Mathcad or Mathematica or finding the integral online at http://integrals.wolfram.com/index.jsp.

Section 38.4

38.38 Calculate the energy needed to change a single ionized helium atom into a double ionized helium atom (that is, change it from He^+ into He^{2+}). Compare it to the energy needed to ionize the hydrogen atom. Assume that both atoms are in their ground state.

38.39 A He^+ ion consists of a nucleus (containing two protons and two neutrons) and a single electron. Find the Bohr radius for this system.

•**38.40** Find the wavelengths of the three lowest-energy lines in the Paschen series of the spectrum of the He^+ ion.

•**38.41** The binding energy of an extra electron added when As atoms are used to dope a Si crystal may be approximately calculated by considering the Bohr model of a hydrogen atom.

a) Express the ground energy of the hydrogen-like atoms in terms of the dielectric constant, the effective mass of an extra electron, and the ground-state energy of a hydrogen atom.

b) Calculate the binding energy of the extra electron in a Si crystal. The dielectric constant of Si is about 10.0, and the effective mass of extra electrons in a Si crystal is about 20.0% of that of free electrons.

•**38.42** What is the wavelength of the first visible line in the spectrum of doubly ionized lithium? Begin by writing the formula for the energy levels of the electron in doubly ionized lithium—then consider energy-level differences that give energies in the appropriate (visible) range. Express the answer as "the transition from state n to state n' produces the first visible line, with wavelength X."

•**38.43** Following the steps used in the text for the hydrogen atom, apply the Bohr model of the atom to derive an expression for

a) the radius of the nth orbit,

b) the speed of the electron in the nth orbit, and

c) the energy levels in a hydrogen-like ionized atom with charge number Z that has lost all of its electrons except for one. Compare the results with the corresponding ones for the hydrogen atom.

•**38.44** Apply the results of Problem 38.43 to determine the maximum and minimum wavelengths of the spectral lines in the Lyman, Balmer, and Paschen series for a singly ionized helium atom (He^+).

Section 38.5

38.45 Consider an electron in a hydrogen atom. If you are able to excite the electron from the $n = 1$ shell to the $n = 2$ shell with laser light of a given wavelength, what wavelength of laser light will excite that electron again from the $n = 2$ to the $n = 3$ shell? Explain.

38.46 A low-power laser has a power of 0.50 mW and a beam diameter of 3.0 mm.

a) Calculate the average light intensity of the laser beam.

b) Compare the value from part (a) to the intensity of light from a 100-W light bulb viewed from 2.0 m.

38.47 A ruby in a laser consists mostly of alumina (Al_2O_3) and a small amount of chromium ions, responsible for its red color. A 3.00-kW ruby laser emits light pulses of duration 10.0 ns and wavelength 694 nm.

a) What is the energy of each of the photons in a pulse?

b) Determine the number of chromium atoms undergoing stimulated emission to produce a pulse.

38.48 You have both a green, 543-nm, 5.00-mW laser and a red, 633-nm, 4.00-mW laser. Which one will produce more photons per second, and why?

Additional Exercises

38.49 What is the shortest possible wavelength in the Lyman series in hydrogen?

38.50 How much energy is required to ionize a hydrogen atom when the electron is in the nth level?

38.51 By what percentage is the mass of the electron changed by using the reduced mass for the hydrogen atom? What would the reduced mass be if the proton had the same mass as the electron?

38.52 Show that the number of different electron states possible for a given value of n is $2n^2$.

38.53 Section 38.2 established that an electron, if observed in the ground state of hydrogen, is expected to have a speed of $0.0073c$. For what nuclear charge Z would an innermost electron have a speed of approximately $0.500c$, when considered classically?

38.54 A collection of hydrogen atoms have all been placed into the $n = 4$ excited state. What wavelengths of photons will be emitted by the hydrogen atoms as they transition back to the ground state?

38.55 Consider a muonic hydrogen atom, in which the electron is replaced by a muon of mass 105.66 MeV/c^2 that orbits the proton. What are the first three energy levels of the muon in this type of atom?

38.56 What is the ionization energy of a hydrogen atom excited to the $n = 2$ state?

38.57 He$^+$ is a helium atom with one electron missing. Treating this ion like a hydrogen atom, what are its first three energy levels?

38.58 What is the energy of a transition capable of producing light of wavelength 10.6 μm? (This is the wavelength of light associated with a commonly available infrared laser.)

38.59 What is the energy of the orbiting electron in a hydrogen atom with a radial quantum number of 45?

•38.60 Find the energy difference between the ground state of hydrogen and the ground state of deuterium (hydrogen with an extra neutron in the nucleus).

•38.61 An excited hydrogen atom, whose electron is in the $n = 4$ state, is motionless. When the electron drops to the ground state, does it set the atom in motion? If so, with what speed?

•38.62 The radius of the $n = 1$ orbit in the hydrogen atom is $a_0 = 0.053$ nm.

a) Compute the radius of the $n = 6$ orbit. How many times larger is this than the $n = 1$ radius?

b) If an electron in the $n = 6$ orbit drops to the $n = 1$ orbit (ground state), what are the frequency and the wavelength of the emitted radiation? What kind of radiation is emitted (visible, infrared, etc.)?

c) How would your answer to part (a) change if the atom was a singly ionized helium atom (He$^+$) instead?

•38.63 An electron in a hydrogen atom is in the ground state (1s). Calculate the probability of finding the electron within a Bohr radius ($a_0 = 0.05295$ nm) of the proton. The ground-state wave function for hydrogen is

$$\psi_{1s}(r) = A_{1s}e^{-r/a_0} = e^{-r/a_0}/\sqrt{\pi a_0^3}.$$

MULTI-VERSION EXERCISES

38.64 A beam of electrons is incident on a system of excited hydrogen atoms. What minimum speed must the electrons have to cause the emission of light due to the transition in the hydrogen atoms from $n = 2$ to $n = 1$? (In the collision of an electron with a hydrogen atom, you may neglect the recoil energy of the hydrogen atom, because it has a mass much greater than that of the electron.)

38.65 A beam of electrons with a speed of 676.01 km/s is incident on a system of excited hydrogen atoms. If an electron hits a hydrogen atom in the $n = 3$ state, what is the highest level n, to which this hydrogen atom can be excited in this collision? (In the collision of the electron with the hydrogen atom, you may neglect the recoil energy of the hydrogen atom, because it has a mass much greater than that of the electron.)

38.66 A beam of electrons with a speed of 378.92 km/s is incident on a system of excited hydrogen atoms. If an electron hits a hydrogen atom and excites it to the $n = 10$ state, what is the lowest level n in which this hydrogen atom could have been before the collision? (In the collision of the electron with the hydrogen atom, you may neglect the recoil energy of the hydrogen atom, because it has a mass much greater than that of the electron.)

38.67 Find the ratio of the number of hydrogen atoms in the $n = 3$ state to the number of hydrogen atoms in the $n = 7$ state at a temperature of 528.3 K.

38.68 At what temperature is the ratio of the number of hydrogen atoms in the $n = 3$ state to the number of hydrogen atoms in the $n = 8$ state equal to $5.1383 \cdot 10^5$?

39

Elementary Particle Physics

FIGURE 39.1 Two galaxies colliding as photographed by the Hubble Space Telescope.

Cosmology is the study of the largest things we know about: the universe and its time evolution. Principally a part of astronomy, cosmology seeks clues about the early universe from studying far-distant galaxies and clusters of galaxies. For example, the collision of two galaxies (Figure 39.1) indicates how galaxies might have grown and evolved billions of years ago. Strangely enough, some of the other major contributions to cosmology come from studying the smallest things we know about: elementary particles and their interactions. Understanding how matter and energy interact at the most basic level gives us some idea of what the earliest moments of the universe must have been like.

This chapter examines the most basic components of matter, the elementary particles that make up atoms and even the particles within atoms. What is known about these basic units is based on quantum mechanics, but we will skip over most of the mathematics involved and emphasize the basic concepts instead. Experimental evidence to back up these ideas is hard to come by, and usually requires analysis of debris formed when particles moving at speeds close to the speed of light are collided in enormous particle accelerators. There has been a continual increase in the energy needed for further research in this field. Today's particle colliders are among the largest and most powerful instruments ever built.

However, the study of the tiniest particles and the highest energies involves some of the most speculative areas of physics. Particle physicists have developed a model of the most basic particles and their interactions, usually referred to as the *standard model*. But even the standard model does not explain everything, and new ideas about how to extend or improve it seem to arise every day. This chapter will look at some of these theories that may become the standard model of the future.

WHAT WE WILL LEARN

- The quest to reduce systems to their components is a main theme in science and has its present culmination in particle physics. While it is equally valid to analyze complexity and emergent structure, this chapter focuses on reductionism in particle physics.

- Substructure is probed using scattering experiments. This chapter defines the concept of a scattering cross section.

- Elementary fermions have spin $\frac{1}{2}\hbar$ and consist of the six quarks (up, down, strange, charm, bottom, top), the electron, muon, and tau leptons, and the electron-, muon-, and tau-neutrinos. Each of these 12 fermions has an antiparticle. Quarks have a noninteger charge of $-\frac{1}{3}e$ or $+\frac{2}{3}e$ and cannot be observed in isolation.

- Elementary bosons are the mediators of the interactions between the fermions. These bosons include the photon (electromagnetic), W and Z bosons (electroweak), gluon (strong), and graviton (gravitational). The graviton is yet to be observed experimentally. Gluons can also interact with other gluons.

- Feynman diagrams represent the interactions of fundamental particles pictorially, and calculations can be based on these diagrams.

- A widely accepted model of particle interactions is the standard model. Extensions of the standard model, which itself is incomplete, are grand unified theories, supersymmetric theories, and string theories.

- Elementary quarks and antiquarks can combine to form particles, which can be observed in isolation. A quark and an antiquark can form a meson (pion, kaon, etc.). Three quarks can form a baryon (proton, neutron, delta baryon, lambda baryon, etc.).

- Particle physics and astrophysics intersect in the study of Big Bang cosmology, its inflationary phase and its electroweak and quark-gluon-plasma phase transitions, all of which happened in the first 3 minutes of the existence of the universe. This chapter also discusses the origin of the cosmic microwave background radiation.

39.1 Reductionism

At least since the times of the ancient Greek philosophers, a central theme has guided humanity's attempts to achieve a quantitative scientific understanding of the world around us. This central theme is **reductionism,** which assumes that a system can be understood in terms of its subsystems. Figure 39.2 illustrates reductionism in particle physics.

Astronomers have established that the universe contains hundreds of billions of galaxies, which are each composed of hundreds of billions of star systems. These star systems can consist of a central star (or sometimes two or more stars), which is orbited by planets, which in turn are orbited by their moons.

Biologists are trying to understand the entire ecosystem as composed of populations of species. Individual organisms of these species contain specialized organs and cells, which have nuclei containing chromosomes consisting of DNA. The DNA, which contains the genetic code that governs all traits passed on from one generation to the next, is understood to be segmented into thousands of genes. Genes are made of usefully readable content, called *exons,* interspersed with unreadable (or at least unread) *introns.* The exons are made of "words"—called codons—each of which contains three amino acid base pairs: each pair consisting of adenine (A) and thymine (T) or cytosine (C) and guanine (G).

Arguably, physicists have led the charge toward reductionism. Physicists now know that matter is composed of molecules, which in turn are made of atoms. The word *atom* derives from the Greek word *átomos* (ατομος, "individual, indivisible"). In particular, the

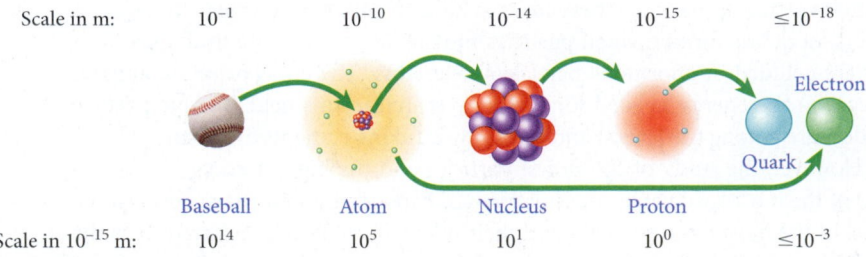

Scale in m:	10^{-1}	10^{-10}	10^{-14}	10^{-15}	$\leq 10^{-18}$

Electron

Quark

Scale in 10^{-15} m:	10^{14}	10^{5}	10^{1}	10^{0}	$\leq 10^{-3}$
	Baseball	Atom	Nucleus	Proton	

FIGURE 39.2 Reductionism in particle physics.

Greek philosopher Democritus was the earliest prominent proponent of the atomistic theory of matter. The four elements of the Greek atomistic theory were earth, air, fire, and water, which were arranged in a two-by-two matrix of hot-cold and wet-dry opposites (see Figure 39.3), and which were thought to be immutable. Aristotle added to these four a fifth element, *quintessence,* which he thought to be the material of which the heavens were made.

The Greek atomistic theory was forgotten until the Renaissance. Then philosophers, physicists, and chemists created a new atomistic view of the world, and it proved to be incredibly successful and productive. By 1869, many different kinds of atoms had been identified as the elementary building blocks of the chemical elements. The Russian chemist Dimitri Mendeleev (1834–1907) took a giant step forward in that year by arranging the atoms of the elements in the periodic table. The current count of the different known elements is 118: the 112 elements from hydrogen through copernicium, plus the yet unnamed elements 113 through 118.

The idea of simplicity is intimately connected with reductionism. According to the rule known as *Occam's razor,* when an observation can be explained by several alternative theories, the simplest explanation is usually preferable. In addition, physical theories are often guided by the search for symmetry and abstract beauty, as well as the more practical demands of consistency and reproducibility. While physics is *the* basic science and is usually concerned with measurable facts, at the level discussed here it is also influenced by concepts from philosophy and even the arts.

The fact that so many kinds of atoms exist led many people to speculate at the end of the 19th century that atoms are not really the most basic fundamental building blocks of matter and that there must be an underlying, more fundamental structure. This more fundamental structure turned out to be a hierarchy of several layers. The rest of this chapter is devoted to peeling back these layers one at a time.

Nuclear physics and elementary particle physics are devoted to searching for this fundamental structure explaining the phenomena observed at the respective size scales. The reductionist quest to discover the most fundamental level is far from complete. For example, nuclear and particle physicists are currently engaged in this quest using the Large Hadron Collider (LHC, Figure 39.4) at CERN (straddling the Swiss-French border, near Geneva), which became operational in 2008.

Cathedral-size caves located approximately 100 m underground house the largest particle detectors ever built. Figure 39.5 shows a construction photo of the partially finished

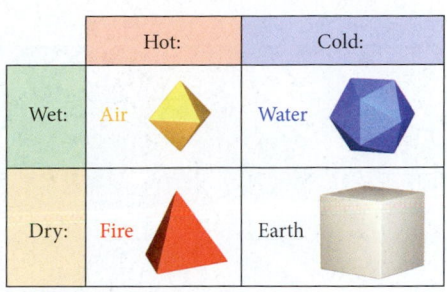

FIGURE 39.3 The four basic elements of ancient Greek philosophy, each shown with the regular polyhedron assigned to it by Plato.

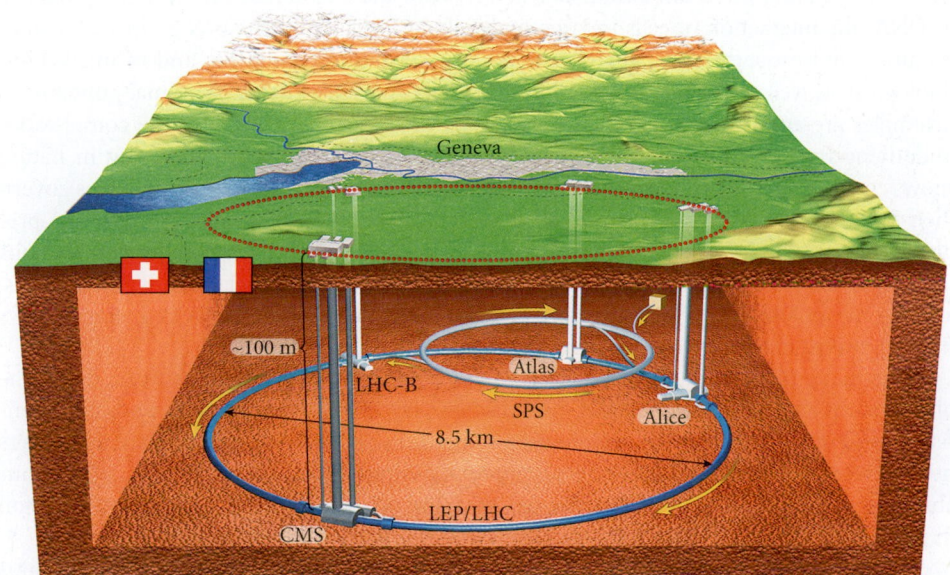

FIGURE 39.4 Schematic layout of the LHC accelerator and detector complex at CERN.

FIGURE 39.5 The ATLAS detector at CERN under construction.

FIGURE 39.6 Construction drawing of the ALICE detector.

ATLAS detector at CERN. Figure 39.6 shows a cutaway construction drawing of the ALICE detector that is also part of the LHC complex at CERN. At the LHC, physicists have discovered the Higgs boson (see Section 39.3) and hope to discover hints about the next more fundamental level in the hierarchy (if it exists!).

Complexity

The remainder of this chapter focuses on atomic nuclei, elementary particles, and smallest constituents, but reductionism is only one part of science research. Just as important is the complementary research goal: finding out how constituents on one level of the size hierarchy work together to create the structure observed on the next-larger level. This is the study of **complexity**—pattern formation, self-organization of systems, and emergence of order out of chaos. This aspect of science is much less mature than the reductionist thread that science has followed for the last few centuries. The study of complexity often relies on computer modeling and is thus not usually covered in introductory textbooks. Nevertheless, many stunning investigations of complex systems are currently ongoing.

Examples of complexity on the atomic scale include pattern formation in complex materials and the self-assembly of nanostructures. Physicists, biochemists, and microbiologists are forming interdisciplinary collaborations to try to understand the mechanism of protein folding in creating three-dimensional structure from the encoding of the genetic sequence in DNA; the interaction of cells in large multicellular organisms; the precise mechanism of evolution and propagation of genetic mutations; and the origins of life and of intelligence (biological, as well as artificial). In biology, the establishment of order in animal populations (examples are anthills and bird-flight formations) is under intense study. The complexities underlying the spread of epidemics need to be understood. Almost everything in nature shows complex patterns, and scientists seek to understand which universal principles govern their formation. In meteorology, the goal is to come closer to being able to describe and predict the complexity of weather patterns. Human interaction has many complex dimensions that can be modeled, such as traffic flow in cities, the rise and fall of social trends and fashions, and the movement of market indicators in economies. Even the universe as a whole shows fractal patterns in the distribution of its galaxy clusters and superclusters.

For many or perhaps most of these fields of research, training in just one discipline of science is not sufficient. Instead, interdisciplinary teams composed of physical scientists, mathematicians, engineers, biologists, computer scientists, social scientists, and others need to work together to obtain a deeper understanding of these areas. In particular, as our ability to shape nature grows faster, the risk of unintended consequences from intervention in poorly understood systems increases, and that risk demands deeper exploration.

But first things first! Let's take things apart, before we put them back together. What is an elementary particle?

39.2 Probing Substructure

For a particle to be considered elementary, it cannot have a substructure of other particles. How is substructure discovered? Visual examination can reveal substructure of systems that are large enough, but as Chapter 38 showed, atomic radii are on the order of 10^{-10} m and thus too small to see. The only means available to explore dimensions that are smaller than this scale is to shoot small particles at atomic targets and observe how the particles are deflected and scattered.

Classical Scattering

Before we proceed to use quantum mechanics, let's extend the classical theory of momentum and collisions, presented in Chapters 7 and 8. In Chapter 7, we considered elastic two-body scattering in two and three spatial dimensions and found that conservation of total energy and momentum did not provide enough constraints to predict the final momenta from the initial ones.

Here we want to examine a very simple case of scattering a very small, low-mass particle off a larger and much higher-mass spherical target. In the limit that the target is much more massive than the projectile, we can consider the target to be stationary. Further, we consider only motion along straight-line trajectories, which implies the absence of any force other than the contact force at the instant when the particles collide.

We define the impact parameter b as the perpendicular distance from the incoming trajectory to the trajectory that would result in a direct head-on collision. What is the deflection angle θ of the trajectory after the collision relative to the trajectory before the collision? A side view is shown in Figure 39.7a. In this figure, the acute angle ϕ, in black, is the angle of the perpendicular to the target at the point at which the small particle makes contact with the target relative to the position vector of the low-mass particle before scattering. It is given from trigonometry (see the blue right triangle in Figure 39.7a) that $\sin\phi = b/R$. We also see from Figure 39.7a that the deflection angle θ is given by

$$\theta = \pi - 2\phi = \pi - 2\sin^{-1}(b/R).$$

Therefore, given the impact parameter and the radius of the target, the deflection angle can be calculated. Alternatively, given the impact parameter and the deflection angle, the radius of the target can be calculated.

In practice, physicists construct macroscopic scattering targets from very thin foils of materials. Because nuclei are so small, each particle hits at most one nucleus in a thin foil. The impact parameter cannot be predicted or measured. However, one can observe the angles at which particles scatter from the foil and make conclusions based on how many particles scatter at a particular angle. This situation is sketched in Figure 39.7b. Because a tiny nucleus will not get hit multiple times, the more realistic situation is shown in Figure 39.7c, which shows a thin foil composed of many very small scattering targets, many of which are not hit, and some of which are hit (but very likely only once). The angular distribution of these scattering events is the same as the one shown in Figure 39.7b.

Instead of a side view of the scattering centers inside the foil, we can sketch a head-on frontal view of the foil as presented to the incoming projectile particles. This situation would look something like Figure 39.8, which shows some unit area A of the foil (gray square) with the scattering centers indicated by the small blue circles. Assume that the projectile particles impinge perpendicular to the page. Each scattering center has an effective area perpendicular to the trajectory of the projectile particles. This effective area is called the **cross section**, denoted by σ. It has the dimension of area, that is, $[\sigma] = \text{m}^2$. However, in nuclear and particle physics, the unit m^2 is not very practical because the relevant cross sections are such small fractions of a square meter. Instead, the (rather ironically named!) units **barn** (b) and millibarn (mb) have been introduced:

$$1\,\text{b} = 10^{-28}\,\text{m}^2$$

$$1\,\text{mb} = 10^{-31}\,\text{m}^2.$$

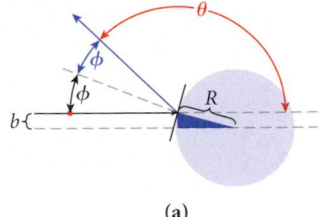

(a)

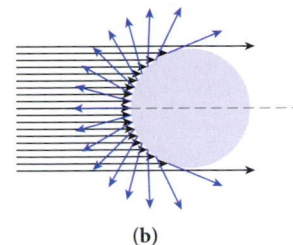

(b)

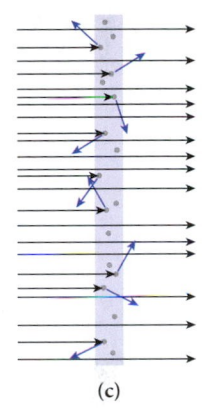

(c)

FIGURE 39.7 Classical scattering of a very small, low-mass particle (red dot) off a larger, high-mass one (blue circle). (a) Scattering at one particular fixed impact parameter; (b) superposition of many scattering events at different impact parameters; (c) superposition of many scattering events of projectile particles off small targets within a thin foil.

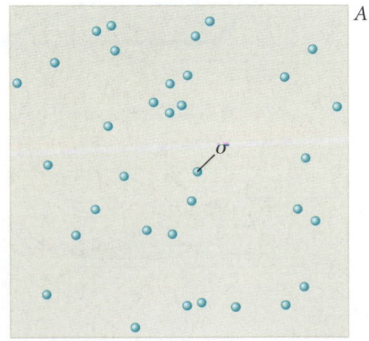

FIGURE 39.8 Frontal view of a target foil composed of many individual, small, randomly distributed scattering centers, as presented to impinging projectile particles.

Assume that the cross section for each scattering center in the target is the same, and that it is small enough that we do not have to worry about the cross sections of different scattering centers overlapping. Then the probability, Π, that any given projectile particle has a scattering event can be stated as the sum of the areas of all scattering centers divided by the area A over which they are spread. Since each scattering center has the same area, σ, the sum of the areas is then $n_t\sigma$, where n_t is the number of scattering centers contained in the target area A. Thus,

$$\Pi = \frac{n_t\sigma}{A}.$$

If a large number n_p of projectile particles fall on this area A, then the total number of reactions N_{pt} is the product of the probability that a projectile particle has a scattering event times the number n_p:

$$N_{pt} = n_p\Pi = \frac{n_p n_t\sigma}{A}.$$

This equation can be solved for the cross section, σ:

$$\sigma = \frac{N_{pt}A}{n_p n_t}.$$

The same basic relationship is also useful for quantum mechanical problems. In general, a cross section is defined as the number of reactions of a particular kind per second divided by the number of projectile particles falling on the target per second and per unit area:

$$\sigma = \frac{\text{Number of reactions per scattering center/s}}{\text{Number of impinging particles/s/m}^2}.$$

A differential cross section, $d\sigma/d\Omega$, can be defined as the number of reactions that lead to a scattering of a particle into some solid angle $d\Omega$ per second, divided by the same denominator as in the definition of σ above:

$$\frac{d\sigma}{d\Omega} = \frac{\text{Number of scatterings into solid angle } d\Omega \text{ per scattering center/s}}{\text{Number of impinging particles/s/m}^2}. \quad (39.1)$$

The units of the differential cross section are $[d\sigma/d\Omega] = \text{mb/sr}$. The abbreviation sr stands for *steradian*, the unit of solid angle. Remember that a sphere subtends a solid angle of 4π sr. Integrating $d\sigma/d\Omega$ over the unit sphere gives the total cross section:

$$\sigma = \int_{4\pi} \left(\frac{d\sigma}{d\Omega}\right) d\Omega.$$

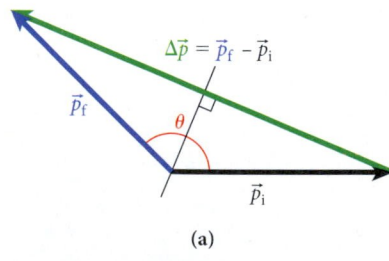

Before we finish with classical considerations, let's examine the impulse (momentum transfer) that the target receives from the projectile. Since the collision is elastic and the target is assumed to be much heavier than the projectile, the lengths of the initial and final momentum vectors of the projectile are the same, $|\vec{p}_i| = |\vec{p}_f|$, and thus the vectors are the equal sides of an isosceles triangle.

Figure 39.9 shows the initial and final momentum vectors for the scattering event depicted in Figure 39.7a, together with the momentum transfer (green vector). If the deflection angle is bisected, the line crosses the momentum transfer vector at a right angle. This forms a right triangle with angles $\frac{1}{2}\theta$, $\frac{1}{2}\pi$, and α. Because the sum of the angles in a triangle is always π, it follows that $\frac{1}{2}\theta + \frac{1}{2}\pi + \alpha = \pi$, and thus $\alpha = \frac{1}{2}\pi - \frac{1}{2}\theta$. The absolute value of the momentum transfer, $\Delta p = |\Delta\vec{p}| \equiv |\vec{p}_f - \vec{p}_i|$, can then be expressed as a function of the deflection angle θ:

$$\Delta p = 2p_i\cos(\alpha) = 2p_i\cos(\tfrac{1}{2}\pi - \tfrac{1}{2}\theta) = 2p_i\sin(\tfrac{1}{2}\theta). \quad (39.2)$$

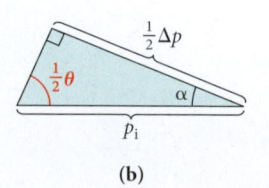

FIGURE 39.9 (a) Momentum transfer for the same deflection angle as in Figure 39.7a. (b) Right triangle formed by the initial momentum and one-half of the momentum transfer.

For backward scattering ($\theta = \pi$), the momentum transfer is at its maximum, with value $2p_i$, and for forward scattering ($\theta = 0$, no collisions), the momentum transfer is zero. Keep in mind that we have assumed that the target is much heavier than the projectile. If this is not the case, then the previous considerations are still true if, instead of using the laboratory frame as was done here, we use the center-of-mass frame to calculate the deflection angle and momentum transfer. In particular, if the projectile is as heavy or heavier than the target, then no backward scattering can be observed in the laboratory frame. (This observation was made in Chapter 7 for one-dimensional collisions of two carts with different masses on a frictionless track.)

Rutherford Scattering

As mentioned in Chapter 38, the experiments of Hans Geiger and Ernest Marsden, and their interpretation by Ernest Rutherford, gave definite proof that the atom consists of an outer layer of electrons surrounding a nucleus. What did Geiger and (then undergraduate student) Marsden actually do?

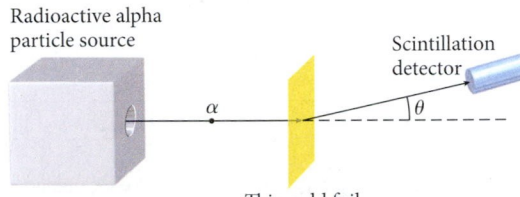

FIGURE 39.10 Basic setup of the Geiger-Marsden experiment in which alpha particles were scattered off gold foil.

Figure 39.10 shows the basic setup of the Geiger-Marsden experiment. They used a radioactive polonium source that emitted alpha particles. (Alpha emission as one possible radioactive decay mechanism will be discussed in Chapter 40.) For now, all you need to know is that the alpha particle is the nucleus of a helium atom and has a charge of $+2e$. The kinetic energy of the alpha particle is typically on the order of 5 MeV. Geiger and Marsden scattered alpha particles off thin gold foil and observed the resulting angular distribution. What can we expect from the scattering in this setup?

Let's first calculate the de Broglie wavelength of a 5-MeV alpha particle. The alpha particle has a mass of 3.73 GeV/c^2, so a kinetic energy of 5 MeV corresponds to a momentum of 193 MeV/c. (Because the mass is very large compared to the kinetic energy in this case, we can use the nonrelativistic approximation, $p = \sqrt{2Em}$.) Thus, the de Broglie wavelength of a 5-MeV alpha particle is

$$\lambda = \frac{h}{p} = \frac{4.136 \cdot 10^{-15} \text{ eV s}}{193 \text{ MeV}/c} = 6.4 \cdot 10^{-15} \text{ m}.$$

The discussion of diffractive scattering of waves in Chapter 34 showed that the angle of the diffraction maximum is approximately $\theta_d \approx \lambda/R$, where R is the radius of the object the waves diffract off—in this case, the atom. At the time, the most prominent model of the atom was the "plum pudding" model, in which negatively charged electrons were thought to be randomly distributed over the volume of the atom, similar to raisins within a plum pudding. The dough of the plum pudding was supposed to be the distribution of positive charge. Geiger and Marsden expected to see diffractive scattering with angles

$$\theta_d \approx 6 \cdot 10^{-15} \text{ m} / 10^{-10} \text{ m} = 6 \cdot 10^{-5} \text{ rad} = 0.003°.$$

In other words, they expected extremely forward-peaked angular distributions. (The limitations imposed by quantum mechanics are discussed in more detail in the next subsection.)

However, to their surprise, Geiger and Marsden observed a significant amount of scattering at large angles, and even some scattering at backward angles. When Ernest Rutherford saw the data, he compared the alpha particles to artillery shells reflected backward by tissue paper. This observation led him to formulate the Rutherford model of the atom, with a very small positively charged nucleus in the center.

Rutherford considered the scattering of a charged particle with (positive) charge $Z_p e$ off a target nucleus with (also positive) charge $Z_t e$. The interaction between these two particles takes place via the Coulomb potential (see Chapter 23), and the potential energy is

$$U = k\frac{Z_p Z_t e^2}{r},$$

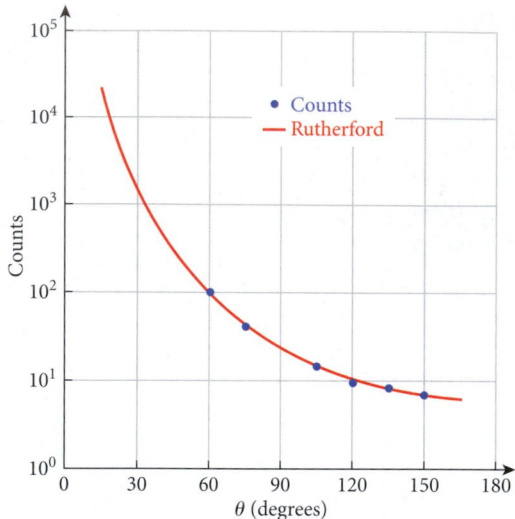

FIGURE 39.11 Comparison of the Rutherford scattering formula (red line) and the angular dependence of the scintillation counts per minute (blue dots), as reported by Geiger and Marsden (*Phil. Mag.* 25 (1913) 604).

where k is Coulomb's constant. Rutherford found that the differential cross section can be written as

$$\frac{d\sigma}{d\Omega} = \left(\frac{kZ_p Z_t e^2}{4K}\right)^2 \frac{1}{\sin^4\left(\frac{1}{2}\theta\right)}, \tag{39.3}$$

where K is the kinetic energy of the projectile particle.

Note that the Rutherford formula (equation 39.3) assumes that projectile and target are both point charges. As Figure 39.11 shows, this formula has a very good fit with the number of scintillation counts per minute (which are proportional to the differential cross section) as a function of the deflection angle.

The relationship between the scattering angle and the momentum transfer that was derived above (equation 39.2) then gives

$$\frac{d\sigma}{d\Omega} = \left(2kZ_p Z_t e^2 m_p\right)^2 \frac{1}{\left(\Delta p\right)^4}, \tag{39.4}$$

where we have used the nonrelativistic approximation $K = p^2/2m_p$, and where m_p is the mass of the projectile. (If we drop the assumption of an infinitely heavy target, then we have to use the reduced mass, $\mu = m_p m_t/(m_p + m_t)$, instead of m_p.)

EXAMPLE 39.1 Backward Scattering of Alpha Particles

We can get a useful upper limit for the size of the atomic nucleus from the observation that 5.00-MeV alpha particles can be scattered in the backward direction.

PROBLEM
How close can a 5.00-MeV alpha particle get to a gold nucleus if the only interaction between them is the Coulomb force?

SOLUTION
We can use a straightforward application of conservation of total mechanical energy. Far from the nucleus, the alpha particle has a kinetic energy of 5.00 MeV and a potential energy of zero. As it approaches the gold nucleus, more and more kinetic energy is converted into potential energy. At the point of closest approach, the alpha particle has zero kinetic energy, and all its initial kinetic energy has been converted to potential energy. The alpha particle has a charge of $+2e$ and the gold nucleus has a charge of $+79e$, so the electrostatic potential energy between them is

$$U(r) = k\frac{(2e)(79e)}{r} = ke^2 \frac{2(79)}{r}.$$

This gives us the equation

$$ke^2 \frac{2(79)}{r_{min}} = 5.00 \text{ MeV},$$

which we need to solve for the minimum distance, r_{min}. The product ke^2 has the dimensions of energy times length. We can evaluate it as follows:

$$ke^2 = (8.9876 \cdot 10^9 \text{ J m/C}^2)(1.602 \cdot 10^{-19} \text{ C})^2$$

$$= 2.307 \cdot 10^{-28} \text{ J m}$$

$$= 1.440 \cdot 10^{-9} \text{ eV m} = 1.44 \text{ MeV fm}.$$

Therefore, we find for the minimum distance:

$$r_{min} = \frac{(1.44 \text{ MeV fm}) \cdot 2 \cdot 79}{5.00 \text{ MeV}} = 45.5 \text{ fm}.$$

The upper bound for the radius of the gold nucleus that we obtain from the backscattering of a 5-MeV alpha particle is $4.55 \cdot 10^{-14}$ m, which is approximately a factor of 3000 smaller than the radius of a gold atom, which is $1.4 \cdot 10^{-10}$ m.

DISCUSSION

The radius of a gold nucleus is approximately 7 fm. Our upper bound is bigger than the real radius of the gold nucleus by a factor of 7, but it is astonishing that such a simple experiment and calculation can yield a result for the radius of the atomic nucleus that has the right order of magnitude.

Why did we use a 5-MeV alpha particle, and not one with 10 MeV kinetic energy, in this example? One reason is that Geiger and Marsden used alpha particles with approximately 5 MeV of kinetic energy. But more importantly, if alpha particles with kinetic energy that is too high are used, the Rutherford scattering approximation breaks down, as we discuss next.

The Rutherford scattering formula (equation 39.3) predicts a $1/K^2$ dependence of the differential cross section on the kinetic energy. However, as the kinetic energy becomes higher, the alpha particle comes closer to the nucleus. At some point, the two nuclei "touch" and the Rutherford formula breaks down. This result is shown very impressively in Figure 39.12, where the differential cross section for alpha particles deflected off lead at a scattering angle of $\theta = 60°$ is shown as a function of the kinetic energy of the alpha particles. Clearly, small kinetic energies show agreement with the Rutherford formula, but above 28 MeV, a clear deviation occurs.

Quantum Limitations

Let's now use what we've learned so far about quantum mechanics to estimate the order of magnitude of the various quantities involved in particle scattering. To resolve structure of size Δx, projectile particles are needed that have a momentum that is larger than the momentum uncertainty Δp_x that is associated with a coordinate space uncertainty Δx. The Heisenberg Uncertainty Relation requires that $\Delta x \cdot \Delta p_x \geq \frac{1}{2}\hbar$ (see Chapter 36), so the minimum momentum needed for a particle to probe a spatial size of Δx is of the order

$$p_{min} \approx \frac{\hbar}{\Delta x}.$$

The minimum particle energy is then

$$E_{min} = \sqrt{p_{min}^2 c^2 + m^2 c^4} = \sqrt{\frac{\hbar^2 c^2}{\Delta x^2} + m^2 c^4}.$$

The product of the constants $\hbar$ and c appears frequently enough that it is worthwhile to know its value:

$$\hbar c = (6.58212 \cdot 10^{-16} \text{ eV s})(2.99792 \cdot 10^8 \text{ m/s}) = 197.327 \text{ MeV fm}.$$

In Example 39.1 we calculated the combination of the Coulomb constant k times the square of the elementary charge quantum e and found that this product also has the units MeV fm. We can take the ratio of these two products that have the same units to find a dimensionless number called the **fine-structure constant**, α, so named because it first appeared in a theory by Arnold Sommerfield to explain the fine structure in spectral lines. His theory was incorrect but the constant is useful, as we will see later, as the coupling constant for electromagnetic interactions:

$$\alpha = \frac{ke^2}{\hbar c} = \frac{e^2}{4\pi\epsilon_0 \hbar c} = \frac{1}{137.036}.$$

The value of α is known to much greater precision than stated here, to about one part in a billion. However, for most purposes it is sufficient to remember that $\alpha = 1/137$.

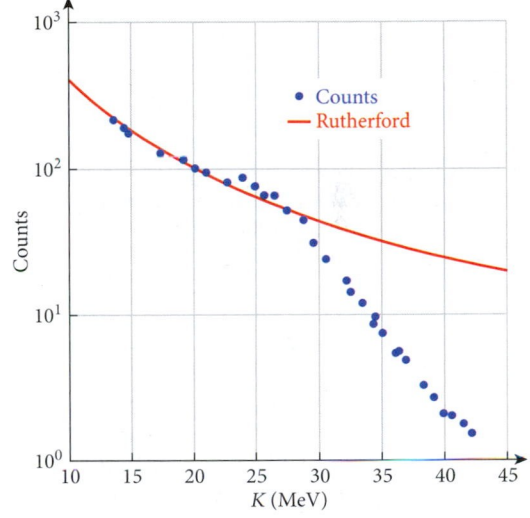

FIGURE 39.12 Dependence of the differential cross section at 60° on the kinetic energy of the alpha particle scattering off lead nuclei (blue circles), as reported by Eisberg and Porter (*Rev. Mod. Phys.* 33 (1961) 190). The red curve shows the expectation values from Rutherford formula.

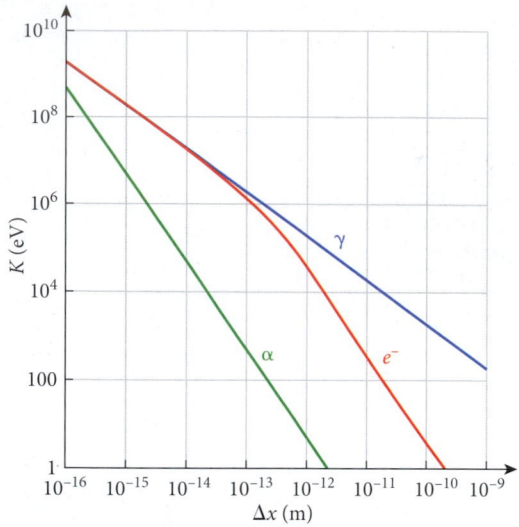

FIGURE 39.13 Minimum kinetic energy required to probe a structure of a given size for electrons (red), photons (blue), and alpha particles (green).

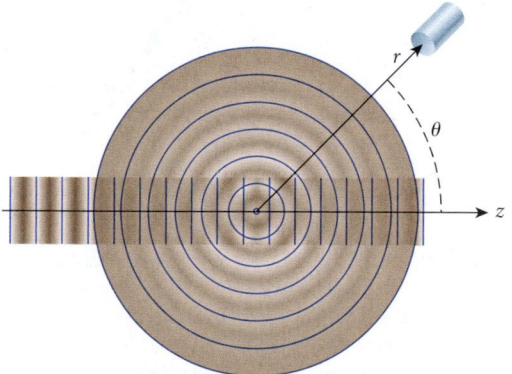

FIGURE 39.14 Scattering of a plane wave, moving from left to right along the z-direction, with a target emitting a spherical wave.

Probing smaller dimensions requires using projectile particles with higher momenta and energies. Figure 39.13 shows the minimum kinetic energies, $K = E - mc^2$, that electrons (mass = 511 keV/c^2, red curve), photons (mass = 0, blue line), and alpha particles (mass = 3.73 GeV/c^2, green line) have to have to probe structure of size Δx. Note that the curve for the electron kinetic energy merges with that of the photon for energies high enough that the mass of the electron becomes negligible compared to its kinetic energy. (The same would happen for the alpha particle when its kinetic energy exceeds its rest mass of 3.73 GeV/c^2, but as we will see later, the alpha particle is a composite particle, so it does not make sense to use it for scattering experiments at these high energies to probe the substructure of some other object.) Chapter 38 established that the atomic size is on the order of 10^{-10} m. Thus, Figure 39.13 shows that probing structure on the atomic scale requires photons with kinetic energies of at least a few thousand electron-volts or electrons with kinetic energies of at least a few tens of electron-volts.

Alternatively, the same order-of-magnitude result can be obtained by using the postulate that the de Broglie wavelength, $\lambda = h/p$, of the probe needs to be smaller than the spatial resolution to be explored. Within a factor of π, the two answers are the same. (We are only interested in the order of magnitude, so a factor of 2 or π is not relevant.)

Quantum Wave Scattering

The Rutherford scattering formula (equation 39.3) was derived with the classical considerations employed above, but it turns out to be correct in the quantum mechanical case as well. For the scattering of quantum waves, consider a plane wave $e^{i\kappa z}$ incident on a target. The target then emits a spherical wave that depends on the polar angle θ, as depicted in Figure 39.14. (Just as for the classical scattering process, we assume no dependence on the azimuthal angle, implying symmetry with respect to rotation about the z-axis.) Then the wave function for the stationary (time-independent) solution of the scattering problem can be written as the sum of the incident plane wave plus the outgoing spherical wave, which is assumed to be emitted from a pointlike distribution without measurable spatial size:

$$\psi_{\text{total}}(\vec{r}) = \psi_{\text{i}}(\vec{r}) + \psi_{\text{f}}(\vec{r}) = N\left(e^{i\kappa z} + f(\theta)\frac{e^{i\kappa r}}{r}\right). \tag{39.5}$$

Here N is a normalization constant and $f(\theta)$ is the scattering amplitude. The physical meaning of the scattering amplitude can be extracted as follows. The number of impinging projectile particles per unit area per unit time used for the definition of the differential cross section (equation 39.1) is the particle density of the incident wave, $|\psi_{\text{i}}(\vec{r})|^2 = |Ne^{i\kappa z}|^2 = |N|^2$. The number of particles per unit time scattered into the area $dA = r^2 d\Omega$ is

$$|\psi_{\text{f}}(\vec{r})|^2 \, dA = \left|N\left(f(\theta)\frac{e^{i\kappa r}}{r}\right)\right|^2 dA$$

$$= |N|^2 |f(\theta)|^2 \frac{1}{r^2} dA$$

$$= |N|^2 |f(\theta)|^2 \frac{1}{r^2} r^2 d\Omega$$

$$= |N|^2 |f(\theta)|^2 \, d\Omega.$$

Therefore, we find from the definition of the differential cross section (equation 39.1) that

$$\frac{d\sigma}{d\Omega} = \frac{|\psi_{\text{f}}(\vec{r})|^2 \, dA}{|\psi_{\text{i}}(\vec{r})|^2 \, d\Omega} = \frac{|N|^2 |f(\theta)|^2 \, d\Omega}{|N|^2 \, d\Omega} = |f(\theta)|^2. \tag{39.6}$$

That is, the differential cross section is the absolute square of the scattering amplitude.

If we relax the condition that the outgoing spherical wave is emitted from a single point, then we can ask how the Rutherford scattering cross section has to be modified. The **form factor**, $F^2(\Delta p)$, is the square of the Fourier transform of the charge distribution, $\rho(\vec{r})$, of the target:

$$F^2(\Delta p) = \left| \frac{1}{e} \int \rho(\vec{r}) e^{i\Delta\vec{p}\cdot\vec{r}/\hbar} dV \right|^2, \qquad (39.7)$$

where the volume integral extends over three-dimensional coordinate space. A Fourier transform is a technique that allows us to express a function of one variable, such as $\vec{r}$, as a function of another, related variable, such as Δp.

The form factor determines how the actual charge (or mass) distribution influences the cross section. In other words, the form factor measures the deviation from the Rutherford pointlike scattering formula (equation 39.4) with the Δp^{-4} dependence on the momentum transfer:

$$\frac{d\sigma}{d\Omega} = \left(\frac{d\sigma}{d\Omega} \right)_{\text{point}} \cdot F^2(\Delta p). \qquad (39.8)$$

In general terms, the scattering of a point particle shows the least-steep fall-off as a function of increasing angle or momentum transfer. Any extended source results in a steeper fall-off of the differential cross section, due to the form factor. If the form factor could be measured at all values of the momentum transfer, then the Fourier transform (equation 39.7) could be inverted directly. However, it is usually not feasible to obtain such measurements with the necessary accuracy. Instead, the usual procedure is to use a charge density distribution with adjustable parameters—for example, the mean radius—and fit these parameters to provide the best match with the experimental scattering data. One example of the use of this procedure is shown in Figure 39.15.

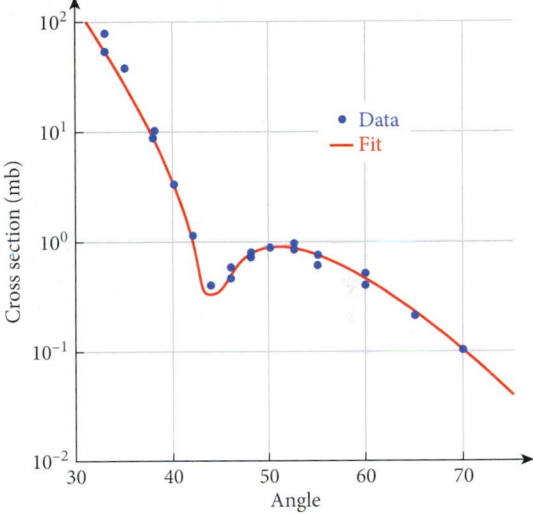

FIGURE 39.15 Scattering of 420-MeV electrons off oxygen. The blue circles represent experimental data; the red curve is the best fit using the form factor, as reported by R. Hofstadter (*Annual Reviews of Nuclear Science* 7 (1957) 231).

39.3 Elementary Particles

A short time after Rutherford's discovery that the nucleus of an atom is smaller than the radius of an atom by a factor of 10,000, it became clear that the atomic nucleus consists of positively charged protons and uncharged neutrons. (Neutrons were discovered in 1932 by the British physicist James Chadwick (1891–1974).)

For a brief period in the 1930s, reductionists were very content with the emerging picture of elementary particles. Only three fundamental indivisible fermionic particles were known—the positively charged proton, the uncharged neutron, and the negatively charged electron. In addition, there was only one known boson, the photon. In 1930, the German physicist Wolfgang Pauli (1900–1958) postulated another fermion, the neutrino, to explain the missing energy in beta decay. (Beta decay is explained later in this chapter.) In addition, in 1935, the Japanese physicist Hideki Yukawa (1907–1981) postulated yet one more boson, the pion, as the particle responsible for the strong interaction (also addressed later in this chapter). Dirac had predicted antiparticles in 1928, and Anderson had identified the positron as the antiparticle to the electron in 1932 (see Section 37.10), so it was generally accepted that each of the elementary particles also has its antiparticle. It seemed that physicists were very close to achieving a satisfactory simple, elegant picture of the layer of truly indivisible elementary particles, similar to the Greek ideal of only four or five elements.

However, small problems began to arise with this picture. In 1936, Anderson discovered another elementary particle, the muon, by examining cosmic rays. At first the muon was thought to be the pion postulated by Yukawa, because it had a mass of approximately 100 MeV/c^2, as demanded by the Yukawa theory. The real pion was discovered 11 years later by Cecil Powell and his collaborators. In 1947, George Rochester and Clifford Butler discovered the kaon in cosmic rays as well. Elementary particles were now grouped into baryons (neutron, proton, and others), mesons (pion, kaon, and others), and leptons (electron, neutrino, muon, and others). Baryons and mesons together are classified as hadrons, which are particles that experience the strong force.

In 1934, Ernest Lawrence (1901–1958) invented the cyclotron, a powerful tool for accelerating particles to higher energies, and the era of accelerator-based physics began. Ever bigger accelerators reaching ever higher energies led to discoveries of more and more "elementary" particles. Eventually, scientists came to believe that this large number of particles meant that another level of substructure existed underneath the layer of what was previously considered elementary. From 1953 to 1957, Robert Hofstadter (1915–1990) and his collaborators used the newly invented linear accelerators to perform a basic variation of Rutherford scattering with high-energy electrons impinging on protons. Their work revealed the charge distributions of nuclei (see, for example, Figure 39.15), but also showed that the proton has a substructure. This means that the proton is composed of even smaller, more fundamental particles. In 1964, Murray Gell-Mann gave this group of particles that make up the proton and other hadrons the name **quarks** (after the phrase "three quarks for Muster Mark" from the novel *Finnegan's Wake* by James Joyce).

This section summarizes the ingredients of the **standard model** of elementary particle physics. It has been in place in its basic form for more than three decades and has withstood a large number of experimental tests. First, we consider the basic constituents, the set of particles currently thought to be fundamental, elementary, or indivisible. These elementary particles have sizes less than 10^{-16} m, which is the approximate experimental limit we can reach, and are probably pointlike—that is, of a size much smaller than the 10^{-16} m that we can resolve. Next, we discuss how these particles interact. Finally, we investigate how the fundamental particles combine to form the next-higher level in the hierarchy of matter.

Elementary Fermions

The elementary fermions are particles with spin $\frac{1}{2}\hbar$, or simply spin $\frac{1}{2}$. (When we say that a particle has a spin that is a number, such as $\frac{1}{2}$, we imply that this number is to be multiplied by $\hbar$.) Chapter 36 showed that a particle's spin is the maximum measurable value for the component of the particle's spin angular momentum in any direction. Fermions include six quarks having names (or *flavors*) of up, down, strange, charm, bottom, and top. These quarks all have a noninteger charge of $-\frac{1}{3}e$ or $+\frac{2}{3}e$. The other fermions are the electron, muon, and tau leptons and the electron-, muon-, and tau-neutrinos. Each of these 12 fermions has an antiparticle.

The original quark flavors of up, down, and strange were introduced by Gell-Mann, and the other three flavors were discovered subsequently. The charm flavor was predicted theoretically in 1970 by Sheldon Glashow, John Iliopoulos, and Luciano Maiani and was discovered experimentally in the form of the J/ψ particle (a meson composed of the charm quark and its antiquark) in 1974 by groups at Stanford Linear Accelerator Center led by Burton Richter and at Brookhaven National Laboratory led by Samuel Ting. The bottom flavor was discovered in 1977 at Fermilab by a collaboration led by Leon Lederman. The top flavor, finally, was discovered simultaneously by two large collaborations, DØ and CDF, at Fermilab in 1995.

Quarks are grouped into three generations, together with the corresponding leptons and neutrinos, as shown in Figure 39.16, where the quarks are indicated by the first letter of their names. Each particle has an antiparticle with the same mass and opposite charge. Table 39.1 lists the charges, masses, and antiparticles for all elementary fermions. The notation for the antiparticle is the same as for the particle, with a horizontal bar added above the symbol. The only exceptions are the antiparticles of the electron, muon, and tau lepton. These antiparticles have a positive sign as superscript instead of the negative sign of their corresponding particles.

Two fundamental conservation laws can be stated at this point: the conservation laws for total lepton number and for total quark number. These laws hold for any reaction or decay process. The **total lepton number,** ℓ_{net}, is simply the number of leptons minus the number of anti-leptons and is constant in time:

Generation:	1	2	3
Charge –1/3 quark:	d	s	b
Charge +2/3 quark:	u	c	t
Lepton:	e^-	μ^-	τ^-
Neutrino:	ν_e	ν_μ	ν_τ

FIGURE 39.16 Elementary fermions of the three generations.

$$\ell_{net} = n_\ell - n_{\bar{\ell}} = \text{constant}. \tag{39.9}$$

Table 39.1	Elementary Fermions					
		Particle			**Antiparticle***	
Name	**Symbol**	**Charge (e)**	**Mass (MeV/c²)**	**Name**		**Symbol**
Down quark	d	–1/3	4.9±0.8	Anti-down		$\bar{d}$
Up quark	u	+2/3	2.4±0.7	Anti-up		$\bar{u}$
Electron	e^-	–1	0.510999	Positron		e^+
Electron-neutrino	ν_e	0	< 0.0000005	Anti-electron-neutrino		$\bar{\nu}_e$
Strange quark	s	–1/3	100±30	Anti-strange		$\bar{s}$
Charm quark	c	+2/3	1290±100	Anti-charm		$\bar{c}$
Muon	μ^-	–1	105.66	Muon		μ^+
Muon-neutrino	ν_μ	0	< 0.19	Anti-muon-neutrino		$\bar{\nu}_\mu$
Bottom quark	b	–1/3	4200±100	Anti-bottom		$\bar{b}$
Top quark	t	+2/3	$(1.729\pm0.06)\cdot10^5$	Anti-top		$\bar{t}$
Tau lepton	τ^-	–1	1776.82±0.16	Tau lepton		τ^+
Tau-neutrino	ν_τ	0	< 18.2	Anti-tau-neutrino		$\bar{\nu}_\tau$

*The antiparticle to a given particle has the same mass, but opposite charge.
Source: Particle Data Group site, 2012, pdg.lbl.gov

An example of the conservation of total lepton number occurs in the decay of the neutron. A neutron decays into a proton, an electron, and an anti-electron-neutrino:

$$n \rightarrow p + e^- + \bar{\nu}_e. \qquad (39.10)$$

The total lepton number before the neutron decays is zero. The total lepton number after the decay is ℓ_{net} = 1 lepton (e^-) – 1 anti-lepton ($\bar{\nu}_e$) = 0, so the total lepton number before the neutron decays is equal to the total lepton number after the decay. The process in equation 39.10 is the prototypical beta-decay process, which we will study in Chapter 40.

Until recently, physicists thought not only that total lepton number is conserved, but also that it is conserved separately for each generation. This would mean that the net number of electron leptons (number of electrons plus number of electron-neutrinos minus number of positrons minus number of anti-electron-neutrinos) is conserved, and separately the net number of muon leptons is conserved, and separately the net number of tau leptons is conserved. Indeed, these separate conservation laws are valid if the standard model of elementary particle physics holds. However, the process of neutrino oscillations, discussed later in this section, invalidates the standard model and the strict conservation laws for each individual lepton species, and only the conservation law for total lepton number in the form of equation 39.9 survives.

In the same way, the **net quark number,** Q_{net}, can be defined as the difference between the total number of quarks and the total number of antiquarks and is also constant in time:

$$Q_{net} = n_q - n_{\bar{q}} = \text{constant}. \qquad (39.11)$$

All physical processes have to obey the fundamental conservation laws of lepton number (equation 39.9) and quark number (equation 39.11), in addition to all of the previously introduced conservation laws of charge (Chapter 21), energy (Chapter 6), momentum (Chapter 7), and angular momentum (Chapter 10). The conservation law of quark number is more often expressed as the conservation law of **baryon number.** This law states that the difference between the number of baryons and the number of anti-baryons is constant in time. It was found before quarks were discovered. Because baryons are composed of quarks, the baryon number conservation law is equivalent to the quark number conservation law stated in equation 39.11.

Elementary particles have a huge range of masses, as shown in Figure 39.17. From the observation of neutrino oscillations, we know that at least one species of neutrino must have a nonzero mass, but presently we can state only upper bounds for the neutrino masses, with the mass of the electron-neutrino being smaller than 0.5 eV/c^2. The light quark masses (u and d) and the lepton mass (e^-) of the first generation are all of the order 1 MeV/c^2. The masses

FIGURE 39.17 Current best values for the masses of the elementary fermions. The widths of the boxes indicate the measurement uncertainty. The arrows indicate that the neutrino masses are upper limits.

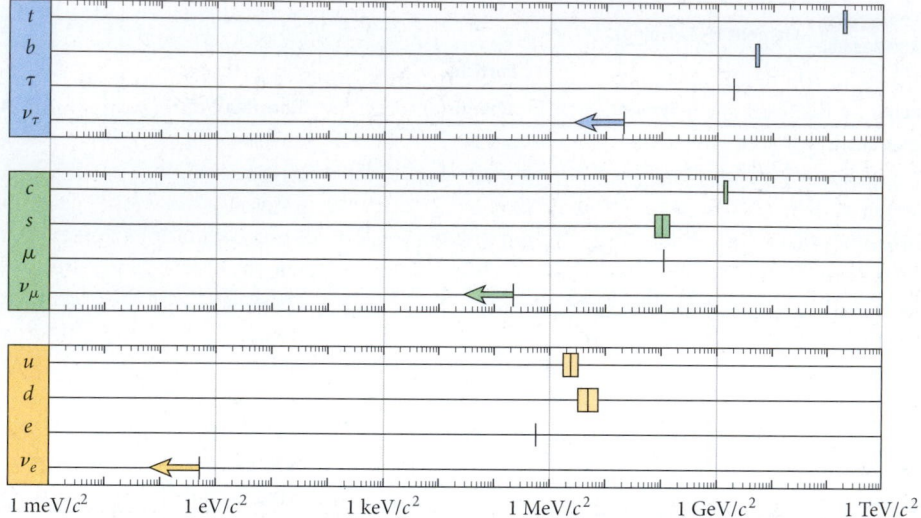

of the second-generation quarks (s and c) and leptons (μ^-) are two to three orders of magnitude larger, in the range 100 MeV/c^2 to GeV/c^2. The masses of the third-generation quarks (b and t) and leptons (τ^-) are in the multi-GeV/c^2 to 100-GeV/c^2 range. Why these particles have the masses they do is currently not known and is an open research topic. It is thought that a particle called the Higgs boson (discussed in the following subsection) is responsible for the generation of the mass of the elementary particles. A linear collider planned for the future, with the working name International Linear Collider, would enable physicists to study the Higgs boson in detail. They hope to then gain an understanding of what sets the scale for the different particle masses.

Is it possible that more than these three generations of fundamental fermions exist? Yes, but we have strong experimental evidence from the decay of W and Z bosons that there are indeed only exactly three generations.

Before we leave the topic of elementary fermions, it is important to note several observations that point to problems in the standard model, possibly hinting at the need for a more fundamental framework.

First, isolated quarks have never been observed, despite a long history of high-precision searches. We will come back to the reason for this when we discuss the interaction of quarks with one another. While we can understand the reasons why it is impossible to isolate a single free quark, it is still legitimate to ask if it is philosophically consistent to speak of a fundamental indivisible particle, if we cannot isolate it.

Second, it is a puzzle why the combined electrical charge of two up quarks and one down quark (we will see below that they combine to make a proton) is exactly the same as the charge of the positron (the opposite of the charge of an electron). Physicists have tested this equality to an incredible precision, better than one part in a trillion, but have never found any violation of it.

Finally, experimental observations have established that the three neutrino species, ν_e, ν_μ, and ν_τ, can change from one species to the other and back again. This phenomenon of neutrino species changes, or **neutrino oscillations,** cannot be explained by the standard model of particle physics.

Fundamental Bosons and Interactions

What about interactions between elementary fermions? How do they exert forces on each other? The answer is that on the fundamental level, all forces are mediated through the exchange of bosons. A full understanding of this concept requires a grasp of quantum field theory, which goes beyond the scope of this book. Some people find it helpful to think of this particle exchange as being similar to two people standing on skateboards and throwing a heavy ball back and forth, thus exerting forces on each other through the exchange of momentum that accompanies the transfer of the ball. However, this analogy is rather crude, because it neglects the wave character of the bosons exchanged. It also can only account for repulsive, and not attractive, interactions.

In classical mechanics or in relativistic mechanics (see Chapter 35), a free particle cannot emit another particle and conserve energy and momentum in the process. However, the Heisenberg uncertainty relation (see Chapter 36) allows a violation of energy conservation of order ΔE over a time $\Delta t < \hbar/\Delta E$. Because this "borrowed" energy has to include the mass energy, $m_b c^2$, of the exchange boson, we find $\Delta t < \hbar/\Delta E < \hbar/m_b c^2$. Also, because the exchanged boson can travel with at most the speed of light, we find $\Delta x \leq c\Delta t$ for the range and thus

$$\Delta x < \frac{\hbar c}{m_b c^2}.$$

Thus, an exchange boson of mass m_b can mediate an interaction only over a range that is inversely proportional to this mass. Infinite-range forces like the Coulomb force thus require an exchange boson that is massless. Table 39.2 lists the elementary bosons of the standard model of particle physics, with their spins (all integers, since they are bosons), their masses, and their charges. You can see that the W^- boson has a mass of more than 80 GeV/c^2. This implies an extremely short-range force of

$$\Delta x(W^-) < \frac{\hbar c}{m_{W^-} c^2} = \frac{197.33 \text{ MeV fm}}{80,399 \text{ MeV}} = 2.5 \cdot 10^{-3} \text{ fm}.$$

Photon

The photon is the massless spin-1 exchange boson of the electromagnetic interaction. It has zero charge. Chapters 21 to 38 examined aspects of the interactions mediated by photons. Since the photon spin is 1, it is called a *vector boson,* as opposed to a *scalar boson* with spin 0.

W Boson

Quarks and leptons can also interact via the so-called **weak interaction.** This interaction can change one quark flavor into another via the emission of a lepton and a neutrino. The weak interaction is the *only* interaction that can change quark flavor. One example of such a reaction is

$$d \rightarrow u + e^- + \bar{\nu}_e. \tag{39.12}$$

The decay of a down quark is similar to the decay of a neutron (equation 39.10), but occurs on the fundamental level. The down quark has a charge of $-\frac{1}{3}e$ and thus needs to emit a particle with charge $-e$ to convert itself to an up quark with charge $+\frac{2}{3}e$. This is accomplished by emitting a W^- boson, which in turn decays into an electron and an anti-electron-neutrino. You can easily convince yourself that both sides of equation 39.12 have the same net charge, quark number (equation 39.11), and total lepton number (equation 39.9).

Table 39.2	Elementary Bosons					
Name	**Symbol**	**Spin ($\hbar$)**	**Mass (GeV/c^2)**	**Charge (e)**	**Interaction***
Photon	γ	1	0	0	Electromagnetic
W boson	W^+	1	80.399±0.023	+1	Weak
W boson	W^-	1	80.399±0.023	−1	Weak
Z boson	Z	1	91.1876±0.0021	0	Weak
Gluon	g	1	0	0	Strong
Higgs boson	H^0	0	125	0	(Mass)
Graviton (?)	G	2	0	0	Gravitation

*Interaction mediated by this particle.
Source: Particle Data Group site, 2012 numbers, pdg.lbl.gov

It is also important to note that the masses of the particles on the right side of equation 39.12 must sum up to a number no greater than the mass on the left side. This is required by the conservation of total energy. Table 39.1 shows that the mass of the down quark is 5 MeV/c^2. This is greater than the mass of the up quark of 2.2 MeV/c^2 plus the mass of the electron of 0.511 MeV/c^2 and the negligible mass of the anti-electron-neutrino. In the intermediate stage, the beta-decay process of equation 39.12 may require the emission of a W^- boson with mass of more than 80 GeV/c^2, but in going from the initial state of this reaction to the final state, energy still must be conserved. The intermediate creation and subsequent decay of the W^- boson are said to be *off-shell* (in the sense of "off the energy shell," that is, not at constant energy), and thus the intermediate W^- boson is called a **virtual particle.**

Figure 39.18 shows a slightly different view of the quark masses than that presented in Figure 39.17. Here the quarks are sorted not by generation but by charge, allowing the diagram to show all possible beta decays. Since all beta decays have to conserve energy, and since the quark mass increases from left to right, the only reactions that are possible have an arrow that points toward the left side, that is, from a higher-mass quark to a lower-mass one. In addition, beta decay involves the emission of a charged lepton, and thus only arrows between quarks of different charge are allowed, that is, no horizontal arrows are possible. The green arrows show all possible β^- decays. One of these was described in equation 39.12. The red arrows show all possible β^+ decays; these are decays that involve creation of a virtual W^+ boson. Beta decays within the same generation of quarks are most common (solid lines in Figure 39.18). Beta decays that involve a jump into a neighboring generation of quarks are much more rare (dashed lines), and beta decays that involve a conversion from the third generation to the first generation (dotted lines) are extremely rare.

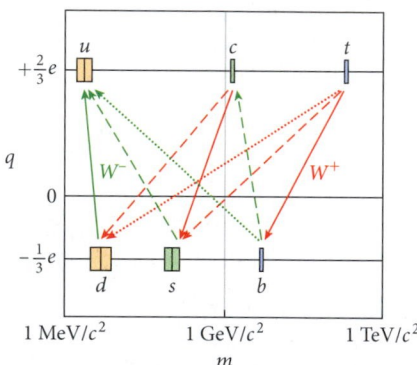

FIGURE 39.18 All possible beta decays of quarks. The red arrows represent β^+ decays, and the green arrows β^- decays. Energy conservation requires that all arrows point only toward the left, never to the right.

Self-Test Opportunity 39.1

Use Figure 39.18 to write reaction equations similar to equation 39.12 for the other two of the most common beta-decay processes—those represented by the solid arrows in the figure.

Z Boson

So far we have only talked about charged leptons and quarks and their electromagnetic and weak interactions. What about uncharged elementary particles? The chapters on electromagnetism showed that uncharged particles do not participate in the electromagnetic interaction. (The neutron has a magnetic moment, but this is due to the fact that it is a composite particle.) Uncharged elementary particles can, however, participate in the weak interaction. The mediator of this interaction needs to be a charge-0 boson, the Z boson. An exchange of an intermediate Z boson is the underlying mechanism of the scattering of a neutrino off an electron.

The W and Z bosons were both postulated in 1968 in a theory developed by the Americans Sheldon Glashow and Steven Weinberg and the Pakistani Abdus Salam. Their formulation provided a unified description of the electromagnetic interaction and the weak interaction, which is now called the *electroweak interaction*. The scattering of neutrinos off electrons, discovered by the Gargamelle collaboration in 1973 at CERN, provided indirect evidence for the existence of the Z boson and convinced the physics community of the validity of the electroweak theory. The W and Z bosons were discovered in early 1983, right after the completion of the CERN SPS accelerator (which is now used as the smaller injector ring for the LHC; see Figure 39.4), by the UA1 and UA2 collaborations. Glashow, Weinberg, and Salam received the 1979 Nobel Prize in Physics for their theoretical work, and the 1984 Nobel Prize in Physics was awarded to the Italian Carlo Rubbia, the leader of the UA1 collaboration, and the Dutch accelerator physicist Simon van der Meer, whose work on stochastic beam cooling made possible the operation of the SPS and the discovery of the W and Z bosons.

Table 39.2 shows that the mass of the Z boson is 91.2 GeV/c^2, which is similar to the mass of a niobium atom. The Z boson can decay into an electron and a positron, which provides a clear experimental signature for this boson, allowing it to be detected and studied directly.

Gluon

The gluon is the mediator of the strong interaction between quarks. To discuss gluons, we need one more detail about quarks. So far, we have told you that quarks come in six flavors.

However, they also have another property, which is whimsically called *color*. Quark colors come in three varieties: red (*r*), blue (*b*), and green (*g*). Antiquarks have anti-colors: anti-red ($\bar{r}$), anti-blue ($\bar{b}$), and anti-green ($\bar{g}$).

A gluon carries a combination of one anti-color and one color. From this information alone, you might conclude that a total of nine different gluons occur. However, a totally color-neutral gluon consisting of equal admixtures of $r\bar{r}$, $b\bar{b}$, and $g\bar{g}$ is not possible, so there are eight gluons. For example, a blue quark can interact with a red quark by exchanging a $b\bar{r}$ gluon, turning the blue quark into a red one and the red into a blue one. Since gluons carry the color charge, they can couple to one another. Thus, gluons can radiate other gluons, which no other exchange boson can do.

Higgs Boson

In 1964, the British theorist Peter Higgs and others postulated a boson that couples to all other elementary particles, including itself, in such a way as to give them mass. This boson is now called the *Higgs boson*. (In Chapter 4 we already briefly mentioned the Higgs boson because of the relationship between force and mass.) The DØ and CDF experiments at Fermilab's Tevatron accelerator complex have been able to extract first possible glimpses of the Higgs boson in their data. With the completion and operation of the LHC at CERN, however, the search for the Higgs boson has now entered a new phase. The first phase of this search came to a successful conclusion on July 4, 2012, when the ATLAS and CMS detector collaborations jointly announced the discovery of the Higgs boson, with a mass of approximately 125 GeV. Figure 39.19 shows two examples of collisions, in which a Higgs boson was produced, and for which characteristic decay signatures were obtained.

The next few years promise to be incredibly exciting for particle physics as we try to understand the properties of this new particle and decode its fundamental importance for the generation of the mass of other particles.

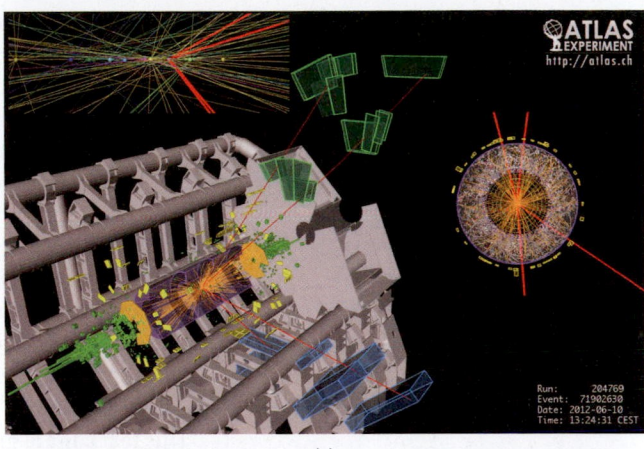

(a)

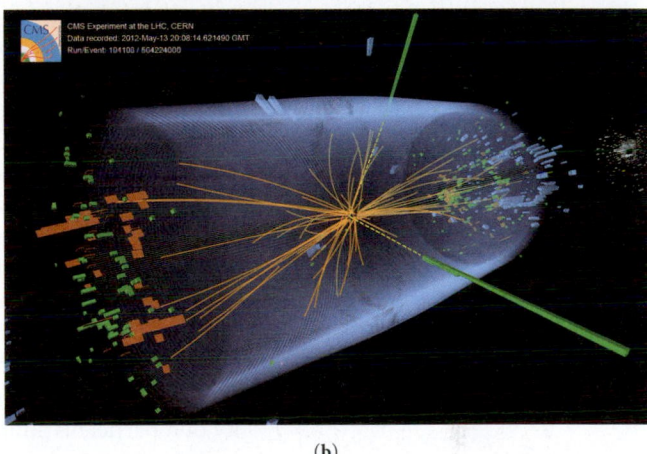

(b)

FIGURE 39.19 (a) Event display from the ATLAS collaboration at CERN, showing the decay of a Higgs boson into four muons (red tracks). (from http://www.atlas.ch/news/2012/latest-results-from-higgs-search.html) (b) Event display from the CMS collaboration at CERN, showing the decay of a Higgs boson into two photons (yellow dashed lines and green towers). (from https://cdsweb.cern.ch/record/1459463)

Graviton

We have not yet talked about gravity in the context of quantum mechanics and elementary particles. Like the electromagnetic force, the gravitational force has infinite range. Thus, the exchange boson must have zero mass. In addition, because of the structure of the gravitational interaction as described by general relativity, the spin of this exchange particle must be 2. However, the standard model of particle physics does not address gravity at all, for two reasons, one experimental and one theoretical. Experimentally, gravity is so weak that it has not been possible to observe a graviton directly. However, a detector called LIGO may detect gravitons of cosmic origin, such as gravitons emitted by colliding galaxies (see Section 15.9). LIGO currently has installations in Louisiana and Washington State. Other similar detectors are under construction or already in operation in other parts of the world. However, much more important is that in its present formulation, the standard model (especially quantum mechanics) and gravity (as described by general relativity) are not compatible theories. This conflict has given rise to a whole new set of theories, including string theories, which we discuss further in Section 39.4.

Feynman Diagrams

If virtual particles can be created and destroyed in intermediate steps of reactions, possibly multiple times, then calculating anything that can be connected to experimentally observable consequences can become very complicated. Calculations might have to take into account infinitely many possible processes. Even the question of how to distinguish between processes that contribute a lot and those that contribute very little to the final result can be an insurmountable challenge.

The American physicist Richard Feynman (1918–1988) found an ingenious way to solve this problem by dividing complex interactions into a few elementary pieces, each of which can be represented by a simple drawing called a **Feynman diagram,** which is created according to rules formulated by Feynman. In this way, he developed quantum electrodynamics, for which he shared the 1965 Nobel Prize in Physics with Sin-Itiro Tomonaga from Japan and the American Julian Schwinger, who made independent contributions to the field. Even though quantum electrodynamics describes only the interactions of photons and charged particles, Feynman diagrams have proven to be incredibly useful for all interactions of all particles.

This text cannot give you a working knowledge of using Feynman diagrams, but simply introduces them to provide an excellent pictorial representation of the underlying physics of particle interactions. We start with freely moving particles propagating through vacuum. Each is simply represented by a line with an arrow at the end, indicating the direction of travel. This line is called a *propagator.* Conventionally, time moves from left to right and the vertical dimension of the diagram is a spatial coordinate (Figure 39.20). (Usually it is assumed that time and space are the standard coordinates, so Feynman diagrams do not show the coordinate axes explicitly. Thus, subsequent figures do not include the axes.) A lepton or a quark is represented by an arrow from left to right, a photon or a W or Z boson by a wavy line, and a gluon by a coiled line. (There is no deeper meaning to the different line styles, which simply follow historical convention.)

Antiparticles move from right to left—that is, backward in time. This statement is not just a computational shortcut; it represents a deep underlying symmetry in the physics that governs the particle-antiparticle relationship: The fundamental theoretical construct of physics is invariant under simultaneous reversal of CPT (charge, parity, time). The parity of a particle is an intrinsic property characterized by the behavior of its wave function under reflection through the origin of coordinates, by which the x, y, and z coordinates are replaced by $-x$, $-y$, and $-z$. If the reflected wave function is identical to the original wave function, the parity is positive. If the reflected wave function is the negative of the original wave function, the parity is negative.

Inverting the charge and parity of an electron gives a positron. In order for this conversion to be a physically valid process, time must be reversed as well. The propagator lines are usually not horizontal, because the spatial coordinate is represented in the vertical direction in a Feynman diagram. However, the angle relative to the horizontal is not a measure for the momentum or velocity, because a Feynman graph is not intended as a plot from which some velocity can be extracted via $\Delta x/\Delta t$-type arguments. The lengths of the lines that represent the different particles also have no physical meaning and are arbitrary.

A Feynman diagram can depict the interaction of different particles. *Any given physical interaction has to have at least two vertices.* A *vertex* is defined as the intersection of three propagator lines. Figure 39.21 shows elementary vertices for six different processes. It is important to realize that none of these vertices shows a physical process that can happen in isolation. For each of these processes, it is impossible to conserve energy and momentum, provided all lines represent real particles. For example, let's look at photon emission from an electron, process (a) in Figure 39.21. In the rest frame of the electron before the emission, all the energy in the system is the rest mass of the electron times c^2, because the electron has no kinetic energy in this frame. After the emission, the photon receives some energy, and the electron experiences recoil, thus also acquiring kinetic energy. Now the total energy in the system in the rest frame is the electron's mass energy plus its kinetic energy plus the photon's energy. Therefore, this process violates energy conservation. From this observation comes an important insight: Each vertex must have at least one virtual particle, which is off-shell, so that $E^2 \neq p^2c^2 + m^2c^4$. Thus, at least two vertices must be combined to create a Feynman diagram that can represent a physically possible process.

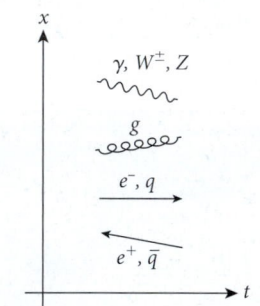

FIGURE 39.20 Feynman diagram showing a freely propagating photon, gluon, electron, positron, quark, antiquark, W boson, and Z boson.

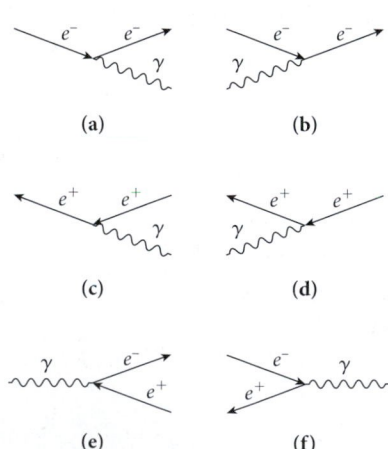

FIGURE 39.21 Elementary Feynman vertices for six different processes: (a) emission of a photon by an electron, (b) absorption of a photon on an electron, (c) emission of a photon by a positron, (d) absorption of a photon by a positron, (e) creation of an electron-positron pair from a photon, (f) electron-positron annihilation into a photon.

The rules that govern the construction of Feynman diagrams representing real processes are as follows:

- All lines that enter or exit a Feynman diagram must represent real particles, for which the energy-momentum relationship $E^2 = p^2c^2 + m^2c^4$ holds.

- Any line that represents a virtual particle, for which $E^2 = p^2c^2 + m^2c^4$ does not hold, must start and end inside the diagram—that is, must have both ends connected to a vertex. This ensures that these virtual particles cannot be observed in experiments.

- At each vertex, energy, momentum, charge, quark number, and all three lepton numbers must be strictly conserved.

From these preliminary considerations, we can produce simple Feynman diagrams. The first physical process we'll consider is the scattering of an electron off a positron (or vice versa).

Figure 39.22 shows four Feynman diagrams, in which an electron and a positron come in and an electron and a positron go out. In all four cases, a virtual photon is exchanged. Thus, all four diagrams represent versions of the same physical process (called *amplitudes*) and need to be added. The cross section for electron-positron scattering can then be calculated by taking the absolute square of this amplitude sum. Note that diagram (a) and diagram (b) in Figure 39.22 differ only in the time ordering of the emission of the photon. In (a) the photon is first emitted by the electron and then absorbed by the positron, and in (b) it is the other way around. Thus, we see that all diagrams that differ only in the internal time ordering represent the same physical process, because there is no way of knowing the internal time ordering of the process. Therefore, both processes (a) and (b) can be written as (c), where the instantaneous exchange of the photon is implied. Process (d) can be interpreted as an electron-positron annihilation into a virtual photon, followed by a pair-creation event.

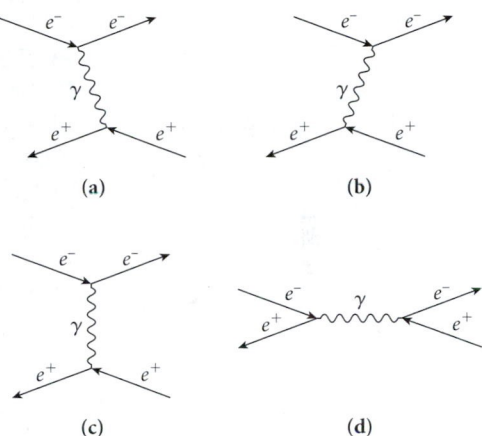

FIGURE 39.22 Four simple Feynman diagrams for electron-positron scattering.

Unfortunately, this is not where the process ends. The electron and positron can also exchange two photons, as shown in Figure 39.23 in one of many possible diagrams, or they can exchange even more photons. For any diagram that involves closed loops of the kind shown in this figure, momentum and energy can travel around a loop without constraint. As a result, we have to integrate over all possible values. Clearly, the summation of diagrams with loops can become arbitrarily complicated very fast.

Here is what saves us: For each virtual photon exchanged, the amplitude for the diagram acquires a multiplicative factor of $\alpha = ke^2/\hbar c = 1/137$. Thus, the contributions of multiphoton diagrams to the overall amplitude sum quickly become smaller, and the series converges.

This description does not amount to an actual calculation, but at least it indicates what is involved in performing this task. In addition, it allows us to draw a quick picture of the lowest-order contributions to a physical process. For example, we can draw a Feynman diagram for the beta decay (equation 39.12) of the down quark (Figure 39.24). With the aid of this diagram, it is straightforward to check that the conservation laws of charge as well as quark and lepton numbers are obeyed in this process, that is, that the Feynman diagram represents a process that can actually happen physically.

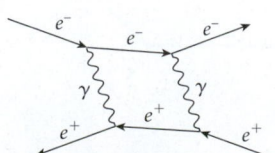

FIGURE 39.23 One possible two-photon exchange diagram for electron-positron scattering.

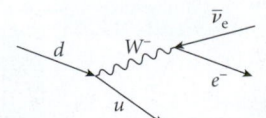

FIGURE 39.24 Feynman diagram for the beta decay of a down quark.

Concept Check 39.1

Which of the following is a valid Feynman diagram for the beta decay of a charm quark?

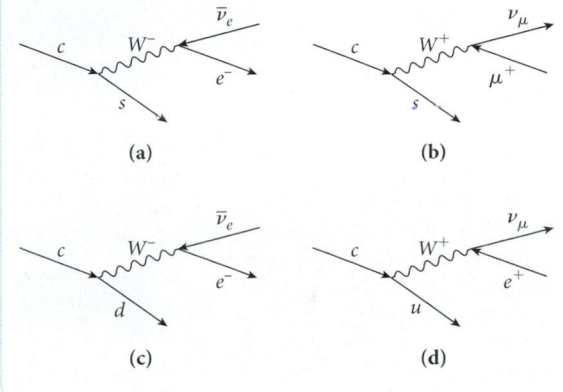

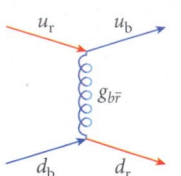

FIGURE 39.25 Simple Feynman graph for a one-gluon exchange in the strong interaction.

The weak and the electromagnetic interactions can be treated beautifully with Feynman diagrams, because the series of amplitudes represented by different graphs converge quickly. We can also construct Feynman diagrams for the strong interaction. One example is shown in Figure 39.25, where a red up quark exchanges a blue/anti-red gluon with a blue down quark. (More on color will follow in Section 39.5.)

However, the situation is much more complicated for the strong interaction, because the coupling constant is not α (which is small compared to 1 in the electromagnetic case), but is approximately 1. Thus, in general, series expansions in terms of Feynman diagrams do not converge. Only for scattering at high momentum are these techniques applicable, as the expansion of the physical process in terms of Feynman diagrams converges quickly enough to be of practical utility. Other methods are needed to perform calculations for the strong interaction.

39.4 Extensions of the Standard Model

Our current understanding is that the interactions in the world around us include three different forces: the strong force, the electroweak force, and gravity. A general guiding principle in the reductionist quest is that it should be possible to show that the different forces are just different aspects of the same force on the next more fundamental level of the reductionist hierarchy. This approach has worked repeatedly to great satisfaction. Figure 39.26 shows a timeline of the unification of forces (previously shown in Chapter 21).

The chapters on electricity and magnetism described these two forces as two aspects of the electromagnetic force, as described by the set of four Maxwell equations. In Section 39.3, we learned that the weak force can also be unified with electromagnetism to form the electroweak force. Consequently, there have been, and continue to be, new attempts to create theories that provide further unification of all known interactions into one.

In addition, although the standard model of particle interactions has been very successful, many scientists are bothered by its seemingly arbitrary values for parameters such as the quark and lepton masses and the coupling constants. Several candidates for a unified theory free of arbitrary parameters have been proposed. These theories are mostly based on arguments from a branch of mathematics called *group theory*. However, two essential features are worth discussing: One is the unification energy scale, and another is the decay of the proton.

Unification Energy Scale

The coupling constant of the electromagnetic interaction is the fine-structure constant, α, which has the value 1/137. The coupling constant of the weak interaction has a value of ~1/30, and the strong interaction has a coupling constant close to 1. The Glashow-

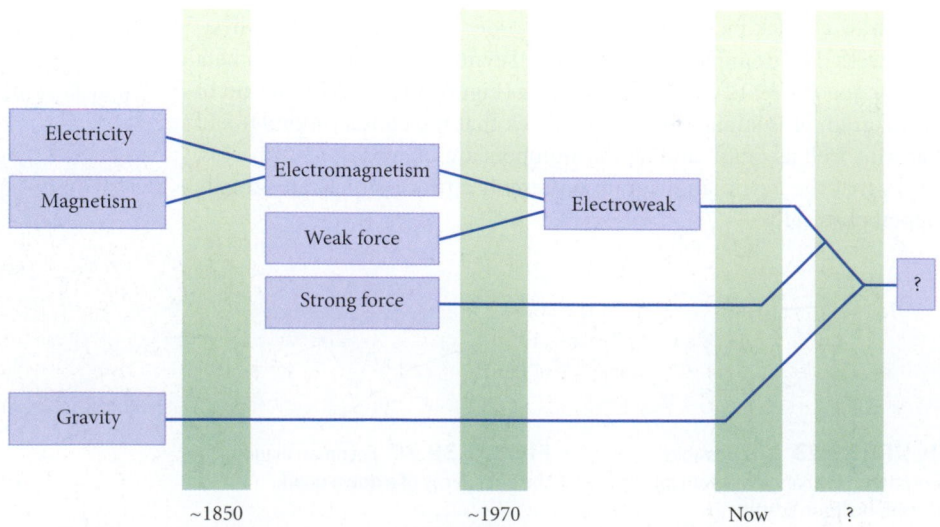

FIGURE 39.26 Timeline of the unification of basic forces.

Weinberg-Salam theory shows that at high energies, the coupling constants for the electromagnetic and weak force merge. Based on this idea, **grand unified theories** have been proposed that merge all three fundamental interactions at very high energies. Thus, if we were able to scatter particles at extremely high momentum transfer, Δp, we would be able to see all coupling constants merge (Figure 39.27). Unfortunately, the kind of momentum transfer required to test these theories is way beyond what can be achieved with present-day accelerators. However, possible hints about grand unified theories may be hidden in the cosmic ray spectrum, which reaches energies above 10^{20} eV.

Proton Decay

Section 39.3 introduced separate absolute conservation laws for the total number of quarks and for the total number of leptons. However, most grand unified theories predict that at the highest energies, quarks and leptons merge and are thus just different aspects of the same fundamental particle, and that the separation between quarks and leptons that we observe is only a manifestation of the low-energy limit of these unified particles. If this is true, then a quark will be able to decay into a lepton, and a proton (composed of two up quarks and one down quark; see Section 39.5) will not be an absolutely stable particle; instead, it will decay exponentially with a very long (but *not* infinite) lifetime on the order of 10^{30} years.

Because the age of the universe is only on the order of 10^{10} years, a very large number of protons must be observed to have a chance to observe even one proton decay event. This is one of the basic purposes for which the Super-Kamiokande (Super-K) detector was built in Japan (Figure 39.28). Super-K is located 1000 m underground and consists of 50,000 tons of purified water contained in a cylindrical steel tank whose diameter and height are both 40 m. Super-K has 11,200 photon detectors inside the tank walls. It can detect the Cherenkov radiation (see Chapter 16) emitted from the motion of high-energy particles through the water. In 1998, the Super-K collaboration was the first to observe the neutrino oscillations, mentioned previously, which imply that at least one of the neutrino species has nonzero mass. However, after several years of running the detector and looking for possible signals, no proton decay events have been reported, thus setting severe limits for entire classes of possible grand unified theories. The present lower limit on the proton's lifetime has been established experimentally as greater than $1.6 \cdot 10^{25}$ years, approximately 100 trillion times the current age of the universe.

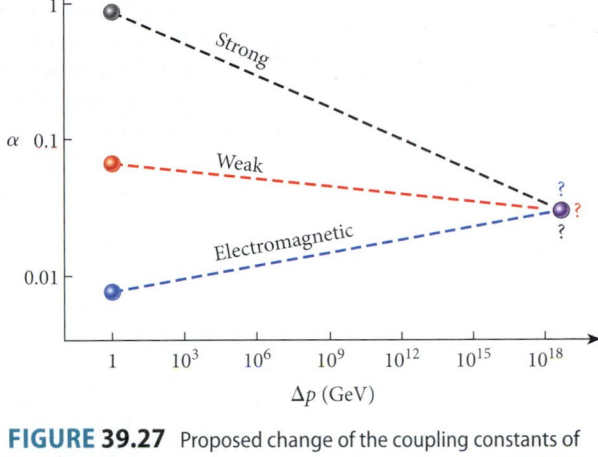

FIGURE 39.27 Proposed change of the coupling constants of fundamental interactions as a function of momentum transfer in a grand unified theory.

FIGURE 39.28 Inside the Super-Kamiokande detector. Normally the detector is filled with purified water, but here the fact that the maintenance crew is at work lets us appreciate the scale of the detector.

EXAMPLE 39.2 | Planck Units

Max Planck proposed a set of natural units that do not depend on the anthropocentric choices made in determining the SI units that we currently use (see Chapter 1). He looked for combinations of the basic universal physics constants of nature that have the dimensions of length, time, mass, and so on.

PROBLEM

Using the universal constants of the speed of light in vacuum, c, Planck's constant divided by 2π, $\hbar$, Coulomb's constant, $k = 1/4\pi\epsilon_0$, the universal gravitational constant, G, and Boltzmann's constant, k_B, what are the values of the Planck length, time, temperature, mass, and charge in SI units?

SOLUTION

The SI units of the speed of light are $[c] = $ m/s, of Planck's constant are $[\hbar] = $ kg m^2/s, of Coulomb's constant are $[k] = $ kg m^3/C^2/s^2, of Boltzmann's constant are $[k_B] = $ kg m^2/s^2/K, and of the gravitational constant are $[G] = $ kg^{-1} m^3/s^2.

– *Continued*

Our first observation is that the ampere (A = C/s) appears only in Coulomb's constant and the kelvin (K) appears only in Boltzmann's constant. Therefore, the kinematics-related quantities of length, time, mass, and energy cannot contain either of these two constants, because they are defined in terms of the MKS base units. If we multiply $\hbar$ by G, the unit kg cancels out and we obtain powers of the units m and s. Multiplication with the appropriate power of the speed of light enables us to cancel either one of these. Thus, we obtain for the Planck length and the Planck time:

$$x_P = \sqrt{\frac{\hbar G}{c^3}}$$

$$t_P = \frac{x_P}{c} = \sqrt{\frac{\hbar G}{c^5}}.$$

For the Planck mass, we must find a combination of $\hbar$ and G that has some power of kg in it, times m/s to some power, which we can cancel out with the appropriate power of the speed of light. This can be done by taking the ratio $\hbar/G$, which has the units $[\hbar/G] = \text{kg}^2 \cdot \text{s/m}$. Then we obtain the Planck mass from

$$m_P = \sqrt{\frac{\hbar c}{G}}.$$

The Planck temperature has to have the Boltzmann constant in the denominator, because this is the only way we can obtain K as the unit. The appropriate powers of the Planck length, time, and mass can then be used to cancel out the non-K units of the Boltzmann constant. This results in

$$T_P = \frac{1}{k_B} \sqrt{\frac{\hbar c^5}{G}}.$$

A similar procedure leads to the Planck charge:

$$q_P = \sqrt{\frac{\hbar c}{k}}.$$

Numerically, these Planck units work out to the following:

$$x_P = 1.61 \cdot 10^{-35} \text{ m}$$
$$t_P = 5.39 \cdot 10^{-44} \text{ s}$$
$$m_P = 2.18 \cdot 10^{-8} \text{ kg}$$
$$T_P = 1.42 \cdot 10^{32} \text{ K}$$
$$q_P = 1.88 \cdot 10^{-18} \text{ C}.$$

The Planck scales for temperature, length, and time are very far from our usual experience and either astronomically high (temperature) or unimaginably small (length and time). The Planck charge, however, is relatively close to the elementary charge quantum e and differs only by a factor of $1/\sqrt{\alpha} = \sqrt{137} = 11.7$.

Supersymmetry and String Theory

The incredible success of the Dirac theory motivated further extension of the standard model of particle physics. Based on symmetries, Dirac predicted that visible matter has undiscovered partners with negative energy states. This led to the discovery of antimatter and to the realization that every particle has a partner in the antiparticle world, with the same mass but opposite charge.

Supersymmetric (abbreviated SUSY) theories propose to do the same for fermions and bosons. **Supersymmetry** postulates for each fermion a supersymmetric bosonic partner, and for each boson a supersymmetric fermionic partner. This idea was originally proposed in 1973 by the Italian-American Bruno Zumino and the Austrian-German Julius Wess.

A severe constraint on SUSY theories is that experimental evidence for these supersymmetric partners has not yet been discovered. This fact excludes all versions of SUSY theories that predict masses below 100 GeV/c^2, the energy-mass region that has been thoroughly explored with existing particle accelerators, for any of the postulated new particles. However, some SUSY extensions of the standard model predict particle masses just above this limit. Thus, there is some hope that the rest-energy regime between 100 GeV and 1 TeV, which is accessible by the Large Hadron Collider at CERN, will reveal discoveries of magnitude comparable to the discovery of the positron.

In addition, many SUSY theories predict the existence of very heavy neutralinos. The neutralinos are thought to be mixtures of zinos (SUSY partner of the Z boson), photinos (SUSY partner of the photon), and higgsinos (SUSY partner of the Higgs boson). The lightest variant of the neutralino is absolutely stable according to these theories and is a leading candidate for the component of cold dark matter. It is called a *WIMP* (weakly interacting massive particle). Chapter 12 introduced the concept of dark matter as the explanation for effects like gravitational lensing and the fact that stars' orbital velocities around their galaxies are in general too large relative to the amount of visible matter.

The branch of particle physics theory that has unquestionably attracted more attention than any other during recent decades is **string theory.** At its most basic level, string theory replaces the elementary (zero-dimensional) point particle as the most basic constituent of matter with one-dimensional strings of very small lengths, possibly on the Planck scale. The elementary particles that are observed experimentally are thought to be strings vibrating at particular resonance frequencies. Strings can merge and split, giving rise to the particle interactions of the standard model discussed in Section 39.3 (see Figure 39.29).

While string theory has been around since the late 1960s, it became dominant after the British theorist Michael Green and the American John Schwartz worked out a version of it in 1984 that was the start of superstring theory for unification of all known forces, including gravity. In the popular press, string theory was called "The Theory of Everything." In the mid-1990s, through the work of Edward Witten and many others, it was discovered that string theory can be thought of as a certain limit of a perhaps more profound theory, which is called *M-theory.*

String theories predict additional spatial dimensions. The most popular versions postulate a 10- or 11-dimensional space-time. Why, then, can we observe only 4 dimensions? The answer given by string theories is that the extra dimensions are compacted to a size so small that we cannot observe them. This can happen at the Planck scale, but it is not excluded that the compactification length scale can be as large as a few micrometers. If this is correct, then we might expect to find distortion of gravity and deviation from the r^{-2} force law on this scale. Almost three decades of high-precision searches, however, have not come up with positive experimental evidence. String theories are very attractive mathematical frameworks in the eyes of particle theorists, but it has not yet been possible to establish any irrefutable connection to experimental observables. In 1997, however, Juan Maldacena from Argentina and the United States proved that string theory defined in a certain space is equivalent to a quantum field theory without gravity on the lower-dimensional boundary of that space. In the end, this may turn out to be the best connection to the standard model of particle physics and thus connect string theory to observable consequences.

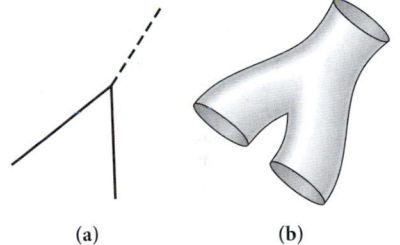

(a)   (b)

FIGURE 39.29 (a) Conventional Feynman diagram of the interaction of point particles; (b) representation of the same process as string fusion in string theory.

39.5 Composite Particles

With a reasonable understanding of the basic fermionic building blocks of matter and their interactions via the exchange of fundamental bosons, we can begin to try to understand how they form composite particles. All composite particles formed by quarks are called **hadrons.** Hadrons are either quark-antiquark pairs, called **mesons,** or groups of three quarks, called **baryons.** Each quark has another trait that can be one of three values. As mentioned in Section 39.3, this trait is arbitrarily called **color,** even though it has nothing to do with the common use of the word as a visual property. All composite particles are color singlets.

What is a color singlet? Quarks carry something called a color charge, coming in three varieties: red (r), green (g), and blue (b). Antiquarks carry anti-colors, which are marked with horizontal bars above the letters. Of course, quarks cannot really be imagined as having a physical color, because what we see as color is only light from a very limited range of

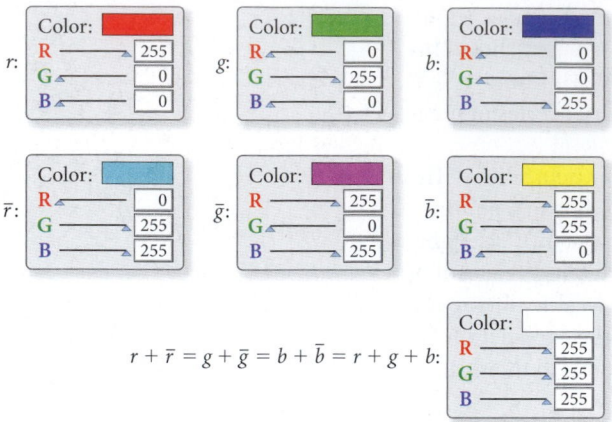

$$r + \bar{r} = g + \bar{g} = b + \bar{b} = r + g + b:$$

FIGURE 39.30 Additive color sliders in a computer graphics program as a model for quark colors (top row), antiquark anti-colors (middle row), and the color singlet (bottom panel).

photon energies, energies that are millions of times smaller than the characteristic scales of the realm of elementary particles. However, the algebra of additive color mixing works as a perfect analogy for the algebra of combining quarks into hadrons.

Figure 39.30 shows RGB color selectors from a typical computer graphics program, where any color is created as a mixture of red, green, and blue by picking numbers between 0 and 255 for each of the three color-components. For example, selecting 255 for red, 0 for green, and 0 for blue produces a pure red (upper left corner). Reversing all three sliders—0 for red, 255 for green, and 255 for blue—produces cyan. In a certain sense, it is the anti-color to red, $\bar{r}$, because if the RGB values of cyan and red are added, we obtain 255 for all three RGB values, which is the "color" white. Pure red, pure green, and pure blue can also be added to obtain white. Thus, white represents the color singlet resulting from a combination of equal amounts of all colors or equal amounts of a color and its anti-color. In particle physics, only color singlets can be physically observable particles. Thus, particles we can observe consist of either three quarks of different colors or a quark of a certain color and an antiquark with the quark's anti-color. The first combination is called a baryon and is a fermion (with either spin $\frac{1}{2}\hbar$ or spin $\frac{3}{2}\hbar$), and the second is a meson, which is always a boson (with either spin 0 or spin $1\hbar$). Three antiquarks with three different anti-colors can also combine to yield a color singlet, and they form the antibaryons.

Mesons

Table 39.3 gives a partial listing of the known mesons, including all mesons with mass less than 1 GeV/c^2 and a few mesons with special properties and higher masses. Hundreds of other mesons are known.

No meson is absolutely stable; they all decay. The lightest mesons are the pions, π^+, π^-, and π^0, all with masses of around 140 MeV/c^2. The neutral pion decays predominantly via the reaction

$$\pi^0 \to 2\gamma. \tag{39.13}$$

This decay proceeds relatively fast, on a time scale of the order of 10^{-16} s, via the electromagnetic interaction. The charged pions, however, cannot decay in this way because of the conservation law of charge. Instead, they have to decay via the weak interaction, predominantly via

$$\pi^+ \to \mu^+ + \nu_\mu$$
$$\pi^- \to \mu^- + \bar{\nu}_\mu. \tag{39.14}$$

In comparison with neutral pion decay, the decay of the charged pions takes practically forever, on the order of 10^{-8} s, or approximately 100 million times longer than the neutral pion decay time.

Let's examine the lifetimes of the other mesons. Figure 39.31 shows three vastly different regions of lifetimes. The area highlighted in yellow shows the decays that proceed via the weak interaction; in white, via the electromagnetic interaction; and in pink, via the strong interaction. Because energy must be conserved in every decay, heavier mesons can decay into lighter mesons but not vice versa. This decay into other mesons happens very fast if it proceeds via the strong interaction. There must be a good reason why the kaons and D and B mesons live such a comparatively long time. From our discussion of quarks, you can understand that it is because they contain strange, charm, and bottom quarks, respectively, which can only decay via the weak interaction into lower-mass quarks of other flavors.

The strangeness quantum number, S, equals the number of antistrange quarks in a hadron minus the number of strange quarks. (It is

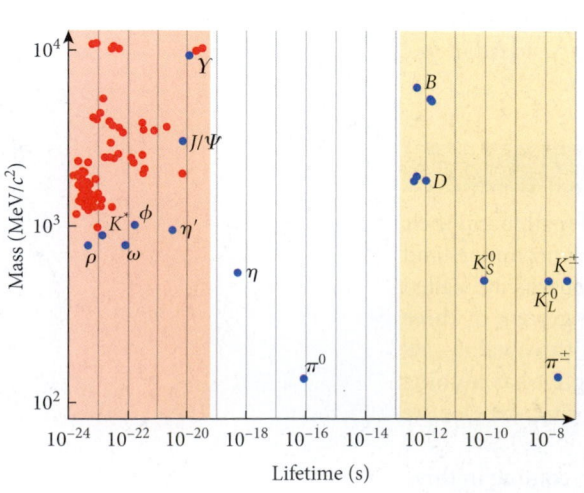

FIGURE 39.31 Plot of the lifetimes and masses for all known mesons. The blue dots represent entries in Table 39.3, and the red dots are other mesons.

Table 39.3		The Most Important Mesons, Including Their Masses, Lifetimes, Quantum Numbers, and Quark Composition							
Name	Symbol	m (MeV/c^2)	Lifetime (s)	s	S	C	B	Quarks	
Pion	π^+	139.570	$2.60 \cdot 10^{-8}$	0	0	0	0	$u\bar{d}$	
Pion	π^-	139.570	$2.60 \cdot 10^{-8}$	0	0	0	0	$\bar{u}d$	
Pion	π^0	134.977	$8.4 \cdot 10^{-17}$	0	0	0	0	$\frac{1}{\sqrt{2}}(u\bar{u}-d\bar{d})$	
Eta	η	547.853	$5.0 \cdot 10^{-19}$	0	0	0	0	$\frac{1}{\sqrt{6}}(u\bar{u}+d\bar{d}-2s\bar{s})$	
Eta prime	η'	957.78	$3.39 \cdot 10^{-21}$	0	0	0	0	$\frac{1}{\sqrt{3}}(u\bar{u}+d\bar{d}+s\bar{s})$	
Rho	ρ^+	775.49	$4.42 \cdot 10^{-24}$	1	0	0	0	$u\bar{d}$	
Rho	ρ^-	775.49	$4.42 \cdot 10^{-24}$	1	0	0	0	$\bar{u}d$	
Rho	ρ^0	775.49	$4.42 \cdot 10^{-24}$	1	0	0	0	$\frac{1}{\sqrt{2}}(u\bar{u}-d\bar{d})$	
Omega	ω	782.65	$7.75 \cdot 10^{-23}$	1	0	0	0	$\frac{1}{\sqrt{2}}(u\bar{u}+d\bar{d})$	
Phi	ϕ	1019.455	$1.55 \cdot 10^{-22}$	1	0	0	0	$s\bar{s}$	
Kaon	K^+	493.677	$1.2380 \cdot 10^{-8}$	0	1	0	0	$u\bar{s}$	
Kaon	K^-	493.677	$1.2380 \cdot 10^{-8}$	0	-1	0	0	$\bar{u}s$	
Kaon	K^0	497.614	$K_S^0{:}8.953 \cdot 10^{-11}$	0	1	0	0	$d\bar{s}$	
Kaon	$\bar{K}^0$	497.614	$K_L^0{:}5.116 \cdot 10^{-8}$	0	-1	0	0	$\bar{d}s$	
K-star	K^{*+}	891.66	$\sim 7.35 \cdot 10^{-20}$	1	1	0	0	$u\bar{s}$	
K-star	K^{*-}	891.66	$\sim 7.35 \cdot 10^{-20}$	1	-1	0	0	$\bar{u}s$	
K-star	K^{*0}	895.94	$\sim 7.346 \cdot 10^{-20}$	1	1	0	0	$d\bar{s}$	
K-star	$\bar{K}^{*0}$	895.94	$\sim 7.346 \cdot 10^{-20}$	1	-1	0	0	$\bar{d}s$	
D meson	D^+	1869.57	$1.040 \cdot 10^{-12}$	0	0	1	0	$c\bar{d}$	
D meson	D^-	1869.54	$1.040 \cdot 10^{-12}$	0	0	-1	0	$\bar{c}d$	
D meson	D^0	1864.80	$4.101 \cdot 10^{-13}$	0	0	1	0	$c\bar{u}$	
D meson	$\bar{D}^0$	1864.80	$4.101 \cdot 10^{-13}$	0	0	-1	0	$\bar{c}u$	
D_s meson	D_s^+	1968.45	$5.0 \cdot 10^{-13}$	0	1	1	0	$c\bar{s}$	
D_s meson	D_s^-	1968.45	$5.0 \cdot 10^{-13}$	0	-1	-1	0	$\bar{c}s$	
J/Psi	J/ψ	3096.9	$7.05 \cdot 10^{-21}$	1	0	0	0	$c\bar{c}$	
B meson	B^+	5279.17	$1.641 \cdot 10^{-12}$	0	0	0	1	$u\bar{b}$	
B meson	B^-	5279.17	$1.641 \cdot 10^{-12}$	0	0	0	-1	$\bar{u}b$	
B meson	B^0	5279.5	$1.519 \cdot 10^{-12}$	0	0	0	1	$d\bar{b}$	
B meson	$\bar{B}^0$	5279.5	$1.519 \cdot 10^{-12}$	0	0	0	-1	$\bar{d}b$	
B_s meson	B_s^0	5366.3	$1.472 \cdot 10^{-12}$	0	-1	0	1	$s\bar{b}$	
B_s meson	$\bar{B}_s^0$	5366.3	$1.472 \cdot 10^{-12}$	0	1	0	-1	$s\bar{b}$	
B_c meson	B_c^+	6277	$4.53 \cdot 10^{-13}$	0	0	1	1	$c\bar{b}$	
B_c meson	B_c^-	6277	$4.53 \cdot 10^{-13}$	0	0	-1	-1	$\bar{c}b$	
Upsilon	Y	9460.30	$1.22 \cdot 10^{-20}$	1	0	0	0	$b\bar{b}$	

K. Nakamura et al. (Particle Data Group), JP G 37, 075021 (2010) and 2011 partial update for the 2012 edition (URL: http://pdg.lbl.gov)
Source: Particle Data Group site 2011 numbers pdg.lbl.gov

a purely historical artifact that anti-strange quarks are counted with positive numbers. For the quantum numbers of charm, C, and bottom, B, which give the net numbers of charm and bottom quarks, the sign is reversed.) Another useful quantum number is the isospin, t, and its projection, t_z. Even though no real angular momentum is involved in this case, the same algebra can be used for isospin as was used for conventional spin. In particular, a given isospin t has $2t + 1$ states of different projection t_z, just as a given spin s has $2s + 1$ states of different projection s_z. The charge number Z, the isospin, and the strangeness quantum number for the low-mass mesons (below the mass threshold where they start to contain quarks other than up, down, and strange) are related via

$$Z = t_z + \tfrac{1}{2}S. \qquad (39.15)$$

Figure 39.32 shows the arrangement of the low-mass mesons with spin 0 (part a) and with spin $1\hbar$ (part b) as a function of isospin and strangeness. They form characteristic nonets,

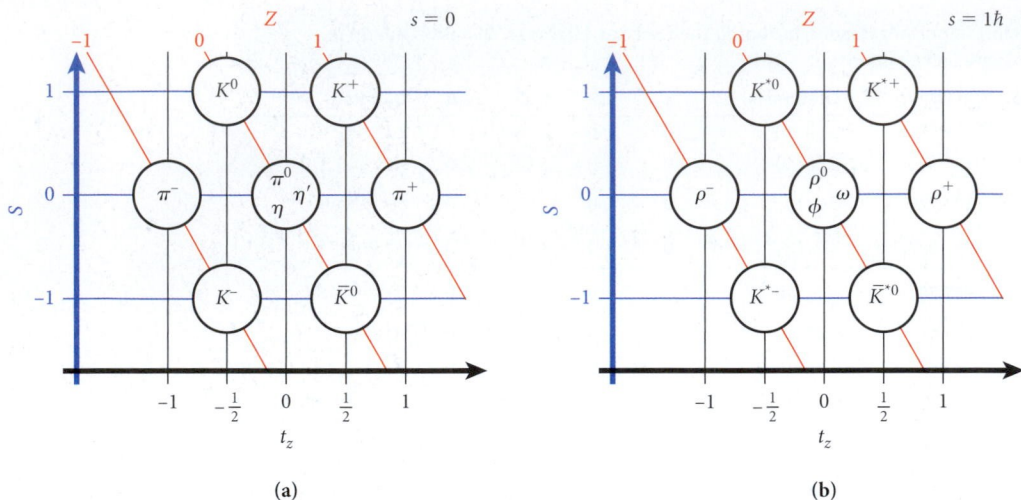

FIGURE 39.32   Meson nonets with (a) spin 0 and (b) spin $1\hbar$.

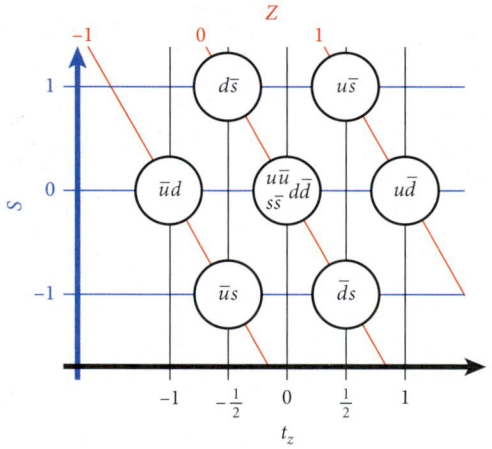

FIGURE 39.33   Quark composition of the two meson nonets.

which the quark model is able to explain beautifully. Figure 39.33 shows the same plot, but in terms of the constituent quarks of the different mesons. In each case, the quark composition of the three neutral mesons with $S = t_z = 0$ is a linear combination of the three quark-antiquark pairs. The quark charges in Table 39.1 confirm that their charges always add up to the corresponding charge of the meson.

The up and down quarks have isospin $\frac{1}{2}$, with projections $t_z = \frac{1}{2}$ for the up quark and $t_z = -\frac{1}{2}$ for the down quark. The isospin for all other quarks and antiquarks (strange, charm, bottom, and top) is 0. Similar to the case of angular momentum and regular spin, if a particle with isospin $\frac{1}{2}$ is combined with a particle with isospin 0, the resulting isospin is $\frac{1}{2}$. If two particles with isospin $\frac{1}{2}$ are combined, the resulting isospin can be 0 (if the two isospin vectors are antiparallel) or 1 (if the two isospin vectors are parallel). If three particles with isospin $\frac{1}{2}$ are combined, the resulting isospin can be $\frac{3}{2}$ (all three isospin vectors parallel) or $\frac{1}{2}$ (one antiparallel to the other two). This "algebra of isospin" allows us to determine the observed isospins of all mesons from the quark isospins, as shown in Figure 39.33.

Concept Check 39.3

From Figure 39.33, the isospin projections, t_z, of the anti-up quark ($\bar{u}$) and the anti-down quark ($\bar{d}$) are

a) $t_z(\bar{u}) = -\frac{1}{2}$, $t_z(\bar{d}) = -\frac{1}{2}$.

b) $t_z(\bar{u}) = -\frac{1}{2}$, $t_z(\bar{d}) = \frac{1}{2}$.

c) $t_z(\bar{u}) = \frac{1}{2}$, $t_z(\bar{d}) = -\frac{1}{2}$.

d) $t_z(\bar{u}) = \frac{1}{2}$, $t_z(\bar{d}) = \frac{1}{2}$.

e) $t_z(\bar{u}) = 0$, $t_z(\bar{d}) = 0$.

Baryons

The most important baryons are listed in Table 39.4. As we have already stated, baryons are composed of three quarks each, with one quark of each color. The only stable baryon is the proton, the lightest baryon. As discussed previously, a neutron can decay into a proton via a beta decay:

$$n \rightarrow p + e^- + \bar{\nu}_e. \tag{39.16}$$

This is the same process as the beta decay of the down quark expressed in equation 39.12, because in this process one of the two down quarks of the neutron decays into an up quark, and the other up and down quarks remain unchanged (Figure 39.34).

The dominant decay processes for all heavier baryons occur via emission of one or more pions. Any of these processes can involve only the strong interaction, in which case lifetimes of the order of 10^{-20} s or shorter are found. If, in addition, a quark needs to change flavor in the process, then much longer lifetimes are seen, typical of the weak interaction.

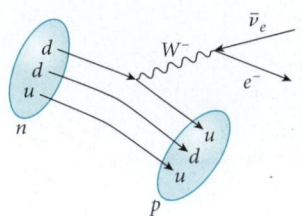

FIGURE 39.34   Feynman diagram for the beta decay of the neutron.

Name	Symbol	m (MeV/c^2)	Lifetime (s)	Z	t_z	s	S	C	B	Quarks
Proton	p	938.272	∞	1	$\frac{1}{2}$	$\frac{1}{2}$	0	0	0	uud
Neutron	n	939.565	881.5	0	$-\frac{1}{2}$	$\frac{1}{2}$	0	0	0	udd
Delta	Δ^{++}	1232	$5.58 \cdot 10^{-24}$	2	$\frac{3}{2}$	$\frac{3}{2}$	0	0	0	uuu
Delta	Δ^{+}	1232	$5.58 \cdot 10^{-24}$	1	$\frac{1}{2}$	$\frac{3}{2}$	0	0	0	uud
Delta	Δ^{-}	1232	$5.58 \cdot 10^{-24}$	−1	$-\frac{1}{2}$	$\frac{3}{2}$	0	0	0	udd
Delta	Δ^{--}	1232	$5.58 \cdot 10^{-24}$	−2	$-\frac{3}{2}$	$\frac{3}{2}$	0	0	0	ddd
Lambda	Λ^{0}	1115.683	$2.632 \cdot 10^{-10}$	0	0	$\frac{1}{2}$	−1	0	0	uds
Sigma	Σ^{+}	1189.37	$8.018 \cdot 10^{-11}$	1	1	$\frac{1}{2}$	−1	0	0	uus
Sigma	Σ^{0}	1192.642	$7.4 \cdot 10^{-20}$	0	0	$\frac{1}{2}$	−1	0	0	uds
Sigma	Σ^{-}	1197.449	$1.48 \cdot 10^{-10}$	−1	−1	$\frac{1}{2}$	−1	0	0	dds
Xi	Ξ^{0}	1314.86	$2.90 \cdot 10^{-10}$	0	$\frac{1}{2}$	$\frac{1}{2}$	−2	0	0	uss
Xi	Ξ^{-}	1321.71	$1.64 \cdot 10^{-10}$	−1	$-\frac{1}{2}$	$\frac{1}{2}$	−2	0	0	dss
Omega	Ω^{-}	1672.45	$8.21 \cdot 10^{-11}$	−1	0	$\frac{3}{2}$	−3	0	0	sss
Lambda-c	Λ_c^{+}	2286.46	$2.0 \cdot 10^{-13}$	1	0	$\frac{1}{2}$	0	1	0	udc
Lambda-b	Λ_b^{0}	5620.2	$1.425 \cdot 10^{-12}$	0	0	$\frac{1}{2}$	0	0	1	udb

Table 39.4 The Most Important Baryons, Including Their Masses, Lifetimes, Quantum Numbers, and Quark Composition

K. Nakamura et al. (Particle Data Group), JP G 37, 075021 (2010) and 2011 partial update for the 2012 edition (URL: http://pdg.lbl.gov).

In Figure 39.35, the lowest-mass baryons with spin $\frac{1}{2}\hbar$ and spin $\frac{3}{2}\hbar$ are arranged as an octet and a decuplet as functions of isospin and strangeness, similar to the diagrams for the meson nonets in Figure 39.32. This arrangement into nonets (for the mesons) and octet and decuplet (for the baryons) derives naturally from the underlying quark structure. In particular, note the omega baryon. Because it has strangeness –3, it is composed of three strange quarks. Its existence was predicted by the quark model, and it was subsequently discovered in 1964. This was the first great success of the quark model.

Self-Test Opportunity 39.3

Construct the quark composition for the baryons of Figure 39.35 in the same way as in Figure 39.33.

Lattice-QCD, Confinement, and Asymptotic Freedom

The sum of the masses of the constituent quarks inside a proton is less than 15 MeV/c^2, which is only 1.5% of the proton's mass. Where does the rest of its mass come from? Answering this question is very difficult and requires mathematical tools that go beyond the scope of this book. However, we can describe a bit of the search for the answer.

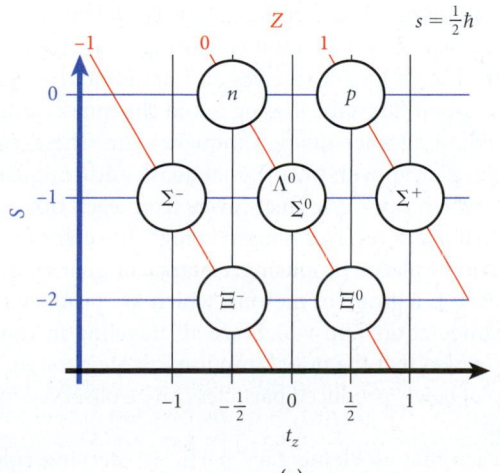

(a)

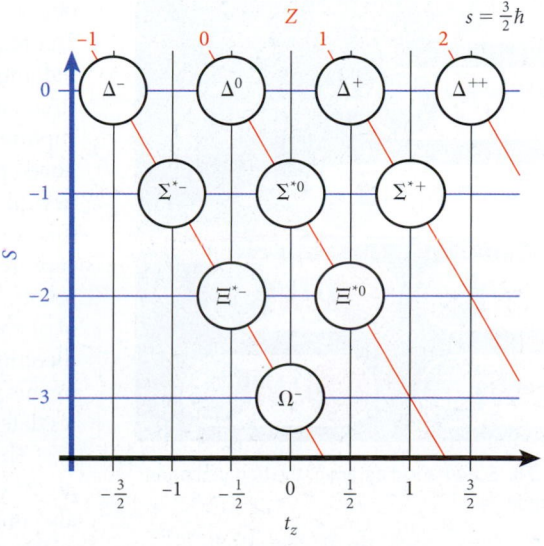

(b)

FIGURE 39.35 (a) Baryon octet for spin $\frac{1}{2}\hbar$ and (b) decuplet for spin $\frac{3}{2}\hbar$.

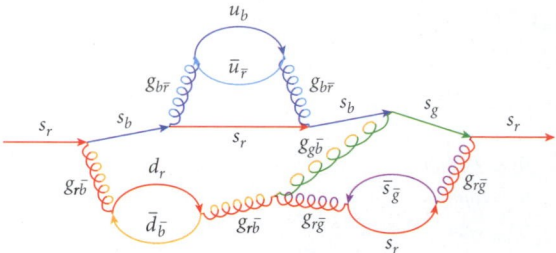

FIGURE 39.36 A Feynman diagram of a red strange quark propagating through the QCD vacuum.

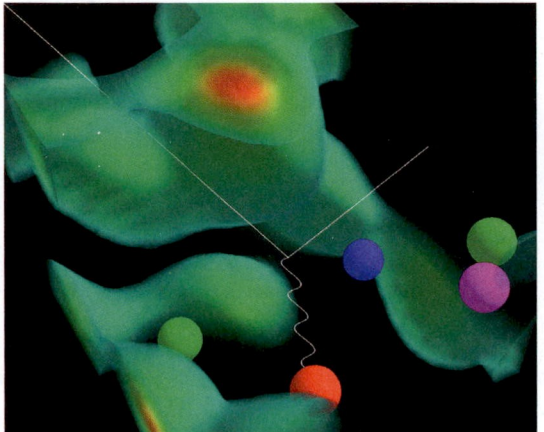

FIGURE 39.37 Artist's rendition of a proton. The red, green, and blue balls represent the locations of the constituent quarks. The green-magenta ("anti-green") pair of balls on the right represents a virtual quark-antiquark pair. The white line represents a high-energy electron that scatters off one of the quarks through exchange of a photon (wavy line). All these objects are displayed on the background of a lattice-QCD calculation of the gluon field, as performed by the CSSM group at the University of Adelaide, Australia.

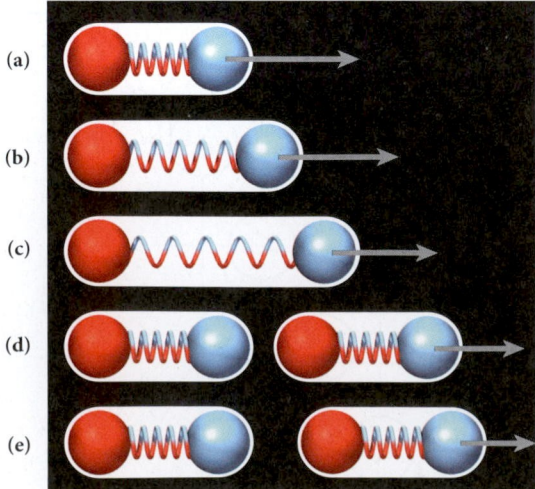

FIGURE 39.38 Schematic diagram of the fragmentation of a color flux-tube.

The fundamental theory, including relativity and quantum physics, that describes the interaction of quarks and gluons with one another, with themselves, and with the vacuum, is called **quantum chromodynamics (QCD).** As an illustrative example of the kinds of processes QCD needs to take into account, Figure 39.36 shows a Feynman diagram of a strange quark propagating through the vacuum, emitting and absorbing gluons and virtual quark-antiquark pairs.

Physicists understand the underlying framework of QCD in principle, but it is very hard to obtain numerical results. To obtain some predictions that can be compared to experimental results, physicists have built customized supercomputers to simulate QCD. Despite the exponential growth of computer power during the last few decades, it is fair to say that progress in computing observables such as hadron masses has been slow.

Solving the equations of QCD is difficult. One method that allows solutions to these equations is **lattice-QCD.** In lattice-QCD, quarks are assumed to reside only on the vertices of a crystal-like lattice and the gluons must travel in straight lines between the quarks. This technique builds in cutoffs that allow calculations to be done. Of course, the real world is not a crystal lattice. However, the aim is to carry out the calculations for smaller and smaller lattice spacings and larger and larger lattices and thus do a better job of approximating physical reality.

Figure 39.37 shows the result of a lattice-QCD calculation of the spatial configuration of the gluon field inside a proton. Overlaid on the numerical plot are drawings of three balls marking the constituent quarks of the proton, as well as two balls marking the position of a temporarily created virtual quark-antiquark pair.

Quarks, antiquarks, and gluons all carry color charge. QCD predicts that only color-neutral objects can exist in isolation. This means that a single quark or gluon cannot be observed by itself. Very sensitive searches for free quarks have been conducted, all of them with negative results.

What happens if we try to remove a quark from a baryon—for example, by shooting an electron with very high energy into a proton, trying to knock out a quark? The quark can acquire a very high momentum after the collision with the electron and move rapidly away from the other two quarks inside the proton. However, it remains tethered to them by a string of gluons called a *flux-tube.* QCD predicts that the energy contained in this flux-tube increases in proportion to its length. Once it is long enough and energetic enough, it can rupture and split into two or more color-neutral objects—mainly pions. This fragmentation is illustrated schematically in Figure 39.38, which shows a meson consisting of a red quark and an anti-red antiquark. The antiquark receives an impulse and moves to the right, stretching the gluon flux-tube joining it and the quark, until the flux-tube ruptures, yielding another quark-antiquark pair. These two newly produced particles pair up with the original quark and antiquark to form two new mesons, which can move freely away from each other. If the impulse that the original quark receives is high enough, the color flux-tube can also rupture in multiple places, producing a quark-antiquark pair at each breakpoint. The result is a group of mesons, which are produced in the process of string fragmentation, and which are all traveling in roughly the same direction. Provided that the initial impulse is large enough, the jet formed by this group of newly produced particles can be observed in particle physics detectors.

This means that all elementary particles carrying color charge—all gluons, quarks, and antiquarks—are absolutely confined. Confinement also implies that any attempt to remove a quark from its color singlet will only produce additional color singlets, that is, hadrons.

The color force acts in such a way that the farther away objects with color charge move from each other, the more strongly they are attached to each other. Does this also mean the opposite, that is, that when color-charged particles inside a singlet are close to each other, they can move freely? Indeed, this is the case. At close distances or high relative momenta, color-charged objects experience *asymptotic freedom*. This was predicted by David Gross, Frank Wilczek, and David Politzer in 1973 and subsequently confirmed experimentally. They shared the 2004 Nobel Prize in Physics for this achievement.

EXAMPLE 39.3 Feynman Diagrams of the Decay of Positive Pions and Positive Muons

When high-energy cosmic rays consisting mainly of protons are incident on the Earth's atmosphere, reactions take place that produce positive pions. A positive pion decays with a mean lifetime of 26 ns into a positive muon and a muon-neutrino:

$$\pi^+ \rightarrow \mu^+ + \nu_\mu.$$

In turn, the positive muon decays with a mean lifetime of 2.2 μs into a positron, an electron-neutrino, and an anti-muon-neutrino

$$\mu^+ \rightarrow e^+ + \nu_e + \bar{\nu}_\mu.$$

PROBLEM
Draw the Feynman diagrams representing these two decays.

SOLUTION
The positive pion is a meson composed of an up quark and an anti-down quark, as shown in Figure 39.39. This quark-antiquark pair $u/\bar{d}$ annihilates to produce a W^+ meson. The W^+ meson then materializes as the lepton/anti-lepton pair ν_μ/μ^+.

The positive muon that is produced in the decay emits a W^+ meson and transforms into an anti-muon-neutrino, as shown in Figure 39.40. This W^+ meson then materializes as the lepton/anti-lepton pair ν_e/e^+.

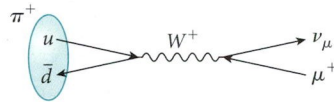

FIGURE 39.39 Feynman diagram of the decay of a positive pion.

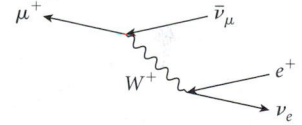

FIGURE 39.40 Feynman diagram of the decay of a positive muon.

DISCUSSION
The decay of positive pions is particularly interesting because these particles are produced at the top of the Earth's atmosphere with relatively high energies via reactions of cosmic ray protons with the nuclei of nitrogen and oxygen atoms. These pions decay into muons, and the resulting muons travel at a large fraction of the speed of light. They should travel at most only 0.66 km before they decay into positrons. However, muons produced by cosmic rays are observed at the surface of the Earth. About 10 muons per second pass through your hand. One reason the muons reach the surface of the Earth is relativistic time dilation, as discussed in Chapter 35 (see Example 35.2). For a speed $\beta = v/c = 0.998$, the time dilation factor is

$$\gamma = \frac{1}{\sqrt{1-\beta^2}} = \frac{1}{\sqrt{1-(0.998)^2}} = 15.8.$$

Thus, the muons will travel a distance of 10 km before they decay, allowing them to reach the surface of the Earth. A second reason the muons reach the Earth's surface is that they can traverse the Earth's atmosphere without interacting with the nuclei in the atoms of atmospheric gases, because a muon interacts only via the weak force and the electromagnetic force, not via the strong force.

39.6 Big Bang Cosmology

The night sky presents an amazing display of light that has traveled for many years from distant stars (Figure 39.41). Since the dawn of civilization, humans have asked where all of this came from (and how, and why). Since the middle of the 20th century, scientists have been reasonably sure that they know the answer, and the picture is becoming clearer and more detailed. Through the interplay of observational astronomy, cosmology, and nuclear and particle physics, the models have been refined.

FIGURE 39.41 View of the universe from Earth in the visible spectrum, using the Aitoff projection in galactic coordinates. This projection shows the view over the entire night sky projected onto an ellipse. The bright stars and clouds in the central horizontal band belong to the Milky Way.

The entire universe came into being through a singular event that we call the *Big Bang*. This event happened $(13.75 \pm 0.11) \cdot 10^9$ years ago. Although the age of the universe cannot be fixed with an uncertainty smaller than 120 million years, physicists are able to state how the universe developed in the first fractions of a second after the Big Bang.

The Planck time is the earliest time after the Big Bang for which it makes any sense to try to deduce a history of the universe (Figure 39.42). (See Example 39.2.) At a time of 10^{-43} s, the temperature was on the order of 10^{32} K. All forces were unified at that time, with gravity separating from the other forces a short time thereafter. The era of grand unification of the strong and electroweak forces is estimated to have lasted to about 10^{-34} s, when the strong force became separated from the electroweak force. At the end of this era, the universe underwent a rapid inflationary expansion. To understand why this process had to have been part of the universe's history, we need to discuss the cosmological concept of inflation.

Inflation

Three important questions have puzzled astronomers for a long time:

1. Why is the cosmic microwave background radiation so incredibly isotropic, which is to say, smooth and very much the same in every direction?
2. Why is the universe made entirely of matter and not of antimatter?
3. Why is the universe so incredibly flat (a term to be explained shortly)?

All three questions deserve some elaboration to show how truly startling these facts are.

First, let's think about the smoothness of the cosmic microwave background. As we saw in Chapter 36, the cosmic microwave background (CMB) radiation shows a perfect blackbody spectrum at a temperature of 2.725 K. CMB radiation was discovered in 1965 by Arno Penzias and Robert Woodrow Wilson, who received the Nobel Prize in Physics for this discovery in 1978. This radiation had been predicted in 1948 by George Gamov, Ralph Alpher, and Robert Herman as a consequence of the Big Bang model, which at that time was still very speculative and only one of several possible models of the time evolution of the universe. Only

FIGURE 39.42 Temperature of the universe as a function of time. The red circle with the cross marks the current universe, and the vertical gray lines mark the approximate boundaries between the different eras. For each era, the basic constituents are shown. Radiation was present at all times and in equilibrium with matter until 10^{13} s, when the universe became transparent to radiation.

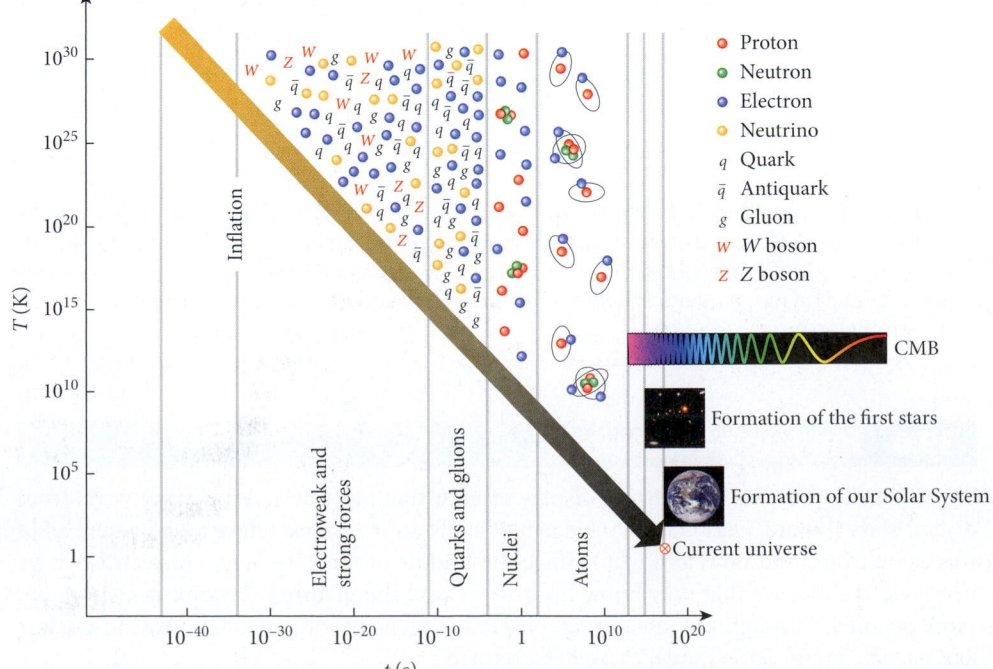

through the COBE and subsequently the WMAP satellite missions was it possible to measure just how isotropic this radiation is, an accomplishment that earned the 2006 Nobel Prize in Physics for George Smoot and John C. Mather. Figure 39.43 shows a picture of the universe at microwave wavelengths. This image was produced by the team that operated the WMAP satellite. It shows that the fluctuations of the temperature of space as a function of direction are only of the order of $|\delta T/T| \approx 10^{-4}$ to 10^{-5}. For comparison, the WMAP team also produced a temperature map of the surface of Earth for June 1992, for which the fluctuations are on the order of $|\delta T/T| \approx 0.2$.

The smoothness of the CMB is so stunning because the radiation comes to us from all directions and has the same value for all points in the sky, even if they could not have been in causal contact during the history of the universe (that is, without inflation).

Second, the universe is currently filled with matter and almost no antimatter. We know that initially matter and antimatter were produced in almost equal amounts and then almost completely annihilated into radiation, which now constitutes the cosmic microwave background. Estimates are that approximately 10^{10} photons presently exist for each hadron. This means that initially there were 10,000,000,001 matter particles for each 10,000,000,000 antimatter particles. Where did this very small asymmetry come from?

Third, after more than 13 billion years, the universe is still expanding, and we still cannot decide whether it will eventually stop expanding or go on expanding forever. The total kinetic energy of the universe is almost exactly equal to the total potential energy, but opposite in sign, so the total energy of the universe must be unbelievably close to zero. We thus say that the universe is *flat*. Imagine shooting off a gun from the surface of Earth. If the bullet has an initial speed that is large compared to the escape speed, then it will easily leave Earth; whereas if the initial speed is smaller than the escape speed, the bullet will fall back down to Earth's surface. Now imagine that after billions of years we still cannot decide if the bullet will eventually fall back down to Earth or will keep going! (As an aside, we note that recent findings, mentioned in Chapter 12, seem to imply that the universe is dominated by dark energy, which is slowly causing the re-acceleration of the universe, presenting yet another mystery. Thus it seems likely that, according to what we know now, that the universe will continue to expand forever.)

A model proposed by Alan Guth in 1980 and later expanded on by others answers all three questions in the same model, one that sounds like science fiction: A very short fraction of time after the Big Bang, the universe went through a very rapid period of inflation. Figure 39.44 depicts this inflation. The idea is that approximately 10^{-34} s after the Big Bang, the universe was at a temperature of around 10^{27} K and in a process of rapid cooling. It then became supercooled. This is analogous to freezing rain still being liquid at a temperature at which water should be frozen. When the freezing rain hits the ground, the relatively small impact shock creates fluctuations that increase exponentially and cause the formation of very smooth ice. The supercooled state of the freezing raindrop is a false ground state, and small fluctuations cause it to make a transition to the true ground state, ice. At the onset of the period of inflation, the universe also entered a false ground state and became supercooled (albeit still at an astronomically high temperature). The transition to the true ground state caused an exponential expansion in size, resulting in the smoothness of the distribution of matter and radiation still seen today in the cosmic microwave background radiation.

After the reheating of the inflationary period, which may have lasted until 10^{-32} s, the universe entered a radiation-dominated era that lasted until approximately 10^{11} s. During this time, the temperature, T, continued to fall gradually and in inverse proportion to the square root of the time, t:

$$T(t) = \frac{1.5 \cdot 10^{10} \text{ K s}^{1/2}}{\sqrt{t}}. \tag{39.17}$$

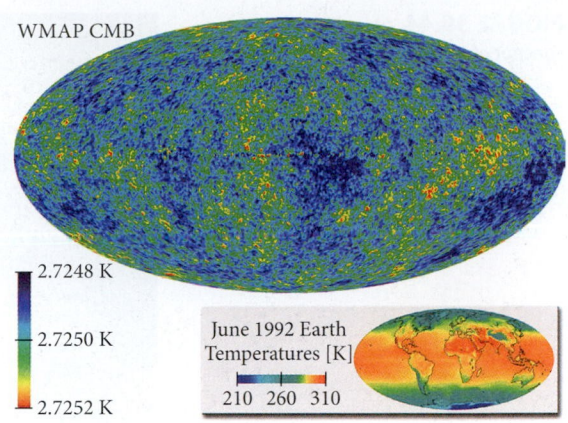

FIGURE 39.43 Temperature map of the cosmic microwave background (CMB) radiation of the universe. The inset shows temperatures on the surface of Earth in June 1992.

The electromagnetic and weak forces parted ways at approximately 10^{-11} s. At that point, the W and Z bosons became massive, through a process called *spontaneous symmetry breaking,* and subsequently decayed away rapidly. (Perhaps it is instructive to look at another example

FIGURE 39.44 Illustration of the concept of inflation of the early universe.

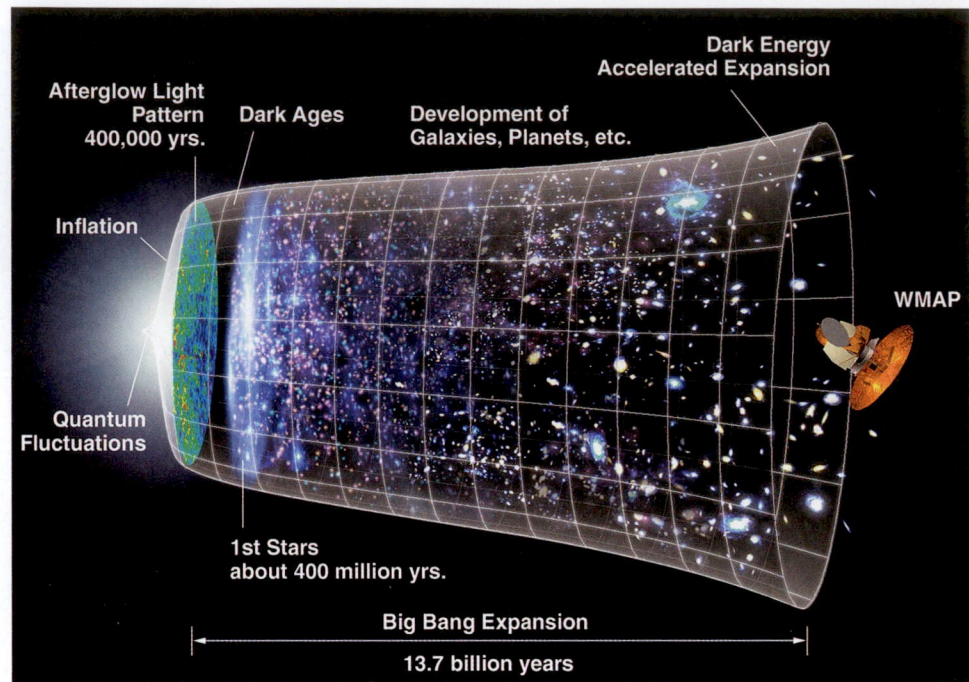

of spontaneous symmetry breaking, to illustrate the concept: In liquid water, the water molecules can have any arrangement and orientation. However, in the process of freezing, this symmetry has to be broken to force the molecules into the crystalline structure of ice.)

Quark-Gluon Plasma

After 10^{-11} s, and lasting until about 10^{-4} s, the universe was a mixture of quarks, gluons, and leptons, forming a plasma. Despite carrying color charges, the quarks and gluons were asymptotically free due to the high temperature. At about 10^{-4} s, the universe had cooled to a temperature of approximately $2.1 \cdot 10^{12}$ K ($k_B T = 180$ MeV). Lattice-QCD calculations indicate that at this temperature the quarks and gluons coalesced into color singlets.

Amazingly, the quark-gluon state of matter that dominated the early universe from 10^{-11} s to 10^{-4} s after the Big Bang can be recreated in accelerator-based experiments today. Experiments at the Relativistic Heavy Ion Collider (RHIC) in Brookhaven, Long Island, use colliding beams of gold nuclei with total kinetic energy of up to 20 TeV each to probe the conditions during the early time evolution of the universe and to produce small volumes of the quark-gluon matter. However, this state of matter lasts less than 10^{-23} s in the laboratory before it explodes into more than 5000 particles, mainly pions. A collision event recorded at RHIC is displayed in Figure 39.45. Since 2010, nuclear physicists also have the Large Hadron Collider (LHC) at CERN at their disposal. It can collide lead ions at an energy of 2.76 TeV per nucleon pair, which is 13 times higher than the energy achieved by the RHIC.

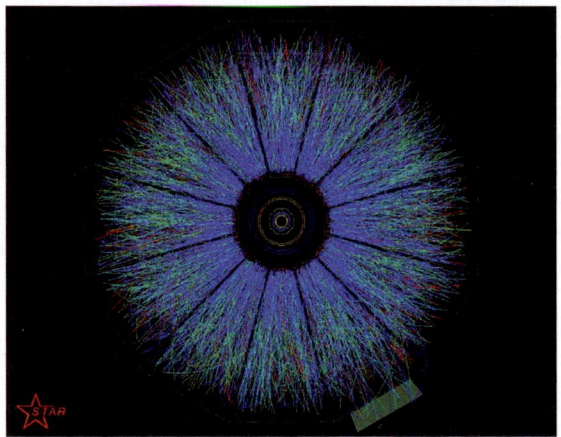

FIGURE 39.45 Event display from a collision of gold nuclei the Relativistic Heavy Ion Collider. Each line represents the track that one of the over 5000 particles resulting from the collision produced in the STAR detector. From these tracks, scientists are trying to determine the properties of the quark-gluon state of matter that was created transiently during the nuclear collision and learn about the history of the universe.

Nucleosynthesis

As the universe emerged from its quark-gluon phase at a time of 10^{-4} s after the Big Bang, it consisted overwhelmingly of photons, electrons, positrons, neutrinos, and anti-neutrinos. A small number of protons and neutrons—the lightest color singlets into which the quarks could condense—were also present. The protons and neutrons were constantly interconverting via the reactions

$$n + \nu_e \rightleftarrows p + e^-$$
$$n + e^+ \rightleftarrows p + \bar{\nu}_e.$$

(39.18)

This interconversion was easily possible because the temperature was still on the order of 10^{11} or 10^{12} K, which corresponds to 10 to 100 MeV, much larger than the mass difference of the neutron and the proton, which is only 1.293 MeV/c^2. This means that protons and neutrons were in equilibrium, with their relative numbers determined by the ratio of their Boltzmann factors:

$$\frac{n_n}{n_p} = e^{(m_n c^2 - m_p c^2)/k_B T}. \tag{39.19}$$

As the universe reached a temperature of approximately 10^{10} K (~0.86 MeV), the reactions of equation 39.18 proceeded too slowly to maintain equilibrium. At this point, the ratio of neutrons to protons froze at the value $e^{-1.293/0.86} = 0.222$. This means that there were then about 22 neutrons for every 100 protons. At this time, the universe was at an age of 1 s. However, the temperature was still too high at this time for nuclei to form. Instead, all neutrons remained free, and thus could beta decay according to equation 39.16, with a 15-min lifetime.

After 100 s, the universe had cooled down to 10^9 K, and due to beta decay, only about 16 neutrons were left for every 100 protons. This was the point where alpha particles (consisting of two neutrons and two protons each) could form, and essentially all available neutrons became bound in them. Small traces of deuterium, tritium, helium-3, and lithium (more on these in Chapter 40) were also formed. Sixteen neutrons and 16 protons can form 8 alpha particles. Therefore, the total mass fraction of nucleons trapped inside alpha particles at that time was 16 + 16 out of 100 + 16: (16+16)/(100 + 16) ≈ 27%. This rough estimate is consistent with the observed primordial mass fraction of 23% helium atoms (each consisting of an alpha particle plus electrons). This matchup is considered one of the greatest successes of the Big Bang theory.

The radiation dominance ended at approximately 10^{11} s (~3000 yr), when the temperature had reached 10^5 K. As the temperature fell further, to below approximately 3000 K (~0.25 eV), electrons were captured by protons and alpha particles to form neutral hydrogen and helium atoms. A photon needs to have at least 10.2 eV of energy to excite a hydrogen atom's electron from the 1s to the 2s state (see Chapter 38), so the photons then had many fewer electrons to scatter from, and thus the opacity of the universe dropped. At that time, photons were able to roam freely through the universe, and radiation decoupled from matter. The photons thus showed a perfect blackbody spectrum with the temperature of the universe at the time of the decoupling. The universe was then about 10^{13} s (~300,000 yr) old. These freely roaming photons are still visible today; they form the cosmic microwave background (CMB).

Why are the CMB photons now showing a temperature of 2.725 K, approximately 1000 times lower than the original temperature? The answer is that the photons have been stretched due to the expansion of the universe. The scale factor of the universe is approximately $R = 1100$ times bigger than it was at the time the CMB decoupled. This means that the wavelengths of these photons are now $\lambda = R\lambda_0$. According to the Wien displacement law (see Chapter 36), the wavelength λ_m at which the blackbody spectrum has its maximum is related to the temperature by $\lambda_m T$ = constant. Consequently, with the expansion of the wavelengths by a factor of 1100, the temperature has to fall by the same factor. (Figure 39.42 indicates this effect schematically by stretching the wavelength of the CMB photons and shifting the color toward the longer wavelengths.)

What followed was a period of darkness in the universe, lasting a few million years perhaps, before the first stars were formed through the gravitational interaction in the gas of hydrogen and helium atoms. Our Solar System formed approximately 4.5 billion years ago, about 9 billion years after the Big Bang. In Figure 39.42, this point in time corresponds to the rightmost gray vertical line, which in this logarithmic scale is only ever so slightly to the left of the present, which is marked by the red circle with a cross at a time of $4.3 \cdot 10^{17}$ s after the Big Bang. Chapter 40 will return to the nuclear physics that went into the formation of the heavy elements of which Earth is mainly composed. Their presence cannot be explained based on Big Bang cosmology alone and requires other events, which we will explore.

From this short narration of Big Bang cosmology, it is clear how particle physics at the smallest imaginable scales is intimately connected to astronomy at the largest possible scales. This amazing connection between the smallest and largest has fascinated physicists for the last 60 years and continues to do so, with new results being discovered each year.

WHAT WE HAVE LEARNED | EXAM STUDY GUIDE

- The reductionist quest to reduce systems to their component parts is a main theme in science and has its present culmination in particle physics. While an equally valid methodology seeks to understand complexity and emergent structure, this chapter focused on reductionism in particle physics.

- Substructure is probed using scattering experiments. The scattering cross section is defined as

$$\sigma = \frac{\text{Number of reactions per scattering center/s}}{\text{Number of impinging particles/s/m}^2}.$$

- The scattering cross section has the physical dimension of area and is measured in the unit barn (b) or millibarn (mb): $1\ \text{b} = 10^{-28}\ \text{m}^2$, $1\ \text{mb} = 10^{-31}\ \text{m}^2$.

- The classical Rutherford cross section for scattering from a pointlike target by the Coulomb interaction is

$$\frac{d\sigma}{d\Omega} = \left(\frac{kZ_p Z_t e^2}{4K}\right)^2 \frac{1}{\sin^4\left(\frac{1}{2}\theta\right)}.$$

- For scattering of a plane wave off a point source, the scattering wave function is

$$\psi_{\text{total}}(\vec{r}) = \psi_i(\vec{r}) + \psi_f(\vec{r}) = N\left(e^{i\kappa z} + f(\theta)\frac{e^{i\kappa r}}{r}\right),$$

and the cross section is related to the scattering amplitude by $\dfrac{d\sigma}{d\Omega} = \left|f(\theta)\right|^2$.

- The form factor is the Fourier transform of the density distribution,

$$F^2(\Delta p) = \left|\frac{1}{e}\int \rho(\vec{r})e^{i\Delta\vec{p}\cdot\vec{r}/\hbar}\,dV\right|^2,$$

and measures the deviation of the scattering cross section from the Rutherford cross section of a pointlike target,

$$\frac{d\sigma}{d\Omega} = \left(\frac{d\sigma}{d\Omega}\right)_{\text{point}} \cdot F^2(\Delta p).$$

- Elementary fermions have spin $\frac{1}{2}\hbar$ and include the six quarks (up, down, strange, charm, bottom, and top), the electron, muon, and tau leptons, and the electron-, muon-, and tau-neutrinos. Each of these 12 fermions has an antiparticle. Quarks all have a noninteger charge of $-\frac{1}{3}e$ or $+\frac{2}{3}e$ and cannot be observed in isolation.

- Elementary bosons are the mediators of the interactions between the fermions. They are the photon (electromagnetic), the W and Z bosons (electroweak), the gluon (strong), and the graviton (gravitational). The graviton is yet to be found experimentally. Gluons can also interact with other gluons.

- The interactions of fundamental particles can be represented pictorially by Feynman diagrams, and calculations can be based on the Feynman diagrams.

- The standard model of particle physics has several extensions, none of which has yet received experimental confirmation. At the Planck scale, a grand unified theory of all forces is postulated, which allows for decay of protons on very long time scales. Supersymmetry is another extension of the standard model, which postulates the existence of a supersymmetric partner for every known fermion and boson. Other particle physics theories are string theories, which hold that the dimensionality of space-time is 10 or 11, with the other dimensions compacted to very small scales.

- Elementary quarks and antiquarks can combine to form color singlets, which are particles that can be observed in isolation. A quark and an antiquark can form a meson (pion, kaon, etc.). Three quarks can form a baryon (proton, neutron, delta baryon, lambda baryon, etc.). The only stable baryon is the proton, and none of the mesons is stable. Lifetimes of the unstable particles vary from 10^{-23} s to 15 min.

- Because gluons carry color charge, they can interact with other gluons. This leads to the confinement of quarks. The theory of the interaction of quarks and gluons, quantum chromodynamics, can only be analyzed in the realm of large momentum transfer and short distances. There the elementary particles show asymptotic freedom. For low-momentum transfers and large distances, the only solution method is lattice-QCD simulated on a computer.

- The history of the early universe at fractions of a second after the Big Bang is described by particle physics. The early inflationary phase left traces in the isotropy of the cosmic microwave background radiation, which originated 300,000 years after the Big Bang at a temperature of 3000 K and cooled down to the presently observed 2.725 K due to the expansion of the universe. The quark-gluon-plasma phase transition of the early universe can be probed in the laboratory with relativistic heavy ion collisions. The primordial fraction of 23% helium in the universe can be explained from the neutron-proton mass difference, which fixes the ratio of proton and neutron numbers to

$$n_n/n_p = e^{(m_n c^2 - m_p c^2)/k_B T}.$$

ANSWERS TO SELF-TEST OPPORTUNITIES

39.1 $c \rightarrow s + \mu^+ + \nu_\mu$ and $t \rightarrow b + \tau^+ + \nu_\tau$

39.2 The Planck energy is most straightforwardly obtained by multiplying the Planck mass with two powers of the speed of light:

$$E_P = m_P c^2 = \sqrt{\frac{\hbar c^5}{G}} = 1.96 \cdot 10^9\ \text{J} = 1.22 \cdot 10^{19}\ \text{GeV}.$$

39.3 Most of these are listed in Table 39.4, with the exception of Σ^* and Ξ^*, which have the same constituent quark composition as Σ and Ξ, respectively.

PROBLEM-SOLVING GUIDELINES

1. The starting point for most calculations involving elementary particles is relativistic kinematics. In particular, $E = mc^2$ is a very useful relationship to remember.

2. Conservation laws for energy and momentum apply to all physical processes. In addition, remember the conservation laws of baryon number, charge, and angular momentum (spin) when solving particle physics problems.

3. Understanding the quark composition of mesons and baryons is important in solving many problems involving decays of these particles.

MULTIPLE-CHOICE QUESTIONS

39.1 According to the text, the de Broglie wavelength, λ, of a 5-MeV alpha particle is 6.4 fm, and the closest distance, $r_{\min}$, to the gold nucleus this alpha particle can get is 45.5 fm (calculated in Example 39.1). Based on the fact that $\lambda \ll r_{\min}$, one can conclude that, for this Rutherford scattering experiment, it is adequate to treat the alpha particle as a

a) particle.   b) wave.

39.2 Which of the following is a composite particle? (select all that apply)

a) electron   c) proton
b) neutrino   d) muon

39.3 Which of the following formed latest in the universe?

a) quarks   d) helium nuclei
b) protons and neutrons e) gluons
c) hydrogen atoms

39.4 An exchange particle for the weak force is the

a) photon.   c) W boson.   e) gluon.
b) meson.   d) graviton.

39.5 Which of the following particles does not have an integer spin?

a) photon   c) ω meson
b) π meson   d) ν_e lepton

39.6 At about what kinetic energy is the length scale probed by an α particle no longer calculated by the classical formula but rather by the relativistic formula?

a) 0.3 GeV   c) 3 GeV
b) 0.03 GeV   d) 30 GeV

39.7 Which of the following experiments proved the existence of the nucleus?

a) the photoelectric effect
b) the Millikan oil-drop experiment
c) the Rutherford scattering experiment
d) the Stern-Gerlach experiment

CONCEPTUAL QUESTIONS

39.8 Which of the following reactions cannot occur, and why?

a) $p \rightarrow \pi^+ + \pi^0$
b) $p\pi^0 \rightarrow n + e^+$
c) $\Lambda^0(1116) \rightarrow p + K^-\pi^+$
d) $\Lambda^0(1450) \rightarrow p + K^- + \pi^+$

39.9 Would neutron scattering or electromagnetic wave scattering (using X-rays or light) be more appropriate for investigating the scattering cross section of an atom as a whole? Which would be more appropriate for investigating the cross section of a nucleus of an atom? Which result will depend on Z, the atomic number?

39.10 Consider a hypothetical force mediated by the exchange of bosons that have the same mass as protons. Approximately what would be the maximum range of such a force? You may assume that the total energy of these particles is simply the rest-mass energy and that they travel close to the speed of light. If you do not make these assumptions and instead use the relativistic expression for total energy, what happens to your estimate of the maximum range of the force?

39.11 Looking at Table 39.3, do the constituent quarks uniquely define the type of meson?

39.12 A free neutron decays into a proton and an electron (and an antineutrino). A free proton has never been observed to decay. Why then do we consider the neutron to be as "fundamental" (at the nuclear level) a particle as the proton? Why do we not consider a neutron to be a proton-electron composite?

39.13 In a positron annihilation experiment, positrons are directed toward a metal. What are we likely to observe in such an experiment, and how might it provide information about the momentum of electrons in the metal?

39.14 If the energy of the virtual photon mediating an electron-proton scattering, $e^- + p \rightarrow e^- + p$, is E, what is the range of this electromagnetic interaction in terms of E?

39.15 Describe the physical processes that the following Feynman diagrams represent:

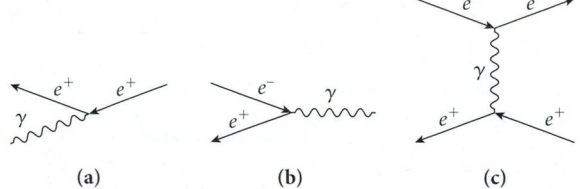

(a)   (b)   (c)

39.16 Figure 39.34 shows a Feynman diagram for the fundamental process involved in the decay of a free neutron: One of the neutron's down quarks converts to an up quark, emitting a virtual W^- boson, which decays into an electron and an anti-electron-neutrino (the only decay energetically possible). Sketch the basic Feynman diagram for the fundamental process involved in each of the following decays:

a) $\mu^- \rightarrow e^- + \nu_\mu + \bar{\nu}_e$
b) $\tau^- \rightarrow \pi^- + \nu_\tau$
c) $\Delta^{++} \rightarrow p + \pi^+$
d) $K^+ \rightarrow \mu^+ + \nu_\mu$
e) $\Lambda^0 \rightarrow p + \pi^-$

39.17 Does the decay process $n \rightarrow p + \pi^-$ violate any conservation rules?

39.18 Consider the decay process $\pi^+ \rightarrow \mu^+ + \nu_\mu + \nu_e$. Can this decay occur?

39.19 Can the reaction $\pi^0 + n \rightarrow K^- + \Sigma^+$ occur?

39.20 How do we know for certain that the scattering process $e^+ + \nu_\mu \to e^+ + \nu_\mu$ proceeds through an intermediate Z boson and cannot proceed through an intermediate charged W boson, while both options are possible for $e^+ + \nu_e \to e^+ + \nu_e$?

39.21 What baryons have the quark composition uds? What is the mass of these baryons?

39.22 In the following Feynman diagram for proton-neutron scattering, what is the virtual particle?

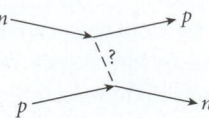

EXERCISES

A blue problem number indicates a worked-out solution is available in the Student Solutions Manual. One • and two •• indicate increasing level of problem difficulty.

Section 39.2

39.23 A 4.50-MeV alpha particle is incident on a platinum nucleus ($Z = 78$). What is the minimum distance of approach, r_{min}?

39.24 A 6.50-MeV alpha particle is incident on a lead nucleus. Because of the Coulomb force between them, the alpha particle will approach the nucleus to a minimum distance, r_{min}.

a) Determine r_{min}.

b) If the kinetic energy of the alpha particle is increased, will the particle's distance of approach increase, decrease, or remain the same? Explain.

39.25 A 6.50-MeV alpha particle scatters at a 60.0° angle off a lead nucleus. Determine the differential cross section of the alpha particle.

39.26 Protons with a kinetic energy of 2.00 MeV scatter off gold nuclei in a foil target. Each gold nucleus contains 79 protons. If both the incoming protons and the gold nuclei can be treated as point objects, what is the differential cross section that will cause the protons to scatter off the gold nuclei at an angle of 30.0° from their initial trajectory?

•**39.27** The de Broglie wavelength, λ, of a 5.00-MeV alpha particle is 6.40 fm, and the closest distance, r_{min}, to the gold nucleus this alpha particle can get is 45.5 fm (calculated in Example 39.1). How does the ratio r_{min}/λ vary with the kinetic energy of the alpha particle?

•**39.28** An experiment similar to the Geiger-Marsden experiment is done by bombarding a 1.00-μm-thick gold foil with 8.00-MeV alpha particles. Calculate the fraction of particles scattered at an angle

a) between 5.00° and 6.00° and

b) between 30.0° and 31.0°.

(The atomic mass number of gold is 197, and its density is 19.3 g/cm³.)

•**39.29** The differential cross section that will cause particles to scatter at an angle 55° off a target is $4.0 \cdot 10^{-18}$ m²/sr. A detector with an area of 1.0 cm² is placed 1.0 m away from the target in order to detect particles that have been scattered at 55°. If $3.0 \cdot 10^{17}$ particles hit the 1.0-mm²-area target every second, how many will strike the detector every second?

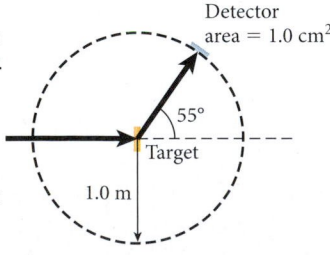

Detector area = 1.0 cm²

55°

Target

1.0 m

••**39.30** Some particle detectors measure the total number of particles integrated over part of a sphere of radius R, where the target is at the center of the sphere. Assuming symmetry about the axis of the incoming particle beam, use the Rutherford scattering formula to obtain the total number of particles detected in an interval of width $d\theta$ as a function of the scattering angle, θ.

••**39.31** Evaluate the form factor and the differential cross section, $d\sigma/d\Omega$, for a beam of electrons scattering off a *uniform-density* charged sphere of total charge Ze and radius R. Describe the scattering pattern.

Section 39.3

39.32 A proton is made of two up quarks and a down quark (uud). Calculate its charge.

39.33 Use the fact that the observed magnetic moment of a proton is $1.4 \cdot 10^{-26}$ A m² to estimate the speed of its quarks. For this estimate, assume that the quarks move in circular orbits of radius 0.80 fm and that they all move at the same speed and direction. Ignore any relativistic effects.

39.34 Determine the approximate probing distance of a photon with an energy of 2.0 keV.

39.35 Draw a Feynman diagram for an electron-proton scattering, $e^- + p \to e^- + p$, mediated by photon exchange.

39.36 Based on the information in Table 39.2, what is the approximate upper bound on the range of a reaction mediated by the Higgs boson?

39.37 Draw Feynman diagrams for the following phenomena:

a) protons scattering off each other

b) a neutron beta decays to a proton: $n \to p + e^- + \bar{\nu}_e$

•**39.38** A proton and a neutron interact via the strong nuclear force. Their interaction is mediated by a meson, much like the interaction between charged particles is mediated by photons—the particles of the electromagnetic field.

a) Perform a rough estimate of the mass of the meson from the uncertainty principle and the known dimensions of a nucleus ($\sim 10^{-15}$ m). Assume that the meson travels at relativistic speed.

b) Use a line of reasoning similar to that in part (a) to prove that the theoretically expected rest mass of the photon is zero.

Section 39.5

39.39 How many fundamental fermions are there in a carbon dioxide molecule (CO_2)?

39.40 Suppose a neutral pion at rest decays into two identical photons.

a) What is the energy of each photon?

b) What is the frequency of each photon?

c) To what part of the electromagnetic spectrum do the photons correspond?

•**39.41** Draw a quark-level Feynman diagram for the decay of a neutral kaon into two charged pions: $K^0 \to \pi^+ + \pi^-$.

Section 39.6

39.42 During the radiation-dominated era of the universe, the temperature was falling gradually according to equation 39.17. Using Stefan's Law, find the time dependence of background-radiation intensity during that era.

39.43 Use equation 39.17 to estimate the age of the universe when protons and neutrons began to form.

39.44 Three hundred thousand years after the Big Bang, the average temperature of the universe was about 3000 K.

a) At what wavelength would the blackbody spectrum peak for this temperature?

b) In what portion of the electromagnetic spectrum is this wavelength found?

•**39.45** At about 10^{-6} s after the Big Bang, the universe had cooled to a temperature of approximately 10^{13} K.

a) Calculate the thermal energy $k_B T$ of the universe at that temperature.

b) Explain what happened to most of the hadrons—protons and neutrons—at that time.

c) Explain what happened to electrons and positrons in terms of temperature and time.

•**39.46** Three hundred thousand years after the Big Bang, the temperature of the universe was 3000 K. Because of expansion, the temperature of the universe is now 2.75 K. Modeling the universe as an ideal gas and assuming that the expansion is adiabatic, calculate how much the volume of the universe has changed. If the process is irreversible, estimate the change in the entropy of the universe based on the change in volume.

•**39.47** The fundamental observation underlying the Big Bang theory of cosmology is Edwin Hubble's 1929 discovery that the arrangement of galaxies throughout space is expanding. Like the photons of the cosmic microwave background, the light from distant galaxies is stretched to longer wavelengths by the expansion of the universe. This is *not* a Doppler shift: Except for their local motions around each other, the galaxies are essentially at rest in space; it is space itself that expands. The ratio of the wavelength of light received at Earth from a galaxy, λ_{rec}, to its wavelength at emission, λ_{emit}, is equal to the ratio of the scale factor (radius of curvature) a of the universe at reception to its value at emission. The redshift, z, of the light—which is what Hubble could measure—is defined by $1 + z = \lambda_{rec}/\lambda_{emit} = a_{rec}/a_{emit}$.

a) *Hubble's Law* states that the redshift, z, of light from a galaxy is proportional to the galaxy's distance from Earth (for reasonably nearby galaxies): $z \cong c^{-1}H\Delta s$, where c is the vacuum speed of light, H is the *Hubble constant*, and Δs is the distance of the galaxy from Earth. Derive this law from the relationships described in the problem statement, and determine the Hubble constant in terms of the scale-factor function $a(t)$.

b) If the Hubble constant currently has the value $H_0 = 72$ (km/s)/Mpc, how far away is a galaxy whose light has the redshift $z = 0.10$? (The megaparsec (Mpc) is a unit of length equal to $3.26 \cdot 10^6$ light-years. For comparison, the Great Nebula in Andromeda is approximately 0.60 Mpc from Earth.)

Additional Exercises

39.48 What is the minimum energy of a photon capable of producing an electron-positron pair? What is the wavelength of this photon?

39.49 a) Calculate the kinetic energy of a neutron that has a de Broglie wavelength of 0.15 nm. Compare this with the energy of an X-ray photon that has the same wavelength.

b) Comment on how this energy difference is relevant to using neutrons or X-rays for investigating biological samples.

39.50 A photon can interact with matter by producing a proton-antiproton pair. What is the minimum energy the photon must have?

39.51 Suppose you had been doing an experiment to probe structure on a scale for which you needed electrons with 100. eV of kinetic energy. Then a neutron beam became available for the experiment. What energy would the neutrons need to have to give you the same resolution?

39.52 What is the de Broglie wavelength of an alpha particle that has a kinetic energy of 100. MeV? According to Figure 39.13, how does this wavelength compare to the size of structure that can be probed with this alpha particle?

39.53 One of the elementary bosons that can mediate electroweak interactions is the Z^0 boson, having the mass of 91.1876 GeV/c^2. Find the order of magnitude of the range of the electroweak interaction.

39.54 What are the wavelengths of the two photons produced when a proton and an antiproton at rest annihilate?

39.55 Estimate the cross section of a Λ^0 particle decay (into $p + \pi^-$, $n + \pi^0$) if the time it takes for this electroweak interaction to occur is $\sim 10^{-10}$ s.

39.56 Determine the classical differential cross section for Rutherford scattering of alpha particles of energy 5.00 MeV projected at uranium atoms and scattered at an angle of 35.0° from the initial trajectory. Assume that both the target and the projectile atoms are pointlike.

39.57 The Geiger-Marsden experiment successfully demonstrated the existence of the nucleus and put limits on its size using the scattering of alpha particles from gold foils. Assume that the alpha particles were fired with a speed about 5.00% of the speed of light.

a) Derive the upper bound of the radius of the nucleus in terms of the speed of the alpha particle that is scattered in the backward direction.

b) Calculate the approximate radius of the gold nucleus using the result from part (a).

•**39.58** An electron-positron pair, traveling toward each other with a speed of 0.99c with respect to their center of mass, collide and annihilate according to $e^- + e^+ \to \gamma + \gamma$. Assuming that the observer is at rest with respect to the center of mass of the electron-positron pair, what is the wavelength of the emitted photons?

•**39.59** Electron and positron beams are collided, and pairs of tau leptons are produced. If the angular distribution of the tau leptons varies as $(1 + \cos^2\theta)$, what fraction of the tau lepton pairs will be captured in a detector that covers only the angles from 60° to 120°?

•**39.60** On July 4, 2012, the discovery of the Higgs boson at the Large Hadron Collider was announced. During the data-taking run, the LHC reached a peak luminosity of $4.00 \cdot 10^{33}$ cm^{-2} s^{-1} (this means that in an area of 1 square centimeter, $4.00 \cdot 10^{33}$ protons collided every second). Assume that the cross section for the production of the Higgs boson in these proton-proton collisions is 1.00 pb (picobarn). If the LHC accelerator ran without interruption for 1.00 yr at this luminosity, how many Higgs bosons would be produced?

••**39.61** Evaluate the form factor and the differential cross section, $d\sigma/d\Omega$, for a beam of electrons scattering off a *thin spherical shell* of total charge Ze and radius a. Could this scattering experiment distinguish between thin-shell and solid-sphere charge distributions? Explain.

MULTI-VERSION EXERCISES

39.62 A neutrino beam with $E = 337$ GeV is passed through a 68.5-cm-thick slab of aluminum-27 (with 27 nucleons in each nucleus). What fraction of the neutrinos will scatter off a nucleon if the cross section is given by $\sigma(E) = (0.68 \cdot 10^{-38}$ cm^2 GeV$^{-1})E$? (Aluminum has a density of 2.77 g/cm^3.)

39.63 A neutrino beam with $E = 143$ GeV is passed through a slab of aluminum-27 (with 27 nucleons in each nucleus). The probability that a neutrino in the beam will scatter off a nucleon in the aluminum slab is $4.19 \cdot 10^{-12}$. The scattering cross section is given by $\sigma(E) = (0.68 \cdot 10^{-38}$ cm^2 GeV$^{-1})E$, and aluminum has a density of 2.77 g/cm^3. How thick is the slab?

39.64 A high-energy neutrino beam is passed through a slab of aluminum-27 (with 27 nucleons in each nucleus) of thickness 71.1 cm. The probability that a neutrino in the beam will scatter off a nucleon in the aluminum slab is $6.00 \cdot 10^{-12}$. The scattering cross section is given by $\sigma(E) = (0.68 \cdot 10^{-38}$ cm^2 GeV$^{-1})E$, and aluminum has a density of 2.77 g/cm^3. What is the energy (in GeV) of the neutrino beam?

39.65 A Geiger-Marsden experiment, in which alpha particles are scattered off a thin gold film, yields an intensity of $I(94.9°) = 853$ counts/s at a scattering angle of 94.9°±0.7°. What is the intensity (in counts/s) at a scattering angle of 60.5°±0.7° if the scattering obeys the Rutherford formula?

39.66 A Geiger-Marsden experiment, in which alpha particles are scattered off a thin gold film, yields an intensity of $I(95.1°) = 1129$ counts/s at a scattering angle of 95.1°±0.4°. At a second scattering angle, the intensity is measured to be 4840 counts/s. Assuming that the scattering obeys the Rutherford formula, what is that second angle (in degrees, to the same uncertainty)?

39.67 A Geiger-Marsden experiment, in which alpha particles are scattered off a thin gold film, is set up with two detectors at $\theta_1 = 85.1°±0.9°$ and $\theta_2 = 62.9°±0.9°$. Assuming that the scattering obeys the Rutherford formula, what is the ratio of the measured intensities, I_1/I_2?

Nuclear Physics

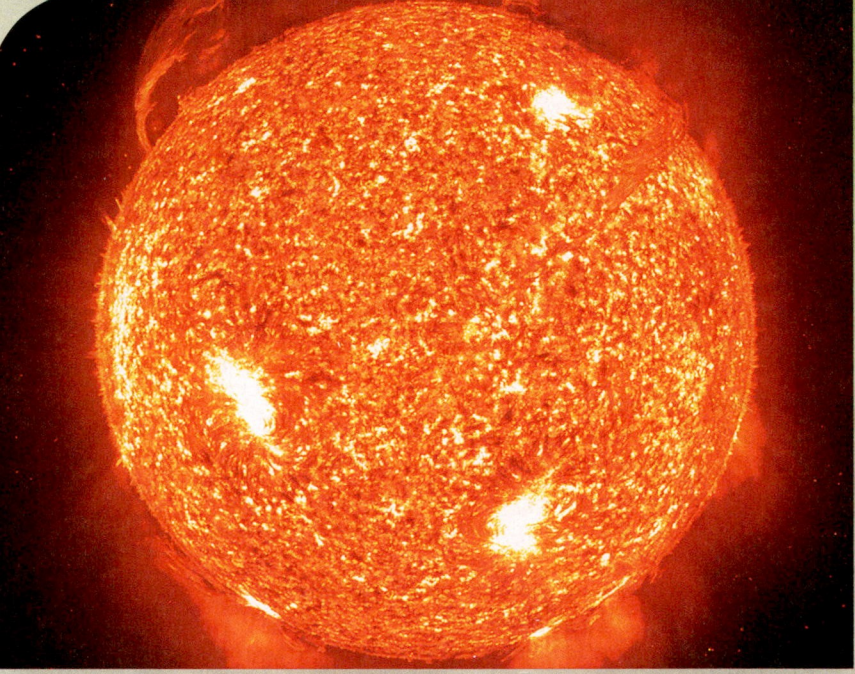

FIGURE **40.1** Our Sun, the giant nuclear fusion reactor that provides almost all of Earth's energy resources, as viewed by the Extreme Ultraviolet Imaging Telescope on the *SOHO* spacecraft using UV light of wavelength 30.4 nm.

Almost all of the energy resources used on Earth trace back to nuclear reactions. The overwhelming majority of these resources are due to our Sun (Figure 40.1), which at its core is a nuclear fusion reactor. This dependence on nuclear reactions makes it important to try to understand them better.

Chapter 39 explained that most of the mass of an atom resides in a small central region called the *nucleus*. This chapter explores the physics of the nucleus, which offers unique insights into the behavior of matter at its most basic level.

We will examine several models that help explain the properties of the nucleus, particularly those that explain the stability of some nuclei and the decay modes of others. Most nuclei are unstable, or radioactive, which has led to many important applications of nuclear physics, including nuclear power, nuclear weapons, and nuclear medicine. In addition, studies of nuclear properties have contributed to our understanding of astronomy and of the origin of the chemical elements in the life cycles of stars.

WHAT WE WILL LEARN

- The atomic nucleus is composed of nearly equal numbers of protons and neutrons (nucleons), densely packed so that the nuclear volume increases approximately linearly with the mass number (which equals the number of nucleons).

- Nucleons interact primarily via the strong interaction, which has a short range, on the order of 10^{-15} m.

- Only 251 stable isotopes are known to exist. More than 2400 unstable isotopes have been discovered, with several thousand more predicted to exist.

- Nuclear masses can be measured very precisely, typically with the precision of 1 part in 100 million. Knowing the masses, we can calculate the binding energies of the isotopes, as well as the Q-values of nuclear reactions. The Q-value determines whether or not a reaction is allowed by energy conservation and is used to find how much energy the reaction releases.

- Several different models of the nucleus can be constructed. One model is based on the assumption that nuclear matter behaves like a liquid drop. This liquid-drop model explains the systematics of the nuclear binding energy. Another model proposes a quantum gas of particles trapped inside the nuclear walls. A third model takes into account the angular momentum quantum numbers and yields a picture of nuclear shells not unlike the electron shells in the atom.

- Unstable isotopes and their excited states can decay in three primary ways. Alpha decay is the emission of a helium-4 nucleus. Beta decay is a weak interaction process and results in either the emission of an anti-neutrino and an electron, the emission of a neutrino and a positron, or the capture of an electron by the nucleus along with the emission of a neutrino. Gamma decay is the emission of a high-energy photon.

- Energy can be released from nuclear reactions in two primary ways—fission and fusion. In fission, a massive nucleus splits into two less massive nuclei and a few neutrons. In fusion, two light nuclei merge into a heavier one.

- Neutron-induced nuclear fission and the ensuing chain reactions can be used for nuclear power plants, but also for nuclear weapons.

- Nuclear fusion powers most stars and is responsible for the light emitted by them as well as the creation of the elements up to iron and nickel.

- Elements heavier than iron are predominantly produced in supernova explosions.

- Nuclear physics has many applications in medicine, both in diagnostics and in the radiation treatment of cancers.

40.1 Nuclear Properties

Isotopes

The atomic nucleus consists of two kinds of **nucleons**: protons and neutrons. These nucleons are held together by the strong interaction. However, not all combinations of protons and neutrons are possible, because of the limitations imposed by the strong interaction, the Coulomb interaction, and quantum mechanics. The number of protons, Z, inside a nucleus determines the element that is formed. Z is the *charge number*, which is sometimes called the *atomic number*. For atoms of a given element, different numbers of neutrons, N, are possible. Nuclei of the same element with different numbers of neutrons are called **isotopes**. The total number of protons and neutrons combined is called the **mass number**, A:

$$A = Z + N. \tag{40.1}$$

Conventionally, a given nucleus is denoted by the symbol that represents its atom in the periodic table, with a superscript for the mass number of the isotope on the left and a subscript for the charge number on the left, and sometimes a subscript for the number of neutrons on the right. For example, the nucleus formed by 8 protons and 9 neutrons is written as $^{17}_{8}O_9$, and the nucleus formed by 92 protons and 146 neutrons is written as $^{238}_{92}U_{146}$. This is the most complete notation for isotopes. However, most of the time the subscript on the right is omitted, because the neutron number is simply $N = A - Z$ and thus can be computed easily from the superscript and subscript on the left. Most practicing nuclear physicists also omit the subscript for the charge number; they simply assume we know that uranium has 92 protons and oxygen has 8 protons. Thus, the above two isotopes are written in many nuclear physics textbooks as ^{17}O (pronounced "oxygen seventeen") and ^{238}U (pronounced "uranium two-thirty-eight"). This book will show the subscript for the charge

FIGURE 40.2 Stable isotopes, marked with blue squares that indicate their neutron and proton numbers. The gray line marks the $N = Z$ line.

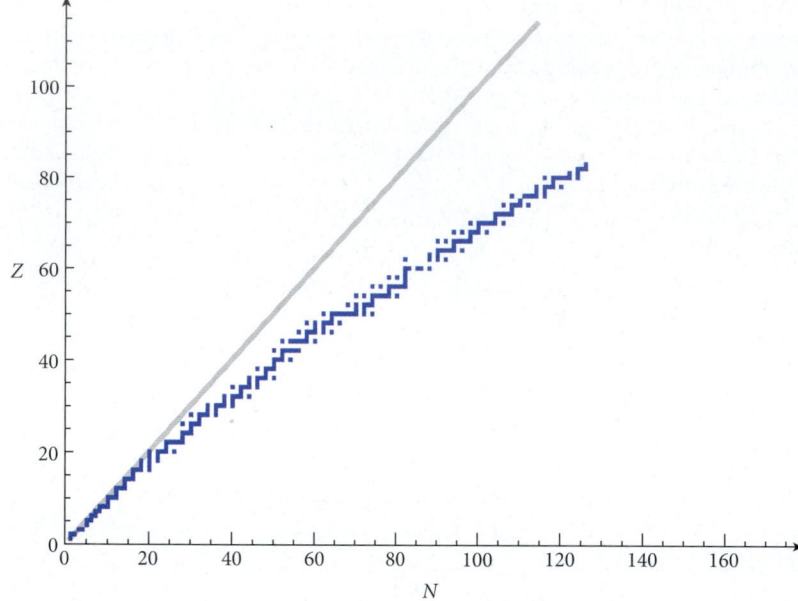

number, but not that for the neutron number. The above isotopes of oxygen and uranium are thus written as $^{17}_{8}\text{O}$ and $^{238}_{92}\text{U}$.

The only two isotopes that have historically received special names are the hydrogen isotopes deuterium ($^{2}_{1}\text{H} \equiv d$, an atom made up of a nucleus with a proton and a neutron, and a single electron) and tritium ($^{3}_{1}\text{H} \equiv t$, an atom made up of a nucleus with a proton and two neutrons, and a single electron).

What kinds of isotopes are found in nature? Answering this question will occupy us several times during this chapter. The overwhelming majority of isotopes decay over time. Only 251 stable isotopes are known, and they are shown in Figure 40.2, where each of them is marked by a blue square that indicates its neutron and proton numbers.

Right away, we can see that there is no stable isotope with $Z > 83$ (bismuth has $Z = 83$). Also, note that there are no stable isotopes for $Z = 43$ (technetium) or for $Z = 61$ (promethium). Further, it is apparent that only a few stable isotopes occur for each Z value, and they are located along a narrow region in this plot, known as the *valley of stability*. For small values of N, this valley follows the $N = Z$ line, which is shown as a diagonal gray line in the plot. However, at $N \approx 20$, the valley of stability begins to veer off this line toward the neutron-rich ($N > Z$) side. This effect is due to the Coulomb interaction among the protons, which also limits the size of the largest nucleus.

Nuclear Interactions

It is not convenient to describe the strong interaction between nucleons as being based on gluon exchange as proposed in QCD, described in Chapter 39. Instead, it is more efficient to describe effective interactions as being based on the exchange of color-singlets between the nucleons (which are also color-singlets). Since the least massive color-singlets are mesons, and the least massive meson is the pion, characterizing the nucleon-nucleon interaction in terms of pion exchange has been very successful. Historically, pion exchange potentials were formulated before the strong interaction was understood in terms of quarks and gluons, but that understanding explains the success of those potentials.

The Japanese physicist Hideki Yukawa (1907–1981) invented the pion exchange potential theory in 1935 and won the 1949 Nobel Prize in Physics for this achievement. The Yukawa potential is conventionally written in the form

$$U(r) = -g^2 \frac{e^{-r/R_\pi}}{r}, \tag{40.2}$$

where g is a real number and is the effective coupling constant in the same way that ke^2 is the coupling constant for the Coulomb potential of the electrostatic interaction. Here R_π is the range of the potential and is given in terms of the pion's mass as

$$R_\pi = \frac{\hbar}{m_\pi c} = \frac{\hbar c}{m_\pi c^2} = \frac{197.3\ \text{MeV fm}}{139.6\ \text{MeV}} = 1.413\ \text{fm}.$$

The derivative of the one-pion exchange potential with respect to r is greater than zero, so the corresponding force, $F(r) = -dU(r)/dr$, is attractive at all distances r.

The one-pion exchange potential is a very good approximation to the nucleon-nucleon potential at large separations. However, at shorter distances, two- and three-pion exchanges dominate nucleon-nucleon interactions. The effective nucleon-nucleon potential is strongly repulsive at short distances, thus preventing nucleons from penetrating each other. Schematically, the nucleon-nucleon potential is plotted in Figure 40.3. It varies depending on the spin and isospin projections (proton or neutron) of the two nucleons involved in the interaction, but has the general shape shown in the plot, with the minimum located at approximately 1 fm.

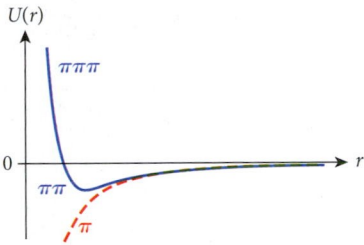

FIGURE 40.3 Schematic plot of the nucleon-nucleon potential (blue). At the point labeled $\pi\pi$, two-pion exchange begins to dominate. At the point labeled $\pi\pi\pi$, three-pion exchange becomes important. The one-pion exchange potential is shown in red.

Nuclear Radius and Nuclear Density

Because the strong nuclear interaction is very short range, what is most important for nuclear physics is the nearest-neighbor interactions between nucleons. The repulsive nature of the potential at close distances prevents the nucleons from penetrating each other, so it has become common to visualize the nucleus as a roughly spherically shaped and densely packed collection of nucleons. This means that the volume of the nucleus should be proportional to the mass number, A. Since the volume of a sphere is proportional to the third power of the radius, the third power of the nuclear radius is proportional to its mass number, or, alternatively:

$$R(A) = R_0 A^{1/3}, \tag{40.3}$$

where the constant $R_0 = 1.12$ fm has been determined experimentally.

EXAMPLE 40.1 Nuclear Density

PROBLEM

What is the nuclear matter density, that is, the mass density inside an atomic nucleus?

SOLUTION

Since the nucleus can be approximated by a sphere with a radius given by equation 40.3, we can calculate its volume. We know the number of nucleons inside the sphere and their mass, so we can find the nuclear density as the ratio of the mass to the volume.

The volume of a nucleus of mass number A is

$$V = \frac{4\pi R(A)^3}{3} = \frac{4\pi R_0^3 A}{3} = (5.88\ \text{fm}^3)A.$$

This tells us that on average a nucleon occupies approximately 5.9 fm³ of space inside a nucleus. We can now find the number density of nucleons inside the nucleus:

$$n = \frac{A}{V} = \frac{A}{(5.88\ \text{fm}^3)A} = 0.170\ \text{fm}^{-3}.$$

Thus, an atomic nucleus has 0.170 nucleons per cubic femtometer. The mass of a nucleon is approximately $1.67 \cdot 10^{-27}$ kg. Multiplying this mass of a single nucleon by the number density of nucleons gives us the mass density of a nucleus:

$$\rho = m_{\text{nucleon}} n = m_{\text{nucleon}} \frac{A}{V} = (1.67 \cdot 10^{-27}\ \text{kg})(0.170\ (10^{-15}\ \text{m})^{-3}) = 2.84 \cdot 10^{17}\ \text{kg/m}^3.$$

For comparison, the density of liquid water is 10^3 kg/m³; thus, nuclear matter density is approximately 280 trillion times higher than the density of liquid water.

FIGURE 40.4 (a) Nuclear density profile as a function of the radial coordinate. (b) Fermi function for three different nuclear radii.

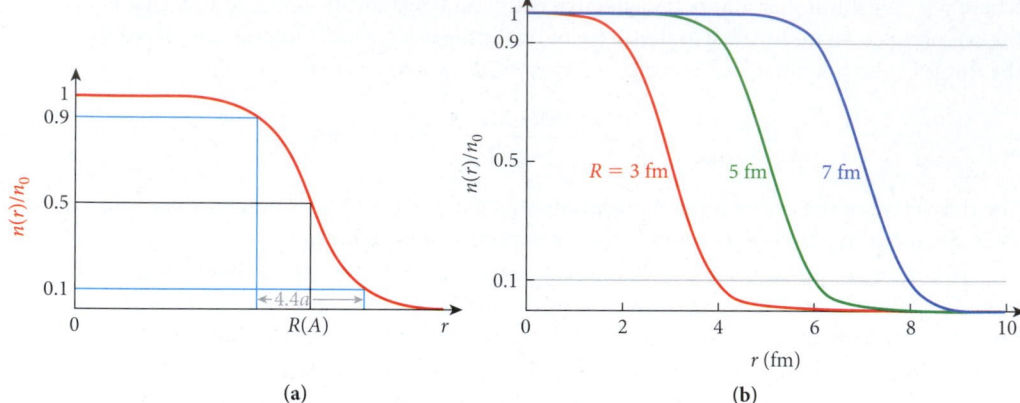

(a)   (b)

Electron-scattering experiments of the kind described in Chapter 39 have established that the density is approximately constant in the interior of a heavy nucleus, and it falls off gradually near the surface. The dependence of the nuclear density on the radial coordinate r can be described by the *Fermi function*:

$$n(r) = \frac{n_0}{1 + e^{(r-R(A))/a}},$$ (40.4)

where $R(A)$ is given by equation 40.3, and the constant a has the value of 0.54 fm (Figure 40.4). The distance over which the density falls from 90% of its central value to 10% of the central value is conventionally defined as the nuclear surface thickness, t. By using equation 40.4, we can show that $t \approx 4.4a$ (see Solved Problem 40.1). Also, with $a = 0.54$ fm, the nuclear surface thickness for large nuclei is found to be approximately $t = 2.4$ fm.

SOLVED PROBLEM 40.1 Nuclear Surface

PROBLEM
Derive a relationship between the nuclear surface thickness, t, and the parameters of the Fermi function (equation 40.4). What are the distances over which the nuclear density falls from 80% to 20% and from 70% to 30% of its central value?

SOLUTION
THINK Equation 40.4 gives the Fermi function for the nuclear density, $n(r)$, as a function of the radial coordinate r. It contains the three parameters n_0, $R(A)$, and a:

$$n(r) = \frac{n_0}{1 + e^{(r-R(A))/a}}.$$

We know that $a = 0.54$ fm, $n_0 = 0.17$ fm^{-3}, and $R(A) = R_0 A^{1/3}$, with $R_0 = 1.12$ fm. Since n_0 is a multiplicative constant in the Fermi function, it cannot have any influence on the thickness of the nuclear surface. However, it is not quite as easy to tell whether $R(A)$ or a determines the surface thickness.

SKETCH Figure 40.4b shows the Fermi function for three different values of the nuclear radius, R. It is clear from this sketch that the nuclear surface thickness, t, is independent of R. Thus, the nuclear surface thickness can only be a function of the parameter a.

RESEARCH Our starting point is the Fermi function, which gives us the nuclear density as a function of the radial coordinate:

$$\frac{n(r)}{n_0} = \frac{1}{1 + e^{(r-R)/a}}.$$ (i)

Here we have simplified the notation by writing R instead of $R(A)$, recognizing from our sketch that the particular value of the nuclear radius will have no influence on the nuclear surface thickness. We need to invert this function to obtain the radial coordinate as a function of the density, $r(n/n_0)$. Then the thickness is defined as

$$t = r(10\%) - r(90\%).$$ (ii)

SIMPLIFY We solve equation (i) for r by taking the inverse of both sides and then subtracting 1 from both sides:

$$\frac{n_0}{n} = 1 + e^{(r-R)/a} \Rightarrow \frac{n_0}{n} - 1 = e^{(r-R)/a}.$$

Now we take the natural log and find

$$\ln\left(\frac{n_0}{n} - 1\right) = (r - R)/a.$$

Multiplication of both sides by a and addition of R then results in

$$r = R + a \ln\left(\frac{n_0}{n} - 1\right),$$

which expresses the radial coordinate as a function of the density, $r(n/n_0)$, as desired. Using this expression with $n = 0.1n_0$ and $n = 0.9n_0$ and equation (ii), we can obtain the surface thickness:

$$t = r(0.1) - r(0.9) = \left[R + a \ln\left(\frac{1}{0.1} - 1\right)\right] - \left[R + a \ln\left(\frac{1}{0.9} - 1\right)\right]$$

$$t = a\left[\ln\left(\frac{1}{0.1} - 1\right) - \ln\left(\frac{1}{0.9} - 1\right)\right] = a\left[\ln 9 - \ln 0.\overline{1}\right] = 2a \ln 9$$

$$t = 4.39445a.$$

CALCULATE We have obtained the desired relationship between the parameter a of the Fermi function and the nuclear surface thickness: $t \approx 4.4a$. Using $a = 0.54$ fm, we find that the surface thickness is 2.4 fm. What remains to be calculated are the thicknesses of the regions over which the density falls from 80% to 20% and from 70% to 30% of its central value:

$$r(0.2) - r(0.8) = (0.54 \text{ fm})\left[\ln\left(\frac{1}{0.2} - 1\right) - \ln\left(\frac{1}{0.8} - 1\right)\right] = 1.4972 \text{ fm}$$

$$r(0.3) - r(0.7) = (0.54 \text{ fm})\left[\ln\left(\frac{1}{0.3} - 1\right) - \ln\left(\frac{1}{0.7} - 1\right)\right] = 0.91508 \text{ fm}.$$

ROUND Since the constant $a = 0.54$ fm is given to only two significant figures, we round our results to

$$r(0.2) - r(0.8) = 1.5 \text{ fm}$$
$$r(0.3) - r(0.7) = 0.92 \text{ fm}.$$

DOUBLE-CHECK As you can see from Figure 40.4b, the Fermi function falls off almost linearly between $n/n_0 = 0.9$ and $n/n_0 = 0.1$. Since the density interval between $n/n_0 = 0.7$ and $n/n_0 = 0.3$ is half of that between 0.9 and 0.1, we expect that the distance over which this fall-off occurs is approximately half of the nuclear surface thickness. (Remember: The surface thickness is defined as the distance over which the density falls off from 0.9 to 0.1 of the central density!) We found that the thickness is $t = 2.4$ fm, so it is not unreasonable to find $r(0.3) - r(0.7) = 0.92$ fm.

Nuclear Lifetimes

The stable isotopes shown in Figure 40.2 are not the only isotopes that can exist. A huge number of unstable isotopes are produced through natural nuclear decays and interactions or in the laboratory. The mean lifetime of an unstable isotope, which is also called its **nuclear lifetime,** is the average time it exists prior to decaying. A quantitative discussion of lifetimes is presented in Section 40.2. Lifetimes vary over an incredible range, from greater than the age of the universe to less than a microsecond. So far, approximately 2400 unstable isotopes are known to exist in addition to the 251 stable ones. Theoretical predictions of the number of isotopes that can possibly exist range up to approximately 6000, so there are many yet to be discovered.

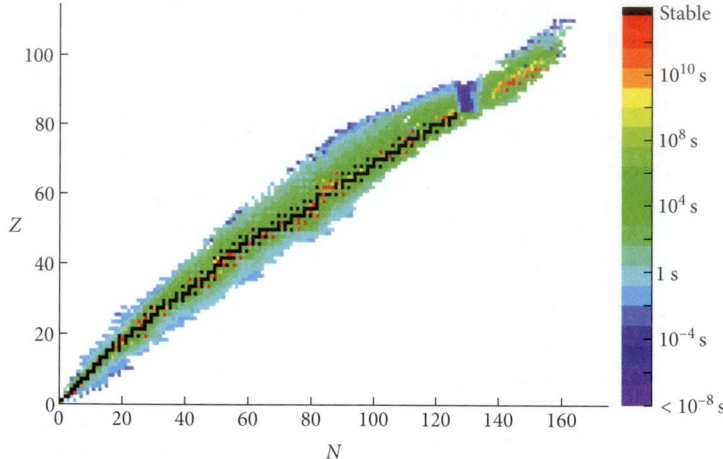

FIGURE 40.5 Measured lifetimes of the known isotopes.

Figure 40.5 shows the isotopes for which lifetimes have been measured. Each square represents an isotope, and the color of the square indicates the isotope's lifetime, according to the scale to the right of the plot. In general, lifetimes are longest for isotopes that are near the stable isotopes, with some lifetimes (shown in dark red) even exceeding the age of the universe (which is approximately $4.3 \cdot 10^{17}$ s). Moving away from the stable isotopes, lifetimes become rapidly shorter. Note in particular the very short lifetimes of all isotopes with neutron numbers around 130 and the longer lifetimes for those with larger neutron numbers and proton numbers around 90. These isotopes with longer lifetimes include the isotopes of the actinides, most notably thorium, uranium, and plutonium. Uranium has no stable isotopes, but the isotopes $^{235}_{92}$U and $^{238}_{92}$U have lifetimes of 700 million years and 4.5 billion years, respectively, and thus live long enough to be found in large quantities on Earth. The following discussion will help you understand the systematic trends of the observed lifetimes.

Nuclear and Atomic Mass

The masses of the proton and the neutron are known to great precision. The values of the mass and the mass-energy equivalent for the proton are

$$m_{\mathrm{p}} = 1.672\,621\,637(83) \cdot 10^{-27} \text{ kg}$$

$$m_{\mathrm{p}}c^2 = 1.503\,277\,359(75) \cdot 10^{-10} \text{ J} = 938.272\,013(23) \text{ MeV}$$

and the values for the neutron are

$$m_{\mathrm{n}} = 1.674\,927\,211(84) \cdot 10^{-27} \text{ kg}$$

$$m_{\mathrm{n}}c^2 = 1.505\,349\,505(75) \cdot 10^{-10} \text{ J} = 939.565\,346(23) \text{ MeV}.$$

(a)

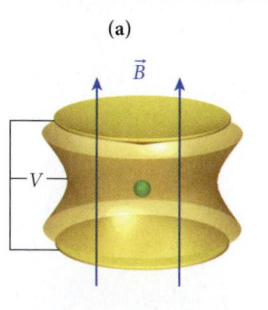

(b)

FIGURE 40.6 (a) Ion trap, which provides mass measurements by measuring an ion's cyclotron frequency. A U.S. silver dollar is shown for a scale comparison. (b) The configuration of the electrodes and magnetic field of the ion trap. The green sphere indicates the region where the ions are trapped.

The parentheses around the last two digits in these numbers indicate that the preceding two digits are uncertain by those amounts; this is a more concise notation for the standard use of $\pm$ to denote uncertainties. For example, the notation 1.672 621 637(83) is equivalent to 1.672 621 637 $\pm$ 0.000 000 083. The stated uncertainties in the proton and neutron masses mean that they are known to approximately 1 part in 10 million. Interestingly, what limits the precision of these masses is the accuracy to which Avogadro's number (N_A; see Chapter 13) is determined or, equivalently, to what precision the SI standard measure of the kilogram is known (Chapter 1).

Mass measurements for nuclei also reach this precision, even if the isotopes in question live only a few seconds or even a fraction of a second. One way to achieve such a mass measurement is to trap a single ion of a given atom in an electromagnetic trap inside a magnetic field (Figure 40.6a) and then measure its cyclotron frequency $\omega = qB/m \Rightarrow m = qB/\omega$ (see Chapter 27). Ion manipulation and storage are made possible by the electrode configurations shown in Figure 40.6b.

The charge of an ion is known precisely, because it is an integer multiple of the electron charge, and frequencies can be measured to essentially arbitrary accuracy by simply counting cycles, so the limit of this type of mass measurement depends on the accuracy with which the strength of the magnetic field, B, can be measured. The magnetic field cannot be measured to the necessary accuracy, so the cyclotron frequency of a known reference atom is measured in the same trap, thereby determining the unknown mass relative to the known one.

The reference mass usually chosen is the isotope $^{12}_{6}$C. Because this isotope consists of 12 nucleons, the **atomic mass unit (u)** is defined as exactly $\frac{1}{12}$ of the mass of a $^{12}_{6}$C atom (mass of the nucleus of $^{12}_{6}$C plus the mass of the six electrons bound to the nucleus via the

Coulomb interaction). The conversion of the atomic mass unit to kilograms and MeV/c^2 is given by

$$1\,u = 1.660538782(83) \cdot 10^{-27}\,kg = 931.494028(23)\,MeV/c^2. \tag{40.5}$$

Note that some older references use "amu" instead of "u," and in chemistry the dalton (Da) is often used instead of the atomic mass unit.

Chapter 13 introduced Avogadro's number as the number of atoms making up exactly 1 mole of a substance; a mole of an element has a mass in grams equal to the atomic mass number of the element. Since the atomic mass number of $^{12}_{6}C$ is 12, Avogadro's number of $^{12}_{6}C$ atoms have a mass of 12 g, and we can see that the relationship between the atomic mass unit and Avogadro's number is given by

$$1\,u = 1\,g/N_A.$$

If masses are stated in terms of the atomic mass unit and the above definition of this unit is used, the masses of other atoms can be measured relative to that of $^{12}_{6}C$. Such a measurement has a high precision because it is not limited by the precision to which Avogadro's number is known. For example, the proton and neutron masses can be specified to 10 significant digits (1 in 10 billion accuracy!):

$$m_p = 1.007\ 276\ 466\ 77(10)\,u$$
$$m_n = 1.008\ 664\ 915\ 97(43)\,u. \tag{40.6}$$

Why use the mass of the neutral atom of $^{12}_{6}C$, including its six electrons, as a reference value, instead of the mass of just the nucleus of $^{12}_{6}C$? This is simply for convenience, because it is very hard to strip all the electrons off an atom. For the same reason, masses of all isotopes are always listed as atomic masses and include the same number of electrons as an isotope's protons.

An ion trap like the one shown in Figure 40.6 can yield a precision of 1 part in 100 million for measurement of the mass of an atom with a lifetime of only 1 s. This precision is equivalent to measuring the mass of a convoy of 10 large 18-wheel trucks, each of mass 20 tons, to the accuracy of the weight of a single dime in the pocket of one of the truck drivers!

The mass of a nucleus is not simply the sum of the masses of the protons and neutrons contained in it. The nucleus is a bound object, and it takes energy to pull it apart into its constituents. Chapter 35 showed that energy is stored in the form of mass, and that energy and mass are related through the famous Einstein formula, $E = mc^2$. Thus, the **binding energy**, $B(N,Z)$, of a nucleus that consists of N neutrons and Z protons can be written as the difference between the mass-energy of the collection of N neutrons plus Z hydrogen atoms (consisting of 1 proton and 1 electron each) and the mass-energy of the atom of mass $m(N,Z)$, consisting of N neutrons, Z protons, and Z electrons:

$$B(N,Z) = Zm(0,1)c^2 + Nm_nc^2 - m(N,Z)c^2. \tag{40.7}$$

Here $m(0,1)$ is the mass of the hydrogen atom, with 0 neutrons, 1 proton, and 1 electron:

$$m(0,1) = 1.007825032\,u.$$

Note that this value is slightly bigger than the mass of a proton given in equation 40.6; the difference is due to the electron's mass.

While equation 40.7 gives an expression for the total binding energy of a nucleus, it is more instructive to examine the binding energy per nucleon:

$$\frac{B(N,Z)}{(N+Z)} = \frac{B(N,Z)}{A}. \tag{40.8}$$

Figure 40.7 shows the binding energy per nucleon (blue dots) for all stable isotopes as a function of the mass number, A. There is a strong increase for small Z, with a spike at $Z = 2$, a data point representing the binding energy of the nucleus of the helium atom—that is, the alpha particle. The value

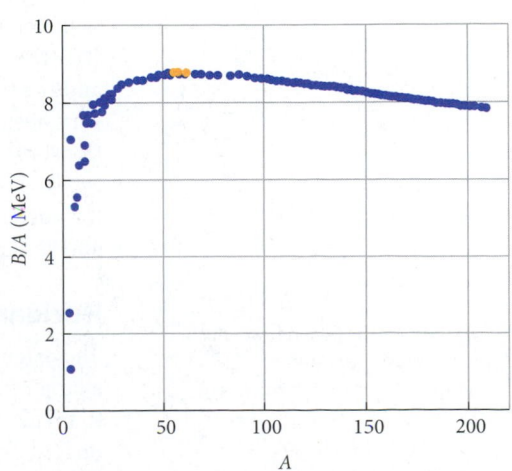

FIGURE 40.7 Binding energy per nucleon as a function of the mass number for all stable isotopes.

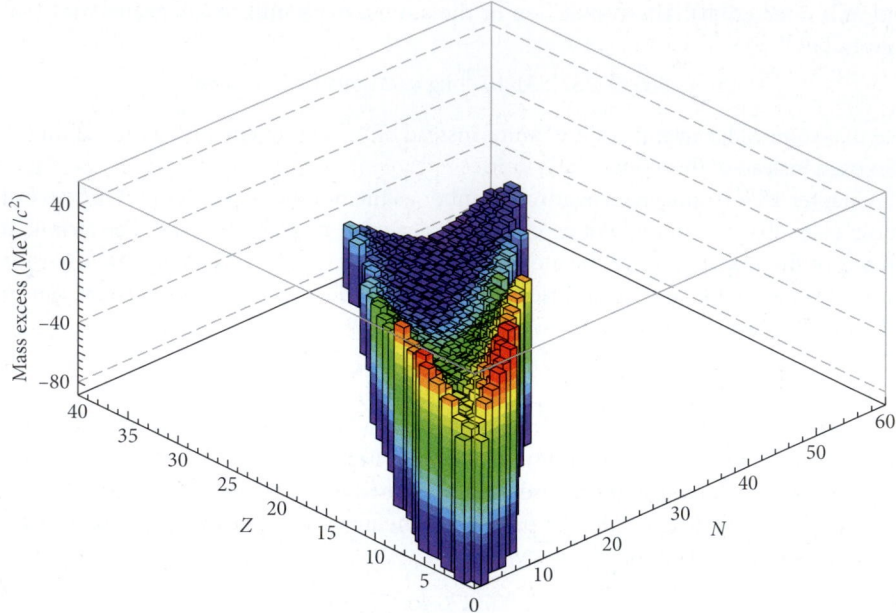

of the binding energy per nucleon of the alpha particle is $B(^4_2\text{He})/A = 7.074$ MeV. The curve reaches a maximum at iron ($Z = 26$) and nickel ($Z = 28$). The highest experimentally measured values of the binding energy per nucleon are $B(^{62}_{28}\text{Ni})/A = 8.795$ MeV, $B(^{58}_{26}\text{Fe})/A = 8.792$ MeV, and $B(^{56}_{26}\text{Fe})/A = 8.790$ MeV (indicated by the yellow dots in Figure 40.7). For $Z > 28$ and $A > 60$, the binding energy per nucleon falls gradually to a value slightly below 8 MeV. For $A > 100$, the binding energy per nucleon falls very nearly linearly with A, with a slope of

$$\left. \frac{\Delta(B/A)}{\Delta A} \right|_{A>100} = -7.1 \cdot 10^{-3} \text{ MeV}.$$

Section 40.3 will present models for the nucleus that will help you understand why the binding energies of nuclei exhibit the trends shown in Figure 40.7.

Another way to express how well a nucleus is bound is the **mass excess,** defined as the difference between the mass of a nucleus and the mass number times the atomic mass unit: mass excess $= m(N,Z) - (A)(1 \text{ u})$.

The mass excess can be expressed in terms of energy units by converting from atomic mass units using equation 40.5. The binding energy is defined in terms of the mass of the neutron and the hydrogen atom, while the mass excess is defined in terms of the mass of $^{12}_6\text{C}$. The mass of $^{12}_6\text{C}$ is defined to be 12 u, so its mass excess is zero. Thus, the mass excess and the binding energy are similar but are not the same. If the mass excess of a nucleus is very negative, the nucleus will have a large binding energy per nucleon, as defined by equations 40.7 and 40.8. For example, one of the most tightly bound nuclei, $^{56}_{26}\text{Fe}$, has a mass excess of -60.6 MeV/c^2 and a binding energy per nucleon of 8.79 MeV/c^2. The mass excesses for nuclei with $Z \leq 40$ are shown in Figure 40.8. The valley of stability described in Section 40.1 is clearly visible in this plot.

Nuclear Reactions and Q-Values

The calculation of the binding energy of a nucleus with equation 40.7 is a special case of a larger class of problems. The binding energy is simply the energy that must be supplied to break up one nucleus consisting of N neutrons and Z protons into its individual nucleons. In general, we can ask about the net energy change due to any rearrangement of an arbitrary group of neutrons and protons from an initial distribution into a final distribution. This rearrangement is called a **nuclear reaction,** in analogy with chemical reactions in which atoms get redistributed among different molecules. In chemical reactions, the number of atoms of a given species on the right-hand side (final state) of a reaction equation is exactly the same as that on the left-hand side (initial state). In nuclear reactions, a similar conservation law is observed: Because baryon number is a conserved quantity, the numbers of nucleons on the left-hand side and the right-hand side are the same. In addition, the number of protons and the number of neutrons are also separately conserved, with reactions that involve the weak force constituting the only exception. (The weak force was discussed in Chapter 39, in the form of the beta decay of the neutron or, equivalently, of the down quark.)

In practically all nuclear reactions, the initial state consists of either one or two nuclei, but not more, because nuclear sizes and therefore nuclear cross sections (see Chapter 39) are so small that the probability of three or more nuclei running into each other simultaneously is negligible.

The energy difference between the initial and final states is conventionally called the **Q-value** of the reaction. If the masses of all isotopes involved are known, then the Q-value is easily computed as the difference between the sum of the masses of the initial state nuclei and the sum of the masses of the final state nuclei. For example, for an initial state composed of a deuteron ($^2_1$H nucleus) and $^{12}_6$C and a final state composed of an isolated proton and $^{13}_6$C, the reaction can be written as $^2_1$H + $^{12}_6$C $\rightarrow$ p + $^{13}_6$C, and the Q-value of this reaction is computed as $Q = m(1,1)c^2 + m(6,6)c^2 - (m(0,1)c^2 + m(7,6)c^2)$, where the mass of a hydrogen atom is $m(0,1)$ and the mass of a deuterium atom is $m(1,1)$. (An often-used alternative notation for the same reaction is $^{12}_6$C($^2_1$H,p)$^{13}_6$C or $^{12}_6$C(d,p)$^{13}_6$C. Such d,p reactions are very popular tools for exploring nuclear structure.)

Why is the Q-value an interesting quantity? The answer is the same as in chemistry: The Q-value indicates whether the reaction is exothermic ($Q > 0$) or endothermic ($Q < 0$)—in other words, whether energy is derived from the reaction or must be put in to make the reaction proceed. Many concepts and applications of nuclear physics depend on the Q-value, and we will repeatedly return to it in this chapter.

If we know the masses of the isotopes, we can also ask how much energy it takes to separate a particular isotope into two parts. In general, for division of a nucleus with N neutrons and Z protons into two smaller nuclei with neutron numbers N_1 and N_2 and proton numbers Z_1 and Z_2, the Q-value can be computed as

$$Q_{12} = m(N,Z)c^2 - m(N_1,Z_1)c^2 - m(N_2,Z_2)c^2, \tag{40.9}$$

where $N_1 + N_2 = N$ and $Z_1 + Z_2 = Z$. The negative of the Q-value associated with this separation process, $-Q_{12}$, is called the **separation energy**, denoted by S; so $S = -Q_{12}$. If $S > 0$, then energy is required to separate the nucleus into the two parts; while if $Q_{12} > 0$, then energy is released when the separation takes place.

In these separation reactions, the numbers of protons and the numbers of neutrons remain the same in the initial and final states, so equation 40.7 can be used and the separation energy can be expressed as the difference in binding energies:

$$S = B(N_1+N_2,Z_1+Z_2) - B(N_1,Z_1) - B(N_2,Z_2). \tag{40.10}$$

In the special case where one of the two nuclei is an alpha particle, this separation energy is usually denoted by the symbol S_α. Other conventionally quoted separation energies are those for proton emission, S_p, and single- and double-neutron emission, S_n and S_{2n}.

Concept Check 40.1

Which isotope X is needed to complete the reaction $n + ^{235}_{92}$U $\rightarrow$ $^{134}_{54}$Xe $+ 2n + X$?

a) $X = ^{100}_{38}$Sr

b) $X = ^{100}_{38}$Xe

c) $X = ^{100}_{62}$Sr

d) $X = ^{102}_{38}$Sr

e) $X = ^{100}_{37}$Rb

Self-Test Opportunity 40.1

Is the reaction $d + ^{12}_6$C $\rightarrow$ $p + ^{13}_6$C exothermic or endothermic?

EXAMPLE 40.2   Separation Energy

PROBLEM
The binding energy per nucleon of the tin isotopes $^{136}_{50}$Sn, $^{134}_{50}$Sn, $^{132}_{50}$Sn, $^{130}_{50}$Sn, $^{128}_{50}$Sn, and $^{126}_{50}$Sn are measured as 8.1991 MeV, 8.2778 MeV, 8.3549 MeV, 8.3868 MeV, 8.4167 MeV, and 8.4435 MeV, respectively. What are the two-neutron separation energies of the first five of these isotopes?

– Continued

SOLUTION

From the given values of the binding energy per nucleon, we can obtain the total binding energy of each isotope by multiplying by its respective number of nucleons. We thus find

$$^{136}_{50}\text{Sn}: \quad B(86,50) = 136 \cdot 8.1991 \text{ MeV} = 1115.08 \text{ MeV}$$

$$^{134}_{50}\text{Sn}: \quad B(84,50) = 134 \cdot 8.2778 \text{ MeV} = 1109.23 \text{ MeV}$$

$$^{132}_{50}\text{Sn}: \quad B(82,50) = 132 \cdot 8.3549 \text{ MeV} = 1102.85 \text{ MeV}$$

$$^{130}_{50}\text{Sn}: \quad B(80,50) = 130 \cdot 8.3868 \text{ MeV} = 1090.28 \text{ MeV}$$

$$^{128}_{50}\text{Sn}: \quad B(78,50) = 128 \cdot 8.4167 \text{ MeV} = 1077.34 \text{ MeV}$$

$$^{126}_{50}\text{Sn}: \quad B(76,50) = 126 \cdot 8.4435 \text{ MeV} = 1063.88 \text{ MeV}.$$

Since two neutrons do not form a bound state, the binding energy of the two neutrons is zero. Thus, in general, we have this simple formula for the two-neutron separation energy:

$$S_{2n}(N,Z) = B(N,Z) - B(N-2,Z). \tag{40.11}$$

Inserting the values for the total binding energies that we have just computed into equation 40.11, we then find

$$S_{2n}(^{136}_{50}\text{Sn}) = (1115.08 - 1109.23) \text{ MeV} = 5.85 \text{ MeV}$$

$$S_{2n}(^{134}_{50}\text{Sn}) = (1109.23 - 1102.85) \text{ MeV} = 6.38 \text{ MeV}$$

$$S_{2n}(^{132}_{50}\text{Sn}) = (1102.85 - 1090.28) \text{ MeV} = 12.57 \text{ MeV}$$

$$S_{2n}(^{130}_{50}\text{Sn}) = (1090.28 - 1077.34) \text{ MeV} = 12.94 \text{ MeV}$$

$$S_{2n}(^{128}_{50}\text{Sn}) = (1077.34 - 1063.88) \text{ MeV} = 13.46 \text{ MeV}.$$

This is a very interesting result. It shows that it suddenly becomes much harder to remove a pair of neutrons from a tin isotope as the neutron number reaches 82. Why is there a big jump in the value of the two-neutron separation energy at this neutron number? This question will be answered in the discussion of the nuclear shell model in Section 40.3.

40.2 Nuclear Decay

As we've noted, not all nuclear isotopes found in nature are stable. An example is uranium, which can be found on Earth in three naturally occurring isotopes: $^{238}_{92}\text{U}$ (99.3% abundance), $^{235}_{92}\text{U}$ (0.7%), and $^{234}_{92}\text{U}$ (trace amounts). They all decay naturally over very long times, and thus are present in appreciable quantities that have survived since the time Earth was formed approximately 4.5 billion years ago. This section looks at the factors that make nuclei unstable and cause them to decay over time. We will discuss alpha, beta, and gamma decays, as well as other decays. These nuclear decays are collectively called **radioactivity.** Radioactivity was discovered in 1896 by Pierre (1859–1906) and Marie Curie (1867–1934) and by Henri Becquerel (1852–1908), for which the three shared the 1903 Nobel Prize in Chemistry.

Exponential Decay Law

Because the laws of quantum mechanics govern atomic nuclei, all decays can be viewed as transitions from one quantum state to another. Thus, decay processes follow quantum mechanical probability rules like those for tunneling, presented in Chapters 36 and 37. It is possible to quantify most decays using quantum mechanics, but we do not need to do that here. All we need to know to understand radioactive decays is that the probability of observing a decay in a given set of atomic nuclei in a given time interval, dt, is proportional to the number of nuclei present. Letting the rate of change of the number of nuclei be dN/dt, we can express this proportionality as

$$dN = -\lambda N dt \Leftrightarrow \frac{dN}{dt} = -\lambda N,$$

where λ is the decay constant. (The negative sign indicates that nuclei are lost as a function of time.) The solution of this differential equation leads to the exponential decay law:

$$N(t) = N_0 e^{-\lambda t}, \qquad (40.12)$$

where N_0 is the initial number of nuclei and $N(t)$ is the number of nuclei that remain as a function of time. Figure 40.9 shows plots of equation 40.12.

The **half-life,** $t_{1/2}$, is defined as the time it takes a quantity of nuclei of a given material to decay to half of the original number, N_0:

$$N(t_{1/2}) = \tfrac{1}{2} N_0. \qquad (40.13)$$

After two half-lives, the population has decreased to one-quarter of its initial value, and after three half-lives, to one-eighth. The decay constant can be related to the half-life by substituting from equation 40.13 into equation 40.12. This results in

$$\tfrac{1}{2} N_0 = N_0 e^{-\lambda t_{1/2}} \Rightarrow$$

$$\tfrac{1}{2} = e^{-\lambda t_{1/2}} \Rightarrow$$

$$\ln \tfrac{1}{2} = -\lambda t_{1/2}$$

$$t_{1/2} = \frac{\ln 2}{\lambda}. \qquad (40.14)$$

It is also common to talk about the **mean lifetime,** τ. This is defined as the average time it takes a nucleus to decay if the population of nuclei obeys the exponential decay law (equation 40.12). The mean lifetime is obtained by integration:

$$\tau = \left\langle N(t) \right\rangle_t = \frac{\displaystyle\int_0^\infty t\, N(t)\, dt}{\displaystyle\int_0^\infty N(t)\, dt} = \frac{\displaystyle\int_0^\infty t\, N_0 e^{-\lambda t}\, dt}{\displaystyle\int_0^\infty N_0 e^{-\lambda t}\, dt} = \frac{N_0(-1/\lambda^2)e^{-\lambda t}(1+\lambda t)\Big|_0^\infty}{N_0(-1/\lambda)e^{-\lambda t}\Big|_0^\infty} = \frac{1}{\lambda}. \qquad (40.15)$$

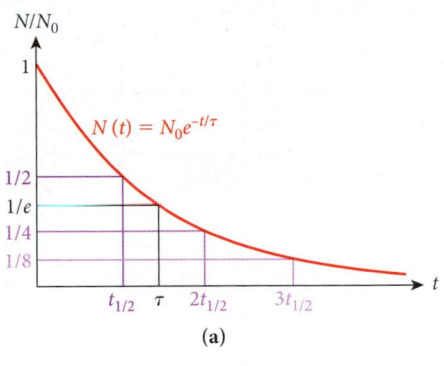

(a)

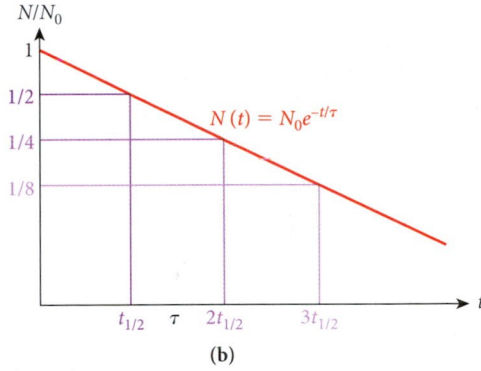

(b)

FIGURE 40.9 Exponential decay over time: (a) linear plot; (b) logarithmic plot.

Thus, the mean lifetime, τ, is simply the inverse of the decay constant, λ. Therefore, as an alternative to equation 40.12, the exponential decay law can be written as

$$N(t) = N_0 e^{-t/\tau}.$$

After one mean lifetime, the population has been reduced by a factor of $1/e$:

$$N(\tau) = N_0 e^{-\tau/\tau} = N_0/e.$$

Finally, combining equations 40.15 and 40.14, we find that the half-life $t_{1/2}$ and the mean lifetime τ are related via

$$t_{1/2} = \frac{\ln 2}{\lambda} = \tau \ln 2.$$

Thus, the half-life is not a half of the lifetime, but a factor $\ln 2 \approx 0.693$ of the lifetime. For example, Chapter 39 gave the lifetime of the neutron as 881.5 s. This means that its half-life is $(881.5 \text{ s})\ln 2 = 611.0$ s.

The mean lifetimes of isotopes are the values displayed in Figure 40.5. What kinds of nuclear decays leading to the lifetimes shown in Figure 40.5 are possible in nature?

The three main nuclear decays are the emission of an alpha particle, the emission of an electron or a positron (or, equivalently, the capture of an electron), and the emission of a photon. These decays constitute the three components of what is commonly called *radioactivity*. Radioactive decays can be very harmful to human health, completely harmless, or in some cases even very helpful in medical diagnostics and treatment. Their effect on health depends on the type of decay, the energy of the decay product, and the radiation dose, that is, the amount of radioactive material present and the amount of radiation

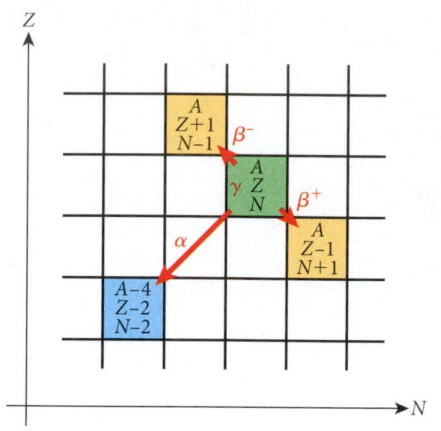

FIGURE 40.10 Nuclear decays in a chart of isotopes.

emitted. Figure 40.10 shows how the different radioactive decays change the nucleus that decays. The following subsections discuss each decay mode in more detail.

In all decays, the decaying nucleus is called the *parent* and the nucleus it decays into is the *daughter*. If the parent and daughter nuclei are different elements, the process is known as **transmutation.** Only those decays that obey the conservation laws, in particular those of energy, charge, and baryon number, are possible.

Alpha Decay

In an **alpha decay** (α decay), the nucleus emits an alpha particle, which is the nucleus of a helium atom, $^4_2$He. This means that the mass number of the parent nucleus decreases by four and the charge number by two:

$$^A_Z \text{Nuc} \rightarrow\ ^4_2\text{He} +\ ^{A-4}_{Z-2}\text{Nuc'}. \tag{40.16}$$

(The notation Nuc' is used to indicate a nucleus that is different in its composition from the initial nucleus, before the decay.) In general, alpha decays are possible when the energy contained in the mass of the alpha particle plus the mass of the daughter nucleus, $m\!\left(^{A-4}_{Z-2}\text{Nuc'}\right)$, is smaller than the mass of the nucleus that undergoes the alpha decay:

$$m\!\left(^A_Z\text{Nuc}\right) > m_\alpha + m\!\left(^{A-4}_{Z-2}\text{Nuc'}\right). \tag{40.17}$$

Since the binding energy per nucleon of the alpha particle (of mass m_α) is very large, 7.074 MeV, the mass of the alpha particle is comparatively low. In addition, as shown in Figure 40.7, the binding energy per nucleon falls very gradually as a function of the mass number, A, for nuclei with large mass numbers. This makes alpha decay possible for almost all unstable isotopes with $A > 150$.

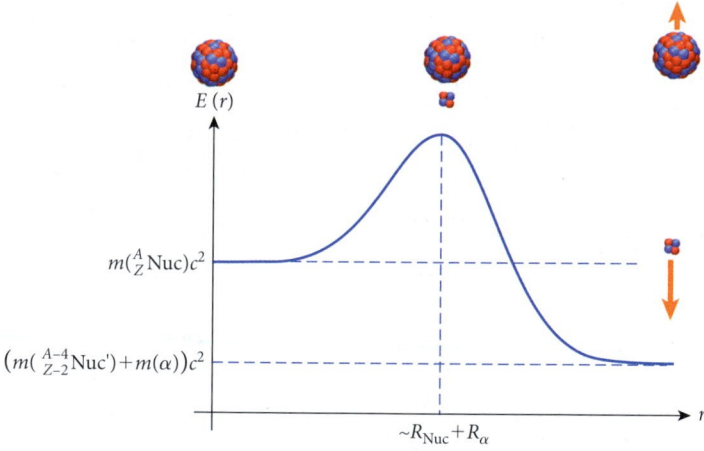

FIGURE 40.11 Plot of the potential energy of the alpha particle and the daughter nucleus, showing the potential barrier (located at the vertical dashed line) for alpha decay.

Chapter 37 mentioned that alpha decay is an example of tunneling—the transmission of the wave function of the alpha particle through a classically forbidden region. This is illustrated in Figure 40.11, where the total energy of the nuclear system is plotted as a function of the separation r between the center of the alpha particle and the center of the daughter nucleus. If the four nucleons that constitute the alpha particle are approximately in the center of the parent nucleus, then the total energy is approximately just the energy contained in the mass of the parent nucleus, $m(^A_Z\text{Nuc})c^2$, as shown in Figure 40.11 at $r = 0$. When the alpha particle and the daughter nucleus are widely separated, $r \rightarrow \infty$, the interaction between the two becomes negligible, and the total energy is the mass-energy of the daughter nucleus plus that of the alpha particle, $\left(m\!\left(^{A-4}_{Z-2}\text{Nuc'}\right)+m(\alpha)\right)c^2$. For many heavy nuclei, this value of the total energy is *lower* than the value at the center. Therefore, the emission of the alpha particle from the parent nucleus is energetically favorable.

However, first the alpha particle has to get out of the parent nucleus. Moving the four nucleons of the alpha particle to one side or the other deforms the nucleus. This deformation increases the excitation of the nucleus, and the potential energy increases relative to the value at $r = 0$. At a configuration where the alpha particle and the daughter nucleus barely touch, indicated by the vertical dashed line in Figure 40.11, the potential energy has a maximum, because of the Coulomb repulsion between the alpha particle and the daughter nucleus. This increase in potential energy forms a potential barrier and prevents spontaneous alpha decay. However, the wave function of the alpha particle can tunnel through this potential barrier, leading to emission of the alpha particle. The tunneling probability and thus the lifetime of the nucleus before alpha decay depend very strongly on the shape (mainly width, but also height) of the barrier.

When the alpha particle escapes from the nucleus, the energy difference between the mass-energies of the parent and daughter nuclei, which is the Q-value of the reaction, is converted into kinetic energy of the alpha particle and the heavy remnant:

$$K_\alpha + K_{\mathrm{Nuc'}} = Q = \left(m\left({}_{Z}^{A}\mathrm{Nuc}\right) - m\left({}_{Z-2}^{A-4}\mathrm{Nuc'}\right) - m_\alpha \right) c^2. \qquad (40.18)$$

Total momentum is conserved, so the momentum of the alpha particle and the momentum of the daughter nucleus must be equal in magnitude and opposite in direction in the rest frame of the parent nucleus, $|\vec{p}_\alpha| = |\vec{p}_{\mathrm{Nuc'}}|$. For the low kinetic energies occurring here, the nonrelativistic approximation $K = p^2/2m$ is sufficient. Momentum conservation then means that the kinetic energy of the daughter nucleus is related to the kinetic energy of the alpha particle via

$$K_{\mathrm{Nuc'}} m\left({}_{Z-2}^{A-4}\mathrm{Nuc'}\right) = K_\alpha m_\alpha \Leftrightarrow K_{\mathrm{Nuc'}} = K_\alpha \frac{m_\alpha}{m\left({}_{Z-2}^{A-4}\mathrm{Nuc'}\right)}.$$

Inserting this result into equation 40.18, we find that the kinetic energy of the alpha particle in the rest frame of the parent nucleus is

$$K_\alpha = \frac{m\left({}_{Z-2}^{A-4}\mathrm{Nuc'}\right)}{m\left({}_{Z-2}^{A-4}\mathrm{Nuc'}\right) + m_\alpha}\left(m\left({}_{Z}^{A}\mathrm{Nuc}\right) - m\left({}_{Z-2}^{A-4}\mathrm{Nuc'}\right) - m_\alpha \right) c^2. \qquad (40.19)$$

EXAMPLE 40.3　Roentgenium Decay

By shooting a beam of ${}_{28}^{64}\mathrm{Ni}$ nuclei at a ${}_{83}^{209}\mathrm{Bi}$ target in December 1994, a group at the GSI national laboratory in Germany was able to produce a new element with 111 protons and 161 neutrons through the reaction ${}_{28}^{64}\mathrm{Ni} + {}_{83}^{209}\mathrm{Bi} \rightarrow {}_{111}^{272}\mathrm{Rg} + n$. Here Rg stands for roentgenium, the official name this new element received in November 2006. The GSI group detected three events in which the new element was produced by observing successive alpha decays of the new element and its known daughter nuclei. One of the three events, with the measured alpha decay times, is shown in Figure 40.12.

PROBLEM
The masses of the nuclei in the decay chain are listed in nuclear data tables (in atomic mass units) as 272.1536, 268.138728, 264.1246, 260.1113, 256.098629, and 252.08656. What energies would you predict for the emitted alpha particles?

SOLUTION
Since the masses of the isotopes are given, we can calculate the Q-value for each decay using equation 40.18. We can then use equation 40.19 to calculate the expected kinetic energy of the alpha particle. This is done in the table below for each of the decays.

The last column of the table compares our predictions with the experimental results that were reported by the GSI group. As you can see, the two sets of numbers are in reasonably good agreement.

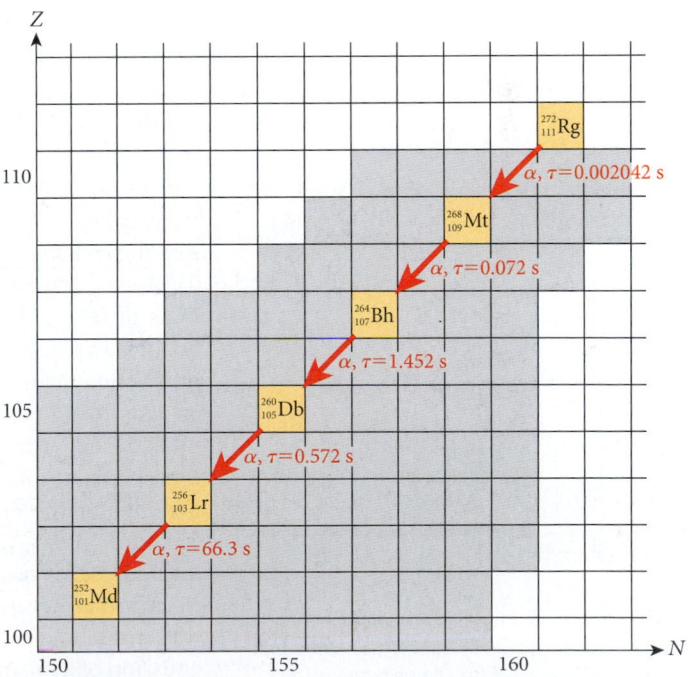

FIGURE 40.12 Roentgenium decay chain as observed on Dec. 17, 1994, at the GSI. The alpha decay chain proceeds through meitnerium (Mt), bohrium (Bh), dubnium (Db), and lawrencium (Lr) to mendelevium (Md). The gray boxes indicate isotopes that have been observed.

DISCUSSION
In practice, the masses of the so-called superheavy elements (those with triple-digit charge numbers) are mostly determined from measurements of the kinetic energies of the alpha particles

– Continued

from the isotopes' decays. The values listed in the table are from one particular event in one particular experiment. The isotope masses are determined from best fits to all available data. For the heaviest elements, these data consist of only a few events, but for charge numbers below 100, millions of events have been recorded. The fact that the observed kinetic energies of the emitted alpha particles and the decay times agreed so well with previously measured data served to convince the GSI group that they had indeed seen the first events of the production of the new element roentgenium.

Name	A	Z	m (u)	$m_\alpha + m$ (u)	Q (MeV)	K_α predicted (MeV)	K_α data (MeV)
Rg	272	111	272.1536				
Mt	268	109	268.138728	272.1413313	11.42826	11.3	10.82
Bh	264	107	264.1246	268.1272033	10.73523	10.6	10.221
Db	260	105	260.1113	264.1139033	9.96395	9.8	9.621
Lr	256	103	256.098629	260.1012323	9.37804	9.2	9.2
Md	252	101	252.08656	256.0891633	8.81729	8.7	8.463

Beta Decay

In a **beta decay** (β decay), the nucleus emits an electron (e^-) or a positron (e^+) or captures one of its own orbiting electrons. Chapter 39 discussed the beta decay of quarks, particularly the β^- decay of a down quark into an up quark: $d \rightarrow u + e^- + \bar{\nu}_e$. Since the neutron is composed of two down quarks and an up quark, one way that the β^- decay of the down quark manifests itself is the β^- decay of the neutron, $n \rightarrow p + e^- + \bar{\nu}_e$, discussed earlier. The general form of a nuclear β^- decay can be written as

$$_Z^A\text{Nuc} \rightarrow \, _{Z+1}^A\text{Nuc'} + e^- + \bar{\nu}_e. \tag{40.20}$$

Note that in nuclear β^- decay, the mass number of the nucleus remains the same, but the charge number increases by 1.

Because nuclei are made of neutrons and protons, it might seem that beta decays should always be possible. However, a beta decay can happen only if it is allowed by energy conservation—that is, the combined mass of the electron and the daughter nucleus must be smaller than the mass of the parent nucleus. It might be expected that this will always be the case because of the mass difference of 1.293 MeV/c^2 (or 0.00139 u) between neutron and proton, which is quite a bit larger than the electron mass of 0.511 MeV/c^2. Although this argument is correct for free neutrons, it is not always true for neutrons bound inside a nucleus. The reason is that the nuclear interaction favors configurations with equal numbers of neutrons and protons. As an example of the effect that this part of the nuclear interaction has, Figure 40.13 plots the experimentally measured masses (horizontal red lines) of all known isotopes with $A = 82$ as a function of the charge number, Z. Clearly it is energetically possible for the bromine ($Z = 35$) isotope with mass number 82 to undergo the β^- decay $_{35}^{82}\text{Br} \rightarrow \, _{36}^{82}\text{Kr} + e^- + \bar{\nu}_e$, but the β^- decay of $_{36}^{82}\text{Kr}$ into $_{37}^{82}\text{Rb}$ is energetically forbidden, because $_{37}^{82}\text{Rb}$ has a greater mass than $_{36}^{82}\text{Kr}$.

Furthermore, it is also possible for beta decays to proceed in the opposite direction inside nuclei. In such a β^+ decay, a proton is converted into a neutron via the emission of a positron and an electron-neutrino: $p \rightarrow n + e^+ + \nu_e$. For a free proton, this process is energetically not possible because the neutron has a higher mass than the proton. However, inside the nucleus, the reaction

$$_Z^A\text{Nuc} \rightarrow \, _{Z-1}^A\text{Nuc'} + e^+ + \nu_e, \tag{40.21}$$

can proceed when the masses of initial and final isotopes are such that the Q-value of the reaction is positive. In addition, a proton inside the nucleus can be converted into a neutron via electron capture: $e^- + p \rightarrow n + \nu_e$. In this β^+ decay, the nucleus captures one of its own orbiting electrons:

$$e^- + \, _Z^A\text{Nuc} \rightarrow \, _{Z-1}^A\text{Nuc'} + \nu_e. \tag{40.22}$$

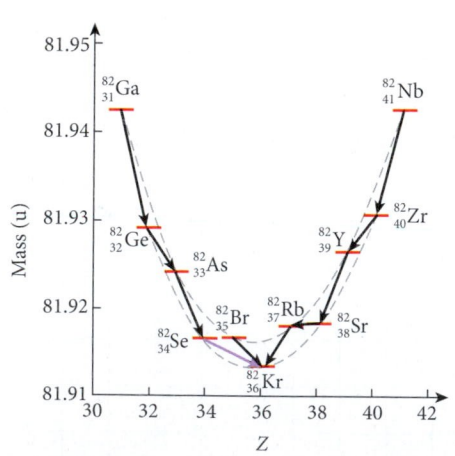

FIGURE 40.13 Nuclei of mass number 82. Their measured masses (in atomic mass units) are shown as a function of the charge number. The black arrows represent beta-decay processes. The purple arrow represents a double beta decay.

We use β^+ decay to refer to both processes, positron emission (equation 40.21) and electron capture (equation 40.22). Theoretically, a third way to convert a proton into a neutron is via anti-neutrino capture: $\bar{\nu}_e + p \rightarrow n + e^+$. However, this process is of negligible importance to us because the cross section is extremely small and because an atom does not have a source of anti-neutrinos present. Consulting Figure 40.13 once again, you see that the nuclei on the right-hand side of the figure can all undergo β^+ decay.

Note that the Q-values for the reactions in equations 40.21 and 40.22 are not the same. The initial state of the reaction in equation 40.21 consists of Z protons, $N = A - Z$ neutrons, and Z electrons, all of which are accounted for in the mass of the initial atom, $m(N,Z)$. The final state consists of an atom with $Z - 1$ protons, $N + 1$ neutrons, and $Z - 1$ electrons, plus one additional electron, plus the newly created positron and neutrino. This atom has a mass of $m(N+1, Z-1)$, and the electron and the positron each have a mass of $m_e = 0.511$ MeV/c^2. Neglecting the binding energy of the electron and the mass of the neutrino, each of which are on the order of 1 eV or less, we then obtain for the Q-value of the positron-emitting β^+ decay (equation 40.21)

$$Q(e^+) = m(N,Z)c^2 - m(N+1, Z-1)c^2 - 2(0.511 \text{ MeV}).$$

The initial state of the reaction in equation 40.22 consists of the same atom with Z protons, $N = A - Z$ neutrons, and Z electrons, and the final state consists of a neutrino and an atom with $Z - 1$ protons, $N + 1$ neutrons, and $Z - 1$ electrons. Thus, in this case, the Q-value is simply

$$Q(ec) = m(N,Z)c^2 - m(N+1, Z-1)c^2.$$

(Remember, the nucleus captured one of its atom's own electrons!) Therefore, the Q-value of the electron capture (*ec*) reaction is always 1.022 MeV larger than that for the same β^+ decay that involves positron emission (e^+). This implies that for some isotopes only electron capture is possible, not positron emission.

For alpha decay, we have seen that the emitted alpha particle has a characteristic energy (equation 40.19), because energy and momentum need to be conserved in the decay. However, in beta decays, the situation is more complicated. In β^- decays as well as in positron-emitting β^+ decays, the final state consists of three particles that can share the decay energy. The emitted neutrinos cannot be observed directly and can carry different amounts of energy. Thus, the electrons or positrons produced by these decays do not have well-defined kinetic energy values, but instead show a continuous distribution of energies.

Gamma Decay

A **gamma decay** (γ decay) is the emission of a photon from a nucleus and is always the product of de-excitation of an excited nuclear state. Gamma decays are qualitatively different from alpha or beta decays in that they do not result in transmutation. Nuclei can attain excited states similarly to how atoms can, by collisions with other objects. The kinetic energy from these processes can be converted into excitation energy, lifting nucleons into higher shells or causing collective vibrations or rotations of the entire nucleus.

The process of gamma decay in nuclei is similar to the emission of photons in the de-excitation of atoms. However, the characteristic electron energies in the atom are on the order of electron-volts, whereas the characteristic energies of nuclear excited states are on the order of millions of electron-volts. Thus, the photons emitted in gamma decays typically are a million times more energetic than the photons emitted in atomic decays.

The initial and final states of the nucleus have well-defined energies, so the photon energy is also, in principle, well defined. However, an excited nuclear state has a finite lifetime, τ. Thus, the uncertainty in the energy of the emitted photon, denoted as the width, Γ, has a lower limit given by the uncertainty relation: $\Gamma > \hbar/\tau$. Thus, the measured energies of gamma rays resulting from the decay between two specific states with finite lifetimes in a nucleus will have a distribution that is not a spike at a single energy, but rather takes the form of a peak in energy with a characteristic width in energy.

Most excited isotopes can de-excite via gamma decay. However, the larger the difference between the angular momentum of the initial state and that of the final state, the

FIGURE 40.14 Gammasphere, the world's most sensitive detector for nuclear gamma rays. Gammasphere consists of 110 gamma-ray detectors cooled by liquid nitrogen and arranged around the point where the nuclear interactions take place. In the photograph, only the support structure and the liquid nitrogen containers are visible.

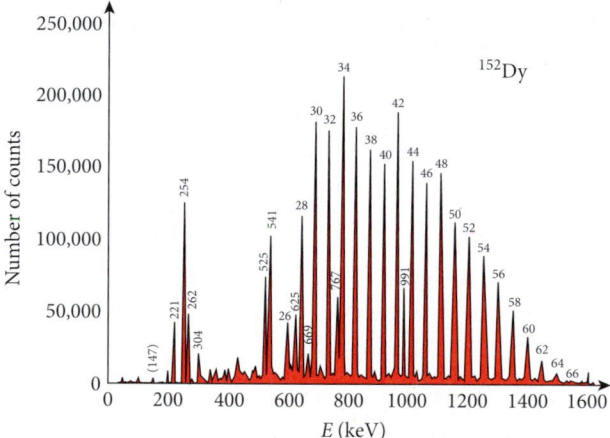

FIGURE 40.15 Gamma-ray spectrum of the superdeformed isotope dysprosium-152. The measured gamma rays are identified either by the energy (in keV, vertical numbers) or by the angular momentum (in units of $\hbar$, horizontal numbers).

Concept Check 40.3

Which of the following types of nuclear decays are possible for the isotope $^{82}_{33}\text{As}$? (*Hint:* The mass of $^{82}_{33}\text{As}$ is 81.925 u, the mass of $^{82}_{32}\text{Ge}$ is 81.930 u, the mass of $^{82}_{34}\text{Se}$ is 81.917 u, the mass of $^{78}_{31}\text{Ga}$ is 77.932 u, and the mass of $^{4}_{2}\text{He}$ is 4.0026u.)

a) alpha decay

b) beta decay

c) positron emission

d) electron capture

e) gamma decay

lower the probability for gamma decay becomes. In some isotopes, the angular momentum of the lowest excited state has a large difference from that of the ground state, and so a gamma decay is very improbable. These isotopes form very long-lived "isomer" states.

Gamma rays emitted from nuclei are very important diagnostic tools for learning about nuclear structure. Since a very large number of states exist in any sample of a given isotope, a large number of gamma rays can be detected at different energies. It requires ingenious detective work and large photon detector arrays (Figure 40.14) to interpret the information on nuclear structure contained in these spectra.

Gamma rays are not only emitted in single-particle transitions in which a single nucleon jumps from an excited state to another state. They can also be emitted from collective vibrations of nuclei, as well as from rotating deformed nuclei. *Superdeformed nuclei*, which are cigar-shaped with an axes ratio of 2:1, were discovered only recently with the aid of gamma-ray spectroscopy. Figure 40.15 shows the gamma-ray spectrum of $^{152}_{66}\text{Dy}$, whose most prominent feature is the sequence of peaks with a very regular spacing of $\Delta E = 47.5$ keV. These peaks result when a rotating nucleus makes transitions between angular momentum states having even multiples of $\hbar$.

Other Decays

While alpha, beta, and gamma decays constitute the overwhelming majority of radioactive decay, we should note a few other decay modes.

Light, very neutron-rich isotopes can decay by emission of a single neutron. Light and very proton-rich isotopes can decay via the emission of a proton. Neutron and proton emitters usually have extremely short lifetimes and are not quite bound nuclei. An example of a neutron emitter is $^{10}_{3}\text{Li}$. The isotope with one more neutron than $^{10}_{3}\text{Li}$, $^{11}_{3}\text{Li}$, has been intensively investigated recently because it consists of three parts $(^{9}_{3}\text{Li} + 2n)$ that form a bound state only if all three are together. There is no bound state of two neutrons, and the neutron emitter $^{10}_{3}\text{Li}$ $(^{9}_{3}\text{Li} + 1n)$ is also not a bound nucleus. Thus, $^{11}_{3}\text{Li}$ is a strange nucleus consisting of a core of $^{9}_{3}\text{Li}$ and a very large "halo" made of two neutrons. The diameter of this halo has been measured to be almost as large as that of a lead nucleus, which consists of 208 nucleons. $^{11}_{3}\text{Li}$ therefore deviates strikingly from the nuclear size law expressed in equation 40.3.

A few isotopes are known to exhibit cluster decays. **Cluster decays** involve the emission of nuclei heavier than helium. Almost all known cluster decays emit carbon nuclei, $^{12}_{6}\text{C}$ or $^{14}_{6}\text{C}$, and in some very rare cases oxygen, neon, or magnesium nuclei.

One very important decay mode of heavy nuclei, fission, will be discussed in Section 40.4.

Many isotopes have two or more possible decay modes. Figure 40.16 presents an overview of the dominant decay modes for all known isotopes. Proton-rich isotopes decay via β^+ decays or proton emission, and neutron-rich nuclei via β^- decays or neutron emission. For heavier nuclei, alpha decay becomes dominant, and the heaviest isotopes often decay predominantly via spontaneous fission.

An extremely rare decay mode of some isotopes is *double beta decay*, which was first observed in 1987 in an isotope of selenium, $^{82}_{34}\text{Se}$, by Michael Moe and colleagues at the University of California–Irvine. Referring back to Figure 40.13, you can see that $^{82}_{34}\text{Se}$ cannot undergo a simple β^- decay to $^{82}_{35}\text{Br}$, because its mass is lower than that of $^{82}_{35}\text{Br}$. Thus, that reaction is forbidden by energy conservation. However, the mass of $^{82}_{36}\text{Kr}$ is lower than that of $^{82}_{34}\text{Se}$. So a nucleus of $^{82}_{34}\text{Se}$ can be converted to a nucleus of $^{82}_{36}\text{Kr}$ via double beta decay:

$$^{82}_{34}\text{Se} \rightarrow {}^{82}_{36}\text{Kr} + 2e^- + 2\bar{\nu}_e. \tag{40.23}$$

Double beta decay is the only reaction that prevents $^{82}_{34}\text{Se}$ from being a stable isotope. This decay is depicted in Figure 40.17. Chapter 39 noted the beta-decay process involves the

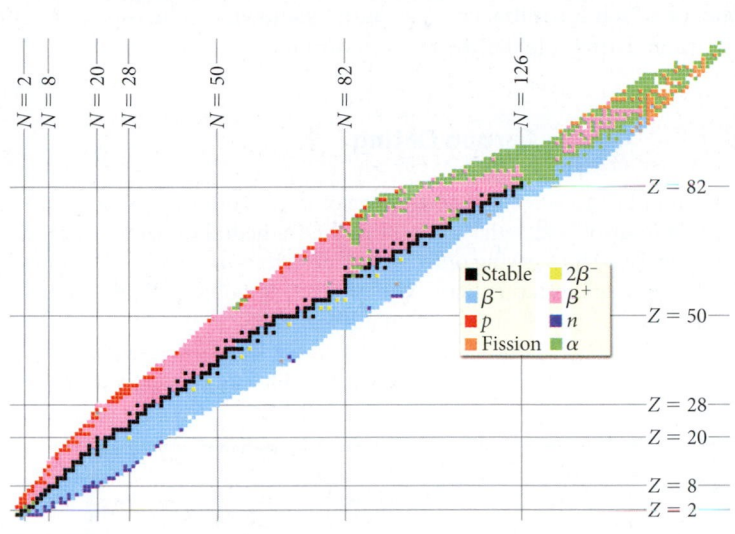

FIGURE 40.16 Dominant decay modes for the known isotopes.

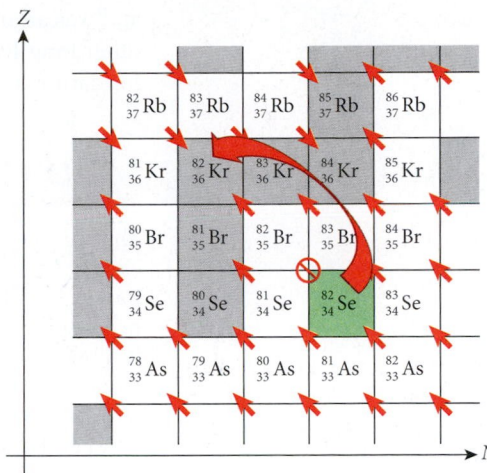

FIGURE 40.17 Chart showing nuclei around mass number 82. The gray shaded squares indicate stable nuclei. The red arrows indicate the directions of beta decays. The green shaded box highlights the isotope selenium-82, which can undergo double beta decay.

exchange of a W boson and thus leads to comparatively long lifetimes. A double beta decay requires two W boson exchanges and thus leads to extremely long lifetimes. The lifetime of $^{82}_{34}$Se has been determined to be 10^{20} years, which is approximately 10 billion times the age of the universe! Only 12 isotopes are known to undergo double beta-decay processes (shown in yellow in Figure 40.16), and their average lifetimes are all of the same order as that of $^{82}_{34}$Se. Thus, double beta-decay event rates are extremely small, making the detection of these decays very difficult experimentally.

Research into double beta decays is flourishing. Double β^+ decays are theoretically possible, but no isotope showing this type of decay has yet been observed. In addition, there are ongoing searches for double beta decays in which no anti-neutrino is emitted. This process could work only if the neutrino is its own antiparticle. If this decay mode were observed, it would violate the standard model of particle physics, introduced in Chapter 39, and would point toward completely new discoveries in physics.

Carbon Dating

Carbon has two stable isotopes: $^{12}_{6}$C at 98.90% abundance and $^{13}_{6}$C at 1.10% abundance. All other carbon isotopes, except $^{14}_{6}$C, decay with half-lives ranging from milliseconds to minutes. The exception, $^{14}_{6}$C, has a half-life of $t_{1/2}\left(^{14}_{6}\text{C}\right) = (5730 \pm 40)$ years. It decays via β^- decay into nitrogen: $^{14}_{6}\text{C} \rightarrow {}^{14}_{7}\text{N} + e^- + \overline{\nu}_e$. The Q-value of this reaction is 156.5 keV.

The $^{14}_{6}$C isotope is produced in the upper atmosphere at a constant rate via the interaction of cosmic rays with atmospheric nitrogen. Because this isotope is produced at a constant rate and decays at a constant rate, its concentration in the atmosphere, and thus the ratio of the number of $^{14}_{6}$C atoms relative to the number of $^{12}_{6}$C atoms, is constant in time, at a value of approximately $1.20 \cdot 10^{-12}$. The plants on Earth's surface consume both carbon isotopes, mainly in CO_2 molecules. Thus, the ratio of $^{14}_{6}\text{C}/^{12}_{6}\text{C}$ is also constant in living plants, and consequently all the way up the food chain as well.

While plants and animals are living and consuming food, their carbon isotope ratio stays at a constant value. However, at the moment of death, the intake of carbon isotopes ceases, and thus the ratio $^{14}_{6}\text{C}/^{12}_{6}\text{C}$ decreases in time as $^{14}_{6}$C decays. Measuring this ratio in a tissue sample can determine how long a plant or animal has been dead. This method can also be used on products made from plants or animals, such as textiles and wooden objects. The procedure of age determination using radioactive decays, called **radiocarbon dating,** or just *carbon dating,* revolutionized the field of archeology. It allows dating of samples up to an age of approximately 10 half-lives of $^{14}_{6}$C, more than 50,000 years. The American physical chemist Willard Frank Libby (1908–1980) discovered the method of carbon dating in 1949 and was awarded the 1960 Nobel Prize in Chemistry for this achievement. In principle, any other long-lived isotope for which the initial concentration is known can be used as a tool for dating objects, but carbon dating is by far the most important.

SOLVED PROBLEM 40.2 | Carbon Dating

The Shroud of Turin, shown in Figure 40.18, is a large piece of linen cloth that some people claim is the burial shroud of Jesus of Nazareth. It is kept in the Cathedral of Saint John the Baptist in Turin, Italy, and displayed for viewing only on very rare occasions. However, some people expressed doubts concerning the authenticity of this object, and other people claimed it to be a hoax of medieval origin. In 1988, the Vatican allowed radiocarbon dating of the shroud by three laboratories, in Zürich, Oxford, and Arizona, and it was shown to be of medieval origin. The testing concluded with a 95% confidence level that the shroud was made between 1260 and 1390.

FIGURE 40.18 Example of the use of carbon dating: the Shroud of Turin.

PROBLEM

If a textile sample contains $(1.08 \pm 0.01) \cdot 10^{-12}$ atoms of $^{14}_{6}\text{C}$ for each atom of $^{12}_{6}\text{C}$, what is its age?

SOLUTION

THINK We know that radioactive decays follow an exponential decay law that determines the number of remaining $^{14}_{6}\text{C}$ atoms as a function of time (equation 40.12). The number of $^{12}_{6}\text{C}$ atoms stays constant in time because this isotope is stable. Therefore, the ratio of the two isotopes, $f_{14/12}(t) = N(^{14}_{6}\text{C},t)/N(^{12}_{6}\text{C})$, follows the same exponential decay law. Because we know the half-life of $^{14}_{6}\text{C}$ and the initial and final fractions of $^{14}_{6}\text{C}$, we can solve the exponential decay law for the time and obtain our answer.

SKETCH Figure 40.19 serves to remind us that the exponential decay law also holds for the fraction of $^{14}_{6}\text{C}$ relative to $^{12}_{6}\text{C}$. Given a certain number for the fraction, we can obtain the age of the sample.

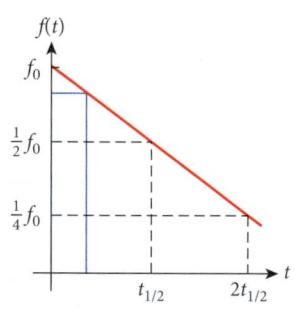

FIGURE 40.19 Carbon-14 fraction as a function of time, plotted on a semilogarithmic scale.

RESEARCH We can use the exponential decay law (equation 40.12) for the number of $^{14}_{6}\text{C}$ atoms as a function of time:

$$N(t) = N_0 e^{-\lambda t}.$$

Since we are given the half-life of the isotope, we need to relate the decay constant λ to the half-life by using the relationship in equation 40.14:

$$t_{1/2} = \ln 2/\lambda.$$

SIMPLIFY First, we write the decay law in terms of the half-life:

$$N(^{14}_{6}\text{C},t) = N_0(^{14}_{6}\text{C})e^{-\lambda t} = N_0(^{14}_{6}\text{C})e^{-t \ln 2/t_{1/2}}.$$

Dividing both sides by the number of $^{12}_{6}\text{C}$ isotopes gives us the fraction of $^{14}_{6}\text{C}$ relative to $^{12}_{6}\text{C}$:

$$f_{14/12}(t) = f_{14/12}(t=0)e^{-t \ln 2/t_{1/2}}.$$

Now we can solve this equation for the time:

$$\frac{f_{14/12}(t)}{f_{14/12}(t=0)} = e^{-t \ln 2/t_{1/2}} \Rightarrow$$

$$\ln\left(\frac{f_{14/12}(t)}{f_{14/12}(t=0)}\right) = -t\frac{\ln 2}{t_{1/2}} \Rightarrow$$

$$t = -\frac{t_{1/2}}{\ln 2}\ln\left(\frac{f_{14/12}(t)}{f_{14/12}(t=0)}\right).$$

CALCULATE   Now we are in a position to insert numbers. We know that the half-life of $^{14}_{6}C$ is $t_{1/2}(^{14}_{6}C) = (5730\pm40)$ years and $f_{14/12}(t=0) = 1.20 \cdot 10^{-12}$. The problem specifies the fraction: $f_{14/12}(t) = (1.08\pm0.01)\cdot 10^{-12}$. Thus, we insert the average value of the fraction into the above equation for the time to find the mean value:

$$t = -\frac{5730 \text{ yr}}{\ln 2}\ln\left(\frac{1.08}{1.20}\right) = 870.978 \text{ yr.}$$

Inserting the value 1.09 in the equation for the time results in an upper limit of uncertainty of 794.787 yr (76 years less), and the value 1.07 yields a lower limit of uncertainty of 947.877 yr (77 years more).

ROUND   From the given uncertainty in the fraction of $^{14}_{6}C$, we have determined the uncertainty in the time. Thus, we round our final answer to a value of 870 yr and quote the uncertainty as ±80 yr:

$$t = (870 \pm 80) \text{ yr.}$$

The textile sample in question was probably produced in the 12th century, but perhaps as early as the 11th or as late as the 13th century.

DOUBLE-CHECK   It is quite reasonable that the time we calculated is small compared to the half-life of $^{14}_{6}C$, because after one half-life the remaining concentration of that isotope would be $0.60 \cdot 10^{-12}$ $(= \frac{1}{2} \cdot 1.20 \cdot 10^{-12})$, and the problem statement indicated that the textile sample had lost only 10% of its $^{14}_{6}C$ content. In fact, for small values of $t/t_{1/2}$ the exponential decay function can be expanded as $e^{-\ln 2 \cdot t/t_{1/2}} \approx 1 - \ln 2 \cdot t/t_{1/2}$. Since 90% of the initial $^{14}_{6}C$ concentration is left over at that time, this linear approximation results in

$$0.9 = 1 - \ln 2 \cdot t/t_{1/2} \Rightarrow t = t_{1/2} \cdot 0.1/\ln 2 = 0.144 t_{1/2} = 830 \text{ yr.}$$

It's worthwhile noting that the assumption that the atmospheric concentration of $^{14}_{6}C$ has always been constant is not quite true. Historic changes in the cosmic-ray flux and other events have resulted in temporal fluctuations. In addition, since the 1950s, the atmospheric concentrations of $^{14}_{6}C$ and other isotopes have been changed by atmospheric tests of nuclear weapons. However, correlating the results of carbon dating with other sources of age determination (ice cores, tree rings, etc.) has allowed calibration curves for the numerical results obtained with carbon dating to be developed.

Units of Radioactivity

How can the radioactivity of a sample be measured? There are two main quantities of interest. First is the intensity of the emission from the sample of radioactive material, that is, the number of decays per unit time. Second is the effect that a given type of radiation has on the human body. (Figure 40.20 is a radioactivity warning sign.)

The SI unit of radioactivity is the **becquerel (Bq)**, named in honor of the French physicist Henri Becquerel (1852–1908), co-discoverer of radioactivity:

$$1 \text{ Bq} = 1 \text{ nuclear decay/s.} \tag{40.24}$$

FIGURE 40.20 The trefoil symbol is internationally recognized as indicating radioactive material.

This is an extremely small unit of radioactivity. A non-SI, but commonly used, unit of radioactivity is the **curie (Ci)**, named in honor of the husband-and-wife team of Frenchman Pierre Curie (1859–1906) and his Polish-born wife Marie Skłodowska-Curie (known as Madame Curie, 1867–1934), who were the other two co-discoverers of radioactivity. Initially, the curie was defined as the activity of 1 gram of the isotope $^{226}_{88}Ra$, which decays via emission of an alpha particle. Currently, the curie is defined in terms of the becquerel as follows:

$$1 \text{ Ci} = 3.7 \cdot 10^{10} \text{ Bq}$$
$$1 \text{ Bq} = 2.7 \cdot 10^{-11} \text{ Ci.} \tag{40.25}$$

Older textbooks used a unit called the rutherford (Rd): 1 Rd = 1 MBq.

As mentioned above, the number of decays alone is not enough to measure the effects of radioactivity. Much more important is the absorbed radiation dose. The SI unit for the absorbed dose is the *gray* (Gy), defined in terms of other SI units as an absorbed energy of 1 joule per kilogram of absorbing material:

$$1 \text{ Gy} = 1 \text{ J/kg}. \tag{40.26}$$

A commonly used non-SI unit is the *rad* (rd). Its conversion to the gray is given by

$$1 \text{ rd} = 0.01 \text{ Gy}$$
$$1 \text{ Gy} = 100 \text{ rd}. \tag{40.27}$$

For X-rays and gamma rays, it is also important to measure the amount of ionization (separation of electrons from their previously neutral atoms). This quantity is called *exposure* and is measured in coulombs per kilogram. A non-SI unit for exposure is the *röntgen* (R), and it is defined when the material is in air as

$$1 \text{ R} = 2.58 \cdot 10^{-4} \text{ C/kg of air}. \tag{40.28}$$

The exposure and absorbed dose are closely related, and for biological tissue 1 röntgen is a close approximation to 1 rad.

The extent of damage to of biological tissue depends not only on the deposited energy (absorbed dose), but also on the type of particle the nuclear decay emitted. Different types of radiation interact in very different ways with matter, in particular with biological tissue, and deposit energy with very different depth profiles. When photons penetrate into matter, they lose energy exponentially as a function of depth into the material. Alpha particles and heavy nuclei, on the other hand, lose a very small fraction of their energy at the entry point, but deposit almost all of their energy at their maximum penetration depth, which is a function of the initial energy. Thus, for example, alpha particles of a given initial energy all penetrate to approximately the same depth in biological tissue and then deposit a very significant fraction of their energy at that depth (Figure 40.21).

Because of the differences in energy deposition as a function of penetration depth, an alpha particle is much more damaging to the cell walls of biological tissue than a photon of the same energy. The *radiation weight factor, w_r,* is 1 for all photons and leptons, independent of their energy. Also, $w_r = 5$ for protons with energy greater than 2 MeV, and for neutrons with energy either greater than 20 MeV or less than 10 keV. In the energy intervals between 10 keV and 100 keV and between 2 MeV and 20 MeV, the weight factor has the value of 10 for neutrons. Neutrons are at their most damaging in the energy interval between 100 keV and 2 MeV, where the weight factor is $w_r = 20$. The same weight factor of 20 is also assigned to alpha particles and heavy nuclei. The weight factor is used to determine the dose equivalent, which is the absorbed dose times the weight factor.

The SI unit for the dose equivalent is the *sievert* (Sv), defined as

$$1 \text{ Sv} = w_r (1 \text{ Gy}). \tag{40.29}$$

Again, a non-SI unit is commonly used, the *rem* (short for "röntgen equivalent, man"). It is defined as

$$1 \text{ rem} = w_r (1 \text{ rd}). \tag{40.30}$$

Comparing equations 40.30, 40.29, and 40.27, we see that

$$1 \text{ Sv} = 100 \text{ rem}. \tag{40.31}$$

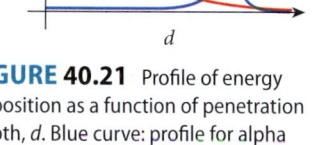

FIGURE 40.21 Profile of energy deposition as a function of penetration depth, *d*. Blue curve: profile for alpha particles and heavy ions. Red curve: profile for photons and leptons.

Radiation Exposure

What is the average dose of radiation that a typical U.S. resident receives each year? To answer this question, we need to consider many sources of radiation. Perhaps the easiest to quantify is the dose we receive from cosmic radiation, or cosmic rays. This dose strongly depends on the elevation above sea level of the city or town in which you live (Figure 40.22). At sea level you receive an annual dose equivalent of 0.26 mSv, whereas in Denver your annual dose equivalent is almost 0.5 mSv. If you build your dream house on a mountaintop, you may get up to 1 mSv per year.

Among other sources of natural radiation, the most important is radon gas. This gas is part of the air you breathe, and it deposits its radiation on the inside of your body, in your lungs. This radiation contributes 2 mSv per year to your dose equivalent. You also ingest radioactive isotopes (mainly $^{14}_{6}C$ and $^{40}_{19}K$) with your food, another 0.4 mSv annually. The Earth itself contributes radiation: 0.16 mSv at the Atlantic Coast, 0.3 mSv in the continental United States, except for the area around Denver, where the dose equivalent is 0.63 mSv per year.

Medical X-rays (dental imaging is 0.01 mSv at the lower end and upper-GI-tract imaging is 2.5 mSv at the upper end) and other procedures add an average of 0.5 mSv to your annual dose equivalent. Each hour of air travel contributes approximately 0.005 mSv to your dose equivalent; a passenger enjoying Gold Elite status (flying more than 50,000 mi/yr) receives an additional 0.5 to 1 mSv. Other minute sources of radiation (watching TV, working at a computer, airport screening, using a gas lantern while camping or a smoke detector, etc.) contribute a combined 0.1 mSv per year. The total annual exposure dose equivalent for the average U.S. resident, the sum of all of the above contributions, is approximately 3.6 mSv and is broken down by source in Figure 40.23.

People who work with radiation sources such as X-ray machines or nuclear reactors are strictly monitored for radiation exposure. Their maximum annual dose equivalent is not allowed to exceed 50 mSv, which is approximately 15 times the dose they receive from the natural environmental sources listed here.

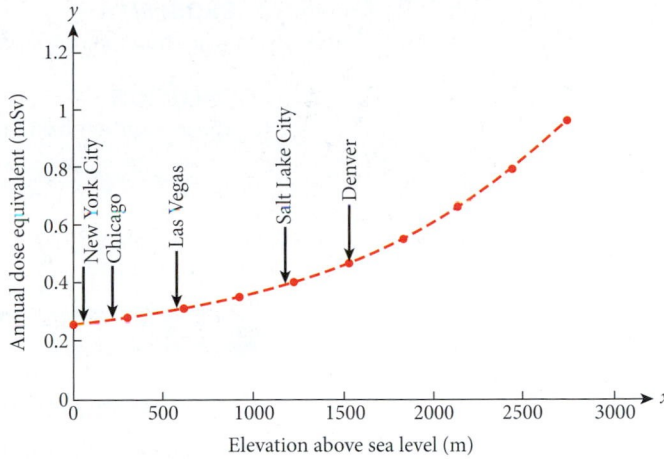

FIGURE 40.22 Annual dose equivalent from cosmic rays for U.S. residents as a function of elevation.

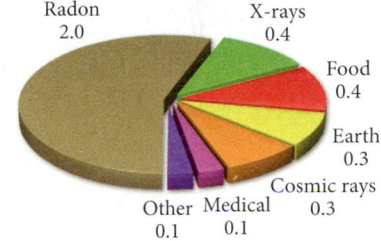

FIGURE 40.23 Average annual radiation exposure dose equivalent for a U.S. resident (in mSv).

EXAMPLE 40.4 | Chest X-Ray

Suppose your last physical exam included a chest X-ray, during which you received a dose of 60. μSv.

PROBLEM 1
What was your dose in millirems (mrem)?

SOLUTION
The conversion factor from sieverts to rems is given by equation 40.31, so you can convert your dose as follows:

$$\left(60.\cdot 10^{-6}\ \text{Sv}\right)\frac{100\ \text{rem}}{1\ \text{Sv}} = 6.0\cdot 10^{-3}\ \text{rem} = 6.0\ \text{mrem}.$$

PROBLEM 2
What was the absorbed dose in micrograys (μGy) and millirads (mrad)?

SOLUTION
The radiation weight factor, w_r, for X-rays is 1. Therefore, the absorbed dose according to equation 40.29 was

$$\frac{60.\ \mu\text{Sv}}{1\ \text{Sv/Gy}} = 60.\ \mu\text{Gy} \qquad \text{or} \qquad \frac{6.0\ \text{mrem}}{1\ \text{rem/rad}} = 6.0\ \text{mrad}.$$

PROBLEM 3
How much energy did you absorb, assuming that the X-rays illuminated 15 kg of your body?

SOLUTION
The gray (equation 40.26) is defined in terms of the energy absorbed divided by the mass over which the energy is absorbed. Thus, the energy you absorbed was

$$\left(60.\ \mu\text{Gy}\right)\left(15\ \text{kg}\right) = \left(60.\cdot 10^{-6}\ \text{J/kg}\right)\left(15\ \text{kg}\right) = 9.0\cdot 10^{-4}\ \text{J}.$$

– Continued

40.3 Nuclear Models

Liquid-Drop Model and Empirical Mass Formula

How can we understand the systematic trends of the binding energy per nucleon, given by equation 40.7 and displayed in Figure 40.7? One model of the nucleus that yields astonishingly good results is the **liquid-drop model.** In this model, the nucleus is treated as a spherical drop of quantum liquid composed of individual nucleons. Because the strong interaction between the nucleons inside the nucleus is attractive, each nucleon in the drop receives a positive contribution to the binding energy. Conventionally, this contribution is called the *volume term* and is proportional to the number of nucleons A:

$$B_v(N,Z) = a_v A = a_v(N+Z). \tag{40.32}$$

Here a_v is a positive constant that is adjusted to the data.

Nucleons at the surface of the liquid drop are not surrounded by other nucleons. Thus, they have fewer nearest-neighbor interactions and therefore are less strongly bound. Accordingly, the model must include a negative term that is proportional to the nuclear surface area. The surface area of a spherical drop is proportional to R^2, the square of its radius. According to equation 40.3, the nuclear radius is proportional to $A^{1/3}$. Therefore, the surface area of the nucleus is proportional to $A^{2/3}$, and the contribution of the nuclear surface to the overall binding energy can be parameterized as

$$B_s(N,Z) = -a_s A^{2/3} = -a_s(N+Z)^{2/3}, \tag{40.33}$$

with a positive fit constant a_s.

Next, we need to take into account that the protons are all positively charged and thus have repulsive Coulomb interactions with one another. Since the Coulomb potential is proportional to the square of the charge number and inversely proportional to the radius, we can write for this term:

$$B_c(N,Z) = -a_c \frac{Z^2}{A^{1/3}} = -a_c \frac{Z^2}{(N+Z)^{1/3}}, \tag{40.34}$$

where we have again made use of $R \propto A^{1/3}$ from equation 40.3. The constant a_c can be calculated if the protons are assumed to be evenly distributed throughout the nucleus. It has the value $a_c = 0.71$ MeV.

The Coulomb interaction alone might make us conclude that it is most advantageous to fill the nucleus with neutrons. Because neutrons do not experience the Coulomb interaction, the term in equation 40.34 does not contribute to their binding energy, so a number of neutrons alone would have a higher binding energy than the same number of neutrons and protons mixed. However, we also have to take into consideration that protons and neutrons are fermions and thus need to respect the Pauli Exclusion Principle (see Chapter 36) separately, for spin-up and spin-down protons and for spin-up and spin-down neutrons. Just as in the case of the electrons filling shells in the atom at higher and higher angular momentum and energy, higher and higher energy levels are filled as more protons and neutrons are added. Simply adding neutrons would become energetically unfavorable at some point. Thus, filling the nucleus approximately equally with protons and neutrons is encouraged by the Pauli Exclusion Principle (Figure 40.24).

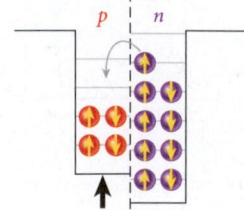

FIGURE 40.24 Illustration for the asymmetry term in the mass formula of the liquid-drop model.

This effect can be parameterized via an *asymmetry term,* which is negative (representing decreased binding) when the number of protons is different from the number of neutrons. This motivates the following expression for the asymmetry term, where $A = Z + N$ in the second equality:

$$B_a(N,Z) = -a_a \frac{\left(Z - \frac{1}{2}A\right)^2}{A} = -\frac{a_a}{4} \frac{(Z-N)^2}{N+Z}. \tag{40.35}$$

Finally, the two-body interaction of nucleon pairs depends on their relative spin. This leads to higher binding energies for nuclei that have all of their protons paired up and all of their neutrons paired up. Thus, an even number of protons in a nucleus results in an additional positive contribution to the binding energy. In the same way, an even number of neutrons creates a higher binding energy as well. Empirically supported results are found with this parameterization for the *pairing term*:

$$B_p(N,Z) = +a_p \frac{(-1)^Z + (-1)^N}{\sqrt{A}} = +a_p \frac{(-1)^Z + (-1)^N}{\sqrt{N+Z}}. \tag{40.36}$$

By combining equations 40.32 to 40.36, we can write an expression for the binding energy as a function of the mass number and charge number of a given nucleus:

$$B(N,Z) = B_v(N,Z) + B_s(N,Z) + B_c(N,Z) + B_a(N,Z) + B_p(N,Z)$$

$$= a_v A - a_s A^{2/3} - a_c \frac{Z^2}{A^{1/3}} - a_a \frac{\left(Z - \frac{1}{2}A\right)^2}{A} + a_p \frac{(-1)^Z + (-1)^N}{\sqrt{A}}.$$

Dividing this expression by the mass number, the binding energy per nucleon is

$$\frac{B(N,Z)}{A} = a_v - a_s A^{-1/3} - a_c \frac{Z^2}{A^{4/3}} - a_a \left(\frac{Z}{A} - \frac{1}{2}\right)^2 + a_p \frac{(-1)^Z + (-1)^N}{A^{3/2}}. \tag{40.37}$$

The constants in this expression can be used as fit parameters to obtain the best agreement with all binding energies for all nuclei. This procedure to explain the systematic trends in the binding energy per nucleon as a function of mass number was first developed by the German physicists Hans Bethe and Carl Friedrich von Weizsäcker in 1935, and the resulting **empirical mass formula** (equation 40.37) carries their name. Several successful fits have been published in the literature. We use the values obtained by Bertulani and Schechter (2002):

$$a_v = 15.85 \text{ MeV}, a_s = 18.34 \text{ MeV}, a_c = 0.71 \text{ MeV}, \\ a_a = 92.86 \text{ MeV}, a_p = 11.46 \text{ MeV}. \tag{40.38}$$

Figure 40.25 displays the same experimental values of the binding energies as Figure 40.7, but only for the nuclei with odd values of A. For these, the last term in equation 40.37 is zero, and so the effect of the other terms in the equation can be compared to the experimental data as a function of the mass number A. (The curve is more complicated for nuclei with even values of A, because they can have either an even number of protons and an even number of neutrons or an odd number of neutrons and an odd number of protons, which changes the sign of the term in equation 40.36 and thus does not produce a single curve like the one shown in Figure 40.7.) The yellow line in Figure 40.25 plots the (constant) volume term. Adding the surface term to it leads to the orange line, which shows very good agreement with the experimental data for low mass numbers. Adding the Coulomb term leads to the red line, which is close to the experimental data, showing that the Coulomb interaction becomes important for large nuclei. Finally, adding the asymmetry term leads to the green line, which gives a very good overall description of all binding energies of all odd-mass nuclei, at least at the resolution of the plot shown here.

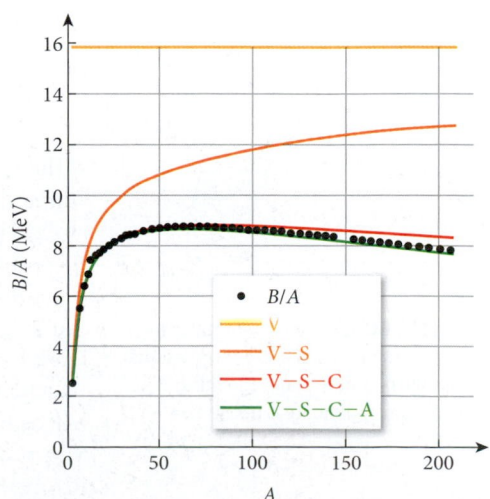

FIGURE 40.25 Binding energy per nucleon for nuclei with odd values of A. Black dots: experimental data. Lines are fits that include different numbers of terms in the Bethe-Weizsäcker formula (V: volume, S: symmetry, C: Coulomb, A: asymmetry).

The success of the Bethe-Weizsäcker formula (equation 40.37) for the binding energy is conventionally interpreted as strong support for the liquid-drop model of the nucleus. However, it is really only a consequence of the approximately spherical shape of nuclei, combined with the fact that the strong interaction is short-range and of the order of the nearest-neighbor spacing of the nucleons inside a nucleus.

Fermi Gas Model

Judging from the success of the liquid-drop model of the nucleus in reproducing the binding energies of nuclei, one might be tempted to think about nucleons as resting in a fixed and approximately spherical arrangement inside the nucleus. However, the Heisenberg uncertainty relation (see Chapter 36) imposes constraints on quantum particles (waves) confined to such a small volume. Since $\Delta p_x \cdot \Delta x \geq \frac{1}{2}\hbar$ and since a nucleon inside a nucleus is localized to within a few femtometers, there is a high uncertainty in momentum, on the order of 100 MeV/c. (If you remember the handy rule $\hbar c = 197.33$ MeV/fm, this is easy to see.) That means that nucleons inside the nucleus cannot be considered to be resting in some fixed configuration, but rather they have fairly high momentum uncertainty and thus momentum.

The **Fermi gas model** of the nucleus (proposed by the Italian-American physicist Enrico Fermi, 1901–1953) takes this momentum into consideration and assumes that nucleons can move freely inside the nucleus like particles of a gas (Figure 40.26).

Chapter 37 showed that a particle with mass m and momentum k confined to a cubic box of width a in three dimensions, but able to move freely inside the box, has a total energy of

$$E = \frac{\hbar^2}{2m}k^2 = \frac{\hbar^2}{2m}\left(k_x^2 + k_y^2 + k_z^2\right) = \frac{\hbar^2\pi^2}{2ma^2}\left(n_x^2 + n_y^2 + n_z^2\right). \qquad (40.39)$$

The Pauli Exclusion Principle allows each quantum state to be occupied by exactly four nucleons (one spin-up and one spin-down proton and one spin-up and one spin-down neutron, for each state).

We can use this information to calculate the density of states, $dN(E)$, which will tell us how many quantum states reside in an interval dE around a given energy E, for the Fermi gas model.

Figure 40.27 illustrates the possible momentum states that can be occupied by the nucleons in the Fermi gas model. Each possible quantum state occupies a volume of $(\pi/a)^3$. Then the number of states, $dN(k)$, between the absolute value of the momentum k and $k + dk$ is given by the ratio of the volume of the spherical shell with radius k and thickness dk and the volume occupied by each quantum state:

$$dN(k) = \frac{\frac{1}{8}4\pi k^2 dk}{(\pi/a)^3} = \frac{a^3 k^2}{2\pi^2}dk. \qquad (40.40)$$

(The factor $\frac{1}{8}$ appears in this equation because we are considering only the octant of the three-dimensional sphere in which all three momentum coordinates are positive. Negative values of the momentum coordinates do not add additional solutions, because the energy is proportional to k^2.)

Now we need to convert the density of momentum states, $dN(k)$, which we just calculated, into a density of states, $dN(E)$. Equation 40.39 shows that $E = \hbar^2 k^2/2m$, and therefore $k = \sqrt{2mE}/\hbar$, so the derivative of the energy with respect to the wave number can be calculated: $dE/dk = \hbar^2 k/m$. The differential dk can be expressed in terms of the differential dE as follows:

$$dk = \frac{dE}{\hbar\sqrt{2E/m}}.$$

Combining this result with equation 40.40 gives the density of states as a function of energy:

$$dN(E) = \frac{a^3 m^{3/2}}{2^{1/2}\pi^2\hbar^3}E^{1/2}dE. \qquad (40.41)$$

Self-Test Opportunity 40.2

Figure 40.13 shows that the masses of the $A = 82$ nuclei with odd charge numbers lie on a parabola (dashed line), and the masses of the $A = 82$ nuclei with even charge numbers lie on another parabola. Explain.

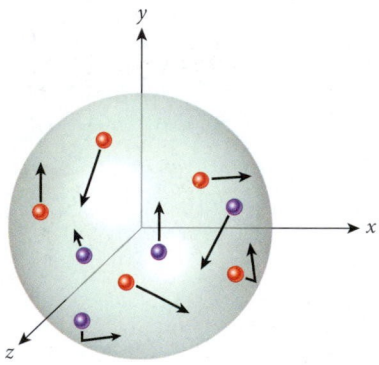

FIGURE 40.26 The Fermi gas model of the nucleus. Protons (purple) and neutrons (red) move independently and freely throughout the interior of the nucleus and are reflected off the nuclear surface that confines them.

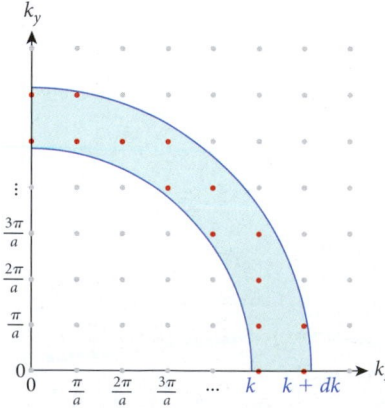

FIGURE 40.27 Density of states in the Fermi gas model. Each dot represents a possible quantum state. The third momentum coordinate is not shown.

The maximum energy required to accommodate all A nucleons in a nucleus according to the Fermi gas model is called the **Fermi energy**, E_F. To find the Fermi energy, we need to integrate the density of states (equation 40.41) up to that energy and set this number equal to $\frac{1}{4}A$. (Remember that each quantum state can be occupied by four nucleons.)

$$\int_0^{n(E_F)} dN(E) = \frac{1}{4}A \Rightarrow$$

$$\int_0^{E_F} \frac{a^3 m^{3/2}}{2^{1/2}\pi^2\hbar^3} E^{1/2} dE = \frac{2^{1/2} a^3 m^{3/2}}{3\pi^2\hbar^3} E_F^{3/2} = \frac{1}{4}A \Rightarrow$$

$$E_F = \frac{\hbar^2}{2m}\left(\frac{3A\pi^2}{2a^3}\right)^{2/3} = \frac{\hbar^2}{2m}\left(\frac{3\pi^2}{2}n\right)^{2/3}, \tag{40.42}$$

where $n = A/a^3$ is the nucleon density inside the nucleus. Using $n = 0.17 \text{ fm}^{-3}$ for the nucleon density (Example 40.1), a value of 938.9 MeV/c^2 as the average of the proton and neutron mass, and $hc = 197.33$ MeV fm, we arrive at the numerical value of the Fermi energy:

$$E_F = \frac{(197.33 \text{ MeV fm}/c)^2}{2(938.9 \text{ MeV}/c^2)}\left(\frac{3\pi^2}{2}(0.17 \text{ fm}^{-3})\right)^{2/3} = 38 \text{ MeV}.$$

The density of the states occupied by nucleons inside the nuclear potential well is sketched in Figure 40.28 as a function of the energy relative to the bottom of the well. You can see that the largest probability of finding a nucleon inside a nucleus corresponds to an energy just below the Fermi energy. Note that the case displayed here is for the temperature $T = 0$ eV; for nonzero temperatures, the vertical drop-off at the Fermi energy is replaced by a more gradual decline.

The *Fermi momentum*, p_F, is the momentum (magnitude) that corresponds to the Fermi energy:

$$p_F = \sqrt{2mE_F} = 270 \text{ MeV}/c$$

$$k_F = p_F/\hbar = 1.36 \text{ fm}^{-1}.$$

Dividing the Fermi momentum by the nucleon mass gives the maximum velocity with which nucleons move around inside the nucleus: $v_F = p_F/m = 0.29c$. That is, nucleons move around inside the nucleus with speeds of up to almost 30% of the speed of light!

Are these nucleon momentum values inside the nucleus real, or are they an artifact of a false interpretation of a quantum result in terms of classical physics? The answer is that they are real. This can be shown via particle production in heavy ion collisions, where the maximum energy of the particles produced can exceed the beam energy, a result that can be traced back directly to the Fermi momentum of the nucleons inside the nucleus.

The picture of the nucleus that emerges in this model can be illustrated as shown in Figure 40.29: Nucleons inside the nucleus are bound by a potential of depth V_0, which is the sum of the Fermi energy and the separation energy of the least-bound nucleon (S_n or S_p). Separation energies are typically on the order of 8 MeV. Thus, the depth of the nuclear well is approximately $V_0 = 8 \text{ MeV} + 38 \text{ MeV} = 46 \text{ MeV}$. This result is valid for all nuclei with A larger than about 12, because from then on the nucleon density is a constant $n = 0.17 \text{ fm}^{-3}$.

This picture can be refined by introducing a separate Fermi energy for neutrons and protons. This gives rise to the asymmetry term (equation 40.35) in the Bethe-Weizsäcker formula (compare also Figure 40.24).

Shell Model

Chapter 38 showed that the electrons in atoms arrange themselves in shells because of the combined effects of their Coulomb interactions with the positively charged nucleus, quantum mechanics, and the Pauli Exclusion Principle. One essential piece of experimental evidence that supports the existence of electron shells is the separation (ionization) energy of

Self-Test Opportunity 40.3

Derive equation 40.41.

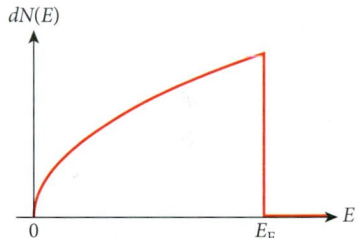

FIGURE 40.28 Density of states as a function of energy in the Fermi gas model. Here the energy is measured relative to the bottom of the potential well.

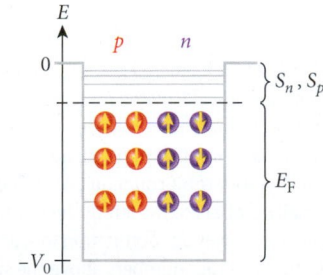

FIGURE 40.29 Approximation for the nuclear potential in the Fermi gas model.

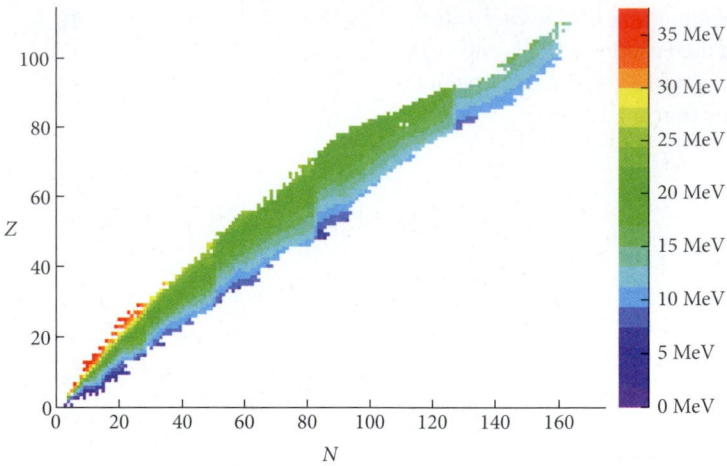

FIGURE 40.30 Experimentally measured two-neutron separation energies.

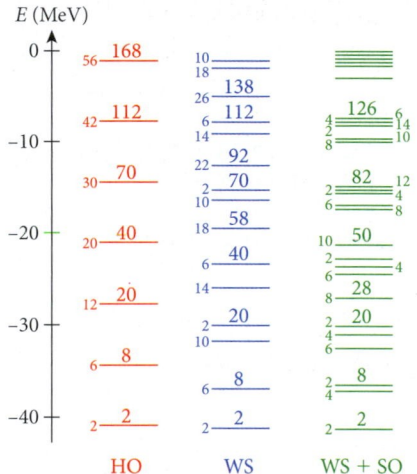

FIGURE 40.31 Energy levels for neutrons in a lead nucleus, calculated for three different potentials: harmonic oscillator (HO), Woods-Saxon (WS), and Woods-Saxon with strong spin-orbit coupling (WS + SO). The small numbers to the left of the levels indicate the maximum occupation for each level, and the larger numbers above the states indicate the cumulative sum of the occupation numbers.

the outermost electron as a function of the charge number, which shows a pronounced maximum for a filled shell. In Example 40.2, we calculated the two-neutron separation energies for a series of tin isotopes and found a pronounced jump of the separation energy at a neutron number of 82. Figure 40.30 shows a compilation of two-neutron separation energies for all known isotopes. It is clear that it becomes harder to remove neutron pairs from nuclei with a given charge number Z and fewer neutrons N. This observation is a consequence of the asymmetry term (equation 40.35) for the liquid-drop model. However, the color coding in Figure 40.30 shows sudden jumps at certain neutron numbers, which are superimposed on the general trend of rising two-neutron separation energies as we move from right to left in a given isotope chain. It is clear from this figure that the neutron numbers 50, 82, and 126 are special, because they exhibit these sudden jumps. When two-proton separation energies are compiled, a similar picture emerges, with sudden jumps at 50 and 82 as we move along lines of constant neutron number. (There is no nucleus with charge number $Z = 126$.)

These and other data indicate that when certain numbers of protons or neutrons occur in a nucleus, a shell is filled. These numbers are called **magic numbers.** The magic numbers indicate that shell closures exist in nuclei just as they exist for the electron configurations in atoms. The magic numbers of either neutrons or protons in nuclei are

$$2, 8, 20, 28, 50, 82, 126.$$

For comparison, the magic numbers for the electron configurations are the charge numbers of the noble gases:

$$2, 10, 18, 36, 54, 86.$$

These sets of numbers are similar, but obviously not identical. Chapter 38 showed that the magic numbers of electrons indicate filled shells. The shells are formed by maximally occupying states with a given orbital angular momentum ℓ, where the maximum occupation of a given state is $2\ell + 1$. In the case of the electrons in an atom, the potential is provided by the Coulomb potential of the nucleus.

The potential for the nucleons inside the nucleus, on the other hand, is provided by the collective action of the other nucleons. Nevertheless, the quantum numbers of the nucleons inside the nuclear potential can also be calculated. Figure 40.31 shows the energy levels of neutrons inside the lead nucleus, calculated for three different potentials. The left side represents the harmonic oscillator potential, for which the solutions were calculated in Chapter 37. These energy levels are equally spaced. Interestingly, this simple harmonic oscillator potential reproduces the three lowest magic numbers. A similar agreement is achieved by using the Woods-Saxon potential, $V(r)$, which has the same dependence on the radial coordinate r as the nucleon density has (equation 40.4):

$$V(r) = -\frac{V_0}{1 + e^{(r-R(A))/a}}. \tag{40.43}$$

Again, the surface thickness constant is $a = 0.54$ fm, and the radius is given by equation 40.3 as $R(A) = R_0 A^{1/3}$. The depth of the potential is $V_0 = 50$ MeV, close to that obtained for the Fermi gas model. The energy levels in this potential cannot be calculated analytically, but only with the aid of a computer. It is still possible to assign angular momentum quantum numbers to these numerical solutions, though, and so the maximum occupation numbers for the Woods-Saxon potential can be calculated. The results of this calculation are shown in blue in Figure 40.31. The maximum occupation number is given for each level. The cumulative occupation numbers are shown only where there is a larger-than-average energy gap to the next level.

The decisive insight into the nuclear shell model was achieved in 1949 by the Germans Maria Goeppert Mayer, Otto Hahn, J. Hans, D. Jensen, and Hans Suess. They realized that an essential part of the nuclear interaction is the coupling between the spin and the orbital angular momentum. (We will not derive this interaction term here, but merely state that it exists; a course on advanced quantum physics will fill this gap.) This spin-orbit coupling is also present for electrons, but it is much larger for nucleons. If a spin-orbit interaction potential with the correct strength is added to the Woods-Saxon potential, we obtain the energy levels shown in green on the right-hand side of Figure 40.31. Now we see that the large energy gaps between the levels occur at the magic numbers observed in experiment.

This simple shell model has been refined repeatedly since it was introduced, and it remains the main workhorse of low-energy nuclear structure calculations. Because the numerical solution of the shell model for heavy nuclei involves inversion of extremely large matrices, it cannot be achieved exactly, even using the largest available computers. Research on the nuclear shell model is thus a vibrant field, focusing on novel approximation techniques and steady refinement of nuclear interactions.

Other Models of the Nucleus

The three models of the nucleus presented so far are by no means the only ones. Nuclei are a very interesting testing ground for many theoretical models, in particular for understanding collisions of nuclei, in which they can be heated and compressed. Models of nuclear collisions involve several subfields of physics: quantum physics, fluid dynamics, thermodynamics, and electromagnetism. Nuclei can even undergo thermodynamic phase transitions between liquid-like and gas-like phases. The ongoing study of these phenomena, their technical applications, and their interdisciplinary relevance promises to yield interesting results in the coming decades.

As one example of a current nuclear model, we show a result from our own research on a nuclear transport model. Figure 40.32 shows the time sequence of a computer simulation of a collision of krypton-86 with niobium-93. The nuclei experience an off-center collision at an impact parameter of 5 fm. The time between two successive frames is 8.0 fm/c = $2.7 \cdot 10^{-23}$ s. Displayed is the nuclear density, which is color-coded according to the scale on the right-hand side of the figure. The compression, deformation, and sideways deflection of the nuclei are clearly visible. From the comparison of these and other model simulations to experimental data, we are able to draw conclusions about the dynamics and thermodynamics of nuclei and the properties of nuclear matter.

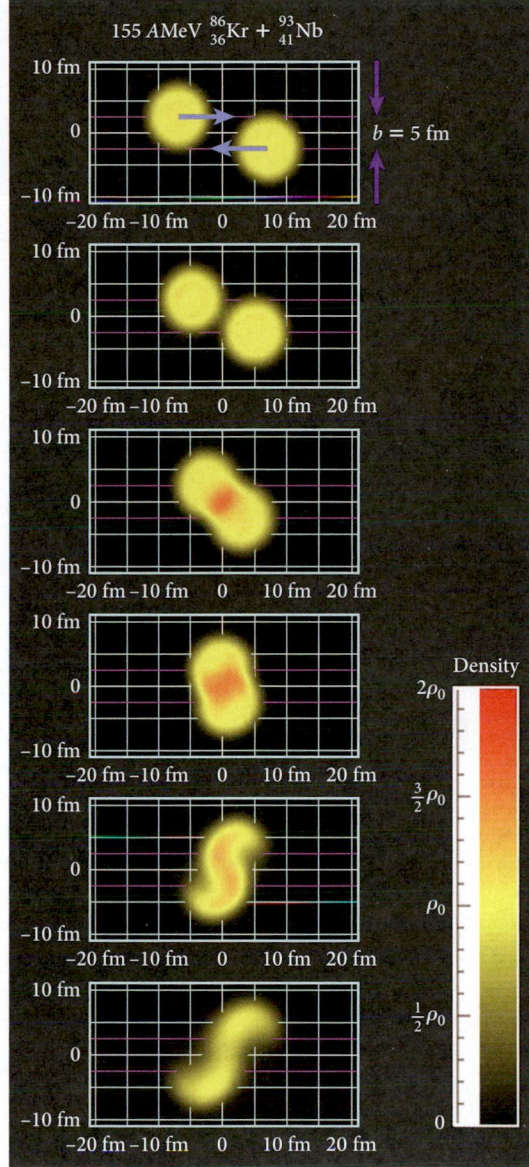

FIGURE 40.32 Time sequence of a nuclear collision, as calculated with a nuclear transport model.

40.4 Nuclear Energy: Fission and Fusion

Let's return once again to Figure 40.7, which shows the binding energy per nucleon as a function of the mass number for stable isotopes. The iron and nickel nuclei have the largest binding energies per nucleon and are thus most strongly bound. Any other arrangement of nucleons into nuclei, lighter or heavier, is less strongly bound. This suggests that very heavy nuclei can be split into two more strongly bound medium-mass nuclei, with the reaction showing a positive Q-value (Figure 40.33). Such a reaction yields useful energy, in a process called **nuclear fission.** In the same way, very light nuclei can be combined into heavier ones in reactions that have positive Q-values and associated useful energy output, in the process of **nuclear fusion.** This is the entire physical basis for the energy production by the Sun and almost all other stars, for fission and fusion reactor power plants, as well as nuclear weapons.

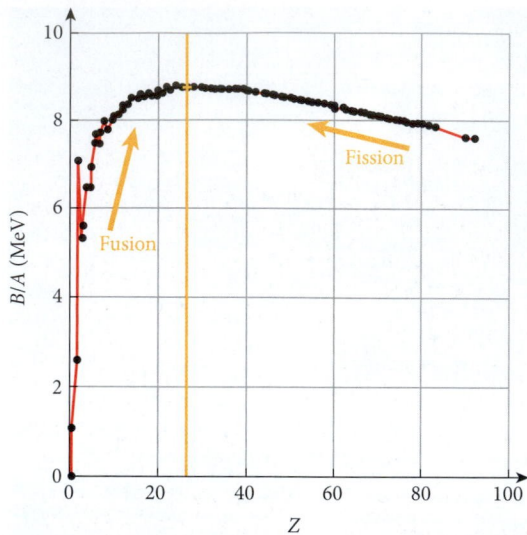

FIGURE 40.33 Energy gain through fission and fusion reactions.

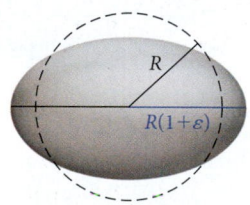

FIGURE 40.34 Nuclear deformation and definition of the deformation parameter.

Nuclear Fission

The nuclear fission process represents the splitting of an atomic nucleus into two smaller nuclei, often with the emission of one or more neutrons. The two smaller nuclei have mass numbers close to half that of the parent nucleus. As we will see, nuclei that undergo fission typically have mass numbers of 230 or greater, so isotopes resulting from nuclear fission typically have mass numbers ranging from 100 to 150. Important examples of fission fragments are the isotopes of krypton through barium and some isotopes of the lanthanides. These fission fragments tend to undergo subsequent radioactive decays, sometimes with very long half-lives. Many of them are easily ingested and accumulated in our bodies, so they pose great health risks. Examples are technetium-99 (more on this isotope later), cesium-137 (30-year half-life), and iodine-131 (8-day half-life), which is stored in the human thyroid.

A nucleus that is undergoing fission has to pass through a deformed configuration not unlike that in alpha decay (see Figure 40.11). A quantitative understanding of the onset of spontaneous fission in heavy nuclei requires a criterion in terms of mass number and charge number. To develop this criterion, we first parameterize the deformation of the nucleus (Figure 40.34) in terms of an ellipsoid with a semimajor axis of length $R(1 + \epsilon)$, where R is the radius of the same nucleus if it were spherical and ϵ is the **deformation parameter.**

When a nucleus is deformed from its spherical configuration, how is the binding energy per nucleon (equation 40.37) affected? The volume term is not changed, because it depends only on the number of nucleons, which stays the same. Since the charge-to-mass ratio is not changed, the symmetry term should also not be affected by deformation. The numbers of neutron pairs and proton pairs, and therefore the pairing term in equation 40.37, also remain constant. However, the surface is slightly increased when the sphere is deformed, so the surface term increases as a function of deformation. In addition, the protons are on average farther apart in the deformed nucleus, so the Coulomb term decreases. Niels Bohr (1885–1962) and John Archibald Wheeler (1911–2008) showed in 1939 that the change in the surface term as a function of deformation is (to leading order) given by $\Delta B_s(\epsilon) = \frac{2}{5}\epsilon^2 B_s(N,Z) = -\frac{2}{5}\epsilon^2 a_s A^{2/3}$, and the change in the Coulomb term by $\Delta B_c(\epsilon) = -\frac{1}{5}\epsilon^2 B_c(N,Z) = \frac{1}{5}\epsilon^2 a_c A^{-1/3} Z^2$. Therefore, the total change in the binding energy is

$$\Delta B(\epsilon) = \Delta B_c(\epsilon) + \Delta B_s(\epsilon)$$
$$= \frac{1}{5}\epsilon^2 a_c A^{-1/3} Z^2 - \frac{2}{5}\epsilon^2 a_s A^{2/3}$$
$$= \frac{1}{5}\epsilon^2 (a_c A^{-1/3} Z^2 - 2a_s A^{2/3}).$$

If $\Delta B(\epsilon) > 0$, energy can be gained by deforming a nucleus, and it can fission instantaneously. This transition point to spontaneous fission is reached when

$$\Delta B(\epsilon) = 0 \Rightarrow a_c A^{-1/3} Z^2 - 2a_s A^{2/3} = 0 \Rightarrow$$
$$\frac{Z^2}{A} = \frac{2a_s}{a_c} = \frac{2(18.34 \text{ MeV})}{0.71 \text{ MeV}} \approx 51.7,$$

using the values given in equation 40.38.

The quantity Z^2/A is called the *fissionability parameter*. Higher fissionability means a higher likelihood of fission and thus a shorter lifetime for the isotope.

In Figure 40.16, you can see that spontaneous fission is the dominant decay mode for some of the heaviest isotopes. However, in almost all cases where spontaneous fission dominates, the half-life of the isotope is so short that it does not exist in nature. Much more important for the use of nuclear fission in nuclear reactors and nuclear weapons is the very long-lived isotope $^{235}_{92}$U, which has a half-life of 700 million years and which can be found in natural uranium deposits at 0.7% abundance. The most abundant uranium isotope is $^{238}_{92}$U, with a natural abundance of 99.3% and a half-life of 4.5 billion years. Both isotopes decay overwhelmingly via emission of an alpha particle and have only a very small probability of

Concept Check 40.4

Which of the following isotopes has the largest fissionability?

a) $^{235}_{92}$U   c) $^{254}_{98}$Cf

b) $^{240}_{94}$Pu   d) $^{258}_{100}$Fm

undergoing spontaneous fission. However, $^{235}_{92}$U fissions almost instantaneously after being hit by a neutron of fairly low energy. This process releases two or three neutrons, which can then trigger further induced fission reactions in other $^{235}_{92}$U nuclei. A typical neutron-induced fission reaction of $^{235}_{92}$U is

$$n + {}^{235}_{92}\text{U} \rightarrow {}^{141}_{56}\text{Ba} + {}^{92}_{36}\text{Kr} + 3n,$$

which releases 173.3 MeV of energy.

If sufficient $^{235}_{92}$U is present, the resulting **chain reaction** will exponentially multiply the number of neutrons in the sample and cause a nuclear explosion. However, for this to happen, the uranium needs to be predominantly composed of $^{235}_{92}$U and not $^{238}_{92}$U (which does not show induced fission when bombarded with low-energy neutrons). Thus, for nuclear weapons, the concentration of $^{235}_{92}$U needs to be *enriched*. Uranium is said to be *weapons-grade* if it contains 80% or more $^{235}_{92}$U. Enrichment can be done by various means, but the dominant technique is to use gas centrifuges for separating the isotopes.

How much weapons-grade uranium is needed to make a nuclear weapon, that is, what is the **critical mass?** The optimum geometrical arrangement of the uranium is in the shape of a sphere. Within the sphere, the neutrons trigger further induced fission reactions, whereas on the surface of the sphere, on average, half of the neutrons escape. The critical mass of a sphere of $^{235}_{92}$U is approximately 50 kg. Since uranium has a mass density 19.2 times that of water, the radius of a sphere of uranium $^{235}_{92}$U with the critical mass is only approximately 8.6 cm. In a nuclear weapon, a sufficient quantity of $^{235}_{92}$U is kept separated in several pieces and then detonated by pushing these pieces together via the use of conventional chemical explosives. The exact details of this process are closely guarded military secrets.

Quantities of $^{235}_{92}$U can also be used for power production in nuclear fission reactors. The uranium needed for this is much less enriched than that used for nuclear weapons. The key to reactor construction is to keep the chain reaction controlled and thus avoid runaway reactions. This is achieved by using moderator materials, which capture neutrons and thus make them unavailable to induce further fission reactions. The disastrous reactor core meltdown at the Ukrainian city of Chernobyl on April 26, 1986, was the result of such a runaway reaction when the moderator system failed. The second biggest nuclear disaster happened on March 11, 2011, when an earthquake and subsequent tsunami caused a series of catastrophic equipment failures at the Fukushima Daiichi nuclear power plant complex in Japan. The resulting nuclear meltdown released significant amounts of radioactive materials into the environment.

A second isotope that is very important for nuclear weapons is plutonium-239. $^{239}_{94}$Pu is produced in reactors from $^{238}_{92}$U via neutron capture and subsequent β^- decays:

$$^{238}_{92}\text{U} + n \rightarrow {}^{239}_{92}\text{U} + \gamma$$
$$^{239}_{92}\text{U} \rightarrow {}^{239}_{93}\text{Np} + e^- + \bar{\nu}_e$$
$$^{239}_{93}\text{Np} \rightarrow {}^{239}_{94}\text{Pu} + e^- + \bar{\nu}_e.$$

The process of producing $^{239}_{94}$Pu is often referred to as **breeding.** $^{239}_{94}$Pu also undergoes neutron-induced fission and gives a higher yield of neutrons than $^{235}_{92}$U. Thus, its critical mass is only 10 kg, and it is much more valuable for nuclear weapons than $^{235}_{92}$U. However, breeding weapons-grade plutonium is not easy because $^{239}_{94}$Pu can easily capture another neutron and turn into $^{240}_{94}$Pu, which does not show spontaneous fission. Too high a fraction of $^{240}_{94}$Pu will lead to precritical detonation (fizzle) of a warhead, so the $^{239}_{94}$Pu needs to be sufficiently pure to be useful in weapons.

Clearly, nuclear weapons pose a great threat, and increasing numbers of countries are acquiring this dangerous technology. After the United States and Russia in the 1950s, China, France, Great Britain, Israel, India, Pakistan, and perhaps others (such as North Korea, Iran, and South Africa) have become nuclear powers. With every additional country that is able to produce nuclear weapons, the danger increases that the weapons could fall into the wrong hands. Small nuclear warheads fit into suitcases, and terrorists have turned their attention to them. Even detonating a bomb of conventional explosives mixed with radioactive material, called a *dirty bomb,* could have devastating consequences.

Equally clearly, the peaceful use of nuclear power from fission reactions is a way to avoid the greenhouse gas emissions associated with the burning of fossil fuels in conventional power plants. However, the storage of radioactive wastes with very long half-lives is still a problem that is not completely solved. Countries such as Germany, for example, have sworn off the use of fission reactors and are turning their attention to alternative carbon-neutral energy sources (wind, biofuels, hydroelectric, photovoltaics, and others) to solve the greenhouse gas problem. However, with global energy consumption on a very steep rise, it remains to be seen if these alternatives to fission power can satisfy the world's energy needs.

EXAMPLE 40.5 | Fission Energy Yield

PROBLEM

What is the energy yield per kilogram of uranium for the neutron-induced fission reaction of $^{235}_{92}$U to form $^{92}_{36}$Kr and three neutrons, if $^{235}_{92}$U is at 20% enrichment in the reactor?

SOLUTION

First, we need to write the equation for this fission reaction. This will tell us what the second fission fragment is. The reactants (a uranium nucleus and a proton) have 92 protons, the charge number of uranium. Since one of the fission fragments is krypton with 36 protons, the other fission fragment has $Z = 92 - 36 = 56$ protons. This means that it is a barium nucleus. For the total mass number for the reactants, we find $A = 236$. Subtracting the mass number total for the three neutrons and the mass number 92 of the krypton isotope, we find the mass number of the barium isotope: $A = 236 - 3 - 92 = 141$. Thus, the induced fission reaction equation is

$$n + {}^{235}_{92}\text{U} \rightarrow {}^{141}_{56}\text{Ba} + {}^{92}_{36}\text{Kr} + 3n.$$

The mass of uranium-235 is 235.0439299 u, the mass of barium-141 is 140.914411 u, the mass of krypton-92 is 91.92615621 u, and the mass of a neutron is 1.008664916 u. Therefore, the mass difference between the reactants and the products is

$$235.0439299 - (140.914411 + 91.92615621 + 2 \cdot 1.008664916) = 0.186033 \text{ u}.$$

In electron-volts, the mass-energy difference between the initial and final states is 173.3 MeV. This is also the sum of the kinetic energies of the three emitted neutrons plus the two fission fragments minus the kinetic energy of the initial neutron.

This energy output can be converted into joules:

$$\Delta E = (173.3 \text{ MeV})(1.602 \cdot 10^{-13} \text{ J/MeV}) = 2.776 \cdot 10^{-11} \text{ J}.$$

A mixture of 80% $^{238}_{92}$U and 20% $^{235}_{92}$U has an average mass number of $0.8(238) + 0.2(235) = 237.4$. Thus, 237.4 g of uranium at the given enrichment has $6.022 \cdot 10^{23}$ uranium atoms (Avogadro's number, see Chapter 13). Of these atoms, 20% are uranium-235 and can undergo neutron-induced fission. Therefore, 237.4 g of the enriched uranium mixture contains $0.2(6.022 \cdot 10^{23}) = 1.2044 \cdot 10^{23}$ uranium-235 atoms, and consequently 1 kg of the mixture contains $1.2044 \cdot 10^{23}/0.2374 = 5.073 \cdot 10^{23}$ uranium-235 atoms.

All that is left to do is to multiply the number of atoms times the energy yield per atom to obtain the total energy yield from fission of a kilogram of the enriched uranium:

$$(2.776 \cdot 10^{-11} \text{ J})(5.073 \cdot 10^{23}) = 14.08 \cdot 10^{12} \text{ J}.$$

Keep in mind that not all of this energy is available for feeding into the electric grid. Some uranium-235 will always be left nonfissioned, and some of the neutrons released will hit uranium-238 and breed it into plutonium.

For comparison, the combustion of 1 kg of gasoline yields a maximum of $4.6 \cdot 10^7$ J. This means that the nuclear fission of enriched uranium yields more than 300,000 times more energy than the combustion of the same mass of gasoline.

Thorium Fission Cycle

Uranium also has the isotope $^{233}_{92}\text{U}$, which can decay via spontaneous fission (or via neutron-induced fission) and has a half-life of 159,200 years. It can be produced from thorium via a neutron-capture reaction:

$$^{232}_{90}\text{Th} + n \rightarrow \ ^{233}_{90}\text{Th} + \gamma$$

$$^{233}_{90}\text{Th} \rightarrow \ ^{233}_{91}\text{Pa} + e^- + \nu_e$$

$$^{233}_{91}\text{Pa} \rightarrow \ ^{233}_{92}\text{U} + e^- + \nu_e.$$

The spontaneous fission of $^{233}_{92}\text{U}$ releases 198 MeV of energy, and this isotope is thus useful as a fuel for fission reactors. The dominant decay mode of $^{233}_{92}\text{U}$ is alpha decay to $^{229}_{90}\text{Th}$. This thorium isotope can then capture three neutrons and become $^{232}_{90}\text{Th}$, and the thorium fission cycle can repeat, until, in principle, all of the fuel has undergone fission. In theory, the fission of $^{233}_{92}\text{U}$ should provide enough neutrons to keep this cycle going as a chain reaction, but in practice even small losses of neutrons prevent this from occurring.

The protactinium isotope $^{233}_{91}\text{Pa}$ is the key to the neutron balance of the thorium fission cycle, because it can capture neutrons before it has a chance to undergo beta decay to $^{233}_{92}\text{U}$ (with a mean lifetime of 27 days). One strategy is to chemically remove the $^{233}_{91}\text{Pa}$ from the reactor core, to prevent it from capturing the neutrons. This can be accomplished using a molten-salt reactor design. Another strategy, implemented by an accelerator-driven reactor design, is to supply additional neutrons to the reactor via an external particle accelerator. Both of these reactor designs are significantly more challenging to develop than designs based on the fission of $^{235}_{92}\text{U}$. However, they have several advantages: less radioactive waste, significantly improved fuel utilization, and increased safety due to the impossibility of runaway reactions like those responsible for the meltdowns at Chernobyl and Fukushima.

Thorium is approximately four times more abundant on Earth than uranium; there are approximately 400,000 tons of minable thorium dioxide in the United States alone. Successful utilization of the thorium fission cycle could therefore satisfy the world's energy needs for the next few thousand years. Several countries, most notably China, have started large-scale efforts to utilize this thorium resource.

Nuclear Fusion

The nuclear fusion process merges light nuclei into heavier ones and thus also increases the binding energy per nucleon, as long as the end product of the fusion process is lighter than iron.

Stellar Fusion

Nuclear fusion is the basic process of energy production in stars. Two main reaction chains, discovered in 1938 by Hans Bethe, accomplish this. The first is the proton-proton chain, and the second is the CNO cycle. Both of these serve to fuse four hydrogen atomic nuclei into one helium nucleus.

The **proton-proton chain** starts with two protons forming a deuteron plus a positron. This is a weak interaction process and takes a long time, approximately 10 billion years on average (half-life). This explains why the Sun has already existed for approximately 5 billion years and did not explode right after it formed. The deuteron from the initial reaction can quickly capture another proton and form helium-3; the positron from the initial reaction annihilates almost instantaneously with an electron:

$$p + p \rightarrow d + e^+ + \nu_e \qquad Q = 0.42 \text{ MeV}$$

$$d + p \rightarrow \ ^3_2\text{He} + \gamma \qquad Q = 5.49 \text{ MeV}$$

$$e^+ + e^- \rightarrow 2\gamma \qquad Q = 1.02 \text{ MeV}.$$

The net result of these three reactions is the fusion of three protons into one helium-3 nucleus plus three photons and one neutrino. The combined Q-value of the three reactions is $Q_{\text{net}} = (0.42 + 5.49 + 1.02) \text{ MeV} = 6.93 \text{ MeV}$.

The next step involves three different reactions through which fusion can proceed toward helium-4. In the Sun, 86% of the time, two helium-3 nuclei undergo a fusion reaction that has a Q-value of 12.86 MeV:

$$\,^3_2\text{He} + \,^3_2\text{He} \rightarrow \,^4_2\text{He} + p + p.$$

Adding the Q-value of this reaction to the Q-values for the formation of the two helium-3 nuclei gives

$$Q_{\text{net}} = (2 \cdot 6.93 + 12.86) \text{ MeV} = 26.7 \text{ MeV}.$$

This is exactly the mass difference between four protons and one helium-4 nucleus, as expected from conservation of energy.

Alternatively, 14% of the time, a helium-3 nucleus fuses with an existing helium-4 nucleus in the Sun:

$$\,^3_2\text{He} + \,^4_2\text{He} \rightarrow \,^7_4\text{Be} + \gamma.$$

This beryllium-7 nucleus then captures an electron to form lithium-7, which in turn undergoes a fusion reaction to form two helium-4 nuclei:

$$\,^7_4\text{Be} + e^- \rightarrow \,^7_3\text{Li} + \nu_e$$

$$\,^7_3\text{Li} + p \rightarrow 2\,^4_2\text{He}.$$

The production of 26.7 MeV of kinetic energy from the fusion of four protons into one helium-4 nucleus is the basic reason why the Sun shines. What happens to this energy? Part of it is carried away by neutrinos. Due to the low cross sections for their interactions, these neutrinos can leave the Sun right away on straight paths. The photons, on the other hand, scatter off the electrons and ions in the Sun, become absorbed and re-emitted in random directions, and take, on average, approximately 50,000 years to reach the Sun's surface.

The name of the **CNO cycle** incorporates the chemical symbols of the elements carbon, nitrogen, and oxygen. This cycle is dominant in stars that are significantly more massive than the Sun. In the CNO cycle, a carbon nucleus serves as a catalyst (meaning that it is regenerated by the end of the cycle) to fuse four protons into one helium-4 nucleus:

$$\,^{12}_6\text{C} + p \rightarrow \,^{13}_7\text{N} + \gamma \qquad\qquad Q = 1.95 \text{ MeV}$$

$$\,^{13}_7\text{N} \rightarrow \,^{13}_6\text{C} + e^+ + \nu_e \qquad\qquad Q = 1.20 \text{ MeV}$$

$$e^+ + e^- \rightarrow 2\gamma \qquad\qquad Q = 1.02 \text{ MeV}$$

$$\,^{13}_6\text{C} + p \rightarrow \,^{14}_7\text{N} + \gamma \qquad\qquad Q = 7.54 \text{ MeV}$$

$$\,^{14}_7\text{N} + p \rightarrow \,^{15}_8\text{O} + \gamma \qquad\qquad Q = 7.35 \text{ MeV}$$

$$\,^{15}_8\text{O} \rightarrow \,^{15}_7\text{N} + e^+ + \nu_e \qquad\qquad Q = 1.73 \text{ MeV}$$

$$e^+ + e^- \rightarrow 2\gamma \qquad\qquad Q = 1.02 \text{ MeV}$$

$$\,^{15}_7\text{N} + p \rightarrow \,^{12}_6\text{C} + \,^4_2\text{He} \qquad\qquad Q = 4.96 \text{ MeV}$$

$$\underline{\text{net:}}\;\; 4p \rightarrow \,^4_2\text{He} + 7\gamma + 2\nu_e \quad Q_{\text{net}} = 26.8 \text{ MeV}$$

These fusion processes need to overcome the Coulomb repulsions between positively charged atomic nuclei. Thus, fusion is restricted to the inner core of the Sun, which extends from the center to about 20% of the solar radius. The core contains about 10% of the mass of the Sun. In the core, the density is up to 150 tons per cubic meter, and the temperature is approximately 13.6 million K. The high temperatures and densities in the Sun's core are sufficient for the proton-proton fusion chain to proceed. By contrast, the CNO cycle requires somewhat higher temperatures than the Sun provides, and so contributes relatively little to the Sun's total energy output. However, as the Sun ages, it will reach stages of stellar evolution when the densities and temperatures in its core will be high enough for the CNO cycle to proceed.

EXAMPLE 40.6 Fusion in the Sun

As noted at the beginning of this chapter, nuclear fusion reactions are responsible for almost all of the energy resources at our disposal on Earth. Let's see how the numbers work out.

PROBLEM

The Sun radiates approximately 1370 W/m^2 onto Earth. How much total energy do fusion reactions inside the solar core need to produce per second, and how many protons does it take per second to generate this much power?

SOLUTION

The Sun is an almost perfect sphere and radiates its power uniformly over the 4π solid angle. Thus, if we calculate the surface area of a sphere with the radius of the Earth's orbit around the Sun, we can calculate the total power output. The orbital radius is 1 astronomical unit, 149.6 million km. Therefore, the area of the surface of the sphere with a radius equal to Earth's orbit is

$$A = 4\pi r^2 = 4\pi (1.496 \cdot 10^{11} \text{ m})^2 = 2.812 \cdot 10^{23} \text{ m}^2.$$

Given that each square meter of this area receives approximately 1370 W of power, the total power output of the Sun is

$$P_{total} = (1.370 \cdot 10^3 \text{ W/m}^2)(2.812 \cdot 10^{23} \text{ m}^2) = 3.85 \cdot 10^{26} \text{ W (385 yottawatts)}.$$

As a reminder, 1 W = 1 J/s and 1 eV = $1.602178 \cdot 10^{-19}$ J. The energy released by the fusion of four protons into one helium-4 nucleus can be converted to joules:

$$26.7 \text{ MeV} = (26.7 \cdot 10^6)(1.602178 \cdot 10^{-19} \text{ J}) = 4.28 \cdot 10^{-12} \text{ J}.$$

One-quarter of this energy is from each proton involved in the fusion cycle, $1.07 \cdot 10^{-12}$ J. Therefore, $(3.85 \cdot 10^{26} \text{ W})/(1.07 \cdot 10^{-12} \text{ J}) = 3.60 \cdot 10^{38}$ protons need to fuse into helium-4 each second. With a proton mass of $1.6726 \cdot 10^{-27}$ kg, this means that a total mass of $(3.60 \cdot 10^{38})$ $(1.6726 \cdot 10^{-27}$ kg$) = 6.02 \cdot 10^{11}$ kg (600 million metric tons) of protons are converted into helium each second!

Concept Check 40.5

With that much hydrogen being converted into helium, how much longer can the Sun keep shining at the current burn rate? (*Hint:* The mass of the Sun is $1.99 \cdot 10^{30}$ kg, the core contains approximately 10% of that mass, and approximately half of the core mass is currently still hydrogen.)

a) 30,000 years

b) 2 million years

c) 100 million years

d) 5 billion years

e) $3.3 \cdot 10^{18}$ years

Terrestrial Fusion

Can the power of fusion be harnessed on Earth? The answer is yes and maybe. In 1952, the first hydrogen bomb was detonated by the United States. Hydrogen bombs use nuclear fusion reactions to achieve maximum energy output. The high temperatures and compressions needed to achieve fusion are obtained by using a nuclear bomb of the fission type as a trigger. So yes, it is possible to achieve nuclear fusion reactions and to obtain a very large energy release. However, the detonation of a hydrogen bomb cannot be considered to be a harnessing, or controlling, of fusion. Controlled fusion as a means of production of useful energy could be the solution to global energy problems. During the past four or five decades, very large sums of money have been dedicated to achieving controlled fusion, so far with little success. The high pressures and temperatures needed for thermonuclear fusion have been an elusive goal.

Presently, two approaches hold promise. One is magnetic confinement technology, which an international collaboration called ITER is using in a thermonuclear fusion reactor being built in France. Figure 40.35 shows the central fusion reactor of the ITER facility. The toroidal vacuum chamber is lined with a shield that

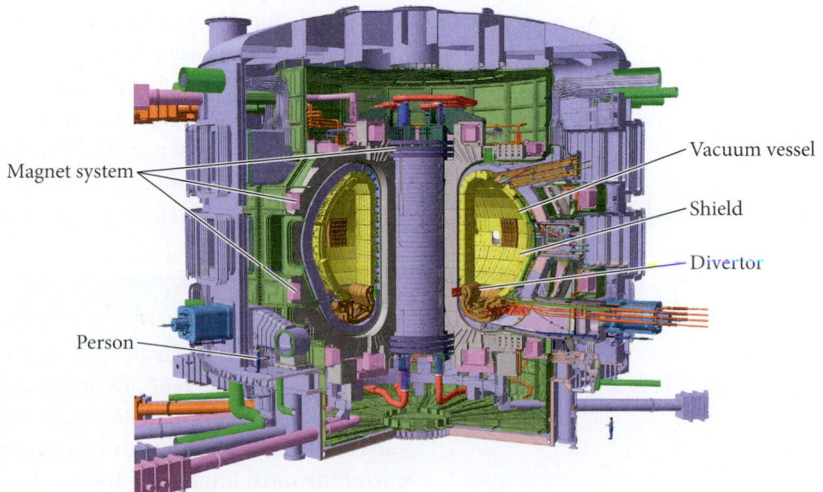

FIGURE 40.35 Cutaway drawing of the central fusion reactor of the ITER facility, now under construction in Cadarache, France.

FIGURE 40.36 (a) The nuclear fusion process at the NIF. (b) A hohlraum cylinder is just a few millimeters wide (less than the size of a dime), with beam entrance holes at either end. It contains the fusion fuel capsule, which is the size of a small pea or a pencil eraser. (c) A technician in the Target Chamber Service System lift checks the target chamber. Technicians must dress in full bodysuits to protect the chamber from any lint or microscopic particles. The target positioner, which holds the target during a shot, is on the right.

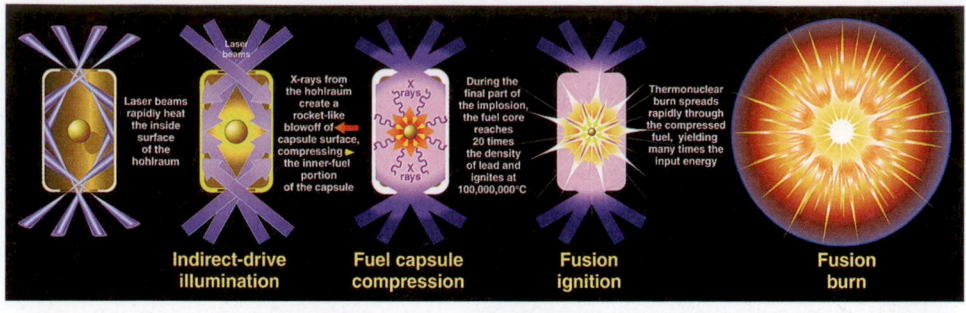

(a)

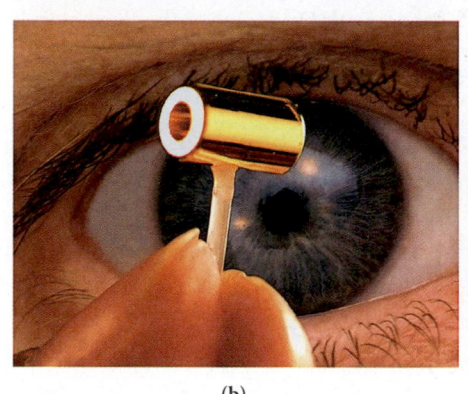

(b)

(c)

absorbs heat and neutrons produced in the fusion reactions. Magnet systems to confine the heated plasma surround the vacuum chamber. There is also a divertor to handle neutral particles that are not contained by the magnetic field.

In 2006, the European Union, the United States, China, Russia, Japan, India, and South Korea agreed to fund this project jointly and selected the construction site, Cadarache, in the south of France. The project cost will be approximately $15 billion, and it will take 10 years to complete. It is expected that the ITER reactor will provide 500 MW of power sustained for approximately 10 minutes. This is an improvement of several orders of magnitude over what has been achieved with earlier magnetic confinement fusion reactors and a big step toward a break-even point—that is, extracting at least as much power as has been put in.

The other very promising approach to fusion is laser fusion. In Livermore, California, the U.S. Department of Energy has constructed the National Ignition Facility (NIF). Here the world's most powerful lasers are shot into small hollow cylinders (called *hohlraum*), and the X-rays released from the interaction of the lasers with the hohlraum walls compress the nuclear fusion fuel in the interiors of the cylinders, causing fusion ignition. (The laser assembly at the NIF was shown in Figures 38.27 and 38.28.) Figure 40.36a presents an overview of the laser-induced fusion process, Figure 40.36b shows one of the hohlraum cylinders, and Figure 40.36c shows the target chamber, in which the target is ignited and the fusion energy that is released is recaptured.

40.5 Nuclear Astrophysics

The interface of nuclear physics, particle physics, and astrophysics is one of the most fascinating areas of current physics research. Laboratory experiments, computer simulations, model studies, and calculations can be performed to answer basic questions about the universe. Several aspects of this interdisciplinary field were discussed in Chapter 39. However, one of the most interesting topics is the origin of the chemical elements. We saw in Chapter 39 that the Big Bang created almost all of the hydrogen and most of the helium in the universe. Fusion reactions in stars, as we have just seen, generate helium as well. On the other hand, the elements that we observe around us are different. The Earth is predominantly composed of iron, silicon, oxygen, aluminum, sodium, and other elements. Elements as heavy as uranium can be found in sizable quantities. Our own bodies consist mainly of oxygen (at 65% of body weight), carbon (18%), hydrogen (3%), calcium (1.5%), phosphorus (1%), and traces of many other elements. Figure 40.37 shows the percent abundance by weight of the elements in the Solar System. Hydrogen and helium dominate, but all stable elements can be found, and even thorium and uranium exist in appreciable quantities.

Where do the heavier elements come from? To answer this question, we have to examine the later stages in a star's life. For example, let's consider a star with a mass about 20 times the mass of the Sun. The combination of nuclear physics and stellar modeling shows that a star this massive uses up its hydrogen fuel much faster than the Sun does, and after only 10 million years, its core is predominantly composed of helium. The helium-4 nuclei then fuse to carbon-12 in what is called *triple-alpha process*:

$$\,^4_2\text{He} + \,^4_2\text{He} \leftrightarrow \,^8_4\text{Be}$$

$$\,^8_4\text{Be} + \,^4_2\text{He} \rightarrow \,^{12}_6\text{C} + \gamma.$$

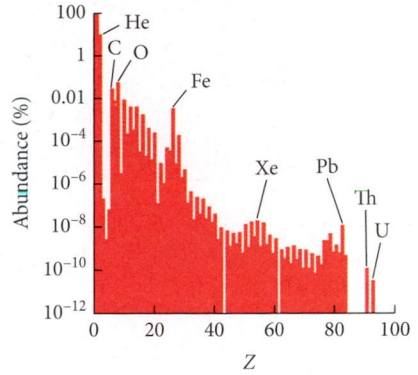

FIGURE 40.37 Abundance of the chemical elements in the Solar System by weight.

This two-step process is necessary because ^8_4Be is not stable. It has a half-life of only $6.7 \cdot 10^{-17}$ s and decays back into two helium-4 nuclei. Therefore, the density of helium-4 needs to be very high for this process to proceed. It then takes approximately 1 million years for much of the helium to be converted into carbon. As the core temperature rises, it eventually becomes sufficient to overcome the Coulomb repulsions between ever heavier nuclei. This leads to the production of nuclei of oxygen, neon, silicon, and then finally iron and nickel.

As we have seen, the stable iron and nickel isotopes have the highest binding energy per nucleon. The production of elements heavier than iron and nickel through fusion processes is thus not possible, which means that the core of the star can no longer produce energy through those processes. This iron core of the 20-solar-mass star has a mass of 1 solar mass. Once the fusion energy output ceases, the thermal pressure due to the fusion photons vanishes, and the core starts collapsing due to the force of gravity. Capture of a significant fraction of the electrons by protons, $p + e^- \rightarrow n + \nu_e$, accelerates the collapse. After only a few milliseconds, the center of the core has been compressed to the density of nuclear matter, and a shock wave begins to propagate radially outward through the core, causing the star to explode and become a **supernova.**

Exactly how the process from core collapse through explosion works in detail is still under intensive investigation. However, observations of supernovas show that the total energy emitted in the form of photons and neutrinos is on the order of 10^{44} J to 10^{45} J, which corresponds to the energy release of more than a billion billion billion (10^{27}) hydrogen bombs ignited simultaneously. What's left behind is a neutron star that is very small (radius of approximately 10 km) and incredibly dense (mass of approximately 1.5 solar masses at nuclear matter density). As the neutron star is formed, angular momentum is conserved, so it rotates very rapidly. As a neutron star spins, it can emit what are known as *lighthouse beams* of radio waves, light, and/or X-rays. Such a rotating neutron star is called a *pulsar*. Pulsars have been observed to rotate as fast as 40 times per second! The rotational speed of the Sun is $4.6 \cdot 10^{-7}$ rotations/s (one rotation every 25 days).

As a result of the supernova explosion, matter is thrown into interstellar space and forms clouds of dust and gas, which are some of the most picturesque objects in the universe. Figure 40.38 shows the famous Crab Nebula, which is the remnant of a supernova that exploded 6500 light-years from Earth and was observed in the year 1054. It was so bright that it could be seen during bright daylight, even though the explosion happened at a distance almost half a billion times greater than the distance from Earth to the Sun.

During a supernova explosion, the isotopes that exist outside the core are bombarded by an incredibly high flux of neutrons. They capture these neutrons very rapidly and also undergo fast β^- decays, thus adding to their neutron and proton numbers. The path of this so-called *r-process* (r stands for "rapid") through the isotope chart is sketched in Figure 40.39. It takes only seconds to form the r-process isotopes, which

FIGURE 40.38 The Crab Nebula, a supernova remnant.

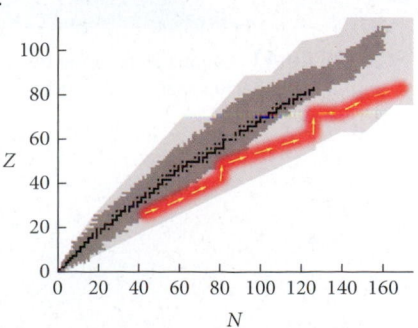

FIGURE 40.39 Path of the r-process (red with yellow arrows) through the isotope chart. Light gray: postulated isotopes. Gray: known isotopes. Black: stable isotopes.

then decay back toward the valley of stability. This process is understood qualitatively, but not in quantitative detail. However, astronomers generally agree that the elemental abundance observed in the Solar System (see Figure 40.37) is the result of a supernova explosion more than 5 billion years ago. The Solar System formed from the ashes of that supernova. The vast majority of the atoms around us and inside our bodies are thus at least 5 billion years old and have been recycled again and again.

40.6 Nuclear Medicine

Nuclear medicine is a flourishing subfield of medicine, where nuclear physics finds direct applications in diagnosis and treatment. Great progress continues to be made by using more precise tools for delivering radiation to the relevant areas of patients' bodies. There has also been steady improvement in detection equipment, resulting in the continual lowering of the minimum radiation doses needed for diagnostic tests. Progress in basic nuclear physics research thus translates directly into medical advances.

Radiation of various kinds is used for cancer treatment. The idea is straightforward: Concentrate the radiation on tumor cells that need to be destroyed, while at the same time trying to avoid irradiating healthy tissues. This treatment option is particularly attractive in cases where the tumor cannot be removed surgically or where the surgery area needs to be treated postoperatively. Brain cancers in particular have been successfully treated with radiation techniques, which are being refined steadily.

The most common radiation treatment of cancers uses gamma rays. In this technique, a very strong radioactive isotope, usually $^{60}_{27}\text{Co}$, β^- decays into an excited state of nickel via $^{60}_{27}\text{Co} \rightarrow {}^{60}_{28}\text{Ni} + e^- + \bar{\nu}_e$, with a Q-value of 318 keV. The half-life for this process is 5.27 years. Once populated, the excited state of $^{60}_{28}\text{Ni}$ then decays promptly into the ground state via the emission of two high-energy gamma rays (Figure 40.40). Thus, the $^{60}_{27}\text{Co}$ isotope delivers two high-energy gamma rays, but still has a very long half-life. Since $^{60}_{27}\text{Co}$ is straightforwardly produced with 99% purity in reactors via neutron capture by the stable isotope $^{59}_{27}\text{Co}$, it is ideal for medical purposes as well as other purposes, such as food irradiation.

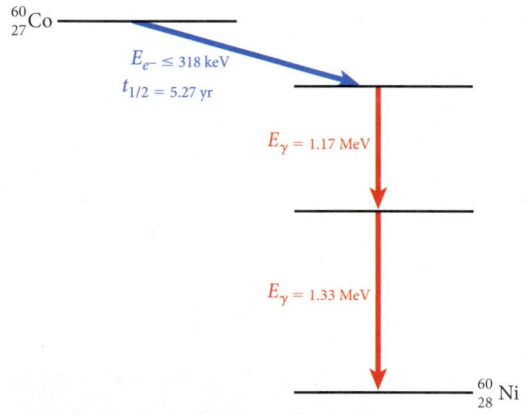

FIGURE 40.40 Decay process for cobalt-60.

The radioactive cobalt is contained behind thick shielding. Narrow channels in the shielding allow photons to escape in well-defined directions to be used as a beam for radiation. This is the basis for the treatment technique called **gamma knife** (invented in 1968), mainly used to treat inoperable brain cancers (Figure 40.41). For each patient, a custom radiation collimator in the shape of a very thick helmet is designed; the many channels through the shielding all point to the point inside the patient's head where the cancer tumor is located.

The last few years have seen a very strong increase in research into cancer treatment with heavy ion beams, which, like alpha particles, deposit most of their energy near their maximum penetration depth and so allow greater depth control of the deposited radiation dose. Great advances in radiation treatment can be expected from the use of this method.

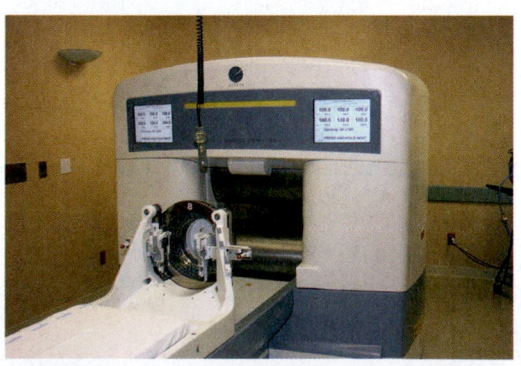

FIGURE 40.41 Gamma-knife cancer treatment facility at Scripps Memorial Hospital, La Jolla, California.

In the field of imaging, the most important technology is based on nuclear magnetic resonance (NMR). The imaging technology based on NMR is called *magnetic resonance imaging* (MRI). How does MRI work? Particles such as protons have an intrinsic magnetic dipole moment, as discussed in Chapter 28. When protons are placed in a strong magnetic field, their spins and therefore their magnetic dipole moments can have only two directions: parallel or antiparallel to the external field. The difference in energy between the two states is the difference in magnetic potential energy, which is $2\mu B$, where μ is the component of the proton's magnetic moment along the direction of the external field. This potential energy difference was discussed in Chapter 27.

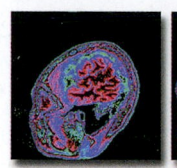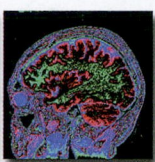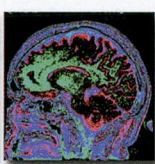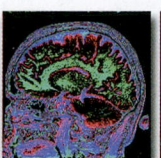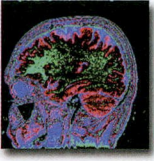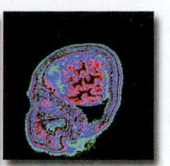

FIGURE 40.42 Image slices through a human head, obtained via MRI.

A time-varying electric field at the proper frequency can induce the dipole moments of some of the protons to flip their direction from parallel to antiparallel to the external field, and the protons thereby gain potential energy. Because the magnetic potential energy can have only two possible values, the energy required to flip the dipole moments is a discrete value, depending on the magnitude of the external field. Chapter 36 showed that the energy delivered by the field is proportional to the frequency. Thus, only one given oscillation frequency will cause the dipole moments to flip. If the time-varying electric field is switched off, the protons in the higher energy state, with dipole moments that are not aligned with the field, will flip to having their dipole moments parallel with the field, emitting photons of a well-defined energy that can be detected.

An MRI device applies this physical principle of nuclear magnetic resonance. Such a device can image the location of the protons in a human body by introducing a time-varying electric field and then varying the magnetic field in a known, precise manner to produce a three-dimensional picture of the distribution of tissue containing hydrogen. The quality of this imaging depends on the strength of the external magnetic field. Figure 40.42 shows results of an MRI scan.

Finally, let's consider an example of the diagnostic use of nuclear radiation. The element technetium ($Z = 43$) has no stable isotopes. The longest-lived technetium isotope is $^{99}_{43}$Tc, with a half-life of 4.2 million years. Its lowest energy levels, their angular momentum values, and their half-lives are shown in Figure 40.43. The state with angular momentum $\frac{1}{2}\hbar$ is an isomeric state, because its angular momentum is more than $1\hbar$ different from either of the two lower-energy states. Thus, it is relatively long-lived and decays with a half-life of 6.02 h into the $j = \frac{7}{2}\hbar$ state via the emission of a photon with an energy of 2.17 keV. This is followed by a very quick decay to the ground state, with a half-life of only 19 ns, via the emission of a 140.5-keV photon. Thus, the decay of $^{99}_{43}$Tc delivers high-energy photons over several hours after the initial preparation of the isotope. Since a 140.5-keV photon easily penetrates biological tissue, this isotope is very valuable for diagnostic purposes.

In a technetium scan, $^{99}_{43}$Tc is injected into the patient. After a few minutes to a few hours, a picture can be taken of the patient with a gamma-ray camera to monitor where in the body the technetium has traveled. Certain tumors result in an increased local concentration of $^{99}_{43}$Tc and appear as black areas in the gamma-ray picture. The patient shown in Figure 40.44 fortunately turned out to be cancer-free. (The black dot near the left elbow is the injection point.) $^{99}_{43}$Tc is thus a great example of the medical use of a specific isotope.

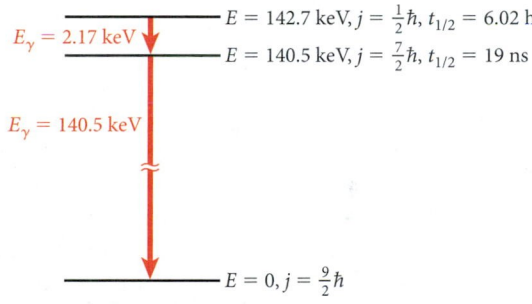

$E_\gamma = 2.17$ keV

$E = 142.7$ keV, $j = \frac{1}{2}\hbar$, $t_{1/2} = 6.02$ h

$E = 140.5$ keV, $j = \frac{7}{2}\hbar$, $t_{1/2} = 19$ ns

$E_\gamma = 140.5$ keV

$E = 0, j = \frac{9}{2}\hbar$

FIGURE 40.43 Lowest energy levels of technetium, their angular momenta in terms of Planck's constant, and their half-lives.

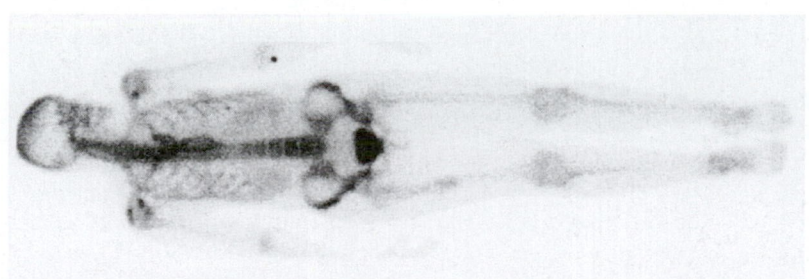

FIGURE 40.44 Technetium scan of a female patient. Dark areas show where there is increased concentration of the technetium-99 isotope in the body.

WHAT WE HAVE LEARNED | EXAM STUDY GUIDE

- The atomic nucleus consists of protons and neutrons, collectively called *nucleons*. Nuclei with different numbers of neutrons but the same number of protons are isotopes of the same element. The mass number of an isotope is the sum of the number of protons and the number of neutrons: $A = Z + N$.

- The interaction between nucleons is short-range and can effectively be described by potentials based on pion exchange.

- Nuclei are approximately spherical in shape, with the radius of the sphere depending on the mass number: $R(A) = R_0 A^{1/3}$, where $R_0 = 1.12$ fm.

- The nucleon density in the interior of a nucleus is $n = 0.17$ fm^{-3}, and the mass density is $\rho = m_{\text{nucleon}} n = 2.8 \cdot 10^{17}$ kg/m^3. The dependence of the density on the radial coordinate is given by the Fermi function: $n(r) = n_0/[1 + e^{(r - R(A))/a}]$, with $a = 0.54$ fm.

- There are 251 stable isotopes and more than 2400 known unstable isotopes, with lifetimes from fractions of seconds to many times the age of the universe.

- Nuclear mass is measured in multiples of the atomic mass unit, $1 \text{ u} = 1.660538782 \cdot 10^{-27}$ kg $= 931.494028$ MeV/c^2, which is defined as being 1/12 of the mass of the $^{12}_{6}$C atom.

- The mass of a nucleus with Z protons and N neutrons is smaller than the sum of the masses of the individual nucleons, and the binding energy is defined as that mass difference times c^2: $B(N,Z) = Z m(0,1)c^2 + N m_n c^2 - m(N,Z)c^2$.

- The mass excess of a nucleus is defined as the difference between the mass of a nucleus expressed in atomic mass units and the mass number: mass excess $= m(N,Z) - (A)(1 \text{ u})$.

- The binding energies of different isotopes can be reproduced well by the Bethe-Weizsäcker formula from the liquid-drop model, as the sum of volume, surface, Coulomb, asymmetry, and pairing contributions:

$$B(N,Z) = a_v A - a_s A^{2/3} - a_c Z^2 A^{-1/3}$$
$$- a_a \left(Z - \tfrac{1}{2}A\right)^2 A^{-1} + a_p \left((-1)^Z + (-1)^N\right) A^{-1/2}.$$

- The Q-value of a given nuclear reaction is the difference between the sum of the mass-energies of the initial nuclei minus that of the final nuclei.

- The Fermi gas model proposes a quantum gas of nucleons that can move freely inside the nucleus but are confined by the nuclear surface. The density of states in the Fermi gas model is $dN(E) = m^{3/2} E^{1/2} dE/(2^{1/2}\pi^2 a^3 \hbar^3)$.

- The Fermi energy is

$$E_F = (\hbar^2/2m)\left(\tfrac{3}{2}\pi^2 n_0\right)^{2/3} = 38 \text{ MeV}.$$

- The nuclear shell model predicts angular momentum shells inside the nucleus similar to the electron shells in the atom. It reproduces the magic numbers of 2, 8, 20, 28, 50, 82, 126.

- Nuclear decays follow an exponential decay law: $N(t) = N_0 e^{-\lambda t}$. The decay constant λ, half-life $t_{1/2}$, and mean lifetime τ are related via $t_{1/2} = \ln 2/\lambda = \tau \ln 2$.

- In alpha decay, a heavier nucleus emits a helium-4 nucleus:

$$^A_Z\text{Nuc} \rightarrow \, ^4_2\text{He} + \, ^{A-4}_{Z-2}\text{Nuc}'.$$

In a β^- decay, an electron and an anti-neutrino are emitted:

$$^A_Z\text{Nuc} \rightarrow \, ^A_{Z+1}\text{Nuc}' + e^- + \bar{\nu}_e.$$

A β^+ decay can proceed via a positron emission:

$$^A_Z\text{Nuc} \rightarrow \, ^A_{Z-1}\text{Nuc}' + e^+ + \nu_e.$$

This type of decay can also occur via electron capture:

$$e^- + \, ^A_Z\text{Nuc} \rightarrow \, ^A_{Z-1}\text{Nuc}' + \nu_e.$$

A gamma decay is the emission of a high-energy photon from an excited nucleus, a process that does not transmute the nucleus.

- The SI unit for radioactivity is the becquerel: 1 Bq = 1 nuclear decay/s. The SI unit for the absorbed dose is the gray: 1 Gy = 1 J/kg. The SI unit for the dose equivalent is the sievert: $1 \text{ Sv} = w_r(1 \text{ Gy})$, where the radiation weight factor, w_r, varies between 1 and 20, depending on the type and energy of the emitted particle.

- Nuclear fission is the splitting of a very heavy nucleus into two medium-mass nuclei, usually with the associated emission of one or a few neutrons. This process is the physical basis for nuclear power plants, as well as nuclear weapons.

- Nuclear fusion is the merging of two light nuclei into a heavier one. Nuclear fusion is the basic process that powers the stars. The two most common fusion chains are the proton-proton chain and the CNO cycle.

- Nuclear astrophysics research is employed to explain, among other things, the abundance of the chemical elements in the Solar System.

- Nuclear medicine applies nuclear physics to medical diagnostics and cancer radiation treatment.

ANSWERS TO SELF-TEST OPPORTUNITIES

40.1

Particle	Mass (u)
d	2.014101778
$^{12}_{6}C$	12
p	1.007825032
$^{13}_{6}C$	13.00573861

$Q = (2.014101778 \text{ u})c^2 + (12 \text{ u})c^2 -$

$\qquad ((1.007825032 \text{ u})c^2 + (13.00573861 \text{ u})c^2) =$

$\qquad (0.000538128 \text{ u})c^2 = 0.501263 \text{ MeV}.$

Q is positive, so the reaction is exothermic.

40.2 In the empirical mass formula (equation 40.37), the nuclei with odd charge numbers have the same values for the volume term, surface term, and pairing term. The asymmetry term and the Coulomb term combine to yield the parabolic shape. The nuclei with even charge numbers form a similar parabola that is offset vertically relative to the other parabola by the pairing term.

40.3

$$dE = \frac{\hbar^2}{m} k\, dk = \frac{\hbar^2}{m} \frac{\sqrt{2mE}}{\hbar} dk = \hbar\sqrt{2E/m}\, dk \Rightarrow \frac{dE}{\hbar\sqrt{2E/m}} = dk$$

$$dN(E) = \frac{a^3 k^2}{2\pi^2} \frac{dE}{\hbar\sqrt{2E/m}} = \frac{a^3}{2\pi^2} \frac{2mE/\hbar^2}{\hbar\sqrt{2E/m}} dE = \frac{a^3 m^{3/2}}{2^{1/2}\pi^2\hbar^3} E^{1/2} dE.$$

PROBLEM-SOLVING GUIDELINES

1. Remember that the nucleus is the prototypical quantum system. The Pauli Exclusion Principle and the uncertainty relation provide essential tools for working out relationships of momenta, energies, and sizes.

2. When calculating binding energies, you should first consider using the Bethe-Weizsäcker formula.

3. To solve a problem involving nuclear decay, first make sure which decay process is involved. You can do this by comparing the initial and final nuclei involved in the decay. Then you need to make sure that the process obeys the conservation laws (lepton number, baryon number, momentum, energy, charge, etc.). Calculating the Q-value is a good idea, because it lets you find out whether a decay is possible.

MULTIPLE-CHOICE QUESTIONS

40.1 Radium-226 decays by emitting an alpha particle. What is the daughter nucleus?

a) Rd b) Rn c) Bi d) Pb

40.2 Which of the following decay modes involves a transition between states of the same nucleus?

a) alpha decay c) gamma decay
b) beta decay d) none of the above

40.3 In neutron stars, which are roughly 90% neutrons and are held together almost entirely by nuclear forces, which of the following terms become(s) relatively dominant for the binding energy, compared to that energy in an ordinary nucleus?

a) the Coulomb term d) all of the above
b) the asymmetry term e) none of the above
c) the pairing term

40.4 When a target nucleus is bombarded by a beam of the appropriate particles, it is possible to produce

a) a less massive nucleus, but not a more massive one.

b) a more massive nucleus, but not a less massive one.

c) a nucleus with a smaller charge number, but not one with a larger charge number.

d) a nucleus with a larger charge number, but not one with a smaller charge number.

e) a nucleus with either a larger or smaller charge number.

40.5 The strong force

a) is only attractive.

b) does not act on electrons.

c) only acts over distances of a few femtometers.

d) All of the above are true.

e) None of the above are true.

40.6 Cobalt has a stable isotope, $^{59}_{27}Co$, and 22 radioactive isotopes. The most stable radioactive isotope is $^{60}_{27}Co$. What is the dominant decay mode of this isotope?

a) β^+

b) β^-

c) proton emission

d) neutron emission

40.7 The mass of an atom (atomic mass) is equal to

a) the sum of the masses of the protons.

b) the sum of the masses of protons and neutrons.

c) the sum of the masses of protons, neutrons, and electrons.

d) the sum of the masses of protons, neutrons, and electrons minus the atom's binding energy.

CONCEPTUAL QUESTIONS

40.8 What is more dangerous, a radioactive material with a short half-life or a long one?

40.9 Apart from fatigue, what is another reason the Federal Aviation Administration limits the number of hours that commercial jet pilots can fly each year?

40.10 Why are there magic numbers in the nuclear shell model?

40.11 The binding energy of $_2^3\text{He}$ is lower than that of $_1^3\text{H}$. Provide a plausible explanation, taking into consideration the Coulomb interaction between two protons in $_2^3\text{He}$.

40.12 Which of the following quantities is conserved during a nuclear reaction, and how?

a) charge

b) the number of nucleons, A

c) mass-energy

d) linear momentum

e) angular momentum

40.13 Some food is treated with gamma radiation to kill bacteria. Why is it not a concern that people who eat such food are ingesting gamma radiation?

40.14 The subsection "Terrestrial Fusion" in Section 40.4 discussed how achieving controlled fusion would be the solution to the world's energy problems and how difficult it is to do this. Why is it so hard? The Sun does it all the time (see the subsection "Stellar Fusion"). Do we need to understand better how the Sun works to build a nuclear fusion reactor?

40.15 Why are atomic nuclei more or less limited in size and in neutron-proton ratio? That is, why are there no stable nuclei with 10 times as many neutrons as protons, and why are there no atomic nuclei the size of marbles?

40.16 A nuclear reaction of the kind $_2^3\text{He} + _6^{12}\text{C} \rightarrow \text{X} + \alpha$ is called a *pick-up reaction*.

a) Why does it have this name, that is, what is picked up, what picked it up, and where did it come from?

b) What is the resulting nucleus, X?

c) What is the Q-value of this reaction?

d) Is this reaction endothermic or exothermic?

40.17 Isospin, or *isotopic spin*, is a quantum variable describing the relationship between protons and neutrons in nuclear and particle physics. (Strictly, it describes the relationship between up and down quarks, as described in Chapter 39, but it was introduced before the advent of the quark model.) It has the same algebraic properties as quantum angular momentum: A proton and a neutron form an iso-doublet of states, with total isospin quantum number $\frac{1}{2}$; the proton is in the $t_z = +\frac{1}{2}$ state, and the neutron is in the $t_z = -\frac{1}{2}$ state, where z refers to a direction in an abstract isospin space.

a) What isospin states can be constructed from two nucleons, that is, two particles with $t = \frac{1}{2}$? To what nuclei do these states correspond?

b) What isospin states can be constructed from three nucleons? To what nuclei do these correspond?

40.18 Before looking it up, predict intrinsic spin (i.e., actual angular momentum) of the deuteron, $_1^2\text{H}$. Explain your reasoning. (*Hint:* Nucleons are fermions.)

40.19 $_{18}^{39}\text{Ar}$ is an isotope with a half-life of 269 yr. If it decays through β^- decay, what isotope will result?

40.20 A neutron star is essentially a gigantic nucleus with mass 1.35 times that of the Sun, or a mass number of order 10^{57}. It consists of approximately 99% neutrons, the rest being protons and an equal number of electrons. Explain the physics that determines these features.

40.21 What is the nuclear configuration of the daughter nucleus associated with the alpha decay of Hf ($A = 157$, $Z = 72$)?

EXERCISES

A blue problem number indicates a worked-out solution is available in the Student Solutions Manual. One • and two •• indicate increasing level of problem difficulty.

Section 40.1

40.22 Estimate the volume of the uranium-235 nucleus.

40.23 Calculate the binding energies of the following nuclei.

a) $_3^7\text{Li}$
b) $_6^{12}\text{C}$
c) $_{26}^{56}\text{Fe}$
d) $_{37}^{85}\text{Rb}$

40.24 Give the numbers of protons, nucleons, neutrons, and electrons in an atom of $_{54}^{134}\text{Xe}$.

•40.25 Using the Fermi function, determine the relative change in density, $(dn(r)/dr)/n_0$, at the nuclear surface, $r = R(A)$.

•40.26 Calculate the binding energy for the following two uranium isotopes (where $u = 1.66 \cdot 10^{-27}$ kg):

a) $_{92}^{238}\text{U}$, which consists of 92 protons, 92 electrons, and 146 neutrons, with a total mass of 238.0507826 u

b) $_{92}^{235}\text{U}$, which consists of 92 protons, 92 electrons, and 143 neutrons, with a total mass of 235.0439299 u

Which isotope is more stable (or less unstable)?

Section 40.2

40.27 Write equations for the β^- decay of the following isotopes:

a) $_{27}^{60}\text{Co}$
b) $_1^3\text{H}$
c) $_6^{14}\text{C}$

40.28 Write equations for the alpha decay of the following isotopes:

a) $_{86}^{212}\text{Rn}$
b) $_{95}^{241}\text{Am}$

40.29 How much energy is released in the beta decay of $_6^{14}\text{C}$?

40.30 A certain radioactive isotope decays to one-eighth of its original amount in 5.0 h.

a) What is its half-life?
b) What is its mean lifetime?

40.31 A certain radioactive isotope decays to one-eighth of its original amount in 5.00 h. How long would it take for 10.0% of it to decay?

40.32 Determine the decay constant of radium-226, which has a half-life of 1600 yr.

40.33 A 1.00-g sample of radioactive thorium-228 decays via β^- decay, and 75 counts are recorded in one day by a detector that has 10.0% efficiency (that is, 10.0% of all events that occur are recorded by the detector). What is the lifetime of this isotope?

40.34 The half-life of a sample of 10^{11} atoms that decay by alpha particle emission is 10 min. How many alpha particles are emitted in the time interval from 100 min to 200 min?

•40.35 The specific activity of a radioactive material is the number of disintegrations per second per gram of radioactive atoms.

a) Given the half-life of $_6^{14}\text{C}$ of 5730 yr, calculate the specific activity of $_6^{14}\text{C}$. Express your result in disintegrations per second per gram, becquerel per gram, and curie per gram.

b) Calculate the initial activity of a 5.00-g piece of wood.

c) How many $^{14}_{6}C$ disintegrations have occurred in a 5.00-g piece of wood that was cut from a tree on January 1, 1700?

•**40.36** During a trip to an excavation site, an archeologist found a piece of charcoal. Analysis of the charcoal found the activity of $^{14}_{6}C$ in the sample to be 0.42 Bq. If the mass of the charcoal is 7.2 g, estimate the approximate age of the site.

•**40.37** In 2008, crime scene investigators discover the bones of a person who appeared to have been the victim of a brutal attack that occurred a long time ago. They would like to know the year when the person was murdered. Using carbon dating, they determine that the rate of change of the $^{14}_{6}C$ is 0.268 Bq per gram of carbon. The rate of change of $^{14}_{6}C$ in the bones of a person who had just died is 0.270 Bq per gram of carbon. What year was the victim killed? The half-life of $^{14}_{6}C$ is $5.73 \cdot 10^{3}$ yr.

•**40.38** Physicists blow stuff up better than anyone else. The measure for gauging the usefulness of blowing something up is the fraction of initial rest mass converted into energy in the process. Looking up the necessary data, calculate this fraction for the following processes:

a) chemical combustion of hydrogen: $2H_2 + O_2 \rightarrow 2H_2O$

b) nuclear fission: $n + {}^{235}_{92}U \rightarrow {}^{89}_{36}Kr + {}^{142}_{56}Ba + 5n$

c) thermonuclear fusion: $^{6}_{3}Li + ^{2}_{1}H \rightarrow ^{7}_{4}Be + n$

d) decay of free neutron: $n \rightarrow p + e^{-} + \bar{\nu}_e$

e) decay of muon: $\mu^{-} \rightarrow e^{-} + \nu_\mu + \bar{\nu}_e$

f) electron-positron annihilation: $e^{-} + e^{+} \rightarrow 2\gamma$

••**40.39** An unstable nucleus A decays to an unstable nucleus B, which in turn decays to a stable nucleus. If at $t = 0$ s there are N_{A0} and N_{B0} nuclei present, derive an expression for N_B, the number of B nuclei present, as a function of time.

••**40.40** In a simple case of *chain* radioactive decay, a *parent* radioactive nucleus, A, decays with a decay constant λ_1 into a *daughter* radioactive nucleus, B, which then decays with a decay constant λ_2 to a *stable* nucleus, C.

a) Write the equations describing the number of nuclei of each of the three types as a function of time, and derive expressions for the number of daughter nuclei, N_2, as a function of time and for the activity of the daughter nuclei, A_2, as a function of time.

b) Discuss the results in the case when $\lambda_2 > \lambda_1$ ($\lambda_2 \approx 10 \lambda_1$) and when $\lambda_2 >> \lambda_1$ ($\lambda_2 \approx 100 \lambda_1$).

Section 40.3

•**40.41** Show that for the case of nuclei with odd mass number, A, the Bethe-Weizsäcker formula can be written as a quadratic in Z—and thus, for any given A, the binding energies of the isotopes having that A take a quadratic form, $B = a + bZ + cZ^2$. Use your result to find the most strongly bound isotope (the most stable one) having $A = 117$.

••**40.42** The *neutron drip line* is defined to be the point at which the neutron separation energy for any isotope of an element is negative. That is, the neutron is unbound. Using the Bethe-Weizsäcker formula, find the neutron drip line for the element Sn. Find this value using S_n and S_{2n}. Plot both S_n and $S_{2n}/2$ as a function of neutron number.

Section 40.4

40.43 A nuclear fission power plant produces about 1.50 GW of electrical power. Assume that the plant has an overall efficiency of 35.0% and that each fission event produces 200. MeV of energy. Calculate the mass of $^{235}_{92}U$ consumed each day.

40.44 a) What is the energy released in the fusion reaction $^{2}_{1}H + ^{2}_{1}H \rightarrow ^{4}_{2}He + Q$?

b) The Earth's oceans have a total mass of water of $1.50 \cdot 10^{16}$ kg, and 0.0300% of this quantity is deuterium, $^{2}_{1}H$. If all the deuterium in the oceans were fused by controlled fusion into $^{4}_{2}He$, how many joules of energy would be released?

c) World power consumption is about $1.00 \cdot 10^{13}$ W. If consumption stayed constant and all problems arising from ocean water consumption (including

those of political, meteorological, and ecological nature) could be avoided, how many years would the energy calculated in part (b) last?

40.45 The Sun radiates energy at the rate of $3.85 \cdot 10^{26}$ W.

a) At what rate, in kilograms per second, is the Sun's mass converted into energy?

b) Why is this result different from the rate calculated in Example 40.6: $6.02 \cdot 10^{11}$ kg of protons being converted into helium each second?

c) Assuming that the current mass of the Sun is $1.99 \cdot 10^{30}$ kg and that it has radiated at the same rate for its entire lifetime of $4.50 \cdot 10^{9}$ yr, what percentage of the Sun's mass has been converted into energy during its entire lifetime?

40.46 Consider the following fusion reaction, through which stars produce progressively heavier elements: $^{3}_{2}He + ^{4}_{2}He \rightarrow ^{7}_{4}Be + \gamma$. The mass of $^{3}_{2}He$ is 3.016029 u, the mass of $^{4}_{2}He$ is 4.002603 u, and the mass of $^{7}_{4}Be$ is 7.0169298 u. The atomic mass unit is 1 u = $1.66 \cdot 10^{-27}$ kg. Assuming that the Be atom is at rest after the reaction and neglecting any potential energy between the atoms and the kinetic energy of the He nuclei, calculate the minimum possible energy and maximum possible wavelength of the photon, γ, that is emitted in this reaction.

•**40.47** Estimate the temperature that would be needed to initiate the fusion reaction $^{3}_{2}He + ^{3}_{2}He \rightarrow ^{4}_{2}He + p + p$.

•**40.48** Consider a hypothetical fission process in which a $^{120}_{52}Te$ nucleus splits into two identical $^{60}_{26}Fe$ nuclei without producing any other particles or radiation. The mass of $^{120}_{52}Te$ is 119.904040 u, and the mass of $^{60}_{26}Fe$ is 59.934078 u. At the moment when the two iron nuclei form, but before they start moving away due to Coulomb repulsion, how far apart are the two nuclei?

•**40.49** The *mass excess* of a nucleus is defined as the difference between the atomic mass (in atomic mass units, u), and the mass number of the nucleus, A. Using the mass-energy conversion 1 u = 931.49 MeV/c^2, this mass excess is usually expressed in kilo-electron-volts (keV). The table below presents the mass excess for several nuclei (from the Berkeley National Lab *NuBase* database):

No.	Nucleus	Mass number, A	Mass excess, Δm (keV/c^2)	Atomic mass (u)
1	$^{1}_{0}n$	1	8,071.3	1.00866491
2	$^{252}_{98}Cf$	252	76,034	
3	$^{256}_{100}Fm$	256	85,496	
4	$^{140}_{56}Ba$	140	−83,271	
5	$^{140}_{54}Xe$	140	−72,990	
6	$^{112}_{46}Pd$	112	−86,336	
7	$^{109}_{42}Mo$	109	−67,250	

a) Calculate the atomic mass (in atomic mass units) for each of the nuclei in the table. For reference, the atomic mass of the neutron is given.

b) Using your results from part (a), determine the mass-energy difference between the initial and final states for the following possible fission reactions:

$^{252}_{98}Cf \rightarrow ^{140}_{56}Ba + ^{109}_{42}Mo + 3n \qquad ^{256}_{100}Fm \rightarrow ^{140}_{54}Xe + ^{112}_{46}Pd + 4n$

c) Will these reactions occur spontaneously?

Section 40.5

40.50 Neutron stars are sometimes approximated to be nothing more than large atomic nuclei (but with many more neutrons). Assuming that a neutron star is as dense as an atomic nucleus, estimate the number of nucleons in a 10.0-km-diameter star.

40.51 What is the average kinetic energy of protons at the center of a star, where the temperature is $1.00 \cdot 10^{7}$ K? What is the average velocity of those protons?

•**40.52** Billions of years ago, the Solar System was created out of the remnants of a supernova explosion. Nuclear scientists believe that two isotopes of uranium, $^{235}_{92}$U and $^{238}_{92}$U, were created in equal amounts at that time. However, today 99.28% of uranium is in the form of $^{238}_{92}$U and only 0.72% is in the form of $^{235}_{92}$U. Assuming a simplified model in which all of the matter in the Solar System originated in a single exploding star, estimate the approximate time of this explosion.

Section 40.6

40.53 A drug containing $^{99}_{43}$Tc ($t_{1/2} = 6.05$ h) with an activity of 1.50 μCi is to be injected into a patient at 9.30 a.m. You are to prepare the drug 2.50 h before the injection (at 7:00 a.m.). What activity should the drug have at the preparation time (7:00 a.m.)?

40.54 A 42.58-MHz photon is needed to produce nuclear magnetic resonance in free protons in a magnetic field of 1.000 T. What is the wavelength of the photon, its energy, and the region of the spectrum in which it lies? Could it be harmful to the human body?

40.55 The radon isotope $^{222}_{86}$Rn, which has a half-life of 3.825 days, is used for medical purposes such as radiotherapy. How long does it take until $^{222}_{86}$Rn decays to 10.00% of its initial quantity?

40.56 Radiation therapy is one of the techniques used for cancer treatment. Based on the approximate mass of a tumor, oncologists can calculate the radiation dose necessary to treat a patient. Suppose a patient has a 50.0-g tumor and needs to receive 0.180 J of energy to kill the cancer cells. What radiation absorbed dose should the patient receive?

Additional Exercises

40.57 The atom of sodium-22 ($^{22}_{11}$Na) has a mass of 21.994435 u. How much work would be needed to take this nucleus completely apart into its constituent pieces (protons, neutrons, and electrons)?

40.58 A Geiger counter initially records 7210 counts/s from a sample of radioactive material. After 45 min, it records 4585 counts/s. Ignore any uncertainty in the counts and find the half-life of the material.

40.59 How close can a 5.00-MeV alpha particle get to a uranium-238 nucleus, assuming that the only interaction is Coulomb?

40.60 $^{239}_{94}$Pu decays with a half-life of 24,100 yr via emission of a 5.25-MeV alpha particle. If you have a 1.00 kg spherical sample of $^{239}_{94}$Pu, find the initial activity in becquerels.

40.61 The activity of a sample of $^{210}_{83}$Bi (with a half-life of 5.01 days) is measured to be 1.000 μCi. What will the activity of this sample be after 1 yr?

40.62 Assuming that carbon makes up 14% of the mass of a human body, calculate the activity of a 75-kg person considering only the beta decays of carbon-14.

40.63 $^{8}_{3}$Li is an isotope that has a lifetime of less than a second. Its mass is 8.022485 u. Calculate its binding energy in MeV.

40.64 What is the total energy released in the decay $n \rightarrow p + e^- + \bar{\nu}_e$?

40.65 A gallon of regular gasoline (density of 737 kg/m^3) contains about 131 MJ of chemical energy. How much energy is contained in the rest mass of this gallon?

40.66 If 10^{30} atoms of a radioactive sample remain after 10 half-lives, how many atoms remain after 20 half-lives?

40.67 Calculate the binding energy per nucleon of

a) $^{4}_{2}$He (4.002603 u).
b) $^{3}_{2}$He (3.016030 u).
c) $^{3}_{1}$H (3.016050 u).
d) $^{2}_{1}$H (2.014102 u).

40.68 The mean lifetime for a radioactive nucleus is 4300 s. What is its half-life?

40.69 $^{214}_{82}$Pb has a half-life of 26.8 min. How many minutes must elapse for 90.0% of a given sample of $^{214}_{82}$Pb atoms to decay?

••40.70 The most common isotope of uranium, $^{238}_{92}$U, produces radon, $^{222}_{86}$Rn, through the following sequence of decays:

$$^{238}_{92}U \rightarrow {}^{234}_{90}Th + \alpha, \quad {}^{234}_{90}Th \rightarrow {}^{234}_{91}Pa + \beta^- + \bar{\nu}_e,$$

$$^{234}_{91}Pa \rightarrow {}^{234}_{92}U + \beta^- + \bar{\nu}_e, \quad {}^{234}_{92}U \rightarrow {}^{230}_{90}Th + \alpha,$$

$$^{230}_{90}Th \rightarrow {}^{226}_{88}Ra + \alpha, \quad {}^{226}_{88}Ra \rightarrow {}^{222}_{86}Rn + \alpha.$$

A sample of $^{238}_{92}$U will build up equilibrium concentrations of its daughter nuclei down to $^{226}_{88}$Ra; the concentrations of each are such that each daughter is produced as fast as it decays. The $^{226}_{88}$Ra decays to $^{222}_{86}$Rn, which escapes as a gas. (The alpha particles also escape, as helium; this is a source of much of the helium found on Earth.) Radon is a health hazard when it occurs in high concentrations in buildings built on soil or foundations containing uranium ores, as it can be inhaled.

a) Look up the necessary data, and calculate the rate at which 1.00 kg of an equilibrium mixture of $^{238}_{92}$U and its first five daughters produces $^{222}_{86}$Rn (mass per unit time).

b) What activity (in curies per unit time) of radon does this represent?

•40.71 After a tree has been chopped down and burned to ash, the carbon isotopes in the ash are found to have a $^{14}_{6}$C to $^{12}_{6}$C ratio of $1.300 \cdot 10^{-12}$. Experimental tests on the $^{14}_{6}$C atoms reveal that $^{14}_{6}$C is a beta emitter with a half-life of 5730 yr. At an archeological excavation, a skeleton is found next to some wood ash from a campfire. If 50.0 g of carbon from the ash emits electrons at a rate of 20.0 per hour, how long ago did the campfire burn?

•40.72 If your mass is 70.0 kg and you have a lifetime of 70.0 yr, how many proton decays will occur in your body during your life (assuming that your body is entirely composed of water)? Use a half-life of $1.00 \cdot 10^{30}$ yr.

•40.73 You have developed a grand unified theory that predicts the following things about the decay of protons: (1) protons never get any older, in the sense that their probability of decay per unit time never changes, and (2) half the protons in any given collection of protons will have decayed in $1.80 \cdot 10^{29}$ yr. You are given experimental facilities to test your theory: a tank containing $1.00 \cdot 10^4$ metric tons of water and sensors to record proton decays. You will be allowed access to this facility for 2 years. How many proton decays will occur in this period if your theory is correct?

•40.74 The precession frequency of the protons in a laboratory NMR spectrometer is 15.35850 MHz. The magnetic dipole moment of the proton is $1.410608 \cdot 10^{-26}$ J/T, while its spin angular momentum is $0.5272863 \cdot 10^{-34}$ J s. Calculate the magnitude of the magnetic field in which the protons are immersed.

••40.75 Two species of radioactive nuclei, A and B, each with an initial population N_0, start decaying. After a time of 100. s, it is observed that $N_A = 100 N_B$. If $\tau_A = 2\tau_B$, find the value of τ_B.

MULTI-VERSION EXERCISES

40.76 A 12.43-g fragment of charcoal is to be carbon dated. Measurements show that it has an activity of 105 decays/min. How many years ago did the tree from which the charcoal was produced die? (*Hint:* The half-life of $^{14}_{6}$C is 5730 yr, and the $^{14}_{6}$C/$^{12}_{6}$C ratio in living organic matter is $1.20 \cdot 10^{-12}$.)

40.77 A fragment of charcoal has been determined by carbon dating to be 4384 years old. Measurements show that it has an activity of 107 decays/min.

What is the mass of the charcoal fragment? (*Hint:* The half-life of $^{14}_{6}$C is 5730 yr, and the $^{14}_{6}$C/$^{12}_{6}$C ratio in living organic matter is $1.20 \cdot 10^{-12}$.)

40.78 A 13.83-g fragment of charcoal has been determined by carbon dating to be 4814 years old. How many decays per minute were measured in the carbon dating process? (*Hint:* The half-life of $^{14}_{6}$C is 5730 yr, and the $^{14}_{6}$C/$^{12}_{6}$C ratio in living organic matter is $1.20 \cdot 10^{-12}$.)

Appendix A

Mathematics Primer

Notation:

The letters a, b, c, x, and y represent real numbers.

The letter n represents integer numbers.

The Greek letters α, β, and γ represent angles, which are measured in radians.

1. Algebra

1.1 Basics

Factors:

$$ax + bx + cx = (a + b + c)x \tag{A.1}$$

$$(a + b)^2 = a^2 + 2ab + b^2 \tag{A.2}$$

$$(a - b)^2 = a^2 - 2ab + b^2 \tag{A.3}$$

$$(a + b)(a - b) = a^2 - b^2 \tag{A.4}$$

Quadratic equation:
An equation of the form

$$ax^2 + bx + c = 0 \tag{A.5}$$

for given values of a, b, and c has the two solutions:

$$x = \frac{-b + \sqrt{b^2 - 4ac}}{2a}$$

and $\tag{A.6}$

$$x = \frac{-b - \sqrt{b^2 - 4ac}}{2a}$$

These solutions of the quadratic equation are called *roots*. The roots are real numbers if $b^2 \geq 4ac$.

1.2 Exponents

If a is a number, a^n is the product of a with itself n times:

$$a^n = \underbrace{a \times a \times a \times \cdots \times a}_{n \text{ factors}} \tag{A.7}$$

The number n is called the *exponent*. However, an exponent does not have to be a positive number or an integer. Any real number x can be used as an exponent.

$$a^{-x} = \frac{1}{a^x} \tag{A.8}$$

$$a^0 = 1 \tag{A.9}$$

$$a^1 = a \tag{A.10}$$

Roots:

$$a^{1/2} = \sqrt{a} \tag{A.11}$$

$$a^{1/n} = \sqrt[n]{a} \tag{A.12}$$

Multiplication and division:

$$a^x a^y = a^{x+y} \tag{A.13}$$

$$\frac{a^x}{a^y} = a^{x-y} \tag{A.14}$$

$$\left(a^x\right)^y = a^{xy} \tag{A.15}$$

1.3 Logarithms

The logarithm is the inverse function of the exponential function:

$$y = a^x \Leftrightarrow x = \log_a y \tag{A.16}$$

The notation $\log_a y$ indicates the logarithm of y with respect to the base a. Since the exponential and logarithmic functions are inverses of each other, we can also write this identity:

$$x = \log_a(a^x) = a^{\log_a x} \quad \text{(for any base } a) \tag{A.17}$$

The two bases most commonly used are base 10, the common logarithm base, and base e, the natural logarithm base. The numerical value of e is

$$e = 2.718281828\ldots \tag{A.18}$$

Base 10:

$$y = 10^x \Leftrightarrow x = \log_{10} y \tag{A.19}$$

Base e:

$$y = e^x \Leftrightarrow x = \ln y \tag{A.20}$$

This book follows the convention of using ln to indicate the logarithm with respect to the base e.

The rules for calculating with logarithms follow from the rules for exponents:

$$\log(ab) = \log a + \log b \tag{A.21}$$

$$\log\left(\frac{a}{b}\right) = \log a - \log b \tag{A.22}$$

$$\log(a^x) = x \log a \tag{A.23}$$

$$\log 1 = 0 \tag{A.24}$$

Since these rules are valid for any base, the subscript indicating the base has been omitted.

1.4 Linear Equations

The general form of a linear equation is

$$y = ax + b \tag{A.25}$$

where a and b are constants. The graph of y versus x is a straight line; a is the slope of this line, and b is the y-intercept. See Figure A.1.

The slope of the line can be calculated by inserting two different values, x_1 and x_2, into the linear equation and calculating the resulting values, y_1 and y_2:

$$a = \frac{y_2 - y_1}{x_2 - x_1} = \frac{\Delta y}{\Delta x} \tag{A.26}$$

If $a = 0$, then the line will be horizontal; if $a > 0$, then the line will rise as x increases, as shown in the example of Figure A.1; if $a < 0$, then the line will fall as x increases.

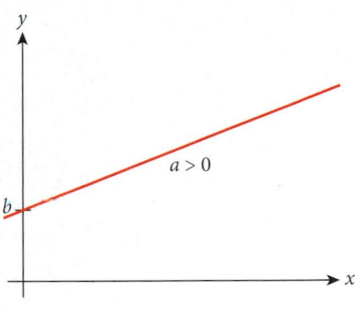

FIGURE A.1 Graphical representation of a linear equation.

2. Geometry

2.1 Geometrical Shapes in Two Dimensions

Figure A.2 gives the area, A, and perimeter length (or circumference), C, of common two-dimensional geometrical shapes.

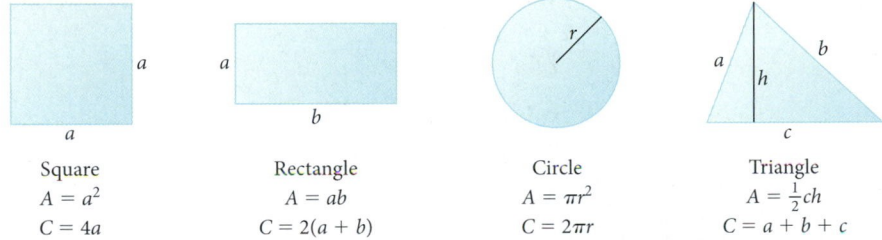

Square
$A = a^2$
$C = 4a$

Rectangle
$A = ab$
$C = 2(a + b)$

Circle
$A = \pi r^2$
$C = 2\pi r$

Triangle
$A = \frac{1}{2}ch$
$C = a + b + c$

FIGURE A.2 Area, A, and perimeter length, C, for square, rectangle, circle, and triangle.

2.2 Geometrical Shapes in Three Dimensions

Figure A.3 gives the volume, V, and surface area, A, of common three-dimensional geometrical shapes.

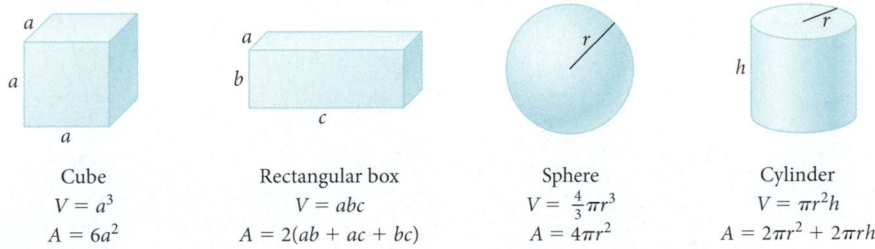

Cube
$V = a^3$
$A = 6a^2$

Rectangular box
$V = abc$
$A = 2(ab + ac + bc)$

Sphere
$V = \frac{4}{3}\pi r^3$
$A = 4\pi r^2$

Cylinder
$V = \pi r^2 h$
$A = 2\pi r^2 + 2\pi rh$

FIGURE A.3 Volume, V, and surface area, A, for cube, rectangular box, sphere, and cylinder.

3. Trigonometry

It is important to note that for the following all angles need to be measured in radians.

3.1 Right Triangles

A right triangle is a triangle for which one of the three angles is a right angle, that is, an angle of exactly 90° ($\pi/2$ rad) (indicated by the small corner mark in Figure A.4). The hypotenuse is the side opposite the 90° angle. Conventionally, the letter c marks the hypotenuse.

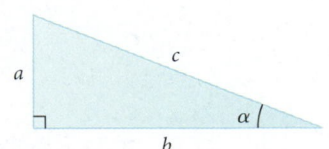

FIGURE A.4 Definition of the side lengths a, b, and c and the angles for the right triangle.

Pythagorean Theorem:

$$a^2 + b^2 = c^2 \tag{A.27}$$

Trigonometric functions (see Figure A.5):

$$\sin \alpha = \frac{a}{c} = \frac{\text{opposite side}}{\text{hypotenuse}} \tag{A.28}$$

$$\cos \alpha = \frac{b}{c} = \frac{\text{adjacent side}}{\text{hypotenuse}} \tag{A.29}$$

$$\tan \alpha = \frac{\sin \alpha}{\cos \alpha} = \frac{a}{b} \tag{A.30}$$

$$\cot \alpha = \frac{\cos \alpha}{\sin \alpha} = \frac{1}{\tan \alpha} = \frac{b}{a} \tag{A.31}$$

$$\csc \alpha = \frac{1}{\sin \alpha} = \frac{c}{a} \tag{A.32}$$

$$\sec \alpha = \frac{1}{\cos \alpha} = \frac{c}{b} \tag{A.33}$$

Inverse trigonometric functions (the notations $\sin^{-1}$, $\cos^{-1}$, etc., are used in this book):

$$\sin^{-1} \frac{a}{c} = \arcsin \frac{a}{c} = \alpha \tag{A.34}$$

$$\cos^{-1} \frac{b}{c} = \arccos \frac{b}{c} = \alpha \tag{A.35}$$

FIGURE A.5 The trigonometric functions sin, cos, tan, and cot.

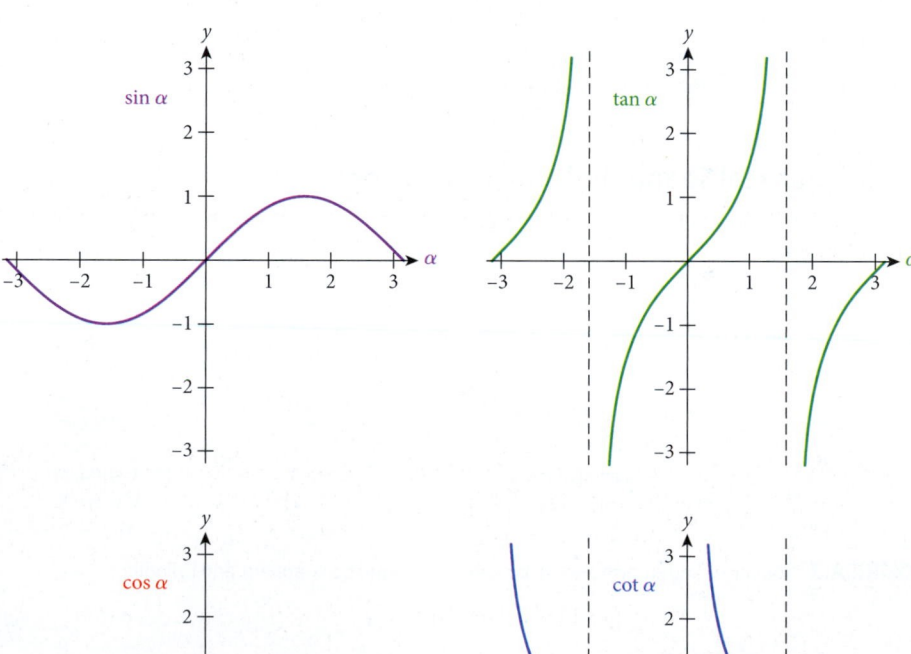

$$\tan^{-1}\frac{a}{b} = \arctan\frac{a}{b} = \alpha \qquad\qquad \text{(A.36)}$$

$$\cot^{-1}\frac{b}{a} = \text{arccot}\,\frac{b}{a} = \alpha \qquad\qquad \text{(A.37)}$$

$$\csc^{-1}\frac{c}{a} = \text{arccsc}\,\frac{c}{a} = \alpha \qquad\qquad \text{(A.38)}$$

$$\sec^{-1}\frac{c}{b} = \text{arcsec}\,\frac{c}{b} = \alpha \qquad\qquad \text{(A.39)}$$

All trigonometric functions are periodic:

$$\sin\left(\alpha + 2\pi\right) = \sin\alpha \qquad\qquad \text{(A.40)}$$

$$\cos\left(\alpha + 2\pi\right) = \cos\alpha \qquad\qquad \text{(A.41)}$$

$$\tan\left(\alpha + \pi\right) = \tan\alpha \qquad\qquad \text{(A.42)}$$

$$\cot\left(\alpha + \pi\right) = \cot\alpha \qquad\qquad \text{(A.43)}$$

Other relations between trigonometric functions:

$$\sin^2\alpha + \cos^2\alpha = 1 \qquad\qquad \text{(A.44)}$$

$$\sin(-\alpha) = -\sin\alpha \qquad\qquad \text{(A.45)}$$

$$\cos(-\alpha) = \cos\alpha \qquad\qquad \text{(A.46)}$$

$$\sin(\alpha \pm \pi/2) = \pm\cos\alpha \qquad\qquad \text{(A.47)}$$

$$\sin(\alpha \pm \pi) = -\sin\alpha \qquad\qquad \text{(A.48)}$$

$$\cos(\alpha \pm \pi/2) = \mp\sin\alpha \qquad\qquad \text{(A.49)}$$

$$\cos(\alpha \pm \pi) = -\cos\alpha \qquad\qquad \text{(A.50)}$$

Addition formulas:

$$\sin(\alpha \pm \beta) = \sin\alpha\cos\beta \pm \cos\alpha\sin\beta \qquad\qquad \text{(A.51)}$$

$$\cos(\alpha \pm \beta) = \cos\alpha\cos\beta \mp \sin\alpha\sin\beta \qquad\qquad \text{(A.52)}$$

Small-angle approximations:

$$\sin\alpha \approx \alpha - \tfrac{1}{6}\alpha^3 + \cdots \quad \left(\text{for } |\alpha| \ll 1\right) \qquad\qquad \text{(A.53)}$$

$$\cos\alpha \approx 1 - \tfrac{1}{2}\alpha^2 + \cdots \quad \left(\text{for } |\alpha| \ll 1\right) \qquad\qquad \text{(A.54)}$$

For small angles, that is, where $|\alpha| \ll 1$, it is often acceptable to use the small-angle approximations $\cos\alpha = 1$ and $\sin\alpha = \tan\alpha = \alpha$.

3.2 General Triangles

The three angles of any triangle add up to π rad (see Figure A.6):

$$\alpha + \beta + \gamma = \pi \qquad\qquad \text{(A.55)}$$

Law of cosines:

$$c^2 = a^2 + b^2 - 2ab\cos\gamma \qquad\qquad \text{(A.56)}$$

(This is a generalization of the Pythagorean Theorem for the case where the angle γ has a value that is not 90°, or $\pi/2$ rads.)

Law of sines:

$$\frac{\sin\alpha}{a} = \frac{\sin\beta}{b} = \frac{\sin\gamma}{c} \qquad\qquad \text{(A.57)}$$

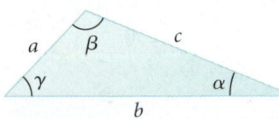

FIGURE A.6 Definition of the sides and angles for a general triangle.

4. Calculus
4.1 Derivatives
Polynomials:

$$\frac{d}{dx}x^n = nx^{n-1} \tag{A.58}$$

Trigonometric functions:

$$\frac{d}{dx}\sin(ax) = a\cos(ax) \tag{A.59}$$

$$\frac{d}{dx}\cos(ax) = -a\sin(ax) \tag{A.60}$$

$$\frac{d}{dx}\tan(ax) = \frac{a}{\cos^2(ax)} \tag{A.61}$$

$$\frac{d}{dx}\cot(ax) = -\frac{a}{\sin^2(ax)} \tag{A.62}$$

Exponentials and logarithms:

$$\frac{d}{dx}e^{ax} = ae^{ax} \tag{A.63}$$

$$\frac{d}{dx}\ln(ax) = \frac{1}{x} \tag{A.64}$$

$$\frac{d}{dx}a^x = a^x \ln a \tag{A.65}$$

Product rule:

$$\frac{d}{dx}\big(f(x)g(x)\big) = \left(\frac{df(x)}{dx}\right)g(x) + f(x)\left(\frac{dg(x)}{dx}\right) \tag{A.66}$$

Chain rule:

$$\frac{dy}{dx} = \frac{dy}{du}\frac{du}{dx} \tag{A.67}$$

4.2 Integrals
All indefinite integrals have an additive integration constant, c.
Polynomials:

$$\int x^n dx = \frac{1}{n+1}x^{n+1} + c \quad \text{(for } n \neq -1) \tag{A.68}$$

$$\int x^{-1}dx = \ln|x| + c \tag{A.69}$$

$$\int \frac{1}{a^2 + x^2}dx = \frac{1}{a}\tan^{-1}\frac{x}{a} + c \tag{A.70}$$

$$\int \frac{1}{\sqrt{a^2 + x^2}}dx = \ln\left|x + \sqrt{a^2 + x^2}\right| + c \tag{A.71}$$

$$\int \frac{1}{\sqrt{a^2 - x^2}}dx = \sin^{-1}\frac{x}{|a|} + c = \tan^{-1}\frac{x}{\sqrt{a^2 - x^2}} + c \tag{A.72}$$

$$\int \frac{1}{\left(a^2 + x^2\right)^{3/2}} dx = \frac{1}{a^2} \frac{x}{\sqrt{a^2 + x^2}} + c \qquad (A.73)$$

$$\int \frac{x}{\left(a^2 + x^2\right)^{3/2}} dx = -\frac{1}{\sqrt{a^2 + x^2}} + c \qquad (A.74)$$

Trigonometric functions:

$$\int \sin(ax) dx = -\frac{1}{a} \cos(ax) + c \qquad (A.75)$$

$$\int \cos(ax) dx = \frac{1}{a} \sin(ax) + c \qquad (A.76)$$

Exponentials:

$$\int e^{ax} dx = \frac{1}{a} e^{ax} + c \qquad (A.77)$$

5. Complex Numbers

We are all familiar with real numbers, which can be sorted along a number line in order of increasing value, from $-\infty$ to $+\infty$. The real numbers are embedded in a much larger set of numbers, called the *complex numbers*. Complex numbers are defined in terms of their real part and their imaginary part. The space of complex numbers is a plane, for which the real parts form one axis, labeled $\Re(z)$ in Figure A.7. The imaginary parts form the other axis, labeled $\Im(z)$ in Figure A.7. (It is conventional to use the Old German script letters $\Re$ and $\Im$ to represent the real and imaginary parts of complex numbers.)

A complex number z is defined in terms of its real part, x, its imaginary part, y, and Euler's constant, i:

$$z = x + iy \qquad (A.78)$$

Euler's constant is defined as follows:

$$i^2 = -1 \qquad (A.79)$$

Both the real part, $x = \Re(z)$, and the imaginary part, $y = \Im(z)$, of a complex number are real numbers. Addition, subtraction, multiplication, and division of complex numbers are defined in analogy to those operations for real numbers, with $i^2 = -1$:

$$(a + ib) + (c + id) = (a + c) + i(b + d) \qquad (A.80)$$

$$(a + ib) - (c + id) = (a - c) + i(b - d) \qquad (A.81)$$

$$(a + ib)(c + id) = (ac - bd) + i(ad + bc) \qquad (A.82)$$

$$\frac{a + ib}{c + id} = \frac{(ac + bd) + i(bc - ad)}{c^2 + d^2} \qquad (A.83)$$

For each complex number z there exists a complex conjugate z^*, which has the same real part but an imaginary part with the opposite sign:

$$z = x + iy \Leftrightarrow z^* = x - iy \qquad (A.84)$$

We can express the real and imaginary parts of a complex number in terms of the number and its complex conjugate:

$$\Re(z) = \tfrac{1}{2}(z + z^*) \qquad (A.85)$$

$$\Im(z) = \tfrac{1}{2}i(z - z^*) \qquad (A.86)$$

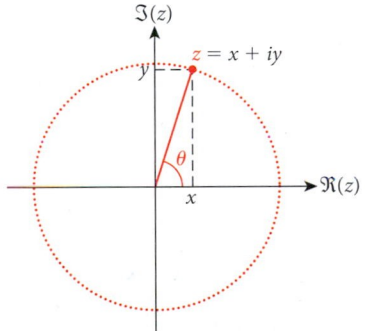

FIGURE A.7 The complex plane. The horizontal axis is formed by the real parts of complex numbers, and the vertical axis by the imaginary parts.

Just like a two-dimensional vector, a complex number $z = x + iy$ has the magnitude $|z|$ as well as an angle θ with respect to the horizontal axis of the complex plane, as indicated in Figure A.7:

$$\left|z\right|^2 = zz^*$$ (A.87)

$$\theta = \tan^{-1}\frac{\Im(z)}{\Re(z)} = \tan^{-1}\frac{i(z-z^*)}{(z+z^*)}$$ (A.88)

We can thus write the complex number $z = x + iy$ in terms of the magnitude and the "phase angle":

$$z = \left|z\right|(\cos\theta + i\sin\theta)$$ (A.89)

An interesting and most useful identity is *Euler's formula*:

$$e^{i\theta} = \cos\theta + i\sin\theta$$ (A.90)

With the aid of this identity, we can write, for any complex number, z,

$$z = \left|z\right|e^{i\theta}$$ (A.91)

We can thus take a complex number z to any power n:

$$z^n = \left|z\right|^n e^{in\theta}$$ (A.92)

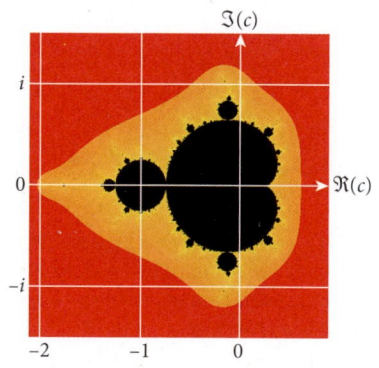

FIGURE A.8 Mandelbrot set in the complex plane.

EXAMPLE A.1 Mandelbrot Set

We can put our knowledge of complex numbers and their multiplication to good use by examining the *Mandelbrot set,* defined as the set of all points c in the complex plane for which the series of iterations

$$z_{n+1} = z_n^2 + c, \qquad \text{with } z_0 = c,$$

does not escape to infinity, that is, for which $|z_n|$ remains finite for all iterations.

This iteration prescription is seemingly simple. For example, we can see that any number for which $|c| > 2$ cannot be part of the Mandelbrot set. However, if we plot the points of the Mandelbrot set in the complex plane, a strangely beautiful object emerges. In Figure A.8, the black points are part of the Mandelbrot set, and the remaining points are color-coded by how fast z_n escapes to infinity.

Appendix B

Element Properties

Z	Charge number (number of protons in the nucleus = number of electrons)
ρ	Mass density at temperature 0 °C (= 273.15 K) and pressure of 1 atmosphere
m	Standard atomic weight (average mass of an atom, abundance-weighted average of the isotope masses)
$T_{melting}$	Temperature of melting point (transition point between solid and liquid phase) at 1 atm pressure
$T_{boiling}$	Temperature of boiling point (transition point between liquid and gas phase) at 1 atm pressure
L_m	Heat of melting/fusion
L_v	Heat of vaporization
E_1	Ionization energy (energy to remove least bound electron)

Z	Sym	Name	Electron Configuration	ρ (g/cm^3)	m (g/mol)	$T_{melting}$ (K)	$T_{boiling}$ (K)	L_m (kJ/mol)	L_v (kJ/mol)	E_1 (eV)
1	H	Hydrogen$_{gas}$	$1s^1$	$8.988 \cdot 10^{-5}$	1.00794	14.01	20.28	0.117	0.904	13.5984
2	He	Helium$_{gas}$	$1s^2$	$1.786 \cdot 10^{-4}$	4.002602	—	4.22	—	0.0829	24.5874
3	Li	Lithium	$[He]2s^1$	0.534	6.941	453.69	1615	3.00	147.1	5.3917
4	Be	Beryllium	$[He]2s^2$	1.85	9.012182	1560	2742	7.895	297	9.3227
5	B	Boron	$[He]2s^2 2p^1$	2.34	10.811	2349	4200	50.2	480	8.2980
6	C	Carbon$_{graphite}$	$[He]2s^2 2p^2$	2.267	12.0107	3800	4300	117	710.9	11.2603
7	N	Nitrogen$_{gas}$	$[He]2s^2 2p^3$	$1.251 \cdot 10^{-3}$	14.0067	63.1526	77.36	0.72	5.56	14.5341
8	O	Oxygen$_{gas}$	$[He]2s^2 2p^4$	$1.429 \cdot 10^{-3}$	15.9994	54.36	90.20	0.444	6.82	13.6181
9	F	Fluorine$_{gas}$	$[He]2s^2 2p^5$	$1.7 \cdot 10^{-3}$	18.998403	53.53	85.03	0.510	6.62	17.4228
10	Ne	Neon$_{gas}$	$[He]2s^2 2p^6$	$9.002 \cdot 10^{-4}$	20.1797	24.56	27.07	0.335	1.71	21.5645
11	Na	Sodium	$[Ne]3s^1$	0.968	22.989770	370.87	1156	2.60	97.42	5.1391
12	Mg	Magnesium	$[Ne]3s^2$	1.738	24.3050	923	1363	8.48	128	7.6462
13	Al	Aluminum	$[Ne]3s^2 3p^1$	2.70	26.981538	933.47	2792	10.71	294.0	5.9858
14	Si	Silicon	$[Ne]3s^2 3p^2$	2.3290	28.0855	1687	3538	50.21	359	8.1517
15	P	Phosphorus$_{white}$	$[Ne]3s^2 3p^3$	1.823	30.973761	317.3	550	0.66	12.4	10.4867
16	S	Sulfur	$[Ne]3s^2 3p^4$	1.92–2.07	32.065	388.36	717.8	1.727	45	10.3600
17	Cl	Chlorine	$[Ne]3s^2 3p^5$	$3.2 \cdot 10^{-3}$	35.453	171.6	239.11	6.406	20.41	12.9676
18	Ar	Argon	$[Ne]3s^2 3p^6$	$1.784 \cdot 10^{-3}$	39.948	83.80	87.30	1.18	6.43	15.7596
19	K	Potassium	$[Ar]4s^1$	0.89	39.0983	336.53	1032	2.4	79.1	4.3407

Z	Sym	Name	Electron Configuration	ρ (g/cm^3)	m (g/mol)	$T_{melting}$ (K)	$T_{boiling}$ (K)	L_m (kJ/mol)	L_v (kJ/mol)	E_1 (eV)
20	Ca	Calcium	[Ar]$4s^2$	1.55	40.078	1115	1757	8.54	154.7	6.1132
21	Sc	Scandium	[Ar]$3d^1\,4s^2$	2.985	44.955910	1814	3109	14.1	332.7	6.5615
22	Ti	Titanium	[Ar]$3d^2\,4s^2$	4.506	47.867	1941	3560	14.15	425	6.8281
23	V	Vanadium	[Ar]$3d^3\,4s^2$	6.0	50.9415	2183	3680	21.5	459	6.7462
24	Cr	Chromium	[Ar]$3d^5\,4s^1$	7.19	51.9961	2180	2944	21.0	339.5	6.7665
25	Mn	Manganese	[Ar]$3d^5\,4s^2$	7.21	54.938049	1519	2334	12.91	221	7.4340
26	Fe	Iron	[Ar]$3d^6\,4s^2$	7.874	55.845	1811	3134	13.81	340	7.9024
27	Co	Cobalt	[Ar]$3d^7\,4s^2$	8.90	58.933200	1768	3200	16.06	377	7.8810
28	Ni	Nickel	[Ar]$3d^8\,4s^2$	8.908	58.6934	1728	3186	17.48	377.5	7.6398
29	Cu	Copper	[Ar]$3d^{10}\,4s^1$	8.94	63.546	1357.77	2835	13.26	300.4	7.7264
30	Zn	Zinc	[Ar]$3d^{10}\,4s^2$	7.14	65.409	692.68	1180	7.32	123.6	9.3942
31	Ga	Gallium	[Ar]$3d^{10}\,4s^2\,4p^1$	5.91	69.723	302.9146	2477	5.59	254	5.9993
32	Ge	Germanium	[Ar]$3d^{10}\,4s^2\,4p^2$	5.323	72.64	1211.40	3106	36.94	334	7.8994
33	As	Arsenic	[Ar]$3d^{10}\,4s^2\,4p^3$	5.727	74.92160	1090	887	24.44	34.76	9.7886
34	Se	Selenium	[Ar]$3d^{10}\,4s^2\,4p^4$	4.28–4.81	78.96	494	958	6.69	95.48	9.7524
35	Br	Bromine$_{liquid}$	[Ar]$3d^{10}\,4s^2\,4p^5$	3.1028	79.904	265.8	332.0	10.571	29.96	11.8138
36	Kr	Krypton$_{gas}$	[Ar]$3d^{10}\,4s^2\,4p^6$	3.749·10^{-3}	83.798	115.79	119.93	1.64	9.08	13.9996
37	Rb	Rubidium	[Kr]$5s^1$	1.532	85.4678	312.46	961	2.19	75.77	4.1771
38	Sr	Strontium	[Kr]$5s^2$	2.64	87.62	1050	1655	7.43	136.9	5.6949
39	Y	Yttrium	[Kr]$4d^1\,5s^2$	4.472	88.90585	1799	3609	11.42	365	6.2173
40	Zr	Zirconium	[Kr]$4d^2\,5s^2$	6.52	91.224	2128	4682	14	573	6.6339
41	Nb	Niobium	[Kr]$4d^4\,5s^1$	8.57	92.90638	2750	5017	30	689.9	6.7589
42	Mo	Molybdenum	[Kr]$4d^5\,5s^1$	10.28	95.94	2896	4912	37.48	617	7.0924
43	Tc	Technetium	[Kr]$4d^5\,5s^2$	11	(98)	2430	4538	33.29	585.2	7.28
44	Ru	Ruthenium	[Kr]$4d^7\,5s^1$	12.45	101.07	2607	4423	38.59	591.6	7.3605
45	Rh	Rhodium	[Kr]$4d^8\,5s^1$	12.41	102.90550	2237	3968	26.59	494	7.4589
46	Pd	Palladium	[Kr]$4d^{10}$	12.023	106.42	1828.05	3236	16.74	362	8.3369
47	Ag	Silver	[Kr]$4d^{10}\,5s^1$	10.49	107.8682	1234.93	2435	11.28	250.58	7.5762
48	Cd	Cadmium	[Kr]$4d^{10}\,5s^2$	8.65	112.411	594.22	1040	6.21	99.87	8.9938
49	In	Indium	[Kr]$4d^{10}\,5s^2\,5p^1$	7.31	114.818	429.7485	2345	3.281	231.8	5.7864
50	Sn	Tin$_{white}$	[Kr]$4d^{10}\,5s^2\,5p^2$	7.365	118.710	505.08	2875	7.03	296.1	7.3439
51	Sb	Antimony	[Kr]$4d^{10}\,5s^2\,5p^3$	6.697	121.760	903.78	1860	19.79	193.43	8.6084
52	Te	Tellurium	[Kr]$4d^{10}\,5s^2\,5p^4$	6.24	127.60	722.66	1261	17.49	114.1	9.0096
53	I	Iodine	[Kr]$4d^{10}\,5s^2\,5p^5$	4.933	126.90447	386.85	457.4	15.52	41.57	10.4513
54	Xe	Xenon$_{gas}$	[Kr]$4d^{10}\,5s^2\,5p^6$	5.894·10^{-3}	131.293	161.4	165.03	2.27	12.64	12.1298
55	Cs	Cesium	[Xe]$6s^1$	1.93	132.90545	301.59	944	2.09	63.9	3.8939
56	Ba	Barium	[Xe]$6s^2$	3.51	137.327	1000	2170	7.12	140.3	5.2117
57	La	Lanthanum	[Xe]$5d^1\,6s^2$	6.162	138.9055	1193	3737	6.20	402.1	5.5769
58	Ce	Cerium	[Xe]$4f^1\,5d^1\,6s^2$	6.770	140.116	1068	3716	5.46	398	5.5387
59	Pr	Praseodymium	[Xe]$4f^3\,6s^2$	6.77	140.90765	1208	3793	6.89	331	5.473
60	Nd	Neodymium	[Xe]$4f^4\,6s^2$	7.01	144.24	1297	3347	7.14	289	5.5250
61	Pm	Promethium	[Xe]$4f^5\,6s^2$	7.26	(145)	1315	3273	7.13	289	5.582

Z	Sym	Name	Electron Configuration	ρ (g/cm^3)	m (g/mol)	$T_{melting}$ (K)	$T_{boiling}$ (K)	L_m (kJ/mol)	L_v (kJ/mol)	E_1 (eV)
62	Sm	Samarium	[Xe]$4f^6\,6s^2$	7.52	150.36	1345	2067	8.62	165	5.6437
63	Eu	Europium	[Xe]$4f^7\,6s^2$	5.264	151.964	1099	1802	9.21	176	5.6704
64	Gd	Gadolinium	[Xe]$4f^7\,5d^1\,6s^2$	7.90	157.25	1585	3546	10.05	301.3	6.1498
65	Tb	Terbium	[Xe]$4f^9\,6s^2$	8.23	158.92534	1629	3503	10.15	293	5.8638
66	Dy	Dysprosium	[Xe]$4f^{10}\,6s^2$	8.540	162.500	1680	2840	11.06	280	5.9389
67	Ho	Holmium	[Xe]$4f^{11}\,6s^2$	8.79	164.93032	1734	2993	17.0	265	6.0215
68	Er	Erbium	[Xe]$4f^{12}\,6s^2$	9.066	167.259	1802	3141	19.90	280	6.1077
69	Tm	Thulium	[Xe]$4f^{13}\,6s^2$	9.32	168.93421	1818	2223	16.84	247	6.1843
70	Yb	Ytterbium	[Xe]$4f^{14}\,6s^2$	6.90	173.04	1097	1469	7.66	159	6.2542
71	Lu	Lutetium	[Xe]$4f^{14}\,5d^1\,6s^2$	9.841	174.967	1925	3675	22	414	5.4259
72	Hf	Hafnium	[Xe]$4f^{14}\,5d^2\,6s^2$	13.31	178.49	2506	4876	27.2	571	6.8251
73	Ta	Tantalum	[Xe]$4f^{14}\,5d^3\,6s^2$	16.69	180.9479	3290	5731	36.57	732.8	7.5496
74	W	Tungsten	[Xe]$4f^{14}\,5d^4\,6s^2$	19.25	183.84	3695	5828	52.31	806.7	7.8640
75	Re	Rhenium	[Xe]$4f^{14}\,5d^5\,6s^2$	21.02	186.207	3459	5869	60.3	704	7.8335
76	Os	Osmium	[Xe]$4f^{14}\,5d^6\,6s^2$	22.61	190.23	3306	5285	57.85	738	8.4382
77	Ir	Iridium	[Xe]$4f^{14}\,5d^7\,6s^2$	22.56	192.217	2739	4701	41.12	563	8.9670
78	Pt	Platinum	[Xe]$4f^{14}\,5d^9\,6s^1$	21.45	195.078	2041.4	4098	22.17	469	8.9588
79	Au	Gold	[Xe]$4f^{14}\,5d^{10}\,6s^1$	19.3	196.96655	1337.33	3129	12.55	324	9.2255
80	Hg	Mercury$_{liquid}$	[Xe]$4f^{14}\,5d^{10}\,6s^2$	13.534	200.59	234.32	629.88	2.29	59.11	10.4375
81	Tl	Thallium	[Xe]$4f^{14}\,5d^{10}\,6s^2\,6p^1$	11.85	204.3833	577	1746	4.14	165	6.1082
82	Pb	Lead	[Xe]$4f^{14}\,5d^{10}\,6s^2\,6p^2$	11.34	207.2	600.61	2022	4.77	179.5	7.4167
83	Bi	Bismuth	[Xe]$4f^{14}\,5d^{10}\,6s^2\,6p^3$	9.78	208.98038	544.7	1837	11.30	151	7.2855
84	Po	Polonium	[Xe]$4f^{14}\,5d^{10}\,6s^2\,6p^4$	9.320	(209)	527	1235	13	102.91	8.414
85	At	Astatine	[Xe]$4f^{14}\,5d^{10}\,6s^2\,6p^5$	?	(210)	?	?	?	?	?
86	Rn	Radon	[Xe]$4f^{14}\,5d^{10}\,6s^2\,6p^6$	$9.73 \cdot 10^{-3}$	(222)	202	211.3	3.247	18.10	10.7485
87	Fr	Francium	[Rn]$7s^1$	1.87	(223)	~300	~950	~2	~65	4.0727
88	Ra	Radium	[Rn]$7s^2$	5.5	(226)	973	2010	8.5	113	5.2784
89	Ac	Actinium	[Rn]$6d^1\,7s^2$	10	(227)	1323	3471	14	400	5.17
90	Th	Thorium	[Rn]$6d^2\,7s^2$	11.7	232.0381	2115	5061	13.81	514	6.3067
91	Pa	Protactinium	[Rn]$5f^2\,6d^1\,7s^2$	15.37	231.03588	1841	~4300	12.34	481	5.89
92	U	Uranium	[Rn]$5f^3\,6d^1\,7s^2$	19.1	238.02891	1405.3	4404	9.14	417.1	6.1941
93	Np	Neptunium	[Rn]$5f^4\,6d^1\,7s^2$	20.45	(237)	910	4273	3.20	336	6.2657
94	Pu	Plutonium	[Rn]$5f^6\,7s^2$	19.816	(244)	912.5	3505	2.82	333.5	6.0260
95	Am	Americium	[Rn]$5f^7\,7s^2$	12	(243)	1449	2880	14.39	238.5	5.9738
96	Cm	Curium	[Rn]$5f^7\,6d^1\,7s^2$	13.51	(247)	1613	3383	~15	?	5.9914
97	Bk	Berkelium	[Rn]$5f^9\,7s^2$	~14	(247)	1259	?	?	?	6.1979
98	Cf	Californium	[Rn]$5f^{10}\,7s^2$	15.1	(251)	1173	1743	?	?	6.2817
99	Es	Einsteinium	[Rn]$5f^{11}\,7s^2$	8.84	(252)	1133	?	?	?	6.42
100	Fm	Fermium	[Rn]$5f^{12}\,7s^2$	?	(257)	1800	?	?	?	6.50
101	Md	Mendelevium	[Rn]$5f^{13}\,7s^2$	?	(258)	1100	?	?	?	6.58
102	No	Nobelium	[Rn]$5f^{14}\,7s^2$	?	(259)	?	?	?	?	6.65

– Continued

Z	Sym	Name	Electron Configuration	ρ (g/cm^3)	m (g/mol)	T_{melting} (K)	T_{boiling} (K)	L_m (kJ/mol)	L_v (kJ/mol)	E_1 (eV)
103	Lr	Lawrencium	[Rn]$5f^{14}\,7s^2\,7p^1$	?	(262)	?	?	?	?	4.9
104	Rf	Rutherfordium	[Rn]$5f^{14}\,6d^2\,7s^2$	?	(263)	?	?	?	?	6
105	Db	Dubnium	[Rn]$5f^{14}\,6d^3\,7s^2$	?	(268)	?	?	?	?	?
106	Sg	Seaborgium	[Rn]$5f^{14}\,6d^4\,7s^2$	?	(271)	?	?	?	?	?
107	Bh	Bohrium	[Rn]$5f^{14}\,6d^5\,7s^2$	?	(270)	?	?	?	?	?
108	Hs	Hassium	[Rn]$5f^{14}\,6d^6\,7s^2$	?	(270)	?	?	?	?	?
109	Mt	Meitnerium	[Rn]$5f^{14}\,6d^7\,7s^2$	?	(278)	?	?	?	?	?
110	Ds	Darmstadtium	*[Rn]$5f^{14}\,6d^9\,7s^1$	?	(281)	?	?	?	?	?
111	Rg	Roentgenium	*[Rn]$5f^{14}\,6d^9\,7s^2$	?	(281)	?	?	?	?	?
112	Cn	Copernicium	*[Rn]$5f^{14}\,6d^{10}\,7s^2$	?	(285)	?	?	?	?	?
113			*[Rn]$5f^{14}\,6d^{10}\,7s^2\,7p^1$	?	(286)	?	?	?	?	?
114	Fl	Flerovium	*[Rn]$5f^{14}\,6d^{10}\,7s^2\,7p^2$	?	(289)	?	?	?	?	?
115			*[Rn]$5f^{14}\,6d^{10}\,7s^2\,7p^3$	?	(289)	?	?	?	?	?
116	Lv	Livermorium	*[Rn]$5f^{14}\,6d^{10}\,7s^2\,7p^4$	?	(293)	?	?	?	?	?
117			*[Rn]$5f^{14}\,6d^{10}\,7s^2\,7p^5$	?	(294)	?	?	?	?	?
118			*[Rn]$5f^{14}\,6d^{10}\,7s^2\,7p^6$	?	(294)	?	?	?	?	?

*Predicted   (longest-lived isotope)

Answers to Selected Questions and Problems

Chapter 1: Overview

Multiple Choice

1.1 c. **1.3** d. **1.5** a. **1.7** b. **1.9** c. **1.11** d. **1.13** c. **1.15** e.

Exercises

1.35 (a) Three. (b) Four. (c) One. (d) Six. (e) One. (f) Two. (g) Three.
1.37 6.34. **1.39** $1 \cdot 10^{-7}$ cm. **1.41** $1.94822 \cdot 10^{6}$ inches. **1.43** $1 \cdot 10^{6}$ mm.
1.45 1 milliPascal. **1.47** 2420 cm². **1.49** (a) 356,000 km = 221,000 miles.
(b) 407,000 km = 253,000 miles. **1.51** $x_{\text{total}} = 5.50 \cdot 10^{-1}$ m; $x_{\text{avg}} =$
$9.17 \cdot 10^{-2}$ m. **1.53** 120 millifurlongs/microfortnight. **1.55** 76 times
the surface area of Earth. **1.57** 39 km. **1.59** 1.56 barrels is equivalent
to $1.51 \cdot 10^{4}$ cubic inches. **1.61** (a) $V_S = 1.41 \times 10^{27}$ m³. (b) $V_E = 1.08$
$\times 10^{21}$ m³. (c) $\rho_S = 1.41 \times 10^{3}$ kg/m³. (d) $\rho_E = 5.52 \times 10^{3}$ kg/m³.
1.63 1.0×10^{2} cm. **1.65** $x = 21.8$ m and $y = 33.5$ m.
1.67 $\vec{A} = 65.0\hat{x} + 37.5\hat{y}$, $\hat{B} = -56.7\hat{x} + 19.5\hat{y}$, $\vec{C} = -15.4\hat{x} - 19.7\hat{y}$,
$\vec{D} = 80.2\hat{x} - 40.9\hat{y}$. **1.69** 3.27 km. **1.71** $\vec{D} = -15\sqrt{2}\hat{x} + \left(32 + 15\sqrt{2}\right)\hat{y} - 3\hat{z}$,
$\left|\vec{D}\right| = 57$ paces. **1.73** $f = 16°$, $\alpha = 41°$, and $\theta = 140°$. **1.75** $2 \cdot 10^{8}$. **1.77** 63.7 m
at $-57.1°$ or 303° (equivalent angles). **1.79** (a) $1.70 \cdot 10^{3}$ at 296°.
(b) $1.61 \cdot 10^{3}$ at 292°. **1.81** $1.00 \cdot 10^{3}$ N. **1.83** (a) 125 miles. (b) 240° or
$-120°$ (from positive x-axis or E). (c) 167 miles. **1.85** 3.79 km at
21.9° W of N. **1.87** $5.62 \cdot 10^{7}$ km. **1.89** 9630. inches. **1.91** $1.4 \cdot 10^{11}$ m,
18° from the Sun.

Multi-Version Exercises

1.93 $\vec{A} = 2.5\hat{x} + 1.5\hat{y}, \vec{B} = 5.5\hat{x} - 1.5\hat{y}, \vec{C} = -6\hat{x} - 3\hat{y}$.
1.95 $(2, -3)$. **1.97** $\vec{D} = 2\hat{x} - 3\hat{y}$. **1.99** $\left|\vec{A}\right| = 58.3$ m, $\left|\vec{B}\right| = 58.3$ m.
1.103 $\vec{A} = 63.3$ at 68.7°; $\vec{B} = 175$ at $-59.0°$. **1.107** d. **1.113** (a) 130.
(b) 11.4. (c) $1.48 \cdot 10^{3}$.

Chapter 2: Motion in a Straight Line

Multiple Choice

2.1 e. **2.3** c. **2.5** e. **2.7** d. **2.9** a. **2.11** b. **2.13** c. **2.15** a.

Exercises

2.29 distance = 66.0 km; displacement = 30.0 km south. **2.31** 0 m/s.
2.33 (a) Between -1 s and $+1$ s; 4.0 m/s. (b) -0.20 m/s. (c) 1.4 m/s. (d) 2 : 1.
(e) $[-5, -4]$, $[1,2]$, and $[4,5]$. **2.35** $x = 0.50$ m. **2.37** (a) 646 m/s.
(b) -0.981 s and 0.663 s. (c) 8.30 m/s².

(d)

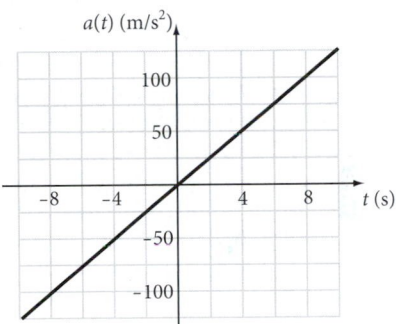

2.39 2.4 m/s² in the backward direction. **2.41** 10.0 m/s².
2.43 $-1.0 \cdot 10^{2}$ m/s². **2.45** (a) 8.14 m/s. (b) 7.43 m/s. (c) 0 m/s².
2.47 $-1.20 \cdot 10^{3}$ cm. **2.49** $x = 23$ m. **2.51** $x = 18$ m. **2.53** (a) At $t = 4.00$ s,
the speed is 20.0 m/s. At $t = 14.0$ s, the speed is 12.0 m/s. (b) 232 m.
2.55 (a) 17.7 s. (b) -1.08 m/s². **2.57** 33.3 m/s. **2.59** 20.0 m.
2.61 (a) 2.50 m/s. (b) 10.0 m. **2.63** (a) 16 s. (b) 0.84 m/s².
2.65 (a) 61.3 m from the first car's start point. (b) 7.83 s. **2.67** (a) 5.1 m/s.
(b) 3.8 m. **2.69** 2.33 s. **2.71** $v_{\frac{1}{2}y} = \sqrt{gy}$. **2.73** 1.46 s. **2.75** 29 m/s.
2.77 (a) 3.52 s. (b) 0.515 s. **2.79** 395 m. **2.81** (a) 6.39 m/s². (b) 56.3 m.
2.83 (a) 33.3 m. (b) -4.17 m/s². **2.85** The trains collide. **2.87** 570 m.
2.89 (a) between $x = 160$ m and $x = 1600$ m. (b) 290 m/s.
2.91 (a) 2.46 m/s². (b) 273 m. **2.93** (a) $v(t) = 1.7\cos(0.46t/\text{s} - 0.31)$ m/s $-$
0.2 m/s, $a(t) = -0.80\sin(0.46t/\text{s} - 0.31)$ m/s². (b) 0.67 s, 7.5 s, 14 s, 21 s,
and 28 s. **2.95** (a) 18 hours.

(b)

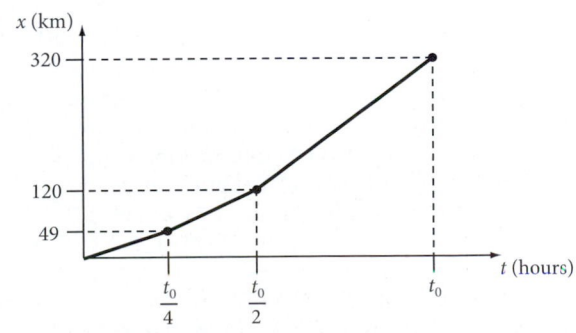

2.97 (a) 37.9 m/s. (b) 26.8 m/s. (c) 1.13 s. **2.99** 693 m.
2.101 (a) $t = \dfrac{\ln 8}{3\alpha}$. (b) $v(t) = \frac{3}{4}\alpha x_0 e^{3\alpha t}$. (c) $a(t) = \frac{9}{4}\alpha^2 x_0 e^{3\alpha t}$. (d) s⁻¹.

Multi-Version Exercises

2.103 2.85 s. **2.106** 1.243 s. **2.108** 9.917 m/s².

Chapter 3: Motion in Two and Three Dimensions

Multiple Choice

3.1 c. **3.3** d. **3.5** c. **3.7** a. **3.9** a. **3.11** a. **3.13** a. **3.15** a.

Exercises

3.35 2.8 m/s. **3.37** 3.06 km 67.5° north of east. **3.39** (a) 174 m from the origin, 60.8° north of west. (b) 21.8 m/s, 44.6° north of west. (c) 1.14 m/s^2, 37.9° north of west. **3.41** 30.0 m/s horizontally and 19.6 m/s vertically. **3.43** 4.69 s. **3.45** 4:1. **3.47** (a) 7.27 m. (b) –9.13 m/s. **3.49** 6.61 m. **3.51** (a) 60.0 m. (b) 75.0°. (c) 31.0 m. **3.53** initial: 24.6 m/s at 47.3°; final: 20.2 m/s at 34.3° downward. **3.55** 81 m/s. **3.57** 3.47 m/s. **3.59** 5. **3.61** (a) 62.0 m/s. (b) 62.3 m/s. **3.63** 14.3 s. **3.65** 3.94 m/s. **3.67** (a) 72.3°. (b) 7.62 s. (c) 90.0°. (d) 7.26 s. (e) $v_{boat} > 5.33$ m/s. **3.69** 95.4 m/s. **3.71** 26.0 m/s. **3.73** 24.8° below the horizontal. **3.75** helicopter: 14.9 m/s; box: 100. m/s. **3.77** 37.7 m/s at 84.1° above the horizontal. **3.79** The eighth floor. **3.81** 9.07 s. **3.83** 1.00 m/s^2. **3.85** 2.69 s. **3.87** (a) 19.1 m. (b) 2.01 s. **3.89** No. After the burglar reaches a horizontal displacement of 5.50 m, he has dropped 8.41 m from the first rooftop and cannot reach the second rooftop. **3.91** (a) Yes. (b) 49.0 m/s at 29.1° above the horizontal. **3.93** 9.20 m/s. **3.95** 8.87 km before the target; 0.180 s window of opportunity. **3.97** (a) 77.4 m/s at 49.7° above the horizontal. (b) 178 m. (c) 63.4 m/s at 38.0° below the horizontal.

Multi-Version Exercises

3.99 12.4 m. **3.103** 349.6°.

Chapter 4: Force

Multiple Choice

4.1 d. **4.3** d. **4.5** a. **4.7** b. **4.9** a. **4.11** b. **4.13** c and d. **4.15** b.

Exercises

4.27 (a) 0.167 N. (b) 0.102 kg. **4.29** 229 lb. **4.31** 4.32 m/s^2. **4.33** 3.62×10^8 N. **4.35** 183 N. **4.37** (a) 1.09 m/s^2. (b) 4.36 N. **4.39** $m_3 = 0.0500$ kg; $\theta = 217°$. **4.41** (a) 441 N. (b) 531 N. **4.43** (a) 2.60 m/s^2. (b) 0.346 N. **4.45** 49.2°. **4.47** (a) 471 N. (b) 589 N. **4.49** Left: 44.4 N; right: 56.6 N. **4.51** 0.69 m/s^2 downward. **4.53** 284 N. **4.55** 807 N. **4.57** 84.9 m. **4.59** 5.84 N. **4.61** (a) 300. N. (b) 500. N. (c) Initially the force of friction is 506 N. After the refrigerator is in motion, the force of friction is the force of kinetic friction, 407 N. **4.63** 17.9 m/s. **4.65** 2.30 m/s^2. **4.67** (a) Block 1 moves up, block 2 moves down. (b) 4.56 m/s^2. **4.69** (a) 4.22 m/s^2. (b) 26.7 m. **4.71** (a) 58.9 N. (b) 76.9 N. **4.73** 2.45 m/s^2. **4.75** (a) 243 N. (b) 46.4 N. (c) 3.05 m/s. **4.77** 1.40 m/s^2. **4.79** 6760 N. **4.81** (a) 29.5 N. (b) 0.754. **4.83** 9.16°. **4.85** (a) $1.69 \cdot 10^{-5}$ kg/m. (b) 0.0274 N. **4.87** (a) 18.6 N. (b) $a_1 = 6.07$ m/s^2; $a_2 = 2.49$ m/s^2. **4.89** 1.72 m/s^2. **4.91** (a) $a_1 = 5.17$ m/s^2; $a_2 = 3.43$ m/s^2. (b) 35.3 N. **4.93** (a) 32.6°. (b) 243 N. **4.95** (a) 3.34 m/s^2. (b) 6.57 m/s^2.

Multi-Version Exercises

4.96 3.312 N. **4.100** 4.997 m/s^2. **4.103** 14.7 N.

Chapter 5: Kinetic Energy, Work, and Power

Multiple Choice

5.1 c. **5.3** a. **5.5** e. **5.7** c. **5.9** b. **5.11** e. **5.13** b.

Exercises

5.19 (a) $4.50 \cdot 10^3$ J. (b) $1.80 \cdot 10^2$ J. (c) $9.00 \cdot 10^2$ J. **5.21** $4.38 \cdot 10^6$ J. **5.23** 12.0 m/s. **5.25** $3.50 \cdot 10^3$ J. **5.27** 0. **5.29** 7.85 J. **5.31** $5.41 \cdot 10^2$ J. **5.33** 1.25 J. **5.35** $\mu = 0.123$; no. **5.37** 44 m/s. **5.39** 17.4 J. **5.41** (a) $W = 1.60 \cdot 10^3$ J. (b) $v_f = 56.6$ m/s. **5.43** $2.40 \cdot 10^4$ N/m. **5.45** 17.6 m/s. **5.47** 3.42 m/s. **5.49** $2.01 \cdot 10^6$ W; The power released by the car. **5.51** 31.8 m/s. **5.53** $3.33 \cdot 10^4$ J. **5.55** 9.12 kJ. **5.57** 42.0 kW = 56.3 hp. **5.59** 62.8 hp. **5.61** 44.2 m/s. **5.63** 5.15 m/s. **5.65** 25.0 N. **5.67** 366 kJ. **5.69** $v_f = 23.9$ m/s in the direction of F_1. **5.71** 35.3°. **5.73** 15.8 J. **5.75** 16.1 hp.

Multi-Version Exercises

5.76 3.899 kJ. **5.79** 3.149 hp. **5.82** 10.72 m. **5.84** 636.7 W.

Chapter 6: Potential Energy and Energy Conservation

Multiple Choice

6.1 a. **6.3** e. **6.5** d. **6.7** d. **6.9** a. **6.11** c. **6.13** b.

Exercises

6.31 29.4 J. **6.33** 0.0869 J. **6.35** $1.93 \cdot 10^6$ J; The net work done on the car is zero. **6.37** 11.6 J. **6.39** (a) $F(y) = 2by - 3ay^2$. (b) $F(y) = -cU_0\cos(cy)$. **6.41** 9.90 m/s. **6.43** 19.1 m/s. **6.45** (a) 8.72 J. (b) 18.3 m/s. **6.47** (a) 28.0 m/s. (b) No. (c) Yes. **6.49** (a) 1.20 J. (b) 0 J. **6.51** (a) 3.89 J. (b) 2.79 m/s. **6.53** 5.37 m/s. **6.55** 16.9 kJ. **6.57** 39.0 kJ. **6.59** 7.65 J. **6.61** (a) 8.93 m/s. (b) 4.08 m/s. (c) –8.52 J. **6.63** $x = 42$ m, $y = 24$ m. **6.65** (a) 14 m/s. (b) 14 m/s. (c) 0.2 m; 6.6 m. **6.67** $41.0 \cdot 10^4$ J. **6.69** $2.0 \cdot 10^8$ J. **6.71** 521 J. **6.73** 1.63 m. **6.75** 3.77 m/s. **6.77** 8.86 m/s. **6.79** $1.27 \cdot 10^2$ m. **6.81** 2.21 kJ. **6.83** (a) $-1.02 \cdot 10^{-1}$ J (lost to friction). (b) 138 N/m. **6.85** (a) 12.5 J. (b) 3.13; 9.38 J. (c) 12.5 J. (d) by a factor of 1/4. (e) by a factor of 1/4. **6.87** $E_{new} = 2.50$ J, $v_{max,2} = 2.24$ m/s, $A_2 = 22.4$ cm. **6.89** (a) $\dfrac{v}{\mu_k g}$. (b) $\dfrac{v^2}{2\mu_k g}$. (c) $mv^2/2$. (d) mv^2. **6.91** (a) 667 J. (b) 667 J. (c) 667 J. (d) 0 J. (e) 0 J.

Multi-Version Exercises

6.93 732.0 m. **6.96** 10.76 m/s. **6.99** 38.70°.

Chapter 7: Momentum and Collisions

Multiple Choice

7.1 b. **7.3** b, d. **7.5** e. **7.7** c. **7.9** c. **7.11** a, b, c. **7.13** a.

Exercises

7.25 (a) 1.0. (b) 0.67. **7.27** $p_x = 3.51$ kg m/s, $p_y = 5.61$ kg m/s. **7.29** 30,500 N, 1.02 s. **7.31** (a) 675 N s opposite to v. (b) 675 N s opposite to v. (c) 136 kg m/s opposite to v. (d) No, the running back will start pushing back with his feet. **7.33** 0.0144 m/s; 2.42 m/s; 10.5 m/s; 762 months. **7.35** (a) $3.15 \cdot 10^9$ m/s. (b) $5.50 \cdot 10^7$ m/s. **7.37** (a) –810. m/s. (b) 43.0 km. **7.39** 4.77 m/s. **7.41** –0.224 m/s. **7.43** 1.26 m/s. **7.45** 21.4 m/s at an angle of 41.5° above the horizontal. **7.47** –34.5 km/s. **7.49** $v_B = 0.433$ m/s, $v_A = 1.30$ m/s.

7.51 Position-Time:

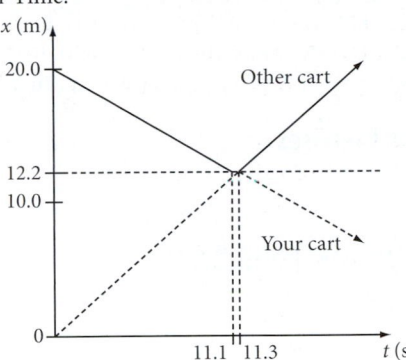

Velocity-Time:

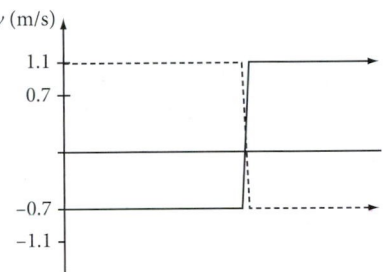

Force-Time:

$$F_0 = \frac{\Delta P}{\Delta t} = \frac{m(v_{i1} - v_{i2})}{\Delta t}$$

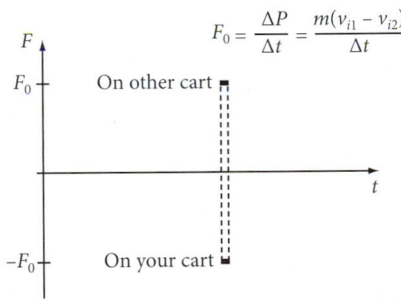

7.53 $3.94 \cdot 10^4$ m/s. **7.55** 0.929 m/s; −23.8°; not elastic. **7.57** $v_{1f} = 582$ m/s in the positive y-direction, and $v_{2f} = 416$ m/s at 36.2° below the positive x-axis. **7.59** Betty: 1.55 kJ; Sally: 649 J; The ratio K_f/K_i is not equal to one, so the collision is inelastic. **7.61** 6.00 m/s. **7.63** smaller car: −48.2g; larger car: 16.1g. **7.65** 42.0 m/s. **7.67** 7.00 m/s.

7.69

Object	H (cm)	h_1 (cm)	ϵ
Range golf ball	85.0	62.6	0.858
Tennis ball	85.0	43.1	0.712
Billiard ball	85.0	54.9	0.804
Hand ball	85.0	48.1	0.752
Wooden ball	85.0	30.9	0.603
Steel ball bearing	85.0	30.3	0.597
Glass marble	85.0	36.8	0.658
Ball of rubber bands	85.0	58.3	0.828
Hollow, hard plastic ball	85.0	40.2	0.688

7.71 40.7°. **7.73** Yes, with exactly 1 m to spare. **7.75** $\epsilon = 0.688$, $K_f/K_i = 0.605$. **7.77** (a) 0.633 m/s. (b) No. **7.79** 1.79 s. **7.81** $2.99 \cdot 10^5$ m/s. **7.83** 0.190 m/s. **7.85** Average force is 1590 N; 2.94g.
7.87 30.0 kg m/s. **7.89** The momentum vector of the formation is 0.0865 kg m/s$\hat{x}$ + 3.05 kg m/s$\hat{y}$. The 115-g bird would have a speed of 26.5 m/s at 1.63° east of north. **7.91** (a) The distance between the first ball and the pallina is 2.0 m and the distance between the second ball and the pallina is 1.75 m. (b) The distance between the first ball and the pallina is 0.98 m and the distance between the second ball and

the pallina is 0.76 m. **7.93** The speed is $\sqrt{17}\,v_i$ in the direction of 14.0° below the horizontal, where v_i is the speed just before breakup. **7.95** 15.9°.

7.97

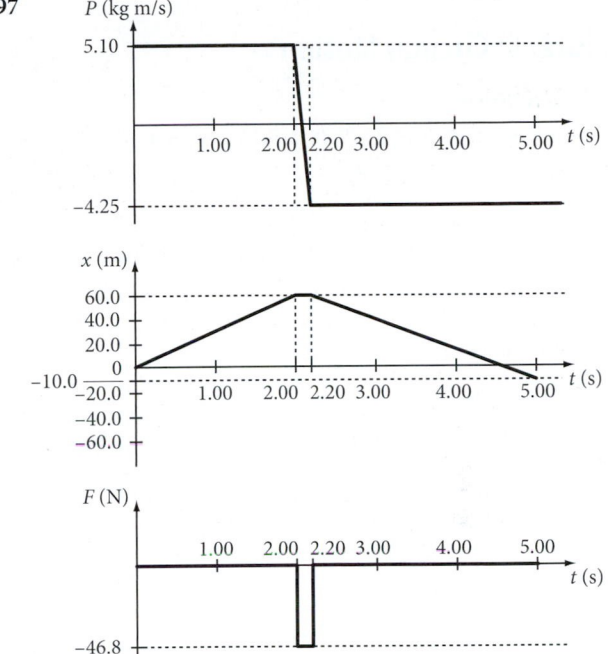

7.99 At least four keys; key ring: 0.12 m/s; phone: 1.33 m/s. **7.101** 22.0 m/s at 0.216° to the right of the initial direction. **7.103** 4.78 pN.
7.105 (a) −(14.9 m/s) $\hat{x}$. (b) $E_{mec,f} = E_{mec,i} = 14.1$ kJ. **7.107** 1.39 m/s.
7.109 $1.16 \cdot 10^{-25}$ kg; germanium.

Multi-Version Exercises

7.110 1.898 m. **7.113** −5.87 m/s. **7.116** 49.20°. **7.119** 0.244 m/s.

Chapter 8: Systems of Particles and Extended Objects

Multiple Choice

8.1 d. **8.3** d. **8.5** e. **8.7** b. **8.9** b. **8.11** a. **8.13** a. **8.15** b.

Exercises

8.29 (a) 4670. km. (b) 742,200 km. **8.31** (−0.500 m,−2.00 m).
8.33 (a) $2.55\hat{x}$ m/s. (b) With respect to the center of mass, before the collision, $\vec{v}_t' = 1.45\hat{x}$ m/s and $\vec{v}_c' = -2.55\hat{x}$ m/s. After the collision, $\vec{v}_t' = -1.45\hat{x}$ m/s and $\vec{v}_c' = 2.55\hat{x}$ m/s. **8.35** (a) −0.769 m/s (to the left).
(b) 0.769 m/s (to the right). (c) −1.50 m/s (to the left). (d) 1.77 m/s
(to the right). **8.37** 0.00603c. **8.39** 126 N in the direction of the water's velocity. **8.41** (a) v/g. (b) $J_{spec, toy} = 81.6$ s, $J_{spec, chem} = 408$ s, $J_{spec, toy} = 0.200J_{spec, chem}$. **8.43** 5.52 h. **8.45** (a) 11,100 kg/s. (b) $1.63 \cdot 10^4$ m/s.
(c) 88.4 m/s². **8.47** (16.9 cm,17.3 cm). **8.49** (3.33 cm,5.77 cm).
8.51 0.286 m. **8.53** $\frac{11x_0}{9}$. **8.55** $6.46 \cdot 10^{-11}$ m. **8.57** 0.1363 ft/s away from the direction the cannons fire. **8.59** (a) 0.866 m/s. (b) 54.5 J.
(c) The chemical energy stored in the muscle tissue of Sam's arms.
8.61 9.09 m/s horizontally. **8.63** 4.08 km/s. **8.65** (a) 2.24 m/s².
(b) 32.4 m/s². (c) 3380 m/s. **8.67** (12.0 cm,5.00 cm). **8.69** $(6a/(8+\pi), (4+3\pi)a/(8+\pi))$. **8.71** 69.3 s. **8.73** (a) $(-5.00\hat{x} + 12.0\hat{y})$ kg m/s.
(b) $(4.00\hat{x} + 6.00\hat{y})$ kg m/s. (c) $(-9.00\hat{x} + 6.00\hat{y})$ kg m/s. **8.75** 88.3 N.

Multi-Version Exercises

8.76 $5.334 \cdot 10^{-7}$ kg/s. **8.79** 309.8 m/s. **8.82** 0.3842 m/s.

Chapter 9: Circular Motion

Multiple Choice

9.1 d. **9.3** b. **9.5** c. **9.7** c. **9.9** a. **9.11** a. **9.13** d. **9.15** d.

Exercises

9.31 $\frac{\pi}{2}$ rad ≈ 1.57 rad. **9.33** $3.07 \cdot 10^6$ m at 13.94° below the surface of the Earth. **9.35** (a) $\alpha = 0.697$ rad/s². (b) $\theta = 8.72$ rad. **9.37** $\Delta d_{12} = 4.76$ m, $\Delta d_{23} = 3.62$ m. **9.39** (a) $v_A = 266.44277$ m/s, $v_B = 266.44396$ m/s, in the direction of the Earth's rotation eastward. (b) $5.97 \cdot 10^{-5}$ rad/s. (c) 29.2 hours. (d) At the Equator there is no difference between the velocities at A and B, so the period is $T_R = \infty$. This means the pendulum does not rotate. **9.41** $\alpha = -6.54$ rad/s². **9.43** -188 rad/s². **9.45** (a) $\omega_3 = 0.209$ rad/s. (b) They all equal 0.209 m/s. (c) $\omega_1 = 2.09$ rad/s; $\omega_2 = 0.419$ rad/s². (d) $\alpha_2 = 2.00 \cdot 10^{-2}$ rad/s²; $\alpha_3 = 1.00 \cdot 10^{-2}$ rad/s². **9.47** (a) 7.00 rad/s. (b) 2.54 rad/s. (c) $-2.47 \cdot 10^{-3}$ rad/s². **9.49** (a) $\Delta\theta = 2.72$ rad. (b) Position of the point: $-0.456\hat{x} + 0.206\hat{y}$; $v = 8.53$ m/s; $a_t = 3.20$ m/s², $a_r = 146$ m/s². **9.51** 6 mg.

9.53 $\Delta t = \sqrt{\dfrac{2g}{\mu d\alpha^2}}$. **9.55** 9.40 m/s. **9.57** 13.2°. **9.59** (a) $v_{\text{zero friction}} = \sqrt{\dfrac{Rg\sin\theta}{\cos\theta}} = \sqrt{Rg\tan\theta}$. (b) v_{max} as in Solved Problem 9.4;

$v_{\text{min}} = \sqrt{\dfrac{Rg(\sin\theta - \mu_s\cos\theta)}{\cos\theta + \mu_s\sin\theta}}$. (c) $v_{\text{zero friction}} = 62.6$ m/s, $v_{\text{min}} = 26.3$ m/s and $v_{\text{max}} = 149$ m/s. **9.61** (a) 0.524 rad/s. (b) $\alpha = -0.0873$ rad/s². (c) $a_t = -0.785$ m/s². **9.63** (a) 54.3 revolutions. (b) 12.1 rev/s. **9.65** $\alpha = -0.105$ rad/s²; $\Delta\theta = 1.88 \cdot 10^4$ rad. **9.67** 4.17 rotations. **9.69** $5.93 \cdot 10^{-3}$ m/s². **9.71** (a) 65.1 s⁻¹. (b) -8.39 s⁻². (c) 139 m. **9.73** $N = 80.3$ N. **9.75** (a) $a_c = 62.5$ m/s²; $F_c = 5.00 \cdot 10^3$ N. (b) $w = 5780$ N. **9.77** 47.5 m. **9.79** (a) $\alpha = 0.471$ rad/s². (b) $a_c = 39.1$ m/s² and $\alpha = 0.471$ rad/s². (c) $a = 39.1$ m/s² at $\theta = 1.90°$ forward of the radius. **9.81** (a) $v = 5.10$ m/s. (b) $T = 736$ N.

Multi-Version Exercises

9.82 16.40 m/s. **9.85** $d = 2.726$ m. **9.87** 15.83 s⁻². **9.90** $7.255 \cdot 10^4$ m/s².

Chapter 10: Rotation

Multiple Choice

10.1 b. **10.3** b. **10.5** c. **10.7** c. **10.9** b. **10.11** b. **10.13** c. **10.15** c. **10.17** a. **10.19** c.

Exercises

10.39 $1.12 \cdot 10^3$ kg m². **10.41** (a) The solid sphere reaches the bottom first. (b) The ice cube travels faster than the solid ball at the base of the incline. (c) 4.91 m/s. **10.45** 5.00 m. **10.47** 8.08 m/s². **10.49** (a) 2.17 N m. (b) 52.7 rad/s. (c) 219 J. **10.51** (a) $7.27 \cdot 10^{-2}$ kg m². (b) 2.28 s. **10.53** (a) $t_{\text{throw}} = 7.68$ s. (b) 5.40 m/s. (c) $t_{\text{throw}} = 7.69$ s, 5.39 m/s. **10.55** 3.31 rad/s². **10.57** (a) 0.577 m. (b) 0.184 m. **10.59** (a) 6.00 rad/s². (b) 150. N m. (c) 1880 rad. (d) 281 kJ. (e) 281 kJ. **10.61** (a) $9.704 \cdot 10^{37}$ kg m². (b) $3.69 \cdot 10^{24}$ kg m². (c) $1.16 \cdot 10^{18}$ kg m². (d) $1.19 \cdot 10^{-20}$.

10.63 (a) $v = \dfrac{J}{M}$, $\omega = \dfrac{5J(h-R)}{2MR^2}$. (b) $h_0 = \dfrac{7}{5}R$.

10.65 (a) 0.150 rad/s. (b) 0.900 m/s **10.67** 0.195 rad/s. **10.69** $E = 1.01 \cdot 10^7$ J, $\tau = 17{,}300$ N m. **10.71** (a) $1.95 \cdot 10^{-46}$ kg m². (b) $2.06 \cdot 10^{-21}$ J. **10.73** 0.487 rad/s. **10.75** 11.0 m/s. **10.77** 0.344 kg m². **10.79** $3.80 \cdot 10^4$ s. **10.81** (a) 259 kg. (b) -15.2 N m. (c) 3.62 revolutions. **10.83** $c = 0.443$.

Multi-Version Exercises

10.85 345 kJ. **10.88** 32.17 J. **10.91** 4.10 m/s².

Chapter 11: Static Equilibrium

Multiple Choice

11.1 c. **11.3** b .**11.5** c. **11.7** a. **11.9** d. **11.11** c.

Exercises

11.27 The tension in the right rope is 600. N and the right rope is attached a distance of $(2/3)L$ from the left edge of the crate. **11.29** 368 N each. **11.31** 42.1 N clockwise. **11.33** (a) $6m_1g$. (b) 7/5. **11.35** 287 N; 939 N. **11.37** The forces are: 743 N on the end farthest from the bricks, 1160 N on the end nearest the bricks. Both forces are upward. **11.39** 777 N downward. **11.41** (a) $T = \dfrac{mg\ell}{\sqrt{L^2 - \ell^2}}$. (b) $F = T = \dfrac{mg\ell}{\sqrt{L^2 - \ell^2}}$. (c) $N = mg$. **11.43** 88.4 N. **11.45** 28.6 N at each hinge. **11.47** $T = 2380$ N; $F_{\text{by}} = 7950$ N; $F_{\text{bx}} = 1190$ N. **11.49** $m = 25.5$ kg. **11.51** 0.25. **11.53** 32.0°. **11.55** (a) The right support applies an upward force of 133 N (29.9 lb). The left support applies an upward force of 267 N (60.0 lb). (b) 2.14 m (7.02 ft) from the left edge of the board. **11.57** (a) 0.206 m. (b) 0.0880 m. **11.59** $d = 3.15$ m. **11.61** unstable: $x = 0$; stable: $x = \pm b$. **11.63** $x = 5L/8$. **11.65** $x_m = 2.54$ m. **11.67** 27.0°. **11.69** 2.74 m from the girl. **11.71** $M_1 = 0.689$ kg. **11.73** (a) $T_r = 196$ N; $T_c = 331$ N. (b) $F_x = 331$ N, $F_y = 687$ N. **11.75** $m_1 = 0.0300$ kg, $m_2 = 0.0300$ kg, $m_3 = 0.0960$ kg. **11.77** 99.6 N in the right chain and 135 N in the left chain. **11.79** (a) 60.5 N. (b) 13.4° below the horizontal.

Multi-Version Exercises

11.81 640.2 N. **11.84** 0.5728. **11.87** 29.95°.

Chapter 12: Gravitation

Multiple Choice

12.1 a. **12.3** b. **12.5** c. **12.7** e. **12.9** a. **12.11** c. **12.13** a. **12.15** a.

Exercises

12.31 3.46 mm.

12.33 $x = \dfrac{L}{1 + \sqrt{\dfrac{M_2}{M_1}}}$.

12.35 $4.36 \cdot 10^{-8}$ N in the positive y-direction. **12.37** This weight of the object on the new planet is $\sqrt[3]{2}$ times the weight of the object on the surface of the Earth. **12.39** (a) 3190 km. (b) 0.055%. **12.41** $3.28 \cdot 10^{-5}$. **12.43** $\sqrt{2}$. **12.45** $v_{\text{min}} = \sqrt{8GM_E/(5R_E)}$. **12.47** 2.5 km. **12.49** (a) $4.72 \cdot 10^{-5}$ m/s. (b) $1.07 \cdot 10^{-7}$ J. **12.51** 7140 s = 1.98 hours.

12.53 (a) $E = \frac{1}{2}mv^2 - \dfrac{GMm}{r}$. (b) $E = \dfrac{L^2}{2mr^2} - \dfrac{GMm}{r}$.

(c) $r_{\text{min}} = -\dfrac{GM}{2(E/m)} + \sqrt{\dfrac{G^2M^2}{(2(E/m))^2} + \dfrac{(L/m)^2}{2(E/m)}}$;

$$r_{max} = -\frac{GM}{2(E/m)} - \sqrt{\frac{G^2M^2}{\left(2(E/m)\right)^2} + \frac{(L/m)^2}{2(E/m)}}.$$

(d) $a = -\dfrac{GM}{2(E/m)}$; $e = \sqrt{1 + \dfrac{(L/m)^2\left(2(E/m)\right)}{G^2M^2}}$.

12.55 7.50 km/s, 5920 s. **12.57** (a) 7770 m/s. (b) 8150 m/s.
12.59 (a) 5.908 km. (b) 2.954 km. (c) 8.872 mm. **12.61** (a) $3.141 \cdot 10^9$ J.
(b) $3.130 \cdot 10^9$ J; there is a slight energy advantage, about 0.4%.
12.63 The force on the Moon due to the Sun is $4.38 \cdot 10^{20}$ N toward
the Sun. The force on the Moon due to the Earth is $1.98 \cdot 10^{20}$ N
toward the Moon. The net force on the Moon is $2.40 \cdot 10^{20}$ N toward
the Sun. The Moon's orbit never curves away from the Sun.
12.65 $2.94 \cdot 10^{-7}$ N. The ratio of the ball forces to the weight of the
small balls is $2.06 \cdot 10^{-8}$:1. **12.67** $2.45 \cdot 10^{10}$ J. **12.69** $2.02 \cdot 10^7$ m.
12.71 (a) $(9.78 \text{ m/s}^2)m$. (b) 29.9° above the horizontal. (c) 45.0°.
12.73 (a) increases by 3.287%. (b) decreases by 6.467%.
12.75 $1.9888 \cdot 10^{30}$ kg. **12.77** (a) $6.28 \cdot 10^4$ m/s, which is 135 times faster
than the speed of a point on the Earth's Equator. (b) $2.66 \cdot 10^{12}$ m/s^2.
(c) $2.71 \cdot 10^{11}$ times bigger than on Earth. (d) $5.50 \cdot 10^4$. (e) $1.89 \cdot 10^6$ m.
12.79 For the new orbit, the distance at perihelion is $6.72 \cdot 10^3$ km,
the distance at aphelion is $1.09 \cdot 10^4$ km, and the orbital period is
$8.23 \cdot 10^3$ s = 2.28 hours.

Multi-Version Exercises

12.81 4.042 km. **12.83** 38.59 AU. **12.86** 263.6 m/s.

Chapter 13: Solids and Fluids

Multiple Choice

13.1 a. **13.3** d. **13.5** $F_2 = F_3 > F_1$. **13.7** d. **13.9** e. **13.11** b. **13.13** a. **13.15** a.

Exercises

13.27 $1.3 \cdot 10^{22}$. **13.29** 0.08 mm. **13.31** 0.318 cm. **13.33** density:
$10.6 \cdot 10^3$ kg/m^3; pressure: $1.10 \cdot 10^8$ Pa; yes, it is a good approximation to
treat the density of seawater as a constant. **13.35** 290. mm.
13.37 (a) 20.8 cm. (b) 21.7 cm. (c) 24.4 cm. **13.39** 6210 m. **13.41** 1.13 m.
13.43 9 pennies. **13.45** (a) 9320 N. (b) 491 N. (c) 49.1 N.
13.47 $1.17 \cdot 10^{-5}$ m^3. **13.49** (a) 49.2 N. (b) 5.02 kg. (c) 0.820 m/s^2.
13.51 (a) $1.089 \cdot 10^6$ N. (b) $9.249 \cdot 10^5$, which means an 18.82% increase
from using hydrogen. **13.53** $9.81 \cdot 10^5$ N/m^2. **13.55** (a) 2.00 m^3/s.
(b) 3750 m/s^2. (c) $3.38 \cdot 10^4$ m/s^2. **13.57** The velocity of the water at the
valve is 4.5 m/s and the velocity of a drop of water from rest is 4.4 m/s.
13.59 (a) $v_2 = 20.5$ m/s.(b) $p_2 = 95.0$ kPa. (c) $h = 0.614$ m.
13.61 $2.15 \cdot 10^{-7}$ m^3. **13.63** 32,700 N. **13.65** (a) 10. N. (b) Yes, because
seawater is denser. **13.67** 12.5 m/s. **13.69** (a) 0.683 g/cm^3.
(b) 0.853 g/cm^3. **13.71** 1.32 MPa. **13.73** (a) 250. m/s. (b) 14.6 kPa.
(c) 585 kN. **13.75** 14.08 m/s. **13.77** 0.39. **13.79** (a) 2.55 m/s.
(b) 176 kPa.

Multi-Version Exercises

13.81 52.12 kN. **13.84** 7.62 kN. **13.87** 0.008697. **13.90** 352 kW.

Chapter 14: Oscillations

Multiple Choice

14.1 c. **14.3** b. **14.5** b. **14.7** a. **14.9** $\omega_a > \omega_b > \omega_c > \omega_d = \omega_e$. **14.11** b.
14.13 c. **14.15** d.

Exercises

14.25 125 N/m. **14.27** 20. Hz. **14.29** (a) $k = 19.5$ N/m. (b) 1.41 Hz.
14.33 $A = 2.35$ cm. **14.35** (a) $T = 2.01$ s. (b) $T = 1.82$ s. (c) $T = 2.26$ s.
(d) There is no period, or, $T = \infty$. **14.37** $T = 1.10$ s. **14.39** (a) $x = 0$

(b) $x = L/\sqrt{12}$. **14.41** (a) $I = \dfrac{43}{30}ML^2$. (b) $T = 2\pi\sqrt{\dfrac{43L}{45g}}$. (c) 1.0 m.

14.43 (a) $x(0) = 1.00$ m; $v(0) = 2.72$ m/s; $a(0) = -2.47$ m/s^2.

(b) $K(t) = (2.5)\pi^2\cos^2\left(\dfrac{\pi}{2}t + \dfrac{\pi}{6}\right)$. (c) $t = 1.67$ s. **14.45** (a) 4.04 J. (b) 1.95 m/s.

14.47 (a) $x_{max} = 2.82$ m. (b) $t = 0.677$ s.

14.49 (a) $r_0 = \left(\dfrac{2A}{B}\right)^{1/6}$. (b) $\omega = \dfrac{3}{\sqrt{m}}\left(\dfrac{4B^7}{A^4}\right)^{1/6}$.

14.51 10.0 s. **14.53** 1.82 s. **14.55** (a) $k = 126$ N/m. (b) $v_{max} = 12.6$ m/s.
(c) $t_f = 34.3$ s. **14.57** (a) 0.154 m. (b) 0.992 m. (c) 0.0385 m.
14.59 The amplitude of the mass's oscillation will be greatest at a
frequency of $f_{d,max} = 2.99$ s^{-1} with an amplitude of 0.430 m. At a
frequency of $f_{d,half} = 2.37$ s^{-1} the amplitude will be half of the maximum.
14.61 $x = \dfrac{x_{max}}{\sqrt{2}}$. **14.63** (a) $x(t) = (1.00 \text{ m})\sin[(1.00 \text{ rad/s})t]$.
(b) $x(t) = (1.12 \text{ m})\sin[(1.00 \text{ rad/s})t + 0.464 \text{ rad}]$. **14.65** 520 N/m.
14.67 8.77 m/s^2. **14.69** $n = 559.6$; $t = 1124$ s. **14.71** $f = 2.23$ Hz.
14.73 $Q = 4.10 \cdot 10^3$. **14.75** 5060 s.

14.77 (a) $T = 2\displaystyle\int_{-A}^{A}\left[\dfrac{c}{2m}\left(A^4 - x^4\right)\right]^{-1/2} dx$. (b) The period is inversely

proportional to A. (c) $T = 2\sqrt{\dfrac{\alpha m}{2\gamma}}A^{\left(1-\frac{\alpha}{2}\right)}\displaystyle\int_{-1}^{1}\left(1 - y^\alpha\right)^{-1/2} dy$. The period

is proportional to $A^{(1-\alpha/2)}$.

Multi-Version Exercises

14.78 7.949 cm. **14.81** 1.404 s. **14.83** 0.7583 m/s.

Chapter 15: Waves

Multiple Choice

15.1 a. **15.3** c. **15.5** d. **15.7** a. **15.9** c.

Exercises

15.23 The resolution time in air is $t_{max} = 0.20$ m/(343 m/s) $= 5.83 \cdot 10^{-4}$ s.
The resolution time in water is $t_{max} = 0.20$ m/(1500 m/s) $= 1.33 \cdot 10^{-4}$ s.
A person can only resolve a time difference of $5.83 \cdot 10^{-4}$ s, so a diver
will not be able to distinguish a time difference of $1.33 \cdot 10^{-4}$ s.
15.25 (a) 0.00200 m. (b) 6.37 waves. (c) 127 cycles. (d) 0.157 m.
(e) 20.0 m/s. **15.27** (a) $\dfrac{d^2\psi_i}{dt^2} = \omega_0^2\left(\psi_{i-1} - 2\psi_i + \psi_{i+1}\right)$.

(b) $\Omega_j = 2\omega_0\sin\left(\dfrac{j\pi}{2(n+1)}\right)$, $1 \le j \le n$.

15.29 $y(x,t) = (8.91 \text{ mm})\sin(10.5 \text{ m}^{-1}x - 100.\pi \text{ rad } t + 2.68 \text{ s}^{-1})$.
15.31 (a) 52.4 m^{-1}. (b) 0.100 s. (c) 62.8 s^{-1}. (d) 1.20 m/s. (e) $\pi/6$.
(f) $y(x,t) = 0.100 \sin(52.4x - 62.8t \pm \pi/6)$. **15.33** 2.28 times longer.
15.35 The sound in the air reaches Alice 0.255 s before the sound from
the wire does.

15.37 (b) $v = 5$ m/s.

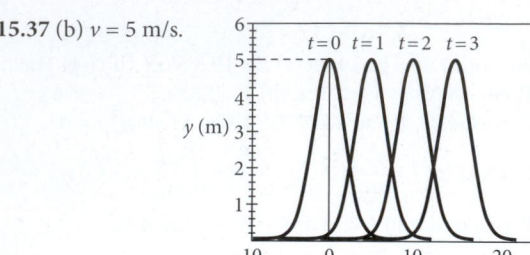

15.39 280 km. **15.41** 360. W. **15.43** (a) 419 Hz. (b) 1.03 m/s^2 upward.
15.45 (a) 173 m/s. (b) 2.00 m, 86.6 Hz. **15.47** 720 N.
15.49 $f(x,t) = (1.00 \text{ cm})\sin((20.0 \text{ m}^{-1})x + (150. \text{ s}^{-1})t)$,
$g(x,t) = (1.00 \text{ cm})\sin((20.0 \text{ m}^{-1})x - (150. \text{ s}^{-1})t)$; 7.50 m/s.
15.51 z (m)

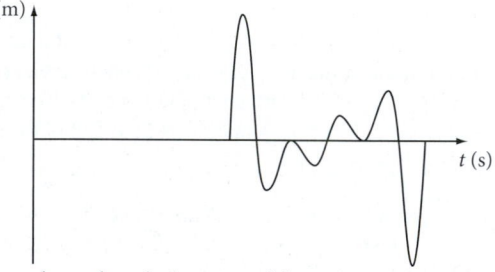

$z(t)$ does not depend on the location of the wave sources at the edges of
the pool. **15.55** (a) 69.2 Hz. (b) 54.7 Hz. **15.57** (a) 80.0 ms. (b) 7.85 m/s.
(c) 617 m/s^2. **15.59** 136 m/s. **15.61** 0.0627. **15.63** $f_2 = f_1$; $v_2 = v_1/\sqrt{3}$;
$\lambda_2 = \lambda_1/\sqrt{3}$. **15.65** (a) 1.26 m, 1.27 Hz. (b) 1.60 m/s. (c) 0.256 N.
15.67 $v = 419.$ m/s; $v_{max} = 3.29$ m/s .
15.69 (a)

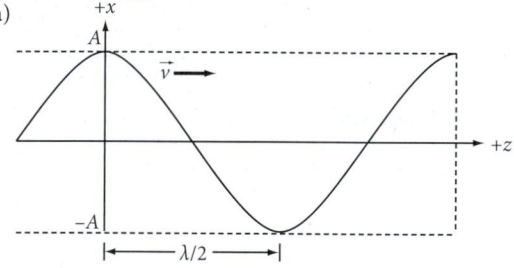

(b) 100. m/s. (c) 31.4 rad/m. (d) 300. N.
(e) $D(z,t) = (0.0300 \text{ m})\cos((10.0\pi \text{ rad/m})x - (1000.\pi \text{ rad/s})t)$.

Multi-Version Exercises

15.71 184.1 Hz. **15.74** $4.512 \cdot 10^3$. **15.77** 43.50 Hz.

Chapter 16: Sound

Multiple Choice

16.1 b. **16.3** c. **16.5** c. **16.7** a. **16.9** b. **16.11** b. **16.13** a.

Exercises

16.25 172 m. **16.27** (a) 343 m/s. (b) 20 °C. **16.29** $1.0 \cdot 10^{20}$ N/m^2 ;
This value is some nine orders of magnitude larger than the actual
value. Light waves are electromagnetic oscillations that do not
require the motion of glass molecules, or the hypothetical ether, for
transmission. **16.31** 6.32 Pa. **16.33** $6.19 \cdot 10^{-8}$ W/m^2. **16.35** 2810 m.
16.37 0.350 m. **16.39** 5.00 m. **16.41** (a) 0.339 W/m^2. (b) 115 dB.
(c) 0.0462 m. **16.43** (a) 2.23°. (b) 10.2°. **16.45** (a) 32.6 m/s. (b) 1292 Hz.
16.47 (a) 50.3°. (b) 1.69 km. **16.49** (a) 900 Hz. (b) 30 m/s. (c) 17 m.
16.51 (a) 220. Hz. (b) 0.780 m. **16.53** 8.19 cm. **16.55** 3.5 kHz.

16.57 425 Hz. **16.59** 2.26 s.
16.61

Note	Frequency (Hz)	Length (m)
G4	392	0.438
A4	440	0.390
B4	494	0.347
F5	698	0.246
C6	1047	0.164

16.63 80.0 dB. **16.65** 6.00 m/s. **16.67** 36.1 Hz. **16.69** 109 Hz.

16.71 (a) In the Young's modulus case, $\delta p(x,t) = Y\dfrac{\partial}{\partial x}\delta x(x,t)$.

In the bulk modulus case, $\delta p(x,t) = B\dfrac{\partial}{\partial x}\delta x(x,t)$.

(b) $\delta p(x,t) = Y\dfrac{\partial}{\partial x}\delta x(x,t) = Y\dfrac{\partial}{\partial x}A\cos(\kappa x - \omega t) = -Y\kappa A\sin(\kappa x - \omega t)$,

$\delta p(x,t) = -B\dfrac{\partial}{\partial x}A\cos(\kappa x - \omega t) = -B\kappa A\sin(\kappa x - \omega t)$;

The pressure amplitude for the Young's modulus case is $\delta p_{max} = -Y\kappa A$
and is $\delta p_{max} = -B\kappa A$ in the bulk modulus case. **16.73** $P_{0.00} = 2.87 \cdot 10^{-5}$ Pa,
$P_{120.} = 28.7$ Pa, $A_{0.00} = 1.11 \cdot 10^{-11}$ m, $A_{120.} = 1.11 \cdot 10^{-5}$ m.

Multi-Version Exercises

16.74 14.3 m/s. **16.77** 23.26 Hz. **16.79** 8726 m/s.

Chapter 17: Temperature

Multiple Choice

17.1 a. **17.3** c. **17.5** d. **17.7** a. **17.9** c. **17.11** b. **17.13** e.

Exercises

17.27 –21.8 °C. **17.29** –89.4 °C. **17.31** (a) 190 K. (b) –110 °F.
17.33 574.59 °F = 574.59 K. **17.35** 7776 kg/m^3. **17.37** 550 °C.
17.39 4.1 mm. **17.41** 116 °C. **17.43** 3.03 m. **17.45** 24 hours and 37
seconds. **17.47** 180 °C. **17.49** $L = 0.16$ m. **17.51** 0.252 cm.
17.53 46.1 μm downward.

17.55 (a) $\beta(T) = \dfrac{-4.52 \cdot 10^{-5} + 11.36 \cdot 10^{-6}T}{1.00016 - 4.52 \cdot 10^{-5}T + 5.68 \cdot 10^{-6}T^2}$.

(b) $\beta(T = 20.0 \text{ °C}) = 1.82 \cdot 10^{-4}$/°C.
17.57 330 cm^3. **17.59** 0.189 L. **17.61** 10. mm. **17.63** 1.45 mm.
17.65 6.8 mm. **17.67** 10.1 °C. **17.69** 0.30%. **17.71** 136 N.
17.73 $\beta = 6.00 \cdot 10^{-6}$ °C. **17.75** (a) 48.6 Hz. (b) 46.8 Hz. (c) 48.6 Hz.

Multi-Version Exercises

17.76 275.8 °C. **17.79** $5.097 \cdot 10^{-3}$ m^2. **17.82** 0.3067 m shorter.

Chapter 18: Heat and the First Law of Thermodynamics

Multiple Choice

18.1 d. **18.3** b. **18.5** c. **18.7** b. **18.9** f. **18.11** a. **18.13** b.

Exercises

18.27 (a) $9.8 \cdot 10^4$ J. (b) If the body converts 100% of food energy into
mechanical energy, then the number of doughnuts needed is 0.094.
The body usually converts only 30% of the energy consumed. This
corresponds to 0.31 of a doughnut. **18.29** 40 J. **18.31** Highest: lead,

22.684 °C; lowest: water, 22.239 °C. **18.33** 10.4 °C. **18.35** They both equal 25.8 kJ. **18.37** 130. J/(kg K); The brick is made of lead. **18.39** 32.9 °C. **18.41** 334 g. **18.43** $2.00 \cdot 10^4$ s; 2090 s. **18.45** (a) None of the water boils away. (b) The aluminum will completely solidify. (c) 44.6 °C. (d) No, it is not possible without the specific heat of aluminum in its liquid phase. **18.47** 291 g. **18.49** 384 W/(m K); copper. **18.51** 5780 K. **18.53** 85.7 s. **18.55** 1.8 kW. **18.57** (a) 592 W/m^2. (b) 226 K. **18.59** (a) $f = (5.88 \cdot 10^{10} \text{ Hz/K})T$. (b) $3.53 \cdot 10^{14}$ Hz. (c) $1.61 \cdot 10^{11}$ Hz. (d) $1.76 \cdot 10^{13}$ Hz. **18.61** 11.6 ft^2 °F h/BTU. **18.63** 5.2 K. **18.65** $2.00 \cdot 10^{-2}$ W/(m K). **18.67** $6.0 \cdot 10^2$ W/m^2. **18.69** (a) 7.07 °C. (b) 0.00 °C. **18.71** $3.98 \cdot 10^{13}$ W. **18.73** 3690:1. **18.75** (a) $4.0 \cdot 10^6$ J. (b) $12.

Multi-Version Exercises

18.77 215 yr. **18.80** 74 W.

Chapter 19: Ideal Gases

Multiple Choice

19.1 a. **19.3** c. **19.5** a. **19.7** b. **19.9** d. **19.11** e. **19.13** b.

Exercises

19.29 342 kPa. **19.31** (a) 259 kPa. (b) 8.84%. **19.33** 374 K. **19.35** 354 m/s. **19.37** 416 kPa. **19.39** 699 L. **19.41** 200. kPa. **19.43** 2.16 cm. **19.45** (a) $3.77 \cdot 10^{-17}$ Pa. (b) 260. m/s. (c) 261 km. **19.47** The rms speed of ^{235}UF$_6$ is 1.0043 times that of ^{238}UF$_6$. **19.49** 27.9 kPa. **19.51** (a) He: $6.69 \cdot 10^{-21}$ J; N: $1.12 \cdot 10^{-20}$ J. (b) He: constant volume: 12.5 J/(mol K); constant pressure: 20.8 J/(mol K); N: constant volume: 20.8 J/(mol K); N: constant pressure: 29.1 J/(mol K). (c) $\gamma_{He} = 5/3$; $\gamma_{N_2} = 7/5$. **19.53** 41.6 J. **19.55** 92.3 J. **19.57** (a) $4.72 \cdot 10^4$ kPa. (b) 189 K. **19.59** 37.6 atm; 826 K. **19.61** $Q_{12} = W_{12} = 2.53$ kJ, $W_{23} = -0.776$ kJ, $Q_{23} = -1.94$ kJ, $W_{31} = -1.16$ kJ, $Q_{31} = 0$.

19.63 (a) $\rho(h) = \rho_0 \left(1 - \frac{\gamma - 1}{\gamma} \cdot \frac{Mgh}{RT_0} \right)^{\frac{1}{\gamma - 1}}$;

$p(h) = p_0 \left(1 - \left(\frac{\gamma - 1}{\gamma} \right) \frac{Mgh}{RT_0} \right)^{\frac{\gamma}{\gamma - 1}}$; $T(h) = T_0 - \frac{\gamma - 1}{\gamma} \cdot \frac{Mgh}{R}$.

(b) 50% sea-level pressure: 5.39 km, 241 K; 50% sea-level density: 7.27 km, 222 K. (c) 5.95 km, 293 K. **19.65** (a) 469 m/s. (b) 1750 m/s.

19.67 $4\pi \left[\frac{m}{2\pi k_B T} \right]^{3/2} \int_{v_S}^{\infty} v^2 e^{-\frac{mv^2}{2k_B T}} dv$; rms speed: 699 m/s; average speed: 644 m/s. **19.69** 83.1 J. **19.71** (a) $8.39 \cdot 10^{23}$ atoms (b) $6.07 \cdot 10^{-21}$ J. (c) 1350 m/s. **19.73** (a) $3.22 \cdot 10^5$ Pa. (b) 605 cm^2. **19.75** 2.02 g/mol, likely hydrogen gas. **19.77** (a) 17.0 kJ. (b) 1.53 kPa. **19.79** $2.07 \cdot 10^{-21}$ J; The energy depends only on temperature, not on the identity of the gas. **19.81** 0.560 L.

Multi-Version Exercises

19.83 -62.47 J. **19.86** $1.204 \cdot 10^6$ Pa.

Chapter 20: The Second Law of Thermodynamics

Multiple Choice

20.1 d. **20.3** a. **20.5** c. **20.7** d. **20.9** a. **20.11** a. **20.13** c.

Exercises

20.27 $\Delta t = 96.5$ s.
20.29 (a)

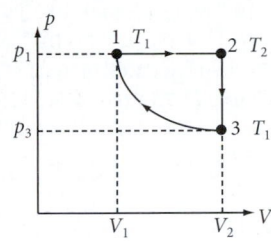

(b) $W_{12} = -90.0$ J, $Q_{12} = -225$ J, $W_{23} = 0$ J, $Q_{23} = 135$ J, $W_{31} = 49.4$ J, and $Q_{31} = 49.4$ J. (c) $\epsilon = 0.180$.
20.31 (a)

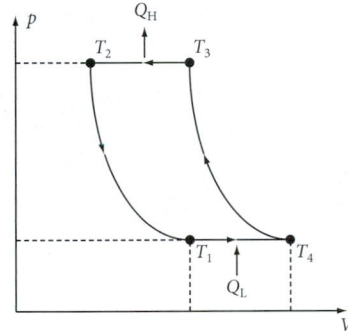

(b) $K = (T_4 - T_1)/(T_3 - T_2 - T_4 + T_1)$. **20.33** (a) 0.7000. (b) 1430 J. (c) $Q_L = 430$. J. **20.35** 2.12%. **20.37** efficiency: 0.33; $T_2 : T_1 = 2 : 3$. **20.39** 48.0 cal. **20.41** $r = 1.75$. **20.43** (a) $p_1 = 101$ kPa, $V_1 = 1.20 \cdot 10^{-3}$ m^3; $T_1 = 586$ K, $p_2 = 507$ kPa, $V_2 = 1.20 \cdot 10^{-3}$ m^3; $T_2 = 2930$ K, $p_3 = 101$ kPa, $V_3 = 6.00 \cdot 10^{-3}$ m^3; $T_3 = 2930$ K. (b) $\epsilon = 0.288$. (c) $\epsilon_{max} = 0.800$. **20.45** (a) $\Delta S_H = -12.1$ J/K, $\Delta S_L = 85.0$ J/K. (b) $\Delta S_{rod} = 0$ J/K (no change in entropy). (c) $\Delta S_{system} = 72.9$ J/K. **20.47** (a) $\epsilon = 0.250$. (b) 0. **20.49** $k_B \ln2$. **20.51** $S_{5\,up} = 0$ (exact); $S_{3\,up} = 3.18 \cdot 10^{-23}$ J/K. **20.53** $-5.94 \cdot 10^{14}$ W/K. **20.55** 0.95. **20.57** 91: 57.5%, 93: 58.5%, 95: 59.9%, 97: 61.0%, % increase = $(60.96\% - 57.5\%)/57.52\% = 6.0\%$. **20.59** $\Delta S = 0.0386$ J/K. **20.61** For two dice the entropy is doubled, and for three dice the entropy is tripled. **20.63** 77.2 K. **20.65** (a) $\epsilon_{max} = 0.471$. (b) $\epsilon = 0.333$. (c) $T_f = 31.4$ °C. **20.67** $\Delta S_{adiabatic} = 0$ (exact), $\Delta S_{isobar} = -6.32$ J/K, $\Delta S_{isotherm} = 6.32$ J/K. **20.69** $\epsilon = 0.253$.

Multi-Version Exercises

20.71 $5.399. **20.74** 164.7 J. **20.77** 0.4356.

Chapter 21: Electrostatics

Multiple Choice

21.1 b. **21.3** b. **21.5** b. **21.7** a. **21.9** c. **21.11** a. **21.13** a.

Exercises

21.31 96,470 C. **21.33** $3.12 \cdot 10^{17}$ electrons. **21.35** 31.75 C. **21.37** (a) $5.00 \cdot 10^{16}$ conduction electrons/cm^3. (b) There are $5.88 \cdot 10^{-7}$ conduction electrons in the doped silicon sample for every conduction electron in the copper sample. **21.39** $1.05 \cdot 10^{-5}$ C; The force is attractive. **21.41** $-2.9 \cdot 10^{-9}$ N. **21.43** 127 N. **21.45** $q = 2.02 \cdot 10^{-5}$ C. **21.47** 3.1 N. **21.49** (a) 0. (b) $\pm a/\sqrt{2}$. **21.51** $\vec{F}_{net, 2} = (-1.22 \cdot 10^8 \text{ N})\hat{x} + (7.25 \cdot 10^7 \text{ N})\hat{y}$; $\left| \vec{F}_{net, 2} \right| = 1.42 \cdot 10^8$ N. **21.53** No; $T = -0.582$ N. **21.55** $-3.66 \cdot 10^{-17}$ C $= -288e$. **21.59** $5.71 \cdot 10^{12}$ C.

21.61 $n = 1$: $F_1 = 8.24 \cdot 10^{-8}$ N; $F_{g,1} = 3.63 \cdot 10^{-47}$ N
$n = 2$: $F_2 = 5.15 \cdot 10^{-9}$ N; $F_{g,2} = 2.27 \cdot 10^{-48}$ N
$n = 3$: $F_3 = 1.02 \cdot 10^{-9}$ N; $F_{g,3} = 4.49 \cdot 10^{-49}$ N
$n = 4$: $F_4 = 3.22 \cdot 10^{-10}$ N; $F_{g,4} = 1.42 \cdot 10^{-49}$ N.
21.63 $4.41 \cdot 10^{-40}$. **21.65** -4.80 N$\hat{y}$. **21.67** (a) 68.2 N. (b) $4.08 \cdot 10^{28}$ m/s^2.
21.69 114 N. **21.71** -65 μC. **21.73** $1.65 \cdot 10^{-7}$ e; $9.39 \cdot 10^{-19}$ kg.
21.75 0.169 μC. **21.77** $m_2 = 50.4$ g. **21.79** $q = 0.105$ μC.
21.81 -24.1 cm. **21.83** 3.04 nC.

Multi-Version Exercises

21.84 1.211 m. **21.87** 2.315 m.

Chapter 22: Electric Fields and Gauss's Law
Multiple Choice

22.1 e. **22.3** a. **22.5** d. **22.7** c. **22.9** a. **22.11** a.

Exercises

22.25 575 N/C. **22.27** 192.5° counterclockwise from the positive
x-axis. **22.29** 4.44 m. **22.31** $E = -kp/[(d/2)^2 + x^2]^{3/2}$. When $x \gg d$ the
magnitude of the electric field reduces to $E = -kp/x^3$.
22.33 $(3.71$ m/s$)\hat{x} + (2.43$ m/s$)\hat{y}$. **22.35** $\vec{E} = (-Q/\pi^2\epsilon_0 R^2)\hat{y}$.

22.37 $\vec{E}(d) = \left(\dfrac{kQ}{d\sqrt{d^2+L^2}}\right)\hat{x} - \left(\dfrac{kQ}{dL} - \dfrac{kQ}{L\sqrt{d^2+L^2}}\right)\hat{y}$. **22.39** 683 N/C.

22.41 $3.46 \cdot 10^{-15}$ N m. **22.43** 0.189 m. **22.45** (a) $v = \sqrt{2h(g - QE/M)}$.
(b) If the value $g - QE/M$ is less than zero the value is non-real and
the body does not fall. **22.47** (a) 0.0141 N/C. (b) $1.35 \cdot 10^6$ m/s^2.
(c) toward the wire. **22.49** $-1.12 \cdot 10^{-8}$ C. **22.51** 60.0 N/C into the face
of the cube. **22.53** Since the radius of the balloon never reaches R, the
charge enclosed is constant and the electric field does not change.
22.55 (a) -54.0 N/C. (b) 0 N/C. (c) 0.360 N/C. (d) $4.97 \cdot 10^{-12}$ C/m^2.
22.57 $-6.77 \cdot 10^5$ C. **22.59** The force on the electron between the plates
has a magnitude of $1.81 \cdot 10^{-14}$ N and is directed from the negative plate
toward the positive plate. The force on the electron outside the plates is
zero. **22.61** $Q = (4/5)\pi AR^5$. **22.63** (a) $4.52 \cdot 10^8$ N/C. (b) $1.95 \cdot 10^8$ N/C.
(c) 0, since it is in the conducting shell. (d) $-1.59 \cdot 10^7$ N/C.
22.65 (a) $1.13 \cdot 10^5$ N/C in the positive x-direction. (b) $4.52 \cdot 10^5$ N/C in
the positive x-direction. **22.67** (a) $\vec{E} = (Qr/4\pi a^3\epsilon_0)\hat{r}$. (b) $\vec{E} = (Q/4\pi\epsilon_0 r^2)\hat{r}$.
(c)

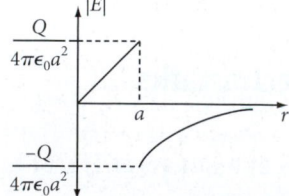

The discontinuity at $r = a$ is due to the surface charge density of the
gold. The charge on the gold layer causes a sudden spike in the total
charge resulting in a discontinuity in the electric fields. **22.69** $E_x =$
$1.83 \cdot 10^5$ N/C, $E_y = 1.34 \cdot 10^5$ N/C. **22.71** 0. **22.73** 145 N/C directed
away from the y-axis. **22.75** (a) -5.00 μC. (b) 0. **22.77** $2.64 \cdot 10^{13}$ m/s^2.
22.79 $4.31 \cdot 10^{-5}$ C/m. **22.81** $3.10 \cdot 10^{10}$ electrons. **22.83** $7.13 \cdot 10^4$ N/C.

22.85 (a)

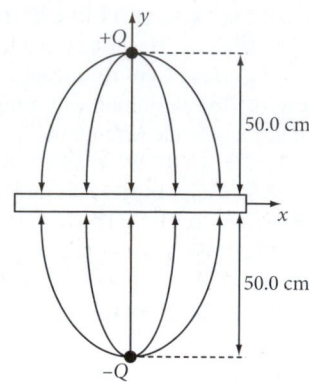

(b) $8.99 \cdot 10^{-3}$ N downward. (c) $E_{\text{total}} = \dfrac{1}{2\pi\epsilon_0} \dfrac{Qa}{\left(a^2 + \rho^2\right)^{3/2}}$.

(d) $\sigma(\rho) = E\epsilon_0 = \dfrac{1}{2\pi} \dfrac{Qa}{\left(a^2 + \rho^2\right)^{3/2}}$.

(e) The total charge induced is equal to the charge in magnitude.

Multi-Version Exercises

22.86 $7.189 \cdot 10^6$ m/s^2. **22.89** $1.438 \cdot 10^4$ N/C.

Chapter 23: Electric Potential
Multiple Choice

23.1 a. **23.3** c. **23.5** a. **23.7** a. **23.9** a. **23.11** c. **23.13** a.

Exercises

23.25 0.734 m. **23.27** $3.84 \cdot 10^{-17}$ J. **23.29** $3.10 \cdot 10^5$ m/s. **23.31** (a) no.
(b) 7.93 cm. (c) 247 km/s. **23.33** (a) 18.0 kV. (b) -71.9 kV.
23.35 $6.95 \cdot 10^{12}$ electrons. **23.37** 1.12 MV. **23.39** 847 V.
23.41 (a) 7.19 kV. (b) 10.8 kV. **23.43** 481 V. **23.45** (a) $x_{\text{max}} = 1$.
(b) $E_0(2e^{-1} - 1)$. **23.47** 70.2 V/m. **23.49** 11.2 m/s^2. **23.51** (a) $10x$ V/m^2.
(b) $3.83 \cdot 10^9$ m/s^2. (c) $2.84 \cdot 10^5$ m/s.

23.53 $\vec{E}(\vec{r}) = \dfrac{kq}{r^3}(x\hat{x} + y\hat{y} + z\hat{z})$.

23.55 (a) $\vec{E}(\vec{r}) = \dfrac{2V_0\vec{r}}{a^2} e^{-r^2/a^2}$. (b) $\rho(\vec{r}) = \dfrac{2\epsilon_0 V_0}{a^2}\left[1 - 2\left(\dfrac{r^2}{a^2}\right)\right]e^{-r^2/a^2}$. (c) 0.

(d)

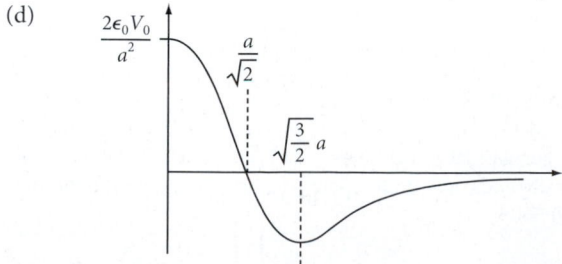

23.57 $2.31 \cdot 10^{-13}$ J = 1.44 MeV. **23.59** 144 keV. **23.61** $v_1 = 0.105$ m/s,
$v_2 = 0.0658$ m/s. **23.63** field: 0; potential: 12 V. **23.65** $1.98 \cdot 10^5$ V.
23.67 $V_A = 2.28 \cdot 10^5$ V; $V_B = V_C = 3.05 \cdot 10^5$ V. **23.69** $5.00 \cdot 10^5$ V;
$1.39 \cdot 10^{-5}$ C. **23.71** (a) 4:1. (b) 33.3 μC. **23.73** 2.02 J. **23.75** first sphere:
$3.00 \cdot 10^3$ N/C; second sphere: $6.00 \cdot 10^3$ N/C. **23.77** (a) 46.8 V.
(b) 7.29 cm. **23.79** (a) $9.44 \cdot 10^4$ V. (b) $0.840 \cdot 10^{-6}$ C,
$E_1 = 1.89 \cdot 10^5$ V/m, $E_2 = 7.55 \cdot 10^5$ V/m.

23.81 (a) $V(x) = \dfrac{1}{4\pi\epsilon_0}\left(\dfrac{q_1}{x - x_1} + \dfrac{q_2}{x - x_2}\right)$ for $x > x_1, x_2$,

$V(x) = \dfrac{1}{4\pi\epsilon_0}\left(\dfrac{q_1}{x_1 - x} + \dfrac{q_2}{x - x_2}\right)$ for $x_1 < x < x_2$, and

$V(x) = \dfrac{1}{4\pi\epsilon_0}\left(\dfrac{q_1}{x_1 - x} + \dfrac{q_2}{x_2 - x}\right)$ for $x < x_1, x_2$. (b) $x = 11.0$ m, $x = -0.250$ m.

(c) $E = \dfrac{1}{4\pi\epsilon_0}\left[\dfrac{q_1}{(x - x_1)^2} + \dfrac{q_2}{(x - x_2)^2}\right]$ for $x > x_1, x_2$,

$E = \dfrac{1}{4\pi\epsilon_0}\left[-\dfrac{q_1}{(x_1 - x)^2} + \dfrac{q_2}{(x - x_2)^2}\right]$ for $x_1 < x < x_2$, and

$E = \dfrac{1}{4\pi\epsilon_0}\left[-\dfrac{q_1}{(x_1 - x)^2} - \dfrac{q_2}{(x_2 - x)^2}\right]$ for $x < x_1, x_2$.

23.83 (a) $V = \dfrac{q}{4\pi\epsilon_0}\left(\dfrac{1}{R}\right)$. (b) $V = \dfrac{q}{4\pi\epsilon_0}\left(\dfrac{1}{R}\right)$.

Multi-Version Exercises

23.85 $6.571 \cdot 10^{-7}$ C. **23.88** $5.182 \cdot 10^5$ V.

Chapter 24: Capacitors

Multiple Choice

24.1 b. **24.3** c. **24.5** c. **24.7** d. **24.9.** a. **24.11** b. **24.13** b.

Exercises

24.29 $1.13 \cdot 10^2$ km^2. **24.31** $8.99 \cdot 10^9$ m. **24.33** $7.089 \cdot 10^{-4}$ F.
24.35 (a) $\sigma = 1.59 \cdot 10^{-8}$ C/m^2, $C' = 64.0$ pF, $\sigma' = 3.19 \cdot 10^{-8}$ C/m^2.
(b) 4.50 V. **24.37** 1.33 pF. **24.39** 4.34 nF. **24.41** (a) 2.700 nF. (b) 3.101 nF.
(c) 1.533 nF. (d) 17.94 nC. (e) 2.330 V. **24.43** $C_1/1275$. **24.45** 0.0360 J.
24.47 $9.96 \cdot 10^{-8}$ J/m^3. **24.49** 1.11 F; $7.10 \cdot 10^{13}$ J. **24.51** (a) 72.0 nJ.
(b) 36.0 nJ. (c) 9.00 nJ. (d) 18.0 nJ is lost; The energy is lost to heat.
24.53 $3.89 \cdot 10^4$. **24.55** $2.2 \cdot 10^{-5}$ C/m^2. **24.57** 70.8 pF.

24.59 $q = \dfrac{\pi\epsilon_0 L(\kappa + 1)V}{2\ln(r_2/r_1)}$; $\dfrac{\kappa + 1}{2}$. **24.61** (a) 12.6 kV/m. (b) 6.00 nC.
(c) 4.89 nC. **24.65** 55 nC. **24.67** 210%. **24.69** 1.50 km. **24.71** $V_A = 0.300$ V.
24.73 $4.03 \cdot 10^{-13}$ F. **24.75** 0.792 nF. **24.77** (a) 5.40. (b) 0.130 μC.
(c) 542. kV/m. (d) 542. kV/m. **24.79** $1.42 \cdot 10^{-8}$ N in the direction
of the motion of the dielectric material. **24.81** (a) $C_0 = 35.4$ pF,
$U_0 = 9.96 \cdot 10^{-8}$ J. (b) $C' = 92.1$ pF, $V' = 28.9$ V, $U' = 3.83 \cdot 10^{-8}$ J. (c) No.
24.83 (a) 0.744 nC. (b) 35.7 nJ. (c) $1.20 \cdot 10^5$ V/m. **24.85** 45.0 μC.
24.87 (b)

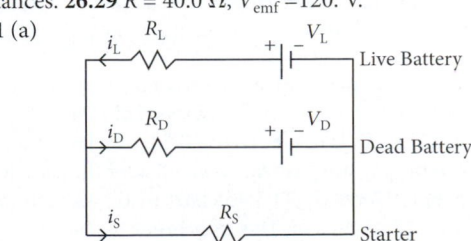

(c) $V = mv_i^2 \sin^2\theta/(2q)$. (d) $v_i = 2.00 \cdot 10^5$ m/s.

Multi-Version Exercises

24.89 7094. **24.92** $6.453 \cdot 10^{-4}$ J.

Chapter 25: Current and Resistance

Multiple Choice

25.1 d. **25.3** c. **25.5** c. **25.7** c. **25.9** c. **25.11** c. **25.13** d.

Exercises

25.29 Current density: 318 A/m^2; Drift speed: $3.30 \cdot 10^{-8}$ m/s.
25.31 (a) $5.85 \cdot 10^{28}$ m^{-3}. (b) 133 A/m^2. (c) $1.42 \cdot 10^{-8}$ m/s.
25.33 $R_B = 1.12$ mm. **25.35** between 9- and 10-gauge wire. **25.37** 3,
independent of material. **25.41** 0.0200 Ω. **25.43** (a) –84%. (b) 530%.
(c) At room temperature, $V_d = 0.43$ mm/s. At 77 K the speed is
$V_d = 2.7$ mm/s. **25.45** 2.570 Ω. **25.47** (a) 1.51 Ω. (b) 1.08 Ω. (c) When
two bulbs are put in series, it is expected that they glow dimmer
than only one bulb. This would mean the one bulb would be hotter and
thus have a larger resistance. **25.49** $R_{eq} = 60.9$ Ω. **25.51** (a) $R_{eq} = 12.00$ Ω.
(b) $i = 1.00$ A. (c) $\Delta V_3 = 1.00$ V. **25.53** (a) $\Delta V_1 = 5.41$ V, $\Delta V_2 = 10.8$ V,
$\Delta V_3 = \Delta V_4 = 2.70$ V, $\Delta V_5 = \Delta V_6 = 1.08$ V. (b) $i_1 = i_2 = 1.08$ A, $i_3 = i_4 =$
$i_5 = i_6 = 0.541$ A. **25.55** 86.0% brighter. **25.57** (a) no. (b) 7.56 Ω.
25.59 (a) 50.0 W. (b) $\Delta V_1 = 80.0$ V and $\Delta V_2 = \Delta V_3 = 40.0$ V.
25.61 $P_1 = 16.0$ W; $P_2 = 18.0$ W; $P_3 = 17.3$ W. **25.63** (a) $R_C = 64.6$ mΩ.
(b) $i = 9.99$ mA. (c) $b = 1.31$ mm. (d) 0.772 s. (e) 4.99 mW. (f) Electric
power is lost via heat. **25.65** 10.0 A. **25.67** $L = 1.00$ m. **25.69** (a) 0.0450 W.
(b) 0.405 W. When the resistors are connected in parallel, the power
delivered to the 200. Ω resistor by the 9.00 V battery is 9.00 times
greater than in the series configuration. **25.71** 25.0 W. **25.73** 9.00 W.
25.75 (a) $3.2 \cdot 10^{-5}$ A. (b) 640 kW. (c) $6.2 \cdot 10^{14}$ Ω. **25.77** (a) $V_{bc} = 55.0$ V.
(b) $i = 27.5$ A. (c) $P = 1.01$ kW. **25.79** 2.73 s. **25.81** (a) 23.8:1.
(b) Copper has a lower resistivity than steel and thus dissipates much
less power in a transmission line. **25.83** (a) EJ. (b) $\rho J^2 = \sigma E^2$.

Multi-Version Exercises

25.84 14.58%. **25.87** 1.73 MJ. **25.89** 1520 °C.

Chapter 26: Direct Current Circuits

Multiple Choice

26.1 a. **26.3** b. **26.5** a. **26.7** c. **26.9** d, e & f. **26.11** e. **26.13** c.

Exercises

26.27 $V_1 = \left(\dfrac{R_1}{R_1 + R_2}\right)\Delta V$, $V_2 = \left(\dfrac{R_2}{R_1 + R_2}\right)\Delta V$; The resistors in series
construct a voltage divider. The voltage ΔV is divided between the
two resistors with potential drop proportional to their respective
resistances. **26.29** $R = 40.0$ Ω, $V_{emf} = 120.$ V.
26.31 (a)

(b) starter: 149.9 A; live battery: 150.4 A; dead battery: 0.4960 A.

26.33 $i_1 = 0.200$ A, $i_2 = 0.200$ A, $i_3 = 0.400$ A, $P_A = 1.20$ W, $P_B = 2.40$ W.

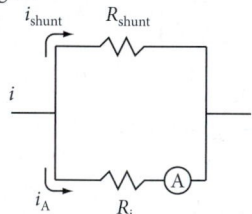

26.35 $i_1 = 0.251$ A to the right, $i_2 = 0.625$ A to the right, $i_3 = 0.251$ A upward, $i_4 = 0.375$ A upward, $i_5 = 0.152$ A to the right, $i_6 = 0.403$ A downward, $i_7 = 0.403$ A to the left; $P(V_{emf,1}) = 1.50$ W, $P(V_{emf,2}) = 23.2$ W.

26.37 $R_L = (1 + \sqrt{3})R$. **26.39** $R_{Shunt} = \dfrac{R_{Ammeter}}{N-1}$; $R_{Shunt} = 10.1$ mΩ; 1/100 through the ammeter and 99/100 through the shunt. **26.41** $V_{ab} = 2.985$ V, $V_{bc} = 3.015$ V. Increasing R_i will reduce the error since the voltmeter will draw less current. **26.43** (a) 6.00 mA each. (b) 12.0 mA. **26.45** 8.99 s. **26.47** $2.40 \cdot 10^{-4}$ C; 41.6 μs. **26.49** 4.35 ms. **26.51** 88.5 s. **26.53** 22.4 kV, 0.0200 μF. **26.55** (a) 2.22 V. (b) 4.94 μJ. (c) 1.41 μJ. **26.57** $(\sqrt{3}-1)C/2$.

26.59 (a) The shunt resistor carries most of the load so that the ammeter is not damaged.

(b) 7.50 mΩ. **26.61** $C = 80.0$ nF, $R = 10.0$ kΩ. **26.63** (a) 225 mA. (b) 18.5 mA. **26.65** 3.81 s. **26.67** $P_1 = 7.44$ mW, $P_2 = 4.30$ W, $P_3 = 7.23$ W. **26.69** (a) $1.22 \cdot 10^{-4}$ s. (b) 68.5 pC. **26.71** (a) 0.693 τ. (b) 1:4. (c) 10.0 μs. (d) 6.93 μs. **26.73** $i_2 = 469$ mA. **26.75** 2C.

Multi-Version Exercises

26.76 12.79 V. **26.79** 0.2933 A. **26.82** 0.1407 A.

Chapter 27: Magnetism

Multiple Choice

27.1 a. **27.3** e. **27.5** a. **27.7** a,c,d,e. **27.9** e. **27.11** d.

Exercises

27.25 $9.36 \cdot 10^{-5}$ T. **27.27** $|\vec{F}| = 7.62 \cdot 10^{-4}$ N, Direction of the force is in the yz-plane, $\theta = 23.2°$ above the negative y-axis. **27.29** $B = 1.44 \cdot 10^{-2}$ T. **27.31** $\dfrac{d\vec{p}}{dt} = -q(\vec{E} + \vec{v} \times \vec{B})$, $\dfrac{dK}{dt} = -q\vec{E} \cdot \vec{v}$. **27.33** $6.82\hat{z}$ T. **27.35** The proton will create a helical path with a velocity of $3.00 \cdot 10^5$ m/s along the z-axis, with the circular motion having a speed of $2.24 \cdot 10^5$ m/s and a radius of 4.67 mm. **27.37** 1:4; The electric field must have magnitude $E = vB$ and point in the positive y-direction in order for the particles to move in a straight line. **27.39** (a) 1.71 T. (b) 4.62 cm. (c) $2.02 \cdot 10^6$ m/s. **27.41** $m = 0.408$ kg. **27.43** 15.0 N; This is the same as the force on a wire of the same length with the same current and magnetic field. **27.45** 917 rad/s^2. **27.47** $\dfrac{iB_0 l^2}{a}\hat{x}$. **27.49** 0.135 T. **27.51** 12.7 rad/s^2.

27.55 (a) $\vec{F}_{ab} = -0.320\hat{y}$ N. (b) 0.619 N directed 14.6° from the x-axis toward the negative z-axis. (c) $F_{net} = 0$. (d) $\tau = 0.0250$ N m and rotates along the y-axis in counterclockwise fashion. (e) The coil rotates in a counterclockwise fashion as seen from above. **27.57** (a) holes.

(b) $2.29 \cdot 10^{24}$ holes/m^3. **27.59** $7.39 \cdot 10^{-2}$ N. **27.61** $2.33 \cdot 10^5$ A. **27.63** (a) $1.89 \cdot 10^{-6}$ T. (b) none. (c) $4.71 \cdot 10^{-6}$ s. **27.65** 0.0311 m, $3.83 \cdot 10^{-7}$s. **27.67** $1.81 \cdot 10^5$ m/s. The velocity increases by a factor of 4/3. **27.69** $B = -4.23 \cdot 10^{-3}\hat{z}$ T. **27.71** 6.97°. **27.73** (a) The trajectory of the proton will be a helix. (b) 10.4 mm. (c) $T = 1.31 \cdot 10^{-7}$ s, $f = 7.62 \cdot 10^6$ Hz. (d) 114 mm.

Multi-Version Exercises

27.74 0.6614 T. **27.77** $3.783 \cdot 10^5$ m/s.

Chapter 28: Magnetic Fields of Moving Charges

Multiple Choice

28.1 b. **28.3** c. **28.5** d. **28.7** c. **28.9** a. **28.11** a. **28.13** d.

Exercises

28.31 $a = 4.2 \cdot 10^{12}$ m/s^2. **28.33** $i = 7.14 \cdot 10^9$ m **28.35** parallel to the x-axis through $(0,4,0)$, carrying a current in the opposite direction of the first wire. **28.37** 7.5° below the x-axis. **28.39** $B = \mu_0 i/(\sqrt{2}\pi d)$, at $-45°$ in the xy plane at a height of b. **28.41** 204 A. **28.43** (a) $-1.89 \cdot 10^{-2}$ N$\hat{y}$. (b) $\vec{F} = 1.89 \cdot 10^{-2}$ N $\hat{x}$. (c) $\vec{F} = 0$ N. **28.45** $9.42 \cdot 10^{-5}$ T at 45.0° from the negative x-direction toward the positive y-axis. **28.47** $B_a = 0$; $B_b = 1.08 \cdot 10^{-6}$ T; $B_c = 2.70 \cdot 10^{-6}$ T; $B_d = 1.69 \cdot 10^{-6}$ T. **28.49** $9.4 \cdot 10^{-5}$ T. **28.51** 4:3. **28.53** (a) $1.3 \cdot 10^{-5}$ T. (b) $1.0 \cdot 10^{-2}$ T. **28.55** $1.9 \cdot 10^{-19}$ kg m/s. **28.57** 0.20 K. **28.59** 55.0. **28.61** $1.12 \cdot 10^{-5}$ Am^2. **28.63** (a) $L = I\omega = (2/5)mR^2\omega$. (b) $\mu = q\omega R^2/5$. (c) $\gamma_e = -e/(2m)$. **28.65** 4.00 mT into the page. **28.67** $4.1 \cdot 10^9$ A. **28.69** 18,000 turns. **28.71** 24.5 mT. **28.73** $q = 4.91 \cdot 10^{-4}$ C. **28.75** $1.25 \cdot 10^{-7}$ N m. **28.77** (a) $-7.28 \cdot 10^{10}$ m/s^2. (b) $-1.19 \cdot 10^{11}$ m/s^2. **28.79** (a) $1.06 \cdot 10^{-4}$ N m. (b) 1.72 rad/s. **28.81** 4.42 A counterclockwise (as viewed from the bar magnet).

28.83 $B = \mu_0 J_0 \left| \dfrac{r}{2} - \dfrac{r^2}{3R} \right|$.

Multi-Version Exercises

28.85 $B = 9.303 \cdot 10^{-7}$ T. **28.88** 18,916.

Chapter 29: Electromagnetic Induction

Multiple Choice

29.1 d. **29.3** a. **29.5** a. **29.7** d. **29.9** a. **29.11** c. **29.13** e.

Exercises

29.29 $1.89 \cdot 10^{-5}$ V. **29.31** 0. **29.33** $\Delta V_{ind} = -\dfrac{\mu_0 \pi a^2 V_0 \omega}{2bR_1}\cos \omega t$; $i = -\dfrac{\mu_0 \pi a^2 V_0 \omega}{2bR_1 R_2}\cos \omega t$. **29.35** (a) 7.07 mA. (b) clockwise. **29.37** 0.558 V. **29.39** $v_{term} = \dfrac{mgR}{w^2 B^2}$. **29.41** $6.24 \cdot 10^{-7}$ V. **29.43** 17.5 Hz. **29.45** (a) 0.370 A. (b) 0.262 A, 103 W. **29.47**

29.49 (a) 1.00 μs. (b) 0; 8.65 μA; 10.0 μA. **29.51** 11.3 V. **29.53** (a) 6.00 A.
(b) 3.00 A. (c) 3.00 A. (d) –18.0 V. (e) –18.0 V. (f) 0. (g) 0.
29.55 $1.01 \cdot 10^3$ m³. **29.57** (a) $6.37 \cdot 10^{26}$ J/m³. (b) $7.07 \cdot 10^9$ kg/m³.
29.59 $4.17 \cdot 10^{-5}$ °C. **29.61** 1:1. **29.63** $7.54 \cdot 10^{-3}$ V. **29.65** It does not
change. **29.67** 0.0866 H. **29.69** 10.0 V; The induced current is
counterclockwise. **29.71** 1.96 m². **29.73** 0.275 H. **29.75** (a) $i_1 = 4.00$ mA;
$i_2 = 2.00$ mA. (b) 2.40 mW. (c) 0.300 mN. **29.77** 3.51 μV.
29.79 (a) 0.257 mN. (b) 25.8 μW. (c) 25.7 μW.

Multi-Version Exercises

29.80 17.3 Ω. **29.83** $V_{ind,max} = 1.21 \cdot 10^{-7}$ V.

Chapter 30: Alternating Current Circuits

Multiple Choice

30.1 d. **30.3** b. **30.5** d. **30.7** d. **30.9** a.

Exercises

30.27 $9.42 \cdot 10^{-4}$ s. **30.29** 7.15 mF. **30.31** 1.19 ms, 3.25 ms, 5.63 ms.
30.33 Drops by 15.2%. **30.35** 251 Hz. **30.37** 500. rad/s.
30.39 (a) 2240 rad/s. (b) 0.400 A. **30.41** 0.549 rad, 117 Ω. **30.43** 537 V.
30.45 (a) $I_{max} = 34.3$ A. (b) $\theta = 0.816$ rad. (c) $C' = 757$ μF, $I'_m = 50.0$ A,
$\phi' = 0$ rad. **30.47** (a) $C = 1.33$ nF; Use a capacitor of capacitance 1.00 nF.
(b) 124 kHz. **30.49** $Q = (1/R)\sqrt{L/C}$. **30.51** (a) 18.4 kHz. (b) 2.25 W.
30.53 (a) 0.392 pF. (b) 11.9 Ω. **30.55** (a) 1.1 V. (b) 0 V. **30.57** 2.03 W.
30.59 (a) 14 V. (b) 9.0 V. **30.61** 2.53 pF. **30.63** (a) 0.77 A. (b) 160 V.
30.65 1.45 Ω. **30.67** (a) 0.101 A. (b) $7.02 \cdot 10^{-4}$ s. **30.69** (a) 10.0 Ω.
(b) 7.50 Ω. (c) $5.97 \cdot 10^{-5}$ H. (d) 26.7 kHz. **30.71** (a) $1.01 \cdot 10^{-3}$ J/m³.
(b) 10.1 A. **30.73** $P = \frac{1}{2}I_R^2 R(1 - \cos(2\omega t))$. **30.75** 377 Hz. **30.77** 1990 Hz.

Multi-Version Exercises

30.78 160. Ω.

Chapter 31: Electromagnetic Waves

Multiple Choice

31.1 c. **31.3** b. **31.5** a. **31.7** c. **31.9** c.

Exercises

31.23 $2.50 \cdot 10^{-5}$ T. **31.25** 10.0 μA.
31.27 $i_d = \epsilon_0 R\left(\dfrac{A}{L}\right)\dfrac{di}{dt} = \epsilon_0 \rho \dfrac{di}{dt}$.
31.29 0.984 ft. **31.31** (a) 0.03 s. (b) 0.24 s; the delay via the cable is
not noticeable. However, via satellite, Alice will receive a response
from her fiancé after 0.5 s, which is quite noticeable. **31.33** $4 \cdot 10^{14}$ Hz
to $8 \cdot 10^{14}$ Hz. **31.35** 3.2 mH. **31.37** (a) 1.3 W/m². (b) 1.2 W/m².
(c) 0.12 W/m². **31.39** $1.70 \cdot 10^6$ V/m. **31.41** $S_{ave} = 13.3$ W/m².
(a) $U = 4.43 \cdot 10^{-8}$ J/m³. (b) $B = 3.33 \cdot 10^{-7}$ T. **31.43** (a) $I = 1.27 \cdot 10^7$ W/m².
The intensity is much larger than the intensity of sunlight.
(b) $E_{rms} = 6.93 \cdot 10^4$ V/m. (c) $S_{ave} = 1.27 \cdot 10^7$ W/m².
(d) $S(x,t) = 2.55 \cdot 10^7$ W/m² $\sin^2(1.22 \cdot 10^7 x - 3.66 \cdot 10^{15} t)$.
(e) $B_{rms} = 2.31 \cdot 10^{-4}$ T. **31.45** (a) $E = 1.03$ kV/m, $B = 3.42$ μT.
(b) $P_r = 3.33$ μPa, $F = 2.50$ μN. **31.47** $3.27 \cdot 10^{-6}$ N. **31.49** 15.1 mW.
31.51 (a) $w = 3.08 \cdot 10^{-9}$ N = 3.08 nN. (b) $I = 1.59$ kW/m²,
$P_r = 5.31$ μN/m². (c) $N = 185$ lasers. **31.53** 2.50 mW.

31.55 $I_2 = 1.13 \cdot 10^4$ W/m², $E = 2.92 \cdot 10^3$ V/m, $B = 9.74 \cdot 10^{-6}$ T.
31.57 $E_{total} = 518$ kWh. **31.59** $3.88 \cdot 10^5$ V/m. **31.61** 136 W.
31.63 100 V/m. **31.65** 6.3 cm. **31.67** (a) $7.07 \cdot 10^{-8}$ T. (b) 37.9 W.
31.69 (a) $E_{rms} = 775$ V/m. (b) $E_{tot} = 1.67 \cdot 10^{-11}$ J. **31.71** $2.07 \cdot 10^7$ m/s².
31.73 (a) $S = 3.32 \cdot 10^{-5}$ W/m². (b) $F_{rms} = 1.27 \cdot 10^{-20}$ N. **31.75** $t = 8.95$ h
(or 8 hours 57 minutes); $N = 4.00 \cdot 10^{23}$.

Multi-Version Exercises

31.77 262.5 V/m. **31.81** 87.9%.

Chapter 32: Geometric Optics

Multiple Choice

32.1 c. **32.3** a. **32.5** b. **32.7** e. **32.9** b. **32.11** a.

Exercises

32.27 –1.00 m. **32.29** 5.66 m. **32.31** –12.5 cm. **32.33** –4.28 m, 0.389.
32.35 –3.0 m. **32.39** $\theta_{c,air} = 41.8°$, $\theta_{c,water} = 62.7°$, $\theta_{c,oil} = 90.0°$. **32.41** 52.4°.
32.43 16.3°. **32.45** 1.35%. **32.49** (a) The image distance is 50.0 cm
behind the mirror. (b) The image has the same height, $h = 2.00$ m.
(c) The image is upright. (d) The image is virtual. **32.51** If the object
is placed at 15.0 cm, the image will be real and at a distance of 30.0 cm
from the mirror. If the object is placed at 5.00 cm, the image will be
virtual and –10.0 cm from the mirror. **32.53** 128.0°. **32.55** The critical
angle in water is 9.03° greater than the critical angle in air. **32.57** 28.8°
with respect to the normal. **32.59** 1.92 m. **32.61** (a) $\rho(t) \rightarrow \rho(-t)$,
$\vec{j}(t) \rightarrow \vec{j}(-t)$, $\vec{E}(\vec{x},t) \rightarrow \vec{E}(\vec{x},-t)$ and $\vec{B}(\vec{x},t) \rightarrow -\vec{B}(\vec{x},-t)$. (b) They do
not transmit light unidirectionally. **32.63** 1.40 rad/s.

Multi-Version Exercises

32.65 11.7 cm. **32.69** 60.61°.

Chapter 33: Lenses and Optical Instruments

Multiple Choice

33.1 b. **33.3** c. **33.5** a. **33.7** b. **33.9** b. **33.11** b. **33.13** d. **33.15** e. **33.17** d

Exercises

33.35 3.0. **33.37** 0.198 m. **33.39** 1.5; inverted. **33.41** (a) 48.0 cm to
the left of the second surface.
(b)

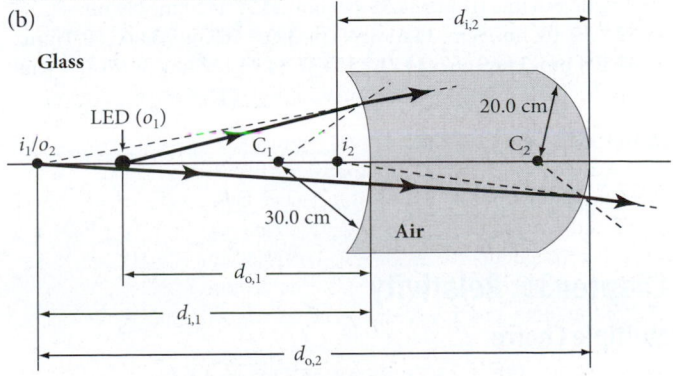

33.43 5.0. **33.45** 6.00 cm. **33.47** 8.3 mm, inverted, virtual.

33.49 (a) at the left focal point of the second lens. (b) –2.
(c)

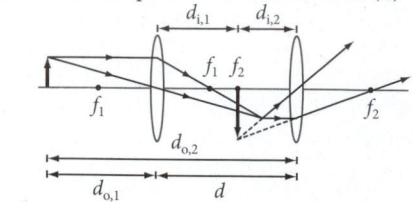

(d) virtual, inverted, larger.
33.51 142 cm to the right of the object. **33.53** (a) 2.50 cm. (b) 2.27 cm.
33.55 (a) 40. cm. (b) –125 cm; The negative indicates the image is on
the same side as the object. (c) 59 cm. (d) 1.7 diopter. (e) converging.
33.57 39 cm; 50. cm. **33.59** 29 cm; 9.9 cm; $m_{norm} = 2.9 \, m_{near}$.
33.61 (a) $3.5 \cdot 10^2$ mm. (b) $2.5 \cdot 10^2$ mm. **33.63** 4.0 times that of the original
lens. **33.65** 2.4 cm. **33.67** 4.00 cm. **33.69** (a) 20. mm. (b) 9.7 mm.
33.71 20.0, inverted. **33.73** –20.0. **33.75** $f_e = 7.2$ mm, $f_o = 1.3$ m.
33.79 –4 diopter. **33.81** 13.3 cm. **33.83** 18 cm from the lens, on the
same side of the lens as the object. **33.85** –1.4 diopter. **33.87** 8.84 cm.
33.89 0.243 mm. **33.91** 185 cm to the right of the object; 5.0 cm.
33.93 nearsighted; 0.12 m. **33.95** 1.3 in; magnification –0.070.
33.97 38.0 cm. **33.99** 8.33 cm. **33.101** 1.0 diopter.

Multi-Version Exercises

33.103 1.736 D. **33.106** 19.05 D.

Chapter 34: Wave Optics

Multiple Choice

34.1 c. **34.3** d. **34.5** a. **34.7** b. **34.9** d. **34.11** b.

Exercises

34.23 (a) 421.9 nm. (b) $1.999 \cdot 10^8$ m/s. **34.25** 87.5 nm. **34.27** 1.0 m.
34.29 1.05 mm. **34.31** 210 nm, 420 nm. **34.33** 139 nm. **34.35** 1.50.
34.37 720.3 nm. **34.39** $17.0 \cdot 10^2$. **34.41** 2370 nm. **34.43** 600. nm.
34.45 $2.7 \cdot 10^{-6}$ degrees. **34.47** Hubble Space Telescope: $1.31 \cdot 10^{-5}$ degrees;
Keck Telescope: $3.15 \cdot 10^{-6}$ degrees; Arecibo radio telescope:
0.0481 degrees. The Arecibo radio telescope is worse than the other
telescopes in terms of angular resolution. The Keck Telescope is better
than the Hubble Space Telescope because of its larger diameter.
34.49 (a) $7.69 \cdot 10^{-3}$ degrees. (b) 11.2 km. **34.51** 31. **34.53** (a) $a = 10\lambda$.
(b) $d = 100\lambda$. (c) $a/d = 1:10$. (d) Without λ, there is insufficient
information to find a or d. **34.55** 449 nm. **34.57** 442 nm, 589 nm.
34.59 $1.25 \cdot 10^4$ lines/cm. **34.61** 0.593 m. **34.63** 665 nm. **34.65** 30.0 mm.
34.67 104 nm. **34.69** 667. **34.71** 2.15 km. **34.73** 122 nm. **34.75** 1.07 μm.

Multi-Version Exercises

34.77 2.36 cm. **34.81** 5.63 mm.

Chapter 35: Relativity

Multiple Choice

35.1 a. **35.3** c. **35.5** a. **35.7** c. **35.9** a. **35.11** c. **35.13** d.

Exercises

35.27 0.984 ft/ns. **35.29** $(1/v)\sqrt{v^2-u^2}$. **35.31** (a) one meter. (b) 0.87 m.
35.33 0.94c. **35.35** 0.600c. (a) $t_s = 4.44 \cdot 10^{-8}$ s. (b) The truck will never
fit from the truck's point of view. **35.37** (a) 0.81c. (b) 20. light-years.
35.39 77.7°.

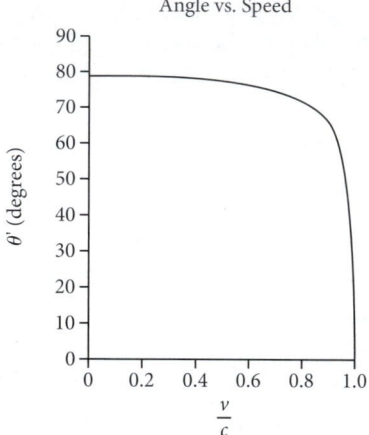

Angle vs. Speed

35.41 0.22c. **35.43** 314 Hz. **35.45** (a) B. (b) $3.73 \cdot 10^{-5}$s. **35.47** –100. m/s.
35.49 (a) 4.94 light-years. (b) 7.60 yr. **35.51** (a) 66.14 m. (b) 0.8514c.
(c) 2.06 μs. (d) 529 m. **35.53** 0.707c. **35.55** $4.02 \cdot 10^6$ GeV. **35.57** 1200 MeV.
35.59 (a) 5.00 keV. (b) 0.139c. (c) The relativistic and nonrelativistic
energies are $E_R = 516$ keV and 5.00 keV, respectively. The relativistic
and nonrelativistic momenta are 71.7 keV/c and 71.5 keV/c, respectively.
35.61 (a) $c/\sqrt{2}$. (b) $\sqrt{17}m$. (c) $c/\sqrt{18}$. **35.65** (a) 2.954 km. (b) $2.485 \cdot 10^{-54}$ m;
if a proton had a nonrelativistic space-time description, it would be a
bizarre, unphysical object, known as *naked singularity*. **35.67** 0.0735 AU.
35.69 First electron: $p_R = 2.44 \cdot 10^{-22}$ kg m/s, 34% different from its
nonrelativistic value. Second electron: $p_R = 1.82 \cdot 10^{-27}$ kg m/s, pretty
much exactly its nonrelativistic value. **35.71** Newtonian: 5.03 keV.
Relativistic: 32.5 MeV. **35.73** 0.697 g. **35.75** 0.882c. **35.77** 0.484 ns, too
small to detect. **35.79** As measured by someone on the spaceship,
$t_1 = 32.9$ yr. As seen by someone on the space station, $t_2 = 105$ yr.
35.81 0.99995c. **35.83** (a) $0.736c = 2.21 \cdot 10^8$ m/s. (b) The photons would
appear to be traveling at the speed of light. **35.85** 18.0 h.

35.87 (a) $dv = g\left(1 - \dfrac{v^2}{c^2}\right)dt$. (b) $v(t) = \dfrac{gt}{\left[1 + \left(gt/c\right)^2\right]^{1/2}}$.

(c) In the time $gt \ll c$, i.e., the Newtonian limit, $v(t) \cong gt$. In the

Einsteinian (relativistic) limit, $\lim\limits_{t \to +\infty} v(t) = c$. (d) $x(t) = \dfrac{c^2}{g}\sqrt{1 + \left(\dfrac{gt}{c}\right)^2}$.

(e) The trajectory is part of a branch of a hyperbola on a Minkowski
diagram. (f) 354 days, 0.401 light-years.

Multi-Version Exercises

35.88 145.89 GeV. **35.91** 12.072 TeV.

Chapter 36: Quantum Physics

Multiple Choice

36.1 b. **36.3** a, c. **36.5** c. **36.7** c.

Exercises

36.19 (a) $5.00 \cdot 10^{-7}$ m. (b) $9.67 \cdot 10^{-6}$ m. **36.21** $3.5 \cdot 10^{-19}$ m; the energy
of the gamma ray is 3700 times greater than the rest energy of a

proton. **36.23** (a) 9.41 μm. (b) 1.02 kW. (c) Your wavelength is not in the visible spectrum. **36.25** (a) $E = 0.243$ J. (b) $n = 1.21 \cdot 10^{19}$ photons per second. (c) $V_T = 1.49 \cdot 10^{-6}$ m³. **36.27** 0.34 eV. **36.29** (a) 2.30 eV. (b) potassium or sodium. **36.31** (a) 0.622 eV. (b) no. (c) 500. nm. **36.33** $2.42 \cdot 10^{-12}$ m. **36.35** 64 eV. **36.37** (a) $1.49 \cdot 10^{-4}$. (b) 9.48 MeV. **36.39** (a) 620. nm. (b) 0.867 nm. **36.41** (a) 47.52 pm. (b) 11 nm. **36.43** (a) 1300 m/s. (b) This speed is much less than the speed of light, so the nonrelativistic approximation is sufficient. (c) 5.0 μeV.

36.45 (a) $\omega(k) = \dfrac{hk^2}{4\pi m}$. (b) $v_P = \dfrac{p}{2m}, v_g = \dfrac{p}{m}$. **36.47** (a) $6.63 \cdot 10^{-35}$ m/s. (b) $5.27 \cdot 10^{-36}$ m/s. **36.49** $6.64 \cdot 10^{-55}$ kg. **36.51** 2.89 m/s.

36.53 $\langle n \rangle = \dfrac{e^{-E/k_B T}}{1 - e^{-E/k_B T}}$.

36.55 $C = N k_B \left(\dfrac{E}{k_B T}\right)^2 \dfrac{e^{E/k_B T}}{\left(e^{E/k_B T} + 1\right)^2}$. For $k_B T \gg 1$, $C \approx \dfrac{N k_B}{4}\left(\dfrac{E}{k_B T}\right)^2$.

For $0 < k_B T \ll 1$, $C \approx N k_B \left(\dfrac{E}{k_B T}\right)^2 e^{-E/k_B T}$.

36.57 0.0233. **36.59** (a) $1.48 \cdot 10^{-34}$ m. (b) $7.63 \cdot 10^{-41}$ m. **36.61** $1.06 \cdot 10^{19}$ s⁻¹. **36.63** 130 nm. **36.65** 0.579 m. **36.67** 61. **36.69** 388 nm. **36.71** $7.62 \cdot 10^{42}$.

Multi-Version Exercises

36.73 110.2 km/s. **36.78** 998.5 MeV.

Chapter 37: Quantum Mechanics

Multiple Choice

37.1 c. **37.3** d. **37.5** b. **37.7** d. **37.9** a.

Exercises

37.23 9.05 fm; Since protons and neutrons have a diameter of about 1.00 fm, they would be useful targets for demonstrating the wave nature of 10.0-MeV neutrons. **37.25** 0.094 eV, 0.38 eV. **37.27** 4.

37.29

$$\psi(x) = \begin{cases} 0 & \text{for } x < -a/2 \text{ and } x > a/2 \\ \sqrt{\dfrac{2}{a}} \sin\left(\dfrac{n\pi x}{a}\right) & \text{for } -a/2 \le x \le a/2 \text{ with even } n \\ \sqrt{\dfrac{2}{a}} \cos\left(\dfrac{n\pi x}{a}\right) & \text{for } -a/2 \le x \le a/2 \text{ with odd } n \end{cases}$$

37.31 No. **37.33** 5.9. **37.35** 6.99 eV. **37.37** 45.5 eV.

37.39 (a) For even parity: $\psi(x) = \begin{cases} Ae^{\gamma x} & \text{for } x \le -a/2 \\ C\cos(\kappa x) & \text{for } -a/2 \le x \le a/2 \\ Ae^{-\gamma x} & \text{for } x \ge a/2 \end{cases}$

For odd parity: $\psi(x) = \begin{cases} Ae^{\gamma x} & \text{for } x \le -a/2 \\ D\sin(\kappa x) & \text{for } -a/2 \le x \le a/2 \\ -Ae^{-\gamma x} & \text{for } x \ge a/2 \end{cases}$

where $E = \dfrac{\hbar^2 \kappa^2}{2m}$. (b) $\eta = \dfrac{\hbar}{\sqrt{2m(U_0 - E)}}$. (c) 467 pm.

37.41 720 nm. **37.43** 101 eV; 85.5 kN/m.

37.45 (a) $A_2 = \dfrac{1}{\sqrt[4]{\pi}\sqrt{b}}$. (b) 0.157. **37.47** (a) $A = \sqrt{\dfrac{2}{L}}$. (b) 0.402.

37.49 0.250. **37.51** $\Psi(\vec{r},t) = Ae^{i(\vec{p}\cdot\vec{r})/\hbar} e^{-ip^2 t/2m\hbar}$; A^2.

37.53 $|\Psi(x,t)|^2 = \dfrac{1}{a}\sin^2\left(\dfrac{\pi x}{a}\right)\left[1 + 4\cos^2\left(\dfrac{\pi x}{a}\right) + 4\cos\left(\dfrac{\pi x}{a}\right)\cos\left(\dfrac{3\hbar\pi^2 t}{2ma^2}\right)\right]$.

37.55 (a) $\Delta t \ge 3.22 \cdot 10^{-22}$ s. (b) $\Delta t \ge 1.75 \cdot 10^{-25}$ s. **37.57** $8.21 \cdot 10^{-4}$ eV.

37.59 (a) $\psi = \left(\sqrt{\dfrac{2}{L}}\right)^3 \sin\dfrac{\pi x}{L}\sin\dfrac{\pi y}{L}\sin\dfrac{\pi z}{L}$; $0 < x, y, z < L$. (b) 14.

37.61 0.54 μm. **37.63** (a) 1.2 eV. (b) 0.19 eV. **37.65** 2.9 MeV. **37.67** $1.10 \cdot 10^3$ nm. **37.69** 38 eV. **37.71** (2,2) and (1,4).

Multi-Version Exercises

37.72 $4.95 \cdot 10^{-15}$. **37.76** $7.93 \cdot 10^{-21}$ J.

Chapter 38: Atomic Physics

Multiple Choice

38.1 d. **38.3** d. **38.5** b. **38.7** b.

Exercises

38.19 91.2 nm. **38.21** 2279 nm, 7458 nm; No. **38.23** −0.378 eV. **38.25** (a) $1.0968 \cdot 10^7$ m⁻¹. (b) $5.4869 \cdot 10^6$ m⁻¹. **38.27** initial: $n = 6$; final: $n = 3$. **38.29** $r = \dfrac{\hbar^2}{GMm^2} n^2$; $2.523 \cdot 10^{74}$. **38.31** $4.716 \cdot 10^{-34}$ J s and 0 J s. **38.33** 65.9°. **38.35** (a) $\dfrac{e^{-1}}{4\pi a_0}$. (b) The probability density includes an r^2 factor. **38.37** 0.03432. **38.39** $a_0/2 = 0.0265$ nm.

38.41 (a) $\dfrac{0.200 E_0}{\kappa^2}$. (b) 27.2 meV.

38.43 (a) $r_n = \dfrac{a_0 n^2}{Z}$. (b) $v_n = \dfrac{Z}{n}\sqrt{\dfrac{ke^2}{\mu a_0}}$. (c) $E' = \dfrac{-Z^2}{n^2} E_0$.

38.45 656 nm; a laser with a wavelength 5.4 times bigger is needed. **38.47** (a) $2.86 \cdot 10^{-19}$ J. (b) $1.05 \cdot 10^{14}$ atoms. **38.49** 91.16 nm. **38.51** 0.05442%, $4.555 \cdot 10^{-31}$ kg. **38.53** $Z = 68.5$, so either erbium (68) or thulium (69). **38.55** −2530 eV, −632 eV, −281 eV. **38.57** −54.4 eV, −13.6 eV, −6.05 eV. **38.59** −6.72 meV. **38.61** Yes. $v = 4.07$ m/s. **38.63** 0.3233.

Multi-Version Exercises

38.64 $1.078 \cdot 10^6$ m/s. **38.67** $5.917 \cdot 10^{11}$.

Chapter 39: Elementary Particle Physics

Multiple Choice

39.1 a. **39.3** c. **39.5** d. **39.7** c.

Exercises

39.23 49.9 fm. **39.25** 13.2 b/sr. **39.27** r_{min}/λ is proportional to $1/\sqrt{K}$. **39.29** 120. particles per second.

39.31 $F^2(\Delta p) = \left| \dfrac{4\pi\rho\hbar^3}{Ze(\Delta p)^3}\left[\sin\left(\dfrac{\Delta p R}{\hbar}\right) - \left(\dfrac{\Delta p R}{\hbar}\right)\cos\left(\dfrac{\Delta p R}{\hbar}\right)\right]\right|^2$;

$\dfrac{d\sigma}{d\Omega} = \dfrac{(2kZ_p Z_t e^2 m_p)^2}{(\Delta p)^4}\left|\dfrac{3\hbar^3}{\Delta p^3 R^3}\left[\sin\left(\dfrac{\Delta p R}{\hbar}\right) - \left(\dfrac{\Delta p R}{\hbar}\right)\cos\left(\dfrac{\Delta p R}{\hbar}\right)\right]\right|^2$.

39.33 $2.2 \cdot 10^8$ m/s.

39.35

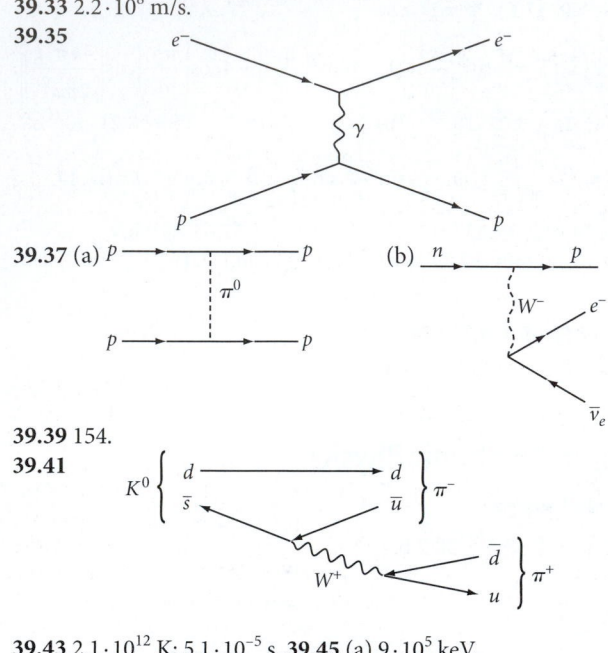

39.37 (a) P ... π^0 ... (b) n ... p ... W^- ... e^- ... $\bar{\nu}_e$

39.39 154.

39.41 K^0 ... d, $\bar{s}$ ... d, $\bar{u}$ $\}\pi^-$ ... W^+ ... $\bar{d}$, u $\}\pi^+$

39.43 $2.1 \cdot 10^{12}$ K; $5.1 \cdot 10^{-5}$ s. **39.45** (a) $9 \cdot 10^5$ keV.

39.47 (a) $H = \left(\dfrac{1}{a}\dfrac{da}{dt}\right)_{rec}$. (b) 420 Mpc. **39.49** (a) 0.036 eV compared to 8.3 keV, over 200,000 times smaller. **39.51** 54. meV. **39.53** 10^{-18} m.

39.55 $9 \cdot 10^{-4}$ m^2. **39.57** (a) $316 \, ke^2/mv^2$. (b) 48.8 fm. **39.59** 13/32.

39.61 $F^2(\Delta p) = \left(\dfrac{\sin[(\Delta p)a/\hbar]}{(\Delta p)a/\hbar}\right)^2$;

$\dfrac{d\sigma}{d\Omega} = \left(\dfrac{2Ze^2 m_e}{4\pi\epsilon_0}\right)^2 \dfrac{1}{(\Delta p)^4}\left(\dfrac{\sin[(\Delta p)a/\hbar]}{(\Delta p)a/\hbar}\right)^2$; Yes.

Multi-Version Exercises

39.62 $9.7 \cdot 10^{-12}$. **39.65** 3900 counts/s.

Chapter 40: Nuclear Physics

Multiple Choice

40.1 b. **40.3** b. **40.5** d. **40.7** d.

Exercises

40.23 (a) 39 MeV. (b) 92 MeV. (c) 493 MeV. (d) 739 MeV.
40.25 -0.46 fm^{-1}. **40.27** (a) $^{60}_{27}\text{Co} \rightarrow {}^{60}_{28}\text{Ni} + {}^{0}_{-1}e^- + \bar{\nu}_e$.
(b) $^{3}_{1}\text{H} \rightarrow {}^{3}_{2}\text{He} + {}^{0}_{-1}e^- + \bar{\nu}_e$. (c) $^{14}_{6}\text{C} \rightarrow {}^{14}_{7}\text{N} + {}^{0}_{-1}e^- + \bar{\nu}_e$.
40.29 0.352 MeV. **40.31** 0.253 h. **40.33** $9.64 \cdot 10^{15}$ yr.
40.35 (a) $1.65 \cdot 10^{11}$ disint/(g s) = $1.65 \cdot 10^{11}$ Bq/g = 4.46 Ci/g.
(b) 1.15 Bq. (c) $1.12 \cdot 10^{10}$ disint. **40.37** 1946 or 1947.

40.39 $N_B(t) = \dfrac{\lambda_A}{\lambda_B - \lambda_A} N_{A0}e^{-\lambda_A t} + \left(N_{B0} - \dfrac{\lambda_A}{\lambda_B - \lambda_A} N_{A0}\right)e^{-\lambda_B t}$.

40.41 $^{117}_{50}\text{Sn}$. **40.43** 4.51 kg. **40.45** (a) $4.28 \cdot 10^9$ kg/s. (c) 0.0305%.
40.47 13.8 GK.
40.49 (a)

No.	Nuclide	Mass number, A	Mass excess, Δm (keV/c^2)	Atomic mass (u)
1	$^{1}_{0}n$	1	8,071.3	1.0087
2	$^{252}_{98}\text{Cf}$	252	76,034	252.08
3	$^{256}_{100}\text{Fm}$	256	85,496	256.09
4	$^{140}_{56}\text{Ba}$	140	$-83,271$	139.91
5	$^{140}_{54}\text{Xe}$	140	$-72,990$	139.92
6	$^{112}_{46}\text{Pd}$	112	$-86,336$	111.91
7	$^{109}_{42}\text{Mo}$	109	$-67,250$	108.93

(b) $\Delta E_{Cf} = 202.34$ MeV and $\Delta E_{Fm} = 212.54$ MeV. (c) Yes.
40.51 $2.07 \cdot 10^{-16}$ J; $4.98 \cdot 10^5$ m/s. **40.53** 2.00 μCi. **40.55** 12.71 days.
40.57 174.147 MeV. **40.59** 53.0 fm. **40.61** $1.13 \cdot 10^{-22}$ μCi.
40.63 41.279 MeV. **40.65** $2.51 \cdot 10^{17}$ J. **40.67** (a) 7.074 MeV.
(b) 2.572 MeV. (c) 2.827 MeV. (d) 1.112 MeV. **40.69** 89.0 min.
40.71 $6.38 \cdot 10^4$ yr. **40.73** $2.57 \cdot 10^4$ decays **40.75** 10.9 s.

Multi-Version Exercises

40.76 4100 yr.

Credits

Photo Credits

About the Authors

Page vi: Photo Courtesy of Okemos Studio of Photography.

A Note from the Authors

Page vii: © Malcolm Fife/Getty Images RF.

The Big Picture

Figure 1a: U.S. Coast Guard photo; **1b:** © DigitalGlobe, Inc.; **2:** © Siemens Press Pictures; **3:** © Lawrence Livermore National Laboratory; **4:** © Volker Lannert/Universität Bonn; **5:** © M. F. Crommie, C. P. Lutz, and D. M. Eigler, IBM Almaden Research Center Visualization Lab, http://www.almaden.ibm.com/vis/stm/images/stm15.jpg. Image reproduced by permission of IBM Research, Almaden Research Center. Unauthorized use not permitted; **6:** This image was made with VMD and is owned by the Theoretical and Computational Biophysics Group, NIH Center for Macromolecular Modeling and Bioinformatics, at the Beckman Institute, University of Illinois at Urbana-Champaign; **7:** © CERN; **8a:** © Mona Schweizer/CERN; **8b:** © CERN; **9:** NASA; **10:** Andrew Fruchter (STScI) et al., WFPC2, HST, NASA.

Chapter 1

Figure 1.1: NASA/JPL-Caltech/L. Allen (Harvard Smithsonian CfA); **1.3:** © Digital Vision/Getty Images RF; **1.4:** © Photograph reproduced with permission of the BIPM; **1.5:** Courtesy NASA/JPL-Caltech; **1.6:** J. Burrus/NIST; **1.7 (top):** Gemini Observatory-GMOS Team; **1.7 (middle):** © BananaStock/PunchStock RF; **1.7 (bottom):** Dr. Fred Murphy, 1975, Centers for Disease Control and Prevention; **1.8(1):** © Hans Gelderblom/Stone/Getty Images; **1.8(2):** © Wolfgang Bauer; **1.8(3):** © Edmond Van Hoorick/Getty Images RF; **1.8(4):** © Digital Vision/Getty Images RF; **1.8(5):** Gemini Observatory-GMOS Team; **1.8(6):** NASA, ESA, and The Hubble Heritage Team (STScI/AURA)/Hubble Space Telescope ACS/STScI-PRC05-20; **1.27:** © The McGraw-Hill Companies, Inc./Mark Dierker, photographer; **p. 36:** NASA.

Chapter 2

Figure 2.1: © Royalty-Free/Corbis; **2.2a-b:** © W. Bauer and G. D. Westfall; **2.8:** © Ted Foxx/Alamy RF; **2.9:** © Ryan McVay/Getty Images RF; **2.17:** © Larry Caruso/WireImage/Getty Images; **2.20-2.23, 2.25:** © W. Bauer and G. D. Westfall; **2.26:** © Royalty-Free/Corbis; **p. 66:** © W. Bauer and G. D. Westfall.

Chapter 3

Figure 3.1: © Terry Oakley/Alamy; **3.3, 3.6, 3.7, 3.9, 3.13:** © W. Bauer and G. D. Westfall; **3.14:** © 2011 Scott Cunningham/Getty Images; **3.17:** © W. Bauer and G. D. Westfall; **3.22-3.23:** © Edmond Van Hoorick/Getty Images RF.

Chapter 4

Figure 4.1: NASA; **4.2a:** © Dex Image/Corbis RF; **4.2b:** © Richard McDowell/Alamy; **4.2c:** Photo by Lynn Betts, USDA Natural Resources Conservation Service; **4.3a:** © Photodisc/Getty Images RF; **4.3b:** © W. Cody/Corbis; **4.3c:** © Boston Globe/Getty Images; **4.4:** © W. Bauer and G. D. Westfall; **4.5a:** © Radius Images/Alamy RF; **4.5c:** © The McGraw-Hill Companies, Inc./Joe De Grandis, photographer; **4.5d:** © W. Bauer and G. D. Westfall; **4.5e:** © Brand X Pictures/PunchStock RF; **4.5f:** © W. Bauer and G. D. Westfall; **4.5g:** R. Stockli, A. Nelson, F. Hasler, NASA/GSFC/NOAA/USGS; **4.6, 4.7a, 4.8:** © W. Bauer and G. D. Westfall; **4.9:** © Tim Graham/Getty Images; **4.11a-b:** © Ryan McVay/Getty Images RF; **4.15, 4.16, 4.20:** © W. Bauer and G. D. Westfall; **4.21a:** © Digital Vision/Getty Images RF; **4.21b:** © Brand X Pictures/Jupiterimages RF; **4.22b:** © Jan Gräser.

Chapter 5

Figure 5.1: NASA/Goddard Space Flight Center Scientific Visualization Studio; **5.2:** © Malcolm Fife/Getty Images RF; **5.3a:** © Earl Roberge/Photo Researchers, Inc.; **5.3b:** © Mike Goldwater/Alamy; **5.3c:** © John Henshall/Alamy; **5.4a:** © Royalty-Free/Corbis; **5.4b:** © Geostock/Getty Images RF; **5.5b:** © Cre8tive Studios/Alamy RF; **5.5c:** © Creatas/PunchStock RF; **5.5d:** © Royalty-Free/Corbis; **5.5e:** © General Motors Corp. Used with permission, GM Media Archives; **5.5f:** Courtesy National Nuclear Security Administration, Nevada Site Office; **5.5g:** © Royalty-Free/Corbis;

5.5h: NASA, ESA, J. Hester and A. Loll (Arizona State University); **5.15a-c:** © W. Bauer and G. D. Westfall; **5.18a:** © Photodisc/Getty Images RF; **5.18b:** © W. Bauer and G. D. Westfall; **5.18c:** NRC File Photo; **5.18d:** © John Henshall/Alamy; **5.18e:** © Royalty-Free/Corbis; **5.18f:** © The McGraw-Hill Companies, Inc./Jill Braaten, photographer; **5.18g:** © Robin Lund/Alamy RF; **5.18h:** © W. Bauer and G. D. Westfall; **5.18i:** © Comstock Images/Getty Images RF; **5.18j:** © izmostock/Alamy; **5.18k:** © Michele Peterson/Alamy; **5.18l:** NASA/Goddard Space Flight Center Scientific Visualization Studio; **5.20:** © W. Bauer and G. D. Westfall; **5.21:** © Thinkstock/Masterfile RF; **5.22:** © W. Bauer and G. D. Westfall; **5.23:** Wind resource map developed by NREL with data from AWS Truepower.

Chapter 6

Figures 6.1, 6.3, 6.11, 6.13, 6.15, 6.19: © W. Bauer and G. D. Westfall; **p. 192:** © The Texas Collection, Baylor University, Waco, Texas.

Chapter 7

Figure 7.1: © Getty Images RF; **7.2(1-4):** © Sandia National Laboratories/Getty Images RF; **7.5:** Reproduced by permission of the News-Gazette, Inc. Permission does not imply endorsement; **7.6:** © Photron; **7.8:** © W. Bauer and G. D. Westfall; **p. 205:** © Photron; **7.10, 7.14:** © W. Bauer and G. D. Westfall; **7.21:** © Delacroix/The Bridgeman Art Library/Getty Images; **p. 225 (golfball):** © Stockdisc/PunchStock RF; **p. 225 (basketball):** © Photolink/Getty Images RF.

Chapter 8

Figure 8.1: NASA; **8.7:** © AP Photo; **8.8a-b:** © W. Bauer and G. D. Westfall; **8.9:** © Boeing; **8.14, 8.22a:** © W. Bauer and G. D. Westfall; **8.22b:** © Comstock Images/Alamy RF; **p. 255 (top right):** © W. Bauer and G. D. Westfall.

Chapter 9

Figure 9.1: NASA; **9.2:** © Royalty-Free/Corbis; **9.7a-b:** © Baokang Bi, Michigan State University; **9.8:** © The McGraw-Hill Companies, Inc./Mark Dierker, photographer; **9.9, 9.14, 9.15:** © W. Bauer and G. D. Westfall; **9.17:** © Royalty-Free/Corbis; **9.20:** © Patrick Reddy/Getty Images.

Chapter 10

Figure 10.1: © DreamPictures/Stone/Getty Images; **10.2a:** © Stock Trek/Getty Images RF; **10.2b:** © Gemini Observatory-GMOS Team; **10.13, 10.14, 10.16:** © W. Bauer and G. D. Westfall; **10.17:** The McGraw-Hill Companies, Inc./Mark Dierker, photographer; **10.24-10.25:** © W. Bauer and G. D. Westfall; **10.27-10.29b:** © The McGraw-Hill Companies, Inc./Mark Dierker, photographer; **10.30:** © Don Farrall/Getty Images RF; **10.31a-c:** © Otto Greule Jr./Allsport/Getty Images; **10.32:** © T. Bollenbach; **p. 317:** © Mark Thompson/Getty Images; **10.35:** © Royalty-Free/Corbis; **10.37:** © W. Bauer and G. D. Westfall.

Chapter 11

Figure 11.1a: © Digital Vision/Alamy RF; **11.1b:** © Guillaume Paumier/Wikimedia Commons; **11.1c:** Copyright © Skyscraper Source Media Inc. http://SkyscraperPage.com; **11.2, 11.4, 11.5:** © W. Bauer and G. D. Westfall; **11.6a:** © Ole Graf/Corbis; **11.8, 11.12:** © W. Bauer and G. D. Westfall; **11.14:** © Scott Olson/Getty Images; **11.15a-b:** © W. Bauer and G. D. Westfall; **11.23:** © Segway, Inc.; **p. 351 (top):** © W. Bauer and G. D. Westfall; **p. 351 (construction worker):** © Creatas Images/Jupiterimages RF; **p. 352 (top)-(bottom):** © W. Bauer and G. D. Westfall.

Chapter 12

Figure 12.1: © JAXA/NHK; **12.2:** © NASA/JPL-Caltech/S. Stolovy (SSC/Caltech); **12.3, 12.7, 12.13:** NASA; **12.14:** U.S. Geological Survey; **12.21:** These images/animations were created by Prof. Andrea Ghez and her research team at UCLA and are from data sets obtained with the W. M. Keck Telescopes; **12.23, p. 378:** NASA; **12.28:** © Vera Rubin; **12.29a:** Courtesy of NASA, ESA, M. J. Jee and H. Ford (Johns Hopkins University); **12.29b:** Courtesy of NASA, ESA, M. J. Jee and H. Ford (Johns Hopkins University). Data from M. J. Jee et al., "Discovery of a Ringlike Dark Matter Structure in the Core of the Galaxy Cluster CL 0024+17," *The Astrophysical Journal*, 661:730, 2007 June 1; **12.30:** X-ray: NASA/CXC/CfA/M. Markevitch et al. Optical: NASA/STScI/Magellan/U. Arizona/D. Clowe et al. Lensing Map: NASA/STScI/ESO WFI/Magellan/U. Arizona/D. Clowe et al.

Chapter 13

Figure 13.1: Horns Rev 1 owned by Vattenfall. Photographer Christian Steiness; **13.2b:** © Royalty-Free/Corbis; **13.2d:** © Image Source/Getty Images RF; **13.5a:** © Photolink/Getty Images RF; **13.5b:** © Don Farrall/Getty Images RF; **13.6a:** C. Borland/Photolink/Getty Images RF; **13.6b:** BananaStock/PunchStock RF; **13.6c:** © Comstock Images/Alamy RF; **13.15:** © W. Bauer and G. D. Westfall; **13.16, 13.20:** © Royalty-Free/Corbis; **13.21:** © W. Bauer and G. D. Westfall; **13.26a:** © Royalty-Free/Corbis; **13.27a-c:** © W. Bauer and G. D. Westfall; **13.28:** © Photolink/Getty Images RF; **13.29, 13.33a:** © W. Bauer and G. D.

Westfall; **13.33b:** © Don Farrall/Getty Images RF; **13.33c, 13.35, 13.39, 13.41:** © W. Bauer and G. D. Westfall; **13.43a-b:** © SHOTFILE/Alamy RF; **13.44:** © The McGraw-Hill Companies, Inc./Ken Karp, photographer; **13.46:** © W. Bauer and G. D. Westfall; **13.50:** © Siede Preis/Getty Images RF; **13.51:** © Royalty-Free/Corbis; **13.52a:** © Kim Steele/Getty Images RF; **13.52b:** © W. Bauer and G. D. Westfall; **13.53:** © *The Astrophysical Journal*, Fryer & Warren 2002 (*Apj*, 574, L65). Reproduced by permission of the AAS.

Chapter 14

Figure 14.1a: © W. Bauer and G. D. Westfall; **14.1b:** Courtesy National Institute of Standards and Technology; **14.2, 14.3, 14.12, 14.15a, 14.19a,b, 14.25:** © W. Bauer and G. D. Westfall; **14.32:** © University of Washington Libraries, Special Collections, UW 21422.

Chapter 15

Figure 15.1: © Matt King/Getty Images; **15.2:** © Don Farrall/Getty Images RF; **15.4:** © W. Bauer and G. D. Westfall; **15.12a:** © Ingram Publishing/Alamy RF; **15.12b:** Image courtesy of Cornell University; **15.18:** NOAA/PMEL/Center for Tsunami Research; **15.27a-d:** © W. Bauer and G. D. Westfall; **p. 483 (top):** Courtesy Vera Sazonova and Paul McEuen; **15.29a, 15.30, p. 485 (middle):** © W. Bauer and G. D. Westfall; **15.31:** M. F. Crommie, C. P. Lutz, and D. M. Eigler, IBM Almaden Research Center Visualization Lab, http://www.almaden.ibm.com/vis/stm/images/stm15.jpg. Image reproduced by permission of IBM Research, Almaden Research Center. Unauthorized use not permitted; **15.32a-d:** Images courtesy of LIGO Laboratory. Supported by the National Science Foundation; **15.33:** Henze/NASA.

Chapter 16

Figure 16.1: © RG4 WENN Photos/Newscom; **16.3:** © Royalty-Free/Corbis; **16.5:** © R. Morley/Photolink/Getty Images RF; **16.7:** © Steve Allen/Getty Images RF; **16.8:** © Arnold Song and Jose Iriarte-Diaz; **16.10:** © McGraw-Hill Companies, Inc.; **16.11:** Photo by Petty Officer 3rd Class John Hyde, U.S. Navy; **16.13, 16.15, 16.16, 16.21:** © W. Bauer and G. D. Westfall; **16.22:** Courtesy Wake Radiology; **16.24:** Courtesy U.S. Nuclear Regulatory Commission; **16.25:** © Photolink/Getty Images RF; **16.26:** © Stockbyte/Getty Images RF; **16.28:** © Annie Reynolds/Photolink/Getty Images RF.

Chapter 17

Figure 17.1: © Antonio Fernandez Sanchez; **17.4 (top left):** © Royalty-Free/Corbis; **17.4 (top middle left):** © W. Bauer and G. D. Westfall; **17.4 (top middle right):** © Digital Vision/Getty Images RF; **17.4 (top right):** The STAR Experiments, Brookhaven National Laboratory; **17.4 (bottom left):** © MSU National Superconducting Cyclotron Laboratory; **17.4 (bottom middle left):** NASA/WMAP Science Team; **17.4 (bottom middle**

right): NOAA; **17.4 (bottom right):** © ITER; **17.6:** © Siede Preis/Getty Images RF; **17.8a-b:** © Janis Research Co., Inc. www.janis.com; **17.9:** © H. Mark Helfer/NIST; **17.10:** © ITER; **17.12:** © Michigan Department of Transportation Photography Unit; **17.14, 17.15, 17.17:** © W. Bauer and G. D. Westfall; **17.20, 17.24:** NASA/WMAP Science Team.

Chapter 18

Figure 18.1: © Kirk Treakle/Alamy; **18.8:** © Hassan Ammar/AP Photo; **18.11:** © W. Bauer and G. D. Westfall; **18.12:** NASA, ESA and J. Hester (ASU); **18.14:** © Skip Brown/National Geographic/Getty Images; **18.16a-b:** Photos used with permission of Owens Corning. Copyright © 2012 Owens Corning. The color PINK is a registered trademark of Owens Corning; **18.18:** NASA; **18.20a-b:** NASA Glenn Research Center; **18.21:** NASA Goddard Space Flight Center, Visible Earth; **18.24:** Courtesy of www.EnergyEfficientSolutions.com; **18.25:** © National Geographic/SuperStock; **18.27a-b:** NASA/IPAC/Caltech; **18.29:** © David J. Green/Alamy RF; **18.31:** INL and SMU; **18.32:** Photo courtesy of Dr. Gretar Ívarsson.

Chapter 19

Figure 19.1: © image 100/PunchStock RF; **19.8-19.9:** © W. Bauer and G. D. Westfall; **19.10:** The Hubble Heritage Team (AURA/STScI/NASA); **19.13:** © W. Bauer and G. D. Westfall; **19.25:** Courtesy of Brookhaven National Laboratory.

Chapter 20

Figure 20.1a: © Frans Lemmens/Getty Images; **20.1b:** © Keith Kent/Photo Researchers, Inc.; **20.4:** © Royalty-Free/Corbis; **20.5:** © Photodisc/PunchStock RF; **20.7, 20.11:** © W. Bauer and G. D. Westfall; **20.13:** © Koichi Kamoshida/Getty Images.

Chapter 21

Figure 21.1a-c: © W. Bauer and G. D. Westfall; **21.2:** © R. Morley/Photolink/Getty Images RF; **21.7a:** © Kim Steele/Getty Images RF; **21.7b:** © Geostock/Getty Images RF; **21.8-21.11, 21.19, 21.21:** © W. Bauer and G. D. Westfall.

Chapter 22

Figure 22.1: © Royalty-Free/Corbis; **22.2:** National Weather Service; **22.17:** Courtesy of Brookhaven National Laboratory; **22.20:** © The McGraw-Hill Companies, Inc./Mark Dierker, photographer; **22.31a-b:** © W. Bauer and G. D. Westfall; **22.32:** © Gerd Kortemeyer; **22.41:** © Royalty-Free/Corbis.

Chapter 23

Figure 23.7a-b: © W. Bauer and G. D. Westfall; **23.9 (left):** © Regis Duvignau/Reuters; **23.9 (right):** © Candy Lab Studios; **23.10:** © 2009 Tesla Motors, Inc. All rights reserved. 'Tesla Motors' and 'Tesla Roadster' are trademarks of Tesla Motors,

Inc. (version 1-5-0296-228); **23.11a-b:** © W. Bauer and G. D. Westfall.

Chapter 24

Figure 24.1, 24.2: © W. Bauer and G. D. Westfall; **24.21:** © Lawrence Livermore National Laboratory; **24.33:** © W. Bauer and G. D. Westfall.

Chapter 25

Figure 25.1: © Image Source/agefotostock RF; **25.2a-f:** © W. Bauer and G. D. Westfall; **25.3a:** © IBM; **25.3c:** © Brand X Pictures/PunchStock RF; **25.3d:** © Don Farrall/Getty Images RF; **25.3e:** © W. Bauer and G. D. Westfall; **25.3f:** © Craig Bickford/International Light Technologies; **25.3g:** © 1000Bulbs.com; **25.3h:** © 2012, Digi-Key Corporation. All rights reserved; **25.3i:** © Thomas Allen/Getty Images RF; **25.3j:** NASA; **25.3k:** © Lawrence Livermore National Laboratory; **25.3l:** NASA artist Werner Heil; **25.4:** © W. Bauer and G. D. Westfall; **25.8a:** Photo courtesy Mike Smith, www.mds975.co.uk; **25.8b:** © 2012, Digi-Key Corporation. All rights reserved; **25.15:** Printed with permission from Mayfield Clinic; **25.21a-b:** Courtesy of ABB; **25.22:** http://en.wikipedia.org/wiki/File:Path_65_P0002014.jpg; **25.24a-b:** © W. Bauer and G. D. Westfall; **25.26:** © Julie Jacobson/AP Photo.

Chapter 26

Figure 26.1: © Royalty-Free/Corbis.

Chapter 27

Figure 27.1: © W. Bauer and G. D. Westfall; **27.4b:** © Steve Cole/Getty Images RF; **27.7:** NASA; **27.8:** NASA/Hubble/Z. Levay and J. Clarke; **27.11:** © W. Bauer and G. D. Westfall; **27.12:** © The McGraw-Hill Companies, Inc./Mark Dierker, photographer; **27.13:** National Geophysical Data Center; **27.15:** © W. Bauer and G. D. Westfall; **27.16:** Lawrence Berkeley National Lab; **27.17a:** Courtesy of Brookhaven National Laboratory; **27.17b:** STAR collaboration/RHIC/Brookhaven National Laboratory; **27.18:** Courtesy of Brookhaven National Laboratory; **27.19:** © Michigan State University; **27.22:** © W. Bauer and G. D. Westfall; **27.24b, 27.28:** © The McGraw-Hill Companies, Inc./Mark Dierker, photographer.

Chapter 28

Figure 28.1: Courtesy of International Business Machines Corporation, © 2012 International Business Machines Corporation; **28.3b, 28.5:** © The McGraw-Hill Companies, Inc./Mark Dierker, photographer; **28.11:** U.S. Navy Photograph by Mr. John F. Williams; **28.20a:** © W. Bauer and G. D. Westfall; **28.24:** © The McGraw-Hill Companies, Inc./Mark Dierker, photographer; **28.28:** © High Field Magnet Laboratory, Radboud University Nijmegen, The Netherlands; **28.31:** © W. Bauer and G. D. Westfall; **28.33:** © Royalty-Free/Corbis; **28.34:** Courtesy of AMSC®. © 2008

American Superconductor Corp.; **28.35(1-4):** © The McGraw-Hill Companies, Inc./Mark Dierker, photographer.

Chapter 29

Figure 29.1: © Photolink/Getty Images RF; **29.11:** © W. Bauer and G. D. Westfall; **29.14a-b:** NASA; **29.18:** © W. Bauer and G. D. Westfall; **29.21:** © General Motors, LLC 2012; **29.31:** © Daisuke Morita/Getty Images RF; **p. 904:** © Royalty-Free/Corbis.

Chapter 30

Figure 30.1, 30.18: © W. Bauer and G. D. Westfall; **30.21:** © The McGraw-Hill Companies, Inc./Mark Dierker, photographer; **30.32a:** © Edmond Van Hoorick/Getty Images RF; **30.32b, 30.33:** © W. Bauer and G. D. Westfall.

Chapter 31

Figure 31.1: © Kevin Foy/Alamy; **31.9a:** © Kim Steele/Getty Images RF; **31.9b:** NASA; **31.9c:** © W. Bauer and G. D. Westfall; **31.10(1):** © C. Borland/Photolink/Getty Images RF; **31.10(2):** © W. Bauer and G. D. Westfall; **31.10(3):** © Don Tremain/Getty Images RF; **31.10(4):** © Photolink/Getty Images RF; **31.10(5):** © Geostock/Getty Images RF; **31.10(6-7):** © Russell Illig/Getty Images RF; **31.10(8):** © Royalty-Free/Corbis; **31.10(9):** © David R. Frazier Photography/Alamy RF; **31.10(10):** © Photodisc/Getty Images RF; **31.10(11):** © W. Bauer and G. D. Westfall; **31.10(12):** © ImageState/Alamy RF; **31.14a:** © John Keating/Photo Researchers, Inc.; **31.14b:** © General Motors Corp. Used with permission, GM Media Archives; **31.26:** © W. Bauer and G. D. Westfall.

Chapter 32

Figure 32.1, 32.7, 32.9, 32.10, 32.18, 32.25, 32.31a: © W. Bauer and G. D. Westfall; **32.33:** © P. K. Chen; **32.34:** © Hank Morgan/Photo Researchers, Inc.; **32.35:** © 2011 Abengoa Solar; **32.37, 32.39, 32.42, 32.43:** © W. Bauer and G. D. Westfall; **32.45a:** Courtesy IT Concepts GmbH; **32.45b:** © David M. Martin, M.D./Photo Researchers, Inc.; **32.47, 32.53:** © W. Bauer and G. D. Westfall.

Chapter 33

Figure 33.1a: NASA and The Hubble Heritage Team (STScI/AURA); **33.1b:** Prof. Gordon T. Taylor, Stony Brook University/NSF Polar Programs; **33.2, 33.3, 33.6, 33.7:** © W. Bauer and G. D. Westfall; **33.17:** © Robert George Young/Getty Images; **33.21a-c:** © W. Bauer and G. D. Westfall; **33.24:** © Scott Bodell/Getty Images RF; **33.30, 33.32, 33.33, 33.35:** © W. Bauer and G. D. Westfall; **33.36 (bottom right):** © Comstock/PunchStock RF; **33.37:** © Meade; **33.39:** © Victor Krabbendam/Southern Astrophysical Research Telescope; **33.41a:** STS-82 Crew/STScI/NASA; **33.41b:** NASA, ESA, STScI, J. Hester and P. Scowen (Arizona State University); **33.42a-b:** NASA,

STScI; **33.43:** NASA; **33.44a:** NASA/CXC/D. Berry; **33.44b:** NASA/CXC/D. Berry & A. Hobart; **33.44c:** NASA/CXC/MIT/F.K. Baganoff et al.

Chapter 34

Figure 34.1: © Jim DeLillo/Shutterstock.com; **34.12a, 34.17, 34.20:** © W. Bauer and G. D. Westfall; **34.22:** © GeoEye satellite image; **34.24, 34.29a, 34.31, 34.32:** © W. Bauer and G. D. Westfall; **34.33:** NASA, 1990; **34.36, 34.39, 34.42, 34.43:** © W. Bauer and G. D. Westfall; **34.52:** © Dr. Bernard Santarsiero, University of Illinois at Chicago; **34.53:** Argonne National Laboratory, managed and operated by UChicago Argonne, LLC, for the U.S. Department of Energy under Contract No. DE-AC02-06CH11357.

Chapter 35

Figure 35.1: NASA, ESA, and STScI; **35.8:** U.S. Air Force photo by Master Sgt Michael A. Kaplan, NASCAR Race; **35.17:** © W. Bauer and G. D. Westfall; **35.19:** F. W. Dyson, A. S. Eddington, and C. Davidson, "A Determination of the Deflection of Light by the Sun's Gravitational Field, from Observations Made at the Total Eclipse of May 29, 1919," *Philosophical Transactions of the Royal Society of London.* Series A, Containing Papers of a Mathematical or Physical Character (1920):291-333, on p. 332; **35.21b:** NASA, Andrew Fruchter and the ERO Team [Sylvia Baggett (STScI), Richard Hook (ST-ECF), Zoltan Levay (STScI)] (STScI); **35.22:** NASA/JPL/Caltech; **35.23:** Image courtesy of Lockheed Martin Corporation.

Chapter 36

Figure 36.1: Courtesy of Brookhaven National Laboratory; **36.2:** © Royalty-Free/Corbis; **36.10a:** © Lencho Guerra; **36.10c:** © Kai Pfaffenbach/Reuters/Corbis; **36.16 (top):** © W. Bauer and G. D. Westfall; **36.17:** Reprinted with permission from P.G. Merli, G.F. Missiroli, G. Pozzi, *American Journal of Physics*, "On the Statistical Aspect of Electron Interference Phenomena," 44(3):306-7, March 1976. © 1976, American Association of Physics Teachers; **36.26:** © Mike Matthews, JILA.

Chapter 37

Figure 37.1a: Photo courtesy of Erik Lucero and Max Hofheinz; **37.9, 37.17b:** Images reproduced by permission of IBM Research, Almaden Research Center. Unauthorized use not permitted; **37.23:** Courtesy of Jeff S. Lundeen, Brandon Sutherland, Aabid Patel, Corey Stewart, and Charles Bamber; **37.29:** From R. Blatt, "Quantum Information Processing: Dream and Realization," *Entangled World*, pp. 235-70, Wiley-VCH, Weinheim 2006. Image courtesy R. Blatt, University of Innsbruck.

Chapter 38

Figure 38.1: © WMKO; **38.2, 38.3, 38.26:** © W. Bauer and G. D. Westfall; **38.27, 38.28:** Images courtesy of University of California,

Lawrence Livermore National Laboratory, and the Department of Energy.

Chapter 39

Figure 39.1: NASA/ESA/STScI/AURA; **39.2a:** © Royalty-Free/Corbis; **39.4-39.6:** © CERN; **39.19a:** ATLAS Experiment © 2012 CERN; **39.19b:** © 2012 CERN, for the benefit of the CMS Collaboration; **39.28:** © Kamioka Observatory, ICRR (Institute for Cosmic Ray Research), The University of Tokyo; **39.37:** Image courtesy of Derek Leinweber, CSSM, University of Adelaide; **39.41:** © Axel Mellinger, University of Potsdam, Germany; **39.42 (stars):** NASA, ESA, and The Hubble Heritage Team (STScI/AURA). Hubble Space Telescope ACS. STScI-PRC05-20; **39.42 (solar system):** © Royalty-Free/Corbis; **39.43-39.44:** NASA/WMAP Science Team; **39.45:** STAR collaboration/RHIC/ Brookhaven National Laboratory.

Chapter 40

Figure 40.1: © Brand X Pictures/PunchStock RF; **40.6a:** © MSU National Superconducting Cyclotron Laboratory; **40.14:** Lawrence Berkeley National Lab; **40.18:** http://en.wikipedia.org/wiki/ File:Shroudofturin.jpg; **40.35:** © ITER; **40.36a-c:** Images courtesy of University of California, Lawrence Livermore National Laboratory, and the U.S. Department of Energy; **40.38:** NASA/ ESA/JPL/Arizona State University; **40.41:** Image courtesy of Steve Goetsch; **40.42, 40.44:** © W. Bauer and G. D. Westfall.

Line Art and Text Credits

Chapter 7

Figure 7.17: Based on information from DZero collaboration and Fermi National Accelerator Laboratory, Education Office.

Chapter 17

Figure 17.4: RHIC figure courtesy of Brookhaven National Laboratory. Cosmic microwave background image credit NASA/WMAP Science Team. ITER machine © ITER Organization; **Figure 17.21:** From data compiled by the National Oceanic and Atmospheric Administration, Brohan et al., J. Geophys. Res., 111, D12106 (2006); **Figure 17.22:** Data from the Vostok ice core, J.R. Petit et al., Nature 399, 429–436 (1999); **Problem 17.52:** Figure and problem based on V.A. Henneken et al., J. Micromech. Microeng. 16 (2006) S107–S115;

Problem 17.53: Figure and problem based on V.A. Henneken et al., J. Micromech. Microeng. 16 (2006) S107–S115.

Chapter 18

Figure 18.28: Concentration of carbon dioxide in the atmosphere from 1832 to 2004. The measurements from 1832 to 1978 were done using ice cores in Antarctica: ride line based on information from D.M. Etheridge, L.P. Steele, R.L. Langenfelds, R.J. Francey, J.-M. Barnola and V.I. Morgan,1998. The measurements from 1959 to 2004 were carried out in the atmosphere on Mauna Loa in Hawaii: blue line based on information from C.D. Keeling and T.P. Whorf, 2005; Historical carbon dioxide records from the Law Dome DE08, DE08-2, and DSS ice cores, *Trends: A Compendium of Data on Global Change,* Carbon Dioxide Information Analysis Center, Oak Ridge National Laboratory, U.S. Department of Energy, Oak Ridge, Tenn., U.S.A. Carbon dioxide concentrations for the past 420,000 years, extracted using ice cores in Antarctica based on information from J.M. Barnola, et. al., 2003.

Chapter 19

Figure 19.25 (b) and (c): Data courtesy of the STAR collaboration.

Chapter 20

Figure 20.14: Image data courtesy of Steve Chu, U.S. Secretary of Energy.

Chapter 25

Table 25.2: Data from American Wire Gauge convention.

Chapter 27

Figure 27.9: Image data from NOAA's National Geophysical Data Center (NGDC), Boulder, Colorado, based on the International Goemagnetic Reference Field (IGRF), Epoch 2000 updated to December 31, 2004. The IGRF is developed by the International Association of Geomagnetism and Aeronomy (IAGA) Division V.

Chapter 36

Figure 36.6: COBE data from NASA / COBE Science Team.

Chapter 39

Figure 39.11: Data as reported by Geiger and Marsden in Phil. Mag. 25 (1913), p. 604; **Figure 39.12:** Data as reported by Eisberg and Porter, Rev. Mod. Phys. 33 (1961), p. 190; **Figure 39.15:** Data as reported by R. Hofstadter in Annual Reviews of Nuclear Science 7 (1957), p. 231; **Table 39.1:** Data source: Particle Data Group site, 2012 edition, (pdg.lbl.gov); **Table 39.2:** Data source: Particle Data Group site, 2011 numbers (pdg.lbl.gov); **Table 39.3:** Data from W.-M. Yao, et al. (Particle Data Group), J. Phys. G 33, 1 (2006), K. Nakamura et al. (Particle Data Group), J. Phys. G 37, 075021 (2012), and 2011 partial update for the 2012 edition (http://pdg.lbl.gov), data source: Particle Data Group site 2011 numbers (http://pdg.lbl.gov); **Figure 39.31:** Data from W.-M. Yao et al. (Particle Data Group), J. Phys. G 33, 1 (2006), (http://pdg. lbl.gov); **Table 39.4:** Data from W.-M. Yao et al. (Particle Data Group), J. Phys. G 33, 1 (2006), K. Nakamura et al. (Particle Data Group), J. Phys. G 37, 075021 (2012), and 2011 partial update for the 2012 edition (http://pdg.lbl.gov).

Chapter 40

Figure 40.12: Data from the GSI National Laboratory in Germany, December 17, 1994; **Figure 40.15:** Figure courtesy of Argonne National Laboratory, adapted from T. Lauritzen et al., Phys. Rev. Lett. 88, 042501 (2002); **Figure 40.31:** Calculations by B. Alex Brown, Michigan State University; **Problem 40.49 Table:** Data per Berkeley National Lab *NuBase* data base.

Appendix B

Data sources: http://physics.nist.gov/PhysRefData/ PerTable/periodic-table.pdf

http://www.wikipedia.org/ and Generalic, Eni. "EniG. Periodic Table of the Elements." 31 Mar. 2008. KTF-Split. <http://www.periodni.com/>.

Inside Front Cover

Fundamental Constants from National Institute of Standards and Technology, http://physics.nist.gov/ constants.

Other Useful Constants from National Institute of Standards and Technology, http://physics.nist.gov/ constants.

Index

Periodic Table of the Elements

Key

6 — Atomic number
C — Symbol
Carbon — Name
12.01 — Average atomic mass
An element

Transitional metals

Metals
Nonmetals
Metalloids

Period number	1A 1	2A 2	3B 3	4B 4	5B 5	6B 6	7B 7	8B 8	8B 9	8B 10	1B 11	2B 12	3A 13	4A 14	5A 15	6A 16	7A 17	8A 18
1	1 H Hydrogen 1.008																	2 He Helium 4.003
2	3 Li Lithium 6.941	4 Be Beryllium 9.012											5 B Boron 10.81	6 C Carbon 12.01	7 N Nitrogen 14.01	8 O Oxygen 16.00	9 F Fluorine 19.00	10 Ne Neon 20.18
3	11 Na Sodium 22.99	12 Mg Magnesium 24.31											13 Al Aluminum 26.98	14 Si Silicon 28.09	15 P Phosphorus 30.97	16 S Sulfur 32.07	17 Cl Chlorine 35.45	18 Ar Argon 39.95
4	19 K Potassium 39.10	20 Ca Calcium 40.08	21 Sc Scandium 44.96	22 Ti Titanium 47.87	23 V Vanadium 50.94	24 Cr Chromium 52.00	25 Mn Manganese 54.94	26 Fe Iron 55.85	27 Co Cobalt 58.93	28 Ni Nickel 58.69	29 Cu Copper 63.55	30 Zn Zinc 65.41	31 Ga Gallium 69.72	32 Ge Germanium 72.64	33 As Arsenic 74.92	34 Se Selenium 78.96	35 Br Bromine 79.90	36 Kr Krypton 83.80
5	37 Rb Rubidium 85.47	38 Sr Strontium 87.62	39 Y Yttrium 88.91	40 Zr Zirconium 91.22	41 Nb Niobium 92.91	42 Mo Molybdenum 95.94	43 Tc Technetium (98)	44 Ru Ruthenium 101.1	45 Rh Rhodium 102.9	46 Pd Palladium 106.4	47 Ag Silver 107.9	48 Cd Cadmium 112.4	49 In Indium 114.8	50 Sn Tin 118.7	51 Sb Antimony 121.8	52 Te Tellurium 127.6	53 I Iodine 126.9	54 Xe Xenon 131.3
6	55 Cs Cesium 132.9	56 Ba Barium 137.3	71 Lu Lutetium 175.0	72 Hf Hafnium 178.5	73 Ta Tantalum 180.9	74 W Tungsten 183.8	75 Re Rhenium 186.2	76 Os Osmium 190.2	77 Ir Iridium 192.2	78 Pt Platinum 195.1	79 Au Gold 197.0	80 Hg Mercury 200.6	81 Tl Thallium 204.4	82 Pb Lead 207.2	83 Bi Bismuth 209.0	84 Po Polonium (209)	85 At Astatine (210)	86 Rn Radon (222)
7	87 Fr Francium (223)	88 Ra Radium (226)	103 Lr Lawrencium (262)	104 Rf Rutherfordium (263)	105 Db Dubnium (268)	106 Sg Seaborgium (271)	107 Bh Bohrium (270)	108 Hs Hassium (270)	109 Mt Meitnerium (278)	110 Ds Darmstadtium (281)	111 Rg Roentgenium (281)	112 Cn Copernicium (285)	113 Uut Ununtrium (286)	114 Fl Flerovium (289)	115 Uup Ununpentium (289)	116 Lv Livermorium (293)	117 Uus Ununseptium (293)	118 Uuo Ununoctium (294)

Lanthanides 6

57 La Lanthanum 138.9	58 Ce Cerium 140.1	59 Pr Praseodymium 140.9	60 Nd Neodymium 144.2	61 Pm Promethium (145)	62 Sm Samarium 150.4	63 Eu Europium 152.0	64 Gd Gadolium 157.3	65 Tb Terbium 158.9	66 Dy Dysprosium 162.5	67 Ho Holmium 164.9	68 Er Erbium 167.3	69 Tm Thulium 168.9	70 Yb Ytterbium 173.0

Actinides 7

89 Ac Actinium (227)	90 Th Thorium 232.0	91 Pa Protactinium 231.0	92 U Uranium 238.0	93 Np Neptunium (237)	94 Pu Plutonium (244)	95 Am Americium (243)	96 Cm Curium (247)	97 Bk Berkelium (247)	98 Cf Californium (251)	99 Es Einsteinium (252)	100 Fm Fermium (257)	101 Md Mendelevium (258)	102 No Nobelium (259)